PLASTICS ADDITIVES

Plastics Additives

Advanced Industrial Analysis

By

Jan C.J. Bart

DSM Research, The Netherlands

IOS Press

Amsterdam • Berlin • Oxford • Tokyo • Washington, DC

The author and the publisher wish to thank Adri Geeve, DSM Coating Resins B.V. (Zwolle, The Netherlands) for providing the cover image 'Analytical Website'.

ISBN 1-58603-533-9
Library of Congress Control Number: 2005931631

Publisher
IOS Press
Nieuwe Hemweg 6B
1013 BG Amsterdam
Netherlands
fax: +31 20 687 0019
e-mail: order@iospress.nl

Distributor in the UK and Ireland
Gazelle Books
Falcon House
Queen Square
Lancaster LA1 1RN
United Kingdom
fax: +44 1524 63232

Distributor in the USA and Canada
IOS Press, Inc.
4502 Rachael Manor Drive
Fairfax, VA 22032
USA
fax: +1 703 323 3668
e-mail: iosbooks@iospress.com

PRINTED IN THE NETHERLANDS

Table of Contents

Preface . . . xi

About the Author . . . xiii

Acknowledgements . . . xv

Chapter 1 **In-Polymer Spectroscopic Analysis of Additives** . . . 1

1.1. Direct Ultraviolet/Visible Spectrophotometry . . . 4
1.1.1. Vapour-phase Ultraviolet Absorption Spectrometry . . . 10
1.2. Solid-state Vibrational Spectroscopies . . . 11
1.2.1. Mid-infrared Spectroscopic Analysis . . . 14
1.2.2. Near-infrared Spectroscopy . . . 34
1.2.3. Raman Spectroscopic Techniques . . . 52
1.3. Photoacoustic Spectroscopy . . . 66
1.4. Emission Spectroscopy . . . 72
1.4.1. Infrared Emission Spectroscopy . . . 72
1.4.2. Molecular Fluorescence Spectroscopy . . . 75
1.4.3. Phosphorescence Spectroscopy . . . 81
1.4.4. Chemiluminescence . . . 82
1.5. Nuclear Spectroscopies . . . 94
1.5.1. Solid-state NMR Spectroscopy . . . 95
1.5.2. Nuclear Quadrupole Resonance . . . 110
1.5.3. Electron Spin Resonance Spectroscopy . . . 112
1.5.4. Mössbauer Spectroscopy . . . 120
1.6. Dielectric Loss Spectroscopy . . . 123
1.7. Ultrasonic Spectroscopy . . . 127
Bibliography . . . 129
General Spectroscopy . . . 129
Direct UV/VIS Spectrophotometry . . . 129
Infrared Spectroscopy . . . 129
Near-infrared Spectroscopy . . . 130
Raman Spectroscopy . . . 130
Photoacoustics . . . 130
Emission Spectroscopy . . . 131
NMR Spectroscopy . . . 131
Electron Spin Resonance Spectroscopy . . . 131
Dielectric Spectroscopy . . . 131
Polymer Characterisation . . . 131
References . . . 132

Chapter 2 **Polymer/Additive Analysis by Thermal Methods** **155**

2.1. Thermal Analysis Techniques 158
2.1.1. Differential Scanning Calorimetry 163
2.1.2. Differential Thermal Analysis 173
2.1.3. Thermogravimetric Analysis 175
2.1.4. Simultaneous Thermal Analysis Methods 189
2.1.5. (Multi)hyphenated Thermal Analysis Techniques 192
2.1.6. Thermal Microscopy 209
2.1.7. Thermoluminescence 213
2.2. Pyrolysis Techniques 214
2.2.1. Pyrolysis–Gas Chromatography 222
2.2.2. Pyrolysis–Mass Spectrometry 235
2.2.3. Pyrolysis–Gas Chromatography–Mass Spectrometry 244
2.2.4. Pyrolysis–Fourier Transform Infrared Spectroscopy 261
2.2.5. Pyrolysis–Gas Chromatography–Fourier Transform Infrared Spectroscopy 263
2.2.6. Pyrolysis–Gas Chromatography–Atomic Emission Detection . . . 264
2.2.7. Temperature-programmed Pyrolysis 266
2.3. Thermal Volatilisation and Desorption Techniques 275
2.3.1. Thermal Separation Techniques 278
2.3.2. Direct Solid Sampling Techniques for Gas Chromatography 282
2.3.3. Thermal Desorption–Mass Spectrometric Techniques 299
Bibliography 300
Thermal Analysis 300
Pyrolysis 301
Thermal Desorption 301
References 301

Chapter 3 **Lasers in Polymer/Additive Analysis** **325**

3.1. Lasers 325
3.2. Laser Ablation 331
3.2.1. Laser Ablation – Plasma Source Spectrometry 335
3.3. Laser Spectroscopy 341
3.3.1. Laser-induced Atomic and Molecular Fluorescence Spectrometry 343
3.3.2. Laser-induced Breakdown Spectroscopy 346
3.4. Laser Desorption/Ionisation Methods 353
3.4.1. Laser Desorption Mass Spectrometry 354
3.4.2. Laser Ionisation 363
3.4.3. Decoupled Laser Desorption/Ionisation 366
3.4.4. Matrix-assisted Laser Desorption/Ionisation 374
3.4.5. Laser Microprobe Mass Spectrometry 381
3.5. Laser Pyrolysis 388
Bibliography 392
Lasers 392
Laser Ablation 392
Laser Spectroscopy/Spectrometry 392
Laser-induced Chemistry 393
Laser Safety 393
References 393

Chapter 4 **Surface Analytical Techniques for Polymer/Additive Formulations** **403**

4.1. Electron Spectroscopy 408
4.1.1. Auger Electron Spectroscopy 409
4.1.2. X-ray Photoelectron Spectroscopy 411
4.2. Surface Mass Spectrometry 420
4.2.1. Secondary Ion Mass Spectrometry 422
4.2.2. Secondary Neutral Mass Spectrometry 439
4.3. Ion Scattering Techniques 441
4.3.1. Low-energy Ion Scattering 443
4.3.2. Rutherford Backscattering Spectroscopy 444
Bibliography 446
Surface Characterisation 446
Electron Spectroscopy 447
Surface Mass Spectrometry 447
Ion Scattering Techniques 447
References 447

Chapter 5 **Microscopy and Microanalysis of Polymer/Additive Formulations** **455**

5.1. Chemical Microanalysis 458
5.2. Microscopy and Imaging Techniques 460
5.3. Light Microscopy 464
5.3.1. Conventional Optical Microscopy 466
5.3.2. Ultraviolet Microscopy 472
5.3.3. Fluorescence Microscopy 475
5.3.4. Confocal and Laser Microscopy 478
5.4. Electron Microscopy 483
5.4.1. Scanning Electron Microscopy 485
5.4.2. Transmission Electron Microscopy 494
5.4.3. Analytical Electron Microscopy 497
5.5. Scanning Probe Microscopy Techniques 501
5.5.1. Atomic Force Microscopy 504
5.5.2. Near-field Scanning Optical Microscopy 511
5.5.3. Scanning Kelvin Microscopy 514
5.6. Microspectroscopic Imaging of Additives 514
5.6.1. UV/Visible Microspectroscopy 519
5.6.2. Infrared Microspectroscopy and Imaging 521
5.6.3. Laser-Raman Microprobe and Microscopy 532
5.6.4. Fluorescence and Luminescence Imaging 541
5.7. Magnetic Resonance Imaging 546
5.7.1. Nuclear Magnetic Resonance Imaging 547
5.7.2. Electron Spin Resonance Imaging 555
5.8. X-ray Microscopy and Microspectroscopy 559
5.8.1. X-ray Microradiography 560
5.8.2. Scanning X-ray Microscopy 561
5.8.3. X-ray Microfluorescence 563
5.8.4. Micro X-ray Photoelectron Spectroscopy 564
5.9. Ion Imaging of Additives 566
5.9.1. Laser-microprobe Mapping 566
5.9.2. Imaging Secondary Ion Mass Spectrometry 567

Bibliography 573
Light Microscopy 573
Electron Microscopy 573
Scanning Probe Microscopy 574
Near-field Optics 574
Microbeam Analysis 574
Microspectroscopy 575
Imaging/Image Analysis 575
Polymer Microscopy 575
General 576
References 576

Chapter 6 **Quantitative Analysis of Additives in Polymers 597**

6.1. Sampling Procedures for Quantitative Analysis of Polymer/Additive Packages 600
6.1.1. Quantitative Analysis of Mineral Filled Engineering Plastics 605
6.1.2. Reverse Engineering of Cured Rubber Compounds 606
6.1.3. Determination of Additive Blends in Polymers 606
6.2. Quantitative Solvent and Thermal Extraction 609
6.2.1. Extraction and Quantification of Polyolefin Additives 613
6.2.2. Supercritical Fluid Extraction 614
6.2.3. Quantification of Antioxidants in Polyolefins 615
6.2.4. Determination of Plasticisers by Solvent and Thermal Extraction . 619
6.2.5. Oil-extended EPDM 623
6.2.6. Migration Rates of Phthalate Esters from Soft PVC Products 624
6.3. Quantitative Chromatographic Methods 624
6.3.1. Quantitative Gas Chromatography 626
6.3.2. Quantitative Liquid Chromatography 628
6.3.3. Quantitative Supercritical Fluid Chromatography 629
6.3.4. Quantitative Thin-layer Chromatography 630
6.4. Quantitative Spectroscopic Techniques 633
6.4.1. Quantitative Ultraviolet/Visible Spectrophotometry 637
6.4.2. Quantitative Fluorescence Spectroscopy 639
6.4.3. Quantitative Infrared Spectroscopy 639
6.4.4. Quantitative Near-infrared Spectroscopy 644
6.4.5. Quantitative Raman Spectroscopy 645
6.4.6. Quantitative Nuclear Magnetic Resonance Methods 646
6.5. Quantitative Mass Spectrometric Techniques 647
6.6. Quantitative Surface Analysis Techniques 651
6.7. Quantitative Microscopy 653
Bibliography 654
General Quantitative Analysis 654
Sampling 654
Chromatography 654
Spectroscopy 655
Mass Spectrometry 655
Surface Analysis 655
Chemometric Techniques 655
References 655

Chapter 7 **Process Analytics** **663**

7.1. In-process Analysers 667
7.2. Process Spectroscopy 675
7.2.1. Remote Spectroscopy 677
7.2.2. Process Electronic Spectroscopy 679
7.2.3. Mid-infrared Process Analysis of Polymer Formulations 683
7.2.4. Near-infrared Spectroscopic Process Analysis 693
7.2.5. Process Raman Spectroscopy 701
7.2.6. Process Nuclear Magnetic Resonance 704
7.2.7. Acoustic Emission Technology 716
7.2.8. Real-time Dielectric Spectroscopy 719
7.3. Process Chromatography 720
7.4. *In Situ* Elemental Analysis 721
Bibliography 722
Process Analytical Chemistry 722
Process Spectroscopy 722
Process Data Analysis 723
References 723

Chapter 8 **Modern Analytical Method Development and Validation** **731**

8.1. Status of Existing Methods for Polymer/Additive Analysis 732
8.2. In-polymer Additive Analysis: Method Development and Optimisation . . 732
8.3. Certified Reference Materials 736
8.4. Analytical Method Validation Approaches 746
8.4.1. Analytical Performance Parameters 751
8.4.2. Interlaboratory Collaborative Studies 755
8.4.3. Validation of Antioxidant Migration Testing 757
8.5. Total Validation Process 757
8.5.1. Software/Hardware Validation/Qualification 758
8.5.2. System Suitability 760
8.6. Rational Step-by-step Method Development and Validation for Polymer/Additive Analysis 760
Bibliography 762
Method Development and Validation 762
Reference Materials 762
References 762

Appendix: List of Symbols **767**

Acronyms of Techniques 767
Chemical Nomenclature 778
Polymers and Products 778
Additives/Chemicals 780
Physical and Mathematical Symbols 785
Physical and Mathematical Greek Symbols 789
General Abbreviations 790

Subject Index **793**

Preface

Modern polymer/additive deformulation is essentially carried out according to three different approaches, in increasing order of sophistication, namely analysis of analytes separated from the polymer (typically an extract), of analytes and polymer in solution, or directly in-polymer (solid state or melt). The current status of conventional, indirect, methods of deformulation of polymer/additive extracts and dissolutions has recently been described in a comprehensive fashion. However, there is an impelling need to tackle polymer/additive deformulations strategically in an ever-increasing order of sophistication in analytical ingenuity, from indirect to direct analysis procedures, from macro to micro, from slow to rapid, from close to remote, from lab to process. Established wet chemical routes for low-molecular-weight additives are frequently no option for analytical problems of considerable complexity (high-molecular-weight additives, grafting, incorporation in the polymer backbone, reactive systems, *etc.*) or in case of surface analysis, microanalysis and spatially resolved analysis. Profiling, process analysis, product safety, quality assurance and industrial troubleshooting all benefit from direct analysis modes.

In recent years, techniques for direct analysis of the non-polymer components have developed apace and it has become increasingly important for scientists, engineers and technicians to have a basic grounding in these methods. This treatise is concerned with the *in situ* characterisation of additives embedded in a broad variety of polymeric matrices and evaluates critically the extensive problem-solving experience and state-of-the-art in the polymer industry. Despite well-deserved attention and considerable efforts direct polymer/additive analysis (without separation) has not yet turned into a great many general and routinely workable concepts. Nevertheless, the future foresees a greater share for in-polymer analysis.

This book, containing an outline of the principles and characteristics of relevant instrumental techniques (without unnecessary detail), provides an in-depth overview of various aspects of direct additive analysis by focusing on a wide array of applications in R&D, production, quality control and technical service. The book describes the fundamental characteristics of the arsenal of techniques utilised industrially in direct relation to application in real-life polymer/additive analysis. Instrumental methods are categorised according to general deformulation principles with emphasis on promoting understanding and on effective problem solving. The chapters are replete with selected and more common applications illustrating why particular additives are analysed by a specific method. The value of the book stays in the applications.

In *Plastics Additives: Advanced Industrial Analysis* the author has attempted to bring together many recent developments in the field in order to provide the reader with valuable insight into current trends and thinking. For each individual technique more excellent textbooks are available, properly referenced, albeit with less focus on the analysis of additives in polymers.

As an alternative to wet chemical routes of analysis, this monograph deals mainly with the direct deformulation of solid polymer/additive compounds. In Chapter 1 in-polymer spectroscopic analysis of additives by means of UV/VIS, FTIR, near-IR, Raman, fluorescence spectroscopy, high-resolution solid-state NMR, ESR, Mössbauer and dielectric resonance spectroscopy is considered with a wide coverage of experimental data. Chapter 2 deals mainly with thermal extraction (as opposed to solvent extraction) of additives and volatiles from polymeric material by means of (hyphenated) thermal analysis, pyrolysis and thermal desorption techniques. Use and applications of various laser-based techniques (ablation, spectroscopy, desorption/ionisation and pyrolysis) to polymer/additive analysis are described in Chapter 3 and are critically evaluated. Chapter 4 gives particular emphasis to the determination of additives on polymeric surfaces. The classical methods of surface analysis (electron spectroscopy, surface mass spectrometry and ion scattering techniques) are applied to practical cases. A variety of options for (surface) microanalysis and spatially resolved analysis by means of microscopy, microspectroscopy, spectromicroscopy, and imaging techniques, as applied to polymer/additive materials, are discussed in Chapter 5. Quantitative analysis (Chapter 6) in an essential part of polymer/additive analysis, in particular in the industrial environment. For quantitation, the separation procedure can be the most important factor for success or failure of the analysis. While this analytical task is recognised to be considerably more difficult than the qualitative analysis of previous chapters, recent round-robins indicate the need for critical self-inspection of the polymer analytical community. In Chapter 7 the various tools for in-process analysis (UV/VIS, mid-IR, near-IR, Raman and low-resolution NMR) are applied to polymer melts. The current status of polymer/additive analytical methodology is described in Chapter 8 and optimisation procedures are outlined. The lack of certified reference materials hampers analytical method validation. A rational step-by-step method development and validation approach to polymer/additive analysis is described.

Each chapter of this monograph is essentially self-contained. The reader may consult any sub-chapter individually. To facilitate rapid scanning the text has been provided with **eye-catchers**. Each chapter concludes with up-to-date references to the primary literature (no patent literature) and a critical list of recommended general reading (books, reviews) for greater insight. The majority of references in the text are from recent publications (1980–2003 and beyond). The book ends with a glossary of symbols and an index compiled with respect to both instrumental methods and analytes. Although every effort has been made to keep the book up-to-date with the latest methodological developments this report represents only work in evolution and contains suggestions for future improvements. In J.R. Thorbecke's words *"De tijd om alles te zeggen is nog niet gekomen"*, or *"Time is not yet ripe to tell everything"*.

Geleen, December 2004

About the Author

Jan C. J. Bart (PhD Structural Chemistry, University of Amsterdam) is a senior scientist with a wide interest in materials characterisation, heterogeneous catalysis and product development who has gained broad industrial experience (Monsanto, Montedison, DSM) in various countries. The contents of this book derive from the author's experience as a previous Head of an Analytical Research Department concerned with polyolefins and engineering plastics at a major plastics producer and are also based on an extensive evaluation of the literature. Dr. Bart has held several teaching assignments (Universities of Amsterdam, Sassari and Pavia), researched extensively in both academic and industrial areas, and authored over 250 scientific papers and chapters in books; he is also author of the related monograph on *Additives in Polymers. Industrial Analysis and Applications*, John Wiley & Sons, Chichester (2005). Dr. Bart has acted as Ramsay Memorial Fellow at the Universities of Leeds (Colour Chemistry) and Oxford (Material Science), visiting scientist at the Institut de Recherches sur la Catalyse (CNRS, Villeurbanne), and Meyerhoff Visiting Professor at the Weizmann Institute of Science (Rehovoth, Israel), and held an Invited Professorship at the University of Science and Technology of China (Hefei, PRC). He is currently a Full Professor of Industrial Chemistry at the University of Messina (Italy).

He is also a member of the Royal Dutch Chemical Society, Royal Society of Chemistry, Society of Plastics Engineers, the Institute of Materials and Associazione Italiana delle Macromolecole.

Acknowledgements

This monograph describes the current state-of-the-art in direct polymer/additive analysis. The high degree of creativity and ingenuity within the international scientific community is both amazing and inspiring. The size of the book shows the high overall productivity in academia and in industry. Yet, only a fraction of the pertinent literature was cited.

The author wishes to thank in particular DSM for actively stimulating the work, for granting permission for publication and financial support. The author thanks colleagues (at DSM Research) and former colleagues (now at SABIC Europe) for reviewing various chapters of the book. Information Services at DSM Research have been crucial in providing much needed access to literature. Each chapter saw many revised versions. Without the expert help and endurance of Mrs. Coba Hendriks, who produced many word-processed issues with endless patience, it would not have been possible to complete this work successfully. The author has not failed to disturb relatives and friends during the many years of preparation of this text, notably in Bucharest and Messina. Without their understanding and hospitality this book would never have been finished.

The author expresses his gratitude to peer reviewers of this project for recommendation to the publisher and thanks editor and members of staff at IOS Press for their professional assistance and guidance from manuscript to printed volume. The kind permission granted by journal publishers, book editors and equipment producers to use illustrations and tables from other sources is gratefully acknowledged. The exact references are given in the figure and table captions. Every effort has been made to contact copyright holders of any material reproduced within the text and the author apologises if any have been overlooked.

Jan C. J. Bart

Geleen, December 2004

Chapter 1

Shining light on obscure matters

In-Polymer Spectroscopic Analysis of Additives

1.1. Direct Ultraviolet/Visible Spectrophotometry . . . 4
1.1.1. Vapour-phase Ultraviolet Absorption Spectrometry . . . 10
1.2. Solid-state Vibrational Spectroscopies . . . 11
1.2.1. Mid-infrared Spectroscopic Analysis . . . 14
1.2.2. Near-infrared Spectroscopy . . . 34
1.2.3. Raman Spectroscopic Techniques . . . 52
1.3. Photoacoustic Spectroscopy . . . 66
1.4. Emission Spectroscopy . . . 72
1.4.1. Infrared Emission Spectroscopy . . . 72
1.4.2. Molecular Fluorescence Spectroscopy . . . 75
1.4.3. Phosphorescence Spectroscopy . . . 81
1.4.4. Chemiluminescence . . . 82
1.5. Nuclear Spectroscopies . . . 94
1.5.1. Solid-state NMR Spectroscopy . . . 95
1.5.2. Nuclear Quadrupole Resonance . . . 110
1.5.3. Electron Spin Resonance Spectroscopy . . . 112
1.5.4. Mössbauer Spectroscopy . . . 120
1.6. Dielectric Loss Spectroscopy . . . 123
1.7. Ultrasonic Spectroscopy . . . 127
Bibliography . . . 129
General Spectroscopy . . . 129
Direct UV/VIS Spectrophotometry . . . 129
Infrared Spectroscopy . . . 129
Near-infrared Spectroscopy . . . 130
Raman Spectroscopy . . . 130
Photoacoustics . . . 130
Emission Spectroscopy . . . 131
NMR Spectroscopy . . . 131
Electron Spin Resonance Spectroscopy . . . 131
Dielectric Spectroscopy . . . 131
Polymer Characterisation . . . 131
References . . . 132

As industrial problem solving requires avoidance of labour intensive procedures *in situ* analytical techniques come to focus (as opposed to methods based on extraction and dissolution), both in a production environment and in a research laboratory. Not only, some classical sample preparation techniques, such as dissolving a sample or forming a melt film in a heated press, may involve volatilisation and degradation of the additives. Other reasons prompting to explore new analytical grounds are the fact that extraction procedures are in principle not the best option in quantitative analysis. Moreover, a wide variety of materials comprising cross-linked polymers, insoluble elastomers, semi-crystalline materials, as well as high-MW or grafted additives are difficult to extract. Traditional sample preparation procedures (Chp. 3 of ref. [1]) often fail in these cases. However, some alternatives were already indicated. As mentioned in ref. [1], additive analysis may be carried out via the examination of extracts or dissolutions of the polymer, (semi) destructive testing by thermal methods, pyrolysis or laser desorption, mainly by examination of volatiles released or non-destructive testing, *i.e.* direct spectroscopic examination of the polymer in the solid or melt. Spectroscopic approaches to the analysis of extracts or chromatographic fractions were discussed already by Bart [1].

Polymers and plastics come in a wide variety of textures. Bulk materials are supplied as chips/granules or powder; fabricated material is sold in sheet, film or fibre form, while speciality products are available as latex, dispersion or emulsion form. Each of these requires particular consideration in sampling technique and approach for off-line analysis, particularly when maintaining sample physical property integrity may be all important. The traditional methods for additive analysis are destructive. Although this may frequently be acceptable, this is not always the case. For example, forensic material, historic and archaeological textiles should best be approached in a non-destructive fashion. Small amounts of sample should not be consumed at the first attempt of analysis. Also, the process of stripping the dye from the fibre destroys the dye-fibre complex, leading to the loss of potentially useful information concerning the distribution of dye(s) within the fibres and thus the dyeing process itself. Consequently, there is considerable scope for the development and use of **alternative non-destructive methods.** Direct methods for polymer/additive analysis are considered to be those in which there is no need to separate the polymer from the additive part for the purpose of analysis. Various factors severely restrict the choice of analytical methods that can be applied to a given polymer compound "as received" without prior separation of the additive from the macromolecular matrix. A selection of practical considerations is:

- Embedding of the additives in a more or less insoluble matrix.
- Low concentration of the additive in the matrix.
- Difference in structure between additive and matrix fragments.
- Fragmentation or thermal stability of the additive.
- Reactions between additive and matrix fragments.

Table 1.1. Main characteristics of *in situ* spectroscopic techniques

Advantages:
- Fast sample analysis turnaround time
- Exclusion of a cost-intensive separation step
- No solvents; safety
- Various sampling modes
- Potentially reliable quantitation of known analytes
- Applicable to intractable solids, artwork, forensic science objects

Disadvantages:
- Interferences (from co-additives and polymeric matrix)
- Lack of specificity
- Poor detection limits
- Limited usefulness
- Restrictive identification of unknown analytes
- Difficult quantitation of multicomponent systems

Considerable progress has been made toward the realisation of direct compound analysis by various forms of spectroscopy. It should not be forgotten, however, that sample preparation in conventional spectroscopy is an important factor, often close to an art.

Spectroscopy of solids is defined as the qualitative or quantitative measurement of the interaction of electromagnetic radiation (emr) with matter in the solid state. The emr interacts as scattering, absorption, emission, fluorescence or diffraction. A variety of spectrometer configurations is used to optimise the measurements of electromagnetic radiation interacting with solid matter in different sampling modes. In this case, scattering is often a requirement for analysis rather than a problem. It is fundamental to diffuse reflectance, a common sample interfacing method used for dedicated applications.

The **main characteristics** of *in situ* spectroscopic methods are given in Table 1.1. Each spectroscopic technique has its own strengths and weaknesses, which determine its utility for studying additives directly in the polymeric matrix. The applicability depends on the identity of the particular additive and polymer matrix, on concentration and amount of sample available, analysis time desired, and need for quantitation. Polymers for which no solvent can be found present analytical difficulties, especially if appreciable amounts of fillers or additives are present. In favourable cases, rapid additive analyses can be carried out without extensive pretreatment steps, *i.e.* without extraction by UV spectrometry [1a], NMR [2] or UV desorption/mass spectrometry [3], but generally these methods suffer from disadvantages due to non-specificity of the tests used. The main disadvantage of direct spectroscopic methods is interference between the variety of groups present and hence lack of specificity. In the direct examination of polymer films by UV or IR, or of the thicker sections of polymer by ATR, the additive is heavily diluted by the matrix. Consequently, detection limits are usually well above the low concentration of additive present (minimum level typically 500 ppm for additives in polyolefins), and the method is only applicable if the additive exhibits strong absorption bands in regions where the polymer shows little or no absorption. The polymer should exhibit a relatively flat absorption curve in the wavelength range used for the quantitative determination of additives. Direct spectroscopic techniques have limited usefulness and generally allow only the quantification of known additives in the polymer batch but not readily the analysis of unknown analytes. It is also generally difficult to obtain both qualitative and quantitative results from a single type of spectroscopy. On the other hand, for well-defined systems (*i.e.* containing a set of known additives in varying concentration) *in situ* spectroscopic techniques are quite useful. In fact, these methods are used mainly for quality control and certification analysis where rapid and cheap methods are available. Direct spectroscopy of polymer films may be very useful for the study of solvent-extraction procedures or stabiliser-ageing processes during simulated processing or end-use conditions. Methods requiring little or no sample preparation are NIRS and laser-Raman spectroscopy.

Despite the fact that direct analysis methods exclude a cost-intensive separation step overall analysis cost may still be high, namely by the need for more sophisticated instrumentation (allowing for a physical rather than chemical separation of components) or extensive application of chemometric techniques. The wide variety of additives that are commercially available and employed complicate spectroscopic data analysis. For multicomponent analysis some kind of **physical separation** of additive signals is often quite helpful, *e.g.* based on mobility (as in LR-NMR or NMRI), diffusion coefficient (as in DOSY NMR), thermal behaviour (as in a thermal analysis and pyrolysis techniques) or mass (as in tandem mass spectrometry). The power of signal processing techniques (such as multi-wavelength techniques, derivative spectrophotometry) is also used to the fullest extent.

Direct UV spectrophotometry is mainly used in favourable cases, namely for the determination of one UV absorber in the absence of other interferences. The technique also finds application in the verification of extraction yields and in migration studies. **Vibrational spectroscopy** holds a prominent place in the routine analysis of additives in polymers. There are three main categories of vibrational spectroscopy that provide useful structural information in the analysis of organic and inorganic molecules: mid-infrared, Raman, and near-infrared spectroscopies. Pre-eminent among these techniques is mid-IR spectroscopy. The advent of the laser has reactivated Raman spectroscopy but the ubiquitous fluorescence of real-life industrial polymers limits application. Vibrational spectroscopy is not an exact technique: rarely, if ever, can the analyst clearly and unambiguously identify a compound using vibrational techniques alone. Nevertheless, information is often obtained not forthcoming from any other analytical technique.

Whereas **NMR spectroscopy** in solution is a highly developed technique for absolute determination of microstructure, solid-state NMR was highly limited until the development of magic-angle spinning, high power decoupling and cross-polarisation. These developments have opened up an entirely new area of structural characterisation as the samples can be examined in their native state. Cross-linked systems and the mechanisms of network formation can be unravelled by s-NMR. However, there are only relatively few *in situ* studies of NMR spectroscopy of polymer/additive formulations due to its low sensitivity. Thus, NMR spectroscopy is used as a standard *ex situ* method for the analysis of reaction products. **ESR spectroscopy** is useful for characterising paramagnetic species both in solution and in the solid state. If the spectra are complicated (hyperfine splitting), or if a mixture of species is produced, higher concentrations or longer lifetimes are required. Due to the fact that most elements have an isotope with finite nuclear spin, the applicability of NMR is much broader than that of ESR spectroscopy. Similarly, **NQR and Mössbauer spectroscopy** show an even more limited applicability. **Chemiluminescence** has recently yielded surprising results in relation to stabilised polymers.

Despite the fact that many spectroscopic techniques are considered mature, many important improvements have gradually been introduced, *e.g.* rapid-scanning Fourier transform infrared (FTIR) spectroscopy, Fourier transform Raman spectroscopy, the more efficient exploitation of the near-infrared region, increased sensitivity leading to breakthrough sampling techniques (*e.g.* PAS, DRIFTS), improved time resolution (allowing for on-line combination with other techniques such as GC, HPLC or thermal analysis) and characterisation of time-dependent phenomena, multivariate data evaluation, optical fibre technology (opening up completely new areas for process control, remote sensing and field-portable instruments), laser and molecular beams, multiphoton spectroscopy, microspectroscopy, miniaturisation, imaging, *etc.* Multiphoton spectroscopy involves excitation of an atom or molecule from one electronic state to another by absorption of two or more photons in contrast to more conventional spectroscopies that involve just a single photon. Lack of intensity is one of the major limitations in many spectroscopic investigations. Consequently, much impetus to the whole field of spectroscopy was given by the introduction of lasers (*cfr.* Chp. 3). Lasers are able to overcome some basic limitations of classical spectroscopy. Recent advances in laser and optical detection instrumentation have allowed the development of **major new spectroscopic techniques**, such as UV resonance Raman spectroscopy [4] and NIR FT-Raman spectroscopy [5]. Time resolution down to the fs range is now possible. Miniature fibre optic spectrometers configured for UV/VIS or NIR applications are now available and measurements can be made in transmission, reflection or absorbance mode. Advances in optical spectroscopy are needed to evaluate the interface between the matrix and the fibre, plate, or particulate filler in composite materials and to improve non-destructive testing and process monitoring [6]. As instruments are increasingly miniaturised, sample sizes will continue to shrink and sample preparation and handling techniques will need to improve.

This Chapter deals with the non-destructive determination of additives in the solid polymeric matrix (bulk) by spectroscopic methods, however without any concern for surface distributions or microanalytical aspects, for which the reader is referred to Chapters 4 and 5. As the additives might be heterogeneously distributed in the polymer, measurements at various positions are recommended. Table 1.2 indicates the main electronic and vibrational spectroscopic techniques currently in use for direct polymer/additive analysis.

For textbooks on polymer analysis, *cfr.* Bibliography.

Table 1.2. Main *in situ* electronic and vibrational spectroscopies for polymer/additive analysis

Spectroscopic technique	Main application modes
Absorption	UV/VIS, FTIR, NIR
Reflectance	UV/VIS, FTIR, NIR
Emission	FTIR, FL, CL
Raman scattering	UV/VIS, NIR

1.1. DIRECT ULTRAVIOLET/VISIBLE SPECTROPHOTOMETRY

Principles and Characteristics

UV/VIS spectrophotometry may be used in the analysis of extracts (*cfr.* Section 5.1 of ref. [1]). One might also wish to measure solid samples for identification and quantitation of the components present. Direct UV/VIS spectrophotometry of a polymeric material without previous extraction or dissolution of the matrix is one of the fastest means for additive analysis. Modern UV spectrophotometers are suitable to investigate efficiently the transmission and/or reflection of polymers either as powders, plates or film. In principle, UV spectrophotometry is an exact tool for the quantitative determination of additives in polymers (primarily stabilisers), **directly in-polymer**. Typical analysable sample quantities amount to about 0.1 to 0.2 mg. Such small samples permit stabiliser contents down to concentrations of 0.03% to be determined with an error of ±10% within 15 min [7]. UV detection can, however, be utilised only in polymer films with a sufficiently low absorbance. Ideally, a blank film sample of the polymer used to make the film is taken as the background. However, as an additive-free matrix is not always available, the blank measurement may be impaired.

Various factors can interfere with accurate and precise measurement of **transparent solid samples**, such as films, glasses or crystals. Direct analysis of additives in film by means of UV spectrophotometry is limited by excessive beam dispersion due to undesired light scattering from the polymer crystalline regions [8]. This crystallinity problem (as in PE) can be eliminated by measurements on molten polymers (*cfr.* Chp. 7.2.2). Additives at low concentrations (0.1%) require a sample thickness such that analysis must be performed in the presence of a high level of light scattering, which may change unpredictably with wavelength. At lower concentration levels and

Table 1.3. Main characteristics of direct UV spectrophotometry

Advantages:
- Routine techniques
- No sample preparation
- No solvents (extraction or dissolution)
- Simple, low cost (rapid QA/QC)
- Fast analysis times (<2 min)
- Various measurement modes
- Safety
- Wide applicability

Disadvantages:
- Poor selectivity (mixtures!)
- Limited qualitative information (unknown additives)
- Interferences (polymer, co-additives, impurities)
- Beam dispersion
- Poor detection limits (matrix dilution effect)
- Not universal (low sensitivity for UV transparent additives)
- Questionable reliability for quantation in mixture analysis (chemometrics required)

correspondingly greater sample thickness, unacceptable signal-to-noise ratios exist. Nevertheless, UV spectrophotometry remains an attractive method for analysis for many additives with high extinction coefficient (aromatics); the principal advantage over IR spectroscopy is the greater sensitivity arising from higher extinction coefficients.

Table 1.3 summarises the **main features** of direct UV spectrophotometry. Direct examination of a polymer film by UV or FTIR spectroscopy or of thicker sections of polymer by ATR has severe limitations in that detection limits are poor because the additive is heavily diluted with polymer; the method is only applicable if the additive has distinct sharp absorption bands in regions where the polymer itself shows little or no absorption. Although UV spectrophotometry is very sensitive and permits direct examination of polymer films, in view of its rather broad absorbencies it may be difficult to resolve those that are the result of additives or degraded species. Direct UV spectroscopy is liable to be in error owing to interference by other highly absorbing impurities that may be present in the sample (*e.g.* fillers or pigments). Interference by such impurities in direct UV spectroscopy may be overcome by selective solvent extraction or by chromatography. The usefulness of in-polymer UV analysis for qualitative and quantitative characterisation is also restricted by the poor selectivity of the technique (many additives show rather similar absorbance bands). This limits the use of unambiguous UV analysis to special cases in which the additive package in a sample is known; the additive concentrations may be determined provided no unknown species are present. Direct UV spectrophotometry cannot easily be used to identify unknown additives and requires multivariate analysis to indicate the presence of more than one absorbing species.

Direct UV spectrophotometry for polymer/additive mixtures is thus mainly applied for UV transparent polymeric matrices (*e.g.* PE) and is not suitable for the detection of additives in polymers absorbing above about 250 nm (*e.g.* styrenics). Only polymer additives which are absorbers of UV radiation (such as light stabilisers and other compounds containing UV-active structural moieties) can be detected. For quantitative determinations the UV approach requires standards or measurement of extinction coefficients.

UV/VIS/NIR measurement modes from 190–2500 nm may be in total transmittance (for turbid liquids, films, emulsions, scattering solids), regular transmittance (liquids), diffuse reflectance (most solids, powders, films, coatings), or specular reflectance (high gloss coatings). While inexpensive UV/VIS spectrometers are typically used for measurement of transmittance of clear solutions, more sophisticated instrumentation with accessories such as absorption-reflection units and 60 or 150 mm integrating spheres has multiple uses, including the characterisation of solid materials in **reflectance mode**. Integrating spheres (wavelength range of 350–980 nm), which function as light collectors, coupled by fibre optics for remote reflectance measurements (FORS) from flat solid surfaces, are used for spectral and colour measurements. Whenever polymer films are slightly opaque and scatter, the use of an integrating sphere is required. Measurement of diffuse reflectance, total hemispherical reflectance, colour (both specular included and excluded), variable incident angle diffuse reflectance, diffuse transmittance and relative specular reflectance can be made using an integrating sphere accessory, suitable for visible and near-IR (*cfr.* Fig. 1.1). Multi-angle spectrophotometers with a wide range of angular viewings (15° to 110°) allow complete and accurate evaluation of the changes exhibited in metallic, pearlescent and special effect finishes.

A 150 mm integrating sphere is the primary instrument used in the characterisation of the reflectance properties of optical materials, *e.g.* for the measurement of UV transmittance of paint films in the automotive field. UV analysis of granulate in reflectance can be carried out by means of an integrating sphere equipped with a special sampling device. This method, which does not require any sample preparation step, is ideal in a plant service environment. More reporting of results is desired here. The reflectance results reveal that at weak to medium absorption bands the bulk of the polymer considerably contributes to the spectrum. Only at strong bands the information is limited to the surface (*viz.*, less than 10 μm). The stronger the absorption bands, the greater the influence of inhomogeneity will be on the accuracy of the prediction.

Table 1.4 opposes double-beam UV/VIS spectrophotometry in solution to the use of a reflectance sphere (standard for industry). Recently also hand-held reflection spectrometers have been introduced.

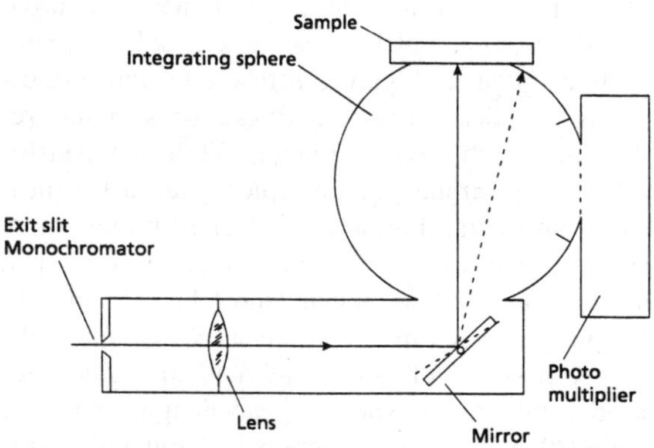

Fig. 1.1. Integrating sphere attachment. After Burgess [9]. Reprinted with permission from *Spectroscopy in Process and Quality Control (SPQ)*, 1998. *Proceedings SPQ-98* is a copyrighted publication of Advanstar Communications Inc. All rights reserved.

In a technique called **solid-phase spectrophotometry** absorption of a colour complex of the analyte sorbed on a solid support is measured without subsequent stripping of the chromogenic species. Solid-phase spectrophotometry offers the advantage of *in situ* preconcentration of the analyte. Therefore, it is (several) orders of magnitude more sensitive than the corresponding conventional spectrophotometric methods [11]. **UV microscopy** (*cfr.* Chp. 5.3.2) may find application in studies aiming at the study of the physical distribution of additives; **UV microspectroscopy** is discussed in Chp. 5.6.1. For UV/VIS reflectance, *cfr.* ref. [12].

Applications

As already noticed elsewhere [1], in-polymer UV analysis of a polyolefin matrix is often a first step in the **deformulation procedure**. In accordance with Scheme 2.12 of ref. [1], UV spectrophotometry also comes into play *after extraction* of a polymer/additive matrix, when the residue is pressed into a thin film to verify removal of all extractables with a chromophoric moiety.

Direct UV/VIS analysis of plastics may be performed on transparent films or compression moulded plaques, with sample thicknesses usually between 50 and 500 μm depending on the absorbances of the analytes and the polymer. For purposes of reproducibility it is advised to press several thin films of various polymer granules. In-polymer UV analysis of a UV transparent polyolefin matrix allows detection of phenolic AOs (at 280 nm) or UVAs (at 330–340 nm) down to 25 ppm level.

An early report on the direct determination of **stabilisers** in pressed polymer films by UV spectrophotometry is due to Drushel *et al.* [13]. The determination of a variety of additives (Santonox R, Ionol, Ionox 330, CAO-5, CAO-6, DPPD, Polygard, Topanol) in the 0.002–1.0% range in ten mils thick

Table 1.4. Reflectance and conventional double-beam UV/VIS spectrophotometry[a]

Feature	Reflectance sphere	Double-beam solution
Measurement time	$t \sim 1$ s	$t \sim 20$–60 s
Resolution	5–10 nm	1 nm
Nature of sample	Opaque, translucent and transparent	Transparent only
Sensitivity limit	0.1% of transmitted light	Highly accurate and linear up to 0.0001% of transmitted light
Measurement of standard and trial	Separate readings	Simultaneous measurement (split beam)

[a] After Shakhnovich and Barren [10]. Reproduced by permission of the Society of Plastics Engineers (SPE).

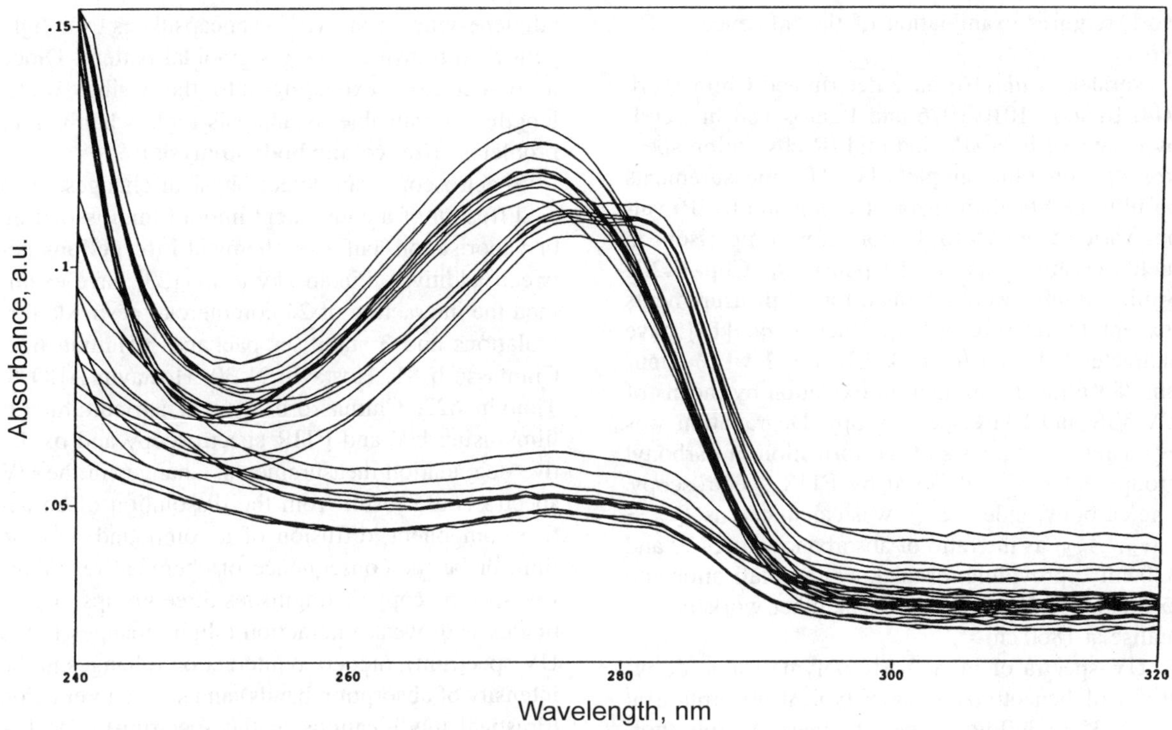

Fig. 1.2. UV absorption spectra (nm) of a calibration set of HDPE/(Irganox 1010/1076, Irgafos 168, oleamide) film samples with variable additive concentrations. After Bremmers and Swagten-Linssen [15]. Reproduced by permission of DSM Research, Geleen.

UV transparent PE film has also been reported long ago [14]. In 1965, Luongo [14] has substantiated that UV spectrophotometry is very sensitive and permits direct examination of hot-pressed polymer films but failed to identify unknown additives or indicate the presence of more than one antioxidant. However, joint use of UV/VIS spectrophotometry and multivariate calibration now facilitates such simultaneous determinations.

With the restrictions given, in-polymer UV spectrophotometry is a very efficient analytical method for the qualitative and above all, quantitative analysis of stabilisers and other substances, directly in solid polymers. In the absence of unknowns the additive concentrations may be determined, as shown by Sehan *et al.* [7] who reported direct analysis of phenolic stabilisers (0.03–0.3%) in very small quantities of solid polymers (<1 mg) and the stabiliser **distribution analysis** of Irganox 1076 in polyolefins by means of UV spectrophotometry using a 0.085 cm diameter pinhole. For this purpose 200 μm thick films were used, where the exact knowledge of the film thickness is a prerequisite for accurate determination of the phenolic stabiliser content (±10%). The analytical method, which requires a calibration curve based on films of known composition, takes only 15 min in contrast to at least 5 h following the extraction route.

Multivariate calibration is particularly useful in case of complex additive packages with overlapping peaks of low intensity. Bremmers *et al.* [15] have examined Irganox 1076 and oleamide in HDPE film (containing Irganox 1010/1076, Irgafos 168 *cq.* 168 phosphate and oleamide) using direct **quantitative determination** by means of both UV (Fig. 1.2) and IR methods and 21 calibration samples with low (200–300 ppm) and high (1200–1500 ppm) Irganox 1076/oleamide concentrations; HPLC was used as a reference method. Film thickness was measured by means of a radioactive source and factor analysis allowed for thickness corrections. Reported standard deviations for Irganox 1076 were *ca.* 25 and 80 ppm in the low (200–300 ppm) and high (1200–1500 ppm) concentration ranges, respectively; oleamide could not be determined by means of UV spectroscopy. Further extension of this work requires examination of the influence of PE type.

Verlaek *et al.* [16] have determined Chimassorb 944, Irganox 1010/1076 and Irgafos 168 in LDPE by means of both UV and mid-IR absorption spectroscopy on film samples. For UV measurements on film the Standard Error of Prediction (SEP) values varied from 15 to 45 ppm (for comparison, in melt measurements *ca.* 10 ppm), *cfr.* Chp. 7.2.2. Similar figures were obtained for IR measurements (except for Irganox 1010). Kaci *et al.* [17] have characterised 80 μm thick LDPE/0.2 wt.% Tinuvin 783 films during thermo-oxidation by means of UV/VIS and FTIR spectroscopy. Degradation was evaluated on the basis of the formation of carbonyl groups detected and dosed by FTIR spectroscopy. The carbonyl index ($I_{c=o}$) was calculated as $I_{c=o} = A_{1720}/A_{720}$ as the ratio of absorbances at 1720 and 720 cm^{-1} providing for thickness normalisation and exclusion of degradation effects; other workers normalise at 1860 cm^{-1}.

UV spectra of monomeric polymerisable derivatives of benzotriazole show typical absorption at λ_{max} 335 to 340 nm. When the benzotriazole moiety is incorporated into polymers, bathochromic and hypsochromic shifts are observed, depending on the polymer and concentration of the stabiliser moiety [18,19]. Using UV spectroscopy and SEC, Pasch *et al.* [19,20] have characterised styrene and methylmethacrylate copolymers containing different UV stabiliser units (benzophenone, phenylbenzotriazole and naphthylbenzotriazole) fixed to the **polymer backbone**. UV spectroscopy is suitable for the determination of the copolymer composition. The chemical heterogeneity of the polymers was evaluated by means of simultaneous refractive index (RI) and UV detection at 313 nm. At 313 nm only the UV stabiliser units absorb, *i.e.*, are "visible" in the detector, whereas the styrene and methylmethacrylate units are transparent at this wavelength. Therefore, the UV profile shows the distribution of the UV stabiliser units along the molar mass axis. Using SEC-RI/UV, it was concluded that the UV stabiliser units are statistically distributed along the different molar mass fractions.

Thai *et al.* [21] have recently studied the dynamics of **vaporisation** and consumption of 2,6-di-*t*-butyl-4-phenylphenol (I) during oxidation of PE containing both (I) and dilauryl thiodipropionate (II) by UV/VIS spectrophotometry, as a function of oxidation time and concentration of (II). Pern [22] noticed that the **loss rate** of the UV absorber Cyasorb UV531 and the progress of discoloration of ethylene-vinylacetate (EVA) encapsulates from light yellow to brown follow a sigmoidal pattern. Diode array detectors extending into the visible wavelengths are valuable as analysis tools when colour problems arise (**colour body analysis**).

Spectroscopy can detect physical changes, such as diffusion of a component into a film, dissolution or vaporisation, but also chemical **interactions** between additives. Pukánszky *et al.* [23] have examined the interaction of 24 commercial pesticide formulations and 3 stabiliser packages (Tinuvin 622, Chimassorb 81; Hostavin N 30, Hostavin ARO 8; Tinuvin 622, Chimassorb 944) in agricultural PE films using UV and FTIR spectroscopy and oxidative degradation measurements. Changes in the UV spectra could result from the dissolution of an active component, diffusion of a compound into the film, or be the consequence of chemical reactions. UV spectroscopy distinguishes three groups of pesticides with weak interaction (slight changes in the UV spectrum), moderate interaction (changes in the intensity of absorption bands) and strong interaction (drastical modifications in the spectrum). The latter occurs mainly for sulfur and organic halogenide compounds, such as Hostavin ARO 8 – Hostavin N 30 and Actellic 50EC.

Albarino [24] has demonstrated the feasibility of quantitative UV analysis of 0.01–0.1 wt.% of Irganox 1010 in **molten polyethylene** when the crystallites, which account for much of the scattering, are eliminated. Greater sample thickness (0.78 cm) and analytical sensitivity are possible compared to analysis of solid samples at room temperature. In-process monitoring using UV/VIS spectrophotometry makes ample use of polymer melts (*cfr.* Table 7.15).

The permanence of UV absorbers in a rubber-modified acrylic film was evaluated by UV/VIS spectrophotometry; the effects of reflectance, light scattering, and matrix absorbance were deconvoluted from the total apparent absorbance, resulting in an absorbance spectrum due to the UVA alone [25]. UV reflectance spectroscopy has been used for the semi-quantitative determination of a benzotriazole and oxanilide UV absorber in two-coat metallics [26]. Berner *et al.* [26] have noticed clearcoat/basecoat migration of UVA in **automotive coatings**. The optical pathway in such systems is not dissimilar from that found in photographic colour prints where a thin coloured gelatine transparency is overlaying a diffusively reflecting support [27].

Gerlock *et al.* [28] have examined curing of various benzotriazole and oxanilide UVA coatings deposited on quartz slides with adjusted film thickness. Carter *et al.* [29] have addressed the evaluation of automotive clearcoats using UV microspectroscopy and other tools (*cfr.* Chp. 5.6.1). Migration of UVAs (Cyagard UV1164 and Tinuvin 384) and HALS (*e.g.* Sanduvor S 3058) in acrylic/melamine clearcoats during cure was studied by microtoming and UV and (subtractive) FTIR additive analysis of thin sections [30]. Compared to UV analysis, IR measurements are more complicated and time consuming. The strong IR bands of the matrix mask the much weaker additive bands. Infrared analysis was mostly used for investigating the distribution profiles of HALS compounds, which in general do not absorb in the UV region of interest.

Micro-UV spectroscopy is a useful tool to determine the distribution of UV light absorbers in paint systems [31]. Figure 1.3 shows micro-transmission UV spectroscopy results for an acrylic/melamine clearcoat containing a benzotriazole UVA [32]. The observed weak gradient in the UV intensity in the non-weathered test specimen suggests UVA volatilisation during cure; upon weathering UVA is depleted from the clearcoat. Spectra were recorded with 5 × 10 μm spot size in 5 μm steps. Photodegradation of currently available benzophenone and benzotriazole type UV screeners, such as Cyasorb UV531/5411/1164, Uvinul N-539 and Sanduvor VSU, occurs at such a rate that most of the screener will be depleted from the surface layers of a coating or in the bulk of a polar polymer after only 3–5 years of direct sun exposure [33].

UV chambers play an important role in comparing and predicting the performance of construction materials (elastomers, plastics, polymeric composites and coatings) and determining the effect of different weathering factors on the performance of a construction material. Recently, an innovative integrating sphere UV chamber design has been proposed for enhanced repeatability and reproducibility of the exposure results [34].

Figure 1.4 shows a calibration curve for UV analysis in reflection (using an integrating sphere) of films of Irganox B blends, based on absorption in the 250–290 nm region in order to account for the total amount of (degraded) Irganox 1076, Irgafos 168 and Irgafos 168 phosphate [35].

Direct UV/VIS spectrophotometry is used in the textile industry for measuring colours, in the paper

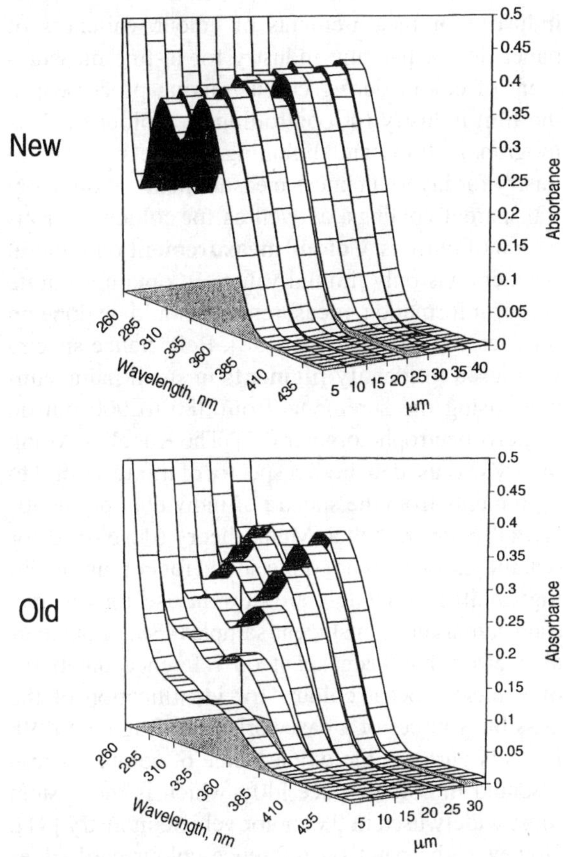

Fig. 1.3. Micro-UV spectra of a benzotriazole UVA containing acrylic/melamine clearcoat before exposure (upper) and after 4 years of Florida weathering. After Gerlock *et al.* [32]. Reprinted with permission from J.L. Gerlock *et al.*, *ACS Symposium Series* **805**, 212–249 (2002). Copyright (2002) American Chemical Society.

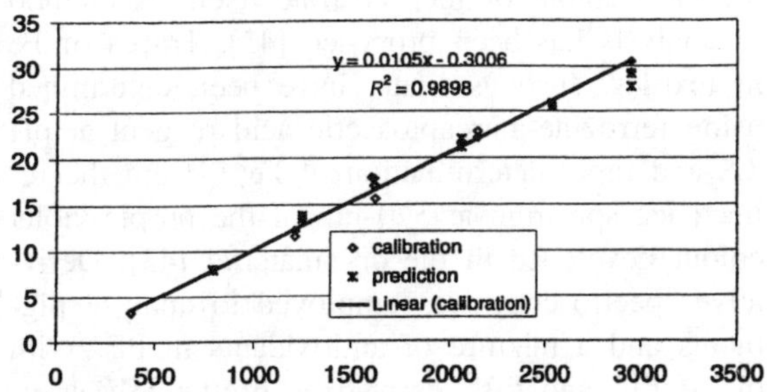

Fig. 1.4. UV integral in the 250–290 nm range in relative absorbance units for analysis of Irganox B 215/220/225/900/921 blends (concentration in ppm). After Knape and Wienke [35]. Reproduced by permission of DSM Research, Geleen.

industry for measurements of colour/whiteness of paper, in the painting industry for in-line measurement of colour during colour mixing processes, in the film industry for controlling the colour of thermographic films and lightning, in the coating industry for layer thickness measurements of optically transparent coatings, as well as for colour determination of plastics. **Colour measurement** on mineral powders has only limited value. To obtain accurate colour information, measurement should be done on product mixed in polymer [36]. Reflectance spectra were used to identify **pigments** in small paint samples, using measurements from 380 to 900 nm on a microspectrophotometer [37]. The Kubelka–Munk theory was used to predict spectra of mixtures of 2 to 3 pigments from the spectra of individual pigments. Actually, the Kubelka–Munk theory (developed for opaque samples) is not quite suited for colour matching; multi-flux models are now in use for opaque, transparent and translucent samples [38]. Considerable effort has been spent on reference databases of vehicle topcoat colours for identification of the possible sources of a casework paint fragment [39]. For this purpose, the whole range of colours is represented in colour space [40], which is the system most widely used in the motor vehicle industry [41]. However, the paint on reference colour cards does not necessarily contain the same pigment mix as that supplied for vehicle use. Also, ageing and weathering of the paint pigment resulting in fading or a colour change is not taken into account.

UV transmittance analysers are used for QA in the textile industry. Identification of **dyes** on textile fibres by assessment of reflectance curves is difficult owing to the dependence of spectral reflectance on concentration and spectral interference due to the base colour of the substrate itself. A method of analysis has been proposed [42]. Traces of Fe in textiles, such as linen, have been determined using ferrozine-mercaptoacetic acid reagent at pH 3.8 and rapid determination of Fe^{2+} from the reflectance spectrum at 570 nm of the purple-violet colour developed in the dry material [43]. Derivative spectroscopy was employed to analyse pigments and a mixture of antioxidants in PE. With the fourth and fifth derivatives of the UV spectra, Irganox 1010 could be determined in PS; the UV absorbance of this polymer makes evaluation of the original spectrum impossible [44]. A variety of chemical derivatisations in UV spectrophotometry have been described.

UV spectrophotometry has also been used to follow up **polymer impregnation** with additives in $scCO_2$ [45]. In another typical UV application the molar absorptivity may be determined, which is an inherent characteristic of a pure compound, as a measure of **purity**.

UV/VIS spectrophotometry is also being applied for *in situ* analysis of separated spots in TLC. The amount of substance required for an interpretable spectrum (typically 0.01–1.0 μg) depends on the chromatographic conditions and the absorption coefficient for the compound.

Fibre optics reflectance spectroscopy (FORS) is a powerful and non-destructive method for the analysis of works of art [46].

1.1.1. Vapour-phase Ultraviolet Absorption Spectrometry

Principles and Characteristics

Vapour-phase or **thermal ultraviolet** (TUV) absorption spectrometry has been proposed by Thompson *et al.* [47] as a rapid analytical technique for the characterisation of organic compounds. Practically, vapour-phase UV spectrometry is carried out by heating a very small amount of sample introduced in a graphite furnace commonly employed for flameless AAS. The vapours evolving from the graphite surface can absorb UV radiation. Vapour-phase profiles of absorbance *vs.* temperature or *vs.* wavelength are obtained according to whether the measurement is performed at fixed wavelength (typically 200 nm) or at constant temperature, respectively. The main **advantages** of the technique are: use of flameless atomic absorption spectrometers without any instrumental modification; rapid performance (1–3 min for each run); wide thermal range (from 150 to 2300°C); good repeatability; *in situ* absorption measurement of the vapours evolved from the graphite surface [48].

Applications

Tittarelli *et al.* [48] have identified some 20 organic and inorganic pigments (used for coloration of polymers, polymer films and coatings) directly without dissolution by means of vapour-phase UV absorption spectrometry. The influence of a polymeric matrix on TUV profiles was not specified. Each pigment has a characteristic thermal UV profile at a particular temperature. Alpha and beta forms of copper phthalocyanine (Pigment Blues 15 and 15.4) produce different TUV profiles. The heating rate plays the main role in the resolution of TUV profiles. The procedure is useful for obtaining qualitative information regarding the type of pigments in polymers. The method is of limited use only.

1.2. SOLID-STATE VIBRATIONAL SPECTROSCOPIES

Vibrational spectroscopies (mid-IR, near-IR, Raman) play an important role in polymer/additive analysis. Optical advances as well as spectacular advances in computing technology and data processing algorithms have greatly impacted vibrational spectroscopy over the past 25 years (*cfr.* Table 1.5). Rapid digital data acquisition is required for FTIR, FT-Raman or CCD-Raman spectroscopy. The raw data obtained from these instruments must always be manipulated before a recognisable spectrum can be displayed.

Although the three spectroscopic techniques are very different in several aspects, their basic physical origin is the same: absorption in mid-IR and NIR, and scattering in Raman, as a consequence of **molecular vibrations**. Due to the different excitation conditions, the relationships between the observed spectral intensities and the chemical nature of the vibrating molecules vary significantly.

Where scanning mid-IR and NIR spectroscopy operate with a polychromatic source from which the sample absorbs specific frequencies corresponding to molecular vibrational transitions, in Raman spectroscopy the sample is irradiated with monochromatic laser light whose frequency may vary from the VIS to the NIR region. Multicomponent analysis can be achieved for samples containing up to ten components through a variety of multivariate statistical algorithms. Table 1.6 compares the **main characteristics** of vibrational spectroscopies. As far as the quantitative evaluation of vibrational spectra is concerned, mid-IR and NIRS follow Beer's law whereas the Raman intensity is directly proportional to the concentration of the compound to be determined.

Near-IR complements mid-IR. Some polymer applications are better suited to the NIR region of the spectrum while other applications are more performing in the mid-IR region. Packaging materials, including laminates and other types of multilayered films, can be analysed intact by NIRS where the light penetrates all layers. Pellets or moulded product can be analysed "as is", without regard to sample thickness. Reinforced thermoplastics or composites are often non-homogeneous and require much averaging for a representative result. Overtone bands of non-hydrogenic bonds of inorganic compounds are very weak and do not absorb appreciably in this region. Therefore, NIR is useful for non-destructive determination of organic compounds in the presence of inorganic fillers, such as percent binder and degree of cure in composites.

Near-infrared reflectance measurements are non-destructive, require no direct contact with the sample analysed (often an important factor in maintaining hygienic processing conditions), and can provide real-time analytical information. They have become possible by combining two fairly recent

Table 1.5. Development of vibrational spectroscopies

Year	Method	Applicability
1968	NIRS	On-line analysis
1980	FTIR	Various sampling techniques (ATR, SR, R-A, DRIFTS)
1980	μRaman	Chemical analysis on 1 μm^2 sections
	Laser excitation	Fluorescence perturbations
1985	NIRS	Laboratory analyses
	FT-Raman	Use of Nd:YAG laser
1986	μFTIR	Layer analysis (20 to 10 μm)
		Analysis on 1000 μm^3 volumes
1990	PA-FTIR	Surface layers of opaque samples
1991	FT-NIR	Optimised quantitative analysis, remote control
1995	μFT-Raman	Chemical analysis on 10–100 μm^3 volumes
		Perturbations due to emission of IR photons by dark samples
1996	PA-FTIR (step scan)	Analysis of deep layers of opaque samples
2002	Raman/FTIR	Complementary information

Table 1.6. Main characteristics of vibrational spectroscopies

Feature	Raman	Mid-infrared	Near-infrared
Frequency range	4000–200 cm^{-1}	4000–200 cm^{-1}	12,820–4000 cm^{-1}
Vibrations	Fundamentals	Fundamentals	Overtone combinations
Excitation condition[a]	$\delta\alpha/\delta q \neq 0$	$\delta\mu/\delta q \neq 0$	$\delta\mu/\delta q \neq 0$ (anharmonicity; $m \ll M$)
Functionalities	Homonuclear	Polar	CH/OH/NH
Structural selectivity	High	High	Low
Intensity	$I_{Raman} \sim c \sim \upsilon_{exc}^4$	$A = \varepsilon \cdot c \cdot l$ Beer's law	$A = \varepsilon \cdot c \cdot l$ Beer's law
Sample preparation	No	Yes (except ATR)	No
Sample volume/thickness	Small (μL, μm)	Small (μL, μm)	Large (up to cm)
Probing	At-line/in-line	ATR	Transmission, transflection, diffuse-reflection
Fibre optics	>100 m	Limited	>100 m

[a] α, polarisability; μ, dipole moment; q, internuclear distance.

developments: *(i)* commercial availability of spectrometers of high precision and reproducibility; and *(ii)* application of sophisticated mathematical methods to extract useful information from complex spectra. The intensities of the absorption bands in NIR are some 10 to 100 times lower than in mid-IR. An advantage of NIR is the use of fast, cheap detectors in combination with quartz-glass optical fibres. In view of the better S/N ratio of NIR signals ($\gg$10,000), as compared to mid-IR absorptions, the use of chemometric techniques for qualitative identity control and quantitative multiple component analysis of complex mixture is favoured. The NIR user is "model" and "statistically" oriented whereas the mid-IR user is more concerned with functional groups.

Classical spectroscopy requires physical separation of the constituent of interest from the matrix, usually by dissolution in a solvent. When considering vibrational spectroscopic analysers, a major component will have numerous wavelengths at which it may be analysed. Minor components require the analyst to seek wavelengths at which they have major absorbances and, almost invariably, use multiple wavelength correlation techniques. In an ideal Beer's law calibration, the matrix is non-absorbing (and non-scattering) and does not interact with the analyte. This is rare in industrial practice. Usually, the matrix will be a major consideration in how analysis is to be performed. By applying **chemometric principles** to NIR spectra, the absorption band due to the constituent of interest can be "mathematically" separated from the absorption bands of the matrix, eliminating the need to physically separate the analyte from the matrix. NIRS has developed strongly over the last 25 years in conjunction with chemometrics. Chemometrics has made NIR analysis different from traditional spectroscopies and is useful not only for quantitative analysis, but also for qualitative information related to unexpected systematic patterns in the data. Although the practical applications of NIR spectroscopy in polymer industries are extensive, the understanding of the basis of analysis has fallen behind the applications. Use of 2D correlation [49] can bring useful information for understanding complicated NIR spectra [50]. Hindle [51] has traced the history of (near-) infrared technology.

Mid-IR absorption and Stokes Raman deal with the same vibrations but are subject to different selection rules (and consequently the spectra differ). IR and RS provide complementary images of molecular vibrations. Vibrations which modulate the molecular dipole moment are visible in the IR spectrum, while those which modulate the polarisability appear in the Raman spectrum. Compositions that do not absorb in the IR range generally give a Raman spectrum and strong IR absorbers will produce a weak spectrum by Raman. Examples of silent Raman vibrational modes are specific point groups (*e.g.* C_6, D_6, C_{6v}, C_{4h}, D_{2h}, D_{3h}, D_{6h}, *etc.*). Other vibrations may be forbidden in both spectra. **Raman spectroscopy** complements IR spectroscopy, particularly for the study of non-polar bonds and functional groups (*e.g.* C=C, C−S, S−S, metal–metal bonds).

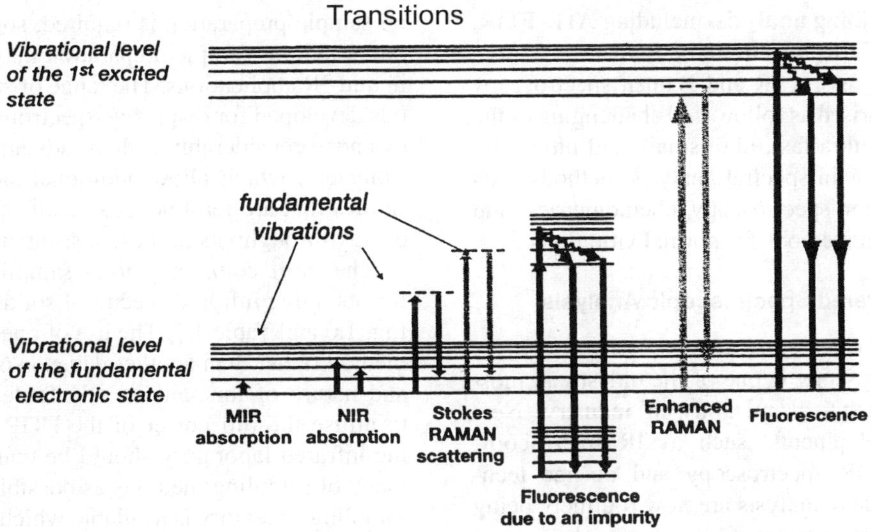

Fig. 1.5. Infrared absorption, Raman scattering and fluorescence. After Zanier [53]. Reprinted with permission from *Spectroscopy in Process and Quality Control (SPQ)*, 1998. *Proceedings SPQ-98* is a copyrighted publication of Advanstar Communications Inc. All rights reserved.

Raman is generally less sensitive than infrared, in particular for oxygenated functional groups, such as −OH, C−O and −COOH. However, the sensitivity of CCD based Raman spectrometers for strongly scattering materials is on a par with FTIR spectrometers for strong IR absorbers (ppm level). Inorganic species often give sharp Raman bands rather than broad features that can mask large regions of the IR spectrum. Raman spectroscopy also provides facile access to the low frequency region (below 400 cm^{-1} Raman shift), an area that is more difficult for IR spectroscopy. However, IR and Raman measurements in combination allow more precise identification of materials.

Raman provides easy sampling, whereas IR spectroscopy frequently needs some form of sample preparation. Materials which are difficult to handle in IR (highly viscous liquids, solids requiring pellets, mulls, or diffuse reflectance) are often easily measured by Raman. Unlike IR reflectance spectra, Raman spectra of solid samples are not affected by sample properties such as particle size. A significant difference with infrared absorption spectroscopy is that the Raman signal is *emitted* from the sample. Consequently, matrix effects are seldom as severe in RS as they are with mid-IR and NIR. Water may be used as a solvent with no loss in signal or resolution. Glass, even tinted, does not interfere with the Raman spectra.

Since the discovery of Raman scattering in 1928 the technique has greatly developed, including surface enhanced Raman spectroscopy (SERS), coherent anti-Stokes Raman spectroscopy (CARS), time-resolved Raman spectroscopy and microspectroscopy. With the development of stable diode lasers (NIR excitation), fibre-optic sample probes, compact optical designs, high quantum efficiency detectors, fast electronics and data elaboration, Raman spectroscopy is moving out of the shadow of IR spectroscopy. It is not expected though that Raman spectroscopy will ever replace FTIR as a simple, laboratory based technique which will most often yield a vibrational spectrum from the majority of samples at much lower cost [52]. However, when applicable, it may well enable measurements to be made which are impossible by other techniques! Areas in which Raman retains key advantages with respect to infrared are microspectrometry, where spectra can be obtained with roughly an order of magnitude better spatial resolution compared with μFTIR, and in remote sampling/*in situ*/on-line analysis.

The inelastic scattering Raman phenomenon is distinct from the relaxed emission denoted fluorescence (Fig. 1.5) because the inelastic scattering is a single event, and a real emitting excited state is never created.

Several techniques in vibrational spectroscopy are available to perform destructive or non-destructive **depth profiling analysis**, including ATR-FTIR, DRIFTS, PA-FTIR, μFTIR and μRaman.

Recent progress in IR and Raman spectroscopy may be summarised as follows: *(i)* challenging of the "ultra" world: ultra-fast, ultra-small, and ultra-thin; and *(ii)* progress in spectral analysis methods such as 2D correlation spectroscopy, chemometrics, and new calculation methods for normal vibrations.

1.2.1. Mid-infrared Spectroscopic Analysis

Principles and Characteristics

Infrared spectroscopy is one of the oldest and most established analytical methods in industry. New technical developments, such as IR microscopy, photoacoustic IR spectroscopy and on-line techniques for process analysis are now routinely being used in many laboratories. Furthermore, chemometric data evaluation, which is very frequently used in near-IR spectroscopy, is often advantageous also in the field of mid-IR spectroscopy and strengthens its outstanding position towards both basic and applied research.

Additive analysis of a polymeric material can be accelerated considerably by omitting the slow extraction or dissolution step. Infrared spectroscopy is suited to direct identification and quantitative determination of additives in polymers in whatever form: film, plates, microtome coupes, powders, flakes, pellets, fibres, rigid parts, *etc.* General principles and characteristics of IR spectroscopy have already been outlined in Section 5.2 of ref. [1]. Here we emphasise the peculiarities of IR spectroscopy as far as solids are concerned.

Infrared spectroscopy has the advantage of relatively simple sample preparation and non-destructive measurement; practically all types of samples (both as regards the state of aggregation and solubility) can be investigated with the aid of a variety of special measuring techniques. Unlike near-IR, where no sample preparation is required, sometimes some rather tedious sample preparation may be necessary in mid-IR applications. The range of sampling methods developed for dispersive spectrometers has been extended considerably with the advent of FTIR spectrometers, which allow additional **sampling techniques** that are feasible as a result of the increased energy throughputs of these instruments.

The most commonly used sampling techniques for obtaining infrared spectra of solids are shown in Fig. 1.6 and Table 1.7. The use of one spectroscopic method rather than another depends on the problem and nature of the sample, *cfr.* Table 1.8. In order to utilise the full power of the FTIR spectrometer, the infrared laboratory should be equipped with as many of sampling methods as possible. A universal sampling accessory is available which is a multipurpose sample compartment for transmission, diffuse reflectance, variable angle specular reflectance, and polarised grazing angle reflectance measurements. Sampling techniques that are inherently surface sensitive may not yield spectra that are characteristic of the sample bulk. As a result of their total thickness or their embossed surfaces samples may not be amenable to direct transmission or surface reflection FTIR.

Table 1.9 summarises the **main features** of *in situ* FTIR spectroscopy as applied to polymer/additive

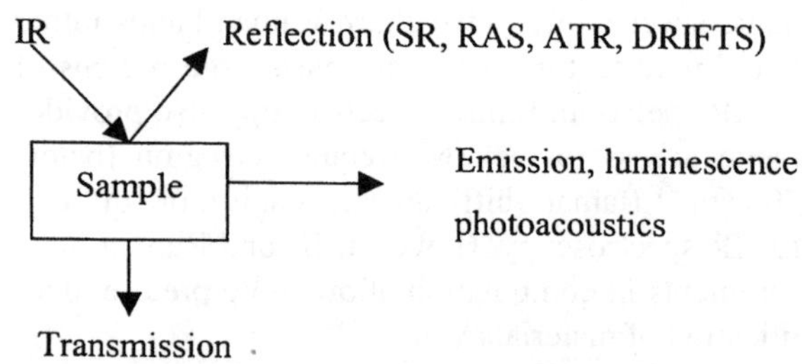

Fig. 1.6. Common methods of FTIR measurements of solids.

Table 1.7. ***In situ*** **infrared sampling methods**

Mode	Techniques	Chapter
Transmission	Ex-solution, cast film, melt, mulls, KBr discs	1.2.1.1, 7.2.3
Reflectance	IRS, ATR, R-A, DRIFTS, SR, abrasion	1.2.1.2–4, 7.2.3
Emission	PAS, FTIES	1.3, 1.4.1
Micro-FTIR	Micro KBr discs (1.5 mm), ATR[a]	5.6.2
NIRS	No sample preparation	1.2.2, 7.2.4
Pyrolysis	PyIR	2.2.4

[a] Golden Gate Single Reflection Diamond ATR.

Table 1.8. Applications of various FTIR accessories

Sample	Sampling mode	Comments
Transparent films and mouldings	Compression moulding	Affects thermal history
	Microtome films	No effect on thermal history
	SR, ATR	
Large mouldings	Abrasion, DRIFTS	Micro destructive
Polymeric powder, reactor fluff	DRIFTS (KBr), HATR	Very low scattered radiation intensity
Granules	DRIFTS (SiC), HATR	
Films on glossy substrate	SR, ATR	
Absorptive surface coatings	ATR	
Opaque and flexible samples	ATR	Ideal for rubbers and plasticised samples
Rigid plastic parts	DRIFTS (SiC)	
Opaque and rigid samples	PA-FTIR	Ideal for dark pigmented samples
Rough surfaces	DRIFTS	
Multilayer samples	ATR	Variation of angle of incidence
Liquid polymers	HATR	
Inclusions in film	μFTIR	Transmission analysis (limit: 10 μm, 1 ng)
Fibres	μFTIR	Transmission mode
	DRIFTS, ATR	(with diamond anvil)
Paint flakes	μFTIR	Reflectance mode
Polymer ash	KBr fused disc	Limit: 0.1 mg
Pigments and solid additives	KBr fused disc	Limit: 0.1 mg

samples. In many industrial analytical problems the samples available are not necessarily in the most suitable form for infrared analysis. Thanks to the differentiated accessory technology, *e.g.* the vertical and horizontal ATR (for powder, films and liquid polymers), diffuse reflection (for powder, granulates, rough surfaces and hard polymers), and regular reflection (for layer systems and layer thickness determination), the main components can be analysed easily and quickly – in a matter of seconds.

Polymer samples can be analysed in all possible textures and excellent spectra can be obtained. FTIR exhibits sensitivity to sample geometry and sample surface. As the additives are heterogeneously distributed in the polymer, measurements at various positions are recommended when necessary. The usefulness for exhaustive IR in-polymer analysis of additive packages containing unknown components (*i.e.* not contained in any reference library) is limited by the inherent characteristics of the method (essentially only functional group identification). Unique identification of unknown components may also be restricted by interference with co-additives and absorption of the polymeric matrix. Spectral subtraction of an appropriate reference polymer may be used to remove matrix interferences and allows tracing of minor components. However, this is not always possible as additive-free material is not always available. IR is limited mostly by the similarity and overlap of many additive absorption bands and by the level of sophistication required to interpret the fingerprint in detail. This presents a major opportunity for qualitative multivariate classification techniques, which can be used to recognise the many subtle details in a polymer formulation. Principle components/Mahalanobis distance Discriminant analysis (PMD) is such a technique designed to classify complex materials into groups or identify unknowns by using n principle components to map data characteristics into an n-space cluster [54].

Infrared spectroscopy has originally mainly been used as a qualitative tool, as opposed to UV spectrophotometry, but this situation is now slowly changing. Quantitation requires a calibration curve and/or multivariate analysis in case of mixtures. In view of the frequently low additive concentrations only the most intense bands (*e.g.* carbonyl bands) can be used for quantitation.

The National Physical Laboratory offers a service for calibrating the transmittance scale of mid-IR spectrophotometers [55]. Excellent wavelength accuracy is an important property of FTIR, making highly accurate spectral subtraction possible.

Many authors [56–61] have recently reviewed sampling techniques in IR spectroscopy. Numerous

Table 1.9. Main characteristics of *in situ* FTIR spectroscopy

Advantages:

- Easy to operate, rapid, reliable, versatile, low cost
- Relatively simple
- Non-destructive
- Fundamental vibration frequencies
- Qualitative and quantitative information
- Specific and characteristic absorption bands
- Excellent reference databases (verification, identification)
- Simultaneous detection of different components of a mixture in one scan
- Identification of polymer and additives (organic, inorganic)
- High absorption coefficients
- Good resolution
- Favourable S/N ratio ($<10^5$)
- Simple, robust quantitative algorithms
- Various measuring modes (differentiated accessory technology)
- Suitable for opaque samples, extremely small sample amounts; few limitations on sample geometry
- Excellent wavenumber accuracy
- Small number of calibration standards; calibration transfer
- Mature technique and instrumentation
- On-line hyphenation
- Wide applicability (including QC)

Disadvantages:

- Some sample preparation needed (grinding, pellet or film pressing)
- Short pathlengths (difficult implementation)
- Low specificity
- Matrix dependency (polymer and co-additives)
- Insensitive to minor components in mixture analysis
- Difficult speciation of components in mixture analysis
- Energy-limited technique
- Low radiation intensity at detector
- Highly dependent on well-characterised calibration standard and sample presentation
- Few commercially available traceable standards
- Interferences (strong water absorption)

books cover the topic of sampling methods in IR spectroscopy [12,62–65]. For databases, *cfr.* Chps. 1 and 5.2 of ref. [1]. FTIR (KBr, nujol and liquid film techniques) spectral libraries of 1124 polymers and polymer additives and 845 dyes and pigments are available [65a].

Table 1.10. Selection of applications of *in situ* infrared techniques

- Identification of polymeric resins, additives and volatiles
- Identification of volatile components in complex mixtures by HS-GC-FTIR
- Analysis of finishes on fibres and fabrics
- Network characterisation (cross-linked systems, rubbers, curing, compositional and degradation studies)
- Quantitative analysis of blends and additives
- Reverse engineering
- Monitoring of chemical changes
- Depth analysis
- Crystallinity and orientation measurements
- Photoacoustic analysis for identification of cured or insoluble materials such as composite materials, thermoplastic parts and inorganic fillers
- Near-surface reflectance analysis for the study of adhesion, coating problems and identification of pliable materials such as elastomers and coated adhesives
- Fire smoke analysis
- Troubleshooting (identification of contaminants, film inclusions, or samples of μg quantities using IR microscopy)
- Quality control
- Chemical imaging

Applications

The scope for IR spectroscopic techniques for direct in-polymer additive analysis is much broader than for extracts. In many real-life cases the form of the sample as presented for analysis is not at all suitable for routine transmission spectroscopy, which would, of course, have been the only method feasible with dispersive IR instruments. Most real-life samples are much too intensely absorbing or scattering for this to be possible. Yet, this does not preclude their routine measurement with Fourier transform spectrometers with the variety of sampling modes. *In situ* infrared analysis has been used for a host of analytical problems, as indicated in Table 1.10.

In the analysis of an **"unknown" plastic**, characterisation into a broad group is usually relatively simple, taking into account the origin of the sample, its use, appearance, and elemental composition. FTIR analysis of intact polymeric materials may be precluded for polymers which themselves have strong infrared absorption. An increasing number of polymers are now compounded with other materials, *e.g.* as composites containing fillers, such as glass fibres, or as coatings, which contain pigments. These additives tend to interfere strongly with IR spectra of the polymers because of their own characteristic absorptions or the scatter of the incident radiation that they cause. Although FT procedures and unconventional sampling methods have improved, the situation with regard to these types of sample is far from satisfactory. Identification of polymers by FTIR is often complicated by the presence of fillers. For example, it was reported that filled semicrystalline polymers of simple chemical structure, such as PE or PTFE, and polyamide, polyester or PC/ABS blends filled with an unknown filler, were difficult to characterise using FTIR due to overlapping spectral regions, low sensitivity to certain bonds or similar repeating units [66]. On the other hand, it was easy to identify the polymer by determining melting peaks or glass transition temperatures using DSC.

IR measurements of **plaques** before and after extraction are a widely used method for evaluation of the amount of antioxidant [67]. Sinclair *et al.* [68] determined PP/DSTDP (0.1–0.7 wt.%) in 0.6 mm thick plaques by means of the carbonyl absorption at 5.75 μm. Using 200 μm thick PS samples an amide wax slip additive was identified in the material using FTIR [69].

The spectroscopic approach of analysing **antioxidants** in the polymer is difficult because of ultra low concentration (100–1000 ppm) and interference of the parent polymer matrix. IR spectroscopy is more specific but less sensitive than UV spectroscopy. Consequently, thicker polymer films (0.2–0.3 mm) have to be used to overcome the disadvantage of lower sensitivity. Another advantage of UV over direct IR spectroscopy in the determination of AOs and light stabilisers in polyolefins is the lack of interfering absorption from the polymer matrix. Thus, UV spectroscopy of thin polyolefin films is able to determine 0.002–1.0% stabilisers. On the other hand, direct IR analysis of additives in POs is limited by excessive beam dispersion due to light scattering from the crystalline regions of the polymer. Depending on additive and polymer *ca.* 500 ppm is usually considered to be a realistic lower limit of detection for IR spectroscopy. Despite reports in the literature [70] that the degree of conversion of phosphite to phosphate can be measured by FTIR, limitations of this method make it unsuitable for quantitative work. The major limitation is that the phosphate P$-$O stretching absorption at 968 cm^{-1} is in the same region as the *trans*-vinylene group absorption in PE. Disappearance of the phosphite P$-$O absorption at 850 cm^{-1} is indicative of complete oxidation of the phosphite. However, changes in the vinyl concentration with processing of LLDPE, which contains a partially degraded phosphite antioxidant, cannot be followed accurately by FTIR. Johnston *et al.* [71] used FTIR spectroscopy to monitor the consumption of Irgafos 168 in LLDPE with progressive processing; samples were studied in which the phosphite was completely oxidised. In a stability study Allen *et al.* [72] have used FTIR analysis of microtomed sections of a PE-X gas pipe and DSC-OIT (200°C) for the evaluation of leaching and consumption of polymer additives at the outer and bore surfaces of the pipe. The observed effect was more prevalent for the AOs than for the UV stabiliser (Cyasorb UV531). Several HALS stabilisers were determined in the presence of other additives in PP using digital spectrafitting techniques [73].

Differences in the IR spectrum resulting from variations in aggregation state have been used in evaluating **additive solubility**. In case of bis(2,2,6,6-tetrametyl-4-piperidinyl)sebacate (Tinuvin 770) in LDPE, a shift of the carbonyl absorption of the ester group has been observed when it is dissolved in the matrix or bloomed at the surface [74]. The concentration of the soluble part was obtained with the usual value of the extinction coefficient $\varepsilon_{solute} = 580$ cm^{-1} mol^{-1} litre for the absorption of ester groups at $\lambda_{max} = 1736$ cm^{-1}, and that of the bloomed part into the surface solid phase with $\varepsilon_{solid} = 945$ cm^{-1} mol^{-1} litre determined for the absorption of the ester groups at $\lambda_{max} = 1718$ cm^{-1}.

IR spectroscopy may be used for detection of **plasticisers** in soft PVC cables [75], but does not distinguish clearly between the many possible dialkylphthalates. With the advent of difference spectroscopy, identification of a plasticiser in a polymer no longer requires isolation of the additive. Identification can readily be made without separation if the polymer is known and a plasticiser-free spectrum is available. This was illustrated for di-2-ethylhexylsebacate in an acrylonitrile–butadiene copolymer [76]. IR can sometimes quantify plasticisers in solid plastic compositions without the need for extraction or dissolution steps. FTIR difference spectroscopy has also been used for quantitative analysis. Another example of difference spectroscopy is the case of two plastic films which differed in printability [77]. Difference mid-IR spectra of the surfaces of the two films in the 1600–1300 cm^{-1} region revealed a stearate (and eventually a free acid, at 1720 cm^{-1}). Surface properties of films are strongly affected by such trace impurities. For example, a low level of fatty acid salt, possibly with free fatty acid, interferes with printability.

Depth analysis of PVC, nitrile rubber and alkyd-melamine coatings by microtome cutting and FTIR microanalysis has enabled measurement of the distribution and migration of additives and evaluation of weathering tests [78]. Murase *et al.* [79] have studied the **migration** of a plasticiser, di-2-ethylhexylphthalate (DEHP), in vinyl chloride resins containing stabilisers, di-*n*-butyltin dilaurate and di-*n*-butyltin maleate (DBTM), under various conditions (heat, accelerated weathering, outdoor exposure, hot water immersion) by depth analysis using FTIR.

With regard to in-polymer analysis of **flame retardants**, it should be considered that their determination in polyesters/polyamides by means of IR spectroscopy differs from the aforementioned work on polyolefins in that: *(i)* IR spectra of polyolefins show few vibrations as opposed to the band rich spectra of polyesters and polyamides, which restricts the useful IR window and limits observation of vibrations of FR analytes; and *(ii)* the molar extinctions of the various IR vibrations of polyesters are considerably higher than those of PE vibrations. Identification of FRs in concentrations below 5% requires other analytical methods than FTIR.

While IR spectroscopy is widely accepted as a method for identification of organic substances, its use may also be extended to identification of inorganic materials, in particular of **fillers** such as silicates, aluminium trihydrate, calcium carbonate, fibre-glass, talc and sulfates. Fillers have characteristic and strong FTIR bands (C−O, Si−O, Al−O and S−O vibrations), which can be easily identified within a spectrum of the polymer. Although these IR bands are good evidence for the presence of certain anions, the nature of the cation is often to be derived from elementary analysis, or eventually from minor features of the IR spectrum. While oxides of light elements, *e.g.*, silica and alumina, show useful spectra in the visible to 15 μm region, heavy metal oxides, metallic sulfides, chlorides and bromides, show no significant bands in this region [80]. Good IR spectra of inorganic materials require a very small particle size, but this is rarely a problem with fillers in plastics compositions. The IR spectra of fillers may be recorded either as alkali halide discs or in paraffin oil mulls. There are several excellent collections of spectra of fillers (*cfr.* Chp. 1 of ref. [1]), so there is generally no difficulty in identification. However, as already mentioned, identification of polymers by FTIR is often complicated by the presence of fillers. Materials with high filler content can be identified quite reliably by difference spectroscopy, as long as the filler is mainly monodisperse. Fillers in a vulcanisate (*e.g.* kaolinite, SiO_2 or carbonate) were easily deduced from the IR spectrum [81]. A procedure was developed for reliable quantification of kaolin in hot-melt EVA polymer [82]. An alternative method for the determination of fillers, *i.e.* ashing, has the disadvantage that it might change the chemical composition of the filler. The alumina phases boehmite, diaspore, gibbsite and bayerite in PE can be distinguished by far-IR spectroscopy (50–400 cm^{-1}) [83]. FTIR was unable to detect 20 wt.% PTFE filler in polyacetal [66]. Generally, IR spectroscopy is also not very good at detecting the presence of halogens. DSC is very successful in detecting Teflon in plastics. Because of the additive package (fillers, pigments, impact modifiers, stabilisers, *etc.*) IR of non-exposed PVC siding panels was reported to be complex and precise identification failed [84].

Fluoropolymer **processing aid** concentrates can be analysed by FTIR to within 0.1% within a few min [85]. Letdown processing aid levels can be determined down to approximately 400 to 500 ppm with an accuracy of ±50 ppm. It is not envisaged that FTIR will be a suitable means for analysing **tracers** for ownership (of defective products). Low concentrations are necessary here in order not to upset materials properties and to avoid confusion with the additive package. This rules out many aromatics, S and P compounds, Si-based materials, Cl and Br compounds, elements found in colourants (Ti, Ba, Ca, *etc.*) as well as other elements (*e.g.* Zn) and various functional groups (COOH, *etc.*).

FTIR is particularly useful in the study of **composite materials** yielding much information about the molecular structure of coupling agents on various substrates, including silica, metals and fillers. Ishida *et al.* [86] used a non-destructive FTIR sampling technique to study glass fibre composite interfaces.

Vibrational spectroscopy is also widely used for the analysis of filled elastomers [87–89] in order to describe the polymer-filler interaction [87,89], interfacial region [89], sulfur and cross-linking chemistry of elastomers and bonding of BHT fragments to the EPDM matrix [88]. In the **rubber industry** the FTIR transmission technique is generally accepted

(ASTM D 3677–90). However, this procedure is preceded by a time-consuming (*i.e.*, several days) and complicated extraction procedure. IR is used extensively to distinguish various types of rubber (*e.g.* SBR, NBR, BR, IR, *etc.*), pyrolysates of rubber, additives, and for QC purposes. Special micro-cut techniques have been developed to allow microtomed IR spectroscopy [90]. Implementation of QA schemes in rubber manufacturing entails controlling the composition of additive mixtures (mainly of AOs and antiozonants) prior to mixing with the vulcanising agents and polymers used for rubber vulcanisation. As a result of the high dipole moments of the polar co-agents, IR spectroscopy is of great value in the study of co-agent-assisted peroxide curing of elastomers [88].

Fourier transform spectroscopy with its speed and improved signal-to-noise ratio has allowed for further applications of IR spectroscopy to polymer research, including **network characterisation** (cross-linking, vulcanised rubber systems, composition and degradation studies). FTIR and ^{13}C s-NMR have been used to characterise the complex of diphenylmethane-4,4-diisocyanate (MDI) with acetone oxime, a potential cross-linking agent for aliphatic polyester/urethane block copolymers [91]. Real-time FTIR is particularly useful for monitoring of the kinetics of UV curing [92] and for determination of the efficiency of **grafting** of functional groups onto polyolefin macromolecules [93]. The latter method is based on liquid extraction of a non-reacted portion from film samples followed by comparison of spectral characteristics of the sample before and after extraction. Bartholin *et al.* [94] used IR to gather evidence for grafting of ester groups on PVC stabilised by Zn and Ca stearates.

IR spectroscopy plays a great role in product analysis under **reverse engineering**. Dormagen [95] and Coz *et al.* [96] reconstructed the formulation of a tyre, based on FTIR, thermal and HPLC analyses.

In an industrial environment, such as the automotive industry, the principal requirement is for rapid analysis of difficult samples by direct means (*i.e.* with a minimum of sample preparation), followed by an informative analysis of the data without the need for extensive spectral interpretation. Quite apart from the obvious need for rapid sampling in such a regime, detailed individual spectral analysis cannot be undertaken, and must be replaced with a quantitative semi-expert system approach. The availability of such software systems for quantitative analysis and qualitative discrimination can solve many seemingly intractable spectroscopic problems. Once calibrated using PLS multivariate techniques, FTIR analysis offers an almost **instantaneous analysis** of multicomponent mixtures. Major applications of FTIR spectroscopy to **automotive industry chemicals** include the analysis of difficult samples in the raw state, semi-automatic quality control and monitoring in use, creation of user-databases to analyse different formulation recipes (*e.g.* PyGC-MS spectral data). Analytical applications of FTIR in the automotive industry have been illustrated with examples covering multicomponent PLS regression calibration of additive packages in commercial motor oil, condition monitoring of used oil, QC discrimination between different formulations of a complex resin-based mixture, and routine analysis of automobile plastics for failure analysis [97]. A mid-IR acousto-optical tuneable filter (AOTF) spectrometer has been described for rapid identification of black plastics in automotive recycling [98]. As mentioned already, mid-IR allows unequivocal identification of polymers albeit with some restrictions. Although FTIR can technically be employed for identification of **automotive scrap**, negative considerations are great sensitivity to sample geometry, sample surface, contaminants, necessary sample preparation, difficult automation, high investment costs and overall unfavourable economic considerations.

FTIR has been used extensively for identification of coatings. Important methods for the study of **paint material** are KBr pellets prepared from scratched off paint material, ATR measurement of coated surfaces and measurements on cross sections of a coating with FTIR microspectroscopy. FTIR is an excellent way of obtaining information quickly about the basic chemical class of a binding material. Samples are placed in a diamond anvil cell and compressed to a thin film; a beam condenser focuses the IR beam to an area of 1.0 mm^2 [99]. Other methods for determining the binder structure of cross-linked systems comprise PyGC.

The possibilities of **quality control** in the polymer industry have been considerably enhanced by the introduction of computer assisted IR and FTIR spectrometers. One of the most frequently used techniques in this field is difference spectroscopy. FTIR is used in purity checking for **QA purposes**. Examples of application of IR spectroscopy in **damage analysis** (often in a combined multianalytical approach) have been reviewed [75,97]. Applications involve analyses of plastics, rubbers, lacquers, lubricant additives, lubricating oil, *etc.*

FTIR analysis was used to monitor **gaseous reaction products** emitted from LDPE/TiO_2, PVC/TiO_2, polyester and rubber samples exposed to UV irradiation [100]. Ranking of the pigments with different photo-activities as protectants or pro-degradants coincided with that obtained from much more time-consuming laboratory testing (chromatography) and field experience. This approach to dynamic monitoring of photooxidation using a specially designed FTIR cell to measure CO_2 emission was also applied to paints.

Infrared spectroscopy and multivariate calibration were used in **quantitative analysis** for additives in HDPE [101] and LDPE [102], *cfr.* Chp. 6.4.3. Difference spectroscopy was reported for quantitation of additive packages in PE and EVA copolymer [103,104]. For quantitative IR analysis, *cfr.* also ref. [105].

FTIR is one of the few analytical techniques which allows **time-resolved studies** of chemical processes associated with film drying and hardening. There are many examples of the use of FTIR to characterise polymer coatings, including curing reactions [106]. Nevertheless, some restrictions are to be noticed. Many coatings are water-based, and cannot be analysed completely successfully using FTIR, due to the strong water absorption bands that obscure large portions of the spectral range. In case of paint films one tends to be restricted to reflectance measurements. Infrared absorption is also used for thickness measurements (within 0.1 μm) of high quality polymer films and coatings.

Spectra (FTIR, ^{1}H NMR, MS) for the identification of some 100 additives in **food packaging** are available [107]. A recent atlas of plastics additives contains 772 FTIR spectra [108] and describes the application of vibrational (FTIR, Raman), electronic and many spectrometries for identification and structural elucidation of plastics additives. Use of FTIR as a versatile technique for real-life samples has been reviewed [109].

1.2.1.1. *Transmission Infrared Spectroscopy*

Principles and Characteristics

While mid-IR spectroscopy is very useful for the quantitative analysis of liquids, it may not automatically be the method of choice for solid products because a homogeneous sample is required. As additives can easily be distributed heterogeneously in the polymer, it is good practice to examine various positions of the solid. Off-line mid-IR evaluation of the plastics composition and concentration of constituents may take from 1 to 2 h depending upon the sampling technique used, *e.g.*, making a thin film (during which polymer properties may be altered). Transmission spectroscopy remains the most commonly used and traditional IR measurement for samples that can be prepared in a transparent form. In transmission measurements the sample is placed in the optical path of the IR beam.

The general principles of transmission IR spectroscopy for solids conform to those for gaseous and liquid samples. **Solid transmission methods** comprise various sampling modes which often involve a modification to the morphology of the sample. Polymer samples can be prepared as mulls or KBr pellets (the most widely used methods) or by casting from solvents and latices, thin film preparation with the microtome, hot pressing into transparent wafers, compression into cold sintered discs and abrasion or grinding in the frozen state. For suitable samples, the transmission technique produces spectra with high signal-to-noise ratios and given the nature of the method, the spectra are quantitative in the IR region. Films can be transilluminated directly. Powders can be studied in the form of a suspension or **mull**. The most commonly used mulling agent is a clear white mineral oil (Nujol™). Preparation of mulls is a low cost method. Another popular technique for solid sampling is pressing of an **alkali halide pellet**. Solids are run more frequently as KBr pellets than as mulls. The preference for pellets probably stems from the fact that KBr is IR transparent over its entire transmission range. However, KBr is a very hygroscopic material. The amount of sample needed for the measurement is *ca.* 5–20 mg.

Transmission techniques for IR spectroscopy further comprise capillary films (layers of a non-volatile liquid). In analogy, in a widely used procedure for obtaining spectra of polymeric materials the sample is first dissolved in a moderate to highly volatile solvent. The solvent is then evaporated, leaving behind a very thin film of the sample adhering to the window. There is only limited control over the thickness of a **cast film**. Yet thickness is important in producing a good spectrum. Thin films for FTIR spectroscopy may also be drawn from polymer solutions using surface tension. If the polymer is thermoplastic, the sample may be run as 5–25 μm thin compression moulded film. Thin sections can also be prepared by standard techniques used in microscopy. These can be studied in a similar way to powders. Special micro-techniques have been developed for studying fibres; parallel winding is also used. Short cut fibres can be studied as powders. Various alternatives to transmission spectroscopy do exist.

A heated film press for preparation of plastic films with reproducible thickness (from 20 to 500 μm) for IR analysis has been described [69]. The actual sample thickness required for analysis depends on factors such as concentration and extinction coefficient of the analyte and opacity of the sample (pigmented or non-pigmented). Film thickness measurements need to be carried out for quantitative measurements. Differences in thickness cause a shift in spectra and methods for spectral normalisation then become necessary [101].

Applications

Applications under review in transmission concern: *(i)* transparent polymeric matrices containing one additive; *(ii)* transparent polymeric matrices containing a multicomponent additive package; *(iii)* additive distribution analysis (using the multi-film stacking method); *(iv)* interactions and degradations in the solid-state; *(v)* quality control; and *(vi)* polymer melts.

Mid-IR spectrometry can detect a high percentage of polyolefin additives by direct transmission measurement of films in a test taking less than 10 minutes (*i.e.* considerably less than extraction/chromatography). Nevertheless, relatively few reports deal with the use of direct spectroscopic methods in transmission on films. This is attributed to problems caused by light scattering and reflection, and polymer-additive interferences. According to the polymer/additive deformulation set-up of Scheme 2.12 of ref. [1], the polymer/additive sample to be examined is routinely first pressed into a thin film for IR analysis to establish the nature of the polymer matrix and to define the extraction solvent. After extraction, the residue is again pressed into a thin film to verify that all extractables have been removed.

Nelissen [110] reported identification of some closely related polybrominated polystyrene **flame retardants**, namely PDBS 80 (Great Lakes), Pyrochek 68PB (Ferro), Pyrochek 68PBI (Ferro) and Saytex HP 7010 (Albemarle) in polyamides by means of FTIR and PyGC-MS. PDBS 80, Pyrochek 68PB or Saytex HP 7010, could not be distinguished in a solid polyamide matrix by means of FTIR transmission spectroscopy due to interference of additive and polymer absorption bands. After extraction, IR spectroscopy suggests that the products differ in the substitution patterns of bromine. Pyrochek 68PB (polystyrene brominated with BrCl; 68 wt.% Br, 0.1 wt.% Cl) and Pyrochek 68PBI (brominated polystyrene, 68 wt.% Br) cannot be differentiated by means of FTIR. **Slip agents**, such as oleamide, in a pressed PE film have also been determined by means of direct IR [80]. Solid-state FTIR analysis of erucamide levels can be hindered because its weak absorption bands can be easily obscured by crystalline bands in the polymer film spectrum [111]. Direct determination by means of IR of dilauryl thiodipropionate (DLTDP) in a pressed PP film [112] and of oleamide in PE have been reported [80].

Miller *et al.* [113] examined IR spectra of **antioxidants** from polymer films taking care to compensate with additive-free polymer in the reference beam. Although IR spectroscopy is more specific than UV spectroscopy, the AO level in polymers is often too low to give suitable spectra [114]. With the low usage level of antioxidants in plastics (250–2000 ppm) a relatively thick sample pathlength is required (0.25–0.75 mm) for the absorption bands of AOs to be visible. Direct film IR spectroscopy has been used for quantitative determination of 0.1 to 1.0 wt.% Cyasorb UV531 in unpigmented HDPE; AOs such as Polygard and Santonox R do not interfere [115]. A calibration curve (absorbance per unit thickness *vs.* concentration) was used for quantitation. Also Tinuvin 326 in PP has been determined [115]. Similarly, Tinuvin 770 and Chimassorb 944 can be determined by in-polymer methods (lower limit: 500 ppm). Tinuvin 770 in PP film is usually quantified on the basis of the 1740 cm^{-1} C=O (ester) vibration, Chimassorb 944 by 1532 cm^{-1} or 1570 cm^{-1} N−H vibrations of the triazine ring. The method is restricted in the presence of additives interfering at these wavenumbers (*e.g.* Ca-stearate with absorptions at 1560 and 1532 cm^{-1}). Crystalline ethylene-bis-stearamide (EBA) in ABS film (0.3–3.0 wt.%) can be quantified on the basis of N−H vibrations in the 3233–3350 cm^{-1} range. Meszlényi *et al.* [104] have described a very simple and rapid qualitative and quantitative IR analysis method of a 180 μm PE film containing light stabilisers (Chimassorb 81/944 or Tinuvin 622) that avoids laborious extraction. As the basis of analysis is absorption of a functional group or moiety in the IR region ester containing stabilisers (Tinuvin 622/770, Irganox 1010/1076, Hostanox O3 and Plastanox STDP) cannot be distinguished. Also Chimassorb 944 and Cyasorb UV3346 interfere (triazine ring) and the determination of Chimassorb 81 can break in on other benzophenone or isocyanurate compounds. The method is thus not generally applicable when the light stabiliser in PE is unknown. Tinuvin 783 (*i.e.* a Tinuvin 622: Chimassorb 944 1:1 blend) was determined quantitatively in 100 μm LDPE film (RSD 10–15% for IR, 1–5% for UV) for QC purposes [116]. Variations in Tinuvin 783 concentration in samples exposed to thermo-oxidation at 90°C up to 98 days were monitored using a calibration curve [17]. Direct measurement of the stabiliser concentration by means of IR spectroscopy has also been used to advantage to determine the rate of evaporation of Irganox 1010/1076/3114 from a 40 μm thick PP film at 150°C in an N_2 flow. Physical loss of stabilisers in 80 μm thick LDPE film for greenhouses was monitored by determination of Chimassorb 944 by UV absorption at 225 nm and of Tinuvin 622 by FTIR ester group absorption at 1734 cm^{-1} [117]. Pukánszky *et al.* [23] have examined the **interactions** between pesticides and stabilisers in agricultural PE films using both in-polymer FTIR and UV measurements. Because of the relatively high concentration of the stabilisers, their presence could easily be detected by FTIR spectroscopy, in spite of the lower sensitivity of this technique in comparison to UV spectroscopy. By means of FTIR strong interaction in the PE film stabilised by Hostavin ARO 8 – Hostavin N 30 with Neviken was observed.

Methods for determination of **diffusion coefficients** of additives in polymers are *in situ* analysis of microtomed sections by means of IR and Raman (imaging) spectroscopy/microscopy, NMRI or radio-tracer techniques. In the so-called multi-film stacking technique a stack of films (top and bottom) is in direct contact with the penetrant; at a predetermined time the films are analysed by IR (directly) or extraction/chromatography (indirectly). The film stacking method is laborious and not very accurate (poor film contact; evaporation during de-stacking, adsorbed penetrant traces at top and bottom). The method can be used for slowly diffusing penetrants at temperatures up to the m.p. of the polymer. Assuming a diffusion model the diffusion coefficient D can be calculated from the concentration profile, as in case of Chimassorb 81 in 40 μm LDPE films [118].

Vigerust *et al.* [102] have reported **quantitative analysis** of additives (silica, erucamide and BHT; 20–1100 ppm) in 80 samples (1 mm thick films) of four different LDPE polymers with varying melt indexes using IR spectroscopy and multivariate calibration; 60 samples were used for model calibration and 20 samples for verification. Results were as follows (correlation coefficient, R^2, and standard error of estimate): SiO_2, 0.91, 30 ppm; BHT, 0.84, 69 ppm; erucamide, 0.91, 72 ppm. External validation indicates that this method for additive analysis has potential for **quality control** of PE. The method is both time and cost effective, at approximately the same standard error as observed for traditional methods. Karstang *et al.* [101] have described IR spectroscopy and multivariate calibration in quantitative analysis of 1 mm thick HDPE/(Irganox 1010, Irgafos 168 phosphate, Ca stearate) films; 55 samples were included in the calibration set. One of the problems in quantification of additives in polymers by IR spectroscopy originates from the sample preparation procedure. The samples should be films of equal thickness (in a typical preparation the thickness varies from 0.9 to 1.1 mm). Calibration models should be used which take thickness variations into account. Similarly, Bremmers *et al.* [15] have examined HDPE/(Irganox 1076, oleamide) films by means of IR (and UV) spectroscopy (Fig. 1.7) and multivariate calibration methods. Using a calibration set of 25 samples for Irganox 1076 and 26

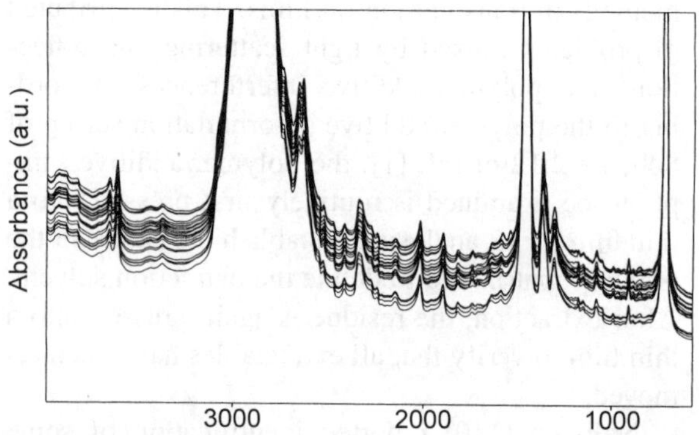

Fig. 1.7. Infrared spectra (cm^{-1}) of a calibration set of HDPE/(Irganox 1010/1076, Irgafos 168, oleamide) film samples with variable additive concentrations. After Bremmers and Swagten-Linssen [15]. Reproduced by permission of DSM Research, Geleen.

samples for oleamide results were ±30 ppm for Irganox 1076 both at low (200–300 ppm) and high (1200–1500 ppm) loadings, ±40 ppm for oleamide at low (250–320 ppm) and ±70 ppm at high loadings (1250–1500 ppm); all results were referred to HPLC reference methods. Film thickness corrections were properly applied, but the influence of PE type was not evaluated. Verlaek *et al.* [16] have determined Chimassorb 944, Irganox 1010/1076 and Irgafos 168 in LDPE film, obtaining standard errors of prediction (SEP) values of 12, 230, 42 and 42 ppm, respectively. The unacceptably high SEP value for Irganox 1010 was ascribed to changes in crystal morphology. For comparison, SEP values of *ca.* 8 ppm were found for mid-IR measurements on melts. Mid-IR was also used for the non-UV absorbers Zn-stearate, Ca-stearate and oleamide (SEP values: 57, 37 and 34 ppm on film, 29, 12 and 49 ppm, respectively, in the melt). Also the quantitative analysis of polyester (0.1–0.3%) and calcium propionate (0.05%) in 500 μm PE films has been reported [69].

Brandolini *et al.* [119] have pointed out some drawbacks to the use of FTIR spectroscopy for quantitative analysis of the extent of phosphite and phosphonite additive degradation in PE. Band positions are not as distinctive as ^{31}P NMR resonances. Consequently, it is difficult to distinguish **degradation products** of similar additives, such as A and B of Fig. 9.3 of ref. [1]. Quantitation in FTIR is also not as straightforward as in NMR. To accurately assess the extent of degradation, appropriate standards would have to be developed. Most importantly, however, the FTIR spectrum contains absorbances from the polymer background and other additives which may obscure the peaks of interest. Spectral subtraction of an appropiate reference polymer can obviate some of this concern, if available. Use of a similar, but not identical, polymer can result in artifacts. In the specific case at hand, the polymer sample was pressed as a 0.1 mm film. An appropriate, secondary oxidant-free reference polymer was available so that spectral subtraction could be performed to remove matrix interferences.

Several laboratory scale mid-IR analyses of **melts** have been reported, both to overcome crystallinity effects in solid polymers and for scaling-up purposes. Palmen *et al.* [120] have used on-line mid-IR (with optical path length of 0.5 mm) for monitoring Irganox B220, Chimassorb 944 and Ca stearate in HDPE melt at 210°C using a mini-extruder (Göttfert; capacity: 10 kg hr^{-1}). The Irganox B220 content (in the 750 to 1800 ppm range) could be predicted with a standard error of prediction (standard deviation σ of difference between mid-IR prediction and a reference value) of 38 ppm (reference: XRF); similar figures for Chimassorb 944 (in 150 to 1000 ppm range) were 32 ppm (reference: N-content analysis) and for Ca stearate (in 950 to 2300 ppm range) 116 ppm (reference: XRF).

It should be noticed that a mid-IR spectrophotometer with tuneable diode lasers has also been used to reduce production waste and to improve quality control. Molten polymer was pumped from a process stream to chilled calendering rolls, producing a film which passed through the IR spectrophotometer and was selectively absorbed when multiplexing the diode lasers. This technology worked well for laboratory "trial" tests but was found unpractical and was too expensive for an on-line polymer production plant.

1.2.1.2. *External Reflectance Techniques*

Principles and Characteristics

There are many types of samples for which the transmission approach is not optimum, desirable or even practicable, *e.g.* urethane foams, polymer laminates, and surface coatings. To obtain spectra from these types of sample it is more usual to employ a reflection technique. Reflection measurements are often also needed when materials are to be measured in their original form, except for thin films. This essentially turns IR spectroscopy into a **surface analysis technique**, but of low sensitivity compared to high vacuum spectroscopic techniques such as XPS, LEED, EELS and SIMS. Since the advent of FTIR spectrometers, infrared sensitivity has so much improved that nowadays a measurable spectrum can be produced from even a single monolayer on a flat surface; interfaces are also commonly examined.

Where the substrate is transparent, IR spectra of the surface layers can be obtained either by transmission or by reflection; when the substrate is non-transmitting, then reflection is normally used. Specialised IR techniques that are suitable for surface analysis are external, internal and diffuse reflectance. When light propagating in a medium of refractive index n_2 reaches a medium of refractive index n_1 radiation is partly reflected and partly refracted and both parts contain information on the material composition. The ratio reflected/refracted radiation depends on n_1, n_2 and the angle of incidence (θ). The choice of methods for obtaining spectra by reflection has expanded significantly in recent years. The purpose of an **integrating sphere detector** system is to provide a collection device for reflected, divergent, and scattered light from a sample. Whenever it is desirable to capture the total reflected light from a sample, the integrating sphere must be used. Photoacoustic spectroscopy (PAS) can also be used for certain surface-analytical problems for powders.

Internal reflectance spectroscopy (IRS; alternatively named attenuated total reflectance, ATR) is a quick and easy non-destructive sampling technique for obtaining the IR spectrum of a material's surface or of material which is either too thick, or strongly absorbing, to be analysed by more traditional transmission methods, *cfr.* Chp. 1.2.1.4. Internal reflection techniques, which require close contact with an internal reflection element (IRE) are unsuitable for rapid screening of plastic materials.

External reflectance spectroscopy (ERS) techniques, which can be combined with IR microscopy are specular reflection spectroscopy (SRS) and reflection–absorption spectroscopy (RAS). These techniques vary in the way the irradiated IR radiation on the surface is reflected from the surface.

Specular reflectance (SR) is defined as reflection in which the angle of incidence θ_i on the sample is exactly equal to the angle of reflection θ_r. The intensity of the reflected beam depends on θ, the surface roughness and absorption by the sample. In general, an SR accessory is simple and easy to use. The amount of light reflected is usually very low, sometimes only a few per cent, thus making this technique more useable with FTIR than with dispersive instruments. Good detectors are a prerequisite for a useful signal-to-noise ratio. There is no control over sample thickness, and the technique works well with coatings in the 0.2 to 20 pm range. Thicker coatings usually produce spectra in which many of the bands show total absorption, and thinner coatings result in very weak spectra. Obviously, the technique provides information on the top layers only. The SR technique is useful for thin film (at least 1 μm thick), solids or liquids on a reflecting substrate. For a good spectrum, the sample surface must be smooth and flat. Interpretation of specular reflectance data is complicated.

Infrared external reflection spectroscopy (IR-ERS) is generally based on single external reflection of IR radiation at the surface of a metal or metal film. This method is also called **reflection–absorption infrared spectroscopy** (RAIRS) or grazing angle reflection, which is actually a form of transmission through an absorbing film on a reflecting substrate. For this purpose specular reflectance accessories are widely used. Reflection–absorption (R-A) measurements are particularly useful for thin film samples a few monolayers thick, adsorbed or cast on a reflecting medium. A typical sample might be the polymer coating usually applied to the inside surface of a food or beverage tin. If the sample is placed, coated side down, on the stage of a specular reflectance accessory, the sample beam penetrates the coating once, reflects from the metal substrate, and passes a second time through the coating before ultimately reaching the detector. The result is a transmission spectrum. In the **multiple external reflection** technique the IR beam is reflected several times between two parallel reflecting plates and spectral information characterises the surface and adsorbates.

For standards for reflectance measurements, *cfr.* ref. [121]. Reflectance spectra are usually measured more easily than emission. Overviews of reflectance spectroscopy have been given [122,123], including ERS of polymer films on metals [124,125].

Applications

The infrared external reflection technique is valuable for surface analysis of all kinds of solid materials and the usefulness of the method for characterisation of polymeric surfaces was shown [126]. Lutz *et al.* [127] have described use of external reflection FTIR spectroscopy in combination with an advanced chemometric procedure (PLS) to determine routinely the qualitative and quantitative composition of **rubber materials** with high carbon-black content (25–50 wt.%). Due to the simplicity and precision of the procedure, the method is very useful in **production control, troubleshooting**, and fast product analysis of CB-filled polymers and other weakly reflecting samples in an industrial environment. Zachmann *et al.* [128] have compared mid-IR and near-IR for fast and reliable identification of **black plastics**. The method of specular reflectance spectroscopy in the mid-IR spectral range between 2 and 20 μm enables identification of a wide range of technical thermoplastics, even those with high filler contents. The comparable method of diffuse reflectance in the NIR range between l and 2 μm fails in case of black material. Polymer identification systems using mid-IR reflectance are being used worldwide (Bruker polymer identification system) for automobile and electronics recycling.

Specular reflectance spectroscopy in mid-IR allows identification of non-coated technical thermoplastics without sample preparation in just a few seconds. First derivative spectra permit to distinguish various **filler** types (talcum, chalk or barium sulfate) in PP. Identification of **flame retardants**, which is important for the sorting process of used computer displays or TV sets, is very complex in view of the wide variety of FRs. It is possible, however, to detect specific FRs in certain polymers by mid-IR reflectance techniques, for example polybrominated diphenylethers (PBDE) in ABS. This technique probably has limited applicability only.

Microscopical RAIRS was shown as a viable technique for analysing the polymer resins contained in dry, black photocopy and printer toners for forensic applications [129,130].

Reflectance spectrophotometry is a means for identification of pigments [131]. For qualitative analysis of **pigments** the log (k/S) profiles of the Kubelka–Munk analysis can be used, as their shapes are independent of concentration. The same principles apply to characterisation of colorants on textile fibres. However, identification of **dyes** on textile fibres by assessment of reflectance curves is difficult owing to the dependence of spectral reflectance on concentration and spectral interference due to the base colour of the substrate itself. Interference due to fibre absorption may be overcome by a differentiation with respect to wavelength, so that the resulting profile generated is a property of the dyestuff alone.

FTIR reflection spectra supply a fast and simple means for determination of foil thickness in polymer analysis (for QC purposes). The layer thickness of film (t) can be determined from the number of the interference waves (N) over a range of wavelengths in case of a known refractive index (n) of the compound.

1.2.1.3. *Diffuse Reflectance Spectroscopy*

Principles and Characteristics

Diffuse reflectance spectroscopy (DRS) is concerned with the efficient collection of diffusively scattered light, the direction of which is unrelated to that of the incident radiation. The technique of DRS enables IR measurements to be made on diffusively scattering solids such as powdered samples without the need for extensive sample preparation. This weak diffuse radiation is collected in a manner

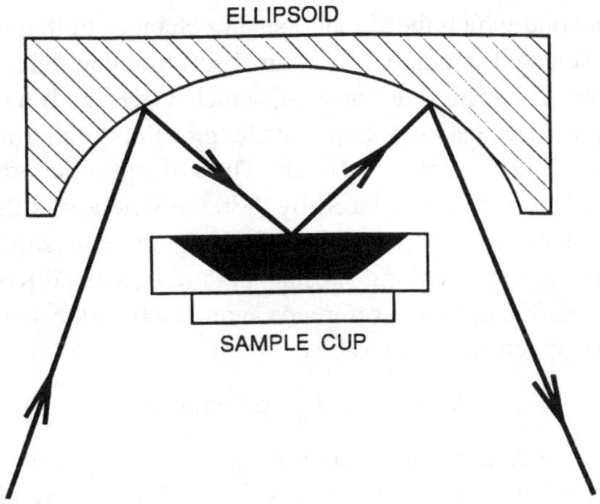

Fig. 1.8. Sample configuration used for diffuse reflectance. After Perkins [56]. Reprinted with permission from W.D. Perkins, in *Practical Sampling Techniques for Infrared Analysis* (P.B. Coleman, ed.), CRC Press, Boca Raton (1993). Copyright CRC Press, Boca Raton, Florida.

which minimises the specular reflectance component. In the visible and near-IR regions of the spectrum, diffuse reflectance measurements have been made for many years using integrating spheres (*cfr.* Fig. 1.1). There have been some attempts to use these devices in the mid-IR, but the very low levels of reflectivity and the energy limitations of dispersive instrumentation make the technique unattractive. Diffuse reflectance spectroscopy requires specially designed cells with hemispherical or ellipsoidal mirrors with high focusing power (Fig. 1.8). The higher throughputs of FTIR spectrometers have made infrared diffuse reflectance feasible, if not common place. The procedure is often referred to as **Diffuse Reflectance Infrared Fourier Transform Spectroscopy (DRIFTS).** A variety of commercial accessories is available on an original design by Griffiths *et al.* [132,133]. As typically only 10 to 15% of the energy throughput is available for DRIFTS analysis, this is an energy loss technique compared to the transmission mode.

Conventionally, diffuse reflectance is analysed in terms of the relationship derived by Kubelka and Munk [134,135]. In practice, the reflectance function may be written in the form of:

$$f(R_\infty) = (1 - R_\infty)^2/2R_\infty = k/S = 2.303 \cdot \varepsilon \cdot c/S \tag{1.1}$$

where R_∞ is the ratio of the diffuse reflectance of the opaque sample at infinite depth (*i.e.* at a depth beyond which the signal does not change) to that of a selected standard; k is as an absorption coefficient, and S a scattering constant, which varies with particle size, packing, and wavelength, thus rendering baseline correction difficult. The absorption coefficient k may be replaced by $2.303\varepsilon c$ where ε is the molar extinction coefficient and c is the concentration of the absorbing substance. Pure powdered KBr or KCl is used as a reference against which the sample spectrum is ratioed, *i.e.*:

$$R_\infty = R(\text{sample})/R_\infty(\text{reference}) \qquad (1.2)$$

where R is the absolute reflectance. The Kubelka–Munk (K-M) model has several limiting conditions [136]. The particles are considered uniformly and randomly distributed, and their diameter smaller than the thickness of the layer. It is conventional to use the K-M function to transform the reflectance spectrum into a spectrum resembling a linear absorbance spectrum of the same sample, except that the relative intensities of the bands will be different. It is possible to obtain good diffuse reflectance mid-IR spectra by utilising Fourier transform instruments but in the mid-IR, unlike near-IR, **solid samples** have to be mixed with a diluent. The best spectra will be obtained if the sample is first ground (10 to 20 μm average particle size is sufficient) and then mixed with the non-absorbing matrix (KBr of KCl); an appropriate concentration of sample in diluent is about 5%. This procedure *must* be followed if the K-M transform is to be carried out and/or if quantitative work is to be attempted. For quantitative analysis of a powdered sample, it is necessary to satisfy the basic requirement of the K-M theory: keep the scattering from the samples constant. This implies that the particle size be kept constant; the preparation should be reproducible. Spectral distortion can occur in diffuse reflectance spectra if the particle size is not uniformly fine.

The most important advantage of the DRIFTS technique is perhaps the ease of **sample preparation.** For powdered samples, no sample preparation (which could change the morphology of the sample) is required. The particle size of a sample for DRIFTS can be much coarser than that needed for preparing a KBr pellet or a mull, where particle sizes need to be 0.5 μm or less to avoid scattering and sloping baselines. The DRIFTS technique is quite powerful for analysis of all types of granular, high surface area powders because of the internal scattering of radiation that occurs. The effective pathlengths of the IR radiation are increased manifold by this scattering. Therefore, it is possible to detect very low concentrations of species in powder samples compared to the standard one-pass transmission method. Sensitivities in the sub-ng range are quoted.

Any sample which scatters IR radiation can be studied using the DRIFT cell. In many situations, diffuse reflectance has superseded KBr disc analysis as the more convenient method. A very useful DRIFTS sample preparation technique, the so-called Si-Carb^{TM} sampling method [137], which extends the utility of DRIFTS for qualitative analysis of almost any solid, involves use of 340 and 400-grit SiC emery paper; the abraded powder on the paper scatters the IR radiation of an abraded sample in all directions so that the basic requirements of the K-M theory are fulfilled. Si-Carb^{TM} sampling is well suited for analysis of solid materials such as paint chips, hard powders and any inflexible sample that can be abraded with SiC paper. This procedure was used by Spragg [137] for really intractable samples, such as cured epoxy resins, too hard and brittle to be easily ground for making a KBr pellet, not readily soluble thus preventing preparation as a cast film, and thermosetting which precluded making a hot pressed film. With data acquisition times of less than 30 s spectra are obtained of more than adequate quality for library searching and identification.

DRIFT spectra can be complex. They are strongly dependent upon the conditions under which they are obtained. They can exhibit both absorbance and reflectance features due to contributions from transmission, internal, and specular reflectance processes, as well as scattering phenomena. DRIFT spectra are affected by the refractive index of the sample, particle size (due to changes in the light scattering coefficient), packing density, homogeneity, concentration and absorption coefficients. Many neat powders absorb far too strongly, and need to be diluted if meaningful DRIFT spectra are to be obtained. Consequently, it is often difficult to achieve good reproducibility. Other disadvantages are that relative band intensities in the raw spectra differ from those of the corresponding transmission spectrum, and that the amount of energy diffusely reflected is very low, often only a few percent and sometimes even less. In order to achieve a good signal-to-noise ratio (SNR), longer data collection times (more scans averaged) are required. A major limitation with the DRIFTS technique arises from functional groups that have high molar absorptivity

Table 1.11. Main characteristics of DRIFT spectroscopy

Advantages:
- Ease of sample preparation
- Suitable for strongly scattering samples and dark solids
- Speed
- Depth profiling
- Wide applicability for powders, granules, fibres (incl. QC)

Disadvantages:
- Very low diffusively reflected radiation intensities
- Relative band intensities differing from corresponding transmission spectra
- Modest reproducibility (particle size, sample packing) and quantitation
- Complex spectra
- Not a real surface technique

coefficients. These species tend to absorb more radiation than they reflect and usually all the detailed spectral information is either lost or greatly distorted in these regions. This is particularly true of silica filler systems which very strongly absorb IR radiation from 1300 to 800 cm^{-1}. Table 1.11 shows the **main assets** of DRIFTS.

Culler [138] has recently reviewed the sampling techniques for qualitative and quantitative analysis of solids by DRIFTS.

Applications
Recently, DRIFTS is gaining popularity as a sensitive technique for the study of a wide range of organic and inorganic samples including powders, crystals, solids with rough surfaces, coarse textured samples such as polymer pellets and fibres. Strongly scattering, or black samples such as coal can be handled by this technique. Bulk solids can be analysed if they reflect enough energy. Although fibres can also be characterised, the fibre orientation will affect the scattering intensity. The method is not indicated for transparent films.

A natural application for DRIFTS is particulate minerals and **fillers** because the nature of the surfaces can easily be determined. Chalk-filled PP was analysed using the diffuse reflectance probe because this material is not transparent. A calibration model with 18 samples using three relevant spectral regions (5307–6275, 6838–7505, 7987–8894 cm^{-1}) was developed for quantification of the filler content [139]. The interfaces of various **organic coatings** (PAA, PMMA, oleic or stearic acid) with ceramic or silica glass surfaces were studied by means of DRIFTS [140]. The coating process of the particulate fillers $Mg(OH)_2$ and $CaCO_3$ with stearic acid and Mg, Ca and Zn stearates was followed with quantitative DRIFTS, XPS and XRD [141]. Another useful application for DRIFTS is the study of silane coupling agent interactions with fillers/fibres used in the manufacture of high strength-reinforced composite materials [138,142,143]. DRIFT and diffuse reflectance UV/VIS spectroscopy were used to study the modifications of various cellulosic materials with different coupling agents [144].

DRIFT spectroscopy of microscopic amounts of dye mixtures extracted from small **textile samples** has been reported; raw and pretreated data matrices were interpreted with the use of chemometrics (PCA, SIMCA, FC) [145]. DRIFTS can readily detect ~200 ng quantities of pure, standard **dyes**. Bridge *et al.* [42] have qualitatively characterised acid dyes (CI Acid Red 17, Red 18, Red 44, Red 88, Blue 45 and Yellow 17) applied to wool and nylon. Near-infrared diffuse reflectance spectroscopy was evaluated for its ability to analyse solid **antioxidant blends** [146]. These opaque materials do not transmit near-IR light. This fast method effectively predicts weight percentage composition with a precision comparable to the currently accepted HPLC method of analysis, and can identify blend types and contaminated materials.

DRIFTS is an easy way to answer questions and solve problems in the product development, quality control, or basic research laboratory [147]. SiC DRIFT sampling was used for destructive **depth profiling** analysis of PE samples [148]. DRIFTS depth profiling provides greater sensitivity than ATR and PA techniques. Simpson [97] has illustrated the use of FTIR spectroscopy as an *in situ* sampling method for **failure analysis**. DRIFTS is here an extremely convenient and rapid sampling technique for polymers such as in-car plastics, especially when used in conjunction with carborundum or diamond abrasive paper sampling. The abrasive pad is first used to measure a background spectrum and is then lightly abraded over the polymer surface before the diffuse reflectance spectrum of the sample is measured. Mid-IR with modified DRIFT cell and in specular reflectance has also been proposed as a method of identification in **plastics recycling** [149, 150].

Fischer *et al.* [151] investigated the simultaneous quantification of several additives in PVC with an in-line diffuse reflectance probe. In cases where conventional transmission spectroscopy is made difficult because of morphological changes occurring as a result of grinding or hot compression moulding procedures adopted during sample preparation, DRIFT is an alternative, as *e.g.* in case of PVC/Ca-stearate [152].

DRIFTS has equally been applied to the identification of HPLC separated fractions [152]. On the other hand, the characterisation of spots on TLC plates is not particularly well suited to DRIFT: some spectral regions are heavily obscured and interactions between the sample and the substrate cause wavenumber shifts. Subtraction of the substrate spectrum from the diffuse reflectance spectrum of 20 μg of Irganox 1076 on a cellulose plate did not reveal useful information [152].

Application of DRIFT to the investigation of polymer surfaces has been reviewed [153].

1.2.1.4. *Attenuated Total Reflection*

Principles and Characteristics

According to ASTM E131-66T *(Nomenclature for Internal Reflection Spectroscopy)* **internal reflection spectroscopy** (IRS) is the technique of recording optical spectra by placing a sample material in contact with a transparent medium of greater refractive index and measuring the (single or multiple) reflectance from the interface, generally at angles of incidence greater than the critical angle. The physical phenomenon on which IRS is based has long been known. An electromagnetic wave incident at the interface between two different media is totally reflected. The transparent optical element used in IRS for establishing the conditions necessary to obtain internal reflection spectra of materials (a crystal of high refractive index) is called the **internal reflection element** (IRE). IRS is a quick and easy non-destructive sampling technique for obtaining the vibrational spectra of a material's surface or of material which is either too thick, or too strongly absorbing, to be analysed by more traditional transmission methods. IRS has been used in IR, Raman and fluorescence modes. Internal reflection spectra are not exact duplicates of normal transmission spectra. Samples examined by IRS generally require minimal, or no, sample preparation.

In recording the spectra of bulk materials via IRS only a thin film of the material near the surface is sampled. This is particularly so when large angles of incidence are employed. The penetration depth d_p depends on the refractive indices of the internal reflectance element (n_0) and sample (n), the angle of incidence θ and the wavelength:

$$d_p = \lambda/2\pi n_0(\sin^2\theta - n^2/n_0^2)^{1/2} \quad (1.3)$$

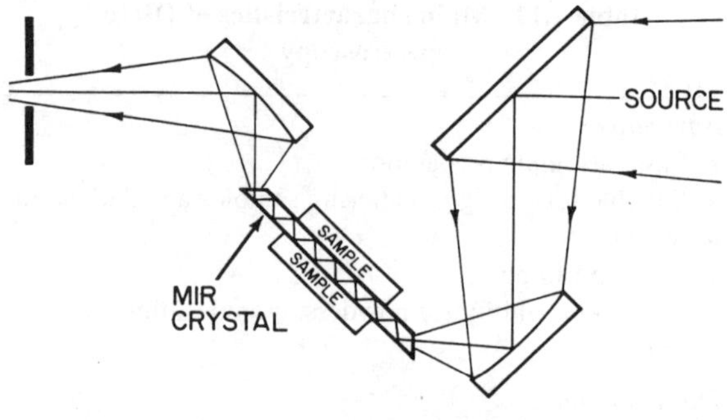

Fig. 1.9. Optical diagram for a typical internal reflectance accessory. After Perkins [56]. Reprinted with permission from W.D. Perkins, in *Practical Sampling Techniques for Infrared Analysis* (P.B. Coleman, ed.), CRC Press, Boca Raton (1993). Copyright CRC Press, Boca Raton, Florida.

The internal reflection spectrum may thus be strongly influenced by the surface, and may differ from the bulk. Whether or not the surface is different from the bulk can be determined from measurements as a function of angle of incidence. The technique was developed by Harrick [154,155] and Fahrenfort [156] and is also being referred to as **attenuated total reflection** (ATR) spectroscopy to account for the fact that reflectance is less than unity. The use of ATR in spectroscopy is based upon the fact that although complete internal reflection occurs at the sample–crystal interface, radiation does in fact penetrate a short distance into the sample. This penetration is termed the evanescent wave. Radiation of selected wavelengths is absorbed by the sample (according to the Beer–Lambert law), which is in contact with the IRE at each point of reflectance. The reflected beam thus contains spectral data from the sample. The resulting reflection is said to be attenuated (loses energy). ATR is observed when the angle of incidence is set and remains above the critical angle and the wavelength is swept through an absorption. ATR is a **surface technique** with a penetration depth of some 0.4–2.0 μm.

Important variables of the ATR technique are the type of crystal material, angle of incidence, wavelength of the radiation, number of reflections (*cfr.* Fig. 1.9), single/double sidedness, contact area, crystal to sample contact and refractive index of the sample. The resultant absorption spectrum, which closely resembles that of a transmission spectrum but is generally weaker, depends on those several parameters [157]. ATR is, insofar as instrumentation is concerned, generally more complicated than conventional transmission, requiring judicious choice in the refractive index of the IRE and/or angle of incidence. A requirement for ATR is a refractive index of the ATR prism which is higher than that of the sample.

Successful use of ATR spectroscopy highly depends on the choice of **IRE crystal material** (ZnS, ZnSe, CdTe, Ge, Si, cubic zirconium and diamond); the standard ZnSe crystal ($d_p = 2\ \mu m$) is useful for routine sampling in survey mode. Consideration must be given to possible undesired corrosive attack of the sample to the IRE. Silicon and Ge IRE ($d_p = 0.66\ \mu m$) perform well in ATR analysis of carbon-filled materials such as black rubbers or "O" rings. Diamond IRE ($d_p = 2\ \mu m$) is highly suitable for ATR spectroscopy in view of its high refractive index, hardness (9000 kg/mm^2), chemical resistance, IR transmission characteristics and thermal conductivity. Optical and physical properties of IRE materials are listed [157,158]. As IR radiation is penetrating only a few μm, it is most important to secure tight optical contact between sample and ATR crystal. A soft sample, such as a polymer film, can be persuaded to contact a large crystal. Solid and powder samples must be pressed on the crystal surface with a sample clamp. The high-pressure diamond ATR-IR accessory is designed for obtaining IR spectra of hard or non-easily deformed materials; it therefore has the ability to make high-pressure contact with the ATR element. As the FTIR absorption intensities vary with the clamping pressure exercised, quantitation is greatly compromised.

Various **ATR devices** are commercially available: single- and multi-bounce ATR, micro-ATR, horizontal ATR (HATR), cylindrical internal reflection, thermal ATR and imaging ATR. HATR has a horizontal sampling surface. In horizontal IRS, the IRE crystal is beneath the sample with a single exposed face, rather than sandwiched between the sample as in traditional vertical accessories. The optical material is usually silver chloride or KRS-5 (42 wt.% TlBr, 58 wt.% TlI). Single reflection systems are capable of generating higher contact efficiencies. Many of the systems used are multiple reflection accessories. In these versions, the crystal is configured to allow the beam of the instrument to reflect many times. The sampling surface is small (0.75 mm for a single-bounce up to 4 mm for the 9-bounce). Single-bounce ATR can be used to analyse discrete areas. The shorter, single-reflection, pathlength produces absorbance values that are within the linearity of the FTIR technique and therefore permits more complete and accurate spectral subtractions. Developments in single reflection systems now allow a viewing capability for micro-samples (typically $500 \times 500\ \mu m$). ATR micro-samplers (sampling area less than 250 μm) can accommodate a wide range of samples, like paint chips, single fibres, films. Micro-ATR mounted on a microscope is useful for measuring small solid samples, which are neither reflective nor transmissive, such as black polymers. Single-bounce HATR is recommended when analysing spectra with strongly absorbing spectral peaks. Multi-bounce HATR is particularly suitable for analysing liquids, pastes, gels, films and soft powders. Cylindrical internal reflection (CIR) is a modification of ATR characterised by using a cylindrical crystal rod.

Depth resolution in ATR spectroscopy depends on the optical parameters such as the refractive index of the crystal element, wavelength and incident angle. The probe thickness is in the range of 1 to 10 μm. Alternative approaches are variable angle ATR [159] and multi-frequency data treatment [160]. Recently, **ATR imaging** has been introduced as an enhancement of IR imaging spectrometers. Diamond ATR devices can be heated up to 200°C (thermal ATR) allowing the study of polymerisation and curing reactions, decompositions, phase transitions, *etc.* Low-temperature diamond ATR extends FTIR measurement of solids and liquids from ambient to near liquid-nitrogen temperatures. A supercritical fluids analyser has been developed as a high-temperature, high-pressure, low-volume FTIR accessory for the study of supercritical fluids. The device, a special version of the diamond ATR system can be used to study the performance of polymer films under extreme conditions. Solvents can be introduced into the sample chamber, at various temperatures and pressures, and the changes in the polymer monitored.

Table 1.12 summarises the **main characteristics** of ATR-FTIR spectroscopy. ATR microsampling eliminates sample preparation and the sample thickness problem and is ideal for non-destructive analysis. A strong advantage of the ATR technique is the very high quality (due to easily obtainable

Table 1.12. Main features of ATR-FTIR spectroscopy

Advantages:
- Direct analysis (little sample preparation)
- No sample thickness concerns
- Non-destructive technique
- High quality data
- Surface analysis (sub-μm range) without requirements for UHV
- Depth profiling
- Suitable for almost all sample types (no gases), incl. non-transparent or intractable samples and aqueous solutions
- Microprobing (single-bounce ATR)
- Simple cleaning

Disadvantages:
- Critical optical contact efficiency
- Less suitable for weakly absorbing systems
- Difficult quantitation (pressure dependent sample contact and FTIR absorption intensities)

Table 1.13. Selection of ATR-FTIR applications

- Analysis of surface active additives
- Blooming, migration of additives towards a polymer surface
- Analysis of fibres, fabrics, coatings
- In-depth distribution analysis
- Analysis of surface impurities and defects (μATR)
- Analysis related to paintability, adhesion, delamination
- Analysis of optically dense or high carbon-black content materials
- Oxidation, degradation of polymer surfaces
- Analysis of multi-layered foils
- Analysis of melt flows
- *In situ* studies; real-time reaction monitoring

high signal-to-noise ratios even with very thin films) and quantity of data which can be acquired over the reasonably lengthy experimental times required. In most cases, industrial analytical laboratories are using reflectance IR spectroscopy as one of their prime tools. The versatility of the technique makes it useful for identification of many condensed phase materials. Conventional ATR spectroscopy of plastics often fails when employed for additive analysis because of the low use concentrations. However, the Fourier transform technique improves the signal-to-noise ratio. Identification of complex mixtures is difficult by means of ATR. As it is also difficult, if not impossible, to generate *a priori* a calibration curve of peak area *vs.* surface concentration ATR-FTIR does not provide a quantitative measure of the surface concentration [161]. But ATR does allow examination of *relative* amounts of surface-segregated additives, provided that the film surfaces are pressed against the crystal reproducibly.

A typical ATR crystal has a pathlength of approximately 10 μm, while the smallest pathlength of a typical transmission cell is 15 μm. This short pathlength makes ATR suitable for samples that absorb strongly and yields better results than transmission. There are certain areas wherein ATR should not at all be employed. Because the total effective thickness that can readily be achieved does not exceed a fraction of a millimetre even with multiple reflections, ATR is less suitable for weakly absorbing systems in which long pathlengths are required. ATR is also not the best tool for examination of brittle materials and should in general be used with caution and suspect when intimate contact between sample and IRE is not assured as relative band intensity data cannot be used reliably. A marked disadvantage of the ATR technique is that it does not measure directly a concentration profile. A calibrated pressure applicator enhances reproducible sample contact.

Internal reflection spectroscopy has been reviewed [123]; several textbooks are available [155,162].

Applications

Internal reflection spectroscopy has found great usefulness in quick qualitative identification of a wide variety of materials as it permits the spectroscopist to obtain IR spectra on many samples with little or no preparation. The application field of ATR spectroscopy covers the full range from identifying micro impurities at the surface of solids to real-time monitoring in production processes; the information gained is characteristic of the top surface (0.5 to 5 μm). Typical applications of ATR-FTIR are shown in Table 1.13.

It appears that vertical and horizontal ATR are ideal for liquids and pastes, and non-destructive sampling of pliable solids such as films on absorbing substrates, rubbers, plasticised plastics, and samples on filter paper, whereas high-pressure ATR qualifies for hard or non-easily deformed materials. Many users are discovering that single-reflection ATR offers a simple way of analysing powders without dilution. The ATR method has been utilised extensively in the analysis of polymeric materials and in interfacial studies. Applications include strongly absorbing probes (including CB-filled polymers), opaque

Table 1.14. Preferred accessory/technique reference chart[a]

Material	Multi-bounce HATR	Single-bounce HATR	DRIFTS
Soft powders	2		1
Hard powders			1
Soft polymers	1	1	
Rigid polymers		2	1[b]
Carbon-filled or black polymer	1[c]	1[c]	
Thin films (free standing)	1	1	
Foams	1		

[a] 1, primary accessory choice; 2, alternative technique.
[b] Si-Carb sampler to abrade sample.
[c] Ge crystal.

solids, single fibres and fabrics, and air-sensitive probes. Potential applications for ATR spectroscopy are enormous [157].

Liquids are one of the easiest classes of materials to study quantitatively via single- or multi-reflection ATR because a well-defined contact surface is obtained. The particular advantage of ATR over conventional transmission for the study of liquids is that the requirements on the liquid cell can be relaxed, especially where small thicknesses are required for transmission measurements. Simpson [97] used HATR-FTIR for the multicomponent analysis of formulated oils. Internal reflection spectrometry can also be used to identify solutes in volatile solvents since the solvent can be readily evaporated, leaving the solute as a thin layer on the surface of the IRE. Repetitive analysis of liquid samples is made easy by the wipe on/wipe off sampling afforded by the horizontal ATR accessory. Table 1.14 lists the preferred accessories for various **solid sample types**. When both single-bounce and multi-bounce are indicated, single-bounce is more suitable for examining the main component; multi-bounce qualifies for lower concentration components or weaker spectral features.

Botros [163] used ATR-FTIR (both isothermal and non-isothermal *in situ* measurements), HPLC and dielectric constant measurements in evaluating **antiblock performance** of fatty amides in EVA copolymer and LDPE in order to optimise the conditions for bulk shipments of these polymers in warm weather conditions. It was observed that two different amides showed quite opposite behaviour. At 50°C (representing the temperature inside a rail car in summer), (the preferred) amide-I gave a maximum concentration on EVA surface while amide-II totally disappeared. LDPE showed the opposite trend. *In situ* measurements of the amide concentration on the EVA copolymer surface were carried out isothermally at 40°C and non-isothermally (30–65°C) on samples with 3000 ppm loadings. The ATR (Ge) technique was used and the apparent angle of incidence of the IR beam was varied between 40–65°C to measure amide concentration at different penetration depths from the polymer surface. Factors affecting antiblock performance of fatty amides in LDPE and EVA are amide concentration, amide type, time, temperature and base resin. The two main mechanisms influencing amide performance are the migration ability to the resin surface and the (in)compatibility with the polymer matrix. Use of ATR-FTIR for **surface measurements** at elevated temperatures has been cited elsewhere [164–166]. An important field of application of (micro) ATR is the study of surfaces (layers, coatings) and surface reactions (oxidation, degradation and blooming) and problems in which spectra of monolayer films must be monitored. For surface analysis of **blooming** phenomena on vulcanisates various IR techniques can be used, such as reflection and ATR; similarly, surface migration of stearates induced in conditions of high temperatures and high humidity is easily detected by ATR-FTIR. Polysulfone membranes were characterised by ATR-FTIR, FAB-MS and XPS [167]; the unexpected observation of N was ascribed to residual solvent DMF rather than an additive. Plasma- and wet chemical-induced surface functionalisation of polyolefins for increased adhesion with fibres was similarly monitored using ATR-

FTIR, AFM and XPS [168]. Surface capable infrared techniques, such as ATR-FTIR and PA-FTIR are also suitable tools for *in situ* analysis of **automotive clearcoats** [29]. ATR-FTIR has further been used to examine skin and core structure of UD-PE film [165]. The ATR optical arrangement is a well understood way of obtaining top layer spectra.

ATR-FTIR spectroscopy has been widely applied as an analytical tool in the (sub)μm range allowing for surface characterisation and **depth profiling** of materials without the need for sectioning of the sample, and subsequent chemical analysis or surface etching, such as sputtering. **Migration** of a plasticiser (DEHP) in PVC containing stabilisers, and of di-*n*-butyltin dilaurate (DBTDL) and di-*n*-butyltin maleate (DBTM) under various conditions (heat, accelerated weathering, outdoor exposure, hot water immersion) was studied by depth analysis using μATR-FTIR and μFTIR on microtomed thin slices [79]. Tatsumi *et al.* [169] used variable angle ATR-FTIR depth profiling (optical microtoming) and ATR-FTIR spectroscopy with sputter etching for the determination of the depth distribution of a chemical additive (a cationic polyacrylamide) within a pulp fibre.

IRS examines only the surface layers of a sample. If the sample is not homogeneous, the spectrum will not be completely representative of the sample as a whole. However, this characteristic can be used to advantage when studying the migration of species to the surface of a polymer (such as antistatics, mould release agents, plasticisers, low-MW pigments, *etc.*). **Surface migration** characteristics of a tackifier additive, polyisobutylene (PIB), in 25 μm thick LLDPE films were investigated by means of ATR-FTIR [170]. Hirt *et al.* [171] have studied the surface concentration of fluorinated additives in HDPE films by means of ATR-FTIR and XPS. ATR-FTIR was also used to determine migration of a phthalate plasticiser to the surface of a PVC article [172]. Surfactant migration in acrylate copolymer coatings was monitored using ATR [173]. Exudation of sodium dioctylsulfosuccinate (SDOSS) surfactant molecules to the film–substrate (F-S) and film-air (F-A) interfaces in styrene/*n*-butyl acrylate latex films in the presence of trimethoxysilylpropylmethacrylate (MSMA) molecules was examined by polarised ATR-FTIR spectroscopy [174]. The distribution of surfactants in latex films can be studied by ATR-FTIR, FT-Raman microscopy and by PA-FTIR (*cfr.* Chp. 1.3). Miller *et al.* [175] have used mid-IR *in situ* IRS with reactive internal reflection elements to quantitatively monitor the adsorption of surfactant species.

Van Alsten *et al.* [166] have used ATR-FTIR spectroscopy for measuring the dissolution of a (polymeric) diffusant into a matrix of another polymer; the method is applicable wherever the components have spectroscopically distinguishable absorption bands. Recently, ATR-FTIR has been used for monitoring small particle **diffusion** in polymer film [170,171,176–180]. Balik *et al.* [180] have described an ATR cell for analysis of diffusion of small molecules, such as amylacetate and limonene in polymer thin films with FTIR. The cell was designed to be used with precast (commercially extruded) polymer films, avoiding the need to cast the film directly onto the ATR crystal and allowing the as-processed transport properties of the film to be assessed. Diffusion coefficients obtained from the ATR cell compare favourably with values obtained gravimetrically for the same polymer and penetrants. ATR-FTIR has also been used to study diffusion of alcohols through sulfonated PS/PIB/PS block copolymers [181]; the challenge liquid was allowed over the sample and the transport process at the polymer/crystal interface was monitored at 3450 cm^{-1}. Yarwood *et al.* [182] used ATR-FTIR experiments (with approximately 40 reflections) in the study of diffusion of silane coupling agents in 10 μm thick PVC film (unplasticised and DHA or Diolpate 7170 plasticised). Fibre optic evanescent wave spectroscopy (FEWS), which is based on ATR, has been used for *in situ* and real-time investigation of the initial stages of diffusion of water and organic compounds in amorphous polymers [183]. Variable temperature ATR-FTIR has been used to investigate the migration of automotive fuel components (methanol, toluene) in a series of high barrier fluoropolymer (PVDF), polyester (PBT) and polyamide (PA12, PEI) based films at various temperatures [176].

In many instances it is of interest to obtain the spectrum of a film on an absorbing substrate without destroying the sample. This includes films on opaque substrates (*e.g.* coatings on metals) as well as films on partially absorbing substrates (*e.g.* protective coatings on fabrics). ATR-FTIR may be used for quality control of **packaging materials**, troubleshooting, competitive product analysis and product authenticity. Micro-ATR-FTIR has been used to determine silicone rubber inclusions in thick acrylonitrile/methyl acrylate copolymer film [184]. ATR-FTIR is a tool to identify analytes on filter **paper**. The use of ATR-FTIR to identify the components separated by paper chromatography has been reported [185].

Fibres and fabrics, among the most difficult materials to handle via transmission spectroscopy, are quite amenable to being studied via IRS. Multiple-bounce ATR-FTIR is of importance in identifying fibres, showing quantitative blend ratios, and in the analysis of fabric additives. ATR-FTIR was used to characterise the surface of graphitised carbon fibres [186].

Dyes present in polymers obviously exhibit very strong UV/VIS absorption and can thus be *detected* at very low concentration. However, *identification* of such a **dye** is generally based on IR absorption (ATR). Although IR spectra are helpful for dye identification, the sensitivity of the ATR-FTIR method is relatively poor. When the dye concentration is only 0.5–1 wt.%, the dye IR absorption bands are simply too weak to detect. Successful use of IR to identify dye functional groups therefore often requires extraction of the dye from its polymer matrix. Although ATR-FTIR cannot be used to discriminate between discolouring of pigments and degradation of PP, the technique can be used to determine the degree of oxidation of the polymer regardless its colour [187].

Carbon-black (CB) filled rubbers are difficult to analyse by IR spectroscopy (even in thin sections) because CB causes large absorption bands in the spectrum which often obscure the region of interest. For such totally absorbing materials good quality spectra may be obtained using micro-ATR equipped with the high refractive index Ge crystal to limit the absorption of the infrared beam in the first microns of the surface [157]. The analysis is obviously characteristic of the surface of the sample. Delor *et al.* [188] have reported a comparison of FTIR techniques (TIR, ATR, PAS) on transparent materials in order to validate the use of horizontal ATR with a germanium crystal (HATR (Ge)) for the study of industrial EPDM, NBR and CR formulations in automotive applications (tyres, hoses, belts, weather-strips, *etc.*). In rubber samples with 20 wt.% carbon-black ATR fails to discern useful structural features. Large CB contents (up to 50 wt.%) in most elastomer formulations limit conventional transmission IR spectroscopy (TIR) to the study of very thin samples (few μm) obtained with a cryogenic microtome. Analyses involving reflection techniques based on the use of crystals with high refractive indexes are suitable alternatives for ageing studies directly on heavily loaded industrial formulations, which cannot be carried out by transmission or photoacoustic methods.

ATR-FTIR was used to assess the amount of erucamide at film surfaces [189]. The disadvantage of ATR-FTIR is that it is difficult, and in many cases impossible, to correlate a specific ATR peak area with a surface concentration. A quantitative analysis is not quite feasible as usually a concentration gradient exists at the surface. **Semiquantitative ATR-FTIR analysis** of high-MW HALS in 150 μm thick LDPE film has been reported [190]. The "surface concentration" of HALS in the LDPE film was derived. ATR-FTIR has also been used to characterise PVC plastisol behaviour [191]. Herres [192] has compared the suitability of various non-destructive methods (FTIR, ATR-FTIR, PA-FTIR and LR-NMR) for quantitative determination of plasticiser content in filled PVC. LR-NMR was found to be more accurate and faster, while the IR techniques provided additional information, *e.g.* on the possible accumulation of plasticiser near the surface. ATR-FTIR is useful also for quantitative determination of a polyacrylamide resin (PAM), which is a dry-strength additive for paper sheets [193]. Coles *et al.* [194] determined quantitatively kaolin clay in polyethylene/vinyl acetate by means of ATR-FTIR and μFTIR. The latter method, while providing better resolution of kaolin, does not have a large enough sampling area and is therefore subject to small shifts in concentration of filler within the sample. ATR is not as sensitive to kaolin as μFTIR, but provides a larger sampling area and more consistent results. The filler content of the polymer was confirmed by ashing. When a polymer is ashed care must be taken as the composition of the filler could change during the process.

The study of printed inks is a challenge for the infrared spectroscopist, especially because the sample typically has to be studied *in situ*. Single-reflection micro-ATR is the only practical sampling method because of the high spectral background associated with the substrate material, which is usually paper or some other cellulosic material. Multiple internal reflection has been used for determining commercial FRs in formulations and as finishes on acrylic fabrics [195]. TiO_2 and Fe_2O_3 photocatalysts immobilised on modified PE films were studied by ATR-FTIR and XPS [196]. Reflection infrared can further be used to determine the **molecular orientation** in biaxially oriented samples.

A particularly important application of ATR-FTIR is the *in situ* study of swelling of polymer O-rings and seals under extreme conditions. *In situ* ATR-FTIR also allows the study of the behaviour of polymers subjected to supercritical fluids [197]. Reliable measurements of the solubilities of CO_2 or any other IR-absorbing gas in polymers at various temperatures and pressures are possible. Kazarian *et al.* [198,199] have examined the interaction of $scCO_2$ with a variety of polymers (PMMA, PVAc, PC, PET, PS, PE) by means of *in situ* ATR-FTIR. The technique is useful for monitoring supercritical fluid processing.

Internal reflection spectroscopy is also ideal for monitoring of reactions in real-time, such as on-line cure monitoring. Compton *et al.* [157] have reported the cure of a viscous adhesive smeared across the IRE surface of a HATR accessory and collected spectra at 30 s intervals. ATR-FTIR spectroscopy also shows strong promise for allowing a combinatorial approach in searching for new and useful polymerisation parameters [200]. Fischer *et al.* [151] have recently used FTIR with a **melt flow cell** and an in-line ATR-dipping probe for quantitative simultaneous multicomponent analysis of several additives in PVC using chemometric methods (PCR and PLS). Relatively small amounts of additives (3%) were detected with an absolute prediction error of 0.3%. The same authors also quantified acrylic monomers in an acrylate-butadiene rubber during the mixing process in an extruder using on-line IR and in-line ATR-dipping. As may be seen from Chp. 7.2.3, many mid-IR process control systems for polymer melts have been developed and are being used in industry.

Applications of IR spectroscopy to investigations of a variety of real surfaces were illustrated elsewhere [123,155]. ASTM E573-96 relates to standard practices for internal reflection spectroscopy [201]. Aldrich/IChem/STJ ATR-FTIR spectral libraries hold 1567 polymers and polymer additives and 949 dyes and pigments [65a].

1.2.2. Near-infrared Spectroscopy

Principles and Characteristics

The near-IR region, which extends from about 780 nm to 2500 nm (or 12,820 to 4000 cm^{-1}) and is located between mid-IR (from 2500 to 50,000 nm or 4000 to 200 cm^{-1}) and visible light, was essentially discovered by Herschel [202]. The region is commonly divided into shortwave near-infrared (SW-NIR: 780–1100 nm) or third overtone region, and longwave near-infrared (LW-NIR: 1100–2500 nm), mainly as a result of detector optimisation. Current interest in near-infrared high overtone spectroscopy (NIRS) arises from a number of hardware developments, including improvements in Fourier transform technology, infrared detectors, monochromators and diode laser sources, the development of fibre optics for the near-infrared range and rapid data acquisition. PC technology in the early 80s became the driving force behind NIRS. NIRS is a secondary analytical technique with results calibrated against reference analytical techniques.

For the ideal harmonic oscillator only the fundamental vibrations are allowed and there would be no NIR spectrum. An important consequence of the anharmonic nature of molecular vibrations is that transitions between more than one energy levels are allowed. These transitions give rise to overtone absorption bands. The near-IR bands result from transitions between the ground state and second or third excited vibrational states. The near-IR region of the spectrum thus contains mainly **overtones and combination bands** of fundamental mid-IR absorption bands (*cfr.* Fig. 1.5). The intensity of the overtones depends on the anharmonicity of the vibration. Near-IR intensities are some 10 to 100 times lower than the corresponding fundamentals in mid-IR; to compensate this, samples are 0.1 to 1 mm thick, which is a large virtue in comparison to mid-IR. There is no special theory of near-IR spectroscopy.

In the NIR region vibrations predominate of light atoms with strong molecular bonds. Typically, strong NIR absorbers include C−H, O−H, N−H, S−H, C=O, =C−H, COOH, and aromatic C−H functionalities. Consequently, nearly all organic analytes (vapour, liquid or solid) have a characteristic NIR spectrum; however, the spectral interpretation in terms of molecular structure is usually rather complex and many band assignments are unresolved. This fact strongly reduces the qualitative power of the spectra, relative to mid-IR. Moreover, as NIR absorption mainly reflects vibrational contributions from very few functional groups, for detailed qualitative analysis it is inferior to mid-IR, which shows all (active) fundamentals. The NIR spectral range (commonly not even shown in collections of IR spectra) is often avoided by organic spectroscopists because of the difficulty of interpreting its spectral features. In fact, considering the theoretically enormously large number of combination and overtone transitions for polyatomic molecules, one might expect that a NIR spectrum consists of so many absorption components as to not being of any practical value for qualitative and quantitative analysis. However, inspection of NIR spectra shows that even larger molecules exhibit only relatively few bands, which is explained by a transition to "local modes" at higher energy [203]. The occurrence of relatively few bands at high NIR wavenumbers may also be understood as a consequence of vibrational intensity being strongly diminished towards higher order combination and overtone modes. Therefore, the near-IR region may profitably be employed for qualitative and quantitative studies. Furthermore, NIRS is suitable for analysis of compounds lacking in UV absorption, so that detection is possible without prior derivatisation.

Some form of **sample preparation** is fundamental in successful NIR analysis. Factors involved in sample preparation comprise the nature of the material itself, including physical size, texture, *etc.*, its composition, amount and type of foreign material present, grinding and other forms of size reduction, blending, *etc.* NIRS can handle thicker samples than mid-IR; the average penetration depth of the NIR radiation is about 10 mm. According to Williams [204], some 30 factors affect the accuracy and precision of NIR analysis which are attributable to samples, sampling and sample presentation. Sampling of **liquids** depends on the viscosity of the liquid. Free-flowing liquids are analysed using flow-through cells. Extremely viscous materials, such as epoxy and other types of resins and opaque slurries, are best analysed by diffuse reflectance. Materials such as rubbers, solid plastics and other materials can be non-sampled and analysed directly. Conventional sampling of these materials must be used for reference analyses. Fibre optics can extend the range of off-line measurement to several meters (*e.g.* in bulk containers). Fibre-optic sampling is an established technique for remote sampling for NIRS (*cfr.* also Chp. 7.2.4). NIR analysis is often simplified by the fact that usually only one or two constituents are to be determined. There is frequently a very strong signal at specific wavelengths, relative to the background, for the constituents to be determined.

In comparison with process liquids, solid samples are much more difficult to be handled by continuous analytical methods. In addition, solid materials, crystalline powders or pelletised plastics are typically inhomogeneous. Moreover, the inhomogeneities of the samples are physical, chemical, or both. Camajani *et al.* [205] have addressed sample handling in NIR analysis of non-homogeneous samples, such as glass-filled polymers. Near-infrared analysis can be directed to the determination of bulk properties and concentrations of such samples. In order to ensure precision of analysis, a large enough number of particles has to be presented to the sample cell. Hirschfeld [206] discussed the relationship of measurement error as a function of sample area geometry and average particle diameter. There are several experimental and commercial **contact solid sampling arrangements** in which the solid materials come in contact with the optical window, including an on-line NIR diffuse reflectance analyser. Solid sampling is needed in many instances, such as inspection of incoming raw materials (batch control) or on-line solid sampling of powdered solids (in continuous processes). Both contact solid analysers and non-contact NIR analysers have been developed. The **non-contact arrangement** is the most suitable in most processes with solid materials. A typical sampling arrangement involves transport of the sample on a conveyer belt. In the non-contact mode the material does not have to be diverted, ground, and wasted after measurement. Factors influencing non-contact analysis have been discussed [207]. In statistical terms the measurement on continuously moving samples is a sampling of continuous random variables. The first use of non-contact NIR analysers was in measuring moisture content of samples moving on conveyors.

Basic **instrument configurations** for NIRS are near-infrared transmittance (NIT) and near-infrared reflectance (NIR). An ideal research instrument would have both capabilities. **Transmission spectroscopy** (subject to Lambert-Beer's law) can analyse non-scattering bulk polymers, molten polymers, and polymer solutions. Transmittance techniques are most useful for measuring large particles. In transmittance measurements, particle size can be small enough to begin to scatter most of the energy striking the sample. For **near-infrared reflectance analysis** (NIRA), the NIR region has some special advantages over mid-IR. The reflectance/absorbance ratio is larger in NIR and calibration plots are therefore more likely to be linear and reproducible [208]. Also, NIR sources have higher energy and detectors have greater sensitivity than those available for mid-IR instrumentation. Consequently, measurement of reflected NIR radiation is inherently more

Identification by DRIFT spectroscopy

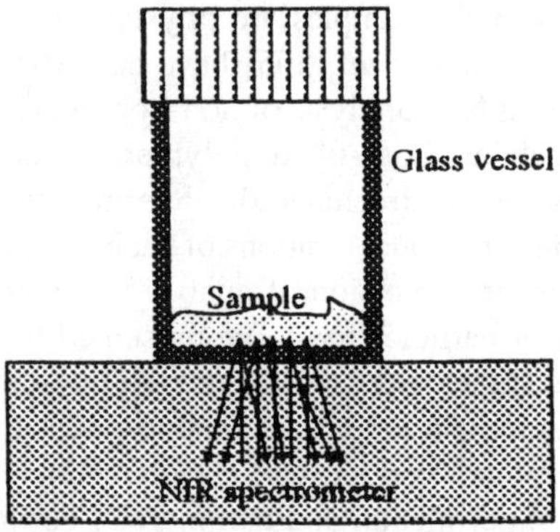

Fig. 1.10. Identification by NIR diffuse reflectance spectroscopy. After Van der Maas [210]. Reprinted with permission from *Spectroscopy in Process and Quality Control (SPQ)*, 1998. *Proceedings SPQ-98* is a copyrighted publication of Advanstar Communications Inc. All rights reserved.

efficient than that of reflected mid-IR radiation. NIR reflectance measurements have a penetration depth of only 1–4 mm. This brings about greater variation when measuring non-homogeneous samples than transmittance techniques. For highly scattering polymer samples, such as semi-crystalline bulk polymers, synthetic fibres, polymers filled with small particles, colloidal suspensions and paints, **NIR diffuse reflectance spectroscopy** can be used. The low absorptivities of NIR bands enables analysis of samples with long "effective pathlengths". As a result, samples with large particle sizes, low crystallinities, or low filler loading can be directly analysed by NIR diffuse reflectance spectroscopy. A top-loading diffuse reflectance accessory that utilises a unique optical focusing system virtually eliminates the specular reflected component. Samples can be analysed directly from plastic bags and most plastic and glass bottles (Fig. 1.10). Spectra of opaque solids can be obtained by utilising light collection devices such as integrating spheres (*cfr.* Fig. 1.1). The diffuse reflectance mode (subject to Kubelka–Munk's theory) exhibits dependency of the scattering coefficient on particle size, sample packing, refractive index effects, *etc.* The finer the powder, the better the scattering. For homogeneous solid sampling quantitative results are excellent; for heterogeneous materials averaging of replicate probing by means of a rotating sampling device is required. Diffuse reflectance spectroscopy is important for NIR analysis and has been used for almost three decades for the analysis of agricultural and industrial products [209]. Most NIR applications involve the diffuse reflectance mode.

Attenuated total reflectance (ATR) can also be performed in the NIR region, especially with FT-NIR instruments, although the spectra are very weak (low extinction coefficient, small wavelength) [175, 211]. The major advantage of NIR spectroscopy over IR spectroscopy in evanescent-field sampling is the availability of optical fibres and ATR crystals (ZnSe and Si) that are non-absorbing in the NIR region. **NIR photoacoustic spectroscopy (PA-NIR)** can also be used to analyse bulk polymer samples [212, 213]. The penetration depth of the NIR beam can be controlled so that spectra can be obtained from a defined region even 1 cm below the surface. Surface chemistry studies in the NIR region are limited.

Near-infrared instruments of the UV-VIS-NIR type have become commercially available about 1955 with applications for agricultural commodities. Instruments designed specifically for measuring NIR energy reflected from solids have been commercially available as from 1971 [214]; the development of these devices was pioneered by Norris [215]. The first successful uses of modern NIRS were in the 1100–2500 nm region. NIR instrumentation is now extremely varied: from UV-VIS-NIR to FTIR instruments, NIR reflectance instruments, PAS technology, on-line and portable analysers.

A practical advantage of the NIR spectral range is that powerful broadband light sources and sufficiently sensitive and stable photodetectors are available in the form of tungsten-halogen lamps and either photomultipliers, solid-state photocells or diode array detectors, respectively. The polychromatic light emitted by the source is commonly separated into monochromatic light by use of diffraction gratings or narrow band pass interference filters. Typical wavelength ranges of commercial NIR spectrometers are 400–2500 nm. No single detector covers the entire 780–2500 nm near-IR range. Detectors used include PbS(Se) photocells (1100–2500 nm), Si (visible, 400–900 nm), Ge, InSb, and recently (extended) InGaAs (*cfr.* ref. [216]). NIR photodiode array detectors using InGaAs are now available in the 900–1650 nm or less sensitive but longer wavelength 1100–2400 nm versions, making it possible to measure an entire NIR spectrum within a time scale of 1 to 2 msec.

Various **NIR technologies** are available from over 50 manufacturers:

(i) *Double-beam filter systems*
(Characteristics: discrete λ, reference measurements, at-line/on-line for solids and pastes, robust, low cost, user's friendly, long life).

(ii) *Grating technology*
(Characteristics: "scanning" system, slow method development, unlimited number of applications, at-line/laboratory, medium cost).

(iii) *FT-NIR technology*
(Characteristics: all wavelengths simultaneously, wavelength accuracy, crystal technology, fibre optics, high spectral resolution, integrating sphere, remote control, mobile product identification, microscope, near-line/at-line, expensive).

(iv) *Acoustic Optical Tuneable Spectrometer (AOTS)*
(Characteristics: crystal technology, no moving parts, ultrafast scanning, robust, glass fibres, probes, process control, multiplexing, application to liquids only, expensive).

(v) *Diode array spectrophotometer*
(Characteristics: real-time monitoring, 2 spectra/ms, 900–2400 nm InGaAs array detectors).

Simple NIR instruments employing interference filters for wavelength selection are widely used for quantitative analysis. Rose [217] demonstrated the potential of NIRS for qualitative analysis on a filter instrument. Filter instrumentation (with insufficient numbers of wavelengths) is most suited for the determination of one component but unsuitable to tackle complicated problems. Dispersive NIR (D-NIR) spectrometers, which use gratings or prisms, are commercial since 1978 and are still the main research instrument. The predictable FT-NIR advantages over typical D-NIR systems are wavelength accuracy and precision, signal intensity accuracy and precision and resolution capability. Peters *et al.* [218] have compared D-NIR and FT-NIR. Silicon photodiode array (PDA) detectors have been in use for chemical analysis since 1982 [219]; extension of this technology for short wavelength near-IR (600–1100 nm) measurements was proposed in 1989 [220,221]. SW-NIR instruments are compact, robust and relatively cheap. They allow identification of only a limited number of plastics and cannot be applied for the identification of black-pigmented parts in recycling applications. Portable NIR analysers are commercially available for use in the polymer, chemical and food industries and are ideal for non-destructive QC applications. Finally, also ATR instruments are available.

Near-IR measurement has four aspects: *(i)* experimental design for sampling and calibration set construction; *(ii)* spectrometry; *(iii)* multivariate calibration; and *(iv)* data analysis with presentation of results. The greatest advantage of NIR spectroscopy is the ease of sample handling (intact sampling). NIR spectroscopy allows various measurement modes: in reflection (for solids, powders, granulate), transflection (for slurries, semi-solids, liquids, films, emulsions) or transmission (for clean liquids). In the NIR region more concentrated solutions (10–100 g/L) are required as compared to the UV region because the vibrational overtones are weak and solvent absorption can be more of a problem. Several of the aforementioned sampling methods of NIR spectroscopy can also be used for polymer analysis.

NIR spectra contain a wealth of information about the physical and chemical properties of molecules, which however cannot easily be extracted from the spectra. The weak near-IR absorption bands are rather broad and overlapping, which reduces the need to use a large range of wavelengths in calibration and analysis but makes spectral analysis intricate. High overtone spectra have a deceptively simple appearance, usually dominated by the overtones of hydrogen stretching vibrations in a diatomic-like pattern. The three major parameters that affect the quality of NIR spectra are: *(i)* stray light; *(ii)* wavelength reproducibility; and *(iii)* instrument noise [222]. The more demanding problems need high quality data, calibration and validation (traceable standards for wavelength accuracy and reproducibility). High-quality NIR spectra are provided by improved conventional and Fourier-transform (FT) spectrometers. A FT-NIR atlas comprising the spectra of approximately 2000 substances in the wavenumber range of 3800 to 10,500 cm^{-1} has been edited [223]. As a result of the complexity of the NIR spectrum it is very difficult to isolate just the band of the material to be measured. This overlapping also makes interpretation of spectra in qualitative terms very difficult. Efforts to analyse complex NIR spectra thoroughly and more clearly have taken several directions: *(i)* systematic studies of NIR spectra of basic molecules; *(ii)* theoretical studies of manipulation of overtones and combination modes for band assignment; and *(iii)* 2D correlation spectroscopy [224]. Two-dimensional (2D) correlation spectroscopy enhances similarities and differences of the variations of individual spectral intensities, accentuating useful information often obscured in the original spectral data set.

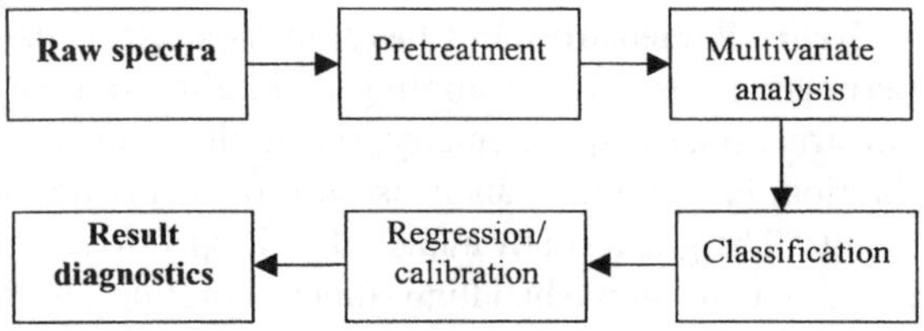

Fig. 1.11. Flow-chart of NIR data analysis.

Several pretreatments are used in **spectral data processing** (scaling, non-linear transformations, *etc.*). Recent advances in spectral matching using various discriminant analysis algorithms have broadened the usage of NIR to include qualitative analysis. In general, three types of spectral regions are identified: *(i)* a region containing information about the analyte; *(ii)* a region containing information about more than one analyte; and *(iii)* a region exhibiting very little spectral information (as reference wavelengths). Because NIR is composed of overlapping overtones and combinations of bands originating in the mid-IR, a chemometric algorithm is needed for all but the simplest chemical systems.

The theory and principles of NIR are not just the physics and electronics. The theory and principles must include **chemometrics** and the reference analysis as part of the total technique. Lack of structural interpretative value has been partly compensated by chemometric evaluation techniques [225–228]. Chemometrics (after 1985) has enabled use of diffuse reflectance techniques on granular solids that produce spectra that could not be treated with a simple one-wavelength Beer's law approach.

Calibration is very important for NIR since classical interpretation of the spectra is nearly impossible. Practically any application calls for calibration of the instrument against a reference method. Nowadays, a NIR instrument without a calibration model is useless. Calibration-free methods for NIR are being researched.

A flow-chart of **NIRS data analysis** is shown in Fig. 1.11. There are two different procedures that are commonly used for NIR data analyses: *(i)* the calibration step, during which (linear combinations of) wavelength responses are selected and related to the property, using calibration samples with known properties; and *(ii)* the validation step, during which the calibration is tested with additional samples that have known properties. In general, calibrations of NIR responses to polymer characteristics can be done for chemical properties (such as composition, monomer ratio, or crystallinity) or for non-chemical properties (such as modulus). For chemical property calibrations that use NIR transmission spectroscopy, in most cases the Beer–Lambert law describes the relationship between the property and the NIR responses. Building of a good, robust **calibration model** with excellent prediction properties is very time consuming. In many uses of NIR instruments and calibration models, long series of measurements are required and one has to be sure that the calibration model works well over long periods of time. Standard practices for the (near) infrared, multivariate and quantitative analyses have been developed [229–231]. Outlined are methods for selecting and analysing the calibration set: collection of near-IR (and IR) spectra; and calibration and validation of the spectrophotometer.

Choosing samples for a **learning set** in correlation spectroscopy is the important first step toward developing a near-IR (chemometric) analytical method. The final objective is to produce an empirical analytical equation including terms for selecting wavelengths that will be "robust" upon future application to all anticipated quantitative tasks for the analyte of interest in the matrix of practical consideration. The significance of a particular calibration can be estimated using several statistics calculated during modelling. Statistical "overfitting" of the data must always be avoided. Conventional methods of sample selection involve identification of all sources of variance likely to be encountered in future analysis such as sample source, composition range of the constituents and parameters to be tested. Important factors that affect the behaviour of solid samples are chemical composition, physical texture, bulk density, and colour. The first three factors mainly affect the diffuse reflectance from the surface. Colour affects the gross reflectance of light energy from the surface. Sufficient samples with reference analyses are to be accumulated to enable the generation of calibration files. The spectral method of sample selection involves selection of a large number of samples strictly on the basis of spectral characteristics. These are often quite similar, and only vary in relatively minor respects. Reference analyses are also carried out on a relatively small number of samples that display the most comprehensive variance in optical data.

Regression analysis can be used to extract information from NIR spectra. Multivariate discriminant analysis enables to separate samples into different classes. There are two mathematical approaches

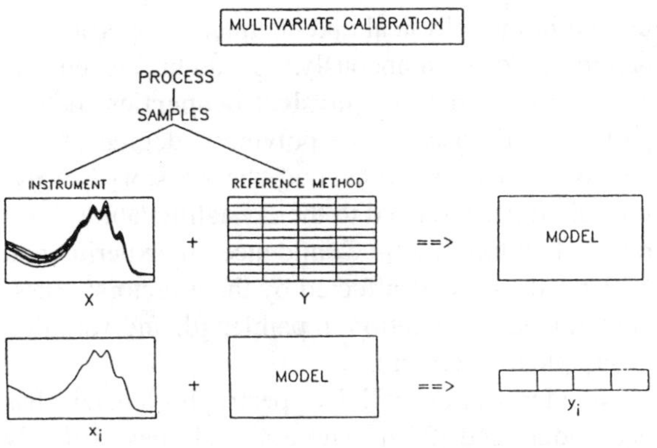

Fig. 1.12. Multivariate calibration.

to both exploratory data analyses and **regression analyses**: the discrete-wavelength and full-spectrum approaches. In the discrete-wavelength approach, only a few wavelength responses (typically six) in the spectrum are used. The wavelengths chosen for regression should be evaluated as to whether they are in fact characteristic of the chemical moieties or components being analysed. With the need of measuring more demanding samples, lower concentration levels, less absorbing analytes, and more complex matrices, more wavelengths are required to gain information about a sample. In the full-spectrum approach, linear combinations of all available wavelength responses in the spectrum are used. Complete spectra are now often used for product identification.

The simplest and most general applicable calibration algorithm used for modelling is multilinear regression (MLR). Generally, MLR is applied to simple systems where unique spectral features are readily isolated for the analyte of interest. In statistical **multivariate calibration** (MVC) methods applied to NIR analysis, a calibration is determined not from band assignments but rather from statistical analyses of spectra of samples in which the property of interest is known from an independent reference method (schematically illustrated in Fig. 1.12). Various full-spectrum calibration methods have been applied to NIR analyses, such as PLS, PCR, PCA, CLS, factor and discriminant analysis [232,233]. Full-spectrum multivariate approaches are employed when unique spectral features can not be isolated, when the absorption bands are strongly influenced by matrix effects or when matrix properties, that depend upon multiple chemical constituents within the matrix, are being correlated to NIR spectra [234]. PLS can be used for the analysis of data where the number of variables exceeds the number of observations. MVC based on PLS or PCR works even in cases when no selective signals exist for any of the constituents, and also in case of "unknown" interferents. The spectra of the pure constituents need not be known. This has caused this approach to be widely used for analysis of complicated samples.

Qualitative analysis is largely used to determine whether a sample is of a desired quality. In many cases the similarity or differences between spectra are subtle and thus cannot be decided by human judgement on a quick, reliable basis. The basic idea of qualitative analysis in NIR is to use spectral measurements to classify objects in groups, as opposed to predicting the value of some quantitative measurement. Chemometric procedures for qualitative analysis require that data for a number of variables should be measured with good precision. Although it has become accepted that complete spectra are required for qualitative analysis, often only a limited number of wavelengths is used. Using a filter instrument Pradhan *et al.* [235] successfully carried out qualitative analysis. Filter instruments are less expensive and generally more robust than scanning instruments and their use in qualitative control systems (such as identity check of raw materials at the in-take point [236,237]) should be given serious consideration.

One of the main purposes of measuring NIR data is the determination of chemical composition or physical properties in a quantitative way. The principle of the measurement procedure for **quantitative analysis** is based on recording the NIR spectra of reference samples (the number depending on the number of components or parameters to be determined) of known composition. The levels of the constituents or the physical parameters are determined by independent, conventional analytical or physical methods. Then the set of reference spectra and the independently determined values of the parameters under investigation are used by a selected statistical method to build a calibration. This enables unknown samples to be evaluated with regard to the individual parameters of interest. The accuracy of the NIR technique depends upon the validity of the calibration data set, which must incorporate the entire range of concentrations that will be determined by the instrument. This set must contain samples with varying ratios of each component. NIR calibrations do not typically extrapolate or interpolate well across concentrations. Typical calibration sets include more than

25 samples. In spectroscopic measurements covering an extended NIR wavenumber range, overtone and combination modes with very different molar absorption coefficients can be recorded simultaneously (*e.g.* water in organics, organics in water). This makes it possible to determine concentrations differing by several orders of magnitude in a single experiment and provides a large dynamic range for concentration measurements. Minor components may thus be analysed together with the primary components [238].

Because many factors influence the response at each wavelength, good **correlations** can often be found even though they may not be associated with any particular spectral feature. Moreover, non-chemical factors can influence the spectra but may carry chemical information at the same time. For example, in reflectance measurements, variations in particle size affect the amount of light reaching the detector. The empirical correlation is quite specific and one which is valid for HDPE is unlikely to apply also to wool (where yet other factors influence the reflectivity) or even to LLDPE. Each method must be calibrated for a particular industrial product, and is vulnerable to distortion when there is an unforeseen change in input materials (*e.g.* through a change in additive used) or in the manufacturing process.

As plastics are "formulated" polymers (containing additives, pigments, dyes, fillers, reinforcing material, *etc.*), which may be blended or coated, and are available in variety of physical forms (chips, powder, film, fibre, sheet, latex, dispersion, emulsion), it is important to use the right calibration model. The concentration of the additives in the "standards" used for model development must span the values of the samples to be analysed successively. Nonetheless, once calibrated, these methods (which are labour intensive in development) can be of great economic value in maintaining consistent product quality, in reducing spoilage, waste, and off-spec material, and in staying as close as possible to regulatory requirements. They find increasing use for quality control in a wide variety of manufacturing processes, including polymer processing, textile finishing and in the pharmaceutical and petrochemical industries. Such rapid, readily automated quantitative analytical methods make possible economical mass production of consistent, high quality.

With a scanning time of 0.1 s and sampling intervals of 2 nm NIR spectrometers are ideal for non-destructive QC applications. Non-destructive analysis of the samples makes NIR spectroscopy also unique in that physical and chemical properties can be measured simultaneously. Spectra have been correlated to a variety of **physical properties** such as particle size, viscosity of polymers, degree of dispersion in paint systems, and "heat history" of nylons, a property related to the crystalline/amorphous ratio of nylon [239]. Confidence in experimental NIR results may be affected by three factors: cross-contamination, variation in pathlength and variation in sample temperature.

Standardisation of NIR spectrophotometers has been addressed [240]. Burgess [241] has critically evaluated the qualification and **validation** of NIRS systems. Key parameters in need of control are: wavelength accuracy and reproducibility, photometric scale linearity, noise, drift, spectral bandwidth and stray light. Krischenko *et al.* [242] described a validation procedure to test the degree of standardisation of NIR spectrometers. For NIR wavelength validation is secured (λ He–Ne laser). The determination of wavelength accuracy in the NIRS region has been described [243]. Developments in calibration methodology and availability of new transfer standards are required in order to ensure "fitness for purpose" and transferability of calibrations and methods. There are currently no commercially available traceable wavelength standards for the 2000–2500 nm region. Freeman [244] has pointed attention to transmittance and reflectance standards and NIR calibration services. Measurement services available from NPL cover transmittance, reflectance, detectors and sources; transmittance standards exist for NIR.

Successful application of NIRS greatly depends on the robustness, specificity, selectivity and transferability of the calibration. In particular, it is highly desirable to be able to support a plant analyser from a laboratory instrument. Sampling-based optical artefacts and the optical characteristics of each instrument, including mechanical and photometric errors, impair transferability of data between instruments. Collaborative NIRS projects with transfer between instruments of calibration files, equations and spectra have been described [245] and interlaboratory collaborative studies of NIRS calibration methodology (for moisture analysis) have been carried out [246]. Blanco *et al.* [247] have compared various calibration methods in NIR diffuse reflectance spectroscopy. An absolute procedure for instrument calibration and standardisation has been presented [248]. A cloning procedure and methods for wavelength and optical corrections were described [249]. Although calibration transfer (from one instrument to another) is not without problems [250], the feasibility for FT-NIR spectrometers was shown recently [251]. Hammond [252] has discussed regulatory acceptance of NIR spectroscopy.

Table 1.15 shows the **main characteristics** of NIRS. Accuracy in the determination of laboratory reference values for use in the development of NIRS calibration equations is a critical component of useful NIR technology. Coates [253] has shown that it is possible for NIR spectroscopic calibration equations to produce predictions that are more accurate than the laboratory reference values used in the calibration set. Despite it being a secondary analytical method, near-IR spectroscopy has tremendous advantages over the reference methods once calibration data sets have been developed. Major advantages of NIRS are speed, easy sample preparation and lack of dependence on extensive use of wet chemistry. The low absorptivity of near-IR energy enables it to penetrate deeply into a sample (up to several centimetres) with absence of material damage. Because of this, samples subjected to NIR analysis can have almost any convenient size or shape (for applications such as the direct analysis of plastic pellets). As NIR analysis can be conducted on relatively large sample sizes (*e.g.* 10–100 g of plastic pellets), any minor variation in sample content homogeneity, such as the degree of dispersion of the antioxidant(s), does not impose any significant precision problems on the analytical result obtained. Care must be exercised, however, to ensure that the particle shape and size distribution do not vary significantly from sample to sample because NIR spectra are easily influenced by different sample morphologies, which can affect the absorption band intensity. A further advantage of the near-IR method is its applicability to insoluble cross-linked systems. The selectivity of near-IR exceeds that of UV/VIS and far-IR spectroscopy. NIRS allows for long pathlength cells (gas cell 40 m; liquid cell 0.5–20 cm). The NIR region has virtually no meaning for structural analysis of unknown polymer samples; spectroscopic experience is less useful. One disadvantage of NIR with respect to mid-IR is that the information it supplies is less detailed. This makes quantitative analysis more difficult, but the information contained in the spectrum can be fully exploited through the use of chemometrics. However, due attention is required for non-casual correlations. The development of a good calibration model is generally very time consuming. Nowadays, the various optimisation steps of method development are greatly automated.

Table 1.15. Main characteristics of near-infrared spectroscopy

Advantages:

- Flexible sample presentation
- All sample types (of any convenient sample size or shape, from transparent to totally opaque)
- Representative sampling (10–100 g of polymer)
- Large dynamic range of sample thicknesses
- Non-invasive, non-destructive
- Small absorption coefficient(s)
- (Ultra)fast measurements (<1 min)
- Various measurement configurations; excellent diffuse reflectance spectra
- Multicomponent analysis
- Medium sensitivity
- Favourable S/N ratio ($10^5 : 1$)
- Mature technology
- Wavelength and ordinate accuracy and precision
- Simple and robust instrumentation (favourable hardware cost)
- High energy sources
- Quartz or cheap glass slides
- Long pathlength cells
- Low maintenance costs
- Qualitative and quantitative applications
- Ideal for QC and manufacturing environment
- Limited operator training needed
- 2D correlation spectroscopy; imaging

Disadvantages:

- Secondary method (requires calibration against reference method)
- Dependence on a large reference set
- Influence of sample morphology
- Slow and costly method development
- Need for quantitative calibration model
- Troublesome calibration transfer
- Strict sample temperature control required
- Spectroscopic complexity (lack of specificity: no characteristic absorption bands)
- Lack in structural interpretative value (difficult to identify unknowns)
- Lack of reference data
- Need for sophisticated data evaluation algorithms (heavy computation load)
- Weak sensitivity to minor components; minimum concentration $>0.1\%$ (no trace analysis)

As all vibrational spectroscopies, NIR is not a trace analytical method. The detection limits are in the low percentage range, in favourable cases in the high per mille range for analyses in liquid media. The advantages of shortwave near-infrared (SW-NIR, 780–1100 nm region) include the ability to use inexpensive silicon detectors, silica-based optics and optical fibres, and inexpensive light sources such as tungsten lamps and light-emitting diodes. Characteristic local mode features in the SW-NIR region can be related to molecular structures. For 2D correlation spectroscopy in the NIR region, *cfr.* Chp. 7.5 of ref. [1]. Fibre optic technologies are now providing sampling opportunities both for in-lab and in-process environments (*cfr.* Chp. 7.2.4).

Official NIRS methods (by AOAC) have been reported [245,246]. For standard practice for near-IR qualitative analysis, *cfr.* also ASTM E1790-96.

Recent reviews deal with various aspects of near-IR spectroscopy [216,238,254–260], including the theory of diffuse reflectance in the NIR region [261]; several books have appeared (*cfr.* Bibliography). Computational methods and chemometrics in NIRS were reviewed [262–264]. There is an explosion in the number of books embracing chemometrics [265–269], *cfr.* also Bibliography of Chp. 6.

Applications

Although the main areas of NIR analysis were agriculture and food industry [270], today all fields of research and quality control seem to be concerned (*cfr.* also ref. [271]). Applications of NIR spectroscopy are now found in many areas: polymers, petrochemicals, textiles, pharmaceuticals, agricultural products, dairy products, packed products, beverages, *etc.* [272]. Some 450 NIR analysers are operative in the Benelux area only.

Near-infrared spectroscopy (NIRS) can be used for product identification, classification and quality control, as well as for the determination of product properties (chemical and physical) and component concentrations in process applications, all with the object of **rapid analysis**. Near-IR analysis was born of a need to solve practical quality control problems rather than the desire to perform high-resolution molecular structure analysis in the laboratory. The samples subjected to NIRS are often very complex mixtures and are studied without any sample preparation. Competitor analysis is not possible.

NIR spectroscopy is capable of determining very different chemical and physical properties of polymers, as shown in Table 1.16. For these purposes transmittance, reflectance, specular and diffuse and transflectance modes are being used. The universal capabilities of this method are based on statistical algorithms (chemometrics), through which it is possible to establish a correlation between properties and spectral information. Consequently, this technique has great meaning in polymer analysis. However, the complex relationship between the chemical and physical structure of polymers can easily lead to misinterpretations. Temperature is a critical parameter, which needs to be controlled as temperature variations can easily deteriorate the reliability of near-IR measurements and calibrations [273,274]. Because polymers are difficult to characterise as "good" or "bad" under normal conditions, a multilevel testing procedure is helpful.

The primary advantage of NIR for industrial applications is the ability to employ robust, relatively inexpensive optical fibres to form a convenient op-

Table 1.16. Applications of NIRS to polymeric materials

Chemical properties:
- Identification of raw material, in-process stock and finished products
- Composition determinations (residual monomer content, molecular weight, copolymer content)
- End-group analysis (carboxyl-, amino-)
- Polymerisation monitoring
- Concentrations of (in)organic chemical constituents (*e.g.* additives; 0.5–100%)
- Monitoring of polymer melts for additive and/or (co)polymer composition
- Volatiles (loss on drying) and moisture content
- Glass fibre content
- Mineral oils
- Finishes on textile fibres
- Remote identification/classification of polymeric materials (recycling)
- Quality control

Physical properties:
- Amount of branching, cross-linking
- Mechanical properties (impact resistance, *etc.*)
- Transparency
- Particle sizes, fibre diameters, layer thickness
- Degree of crystallinity, orientation
- Isotacticity index
- Density
- Viscosity, MFI
- Colour, yellowing index
- Dyeability
- OH/I/acidity number
- Sensorial parameters

tical probe for gathering the spectral data on which the **material identification** is based. NIR is used not only for identification of materials anywhere in the formulation process, from raw materials to intermediates and finished products, but also for quantitative analysis. **Quantitative analysis** of materials is inherently more complex than identification and requires a more elaborate calibration set on which to base the analysis of new samples.

NIR spectroscopy has been used as an **off-line technique** since the 1950s for the analysis of polymers. Already in 1968, Greive *et al.* [275] used mid-IR and NIRS to determine the amount of ABS in PVC. Each standard film contained 3 wt.% stabiliser (dibutyltin thioglycolate), which was used as an internal standard in quantification by means of IR; calibration standards of PVC containing various amounts of ABS were used for quantification by means of NIRS. Either procedure, which requires about 30 min each, can be used as a plant control method but lacks the required precision. Since then, NIR has evolved vigorously, both in terms of hardware and software. Whereas the 1980–1995 period has been characterised mainly by instrumental and methodological design developments, more recently more chemical industrial applications (petroleum, polymers, textiles) are being reported. NIRS is one of the few analytical methods that can non-destructively analyse the many **different sample forms** that are commonly encountered in synthetic polymer chemistry (*e.g.* polymer pellets without grinding [205]). Many polymers are not very soluble and need to be examined as pressed films. Thin (1–25 mils) films for mid-IR are difficult to prepare and to measure accurately in thickness, while NIR analyses much thicker samples. Reflectance and photoacoustic techniques also provide excellent means for examining polymers in the NIR and may be developed into quantitative analyses with negligible sample treatment. Both of these methods are well suited to NIRS due to high-energy throughput and low sample absorbance.

Haanstra *et al.* [276] have determined the **information depth** (defined as the thickness of material that results in 50% loss of light) of VIS-NIR in 0.2 to 8.46 mm thick LDPE films using transmission (400 to 2500 nm) and reflectance (400 to 2200 nm) measurements. For transmission, the information depth varies between ~100 μm (at 2300 nm) and 830 μm (at 1600 nm). For reflectance, the information depth varies between 0 and 2.5 mm. The reflectance geometry proves to be the better way of measuring through a PE film.

NIR spectroscopy has been used for analysis of synthetic polymers for many years. In a review, Miller [234] has summarised various applications. Information concerning molecular weight, residual monomer content, copolymer or blend composition, and crystallinity effects has been obtained. Determinations of the heatset temperature of nylons and analysis of plastic laminates by NIR have been studied. Davies *et al.* [277] have reported the use of NIR spectrometry for analysis of a food **packaging laminate** consisting of cast PP, LLDPE and nylon. The absence of any of the laminates was detected using a semi-quantitative method with a simple linear regression technique. The method was expanded to measure not only the presence or absence of a given layer but also the thickness of each layer with a precision of 10% relative for a single determination. **Composition determinations** have concerned polyolefins, *N*-containing polymers, diene polymers, polyesters/polyethers, fibres/textiles, cellulosics, *etc.* Sensitivity to compositional properties, such as monomer ratios in copolymers, compositions of polymer blends, and concentrations of additives, depends on the ability of NIRS to detect different functional groups in a polymer.

In combination with full-spectrum multivariate analysis methods and developments in fibre-optic technology NIR has gained great importance especially for chemical quality assurance but also for automatic reaction process control of polymers, in a rational and economical manner [272,278]. Although multiple component quantitations are now routinely being performed, NIRS is not an easy to use technique. Each specific application needs to be calibrated. The complex relation between chemical and physical structure of polymers can easily lead to misinterpretations by uncritical use. Clearly, NIRS is not a technique suitable for analysis of competitor products beyond the training set.

Quality assurance:
Near-IR light-fibre spectroscopy is particularly well suited for assessing product quality of **polymeric materials**, with minimal or no sample preparation. This is actually one of the most active fields in the application of NIRS to polymers. However, it is necessary to have reasonably accurate training sets of the types of polymer to be analysed. NIRS has replaced many conventional methods in polymer analysis. Methods exist for the determination of OH number, acid value, chain length and cross-linking, methyl and methylene end-groups, primary

Table 1.17. Quality assurance applications of NIRS

Matrix	Analyte(s)	Reference(s)	Year
Colloidal suspensions	Silane coupling agents	[284]	1989
Polyolefin (powder, moulded part)	Antioxidants, stabilisers	[280,285]	1991
PP pellets	Unknown additives	[286]	1993
	Neat additives	[237]	1994
PP granulate	Tinuvin 770 (0.05–0.4 wt.%) Irganox 225 (0.1–0.45 wt.%)	[287]	1995
PVC dry blends	Loxiol G 60, Loxiol G 21 Stabilox CZE 2040, chalk	[288]	1997
Lubricant	Viscosity improvers	[289]	1998
Polymer yarn	Oil finish	[290]	1998
	Couplers	[291]	1998

and secondary amines, degree of unsaturation, residual monomer content [279], additive content [275, 280], percent components in block and random copolymers as well as mixtures, hydration number, rate or degree of cure. Starting materials in a polymerisation can be checked for correct material levels and the polymerisation itself can be followed. Degree of crystallinity and orientation [281], melting range, intrinsic viscosity, and elasticity of rubber are some of the "physical" properties measured by NIRS. Molecular weight determination by end-group analysis has been performed by measuring OH or NH absorption in NIRS [279,282]. Quantitative analysis of copolymers is another important area in which NIRS can be useful. Styrene content in styrene–acrylic copolymers can be determined using the 2100 nm combination band attributed to aromatic C—H stretch [283]. Current applications are mainly qualitative and concern fingerprinting, unique on-line identification of raw materials, verification of sources of supply, identification of plastic packaging materials (PE, PET, PP, PS, PVC, *etc.*), check for contaminants, *etc.*

Other reported NIRS applications are the determination of micro-additives in PP pellets [260], of additive levels in masterbatches or shipments, of plasticisers in PVC, of moisture content in polyalkylene glycol ethers [292], of rest monomer in polymers (*e.g.* PPO) [278]. On-line monitoring of the moisture and lubricant levels on polyacetate fibre film using NIR reflectance measurement was reported [293]. NIRS allows rapid identification of polymer dispersions and an accurate water content determination (±0.2%). The method replaces the tedious gravimetric determination of the non-volatile solid content of dispersions according to DIN 53189.

At-line NIRS measurements of the **resin content** in GFR PS materials has been reported; silica does not have a significant absorbance in the NIR region [294]. Quantitative at-line process control is still underdeveloped. Table 1.17 shows a selection of QA applications of NIRS, mostly in diffuse reflectance mode. Near-IR spectroscopy has also been used for the qualification of solid raw materials, such as batches of photographic couplers [291]. The NIR method takes 1–2 min, as compared to 10–20 min for HPLC. NIRS can predict coupler concentrations as accurately as HPLC but with a higher precision. Interaction of polymers with surfaces is an important process that affects several materials properties, such as adhesion and colloidal stability. NIR diffuse reflectance spectroscopy allows studying adsorption onto particle surfaces. Krysztafkiewicz [284] has used NIRS to evaluate silane **coupling agents** to silica fillers in elastomers using silanol bands at 7326 cm^{-1} and 3748 cm^{-1}. Timm *et al.* [289] used NIRS for qualification of **lubricants** and other mineral oil products for QA and examined 22 samples of 6 different additive packages (viscosity improvers, polymethacrylate) in transmission mode (pathlength: 0.5 cm). Other applications concerned discrimination of hydraulic oils, quantification of the oil content in water-soluble cooling lubricants, total oil content, impurities (foreign oil). For QC purposes in the paint industry NIRS is used for purity control of solvents. For example, the water and alcohol content of butylacetate are relevant in relation to the production process.

Full NIR spectra of a set of 74 neat polymer additives, measured with an optical fibre probe and taken in diffuse reflectance mode, have been used for **qual-**

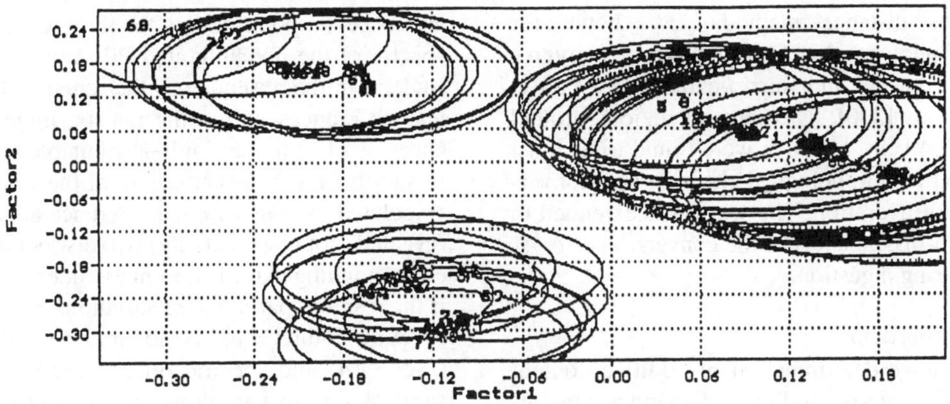

Fig. 1.13. Two-factor plot of the NIR spectral series of calcium stearate additives of different morphological forms (crude crystals, fine powder and normal powder). After Molt and Ihlbrock [237]. Reproduced from K. Molt and D. Ihlbrock, *Fresenius J. Anal. Chem.* **348**, 523–529 (1994), by permission of Springer-Verlag, Copyright (1994).

ity control of incoming raw materials [237]. Differentiated calibration was necessary for the group of products that are chemically and spectrally distinctly different and those very similar products with few, generally weak distinctive spectral features. The calibration procedure must provide enough sensitivity with respect to small but significant differences between the spectra within such a group. Discriminant and cluster analysis in factor space are powerful tools for calibration of spectral libraries for the purpose of automatic quality control. Discriminant analysis, which has high selectivity and robustness against spectral disturbances but low sensitivity, was used for calibration of the chemically distinct different compounds. Cluster analysis, due to its high sensitivity, was used for calibration of groups of rather similar products. Yet, unexpected spectral disturbances may cause false positive results. Reasons why spectra of different products may be very similar are: *(i)* similar chemical structures (*e.g.* Tinuvin 326/327; Loxamid OHA/OP); *(ii)* similar composition of multicomponent products (*e.g.* Irganox B215/B220); *(iii)* different morphologies (*e.g.* calcium stearates CA 710/720, CA 740, CA 860); or *(iv)* polymorphism (*e.g.* α- and β-Irganox 1076). When generating a spectral library, therefore, every existing polymorphous modification of a product should be known and treated as a separate product. At the same time, this opens the opportunity of including polymorphism into quality control. Complicated situations arise if a product is composed of several chemical compounds each with the property of polymorphism, *e.g.* Irganox 1010/1076, which are components of Irganox B731 and B991. Figure 1.13 shows a 2D plot of factors relevant for clustering of calibration spectra of three calcium stearates of different grain size. An unknown spectrum which needs to be assigned to one of the series of chemically substantially different calibration spectra may be judged on the basis of distance criteria (Mahalanobis distance [295]). Reliable quality control of incoming product was shown. Using NIRS the risk of using off-spec raw materials is much reduced.

Differences in polymorphic states may result in significant differences in properties and performance, *e.g.* in photographic and pharmaceutical applications. In photographic applications morphology needs to be carefully controlled during production in view of its influence on properties such as visible absorption characteristics, hygroscopicity, stability, *etc.* Consequently, the exact morphology of a given dye is critical in photographic products. Analytical techniques used to identify and classify crystal polymorphs comprise XRD, DSC, microscopy, FTIR, NIRS and Raman [296].

Manufacturers rely on the ability of NIRS for off-line identification and classification purposes [297, 298] and on-line for process analysis [299]. In particular, the use of NIR fibre optic probes for diffuse reflectance measurements and the minimal requirement for sample preparation minimises the likelihood of additional polymorphic forms being introduced during measurements.

FT-NIRS using transflectance and diffuse reflectance probes was particularly successful in **morphology identification** and differentiation (using

Mahalanobis distance methods), as shown for batches of photographic formulations of a hygroscopic dye (powder and slurry) exhibiting 17 polymorphic forms [300]. The spectral regions around 5240 and 7040 cm^{-1}, which were found to be most sensitive to the morphological changes, suggest various levels of hydration. The work was extended to real-time, in-line analysis of the conversion of crystal types during digestion.

Additives in polymers:
The quantitative determination of additive content in plastics frequently requires a demanding analytical approach. Hall *et al.* [286] have given a good example of **model building** for quantification of (unknown) additive levels in PP pellets obtained from two process extruders/pelletisers using NIR reflectance spectroscopy and multiple linear least-squares regression (MLR). While designed to produce identical product, subtle differences in pellet size, extrusion temperature and composition of the product from different processing lines must be accounted for in a spectroscopic procedure. Figure 1.14 shows absorption of the additive at about 2165 nm, where the PP matrix is relatively non-absorbing. The analytical wavelength most sensitive to additive concentration appears to be at 2172 nm, where the additive exhibits strong absorption and an isosbestic point occurs for the second-derivative spectra of the unadulterated PP samples of two extruders. In the absence of other chemical and physical differences within the matrix, intensity differences at 2172 nm are only attributable to variations in additive level. A scatter plot of the additive levels calculated by NIRS *vs.* the reference additive levels revealed (not unexpectedly) two distinct populations corresponding to the different extruder samples. This is the result of particle size variations. Since a unique spectral feature could be identified for the additive (Fig. 1.14) and spectral differences in the process samples due to variations in the additive level occur in this same spectral region (Fig. 1.15), an MLR approach was indicated. In order to minimise multiplicative scatter effects that have not been compensated for by derivative techniques, the ratio of the second-derivative intensity at two wavelengths was used [301]. In this case, by inclusion of a secondary reference wavelength (1946 nm) that mimics the entire matrix and inherent spectral variations, a single spectroscopic model was implemented for polymer pellets originating from multiple extruders, pelletisers or process lines. This enables accurate monitoring of the additive level to optimise product performance and quality with minimal concern for process variations that affect the spectroscopic measurement.

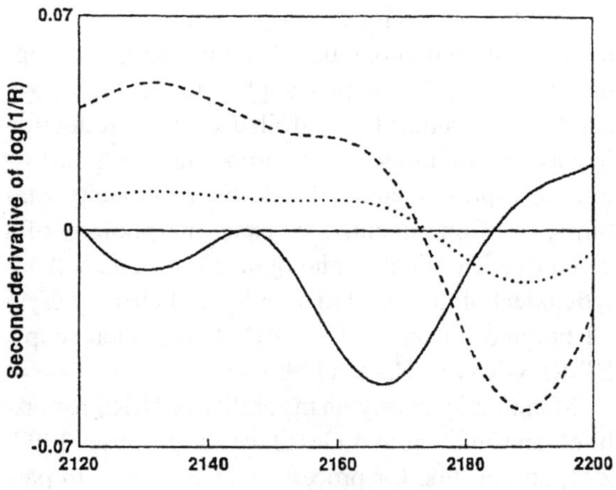

Fig. 1.14. Expanded second-derivative NIR spectra of pure additive (—) powder and extruder A (. . . .) and extruder B (- - -) pure PP pellets. Region suitable for additive level determination. After Hall *et al.* [286]. Reprinted with permission from *Journal of Near Infrared Spectroscopy*, Vol. 1, 55–62 (1993). *J. Near Infrared Spectr.* is a copyrighted publication of NIR Publications, Chichester.

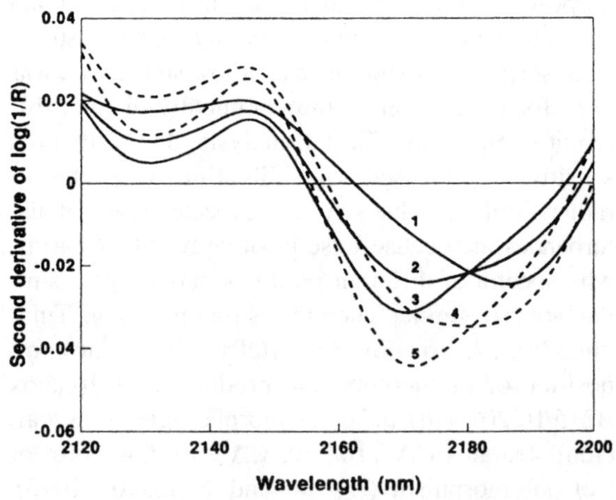

Fig. 1.15. Expanded second-derivative spectra of PP pellets from two extruders with varying additive levels; (—) extruder A and (- - -) extruder B. Additive levels (%): (1) 5.1; (2) 9.7; (3) 12.2; (4) 7.2; and (5) 10.4. After Hall *et al.* [286]. Reprinted with permission from *Journal of Near Infrared Spectroscopy*, Vol. 1, 55–62 (1993). *J. Near Infrared Spectr.* is a copyrighted publication of NIR Publications, Chichester.

Cost-effective, non-invasive, quantitative and simultaneous determination of low level Tinuvin 770 (0.05 to 0.4 wt.%) and Irganox 225 (0.1 to 0.45 wt.%) contents in PP pellets by NIRS in diffuse reflection mode using MLWR and PLS spectroscopic models has been reported [287]. Seven samples were used for calibration and two for validation. Spectral bands attributable to Tinuvin 770 and Irganox 225 appear at 1560 nm and 1390 nm, respectively. For Tinuvin 770 a two factor PLS model from 1500 to 1600 nm was developed; for Irganox 225 a four factor PLS model in the 1360 to 1460 nm region. A quotient-term multiple linear least-squares spectroscopic model was derived that characterises analyte concentration and corrects for spectroscopic differences within the matrix due to the extruder/pelletisers. Reported standard deviations of *ca.* 25 ppm for Tinuvin 770 and *ca.* 80 ppm for Irganox 225, or relative standard errors of 0.01 wt.% for Tinuvin 770 and 0.03 wt.% for Irganox 225, approach the accuracy of the reference analytical method.

The greatest potential for near-infrared reflectance analysis (NIRA) is in the statistical process analysis of manufacturing processes. The speed and non-destructive nature of this technique make it ideally suited to continuous material control. Several resin producers have considered programs for the introduction of on-line NIR analysis into their production facilities, *i.e.* **process or product control** of additive dosage. Herman *et al.* [302] have reported NIR determinations of additives in 10 μm PE film. Spatafore *et al.* [280,285] have described NIRA as an effective QC tool for quantification of thermal and light stabilisers in polyolefin processing. The chemical moieties found in organic additives, such as AOs and UVAs all exhibit characteristic absorbance bands in the near-IR region of the spectrum. Thirty-three formulations of Himont Profax 6301 resin containing various ratios of a primary AO (Irganox 1010), secondary AO (Irgafos 168), and UVA (Tinuvin 770) were extruded as a calibration set. The concentration of each of the additives ranged from 0 to 1%. Twenty-two formulations were utilised for calibration. The remainder were then used as a validation set to obtain estimates of the standard error of prediction. Samples were scanned from 1200–2400 nm (5 scans/s). Fifty rapid scans were averaged to produce each scan sequence. All spectra were collected in the transflectance mode. Calibration was performed using the first derivative of the absorbance spectra. First derivative transforms were used to remove baseline offsets due to particle size, colour, and minor thickness differences. Multiple wavelength regressions were carried out on the data; in each case the wavelengths were chosen to be characteristic of functional groups present in each of the additives (typically 2032 nm and 1546 nm for O−H and N−H bands, respectively). The work shows that NIRA can effectively be utilised to quantitate AOs and UVAs in polyolefins up to 1 wt.% with an accuracy close to that achieved by more time-consuming methods (GC and HPLC), with the additional advantages of simple sample preparation and ease of operation. Results for multiple constituents can be obtained in 10–20 seconds. The speed of analysis allows determination of the concentration of additives in a powdered matrix or moulded parts in time to make process changes, or to rapidly respond to customers experiencing product failure.

NIRS and acoustic emission have also been used for monitoring powder blend homogeneity in a mixing process of drug material; it enables optimisation of both the mixing quality and duration of a mixing process [303,304]. The blending process was also monitored with an InSb imaging camera. The same approach can be extended to additive blends. NIR diffuse reflectance spectroscopy has been used for analysis of granular **antioxidant blends** [146]. Powdered AO blends are opaque materials that will not transmit near-IR light. NIR limitations were identified with respect to particle size variation (from powder to granular) and sample presentation. Apparently, the spectral data contain information related to the particle size of the sample. However, different crystalline forms did not affect the results. The example demonstrates again the importance of using a calibration set (here: 41 samples) with physical properties similar to the unknown samples. The method developed effectively predicts weight percent composition for a series of 50:50 blends with a precision comparable to currently accepted HPLC analysis methods. It is very fast and can identify blend types and contaminated materials for quality control. Also, a calibration model based on PLS regression was reported for HDPE/(Irganox 1010, Irgafos 168) pellets valid for concentrations up to 4500 ppm [305]. Data pretreatment of the raw FT-NIR data, measured in the diffuse reflectance mode, was necessary to eliminate the physical differences of the samples, *e.g.* size and shape. Multiple scattering correction (MSC) and second derivative of the

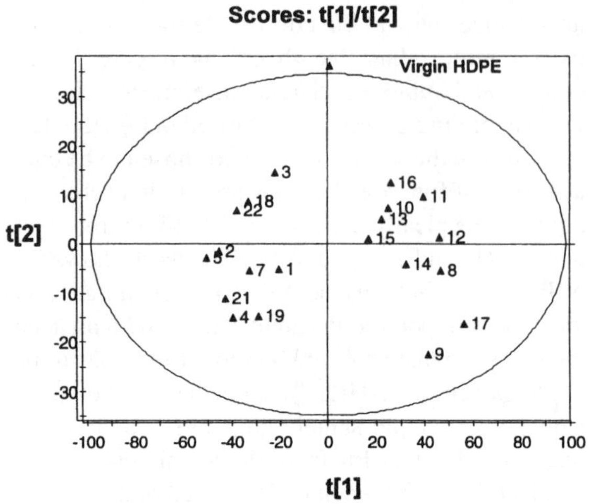

Fig. 1.16. Score plot of PCA factor 1 and factor 2 for virgin HDPE and HDPE/(Irganox 1010, Irgafos 168) calibration samples based on NIR spectra in the 5000–9000 cm^{-1} region. After Camacho and Karlsson [305]. Reproduced from W. Camacho and S. Karlsson, *Int. J. Polym. Anal. Charact.* **7**, 41–51 (2002), by permission of Taylor & Francis Ltd. (http://www.tandf.co.uk/journals).

raw data were used for this purpose. Characteristic regions for Irganox 1010 (6850 to 7350 cm^{-1}; phenolics) and Irgafos 168 (4694 to 5230 cm^{-1}; P—O—Aryl) were identified. Figure 1.16 shows a PCA score plot on the NIR data with two clusters of samples, those with total AO concentrations below 2200 ppm and above 2500 ppm, whereas the (partially oxidised) virgin sample differs from the rest. Root-mean-square errors of prediction (RMSEP) for Irganox 1010 and Irgafos 168 were reported as 45 and 97 ppm, respectively, in close comparison to 35 and 68 ppm for the error of wet methods, *i.e.* extraction (MAE or US) plus HPLC. The inaccuracy in the quantification of Irgafos 168 is due to degradation during polymer processing.

Stabilisers in pigmented plastics can also be measured using NIRS. Transflectance NIR analysis of stabilised PS pellets pressed into discs allows detection of 100 ppm additive [294]. Quantification of **UVAs** containing aromatic CH absorptions in PP is also possible to a detection limit of 100 ppm [306].

For NIR analysis of flexible PVC/(24.75–29.50% DOP; 5% epoxidised soybean oil) in reflectance mode SEP values of 0.2% (DOP) and 0.1% (soybean oil) were reported [306]. Using the wavelengths 1680, 1722 and 2336 nm ($-C-H_n$) stretching) and 1982 nm (second overtone of C=O) 9.8 to 23.1% **plasticiser** in cellulose acetate was determined by a fixed filter analyser with a standard error of 0.4% [294]. Pahl *et al.* [288] have used NIRS for analysis of *all* the components of **dryblends** for PVC window profiles containing chalk, internal lubricants (Loxiol G 60/G 21) and a Ca/Zn stabiliser (Stabilox CZE 2040). Unlike conventional methods for analysis of mixtures, this technique is not limited to the determination of one "critical" component (to reduce analysis costs). Surprisingly, and contrary to common wisdom, the stabiliser is most readily dispersed in the mixture and is therefore not the "critical" component, as previously assumed. A neural-network analysis [307] was successfully applied to discriminate 123 second-derivative NIR spectra of 41 PVC samples with different plasticisers (DINP, DOP, DOA, TOTM and polyester) in concentrations up to 49%; a calibration model allowed prediction of the content of plasticisers with high correlation coefficients (0.999–1.000) and RMSEP of 0.41 wt.% for DINP [308]. The non-linear neural-network analysis gave better results than a linear PLS regression analysis. The procedure is useful for practical PVC sorting according to plasticiser type in plastic recycling.

NIRS in internal reflection mode has also been used to study *in situ* low-level **surfactant** adsorption reactions (sub-monolayer coverage) using reactive internal reflection elements [175]. Standard errors of 0.04% were reported for the determination of 0.25–1.25% additives in nylon (cubes and films) [294]. Turley *et al.* [309] used NIRS to determine ethylene oxide content and glycerin additive concentrations in ethylene oxide/propylene oxide copolymers. NIR diffuse reflectance spectroscopy was used to analyse up to 10% paper (cellulose) in agglomerate plastic waste (PE 60%, PS 20%, PP 15%, PVC 4%) [161]. For **pigment applications**, *cfr.* ref. [310].

The major polymer manufacturers should actively be pursuing the introduction of NIR into their production facilities [311].

Textiles:
NIRS has been used for qualitative identification of textile fibres, polymer microstructure and composition studies, determination of finishes on textile fibres and colour deviations in dye batches. As shown in Table 1.18, near-infrared reflectance analysis (NIRA) is useful for characterising textile raw materials, fibres, yarns, and fabrics and is an excellent means to obtain real-time process/product information in textile manufacturing [312]. The non-destructive quantitative analysis is simple to use and

Table 1.18. Some applications of NIRA in textile research

Textile specimen	Analyte(s)	Reference(s)	Year(s)
Nylon fabric	Moisture, finish-on-fibre	[313]	1984
Cotton	Sugar	[314]	1988
PA6.6	Moisture, finish-on-fibre	[315]	1992
Textured PA6	Dyes	[316]	1992
Aramid knitted fabrics	Sizing agents	[317]	1997
Acrylic fibres	Finishing oils	[318,319]	1997, 1998
Wool, cotton yarns	Fibrillated HDPE/PP contaminants, extractables	[236]	1998

allows rapid simultaneous measurement of multiple components in a sample. Many innovative mathematical treatments, *e.g.* discriminant analysis and spectral reconstruction, have been developed for quantitative analysis of such data, and for morphological investigations of fibres and yarns and the evaluation of surface properties of textiles. On-line NIR sensors are being used in textile fibre installations.

Textile **fibre quality** measurements (fibre types, sizings, moisture), particularly for natural fibres such as cotton and wool, utilise NIR technology replacing less precise and more labour intensive methods. NIRA performs a quantitative polyester/cotton blend analysis within 2 minutes [315]. **Moisture content** in synthetic fibres, yarns, and fabrics is a critical variable with significant impact on physical properties, processing behaviour, and manufacturing productivity. Several methods can be used to measure the moisture content of nylon materials including the Karl Fischer reagent (KFR) titration method. Although the KFR method is very accurate, it is time consuming, requires careful laboratory and sampling techniques, and normally uses odorous chemicals. The determination of water in textiles and polymers was the subject of several early NIR investigations, *cfr.* ref. [320]; systems studied comprised cotton, wool, PVAL, polyols, oligoether acrylates, polyurethanes, PET and PA6 (at 5150 cm^{-1} for nylon and 5240 cm^{-1} for PET). NIRA allows analysis on yarns in less than 5 min in a textile laboratory and without using odorous chemicals. A comparison of the moisture content on nylon 6.6 spun yarn by KFR and NIRA (three-wavelength model) has been reported [313]. Moisture measurements dominate NIRS applications, *e.g.* in thin cotton fibre layers using transflectance [321]. Hammersley *et al.* [322] used NIR to measure moisture and residual grease in wool scouring. On-line monitoring of moisture and lubricant levels on a polyacetate fibre has also been reported [293].

It is common practice in manufacturing of synthetic fibres to add a small amount of lubricating material (“finish”) to the fibre to assist its performance and runnability in downstream textile manufacturing processes. The primary functions of the finish are lubrication, static protection, and filament cohesion. The amount of finish on the fibre surface, called **finish-on-fibre** (FOF), is of critical importance in textile manufacturing processes. Traditional methods for measuring FOF of synthetic fibres consist of extraction of the finish oils from the fibre with either hot or cold organic solvents (*e.g.* CCl_4), followed by gravimetric or IR analysis. Blanco *et al.* [318] have reported QC analyses of finishing oils in acrylic fibres using NIR diffuse reflectance spectroscopy and PLS regression methods. Sizing agents on sized fibre reinforcements can easily be analysed *in situ* by means of NIR light-fibre optics spectroscopy. Ozaki *et al.* [317] have reported *in situ* analysis of bisphenol-A epoxy dispersion-based and PE type dispersion-based sizing agents on aramid knitted fabrics and on glass strands. For sized aramid knitted fabrics, characteristic NIR bands of epoxy and PE types of sizing agents were superimposed on the spectrum of non-treated aramid fabrics. The spectra of non-treated aramid knitted fabrics and aramid fabrics with two different sizing agents were clearly distinguished without chemometric analysis. Quantitative analysis of the epoxy sizing agent is possible. FTIR spectroscopy is not always suitable for *in situ* analysis of aramid fibres because of strong absorptions in the mid-IR region. Formation of an amide bond between a maleic anhydride-modified polypropylene sizing agent and the amino silane coupling agent was revealed by NIR spectral measurements of the sizing glass strand. NIRA is a widely accepted quantitative tool for desize testing. Lemere [323] used NIRS to make size add-on determinations of blends involving cotton-polyester yarns. NIR spectroscopy was also used to determine the amount of processing oil in polyester yarns [324]. The application of silicone lubricants on textile fibres is frequently measured by NIRA techniques.

NIRS in the 2250–2400 nm region in combination with PCA was demonstrated to be also a viable technique for detection of polymeric contaminations (HDPE and PP) in both wool and cotton yarns. Moreover, wool researchers have shown strong interest in NIRS for the determination of composition (extractables) [236]. The heatset temperature of carpet yarns, amount of dye(s) in yarn(s), and quality of cotton and wool [325] are also routinely measured.

NIRA was used for the analysis of the **dyeability** of textured PA6 carpet yarns (one quality type) with C.I. Blue 127:1 and luminosity measurements of partially oriented polyester-yarns [316]. NIR diffuse reflectance spectroscopy has been used to investigate dye uptake potential of PET fibres [326]. NIRS has been used for evaluating colour deviations in different production batches of the acid dye Tectilon Red 2B [327]. If an adequate number of samples are available to generate robust regression models, then NIRS can be used as a QC tool to evaluate the dye using only drying and grinding for NIRS sample preparation.

Polymer identification:
A major application of NIR in the polymer industry is compositional analysis. Identification of different polymers, their properties, and their morphological differences normally requires extensive testing and complicated, time-consuming analysis. A difficulty with polymer applications is that it is not easy to create a stable calibration with stepwise multiple linear regression analysis because the constituent of interest may show little variation and, as a result, regression coefficients tend to have small values.

Neural-network analysis methods were used to discriminate more than 50 different kinds of plastic patterns [328] and to separate **PE grades** [329]. Shimoyama *et al.* [330] have recently reported discrimination of five HDPE, six LLDPE and seven LDPE grades by NIRS and chemometric analysis. PLS regression has enabled to propose good calibration models which predict the density, crystallinity and melting points of PE by use of NIR spectra. This allows use of vibrational spectroscopy for QC of polymers. NIR diffuse reflectance bands at 1192, 1376 and 1698 nm due to the overtone and combination modes of CH_3 groups play important roles in the prediction of density (SEP of 0.001 $g\,cm^{-3}$) of LLDPE pellets with densities in the 0.911–0.925 $g\,cm^{-3}$ range [224]. Similarly, NIRS has been used for in-line monitoring of LDPE density (based on I_{1170}/I_{1213} ratio) [331].

Miller [332] has shown that NIR transflectance is useful for polymer characterisation of **food packaging materials**. The method was applied to PE film (before and after stretching and heat-sealing), nylon/EVOH/nylon coextruded sheets (before and after annealing) and PE/nylon laminate (before and after pasteurisation).

Recycling of plastics (packaging waste, textiles, electronic and automotive waste) constitutes a real challenge for optical recognition technology. Remote sensor systems for the automated identification of plastics have been developed [333]. Since most plastics are opaque, measuring remotely can only be done in diffuse reflectance, where large light losses occur. NIRS allows fast on-line mobile product identification of plastics for use in **recycling processes** [334–336] and additive identification within seconds. Miller [337] has used the rapid classification capability of NIRS to distinguish various classes of PUR foams. Most consumer packages (PE, PET, PP, PS) can be classified with an integration time of 6.3 msec per sample with an identification rate of better than 98% by means of NIR transflectance spectroscopy [338]. Rohe *et al.* [334] have considered application of NIRS to plastic sorting problems, investigating bottles, cups, containers, foils, *etc.*, from household waste as well as mass consumer technical products, cases and parts of computers, keyboards, monitors, printers, plotters, *etc.* and cases of tools for drilling, sawing, *etc.* The latter parts often contained glass fibre fillers and dyes. Fillers, plasticisers, dyes and additives, contained in mass consumer products strongly influence the NIRS spectra. Neural-networks have also been used in the identification of plastic waste [339]. Kulcke *et al.* [340] have described the potential for fully-automated industrial polymer waste sorters for waste recycling, based on NIR spectral imaging. Although NIRS is being used for plastic waste sorting since 1993, additives and fillers disturb polymer identification [341], or make it even impossible (when loaded with black pigments, *cfr.* Fig. 1.17). For recycling purposes NIRS

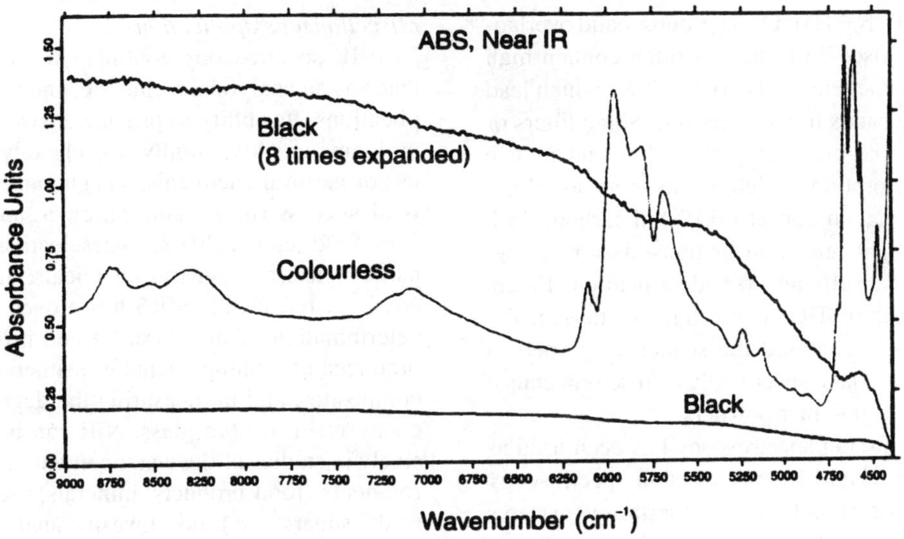

Fig. 1.17. NIR spectra of colourless and black ABS. The spectrum of the black material lacks information of sufficient detail allowing reliable material identification. After Zachmann and Turner [128]. Reprinted from G. Zachmann and P. Turner, *Spectroscopy Europe* **9**, 18–22 (1997). Copyright 1997 © Wiley-VCH, Weinheim. Reproduced with permission.

is not yet capable of making accurate distinctions between dark and coloured plastics with different compositions [149]. Household waste gives spectra of good quality so that the 1600 to 1800 nm range can be scanned in 1 ms or less. GFR materials of technical products need longer scan times. A NIR-AOTF spectrometer is capable of identifying the most common recycling plastic materials in very short times (<1 s) [334]. Rapid NIR identifiers for household plastic waste sorting do not cover the NIR spectral range in which discrimination between PA6 and PA6.6 is possible.

The statistical analysis method of discriminant analysis [342] has been combined with NIRA to identify dissimilar textile products. Most textile fibres, yarns, and fabrics have chemical structures which yield complex NIR spectra, and as such these species normally require three or more wavelengths to classify the material. Discriminant analysis is simple to use, rapid, and does not require extensive, time-consuming sample preparation and analysis. Polyester staple fibres of different tenacity levels have different fabric dyeing properties. NIRA method with discriminant analysis successfully identifies and classifies the polyester staple samples by tenacity level and thus provides a quick technique for identification of polyester fibre anticipating quality problems [315]. Mitchell *et al.* [343] used NIRS to determine the acetyl content in cellulose acetate.

Various **fibre identification systems** for carpet recycling have been developed. The Bruker FT-NIR integrating sphere carpet fibre identification system, pre-programmed to identify PA6, PA6.6, PET, PP, acrylic and wool, is fast (2 s) and can account for inclusion of undesired material in the form of calcium carbonate and other contaminants. Mixtures of known percentages of PA6 (70–100 wt.%), PP (0–12 wt.%) and $CaCO_3$ (0–18 wt.%) were quantified by means of FT-NIR spectroscopy to an accuracy of ±2% [335]. For this purpose the spectra were collected by means of an integrating sphere for sampling of a large area (1 inch in diameter), providing some spatial averaging of the inherently heterogeneous mixture of material. Kip *et al.* [344] have developed commercial ID/sorting technology for generating monostreams of post-consumer carpet waste feedstock according to face fibre type (PP, PA6, PA6.6, PET or wool) using NIR-AOTF technology. The various face fibre types can be identified in 50 ms. A carpet sample library was build up from some 600 US and 2000 European carpets (both new and post-consumer). Portable near-IR devices allow identification of at least 20 types of polymers, blends, or additives (in 30 s).

Some problems are encountered in polymer identification by means of NIRS. If a plastic component contains C−H, N−H or O−H groups, band overlapping occurs. Also, E&E plastics often contain high amounts of flame retardants (10 to 30%), which lead to observable bands in the spectrum. Some fillers or dyes behave like black or grey bodies and induce broadband absorption, which results in smaller light intensities. Small amounts (>0.1%) of carbon-black reduce NIR light reflection or transmission to levels which are insufficient for identification. Fillers like glass fibres or TiO_2 cause light scattering. Papini *et al.* [345] analysed the scattering properties of granular materials, specifically diffuse reflectance spectroscopy of polymer powders.

NIR transmission spectroscopy has been used as an alternative to mid-IR transmission spectroscopy for *in situ* cure monitoring of thermoset systems [346]. NIR has also been used to evaluate the transparency of acrylic adhesives [347].

The applications of NIRS to synthetic polymers were reviewed [234,294].

Paper industry:
NIR methods have been described for control of incoming materials in the paper industry and in the analysis of paper related goods. Quantitative models for paper components such as mechanical pulp (lignin containing) or chemical pulp (cellulose based) and the various detectable filler materials (clay, talc, chalk) or coatings thereof have also been established. Accuracy of the determination of clay (filler) is better than 1%. Much more sophisticated applications involve the measurement of additives in papers, such as sizing agents and pigments, and wet or dry strengths of materials. Characterisation of papers by their physical properties (hydrophobicity, ash content, basic weight, thickness, tear index and tearing strength, bursting strength, tensile stretch and strength, fibre length, debonding energy, wettability and printability, *etc.*) can also be investigated by NIRS. Other applications for the paper industry are hardwood/softwood ratio, coating levels, waxes, resin uptake, kaolin modifications, rag content, super absorbent treatment and filter paper parameters. Also alkoxylates of various types have been analysed by NIRA, including alcohol and nonylphenyl ethoxylates, PEG and PPG [309,348]; ethoxylates are used for paper manufacture and textile processing. NIRS has also been used in measuring the concentration of residual ink in recycled newsprint [349].

Miscellaneous applications:
Near-IR spectroscopy continues to grow in importance as an analytical technique, and to find new applications. Its ability to provide answers to practical problems quickly, simply and cheaply, without the use of harmful chemicals, has guaranteed its **industrial success** (in agriculture, environmental chemistry, food science, life sciences, leather, textiles and paper industries, general chemicals, and polymers, *etc.*), *cfr.* ref. [259]. NIRS has proved useful in the determination of moisture, fat and protein in dairy products, of tanning agents in leather, of binders in composites, and in measuring the degree of cure of epoxy resin in fibre glass. NIR can be used extensively in studies of the water content in various environments (food products, minerals, proteins, amino acids, sugars, *etc.*), non-invasive analysis of natural resin extracts, *etc.* NIRS is a widely applicable analytical technique for the quantitative study of liquid and compressed gaseous systems, including fluid states, up to high pressures and temperatures. Typical applications include monitoring feed gas composition and multicomponent analysis of liquids (reaction products) and solids. Other applications of FT-(N)IR are reported for the qualitative analysis of raw material, determination of impurities ("trace" analysis), and fruit analysis (taste analyser, ripeness). NIR can be used for on-line measurement of film thickness, which is useful in analysis of food packaging materials.

Ciurczak [350] has reviewed general analytical applications of NIRS. Applications of NIR spectroscopy to polymers [259,294] and textiles [315] have also been described. Siesler [278] has recently reviewed the quantitative determination of the additive content in plastics by means of NIR in diffuse reflection mode.

1.2.3. Raman Spectroscopic Techniques

Principles and Characteristics
Raman light scattering was first described in 1928 [351]. Raman spectroscopy has historically been neglected in view of fluorescence, difficulty in using the equipment, and lack of sensitivity. It was not until the mid to late 1980s, when the first FT-Raman experiments were carried out, that Raman truly began its renaissance. Meanwhile also the problem of efficiently filtering the Raman scattered light from the Rayleigh scattering has been solved and techniques to enhance sensitivity, such as resonance Raman spectroscopy [352] and surface-enhanced Raman spectroscopy [353], have been developed.

Raman and infrared spectroscopy both excite vibrational states, but by different mechanisms. Vibrating molecules generate changes in the polarisability and/or in the magnitude of the dipole associated with the molecule. These changes give rise to the spectroscopic effects of **Raman (inelastic) scattering and infrared absorption**, respectively, which are the two most valuable techniques available for the measurement of the vibrational and rotational characteristics of molecules.

A molecule can be polarised by the application of a field which induces a dipole. Induction and subsequent relaxation of a dipole results in absorption and then scattering of radiation. More precisely, considering a monochromatic light wave of frequency ν_0 in an electric field E, and a molecule vibrating with a frequency ν_{vib}, a dipole will be induced (and hence scattering will occur) at the same frequency as the source of radiation ν_0 (*elastic* or **Rayleigh scatter**), but also at the *shifted* frequencies $\nu_0 \pm \nu_{\text{vib}}$ (*inelastic* or **Raman scatter**). The inelastic light scatter with frequency $(\nu_0 + \nu_{\text{vib}})$ is known as anti-Stokes Raman scatter, and scatter at frequency $(\nu_0 - \nu_{\text{vib}})$ as Stokes Raman scatter. Inelastic Raman scattering, which leads to frequency shifts, is an **emission** technique but phenomenologically distinct from relaxed emission denoted as fluorescence, *cfr.* Fig. 1.5.

Raman spectra are **excited** by monochromatic radiation, emitted by different lasers in the ultraviolet (UV), visible (VIS) or near-infrared (NIR) range. The frequency of the incident radiation can be chosen to interact predominantly with specific electrons of a sample. When the frequency of a light wave does not match an energy change in a molecule, the light wave is scattered instead of being absorbed. Molecules emit Raman lines with a frequency difference $(\Delta\nu)$ to that of the exciting frequency (ν_0) between 0 and +3700 or −3700 cm^{-1}. Usually only the Raman spectrum which is shifted to smaller wavenumbers, the "Stokes" Raman spectrum, is recorded. The intensity of the Stokes scatter is invariably greater than the anti-Stokes equivalent bands. Most of the light undergoes elastic (Rayleigh) scattering (no change in frequency) but a small fraction changes frequency through interaction with the vibrational states of the molecule:

$$I_{\text{Raman}} \approx 10^{-4} I_{\text{Rayleigh}} \approx 10^{-8} I_{\text{source}} \qquad (1.4)$$

Not surprisingly, therefore, modern RS relies on the availability of lasers as intense and stable light sources.

All materials possess "polarisability", *i.e.* if placed in an oscillating electric field of magnitude $\mathbf{E}$ (V m^{-1}) an electric dipole μ will be induced. The proportionality constant or **polarisability** α (a tensor quantity) in

$$\mu = \alpha \mathbf{E} \qquad (1.5)$$

stands for the ease with which the electron cloud can be distorted by the applied potential field. Molecular vibrations modulate the molecular polarisability. The vibrational modes must satisfy specific "selection rules". Infrared and Raman activity are related to variations in dipole (μ) and polarisability (α) as a result of atomic displacements (internuclear distance q). Infrared spectroscopy only observes vibrations which have non-zero values of $(\delta\mu/\delta\text{q})_0$, whilst for Raman activity $(\delta\alpha/\delta\text{q})_0$ must have non-zero values. In Raman spectroscopy it is the absorption of photons related to a change in polarisability of chemical bonds that determines activity, but the absorbed light is then reemitted rather than being permanently absorbed. Raman and infrared spectroscopy are complementary techniques.

Raman-shifted frequencies can be used in the same manner as IR vibrational absorption spectroscopy. Fundamental vibrational frequencies give information on analyte structure and dynamics: the frequencies intimately depend upon the bond force constants and atom connectivities. Structural information can be extracted from Raman spectra using functional group frequencies. For vibrating molecules the potential energy *vs.* interatomic separation for a diatomic molecule is described by the classical Morse curve. The potential function is *not* parabolic and the motion is not simply harmonic but *anharmonic*. As with all forms of molecular energy, vibrational energy is quantised. A non-linear molecule consisting of N atoms with fixed position and molecular orientation can possess up to $3N - 6$ **fundamental modes of vibration**, depending upon symmetry (point group) [354].

The Raman cross-section is proportional to ν^4 and to α^2. The **molecular cross-section** of Raman spectroscopy is some ten orders of magnitude smaller than that of IR spectroscopy. However, for a typical "classical" detector of Raman spectra, the photomultiplier (PMT), the number of photons to produce the noise equivalent power (NEP), which is the light flux necessary to produce a signal of the same magnitude as the noise, is up to ten orders of magnitude smaller than those of detectors employed

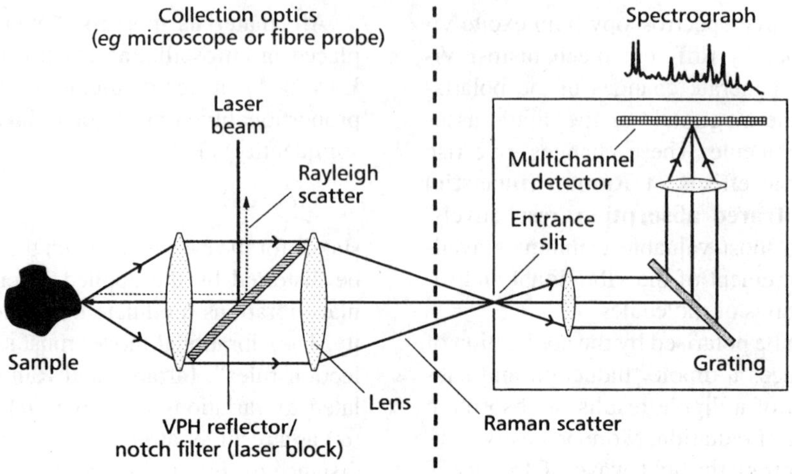

Fig. 1.18. Schematic of a typical modern dispersive Raman spectrometer. After Everall *et al.* [356]. From N. Everall *et al.*, *Chemistry in Britain* **36** (7), 40–43 (2000). Reproduced by permission of The Royal Society of Chemistry.

for infrared spectroscopy. This fully compensates the smaller cross-section in Raman spectroscopy. Consequently, IR and Raman spectra have a limit of detection of the same order of magnitude [355]. Major gains in Raman sensitivity are achievable by application of resonance Raman spectroscopy (RRS) and surface-enhanced Raman spectroscopy (SERS).

The Raman **sampling technique** is considerably easier than for IR and the information supplied can be more significant. Whereas IR requires use of thin film samples (unless reflectance), a Raman cell can easily accommodate a bulk specimen, fibre, powder, film or solution with minimum sample thickness of the order of several mm, which can easily be heated or cooled, pressurised or irradiated, stretched or bent.

A **Raman system** is an integrated package which generally includes the sample interface, Raman instrument, a computer and software. The sample interface is the device that produces and delivers illumination to the sample (laser excitation), collects light scattered from the sample (collection optics) and passes the collected light on to the Raman spectrograph. In this system ease of sampling should be preserved. Raman measurements can usually be made in a non-invasive manner. Basic **instrument types** are: *(i)* filter instruments; *(ii)* spectrographs with array detectors; *(iii)* wavelength scanning with array detection; *(iv)* interferometers with single-element detectors (FT-Raman spectrometers); and *(v)* interferometers with array detectors (multi-channel Fourier transform or MCFT technique). As Raman scattering is very weak a Raman instrument needs to detect as many Raman photons from the sample as possible. Interfering light sources, such as elastically scattered (Rayleigh) photons, laser-induced fluorescence, room lighting and sunlight, should be excluded. In particular, unless the Rayleigh scatter is blocked, the Raman photons are completely swamped. By using Volume Phase Holography (VPH) very efficient laser-blocking "notch" filters have become available and compact Raman systems can now be designed. Silicon CCD (charge-coupled device) detectors are nowadays the detectors of choice for most Raman measurements at wavelengths shorter than about 1000 nm. Raman spectrographs are either one-dimensional (*e.g.* Czerny-Turner) or two-dimensional (Echelle). Thanks to the development of CCD detectors, air-cooled lasers and holographic notch filters (HNFs) to remove the scattered laser light from the collected Raman emission, and fibre optics, spectra acquisition is now extremely rapid. Raman technology is still improving. With continuing improvements in spectral resolution, the MCFT Raman instrument could become competitive with any of today's Raman instruments.

At present two Raman instrument types are in major use: *(i)* VIS to NIR excitation with monochromatisation of the scattered radiation by a holographic grating and simultaneous detection of the dispersed, narrow frequency range by a CCD detector (dispersive Raman, D-Raman), *cfr.* Fig. 1.18; and *(ii)* NIR laser excitation and measurement in

Table 1.19. Dispersive *vs.* FT-Raman spectroscopy

Dispersive Raman	FT-Raman
• Array detection and imaging	• Higher and continually variable resolution
• Better S/N performance	• High frequency precision
• Smaller spot size for microscopy	• True linearity in wavenumber
• Visible optics and alignment	• Fluorescence rejection (1064 nm excitation)
Applications:	• Use of a single detector
• Aqueous and dark samples	*Applications:*
• Laminates and paint samples	• Forensics
• Minor component analysis	• Polymers, pulp and paper, textiles
• On-line analysis	• Raw materials identification

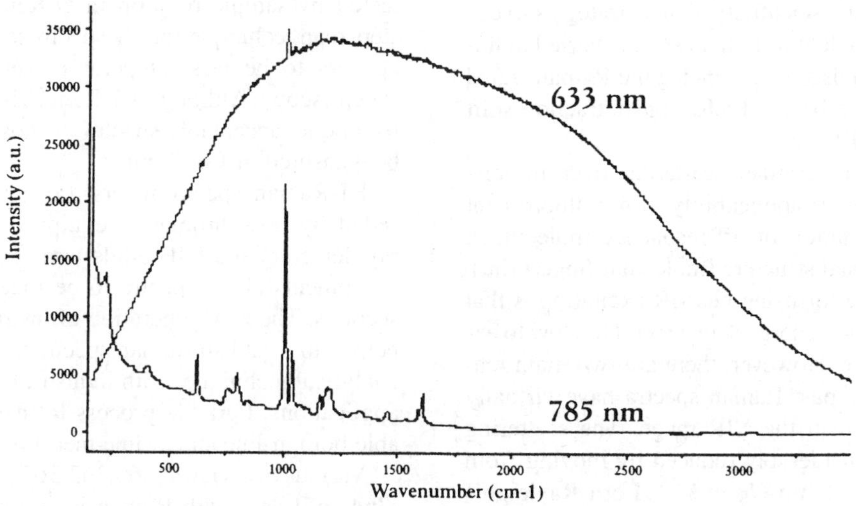

Fig. 1.19. Raman spectra of a coloured polymer highlighting the fluorescence background difference when recorded at 633 and 785 nm. Reproduced by permission of Jobin Yvon S.A., Villeneuve d'Ascq, France.

a Fourier transform spectrometer (FT-Raman). Table 1.19 compares these technologies. Despite the fact that interferometers (in combination with the use of near-infrared) are superior to grating spectrometers in various respects, there has been considerable interest in the NIR performance of **dispersive Raman spectrometers** with grating monochromators, initially using scanning spectrometers. The scanning technique is more valuable for large amounts of sample but less so for the study of μm size samples. For a large proportion of samples, irradiation with visible light causes strong fluorescence by additives or impurities (or by the sample itself). Dispersive Raman analysis complements FT-Raman techniques. There are a significant number of materials that *can* be studied by D-Raman but *cannot* be studied by FT-Raman, including inorganics (in particular oxides): fluorescence; carbon and other black materials: heating and black-body radiation; and aqueous solutions: YAG laser incident with 2nd overtone of water.

Sensitivity depends on the relative intensities of the analyte Raman bands compared with overlapping, interfering Raman bands and emissions from the sample. Raman analysis is often hindered by fluorescence by the sample or impurities with the laser excitation line being used. Fluorescence occurs when the excitation line is partially absorbed and reemitted. The quantum yield of the fluorescence process is often several orders of magnitude higher than that of the Raman process, and thus any useful spectroscopic information is lost, (*cfr.* Fig. 1.19). **Fluorescence interference** does not normally occur in condensed phases with UV excitation wavelengths below 260 nm [357]. There is no single solution to the fluorescence problem in Raman spectroscopy. In favourable cases photobleaching (decay of fluorescence in time) solves the problem. Currently the most widely used strategy is to use an excitation wavelength where the fluorescent molecules do not absorb (no absorption, then no fluorescence). FT-Raman with a laser wavelength of 1064 nm is the most successful and potential practical solution. Near-IR excitation is below the electronic absorption process that leads to fluorescence for most organic substances. However, shift to higher wavelength sacrifices sensitivity due to the ν^{-4} decrease of scattering efficiency. Of course, if there is no fluorescence problem then a visible laser and CCD detector give maximum sensitivity. Other strategies to circumvent the problem of Raman signals buried in fluorescence and noise are enhancing the Raman signal (resonance RS or SERS/SERRS) or subtraction-shift procedures (SSRS) [358].

Advantages of Raman scattering over fluorescence comprise: *(i)* applicability to non-fluorescent molecules (excitation of off-resonance molecules); and *(ii)* vibrational structure (molecular fingerprint).

The rationale for using near-IR excitation is that the laser excitation energy is in general too low to excite fluorescence. However, there are two main reasons why in the past Raman spectra have virtually not been excited in the NIR range. The scattering efficiency is considerably reduced by moving from 488 nm to 1064.1 nm ($I_R \sim \lambda^{-4}$, Lord Rayleigh's electron resonant scattering law). Thus, Raman lines ($0 \ldots 3700$ cm^{-1}) excited by a Nd:YAG laser would have only a fraction (1/23 to 1/75) of the intensity of those excited by an Ar^+ laser at 488 nm. Moreover, near-IR solid-state detectors are orders of magnitude less sensitive than the photomultiplier tubes used in conventional Raman spectroscopy. Both effects cannot be compensated by an increase in laser power since this would destroy the sample. An increase of the signal/noise ratio by increasing the time constant would lead to unacceptably long recording times.

Nowadays it is possible to use **FT-Raman spectroscopy** with a CW Nd:YAG laser (1064.1 nm) and an FTIR spectrometer instead of the conventional D-Raman spectrometer [5,359]. In this experimental set-up it is possible to measure typical samples various orders of magnitude more efficiently than with "classical" Raman spectrometers [360]. The intensity of the Raman spectrum may be increased (roughly 300 times) by making use of the Jacquinot advantage [361]. In FT-Raman also the multiplex advantage of the Michelson interferometer is used to increase the spectral S/N that is limited by IR detector background noise. Interferometers of modern Raman spectrometers have a 300 fold optical conductance compared to grating spectrometers, a tolerant imaging system, and permit easy sampling. Also, fibre optics, developed for telecommunication, shows highest transmittance in the NIR range (80% per km). Since the cross-section of Raman scattering is very small, the efficiency of the sample arrangement needs to be as large as possible.

A great variety of thermally sensitive samples have become accessible to FT-Raman, including polymers. Thermally sensitive samples may be protected by sample rotation in D-Raman and by the step-scan technique for FT spectroscopy. FT-Raman appears to be less temperature sensitive than IR spectroscopy. Although FT-Raman is not a sensitive technique, acceptable spectra of most polymers can be measured in 1 or 2 min.

FT-Raman spectrometers are now being superseded by polychromators equipped with CCD array detectors and NIR diode laser excitation. These instruments allow spectra to be measured in a few seconds. The new generation of instrumentation has helped to establish Raman spectroscopy as a routine analytical technique, with industrial process-control applications. Portable process Raman analysers enable both in-line and at-line measurements.

Various overviews [360,362,363,363a] and books [364,365] deal with Raman instrumentation. A variety of special issues of *Spectrochim. Acta A* have been devoted to this spectroscopy [366].

Raman spectra of condensed-phase samples consist of bands, typically 5–20 cm^{-1} wide, within a spectral region, typically between 70 cm^{-1} and 3700 cm^{-1}. Analytical information can be extracted not only from band position, but also from band shape and band intensity. Manual **qualitative analysis** of Raman spectra is often time-consuming and requires an experienced analyst. Raman spectral library searching is currently limited to specialised applications where small libraries can be effective [367]. Large, general purpose and specific Raman spectral libraries are being developed [368]. For interpretation of IR, Raman and NMR spectra, *cfr.* ref. [369]. Qualitative analysis can avail itself of chemometric methods [370]. There is no agreed standard approach for correcting the instrument response as a function of wavelength. Accurate corrections will be needed to enable useful Raman polymer libraries to be constructed and to allow calibration transfer between instruments.

Unlike IR absorption spectroscopy, Raman is a single-beam technique and not intrinsically quantitative. Although absolute Raman intensities are notoriously difficult to measure, Raman spectroscopy can be used for **quantitation** since, to a first approximation, the intensity of a Raman band is proportional to the concentration of the vibrating species. The absolute intensity is also dependent on factors such as laser power (total number of photons delivered to the sample during the measurement interval), sample refractive index, and changes in scattering cross-section due to intermolecular interactions, optical sampling geometry, *etc.* As the observed Raman intensity of particulate solids can depend upon particle size distribution, a calibration set should accurately reflect the particle size of the samples of interest. Calibrations are often non-linear due to the intrinsic scattering efficiency varying with concentration. To avoid compensation problems, in most cases quantitative Raman spectroscopy is normalised relative to an internal standard band in the vicinity of the analytical absorption band (*e.g.* a solvent peak). However, in industrial practice the use of an *external* reference is usually preferred. Hendra *et al.* [371] have addressed the use of external reference materials in quantitative analysis to standardise the intensities of FT-Raman scattering spectra. Pelletier [372] and others [373] have recently discussed quantitative analysis using Raman spectroscopy.

Table 1.20 shows the **main characteristics** of Raman spectroscopy. Raman spectroscopy is characterised by simplicity of sampling (unlike IR). Its versatility via macro optical arrangements and microscopes, allied to the ability to sample through glass and other transparent packaging media in non-contact mode and ease of sampling via coupling to fibre optics, has been used to study species in all physical forms (size, shape, transparency), including liquids (aqueous and other solutions) and gases, without the risk of sample contamination. Raman spectroscopy is highly suitable for analysis of solids and allows high-throughput screening; water signals are very weak. RS gives access to the low frequency vibrations below 400 cm^{-1} (lattice vibrations, concerted motions in polymers, catalyst supports, metal-organic bonds, sample morphology). High temperature studies can be performed easily. A disadvantage of Raman scattering is that it is a very weak effect (some six orders of magnitude weaker than fluorescence). Enhancement of 10^{14} fold is necessary for single molecule detection. RS is not a trace element technique. Liquids or diluted solids show low sensitivity (no effect of increasing pathlength). The inherent problems associated with the technique, such as fluorescence and lack of sensitivity, have been addressed and can be overcome. The small laser spot sizes on the sample (1 mm–1 μm) can result in non-representative spectra of inhomogeneous samples and may determine unsatisfactory repro-

Table 1.20. Main characteristics of Raman spectroscopy

Advantages:

- Very flexible sampling ("as received")
- Non-destructive, non-invasive
- Small sample size (microscopy: 1 μm^3)
- Remote sampling (optical fibres over >100 m); process analysis
- *In situ* measurements
- Broad spectral range (Raman shift values from 70 cm^{-1} to over 3500 cm^{-1})
- Highly selective (RRS, SERS)
- Relatively high sensitivity (ppm)
- Very accurate peak positions
- Well resolved spectra with high information content (vibrational frequencies of chemical bonds)
- Fast material identification (database dependent)
- Chemometrics for complex analysis
- High spatial resolution (μRS: 1 μm)
- Imaging
- Well-developed theory
- Applicable to almost any chemical substance (more universal than UV/VIS or F)

Drawbacks:

- Very small scattering cross-section ($\sim 10^{-30}$ cm^2/molecule)
- High fluorescence quantum yield for certain molecular systems
- Poor Raman scattering of certain substance classes
- Limited variation in pathlength
- Non-representative spectra; unsatisfactory reproducibility
- Difficult quantitation (calibration needed); usually qualitative only
- Depth profiling limited to transparent materials
- Risk of sample degradation (UV; laser damage)
- Limited reference libraries (databases up to 15,000 compounds)
- Validation
- Most applications limited to percentage range
- Relatively high instrument cost
- Safety (use of lasers)

ducibility. Samples can degrade in the laser beam, or change morphology, or simply heat up and incandesce. The limited availability of digitised specific Raman libraries restricts widespread use of the technique. Quantitation is relatively inaccurate in view of the low intensities. Transferability and validation require improvements.

Raman spectroscopy has gained importance by introducing **lasers** as a light source. Lasers provide a coherent, single-frequency, high-power, small-beam source ($\sim$100 μm) that is nearly ideal for Raman spectroscopy. Improvements in laser technology have resulted in a large array of available frequencies ranging from UV to IR (*cfr.* Chp. 3.1), and Raman spectroscopy has been the beneficiary of these advances. The majority of lasers used for Raman spectroscopy have visible or near-visible emission frequencies. UV exitations are also used for specialist applications. Some popular lasers are HeCd (325, 354, 442 nm), Ar^+ (488, 514 nm), HeNe (633 nm) or diode (785 nm). Intracavity frequency-doubled Ar^+ lasers (257, 248, 244, 238, 229 nm) and Kr^+ lasers (234, 206.5 nm) give the desired continuous-wave (CW) excitation while the Nd:YAG lasers (1064, 532, 355, 266, 213, 204, 200, 184 nm) and XeCl excimer lasers (308, 208–950 nm) are low duty-cycle $\sim$3–15-ns sources. The benefits of using a laser system capable of providing high average powers with low peak power have been clearly demonstrated. The intercavity doubled Ar^+ laser makes the UV-Raman measurement comparable in difficulty to the typical visible-wavelength Raman measurement. The choice of laser excitation frequency, ν, depends on the type of sample being examined. In most cases, the laser wavelength is chosen to avoid any absorption by the sample as it may be destroyed by photodecomposition. Since the Raman scattering cross-section varies as ν^4 the wavelength of the source should be as short as possible to increase the probability of Raman-scattered photons. The excitation region covered by Ar^+ lasers (between 450 and 520 nm) is unfortunately especially prone to interference from fluorescent impurities. Taking into account the fluorescence problem, the most practical laser of choice is the Nd:YAG system, lasing at 1.064 μm (9395 cm^{-1}).

Despite its potential abilities Raman spectroscopy has until recently not been used substantially in analytical laboratories, but has been applied mainly to academic problems as a major tool for fundamental studies in physics and physical chemistry. This finds its origin in the fact that for classical Raman spectroscopy photons of the visible spectrum were usually employed. Fluorescence phenomena limit the **applicability** of classical Raman spectroscopy to highly purified materials, as opposed to real-life samples. Other factors, such as: (*i*) high cost of the equipment; (*ii*) need for highly skilled operators; (*iii*) slow data-acquisition rate; and (*iv*) lack of extensive databases have further contributed to the perception that Raman spectroscopy is inferior to IR spectroscopy for applied analysis of polymers in an industrial laboratory. However, this picture is now changing. Today, Raman spectroscopists have at their disposal both more efficient grating monochromators and CCDs for detection (dispersive Raman spectroscopy), Fourier transform technology and high-power lasers for excitation.

Modern Raman systems are ideally suited for at- or near-line analysis. Fibre-optic probes, which can be interfaced to CCD-Raman spectrometers with greater ease than to FT-Raman instruments, have greatly expanded the utility of Raman spectroscopy by taking the measurement capability to the sample [374]. It is also relatively simple to interface Raman spectrometers to other techniques, such as chromatography, light scattering, XRD, DSC, *etc.* but this is not yet an active area of research. Everall [375] has reported off-line LC-Raman (LC-Transform) interfacing.

If Raman is to become a routine analytical technique, then it is clear that calibration and transferability issues will have to be addressed along with the introduction of traceable reference standards.

Various aspects of Raman spectroscopy have been reviewed [376–382]; several books have appeared (*cfr.* Bibliography). Brookes [383] and Adar [363a] have addressed the prospects of Raman spectroscopy.

Applications

As it is common in the Raman scattering process to observe Raman band intensities of *ca.* 10^{-9} of the incident photons (UV, VIS, NIR) provided by a monochromatic laser source, Raman spectroscopy is an inherently insensitive analytical method that usually requires molecular concentrations of >0.01 M. Raman spectroscopy probably represents the single largest application of laser spectroscopy in industrial analysis and is being used in industry only as from the 1980s for the analysis of a wide range of materials, mainly solids. Raman spectroscopy is sensitive to molecular and crystal structure and can be used for identification purposes using a collection of fingerprint spectra, *i.e.* for confirming incoming product (QC), monitoring products, speciation, molecular identification (impurities or components in mixtures), microspectroscopy (*cfr.* Chp. 5.6.3), polymer morphology, investigations of fibres and films, reaction monitoring and on-line process control (*cfr.* Chp. 7.2.5). Some cases where Raman generally works particularly well relative to IR are inorganic materials (especially those with bands below 400 cm^{-1}), unsaturated compounds, aqueous solutions, and irregularly shaped objects or containers, where the ability to measure spectra without contacting the sample can be used effectively. Raman analysis is hindered by samples that fluoresce with the laser excitation line being used, are weak Raman scatterers, or decompose or burn under the laser light. Raman spectroscopy is also less effective than IR for samples dissolved in solvent. With Raman there is no simple way to increase the pathlength of the measurement and sensitivity for the materials of interest is often lower when a solvent is present.

Polymerisation reactions of unsaturated **monomers** (*e.g.* vinyl chloride, styrene, various acrylates/methacrylates), which involve loss of a C=C double bond, are easily followed by *in situ* Raman spectroscopy in view of the very strong monomer Raman band [356]. For example, the styrene monomer concentration was determined from the C=C stretch near 1640 cm^{-1} in on-line Raman spectra obtained during production of syndiotactic polystyrene [384].

Applications of Raman to **polymer/additive deformulation** are still rather few, especially if compared to IR methods (*cfr.* Chp. 1.2.1). Hummel [108] has attributed the general lack of applications of RS in the field of plastics additives to poor Raman scattering of certain substance categories, unsatisfactory reproducibility of the spectra and scarcity of specific Raman libraries [385,386]. Polymer/additive analysis by means of Raman spectroscopy is mainly restricted to fillers, pigments and dyes; the major usefulness comes from NIR FT-Raman, which greatly overcomes the fluorescence problem. The ion-pair dissociation effect of the 2-keto-4-(2,5,8,11-tetraoxadodecyl)-1,3-dioxolane modified carbonate (MC3) plasticiser in poly(ethylene oxide) (PEO) was studied by means of Raman, FTIR and EXAFS [387]. Another study established the feasibility of using Raman spectroscopy to quantify levels of melamine and melamine cyanurate in nylons [388]. In principle, grafted chromophore-containing additives can be determined spectroscopically.

Heavily **filled polymer composites** may be very difficult to analyse using IR spectroscopy because of broad and strong Si−O absorptions of fillers such as glass, clay and silica, but these fillers are poor Raman scatterers, and therefore the Raman spectrum of the polymer is obtainable without removal of the filler [389]. An illustrative example is the IR spectrum of PP/(DBDPE, Sb_2O_3, talc), which was greatly obscured by strong silicate bands at 9.8 and 14.9 μm, with only weak features at 13.4 μm (Sb_2O_3) and 7.3–7.7 μm (DBDPE). On the other hand, Raman spectra showed strongest bands for Sb_2O_3 (250 and 185 cm^{-1} shift), medium bands for DBDPE (140 and 220 cm^{-1} shift) and for PP. The silicate bands that obscured the regions of the IR spectrum were not observed in the Raman spectrum [389a]. Many fillers actually give much sharper Raman than IR bands, simplifying identification of the filler itself. It is trivial to distinguish the anatase and rutile forms of TiO_2 fillers from their Raman spectra.

Although Raman spectroscopy is very useful for identification and quantitation of carbonaceous species in various matrices, carbon is the most problematical filler. Common **carbon fillers** (amorphous coke or graphite) are strong Raman scatterers, but they also strongly absorb the Raman scattered light from the polymer. Thus, a carbon-filled polymer often displays only the spectrum of carbon, or if excessive laser power is used, the sample is burnt by laser absorption, When using 1064 nm excitation (FT-Raman) carbon-filled samples are strongly heated and will incandesce.

UV/VIS laser excitation of most organic **pigments**, which are aromatic *cq.* condensed, produces strong fluorescence. Reasonable RS may be obtained using red (785 nm) or near-IR (1064 nm) excitation. Generally, IR spectroscopy is faster, cheaper and more specific than RS in the identification of organic pigments. On the other hand, Raman spectroscopy is frequently used for (inorganic) pigment analysis of artworks [390,391]. Most common **dyes** fluoresce strongly and intrinsically when exposed to visible light. It is therefore not surprising to find no direct in-polymer Raman analysis of some main classes of additives (colorants, dyestuffs, pigments, *etc.*). NIR FT-Raman spectroscopy is here a more obvious analytical tool [392]. Dye spectra show very clearly in the presence of cellulose, which is a weak Raman scatterer.

Raman spectroscopy is extremely useful in the analysis of **surfactants**, particularly those in which the hydrophile is inorganic (sulfate, carbonate, phosphate, *etc.*). Infrared and Raman spectroscopy of surfactants were reviewed [393].

Most polymers can be analysed as received, as pellets, powders, films, fibres, in solution, or even as whole articles such as mouldings. Fine fibres can present some difficulties if a Raman microscope is not available. Raman spectroscopy has found applications in the identification of polymers in which additives obscure the polymer peaks in the IR spectrum. Reclaimed polymer is more prone to fluorescence than virgin material, causing problems for Raman analysis [394]. Laser-Raman spectroscopy often allows polymer identification (*e.g.* in recycled material) only in conjunction with IR spectroscopy. Raman spectroscopy has been used to examine weathered PVC plasticised with DOP, DOA and BBP for dehydrochlorination [395]. Laser-Raman spectroscopy has also been proposed as a suitable method for precise detection of ageing deterioration of vinyl chloride resins containing plasticisers and fillers used as electrical wire and cable coatings [396].

Laser-Raman spectroscopy is an ideal technique for **contactless monitoring** of extruded films, sheets, and moving fibres for the evaluation of crystallinity. These are perhaps ideal samples since they can have a relatively smooth surface, which can be held at the focus of the laser beam. A difficult sampling problem is that of a rough surface such as a bed of polymer pellets, when the roughness exceeds the depth of focus of the Raman collection lens. One solution is to grind the sample to produce a fine powder.

As a result of the high polarisability of C−S and S−S bonds, Raman spectroscopy is especially suitable for studying sulfur vulcanisation of **elastomers**. However, conventional Raman studies of elastomers are limited on account of sample fluorescence (often due to impurities).

Highly coloured samples (either pigmented or degraded/contaminated) often tend to burn in the laser beam, to fluoresce, or to heat up and incandesce. Other **difficult samples** or problems for Raman include: analysis of carbon-filled materials, measurement of trace ($\ll 1\%$) levels of additives or components in the polymer (unless subject to resonance enhancement), estimation of non-unsaturated end-groups in high polymers, analysis of degradation and measurement of thin ($\ll 1\ \mu m$) surface coatings or treatments on bulk polymers. Samples difficult by FT-Raman are dark specimens, some inorganic materials, dilute aqueous solutions, fragile or thermally sensitive samples. Raman spectroscopy plays also only a minor role in the hyphenation to separation techniques, such as TLC [397].

Although **FT-Raman** has determined an improvement in the performance of classical Raman spectroscopy of highly fluorescing polymeric specimens (blends, degraded samples, heat treated samples, vulcanisates, fully formulated oils, additives and coloured materials), it is far from true to state that the technique is entirely fluorescence free. NIR FT-Raman has been proved useful in the identification of polymers, end-group analysis, examination of vulcanisates, observation of dyestuffs in polymeric materials, morphological studies, kinetic measurements, and in the investigation of mechanical changes and degradation of polymers. The optimal sample thickness for FT-Raman analysis of PE, PET and cellulose was determined [398]. As Raman spectroscopy is ideal for the study of changes occurring in the C=C moiety of polymers, it is of great use in the study of polybutadiene rubbers [399], where results obtained by FT-Raman spectroscopy are more reliable than those derived from NMR spectroscopy.

FT-Raman has been used as an alternative to TG techniques to determine filler content in HDPE/$CaCO_3$ composites and provides comparable results [400]. As most pigments (apart from carbon-black) and glass are poor Raman scatterers, in principle Raman spectra are obtainable from these samples without removal of the fillers or difficult sample preparation. Conventional visible Raman spectroscopy has failed in attempting to analyse dyestuffs. Conventional Raman spectra of dyed textiles tend to be dominated by the (fluorescent) spectrum of the dye [401]. Consequently, FT-Raman spectroscopy may be a more useful tool for direct observation of low levels of **dyestuffs** in polymeric materials. Indeed, by using NIR excitation dramatic improvements in the Raman spectra of these dyes can be achieved [392]. FT-Raman was quite useful for the discrimination of differently dyed cotton-cellulose fabrics with the bifunctional reactive dye Cibacron C, provided that the interpretation was facilitated by chemometrics [402]. Schrader *et al.* [403] have used FT-Raman spectra to distinguish non-destructively the main dye components in historical textiles. Bourgeois *et al.* [401] have successfully used FT-Raman in the characterisation of low levels (1–2%) of dyestuffs in acrylic fibres. Unlike Raman data, DRIFT spectra are essentially of the acrylic fibres and yield no information as to the nature of the dye.

In situ Raman spectroscopy of the decomposition of *t*-butyl peroxy pivalate (TBPP) in *n*-hexane at 1900 bar and 100°C was reported [404].

Whereas conventional Raman studies of elastomers have been severely limited due to sample fluorescence (only highly purified and non-vulcanised samples could be studied), vulcanised systems can now be investigated quickly and with ease using NIR FT-Raman spectroscopy. As shown by Hendra *et al.* [386] even a **black oil-extended natural rubber** containing a significant quantity of fluorescent material can give recognisable spectra with no sample treatment. FT-Raman spectroscopy is also proving to be an excellent tool in the examination of cross-linked materials, because the S−S bond gives a prominent band in the Raman spectrum near 480 cm^{-1}. Also information about composition, crystallinity and orientation is contained in Raman spectra of polymers. The only additive to date to prevent acquisition of useful FT-Raman spectra is carbon-black.

The FT-Raman remote sensing probe was used to discriminate ivory specimens [405]. FT-Raman should not be used to study catalysts, carbons and emulsion polymerisation, where D-Raman can provide very useful spectra.

Hendra *et al.* [386] have recently reviewed the use of NIR FT-Raman spectroscopy in the study of many (co)polymers and blends, both qualitatively and quantitatively. For an overview of FT-Raman of elastomers, *cfr.* ref. [406].

Polymer applications in Raman spectroscopy were reviewed [375,407,408], as well as general applications in the chemical industry [52,384,409]. For Raman spectroscopy of synthetic polymers, *cfr.* ref. [394]. The use of Raman spectroscopy in art analysis has recently been reviewed [410,410a]. For applications of non-classical Raman spectroscopy, *cfr.* ref. [411] and for FT-Raman spectroscopy, *cfr.* also ref. [412]. A textbook is available [394].

1.2.3.1. *Specialised Raman Techniques*

Principles and Characteristics

In general, Raman spectroscopy suffers from low sensitivity, so that Raman analysis is typically performed on not or fairly concentrated samples. Many instrumental developments have greatly extended the potential usefulness of Raman spectroscopy to industrial problem solving. Several techniques have emerged which enhance the sensitivity of certain applications, such as **resonance Raman spectroscopy** (RRS) [352] and **surface-enhanced Raman spectroscopy** (SERS) [353]. The goal of **time-resolved Raman scattering** is to measure the transition condition of a sample (with time intervals ranging typically from ps to sec), *e.g.* for monitoring a chemical reaction. These more specialised Raman techniques are applied in important niches, but generally not yet routinely for problems in the chemical industry. There are unresolved questions concerning the quantitative nature of these methods.

Applications

Surface Raman techniques have been used in applications such as *in situ* ink analysis (*cfr.* also Chps. 1.2.3.1.1–2). Nanosecond laser flash photolysis and time-resolved resonance Raman spectroscopy have been used to study reactions between the AOs α-tocopherol and ascorbate and the triplet excited states of duroquinone (DQ) and ubiquinone (UQ).

1.2.3.1.1. *Resonance Raman Spectroscopy*

Principles and Characteristics

The spontaneous Raman effect can be initiated by a photon with sufficient energy to raise a molecule to a virtual state, which exists long enough to emit the Stokes or anti-Stokes photon in an inelastic manner. When the incident light photon's energy matches the energy necessary to reach an excited but stable electronic state of a molecule the process is called resonant Raman (RR). In **resonance excitation** conditions of a chromophore the induced dipole moment becomes much larger, causing a large increase in intensity of the Raman scattering [413]. The increase, by as much as 10^8 times over non-resonance conditions (*i.e.* about as strong as fluorescence), means that vibrational Raman spectra of dilute samples (in sub-mmolar concentrations) can then be studied quite easily. The dramatic increase in sensitivity happens for only a few of the molecule's vibrations, giving resonance Raman much greater **specificity** than normal spontaneous Raman scattering. In principle, resonance enhancement of the Raman scattered intensity can be used to increase the sensitivity of almost any type of Raman process. Sensitivity depends on the relative intensities of the analyte Raman bands compared with overlapping, interfering Raman bands and emissions from the sample. For the study of resonance Raman phenomena tuneable lasers (dye or Ti-sapphire) are mainly used. Different Raman spectra are observed with excitation in resonance *vs.* not in resonance.

Resonance Raman spectroscopy (RRS) leads to increased selectivity in Raman spectral measurements. The Raman spectrum of individual components in a complex mixture can be selectively enhanced by a judicious choice of laser wavelength. Only the Raman bands of the chromophore which is in resonance at the wavelength of excitation are significantly enhanced. Raman bands of non-absorbing species are not enhanced and do not interfere with those of the chromophore. Clearly, resonance Raman is a very sensitive analytical tool capable of providing detailed molecular vibrational information.

In principle, no special Raman instrumentation is needed to perform RRS because RR spectra can be obtained with conventional Raman spectrometers, if only the suitable excitation wavelength is applied. However, resonance Raman scattering is experimentally more difficult to implement than normal spontaneous Raman scattering. The excitation wavelength must be made to match the absorption band of the electronic chromophore of interest. The absorption band makes both the excitation intensity and Raman scattered intensity dependent on sample thickness, complicating quantitative analysis. Absorption of the excitation intensity can damage the sample due to heating and/or photochemistry.

The **advantages** of resonance Raman spectroscopy in molecular studies can be summarised as follows: low detection limits of chromophores ($<10^{-6}$ M), ability to excite particular species, structural sensitivity with high resolution, and lack of interference from non-chromophoric species [413,414].

Prior to the introduction of commercially available tuneable UV lasers, the major constraint for exploitation of RRS in industry has been the restricted number of systems which exhibited visible chromophores and could be conveniently probed. Clearly a more widespread use of **visible resonance Raman spectroscopy** would be possible except that in practice: *(i)* only a few chromophores absorb in the visible region; and *(ii)* fluorescence interference is nearly ubiquitous in "real life" samples.

Frequency doubled Ar^+ lasers with excitation in the UV region (257, 248, 244, 238 and 228.9 nm) allows UV-Raman studies of solid absorbing samples. At these wavelengths probing depth is about 1 μm. A major advantage of UV Raman over visible Raman spectroscopy is the fact that the majority of molecules have UV absorption bands. Factors limiting application of **UV resonance Raman spectroscopy** (UVRRS) are instrument cost, additive degradation and lack of extensive databases of UVRR reference spectra for different excitation wavelengths.

Resonance Raman spectroscopy has been reviewed [352,415]. Asher *et al.* [413,414,416] have recently reviewed UV resonance Raman spectroscopy.

Applications

Resonance Raman has been applied to the determination of **dyes** [417]. Resonance Raman spectroscopy and ESR have identified the key chromophores in ultramarine blue as S_3^- and S_2^- species [418]. Bell *et al.* [358] have overcome a fluorescence problem in a resonance Raman study (λ_{ex} = 363.8 nm) of ancient Chinese documents containing the yellow dye compound berberine by means of subtraction of shifted spectra (shifted-subtracted Raman spectroscopy, SSRS). The procedure may find wider use.

Until recently, RRS was limited to the small subset of compounds absorbing in the visible and near-UV spectral regions where laser sources typically have been available. UV resonance Raman spectroscopy (UVRRS) is highly sensitive and selective for studying the vibrational spectra of molecules with electronic transitions in the 180 to 300 nm region. The ability of UVRRS to selectively examine the vibrational behaviour of particular species in low concentrations in complex mixtures makes the technique unique for analytical applications. For example, the determination of low residual levels of **olefin monomers** in polymers is difficult by chemical or by spectroscopic methods, such as Iodine Number, ^{1}H NMR, ^{13}C NMR, IR or Raman. Although the C=C bond is a strong Raman scatterer, the normal Raman effect is weak, and it is usually difficult to detect concentrations below 10,000 to 5000 ppm. By resonance enhancement of the π–π^* transition it is possible to overcome the concentration problems associated with normal Raman spectroscopy while maintaining vibrational specificity. UVRRS near 220 nm has been used to probe low concentrations of unsaturated species in polypropylene [419]. A suitable internal standard is required for quantitative comparisons of olefin content. Because of the high energy carried per photon in the UV, care must be taken to prevent degradation of the sample.

Raman spectroscopy would not normally be considered a good tool for studying polymer degradation, since the latter often proceeds via an oxidative route, and oxygenated species are not usually strong Raman scatterers. Furthermore, degradation often yields discoloured materials that are prone to fluorescence, making Raman measurements difficult. Resonance Raman spectroscopy has been put to advantage in the study of **PVC degradation**, where low levels of conjugated polyene (*ca.* 0.0001%) are sufficient for discoloration as a result of dehydrochlorination [376,420]. Similarly, low levels of dehydrochlorination in natural weathering studies can be detected at such an early stage. Uncoloured PVC films give no resonance Raman spectra. It is also well known that failures arise when PVC comes into contact with polyurethane foams (which contain amines as catalysts for foaming). GC and IR analyses indicate plasticiser loss. Laser Raman associates this effect with degradation of PVC, namely chemical dehydrochlorination under the influence of bases, followed by cross-linking, which results in plasticiser incompatibility and consequent mass loss.

Williams *et al.* [421] have assessed the potential of UVRRS as a general analytical tool, both to distinguish between molecules with similar electronic absorptions, and to wavelength tune the laser to enhance selectively the Raman spectrum of individual components in a complex mixture. Raman spectroscopy provides a method whereby the distribution of polyene sequence lengths can be monitored. Not surprisingly, resonance Raman is also a major tool for characterising conducting polymers [422].

Asher *et al.* [423] have indicated that UVRRS may be used as a technique for **speciation** of aromatic ring systems in complex matrices of industrial interest. This result might suggest an extension to direct deformulation of a complex mixture of additives in polymers provided that excitation can selectively occur within a discrete absorption band of each analyte. However, at the same time this requirement may easily turn out to be the major limitation in practical applicability as virtually identical UV spectra (or λ_{max}) are often observed for a variety of stabilisers (*cfr.* Chp. 1.1). Klenerman *et al.* [424] have examined Chimassorb 944/966, ethoxylamine, glycerol monostearate, Irganox B215/1076/PS 802, Tinuvin 770 and Kemamide by means of UVRRS at 244 nm excitation observing the lack of a major resonance Raman enhancement for the antioxidants and no specificity. Sample degradation is a serious problem in UV-Raman measurements of UVAs and AOs. Fluorescence life-time measurements, based on the principle of time correlated single-photon counting, seem very promising but require further development. Fluorescence decay measurements only use laser power in the μW range, as opposed to conventional macro-Raman measurements (mW range).

Raman spectroscopy is one of the optical molecular spectroscopic techniques capable of giving quantitative information about molecular orientation in polymers. A resonant Raman-active agent and/or highly anisotropic rigid rod polymeric substance incorporated into polymers can be easily detected at low concentration levels and used as an indicator of the **molecular orientation** of the processed polymer itself [425]. It is possible to apply resonance Raman spectroscopy to many more problems. The combined application of UV/VIS and near-IR Raman excitation may be advantageous. UV excitation selects for a small number of resonance-enhanced bands of an analyte and the measurements are made with high sensitivity and selectivity. In contrast, with non-resonance visible and near-IR excitation, numerous Raman bands occur with similar intensities for all components in the sample in proportion to their concentrations. Consequently, one obtains both average and specific information on sample composition [419].

UV-Raman spectroscopy is, as yet, used in few analytical studies not in the last place because the technique is confined to a few laboratories only. Asher *et al.* [4,419,426] have reported applications of UVRRS.

1.2.3.1.2. *Surface-enhanced Raman Spectroscopy*

Principles and Characteristics

In a classical picture, Raman scattering of surface species can be separated into two processes: *(i)* excitation induced by the electromagnetic field at the surface from the vibrational ground state in the fundamental electronic state to an excited virtual state; and *(ii)* subsequent transition from the excited state to an excited vibrational level in the electronic ground state accompanied by spontaneous emission of Raman shifted light. The probability of the induced transition to the excited state increases with the strength of the electromagnetic field at the surface. It follows that the intensity of Raman scattered light increases with the enhancement of the surface electromagnetic field, which is brought about by incidence of the exciting laser beam. The effect is called **surface-enhanced Raman spectroscopy** (SERS) [427]. The Raman cross-section of a material may be increased by a factor of 10^7 or more by the presence of metal colloids or a roughened metal surface (Ag, Au, Cu); surface roughness of the order of *ca.* 20–200 nm is required. The increase occurs only for material directly in contact with the active metal surface. While resonance Raman spectroscopy is invariably more useful with UV or VIS laser excitation, SERS is equally applicable in any electromagnetic region of radiation. NIR SERS has all the advantages of NIR Raman and the added advantage of high sensitivity. SERS may be induced using a conventional Raman set-up.

Surface-enhanced Raman scattering was first demonstrated in 1974 [428]. The nature of the mechanism that produces SERS is still subject of debate: electromagnetic (EM) enhancement, which does not require a chemical bond between adsorbate and metal surface and chemical or charge transfer (CT) enhancement, which favours such specific bonding [429]. In any case, the Raman enhancement due to the surface effect decreases very quickly as a function of distance, and little enhancement is obtained for molecules a few monolayers away from the surface. As a result, SERS is surface-selective. For intensity enhancement the SERS technique is particularly attractive because the excitation wavelength does not have to coincide with absorbance of a molecule, as opposed to resonance Raman. The excitation wavelength must be one that can launch a surface electromagnetic wave (surface plasmon) on the metal surface. However, plasmon resonance is very broad and peaks in the far visible and NIR for the active metals mentioned above. Therefore, red and NIR wavelength lasers provide highest efficiency.

SERS incorporates most of the **advantages** of Raman spectroscopy. The greatest benefits are enhanced sensitivity (10^{-9} M, ng level), selectivity and surface specificity. However, the great analytical potential for SERS is limited by several factors, amongst which the need for adsorbates on a limited number of metal surfaces [430]. Quantitative applications of SERS are difficult [431]. For SERS to become a useful method for real-life applications, stable and reproducible substrates must be manufactured (nanotechnology), *e.g.* Klarite®. Recently, stable Raman enhancing reagents (*e.g.* Au hydrosol) were reported [432].

Surface-enhanced resonance Raman scattering (SERRS) is obtained when a molecule with a chromophore is adsorbed onto or is in close proximity to a suitable metal surface and the excitation wavelength is tuned to the molecular frequency of the analyte. By combining SERS with resonance enhancement, Raman cross-sections have been increased by a factor of 10^{14} to 10^{15}, *i.e.* much more than that of either resonance Raman or SERS. The technique enables very sensitive analysis and low detection limits to be achieved, making single molecule Raman spectroscopy possible [433]. Signal assignment in SERRS is simple and reliable, as opposed to the difficulty in SERS. In using SERRS, fluorescence is usually quenched. Further, the technique requires very low laser powers and consequently photodegradation – common in SERS – is seldom a problem.

SERS was reviewed [431,434] and described in a specific textbook [427]; a monograph on surface Raman spectroscopy is available [123].

Applications

The advent of surface-enhanced Raman spectroscopy [435] allows studying extremely low concentrations of molecules on surfaces. Yet SERS is still a rarely applied vibrational technique. Because SERS provides both rich spectroscopic information and high sensitivity as a result of the large enhancement effect, it is an ideal tool for **trace analysis** as well as for *in situ* investigations of interfacial phenomena. A number of investigations has explored the possibility of using SERS for direct analysis of species separated by TLC, HPLC and GC. Tran [436] reported sub-ng detection of dyes on filter paper by SERS.

The ability to probe surfaces using *in situ* SERS can be exploited in polymer chemistry to characterise the surface of polymers for comparison with the bulk properties and to study polymer-metal composites such as adhesives and coatings. Interactions between adhesives and metal polymeric surfaces have been investigated [437]. The applicability of SERS to **polymer surfaces** has also been reported in a study on poly-*p*-nitrostyrene [438], and in another study on the effects of chromic/sulfuric acid etching on PE films [439]. The desired surface sensitivity

Table 1.21. *In situ* probing by Raman scattering techniques of ink jet dyes printed on paper

Feature	SERRS	NIR-FTRS
Sample preparation	Silver-colloid addition	Non-invasive
Excitation range	Visible	Near-infrared
Resonance enhancement	Yes	No
Laser power	Low (<1 mW)	High (>200 mW)
Accumulation time	Low (1–10 s)	High (2–30 min)
Volume sampled	Low (surface sampling)	High (bulk sampling)
Paper/filler identification	No	Yes

was obtained, but sample preparation is complex. Of particular interest in SERS is that monolayers give spectra influenced by the surface selection rules. As a consequence, the orientation of chemical groups in polymers relative to the substrate metal surface can be estimated. SERS can also be exploited for the study of interdiffusion of polymers. Applications of SERS were reviewed [353].

Characteristic spectra routinely observed with SERRS permit identification of mixtures without the need for preseparation. Since a SERRS spectrum is characteristic of the molecule the technique can be used to discriminate between dyes and identify dyes in mixtures, even when the **dyes** have very similar chemical structures. Recently, an *in situ* SERRS method has been reported for the detection of a reactive dye covalently bound to cotton, whereby a fibre was treated with colloid [440]. The analysis of 20 similar monoazo dyes, all of which produced unique characteristic SERRS spectra, has been mentioned [441]. An RSD of 5% was routinely obtained.

Smith *et al.* [442] have reported a comparison between two Raman scattering techniques, SERRS and NIR-FTRS, for the characterisation of ink jet dyes and inks printed onto paper surfaces. Table 1.21 shows the main differences in performance. At variance to SERRS, which identifies only the dye chromophore, NIR-FTRS observes a number of bands that arise from the paper and ink systems. Combination of the two techniques provides some information on the electronic as well as the vibrational properties of the dyes *in situ*.

The sensitivity and selectivity of SERRS can be exploited in **forensic science** by determining the nature of the dye mixture *in situ* from a single fibre, or from ink. SERRS was also used successfully for *in situ* analysis of chromophores in lipstick smears on glass and cotton surfaces, without any preseparation [443]. Reproducible SERS/SERRS substrates are still in search of high-value applications [444].

Applications of Raman spectroscopy to investigations of a variety of real surfaces were reported [123].

1.2.3.1.3. *Low-resolution Raman Spectroscopy*

Principles and Characteristics

The main drawback to Raman has been its high price tag relative to mid-IR and near-IR, although the price barrier to Raman is lessening. Nevertheless, a main issue is still the laser system required to produce quality high-resolution spectra. Clarke *et al.* [445] have recently introduced the concept of low-resolution Raman spectroscopy (LRRS) as a potentially highly useful, low-cost approach to organic identification and analysis.

In a typical LRRS application the need for feature separation is much like that encountered in mid-IR spectroscopy. One rarely requires single wavenumber resolution to find the fingerprint feature that allows identification and quantification of the system under analysis. Therefore, a broader band laser source may often suffice for the Raman analysis. Simple multi-mode, solid-state laser sources are both sufficient for the task and extremely cost-effective. The complete LRRS system thus consists of an inexpensive multi-mode laser diode, a low-resolution monochromator matched to a simple CCD detector, and a Rayleigh filtering device. Even though all spectral features are not necessarily cleanly resolved with either LRRS or near-IR, the ability to use broad vibrational bands as fundamentals gives LRRS an inherent advantage over near-IR. In essence, the LRRS approach relies on the fact that only certain spectral features are required to be fully resolved to identify the components in an organic sample.

Applications
Clarke *et al.* [445] described the application of LRRS for mixtures of organic liquids, and petroleum products (in particular aromatic ring structures in aliphatic admixtures). The aromatic composition of fuels is characterisable by Raman spectroscopy on the basis of the identifying ring vibrational modes that are strong scatterers in the Raman spectrum. LRRS also finds medical applications, *e.g.* the analysis of skin creams in which the lubricating ingredient is a petroleum-based component (benzoylperoxide), characterised by aromatic peaks and hosted in an olefinic. The analogy to aromatic additives in polyolefins is clear.

1.3. PHOTOACOUSTIC SPECTROSCOPY

Principles and Characteristics
Absorbance methods can be separated into two groups: methods that measure transmission, including conventional spectrophotometry, and those that measure the power absorbed by the sample, the so-called calorimetric techniques, including photoacoustic methods. Photoacoustic (PA) spectroscopy is an **emission technique** for examining (the surfaces of) solid materials. There is broad general agreement on the fundamental processes which account for the photoacoustic effect in solids [446], first described by Bell [447], but applied in spectroscopy only since 1968.

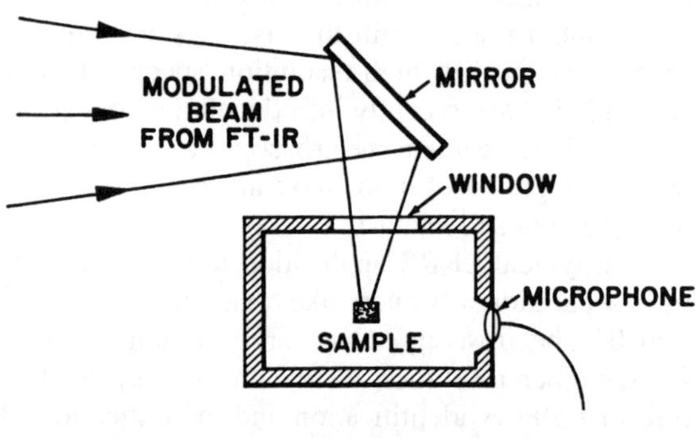

Fig. 1.20. Schematic diagram of a photoacoustic detector. After Perkins [56]. Reprinted with permission from W.D. Perkins, in *Practical Sampling Techniques for Infrared Analysis* (P.B. Coleman, ed.), CRC Press, Boca Raton (1993). Copyright CRC Press, Boca Raton, Florida.

Photoacoustic measurements are unique in that they depend directly on the energy absorbed by the sample, rather than on what is transmitted or reflected. The PA effect is based on detection of **acoustic waves** generated by radiationless relaxations of an absorption process initiated by a non-stationary (modulated or pulsed) light source (Rosencwaig–Gersho or R-G theory). Experimentally, a sample is placed in a metal sample cup (∅10 mm, 3 mm deep) in an acoustically isolated closed chamber with a KBr window (*cfr.* Fig. 1.20) and the cell is filled with an IR transparent coupling gas (He, Ar or air). Periodic illumination of the sample at a wavelength at which it absorbs heats it. The complex PA effect consists of three steps, namely from a modulated electromagnetic wave to modulated heat and modulated pressure (sound), as follows: *(i)* heat release in the sample material due to optical absorption; *(ii)* acoustic and thermal wave generation in the sample material and surrounding gas phase; and *(iii)* determination of gas pressure fluctuations in a PA detector (Fig. 1.21). The increase in temperature can be detected either with a sensitive microphone (for solids) or with a piezoelectric transducer (for liquids), in which the heat-induced stress generates an electric signal. The microphone *listens* to a sample becoming warm at its characteristic absorbing wavelengths. Measuring the signal as a function of wavelength gives the absorption spectrum of the analyte. The magnitude of the acoustic signal corresponds to the amount of light absorbed by the sample. Consequently, a photoacoustic spectrum resembles an optical absorption

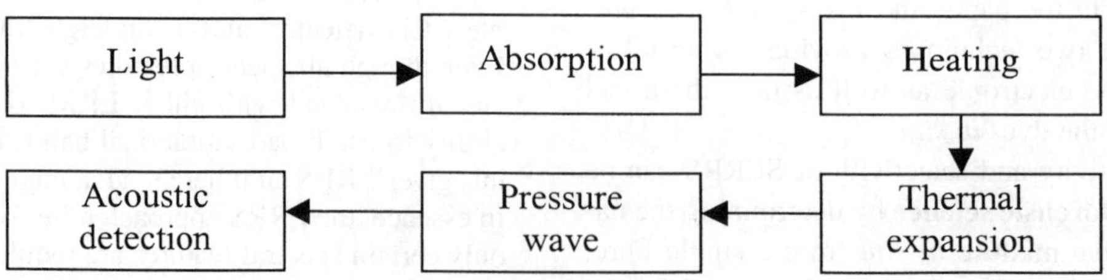

Fig. 1.21. Principle of a photoacoustic experiment.

spectrum. Since photoacoustics measures the transient internal heating of the sample, it is clearly a form of calorimetry as well as a form of optical spectroscopy. Although the name photoacoustic spectroscopy (PAS) is well established, a more descriptive name would be **photothermal spectroscopy**.

Photoacoustic spectroscopy is the general term to describe spectroscopic measurements taken by detecting an acoustic signal generated by the absorption within the sample of an amplitude-modulated beam of energetic radiation, including electromagnetic radiation from radiofrequency to X-rays, electrons, protons, ions and other particles. Most of the early PA experiments were in the **UV/VIS region** of the spectrum, because of the dependence of PA intensity on the source intensity [448]. PA-UV makes use of a high intensity xenon arc source and a dispersive spectrometer. PAS was not used in the **mid-IR region** until the advent of FTIR. The increased signal intensity available in the FTIR experiment coupled with PA detection has now provided a method which is extremely versatile for various types of solid samples, which are difficult to sample otherwise. PA-FTIR differs from most infrared techniques in that it is an emission rather than an absorption technique. FTIR photoacoustic spectra are measured with a photoacoustic cell accessory, which mounts in the sample compartment of the FTIR. A microphone is connected to the sample chamber for detection of the photoacoustic signal. Spectra are obtained by Fourier transforming the PA-FTIR interferogram and appear as absorbance spectra without further computation. Most FTIR instruments cover the range of 4000 cm^{-1} to 400 cm^{-1}, but some sources also cover the near and/or far IR regions. Since **NIR sources** have high energy, the NIR region is particularly well suited to photoacoustic experiments. As NIR penetrates more than mid-IR, the sample depth of PA-NIR exceeds that of PA-FTIR. **Raman photoacoustic spectroscopy** was also described [449]. The advent of lasers and sensitive microphones has led to an extensive re-examination of the technique. Since the generated PAS signal is proportional to the absorbed (and thus to the incident) radiation power, powerful radiation sources, particularly lasers offering high spectral brightness, are advantageous for achieving high detection sensitivity and selectivity. In the UV/VIS spectral range, excimer and dye lasers have been employed, whereas in the mid-IR wavelength range line-tuneable CO_2 and CO lasers dominate the applications. Photoacoustic detection modes are absorbance, diffuse reflectance and transmittance.

The **theory** of PAS is sufficiently complicated [446]. PAS has some limitations insofar as the frequency positions of absorption bands are reproduced but the band intensities cannot be interpreted according to the conventional rules of transmission spectroscopy. The intensity of a PA signal depends on the coefficients of optical absorption and thermal diffusion.

Photoacoustic spectroscopy is the only **sampling method** that is non-destructive, non-contact, and produces spectra directly without sample preparation or alteration. The absence of sample preparation is particularly useful with highly air- or moisture-sensitive polymers, such as polyacetylene for which conventional absorption spectra are difficult to obtain. An advantage of PAS is that spectra can be obtained from strongly absorbing samples and for materials that cannot be ground to a fine powder or be prepared with a flat surface. The main limitation to sampling in PAS is that the samples be small enough to fit into the sample cup. PA-FTIR allows a wide variety of sample forms: polymer chips, pellets, low-surface-area samples (*e.g.* plugs excised from a moulding), extruded film, coatings, fibres, blown foam, gels, pastes, viscous liquids and powders without crushing, grinding or diluting. Opaque, hard and insoluble solids may be handled without abrasion or milling. Because of the energy-conversion process (light absorption-emission of acoustic waves), PAS detection is a valuable tool when the optical absorption is so strong that it prevents light passage through the sample. Consequently, PA-FTIR is actually the most broadly applicable mode to overcome the opacity problem in the infrared spectral region (other modes are NIRS and reflectance). Ideal samples are loosely packed powders or open-cell polymer foams. Since the photoacoustic signal is proportional to the absorbed energy, even spectra of strongly scattering samples, *e.g.* powders, can easily be measured. However, PA studies on powders are usually only qualitative. Samples containing low levels of elemental carbon also give an increased signal, due to the absorption coefficient of carbon. Samples which are pigmented black with carbon can sometimes also yield strong spectra, but heavily carbon-filled samples suffer from too much absorption. While PAS can be applied to bulk solids, powdered samples give the best

PA spectra because their increased surface area facilitates heat transfer between sample and surrounding gas.

When neither transmission nor reflection infrared spectra are satisfactory, spectra can generally be obtained by photoacoustic spectroscopy. The IR spectrum that is obtained is ratioed against an IR spectrum of carbon-black. **PAS spectra** resemble normal IR spectra with the same absorbance peak wavenumber locations as classic transmission spectra. Because the signal-to-noise ratio of PAS is very low, several thousand scans are needed to obtain a spectrum. The nature and interpretation of the spectra are markedly influenced by the thermal and acoustic properties of the sample.

PAS is essentially a **surface sampling technique** in which the penetration depth is related to the interferometer frequency by the following equation:

$$d = (\alpha / w)^{1/2} \quad (1.6)$$

where w is the modulation frequency of the incident radiation and α represents the thermal diffusivity of the sample. This is the basis of PAS **depth profiling**, which is an important application of the technique [450,451]. Depth resolution in PAS is thus rather complex and depends on the optical and thermal properties of the material [452]. The most useful technique for depth profiling is the phase modulated (PM) mode. Typical sampling depths for polymer samples in PA-FTIR spectra range from a few to a hundred μm. This allows measurement of spectra with either high surface specificity to analyse a coating, or with bulk specificity to observe the absorbance bands of a substrate. Samples with compositional variations due to layers or gradients may be examined. PAS depth profiling is especially a viable approach when the top layer is either thin or not strongly absorbing in key regions of the IR spectrum where deeper layers absorb [452a]. However, measuring a concentration quantitatively as a function of depth has not been achieved.

The **main characteristics** of PAS are collected in Table 1.22. The major benefit of PAS is that, to a large extent, the general features of the spectrum are independent of the sample morphology. This is in contrast to other IR sampling techniques for solids such as diffuse reflectance, for which particle size and shape of the sample are critical.

Table 1.22. Main characteristics of photoacoustic spectroscopy

Advantages:

- No elaborate sample preparation (samples to be sealed in a cell)
- Rapid, non-destructive, non-contact
- High selectivity
- Relatively high sensitivity
- Any sample morphology allowed
- UV/VIS, mid-IR and NIR applications
- Suitable for samples in any aggregation state (macro- and micro-size)
- Various detection modes (absorbance, diffuse reflectance, transmittance)
- Suitable for optically opaque materials
- Artefact-free, library searchable spectra
- Quantitation feasible
- Surface characterisation tool
- Depth profiling (compositional gradients and layers), μm range
- Very stable instrumentation
- Spectroscopic and non-spectroscopic applications

Disadvantages:

- Complicated theory
- Relatively high cost of the accessory
- Low signal-to-noise ratio

Quantitative analysis of the PA signal is possible when all the steps of the process can be described quantitatively. This has been achieved only in a few special cases. Application of factor analysis of processing PA-FTIR spectra enables quantitative analyses to be performed with standard error of prediction (SEP) values below 1%. Both PCR and PLC factor analysis tolerate non-linearity in spectra and allow concentrations of multicomponent systems to be determined. The conditions for quantitative PA-FTIR analysis were discussed [453]. Parker [454] has recently demonstrated that quantitative PA-FTIR analysis of rubbers and polymers is generally applicable.

Gardella *et al.* [451] have reported a comparison between **ATR and PA-FTIR sampling** for surface analysis of polymer mixtures. ATR provides an IR spectrum representative of surface regions of different depths, giving a non-destructive depth profile over a fairly large sample depth, in comparison to particle ejection techniques. Since PA-FTIR requires no contact with the sample, sample shape and surface roughness do not matter, as opposed to ATR techniques. The wavelength range covered is limited only by the FTIR instrumentation and PA detector window and not also by the ATR crystal. PA-FTIR has a distinct advantage over ATR spectroscopy because of its higher sensitivity in the high-wavenumber regions of the IR spectrum. The depth being sampled by PA-FTIR in rubbers is typically 3 to 11 μm at 2000 cm^{-1}, which is an order of magnitude greater than that sampled by ATR techniques. Step-scan PA-FTIR improves surface layer discrimination and resolves a layer thickness of less than 1 μm, *i.e.* better surface specificity than IRS [455]. For smooth surfaces, ATR is fast but suffers from variable background arising from imperfect sample IRE contact, which can affect apparent photometric accuracy. PAS methods are slow; having low signal intensity for smooth, low surface area solids, but under specific conditions can have a shallower sampling depth than ATR. PAS methods suffer from the variable signal level across the spectrum, which creates photometric inaccuracies, but PAS is more sensitive to surface impurities. For higher surface area samples, or air-sensitive samples, PAS can provide a simple means of obtaining a vibrational spectrum, without the surface disruption which could occur when pressing a sample against an IRE for ATR analysis. Thus powdered samples and rough surface morphologies are more favourable for PAS sampling whereas smooth morphologies are needed for ATR (for good contact between crystal and sample), because inefficient scattering of the excitation source does not greatly affect the resulting PA spectrum [451]. In the infrared region, there is little doubt that ATR and FFT interferometry provide information as easily and with greater sensitivity and resolution. In the visible region there is more to be said in favour of PAS *vs.* diffuse reflectance, but it is a region of less general analytical significance than the infrared. PAS can handle samples which cannot be measured by diffuse reflectance. Both ATR and PAS methods for surface analysis of polymers by FTIR have distinctive advantages and disadvantages, which make conjunctive use a viable means to increase structural information.

Photoacoustic spectroscopy has also been applied to chromatographic analysis. Thin-layer chromatograms have been analysed quantitatively by removal of plate sections and also directly on the plate. A PAS cell has also been developed as a detector for liquid chromatography based on single wavelength measurement. Phase-resolved PAS (PR-PAS) is a new non-destructive and quick analysis tool for the characterisation of multilayered samples or for materials that have a changing composition going from surface to deeper within. Future developments foresee micro-PAS detection with focusing capabilities allowing for better sensitivity for single particles, as well as time-resolved PAS for diffusion studies.

PA-FTIR has been reviewed [453,456–462], as well as PA-UV/VIS/NIR (250–2500 nm) [459]; a monograph has appeared [448].

Applications

Photoacoustic spectroscopy is utilised as a non-destructive method for characterisation of solids and other materials which are difficult to study by other methods. PAS is a valuable tool in food science. Applications in polymer research range from surface analysis to depth profiling, determination of stratification and degradation of polymers, non-equilibrium processes, time-dependent phenomena, *etc.* Qualitative PA-FTIR analysis has concerned **macrosamples**, *e.g.* polymer identification (by computer search after converting the spectrum to transmission) and adhesive analysis (using a spectral subtraction approach) [453]. On the other hand, PA-FTIR offers some unique advantages for **microsample analysis** over FTIR microscopy, such as no requirement for pressing to reduce optical density, no need for delicate optical alignment and an extended spectral range. PAS is suitable for materials which cannot easily be handled by FTIR such as highly cross-linked materials (rubbers), oxidation sensitive materials (*e.g.* polyacetylenes), multilayer materials, opaque and rigid samples [463]. FTIR is used more frequently in PAS mode than either UV/VIS or NIR.

Photoacoustic UV spectroscopy (PA-UV) has been used for evaluation of UV-absorbing paint additives in clear intact paint layers [32,464]. The technique enables quantitative analysis of additive concentration and aids in determining the effects due to paint processes and substrate composition changes on additive concentration. PA-UV is primarily a top-layer detection tool, enabling the analyst to evaluate the outer regions of a sample for additive disposition without requiring layer removal. Any loss of UVA additive to the substrate can be clearly seen from PA-UV spectra. Analysis depth is determined by the modulation frequency. PA-UV analysis is readily applicable to commercial paint systems and real components with some curvature. This is a distinct advantage over methods requiring removal of paint from a plastic substrate or good contact, as in case of ATR. Photoacoustic absorption measurements **(PA-VIS)** have been used for on-line monitoring of changes in dyeing processes involving concentrated dyestuff (5 g L^{-1}) with extremely high absorption and scattering particles in the dye solution, which make application of classical transmission spectroscopic analysis impossible [465]. Instead of a modulated light source, in this case a pulsed Nd:YAG laser (532 nm) was used.

Photoacoustic and diffuse reflectance techniques in the NIR region were used to determine the amount of combined vinyl acetate in powdered PVC/PVAc copolymer [213]. In **PA-NIR** careful control of the particle size range, distribution of the sample and multiple wavelength measurements are required for quantitative studies; no pre-weighing of the sample is required. PAS is a depth-sensitive technique. When used in polymer analysis, attention should be paid to sample characteristics which may affect surface layer composition and thermal properties: additives, segregated low-MW material, contaminants, *etc.* PA-NIR spectra of LDPE taken at various modulation frequencies show strong differences at high and low modulation frequencies of the light used [212]. For a thermal diffusivity $\alpha = 0.001\ cm^2\ s^{-1}$ the sampling depth μ in LDPE as a function of the light modulation frequency ω varies from 56 μm at 10 Hz to 11 μm at 240 Hz. The observed increase in concentration of methyl and vinyl groups from the polymer bulk to the surface is assumed relating to the presence of low-MW, waxy, non-crystalline material at the polymer surface. The higher concentration of hydroxyl groups at the surface layers is due to polymer oxidation. PA-NIR has also been used in the determination of moisture [466].

PA-FTIR can be used for many different types of polymer analyses, from a simple qualitative identification to more advanced variable depth sampling measurements. In **qualitative PA-FTIR analyses**, polymers in pellet or other forms can be identified within a few seconds using rapid computerised search and spectral libraries commonly available with FTIR systems. Polymer degradation chemistry, due to weathering, can be studied by PA-FTIR by measuring spectra as a function of exposure time. Ashworth *et al.* [467] have used PAS of cryogenically ground cloth to monitor the content of polyester-cotton blends. Other applications of PAS are identification of polymer-coatings on a metallic substrate, determination of the thickness of polymer coatings, identification of the (metal) substrate of the laminate and characterisation of the adhesion between polymer coating and substrate [468]. In characterisation studies of polymer coatings on metals, the quality of thermal contact between polymer coating and metal substrate influences the photoacoustic phase spectrum [469]. PA-FTIR provides a ready means to examine the chemical composition of a clearcoat's surface in complete paint systems [32].

PA-FTIR spectroscopy is an excellent technique for comparison of the surface-to-bulk composition of polymer samples, since surface and bulk sample spectra are easily obtained. The sampling depth varies as a function of wavenumber. PA-FTIR is frequently used for **depth profile studies** in polymer films, *i.e.* to roughly 50 μm in depth. Frequency-, phase- and time-resolved step-scan PA-FTIR approaches have been used in depth profiling chemical analysis of layered polymeric samples [470]. Depth profiling may be applied to studies of blooming, photooxidation, laminates, *etc.* Carter *et al.* [471] examined **bloom** of dimorpholinyl thione and zinc stearate from ingredients used during the vulcanisation of a NR and silica-filled SBR sample. Using the depth profiling capability of PA-FTIR it has been established that the concentrations of polyurethane **sizing agent** are higher at the cotton yarn surface than in the bulk [472]. PAS depth profiling of a polymer surface from *ca.* 6 to 25 μm was used for surface mapping of lauric diethanolamide in PE masterbatch pellets [473]. The mapping technique showed a lower concentration of the additive on the surface of the pellet and a higher concentration towards the core, *i.e.* a reservoir of additive concentration at higher depth. Polymer surfaces can be characterised by PA-FTIR in terms of surface segregation of additives, chemical treatments or fluorination [474]. Also surfaces of vulcanisates have been analysed by PA-FTIR [475]. Carter *et al.* [476] used PA-FTIR in depth profiling of paints and carbon-black filled rubbers.

PA-FTIR depth profiling results are consistent with the known layer structure of a packaging laminate film and an adhesive label [477]. Double-layered PET/PET, PP/PET and PET/PP laminates were studied by PA-FTIR [478]. PA-FTIR and DSC identified a skin layer and a core in injection moulded PET plates [479]. Plastic-coated paper was analysed by both PA-FTIR and DRIFTS allowing for shallow- and deep-sampling, respectively [453]. PA-FTIR is also a suitable tool for the analysis of **polymer films** used as a barrier coating on beverage and food containers; at variance to specular reflectance measurements surface flatness is not critical.

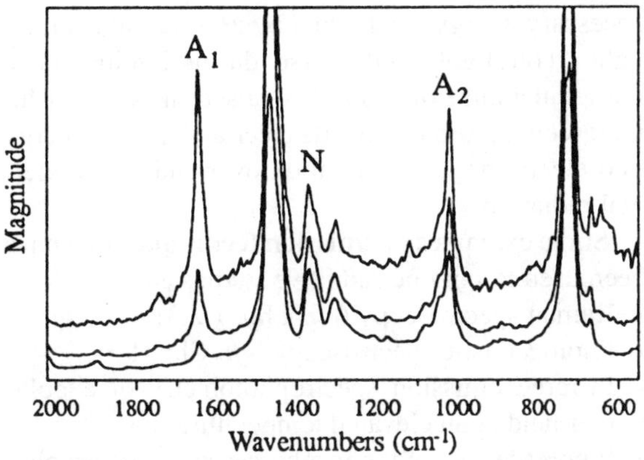

Fig. 1.22. PA-FTIR analysis of additive distributions in polyethylene. After McClelland *et al.* [481]. Reproduced by permission of the Society of Plastics Engineers (SPE).

PA-FTIR has been used to identify additives and to determine their distribution in surface layers. Urban *et al.* [480] have described surface imaging and depth profiling of 50%/50% styrene-*n*-butylacrylate latex film cast on a PTFE substrate by means of various vibrational surface sensitive measurements, including PA-FTIR and microscopic ATR-FTIR/FT-Raman microscopy to study the distribution of **surfactants** in latex films. Surfactant aggregates were detected at the film–air interface at significantly higher concentrations than at the film–substrate interface. PA-FTIR spectroscopy allows monitoring of the surfactant distribution across the film thickness. Exudation phenomena of surfactant molecules in latex films have been observed. The distribution of surfactants in latex films can be affected by surfactant-latex compatibility, interfacial surface tension difference, temperature and humidity, and other factors occurring during latex coalescence. The **distribution of additives** in PE has been determined using spectra measured at different modulation frequencies. Figure 1.22 shows PA-FTIR spectra for A_1 (A_2) additive bands of 39 μm, 11 μm, and 3.8 μm (resp. 30 μm, 8.6 μm, and 3.0 μm) from lower to higher on the plot (scaled to equal height of the PE band labelled N). The information gathered with the variable frequency data indicates that the A_1 additive gradient is sharper than that of the A_2 additive [481].

PLS factor analysis supplied with FTIR systems has been applied to determine vinyl acetate concentrations in PE copolymer pellets of varying size using PA-FTIR without sample preparation [453]. Herres [192] has compared the suitability of various non-destructive methods (FTIR, ATR-FTIR, PA-FTIR and pulsed NMR) for quantitative determination of **plasticiser content** in filled PVC. PAS was also instrumental in detecting the presence of absorbed water and **filler modifiers**, such as calcium carbonate treated with stearic acid, in aged silicone based sealants [482].

Very black materials, such as tyres and graphite fibre composites, are very difficult samples for which PA-FTIR can obtain results, although with more difficulty than for non-carbon filled materials. PA-FTIR is probably the best technique for acquiring IR spectra of **carbon-black-filled polymers** [454]. In fact, the standard reference material used in PA-FTIR spectroscopy is a 60 wt.% carbon-black-filled natural rubber. However, even in PA-FTIR carbon-black can dominate the thermal response of the sample at high loading levels. Carbon fibre/epoxy prepregs and automobile tyres, which are too opaque for transmission spectroscopy, yield good PA-FTIR spectra [453]. PA-FTIR has been used for compositional analysis of off-road tyre treads based on *cis*-1,4-polyisoprene [483]. An excellent PA-FTIR spectrum may be obtained from a sheet of carbon-black-filled PC, at variance to FT-Raman spectroscopy [484]. PAS has been used to identify black contamination of ABS chips as PE on the basis of excellent spectra over the full range of 4000–400 cm^{-1}, where FTIR microscopy failed [109].

As PA-FTIR requires minimal or no sample preparation, it is a well-suited method for non-destructive analysis of **fibres**. PA-FTIR has been used for single fibre sampling (forensic application) and as a tool for identifying surface coatings on fibres [485]. The PAS method allows investigation of the surface treatment of wool fibres [486]. PA-FTIR polarised light measurements can be used to study molecular orientation in drawn fibres and films.

PA-FTIR has also been applied for *in situ* **detection in TLC** [487]. Cure chemistry is readily studied by PA-FTIR. The use of PA-FTIR spectroscopy for *in situ* studies of cross-linking can provide further insight into chemical and physical processes. PA-FTIR has sufficient sensitivity to monitor phase transitions during cross-linking processes. Also on-line headspace analysis of volatile organic compounds in wastewater using PA-FTIR has been reported [488].

Photoacoustics is also widely used for numerous **non-spectroscopic applications**, such as the determination of thermal diffusivity, non-destructive testing of materials (in particular probing of sub-surface defects) by thermal wave imaging, time-resolved studies of de-excitation processes, studies of phase transitions, *etc.*

1.4. EMISSION SPECTROSCOPY

Various competitive routes are available for dissipation of absorbed radiant energy. These include both non-radiative transitions and radiative photophysical processes, such as fluorescence and phosphorescence. Energy can be transferred directly to other molecules by a process known as quenching dissipated through the vibrational motion of the molecule. Quenching depends on collisions between molecules; internal transfer of energy as a result of which a molecule passes over into a lower-lying electronic state facilitates the vibrational process.

Radiative processes result in the emission of one or more wavelengths of radiation as excited-state electrons with differing energies participate in these processes (Fig. 1.23). Emission of UV/VIS radiation by matter is due to spontaneous radiative decay of electronically excited atoms or molecules. Emission entails that the emitting matter loses energy in the form of electromagnetic radiation and therefore some form of energy must be supplied for a continuous emission of radiation. Classical emission spectroscopy is based on excitation of atoms or molecules into higher electronic states by electron impact, photon absorption or thermal excitation at high temperatures. **Incandescence** is the process in which the emission of radiation is sustained by simple heating, *e.g.* a hot body emitting light ("hot light"). All other forms of light emission by matter, not involving high temperature, can be included under the more general term of **luminescence** and the most common, and analytically most useful, type of luminescence is photoluminescence, in which the

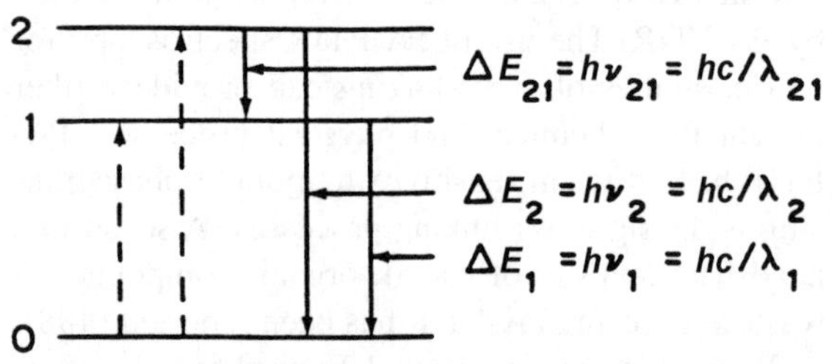

Fig. 1.23. Emission. Addition of thermal, electrical or chemical energy causes non-radiational excitation of the analyte and emission of radiation.

necessary energy is supplied by an external exciting light ("cold light"). Of course, during luminescence the temperature of a solid is raised, however a luminescent material converts a certain type of energy into electromagnetic radiation over and above thermal radiation.

Some examples of **emission techniques** that have been used to polymer/additive studies are:

- Raman spectroscopy (*cfr.* Chp. 1.2.3).
- Photoacoustic spectroscopy (*cfr.* Chp. 1.3).
- Infrared emission spectrophotometry of a polymer held at an elevated temperature.
- Fluorescence and phosphorescence spectrophotometry to detect conjugated chromophores with sensitivity 10–100 times that of absorption spectrophotometry.
- Oxyluminescence (OL), *i.e.* weak visible chemiluminescence (CL) emitted during oxidation.
- X-ray photoelectron spectroscopy (XPS) for probing the outer surface layer (*ca.* 5 nm).

Photothermal spectroscopy is strictly related to photoacoustic spectroscopy [489]. In this case, the energy of the incident electromagnetic radiation is converted to thermal energy. The latter induces radiative emission at the surface, which is detected.

Emission spectroscopic techniques are more sensitive than absorption or reflectance spectrophotometry. Excitation by narrow-band lasers may result in the selective population of wanted levels, which emit their excitation energy as fluorescence photons. A laser-induced fluorescence (LIF) spectrum is much simpler than the emission spectrum of a gas discharge, where the superposition of fluorescence from many emitting levels is observed.

1.4.1. Infrared Emission Spectroscopy

Principles and Characteristics

In all ranges of the infrared (NIR, mid-IR, FIR) mainly *absorption* of radiation by a sample is used as an analytical tool. *Emission* spectra are rarely recorded, even though they are powerful for problems which cannot be investigated by other methods.

In order to observe emission it is necessary to populate a higher lying unoccupied quantised state. A molecule in a vibrationally excited state has a certain probability of emitting IR radiation in the presence or absence of incident electromagnetic radiation, resulting in induced and spontaneous emission, respectively. At r.t. the number of molecules in a first excited state is less than 1% of the population in the ground state, when the separation of energy levels is about 1000 cm^{-1}, typical in the infrared. Induced IR emission is much weaker than (induced) absorption. On the other hand, at moderately elevated temperatures the IR emission spectrum of a thin film on a solid surface may easily be observed. Solids, as black body radiators, emit light that can be characterised by the radiated power, spectral profile, and photon flux.

If a sample absorbs IR radiation at characteristic wavenumbers, it is capable of emitting radiation at these wavenumbers. A thin sample of a material will emit radiation with a spectrum very similar to its absorption spectrum. By ratioing the emitted radiation from the thin film to that from a black body at the same temperature, an **emissivity spectrum** is obtained which generally has the appearance of an inverted transmission spectrum. Emission spectra used to be collected from samples heated well above r.t., typically to 40–100°C (to minimise sample degradation), with a black-body source (*e.g.* graphite) at the same temperature as a reference. With FTIR instruments, emission spectra can also be recorded at room temperature.

Fourier transform infrared emission spectroscopy (FTIES) is a single beam technique in which IR radiation emitted from heated materials is directly analysed by a highly sensitive FTIR spectrometer. Commercial IR spectrometers modified to measure emission spectra have been described [109,490]. An emission spectrum usually takes the form of a plot of the relative power of the emitted radiation as a function of wavelength or frequency. In FTIES the sample (usually as a thin film) is placed in intimate contact with a miniature platinum hotplate, heated in either nitrogen or oxygen flowing directly over the hotplate, and the emitted radiation is gathered (Fig. 1.24). It is necessary to collect a background spectrum at each temperature of interest to allow for variation in emission intensity at different temperatures. Precise temperature control is strictly necessary for quantitative analysis. The physical background and approaches to quantitative FTIES analysis have been reviewed [491].

Theory of infrared emission is more complex and not as well established as that of conventional absorption spectroscopy. As the same vibrational modes are probed in emission and absorbance measurements, the structural information content is the same in both cases. For the majority of analyses, there is little or no advantage to be gained in recording the emission rather than the absorbance spectrum

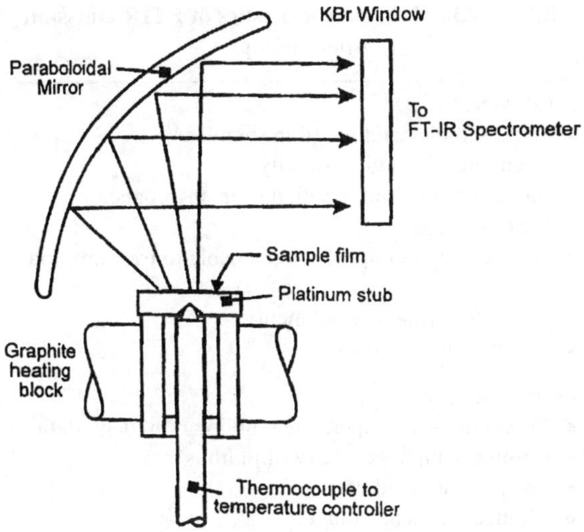

Fig. 1.24. Schematic diagram of sampling arrangement for FTIR emission spectroscopy. After Rintoul *et al.* [109]. From L. Rintoul *et al.*, *Analyst* **123**, 571–577 (1998). Reproduced by permission of The Royal Society of Chemistry.

of a condensed phase sample. Modern reflectance and FTIR microscopy techniques restrict the application area. In some circumstances, however, it may prove convenient and practicable. Emission spectroscopy is a very attractive option for use in on-line IR spectroscopy [492]. However, the conditions under which it may be applied are limited to thin samples or to cases with a suitable thermal gradient. In the IR region of the spectrum, the vibrational energy levels may be populated at temperatures at which polymer oxidation occurs.

Table 1.23 summarises the **main features** of FTIES. Advantages of FTIES are the ease with which spectra of samples at elevated temperature are obtained *in situ* and the high intensity. Another advantage is that the radiation is produced by the sample; no additional source or probe is needed and no mechanical contact is required. Infrared emission has an appreciable intensity, particularly in the long wavelength region when the temperature of the sample rises beyond room temperature. Emission techniques are inherently more sensitive than other IR techniques because the small signal of the emitted photons is measured directly rather than requiring detection of small differences between large signals, as in transmittance or reflectance measurements. IR emission spectra can be routinely obtained on μg levels of material and for samples not amenable to transmission techniques. Most limitations in IR

Table 1.23. Main characteristics of FTIR emission spectroscopy

Advantages:
- Little sample handling (thin samples)
- High intensity and sensitivity
- Same structural information as in absorbance measurements
- Applicable to samples not amenable to transmission techniques
- Non interfering measurement
- Mainly qualitative tool

Disadvantages:
- Maximum use temperature (sublimation, degradation)
- Limited sample geometry (thin films)
- Temperature gradients
- Limited frequency range
- Difficult quantitation
- Limited applicability

emission spectroscopy are obvious. Thermal emission is limited due to sample heating requirements and to self-absorption of emitted radiation, which washes out spectral features. A large temperature differential is desired to increase the radiant flux from the sample, but sublimation and decomposition define a maximum use temperature for many samples, Temperature gradients across a sample can pose severe problems and are a major reason why quantitative IR emission spectroscopy is difficult. The frequency range is limited because of the low intensity of emitted radiation above 2000 cm^{-1} for samples near ambient temperature. A further problem involves multiple passing of radiation through the modulator leading to spectral artefacts. If the sample is at uniform temperature and thicker than a few μm, the processes of absorption and emission will result in the loss of spectral information and an uninformative black-body spectrum will be all that remains. Spectral information can be obtained if the sample has a sufficiently high thermal gradient at the surface. The condition of thin samples is met by materials such as packaging films (*e.g.* a 10 μm thick PVC film). Under such circumstances the thermal gradients arising during processing may be sufficient to allow observation of emission spectra. Another circumstance in which the black-body spectrum is not observed is when the emitting material is transparent in the wavelength region of interest. For an opaque sample the information provided by emission spectra is identical with that in the reflection spectra.

Table 1.24. General applications of FTIR emission spectroscopy

- Thermal degradation of polymers
- Oxidation studies
- Evaluation of stabiliser packages
- Cure analysis
- Characterisation of textile fibres
- Investigation of surface layers
- Analysis of heated food materials

The data obtained by FTIES is not quantitative due to such interferences as sample reflectivity and re-absorbance in thicker samples. Pell *et al.* [493] have suggested an approximate relationship between absorbance and emittance. Infrared emission spectroscopy was reviewed [123,494].

Applications

Table 1.24 lists some applications of FTIES related to polymers. Important industrial applications are found outside the laboratory in remote sensing (environmental analysis, astronomy).

Emission is very useful to obtain spectra from thin coatings on uneven metal surfaces when diffuse reflection measurements would require a small sample. Infrared emission spectroscopy enables easy acquisition of spectroscopic information from samples undergoing degradation at elevated temperatures, by simply detecting the time-dependent IR emission originating from the samples as they degrade. It is therefore promising for applications in **polymer degradation** and characterisation, where spectral information at higher temperatures is often needed to elucidate polymer reactions. The capability of identifying and determining degradation intermediates and products is attractive, compared with the traditional thermal analysis method. Despite this potential, the technique is rather neglected. The thermal degradation of PVC/PVP blends was studied by means of FTIES [495].

Infrared emission has the potential to both identify and determine the relative concentration of **oxidation products** at the oxidation temperature. Emission spectroscopy is generally regarded as one of the most sensitive methods for detecting certain oxidation products and enables an easy way of obtaining real-time spectral information. This is particularly important when investigating the oxidation product distribution at the very early stages of polymer degradation. The rate of formation of carbonyl and γ-lactone oxidation products of single reactor particles of unstabilised PP at 150°C has been followed by FTIES [496]. Quantitative FTIES was applied to investigate the real-time thermal oxidation of thin polyolefin (HDPE, XLPE, PP) samples [490]. George *et al.* [496] have established the relationship between single particle CL and FTIR emission of pressed polyolefin particles. FTIES contributes to the ongoing efforts to evaluate **stabiliser packages**, as an alternative technique to determine induction periods and to investigate the performance of PVC formulations [497]. Wheals [498] has compared emission spectrometry and PyGC in the analysis of 190 **dyes**; 53 dyes were identified by emission spectrometry and 141 by PyGC.

Carbon-black-filled materials defy spectral analysis, since featureless black-body emission is observed; however, other highly filled polymers containing inorganic pigments and fillers can easily be analysed. FTIES has been applied to studies of *in situ* polymer degradation of carbon-black-free EPDM and Buna-N (nitrile) **rubbers** as 10 μm thick microtome cuttings [497]. FTIR emission has also been used for the characterisation of **textile fibres** [499]; no special sample preparation is required as in case of transmission spectroscopy. It would be useful to collect a library of emission spectra of fibres.

Cure analysis of thermoset thin films by FTIR emission has also been reported [500]. FTIES was used to study **cure monitoring** of aerospace epoxy resins and prepregs and provides a rapid method for prepreg quality acceptance [501]. Also curing of commercial polyurethane two-component paints has been carried out by FTIR emission [493].

The potential of mid-IR emission spectroscopy for **on-line analysis** of heated food (starch and other materials) has been demonstrated [492]. A related application is **infrared heating**, which relies on the fact that specific components of the emitted IR radiation spectrum coincide with the wavelength of the molecular oscillation in the material to be heated. The material absorbs radiation at these wavelengths and this absorption involves transferring energy to the appropriate molecules so that the material is heated. Infrared emitters have been specially developed for applications in the plastics industry and are designed to precisely match the requirements of individual heating processes. Emitters are available for various processing operations including surface heating for adhesion processes or for drying, surface heating of foils and thin films and uniform volumetric heating of large components, such as billets and plates. **IR radiometry** can yield accurate resin temperatures for transparent resins of known emissivity, but problems exist for the calibration of this instrument should the resin emissivity change such as with a filled material.

An overview of the application of infrared emission spectroscopy to solid materials is available [502].

1.4.2. Molecular Fluorescence Spectroscopy

Principles and Characteristics

Absorption measures the loss of photons, but makes no statement about their fate. Absorption processes usually induce excited states. When molecules absorb radiation in electronic transitions to form excited states, the latter may lose the acquired energy via several mechanisms. If the energy loss occurs through emission of radiation, the process is called **luminescence**. Luminescence is defined as non-equilibrium radiation that is in excess over and above the thermal radiation background and arises in the presence of intermediate processes of energy transformation between absorption and emission. Luminescence is generally known as cold emitted light in contrast to incandescence, which is the light emitted by hot bodies. The wavelength of luminescence is invariably greater than that of the exciting radiation (Stokes' law [503]). For a classification of luminescence, *cfr.* Table 5.11 of ref. [1]. The theory of luminescence was reviewed [504].

Fluorescence is a form of luminescence and is simply the electric dipole transition from an excited electronic state to a lower state, usually the ground state, of the same multiplicity, as shown in Fig. 1.25 for the lowest excited single state (S_1) to the singlet ground state (S_0). The process has considerable similarity to Raman spectroscopy. However, the wavelength of the emitted light is different from that produced by fluorescence. Moreover, the narrow Raman spectral bands carry a great deal of information on molecular structure, in contrast to the broad fluorescence emission. The rules governing the emission process for an excited fluorescent molecule are complex. Like molecular absorption bands, molecular fluorescence bands are made up of a multitude of closely spaced lines (that are often difficult to resolve).

Generally, the lifetime of an excited species is short because there are several ways an excited molecule can give up the excess energy. Apart from

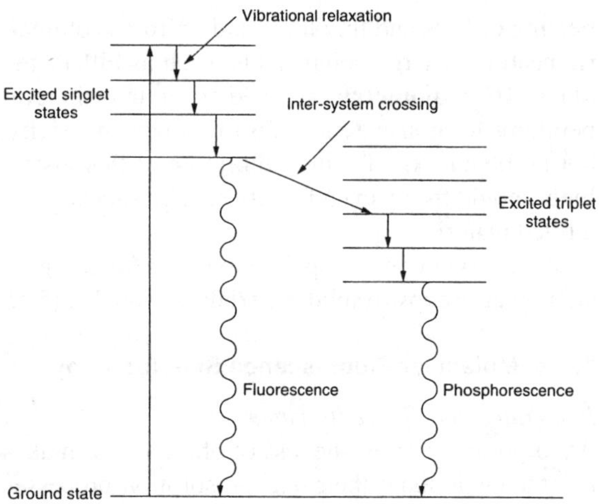

Fig. 1.25. Typical transitions for fluorescence and phosphorescence processes. After Stuart [505]. Reprinted from B. Stuart, *Polymer Analysis*, Copyright © 2002 John Wiley & Sons, Ltd. Reproduced with permission.

the radiative transition (fluorescence) the molecule may undergo a chemical reaction resulting in photodestruction or photobleaching of the fluorescent molecule or returns to the ground state by nonradiative deactivation. Excited state bimolecular reactions and collisional quenching complicate the mechanism of fluorescence in solid polymers. Since fluorescence transitions are spin-allowed, they occur very rapidly and the average lifetimes of the excited states responsible for fluorescence are typically 10^{-8}–10^{-10} s. Fluorescence ceases virtually immediately after the excitation radiation is removed. The fluorescence signal decays as $I = I_0 \exp(-t/\tau)$, where τ is a characteristic property of the fluorescent molecule.

The fluorescence process is described by the excitation wavelength that generates the excited state of the fluorophore and an emission wavelength at which the fluorescence signal is detected. Generally, the emission wavelength is shifted to lower energies relative to the absorption wavelength. The origin of this effect is the vibrational relaxation (*i.e.*, the nonradiative transition to the lowest vibrational level of the excited electronic state) that occurs before fluorescence is emitted (with minute heat dissipation), causing the emitted photon to be "red-shifted" to longer wavelengths (Stokes-shifted fluorescence), *cfr.* Fig. 1.5. Typical excitation wavelengths for fluorescence range from 230 to 600 nm with emission from 250 to 800 nm. Resonance fluorescence has an identical wavelength to the radiation that caused the fluorescence.

Any fluorescent molecule has two characteristic spectra: an *excitation spectrum*, depicting the dependence on wavelength of the exciting light of the fluorescence intensity at a fixed emission wavelength, and an *emission spectrum*. In principle, fluorescent substances can be characterised on the basis of their combined excitation (ex) and emission (em) spectra. Consequently, in comparison to absorption spectrometry, fluorescence spectrometry generally offers enhanced selectivity because for each measurement two different wavelengths (ex and em) are involved. Fluorescence detection is much more sensitive compared with UV/VIS reflectance densitometry, yielding detection limits in the low-pg range [506]. Fluorescence based identification of related analytes in a mixture often requires a chromatographic separation of the components prior to detection. Nonfluorescent UV-absorbing compounds are not detected.

Fluorescence measurements are made with (spectro)fluorimeters, and, for chromatographic applications, with fluorescence detectors. **Fluorescence spectrometers** are either lifetime or steady-state instruments, depending on whether they resolve the temporal behaviour of the emission (or more correctly the excited state), or not, respectively. In both cases there are strong similarities with single beam UV absorption instruments. However, the levels of photons detected in fluorescence (or equally phosphorescence) are typically much lower than those in absorbance. As a consequence, certain features are optimised differently for fluorescence. Fluorescence is detected orthogonal to the direction of the ex beam incident on the sample, so as to delineate the em photons from those of the ex beam and minimise those from Rayleigh and Raman scattering (*cfr.* Fig. 5.7 of ref. [1]). The selection of ex wavelength and detected em wavelength may be controlled independently. Scanning of the em wavelength at fixed ex wavelength gives the **emission spectrum**, or *vice versa* the **excitation spectrum** (which is identical with the absorption spectrum). It is possible to automatically scan both ex and em wavelengths (from near to the ex wavelength up to longer wavelengths) to give a high-resolution excitation–emission map **(2D fluorescence)**. The result of the fluorescence scan can be viewed in a 3D plot showing ex wavelength, em wavelength and fluorescence intensity on the axes. The development of 2D fluorescence makes it possible to monitor several components simultaneously.

The magnitude of the fluorescence signal (F) is given by:

$$F = f(\theta)g(\lambda)I_0\theta_f\varepsilon lc \qquad (1.7)$$

where $f(\theta)$ is a geometry factor related to the positioning in the detector, $g(\lambda)$ is the wavelength response characteristics of the detector, I_0 is the intensity of the excitation source, θ_f is the quantum yield of the analyte molecule, ε is the molar absorptivity of the analyte, l is the optical pathlength and c is the molar concentration of the analyte. Fluorescence detection is subject to Beer's law (dilute solutions: $\varepsilon cl < 0.01$). Since all terms in eq. (1.7) are constant, or fixed by experiment, fluorescence emission is linearly dependent on the sample amount. As fluorescence is also directly proportional to the number of photons absorbed by the sample, it is advantageous to employ very high intensity **light sources**. Although there are several major differences between fluorescence and absorption instruments, a most important difference concerns the source. The use of lasers as excitation sources instead of conventional lamps provides several advantages, amongst which the increase in sensitivity. Key to every technique that uses laser-induced fluorescence for the ultimate single-molecule spectroscopy is minimising background noise.

Modern fluorescence spectrometry combines a fast tuneable excitation light source (deuterium lamp, low-pressure mercury lamp, or xenon discharge lamp) with a scanning monochromator that can quickly be tuned at high speed – up to 140 nm/ms – to the different wavelengths, an extremely fast and sensitive diode array spectrometer as a detector (able to register a complete spectrum in less than a msec) and UV optical fibre technology down to 200 nm to connect sample area and detector. Excitation wavelengths are typically from UV to NIR, with emission wavelengths of 300 to 1100 nm. Moreover, two wavelength selectors are required (for ex and em radiation), either filters (filter fluorimeters) or monochromators (spectrofluorimeters). Fluorimeters are usually double-beam in order to compensate for fluctuations in the power of the source.

A substance must absorb light in order to fluoresce; as a result, any substance that can be measured fluorimetrically can also be determined spectrophotometrically. As fluorescence is one of several mechanisms by which a molecule returns to the ground state after excitation by absorption of radiation, in principle all absorbing molecules have the potential to fluoresce. Most do not, however, because their structure provides radiationless pathways by which relaxation can occur *at greater rate* than fluorescent emission. Only an estimated 5 to 8% of absorbing compounds possess the structural features necessary for natural fluorescence. Such chromophores are called **fluorophores**. Fluorescence is expected in molecules that are aromatic or contain multiple conjugated double bonds with a high degree of resonance stability. Electron-donating groups tend to enhance fluorescence, whereas electron-withdrawing groups promote quenching. Molecular rigidity is also important for fluorescence. In general, aromatic compounds that are the most planar, rigid, and sterically uncrowded are the most fluorescent, such as anthracene and pyrene, and other highly conjugated ring systems such as fluorescein, rhodamine, dansyl chloride, and their derivatives. Fluorescence spectroscopy is thus limited to molecules with the required properties, but compounds that do not fluoresce can be chemically modified with a fluorescent molecular group before analysis/detection (**fluorescence derivatisation**). Fluorescing compounds at high wavelengths (500–600 nm) are available.

In conventional fluorescence analysis the majority of **quantitative measurements** is made using fixed wavelengths for excitation and emission. At low concentrations, a plot of the fluorescent power of a solution *vs.* the concentration of the emitting species ordinarily is linear. As to the **sensitivity** of fluorescence procedures in analysis, it is convenient to separate the contributions from the properties of the fluorescent molecule itself (*absolute sensitivity*), the performance of the instrument (*instrumental sensitivity*) and the chemistry involved in the preparation of the sample (*method sensitivity*). The absolute sensitivity is determined chiefly by the molar absorptivity and the fluorescence efficiency of the analyte molecule itself. High fluorescence efficiency is usually associated with some rigidity. The method sensitivity takes account of pre-concentration steps in the preparation of the sample on the one hand and the limitations imposed by the fluorescence of the blank on the other. The sensitivities and selectivities attained by fluorescence, phosphorescence and chemiluminescence are hardly paralleled by other techniques. In many cases it is not required that the analyte be isolated from the matrix.

Table 1.25. Main characteristics of fluorescence spectroscopy

Advantages:
- Excellent sensitivity (10 fg/μL) for lowest detection limits; trace analysis
- Wide dynamic range (10–4 × 10^5 ng/mL)
- High scan speeds (0.6 s/spectrum: 200–400 nm, 10 nm step)
- Increased selectivity compared with UV/VIS spectrophotometry
- Multiwavelength detection
- Spectral library search for peak confirmation
- Non-destructive
- Quantitation at ultratrace level (with appropriate calibration)
- Off-line, on-line

Disadvantages:
- Low-light-level phenomenon
- Limited compound classes and applicability
- Limited spectroscopic identification power
- Method development required (FLD)

Wavelength calibration of fluorimeters is highly important; a **calibration** sample is to be used. Fluorescence intensity samples such as Rhodamine B are routinely measured to calibrate and monitor the performance of fluorescence spectrophotometers [507]. Also PMMA/fluorescent materials are used to check instrumental stability resolution and wavelength precision (Anadis Instruments/Malden).

Table 1.25 lists the **main features** of fluorescence spectroscopy. One of the most attractive features of fluorescence detection is the fact that near-zero background and direct proportionality between excitation power and emission signal intensity render fluorimetry a very sensitive detection technique, often one to three orders of magnitude better than UV absorption spectroscopy. Fluorimetry shows exceptional limits of detection accessible in favourable circumstances (low pg range) [506]. Another advantage of fluorescence methods is the large linear range, which is often significantly greater than that encountered in absorption spectroscopy. Moreover, fluorescence spectra often show more fine structure than UV spectra and are consequently more reliable for identification purposes by spectra matching. In addition, for a given compound, fluorescence spectra can be obtained over a large number of different excitation wavelengths, each providing a unique spectrum that improves the confidence of identification. In principle, fluorescent substances can be characterised on the basis of their combined ex/em spectra. One of the major advantages of fluorescence investigations over absorption studies is a greater selectivity in the analysis of multicomponent samples.

Fluorescence spectra suffer from the same limitations as UV absorption spectra in not being very characteristic of the compound from which they are obtained and in general it is not possible to identify a completely "unknown" compound from ex/em spectra. In fact, both techniques depend upon transitions between electronic energy levels. The actual energy values are determined by the π-system and so the bands observed for all compounds containing the same basic π-system (C=C, C=C−C=O, the benzenoid ring, *etc.*) occur in the same narrow region of the spectrum. Fluorescence spectra can provide a positive means of identification when bands show vibrational fine structure. However, fluorescence spectroscopy is clearly applicable to a more limited range of chemical systems than absorption spectroscopy, due to the limited number of fluorophores. Non-fluorescent UV-absorbing compounds are not detected. Infrared and NMR spectra are usually much more powerful for identification purposes. Background emission from the blank is a serious problem in fluorescence work and often determines the lower limit of concentration that can be reached.

Fluorescence lifetime imaging (FLIM) can be used to distinguish between different fluorophores in a field of view. Confocal scanning can be used in combination with fluorescence imaging to acquire 3D spectroscopic images and to map the distribution of different fluorophores present (*cfr.* Chp. 5.6.4). Absorption and **fluorescence imaging** instruments are beginning to be developed that are able to produce a spectroscopic image of a sample, each pixel of the image being a complete spectrum. Such instruments will find growing use in the investigation of heterogeneous material for which traditional methods are only able to give spatially averaged results. The reader is referred to other sections for **alternative fluorescence techniques**, such as LEAFS (Chp. 3.3.1) and XRF (Section 8.4.1 of ref. [1]), which both provide information on the atomic rather than the molecular level.

Fluorescence spectroscopy has been reviewed [508]. Various monographs deal with fluorescence [509,510], *cfr.* also Bibliography; for fluorescent probes, *cfr.* ref. [511], and for standards in fluorescence spectroscopy ref. [512].

Applications

Fluorescence spectroscopy, through the analysis of the spectra and of the lifetime of excited states, enables the study of the electronic states of the species present. The wavelengths used in fluorescence studies are in the 300–700 nm range; a layer with a thickness of hundredths or tenths of a μm can be observed.

Direct fluorescence, phosphorescence and X-ray fluorescence spectroscopy for polymer/additive analysis have been reported [513]. In commercial polymers, additives having electronic absorption bands in the visible and near-UV wavelength regions may fluoresce and give rise to composite spectra. Some **general applications** of fluorescence spectroscopic analysis for polymeric materials relate to: *(i)* the direct determination of additives in polymers; *(ii)* impurity detection; *(iii)* ageing studies of polymers; *(iv)* interactions between polymer and additives; *(v)* tracer systems (markers); and *(vi) in situ* monitoring capability.

Fluorescence characteristics of some common polymers have been listed [505]. Fluorescence analysis of **polyolefins** has been the subject of much controversy but is generally now considered to be associated with the presence of low levels of cyclic α, β-unsaturated carbonyl compounds of the enone or enal type [514]. Fluorescence and phosphorescence excitation and emission of LLDPE, HDPE and mPE were reported [515].

In a very early report of direct determination of **stabilisers** in polymers by luminescence techniques by Drushel *et al.* [13] the fluorescence of EPR/Age Rite D (trimethyldihydroquinoline) and of EPR/Santonox R were examined. Lack of interference by other polymer additives and polymerisation catalyst residues was emphasised. Age Rite D concentrations can be measured directly in pressed EPR films ($<$0.01 cm thickness) by fluorescence at levels below 0.1–0.2 wt.% in order to prevent concentration quenching. In the fluorescence emission spectra of Irgafos 168 the fluorescence quantum yield of the phosphate is much greater than that of the phosphite [516]. This difference enables quantification of the phosphate concentration. Although direct quantitative determination of UV stabilisers in extruded polyolefins by means of fluorescence spectroscopy (ex, 370 nm; em, 390–550 nm) has been described [517], this is certainly not a universally applicable technique (being additive and matrix dependent). The effects of additives (AOs and cross-linking agent by-products) on electroluminescence of LDPE and XPE were evaluated [518]. Figure 1.26 shows fluorescence emission spectra of benzotriazole UVAs in a model acrylic melamine clearcoat (as film) [519]. A significant emission is only obtained with 3′ unsubstituted compounds.

Provorov *et al.* [520] have studied a rather extensive group of elastomer additives (accelerators, stabilisers, softeners, fillers, and other ingredients) for possible analysis by fluorescence techniques. No fluorescence lifetime measurements have been applied for discriminating stabilisers in polymers. UV microscopy is another means of measuring the concentration (and distribution) of UV absorbing or fluorescent additives in plastics (*cfr.* Chp. 5.3.2).

Fluorescence is also commonly being used in **pigment identification** (using UV excimer laser for accurate removal of varnishes and over-paintings). Pigments and dyes exhibit their own characteristic emission spectra. Different crystalline forms of TiO_2 may be differentiated by their characteristic emissions.

In situ measurements on **HPTLC plates** can be made by fluorescence and fluorescence quenching [521]. Because many organic compounds are intrinsically fluorescent, they are readily determined without requiring chemical derivatisation. Calibration curves in fluorescence are usually linear over two or three orders of magnitude. The fluorescence response for substances in TLC may differ considerably from measurements made in solution; this is ascribed to adsorption onto the sorbent layer, which provides additional non-radiative pathways for the dissipation of the excitation energy. TLC applications with fluorodensitometric detection have been reviewed [506].

Fluorimetry has also been used for the detection of **metal traces** in polymers, such as Al, Ti (catalyst residues), Fe and Ca in PE, Fe in PVC, and Cu in PA (after extraction) [522]. Chelate-forming reagents used were 8-hydroxychinoline (for Al), fluorexon (for Ca), stilbexon (for Fe) and a ZnS(Ag) phosphor (for Ca).

Analysis of thermally oxidised PET by fluorescence spectroscopy has shown the presence of hydroxylated terephthalate units [523]. These groups are formed by hydroxyl radical attack on the terephthalate units and then undergo further oxidation to produce stilbene-quinone units causing discoloration of the polymer. Fluorescence analysis has also been used to study photobleaching and photoinduced discoloration reactions in EVA formulations [22]. Fluorescence spectroscopy offers a

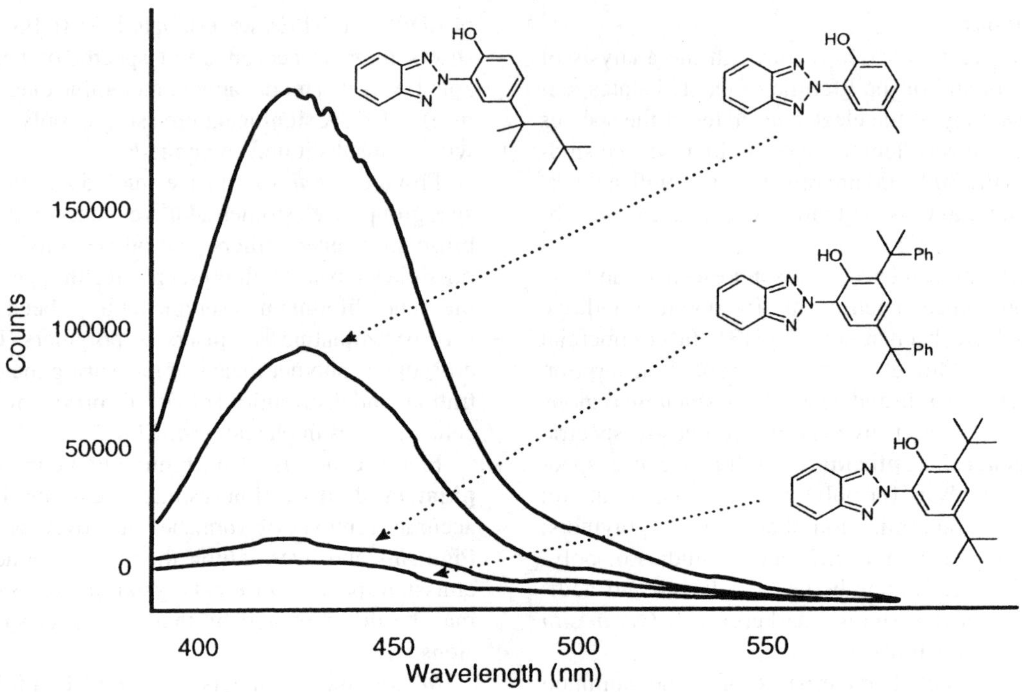

Fig. 1.26. Fluorescence emission spectra of benzotriazole UVAs in a model acrylic melamine clearcoat (excitation wavelength: 330 nm). After DeBellis *et al.* [519]. Reprinted with permission from A.D. DeBellis *et al.*, *ACS Symposium Series* **805**, 453–467 (2002). Copyright (2002) American Chemical Society.

straightforward non-destructive means of studying the properties of solid polymers and can be utilised to identify the chromophores and to study their roles in the process of **weathering** in PVC. The carbonyl groups and DOP are the two major chromophoric units in plasticised PVC. Polymeric materials under physico-chemical stresses ($h\nu$, O_2, $\Delta, \ldots$) may also lead to fluorescent products and yellowing. Fluorescence microscopy (0.5 μm resolution) can be used for **diffusion** problems.

The applicability of fluorescent tagging for coding of plastics to aid waste separation techniques prior to recycling has been investigated [524–527]. As reported by Simmons *et al.* [528], a **fluorescent tracer system** for automatic identification and sorting of waste plastics has been developed, which is capable of identifying different plastics with high precision and high speed. The identification system relies on incorporation of very low levels (1 to 10 ppm) of fluorescent substances ("tracers") into the plastics requiring to be identified. Because of the highly diverse nature of polymer types it is not practical to have a different tracer for each potential variant. To overcome this, combinations of tracers are used to identify materials. Seven tracers are sufficient to identify up to 63 different material variants. Identification is carried out by illuminating the plastics with UV light. Each combination of fluorescing tracers shows a unique fingerprint. A fluorosensor with associated data processing system detects the fluorescent light emissions and identifies the nature of the host material. Within the framework of a Brite-Euram program some 60 fluorescent compounds from various sources were evaluated on their properties:

- essentially colourless under normal lighting conditions;
- stable in plastics under normal processing and use conditions;
- compatible and photophysically non-interactive with the host plastic materials;
- toxicologically compatible with the processing and application requirements;
- common excitation wavelength band for optimum fluorescence;
- high fluorescence quantum yield; and
- narrow and well separated emission bands.

Inorganic tracers frequently exhibit line emissions rather than the usual broad emission signals provided by organic fluorophores. The fluorosensor system identified container types with an accuracy of 98–100% at rates of up to 5 items/s, and can be used with some pigmented materials. Fluorescence in fibre recycling was also discussed [529].

In a completely different application, Shimoyama *et al.* [530] have reported a 3D fluorescence method using quartz fibre optics for the non-destructive determination of colorants on woodblock prints. Cleve *et al.* [531] have developed an **on-line fluorescence** spectroscopic method to measure sizing effects and optical brightening agents in polyamide woven fabrics. According to Allen [532] the *in situ* capability of fluorescence is very viable as a monitoring probe via a fibre optic technique.

Films of PBMA containing the fluorescent dye Nile Red display both spontaneous and stimulated luminescence emission [533]. The spontaneous luminescence from the dye–polymer combination is inferred to arise largely from impurities, including benzoyl peroxide present as a residual polymerisation initiator. **Fluorescent dyes** may be used for real-time measurements of temperature and shear gradients within flowing polymers [534,535]; fluorescence anisotropy of fluorescent dyes (at 10 ppm concentration level) has been used to determine molecular orientation during polymer processing [536]. Temperature sensitive fluorescent dyes, such as the excimer producing mobility dye bispyrene propane (BPP) and the fluorescence band broadening dye perylene, doped into polymer resins (PC, PMMA, PS) as process probes, were used to monitor the true resin temperature during extrusion processing [537,538]. Non-contact temperature monitoring during capillary rheometry testing may be based on temperature variations in the spectrum of perylene used as a molecular probe (in 10^{-6} mass fraction) in PE (I_{464}/I_{473}); standard uncertainty is 2°C [539]. Fluorescence spectroscopy can be used for monitoring composite curing reactions [540], and for concentration measurements in flows (using fluorescent dyes). Failure studies of **adhesion** of PUR on epoxy-coated steel and PET/PE by UV reflection and fluorescence techniques were reported [541].

UV fluorescent sulfur detection and chemiluminescent nitrogen methods have been developed for total sulfur only, nitrogen only or simultaneous sulfur/nitrogen analysis. These systems provide quick automated analysis to estimate the type and amount of sulfur and nitrogen-containing additives present in a polyolefin sample, *i.e.* slip agents, antistatics, AOs and UVAs. Samples can be analysed in pellet form without sample preparation. Such analysers are particularly suitable to monitor product quality and control, analysis of bulk additive purity and regular checks of the amounts of additives in resin batches.

Plitt *et al.* [542] have surveyed the literature covering the fluorescence of fibres, rubber, cellulose, polymers, and plastics long ago. On the whole, fluorescence and phosphorescence techniques find restricted practical application for polymer/additive analysis. There is also little information in the literature on the quantitative aspects of the direct examination of polymer films by luminescence techniques.

Fluorescence is not only useful simply for chemical analysis. **Fluorescent pigments** are incorporated in protected objects via ink, fibres or paper pulp [543]. The most widespread types of fluorescent pigments used in protected documents are organic pigments with large Stokes shift fluorescing green and red under UV excitation (365 nm), *e.g.* Luminor 540T. However, also anti-Stokes inorganic pigments having a green and red luminescence under IR excitation at 980 nm are available [543]. Such taggants embedded in materials can be analysed chemically as well as by optical methods (laser visualisation). Fluorescent additives that stick to textile fibres are added to laundry soap in order to make clothing appear "whiter". Fluorescence is also encountered in cathode tube lighting where the internal walls are covered with mineral salts (luminophores) that emit in the visible region due to excitation by electrons. Smartlight RL 1000 (Ciba Specialty Chemicals), a red luminescent additive for agricultural films, shifts light from the UV part of the spectrum to visible light and improves crop quality and productivity.

1.4.3. Phosphorescence Spectroscopy

Principles and Characteristics

Phosphorescence is another form of luminescence and refers to emission of light associated with a radiative transition from an excited electronic state that has a different spin multiplicity from that of the ground state (*cfr.* Fig. 1.25). Such a process involves inter-system crossing between an excited singlet state and an excited triplet state. The process takes much longer than fluorescence since the transition from triplet to singlet states involves a change of electronic spin. Phosphorescence has a longer wavelength than fluorescence and has also a longer lifetime, ranging from ms to hours after the excitation radiation is removed (slow decay or "afterglow"). The delineation into fluorescence and phosphorescence is an arbitrary one.

Phosphorescence is more difficult to observe than fluorescence because of the possibility of quenching by impurities present in a sample. Phosphorescence is also less probable than fluorescence. On the other hand, many compounds that do not fluoresce do exhibit phosphorescence. Phosphorescence offers sometimes advantages with respect to UV spectrometry as to specificity; however, in practice the sensitivity of phosphorescence usually does not exceed that of UV spectroscopy [389a]. Fluorescence is considerably more important than phosphorescence in analytical chemistry. Although application of phosphorescence to the analysis of additives in polymers is restricted, the direct determination is an asset.

Various textbooks describe phosphorescence spectroscopy (*cfr.* Bibliography).

Applications

Some additives show phosphorescence [544]. Phosphorescent plastics having long decay times at room temperature were once of interest for mail-coding applications [545]. Conventional phosphorescent additives (*e.g.* radioactive paints, ZnS:Cu) and new metal-oxide based super-phosphorescent additives [546] with substantially longer glow time at considerably lower loadings (1–10%) are used for plastic products such as backlit panels for safety signs and sporting goods.

Since the determination of inhibitors in polymers by Drushel *et al.* [13] little information has been added to the literature on the quantitative aspects of the direct examination of polymer films by phosphorescence spectroscopy. These authors examined phosphorescence (at liquid-nitrogen temperature) of thin EPR films containing Santonox R (2,2′-dimethyl-5,5′-di-*t*-butyl-4,4′-dihydroxydiphenyl sulfide) and *N*-phenyl-2-naphthyl-amine (PBN). The rather intense phosphorescence of PBN may be used to advantage when other additives interfere in the UV absorption method. As to quantitative phosphorescence analysis, several factors, *e.g.* film thickness, concentration quenching, and background absorption, *etc.*, affect the linearity of the analytical working curves and precision of the measurements [13]. The reliability of a correlation between stabiliser concentration in the film and phosphorescence intensity at 77 K is also influenced by the degree of crystallinity [544].

Oxygen quenches phosphorescence of **aromatic hydrocarbons** in plastics at r.t. but not at liquid nitrogen temperature [547]. Luminescence spectroscopy (phosphorescence at 77 K and fluorescence at r.t.) may be used to evaluate oxidation processes in plastic materials, *e.g.* in LDPE films [548]. Recently, Allen *et al.* [549] have reported that prolonged melt oxidation of PET results in extensive discoloration and the formation of highly fluorescent hydroxylated terephthalate units which exist in equilibrium with highly phosphorescent stilbenequinone units. Phosphorescence characteristics of some common polymers are available [505].

Phosphorescence of **dyes** in PVAL can be observed at room temperature without prior evacuation to remove oxygen, which often quenches the emission in other systems [550]. Also phosphorescence of dyes in PVC has been observed.

It is clear that phosphorescence spectroscopy is only sporadically being used for polymer/additive analysis.

1.4.4. Chemiluminescence

Principles and Characteristics

In photochemical processes light may be absorbed by matter to promote photochemical reactions. There are chemical processes which give the opposite result, that is the production of light from a thermal reaction. When exothermic chemical reactions occur, the product species are usually in their ground electronic states. However, some thermal (dark) reactions (in whatever aggregation state) are so exothermic ($\Delta H > 40$ kcal/mol) that the energy exceeds that of the electronically excited state of one of the product molecules and visible light (400–700 nm) may be emitted upon relaxation. This phenomenon is called chemiluminescence (CL). The most effective CL reactions involve electron transfer, singlet oxygen or peroxide decomposition. Decomposition of cyclic peroxides is at the basis of most bioluminescence processes. There are some other physicochemical processes which can lead to the formation of excited states and thereby to the emission of light; these are based on the bimolecular recombination of high-energy species such as free radicals and radical ions [551]. Depending on the mode of formation of the excited molecule, the origin of luminescence may also be referred to as "photoluminescence" or "electroluminescence" (*cfr.* also Table 5.11 of ref. [1]). The process of chemiluminescence is widespread in nature, where it is known as "bioluminescence"; for example the emission of VIS light by fireflies and glow-worms.

CL was already observed in the 19th century [552] and has been known as an analytical tool for a long time. The applications of CL to sensitive analyses in complex samples is advantageous for various reasons: *(i)* CL is observed against a dark, low-noise background; *(ii)* the reaction generating emission occurs between a limited number of compounds; and *(iii)* the measurement of specific emissions can be selected using optical filters.

Light emitted by oxidation reactions with ozone leads to the most numerous applications of chemiluminescence. A common example of CL in the gas phase is the reaction of nitric oxide with ozone:

$$NO + O_3 \rightarrow NO_2^* + O_2 \rightarrow NO_2 + O_2 + h\nu \text{ (600–900 nm)} \quad (1.8)$$

Here, the product species NO_2 is produced in an excited electronic state and emits light in the VIS/NIR region. The intensity of chemiluminescence is proportional to the concentration of NO in the ppm-ppb range. Equation 1.8 can be used as the basis for development of a chemical sensor for NO. The method is used for quantification of non-extractable nitrogen-containing additives. In **nitrogen pyrochemiluminescence** the polymer (in pellet form up to 500 mg; no sample preparation) is combusted in a stream of oxygen to give NO, as follows:

$$R_3N + O_2 \xrightarrow{1000°C} CO_2 + H_2O + NO \quad (1.9)$$

The total elemental nitrogen analysers [553] and various chromatographic **chemiluminescent nitrogen detection** (CLND) techniques [554] collectively are based on the detection mechanism of eqs. 1.8 and 1.9. A test method for the use of nitrogen pyrochemiluminescence as an analytical method is described in ASTM D 4629-86. Pyrochemiluminescence gives the true value of chemically bound nitrogen (ppm to 17%) and delivers equimolar response in 5–15 min. The data correlate very well with values by the Kjeldahl method. In GC-CLND, described by Fujinari [554], the sample components are eluted from the column and then oxidised at high temperatures (1000–1100°C). The chemiluminescence detected by the PMT detector is proportional to each nitrogen containing compound eluting from the chromatographic column. CLND is also a truly equimolar HPLC detector (no response to N_2). True equimolar response for all nitrogen containing samples regardless of compound structure allows a single standard to quantitate multiple and complex samples. CLND provides easy quantitation (sensitivity: <0.3 ng N; linearity: 10^5). When used in tandem with a mass spectrometer a powerful detection and identification system is obtained.

Similarly, the **sulfur chemiluminescence detectors** (SCD), namely fluorine-induced SCD (FSCD), ozone-induced SCD (O-SCD) and redox CD (RCD), work by first oxidising the organosulfur compound to give a species which may either react with fluorine or ozone to form a chemiluminescence species (HF^* and SO_2^*, respectively). Reaction equations for O-SCD at 800–1100°C, developed by ref. [555], are:

$$RS + O \rightarrow SO + \text{combustion products} \quad (1.10)$$

$$SO + O_3 \rightarrow SO_2^* + O_2 \rightarrow SO_2 + O_2 + h\nu \text{ (300–400 nm)} \quad (1.11)$$

O-SCD is selective for sulfur, to which it responds quantitatively, irrespective of the compound. Other products of combustion do not chemiluminesce with ozone. O-SCD shows high sensitivity (<3 ng S), equimolar and linear response (>4 orders), high selectivity (>10^7), and absence of quenching (reliable quantitation in ppm to 40% range); only a single component calibration blend is required. Samples up to 500 mg may be handled. GC-SCD is suitable for odorant analysis. In the case of RCD it is the NO, produced from oxidation of the sulfur compound by NO_2, which is reacted with O_3 to form the chemiluminescent NO_2^*; the RCD responds to oxidisable species and not selectively to sulfur. Simultaneous nitrogen/sulfur chemiluminescence (GC-SCD/CLND) is possible [556]. The use of SCDs was reviewed [557].

Sulfur analysis is demanding as sulfur compounds are inherently difficult to measure because they are polar, reactive and often present at trace levels. Preparation of standards, especially gases, is also difficult. Sulfur-selective detection in FPD is based on combustion of sulfur-containing compounds in a hydrogen-rich/air flame to produce S_2^*. The emission from S_2^* is monitored using a PMT positioned near the flame. The problems of FPD for sulfur detection are well documented [558].

In 1961, Ashby [559] first observed chemiluminescence during the oxidation of polymers. **Oxychemiluminescence** or polymer chemiluminescence is weak visible chemiluminescence. The spectral

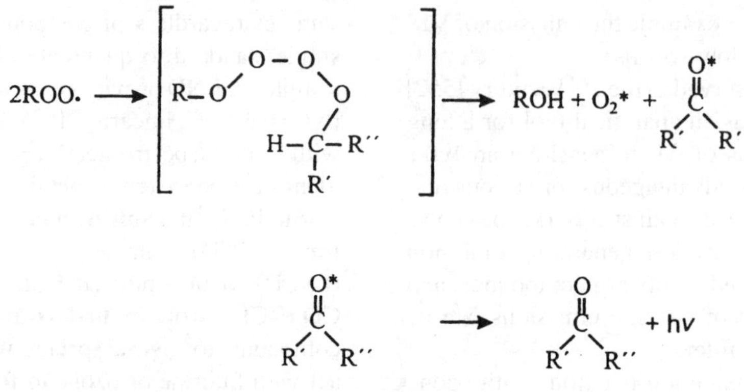

Scheme 1.1. The Russell mechanism for CL emission.

range varies with the polymer type but is generally in the blue-violet region (*ca.* 350–450 nm). Chemiluminescence provides a very sensitive method for determining the rate of peroxidation of polymers due to the ability of modern photodetection techniques to detect extremely low luminescence emissions (10^{-8} to 10^{-10} lumens). Only a few of the reactions involved in oxidation are potentially capable of light emission, which requires that the reaction be sufficiently exoenergetic to produce a product in an excited state and that the excited product should have a finite probability of photon emission rather than radiationless deactivation. The CL spectra exhibited in the oxidation of polymers, as well as of hydrocarbons, are in most cases consistent with the phosphorescence emission from carbonyl chromophores. The most widely accepted **mechanism** is the Russell mechanism (Scheme 1.1): a self-termination reaction between two secondary alkylperoxy radicals that, through an intermediate tetroxide, yields a triplet carbonyl species together with an alcohol group and a molecule of oxygen [560–562]. Luminescence is observed due to the relaxation of this excited state. Peroxy radicals are formed in the propagation reactions during thermooxidation. However, peroxy radicals may also be formed by molecular decomposition of hydroperoxides, explaining the CL emission commonly observed in N_2 atmosphere. It seems that the peroxy radicals termination proposed by Lee and Mendenhall [563] is more appropriate to explain the origin of light from the termination reaction of oxidation of polyamides [564]. Most chemiluminescence in polyamides is produced by the reactions of foreign groups in polyamides, such as end-groups (carbonyls, primary amines), as well as ketone and imine structures. It thus appears that polymer CL profiles probe different species depending on composition and purity (polyolefins, polyamides). Potential sources for light emission by polymers were reviewed [565].

It has been pointed out that PP, which predominantly has tertiary radicals, still gives rise to a relatively intense CL. This observation has been used as an argument against the Russell mechanism, which requires that at least one of the two reacting peroxy radicals is primary or secondary [565a]. However, the alkoxy radicals formed in PP may undergo beta-scissions, forming a ketone and an alkyl radical; in the presence of oxygen, the latter will react to a primary or secondary peroxy radical that may react with a tertiary peroxy radical and, via the Russell mechanism, may give rise to CL. The Russell mechanism (in solution) has recently been challenged [566,567] and is probably inappropriate in the solid. For several reasons, oxidation of polymers in the solid state, especially semi-crystalline polymers must be heterogeneous in nature [568]. Knowledge of the mechanisms of chemiluminescence reactions in the solid state remains insatisfactory in some important details.

Polymer luminescence varies from strong (*e.g.* polyolefins, polyamides) to medium (*e.g.* cellulose, poly(2,6-dimethyl-*p*-phenylene oxide)) and weak (*e.g.* PS, PMMA). Almost all polymers give CL under oxidation but most studies have been limited to polyolefins, polyamides and elastomers. In the past few years, oxyluminescence has been gaining acceptance as a sensitive method of studying oxidative polymer degradation. Chemiluminescence can be used in an analogous manner to that of oxygen absorption and IR carbonyl measurements to determine the degree of inhibition as a function of time, temperature, and the oxygen content and pressure of the gas phase surrounding the polymer sample or composition. Classical kinetic parameters cannot be determined from oxidative product formation (carbonyl index or O_2 uptake). Schard and Russell [569] proposed oxyluminescence as a way of looking at the performance of antioxidants.

The instrumental CL assembly has four major functions: *(i)* temperature and atmosphere control and monitoring; *(ii)* photon emission detection; *(iii)* pulse counting and data storage; and *(iv)* data retrieval and analysis [562,570,571]. **CL instruments** used in polymer studies essentially consist of a sensitive light sensor linked to a light-tight oven with temperature adjustable sample chamber and gas exchange facility. Conventional CL detector techniques were once based upon photon multiplier tubes (PMT) and did not allow monitoring the gradual development of oxidation profiles [572]. Meanwhile, CL apparatus has developed from the use of continuous electrometer electronics with poor sensitivity in the 1960s to the present use of a photon counting device with a resistive anode encoder (RAE), CCD camera or image intensified device. Sensitive CCD detection, which also allows spatial resolution, as opposed to PMT, has greatly simplified CL experiments [573], *cfr.* Fig. 5.18. Many authors use proprietary apparatus designs to follow the degradation of unstabilised and stabilised polymers. CL technology has been slow in gaining acceptance as a test method in the industrial environment, which is mainly due to the fact that the very low quantum yield (*ca.* 10^{-9}) of CL reactions requires highly efficient photon counting technologies. The availability of commercial (multi-sample) equipment [574,575] has changed the picture resulting in renewed interest in the application of chemiluminescence to oxidising polymers by the international scientific community.

In a chemiluminograph a very small amount of sample (*e.g.* mg) of polymer pellet, flake, film, powder or liquid is placed in a clean pure aluminium or stainless steel cuvette, or sample holder, which is then entered in a carefully temperature-controlled detector cell. **Chemiluminescence experiments** are usually carried out in inert atmosphere either isothermally or in temperature ramping up to 250°C (*e.g.* after photooxidation), in oxidising atmosphere isothermally (*e.g.* during thermo-oxidation), or by switching the atmosphere from oxygen to nitrogen isothermally. Polymer samples are generally heated to increase the emission. In case of evaluating the degree of oxidation from an exposed (*e.g.* irradiated or weathered) sample, the test would only use nitrogen as to avoid further degradation of the sample. By gently ramping up the temperature the maximum CL signal is used to indicate the relative performance of different samples. Use of a temperature gradient also serves in "scouting" for the isothermal temperature for thermo-oxidation analysis, which is usually carried out below T_g. Pure oxygen is then admitted to the sample after a conditioning phase. Reproducible results are obtained for homogeneous samples with the same thermal history that are kept in close contact to the heat cell.

The fundamental influencing **parameters** of the CL analysis technique have been described [576]. CL measurements are influenced by a great many parameters and depend on: *(i)* geometric factors (foil, powder, granule, fibre); *(ii)* sample quantity; *(iii)* sample cutting, breaking, punching (active surface); *(iv)* atmospheric conditions; *(v)* organic pollution; *(vi)* shape of sample holder (for liquids); *(vii)* light emission of sample holder; *(viii)* self-absorption of light by the sample; and *(ix)* contact *cq.* heat conduction between sample holder and sample. Standardisation of the operating procedure is thus strongly recommended. Calibration of both radiation and temperature measurement are required. Besides geometrical factors, molecular sizes and the chemical nature of AOs influence the CL induction time: low-mobility stabilisers such as oligomeric HAS cause no or minor variation of CL-OIT in contrast to low-MW stabilisers with high mobility. With so many factors affecting the generation of light during thermooxidation of a polymer, in particular non-oxidative processes, the usefulness of the tool in understanding polymeric oxidation has been questioned [577].

When chemiluminescence is used to monitor oxidation reactions in a polymer composition as it is heated or held isothermally in air or oxygen, the testing is sometimes referred to as **thermal oxyluminescence** (TOL), if applied in a thermal analysis mode. In CL experiments run in oxygen-containing atmospheres, oxidation processes are fed with fresh oxygen during the test. The isothermal experiment in **oxidising atmosphere** is related to conventional oxidation measurements. The CL signal has a typical S-shape with an induction period, which is interpreted as the time taken for isolated zones with very high extents of oxidation to spread beyond their initial volume. Under these conditions the total CL integrated over time from samples of natural rubber or

PP correlates well with the oxygen uptake and the CL intensity is assumed to be directly proportional to the rate of oxidation [578]. The intensity of emitted light is proportional to the rate of termination for the reactions involved in the Russell mechanism. Chemiluminescence reveals a difficulty in the definition of the induction period and the steady state of oxidation; as the sensitivity of the CL analysis is increased, the apparent induction time and the limiting rate decrease.

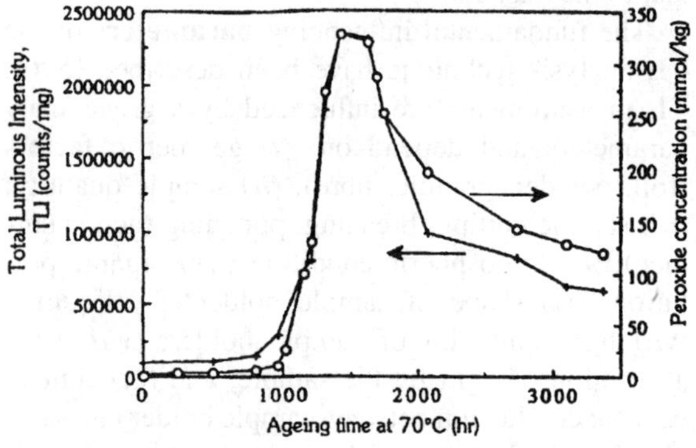

Fig. 1.27. Total luminous intensity and peroxide concentration for PP powder aged at 70°C. Reproduced by permission of G. Ahlblad, Royal Institute of Technology, Stockholm.

Chemiluminescence emission may also occur when an oxidised polymer is heated in an inert atmosphere. The integrated CL emission from a sample heated in inert atmosphere, which is frequently denoted TLI (total luminescence intensity), has for PP and natural rubber been found to be proportional to the concentration of hydroperoxides [577,579–582]. As shown in Fig. 1.27, the TLI curve mimics the iodometrically measured peroxide concentration closely [583] and provides an additional method of following the induction time (τ) to peroxidation. The recorded intensity may be regarded as a measure of the degree of oxidation [578]. Luminescence in nitrogen can be measured isothermally or by ramping. Both experiments allow TLI measurement, *cfr.* Fig. 1.28. Linear relations between oxygen uptake of pre-oxidised polymers and TLI measured in **ramp experiments in nitrogen** are polymer-, temperature- and sample geometry (powder, film, granule) dependent [584]. Thus ramp CL experiments are less satisfactory for kinetic analysis and cannot be used to quantify the degradation of a sample without a previous correlation curve under the same ageing conditions being available. However, ramp experiments can be used to determine differences in the early oxidation state of a sample of comparable shape and with a comparable degradation history. Ramping experiments under N_2 tend to show higher sensitivity and show different kinds of peroxides (for PP).

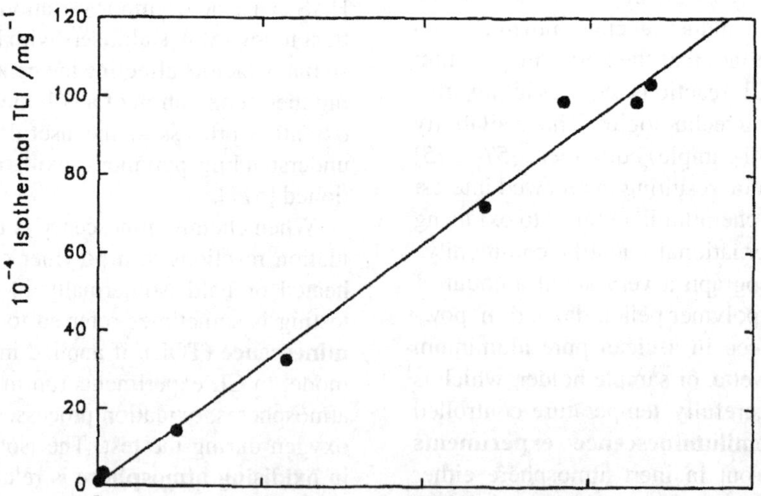

Fig. 1.28. Relationship of TLI measured isothermally at 100°C and by the temperature-ramp method for PP powder samples with the same geometry. After Billingham *et al.* [579]. Reprinted from *Polymer Degradation and Stability* **34**, N.C. Billingham *et al.*, 263–277, Copyright (1991), with permission of Elsevier.

The emission intensity is directly proportional to the resulting carbonyl concentration [585]. The relation between **chemiluminescence intensity** and the rate of oxidation is complicated. Whatever the process for CL emission, the intensity I can be expressed in the form:

$$I = c \cdot m \cdot G \cdot \Phi \cdot \tau \cdot R(t) \tag{1.12}$$

where the instrumental constant c, mass m and geometry G are sample specific terms whereas Φ (chemiluminescence yield), τ (transmittance) and $R(t)$ (luminescent reaction rate) are material specific. The instrumental and geometrical terms must be kept constant for comparative measurements; Φ and τ reflect the (low) quantum efficiency. In principle, G can be calculated for any given instrument, being a function of the detection efficiency of the photomultiplier and the geometry of the detector, though it also depends on the form of the sample. Determination of Φ is more problematic, but values around 10^{-9}–10^{-11} are typical of carbonyl emissions in solid polymers. By using reference samples standardised by a chemical method, CL can be used for the quantitative determination of ROOH with the advantage of simplicity and rapidity. It is assumed that the measured integrated CL in isothermal or temperature ramp experiments is proportional to the amount of peroxides formed during previous ageing [579]. CL can be used to calculate the total luminescence intensity of pre-aged specimens. The TLI value is proportional to the amount of hydroperoxides in a sample and gives a measure of the degree of oxidation [580].

Table 1.26 shows the **main characteristics** of oxyluminescence for polymer oxidation studies. The most attractive feature of the CL technique is high sensitivity. Unlike other methods for studying the oxidation of polymers such as oxygen uptake or FTIR carbonyl index, which need relatively large samples, CL is sensitive enough to handle very small ($<$20 μg) polymer samples despite the low quantum yield; oxidative degradation in single powder particles can be monitored. The sensitivity of CL measurements enables the early stages of both thermal and photooxidation and small changes to be studied in detail under actual conditions and offers interesting prospects for industrial application. The large dynamic sensitivity range enables continuous monitoring of oxidation from the very weak early stages to the main oxidation of a material. Chemiluminescence is the only method that permits direct measurement of thermo-oxidative polymer decomposition. CL measurements are particularly suitable for quality control and for comparisons between various stabilising systems, given the same basic polymer material.

Table 1.26. Main features of oxyluminescence for polymer studies

Advantages:
- Little sample preparation
- Micro sample sizes ($>$10 μg)
- Extreme sensitivity; measurements at low temperature; large dynamic range
- Speed, simplicity
- Discrimination of *low* stabiliser concentrations
- Early detection of sample defects
- Low volatilisation of additives (applicable to volatile samples)
- Accommodates wide range of sample forms (film, pellet, fibre, powder, liquid)
- Discrimination of various sample geometrics
- Measurement of peroxide concentration (TLI, N_2 atmosphere)
- Quantitation
- Acceleration *vs.* oven aging: 10–20×
- Very sensitive for OIT measurements (superior to DSC)
- Commercial equipment; automation
- Applicable for industrial purposes (QC; efficient screening and ranking of oxidative stability)

Disadvantages:
- Strong geometry dependence
- Lack of standards (wavelength, S/N ratio)
- No standardised testing procedures
- Poor reproducibility (improving with commercial equipment)
- "New" test method (limited experience in industry)
- Not equally applicable to all polymer systems
- Not applicable for black samples
- Sampling position dependent for heterogeneous materials *cq.* phenomena (repeatability)
- Not suitable for kinetic evaluations of polymer oxidation
- Bulk rather than surface technique
- Theoretically still debated

Chemiluminescence (CL) is not an absolute method; for analytical and even kinetic measurements it requires the calibration by another technique. The technique has to be used with care, particularly in comparing samples with different forms (film, powder, plaque, *etc.*) or from different age-

Table 1.27. General applications of chemiluminescence

- Ageing studies (real-time)
- Effect of processing and storage on polymer degradation
- Fundamental studies of oxidation of polymers
- Industrial test method for assessment of stabiliser efficiency

ing temperatures. The analysis of CL data is usually not quite straightforward, as a detailed knowledge of mechanisms leading to light emission, and an advanced knowledge of chemical kinetics for the derivation of equations describing the CL phenomena are needed. It is difficult to discriminate contributions for chemoluminescence and thermoluminescence (TL). In view of heterogeneous oxidation CL results may depend upon the sampling position. This problem is overcome in chemiluminescence imaging (ICL). As chemiluminescence measures light, the detected CL intensity may be reduced by self-absorption in opaque materials (*e.g.* filled polymers or rubbers). The technique is not entirely suitable for black particles. Additives like carbon-black absorb all light except the photons created in a thin zone close to the surface. Another drawback of chemiluminescence is that it gives information on the oxidation in the bulk of the polymer whereas the performance of a coating is usually determined by a physical property of the surface. There is still a lack of understanding on their mutual correlation [586]. CL measurements on commercial polymeric materials are often hard to interpret. Chemiluminescence offers poor reproducibility conditions with the use of proprietary equipment with different sensitivity limits. This is set to improve with the availability of commercial apparatus. Standardised testing procedures are in the development phase.

Table 1.27 lists some **general applications** of chemiluminescence. Early CL studies were limited by a low S/N ratio. Thus, oxidation could only be observed at high temperatures where other techniques provide more direct results. With the development of fast pulse analysers low light levels can now be measured with great sensitivity (a few photons/s). Consequently, it has been claimed that low temperature luminescence measurements are useful in lifetime prediction [570]. Some major applications of the CL technique are identified as isothermal measurements in oxygen to study progress of oxidation, and in nitrogen to estimate the amounts of peroxides present in a preoxidised sample. CL has been used principally in fundamental studies of unstabilised polymers to determine the degree of oxidation [587,588], to measure the kinetics of oxidation [562], to estimate the peroxide content [579], or to assess the relative stabilisation efficiency [573,589]. The early stages of oxidation of polymers may be followed up. CL monitors peroxide formation very well for the lower concentrations which are relevant to all practical ageing conditions. Nevertheless, the low quantum yield of CL and the small fraction of the polymer initially oxidising require long integration times at low temperatures. CL has widely been used for the characterisation of the thermo-oxidative degradation of polyolefins (LDPE, LLDPE, HDPE, iPP). In an inert atmosphere CL is a valuable tool for the assessment of small degradation effects during processing and storage [590].

In order to prove the **validity** of chemiluminescence, a comparison of CL with other analyses techniques was carried out. In comparison with other techniques, such as oven ageing testing, iodometric peroxide determination, UV and FTIR spectroscopy, XPS, x-ray tomography, density and modulus profiling, chemiluminescence offers higher sensitivity, simplicity and quickness. However, care should be exercised in comparing samples with different forms and ageing times.

Billingham *et al.* [591,592] have compared the use of CL and DSC in measuring **oxidation induction times** (OITs) and temperatures for various polyolefin and PET samples with varying stabiliser concentrations and packages. Direct comparison of CL (Fig. 1.29) and DSC (not shown) indicated essentially identical data over a range of temperatures. For long OITs CL is more sensitive and reliable than a standard DSC test, allowing measurements at low temperatures closer to that of real degradation conditions. A further advantage of the CL techniques is its selectivity towards the oxidation reaction. A DSC curve in an oxidising atmosphere is the sum of many different exothermic reactions, sometimes rendering determination of OIT difficult. A significant problem in using DSC to measure OITs at high temperature is the formation of volatiles. This effect can be minimised at elevated pressures, but the design of high-pressure DSC is difficult; CL measurements are unaffected by high pressures. Unlike DSC-OIT, CL-OIT is not standardised. The OIT test appears to be more suitable for assessing processing stabilisers than for lifetime prediction at service temperature.

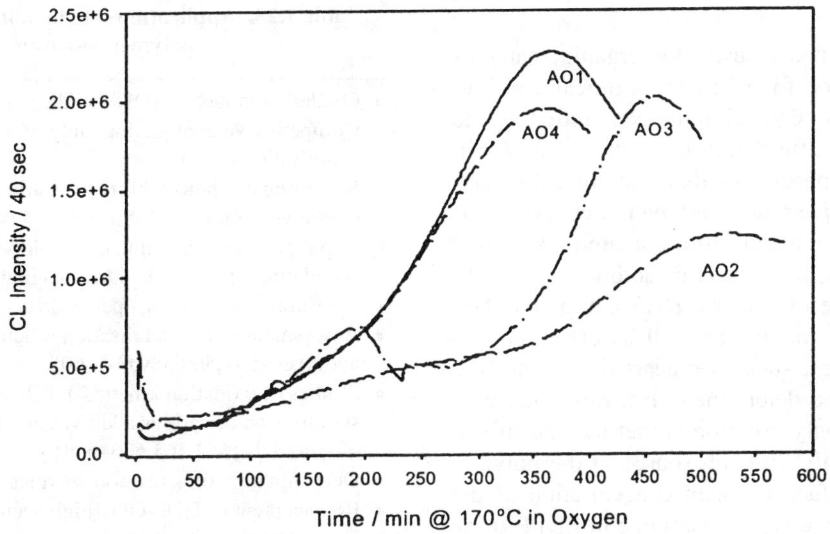

Fig. 1.29. CL-OIT for LDPE with four antioxidant formulations. Data obtained using four separate experiments. After Fearon *et al.* [591]. Reproduced by permission of N.C. Billingham, University of Sussex.

The OIT technique is not recommended for gauging the effectiveness of long-term AOs since at the high test temperatures employed, they exhibit high solubility and mobility in the polymer melt, often in contrast to the poor solubility in the polymer at room temperature (an example of this behaviour is Santonox® in PE). The induction times obtained for process stabilisers by CL generally correlate well with those obtained by DSC. The CL-OIT test gives a measure of polymer process stabilisation efficiency in PP samples subject to multipass extrusion.

Oxidation of polymers is generally accompanied by an exothermic heat flow as well as weak chemiluminescence. Any of these quantities may be used to evaluate the oxidative stability since they are all related to the oxidation rate. Not surprisingly, therefore, simultaneous measurement of oxyluminescence with DTA and DSC has been reported already long ago [593,594]. Commercial **DSC-CL** equipment, which consists of a calorimeter equipped with a single photon counting detector (PMT), is now available and allows simultaneous recording of both enthalpic changes and CL-OIT data of polymer samples [595]. The DSC-CL technique is highly reliable for determining OIT values [596]. CL is also good replacement for PDSC [597]. The use of CL in combination with thermal analysis has enabled a more satisfactory interpretation of CL data to be made.

Other possible simultaneous techniques in combination with CL are μFTIR, FTIES and oxygen uptake. **Simultaneous FTIES-CL** analysis (both emission spectroscopies) has been used to evaluate oxidation models for polypropylene [598]. In a further instrumental advancement, imaging chemiluminescence (ICL) has become available (*cfr.* Chp. 5.6.4.1). Early stages of polymer oxidation can also be studied using $^{18}O_2$ exposure and ToF-SIMS analysis [599,600].

Undoubtedly, the CL technique has a large potential as a method to study the polymer degradation and is close to being recognised as a standard method. (CL is under discussion in ISO/TC35 for paints and varnishes to be standardised as an analytical test method). However, it needs to be further improved to reach the level of confidence required by industry. Optimisation of technical components (optics, oven and atmosphere control), the standardisation of testing procedures and a better understanding of the CL reaction are some key points for industrial acceptance. Presently, CL cannot match the cost/performance ratio of the simple oven-ageing test. Expectations are though that the CL technique will make it possible to determine the efficiency of antidegradants more quickly than with current available techniques such as FTIR and DSC-OIT.

Recent books dealing with (chemi)luminescence are few (*cfr.* Bibliography). For thermoluminescence, *cfr.* Chp. 2.1.7.

Applications

Chemiluminescence is used for organic materials such as polymers, foods, pharmaceutical and biological materials. Chemiluminescent nitrogen and UV fluorescent sulfur detectors are particularly suitable to monitor product quality and control, analyses of bulk additive purity and regular checks of the amounts of nitrogen and sulfur containing additives in resin batches. Low levels of additives (*i.e.* <1–2%), such as the oligomeric HAS compound Tinuvin 622 in PE and PP, as well as otherwise extractable additives, such as erucamide and antistatic agents, can be determined on a rapid and routine basis. The only criterion is that the additive of interest is the only nitrogen source in the polymer sample. The surface nitrogen concentration of the slip agents oleamide and stearamide in LDPE/0.3% Armoslip CP (oleamide) and LDPE/0.3% Armoslip 18LF (stearamide) was monitored to quantify the migration process [473]. When using HPLC-CLND acetonitrile obviously is not the most suitable eluent.

Progress in oxidation of polymers may be studied by various means: FTIR, DSC, TG, SEC, CL, ESR, MALDI-ToFMS, oxygen uptake, *etc.* Oxyluminescence is an effective tool for determining the extent and nature of the oxidation process of polymers at very early stages (not yet detectable by macroscopic measurements) and under conditions similar to those during service. The opportunity to study oxidation quantitatively has stimulated the application of chemiluminescence for evaluation of a broad range of materials. Applications of CL in polymer research are shown in Table 1.28.

Table 1.28. Applications of chemiluminescence in polymer research

- Oxidation induction (OIT) studies of polymers [601]
- Competitive technique for study of degradation and stabilisation of polymers
- Screening of photostability of coatings and of oxidative stability of lubricants
- Rapid property screening of developed materials (by correlation of CL-OIT *vs.* oven aging time)
- Measurement of (hydro)peroxide content [579,602]
- Assessment of the stabilisation efficiency of (new) additives [537,569,589,603–605]
- Studies of oxidation kinetics, mechanism of stabilisation of various additives and long term stability of materials [562,565,606–609]
- Determination of activation energies
- Replacement of DSC-OIT (high accuracy, faster)
- Employment for rapid routine and release inspections of processed material (QC)
- Means of establishing stabiliser distributions (ICL)
- Oxidation profiling [602]
- Homogeneity testing [602]
- Damage analysis

Oxyluminescence is now widely used to study the rate of **oxidation of polymer particles**, in particular for rapid quantitative evaluation of the thermal oxidative stability of raw and stabilised polymers, the efficacy of additives, the effects of processing or service temperatures, storage conditions, and the effects of weathering and ageing, ozonisation, electron beam and γ-irradiation. The technique is also useful in evaluating the photooxidative stability of exposed polymers, specifically by quantifying the amount of hydroperoxides formed during the early stages of degradation. CL is ideal for a variety of applications, including QC, routine QA testing and product formulation research. Luminescence is a suitable means of detecting pre-oxidation of materials such as oleamide, erucamide, Ca-stearate, *etc.*, mechanical stress [610], and laser radiation perturbation [561] and allows evaluation of sources of supply. Yang [611] has reported a CL study of sodium benzoate as a nucleating agent. In this case, benzoic acid acted as an interfering chemiluminescence substance. Since the first studies on the efficiency of stabilisers by CL [569] the technique has been applied extensively to PP [559,609], PE [584,612–614], PVC, acrylics [615,616], rubbers [578,617, 618], polyamides [584,619–622], polyesters [584] and polymer blends [623].

Polypropylene has been the most studied polymer by means of chemiluminescence. CL has been applied to study the oxidation of PP under diverse conditions, namely after photooxidation [624], in the presence of metal ion prodegradants [625], after storage oxidation [626] as well as under high temperature ageing [627]. The sensitivity of CL has been demonstrated by measurement of complete oxidation curves of single reactor powder particles of PP of 30 μg mass [608] and by imaging chemiluminescence with a spatial resolution of 20 μm. CL emission from the isothermal oxidation of PP is fundamentally related to the formation of an oxidation product (excited carbonyl). CL is a very attractive method for studying peroxidation in PP because the measurement is relatively simple and needs little sample preparation, in marked contrast with the conventional iodometric approach. CL has been applied as a technique to measure the hydroperoxide formation during photooxidation of PP film samples [624]. A correlation between the development of CL emission and formation of hydrogen peroxide or related species during thermal oxidation of PP was reported [565]; these species may function as mediators for the gas-phase spreading [628]. The hydroperoxide content determined with the integrated total chemiluminescence of various samples could not be detected by ATR-FTIR or XPS analysis [624]. Similarly, very minor defects in iPP film in the very

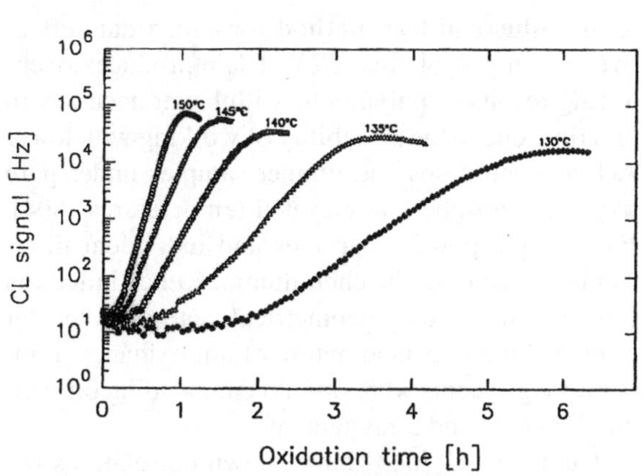

Fig. 1.30. Typical chemiluminescence signals from the oxidation of unstabilised PP powder samples at different temperatures under oxygen. After Celina and George [568]. Reprinted from *Polymer Degradation and Stability* **40**, 323, M. Celina and G.A. George, Copyright (1993), with permission of Elsevier.

early stages of photoirradiation could not be detected by IR and UV/VIS spectroscopies [629].

The oxidation-time profile of polymers is usually interpreted on the basis of the "classical" homogeneous kinetic analysis. There is however considerable evidence for **"infectious spreading"** of the oxidation zone from a small number of sites, such as catalyst residues [630–632], both for single particles and aggregates of PP particles, as shown by CL–time curves [628], and oxidation product–time curves obtained by IR spectrophotometry [632]. The heterogeneous nature of degradation during thermal oxidation of polymers, with formation of highly oxidised zones in the amorphous region, is now well accepted. The physical spreading model [568,631], proposed to explain the development of oxidation, was mainly based on CL experiments performed on films. Oxidation of solid PP measured by CL has been interpreted as involving heterogeneous initiation that leads to high oxidation rates in localised zones and is followed by the physical spreading of oxidation [633]. Figure 1.30 shows typical CL signals from the oxidation of unstabilised PP powder samples at different temperatures under O_2. The interpretation of PP oxidation should not be extended to other polymers.

The semi-quantitative CL analysis of Fig. 1.31 was taken as evidence that the antioxidant is heterogeneously distributed in the individual PP/75 ppm Hostanox O-10 particles [605].

Chemiluminescence of PE is much less intense than in case of PP but CL is still a sensitive dry

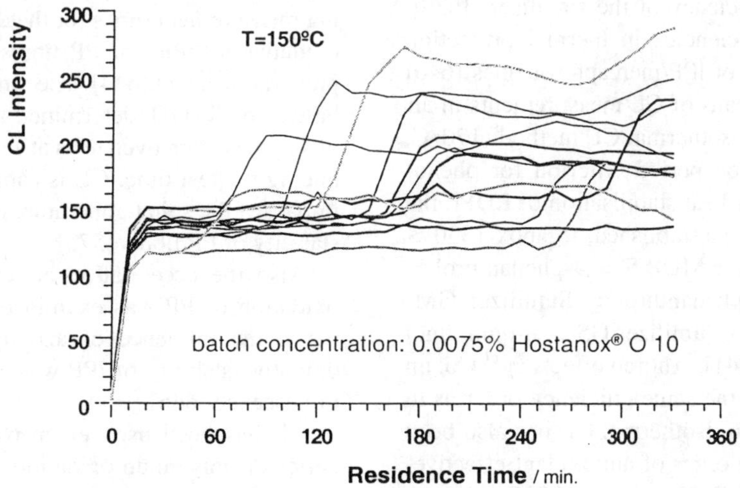

Fig. 1.31. CL curves of 300 μm PP/Hostanox O-10 powder particles. After Kröhnke [605]. Reproduced by permission of Atlas MTT GmbH, Linsengericht, Germany.

technique for the detection of residual peroxides in PE [634]. In a study on radiation-induced oxidation of HDPE/Irganox 1010 it was noticed that the bloomed antioxidant makes a significant contribution to the observed high CL emission [635]. Isothermal CL measurements have been used to evaluate the stability of EVA/PET/XLPE I&C cables [636]. CL (in air at 130°C, and in N_2) has been used for quality control of batches of slip agents (oleamide and erucamide) from different suppliers for LDPE [637]. A relation was established with complaint samples. Small amounts of degraded erucamide impair the stability of the end product. A different method for the determination of degradation products is SEC-UV.

Typical applications of CL are also isothermal experiments under O_2 for **ranking of stabilisers** based on the induction period as a criterion. Additives should be extracted before carrying out the CL experiments to avoid contributions of peroxides and stabilisers to the observed CL emission. However, extraction may change the peroxide concentration. The effect of different stabilisers (Cyasorb UV531, Irganox 1010, Tinuvin 770, Irgafos 168, Hostanox SE2) on the formation and photolysis of hydroperoxides in the very early stages of PP photooxidation has been studied by CL [603]. Several stabilisers interfered with the chemiluminescence experiment, such as hindered phenols. Hindered piperidine additives quench the emission only weakly. CL of photo-irradiated iPP containing HALS and a mercapto-1,3,5-triazine phenolic antioxidant was used to rank the efficiency of the stabilisers [638]. The antioxidant efficiencies in thermal protection of iPP/Se [639] and of iPP/mercaptotriazines [640] were assessed by means of CL measurements in air at 180°C. Using the isothermal CL method (170°C, O_2) and the induction period criterion for phenolic AO evaluation for heat stabilisation of LDPE the following ranking was established: Irganox 1330 > 1-pyrenol > Sumilizer MDP-S > 4-phenanthrol > 1-phenanthrol > 3-phenanthrol > Sumilizer GM-S > 2-phenanthrol > Sumilizer GS > 8-quinolinol > Sumilizer BHT [641]. Antirad effects (γ^{60}Co, up to 10 Mrad/h) show the same efficiency order as in the thermal oxidation. Isothermal CL has also been used to establish the order of antioxidant effectiveness in HDPE and LDPE films at 190°C in air as Irganox 1010 $\gg$ Ethanox 330 > Irganox 1076 > Topanol OC [642]. Chemiluminescence can be used as an **industrial test method** for antioxidant effectiveness in polyolefins [573]. CL reproduces oven-ageing results of polyolefins with better accuracy in less time due to the possibility of working with lower AO concentrations and thinner samples under pure oxygen atmosphere at elevated temperatures [604]. Even single powder particles and individual fibres can be measured. The chemiluminescence induction time is influenced by geometrical factors, molecular sizes and the chemical nature of antioxidants. Temperature gradients were observed depending on sample thickness and arrangement.

Table 1.29. Comparison of CL-isothermal and oven tests on PP fibres with spin preparation

PP fibre	OIT[a] (h)	Oven ageing[b] (h)
A	6.8	750
B	30.6	1300
C	36.9	3250

[a]Temperature: 150°C.
[b]Temperature: 130°C.

Dudler *et al.* [643] have shown correlations between CL and oven aging data of stabilised PP. CL was found to accelerate testing times by a factor of 4 to 12; this reduction is most likely caused by the fact that CL measurements are usually carried out in pure oxygen. CL testing is more likely to be a back-up to oven-ageing tests for the determination of stabiliser effectiveness rather than a replacement. Table 1.29 compares typical CL and oven ageing measurement times for the evaluation of thermo-oxidative stability of PP fibres treated with a spin preparation agent [644]. The ranking of the materials based on CL-OIT determined at 150°C equals that of samples from oven tests at 130°C, thus greatly reducing the test time. CL is capable of detecting the impact of the spin temperature on the long term heat stability of PP fibres [573].

Also the acceleration effect of stearic acid on oxidation of PP was examined by CL [645]. Similarly, the influence of azo dyes on the thermo-oxidative stability of iPP was assessed by chemiluminescence [646].

CL has been used extensively to study the kinetics of polyamide oxidation. Chemiluminescence cannot be used to describe the oxidation rate of **polyamides** [619]. CL should be used only to evaluate the oxidation states of polyamides. Forsström *et al.* [647] have investigated the effect of two commercial stabilisers, *i.e.* Irganox 1098 and B1171 on the oxidative stability of 40 μm thin PA6 films in air and oxygen in the temperature range of 100–140°C. Interpretation of the time profile of CL from oxidation of polyamides, polyethers and hardened epoxies remains an unsolved problem. In polyamides the content and ratio of carboxylic acid and amine end-groups plays a role [648,649].

The CL emission of poly(ethylene-*co*-1,4-cyclohexane-dimethylene terephthalate) (PECT) is highly dependent upon the thermal and UV oxidative history of the material [650]. Thermal oxidation of the polymer as measured by hydroperoxide concentration is directly related to CL intensity and can predict the behaviour of antioxidants.

Mattson *et al.* [651] used various techniques (CL, density profiling, computed x-ray tomography and modulus profiling) to assess the ageing of CB-filled EPDM cable materials. CL showed the highest sensitivity at low temperatures and/or over short time intervals. However, caution is warranted when interpreting CL data. The other three techniques (DP, CT and MP) were more easily connected to changes in macroscopic mechanical properties and are helpful in monitoring and understanding heterogeneous ageing phenomena such as diffusion-limited oxidation. Proportionality between the TLI values and the peroxide concentration has been found, but needs confirmation.

Chemiluminescence can be used to evaluate the effect of various compounding and processing variables on the elastomer thermooxidative stability. Variations in mixing, polymer type, cure state and stabilisers can be characterised in terms of induction period, oxidation rate constant and durability. The usefulness of the technique has been demonstrated for a variety of **elastomeric systems**: unvulcanised and vulcanised compositions as well as formulations with fillers and antioxidants. Fillers, especially carbon-black markedly reduce the level of light emitted during oxidation. CL can be efficiently employed, even in evaluation of such low-emitting compounds as vulcanisates containing 40 phr of carbon-black [617]. Chemiluminescence emission from a hydroxyl-terminated polybutadiene (HTPB) rubber was measured during isothermal oxidation from 70 to 130°C [618].

Chemiluminescence can be used as an alternative to the determination of thermal stability and AO performance by means of DSC-OIT. Figure 1.32 shows

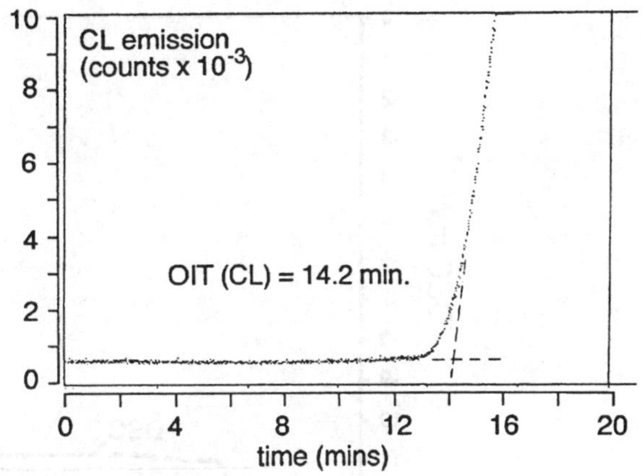

Fig. 1.32. CL-OIT data for PP/(Irganox 1010, Irgafos 168). After Scheirs *et al.* [575]. Reproduced by permission of J. Scheirs, ExcelPlas Australia, Edithvale, Victoria.

typical OIT data as obtained by CL for PP/(Irganox 1010, Irgafos 168). CL offers many advantages over DSC, such as much higher sensitivity that enables measurements at more realistic use temperatures (below T_g) closer to realistic degradation conditions, sharp onset time/temperature, and needs small samples only (10 μg). A significant problem in using DSC to measure OITs at high temperatures is that the sample may be too volatile, or may produce volatile oxidation products. CL detection is potentially a very valuable method for studying volatile samples. The correlation between CL and DSC data is generally satisfactory [614,617]. The CL technique provides more information on the oxidation process than DSC. Figure 1.33 shows simultaneous DSC-CL measurements on a highly stabilised PP plaque at 150°C in O_2. In this case the CL detector determined OIT of 66 h whereas DSC was too insensitive to detect the onset of oxidation [614]. The oxidation of 100 μm thick PP/Irganox 1010 films was studied by means of simultaneous DSC-CL over a wide range of oxygen pressures (1–25 bar) in order to lower testing temperatures (130–150°C) [597]. CL turned out to be good replacement of expensive pressurised DSC equipment. A simplified approach to quantitatively assessing the effects of polymer additives has been applied to DSC-CL data for LDPE/(Chimassorb 944, DCP) based formulations and DSC-OIT data for MDPE/(CB, Irgafos 168, Irganox 1010) [652].

Forsström [653] has reported simultaneous detection of heat flow (using a microcalorimeter) and light emission (CL) during oxidation of unstabilised PP: a time shift between both techniques

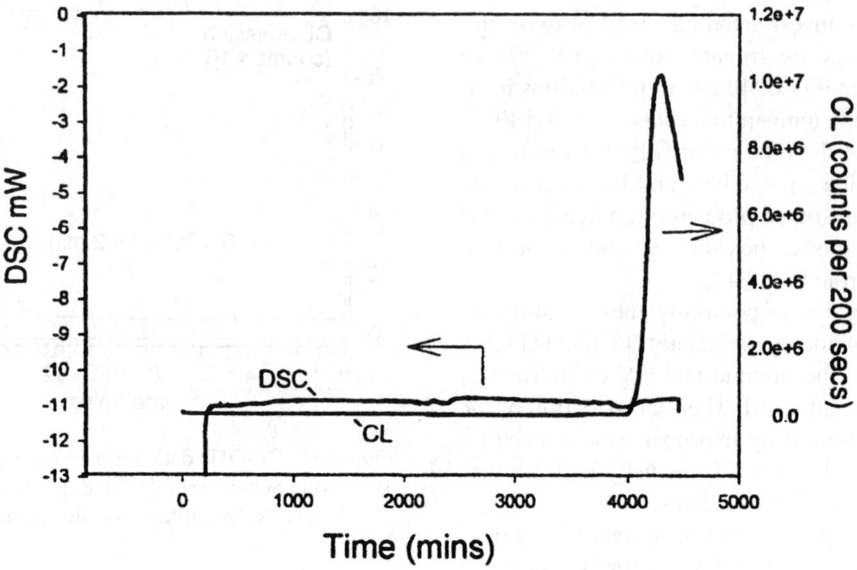

Fig. 1.33. Simultaneous DSC-CL of a highly stabilised PP plaque. After Billingham *et al.* [614]. Reproduced by permission of Rapra Technology Ltd., Shawbury.

was observed. George *et al.* [496] have established the relationship between single particle CL and FTIES of pressed polyolefin particles. FTIR emission spectroscopy may contribute to the ongoing efforts to evaluate stabiliser packages, as an alternative technique to determine induction periods and to investigate the performance of PVC formulations [497].

The CL-technique is also capable as a short-term test to predict the tendency of spontaneous ignition (not necessarily caused by a CL process) of pigments and/or additive concentrates when added to the polymer, *e.g.* during extrusion at high temperature [644].

Chemiluminescence has also been proposed as a novel tool in **paper conservation** studies [654,655]. CL phenomena can be used for assessing the thermal and oxidative degradation pathways of paper-based historical documents. In contrast with the usual accelerated degradation experiments in climatic chambers, measurement of isothermal CL is quick. The influence of all paper components (alkalinity, metal content, cellulose peroxides and carbonyl groups, moisture) and exposure to light will be investigated in the framework of the PAPYLUM project (ending October 2004).

Chemiluminescence applied to oxidation and degradation of polyolefins was reviewed [656,657].

1.5. NUCLEAR SPECTROSCOPIES

Nuclear spectroscopic studies in polymer/additive research comprise nuclear magnetic resonance (NMR), nuclear quadropole resonance (NQR), electron spin resonance (ESR) and Mössbauer (absorption/emission) spectroscopy (MAS, MES). When everything else has failed in elucidating difficult problems a safe, almost universally valid advice is to try magnetic resonance techniques, NMR and ESR, in this order.

The **magnetic spectroscopies** exploit the effect of a strong magnetic field on the interactions of matter with electromagnetic radiation. The magnetic field can induce small energy differences as a consequence of the magnetic properties of electrons and of some (though not all) atomic nuclei. In NMR and NQR radiation is absorbed in the **radiofrequency region** by the same fundamental process as at all other wavelengths, but the energy of quanta at these frequencies is very small (typically 100 neV). The small splittings necessary to produce absorption in the rf region are those normally associated with the hyperfine structure of electronic spectra. Both NMR and NQR involve the coupling of rf radiation with a nuclear magnetic moment to bring about transitions between nuclear orientations of different energies. The difference between the two lies in the origin of the external nuclear energy levels. In the case of NMR, the energy levels are governed by the interaction of the nuclear magnetic dipole moment with an externally applied magnetic induction, whereas in NQR the levels are governed by an interaction of the nuclear electric quadrupole moment with the electric field gradient produced at the nucleus by the charge distribution to its environment.

ESR and NMR share the same basic theory. Whereas NMR deals with nuclei having magnetic moments, ESR refers to electrons, but these must be unpaired. There are many chemical compounds which have odd numbers of electrons. ESR finds application in the study of paramagnetic transition metal complexes, organic free radicals, free radicals formed when most materials are subjected to ionising radiation, *etc.* The ESR phenomenon is due to absorption of energy by a "spin" system from electromagnetic radiation with frequency ranging from that for **microwaves** to sub-mm waves. The fact that the population difference between spin states is greater for electrons than for nuclei means that ESR spectroscopy is much more sensitive than NMR. This higher sensitivity stands in relation to the frequency range (microwave *vs.* radiofrequency) and short lifetimes of the excited states. Because ESR involves frequencies on the order of 10^9 Hz (and a resulting time scale of 10^{-9} s), it takes a much faster "snapshot" of dynamic systems than does NMR. Consequently, ESR can generate information about chemical processes that are too fast to study by NMR.

NMR, NQR and ESR depend for their chemical significance on the nuclear moments of the isotopes present in the species under study. **Magnetic resonance spectroscopies** may be used for the determination of the chemical structure as well as for the dynamics of polymer chains. Questions regarding the presence of additives, cross-linking and the dynamic behaviour of matter may be tackled.

Also Mössbauer spectroscopy is a resonance phenomenon, but involves **γ-rays**. Mössbauer spectroscopy is a probe of short and medium range structure. Mössbauer nuclei of interest to additives in polymers are rather few. Other nuclear methods in the research of chemical structure, such as position annihilation spectroscopy and nuclear resonant scattering of synchrotron radiation (an extension of conventional Mössbauer spectroscopy) are emerging techniques with no reported applications in the field of polymer/additive analysis yet.

1.5.1. Solid-state NMR Spectroscopy

Principles and Characteristics

As a non-destructive technique probing the magnetic interactions of atomic nuclei, nuclear magnetic resonance spectroscopy is one of the most powerful structural information tools for almost all additive classes, including highly polar, ionic and thermo-labile compounds. Solution NMR is exceptionally useful to chemists because the high resolution achieved (with line widths for ^1H less than 1 Hz) allows small but important effects (*i.e.* chemical shifts and splittings due to coupling constants) to be observed and structural assignments to be made. Solution NMR analysis of the products extracted from polymeric matrices and for dissolved polymer/additive systems has been described elsewhere [1].

NMR experiments are not restricted to solutions but can also be conducted directly in the solid state. NMR was first observed in solids in 1945. Solid-state NMR has gained momentum since the introduction of the Fourier transform principle (after 1975). Recently, 750 MHz s-NMR instruments have been introduced. In order to obtain high-resolution s-NMR spectra, special techniques and spectrometer designs are employed. Although it is possible to use the same **spectrometer** for both solution and solid-state studies (and manufacturers are developing systems which can be modulated for any technique like l-NMR, s-NMR, NMR Imaging, NMR Microscopy, or Localised NMR spectroscopy), usually each customer configures a particular spectrometer for only one experimental technique.

There are compelling reasons why it may be preferable to characterise **solid-phase samples**, *e.g.*:

- many high-value products produced by the chemical industry are solids;
- many samples cannot be dissolved (*e.g.* highly cross-linked and filled polymer systems) or are altered by dissolution;
- phenomena inherent only to the solid phase (*e.g.* entanglements);
- intermolecular interactions;
- chemical and physical processes in the solid state; and
- molecular motions.

Still, solid-state NMR has attracted much less attention than NMR of liquids. An early impetus for the development of s-NMR was the study of polymers. The technique allows to investigate structure, dynamics and order of the polymeric solid state.

Fourier spectroscopy has unified solid- and liquid-state NMR in an unprecedented manner. Despite the fact that the principles of the techniques are the same, there are several factors causing significant differences between spectra of solids and liquids. The Hamiltonian, $\boldsymbol{H}$, *i.e.* the quantum mechanical description of the various interactions experienced by a nuclear spin system, is given by

$$\boldsymbol{H} = \boldsymbol{H}_{\mathrm{Z}} + \boldsymbol{H}_{\mathrm{D}} + \boldsymbol{H}_{\mathrm{CS}} + \boldsymbol{H}_{\mathrm{J}} + \boldsymbol{H}_{\mathrm{Q}} \qquad (1.13)$$

where the various terms represent Zeeman interaction ($\boldsymbol{H}_{\mathrm{Z}}$), dipolar interactions ($\boldsymbol{H}_{\mathrm{D}}$), chemical shift ($\boldsymbol{H}_{\mathrm{CS}}$), nuclear–nuclear interactions ($\boldsymbol{H}_{\mathrm{J}}$) and quadrupolar interaction ($\boldsymbol{H}_{\mathrm{Q}}$). Each interaction is mathematically described as a second-rank tensor. Tensors may be isotropic (no orientative dependence), axial or asymmetric. The narrow line widths of resonances in solution spectra are a direct result of the rapid molecular motion, which averages out strong dipolar interactions between spins; in this case, the isotropic parts of $\boldsymbol{H}$ dominate. On the other hand, the **anisotropic interactions** in solids broaden NMR resonances to such an extent that chemical shifts and indirect spin–spin couplings are no longer resolved. The maximum coupling that can exist between a pair of protons can be on the order of 50 kHz, far larger than the few Hertz found in solution spectra. Consequently, different ways of handling s-NMR problems have been developed. The main features of the solid state that make the NMR spectra look different for the liquid state are: *(i)* dipolar interactions: *(ii)* broadening due to chemical shift anisotropy (reduceable using MAS techniques); and *(iii)* relaxation times (reduceable using cross-polarisation). The primary difference between solid-state and liquid-state NMR is one of timescale. s-NMR is characterised by inefficient spin–lattice relaxation (long T_1's) and extremely efficient spin–spin relaxation (short T_2's).

NMR studies of solids can generally be classified into **three categories** based on: *(i)* high-resolution spectra; *(ii)* relaxation times; and *(iii)* broadline spectra. Of course, spin–spin (transverse) relaxation directly affects the observed signal (FID) from pulsed NMR operation, which is Fourier transformed to yield the spectrum, so that the three areas are not totally distinct. Major problems encountered in **high-resolution s-NMR techniques** are line-broadening and low sensitive nuclei. In solid samples, which present a complete range of molecular orientations in the applied magnetic field $\mathbf{B}_0$, all or part of the anisotropic interactions of nuclear spins remain static, leading to complex spectra and substantial **line-broadening**, typically 10 kHz. The most commonly encountered broadening interactions in solids are chemical shift anisotropy, direct dipole–dipole interaction, and quadrupolar interaction (often dominating). Spin–spin interactions and dipolar line broadening are closely related phenomena, but not identical. Spin–spin coupling is an intramolecular phenomenon, where neighbouring molecules are not involved. Because each spin possesses a magnetic moment μ, each is surrounded by a magnetic field that is experienced by the others. This is the direct, or through-space, dipole–dipole (or dipolar) coupling. The direct splitting $\Delta\nu$ (in Hz) in the spectrum is:

$$\Delta\nu = \frac{3\mu^2}{hr^3}(3\cos^2\theta - 1) \qquad (1.14)$$

The splitting depends very strongly upon the distance r between the nuclei and is a function of the angle θ between the internuclear vector with the static field $\mathbf{B}_0$. In solids the through-space dipolar coupling between magnetic nuclei is not averaged to zero (as for liquids) and gives rise to characteristic splitting patterns. It is thus not surprising that featureless broad bands are observed in s-NMR spectra unless the dipolar coupling is minimum (*i.e.* zero) for $\theta = 54.74°$.

The effects of direct dipole–dipole coupling on solid-state spectra need to be reduced in order to resolve the chemical shifts. A major goal in NMR has thus been to develop various techniques for **line-narrowing** of the solid-state resonance spectra: high-power dipolar decoupling (DEC), magic-angle spinning (MAS), dynamic-angle spinning (DAS), double rotation (DOR) or multiple-quantum magic-angle spinning (MQMAS). A detailed treatment of these techniques is beyond the scope of this text. **Magic-angle spinning** is by far the most powerful tool in s-NMR. This rapid mechanical spinning technique averages anisotropic interactions by acting on the factor $(3\cos^2\theta - 1)$ in the Hamiltonians, which in solids is not averaged to zero by rapid molecular motion (*cfr.* eq. (1.14)). It is possible to convert solid-state spectra to something akin to those of fluids, namely spectra containing sharp resonances with one resonance per distinguishable nuclear site by rapid spinning (6–35 kHz) at the magic angle ($\theta = 54.74°$), which imposes motional averaging.

MAS affects line broadenings from dipolar interaction, chemical shift anisotropy and quadrupolar interaction, which all contain the angular dependence. By spinning at the magic angle ($3\cos^2\theta - 1 = 0$) the dipolar interaction (^{1}H$-^{13}$C, ^{1}H$-$O$-^{29}$Si) vanishes, the chemical shift anisotropy is averaged to the isotropic value (^{13}C, ^{29}Si) and the first order quadrupole interaction vanishes, while the second order is reduced (^{27}Al). Although homonuclear dipolar couplings are in principle removable by MAS alone, with abundant nuclei they are often very strong. The alternative to MAS is to manipulate the nuclear spins themselves using multiple-pulse line narrowing so as to average the dipolar interaction. When dilute spins, such as ^{13}C, interact via the dipole interaction with ^{1}H or other abundant nuclei, the large heteronuclear broadening of an already low-intensity spectrum is a considerable problem. To obtain spectra free of heteronuclear couplings a strong continuous rf may be applied at the given nuclear resonance frequencies (*e.g.* proton decoupling for ^{13}C).

While MAS can provide significant resolution enhancement, it enhances sensitivity only insofar as the signal from broad resonances is concentrated into narrower resonances. For naturally low-abundance nuclei like ^{13}C (1% naturally occurring), this increase may be insufficient. Other techniques have emerged which substantially increase the **NMR sensitivity**. In fact, modern s-NMR is capable of producing high-quality high-resolution spectra of dilute spins such as ^{13}C and ^{15}N in solid samples in a relatively short time. Dilute and abundant nuclei are often in close proximity, and coupled via dipolar interaction. Dilute nuclei are more difficult to observe than abundant nuclei, such as ^{1}H or ^{31}P, particularly those with a low gyromagnetic ratio. Possible solutions to the problem of low NMR sensitive nuclei (*e.g.* ^{13}C) in solids are isotope enrichment (expensive) and **polarisation-transfer techniques** [658]. The latter techniques are based on the fact that it is possible to alter the polarisation, and hence the strength of the NMR signal, of certain spin species (typically low abundance and low γ-nuclei) by manipulating the polarisation of other spin species (*e.g.* high abundance and high γ-nuclei). Several such polarisation-transfer techniques exist. The most well known is **cross-polarisation** (CP), usually applied to measure NMR of rare spins (*e.g.* ^{13}C, ^{15}N, ^{29}Si, ^{31}P) in solid materials containing abundant spins (*e.g.* ^{1}H, ^{19}F) as well. Cross-polarisation is based on an indirect excitation of dilute spins S by rf-mediated polarisation transfer of magnetisation from abundant spins I. The exchange requires an energy match and a coupling interaction for polarisation to be transferred; on CP, the rf amplitudes are adjusted so that the Larmor precession frequencies of the abundant and rare nuclear spin species are equal (Hartmann-Hahn condition). The CP efficiency depends on the strength of the I-S dipolar interaction, *i.e.* on the distance between I and S nuclei (^{1}H$-^{13}$C, ^{1}H$-$O$-^{29}$Si, *etc.*). The ^{1}H$-^{13}$C cross-polarisation pulse sequence has become the standard for s-NMR, and has made ^{13}C s-NMR practical for the first time. Cross-polarisation overcomes two serious problems: low sensitivity and long spin–lattice relaxation times of spin-½ nuclei. In a typical organic solid it is not unusual to have proton T_1 values of a few seconds and carbon T_1 values of minutes or hours. Cross-polarisation overcomes long T_1's. The ability to recycle at the proton T_1 rather than at the carbon T_1 represents a dramatic sensitivity enhancement. CP can also be used to detect whether I and S spins are physically near each other. Cross-polarisation is usually combined with magic-angle spinning in the most frequently encountered CP/MAS s-NMR experiment. CP/MAS NMR provides structural and dynamic information on the molecular level for solid polymeric materials. Although the line widths in such high-resolution spectra are still greater than those in liquids, the various non-equivalent nuclei can usually be resolved. Another polarisation-transfer technique is the **nuclear Overhauser effect** (NOE), in which, as in liquids, polarisation changes are obtained through mutual relaxation transitions.

High-resolution ^{13}C spectra of solid polymers can principally be obtained by two ways: from normal Bloch decays (SPE: **single-pulse excitation**) of the carbon magnetisation, just as in l-NMR, or from cross-polarisation. These techniques are complementary. Discriminating experiments may consist of comparing CP/MAS and SPE spectra (the latter obtained without cross-polarisation). Whereas the former depends on *proton* relaxation, the latter is affected only by *carbon* relaxation. Because of the great segmental mobility in elastomers, these systems have shorter spin–lattice relaxation times (in the order of seconds), which makes SPE feasible.

Table 1.30 mentions the most important techniques which are often applied in modern solid-state NMR. Owing to the successful combination of MAS with sensitivity enhancement pulse sequences (most notably cross-polarisation from abundant to dilute

Table 1.30. Some solid-state NMR techniques and their major effects[a]

Technique	Major effects
Line-narrowing techniques:	
Magic-angle spinning (MAS)	Eliminates all anisotropies to first order
Decoupling	Eliminates heteronuclear dipolar interactions, heteronuclear J-coupling
Multiple pulse decoupling	Eliminates homonuclear dipolar interactions
Polarisation-transfer techniques:	
Cross-polarisation (CP)	Rare-spin NMR with increased sensitivity Connectivity between cross-polarising spins
Nuclear Overhauser effect (NOE)	Connectivity between mutually relaxing spins
Two-dimensional (2-D) NMR:	
Homo- and heteronuclear 2-D spectroscopy	Connectivity between nuclear spins of the same or different species
Relaxation measurements:	
Zeeman relaxation	Study of molecular motions with correlation times of the order of 10^{-7}–10^{-10} s
Rotating-frame relaxation	Study of molecular motions with correlation times of the order of 10^{-4}–10^{-6} s

[a] After Wind [659]. Reprinted from *Encyclopedia of Analytical Science* (A. Townshend, ed.), R.A. Wind, pp. 3477–3485. Copyright (1995), with permission from Elsevier.

spins), s-NMR has evolved into a technique with sensitivity and resolution comparable to its solution counterpart. Nevertheless, sensitivity is still a limiting factor and makes it difficult to obtain spectra from isolated thin films or from surfaces in low-surface-area materials.

The range of useful **s-NMR nuclei** is limited both by the technique and by the characteristics of the materials (additives). In spite of the fact that ^{1}H and ^{19}F are very sensitive, few s-NMR applications are possible because of the broad line widths (strong dipolar coupling). Where deuterium labelling at a specific location of a component molecule is used, this allows selective experiments at quite a high level of sensitivity and reasonable ease of interpretation. Useful solid-state NMR nuclei are in particular ^{13}C, ^{23}Na, ^{29}Si, ^{31}P and to a lesser extent ^{11}B, ^{25}Mg and ^{27}Al.

Important objectives of s-NMR spectroscopy are the determination of **molecular structure**, micromorphology and molecular mobility. Studies of molecular structure require high resolution so that individual chemical shifts are revealed free of overlap from other interactions as well as the anisotropy of the magnetic shielding. By measuring the splitting caused by direct dipole–dipole coupling, internuclear distances can be measured with great accuracy. Obvious applications of s-NMR are conformational studies. s-NMR allows the solid-state identification of insoluble polymers and of additives therein contained, the study of additive degradation and reactions in the polymer matrix, stabilisation studies and examination of systems which are difficult to approach for l-NMR, such as the analysis of Na benzoate or grafted polymers. However, s-NMR lacks the sensitivity to readily determine the presence of smaller amounts of additives.

The general advantage of NMR is its high specificity. The method measures volume average particles. Consequently, errors due to the heterogeneity of the sample are negligible. Although averaging is macroscopic, the answer is on a nm scale. Any sample type and shape can be analysed. Typical detection limits are *ca.* 10^{18}–10^{20} atoms of the nuclear isotope studied; for ^{13}C at least 0.5 wt.% additives should be present. Whereas ^{1}H and ^{13}C l-NMR are both easily used for quantitative purposes, the same is not true in s-NMR where proton NMR is hampered by a resolution problem. In general, s-NMR is quantifiable only for those nuclei which do not require CP/MAS (which upsets intensity ratios); quantification by means of ^{13}C s-NMR is therefore difficult, but feasible for ^{29}Si and ^{31}P provided that an internal standard is used. Although in principle FID height F_0 immediately after a single rf pulse is a faithful relative measure of molecular concentration,

Table 1.31. Main characteristics of high-resolution s-NMR spectroscopy

Advantages:
- Multi-nuclear detectability
- Non-destructive, non-invasive bulk probe
- Sample form (powder, single crystal, randomly oriented or aligned film)
- High specificity
- Relatively high spectral resolution
- Ease of manipulation of nuclear spin Hamiltonians (spectral simplification)
- Structure/dynamics-property relationship
- Micromorphological information (relaxation measurements)
- Multidimensionality
- Applicable to all additive classes (non-polar, highly polar, ionic, thermolabile)

Disadvantages:
- Relatively insensitive
- Typical sample size: 10–500 mg (nucleus dependent)
- Fairly long data acquisition times (nucleus dependent)
- No separation involved
- Difficult quantitation (nucleus dependent)
- Expensive equipment
- Laboratory-based technique
- Need for skilled operator

practical spectral analysis is unfortunately prone to error and an internal standard is useful. The **main features** of s-NMR are shown in Table 1.31.

High-resolution s-NMR has some obvious advantages over standard liquid-phase high-resolution FT-NMR: it provides qualitative and quantitative information about the less mobile constituents of a sample *in situ*, without lengthy sample preparation. s-NMR is highly sensitive to molecular mobilities, with respect to relaxation and line-broadening. The presence of anisotropic broadenings provides extra information about structure and dynamics. Therefore, by using techniques in which specific broadenings are retained and/or by using spin-labelled samples in which specific broadenings are selected, s-NMR can provide complementary information to l-NMR.

Each nucleus has its own (dis)advantages. In case of ^{13}C NMR ^{1}H-decoupled spectra are advantageous since there is only one line for each carbon. Moreover, inorganic components do not interfere if they do not contain carbon (*e.g.* glass, inorganic flame retardants, *etc.*). Several disadvantages may be noted: *(i)* the relatively low sensitivity of ^{13}C s-NMR (requiring typically *ca.* 200 mg sample and 10–16 h accumulation time for a 400 MHz NMR); *(ii)* polymeric matrix interference; *(iii)* pronounced differences in ^{13}C spin–lattice relaxation times; and *(iv)* difficult quantitation.

Most polymers of technical importance are heterogeneous in many respects: chemical structure (*e.g.* block copolymers), segregation of hard and soft segments (*e.g.* in polyurethanes), crystallinity, macrostructure (*e.g.* impact-modified polymers). s-NMR is most appropriate to characterise heterogeneous polymer systems and to correlate chemical structure and dynamics. For the characterisation of heterogeneous systems a wide range of NMR tools is available, ranging from high-resolution s-NMR with magic-angle spinning to low-resolution bench-top NMR [660]. Magic-angle spinning of non-solids will benefit all heterogeneous samples, such as polymers in suspension, gels, viscous liquids, *etc.* Information on distances involved, such as the size of the domains, may be obtained from the effects of spin diffusion, a transport of magnetisation through space without particle motion, which covers the range from 1 nm to 100 nm [660]. Distance information can also be obtained from NMR experiments which exploit the dipolar coupling between nuclear spins. Spin diffusion measurements have proved to be very effective to study **micromorphology** of blends (miscibility, phase separation, amorphous content).

For detecting microheterogeneities, one of the more generally applicable and molecularly specific techniques in the nm range is **relaxation measurements** using broadline NMR. Molecular miscibility is measured here by means of T_{1r} (^{1}H) NMR relaxation times. It is the strength of s-NMR that it is possible to view rigid and more mobile parts of the polymeric material separately. The family of so-called "solids NMR" techniques can probe molecular order and dynamics in a lattice, and are sensitive to the proximity between magnetically active nuclei. Typically, a variety of NMR methods may be used to characterise various aspects of copolymers, such as ^{13}C single pulse experiments (for crystallinity), ^{1}H and CP/MAS (for characterisation of the composition and molecular mobility in the crystalline domain), and spin-diffusion, T_1 and $T_{1\rho}$ measurements (for the determination of lamellas thickness) [661]. Molecular mixing of an additive with the matrix material may as well be distinguished from segregation into a separate phase.

Many solid-state NMR techniques enhance the power of this technology, such as:

Table 1.32. Properties of polymeric solids, studied by means of solid-state NMR[a]

Material	Properties studied using s-NMR
Organic polymers Inorganic polymers Copolymers/ polymer blends/ composites Amorphous polymers Polymer conductors Resins Fibres	Characterisation of amorphous, crystalline and reinforced phases; determination of insoluble polymers; melts; additives, miscibility of additives, polymer-additive interactions; grafting; dopants; morphology; domain sizes; interfacial regions; chain diffusion effects; homopolymer tacticity; copolymer sequence distribution; chain branching; network characterisation, cross-link density; heterogeneities; defect structures; structural changes due to oxidation, hydration, irradiation, pyrolysis, cross-linking and curing processes; dynamics of small molecules dissolved in a polymer matrix; polymer dynamics (spin relaxation); molecular motions; characterisation of dangling bonds; conductivity changes due to dopants, distribution of conducting electrons; phase transitions; determination of the order parameter

[a] After Wind [659]. Reprinted from *Encyclopedia of Analytical Science* (A. Townshend, ed.), R.A. Wind, pp. 3477–3485. Copyright (1995), with permission from Elsevier.

- *Multidimensional NMR:* 2D NMR experiments can be used for connectivity studies, can detect interactions between polymer chains and determine chain conformation and packing.
- *Solids micro-imaging:* This technique can be used to study voids and cracks in solids, diffusion processes, heterogeneous distribution and *in situ* localised dynamics such as chemical reactions. The main limitation is the spatial resolution, which is currently 30–100 μm, whereas for many applications a resolution of 1 μm or less is required (see Chp. 5.7.1).
- *Solid-state NMR as a process-control technique:* Relatively simple spectrometers, capable of a limited amount of experiments only, are gradually being introduced in industrial plants in order to control and optimise processes such as the production of polymeric materials, catalytic processes and combustion (see Chp. 7.2.6).

In the area of high-resolution s-NMR new developments and applications are mainly taking place in the field of multidimensional NMR spectroscopy (*e.g.* domain studies in polymers and polymer blends). Thin-layer chromatography combined with HRMAS s-NMR can be used for compound identification without the need for substance elution from the stationary phase [662].

A collection of ^{13}C CP/MAS NMR spectra of common polymers is available [663].

Solid-state NMR has been reviewed [659,664, 665] and several books have appeared [666–668]. Also solid-state NMR of polymers has been dealt with [669,670]; *cfr.* also Bibliography. Cross-polarisation has been reviewed [671,672].

Applications

Solid-state NMR is used to study both *structure* and *dynamics* in materials. Since NMR is a probe that is sensitive in dimensions where dipolar interactions are active, it can yield information about the near environment of a nucleus, and hence about miscibility of a polymer system on the molecular scale, provided however that the concentration of spins is high enough. Both high- and low-resolution s-NMR find applications in polymer analysis.

Table 1.32 emphasises the wide **scope of s-NMR** of polymers and gives examples of the structural and dynamical information that can be obtained. The ultimate use of most polymers is in the solid state, and it is therefore desirable to characterise the properties of this state, in particular the chemical microstructure, micromorphology and molecular-level dynamics. Hence, polymer chemists should have strong interest in s-NMR. However, access to s-NMR equipment seems to be diminishing.

In crystalline solids NMR is complementary to XRD for **structure determination** (even with remarkable results: $C{-}H_{aliph} = 0.95 \pm 0.01$ Å, $C{-}H_{arom} = 1.05 \pm 0.01$ Å). In non-crystalline solids NMR and x-ray absorption spectroscopy (XAS) are amongst the most important tools to investigate structure on a molecular level. It is advantageous to undertake comprehensive studies using both ^{13}C and ^{1}H nuclei, with measurements of both spectra and relaxation times. Techniques which are especially powerful for the analysis of cross-linked network polymers are s-NMR and FTIR spectroscopy. Infrared is often a strong competitor for high-resolution s-NMR. Like vibrational spectroscopy, CP/DD/MAS NMR is similarly rather insensitive to microstructural issues within the crystalline and amorphous states. Other materials which are often studied by s-NMR are melts, swollen gels, foams, emulsions or suspensions. Mineral fillers in powder or granulate generally do not disturb. Although **high-resolution ^{1}H s-NMR spectroscopy** is possible, most applications have focused on other nuclei such as ^{13}C. Grossman [673, 674] used both high-resolution ^{1}H and ^{13}C MAS NMR spectra to demonstrate that the lead-based **heat stabilisers** mono-, tri- and tetrabasic lead sulfate, dibasic lead phosphite, dibasic lead phthalate, tribasic lead maleate and tetrabasic lead fumarate, are unique compounds rather than double salts of lead oxide, such as $3PbO{\cdot}PbSO_4{\cdot}H_2O$, $2PbO{\cdot}Pb[C_6H_4(CO_2)_2]$, $2PbO{\cdot}PbHPO_3{\cdot}½H_2O$ and $2PbO{\cdot}Pb(C_{17}H_{35}CO_2)_2$. The crystal structure of tribasic lead sulfate, $3PbO{\cdot}PbSO_4{\cdot}H_2O$, the largest volume stabiliser worldwide for PVC, is more accurately designed as $4PbO{\cdot}H_2SO_4$, to emphasise that H_2O is not present in the structure [675].

Solid-state ^{13}C NMR has been widely employed for problems related to flame retardants, impact modifiers, plasticisers (and plasticiser motion), fillers (including polymer-filler interactions), co-polymers, grafting, elastomers and filled vulcanisates, molecular symmetry and heterogeneity, *etc.* Use of ^{13}C NMR is recommended particularly for insoluble components (such as high-MW species) at high levels (typically $>1\%$). Obviously, direct ^{13}C NMR of polymers suffers from matrix interference of the polymer carbon backbone yielding complex spectra. Therefore, studies on polyolefins and PVC are relatively favoured, whereas polyacrylates are unfavoured. ^{13}C (SPE and CP/MAS) NMR and *in situ* ^{1}H NMR were used in a study of PU/melamine [676].

^{1}H and ^{13}C s-NMR, in conjunction with DSC, DMA and x-ray scattering, have been used to study the solubilisation of various **flame retardants** in HIPS [677]. As many FRs (in particular Br-containing) do not dissolve in common NMR solvents such as $CDCl_3$ and tetrachloroethane, use of s-NMR is ideal. Moreover, FRs are often aromatic compounds which reduces matrix effects of polyolefins and PVC. Van der Velden *et al.* [678] have analysed the low-MW perbrominated FR decabromodiphenyl (Adine 0102, ATO®), the partially brominated FR 1,2-pentabromophenylethane (Saytex 8010, Albemarle®), 2,4,6-tribromophenyl terminated tetrabromobisphenol A-carbonate oligomer (BC-58, Great Lakes®), a tetrabromobisphenol A-based epoxy resin (F 2400, Great Lakes®) and the polymeric FR polypentabromobenzylacrylate (FR 1025, Ameribrom®) in PBT (containing 6–14% FR, 1% Teflon, 15–30% GF, 4–7% Sb_2O_3) by means of ^{13}C s-NMR. The resonances of the partially brominated FRs BC-58, FR 1025 and F 2400 are quite distinct from those of PBT and these additives can readily be identified in PBT via ^{13}C s-NMR techniques. In the ^{13}C SPE/MAS NMR spectrum the resonances of Adine 0102 coincide with those of the aromatic C atoms of PBT. The ^{13}C resonance position of the ethyl fragment of Saytex 8010 in the ^{13}C CP/MAS NMR spectrum (not interfering with PBT) is not highly specific and may coincide with the resonances of the main chain C atoms of impact modifiers and polymeric FRs (such as polypentabromobenzylalcohol and polystyrenes). This renders unambiguous identification of Saytex 8010 in PBT via ^{13}C s-NMR impossible. Advantages of ^{13}C s-NMR in the determination of FRs in polyester are: *(i)* no interference of inorganic components (such as glass fibres, wollastonite, Sb_2O_3, etc.); *(ii)* no disturbance of fluoro copolymers (used in PBT) in ^{13}C CP/MAS experiments; *(iii)* simultaneous generation of structural information on FR and polyester; *(iv)* no interference of impact modifiers; and *(v)* no sample preparation. Disadvantages are: *(i)* the relatively low sensitivity of ^{13}C NMR (requiring *ca.* 200 mg sample and 10–16 h measuring time on a 400 MHz instrument for PBT/10 wt.% FR); *(ii)* failure of the standard ^{13}C CP/MAS NMR technique for perbrominated or proton-poor FRs (for such FRs SPE NMR is needed, which requires very long pulse cycle times, up to 120 s); and *(iii)* difficult quantitation.

Hydroperoxides, which play a key role in the oxidative degradation of many polyolefins were studied by ^{13}C NMR and ESR in γ-irradiated ^{13}C-polyethylene [679]. It is possible to identify and quantify aromatic additives in PE directly by PE signal suppression [680], but ^{1}H l-NMR serves the purpose as well. Solid-state ^{13}C CP/MAS NMR was used to quantify starch in PE [681].

^{13}C MAS NMR is also an efficient technique for the direct identification of (insoluble) **impact modifiers** (IMs), such as polar LDPE co- and terpolymers (*e.g.* ethylene acrylates), "all acrylic" core–shell rubbers (*e.g.* PBA core/PMMA shell), MBS core–shell rubber (butadiene rubber core/S-MMA co-polymer shell). Sufficient sensitivity derives from high IM loadings (typically 20 wt.%). Van der Velden *et al.* [678,682] have used ^{13}C SPE NMR and ^{13}C CP/MAS NMR techniques for the study of three different types of IM in toughened PBT, namely E-MA-GMA Lotader AX 8900 (ATO®), the PB/PMMA core shell product Palaroid EXL 3361 (Röhm & Haas®) and the PB-SMMA core shell product Kane Ace M 511P (Kaneka®). In general, the ^{13}C NMR resonances of the rubbery-like materials can be clearly visualised by using SPE with short cycle delays. For the core–shell products additional CP experiments had to be performed for identification of the rigid shell. With these techniques the type of acrylate monomer (MA, EA, BA) present in ATO's Lotader series could be identified. In addition, small amounts of glycidyl-methacrylate (GMA) (0.3–1 wt.%) were detected. Styrene blocks in the M/S core shell products in the PBT compound could not be detected. Also identification of IMs in nylons (*e.g.* Zytel ST 801®) is a rather complex matter [683]. For the analysis of complex impact modifiers s-NMR is usually part of a multidisciplinary approach (^{13}C NMR, IR, Raman, PyGC-MS, 2D FTIR); quantitation often requires ^{1}H l-NMR and PyGC-MS.

NMR is a powerful tool to investigate molecular structure and motion and to obtain information about the range of certain interactions. Modern s-NMR techniques allow to analyse the effects of polymer-polymer (*i.e.* PP-EPDM) and polymer-filler interactions and to detect cause for the properties of a composite. **Polymer-filler interactions** may result in formation of an interphase connecting two incompatible polymer phases or a polymer and a filler phase. Lipatov's model [684] consists of the rigid filler particle encapsulated by the layer of the interphase. This structure is embedded in the bulk polymer. However, it is difficult to obtain information about the properties of these interphases: the layer-thickness of such phases on a filler does not exceed a few nanometers. The study of filled polymers is in development due to improvements in the methods of analysis [685,686]. Legrand [686] has discussed the application of magnetic resonance spectroscopies to the characterisation of elastomer/filler interface systems, in particular the dynamic behaviour of a polymer in the vicinity of the filler. Veeman *et al.* [687] used ^{13}C CP/MAS NMR in the study of polymer-filler interactions using ternary systems consisting of PP, EPDM and different types of inorganic fillers (kaolin, $BaSO_4$, lithopone, ZnS). Kaolin is a filler with strong interactions, $BaSO_4$ and ZnS show weak interactions, with lithopone occupying an intermediate position. The molecular details of **polymer-surfactant interaction** have also been investigated, using a large family of modern pulsed NMR techniques [688].

Solid-state NMR is widely used for the characterisation of **elastomers** and rubber compounds. Kelm [689] has published a catalogue of interpreted high-resolution carbon- and proton NMR spectra for the determination of (filled) elastomers, blends and thermoplastic elastomers. ^{13}C CP/MAS NMR and SPE MAS have been used for the compositional study of a series of E-VA, E-GMA, E-VA-GMA co- and terpolymers [690]. High-resolution s-NMR is also a powerful technique for studying the morphology and microphase structure of block co-poly(ether esters), such as those consisting of poly(tetramethylene oxide) (PTMO) "soft" segments and poly(butyleneterephthalate) (PBT) "hard" segments [691]. Quantitative ^{13}C MAS spectra were used to estimate the soft component fraction. Van der Velden *et al.* [683] have examined various fractions of the heterogeneous polymer system (EPDM-*g*-MA)-*g*-PA6.6 using high-resolution ^{13}C s-NMR, including single-pulse excitation (SPE or Block decay), and IR techniques. ^{13}C s-NMR has also been used for elucidation of other graft structures [692] and indeed could be a useful tool for characterisation of **additives grafted** on polymers (Pol-*g*-Add). s-NMR is useful to follow the fate of **accelerators** and stabilisers in rubber vulcanisates [693]. Both CIMS and s-NMR are ideal tools for studying the accelerator breakdown process during rubber vulcanisation. In case of the sulfenamide accelerator *N*-cyclohexyl-2-benzothiazole sulfenamide (CBS) it is supposed that mercaptobenzothiazole (MBT) and cyclohexylamine moieties are formed. In order to confirm that the majority of the amine remains polymer-bound in the cured rubber polyisoprene/^{15}N labelled CBS (labelling in the cyclohexylamine moiety) ^{15}N s-NMR was used [693]. Similarly NR/^{13}C labelled IPPD was studied by ^{13}C NMR. Shortly after heat ageing the degree of polymerisation is low allowing migration and extraction of the antidegradant. After a few months of storage at r.t. antidegradant reaction products become non-extractable.

Solution-state NMR and solid ^{13}C NMR are frequently used for the characterisation of the elastomeric components of **filled vulcanisates** [694].

Fillers can influence the linewidth of the spectra by introduction of microscopic inhomogeneities by susceptibility variations and by reduction of the molecular mobility of the polymer chains. However, both effects can be averaged out by magic-angle spinning and high power decoupling. Consequently, fillers do not have a detrimental effect on the resolution of the solid-state elastomer spectra. Therefore, no decomposition procedures are necessary prior to NMR investigation of rubbers. Van der Velden *et al.* [695] have studied a complex vulcanised di-blend (SBR/EPDM) *via* s-NMR methods mainly to quantify the various microstructures present. The spectra show relatively broad lines, with typical line-widths of 100–500 Hz, which are roughly 10 to 100 times larger than in the liquid state. At variance to thermoplastics and thermosets, the NMR spectra of elastomers have a much better resolution if the measurements are performed well above T_g (SBR, $-100°C$). Above T_g segmental motions, comparable to the liquid phase, average out line-broadening effects. In comparison with other methods like IR and PyGC-IR, ^{13}C s-NMR appears to be advantageous for structure analysis and/or identification. Komoroski [696] has used ^{13}C MAS NMR in the study of filled *cis*-BR/SBR and NR/*cis*-BR/SBR vulcanisates as an alternative to IR spectroscopy or l-NMR for the characterisation of the elastomeric components of filled vulcanisates. ^{13}C MAS NMR spectra are of sufficient quality for polymer identification in simple filled vulcanisates [694]. The MAS spectra are usable for *direct* **quantitative analysis** of the polymeric components without prior sample work-up. This has been demonstrated also for simple di-blends and tri-blends [696]. The accuracy of the method is comparable to IR. The method does not need to rely on calibration curves derived from standard blends. However, as demonstrated for NR/BR/SBR analysis, a standard curve can be used for part of the analysis with improvement in accuracy.

In the presence of large amounts of carbon-black in technical rubber goods, ^{13}C s-NMR and ^{1}H and ^{13}C s-NMR relaxation-time experiments are often better analytical tools than either IR or Raman spectroscopy. However, the sensitivity of ^{13}C s-NMR is not as high as that of IR and Raman spectroscopy. For instance, ^{13}C s-NMR of sulfur-vulcanised EPDM could only be performed when the ENB unsaturation of EPDM was fully isotopically enriched [697].

In two situations in particular ^{13}C MAS NMR has a strong edge over IR or ^{13}C NMR with solubilisation [696]. The first involves highly cured samples or samples where solubilisation of the elastomer component is difficult or impossible. For example, peroxide cured rubber is difficult to devulcanise using ODCB. Here, ^{13}C MAS NMR with or without CP, as appropriate, provides spectra of equal quality as for samples cured to a lesser degree. The problems of incomplete or selective solubilisation of elastomeric components can be avoided. ^{13}C MAS NMR may be the method of choice for peroxide-cured rubber. The second application for which ^{13}C MAS NMR is well suited is the aforementioned analysis of relatively small amounts of NR or synthetic *cis*-polyisoprene in filled vulcanisates [696].

Barendswaard *et al.* [698] analysed various polymer stabilisers (Irganox 1010/B225 and Irgafos 168) by means of ^{13}C CP/MAS NMR to gain information on **molecular symmetry**. Equivalent molecular positions in solution can lead to several signals in the solid state when molecules are situated at non-equivalent positions within a crystal or if the symmetry of the lattice is less than that of the molecule. NMR spectra of crystalline stabilisers show a strong influence of the crystalline surroundings on resonance positions. Whereas Irganox 1010 exhibits different crystalline modifications, CP/MAS NMR experiments suggest molecules devoid of any symmetry once embedded in LLDPE. Barendswaard *et al.* [698] also used $T_{1\rho}$ (^{1}H) relaxation time measurements of stabiliser and polymer matrix to detect the **molecular heterogeneity/homogeneity** of the low-MW additive Zn/Ca stearate in a cast PVC film in the nm range with ^{13}C detection via cross-polarisation. The scale of heterogeneity of the stearate in the PVC film is larger than about 50 nm. The bulk of the stearate is clearly not molecularly distributed in the PVC matrix.

PA6/montmorillonite clay nanocomposites were characterised by ^{13}C l-NMR and ^{15}N CP/MAS NMR spectroscopy [699]. Indications from the latter technique are that the nanocomposite thermal history dictates the ratio and type of crystallites formed. ^{15}N CP/MAS NMR has also been used to follow ^{15}N-labelled HALS in automotive painting [32].

Determination of minor inorganic components in polymers such as polyesters is industrially relevant. It would appear that identification of Na-containing additives in a polymeric matrix by means of **^{23}Na s-NMR** is not a trivial matter in the presence of other sodium sources, such as glass fibres, $NaSbO_3$, metal salts, a sodium-containing PE ionomer (Surlyn), *etc.* The flame retardant $NaSbO_3$, dispersed in a polymeric solid, can fairly easily be identified on the basis of the ^{23}Na chemical shift (11.0 ppm), which differs significantly from other sodium sources such as Pyrochek AM-595 (1.0 and −6.4 ppm), Na-stearate (−8.4 ppm) or a glass fibre reinforced (Na-containing) polymer (10.5 ppm). The ^{23}Na NMR pattern of Na_2HPO_4 is complex and therefore highly specific for this compound, allowing easy identification [700].

Silica and silanes can be examined through the ^{29}Si nucleus. **^{29}Si s-NMR** has been used to study the deposition of amine functional silanes, such as isocyanurate silane and ureidosilane, onto E-glass fibres [701]. Derouet *et al.* [702] used ^{13}C and ^{29}Si CP/MAS NMR for the characterisation of alkenyltrialkoxysilane and trialkoxysilyl terminated polyisoprene grafting onto silica micro-particles. CP/MAS NMR spectroscopy is also a useful technique to detect and identify polymeric structures chemically grafted onto a silica surface. Polymer-grafted silica gels are used for rubber reinforcement, as a stationary phase in chromatography, *etc.* ^{13}C and ^{29}Si CP/MAS NMR and proton spin–lattice relaxation time measurements were used to study polycarbonate oligomer grafting onto the surface of amorphous silica [703]. Various Si environments in the interfacial region of glass-filled PA6/γAPS were identified using ^{29}Si CP/MAS spectra [703a]. On the whole, it appears that there is limited scope for ^{29}Si s-NMR studies of additives in polymers. Apart from antiblocking agents (100 ppm level), which cannot easily be detected, Si-containing fillers (glass-fibre, mica, wollastonite, *etc.*) are usually determined by other techniques (*e.g.* IR, XRD, *etc.*).

Although it would be interesting to study **^{33}S s-NMR** for rubber vulcanisates, this nucleus has such low abundance and sensitivity that it is now not possible. On the other hand, **^{31}P s-NMR** is of more interest because of the sensitivity of the nucleus and lack of polymeric matrix interference; the spectra can usually be acquired in a relatively short time. The main applications in polymer/additive deformulation are found in the analysis of phosphorous containing additives such as secondary antioxidants (*e.g.* Irgafos 168 and Sandostab P-EPQ), flame retardants and transesterification suppressants, as well as in quantitative determinations. ^{31}P s-NMR is an efficient tool for the structural analysis of insoluble polyphosphates and melamine phosphates. The effect of the intumescent FR melamine pyrophosphate on the thermal degradation of PA6 and PA6.6 was studied by means of ^{31}P NMR, ^{13}C NMR and XRD [704]. ^{31}P MAS NMR has unravelled the phosphorous-based chemistry associated with the **phosphite stabiliser** Ultranox 626 or bis(2,4-di-*t*-butylphenyl) pentaerythritol diphosphite (BTBP) in polymer blends during extrusion at 280–300°C [705]. ^{31}P CP/MAS NMR and l-NMR experiments were used to identify BTBP in polycarbonate; BTBP undergoes a complex process of hydrolysis leading to various new phosphorous species [706]. It was also demonstrated by ^{31}P CP/MAS NMR that conversion of the phosphite group of BTBP to a phosphonate moiety is a prerequisite for effective inhibition of transesterification in PC/PET/PAR blends [707,708]. Klender [709] has reported extensive ^{31}P NMR work on fluorophosphonites as co-stabilisers in stabilisation of polyolefins. Sultany [710] has determined the miscibility of phosphorous additives (Ultranox 626 and Irgafos 168) in **masterbatches** in LLDPE by high-resolution s-NMR using both chemical shifts and relaxation studies. In case of extensive intermixing of two components at the nm level the proton $T_{1\rho}$ values (proton decay rate constants in the rotating frame) of two blended materials are averaged to a single value by spin diffusion. Thus if two materials are highly miscible, they will both have similar $T_{1\rho}$ values in a blend as measured by ^{1}H s-NMR relaxation studies. With high-resolution s-NMR using cross-polarisation techniques, the proton $T_{1\rho}$ decays can be monitored indirectly through other nuclei (*e.g.* ^{31}P or ^{13}C) in the vicinity of the protons. In 5% masterbatches, the observed proton $T_{1\rho}$ value for Ultranox 626 has become quite close to that of LLDPE reflecting good compatibility with the polymer, quite opposite to Irgafos 168, which shows a large difference in chemical shift between solid (−154 ppm; reference $CaH_4(PO_4)_2{\cdot}H_2O$) and solution (−131 ppm; reference 85% H_3PO_4) ^{31}P NMR. Results from both chemical shifts and relaxation studies indicate a difference in miscibility of Ultranox 626 and Irgafos 168 in 5% masterbatches in LLDPE with Ultranox 626 forming a homogeneous dispersion and Irgafos 168 segregating into domains of pure and dissolved Irgafos 168. The results are indicative that a 5% loading exceeds the solubility of Irgafos 168 in LLDPE. This method shows promise in examining the relative dispersion of phosphorous containing additives in polymer matrices.

Multinuclear (^{13}C, ^{23}Na and ^{31}P) s-NMR of FR Pyrochek AM-595 shows very specific ^{23}Na or ^{31}P NMR signals for this 3:1 mixture of Na_2HPO_4 and barium-alkylphosphate. However, as the complex ^{23}Na pattern for the Na_2HPO_4 part of the Pyrochek mixture overlaps severely with Na-signals in glass fibres, Pyrochek AM-595 is difficult to detect in a neat GFR polymer sample (*e.g.* PCT) using s-NMR. ^{13}C single pulse NMR indicates the presence of several branched alkyl residues in Ba (alkyl)phosphates. Bourbigot *et al.* [711–713] studied the synergetic action of zinc borates, $4ZnO{\cdot}B_2O_3{\cdot}H_2O$ (Firebrake 415) and $2ZnO{\cdot}3B_2O_5{\cdot}3{\cdot}5H_2O$ (Firebrake ZB), with metal hydroxides ($Mg(OH)_2$ and $Al(OH)_3$) in EVA-copolymers by means of **multinuclear ^{11}B, ^{13}C, ^{25}Mg, ^{27}Al NMR** to characterise samples after compounding and to show polymer/filler interactions. Measurement by ^{13}C CP/DD/MAS NMR of the spin lattice times indicated structural modifications of the polymeric matrix suggesting that $4ZnO{\cdot}B_2O_3{\cdot}H_2O$ shows poor compatibility with the polymeric matrix. ^{11}B, ^{25}Mg and ^{27}Al s-NMR were used to determine the modifications of the fillers. Bourbigot *et al.* [714] also examined FR EVA-based materials containing a PA6 (exfoliated montmorillonite) clay nanocomposite hybrid (PA6-nano) as a charring agent. ^{13}C CP/DD/MAS NMR, ^{31}P DD/MAS NMR and ^{27}Al MAS NMR were used to characterise ammonium polyphosphate, $(NH_4PO_3)_n$ (APP), EVA/PA6 formulations. The ^{27}Al MAS NMR spectra showed interaction of the clay with APP to form aluminaphosphates above 310°C; at higher temperatures the aluminaphosphate structure collapses. Multinuclear (^{1}H, ^{13}C, ^{29}Si) s-NMR was used to determine that only the PEO segments of PS-*b*-PEO copolymers are intercalated in the silicate galleries of hectorite nanocomposites [715].

In conclusion, the main applications of s-NMR concern the ^{13}C and ^{31}P nuclei. However, publications in the open literature are scarce and prospects are obscure.

1.5.1.1. *Dynamics in Solids*

Principles and Characteristics

The largest areas of interest for NMR in polymer science are structural and molecular dynamics studies. High-resolution ^{13}C NMR is a most powerful tool for investigating **local dynamics** in polymers. Unlike other methods, such as ESR or fluorescence anisotropy, it does not require any labelling and yields direct information on the compound under study. What is specific to ^{13}C NMR is high selectivity allied with the natural abundance of ^{13}C nuclei. As a selective technique, ^{13}C s-NMR allows the observation of one signal per magnetically inequivalent carbon, and therefore the dynamic behaviour of each part of a molecule can be followed independently. Moreover, many NMR parameters are sensitive to molecular motions. These include the relaxation times and line widths, strength of ^{1}H–^{13}C dipolar interactions and chemical shift anisotropies (2D NMR techniques). These parameters differ in the information they carry. The available spectral windows depend on the type of measurement and range from 10^{-1} Hz for slow processes to several hundreds of MHz for very fast modes. For bulk polymers at $T > T_g$, the fast processes of the local dynamics can be investigated by determining the spin–lattice relaxation time, T_1 (^{13}C), and the nuclear Overhauser enhancement. Line shape analysis and measurements of the tensorial interactions, line widths and $T_{1\rho}$ (^{13}C) relaxation times are more appropriate for probing slower motions in glassy state investigations. The spectrum line shape is strongly dependent on the rate of motion in the range of 10^{-1} to 10^6 Hz. Lauprêtre [716] has considered the sensitivity of the different NMR parameters to molecular motions.

Diffusivity is no longer a phenomenological coefficient and very firm validation from molecular theories now exists for Fick's law. Molecular dynamics (MD) simulation has contributed significantly to the understanding of liquid and solid behaviour, in particular as to diffusion in rubbery polymers. Most of the atomistic MD simulation work has focused on chemically simple penetrants (He, H_2, O_2, N_2, CH_4) and polymers (PE, PP, PIB) in systems that, macroscopically, exhibit Fickian behaviour. **Small-molecule mobility** in macromolecular materials dictates physical and chemical characteristics of the polymer produced. The investigation of diffusion phenomena is an important topic in both fundamental research and industrial application. Real-life applications often have to do with large, complex, or strongly interacting solvent or plasticiser molecules, whose thermodynamic and transport behaviour have not been investigated sufficiently with molecular modelling. Polymer processing operations affected by molecular transport include devolatilisation, mixing of plasticisers or other additives, and formation of films, coatings and foams. Distinctive molecular diffusion behaviour is essential for miscellaneous polymer products such as barrier materials, controlled drug delivery systems, and membranes for separation processes. The fundamental physical property required to design and optimise processing operations is the mutual **diffusion coefficient**, D (typically widely ranging from 10^{-16} to 10^{-5} cm^2/s). In addition to temperature and composition, diffusion in polymers is controlled by morphological features such as crystallinity and cross-linking, both of which tend to reduce molecular mobility [717].

While polymer scientists have many excellent tools at their disposal with which to study polymeric materials at both the micro- and macrostructural levels, the choice is more restricted when it comes to analysing *dynamic* structural changes. Studies of molecular mobility cover a wide range of techniques, depending on the characteristic **time scale** of the motion. The time scale accessible by NMR is limited on one end by the fast and unrestricted segmental motion and at the other end by the spin–lattice relaxation time. Thus molecular dynamics can be investigated within a range of 10^{-12} s to some 100 s. High-resolution NMR is often used for studying fast molecular motion, and wide-line NMR for slow molecular motion. Wide-line spectra can provide detailed information about type and time scale of reorientational processes. NMR cannot yet sense molecular translation on a molecular distance scale, but on a larger scale in the range of 0.1 μm up to about 10 μm by measuring the particle diffusion in magnetic field gradients. Magnetic resonance imaging (MRI) plays a much more modest role in comparison to areas such as food science.

Numerous **non-NMR methods** exist for measuring diffusion such as light and neutron scattering, forced Rayleigh scattering, fluorescence and centrifuge methods, sorption, permeation and radioactive tracing, but they are generally of limited application (*e.g.* concentration range) or are invasive in nature. NMR has gained a most decisive role for diffusion studies with fluids, in particular through the application of the NMR pulse field gradient technique. NMR is valuable because of its non-invasive nature; no optical labelling of the probe species is required. With this technique a direct measure of the **self-diffusion coefficient** of the penetrant is achieved by observing the molecules microscopically, while other methods (*e.g.* sorption) indirectly determine the self-diffusion coefficient from macroscopic measurements. By using NMR techniques diffusion may be studied in the absence of a concentration gradient. The strong concentration dependence of the diffusion coefficient in polymers presents difficulties for experimental diffusion studies. While structural NMR studies often have to compete with powerful scattering techniques, multidimensional exchange NMR in solids is without rival in providing details about polymer dynamics on a molecular level. NMR can be used to measure molecular motion in aggregates of polymer molecules such as solutions, melts, and entangled or cross-linked networks. As most polymers of technical importance are heterogeneous it is not surprising that molecular dynamics is also heterogeneous.

NMR techniques for measuring translational diffusion can be separated into two classes: *(i)* relaxation-based; and *(ii)* gradient-based. Because the NMR signal is observed only after the nuclear magnetisation has been perturbed from its equilibrium state, relaxation is a standard feature of all NMR experiments. NMR **relaxation measurements** provide a powerful tool for investigating molecular dynamics. Two primary relaxation processes are usually identificable: spin–lattice relaxation times T_1 or $T_{1\rho}$ and spin–spin relaxation times T_2. In solids T_1 ranges typically from 10^{-3} to 10^3 s, and T_2 from 10^{-4} to 10^{-2} s. Therefore, measurements of relaxation times are indirect probes of the dynamics in the solid. Proton T_1 is a parameter associated with high frequencies while proton $T_{1\rho}$ is attributed to low frequencies. In order words, the response obtained from T_1 and $T_{1\rho}$ from protons is related to distinct regions of molecular mobility. The relaxation method necessarily reports on motions that occur on an extremely short time scale. NMR (by means of relaxation times) determines molecular dynamics or mobility of a component in the amorphous fraction of a polymer. Phases with different motional characteristics can be easily differentiated using NMR techniques. Rigid solids tend to have long spin lattice relaxation times and very broad lines, as large as 40 kHz. They also cross polarise very effectively, due to the static dipolar interactions. Rubbery solids, on the other hand, possess much shorter spin lattice relaxation times, narrower lines and do not cross polarise well. Since relaxation times are related to mobility, temperature and phase strongly influence the observed values [669]. An easier qualitative assessment of dynamics can often be obtained from resonance line shapes. Relaxation times and line shapes characterise molecular mobility in various phases (0.5–500 nm). More detailed information on dynamics is available from so-called exchange experiments. By relaxation measurements, line-shape studies, and 2D exchange experiments, correlation times between 10^{-10}–10^{-2}, 10^{-5}–10^{-1}, and 10^{-3}–10^{2} seconds can be determined, respectively.

Linear magnetic field gradients can also be used for the detection of transport phenomena such as diffusion and flow. The traditional and most widespread NMR method for measuring diffusion is based on the Hahn **spin-echo experiment** [719] in such a field gradient (FGSE). Originally the concepts and experiments were developed and performed in static magnetic field gradients (hence the notation SGSE-static gradient spin-echo). Because field-gradient spin-echo measurements of D depend on no driving force such as a concentration, temperature, or velocity gradient, *etc.*, they reflect Brownian motion of the molecules and are usually referred to as self-diffusion. In field-gradient spin-echo (FGSE) methods of measuring self-diffusion, a set of measurements of the magnitude of the spin-echo as a function of the magnitude and duration of the calibrated field gradient yields the diffusion coefficient D of the species at resonance. The only severe limitation of the method is the relatively modest lower limit for the measurable diffusivity; no more than another order of magnitude (to $D \geq 10^{-11}$ cm^2 s^{-1}) can be reasonably expected to be gained in optimal cases through the use of pulse sequences which elicit spin echoes at long diffusion times. In polymers, the FGSE methods of measuring self-diffusion have been useful in three more or less distinct areas, the diffusion of polymers in the melt, in concentrated, dilute and semidilute solutions, and the diffusion of penetrants and diluents in polymer hosts.

The **pulsed gradient spin-echo** (PGSE) was suggested in 1965 [720]. PGSEs are now the overwhelmingly dominant modes for measuring self-diffusion by NMR [721]. SGSE and PGSE diffusion measurements require a pulsed NMR spectrometer with a provision for creating a uniform calibrated magnetic field gradient in the region of the sample. Using the pulsed field gradient or Stejskal and Tanner (S-T) sequence, consisting of a modified Hahn spin-echo sequence, two equal rectangular pulsed gradients of strength g and duration δ are applied into each τ period a time Δ apart (*cfr.* Fig. 1.34), from which the self-diffusivity of mobile species within a material may be obtained [722–724]. Translation diffusion in the phase evolution time interval Δ between the gradient pulses results in attenuation of the spin-echo, as given by the S-T factor $\exp[Dq^2(\Delta - \delta/3)]$; $q = \gamma g\delta$, where γ is the gyromagnetic ratio, g and δ are the gradient pulse and duration, respectively, q is the area of the gradient pulse, and D is a self-diffusion coefficient. The spin-echo is attenuated not only by diffusion but also by relaxation. There are many sequences other than the S-T sequence (*cfr.* ref. [725]). ^{1}H NMR is generally used for diffusion measurements in polymers since protons tend to be abundant and offer large NMR signal strength. The main advantage of the spin-echo method for measurement of the diffusion coefficient of small molecules in a semicrystalline polymer is its independence of large-scale morphological features. Present-day PGSE instrumentation is often capable of producing high-resolution FT-PGSE spectra at maximum gradient settings of 100–1000 Gauss cm^{-1}. It has recently become popular to present results in a 2D manner, with spectral information on the x and z axes (frequency and intensity, respectively) and diffusion information on the y axis. Of particular interest in practical polymer work are cases where several substances diffuse simultaneously, or where diffusion is anisotropic or inhomogeneous, as in partially crystalline or filled rubbery polymers. For such cases PGSE measurements offer their greatest advantages. Some variants of the original (static gradient) spin-echo experiments are useful in cases of very slow diffusion (*e.g.* stray field spin-echo or STRAFI).

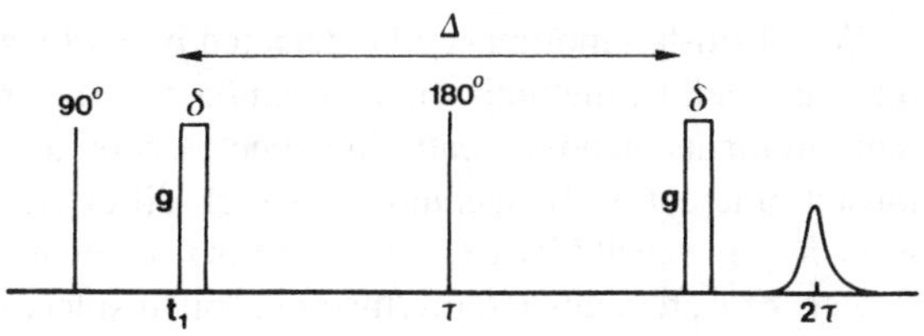

Fig. 1.34. Basic Stejskal–Tanner pulsed gradient spin-echo (PGSE) pulse sequence $\pi/2$–g(δ)–π–g(δ)–echo used for displacement spectroscopy. The echo time TE is 2τ and the displacement time is Δ. After Hills [718]. Reprinted from B. Hills, *Magnetic Resonance Imaging in Food Science*, John Wiley & Sons, Inc., New York, NY. Copyright © (1998, John Wiley & Sons, Inc.). This material is used by permission of John Wiley & Sons, Inc.

Self-diffusion motion can be detected by various nuclear labelling methods, such as radioactive tracer measurements, neutron scattering spectroscopy and pulsed gradient NMR techniques, which differ significantly in sensitivity to molecular displacements. Tracer measurements require macroscopic displacements on the mm scale and are applicable only to rapidly diffusing molecules, Neutron scattering is sensitive to nuclear position correlations over a few Ångstroms. Pulsed gradient NMR bridges the gap between the macroscopic and microscopic domains and detects molecular self-displacements in excess of a few hundred Ångstroms.

Measurement of diffusion using **pulsed field gradient NMR** (PFG-NMR) is a powerful analytical tool because it combines the high specificity and information content of NMR spectroscopy with the size selectivity of diffusion coefficients. PFG-NMR employs timescales of tens of ms and has a displacement sensitivity of the order of 100 nm. PFG-NMR can determine molecular self-diffusion coefficients in liquid phases down to a lower limit of 10^{-14} $m^2 s^{-1}$. Due to the combination of experimental convenience and straightforward interpretation, PFG-NMR has become the method of choice for studying translational diffusion. PFG-NMR experiments have been reported using ^1H, ^2H, ^7Li, ^{13}C, ^{19}F and other nuclei. The time Δ over which PFG-NMR measurements are possible is limited.

An advantage of PFG-NMR is that it can be employed to simplify complex NMR spectra. Pulsed field gradients find application in numerous 1D and multidimensional NMR techniques as a means of selecting those signals deemed interesting and suppressing those which are not. The simplification is achieved by attenuation of resonances based on the differential diffusion properties of components present in the mixture. One of the more obvious and useful applications of this approach is the use of PFG-NMR for suppression of the solvent resonance in the ^1H NMR spectra of solutions. PFG-NMR is also a useful tool for the spectral analysis of mixtures of polymer additives with different diffusion coefficients [726]. Diffusion provides a criterion by which to separate mixtures of species according to size and shape. Diffusion-ordered spectroscopy (DOSY) is one of the elaborate methods for separating complex mixtures, *cfr.* Section 5.4.1.1 of ref. [1]. Other NMR applications of gradients include NMR imaging and microscopy.

Diffusion studied by NMR was recently reviewed [727]. Various reviews deal with gradient-based NMR diffusion measurements [722–725,728]. The literature on diffusion is vast and highly mathematical [729–731].

Applications

Diffusion of small molecules in rubbers is of both theoretical and practical importance. Self-diffusion of small molecules must be understood in relation to applications of rubbers as seals in contact with solvents, and for diffusion of plasticisers and other small molecules. NMR studies provide a first insight into the interactions on the molecular scale by observation of molecular mobilities. Examples of dynamic processes which can be investigated using NMR are overall and local molecular motions and kinetics of processes, such as chemical exchange phenomena and chemical reactions.

Pulsed field gradient NMR (PFG-NMR) has been used to analyse mixtures of polymer additives and simple polymer solutions. PFG-NMR experiments were utilised to determine diffusion coefficients of the individual components of a mixture and in this way facilitate resonance assignments [726]. PFG-NMR was used to edit the NMR spectra of polymer solutions by eliminating the resonances of fast-diffusing components, such as low-MW additives or residual solvent. PFG-NMR is ideal to study anomalous diffusion (time-dependent diffusion coefficient, as in semi-crystalline polymers), when at least the diffusing molecule can be identified by NMR (*e.g.* Xe). A number of field-gradient spin-echo investigations has reported on transport and migration of molecules dissolved in polymers near and above T_g (Table 1.33). PGSE-NMR is well established in self-diffusion studies of surfactant solutions and polymer-surfactant interactions [732]. Fleischer [733] measured the diffusion of each component in benzene-cyclohexane and benzene–toluene mixtures in LDPE with deuterated and protonated diffusants.

Film formation of latexes can be followed by s-NMR experiments. Three different kinds of **water** were found in poly(butylacrylate)/polystyrene/poly(acrylic acid) latex films: free water, mobile water bound to the polymer and immobilised water inside the polymer [750]. The effects of water and DEP plasticiser on the molecular motion of cellulose acetate (CA) have been examined by ^1H, ^{13}C and CP/MAS NMR [751]. ^{13}C l-NMR relaxation

Table 1.33. Field gradient spin-echo NMR diffusion measurements

Method[a]	Nucleus	Polymer	Diffusant(s)	Reference(s)
SGSE	^1H	PIB	Cyclohexane	[734]
SGSE	^1H	Cross-linked rubber	Benzene	[735]
PGSE	^1H	PEO, PDMS	Benzene, $CHCl_3$	[736]
PGSE	^1H	PIB	Benzene	[737]
PGSE	^1H	PVC, PS	DMP, DBP, DOP	[738,739]
PGSE	^1H	LDPE	Butane	[740]
PGSE	^{13}C	PS	*Trans*-decaline	[741]
PGSE	^{13}C	PS	CH_2Cl_2, cyclopentane	[742]
PGSE	^{19}F, ^1H	PBD	C_6F_6, *n*-dodecane, *n*-hexatricontane	[743]
PGSE	^{19}F, ^1H	PBD	1,3-diadamantane (DMA); DMA + C_6F_6	[744]
PGSE	^1H	PIB	Toluene	[745]
PGSE	^1H	PBD	Cyclohexane	[746]
PGSE	^1H	*Cis*-PIP	*n*-Paraffins (C_8–C_{36})	[747]
PGSE	^1H	PEs	*n*-Alkanes	[748]
PGSE	^1H	LDPE	Benzene-cyclohexane, benzene-toluene	[723]
PGSE	^1H	PIP	Benzene-cyclohexane	[749]

[a] SGSE, static gradient spin-echo; PGSE, pulsed gradient spin-echo.

and CP/MAS NMR measurements have also been used to compare the motional characteristics of di-*n*-hexyl adipate (DHA) in solution and in the solid state of a poly(vinylbutyral-*co*-vinyl alcohol) (PVB) matrix [752]. Plasticiser molecules would be expected to exhibit high levels of mobility even in the polymer matrix. The morphologies of plasticised polymers like PVB/DHA are complex but can nevertheless be evaluated with NMR techniques. s-NMR studies of plasticised polymers have revealed that these systems are not simple homogeneous blends but rather complex multiphased matrices with concentration gradients ranging from plasticiser pools to rigid polymer domains. The results indicate that the DHA molecules exist in separate liquid and solid type environments in the PVB/DHA matrix.

^{31}P line shapes have been used to study the motion of a phosphate ester in BPA-PC and in a blend of PS and PPO [753,754]. One-dimensional solid echo ^{31}P chemical shift anisotropy line shapes are an effective means of determining rate and amplitude of **ester motion**. ^{31}P Hahn echo spectra of 5 to 20 wt.% tris(2-ethylhexyl)phosphate (TEHP) in tetramethylpolycarbonate (TMBPA-PC) were the basis of a study of diluent dynamics [755].

Harris *et al.* [756] have studied thick PVC films plasticised with up to 180 pph DIBP and DEHP by solid-state ^1H and ^{13}C spectroscopies, T_1 and $T_{1\rho}$ relaxation times and ^{13}C CP/HPHD/MAS spectra. ^{13}C chemical shifts give information about any possible interaction between the PVC matrix and plasticiser molecules. The data were considered in terms of the domain structure of the samples at the microscopic level and of the role of the **plasticiser**. The very mobile plasticiser cross-polarises badly and gives intense peaks only at long contact times (Fig. 1.35). Below 50 phr plasticiser molecules are intimately involved with the PVC chains; at higher concentrations they agglomerate to form highly mobile domains. NMR measurements (^1H NMR relaxation times, T_1, $T_{1\rho}$ and T_2, high-resolution ^{13}C NMR) have equally given evidence that highly plasticised PVC (35 wt.% or 80 pph) has a rather homogeneous morphology involving a molecular level distribution of DIDP plasticiser molecules without any significant domains of plasticiser and only small domains of ordered PVC, which remain free of plasticiser [757]. ^{13}C CP/MAS NMR experiments of PVC/50 wt.% DOP at $T > 60°C$ have given evidence for a multiphase system: *(i)* a DOP rich PVC phase (relatively narrow PVC and narrow DOP resonances); *(ii)* a more rigid PVC/DOP phase (broad components of the PVC and DOP signals in the proton dimension); and *(iii)* a pure DOP phase (narrow DOP resonances, high mobility on the NMR timescale). Harris *et al.* [758] have also investigated the interactions between PVC and aliphatic ketones

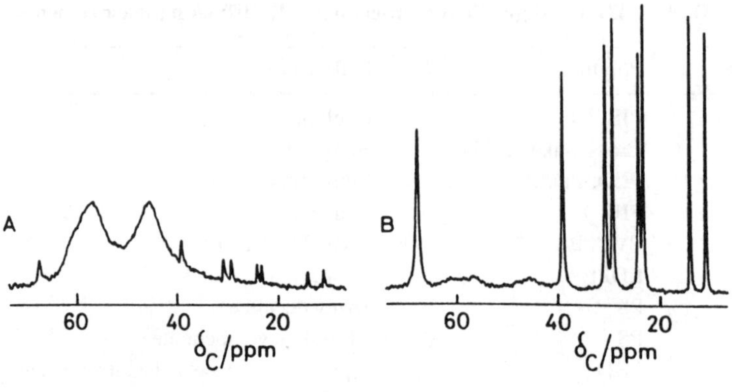

Fig. 1.35. Discrimination by contact time for PVC/180 pph DEHP. 300 MHz ^{13}C/HPHD/MAS spectra: A, contact time 200 μs; B, contact time 5 ms. The broad peaks arise from PVC and the sharp ones from the plasticiser. After Harris [669]. Reprinted from R.K. Harris, in *Polymer Spectroscopy* (A.H. Fawcett, ed.). Copyright © 1996 John Wiley & Sons, Ltd. Reproduced with permission.

by nuclear relaxation times such as ^{1}H and ^{13}C spin–lattice relaxation time (T_1) and proton-lattice relaxation time in the rotating frame ($T_{1\rho}$). Similarly, the influence of polyols as plasticisers on the starch molecular organisation was studied by s-NMR techniques (CP/MAS and HP/DEC) [759]. NMR data have indicated that polyol chains in flexible and water-blown flame retarded **polyurethane foams** retain significant mobility during thermal degradation [760].

EPDM is known to provide solution-like high-resolution s-NMR spectra, as a result of fast local motion occurring at temperatures of use much higher than T_g. Gelfer *et al.* [661] have described the morphology and molecular mobility in ethylene-hexene copolymers by s-NMR methods. ^{13}C MAS single pulse experiments were used to determine crystallinity; ^{1}H CP/MAS, T_1 and $T_{1\rho}$ data characterised the molecular mobility, whereas the crystalline-amorphous interface was investigated using a combination of ^{1}H spin-diffusion and relaxation measurements.

Smith *et al.* [761] have used ^{1}H MAS NMR (200 MHz) in the determination of the **phase partitioning** of 2,6-di-*t*-butyl-4-methylphenol (Ionol) between rigid PS and polybutadiene (PBD) rubber in HIPS/(9 wt.% PBD; Irganox 1076, ZnSt, Ionol). The NMR method to quantify partitioning is based on the fact that the rubber phase and molecules dissolved therein can be easily distinguished due to this phase's enhanced molecular motional characteristics. NMR is useful when the phases composing the blend have very different T_g values. Standard rubber-Ionol blends were used for calibration. The level of Ionol in the rubber phase was determined by ^{1}H s-NMR and the total amount in HIPS was derived from LC. Ionol was found to preferentially partition into the rubber phase with a partition coefficient of about 2. Similarly, Tinuvin P and Tinuvin 770 in SAN-EPDM (23 wt.%) were determined with ^{13}C s-NMR (75 MHz) at 110°C [761].

Multidimensional s-NMR spectroscopy has yielded ample molecular-scale information on reorientational and translational dynamics in semicrystalline and amorphous polymers, on their chemical and phase structure, and on orientational order. The dynamics and structure of amorphous polymers studied by multidimensional solid-state ^{13}C exchange NMR spectroscopy has been reviewed [762].

1.5.2. Nuclear Quadrupole Resonance

Principles and Characteristics

An interaction that is never directly seen in liquid spectra but that, if present, always dominates solid-state spectra is quadrupole interaction. Nuclei with $I > ½$ have an electric quadrupole moment Q that is a measure of the deviation of the nuclear charge distribution from spherical symmetry. Nuclei with $I = 0, ½$ do not care about electric field gradients: their charge distribution is spherical. Some 74% of all NMR-active nuclei have $I > ½$, as listed elsewhere [763]. The nuclear electric quadrupole moment, Q, of an $I \geq 1$ nucleus can interact with the electronic environment near that nucleus to affect the nuclear spin angular momentum energy levels, even in zero magnetic field. Quadrupole interactions can get quite large, and in most cases they will dominate the chemical shift spectrum. Magic-angle spinning can be used on quadrupole couplings as well as on the other interactions.

Nuclear quadrupole resonance (NQR) is concerned with the absence of magnetic induction ("zero field"); there is no magnetic interaction and unperturbed or "pure" resonance lines are observed. When the quadrupole interaction is dominant, the transition frequencies between the energy levels are largely determined by the electric field gradients at the nucleus. In an electric field of inhomogeneous charge distribution Q interacts with the electric field gradient to produce a set of orientation dependent energy levels. NQR involves coupling of radiofrequency radiation with a nuclear magnetic moment to bring about transitions between nuclear orientations of different energy. NQR is a powerful tool for studying the electronic structure and molecular dynamics of matter.

The **fundamental requirements** of NQR spectroscopy [764] are:

(i) a nucleus with a quadrupole moment ($I > ½$) in an asymmetrical environment;
(ii) solid-state effect only;
(iii) reasonably high natural abundance of the nuclear isotope of choice; and
(iv) sensitive RF detection with variable operating frequency.

With NQR the electric induction gradient is a molecular or solid-state property and is considerably larger than any practical externally applied field gradient. This implies that a variable-frequency detection system must be used. The NQR frequencies for the various nuclei vary from 100 kHz up to 1 GHz, making detection by a single spectrometer very difficult. Their values depend on quadrupole moments of the nucleus, the valence electrons state and the type of chemical bonds in which the studied atom participate.

NQR spectroscopy uses **instrumentation** and techniques similar to NMR spectroscopy to probe the electronic environment near a quadrupolar nucleus. However, in contrast to NMR, NQR can operate without a strong external DC magnetic field. There are various methods for NQR detection [764]. **Direct NQR detection** techniques are either continuous wave (CW) or pulsed methods. Pulsed techniques are most widely used and employ the latest signal processing methods, including fast Fourier transform and others. The essence of the pulse method approach consists of irradiating the spin-system by RF pulses with frequencies equal or close to the NQR transition frequency. This determines a variation in spin state. Relaxation from the excited state is accompanied by emission of photon energy, characteristic of the nucleus. Multipulse sequences, widely used in magnetic resonance, are also very common in NQR spectroscopy. They are effectively used for increasing sensitivity, reducing the duration of the experiment, and for measuring relaxation times in the sample. Sensitivity of the NQR spectrometer is important, as the intensity of NQR signals is very low. Besides, **indirect NQR detection** methods have also been developed, which are mainly used at low frequencies or in cases when the concentration of quadrupolar nuclei is not high. Indirect NQR detection permits high sensitivity for detecting many light elements.

The main **spectral parameters** in NQR experiments are the transition frequencies of the nucleus and the line width Δf. Pulsed NQR produces (nearly) single peak signals at specific frequencies that depend on the local structure around the observed atom and its chemical bonding, usually in a crystalline solid. Because the resonance frequency is almost unique to each compound, NQR exhibits great specificity for various analytes, notably (^{14}N containing) explosives and narcotics. The most useful elements to monitor by NQR are ^{14}N, ^{35}Cl and ^{37}Cl. Since the NQR frequency depends on the electric field gradient at the nucleus under study, NQR data provides valuable information about the electronic structure of the molecules in the solid state.

Pulsed NQR methods are very useful for structure determination [765,766]. When applied to **structural investigations**, NQR spectra may prove an effective tool for the preliminary study of crystal structure in the absence of detailed x-ray data. Differences between chemically non-equivalent atomic positions are readily revealed by NQR spectroscopy; splitting may be utilised to identify geometric isomers.

NQR is a well established spectroscopic method that has, however, a minor place in performing structural studies of polymeric materials. One of the major problems with NQR in the examination of polymers is that line widths are generally broad and that individual lines that can be assigned to separate structures are rarely observed. With pulse methods some of these disadvantages can be overcome [767].

Table 1.34 summarises the **main features** of NQR spectroscopy. The non-invasive nature of NQR

Table 1.34. Main characteristics of NQR spectroscopy

Advantages:
- Non-destructive, non-invasive
- Speed of measurements
- Compound specific
- High spectral resolution
- Local probe (structure determination)
- Phase identification and quantification
- Well-established bulk technique
- Mixture analysis

Disadvantages:
- Solids probe only
- Limited to $I > ½$ nuclei
- Low NQR signal intensities
- Sample size (2 g of polycrystalline material)
- Less flexible than NMR
- Lack of sufficiently sophisticated equipment

(closely connected with the absence of magnets) gives it some advantages over other methods. NQR nuclei of interest in polymer/additive analysis are ^{14}N, ^{35}Cl, ^{37}Cl, ^{79}Br, ^{81}Br, ^{121}Sb, ^{123}Sb. Because NQR is so compound-specific, other additives do not interfere with the signal for a target compound; consequently, NQR can be used for direct identification of additives in mixtures. Liquids and polymers are too disordered to give an NQR signal. NQR is not as extensively useful as NMR spectroscopy and inherently less flexible but when it works it is extremely attractive because of its specificity. NQR can work with slurries, aggregates and possibly even emulsions, as long as the molecular dynamics are slower than the NQR method time scale (MHz range).

NQR was repeatedly reviewed [764,768–772] and was also the topic of several books [773,774].

Applications
The main uses of NQR are: *(i)* information about chemical bonding in the solid state; *(ii)* molecular structure information; *(iii)* characterisation of molecular or ionic species (fingerprinting); *(iv)* crystallographic and molecular symmetry information; *(v)* solid-state molecular motion studies; *(vi)* phase transitions; and *(vii)* studies of impurities. The reason for the relatively limited practical application of NQR seems to lie in the scarcity of sufficiently sophisticated equipment.

Brame [767] has used ^{35}Cl NQR for the study of polychloroprene (Neoprene W) **rubbers** at different states of cure (ordered and disordered fraction). ^{35}Cl NQR can be used for product quality control verifying the microstructure of different rubbers. The microstructures of some chloroprene rubbers, chloroprene-styrene copolymer and chloroprene–dichlorobutadiene copolymer have been examined by NQR [766].

Bromine NQR poses many challenges, most notably the very wide frequency range over which transitions may occur. The dispersion of brominated **flame retardants** (Saytex 102/BT-93/RB-49, 1,3,5-tribromobenzene, 1-bromo-4-(4-bromophenoxybenzene) and 1,2,4,5-tetrabromobenzene) in polymer blends has been monitored with pulsed ^{79,81}Br NQR spectroscopy exploiting the transition frequency dependence on intermolecular contacts [775]. The **degree of dispersion** may be derived from a line width analysis of ^{81}Br NQR resonances. Dispersion yields NQR resonances inhomogeneously broadened relative to the pure crystalline material by factors of 4- to 20-fold. The ^{81}Br NQR spectra of Saytex BT-93, pure and in HIPS, are shown in Fig. 1.36. For these FRs the line widths are the most informative features and indicate changes in the range of intermolecular Br···Br contacts. Saytex BT-93 in HIPS shows a substantially broader ^{81}Br NQR transition than the pure material: 799 kHz *vs.* 214 kHz. This denotes different environments at the bromine sites.

^{81}Br NQR transition frequencies can be partially correlated with molecular structure. Small frequency shifts can be attributed to lattice packing. Since the crystallographic differences in the bromine sites are retained in the 1,3,5-tribromobenzene/polyester mixtures, the ^{81}Br NQR spectrum is taken as evidence that 1,3,5-tribromobenzene has not dissolved. Chang *et al.* [776] have discussed **interaction** of additives with a polymer matrix. The higher the melting point of the additive in relation to the processing temperature of the plastic, the greater the chance that the additive will phase separate, creating a heterogeneous additive/polymer mixture. The NQR analysis of FRs in HIPS [775] is consistent with Chang's results. Quadrupole interactions of ^{14}N in benzotriazole have also been examined [777]. Applications of NQR were reviewed [778].

1.5.3. Electron Spin Resonance Spectroscopy

Principles and Characteristics
Electron spin resonance (ESR) or electron paramagnetic resonance (EPR) is meant to characterise paramagnetic ions and radicals because of its ability

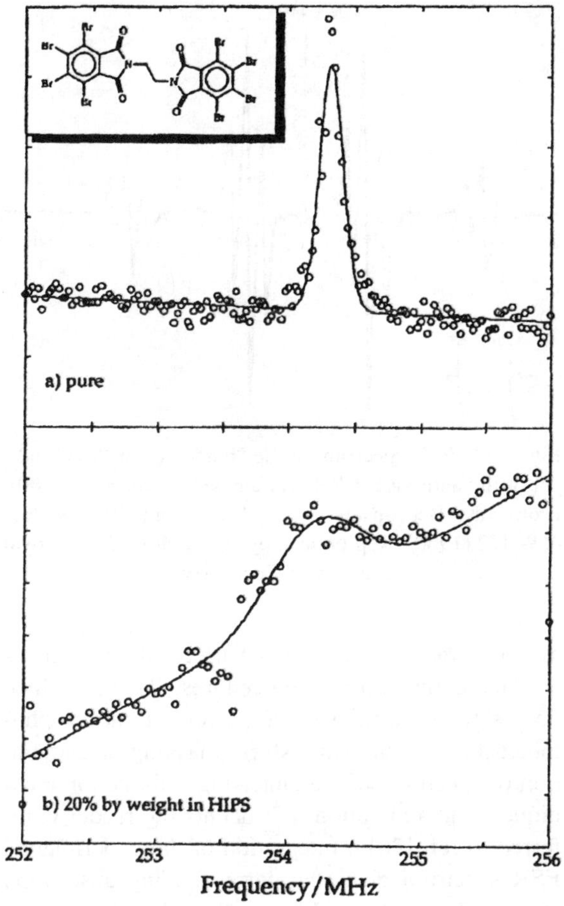

Fig. 1.36. ^{81}Br NQR spectra of 3,3′,4,4′,5,5′,6,6′-octabromo-*N,N′*-ethylenediphthalimide (Saytex BT-93), pure and in high impact polystyrene. The frequency-dependent baselines derive from changes in probe tuning over the scan range. After Mrse *et al.* [775]. Reprinted with permission from A.A. Mrse *et al., Chem. Mater.* **10**, 1291–1300 (1998). Copyright (1998) American Chemical Society.

to detect unpaired electrons. In ESR experiments, a solid sample is placed in an external magnetic field of constant strength, $\mathbf{H}_0$, that splits the energy levels (allowed spin states) of atoms, atomic groups or molecules containing unpaired electrons. Such species are described as **paramagnetic**. The few organic molecules that do posses an unpaired electron and are paramagnetic are called **free radicals**. Organic free radicals are usually encountered as intermediates in chemical reactions, such as one-electron oxidation or reduction reactions, irradiation processes or homolytic cleavage of a chemical bond.

Under the effect of radiofrequency electromagnetic radiation, the spin moments become aligned with the field; the two spin orientations correspond to two energy levels $E_\pm = \pm\frac{1}{2}g\beta G$, where g is a dimensionless proportionality constant called the electron Zeeman or g factor, β the magnetic moment of the electron or Bohr magneton and G the magnetic induction. Values for g factors of common organic radicals, which depend on the exact structure of the free radical possessing the unpaired electron, are now well established. The transition between the two levels corresponds to spin inversion and is accompanied by absorption or emission of photon energy

$$h\nu_r = E_+ + E_- = g\beta\mathbf{H}_r \qquad (1.15)$$

where $\mathbf{H}_r$ is the applied magnetic field strength. This fundamental equation expresses the **resonance condition** in ESR spectroscopy. The probability of transition from lower to higher spin state is identical to the inverse transition. Consequently, energy absorption in the resonance condition is only different from zero if there is a difference in population between the two levels, and in particular if the lower level is more highly populated.

ESR experiments in commercial **spectrometers** consist in exposing a sample containing paramagnetic species to the combined action of a flux of microwaves at constant frequency and a magnetic field of about 3300 G which is varied in order to satisfy the resonance condition. Operating frequencies of the microwave generator (klystron) are in the range of 1–100 GHz (X band: 9.5 GHz, λ 3.2 cm; K band: 24 GHz, λ 1.25 cm; Q band: 35 GHz, λ 0.85 cm).

ESR spectroscopy has developed significantly since its introduction to chemical applications in the 1950s [779], with major advances in the stability of the magnetic field, in the sensitivity to low radical concentrations, in data collection and manipulation. ESR spectroscopy enables both identification of radicals and measurement of their concentration. It is a non-destructive technique and spectra can be recorded during polymerisation, and, in suitable circumstances, during degradation of polymers. A number of characteristics of the spectrum of a radical can be predicted from its structure and used to identify the presence of the radical in an ESR spectrum.

ESR spectra are obtained as first-derivative spectra of signal intensity *vs.* magnetic field because of the method of observation of the absorption of microwave power. The main parameters of an **ESR spectrum** are:

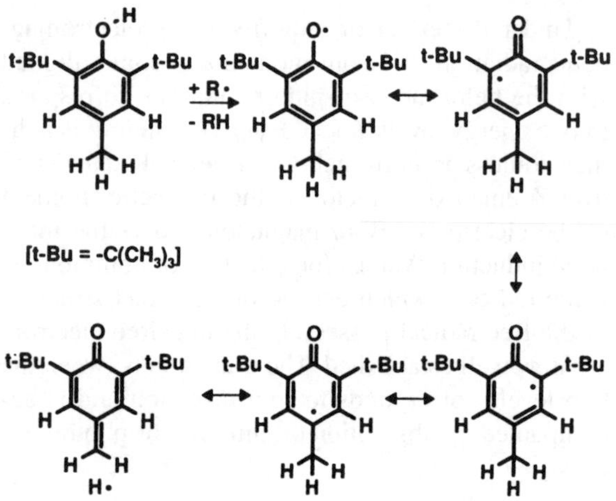

Fig. 1.37. The common antioxidant BHT and the principle resonance structures of its phenoxy radical. After Becconsall *et al.* [780]. From J.K. Becconsall *et al.*, *Trans. Faraday Soc.* **56**, 459–472 (1960). Reproduced by permission of The Royal Society of Chemistry.

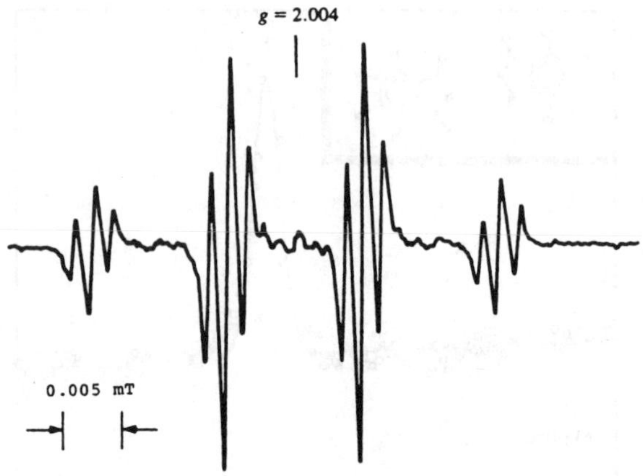

Fig. 1.38. ESR spectrum of the "hindered" aryloxyl radical of the antioxidant BHT. After Becconsall *et al.* [780]. From J.K. Becconsall *et al.*, *Trans. Faraday Soc.* **56**, 459–472 (1960). Reproduced by permission of The Royal Society of Chemistry.

(i) *g* value – or position parameter – corresponding to the proportionality between magnetic field **H** and microwave frequency, expressed in the resonance relationship of eq. (1.15); the *g* factor is determined by accurate measurement of the frequency and magnetic field strength in the resonance condition and is similar in some respects to the gyromagnetic ratio (γ) used in NMR spectroscopy;
(ii) number of lines in the spectrum, resulting from interactions between the unpaired electron spin on the radical and the nuclear spins of adjacent atoms;
(iii) relative intensities of the component lines of the spectrum of the radical;
(iv) hyperfine splitting (hfs) between the lines, which depends on the electron spin on the radical site, the magnitude of interacting nuclear spins and conformation of the radical;
(v) line widths; and
(vi) line shape, usually represented by a Gaussian or Lorentzian expression, reflecting the environment of the radical.

Figure 1.37 shows the structure of the phenoxy radical of BHT, existing as a hybrid of five principle resonance structures; Fig. 1.38 shows the ESR spectrum of this phenoxy radical [780]. ESR signals are usually detected and displayed in the dispersion mode.

The assignment of ESR spectra to component radicals and the measurement of the concentrations of these radicals require a variety of experimental and computational procedures. These include dose saturation, microwave power saturation, photobleaching, Fourier transform masking, accumulation of spectra, thermal annealing, subtraction techniques, and simulation. For details the reader is referred to ref. [781]. Integration of the experimental ESR spectrum gives the corresponding absorption spectrum and a second integration gives the area of the spectrum, which is proportional to the number of unpaired electrons provided that microwave power saturation is avoided.

As ESR can only be applied to atoms or molecules containing at least an unpaired electron, this specific spectroscopic technique can be used for applications in the chemistry of labile paramagnetic intermediates, for the study of reaction mechanisms and of molecular mobility of paramagnetic particles. The main monitored parameter is the line width in the ESR spectrum, which reflects molecular motion of a radical in a condensed medium. Analysis of change of ESR line width forms a basis for determination of dynamic parameters [782]. At high concentration of paramagnetic particles the broadening of the ESR lines is determined by interradical dipole and exchange interactions of unpaired electrons.

Table 1.35 shows the **main characteristics** of ESR. The technique provides information (usually at ambient pressure and temperature) about the nature of paramagnetic defects (organic radicals or transition metal radicals), spin-state, valence state and

Table 1.35. Main features of electron spin resonance spectroscopy

Advantages:
- Highly sensitive and specific
- Non-destructive
- Detection of the electronic state of the local site near an unpaired electron
- Element selective
- Quantitative
- Imaging capabilities (ESRI)

Disadvantages:
- Limited to few ions and organic free radicals
- Applicable only to isolated paramagnetic species in a diamagnetic matrix
- Relatively high cost

site symmetry, (sometimes) first shell coordination geometry and type of ligands. The method can be applied to crystalline as well as to amorphous materials: single crystals, powders, gels, and solutions. The maximum information from ESR spectra is obtained usually from solid-state rather than liquid solution samples and especially from oriented single crystals. ESR is representative of bulk properties but provides also surface information of adsorbed species. ESR is one of the most sensitive spectroscopic techniques with a lower limit of sensitivity of $\approx 10^{-7}$ M or 10^{11} spins (typical sample size: 10 mg to several g). All elements possessing an unpaired electron may be detected. The majority of ESR investigations deal with a few ions only: Mn^{2+}, Fe^{3+}, Cr^{3+}, VO^{2+}, Cu^{2+}, radiation defects ("colour centres"). A limitation of the technique is that it is applicable only to isolated paramagnetic species. **Electron spin-imaging** (ESRI) using a spin-echo spectrometer is described in Chp. 5.7.2.

The theory of ESR was recently reviewed [781, 783,784]; several books are available [785–787], *cfr.* also Bibliography.

Applications

Electron spin resonance is a powerful tool for free radical studies. Applications of ESR spectroscopy to polymers are specific and often almost exclusive in various sectors of the physico-chemical characterisation of polymers and processes (Table 1.36). ESR spectroscopy offers a unique technique to study the role of radical species as intermediates in both **polymerisation** and polymer degradation processes. In particular, ESR spectroscopy enables measurement of radical concentrations [781] and is therefore a powerful technique for developing a fundamental understanding of the mechanism and kinetics of free radical polymerisation.

Table 1.36. Free radical studies related to polymers

- Polymerisation and cross-linking reactions
- Grafting processes
- Oxidative and radiation degradation of organic polymers
- Mechanical fracture of polymers
- Kinetics of radical reactions
- Initiation reactions (using photons or high-energy radiation)
- Free radical intermediates *cq.* mechanisms
- Mechanisms of photolysis and thermolysis (pyrolysis)
- Molecular dynamics of polymers
- Action of stabilisers
- Additive migration

Although ESR spectroscopy may be applied to both solutions and the solid state, topics related to polymer/additive analysis are confined almost exclusively to in-polymer analysis. Zhou *et al.* [788] have described an on-line ESR study of peroxide-induced **cross-linking** of HDPE. Peroxides were used to provide primary radicals upon thermal decomposition at elevated temperatures for the generation of polymer backbone radicals. ESR spectra showed that some backbone radicals were trapped into the cross-linked polymer network and were still detectable after several months. The termination of backbone radicals is diffusion controlled. An ESR study of chemical cross-linking of PE with dicumyl peroxide (DCP) at high temperature has confirmed that the radicals originated from DCP decomposition react with amine type AOs to produce nitroxyl radicals; the antioxidants retard the initiation reaction of the PE cross-linking process [789]. Sulfur and phosphorous AOs also react with radicals yielded by decomposed DCP; 2-phenylisopropyl radicals were observed [790]. The role of polymer texture (crystallite size) on peroxide (t-butyl peroxylbenzoate) distribution (or solubility) in various PPs was studied by ESR at 145°C [791].

ESR is a suitable means for studying **polymer degradation** by external forces (fracture processes), UV radiation (photolysis, weathering) or exposure to other high-energy radiation (γ- or x-rays) or high-energy particles (*e.g.* fast electrons). Degradation of polymers is often understood from a practical viewpoint as deterioration in the properties of polymer materials leading to failure in service. The degradation reactions usually involve free radical intermediates, and therefore ESR spectroscopy is a valuable technique for investigating the chemical mechanism of degradation. Sommer *et al.* [792] have proposed to apply ultra fast *in situ* weathering of samples and directly measure the evolution of radicals by ESR; correlation with outdoor results were not presented and need to be demonstrated.

ESR has been used since 1960 to observe **radiation degradation** of polymers, and hence to provide evidence for intermediate species in radiolysis. The technique is suitable in identifying the free radicals produced at the earliest stage by UV and high-energy irradiation of PE, PP, PTFE, PMMA, PS and other polymers [793,794]. ESR spectra of alkyl radical pairs in e-beam irradiated PE were reported [795]. In-source and post-irradiation oxidation of PP/HALS films has been investigated by ESR and product analysis [796]. Concentration gradients of peroxy radicals, nitroxyl radicals, hydroperoxides, alcohols and carbonyl compounds have been determined with the multilayer technique up to a depth of 250 μm. The loss of a phenol group and formation of oxidation products in γ-irradiated HDPE/Irganox 1010 have been followed by direct use of ESR and FTIR [797]. Grafting through a peroxide link to the HDPE backbone, leaving three phenolic groups potentially active, was considered as the reason for poor antioxidant activity in γ-irradiated HDPE.

ESR was also used to study γ-radiation effects on an amine antioxidant in an ethylene–propylene copolymer [798]; free radicals in the polymer interacted with the AO leading to stable nitroxyl $R{-}NO^{\bullet}$ radicals. The signals of samples loaded with the AO recorded after irradiation in air are a superposition of two signals, namely antioxidant $R{-}NO^{\bullet}$ radicals and polymer peroxy radicals. The extractable AO levels decreased to nihil as the total dose increased to 400 kGy. ESR and extraction results are rationalised on the basis of the following simplified reaction scheme:

$$POO^{\bullet} + R{-}NH \rightarrow POOH + R{-}N^{\bullet} \qquad (1.16)$$

$$POO^{\bullet} + R{-}N^{\bullet} \rightarrow PO^{\bullet} + R{-}NO^{\bullet} \qquad (1.17)$$

$$R{-}NO^{\bullet} + P^{\bullet} \rightarrow R{-}NOP \qquad (1.18)$$

Simulation analysis of the ESR spectrum of the benzophenone (BP)-UV photoinitiated reaction of LDPE/alkylfullerene (C_{60}) in the molten state has given evidence for C_{60}-bound LDPE materials [799]. Time-resolved ESR (TREPR) and laser flash photolysis were used to characterise fullerene derivatives in PMMA; the fullerene adduct was cross-linked to the polymer chains [800].

ESR has been useful in studying the influence of dissolved gases on **polymer mobility** [801]. Stable nitroxyl radicals, such as 2,2,6,6-tetramethylpiperidin-1-oxyl (TEMPO) are widely employed as spectroscopic probes for observing binding sites and molecular motion of macromolecules [802]. ESR spectra of the TEMPO free radical in PC film at various temperature and in solution were reported [795]. The TEMPO spin probe method was also used to study diisooctylphthalate (DIOP) **plasticiser diffusion** in suspension polymerised PVC particles [803]. Similarly, the compatibility limit of PVAc and dinonylphthalate (DNP) was studied by means of 2,2-di-*n*-nonyl-5,5-dimethyl-3-oxazolidinyloxy spin probe ESR measurements and DSC [804]; DNP is an effective plasticiser for PVAc for concentrations not exceeding 17 wt.%. According to ESR evidence BBP in PVC forms radicals more easily than DOP [805].

It is well known that constituents of plastic packages can migrate towards foodstuffs in contact with them, leading to possible organoleptic and toxic consequences. The main factors determining **migration** from polymers to food are, *inter alia*: *(i)* mobility of the migrant in the plastic; *(ii)* penetration of food constituents or simulant into the polymeric network; and *(iii)* affinity of the migrant for the food simulant.

There exists considerable interest in quick methods to control compliance of plastic materials with food packaging regulations [806]. Food-polymer packaging interactions have been mainly demonstrated indirectly, by monitoring migration of residual monomers or technological additives into food [807,808]. Penetration of food into packaging has been demonstrated by a variety of techniques amongst which ESR [809–811]. Feigenbaum *et al.* [810] have recently shown that ESR allows evaluation of the influence of factors *(i)* and *(ii)* in the case of paramagnetic adjuvants (150 ppm DOXYL, TEMPO and BHT derivatives) in rigid PVC in contact with aqueous and fatty simulants. The ESR method has also been used to study the influence of chain length of fatty esters on their penetration into PVC and on migration of additives from PVC to these media [812]. In particular, attention was paid to migration of the paramagnetic additives 5-DOXYL methyl stearate, 16-DOXYL methyl stearate and 4-amino-TEMPO from rigid PVC to pure or mixed fatty esters used as food simulators. Feigenbaum *et al.* [813] used ESR also in a study of varnish-food simulant interactions, namely the behaviour of amino-oxyls added as probes to epoxyphenolic and PVC resins, constituents of a can coating, in contact with food simulants. Sawada *et al.* [814] have reported a DOXYL spin-label investigation of the dynamic behaviour of stearic acid additives in PVC/DOP.

ESR can equally be used for detection of radicals in **masticated rubber**; their identification in relation to the chemical structure might be approached with specific techniques such as electron nuclear double resonance (ENDOR). ESR studies also contribute to the understanding of the char forming process of various polymers [815], to the study of mechanical fracture, which produces free radicals, grafting reactions, *etc.* Pedulli *et al.* [816,817] have determined the bond dissociation enthalpies of α-tocopherol and other phenolic AOs by means of ESR. The determination of the O−H bond dissociation enthalpies of phenolic molecules is of considerable practical interest since this class of chemical compounds includes most of the synthetic and naturally occurring antioxidants which exert their action via an initial hydrogen transfer reaction whose rate constant depends on the strength of the O−H bond.

ESR spectroscopy has widely been used for the study of **stabilisers** which act as inhibitors in radical processes. Amongst these are phenolics, which show a mechanism involving the transformation of hydroperoxide chain propagation radicals into less reactive phenoxy radicals. Scott *et al.* [780] have identified the first hindered aryloxyl radical from the well-known antioxidant BHT (2,6-di-t-butyl-4-methylphenol) to be unequivocally identified by ESR (*cfr.* Fig. 1.37). The proposed mode of action of HALS (as deduced from investigations on polyolefins) is given by the Denisov cycle and involves nitroxyl radicals which can profitably be studied by means of ESR spectroscopy. Fully hindered amines show excellent UV stability on account of their ability to form stable nitroxyl radicals which function as chain breaking electron acceptors but not as chain breaking hydrogen atom donors in the free radical oxidative process. According to the ESR study of Ganem [802], N-oxyl radicals can oxidise aliphatic alcohols to ketones. Similarly, interaction between an N-oxyl radical and Irganox 1010 gives a resonance-stabilised quinone radical and a hydroxylamine [818,819]. This quinone is photoactive, and sensitises the photooxidation of the polymer via hydrogen abstraction or hydroperoxide formation.

ESR is a widely used spin probe technique for the study of nitroxide radicals in macromolecular systems. The structure of stable nitroxide radicals is rather diverse, although all of them contain a paramagnetic fragment $\geq$N−O$^\bullet$ as a structural element. Hundreds of these radicals have been synthesised. The following properties make nitroxide radicals ideal subjects in polymer studies:

- resistance to relatively high temperatures (100–200°C);
- structural variety, which allows modelling of a distinct organic compound; and
- paramagnetism, which allows using standard ESR for determining the dynamic parameters of the particles, introduced in trace amounts (10^{-4}–10^{-2} mol/kg).

ESR studies of free radicals formed under UV-irradiation were reported for hindered piperidine photostabilisers and antioxidants [820]. Kelen *et al.* [819] reported an ESR study of hindered piperidine derivatives in a chalk filled PP matrix in the presence of other additives (Irganox 1010, Tinuvin 770/622), with particular emphasis on concentration changes of N-oxyl radicals and interaction between a HALS compound and a hindered phenol. Other additives present in the polymer influence the concentration of the N-oxyl radicals. Lattimer *et al.* [821] studied oxidation of the partially hindered bicyclic amine 3,3-dialkyldecahydroquinoxalin-2-ones (excellent UV stabiliser and thermal antioxidant) with m-chloroperbenzoic acid by means of ESR and reported some extremely stable radical derivatives (over 231 days of stability). ESR was also used to measure the piperidinoxyl radical concentration, and hence the HALS content in LDPE/(Chimassorb 944, Tinuvin 622) agricultural film during use. Evidence was reported for polymer-bound radicals [117].

ESR experiments have also allowed insight into the mechanistic aspects of **benzofuranone (lactone) stabilisation**. Upon oxidation, lactones form C-centred radicals (Fig. 1.39). Formation of the lactone radical results in generation of H$^\bullet$, which functions as a carbon centred radical trap. The intensity of the lactone-radical ESR signal (Fig. 1.40) at about 200°C in polyolefins including the lactone and hindered phenol is much higher than compared to compositions containing the lactone alone. This denotes the capability of the lactone to efficiently reduce phenoxy radicals into the corresponding phenols, *i.e.* regenerating the phenolic antioxidant [823].

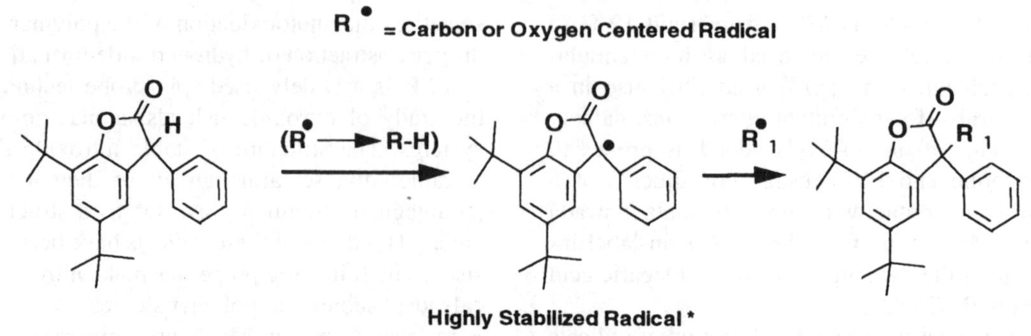

Fig. 1.39. Benzofuranonyl radical. After Kenny [822]. Reproduced by permission of Rapra Technology Ltd.

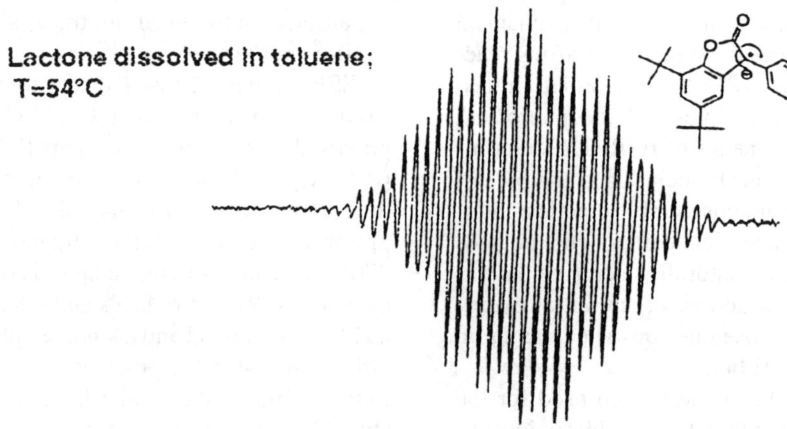

Fig. 1.40. ESR spectrum of C-centred lactone radical. After Kröhnke [823]. Reproduced by permission of the Society of Plastics Engineers (SPE).

ESR of paramagnetic free radicals can be used to check the efficacy of AOs and other stabilisers. ESR was used in the study of phenothiazines as antioxidants in PP; aromatic secondary amines can retard polymer oxidation by reacting with alkylperoxy radicals [824]. Tkáč [825] has described hydrogen and electron transfer reactions of AOs by ESR and has shown the efficiency of the ESR technique in elucidating the relationship between structure and reactivity of radicals formed from antioxidants possessing different H- and e-donor functional groups, including (hindered) phenols, amines, *etc.*

Automotive paint weathering research is based on measurement of chemical changes by means of FTIR (all coating layers), transmittance UV (clearcoat only) and ESR (determination of "active HALS" content of clearcoat and basecoat slices from weathered test panels) [826]. Gerlock *et al.* [827, 828] have determined the photooxidative stability of organic coatings by doping with a nitroxide and ESR monitoring of its concentration, as free radicals produced in the coating by photolysis are scavenged. ESR was also used to quantify the steady-state concentration of HALS-based nitroxyl radicals and the concentration of nitroxyl radicals produced when HALS and its inhibition cycle products are oxidised with peracid for various clearcoat/basecoat paint systems [31]. ESR has also been used to monitor the kinetics of nitroxide formation and decay during UV photodegradation of acrylic/melamine coatings doped with either a HALS (Tinuvin 770) or a hindered amine based nitroxide [829]. The nitroxide level *vs.* exposure time for these coatings has been measured as a function of light intensity, humidity and HALS dopant level. In the nitroxide doped coatings, the nitroxide decreases as it scavenges radi-

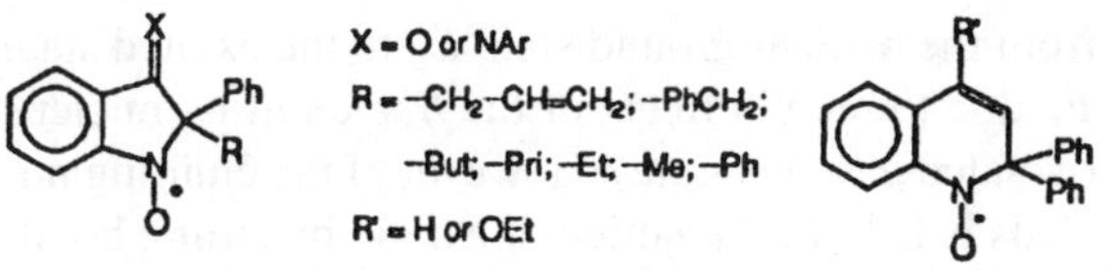

Scheme 1.2. Indolinonic and quinolinic aminoxyls. After Greci *et al.* [831]. Reproduced by permission of L. Greci, Università Politecnica delle Marche, Ancona.

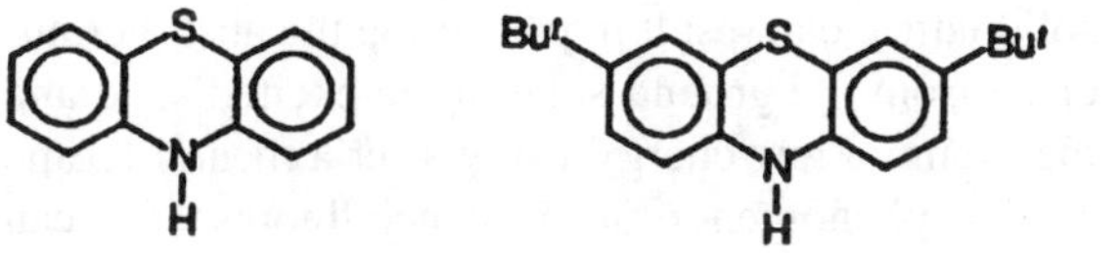

(a)

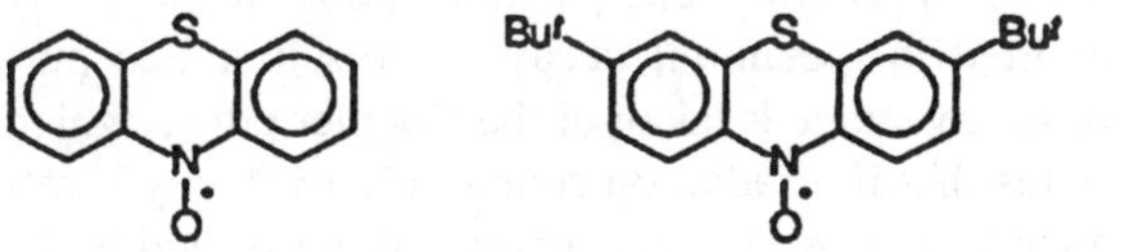

(b)

Scheme 1.3. Phenothiazines (a) and corresponding aminoxyls (b).

cals produced in the coating. The formation rates in acrylic/urethane coatings are much lower than those in an acrylic/melamine coating under the same conditions. Also the effect of pigments on the coating degradation was assessed by ESR [830].

Faucitano *et al.* [832] have reported ESR evidence for the existence of an N-peroxyl radical intermediate in the conversion of the 2,2,6,6-tetramethylpiperidinaminyl radical to the corresponding nitroxide in isotactic PP films. Faucitano *et al.* [831] have also examined the role of indolinonic and quinolinic aminoxyls (Scheme 1.2) in PP processing by means of ESR, measuring the concentration *vs.* number of extrusions. By extracting phenothiazines (Scheme 1.3a) containing PP after thermal oxidation at 160°C, very intense ESR signals were recorded, different from those of the aminoxyls (Scheme 1.3b). Geuskens *et al.* [833] have monitored oxidation of Tinuvin 770 to nitroxy radicals by ESR spectroscopy in an ethylene–propylene random copolymer (EPM), a styrene–butadiene–styrene block copolymer (SBS) and the same block copolymer previously hydroperoxidised by reaction with singlet oxygen. In the photooxidation of all three polymers, HALS is oxidised to nitroxy radicals by peroxy radicals generated photochemically but these can also originate from the thermal decomposition of clustered hydroperoxides in the dark.

Lacoste *et al.* [834] have recently proposed a novel ESR method for *in situ* checking of the consumption of total piperidyl species (intact HAS and all of its byproducts) in PP films through photooxidation. The concentration of nitroxyl radicals produced upon irradiation in stabilised PP has first been measured by **direct ESR analysis**. Several authors have used direct ESR measurements to monitor the concentration of nitroxyl-free radicals in HAS doped polymer films as a function of exposure time to oxidation [833,835]. However, direct ESR is not an ideal method to follow HAS consumption in PP through oxidation as 2,2,6,6-tetramethylpiperidine-based additives convert through a series of oxidation products, several of which are themselves stabilisers. Consequently, it is necessary to monitor *all species* involved in stabilisation of the polymer throughout its oxidative lifetime. Therefore, the change of concentration of the overall stabilising species has been detected by **indirect ESR**, after conversion of the overall HAS derivatives into nitroxyl-free radicals by exposure of photooxidised PP to peracetic acid vapour. The proposed indirect ESR technique is easier, faster, accurate, and a very sensitive method which avoids questionable extraction procedures.

Experimental ESR evidence obtained in *solution* [836] indicates that various N-substituted-2,2,6,6-tetramethyl-4-piperidinyl derivatives are oxidised to nitroxy radicals by peroxy and acylperoxy radicals:

$$\geq NX + POO^{\bullet} \rightarrow \geq NO^{\bullet} + \text{products} \qquad (1.19)$$

As stabilisers are often used in combination **interactions** are possible. ESR studies in the liquid state have been used to elucidate such interactions, *e.g.* with HALS/phenol mixtures it is possible to obtain information about the interactions between nitroxyl radicals and phenols, nitroxide radicals and phenoxy radicals, between phosphites, nitroxyl and phenoxy radicals in phosphite/phenol and phosphite/HALS mixtures. The results are useful for optimisation of additive formulations.

The key chromophore in **ultramarine blue** (lapis lazuli), $Na_6(Al_6Si_6O_{24}){\cdot}2NaS_3$ with sodalite type structure, has been identified by ESR (Fig. 1.41) and resonance Raman spectroscopy as the paramagnetic S_3^- species. ESR offers a non-destructive method for identification of ultramarine in PVC at a detection limit of 50 ppm for ultramarine blue and

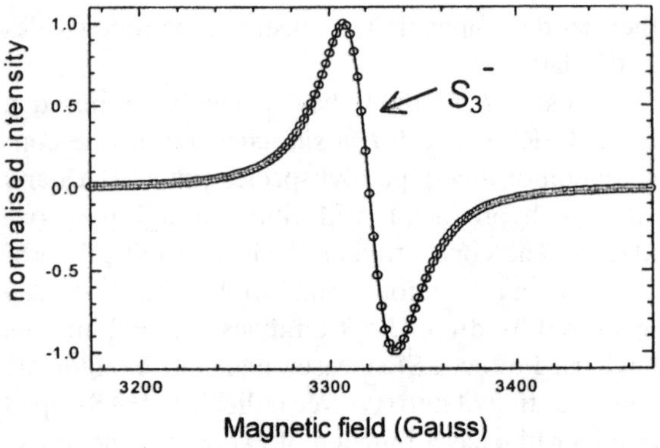

Fig. 1.41. ESR spectrum of ultramarine blue pigment. After Duhayon [837]. Reproduced by permission of the Society of Plastics Engineers (SPE).

100 ppm for ultramarine violet [837]. Clear ultramarine tinted bisphenol-A polycarbonate (BPA-PC) discolours when processed at too high a temperature. ESR has been used to reveal an interaction between pigment, stabiliser and resin [838].

Kawaguchi *et al.* [839] have reviewed the application of ESR for studies of **reaction mechanisms** of polymer additives (light stabilisers, antioxidants, carbon-black/rubber coupling agent), and of molecular motions of polymers. More recently, more general ESR applications have been reviewed [840]. Various books deal with applications of ESR [841], in particular also in relation to polymer research [842].

1.5.4. Mössbauer Spectroscopy

Principles and Characteristics

Mössbauer spectroscopy or nuclear gamma resonance fluorescence is a peculiar nuclear phenomenon, namely the recoil-free γ-ray resonance emission and absorption in solids, which analyses the energy levels of the nucleus with extremely high accuracy [843]. The fundamental physics of this effect involves transition (decay) of a nucleus from an excited state of energy E_e to a ground state of energy E_g with the emission of a γ-ray of energy E_γ (typically 10–100 keV). If the emitting nucleus is free to recoil the emitting γ-ray energy is $E_\gamma = (E_e - E_g) - E_r$, where E_r is the recoil energy of the nucleus. The magnitude of E_r is given classically by the relationship $E_r = E_\gamma^2/2mc^2$, where m is the mass of the recoiling atom. It follows that $E_\gamma < (E_e - E_g)$ and absorption of the emitted γ-photon by a nucleus of the same species will fail to promote transition from the nuclear ground state E_g to the excited state E_e due to recoil effects of the *free* emitting nucleus (isolated gaseous state). However, if the emitting nucleus is held in the lattice of a solid by strong bonding forces the recoil energy is taken up by the lattice and the mass in the recoil energy equation corresponds to that of some $10^{10} - 10^{20}$ atoms, leading to $E_r \approx 0$ or $E_\gamma = E_e - E_g$. Consequently, recoil-free absorption of a γ-ray by a nucleus bound to a solid lattice can result in promoting the absorber nucleus from the ground state to the excited state and may remit a low energy γ-ray after a mean lifetime τ. This phenomenon of resonance fluorescence can be turned into a **spectroscopic technique** by applying an appropriate energy modulation of the γ-ray emitted in the initial decay process. For this purpose advantage is taken of the Doppler effect, which states that if a radiation source has a velocity ν relative to an observer, its energy will be shifted by an amount of energy $E = (\nu/c)E_\gamma$. This can be used to modulate the γ-ray emitted in a typical Mössbauer transition, that is, to "sweep through" the energy width of the nuclear transition. Nuclear levels exhibit a discrete fine structure (hyperfine structure), which arises from the environmental electronic configurations. For the study of these shifts and splits the incident γ-ray energy may be controlled by using the Doppler effect. Although Doppler motion is unnecessary to compensate the recoil energy, the Doppler velocity is indispensable for spectroscopy.

Mössbauer spectroscopy is thus based on the resonant, recoil-free absorption of nuclear γ-radiation. Conditions for the observation of the Mössbauer effect are:

- Nuclei in the excited state as a source of γ-photons.
- Emitting and absorbing atoms in rigid lattices.
- Recoil-free events.

The Mössbauer apparatus consists of an emitter, an absorber, and a γ-ray detector. In a typical **Mössbauer experiment**, which can be performed either in transmission or in backscattering mode, a radioactive source is mounted on a velocity transducer which imparts a smoothly varying motion to the source of the γ-rays (relative to the absorber, which is held stationary), up to a maximum of several cm/s (Fig. 1.42). In practice, a source is needed which decays to the excited state of the nucleus of interest with a sufficiently long lifetime such that experiments are practical. The source usually consists of nuclei in the excited state which are obtained

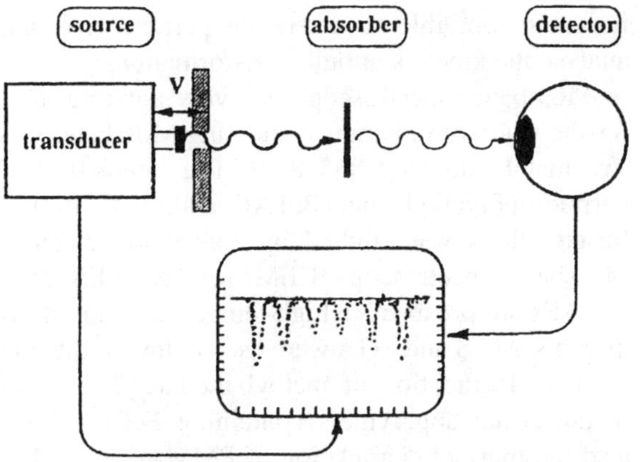

Fig. 1.42. Experimental arrangement for performing Mössbauer effect spectroscopy. After Fujita [844]. Reproduced from F.E. Fujita, *Contemp. Phys.* **40**, 323–337 (1999), by permission of Taylor & Francis Ltd. (http://www.tandf.co.uk/journals).

Table 1.37. Mössbauer nuclei, sources, half-life times and energies

Isotope	Source	Half-life	Energy (keV)
^{57}Fe	^{57}Co	270 d	14.4
^{119}Sn	^{119m}Sn	245 d	23.9
^{121}Sb	^{121m}Sn	75 y	37.2

from radioactive isotopes. Decay of the excited state to the ground state leads to emission of a γ-quantum with an extremely narrow linewidth (neV). Only a limited number of elements satisfy the experimental conditions. Mössbauer nuclei of interest to additives in polymers are given in Table 1.37.

Because the nucleus is coupled to its environment through hyperfine interactions, nuclear levels in an absorber have slightly different energy than in an emitter in a different chemical environment. The Mössbauer effect will then not be observed because the energy of the emitted γ-quantum does not match the energy difference between the levels in the absorber. The Doppler effect is used to vary the energy of the radiation within a narrow energy window of at most 500 neV. Resonant absorption will take place only when the $(E_e - E_g)$ separations in emitter and absorber are precisely matched. A gamma ray detector is used to register a spectrum with one or several absorption peaks at different velocities. A **Mössbauer spectrum** is a plot of the γ-ray intensity transmitted by the sample against the displacement of the radioactive source relative to the sample. Mössbauer parameters are the position δ of the resonance maximum, the line width Γ, and the resonance effect magnitude ε corresponding to the total area A under the resonance curve. The following information can be extracted from the absorption spectrum: *(i)* characterisation of the electronic charge density at the nucleus of the resonant atom, through the isomer shift; *(ii)* local symmetry of the site of the resonant atom, through the quadrupole splitting; *(iii)* dynamic properties of the lattice in which the resonant atom is bound, through the recoil-free fraction f; and *(iv)* nature of magnetic interactions between ions, through the hyperfine splitting (Zeeman effect) [845]. Mössbauer spectroscopy is a probe of short and medium range structure, a local probe of the vibrational density of states. Hyperfine interactions couple the nucleus to its surroundings and make it a sensitive probe for the state of the absorber. The very narrow line width of Mössbauer γ-radiation allows very small perturbations in the sample environment to be measured. All hyperfine interactions can occur simultaneously. The intensity of a Mössbauer spectrum depends not only on the recoil-free fractions of the source and the absorber and on the number of absorbing nuclei, but also on the line width of the absorption lines and saturation effects. Using Mössbauer derived information one can investigate the local electronic and structural properties of solid materials, in particular with regard to oxidation states, magnetic properties of the nucleus and lattice symmetry of selected elements [844]. Mössbauer spectroscopy can also quantitatively analyse phases (phase distributions), structures, chemical bonds, valences, lattice distortions and vibrations, impurities, defects and atomic jumps in solids, including polymers.

Table 1.38 summarises the **main features** of Mössbauer spectroscopy. The great advantage of Mössbauer spectroscopy for in-polymer additive analysis is that it provides *in situ* information. An economic advantage is that the technique is relatively inexpensive in comparison to electron microscopy or XPS. The technique is limited to those isotopes that exhibit the Mössbauer effect. The detection limit is $\sim 10^{18}$ atoms of the nuclear isotope studied. Through the Mössbauer effect in iron, it is possible to obtain information on the state of cobalt. Whereas in Mössbauer absorption spectroscopy (MAS) a single-line source is moved and

Table 1.38. Main characteristics of Mössbauer spectroscopy

Advantages:
- Element selective
- Speed of measurements
- *In situ*
- Non-destructive
- High spectral resolution
- Local probe (structure, valence state, spin-state, magnetic state)
- Phase identification and quantification (distribution)
- Structural characterisation of disordered states
- Bulk technique (0.1–10 μm); surface information for highly dispersed systems

Disadvantages:
- Limited to relatively few isotopes
- Not suitable for gases or liquids
- Sample size (500 mg–g)

the absorbing sample is in fixed position, it is also possible to fix the ^{57}Co-containing source and move the single-line ^{57}Fe absorber, in order to investigate cobalt-containing additives (Mössbauer emission spectroscopy, MES).

New methodological developments in Mössbauer spectroscopy are the use of monochromatic synchrotron radiation and Coulomb excitation instead of radioisotope sources, the simultaneous detection of Mössbauer γ-rays, internal conversion electrons and x-rays from different depths of one specimen [844]. A competitor technique yielding similar information on chemical order is EXAFS.

Mössbauer spectroscopy is one of the techniques that is not frequently used in in-polymer additive analysis. Nevertheless it may yield very useful information on a number of important additives (mainly stabilisers, flame retardants and plasticisers) using Mössbauer isotopes such as ^{57}Fe(Co), ^{119}Sn, and ^{121}Sb.

Mössbauer spectroscopy has recently been reviewed [846–849]. Several books on Mössbauer spectroscopy are available [850–854].

Applications

Applications of Mössbauer spectroscopy in additive analysis are rather few and fall in one of the following categories:
- identification of interaction products
- determination *cq.* verification of oxidation states
- structure information.

It is also possible to determine particle size and analyse the kinetics of bulk transformations.

Mössbauer spectroscopy is a very powerful tool for the study of polymers containing Mössbauer active metal ions [845,855–857]. The interaction of perfluoropolyalkyl ether (PFPAE) additives with **Fe-based** alloys was studied by conversion electron Mössbauer spectroscopy (CEMS) and XANES [858]; PFPAEs are prospective high-temperature liquid lubricants. MAS and TGA were used to investigate the thermal degradation of methyl methacrylate–ethyl methacrylate copolymers containing $FeCl_3$ [859]; also the thermal degradation of PMMA-*co*-*n*BMA/$FeCl_3$ was studied by means of MAS using a 25 mCi ^{57}Co(Cu) source [860]. Similarly, PMMA, PEMA and PBMA containing $FeCl_3$ and $FeSO_4$ as stabilisers were examined by means of Mössbauer spectroscopy [861]. Quadrupole splitting values quite different from those for pure ferrous sulphate indicate that the environment of the Fe^{2+} moiety changes in the polymer. The isomer shift values denote that no reduction of Fe^{3+} takes place during free radical polymerisation. Recently, a Mössbauer study of metal-filled composites based on porous PE matrices prepared by reduction of Mohr's salt with $LiBH_4$ with formation of supermagnetic nanoclusters of Fe(0) was reported [862]. Mössbauer spectroscopy was also used to study interaction of the heat stabiliser Fe(III) formate with poly(phenylmethylhydrosiloxane) films during degradation below 450°C [863]. Gol'danskii *et al.* [864] have studied ion containing polymers in the solid state by means of Mössbauer spectroscopy. The technique has also been used for Nafion perfluorinated acid membranes exchanged with Fe^{2+}, Fe^{3+} and Eu^{3+} [845].

Mössbauer spectroscopy has equally been used to study the structure and reactivity of organotin derivatives in PE, and the mechanism of polymer stabilisation by organotin compounds, Sn chlorides and $FeCl_3$ [865]. **^{119m}Sn Mössbauer studies** have been reported of the thermal [866] and photochemical [867] degradation of organotin stabilised PVC, as well as after γ-irradiation [868]. In an *in situ* study of the reactions undergone by the **organotin stabilisers** $R_2Sn(SCH_2CO_2C_8H_{17})_2$ or $R_2Sn(IOTG)_2$, where R = butyl or octyl, and $Bu_2Sn(O_2C{-}CH{=}CH{-}CO_2C_8H_{17})_2$ or $Bu_2Sn(IOM)_2$, during thermal degradation of PVC at 185°C, it was noticed that the stabiliser was converted into the mixed halomercaptide $R_2SnCl(X)$

(X = IOTG or IOM) [866,869] and not into R_2SnCl_2, as suggested earlier [870]. Similarly, reactions undergone by the stabilisers $Bu_2Sn(IOTG)_2$ and $Bu_2Sn(IOM)_2$ during UV degradation of the polymer in air at 25°C were studied [867]. The thioglycollate is rapidly converted to the monochloroester, $Bu_2SnCl(IOTG)$. Prolonged exposure of $Bu_2Sn(IOM)_2$ stabilised PVC leads to formation of $SnOCl_2$. The maleate stabiliser remains chemically unaltered after considerable irradiation. No evidence was found for coordinative interactions between the chlorine atoms of the polymer and the tin atom. Owing to the relatively low ^{119}Sn levels in the PVC samples (1.2 to 2% stabiliser), long runtimes were necessary. ^{119m}Sn Mössbauer spectroscopy has also been used to study the chemical changes undergone by a range of other tin-containing stabilisers (dialkyltin dilaurates, dialkyltin bis(ethylcysteinates), stannous stearate and stannous cysteinate) during thermal degradation of PVC at 185°C [871]. Mössbauer parameters indicate substantial changes on incorporation of these compounds into PVC by hot milling. Stannous stearate undergoes almost complete conversion to stannic oxide on milling. Stannous cysteinate withstands hot-milling better than the related stearate. Attempts to trace intermediate monochlorotin derivatives in PVC in solution stabilised with dialkyltin dilaurates and maleates by means of Mössbauer spectroscopy were inconclusive [872]. Also PVC stabilised with lauroyltributyltin, dibutyltin dicaproate or tetraphenyltin was examined by means of Mössbauer spectroscopy [873].

There is little published work on the packaging aspects of radiation sterilisation. ^{119m}Sn Mössbauer spectroscopy (15 mCi ^{119m}Sn barium stannate source) has been used to study the changes occuring in the organotin stabilisers $Oct_2Sn(IOTG)_2$, $Bu_2Sn(IOTG)_2$ and $Bu_2Sn(IOM)_2$ within a PVC matrix when exposed to γ-radiation from a ^{60}Co source up to 20 Mrad [868]. The final degradation product for all three stabilisers is $SnCl_4$. The maleate $Bu_2Sn(IOM)_2$ is the most stable of the three stabilisers studied, up to 10 Mrad doses. In case of the $Bu_2Sn(IOTG)_2$, evidence was found for the existence of $Bu_2SnCl(IOTG)_2$ and Bu_2SnCl_2 as intermediate degradation products.

Neoprene GW-DuPont (formulation: polychloroprene 100, MgO 4, ZnO 5, stearic acid 0.5 phr), modified with 1 to 5 phr SnO_2 and $ZnSn(OH)_6$ and 50 phr chlorinated paraffin for increased **flame retardancy** and reduced rates of smoke evolution rates, was studied by thermal analysis techniques and ^{119m}Sn Mössbauer spectroscopy (10 mCi $Ca^{119m}SnO_3$ source) to elucidate the role of the tin compounds and to investigate the chemical changes which occur during thermal degradation and combustion [874].

Conversion electron Mössbauer spectroscopy (CEMS; 30 mCi $^{57}Co(Rh)$) was used for the quantitation of Fe^{2+}/Fe^{3+} in ancient manuscripts written with iron-gall ink [875].

The use of Mössbauer spectroscopy for the study of polymerisation catalysts is feasible. Mössbauer spectroscopy is equally a very useful tool for investigating aggregation and coupling between metal ions and host lattices. Mössbauer emission spectroscopy has not been applied to the study of additives in polymers. Applications in Mössbauer spectroscopy have been collected in refs. [854,855].

1.6. DIELECTRIC LOSS SPECTROSCOPY

Principles and Characteristics

Dielectric loss spectroscopy (DIES), also named dielectric relaxation spectroscopy (DRS), dielectric analysis (DEA), or dielectrometry, is a method by which the behaviour of (polar) molecules or the mobility of charged sites in a material in an electric field can be observed. The foundation of dielectric sensing is its ability to measure the changes at the molecular level in the translational mobility of ions and changes in the rotational mobility of dipoles in the presence of a force created by an electric field. DIES measures the electrical polarisation and conduction properties of a sample subjected to a time varying electric field. This technique has long been known for studying dynamic properties, charged transport, molecular structures, and morphology of polymeric materials.

When a (polar) molecule is placed in an electric field, two types of molecule/field interactions take place, namely reversible storage and irreversible dissipation of field energy. The first interaction is a capacitive effect, caused by the polarisability of a molecule. Molecules placed in an electric field are polarised. Various polarisation mechanisms are distinguished (atomic, electrical and macroscopic polarisation, and dipole orientation). When the electric field is removed the molecules will return to their original state and the energy is reversibly released. The second interaction results from two separate mechanisms by which electric field energy is dissipated, namely the electrical conductivity of the material and friction energy. This dissipation of field energy is an irreversible process.

The polarisability of a material is given by its relative dielectric constant ε_r, which is the ratio between the permittivity of the examined material and the permittivity of free space ε_0 ($\varepsilon_0 = 8.85\ \text{pF m}^{-1}$). To describe both storage and dissipation dielectric properties, this relative dielectric constant is expressed in its complex form

$$\varepsilon^*(\omega, T) = \varepsilon'(\omega, T) - i\varepsilon''(\omega, T) \qquad (1.20)$$

with $\varepsilon^*(\omega, T)$ being the complex dielectric constant, and $\varepsilon'(\omega, T)$ and $\varepsilon''(\omega, T)$ the real and imaginary part, respectively. Dielectric relaxation arises from the frequency (ω) dependence of the complex permittivity by monitoring the changes in its real and imaginary parts. The real part of the dielectric constant (ε') is a measure for the capacitive nature of the material and is normally simply called the **dielectric constant**. The imaginary part ε'' is a measure for the dielectric losses, called the loss index. The **dielectric loss tangent** is given by $\tan\delta = \varepsilon''/\varepsilon'$. Capacitance, or the ability to store electrical charge, is proportional to the relative permittivity (ε'), which is a measure of the alignment and the number of dipoles in the sample. Conductance is the ability to transfer electric charge and is proportional to the dielectric loss factor (ε''). With the use of a dielectric spectrometer, the complex dielectric constant of a material can be measured as a function of temperature (T) and frequency of the field and the fundamental electrical characteristics of a material, conductance and capacity, can be studied as a function of temperature, time, and frequency. For non-polar thermoplasts and thermosets typical values are $\varepsilon'' \leq 10^{-3}$ and $\tan\delta \leq 10^{-4}$; for polar thermoplasts ($T < T_g$) $\varepsilon'' \leq 10^{-2}$ and $\tan\delta \leq 10^{-3}$.

In its modern form DIES is broadband in frequency and covers the range from 10^{-6} to 10^{12} Hz, thus making possible the study of both fast processes and slow relaxations. To span this huge frequency window a variety of different measurement techniques have to be combined. In practice, DIES broadly breaks down into studies below and above $\sim 10^7$ Hz. The dielectric dispersion and absorption features for solutes (*e.g.* polymers in solution) occur in the microwave region (10^8–10^{11} Hz). For the low frequency range ($< 10^7$ Hz) a vast body of accurate data has now been accumulated that describes the dielectric dispersion behaviour of amorphous and crystalline polymers and of rod-like polymers in solution. Applying a sinusoidal voltage to the sample and measuring the current the mobility of ions and dipoles is derived. A wide frequency range is scanned and the desired dielectric properties are calculated from the loss factor data. With the introduction of modern frequency and impedance analysers, which allow measurements over a wide frequency range, this technique has become more generally applicable. **Dielectrics equipment** is commercially available [876]. The speed of operation of modern measuring equipment now permits real-time measurements of $\varepsilon(\omega)$ as a system undergoes chemical or physical transformations such as polymerisation or crystallisation, respectively. Dielectric sensing techniques are applied both in the laboratory and on-line *in situ* in production facilities.

Table 1.39. Main characteristics of dielectric spectroscopy

Advantages:
- Relatively known and cheap technique
- Extraordinary width of frequency range (μHz to THz)
- Rapid measurements, ease of interpretation
- Qualitative monitoring and quantitative measurements
- Analysis of bulk and surface properties
- Small samples (mg)
- Simple, commercial equipment and software
- Rugged
- Reusable sensors
- On-line sensing
- Insight in dynamic properties of materials
- Applicable to molecular liquids, solutions, solids

Disadvantages:
- Characterisation of a macroscopic property (conductivity) only
- Limited access to the high frequency microwave region ($> 10^7$ Hz; MDS)

Table 1.39 lists the **main features** of DIES. The merits of DIES include small sample size (typically 1 cm^2 × 50 μm), wide frequency range (usually 10^{-3} to 10^7 Hz) and sound theoretical basis both in phenomenological and molecular terms [877,878]. Reduction of the required amount of sample material to the mg level have made it an attractive spectroscopic tool for samples that are available in small quantities only. Difficulties with DIES include: *(i)* low frequency conductivity-related processes, which may obscure the dipole relaxation processes; and *(ii)* limited access to high frequency range. Detectable effects of additives require concentrations exceeding 0.5% (system dependent).

The connection between dielectric permittivity (dielectric constant) and molecular dipole moments provides a means of determining molecular structure. Dielectric sensing allows monitoring of the changes in transitional mobility of ions and in the rotational mobility of dipoles in the presence of an electric field. DIES has been used for nearly sixty years as a leading method for studying the orientational **motions** of molecules in the liquid, amorphous solid, crystalline and liquid-crystalline states [879,880]. The variations in molecular position are a probe for monitoring changes in macroscopic mechanical properties such as viscosity, modulus, T_g, and degree of cure.

DIES can be used both for qualitative monitoring of chemical reactions in organic materials (*e.g.* curing, drying) and for quantitative measurements (*e.g.* determination of the concentration of polar liquids in materials such as water content in polymers). DIES can be combined with other techniques, such as FTIR, to gain specific molecular information on reactions that take place simultaneously and monitor these reactions. However, only conductivity, a macroscopic property, is measured. Consequently, molecular differentiation between combined reactions cannot be made. A lab-scale experiment in combination with more specific techniques (*e.g.* FTIR) is necessary to determine quantitatively the specific reactions. Dielectric analysis also measures changes in the properties of a polymer as it is subjected to a periodic field. A general problem in interpretation of dielectric and conductive methods is that they are not specific and are affected by many sources of interferences. These factors may explain the relatively slow introduction of this technique in characterisation of elastomer systems.

Schreyer *et al.* [881] have reviewed the theoretical principles of dielectric behaviour (*e.g.* polarisation, relaxation, relaxation time and spectrum, frequency and temperature variation, activation energy); several books are also available [878–880]. Monographs on dielectric spectroscopy of polymeric materials have appeared [877,882].

Applications

Dielectric measurements find an application in the testing of polymers which are to be employed in electrotechnical fields. However, dielectric relaxation spectroscopy serves primarily to elucidate the (supra)molecular structure of polymers and in particular the mechanisms of motion (Table 1.40). Certain polymer properties are invariably impaired as a result of chemical and physical additive interactions; notably among these are the dielectric properties [883]. Additives may increase dielectric losses either because of their intrinsic ionic and/or polar nature or because they may absorb water, which increases further the dipolar and ionic constituents of the system.

Table 1.40. General applications of dielectric spectroscopy

- Molecular dynamics in polymers [885–888]
- Fundamental studies on molecular relaxation processes
- Monitoring of cure of coatings, films and adhesives [882]
- Interface studies in heterogeneous materials [884,889]
- Influence of additives (incl. moisture) [884,890]
- On-line monitoring of additive concentrations [891–893]
- Polymer degradation studies [894,895]
- Measurements of chemical concentrations in opaque liquids
- Automatic quality monitoring [891–893]

The dielectric technique is a very powerful tool in studying heterogeneous materials. Dielectric measurements enable to ascertain whether a homogeneous or multiphase system is present in polymer mixtures. If a heterogeneous material contains components with a different electrical conductivity very strong dielectric effects are detected, due to interfacial polarisation caused by blocking of charge carrier transport at the boundaries between the constituents. For composite materials of non-polar components (fillers and fibres) **interfacial water**, adsorbed on the filler surface, can be readily detected by an increase of the dielectric constant ε' and loss index ε''. The theoretically predicted dielectric loss effects due to a conductive water interlayer at the filler/matrix interface are dominant especially at low frequencies. Steeman [884] has described a DIES study on water absorption of glass-bead filled HDPE composites. Water uptake by such composites is only due to adsorption of water molecules at the glass-sphere surface. DIES can equally be used to investigate the effects of moisture in polyesters and polyamides (in competition with LR-NMR) and the water concentration in polyurethanes (±0.003%). It may be envisaged that the interface of non-dissolved **flame retardant particles** in a polymeric matrix can also be studied by means of DIES.

Botros [163] used ATR-FTIR, HPLC and dielectric constant measurements to gain insight in the antiblock performance of some **fatty amides** in EVA copolymer and LDPE. Incompatibility of the amide with the EVA matrix is an important factor influencing antiblock performance; this is in addition to the rate of migration to the polymer surface. Certain fatty amides should therefore best be used with polar polymers and others with non-polar polymers. The dielectric properties of ion-selective PVC membranes highly **plasticised** with various citrates, sebacates, azelates and adipates were studied [896]; dielectric measurements of PVC/DOS were also reported [897].

Automated short-time dielectric breakdown tests were used to evaluate the dielectric strength of HDPE insulating materials for medium voltage **cables** as a function of additive composition and levels (TiO_2, carbon-black, Irganox B215, Tinuvin 111) [898,899]. **Carbon-black** is the component that affects dielectric strength most. The results were used to evaluate additive mixing levels in the compounds. The weakest point for formation of the rupture channel is on the carbon-black agglomerate. The dielectric behaviour of ac aged (25 years) XLPE cables was also reported [900] and DIES (in the microwave region) of poly(ethylene glycol adipate) containing a binary filler composed of graphite and pyrogenic silica was described [901]. Microwave (10 MHz–20 GHz) dielectric loss spectroscopy (MDS) has potential as a tool for the measurement of natural rubber/carbon-black interactions [902].

Dielectric analysis can determine concentrations of ingredients in mixtures based on differences in the electrical properties. **Mixing rules** describe how dielectric constant varies with concentration. For many materials, the relative permittivity ε of a mixture containing volume fraction ϕ_A of non-polar polymer A with relative permittivity ε_A and volume fraction ϕ_B of additive material B with relative permittivity ε_B is given by

$$\varepsilon^{1/3} = \phi_A \varepsilon_A^{1/3} + \phi_B \varepsilon_B^{1/3} \tag{1.21}$$

In two-component mixtures, the volume fraction concentrations add to one and the additive concentration ϕ_B is proportional to the cube root of the permittivity. Uncertainty in concentration determinations depends on the contrast in permittivity between matrix and additive (including fillers, metals, solvents, water, air, *etc.*). The concentration of a polar additive like water, with a large dielectric constant (~80), is easily measured in a non-polar material like oil with a low dielectric constant (~2–3).

Dielectric spectroscopy is also a very powerful tool to characterise the effects of (relatively) small quantities of a polar additive in a polymer foaming agent. Addition of low-MW additives (typically fatty acids) is essential in the production of dimensionally stable polymeric foams. The working mechanism of these additives in based on an **interfacial effect**. Common techniques like electron microscopy, IR and XRD fail to characterise these additives inside polymeric foams which constitute an extreme case of a heterogeneous system. However, the low-MW additives can form a thin conducting layer (10–20 μm) between polymer and gas phase. This results in interfacial polarisation at low field frequencies. Using LDPE with stearyl stearamide, GMS and ethoxylated-$C_{14/16}$-amines as polar **cell stabilising additives** in the blowing agent, it was shown that the dielectric constant of a foam depends only on the dielectric constant of the filler and the filler volume fraction, φ_{filler} [889]:

$$\varepsilon'_{\text{foam}} = \varepsilon''_{\text{polymer}}(1 - \varphi_{\text{filler}}^{1/3})^{-1} \tag{1.22}$$

Dielectric analysis has also been applied to study polymer thermal ageing [894], *e.g.* of adhesive bonded structures [895]. Dielectric loss is a means for detecting early steps in **polymer degradation** by oxidation, although nowadays CL is a strong competitor. The dielectric behaviour of post-irradiation oxidised PP/HALS was also reported [796]. Dielectric analysis has not been extended to many elastomer applications. Dissociation and mobility of salt complexes as **heat stabilisers** in PA4.6 have been investigated by means of DIES; the technique was also used to study thiourea-doped PVAL [903].

Frequency dependent dielectric measurements made over many decades of frequency (Hz to MHz) provide a sensitive, convenient means for characterising **processing properties** of thermosets and thermoplastics [882]. DIES contributes to the understanding of the dynamics of complex solid polymer systems such as blends, of polymer solutions, and of polymerisation and curing or drying reactions. On-line *in situ* dielectric sensing is applied in monitoring the polymerisation step in the production of cast nylon and the molecular mobility in curing reactions of thermoset systems such as (S)-RIM, and to control manufacturing processes. Dielectric sensing provides valuable insight in observing the state of the resin during the process, verifying and reducing the time in developing a cure process, as well as providing an automated self-correcting intelligent control system. Cure monitoring of thick HSPE/vinyl ester composites was DIES controlled [904]. DIES is actively being used for in-line measurement of the melting point of a polyester during reactive extrusion, for phase inversion detection in water/oil systems and in moisture level detection.

Relative permittivity measurements in melts can be used for the quantitative determination of individual or total additive concentrations in polymers, including multicomponent resins and compounds in which one additive dominates in concentration of permittivity, and materials with additives or primary materials that are mixtures (such as masterbatches or copolymers). Dielectric spectroscopy has been used in **process control** by determining the concentration of ionic impurities, metal hydride catalysts and stabilisers (such as boric acid, alkylaryl phosphites and epoxides included in the monomer mixture) and other process variables in a continuous polymerisation process of polycarbonates and in batchwise production of polyetherimides [891]. **In-line dielectric sensors** enable successful measurement of concentrations of single additives and co-monomers in both polar and non-polar polymers during extruder processing. Applications for different ethylene copolymers and PE and PS melts containing graduated concentrations of calcium carbonate and alumina powder were reported [892]; also PA6.6/5 wt.% nanoclay was measured [893]. For PS/12 vol.% Al_2O_3 a calculated uncertainty in the filler volume fraction of 0.4% was achieved [893]. The feasibility of a prospective in-line application can be predicted by calculation. Theory and experiments indicate that dynamic dielectric measurements in melts can quantitatively determine individual or total additive concentrations with standard uncertainties that depend on electrical contrast and can be smaller than 0.1 vol.% with polymers and compounds containing one or possibly two dominant additives. In-line dielectric sensors provide measurements that enable resin producers and compounders to automatically control the concentration of additives, to promptly correct for any process disturbances and make faster product transitions. The method has been extended to simultaneously measure three individual concentrations in three-component mixtures.

Electrical measurements can thus be used in input for on-line intelligent closed-loop process control and automatic quality monitoring, even in multicomponent mixtures.

1.7. ULTRASONIC SPECTROSCOPY

Principles and Characteristics

Ultrasonic spectroscopy is simply spectroscopy employing sound waves. In particular, it uses high-frequency acoustical waves (typically 0.5–20 MHz). The waves probe intermolecular forces in materials. Oscillating compression (and decompression) in ultrasound (US) waves cause oscillation of molecular arrangements in the sample, which responds by intermolecular attraction or repulsion. In an ultrasonic wave, oscillating pressure (stress in general terms) causes the oscillation of compressions (mechanical deformation), and therefore, by its nature, is a rheological wave. Hence, ultrasonic parameters have the simple meaning of elasticity and viscosity. However, while classical rheology deals with slow or low-frequency (typically below 1 kHz) deformations, ultrasound involves fast or high-frequency deformations (above 100 kHz).

Sound waves travel slowest through gases, faster through liquids and fastest through solids. **Active acoustics** or acoustic spectroscopy studies the attenuation and/or changes in velocity of ultrasound passed into a system. Amplitudes of deformations in the ultrasound waves employed in analytical US are extremely small, making ultrasound analysis a non-destructive technique. Although ultrasonic techniques have been used in non-destructive testing and imaging for decades, application of US to material's analyses has been held back by problems with ultrasonic design, electronics, sample handling, complex measuring procedures and limited resolution. Recently, high-resolution ultrasonic (HR-US) spectrometers have become available.

The general principles of **HR-US measurements** are simple. A generated electronic signal is transformed by a piezotransducer into the ultrasonic wave travelling through the sample in a 30 μL to 4 mL cell. Another piezotransducer transforms the received ultrasonic wave into an electronic signal

Table 1.41. Main characteristics of high-resolution ultrasonic spectroscopy

Advantages:
- Non-destructive, easy sample handling
- Small sample size
- Broad temperature range (−20 to 140°C)
- Fast analysis (s)
- Ease of wavelength variation
- Applicable to opaque samples (liquids, solutions, solids)
- Large dynamic range (down to 0.3 ppm)
- High resolution and high sensitivity
- No need for optical marker
- Robust, multipurpose commercial instruments (R&D, QC, process control)

Disadvantages:
- Emerging technology
- Temperature control needed
- Validation unsettled

for subsequent analysis. The two **major parameters** measured in HR-US are attenuation (resolution: 0.1%) and velocity of the waves. When an ultrasonic wave propagates through a material, it loses part of its energy. Attenuation of the US wave is determined by scattering of ultrasonic waves on particles and by fast chemical relaxation. Therefore, measurements of ultrasonic attenuation are a powerful tool for analysis of structure of materials and their chemical dynamics. Attenuation measurements do not require high temperature stability and can be performed on large samples. The longitudinal ultrasonic velocity c_l is determined by the compressibility or elasticity E and density (ρ) of the medium.

$$c_l = (E/\rho)^{1/2} \quad (1.23)$$

and is extremely sensitive to the molecular organisation and intermolecular interactions in the sample. Measurement of velocity requires high resolution (down to 10^{-5}%) and accurate temperature control (small samples). High-resolution measurement of ultrasound waves propagating through test materials is often superior to other analytical techniques that utilise electromagnetic waves or other measuring principles.

Table 1.41 lists the **main characteristics** of HR-US. As ultrasonic waves are synthesised electronically, it is easy to change their wavelength, quite unlike optical techniques. Ultrasonic velocity and attenuation can be measured simultaneously at different wavelengths as a function of time. US analysis allows a broad variety of sample types, chemical reactions and processes. HR-US can measure concentrations of components, transition temperatures and temperature intervals, characterise temperature stability and shelf-life of materials, analyse particle sizes in suspensions and emulsions. It is possible to make ultrasonic measurements in the temperature ramp regime for analysis of heat stability, phase and conformational transitions. Fast measurements allow analysis of flowing samples (HPLC). HR-US can analyse chemical reactions, transitions and processes as fast as 10^{-5} to 10^{-7} s.

The analytical power of ultrasound spectroscopy was recently illustrated [905]. For ultrasound imaging, *cfr.* refs. [906,907]. Atomic force acoustic microscopy (AFAM) is a new SPM technique that enables measurement of qualitative and quantitative local elastic properties of different materials and is awaiting application.

Applications

Ultrasonic spectroscopy allows measuring a wide variety of liquid systems, from dilute to concentrated solutions, and can be used to monitor processes such as molecular structural changes, thermal transitions, chemical reactions, aggregation formation, crystallisation, *etc.* Attenuation measurements are used for particle sizing in emulsions and suspensions and for kinetics of fast chemical reactions.

US velocity and attenuation measurements can also be used to determine solid-state material properties such as concentration and dispersion of **fillers** [908,909]. The active level of the acoustic emission signal of PP/talc composites was related to the degree of dispersion of the filler in the matrix [910]. LDPE/28–32 wt.% $Mg(OH)_2$ and HDPE/0–10 wt.% $Mg(OH)_2$ samples were examined over a wide range of temperatures (160–200°C) and pressures (up to 60 bar) to determine the effect of melt T, p and filler concentration on US velocity and attenuation in the melt [911]. Ultrasound velocity is affected by melt T, p and material density (filler content); US attenuation increases with increasing filler content. As US calculated filler concentrations deviated consistently from off-line TGA measured values (Fig. 1.43) it is obvious that further validation is required. In principle, extrusion processing data can be used to predict filler concentration. Accurate determination of filler concentration in real time is potentially useful to reduce excess and unnecessary usage of filler, to reduce scrap product and save production costs.

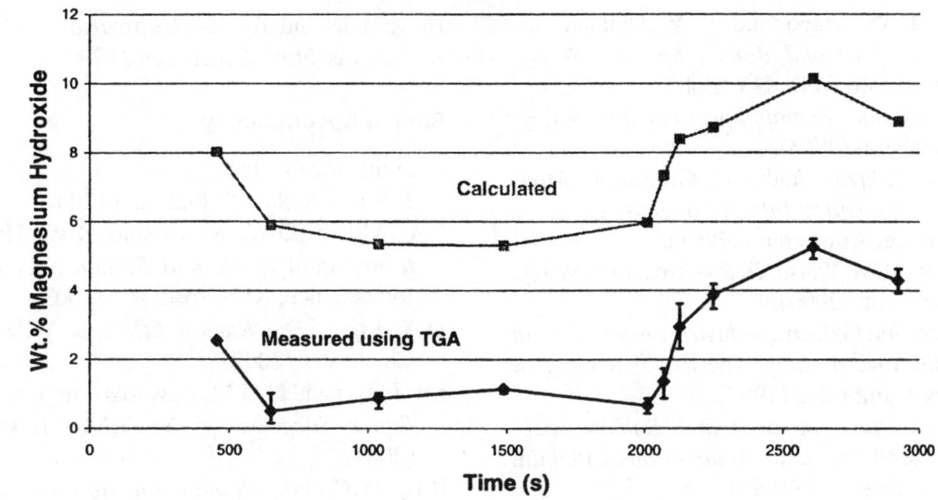

Fig. 1.43. Comparison between calculated (US process data) and measured (off-line TGA) $Mg(OH)_2$ contents in LDPE during twin screw extrusion. After Smith *et al.* [911]. Reproduced by permission of the Society of Plastics Engineers (SPE).

Hull *et al.* [912] investigated the application of ultrasound for **polymer identification**. Polymers were identified from differences in their attenuation coefficients at varying frequencies. Ultrasonic spectroscopy also detects defects and variations in material properties; it has been applied to failed adhesive joints.

Photoacoustic sensors are usually based upon direct absorption of the excitation energy by the sample system; as a result, such measurements are wavelength specific or dependent. Wan *et al.* [913] have described wavelength independent photosensitised measurements for detection of pulsed laser (Nd:YAG, 532 nm) induced acoustic signals with the aim of on-line sorting of **waste plastics**. The sound velocity is expected to vary with composition, density, crystallinity and degree of orientation of the sample. Application of a sensitiser removes the problems related to light absorption by tested samples. The technique is not limited by surface modifications or colour, printed labels, *etc.*

BIBLIOGRAPHY

General Spectroscopy

H.-H. Perkampus and N.P. Ernsting, *Encyclopedia of Spectroscopy*, J. Wiley & Sons, Chichester (2004).

J.M. Hollas, *Modern Spectroscopy*, J. Wiley & Sons, Chichester (2004).

R.A. Nyquist, *Interpreting Infrared, Raman and Nuclear Magnetic Resonance Spectra*, Academic Press, San Diego, CA (2001).

J. Workman and A. Springsteen (eds.), *Applied Spectroscopy: A Compact Reference for Practitioners*, Academic Press, San Diego, CA (1998).

C. Burgess and D.G. Jones (eds.), *Spectrophotometry, Luminescence and Colour; Science and Compliance*, Elsevier, Amsterdam (1995).

G. Talsky, *Derivative Spectrophotometry: Low and High Order*, VCH, Weinheim (1994).

W.O. George and H.A. Willis (eds.), *Computer Methods in UV, Visible, and IR Spectroscopy*, The Royal Society of Chemistry, Cambridge (1990).

R.J.H. Clark and R.E. Hester (eds.), *Spectroscopy of Surfaces*, J. Wiley & Sons, New York, NY (1988).

Direct UV/VIS Spectrophotometry

H.G. Völz, *Industrial Color Testing*, Wiley-VCH, Weinheim (2001).

H.-H. Perkampus, *UV-VIS Spectroscopy and Its Applications*, Springer-Verlag, Berlin (1992).

G. Wyszecki and W.S. Stiles, *Colour Science*, J. Wiley & Sons, Chichester (1982).

Infrared Spectroscopy

B.H. Stuart, *Infrared Spectroscopy, Fundamentals and Applications*, J. Wiley & Sons, Chichester (2004).

K.H. Michaelian, *Photoacoustic Infrared Spectroscopy*, Wiley-Interscience, New York, NY (2003).

F.A. Miller, D.W. Mayo and R.W. Hannah, *Interpretation of IR and Raman Spectra*, Wiley-Interscience, New York, NY (2003).

S.Hartewig, *IR and Raman Spectroscopy*, Wiley-VCH, Weinheim (2003).

A.A. Christy, Y. Ozaki and V.G. Gregoriou, *Modern Fourier Transform Infrared Spectroscopy*, Elsevier Science, Amsterdam (2001).

H. Günzler and H.M. Weise, *IR Spectroscopy*, Wiley-VCH, Weinheim (2000).

J.M. Chalmers and G. Dent, *Industrial Analysis with Vibrational Spectroscopy*, The Royal Society of Chemistry, Cambridge (1997).

W. Suëtaka, *Surface Infrared and Raman Spectroscopy. Methods and Applications*, Plenum Press, New York, NY (1995).

F.M. Mirabella (ed.), *Internal Reflection Spectroscopy. Theory and Applications*, M. Dekker, New York, NY (1993).

P.B. Coleman (ed.), *Practical Sampling Techniques for Infrared Analysis*, CRC Press, Boca Raton, FL (1993).

N.J. Harrick, *Internal Reflection Spectroscopy*, Interscience Publishers, New York, NY (1967).

Near-infrared Spectroscopy

H.W. Siesler, Y. Ozaki, S. Kawata and H.M. Heise (eds.), *Near-Infrared Spectroscopy, Principles, Instruments, Applications*, Wiley-VCH, Weinheim (2002).

T. Næs, T. Isaksson, T. Fearn and T. Davies, *A User-Friendly Guide to Multivariate Calibration and Classification*, NIR Publications, Chichester (2001).

D.A. Burns and E.W. Ciurczak (eds.), *Handbook of Near-Infrared Analysis*, M. Dekker, New York, NY (2001).

A. Davies and P. Williams (eds.), *Near Infrared Spectroscopy: The Future Waves*, NIR Publications, Chichester (1996).

B.G. Osborne, T. Fearn and P.H. Hindle, *Practical Near Infrared Spectroscopy with Applications in Food and Beverage Analysis*, Longman Scientific & Technical, Singapore (1993).

M. Buback and H.P. Vögele, *FT-NIR Atlas*, Wiley-VCH, Weinheim (1993).

I. Murray and I.A. Cowe (eds.), *Making Light Work: Advances in Near Infrared Spectroscopy*, VCH, Weinheim (1992).

K.I. Hildrum, T. Isaksson, T. Næs and A. Tandberg, *Near Infrared Spectroscopy*, Ellis Horwood, Chichester (1992).

H. Martens and T. Næs, *Multivariate Calibration*, J. Wiley & Sons, Chichester (1989).

Raman Spectroscopy

E. Smith and G. Dent, *Modern Raman Spectroscopy*, J. Wiley & Sons, Chichester (2004).

F.A. Miller, D.W. Mayo and R.W. Hannah, *Interpretation of IR and Raman Spectra*, Wiley-Interscience, New York, NY (2003).

D.A. Long, *The Raman Effect*, J. Wiley & Sons, Chichester (2002).

I.R. Lewis and H.G.M. Edwards (eds.), *Handbook of Raman Spectroscopy*, M. Dekker, New York, NY (2001).

R.L. McCreery, *Raman Spectroscopy for Chemical Analysis*, Wiley-Interscience, New York, NY (2000).

M.J. Pelletier (ed.), *Analytical Applications of Raman Spectroscopy*, Blackwell Science, Oxford (1999).

J.J. Laserna (ed.), *Modern Techniques in Raman Spectroscopy*, J. Wiley & Sons, New York, NY (1996).

B. Schrader (ed.), *Infrared and Raman Spectroscopy. Methods and Applications*, Wiley-VCH, Weinheim (1995).

J.R. Ferraro and K. Nakamoto, *Introductory Raman Spectroscopy*, Academic Press, San Diego, CA (1994).

D.B. Chase and J.F. Rabolt (eds.), *Fourier Transform Raman Spectroscopy; from Concept to Experiment*, Academic Press, San Diego, CA (1994).

P. Hendra, C. Jones and G. Warnes, *Fourier Transform Raman Spectroscopy: Instrumentation and Chemical Applications*, Ellis Horwood, New York, NY (1991).

J.G. Grasselli and B.J. Bulkin (eds.), *Analytical Raman Spectroscopy*, J. Wiley & Sons, New York, NY (1991).

N.B. Colthup, L.H. Daly and S.E. Wiberley, *Introduction to Infrared and Raman Spectroscopy*, Academic Press, San Diego, CA (1990).

D.J. Gardiner and P.R. Graves (eds.), *Practical Raman Spectroscopy*, Springer-Verlag, Berlin (1989).

Photoacoustics

A. Mandelis (ed.), *Principles and Perspectives of Photothermal and Photoacoustic Phenomena*, Elsevier, Amsterdam (1992).

A.C. Tam, *Photoacoustic and Photothermal Phenomena III, Springer Serreson Optics Sciences, Vol. 69*, Springer, New York, NY (1992).

A. Rosencwaig, *Photoacoustics and Photoacoustic Spectroscopy*, J. Wiley & Sons, New York, NY (1980).

Emission Spectroscopy

R. Kraayenhof, A.J.W.G. Visser and H.C. Gerritsen (eds.), *Fluorescence Spectroscopy, Imaging and Probes*, Springer-Verlag, Berlin (2002).

A.M. Garcia-Campana, *Chemiluminescence in Analytical Chemistry*, CRC Press, Boca Raton, FL (2001).

A. Sharma and S.G. Schulman, *Introduction to Fluorescence Spectroscopy*, Wiley-Interscience, New York, NY (1999).

J.R. Lakowicz (ed.), *Principles of Fluorescence Spectroscopy*, Kluwer Academic Publishers, Dordrecht (1999).

D.R. Vij (ed.), *Luminescence of Solids*, Plenum Press, New York, NY (1998).

S.G. Burnay, T.L. Williams and C.H. Jones (eds.), *Applications of Thermal Imaging*, A. Hilger, Bristol (1998).

D. Stiles, A. Calokerinos and A. Townshend (eds.), *Flame Chemiluminescence Analysis by Molecular Emission Cavity Detection*, J. Wiley & Sons, Chichester (1994).

R.C. Ropp, *Luminescence and The Solid State*, Elsevier Science, Amsterdam (1991).

D. Rendell, *Fluorescence and Phosphorescence Spectroscopy*, J. Wiley & Sons, Chichester (1987).

NMR Spectroscopy

B. Blümich, *Essential NMR*, Springer-Verlag, Berlin (2005).

K.J.D. MacKenzie and M.E. Smith, *Multinuclear Solid-State NMR of Inorganic Materials*, Pergamon, Oxford (2002).

M.J. Duer (ed.), *Solid State NMR Spectroscopy*, Blackwell Science, Oxford (2001).

E.D. Becker, *High Resolution NMR*, Academic Press, San Diego, CA (1999).

J. Klinowski and W. Kolodziejski (eds.), *Solid-State NMR Techniques*, World Scientific, Singapore (1998).

B. Blümich, P. Blümler and B.F. Chmelka (eds.), *Solid State NMR I: Methods*, Springer, Berlin (1994).

B. Blümich, H.W. Beckham and F. Engelke (eds.), *Solid State NMR III: Organic Matter*, Springer, Berlin (1994).

B. Blümich, A.E. Bennett and B.F. Chmelka (eds.), *Solid State NMR IV: Methods and Applications of Solid State NMR*, Springer, Berlin (1994).

Electron Spin Resonance Spectroscopy

F. Gerson and W. Huber, *Electron Spin Resonance Spectroscopy for Organic Chemists*, J. Wiley & Sons, Chichester (2003).

J.A. Weil, J.R. Bolton and J.E. Wertz, *Electron Paramagnetic Resonance*, Wiley-Interscience, New York, NY (1994).

N.M. Atherton, *Principles of Electron Spin Resonance*, Prentice-Hall, Englewood Cliff, NJ (1993).

M. Ikeya, *New Applications of Electron Spin Resonance*, World Scientific, London (1992).

L.Kevan and M.K. Bowman, *Modern Pulsed and Continuous-Wave Electron Spin Resonance*, Wiley-Interscience, New York, NY (1990).

Dielectric Spectroscopy

F. Kremer and A. Schönhals (eds.), *Broadband Dielectric Spectroscopy*, Springer-Verlag, Berlin (2002).

J. Runt and J. Fitzgerald (eds.), *Dielectric Spectroscopy of Polymeric Materials*, American Chemical Society, Washington, DC (1997).

S.J. Havriliak and S. Havriliak, *Dielectric and Mechanical Relaxation in Materials*, Hanser Gardner Publishers, Cincinnati, OH (1996).

Polymer Characterisation

K. Hatada and T. Kitayama, *NMR Spectroscopy of Polymers*, Springer-Verlag, Berlin (2004).

V.M. Litvinov and P.P. De (eds.), *Spectroscopy of Rubbers and Rubbery Materials*, Rapra Technology Ltd., Shawbury (2002).

P.A. Mirau, *A Practical Guide to Understanding the NMR of Polymers*, Wiley-Interscience, New York, NY (2002).

J.L. Koenig, *Infrared and Raman Spectroscopy of Polymers, Rapra Review Report No. 134*, Rapra Technology Ltd., Shawbury (2001).

P.A. Mirau, *Solid-State NMR of Polymers, Rapra Review Report No. 128*, Rapra Technology Ltd., Shawbury (2001).

J.L. Koenig, *Spectroscopy of Polymers*, Elsevier Science, Amsterdam (1999).

R.A. Pethrick and J.V. Dawkins (eds.), *Modern Techniques for Polymer Characterization*, J. Wiley & Sons, Chichester (1999).

I. Ando and T. Asakura, *Solid State NMR of Polymers*, Elsevier Science, Amsterdam (1998).

A.H. Kuptsov and G.N. Zhizhin, *Handbook of Fourier Transform Raman and Infrared Spectra of Polymers*, Elsevier, Amsterdam (1998).

A.H. Fawcett, *Polymer Spectroscopy*, J. Wiley & Sons, Chichester (1996).

M.W. Urban (ed.), *Attenuated Total Reflection Spectroscopy of Polymers: Theory and Practice*, American Chemical Society, Washington, DC (1996).

F.A. Bovey and P.A. Mirau, *NMR of Polymers*, Academic Press, San Diego, CA (1996).

K. Schmidt-Rohr and H.W. Spiess, *Multidimensional Solid-State NMR and Polymers*, Academic Press, San Diego, CA (1994).

P.J. Hendra and J.K. Agbenyega, *The Raman Spectra of Polymers*, J. Wiley & Sons, Chichester (1994).

F. Garbassi, M. Morra and E. Occhiello, *Polymer Surfaces from Physics to Technology*, Wiley-Interscience, Chichester (1994).

B.J. Hunt and M.I. James (eds.), *Polymer Characterization*, Blackie, London (1993).

R.N. Ibbett (ed.), *NMR Spectroscopy of Polymers*, Blackie, Glasgow (1993).

V.J. McBrierty and K.J. Packer, *Nuclear Magnetic Resonance in Solid Polymers*, Cambridge University Press, Cambridge (1993).

D.I. Bower and W.F. Maddams, *The Vibrational Spectroscopy of Polymers*, Cambridge University Press, Cambridge (1992).

A. Garton, *Infrared Spectroscopy of Polymer Blends, Composites and Surfaces*, Hanser Gardner, Cincinnati, OH (1992).

G. Bodor, *Structural Investigation of Polymers*, Ellis Horwood, New York, NY (1991).

J.L. Zlatkevich (ed.), *Luminescence Techniques in Solid State Polymer Research*, M. Dekker, New York, NY (1989).

I. Perepechko, *Acoustic Methods for Investigating Polymers*, Mir Publishers, Moscow (1975).

REFERENCES

[1] J.C.J. Bart, *Additives in Polymers. Industrial Analysis and Applications*, J. Wiley & Sons, Chichester (2005).

[1a] B. Brauer, T. Funke and H. Schulenberg-Schell, *Dtsch. Lebensm. Rdsch.* **91**, 381–5 (1995).

[2] F.C. Schilling and V.J. Kuck, *Polym. Degr. Stabil.* **31**, 141–52 (1991).

[3] S.J. Wright, M.J. Dale, P.R.R. Langridge-Smith, Q. Zhan and R. Zenobi, *Anal. Chem.* **68**, 3585–94 (1996).

[4] C.M. Jones, T. Naim, M. Ludwig, J. Murtaugh, P.L. Flaugh, J.M. Dudik, C.R. Johnson and S.A. Asher, *Trends Anal. Chem.* **4**, 75–80 (1985).

[5] T. Hirschfeld and D.B. Chase, *Appl. Spectrosc.* **40**, 133 (1986).

[6] S. White, in *Applied Spectroscopy* (J. Workman and A.W. Springsteen, eds.), Academic Press, San Diego, CA (1998), pp. 389–97.

[7] R. Sehan and W. Kimmer, *Jena Rev.* **35** (1), 31–3 (1990).

[8] R.V. Albarino, *Appl. Spectrosc.* **27**, 46–50 (1973).

[9] C. Burgess, *Proceedings SPQ-98 (Spectroscopy in Process and Quality Control)*, Advanstar, London (1998).

[10] A. Shakhnovich and J. Barren, *Proceedings SPE ANTEC 2002*, San Francisco, CA (2002), pp. 2512–6.

[11] R. Łobiński and Z. Marczenko, *Crit. Rev. Anal. Chem.* **23**, 55–111 (1992).

[12] J. Workman and A.W. Springsteen, *Applied Spectroscopy. A Compact Reference for Practitioners*, Academic Press, San Diego, CA (1998).

[13] H.V. Drushel and A.L. Sommers, *Anal. Chem.* **36**, 836–40 (1964).

[14] J.P. Luongo, *Appl. Spectrosc.* **19**, 117–9 (1965).

[15] S. Bremmers and J. Swagten-Linssen, *Techn. Rept. DSM Research*, Geleen (1994).

[16] R.L.H.M. Verlaek, E.M.F.J. Verdurmen and B.J. Kip, *Techn. Rept. DSM Research*, Geleen (1994).

[17] M. Kaci, F. Rezgui and T. Sadoun, *Macromol. Symp.* **127**, 123–30 (1998).

[18] J. Pospíšil, *Adv. Polym. Sci.* **101**, 65–167 (1991).

[19] H. Pasch, K.F. Shuhaibar and S. Attari, *J. Appl. Polym. Sci.* **36**, 263–74 (1991).

[20] H. Pasch, A. Al-Mobasher, S. Attari, K.F. Shuhaibar and F.A. Rasoul, *J. Appl. Polym. Sci.: Appl. Polym. Symp.* **45**, 209–26 (1990).

[21] H. Thai and N.K. Tuyleneva, *Tap Chi Hoa Hoc* **36** (2), 50–3 (1998).

[22] F.J. Pern, *Angew. Makromol. Chem.* **252**, 195–216 (1997).

[23] E. Epacher and B. Pukánszky, *Proceedings SPE ANTEC '99*, New York, NY (1999), pp. 3785–90.

[24] R.V. Albarino, *Appl. Spectrosc.* **27**, 46–50 (1973).

[25] J.E. Bonekamp and N.L. Maecker, *J. Appl. Polym. Sci.* **54** (11), 1593–604 (1994).

[26] M. Rembold and G. Berner, *ACS Org. Coatings Proc.* **47**, 624–9 (1982).

[27] F.C. Williams and F.R. Clapper, *J. Opt. Soc. Amer.* **43** (7), 505–9 (1953).

[28] J.L. Gerlock, C.A. Smith, V.A. Cooper, T.G. Dusbiber and W.H. Weeber, *Polym. Degr. Stabil.* **62**, 225–34 (1998).

[29] R.O. Carter III, J.L. Gerlock and C.A. Smith, *Proceedings SPE ANTEC '99*, New York, NY (1999), pp. 2480–4.

[30] G. Haacke, F.F. Andrawes and B.H. Campbell, *J. Coat. Technol.* **68**, 57–62 (1996).

[31] J.L. Gerlock, A.V. Kucherov and C.A. Smith, *Polym. Degr. Stabil.* **73**, 201–10 (2001).

[32] J.L. Gerlock, C.A. Smith, V.A. Cooper, S.A. Kaberline, T.J. Prater, R.O. Carter III, A.V. Kucherov, T. Misovski and M.E. Nichols, *ACS Symp. Ser.* **805**, 212–49 (2002).

[33] J.B. Pickett and J.E. Moore, *Polym. Preprints* **34**, 153–4 (1993).

[34] J.W. Chin, W.E. Byrd, E. Embree and J. Martin, in *Service Life Prediction* (J.W. Martin and D.R. Bauer, eds.), *ACS Symp. Ser.* **805**, 144–60 (2002).

[35] P. Knape and D. Wienke, *Techn. Rept. DSM Research*, Geleen (1999).

[36] L. van Aken, *Proceedings 3rd European Additives & Colors Conference* (J.M. Liégois, ed.), SPE Benelux, Antwerp (2003), pp. 16–24.

[37] D.R. Cousins, C.R. Platoni and L.W. Russell, *Forensic Sci. Intl.* **24** (3), 183–96 (1984).

[38] T.H.J. Rottiers, *Proceedings 3rd European Additives & Colors Conference* (J.M. Liégois, ed.), SPE Benelux, Antwerp (2003), pp. 98–9.

[39] J. Locke, J.M. Wilkinson and T.J. Hanford, *Forensic Sci. Intl.* **37**, 177–87 (1988).

[40] G. Wyszecki and W.S. Stiles, *Colour Science*, J. Wiley & Sons, Chichester (1981).

[41] H. Terstiege, *J. Coat. Technol.* **58**, 37–41 (1986).

[42] T.P. Bridge, R.H. Wardman and A.F. Fell, *Analyst* **110**, 1307–12 (1985).

[43] K. Moskaliuk and L. Bokic, *Textilveredlung* **19** (11), 26–8 (1984).

[44] O. Nuyken and G. Talsky, *Polym. Bull.* **2**, 719 (1980).

[45] P. Alessi, A. Cortesi and I. Kikic, *Proceedings GRICU (Gruppo Ricercatori Italiani di Ingegneria Chimica dell'Università)*, Ferrara (1998), pp. 379–82.

[46] L. Wang, G. Dang, L. Zheng and X. Wang, *Anal. Lett.* **34** (13), 2403–14 (2001).

[47] K.C. Thompson and K. Wagstaff, *Analyst (London)* **104**, 668–79 (1979).

[48] P. Tittarelli, P. Zerlia, A. Colli and G. Ferrari, *Anal. Chem.* **55**, 220–4 (1983).

[49] I. Noda, *Appl. Spectrosc.* **47**, 1329 (1993).

[50] Y. Ozaki, Y. Lui, M. Czarnecki and I. Noda, *Macromol. Symp.* **94**, 51 (1995).

[51] P.H. Hindle, in *Near Infrared Spectroscopy* (A.M.C. Davies and R. Giangiacomo, eds.), NIR Publications, Chichester (2000), pp. 1–6.

[52] N.J. Everall, in *An Introduction to Laser Spectroscopy* (D.L. Andrews and A.A. Demidov, eds.), Plenum Press, New York, NY (1995), pp. 115–31.

[53] N. Zanier, *Proceedings SPQ-98 (Spectroscopy in Process and Quality Control)*, Advanstar, London (1998).

[54] P.J. Gemperline and N.K. Shah, *Anal. Chem.* **62**, 465 (1990).

[55] F.J. Clarke, *Proceedings SPIE – Intl. Soc. Opt. Eng.* **2775**, 6–17 (1996).

[56] W.D. Perkins, in *Practical Sampling Techniques for Infrared Analysis* (P.B. Coleman, ed.), CRC Press, Boca Raton, FL (1993), pp. 11–53.

[57] A.A. Ismail, F.R. van de Voort and J. Sedman, in *Instrumental Methods in Food Analysis* (J.R.J. Paré and J.M.R. Bélanger, eds.), Elsevier Science, Oxford (1997), pp. 93–139.

[58] A.M. Zaper and J.L. Koenig, in *Determination of Molecular Weight* (A.R. Cooper, ed.), J. Wiley & Sons, New York, NY (1989), pp. 432–85.

[59] J. Coates, in *Applied Spectroscopy* (J. Workman and A.W. Springsteen, eds.), Academic Press, San Diego, CA (1998), pp. 49–91.

[60] S.V. Compton and D.A.C. Compton, in *Practical Sampling Techniques for Infrared Analysis* (P.B. Coleman, ed.), CRC Press, Boca Raton, FL (1993), pp. 217–54.

[61] J.O. Leppinen, in *Surface Characterization. A User's Sourcebook* (D. Brune, R. Hellborg, H.J. Whitlow and O. Hunderi, eds.), Wiley-VCH, Weinheim (1997), Chp. 22.

[62] P.B. Coleman (ed.), *Practical Sampling Techniques for Infrared Analysis*, CRC Press, Boca Raton, FL (1993).

[63] J.R. Ferraro and K. Krishnan (eds.), *Practical FT-IR Spectroscopy: Industrial and Laboratory Chemical Analysis*, Academic Press, New York, NY (1990).

[64] J.M. Chalmers and G. Dent, *Industrial Analysis with Vibrational Spectroscopy*, The Royal Society of Chemistry, Cambridge (1997).

[65] N.B. Colthup, L.H. Daly and S.E. Wiberley, *Introduction to Infrared and Raman Spectroscopy*, Academic Press, San Diego, CA (1990).

[65a] Available from Anadis Instruments, Malden, The Netherlands.

[66] J.J. He and B. Gilpatrick, *Proceedings SPE ANTEC '99*, New York, NY (1999), pp. 2314–8.

[67] R.L. Gray, R.E. Lee and C. Neri, *Proceedings Polyolefins X* (SPE, ed.), Houston, TX (1997), pp. 599–613.

[68] J.W. Sinclair, L. Schall and N.T. Crabb, *J. Chromatogr. Sci.* **18**, 30–4 (1980).

[69] R. Osland, *Lab. Pract.* **37** (2), 73 (1988).

[70] R.C. Wieboldt, K.D. Kempfert and D.L. Dalrymple, *Appl. Spectrosc.* **44**, 1028 (1990).

[71] R.T. Johnston and E.J. Slone, *Proceedings Polyolefins VII* (SPE, ed.), Houston, TX (1991), p. 207.
[72] N.S. Allen, S.J. Palmer, G.P. Marshall and J.L. Gardette, *Polym. Degr. Stabil.* **56** (3), 265–74 (1997).
[73] P. Li, G. Wu, Y. Wang and X. Yang, *Fenxi Huaxue* **13**, 664 (1985); *C.A.* **104**, 51459e (1986).
[74] S. Zehnacker and J. Marchal, *Polym. Degr. Stabil.* **45**, 435–9 (1994).
[75] H. Grupp, *Maschinenschaden* **44** (1), 28–32 (1971).
[76] E.G. Bartick, *Perkin-Elmer Infrared Bulletin 70*, Perkin-Elmer, Norwalk, CT (1979).
[77] S.C. Pattacini and R.E. Anacreon, *Perkin-Elmer Infrared Bulletin 74*, Perkin-Elmer, Norwalk, CT (1980).
[78] A. Murase, Y. Esaki, M. Sugiura and T. Araga, *Anal. Sci.* **7**, 1597–600 (1991).
[79] A. Murase, M. Sugiura and T. Araga, *Polym. Degr. Stabil.* **43** (3), 415–22 (1994).
[80] J. Haslam, H.A. Willis and D.C.M. Squirrell, *Identification and Analysis of Plastics*, J. Wiley & Sons, Chichester (1983).
[81] D. Brück, *Fortbildungsseminar Moderne Prüfmethoden – von Rohstoffen über Mischungen bis zum Produkte*, DIK, Hannover (May 2000).
[82] B.J. Coles and C.J. Hall, *Proceedings SPE ANTEC '98*, Atlanta, GA (1998), pp. 2215–8.
[83] H.D. Ruan, R.L. Frost, J.T. Kloprogge and L. Duong, *Spectrochim. Acta* **58A** (2), 265–72 (2002).
[84] D.J. Carlsson, M. Krzymien, D.J. Worsfold and M. Day, *J. Vinyl Addit. Technol.* **3**, 100–6 (1997).
[85] S.E. Amos, G.M. Giacoletto, J.H. Horns, C. Lavallée and S.S. Woods, in *Plastics Additives Handbook* (H. Zweifel, ed.), Hanser Publishers, Munich (2000), pp. 553–84.
[86] S.R. Culler, H. Ishida and J.L. Koenig, *Appl. Spectrosc.* **38**, 1–7 (1984).
[87] P.P. De, in *Spectroscopy of Rubbers and Rubbery Materials* (V.M. Litvinov and P.P. De, eds.), Rapra Technology Ltd., Shawbury (2002), pp. 77–124.
[88] H.G. Dikland and M. van Duin, in *Spectroscopy of Rubbers and Rubbery Materials* (V.M. Litvinov and P.P. De, eds.), Rapra Technology Ltd., Shawbury (2002), pp. 207–46.
[89] L. Bokobza, *Anal. Sci.* **17**, i675–8 (2001).
[90] C.Y. Chu, *Kautsch. Gummi Kunstst.* **41** (1), 33–9 (1988).
[91] A.L. Cholli, J.L. Koenig, T. Sun and H. Zhou, *J. Appl. Polym. Sci.* **28**, 3497–502 (1983).
[92] B.-S. Chiou and S.A. Khan, *Proceedings 50th IS&T (Imaging Science and Technology) Annu. Conf.*, Cambridge, MA (1997), pp. 570–4.
[93] S.S. Pesetskii, A.I. Kuzakov, O.M. Kasperovich and Yu.M. Krivoguz, *Vest. Akad. Navuk Belarusi, Ser. Khim. Navuk* (4), 15–9 (1997).
[94] M. Bartholin, N. Bensemra, T. Van Hoang and A. Guyot, *Polym. Bull. (Berlin)* **23**, 425–30 (1990).
[95] D.B. Dormagen, *Kautsch. Gummi Kunstst.* **39** (12), 1165 (1986).
[96] D. Coz and K. Baranwal, *Rubber World*, 30 (Jan. 1999).
[97] M.B. Simpson, *Spec. Publ.-Roy. Soc. Chem.* **93**, 185–97 (1991).
[98] F. Kowol, M. Oleimeulen, H. Freitag and T. Huth-Fehre, *J. Near Infrared Spectrosc.* **6** (1–4), A149–51 (1998).
[99] T. Learner, in *Resins, Ancient and Modern (SSCR Conference Preprints)*, Aberdeen (1995), pp. 76–84.
[100] C. Jin, T.A. Egerton, J.R. White, P.A. Christensen and N. MacDonald, *Proceedings SPE ANTEC 2003*, Nashville, TN (2003) pp. 1976–80.
[101] T.V. Karstang and A. Henriksen, *Chemom. Intell. Lab. Syst.* **14** (1–3), 331–9 (1992).
[102] B. Vigerust, K. Kolset, S. Nordenson, A. Henriksen and K. Kleveland, *Appl. Spectrosc.* **45** (2), 173–7 (1991).
[103] H. Winterberg, *Plaste Kautsch.* **37**, 185 (1990).
[104] G. Meszlényi, I. Lovas and Z. Nyitrai, *J. Appl. Polym. Sci.: Appl. Polym. Symp.* **48**, 411–5 (1991).
[105] R.A. Crocombe, M.L. Olson and S.L. Hill, *Computerized Quantitative Infrared Analysis, ASTM STP 934* (G.L. McClure, ed.), ASTM, Philadelphia, PA (1987), pp. 95–130.
[106] E.M. Salazar-Rojas and M.W. Urban, *Progr. Org. Coatings* **1b**, 371 (1989).
[107] B. van Lierop, L. Castle, A. Feigenbaum and A. Boenke, *Spectra for the Identification of Additives in Food Packaging*, Kluwer Academic Publishers, Dordrecht (1998).
[108] D.O. Hummel, *Atlas of Plastics Additives: Analysis by Spectrometric Methods*, Springer-Verlag, Berlin (2002).
[109] L. Rintoul, H. Panayiotou, S. Kokot, G. George, G. Cash, R. Frost, T. Bui and P. Fredericks, *Analyst* **123** (4), 571–7 (1998).
[110] H. Nelissen, *Techn. Rept. DSM Research*, Geleen (1998).
[111] R.A. Fidler, *Plast. Compd.* **15** (6), 47–50 (1992).
[112] J. Major and V. Kocmanova, *Chem. Prum.* **17**, 372 (1967).
[113] R.G.J. Miller and H.A. Willis, *Spectrochim. Acta* **14**, 119 (1959).
[114] G. Guiochon and J. Henniker, *Br. Plast.* **37**, 74 (1964).
[115] T.R. Crompton, *Manual of Plastics Analysis*, Plenum Press, New York, NY (1998).

[116] G. Meszlényi, G. Kortvelyessy and M. Sipos, *ACH-Models Chem.* **136** (4), 407–13 (1999).
[117] M. Scoponi, S. Cimmino and M. Kaci, *Polymer* **41**, 7969–80 (2000).
[118] J. Sampers, *Techn. Rept. DSM Research*, Geleen (1991).
[119] A.J. Brandolini, J.M. Garcia and R.E. Truitt, *Spectroscopy* **7** (3), 34–9 (1992).
[120] P. Palmen and G. Meyer, *Techn. Rept. DSM Research*, Geleen (1996).
[121] A.W. Springsteen, in *Applied Spectroscopy* (J. Workman and A.W. Springsteen), Academic Press, San Diego, CA (1997), pp. 247–67.
[122] A.W. Springsteen, in *Applied Spectroscopy* (J. Workman and A.W. Springsteen), Academic Press, San Diego, CA (1997), pp. 193–224.
[123] W. Suĕtaka, *Surface Infrared and Raman Spectroscopy, Methods and Applications*, Plenum Press, New York, NY (1995).
[124] D.L. Allara, in *Characterization of Metal and Polymer Surfaces* (L.H. Lee, ed.), Academic Press, London (1977).
[125] H.G. Tompkins, in *Methods of Surface Analysis* (A.W. Czanderna, ed.), Elsevier, Amsterdam (1975).
[126] J.A.J. Jansen, J.H. van der Maas and A. Posthuma de Boer, *Appl. Spectrosc.* **45**, 1149 (1991).
[127] E.T.G. Lutz, H.J. Luinge, J.H. van der Maas and R. van Agen, *Appl. Spectrosc.* **48**, 1021–5 (1994).
[128] G. Zachmann and P. Turner, *Spectrosc. Europe* **9** (1), 18–22 (1997).
[129] R.A. Merrill, E.G. Bartick and J.H. Taylor III, *Anal. Bioanal. Chem.* **376**, 1272–8 (2003).
[130] W.J. Egan, S.L. Morgan, E.G. Bartick, R.A. Merrill and H.J. Taylor III, *Anal. Bioanal. Chem.* **376**, 1279–85 (2003).
[131] D.L. MacAdam, *Colour Measurement*, Springer-Verlag, Berlin (1981), pp. 106–28.
[132] M.P. Fuller and P.R. Griffiths, *Anal. Chem.* **50**, 1906 (1978).
[133] M.P. Fuller and P.R. Griffiths, *Appl. Spectrosc.* **34**, 533 (1980).
[134] P. Kubelka and F. Munk, *Z. Tech. Phys.* **12**, 593 (1931).
[135] P. Kubelka and F. Munk, *J. Opt. Soc. Am.* **38**, 448 (1948).
[136] F. Garbassi, M. Morra and E. Occhiello, *Polymer Surfaces. From Physics to Technology*, Wiley-Interscience, Chichester (1994).
[137] R.A. Spragg, *Appl. Spectrosc.* **38**, 604 (1984).
[138] S.R. Culler, in *Practical Sampling Techniques for Infrared Analysis* (P.B. Coleman, ed.), CRC Press, Boca Raton, FL (1993), pp. 93–105.
[139] D. Fischer, T. Bayer, K.-J. Eichhorn and M. Otto, *Fresenius J. Anal. Chem.* **359**, 74–7 (1997).
[140] R.A. Condrate and D.H. Lee, *Proceedings Conf. Sci. Whitewares II* (W.M. Carty and C.W. Sinton, eds.), American Ceramic Society, Westerville, OH (2000), pp. 131–9.
[141] M. Gilbert, P. Petiraksakul and I. Mathieson, *Mater. Sci. Technol.* **17** (11), 1472–8 (2001).
[142] R.T. Graf, J.L. Koenig and H. Ishida, *Anal. Chem.* **56**, 773–8 (1984).
[143] L. Vågberg, P. De Potocki and P. Stenius, *Appl. Spectrosc.* **43**, 1240–8 (1989).
[144] M.C. Matias, M.U. De La Orden, C. Gonzalez Sánchez and J. Martinez Urreaga, *J. Appl. Polym. Sci.* **75** (2), 256–66 (2000).
[145] S. Kokot, S. Carswell and D.L. Massart, *Appl. Spectrosc.* **46**, 1393–9 (1992).
[146] M.S. Weeks and W. Mockel, *AT-Process* **3** (1, 2), 33–42 (1997).
[147] J.P. Coates, J.M. D'Agostino and C.R. Friedman, *Am. Lab.* **18** (11), 82–6 (1986).
[148] N. Dumont and C. Depecker, *Vibr. Spectrosc.* **20** (1), 5–14 (1999).
[149] P.E.R. Mucci, in *Chemical Aspects of Plastics Recycling* (W. Hoyle and D.R. Karsa, eds.), The Royal Society of Chemistry, Cambridge (1997), pp. 54–70.
[150] G. Zachmann and P. Turner, in *Chemical Aspects of Plastics Recycling* (W. Hoyle and D.R. Karsa, eds.), The Royal Society of Chemistry, Cambridge (1997), pp. 71–6.
[151] D. Fischer, E. Henßge and E. Pigorsch, *Abstracts Advanced Methods of Polymer Characterization: New Developments and Applications in Industry*, Mainz (1999), Paper O11.
[152] J.M. Chalmers and M.W. Mckenzie, *Appl. Spectrosc.* **39**, 634–41 (1985).
[153] M.B. Mitchell, *Adv. Chem. Ser.* **236**, 351 (1993).
[154] N.J. Harrick, *J. Opt. Soc. Amer.* **49**, 376 (1959).
[155] N.J. Harrick, *Internal Reflectance Spectroscopy*, Wiley-Interscience, New York, NY (1967).
[156] J. Fahrenfort, *Spectrochim. Acta* **17**, 698 (1961).
[157] S.V. Compton and D.A.C. Compton, in *Practical Sampling Techniques for Infrared Analysis* (P.B. Coleman, ed.), CRC Press, Boca Raton, FL (1993), pp. 55–92.
[158] J. Coates and A. Sanders, *Spectrosc. Europe* **12** (5), 12–22 (2000).
[159] W.P. Rothwell, D.R. Holecek and J.A. Vershaw, *J. Polym. Sci. Lett. Ed.* **22**, 241 (1984).
[160] S. Ekgasit and H. Ishida, *Appl. Spectrosc.* **51**, 1488 (1997).
[161] M.W. Urban, *Proceedings ACS Div. Polym. Mat. Sci. & Engng.* **78**, 18–9 (1998).
[162] M.W. Urban, *Attenuated Total Spectroscopy of Polymers: Theory and Practice*, American Chemical Society, Washington, DC (1996).

[163] M.G. Botros, *J. Plast. Film. Shtg.* **11**, 326–37 (1995).

[164] K. Fujimoto and K. Wataya, *J. Appl. Polym. Sci.* **13**, 2513–26 (1969).

[165] E. Agosti, G. Zerbi and I.M. Ward, *Polymer* **33** (20), 4219–29 (1992).

[166] J.G. van Alsten and S.R. Lusting, *Macromolecules* **25**, 5069–73 (1992).

[167] M. Fontijn, K. Van 't Riet and B.H. Bijsterbosch, *Colloids Surf.* **54** (3–4), 331–47 (1991).

[168] D. Feldman, F. Denes, Z. Zeng, A.R. Denes and D. Banu, *J. Adhes. Sci. Technol.* **14** (13), 1705–21 (2000).

[169] D. Tatsumi and T. Yamauchi, *J. Appl. Polym. Sci.* **69** (3), 461–8 (1998).

[170] C.M. Small, G.M. McNally, A. Marks and W.R. Murphy, *Proceedings SPE ANTEC 2002*, San Francisco, CA (2002), pp. 2882–6.

[171] K.B. Walters and D.E. Hirt, *Proceedings SPE ANTEC 2001*, Dallas (2001), pp. 2644–8.

[172] P.K. Dutta and K.R. Graf, *J. Appl. Polym. Sci.* **29**, 2247 (1984).

[173] T. Thorstenson, L. Tebelius and M.W. Urban, *Polym. Mater. Sci. Eng.* **68**, 203–4 (1993).

[174] B.-J. Niu and M.W. Urban, *J. Appl. Polym. Sci.* **60**, 389–97 (1996).

[175] J.J. Kellar, W.M. Cross and J.D. Miller, *Appl. Spectrosc.* **44**, 1508–12 (1990).

[176] M.P. McCourt, G.M. McNally, C. McGlory, T. McNally and W.R. Murphy, *Proceedings SPE ANTEC 2002*, San Francisco, CA (2002), pp. 2563–7.

[177] N.B. Joshi and D.E. Hirt, *Proceedings SPE ANTEC '98*, Atlanta, GA (1998), pp. 2832–6.

[178] G.T. Fieldson and T.A. Barbari, *Polymer* **34** (6), 1146–53 (1993).

[179] G.T. Fieldson and T.A. Barbari, *AICHE J.* **41** (4), 795–805 (1995).

[180] C.M. Balik and W.H. Simendinger III, *Polymer* **39** (20), 4723–8 (1998).

[181] G. Napadensky, J.M. Sloan, N.B. Tan, D. Crawford, D.A. Mountz and K.A. Mauritz, *Proceedings ACS Div. Polym. Mater., Sci. & Engng.* **83**, 426–7 (2000).

[182] P. Eaton, P. Holmes and J. Yarwood, *Appl. Spectrosc.* **54** (4), 508–16 (2000).

[183] E. Bormashenko, J. Reichlin, R. Pogreb, I. Wassermann and A. Katzir, *Proceedings SPE ANTEC 2000*, Orlando, FL (2000), pp. 2069–73.

[184] J.G. Grasselli, S.E. Mocadlo and J.R. Mooney, in *Applied Polymer Analysis and Characterization* (J. Mitchell, ed.), Hanser Publishers, Munich (1987), pp. 315–28.

[185] K. Sreenivasan, *Chromatographia* **22** (1–6), 199–200 (1986).

[186] C. Sellitti, J.L. Koenig and H. Ishida, in *Interfaces in Polymer, Ceramic and Metal Matrix Composites* (H. Ishida, ed.), Elsevier, New York, NY (1988), p. 163.

[187] G. Meyers, P. Gijsman and J. Sampers, *Techn. Rept. DSM Research*, Geleen (2000).

[188] F. Delor, N. Barrois-Oudin, X. Duteurtre, C. Cardinet, J. Lemaire and J. Lacoste, *Polym. Degr. Stabil.* **62**, 395–401 (1998).

[189] B.H. Schumann and J.J. Wooster, *Proceedings SPE ANTEC '99*, New York, NY (1999), pp. 46–51.

[190] M. Vitali, *Abstracts First Intl. Conference on Polymer Modification, Degradation and Stabilisation (MoDeSt)*, Palermo (2000).

[191] A. Marcilla, M. Beltrán, J.C. García and D. Mang, *J. Vinyl Addit. Technol.* **1**, 10–4 (1995).

[192] W. Herres, *Kunststoffe* **76** (7), 591–4 (1986).

[193] D. Tatsumi, T. Yamauchi and K. Murakami, *Nord. Pulp. Pap. Res. J.* **10**, 94 (1995).

[194] B.J. Coles and C.J. Hall, *Proceedings SPE ANTEC '98*, Atlanta, GA (1998), pp. 2215–8.

[195] P. Pages and J. Ros, *J. Soc. Dyers Color.* **99** (11), 333–5 (1983).

[196] M.R. Dhananjeyan, E. Mielczarski, K.R. Thampi, Ph. Buffat, M. Bensimon, A. Kulik, J. Mielczarski and J. Kiwi, *J. Phys. Chem.* **B105** (48), 12046–55 (2001).

[197] S.G. Kazarian, N.M.B. Flichy, D. Coombs and G. Poulter, *Intl. Lab.* **32** (2), 28–32 (2002).

[198] S.G. Kazarian, M.F. Vincent, F.V. Bright, C.L. Liotta and C.E. Eckert, *J. Am. Chem. Soc.* **118**, 1729–36 (1996).

[199] S.G. Kazarian, *Proceedings SPE ANTEC '99*, New York, NY (1999), pp. 2433–5.

[200] A. Tuchbreiter, J. Marquardt, J. Zimmermann, P. Walter, R. Mülhaupt, B. Kappler, D. Faller, T. Roths and J. Honerkamp, *J. Combin. Chem.* **3** (6), 598–603 (2001).

[201] ASTM E 573-90, *Standard Practices for Internal Reflection Spectroscopy, Annual Book of ASTM Standards*, American Society for Testing and Materials, Philadelphia, PA (1995), Vol. 03.06.

[202] W. Herschel, *Phil. Trans. Roy. Soc. (London)* **284**, 293, 538 (1800).

[203] I.M. Mills, in *Making Light Work: Advances in Near Infrared Spectroscopy* (I. Murray and I.A. Cowe, eds.), VCH, Weinheim (1992).

[204] P.C. Williams, in *Handbook of Near Infrared Analysis* (D.A. Burns and E.W. Ciurczak, eds.), M. Dekker, New York, NY (1992), pp. 281–315.

[205] N. Camajani and C. Kradjel, in *Making Light Work: Advances in Near Infrared Spectroscopy* (I. Murray and I.A. Cowe, eds.), VCH, Weinheim (1992), pp. 461–7.

[206] T. Hirschfeld, *Appl. Spectrosc.* **39** (6), 1085–6 (1985).

[207] G.J. Kemeny, in *Handbook of Near-Infrared Analysis* (D.A. Burns and E.W. Ciurczak, eds.), M. Dekker, New York, NY (1992), pp. 53–105.

[208] M.S. Day and F.R.B. Fearn, *Lab. Pract.*, 328–30 (Apr. 1982).

[209] B.G. Osborne and T. Fearn, *Near Infrared Spectroscopy in Food Analysis*, J. Wiley & Sons, New York, NY (1986).

[210] J. van der Maas, *Proceedings SPQ-98 (Spectroscopy in Process and Quality Control)*, Advanstar, London (1998).

[211] E. Nishio, M. Morimoto and K. Nishikida, *Appl. Spectrosc.* **44**, 1639 (1990).

[212] M. Ganzarolli de Oliveira, O. Pessoa, H. Vargas and F. Galembeck, *J. Appl. Polym. Sci.* **35**, 1791 (1988).

[213] G.F. Kirkbright and K.R. Menon, *Anal. Chim. Acta* **136**, 373 (1982).

[214] I.A. Cowe, *Anal. Proc. (London)* **20** (2), 65–8 (1983).

[215] K.H. Norris, *J. Near Infrared Spectrosc.* **4**, 31–7 (1996).

[216] R.A. Shaw and H.H. Mantsch, in *Encyclopedia of Spectroscopy and Spectrometry* (J.C. Lindon, ed.), Academic Press, San Diego, CA (2000), pp. 1451–61.

[217] J. Rose, *Pittsburgh Conference on Analytical Chemistry & Applied Spectroscopy (PittCon 82)*, Atlantic City, NJ (1982), Paper No. 182.

[218] D. Peters and J. Shearer, *Proceedings SPIE – Intl. Soc. Opt. Eng.* **2069**, 42–9 (1993).

[219] Y. Talmi, *Appl. Spectrosc.* **36**, 1 (1982).

[220] D.M. Mayes and J.B. Callis, *Appl. Spectrosc.* **43**, 27 (1989).

[221] J.J. Kelly, C.H. Barlow, T.M. Jinguji and J.B. Callis, *Anal. Chem.* **61**, 313 (1989).

[222] P.J. Brimmer and F.A. DeThomas, in *Making Light Work: Advances in Near Infrared Spectroscopy* (I. Murray and I.A. Cowe, eds.), VCH Publishers, Weinheim (1992), pp. 35–40.

[223] M. Buback and H.P. Vögele, *FT-NIR Atlas*, VCH, Weinheim (1993).

[224] M. Shimoyama, T. Ninomiya, K. Sano, Y. Ozaki, H. Higashiyama, M. Watari and M. Tomo, *J. Near Infrared Spectrosc.* **6**, 317–24 (1998).

[225] J.J. Workman, P.R. Mobley and B.R. Kowalski, *Appl. Spectrosc. Rev.* **31** (1/2), 73 (1996).

[226] S.D. Brown, T.S. Sum, F. Despagne and B.K. Lavine, *Anal. Chem.* **68**, 21R (1996).

[227] R. Bro, J.J. Workman, P.R. Mobley and B.R. Kowalski, *Appl. Spectrosc. Rev.* **32**, 237 (1997).

[228] P.R. Mobley, B.R. Kowalski, J.J. Workman and R. Bro, *Appl. Spectrosc. Rev.* **31**, 347 (1996).

[229] ASTM E 1655-97, *Practices for Infrared, Multivariate, Quantitative Analysis, Annual Book of ASTM Standards*, ASTM, West Conshohocken, PA (1997), Vol. 03.06.

[230] J. Workman and J. Brown, *Spectroscopy* **11** (2), 48 (1996); **11** (9), 24 (1996).

[231] *AOAC Official Methods of Analysis* (15th ed.), AOAC, Arlington, VA (1990), pp. 74–6.

[232] P. Robert, M.-F. Devaux and D. Bertrand, *J. Near Infrared Spectrosc.* **4**, 75–84 (1996).

[233] S.M. Buco, in *Making Light Work: Advances in Near Infrared Spectroscopy* (I. Murray and I.A. Cowe, eds.), VCH Publishers, Weinheim (1992), pp. 124–33.

[234] C.E. Miller, *Appl. Spectrosc. Rev.* **26**, 277–339 (1991).

[235] H.K. Pradhan, A.M.C. Davies and R. Grinter, in *Making Light Work: Advances in Near Infrared Spectroscopy* (I. Murray and I.A. Cowe, eds.), VCH Publishers, Weinheim (1992), pp. 55–60.

[236] J.S. Church, J.A. O'Neill and A.L. Woodhead, *Appl. Spectrosc.* **52** (8), 1039–46 (1998).

[237] K. Molt and D. Ihlbrock, *Fresenius J. Anal. Chem.* **348**, 523–9 (1994).

[238] M. Buback, in *Infrared and Raman Spectroscopy* (B. Schrader, ed.), VCH, Weinheim (1995).

[239] C. Kradjel and L. McDermott, in *Handbook of Near Infrared Analysis* (D.A. Burns and E.W. Ciurczak, eds.), M. Dekker, New York, NY (1992), pp. 565–608.

[240] D.B. Funk, *J. Near Infrared Spectrosc.* **4**, 101–6 (1996).

[241] C. Burgess, in *Near-Infrared Spectroscopy* (A.M.C. Davies and R. Giangiacomo, eds.), NIR Publications, Chichester (2000), pp. 101–7.

[242] V.P. Krischenko, V.G. Efremtsev and M.A. Gundaev, in *Making Light Work: Advances in Near Infrared Spectroscopy* (I. Murray and I.A. Cowe, eds.), VCH Publishers, Weinheim (1992), pp. 229–32.

[243] S.R. Lowry, J. Hyatt and W.J. McCarthy, *Appl. Spectrosc.* **54** (3), 450–5 (2000).

[244] G.H.C. Freeman, in *Making Light Work: Advances in Near Infrared Spectroscopy* (I. Murray and I.A. Cowe, eds.), VCH Publishers, Weinheim (1992), pp. 115–23.

[245] F.E. Barton II and W.R. Windham, *J. AOAC* **71**, 1162–7 (1988).

[246] W.R. Windham and F.E. Barton II, *J. AOAC* **74**, 324–31 (1991).

[247] M. Blanco, J. Coello, H. Itturiaga, S. Maspoch and C. de la Pezuela, *J. Near Infrared Spectrosc.* **5**, 67–75 (1997).

[248] V. Grassi, A. Granz and J. Coates, in *Abstracts European Seminar on Infrared Spectroscopy*, Lyon (1995), p. 29.

[249] P. Dardenne, in *Abstracts European Seminar on Infrared Spectroscopy*, Lyon (1995), p. 23.
[250] Q. Wang, S. DeJesus, J.-P. Conzen, A. Schmidt and H. Weiler, *J. Near Infrared Spectrosc.* **6**, A201–5 (1998).
[251] J.-P. Conzen, T. Stadelmann, A. Schmidt, H. Weiler, T. Droz-Georget and A. Zillan, in *Near Infrared Spectroscopy* (A.C.M. Davies and R. Giangiacomo, eds.), NIR Publications, Chichester (2000), pp. 221–4.
[252] S.V. Hammond, *Proceedings SPQ-98 (Spectroscopy in Process and Quality Control)*, Advanstar, London (1998).
[253] D.B. Coates, *Spectrosc. Europe* **14** (4), 24–6 (2002).
[254] J.J. Kellar, W.M. Cross and J.D. Miller, *Appl. Spectrosc.* **43**, 1456 (1989).
[255] L. Bokobza, *J. Near Infrared Spectrosc.* **6**, 3–17 (1998).
[256] W.F. McClure, *Anal. Chem.* **66** (1), 43A (1994).
[257] D.A. Burns and M. Margoshes, in *Handbook of Near-Infrared Analysis* (D.A. Burns and E.W. Ciurczak, eds.), M. Dekker, New York, NY (1992), pp. 1–5.
[258] J.J. Workman and D.A. Burns, in *Handbook of Near-Infrared Analysis* (D.A. Burns and E.W. Ciurczak, eds.), M. Dekker, New York, NY (1992), pp. 37–51.
[259] L.G. Weyer, *Appl. Spectrosc. Rev.* **21**, 1–43 (1985).
[260] K. Tatsubayashi and F.A. DeThomas, *Purasuchikkusu Eji* **40** (6), 157–62 (1994).
[261] J.M. Olinger and P.R. Griffiths, in *Handbook of Near-Infrared Analysis* (D.A. Burns and E.W. Ciurczak, eds.), M. Dekker, New York, NY (1992), pp. 13–35.
[262] P. Geladi and E. Dåbakk, in *Encyclopedia of Spectroscopy and Spectrometry* (J.C. Lindon, ed.), Academic Press, San Diego, CA (2000), pp. 343–9.
[263] P. Geladi and E. Dåbakk, *J. Near Infrared Spectrosc.* **3**, 119–32 (1995).
[264] P. Rees, *Scient. Comput. World* **38** (5), 18–9 (1998).
[265] T. Næs, T. Isaksson, T. Fearn and T. Davies, *A User-Friendly Guide to Multivariate Calibration and Classification*, NIR Publications, Chichester (2002).
[266] K. Kemsley, *Discriminant Analysis in Spectroscopy*, J. Wiley & Sons, New York, NY (1998).
[267] W.P. Gardiner, *Statistical Analysis Methods for Chemists*, The Royal Society of Chemistry, Cambridge (1997).
[268] G.J. McLachlan, *Discriminant Analysis and Statistical Pattern Recognition*, J. Wiley & Sons, Chichester (1992).
[269] W.J. Krzanowski, *Principles of Multivariate Analysis*, Oxford University Press, Oxford (1988).
[270] P. Williams and K.H. Norris (eds.), *Near-Infrared Technology in the Agricultural and Food Industries*, American Association of Cereal Chemists, St. Paul, MN (1990).
[271] G. Lachenal, A. Pierre and N. Poisson, *Micron* **27**, 329–34 (1996).
[272] D.A. Burns and E.W. Ciurczak (eds.), *Handbook of Near-Infrared Analysis*, M. Dekker, New York, NY (1992).
[273] A.K. Smilde, *Anal. Chem.* **70**, 1761–7 (1998).
[274] A.K. Smilde, *Proceedings SPQ-98 (Spectroscopy in Process and Quality Control)*, Advanstar, London (1998).
[275] W.H. Greive and D.D. Doepken, *Polym. Engng. Sci.* **8**, 19–23 (1968).
[276] W.G. Haanstra, W. Hansen, M.J.G. Huys, B.J. Kip, P. Palmen, J. Roumen, M. Snieder, T. v.d. Weerdhof, S. Wiedemann and V.A.L. Wortel, *Appl. Spectrosc.* **52** (6), 863–8 (1998).
[277] A.M.C. Davies, A. Grant, G.M. Gavrel and R.V. Steeper, *Analyst* **110** (6), 643–7 (1985).
[278] H.W. Siesler, *Kunststoffe* **86** (7), 1007–10 (1996).
[279] R.G.J. Miller and H.A. Willis, *J. Appl. Chem.* **6**, 385 (1956).
[280] R. Spatafore and L. McDermott, *Proceedings SPE ANTEC '91*, Montréal (1991), pp. 1942–5.
[281] M. Zahedi, *Iran. J. Polym. Sci. Technol.* **11** (1), 31 (1998).
[282] R.J.W. Le Fevre, G.M. Parkins and R. Roper, *Aust. J. Chem.* **13**, 169 (1960).
[283] A. Giammarise, *Anal. Lett.* **2** (3), 117–21 (1969).
[284] A. Krysztafkiewicz, *Colloid Polym. Sci.* **267**, 399 (1989).
[285] R. Spatafore and L. McDermott, *Plast. Compd.* **14** (6), 68–71 (1991).
[286] J.W. Hall, D.E. Grzybowski and S.L. Monfre, *J. Near Infrared Spectrosc.* **1**, 55–62 (1993).
[287] F.A. DeThomas, J.W. Hall and D.E. Grzybowski, in *Leaping Ahead Near Infrared Spectroscopy, Proceedings 6th Intl. Conference on Near Infrared Spectroscopy* (G.D. Batten, ed.), Near Infrared Spectroscopy Group, North Melbourne (1995), pp. 364–6.
[288] M.H. Pahl and M. Grosse-Aschhoff, *Kunststoffe* **87**, 872–4 (1997).
[289] D. Timm, C. Kurowski, U. Grummisch, U. Meyhack and H. Grunewald, *J. Near Infrared Spectrosc.* **6**, A243–6 (1998).
[290] K. Michalski and W.B. Mroczyk, *J. Near Infrared Spectrosc.* **6**, A247–51 (1998).
[291] R. Cinier and J. Guilment, *J. Near Infrared Spectrosc.* **6**, A231–7 (1998).
[292] D. Coombs, G. Poulter and A.I. Thomson, *Spectrosc. Europe* **10** (5), 10–4 (1998).

[293] S.L. Monfre and F.A. DeThomas, in *Near Infrared Spectroscopy* (K.I. Hildrum, T. Isaksson, T. Næs and A. Tandberg, eds.), Ellis Horwood, Chichester (1992), pp. 447–51.

[294] C. Kradjel and L. McDermott, in *Handbook of Near Infrared Analysis* (D.A. Burns and E.W. Ciurczak, eds.), M. Dekker, New York, NY (1992), pp. 565–608.

[295] R. Gnanadesikan, *Methods for Statistical Data Analysis of Multivariate Observations*, J. Wiley & Sons, New York, NY (1977).

[296] T.L. Threlfall, *Analyst* **120**, 2435 (1995).

[297] P. Corti, L. Savini, E. Dreassi, S. Perticoni, R. Genga, L. Montecchi and S. Lonardi, *Process Control Qual.* **2**, 131 (1992).

[298] P.K. Aldridge, C.L. Evans, H.W. Ward, S.T. Colgan, N. Boyer and P.J. Gemperline, *Anal. Chem.* **68**, 997 (1996).

[299] K. De Braekeleer, R. De Maesschalck, P.A. Hailey, D.C.A. Sharp and D.L. Massart, *Chemom. Intell. Lab. Syst.* **46**, 103 (1999).

[300] S.A. Yeboah and M.J. Robbins, in *Near Infrared Spectroscopy* (A.M.C. Davies and R. Giangiacomo, eds.), NIR Publications, Chichester (2000), pp. 45–51.

[301] K.H. Norris, in *Food Research and Data Analysis* (H. Martens and H. Russwarm, eds.), Applied Science Publishers, New York, NY (1985), pp. 95–114.

[302] H. Herman, K. Elliott, P. Hope and J. Hunter, *Proceedings SPQ-98 (Spectroscopy in Process and Quality Control)*, Advanstar, London (1998).

[303] L. Bellamy, A. Nordon and D. Littlejohn, *Abstracts Colloquium Spectroscopicum Internationale XXXIII*, Granada (2003), p. 129.

[304] A.S. El-Hagrasy, H.R. Morris, F. D'Amico, R.A. Lodder and J.K. Drennen, *J. Pharm. Sci.* **90**, 1298–307; 1915 (2001).

[305] W. Camacho and S. Karlsson, *Int. J. Polym. Anal. Charact.* **7**, 41–51 (2002).

[306] L. Weyer, *Appl. Spectrosc. Rev.* **21** (1–2), 1 (1985).

[307] J. Zupan and J. Gasteiger, *Neural Networks for Chemists*, VCH Publishers, Weinheim (1993).

[308] K. Saeki, K. Funatsu and K. Tanabe, *Anal. Sci.* **19**, 309–12 (2003).

[309] P.A. Turley and A. Pietrantonio, *J. Cellular Plastics* **20**, 274–7 (1984).

[310] D. Desilets, P.M. Kazmaier and R.A. Burt, *Can. J. Chem.* **73**, 319 (1995).

[311] M.P. Thomas, *J. Vinyl Addit. Technol.* **2** (4), 330–8 (1996).

[312] S. Ghosh, in *Near Infrared Spectroscopy: The Future Waves* (A.M.C. Davies and P. Williams, eds.), NIR Publications, Chichester (1996), pp. 300–5.

[313] J. Rodgers, *Proceedings 7th Intl. NIRA Symposium*, Tarrytown, NY (1984).

[314] S. Ghosh and R. Roy, *J. Textile Inst.* **79**, 504 (1988).

[315] S. Ghosh and J. Rodgers, in *Handbook of Near Infrared Analysis* (D.A. Burns and E.W. Ciurczak, eds.), M. Dekker, New York, NY (1992), pp. 495–526.

[316] T. Malfait and L. Ruys, in *Near Infrared Spectroscopy* (K.I. Hildrum, T. Isaksson, T. Næs and A. Tandberg, eds.), Ellis Horwood, Chichester (1992), pp. 423–7.

[317] K. Kitagawa, S. Hayasaki and Y. Ozaki, *Vibr. Spectrosc.* **15**, 43–51 (1997).

[318] M. Blanco, J. Coello, J.M. Garcia Fraga, H. Iturriaga, S. Maspoch and J. Pages, *Analyst* **122** (8), 777–81 (1997).

[319] M. Blanco, J. Coello, H. Iturriaga, S. Maspoch and J. Pages, *Proceedings 6th Symposium on Analytical Sciences*, Deauville (1998).

[320] W.A.P. Luck, in *Water in Polymers* (S.P. Rowland, ed.), American Chemical Society, Washington, DC (1980).

[321] R.A. Taylor, in *Near Infrared Spectroscopy: The Future Waves*, NIR Publications, Chichester (1996), pp. 677–83.

[322] M. Hammersley and P. Wilkins, *Automation and Control* (10), 13–6 (1989).

[323] L.F. Lemere, *Text. Chem. Color.* **21**, 19 (1989).

[324] G.D. Ogilvie, C. Slater and A. Ferguson, *ICI Fibres Rept.*, Harrogate (1980).

[325] J. Connel and G. Brown, *J. Text. Inst.* **69**, 357 (1978).

[326] S. Ghosh and J.R. Hassick, in *Analytical Applications of Spectroscopy* (C.S. Creaser and A.M.C. Davies, eds.), The Royal Society of Chemistry, London (1988), pp. 147–57.

[327] M.J. Drews and M. Cartree, *Text. Chem. Colorist* **26**, 23–8 (1994).

[328] T. Matsumoto, K. Tanabe, K. Saeki, T. Amano and H. Uesaka, *Bunseki Kagaku* **48**, 483 (1999).

[329] K. Saeki, T. Matsumoto, K. Tanabe and T. Amano, *Bunseki Kagaku* **50**, 223 (2001).

[330] M. Shimoyama, T. Amari, T. Kamiya, H. Sato, T. Ninomiya and Y. Ozaki, *IUPAC Polymer Conference on the Mission and Challenges of Polymer Science and Technology (IUPAC-PC 2002)*, Kyoto (2002), p. 536.

[331] T. Nagata, M. Oshima and M. Tanigaki, *Polym. Engng. Sci.* **40** (5), 1107–13 (2000).

[332] C.E. Miller, in *Near Infrared Spectroscopy* (K.I. Hildrum, T. Isaksson, T. Næs and A. Tandberg, eds.), Ellis Horwood, Chichester (1992), pp. 413–22.

[333] T. Huth-Fehre, R. Feldhoff, F. Kowol, H. Freitag, S. Kuttler, B. Lohwasser and M. Oleimeulen, *J. Near Infrared Spectrosc.* **6**, A7–A11 (1998).

[334] N. Eisenreich, J. Herz, H. Kull, W. Mayer and T. Rohe, *Proceedings SPE ANTEC '96*, Indianapolis, IN (1996), pp. 3131–5.
[335] S.T. DeJesus and Q. Wang, *J. Near Infrared Spectrosc.* **6**, A259–63 (1998).
[336] R. Feldhoff, *Dissertation*, Münster (1996).
[337] C.E. Miller, in *Making Light Work: Advances in Near Infrared Spectroscopy* (I. Murray and I.A. Cowe, eds.), VCH Verlag, Weinheim (1992), pp. 519–24.
[338] R. Feldhoff, T. Huth-Fehre, T. Kantimm, L. Quick, K. Cammann, W. v.d. Broek, D. Wienke and H. Fuchs, *J. Near Infrared Spectrosc.* **3**, 3–9 (1995).
[339] D.M. Scott and R.L. Waterland, *Polym. Engng. Sci.* **35**, 1011–5 (1995).
[340] A. Kulcke, C. Gurschler, G. Spock, R. Leitner and M. Kraft, *J. Near Infrared Spectrosc.* **11**, 71–81 (2003).
[341] N. Eisenreich, *Identifizieren der Kunststoffe in Verpackungsabfällen. Schnelles Identifizieren mit Nah-Infrarot-Spektroskopie. Information Bulletin*, Fraunhofer Institut f. Chemische Technologie, Kunststoff-Messe K'92, Düsseldorf (1992).
[342] H. Mark and D. Tunnel, *Anal. Chem.* **57**, 1449 (1985).
[343] J.A. Mitchell, C.D. Bockman and A.V. Lee, *Anal. Chem.* **29**, 499 (1957).
[344] B. Kip, M. Huys and T. Berghmans, *Techn. Rept. DSM Research*, Geleen (1998).
[345] C. Ventura and M. Papini, *J. Near Infrared Spectrosc.* **5**, 123–33 (1997).
[346] L. Rey, J. Galy, H. Sautereau, G. Lachenal, D. Henry and J. Vial, *Appl. Spectrosc.* **54** (1), 39–43 (2000).
[347] M. Snir, S. Wasserman, H. Dodiuk and S. Kenig, *J. Adhes.* **27**, 175 (1989).
[348] W. Campbell, *Proceedings NIRS Intl. Symposium*, Scheveningen (1986).
[349] B.D. Jordan and S.J. Popson, *J. Pulp Pap. Sci.* **20**, 161 (1994).
[350] E.W. Ciurczak, *ChemTech* (6), 374–80 (1992).
[351] C.V. Raman and K.S. Krishnan, *Nature* **12** (3048), 501 (1928).
[352] T.J. Vickers, C.K. Mann, J. Zhu and C.K. Chong, *Appl. Spectrosc. Rev.* **26** (4), 339–73 (1991).
[353] I. Nabiev and M. Manfait, *Rev. Inst. Fr. Petr.* **48** (3), 261–85 (1993).
[354] A. Fadini and F.M. Schnepel, *Vibrational Spectroscopy: Methods and Applications*, Ellis Horwood, Chichester (1989).
[355] B. Schrader (ed.), *Infrared and Raman Spectroscopy. Methods and Applications*, VCH Publishers, Weinheim (1995).
[356] N. Everall, B. King and I. Clegg, *Chem. Br.* **36** (7), 40–3 (2000).
[357] S.A. Asher and C.R. Johnson, *Science* **225**, 311–3 (1984).
[358] S.E.J. Bell, E.S.O. Bourguignon, A. O'Grady, J. Villaumie and A.C. Dennis, *Spectrosc. Europe* **14** (6), 17–20 (2002).
[359] B. Chase, *Anal. Chem.* **59**, 881A (1987).
[360] B. Schrader, *Fresenius J. Anal. Chem.* **355**, 233–9 (1996).
[361] P. Jacquinot, *J. Opt. Soc. Am.* **44**, 761 (1954).
[362] M.J. Pelletier, in *Analytical Applications of Raman Spectroscopy* (M.J. Pelletier, ed.), Blackwell Science, Oxford (1999), pp. 53–105.
[363] B. Schrader, in *Encyclopedia of Spectroscopy and Spectrometry* (J.C. Lindon, ed.), Academic Press, San Diego, CA (2000), pp. 1986–92.
[363a] F. Adar, *LabPlus Intl.* **18** (1), 18–21 (2004).
[364] D.B. Chase and J.F. Rabolt (eds.), *Fourier Transform Raman Spectroscopy*, Academic Press, San Diego, CA (1994).
[365] P. Hendra, C. Jones and G. Warnes, *Fourier Transform Raman Spectroscopy: Instrumentation and Chemical Applications*, Prentice Hall, New York, NY (1991).
[366] Special issues of *Spectrochim. Acta A* devoted to FT-Raman spectroscopy: **46A** (2) (1990); **47A** (9/10) (1991); **49A** (5/6) (1993); **50A** (11) (1994); **51A** (12) (1995); **53A** (13) (1997).
[367] R.A. Nyquist, C.L. Putzig and M.A. Leugers, *Infrared and Raman Spectral Atlas of Inorganic Compounds and Organic Salts: Raman Spectra*, Academic Press, San Diego, CA (1997).
[368] Nicolet Instrument Corp. (Madison, WI), 15000 spectra of common chemical compounds; Sadtler (BioRad Laboratories, Cambridge, MA), 1650 spectra of monomers and polymers; AIST/STJ (Japan), 735 polymers and polymer additives and 125 dyes and pigments.
[369] R.A. Nyquist, *Interpreting Infrared, Raman and Nuclear Magnetic Resonance Spectra*, Academic Press, Orlando, FL (2001).
[370] M.J. Adams, *Chemometrics in Analytical Spectroscopy*, The Royal Society of Chemistry, Cambridge (1995).
[371] A.T.G. De Paepe, J.M. Dyke, P.J. Hendra and F.W. Langkilde, *Spectrochim. Acta* **A53**, 2267–73 (1997).
[372] M.J. Pelletier, *Appl. Spectrosc.* **57** (1), 20–39A (2003).
[373] T.J. Vickers and C.K. Mann, in *Analytical Raman Spectroscopy* (J.G. Grasselli and B.J. Bulkin, eds.), Wiley-Interscience, New York, NY (1991), pp. 107–35.
[374] I.R. Lewis and P.R. Griffiths, *Appl. Spectrosc.* **50**, 12–30A (1996).

[375] N. Everall, in *Analytical Applications of Raman Spectroscopy* (M.J. Pelletier, ed.), Blackwell Science, Oxford (1999), pp. 127–92.

[376] D. Wolverson, in *An Introduction to Laser Spectroscopy* (D.L. Andrews and A.A. Demidov, eds.), Plenum Press, New York, NY (1995), pp. 91–114.

[377] D.L. Andrews, in *Encyclopedia of Spectroscopy and Spectrometry* (J.C. Lindon, ed.), Academic Press, San Diego, CA (2000), pp. 1993–2000.

[378] D. Bougeard, H. Hamaguchi and L.D. Ziegler, *J. Raman Spectrosc.* **34** (2), 97–9 (2003).

[379] D.L. Gerrard and H.J. Bowley, *Anal. Chem.* **58**, 6R–13R (1986).

[380] P.J. Hendra, in *Modern Techniques in Raman Spectroscopy* (J.J. Laserna, ed.), J. Wiley & Sons, Chichester (1996), pp. 73–108.

[381] R.L. McCreery, in *Modern Techniques in Raman Spectroscopy* (J.J. Laserna, ed.), J. Wiley & Sons, Chichester (1996), pp. 41–71.

[382] B.P. Lenain, *Analysis* **28**, 11–4 (2000).

[383] A. Brookes, *Spectrosc. Europe* **11** (2), 10–4 (1999).

[384] E.D. Lipp and M.A. Leugers, in *Analytical Applications of Raman Spectroscopy* (M.J. Pelletier, ed.), Blackwell Science, Oxford (1999), pp. 106–26.

[385] D.O. Hummel, *Atlas of Polymer and Plastics Analysis*, C. Hanser Verlag, Munich (1990), Vol. 2.

[386] P.J. Hendra and J. Agbenyega, *The Raman Spectroscopy of Polymers*, J. Wiley & Sons, Chichester (1993).

[387] X.Q. Yang, H.S. Lee, L. Hanson, J. McBreen and Y. Okamoto, *J. Power Sources* **54** (2), 198–204 (1995).

[388] C. Tseng, C. Mann and T. Vickers, *Appl. Spectrosc.* **48**, 535–7 (1994).

[389] W.F. Maddams, *Spectrochim. Acta* **50A** (11), 1967–86 (1994).

[389a] T.R. Crompton, *Practical Polymer Analysis*, Plenum Press, New York, NY (1993).

[390] M.J. Manzaneda, J.M. Yúfera, S. Ruiz-Moreno, M.J. Someira, P. Morillo and M. Breitman, in *Lasers in the Conservation of Artworks* (W. Kautek and E. König, eds.), Verlag Mayer & Comp., Klosterneuburg and Vienna (1997), pp. 61–7.

[391] Special Issue – *Raman Spectroscopy in Art, Medicine and Archaeology, J. Raman Spectrosc.* **28**, 77–197 (1997).

[392] V.M. Hallmark, C.G. Zimba, J.D. Swalen and J.F. Rabolt, *Spectroscopy* **2**, 40 (1987).

[393] T.M. Hancewicz, in *Anionic Surfactants* (J. Cross, ed.), M. Dekker, New York, NY (1998), pp. 125–67.

[394] M.J. Pelletier (ed.), *Analytical Applications of Raman Spectroscopy*, Blackwell Science, Oxford (1999).

[395] H.J. Bowley, D.L. Gerrard and I.S. Biggin, *Polym. Degr. Stabil.* **20**, 257–69 (1988).

[396] T. Mitsui, S. Hanai and H. Umeda (to Densen Sogo Gijutsu Sentaa), *Jpn. Kokai Tokkyo Koho* JP 1993–65746 (March 24, 1993).

[397] N.J. Everall, M. Chalmers and I.D. Newton, *Appl. Spectrosc.* **40**, 597–601 (1992).

[398] B.A. Barrera and A.J. Sommer, *Appl. Spectrosc.* **52** (11), 1483–7 (1998).

[399] C.H. Jones, *Spectrochim. Acta* **47A** (9/10), 1313 (1990).

[400] C. Pakjamsai and J. Suwanprateeb, *J. Appl. Polym. Sci.* **78** (11), 1947–54 (2000).

[401] D. Bourgeois and S.P. Church, *Spectrochim. Acta* **46A**, 295–301 (1990).

[402] S. Kokot, N.A. Tuan and L. Rintoul, *Appl. Spectrosc.* **51**, 387–95 (1997).

[403] G.N. Andreev, B. Schrader, H. Schulz, R. Fuchs, S. Popov and N. Handjieva, *Fresenius J. Anal. Chem.* **371** (7), 1009–17 (2001).

[404] D. Freitag, C. Götz, G. Luft, A. van der Pol and K. Wierda, *Appl. Spectrosc.* **55** (2), 136–41 (2001).

[405] H.G.M. Edwards, D.W. Farwell, J.M. Holden and E.M. Lawson, *Studies Conserv.* **43**, 9–16 (1998).

[406] K.D.O. Jackson, M.J.R. Loadman, C.H. Jones and G. Ellis, *Spectrochim. Acta* **46A** (2), 217–26 (1990).

[407] C.M. Snively and J.L. Koenig, in *Encyclopedia of Spectroscopy and Spectrometry* (J.C. Lindon, ed.), Academic Press, San Diego, CA (2000), pp. 1858–64.

[408] D.L. Gerrard and W.F. Maddams, *Appl. Spectrosc. Rev.* **22**, 251–334 (1986).

[409] F.J. Bergin and H.F. Shurvell, *Anal. Proc. (London)* **26** (7), 263–4 (1989).

[410] G.D. Smith, *Revs. Conserv.* **2**, 92–106 (2001).

[410a] H.G.M. Edwards and J.M. Chalmers (eds.), *Raman Spectroscopy in Archaeology and Art History*, The Royal Society of Chemistry, Cambridge (2005).

[411] W. Kiefer and M. Spiekermann, in *Infrared and Raman Spectroscopy* (B. Schrader, ed.), VCH, Weinheim (1995), Chp. 6.

[412] R.H. Brody, E.A. Carter, H.G.M. Edwards and A.M. Pollard, in *Encyclopedia of Spectroscopy and Spectrometry* (J.C. Lindon, ed.), Academic Press, San Diego, CA (2000), pp. 649–57.

[413] S. Asher, *Anal. Chem.* **65**, 59–66A (1993).

[414] S.A. Asher, C.H. Munro and Z. Chi, *Laser Focus World* **33** (7), 99–109 (1997).

[415] J.M. Hollas, *High Resolution Spectroscopy*, J. Wiley & Sons, Chichester (1998), pp. 468–710.

[416] V. Pajcini, C.H. Munro, R.W. Bormett, R.E. Witkowski and S.A. Asher, *Appl. Spectrosc.* **51** (1), 81–6 (1997).

[417] L. Van Haverbeke, P.F. Lynch and C.W. Brown, *Anal. Chem.* **50**, 315–7 (1978).

[418] R.J.H. Clark, T.J. Dines and M. Kurmoo, *Inorg. Chem.* **22** (19), 2766–72 (1983).

[419] P.M. Killough, V.L. DeVito and S.A. Asher, *Appl. Spectrosc.* **45** (7), 1067–9 (1991).

[420] D.A. Long, *Raman Spectroscopy*, McGraw-Hill, New York, NY (1977).

[421] K.P.J. Williams and D. Klenerman, *J. Raman Spectrosc.* **23** (4), 191–6 (1992).

[422] Y. Furukawa, H. Ohta, A. Sakamoto and M. Tasumi, *Spectrochim. Acta* **47A**, 1367 (1991).

[423] S.A. Asher and C.M. Jones, *Prepr. Pap. – Am. Chem. Soc., Div. Fuel Chem.* 31 (1), 170–80 (1986).

[424] D. Klenerman and A. v.d. Pol, unpubl. results (1999).

[425] G.A. Voyiatzis and K.S. Andrikopoulos, *Proceedings SPQ-98 (Spectroscopy in Process and Quality Control)*, Advanstar, London (1998).

[426] S.A. Asher, *Anal. Chem.* **65** (4), 201A–10A (1993).

[427] R. Chang and T.E. Furtak, *Surface Enhanced Raman Spectroscopy*, Plenum Press, New York, NY (1982).

[428] M. Fleischmann, P.J. Hendra and A.J. McQuillan, *Chem. Phys. Lett.* **26**, 163–6 (1974).

[429] A. Otto, *J. Raman Spectrosc.* **22**, 743 (1991).

[430] M.J. Pelletier, in *Analytical Applications of Raman Spectroscopy* (M.J. Pelletier, ed.), Blackwell Science, Oxford (1999), pp. 1–52.

[431] A. Rupérez and J.J. Laserna, in *Modern Techniques in Raman Spectroscopy* (J.J. Laserna, ed.), J. Wiley & Sons, Chichester (1996), pp. 227–64.

[432] T. Fukuoka, K. Nakamura, Y. Mori, Y. Yoshimura and K. Matsukawa, *Proceedings IUPAC Intl. Congress on Analytical Sciences 2001*, Tokyo (2001), Poster P4071.

[433] S. Nie and S.R. Emory, *Science* **275**, 1102–6 (1997).

[434] J.J. Laserna, *Anal. Chem. Acta* **283**, 607 (1993).

[435] R.L. Garrell, *Anal. Chem.* **61**, 401A (1989).

[436] C.D. Tran, *Anal. Chem.* **56**, 824 (1984).

[437] F.J. Boerio, P.P. Hong, P.J. Clark and Y. Okamoto, *Langmuir* **6**, 721 (1990).

[438] D.L. Allara, C.A. Murray and S. Bodoff, in *Physicochemical Aspects of Polymer Surfaces* (K.L. Mittal, ed.), Plenum Press, New York, NY (1983), Vol. 1

[439] L.M. Siperko, *Appl. Spectrosc.* **43**, 226 (1989).

[440] C.H. Munro, W.E. Smith, M. Garner, J. Clarkson and P.C. White, *Langmuir* **11**, 3712 (1995).

[441] W.E. Smith and C. Rodger, in *Encyclopedia of Spectroscopy and Spectrometry* (J.C. Lindon, ed.), Academic Press, San Diego, CA (2000), pp. 2329–34.

[442] C. Rodger, G. Dent, J. Watkinson and W.E. Smith, *Appl. Spectrosc.* **54** (11), 1567–76 (2000).

[443] C. Rodger and D. Broughton, *Analyst* (Cambridge) **123** (9), 1823–6 (1998).

[444] D. Filmore, *Today's Chemist at Work*, 15–8 (2002).

[445] R.H. Clarke, S. Londhe and M.E. Womble, *Spectroscopy* **13** (10), 29–35 (1998).

[446] A. Rosencwaig and A. Gersho, *J. Appl. Phys.* **47**, 64–9 (1976).

[447] A.G. Bell, *Proc. Am. Assoc. Adv. Sci.* **29**, 115 (1880).

[448] A. Rosencwaig, *Photoacoustics and Photoacoustic Spectroscopy*, Wiley-Interscience, New York, NY (1980).

[449] G.A. West and J.J. Barrett, *Opt. Lett.* **4**, 395 (1979).

[450] M.J. Adams, A.A. King and J.F. Kirkbright, *Analyst* **101**, 733 (1976).

[451] J.A. Gardella, G.L. Grobe III, W.L. Hopson and E.M. Eyring, *Anal. Chem.* **56**, 1169–77 (1984).

[452] M. Urban, *J. Adhesion Sci. Technol.* **7**, 1 (1993).

[452a] G.M. Story and C. Marcott, *AIP Conf. Proceedings (Fourier Transform Spectroscopy)* **430**, 513–5 (1998).

[453] J.F. McClelland, R.W. Jones, S. Luo and L.M. Seaverson, in *Practical Sampling Techniques for Infrared Analysis* (P.B. Coleman, ed.), CRC Press, Boca Raton, FL (1993), Chp. 5.

[454] J.R. Parker, in *Spectroscopy of Rubbers and Rubbery Materials* (V.M. Litvinov and P.P. De, eds.), Rapra Technology Ltd., Shawbury (2002), pp. 49–76.

[455] J.M. Chalmers, N.J. Everall, K. Hewitson and A. Grady, *Proceedings ACS Div. Polym. Mater.: Sci. Engng.* **82**, 402–3 (2000).

[456] M.W. Sigrist, in *Encyclopedia of Spectroscopy and Spectrometry* (J.C. Lindon, ed.), Academic Press, San Diego, CA (2000), pp. 1810–4.

[457] A. Miklós, S. Schäfer and P. Hess, in *Encyclopedia of Spectroscopy and Spectrometry* (J.C. Lindon, ed.), Academic Press, San Diego, CA (2000), pp. 1815–22.

[458] J.F. McClelland, S.J. Bajic, R.W. Jones and L.M. Seaverson, in *Modern Techniques in Applied Molecular Spectroscopy* (F.M. Mirabella, ed.), J. Wiley & Sons, New York, NY (1998), Chp. 6.

[459] D.E.M. Spillane, in *Encyclopedia of Analytical Science* (A. Townshend, ed.), Academic Press, London (1995), pp. 3971–6.

[460] D. Betteridge and P.J. Meylor, *Crit. Rev. Anal. Chem.* **14** (4), 267–95 (1984).

[461] R.M. Leblanc, in *Encyclopedia of Physical Science and Technology* (R.A. Meyers, ed.), Academic Press, San Diego, CA (2001), Vol. 12.
[462] D.W. Vidrine, in *Fourier Transform Infrared Spectroscopy: Techniques in Fourier Transform Interferometry* (J.R. Ferraro and L.J. Basile, eds.), Academic Press, New York, NY (1982), Vol. 3, Chp. 4.
[463] D.O. Hummel and F. Scholl, *Atlas der Polymer- und Kunststoffanalyse*, C. Hanser Verlag and VCH Verlagsgesellschaft, Munich (1988), Vol. 2.
[464] R.O. Carter III, *Opt. Eng.* **36**, 326–31 (1997).
[465] C. Haisch and R. Niessner, *Spectrosc. Europe* **14** (5), 10–5 (2002).
[466] Q. Jin, G.F. Kirkbright and D.E.M. Spillane, *Appl. Spectrosc.* **36** (2), 120–4 (1982).
[467] C.M. Ashworth, G.F. Kirkbright and D.E.M. Spillane, *Analyst* **108**, 1481–4 (1983).
[468] M.W.C. Wahls and J.C. Leyte, *Proceedings 5th European Conference on Advanced Materials, Processes & Applications (Euromat-97)* (L.A.J.L. Sarton and H.B. Zeedijk, eds.), Maastricht (1997), Vol. 3, pp. 307–10.
[469] M.W.C. Wahls and J.C. Leyte, *Appl. Spectrosc.* **52** (1), 123–7 (1998).
[470] E.Y. Jiang, D.L. Drapcho, W.J. McCarthy and R.A. Crocombe, *AIP Conference Proceedings (Fourier Transform Spectroscopy)*, Vol. 430, pp. 381–4 (1998).
[471] M.C. Paputa Peck, M.A. Samus, P.C. Kilgoar and R.O. Carter III, *Rubber Chem. Technol.* **64** (4), 610 (1991).
[472] C.Q. Yang, R.R. Bresce and W.G. Fateley, *Appl. Spectrosc.* **41** (5), 889–96 (1987).
[473] A.H. Sharma, F. Mozayeni and J. Alberts, *Proceedings Polyolefins XI* (SPE, ed.), Houston, TX (1999), pp. 679–703.
[474] A.H. Sharma and B.C. Beard, *Communication at SPE Regional Technical Conference*, Atlantic City, NJ (1996).
[475] D. Brück, *Kautsch. Gummi Kunstst.* **39** (12), 1165–74 (1986).
[476] R.O. Carter, R.A. Palmen, R.M. Dittmar, C.J. Manning, S.S. Bains and J.L. Chao, *Proceedings SPIE – Intl. Soc. Opt. Eng.* **1145**, 362–3 (1989).
[477] G.M. Story and C. Marcott, *AIP Conference Proceedings (Fourier Transform Spectroscopy)*, Vol. 430, pp. 513–5 (1998).
[478] M.W.C. Wahls and J.C. Leyte, *J. Appl. Phys.* **83** (1), 504–9 (1998).
[479] L. Quintanilla, J.C. Rodriguez-Cabello, T. Jawhari and J.M. Pastor, *Polymer* **34** (18), 3787–95 (1993).
[480] Y. Zhao and M.W. Urban, *Proceedings A.C.S. Div. Polym. Mater.: Sci. & Engng.* **78**, 25–6 (1998).
[481] J.F. McClelland, R. Jones and S. Bajic, *Proceedings SPE ANTEC '99*, New York, NY (1999), pp. 2424–8.
[482] M.A. Lacasse and R.M. Paroli, *ASTM Spec. Tech. Publ., STP* **1243** (1995), pp. 29–48.
[483] W.H. Waddell and J.R. Parker, *Rubber Chem. Technol.* **65**, 836 (1992).
[484] R.I. Grose, S. Hvilsted and H.W. Siesler, *Makromol. Chem., Macromol. Symp.* **52**, 175 (1991).
[485] M.W. Urban, E.G. Chatzi, B.C. Perry and J.L. Koenig, *Appl. Spectrosc.* **40** (8), 1103 (1986).
[486] E.A. Carter, P.M. Fredericks and J.S. Church, *Text. Res. J.* **66**, 787 (1996).
[487] L.B. Lloyd, R.C. Yeates and E.M. Eyring, *Anal. Chem.* **54**, 549 (1982).
[488] I.L.B. Olsen and M. Olsen, *Chemical Plants & Process.* (3), 48 (2000).
[489] S.O. Karstad and P.E. Nordell, *Appl. Surf. Sci.* **6**, 372 (1980).
[490] G.A. George, M. Celina, A.M. Vassallo and P.A. Cole-Clarke, *Polym. Degr. Stabil.* **48** (2), 199–210 (1995).
[491] F.J. DeBlase and S. Compton, *Appl. Spectrosc.* **454**, 611 (1991).
[492] P.S. Belton, E.K. Kemsley, J. Potter and R.H. Wilson, *Proceedings 8th Eur. Conference on Food Chemistry* (G. Sontag and W. Pfannhauser, eds.), Vienna (1995), Vol. 1, pp. 164–7.
[493] R.J. Pell, J.B. Callis and B.R. Kowalski, *Appl. Spectrosc.* **45**, 808–18 (1991).
[494] N. Sheppard, in *Analytical Application of FT-IR to Molecular and Biological Systems* (J.R. Durig, ed.), Reidel, Dordrecht (1980), p. 125.
[495] J. Dong, P.M. Fredericks and G.A. George, *Polym. Degr. Stabil.* **58** (1–2), 159–69 (1997).
[496] G.A. George, A.M. Vassallo, P.A. Cole-Clarke and M. Celina, *Angew. Makromol. Chem.* **232**, 105–18 (1995).
[497] M. Celina, D.K. Ottesen, K.T. Gillen and R.L. Clough, *Polym. Degr. Stabil.* **58**, 15–31 (1997).
[498] B.B. Wheals, in *Analytical Pyrolysis* (C.E.R. Jones and C.A. Cramers, eds.), Elsevier, Amsterdam (1997), p. 89.
[499] M. Friedrich and D.R.T. Zahn, *Appl. Spectrosc.* **52** (12), 1530–5 (1998).
[500] G.A. George, G.A. Cash, H. Liu and T. Vassallo, *Polym. Mater. Sci. Eng.* **74**, 90–1 (1996).
[501] G.A. George, G.A. Cash and L. Rintoul, *Polym. Intl.* **41** (2), 169–82 (1996).
[502] J.B. Bates, in *Fourier Transform Infrared Spectroscopy* (J.R. Ferraro and L.J. Basile, eds.), Academic Press, New York, NY (1978), Chp. 3.
[503] G. Stokes, *Phil. Trans.* **143**, 463 (1852).
[504] V.P. Gribkovskii, in *Luminescence of Solids* (D.R. Vij, ed.), Plenum Press, New York, NY (1998), pp. 1–43.
[505] B. Stuart, *Polymer Analysis*, J. Wiley & Sons, Chichester (2002).

[506] W.R.G. Baeyens and B. Lin Ling, *J. Planar Chromatogr.* **1**, 198 (1988).
[507] Starna Brand Fluorescence Intensity Samples.
[508] S.G. Schulman, Q.Q. Di and J. Jochum, in *Encyclopedia of Spectroscopy and Spectrometry* (J.C. Lindon, ed.), Academic Press, San Diego, CA (2000), pp. 1718–25.
[509] R.S. Becker, *Theory and Interpretation of Fluorescence and Phosphorescence*, Wiley-Interscience, New York, NY (1969).
[510] F.C. De Schrijver and D. De Keukeleire, in *Luminescence Techniques in Chemical and Biochemical Analysis* (W.R.G. Baeyens, D. De Keukeleire and K. Korkidis, eds.), M. Dekker, New York, NY (1991), pp. 13–20.
[511] R.P. Haugland, *Molecular Probes Handbook of Fluorescent Probes and Research Chemicals*, Molecular Probes, Eugene, OR (1992).
[512] J.N. Miller, *Standards in Fluorescence Spectrometry*, Chapman & Hall, London (1981).
[513] G.R. Goe, *Proceedings SPE ANTEC '77*, Montreal (1977), p. 496.
[514] N.S. Allen, *Polym. Degr. Stabil.* **6**, 193 (1984).
[515] N.S. Allen, M. Edge, D. Holdsworth, A. Rahman, F. Catalina, E. Fontan, A.M. Escalona and F.F. Sibon, *Polym. Degr. Stabil.* **67**, 57–67 (2000).
[516] J. Scheirs, J. Pospíšil, M.J. O'Connor and S.W. Bigger, *Adv. Chem. Ser.* **249**, 359–74 (1996).
[517] *Application Note FLA-34, Fluorescence Applications*, Perkin-Elmer, Norwalk, CT (n.d.).
[518] F. Zong-huai, H. Miyata, T. Takahashi, T. Itou and T. Nakiri, *Proceedings 2nd Intl. Symposium on Electrical Insulating Materials*, Toyohashi (1998), pp. 211–4.
[519] A.D. DeBellis, R. Iyengar, N.A. Kaprinidis, R.K. Rodebaugh and J. Suhadolnik, *ACS Symp. Ser.* **805**, 453–67 (2002).
[520] V.N. Provorov and V.D. Zaitseva, *Metody Analiza Syr'ya i Materialov, Primenyaemykh v Rezino Prom.*, Moscow (1959), 49–53; *C.A.* **56**, 2539c (1962).
[521] C.F. Poole and S. Khatib, in *Quantitative Analysis Using Chromatographic Techniques* (E. Katz, ed.), J. Wiley & Sons, Chichester (1987), pp. 193–270.
[522] W. Mische, E. Schröder and K. Thinius, *Plaste Kautsch.* **19** (12), 882–6 (1972).
[523] N.S. Allen, M. Edge, J. Daniels and D. Royall, *Abstracts 37th IUPAC Intl. Symposium on Macromolecules*, Gold Coast (1998), p. 956.
[524] E.C. Corbett, J.G. Frey, R.I. Grose, P.J. Hendra and T. Taylor-Brown, *Plast. Rubber Compos. Process. Appl.* **21** (1), 5–11 (1994).
[525] B. Simmons, R. Ahmad and B. Overton, *Br. Plast. Rubber*, 4–10 (June 1998).
[526] J.J. Krutak, M.R. Cushman, C.A. Coates, W.W. Parham, M.A. Weaver and G. Patonay (to Eastman Co.), *U.S. Pat.* 5,397,819 (March 3, 1995).
[527] J. Slavík (ed.), *Fluorescence Microscopy and Fluorescent Probes*, Plenum Press, New York, NY (1996), Vol. 2.
[528] B. Simmons, R. Ahmad and B. Overton, *Br. Plast. Rubber* (6), 4–10 (1998).
[529] M. Dubreuil, *Progr. Pap. Recycl.* **4**, 98–108 (1995).
[530] S. Shimoyama, Y. Noda and S. Katsuhara, *Bunseki Kagaku* **47** (2), 93–100 (1998).
[531] E. Cleve and E. Schollmeyer, *Textilveredl.* **30**, 18–21 (1995).
[532] N.S. Allen, *Photochem.* **28**, 381–452 (1997).
[533] G.D. Mendenhall, *Adv. Chem. Ser.* **249**, 213–21 (1996).
[534] A.J. Bur, R.E. Lowry, S.C. Roth, C.L. Thomas and F.W. Wang, *Macromolecules* **25**, 3503 (1992).
[535] A.J. Bur, K. Migler, M.G. Vangel and D.S. Johnsonbaugh, in *CAE and Intelligent Processing of Polymeric Materials* (Am. Soc. Mech. Eng. Mater. Div., ed.), Vol. 97, pp. 199–211 (1997).
[536] A.J. Bur and S.C. Roth, *Proceedings SPE ANTEC '98*, Atlanta, GA (1998), pp. 2090–4.
[537] A.J. Bur and S.C. Roth, *Proceedings SPE ANTEC 2000*, Orlando, FL (2000), pp. 2058–63.
[538] A.J. Bur, S.C. Roth, D.W.Baugh, K.A. Koppi, M.A. Spalding, J. Gunderson and W.C. Buzanowski, *Proceedings SPE ANTEC 2003*, Nashville, TN (2003), pp. 3341–5.
[539] A.J. Bur, S.C. Roth and H. Lobo, *Proceedings SPE ANTEC 2001*, Dallas (2001), pp. 3071–5.
[540] J. Dunkers, D.L. Woerdeman and R. Parnas, *Proceedings 10th Annu. ASM/ESD Adv. Comp. Conf.*, Dearborn, MI (1994), p. 257.
[541] N.-J. Kim, C.N. Jeong, C.S.P. Sung and N.-H. Sung, *Polym. Mater. Sci. Eng.* **81**, 387–8 (1999).
[542] K.F. Plitt and S.D. Toner, *J. Appl. Polym. Sci.* **5**, 534 (1961).
[543] A. Gorelenko, L. Korochkin and S. Pliska, *Proceedings SPIE – Intl. Soc. Opt. Eng.* **4535**, 98–110 (2001).
[544] G.F. Kirkbright, R. Narayanaswamy and T.S. West, *Anal. Chim. Acta* **52**, 237–46 (1970).
[545] C.F. Forster and E.F. Rickard, *Nature* **197**, 1199 (1963).
[546] J.A. Bodi and A. Burgess, *Proceedings SPE ANTEC 2000*, Orlando, FL (2000), pp. 2611–5.
[547] G. Oster, N. Geacintov and A.U. Khan, *Nature* **197**, 1089 (1962).
[548] W.G. Charlston, D.M. Ryder, C.J. Olliff, C. Laurent, F. Massines, G. Teyssedre and P. Tiemblo, *Proceedings IEEC Annu. Rep.*, Minneapolis, MN (1997), Vol. 1, pp. 250–3.

[549] N.S. Allen, M. Edge, J. Daniels and D. Royall, *Polym. Degr. Stabil.* **62**, 373–83 (1998).

[550] E.V. Anufrieva and A.D. Saitzeva, *Izv. Akad. Nauk SSSR, Ser. Fiz.* **24**, 755 (1960).

[551] P. Suppan, *Chemistry and Light*, The Royal Society of Chemistry, Cambridge (1994).

[552] B. Radziszewski, *Ber. Dtsch. Chem. Ges.* **10**, 70, 321 (1877).

[553] J. Crnko, B.C. Kibler and E.M. Fujinari, in *Instrumental Methods in Food and Beverage Analysis* (D.L.B. Wetzel and G. Charalambous, eds.), Elsevier, Amsterdam (1998), p. 379.

[554] E.M. Fujinari, in *Instrumental Methods in Food and Beverage Analysis* (D.L.B. Wetzel and G. Charalambous, eds.), Elsevier, Amsterdam (1998), pp. 385–423.

[555] R.L. Benner and D.H. Stedman, *Anal. Chem.* **61**, 1268 (1989).

[556] J.-F. Borny and J. Werner, *Abstracts 6th Intl. Symposium on Hyphenated Techniques in Chromatography and Hyphenated Chromatographic Analyzers (HTC-6)*, Bruges (2000), P13.

[557] R.S. Hutte, *J. Chromatogr. Libr.* **56**, 201–29 (1995).

[558] S.O. Farwell and C.J. Barinaga, *J. Chromatogr. Sci.* **24**, 483 (1986).

[559] G.E. Ashby, *J. Polym. Sci.* **50**, 99 (1961).

[560] G.A. Russell, *Chem. Ind. (London)*, 1483 (1956).

[561] G.A. Russell, *J. Am. Chem. Soc.* **79**, 3871 (1957).

[562] G.A. George, in *Developments in Polymer Degradation-3* (N. Grassie, ed.), Applied Science Publishers, London (1981), p. 173.

[563] S.H. Lee and G.D. Mendenhall, *J. Am. Chem. Soc.* **110**, 4318 (1989).

[564] B. Lánská, *Polym. Degr. Stabil.* **61**, 27–31 (1998).

[565] L. Matisová-Rychlá and J. Rychlý, *Adv. Chem. Ser.* **249**, 175–93 (1996).

[565a] G.A. George, in *Luminescence Techniques in Solid State Polymer Research* (L. Zlatkevich, ed.), M. Dekker, New York, NY (1989), Chp. 3.

[566] I. Blakey and G.A. George, *Macromolecules* **34** (6), 1873–80 (2001).

[567] G. George, *Abstracts First Intl. Conference on Polymer Modification, Degradation and Stabilisation (MoDeSt)*, Palermo (2000).

[568] M. Celina and G.A. George, *Polym. Degr. Stabil.* **40**, 323 (1993).

[569] M.P. Schard and C.A. Russell, *J. Appl. Polym. Sci.* **8**, 985, 997 (1964).

[570] G.D. Mendenhall, *Angew. Chem. Int. Ed.* **16**, 225 (1977).

[571] E.S. O'Keefe and N.C. Billingham, *Polym. Degr. Stabil.* **10**, 137 (1985).

[572] L. Zlatkevich (ed.), *Luminescence Techniques in the Solid State Polymer Research*, M. Dekker, New York, NY (1989), p. 135.

[573] D.R. Kohler and C. Kröhnke, *Polym. Degr. Stabil.* **62**, 385–93 (1998).

[574] Atlas Electric Devices Co., Chicago, USA; Tohoku Electronic Industrial Co., Ltd., Japan; Slovak Academy of Sciences, Bratislava, Slovak Republic.

[575] J. Scheirs, H.C. Ashton and P.S. Gardner, in *Proceedings XVIIth Intl. Conference on Advances in the Stabilization and Degradation of Polymers* (A.V. Patsis, ed.), Luzern (1995), pp. 275–96.

[576] D. Forsström, A. Kron, B. Stenberg, B. Terselius and T. Reitberger, *Polym. Degr. Stabil.* **43**, 277 (1994).

[577] N.C. Billingham, *Polym. Prepr., Am. Chem. Soc., Div. Polym. Chem.* **34** (2), 237–42 (1993).

[578] B. Mattson and B. Stenberg, *Adv. Chem. Ser.* **249**, 235–48 (1996).

[579] N.C. Billingham, E.T.H. Then and P.J. Gijsman, *Polym. Degr. Stabil.* **34**, 263–77 (1991).

[580] N.C. Billingham, J.W. Burdon, I.W. Kaluska, E.S. O'Keefe and E.T.H. Then, *Proceedings Tenth Intl. Conference on Advances in the Stabilization and Degradation of Polymers* (A.V. Patsis, ed.), Luzern (1988).

[581] A. Kron, B. Stenberg, T. Reitberger and N.C. Billingham, *Polym. Degr. Stabil.* **53**, 119–27 (1996).

[582] L. Matisová-Rychlá, Zs. Fodor, J. Rychlý and M. Iring, *Polym. Degr. Stabil.* **3**, 371 (1981).

[583] A. Kron, *Ph. D. Thesis*, Royal Institute of Technology, Stockholm (1996).

[584] P. Gijsman and F. Verdun, *Polym. Degr. Stabil.* **74** (3), 533–42 (2001).

[585] L. Reich and S.S. Stivala, *Autooxidation of Hydrocarbons and Polyolefins*, M. Dekker, New York, NY (1969), Chp. 5.

[586] V. Dudler, Th. Bolle and G. Rytz, *Polym. Degr. Stabil.* **60**, 351 (1998).

[587] Z. Osawa, S. Wu and F. Xonoma, *Polym. Degr. Stabil.* **22**, 97–107 (1988).

[588] J.M. Gromek and M.R. Derrick, *Polym. Degr. Stabil.* **39**, 261–8 (1993).

[589] V. Dudler, D.J. Lacey and C. Kröhnke, *Polym. Degr. Stabil.* **51**, 115–24 (1996).

[590] R. Setnescu, S. Jipa and Z. Osawa, *Polym. Degr. Stabil.* **60**, 377–83 (1998).

[591] P.K. Fearon, D.J. Whiteman, S.W. Bigger and N.C. Billingham, *Abstracts First Intl. Conference on Polymer Modification, Degradation and Stabilisation (MoDeSt)*, Palermo (2000).

[592] D.J. Whiteman, P.K. Fearon and N.C. Billingham, *Abstracts First Intl. Conference on Polymer Modification, Degradation and Stabilisation (MoDeSt)*, Palermo (2000).

[593] D.J. David, *Thermochim. Acta* **3**, 277 (1972).

[594] A.M. Wynne and W.W. Wendtland, *Thermochim. Acta* **13**, 393 (1975); **14**, 61 (1975).

[595] J. Scheirs, S.W. Bigger and N.C. Billingham, *J. Polym. Sci., Polym. Chem. Edn.* **30**, 1873 (1992).

[596] N.C. Billingham, P.K. Fearon, D.J. Whiteman and V. Dudler, *Proceedings Fifth AddCon World Conference*, Prague (1999), Paper 23.

[597] M.N. Grigg and N.C. Billingham, *Proceedings 2nd Intl. Conference on Polymer Modification, Degradation and Stabilisation (MoDeSt2)*, Budapest (2002).

[598] I. Blakey and G.A. George, *Preprints 37th Intl. Symposium on Macromolecules*, Gold Coast (1998), p. 423.

[599] J. Gerlock, T. Prater, S. Kaberline and J. deVries, *Polym. Degr. Stabil.* **47**, 405–11 (1995).

[600] K. Möller, A. Jansson and P. Sjövall, *Proceedings 2nd Intl. Conference on Polymer Modification, Degradation and Stabilisation (MoDeSt2)*, Budapest (2002).

[601] N.C. Billingham, in *Symposium on Chemiluminescence 2000*, Atlas, Leipzig (2000).

[602] B. Stenberg, T. Reitberger and B. Terselius, in *Abstracts Intl. Workshop on Practical Applications of Chemiluminescence at the Oxidation of Chemical Systems*, Smolenice Castle (1998), p. 8.

[603] G.A. George and M. Ghaemy, *Polym. Degr. Stabil.* **34**, 37 (1991).

[604] D.R. Kohler and C. Kröhnke, *Polym. Degr. Stabil.* **63**, 165–73 (1999).

[605] C. Kröhnke, in *Symposium on Chemiluminescence 2000*, Atlas, Leipzig (2000).

[606] M. Celina, G.A. George, and N.C. Billingham, *Polym. Degr. Stabil.* **42**, 335 (1993).

[607] L. Zlatkevich, *Polym. Degr. Stabil.* **50**, 83 (1995).

[608] M. Celina, G.A. George, D.J. Lacey and N.C. Billingham, *Polym. Degr. Stabil.* **47**, 311 (1995).

[609] L. Matisová-Rychlá, J. Rychlý, J. Verdu, L. Audouin and K. Csomorova, *Polym. Degr. Stabil.* **49**, 51 (1995).

[610] G.D. Mendenhall, *Angew. Chem. Int. Ed.* **29**, 362 (1990).

[611] T. Yang, *Polym. Prepr. (Am. Chem. Soc., Div. Polym. Chem.)* **39** (2), 693–4 (1998).

[612] L. Audouin-Jirackova and J. Verdu, *J. Polym. Sci., Polym. Chem. Ed.* **25**, 1205 (1987).

[613] P.V. Zamotaev, Z. Strel'tsova, L. Matisová-Rychlá and I. Chodak, *Polym. Degr. Stabil.* **42**, 167 (1993).

[614] N.C. Billingham, P. Fearon, D.J. Whiteman, N. Marshall and S.P. Bigger, *Proceedings AddCon World 2000*, Basel (2000), Paper 7.

[615] S.S. Parikh and L. Zlatkevich, *Proceedings SPE ANTEC '94*, San Francisco, CA (1994), pp. 2546.

[616] S.S. Parikh and L. Zlatkevich, *Angew. Makromol. Chem.* **247**, 255–68 (1997).

[617] L. Zlatkevich and D.J. Burlett, *Polym. Degr. Stabil.* **65**, 53–8 (1999).

[618] G. Ahlblad, T. Reitberger, B. Terselius and B. Stenberg, *Polym. Degr. Stabil.* **65**, 179–84 (1999).

[619] B. Lánská, D. Doskocilová, L. Matisová-Rychlá, R. Puffr and J. Rychlý, *Polym. Degr. Stabil.* **63**, 469–79 (1999).

[620] G.A. George, *Polym. Degr. Stabil.* **1**, 217 (1979).

[621] L. Matisová-Rychlá, B. Lánská, J. Rychlý and N.C. Billingham, *Polym. Degr. Stabil.* **43**, 131–9 (1994).

[622] L. Zlatkevich, *ACS Symp. Ser.* **280**, 387–409 (1985).

[623] S.W. Bigger, M.J. Cran, P.K. Fearon and N.C. Billingham, *Proceedings 11th Intl. Conference Additives 2002*, Clearwater Beach, FL (2002).

[624] G.A. George and M. Ghaemy, *Polym. Degr. Stabil.* **33**, 411 (1991).

[625] Z. Osawa and K. Tsurumi, *Polym. Degr. Stabil.* **26**, 151 (1989).

[626] Z. Osawa, *Polym. Degr. Stabil.* **20**, 203 (1988).

[627] L. Zlatkevich, *J. Polym. Sci. Phys.* **23**, 1691 (1985).

[628] G.A. George, M. Celina, C. Lerf, G. Cash and D. Wedell, *Macromol. Symp.* **115**, 69–92 (1997).

[629] Z. Osawa, M. Someya, Y.S. Fu and F. Konoma, *Polym. Degr. Stabil.* **43**, 461 (1994).

[630] B.G.S. Goss, M.D. Barry, D. Birtwhistle and G.A. George, *Polym. Degr. Stabil.* **74** (2), 271–84 (2001).

[631] M. Celina and G.A. George, *Polym. Degr. Stabil.* **50**, 89–99 (1995).

[632] F. Gugumus, *Polym. Degr. Stabil.* **52**, 161–70 (1996).

[633] M. Celina, G.A. George and N.C. Billingham, *Adv. Chem. Ser.* **249**, 159–74 (1996).

[634] P.K. Fearon, S.W. Bigger and N.C. Billingham, *Polym. Prepr. (ACS Div. Polym. Chem.)* **42** (1), 352–3 (2001).

[635] X. Zhong, F. Yoshii, T. Sasaki, T. Yagi and K. Makuuchi, *Polym. Degr. Stabil.* **51**, 159–65 (1996).

[636] R. Setnescu, G. Marek, S. Jipa, T. Setnescu, C. Georgian and I. Mihalcea, *Proceedings First Intl. Conference on Polymer Modification, Degradation and Stabilisation (MoDeSt)*, Palermo (2000).

[637] B.J. Kip and R. Tabaksblat, *Tech. Rept. DSM Research*, Geleen (1992).

[638] S. Jipa, R. Setnescu, T. Zaharescu and I. Mihalcea, *Mater. Plast. (Bucharest)* **36** (4), 211–16 (1999).

[639] S. Jipa, T. Zaharescu, R. Setnescu, T. Setnescu, D. Wurm and M. Dumitru, *Mater. Plast. (Bucharest)* **37** (1), 29–33 (2000).

[640] S. Jipa, R. Setnescu, T. Setnescu and T. Zaharescu, *Polym. Degr. Stabil.* **68**, 159–64 (2000).

[641] S. Jipa, Z. Osawa, H. Otsuki and M. Nishimoto, *Polym. Degr. Stabil.* **56**, 45 (1997).

[642] S. Jipa, R. Setnescu, T. Setnescu, C. Podina and I. Mihalcea, *South. Braz. J. Chem.* **6** (6), 71–6 (1998).

[643] V. Dudler and D. Lacey, *Proceedings 17th Int. Conference on Advances in the Stabilization and Degradation of Polymers* (A.V. Patsis, ed.), Luzern (1995), p. 57.

[644] J. Boxhammer, *Symposium on Chemiluminescence 2000*, Atlas, Leipzig (2000).

[645] R. Broska, J. Rychlý and K. Csomorova, *Polym. Degr. Stabil.* **63** (2), 231–6 (1999).

[646] C. Dumitrescu, L.M. Georghiu, R.L. Olteanu, S. Jipa and C.M. Tarabasanu, *Mater. Plast. (Bucharest)* **38** (2), 88–92 (2001).

[647] D. Forsström, B. Terselius and L.-G. Svensson, in *Abstracts Intl. Workshop on Practical Applications of Chemiluminescence at the Oxidation of Chemical Systems*, Smolenice Castle (1998), p. 11.

[648] B. Lánská, M. Matisová-Rychlá, J. Brozek and J. Rychlý, *Polym. Degr. Stabil.* **66**, 433 (1999).

[649] M. Grigg, N.C. Billingham and G.A. George, *Proceedings 2nd Intl. Conference on Polymer Modification, Degradation and Stabilisation (MoDeSt2)*, Budapest (2002).

[650] N.S. Allen, G. Rivalle, M. Edge, T. Corrales and F. Catalina, *Polym. Degr. Stabil.* **75** (2), 237–46 (2002).

[651] B. Mattson, B. Stenberg, K.T. Gillen, R.L. Clough and E. Östman, *Polym. Degr. Stabil.* **41**, 211–21 (1993).

[652] P.K. Fearon, T.L. Phease, N.C. Billingham and S.W. Bigger, *Polymer* **43**, 4611–8 (2002).

[653] D. Forsström, *Proceedings 2nd Intl. Conference on Polymer Modification, Degradation and Stabilisation (MoDeSt2)*, Budapest (2002).

[654] M. Strlič, J. Rychlý, B. Pihlar, J. Kolar, J.L. Pedersoli, J. Lemaire, L. Rychlá and D. Kočar, *Proceedings 2nd Intl. Conference on Polymer Modification, Degradation and Stabilisation (MoDeSt2)*, Budapest (2002).

[655] M. Strlič, J. Rychlý, B. Pihlar, J. Kolar, J.L. Pedersoli, D. Fromageot, L. Rychlá and D. Kočar, *PAPYLUM Project EEC Contract No. EVK4-CT-2000-0038.*

[656] A. Kron, B. Stenberg and T. Reitberger, *Progr. Rubber Plastics Technol.* **13** (2), 81 (1997).

[657] J.M. Gomez-Elvira and P. Tiemblo, *Rev. Plast. Mod.* **75** (502), 384–90 (1998).

[658] A. Pines, M.G. Gibby and J.S. Waugh, *J. Chem. Phys.* **59**, 569 (1973).

[659] R.A. Wind, in *Encyclopedia of Analytical Science* (A. Townshend, ed.), Academic Press, London (1995), pp. 3477–85.

[660] R. Voelkel, *Abstracts Advanced Methods of Polymer Characterization: New Developments and Applications in Industry*, Mainz (1999), Paper M4.

[661] M.Y. Gelfer and K. Schmidt-Rohr, *Proceedings ACS Div. Polym. Mater.: Sci. & Engng.* **80**, 380–1 (1999).

[662] I.D. Wilson, M. Spraul and E. Humpfer, *J. Planar Chromatogr.* **10**, 217–9 (1997).

[663] M. Wilhelm, M. Neidhoefer, S. Spiegel and H.W. Spiess, *Macromol. Chem. Phys.* **200** (10), 2205–7 (1999).

[664] J.L. Ackerman, in *A Complete Introduction to Modern NMR Spectroscopy* (R.S. Macomber, ed.), J. Wiley & Sons, New York, NY (1998), pp. 283–305.

[665] D.J. O'Donnel, *ACS Symp. Series* **247**, 21–41 (1984).

[666] C.A. Fyfe, *Solid State NMR for Chemists*, CFC Press, Guelph (1983).

[667] M. Mehring, *High Resolution NMR in Solids*, Springer Verlag, New York, NY (1983).

[668] M.C. Gerstein and C.R. Dybrowski, *Transient Techniques in NMR of Solids*, Academic Press, Orlando, FL (1985).

[669] R.K. Harris, in *Polymer Spectroscopy* (A.H. Fawcett, ed.), J. Wiley & Sons, Chichester (1996).

[670] L. Mathias (ed.), *Solid State NMR of Polymers*, Plenum Press, New York, NY (1991).

[671] D. Michel and F. Engelke, in *NMR. Basic Principles and Progress* (P. Diehl, E. Flück, H. Günther, R. Kosfeld and J. Seelig, eds.), Springer-Verlag, Berlin (1994), pp. 69–125.

[672] M. Heidenreich, A. Spyros, W. Köckenberger, N. Chandrakumar, R. Bowtell and R. Kimmich, in *Spatially Resolved Magnetic Resonance* (P. Blümler, B. Blümich, R. Botto and E. Fukushima, eds.), Wiley-VCH, Weinheim (1998), pp. 21–52.

[673] R.F. Grossman, *Proceedings SPE ANTEC '99*, New York, NY (1999), pp. 3095–100.

[674] R.F. Grossman, *J. Vinyl Addit. Technol.* **5** (3), 148–58 (1999).

[675] I.M. Steele, J.J. Pluth and J.W. Richardson, *J. Solid State Chem.* **132**, 173–81 (1997).

[676] C.M. Dick, J.J. Liggatt, C.E. Snape, S.C. Martin, C. Denecker, G. Seeley, B. Eling, C. Lindsay and P. Chaffanjon, *Proceedings A.C.S. Div. Polym. Mater.: Sci. & Engng.* **83**, 44 (2000).

[677] D. Radloff, H.W. Spiess, J.T. Books and K.C. Dowling, *J. Appl. Polym. Sci.* **60**, 715–20 (1996).

[678] G. v.d. Velden, W. Gabriëlse and V. Litvinov, *Techn. Rept. DSM Research*, Geleen (1997).

[679] R.A. Assink, M. Celina, T.D. Dunbar, T.M. Alam, R.L. Clough and K.T. Gillen, *Proceedings A.C.S.*

Div. Polym. Mater.: Sci. & Engng. **83**, 176–8 (2000).
[680] N.K. de Vries and J. Beulen, *Techn. Rept. DSM Research*, Geleen (1992).
[681] T. Tikuisis, D.E. Axelson and A. Sharma, *Polym. Eng. Sci.* **33**, 26–31 (1993).
[682] G. v.d. Velden, H. Nelissen and J. Bremmers, *Techn. Rept. DSM Research*, Geleen (1998).
[683] G. v.d. Velden, H. Nelissen and T. Veermans, *Acta Polym.* **47**, 161–9 (1996).
[684] Y.S. Lipatov, *Adv. Polym. Sci.* **22**, 1 (1977).
[685] J.P. Cohen Addad, *Progr. NMR Spectrosc.* **25**, 1–316 (1993).
[686] A.P. Legrand, *Proceedings Eurofillers '95, Macromol. Symp.* **108**, 81–96 (1996).
[687] M. Hess, W. Veeman and P. Magusin, *Proceedings SPE ANTEC '96*, Indianapolis, IN (1996), pp. 3682–6.
[688] P. Stilbs, in *Polymer–Surfactant Systems* (J.C.T. Kwak, ed.), M. Dekker, New York, NY (1998), pp. 239–66.
[689] J. Kelm, *Forschungsbericht 213. High Resolution NMR Spectroscopy for the Determination of Elastomers, Blends and Thermoplastic Elastomers – Carbon and Proton NMR Spectra Catalogue*, BAM, Berlin (1995).
[690] M.F. Grenier-Loustalot, *Polym. Comm.* **31**, 329–32 (1990).
[691] W. Gabriëlse, personal comunication (1998).
[692] W. Heinen, M. v. Duin, C.H. Rosenmöller, C.B. Wenzel, H.J.M. de Groot and J. Lugtenburg, *Macromol. Symp.* **129**, 119–25 (1998).
[693] A.J. Aarts and K.M. Baker, *Kautsch. Gummi Kunstst.* **37**, 497–500 (1984).
[694] F. Mersch and R. Zimmer, *Kautsch. Gummi Kunstst.* **39**, 427–32 (1986).
[695] G.P.M. v.d. Velden and J. Kelm, *Rubber Chem. Technol.* **63**, 215–22 (1990).
[696] R.A. Komoroski, *Rubber Chem. Technol.* **56**, 959–70 (1983).
[697] R. Winters, W. Heinen, M.A.L. Verbruggen, J. Lugtenburg, M. van Duin and H.J.M. de Groot, *Macromolecules* **35** (5), 1958–66 (2002).
[698] W. Barendswaard, J. Moonen and M. Neilen, *Anal. Chim. Acta* **283**, 1007–24 (1993).
[699] R.D. Davis, W.L. Jarrett and L.J. Mathias, *ACS Symp. Ser.* **804**, 117–26 (2002).
[700] H. Nelissen and G. v.d. Velden, *Techn. Rept. DSM Research*, Geleen (1998).
[701] H. Watson, A.E.E. Norstrom, J.G. Matisons, A. Root and J.B. Rosenholm, *J. Adhes. Sci. Technol.* **15** (9), 1103–17 (2001).
[702] D. Derouet, S. Forgeard, J.-C. Brosse, J. Emery and J.-Y. Buzare, *Extended Abstracts 2nd Intl. Conference on Polymer Fillers (Eurofillers '97)*, Manchester (1997), pp. 123–6.
[703] X.-Q. Zie, S.V. Ramade and A.T. DiBenedetto, *Polymer* **40** (23), 6297–6306 (1999).
[703a] T.L. Weeding, W.S. Veeman, L.W. Jenneskens, H. Angad Gaur, H.E.C. Schuurs and W.G.B. Huysmans, *Macromolecules* **22**, 706–14 (1989).
[704] S. Jahromi, W. Gabriëlse and A. Braam, *Proceedings 11th Intl. Conference Additives 2002*, Clearwater Beach, FL (2002).
[705] K.R. Carduner and R.O. Carter III, *J. Magn. Reson.* **84**, 361–6 (1989).
[706] K.R. Carduner, R.O. Carter III, M.F. Cheung, A. Golovoy and H. Van Oene, *J. Appl. Polym. Sci.* **40**, 963–75 (1990).
[707] A. Golovoy, M.-F. Cheung, K.R. Carduner and M.J. Rokosz, *Polym. Bull.* **21**, 327–34 (1989).
[708] M.F. Cheung, K.R. Carduner, A. Golovoy and H. Van Oene, *J. Appl. Polym. Sci.* **40**, 977–87 (1990).
[709] G.J. Klender, *Adv. Chem. Ser.* **249**, 397–423 (1996).
[710] C.M. Sultany, *Polym. Bull.* **20**, 463–6 (1988).
[711] S. Bourbigot, F. Carpentier, M. Le Bras, C. Fernandez, J.-P. Amoureux and R. Delobel, *Extended Abstracts 2nd Intl. Conference on Polymer Fillers (Eurofillers '97)*, Manchester (1997), pp. 419–23.
[712] M. Le Bras, S. Bourbigot, F. Carpentier, R.M. Leeuwendal and D. Schubert, *Gummi Fasern Kunstst.* **51** (12), 972–82 (1998).
[713] S. Bourbigot, M. Le Bras, R.M. Leeuwendal, K.K. Shen and D. Schubert, *Polym. Degr. Stabil.* **64**, 419–25 (1999).
[714] S. Bourbigot, M. Le Bras, F. Dabrowski, J.F. Gilman and T. Kashiwagi, *Proceedings 10th BBC Annu. Conference on Flame Retardancy*, Stamford, CT (1999).
[715] S.-S. Hou, T.J. Bonagamba, F.L. Beyer and K. Schmidt-Rohr, *Polym. Prepr.* **43** (2), 1281–2 (2002).
[716] F. Lauprêtre, in *Solid-State NMR. I. Methods* (B. Blümich, ed.), Springer-Verlag, Berlin (1994), pp. 64–109.
[717] J.L. Duda and J.M. Zielinski, in *Diffusion in Polymers* (P. Neogi, ed.), M. Dekker, New York, NY (1996), pp. 143–71.
[718] B. Hills, *Magnetic Resonance Imaging in Food Science*, J. Wiley & Sons, New York, NY (1998).
[719] E.L. Hahn, *Phys. Rev.* **80**, 580 (1950).
[720] E.O. Stejskal and J.E. Tanner, *J. Chem. Phys.* **42**, 288–92 (1965).
[721] R.W. Korsmeyer, E. von Meerwall and N.A. Peppas, *J. Polym. Sci., Polym. Phys. Edn.* **24**, 409–34 (1986).
[722] E.D. von Meerwall, *Rubber Chem. Technol.* **58**, 527 (1985).
[723] P. Stilbs, *Progr. NMR Spectrosc.* **19**, 1 (1987).

[724] J. Kärger, H. Pfeiffer and W. Heink, *Adv. Magn. Reson.* **12**, 1 (1988).

[725] W.S. Price, in *New Advances in Analytical Chemistry* (Atta-ur-Rahman, ed.), Harwood Academic Publishers, Amsterdam (2000), pp. 31–72.

[726] D.A. Jayawickrama, C.K. Larive, E.F. McCord and D.C. Roe, *Magn. Res. Chem.* **36**, 755–60 (1998).

[727] P. Stilbs, in *Encyclopedia of Spectroscopy and Spectrometry* (J.C. Lindon, ed.), Academic Press, San Diego, CA (2000), pp. 369–75.

[728] E. von Meerwall, *Adv. Polym. Sci.* **54**, 1 (1984).

[729] J. Crank, *The Mathematics of Diffusion*, Clarendon Press, Oxford (1975).

[730] J. Crank, *Free and Moving Boundary Problems*, Clarendon Press, Oxford (1984).

[731] P. Neogi, *Diffusion in Polymers*, M. Dekker, New York, NY (1996).

[732] A. Carlsson, G. Karlström and B. Lindman, *J. Phys. Chem.* **93**, 3673–7 (1989).

[733] G. Fleischer, *Polym. Commun.* **26**, 20–2 (1985).

[734] D.W. McCall, D.C. Douglas and E.W. Anderson, *J. Polym. Sci.* **A1**, 1709–20 (1963).

[735] D.E. Woessner, *J. Phys. Chem.* **67**, 1365 (1983).

[736] J.E. Tanner, *Macromolecules* **4**, 748 (1971).

[737] B.D. Boss, E.O. Stejskal and J.D. Ferry, *J. Phys. Chem.* **71**, 1501 (1967).

[738] A.I. Maklakov, A.G. Smechko and A.A. Maklalov, *Macromol. Sci. (Moscow)* **19**, 2611 (1977).

[739] N.M. Azancheev and A.I. Maklakov, *Macromol. Sci. (Moscow)* **21**, 1574 (1979).

[740] I. Zupančič, G. Lahajnar, R. Blinc, D.H. Reneker and A. Peterlin, *J. Polym. Sci.: Polym. Phys.* **16**, 1399–407 (1978).

[741] M.E. Moseley and P. Stilbs, *Chem. Scripta* **16**, 114 (1980).

[742] B. Nyström, M.E. Moseley, P. Stilbs and J. Roots, *Polymer* **22**, 218 (1981).

[743] R.D. Ferguson and E. von Meerwall, *J. Polym. Sci.: Polym. Phys. Ed.* **18**, 1285 (1980).

[744] E. von Meerwall and R. van Antwerp, *Macromolecules* **15**, 1115 (1982).

[745] A. Banis, P.T. Inglefield, A.A. Jones and W.Y. Wen, *J. Polym. Sci.: Polym. Phys. Edn.* **33**, 1495–504, 1505–14, 1515–22 (1995).

[746] A. Guillermo, M. Todica and J.P. Cohen-Addad, *Macromolecules* **26**, 3946–52 (1993).

[747] E. von Meerwall and R.D. Ferguson, *J. Appl. Polym. Sci.* **23**, 3657 (1979).

[748] G. Fleischer and P. Holstein, *Acta Polym.* **35**, 738 (1984).

[749] S. Schlick, Z. Gao, S. Matsukawa, I. Ando, E. Fead and G. Rossi, *Macromolecules* **31**, 8124–33 (1998).

[750] K. Landfester, F. Mellinger, J. Rottstegge and H.W. Spiess, *Preprints 37th IUPAC Intl. Symposium on Macromolecules*, Gold Coast (1998), p. 117.

[751] C.M. Keely, X. Zhang and V.J. McBrierty, *J. Mol. Struct.* **355**, 33–46 (1995).

[752] A.A. Parker, D.P. Hedrick and W.M. Ritchey, *Macromolecules* **25** (13), 3365–8 (1992).

[753] B.J. Cauley, C. Cipriani, K. Ellis, A.K. Roy, A.A. Jones, P.T. Inglefield, B.J. McKinley and R.P. Kambour, *Macromolecules* **24**, 403 (1991).

[754] Y. Liu, P.T. Inglefield, A.A. Jones and R.P. Kambour, *Magn. Reson. Chem.* **32**, 518 (1994).

[755] P. Bergquist, Y. Zhu, A.A. Jones and P.T. Inglefield, *Macromolecules* **32**, 7925–31 (1991).

[756] M.I.B. Tavares, E.E.C. Monteiro, R.K. Harris and A.M. Kenwright, *Eur. Polym. J.* **30** (9), 1089–95 (1994).

[757] N.J. Clayden and C. Howick, *Polymer* **34**, 2508–15 (1993).

[758] M.I.B. Tavares, E.E.C. Monteiro, A. Kenwright and R. Harris, *Polym. Bull.* (Berlin) **30** (6), 677–84 (1993).

[759] A.L.M. Smits, P.H. Kruiskamp, J.J.G. van Soest and J.F.G. Vliegenthart, *Abstracts First Intl. Conference on Polymer Modification, Degradation and Stabilisation (MoDeSt)*, Palermo (2000).

[760] J.J. Liggar, C. Denecker, C. Dick, S.C. Martin, C.E. Snape, M. Mohammed, G. Seeley, B. Eling, C. Lindsay and P. Chaffanjon, *Polym. Mater. Sci. Eng.* **82**, 186 (2000).

[761] P.B. Smith, W.C. Buzanowski, J.J. Gunderson, D.B. Priddy and L. Pfenninger, *Proceedings SPE ANTEC 2002*, San Francisco, CA (2002), pp. 2030–4.

[762] K. Schmidt-Rohr, *ACS Symp. Ser.* **598**, 191–214 (1995).

[763] J. Selinger, in *Encyclopedia of Spectroscopy and Spectrometry* (J.C. Lindon, ed.), Academic Press, San Diego, CA (2000), pp. 1672–80.

[764] B.P. Straughan and S. Walker (eds.), *Spectroscopy*, Chapman & Hall, London (1976), pp. 175–208.

[765] T.A. Babushkina and G.K. Semin, *Bull. Acad. Sci.: Phys. Ser.* **45** (3), 508 (1981).

[766] T.A. Babushkina, L.N. Grozdeva and G.K. Semin, *J. Mol. Struct.* **73**, 215–22 (1981).

[767] E.G. Brame, *Makromol. Chem., Macromol. Symp.* **72**, 1–9 (1993).

[768] M. Mackowiak, in *New Advances in Analytical Chemistry* (Atta-ur-Rahman, ed.), Harwood Academic Publishers, Amsterdam (2000), pp. 433–86.

[769] A.C. Olivieri, in *Encyclopedia of Spectroscopy and Spectrometry* (J.C. Lindon, ed.), Academic Press, San Diego, CA (2000), pp. 2116–27.

[770] T.N. Rudakov, in *Encyclopedia of Spectroscopy and Spectrometry* (J.C. Lindon, ed.), Academic Press, San Diego, CA (2000), pp. 1663–71.
[771] M. Geoffrey and E.A.C. Lucken, in *The Chemistry of Halides, Pseudo-Halides and Azides* (S. Patai and Z. Rappoport, eds.), J. Wiley & Sons, Chichester (1995), Pt. 1, pp. 289–335.
[772] R. Ikeda and K. Horiuchi, *Jikken Kagaku Koza* **8**, 563 (1993); *C.A.* **123**, 356973 (1995).
[773] J.A. Buslaev, L. Kolditz and E.A. Kravchenko, *Nuclear Quadrupole Resonance in Inorganic Chemistry*, VEB Deutsche Verlag der Wissenschaften, Berlin (1987).
[774] G.K. Semin, T.A. Babushkina and G.G. Yakobson, *Nuclear Quadrupole Resonance in Chemistry*, J. Wiley & Sons, New York, NY (1975).
[775] A.A. Mrse, Y. Lee, P.L. Bryant, F.R. Fronczek, L.G. Butler and L.S. Simeral, *Chem. Mater.* **10** (5), 1291–300 (1998).
[776] E.P. Chang, R. Kirsten and J. Slagowski, *J. Appl. Polym. Sci.* **21**, 2167–80 (1977).
[777] M.H. Palmer, M.M.P. Khurshid, T.J. Rayner and J.A.S. Smith, *Chem. Phys.* **182**, 27 (1994).
[778] O. Kh. Poleshchuk and J.N. Latosińska, in *Encyclopedia of Spectroscopy and Spectrometry* (J.C. Lindon, ed.), Academic Press, San Diego, CA (2000), pp. 1653–62.
[779] P.B. Ayscough, *Electron Spin Resonance in Chemistry*, Methuen, London (1967).
[780] J.K. Becconsall, S. Clough and G. Scott, *Trans. Faraday Soc.* **56**, 459–72 (1960).
[781] T.G. Carswell, R.W. Garrett, D.J.T. Hill, J.H. O'Donnell, P.J. Pomery and C.L. Winzor, in *Polymer Spectroscopy* (A.H. Fawcett, ed.), J. Wiley & Sons, Chichester (1996), pp. 253–74.
[782] A.L. Kovarski, *Molecular Dynamics of Additives in Polymers*, VSP, Utrecht (1997).
[783] C.C. Rowlands and D.M. Murphy, *Encyclopedia of Spectroscopy and Spectrometry* (J.C. Lindon, ed.), Academic Press, San Diego, CA (2000), pp. 445–56.
[784] L. Kevan, in *Encyclopedia of Physical Science and Technology* (R.A. Meyers, ed.), Academic Press, San Diego, CA (2001), Vol. 5.
[785] J.E. Wertz and J.R. Bolton, *Electron Spin Resonance: Elementary Theory and Practical Applications*, Chapman & Hall, New York, NY (1986).
[786] B. Rånby and J.F. Rabek, *ESR Spectroscopy in Polymer Research*, Springer-Verlag, Berlin (1977).
[787] E.G. Rozantsev, *Free Nitroxyl Radicals*, Plenum Press, New York, NY (1970).
[788] W. Zhou and S. Zhu, *Macromolecules* **31** (13), 4335–41 (1998).
[789] T. Yamazaki and T. Seguchi, *J. Polym. Sci., Part A: Polym. Chem.* **37** (3), 349–56 (1999).
[790] T. Yamazaki and T. Seguchi, *J. Polym. Sci., Part A: Polym. Chem.* **38** (17), 3092–9 (2000).
[791] Z. Laszlo-Hedvig, E. Dobai, L. Korecz, P. Nagy, A. Hesse, M. Ratzsch and F. Tudos, *Macromol. Chem. Phys.* **202** (13), 2696–701 (2001).
[792] A. Sommer, E. Zirngiebl, L. Kahl and M. Schonfelder, *Progr. Org. Coatings* **19** (1), 79–87 (1991).
[793] Z. Joffe and B. Rånby, in *ESR Applications to Polymer Research* (P.-O. Kinell, B. Rånby and V. Runnström-Reio, eds.), J. Wiley & Sons, New York, NY (1973), pp. 171–85.
[794] K. Tsuji and S. Okamura, in *ESR Applications to Polymer Research* (P.-O. Kinell, B. Rånby and V. Runnström-Reio, eds.), J. Wiley & Sons, New York, NY (1973), pp. 187–99.
[795] M. Narayana, in *Encyclopedia of Analytical Science* (A. Townshend, ed.), Academic Press, London (1995), pp. 1022–31.
[796] H.M. Banford, R.A. Fouracre, A. Faucitano, A. Buttafava and F. Martinotti, *IEEE Trans. Dielectr. Electr. Insul.* **3** (4), 594–8 (1996).
[797] J. Mallégol, D. Carlsson and L. Deschênes, *Polym. Degr. Stabil.* **73**, 259–67 (2001).
[798] P. Anelli, S. Baccaro and C. Casadio, *Radiat. Phys. Chem.* **52** (1–6), 183–6 (1998).
[799] B. Qu, S.M. Chen and L.M. Dai, *Appl. Magn. Reson.* **19** (1), 59–67 (2000).
[800] G. Agostini, L. Pasimeni, M. Ruzzi, S. Monti, M. Maggini, M. Prato, I. Lamparth and A. Hirsch, *Chem. Phys.* **253** (1), 105–13 (2000).
[801] E.J. Harbron, W.C.Bunyard, J.M. DeSimone and M.D.E. Forbes, *Polym. Prepr. ACS, Div. Polym. Chem.* **39** (2), 735–6 (1998).
[802] B. Ganem, *J. Org. Chem.* **40**, 1998–2000 (1975).
[803] M. Gyor, A. Rockenbauer, L. Jokay and F. Tudos, *Polym. Bull.* **15** (6), 525–30 (1986).
[804] A.T. Bullock, C.G. Cameron and I.S. Miles, *Polym. Comm.* **24** (1), 22–4 (1983).
[805] H.J. Bowley, D.L. Gerrard and I.S. Biggin, *Polym. Degr. Stabil.* **20**, 257–69 (1988).
[806] J. Bouquant, M. Hamdani, P. Métois, A.M. Riquet and D. Scholler, *Food Addit. Contam.* **14** (6–7), 571–82 (1997).
[807] P.A. Tice, *Food Addit. Contam.* **11** (2), 187 (1994).
[808] L.H. Adcock, *Plast. Rubber Appl.* **4**, 53 (1984).
[809] A.M. Riquet and O. Akermann, *Food Chem.* **26**, 271 (1987).
[810] A.M. Riquet, V. Sandray, O. Akermann and A. Feigenbaum, *Sci. Aliments* **11**, 341–59 (1991).
[811] A.M. Riquet, M. Hamdani and A. Feigenbaum, *J. Polym. Eng.* **15**, 1 (1995/96).
[812] A.M. Riquet, N. Wolff and A. Feigenbaum, *Packag. Technol. Sci.* **7** (4), 175–85 (1994).
[813] S. Cottier, A.M. Riquet, A. Feigenbaum and P. Mortreuil, *Polym. Intl.* **43**, 353–8 (1997).

[814] K. Sawada, S. Onodera, H. Enomoto, S. Nagaoka and K. Mukai, *Kobunshi Ronbushu* **56** (5), 298–306 (1999).

[815] A. Factor, *ACS Symp. Ser.* **425**, 274–87 (1989).

[816] M. Lucarini, G.F. Pedulli and M. Cipollone, *J. Org. Chem.* **59**, 5063–70 (1994).

[817] F. Magnaterra, P. Pedrielli and G.F. Pedulli, *Gazz. Chim. Ital.* **126** (10), 673–7 (1996).

[818] N.S. Allen, *Polym. Photochem.* **1**, 234 (1981).

[819] T. Kelen, F. Tüdös, G. Bálint and A. Rockenbauer, *ACS Symp. Ser.* **280**, 109–17 (1985).

[820] J. Lucki, J.F. Rabek and B. Rånby, *Polym. Prepr.* **25** (1), 38–9 (1984).

[821] R.W. Layer, J.T. Lai, R.P. Lattimer and J.C. Westfahl, *ACS Symp. Ser.* **280**, 99–107 (1985).

[822] J. Kenny, *Proceedings 5th AddCon World '99*, Prague (1999), Paper 3.

[823] C. Kröhnke, *Proceedings Polyolefins X* (SPE, ed.), Houston, TX (1997), pp. 555–69.

[824] L. Greci, A. Mar'in, P. Stipa and P. Carloni, *Polym. Degr. Stabil.* **50**, 305–12 (1995).

[825] A. Tkáč, in *Developments in Polymer Stabilisation-8* (G. Scott, ed.), Elsevier Applied Science, London (1987), pp. 61–178.

[826] J.L. Gerlock, A.V. Kucherov and C.A. Smith, *Atlas SunSpots (Material Testing Product and Technology News)* **31/66**, 1–11 (2001).

[827] D.R. Bauer, L.M. Briggs and J.L. Gerlock, *J. Polym. Sci. Part B: Polym. Phys.* **24**, 1651 (1986).

[828] M.V. Motyakin, J.L. Gerlock and S. Schlick, *Macromolecules* **32**, 5463 (1999).

[829] J.L. Gerlock, D.R. Bauer and L.M. Briggs, *Polym. Prepr.* **25** (1), 30–1 (1984).

[830] S. Scierka and N. Blough, *Proceedings ACS Div. Polym. Mater.: Sci. & Engng.* **83**, 338–9 (2000).

[831] L. Greci, A. Mar'in, P. Stipa, P. Carloni, E. Damiani and A. Faucitano, in *Proceedings 17th Int. Conference on Advances in the Stabilization and Degradation of Polymers* (A.V. Patsis, ed.), Luzern (1995), pp. 109–24.

[832] A. Faucitano, A. Buttafava, F. Martinotti and P. Bortolus, *J. Phys. Chem.* **88**, 1187–90 (1984).

[833] G. Geuskens, M.N. Kanda and G. Nedelkos, *Bull. Soc. Chim. Belg.* **99**, 1085–1100 (1990).

[834] S. Commereuc, S. Scheirs, V. Verney and J. Lacoste, *J. Appl. Polym. Sci.* **69** (6), 1107–14 (1998).

[835] J.L. Gerlock, D.R. Bauer and L.M. Briggs, *Polym. Degr. Stabil.* **14**, 53–71 (1986).

[836] P.P. Klemchuk, M.E. Gande and E. Cordola, *Polym. Degr. Stabil.* **27**, 65 (1990).

[837] C. Duhayon, *Abstracts SPE Benelux Symposium on Pigments, Dyes & Polymer Additives*, Oostende (2000).

[838] A. Factor, D.L. Tomaja and M.L. Chu, *Angew. Makromol. Chem.* **261–2**, 55–64 (1998).

[839] T. Kawaguchi, K. Nakai and K. Ohashi, *Sumitomo Kagaku (Osaka)* (1), 48–60 (1989).

[840] A.B. Vassilikou-Dova, in *New Advances in Analytical Chemistry* (Atta-ur-Rahman, ed.), Harwood Academic Publishers, Amsterdam (2000), pp. 405–40.

[841] C.Z. Rudowicz (ed.), *Modern Applications of EPR/ESR: From Biophysics to Materials Science*, World Science, Singapore (1998).

[842] P.-O. Kinell, B. Rånby and V. Runnström-Reio (eds.), *ESR Applications to Polymer Research*, J. Wiley & Sons, New York, NY (1973).

[843] R.L. Mössbauer, *Naturwissensch.* **45**, 538 (1958); *Z. Phys.* **151**, 124 (1958).

[844] F.E. Fujita, *Contemp. Phys.* **40** (5), 323–37 (1999).

[845] B. Rodmacq, M. Pineri, J.M.D. Coey and A. Meagher, *J. Polym. Sci., Polym. Chem. Ed.* **20**, 603–21 (1982).

[846] J.C. Waerenborgh, *Quimica (Lisboa)* **74**, 22–33 (1999).

[847] P. Gutlich and J. Ensling, in *Inorganic Electronic Structure and Spectroscopy* (E.I. Solomon and A.B.P. Lever, eds.), J. Wiley & Sons, New York, NY (1999), Vol. 1, pp. 161–211.

[848] G.N. Belozerski, in *Encyclopedia of Spectroscopy and Spectrometry* (J.C. Lindon, ed.), Academic Press, San Diego, CA (2000), pp. 1315–23.

[849] G.R. Hoy, in *Encyclopedia of Physical Science and Technology* (R.A. Meyers, ed.), Academic Press, San Diego, CA (2001), Vol. 10.

[850] D.P. Dickson and F.J. Berry, *Mössbauer Spectroscopy*, Cambridge University Press, Cambridge (1986).

[851] T.E. Cranshaw, B.W. Dale, G.O. Longworth and C.E. Johnson, *Mössbauer Spectroscopy and its Applications*, Cambridge University Press, Cambridge (1985).

[852] R.H. Herber (ed.), *Chemical Mössbauer Spectroscopy*, Plenum Press, New York, NY (1984).

[853] U. Gonser (ed.), *Mössbauer Spectroscopy, Topics in Applied Physics*, Springer-Verlag, Berlin (1981), Vol. 5.

[854] A.G. Maddock, *Mössbauer Spectroscopy*, Horwood Publishing, Amsterdam (1997).

[855] R.L. Cohen (ed.), *Applications of Mössbauer Spectroscopy*, Academic Press, New York, NY (1976).

[856] F.J. Litterest, *Nucl. Instrum. Meth.* **199**, 87–97 (1982).

[857] I.P. Suzdalev and V.K. Imshennik, *Nucl. Instrum. Meth.* **199**, 125–32 (1982).

[858] J.N. Cutler, J.H. Sanders and G. John, *Tribol. Lett.* **4** (2), 149–54 (1998).

[859] G.S. Kapur and A.S. Brar, *Hyperfine Interactions* **56**, 1695–700 (1990).

[860] G.S. Kapur and A.S. Brar, *J. Radioanal. Nucl. Chem. Lett.* **136**, 169–75 (1989).

[861] A.S. Brar and G.S. Kapur, *Hyperfine Interactions* **45**, 323–9 (1989).

[862] V.Ya. Rochev, V.G. Bekeshev, N.N. Savvateev and N.K. Kivrina, *Izv. Akad. Nauk, Ser. Fiz.* **65** (7), 1032–4 (2001).

[863] M.P. Glazunov, E.K. Kondrashov, N.P. Sokolova, M.S. Grigor'ev, Yu.V. Golenko and R.A. Bulgakova, *Vysokomol. Soedin., Ser. B* **30** (6), 415–7 (1988).

[864] V.K. Gol'danskii and L.A. Korytko, in *Applications of Mössbauer Spectroscopy* (R.L. Cohen, ed.), Academic Press, New York, NY (1976), Vol. 1, p. 287.

[865] V.I. Gol'danskii, *Vysokomol. Soedin., Ser. A* **13** (2), 311–24 (1971).

[866] D.W. Allen, J.S. Brooks, R.W. Clarkson, M.T.J. Mellor and A.G. Williamson, *J. Organomet. Chem.* **199**, 299–310 (1980).

[867] J.S. Brooks, R.W. Clarkson, D.W. Allen, M.T.J. Mellor and A.G. Williamson, *Polym. Degr. Stabil.* **4**, 359–63 (1982).

[868] J.S. Brooks, D.W. Allen and J. Unwin, *Polym. Degr. Stabil.* **10**, 79–94 (1985).

[869] D.W. Allen, J.S. Brooks, R.W. Clarkson, M.T.J. Mellor and A.G. Williamson, *Chem. Ind.*, 663–4 (1979).

[870] P.G. Harrison, T.J. King and M.A. Healy, *J. Organomet. Chem.* **182**, 17 (1979).

[871] D.W. Allen, J.S. Brooks, R.W. Clarkson, J. Unwin and P.J. Smith, *Polym. Degr. Stabil.* **13** (3), 191–200 (1985).

[872] C. Garrigues, A. Guyot, V.H. Tran and M. Thomalla, *Polym. Degr. Stabil.* **43** (2), 299–306 (1994).

[873] A.Yu. Aleksandrov, V.I. Gol'danskii, T.B. Zavarova and L.A. Korytko, *Vysokomol. Soedin., Ser. B* **13** (10), 784–6 (1971).

[874] P.R. Hornsby, P.A. Mitchell and P.A. Cusack, *Polym. Degr. Stabil.* **32** (3), 299–312 (1991).

[875] E. Bulska, B. Wagner, B. Stahl, M. Heck and H.M. Ortner, unpubl. results (2001).

[876] J.M. Pochan, J.J. Fitzgerald and G. Williams, in *Determination of Electronic and Optical Properties* (B.W. Rossitter and R.C. Baetzold, eds.), Wiley-Interscience, New York, NY (1993), Vol. VIII.

[877] J. Runt and J. Fitzgerald (eds.), *Dielectric Spectroscopy of Polymeric Materials*, American Chemical Society, Washington, DC (1997).

[878] N.E. Hill, W. Vaughan, A.H. Price and M. Davies, *Dielectric Properties and Molecular Behaviour*, Van Nostrand, London (1969).

[879] A.K. Jonscher, *Dielectric Relaxation in Solids*, Chelsea Dielectrics Press, London (1983).

[880] C.J.F. Böttcher and P. Bordewijk, *Theory of Electric Polarization*, Elsevier, Amsterdam (1978), Vol. 2.

[881] G.W. Schreyer and M. May, *Wiss. Z. Tech. Hochsch. "Carl Schorlemmer" Leuna-Merseburg* **26** (1), 128–46 (1984).

[882] P. Hedvig, *Dielectric Spectroscopy of Polymers*, J. Wiley & Sons, New York, NY (1977).

[883] L. Mascia, *The Role of Additives in Plastics*, Edward Arnold Publishers, London (1974).

[884] P.A.M. Steeman, *Interfacial Phenomena in Polymer Systems, Ph. D. Thesis*, Delft Technical University (1992); *cfr.* also P.A.M. Steeman and J. van Turnhout, in *Broadband Dielectric Spectroscopy* (F. Kremer and A. Schönhals), Springer-Verlag, Berlin (2002), pp. 495–522.

[885] O. Urakawa, K. Adachi and T. Kotaka, *Macromolecules* **26**, 2036 (1993).

[886] A. Schönhals, *Macromolecules* **26**, 1309 (1993).

[887] M. Müller, K. Kremer, R. Stadler, E.W. Fischer and U. Seidel, *J. Colloid Polym. Sci.* **273**, 38 (1995).

[888] G. Williams, C. Duck, J. Fournier and J.R. Hayden, in *Polymer Spectroscopy* (A.H. Fawcett, ed.), J. Wiley & Sons, Chichester (1996), pp. 275–96.

[889] J.N. Barsema, *M. Sc. Thesis*, Technical University of Twente (1999).

[890] L. Boden, M. Lundgren, K.-E. Stensio and M. Gorzynski, *J. Chromatogr.* **A788** (1/2), 195–203 (1997).

[891] J.C. Golba and M.G. Hansen (to General Electric), *U.S. Pat.* 4,448,943 (May 15, 1984).

[892] M. McBrearty, A. Bur and S. Perusich, *Proceedings SPE ANTEC '99*, New York, NY (1999), pp. 2580–4.

[893] M. McBrearty, A. Bur and S. Roth, *Proceedings SPE ANTEC 2000*, Orlando, FL (2000), pp. 287–91.

[894] G. Seytre, G. Boiteux, J.F. Chailan, J. Chauchard and B. Pinel, in *Proceedings 17th Intl. Conference on Advances in the Stabilization and Degradation of Polymers* (A.V. Patzis, ed.), Luzern (1995), pp. 315–26.

[895] P.A. Pethrick, S.B. Joshi, D. Hayward, Z.-C. Li, S. Halliday, W.M. Banks, R. Gilmore and L.W. Yates, *Mater. Res. Soc. Symp. Proc.* **503**, 69–74 (1998).

[896] W.S. Gibbons and R.P. Kusy, *Polymer* **39** (14), 3167–78 (1998).

[897] S. Mahrous and M.S. Sobhy, *Int. J. Polym. Mater.* **44** (1–2), 171–8 (1999).

[898] M.M. Ueki and M. Zanin, *Polim.: Cienc. Tecnol.* **7** (4), 42–50 (1997).

[899] M.M. Ueki and M. Zanin, *IEEE Trans. Dielectr. Electr. Insul.* **6** (6), 876–81 (1999).

[900] E.L. Leguenza, G.C. Silva, J.V. Gulmine, P.C.N. Scarpa and D.K. Das-Gupta, *IEE Conf. Publ.* **473** (Dielectric Materials, Measurements and Applications), 241–6 (2000).

[901] L.V. Dubrovina, V.M. Ogenko, S.N. Makhno and A.A. Chuiko, *Vysokomol. Soedin. A/B* **43** (9), 1535–9 (2001).

[902] L. Lucchese, C.M. Liauw, N.S. Allen, M. Edge, F. Thompson and R.S. Whitehouse, *Polym. Bull.* **44**, 187–94 (2000).

[903] T. Fahmy, *Int. J. Polym. Mater.* **50** (1), 109–27 (2001).

[904] S.-G. Lee, D.-C. Kim and C.W. Joo, *J. Korean Fiber Soc.* **38** (5), 214–21 (2001).

[905] V. Buckin, B.O. Driscoll, C. Smyth, A.C. Alting and R.W. Visschers, *Spectrosc. Europe* **15** (1), 20–5 (2003).

[906] A.R. Clarke and C.N. Eberhardt, *Microscopy Techniques for Materials Science*, Woodhead Publishing Ltd., Cambridge (2002).

[907] A. Briggs, *An Introduction to Scanning Acoustic Microscopy*, Oxford University Press, Oxford (1985).

[908] B. Bridge, *Br. J. Non Destruct. Testing*, 326–31 (Sept. 1987).

[909] D. Roberts, *Materials World*, 12–4 (Jan. 1996).

[910] T. Xu, H. Lei and C.S. Xie, *Polym. Testing* **21** (3), 319–24 (2002).

[911] G.D. Smith, E.C. Brown and P.D. Coates, *Proceedings SPE ANTEC 2001*, Dallas, TX (2001), pp. 3105–9.

[912] J.B. Hull, C.M. Langton and A.R. Jones, *Macromol. Repts.* **A31**, 1191–9 (1994).

[913] J.K.S. Wan, K.P. Vepsäläinen and M.S. Ioffe, *J. Appl. Polym. Sci.* **54** (1), 25–31 (1994).

Chapter 2

Thinking small

Polymer/Additive Analysis by Thermal Methods

2.1. Thermal Analysis Techniques . . . 158
  2.1.1. Differential Scanning Calorimetry . . . 163
  2.1.2. Differential Thermal Analysis . . . 173
  2.1.3. Thermogravimetric Analysis . . . 175
  2.1.4. Simultaneous Thermal Analysis Methods . . . 189
  2.1.5. (Multi)hyphenated Thermal Analysis Techniques . . . 192
  2.1.6. Thermal Microscopy . . . 209
  2.1.7. Thermoluminescence . . . 213
2.2. Pyrolysis Techniques . . . 214
  2.2.1. Pyrolysis–Gas Chromatography . . . 222
  2.2.2. Pyrolysis–Mass Spectrometry . . . 235
  2.2.3. Pyrolysis–Gas Chromatography–Mass Spectrometry . . . 244
  2.2.4. Pyrolysis–Fourier Transform Infrared Spectroscopy . . . 261
  2.2.5. Pyrolysis–Gas Chromatography–Fourier Transform Infrared Spectroscopy . . . 263
  2.2.6. Pyrolysis–Gas Chromatography–Atomic Emission Detection . . . 264
  2.2.7. Temperature-programmed Pyrolysis . . . 266
2.3. Thermal Volatilisation and Desorption Techniques . . . 275
  2.3.1. Thermal Separation Techniques . . . 278
  2.3.2. Direct Solid Sampling Techniques for Gas Chromatography . . . 282
  2.3.3. Thermal Desorption–Mass Spectrometric Techniques . . . 299
  Bibliography . . . 300
    Thermal Analysis . . . 300
    Pyrolysis . . . 301
    Thermal Desorption . . . 301
  References . . . 301

Additive analysis may be carried out by examination of extracts or dissolutions of the polymer, by non-destructive spectroscopic (in-polymer) testing of solid or melt, or by degradative testing using thermal methods mainly through the examination of volatiles released (**"thermal extraction"**). In this Chapter we consider thermo-analytical and pyrolysis methods applied to polymer/additive formulations "as received"; Chp. 3 deals with laser desorption techniques.

Thermoanalytical methods are especially suitable when liquid or gas extraction fails and for characterising intrinsically insoluble polymers, *e.g.* cross-linked materials such as vulcanised rubbers. Other materials (such as drying oils, lacquers, and some synthetic polymers) become insoluble and non-volatile on ageing. This renders them unsuitable for conventional analysis requiring solubility (*e.g.* HPLC) or high volatility (*e.g.* GC). Analysis of non-volatile polymeric organic samples is often compromised by their intractability. Heating polymeric materials may lead to post-cross-linking, release of gaseous products (rest monomers and solvents, plasticisers and other low-MW additives) and toxic gases (from thermal decomposition or interaction), bubble formation (caused by outgasing) and decomposition. The non-volatility of macromolecules is exploited to advantage in polymer/additive analysis.

Table 2.1. Tools for materials analysis by thermal methods

Heating rate (dT/dt)	Atmosphere	
	Air	Inert
Quasi-isothermal	MTDSC	MTDSC
$>10°C/min$	TG, DSC	TG, TPPy[a]
$>100°C/s$	Flash TG, fast thermolysis, oxidative pyrolysis, HPer DSC	Flash TG, fast thermolysis, flash pyrolysis[b], HPer DSC
$\gg 1000°C/s$	Combustion, laser desorption	Laser pyrolysis

[a] Typically 1°C/min to 600°C.
[b] Typically 0.2 s to the pyrolysis temperature (500 to 800°C).

In principle, heating a material to desorb the volatile components (thermal desorption) is the most direct way to analyse for organic additives in a compounded polymer without interference of the matrix. In this context a volatile compound is considered being one having a vapour pressure high enough so that at least some of it can be vaporised by heating at a temperature lower than the thermal decomposition point of the polymeric component, which is the case for numerous organic additives for polymers. These additives can be selectively volatilised and identified. The efficiency of **volatile removal** from a polymer matrix is influenced by several factors [1,2]:

1. *Particle size.* Volatiles are removed more efficiently from small particles (powder) than from larger ones.
2. *Temperature.* Higher temperatures are generally most effective. Diffusion rates are markedly higher above T_g of the polymer. Temperatures in the range of 100–300°C are typically used for desorption of volatiles from polymers.
3. *Vapour pressure.* Most polymer additives are solids at room temperature and exhibit low vapour pressures. Detection of the maximum number of additives may require heating of the sample rather close to the decomposition point of the polymer.
4. *Residence time.* Volatiles will be desorbed more completely from the polymer if they are removed from the heating zone as they evolve. Fast removal of desorbed species is accomplished either by heating in (high) vacuum (*e.g.* DI-MS, vacuum TG-MS) or by use of a continuous flow (*e.g.* thermal desorbers, DHS-GC-MS).

As shown in Table 2.1 there are thermoanalytical techniques, such as **thermogravimetric analysis** (TGA) or temperature programmed pyrolysis (TPPy), in which slow heating profiles are taken to advantage, in particular in combination with appropriate detection modes (*e.g.* TG-MS, TG-FTIR, TPPy-MS, TD-MS). In these volatile removal techniques, the additives are generally all detected at temperatures below the decomposition temperature of the polymer. However, it is also possible to gain information on additives from flash pyrolysis experiments. Fast thermolysis/FT-IR is to be positioned between conventional thermogravimetry and fast pyrolysis. Thermal studies of polymers and polymer formulations may be classified according to the amount of energy provided to the system (Table 2.2).

The partial pressure of some polymer additives and auxiliary agents is so low that these cannot be introduced into a GC-MS system using the classic method without undergoing decomposition. Such compounds with molecular masses >1000 Da are often low-MW oligomeric additives and can only be analysed using GC-MS by means of pyrolysis, *i.e.* when fragmented. Thermal desorption and PyGC-MS are uniquely sensitive and versatile methods of analysis. Whereas TG-MS is more suitable for volatile compounds, PyGC is widely used in analysis of non-volatile compounds. The present power of TG-MS, TG-FTIR and (microfurnace) PyGC-MS is typically illustrated in the thermal decomposition of sodium ethyl xanthate (SEX), which leads to a complex mixture with carbon disulfide, diethyl disulfide, ethanol and carbonyl sulfide as major gases [3]. PyGC-MS was the only technique that permitted unambiguous identification of all the evolved gases. Interpretation of the TG-MS data was reliant on the PyGC-MS data. The overlapping of molecular ion signals with isotope and/or fragment ion signals posed a significant problem in determining the amounts of each gas produced. TG-FTIR was limited to identifying gases with very characteristic vibration frequencies, such as CS_2 and carbonyl sulfide, and monitoring of functional groups. TG-MS

Table 2.2. Classification of thermal studies of polymeric materials

Energy provided	Effect	Structural information
Very high	Complete pyrolysis *cq.* combustion (CO_2, H_2O)	Elementary composition
High	Pyrolysis, thermolysis (Release of structurally significant fragments)	Polymer structure, additive structure[a]
Moderate	Thermal degradation (Release of residual volatiles)	Volatiles
Low	Thermal desorption (Release of volatiles)	Monomers, oligomers, additives, *etc.*

[a] Data reduction required.

Table 2.3. Comparison of some thermal decomposition techniques

Feature	DP-MS[a]	TVA[b]	Flash PyGC-MS
Sample size	μg-mg	50 mg	mg
Residence time in hot zone	μsec-msec	sec	<1 sec
Transport time[c]	μsec-msec	sec	10^1–10^2 min
Fragment masses	high	stable[d]	ionisation mode dependent
Probability of secondary reactions	low	high	low

[a] Direct probe mass spectrometry.
[b] Thermal volatilisation analysis.
[c] Time elapsing from product formation to detection.
[d] Thermally labile products may escape detection.

and TG-FTIR may be insufficient for the identification and monitoring of gases in a complex mixture because of lack of a separating medium.

Pyrolysis is essentially the cleavage of chemical bonds by use of thermal energy only (in inert atmosphere). Analytical pyrolysis is the technique of studying molecules either by observing their behaviour during pyrolysis or by studying the resulting molecular fragments. Commercially available pyrolysis instruments are capable of heating filaments to temperatures in excess of 800°C in milliseconds, producing rapid thermal degradation of small samples. To obtain reproducible pyrolysis many experimental parameters need to be optimised and carefully controlled. Continuous-mode and pulse-mode pyrolysers are commercially available to control the necessary parameters needed to give reproducible pyrolysis. Analytical pyrolysis represents an extensive family of techniques with very little of interlaboratory comparison.

It should be stressed that the (micro) destructive methods described in this Chapter should not be considered as in-polymer techniques. In fact, generally no properties of the polymeric matrix are being measured directly, but just off-gases (desorption) or fragments (pyrolysis).

It is quite obvious that as with every comparative technique, analytical pyrolysis benefits from an extensive reference library of pyrograms; similarly, specific mass spectral databases of additives are very useful for PyGC-MS and TG-MS experiments. Analytical pyrolysis and thermogravimetric analysis are closely related. Temperature-programmed pyrolysers (TPPy) and TGA have similar features, but TPPy lacks any weight information. Vacuum TG-MS and in-source direct pyrolysis-MS (DP-MS) have a similar relationship. Nevertheless important differences should be noted (Table 2.3; *cfr.* also Table 2.40). In fact, these techniques may see different products. In case of DP-MS very small sample weights, 0.1–1 μg, are allowed, as opposed to some 10 mg for TG-MS. Apart from handling problems of solid material this raises the important question of **sample representativity**. For example, Sinclair *et al.* [4] noticed that pyrolysis of solid PP/DSTDP led to too much scatter in the results of repetitive analyses for the approach to give quantitative data. The sample sizes were on the order of 100 μg and the origin of the scatter was attributed to inhomogeneities in additive dispersal (granule-to-granule variations).

Insuring that a solid sample of only a few micrograms is homogeneous and therefore representative of the material from which it was taken presents a constant problem to the methods described in this Chapter. Especially the extremely small sample sizes used in the very sensitive technique which is DP-MS gives rise to concern. In many cases larger sample capacity is more an advantage than an obstacle. This certainly holds true for the analytical investigation of rubbers. Analysts sampling materials for which homogeneity is a concern have devised several methods to deal with the problem. If possible, the sample material may be ground to a fine powder from which small portions are taken for analysis.

The various techniques in the realm of thermal analysis have a variety of uses in quality control testing, R&D, and failure investigations of insoluble or carbon-black containing polymeric materials, such as rubbers.

2.1. THERMAL ANALYSIS TECHNIQUES

Principles and Characteristics

Thermal analysis (TA) is the general denomination of methods in which bulk **physical property changes** of a material, a mixture of substances or a reaction mixture are measured in response to programmed changes in temperature in a specified atmosphere [5,6]. The main physical properties measured are transition temperatures, enthalpy, dimensional changes, viscoelastic properties, dielectric properties and mass changes (Table 2.4).

Some 12 major thermal analysis techniques do exist, including thermometry which provides the standard of temperature measurements. The most widely used instruments in thermal analysis (and their primary output signals) are DTA (T), DSC (T, heat flow-rate, enthalpy), TG (T, mass), dilatometry (T, length), TMA (T, length, force) and DMA (T, length, force, frequency). As none of these measurands is measured using an absolute technique, all instruments need to be calibrated. In dynamical mechanical analysis (DMA) a sample is subjected to sinusoidal mechanical deformation of frequency, f, and the corresponding forces measured. Conversely, the sample can be subjected to a defined force amplitude and the resulting deformation measured. DMA measurements provide an insight into temperature- and frequency-related molecular movement, and supply information regarding elasticity and stiffness. In general, DSC measurements aid the interpretation of DMA curves (and *vice versa*). There are only restricted links to polymer/additive analysis (identification/quantification).

Table 2.4. Main thermal analysis methods

Basis of method	Thermal analysis techniques
Mass change	TG, DTG, STA, TG-DTA, TG-DSC
Temperature change	DTA, STA
Energy change	DSC, PDSC, MTDSC, HPer DSC
Dimensional change	TMA
Mechanical properties	DMA, DETA
Optometry	Thermomicroscopy, TOL

When the physical property measured is light energy, the technique is thermoptometry, *i.e.* a family of techniques in which an optical property of a sample is monitored against time or temperature, while the temperature of the sample, in a specified atmosphere, is programmed. Two examples of thermoptometry are: thermomicroscopy (Chp. 2.1.6), where the sample is observed directly under a microscope; and thermoluminescence (Chp. 2.1.7), where the luminescence intensity of a sample is monitored as a function of temperature.

Classic thermal analysis observes property changes in isothermal runs, including stepwise heating and cooling, or constant rate heating or cooling. However, thermoanalytical methods are generally not equilibrium methods. More recently, other **temperature control modes** have been introduced such as sample controlled thermal analysis (SCTA or controlled-rate thermal analysis, CRTA) [7], temperature jump [8] and rate jump [9], temperature modulation [10–12], and repeated temperature scanning [13]. Each mode has its suitable applications, advantages and drawbacks [14]. Differential analytical techniques may be used to resolve overlapping thermal effects.

In contrast to analytical pyrolysis, thermal analysis techniques are not usually concerned with the chemical nature of the reaction products during heating. However, during such events, analysis of the decomposition products can be done (evolved gas analysis). Most TA experiments are conducted at atmospheric pressure, with many applications requiring the use of a defined gas atmosphere (oxidising, reducing, inert). For desorption experiments (high) vacuum methods are desirable.

Definitions, nomenclature, terms and sources of information in thermal analysis are to be found in refs. [15,16]. The basis of thermal analysis has recently been reviewed by Wunderlich [6], thermoanalytical instrumentation, techniques and methodology by Gallagher [17]; the history of thermal analysis was traced by Mackenzie [18]. Thermal analysis of polymers is described in various books [19–23] and reviews [24–28]. Thermal analysis is a powerful **secondary technique**.

Thermal analysis can offer advantages over other analytical techniques, including variability with respect to application of thermal energy (temperature control mode), sample size (from 0.1 μg to 500 g) and form (gel, liquid, glass, solid), ease of variability and control of sample preparation, atmosphere of choice, relatively rapid and moderately priced instrumentation. Various TA techniques are particularly suited to the study of polymeric materials, reflecting structural changes unique to substances composed of large extended chain molecules, and have been in use for several decades. The advantages of (combined) TA techniques for the study of polymer formulations are: a wide accessible temperature range, the ability to vary the atmosphere during thermal treatment, monitoring of evolution behaviour in real time, concentration of volatiles by multicomponent organic sorbent traps, excellent GC performance and powerful analysis of components by GC-MS techniques.

In this Chapter we focus on the analysis by thermal methods of the low-MW ingredients in polymeric materials. The main techniques for this purpose are differential thermal analyses (DTA), differential scanning calorimetry (DSC), thermogravimetry (TG), and corresponding hyphenated techniques. In recent years, DSC has largely supplanted DTA. It is often possible to identify substances by reference to a characteristic transition temperature. By investigating the changes in the measured property (*e.g.* enthalpy, mass, length, stiffness, *etc.*) with temperature, one may be able to quantify filler content but also the degree of crystallinity or cross-link density. TG is a technique that, although limited in scope to those reactions taking place with a change in weight, gives intrinsically quantitative results. DSC and DTA are essentially more versatile techniques which can detect *any* reaction taking place with a change in energy. However, a single thermoanalytical technique is seldom adequate to answer completely and unequivocally a specific problem. It is common, therefore, to use several thermal and other analytical techniques in an investigation. **Simultaneous techniques** incorporate various thermoanalytical measurements, *e.g.* TG, DTG, DTA [29]. A typical modern assembly is TG-DSC (or "STA"). The coupling of thermal analysis methods with other methods of instrumental analysis (*e.g.* FTIR, MS or GC) is becoming increasingly important, in particular for **evolved gas analysis** (EGA). Analysis of volatiles may be carried out off-line (using adsorbent tubes) or on-line.

Like all analytical instruments, thermal analysis equipment offers automation. The instruments are installed, qualified and calibrated (IQ, OQ) and their performance should periodically be checked by the user (PQ) according to ISO-norms and current GMP. In common with a wide range of analytical techniques, there are many difficulties associated with obtaining "equivalent data" using thermal analysis. The limitations identified can be minimised by tight control of specimen size, preparation and evaluation methods employed within laboratories. **Quality assurance** methods in the TA laboratory have been addressed [30–32]. A procedure for the temperature calibration of thermobalances is under elaboration by a working group of GEFTA (see also ASTM E 1582-00). The effect of various parameters on the results is considered to improve the metrological quality beyond the level of previous proposals. An increased need for **standardisation** is felt, both in terms of standard method development and standard materials, *e.g.* a wider range of Standard Reference Materials (NIST, LGC-ORM, ICTAC, *etc.*), and software control, as pointed out by Sarge *et al.* [15]. The COMAR database for reference materials (RMs), consisting of 10,285 RMs (as of June 1998) lists only 19 RMs for temperature, 12 for heat of fusion and 1 for heat capacity. A compilation of RMs for calorimetry and thermal analysis is available [33]. Appropriate and well characterised polymers are needed as ASTM and industrial standards for a variety of applications, including oxidative stability of engineering plastics, composites, coatings and elastomers, quality assurance protocols and control charting. Riga *et al.* [34] have evaluated the thermal oxidative behaviour of a set of reference polymers by means of TG, DTA, DSC and PDSC (pressure DSC). There is good correlation between measured stability by PDSC in oxygen and DTA in air. Affolter *et al.* [35] have recently described **interlaboratory tests** according to ISO 5725 with polymeric materials using thermal analysis, namely the determination of the contents of carbon-black (ISO 9924-1 or ASTM 4218) and ash, of vinylacetate in EVA copolymers, of plasticisers in thermoplastics (all with TGA), as well as the crystallinity of thermoplastic materials, curing of epoxy resins and OIT of polyolefins (all by DSC). The comparisons between TGA and standardised methods reveal that TGA can be an alternative, which is time effective and produces at least equal or better results. For standardisation of thermal analysis, calibration, and (certified) reference materials, *cfr.* ref. [36].

Apart from combined TA techniques (on-line or not) the actual **trends** in thermal analysis are the introduction of modulated and high-resolution techniques, hyphenated thermal analysis methods (*e.g.* TG-FTIR, TG-MS, DSC-XRD, etc.), alternative heating modes, microthermal analysis methods, industrial standardisation and quality control. Modulation means a periodic perturbation of a temperature program. **Temperature modulation** finds application in DSC, TG, DETA, TMA and μTA. Temperature-modulated techniques, such as Modulated DSC (MDSC™) and Modulated TGA (MTGA™), broaden the insight into the material properties. The use of modulated temperature programs in thermal methods has been reviewed [37,37a].

In **microwave thermal analysis (MWTA)** microwaves heat the sample by direct interaction (rather than by conduction of heat as in conventional thermal analysis) and can penetrate up to about 2 cm [38]. This reduces temperature gradients in masses up to 0.5 g. The method allows the study of thermal and dielectric properties of materials. **Microthermal analysis methods** allow to reduce the amount of material observed and considerably increase the range of problems that can be tackled (*cfr.* Chp. 2.1.6.1).

A measurement good practice guide on thermal analysis is available [39].

Applications
The most important **users** of thermal analysis techniques undoubtedly include the plastics processing and manufacturing industries, the pharmaceutical and the chemical industry in general, as well as the food and petrochemical industries. Even with an extremely small amount of sample DTA, DSC, TG/DTG, TMA, and DMA are broadly applicable in all phases of the polymer industry, including raw material and quality control, compound design, processing, vulcanisation, product evaluation, failure analysis, environmental stability, and research and development. Thermal methods of analysis are important for polymer characterisation, mainly because they are relatively immune to the difficulties associated with the non-volatile and insoluble nature of high polymers. The polymers that have been most extensively studied by thermochemical analysis are PVC, PS, styrene-acrylonitrile copolymers, PE, PP, polyacrylates and copolymers, PET, polyphenylenes, and polyphenylene oxides and sulfides. In most instances, thermochemical analysis is performed under inert atmosphere to avoid the production of secondary oxidation products.

Table 2.5 summarises the **main applications** of thermal analysis and combined techniques for polymeric materials. Of these, thermomechanical analysis (TMA) and dynamic mechanical analysis (DMA) provide only physical properties of a very specific nature and yield very little chemical information. DMA was used to study the interaction of fillers with rubber host systems [40]. Thermomechanical analysis (TMA) measures the dimensional changes of a sample as a function of temperature. Relevant applications are reported for on-line TMA-MS (*cfr.* Chp. 2.1.5); μTMA offers opportunities (*cfr.* Chp. 2.1.6.1). The primary TA techniques for certifying **product quality** are DSC and TG (Table 2.6). Specific tests for which these techniques are used in quality testing vary depending upon the type of material and industry. Applications of modulated temperature programme are: *(i)* study of kinetics; *(ii)* AC calorimetry; *(iii)* separation of sample responses (in conjunction with deconvolution algorithms); and *(iv)* microthermal analysis.

Thermal methods are a viable option for **polymer/additive analysis** [41]. Thermal analysis of additives in polymers has recently extensively been reviewed by Bair [27], in particular with regard to protective agents (antioxidants, light and heat stabilisers), plasticisers, flame retardants, nucleating and blowing agents, processing aids (mould lubricants), carbon-black, fillers, photo-initiators, coupling agents and polymeric impact modifiers. Whereas Chiu *et al.* [42] have described thermal evolution techniques for the determination of additives

Table 2.5. Main applications of thermal analysis and combined techniques in polymer development

Application	Techniques
Polymorphism	
Raw materials; characterisation, control of crystallisation	DSC, solution calorimetry, microcalorimetry, sub-ambient DSC, DSC-spectroscopy, DSC-X-ray, thermomicroscopy
Raw materials; storage conditions	DSC, TG, adsorption isotherms
Process control	DSC, TG, DSC-IR
Amorphous state	
Glass transitions (T_g); influence of moisture, plasticisers	DSC, MTDSC
Study of (co)polymers, blends	DSC, MTDSC
Optimisation of formulations	DSC, TG, MTDSC
Quantitation	DSC, microcalorimetry, TG
Polymer identification	
Reverse engineering, process control; supplier monitoring	DSC, TG, TG–MS, TG–FTIR
Polymeric materials development	
Physical interactions, phase diagrams	DSC
Melting point	DSC
Identification, quantitation	DSC, TG
R&D development work	DSC, TG, TG-MS, TG-FTIR
Combustion properties	TG, TG-MS, TG-FTIR
Flame retardant properties	TG, TG-MS, TG-FTIR
Process optimisation	DSC, DSC-spectroscopy
Interaction	
Determination of bound water	DSC-TG, sub-ambient DSC
Plasticisation by gases	HPDSC
Reactivity/reaction monitoring	
Curing, cross-linking	DSC, DTA, CRTA
Sample composition	
Competitive analysis	DSC, TG, TG-MS, TG-FTIR
Volatiles, polymer, fillers	TG, TG-MS, TG-FTIR
Impact modifiers	DSC
New formulations	DSC, TG
Quantitation	
Gas solid reactions, calibration	PTA, TG-MS
Thermal history	
Crystallinity, melting	DSC
Thermal stability	
Thermal decomposition, kinetics	DSC, DTA, TG, TG-MS, TG-IR
Thermo-oxidative degradation (OIT)	DSC, DTA, TG
Evaluation of stabilisers	DSC, DTA, TG
Compatibility of blends	Microcalorimetry
Processing behaviour	DSC, DTA, TG
Product lifetime	DSC, TG
Evolved gas analysis	
Decomposition, stability	TG-MS, TG-FTIR, TG-GC-MS, TMA-MS
Environmental protection	TG-DTA/GC-MS
Raw material control	
Purity	DSC, TG
Automated batch analysis; pass/fail tests, quality assurance	DSC

Table 2.5. (Continued)

Application	Techniques
Product QC/QA	
Determination of T_m, T_g; degree of cure; plasticiser efficiency	DSC
Decomposition temperature	DSC, TG
Component quantification	TG
Failure analysis/troubleshooting	
Polymer identification	DSC, MTDSC, TG
Localised thermal analysis	
Fracture	μTA
Surface analysis	MTDSC
Imaging	
Subsurface imaging	SThM

Table 2.6. Applications of differential scanning calorimetry (DSC) and thermogravimetry (TG)

Application	DSC	TG
Physical properties		
Specific heat capacity	•	
Physical transformations		
Heat of fusion, specific heat ΔH_f, Δc_p	•	
Crystallinity	•	
Evaporation, sublimation, desorption	•	•
Polymorphism	•	
T_g, softening, T_m	•	
Chemical properties		
Purity	•	•
Chemical reactions		
Decomposition, pyrolysis, depolymerisation, thermal stability	•	•
Oxidative decomposition/stability	•	•
Degree of cure, vulcanisation	•	•
Reaction profile, kinetics	•	•
Reaction heat ΔH_r	•	
Safety aspects	•	
Chemical composition		
Compositional analysis (*e.g.* moisture, liquid components, carbon-black, ash, fillers, polymer)	(•)	•

and Pearce *et al.* [43] have emphasised the usefulness of EVA techniques for investigations on flame retardants, Bair [27] has stressed the fact that direct quantitative analysis in thermal analysis is underutilised by many TA practitioners. Often, however, the amount of additive is so small that conventional thermal analysis techniques cannot easily detect it, as in case of a coupling agent in a resin. Similarly, the amount of antioxidant (AO) added to a polyolefin is usually so small (typically below 0.5 wt.%) that it cannot easily be analysed directly by any known thermal analysis method; however, at elevated temperatures in an oxidising atmosphere **indirect thermal analysis methods** can be used to determine AO concentrations to below 0.01 wt.%.

On the other hand, a variety of hyphenated techniques (such as TG-MS, in particular when coupled to powerful data evaluation methods, such as PCA) are quite suitable for the measurement of fairly small concentrations of additives in polymeric matrices. Quite obviously, at very low additive concentrations using small sample sizes, it is necessary to ascertain homogeneity of the samples. When the concentration of an additive exceeds a few weight percent, it is often possible to assay additives calorimetrically in a commercial resin without extraction. If the additive is incompatible with the resin, it can be detected in a separate crystalline of glassy phase by either its melting temperature (T_m) or its glass transition temperature (T_g) and measured quantitatively from heat of fusion (ΔH_f) determinations at T_m or the change in heat capacity (Δc_p) at T_g. Conversely, when an additive is soluble in a polymer, its concentration can be estimated from shifts in T_m or T_g of the resin. Many non-polymeric additives, such as plasticisers, can be vaporised quantitatively by thermogravimetric techniques at temperatures that are well below the degradation temperature of the host polymer and identified by IR or MS analysis [27]. For the effects of temperature for the purpose of evaluating plastics, *cfr.* ref. [44].

Modern thermal analysis, microcalorimetry and new emerging combined techniques, which deliver spectroscopic, microscopic and calorimetric data, are powerful tools for the study of **elastomers** [45]. These techniques are very useful in all steps of new product development and for product quality control. Gaddy *et al.* [46] applied thermal analysis, such as TGA, DSC and DMA to characterise **aged materials**. Both black and non-black filled EPDM vulcanisates were aged according to ASTM D 4637 (artificial heat, ozone and UV light exposure) as well as by outdoor exposure. It was found that thermal analytical methods do show only marginal changes due to ageing. A variety of procedures other than thermal analysis are reported for analysing vulcanisates, but most are too lengthy to use as routine procedures. A summary of many of the wet chemical procedures is presented in ASTM D 297–93 [47]. These procedures are remarkable because of their complexity.

Thermal analytical methods have proven to be useful in the evaluation of a number of flammability parameters (DSC: rate of heat release, solubility; TG: FR activation temperature, quantification, weight loss). Thermal stability is an important criterion for the development of new **flame retardant formulations**, and thermal analysis is widely used in the evaluation of the residual char, the study of the mechanism of action, the interaction between the components, *etc.* [48]. Flame retardants are required to survive various processing steps as, *e.g.* compounding, injection moulding. Thermal analysis, dynamic real-time injection moulding experiments and capillary rheometry supply the necessary important information on the thermostability of flame retardant compounds [49]. Typically, brominated flame retardant systems studied recently were HIPS/(Saytex HBCD, DBTM), PA6.6/(Saytex 8010, Sb_2O_3), GFR PET/(Saytex 8010, $Na_2Sb_2O_6$), GFR PET/(Saytex 120, $Na_2Sb_2O_6$) and GFR PET/(Saytex BT93W, $Na_2Sb_2O_6$) [49]. Similarly, it is quite obvious that stabilisers should not decompose during the different thermal treatments of a polymer; temperatures up to 300°C and more may occasionally be applied for rather short periods. Most commercial antioxidants satisfy this requirement.

In product and process control **quality assurance** of polymeric materials cannot be imagined without thermal analysis [25]. TA is widely used for automated batch analysis of incoming raw material in production laboratories [50] and for QC of engineering thermoplastics. Method development includes the automated calculations necessary for determination of pass/fail criteria. Increasing confidence in thermal analysis data has resulted in many ASTM, DIN and ISO standards. ASTM International Committee E37 on Thermal Measurements has jurisdiction over some 40 standards covering all aspects of thermal analytical techniques and thermophysical properties [51,52]. A detailed list of established **standard methods** and practices, relevant to additives in elastomers and plastics, formulated on thermal measurements, is available [53].

Applications volumes for thermal analysis of polymers and paints are available [54,55].

2.1.1. Differential Scanning Calorimetry

Principles and Characteristics

Calorimetry is the name given to any experiment that is used to measure the transfer of heat in any of its manifestations. Differential scanning calorimetry (DSC) is defined by ICTAC as: "A technique in which the difference in energy inputs into a substance and a thermally inert reference material is measured as a function of temperature, while the substance and reference material are subjected to a controlled temperature programme". DSC is thus designed to measure the actual amount of power (heat flow-rate, in watts) involved directly with the associated thermal event, rather than its indirect effect, the simple change in temperature of the sample. The DSC thermogram is a plot of the differential **heat flow** *versus* temperature (typically 10°C/min) or time. Integration of peaks gives the enthalpy change of the specimen. When the sample absorbs energy the enthalpy change is called endothermal; when the sample releases energy it is called exothermal. Exothermal transitions are typically crystallisation, curing, cross-linking, oxidative degradation, whereas endothermal behaviour is found in the melting transition and vaporisation. Any transition in a material that involves a change in the heat content of the material can be detected and measured by DSC. One rarely starts a DSC run without first running a TGA to determine the degradation point of a material.

Two main types of commercial **DSC instruments** are in use, namely "heat flux" (hf) and "power compensation" (pc) instruments (*cfr.* ref. [21]). The "power compensating" version, originally developed by Perkin-Elmer Co. [56], employs two different ovens. DSC in the "heat flux" mode with one oven is similar in operation to a conventional

Table 2.7. Comparison of DSC instrumentation

Power compensation DSC	Heat flux DSC
+ One point calibration of caloric sensitivity	− Calibration of caloric sensitivity
+ High resolution	− Lower resolution/overall precision
− Baseline correction necessary	+ Baseline stability
− Difficult temperature calibration	+ Easy temperature calibration
− Long stabilisation time in low T range	+ Higher operational temperature ranges
	+ Easy handling

DTA, except that the quantitative compensation for the problem areas, such as the temperature dependence of thermal transport and sensor sensitivity, is built into the associated hardware and software. In an hfDSC instrument, ΔT between sample and reference is recorded, after suitable calorimetric calibration, as a direct measure of the difference in heat flow-rate or the difference in power. In a pcDSC instrument the difference in power supplied to sample and reference, in order to keep their temperatures as nearly the same as possible, is measured directly. In its most refined form the DSC apparatus closely approaches an ideal isothermal calorimeter. Table 2.7 compares conventional DSC instrumentation. The performance of a DSC is mainly dependent on baseline stability and reproducibility, sensitivity and resolution. In the newest designs the advantages of pcDSC and hfDSC are combined (high resolution, baseline stability and high heating/cooling rates) [57]. Advanced hfDSC technology provides a fundamentally more accurate way of measuring heat flow than in the past, with better resolution than pcDSC. A comparative test of DSC, with emphasis on resolution and sensitivity, was reported recently using 4,4′-azoxyanisole; 22 different models of 8 manufacturers were involved [58]. It is distressing to notice that more than half of the contributions had to be corrected because of incorrect interpretation of the results! Yet, in general DSC is more readily interpreted than DTA and yields more reliable values for calorimetric quantities but it can only be applied over a limited temperature range compared with DTA.

For DSC measurements the sample is contained in a metal pan and the reference is an empty pan of the same material (usually aluminium). In a typical DSC, a sample may have a mass of 20 mg and shows a heat capacity of about 50 mJ/K. DSC can examine materials between −170°C and +750°C. As heat is

Table 2.8. Main characteristics of DSC

Advantages:

- Simplicity and ease of use
- Short analysis time
- Small sample quantities required (<10 mg)
- Reproducibility and precision; quantitation
- Mature technology
- Combined techniques
- New developments (PDSC, HPer DSC, TMDSC)
- Wide applicability (QC tool)

Disadvantages:

- Physical information only
- Experimental conditions far removed from industrial operation (no stirring, *etc.*)

transferred through the thermoelectric disc, the differential heat flow to sample and reference is measured. A purge gas (typically N_2, Ar or He) is introduced to the sample chamber. Oxidising gases such as air or oxygen can also be used to observe specific chemical reactions.

Accurate temperature calibration using NIST-ICTA melting point standards (indium, tin, lead, aluminium, zinc of purity >99.999%; accuracy to 0.1°C or better [59]) is essential. Richardson [60] has critically described **standardisation** and quality assurance of DSC. Much work has been done to implement peak separation software techniques.

Table 2.8 shows the **main features** of DSC. Differential scanning calorimetry is the workhorse of the thermal analysis laboratory when it comes to measurements of changes in the heat capacity of a material with temperature. This enables detection and quantification of a wide variety of physical and chemical phenomena, as indicated in Table 2.5. DSC analysis may be applied to polymer products ranging from granules, powders, fibres and films to all kinds of injection-moulded parts. DSC is also invaluable in the characterisation of blends and copolymers and can also be used to study ageing and degradation as well as curing and cross-linking reactions. The improvement of the sensitivity and robustness of microcalorimetry allow a strong increase of the use of this technique for stability studies, quantitation of amorphous content, *etc.* DSC is a relative technique – quantitative data are obtained by comparison of signals from known and unknown, *i.e.* calibrant and sample. Calibration substances for temperature and heat calibration of DSC are commercially available. Portable thermal analysers (DSC) have shortly been introduced for quality control and other routine analysis.

Temperature-modulated differential scanning calorimetry (MTDSC) has recently commercially been introduced as an extension of DSC in which the usually linear or isothermal temperature program is overlaid by some type of temperature perturbation (*e.g.* sinusoidal) [10,11,61]. MTDSC determines a dynamic heat capacity from the relationship between the modulation components of temperature and heat flow. In MTDSC the heat capacity component, which scales directly with heating rate, can be separated from other processes. MTDSC distinguishes between kinetic and thermodynamic processes, thereby revealing information which cannot be obtained by standard DSC. Benefits include increased precision, higher sensitivity for weak transitions, simultaneous optimisation of sensitivity and resolution, improved interpretation of complex transitions, and measurement of heat capacity and thermal conductivity. MTDSC is to be regarded as a valuable extension of DSC, which is primarily advantageous in case of overlapping processes involving excess processes that are not susceptible to the temperature modulation such as curing [62]. MTDSC has been developed into CASM (Calorimetric Analysis with Scanning Microscopy – a hybrid of AFM), also called μMTDSC.

As opposed to the quasi-isothermal scanning rate of MTDSC experiments **high-performance DSC** (HPer DSC) is a form of pcDSC which allows quantitative measurement at very high heating and cooling rates (high speed DSC) – from 50°C/min up to 500°C/min – of 10–100 μg samples [63,64]. The increased scan rate gives significantly higher sensitivity because it leads to higher heat flow. Quantitative results are obtained as instrumental drift is minimal during the very short times of measurement. HPer DSC facilitates analysis of rate-dependent phenomena in real time (process simulation). HPer DSC can profitably be operated in a SEC-FTIR-HPer DSC arrangement [64]. Combined methods are DSC-IR and DSC-XRD. Simultaneous DSC-XRD has been reported for the study of phase transitions [65,66].

DSC thermograms have been published [67]; the reader is referred to a recent book [68] and reviews on DTA and DSC [69–71] for further details. Quantitative DSC has been reviewed [72]. Mathot [73] has reviewed recent advances in DSC, including (very) high pressure DSC (up to 550 MPa). Modulated DSC has been reviewed by ref. [21] and its practical applicability to polymeric systems has been described [74]. Reading *et al.* [75] have discussed origin and interpretation of MTDSC. Special issues on temperature modulated calorimetry [76] and MTDSC [77,78] have appeared.

Applications

Polymer applications of DSC are numerous and concern the determination of T_m (ASTM E 794), T_g (ASTM E 1326-03, ISO/FDIS 11357-2), specific heat capacity of a material (ASTM E 1269, ASTM D 4816), crystallisation temperature upon cooling (ASTM E 794), transition temperatures (ASTM D 3418, ASTM D 4419, ASTM D 4591), purity of a material [79,80], contamination outgassing (ASTM E 1559), reaction rates, sample composition, reaction kinetic constants (ASTM E 698), reaction mechanisms, thermal stability (ASTM E 537), minimum processing temperatures, heat of fusion and crystallisation (ASTM D 3417), heat of crystallisation (ASTM E 793), additive effects on a material, quality control of raw materials [25], discrimination between materials, detection of polymorphism [81], characterisation of thermally and UV cured materials (cure state, degree of cure) (ASTM D 2471, ASTM D 5028), oxidative stability testing, OIT (ASTM D 3895, ASTM D 3012, ASTM E 1858-03), *etc.*

DSC has been used in the determination of an internal moulding lubricant (low-MW PE) in a polyphenylene oxide (PPO) based resin, where the area under the melting peak is proportional to the amount of PE [82]. De *et al.* [83] have used thermal analysis methods (DSC, TG) for characterisation of polyurethane-mica composites; DSC of TPU-mica was carried out to determine the T_g of composites. Examination of T_g by DSC reveals the plasticising effect of small molecules on the **glass transition** behaviour of polymers. The addition of plasticisers lowers T_g of NBR [84]. The plasticiser content in

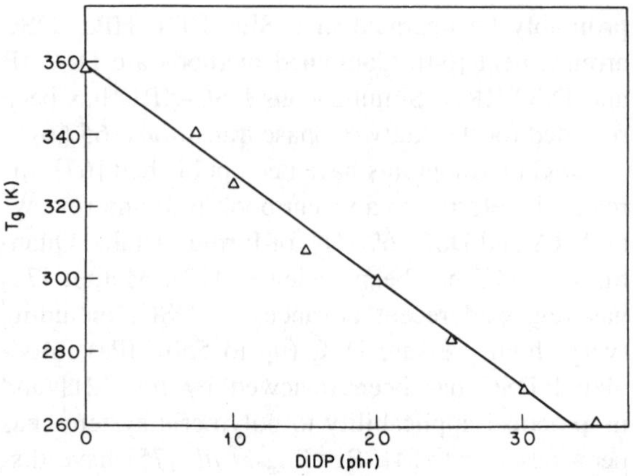

Fig. 2.1. T_g plotted *versus* concentration of diisodecyl phthalate in PVC. After Bair [27]. Reprinted from H.E. Bair, in *Thermal Characterization of Polymeric Materials* (E.A. Turi, ed.), Academic Press, San Diego (1997), pp. 2263–2420. Copyright (1997), with permission from Elsevier.

PVAc has been determined according to equation $T_g = 18.4 - 6.77C + 0.2197C^2$ (plasticiser content C in %; validity range: 0–12.5%; correlation coefficient 0.9986) [85]. Whereas further chlorinating (PVCC) increases T_g from approximately 86°C towards 100°C, it can be lowered to almost any value by addition of plasticisers (−40° to +90°C), as indeed is true also for PVAc. When the additive concentration in a resin exceeds a few weight percent, it is often possible to assay the additive calorimetrically without extracting it. If the additive is incompatible with the resin, it can be detected in a separate crystalline or glassy phase by either its T_m or T_g and measured quantitatively from ΔH_f determinations at T_m or Δc_p measurements at T_g. When an additive is soluble in a polymer, its concentration can be estimated from shifts in T_m or T_g of the resin. Once a master curve of T_g *vs.* plasticiser concentration has been prepared for a particular PVC composition, it can be used to determine the amount of plasticiser in an unknown formulation of the particular plasticiser, *e.g.* PVC/DIDP (*cfr.* Fig. 2.1) [27]. For PVC/DOP a linear relationship from zero to 45 wt.% plasticiser has been reported [86]. DSC was also used in **miscibility studies** of erucamide and PA12 [87].

A round-robin determination of T_g of amorphous thermoplasts (PMMA, PA61/6T, PC, PSU) by means of DSC has indicated an intralaboratory confidence level r between 1.0° and 1.5°C for T_g of non-hygroscopic polymers and an interlaboratory confidence level R of about 3.0°C [88]. On the other hand, the standard deviation of the mean interlaboratory spread for heat capacity (Δc_p) measurements is high (>30%).

Important information on **stabiliser distribution** in PVC can be derived by DSC [26]. As stabiliser impregnation decreases T_g, this can be taken as an indication of the uniformity within PVC granules [89]. The organotin stabilisers dibutyltin tris(isooctylthioglycolate) and dibutyltin bis (dodecylmercaptide) dry-blended into a PVC powder do not mix homogeneously at the molecular level until the polymer is processed. Additives also influence the **heat capacity**, c_p, values, as measured by DSC, as in case of PVC/DIDP [90]. In impact strength PS the styrene/butadiene ratio in the polyblend can readily be determined by the application of DSC on the basis of Δc_p values [82]. Also the content of slip additive (proportional to the determined heat of fusion) in a blend of PBT, PC and EPDM with approximately 5% glass fibres has been determined [85].

Various product forms may occur during storage (typical for fatty acid derivatives) or production. DSC has been used for detection of **polymorphism** of butylated hydroxyanisole (BHA) [91]. Similarly, DSC has allowed to detect various product forms of the hindered phenolic antioxidant octadecyl 3-(3′,5′-di-*t*-butyl-4′-hydroxyphenyl) propionate (Anox PP 18), which caused handling problems during production [92]. The DSC method for **purity determination** as used for curatives, such as 2,2′-benzothiazyldisulfide (MBTS) [93], and for sulfur and accelerators [79], is also applicable to other additives, such as antioxidants and antiozonants. DSC and TGA have been used to establish the oxidation and weight loss characteristics of commercially available triaryl, trialkyl and alkyl-aryl phosphate esters, which are widely used as plasticisers and flame retardants in the polymer industry [94].

Plasticiser efficiency in PVC can be evaluated by a number of semi-empirical tests, such as lowering of T_g as the level of plasticiser is raised. Addition of 20% DOP decreases T_g of PVC from 85°C to 30°C. On a routine basis, a moulder can check incoming materials by monitoring T_g [82]. DSC has also been used to determine the effectiveness of various lubricant additives for PVC [95]. Depending on the processing period, and therefore the temperature profile, each lubricant underwent a change in the internal *vs.* external nature of its behaviour. Monitoring of T_g produced evidence for the effectiveness of additives as internal lubricants. DSC has also been used to determine the effect of plasticisers on the melting point of PA11. A graph of percent plasticiser *vs.* melting point can be used in quality control or to evaluate plasticiser efficiency [82]. Other additives may be determined as well.

Analysis of rubbers and rubber compounds containing curing agents, fillers, accelerators and other additives, often involves DSC, TGA, NMR or MS. The **sulfur concentration** during vulcanisation can be determined by means of DSC, where the enthalpy associated with the melting process (ΔH_m) often correlates with the sulfur content (in a limited sulfur range) [96]. Brazier *et al.* [97] have reported a relation between heat of vulcanisation, as determined by DSC analysis, and sulfur and accelerator content of a fully accelerated natural rubber-polybutadiene blend. A similar relationship between enthalpy of cure and **peroxide content** has been shown for various elastomer systems [98]. Except for hydroperoxides, where the half life ($t_{1/2}$) in monochlorobenzene is determined titrimetrically by measuring the active oxygen content in time, $t_{1/2}$ of peroxides is usually determined by DSC of a dilute solution of an initiator in monochlorobenzene [99].

DSC can be used to study additive **nucleating activity** and has revealed the effect of nucleating agents and pigments on the crystallisation of iPP [100]. Van Every *et al.* [101] have used DSC and factor analysis to detect trace amounts (up to 250 ppm) of the nucleator sodium benzoate (NaBz) in PP formulations. Also other authors [102] have used DSC to study crystal nucleating activity (effect of copper deactivator on ageing life of PP).

Hassel [103] has compared DSC, TG, thermal evolution analysis, TMA and DMA in evaluating **flame retardant** textiles based on different polyester fibres. Also the thermoanalytical analysis (DSC, TGA) of a sisal reinforced flame retardant polyester/(DBDPO, Sb_2O_3) formulation has been described [104]. Larcey *et al.* [105] have reported use of a simultaneous TG-DSC system (STA) to investigate the suitability of using magnesium hydroxide as a flame retardant and **smoke suppressant** in PP formulations.

DSC can be exploited for quantitative analysis of **chemical blowing agents** (CBAs) in commercial foam formulations [106]. The rationale behind this is that decomposition of azodicarbonamide is an exothermic process and that the heat of decomposition, ΔH_d, can be measured quantitatively by DSC. Prasad *et al.* [106] reported a linear relationship between ΔH_d (0–425 J/g) and azodicarbonamide content (0–36%). DSC thus allows detection of the level of undecomposed CBAs present in processed foam products and establishes the onset temperature for the decomposition. Advantages of DSC over EGA techniques are ease of operation, shorter analysis time, and detection of azodicarbonamide concentrations as low as 1%. Dixon *et al.* [107] have correlated thermal analysis data (DSC, TGA) of a variety of CBAs with cell morphology of extruded, expanded PP rod samples. CBAs with a higher temperature and rate of gas evolution lead to foams displaying a finer cell size structure and higher cell density.

As DSC measures the amount of heat flow into or out of the sample as a function of the given materials temperature, it is very useful in determining reaction kinetics or the state of cure (*i.e.* degree of vulcanisation) in **rubber** compounds. The effect of additives on curing reaction of rubber may also be detected by DSC methods. Schnecko *et al.* [108] have considered the effectiveness of thermal analysis methods. Faults that have actually occurred in industrial rubber compounds are often analysed by means of DSC and TGA [109]. Schindbauer *et al.* [110] have reported quantitative investigations on the curing behaviour of phenoplasts by means of DSC measurements. **Thermosets** may be characterised by various thermoanalytical methods [111] such as epoxy curing via T_g measurements (DSC); determination of the rate and degree of cure (TG); uniformity of filler in moulded part (TG); and spot-to-spot or batch-to-batch uniformity of the degree of cure (TMA).

DSC is also used in **plastic identification**. Using DSC, LDPE, HDPE and LLDPE samples can be distinguished from each other without any difficulty. The method is very suitable for rapid QC purposes. Similarly, black coloured (IR absorbing) ABS/PA6 blends may be identified [85]. Generally, however, when a melting peak or a glass transition region of an unknown plastic is obtained from DSC measurements, it requires the assistance of FTIR for plastic identification because of the overlapping range of melting and glass transition temperatures of different plastics. Off-line DSC-FTIR (including microspectroscopy) is often used for plastic identification [112].

Wherever FTIR has difficulties in accurately identifying filled polymers, blends, and polymer families, such as polyamides and polyesters, DSC can assist in determining the unknown by providing information on physical properties.

Thermal oxidation of plastics can be assessed by various methods, amongst which heat measurement (DSC and DTA). Accelerated methods such as DTA and DSC and oxygen uptake measurements have been used quite extensively in studies of thermal oxidative stability of plastics [113]. The definition of thermal stability is very vague and is interpreted differently. Nikulicheva *et al.* [114] have summarised the diversity of methods of thermal stability determination using TA methods. Problems associated with the use of thermal analysis to determine the thermal stability of plastics have been discussed in detail [115,116].

Oxidative degradation is a process easily detected by DSC. Many industrial test specifications exist such as the DSC based Underwriters Laboratory test [118]. **Oxidative induction time** (OIT) is defined as the time to the onset of oxidation of a test specimen, exposed to an oxidising gas at an elevated isothermal test temperature. Bair [119] has described the details of the technique, using DSC, DTA, and TG. A sample is brought to the preselected isothermal (preferably in a N_2 stream), the atmosphere is changed to O_2 at the same flow-rate (zero time of the experiment), and the delay before the oxidation starts (detected as an abrupt departure from the baseline) then serves as an indication of the relative oxidisability of the polymer (Fig. 2.2). It is recommended that OIT experimental conditions are selected so that OIT values are between 15 and 100 min. In dynamic DSC scans the onset temperature of the exotherm transition (T_{onset}) is obtained. Dynamic OIT* (temperature) or OOT (oxidation onset temperature) is quicker.

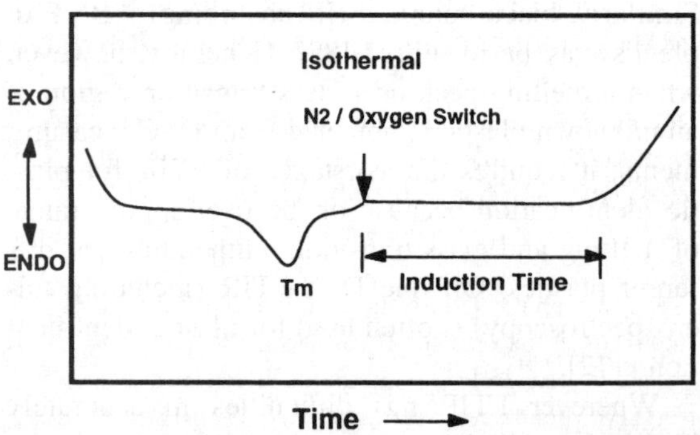

Fig. 2.2. Oxidative induction time tracing from DSC. After Woo *et al.* [117]. Reproduced by permission of the Society of Plastics Engineers (SPE).

OIT is a widely used screening parameter for the oxidative stability of polymers, edible oils, and lubricants, which is typically used as a quality control tool to rank the effectiveness of various oxidation inhibitors. It is a kinetic parameter (*i.e.* dependent on both time and temperature) and not a thermodynamic property. As a parameter dependent on test time and temperature, the OIT* value appears to be decreasing with time but in a well-behaved and predictable manner. OIT is either a measure of the amount of antioxidant present in the polymer or the effectiveness of the particular AO used. If the amount of AO in the polymer is known, then OITime or IOTemperature allow monitoring residual AO contents and calculation of the linear rate of AO consumption. A major limitation of DSC-OIT is that if the isothermal test temperature is lowered below the standard 200°C temperature to reveal small differences in AO concentration at low levels, the polymer's exothermic oxidation rate may decrease below the limits of DSC detectability. Lugão *et al.* [120] have recently introduced a temperature dependent oxidative induction time (TOIT) in order to cope with some limitations of the traditional OIT method.

Various authors [121–124] describe the parameters that may affect reproducibility of the OIT test and consequently the intra- and **interlaboratory precision**. They may be categorised as follows: influences by the sample material itself (additives, fillers, pigments, inhibitors, metal catalysts) and experimental parameters (temperature, pressure, reactant gas, oxygen flow-rate, single and multistage oxidation, specimen mass and surface area, metallic impurities in DSC pans, evaluation procedures, *etc.*). In DSC or DTA oxidative induction testing, the sample thickness used in oxygen absorption testing is between 100 and 250 μm. This requirement minimises diffusion-controlled reactions.

It has been observed that the effectiveness of antioxidants, as measured by OIT at high temperatures, may differ as a function of temperature. OIT test temperatures should preferably be close to actual use temperature. At ambient pressures, significant oxidation is often not detected until the polymer is above the melting point. At these temperatures, essential ingredients in the polymer formulation can be lost. Also, tests performed above the melting point cannot be extrapolated reliably to temperatures below the melting point. Pressure DSC suppresses volatilisation of additives and degradation by-products, an event which thermally competes with the oxidation exotherm. Moreover, the

Table 2.9. Oxidative induction testing methodology[a]

Parameter	ASTM D 3895-95	Generalised method[a]
Temperature mode	Isothermal	Isothermal, scanning
Temperature range	About 200°C	As low as 100°C
Time range	Minutes	Up to 1 wk
Reaction type	Exothermic	Endo- and exothermic
Atmosphere	Oxygen	Various O_2 concentrations
Pressure	Ambient	Up to 68 bars
Polymer type	Polyolefins	Polyolefins, olefin based TPE's, flexible PVC, polyesterether TPE, polyurethanes, natural rubber latex, and natural rubber compounds

[a] After Woo *et al.* [117]. Reproduced by permission of the Society of Plastics Engineers (SPE).

more saturated atmosphere allows for lower test temperatures and shorter measurement times (especially relevant in case of improved additive packages) [125]. Resolution of the oxidation exotherm can be improved by providing a pressurised environment to the sample. High-pressure DSC (HPDSC) cells operate up to 2200 psi. Cassel *et al.* [126] have compared OIT tests using pcDSC, pressurised cell pcDSC and hfDSC for HDPE taking the ASTM E 37.01.10 Task Group Interlaboratory Study as a basis.

It is not uncommon for OIT results to vary widely between labs testing the same material [126a]. In order to overcome this intolerable situation a Standard Reference Material for DSC-OIT testing has been selected by ASTM Committee D9, based on nine interlaboratory test programs, namely a 0.22 mm translucent HDPE/Irganox 1010 film sample [127]. This reference material is statistically homogeneous on a DSC scale, which is a necessary condition for a reference material. Despite its instability the material described by Blaine *et al.* [127] is considered the best available material for OIT testing.

ASTM committees have dealt with **standard test methods** to determine the oxidative properties of materials (*cfr.* ASTM D 3350, 3895, 4565, 5483, 5885 and E 1858). ASTM E 1858-97 (DSC-OIT) can be used to determine the oxidative behaviour of polyolefins (HDPE) and hydrocarbon oils (diluted engine pass oil blend). This method is precise. There is a good correlation between DSC/TGA-OOT in air/O_2 (ASTM E 2009-99) and PDSC-OIT (ASTM E 1858-97) under high-pressure oxygen for polyolefins. The ASTM method D 3895-92 for DSC-OIT and DSC-OIT* has recently been modified into a generalised technique with considerably expanded applications to polymer systems in addition to polyolefins (Table 2.9) [117]. For polyolefins, a high temperature oxidation acceleration was originated from the volatilisation of antioxidants. An **interlaboratory test** (ILT) program for DSC-OIT to determine precision/reproducibility and repeatability has recently been completed (ASTM Committee E 37.01.10). Also dynamic DSC-OIT* has been subject of interlaboratory tests [128] and is standardised. Affolter *et al.* [88,129] have described two interlaboratory tests for determination of the thermal stability of polyolefins in oxygen: *(i)* a static procedure (according to EN 728) at a fixed temperature (210°C) to determine the oxygen induction time (OIT); and *(ii)* a dynamic method with continuous heating the sample with a rate of 10°C/min to determine the oxygen induction temperature (OIT*). The results of the OIT determination are tainted with a considerable uncertainty of measurement and cast doubt on the predictive value for purposes of quality control. Especially for low OIT values (low stabilised plastic materials) the dynamic method (OIT*) seems to be an attractive alternative [126a]; however, differentiation between samples decreases rapidly for higher OIT* values. Bair [27] has demonstrated the efficient use of OIT measurements for evaluating additives under simulated processing.

The OIT test measures the **intrinsic thermal stability** of a material as well as the amount of stabilisers in the material. DSC is a convenient method for measuring concentrations of hindered phenolic antioxidants in polyolefins. The AO concentration-OIT relationship is linear for the most part of nonvolatile AOs [131–134], *cfr.* Fig. 2.3. DSC-OIT has also been used in determination of the oxidative stability of HDPE film (isothermal at 200°C in O_2, ac-

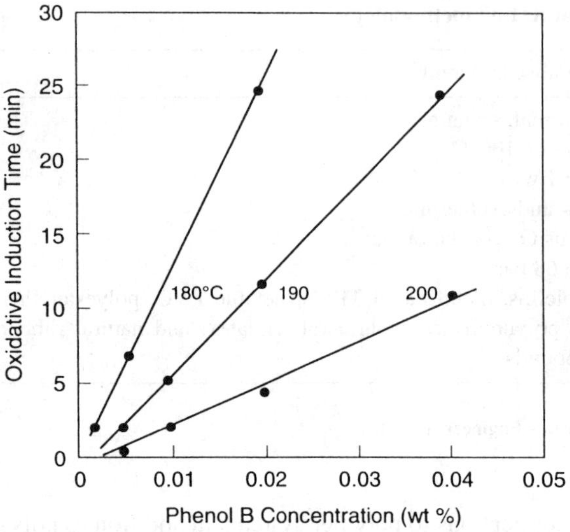

Fig. 2.3. Oxidative induction time *versus* concentration of Irganox 1010 (phenol B) in LDPE as measured by isothermal DSC at 180, 190 and 200°C. After Foster [130]. Reprinted with permission from G.N. Foster, in *Oxidation Inhibition in Organic Materials* (J. Pospíšil and P.P. Klemchuk, eds.), CRC Press, Boca Raton, pp. 299–347 (1990). Copyright CRC Press, Boca Raton, Florida.

cording to PR EN 728, provisional European standard) [85]. A plot of oxidative induction time *versus* AO concentration for Irganox 1010 in HDPE is linear over a range from 50 to 1000 ppm [135]. Stability parameter mapping and stability vector analysis have been applied to DSC-OIT data for MDPE/(CB, Irgafos 168, Irganox 1010) [136]. Woo *et al.* [137] found DSC-OIT particularly useful in aiding the development of stabiliser packages for medical plastics (PVC, PP, EVA, PMMA). A linear relationship of PVC stability *vs.* epoxidised oil content (5–15%) was reported (*cfr.* also ref. [138]).

Both the standard OIT (Std-OIT, according to ASTM D-3895) and high-pressure oxidative induction time (HP-OIT, according to ASTM D-5885) tests can effectively monitor the overall amount of oxidants present in a **geomembrane**. A manufacturing QC specification for HDPE geomembranes, evaluating antioxidant packages, is based on Std-OIT and HP-OIT [139]. Using Std-OIT and HPDSC-OIT tests Hsuan *et al.* [140–142] have noticed depletion of AOs (hindered phenols and phosphites) during thermal oxidation of HDPE. The situation becomes more complicated in blends of AOs and/or antiozonants, because different antioxidants volatilise at different temperatures and rates. For AO packages that contain thiosynergists or hindered amines, HP-OIT is the appropriate test.

Thermoanalytical techniques are a quick way for assessing the **relative performance of AOs** in polymers, rubbers, lubricants, *etc.* and have been widely used. DSC-OIT is used to study base polymer stability, optimum additive level, the degree of material deterioration during processing and the effect of multiple shear histories while reprocessing. DSC is also useful to determine the effective AO concentrations among all the transformation products present in a polyolefin formulation.

Determination of OIT as a technique for evaluating **polymer-ageing** has been gaining popularity. Originally developed by Rudin *et al.* [143], Bair [144] and others, the method has been widely used to evaluate ageing in polyolefins [134,145–147] and in some unvulcanised elastomers. DSC and DTA have been used to evaluate the effectiveness of AOs for many years and were the subject of an early ASTM quality control test (ASTM D3350). In particular, Gilroy *et al.* [148] used the OIT as a test procedure to screen polyethylene insulation used in telephone wire and cable for oxidation resistance in pedestals. The method later became available as ASTM Test Method for Copper Induced Oxidative Induction Time of Polyolefins [149]. Information from the DSC/DTA test can be applied to prevent degradation during processing, to assess the effect of altering process conditions of an actual wire sample after extrusion, or as a routine QC check of the finished product [150].

DSC is specified in USP for the physical testing of PE containers; the quality of packaging material is of decisive importance for the protection of raw materials and end products, such as primary packaging material. As shown in Fig. 2.4, unprotected PE samples decompose almost immediately at the test temperature. However, a PE sample containing 0.04% stabiliser remains protected for approximately 16 min at the test temperature, whereas the PE sample containing 0.055% stabiliser is protected for 25 min [151]. The DSC test thus provides a rapid method of screening for the proper AO levels in a polymer. Bharel *et al.* [152] have reported DSC-OIT for performance evaluation of two diamide antioxidants in HDPE and Hakani *et al.* [153] for the evaluation of oxidative stability of flexible polyolefins (FPO) with the biological γ-oryzanol and α-tocopherol antioxidants for food and medical applications.

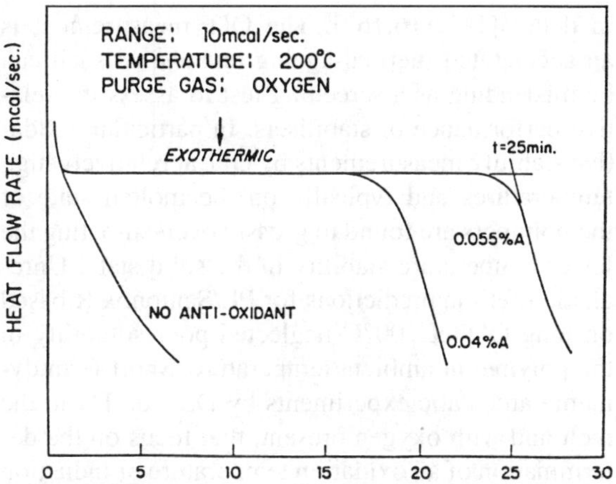

Fig. 2.4. DSC-OIT of polyethylene. After Gibbons [151]. Reproduced by permission of International Scientific Communications, Inc.

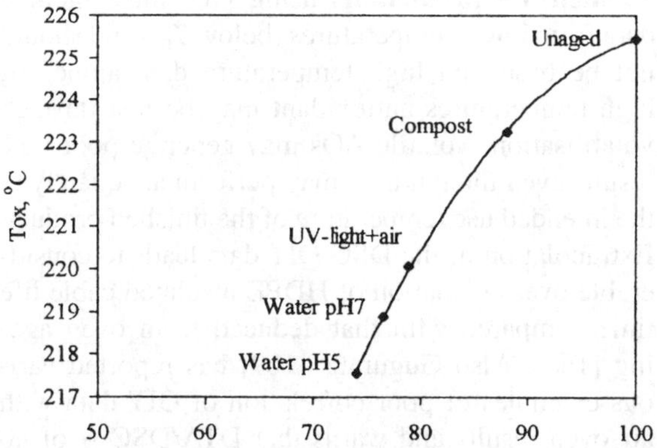

Fig. 2.5. Effect of the residual amount of Chimassorb 944 on T_{OX} of an LDPE film. After Haider and Karlsson [154]. Reprinted from *Polymer Degradation and Stability* **74**, N. Haider and S. Karlsson, 103–112, Copyright (2001), with permission from Elsevier.

Figure 2.5 shows the effect of the residual amount of stabiliser on the thermo-oxidative stability of LDPE/Chimassorb 944 exposed to various testing environments [154]. The differences in the oxidation behaviour of the polymeric matrix (as measured by DSC) are related to the differences in the consumption/migration rate of the stabiliser and the amount of stabiliser remaining in the polymeric matrix (as measured by UV spectroscopy).

Although OIT is a specification for many additive suppliers (product control), in general static DSC-OIT shows considerable uncertainty of measurement and the benefit of these measurements with regard to quality control or life-time prediction for polyolefin component parts is rated very low. Pauquet *et al.* [132] have described limitations and applications of DSC-OIT to QC of polyolefins. Blaine *et al.* [121] have recently reviewed DSC-OIT of polyolefins. Dynamic DSC-OIT* for HDPE leads to essential higher reproducibilities [128]. The aforementioned interlaboratory DSC-OIT* test for the determination of carbon-black content revealed an inhomogeneous distribution in commercial raw polyolefins with 2–3% CB [155]. This requires reprocessing for quality control purposes. Cooney *et al.* [156] used both OIT and OIT* to evaluate the thermal oxidative stability of high-impact polypropylene copolymer. DSC-OIT has also been applied for measuring the thermal stability of PB [157] and iPP [134] with different antioxidant concentrations. hfDSC-OIT was used to compare onset temperature, enthalpy and oxidation rate of various NB/BR compounds containing TMQ and 6PPD as antioxidants [84]. DSC-OIT (ASTM E 537) was used to determine the oxidation characteristics of commercial phosphate esters (flame retardants, lubricants, plasticisers) [94].

Studies applying DSC or DTA techniques for **elastomer ageing** and antioxidant evaluation use various approaches, which depend on the determination of *(i)* enthalpy; *(ii)* onset temperature; *(iii)* isothermal induction time; *(iv)* energy of activation; and *(v)* oxidation peak temperature. Stenberg *et al.* [158] have reported a DSC analysis of the variation of AO concentrations with ageing time at different depths of thick-walled natural rubber samples. In this case, as indeed very often, calibration curves correlating AO concentration to OIT (at atmospheric pressure) are curvilinear. The observed non-linearity of the calibration curve for IPPD (*N*-isopropyl-*N*′-phenyl-*p*-phenylene diamine) concentration in TMTD/ZnO-cured NR *vs.* OIT (Fig. 2.6) was ascribed to the simultaneous loss of AOs by two mechanisms: evaporation and consumption of AOs by oxidation [158]. The decrease in OIT is most rapid at the outer oxygen-exposed parts of the samples. Diffusion of IPPD from the interior of the samples prolongs the OIT at a distance of 12 mm from the centre. No such affect was found with DENA.

Published information about OIT in elastomer systems is relatively scarce. González [159] reported the relative efficiencies of seven AOs in guayule rubber. Savasçi *et al.* [160] used DSC-OIT at 150°C

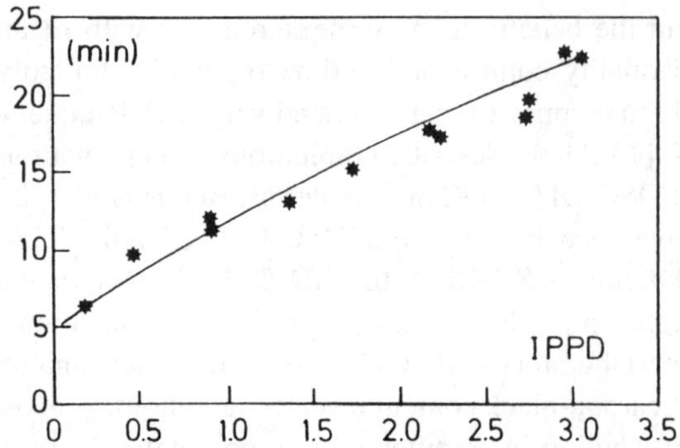

Fig. 2.6. DSC-OIT dependence on concentration of IPPD (%) in natural rubber. After Stenberg and Björk [158]. Reprinted from B. Stenberg and F. Björk, *J. Appl. Polym. Sci.* **31**, 487–492 (1986), John Wiley & Sons, Inc., New York, NY, Copyright © (1986, John Wiley & Sons, Inc.). This material is used by permission of John Wiley & Sons, Inc.

in air to evaluate the effectiveness of 2,6-di-*t*-butylcatechol (Dnx) and tri(mono- and dinonylphenol mixture) phosphite (Plg) and their mixtures in *cis*-BR, whereas Šimon *et al.* [161] have indicated that DSC enables analysis of the induction period in the vulcanisation of rubber compounds. Smith *et al.* [162] have used DSC-OIT to evaluate the effects of different AOs in unvulcanised rubbers and Berg *et al.* [163] used OIT to compare a phenolic AO and a triazine-type AO in hydroxy-terminated polybutadiene elastomer (OHBR). A feasibility study of several antiozonants in different elastomers was reported by Burlett [164], showing potential of the OIT technique for screening AOs and antiozonants in technical compounds. For epoxy curing with different accelerators DSC and conversions calculated immediately indicate the most efficient accelerator. DSC-OIT has also been used for the determination of the oxidation stability of oils [165].

Despite useful DSC-OIT results a word of caution is necessary. Direct comparison between two single OIT values may be dangerous. Determination of the oxidative stability by DSC is fast and easy. It is especially recommended for quality assurance of demanding long term goods, such as electrical cables, medical devices and hot-water PE pipes [166]. Each lot of the raw material should be investigated. There are, however, problems in correlating the results obtained from such studies with those obtained by using oven ageing or a multiple extrusion technique. Problems associated with the use of thermal analysis to assess the stability of plastics have been discussed in detail [115,116,167]. The OIT measurement is an accelerated thermal-ageing test and as such can be misleading as a screening test to assess the relative performance of stabilisers. In particular, oxidative stability measurements by OIT at relatively high temperatures and typically on the molten state of the polymers are found to grossly overestimating the lower temperature stability in the solid state. Unrealistic lifetime predictions for PE/Santonox R based on long OIT at 200°C neglected poor solubility in the polymer at ambient temperature. Short-term dynamic and static experiments by DSC or TG in the melt and with oxygen present, that focus on the determination of an oxidation temperature or induction time, are well suited to facilitating the initial screening of AO systems for various polymers that degrade via a free radical-type mechanism. However, OITs for polyolefins that are acquired rapidly in the melt do not obey a simple Arrhenius relationship.

Shelf-life predictions using OIT must include data from lower temperatures (below T_m) and should not be based on high temperature data alone. At high temperatures antioxidant may be lost through volatilisation. Volatile AOs may generate poor OIT results even though they may perform adequately at the intended use temperature of the finished product. Extrapolation of the DSC-OIT data leads to considerable over-estimation of HDPE insulated cable life time compared with that deduced from oven ageing [168]. Also Gugumus [169] has reported various examples of poor correlation of OIT data with air oven results and warns that DTA/DSC is of no value in the prediction of oven ageing in the solid state even though it is excellent for QC purposes. The use of DSC-OIT, DTA-OIT and CL for thermal life time prediction has recently critically been evaluated [170].

DSC-OIT and DSC-OIT* are commonly used methods to determine if failure is due to oxidative degradation. Ezrin *et al.* [171] have reported several examples. For analytical methods applied to the testing of oxidation inhibition, *cfr.* also Foster [130].

In summary, DSC-OIT is very successful for the determination of activation energy of oxidative degradation, antioxidant effects, optimal processing parameters, and correlation of product performance if oxidation is the primary governing parameter.

Similar to DSC, microcalorimetry may be used to measure the efficiency of stabilisers in polymers [172]. Microcalorimetry appears to be a highly sensitive technique to detect oxidation, also during the initial stages of oxidation.

High-pressure DSC has been used for *in situ* measurements of the plasticisation of polymers by blowing agents (*e.g.* PVC-CO_2, PS-CO_2, PS-HFC134a) [173]. From the T_g–p profiles the plasticising effects induced by dissolved solvents were derived and differences in cellular morphology were related to differences in diffusivities. High-pressure DSC has been used by Sepe [125] to measure the oxidation induction time of virgin and reclaimed PP samples. The oxidative stability of recycled materials, the assessment of useful product lifetimes and the effects of injection moulding on oxidative stability were discussed. PDSC-OIT has also been used to assess the oxidative stability of motor oils. Riga *et al.* [174] have developed a standard test method for determining OIT of hydrocarbons by DSC and HPDSC.

DSC is thus a quick and reliable method of analysis, not only in material development, but primarily in the areas of **quality assurance, raw material control and failure analysis**. DSC is used for identification of incoming plastic materials *e.g.* HDPE/PA6 and LDPE/EVAL/PA6 composite film. DSC can not only identify the major components of polymers, but can also detect minor components such as adhesives, if these have a melting behaviour which differs from that of the polymers. Quality control of packaging film without sample preparation is based on the measurement of the solid/liquid phase transition of melting by means of DSC. Sass [175] has given various examples of quality assurance and defect analysis of plastics by DSC. There is increased demand for sensitivity and capability because of the growing complexity of materials.

For other applications of DSC to studies of polymers, *cfr.* also Crompton [176].

2.1.2. Differential Thermal Analysis

Principles and Characteristics

Differential thermal analysis (DTA) is defined by ICTAC as: "A technique in which the temperature difference between a substance and a thermally inert reference material is measured as a function of temperature, while the substance and reference material are subjected to a controlled temperature programme". For the determination of the differential temperature ΔT temperature sensors, generally thermocouples, are used which are in direct contact with the materials or their containers. The output of the instrument is the difference between the two thermocouple voltages. In a "differential" type measurement the investigated sample and a reference material are treated with the same temperature programme. A thermally inert substance (*e.g.* Al_2O_3), which has no phase change in the temperature range of the experiment, is used as a reference material. DTA apparatus is most properly described as an adiabatic calorimeter with some thermal leakage.

DTA techniques permit study of the thermal behaviour of materials as they undergo transformations as a function of temperature. When the sample undergoes a phase change, or a chemical reaction, energy is absorbed or released, and a ΔT between sample and reference is detected. If the output is positive there is an exothermic reaction, whereas a negative voltage shows an endothermic reaction. When there are no thermal transformations this output voltage is zero. The **main use** of DTA is to detect the initial temperatures of thermal processes and qualitatively characterise them as endothermic or exothermic, reversible or irreversible, first- or higher-order transition, *etc.* This information, and the dependence upon the specific atmosphere, makes DTA particularly valuable for determination of phase diagrams [17]. Ideally, the area under the DTA peak should be proportional to the heat of the process originating the peak. However, many factors influence the curve and are not compensated in the traditional simple DTA plot. Changes in thermal transport properties of the system, detector sensitivity with temperature, *etc.*, will generally diminish the response of DTA with increasing temperature. DTA yields calorimetric information when calibration permits the quantitative conversion of temperature difference to heat flow and ultimately heat of transition or heat capacity.

DTA may be more precise than standard calorimetry in fixing transition temperatures. DTA is cheap and simple and has been widely applied to the study of stabilised polyolefins. Measurements are carried out either isothermally or dynamically. In the isothermal mode, the induction times to the start or the maximum of the exothermic peak are determined.

In **dynamic DTA** the temperature of the start of an exothermal oxidation reaction (T_{ox}) is measured during a constant heating rate experiment performed in an oxygen atmosphere. The attraction of this method, which is much less used than the isothermal method for evaluating the stability of polymer samples by DTA/DSC [177–179], are simplicity and speed (20 min).

DTA and DSC are related techniques that measure the same thermal events with different methods. Whereas DTA in the traditional use of the technique measures a difference in temperature, DSC monitors the difference in heat flow between a sample and a reference material as the material is heated or cooled (*cfr.* Chp. 2.1.1). Degradation processes may occur in a polymer which are not associated with the loss of volatiles. It is here that both DTA and DSC techniques are useful as they show whether any reactions are occurring which involve either heat evolution or absorption.

For recent reviews of DTA/DSC, *cfr.* refs. [69, 180].

Applications
DTA has been widely used as a screening test and for **quality control** purposes of polymer formulations, especially in the wire and cable industries. Most of the work dealing with DTA and DSC for studying polymer oxidation has been performed under isothermal conditions at elevated temperatures well above the melting point, *e.g.* for iPP stabilised with simple phenolic additives (Topanol O/CA, Irganox 1010/1076, Irgastab 2002, Ionox 330, Goodrite 3114/3125, Santowhite Powder, Plastanox 2246/425). Billingham *et al.* [116] have critically reviewed the application of the technique to **oxidation and stabilisation studies** of polymers. Figure 2.7 shows a typical concentration dependence of the induction period (corresponding to the time required to consume all of the additive) for 0.05–0.50 wt.% Irganox 1010. For most of the other additives similar linear curves were obtained, although curvature is sometimes observed. These curves can be used to predict values at other concentrations. It was pointed out [116] that ranking of relative efficiencies of antioxidants is sensitive to the isothermal temperature chosen (effect of activation energies). Where no correlation is apparent between AO efficiency and molecular size the additive mobility is not an important factor. It also appears that impurities in the polymer are very important in determining the efficiency of phenolic stabilisers, which implies that AOs should be compared only by means of DTA in the polymer in which they are to be used. Therefore, the DTA method, although attractive in many ways, should be used only with extreme caution.

As polymers are usually processed under conditions of low oxygen concentration, as in injection

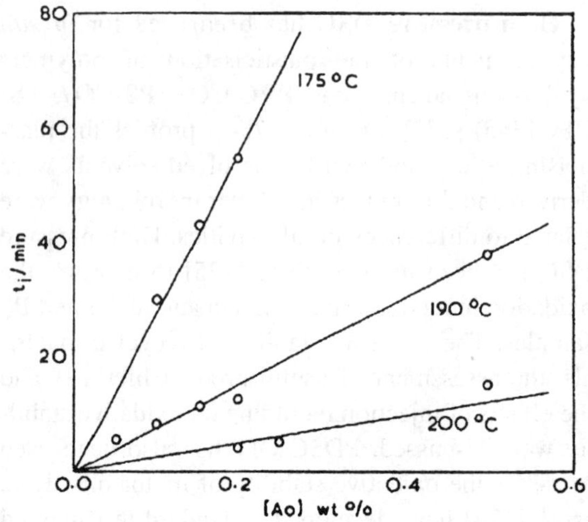

Fig. 2.7. Concentration dependence of the induction period (DTA-OIT) for PP/Irganox 1010 at various temperatures. After Billingham *et al.* [116]. Reprinted from N.C. Billingham *et al.*, in *Developments in Polymer Degradation* (N. Grassie, ed.), Applied Science Publishers, London, Copyright (1981), with permission from Elsevier.

moulding or extrusion operations, DTA measurements in air may be irrelevant to processing conditions. Moreover, in extrapolation of DTA data for stabilised polyolefins (usually in the range of 150–200°C in pure oxygen) to service use temperature, it should also be considered that the polymer passing through its melting range becomes a semicrystalline solid, which causes unpredictable distortions in the Arrhenius plot; besides, the solubility of the antioxidant may be exceeded so that it becomes supersaturated in the polymer and loss of additive may result. Consequently, extrapolation of DTA data to temperatures below the polymer melting point is generally considered to be invalid [116].

The main reason for using induction time data for the determination of **antioxidant concentration** in polymers is the frequently observed linear relationship between induction time and antioxidant concentration [131]. In view of the aforementioned considerations great caution should be exercised in quantitative estimation of antioxidant levels in polymers. Wight [181] and others [143] have used quantitative differential thermal analysis (QDTA), in particular for determining the degree of oxidative stability of polyolefins for QC purposes in the wire and cable industry in lieu of a direct antioxidant analysis. Application of the basic purpose of a QC test (assurance that the raw material is indeed the material specified and that the finished product will perform adequately for its lifetime) to the determination of oxidative stability requires determining that the proper stabiliser package is present in the required concentration and that the finished product has not been unduly degraded during manufacture. These conditions are hard to meet with DTA. In particular, the use of QDTA to selectively determine the presence or absence of specific components in a stabiliser package is slippery ground [181]. Also, the sample size highlights inhomogeneities in the sample and may easily lead to apparent irreproducibilities [131]. While DTA-OIT does provide a measure of the total oxidative stability of the polymer, it does usually not establish the concentration of *individual* stabilisers. The presence of a primary AO and a copper inhibitor in combination could be detected separately by comparing OITs in copper and aluminum pans. However, the presence of the thioester synergist DSTDP interferes with the determination of the effective level of copper inhibitor [181]. Degradation products of polyolefins lower the observed stability, yielding suppressed antioxidant values.

Although DTA-OIT may be a useful tool in quality control since comparison of a stabilised and an unstabilised sample of polymer will certainly show a difference, it need not bear any significant relationship to the actual life expectancy of a finished product. QDTA can only determine a relative degree of stability by comparing a measured OIT against a value for a known material with the same stabiliser package. Some misuses of thermal methods for the measurement of **polymer durability** have been reported by Gugumus [169]. For example, DTA/DSC should definitely not be used for prediction of oven ageing in the solid state. In fact, DTA-OIT data do not correlate with oven ageing for HDPE insulation [182] or moulded PP plaques [183]. Numerous publications have also been devoted to **rubber oxidation measurements** by DTA/DSC techniques but the correlation between DTA test data and antioxidant activity is poor.

Gedde *et al.* [145] have recently used **dynamic DTA** on PE film containing known concentrations (up to wt.%) of different types of primary and secondary AOs (Irganox 1010, Naugard 445/TNPP, Varox DSTDP) as an analytical tool by which the antioxidant content can be determined. The data obtained was consistent with the data from the isothermal method; similar kinetics (ΔE and k_0) were derived.

Quantitative DTA methods for untreated cotton and fabric treated with P- and N-containing **flame retardants** were suitable for determining the efficiency of FRs and provided data that correlated with oxygen index values [184]. Childress *et al.* [185] described DTA, DSC and TG studies on brominated phosphite and phosphate flame retardants. Nara *et al.* [186] have studied pyrolysis of tetrabrominated epoxy resin and its fire retardant mechanism. Pyrolysis of DER 542 (brominated epoxy resin) and Epikote 1001 (non-brominated epoxy resin) was investigated by DTA en TG. Bhatnagar *et al.* [187] have reviewed DTA and DSC studies on flame retardant polymers. Carroll-Porczynski [188] described the applications of simultaneous TG and DTA and DTA/MS analysis for predicting the flame retardancy of composite textile fabrics and polymers. The use of DTA to identify **mineral fillers** in rubber formulations is as old as the technique itself [189].

Chan [190] has compared the evaluation of **metal deactivators** by means of thermal analysis, oxygen absorption and oven-ageing, emphasising that the high test temperatures used in DTA and DSC can give misleading results. The physicochemical changes of the **foaming agent** OBSH (4,4′-oxybis(benzenesulfonyl hydrazide)) during heating were studied by using DTA [191].

DTA was also used to study the diffusion of Irganox 1330 through iPP. The technique has the advantage of being sensitive to low levels of stabiliser. The diffusion values obtained were in good agreement with those predicted by Fick's law [134]. For other applications of DTA to the examination of polymers, *cfr.* Crompton [176].

2.1.3. Thermogravimetric Analysis

Principles and Characteristics

Thermogravimetry (TG) or thermogravimetric analysis (TGA) is a technique in which the mass of a substance is monitored as a function of temperature or time as the specimen is subjected to a controlled temperature program in a controlled atmosphere. Thermogravimetric measurements require a thermobalance. There are many different types of TG analysers, varying in furnace (size, design and positioning), temperature range, size of sample holder, sensitivity, degree of microcomputer control of the hardware, capabilities of the software, *etc.* TGA instruments are essentially of two basic configurations: one positions the sample horizontally with respect to the gas flow through the instrument, while the other makes use of vertical positioning of the sample (bottom- or top-loading). Depending on the problem a specific instrument may be preferred.

The **basic TG experiment** consists of recording the weight of a sample as it is heated in a defined environment (inert or oxidising) either isothermally (iso-TG) or at a controlled heating rate (CRTG). The experimental record is a plot (thermal curve) of some form of the weight change (*e.g.*, actual weight or percent lost) *vs.* time or temperature of the sample. The simple additional step of using the derivative of the primary weight change curve (DTG) extends the capability and scope of the analysis. TGA examines materials between ambient and +1500°C. For the plastics industry the most common temperature range is from ambient to 800°C. The variables affecting resolution for a specific hardware design are typically sample size, heating/cooling rate, purge gas composition, flow-rate, *etc.* Generally, smaller sample sizes, slower heating/cooling rates, and high thermal conductivity purge gases (*e.g.* helium) result in improved resolution. It is recommended to use as small a sample as possible within the limits of resolution of the microbalance (typically 5–10 mg). The homogeneity of a sample can sometimes limit the sample size (*e.g.* in case of polymer blends). Powdered samples, of small particulate size, have the ideal form for TG studies. However, in polymer science samples are often films, fibres, sheets, pellets, granules or blocks. The packing density should be as uniform as possible. **Temperature calibration** is usually carried out with ICTAC/TAI Curie-point materials (accuracy *ca.* 2°C) according to ASTM E1582-00, mass scale calibration according to ASTM E2040-03.

Goals of TGA separation are accuracy, reliability, completeness of separation and minimum turnaround time. Mass changes as small as 50–100 μg can nowadays be detected. In developing an efficient test one needs to balance the needs for resolution, accuracy and test time.

It should be noted that TGA will not always be accurate because various components in polymeric formulations are not observed as independent weight loss in TG curves (*e.g.* sulfur, accelerators, antioxidants and antidegradants in elastomers) and may undergo weight loss over a large temperature range. Low-MW volatile products (*e.g.* oils, waxes, plasticisers and resins) tend to overlap with polymer decomposition for most choices of method parameters. In the presence of multiple decomposition processes such **overlap of thermal events** is thus a major problem [192]. Consequently, there are practical limits to the kind and degree of information that can be extracted by TG analyses of unknown polymer compositions.

Constant heating rate methods are simple and allow separation of overlapping weight losses by the derivative of mass change (DTG) analysis. Disadvantages of linear heating are: *(i)* relatively poor resolution; *(ii)* non-uniform reaction conditions throughout the sample; *(iii)* results affected by experimental conditions (*e.g.* heating rate, gas flow-rate, sample mass); *(iv)* poor sample temperature measurement (heat distribution); and *(v)* little kinetic information. Various methods have been devised to increase resolution. Possible solutions are use of a multiple step temperature program, or of derivative weight loss criteria. Overlapping decompositions may be separated experimentally by very fast heating (infrared furnace, microwave TA), by "event-controlled" thermal analysis or by means of chemometric data evaluation, such as Principle Component Analysis [193] and factor analysis [194]. Various modifications of conventional thermal analysis have been proposed which are based on monitoring the course of gas-solid interactions, such as controlled-rate analysis and pulsed thermal analysis.

In the last decades several **high-resolution techniques** have been introduced. These techniques are **"event-controlled"**, *i.e.* when a thermal event (decomposition, evaporation, oxidation, *etc.*) occurs a change in measuring condition is introduced. Such event-controlled techniques are termed "controlled rate thermal analysis" (CRTA) [7] or "reaction-controlled thermal analysis" (RCTA) [195]. Nomenclature in the pertinent literature is confusing [7, 196]. Scheme 2.1 gives an overview of the relations between the methods which all aim at increasing the resolution of closely occurring thermal events.

In **controlled transformation rate thermal analysis** (CRTA), instead of controlling the temperature (as in conventional thermal analysis (Fig. 2.8a)), some other physical or chemical property X is modified, which is made to follow a pre-determined programme $X = f(t)$ under the appropriate action of temperature (Fig. 2.8b) [7]. Heating of the sample may be controlled by any parameter linked to the rate of thermally activated transformations, such as total gas flow (EGD control; constant decomposition rate thermal analysis [199]), partial gas flow (EGA

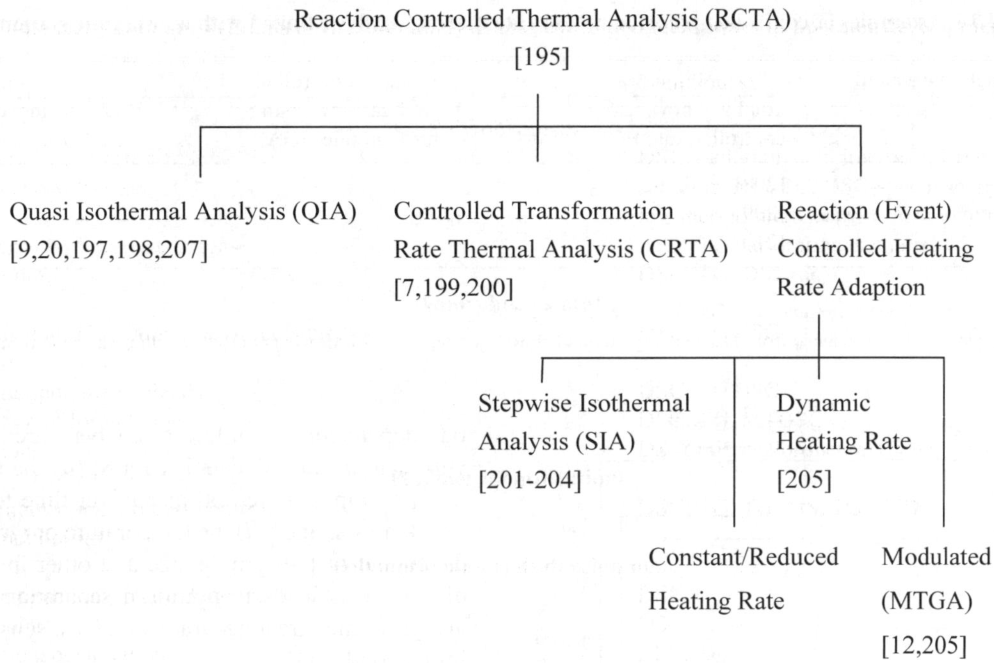

Scheme 2.1. "Event-controlled" thermal analysis techniques. After ref. [195].

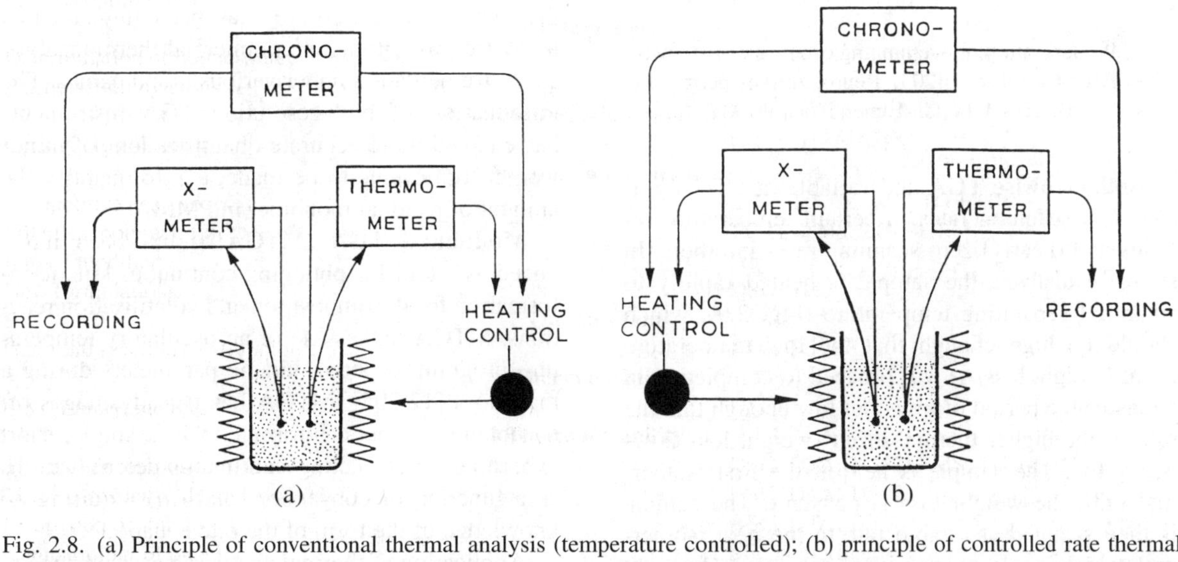

Fig. 2.8. (a) Principle of conventional thermal analysis (temperature controlled); (b) principle of controlled rate thermal analysis (*X*-controlled). After Rouquerol [7]. Reprinted from *Thermochimica Acta* **144**, J. Rouquerol, 209–224 (1989), with permission from Elsevier.

control [206]), mass (DTG control, derivative thermogravimetry; stepwise isothermal heating [207]), length (TD control) or heat flow (DTA, DSC control). Many other possibilities may be envisaged [7]. An even more rewarding way to use CRTA is in combination with simultaneous measurement of a second parameter, *e.g.* mass flow of evolved gas, composition of evolved gas, x-ray analysis, IR absorption, length, heat flow, *etc.* (Table 2.10). For additive analysis a useful approach is EGD or EGA rate control in combination with simultaneous mass measurement (in CRTA-MS configuration).

Table 2.10. Examples of controlled transformation rate thermal analysis linked with another measurement[a,b]

2nd parameter measured	Parameter controlled		
	Total gas flow, using controlled rate EGD	Partial gas flow, using controlled rate EGA	Mass, using controlled rate TG
Mass	[208]	–	–
Composition of evolved gas	[209]	[206]	[211]
Heat flow	[210]	–	–

[a] Examples with a reference in square brackets have been investigated.
[b] After Rouquerol [7]. Reproduced from *Thermochim. Acta* **44**, J. Rouquerol, 209–224. Copyright (1989), with permission from Elsevier.

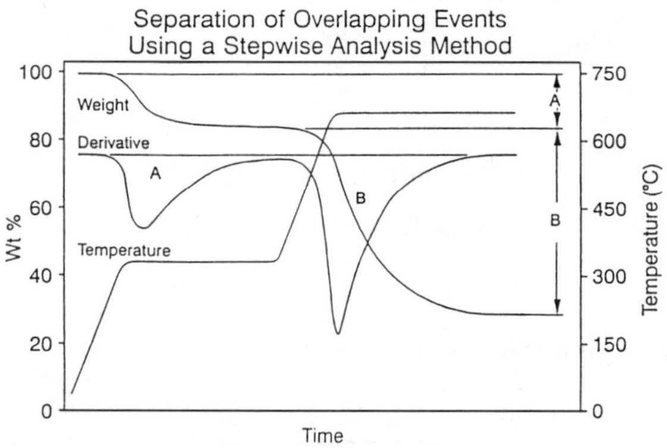

Fig. 2.9. Separation of overlapping events using stepwise TGA. After Cassel *et al.* [203]. Reproduced by permission of B. Cassel, Perkin-Elmer, Norwalk, CT.

Both **stepwise TGA** and variable rate TGA employ fast scanning rates in certain temperature regions and (nearly) zero scanning rates in others. In stepwise analysis the sample is heated rapidly to an initial separating temperature (Fig. 2.9), which should be high enough that the low temperature event (weight loss A) will proceed to completion in a reasonable period of time, but low enough that the rate of the higher thermal event (weight loss B) is negligible. The sample is held at the first isothermal until the weight loss is constant. The sample is then scanned at a rapid rate to the next isothermal, which is selected to optimise the second weight loss. Each temperature is selected to optimise the weight loss of each component in the presence of the others. Cassel *et al.* [203] have compared stepwise TGA with constant rate methods and the rate-dependent, variable rate method. In the latter, the temperature program depends on the rate of weight loss. Hence, separation may depend on initial conditions (sample size, surface area, purge rate and feedback parameters). Some advantages of stepwise to rate-dependent, variable rate analysis are: *(i)* stepwise can use faster scanning rates; *(ii)* the temperature program can be optimised over time for a routine analysis; and *(iii)* the temperature program is independent of the sample size and other initial conditions. This leads to optimum separation at short analysis time, great accuracy and least sensitivity to initial conditions.

"Event-controlled" thermal analysis techniques have repeatedly been reviewed [195,196]. Rouquerol [7] has traced the historical development of the method. Event-control has been implemented in control algorithms in commercial thermoanalytical instrumentation under various brand names. The introduction of high-resolution TGA instruments has enabled more accurate quantifications of minor weight loss events to be made, *e.g.* to quantify the amount of residual monomer in PMMA.

Modulated TGA (MTGA™) has been introduced as a tool for obtaining continuous kinetic information for decomposition and volatilisation reactions. MTGA makes use of an oscillatory temperature program to obtain kinetic parameters during a mass loss [12,205]. MTGA has the advantages of: *(i)* obtaining kinetic information in a single, short experiment; *(ii)* making continuous determinations as a function of conversion; and *(iii)* requiring no knowledge of the form of the rate equation.

Application of thermal analysis has also been extended by the development of **pulse thermal analysis** (PTA). This method is based on injection of a specific amount of the gaseous reactant(s) into an inert carrier gas stream at any temperature (non-isothermal) and/or time (isothermal mode) and monitoring of changes in mass, enthalpy and gas composition resulting from an incremental reaction extent [212]. The method is suitable for the quantification of the evolved gas by MS or FTIR due to the injection of a well-known amount of the chosen gas to the system, which can be used for calibration. PTA provides the following advantages compared to conventional TA: *(i)* quantitative calibration of mass spectrometric signals increasing the sensitivity of TA measurements; *(ii)* monitoring of gas-solid processes with defined extent of reaction (*i.e.* the reaction can be stopped at any point between pulses); and *(iii)* simultaneous monitoring of changes in mass, thermal effects, composition and amount of gaseous reactants and products under pulse conditions [213].

Some **other developments** concern: *(i)* enlarging sample volumes; *(ii)* separation of complex mixtures and identification of individual compounds; *(iii)* hyphenation; *(iv)* alternative heating modes (*e.g.* IR heating up to 500°C/min); and *(v)* factor analysis. Microwave thermal analysis (MWTA) also enables uniform application of heat to large samples (*ca.* 500 mg) [38], but is restricted to samples allowing a change in dielectric properties, *cfr.* Section 3.4.4 of ref. [213a]. The real power of the use of factor analytical methods in the analysis of complex chemical phenomena, such as thermal analysis or pyrolysis of rubber blends, lies in the ability to gain molecular chemical insights that might otherwise be obscured. Using TG and chemometrics allows to gain a good deal of information about the structure of rubber blends [194].

Table 2.11 summarises the **main characteristics** of TGA. A macro-scale TG/DTG-DTA (STA) has been developed (sample size up to 500 g) for

Table 2.11. Main characteristics of thermogravimetric analysis

Advantages:
- Small sample size (*ca.* 10–20 mg)
- No sample preparation
- Rapid
- Quantitative
- High sensitivity
- Various temperature control modes
- Mature technology
- Wide applicability (including QC)

Disadvantages:
- No identification power (unless hyphenated)
- Limited resolution (but HRTGA by rate adjustment)
- Reproducibility (for small sample sizes of heterogeneous materials)

ecotoxicological testing, environmental protection, waste investigations, construction industry, geological samples, *etc.* The major deficiency of TGA is its inability to provide any qualitative support for the analysis. Some type of spectroscopy (FTIR, MS) is required to identify the various components. Commercial instruments are also available that perform DSC and TGA testing simultaneously on the same sample. This allows identification of transitions as either related to or independent of chemical reactions and decomposition processes.

For further information the reader is referred to some recent reviews on thermogravimetry [80,214, 215], in particular related to polymers [216], and on controlled rate thermal analysis and related techniques [195,196]; many textbooks are available (*cfr.* Bibliography).

Applications

The primary application of TGA is to characterise a material's weight loss *vs.* time at a given temperature or within a certain temperature range. The thermoanalytical technique is used for the structural characterisation of homopolymers, copolymers, polymeric blends, composites and rubbers and finds application in the detection of monomeric residuals, solvents, additives, toxic degradation products, ash content, *etc.*, and for measurements related to thermal stability, volatilisation and evaporation. In order to elucidate the structure of complex polymeric materials, it is important to separate the constituting components. This can be done in several ways, such as by admission of air after initial heating in inert atmosphere.

Compositional Analysis:

Through examination of the various steps in the weight loss process TG has considerable potential to provide an effective and relatively rapid analysis of the "basic composition", namely the content of highly volatile matter (*e.g.* moisture, solvents, oil), polymer content, carbon-black or carbon fibre content, ash or filler content. The derivative is used in this process to highlight the different weight loss steps. CRTG enhances the resolution. A standard test is available for composition analysis of polymeric formulations by means of TG [217]. Many of the compositional analysis applications involving TG have focused on the **quantitative determination** of concentrations of one or several additives to a polymeric matrix [218].

In general, TGA provides information about the temperature and course of decomposition reactions in inert atmosphere (essentially a form of controlled pyrolysis), as well as burning profiles in air or oxygen (in conjunction with EGA). Certain classes of additives may require a more reactive atmosphere (such as oxygen) to decompose than the usual nitrogen gas purge, but much useful data can be collected based on the use of the process of elimination by subtracting reactive substances from the inert substrates. TGA results are significantly affected by the choice of atmosphere. In an inert atmosphere the onset of decomposition is delayed and the shape of the entire thermogram is completely different from that in air or oxygen. In relation to compounding and processing it is often necessary to study the decomposition behaviour and stability of additives, *e.g.* of copper-based additives, which were studied by TGA under N_2 [219]. A great deal of information regarding structure can be derived from a prescan which pyrolyses the polymer in an inert atmosphere and then burns off the resulting carbon in an oxidising atmosphere. For example, the amount of carbon formed by pyrolysis may be indicative of the presence of certain flame retardant additives in flammable materials such a polyolefins and styrenic polymers. Thermal analysis (TG and DSC) also offers a rapid means of testing both polymers and antiozonants for ozone reactivity [164].

TG is frequently used for analysing the composition of adhesives by quantifying the amount of **moisture** which is present and the amount of **volatiles** associated with a reaction. Fast heating rate TG allows detection of very low levels of volatiles in small samples. TG is also used for the quantitative determination of **solvents** in polymeric additives used as pour-point depressants and flow improvers [220]. PET moisture analysis by means of TG can be carried out at ppm level [221]. Thermogravimetry (eventually combined with GC or IR and subambient DSC) is very useful for the determination of residual solvents or for the study of interactions of water with polymers (important for modified release formulations for which swelling or gel formation of polymeric excipients is relevant). TGA has also been employed to measure the continuous desorption of sorbed $scCO_2$ in polymeric materials [222].

Thermal methods of analysis are widely used to investigate the process of **additive loss** from polymers. According to several authors [223–225] the volatility of low-MW additives (plasticisers, antioxidants, light stabilisers, accelerators, *etc.*) proceeds according to first-order kinetics. Various interferences have been noticed in these analyses [226]. A common use of TG is to determine the volatility of additives either neat or from polymer composition [130]. Price [227] has determined vapour pressures of plasticisers and UV absorbers by means of TG.

TGA also allows determination of volatile organic additives such as dioctylphthalate (DOP) **plasticisers** in vinyl plastics (*e.g.* in infant teethers). Determination of DOP is simple and quantitative, although it is really a test of total volatile organics, and is not specific of any one additive [228,229]. Efficient PVC/DOP analysis by TG consists in using a heating rate of 20°C/min to 190°C and an isothermal dwell time (*ca.* 10 min) in N_2 to allow volatilisation of the additive, followed by 20°C/min heating through the decomposition region.

Affolter [230] has discussed methods of characterisation and identification of polyester plasticisers. Polymeric and monomeric plasticisers were distinguished on the basis of molecular weight determination, TG, and TLC, and chemically identified by IR spectroscopy, and by the determination of monomeric units by saponification. These methods use sample sizes of about 1 g. Marcilla *et al.* [231] have studied the thermal degradation behaviour of ten commercial PVC resins by TG. TG was also used to study eight commercial phthalate and adipate plasticisers. Different kinetic models were suggested for the correlation of weight loss data at four heating rates for two resins and three plasticisers.

TG/DTG appears as a traditional and effective analytical technique for compositional analysis of **compounded elastomers**, which are complex mixtures of polymer, oil, carbon-black, or mineral filler, curatives, plasticisers, and other ingredients [108, 232–235]. Swarin *et al.* [236] were able to separate volatilisation events of mixed plasticisers in NBR vulcanisates. Ten commercial NBR samples were analysed for plasticiser type using both an extraction/GC procedure and TG/DTG. The correlation between relative retention time of each plasticiser and the DTG peak temperature for volatilisation was excellent. Thus, TG/DTG can be used to identify single plasticisers in NBR formulations. Also oils could be distinguished from one another on the basis of DTG volatilisation data.

A major challenge in TG analysis of elastomer vulcanisates is to accurately separate **oil/plasticiser** and elastomer regions, which often overlap. Most of these materials have volatilisation ranges rather than discrete volatilisation points because they are chemical blends of components of various molecular weights and volatilities. Overlapping of oil and elastomer TG curves is therefore quite common, especially if the oil is of the less volatile paraffinic type. Overlapping is also expected for many other process oils, plasticisers, and processing aids, which decompose in the same temperature region as the elastomers. Various methods for graphical resolution of oil and polymer weight loss have been described [236]. Zeyen [237] observed that analytical data for routine oil/plasticiser production samples obtained by multistep iso-TG in N_2 and O_2 correlate better with known values than those determined according to ASTM D 297-81. However, iso-TG does not work well with paraffinic oils used primarily in moulded rubber goods (particularly in EPDM compounds), which volatilise at a higher temperature. Zeyen [237] has listed volatilisation/oxidation temperatures of different components in rubber formulations in TG/DTG experiments. High resolution or reduced pressure methods are frequently used. Reduced pressure methods (typically 10 mbar) alter the volatilisation temperature of oil and separate it from the polymer. Similarly, if the pyrolytic decomposition of the rubber component is overridden by the release of plasticiser with a high boiling point, exact determination of the plasticiser content can be made by measuring in vacuum, as shown in Fig. 2.10. However, at very low pressures, volatile substances already start to evaporate at room temperature. Sichina [238] has illustrated the usefulness of the auto-stepwise TG for some unidentified rubber/oil samples.

Möhler *et al.* [234] examined a carbon-black (N 550) loaded NR/EPDM with a low-boiling adipic acid ester plasticiser by means of TG/DTG. In the temperature range of volatilisation of the plasticiser also residuals of the vulcanisation and accelerator system and antioxidants or antiozonants evolve. The same authors reported also TG/DTG measurements of EPDM containing a high-boiling paraffinic mineral oil plasticiser, of NR/EPDM and SBR/EPDM with low-boiling adipic ester plasticiser, and the separation of various active CBs (N 220 and N 762) in EPDM compounds containing the low-boiling adipic acid ester plasticiser. Without high-resolution facilities TG/DTG does not allow the qualitative separation of the two carbon-blacks. Carbon-black analysis is also difficult in the presence of chalk.

TG/DTG has gained wide acceptance as a method for compositional analysis of polymer/oil/CB masterbatches and of compounded rubber and vulcanisates as evidenced by the ASTM E 1131-03 test method on "Compositional Analysis by Thermogravimetry". The standard test method for compositional analysis of elastomers by TG [239] describes a general procedure to determine the quantity of four arbitrarily defined components:

(*i*) highly volatile matter (low-boiling components – 300°C and lower – such as moisture, rest monomer, processing oils and extenders, plasticisers, curatives, antioxidants);

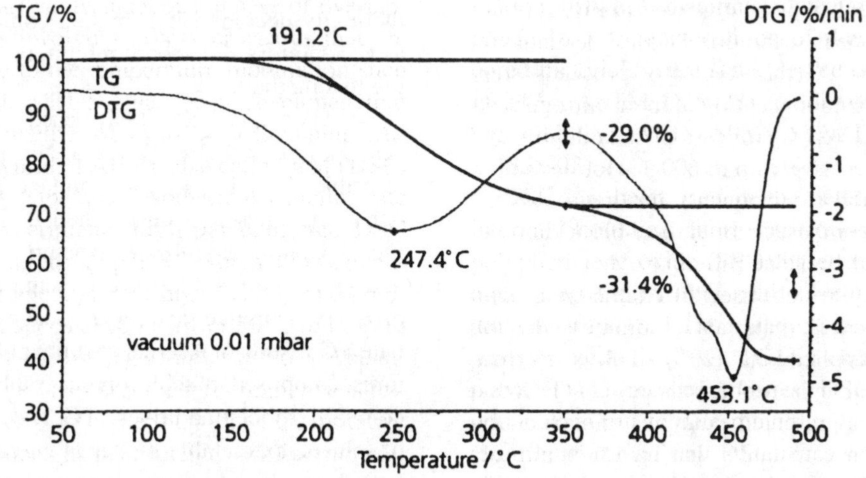

Fig. 2.10. Exact determination of plasticiser content (29%) of SBR rubber by means of vacuum TG. Reproduced by permission of Netzsch-Gerätebau GmbH, Selb (TG 209 Technical Data Sheet).

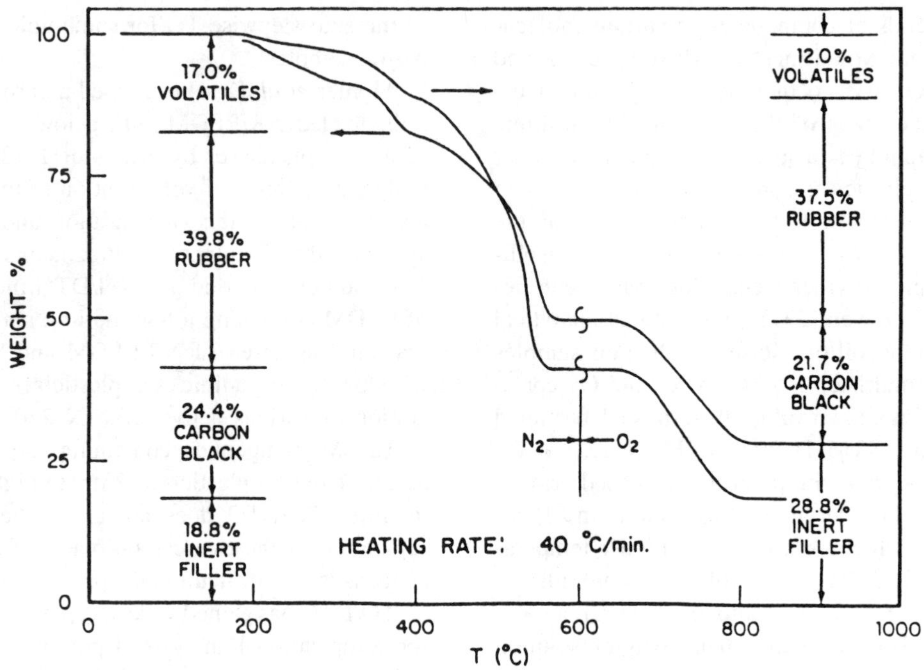

Fig. 2.11. Analysis of automotive V-belt composition. After Gibbons [151]. Reproduced by permission of International Scientific Communications Inc.

(ii) medium volatile matter (materials which degrade at 300 to 750°C, such as processing oil/aid, curing agent, *etc.*, including the elastomer portion of the compound);

(iii) combustible material (oxidisable, non-volatile material at 750°C, *e.g.* CB, graphite); and

(iv) ash (non-volatile residues in an oxidising atmosphere, such as metallic oxides and fillers).

These components may be observed in Fig. 2.11.

Multicomponent separation of a rubber material performed with TG then typically proceeds stepwise, as follows: rapid heating in inert (nitrogen) atmosphere up to 100°C (for loss of volatile oils and extenders), successively up to 600°C (for decomposition of the rubber component), heating in oxygen to 950°C (for combustion of carbon-black) and determination of the residue (fillers).

TG has been widely used to characterise compounded elastomeric materials in commercial [240] and military applications [235]. TG is a **troubleshooting tool** in the rubber industry [241]. Ohtake *et al.* [109] have presented examples of such analyses with reference to faults that have actually occurred in industrial rubber components. Ramirez *et al.* [242] have described TG studies of a wide range of (un)vulcanised elastomers and blends. In most cases it was possible to determine characteristic TG curves for each material, allowing the characterisation of polymers as well as additives, such as fillers and oils. Soos *et al.* [243] have reported a rapid method for the determination of moisture levels in additives used in the rubber industry. The inverse thermometric method of moisture determination was used for powdered additives. Besides mineral fillers, thermally decomposing organic combinations such as accelerators and scorch inhibitors were tested using this method.

Macaione *et al.* [235] have used TG for the characterisation of SBR, BR and NR in mono-, di-, or triblend rubber systems and carbon-filled rubber composites and determined the percentage of highly volatile organics, elastomer(s), carbon-black, and inorganic residue for each sample. Lochmüller *et al.* [194] applied factor analytical methods to evaluate TG results of a series of rubber blends and mixtures composed of chloroprene rubber, NBR, and common rubber additives. TG and measurements of toluene extractable matter of cured siloxane rubbers thermally aged in inert gas atmosphere at 80°C showed a build-up of low-MW fragments in the rubber network with age [244].

Sircar [192] has reviewed the analysis of elastomer vulcanisate compositions by TG/DTG. DTG may serve as an identifier of elastomer type in a compounded formulation. The problem of the determination of the elastomer-carbon residue and added carbon-black in the compounds, which often oxidise together, has not been fully resolved. TG has gained itself wide acceptance as a method for compositional analysis of vulcanisates (ASTM E 1131-03), despite some restrictions. It provides reasonably accurate data, is faster than the classical extraction method, and is an excellent QC tool. The classical ASTM method (D 297-81) is too lengthy to be of much practical use on a routine basis, often requires preliminary identification of the polymer and is costly. However, a 100% materials balance in TG is not always achieved. This may be due to overlap of low-MW volatile material with polymer decomposition products, formation of char which decomposes in the region assigned to carbon-black, or carry-over of early stage decomposition products to the ash (residue) region. Even though accuracy is not always high, precision is still good. Thus, TG/DTG remains the method of choice for compositional analysis of uncured and cured elastomer compounds.

Yang *et al.* [245] have used TG for the study of the thermal weight loss of low-MW surfactants, used as **antistatic agents** in HDPE containers. In a typical example of product development Ward *et al.* [246] have reported the use of TG in combination with static decay and optical measurements for evaluation of the effectiveness of some 13 internal antistatics. With fail/pass criteria of a weight loss of 5% (up to 250°C) and a static decay time of less than 0.5 s at 70% r.h., none of the commercially available surfactants did meet all critical criteria; developmental PMMA antistats were reported.

Thermal analysis is widely used to study the efficiency of **antioxidants** (stabilisers) for polyolefins. Woo *et al.* [117] observed that high temperature oxidation acceleration for HDPE originates from the volatilisation of Irganox 1076. This is in accordance with the rate of volatilisation of Irganox 1010, as determined by TG [119]. Wang *et al.* [247] have used TGA to evaluate the thermostability of various low and high-MW HALS.

Gray *et al.* [248] have used TGA in product development to evaluate antioxidants that volatilise significantly less during foam cure (Lowinox DBNP, Anox BF, Anox PP/Irganox 1076, Anox 20/Irganox 1010) or are not expected to be lost under typical curing conditions (Lowinox OS 330, DTDTDP).

Thermal analysis methods (TG, DSC, cone calorimeter, pyrolysis-combustion flow calorimeter) play a key role in **flame retardancy studies**. Some typical applications of TGA are weight loss/gain, reactivity with atmospheres, oxidative degradation, drying rate, reaction kinetics, volatilisation analysis, compound composition and stabiliser effectiveness. TG and DSC are frequently used for testing of FR materials to verify excellent thermal stability and high onset decomposition temperatures [249]. Benbow *et al.* [250] have carried out TG studies of the thermally stable FRs DBBP and DBDPE. Isothermal studies comparing FR formulations to their general-purpose analogues can also help to determine the effectiveness of the additive system and the weight loss observed under such conditions can be used to quantify the amount of the FR additive. Figure 2.12a compares the weight loss process for a general purpose and a flame retardant ABS, while Fig. 2.12b shows the derivative curves. In this case evidence of the flame retardant additive is seen in the lower temperature of initial decomposition, in the two-phase weight loss of the polymer, and in the presence of a significant amount of carbon that forms during pyrolysis and then burns off in air at the end of the test [24]. DTG was also used to study the influence of BFRs on thermal degradation of polymer blends in air and inert argon atmosphere [251]. Although TG can easily provide the whole weight loss behaviour of the FR system, it cannot provide unequivocal information on the detailed thermal degradation mechanisms.

TG-DTA data of an APP/melamine binary mixture showed interaction with an increase in thermal stability [252]. Learmonth *et al.* [253] have described reaction between Sb_2O_3 and the organic HFRs Cereclor, perchloropentacyclodecane (Dechlorane 4070), tetrakis (pentabromophenoxy) silane (Flammex 4BS) and pentabromotoluene (Flammex 5BT) in a cross-linked polyester resin. Weight loss plots indicated when reaction took place. Quantitative analysis of volatile reaction products from Cereclor–Sb_2O_3 and Sb_2O_3–PVC (Corvic P65-50) mixtures showed $SbCl_3$ as the main product. The main limitation of TG studies of FR polymers is of course that they give little information about reactions resulting in the production of new species, which may exert an inhibiting action on the combustion of the organic polymers by virtue of reactions occurring purely in the condensed phase (*e.g.* charring, interactions between FR and polymer). In flame

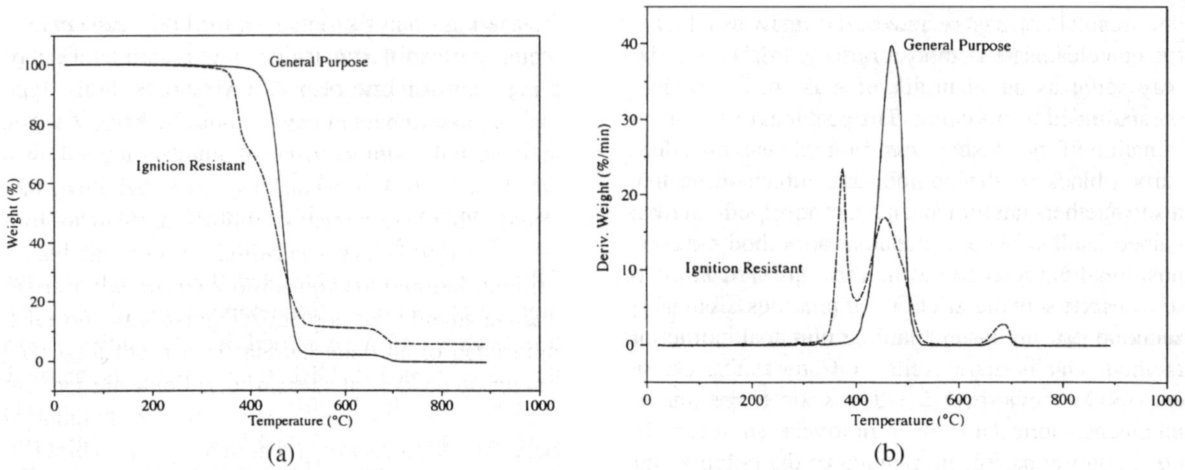

Fig. 2.12. Comparative TGA showing weight loss (a) and the derivative of weight loss (b) for general purpose and ignition-resistant ABS. After Sepe [24]. Reproduced by permission of Rapra Technology Ltd.

retardancy studies thermal analysis (TG, DSC) is therefore more efficient in combination with surface analysis (XPS, ToF-SIMS, AFM) studies which allow determination of the surface composition of FR materials by physical and chemical mapping. Redfern [48] has reviewed the use of thermal analysis for the evaluation of flame retardants.

Thermogravimetric data were also used to evaluate kinetic parameters for thermo-oxidative degradation of some flame retardant PP materials [254]. In addition, isothermal evaluations at normal processing temperatures can be used to evaluate the tendency of materials to produce condensed volatiles. These deposits, known as **plate-out**, negatively influence the acceptability of the manufactured product and also determine increased mould maintenance. Ezrin *et al.* [255] have reported TG in combination with a pH test in screening flame retardant thermoplastics for moulding safety. The acidic nature of FR decomposition products may cause corrosion of moulding equipment, unacceptable moulded parts and also constitutes a potential hazard from the industrial hygiene point of view. TG is well suited to establish the temperature range at which a FR material can be processed without decomposition. The problem is most severe with plastics requiring high moulding temperature due to high melting point, such as PA66 and PBT. Nowadays, the TGA/pH test would probably be replaced by a TG-FTIR or TG-MS analysis (*cfr.* ref. [256]).

Incorporation of **fillers** into a resin generally modifies mechanical, electrical or optical properties, the resin's appearance, or produces a delayed release. Examples of such fillers are carbon-black (pigment; opacity), TiO_2 and $CaCO_3$ (brighteners), and silicone oil (lubricant). TG has frequently been the method of choice for the compositional analysis of filled resin systems. With a typical specimen size of *ca.* 20 mg, TG is used extensively in investigative work to study homogeneity, carbon-black contents and glass fibre levels and to characterise fire retardant polymers. Consequently, TG finds wide use in the composition analysis of filled polymeric resins for structural applications [257]. Actually, TG is frequently the method of choice for composition analysis of filled resin systems as it offers the potential for rapid quantitative detection of multiple components in a single analysis with good precision and accuracy for concentrations down to approximately 1 wt.%. The concentration of carbon-black in a resin, added to the plastic to improve its resistance to thermal and photoinduced degradation, can easily be determined by TG [258]. Weight losses in air at temperatures exceeding 600°C have been used to distinguish between different types of colorant systems and fillers in elastomers [259]. Large amounts of inorganic filler (*e.g.* 70 wt.% of fused silica in an epoxy composite) can be analysed by thermogravimetrically pyrolysing the organic components away and identifying the remaining residue by XRF.

Ostromow [260] has described the analysis of mineral fillers by dry ashing (according to DIN 53568, BS 903 (1950) or ASTM D 297-59T (1960)). Determination of the **ash content** in polymeric compounds can be performed with standard methods (*i.e.*

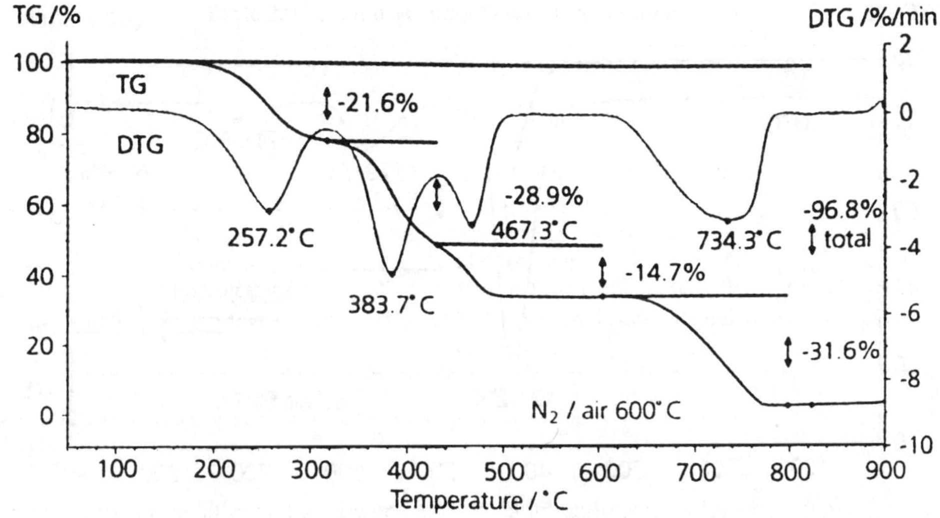

Fig. 2.13. TGA of an NR/EPDM rubber mixture showing release of plasticiser, residue of the vulcanisation system and of the antioxidant (21.6%), decomposition of natural rubber (28.9%) and of EPDM (14.7%), combustion of carbon-black (31.6%) after switching from inert atmosphere to air, and residual ash (3.2%). Reproduced by permission of Netzsch-Gerätebau GmbH, Selb, Germany (TG209 Technical Data Sheet).

ISO 247) and also with TG (following ISO 9924-1), *cfr.* Fig. 2.13. Comparison between both methods reveals that for ash contents over 10% TG is as efficient and precise as conventional methods. Smaller contents lead to a higher uncertainty of measurement in case of TG [155]. Not all fillers are equally stable: glass fibres, quartz and talc do not decompose below 900°C, whereas chalk loses CO_2, kaolin H_2O and aramid fibres pyrolyse; some fillers are unstable in oxygen atmosphere such as carbon-black and carbon fibres. A unique advantage of TGA is the capability to separate most inorganic fillers from carbon-black by first running the sample under a non-oxidising atmosphere and then switching to an oxidising environment to burn off the carbon-black. (Carbonate fillers present difficulty due to liberation of carbon dioxide.)

With TG it is also possible to determine **glass fibres** in polymer systems. Fava [261] recorded TG/DTG curves of PP filled with carbonate and fibreglass. TG is an ideal analytical tool for the control of the glass fibre content in composite materials. Since the glass fibre is thermally inert, there is no problem resolving the weight from the resin (by simple subtraction from 100%). Gibbons [151] has analysed additives such as plasticisers, antioxidants, fillers, and reinforcements for PA11, PE, PP and epoxy resins both qualitatively and quantitatively by DSC and thermomechanical analysis. Figure 2.11 shows a TG analysis of an automotive V-belt for the composition of its various components. Carbon-black is added here for conduction to dissipate the static electricity charge that accumulates in use, improves tear resistance of the belt and aids in allowing for longer trouble-free service. The inert filler minimises the expansion coefficient of the rubber and prevents the belt from stretching out of shape during use. Plasticisers and rubber content can be determined in N_2 atmosphere, whereas CB and any fillers are determined in the presence of oxygen. Subtle differences in composition of the belt compounds can easily be determined by TG [151]. Also compositional analysis of PA6 (polymer, moisture and glass fibre content) by means of TGA has been reported [85]. Determination of glass fibre in nylons is particularly useful when examining a stressed or broken moulded part to insure that the area of failure has the proper nylon–glass ratio [82]. Figure 2.14 shows the simultaneous determination of blend composition (12.3% PTFE) and glass fibre content (30.1%) in a GFR PBT/PTFE blend. Should the filler be unknown, it is also possible to take this residue and identify it by other analytical techniques, such as infrared analysis.

TG can also be used for the evaluation of the thermal stability of **organic and inorganic pigments** and pigmented polymeric samples [262]. Taking advantage of the chemistry of filler components, Gill-

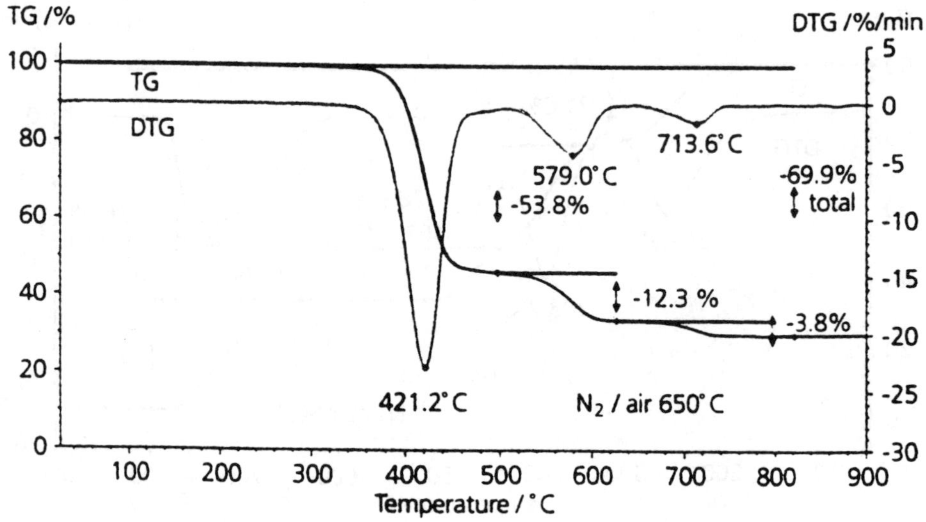

Fig. 2.14. Determination of blend composition and glass fibre content of a GFR PBT/PTFE blend. Reproduced by permission of Netzsch-Gerätebau GmbH, Selb, Germany (TG209 Technical Data Sheet).

mor *et al.* [257] have distinguished four inorganic ingredients ($CaCO_3$, TiO_2, silicone oil and carbon-black) within the pyrolysis ash of a butadiene modified polystyrene matrix in a single TG analysis with appropriate gas switching. Brennan [82] has described the determination of the **lubricant** MoS_2 in PA6.6 by complete degradation of the polymer component in air. Nakatsuka *et al.* [263] have determined 0–12 wt.% **starch** in starch-LDPE blend films by means of TG. Direct FTIR analysis on the basis of the 980 cm^{-1} (C−O stretching)/1460 cm^{-1} (CH_2 bending in LDPE) peak ratio can be used to determine starch levels (up to 40%) in LDPE/starch blend films [264]. However, FTIR analysis is difficult for thick films, particularly when exposed to a soil environment. TG analysis is then more appropriate [264]. The percent weight loss over a specified temperature range, at constant heating rate as determined by TG, correlated well with the starch content of films (in the range of 0 to 12 wt.% starch), as determined by chemical analysis. The method fails for samples exposed for longer periods of time due to formation of low-MW oxidation products of LDPE, which volatilise in the temperature range when starch degrades. Also the filler-content determination of wood-based composites by TGA has been reported [265]. Oil-palm wood flour (OPWF) was investigated as a new type of **wood-based filler** for PP. Characterisation of OPWF composites requires checking for the actual filler content and filler distribution within the matrix. The organic OPWF filler degrades before the PP matrix when subjected to high temperature. Ahmad Fuad *et al.* [265] have described an analytical technique for computation of the OPWF content in composites based on a simple expression derived from TG analysis. The technique has shown good agreement and consistency between determined and actual filler contents.

Techniques based on TG analysis have made it possible to readily and accurately measure the **carbon-black content** in commercial polymer formulations, such as in rubbers, at levels as far apart as 0.1% and 30%. The typical procedure is shown in Fig. 2.15 (sensitivity of the TG scan is 100 wt.% full scale) for a polyethylene masterbatch formulation, which was initially heated in N_2 at a rate of 160°C/min. to about 550°C. Pyrolytic decomposition to gaseous products resulted in a 75% weight loss. After changing to O_2 atmosphere the carbon-black is then oxidised [151]. The precision of the determination in the PE/CB masterbatch formulation is about 0.05 to 0.1% carbon (absolute). The TG method is fast, *i.e.* 6 min at 160°C/min, as compared to 2 h for ASTM D 1063 [266] without TG, thus providing substantial time savings. The compositional analysis (polymer and CB content) of LDPE has been reported [85]; Affolter *et al.* [155] have determined the content of carbon-black in polyolefins (2–3% CB) by TG following ISO 9924-1 and have noticed an inhomogeneous distribution in commercial raw materials (LDPE). The relative oxidation characteristics of the carbon residue and carbon-black control the peak resolution obtainable by DTG. By proper choice of isothermal conditions and dilute oxygen atmosphere, DTG oxidation peaks of most blacks can be separated from the char and their quantity can be estimated by TG/DTG. In addition to quantitative determination, TG can be used to distinguish between different **carbon-black grades**, including medium particle-size reinforcing blacks (N 550, N 660, *etc.*), both in the free form and when incorporated into a rubber formulation. As carbon-blacks oxidise at different temperatures depending on their surface areas the method is based on a linear relationship between specific surface area and temperature at which 15% CB has been oxidised (T_{15} value). Charsley *et al.* [267] have examined the variables which affect T_{15} measurements with a view to optimising the experimental procedure. Using this method, the relationship between T_{15} and surface area for a wide range of free CBs of different surface areas (such as MT, SRF, GPF, FEF, HAF, SAF and channel black types) and compounded CBs has been investigated. The technique is not suitable for the identification of a CB type in unknown formulations. It can be used, however, as a routine quality control check on batch rubbers. Pautrat *et al.* [268] have described quantitative analysis of HAF, SRF, and MT carbon-blacks in EPDM, IIR, and NR, as well as HAF in SBR according to ASTM D 297. Knappe *et al.* [84] have compared CB types N 234 and N 660 by means of TG stressing the fact that this technique is highly suitable for investigating the activity of different types of carbon-black.

Direct TG analysis of carbon-black in impact modified GFR PA6 at low CB concentrations (<0.5 wt.%) may be difficult in view of the heterogeneity problems. In such cases, it is advised to degrade the polymer (in HCl) to form the water-soluble ε-amino caproic acid and to remove the glass fibre with HF, obtaining CB as an insoluble fraction, which is then subjected to TG [269]. Even small amounts of oxygen present during pyrolysis can produce significant errors in the determination of high temperature volatiles, polymer and carbon. To detect this problem carbon-black, toner or charcoal may be run in TGA to verify for constant weight (*i.e.* no oxidation at 600°C).

Two SBR 1502 mixes containing carbon-black and calcium carbonate were analysed by Casa *et al.* [270] using TG and the analytical results were

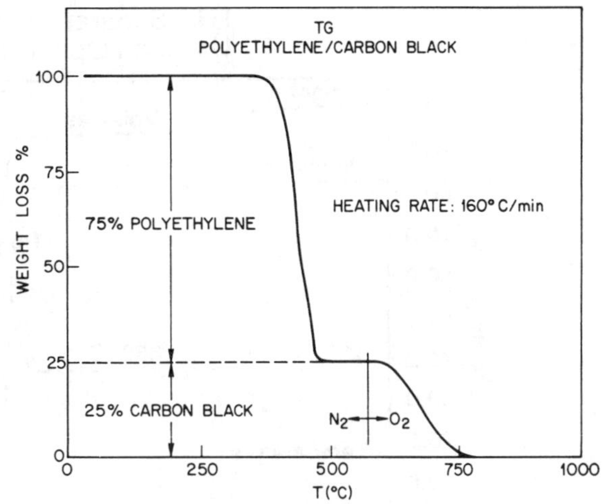

Fig. 2.15. TG of polyethylene containing 25 wt.% carbon-black. After Brennan [258]. Reprinted from *Thermochimica Acta* **18**, W.P. Brennan, 101–111, Copyright (1977), with permission from Elsevier.

compared with the known composition of the mixes. The technique was practically useful, and is probably applicable to the determination of other mineral fillers in polymers, such as hydrated aluminum oxides.

As noticed above, the exact separation of rubber ingredients such as curatives, emulsifiers and antioxidants from the oil-loss curve is hardly ever possible by TG, even with the use of DTG [25]. However, HRTGA, mass detection and multivariate analysis are means which still need to be explored. The effect of high resolution was clearly noticed in a (nitrile rubber, PVC)/plasticiser sample, which shows simultaneous evolution of plasticiser and decomposition of PVC in a linear programmed heating mode, better separation in vacuum TG conditions with constant heating, and plasticiser evolution before PVC decomposition in a controlled rate heating mode [271].

Figure 2.16 shows the TG separation of an LDPE formulation containing a lubricant, carbon-black and inert filler using an optimised stepwise analysis mode. The percentage lubricant, polymer, carbon, and inert filler can quantitatively be determined [272]. Lever *et al.* [273] have shown that HRTGA gives a much cleaner resolution of mass losses than conventional TG of a polymeric derivative used as an oil additive. Also the high resolution TG analysis of DOP in vinyl plastics has been reported [229].

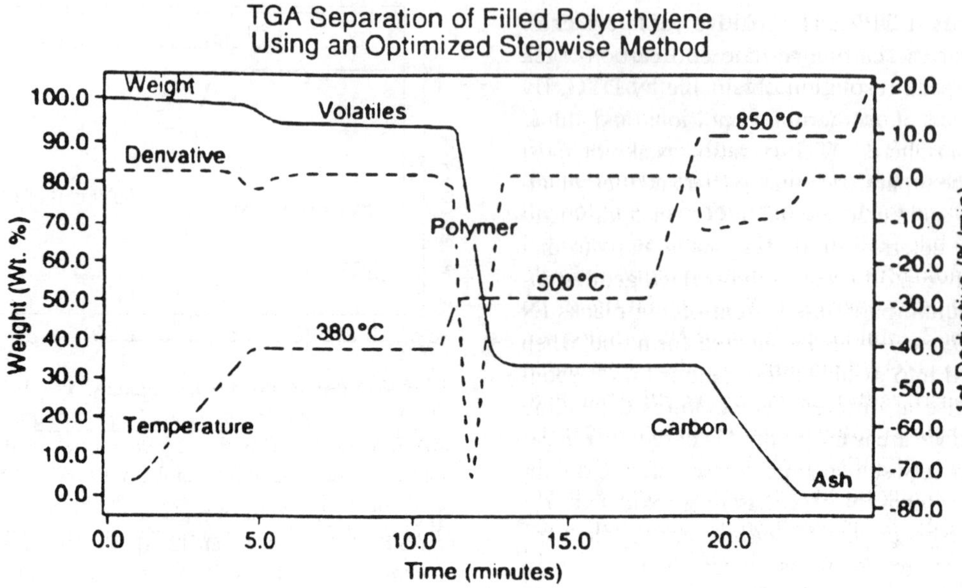

Fig. 2.16. TGA separation of filled polyethylene using an optimised stepwise method. After ref. [272]. Reproduced by permission of Perkin-Elmer.

Compositional analysis by TG also has its limitations, as stressed by Gillmor *et al.* [257]. The method involves removal of the resin matrix through pyrolysis with subsequent evaluation of the remaining ash or residue. Actually, this procedure is like burning a whole haystack in order to find the needle. The best results are obtained when several critical conditions are established in the TG analysis. It is imperative that residual volatiles like moisture or solvent be removed to establish a stable "dry" weight before the onset of pyrolysis of the matrix or evolution of a system component. Pyrolysis of the polymeric matrix should be complete and therefore not contribute carbonaceous residue to the ash. Polymeric matrices that decompose by 'unzipping' during pyrolysis are best suited for compositional analysis studies using TG since they contribute negligibly to the pyrolysis ash. Although TG is an excellent technique for the compositional analysis of compounded elastomers [108], it does not reveal the extent of cure. DSC is required for that purpose. A shortcoming to TG is that data on small amounts of organic additives are difficult to discern. Yet, with the increased TGA sensitivity and accuracy, product quality and product development of plastics will improve. Leyden *et al.* [274] have discussed various test methods in use for quality control; Stroh [275] has reviewed the role of TG in QC of elastomers.

TG may be used in **raw material monitoring**. For example, commercial magnesium stearate can comprise various components. In addition to stearate, these primarily include palmitate, their hydrates and free fatty acid. Both TG and DSC curves may be used for characterisation of such samples.

Thermogravimetry is used for **quality control** and to quantitatively determine the various components in a polymer or elastomer formulation by separating them on the basis of their relative thermal stability for the purpose of full compound analysis and for reverse engineering. This is a considerable challenge. TG has great potential for QC of filler quantity [276]. The importance of filler dispersion in the polymer matrix when working with filled polymers has been described [277]. Consistency of dispersion, mould design, and injection/process parameters are very important to the process engineer and for process validation. Reddy [278] used TG to measure and control the quality of filler dispersion in an injection moulding grade polyolefin compound with 8 wt.% (nominal) filler loading. Samples of about 20 mg were analysed continuously (each 24 min) for polymer and filler concentrations by means of robotic TG. Tighter monitoring and control of filler dispersion throughout the compounding run achieved large process improvements. TG has also been used to determine the composition of the elastomer masterbatches. Harris [279] has presented data for carbon-black determined by rapid TG analysis for quality control of oil- and carbon black-loaded BR and SBR masterbatches.

Thermal Stability Determinations:
Oxygen uptake in a polymer sample can be used to signal the onset of oxidation. In TG experiments at first a weight gain, corresponding to oxygen absorption by the polymer, is observed. Subsequently, a weight loss occurs corresponding to chain scission. The onset of weight gain is delayed by the antioxidants as in DSC-OIT experiments. Under isothermal conditions, the induction period is found to be proportional to the AO concentration in the polymer. In practice, there can be a temperature region of transition where weight gain (oxidation) can be offset by weight loss processes so that there is no detectable sign of oxidation. Concerns such as these have led to a greater acceptance of DSC than of TG for OITs. Bair [27] has discussed the relative merits of the two techniques and experimental details for ageing evaluations by TG.

By measuring the induction period of samples containing known amounts of AO, a calibration curve can be constructed. On this basis TG measurements are then capable of detecting AOs at concentrations above 0.001 wt.%. Increased analytical sensitivity can be gained by simply lowering the test temperature, which will extend the time scale and thus increases the difference in induction times between two samples with different AO concentrations. OIT is a most commonly used measure of additive loading in a polymer. One shortcoming in using OITimes to measure AO content in a polymer is that artificially low levels can be determined in the presence of pigments or transition metals. Bair *et al.* [280] have given an example of the ability of TG to determine quantitatively the concentration of an antioxidant in aged PE samples using OIT measurements. Ohtaki *et al.* [109] used TG-OIT and DSC-OIT on rubber samples.

Although oxidative stability studies are often undertaken in relation to **service life**, it has been pointed out by Gugumus [169] that TG data, while providing some indications concerning processing stability, do not permit any conclusions with respect to thermo-oxidative stability at lower temperatures. The observed dramatic change in slope occurring at point A in Fig. 2.17, which relates TG-OIT data with chemical analysis, indicates that small increases in

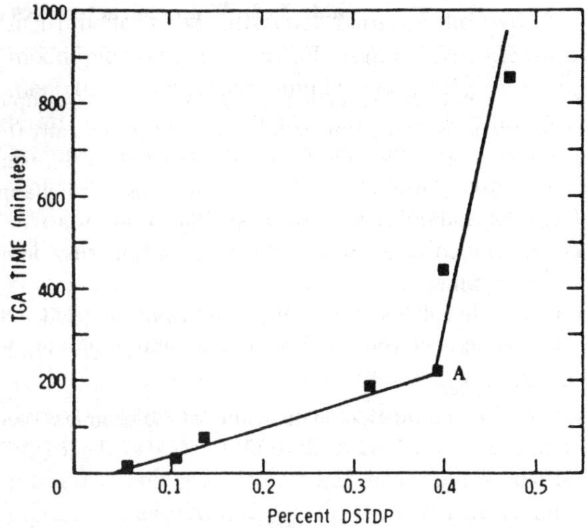

Fig. 2.17. Correlation of TG-OIT at 190°C and DSTDP concentration in PP. After Wims and Swarin [281]. Reprinted from A.M. Wims and S.J. Swarin, *J. Appl. Polym. Sci.* **19**, 1243–1256 (1975), John Wiley & Sons, Inc., New York, NY, Copyright © (1975, John Wiley & Sons, Inc.). This material is used by permission of John Wiley & Sons, Inc.

DSTDP concentration above 0.4% extend the life of the moulded PP part to a much greater degree than do similar increases below 0.4% [281].

Recent work has been performed, which shows that TG can be used to demonstrate change in the thermo-oxidative stability of HDPE due to oven ageing. A variety of ASTM **test methods** for heat ageing and thermal stability measurements are based on the use of TGA (*e.g.* D 1870, D 2126, D 3045, D 4202, D 5510 and E 1641).

2.1.4. Simultaneous Thermal Analysis Methods

Simultaneous thermal analysis (STA) refers to the simultaneous application of two or more thermoanalytical methods on one sample at the same time, such as DTA and thermoconductivity. In practice, however, this term is mostly used for simultaneous measurement of the mass changes and caloric effects on a sample under thermal treatment. The benefits are: *(i)* information on transformation energetics and mass change in one run, under identical conditions; *(ii)* time saving; and *(iii)* no differences in sample composition for the various thermal measurements – important for non-homogeneous sample materials. Although TG-DSC and TG-DTA are the most widely used of the simultaneous techniques due to

Table 2.12. Main characteristics of simultaneous TG-DSC (or TG-DTA)

Advantages:
- Greater efficiency (sample preparation time, run set-up time, instrument time)
- Higher accuracy of temperature calibration (typically 0.1°C for DSC, as compared to only 2°C for stand-alone TGA)
- Weight measurements validate quantitative DSC measurements
- Identical experimental sampling conditions eliminate interpretative uncertainties due to sample geometry and inhomogeneity (a single sectioning of the original material)
- Uniform perturbation of results due to sample environment (thermal history, orientation effects, heat treatment, pressure during cutting, *etc.*)
- Correlation of observed effects (simplifies interpretation)
- Detection of moisture and determination of *in situ* dry weight

Disadvantage:
- No *direct* information on the nature of the chemical species involved

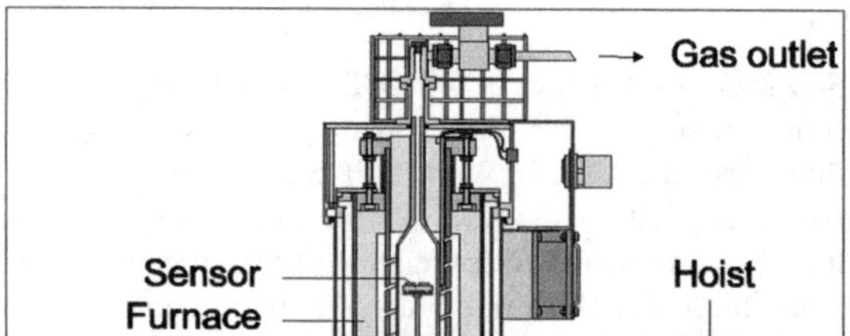

Fig. 2.18. Schematic design of simultaneous TG-DSC/DTA. After Blumm [287]. Reproduced by permission of Reed Elsevier.

their complementary nature, for complex reactions with several overlapping stages, it may be difficult to match the energy changes and weight losses.

Short overviews of thermogravimetric analytical techniques (simultaneous, non-simultaneous, (multi)hyphenated) are available [282,283].

2.1.4.1. *Thermogravimetry–Differential Scanning Calorimetry*

Principles and Characteristics

TG-DSC allows simultaneous measurements for the determination of mass change (TG) and energetic changes (DSC) on one sample under identical test conditions. As with this method all factors which influence the measurement signals (*e.g.* atmosphere, sample structure, temperature gradient, diffusion paths and packing density) are identical, TG and DSC results can be correlated and interpreted more easily. A typical modern assembly for simultaneous TG-DSC (or "STA") with user-exchangeable TG, TG-DTA and TG-DSC sample holders (up to 5 g, −120°C to 1650°C, 10^{-4} mbar) is shown in Fig. 2.18. For **interface design**, *cfr.* refs. [283a,284, 285]. In a recently described macro-STA a totally new concept has been introduced for recording the sample temperature and contact between the purge gas and the sample (a portion of the gas is conveyed *directly through* the sample) [286]. With sample volumes 2000 times higher than is possible with standard TG-DSC/DTA systems representative results are guaranteed when testing inhomogeneous or highly diluted materials such as household or industrial waste.

The **advantages** of single sample simultaneous TG-DSC (or TG-DTA) have been summarised by Redfern [288] and others (Table 2.12). The technique has recently been reviewed [285,289].

Applications

Typical applications that are ideal for TG-DSC are temperature stability, decomposition behaviour, drying and firing processes, transition and reaction temperatures, melting and crystallisation processes. Redfern [290] has reviewed single sample simultaneous thermal analysis, *i.e.* TG-DSC and TG-DTA studies of polymers, and has reported TG-DSC of an uncured polyimide resin in which a more accurate determination of the quantitative measurement of the heat of cure is made possible by the simultaneous technique. Other studies have concerned inert filler content in PE/CB, epoxy/CB and moisture in Kevlar [283a].

Kodama *et al.* [291] have reported TG-DSC curves for the analysis of the interaction between **vulcanisation accelerators** (tetramethylthiuram disulfide, dibenzothiazolyl disulfide, diphenylguanidine and *N*-cyclohexyl-2-benzothiazolylsulfenamide) and **fillers** (CB, hard clay and $CaCO_3$). The initial m.p. of the accelerators was largely influenced by the fillers. Emmott *et al.* [292] have investigated the complex reaction between $Sr(NO_3)_2$ and the binder Alloprene (a pyrotechnic system) at about 300°C by simultaneous TG-DSC and TG-DTA-MS. The same techniques were used to examine the Ti–$NaNO_3$–Alloprene and Mg–$NaNO_3$–Alloprene systems [293–295].

Simultaneous TG-DSC has also been used to study the behaviour of various particle sizes and coatings of magnesium hydroxide as a **flame retardant** and smoke suppressant in PP formulations [296].

2.1.4.2. *Thermogravimetry–Differential Thermal Analysis*

Principles and Characteristics

Since TG and DTA complement each other, it is an obvious move to attempt both investigations simultaneously [297]. TG-DTA measures mass and energy changes as a function of temperature of time. Depending on the atmospheric conditions (vacuum, inert or air conditions) thermal or oxidative stability is measured. The main use of DTA is to detect the initial temperatures of thermal processes and qualitatively characterise them as endothermic or exothermic, reversible or irreversible, *etc.* Ideally, the area under the DTA peak should be proportional to the heat of the process that gave rise to the peak. Simultaneous TG-DTA results improve the interpretation of thermal events. Validation is an important issue for the user; in this relation simultaneous TG-DTA plays a role. Hyphenated TG-DTA provides valuable information even in materials when no weight changes occur over the temperature range studied.

The method of DTG was elaborated in 1954 [298]. This early technique was the first to really measure the rate of mass change. Nowadays only computerised derivation is used. Paulik *et al.* [299] devised simultaneous TG/DTG-DTA (the **derivatograph**). TG/DTG-DTA and TG-DTA-MS instruments are commercially available. In TG-DTA-MS due to the missing separation technique, a positive identification of individual substances is not possible. An additional, off-line identification step by adsorption techniques followed by GC-MS will overcome this restriction. For a description of the TG-DTA interface the reader is referred to ref. [285]. **Advantages and disadvantages** of single sample TG-DTA are as given for TG-DSC (*cfr.* Chp. 2.1.4.1).

TG-DTA has recently been reviewed [285,289].

Applications

Typical TG-DTA applications are thermal and oxidative stability, determination of relative components, decomposition temperatures and thermal decay reactions, action of heat stabilisers, thermal ageing. TG-DTA has been used to screen candidate automotive engineering plastics, elastomeric seals and lubricant additives to establish **quality** and understand **field failures** [300]. The organometallic chemicals used as lubricant additives were employed to increase the thermal and/or oxidative stability of passenger car and heavy-duty diesel oils.

Negri *et al.* [301] have applied TG-DTA to the characterisation of different types of carbon-black in NR **vulcanisates**. The method allowed determination of the overall CB content, but where combinations of different blacks were present it was not possible to determine the proportion of each type. TG-DTA has also been used to correlate TGA in airflow and N_2 gas flow and some other micro-scale flammability tests (*i.e.*, oxygen index, hot-plate ignition and drum friction tests) on covers of different flame-resistant and non flame-resistant rubber conveyer belts [302]. The minimum temperatures at which rapid weight loss of each sample began to appear were determined and compared with the results from the micro-scale flammability tests.

Paulik [20] has described simultaneous TG-DTA of flame-retarded PE. A peak observed at 360°C in the DTG curve was indicative of reaction between the **flame-retardant** components (Sb_3O_3 and a halide), in which $SbCl_3$ is formed. The decomposition of flame-retarded PP/PE copolymers (FR = $Mg(OH)_2$; brominated trimethylphenylindane/Sb_2O_3) was investigated by means of TG/DTA-FTIR [303]. It was pointed out that the results might differ from tests performed on larger specimens.

Koch [304] has applied TG-DTA for quantitative determination of the polymer and rubber phase in

ABS graft polymers and of a polymer/softener/soot/mineral filler mixture. TG-DTA curves have also been used to evaluate antioxidation activities of various types of **antioxidants** [305]. TG-DTA and PDSC are suitable for product quality control as exemplified by OIT measurements for commercial engineering plastics, polyolefins and elastomers [306]. Applications of TG-DTA outnumber those of TG-DSC.

2.1.5. (Multi)hyphenated Thermal Analysis Techniques

Principles and Characteristics
Although thermal analysis techniques (DSC, DTA, TG) provide accurate information regarding macroscopic property changes with temperature, structural information cannot be obtained by these techniques. Hyphenation is mainly directed to providing this much needed chemical and morphological (microscopical) information [307], *cfr.* Table 2.13. A different way of considering hyphenated thermoanalytical techniques is to distinguish methods which aim at analysis of the **evolved decomposition products** or of the residue. Coupled instruments each need to operate under optimum conditions. Key elements in performance are the interface system and integrated software package.

Evolved gas analysis (EGA) coupled to TGA experiments may be carried out on-line (*e.g.* TG-MS, TG-FTIR) or off-line (volatile collection followed by thermal desorption: TD-GC-MS). With the off-line approach complex mixtures, which are often difficult to interpret with the on-line method, can be easily analysed. Less well known is the direct combination of TGA and AAS for simultaneous detection of atomic vapour in thermal analysis [308, 309]. In most EGA designs the substances evolved in the furnace of the TG device are carried to the spectrometer by a carrier gas, but not necessarily so. In the TG-AAS design of Kántor *et al.* [309] the light path of the AAS source passes directly through the furnace of the TG device. Similarly, in the "on-the-spot" TG-FTIR technique [310] the radiation is brought to the thermogravimetric analyser.

Redfern [290] has defined the minimum requirements for good interface design of a thermal analyser linked to an evolved gas analyser. Hyphenated, simultaneous techniques offer obviously considerable advantages over sequential methods in terms of reduced sample handling, uniqueness of the sample, absence of retention times, speed, *etc.* Combined techniques highly increase the domains of applications and the robustness of the information delivered. While many problems are solved by a combination of different TA measurement techniques, in a modern TA system great importance is attached to the software as this alone can determine an enormous increase in efficiency. The notable advancements in the last 15 years have enhanced the sensitivity and resolution capability of the equipment, with an attendant improvement in analysing complex mixtures. Both TG-MS and TG-IR are growth areas, as shown by the increase in research papers, namely totalling some 15-30-80 and 10-70-140, respectively, for the 1990–1995–2000 period, in line with earlier predictions [311]. Combined TG-EGA was reviewed [216].

Apart from the simultaneous coupled thermoanalytical techniques (TG-DSC and TG-DTA), **residue analysis** appears to have attracted fewer experimentalists than evolved gas analysis. Nevertheless, the ability to observe a sample by thermooptical methods, such as DSC-thermomicroscopy (or optical DSC), DTA-photometry or video microscopy imaging-TG (VMI-TG), as it is heated under conditions of controlled atmosphere and heating rate, provides a valuable supplement to thermal analysis techniques. In fact, DSC is non-specific and cannot distinguish between a phase change and a fusion reaction. Using thermomicroscopic methods

Table 2.13. Hyphenated thermoanalytical techniques

Scope	Means	Instrumental tools
Evolved gas analysis	Titrimetry, FTIR, MS, GC, ThGC, AAS	TG-FTIR, TG-MS, TG-GC-IR, TG-GC-MS, TG-AAS, DTA-MS, DTA-GC, ThGC
Residue analysis (morphology)	Photometry, microscopy	DSC-thermomicroscopy, VMI-TG-MS, SThM
Residue analysis (structure)	XRD	DSC-XRD, TG-XRD, TG-XRD-MS
Residue analysis (stability)	OIT, CL	DSC-OIT, DSC-CL, DTA-OIT, TG-OIT

it is possible to observe directly such phenomena as phase changes, fusion, decomposition reactions and processes of sintering, decrepitation, creeping of a liquid melt and foaming or bubbling reactions, which can often complicate the interpretation of thermoanalytical data. In DSC-thermomicroscopy the sample is observed by a photovisual microscopy system and differences between the heat flows of the sample and a thermal inert reference are measured simultaneously [312].

VMI-TG allows a direct visualisation of changes in **morphology** and texture in a substrate during thermal processing. VMI-TGA is to be considered as a valuable tool for studying gas-solid or thermal decomposition reactions since it combines simultaneous monitoring of reaction rates (by sample weight measurements) and direct viewing (or video taping) of the structural transformations that may accompany the heterogeneous reactions. This useful analytical tool is grossly under-utilised for polymer thermal decomposition and flammability studies. VMI-TG-MS can be used to obtain complete time histories of samples undergoing thermal treatment, pyrolysis or combustion, all processes characterised by complex chemical and structural transformations [282,313]. Weight losses (TG), evolved gas analysis (MS) and video records of structural transformation all concord in the evaluation of the heat effects.

Combinations of TA with scanning probe microscopy and x-ray analysis allow characterisation of the morphology of a material *in situ* and in real-time.

For residue analysis various other measurements (such as XRD) may be performed off-line, *i.e.* after removing the sample from the thermoanalytical equipment at different temperatures. It is obviously more desirable to carry out XRD measurements (powder goniometer or x-ray film camera) simultaneously with thermal analysis [314,315]. However, this technique is bedevilled with considerably experimental difficulties (geometrical and focusing problems) and simultaneous TG-XRD [316,317] has never reached commercial implementation. Wiedemann *et al.* [315] have used a simultaneous TG-XRD-MS system (thermomolecular beam analysis, TMBA), in which the weight changes were followed by measuring the gaseous reaction products, and by x-ray powder analysis of the residue. Position-sensitive detectors for fast data acquisition are also used in time-resolved high temperature XRD techniques [318]. Also simultaneous DSC-XRD systems, eventually equipped with a CCD detector, have been reported [319–322]. Small-angle XRD profiles can be obtained with DSC-SRXRD. Combination with other *in situ* temperature-controlled experiments, such as diffraction, scattering, microscopy, and spectroscopy are expected to further develop in the next future in relation to investigations of the structure and dynamics of materials. Quite obviously, residue analysis in combination with evolved gas analysis provides a considerable amount of additional information.

The DSC curve in oxidising atmosphere is the sum of many different exothermic reactions and is sometimes complex, making the graphical determination of OIT difficult. In contrast, the chemiluminescence (CL) signal is related to one reaction only. CL emission is related to the hydroperoxide (ROOH) content of a polymer. The CL curve is always well defined with a sharp onset and the OIT is easy to determine. CL offers many advantages over DSC for the study of polymer oxidation. **Simultaneous DSC-CL**, *i.e.* recording of enthalpic and photochemical signals, is now available in a commercial DSC with added single photon counting detector (PMT). The much higher sensitivity of CL means that it is possible to make OIT measurements at lower temperature, closer to real degradation conditions. In DSC-CL the OIT is taken at the time corresponding to the point at which the extrapolated isotherm or the CL signal intersects the extended baseline. During accelerated testing at elevated temperatures the effects of additive volatility are not usually taken into account. The DSC-CL technique is highly reliable for determining OIT values [323].

Applications

Bigger *et al.* [136] have applied stability parameter mapping and stability vector analysis for OIT data measured by means of simultaneous DSC-CL for a variety of LDPE/(DCP, Chimassorb 944, AOs) samples. Billingham *et al.* [323] have reported various simultaneous DSC-CL measurements of stabilised PET and PP samples.

TMA combined with a gas analyser (MS or FTIR) allows the dimensional changes caused by decomposition processes to be rapidly investigated or foaming processes to be optimised. On-line TMA-MS was used to investigate **delamination** of printed circuit boards (woven fibreglass embedded in an epoxy resin matrix) to determine the temperature at which particular decomposition products are formed. The sudden dimensional change in the *z*-direction of a printed circuit broad at 320°C, typical of lamination, was monitored simultaneously with MS by measuring the intensities of m/z 79 and 94 fragment ions (characteristic of Br and CH_3Br, *i.e.* decomposition products of TBBA) [324]. Similarly, also the expansion behaviour of a vinylidene chloride-styrene copolymer was followed by on-line TMA-MS. **Foaming** at 142°C, with a maximum volume increase of about 6000%, was accompanied by isopentane evolution (m/z 43 fragment ion). The polymer was identified at higher temperatures (HCl m/z 36, benzene m/z 78) [324].

For thermal–spectroscopic and other hyphenated techniques in polymer characterisation, *cfr.* ref. [324a].

2.1.5.1. *Thermogravimetry – Fourier Transform Infrared Spectroscopy*

Principles and Characteristics

For identification purposes the classical technique of infrared spectroscopy is highly suited to hyphenation to thermogravimetry. Compared with dispersive (prism or grating) IR equipment, Fourier transform infrared has the advantage that the use of an *interferometer* allows the whole IR spectrum (from 400 to 4000 cm^{-1}) to be scanned several times per second. This real-time performance is an obvious asset in analysing gases released during a TG experiment. Consequently, TG-FTIR is an important tool for materials characterisation, in particular for polymer analysts concerned with structure/mixture compositions, and degradation/reaction mechanism studies.

Experimental coupling of TGA and FTIR spectroscopy was reported in the literature [325] as early as 1980, but dedicated instruments were not available until 1987 [326–328]. Up to that time most work on the identification of evolved gases from TG had been in the TG-MS combination and reports on polymer studies using hyphenated TG-FTIR were relatively scarce [216,329]. Several approaches to coupling TG and FTIR components have been reported [290,310,330–337]. In conventional commercially available TG-FTIR systems, the evolved gases are led from the TG system to the spectrometer via the shortest possible **heated transfer line** (typically at 250°C) by a carrier gas flow [330–333]. Kaisersberger *et al.* [338] have discussed hyphenation of a thermobalance or an STA instrument to IR spectrometers. Intensive contact between the IR radiation with the gas-flow including the evolved gases is achieved in specially designed gas-measuring cells. Some commercial designs are based on total flow, *i.e.* all of the gases which evolve at atmospheric pressure together with the purge gas flow into the heated IR gas cell. Flow gases normally used for TG (N_2 or dry CO_2-free air or an inert gas) are IR transparent, and ensure fast transportation between TG and FTIR without time gap between release and detection. A long pathlength through the gas is required, since the concentrations are low. As opposed to coupling of mass spectrometers, the whole gas flow from the thermobalance should pass through the gas cell of the IR spectrometer. Normally, the transfer time for the gas is in the range of a few seconds and the flow profile is laminar. A fast detector such as a liquid nitrogen cooled MCT (Hg-Cd-Te 600–4800 cm^{-1}) detector is often used.

Other designs make use of a **sniffler** tube that extends into the sample cup and removes some of the evolved gases along with a portion of the inert gas purge [339]. Problems arise both with on- and off-line TG-FTIR systems for high-MW components, which may deposit on cold spots in the TG equipment or in the transfer line. Yet another approach to TG-FTIR coupling is use of He carrier gas at high flow-rates, leading to the formation of an **aerosol** of the evolved components, which is then introduced into the spectrometer [335,340]. This system performs quantitative measurements and preserves and monitors very high-MW condensibles. In an **"on-the-spot" TG-FTIR** technique the radiation is brought to the TG system, as opposed to bringing the evolved components to the spectrometer [310, 341]. The IR beam is led directly into the TG system, where it is reflected by a mirror mounted inside the TG equipment and is subsequently detected by the standard FTIR detector. Compared to FTIR methods which employ heatable gas cells (*e.g.* fast thermolysis FTIR [342]) the on-the-spot TG-FTIR technique avoids transfer lines and monitors the gaseous atmosphere directly above the sample pan; spectral information is obtained, which is directly correlated to the recorded mass change as a function of time and temperature. The on-the-spot technique also allows detection of higher-MW components than systems based on a heated transfer line [310]. Due to the direct IR detection, there is no loss of evolved components by cold spots or discrimination of high-MW.

At the end of a TG experiment software allows contour plots (scan time/temperature *vs.* wavenumber) and 3D stacked plots (*cfr.* Figs. 2.19 and 2.20)

Table 2.14. Main characteristics of TG-FTIR

Advantages:
- Functional group indentification/specific compound analysis
- Reference vapour-phase spectral libraries [344]
- Suitable for higher-MW fractions (up to 800 Da)
- Analysis of structural isomers
- Real-time analysis (continuous effluent scanning)
- Quantitative (±10%; with appropriate calibration)
- Non-destructive
- Cost-effective
- Commercial equipment

Disadvantages:
- Relatively low sensitivity (sub-μg range)
- Difficult mixture analysis

containing information about the kind (wavenumber) and amount (absorbance units) of the released gases as a function of temperature or time at which they are released. **Evolved Gas Profiles** (EGP) can be reconstructed from the stored interferograms according to Gram-Schmidt [343]. Each point in this EGP corresponds to an IR spectrum of the evolved components in the TG equipment. Specific Gas Profiles (SGP) and Functional Group Profiles (FGP) can also be reconstructed from the stored interferograms in selected wavenumber windows to detect components with specific group frequencies. Comparison of EGP as a function of time with the DTG curve yields a direct comparison of TGA and spectroscopic data. The detection limits are in the sub-μg/sec range and dependent on the extinction coefficient of the evolved components [310]. In general, however, detection limits for components in the condensed phase are a decade lower than those in the gas phase.

Table 2.14 lists the **main characteristics** of TG-FTIR. The on-line combination TG-FTIR makes it possible to identify all molecules or bonds with an oscillating *dipole*. The technique is especially useful for smaller molecules where the high specificity of strong IR absorption bands makes up for the relatively low sensitivity of IR detection. It is rather difficult to use IR to analyse mixtures of compounds with similar functional groups or mixtures of weak IR absorbers in the presence of strong absorbers. In addition to the chemical composition of the evolved components, the technique also provides information on the sequence and kinetics of the mass-loss process, which may not show up in the mass-loss curve. The TG-FTIR user may also take advantage of the non-destructive nature of FTIR analysis. After FTIR detection the gases may be "cold trapped" for further analysis using complementary methods (GC, GC-MS, *etc.*). For TG-FTIR typically a few mg are sufficient.

If the calculated weight loss of observed gases is lower than that measured by TGA, then it can be inferred that other gases are being evolved that are FTIR blind. In fact, a limitation of FTIR lies in detecting only non-symmetrical gas molecules. Gases without IR absorbance (*e.g.* O_2, N_2) cannot be detected and FTIR does not readily distinguish hydrocarbons above C_3H_8.

Another limitation of the technique is that only the vapour phase is being sampled, not the solid state; it is not possible to discern reactions that occur in the solid except by inference from the volatiles that desorb from the sample pan. Where necessary, it is important to analyse also the solid residue at several temperatures in order to ascertain the correlation between the evolved gases and rearrangements which occur in the solid, which permit this evolution.

Although spectral subtraction and spectral search can aid in the identification of evolved gases, which are often a mixture of products, for unambiguous identification of unknown volatiles more powerful methods are required, such as incorporation of a parallel mass spectrometer onto the FTIR stage of a TG-FTIR. The thermal decomposition products of TGA are then often collected in a Tenax (adsorbent charcoal) trap. After desorption, the products are separated on a GC and the sample split, with most going to an IR spectrometer and a much smaller fraction to a mass spectrometer [345]. Also other experimental schemes may be envisaged [346], *cfr.* Schemes 2.2 and 2.3. Such complex systems are neither inexpensive nor can be used routinely.

TG-FTIR allows **quantitative analysis** even when more than one component of interest pyrolyses during a single weight loss [347,348]. However, there are two major difficulties in this area: *(i)* non-linear absorbance against concentration (because of low data resolution); and *(ii)* measurement of changing concentration profiles with possible overlap of co-added scan sets. TG-FTIR in polymer degradation is described in refs. [335,349–352] and has been reviewed by Mittleman *et al.* [353] and Mullens [346].

FTIR uses much lower excitation energy than MS and can therefore detect larger functional groups in

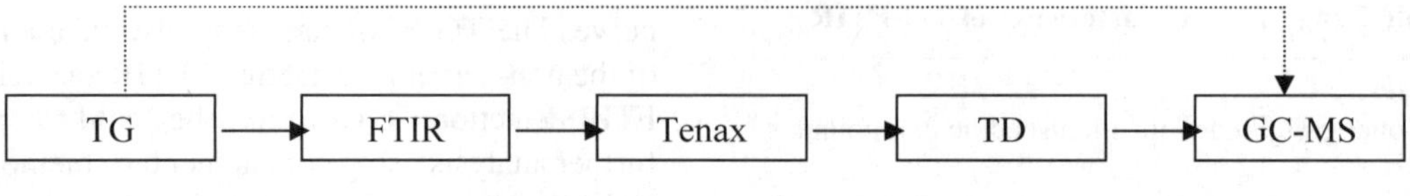

Scheme 2.2. Flow-chart of a combination of techniques (— on-line; · · · off-line).

evolved gases from TG experiments, such as high-boiling oligomers and heavy tar products, which can be analysed as fine aerosols in a gas stream [328, 354]. TG-FTIR also distinguishes structural isomers [355]. At variance to TG-FTIR, TG-MS requires special high-vacuum capabilities for MS and more stringent operating conditions but TG-MS exhibits detection levels which are several orders of magnitude lower than FTIR (pg and sub-μg ranges, respectively). In some conditions, MS results can be misleading because of secondary products resulting from ion fragmentation [356]. Like IR, mass spectrometry has the capability of analysing simultaneously and independently a number of volatile components from a weight loss step. Yet, MS identifies each individual compound and not a class of compounds of the same functional group characteristics. Both MS and FTIR need the support of spectral libraries.

TG-FTIR emission spectroscopy may be used to study the chemical nature of the surface of the sample.

Applications
TG-FTIR has become quite a popular, versatile, cost-effective and informative instrument for modern polymer analysts concerned with thermal decompositions, oxidation processes, desorption behaviour, effectiveness of additives, aging processes, characterisation of raw materials and detection of residues. The growth rate of TG-FTIR instrumentation currently exceeds that of TG-MS.

Some more specific polymer chemistry applications for TG-FTIR are solvent and water retention, curing and vulcanisation reactions, isothermal ageing, product stability, identification of base polymer type and additives (plasticisers, mould lubricants, blowing agents, antioxidants, flame retardants, processing aids, *etc.*) and safety concerns (processing, product safety, product liability, fire hazards) [357]. A wide variety of polymers and elastomers has been studied by TG-FTIR [353,358,359]. The potential applications of an integrated TG-FTIR system were discussed by various authors [346,357].

During the early stages of developmental **polymer processing operations**, various additive packages and processing aids may be explored and evaluated. The exact identity of potential VOCs may not be known. By combining TG with some form of gas analysis, such as IR or MS, the composition and the relative amounts of volatiles evolved under typical processing conditions can be determined.

Used as an **evolved gas analysis technique**, TG-FTIR permits limited identification of neutrals desorbed from a matrix subjected to a TG regime on the basis of functional group recognition. TG-FTIR is used to a great extent to identify the off-gases of a polymer at different stages in the decomposition process. The nature of the volatile products is especially important from an environmental point of view. TG-FTIR is capable of identifying and measuring **solvents** that might be present in polymer samples. TG-FTIR examination of a polybutadiene sample with a high proportion of inorganic fillers and spectral subtraction procedures identified water and a plasticiser at 200°C, and CO_2, CO, H_2O, methane, ethylene, *n*-butane, *n*-pentane and cyclic hydrocarbons at 500°C [290]. The results indicate that a single sample weight loss may well correspond to a very complex mixture of evolved gases. Spectral subtraction and spectral search aid the identification of evolved gases.

Wilkie [339] has recently reviewed the use of TG-FTIR for studying polymer degradation. A significant amount of work has been carried out on the interaction of PMMA with **additives**, including red phosphorous, Wilkinson's salt, $(PPh_3)_3RhCl$, Ph_xSnCl_{4-x} ($x = 0$–4), Ph_2S_2, Nafions, various transition metal halides and copolymers of MMA with 2-sulfoethylmethacrylate [339]. Additive-polymer interactions were spotted in thermal degradation of PMMA/$MnCl_2$ [359a]. Wilkie *et al.* [360, 361] have also used TG-FTIR to examine **copolymers**, which may give a high yield of non-volatile compounds on thermolysis, with the object of developing new flame retardants. The thermal stability of grafts of char-forming monomers, such as

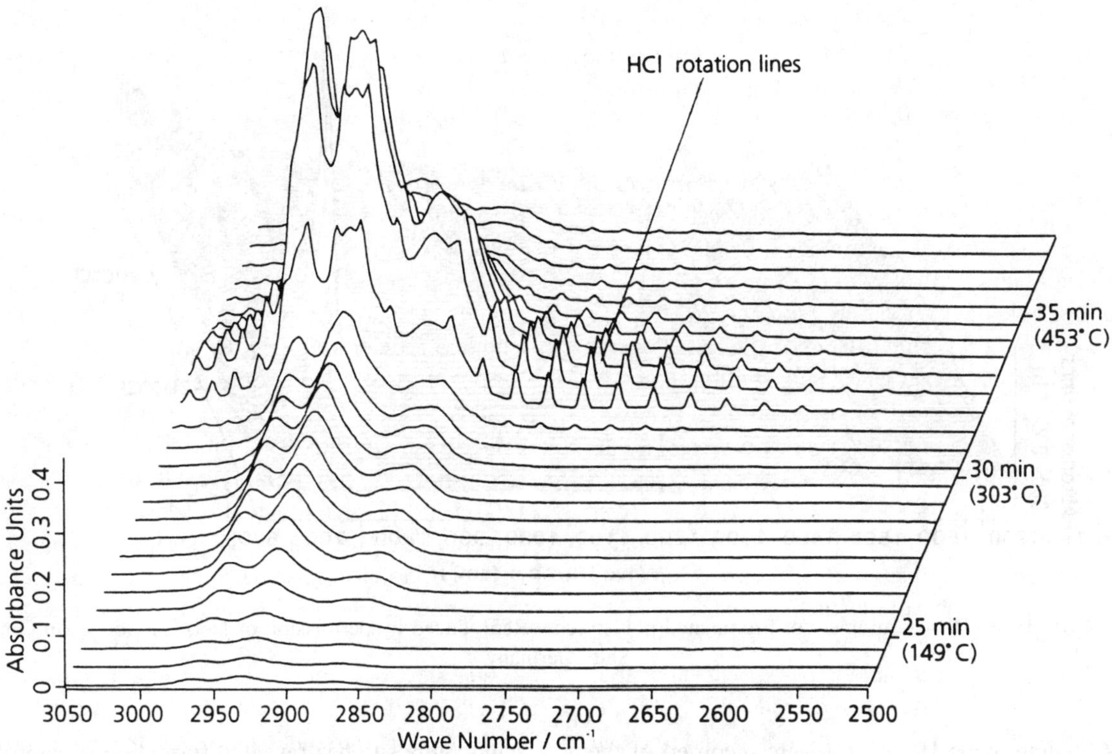

Fig. 2.19. HCl rotation lines during PVC decomposition. After Post *et al.* [362]. Reproduced from *Thermochimica Acta* **263**, E. Post *et al.*, 1–6 (1995), with permission from Elsevier.

sodium methacrylate or acrylonitrile onto butadiene-containing polymers, polystyrene and polyamide-6, was equally assessed by means of TG-FTIR. Similar analysis of ABS grafted with methacrylic acid showed evolution of butadiene and aromatics from the graft copolymer some 90–100°C higher than in virgin ABS [360].

TG-FTIR has also been used to study the **stabilising action** of 3-(2,4-dibromophenylazo)-9-(2,3-epoxypropane) carbazole on the degradation of PVC [363]. Post *et al.* [362] have reported TG-FTIR measurements of recyclable polymer automobile undercoatings containing PVC. Rapid decomposition of PVC begins abruptly at 300°C (extrapolated onset), at lower temperature for the undercoating materials. Figure 2.19 is a 3D representation of the FTIR spectra in the wavenumber range of 3050–2500 cm^{-1} during this TG step. Integration of the rotation lines gives information about the relative HCl emission of different samples. TG-FTIR was also applied to plasticised PVC [364]. As to **flame retardant** applications, Fig. 2.20 shows evolution of bromobenzene at 200°C in TG-FTIR measurement (792 cm^{-1}) of a brominated polystyrene sample [365]. On-the-spot TG-FTIR of PBT/octabromodiphenyl ether (MW 801 Da) detected the brominated diphenylether flame retardant at 275°C and terephthalic acid (the starting monomer of PBT) at 425°C [310]. Similar high-MW species have never been reported in TG-MS experiments; the flame retardant was not observed in off-line TG-GC-FTIR-MS analysis. In an examination of an ABS/PC blend with 8% triphenylphosphate (TPP), in addition to the EGP, the SGPs for the specific wavenumber windows of TPP (900–1200 cm^{-1}), aromatic compounds (3000–3100 cm^{-1}), and carbon oxides originating from PC (2200–2300 cm^{-1}), were obtained. TPP evolving first was detected at about 150°C (detection limit 0.5 μg/s) [310]. Anthony [366] has used FT-IR spectroscopy to examine TG residues and diffuse reflectance as the means of sample preparation for the study of interactions between pyromellitate polyesters (**smoke suppressants**) and polyurethane foams. This was achieved by interrupting the thermal analysis at selected points on the TG curve. In

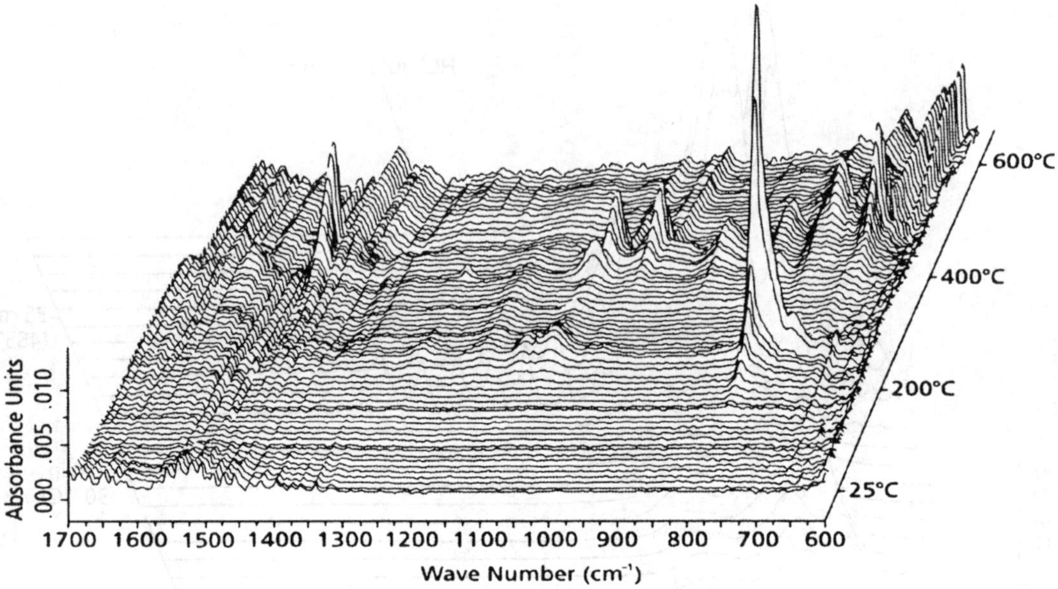

Fig. 2.20. TG-FTIR measurement of brominated polystyrene. Reproduced by permission of Netzsch-Gerätebau GmbH, Selb, Germany.

this off-line mode IR spectra were recorded of the residues at progressive stages of thermal-oxidative degradation. Reaction between the liberated aromatic amine with the pyromellitate ester forms a thermally stable polyimide. The pyromellitimide structure stabilises the main smoke and toxic gas precursors formed during combustion.

Post *et al.* [367] have used TG-FTIR to study outgassing of a plasticiser (type and amount) from an EPDM compound. The plasticiser emerged in the first mass-loss step at 285.1°C, which was identified as diisobutyl adipate by on-line infrared. Jansen *et al.* [310] described on-the-spot TG-FTIR of a masterbatch of 20% **silicone oil** in PS and the analysis of DEHP plasticised PVC; DEHP was clearly revealed by the IR spectrum at 275°C [341]. Polystyrene containing the **blowing agent** azodicarbonamide releases CO, CO_2, NH_3 and formamide at 225°C. The main gaseous product, N_2, cannot be detected, as it does not absorb in IR. The same authors [310] used the technique for the study of a CB-filled styrene–butadiene rubber (SBR) providing information about the composition, including organic additives, polymers, carbon-black and inorganic fillers. At 250°C water, CS_2 and morpholine were detected. The latter two components are degradation products of the **accelerator** used, 2-(morpholinothio) benzothiazole. TG-FTIR examination of PA6/clay nanocomposite has revealed formation of caprolactam, hydrocarbons, CO_2, CO, NH_3 and H_2O [368]. TG-FTIR has also been used to analyse the emission of volatiles during PUR powder paint drying [369] and to study the thermal degradation behaviour of polyurethanes blended with the flame retardant poly(bispropoxyphosphazene) [370].

TG-FTIR has been used to study talc–Irganox 1010 interactions [371]. According to the nature of talc three states (free, surface and adsorbed) of Irganox 1010 molecules could be identified in the presence of the filler. TG-FTIR has also been employed for the study of zinc stearate [372], of wood as a **filler** to thermoset plastics as well as for outgassing, which leads to bubbles in painted plastic surfaces and potentially toxic gases. Ezrin *et al.* [255] have reported **troubleshooting** by means of TG and (off-line) FTIR. For other TG-FTIR and TG-MS applications, *cfr.* ref. [373]. TG-FTIR emission spectroscopy can be used to study the chemical nature of the surface of the sample. Mullens *et al.* [359] have recently reviewed TG-FTIR applications.

2.1.5.2. *Thermolysis – Fourier Transform Infrared Spectroscopy*

Principles and Characteristics

For FTIR spectroscopic monitoring of evolved gases other straightforward instrumentation and methodology may be used, namely **slow thermolysis-FTIR** [374]. In this experimental set-up, the sample is placed in stainless steel tubing, connected to an empty GC column in a GC oven with temperature programming. The gases evolved from the sample are passed through a gold-coated lightpipe and monitored with a MCT detector and the IR data are collected with standard GC-FTIR software. At variance to TG-FTIR no weight information is gathered.

Brill [342,375] has developed **fast thermolysis/FTIR** as a new combination of thermal analysis and spectroscopy, which is to be positioned between TG-FTIR and fast pyrolysis/FTIR. Conventional pyrolysis, as a technique for introducing solid or non-volatile samples to gas chromatography, does not permit real time *in situ* analysis of thermal decomposition gases that exist during ignition, combustion and explosion of a bulk material. By imposing a fast heating rate (up to 200°C/s) a new dimension in hyphenated thermal analysis and spectroscopy is gained. The heating rates of fast thermolysis/FTIR fall between those of conventional thermal analysis and combustion (thousands of degrees per second) (Table 2.1).

In thermolysis FTIR the sample (typically 200 μg) is loaded onto a quartz boat, which is inserted straight into a platinum coil filament. With the beam focused several mm above the filament surface, the IR-active gas products from the fast heated sample can be detected in near real-time. Fast thermolysis/FTIR spectroscopy combines rapid-scan FTIR (20 scans/s) with pyrolysis of a material and real-time measurement of the gas spectra [376]. Temperature, mass changes and spectral data of IR active gases are thus measured simultaneously as a function of time during the rapid heating phase. High-resolution vapour phase libraries are used for identification.

Thermolysis/FTIR is usually carried out in two **measurement modes**, with moderate heating rate (20°C/min; TGA mode) or fast heating rate (exceeding 100°C/s, rapid scan mode) [342,377]. The Brill IR cell (kept in Ar atmoshpere and heated to prevent condensation) allows the same kind of analysis as TG-FTIR except that it produces a total degradation product, which is a complete break-up of the polymeric material. Fast thermolysis/FTIR provides insight into chemical and physical processes where a high heating rate exists, as during ignition, combustion, or explosion of a bulk material [376]. While fast thermolysis/FTIR is the rapid-heating complement to conventional thermal analysis techniques, such as TGA, of course the quantitative aspects of TGA depend on maintaining quasi-equilibrium heat transfer. As this is difficult to achieve at a high heating rate, fast thermolysis/FTIR cannot act as a quantitative analytical tool.

Variations on the theme of fast thermolysis/FTIR spectroscopy include temperature profiling/FTIR spectroscopy, in which the temperature changes of the condensed phase are measured simultaneously with the gas evolution; fast-heat-and-hold/FTIR spectroscopy [378], in which isothermal decomposition is studied following rapid heating to a selected temperature and Simultaneous Mass and Temperature Change (SMATCH)/FTIR spectroscopy [379], which has clearly established the connection between the microscale fast thermolysis approach and steady-state combustion of the bulk material. In T-jump/FTIR spectroscopy the thermal decomposition of a material can be studied isothermally after heating at 2000°C/s [376].

Advantages of the real-time/fast heating approach are *in situ* analysis of dynamically changing processes and detection of some relatively reactive molecules that are lost to side reactions at slower heating rates or with time delays in the detection step. Thus reaction schemes different from those occurring with slow heating can be studied. Pressure and composition of the atmosphere can be set as desired to gain an additional variable. The small sample size permits studies to be performed safely. A **drawback** of fast thermolysis techniques is that the sample decomposes under non-isothermal conditions.

Applications

Provder *et al.* [374] have applied evolved gas analysis-FTIR (essentially slow thermolysis-FTIR) to **chemical cure**. Fast thermolysis/FTIR studies of the intumescent **flame retardant** melamine cyanurate (MC) and PA6.6/10 wt.% MC have been reported [377]. In order to provide a basis for understanding the combustion behaviour of flame retarded polymers a study of the decomposition products at various heating rates is quite useful. Fast thermolysis/FTIR reveals a considerable difference in the components of decomposed gases produced at different heating rates.

Brill *et al.* [376,380] have illustrated the application of T-jump/FTIR spectroscopy with rapid thermolysis of various organoazide polymers and hydroxyl-terminated polybutadiene with and without TiO_2 and melamine additives. The T-jump/FTIR technique determines the chemistry of fast pyrolysis.

2.1.5.3. *Thermogravimetry – Mass Spectrometry*

Principles and Characteristics

Thermal events may bring about a change in the mass of a sample. The need then often arises to correlate thermal behaviour with the underlying chemistry (outgassing, thermostability, degradation) and physics (change in colour, blooming, cracking, foaming, migration) in a simultaneous mode. Coupling with MS adds the **chemical analytical features** allowing the chemist to assign the detected weight losses to specific evolved gases, thereby correlating chemical information with the thermal event. However, TG-MS is also an excellent starting point for endoscopic, audiometric and magnetometric extensions. Simultaneous TG-MS is therefore a very powerful hyphenated technique combining the direct measurement of weight loss as a function of temperature with the use of a sensitive spectrometric detector. In addition to the weight loss information, mass spectrometry permits temporal resolution of the gases that are evolved during thermal or thermo-oxidative degradation of a polymer in controlled atmospheric conditions. As in TG-MS the products of degradation are flushed out with the purge gas, this greatly reduces the possibility of recombination, as opposed to cold trap experiments.

While on the one hand interpretation of TG data is facilitated by the mass spectrometric information, on the other hand TG data ease quantification of MS results. Molecular weight information is collected concerning the **evolved gases**, which are responsible for the detected weight losses. Evolved gases can be identified in sequential order and a specific component may be associated with a specific weight loss. TG-MS involves two distinct axes of information with significantly different time frames. For each sampled point along the TG axis an entire mass spectrum is acquired. The rate of data collection along the MS axis far exceeds the sampling rate along the TG axis. Variables for the TG-MS experimentalist are: *(i)* nature of the thermogravimetric equipment (*cfr.* Chp. 2.1.3); *(ii)* interface; *(iii)* mass analyser type; and *(iv)* ionisation mode [381]. These variables give rise to a multitude of hardware solutions, developed since the usefulness of coupling MS to TG was suggested first in 1965 [382,383].

Bart *et al.* [311] have reviewed the essential **design criteria** for TG-MS allowing routine application for polymers. As to hardware, in TG-MS couplings both vacuum and gas atmosphere mass flow **thermobalances** have been used. Various essential design principles are to be respected in coupling to a mass spectrometer. Kaisersberger *et al.* [338] have described the basic features of coupling systems for TG-MS, TG-FTIR and TG-GC.

Essentially three types of **mass spectrometers** have been used in combination with TG, namely ToF, QMS (mostly) and magnetic sector instruments [381], but no ion traps. Also the type of **ionisation mode** has varied in combination to coupling to TG, usually electron impact (EI) or (atmospheric pressure) chemical ionisation (CI). The most common ionisation method in TG-MS is EI using high-energy electrons (70–90 eV). TG-EIMS is characterised by complex fragmentation and difficult mixture analysis; chemometrics is wanted. Although lower ionising energies (10–30 eV) enhance the relative intensity of molecular ion peaks and reduce the number and relative abundancies of the lower-MW fragment ions as well as the fragmentation, the trade-off is a marked decrease in sensitivity with decreasing electron energy.

The use of CI overcomes some of the limitations of EIMS in case of co-evolution. The feasibility of characterising evolved volatiles by TG-CIMS has been examined [384–386]. CIMS has the advantage of ease of interpretation (due to better control on the complexity of the spectral fragmentation pattern) and of being able to operate at higher input pressures. A TG-CIMS system in which the thermobalance works under normal pressure, connected to a mass spectrometer working at elevated pressure ($p = 1$ mbar), offers the possibility of using the purge gas of the thermobalance as a reaction gas in the chemical ionisation source [385]. This reduces interface problems and restricts the fragmentation of released volatile compounds. In turn, this leads to simple cracking patterns, intensive (quasi) molecular ions and therefore easy-to-interpret spectra, especially useful in mixture analysis. However, molecular mass information alone is insufficient for structural assignment. TG-CIMS has found limited use so far and appears to be restricted to specific cases. Contrary to PyMS [387] no comparative study of EI and CI techniques in TG-MS has been carried out.

In a recent development Lindinger *et al.* [388] and Bassi *et al.* [389] have described a **soft ionisation** (SI) gas analyser mass spectrometer (up to 500 Da), which seems eminently suited for coupling to TG and solving some of the aforementioned problems in mixture analysis, as occur in direct polymer/additive deformulation by means of TG-MS.

IMR-MS with interchangeable ionisation modes is characterised by specific fragmentation, molecular ion peaks (and some secondary peaks), direct molecular identification and 100 ppb sensitivity (outperforming TG-EIMS). TG-SI/EIMS would be a high performance, relatively low-cost TG-MS instrument with excellent evolved gas separation capabilities (partly within the TG and partly within the MS part of the hyphenated instrument) and identifying power. It would also appear that TG-ToFMS [390–392] might soon enjoy a comeback.

It is obvious from the history of TG-MS [393] that the **interface** is of crucial importance, fulfils several functions and poses several problems. It operates simultaneously as a gas-input system for the mass analyser and (usually) as a pressure reduction system. Within the interface, conditions are converted from the high temperature and (usually) atmospheric pressure of TG to the room temperature and (usually) high-vacuum conditions in the mass analyser. Both the temperature and geometry of the interface region influence the coupling. The main aspects of the flow in the thermobalance, relevant to hyphenated techniques, are understood.

Mass spectrometry coupling can be achieved by connecting a heated **capillary** at the end of the gas-flow system of a thermobalance or by means of a direct **nozzle/skimmer coupling** integrated into the furnace of the thermal analyser. The high sensitivity of the skimmer coupling as compared to the capillary coupling of mass spectrometers is ascribed to the perfect gas flow conditions (molecular beam), short transfer path and elimination of condensation effects [338]. In the latest design (Fig. 2.21), the thermobalance furnace stands in direct contact with two chambers (at 10 Pa and 10^{-3} Pa) via a small diameter-sampling orifice (several dozen μm) through which a molecular flow reaches supersonic speed within a few μsec and is directed straight to the mass spectrometer. The system allows high resolution and sensitivity for masses up to 1024 Da [394]. Capillary couplings are usually restricted to some 200°C (interface temperature) [367]. However, upon proper design of both furnace tube and interface condensation memory effects can be overcome [395,396]. Low volatility compounds are the least favoured in reaching the mass spectrometer. Condensation in the thermobalance is the main potential source for fouling and "memory" effects [395]. In the past, high

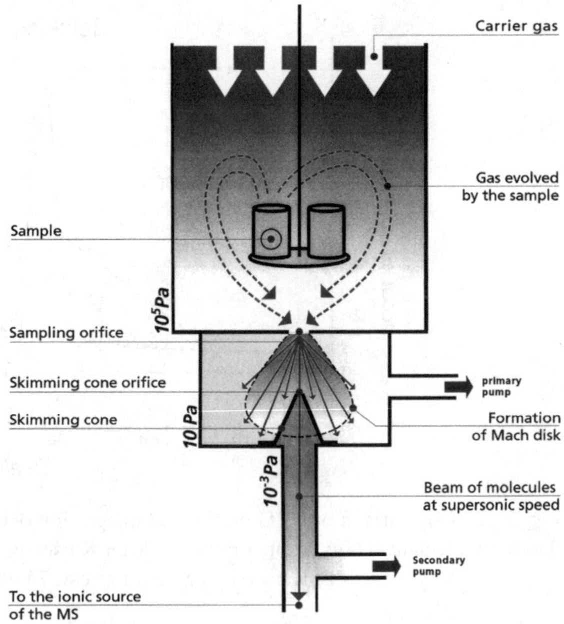

Fig. 2.21. Schematic diagram of the SuperSonic system for coupling a high-temperature microbalance to a mass spectrometer. Reproduced by permission of Setaram, Caluire, France.

masses ($m/z > 200$) were seldom studied by TG-MS (mass range of 1–800 Da is currently felt appropriate) since molecules of higher molecular masses are usually not volatile under atmospheric conditions. This limits the usefulness of TG-MS for direct polymer/additive deformulation. It has been reported that large fragments (*e.g.* $m/z = 312$) are lost in capillary couplings but are easily observed in STA-QMS skimmer couplings. However, using an appropriate capillary interface Wenz *et al.* [397,398] have successfully detected the parent ion ($m/z = 447$) of Tinuvin 234 by TG-MS.

Software allows qualitative comparison of the shape of the DTG curve with integration over all detected mass numbers (total-ion curve) [338]. This can help to assure that the selected mass range for measurement is sufficient to describe all weight loss effects by corresponding MS signals (*i.e.* DTG and total-ion curve show parallel shape), and can be used to observe insufficient mass range, retention and condensation effects (*i.e.* non-parallel profiles of DTG and total-ion curve) (Fig. 2.22).

In TG analysis of polymers handling of large quantities of material released during sample decomposition calls special attention. Typically, in

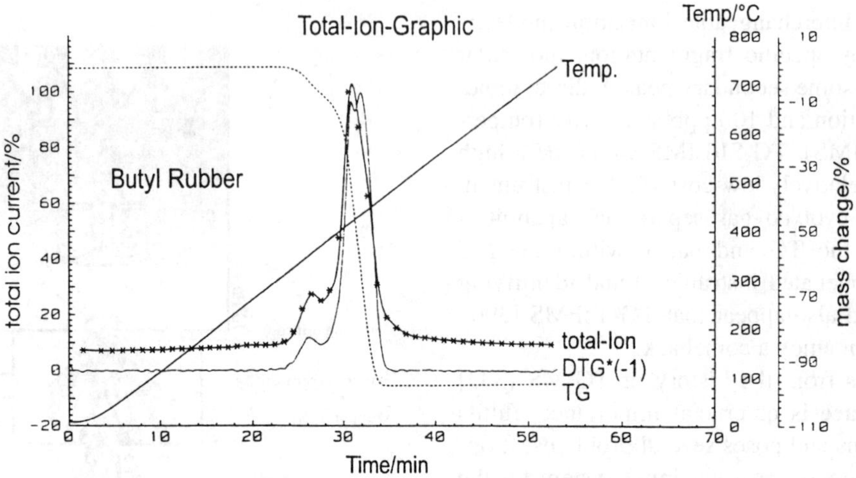

Fig. 2.22. Comparison of DTG profile and integration over all detected mass numbers (total ion curve) for thermal degradation of a technical butyl rubber mixture. After Kaisersberger and Post [338]. Reproduced from *Thermochimica Acta* **295**, E. Kaisersberger and E. Post, 75–93 (1997), with permission from Elsevier.

problems of outgassing, the components of interest are low-MW compounds (entrained solvents and plasticisers) in low concentrations, which evolve before the sample reaches its own anaerobic decomposition temperature. In other processes, large volumes of materials may be expelled as a mixture of decomposition products and particulates. A properly designed TG-MS interface cell for routine purposes must therefore be capable of handling both trace components as well as any large quantities of material released during sample decomposition. Loss of gas by condensation at cold spots, low detection sensitivity because of heavy dilution with purge gas, low time and temperature resolution because of long transfer times and mixing with the purge gas by diffusion and by uncontrolled flow conditions, and variation in gas composition in the coupling interface should be avoided.

It is important to realise that mass spectrometric measurements in TG-MS are not performed directly on the polymer but only evolved gases are detected and identified. Factors influencing component loss from polymeric matrices are volatility, rate of diffusion, solubility in the polymeric matrix, flow-rate, temperature, ΔT, sample thickness, *etc.* Therefore, information about the polymeric matrix is obtained in an *indirect* way, and concerns especially the thermal stability, degradation mechanism and kinetics, performance behaviour, reactivity, and **analysis of volatile additives**, residuals, monomer occlusions

Table 2.15. Main features of TG-MS[a]

Advantages:
- Minimal sample preparation
- Short analysis time
- High detection sensitivity
- Discrimination between various weight change processes
- Quantitation
- Evolved gas analysis (trapped solvents, unreacted reagents, degradation products)
- Wide applicability

Disadvantages:
- Limited identification of evolved gas and residuals
- Small sample size (inhomogeneities)
- Lack of standardisation
- Experimentally vulnerable
- Insufficient interlaboratory reproducibility
- Dependency on gas flow-rate, sample size and heating time
- Vapour fractionation and condensation
- High cost of interface

[a] After Raemaekers and Bart [311]. Reproduced from *Thermochim. Acta* **295**, K.G.H. Raemaekers and J.C.J. Bart, 1–58. Copyright (1997), with permission from Elsevier.

and trace impurities. Attempts to gather information simultaneously about evolved gases and residue have already been mentioned (Chp. 2.1.5).

Table 2.15 shows the **main characteristics** of TG-MS. TG-MS provides direct physical and chemical information simultaneously as a function of temperature, in dynamic mode (as opposed to techniques in static mode). Experimental TG-MS conditions for the examination of a material can be varied (high vacuum to high pressure), at difference to more restricted options in pyrolysis. Proven performance and complexity of tasks in the characterisation of (commercial) plastics, fibres, paints and other polymeric materials have made TG-MS a desirable analytical tool, in competition with methods such as PyGC [399] and other techniques. TG-MS is especially useful for samples which cannot easily be studied by spectroscopic means, such as CB-filled elastomers. Although TG-MS is experimentally vulnerable (*e.g.* O_2 leakage) the presence of MS is an autocheck on proper operation.

Major drawbacks of TG-MS are cost and method standardisation. Although one cannot properly speak of a standardised TG-MS coupling technique this does not necessary constitute a problem. As in case of PyMS there are good reasons to expect that a variety of TG-MS couplings have a future.

Both TG-MS and PyMS are subject to fouling of the detector, which may impair **quantification**. This is less serious in case of TG-MS, where the mass spectrometer is only required to yield correct relative data (quantification via TG), at variance to PyMS where absolute mass spectra data are necessary for quantification. Courtault [400] has described quantitative aspects of TG-MS coupling, which is still difficult matter. Quantitation of TG-MS data requires calibration of the system, *i.e.* determination of the relationship between observed intensities of the ion currents and the amount of the analysed species. Quantatitive work with MS couplings has recently been treated very clearly by Maciejewski *et al.* [212, 401], also introducing a new experimental technology, **Pulse Thermal Analysis** (PTA). PTA enables the introduction of a well-defined amount of a gas (including oxygen) to the system at any temperature (non-isothermal) and/or time (isothermal mode). Injected pulses can be used as a reference for the quantification of the signals originating from the evolution of gas(es) formed during decomposition of solids. A linear dependence between the amount of injected gas and the intensity of m.s. signals enables quantification of mass spectroscopic data with an accuracy for evolved species below 0.01 wt.%. The possibility of exact **calibration** of the MS signal by means of PTA increases significantly the potential of coupled TA-MS methods. Calibration and **interlaboratory reproducibility** are issues which require further attention. Statheropoulos *et al.* [402] have proposed a procedure for evaluation of the performance of a TG-MS system, and for interlaboratory comparisons. The proposed quantitative evaluation procedure includes measurements of mass-flow stability, evolved-gas transfer delay and the evolved gas condensation effect.

Given the limited component separation capability of thermal methods, **single-stage TG-MS** instrumentation is in principle not suited to identify, although it has this capability in simple cases (evolution of low-MW gases, such as CO_2, CO, formaldehyde, *etc.*). Consequently, in its basic form, the technique is more fit to degradation studies than to characterisation of higher-MW species (volatile oligomers, *etc.*). In order to *unambiguously* identify a component in a mixture without forgoing direct TG-mass spectral integrity, **MS/MS techniques** are an obvious choice. Shushan *et al.* [403,404] have described a TG-APCI-MS/MS system for evolved gas analysis in which the soft ionisation mode minimises further fragmentation of gases evolved by thermal degradation. However, this solution adds considerably to the cost of the analysis. Alders *et al.* [381] have argued that HRTG-EI/SI-QMS extended with multivariate data analysis is a desirable option for the near future. Tas *et al.* [193] have recently shown successfully PCA analysis of TG-MS data. Other **new developments** are video-imaging (VMI) TG-MS [282] and high vacuum (10^{-5} mbar) TG-MS. Already Affolter *et al.* [405] have shown the beneficial effects of slightly reduced pressure (1 mbar) on desorption.

A comparison between TG-MS and other EGA techniques been described [311,346]. Kaisersberger *et al.* [338] have compared **TG-MS and TG-FTIR** for evolved gas analysis. For the detection of trace amounts of volatiles, mass spectrometry (in particular ToF-MS) shows, in general, the higher sensitivity with detection limits in the ppb range. TG-FTIR/MS (parallel coupling) was also described [405a].

Applications

The complexity of thermal degradative processes and the great variety of additives present in polymer formulations benefit from the combination of TG with other analytical techniques. This is particularly true in coupling to an identifying technique, such as MS or IR. TG-MS has been used in a wide variety of qualitative and quantitative industrial problem-solving cases (Table 2.16).

Table 2.16. Problem solving areas for TG-MS[a]

(a)	*Thermal stability and degradation studies:*
	testing of thermal and thermo-oxidative degradation of polymers;
	testing close-response relationships of additives (stabilisers);
	identification of degradation products.
(b)	*Structural characterisation and chemical analysis:*
	identity, equivalency and structure of polymeric materials;
	fingerprint identification;
	compositional analysis for identification of components in blends of additives, *etc.*
(c)	*Analysis of evolved gases during synthesis, processing and recycling:*
	outgassing phenomena;
	analysis of additives or processing agents;
	determination of the effect of stabilisers;
	environmental impact of polymer degradation;
	health protection studies;
	product safety studies.

[a] After Raemaekers and Bart [311]. Reproduced from *Thermochim. Acta* **295**, K.G.H. Raemaekers and J.C.J. Bart, 1–58. Copyright (1997), with permission from Elsevier.

Materials for TG-MS may take various forms (powder, granulate, film, fibre, *etc.*). Sample size in the classical TG-MS is about 10–20 mg, but a macro TA-MS/GC-MS can handle up to 500 g.

Product development:
The **industrial problem-solving capability** of TG-MS is highly valued. Kleineberg *et al.* [391] have reported early application of TG-ToFMS for the evaluation of the toxicity potential in normal use and catastrophic situations of some 300 flame retardants materials employed in interiors of passenger and cargo aircraft. Advantage was taken of the inherent high speed scanning capability of ToF-MS. Collection of a complete history of the evolved material from the sample at distinct points of weight loss and temperature enabled the toxicologist to relate conventional TG information to the unequivocal identification of potentially toxic thermal decomposition products. TG-ToFMS of a carboxynitroso rubber showed abrupt, complete decomposition at 292°C. The mass spectrum was interpreted on the basis of two primary decomposition products, namely carbonyl fluoride (m/z 66, 47, 50, 31, 19) and perfluoro-N-methylmethylenimine (m/z 133, 69, 114, 31, 50, 45, 26, 57, 64, 12, 19), the secondary reaction products CO_2 (m/z 44) and trifluoromethylisocyanate (m/z 111) and corrosion products (HF, SiF_4, $(CF_3)_2NH$; m/z 85, *etc.*). Quantitative determination was achieved through correlation of MS and TG data.

Holzapfel [406] has used TG-MS to define moulding conditions (T, t) for polymeric material in order to minimise degradation during processing of both the polymer and the added cross-linking agent triallylisocyanurate (TAIC) in a **toner** for high performance laser printers. TG-MS has also been applied to characterise polymer derivatives as **fuel oil additives** with respect to the propensity to volatilise or oxidise under end-use conditions.

Lehrle *et al.* [407] have studied **controlled release** of the volatile antioxidant butylated hydroxytoluene (BHT) from cross-linked alginate matrix particles. TG-MS results demonstrate that controlled release can be successfully achieved (*i.e.* BHT is retained beyond its normal evolution temperature); polyisoprene rubber is more resistant to oxidation when protected in this way than by the equivalent concentration of unencapsulated antioxidant. Tsuneto *et al.* [386] have analysed evolved gases in a process for removing **binder** polymer (PBMA and LLDPE) from ceramics obtained by injection moulding.

Also several EPDM products were studied by means of TG-MS [311]. TG/DTG of an EPDM without filler and plasticiser shows that during the maximum weight loss phenomenon ENB ($m/z = 66$, 91, 105), aliphatics ($m/z = 43$, 56, 69) and olefins (ethene: $m/z = 26$, 27; propene: $m/z = 40$, 41, 42) are detected. The dynamic DTG and MS curves in inert atmosphere of an EPDM compound charged with oil, filler and carbon-black, indicate loss of oil (max. at 336°C), thermal stability of the polymer up to about 420°C (maximum decomposition at 485°C), and decarboxylation of the filler at 730°C (CO_2: $m/z = 12$, 44); finally, above 900°C in O_2 atmosphere carbon-black is detected. The same authors [311] have reported a TG-MS study of EPDM-SBR blends.

Analysis of additives and volatiles:
Knowledge of compounding ingredients is needed for a number of applications [2]: *(i)* verification of ingredients in compounded stocks; *(ii)* reconstruction of formulations in unknown materials; *(iii)* investigation of manufacturing problems; *(iv)* identification of odorants or irritants evolving from polymeric materials during processing or use; and *(v)*

product quality studies. Identification of these ingredients in a compounded polymer by means of TG-MS is a difficult analytical task, which is made complex by a number of factors, in particular: *(i)* wide variety of additive types, varying greatly in molecular weight, volatility and polarity; *(ii)* lability of many additives; *(iii)* compounding of complex mixture of additives; *(iv)* low organic additive concentrations (<1–5%); and *(v)* experimental limitations of TG-MS (maximum MW 500 Da).

As a result of their limited volatility, identification of organic additives in polymers by using TG-MS is considerably more difficult than that of residual volatiles (such as rest monomers and solvents). Actually, TG-MS is an ideal technique for identifying residual volatiles in polymers. The detection of such volatiles (and of other impurities) can often yield clues as to manufacturing processes.

During processing of polymeric materials, especially at elevated temperatures, **outgassing** of low-MW products, possibly accompanied by additives (notably plasticisers) or degradation products, may occur, causing deterioration of the properties of the material. Control of outgassing phenomena is also important in relation to mould contamination and reprocessability, suitability for finishing processes (*e.g.* gluing, welding, lacquering and plating), to product lifetime, admissible temperatures for use, contact and environmental contamination in product applications, or may concern toxicological and aesthetic aspects [345,408]. The study of outgassing phenomena has been very useful for identification of components that may be held responsible for environmental stress cracking (ESC) of plastic products. Through outgassing experiments using TG-MS and related techniques, harmful components causing ESC can be detected and identified [409]. Also problems such as surface crazing, bubble formation, and chemical degradation of the polymer can sometimes be caused by residuals. TG-MS and purge-and-trap (PT) have been used for the quantitative determination of halogenated hydrocarbons (bubbling agents) in polymeric foam insulation materials [410]. TG-MS was found to be more effective than PT GC-MS; in fact, PT measures only the volatile compounds contained in the open elements and not in the closed pores of the polymeric foam.

TG-MS couplings are increasingly used by the **rubber industry**, especially since aromatic plasticisers are toxicologically suspect. Kaisersberger *et al.* [411] have reported detection of nitrosamine precursory compounds during rubber vulcanisation (originating from vulcanisation agents) and the determination of toxic or environmentally damaging exhaust gases during technical burning processes (polycyclic aromatic compounds, PCBs, *etc.*). Post *et al.* [367] used the skimmer-MS coupling in TG-MS measurements to study the outgassing of a plasticiser from an EPDM compound.

In many cases, such as in the determination of highly volatile materials (*e.g.* moisture in nylons or in polysaccharides [412]), or of residual solvents or plasticisers (as in PVB) [413], use of TG-MS is requested. Specifically, there are reports on the entrapment of curing volatiles in bismaleimide laminates [414] and elastomers [415], of plasticisers such as bambuterol hydrochloride [416] or triphenyl phosphate and diethylterephthalate in cellulose acetate [417], on solvent extraction and formaldehyde loss in phenolic resins [418,419]. Carraher *et al.* [420] have used TG-MS to study the degradation of a fluorescent titanium polyether ester dye.

Monitoring of halogen-free FR **thermosetting plastics** materials for the electronic industry (printed circuit boards), identification and quantification of pyrolysis products from new thermosetting plastics are problem-solving areas for TG-MS. The technique has also been applied in relation to the evolution of toxic compounds from PVC and polyurethane foams [421]. Simultaneous and off-line TA-MS were used for the study of various FR polyurethane foams [422,423]. PU foam containing the **flame retardant** tetrakis(2-chloroethyl)ethylenediphosphate decomposes in an oxidative atmosphere at standard pressure in one rapid reaction whereby several highly toxic species are formed; the TG-MS detection limit of this flame retardant was determined by measuring vinylchloride m/z 64 cleavage [424]. In a study of decomposition of phosphate flame retardants three TG-MS methods were compared; differing results were ascribed to discriminating transport phenomena through the capillary interface and secondary reactions [425].

Mullens *et al.* [256] also addressed the determination of gases released during heating of the flame retardant HET-acid (1,4,5,6,7,7-hexachlorobicyclo-[2.2.1]hept-5-en-2,3-dicarboxylic acid) by means of on-line TG-MS, TG-FTIR and (off-line) TG-Tenax-TD-GC-MS. The combination of techniques offers unambiguous identification of the evolved products (CO_2, H_2O, Cl_2, HCl, C_2Cl_4, maleic acid anhydride,

HET-acid anhydride, chlorinated cyclic hydrocarbons and chlorinated unsaturated linear hydrocarbons) as a function of temperature. Some 18 volatile components adsorbed on Tenax during the TG-Tenax-TD-GC-MS experiments of HET-acid were identified. This combined effort in problem solving is a good example of the possibilities of complementary analytical techniques for unambiguous identification of gaseous decomposition products as a function of temperature. TG-MS was also used for ecotoxicological tests (*e.g.* of incineration products from the London Underground and Düsseldorf Airport fires). Comparison of the thermal stability of various polyurethanes illustrates the use of TG-MS in evaluation of similar materials for heat sensitive applications. In other TG-MS applications the pattern of evolution of styrene, butadiene and acrylonitrile as a function of temperature has provided a unique way for classifying different ABS types. Loss of the **antioxidant** BHT has been detected by MS preceding ethylene vinyl acetate copolymer degradation [426]; BHT could be identified at a concentration level of 20 ppm. The power of TG-MS is further illustrated by identification of fifteen volatile products in polyimide resins [427], amongst which unwanted solvents, indicating shortcomings in the synthesis route.

In spite of the reported use of TG-MS in additive identification in (competitor) products (quantitative), analysis of additive packages is usually carried out with procedures not routinely including TG-MS [428]. This is on account of the low additive concentrations on the one hand, the limited resolution ability of thermal methods, the limited molar masses transiting through the (heated) TG-MS interfaces, and the availability of a broad variety of performing alternative techniques on the other hand. In view of their low concentrations, analysis of additives using TG-MS equipment is often carried out with a condensation trap [429], in which there is no dilution of the evolved samples.

In comparison with the widespread use of single-stage MS in chemical analysis and in polymer analysis in particular [430], there has been little use of TG-MS/MS. For this analytical tool three specific areas can be considered: *(i)* identification of unknown organic additives in compounded polymers [431]; *(ii)* identification of volatile pyrolysates in polymer pyrolysis studies [403,404]; and *(iii)* characterisation of individual oligomers in low-MW polymers [432].

TG-MS applications were reviewed [311,346].

2.1.5.4. *Multihyphenated Thermogravimetry–Differential Scanning Calorimetry Techniques*

Principles and Characteristics

Simultaneous thermal analysis techniques, such as TG-DSC/DTA offer vital information on polymer structure based on heat flow behaviour and mass change [290], but little direct information on the composition of evolved gas products. A more complete thermal profile is provided when a thermal analyser is coupled to an identification tool. Henderson *et al.* [433] have recently described TG-DSC/DTA with evolved gas analysers (MS and FTIR). The skimmer coupling is the most advanced commercial way of combining a thermobalance or simultaneous TG-DSC/DTA instrument with a quadrupole mass spectrometer [338]. For descriptions of interface techniques in this coupled instrumentation, *cfr.* ref. [411]. Simultaneous TG-DSC-MS is capable of operation up to 2000°C [434].

Applications

Möhler *et al.* [234] have reported TG-DSC-MS of the thermal decomposition of the **vulcanisation accelerator** tetramethylthiuramdisulfide (TMTD) in rubber; degradation of TMTD starts at about 155°C, as evidenced by m/z 76 (CS_2) and 44 (radical of the secondary dimethylamine).

The high sensitivity of the instrumental combination was demonstrated by Kaisersberger *et al.* [435], who published TG-DSC-MS data for EPDM showing cumyloxy radicals (m/z 135, 136) from the dicumylperoxide (DCP) system. Without the MS data, the mass loss in the range from 240°C to 400°C would only have been attributed to the plasticiser content. Hyphenation prevents both a misinterpretation of the results and permits optimisation of the process by adjusting the amount of DCP added to the elastomer prior to vulcanisation. TG-DSC-MS was also used to recognise epoxy resin fragments (m/z 58, 92, 135) in electronic scrap from the automobile industry and bromine **flame retardants** in electronic waste [435]. Simultaneous TG-DSC-QMS of polystyrene decomposition at 400°C revealed monomers, trimers and fragments (trimer at m/z 312) [433].

Redfern [284] has reviewed the application of STA (*i.e.* TGA-DSC), STA-MS and STA-FTIR to the degradation of polymers, describing in particular TG-DSC-FTIR of zinc stearate.

2.1.5.5. *Multihyphenated Thermogravimetry–Differential Thermal Analysis Techniques*

Principles and Characteristics

If the TG-DTA approach offers advantages, then even greater synergy is achieved by addition of an Evolved Gas Analyser (EGA). Many of the TG or DTA curves are in fact much more complex than might appear at first sight. Simultaneous TG-DTA-MS yields such supplementary information useful for the interpretation of complex thermoanalytical curves. Commercial TG/DTG-DTA-MS instrumentation (sample size up to 5 g) has been described [436,437]. Kettrup *et al.* [438] have described a macro TG-DTA-MS system (sample volume 170 mL, balance capacity 500 g, T_{max} 1200°C). The furnace is designed for homogeneous heating of this large sample volume and allows examination of heterogeneous materials and trace analysis.

In TG-DTA-MS due to the missing separation technique, a positive identification of individual substances is not always possible. An additional, off-line identification step by adsorption techniques followed by GC-MS overcomes this restriction.

In technical polymer formulations (as for flame retarded compositions) the complexity of TG-DTA spectra often renders interpretation tentative and coupling to FTIR is another way to remove doubts about these interpretations. Gibert *et al.* [303] have described this system. The low sample volume used in DTA/TG-FTIR experiments limits the effect connected to the existence of mass or temperature gradients and mass or heat transfer. Consequently, results obtained with this coupling of techniques may differ from tests (such as fire tests) performed on larger specimens.

Applications

Manley [28] examined a cured **phenolic formaldehyde resin** (PF) by means of TG-DTA-MS observing a lower sensitivity of TG relative to DTA. (However, new TGA instrumental developments have been reported since.) The TG curve shows loss of phenol (m/z 94); DTA observed water (m/z 18), ammonia (m/z 17) and formaldehyde (m/z 29), indicating disrupture of cross-links greatly affecting the mechanical properties of moulded PF compounds. The MS traces show catastrophic deterioration of PF resins at 200°C. The DTA trace also signals a change around 200°C. DTA is thus a useful indicator of temperatures at which engineering properties may change but MS shows clearly why these changes occur.

A TG/DTG-DTA-MS study of PVC **phone cable coatings** containing di-(2-ethylhexyl)phthalate (DEHP) as a plasticiser has given insight in combustion technology [439]. TG-DTA-MS in combination with TG-DSC has been used to study interactions in **pyrotechnic compositions** (*cfr.* Chp. 2.1.4.1).

Gibert *et al.* [303] have reported a study of talc filled PP/PE copolymers flame retarded with $Mg(OH)_2$ or brominated trimethylphenylindane/Sb_2O_3 (and in combination), using TG-DTA on-line coupled with FTIR. A good correlation was found between the maxima of Gram-Schmidt curves and DTG, and between DTA/TG-FTIR conclusions and fire resistance tests. The DTA/TG-FTIR coupling also showed the limitation of use of $Mg(OH)_2$ as a flame retardant.

2.1.5.6. *(Multi)hyphenated Thermogravimetry–Gas Chromatography Techniques*

Principles and Characteristics

Despite the utility of TG-FTIR and TG-MS techniques, a distinct disadvantage is that the presence of components at very low concentrations may be masked by higher concentration interferants. Also, if a pattern of complex species is evolved during heating (as is frequently the case for polymers), it is advantageous to achieve separation prior to entering the final phase of the mass spectrometer. Consequently, it is useful to incorporate the separation power of GC by collecting products in a trap or on the head of a capillary column for all or part of the TG run [440,441]. However, these methods necessarily result in the loss of the time/temperature evolution data for the products analysed.

Coupling of GC is still not so common because of the intermittent sampling of chromatographic analysis, as opposed to the need for continuous analysis of an effluent stream from a thermal analyser. Separation of components between the thermal analysis unit and the mass spectrometer by GC has been achieved either by **trapping** (cold trap, solution or sorbent tube) [345,442] or **directly** [443,444], *cfr.* Scheme 2.3. Kaisersberger *et al.* [338] have reviewed GC couplings with thermal analytical instrumentation from a practical point of view. The gas-flow conditions in thermobalances, design of coupling interfaces and features of the gas analysers relevant to the coupling were discussed. Jansen [445] has reviewed TG-GC techniques.

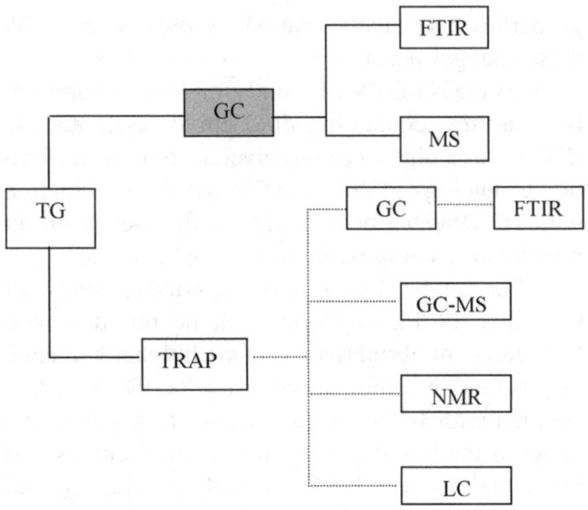

Scheme 2.3. On-line and off-line combinations of thermogravimetry.

McClennen *et al.* [354] have described **on-line TG-GC-IR** and **TG-GC-MS** by approaching the time requirements of TG and GC. Combination of a pulsed automated vapour sampling inlet and transfer line type GC column permits high-speed GC identification of individual TG products while maintaining sufficiently high temporal resolution with a *ca.* 1 min vapour sampling interval. Short capillary GC columns were selected to provide short retention times (<60 s) consistent with time-resolved profiles of the TG curve for comparison to the DTG curves. The classic chromatographic trade-off between efficiency and analytical time is thus balanced in such a way as to provide both real-time thermal evolution profiles of multiple components and separation sufficient to allow a significant degree of component identification by means of TG-GC-IR and TG-GC-MS. **TG-ultrafast GC-ToFMS** is also a desirable (but costly) option.

Meuzelaar *et al.* [446] have described an on-line high-pressure TG-GC-MS system, which requires small amounts (10–100 mg) of sample and can be operated at high pressure under different atmosphere (N_2, He, H_2, *etc.*).

Arii *et al.* [447] have used an integrated simultaneous **TG-DTA/GC-MS system** with dynamic rate control (DRC), a form of a high-resolution TG technique. In this DRC method the heating rate is controlled in such a way that the absolute value of a sample's temperature decrease rate is expressed as a monotonous function of the sample's weight decrease or decomposition rate. This feature improves resolution, identification and quantification. For the purpose of very fast heating response, an IR image furnace is used instead of a conventional type resistivity heater. The TG-DTA/GC-MS system can be used in continuous sampling or direct coupling mode and intermittent sampling or trap coupling mode. In direct coupling mode the GC is bypassed. Intermittent sampling is "off-line". With conventional TG-GC-MS, analysis of evolved gases is complicated since the sample is exposed to higher temperature than that required for decomposition to each component. In **high-resolution TG-GC-MS**, however, analytical treatment becomes easier and improvement of accuracy in identification and quantification is accomplished since the decomposition components on clearly separated steps are expressed as a simple profile. HRTG-GC-MS provides various advantages: *(i)* the heating rate of the sample is dynamically and continuously varied in response to changes in the decomposition rate of the sample; *(ii)* each component of the sample decomposes separately and completely without being affected by decomposition of others; *(iii)* secondary reactions between decomposed gaseous products are limited; and *(iv)* analytical treatment becomes easier by reducing the number of peaks in HRTG-GC-MS profiles.

Combination of a separation technique (usually GC) with TG requires either **successive trapping** at different temperatures or times to obtain temperature or time-resolved separation of the gases. The use of different solvents or adsorption on a suitable solid (Tenax™, charcoal, Chromosorb, *etc.*), followed by thermal desorption (TD), are other possibilities for batch-wise experimentation. In **TG-CT-GC-MS** mode, information is obtained regarding the composition at each specific point of the TG curve at a time. Using the cold trap technique chemical interactions are not excluded (residence time, heating up, catalytic reactions). Similarly, in **trapping TG-GC-IR** the effluent from TG is commonly captured on a trap constructed from a GC capillary injector liner with Tenax™ solid-phase adsorbent and analysed by GC-IR.

As already mentioned before, a macro STA-MS/off-line GC-MS system (sample size 170 g) (for risk assessment) has been described, which allows examination of heterogeneous materials [286].

Significant disadvantages of multihyphenated systems are complexity, cost, and need for a trained operator. It is therefore not at all certain that such techniques hold the future.

Applications
Volatile additives for (un)vulcanised rubbers can be accurately identified by TG or by controlled heating of a test sample in a sealed vial equipped with an overhead collecting headspace, transferring the heated volatiles to a chromatographic column and analysing the separated volatile components emerging from the chromatograph column by various selective analytical detectors. Despite significant instrumental developments polymer/additive analysis by means of hyphenated TG-GC methods is not widely practised. McGrattan [448] examined the decomposition of EVA copolymers using TG-GC-IR. Rau *et al.* [449] have investigated the thermally induced decomposition of polymeric material by a combination of TG-FTIR and successive GC-FTIR measurements of co-evoluted trapped gases.

Arii *et al.* [447] have combined GC-MS with HRTG to study decomposition of a graphite loaded resin. The oxidative degradation products of a flame retardant for polymers (HET-acid, *i.e.* 1,4,5,6,7,7-hexachlorobicyclo[2.2.1]hept-5-en-2,3-dicarboxylic acid) were identified by on-line TG-FTIR and off-line TG-GC-MS (using Tenax and thermal desorption) [256]. Thermal behaviour of flame-retarded polyurethane foams has been investigated using on-line TG-MS and off-line TG-GC-MS (using XAD resin as an adsorbent) [422]. With the same techniques, adhesives used in the automobile industry have been investigated [450]. Mullens *et al.* [451] have described the oxidative degradation of PS by means TG-Tenax-TD-GC-MS. Temperature-programmed reduction (TPR) of PP was studied by simultaneous use of *in situ* FTIR (transmission and DRIFTS) and MS with GC-FID analysis at the point of maximum product formation (at 553 K) [452].

Meuzelaar *et al.* [453] have reported experiments with non-vulcanised styrene-butadiene rubber (SBR) in the presence of various catalysts and co-processing runs of coal and lower grade post-consumer polymers (coloured PE and PS, waste rubber tyres, commingled plastic mixture) in a high pressure TG-GC-MS system at a hydrogen pressure of 900 psi. This system is essentially designed for applied rather than analytical TG-GC-MS work.

Combustion of polymers in horizontal or vertical furnaces and subsequent off-line HRGC-MS of pyrolysis products is suitable for simulation of burning processes [454]. Kettrup *et al.* [455] use a macro-scale TG/DTG-DTA-CT-GC-MS system for enlarged sample capacity, simultaneous TG-MS and GC-MS coupling, identification of trace hazards and separation of complex gas mixtures for identification of individual compounds. The system has been used for thermal decomposition of **flame retarded polymers** both by on-line methods (TA-MS) and off-line techniques (collection of products on Tenax, desorption and GC-MS or HPLC-MS/MS analysis).

Simultaneous multiply hyphenated techniques such as TG-GC-IR-MS [354] and TG-CT-GC-IR-MS have been used for **identification of VOCs** from polymer processing. McGuire [456] has examined VOCs from polyolefin processing by means of TG-CT-GC-IR-MS detecting residual monomers, dimers and trimers, and potential VOCs generated during manufacturing of several synthetic fibre spin finishes, and identified nonanoic acid. These techniques, especially when coupled with a gas chromatograph, are effective for determining thermal stability, product distribution and product evolution as a function of temperature.

2.1.6. Thermal Microscopy

Principles and Characteristics
Thermomicroscopy is a method in which a sample following a temperature program is observed by microscope. Thermomicroscopy is commonly carried out in either reflection or transmission (normally under polarising conditions). In an ordinary hot-stage unit the sample, heated in a DSC pan, can be observed using a stereomicroscope [457]. Other related approaches have been reported [458]. Additional structural information may be obtained when the sample is viewed in transmitted light, using a conventional, modified Kofler light hot-stage unit [459–461]. Any change in structure of the material on heating will result in changes in the recorded light intensity. Commercial equipment is available enabling simultaneous DSC and thermomicroscopy measurements [462]. Kagemoto *et al.* [463] have reported the development of a DTA apparatus equipped with a laser.

Zygourakis *et al.* [464] have first reported visual observation of reacting samples in a TG pan and have developed a custom modified thermogravimetric reactor with *in situ* video microscopy imaging (VMI) capabilities (applied to studying coal pyrolysis and combustion). Video-imaging TG is a direct visualisation of the morphological and textural changes in the sample during thermal processing. Raemaekers *et al.* [282,313] have recently described video-imaging TG-MS. By continuously monitoring the weight loss of the sample, the instantaneous volatile release rates are obtained and may be analysed to elucidate the effects of heating rate. It is clearly important that one is able to monitor and quantify structural changes occurring during thermal treatment since these transformations determine the micropore and macropore structure of the solid. VMI-TG is particularly well suited for recording time histories of thermal processes, because of the easily observable changes in particle size and shape. By monitoring the macroscopic changes of thermally stimulated particles, VMI-TG not only provides important information on the development of internal pore structure and mode of volatile release, but also on the rheological properties of a material and the tendency to agglomerate. VMI-TG-MS also allows distinguishing experimental artefacts from real chemical and physical phenomena and permits total insight in TGA events.

Applications
VMI-TG-MS combines chemical and physical information with direct viewing and video taping for documentation. There are many possible applications of on-line VMI-TG-MS to research in polymer science. Typical experiments relate to solvent action, crystallisation, melting and solidification cycles, decompositions, foaming, paint drying, and topographical changes of polymers during curing. VMI-TG-MS allows insight in the relation between physical measurements (weight change Δw; rate change DTG), chemical observations (identification/verification), and direct viewing of morphological transformations, such as phase changes, particle swelling, bubbling, cracking, blistering, agglomeration, sintering, condensation, and sublimation, crystallisation, plasticisation and melting, observation of the mode of volatile release (foaming, decomposition, migration), detection of colour changes in relation to chemical (MS) and mass loss characteristics (TG), *e.g.* fogging, blooming and iridescence, morphometric analysis (changes in particle size, shape and colour), observation of rheological properties of heat-treated particles, ignition detection and characterisation, recording oxidation reactions, *etc.* The video-microscopy capabilities of a TG-MS allow correlation with reactivity data, better insight in chemical processes and reduce wrong interpretations of deceptively simple TGA results. Raemaekers *et al.* [313] have presented the sublimation dynamics of the flame retardant melamine cyanurate, thermal degradation of the blowing agent azodicarbonamide, thermal degradation of PVC and troubleshooting of black nitrile rubber seals as illustrations of the merits of VMI-TG-MS. In all cases the added power *versus* regular TG-MS was apparent.

The increasing need to correlate thermal behaviour *simultaneously* with the underlying chemistry *and* the accompanying physical phenomena determines the usefulness of other extensions (audiometric and magnetometric).

2.1.6.1. *Microthermal Analysis Methods*

Principles and Characteristics
With traditional methods of thermal analysis the determination of the spatial distribution of single phases, the thermal characterisation of their interfaces and the study of the relation between morphological features and thermal properties is not possible. Conventional thermal methods only yield a sample-averaged response. In order to obtain spatially resolved information about a sample, one must resort to microscopy and imaging techniques (*cfr.* refs. [313,464]). Although IR and Raman microspectrometry may be used to investigate chemical composition on a local scale, spatial and structural resolution is often poor. Recently, the three complementary technologies of thermal analysis (properties), microscopy (structure) and spectroscopy (composition) for bulk analysis have become available for combined problem solving at the micro level (*cfr.* Table 2.17). These three areas are even totally integrated at the micro level by using general scanning (visualisation), local thermal analysis (characterisation) and ablation (analysis), in a μTA-EGC

Table 2.17. Complementary macro- and microanalysis techniques

Analytical technologies		Macro level	Micro level
Thermal analysis	Microscopy	Hot-stage microscopy, VMI-TG-MS	LTA, SthM, CASM
Thermal analysis	Spectroscopy	TG-FTIR	Localised photothermal FTIR
Spectroscopy	Microscopy	Fluorescence microscopy	μFTIR, NSOM

approach. Similarly, interfacing an FTIR microscope with a temperature-programmable hot stage enables simultaneous acquisition of thermograms and IR spectra. Coupling of pulsed laser radiation to NSOM tips permits spatially resolved ($<$100 nm) thermal desorption from molecular surfaces. Also a combination of FTIR spectroscopy and scanning thermal microscopy has been described. Photothermal FTIR spectroscopy using a proximal probe opens the way for IR microscopy at a spatial resolution well below the diffraction limit, but determined instead by the size of the contact between probe and sample (at present on the order of a few hundred nm, ultimately at a scale of 20–30 nm) [465,466]. Modern microthermal analysis techniques and new emerging combined methods, which deliver thermal, microscopic and spectroscopic data, offer powerful analytical tools for the study of polymeric materials.

Temperature and temperature control have always played a major role in plastics processing. Thermal imaging is a powerful tool for solving problems in production and development [467]. Various **thermographic imaging techniques** have been developed [468,468a]. Phase techniques are based on the fact that any heated body emits secondary radiations that can be monitored optically. The amount of radiation detected can be related to the material emissivity, wavelength of detection and to the temperature of the part. Thermography is a non-destructive, non-contact characterisation technique. Infrared thermal imaging equipment is purpose-designed for real-time recording precise temperature over specific periods of time, *e.g.* for the determination of the processing window of a plastic run. Thermographs have been used to infer the parts' thickness distributions as well as material behaviour during blow moulding [468]. Thermal imaging, through the appearance of hot spots, can be used to detect shrinkage and poor adhesion to the mould surface.

Microthermal analysers afford images based on thermal properties such as surface thermal conductivity (10 μm deep) and thermal diffusivity (with modulated frequencies for depth probing) and permit thermal analysis on samples of μm^2 area by combining the imaging ability of AFM and the thermal characterisation ability of temperature modulated DSC (μMTDSC). As only measurements made on small samples can follow the temperature modulation, modulated temperature experiments are particularly indicated for small samples.

Scanning thermal microscopy (SThM) is a new near-surface technology that combines in a single instrument the high-resolution visualisation and positioning methods of atomic scanning force microscopy (AFM) with the more quantitative techniques of thermal analysis [469–471]. In SThM rather than heating an entire specimen, one raises precisely defined regions of the sample through the temperature range of interest, the heat being supplied by an ultra-miniature heat source and thermal sensor replacing the conventional passive AFM tip. The power required to maintain the tip at a constant temperature can be monitored as it is scanned across the specimen and used to build up an image based on the variation in apparent thermal conductivity and diffusivity (concurrent with topographic imaging) [472]. While scanning the surface the temperature of the probe can be modulated by a few degrees at frequencies in the kHz range. Controlling the frequency of temperature modulation provides the ability to vary the **depth of analysis**. As the depth of view depends on the temperature modulation frequency, sub-surface imaging is possible, quite unlike almost all other variants of scanning probe microscopy [473]. This allows analysis of heterogeneous samples. By comparing the sub-surface images with those obtained from the surface, effects of wear, oxidation and degradation can be determined.

As is usually the case in AFM, an area of the surface (up to 100 μm by 100 μm) is first scanned (in contact mode) to image the topography with sub-μm resolution. The microthermal system is capable of providing various images or views of the surface of the sample, such as thermal conductivity/diffusivity, phase transition temperatures and thermal expansion rates for small sample volumes. These images provide information about the size of features, domains or contaminants that may be present. Subsequently, any specific location on the sample (as small as 2×2 μm) can be selected and scanned in temperature to detect glass transitions, crystallisation and melting by local thermal analysis (LTA) using μTMA, and μMTDTA techniques, which are comparable with their "macro" counterparts [474]. The resulting simultaneous measurements of thermal data and spatial detail allow to obtain useful data on the structure of polymer blends, phase separation and buried interfaces [470]. This technique is also known as **calorimetric analysis with scanning microscopy** (CASM). However, as the sample mass involved is unknown, measurements are qualitative only.

Microthermal analysis

Local thermal analysis

μMTDTA, μTMA, L-FTIR, L-Py

Imaging

SThM, CASM

Scheme 2.4. Microthermal analysis family tree.

Table 2.18. Main features of microthermal analysis

Advantages:

- Rapid experimentation (fast heating rates)
- Minute sample size/area
- High spatial resolution
- Surface visualisation (topographical and thermal property mapping)
- Local thermal analysis

Disadvantages:

- Surface/near-surface limited
- Not quantitative
- Temperature calibration
- Few real-life applications

With some knowledge of the components in the sample and their likely thermal responses (*e.g.* melting point, softening temperature, *etc.*) it is possible to use the results from SThM experiments to elucidate nature and distribution of different phases in the bulk. In analogy to TA, μTA cannot be used by itself to obtain detailed local chemical information. This requires coupling, *e.g.* μTG-CT-GC-MS. Where the chemistry of the sample is unknown, the microthermal analyser can be used as a resistively heated probe for locally ($<10 \times 10\ \mu m^2$) desorbing, volatilising or ablating material from the surface. Characterisation of the evolved gases from a selected micro pyrolysis crater resulting from rapidly heating the probe to 600–800°C can be carried out by trapping in a suitable sorbent in a hot micro collection tube, followed by TD-GC-MS (μTA-EGC or local PyGC-MS analysis) [475,476]. This combined approach presents the possibility of visualising a specimen's surface, characterising its thermal properties and then analysing its chemical composition.

Table 2.18 shows the **main features** of μTA. The technique allows fast heating rates (500°C/min) because of the small sample size and low thermal mass of the thermal probe. This provides the ability to make numerous measurements in a few minutes. Temperature calibration is more challenging with microthermal analysis techniques than with the macro relatives [477]. Scheme 2.4 shows the family relationships in micro-sampling and thermal imaging (*cfr.* also http://www.anasys.co.uk/microta).

Microthermal analysis methods overcome some of the problems of ordinary thermal methods, namely long experimental time, sampling (especially for samples which are too small, embedded or difficult to extract), and lack of spatial resolution. Microthermal analysis thus provides:

- *surface visualisation:* image contrast based on surface topography, thermal conductivity and thermal diffusivity of the near-surface region;
- *surface characterisation:* identification of phases by measuring the material's thermal properties;
- *spatial distribution of phases:* size and distribution as a function of temperature;
- *thermal properties of small samples/areas:* differentiation between surface and bulk properties.

Beside its analytical application the scanning thermal microscope can also be used for **thermal micro structuring** or micro patterning of polymer surfaces [478]. Scanning conditions can influence depth and width of thermally generated microstructures.

Thermomicroscopy has been reviewed [479]; a monograph on IR thermography is available [468].

Applications

Thermomicroscopy allows the visual observation of subtle changes in the polymer structure, such as surface melting, degradation, cracking, colour changes, *etc.*, as the temperature is increased under a controlled temperature programme. Thermomicroscopy and thermomicrophotometry (TMP) are not yet widely used in polymer characterisation.

Microthermal analysis constitutes a comprehensive micro characterisation tool for small areas or small polymer samples. This new technology has a wide range of potential applications (Table 2.19). In

Table 2.19. Potential applications of microthermal analysis

- Characterisation of micro-phases, domains, grains and interface surfaces of heterogeneous polymers (*e.g.* multilayer packaging materials, blends or composites) at nano-level [481]
- Measurement of T_g of a polymer in the bulk and in the proximity of an inorganic filler
- Characterisation of gradients in properties in "homogeneous" materials
- Investigations of phase miscibility
- Phase identification on the basis of thermal contrast
- Identification of small polymer film imperfections (gel formations, contamination) or other spatially resolved components [481]
- Investigation of phase transitions
- Weld joint analysis
- Identification of traces
- Depth profiling

polymer science μTA may be used to identify phases in copolymers and polymer blends. Components of composite materials, films and surface coatings can be identified and interface regions can be studied. In addition, by melting surface coatings, coating thickness can be determined. The technique can also be used to characterise surface crystallinity and cure and to identify different morphic forms. An insight, hitherto unobtainable by microscopy, into the size, shape and distribution of phases can be obtained from images constructed from spatial variations in surface adhesion properties [480].

In this area the technological capabilities are more highly advanced than the need in polymer/additive analysis. Microthermal analysis should be able to gain direct (spatially resolved) thermal information on the surface components of a polymeric material, which may be of interest in blooming and plate-out studies. The analysis of glass-filled PP using μTA was reported [482]. Microthermal conductivity can be applied to toner particles embedded in a polymer matrix. In particular, Zur Mühlen [474] has reported the measurement of T_g at the interface between toner particles and a low viscosity film and the analysis of the cross-section of food packaging material by means of μTA. The finding that TGA-EGA and TD-GC-MS both gave evidence for the presence of BHT in a paint film, as opposed to μTG-EGA suggests that the antioxidant has been consumed in use or has been prevented from reaching the surface of the film [475].

Microspectroscopic techniques provide alternative means for the characterisation of laminates. *In situ* μTMA analysis has been used for the identification of all layers of a multilayer **packaging material** on the basis of the softening temperatures [481,483]. Phases in fisheyes in PE film can be identified using μTA-EGC analysis. Gel analysis using hot-stage microscopy was reviewed [484]. The use of a processing aid additive demonstrated significant improvement in reducing gels generated in extruders. Wax coated surfaces were examined by localised thermal analysis [471]. Scanning thermal microscopy is also used to analyse surface induced structure formation on defined heterogenised surfaces achieved by micro printing [478].

Applications of μTA in material science were reviewed [485]. Wunderlich *et al.* [486] described the principles of microthermal analysis and the application to the study of macromolecules. A monograph describes thermal imaging and its applications [487].

2.1.7. Thermoluminescence

Principles and Characteristics
Thermoluminescence (TL) or thermally stimulated luminescence (TSL) is one of a considerable number of thermal analysis methods where a physical property of a substance is measured as a function of temperature, while the substance is subjected to a controlled temperature programme. TSL is a variant of emission spectroscopy, which determines the release of UV/VIS photons when a sample, having been exposed to some form of ionising radiation, is heated. TSL is not the same as the light given off when a substance is heated to incandescence, but is the thermally stimulated emission of light following the previous absorption of energy from radiation. As such, a sample exhibiting TSL cannot be made to emit TSL again by reheating.

For **TSL experiments** the sample, under temperature control, is mounted in a vacuum light box and exposed to UV light (mercury source) or (optionally) to an electron beam. The spectral distribution of TSL is measured with CCD equipment. Temperature ranges for TSL determinations fall into two groups, below and above ambient temperature, with the general range −100 to +600°C. Facilities may also be required for heating the sample under test in vacuum or in oxygen (where, for the latter the term oxyluminescence is used). TSL measurements are very sensitive, *i.e.* very low light levels can be measured and very small concentrations of trapped electrons and impurity luminescence centres can be detected. The measurements may properly be considered as a form of thermal analysis as the phenomena of interest are triggered by heating: TSL monitors photons during a thermal scan (**glow curve**). These photons result from radiative transitions of free electrons which are released from traps to recombination centres (luminescence centres). The electrons are trapped by structural or chemical defects during the low-temperature irradiation of the polymers. Position and intensity of TSL peaks of organic solids are connected with structural and phase changes or with temperatures at which relaxation processes occur. Furthermore, the intensity is influenced by **foreign impurities**, *e.g.* stabilisers. As the glow curve light is often coloured, further data may be acquired by obtaining a series of TSL glow curves recorded for different wavelengths.

TSL is observable in most dielectrics; in polymers the sample is commonly irradiated at liquid nitrogen temperature and heated to room temperature at a rate of approximately 3°C/min. TSL emission in many commercial polymers is negligible above room temperature and the information, which can be extracted from a single TSL measurement on the molecular environment of the trapped electrons, is not as precise as from ESR. The TSL spectrum of a polymer may contain both fluorescent and phosphorescent components. TSL provides information about **ageing processes** and can be used as a method for early recognition of damage in polymers. Fleming [488] has reviewed thermally stimulated luminescence (TSL) for the analysis of polymers.

TSL should not be confused with chemiluminescence (*cfr.* Chp. 1.4.4), which is the emission of light originating in a chemical reaction. Thermoluminescence was reviewed [489]. For a book on thermoluminescence of solids the reader is referred to McKeever [490].

Applications

Kunze *et al.* [491] have described TSL in polymer studies. Thermoluminescence has been proposed for early diagnosis of ageing of greenhouse coverings and analysis of the efficiency of stabilisers in agricultural films [492]. In particular, a TSL study of 50–100 μm thick LLDPE film and HDPE agricultural film after various annealing, stressing and weathering processes (TSL: UV or *e*-excitation) was reported [493]. The temperatures (110 to 230 K) were chosen in connection with different relaxation processes in the polymers. The TSL-glow curves show significant changes in an early stage of ageing *cq.* damage to the polymer, as caused following exposure. The main luminescence centres, which are formed by the **stabiliser substances**, are destroyed during weathering, especially in the amorphous regions. TSL is also used as a detection test of changes induced by food irradiation.

No additive analysis applications were mentioned in Fleming's review on TSL [488].

2.2. PYROLYSIS TECHNIQUES

Principles and Characteristics

Pyrolysis is a chemical degradation reaction that is induced by thermal energy (alone) and generally refers to an inert atmosphere [494]. The sample is subjected to a short burst of intense heat that initiates thermal fragmentation and the production of a range of smaller molecular species that are related to the original sample composition. Pyrolysis of polymeric materials is performed either for analytical purposes or for producing useful materials. **Applied polymer pyrolysis** is used as a process for transformation of polymers or polymer-containing materials into gases, liquids or solids. **Analytical pyrolysis** is one of several commercially available thermal degradative techniques used routinely for characterisation of synthetic polymers. Progress in pyrolysis as an analytical tool is well documented [495, 496]. The techniques of analytical pyrolysis were largely developed in the 1970s, and books by Irwin [495]. Liebman and Levy [497], Wampler [498] and Moldoveanu [499] have dealt with the basic subjects of polymer analysis by pyrolytic methods. Analytical pyrolysis involves an integrated analysis system, which is carefully controlled to produce reproducible results, and which uses small amounts of sample (often in the μg range, up to 100 mg). The small, characteristic volatiles (desorption) and molecular fragments (degradation), which are generated, are used to qualitatively identify the structure of the original polymeric matrix and to determine quantitative information on composition.

Pyrolytic decomposition:
Energy taken up by a molecule is quickly distributed over the molecular structure. Energy dissipation in the form of rotation or translation is not possible in the solid state. Thermally excited *macromolecular* systems are even more complex because of collision interactions existing between macromolecules. Factors influencing macromolecular decomposition mechanisms are: *(i)* type of bonds and bond energy (single and multiple bonds, C−C and C−heteroatom bonds); *(ii)* number and distribution of bonds in the polymer; *(iii)* kinetics and thermodynamics in radical formation; and *(iv)* reactivity and influence of radicals on adjacent bonds in the polymer. The decomposition mechanism is affected by the pyrolysis conditions such as maximum temperature, heating rate, external pressure, *etc.* Hence, it is very difficult to predict the pyrolysis products and their distribution, although some progress has recently been made [500]. The probability of bond breaking increases for all bonds with an increasing maximum temperature of pyrolysis. Bond strengths decrease in the order of C−H, C−C, C−Cl and from secondary to quaternary carbon atoms. Further potential causes of rupture in polymer chains are impurities introduced during manufacture, or partial oxidation due to ageing processes.

The thermal decomposition behaviour of polymers has been investigated and characterised since the sixties. Pyrolysis of macromolecules does not produce a random mixture of non-typical fragments but specific patterns. This fact is used for identification of polymers from their pyrolysates. Hummel *et al.* [501] classified the **major decomposition mechanisms** in polymers into four categories with differing consecutive and parallel reactions, as follows:

1. Retropolymerisation starting with the chain end, with predominant monomer formation (*e.g.* POM, PMMA, poly-α-methylstyrene).
2. Statistical chain scission, followed by: *(i)* retropolymerisation starting from the radical chain ends (*e.g.* PIB, PS); *(ii)* radical transfer and disproportionation (*e.g.* PE, iPP); and *(iii)* stabilisation of the fragments, *e.g.* by cyclisation (PDMS).
3. Elimination of thermally labile side groups: *(i)* followed by fragmentation and cyclisation of the main chain (*e.g.* PVC); and *(ii)* statistically, followed by fragmentation of the main chain or interchain condensation reactions (*e.g.* PAA, PAN).
4. Condensation reactions between the chains by elimination of small molecules (*e.g.* phenol-formaldehyde resins).

For practical use of pyrolysis in the analysis of polymers unequivocal conclusions as to the original polymer are only possible when the pyrolysis products can be attributed unambiguously to a polymer of known composition, for example by means of a standard or reference specimen.

Pyrolysis product formation also depends on **experimental variables** including pyrolyser type. If the energy parameters (T, t and heating rate) are controlled in a reproducible way, the fragmentation is characteristic of the original molecule, based on the relative strengths of the atomic bonds. Reproducibility of pyrolysis data depends on compliance with the conditions prevailing during pyrolysis, GC separation and detection in the mass spectrometer. These parameters must be optimised and carefully controlled. Information on the pyrolysed sample is most complete if the entire spectrum of the pyrolysis products is used. The specificity of the pyrolysis products of polymers increases with their molecular weight. Heavy products are more adequately representatives of the test sample fragments than the less specific light fragments whose formation is also strongly influenced by secondary reactions.

Pyrolysis may be carried out after solvent extraction or, more commonly, directly either in desorption mode (by temperature ramping; TPPy) or by flash pyrolysis.

Sample preparation technique:
The production of a variety of smaller molecules from some larger original molecules has fostered the use of pyrolysis as a (destructive) sample preparation technique. As a result, capillary GC, MS, and FTIR spectroscopy may be used routinely for analysis of synthetic polymers, composites and other complex industrial materials. In pyrolysis experiments sample size and shape, homogeneity and contamination are important issues. Generally, 10–50 μg of sample is desirable for direct PyGC, and about twice that for direct PyFTIR.

Pyrolysis as a sample preparation technique, coupled to GC for separation of the multitude of decomposition products and to an identification technique

Table 2.20. Variables in experimental design of analytical pyrolysis techniques

Pyrolyser type:
Resistively heated devices (filaments, coils, and ribbons), inductively heated devices (Curie-point), microfurnaces, laser, direct pyrolysis (DPMS), in-column, PTV
Heating mode:
Continuous or pulse mode (flash pyrolysis); step mode (temperature-programmed pyrolysis)
Atmosphere:
Inert or oxidative
Gas chromatographic separation:
Column (packed, capillary) and detection (FID, FPD, ECD, NPD, AED, TSD, SCD, PID, MS, IMS and FTIR); fast GC mode
Mass spectrometry type:
QMS, QQQ, ToF-MS, QITMS, and FTICR-MS
Ionisation mode:
EI, LVEI, CI, APCI, ECNI, FI, FD, PI, FAB and MAB
Interfacing:
In ion-source (DI), near ion-source or outside ion-source
Derivatisation:
Simultaneous pyrolysis methylation (SPM)-GC

(MS or FTIR), leads to an extensive analytical pyrolysis family tree [282,381]. Table 2.20 summarises the variables in experimental design.

It follows that standardisation of analytical pyrolysis is not easily achievable. With the manifold in variations in pyrolysis techniques, such as PyGC-MS *vs.* direct PyMS, flash pyrolysis *vs.* temperature-controlled pyrolysis, pyrolysis inside the ion source *vs.* pyrolysis outside the ion source, *etc.*, development of a standard library for pyrolysis mass spectrometry is arduous. The more experimental variables of a technique, the more complicated it is to control the reproducibility. In the field of direct polymer/additive analysis an attempt has been made to set a standard (VW/Shimadzu PyGC-MS standardised additive application and MS library) [502]. Satisfactory analysis may be carried out by high-resolution gas chromatographic separation (PyGC). Improved mass spectrometric separation can be achieved either by soft ionisation methods or by working with tandem MS (MS/MS) or high-resolution (HR) MS devices. In time-resolved PyMS the separation of the products evolved at different temperature intervals helps in determining the origin of the various compounds. Addition of a pyrolyser to separation and identification equipment is not straightforward; in fact, all three components of the on-line equipment have to be optimised for reproducible results.

Pyrolysers:
There are many different types of pyrolyser varying in design, temperature range, sample quantity, sensitivity, *etc.* Depending on the problem a specific design may be preferred. Historically, pyrolysers are classified as **continuous-mode** and **pulse-mode** systems, or **static** (*i.e.* enclosed) and **dynamic** (*i.e.* in continuous-flow). In a static pyrolyser the sample is heated in an enclosed volume for a given period of time, whereas in dynamic pyrolysis systems the sample is rapidly heated in a steady carrier gas flow [503]. In continuous type pyrolysers the sample is supplied to a furnace preheated to the final temperature; in pulse mode reactors the sample is introduced into a cold furnace, which is then heated to the final pyrolysis temperature. Helium is often the preferred atmosphere because the high thermal conductivity facilitates heat transfer from the sample to minimise secondary degradation reactions and helium is a common GC carrier gas. The time required to raise the temperature of the sample from the initial temperature to the final pyrolysis temperature is called the temperature rise time (TRT) and the total time required to raise the sample temperature and pyrolyse it at the final temperature is the pyrolysis interval or total heating (THT) time. Fast temperature rise times (0.02–0.1 s) and stable filament temperatures (better than $\pm 1°C$) are possible. The pyrolysis temperature is a parameter which is not, on its own, sufficient for describing the pyrolysis process. The thermal energy deposited on the sample in a given time is a function of thermal capacity (C) and of the rate of heat transfer. In turn, C depends on the mass and specific heat capacity, while heat transfer is a function of thermal conductivity, mass and morphology of the sample.

The most common dynamic systems distinguish: *(i)* pyrolysers in which the pyrolysis chamber wall temperature is much lower than the pyrolysis temperature; and *(ii)* pyrolysers with a pyrolysis chamber of the tube-furnace type whose walls are heated to the pyrolysis temperature.

In case of expected and inherent inhomogeneity of the sample large sample holders may be desirable although problematic from the point of view of heat conduction and secondary reactions. **Requirements** of a good analytical pyrolyser are production

Table 2.21. Comparison of the main characteristics of several pyrolysers[a]

Property	Curie-point	Heated filament	Microfurnace	Laser
Temperature limit (°C)	1128	1100–1400	1500	High
Temperature control	Discrete	Continuous	Continuous	Uncontrolled
Use of temperature gradients	Not possible	Possible	Common	Possible
Minimum TRT	70 ms	100 μs	0.2 s–1 min	10 μs
Sample size (μg)	10–1000	10–1000	50–5000	20–500
Reproducibility	Very good	Very good	Good	Poor
Catalytic reactions	Some	Low	Low	Very low
Use with analytical instruments	On-line/off-line	On-line/off-line	On-line/off-line	On-line/off-line

[a] After Moldoveanu [499]. Reprinted from S.C. Moldoveanu, *Analytical Pyrolysis of Natural Organic Polymers.* Copyright (1998), with permission of Elsevier.

of degradation products that are: *(i)* as nearly unique to the sample as possible; *(ii)* reproducible; *(iii)* capable of successful separation and elution in GC. These requirements may conflict with some substances.

Current methods to pyrolyse samples rapidly for analysis by GC, MS or FTIR are largely based upon induction-heated filaments or foils (Curie-point pyrolysers) [504,505] and galvanically heated (resistive) filaments [506], both classified as pulse-mode pyrolysers [507], and continuous-mode furnace pyrolysers [508]. Apart from these conventional pyrolysers, other pyrolysis devices are in use: direct probe, laser pyrolysers (*cfr.* Chp. 3.5), programmed temperature vaporisers (PTV, *cfr.* Chp. 2.2.7) and in-column pyrolysers. In the latter device, pyrolysis is carried out in a segment of deactivated stainless steel tubing by passing a pulse of electrical current from a capacitive discharge power supply through the tubing. On-line pyrolysis collects a high fraction of the structurally most significant high-MW fragments [509]. Liebman *et al.* [510] have carefully considered instrumental and standardisation methods. Pyrolyser design has recently been compared by refs. [499,511] and was discussed earlier [496,512, 513]. The essential requirement of the pyrolysis unit is that of reproducibility, both in product formation and in migration of these products from the pyrolysis zone to the analytical device. Pyrolyser design requires: *(i)* minimisation of dead volume; *(ii)* short flow line from pyrolyser to column with heat insulation to prevent condensation of high boiling point pyrolysis products; *(iii)* rapid sample heating; and *(iv)* allowance for any sample state (fibre, pellets, *etc.*). **Advantages/disadvantages** of different types of pyrolysis apparatus have been discussed [495, 497,499,508,514]. Table 2.21 compares several pyrolysers.

A. Furnace-type pyrolysers:
(Micro)furnace pyrolysers, which are preheated to the desired final pyrolysis temperature before introduction of the sample, are categorised as continuous-mode pyrolysers. In such devices, the sample is either moved into a preheated pyrolysis chamber (isothermal mode) or heated rapidly from ambient to pyrolysis temperature (programmable mode). However, furnace pyrolysers are generally held isothermally at the desired pyrolysis temperature, and the samples are introduced into the hot volume.

Solid samples are either dissolved or introduced by means of solid-injection syringes. Care must be taken to introduce the sample for pyrolysis into the furnace without admitting air, since the pyrolysis zone is already hot and degradation starts immediately. The pyrolysis products are then swept into the analytical device by the carrier gas [508]. In furnace pyrolysis problems include long temperature rise times and lack of control over the duration of the pyrolysis.

Hu [516] described a single-chamber two-stage pyrolysis technique which could be used to discriminate between volatile ingredients and other additives, and polymer degradation products. Tsuge *et al.* [508] have described an improved two-stage (TD and Py) vertical microfurnace-type design with two independent temperature-controlled ovens (Fig. 2.23), which leads to highly reproducible and characteristic pyrograms for any form of polymer samples without any particular skills. The vertical furnace-type pyrolyser features characteristics similar to those of pulse filament-type pyrolysers, as developed by Lehrle *et al.* [517]. It appears that certainly in the cases of polymers such as polyolefins, where complex pyrograms are produced, the thermocouple-feedback filament pyrol-

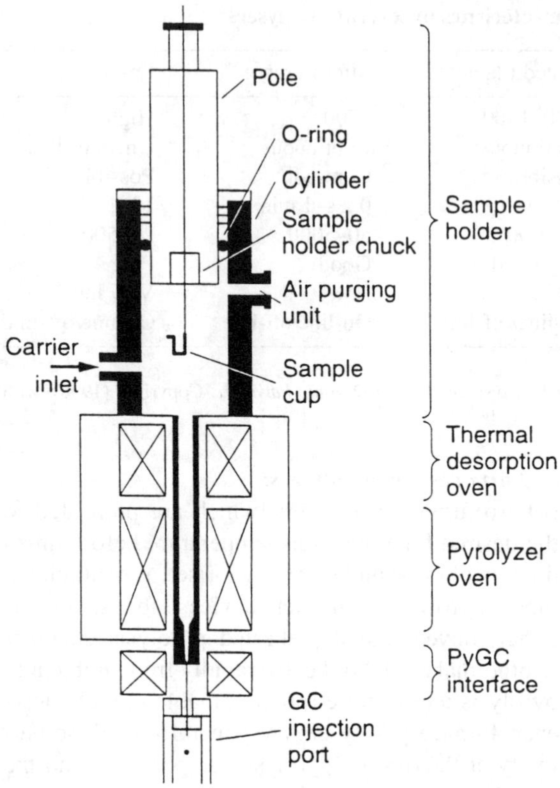

Fig. 2.23. Schematic of a two-stage microfurnace pyrolyser. After Watanabe *et al.* [515]. Reprinted from O. Watanabe *et al.*, *J. High Resolut. Chromatogr.* **14**, 269–272 (1991). Copyright 1991 © Wiley-VCH. Reproduced by permission.

yser [511,518,519] and the vertical furnace-type pyrolyser [508] are the preferred methods.

Amongst the advantages of furnace pyrolysers we mention simple construction, relatively easy operation, and low cost. A disadvantage of isothermal furnaces is that in order to insure thermal stability the furnace tube is usually considerably larger than the sample. This produces a relatively large volume through which the sample must pass before entering the analytical device, *i.e.* a high residence time with the possibility of undesired secondary reactions. Therefore, furnaces are almost always operated with a high flow-rate through the tube (*e.g.* 100 mL/min), generally necessitating split capillary analysis. This high flow-rate reduces the residence time for the sample inside the hot zone.

B. Resistively heated filament pyrolysers:
Whereas isothermal furnaces achieve a fairly fast sample heating by keeping the pyrolysis instrument hot and injecting samples into it, heated filament pyrolysers take the opposite approach in that the sample is placed directly onto the cold heater, which is then rapidly heated to pyrolysis temperature (within *ca.* 20 ms). Either resistance or inductive heating is used. In commercially available filament pyrolysers the sample is deposited on a high-resistance wire or ribbon. Soluble materials may be deposited from a solvent, which is then dried before pyrolysis. Insoluble materials may be melted in place to secure them before pyrolysis. Whenever the filament is a flat ribbon or contains a grooved surface, placement of solid material is facilitated. Upon heating the filament, the material is pyrolysed, the pyrolysis products are swept onto the column by the carrier gas and are separated. Modern resistively heated filament pyrolysers produce a highly predictable temperature-time profile for the filament and also provide a means of varying the heating rate linearly over the initial temperature rise period (ramp control).

The main advantage of a resistively heated pyrolyser is that the filament may be heated to any temperature over its usable range, at a variety of rates. This allows duplicating processes such as TGA. It also permits interfacing to spectroscopic techniques with constant scanning for time-resolved thermal processing. A sample may be inserted directly into the ion source of a mass spectrometer, or placed in the light path of an FTIR [520], and the products are monitored in real-time throughout the heating process. Among the definite advantages of the resistively heated pyrolysers over Curie-point pyrolysers one should further mention the absence of solvent or grinding for sample introduction, ease in weighing the sample, and good repeatability. It is also hard to overestimate the ability of resistively heated pyrolysers to carry out so-called sequential pyrolyses [521], *i.e.*, the pyrolysis temperature and time is chosen in a way that each pyrolysis affords only fractional decomposition of the sample. Secondary reactions are minimised.

Disadvantages of resistively heated filament pyrolysers are difficult automation and the fact that the pyrolysis temperature is difficult to control, as the thermal properties of the sample and filament vary with sample size and filament ageing. Consequently, in spite of constant energy supply to the filament, the temperature attained by the sample during the transient period of pyrolysis is not accurately fixed. The temperature of the surface, which may act catalytically, is difficult to measure.

C. Inductively heated filaments: Curie-point pyrolysers:

The Curie-point flash pyrolyser was originated by Szymanski *et al.* [522], initially developed by Simon *et al.* [523] and later improved [511]. A Curie-point system (Fig. 2.24) can heat a ferromagnetic metal wire inductively with radio frequencies to the pyrolysis temperature in milliseconds. The final temperature is well characterised and reproducible. The alloy of the ferromagnetic material used achieves control of the pyrolysis temperature in a Curie-point instrument. Curie-point reference values are: alumel 154.2°C, nickel 355.3°C, Perkalloy 596°C, iron 780°C, Hisat-50 1000°C. A set of six certified and traceable Curie temperature materials is available (ICTAC/TAI).

This system offers a wide choice of sample holder shape (wire, filament, boat, tube, and folded foil). The technique is suited to the analysis of samples which may be coated onto the filament as a very thin layer. Soluble materials may be dissolved in an appropriate solvent and the wire dipped into the solution. Pyrolysis samples that are not soluble must be applied to the wires in some other fashion [524]. Finely ground samples may be deposited onto the wire from a suspension, which is then dried to leave a coating of particles on the wire. Another approach is to apply the sample as a melt, which then solidifies onto the wire.

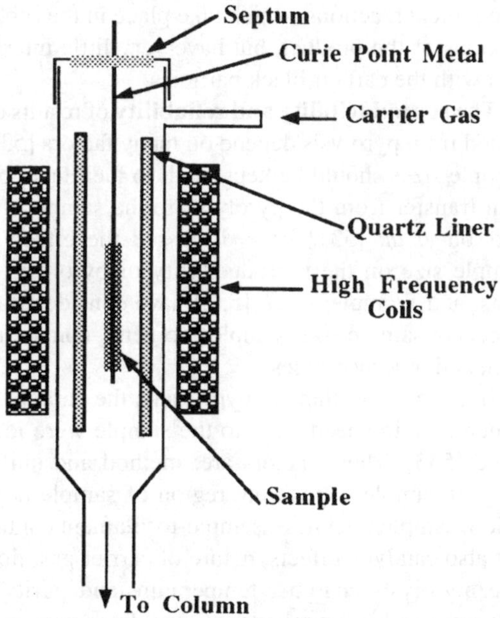

Fig. 2.24. Curie-point pyrolyser. After Scott [525]. Reprinted from R.P.W. Scott, *Introduction to Analytical Gas Chromatography*, Marcel Dekker Inc., New York (1998), by courtesy of Marcel Dekker Inc.

An advantage of Curie-point systems is that there is no temperature calibration to perform since there is no temperature control setting. Curie-point devices are considered to represent the most reproducible pyrolysis method. Disadvantages of Curie-point systems derive from the fact that the temperature of pyrolysis is a function of the Curie-point wire alloy composition. Consequently, the sample may be heated to discrete temperatures only. However, also in this case the temperature of the sample during pyrolysis will still depend to some extent on the size of the sample. With a Curie-point system it is not possible to optimise the pyrolysis temperature by placing the sample into the instrument and increasing the temperature in a stepwise fashion, observing the pyrolysis products after each heating. The catalytic effect that was of some concern with filament pyrolysers is of even greater concern with Curie-point wires. Table 2.21 summarises the main characteristics.

Identification power:

Pyrolysis followed by separation and identification of the pyrolysis products has proved to be particularly useful in polymer/additive analysis. Apart from the absolute concentration of the additive in the polymer, the degree of fragmentation is decisive for identification of an additive in the polymer. The **degree of fragmentation** depends on the temperature selected for pyrolysis. At lower pyrolysis temperature, smaller fragmentation of an additive may be expected and more structural information about the original molecule is contained in the fragment. In that respect a pyrolysis temperature of 450°C would be highly desirable. However, at this temperature the polymer degrades into high-MW fragments (oligomers) which foul the GC system and give rise to considerable memory effects. The great excess of high-MW polymer fragments may severely interfere with detection of characteristic additive fragments. In polymer/additive analysis the highest possible fragmentation of the polymer is beneficial because here the polymer fragments are not of analytical interest. To ensure adequate fragmentation of the polymer matrix a pyrolysis temperature of at least 550°C is required. This working pyrolysis temperature thus generally constitutes a practical overall compromise between high fragmentation of the polymer matrix and sufficiently low fragmentation of the additive (Fig. 2.25). At the same

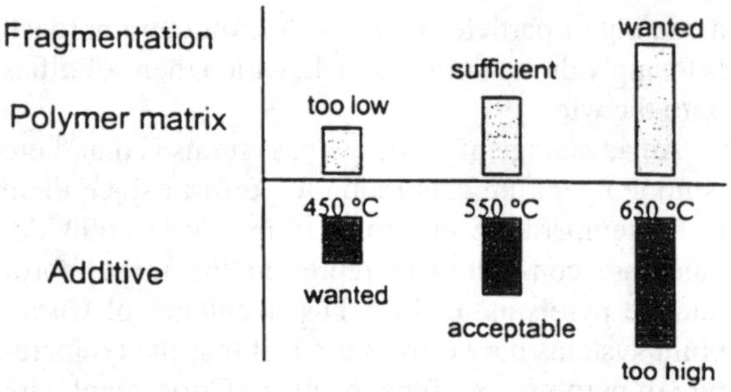

Fig. 2.25. Fragmentation at differing temperatures of the polymer matrix and additives.

time, the choice of this experimental parameter sets the maximum sample size for the chromatographic conditions employed, such as split ratio, flow, and film thickness. A maximum polymer size is about 300 μg unless the concentration of additives is less than 0.1%.

The use of pyrograms as a rapid identification tool in analysis of polymers has been slow to develop partly because of the difficulty in applying pattern recognition techniques to the pyrograms produced. Although many analysts can identify pyrograms of standard polymers on sight an instrumental approach presents significant problems. Qian *et al.* [526] have suggested that a robust standard polymer library, independent of laboratory and mass spectrometer type, may be developed. However, rigorous interlaboratory tests are needed to achieve this goal. For reproducibility a great many experimental variables need to be controlled. Table 2.22 shows the main features of analytical pyrolysis.

Pyrolysis techniques are particularly suited for the more difficult **polymer/additive analysis** problems on account of intricate architecture and morphological features, *e.g.* in case of: *(i)* polymer-bound additive functionalities (AOs, FRs); *(ii)* impact modifiers such as terpolymers (*e.g.* styrene-hydrogenated butadiene-styrene), graft polymers (*e.g.* EPM-*g*-PBT) and an internal rubbery phase in core/shell polymers (*e.g.* acrylate-based cross-linked polymer) [527]; and *(iii)* interfacial agents (*e.g.* graft copolymers, sizings).

Most spectroscopic methods (with the exception of s-NMR) often encounter severe difficulties in the analysis of intractable samples, such as insoluble vulcanised rubbers, which usually also contain a plethora of additives. On the other hand, high-resolution PyGC methods are easily applied for the structural characterisation of EPDM [528]. Some

Table 2.22. Characteristics of analytical pyrolysis

Advantages:
- High sensitivity (detection at 100 ppm level for sufficiently large samples)
- Trace analysis of all organic compounds in liquid or solid state
- Minimal sample preparation (no extraction or enrichments)
- Analysis of intractable samples
- Simultaneous identification and quantification of various additives in one experimental run
- Short analysis time (*ca.* 1 h for GC, <5 min for fast GC and MS)
- Small sample size (10–300 μg)
- Micro destructive only

Disadvantages:
- Difficult reproducibility
- Changes in thermal characteristics of filament affect quality of pyrolysis data
- Inorganic filled polymers may pyrolyse differently from unfilled polymers (catalytic effects)
- Data processing

fillers produce little interference, as shown for the effect of carbon-black on the pyrolysis products from rubber goods [529]. This is mostly because pyrolysis degradation mechanisms are largely intramolecular free radical reactions, which take place in the rubber sections of the product, but have very little interaction with the carbon-black particles.

The **reproducibility and reliability** of results obtained from pyrolysis depend on many factors [530]. Sample sizes should be kept small to facilitate good heat transfer from the pyrolyser to the sample. Andersson *et al.* [531] have discussed the effects of sample size on the reproducibility of pyrolysis results, and Wampler *et al.* [532] have considered the effects of sample size, sample geometry, contamination, and other variables.

In the case of filament pyrolysers, the parameters which pertain specifically to the sample were identified [533]. These factors are: method and uniformity of sample deposition, region of sample deposition, sample thickness, sample-to-filament contact, but also catalytic effects, nature of carrier gas, flowrate, pyrolysis chamber temperature and purity of solvents used in sample deposition. Important parameters are also the temperature-time profile (TTP), which depends upon TRT, THT as well as T_{eq}. Reproducibility is enhanced if the entire sample experiences the same TTP and if the primary products are allowed to migrate rapidly to a cooler zone to prevent further reaction. Analytical pyrolysis should preferably be conducted under mild conditions in order to avoid secondary reactions [534]. Using a ribbon filament at relatively low temperatures can provide mild conditions for pyrolysis.

Various factors affecting thermal degradation of high polymer samples for PyGC were also investigated using a microfurnace-type pyrolyser. It proved that many experimental parameters such as nature and linear velocity of carrier gases, form and quantity of polymer samples and surface-state of the sample-holder were responsible for reproducibility and reliability of the resulting pyrograms [535]. Pyrolysis using a contaminated surface can result in catalytic effects that can drastically alter type and amount of pyrolysis products.

Interlaboratory reproducibility:
The diversity of pyrolysis equipment (laser, resistive-heated boat, ribbon probe, Curie-point foil wire, coil probe with or without insert) and experimental conditions make interlaboratory comparison arduous [536]. Moreover, the analytical instrument at the end of the pyrolyser may influence the quality of the data. Pyrograms obtained on the same equipment have usually proven to be quite reproducible. Multi-user round-robin examinations have been held [537, 538]. The standard procedure [537] using Kraton® 1107 or Cariflex® as a reference copolymer provides a direct method for correlating pyrolysis data from a variety of users. In order to achieve reproducible fragmentation of a polymer each parameter which influences the degradation should be kept constant, such as end temperature, temperature ramp in sample, flow conditions in the pyrolyser or the presence of reactive gases (oxygen), sample weight and geometry. Total system reliability, pyrolyser, rate of temperature rise, interface, detector, data handling, is under constant scrutiny. A serious consideration when comparing results from different laboratories is the applied temperature (gradient) of pyrolysis. It must be recognised that the temperature of pyrolysis differs dramatically depending on the filament used, *e.g.* coil filament with a quartz interior or a platinum ribbon. With reasonable care in sample manipulation and experimental factors, comparable data can be obtained from pyrolytic devices [537].

To promote interlaboratory data comparison in analytical pyrolysis standardisation and reliable compilation of a database for various series of standard samples are among the most important factors.

Hyphenated pyrolysis techniques:
The development of analytical pyrolysis methods is closely related to the advances in instrumental chemical analysis and hyphenation. Pyrolysis has been integrated with various hyphenated techniques. Nowadays, analytical pyrolysis is extensively practised using PyMS (since 1953) and PyGC (since 1959) or PyGC-MS (since 1966), where the characterisation of the original samples is carried out through on-line separative analysis of the resulting complex pyrolysates.

Analytical pyrolysis has been described in books ([512,539], *cfr.* also Bibliography), in many reviews [430,510,517,540–545], in bibliographies [530], and sustains a dedicated journal [546]. A special issue on analytical pyrolysis of synthetic and natural polymeric materials has just appeared [547]. Wampler [514] and Moldoveanu [499] have recently reviewed pyrolysis instrumentation and analysis. Several authors [530,540,548–561] reviewed analytical pyrolysis in polymer studies. The determination of in-polymer additives by flash pyrolysis techniques was reviewed [562,563]. Blazsó [564] has recently reviewed development in analytical and applied pyrolysis. Yearly some 400 pyrolysis-related papers appear.

Applications
General use of analytical pyrolysis is given in Table 2.23. The earliest application of analytical pyrolysis was the identification of the isoprene unit in rubber in 1860 [565]. Analytical pyrolysis is now extensively applied for the analysis of natural and synthetic polymers, textile fibres, wood products, foods, leather, paints, varnishes, adhesives, paper, biopolymers (proteins, polysaccharides), *etc.*, and allows the study of a broad variety of materials including carpets, clothing, electronic components, upholstery, plastic recyclates, fuel sources, oil paintings, *etc.*

Perhaps the widest application of analytical pyrolysis is in the analysis of **synthetic polymers**, both from the standpoints of product analysis and quality control as well as polymer longevity, degradation dynamics, and thermal stability. Nature, composition and structure of macromolecules are elucidated through analysis of volatile pyrolysis products that are large enough to possess the substantial structural elements of the original polymer. When identifying unknowns, a reference library of pyrolysates may be consulted. Pyrolysis finds its greatest utility when dealing with heavily carbon-filled polymers, which

Table 2.23. General use of analytical pyrolysis

- Identification of low-level polymer additives
- Qualitative identification of composition of copolymer or polymer blends
- Characterisation of copolymer sequencing
- Differentiation between copolymers and physical blends of homopolymers
- Determination of monomer ratios in copolymers
- Stereochemistry
- Investigation of defect structures, branching, head-to-head or tail-to-tail linking, extraneous substitutions.
- Polymer kinetics and degradation mechanisms
- Determination of efficiency of curing

are often quite inaccessible by other sampling techniques. At the same time, residual monomers and oligomers, as well as various **additives** that may be present in the polymer matrix, evaporate or decompose, and appear among the volatile pyrolysis products. Analytical pyrolysis is a key tool for the analysis of rubbers and vulcanisates (*cfr.* Schemes 2.1, 2.3, 2.5, 2.6, 2.7 and 2.8 of ref. [213a]). The procedures differ according to the need for characterising the polymer or the additive part. Various additives, catalysts and residual oligomers were analysed in plastics and the emission of toxic compounds under pyrolysis and combustion were monitored. Some common analytical applications of polymer pyrolysis are given in Table 2.24. A variety of pyrolysed-polymer databases is available from various sources.

In polymer/additive analysis it should be considered that the conditions for pyrolysis of a single neat additive are very different from those when embedded in the polymer matrix. A prerequisite for the fragments from a single additive forming in the same way as during pyrolysis of a polymer is absence of interaction with the polymer matrix fragments. This condition is not fulfilled in case of substances used for cross-linking polymer chains because the agent is chemically bound. In principle, the possibility is given that the use of several additives and auxiliary agents in a polymer can lead to interactions during pyrolysis. However, this process is not favoured by the low concentration of additives and their distribution within the polymer.

Frequently, samples like paint flakes present a problem to the analytical lab because they are small, non-volatile and opaque with **inorganic pigments**. Since pyrolysis prepares a volatile organic sample from a polymer or composite, it offers the ability to introduce these organics to an analytical instrument

Table 2.24. Typical analytical applications of polymer pyrolysis

- Polymer identification and analysis of volatiles in polymers
- Determination of toxic compounds among polymer pyrolysis products
- Fingerprint comparison of pyrograms with standard pyrograms to identify major components of copolymers or polymer blends
- Analysis of end-groups and minor copolymer constituents
- Determination of copolymer composition and microstructure
- Triad sequencing in vinyl chloride-vinylidene chloride copolymers
- Forensic identification of paints, fibres, textiles, adhesives, and plastics
- Analysis of coating additives
- Ink and toner identification
- Identification of natural materials and biopolymers
- Authentication and conservation of art materials
- Quality control, *etc.*

separate from the inorganics, using only a few micrograms of sample. This extends the use of analytical techniques such as MS and FTIR to the investigation of small complex samples. Although the more obvious use of analytical pyrolysis is directly on the solid sample, occasionally **extracts** are being pyrolysed as well.

Washall [566] has reviewed analytical pyrolysis of cationic alkylammonium halide surfactants and has shown that analytical pyrolysis is a technique that works well even with **trace quantities** (low ppm level). For applications which require protection of sample integrity, as in forensic science and in art conservation, analytical pyrolysis is an obvious analytical tool. Bart [563] has reviewed polymer/additive analysis by flash pyrolysis techniques and Challinor [567] the applications of analytical pyrolysis in forensic science. Paints, varnishes, glues, pigments, waxes, organic binder formulations have been studied from the aspects of both conservancy and authentication.

2.2.1. Pyrolysis–Gas Chromatography

Principles and Characteristics

The utility of GC has prompted analysts to devise ways to introduce samples by means other than syringe to meet the needs of specific applications, including vapour-phase sample loops, heated headspace injections and thermal desorption of compounds from a solid matrix. Gas chromatography is simple, inexpensive and a popular tool for the investigation of organic materials, provided they are volatile within the working temperature range, generally up to about 300°C. This limitation excludes polymers but pyrolysis extends the applicability of GC to polymeric materials.

PyGC procedures have generally followed one of four basic patterns. The **off-line** procedure, first reported by Davison *et al.* [568] shortly after introduction of GC [569], involves heating of the polymer in a separate enclosure, trapping the off-gases and admitting the pyrolysate to the chromatograph after a given collection interval. An alternative procedure is flash pyrolysis, as described in this paragraph. Wang *et al.* [570] have reported a technique that combines pyrolysis of a polymer with trapping of the pyrolysis products in a solvent followed by GC or HPLC analysis. This procedure adds flexibility to further analysis and also performs the first selection process (dissolve or not dissolve) for the pyrolysis products. In temperature-programmed mode (TPPy-GC) the aim is to separate thermal desorption of additives from pyrolysis of the polymeric matrix in order to increase the information content op the experiment (*cfr.* Chp. 2.2.7). The structural information obtained by PyGC is sometimes unique and complementary to that obtained by the conventional spectroscopic methods, such as IR and NMR.

After the initial introduction of on-line PyGC systems in 1959 [571–573], improvement in pyrolysis technology in combination with high resolution, chemically inert and thermostable capillary GC separation columns, has led to the development of an extended **family of PyGC techniques** for chemical structure determination of polymeric materials with on-line postchromatographic detection, such as FID, rapid scanning MS, FTIR and AED in the presence or absence of reactive reagents and/or catalysts. Most commercially available pyrolysers are simple add-on modifications to the injection systems of standard gas chromatographs or mass spectrometers. Pyrolysers in use for PyGC are flash-filament, Curie-point and horizontal and vertical furnace-type. Pyrolyser operation does not require special training, and the techniques for efficient loading of samples onto pyrolysis probes are an easily acquired skill. To get the most from the PyGC technique, users must be proficient in GC techniques. Production of reproducible experimental results is not trivial.

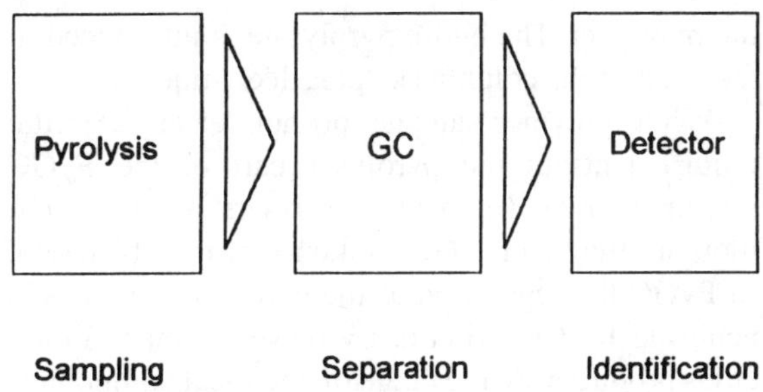

Fig. 2.26. Integrated PyGC system.

Pyrolysis is commonly performed in a flow of an inert gas, and this can be used as a carrier gas for chromatography. Organic samples pyrolysed in the inlet of a GC must be consistent in size with the capacity of the GC so as not to overload column or detector, *i.e.* typically between 10 and 1000 mg. Pyrolysis takes place over a wide range of temperatures, and degradation may begin as low as 300–350°C. The usual working range for flash pyrolysis, however, is between 550 and 800°C, at which temperatures the polymers degrade quickly (in seconds) and the products may be efficiently introduced to a GC column. Pyrolysis generally takes place within a matter of milliseconds.

In an efficient, integrated PyGC system (Fig. 2.26) all essential functions should be well tuned. These are: pyrolysis (sampling); interface; separation (resolution, sensitivity, sample load, column efficiency, reproducibility, standardisation, *etc.*); detection (FID, ECD, FPD, NPD, PID, AED, *etc.*); identification (peak selection, resolution, retention times, internal or external markers, intra-sample/inter-sample variations, interlaboratory reproducibility, techniques assessing correlations between pyrograms, chemometric methods, non-linear mapping, libraries, *etc.*); quantitation (detector type, ...); and data handling (fingerprint comparison, statistical analysis, *etc.*).

PyGC systems require **interfacing** the pyrolyser with the injection port of the gas chromatograph. The interface is critical to successful, reproducible pyrolysis experiments. In PyGC couplings various essential design principles are to be fulfilled. For almost all thermoanalytical experiments an exact control of the sample atmosphere is required. This is most desirable in case of polymer characterisation. The PyGC interface should ensure efficient transfer of pyrolysis products. In fact, limitations on the size and polarity of the pyrolysis products are not just due to chromatographic filtering, but also to trapping in the pyrolyser. The entire pyrolysate is introduced to the head of the column as "plug-like" injection.

While polymer samples do not set any specific requirements to the pyrolyser part of the PyGC instrumentation (filament pyrolysers are preferred though), they do for GC (**polarity** of the off-gases). In PyGC the separation of the pyrolysis products is achieved by GC criteria. Pyrolysis of macromolecules produces a wide range of chemical compounds ranging from non-polar (*e.g.* alkanes and alkenes) to highly polar (*e.g.* alcohols and carboxylic acids). This sets limitations to the GC column choice. Polar compounds, which give useful diagnostic information about the structure of the material, are normally difficult to measure by PyGC due to their partial or complete adsorption in the pyrolysis zone, injection system or capillary column. Polar pyrolysates often show peak tailing characteristics, poor reproducibility, and long elution times. Consequently, a disadvantage of PyGC is the selectivity for apolar and medium polarity products. Without appropriate measures, PyGC is unsuitable for very polar and high-MW pyrolysis products (memory effects) and only thermally stable and relatively low-MW compounds are eluted from a GC column.

In in-column PyGC pyrolysis is carried out in a segment of deactivated stainless steel tubing, connected to a precolumn followed by a GC column [509]. Pyrolysis is carried out by passing a pulse of electrical current from a capacitive discharge power supply through the tubing. In comparison to conventional pyrolysis the technique extends much further toward high-MW fragments carrying more significant structural information.

The state-of-the art in GC is reflected in PyGC; the quality of PyGC data greatly depends on that of the chromatographic system used. The appearance of the pyrogram depends upon the columns (packed or capillary), their length and nature and loading of the stationary phase, nature and linear velocity of the carrier gas, temperature of the analysis and detector response. The information gained in pyrolysis studies is only as good as the degree and type of separation achieved on the column and, certainly in the early stages of investigation work, a variety of columns should be studied. In the past, PyGC utilised almost exclusively packed Carbowax phase GC columns and isothermal operation, which is adequate for "fingerprints"; Kolb *et al.* [574] applied linearly programmed temperature packed and open-tubular columns. However, packed columns have a limited life span and are unsatisfactory for chromatographing very polar and higher-MW compounds, which are often highly diagnostic for many polymers. The limitations of packed column GC prompted the use of capillary GC columns, particularly high-resolution vitreous (fused) silica types [575,576]. Pyrolysis capillary column GC has become the standard adopted in most laboratories. Although the use of a capillary column greatly improves the chromatograms with respect to packed columns, this is at the expense of some difficulty in gas flow control, as the small gas flow possible through a capillary column would not remove the volatile products from the furnace quickly enough to avoid secondary reactions [577]. Challinor [578] has described a simple system for interfacing a Curie-point pyrolyser to a GC equipped with a medium polarity phase capillary column and has compared the results with those obtained from a packed column. Wang *et al.* [579] have reported development of **Py-fast GC** (without cryogenic focusing) for analysis of synthetic polymers, achieving a reduction in retention time from 50 to 5 min. The GC step is then no longer the limiting step in the PyGC operation.

In early PyGC work large **sample sizes** (20–30 mg) were required which favoured occurrence of consecutive and side-reactions of primary pyrolysis products [580,581]. Conventional thermal conductivity detectors (TCDs) were typically used to detect the pyrolysis products separated by means of (the now largely superseded) packed column chromatography; sensitive flame ionisation detectors (FIDs) are applied for capillary column work, which allow operation with much smaller sample size (1–2 mg), minimising secondary reactions [582,583]. This experimental improvement has greatly contributed to a more reliable picture of the primary pyrolysis reactions. **Analyte volumes** can be manipulated in various ways. For example, large-volume injection techniques, designed to introduce more sample in the GC system, eliminate the highly volatile components from the sample (usually the solvent). However, this is not possible where there is no solvent, as in case of on-line PyGC, but can be practised for off-line pyrolysis. On the other hand, PyGC systems with gradient (or step) heating of the pyrolyser allow sending only a fraction (of interest) of the whole pyrolysate into the chromatographic column. Pyrolysis commonly generates enough material for a GC analysis even when a very small amount of sample is taken for analysis. For this reason, concentration techniques, which are used in GC for trace analysis, are not frequently associated with PyGC [584].

For PyGC experiments, the limiting factor is the chromatographic time needed to resolve all of the pyrolysis products. Characterisation of the chromatographic peaks in the pyrogram may be achieved if a range of reference compounds is available for retention time measurements. Interpretation time for comparing pyrograms varies from several minutes to hours depending on the complexity of the pyrolysis data.

The value of PyGC data is highly dependent upon the **GC detector** employed. The most common detectors are the non-selective FID without capability of structural identification (but taking advantage of retention time matching), and MS with qualitative identification capability. FIDs are about three orders of magnitude more sensitive than TCDs, and are ideal for most PyGC kinetic studies of polymers. The detection limit of PyGC-FID of loss of HCl from PVC is smaller than 50 ng (1.4 nmoles) [585]. Nonetheless, in some cases MS has to be used, namely when gaseous components are formed which are FID blind, such as HCl, CO_2, NH_3, *etc*. Lehrle *et al.* [586] have compared PyGC-FID and PyGC-MS and noticed that spurious pyrolysis peaks (due to air gases) can arise in MS detection, as opposed to flame ionisation detection.

To increase selectivity a wide range of selective detectors is in use in successful product identification. HRGC with capillary columns has been reported in combination with such specific detectors as ECD, FPD, NPD, AED, SCD, *etc.* Electron capture or flame photometric detectors are used to obtain specificity for specific elements (such as Cl, S and N) in pyrolysis products. PyGC-FID/TSD has been used for trace determinations (down to 0.2 ppm) of PVP in complex mixtures. Apart from the aforementioned selective detectors successful identification of pyrolysis products is often achieved by means of on-line mass spectrometric facilities, such as PyGC-MS, PyMS/MS, or spectroscopically as in PyFTIR and PyGC-IR. During the experiment particular attention should be paid to analysis and identification of heavy fragments, which give a more complete picture of the sample structure.

The reliability of simultaneous detection of minor flash-pyrolysis products is often quite insufficient. In composite materials polymer and mineral fillers are often contained in large amounts and the concentration of other components is not high, typically 0.5%. For this reason pyrolysis fragments from minority ingredients are few in comparison with the polymer fragments and can be missed in single-step heating of the specimen in the pyrolysis cell. The analyst stands a better chance in detecting low-MW additives by **temperature-programmed pyrolysis** (TPPy), *cfr.* Chp. 2.2.7. This procedure enables fractional separation of the specimen: volatiles, high-boiling non-polymer additives, base polymer, and mineral constituents. The mineral constituents are mechanically removed from the pyrolysis unit after the analysis of the organic portion and can be analysed by independent methods.

Walker [538] has recommended a procedure for calibrating PyGC instrumentation. Obviously, good chromatographic techniques alone will not permit the analyst to solve pyrolysis standardisation problems. High quality PyGC standardisation requires dual calibration (calibrated equipment and calibrated polymer). Standard polymers are Kraton® 1107 and 1108. An atlas of standard polymer pyrograms (in given conditions) is available [587].

Condensation of certain fractions of the pyrolysate, different methods of column interfacing and chromatographic columns are important factors which influence **interlaboratory reproducibility** of PyGC. Several correlation trials assessing the degree of interlaboratory reproducibility of PyGC have been reported [538,588–590]. The use of an isoprene–styrene co-polymer demonstrates clearly the effect of temperature on product distribution [538] and has led to a calibration procedure for PyGC to ensure superior quantitative laboratory reproducibility. Windig *et al.* [591] have recommended a set of standard pyrolysis conditions producing a reasonable degree of interlaboratory reproducibility for Curie-point pyrolysers.

A prerequisite for any meaningful PyGC analysis is to define optimal experimental conditions. To obtain reproducible experimental results equal attention should be given to all stages of sample preparation, pyrolysis and GC analysis. A manifold of pyrolysis conditions must be precisely defined and accurately controlled for quantitative and reliable polymer characterisation work, such as sample handling (size, weight, homogeneity, sample-pyrolyser contact, solvent removal, sample holder shape, cleanliness), pyrolysis (temperature-time profile or temperature rise time TRT, total heating time THT, equilibrium temperature T_{eq}, final and optimum pyrolysis temperature, filament temperature, reproducible heat input), cooling transference (adsorptive loss, condensation, secondary reactions), separation, detection, quantification, identification, and data handling. These aspects have been dealt with extensively [495,497,512,539,542,592]. Reproducible results may be obtained as a result of extensive instrument development [508,511,593,594]. Adequate performance is ensured with relative standard deviations of $<1\%$.

Quantitative analysis of certain constituents requires a rigorous approach to the selection of the conditions experimental [595]. As to pyrolysis, the factors to be considered fall under the following four headings: *(i)* sample deposition effects; *(ii)* sample thickness effects; *(iii)* temperature *vs.* time profiles; and *(iv)* design of the pyrolysis chamber. Lehrle *et al.* [517,518,593,596–598] consider that ribbon filaments give the most reproducible results. In any quantitative study of polymer pyrolysis it is essential to eliminate any anomalous effect of sample thickness because of low thermal conductivities. It is therefore recommended to work with samples as small as 5×10^{-6} g. Equally important are the state of subdivision (powder *vs.* pellet), and mode of packing in the sample tube (ease of pyrolysate egress to reproduce any secondary reactions). Fast temperature-rise times (*e.g.* 0.02 s) and stable filament temperatures (better than $\pm 1°C$) are necessary to ensure that the pyrolysis products are characteristic of a chosen, precisely defined pyrolysis temperature. The design of the pyrolysis chamber is very important as it provides the interface between pyrolyser and analysis system.

The most critical step in quantitative operation of PyGC systems is in mounting the sample for pyrolysis and the design of the pyrolysis unit itself. It is good practice to define the optimal operating temperature (using either TG or pyrolysis). In quantitative GC analysis it is often assumed that the molar response to a component is proportional to the number of carbon atoms of that component, in which case calibration of response is not necessary. However, it is more advantageous to calibrate the response of the GC detector for all characteristic components of interest. This may require that reference materials be available. Just as in other methods an external or internal standard is needed, but a **"pyrolytical standard"** (related to a polymer fragment) may be preferred [599]. In PyGC, FID is frequently used for both qualitative information (based on retention times) and quantitative data (based on peak areas). If no calibration of peak areas *vs.* component size has been performed for all components of interest, the yield ratios of the products are frequently taken as their peak area ratios, tacitly excluding specific sensitivity discriminations with detectors. Lehrle *et al.* [586] have drawn attention to the fact that the ratios obtained by MS detection correspond to number (*i.e.* molar) ratios, whereas those from an FID will correspond to weight ratios. It follows that the mass spectrometer will appear to be relatively less sensitive than FID to molecules of higher molecular weight and that the apparent ratios of the components, as deduced from their peak areas, will be different when FID and MS results are compared. Although PyGC is a well-established analytical technique and involves little sample preparation, it finds limited application in quantitative analysis of additives in polymers.

It is the case to notice that principal component analysis (PCA) has frequently been used to extract maximum information from data matrices of considerable dimensions [600]. Quantitative analysis using PCA has two different features from the conventional calibration curve method. First, all the peaks obtained by PyGC are used for the determination with PCA. Second, the resulting principal component scores (PCSs), which correspond to the concentrations of the components in the sample, are calculated using several known samples (standard materials) and one unknown sample simultaneously; the PCSs of the known samples change with the unknown sample. Thus quantitative results calculated on the basis of many peaks obtained by use of PyGC are often better than those obtained by the conventional calibration curve method [601]. Mitsui *et al.* [602] analysed PyGC data using principal component analysis.

Lehrle *et al.* [597] have reviewed the study of polymer pyrolysis by PyGC with special reference to the objective of obtaining results with quantitative significance. Since polymer decomposition is usually quite sensitive to relatively minor changes in pyrolysis conditions, quantitative analysis imposes more stringent control requirements than are necessary in the purely qualitative approach. Also Berezkin [503] has paid attention to various aspects of quantitative analysis by means of PyGC and has pointed out that it is difficult to predict the quantitative composition of the volatile decomposition products formed in pyrolysis on the basis of sample structure and pyrolysis conditions. By quantitative modelling the detailed pyrolysis mechanism such predictions are now feasible [500]. Quantitative aspects of pyrolysis experiments include calculation of copolymer composition, microstructure measurements and determination of kinetic and degradation parameters [544].

Quantitative analytical schemes (using calibration curves, internal standardisation) have been devised for selected (copolymeric) systems [540]. An example is the determination of the composition of styrene-methacrylate by PyGC [603]. Other applications include the quantification of rubber components [604,605], cellulose esters [606], isocyanate components of polyurethanes [607], nylon [608], plasticisers [609], aliphatic sulfur-containing additives [4]. Standardisation for quantitative analysis based on PyGC data requires careful choice of reference polymers. Use of an **internal standard** may improve the precision when the concentration of a single polymeric component of a blend or copolymer is calculated on the basis of PyGC data [610].

Pyrolysis–gas chromatography is also useful in quantitative kinetic work, where two experimental requirements are of special importance:

(i) uniform deposition of the sample and small enough size to exclude a temperature gradient across the sample; and
(ii) a temperature/time profile which is as close to rectangular as possible [598].

Some qualifying **advantages** of PyGC appealing especially in industrial applications are given in Table 2.25. The strength of PyGC depends greatly on the detection system (FID, FPD, MS, FTIR, *etc.*). Low additive fragment levels can easily be detected (the contributions of various isomers are summed). **Limitations** are that PyGC is an indirect method of investigation, which requires strict standardisation of all experimental conditions. Chemical reactions during pyrolysis are complex, which makes it difficult to establish correlations between structure of the pyrolysate and end products of pyrolysis. The pyrolysis products must have sufficient vapour pressures. On-line derivatisation during pyrolysis [611, 612] may remove restrictions on the detection of polar pyrolysis products. On the other hand, very short column PyGC with almost complete sacrifice of resolution and peak separation has been shown to pass relatively polar large molecules.

The characteristics and relative merits of the alternative PyGC systems were described by Lehrle *et al.* [511,517] and others. PyGC has been the specific subject of several books [497,499,512,613] and many reviews [496,503,510,517,530,541,567, 614,615], some applied to polymers [497,616–618].

Table 2.25. Characteristics of PyGC

Advantages:
- Small sample size (about 100 μg)
- Minimal sample preparation
- High sensitivity
- Short analysis time (5 min using fast GC columns)
- Automation
- Standard chromatographic equipment
- High efficiency of separation
- Rapid information about structural units
- No disturbance of inorganic components, such as fillers (*e.g.* direct examination of pigmented samples)
- Relatively low cost
- Identification power (with MS, IR coupling)
- Principal component analysis
- Universality (wide field of application)

Disadvantages:
- Indirect analysis method
- Strict standardisation of all experimental conditions
- Composition of pyrolysate dependent on experimental conditions
- Limited to determination of volatiles only
- Need for sufficient vapour pressure of pyrolysis products
- Restrictions on polar and high MW pyrolysis products

Comparison of PyGC to related evolved gas techniques:

PyGC allows ready identification of the polymer, but identification of low concentration additives is usually more difficult. Chromatographic methods, such as PyGC and TG-GC-MS *etc.*, are obviously inherently slower than PyMS. The possibilities are improving, though, by using fast columns (*cfr.* ref. [354]) and accurate fast sampling valves. However, only the fragments that are non-reactive, thermally stable, and volatile can be analysed by GC, whereas MS has none of these limitations. As to the ability to identify evolved gaseous species, the **specificity** decreases from unambiguous to not highly specific in the order PyMS/MS > PyGC-MS > PyMS > PyGC. In general, GC-MS is preferable to MS/MS when a large number of unknown components in a mixture are to be identified. Hyphenated PyGC techniques are more powerful in identifying many of the components such as initiators, plasticisers, and flame retardants. In these techniques structural assignment of components is greatly aided by the RRT values. As compared to thermogravimetric methods (such as TG-MS), pyrolysis techniques do not reveal weight changes and in that respect are less favoured for quantitative analysis.

Thermally-assisted hydrolysis and methylation: Although GC separation facilitates identification of the pyrolysate, it prevents detection of certain pyrolysis products, notably polar and high-boiling compounds, which contain particularly useful diagnostic information about the structure of a material. In conventional GC this problem may be remedied by derivatisation of the polar compounds externally or by co-injection to give compounds which may be efficiently separated. Similarly, in order to realise the full potential of on-line PyGC high-temperature *in situ* derivatisation reactions may be carried out.

Thermally-assisted hydrolysis and methylation (THM) using organic alkaline reagents is widely utilised for reliable and informative characterisation of various condensation-type polymers that are often intractable for the conventional pyrolysis techniques [619]. Wang [618] has extended the derivatisation concept and distinguishes "pre-pyrolysis" and "post-pyrolysis" (*i.e.* "pre-column") derivatisation reactions. The purpose of "pre-pyrolysis" derivatisation is to secure a favourable thermal degradation pathway during pyrolysis.

Challinor [611,620] has reported the use of "pyrolysis derivatisation" techniques, simultaneous pyrolysis methylation (SPM), *cq.* THM. In Py-THM-GC the sample (about 5 μg) is typically placed in the hollow of a flattened Curie-point pyrolysis wire with approximately 0.5 μL tetramethyl ammonium hydroxide (TMAH) (25 wt.% aqueous solution) or tetramethylsulfonium hydroxide (TMSH). The prepared wire is then immediately located in the pyrolyser without allowing aqueous TMAH to evaporate and pyrolysis is carried out at the predetermined temperature. Special injectors for **chemolysis** (*e.g.* PTV injector) allow THM also for furnace PyGC experiments. Moldoveanu [499] has listed other common derivatisations utilised in GC analysis.

THM-GC has **advantages** compared to chemical degradation and PyGC in that more structural information about the polar components of some polymers can be obtained with minimal sample manipulation. Further, the technique is more sensitive than existing methods and has the advantage that comparatively low-cost instrumentation is utilised.

Pyrolysis in the presence of reactive reagents has extended the application field of PyGC. THM-GC is applicable to a wide variety of oxygen-containing macromolecular materials amenable to hydrolysis and subsequent alkylation, particularly polyesters, alkyd resins, styrenated unsaturated polyesters, phenolics, aromatic polyesters (LCPs), polyaramides, polycarbonates, polyacrylamides, polycarboxylates, fatty acids, *etc.*

Table 2.26. Examples from the spectrum of samples for pyrolysis gas chromatography

Plastics	Paints	Oils
Rubbers	Films	Bitumen
Caoutchouc	Foams	Polymer additives
Adhesives	Packaging materials	Processing agents
Coatings	Fibres	Cross-linking agents
Dyes	Textiles	
	Cellulose	

Applications

In principle, PyGC allows investigation of all substances that either vaporise without decomposing or can be cleaved thermally into small fragments. The materials mentioned in Table 2.26 originate from kitchen appliances, textile fibres, automotive paints, adhesives, tyre rubber, stationary items, computer parts and artworks from museums, *etc.* Yet, recourse to pyrolysis methods can be taken for granted more easily for some materials than for others. In view of their inert structure it is not surprising that analytical pyrolysis is a frequently used technique to determine the structure of polyolefins. Condensation polymers (polyesters, -amides, -ethers, -carbonates, -urethanes) have been studied much less extensively by PyGC than polyolefins or vinylpolymers (*e.g.* PVC). This is partly due to the fact that condensation polymers can be chemically degraded and are, therefore, more readily studied by conventional analytical techniques. PyGC is a powerful technique for studying the chemical structures of **intractable polymers**, such as polyacetylene [621] and styrene-divinylbenzene (ST-DVB) copolymers [622].

Pyrolysis techniques are among the oldest approaches to the study of the structure of polymeric systems [540,565]. The applications of PyGC to polymer analysis are summarised in Table 2.27.

Additives:

The analysis of additives is a restricted application for PyGC, even though the technique has been applied for the (quantitative) determination of a surprisingly wide variety of additives in polymers, such

Table 2.27. Applications of PyGC to polymer analysis

- Analysis of occluded volatiles, additives and volatile pyrolysis products
- Fingerprint identification of polymers, microorganisms and solids (*e.g.* in forensic science)
- Product quality control
- Quantitation
- Determination of the (micro)structure of polymers (branching, compositional analysis of copolymers and blends, comonomer ratios, sequence distributions, analysis of end-groups)
- Determination of the polymer steric structure (stereoregularity, tacticity, steric block length, and chemical inversions)
- Polymer pyrolysis mechanisms
- Evaluation of the stability of polymers and reaction mechanisms of thermal degradation
- Kinetic studies

as (HALS) stabilisers, plasticisers, flame retardants, fillers, pigments, lubricants, cross-linking agents, coupling agents, sulfur-containing compounds, unreacted monomers, residual process solvents, as well as flavours, taints and odours, binders in paint, toners on paper, *etc.*, using various detectors (FID, AED, NPD, FPD). These analyses have been carried out both on solid samples and on extracts.

Perlstein *et al.* [623] have used CuPyGC using packed columns for the identification and semi-quantitative determination of various low-MW light stabilisers (Tinuvin 144/770, Hostavin TM N20) and polymeric HALS (Tinuvin 622, Chimassorb 944) in LDPE and PP **extracts**. The method involved dissolution of the polymer, treatment with sulfuric acid, neutralisation, extraction of HALS with methylene chloride (procedure Freitag [624]), evaporation to dryness, weighing and pyrolysing for identification. By using two different stationary phases (Porapak QS for Tinuvin144/622, Chimassorb 944, and Carbowax 20M for the same AOs and Tinuvin 770 and Hostavin TM N20) it was possible to differentiate between all the aforementioned HALS. Hostavin TM N20 gives no pyrolysis products detectable on Porapak QS. Complicated sample preparation, high retention times, overlapping characteristic peaks and low recovery (72–94%) limit the practicality of this method. Also Sinclair *et al.* [4] have practised the use of extracts in PyGC analysis of DSTDP and DLTDP in PP by means of PyGC-FPD and achieved a precision of ±10% for 0.1 to 0.7 wt.% DSTDP in PP. Evolved H_2S was detected with the FPD operated in the sulfur mode. Aromatic thioethers do not produce H_2S under the same conditions. Wang *et al.* [625] examined lubricants (fatty acids and their related esters, amides and metallic salts, waxes) after extraction from a PE grade (Spartech 14575). Wang [626] also examined various AOs (Irganox 259/565/1010/1035/1076/MD 1024/1425/3114, Irgafos 168) and an additive extract of GE Cycoloy C 3600 by means of PyGC and PyGC-MS. In PyGC analysis of a polymeric resin used as an antioxidant in SBR a **rubber extract** was taken; the 2-phenylhydroxyphenylpropane fragment, produced in greatest yield, served as diagnostic component and a relative standard deviation of 3% was reported [627].

Analysis of **high-MW stabilisers** may be limited by thermal instability. For example, GC techniques are not feasible for molecules with MW > 800 Da. In pyrolysis the thermal instability may be utilised in generating characteristic fragments. Successful separation of complex stabilisers requires a well-balanced compromise between resolution, separation temperature and analysis time. Roberson *et al.* [628] have presented a PyGC method for quantitative determination of complex stabilisers (Chimassorb 944, Tinuvin 622 and Sandostab P-EPQ) in PP after a one-step extraction procedure by dissolution (toluene)/precipitation with a recovery rate of 89.9–99.4%. The pyrolysis temperature was optimised to produce N- or P-containing characteristic fragments, which were separated on a capillary column and detected by NPD, thus decreasing interference from fragments that do not contain these elements. In this case of very complex fragmentation, FID chromatograms only showed limited usefulness. For quantitative analysis by the standard addition calibration curves in the 0–5000 ppm range were obtained and LODs <50 ppm were achieved. The method resulted in a wide dynamic range with no significant deviation observed from linearity up to 10,000 ppm. The results illustrate the capability of capillary PyGC-NPD for rapid analysis of complex polymer additives. The same technique has been used for identification of various **rubber accelerators** (CZ, NS, MOR, DZ and MBT) in vulcanised elastomers (formulation: SBR 100, CB 25, stearic acid 1, ZnO 4, sulfur 1.5, accelerator 2) on the basis of their decomposition residues (cyclohexylamine, *t*-butylamine, morpholine, dicyclohexylamine, and benzothiazole, respectively) [629].

The frequent use of PyGC rather than GC for extracts of polymer/additive formulations may be dictated by the high-MW nature of the additives to be determined. However, for such additives extraction is usually not ideal. On the other hand, direct PyGC analysis on solids containing a low additive content (such as AOs) may be troublesome in view of the prevailing polymer matrix contributions in the pyrolysate.

Use of PyGC in identification and quantitative analysis of individual polymers or simple polymer mixtures in polymer-containing specimens with a limited number of components is relatively simple [530,541]. Pyrograms of composite materials are much more complex since they result from the superposition of the pyrograms of individual high-MW compounds, the peaks of the liberated thermally stable volatile additives (stabilisers, plasticisers, modifying agents, *etc.*), the peaks of the compounds arising from chemical interactions in the processing of the constituents or in the synthesis of composite materials from raw mixes. However, as the experimental technique undergoes improvement, new applications are possible.

Alexeeva *et al.* [595] have indicated that application of multistep heating of a specimen in the pyrolysis cell followed by chromatographic analysis of the liberated fractions can give much more information about the specimen than that provided by single-step pyrolysis GC. The technique enables determination of polymers in any proportions, high-boiling organic ingredients (stabilisers, plasticisers, *etc.*) and also volatile organic compounds present in the specimen. Analysis of various **commercial rubbers** by means of step-wise filament-type, packed-column PyGC-FID shows the possibility of a simultaneous determination of the type of polymer (butadiene-acrylonitrile rubber), stabiliser (Neozone D) and phthalate plasticisers (DBP and DOP). In isoprene rubber the phenol-type stabiliser Ionol was observed [595]. Thus, a single experiment can detect volatile impurities in rubbers of the same type (natural rubber, and synthetic isoprene rubbers prepared by different manufacturers), the type of polymer and non-polymer additives. It is possible to distinguish materials of different nature and also materials of the same type but with different properties. The method can be used commercially for quality assessment. Hirayanagi *et al.* [630] reported quantitative analysis of various rubber blends using PyGC.

Figure 2.27 shows the use of pyrolysis in the examination of very small samples (1 mg) of cured **adhesives** and the quantitative determination of the antioxidant 2,6-di-*tert*-methylphenol (BHT). The py-

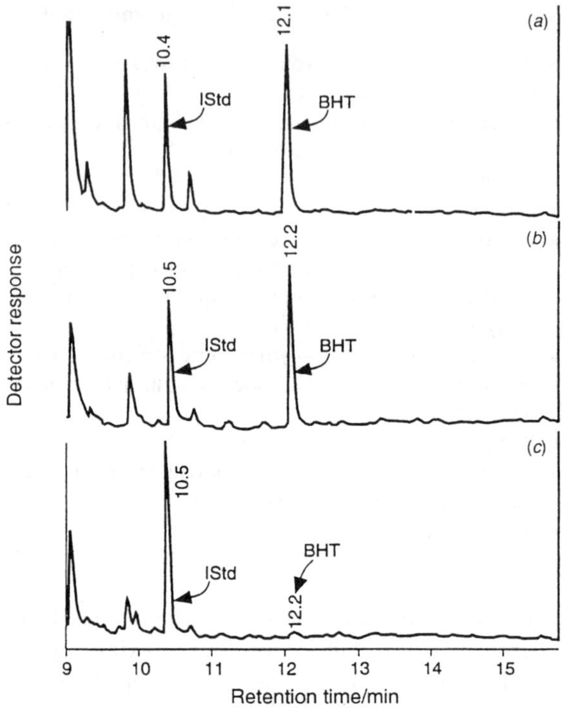

Fig. 2.27. Pyrolysis–gas chromatograms of newly formulated adhesive (*a*), a well-performing adhesive (*b*) and a failed adhesive (*c*) showing resolution of internal standard and BHT peaks. After Franich *et al.* [631]. From R.A. Franich *et al.*, *Analyst* **120**, 1927–1931 (1995). Reproduced by permission of The Royal Society of Chemistry.

rolysis technique gave errors as large as 10% as compared to only 1% with an extraction method employing 100 mg. Despite the lower accuracy, the convenience and the sensitivity of the pyrolysis technique make it a valuable analytical tool for this particular application. PyGC and extraction methods were compared for the quantitative analysis of BHT antioxidant in liquid adhesives and in cured polychloroprene adhesive films [631].

Applications have been reported for mild PyGC analysis of **plasticisers** (at a temperature which is not too high to minimise the pyrolysis of the plasticisers and/or other components of the plastic material) [609,632,633]. Challinor [567] detected butylbenzylphthalate in PVC by means of PyGC. Intraclass discrimination is an important factor in (forensic) examination of materials, and PyGC may be used to distinguish closely related polymers. Wampler *et al.* [634] reported PyGC of vinyl sheeting containing DEHP. This phthalate is not a pyroly-

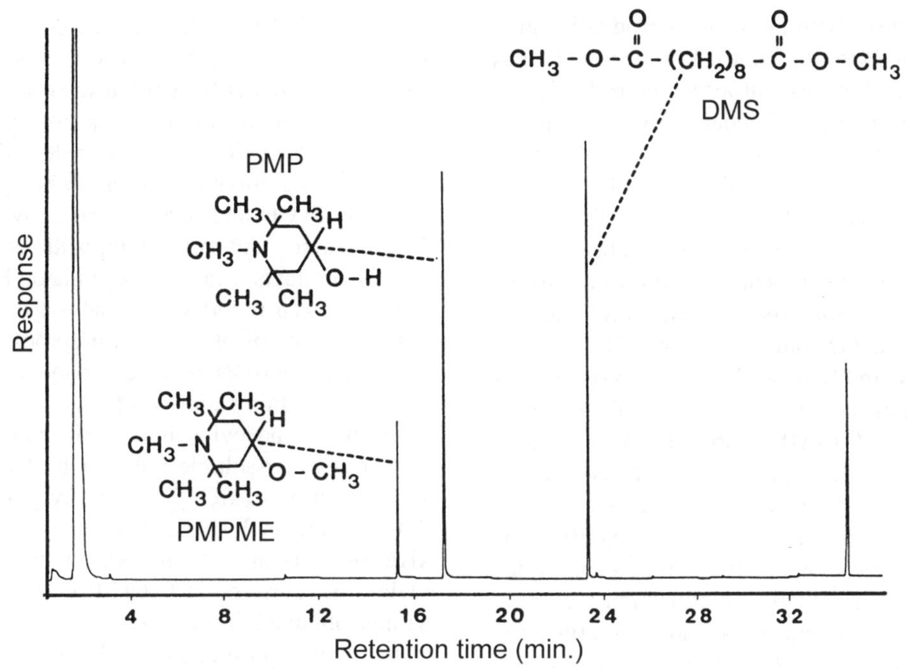

Fig. 2.28. SPM-GC of Tinuvin 292. DMS, dimethylsebacate; PMP, pentamethyl piperidol; PMPME, pentamethyl piperidol methyl ether. After Challinor [636]. Reproduced from *J. Anal. Appl. Pyrol.* **20**, J.M. Challinor, 15–24 (1991), with permission from Elsevier.

sis product *per se* but rather was volatilised from the plastic before it was pyrolysed.

PyGC is also an effective method for identifying and differentiating the organic binder of paint. In some cases, **paint additives** may readily be detected and identified. Automotive paint binder types can be identified on mg-sized samples of topcoat. Challinor [567] has evidenced various phthalates (DBP, BBP, BCP) in a methyl methacrylate (MMA)–butyl methacrylate (BMA) copolymer, BBP in a MMA–BMA–MA (methacrylic acid) terpolymer, DBP and BCP in MMA–MA copolymer and BBP in MMA–EA (ethylacrylate) copolymer. Paint additives may also be identified in architectural paints. Dimethyl orthophthalate (DMOP) was detected in an architectural alkyd enamel which had been subjected to simultaneous pyrolysis methylation (SPM) [635]. Industrial finishes on a beverage can also contain a variety of plasticisers (DMA, DMOP, DMIP), as also determined by SMP-GC [567].

Pyrolysis alkylation gas chromatography has been applied for the characterisation of additives in **surface coating formulations** [576,611]. Some of the hindered amine type UV absorbers used in surface coatings and plastics are complex esters. When subjected to SPM, Tinuvin 292 forms octanedioic acid dimethyl ester (dimethyl sebacate), pentamethyl piperidol and its methyl ether (Fig. 2.28) [636].

For the characterisation of unknown, cured epoxy resins, basic information is required in respect to minor components, such as **coupling agents** or catalysts. These are added in the low percent range and are later not accessible to identification as free compounds due to reactions with the resin. Structural assignment takes place by focusing on those fragments which allow unequivocal identification of the original agent used. Tsuge *et al.* [637] have used PyGC-FID to determine cross-linking reagents of chloromethylated ST-DVB copolymers. PyGC has rarely been used for analysis of composites; an example is the determination of the **hardener** 4,4′-diaminodiphenylsulfone (DDS) in a matrix containing polyfunctional *N,N,N′,N′*-tetraglycidyl-4,4′-diaminodiphenylmethane (TGDDM) resin [638].

Separation of **flame retardants** (usually high boiling point products) from engineering thermoplastics matrices targeted for high-temperature applications is difficult as most of these polymers hardly dissolve in any solvent at room temperature. Therefore, it is not surprising that only few reports have dealt with the analytical identification of FRs in polymers [639,640]. Pyrolysis is a suitable technique for this purpose. Oguri [641] examined an electric circuit board (epoxy resin) by the combustion-GC method and detected BFRs. Wang [642] analysed various BFRs in polyesters and polyamides by PyGC-AED and PyGC-MS. In a different approach to the same problem, Nelissen [269] has used wet chemical means to gain access to the flame retardants. Analysis of BFRs may endanger the chromatographic column (loss of HBr).

Machalkova [643] has described analysis of polymer composites and rubber blends with emphasis on separation of **low-MW additives** by instrumental methods. Examples refer to analysis of inorganic filler- or synthetic fibre-reinforced plastics and laminated plastic films using PyGC and IR. The versatility of PyGC has further been exemplified by Jones [633] as a thermovolatilisation technique for direct determination of occluded volatiles and low-MW additives in lube oil, novolac resins and HDPE, of plasticisers and vinylchloride in PVC, and of solvent residues in paints and bitumens, *etc.* Dicumylperoxide (DCP) in LDPE was identified through detection of three main by-products of reaction, acetophenone, α-methylstyrene and 2-phenylpropan-2-ol [633].

Fillers such as glass, asbestos, graphite, molybdenum sulfide, bronze and lead, were determined by pyrolysis at 700°C under N_2 followed by weighing [644], as well as various **additives in paper** samples [558,645,646]. Crockett *et al.* [647] reported analysis of polymer additives in paper by stepwise PyGC.

Other applications of PyGC to synthetic resins include identification of **inks** and **photocopy toner** in paper [648–650] and surface coatings on currency notes [634]. Toner materials used in photocopiers and laser printers are usually a combination of polymers (such as a styrene/butyl acrylate copolymer) and inorganic colorants, but the pyrograms are simplified since the inorganics are left behind. Wheals [651] has compared emission spectrometry and PyGC in the analysis of 190 **dyes**; 53 dyes were identified by emission spectrometry and 141 by PyGC. Sonoda *et al.* [652] have analysed **organic synthetic pigments** by means of PyGC and XRD.

Lehrle *et al.* [653] analysed polymeric additives, used as **viscosity index** (VI) **improvers**, in a high-grade oil lubricant by means of PyGC-FID.

Use of PyGC with selective detectors and ion-exchange chromatography (IEC) permits determination of the **elemental composition of additives** in polymers from the products of pyrolysis or oxidative thermal degradation. The lower detection limit for additives in a polymer appears to be approximately 0.1% by PyGC and 0.0001–0.01% by IEC. In IEC both the anion and cation compositions of the degradation products can be determined. Experimental data has been reported for analysis of PC containing tetrabromobisphenol as a fireproofing agent and for ethylene–propene–diene copolymer with N-, S-, and Br-containing additives [654].

Sodium dodecyl sulfate surfactant was quantified in a water-soluble polymer by PyGC-FPD via determination of SO_2 liberated by pyrolysis [655]. Multivariate analysis has been applied to **quantitative analysis** of sodium dodecylbenzene sulfonate (DBS) by Curie-point PyGC-FID [601]. The pyrograms obtained from mixtures of DBS and polyoxyethylene lauryl ether (PEG) were normalised for peak heights and areas of *several* internal standards, which appeared in every pyrogram, against the characteristic peaks for DBS and PEG. This normalisation method reduces experimental error compared with a single internal standard. The resultant values were used for multivariate analysis. Principal component scores and a calibration curve were used to determine the DBS content. Calculated values were in fair agreement with theoretical values. The reproducibility was 1.5% using 50 wt.% DBS as opposed to 8% in case of the conventional method using calibration curves based on one peak. CuPyGC was used for quantitative analysis of PMMA adsorbent impurities in drugs (<0.15%) [656].

In conclusion, there exists a fair amount of evidence that PyGC can be used to detect additives and rest monomer, qualitatively and quantitatively directly in the polymeric matrix, without previous separation. Table 2.28 summarises the (few) reported quantitative additive analyses by means of flash PyGC. Relative standard deviations up to 10% are quoted regularly.

Polymers:
The main potential of PyGC (and PyHGC) is to be found in the study of complex polymeric materials. For the identification of polymers by PyGC an ISO standard is in preparation (ISO/DIS 7270). Polymer analysis requires information as given in Table 2.29.

In practice no one hyphenated technique can provide an answer to all these topics. Methods for

Table 2.28. (Semi)quantitative additive analysis by flash PyGC

Analyte(s)[a]	Matrix	Method	RSD[b]	Reference
PAAE	Paper	PyGC-FID	5%	[645]
PAM, PAAE	Paper	PyGC	–	[646]
Tinuvin 144, Tinuvin 770, Hostavin TM N 20, Tinuvin 622, Chimassorb 944	LDPE, PP extracts	CuPyGC	–	[657]
DSTDP	PP extract	PyGC-FPD	10%	[4]
Alurofen (polymeric antioxidant)	SBR extract	PyGC	3%	[627]
Chimassorb 944, Tinuvin 622, Sandostab P-EPQ	PP extract	PyGC-NPD	–	[628]
BHT	Polychloropene adhesives extract	PyGC	10%	[631]
DBS	PEG	PyGC-FID + MVA	1.5 to 8%	[601]
Alkyl methacrylates	Oil lubricant	PyGC-FID	–	[653]

[a]BHT, butylated hydroxytoluene; DBS, sodium dodecylbenzene sulfonate; DSTDP, distearylthiodipropionate; PAAE, polyamide amine epichlorohydrin; PAM, polyacryl(methacrylate); PEG, polyoxyethylene lauryl ether; SBR, styrene–butadiene rubber.
[b]RSD, relative standard deviation.

Table 2.29. Analysis of polymeric materials

- Molecular weight distribution and average molecular weight (1)
- Polymer identification (fingerprinting) (2)
- Compositional analysis (sequence of monomeric units, length distribution and homogeneity) (3)
- Branching, cross-linking, or other side-chain substitutions (4)
- Copolymer structures or variations in the polymeric system; blends and compounds (5)
- Identification of degradation products; outgassing phenomena (6)
- Additives or impurities present (7)

analysing change in a degrading polymeric material may be divided into those which measure properties in the bulk phase of the degrading polymer substrate and those which measure volatile fragments. PyGC belongs to the second category, together with PyMS, PyGC-MS, laser Py-MS, TG-FTIR, TG-MS, TG-GC-MS and others. No measurements are performed directly on the polymer; only evolved gases are detected and identified. PyGC showed that high-MW polyisobutylene can degrade thermally to give more than 60 volatile products [500]. Depending upon the ancillary separation method, such as GC, HPLC, Py, TG and/or chemical modification steps, either off-line or in tandem, a specific problem is addressed. In case of PyGC mainly items (2)–(7) of Table 2.29 may be investigated, whereas item (1) is the domain of SEC and (3) and (4) of NMR. Therefore, information about a polymeric matrix is obtained in an indirect way, and concerns especially microstructure, thermal stability, degradation mechanism, performance behaviour, and volatile additives, residuals and monomer occlusions. Under strict control of the experimental conditions (such as pyrolysis temperature and temperature rise time) it is possible to make kinetic studies of polymer decomposition by PyGC [521]. PyGC has been used for molecular characterisation of acrylic resins, cellulosic materials, fluorocarbon and vinyl polymers, nylon type polymers, polyesters, polyolefins, rubbers, silicones and thermosetting resins. The method has a record of proven performance in the characterisation of (commercial) plastics, man-made homopolymer and copolymer fibres, natural fibres, fibre blends, or partly degraded fibres. PyGC-SCD has been used to detect sulfur-containing products in photodegradated silk samples [658]. Pyrolysis with solvent trapping was used to identify two vinyl acids (acrylic and methacrylic acid) as low-level additives polymerised into emulsion polymer chains [570]. Fast PyGC has been used for determination of low-level monomeric units in polymer analysis [570]. Paints, rubbers, and other polymeric materials have made PyGC a quality control technique.

Fingerprint identification:
Establishing structure and composition of an analysed substance from pyrolysis products is quite difficult. Therefore, in practice empirical correlations between structure of the substance of interest and the range of pyrolysis products prevail (pattern recognition as "fingerprints"). Packed gas chromatography

column pyrolysis is adequate for general fingerprinting but minor details are discernible using Py-CGC. However, much of the qualitative work has been performed with such widely varying conditions of pyrolysis and GC as to make comparison impossible. The most widely used, but crude, method for establishing the nature of a polymer from PyGC data is to pyrolyse all samples in the same standard conditions and to record the chromatogram as a fingerprint of the polymer [596]. In these conditions, PyGC is excellent as a comparative technique [399].

Pyrolysis has found widespread application in the identification of **polymers**, for which standard databases of model pyrograms of known polymers have been described [659]. Tsuge *et al.* [660] have compiled a trial standard database for 135 kinds of typical polymers pyrolysed under the same conditions using PyGC. However, in the literature there is no consensus on the optimal temperature of pyrolysis: temperatures used range from 400 to 1000°C, although a temperature around 600°C is considered best. Morever, the choice of GC column is not obvious as pyrolysis products of polymers for identification are, as a rule, multicomponent mixtures of compounds differing in polarity with different functional groups and boiling points.

The use of PyGC fingerprinting to correlate pyrolysis fragmentation patterns to structural information may be conducted at several levels of sophistication [596,661]. Forensic applications are largely concerned with fingerprint identification of samples (paints, fibres, and adhesives). Other fields of application are in art (identification of forgeries), identification of paints, lacquers and varnishes. In all cases emphasis is placed upon the reliability of comparisons.

PyGC is particularly amenable to use for **quality control** in manufacturing plant situations, where only the patterns of pyrograms are compared [529]. Blend ratios can be determined on the basis of known calibration blends [662]. Stepwise PyGC has also been indicated for product quality control [595].

PyGC finds wide application in the analysis of **rubbers** [663]. Ostromow [260] and others [664] have described the use of off-line PyGC for the identification of rubbers and elastomers, quantitative determination of volatile and polymeric components and carbon-black with the inorganic residue being recovered for subsequent analysis [516]. The properties of elastomer blends can be significantly altered by changing the distribution of carbon-black between the individual phases. Very few methods are available for determination of the CB distribution in polymer blends. Cotten *et al.* [665] have used PyGC for this purpose in SBR/BR blend. Wampler *et al.* [663] have described analysis of rubber materials (automobile rubbers, sealants, gloves, foams) by PyGC. Coulter *et al.* [666] have reported a Curie-point/packed GC column application to analysis of rubber materials. Also Tojo *et al.* [667] reported a series of analytical procedures applied to unknown rubber (RS) and plastic samples (PS) to identify their components and additives. The components were examined by PyGC for RS and by IR for PS. Soxhlet extraction of RS and PS followed by IR and GC-MS effected the determination of the additives, *e.g.* dialkyl phthalates, fatty esters and higher aliphatic alcohols.

In cross-linked elastomers chemical determination of S gives a first impression of the vulcanisation system (peroxide- or sulfur-cross-linked). A more precise characterisation can be obtained by chromatographic methods, *i.e.* by identifying the products which originate from thermal degradation of the cross-linking reagents (S, S donors, peroxides). The vulcanisation additives used (accelerators, retarders) can be determined on the basis of the degradation products of S-linked material. EGA and PyGC with adequate detectors (MS, FID) are suitable degradation chemical analytical methods for such investigations [668]. Analysis of elastomers in rubber products such as tyres is very critical in QC and **failure mode analysis**. Structure and blend ratio of elastomers and uncured stocks can be analysed by FTIR and NMR. However, it is very difficult to analyse the structure of cured rubber due to formation of cross-linking and incorporation of carbon-black. In addition, since the elastomers used in tyres to improve performance have various microstructures, their analysis is quite complex.

PyGC cannot be fully exploited for identification of unknown compounds in complex matrices, such as cured epoxy resins. It is impossible to identify unknown resins by pattern recognition. In those cases identification of pyrolysis products requires postchromatographic detection (MS, FTIR, AED) to collect structural information.

Challinor [611] has applied Py-THM-GC to structure determination of unsaturated **polyester resins**, phenolic resins, oils and fats, surface coating additives, *etc.* In contrast to PyGC, it is possible to identify the polybasic acid and polyhydric alcohol components of polyester resins by means of THM-GC. Pyrolysis methylation of phenolic polymers results in the formation of the respective phenol methyl ethers in contrast to phenolic compounds on conventional pyrolysis. For example, epoxy resins, which give phenol, isopropenyl phenol and bisphenol-A by PyGC, result in phenol methyl ether, isopropenyl phenol methyl ether and bisphenol-A dimethyl ether on pyrolysis methylation. Phenol formaldehyde resins give corresponding phenol methyl esters.

Vegetable **oils and fats** are usually identified as their fatty acid derivatives or their triglycerides. The well-established methods usually depend on GC as a means of identification. Pyrolysis derivatisation procedures (THM-GC) developed more recently [636] provide a method for characterising these materials. Microgram quantities of the triglycerides are reacted with tetramethylammonium hydroxide (TMAH) at high temperature to yield fatty acid methyl esters without employing multistep procedures.

Shedrinsky *et al.* [669] have described the application of analytical pyrolysis (PyGC, PyMS, PyGC-MS, PyGC-FTIR) in **conservation science** for the study of art materials, which are mostly non-volatile and insoluble.

A compilation of gas chromatographic applications of pyrolysis is found in Liebman and Levy [497].

2.2.2. Pyrolysis–Mass Spectrometry

Principles and Characteristics

In the physical analytical approach of pyrolysis solid organic matter is exposed to thermal energy in an inert atmosphere or *in vacuo* such that structurally significant fragments are obtained. Flash pyrolysis of organic material dictates small sample size (μg or, better, ng), fast heating rates (non-isothermal conditions in the sample) and an open low pressure system in which the analytical pyrolysis takes place. PyGC offers suitable capability for detecting low-MW pyrolysis products, but does not offer sufficient potential to allow obtaining information on larger fragments from the polymeric components, and especially many of the additives. PyGC is also quite time-consuming.

Polymer pyrolysates may be rather complex: for example, the pyrogram of 1,4-polybutadiene contains some 500 components [670], complicating considerably the application of PyGC (unless comprehensive) as well as PyMS techniques. To ease the interpretation of the spectral information obtained from pyrolysates various approaches have been described, such as selective (chemical or physical) removal of certain classes of compounds from the original mixture, catalytic alteration (hydrogenation) leading to reduction in the number of components, and application of soft ionisation techniques (*e.g.* FIMS) to ensure minimal MS fragmentation.

PyMS is a mass spectrometric technique in which a flash pyrolysis device is coupled directly or indirectly to a mass spectrometer. Total PyMS experiments can be performed in a few minutes. Off-line PyMS of polymers was first reported in 1948 [671, 672] and on-line PyMS of polymers in 1953 [673]. In the ideal experimental design the pyrolytic fragments of macromolecules are generated under non-isothermal conditions, escape sufficiently fast from the dissociating matrix so that overheating and further rearrangement of the pyrolysis products are prevented, and are analysed without further wall contact by soft ionisation MS techniques. The ideal conditions are most closely met when pyrolysis takes place inside the ionisation chamber, but in practice the analytical PyMS conditions are often quite different.

As will be apparent, PyMS is far from an unambiguous process. A large number of **PyMS configurations** can be designed by varying pyrolyser-MS coupling (off-line; on-line), pyrolyser type (resistively and inductively heated devices, micro-furnace, laser pyrolysers), on-line mode (DP, in-source, near-source, in front of the source), heating mode (flash *vs.* temperature resolved pyrolysis), mass spectrometer type (QMS, QQQ, ToFMS, FTICR-MS), and ionisation mode (EI [674–676], LVEI [677–680], CI [640,675,681–684], DCI [685–687], APCI [688], ECNI [674], FI [387,675,679–681,689–704], FD [675], PI [705–707], FAB [675] and MAB). Application of various ionisation techniques gives PyMS a broad chemical compound sensitivity. It is immediately obvious that soft ionisation techniques are preferred for PyMS. An important aspect of analytical pyrolysis is production and detection of thermal fragments which contain essential structural information. It is equally clear that the technique is quite the opposite from instrument standardisation. Pyrolysis mass spectrometry may benefit from a third dimension: temperature (linear programming), spectrum of product ions (tandem MS), or spatial resolution (laser microprobe).

PyMS systems eliminate some of the problems associated with transfer of pyrolysis products from an external pyrolyser to a gas chromatograph. **Off-line** microscale open system **flash pyrolysis** into a cold trap can be a very useful alternative especially for structural studies because the pyrolysis conditions, as well as the conditions for chemical structure analysis, can be optimised independently. After pyrolysis, the pyrolysate is eventually derivatised, separated by chromatographic techniques and identified using MS (*e.g.* FAB-MS or high-resolution EI-MS) or NMR. As pointed out by Boon [708], **on-line PyMS** without a chromatographic interface is performed in a number of ways: *i.e.* in front of, near, or inside the ion source. Examples of pyrolysis devices **in front of the ion source** are a crucible in a flame close to a molecular beam instrument [709,710] or a thermobalance exiting into an API source [404]. Pyrolysis **near the ion source** utilises an expansion chamber, an extended empty tube inlet, a heated glass liner, an all glass-heated inlet system (AGHIS) or "direct" probe distillation from a glass capillary tube. Curie-point systems are commercially available that can be interfaced with mass spectrometers so that pyrolysis occurs near the ion source. Pyrolysis inside the ionisation chamber is employed in a modified "DCI" (direct chemical ionisation) approach in which pyrolysis takes place on a resistively heated Pt filament ($T_{max} \approx 1000°C$). **In-source direct pyrolysis mass spectrometry** is a wall-free pyrolysis process, which allows temperature-resolved analysis (*cfr.* Chp. 2.2.7). Whereas in flash pyrolysis (where the temperature is rapidly increased to a fixed value), the final pyrolysis temperature affects the mass spectrum, this variable is absent with in-source PyMS, where the pyrolysis mass spectrum is collected over the entire pyrolysis process and the mass spectrum is averaged across a fixed pyrolysis region. Variations in heating rate, sample loading, ion source temperature, and polymer molecular weight then have only minor effects on the relative abundances of the pyrolysis products. With in-source pyrolysis, secondary reactions and condensation of pyrolysate can be largely avoided because of the high-vacuum environment and the very short time gap between the pyrolysis and ionisation events. The mass spectra thus generated are simpler and more reproducible and have fewer experimental variables. Low sample loadings are a disadvantage for heterogeneous materials. Qian *et al.* [526] have developed an in-source PyMS spectral library for quick identification of industrial polymers (over 150 entries of standard and specialty polymers, copolymers and terpolymers; available on request).

Table 2.30. Main characteristics of in-source direct pyrolysis mass spectrometry

Advantages:
- No prior separation needed
- Short analysis times (2 min)
- Few experimental variables
- Reproducible mass spectra for fingerprinting (verification) and pattern recognition
- Provides data on molecular species not passing GC columns
- High confidence level
- Standard spectral reference library
- Allows information at different stages of thermal degradation
- Low matrix interference

Disadvantages:
- Low sample loadings (μg level)
- Quantitation challenging
- Rigorous interlaboratory testing needed
- Fouling of ion source

Table 2.30 summarises the **main features** of in-source polymer mass spectrometry. An important advantage of the in-source PyMS technique for polymer analysis is its low matrix interference, because non-polymeric materials are vaporised prior to polymer pyrolysis. In-source filament pyrolysis can be used for high-speed, broadband screening of additives in polymer blends and is suitable for high-resolution MS and MS^n studies [711]. Despite small sample sizes (about 1 μg or less), contamination of the ionisation chamber is a disadvantage of this approach and frequent cleaning is necessary to avoid ion optical problems. Low sample loadings are an obvious problem for heterogeneous materials. At present, most PyMS instruments are not equipped for pyrolysis inside the ionisation chamber. The pyrolysis chambers and analytical systems are often interfaced in a "near-source mode".

Several **advanced PyMS configurations** have been described. Boon *et al.* [712] have presented a multi-purpose external ion source FTICR mass spectrometer for rapid microscale analysis of complex mixtures. External source DT-FTICR-MS allows obtaining nominal mass spectra, temperature windows, HRMS data and exact elemental composition and MS/MS data on selected ions. For more detailed structural analysis of the more volatile part of the pyrolysate PyGC-MS and PyGC-HRMS are frequently applied. Laser pyrolysis experiments benefit from time-of-flight mass spectrometers in the LPy-EI-ToFMS mode [713].

In the use of PyMS to the study of complex materials, such as polymers containing organic additives, the choice of **ionisation method** is of major interest. The application of electron impact (EI), chemical ionisation (CI), field ionisation (FI), field desorption (FD), and fast atom bombardment (FAB) for the analysis of rubber additives has been described in detail [675]. Although for reference to commercial mass spectral libraries EI ionisation would be recommended, it is yet advantageous to use weak ionisation modes. In fact, in most cases identification by means of spectral libraries is not possible. In the EI-MS approach a number of difficulties are encountered: *(i)* it may not be possible to distinguish which ions arise from fragments and which from molecular species; *(ii)* fragmentation will vary considerably depending on the instrumental conditions; and *(iii)* normally only low mass ions (less than m/z 300) are observed as opposed to FIMS. Therefore EI ionisation is less useful, in particular for mixtures. However, one stands a better chance for reliable identification of pyrolysis products in combination with elemental composition data of mass signals from high-resolution mass spectrometry. In general, the most characteristic information is obtained from molecular ions of primary, high mass, thermal degradation products. Therefore, high yields and good detection sensitivity for high mass ions are required. Various experimental procedures have been developed to reduce the complexity of the PyMS spectrum. One such procedure is the use of low-voltage EI ionisation (LVEIMS), where the electron beam has energies of 14–15 eV (instead of 70 eV) [677,678]. These energy values are only a few eV above the ionisation potential of most molecules and fragmentation at these voltages is lower (but not absent). While secondary fragmentations are reduced a major portion of the ions in the mass pyrogram are intensive molecular ions. In order to obtain even simpler MS spectra, soft ionisation modes can be applied, such as CI, FI, FD and PI. A comparative study of Py-EIMS and Py-CIMS techniques has been reported [387]. The information content of Py mass spectra can be markedly improved by soft ionisation modes. In single-stage Py-CIMS in principle only the molecular weight is easily derived [683]. Whereas structural assignment on the basis of molecular weight only can be a tenuous proposition for unambiguous identification of a component, PyMS/MS is an obvious choice. Rapid identification of co-evolving compounds by MS/MS is especially appealing in combination with pyrolysis [393,714,715]. Tandem MS techniques overcome at least in part the need of a separation. Snyder [688] has emphasised the advantages of Py-APCI-MS/MS for analyte detection in complex solid matrices. Under APCI conditions the probe usually needs to be cleaned, but not the ionisation source. In contrast to EI, APCI is an inherently selective process such that in the positive ion mode only compounds containing organic functional groups will be ionised. APCI is also a soft process because protonated molecules are usually formed, thus placing the analyte signal mainly in one peak as opposed to a set of peaks, as commonly observed in EI mass spectra. APCI is highly sensitive. Whereas EI is a fairly linear process with respect to the concentration of analyte in a mixture *vs.* intensity, APCI is usually non-linear in comparison of MS signal intensity and concentration. As a rapid screening method for many samples, Py-APCI-MS/MS appears to have promise in that sample extraction and clean-up procedures may be avoided. Analytical Py-APCI-MS/MS has a beneficial impact on detection, characterisation and identification of specific analytes in extremely complex solid matrices.

Other soft techniques, like FI and FD, prevent fragmentation and generate mainly molecular ions, which simplifies interpretation of unknown polymeric systems for which no library entries are available. FIMS is particularly advantageous since it normally yields almost exclusively molecular ions of the various components of the pyrolysate, for fairly high-MW pyrolysis products (up to m/z 2000) [697]. Mass pyrograms obtained by Py-FIMS are comparable to those of Py-LVEIMS [679,680]. Py-FIMS is actually a compromise between LVEIMS of pyrolysates and FDMS; it produces less secondary fragmentation than Py-EIMS (though more than FDMS), and it can be applied less selectively than FDMS. The disadvantages of Py-FIMS are low sensitivity (low ion/neutral ratio) and other common problems in PyMS, such as condensations and losses of the labile fragments; the emitter is not commercially available and is most suitable for R&D purposes on sector mass spectrometers.

Hummel [691] has given an excellent account of Py-FIMS in polymer studies, which contains much information of mechanistic and structural value, in spite of the simplicity of the spectra. The use of in-source pyrolysis coupled with FIMS for studies of rubber vulcanisates has been described [675]. Py-FIMS is a very effective technique for direct rubber compound analysis. The sample can be examined directly, without pretreatment, and both organic additives and the rubber components can be identified in the same experiment. Time/temperature-resolved Py-FIMS allows for identification of rubber components and organic additives in pyrolysates of rubber compounds. Recent **Py-FIMS** studies of polyolefins and copolymers were reported by Lattimer [693,694] and others [701], who used field ionisation as a "survey" or screening technique to obtain an overview of the chemical composition, *i.e.* to determine the number of components and their molecular weights on the basis of the observed molecular ions for the various components. In some cases, the supplemental techniques of MS/MS and AC-MS were used in EI and/or CI mode to obtain detailed chemical structure information.

Also another soft ionisation mode, vacuum ultraviolet (VUV) photoionisation (PI) has been proposed for structural characterisation of polymeric materials [707]. Photoionisation is not a common technique in mass spectrometry but has been utilised for both PyMS [705] and PyGC-MS [705]. A photoionisation system usually consists of a windowless, differentially pumped rare gas resonance lamp coupled with the ionisation chamber of the mass spectrometer. Argon, krypton, or other inert gases are used in the lamp. Argon produces energies of 11.6 and 11.8 (Ar I) eV, and krypton 10.0 and 10.6 (Kr I) eV. The pressure inside the ion source is usually about 10^{-2} Torr. In a typical pyrolysis-photoionisation mass spectrometry **(Py-PIMS)** experiment solid samples ($\sim$10 μg) are pyrolysed on a heated probe in the source region of a reflectron time-of-flight mass spectrometer using a temperature-programming system [706]. The pyrolysates are softly ionised by absorption of a single VUV photon with energy just above the ionisation threshold. Unlike FIMS, which may produce significant amounts of both protonated molecules (MH^+) and radical cations ($MH^{\bullet+}$) during ionisation, photoionisation results only in the formation of radical cations with little subsequent fragmentation [716–718]. Therefore, the ion distribution observed in the mass spectrum may match more closely the product distribution produced by thermal degradation. The different ion distributions detected by FI and PI may be attributed to two factors. Field ionisation favours production of ions in the range of 400–600 m/z [697], while VUV photoionisation generally favours production of ions below 300 m/z. Moreover, a sample size effect has been invoked [706]. Py-PIMS has many advantages, including rapid analysis (a few minutes), small sample sizes (10 μg or less), and analysis of solid samples. The technique presents also several advantages over EI, namely intense molecular ions providing additional information in cases when the EI mass spectrum is not diagnostic, perfectly stable energies, efficient photoionisation at low ionisation energies, availability of different photon energies by use of different rare gases in the resonance lamp, selective detection of classes of compounds depending on their ionisation potentials and fragmentation reduction by choosing a photon energy close to the ionisation threshold of the compound of interest [499]. Compared to EI spectra a disadvantage of PI spectra is that they are not standard and therefore not library searchable. An alternative way to reduce the complexity of the mass spectra of the pyrolysate is application of temperature-resolved in-source pyrolysis-mass spectrometry, which allows to separate the processes of volatilisation and degradation of plastics additives and matrices (*cfr.* Chp. 2.2.7).

The basic steps used to obtain structural information about complex samples by PyMS are summarised in Scheme 2.5. Because of the mass spectral complexity, PyMS has often been used in connection with data analysis techniques [687]. In general, a mass range of about 200 masses is acquired and has to be reduced to two or three dimensions for the benefit of visualisation. Multivariate analysis techniques are often used in the evaluation of the resulting complex profiles [720]. Moldoveanu [499] has discussed the application of factor analysis to PyMS data. As a fingerprint analysis method PyMS is rapid and ideally suited to computerised pattern recognition techniques, which have intensively been used for characterisation and differentiation of complex samples [698,721,722]. Data analysis of complex mass spectrometric data, such as those obtained in PyMS experiments, is described in Chp. 6.4 of ref. [213a]. When these techniques are applied to Py-FIMS spectra, which focus the information necessary for sample differentiation, new insight is gained about composition and structure [692,696,698]. Py-FIMS extended with chemometrical evaluation is particularly useful in a **quality control laboratory** where the number of samples is so high that time-consuming visual spectra interpretation is only possible for special cases.

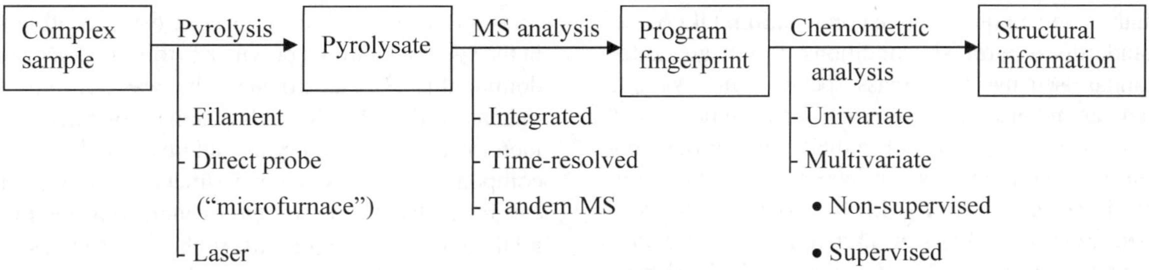

Scheme 2.5. Basic approach options to computerised pyrolysis mass spectrometry After Dworzanski and Meuzelaar [719]. Reprinted from P. Dworzanski and H.L.C. Meuzelaar, in *Encyclopedia of Spectroscopy and Spectrometry*, Academic Press, J.C. Lindon (ed.), pp. 1906–1919, Copyright (2000), with permission from Elsevier.

Table 2.31. General characteristics of flash PyMS

Advantages:

- Small sample amount (generally in μg)
- High-speed analyses (<2 min/sample), temperature resolved
- High sample throughput, automation
- High transfer efficiency of sample to mass spectrometer
- High sensitivity
- Molecular weight of pyrolysis products directly derivable from mass spectrum through (quasi)molecular ions
- Molecular structure determination based on mass spectral fragmentation patterns
- Direct complex mixture analysis
- Broad chemical compound sensitivity
- Spectrum interpretation supported by automatic library search (ionisation mode dependent), spectra subtraction, *etc.*
- Fingerprinting (for fast quality control)
- Chemometrical evaluation techniques for classification and differentiation
- Elimination of the variables associated with gas chromatography

Disadvantages:

- Instrumental reproducibility (condensation, cleanliness of ion source)
- Sample related reproducibility (homogeneity, preparation, deposition)
- Very low sample load (2–20 μg)
- Complex interpretation in Py-EIMS
- No positive identification of thermally evolved products (as in PyGC-MS or DHS-GC-MS)
- Limited quantitative potential

PyMS has been developed as a practical technique in analytical pyrolysis. **Advantages** of PyMS for polymers, partially already given by Schulten *et al.* [692], are as summarised in Table 2.31. Whether PyMS actually represents an improvement over PyGC in terms of reproducibility and discrimination are key questions yet to be addressed. As all techniques, PyMS also has some **limitations**. A major concern is reproducibility. Although in constant experimental conditions reproducibility of PyMS is good, various factors may produce significant variations in the mass spectral information obtained by PyMS. Apart from the multiplicity of instrumental designs (ionisation mode, ion source pressure, mass spectrometer type), there are various intrinsic parameters in the pyrolysis process itself (heating rate, temperature, presence of oxygen during pyrolysis), which may cause deviation from ideal behaviour and make interlaboratory comparability and reproducibility difficult. Boon [708] has discussed some instrumental factors which influence the interlaboratory comparability of the data. Instrumentation reproducibility determined by factors such as cleanliness of ionisation source (especially at high analysis rates), pyrolysate transferability and condensation on cool parts, *etc.*, should be considered [499]. Other affecting factors are sample related, such as sample preparation, sample homogeneity, sample load, sample matrix and type of deposition on the heating device [723]. The low amount of sample (such as 2 to 20 μg) utilised in PyMS is not easy to apply consistently. Without adequate precaution this can easily lead to random results for heterogeneous materials. All these factors must be kept constant to obtain reproducible PyMS results and to guarantee result transferability from one experimental set-up to another. Windig *et al.* [591] have recommended a set of standard pyrolysis conditions for CuPy-QMS (wire cleaning, suspending liquid, sample size, T_{eq}, TRT, total heating time, inlet temperature) for production of a pyrolysate which affords a reasonable degree of interlaboratory reproducibility with respect to its qualitative and quantitative composition. Of course, standardised pyrolysis conditions do not guarantee standardised pyrolysis mass spectra. Until sample transfer and mass analysis conditions can be defined more accurately, these will remain a major potential cause of variation between laboratories. This problem is not unique to PyMS, but holds for all mass spectrometric analyses of multicomponent mixtures of organic compounds. Wampler *et al.* [724] have proposed the use of a library of averaged spectra from pyrograms of polymers (simulated PyMS spectra), which is largely GC parameter independent.

A major disadvantage of PyMS is that a complex mixture is produced by a combination of pyrolysis and electron-impact fragmentation, which makes a mass pyrogram more difficult to interpret than chromatograms produced in PyGC, in which only pyrolytic breakdown is involved. Therefore, PyMS does not easily provide positive identification of thermally evolved products (as can DHS-GC-MS and PyGC-MS). PyMS has also not often been used for quantitative purposes. The present **poor quantitative potential** of the technique still requires considerable development and certainly limits acceptance of PyMS in industry. Some major obstacles to quantitative studies exist, such as char formation, which can preferentially absorb certain compound classes. The importance of such factors is still obscure. In specific configurations, such as in-source Py-EIMS, quantitation has been achieved [725], *cfr.* also Chp. 2.2.7.

Some comparisons between PyMS, PyGC, and PyGC-MS are shown in Table 2.32. For *direct* compound analysis, PyMS offers a way to bypass time-consuming and costly separations, at least in part. Polymer blends and copolymers are readily identified along with certain additives. Temperature-programmed PyMS assists in separating the more volatile components (non-polymeric oils and organic additives) for easier detection.

Boerboom [725a] has reviewed PyMS (direct probe, filament and laser). For recent reviews on PyMS the reader is referred to refs. [719,726]. Boon [708] and others [727] have dealt with analytical polymer PyMS in particular.

Applications

Two types of PyMS techniques are used in industrial laboratories for polymer analysis, in-source PyMS and PyGC-MS. In an early study to assess the applicability of CuPyMS for the determination of *polymer* components in some cases ions due to additives (plasticisers and stabilisers) were clearly identifiable in the spectra although polymer pyrolysate ions were dominant [728]. Nevertheless, this was felt to be an indication that PyMS holds promise for fingerprint identification of both organic additives and polymer components in the same experiment. General applications of PyMS in the field of industrial polymers relate mainly to polyolefins, rubbers, adhesives and paints. The problems addressed vary in degree of sophistication and range from fingerprinting and QC to identification of the pyrolysis products, determination of polymer structure and pyrolysis mechanism, observation of degradation processes as a function of temperature and analysis of oligomers and additives (mainly AOs, FRs, vulcanisation accelerators, plasticisers, tackifiers). PyMS is an excellent tool for comparison, but less useful for obtaining structural information or for studying the presence of impurities in a polymer. PyMS has also been used to study decomposition products, with environmental concern. Moreover, PyMS has found many applications in other areas, such as microbiology, geochemistry, soil science, forensic science, environmental science, art conservation, *etc.*

Whereas non-destructive methods like UV, IR and NMR spectroscopy, as discussed in Chp. 1, offer information about functional groups and structural elements, PyMS enables recording of large sequences of the polymeric chain, in particular in direct probe conditions [729]. In addition to intact monomers and oligomers, which are thermally cleaved during high temperature pyrolysis, PyMS detects additives and trapped contaminants at low temperature. Additives frequently distill out intact in a PyMS experiment.

While PyMS is essentially most suited for direct analysis of solid matter some reports deal with polymer or rubber **extracts**. For example, Hummel *et al.* [699] have studied Py-FIMS of the vulcanisation accelerator 1,3-diphenylguanidine (DPG) and 1,3-di-*o*-tolylguanidine (DOTG) in extracted vulcanisates. The Py-FIMS spectrum of DOTG at $T_{max} = 563$ K (Fig. 2.29) shows the parent peak (m/z 239) as the strongest one (26.4%), followed by fragments at m/z 107 (*o*-toluidine), 132 (*o*-tolylcarbodiimide), and 222 (1,3-di-*o*-tolylcarbodiimide). DOTG is a rather stable molecule as compared with DPG.

In a comprehensive study Curry [730] examined small samples (<10 μg) of 94 household **adhesives** commercially available in the UK. Modern adhesives are quite intractable (insoluble, rubbery, *etc.*)

Table 2.32. Comparison between PyMS, PyGC and PyGC-MS[a]

Parameter	PyMS	PyGC	PyGC-MS
Transfer efficiency	Good	Average (poor for some compounds)	Same as PyGC
Analysis time	Short (1 min)	Long (hours)	Long (hours)
Reproducibility	Very good (not for direct probe)	Good	Good
Fingerprint capability	Yes	Yes	Yes
Molecular information	Poor	Possible with standards	Excellent
Quantitation	Poor	Good	Good
Automation capability	Yes	Yes	Yes

[a] After Moldoveanu [499]. Reprinted from S.C. Moldoveanu, *Analytical Pyrolysis of Natural Organic Polymers*. Copyright (1998), with permission of Elsevier.

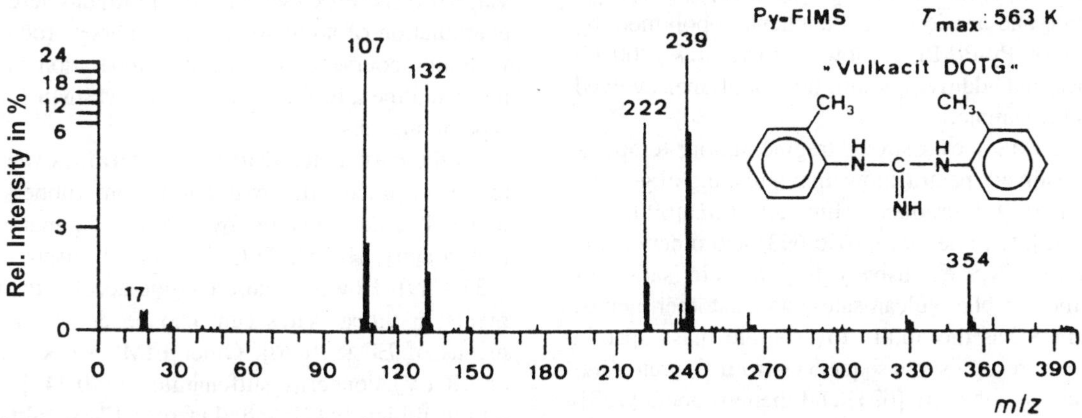

Fig. 2.29. Py-FIMS spectrum of 1,3-di-*o*-tolylguanidine (Vulkacit DOTG, Bayer AG) at $T_{max} = 563$ K; relative intensity in % of the sum of all heights of the recorded peaks in the mass spectrum taken as 100). After Hummel *et al.* [699]. Reprinted from D. Hummel *et al.*, *Makromol. Chem., Rapid Commun.* **3**, 335–341 (1982). Copyright 1982 © Wiley-VCH. Reproduced with permission.

and extremely complex systems, which may contain compounding ingredients such as plasticisers, tackifying resins, AOs, UVAs, processing aids, *etc.*, in addition to the basic polymer resin (usually acrylic, carbohydrate, cyanoacrylate, epoxy, natural rubber, neoprene, nitrile, PS, PUR, PVAc, silicone or SBR). Some of these materials are of considerable importance in art and conservation, such as PVAc, PVAc copolymers, acrylics, silicones, carbohydrates, and epoxy-based adhesives. PyMS identified rosin esters (m/z 239, 197, 195), *tert*-butylphenols (m/z 163, 179, 135, 149) and a phthalate plasticiser (m/z 149) in neoprene rubber adhesives, an adipate plasticiser (m/z 129) in polyurethanes, 2-ethylhexyl acrylate (m/z 57, 55, 70 and 112) in PVAc copolymers and cyanoacrylate "superglues", *etc.* [730]. In cases where complete identification was not possible the spectra could be used as reliable fingerprints. Such fingerprinting studies are clearly very useful also for characterisation and analysis of other synthetic polymers. CuPy-LVEIMS has been applied as a fast primary screening technique for fingerprinting and characterisation of geopolymeric materials [731].

Analytical pyrolysis coupled to mass spectrometry is widely used for identification of thermal decomposition products of **polymers**, and pyrolysis reaction mechanisms are frequently postulated on the basis of the products obtained. PyMS, in its simplest form, compared to other more sophisticated techniques, such as direct FIMS, laser ablation-MS, and FAB, already provides data which, with sufficiently wide databases, can give useful information about structure. Zoller *et al.* [706] analysed the polymer structure of LLDPE/30 wt.% CB and LLDPE/20 wt.% silica by means of Py-PIMS. The mass spectrum of LLDPE/CB was indistinguishable from that of virgin LLDPE. Mass peaks due to carbon-black (a series of peaks separated by 12 Da) are only detected near the end of the heating ramp (>400°C) and not during polymer degradation. The presence of silica prevents microstructural analysis using conventional IR spectroscopy. No mass peaks attributable to silica were found using Py-PIMS. This technique can be used to rapidly quantitatively analyse the microstructure of a wide variety of polyolefins; Py-PIMS composition values are in good agreement with ^{13}C NMR studies. Small sample size (~10 μg) and fast heating ramps (>1°C/s) are needed to study polymer microstructure. Zoller *et al.* [707] also examined polybutadiene, polyacrylonitrile and acrylonitrile–butadiene copolymers by means of Py-PIMS. At low temperatures (200°C) unspecified additives within the copolymer evolved from the sample.

Py-EIMS shows strong fragmentation; temperature-resolved spectra allow distinguishing oligomers from pyrolysis products. Hummel *et al.* [690,699–704,732], Lattimer *et al.* [675,693] and others [697] used Py-FIMS extensively for characterisation of polymers, rubber vulcanisates and elastomer blends. Py-FIMS has the ability to generate mass spectra for long, recognisable sequences from saturated hydrocarbon polymers [693] and diene rubbers [733]. While with Py-EIMS typically molecular weights up to m/z 450 are found and several low-MW fragments are generated due to the hard ionisation, with Py-FIMS molecular ions up to m/z 850 are detectable [734]. As expected, PyMS is also widely used for direct compound analysis of **rubbers** [735]. PyMS gives useful information about S-containing fragments in elastomers [704]. Pausch *et al.* [678] have analysed cured rubber compounds, including commercial hose products, to show the feasibility of determining both polymers and selected additives in the same experiment. CuPy-LVEIMS analysis (11 eV) of an SBR compound containing polymerised 2,2,4-trimethyl-1,2-dihydroquinoline (TMDQ), dioctyl-*p*-phenylenediamine (DOPPD), fatty acids, sulfur, and typical SBR curatives, confirmed SBR and identified poly-TMDQ; DOPPD, stearic acid, sulfur, oil or fillers were not observed. There was no evidence for curatives in any sample examined, apparently because vulcanisation destroys or volatilises these species.

Hummel *et al.* [704] took another approach to manipulate the ionisation mode. Rubbers as well as their vulcanisates on the basis of poly(*cis*-butenylene) (BR), poly(2-methyl-*cis*-butenylene) (NR), poly(2-chlorobutenylene) (CR), poly(butadiene-*co*-styrene) (SBR) and poly(butadiene-*co*-acrylonitrile) (NBR) were investigated with Py-FIMS and thermal decomposition was studied as a function of pyrolysis temperature (773 K to 1173 K). Similarly, Lattimer *et al.* [682] have described direct analysis of elastomer compounds by soft ionisation, tandem (MS/MS) and high-resolution (AC-MS) mass spectrometry. Organic additives, including curatives, could be identified via thermal desorption methods in a commercial EPDM bearing as well as a diene rubber V-belt. The composition of commercial thermoplastic polyurethane was determined via pyrolysis (Py-CIMS). Moore [736] has described examination of additives used in rubber production without recourse to prior sample preparation and has indicated the advantages of tandem MS/MS for this type of analysis.

In the past, **vulcanisation accelerators** were often determined after extraction from rubbers followed by identification by means of paper chromatography, HPLC, TLC, UV, or IR spectroscopy [737–739]. However, sulfenamide accelerator fragments in vulcanisates can also be determined by means of EIMS [676]. Other EIMS work (70 eV or 18 eV) concerns sulfenamide [740,741], thiuramdisulfides [742] and thiazoles [743], although not in a vulcanisate.

Hummel *et al.* [704] have used Py-FIMS to examine a great variety of vulcanisation accelerators of various classes, namely guanidine derivatives (DPG, DOTG), thiourea derivatives (ETU), thiuramsulfides (TMTD, TMTM, TET, DPMTT), dithiocarbamates (ZDMC, ZDEC, ZEPC, Z5MC), thiazoles (MBT, MBTS, ZMBT) and sulfenamides (TBBS, CBS, DCBS, MBS), at 563 K and 873 K within the vacuum of the high temperature inlet system of a mass spectrometer equipped with a combined EI/FI ion source. The decomposition behaviour of two accelerators (tetraethyl thiuramdisulfide and benzothiazyl-2-cyclohexylsulfenamide) was examined in detail. In field ionisation the molecular ion is most intense; mainly molecular ions of the volatile degradation products were observed. In addition to the accelerators, standard vulcanisates on the basis of rubber, carbon-black ("Corax A, N 555"), sulfur and benzothiazyl-2-cyclohexylsulfenamide ("Vulkacit CZ", CBS), were investigated. Py-FIMS spectra of rubbers were compared with spectra of non-extracted and extracted vulcanisates of BR, NR, IR, SBR and NBR. The aforementioned determinations of accelerators in rubbers by means of PyMS were all carried out already some twenty years ago. In case of fairly simple rubber formulations, PyMS can recognise (rest) accelerators. Vulcanised rubber pyrograms often show benzothiazole, generated by the pyrolysis of additives.

More recently, Lattimer *et al.* [2,678] have used direct methods of qualitative rubber analysis. Various direct single-stage mass spectrometric methods were found to be effective for identifying organic additives in rubber. Tandem mass spectrometry (MS/MS) increases the specificity and sensitivity of detection and identification of additives in **direct rubber compound analysis** [431,744]. Again, Py-FIMS turned out to be a good technique for analysis of both the organic additives and rubber components in the same experiment [675]. Results of these studies have been summarised in a review paper on rubber compound analysis [745].

Py-LVEIMS and PyHRMS were used in the determination of structure and composition of clinically important polyurethanes, PEUUs (Biomer, Lycra Spandex, Tecoflex and Pellethane) [746]. The antioxidants found in Biomer and Lycra Spandex were identified as well as an AO in Pellethane and an antistatic agent or residual catalyst in Biomer and Lycra Spandex. These tools are valuable for QC of implant material. PyMS is widely used in **quality control** of industrial products, such as foils and packaging materials [747], silicone and nitrile rubbers [748], can coatings and rubber sealing and tobacco [749]. In particular, it appears that Py-FIMS finds application not only for research in structure determination of non-volatile polymers, but also in product control. Schulten *et al.* [692] have described the application of Py-FIMS to **paints** (commercial can coatings), **epoxy and polyester resins**. Polyamides, acrylic and methacrylic resins, have been analysed in the same way [750–753]. Py-FIMS is well suited for these investigations and allows identification of different polymer sub-units, such as monomers, dimers, backbone fragments, *etc.* Moreover, the technique enables differentiation of the examined compounds, which is of interest to industrial quality control.

Luyk [251] has examined formation of polybrominated dibenzo-*p*-dioxins (PBDD) and dibenzofurans (PBDF) during **thermal processing of polymers** containing polybrominated diphenyl ethers. The formation of PBDDs and PBDFs is a result of thermal and mechanical stress in the melt phase or condensed phase. The yield depends on temperature and residence time in the extruder. After repeated extrusion cycles ($n = 4$) of HIPS/(DBDPO, Sb_2O_3) at 275°C, a significant increase in the yield of PBDFs was observed. Consequently, the contamination of flame retardant polymer blends with PBDFs is explained by their formation during industrial compounding of decabromodiphenyl ether in the HIPS matrix. PyMS analysis of HIPS/(DBDPO, Sb_2O_3) reveals several processes taking place during degradation of this polymer blend, such as debromination of DBDPO into less brominated diphenyl ethers, bromination of polystyrene, formation of PBDFs, antimony(III) bromides and oxybromides. The formation of PBDFs takes place in the temperature range in which the polymer blend degrades (350–400°C) [754]. The results of TPPy-MS and PyGC-MS analyses show that PBDFs are formed during Curie-point pyrolysis of HIPS/(DBDPO, Sb_2O_3) at 510°C. RPLC analysis was used to describe the relative increase of hepta- and octabromodibenzofuran.

Direct Py-CIMS (reagent: methane) has been used for the study of degradation products of PA6.6/poly(pentabromobenzylacrylate) (FR-1025) and PA6.6/(FR-1025, Sb_2O_3) [640], and of PVC/Sb_2O_3 [755–757] and PVC/MoO_3 [755] flame retardant formulations. Gas phase interaction with formation of $SbCl_2$ was reported for PVC/Sb_2O_3 [684]. PyMS was also used for characterisation of nanocomposite FR formulations [758].

McGuire *et al.* [759] reported that direct PyMS could be used routinely for **quantitative analysis** of carbon-black filled poly(epichlorohydrin-*co*-methylene oxide).

Additive analysis by means of flash PyMS both on extracts and directly on solid polymeric formulations is mainly confined to rubbery materials, is based on the employment of soft ionisation techniques (FI, CI, PI, LVEI in this order), cannot claim an extensive record for quantitation and finds only limited application for fingerprinting and industrial quality control. The underlying reason is probably that although flash pyrolysis is frequently used to create fragment ions reproducibly, it provides no molecular weight information for thermally unstable compounds and lacks separation quality.

Whereas PyMS in the analysis of polymers has recently been reviewed by Boon [708], a good overview of fibre analysis by PyMS was published by Hughes *et al.* [760]. Saferstein *et al.* [683] described PyMS as a forensic tool. In-house PyMS libraries containing standard spectra of polymers, oils and their additives have been reported [761].

2.2.3. Pyrolysis–Gas Chromatography–Mass Spectrometry

Principles and Characteristics

As already noticed, direct additive analysis by flash pyrolytic decomposition is usually not easy for complex real life polymeric systems containing a *range of additives* and more elaborate procedures must precede mass spectral detection. The wide varieties of additives that are commercially available complicate data interpretation. In view of these limitations, separation of additives or additive fragments contained in the polymer matrix is usually necessary (*e.g.* by means of temperature-programming PyGC-MS or MS^n techniques). PyGC-MS does not require any pretreatment of the solid sample, which is thermally destroyed under helium atmosphere and carried over in the gas phase; the fragments formed are separated in GC and detected in MS. As opposed to direct desorption techniques (*cfr.* Chp. 2.3), this method thus comprises a chromatographic separation step. With the aid of GC-MS nature and amount of the pyrolytical products can be established. The data are reduced to an additive (-fragment) specific ion chromatogram. Ion chromatogram peaks are identified by means of a search of library records and retention time information. Information on all additive fragments is then grouped together to give a complete determination. Figure 7.13 of ref. [213a] shows schematically a PyGC-MS analysis as a total ion chromatogram (TIC); the mass spectrum indicates the components of the chromatographic peak.

Not unlike TG-MS (*cfr.* Chp. 2.1.5.3), PyGC (*cfr.* Chp. 2.2.1) and PyMS (*cfr.* Chp. 2.2.2), PyGC-MS represents a vast number of **instrumental configurations** differing in pyrolyser type, PyGC interface, GC characteristics (column type), MS characteristics (including ionisation mode), operational variables (oxidative pyrolysis, simultaneous alkylation pyrolysis, temperature-resolved pyrolysis, *etc.*) and data handling procedures (search library, chemometrics). Standardisation in this area is still far off.

On- and off-line PyGC-MS approaches were discussed by Boon [708]. The first directly coupled PyGC-MS system, using a Curie-point pyrolyser, was described by Simon *et al.* [762]. The use of flash pyrolysis has increased dramatically with introduction of fused silica GC columns. In PyGC-MS the type of **ionisation mode** is usually either EI or CI. Electron impact ionisation at the normal ionising voltage (70 eV) causes extensive fragmentation. Thus much information is lost by such MS detection, as many small additive fragments are not specific. Identification of individual compounds is even more difficult when several complex compounds coevolve, *e.g.* during polymer degradation. Soft ionisation techniques allow conservation of more information about structure and molecular mass. However, the use of GC as a separating device sets upper limits to the detectable molecular mass. This restriction is absent in PyMS. Low ionising energies (10–30 eV) enhance the relative intensity of molecular ion peaks and reduce the number and relative abundancies of the lower-MW fragment ions as well as the fragmentation, at the expense, however, of a marked decrease in sensitivity. CIMS has the advantage of ease of interpretation (due to better control on the complexity of the spectral fragmentation pattern) and is able to operate at higher impact pressures. Reported experience with PyGC-CIMS is limited [763].

Additive detection with PyGC-MS is influenced by: *(i)* fragmentation or thermal stability of the additive; *(ii)* concentration of the additive in the matrix; *(iii)* structure (mass) of additive and polymer fragments (specificity); and *(iv)* reactions of additive and polymer fragments.

In PyGC-MS of polymer/additive matrices, polymer fragments in high concentrations are superimposed on the fragments from the additives. This can cause detection problems for additive fragments. A prerequisite for "filtering out" an additive fragment from the background of the polymer matrix is that the mass spectra of the additive or its fragments differ significantly from those of the polymer fragments. It is advantageous when the decomposition behaviour of an additive is known. Degree of fragmentation and absolute concentration of the additive in the polymer are decisive for identification of an additive. The former increases with the pyrolysis temperature. The amount of structural information contained in individual fragments of the original molecule decreases with fragmentation. This is a reason why low additive fragmentation is desirable. PyGC-MS is of limited use for additive analysis of thermally labile, low-volatility products, which give extensive fragmentation. The ideal case of an additive which is not fragmenting at a given pyrolysis temperature remains an exception. Figures 2.30 and 2.31 show some additives with/without significant fragmentation at 550°C. For example, benzophenones, benzotriazoles and phosphates do not fragment easily, while FRs are designed for heat stability; on the other hand, phenolic AOs and HALS

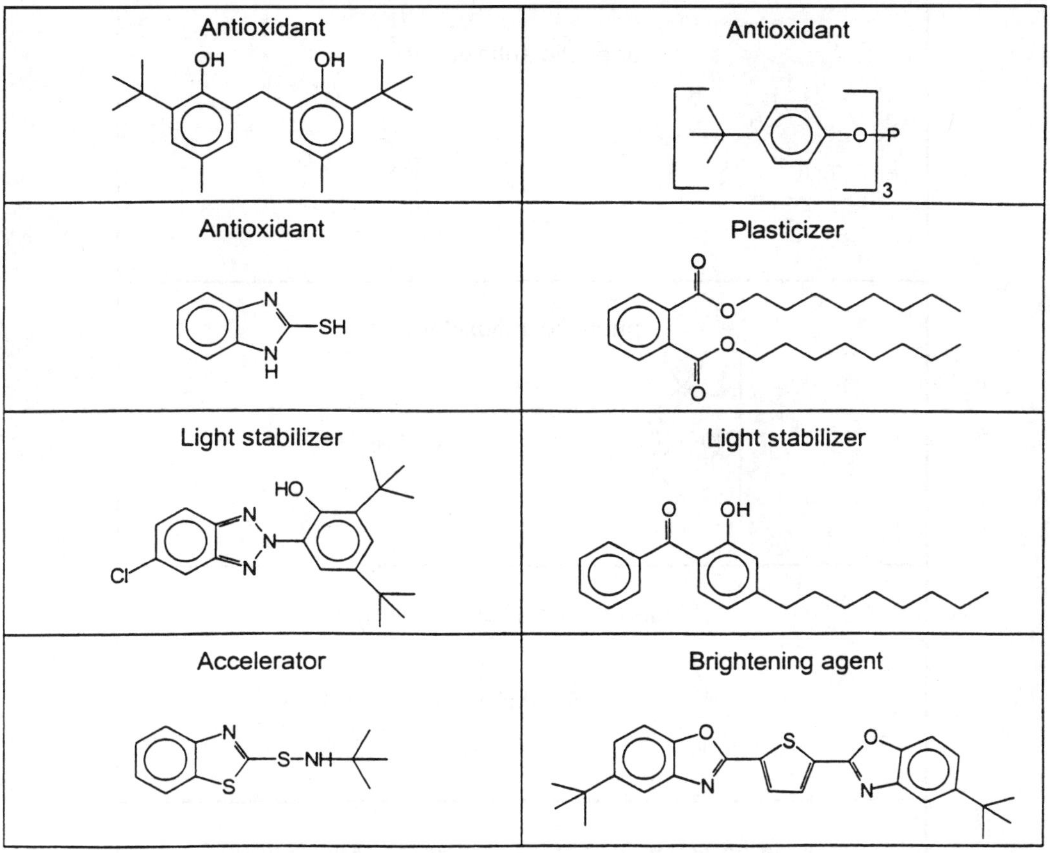

Fig. 2.30. Examples of additives with almost no fragmentation during flash pyrolysis at 550°C. After Kuch [502]. Reproduced by permission of Shimadzu Corporation, Japan.

do fragment at 550°C. In polymer/additive analysis, high fragmentation of the polymer is beneficial because the polymer fragments are not normally of analytical interest when the additives are being investigated.

The concentration of the additive in the polymer is obviously of decisive importance in identification by using PyGC-MS. When considering a major pyrolysis fragment to identify an additive with a concentration of 0.2% having a mass proportion of 20%, then the initial additive concentration corresponds to 0.04% in comparison to a non-fragmenting substance. For a sample weight of 300 μg and a split ratio of 1:30, this means an absolute quantity of 0.004 μg (or 4 ng). At its highest sensitivity, a modern GC-MS is capable of detecting 0.1 ng (or 100 pg) of methyl stearate at m/z 298 for S/N > 30. This exceeds by a factor of 40 the requirements for a concentration of 0.2% and a sample weight of 300 μg. Consequently, in general detection should be straightforward for an additive that completely reaches the GC-MS system without being fragmented. It obviously also helps if the fragments of the additive do not react with those of the polymer matrix to give further products. In order to rule out follow-up reactions between additive and polymer fragments during pyrolysis the TIC of the pure polymer should be compared with that of the polymer plus additive. Conditions such as heating rate for pyrolysis of a pure additive may differ from those in the polymer matrix. Consequently, the calibration curve ("standardisation curve") for a given additive, the so-called external calibration, may not be identical to that for the same additive in the polymer.

Before analysing polymer samples it is useful to examine first the pyrolysis behaviour of the **pure additives**. For example, pyrolysis of N-isopropyl-N'-phenyl-p-phenylenediamine (IPPD) at 550°C leads to the formation of N-phenyl-p-phenylenediamine

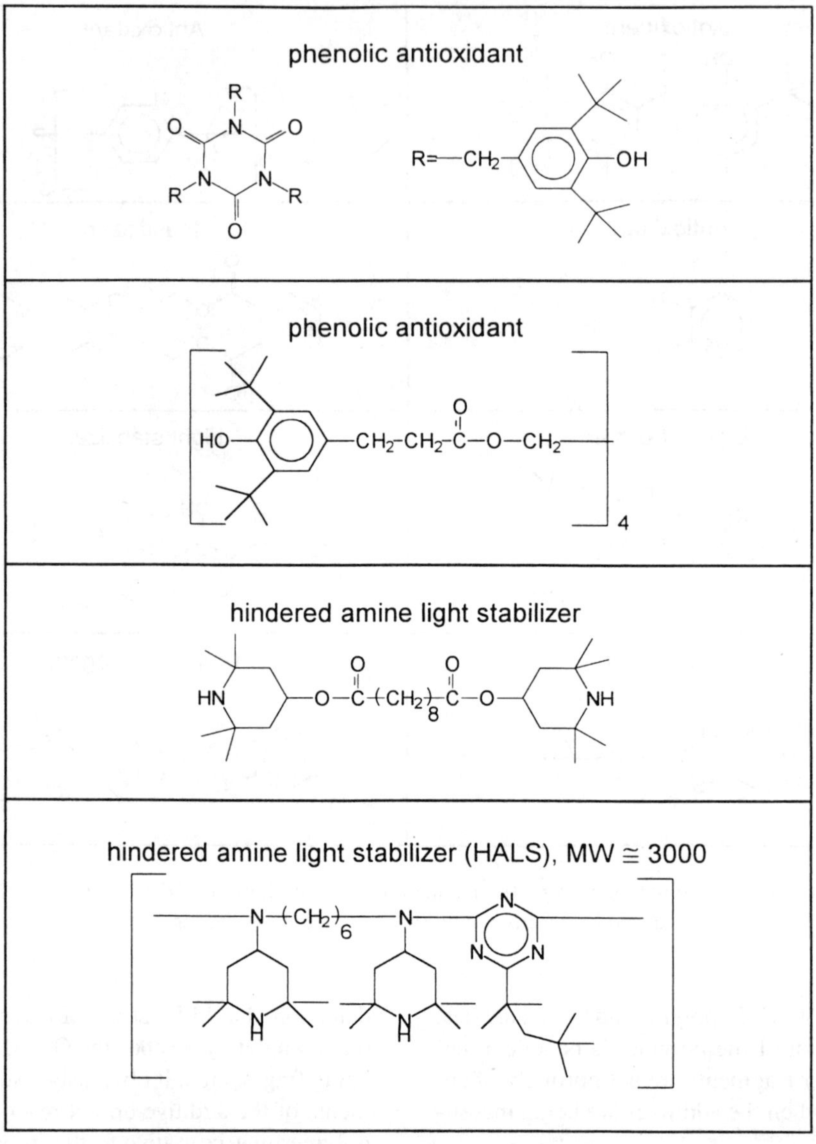

Fig. 2.31. Examples of additives with extensive fragmentation during flash pyrolysis at 550°C. After Kuch [502]. Reproduced by permission of Shimadzu Corporation, Japan.

(PPD) by cleavage of an isopropyl group of IPPD. From the fact that IPPD concentrations of 1% and less in vulcanised NBR, polychloroprene (CR), EPDM, polyisoprene (NR) and styrene–butadiene copolymers (SBR) can be detected, it is concluded that in these cases the fragmentation of IPPD takes place apparently without any influence by decomposition of the polymer matrix.

In case of great **similarity in chemical structures** of additive and polymer fragments it is no longer possible to identify the additive fragment among the polymer fragments. This has been observed for 0.2 phr thiodipropionate in PP [502]. The spectra of the major degradation products of the antioxidant, being 1-octadecene and octylacrylate, are not very specific. In fact, selecting m/z 97 from the mass spectrum of octylacrylate and examination of the TIC of PP shows that this mass is included in each peak for PP in the range of the retention time for the octylacrylate (Fig. 2.32). Direct detection of thiodiproprionate in PP is thus not possible.

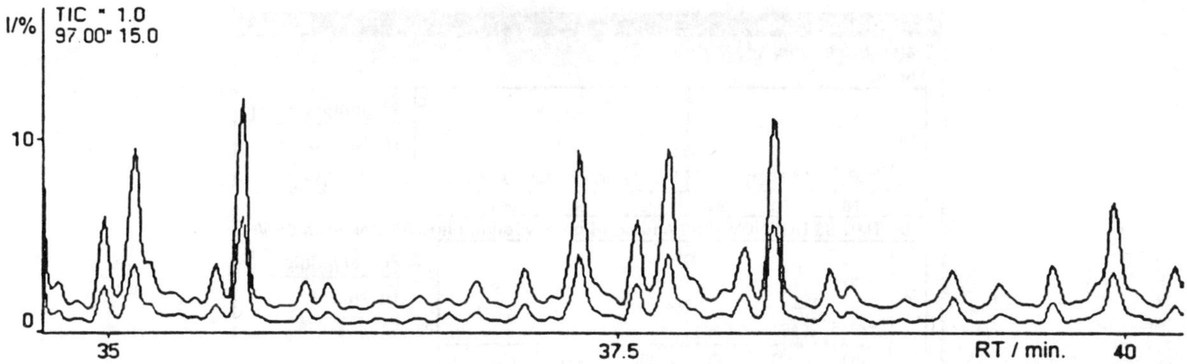

Fig. 2.32. Total ion chromatogram of 0.2 phr thiodipropionate in PP and the mass trace at m/z 97. After Kuch [502]. Reproduced by permission of Shimadzu Corporation, Japan.

GC enables separation of pyrolysis fragments of an additive in an unknown polymer sample. To avoid overloading of the GC column less than some 10 μg of the constituents from a sample capable of pyrolysis should reach the column directly. Therefore, sample sizes of 100 μg to 300 μg are recommended. The maximum sample mass is of the order of magnitude of the inhomogeneities of technical polymeric materials, such as polymer blends and compounds. It follows that in those cases a statistical approach would require multiple determinations and this would benefit from an autosampler.

In order to set an internal company standard for quality control, Volkswagen AG Wolfsburg has developed a **PyGC-MS protocol for additive analysis** in co-operation with Shimadzu. The system consists of a furnace type pyrolyser, capillary GC column, and is equipped with a unique, extensive library, including some 500 additive EIMS fragment spectra from 300 additives (comprising antioxidants, metal deactivators, antiozonants, light stabilisers, processing and cross-linking agents, heat stabilisers, blowing agents, plasticisers, antistatics and flame retardants) collected in standard operating conditions [764–766]. The "VW/Shimadzu Additive – MS library" allows additive identification on the basis of retention times and EI mass spectra. Figure 2.33 shows a typical entry to this library. Each entry contains information about number and type of fragments of an additive, concentration and significance of each fragment. In the standardised VW/Shimadzu procedure the pyrolysis temperature for application of the MS library is set at 550°C (and use is indeed restricted to this temperature). At 550°C the polymer is sufficiently fragmented and the GC system is not too strongly loaded with involatile, high-MW pyrolysis products, which might cause memory effects. In addition, fragmentation of polymer additives and agents is sufficiently low at 550°C so as to allow easy identification. As shown elsewhere (*cfr.* Fig. 2.25), 550°C is a compromise choice for the purpose of additive analysis in polymers and rubbers and not an optimum choice for a given system (consider Figs. 2.30 and 2.31). Other workers have built up similar libraries at 450°C and 800°C [767].

In order to account for column degradation reference substances are often added to the unknown sample and so-called relative retention times (RRT) are used instead of the absolute retention times (RT) for the analyte. The retention times of the analytes are thereby calculated in relation to the retention time interval for the standards. Requirements placed on RRT standards are identical to those for internal standards for the quantitative GC analysis. The procedure adopted is based on the assumption that retention times for the standards are affected by column ageing in the same way as those for the analytes, thus ensuring retention time stability. In addition, standards must have sufficient thermal stability at 550°C and should not undergo any reactions with pyrolysis products of the sample. They should also exhibit characteristic mass spectra that permit unambiguous identification. If standards are to be used for quantitative evaluation on the basis of peak areas, then it must be assured that these are not also being formed by pyrolysis of the polymer matrix. A typical RRT reference solution is made up by dissolving 2.0 mg anthracene and 10.0 mg dibenzo[a,h] anthracene (DBA) in 20 mL dichloromethane.

Analysis by means of the VW/Shimadzu protocol is thus performed in a **standardised fashion:** *(i)*

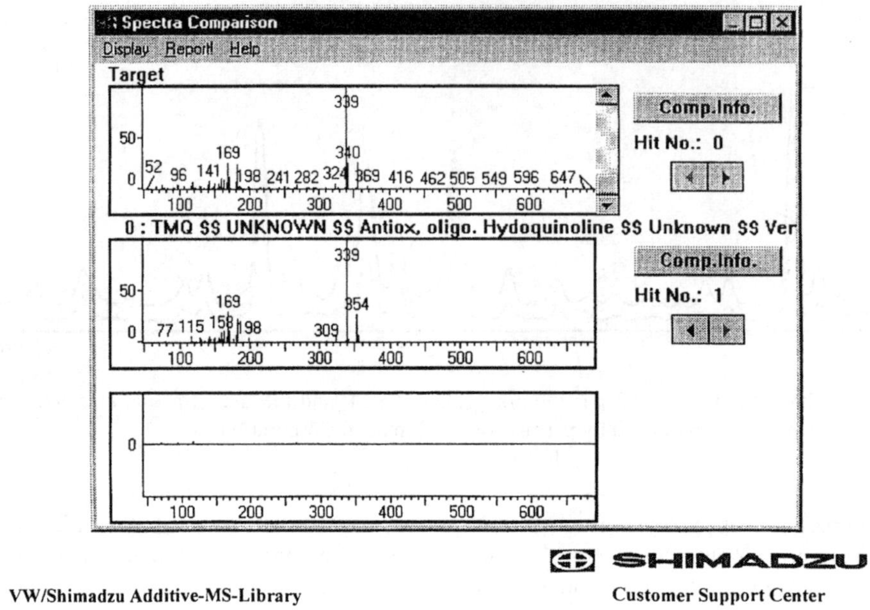

Fig. 2.33. Matching of an unknown component using the VW/Shimadzu Additive–MS Library. After Kuch [502]. Reproduced by permission of Shimadzu Corporation, Japan.

for each pyrolysis fragment of an additive the relative retention time (RRT) is stored in a data library; *(ii)* only fragments with RRT identical to the stored RRT in a database are examined; and *(iii)* identification of specific compounds is based on several characteristic mass ions using mass spectra of additive fragments stored in the library. For unambiguous identification sample and library spectra must agree and retention times must be respected. Only standardisation of the analysis conditions may lead to secure identification by means of target compound software. The VW/Shimadzu Additive – MS library can easily be extended as new spectra can be stored by the user and the method files can be modified (*e.g.* entry of new masses for different additives).

As to its role in identification of additives in polymeric materials it should be emphasised that the **screening power** of PyGC-MS in the aforementioned standard conditions is limited. This is easily appreciated by considering Figs. 2.30 and 2.31; not all additives fragment at 550°C in a selective way. The restriction of PyGC-MS as a screening device is due to the differences in optimum fragmentation temperature for various additives. The damage caused by overfragmentation cannot be restored by the choice of a "good" ionisation mode. Consequently, PyGC-MS is not to be considered a direct full-proof screening device. Broadband screening by means of TLC is essentially more reliable. Moreover, there is a difference between pyrolysis of pure additives and in-polymer additives (detection limit). Nevertheless, PyGC-MS is highly suitable for comparison of *good* and *bad* quality (fingerprinting).

Absolute quantitation in analytical pyrolysis is seldom used. Recently, a few experiments have reported absolute quantitative results using an off-line system, which requires trapping of the pyrolysis products and their solubilisation with a known amount of solvent [768]. An internal standard was added to the solution for further GC analysis. This rather complicated procedure eliminates the most attractive aspect of PyGC-MS, namely minimal sample work-up. Later, Bocchini *et al.* [769] have proposed a simple method to obtain lignin absolute quantitative results by on-line PyGC-MS using 1,3,5-tri-*t*-butylbenzene, 1,2,4-tribenzoic acid trimethyl ester and 1,3,5-trimethoxybenzene as internal standards. Quantitation requires control and optimisation of the many parameters characterising the method (Py, GC, MS). This is not a trivial matter in a dynamic system with variations from a flow of inert gas (PyGC) to vacuum conditions (MS). Flame ionisation detection is one of the most frequently used detection methods for quantitative analysis of pyrolysates. However, few PyGC-MS systems are also equipped with an FID or ECD detector. For the

Table 2.33. Qualifying features of PyGC-MS in polymer/additive analysis

- Identification of the base polymer
- Unequivocal identification of polymer additives by comparison with additive library mass spectra
- Fractionation of unknown compounds into their organic constituents
- Quantification of additives, though exacting
- Direct determination of single additives in the solid matrix down to 0.1 phr
- Identification of interfering compounds and availability of alternative procedures
- Allowance for investigation of the migration behaviour of additives

EI ion source, the generated total ion stream is directly proportional to the gas pressure in the impact field, which provides a basic condition for quantitative analysis. Quantitation is based on the fact that degradation is ion specific, *i.e.* a given substance always produces the same percentage of fragment ions. The additive content may be determined from the total mass and the integrated fragmentation pattern [770]. Internal standards (*e.g.* chrysene or a polymer peak) are frequently used and SPC standards for highest and lowest concentration, and multiple measurements of each concentration or sample (usually 3 times) are recommended. A complete quantification requires considerable time and effort. Multivariate calibration models are in use for quantification [771]. Quantification of additives by means of PyGC-MS is possible for additives depending on their nature (fragmentation behaviour). Relatively few such reports have appeared (*cfr.* Table 2.38). RSD values of about 5–10% are quoted.

From the above, some important features of pyrolysis GC-MS emerge, as given in Table 2.33. On-line flash pyrolysis GC-MS, with Curie-point, resistively-heated filament or furnace pyrolysers, is very widely utilised for identification of pyrolysis products from synthetic polymers. The **main characteristics** of PyGC-MS of polymers, as given by Schulten *et al.* [692], are shown in Table 2.34. PyGC-MS is an excellent tool for fast product quality control; for R&D purposes full control of the (many) experimental parameters is needed. Polymer standards (*e.g.* SEC standards) can be used to determine sensitivity and precision of PyGC-MS.

The small sample size is particularly advantageous for damage cases. Changes in composition

Table 2.34. Main characteristics of PyGC-MS

Advantages:

- Short sample preparation time (no extraction or dissolution required)
- Identification (and quantification) of (non)-volatile components without prior separation
- Small sample quantities (<0.5 mg) (*e.g.* damage cases)
- Molecular weight of pyrolysis products directly derivable from the (quasi)molecular ions in mass spectra
- Elemental compositions of mass signals from high-resolution mass spectrometry
- Molecular structural information from the mass spectral fragmentation pattern (comprehensive EIMS libraries commercially available)
- Spectrum interpretation supported by automatic library search
- Direct analysis of complex mixtures
- *In situ* derivatisation
- Mass spectra of mixtures as fingerprints for fast quality control
- Chemometrical evaluation techniques (PCA, cluster analysis) for classification and differentiation
- Wide application range for organic material fragmenting into volatile products (up to 800 Da)
- High information content
- Short analysis time (5 min using fast GC columns)
- Automation

Disadvantages:

- Sampling problems for inhomogeneous materials
- Multiple measurements needed for statistical evaluations
- Limited sensitivity (LOD 0.1%)
- Influence of sample geometry
- Analysis time (usually about 1 h for standard GC)
- Unsuitable for very polar and high-MW pyrolysis products
- No inorganic probing
- Corrosive fragments (*e.g.* HCl, HF) undesired
- Extensive method development required
- No standard quantitative analysis

within ng or even pg ranges can be detected. Comprehensive EIMS libraries are commercially available; spectrum interpretation is supported by automatic library search. All organic material which fragments into volatile products at up to 800°C or evaporates without fragmentation can be analysed. The selectivity for apolar and medium polarity pyrolysis products depends on the choice of the chromatographic column. On-line derivatisation during pyrolysis [611,612] reduces this drawback to some extent. Purely inorganic additives, such as oxides, chalk or talc are not volatile under GC conditions and are inaccessible to PyGC-MS analysis and residues have to be analysed by other methods like AAS, AES or IR. Organometallic and complex compounds can only be identified by their organic residual substances. Samples that liberate acidic or corrosive products such as halogenated hydrocarbons, HCl, HF or SO_3 are less suitable for PyGC-MS. This holds for PVC, CPE, CR and flame retardants by dehydrohalogenation. Nevertheless, sometimes PyGC-MS is being used in such unwanted circumstances [772]. As all hyphenated techniques, PyGC-MS is not of easy operation and a qualified operator is wanted.

Despite interfacing and chromatographic filtering problems, PyGC-MS is an attractive on-line method for structural characterisation of monomeric and sub-monomeric products of pyrolysis and for the analysis of thermally extractable compounds desorbed from complex matrices. Condensation of certain fractions of the pyrolysate, different methods of column interfacing and the chromatographic columns are important factors which reduce the **interlaboratory reproducibility** of PyGC-MS. In particular, the stability and accuracy of the system are strongly reduced by fouling of the MS and GC-MS interface. Column switching, in which the high-MW fragments are backflushed, can strongly reduce fouling of the ion source. Stankiewicz *et al.* [773] have reported a comparison of the analytical performance of filament and Curie-point pyrolysis devices in combination with GC-MS and concluded to a high degree of comparability provided that the analytical variables, *e.g.* sample size, GC column stationary phase, carrier gas, *etc.* are strictly controlled. With sufficient care regarding the pyrolysis process and in maintaining good chromatographic practices reproducibility of PyGC-MS is satisfactory. Otherwise differences in performance are noticed [774]. Method development is required, as illustrated in Scheme 8.1. Occasional comparisons of PyGC-MS with other techniques for polymer/additive analysis have been reported (*cfr.* Chp. 6.2.3 and 6.2.4).

Essential conditions for application of PyGC-MS as an **industrial analysis tool** are given in Table 2.35. In industrial research laboratories the drive towards cost reduction is high. Analysis costs of PyGC-MS can be cut by a variety of options: *(i)* Programmed Temperature Vaporisation (PTV)-injector as a simple, inexpensive and versatile multi-step thermal desorption/programmed pyrolysis system; *(ii)* autosampling system; *(iii)* fast-GC (*cfr.* Table 2.36); and/or *(iv)* fast column exchange. During

Table 2.35. Requirements for application of PyGC-MS in an industrial analytical laboratory

- High stability of the mass spectrometer (tuning, temperature, pressure)
- No contamination with oxygen
- Calibrated pyrolyser, GC temperature and MS parameters
- Optimal column selection (preferably fast GC)
- Absence of "memory effects" for pyrolyser and GC
- Autosampler for speed and statistical data analysis
- Availability of user defined mass spectral databases (both for polymers and additives)
- Short maintenance time
- Quantitative potential

pyrolysis of polymeric materials at 500 to 600°C some fairly high-MW fragments of the polymer matrix are produced without any relevant information as to the additive package. Generally rather long GC run times are necessary to get rid of these unwanted products. Fast-GC methods employing a narrow-bore column of 20 m × 0.15 mm with a high pressure unit and high split ratio are then highly desirable. Not only does total analysis time decrease by about 70% without loss of separation performance, but less polymer is fouling the GC column and MS detector at high split ratios. The much lower concentrations passing GC lead to less contamination, higher column stability and reduced service needs for MS.

Being in competition with other industrial analytical techniques, PyGC-MS might lose its competitive edge if not further developed. Fast GC-MS analysis of polymer extracts is already available. **Py-fast GC-MS** using 0.2 mm i.d. columns has recently been reported [775]. With robotic systems in particular, sampling problems (arising from heterogeneity of polymer compounds) can be tackled more easily. Results can further be improved using soft ionisation (CI, SI), direct inlet (DI) and multivariate data analysis (MVA).

Some **operational PyGC-MS variants** are available: *(i)* thermally assisted alkylation; *(ii)* oxidative pyrolysis; and *(iii)* temperature-resolved PyGC-MS (*cfr.* Chp. 2.2.7). *In situ* thermally assisted methylation GC-MS with tetramethylammonium hydroxyde (TMAH) (Py-TMAH-GC-MS) has been reported as a useful analysis technique for analysis of polar analytes [636]. In case of normal PyGC analysis, pyrolysis is performed in inert gas (N_2, He or H_2) as a carrier gas. Oguri *et al.* [584] have recently described a combustion gas analyser composed of a

Table 2.36. Typical experimental parameters for PyGC-MS and fast PyGC-MS

Parameter	Standard GC	Fast GC
Column geometry	60 m × 0.32 mm i.d.	20 m × 0.15 mm i.d.
Film thickness	1.0 μm	0.15 μm
Stationary phase	PDMS + 5% phenyl	PDMS + 5% phenyl
Carrier gas	Helium	Helium
Column flow	1.2 mL/min	0.5 mL/min
Split radio	1:30	1:100
Pyrolysis temperature	550°C	550°C
Sample weight	50 μg	160 μg
Mass range	45–900 amu	45–900 amu
Scan speed	1000 amu/s	4000 amu/s

Curie-point pyrolysis/combustion chamber in which sub-μgs of plastic are burned in air atmosphere in the chamber; the exhausted combustion gas is then cryofocused by a purge-and-trap (PT) device and analysed by GC-MS. In this way air-pyrograms are easily obtained. Lehrle *et al.* [598] have proposed PyGC-MS in **oxidative pyrolysis conditions**. Pyrolysis is then performed whilst air still remains within the pyrolysis chamber (before complete evacuation). This approach offers advantages over the so-called Enclosed Curie-Point (ECP) pyrolysis method for polymer and oil oxidation studies [511]. The latter procedure suffers from the limitation that secondary reactions may arise due to the fact that it is effectively a two-stage process, in which the analysis follows the oxidation stage.

Frequently used competitive techniques for PyGC-MS are NMR, IR and MALDI-ToFMS. The subject has been covered elsewhere [497–499].

Applications
Pyrolysis GC-MS finds wide application for a variety of purposes, such as fingerprinting of polymeric materials, screening of additives, deformulations of solid polymer/additives, testing of raw materials, quality control of organic and polymeric products, handling of customer complaints, analysis of damage cases involving polymers (comparison to reference samples), determination of contaminations (impurities, residues), quantitation, study of additive degradation (environmental issues), and forensic applications. The small minimum sample size (approximately 0.05 mg) favours many of these applications. PyGC-MS permits simple identification of monomers. For example, polar monomers in polyacrylate dispersions have been determined after chemical derivatisation (D-PyGC-MS) with BSA and BSTFA to protect the polar groups [776]. PyGC-MS also allows structural analysis of specific polymers used in finished goods – which include the polymeric matrix, but also additives, fillers, fibres, colorants, *etc.*, regardless of the presence of such additives. *Vice-versa*, additives may be determined regardless of the polymer albeit with some limitations as to concentration and type of additive. No molecular weight information is gained.

Pyrolysis GC-MS is frequently used for **compositional studies** of (co)polymers, modified polymers (*e.g.* grafted or impact modified) and characterisation of rubbers (such as BR, SBR, NBR, IIR, IR, NR, EPDM, *etc.*), often replacing the older pyrolysis hydrogenation gas chromatography (PyHGC) technique. For example, the chemical composition of EPDM in terms of the ethene/propene ratio and amounts of dicyclopentadiene (DCPD) and ethylidene norbornene (ENB) can be determined by means of PyGC-MS with similar accuracy as with PyHGC-FID (average deviation on C_2 content of ±2–3%, on ENB and DCPD content of ±0.2–0.3%; internal standard: chrysene). Blazsó [777] examined polysilane copolymers and phenol-formaldehyde polycondensates by means of PyGC-MS. The technique was also used to analyse polyurethane-based consumer products [778]. Lehrle *et al.* [585] have used PyGC-MS to assess the thermal behaviour degradation behaviour of PVC. Starch has been characterised by PyGC-MS and multivariate data analysis [779].

Analysis methods such as PyGC-MS are necessary since in some cases they are the only characterisation method that can be used. Occasionally such small amounts of a material are available only that no film can be made for FTIR measurements. The use of FTIR microspectroscopy determinations of

Table 2.37. Examples for PyGC-MS samples

Chemical	Technological		Damage cases
Monomers	Plastics	Fibres	Crack formation
Oligomers	Rubbers	Textiles	Fracture
Polymers	Adhesives	Paper	Discolorations
Copolymers	Resins	Cellulose	Change in gloss
Blends	Seals	Humic acids	Scaling
Thermoplastics	Coatings	Viscous fluids	Layer deposition
Thermosets	Paints	Dispersions	Contaminations
Elastomers	Lacquers	Oils	Agglomeration
Biopolymers	Films	Inks	Degradation
	Foils	Bitumen	Environmental
	Foams	Additives	Forensic
	Packaging materials		

ENB and DCPD in the aforementioned example is also quite inaccurate due to the low concentrations. Other difficult samples are cross-linked gels or compounds incorporating carbon-black and oil. Direct analysis by means of PyGC-MS is particularly indicated for highly **intractable solids**, such as cross-linked polymers (*e.g.* vulcanisates, resins, grafts), but also polymer blends, samples heavily charged with inorganic parts (*e.g.* glass fibres, fillers, carbon-black, metal stearates) or impact modifiers, extraction resistant oligomeric, polymeric or grafted additives (*e.g.* AOs, FRs). Low levels of polyvinylpyrrolidinone (PVP) in polysulfone (PSU) were determined by PyGC-MS and FTIR [780]. PyGC-MS may also act as a direct inlet for TLC spots. Not surprisingly, PyGC-MS has found early application for rubbers, but lately also for synthetic polymers, paper, paints, forensic textiles, *etc.* Determination of rubber additives (such as IPPD, TMQ and aromatic peroxides), which are usually present in relatively high amount (percentage level), is easier than polymer/additive analysis, where the concentrations are generally lower (hundreds of ppm). The influence of carbon-black on pyrolysis has been studied [665]. As may be noticed, reported applications in polymer/additive analysis mainly concern flame retardants (which are thermally stable compounds, in high concentration), impact modifiers (well identifiable, high concentrations), various phosphorous containing compounds (thermally stabile, volatile) and aromatics (such as UVAs). Identification of impact modifiers by means of PyGC-MS is not trivial, as details in the pyrogram need to be considered for positive identification. Table 2.37 lists some examples of samples which have been subjected to PyGC-MS analysis.

Because of the great variety of polymer types and additives, of special interest are those analytical techniques which allow fast identification of all compounds in a kind of **"screening"** mode. PyGC-MS may be used for that purpose, being fast, specific and sensitive. In the standard VW/Shimadzu protocol [781] for qualitative additive analysis 10 μL of a reference solution consisting of 0.1 mg/mL anthracene and 0.5 mg/mL dibenzo [*a,h*] anthracene (defining a relative retention time scale) are added to 10–300 μg of the sample. Gruber *et al.* [782] mentioned that the minimum amount of additive needed for identification is 0.2 μg (or 0.1 wt.% for a 200 μg sample). More additive is needed if fragmentation leads to many fragments (*e.g.* Wingstay L) or to very large fragments with low intensities (*e.g.* Saytex 8010). Groupswise additive identification is very successful, provided that the additive is incorporated in the mass spectral library and that the mass fragments are sufficiently specific. However, as pointed out before, screening by means of PyGC-MS is not waterproof. Various classes of compounds cannot be observed by PyGC-MS (*e.g.* non-aromatic peroxides in rubbers) or cannot be discriminated (*e.g.* HALS compounds). To fully exploit the possibilities of PyGC-MS more background information about standard polymers and additives is highly desirable.

PyGC-MS permits direct deformulation of **solid polymer/additive matrices**, independently of cross-linking, filler or pigmentation type. The technique allows the determination of the composition of polymer and additives. Liebmann *et al.* [783] used

Py-APCI-MS, PyGC-MS/MS, GC-FTIR and SFC to analyse and classify such difficult systems as core/shell formulation products (micro-encapsulated materials), including thickeners, additives and carriers. Because of the low level of **antioxidants** normally used, they cannot be analysed directly by most common spectroscopic or thermal chemical techniques. Lichtenstein *et al.* [784] reported qualitative and quantitative analysis of 2,6-di-*t*-butyl-*p*-cresol by means of on-line CuPyGC-MS in butadiene/styrene copolymer. A coefficient of variation of 10% was achieved with a detection limit of the additive of 1 μg. The selectivity was enhanced by using single ion monitoring. Wang [626] examined Irganox 1010/1076/1035/MD 1024/259/3114/1425/565 and Irgafos 168 by means of PyGC-MS at 950°C and has identified Irganox 1076 and Irgafos 168 in an extract of GE Cycoloy C 3600 (ABS/PC/PMMA blend). Pasch *et al.* [785] described PyGC-MS analysis of a variety of antioxidants, such as Tinuvin 320 (MW 323.43), Tinuvin 571 (MW 393.57), Irganox 3114 (MW 784.10), Irganox 3052FF (MW 394.55), Hostavin N20 (MW 364.57), the oligomeric HALS Hostavin N30 (MW $\sim$ 1500), Hostanox O3 (MW 795.06), and analysed various polymer formulations such as PA6/0.3 wt.% Tinuvin 320 (m/z 308, 323), PMMA/1 wt.% Irganox 3052FF and PP/1.05 wt.% Hostanox O3. The main peak in the TIC diagram of PA6/Tinuvin 320 is ε-caprolactam; the calibration curve is linear up to 1.0 wt.% [786]. Quantitative determination of PP/(0.3–1.0 wt.% Irganox 3114) used 7-methyl-1-undecene as the internal standard. At variance to MALDI-ToFMS, PyGC-MS discriminates a mixture of the structural isomers Tinuvin 320, Tinuvin 350 and Tinuvin 329 (all $C_{20}H_{25}N_3O$) [785], *cfr.* Figs. 2.34 and 2.35. Direct identification of an additive in a polymer is difficult or impossible when the additive and polymer fragment, *cq.* when their mass spectra, are very similar.

PyGC-MS has also been used for analysis of Irganox 1010 in PE and PBT in the presence of TMAH [787]; reported relative standard deviations of 7% for PE and 3% for PBT samples. It is possible to distinguish various pure high-MW HALS compounds (Chimassorb 16/119/944, Cyasorb UV3346/CEC-3529 and Uvasorb HH 88) by means of PyGC-MS. Blazsó [788] has examined fast pyrolysis from 300 to 900°C of Tinuvin 144 and HM-HALS products (Tinuvin 622, Chimassorb 119/944) in the absence and presence of PVC. Direct analysis of AOs in the polymer by PyGC is considered difficult because of the low additive concentration and possible interference from the original polymer matrix. Antioxidants can be qualitatively and quantitatively analysed by PyGC more easily after separating the polymer and additives.

Recently, Dow Chemical has put much emphasis on the use of PyGC-MS, but mainly for qualitative purposes [625,626,632,789]. **Lubricants** are another class of low-concentration additives. Analysis of lubricants (rather large molecules with polar functional groups) usually relies on polymer-additive separation followed by GC or LC and identification. Awareness of the level of purity of the technical-grade lubricants and composition of the common fatty acid in lubricants is important. Wang *et al.* [625] have examined PyGC-MS for analysis of various lubricants in polymer extracts: low-MW PE wax, paraffin wax, earth wax, stearic acid, technical-grade butyl stearate, zinc stearate, butyl oleate, butyl palmitate, *N,N′*-ethylenebisstearamide (EBS or AcraWax C), and stearamide. PyGC-MS of LDPE/1% oleamide masterbatch (without pretreatment) has identified 9-octadecenamide as a desorption product among the pyrolysis fragments of the polymeric matrix (α,ω-alkanedienes, α-alkenes and alkanes) [790]. Py-GC analysis of lubricants in a polymer requires a preseparation of the additives and polymer because the lubricant level is usually low. Auxiliary techniques may be required for complete identification.

In comparison to analysis of the low concentration lubricants and antioxidants, PyGC-MS stands better chances for plasticisers and flame retardants, which are usually present in relatively high amounts. **Plasticisers** can be qualitatively and quantitatively analysed by PyGC simultaneously with the polymer composition and microstructure. Wang [632] examined various systems by means of PyGC-MS at 700°C: PVC/DEHP flexible tubing, cellulose propionate/DOA, vinyl chloride–vinylidene chloride/(DBS, TBAC) food packaging film, PC-HIPS/TPP, and PU/(DDP mixture) sealant. Polymeric plasticisers as in PVC/butadiene–acrylonitrile and styrene–butyl acrylate copolymers can be resolved by their pyrolysate pattern. The plasticiser that exists as an internally modified polymer backbone can be investigated by analysis of the copolymer composition.

PyGC-MS is amenable to analysis of **organic flame retardants** such as halogenated organics and

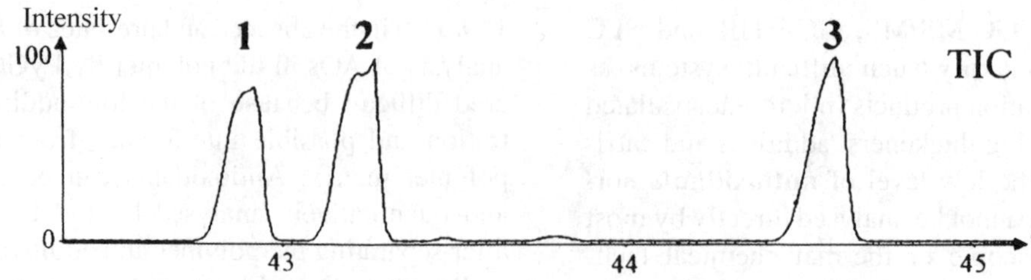

Fig. 2.34. Gas chromatographic separation of the isomers Tinuvin 320 (1), Tinuvin 350 (2) and Tinuvin 329 (3). After Meyer-Dulheuer *et al.* [785]. Reproduced by permission of Hüthig GmbH.

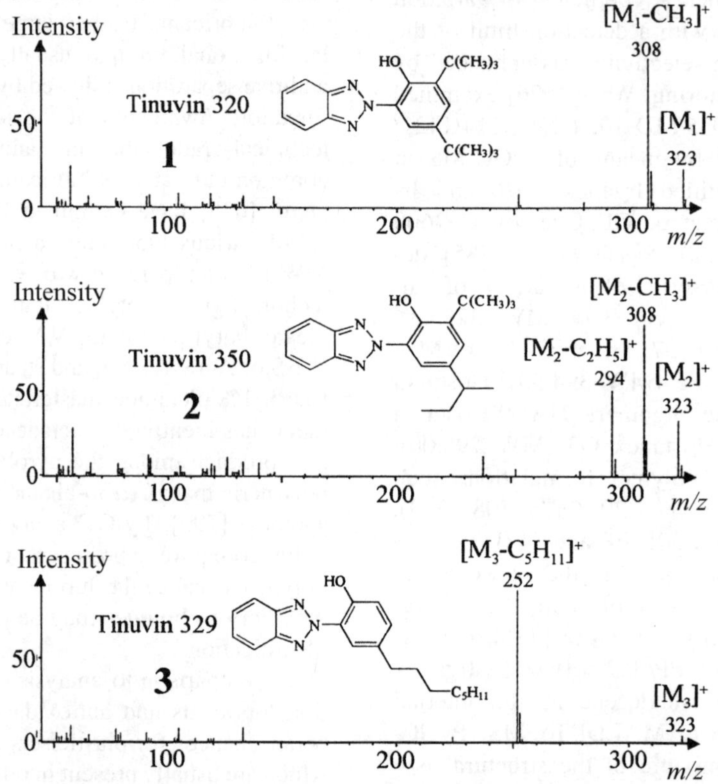

Fig. 2.35. Identification of the UV absorbers Tinuvin 320 (1), Tinuvin 350 (2) and Tinuvin 329 (3) by mass spectrometry. After Meyer-Dulheuer *et al.* [785]. Reproduced by permission of Hüthig GmbH.

halogenated or non-halogenated phosphate esters, which can be analysed qualitatively and quantitatively simultaneously with the polymer composition. Wang [789] examined a cross-linked epoxy resin/brominated bisphenol-A printed circuit board, a poly(diallylphthalate)/Dechlorane Plus connector, a PBT/brominated phenol computer insert card socket, GFR PBT/brominated polystyrene, ABS/PVC, ABS/OBDPO, ABS-PVC/brominated phenol, and an ABS-PVC/TPP instrument panel by means of PyGC-MS at 950°C. Without knowing the identity of the flame retardant, there is no simple way to predict which fragment ion/mass to monitor. Even though many flame retardants contain halogen elements, there is no good way to know the identity of the aliphatic or aromatic counterpart to which the halogen element is attached. In such cases, AED may be superior to MS for detection of specific-element-containing fragments and their relative intensity pattern. Identification of unknown flame retardants may be complex matter. Scheme 2.6 summarises the approach for brominated FRs in polyesters [269].

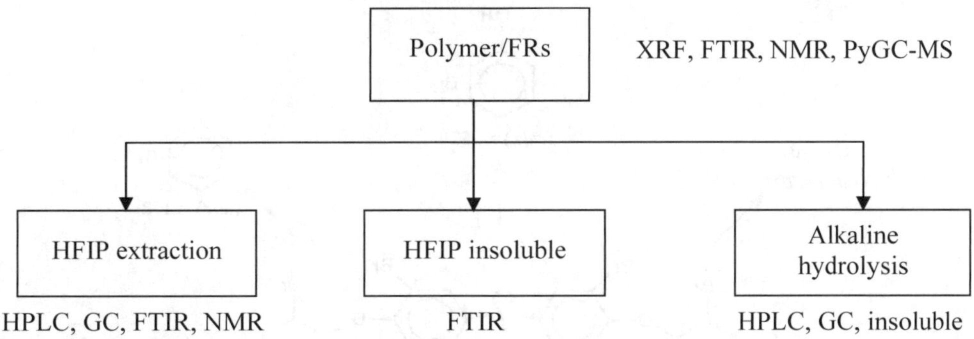

Scheme 2.6. Analysis for identification of flame retardants in polyesters. After Nelissen [269]. Reproduced by permission of DSM Research, Geleen.

PyGC-MS enables differentiation between various brominated flame retardants. Pyrograms of the reference materials (pure FRs) need to be compared with that of the sample to be examined in order to identify the flame retardant class. Selection of the pyrolysis temperature is most important. A compromise between mobilisation of the flame retardants and minimisation of thermal reaction products has to be found. Flame retardants were identified in EPR/TBBA, ABS/TBBA, PBT/TBBA, PP/PBDE, HIPS/PBB, using an optimised pyrolysis temperature (430°C) for these systems [791]. PyGC-MS was also used for polymer and additive (FR) characterisation of a Japanese TV cabinet [792]. Figure 2.36 shows the additive fragments isolated, together with a proposed (sub)structure, sufficient to identify the flame retardant as tetrabromobisphenol-*S*-bis-(2,3-dibromopropylether) (TBBP-S) on the basis of patent search.

Van Eldik *et al.* [793,794] analysed flame retardant recycling materials (television and computer housing material), mainly by means of FTIR and TGA (for polymer identification), ED-XRF (for identification of halogen containing samples), PyGC-MS (for determination of FR class) and HRGC-MS (for the quantification of polybrominated dioxins and furans, PBDD/F). In these plastics for electrotechnical applications, brominated bisphenols (tetrabromobisphenol-A, TBBA) and polybrominated diphenylethers (PBDE) were determined by means of PyGC-MS and a suitable clean-up method was developed for quantification of the PBDD/PBDF content in polymer extracts containing high concentrations of polybrominated diphenylethers. The concentrations of PBDD/F and of selected FRs (PBDE, PBB, TBPE) were monitored during the recycling process in order to characterise the reaction behaviour of the flame retardants. During recycling FRs and PBDD/F tend to form lower brominated products. The main decomposition products of decabromodiphenylether are hexabromobenzene and pentabromophenol. TBBA reacts by elimination of bromine under formation of lower brominated products. PyGC-MS of electronic scrap containing brominated epoxy resin and brominated polystyrene flame retardants was reported [795]. PyGC-MS of a halogen-free flame retarded (PD 3710) epoxy composite did not reveal formation of dioxines [796].

Nelissen [269] examined the identification of the closely related flame retardants PDBS 80 (Great Lakes; polymerised dibromostyrene), Pyrochek 68 PB (Ferro, halogenated polystyrene; 68 wt.% Br, 0.1 wt.% Cl), Pyrochek 68 PBI (Ferro, brominated polystyrene, 68 wt.% Br) and Saytex HP 7010 (Albemarle, brominated polystyrene, 69 wt.% Br) in polyamides by means of FTIR and PyGC-MS. As a result of interference of heavy polymer absorption bands it is not possible to distinguish brominated polystyrene FRs such as PDBS 80, Pyrochek 68 PB or Saytex HP 7010 in polyamide matrices by means of direct FTIR transmission spectroscopy. Extraction is required for that purpose. Identification of both FR and polyamide is possible by means of PyGC-MS without extraction and is less time-consuming. Pyrolysis of PDBS 80 leads to dibromostyrenes as a main product, as opposed to tribromostyrenes for Saytex 7010 [797]. PyGC-MS is thus particularly valuable when no alternative standard procedures are available as in case of brominated polystyrenes, used as flame retardants in polyamides. Problems

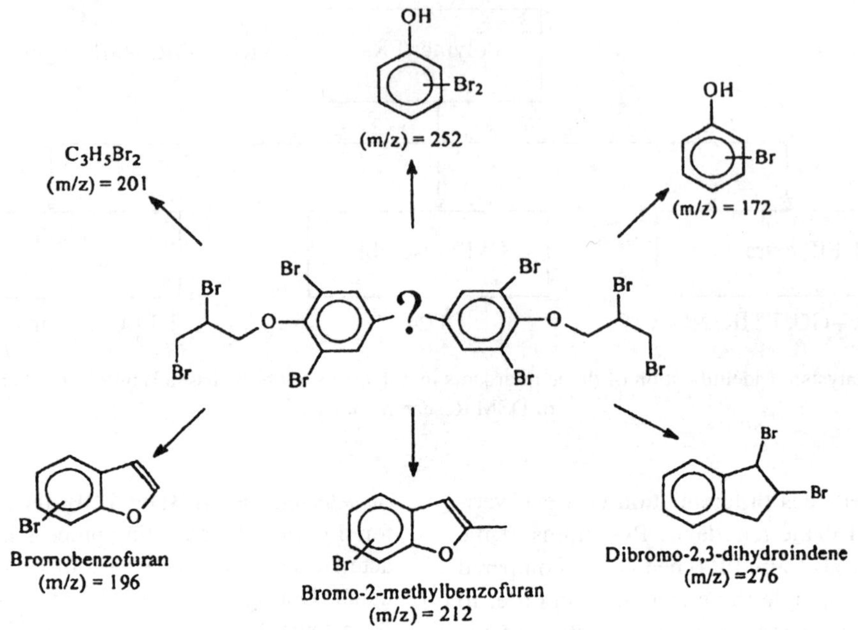

Fig. 2.36. Pyrolysis products of 1,3-dibromo-2-(2,3-dibromopropoxy) benzene. After Dettmer *et al.* [792]. Reproduced from *Chemosphere* **39**, F.T. Dettmer *et al.*, 1523–1532 (1999), with permission from Elsevier.

may occur when additives break up in frequently encountered fragments. Quantification is possible after appropriate calibration. None of the reported techniques allows distinguishing between Pyrochek 68 PB and Pyrochek 68 PBI.

PyGC-MS has also been used to investigate cotton fabrics [798,799], in particular the yields of volatile products and char from FR treated **cellulosic (cotton) fabrics**. As expected, the vapour phase active materials APP-ammonium bromide (Amgard CD) and an antimony(III) oxide-aliphatic bromide (Flacavon H14/587) resulted in low char yields and high yields in volatiles and CO. The condensed phase flame retardants APP (Amgard TR), a phosphonium salt-urea-polycondensate (Proban CC) and a phosphonopropionamide (Pyrovatex CP) produced large amounts of char [798]. Zaikov *et al.* [800] examined the thermal degradation of the polymer FRs triphenylphosphine (TPP) and modified, kaolin intercalated, triphenylphosphine (TPP-i) by means of DSC and oxidative PyGC-MS. DSC study indicated the presence of an exothermal reaction of degradation (char cross-linking) above 320°C for TPP-i, as opposed to the sharp endothermal evaporation (sublimation) for neat TPP. Flame retardant copolymers of styrene and methylmethacrylate with various phosphorous containing monomers were characterised by PyGC-MS [801]. PyGC-MS was also used to establish the nature of the reaction products of triglycidyl isocyanurate (TGI) with ortho- and polyphosphoric acid as flame retarding agents [802]. The mechanism of action of aromatic sulfonates as FRs on PC was investigated by means of PyGC-MS [803]. The technique was also used to study the smoke suppressing effect of FeOOH in CPVC-65% Cl containing the commercially available plasticisers DOP or Santicizer 2148 [804].

Determination of **additives in rubbers** by means of PyGC-MS is actively being pursued. Antioxidants in vulcanised SBR compounds have successfully been analysed using heat-desorption by means of a double-shot-pyrolyser (*cfr.* Fig. 2.23), followed by GC-MS analysis [805]. Kim *et al.* [806] have used PyGC-MS in the identification of organic additives in cured rubber without any sample pretreatment. Takahashi *et al.* [763] have reported use of PyGC-MS (fused silica capillary column, EI/CI mass spectra) to determine both the type of base rubber and additives for tyre rubbers. Polymers used as additives in glass mats could also be determined. Using essentially the same system, Geissler [764] has described PyGC-MS analysis of additives in rubbers and plastics by comparing the pyrolysis fragments with the aforementioned (VW/Shimadzu) additive spectrum library. The technique was also applied for the identification of the antioxidant 2,2,4-trimethyl-1,2-dihydroquinoline in SBR and to the quantitative analysis of *N*-isopropyl-*N*′-phenyl-*p*-phenylene diamine (IPPD) in natural rubber (linear calibration curve for areas of m/z 211 and 226 *vs.* IPPD concentration up to 1%) with a deviation of the peak areas of about 7% using SIM mode [807]. Inconsistencies observed with various polymer lots were ascribed to production problems. The exponential decay of IPPD in a NR vulcanisate during storage at 90°C for up to 600 hrs could be followed quantitatively by means of PyGC-MS using the aforementioned calibration curve [807]. With this method concentration levels as low as 0.1% can be detected.

Sulfidic cross-linking systems usually consist of molecular sulfur, acidic accelerators (mercapto- and dithiocarbamate derivatives), basic secondary accelerators and activators (zinc oxide and fatty acids). **Vulcanisate analysis** aims at identification of vulcanisers, quality assurance and analysis of failures and troubleshooting in production. Evolved gas analysis (EGA) and PyGC with adequate detectors (MS, FID, AED) are suitable chemical-analytical methods for such investigations. Köbisch [808] has described PyGC-MS analysis of various sulfur vulcanised elastomers and examined several zinc-dithiocarbamate and thiuram accelerators (TMTD, ZDMC, ZDEC, ZDBC, TBzTD, TMTM, MPTD), mercapto accelerators (2-MBT, MBTS, ZMBT), sulfenamides (OTOS, MBS, CBS), secondary accelerators (HMTA, DETU, DPTU, ETU, DPG, OTBG, DOTG), sulfur donors (DTDM, DPTT, CLD), peroxidic cross-linking agents and vulcanisation retarders (CTP). In view of the inhomogeneities in elastomers only semi-quantitative analysis of accelerators in a vulcanisate is useful. Using chrysene as an internal standard (added in CH_2Cl_2 solution to the sample) Köbisch [808] reported quantification of several accelerators in vulcanisates on the basis of a specific fragment (pyrolysis at 450°C): TMTD (dimethyldithiocarbamate, m/z 88), MBT (benzothiazole, m/z 135), and MPTD (*N*-methylbenzothiazolthione, m/z 181). MBTS, MBS, CBS and OTOS were determined semi-quantitatively after pyrolysis at 350°C as a higher pyrolysis temperature leads to excessive fragmentation. Cross-linking agents are only identified at low pyrolysis temperatures; peroxidic cross-linking agents can only be analysed qualitatively on the basis of their relatively volatile decomposition products [808]. Herrmann [809] determined 2-MBT in vulcanisates semiquantitatively. The accelerator zinc-*N*-dimethyldithiocarbamate (ZDMC) cannot be detected by PyGC-MS analysis at 550°C in the unfragmented state because of its low thermal stability [502]. However, ZDMC in vulcanised natural rubber (NR) could unambiguously be identified by DI-MS on the basis of the peak spectrum of the molecular mass trace m/z 304 (Fig. 2.37).

The analysis of paint fragments is of interest both in the automotive industry and for forensic purposes (30,000 original samples in EUCAP). Peaks appearing in any pyrogram of an **automotive paint** may be a complex mixture of polymer pyrolysate, additives, plasticisers and other ingredients, each of which has a specific function in the performance of the paint product. In a recent review [810], it has been stressed that the use of PyGC-MS in conjunction with FTIR is an excellent method for the analysis of several types of organic paints and coatings and permits detection of some of the minor ingredients or additives in modern **paint and varnish formulations**. PyGC-MS offers two major advantages over FTIR for most paints. Because it separates out the various components, it can identify mixed media and copolymers more easily. Moreover, the nature of the pigments generally does not interfere with the identification of the binding medium (except for pigments containing a carbonate group such as chalk or lead white, which produce large quantities of CO_2). Using CuPyGC-MS phthalates in PVA **emulsion paints** were identified by Learner [810] by very sharp peaks in the pyrograms and a mass fragment m/z of 149. The very different retention times of dibutyl-, dioctyl- and butylbenzylphthalate differentiate between them.

PyGC-MS can also be used in the characterisation of bituminous pigments in paints [811]. Analysis of asphaltic materials, such as bitumen (mixture of heavy hydrocarbons) or asphalts (distillation residues from heavy crudes) is complicated matter for a geochemist even when large amounts of sample are available. The analytical problem is rendered still more difficult when small amounts of asphalt are contained in oil paint. Late 18th and mid-19th Century artists have used asphaltic materials as pigments in paints. Boon *et al.* [812] is developing a complete analytical strategy for the detection of bitumen/asphalt in **oil paint** samples using DTMS, PyGC-MS and imaging FTIR techniques. As in the course of the polymerisation of the oil paint the asphaltic polymer phase may be affected (oxidised) as well, the analytical strategy adopted

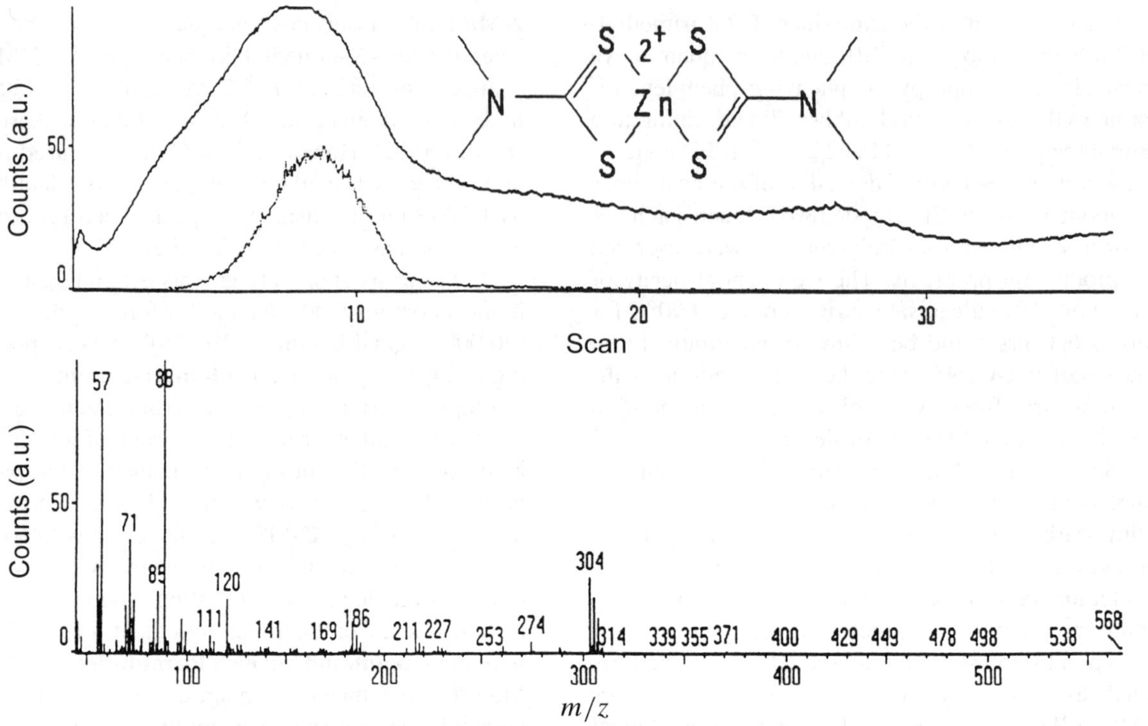

Fig. 2.37. DI-MS of NR/ZDMC with the m/z 304 trace and associated mass spectrum. After Kuch [502]. Reproduced by permission of Shimadzu Corporation, Japan.

by ref. [812] for the identification of asphaltic pigments in paint consists of a search for molecular markers (*e.g.* asphaltenes, hopanes, steranes and porphyrins). PyGC-MS has also been used for the identification of azo dyes (*e.g.* methyl orange) in textile effluents [813].

Various groups [645,646,814–817] have used PyGC and PyGC-MS in the determination of **additives in paper**. The potential of analytical pyrolysis for the qualitative and quantitative analysis of additives applied in paper products and within the paper manufacturing process is demonstrated for the polyamide amine epichlorohydrin (PAAE) **wet strength agent**. Tsuge *et al.* [645] quantified the difficult to extract additive PAAE in paper by means of PyGC-FID on the basis of cyclopentanone. Munakata *et al.* [646] quantified polyacryl(methacrylate) (PAM) and PAAE in paper by means of PyGC, using methylpyrazine as a more reliable key substance for quantification of PAAE. Nevertheless, the results were still unsatisfactory. Odermatt *et al.* [815] finally improved the sensitivity, and by focusing the mass-selective detector on a single key ion (methylpyrazine) decreased the matrix influence. A detection limit of 0.02% was achieved in the proposed absolute and quantitative method. Kleen *et al.* [771] used PyGC-MS in combination with PCA and PLS for quantification of major and minor components in softwood kraft pulps with significant improvement with respect to the traditional method using acid hydrolysis/derivatisation GC analysis. Py-MS and TD-GC-MS were also applied to the analysis of silicone polymer release liners as paper contaminants [818].

As shown in Table 2.38, PyGC-MS has been used for **(semi)-quantitative additive analysis** in a restricted number of cases. For quantitation by pyrolysis methods, calibration must be performed with different concentrations of additive standards to ensure the same pyrolysis efficiency and linearity of the signal intensity. For quantitative additive analysis either an internal standard method (using a polymer fragment peak) or an external standard may be used. In the latter case, normally 2 μL of a standard solution (0.1 mg/mL chrysene in CH_2Cl_2) are added to the solid sample. Obviously, a calibration curve should be batch independent. Reported standard deviations vary considerably with the more reliable re-

Table 2.38. Quantitative additive analysis by flash PyGC-MS

Analyte[a]	Matrix	RSD/R^{2b}	Reference
2,6-di-*t*-butyl-*p*-cresol	Butadiene/styrene copolymer	10%	[784]
Tinuvin 320	PA6	5%	[819]
Irganox 3114	PP	–	[820]
Irganox 1010	PE	7%	[787]
Irganox 1010	PET	3%	[787]
PBDD/PBDF	Polymer extract	–	[793]
IPPD	NR	7%	[807]
DOP	NBR	3%	[819]
N-butylbenzene sulfonamide	PA12	1%	[819]
Chimassorb 944 (MW ∼ 3500)	–	10%	[797]
PDBS 80	PBT	0.9994	[819]
Accelerators	Vulcanisates	0.9218–0.9802	[808]
PAAE	Paper	0.9998	[815]
Unspecified stabiliser(s)	Polymers	0.993–1.000	[820a]
Irgafos 168	(Calibration curve)	0.9995	[819]
Irganox 110	(Calibration curve)	0.9960	[819]
Uvasorb HA88 (MW ∼ 3000)	(Calibration curve)	0.9927	[819]

[a]DOP, dioctylphthalate; IPPD, *N*-isopropyl-*N*′-phenyl-*p*-phenylene diamine; NBR, acrylonitrile–butadiene rubber; NR, natural rubber; PA, polyamide; PAAE, polyamide amine epichlorohydrin; PBDD, polybrominated dibenzo-*p*-dioxins; PBDF, polybrominated dibenzofurans; PBT, polybutylene terephthalate; PDBS, polydibromostyrene.
[b]RDS, relative standard deviation (%); R^2 correlation coefficient.

sults conforming to 1–5% standard deviation. The accuracy of thermal extraction should be compared with that of other techniques (*e.g.* based on spectroscopic methods), which however usually follow solvent extraction. In those cases a major uncertainty is contained in this wet chemical step. A recent interlaboratory test for the quantitative determination of Irganox 1010 in PE by various methods [88] has indicated the strength of the technique.

When making quantitative analyses it is necessary to find one or more additive specific pyrolysis products from the actual substance. It is advised to investigate the influence of pyrolysis temperature on fragmentation. Quantification of additives with more specific mass fragments and fewer fragments might lead to better quantitative results. Lower temperature may result in fewer fragments. It is therefore favourable to optimise the pyrolysis temperature.

From the reported quantitative determinations of additives in polymeric matrices by means of PyGC-MS a few crucial influencing factors appear, namely *(i)* fragmentation behaviour; *(ii)* tuning file and ion chamber conditions; *(iii)* matrix effects on pyrolysis; *(iv)* evaporation before analysis; *(v)* weighing errors due to low concentration; and *(vi)* sample heterogeneity. The **fragmentation behaviour** of the analyte has great influence on both identification and quantitation by means of PyGC-MS, as illustrated by Kuch [797] in the determination of Chimassorb 944. This oligomeric HALS stabiliser with molar mass distribution is relatively difficult to analyse with traditional "wet chemical" analysis methods. Quantitative determination by means of flash PyGC-MS is equally quite difficult on account of the fragmentation behaviour into non-highly specific fragments, which impairs accurate quantification. For quantification of high Chimassorb 944 concentrations the multiple ion chromatogram peaks at $m/z = 91$ (Chimassorb 944) and $m/z = 114$ (chrysene) can be used and for low amounts (up to 1 μg) the mass fragments at $m/z = 56$–57 (diisobutene from Chimassorb 944) and $m/z = 227$–229 (chrysene). It is noticed that Lattimer [681] has described the identification of Chimassorb 944 by means of FD-MS and FI-MS and various authors [628,657] have reported semi-quantitative PyGC analysis of Chimassorb 944 in LDPE and PP extracts (Table 2.28). In more favourable fragmentation conditions, as in the determination of ENB and DCPD in EPDM, more reliable and useful results are obtained more easily.

As mass detectors easily suffer from fouling, it is advantageous to use new tuning files for each series of measurements. For optimal quantification measurement of a complete calibration line for each series of measurements is recommended. Moreover, since polymer matrix effects have shown to influence the fragmentation behaviour during pyrolysis, separate calibration lines are required for each different polymeric matrix. Since errors in sample preparation are always larger for small amounts of product (additive, chrysene, and polymer), part of the observed deviations on the calibration curves can possibly be ascribed to sample preparation. This effect should be more pronounced for stabilisers (present in ppm amounts) than for flame retardants (in percent range). The effects of evaporation of additives in thermal extraction methods should be verified. Kuch [797] considers (semi)-quantitative analysis by means of PyGC-MS not a routine operation in view of the need for calibration standards with the same polymer matrix, time-consuming multiple measurements, inhomogeneous technical samples and alternative standard procedures. A dependency on lot number has been observed (sampling *cq.* production problems), which is not surprising when the inhomogeneities of elastomers and technical polymers (especially in blends and compounds) often exceed the maximum quantities of 0.5 mg which can be handled by PyGC-MS. Rubbers are heterogeneous materials by excellence with antiozonants (such as IPPD) migrating to the surface. Therefore, for these materials PyGC-MS usually cannot pretend to provide better than semi-quantitative analysis. Results may be improved by **multiple sampling**. In order to yield "average" values, the PyGC-MS technique greatly benefits from unattended robotic operation allowing the operator to examine various samples from a batch for the sake of statistics.

It appears that PyGC-MS is being used most as a **quality control** (QC) tool for comparison of *good* and *bad* quality (fingerprinting). As such the method is in use in the automotive industry for the evaluation of incoming materials (rubbers and synthetic polymers) [807]. Wilcken *et al.* [821] use furnace PyGC-MS and principal component analysis (PCA) to differentiate resin-modified paints as a tool for QC of solvent-based can coatings. The method offers detailed information about resin ingredients and network fragments. Especially qualitative modifications, like inadvertent exchanges of resins in these complex technical polymer systems, are clearly detected and identified. Also in this case examination of quantitative modifications is more difficult. PyGC-MS has also been used as a tool for product control in the electronic industry to establish the chemical equivalency of polycarbonates from various sources, to distinguish between PTFE and HFP-TFE copolymer, to determine compositions of PS-PPO blends containing Ph_3PO_4 as a fire retardant, and effluent gases from nigrosine dyes [822]. Stepwise PyGC is also indicated for product quality control [595]. Work by Bradna *et al.* [823] aimed at using filament PyGC-MS and PyGC-HRMS methods for QC purposes of carbon-fibre composites, especially for testing their ability to identify the components of some epoxide matrices, namely N,N,N',N'-tetraglycidyl-4,4′-diaminodiphenylmethane (TGDDM) and a diglycidylether of biphenyl A (DGEBA) type resin. Whereas the accelerator 3-(3,4-dichlorophenyl)1,1-dimethylurea (DIURON) was identified, absence of characteristic pyrolysis products of resin hardeners, such as dicyanodiamide (DICY) and 4,4′-diaminodiphenylsulfone (DDS), prevented their unambiguous detection by this method. As a consequence of insolubility of the matrix and high carbon-fibre content (up to 70 vol.%), chemical analysis of such composites is very difficult, as most chromatographic (GC, HPLC and GPC) and spectral (IR) methods, useful for the analysis of uncured binders of carbon prepregs, cannot be used.

Failures can be examined by comparing fresh and failed parts by PyGC-MS under standard conditions (*cfr.* also Fig. 2.27). In **damage cases** it is often an advantage of the technique that only very small sample quantities (<0.1 mg) are needed from the damage areas, such as cracked and ruptured regions, parts with changes of gloss or coloration and deposits of mechanically moving parts. This requires utmost care in sample preparation. The technique has been used to compare an engineering part composed of an elastomeric blend of NR and BR (as deduced by the observed pyrolysis products butadiene, isoprene, vinylcyclohexene, dipentene and IPPD) and a failed part, composed of NR (as evident from isoprene and dipentene), which contained only a minor fraction of IPPD. In this case, excessive dynamic and thermal stress had apparently caused degradation of both polyisoprene and IPPD [807].

Hardell [824] has characterised **organic impurities** in pulp and paper products using PyGC-MS with the SPM technique. Frequent problems arise from resin and sizing agents which contain polar groups (carboxylic acid and alcohols). These products can be methylated directly on the pyrolysis filament by addition of TMAH. It is possible to destinguish between softwood and hardwood resin and various sizing agents including (fortified) resin and alkylketene dimers (AKD). Typical samples of detrimental substances found in spots, specks and deposits in pulp and paper mills are resin, lignin, shivers (from wood); rozin size, polystyrenes, polyacrylates (from sizing and coating agents); PE, PVC, nylons, polyesters (from synthetic polymers); polyisoprene, polyacrylates (from adhesives, tapes, labels); fatty acid derivatives (anti-foaming agents). **Troubleshooting** includes the analysis of spots in paper, of specks in paperboard, and cratering in wallpaper. Using PyGC-MS, Geissler *et al.* [825] identified fatty acid inclusions in technical drawing paper; the analysis allowed optimisation of the production process.

Chemical recycling of plastic materials often involves pyrolysis of thermally assisted reactions. One of the potential hazards of high-temperature **waste management techniques** is formation of toxic chlorinated aromatic hydrocarbons. Additives, such as bromodiphenyl ether-based FRs, are transformed to bromodioxin under these conditions, as verified by means of various pyrolysis techniques [251]. Blazsó *et al.* [772,826,826a] have studied formation of halogenated products in pyrolysis of polymers and plastics additives. GC-MS results showed that chlorination of phenolic thermal decomposition products (*e.g.* from Irganox 245) occurs when cupric or ferric chloride is present during pyrolysis. Munson [827] has described other environmental applications of pyrolysis.

The fact that very small sample sizes can be investigated is particularly beneficial when only small amounts of sample are available, for instance in **forensic science** [567,767] and in the application to the study of cultural materials [669]. Current applications include analysis of trace evidence samples in forensic laboratories, evaluation of new composite formulations and authentication and conservation of artworks.

It appears that PyGC-MS and PyGC are being used more widely than PyMS for polymer/additive analysis. A major advantage of PyGC-MS over PyMS is the provision for separation of the various components of commercial additive packages. Flash PyGC-MS is used more extensively than flash PyGC for quantitative additive analyses of solid polymer formulations; PyMS finds hardly followers in this field. However, generalised replacement of the slow solvent extraction additive analysis procedures by thermal extraction using pyrolytic methods is still to be achieved. In particular, quantitative analysis by PyGC-MS is feasible but by no means routine. Continuous attention is warranted for further development of the PyGC-MS database for identification purposes.

PyGC-MS of polymers was recently reviewed [828].

2.2.4. Pyrolysis–Fourier Transform Infrared Spectroscopy

Principles and Characteristics
Pyrolysis as a sampling technique in infrared spectroscopy is not new [829] and has been reported for many intractable polymeric materials which make either the traditional techniques of solvent casting or film pressing impossible. Originally the sample was first pyrolysed (often in a test tube) and the pyrolysate was condensed onto an infrared transmitting window material [829–831]. This procedure frequently involved recombination of pyrolysis products, which did not give an accurate picture of the pyrolysis process. Other methods make use of a heated transfer line to transport the volatiles into the IR beam [331]. This causes dilution effects that greatly reduce sensitivity. Liebman *et al.* [832] in 1976 have first described fast pyrolysis/FTIR spectroscopy using filament heating.

Modern PyFTIR equipment allows thermal evolution, vaporisation and pyrolysis directly in the FTIR. In **direct PyFTIR** the sample is located <3 mm below the beam [833,834]. Washall *et al.* [833] have described a cylindrical interface equipped with KBr windows, for connection of a ribbon filament pyrolyser to FTIR. Also sample cells with ZnSe windows are available for insertion into the light path of a Fourier transform infrared spectrometer for direct FTIR measurement of intricate solids.

Library search requires gas phase spectral databases (*e.g.* NIH/EPA) and archived spectra. For identification purposes temperature-programmed PyFTIR with appropriate data processing (differentiation, profile subtraction, *etc.*) has been used [835].

Advantages of pyrolysis directly in the IR beam are that spectra are obtained before side reactions and condensation occur. Direct PyFTIR provides a cleaner, simpler method for polymer analysis and also eliminates the need to transfer the gas phase sample to the IR sampling compartment, resulting in less dilution, analysis of smaller samples, and the possibility of monitoring the earliest species.

The method requires no sample preparation. Complex, non-volatile and polymeric materials, which are rendered opaque or completely non-transparent by the presence of pigments or fillers, may readily be analysed by FTIR using a specially designed PyFTIR Brill cell. PyFTIR analysis can be performed in a fraction of the time of a PyGC run. Unlike GC, FTIR is a direct probe of the molecular structure. PyFTIR can also obtain significant quantitative information; IR calibration curves are required [836]. The method needs only small amounts of sample (<1 mg). Under optimal conditions, sample sizes as low as 50 μg can be used, as described by Luigart [837] for a micro method for identification of vulcanisates and filled plastics. Washall *et al.* [834] expressed a high confidence level for reproducibility of direct PyFTIR.

Although FTIR can readily be utilised for the analysis of pyrolysates, and has some advantages over PyMS and TVA, a **disadvantage** of PyFTIR is the lower sensitivity relative to mass spectrometry. This explains the limited usage of this complementary technique. The sensitivity of pyrolysis–IR spectroscopy is surpassed by pyrolysis–laser photoacoustic spectroscopy, a combination of filament pyrolysis and CO_2 laser photoacoustic detection [838].

Reviews dealing with pyrolysis as a **sampling technique** for IR spectroscopy and for the determination of the microstructure of synthetic polymers are few and dated [557,561,839,840]. In a standard treatise on qualitative and quantitative analysis of rubbers and elastomers (Bayer technology, 1981) Ostromow [260] ranks off-line PyIR still amongst the main techniques utilised.

Applications

Smith [841] has discussed applications of pyrolysis techniques for polymeric systems with emphasis on the qualitative identification of components in a copolymer or polymer blend, identification of low-level polymer contaminants, characterisation of copolymer sequencing, differentiation between copolymers and physical blends of homopolymers, determination of monomer ratios in copolymers, and the study of polymer kinetics and degradation mechanisms. Pyrolysis destroys the stereostructure of the polymers. Gaseous components generated from pyrolysis of a wide variety of polymers have been analysed both off-line and on-line by IR spectroscopy to determine (quantitatively) the major components of the parent **resin**, *e.g.* rubbers and elastomers [260,842], cured epoxy and polyester resins [843], electronic moulding compounds (EMC) [844], plastic for automotive parts [845], polystyrene, nylons, PMMA and PVC [833] and phenol-formaldehyde resins, polycaprolactam, polyacrylonitrile, polyolefins, polyurethanes, copolymers, blends, ligroin, *etc.*

Ishiguro *et al.* [846] have used off-line pyrolysis-infrared spectroscopy for the analysis of polymers in various kinds of plastic materials. The pyrolysis products were obtained by heating small amounts of the plastic samples on a gas burner in middle size test tubes; the IR spectra of the pyrolysates were measured by a KBr sandwich method. This method was found to be simple, speedy and useful for the analysis of complex mixtures consisting of polymers and various kinds of additives in plastic materials. Washall *et al.* [834] used direct PyFTIR for polymer analysis.

The main application of PyFTIR is for **fingerprint identification** of polymers, polymer blends and vulcanisates, in particular in those instances where transmission or ATR spectra are too weak to allow interpretation, as in the presence of fillers such as carbon-black and inorganics. Interfacing of a programmable pyrolyser to a lightpipe of an FTIR provides **off-gas analysis** and details on degradation mechanisms. Liebman *et al.* [832] reported pyrolysis FTIR products of PVC using a heated lightpipe. Because of volume considerations, sensitivity was reduced as a result of dilution effects. Luigart [837] used PyIR to study **damage cases** of NR/SBR materials. Direct-pyrolysis FTIR can be used as a **quality control** tool and for polymer identification purposes in a QC laboratory [833]. Davidson [847] applied TPPy-FTIR for the identification of many components of polyurethanes from the composition of the evolved gases. The ease with which **carbon-filled rubbers** (with high loadings, 35%) can be analysed by PyIR makes it a preferred technique for the initial determination of the polymer content of these materials. An experienced spectroscopist can determine at a glance after a 30 s scan which polymer is present, as illustrated for carbon-filled natural rubber and synthetic styrene–butadiene rubber by Matheson *et al.* [529].

May *et al.* [848] has discriminated 31 household gloss **paints** (basically pentaerythritol-*o*-phthalate alkyds) by means of PyIR and six other common techniques. All the paints gave spectra characteristic of an alkyd paint incorporating phthalic anhydride as the dibasic acid. PyIR is largely insensitive to pigment variations. For characterisation of

179 glues and acrylic, cellulose, epoxy, polyester, rubber, polystyrene, poly(vinyl acetate) and urea-formaldehyde resin **adhesives** by means of pyrolysis, IR spectroscopic detection was applicable only to unfilled samples, but could not readily be applied to filled adhesives, which constitute a fair share of all commercial products [849]. On the other hand, PyGC can analyse all adhesives; for a few classes of adhesive the pyrograms are not very distinctive. Similarly, Fuchslueger *et al.* [850] have compared PyGC with MS and FTIR detection for the identification of epoxy resins.

Werner [851] has used Py-ATR-IR for fingerprint identification of **sizes, finishes** or low-level surface treatments on glass fibres. The procedure allows quality specification and control of finishes by manufacturers and users of glass reinforced plastic products. PyFTIR has been used only sporadically for determination of **additives** in polymeric materials. Truett [839] described application of PyIR for black-pigmented polymers, cross-linked polymers which cannot be pressed into films, complex copolymers for minor component identification, detection of additives and identification of (toxic) gases from burning polymers (materials for aircraft interiors). Since additives are often minor components, they will not be detected by conventional infrared techniques. Using the pyrolyser as sublimer can identify the additives. Peschel *et al.* [852] have reported the determination of **textile auxiliaries**, such as oxyalkylated fatty alcohols, isoalkylphenols, ester oils, and sulfobetaines, on synthetic polyester fibres.

On-line pyrolysis-FTIR studies of evolved degradation products from polymerics provide rapid, unique information that is useful in formulating **fire retardant materials**. PyFTIR studies of flame retarded cotton fabrics [853] have been used to retrieve information on evolved gases and condensibles in dependence on the pyrolysis temperature.

According to Hummel *et al.* [854], who described linear temperature-programmed pyrolysis of several thermo-resistant polymers (Twaron, Kapton and Pyrolin PI-2555) using FTIR with a heatable cell and LVEIMS (18 eV) for evolved gas analysis, FTIR is superior for the analysis of light fragments, whereas EIMS is more sensitive than FTIR. PyFTIR has mainly been used in the '80s and now appears to be declining.

Wang *et al.* [838] have reported the use of pyrolysis–laser photoacoustic spectroscopy (PLPAS) of polymers (PE, PTFE, nylon-1010) using CO_2 laser photoacoustic detection.

2.2.5. Pyrolysis–Gas Chromatography–Fourier Transform Infrared Spectroscopy

Principles and Characteristics

Infrared spectrometers, particularly Fourier transform infrared (FTIR) instruments, have been used as detectors in gas chromatography [855] offering the capability of compound quantitation and identification. Spectral search requires use of the NIH/EPA library of gas phase spectra.

The first papers on PyGC-FTIR have appeared in the mid 80s [856,857]. When a pyrolyser is used at the front end of the chromatograph, it usually acts just as a convenient way of sample transformation/injection into the GC-FTIR. Yet, addition of a pyrolyser to GC-FTIR equipment is not trivial; in fact, all three components of the on-line equipment have to be optimised in order to generate reliable results. Interface designs comprise a lightpipe [858]. The commercial availability of FTIR systems capable of highly sensitive detection, using a mercury-cadmium-telluride (MCT) detector (liquid N_2 cooled) and completely automated sampling and data manipulations, have brought PyGC-FTIR effluent analysis all the advantages of classic IR spectroscopic interpretation. Modern data systems permit analysts to view on-the-fly real-time absorption spectra of the eluting selected peaks.

The **advantages** of FTIR detection are as usual: functional group identifications and specific compound qualitative analysis; simultaneous spectral information on many species; continuous scanning of effluent either from direct thermal processing/pyrolysis or from the GC separation process; quantitative analysis using proper calibration from well-known absorption coefficient information for most all IR-absorbing organics and inorganics; reference spectral libraries on database accessible files with efficient search routines. **Drawbacks** are difficult handling and a limited transfer temperature range for the lightpipe IR technique (T_{max} 250°C; at 300°C background noise renders analysis impossible). Some additives, such as flame retardants, tend to crystallise out in the relatively cool lightpipe. With the relative **insensitivity** of GC-FTIR couplings as compared to GC-MS hyphenation, the former technique finds less application. A similar situation exists for PyGC-FTIR as compared to PyGC-MS.

Identification of additives by means of PyGC-FTIR is usually difficult. The method is only suitable for polymer formulations with **high additive**

loadings (% rather than ‰ range). PyGC-vapour phase IR, which is not to be considered a key method for polymer/additive analysis, was reviewed in the past [561,859].

Applications
As IR detectors are not as popular as MS, PyGC-FTIR has only occasionally been used in polymer analysis. Such applications have been commonly related to analysis of certain gases such as CO_2, CO, CH_4, NH_3, *etc*., where the MS analysis is less successful. PyGC-FTIR has been applied in several studies of synthetic polymer analyses [860]. Reddmann *et al.* [861] used the technique to analyse technical **rubber** samples, *e.g.* for automobile manufacturing. Cleaved products from isoprene, NBR, SBR and EPDM were identified. The authors argue that structural information was obtained more easily from IR spectra than from mass spectra. PyGC-FTIR can be used in **quality control**. Hummel *et al.* [732] have reported a comparison of PyGC-dispersive IR and PyGC-FTIR in the analysis of **acrylic resins**. The latter method was more sensitive. Compared with Py-FIMS, PyGC-FTIR has the advantage of giving direct evidence of the chemical nature of the fragments. Hummel *et al.* [862] also described the use of CuPyGC-FTIR for the study of three industrial poly(ester urethane) elastomers, namely Urepan 600/641 (Bayer) and Elastollan C78A (BASF/Elastogran) in 1 mg sample size. Interference was observed on account of isocyanates in the pyrolysates as a result of a diisocyanate component in the elastomer and the presence of the antihydrolysis additive Stabaxol, bis (2,6-diisopropyl) carbodiimide.

Nishio *et al.* [863] have reported qualitative analysis of **silane coupling agents** on E-glass fibres using PyGC-FTIR. The thermal decomposition products are fractionated organofunctional groups from the coupling agent, which can be identified by FTIR. Analysis for coupling agents on glass fibres using FTIR has also been reported [864]. However, non-destructive techniques using FTIR are limited by optical interference in the IR region.

The complementary nature of IR and MS has been utilised by Duncan [865] in a **PyGC-FTIR-MS** system with library search capability in the study of the pyrolysis products of polybutadiene and the antioxidant 2,6-di-*t*-butyl-4-methylphenol. Oguchi *et al.* [860] have reported a similar instrumental set-up for analysis of a methylmethacrylate–butadiene–styrene copolymer. A vapour-phase FTIR-MS system is a valuable analytical tool in the characterisation of polymeric materials. With on-column quantities of approximately 5 ng for strong IR absorbers and 30 ng for weak absorbers, it is possible to obtain both IR and MS data with a single injection. These two completely independent principles of molecular spectroscopy provide unknown identifications with a high level of confidence.

2.2.6. Pyrolysis–Gas Chromatography–Atomic Emission Detection

Principles and Characteristics
Linking PyGC to AED was first reported by Ou *et al.* [866], who also compared the merits of AED over other GC detectors such as FID, ECD, FPD, and PID. The multielement AED detector is capable of identifying up to four elements simultaneously from a single injection and can define the empirical formula of a compound containing common elements. AED has different sensitivities and selectivities for different elements (*cfr.* Table 8.25 of ref. [213a]). For example, detection of nitrogen and oxygen species in pyrolysis gases is difficult compared to that of sulfur. Increasing the sample size injected can overcome sensitivity problems as long as the element of interest has a high selectivity over carbon. The potential for application of **on-line PyGC-AED** in polymer characterisation is high, in particular for effective monitoring of halogen and phosphorous-containing pyrolysates.

The advantages and disadvantages of using AED have been discussed [642]. The **main features** of PyGC-AED are summarised in Table 2.39. Sample enrichment, a normal necessity for AED before detection of many elements is not possible in direct linkage of PyGC without complicated trapping arrangements. As a consequence, some elements are

Table 2.39. Main characteristics of PyGC-AED

Advantages:
- Multi-element detection
- High selectivity
- Definition of empirical formula

Disadvantages:
- Need for sample enrichment
- Element-dependent sensitivities and selectivities
- High sensitivity to moisture and air

difficult to detect using this technique. The method is also highly sensitive to moisture and air entering the system.

Off-line Py-PTV-GC-AED, using an adsorbent tube assembly inserted into the PTV injector, enables a more concentrated sample to be prepared, thus eliminating reproducibility problems with polymer heterogeneity, better control of the sample volumes injected, and different molecular weight ranges to be injected, thus avoiding carbon breakthrough [867]. The system allows for selective trapping of volatile organic material, which can subsequently be released to the analytical column by temperature-programmed desorption after refocusing, producing sharp and well-defined peaks. Using this technique, prior enrichment of the pyrolysis gases is possible either directly on the adsorbent trap or using a preliminary separation mechanism. Off-line Py-PTV-GC-AED allows use of different adsorbents for selective enrichment (*e.g.* C_6–C_{12} on Tenax TA) and concentration of different species increasing the sensitivity for different elements. Larger samples can be pyrolysed for heterogeneous materials, such as polymers or coal, and large gas volumes (up to 1 L) can be loaded onto the trap.

Applications
For full exploitation of PyGC for identification of unknown components in complex matrices a range of detectors (MS, FTIR, AED) is necessary. AED possesses unique elemental selectivity and can sometimes enable analysis of samples which exhibit severe matrix interference by other technology. The identification of 1 wt.% of **polyacrylamide additive** in PVAL by PyGC-AED (nitrogen trace pyrogram) and PyGC-MS (peak assignment) is a very good example [868]. At such low polyacrylamide levels, most of the non-destructive spectroscopic methods suffer from lack of sensitivity. It is also very difficult to determine low-level additives using PyGC-FID and a more selective detection method is required. While the carbon trace obtained in GC-AED, which is very similar to the GC-FID chromatogram, could not detect 1.0% polyacrylamide in PVA, the nitrogen trace was discriminative. The nitrogen containing peaks, matched with a polyacrylamide standard have been further identified by DI-MS. In this application, AED was far more effective than MS or FID.

Fuchslueger *et al.* [850] have examined cured **epoxy resins** and their minor components combining PyGC-MS and PyGC-AED, and identified ethyltrimethoxysilane as a fragment of the **coupling agent** γ-glycidoxypropoxytrimethoxysilane in the industrial system bisphenol-A-DGE/(4,4′-diaminodiphenylmethane (hardener), SILAN A-187 (bonding agent/filler)). Oguchi *et al.* [860] have reported use of PyGC-AED in combination with PyGC-FTIR-MS.

The types of organic **flame retardants** used in polymers (*e.g.* halogen- and phosphorous containing) are easily recognised by AED element tracing of pyrolysates after PyGC [789]. Wang [642] analysed various BFRs in polyesters and polyamides by PyGC-AED and PyGC-MS, namely poly(dibromostyrene) (PDBS-80), brominated polystyrene (Pyrochek PB 68), pentabromobenzyl polyacrylate (FR-1025P) and brominated epoxy (tetrabromobisphenol A-diglycidal ether) (F-3020) in PBT, SAN, PA4.6, PA6.6 and PA6.9. The ability to detect specific-element-containing pyrolysates, while at the same time discriminating any other possible complications, is the key strength of AED. In this case the atomic emission lines used for detection of C, H and Br were at 496, 486 and 478 nm. On the basis of the AED halogen element trace, peak pattern matching with pure flame retardant identified the type of polymeric FR used. After generation of the pyrogram by PyGC-AED, pyrolysates produced from polymeric FRs were identified by MS analysis of a complex total ion chromatogram (TIC), which was mainly generated by the thermoplastic polymer matrix. Because chlorine and bromine possess characteristic pairs of isotopes with well-known ratios, it is generally relatively easy to detect these elements in a component through a mass spectrum. This facilitates PyGC-MS analysis of HFRs. Most of the commercially available BFRs show two major fragment families after pyrolysis: brominated styrene and phenol. A fairly complete identification of all fragments and correct relative peak intensity pattern are necessary to ensure unambiguous identification of a BFR. In this situation, AED may be superior to MS for detection of specific-element-containing fragments and their relative intensity pattern. Unlike MS detection of the components, AED does not have direct identification capability. However, AED can be used as an identification tool by the peak pattern recognition approach [642]. Moreover, quantitative analysis of polymeric FRs may be achieved by PyGC-AED if a set of polymeric FR standards of varying concentrations are available or can be prepared in similar thermoplastic resin matrices.

On-line PyGC-AED and off-line Py-PTV-GC-AED have been applied to **waste tyres** using the emission lines for C, S, N, O and Cl at 193/495, 181, 174, 777 and 479 nm [867]. In the off-line mode 10 mg of tyre crumb were pyrolysed at 1000°C with injection of 40 mL for carbon and sulfur, and 100 mL for oxygen.

2.2.7. Temperature-programmed Pyrolysis

Principles and Characteristics

The most commonly used pyrolysis-mode is the pulse mode (flash pyrolysis), in which a rapid temperature change is applied for a short period of time (seconds). Application of programmed heating techniques (stepwise, sequential or fractionated) has a main advantage over pulsed heating in that it provides temperature-resolved data of classes of compounds with different thermal stability and desorption characteristics. Controlled heating may be used to simulate thermal processes or to analyse degradation products. This produces a **time-resolved** picture of the generation of specific products. In the programmed-mode the heating rates are more typical of conventional thermal analysers. Performing the pyrolysis inside the ionisation chamber allows temperature-resolved analysis at temperature ramps of up to $20°C\,s^{-1}$. Slow pyrolysis is generally possible using either a programmable furnace or resistively heated filament pyrolyser. Hu [516] has reported a three-step analysis with a first step of programming the pyrolyser to 300°C, a second step to 1000°C, followed by a cleaning step again at 1000°C using a conventional pyrolyser.

Due to the heating applied, **desorption and pyrolysis** generally are competing processes. In order to obtain structure-specific fragments, thermal degradation must be limited as much as possible. Temperature-programmed pyrolysis (TPPy) is a process in which the sample is heated at a controlled rate over a range of temperatures during which pyrolysis occurs. As expected, TPPy proves quite useful in separating organic additives for easier identification and allows characterisation of both organic additives and polymer components in one experiment. The volatiles and additives vaporise at lower temperatures (200 to 400°C), while pyrolysis fragments from polymers are formed at higher temperatures together with evaporation of metallic components. For polymer/additive formulations containing volatile material temperature-programmed or fractionated pyrolysis, which allows for sequential analysis, namely thermal desorption prior to pyrolysis, is often more suitable than single-step flash pyrolysis. This is illustrated in Fig. 2.38, which shows the total ion current (TIC) *vs.* time profile for the direct FI-MS analysis of an uncured rubber with evaporation of organic additives between 50 and 400°C and evolution of rubber thermal decomposition products (pyrolysates) at 400–750°C [745]. The use of a thermal desorption step can profitably be combined with recent developments in the field of GC analysis of polymers.

Several analytical schemes are in use for programmable pyrolysis of materials. The temperature can be raised: *(i)* in a linear programming mode; *(ii)* in a stepwise heating mode; *(iii)* by applying a train of energy pulses to the thermo-element, which warms up in the same manner after each pulse (sequential pyrolysis); and *(iv)* by applying increasing energy pulses to the thermo-element so that each step results in heating of the specimen to a higher temperature than achieved in the previous step (fractionated pyrolysis). The evolution profiles of the products of linear TPPy of polymers contain much useful information about composition and thermal decomposition reactions. Appropriate data processing enhances effective retrieval of this information.

Various **hyphenated temperature-programmed analytical pyrolysis techniques** are in use, such as TPPy-MS, TPPy-GC, TPPy-AED and TPPy-GC-MS [869]. Among these techniques, TPPy-MS can provide specific information on the degradation products, TPPy-AED allows monitoring of the evolution behaviour on the basis of constituent elements in the products, and TPPy-GC combined with MS or AED is a powerful tool for identifying the various degradation products. TPPy-GC equipped with two columns (for low and high volatility components) assures better performance and protects the column lifetime. TPPy-GC-MS is essentially a combined sequential TD-GC-MS and PyGC-MS methodology to conveniently study additives and polymer matrix. A major advantage of using this approach is that the method utilises small sample sizes (*ca.* 0.1 mg), which allow: *(i)* rapid screening of a larger number of samples; and *(ii)* microscale analysis. Stepwise heating of the specimen in a filament-type pyrolysis cell may be employed in order to isolate successively the constituents in combination with chromatographic separation of the liberated compounds after each heating step. **Multi-step heating** can give much more information about a specimen than single-step pyrolysis

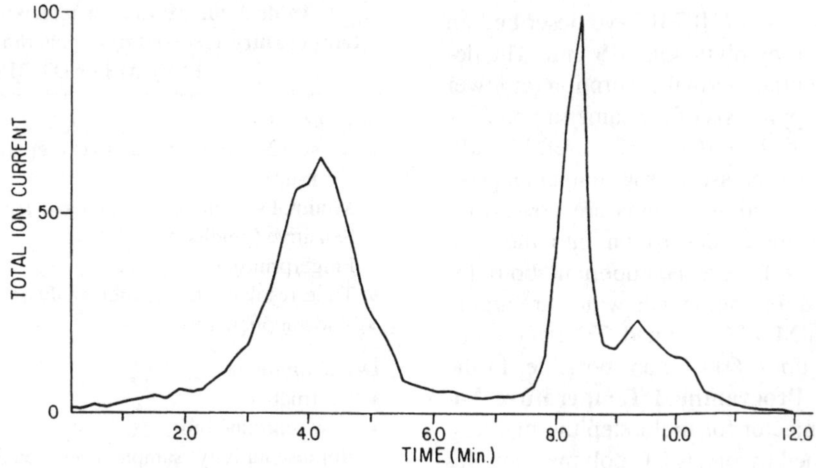

Fig. 2.38. Total ion current (TIC) *versus* time (temperature) profile for direct analysis of an uncured rubber by FI-MS. After Lattimer and Harris [745]. Reprinted with permission from *Rubber Chem. Technol.* **62**, 548–567 (1989). Copyright © (1989), Rubber Division, American Chemical Society, Inc.

GC. In TD/PyGC-MS cryogenic refocusing is necessary because thermal desorption is a relatively slow process. Heating of a specimen in the pyrolysis cell of a chromatograph causes desorption and evaporation of volatile compounds at temperatures close to the boiling point. Further heating of the specimen to the pyrolysis temperature leads to disintegration of the non-volatile organic portion so that finally the mineral constituents remain in the pyrolysis unit. Stepwise heating of a specimen in a pyrolysis cell thus enables separation of the specimen into several fractions: volatile impurities (monomers, solvents), high-boiling non-polymer additives, base polymers, special polymer additives or contaminating polymer impurities, and mineral constituents. Some compounds formed on pyrolysis can undergo partial vapour-phase pyrolysis and give rise to compounds with other structures. The technique of stepwise heating of the specimen in the pyrolysis cell of a chromatograph does not call for any special pretreatment of the specimen, such as isolation of the polymer, solvent removal, isolation of ingredients, inorganic fillers, *etc.* [595]. Andersson *et al.* [870] have discussed in considerable detail the determination of the temperature-time profile for a sample in PyGC, *i.e.* the dependence of the sample temperature on sample size and pyrolysis time.

Several **instrumental designs** have been reported for sequential desorption pyrolysis analysis. Tsuge *et al.* [515] reported a two-stage analyser with two separated ovens (for desorption and pyrolysis), *cfr.* Fig. 2.23. In a commercial **double-shot pyrolyser**, essentially TD/PyGC, a thermal desorption process, consisting of gentle heating of the sample to a thermal desorption temperature, which protects the sample from rapid thermal degradation and decomposition, is followed by instant pyrolysis of the sample with a gravitational free-fall mechanism, providing two different sets of information for a single sample [871]. Using a double-shot pyrolyser evolved gas analysis (EGA) can be carried out with the provision of selective sampling over any desired temperature interval for GC analysis [872]. EGA can provide information reflecting thermal properties of a polymer and is considered as equally effective as TGA. By analysing only selected portions of evolved gases the analysis time is reduced. Combination of EGA, trap-and-purge of desired portions of the evolved fraction (heartcutting) and MS or GC-MS is a very powerful method to investigate complex polymeric materials. The temperature ranges for heartcutting can be determined by means of TPPy-MS in which evolved gas is detected without any separation (using a short length deactivated stainless capillary tube). In automatic heartcut EGA analysis, each zone is cryo-trapped by a micro cryo-trapping device. The contents of the trap are then quickly discharged into the GC oven and identified by MS library search using a pyrogram library comprising some 136 polymers. In this way, the polymeric formulation can be comprehensively characterised with respect to additives and substrate polymer(s). In analogy to the original TD-GC-MS device allowing thermal desorption in the injection port, Ezrin *et al.* [873] have described an injection head for a pyrolytic GC-MS unit. The device permits sequential thermal desorption (at lower temperatures) and pyrolysis of the same sample.

As indicated elsewhere (Chp. 4.2 of ref. [213a]), fully temperature and pressure programmable, powerful multi-mode injection systems are now available which allow direct desorption into the capillary column. This feature is important both for solid sampling and for compounds which are highly active or of high-MW. Combined TD-Py options ($\Delta T = 16°\mathrm{C\,s^{-1}}$ up to 600°C) are possible. In the high temperature **Programmed Temperature Vaporising** (PTV) injector for multi-step thermal desorption/programmed pyrolysis GC polymer samples (typically 5 mg) are loaded directly in the liner of the injector [752]. For proper setting of desorption and pyrolysis temperatures preliminary TGA plots are useful. The PTV device allows analysing residual monomers at the lowest selected temperature and additives (such as release agents) at intermediate injector temperatures; at higher temperature levels macromolecular structural information may be gathered. The results are in good agreement with other pyrolysis systems [874]. For quantitation a calibration standard is injected directly after the TD-Py experiment. Identification of polymer additives in a **PTV-CT-GC-FID** set-up is limited to verification on the basis of retention times. The main advantages of the technique are simplicity, versatility, acceptable reproducibility, and relatively low cost in comparison with dedicated thermal desorption and pyrolysis instruments. As the device does not require a heated transfer line it is also applicable for the analysis of high-MW analytes, such as AOs, UVAs and other high-MW pyrolysis products that otherwise are easily lost. It appears that the lower temperature stages in the multi-step TD-Py method offer a good alternative for time-consuming extractions. The method is useful mainly for qualitative analysis and rapid screening. For quantitative analysis the reproducibility (now within about 20%) needs to be improved. In general terms, quantification by means of TD-GC-MS techniques is a doubtful exercise because total desorption of the analyte(s) at a given temperature is not assured, internal standards are difficult to use and mass spectrometry is not exactly well known for its quantitative excellence.

Direct temperature-resolved pyrolysis-mass spectrometry (DT-MS or TPPy-MS) performed close to the ion source is another rapid and sensitive technique in the characterisation of involatile

Table 2.40. Main characteristics of direct temperature-resolved pyrolysis mass spectrometry (TPPy-MS or DT-MS)

Advantages:

- Rapid (2–5 min; no sample preparation, no chemical work-up)
- Minimal sample requirements (1 μg)
- Sensitive (pmole range)
- Fingerprinting
- Time-resolved degradation analysis
- Wide applicability

Disadvantages:

- Destructive
- Low reproducibility
- Representativity (sample heterogeneity)
- Rapid fouling of ion source
- No separation at molecular level
- Data analysis
- Not easily made quantitative

organic material using powerful mass spectrometric analysis. Temperature-programmed pyrolysis (with supply of ions in time) yields much more information than at constant temperature. The technique requires little sample preparation, but inhomogeneous samples require more care. The sample is usually taken up in a drop of liquid, which is deposited on the filament. Reproducibility is generally quite low; the geometry of the filament is important for temperature control. By means of a pyroprobe (a coiled filament) it is possible to heat 100–200 μg solid matter in a temperature-programmed mode. Small sample sizes are necessary because of the high sensitivity of the mass spectrometer and the need to prevent temperature gradients within the sample. In principle, DT-MS allows separation of the components of a mixture without recourse to more time-consuming and complex separation methods. When needed for identification of molecular fragments present in a DT-MS spectrum, GC-MS can be used to achieve molecular separation. Temperature and time resolution with TPPy-MS allow consecutive stages of the degradation reaction to be followed. Without removal of volatiles the identity of the polymeric matrix might easily be masked. Multivariate mapping from principle component analysis of DT-MS "fingerprint" spectra may be used to classify and describe the (dis)similarity of samples [875–877].

The **main characteristics** of TPPy-MS (or DT-MS) are given in Table 2.40. For quantitative results an internal standard is required; it is also necessary that no material is lost in the vacuum of the ion source or during heating. By introduction of a larger sample size (100–200 μg) in TPPy-MS than in in-source pyrolysis (*ca.* 1 μg or 0.1 ng additive at a 100 ppm content) pyroprobe analysis achieves a lower detection limit. In-source pyrolysis in a QMS (with a typical detection limit of 1 ng in full scan mode) is limited to additive concentration levels of at least 0.1%, such as flame retardants. Analysis of additives at lower concentration levels is theoretically possible by means of pyroprobe MS.

The desorption products can be ionised (EI, CI, MAB, *etc.*) and detected. Chemical ionisation techniques appear to be especially suitable for the analysis of oligomeric fragments released in the early stages of pyrolysis of polymer systems. Although EI ionisation is also useful for TPPy, the extensive fragmentation caused by EI may lead to complex spectra for some materials. Lower electron energies can be used to reduce fragmentation but decreased ion intensities will result in certain species not being ionised at all. Variation of this parameter is a major cause of non-reproducible results in PyMS. Time-temperature resolution obtainable *via* Py-FIMS allows one to distinguish volatile additives [878] and residual monomers [879] from pyrolysis products of the polymer backbone.

Westall *et al.* [761] performed TPPy-MS using a magnetic sector in EI-mode near the ion source. Boon [708] has recently reviewed analytical PyMS, including temperature-resolved in-source PyMS. Analytical pyrolysis inside the ionisation chamber of a mass spectrometer (*i.e.* in-source PyMS) gives a complete inventory of the pyrolysis products evolved from a solid sample. In-source PyMS is valuable for structural investigation of synthetic polymers, blends and compounds and natural macromolecules as the constituents of polymers remain recognisable in this relatively mild thermal degradation mode and are detected by MS [708,754]. Under EI conditions *in vacuo* volatile additives evaporate in the ion source before the polymer matrix starts to degrade. The large amount of data produced by TPPy-MS requires a data system for efficient processing of the results and factor analysis to deconvolute the overlapping degradation and pyrolysis processes, as demonstrated by Windig *et al.* [880,881] for a bipolymer mixture, a wood sample, and an uncured rubber compound. In each of these examples, factor analysis technique was able to deconvolute the pyrolysis curves into the mixture components without prior information on peak shapes and locations. Thus, by resolving "pure" component curves, it is possible to monitor degradation mechanisms and to elucidate how additives are linked into the polymer matrix.

Except for TG, programmable furnaces are rarely interfaced directly to spectroscopic techniques. Davidson [835] has indicated several data processing schemes to extract information about composition and overlapping thermal decomposition reactions of evolved gaseous reaction products of polymers subjected to **linear temperature-programmed pyrolysis–infrared spectroscopy**.

Applications
Temperature-programmed analysis techniques have successfully been applied to the characterisation of polymer blends and compounds yielding information about thermal stability and the **successive stages of degradation and volatilisation** [882]. A variety of materials has been studied, comprising food wraps, flame retardant polymers, organic pigments, additives in poly(etherurethane ureas) and epoxy resins, rubber vulcanisates, polymer structure, oil paintings, *etc.* Figure 2.39 shows an EGA experimental plot from 100°C to 700°C for a HIPS-PC blend. Triphenyl phosphate (TPP; plasticiser, flame retardant) evolves first (from 170°C to 330°C), followed by decomposition of HIPS (from 350°C to 540°C) and PC (from 440°C to 710°C) [632].

TPPy-MS was used for rapid structural investigation of "sheen" on antistatic matting [883]. Thermal desorption (195°C, 10 min) in combination with filament pulse pyrolysis (700°C, 2 s) has been used to study release of the alkenediketene (AKD) sizing agent from paper (cellulose) [884]. Heating-coil TPPy-GC-MS at 175°C was used for determination of **plasticiser** concentration profiles in layered, nitrocellulose-based propellant sheets (total thickness 2–4 mm) [885]. Using microtomed cross-sectional slices of 200 μm evidence was obtained for rapid migration between laminated sheets of nitrocellulose-based propellant containing different amounts of nitrate ester plasticisers. TPPy-CT-GC-MS was used for determination of additives and polymer in food wraps [886]; the major organic compounds from PVC were C_9–C_{20}, epoxide soybean oil, DOA and alkyl phenols, while C_{14}–C_{20} and phthalate esters were detected from PE food wrap. Using the two-stage TD/PyGC analyser, Tsuge *et al.* [515] showed release of various additives, such as dioctyladipate and dioctylphthalate, from an

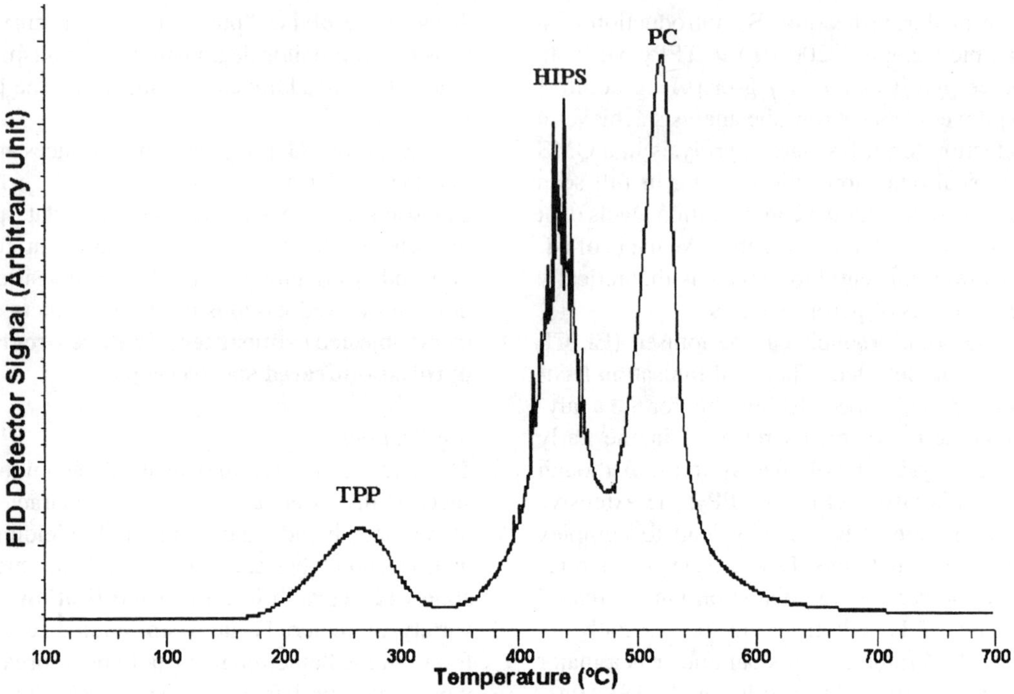

Fig. 2.39. An evolved gas analysis experimental plot (FID signal intensity *vs.* temperature) from 100°C to 700°C for HIPS-PC/TPP. After Wang [632]. Reproduced from *Journal of Chromatography* **A883**, F.C.-Y. Wang, 199–210 (2000), with permission from Elsevier.

acrylonitrile–butadiene rubber in a first desorption step at 300°C. Pyrolysis of the remaining polymer fraction then gave detailed information about the polymer structure. TD/PyGC-FID and PyGC-FID for acrylonitrile–butadiene containing DOA, DOP, DOS, TCP have been compared [871]. Watanabe's selective sample introduction device for evolved gas analysis (EGA-direct FID, EGA-direct MS, EGA-GC-FID, EGA-GC-MS) of polymeric materials has been applied to the analysis of paper and polymeric materials [872]. In the examination of an unknown rubber by means of TPPy with the double-shot furnace pyrolyser a first small peak eluting at 80 to 280°C was composed of benzothiazoles (such as 2-methyl- and 2-mercaptothiazoles), DOP, Nocrac 6C and free fatty acid; the second zone at 300 to 550°C originated from rubber decomposition [887]. Ezrin *et al.* [873] have detected a chlorinated flame retardant using TD as a first step before pyrolysis.

In complex formulations it is usually found that TPPy-GC-MS allows separation of additives from polymer pyrolysis products and their identification is facilitated. Thus both additives and polymer components were identified in commercial blends of PS and PPO, and protocols for characterising intractable rubbers have been published. Programmed direct probe heating of PP compounds of known additive composition (all containing 0.08–0.40 wt.% Irganox 1076/3114, Ultranox 626, Ca-stearate, and Tinuvin 144 or Tinuvin 622 or Tinuvin 770 or Chimassorb 622 or GoodRite 3150) between 200–400°C with EI-MS, isobutane CI-MS and FI-MS analysis has been reported [681]. Residual volatile chemicals and most organic additives were thermally desorbed at lower temperatures (below about 300°C), while the polymeric components were thermally decomposed (pyrolysed) above 300°C. Pyrolysis products from the PP compounds were observed at every carbon number to masses well above 1000 Da. Overall, direct mass spectral analysis is very effective for detection and identification of various organic additives and polymeric compounds. Multistep thermal desorption/programmed pyrolysis for gas chromatography with PTV injection (TD/Py-PTV-GC) can be used to characterise **complex mixtures of several polymers and additives**, as shown by Cramer *et al.* [752], who have selected four temperature levels for the study of an (ABS) impact modified PC/PBT blend (composition: PC, PBT (26, 53%)/(0.15% Irganox 1076, 0.2% AO 2246, 0.25% PETS, 20% ABS)). The total analysis effort therefore consists in four separate GC runs for the blend and each of its constituents (Irganox 1076, antioxidant 2,2′-methylene-bis(4-methyl-6-*t*-butylphenol), release agent pentaerythritoltetrastearate (PETS), PC, PBT, impact modifier ABS). The individual chromatograms of the various constituents of the polymeric sample were correlated with those of the final material in order to identify additives (thermal desorption) and degradation products (pyrolysis). For this blend, residual monomers, process solvents and highly volatile additives were determined at $T_{PTV} = 200°C$, reaction products formed in transesterification between PC and PBT (such as butanediol and THF), less volatile additives and stabiliser residues at 320°C, polymer blend degradation products at 500°C and other pyrolysis products at 600°C. Transfer to the column for the additives is already complete at the lowest desorption temperature ($T = 200°C$). The elution temperature of the release agent PETS in the GC run is 425°C, which illustrates the high-MW nature of the PETS components. In the absence of a heated transfer line even very high-MW components can be transferred to the GC column without losses. Most of the additives found in the thermal desorption of the blend originate from the ABS component in the blend. Further identification of (unknown) additives contained in the blend components, namely Ionol CP, Dressinate and Irganox PS 800 (and cyclic PBT trimer), could not be achieved with PTV-GC-FID. For this purpose mass spectrometric detection would have been required but combination of a high-temperature GC with MS is technically challenging. It is also of interest to notice that HT-GC with on-column injection and FID detection of an SFE extract indicated extraction of PETS from the sample. These high-boiling components could not be analysed with conventional GC-MS. Clearly therefore, the first step in the multi-step thermal desorption programmed pyrolysis method constitutes a good alternative for time-consuming extractions.

Temperature-resolved in-source PyMS is quite suitable for qualitative and quantitative determination of **flame retardants** in polymeric materials (validation with XRF or NAA). On the other hand, the detection limit (1 ng analyte) may not be easily reached for additives such as stabilisers. Tsuge *et al.* [869] have investigated thermal degradation of PBT containing a synergistic flame retardant system based on brominated polycarbonate (Br-PC) and Sb_2O_3 by means of various temperature-programmed analytical pyrolysis techniques, namely TPPy-MS, TPPy-AED and TPPy-GC-MS, with the object of understanding the synergistic flame retardancy of the halogenated organic compounds/Sb_2O_3 system. TPPy-GC-MS measurements were used to identify the thermal degradation products formed during heating from 60°C up to 700°C at a rate of 10°C/min. As might be expected for the synergistic reaction between Br-PC and Sb_2O_3, a prominent peak of the flame poisoner $SbBr_3$ (m/z 362) was identified. TPPy-AED was applied to monitor the evolution profiles on the basis of constituent elements (C, Br and Sb) in the evolved components. The emission curves indicate that thermal degradation of FR-PBT takes place in at least two stages with maxima at *ca.* 330°C and *ca.* 380°C. On the basis of the specific evolution behaviours of the volatile products measured by means of TPPy-MS, TPPy-AED and TPPy-GC-MS thermal degradation mechanisms of the Br-PC/Sb_2O_3 flame retardant system in FR-PBT were suggested. Similarly, TPPy-GC-MS was used for the study of FR-PET [828]. Luyk *et al.* [674] have described the characterisation of BFR polymer blends by TPPy-MS (up to 800°C, heating rate 16.5°C/s) using both EI and ECNI to identify the thermal degradation products. PyMS in EI mode offers a sensitive tool for fast analysis of unknown mixtures of polymers and additives, whereas electron capture negative ionisation is a very soft ionisation technique, which is used to selectively ionise electron-accepting molecules such as brominated compounds. The following systems were investigated: HIPS/(Br_{10}DPO, Sb_2O_3), *p*-BrPS, PS, PBT/(Br_{10}BB, Sb_2O_3), PS/(Br_{10}BB, Sb_2O_3), ABS/(TBBP-A, Sb_2O_3), PS/HBCD, Br_{10}DPO, Br_{10}BB, TBBP-A, HBCD. High-MW pyrolysis products in the m/z range of 1000–2000 Da were detected for *p*-bromopolystyrene and for a compound of HIPS with the FR system Br_{10}DPO/Sb_2O_3. The bromine chemistry in polystyrene spiked with Sb_2O_3 and flame-retarding Br_{10}DPO was also studied by in-source TPPy-MS using electron attachment reactions in Ar atmosphere [754]. Because the polystyrene pyrolysis products are not ionised under these conditions, debromination of the fire retardant and formation of polybrominated dibenzofuran, antimonybromide and antimonyoxybromides and brominated styrene

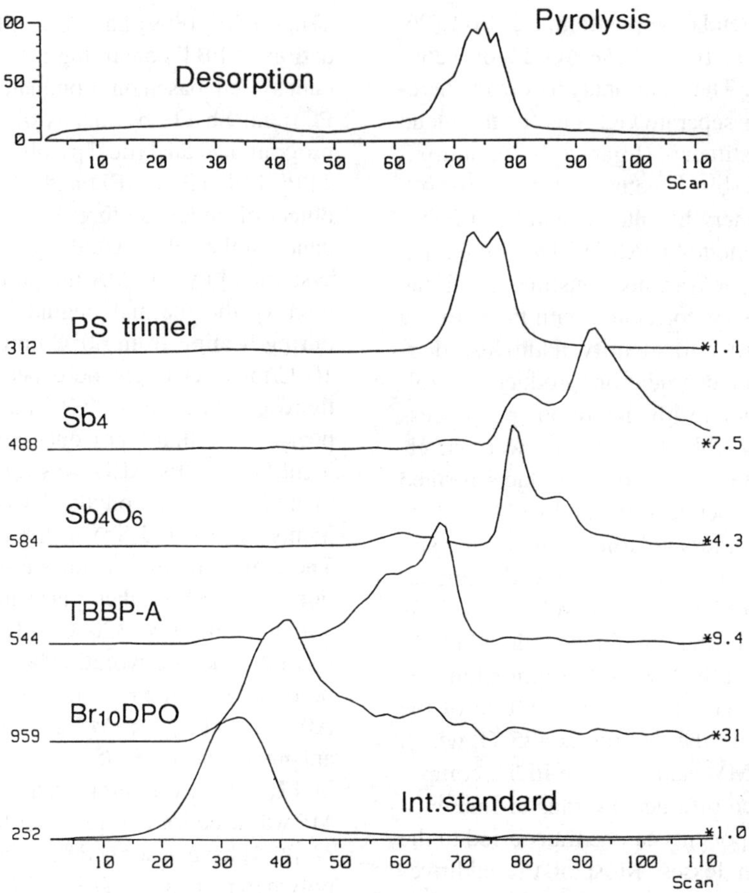

Fig. 2.40. Direct temperature-resolved (in-source) PyMS of PS/(Br_xDPO, TBBP-A, Sb_xO_y); total ion current and mass chromatograms. After De Koster and Boon [725]. Reproduced by permission of Consumentenbond, The Hague.

oligomers up to a degree of polymerisation (DP) 15 can be very clearly observed in the negative ion mass spectra.

De Koster *et al.* [725] have reported CuPyGC-MS (358°C) and TPPy-MS case studies of a variety of homogenised flame retarded systems, such as SAN/(TBBP-A, Sb_xO_y), PS/(Br_xDPO, Sb_xO_y), PS/(TBE, Sb_xO_y), SAN/(Br_xBB, Br_xDPO, Sb_xO_y), PS/(Dechlorane Plus 25, Sb_xO_y), SAN-PC/DPB, PS/(TPP, Sb_xO_y), SAN/(Br_xDPO, Sb_xO_y), *etc.* Figure 2.40 shows the temperature-resolved mass spectrum of a toluene suspension of 1–2 μL of PS/ (Br_xDPO, TBBP-A, Sb_xO_y) spiked with an internal standard (perylene). The total ion current plot recorded on a double focusing (BE) mass spectrometer gives evidence for various desorption and pyrolysis events. Compounds of lower polarity appear in the lower scan range, and more polar compounds, such as TBBP-A at higher temperatures. The desorption behaviour of the internal standard (m/z 252) does not interfere with the analytes. The commercial availability of TBBP-A and Br_{10}DPO enabled development of a quantitative analysis of these compounds in the polymer matrix (based on peak height ratios such as I_{959}/I_{252} for Br_{10}DPO). Extension of the internal standard peak ratio method to the analysis of other BFRs requires the availability of pure reference compounds. The procedure shows quantitation with in-source PyMS. Some 500 complex polymer matrices (consumer electronics and home appliances) were examined.

In analogy to Luyk *et al.* [674], also Blazsó [888] has been driven by environmental concerns in a PyGC-MS and in-source TPPy-MS study to identify the nature and to monitor the evolution of chlorine-containing volatile thermal decomposition products from three **organic pigments** dispersed in synthetic

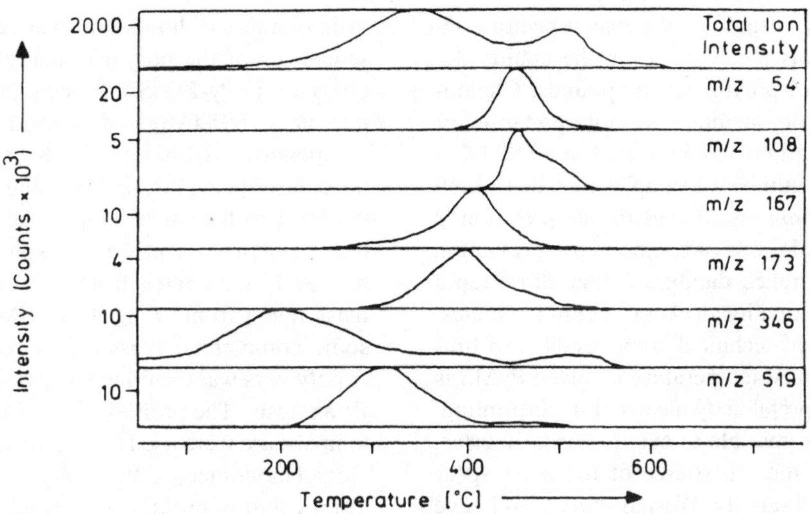

Fig. 2.41. TPPy-FIMS of butadiene rubber: temperature dependency of the total ion intensity and formation of butadiene monomer (*m/z* 54) and dimer (*m/z* 108), mercaptobenzothiazole (*m/z* 167) and TMDQ monomer, dimer, and trimer (*m/z* 173, 346, 519). After Schulten *et al.* [675]. Reprinted with permission from *Rubber Chem. Technol.* **62**, 698 (1989). Copyright © (1989), Rubber Division, American Chemical Society, Inc.

polymers. The systems studied were PE/dioxazine (Violet 23), PE/tetrachloro-isoindolinone (Microlen Yellow 110) and poly(vinylacetate-*co*-vinylchloride)/copper phthalocyanine (Microlith Green). PyMS, PyMS/MS and TPPy-MS were used to study the structure of additives in Biomer and Lycra Spandex **poly(ether urethane urea)s** of identical composition [888a]. Blazsó [777] has also investigated **epoxy resins** synthesised from diglycidyl ether of bisphenol-A (DGEBA) and alkanediol, cured with dimethylbenzylamine accelerator. The evaporation of dimethylbenzylamine, unreacted diol and hydroxyethers of DGEBA and diol was monitored by TPPy-MS and the presence of monomer and oligomer residues and of volatile additives was revealed.

Kodama *et al.* [805] have analysed antioxidants in vulcanised SBR compounds using heat-desorption by means of a double-shot pyrolyser followed by GC-MS analysis. Schulten *et al.* [675] performed TPPy-FIMS experiments using direct introduction. **Rubber vulcanisates** (BR, NR, SBR) were heated without any pretreatment from 50°C to 750°C in high vacuum with a heating rate of 1.2°C s^{-1}; mass range recorded 50–1500 Da. Figure 2.41 shows the thermogram for some pyrolysis products for BR. The counts (in arbitrary units) for the total ion current, butadiene (*m/z* 54), butadiene dimer (*m/z* 108), mercaptobenzothiazole (MBT) (*m/z* 167), 2,2,4-trimethyl-1,2-dihydroquinoline (TMDQ) (*m/z* 173), TMDQ dimer (*m/z* 346), and TMDQ trimer (*m/z* 519) are plotted *vs.* the probe temperature. MBT is a fragment from the accelerator *N*-*t*-butyl-2-benzothiazyl sulfenamide (TBBS). In addition to processing oil, also the antioxidant *N*-(1,3-dimethylbutyl)-*N*′-phenyl-*p*-phenylenediamine (HPPD) was observed (*m/z* 268), as well as stearic acid ($M^{+\bullet}$ *m/z* 284). Schulten and Wilcken [692,734, 821,889] examined commercial can coatings composed of filled and plasticised polyisoprene–styrene copolymers, and polyester resins containing melamine copolymer (methyl/butyl) as cross-linking agents, acrylic and epoxide resin components, *p*-toluenesulfonic acid (catalyst), wax (lubricants), TiO_2 (pigment) and organic solvents. Resin and additive modified laquers were examined by means of CuPyGC-MS, direct TPPy-EIMS, in-source TPPy-FIMS (magnetic sector) using PCA and static HS-GC-MS. All three pyrolysis measurement modes demonstrated a high discrimination power and can be used for **quality control** purposes. After optimisation of the temperature program direct pyrolysis requires a measurement time of 0.5 h. The most pronounced effects on the mass spectra were due to qualitative changes in the cross-linking agent (hardener); minor effects have been assigned to qualitative and quantitative modifications of co-resins (acrylic of epoxy resin). As the mass spectra contain marker signals for every resin, the quality decision includes all polymeric compounds. Chemometrical evaluation methods for interpretation of PyMS data are particularly indicated in a QC laboratory, where the number of samples may be so high that time-consuming visual spectra interpretation is only rarely possible. Use of chemometric evaluation routines and reference databases thus allow rapid good/poor batch quality decisions within minutes. Direct Py-EIMS of technical compounds is of limited application only, temperature-resolved PyMS is more easily interpretable by means of fingerprinting. In TPPy-MS it is possible to avoid solvent interferences. However, the ion source of the mass spectrometer is fouled rapidly. Westall *et al.* [761] have described applications of TPPy-MS to the analysis of materials of interest in defence quality assurance and have demonstrated the advantages of temporal resolution in the analysis of samples such as polymers, lubricants and their additives.

Techniques such as DT-MS, TPPy-HRMS and Py-TMAH-GC-MS are also very well suited for the study of complex and minute (10 μg) samples from paintings to identify the materials used and the degradation processes that have occurred during the lifetime of such **art objects**. Information is obtained on the volatile physically absorbed fractions such as oils and waxes, on the constituents of the macromolecular network such as binding media, gums, varnishes, resins, and on some inorganic components and metals. All this information is contained in one experiment with a sample requirement on the μg level. Boon *et al.* [890] examined a Dutch oil painting (of the 17th century artist Ferdinand Bol) by means of TPPy-MS and revealed the degree of impregnation of the paint film with waxes and varnishes. Reactive PyMS under direct transalkylation conditions was used to study the oxidation state of the oil paint film and varnish. The advantages of the reactive PyMS approach are speed and sensitivity. Less than 10 μg of paint sample is sufficient for analysis. Several analysis conditions (EI, CI and different T-ramps) were investigated. Quantitative results of fatty acids and diacids from the oil paint network were compared with wet chemical data obtained by GC-MS.

Hummel *et al.* [854,891,892] have used linear temperature-programmed PyMS for the study of a variety of **polymers**. Montaudo *et al.* [893] used direct probe PyMS to follow the degradation processes over minutes to hours and were able to determine sequences of the polymers analysed. The methodology of TPPy-FIMS has been applied to synthetic polymers [695,696] and is used for direct rubber compounds analysis [675]. Schulten *et al.* [697] have described TPPy-FIMS using a direct introduction system for small samples (10–400 μg) of synthetic polymers and quasi-linear heating from 50 to 800°C with rates from 0.2 to 10°C s^{-1} and a mass range from m/z 70 to 2100. Linear TPPy of some commercial cross-linked methylmethacrylate copolymers was examined by pyrolysis-evolved gas-IR analysis. The profiles of rate of evolution against temperature were used to identify the presence of occluded monomer and thermally unstable end-groups. The evolution profiles were used to distinguish between samples from different manufacturers, and to identify material in aircraft canopies [894].

2.2.7.1. *Thermochromatography*

Principles and Characteristics

Thermochromatography (ThGC) equipment is composed of an PyGC-FID with a sampling valve [895]. In ThGC a sample is heated under a controlled temperature gradient and the headspace gases are sampled at predetermined temperatures. Heating of the pyrolysis oven and the timed sampling of the evolved gases in the pyrolyser tube are computer-controlled. Thermochromatography is thus essentially a temperature-resolved, multiple-injection GC technique [896,897]. Consecutive chromatograms are collected during pyrolysis, each at known temperature and representing the headspace gas component distribution above the pyrolysing material. Heating rates typically used in TG are also applicable in ThGC. Each isothermal GC run is completed between two sampling events (typically one sample every two minutes). Tens of chromatograms are produced during a pyrolysis process in ThGC. As only a short analysis time is available for each chromatogram they must be run isothermally. This is a limitation as compared to flash PyGC methods. However, it is possible to analyse the important parts of the pyrolysates from most low boiling products up to oligomers by using ThGC.

ThGC is a **multidimensional technique**, where output data is in the form of a 2D response surface in the co-ordinates of pyrolysis temperature and chromatographic run time. Chemometric techniques (PCA, PCR, PLSR and EFA) decompose ThGC data sets into "chromatograms" and their corresponding temperature-profile "thermograms" of the evolving headspace gases [898–900]. ThGC allows constructing a mass-loss curve related to detectable volatile product evolution. The method unites some of the advantages of PyGC and TGA. This kind of evolved gas analysis gives more information about the thermal decomposition than, for example, TG alone. The technique is not in common use. A review has appeared [897].

Applications
ThGC provides information about the number of stages of gas evolution and related headspace gas composition. During thermal programming both thermal vaporisation (desorption) and multi-stage degradations are observed. The technique has potential as characteristic fingerprint plots in polymer studies [895,900]. The usefulness of evolving factor analysis (EFA) in thermochromatographic analysis of polymers has been demonstrated using EPDM rubber containing a flame retardant as an illustrative example [901]. EFA made it possible to recognise the thermal steps during pyrolysis and to identify which evolving compounds dominate at a given temperature. ThGC has also been used to study the effect of ammonium polyphosphate (APP) on thermal and thermo-oxidative degradation of polystyrene. ThGC can be used for quality control of complex polymeric materials.

2.3. THERMAL VOLATILISATION AND DESORPTION TECHNIQUES

Principles and Characteristics
Thermal studies of polymers and polymer formulations may be classified in the first place according to the amount of energy provided to the system. By introducing sufficient energy the additional components present in a polymer formulation, such as additives, modifiers, residual monomers and oligomers may be released. In the extreme case so much energy is provided that the polymer is reduced to a series of small, often structurally important, fragments (pyrolysis products). In the intermediate range residual volatiles may still be released but, with increasing energy, thermal degradation of the polymer rather than complete pyrolysis will occur. Other factors distinguishing various thermal methods are pressure (from vacuum to atmospheric) and flow conditions (from static in vacuum TG, SHS, TVA and DI-MS to dynamic in DHS, TD, TGA and TPPy).

The rate of loss of additives from polymers is determined by migration and volatilisation. Physical loss of additives from polymers is increasingly important in food applications and medical uses of polymers, but also other situations lead to a rapid decrease in antioxidants, such as **mechanical extraction** of tyres (at temperatures up to 100°C at high speed conditions) or simple solvent extraction of seals and hoses, *etc.*

The classical method for separating additives from polymers, namely by means of **solvent extraction**, suffers from several disadvantages, in particular at low levels of the additives (dilution effect), *cfr.* Chp. 3 of ref. [213a]. In principle, more sensitivity (typically 1000 times) may be expected for **heat or thermal extraction**. However, without adequate measures resolution is generally poor, *e.g.* it is often difficult to separate oligomers from additives and oil. Moreover, alternative techniques such as TGA and (temperature-programmed) pyrolysis will volatilise thermostable additives but may easily also induce thermal fragmentation. In other cases, as for thermosetting rubber mixtures, a further problem arises from the many additives which are already fragmented due to the curing and cross-linking process. Removal of volatiles from polymeric materials is a form of distillation. This is actually usually not performed for analytical purposes but constitutes a prerequisite in the design of many consumer products. Polymer devolatilisation as a processing step has recently been reviewed [902].

Table 2.41 lists the **general features** of thermal extraction. The field is characterised by many rather

Table 2.41. Main characteristics of thermal extraction

Advantages:
- No sample preparation (no contamination problems)
- High sensitivity (no dilution effect)
- No analytical interference from (ultra-pure) solvent or solvent artefacts
- Time efficiency
- Desorption efficiency >99%
- Environmental friendly (no solvent disposal)
- Cost effective

Disadvantages:
- Strong dependency on volatility
- Solvent favours extraction process (by swelling)
- Low resolution
- Complex analysis
- Difficult quantitation
- Unsuitable for thermolabile additives

similar methods and devices. Various solvent-free or thermal extractions are available to the experimentalist. As with supercritical fluid extraction, gas extraction of analytes from solids is influenced by the nature of the analyte-matrix interaction. For systems behaving as though volatiles are "dissolved" in the solid (*e.g.* monomer-in-polymer systems), equilibrium headspace is an appropriate method, but for adsorption systems in which the analytes are firmly bound to the matrix (*e.g.* water in polyamides) thermal desorption is favoured. Volatile extraction avoids some of the disadvantages associated with solvent extraction, and, thanks to its suitability for analysis of wide boiling point range samples it complements SFE.

Both static thermal volatilisation and dynamic thermal desorption techniques are available as **sampling methods** for direct polymer/additive analysis (*cfr.* Table 2.42). A group of thermal techniques is based on detection and analysis of materials thermally evolved from a sample upon controlled heating in a **static** fashion. These comprise differential thermal distillation, sublimation, thermal evolution methods based on pressure changes (TVA, SATVA with differential condensation of products), static headspace techniques, direct-probe mass spectrometry, temperature-programmed pyrolysis, programmable temperature vaporisation, *etc.* Various hyphenated *cq.* multiple detection modes are reported for separation and identification of the volatiles (GC, FID, MS, UV, FTIR). Thermal volatilisation techniques provide cost-effective sample preparation for analysis of (semi)-volatile organics in a wide range of matrices. They offer significant productivity and sensitivity benefits compared to solvent extraction, steam distillation or other labour-intensive techniques and allow a broad volatility range and ppt to % level analyses. Applications include the determination of additives and contaminants, vapour pressure measurements, determination of odorous materials in polymer systems, as well as thermal degradation studies and polymer composition and structure identifications [42].

Thermal desorption (TD), as the name implies, is a thermal extraction technique and is essentially a **dynamic process**. As with liquids, solid samples are heated under a continuous flow of inert gas such that (semi)-volatile organics are extracted from the sample matrix into the gas stream and transferred in the vapour phase to an analyser. The terminology is not always strictly adhered to and thermal desorption is

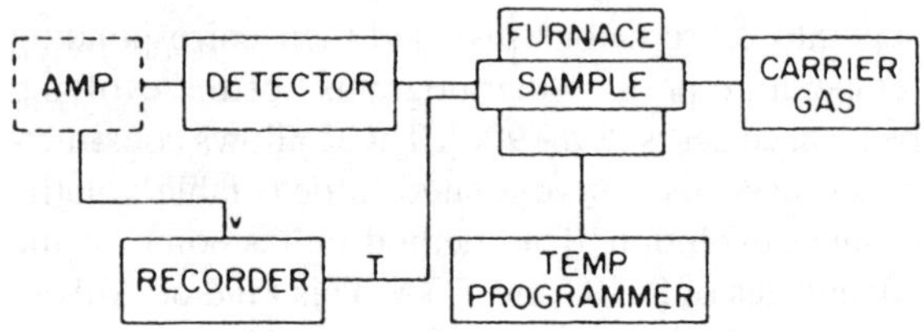

Fig. 2.42. Definition of thermal evolution analysis. After Chiu and Palermo [42]. Reproduced from *Analytica Chimica Acta* **81**, J. Chiu and E.P. Palermo, 1–19 (1976), with permission from Elsevier.

usually also meant to describe processes related to in-source mass spectrometry where thermal extraction occurs in high vacuum conditions. Development of standardised thermal desorption methodologies is still an issue.

Chiu *et al.* [42] have coned the terminology of **thermal evolution analysis** (TEA). TEA is not just a specific heating device to produce volatiles, but rather a family of techniques which provides qualitative and quantitative information on the sample (Fig. 2.42). A typical TEA technique is thermal volatilisation analysis (TVA) with or without differential condensation of products [903,904], which allows information to be obtained on the nature of the products and on compositional changes on heating. Besides detection of gas evolution by pressure change at constant volume (as in TVA), thermal decomposition can also be monitored by change in volume at constant pressure. A constant pressure–variable volume gas detection apparatus has been described [905]. In general terms, TEA techniques may be based on pressure/volume changes [903–905], flame ionisation [906], or on thermal conductivity in combination to MS [382]; other systems described are TEA-IR and TEA-CT-GC [907]. Gas chromatographs equipped with either FID or thermal conductivity detectors can readily be adapted to perform thermal evolution analysis. An ion-selective electrode has been used to monitor HF evolution [908]. Other detectors such as gas density balance, photocells for light scattering, and radioactivity, *etc.*, have been demonstrated as tools for monitoring thermal evolution and should be potentially useful for polymer degradation studies.

TEA techniques thus allow continuous monitoring of materials thermally evolved from a sample on controlled heating. Stepwise detection of such volatiles as a function of temperature or time, and quantitative measurement and identification of these materials provide very useful information.

Table 2.42. Characterisation of thermal volatilisation and desorption techniques

Technique	Experimental conditions	EGA
a. Thermal volatilisation techniques (static)		
Thermal distillation	Vacuum[a]/atm.	UV, TLC
Vacuum TG	Vacuum[b]	MS, FTIR
Sublimation	Atmospheric	–
TVA, SATVA	Vacuum, inert	GC, GC-MS, FTIR, MS
SHS	Atmospheric	GC, GC-MS
HS-SPME	Atmospheric	GC, GC-FID/MS
TD	Vacuum[c], inert	GC, MS, GC-MS/FTIR
LD	Vacuum	MS
b. Thermal desorption techniques (dynamic)		
PT[d]	Inert	GC, GC-MS
DHS	Inert	GC, GC-MS
TD	Inert	CT-GC, CT-GC-MS
TPPy[e]	Inert	GC, FTIR, MS, GC-MS, GC-FTIR
ThGC	Inert	GC

[a] 1 mbar.
[b] Usually 10^{-2}–10^{-5} mbar.
[c] 10^{-5}–10^{-7} mbar.
[d] Room temperature.
[e] Static and dynamic modes.

Table 2.43. Thermal evolution techniques[a]

Thermal evolution	EGD	EGA
Furnace (TEA)	Thermal conductivity	GC
	Flame ionisation	IR
DTA	Pressure change	UV/VIS
TG	Volume change	MS
	Gas density	Chemical
	Electrochemical	
	Photometric	
	Radioactivity	

[a] After Chiu and Palermo [42]. Reproduced from *Analytica Chimica Acta* **81**, J. Chiu *et al.*, 1–19. Copyright (1976), with permission from Elsevier.

The capability of a thermal technique for materials characterisation is greatly increased by hyphenation to identify further either the residue or the effluence (preferably both) during a certain thermal event. Detectors offering the best combination of sensitivity and versatility are MS and FTIR. They enable the simultaneous identification of the gaseous species emitted from a sample, according to their mass or their vibrational spectra. Desorption methods are most commonly used for identification purposes, less so for quantitation.

Volatiles from a well-controlled furnace, a thermogravimetric or differential thermal analyser, *etc.*, can be monitored by a variety of **evolved gas detectors** (EGD) (Table 2.43). Evolved gas detection refers to methods for sensing gases emanating from organic or polymer compositions as a function of temperature or or time, but without identifying their specific chemical natures. On the other hand, according to the ICTAC nomenclature, **evolved gas analysis** (EGA) then includes any technique of determining specifically the composition (nature and amount) of volatile product(s) formed during a thermal analysis event. This definition includes TG-MS, TG-FTIR, and also TPR (Temperature-Programmed Reduction) and TPD (Temperature-Programmed Desorption) coupled to a detection system, and in general all other techniques by which gases are released and detected either directly or indirectly by using an adsorbent or a solvent. Evolved gases may be analysed simultaneously or discontinuously, with or without generating weight loss data. Apart

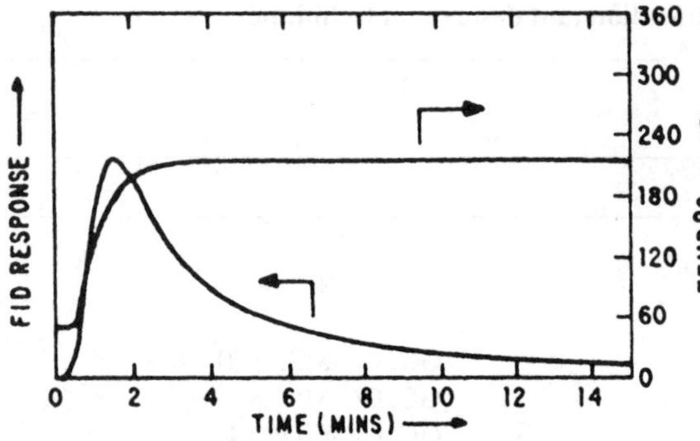

Fig. 2.43. Thermal evolution analysis of antioxidant in polyethylene. After Gill [923]. Reproduced by permission of Du Pont.

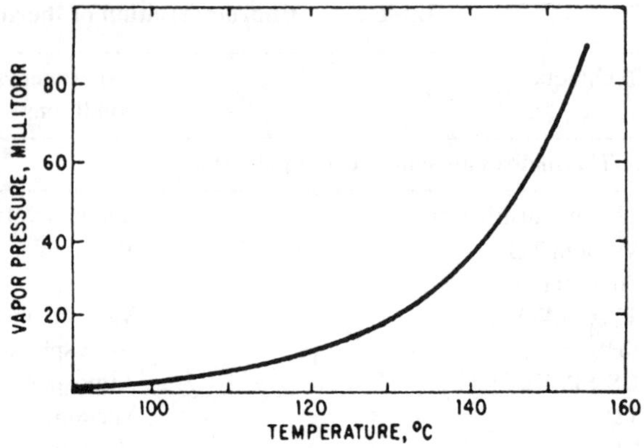

Fig. 2.44. Vapour pressure of di(2-ethylhexyl)phthalate by thermal evolution analysis. After Blaine [924]. Reproduced by permission of the author.

from the aforementioned TVA [909,910], other examples are linear programmed thermal degradation mass spectrometry (LPTD-MS) [882,911,912], temperature-programmed pyrolysis mass spectrometry (TPPy-MS) [761,883,913], and such closely resembling probe MS methods as thermal analysis mass spectrometry (TA-MS) [914] and mass spectrometric thermal analysis (MTA) [915–921]. Thermogravimetry can be uniquely used for qualitative and quantitative evolved gas analysis based on prior knowledge of the sample.

Evolved gas analysis techniques have recently been reviewed [922].

Applications
Thermal evolution analysis is an excellent tool for polymer studies complementary to other thermal techniques such as DTA, TG and pyrolysis. Its applications include thermal degradation studies, determination of additives and contaminants, polymer composition and structure identifications. With small variations, the apparatus can also be used for vapour pressure measurements, and for determination of odorous materials in polymer systems. Coupling of TEA to GC for the identification of effluents is practicable and useful. TEA-CT-GC was used for the analysis of **volatiles** from ABS: 10 ppb of styrene but negligible acrylonitrile was detected in the headspace of a typical ABS resin [42].

Low levels of **antioxidants**, such as 2,6-di-*t*-butyl-*p*-cresol in μg amounts of PE, have been determined by the first commercial thermal evolution analysis equipment based on the design by Eggertsen *et al.* [906] with flame ionisation detection (Fig. 2.43) [923]. The high sensitivity of FID can also be utilised for **vapour pressure measurements** (as for DEHP) (Fig. 2.44) [924]. Recently, Kiran *et al.* [925] coupled molecular weight chromatography to TEA for identification of decomposition products of polymers.

With some restrictions, EGA is applicable to the study of stability and degradation, determination of residual solvent, monomer and additives, evaluation of contamination, compositional analysis, quality control, investigation of kinetics and mechanisms, studies on toxicity, and process control. In the discontinuous mode, thermal desorption methods often use adsorbents or cold traps in TD-CT-GC or TD-CT-MS configuration.

Chiu *et al.* [42] have reviewed polymer characterisation by thermal evolution techniques.

2.3.1. Thermal Separation Techniques

Chances for successful identification and quantification are considerably enhanced when analytes are separated. For solutions, chromatography is the supreme tool, whereas for solids some form of thermal treatment may achieve fractionation of matter according to volatility. Vapour evolution from polymers may be controlled and studied by various means, such as sublimation, thermal distillation, vacuum TG-MS, thermal evolution analysis (TEA) including TVA, headspace techniques or thermal desorption. It is obviously desirable that "evaporation" of the additives takes place below the decomposition temperature of the polymer. In principle, this can also be realised in thermal-programmed pyrolysis (dry distillation in vacuum). Desorption processes are controlled by diffusion.

Table 2.44. Volatility of antioxidants[a]

Antioxidant	Vapour pressure (mm Hg)	Loss of weight (1%) at 150°C
2,6-Di-*t*-butyl-*p*-cresol	22.15	100
2-Benzyl-6-*t*-butyl-*p*-cresol	1.83	100
2,2′-Methylene-bis-6-*t*-butyl-*p*-cresol	0.169	19–28
Diphenylamine	7.52	100
N-isopropyl-*N*′-phenyl-*p*-phenylenediamine	0.59	40–53
N,*N*′-diphenyl-*p*-phenylenediamine	0.032	2–3

[a] After Schröder [926]. Reproduced by permission of IUPAC.

2.3.1.1. *Vacuum Sublimation*

Principles and Characteristics
Sublimation is a direct phase transition from the solid state into the gas phase. The sublimation procedure is not time-consuming and eliminates many conventional clean-up procedures. The technique has only found limited application for analytical purposes. The results of vacuum sublimation depend upon the physical form of the sample (powder, pellet). Some of the antioxidants listed in Table 2.44 are so volatile that direct determination by sublimation is possible.

Applications
Fazio *et al.* [927] have described a multideterminative procedure for various **antioxidants** (DLTDP, BHT, Ionox 100, PG, BHA) in food by means of a (10^{-4} mm Hg) vacuum sublimation technique (with GC and confirmed by UV, IR or MS) and achieved recoveries exceeding 85%. Although no indications were given that this isolation procedure would be equally successful for the quantitative determination of these indirect food additives when incorporated in a polymeric matrix, Shlyapnikov *et al.* [928] have reported the direct determination of AOs in PE by vacuum sublimation.

2.3.1.2. *Thermal Distillation*

Principles and Characteristics
When a sample of a polymeric material is heated *in vacuo* of 10^{-1}–10^{-2} Torr at 100–250°C the low-MW compounds contained in the sample are evaporated and may be condensed in a cold vessel. Shlyapnikov *et al.* [929] have reported a special apparatus for extraction of low-MW compounds, followed by UV detection or TLC separation [930]. Distillation of low-MW compounds from polymers has been studied extensively [928,931]. For each compound there is a temperature (T_1) below which the compound practically does not evaporate (<1%) and a temperature (T_2) above which the compound is run off almost completely (>95%). The span between these two temperatures is about 50–100°C. This allows separating mixtures by varying distillation time and temperature. The quantity of volatile products increases with the evaporation temperature.

Affolter *et al.* [405] have described vacuum TG ($p = 1$ mbar) as a device for thermal distillation of relatively small quantities of sample (up to 30 μL). Thermal separation at moderate temperatures under modest 1 mbar vacuum in TG is often superior to chromatography and cost effective. However, some thermal pyrolysis cannot be excluded.

Recently a new technique for thermal extraction was described, which consists in high-frequency heating of a sample at the bottom of a small sealed glass tube vessel [932]. After cooling the opposite end of the tube the condensed low-MW compounds, which vaporise during the heating process, are collected. The method is used for recovering fatty acids, UVAs, AOs and plasticisers from resins.

Criado *et al.* [933] have reported the use of a conventional high vacuum TG device, operating at 10^{-2} mbar. Rouquerol *et al.* [934] and Raemaekers *et al.* [935] used a *high* vacuum TG device (operating at 10^{-5} mbar). Vacuum TG is to be considered as a combination of TG and TD.

For thermal distillative separation of larger samples (up to 1 mL) a special macroscopic thermogravimetric cell has been developed, which is used for industrial applications (*e.g.* analysis of PVC packaging material) [405].

Distillation in vacuo is free from the setbacks of solvent extraction, such as migration of the solvent into the polymeric matrix and solubility of the additive in the solvent [928,931].

Applications
Vacuum/thermal/displacement extraction procedures are used for the direct isolation or release of volatile

components from a polymeric matrix and may involve combined use of vacuum and heat, as during dry vacuum distillation or in direct insertion probe mass spectrometry. A typical example is gas stripping of a roll of packaging film with an odour complaint making use of an adsorption tube [936]. Literature gives only fragmentary indications on use.

Shlyapnikov *et al.* [929] have reported distillation *in vacuo* at different temperatures (from 20° to 200°C), with indications of the degree of extraction from 0.02–0.1 g PE for 0.2–2.0% diphenylamine, phenyl-α-naphthylamine (Neozon A), phenyl-β-naphthylamine (Neozon D), N-phenyl-N'-cyclohexyl-p-phenylenediamine, N,N'-di-β-naphthyl-p-phenylenediamine, Ionol, 2,4,6-tri-t-butylphenol, propyl gallate, α-naphthene, 2,2′-methylenebis(4-methyl-6-t-butylphenol), 2,2′-thiobis(4-methyl-6-t-butylphenol), 2,2′-methylenebis(4-chloro-6-t-butylphenol), 2,6-bis(2-hydroxy-3-t-butyl-5′-methylbenzyl)-4-methylphenol, and Tinuvin 326; reported accuracy of 1–2% using derivative ($n = 2$) UV spectrophotometry.

Affolter *et al.* [405] have analysed blends of monomeric and polymeric plasticisers in complex plastic materials using extraction (Soxhlet, SFE) and thermal methods (TGA with fixed heating rate, isothermal TGA, vacuum TGA at $p = 1$ mbar) rather than chromatographic techniques. Obviously, entrainment of monomers and rest solvents can never be excluded. Greater amounts of diisodecylphthalate (DIDP) and poly(adipic acid ester) (PAE, MW $\sim$ 4500) were separated by thermal distillative separation at $T(p) = 300°C$ (1 bar) or 220°C (10 mbar). Plasticisers can easily be distinguished by means of TGA [230,405], such as dibutylphthalate in PVC packaging material. Reduction in boiling points in vacuum TG and vacuum TG-DSC-MS were reported by Affolter *et al.* [405] and Henderson *et al.* [433], respectively, for the vaporisation of plasticisers. Use of high-vacuum TGA (10^{-5} mbar) for polymer/additive analysis in combination with soft ionisation MS seems attractive and deserves further attention, in particular in view of the high sensitivity of modern instrumentation.

FAAM [937] contains an analytical method based on vacuum distillation applicable to several regulated **antioxidants**. The analytical method measures only the presence of unoxidised substance. Analytical results may not necessarily indicate or reflect the amount of antioxidant that was originally added to the polymer.

2.3.1.3. *Thermal Volatilisation Analysis*

Principles and Characteristics

In thermal volatilisation analysis (TVA) the volatile products are passed from a heated sample in a continuously evacuated system to the cold surface of a trap some distance away. A small pressure develops which varies with the rate of volatilisation of the sample. If the pressure is recorded as the sample temperature is increased in a linear manner, a thermal volatilisation analysis (TVA) thermogram is obtained showing one or more peaks. The thermogram indicates the variation of rate of volatilisation of the sample with temperature. TVA is essentially a pyrolysis technique, which records the pressure of the volatile pyrolysates. The TVA trace is somewhat dependent on heating rate, as in case of TG, and should therefore be standardised.

In McNeill's TVA design [903], which consists in a **trap system** allowing products to pass selectively to a particular trap by distillation, product fractionation is achieved. The products are separated into non-condensable gases, condensable gases and volatile liquids, a cold ring fraction (tars and waxes) and a solid residue (if any), which may be examined by IR and MS. TVA may thus be used to good effect to determine the volatility distribution of degradation products as a function of oven temperature during the course of a programmed thermal degradation experiment (rate profiling). A low temperature peak is indicative of unpolymerised material and may result from a combination of such ingredients as monomers, absorbed gases, plasticisers, cross-linking agents, antioxidants and organic pigments.

Variants of TVA are differential condensation TVA [903] and **sub-ambient TVA** or SATVA [910, 938]. In SATVA the condensable gases and volatile liquids are further fractionated: volatile components are collected in gas cells and less volatile liquid fractions are separated and characterised by GC-MS. SATVA consists in the slow, controlled, distillation of the volatile products from the −196°C trap (where they have been collected) as it is heated up to room temperature. As they distil, MS or IR identifies the volatiles and at the same time they are collected in narrow fractions for further analysis. Continuous pressure monitoring during distillation is recorded as a SATVA curve [939]. McGill *et al.* [940,941] have indicated that the very simple technique of trap-to-trap distillation of trace amounts of liquid-nitrogen condensable volatiles can be used for qualitative and quantitative analysis. If a cooled trap containing volatiles from a degradation study (*e.g.* following TVA) is allowed to warm up slowly in a closed system the pressure will increase in a stepwise manner as each component evaporates at the appropriate temperature. When the evaporating components are recondensed in a second trap a more readily interpreted differential pressure curve is obtained. This trap-to-trap distillation is called **Differential Distillation Analysis** by McGill [941] and sub-ambient thermal volatilisation analysis (SATVA) by McNeill *et al.* [910,942]. The former term is more descriptive. It also avoids possible confusion with the TVA technique, which employs a series of sub-ambient traps to monitor evolution of volatiles during thermal degradation.

Advantages of the non-destructive TVA techniques are that various product fractions are isolated (on the basis of volatility under high-vacuum conditions), which are available for subsequent spectroscopic analysis. McNeill [943] has described on-line TVA-SATVA-MS and the use of gas-phase IR spectroscopy for identification of the volatiles. IR and NMR may examine the cold ring fraction. Direct weighing allows quantitative measurements of residue and cold ring fraction. Stevenson *et al.* [944] have developed on-line TVA-FTIR with a vacuum-tight long-path gas IR cell for the study of polymer degradation phenomena. TVA thus gives access to all **major product fractions** of polymer degradation (from non-condensable gases to the insoluble residue) and yields information on stages of breakdown and threshold temperatures. In a TVA thermogram, as in a DTG trace, peaks represent rate maxima for degradation processes. Some of the information obtained by (vacuum) TVA may also be found by (vacuum) TG and DTA, but these methods are best regarded as complementary to each other. It is possible to measure quantitatively by TG all changes which result in weight loss, but volatile and "cold ring" fractions are not distinguished. TVA reveals only processes producing some volatile products, but can distinguish between products condensable and non-condensable at -190°C. As to the **disadvantages** of the technique, because of differences in thermal conductivity, it is difficult to make quantitative deductions from peak heights in TVA thermograms. Suitable calibration procedures are necessary. TVA equipment appears to have had limited distribution only; the apparatus is non-commercial and non-standardised. Ideal samples for TVA are films, but also thin layers of powder may be examined; typical sample weights are 50–100 mg, which is a considerable advantage over TGA and for heterogeneous materials.

Applications

As few experimental TVA set-ups have been built, the method has found restricted application, mainly for identification of rubbers, vulcanisation of natural rubber, determination of **total and non-condensable volatiles** in polymers and degradation studies. TVA has proved useful for testing a wide range of polymers (including PS, poly-α-methylstyrene, styrene–butadiene copolymers, PVC, polyisobutene, butyl and chlorobutyl rubber) on the presence of trapped solvents, monomers, *etc.*

TVA and Differential Distillation Analysis (or SATVA) have been applied to studies on the effect of chlorinated hydrocarbon **fire retardants** [945, 946]. Rigby [947] has studied vapour evolution from LDPE cable insulation material (20–60 mg) by means of TVA-ToFMS, a form of in-source TD-MS, and identified Santonox R and traces of the **cross-linking agent** dicumyl peroxide in the low-temperature peak of the TVA curve. The TVA curves of commercial LDPE and of freshly prepared LDPE/Santonox suggested that in commercial samples Santonox was distributed closer to the surface than in freshly prepared ones. The detection limit (using ToF-MS) for Santonox bulk-distributed in LDPE was about 0.01%. Chiantore *et al.* [948] have characterised a phenol-formaldehyde (PF) resin by means of TVA (to isolate the lower-MW components) and SEC (both on volatile fractions and residues).

TVA has been used mainly for the study of the basic degradation patterns of depolymerisation [910,942]. McNeill *et al.* [949] have studied thermal degradation of PS and PS/IDBP by means of TVA, SATVA and GC-MS. In the presence of 4,4′-isopropylidene-bis-(2,6-dibromophenol) (IDBP), the main products were similar, except for propiophenone and phenylpropanoyl bromide in the presence of IDBP. Similarly, TG and TVA have been used to study the thermal stabilities of PET, PBT and PDMT [950]. In this case, the amounts of the main product fractions (residue, cold ring fraction, volatile products) have been determined quantitatively and the various materials present in the volatile and cold ring fractions have been separated and identified. McNeill *et al.* [939] also fractionated PVC/DOP by

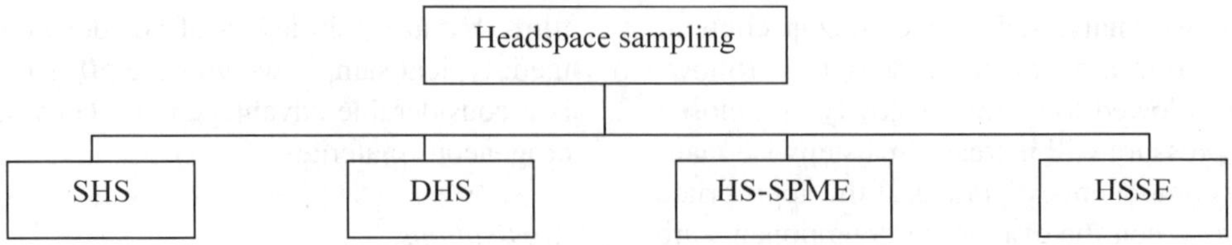

Scheme 2.7. Headspace sampling techniques.

means of TVA and SATVA and analysed by GC-MS and FTIR. Beside the products of the pure components, new products did originate from reactions of free radicals formed by the individual components. PVC/DOP exhibits retarded dehydrochlorination during a TVA experiment. Also the thermal stability of blends of PE, polyethyl acrylate and ethylene–ethyl acrylate copolymer with PDMS was investigated in inert atmosphere using TG/DTG and TVA [951]. The condensable volatile degradation products from the TVA experiments were separated by sub-ambient TVA and investigated by FTIR, GC, MS and GC-MS techniques. The cold ring fraction was characterised by FTIR spectroscopy and GC. Most of the degradation products from the blends were similar to the degradation products from PDMS and the corresponding polyolefin when degraded alone, but the presence of some additional products indicated interactions during degradations as a result of blending.

TVA-FTIR allows method development for online detection and quantification of degradation products, as they are formed in the TVA experiment. Stevenson *et al.* [944] have used TVA and SATVA as a platform for spectroscopic investigations of polymer degradation processes, as in case of PMMA and poly(bisphenol A, 2-hydroxypropylether). The kinetics of thermal and thermo-oxidative degradation and characterisation of the various degradation products of polystyrene alone and in the presence [949] of the flame retardant IDBP were investigated using TVA, FTIR and GC-MS. Other applications of TVA in problems in polymer chemistry have been reported [943,952]. The method is now near defunct.

2.3.2. Direct Solid Sampling Techniques for Gas Chromatography

For reasons of health, safety and environmental concern, it is often important to know the nature and amount of volatile components which can evolve during processing, storage and use. For outgassing problems and the characterisation of odour-active compounds thermal treatment hyphenated with gas chromatography is a versatile tool. Thermal desorption as a clean, labour-saving technique enables and simplifies a wide range of GC applications.

Heat extraction techniques for solid sample preparation in GC are static and dynamic headspace analysis (SHS, DHS, HS-SPME and HSSE), thermal desorption (TD-GC, TD-GC-MS), pyrolysis and thermochromatography. Nomenclature is not unambiguous as to DHS, TD and PT. The terminology purge-and-trap is usually preferred for the simplest dynamic technique in which it is not necessary to subject the sample to either solvents or elevated temperatures. Scheme 2.7 shows the family of **headspace sampling** techniques. Headspace sorptive extraction (HSSE) and HS-SPME represent high capacity static headspace.

Thermal desorption is commonly used in combination with a GC analyser. In the process of thermal desorption, heat and a flow of inert gas are used to extract (semi)volatile organics retained in a sample matrix or on a sorbent bed. **Trapping tube sorbents** (such as Tenax or graphitised carbon) may be selected to quantitatively retain every compound of interest to provide a representative analytical profile of the 'headspace' of (odorous) materials [953]. However, conventional thermal desorption in its most simple single-stage form is of limited application for packed column chromatography (broad component bands) and cannot be used at all for capillary column GC. Most samples do require refocusing. Analytes loaded into tubes with an i.d. above 2 mm simply cannot be desorbed quickly enough to produce capillary GC-compatible peaks. Most commercial thermal desorbers are 'two-stage', *i.e.* they contain a focusing mechanism (capillary cryofocusing or cold adsorbent trapping) for concentrating analytes desorbed from the sample tube before releasing them into the analytical system in as small a volume of vapour as possible. Both procedures do produce excellent, capillary-compatible chromatography, but capillary cryofocusing is quite costly in terms of cryogen consumption. Moreover, the volatility range of capillary cryofocusing devices is limited. Nowadays, refocusing on a small electrically cooled, packed cold trap, which can then be heated rapidly (>60°C/s) to desorb 99% of analytes within 10 seconds, is invariably the technique of choice for thermal desorption. No liquid cryogen is required.

In **purge-and-trap** (PT) a purge gas source is blown through a sample and purge gas and sample volatiles are trapped in an adsorption tube. Besides on-line analysis of the volatile compounds of samples at elevated temperatures, it is possible to proceed off-line by using Tenax adsorption cartridges, thus allowing collection for longer times, *e.g.* 24 hours. Off-line sampling with pre-concentration of working place atmospheres is commonplace. "On-line" and "off-line" PT techniques are of limited use in polymer/additive analysis. In off-line PT there is less restriction on the quantity of sample taken to obtain a satisfactory concentration of the volatiles for analysis. A purging vessel suitable for using up to 200 g quantities may be selected with a supply of filtered high-purity purge gas (He or N_2). Usually the adsorption tube used for cold trapping the volatiles is of dual-packing design (with a weak and a stronger adsorbent). This poses the problem of adequate choice of adsorbents to collect a wide range of volatile components. Moisture can sometimes cause failure of a cryo-trapping system due to icing up and poor chromatography. In dynamic headspace, which employs heat, concentrator technology is used in a typical TD-CT-GC-MS configuration.

Although thermal extracts are often comparable with standard solvent extracts for many GC or GC-MS material-testing applications this is not necessarily always the case. For example, it has been reported that many more products were identified after SPME-GC-MS than with direct HS-GC-MS [954]. Nevertheless, direct desorption of (ultra)volatile organics from samples weighed straight into empty desorption tubes or appropriate tube liners is probably the most straightforward and cost-effective sampling procedure for otherwise difficult materials. A Programmed Temperature Vaporising (PTV) injector can be used both as a thermo-desorption unit and as a programmed pyrolyser (up to 600°C) [752]. TD can yield either a complete, quantitative extraction of specific target analytes or just a characteristic analyte profile. The method combines sample clean-up, analyte extraction, and sample injection or introduction into one automated operation. Temperature-programmed pyrolysis (TPPy) in combination with a capillary GC column and spectral detection methods is another useful tool for the analysis of polymeric materials (*cfr.* Chp. 2.2.7).

TD-EGA techniques have been reviewed [445], as well as direct solid sampling methods for GC analysis of polymer/additive formulations [955].

2.3.2.1. *Solid Static Headspace Sampling*

Principles and Characteristics

Solution headspace gas chromatographic sampling has a counterpart in a solvent-free, direct method for the rapid determination of volatile components in solid samples. Volatile and semi-volatile components can be desorbed directly from sample matrices or from sorbent or cryogenic traps without any significant sample preparation.

The simplest method of chemical analysis of volatiles involves heating the polymer in a closed chamber and directly injecting a sample of the headspace gas onto a chromatographic column (ASTM Method D 4526). This technique, known as the **"hot jar" method**, was originally developed for analysis of residual monomer in PVC [956]. In a variation of the method a mixture of polymer and water is placed in a vial and heated for a period of time [957]. The water facilitates removal of volatiles by a steam-stripper mechanism. Crompton [176] has reported another simple and inexpensive device (**heated copper block**) for liberating both existing volatiles in polymers and those produced by thermal degradation from polymers by heating at temperatures up to 300°C, in the absence of solvents.

In static headspace sampling (SHS), which relies totally on volatilisation to separate analytes from a sample matrix, important factors are related to diffusion and surface area. Precise thermal conditions are needed to determine occluded solvents, residual monomers, and additives in polymers and their composites. In particular, for accurate quantitative analysis in a static headspace experiment, the temperature/pressure conditions of the sample vessel are critical. In SHS-GC an aliquot of the headspace vapour in thermodynamic equilibrium with a solid sample (of known weight) is transferred to a GC or GC-MS for separation and identification. HS-GC is characterised by a relatively long thermostating time (up to 25 min) and short analysis time (2 min). With

Table 2.45. Main characteristics of solid headspace sampling

Advantages:
- Direct analysis
- Automated sampling, high sample throughput
- Higher headspace sensitivity than in solution approach
- Detection levels: sub-ppb
- No loss of residual chemicals as would result from incomplete dissolution or extraction
- No solvent interference
- Minimal sample preparation time
- No sample extraction, clean-up and preconcentration necessary
- Maintenance of a cleaner chromatographic system (no injection of polymer solution)
- Quantitative technique
- Highly reproducible
- High precision ($<1\%$ RSD)
- Ease of use
- Applicable to wide variety of polymers
- Higher sample representativity than in most thermoanalytical techniques (*e.g.* TGA)

Disadvantages:
- Need for separate calibration runs for different polymers (different response factors)
- Restrictions on analysable chemicals (preferably b.p. <130–$150°C$, not extremely polar)
- Equilibrium T, t between headspace and polymer to be established before sampling
- High equilibration times
- Polymer decomposition to be avoided ($T < 90$–$160°C$)

overlapping thermostating for up to 12 samples the period from injection to injection (PII) is 2.1 min, which guarantees high productivity.

The characteristics of solid headspace sampling with an internal standard for the determination of residual volatiles in polymers are given in Table 2.45. Generation of a headspace sample is an equilibrium process that limits the amount of a specific component available for analysis within the practical restraints of time and temperature. Static headspace sampling in atmospheric conditions is limited to about 210°C (oxidation and thermal decomposition of polymers); an alternative is thermo desorption in inert conditions. Sensitivity is enhanced by 100 times using LVI with on-column cryofocusing. Solid headspace provides about 10-fold more sensitivity than solution headspace. HS-GC does not suffer from interference from the solvent peak or from impurities. Typical detectors used in SHS-GC are FID, ECD and MS. Determination of trace components of relatively high-MW is very difficult, mainly because of utilisation of indirect syringe sampling. Chemical interactions are possible.

In order to obtain quantitative results by HS-GC, the system must be calibrated. Absolute quantitation is not possible. **Quantification** can be done by the conventional external calibration method with liquids containing the analytes concerned in known concentrations or by means of standard addition. Pausch *et al.* [958] have developed an internal standard method for solid headspace analysis of residuals in polymers in order to overcome the limitations of external standardisation (*cfr.* Chp. 4.2.2 of ref. [213a]). Use of an internal standard works quite well, as shown in case of the determination of residual hydrocarbon solvent in poly(acrylic acid) using the solid HS-GC-FID approach [959]. In the comparison made by Lattimer *et al.* [959] the concentrations determined by solid HS-GC exceeded those from either solution GC or extraction UV methods. Solid HS-GC-FID allows sub-ppm detection. For quantitative analysis, both in equilibrium and non-equilibrium conditions, *cfr.* ref. [960]. Multiple headspace extraction (MHE) has the advantage that by extracting the *whole* amount of the analyte, any effect of the sample matrix is eliminated; the technique is normally used only for method development and validation.

Headspace analysis has been described in several reviews [561] and books [960–962]. Solid polymer HS-GC methods were reviewed [176].

Applications

Equilibrium HS-GC is used for analysing a wide range of solids (polymers, resins, powders, film, granulate, soils) and – especially – liquids (aqueous samples, oils, emulsions, gels, ointments, *etc.*). Although HS-GC may also be used in qualitative analysis, its main application is for the quantitative determination of volatile components (residual monomers, solvents, impurities) in samples. For identification and/or determination of **residual solvents** in polymers it is mandatory to use solventless methods of analysis, *i.e.* there must be no risk of confusing solvents in which the sample is dissolved for analysis with residual solvents in the sample. Most methods for determination of residual solvents are based on headspace techniques. Chromatographers can routinely determine typical **volatile compounds** at sub-ppm levels. Examples of the use of solid HS-GC sampling applications include fingerprinting, qualitative and/or quantitative analysis of residual monomers in polymers, *e.g.* ACN (ASTM 4322-83), 1,3-butadiene [963], vinylchloride in PVC (ASTM 3749-87), ethylene in PE, or acrylates in polyacrylates, and of volatile residual solvents, such as propanol in polyolefin film [964] and hydrocarbon solvent in poly(acrylic acid) [959], residual volatiles in manufactured goods, *e.g.* acetaldehyde in PET bottles, ethylene oxide in various materials [965], epichlorohydrin in epoxy resins [966], or residual printing ink solvents in packaging materials. Acrylonitrile, α-methylstyrene and styrene monomers were determined by SHS-GC over heated solid polymer samples [967]. Crompton [176] has reported the **quantitative determination** of toluene and ethylacetate on adjacent pieces of a PE adhesive laminated to PP double-coated with saran. The same author has applied SHS-GC-FID also to the determination of styrene monomer in polystyrene and its copolymers, using α-xylene as an internal standard. Similar methods have been used for the quantitative determination of volatiles in styrene-butadiene, PVC, α-methylstyrene, polycarbonates, polyolefins and rubber adhesives. Headspace simplifies the determination of the water content in many products including detergents, lubricating and hydraulic oils and explosives.

In headspace GC analysis of volatiles in polystyrene organics (aromatics, n- and isopentane) may be detected by FID, organic halogen compounds with ECD and inorganics with TCD [968,969]; LOD 0.001% with an accuracy of $\pm 5\%$. For measurement of volatiles in polymers the equilibrium is reached for some plastics fairly rapidly, for example vinylchloride in PVC reaches equilibrium in about 10 min at 100°C. However, styrene equilibration in PS takes much longer than is practical, and solution headspace is then preferred. With heated SHS recovery is biased towards the more volatile compounds. SPME tends to yield a higher recovery with relatively non-volatile compounds than SHS. Karlsson *et al.* [954] have compared headspace GC with SPME methods for polymers.

Oguri [970] has described the use of HS-GC in the analysis of **additives** in vulcanised rubbers and reinforcing materials. Accelerator fragments and AOs in vulcanised rubbers have been determined by SHS-GC-MS/FTIR/FID [971]. HS-GC is also used to determine volatile decomposition products of peroxide initiators in final resins. Tsuge *et al.* [972] have applied HS-GC to the determination of ester plasticisers and phenolic and amine AOs in a butadiene rubber.

Headspace (as a sample preparation technique) in combination with chemosensor technology is of great interest for **odour analysis** [973]. Ezrin *et al.* [974] have given an account of studies using HS-GC-MS of contaminants present in HDPE recycled from food packaging containers. SHS-GC was also used for contaminant diffusivity measurements through LDPE film [975].

A European collection of 50 recycled food contact PET samples was recently analysed by high-temperature (180°C, 1 h) SHS-GC-MS using three substances (limonene, benzophenone, methyl stearate; 1–50 ppm) as external standards for quantification of the contaminants [976]. The results are shown in Fig. 2.45. Maximum contamination levels of 4 ppb were observed. The headspace method was verified by liquid extraction GC-MS. HS-GC-MS showed limits for the extraction of lesser volatiles for which liquid extraction should be applied.

SHS-GC can also be used for the determination of airborne emission profiles for new and reformulated products for the plastics, adhesives, and coatings industries as well as for studies related to fogging of semivolatiles and phthalates in the automotive industry. HS-GC was also used for the analysis of foaming agents in expanded polystyrene [977].

2.3.2.2. *Dynamic Headspace Sampling*

Principles and Characteristics

Dynamic headspace sampling (DHS) is a solvent-free, highly reproducible, automated extraction procedure of volatiles from almost any matrix for quantitative and qualitative determinations, which extends the headspace method and uses **concentrator technology** to achieve highly sensitive detection limits. Table 2.46 summarises the characteristics of concentrator technology using thermal desorption methods.

Unlike static headspace, DHS involves sweeping the sample with a continuous stream of inert (*e.g.* high-purity helium) carrier gas during the sampling period. Sampling may be conducted at ambient or elevated temperatures, but below the degradation temperature region for the material. Carrier gas is adjusted for suitable time and flow-rates that are appropriate to the analysis. Dry samples, such as polymers, construction materials, or environmental matrices can be purged for dynamic headspace analysis. The result of DHS is the generation of a headspace of considerable size (volumes of several cm^3 to litres). There are two general methods

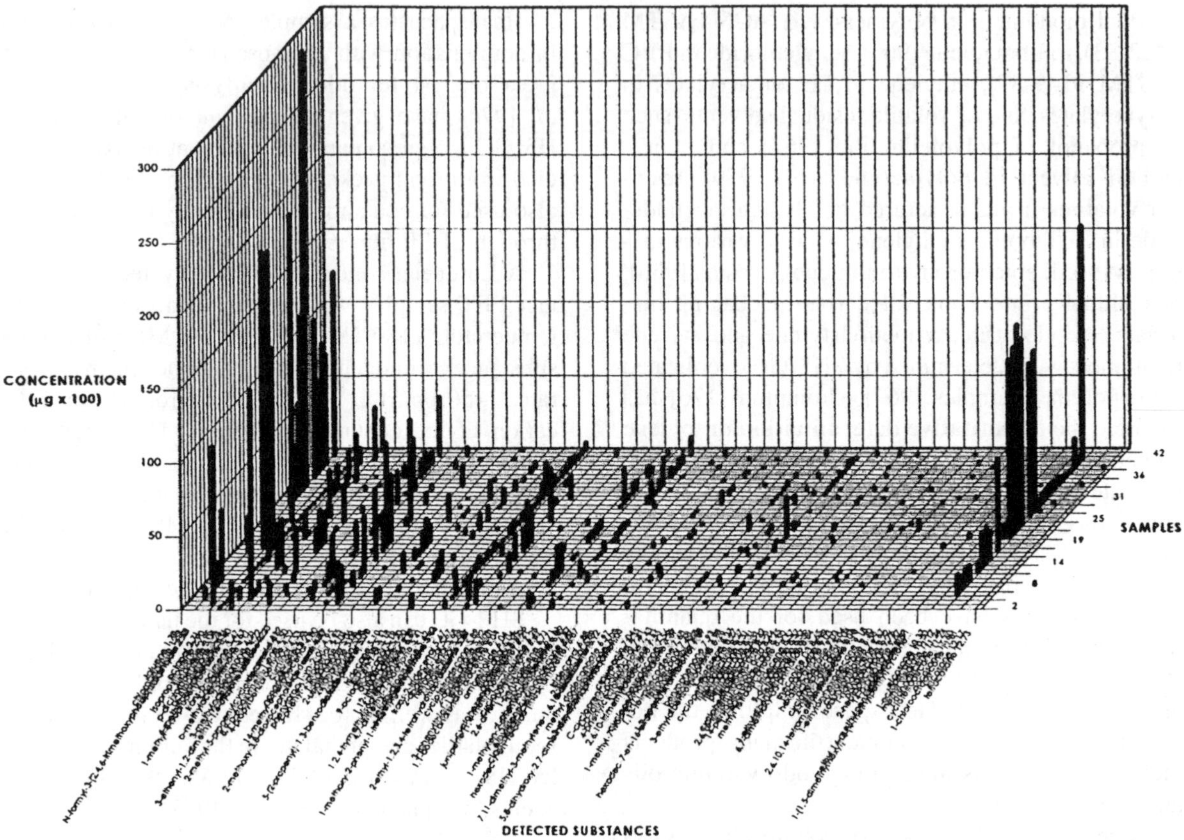

Fig. 2.45. PET headspace extraction of "dirty" samples. After Raffael and Simoneau [976]. Reproduced by permission of B. Raffael, Joint Research Centre, Ispra.

Table 2.46. Characteristics of concentrator technology using thermal desorption[a]

Advantages:

- Useful for gas, liquid, and solid materials
- Minimal sample handling
- Controlled and reproducible operations
- Wide range of sorbent selectivities and capacities, thermal desorption times and temperatures, and sequencing of specific sorption/desorption stages
- High concentration enhancement potential ($>10^3$)
- Fully automated methods for ppb-level analyses
- Automated interfacing to GC, GC-MS, GC-FTIR

Disadvantages:

- Thermally labile compounds difficult
- Requires appropriate internal standardisation

[a] After Liebman and Wampler [561]. Reprinted from S.A. Liebman *et al.*, in *Pyrolysis and GC in Polymer Analysis* (S.A. Liebman *et al.*, eds.), Marcel Dekker Inc., New York, NY (1985), by courtesy of Marcel Dekker Inc.

for retaining the volatiles of interest: cryogenics and solid-phase adsorbent trapping (at 20°C). Because the process requires several minutes to complete, the purged analytes need to be refocused by cryofocusing them on the column or trapping before GC analysis can take place; carrier gas is vented [978, 979]. The stripped compounds, which are trapped, are successively to be desorbed on a GC column. It is essential that analyte molecules be desorbed intact from a matrix. The DHS technique is often referred to as **"purge-and-trap"** (PT) or sample concentration method and is described in ASTM D4526-85. In DHS-GC analyses, care should be taken to avoid introducing extraneous contaminants such as bleed phthalates from rubber septa used for the GC apparatus.

Sample concentrators are of major advantage in the study of volatiles in manufactured products using the DHS approach. Residual solvents, additives, and monomers may be sampled and efficiently analysed

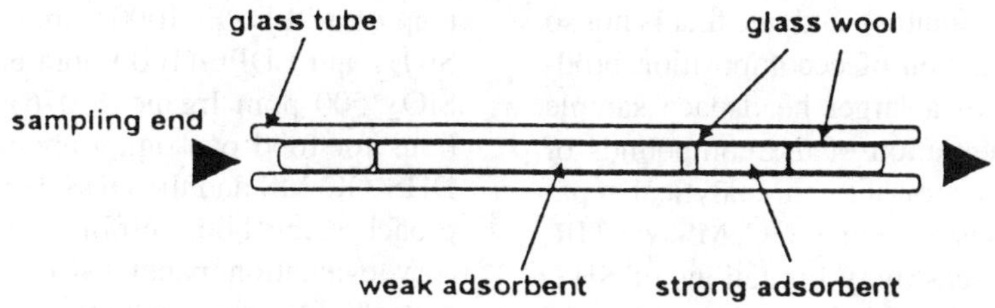

Fig. 2.46. Dual packed adsorbent bed sampling tube.

at trace level in a relatively routine manner. The most cost-effective systems are based on cartridges or on-line traps filled with appropriate sorbents. Since this method increases the sensitivity over static headspace, generally small amounts of sample (typically *ca.* 10 mg) are needed. The ability of solid sorbents to effectively trap compounds at ambient temperatures and to release them at elevated temperatures permits thermal desorption to be used instead of solvent stripping. Because of the ability to sorb compounds reversibly, many of the sorbents are column packings used in GC or HPLC. Different **sorbents** (*e.g.* activated charcoal, Tenax-GC, Chromosorb porous polymers, *etc.*) vary in their specificity and efficiency for retaining different species. The varied retentive properties for different sorbents have resulted in use of multi-layered sorbents within a single trap (or several different sorbent traps in series), *cfr.* Fig. 2.46.

Thermal desorption requires a means to apply heat to the trap while carrier gas is flowing through and to do so in a pulse mode, so as to simulate a typical syringe injection into a GC system. Selection of thermal desorption parameters depends on the thermal stability of sorbent and sorbate, the energy required to desorb the latter (related to its adsorption enthalpy), and close control of the chosen desorption temperature (particularly the risetime to achieve a rapid pulse) [561].

In modern DHS instrumentation the two-stage thermal desorption process consists of tube desorption onto a Peltier-cooled trap, followed by trap desorption of the heated trap. Peltier-**cooled focusing traps** concentrate volatile organics without liquid cryogen. Rapid thermal desorption of the focusing trap (40°C/s) produces component bands about 1 s wide for uncompromised narrow-bore CGC. Flexible split-ratio selection facilitates a wide range of applications from trace level to percent levels of volatiles in solids. An alternative is to trap analytes in individual **portable traps**, which are subsequently desorbed off-line. Advantages of this system are provision for long storage periods, avoidance of cross contamination of in-line systems, remote sampling and automation (high sample throughput).

Ezrin *et al.* [980] developed a **direct dynamic headspace** device for use in GC-MS in which the sample (5–25 mg) is placed in the hot zone directly on the GC column head. The method is adequate even for very high boiling compounds. The design of the **DHS-GC-MS** device probably contributes to this capability, there being no transfer lines and the sample tube being located directly at the head of the GC column. It is capable of isolating trace level compounds that would have been much more difficult to determine by extraction methods. Analysis time is much shorter than by extraction. Identification of compounds is based on GC retention times, mass spectrometry and reference compounds. Alternate methods of analysis, such as SFE and SFC, which use CO_2 for extraction and as the carrier in SFC, operate at considerably lower temperature than in DHS-GC-MS. Possibly the aromatics can be extracted more readily and completely by SFE than by heat alone (headspace).

The **advantages** of DHS and of another close variant of dynamic extraction, namely thermal desorption, depend on the detection limits required, the prevailing legislation or trade practices, and, in particular, the nature of the sample matrix. DHS is applicable to a very broad boiling range of components; nearly 100% recovery of volatiles is possible. Advantages of DHS-GC include high sensitivity, good resolution between compounds, qualitative and quantitative determination of composition. In accurate quantitative analysis of compounds that are difficult to remove from the host polymer with heat alone, a compromise headspace temperature is used between one that will remove most of the compound in a reasonable time (minutes), and one that is not so high as to cause formation of decomposition products. By generation of a larger headspace sample and refocusing/concentration of the compounds of interest before introduction into an analytical separation/identification system, either GC, MS or FTIR, DHS overcomes the sensitivity limitations of SHS. The two-step desorption procedure achieves a sample enrichment up to a factor of 10^3. A **disadvantage** is that not all compounds are sufficiently volatile to be analysed in this way (*e.g.* Irganox 1010, DSTDP, calcium stearate). DHS is also more cumbersome to use than SHS, and usually for quantification it is necessary to completely strip the analyte from the matrix, which could be time-consuming. Thermal desorption analyses are normally successfully **calibrated** using an external standard method. However, for additional confidence internal standard introduction is possible via a sample valve accessory. A theoretical model for quantitative determination of volatile compounds in polymers by DHS has been developed [981].

DHS-GC can be applied to solids, films, powders, semi-solids, and liquids in a direct manner with minimal sample preparation. DHS-GC can be used to analyse less volatile trace components in solid samples. The basic capability of PyGC coupled with sample concentrators for DHS analysis provides versatile instrumentation for polymer research or quality control situations.

A Ph.D. thesis covers dynamic headspace analysis, in particular of volatile compounds in polymers [982].

Applications

Applications of DHS-GC-MS are industrial, including the determination of residual volatiles, semivolatiles and degradation products in polymers, but mainly in food chemistry (flavour and fragrance analysis), in environmental science (pesticides), toxicology (biological fluids), and forensic science. DHS is used for quality control analyses.

The relatively low-temperature headspace sampling/thermal desorption techniques are successfully utilised in particular to study **odorous compounds** in polymers. Although headspace methods offer quick analysis by simple means, they suffer from the fact that odorous volatiles may be present in such small amounts that a concentration step may be required before detection is possible. Bigger *et al.* [973] produced an odour map of 87 LDPE/(1000 ppm erucamide, 1000 ppm SiO_2) and LDPE/(1000 ppm erucamide, 1000 ppm SiO_2, 500 ppm Irganox 1076) film-grade formulations for food packaging applications by means of DHS-GC-MS and the DHS-Tenax-GC-olfactory approach using both sniffing port analysis by a sensory evaluation panel (SEP panel) and an odour meter/SnO_2 semiconductor device. Unsaturated C_6 aldehydes, such as 2-hexenal and 2-methyl-2-pentenal (both odour-active compounds), are produced in the polymer as a result of undesirable degradation reactions involving the polymer, the slip agent erucamide, or the diatomaceous silica antiblock agent. The results suggest that SEPs can be effectively replaced by odour meters [983] for the evaluation of the level of odour in LDPE resins. A DHS technique has also been described for detecting C_6–C_{14} VOCs in LDPE consisting of purging with N_2, trapping at ambient temperature on Tenax-GC, and identification by GC-MS [984,984a]; odoriphores such as C_6 to C_{14} straight-chained aldehydes, alcohols, branched and unbranched alkanes and alkenes were identified.

Experimental variables (sampling volume, desorption time, desorption temperature and headspace pressure) were mathematically modelled and quantitatively assessed with a view to optimisation. Van Eldik *et al.* [985] applied solid polymer PT analysis for the screening of the **outgassing behaviour** of 61 styrenic samples (ABS, HIPS, PPO/PS), containing a variety of BFRs (such as DBBP, DBDPO, OBB, TBBP-A, TBPE, *etc.*), taken from used TV sets, computer housings and printers, and 13 polyamides (PA6, PA6.6) for electrotechnical applications in relation to health and safety for the user. The analytes were collected on a Tenax tube, cryofocused and analysed by means of TD-GC-MS.

Tsuge *et al.* [986] have reported DHS-GC-FID (wide-bore capillary column) in the study of **paint coatings** (acrylic/melamine resin/(LS-440 or LS-292 (HALS), Tinuvin 1130 or Tinuvin 900, surface modifier, organic solvent)). DHS-GC provides a sensitive technique for quantification of HALS and UVA occluded in cured polymeric materials such as coating paints. The optimum thermal desorption temperature was decided empirically. The system permitted fairly accurate determination of HALS (LS-440 and LS-292), which were almost quantitatively recovered, within an analysis time of 1.5 h. Reproducibility for Tinuvin 900 was less satisfactory, whereas Tinuvin 1130 was not observed because of its extremely low volatility.

Other reported applications concerning volatiles in polymers and packaging concern acrylonitrile and styrene in ABS terpolymers, residual solvents in adhesive tape, ethylene oxide in sterilised medical products, acetaldehyde in polyethylene terephthalate, and epichlorohydrin in epoxy resins and volatiles in food packaging film. Ezrin *et al.* [974, 980,987,988] have illustrated the versatility and wide applicability of direct dynamic HS-GC-MS to problems of plastics compositional and failure analysis, in particular for identifying rather low volatility compounds. Examples of such analyses are the identification of the oil source in oil-contaminated electrical systems, wax-like exudate at soldered locations, electrical power cable suspected as a cause of worker health problems, compounds adsorbed on reinforcing fibres, analysis of volatiles collected from a large quantity of material, identification of a high-boiling trace compound, and identification of a polymer not readily identifiable by other methods due to interference from additives. Headspace GC-MS has been developed here as a prime method of analysis of formulations containing compounds sufficiently volatile to be transferred into a GC at headspace temperatures up to 300–350°C. This includes many AOs, FRs, plasticisers, lubricants, *etc.* For example, the antioxidant BHT could be analysed quantitatively at 300–325°C in 10–20 min or less. **Failure analysis** often combines DHS-GC-MS and micro-infrared spectroscopy. Examples reported by Ezrin [989] are exudation of antioxidants, analysis of coupling agents absorbed on reinforcing fibres, DOP on carbon fibres, triphenylphosphine (TPP) on impregnated fabric, toluene sulfonamide in cables, and quantification and control of contaminants in recycled HDPE.

Raisanen [990] has described a device consisting of a cylindrical sample bottle holder, a dry carrier gas flow system, and a moisture transducer as a non-toxic replacement for Karl Fischer analysis of plastics. The sample of test material contained in a 20 mL septum bottle is heated to a preset temperature, usually just below the softening point of the resin. Evolved volatiles in the bottle headspace are passed through a cold trap filter to an analysis cell where the moisture content of the flowing gas is measured. “High boilers” are filtered out. A sophisticated algorithm, which makes use of the fact that as the sample approaches total dryness the rate of evolution of water is proportional to the water remaining in the sample, allows accurate determination of the moisture content. The method overcomes many of the limitations of KF and LOD methods. The **DHS moisture transducer** device has a detection limit of 10 μg of water, gives reliable results down to 10 ppm moisture on a representative large mass of pellets, is simple and quick to operate (5 min), is rugged enough for installation on the molding floor and is economic in use. Calibration and standardisation are quick and easy (10 min). The results correlate well with the reference standard (Karl Fischer coulometry). Validation tests show moisture determination accuracy of better than 2% standard error, and precision (coefficient of variation) of 0.5 to 10%. The range of applicability is 10 ppm to 10% moisture.

2.3.2.3. *Headspace Solid-Phase Microextraction*

Principles and Characteristics

Normally, analysis of solid materials prior to chromatographic separation and detection requires some form of extraction with organic solvents, either by heating (Soxhlet, Soxtec, *etc*), agitation (sonication or shake-flask extraction) of the organic solvent-solid mixture, or by more recently introduced techniques (MAE, SFE, ASE®). In particular the latter approaches are costly in terms of equipment. It has been shown that solid-phase microextraction (SPME) can also be utilised for direct analysis of solids [991].

In SPME analytes can be adsorbed from a liquid sample, by immersion or headspace extraction, or from a solid sample, by headspace extraction using a polymer-coated fused silica fibre. Headspace solid-phase microextraction (HS-SPME) shortens the extraction time and facilitates analysis of solid samples, provided that the analytes are volatile. Pawliszyn *et al.* [992] have reported initial work on HS-SPME in 1993. These authors have evaluated theoretically and experimentally the equilibrium and kinetics of HS-SPME.

Three phases are involved in headspace extraction, namely the condensed phase, its headspace and the SPME polymer. In the sampling process, the SPME fibre acts as a “chemical pump”, forcing compounds out of the headspace of a (liquid or) solid phase into a phase-coated fibre. For headspace sampling of volatiles the vapour phase should be in equilibrium with the sample. Sample/air/fibre partitioning of volatiles depends on many factors, including the nature of the sample matrix, presence of interfering compounds, sample and headspace volumes,

Table 2.47. Main characteristics of HS-SPME solid sampling

Advantages:
- Short equilibration times (2–15 min)
- Simple device; portability (suitable for on-site analysis and process monitoring)
- Not traumatic to analytes
- Low cost
- Concentrator method
- Theoretical principles available (but complex)
- Very low detection limits (ppt)
- Analysis of solid samples

Disadvantages:
- Method development required
- New technology
- Fragile fibres (selected suppliers only)
- Quantitatively exacting
- Difficult calibration of instrumental response
- Equilibrium process
- Overall extraction lower than in direct-immersion
- Only suitable for volatile analytes

temperature, fibre selectivity, and choice of standards (external *vs.* internal standardisation) [993]. As SPME is essentially an equilibration analytical method, the equilibrium of extraction has been reached when the concentration of the analyte is homogeneous within each of the three phases. Theoretical treatment of the kinetics of the diffusion process from the condensed phase to the SPME polymer film through the headspace is very complicated [994].

Table 2.47 lists the **main characteristics** of HS-SPME solid sampling. Methods such as headspace, purge-and-trap, liquid–liquid extraction, liquid phase extraction and thermal desorption present several disadvantages with respect to SPME, such as relatively long sample preparation time and high-boiling solvents. SPME is often a simpler and less expensive alternative method to DHS. In terms of precision, linearity and sensitivity, SPME equals HS methods. Detection limits of HS-SPME are at ppt level when ITMS is used as a detector and very similar to that of direct SPME. Thus, compounds such as phenols are detected by HS-SPME with far greater sensitivity than would be obtained with traditional heated SHS. Relative standard deviations of highly volatile components are 1–5%, for less volatile analytes 5–15% [995]. HS-SPME has potential to extract a wide range of organic compounds, volatile or semi-volatile from various matrices, both liquid and solid phases.

The SPME device not only combines extraction and concentration but also directly transfers the absorbed compounds into a GC injector. These features of HS-SPME provide major advantages over previous headspace techniques. Coupling to GC, GC-MS (including ion-trap), split/splitless and on-column injection or desorption of the analytes in an SPME-HPLC interface have been described. A significant difference in sensitivity between direct and headspace sampling can occur only for very volatile analytes. HS-SPME introduces some selectivity into the extraction technique as only analytes with sufficient vapour pressure at room temperature are detected. An obvious drawback of HS-SPME is that semi- and non-volatiles will not be present in detectable amounts in the headspace. In combination with GC this is actually advantageous and enables faster equilibration than sampling from liquid [992].

Quantification of analytes within solid samples by HS-SPME is not trivial [993]. As each volatile substance has a different equilibrium partition constant between the headspace and SPME fibre, the relative GC peak areas do not reflect the true proportion of these analytes in the headspace. Other factors, such as surface area, sampling time and temperature, will also affect the quantification. A problem in using SPME for analysis of solid samples is therefore calibration of the instrumental response [996]. Further advances in quantitative SPME of volatiles within solid samples are desirable.

Because liquid and headspace sampling methods differ in kinetics, the two approaches are complementary. Equilibrium is attained more rapidly in headspace SPME than in **direct-immersion SPME**, because there is no liquid to hinder diffusion of the analyte onto the stationary phase. For a given sampling time, immersion SPME is more sensitive than HS-SPME for analytes predominantly present in the liquid. The reverse is true for analytes that are primarily in the headspace. Several additional factors can affect SPME and do influence the choice between immersion and headspace sampling [997]. Overall extraction with HS-SPME is apt to be lower than in direct-immersion because transfer of analytes from the sample to the gas phase seldom is quantitative. HS-SPME was compared with PT [998] and HS-GC-MS [954,999]. Application of HS-SPME eliminates many problems of other headspace techniques and extends headspace sampling to less volatile compounds due to the concentration effect at the fibre coating. Thermal desorption of organic compounds from polymers and contaminated soil is quite effective for analysing this difficult category of samples by HS-SPME methods.

Cfr. also Chp. 7.1.1.3 of ref. [213a] for on-line SPME-GC coupling.

Applications

Some typical industrial applications of HS-SPME are analysis of trace impurities in polymers and solid samples, the determination of solvent residues in raw materials, ppt odour analysis, organics in water, *etc.* SPME can also be used for qualitative headspace sampling of fruit volatiles [1000]. Since equilibrium rather than exhaustive extraction occurs in the micro-extraction methods, SPME is suitable for field monitoring.

Karlsson *et al.* [954] have developed an SPME technique which was applied to the complex patterns of UV-initiated thermal degradation products in polymers. The extracted products from Scott-Gilead LDPE films containing photosensitising additives were analysed by GC-MS and compared to those obtained by direct headspace GC-MS. The degradation products were extracted with non-polar PDMS and polar carbowax/divinylbenzene SPME fibres or directly subjected to HS-GC-MS. The polarity of the fibre material has a large influence on the extraction efficiency of the different products. A polar fibre is more efficient in extracting polar compounds and a non-polar fibre for non-polar compounds. The SPME method allowed the identification of homologous series of carboxylic acids, ketones, and furanones, while direct headspace GC-MS identified only a few carboxylic acids (C_1–C_6) and small amounts of some ketones and furanones. The absolute amount of each product was not determined because of the difficulties involved in the exact quantification of a large number of products with different polarities and volatilities. The number of products observed in HS-GC-MS chromatograms was considerably smaller than after SPME fibre extractions followed by GC-MS. Only the most volatile products were observed in HS-GC-MS chromatograms, while SPME was more effective in extracting even less volatile products [999]. The polar carbowax fibre identified also C_7–C_{12} carboxylic acids and 4-oxopentanoic acid. The difference between the SPME method and traditional HS-GC-MS analysis is that in SPME the compounds are constantly removed from the vapour phase due to the absorption into the fibre. This leads to more compounds being volatilised from the polymer matrix to obtain a new equilibrium. The headspace method is less sensitive because the compounds are not removed from the vapour phase before the injection, and after equilibrium is reached no more products are released from the polymer matrix. SPME is therefore a valuable sample preparation technique to be used as a tool to isolate and identify complex series of degradation products in polymers. Karlsson *et al.* [1001] have also reported HS-SPME to extract photoproducts from the gas phase above UV exposed, enhanced degradable LDPE (with photosensitisers and biodegradable fillers). For HS-SPME applied to polymers grinding is necessary.

Residual solvents and monomers are normally monitored using SHS-GC. Penton [1002] has compared the determination of residual solvents and monomers in polystyrene with SPME and SHS. With heated SHS recovery is biased towards the more volatile compounds (such as BHT). This agrees with other studies. Page [1003] has described the quantitative determination of volatiles in solids (food) by means of HS-SPME.

2.3.2.4. *Thermal Desorption–Gas Chromatographic Techniques*

Principles and Characteristics

On-line thermal desorption (TD) is the use of heat with a flow of inert gas to extract volatile chemicals from a solid matrix transfer (often followed by transfer to GC). TD was originally developed as an off-line sampling method with pre-concentration of workplace atmosphere by pumping air through a solid adsorbent material, such as charcoal or Tenax. In this field thermal desorption has gained regulatory acceptance. For a 10 L gas sample a typical detection limit is *ca.* $10\ \mu g\ m^{-3}$.

Conventional thermal desorption systems usually consist of a desorption unit, an intermediate cold trap and a GC inlet. The sample passes through at least one intermediate trapping-desorption cycle (Fig. 2.47) with opportunities to lose active and/or high-MW components; the system requires optimisation of multiple parameters. Dynamic trapping of thermally desorbed organic components from a given sample and subsequent stripping into the GC column enables rapid and sensitive determinations of the trace organic components without sample loss and contamination. In some cases reactive thermal

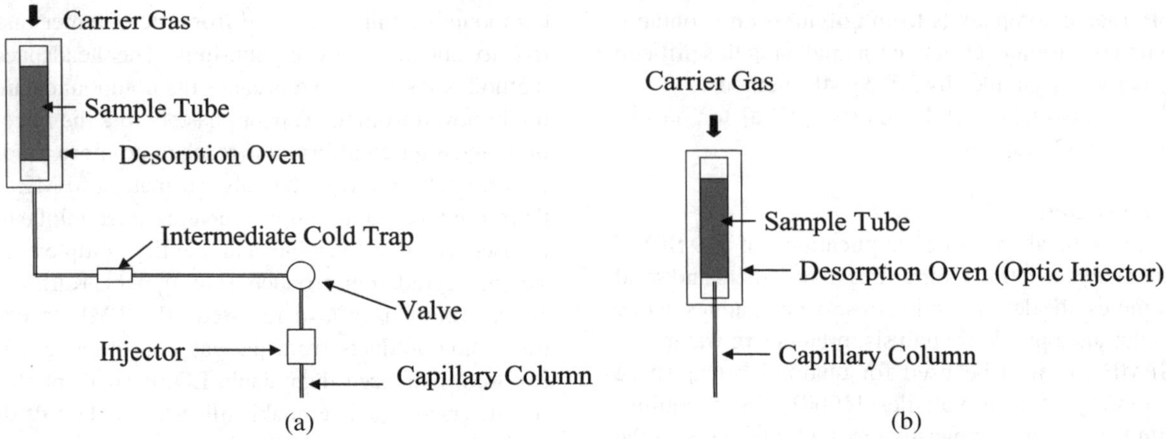

Fig. 2.47. (a) Conventional thermal desorber; (b) direct thermal desorption. Reproduced by permission of ATAS GL International, B.V., Veldhoven. The Netherlands.

desorption-GC (RTD-GC) is appropriate. The TD-GC interface should be uniformly heated. Advantages are high speed, cleanliness (no solvent artefacts), and simplicity (no sample preparation).

Thermal desorption techniques for GC sample preparation now are several [1004]:

(i) Direct interfacing to GC, usually resulting in poor resolution unless the sample is small enough or can be desorbed rapidly (principle of SPME).
(ii) Cryogenic focusing: good resolution, but costly.
(iii) Ambient trapping by means of an intermediate trap, which may be used to enhance selectivity by proper choice of sorbent.
(iv) Cryogenic traps; same rationale as ambient trapping, but the desorbed organics are condensed in the trap rather than adsorbed.
(v) Adsorption cartridges: for air monitoring of organic volatiles.
(vi) Peltier-cooled trapping (cryogen-free operation down to −30°C).

Depending on the nature of the material being tested, samples may be either weighed into empty TD tubes or tube liners for direct desorption (*e.g.* dry materials as polymers, resins, packaging materials) or purged off-line into tubes packed with a sorbent bed (useful for typically non-homogeneous and high-humidity samples). The combination of purge-and-trap with TD not only improves productivity but also facilitates selectivity. Trapping-tube sorbent may be selected to quantitatively retain all volatilised analytes. Alternatively, sorbents may be used that selectively retain certain specific analytes while allowing bulk ingredients to break through [1005].

Direct desorption of (semi)-volatile organics from samples weighed straight into empty desorption tubes or appropriate tube liners is a highly cost-effective sampling procedure for difficult materials. This procedure allows a characteristic odour profile or a complete, quantitative extraction of specific target analytes to be achieved. Sample clean-up, analyte extraction, and sample introduction are combined into one automated operation. Conditions for direct TD are: *(i)* high surface area solid materials: powders, granules (particle size <1 mm^3), fibres or films; *(ii)* unrestricted flow of gas through the sample tube; *(iii)* sample to be placed well within the heated zone of the thermal desorber; and *(iv)* molecules should be desorbed intact from the matrix. Direct thermal desorption (DTD) *cq.* extraction is an alternative to solvent extraction of solid samples, and is compatible with all GCs and GC-MSs. Direct TD is appropriate only if the desired extraction takes place at a temperature below the decomposition point of other materials in the sample matrix and the relatively small sample size that can be measured in a TD tube is representative of the sample as a whole. Fully automated **thermal desorption GC systems** have been reported for the analysis of volatile components in solid samples [1006], some being commercial. Gorman [971] has proposed a controlled thermal desorption and concentration method (essentially headspace) for separating volatile additives from vulcanisable rubber, in a HS-GC-MS configuration without the need for prior sample preparation, such as milling, extraction or pyrolysis. The procedure comprises sealing the test sample (typically 2–10 g of (un)vulcanised rubber) in a glass vial containing a controlled atmosphere and overhead headspace, followed by heating for complete desorption. The volatilised gases (composed of accelerator fragments, AOs, *etc.*) are then swept through a capillary GC column and analysed by MS, FTIR and FID using column splitters. While in-source TD-MS of a rubber vulcanisate shows no separation of components, the proposed method of ref. [971] achieves such separation. Less volatile components such as process oils remain in the vial as a liquid or solid. Coupling of thermal desorption and identification techniques constitutes a powerful means for detailed characterisation of outgassing processes with many potential applications in the field of rubbers.

Thermal desorption GC-MS is usually carried out in equipment which consists of a gas-tight precision oven at precisely controlled temperature through which a carrier gas is passed; the effluent of organic species is trapped in a cartridge (typically Tenax and charcoal). The oven may be set at a temperature at which a given material normally softens. Scrivens *et al.* [1007] have proposed the use of a **TD-GC-MS/FID** set-up composed of a programmable furnace (r.t.–800°C), a wide- and narrow-bore trap filled with sorbent materials of increasing tenacity (glass beads, Tenax, silica gel and Ambersorb XE 304) for trapping of volatiles according to polarity, and a double-focusing mass spectrometer. Gas passing continuously over the sample could be varied to mimic different atmospheres for thermal treatment. The concentrator gas flow is split to an FID detector for real-time monitoring of the evolution of volatiles and measurement of the total amount of sample trapped and to a sorbent trap kept at room temperature for further GC-MS processing. Quantitation is facilitated by the ability to inject an external standard into the wide-bore trap during the collection of volatiles. Quantification of thermally extracted samples depends on the matrix (adsorbent effect).

In direct thermal desorption (DTD) a few mg of a solid (typically <10 mg) are loaded into the cold injector, the carrier gas is temporarily halted, the injector is rapidly heated to the desired temperature (usually 50 to 200°C for polymer analysis), the carrier gas is resumed and the thermally extracted components are swept onto the column (Fig. 2.47b). The flow-path is simplified with few possibilities for sample loss and few parameters to optimise. The system is suitable for labile and high-MW compounds; cryogenic cooling is often unnecessary.

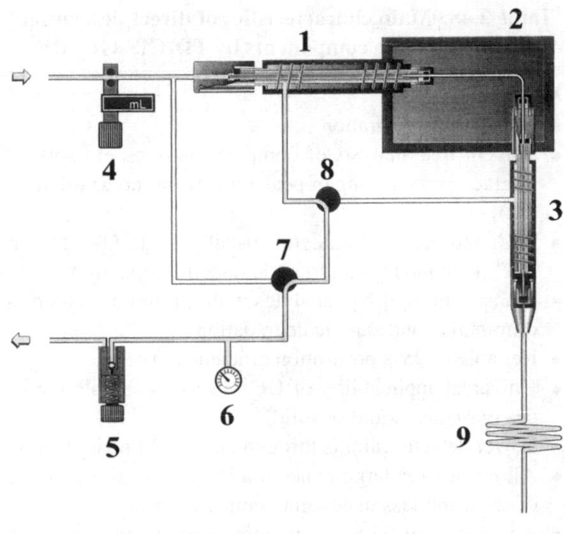

Fig. 2.48. Schematic of a thermodesorption (TDS)-temperature programmable cooled injection system (CIS). Legend: 1, TDS; 2, temperaturised transfer capillary; 3, CIS; 4, mass flow controller; 5, back-pressure regulator; 6, pressure gauge; 7, split/splitless valve; 8, TDS/CIS splitflow switch; 9, analytical column, connected to MSD. Reproduced by permission of Gerstel GmbH & Co. KG, Mülheim a.d. R., Germany.

Recently, another commercial direct thermodesorption system (TDS) with a temperature programmable cooled injection system (CIS) and GC-MS for identification has been introduced, which is suitable for the analysis of both prepacked adsorbent tubes and for direct analysis of solids, liquids and gels. Gas samples are prepared for analysis by passing the sample through a desorption tube containing an appropriate adsorbent. All other sample types (maximum of 200 mg adsorbed on a carrier or as a solid), placed in an empty tube without further preparation, can be inserted directly into the (horizontal) desorption chamber held at 20°C (initial temperature). After purging with the carrier gas and heating to the desired temperature the volatiles are transferred in either split- or splitless-mode via a heated transfer capillary into the CIS for cryofocussing (−150°C). After complete desorption the CIS liner is then heated up to the desired temperature to allow split or splitless transfer of the trapped volatiles to the analytical column, *cfr.* Fig. 2.48 [1008,1009]. In **TD-CIS-GC-MS** the CIS conditions (temperature program) play an important role. Using a cooled injection system – CIS/PTV technology – analytes are focused

Table 2.48. Main characteristics of direct desorption of volatile trace components by TD-CIS-GC-MS

Advantages:
- No sample preparation
- Solvent-free analysis of complex matrices (no solvent artefacts; no masking of peak by solvent; no health hazard)
- Wide b.p. range of analytes (usually C_2 to C_{40}, but up to C_{100} in modern multimode injection systems)
- Transfer of high-b.p. analytes with minimum mass discrimination and sample degradation
- Reliable (>95% desorption efficiencies)
- Universal applicability of GC-MS to all sample matrices (gaseous, liquid or solid)
- Lower detection limits through large-volume injection
- Allowance for large concentration ranges through use of split, splitless or solvent venting modes
- Avoidance of cross contaminations (frequently observed in extractions)
- Reduced sample matrix contribution
- Quantitative (preparation of standards and samples by spiking solutions onto the desorption tube)
- Ease of method development
- Autosampler capability, cost-effective sampling
- High desorption flow allowing short analysis times

Disadvantages:
- Unsuitable for thermolabile additives (and GC incompatible compounds)
- Dependency on volatility

in the inlet liner, not the column, before being transferred onto the analytical column as a narrow band. Few reports are available on the use of direct TD-CIS-GC-MS for quantification purposes. To that end the use of FID detection has considerable advantages (*cfr.* ref. [1007]). The main features of direct desorption of volatile trace compounds by TD-CIS-GC-MS are given in Table 2.48. Detection limits (ppt range) are enhanced by 10^3 as compared to Soxhlet extraction. Thermal desorption is probably best used for compounds of reasonable volatility, which can be desorbed intact and are present at fairly low concentration. Handling of large amounts of sample (up to 200 mg) is a great advantage compared to temperature-programmed pyrolysis methods (up to 1 mg). TGA may be used to define optimum thermal desorption conditions.

In on-line TD-GC-FTIR-FID a thermal desorption cold trap injector is used as an oven for temperature-controlled outgassing of polymeric materials with a maximum temperature of 350°C. The volatile components from the sample are transferred to the cold trap by the carrier gas and preconcentrated. After completion of the outgassing process the cold trap is heated very quickly, causing on-column injection of the trapped components onto the gas chromatograph. In the configuration described by Jansen *et al.* [408], the GC-FTIR interface is provided with a gold-covered capillary lightpipe and a MCT detector. The lightpipe components are detected with a conventional flame ionisation detector. An interpretable IR spectrum is obtained from *ca.* 10^{-7} g of a component depending upon the IR sensitivity. This sets detection limits of outgassing for 0.1 g samples in the $\mu g\ g^{-1}$ range. Quantitative analyses can be made on the basis of extinction coefficients measured on standards.

Off-line sampling with preconcentration of offgases has also been coupled to **on-line TD-GC-FTIR-MS** (ion-trap in parallel configuration) with a thermal desorption unit capable of accommodating Tenax trapping tubes [345]. Thermal desorption GC-FTIR-MS can be operated both in parallel and tandem FTIR-MS configuration, where parallel FTIR-MS operation (Fig. 2.49) is preferred. Compared to FTIR alone, the parallel configuration enhances and facilitates elucidation of the evolved species and furthermore lowers the detection limits from ppm to (sub)ppb level [345].

Advantages of the aforementioned thermal analysis techniques for analysis of polymer formulations include a wide accessible temperature range, ability to vary the atmosphere during thermal treatment, monitoring of evolution behaviour in real time, concentration of volatiles by various multicomponent organic sorbent traps, excellent GC performance and powerful analysis of components by GC-MS techniques. Apart from identification of components in formulations, quantitation of known components has been performed. A major disadvantage in routine analysis by these techniques is the throughput of samples; the longest retained compound determines the analysis time. Other significant disadvantages of *multi*hyphenated systems are complexity, cost, and need for a trained operator.

The subtilities of various forms of TD and DHS are food for connoisseurs: TD-CT-GC-MS exposes analytes thermally, whereas DHS-CT-GC-MS in origin is essentially a room-temperature method.

Applications

General applications of TD are many: industrial emissions, air monitoring, occupational hygiene,

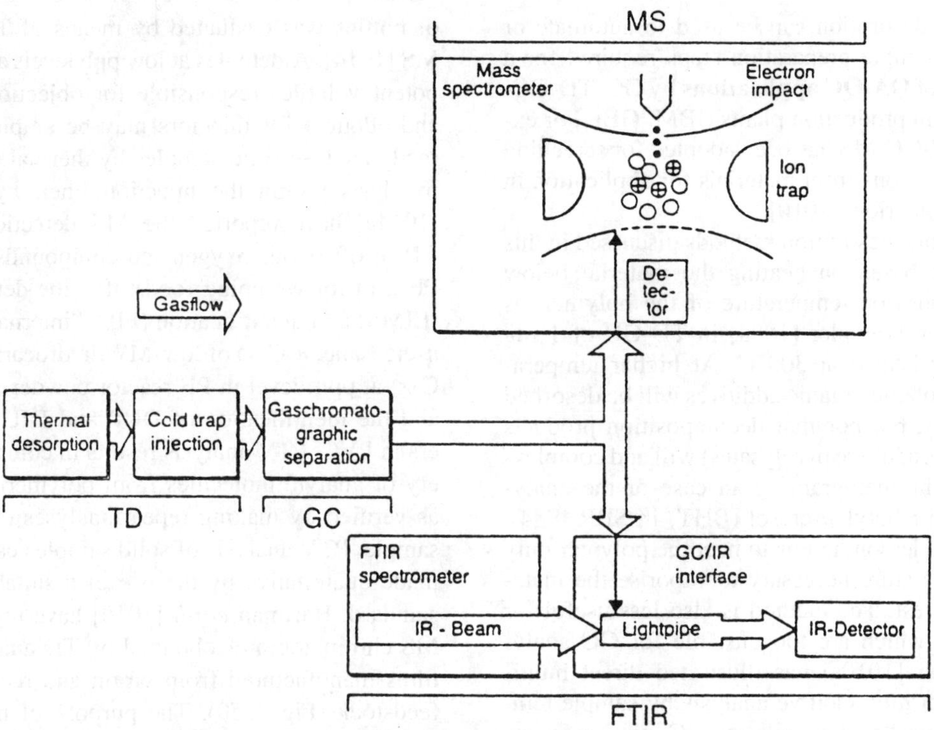

Fig. 2.49. Flow-chart of a parallel TD-GC-FTIR-MS system. After Jansen *et al.* [345]. Reproduced from *Thermochimica Acta* **134**, J.A.J. Jansen *et al.*, 307–312 (1988), with permission from Elsevier.

materials outgassing, VOC analysis, *etc.* During processing and in product applications of polymeric materials, especially at elevated temperatures, evaporation of low-MW products, possibly accompanied by degradation products, may occur and cause deterioration of material properties. In addition, outgassing phenomena may be related to contact and environmental contamination in product applications, toxicological and aesthetic aspects, suitability of plastic products for finishing processes (*e.g.*, glueing, welding, lacquering and plating), admissible temperatures for processing and use, mould contamination and reprocessability. Therefore, evaluation of the type and amounts of volatile components is of considerable practical interest, and is also an analytical challenge. To determine the contents of volatile compounds, various evolved gas analysis methods are in use, such as TG-MS and TG-IR. Thermal desorption is used for **materials emissions testing**, materials QA/QC, purge-and-trap analysis, workplace and ambient air monitoring, breath analysis, *etc.* It offers great versatility with regard to analyte concentration (from ppb to %). Applications range from identification of odoriphores in printed PE film to identification and quantification (at ppm level) of low-MW hydrocarbons in PE, solvents in paint flakes, rubber floorings and characterisation of phthalate esters in PVC. Thermal desorption can be used for almost every organic compound analysed by GC. Coupling of thermal desorption and identification techniques constitutes powerful means for detailed characterisation of outgassing processes. For example, stringent demands are made on outgassing of silicone rubber products in some applications (*e.g.* sealing rings, computer keyboards *etc.*). Jansen *et al.* [408] examined a great variety of polymer samples, including a peroxide vulcanised dimethylsiloxane rubber mixture by means of TD-GC-FTIR-FID, observing CO_2 and 2,4-dichlorobenzene as the degradation products of the peroxide and low-MW dimethylsiloxanes. In a batch of polyacetol components causing smell and headache complaints trioxane, the cyclic trimer of formaldehyde, was detected. Jansen *et al.* [345] also evaluated temperature-controlled outgassing processes of plastics and rubbers using both on-line and off-line TD-GC-FTIR-MS.

Thermal desorption can be used to automate or reduce the sample preparation steps required for a wide range of **QA/QC applications** by GC. TD-GC-MS is used in production plants (IBM, GE). For example, TDS-GC-MS has been adopted for screening of VOC emissions from materials for application in automotive interiors [1010].

The thermal desorption methods discussed in this Chapter are based on heating the material below the decomposition temperature of the polymer, as illustrated by Wampler [1004] in TD-GC of 1 mg PVC/DEHP heated at 300°C. At higher temperatures, less volatile organic additives will be desorbed more readily, but polymer decomposition products (and perhaps additive pyrolysates) will add complexity to the chromatogram, as in case of the analysis of 2,6-di-*t*-butyl-*p*-cresol (BHT) in SBR [784]. The obvious lesson here is to heat the polymer only to the temperature necessary to vaporise the materials of interest. The method is also less useful for compounds which are too unstable for GC analysis. Wampler [1010a] has illustrated direct multi-step polymer *and* additive analysis by sample temperature control on-line with the GC injector for removal of semivolatiles followed by polymer identification by PyGC-MS.

Purge-and-trap screening followed by off-line TD-GC-MS analysis of the collected adsorbent tubes was used to determine emissions from flame retarded polymers (in TV sets) [1011]. Volatile transformation products from additives in γ-irradiated HDPE packaging were analysed by means of TD-GC-MS [1012]. Off-gassed C_5–C_{30} polymer fractions can readily be GC analysed using a Carbotrap™ 370 thermal desorption tube and a TD unit.

Thermal desorption GC-MS methods (usually limited to desorption below 300°C) are ideal for identifying **residual volatiles** in polymers, which can often yield clues as to manufacturing processes [1]. By using principal component analysis (PCA) or TD-GC-MS data it was possible to characterise PP formulations in terms of manufacturer, site of manufacture and additive package used [1013]. Woolfenden [1005] has given some examples of automatic on-line thermal extraction. Using 40 mg of an epoxy resin, the toxic volatile monomer, epichlorohydrin, was extracted efficiently (*ca.* 95%) from a PTFE tube liner in 10 min at 175°C, well below the polymer degradation temperature. For direct TD of paint small aliquots ($\sim$5 μL) are sufficient. Sensory performance of α-tocopherol (Vitamin E) as an antioxidant for LDPE extrusion coating polymers and for HDPE extrusion blow-molding of bottles was evaluated by means of TD-CT-GC-MS [1014]. Aldehydes at low ppb levels are the most potent volatiles responsible for objectionable taste and odour. PTV injectors may be employed to directly analyse solid samples by thermal desorption-pyrolysis within the injection liner. Lynch *et al.* [1014a] have reported the MS detection (EI and CI) of off-odour oxygenated compounds in printed PE film for wrapping foodstuffs, the determination (EIMS) and quantification (FID, "internal" standard *n*-eicosane, *n*-C_{21}) of low-MW hydrocarbons (C_8 to C_{24}) at ppm level in PE reactor powder and pellets, and the identification of DUP and DIUP plasticisers in PVC. PTV analysis results in complete recovery of analyte molecules from polymeric matrices, as verified by making repeat analyses on the same sample. PTV analysis of solid samples can easily be made quantitative by the use of a suitable internal standard. Hartman *et al.* [1015] have reported GC-MS chromatograms obtained by TD analysis of PP films manufactured from virgin and recycled resin feedstock (Fig. 2.50). The purpose of the analysis was to investigate the basis of off-odour complaints associated with recycled product.

Identification of organic **additives** in polymers by using TD-GC-MS is considerably more difficult than that of residual volatiles. The reason for this is simply that most organic additives are not very volatile in the GC sense. The use of short GC columns, low stationary phase loadings (for packed columns), and high-temperature packing materials extends the applicability of GC-MS to some higher-MW materials. Hindered phenol AOs in polymers, such as Irganox 1010 in PBT, were analysed quantitatively by TMAH-TD-GC-MS [1016]. Tsuge *et al.* [1017] have used in-line RTD-GC for direct determination of trace amounts of polymeric HALS, Adekastab LA-68LD (MW $\approx$ 1.900 Da) in PP without any preseparation. The formulated PP/(Adekastab LA-68LD, Irganox 1010, Irgafos 168) sample (0.1 mg) was mixed with 2 μL of TMAH 25 wt.% methanol solution and introduced in a furnace pyrolyser at 300°C; thermodegradation products were then analysed quantitatively by GC. This procedure is an alternative to time-consuming solvent extraction followed by LC. On-line derivatisation can also be carried out with a PTV injector.

Knobel [1018] has compared TD-GC-MS with extraction followed by LC-PDA or GC-MS for the detection of an UV photo-initiator in resins used

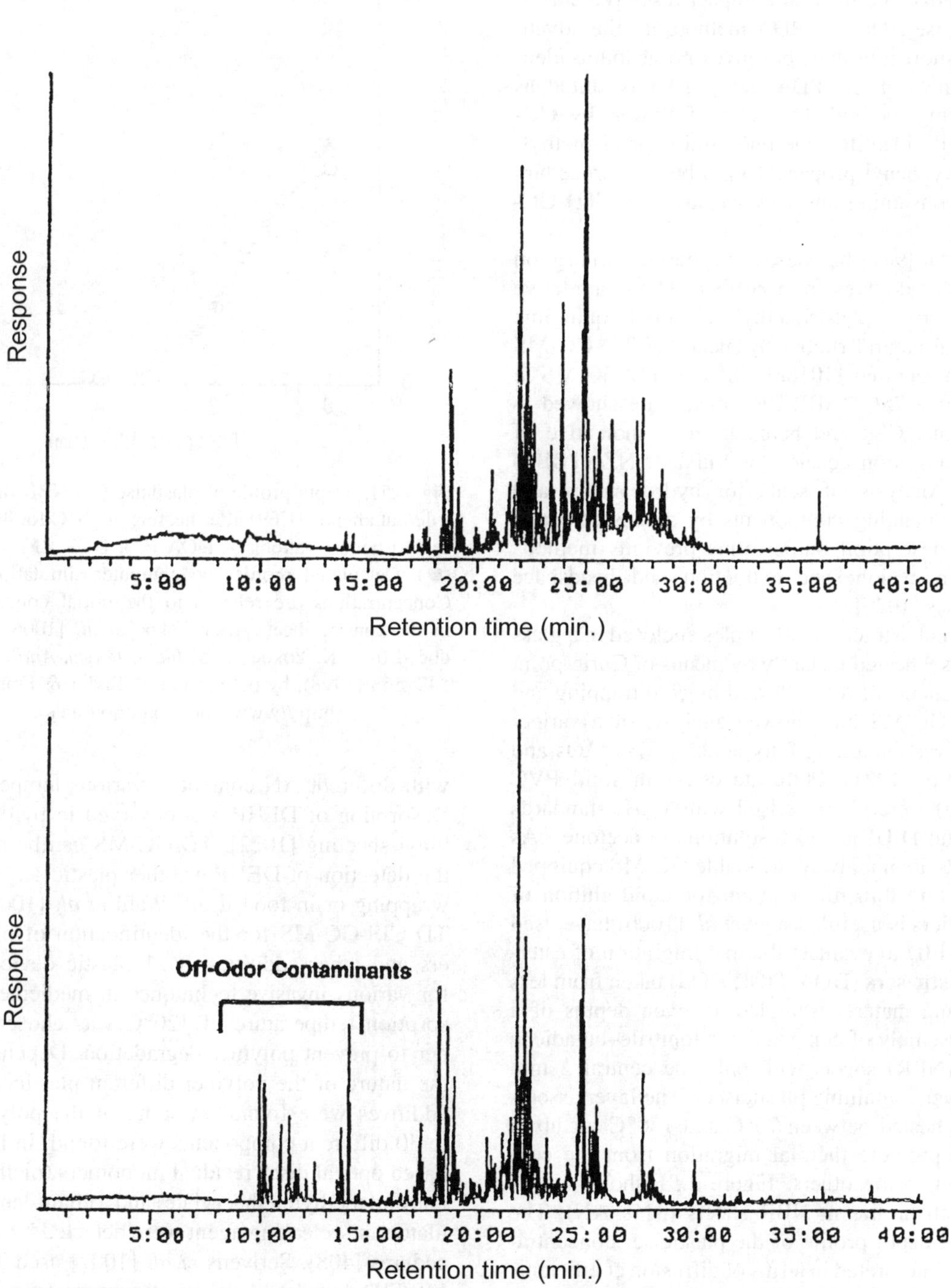

Fig. 2.50. Chromatogram obtained from TD-GC-MS analysis of PP films manufactured from virgin (upper trace) and recycled (lower trace) feedstocks. Off-odour components are evident in the recycled film. After Hartman *et al.* [1015]. Reprinted from T.G. Hartman *et al.*, in *Flavor Measurement* (C.-T. Ho and C.H. Manley, eds.), Marcel Dekker Inc., New York, NY (1993), by courtesy of Marcel Dekker Inc.

as protective coatings on compact discs (patent violation case). The LC-PDA method has the advantage of short runtimes, but gives no absolute identity because of the PDA detector; it is suited as a screening method. Extraction followed by GC-MS readily identifies the photo-initiator (2-methyl-2-hydroxyphenyl-propane-1-one) but the procedure is time-consuming and less sensitive than TD-GC-MS.

Gorman [971] has described thermal desorption of volatile additives from **rubber**. The quantitative analysis of 2,2,4-trimethyl-1,2-dihydroquinoline (TMQ) in natural rubber by means of TD-GC-MS has been reported [1018a]. Off-line TD-GC-MS at 180°C of a 75/25 SBR/BR vulcanisate showed *t*-butylamine, CS_2 and benzothiazole, indicative of the vulcanisation accelerator Vulkacit NZ (TBBS) [1019]. Analysis of seals for hydrocarbons and silicon-containing components by means of direct thermal desorption outperforms previous methods based on cyclohexane extraction and headspace techniques [1020].

Thermal extraction of samples enclosed in a glass tube vessel heated instantly by means of Curie-point ferromagnetic alloys, followed by cold trapping and off-line GC-MS has allowed analysis of a variety of additives, including fatty acids, UVAs, AOs and plasticisers [1021]. Phthalate esters in solid PVC (1–2 mg) were characterised with 1 μL standards (DUP and DIUP as 30% solutions in acetone). As standards are not always available GC-MS equipped with a short thin film column for rapid elution of high boilers is useful. Tsuge *et al.* [1006] have used TD-GC-FID to evaluate thermal migration of ester-type **plasticisers** (DOA, DOP, DOS) taken from less than 1 mg material sampled at given depths of a 3-1-3 assembly of contacted acrylonitrile–butadiene rubber (NBR) sheets with only the central 2 mm thick layer containing plasticisers. The layer assembly was heated between 70°C and 150°C for up to 72 h to promote thermal migration from the central sheet to the others. Figure 2.51 shows a typical depth profile in NBR sheets (31% AN). The observed depth profile of the plasticiser concentration was interpreted in terms of diffusion coefficients of the plasticisers in NBR on the basis of Fick's laws. Total time for all analytical steps, including baking-out for the next run, took about 1 h. This rapid and highly sensitive method thus enabled not only depth profiling of the plasticisers in rubber samples without performing any preliminary sample pretreatment, but also estimation of the diffusion coefficients for the plasticisers in various rubber samples

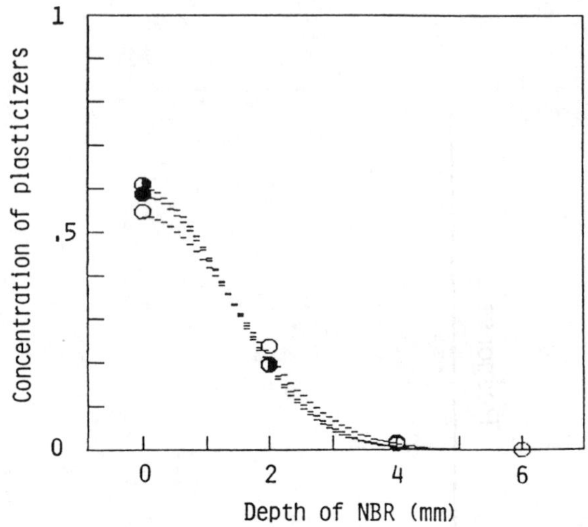

Fig. 2.51. Depth profile of plasticisers in NBR (acrylonitrile/butadiene, 31/69) after heating at 70°C for 48 h. Observed concentrations of DOA (○), DOP (◑) and DOS (●). Calculated profiles by computer simulation (- - -). Concentrations are relative to the initial concentration in the central sheet. After Yokoe *et al.* [1006]. Reproduced from K. Yokoe *et al.*, *Int. J. Polym. Anal. Char.* **4**, 547–563 (1998), by permission of Taylor & Francis Ltd. (http://www.tandf.co.uk/journals).

with different AN content at various temperatures. Desorption of DEHP was observed in pyrolysis of vinyl sheeting [1022]. TD-GC-MS can be used for the detection of DEHP or other plasticisers in food wrapping or in food itself. Wahl *et al.* [1009] used TD-CIS-GC-MS for the identification of plasticisers and other additives in 21 plastic devices used for various invasive techniques in medicine. A desorption temperature of 120°C was chosen in order to prevent polymer degradation. Depending on the nature of the polymer different plasticisers and additives were found. In some of the polymer up to 30 different components were found. In PMMA-based optical discs residual monomers (methylacrylate and methylmethacrylate) and a chain-length regulator and releasing agent were detected by TD procedures [408]. Scrivens *et al.* [1007] used TD-GC-MS/FID for detection of **oligomers** (general formula $Cl\phi SO_2(\phi S\phi SO_2)_n\phi Cl$, $\phi = para$-substituted aromatic ring C_6H_4) and demulsifiers (octyl phenyl formaldehyde resin, MW 900 Da; *t*-butyl phenol formaldehyde resin, MW 1000 Da), materials which usually pose a difficult analytical problem.

Direct thermal extraction is a useful **troubleshooting** tool in polymer analysis. Kenion *et al.* [1023]

have used TD-GC-MS to determine the source and chemical identity of brown **discolorants** in a PE laminate made with a polyurethane adhesive. Polyethylene that had undergone oxidation presented saturated carboxylic acids as major components in a TD-GC-MS analysis, as opposed to PE that had not been oxidised, which did not off-gas fatty acids. TD-GC-MS is also a rapid and direct method in identification of 'free' biomarkers in a broad range of organic materials [1024]. Jansen *et al.* [345] have described detection of PCBs in 2,4-dichlorobenzoylperoxide cured silicone rubbers after outgassing products of a rubber silicone part, obtained after desorption for 10 minutes at 200°C in the thermal desorption cold-trap and subsequent analysis by means of TD-GC-MS. Using a mass range of 290–294 Da MS can be used as a selective detector for these substances. Direct thermal desorption (at 125°C) was applied to identify the **off-odour** of printed PE when heat-sealed. Oxygenated compounds not present in the unprinted film were thought to be responsible for the smell. Low-MW hydrocarbons (C_8–C_{24}) in PE were quantified at ppm levels after desorption at 120°C. Printing and coating solvents in plastic bag headspace have been identified by GC-MS as causes of odour or taint [936]; 2-ethyl-hexadecanoic acid was identified in stained synthetic leather using TD-GC-MS. Decomposition of polyesterurethanes was evaluated by means of TG-Tenax off-line sampling followed by TD-GC-FTIR-MS and revealed CO_2, H_2O, THF, cyclopentanone, dicarbonic acid and aliphatic diols and esters [345]. When still sorbed in the polymer after ESC failure has occurred, a **stress-cracking** agent may be identified by outgassing experiments using TD-GC-FTIR-MS. Ethyl lactate was thus found to have caused stress-cracking of moulded ABS after a lacquering step [1025].

Combined direct TD-CIS can also be very helpful in the chemical **risk assessment** in plastic processing due to polymer compounds and/or additives degradation. The temperature of the TD module needs only have to be set to the processing temperature under investigation or to the degradation temperature of the particular polymer.

2.3.3. Thermal Desorption–Mass Spectrometric Techniques

Principles and Characteristics

In thermal desorption mass spectrometry (TD-MS) polymer samples are introduced into the ion source of a mass spectrometer by means of a direct inlet for solid samples. In the "direct evaporation mode" the procedure consists in placing the solid in a sample tube fitted onto the inlet of a mass spectrometer (DI-MS), heating the cup for a given time (typically 15–20 min) and temperature (150–200°C) in vacuum and recording the mass spectra of the volatiles released in an ionisation mode of choice. Alternatively, thermal degradation may be carried out in a linear temperature program up to about 800°C or by using the direct probe in a temperature-programmable fashion as a fractional distillation device (DT-MS). During these (conceptually very similar) processes components desorb in relation to their volatility and the volatile products are ionised and detected immediately by repeated mass scans. Low-resolution "survey" mass scans may be obtained by various ionisation methods, such as EI, CI, FI, FD or FAB-MS. Some advantages are the absence of sample manipulation and fast *in vacuo* ionisation reducing secondary reactions. Disadvantages are the need for small sample size, which renders it difficult to detect the typical low concentrations of volatile additives in conventional (rubber) compounds, and fouling of the ion source by deposition of non-volatile material. A further problem relates to inefficient extraction and isolation of the organic additives from the inorganic additives in carbon-black matrices. As the thermal separation is often quite inadequate (*e.g.* in selective separation of process oils and volatile fragments) complex spectra are generated. Consequently, the added value of TD-MS identification of additives, namely minimal sample preparation (no extraction or dissolution step) and high selectivity of the mass spectrometer, is frequently off-set by the tedious interpretation of MS spectra requiring expert knowledge. Therefore, it is generally doubtful that single-stage TD-MS techniques will be able to make significant contributions to quantitative additive analysis for unknown (real-life) samples with low additive concentrations.

Cotter [1026] has reviewed thermal desorption/volatilisation for volatility enhancement. Cotter's authoritative coverage includes desorption mechanisms, inert probe desorption and the various interrelations of evaporation of intact neutrals with thermal decomposition.

Applications

Thermal desorption mass spectrometry allows rapid qualitative scanning (2 min) of additive packages [2,

790,1027,1028]. As most common additives are removed quite readily from **rubbers**, TD-MS is therefore often the first and preferred approach for qualitative analysis of untreated rubber [745].

TD-MS allows differentiation of very similar **flame retardants**. Verdurmen *et al.* [790] have discriminated the tetrabromobisphenol-A-carbonate oligomers BC 52 (MW 2494; phenol terminated) and BC 58 (unspecified MW; tribromophenol terminated) in PBT by DIP-MS observing m/z 544 (tetrabromobisphenol A molecular ion), m/z 322 (tribromophenol radical cation) and m/z 369, 325, 203, 149, 105 (pyrolysis products of PBT) but not the characteristic fragment ions m/z 605 and 664 of BC 52. However, without prior knowledge the m/z 322 component would probably have gone unnoticed. Similarly, in dissolution of nylon/(Irgafos 168, 2,4-di-*t*-butylphenol) the latter component could hardly be detected by TD-MS. While it is possible to identify stabilisers in polyamides, TD-MS is not the ideal technique for this purpose. Although TD-MS has been reported to yield quantitative information after internal calibration and has been used to analyse additive packages in electronic goods both qualitatively and quantitatively [1029], quantitation is difficult and costly [1030].

Pleshkova *et al.* [1031] have used TD-MS to determine the composition of a number of commercial bisphenol A-based polycarbonates. PhOH, *p*-*t*-butylphenol, and *p*-isooctylphenol were determined as the main chain-transfer agents for these polycarbonates. Up to 10 additives with concentrations exceeding 0.05% could be determined in a sample. Mould lubricants of various chemical nature were the main additives found.

BIBLIOGRAPHY

Thermal Analysis

B. Wunderlich, *Thermal Analysis of Polymeric Materials*, Springer-Verlag, Berlin (2005).

M. Sorai (JSCTA, ed.), *Comprehensive Handbook of Calorimetry and Thermal Analysis*, J. Wiley & Sons, New York, NY (2004).

G.W.H. Höhne, W.F. Hemminger and H.-J. Flammersheim, *Differential Scanning Calorimetry; An Introduction for Practitioners*, Springer-Verlag, Berlin (2003).

W.M. Groenewoud, *Characterisation of Polymers by Thermal Analysis*, Elsevier Science, Amsterdam (2001).

A.T. Riga and L. Judovits (eds.), *Materials Characterization by Dynamic and Modulated Thermal Analytical Techniques, ASTM Spec. Tech. Publ. STP* **1402** ASTM, West Conshohocken, PA (2001).

D. Mulligan, S. Gnaniah and G. Sims, *Thermal Analysis Techniques for Composites and Adhesives*, National Physics Laboratory, Teddington (2000).

T. Hatakeyama and F.X. Quinn, *Thermal Analysis: Fundamentals and Applications to Polymer Science*, J. Wiley & Sons, New York, NY (1999).

T. Hatakeyama and Z. Liu (eds.), *Handbook of Thermal Analysis*, J. Wiley & Sons, New York, NY (1999).

S. Knappe, H. Möhler, J. Opfermann and H. Walter, *Focus on Thermal Analysis for Paints*, Netzsch-Gerätebau/Fachhochschule Würzburg-Schweinfurt-Aschaffenburg, Selb (1999).

H. Möhler and S. Knappe, *Focus on Thermal Analysis for Polymers*, Netzsch-Gerätebau/ Fachhochschule Würzburg-Schweinfurt-Aschaffenburg, Selb (1998).

M.E. Brown (ed.), *Handbook of Thermal Analysis and Calorimetry*, Vol. 1, Elsevier, Amsterdam (1998), 3 vols.

P.K. Gallagher (ed.), *Handbook of Thermal Analysis and Calorimetry*, Elsevier Science, Amsterdam (1998), 3 vols.

Th. Nitzschke and J. Vogel (eds.), *Thermische Analyse an Kunststoffen – Methoden und Anwendungen*, Ecomed Verlagsgesellschaft, Landsberg (1998).

A.T. Riga and G.H. Patterson (eds.), *Oxidative Behavior of Materials by Thermal Analytical Techniques, ASTM Spec. Tech. Publ. STP* **1326**, ASTM, West Conshohocken, PA (1997).

E.A. Turi (ed.), *Thermal Characterization of Polymeric Materials*, Academic Press, San Diego, CA (1997).

P.J. Haines, *Thermal Methods of Analysis: Principles, Applications and Problems*, Blackie A&P, London (1995).

M.P. Sepe, *Thermal Analysis of Polymers, Rapra Review Report No. 95*, Rapra Technology Ltd., Shawbury (1995).

F. Paulik, *Special Trends in Thermal Analysis*, J. Wiley & Sons, Chichester (1995).

V.B.F. Mathot (ed.), *Calorimetry and Thermal Analysis of Polymers*, C. Hanser Verlag, Munich (1994).

E. Kaisersberger, S. Knappe and H. Möhler, *TA for Polymer Engineering*, Netzsch GmbH, Selb-Würzburg (1993).

E.L. Charsley and S.B. Warrington (eds.), *Thermal Analysis: Techniques and Applications*, The Royal Society of Chemistry, Cambridge (1992).

A.T. Riga and C.M. Neag (eds.), *Materials Characterization by Thermomechanical Analysis, ASTM Spec. Tech. Publ. STP* **1136**, ASTM, West Conshohocken, PA (1991).

B. Wunderlich, *Thermal Analysis*, Academic Press, New York, NY (1990).

W.F. Hemminger and H.K. Cammenga, *Methods of Thermal Analysis*, Springer-Verlag, Heidelberg (1989).

C.M. Earnest (ed.), *Compositional Analysis of Thermogravimetry, ASTM Spec. Tech. Publ. STP* **997**, American Society for Testing and Materials, Philadelphia, PA (1988).

Pyrolysis

S.C. Moldoveanu, *Analytical Pyrolysis of Natural Organic Polymers*, Elsevier, Amsterdam (1998).

T.P. Wampler (ed.), *Applied Pyrolysis Handbook*, M. Dekker, New York, NY (1995).

S. Tsuge and H. Ohtani, *Pyrolysis–Gas Chromatography of High Polymers – Fundamentals and Data Compilation*, Techno System Co., Tokyo (1989); Chinese translation: X. Jin and Y.F. Luo, Chinese Science and Technology Publ. Co., Beijing (1992).

G. Allen and J.C. Bevington (eds.), *Pyrolysis–Gas Chromatography,* Pergamon Press, Oxford (1989).

S.A. Liebman and E.J. Levy, *Pyrolysis and GC in Polymer Analysis*, M. Dekker, New York, NY (1985).

K.J. Voorhees, *Analytical Pyrolysis – Techniques and Applications*, Butterworths, London (1984).

W.J. Irwin, *Analytical Pyrolysis: A Comprehensive Guide*, M. Dekker, New York, NY (1982).

R.W. May, E.F. Pearson and D. Scothern, *Pyrolysis Gas Chromatography*, Chemical Society, London (1977).

Thermal Desorption

B. Kolb and L.S. Ettre, *Static Headspace-Gas Chromatography, Theory and Practice*, Wiley-VCH, New York, NY (1997).

A. Hagman, *Dynamic Headspace. A Versatile Method for Analysing Volatile Compounds in Polymers, Ph.D. Thesis*, University of Stockholm (1988).

REFERENCES

[1] W.V. Ligon and M.C. George, *J. Polym. Sci.: Polym. Chem. Ed.* **16**, 2703–9 (1978).

[2] R.P. Lattimer and R.E. Harris, *Mass Spectrom. Revs.* **4**, 369–90 (1985).

[3] N.G. Fisher and J.G. Dunn, *J. Therm. Anal. Calorim.* **56**, 43–9 (1999).

[4] J.W. Sinclair, L. Schall and N.T. Crabb, *J. Chromatogr. Sci.* **18**, 30–4 (1980).

[5] DIN 51005: *Thermische Analyse (TA). Begriffe.* Beuth-Verlag, Berlin (1983).

[6] B. Wunderlich, *Thermal Characterization of Polymeric Materials* (E.A. Turi, ed.), Academic Press, New York, NY (1997), pp. 206–482.

[7] J. Rouquerol, *Thermochim. Acta* **144**, 209–24 (1989).

[8] B. Dickens, *J. Polym. Sci. – Polym. Chem.* **20**, 1065 (1982).

[9] J. Rouquerol, *J. Therm. Anal.* **5**, 203 (1973).

[10] M. Reading, E. Elliot and V.L. Hill, *J. Therm. Anal.* **40**, 949 (1993).

[11] P.S. Gill, S.R. Sauerbrunn and M. Reading, *J. Therm. Anal.* **40**, 931 (1993).

[12] R.L. Blaine and B.K. Hahn, *J. Therm. Anal. Calorim.* **54**, 695–704 (1998).

[13] T. Ozawa, *Thermochim. Acta* **356**, 173 (2000).

[14] T. Ozawa, *J. Therm. Anal. Calorim.* **64**, 109–26 (2001).

[15] W. Hemminger and S.M. Sarge, in *Handbook of Thermal Analysis and Calorimetry* (M.E. Brown, ed.), Elsevier, Amsterdam, Vol. 1, pp. 1–74 (1998).

[16] E.L. Charsley, in *Thermal Analysis – Techniques and Applications* (E.L. Charsley and S.B. Warrington, eds.), The Royal Society of Chemistry, Cambridge (1992), pp. 284–91.

[17] P.K. Gallagher, in *Thermal Characterization of Polymeric Materials* (E.A. Turi, ed.), Academic Press, San Diego, CA (1997), pp. 1–203.

[18] R.W. Mackenzie, *Thermochim. Acta* **73** (3), 1–124 (1984).

[19] E.A. Turi (ed.), *Thermal Characterization of Polymeric Materials*, Academic Press, San Diego, CA (1997).

[20] F. Paulik, *Special Trends in Thermal Analysis*, J. Wiley & Sons, Chichester (1995).

[21] T. Hatakeyama and F.X. Quinn, *Thermal Analysis. Fundamentals and Applications to Polymer Science*, J. Wiley & Sons, New York, NY (1999).

[22] J. Scheirs, *Compositional and Failure Analysis of Polymers. A Practical Approach*, J. Wiley & Sons, Chichester (2000).

[23] V.B.F. Mathot (ed.), *Calorimetry and Thermal Analysis of Polymers*, Hanser Publishers, Munich (1994).

[24] M.P. Sepe, *Thermal Analysis of Polymers, Rapra Review Rept. No. 95*, Rapra Technology Ltd., Shawbury (1997).

[25] A.K. Sircar, in *Thermal Characterization of Polymeric Materials* (E.A. Turi, ed.), Academic Press, New York, NY (1997), pp. 888–1378.

[26] R.P. Chartoff, in *Thermal Characterization of Polymeric Materials* (E.A. Turi, ed.), Academic Press, San Diego, CA (1997), pp. 484–743.

[27] H.E. Bair, in *Thermal Characterization of Polymeric Materials* (E.A. Turi, ed.), Academic Press, San Diego, CA (1997), pp. 2263–420.

[28] T.R. Manley, *Progr. Rubber Plast. Technol.* **5** (4), 253–88 (1989).

[29] F. Paulik, J. Paulik and L. Erdey, *Z. Anal. Chem.* **160**, 241 (1958).

[30] P.H. Willcocks, in *Thermal Analysis – Techniques and Applications* (E.L. Charsley and S.B. Warrington, eds.), The Royal Society of Chemistry, Cambridge (1992), pp. 265–77.

[31] W. Hemminger, in *Proceedings GSF Symposium on Analytical Chemical Quality Assurance* (A. Kettrup and E. Flammenkamp, eds.), Ecomed Verlagsgesellschaft, Landsberg (1995), pp. 155–63.

[32] P.H. Willcocks, N.R. Bosley, A. Broadhurst and I.D. Luscombe, *J. Therm. Anal. Calorim.* **56**, 1413–8 (1999).

[33] R. Sabbah (ed.), *Reference Materials for Calorimetry and Thermal Analysis*, Special issue in *Thermochim. Acta* **331/2** (1999).

[34] A.T. Riga, R.M. Collins and G.H. Patterson, *ASTM Spec. Techn. Publ., STP* **1326** *(Oxidative Behavior of Materials by Thermal Analytical Techniques)*, A.T. Riga and G.H. Patterson (eds.), ASTM, West Conshohocken, PA (1997), pp. 91–101.

[35] S. Affolter and M. Schmid, in *Developments in Polymer Analysis and Characterisation* (Rapra Technology Ltd., ed.), Rapra Technology Ltd., Shawbury (1999), Paper 5.

[36] Special Issue *J. Thermal. Anal.* **40**, 1379–463 (1993).

[37] M. Reading, *J. Therm. Anal. Calorim.* **64**, 7–14 (2001).

[37a] J.O. Hill, in *Encyclopedia of Analytical Science* (A. Townshend, ed.), Elsevier, Oxford (2005), Vol. 9, pp. 17–50.

[38] G.M.B. Parkes, P.A. Barnes, E.L. Charsley and G. Bond, *J. Therm. Anal. Calorim.* **56**, 723–31 (1999).

[39] Anon., *Thermal Analysis of Composites and Adhesives*, National Physical Laboratory, Teddington (2000).

[40] W. Randall and J. Riedel, *Rubber World* **207** (4), 23–31 (1993).

[41] T.R. Crompton, *Practical Polymer Analysis*, Plenum Press, New York, NY (1993).

[42] J. Chiu and E.F. Palermo, *Anal. Chim. Acta* **81** (1), 1–19 (1976).

[43] E.M. Pearce, Y.P. Khanna and D. Raucher, in *Thermal Characterization of Polymeric Materials* (E.A. Turi, ed.), Academic Press, New York, NY (1980), Chp. 8.

[44] Plastics Design Laboratory (ed.), *The Effect of Temperature and Other Factors on Polymers*, PDL, Norwich, NY (1991).

[45] D.W. Brazier, *Proceedings Workshop on State-of-the Art of Thermal Analysis*, NBS, Gaithersburg, MD, *NBS Special Publ.* **580**, 55–88 (1980).

[46] G.D. Gaddy, W.J. Rossiter and R.K. Eby, *ASTM Spec. Techn. Publ., STP* **1088** (*Roofing Research and Standards Development*), ASTM, Philadelphia, PA (1990), pp. 37–52.

[47] ASTM D 297-93, *Test Methods for Rubber Products – Chemical Analysis, Annual Book of ASTM Standards*, ASTM, West Conshohocken, PA (1998), Vol. 09.01.

[48] J.P. Redfern, *Polym. Degr. Stabil.* **64**, 561–72 (1999).

[49] L. van Wabeeke, in *Flame Retardants '96*, Interscience Communications Ltd., London (1996), pp. 91–105.

[50] M.P. DiVito, K.J. Fielder, G.H. Curran and M.S. Feder, *Am. Lab.* **24** (1), 30–6 (1992).

[51] American Society for Testing of Materials, *Annual Book of ASTM Standards*, ASTM, West Conshohocken, PA (1998), Vol. 14.02.

[52] A. Riga, *ICTAC News* **35** (2), 91–5 (Dec. 2002).

[53] R.G. Ferillo, *ASTM Standard Procedures for Thermal Measurements, Educ. Symp. Rubber Div.*, American Chemical Society, Toronto (1991).

[54] H. Möhler and S. Knappe, *Focus on Thermal Analysis for Polymers*, Netzsch-Gerätebau/Fachhochschule Würzburg-Schweinfurt-Aschaffenburg, Selb (1998).

[55] S. Knappe, H. Möhler, J. Opfermann and H. Walter, *Focus on Thermal Analysis for Paints*, Netzsch-Gerätebau/Fachhochschule Würzburg-Schweinfurt-Aschaffenburg, Selb (1999).

[56] M.J. O'Neill, *Anal. Chem.* **36**, 1238–44 (1964).

[57] J. Koch and Y. Teramoto, *Abstracts 12th Intl. Congress on Thermal Analysis and Calorimetry (ICTAC)*, Copenhagen (2000), p. 93.

[58] P.J. van Ekeren, C.M. Holl and A.J. Witteveen, *J. Therm. Anal.* **49**, 1105–14 (1997).

[59] NIST-ICTA *Standard Materials* GM757-760.

[60] M.J. Richardson, *J. Thermal Anal. Calorim.* **56**, 1401–4 (1999).

[61] M. Reading, *Trends Polym. Sci.* **1** (8), 248–53 (1993).

[62] R. Scherrenberg, V.B.F. Mathot and P. Steeman, *J. Thermal Anal.* **54**, 477–99 (1998).

[63] T.F.J. Pijpers, V.B.F. Mathot and R.B. Cassel, *Proceedings 28th Annu. Conf. North Am. Therm. Anal. Soc. (NATAS)*, Orlando, FL (2000), p. 860.

[64] T.F.J. Pijpers, V.B.F. Mathot, B. Goderis and E. van der Vegte, *Proceedings 28th Annu. Conf. North Am. Therm. Anal. Soc. (NATAS)*, Orlando, FL (2000), pp. 32–7.

[65] H. Yoshida, R. Kinoshita and Y. Teramoto, *Thermochim. Acta* **264**, 173 (1994).

[66] W. Bras, G.E. Derbyshire, A. Devine, S.M. Clark, J. Cooke, B.E. Komanschek and A.J. Ryan, *J. Appl. Crystallogr.* **28**, 26 (1995).

[67] C.D. Price, *The Rapra Collection of DSC Thermograms of Semi-Crystalline Thermoplastic Materials*, Rapra Technology Ltd., Shawbury (1997).

[68] G.W.H. Höhne, W.F. Hemminger and H.-J. Flammersheim, *Differential Scanning Calorimetry: An Introduction to Practitioners*, Springer-Verlag, Berlin (2003).

[69] P.J. Haines, M. Reading and F.W. Wilburn, in *Handbook of Thermal Analysis and Calorimetry* (M.E. Brown, ed.), Elsevier, Amsterdam (1998), Vol. 1, Chp. 5.

[70] V.J. Griffin and P.G. Laye, in *Thermal Analysis – Techniques and Applications* (E.L. Charsley and S.B. Warrington, eds.), The Royal Society of Chemistry, Cambridge (1992), pp. 17–30.

[71] J.O. Hill, in *Encyclopedia of Analytical Science* (A. Townshend, ed.), Academic Press, San Diego, CA (1995), pp. 5155–60.

[72] M.J. Richardson, in *Developments in Polymer Characterization* (J.V. Dawkins, ed.), Applied Science Publishers, London (1978), Vol. 1, pp. 205–44.

[73] V.B.F. Mathot, *J. Therm. Anal. Calorim.* **64**, 15–35 (2001).

[74] R. Scherrenberg, V.B.F. Mathot and A. van Hemelrijck, *Thermochim. Acta* **330** (1–2), 3–19 (1999).

[75] K.J. Jones, I. Kinshott, M. Reading, A.A. Lacey, C. Nikolopoulos and H.M. Pollock, *Thermochim. Acta* **304/305**, 187–99 (1997).

[76] C. Schick and G.W.H. Höhne (eds.), Special Issue on *Temperature Modulated Calorimetry, Thermochim. Acta* **304/5** (1997).

[77] L.H. Judovits and J.D. Menezel (eds.), Special Issue on *Temperature Modulated DSC, J. Therm. Anal. Calorim.* **54** (1998).

[78] T. Ozawa and K. Kanari, *J. Therm. Anal.* **49** (2), 979–89 (1997).

[79] G. Widmann and O. Scherrer, *J. Therm. Anal.* **37**, 1957–64 (1991).

[80] P.K. Gallagher, in *Handbook of Thermal Analysis and Calorimetry* (M.E. Brown, ed.), Elsevier, Amsterdam (1998), Vol. 1, Chp. 4.

[81] B. Perrenot and G. Widmann, *Thermochim. Acta* **234**, 31–9 (1994).

[82] W.P. Brennan, *Perkin-Elmer Thermal Analysis Application Study* 22 (1977).

[83] D. Baral, P.P. De and G.B. Nando, *Polym. Degr. Stabil.* **65**, 47–51 (1999).

[84] S. Knappe and C. Urso, *Thermochim. Acta* **227**, 35–42 (1993).

[85] M.Z. Jandali and G. Widmann (Mettler-Toledo Analytical), *Collected Applications Thermal Analysis (Thermoplastics)*, Mettler-Toledo, Schwerzenbach (1997).

[86] W.P. Brennan, *Perkin-Elmer Thermal Analysis Application Study No. 22*, p. 5 (1977).

[87] M.F. de Velasco-Ruiz, I. Quijada-Garrido, R. Benavente and J.M. Barrales-Rienda, *Polymer* **41** (15), 5819–28 (2000).

[88] S. Affolter, A. Ritter and M. Schmid, *Ringversuche an polymeren Werkstoffen*, EMPA, St. Gallen (2000).

[89] C.Y. Lin, *M.Sc. Thesis*, University of Cincinnati, Cincinnati, OH (1978).

[90] H.E. Bair, *ASTM Spec. Techn. Publ., STP* **1249** *(Assignment of the Glass Transition)*, R.J. Seyler (ed.), ASTM, Philadelphia, PA (1994), pp. 50–74.

[91] Mettler-Toledo Analytical, *Collected Applications Thermal Analysis (Pharmaceuticals)*, Mettler-Toledo, Schwerzenbach (n.d.).

[92] R.E. Lee, S. Narayan, L. Pallini and J.M. Zenner, *Proceedings SPE ANTEC '99*, New York, NY (1999), pp. 2308–11.

[93] D.W. Brazier, *NBS Spec. Publ. (U.S.)* **580**, 55–88 (1979).

[94] S.G. Shankwalkar and D.G. Placek, *I&E Res.* **31** (7), 1810–3 (1992).

[95] P.S. Holsopple, *J. Vinyl Technol.* **9** (4), 183–90 (1987).

[96] J.J. Maurer and D.W. Brazier, *Kautsch. Gummi Kunstst.* **36**, 37 (1983).

[97] D.W. Brazier and G.H. Nickel, *Rubber Chem. Technol.* **48**, 26 (1975).

[98] D.W. Brazier and N.V. Schwartz, *Thermochim. Acta* **39**, 7 (1980).

[99] A. Maltha, *Proceedings Organic Peroxides Symposium*, Apeldoorn (1998).

[100] Y. Mubarak, E. Harkin-Jones, P. Martin and M. Ahmad, *Proceedings SPE ANTEC '99*, New York, NY (1999), pp. 3796–3800.
[101] K.W. Van Every and M.J. Elder, *Proceedings SPE ANTEC '96*, Indianapolis, IN (1996), pp. 2400–5.
[102] T. Yoshikawa and K. Kimura, *Materiaru Raifu* **10** (3), 143–8 (1998).
[103] R.H. Hassel, *Am. Lab.* **9**, 35 (1977).
[104] V.J. Fernandes, A.S. Araujo and N.S. Fernandes, *Proceedings 7th European Symposium on Thermal Analysis and Calorimetry (ESTAC-7)*, Balatonfüred (1998), p. 262.
[105] P.A. Larcey, J.P. Redfern and G.M. Bell, *Fire & Mater.* **19** (6), 283–5 (1995).
[106] A. Prasad and M. Shanker, *Proceedings SPE ANTEC '98*, Atlanta, GA (1998), pp. 1860–4; *Cell Polym.* **18** (1), 35–51 (1999).
[107] D. Dixon, E.M.A. Harkin-Jones and P.J. Martin, *Proceedings SPE ANTEC '99*, New York, NY (1999), pp. 2035–9.
[108] H. Schnecko and G. Angerer, *Kautsch. Gummi Kunstst.* **41** (2), 149–53 (1988).
[109] Y. Ohtake and M. Furukawa, *Intl. Polym. Sci. Technol.* **22** (3), T74–T83 (1995).
[110] H. Schindbauer, G. Henkel, J. Weiss and W. Eichberger, *Angew. Makromol. Chem.* **49**, 115 (1976).
[111] R.B. Cassel, *Paper presented at 28th Pittsburgh Conference, Cleveland, OH* (1977); *Perkin Elmer Therm. Anal. Appl. Study* 19, 9 pp. (1977).
[112] J.J. He and B. Gilpatrick, *Proceedings SPE ANTEC '99*, New York, NY (1999), pp. 2314–8.
[113] D. Munteanu, in *Developments in Polymer Stabilization* (G. Scott, ed.), Elsevier Applied Science, London (1987), Vol. 8, pp. 179–208.
[114] O.N. Nikulicheva, V.P. Fadeeva and V.A. Logvinenko, *J. Therm. Anal.* **44**, 329–37 (1995).
[115] R.H. Still, in *Developments in Polymer Degradation* (N. Grassie. ed.), Applied Science Publishers, London (1977), Vol. 1, p. 1.
[116] N.C. Billingham, D.C. Bott and A.S. Manke, in *Developments in Polymer Degradation* (N. Grassie, ed.), Applied Science Publishers, London (1981), Vol. 3, p. 63.
[117] L. Woo, M.T.K. Ling, S.P. Westphal and S.Y. Ding, *Proceedings SPE ANTEC '97*, Toronto (1997), pp. 3606–10.
[118] S.L. Rosen, *Fundamental Principles of Polymeric Materials*, Wiley-Interscience, New York, NY (1982).
[119] H.E. Bair, in *Thermal Characterization of Polymeric Materials* (E.A. Turi, ed.), Academic Press, New York, NY (1981), pp. 845–909.
[120] A.B. Lugão, E.C.L. Cardoso, B. Hutzler, L.D.B. Machado and R.N. Conceição, *Radiat. Phys. Chem.* **63**, 489–92 (2002).
[121] R.L. Blaine, C.J. Lundgren and M.B. Harris, *ASTM Spec. Techn. Publ., STP* **1326** *(Oxidative Behavior of Materials by Thermal Analytical Techniques)*, A.T. Riga and G.H. Patterson (eds.), ASTM, Philadelphia, PA (1997), pp. 3–15.
[122] R. Truttmann, K. Schiano and R. Riesen, *ASTM Spec. Techn. Publ., STP* **1326** *(Oxidative Behavior of Materials by Thermal Analytical Techniques)*, A.T. Riga and G.H. Patterson (eds.), ASTM, Philadelphia, PA (1997), pp. 184–92.
[123] C.J. Lundgren and R.L. Blaine, *Proceedings 22nd Annu. Conf. North Am. Therm. Anal. Soc. (NATAS)*, Denver, CO (1993), pp. 502–7.
[124] D.S. Rosa, J. Sarti, L.H.I. Mei, M.M. Filho and S. Silveira, *Polym. Testing* **19** (5), 523–31 (2000).
[125] M.P. Sepe, *Proceedings SPE ANTEC '95*, Boston, MA (1995), pp. 2468–74.
[126] R.B. Cassel, A.W. Salamon, G. Curran and A.T. Riga, *ASTM Spec. Techn. Publ., STP* **1326** *(Oxidative Behavior of Materials by Thermal Analytical Techniques)*, A.T. Riga and G.H. Patterson (eds.), ASTM, Philadelpha, PA (1997), pp. 164–71.
[126a] M. Schmid and S. Affolter, *Polymer Testing* **22**, 419–26 (2003).
[127] R.L. Blaine and M.B. Harris, *ASTM Spec. Techn. Publ., STP* **1326** *(Oxidative Behavior of Materials by Thermal Analytical Techniques)*, A.T. Riga and G.H. Patterson (eds.), ASTM, Philadelphia, PA (1997), pp. 193–204.
[128] L. Koski and K. Saarela, *J. Therm. Anal.* **25**, 167 (1982).
[129] S. Affolter, M. Schmid and B. Wampfler, *Kautsch. Gummi Kunstst.* **52**, 519–28 (1999).
[130] G.N. Foster, in *Oxidation Inhibition in Organic Materials* (J. Pospíšil and P.P. Klemchuk, eds.), CRC Press, Boca Raton, FL (1990), Vol. 2, pp. 299–347.
[131] J.B. Howard, *Polym. Engng. Sci.* **13** (6), 429–34 (1973).
[132] J.R. Pauquet, R.V. Todesco and W.O. Drake, *Proceedings 42nd Intl. Wire Cable Symposium*, St. Louis, MO (1993), pp. 77–86.
[133] V. Kuck, *Proceedings VIth Intl. Conference on Plastics in Telecommunications* (Institute of Materials), London, Vol. 3, pp. 1–10 (1992).
[134] T. Schwarz, G. Steiner and J. Koppelmann, *J. Appl. Polym. Sci.* **38**, 1–7 (1989).
[135] P.P. Klemchuk and P.L. Horng, *Polym. Degr. Stabil.* **34**, 333–46 (1991).
[136] P.K. Fearon, T.L. Phease, N.C. Billingham and S.W. Bigger, *Polymer* **43**, 4611–8 (2002).
[137] L. Woo, M.T.K. Ling and E. Chan, *J. Vinyl Technol.* **13** (4), 199–203 (1991).
[138] Y.S. Ding, C. Qin, M.T.K. Ling, S. Westphal and L. Woo, *Proceedings SPE ANTEC '97*, Toronto (1997), pp. 3598–602.

[139] GRI-GM 13, *Standard Specification for Test Properties, Testing Frequency and Recommended Warranty for High Density Polyethylene (HDPE) Smooth and Textured Geomembranes*, Geosynthetic Institute, Philadelphia, PA (1997).

[140] Y.G. Hsuan and Z. Guan, *Proceedings 6th Intl. Conf. Geosynthetics* (R.K. Rowe, ed.), Atlanta, GA (1998), Vol. I, pp. 375–80.

[141] Y.G. Hsuan and Z. Guan, *ASTM Spec. Techn. Publ., STP* **1326** *(Oxidative Behavior of Materials by Thermal Analytical Techniques)*, A.T. Riga and G.H. Patterson (eds.), ASTM, West Conshohocken, PA (1997), pp. 76–90.

[142] Y.G. Hsuan and R.M. Koerner, *J. Geotech. Geoenviron. Eng.* **124** (6), 532–41 (1998).

[143] A. Rudin, H.P. Schreiber and M.H. Waldman, *I&EC* **53**, 137 (1961).

[144] H.E. Bair, *Polym. Eng. Sci.* **13**, 435–9 (1973).

[145] K. Karlsson, C. Assargren and U.W. Gedde, *Polym. Testing* **9**, 421–31 (1990).

[146] T. Schwarz, G. Steiner and J. Koppelmann, *J. Therm. Anal.* **35**, 481–96 (1989).

[147] T. Schwarz, G. Steiner and J. Koppelmann, *J. Appl. Polym. Sci.* **37**, 3335–41 (1989).

[148] J.B. Howard and H.M. Gilroy, *Polym. Eng. Sci.* **15**, 268–71 (1975).

[149] ASTM D 3895-98, *Test Method for Oxidative-Induction Time of Polyolefins by DSC, Annual Book of ASTM Standards*, ASTM, West Conshohocken, PA (1999), Vol. 08.02.

[150] B.L. Treherne, *Elastomerics* **114**, 25–8 (1982).

[151] J.J. Gibbons, *Am. Lab.* **19**, 33–9 (1987).

[152] R. Bharel, R.C. Anand, V. Choudhary and I.K. Varma, *Polym. Degr. Stabil.* **38**, 107–12 (1992).

[153] S. Hakani, S. Mehta, P. Canale and S. McCarthy, *Proceedings SPE ANTEC '99*, New York, NY (1999), pp. 3113–8.

[154] N. Haider and S. Karlsson, *Polym. Degr. Stabil.* **74**, 103–12 (2001).

[155] S. Affolter and M. Schmid, *Proceedings 12th Intl. Symposium on Polymer Analysis and Characterization*, La Rochelle (1999), P53.

[156] D. Cooney, M. Day, C. Klein and J. Fox, *Proceedings 22nd Annu. Conf. North Am. Therm. Anal. Soc. (NATAS)*, Denver, CO (1993), p. 372.

[157] E. Kramer and J. Koppelmann, *Polym. Eng. Sci.* **27**, 945 (1987).

[158] B. Stenberg and F. Björk, *J. Appl. Polym. Sci.* **31**, 487–92 (1986).

[159] V. González, *Rubber Chem. Technol.* **53**, 378 (1980).

[160] O.T. Savaşçi and S.M. Baysal, *J. Appl. Polym. Sci.* **31**, 2157–69 (1986).

[161] P. Šimon and. A. Kučma, *J. Therm. Anal. Calorim.* **56**, 1107–13 (1999).

[162] R.C. Smith and H.L. Stephens, *J. Elastomers Plast.* **7**, 156 (1975).

[163] H. Berg and B. Stenberg, *Plast. Rubber Process. Appl.* **12**, 235–9 (1989).

[164] D.J. Burlett, *Proceedings 17th NATAS Annu. Conf. North Am. Therm. Anal. Soc.* (1988), pp. 654–71.

[165] S. Nakatsuka and A.L. Andrady, *J. Appl. Polym. Sci.* **45**, 1881–7 (1992).

[166] J. Viebke and U.W. Gedde, *Polym. Engng. Sci.* **38** (8), 1244–50 (1998).

[167] F. Gugumus, in *Developments in Polymer Stabilization* (G. Scott, ed.), Applied Science Publishers, London (1979), Vol. 1, p. 261.

[168] B.S. Bernstein and P.-N. Lee, *Proceedings 24th Intl. Wire & Cable Symposium*, NTIS (ed.), Springfield, VA (1975), p. 202.

[169] F. Gugumus, in *Developments in Polymer Stabilization* (G. Scott, ed.), Elsevier Applied Science, London (1987), Vol. 8, pp. 239–89.

[170] P. Budrugeac, *Polym. Degr. Stabil.* **68**, 289–93 (2000).

[171] M. Ezrin, A. Zepke, J. Helwig, G. Lavigne and M. Dudley, *Proceedings SPE ANTEC 2000*, Orlando, FL (2000), pp. 3108–12.

[172] D. Forsström, *Thermometric Application Note 22027*, Järfälla (1999).

[173] Z. Zhang and Y.P. Handa, *Proceedings SPE ANTEC '97*, Toronto (1997), pp. 2051–4.

[174] A.T. Riga and G.H. Patterson, *ASTM Spec. Techn. Publ., STP* **1326** *(Oxidative Behavior of Materials by Thermal Analytical Techniques)*, A.T. Riga and G.H. Patterson (eds.), ASTM, Philadelphia, PA (1997), pp. 151–63.

[175] A. Sass, *Kunststoffe* **88** (2), 231–3 (1998).

[176] T.R. Crompton, *Analysis of Polymers. An Introduction*, Pergamon Press, Oxford (1989).

[177] S.H. Goh, *Thermochim. Acta* **80**, 75 (1984).

[178] L. Peeva, *Angew. Makromol. Chem.* **158/159**, 351 (1988).

[179] U.W. Gedde and M. Ifwarson, *Polym. Engng. Sci.* **30**, 202 (1990).

[180] J.O. Hill, in *Encyclopedia of Analytical Science* (A. Townshend, ed.), Academic Press, San Diego, CA (1995), pp. 5148–55.

[181] F.R. Wight, *Polym. Engng. Sci.* **16**, 652–4 (1976).

[182] D.D. O'Rell and A. Patel, *Proceedings 24th Intl. Wire & Cable Symposium*, NTIS (ed.), Springfield, VA (1975), p. 231.

[183] D.A. Gordon, *Adv. Chem. Ser.* **85**, 224 (1968).

[184] O. Egyed and J. Simon, *J. Therm. Anal.* **16**, 321 (1979).

[185] B.C. Childress and R.H. Barker, *J. Fire Retardant Chem.* **8**, 65 (1981).

[186] S. Nara, T. Kimura and K. Matsuyama, *Rev. Elec. Commun. Lab.* **20**, 159 (1972).

[187] V.M. Bhatnagar and J.M. Vergnaud, *J. Therm. Anal.* **27** (1), 159–200 (1983).
[188] C.Z. Carroll-Porczynski, *Composites* **4**, 9 (1973).
[189] R.C. Mackenzie, *The Differential Thermal Investigation of Clays*, Mineralogical Society, London (1957).
[190] M.G. Chan, in *Oxidation Inhibition in Organic Materials* (J. Pospíšil and P.P. Klemchuk, eds.), CRC Press, Boca Raton, FL (1989), Vol. I, pp. 225–46.
[191] X. Zhu and S. Chen, *Tianranqi Huagong* **22** (5), 45–8 (1997).
[192] A.K. Sircar, *Rubber Chem. Technol.* **65**, 503–25 (1992).
[193] A. Tas, F. Wülfert, K.G.H. Raemaekers and J.C.J. Bart, unpubl. results (2000).
[194] C.H. Lochmüller, S.J. Breiner, M.N. Koel and M.A. Elomaa, *Anal. Chem.* **63** (17), 1685–91 (1991).
[195] O. Toft Sørensen, *J. Therm. Anal. Calorim.* **56**, 17–26 (1999).
[196] M. Reading, in *Handbook of Thermal Analysis and Calorimetry* (M.E. Brown, ed.), Elsevier, Amsterdam (1998), Vol. 1. Chp. 8.
[197] F. Paulik and J. Paulik, *Anal. Chim. Acta* **67**, 203 (1973).
[198] J. Paulik and F. Paulik, *Anal. Chim. Acta* **56**, 329 (1971).
[199] J. Rouquerol, *Bull. Soc. Chim. Fr.*, 31 (1964).
[200] A. Ortega, L.A. Pérez-Maquéda and J.M. Criado, *J. Therm. Anal.* **42**, 551 (1994).
[201] O. Toft Sørensen, *J. Therm. Anal.* **38**, 213 (1992).
[202] O. Toft Sørensen, *J. Therm. Anal.* **13**, 429 (1978).
[203] R.B. Cassel, R.L. Fyans and G.H. Curran, *Proceedings 19th NATAS Annu. Conf. North Am. Therm. Anal. Soc.* (1991), Paper 55.
[204] O. Toft Sørensen, *Thermochim. Acta* **138**, 107 (1989).
[205] R. Blaine, *Am. Lab.* (1), 21–3 (1998).
[206] J. Rouquerol, *Pure Appl. Chem.* **57** (1), 69 (1985).
[207] L. Erdey, F. Paulik and J. Paulik, *Hung. Pat.* 152,197 (Oct. 31, 1962).
[208] J. Rouquerol and L. Davy, *Thermochim. Acta* **24**, 391–7 (1978).
[209] G. Goujon, *Thesis nr. A.O. 11021*, University of Marseille (1975), p. 107.
[210] M. Ganteaume and J. Rouquerol, *J. Therm. Anal.* **3**, 413–20 (1971).
[211] F. Paulik and J. Paulik, *Anal. Chim. Acta* **67**, 437–43 (1973).
[212] M. Maciejewski, C.A. Müller, R. Tschan, W.D. Emmerich and A. Baiker, *Thermochim. Acta* **295**, 167–82 (1997).
[213] M. Maciejewski, W.D. Emmerich and A. Baiker, *J. Therm. Anal. Calorim.* **56**, 627–37 (1999).
[213a] J.C.J. Bart, *Additives in Polymers. Industrial Analysis and Applications*, J. Wiley & Sons, Chichester (2005).
[214] D. Dollimore, in *Thermal Analysis – Techniques and Applications* (E.L. Charsley and S.B. Warrington, eds.), The Royal Society of Chemistry, Cambridge (1992), pp. 31–58.
[215] E. Robens and K.K. Unger, *ICTAC News* **35** (1), 43–51 (June 2002).
[216] J. Chiu, in *Applied Polymer Analysis and Characterization* (J. Mitchell, ed.), Hanser Publishers, Munich (1987), pp. 175–215.
[217] ASTM E 1131-98, *Test Method for Compositional Analysis by Thermogravimetry, Annual Book of ASTM Standards*, ASTM, West Conshohocken, PA (1999), Vol. 14.02.
[218] C.M. Earnest, *Anal. Chem.* **56**, 1471A (1984).
[219] R.D. Pike, W.H. Starnes, J.P. Jeng, W.S. Bryant, P. Kourtesis, C.W. Adams, S.D. Bunge, Y.M. Kang, A.S. Kim, J.H. Kim, J.A. Macko and C.P. O'Brien, *Macromolecules* **30**, 6957–79 (1997).
[220] D. Al-Sammerrai, H. Al-Najjar and W. Selim, *Analyst (London)* **110**, 1267–8 (1985).
[221] E.W. Salamon and R.B. Cassel, *Proceedings SPE ANTEC '99*, New York, NY (1999), pp. 3836–8.
[222] W.J. Shuely and B.S. Ince, *Polym. Mat. Sci. Engng.* **84**, 625–6 (2001).
[223] Z. Mrklić and T. Kovačić, *Thermochim. Acta* **322**, 129–35 (1998).
[224] J. Luston, V. Pastušakova and F. Vašš, *J. Appl. Polym. Sci.* **47**, 555 (1993); **48**, 219 (1993).
[225] S. Matsumoto, *J. Polym. Sci. Polym. Chem.* **21**, 557 (1983).
[226] J. Joroszynska and T. Kleps, *Intl. Polym. Sci. Technol.* **5** (2), T125–8 (1978).
[227] D.M. Price, *Thermochim. Acta* **367/8**, 253–62 (2001).
[228] P. Mapleston, *Modern Plast.*, 35 (1998).
[229] R.B. Cassel and B. Twombly, *Am. Lab.* **31** (1), 22–7 (1999).
[230] S. Affolter, *Kautsch. Gummi Kunstst.* **44** (11), 1026–8 (1991).
[231] A. Marcilla and M. Beltram, *Polym. Degr. Stabil.* **53** (2), 251–60 (1996).
[232] A. Coats and J.P. Redfern, *Analyst* **88**, 905 (1963).
[233] J.J. Maurer, *Rubber Chem. Technol.* **42**, 110 (1969).
[234] H. Möhler, A. Stegmayer and E. Kaisersberger, *Kautsch. Gummi Kunstst.* **44** (4), 369–80 (1991).
[235] D.P. Macaione, R.E. Sacher and R.E. Singler, *ASTM Spec. Techn. Publ., STP* **997** (*Compositional Analysis by Thermogravimetry*), C.M. Earnest (ed.), ASTM, Philadelphia, PA (1988), pp. 59–69.
[236] S.J. Swarin and A.M. Wims, *Rubber Chem. Technol.* **47**, 1193–205 (1974).

[237] R.L. Zeyen, *Rubber World* **199** (4), 14–20 (1989).
[238] W.J. Sichina, *Am. Lab.* **25**, 45–50 (1993).
[239] D.E. Larkin, *ASTM Spec. Techn. Publ., STP* **997** *(Compositional Analysis by Thermogravimetry)*, C.E. Earnest (ed.), ASTM, Philadelphia, PA (1988), pp. 28–37.
[240] J.J. Maurer, in *Thermal Characterisation of Polymeric Materials* (E.A. Turi, ed.), Academic Press, New York, NY (1981), pp. 571–708.
[241] L.G. Britcher, P.D. Folland and J.G. Matisons, *Preprints 37th IUPAC Intl. Symposium on Macromolecules*, Gold Coast (1998), p. 125.
[242] E. Ramirez, B.M. Huerta and A. Ledezma, *Ind. Gomma* **38** (11), 43–8 (1994).
[243] I. Soos and P. Marik Korda, *Muanyag Gumi* **23** (7), 193–5 (1986).
[244] M. Patel and A.R. Skinner, *Abstracts Intl. Conference on Ageing Studies and Lifetime Extension of Materials*, Oxford (1999).
[245] Y.-W. Yang, W.-H. Hwu, M.-D. Ger, A.-H. Chien and C.C. Huang, *Huoyao Jishu* **14** (2), 77–85 (1998).
[246] J. Ward, R. Simmons and P. Chatham, *Proceedings SPE ANTEC '98*, Atlanta, GA (1998), pp. 1782–6.
[247] H. Wang and W. Chen, *Beijing Shifan Daxue Xuebao, Ziran Kexueban* **34** (2), 219–23 (1998).
[248] R.L. Gray and R.E. Lee, in *Plastics Additives. An A–Z Reference* (G. Pritchard, ed.), Chapman & Hall, London (1998), pp. 567–75.
[249] A.B. Morgan and J.M. Tour, *J. Appl. Polym. Sci.* **73** (5), 707–18 (1999).
[250] A.W. Benbow and C.F. Cullis, *Eur. Polym. J.* **11**, 723 (1975).
[251] R. Luyk, *Ph.D. Thesis*, University of Amsterdam (1993).
[252] A.P. Taylor and F.R. Sale, *Makromol. Chem., Macromol. Symp.* **74**, 85–93 (1993).
[253] G.S. Learmonth and D.G. Thwaite, *Br. Polym. J.* **2**, 104 (1970).
[254] R.M. Aseeva, L.V. Ruban and G.E. Zaikov, *Intl. J. Polym. Mat.* **38** (3–4), 297–313 (1997).
[255] M. Ezrin and J. Harten, *Proceedings SPE ANTEC '81*, Boston, MA (1981), pp. 188–9.
[256] J. Mullens, G. Reggers, M. Ruysen, R. Carleer, J.Yperman, D. Franco and L.C. Van Poucke, *J. Therm. Anal.* **49**, 1061–7 (1997).
[257] J. Gillmor and R.J. Seyler, *ASTM Spec. Techn. Publ., STP* **997** *(Compositional Analysis by Thermogravimetry)*, C.E. Earnest (ed.), ASTM, Philadelphia, PA (1988), pp. 38–47.
[258] W.P. Brennan, *Thermochim. Acta* **18**, 101–11 (1977).
[259] M. Sepe, in *Elastomer Technology Handbook* (N.P. Cheremisinoff, ed.), CRC Press, Boca Raton, FL (1993), pp. 105–258.
[260] H. Ostromow, *Analyse von Kautschuken und Elastomeren*, Springer-Verlag, Berlin (1981).
[261] R.A. Fava, *Proceedings 22nd Annu. Conf. North Am. Therm. Anal. Soc. (NATAS)*, Denver, CO (1993), pp. 564–7.
[262] B.R. Carson, G. Liptay, G. Kenessey, J.R. Allen and T. Wadsten, *Proceedings 7th European Symposium on Thermal Analysis and Calorimetry (ESTAC-7)*, Balatonfüred (1998), p. 57.
[263] S. Nakatsuka and A.L. Andrady, *J. Appl. Polym. Sci.* **45** (11), 1881–7 (1992).
[264] S.M. Goheen and R.P. Wool, *J. Appl. Polym. Sci.* **41**, 2691–701 (1991).
[265] M.Y. Ahmad Fuad, M.J. Zaini, M. Jamaludin and R. Ridzuan, *Polym. Testing* **13** (1), 15–24 (1994).
[266] ASTM D 1063, *Standard Test Method for Ash in Rosin, Annual Book of ASTM Standards*, ASTM, Philadelpha, PA (1981), Vol. 06.03.
[267] E.L. Charsley and D.G. Dunn, *Rubber Chem. Technol.* **55**, 382–90 (1982).
[268] R. Pautrat, B. Metavier and J. Marteau, *Rev. Gen. Caoutch.* **52** (4), 273 (1970); *Rubber Chem. Technol.* **49**, 1060–7 (1976).
[269] H. Nelissen, *Techn. Rept. DSM Research*, Geleen (1998).
[270] F. Casa, A. Giacometti and P. Massimo, *Ind. Gomma* **32** (9), 22–8 (1988).
[271] R. Riesen and B. Schenker, *Abstracts 7th European Symposium on Thermal Analysis and Calorimetry (ESTAC-7)*, Balafonfüred (1998).
[272] *Perkin-Elmer Thermal Analysis Newsletter* (PETAN-57).
[273] T.J. Lever and A. Sutkowski, *J. Therm. Anal.* **40**, 257–62 (1993).
[274] J.J. Leyden and J.M. Raab, *Elastomerics* **112** (4), 34–7 (1980).
[275] M. Stroh, in *Thermische Analyse an Kunststoffen – Methoden und Anwendungen* (T. Nitschke and J. Vogel, eds.), Ecomed Verlagsgesellschaft, Landsberg (1998), pp. 53–63.
[276] R. Fyans, *Adhes. Age*, 19 (March 1975).
[277] R.K. Chohan, B. David, A. Nir and Z. Tadmor, *Intl. Polym. Process.* **2**, 13 (1987).
[278] V.N. Reddy, *Proceedings SPE ANTEC '99*, New York, NY (1999), pp. 2969–72.
[279] J. Harris, *Synthesis* **8**, 20 (1977).
[280] H.E. Bair, G.E. Johnson, E.W. Anderson and S. Matsuoka, *Polym. Eng. Sci.* **21** (3), 930–5 (1981).
[281] A.M. Wims and S.J. Swarin, *J. Appl. Polym. Sci.* **19**, 1243–56 (1975).
[282] K.G.H. Raemaekers and J.C.J. Bart, *Proceedings SPE ANTEC '99*, New York, NY (1999), pp. 2599–602.

[283] J.O. Hill, in *Encyclopedia of Analytical Science* (A. Townshend, ed.), Academic Press, San Diego, CA (1995), pp. 5141–8.

[283a] P. Le Parlouër, *Thermochim. Acta* **121**, 307–22 (1987).

[284] J.P. Redfern, *J. Appl. Polym. Sci.: Appl. Polym. Symp.* **55**, 65–86 (1994).

[285] J. van Humbeeck, in *Handbook of Thermal Analysis and Calorimetry* (M.E. Brown, ed.), Elsevier, Amsterdam (1998), Vol. 1, Chp. 11.

[286] Netzsch Gerätebau, *Technical Data Sheet Coupling Systems*, Selb (1996).

[287] J. Blumm, *LabPlus Intl.* **15** (1), 23 (2001).

[288] J.P. Redfern, *Intl. Labmate* **11** (1), 19 (1986).

[289] E.L. Charsley, in *Thermal Analysis – Techniques and Applications* (E.L. Charsley and S.B. Warrington, eds.), The Royal Society of Chemistry, Cambridge (1992), pp. 59–83.

[290] J.P. Redfern, *Polym. Intl.* **26**, 51–8 (1991).

[291] S. Kodama, H. Kawasaki, K. Itatani, F. Kusano and T. Nakatsuka, *Okayama-ken Kogyo Gijutsu Senta Hokoku* **7**, 1 (1981).

[292] P. Emmott, T.T. Griffiths, J. Queay, E.L. Charsley and S.B. Warrington, *Proceedings 16th Intl. Pyrotech. Semin.*, Jönköping (1991), p. 937.

[293] P. Emmott, T.T. Griffiths, J. Queay, E.L. Charsley and S.B. Warrington, *Proceedings 18th Intl. Pyrotech. Semin.*, Breckenridge, OH (1992), p. 221.

[294] E.L. Charsley, S.B. Warrington, P. Emmott, T.T. Griffiths and J. Queay, *J. Therm. Anal.* **38**, 641 (1992).

[295] T. Griffiths, J. Queay, E.L. Charsley and S.B. Warrington, *Proceedings 19th Intl. Pyrotech. Semin.*, Christchurch (1994), p. 716.

[296] P.A. Larcey, J.P. Redfern and G.M. Bell, *Fire & Mater.* **19**, 283–5 (1995).

[297] D.A. Powell, *J. Sci. Instrum.* **34**, 225 (1957).

[298] F. Paulik, J. Paulik and L. Erdey, *Hung. Pat.* 144,548 (1954).

[299] F. Paulik, J. Paulik and L. Erdey, *Hung. Pat.* 145,369 (1955).

[300] A. Riga, *Proceedings 26th Annu. Conf. North Am. Therm. Anal. Soc.* (K.R. Williams, ed.), Omnipress, Madison, WI (1998), pp. 648–53.

[301] M. Negri and F. Alarcon-Lorca, *Rev. Gen. Caoutch. Plast.* **61**, 55–9 (1984).

[302] Y. Nakagawa and T. Komai, *J. Fire Sci.* **8** (6), 455–76 (1990).

[303] J.-P. Gibert, J.-M. Lopez Cuesta, A. Bergeret and A. Crespy, *Polym. Degr. Stabil.* **67** (3), 437–7 (2000).

[304] J. Koch, *LaborPraxis* **18** (10), 58–62 (1994).

[305] S. Otomo, H. Shimizu and Y. Yamamoto, *Nippon Gomu Kyokaishi* **67** (4), 306–13 (1994).

[306] A. Riga, R. Collins and G. Mlachak, *Thermochim. Acta* **324** (1–2), 135 (1998).

[307] T. Ozawa, *Thermochim. Acta* **355**, 35 (2000).

[308] N. Nagayama and T. Takeda, *Thermochim. Acta* **156**, 11 (1989).

[309] T. Kántor and L. Benzúr, *J. Anal. At. Spectrosc.* **1**, 9 (1986).

[310] J.A.J. Jansen, J.H. v.d. Maas and A. Posthuma de Boer, *Appl. Spectrosc.* **46** (1), 88–92 (1992).

[311] K.G.H. Raemaekers and J.C.J. Bart, *Thermochim. Acta* **295**, 1–58 (1997).

[312] K. Pommerenke, *Intl. Lab.* **31** (1), 15–8 (2001).

[313] G. Bongaerts, K. Rabel, K.G.H. Raemaekers and J.C.J. Bart, *Abstracts 12th Intl. Congress on Thermal Analysis and Calorimetry (ICTAC-12)*, Copenhagen (2000).

[314] N. Gerard, *J. Physics* **E7**, 509 (1974).

[315] H.G. Wiedemann and G. Bayer, *Z. Anal. Chem.* **266**, 97–109 (1973).

[316] P. Barret, in *Vacuum Microbalance Technology* (C. Eyraud and M. Escoubes, eds.), Heyden, London (1995), Vol. 3, p. 205.

[317] J.C.J. Bart, *Techn. Rept. DSM Research*, Geleen (1995).

[318] J.P. Auffredic, J. Plevert and D. Louer, *J. Therm. Anal.* **37**, 1727 (1991).

[319] T.G. Fawcett, C.E. Crowther, L.F. Whiting, J.C. Tou, W.F. Scott, R.A. Newman, W.C. Harris, F.J. Knoll and V.J. Caldecourt, *Adv. X-Ray Anal.* **28**, 227 (1985).

[320] T.P. Russell and J.T. Koberstein, *J. Polym. Sci.* **23**, 1109 (1985).

[321] G.Z. Zhang and H. Yoshida, *Abstracts 8th European Symposium on Thermal Analysis and Calorimetry (ESTAC-8)*, Barcelona (2002), Paper 1O-6.

[322] T. Arii, A. Kishi and Y. Kobayashi, *Thermochim. Acta* **325**, 151–6 (1999).

[323] N.C. Billingham, P.K. Fearon, D.J. Whiteman and V. Dudler, *Proceedings 5th AddCon World Conference*, Prague (1999), Paper 23.

[324] C. Darribère, *Mettler-Toledo UserCom* **1**, 21–2 (2002).

[324a] T. Provder, M.W. Urban and H.G. Barth (eds.), *ACS Symp. Ser.* **581** (1994).

[325] J.O. Lephardt and R.A. Fenner, *Appl. Spectrosc.* **34**, 174 (1980).

[326] D.A.C. Compton, *Intl. Labmate* **12** (4), 37 (1987).

[327] R.C. Wieboldt, G.E. Adams, S.R. Lowry and R.J. Rosenthal, *Am. Lab.* **20** (1), 70 (1988).

[328] P.R. Solomon, M.A. Serio, R.M. Carangelo, R. Bassilakis, Z.Z. Yu, S. Charpenay and J. Whelan, *J. Anal. Appl. Pyrol.* **19**, 1–14 (1991).

[329] D.J. McEwen, W.R. Lee and S.J. Swarin, *Thermochim. Acta* **86**, 251–6 (1985).

[330] C.A. Cody, L. DiCarlo and B.K. Faulseit, *Am. Lab.* **13** (1), 93 (1981).

[331] J.O. Lephart, *Appl. Spectrosc. Rev.* **18**, 265 (1982).

[332] P.B. Roush, J.M. Luce and G.A. Totten, *Am. Lab.* **15** (10), 90 (1983).

[333] A.G. Nerheim, in *Fourier Transform Infrared Spectroscopy, Vol. 4. Applications to Chemical Systems* (J.R. Ferraro and L.J. Basile, eds.), Academic Press, New York, NY (1985), p. 147.

[334] M.L. Mittleman, D.A. Compton and P. Engle, *Proceedings 13th Annu. Conf. North Am. Therm. Anal. Soc. (NATAS)*, Philadelphia, PA (1984), p. 410.

[335] R.M. Carangelo, P.R. Solomon, R. Bassilakis, D. Gravel, M. Baillargeon, F. Baudais, G. Vail and J.K. Whelan, *Am. Lab.* **22** (6), 51–60 (1990).

[336] D.A.C. Compton, S.L. Hill, N.A. Wright, M.A. Druy, J. Piche, W.A. Stevenson and D.W. Vidrine, *Appl. Spectrosc.* **42**, 972–9 (1988).

[337] E. Post, S. Rahner, H. Möhler and A. Rager, *Thermochim. Acta* **263**, 1–6 (1995).

[338] E. Kaisersberger and E. Post, *Thermochim. Acta* **295**, 75–93 (1997).

[339] C.A. Wilkie, *Polym. Degr. Stabil.* **66**, 301–6 (1999).

[340] J.K. Whelan, P.R. Solomon, G.V. Deshpande and R.M. Carangelo, *Energy Fuels* **2**, 65 (1988).

[341] J.A.J. Jansen, J.H. van der Maas and A. Posthuma De Boer, in *Integration of Fundamental Polymer Science and Technology-5* (P.J. Lemstra, ed.), Elsevier Applied Sciences, Oxford (1991), pp. 316–22.

[342] T.B. Brill, *Anal. Chem.* **61**, 897A–906A (1989).

[343] W. Herres, *HRGC-FTIR: Capillary Gas Chromatography–Fourier Transform Infrared Spectroscopy, Theory and Applications*, Hüthig Publishers, Heidelberg (1987).

[344] R.A. Nyquist, *The Interpretation of Vapor-Phase Infrared Spectra: Group Frequency Data*, Sadtler Research Laboratories, Philadelphia, PA (1984), pp. 87–96.

[345] J.A.J. Jansen, W.E. Haas, H.G.M. Neutkens and A.J.H. Leenen, *Thermochim. Acta* **134**, 307–12 (1988).

[346] J. Mullens, in *Handbook of Thermal Analysis and Calorimetry* (M.E. Brown, ed.), Elsevier, Amsterdam (1998), pp. 509–46.

[347] D.J. Johnson and D.A.C. Compton, *Am. Lab.* **23** (1), 37 (1991).

[348] M.L. Mittleman, *Thermochim. Acta* **166**, 301 (1990).

[349] J.A.J. Jansen, *Ph.D. Thesis*, University of Utrecht (1992).

[350] C.A. Wilkie and M.L. Mittleman, *ACS Symp. Ser.* **581**, 116 (1994).

[351] D.J. Johnson, P.J. Stout, S.L. Hill and K. Krishnan, *ACS Symp. Ser.* **581**, 149 (1994).

[352] M.L. Mittleman, J.R. Thomsen and C.A. Wilkie, *Polym. Mater. Sci. Eng.* **63**, 957 (1990).

[353] M.L. Mittleman, D. Johnson and C.A. Wilkie, *Trends Polym. Sci.* **2** (11), 391–8 (1994).

[354] W.H. McClennen, R.M. Buchanan, N.S. Arnold, J.P. Dworzanski and H.L.C. Meuzelaar, *Anal. Chem.* **65**, 2819–23 (1993).

[355] L. Meublat and P. Le Parlouër, *Spectra 2000* **161**, 59 (1991).

[356] J.P. Redfern, P.H. Newbatt and P. Larcey, *Polym. Mater. Sci. Eng.* **69**, 144 (1993).

[357] D.A.C. Compton, D.J. Johnson and M.L. Mittleman, *Res. Dev.* **31** (4), 68–73 (1989).

[358] J.C.J. Bart and K.G.H. Raemaekers, in *Spectroscopy of Rubbers and Rubbery Materials* (V. Litvinov and P.P. De, eds.), Rapra Technology Ltd., Shawbury (2002), pp. 1–47.

[359] J. Mullens, R. Carleer, G. Reggers, J. Yperman, J. Vanhees and L.C. van Poucke, *Thermochim. Acta* **212**, 219 (1992).

[359a] M.L. Mittleman and C.A. Wilkie, *Polym. Mater. Sci. Engng.* **64**, 90–1 (1991).

[360] D.D. Jiang, Q. Yao, M.A. McKinney and C.A. Wilkie, *Polym. Degr. Stabil.* **63**, 423–34 (1999).

[361] C.A. Wilkie, in *Chemistry and Technology of Polymer Additives* (S. Al-Malaika, A. Golovoy and C.A. Wilkie, eds.), Blackwell Science, Oxford (1999), pp. 218–34.

[362] E. Post, S. Rahner, H. Möhler and A. Rager, *Thermochim. Acta* **263**, 1–6 (1995).

[363] K. Pielichowski, *J. Therm. Anal. Calorim.* **55**, 559–63 (1999).

[364] R.C. Wieboldt, S.R. Lowry and R.J. Rosenthal, *Mikrochim. Acta (Wien)* **1**, 179–82 (1988).

[365] Netzsch Gerätebau GmbH, *Application Note TG-DTA/DSC, MS, FTIR Coupling Systems*, Selb (n.d.).

[366] G.M. Anthony, *Makromol. Chem., Macromol. Symp.* **74**, 117–23 (1993).

[367] E. Post, S. Rahner and F. Giblin, *Proceedings SPE ANTEC '97*, Toronto (1997), pp. 2300–4.

[368] K.P. Pramoda, T.X. Liu, Z.H. Liu, C.B. He and H.-J. Sue, *Polymer Preprints* **43** (2), 1217 (2002).

[369] H. Möhler, H. Walter and S. Knappe, in *Coupling of Thermal Analysis and Gas Analysis Techniques and Applications* (E. Kaisersberger, E. Kapsch, E. Post and A. Schranner, eds.), Netzsch, Selb (2001), pp. 32–40.

[370] P.-S. Wang, W.-Y. Chiu, L.-W. Chen, B.-L. Denq, T.-M. Don and Y.-S. Chiu, *Polym. Degr. Stabil.* **66** (3), 307–15 (1999).

[371] D. Bertin, B. Boutevin, G. Rigal and G. Fourty, *Eur. Polym. J.* **34** (2), 163–70 (1998).

[372] J.P. Redfern and J. Powell, in *Hyphenated Techniques in Polymer Characterisation* (T. Provder, M.W. Urban and H.G. Barth, eds.), *ACS Symp. Sci.* **581**, 90 (1994).

[373] *Mettler Toledo Application Booklet 51725056, Evolved Gas Analysis*, Mettler Toledo, Schwerzenbach (2002).
[374] C.-Y. Kuo and T. Provder, *Adv. Chem. Ser.* **227**, 343–55 (1990).
[375] Y. Ogumi and T.B. Brill, *Combust. Flame* **62**, 213–24 (1985).
[376] T.B. Brill, P.J. Brush, K.J. James, J.E. Shepherd and K.J. Pfeiffer, *Appl. Spectrosc.* **46**, 900–11 (1992).
[377] Y. Nagasawa, M. Hotta and K. Ozawa, *J. Anal. Appl. Pyrol.* **33**, 253–67 (1995).
[378] T.B. Brill and P.J. Brush, *Proceedings 9th Intl. Symposium on Detonation*, Portland, OR (1989).
[379] M.D. Timken, J.-K. Chen and T.B. Brill, *Appl. Spectrosc.* **44**, 701–6 (1990).
[380] T.B. Brill and B.T. Budenz, *CPIA Publ.* **680**, 617–28 (1998).
[381] A.F. Alders, K.G.H. Raemaekers and J.C.J. Bart, *Proceedings SPE ANTEC '99*, New York, NY (1999), pp. 3761–4.
[382] W.W. Wendlandt and T.M. Southern, *Anal. Chim. Acta* **32**, 405 (1965).
[383] R.S. Gohlke and H.G. Langer, *Anal. Chem.* **37** (10), 25A (1965).
[384] S.M. Dyszel, *Thermochim. Acta* **61**, 169 (1983).
[385] E. Baumgartner and E. Nachbaur, *Thermochim. Acta* **19**, 3 (1977).
[386] T. Tsuneto, I. Murasawa, M. Nagata and Y. Kubota, *J. Anal. Appl. Pyrol.* **33**, 139 (1995).
[387] Y. Shimizu and B. Munson, *J. Polym. Sci.: Polym. Chem. Ed.* **17**, 1991 (1979).
[388] W. Lindinger, J. Hirber and H. Paretzke, *Intl. J. Mass Spectrom. Ion Process.* **129**, 79–88 (1993).
[389] D. Bassi. P. Tosi and R. Schögl, *J. Vac. Sci. Technol.* **A16** (1), 114–22 (1998).
[390] F. Zitomer, *Anal. Chem.* **40**, 1091 (1968).
[391] G.A. Kleineberg and D.L. Geiger, *Proceedings 3rd Intl. Congress on Thermal Analysis (ICTA)*, Davos (1971), pp. 325–36.
[392] D. Price, D. Dollimore, N.S. Fatemi and R. Whitehead, *Thermochim. Acta* **42**, 323–32 (1980).
[393] A. Ballistreri, M. Garozzo, M. Giuffrida and G. Montaudo, *Polym. Degr. Stabil.* **16**, 337 (1986).
[394] G. Mathonat, G. Hurel and S. Moreau, *Abstracts 8th European Symposium on Thermal Analysis and Calorimetry (ESTAC-8)*, Barcelona (2000), Poster P-11.
[395] P.R. Dufour, K.G.H. Raemaekers and J.C.J. Bart, *Thermochim. Acta* **175**, 263–79 (1991).
[396] K.G.H. Raemaekers, *Fachtagung Kopplungen der instrumentellen Analytik (TA, IR, MS) für die Kunststoff- und Kautschukindustrie*, Würzburg (Sept. 1991).
[397] W. Schwanebeck and H.W. Wenz, *Fresenius Z. Anal. Chem.* **331**, 61 (1988).
[398] H. Wenz and H. Rohrbach, *Fachtagung Kopplungen der instrumentellen Analytik (TA, IR, MS) für die Kunststoff- und Kautschukindustrie*, Würzburg (Sept. 1991).
[399] R. Saferstein, in *Pyrolysis and GC in Polymer Analysis* (S.A. Liebman and E.J. Levy, eds.), M. Dekker, New York, NY (1985), p. 339.
[400] B. Courtauld, *Analysis* **7**, 481 (1979).
[401] M. Maciejewski and A. Baiker, *Thermochim. Acta* **295**, 95–105 (1997).
[402] M. Statheropoulos, S. Kyriakou and N. Tzamtzis, *Thermochim. Acta* **322**, 167–73 (1998).
[403] B. Shushan, B. Davidson and R.B. Prime, *Anal. Calorim.* **5**, 105 (1984).
[404] R.B. Prime and B. Shushan, *Anal. Chem.* **61**, 1195–201 (1989).
[405] S. Affolter and S. Hofstetter, *Kautsch. Gummi Kunstst.* **48**, 173–9 (1995).
[405a] M.W. Groenewoud and W. de Jong, *Thermochim. Acta* **286**, 341–54 (1996).
[406] W. Holzapfel, in *Fachtagung Kopplungen der instrumentellen Analytik (TA, IR, MS) für die Kunststoff- und Kautschukindustrie*, Würzburg (Sept. 1991).
[407] F.E. Keen, R.S. Lehrle, E. Jacab and T. Szekely, *Polym. Degr. Stabil.* **38**, 219 (1992).
[408] J.A.J. Jansen and W.E. Haas, *Anal. Chim. Acta* **196**, 69–74 (1987).
[409] J.A.J. Jansen, in *Calorimetry and Thermal Analysis of Polymers* (V.B.F. Mathot, ed.), Hanser Publishers, Munich (1994), p. 335.
[410] R. Kunze, D. Neubert, K. Brademann-Jock and U. Erhardt, *J. Therm. Anal.* **53**, 27–33 (1998).
[411] E. Kaisersberger, E. Post and J. Janoschek, *ACS Symp. Ser.* **581**, 74 (1994).
[412] J.C. May, A. Del Grosso and R. Wheeler, *Thermochim. Acta* **115**, 289 (1987).
[413] J. Chiu, *Appl. Polym. Symp.* **2**, 25 (1966).
[414] M. Simpson, P.M. Jacobs and F.R. Jones, *Composites* **22**, 105 (1991).
[415] A.R. Cooper, M.M. Steiner, G.B. McCauley, G.R. Kwiatkowski, J.G. Moncur and E.H. Kawasaki, *Proceedings SPE ANTEC '89*, New York, NY (1989), p. 1137.
[416] M. Wingfield, in *Fachtagung Kopplungen der instrumentellen Analytik (TA, IR, MS) für die Kunststoff- und Kautschukindustrie*, Würzburg (Sept. 1991).
[417] R.G. Beimer, *Proceedings ACS Div. Org. Coat. Plast. Chem., Pap.* **35**, 428–33 (1975).
[418] C. Chang and J.R. Tackett, *Thermochim. Acta* **192**, 181 (1991).
[419] E. Kaisersberger, in *Fachtagung Kopplungen der instrumentellen Analytik (TA, IR, MS) für die Kunststoff- und Kautschukindustrie*, Würzburg (Sept. 1991).

[420] C.E. Carraher, H.M. Molloy, J.W. Louda and E. Baker, *Proceedings ACS Div. Polym. Mater. Sci. & Engng.* **78**, 102–3 (1998).

[421] A. Kettrup and K.H. Ohrbach, *Proceedings 8th Intl. Congress on Thermal Analysis* (A. Blazek, ed.), Alfa, Bratislava (1985), Vol. 2, p. 629.

[422] G. Matuschek, *Thermochim. Acta* **263**, 59–71 (1995).

[423] G. Matuschek, in *Thermische Analyse an Kunststoffen – Methoden und Anwendungen* (Th. Nitschke and J. Vogel, eds.), Ecomed Verlagsgesellschaft, Landsberg (1998), pp. 87–98.

[424] H.-K. Ohrbach and A. Kettrup, *Polym. Degr. Stabil.* **13**, 99 (1985).

[425] H.G. Langer, J.D. Fellmann and C.D. Wood, *Proceedings 7th Intl. Congress on Thermal Analysis* (B. Miller, ed.), J. Wiley & Sons, Chichester (1982), Vol. 2, pp. 1350–5.

[426] H.K. Yuen, G.W. Mappes and W.A. Grote, *Thermochim. Acta* **52**, 143 (1982).

[427] E.G. Jones, D.L. Pedrick and I.J. Goldfarb, *Polym. Eng. Sci.* **28**, 1046 (1988).

[428] E.M. Verdurmen, R. Verstappen, J. Swagten, H. Nelissen, G. Heemels and J.C.J. Bart, *Abstracts 4th Intl. Symposium on Hyphenated Techniques in Chromatography and Hyphenated Chromatographic Analyzers (HTC-4)*, Bruges (1996).

[429] J. Chiu and A.J. Beattie, *Thermochim. Acta* **21**, 263 (1977); **40**, 251 (1980).

[430] H.-R. Schulten and R.P. Lattimer, *Mass Spectrom. Revs.* **3**, 231 (1984).

[431] R.P. Lattimer, *Rubber Chem. Technol.* **61**, 658 (1988).

[432] R.P. Lattimer, H. Muenster and H. Budzikiewicz, *Intl. J. Mass Spectrom. Ion Process.* **90**, 119 (1989).

[433] J.B. Henderson, E. Post, E. Treser and B. Fidler, *Proceedings SPE ANTEC '99*, New York, NY (1999), pp. 2589–93.

[434] F. Giblin, E. Post, W.-D. Emmerich and G. Bräuer, *Proceedings 23rd Annu. Conf. North Am. Therm. Anal. Soc. (NATAS)*, Toronto (1994).

[435] J. Janoschek, E. Kaisersberger and E. Post, *Polym. Mater. Sci. Eng.* **69**, 458–9 (1993).

[436] W.D. Emmerich and E. Kaisersberger, *J. Therm. Anal.* **17**, 197 (1979).

[437] I. Berecz, S. Bohátka, G. Langer and G. Szöör, *Intl. J. Mass Spectrom. Ion Phys.* **47**, 273 (1983).

[438] G. Matuschek, H. Utschick, Ch. Namendorf and A. Kettrup, *Abstracts 13th Intl. Symposium on Analytical and Applied Pyrolysis Abstracts (Pyrolysis '98)*, Neuherberg/München (1998), p. 37.

[439] K.-H. Ohrbach, G. Radhoff and A. Kettrup, *Thermochim. Acta* **85**, 403 (1985).

[440] A.E. Pavlath, K.S. Gregorski and R. Young, *Thermochim. Acta* **92**, 383–4 (1985).

[441] H.K. Yuen and G.W. Mappes, *Thermochim. Acta* **70**, 269–81 (1983).

[442] J. Chiu, *Thermochim. Acta* **1**, 231 (1970).

[443] L.F. Whiting and P.W. Langvardt, *Anal. Chem.* **56**, 1755 (1984).

[444] P.A. Barnes, G. Stevenson and S.B. Warrington, *Proceedings 2nd ESTA*, Heyden, London (1981), p. 47.

[445] J.A.J. Jansen, in *Calorimetry and Thermal Analysis of Polymers* (V.B.F. Mathot, ed.), Hanser Publishers, Munich (1994), pp. 335–52.

[446] K. Liu, E. Jakab, W.H. McClennen and H.L.C. Meuzelaar, *Prepr. Pap. Am. Chem. Soc., Div. Fuel Chem.* **38**, 823 (1993).

[447] T. Arii, T. Senda and N. Fuji, *Thermochim. Acta* **267**, 209–21 (1995).

[448] B.J. McGrattan, *Appl. Spectrosc.* **48** (12), 1472–6 (1994).

[449] A. Rau and B.J. McGrattan, *Abstracts 13th Intl. Symposium on Analytical and Applied Pyrolysis (Pyrolysis '98)*, Neuherberg/München (1998), p. 79.

[450] G. Lörinci, G. Matuschek, J. Fekete, I. Gebefügi and A. Kettrup, *Thermochim. Acta* **263**, 73 (1995).

[451] J. Mullens, B. Carleer, G. Reggers, M. Ruysen, J. Yperman and L.C. van Poucke, *Bull. Soc. Chim. Belg.* **101**, 267 (1992).

[452] S.A. Hedrick and S.S.C. Chuang, *Thermochim. Acta* **315**, 159–68 (1998).

[453] K. Liu, E. Jakab, W. Zmierczak, J. Shabtai and H.L.C. Meuzelaar, *Prepr. Pap. Am. Chem. Soc., Div. Fuel Chem.* **39** (2), 576–80 (1994).

[454] A. Kettrup, G. Matuschek, J. Cyrys and K.-H. Ohrbach, *Makromol. Chem., Macromol. Symp.* **74**, 219–24 (1993).

[455] A. Kettrup, G. Matuschek and D. Lenoir, *Proceedings 11th Intl. Symposium on Analytical and Applied Pyrolysis (Pyrolysis '94)*, Nagoya (1994), Paper L21.

[456] J.M. McGuire, *Proceedings SPE ANTEC '97*, Toronto (1997), pp. 2296–9.

[457] E.L. Charsley and A.C.F. Kamp, in *Thermal Analysis* (H.G. Wiedemann, ed.), Birkhauser Verlag, Basel (1972), Vol. 1, p. 499.

[458] P.J. Haines and G.A. Skinner, *Thermochim. Acta* **134**, 201 (1988).

[459] L. Kofler, A. Kofler and M. Brandstätter, *Thermo-Mikromethoden,* Verlag Wagner, Innsbruck (1954).

[460] W. McCrone, *Fusion Methods in Chemical Microscopy*, Interscience, New York, NY (1957).

[461] E.L. Charsley, A.C.F. Kamp and J.A. Rumsey, in *Thermal Analysis* (H.G. Wiedemann, ed.), Birkhauser Verlag, Basel (1980), p. 285.

[462] H.G. Wiedemann and G. Bayer, *Thermochim. Acta* **92**, 371 (1985).

[463] A. Kagemoto, Y. Okado, H. Iree, M. Oka and Y. Baba, *Thermochim. Acta* **176**, 1 (1991).

[464] A.N. Matzakos and K. Zygourakis, *Rev. Sci. Instrum.* **64** (6), 1541–8 (1993).

[465] A. Hammiche, H.M. Pollock, M. Reading, M. Claybourn, P.H. Turner and K. Jewkes, *Appl. Spectrosc.* **53**, 810–5 (1999).

[466] L. Bozec, A. Hammiche, H.M. Pollock, M. Conroy, J.M. Chalmers, N.J. Everall and L. Turin, *J. Appl. Phys.* **90** (10), 5159–65 (2001).

[467] M.S. Mitchell, *Proceedings SPE ANTEC '97*, Toronto (1997), pp. 862–5.

[468] X.P.V. Maldague, *Non Destructive Evaluation of Materials by Infrared Thermography*, Springer-Verlag, Berlin (1995).

[468a] S.R. Williams, *Petro Industry News* **6** (3), 24–25 (2005).

[469] M. Reading, D.J. Hourston, M. Song, H. Pollock and A. Hammiche, *Proceedings ACS Div. Polym. Mat.: Sci. & Engng.* **78**, 122 (1998).

[470] H.M. Pollock, A. Hammiche, M. Song, D.J. Hourston and M. Reading, *J. Adhes.* **67** (1–4), 217–34 (1998).

[471] D.M. Price, M. Reading and T.J. Lever, *J. Therm. Anal. Calorim.* **56**, 673–9 (1999).

[472] C.C. Williams and H.K. Wickramasinghe, *Appl. Phys. Lett.* **49**, 1587–9 (1986).

[473] A. Hammiche, H.M. Pollock, M. Song and D.J. Hourston, *Measurement Sci. Technol.* **7**, 142–50 (1996).

[474] E. Zur Mühlen, in *Developments in Polymer Analysis and Characterisation* (Rapra Technology Ltd., ed.), Rapra Technology Ltd., Shawbury (1999), Paper 2.

[475] R.M. Smith, D.J. Hourston, D.M. Price, M. Reading, A. Hammiche and H.M. Pollock, *Proceedings 6th Intl. Symposium on Hyphenated Techniques in Chromatography and Hyphenated Chromatographic Analyzers (HTC-6)*, Bruges (2000), P69.

[476] D.M. Price, M. Reading, R.M. Smith, H.M. Pollock and A. Hammiche, *J. Therm. Anal. Calorim.* **64**, 309–14 (2001).

[477] R.L. Blaine, C.G. Slough and D.M. Price, *Abstracts 12th Intl. Congress on Thermal Analysis and Calorimetry (ICTAC-12)*, Copenhagen (2000), p. 110.

[478] E. Meyer and H.-G. Braun, *Abstracts Advanced Methods of Polymer Characterization: New Developments and Applications in Industry*, Deutsche Bunsen-Gesellschaft f. Physikalische Chemie, Mainz (1999), P1.32.

[479] H.G. Wiedemann and S. Felder-Casagrande, in *Handbook of Thermal Analysis and Calorimetry* (M.E. Brown, ed.), Elsevier, Amsterdam (1998), Vol. 1, Chp. 10.

[480] D.B. Grandy, D.J. Hourston, D.M. Price, M. Reading, M. Song and P.A. Sykes, *Abstracts 12th Intl. Congress on Thermal Analysis and Calorimetry (ICTAC-12)*, Copenhagen (2000), p. 119.

[481] T.J. Lever and D.M. Price, *Am. Lab.* **30** (16), 15–8 (1998).

[482] W. Xie, B. Sisk, P. Zongming and W.-P. Pan, *Proceedings 28th NATAS Annu. Conf. Therm. Anal. Soc. (NATAS)*, Orlando, FL (2000), pp. 45–51.

[483] T. Lever, J. Leckenby, D. Price and M. Branch, *The TA Hotline (TA Instruments)* **1**, 6–7 (1998).

[484] T.I. Butler and S.E. Pirtle, *Proceedings Polym., Laminations Coat. Conf.*, Tappi Press, Vol. 2, 601–7 (1996).

[485] P. Hack, W. Xie, B. Sisk and W.-P. Pan, *Proceedings 29th Annu. Conf. North Am. Therm. Anal. Soc. (NATAS)*, St. Louis, MO (2001), pp. 580–5.

[486] I. Moon, R. Androsch, W. Chen and B. Wunderlich, *J. Therm. Anal. Calorim.* **59** (1–2), 187–203 (2000).

[487] J.M.B. Webber, in *Applications of Thermal Imaging* (S.G. Barnay, T.L. Williams and C.H. Jones, eds.), Adam Hilger, Bristol (1988).

[488] R.J. Fleming, *J. Therm. Anal.* **36**, 331–59 (1990).

[489] D.R. Vij, in *Luminescence of Solids* (D.R. Vij, ed.), Plenum Press, New York, NY (1998), pp. 271–308.

[490] S.W.S. McKeever, *Thermoluminescence of Solids*, Cambridge University Press, Cambridge (1985).

[491] R. Kunze and K. Brademann-Jock, *Thermolumineszenz in Polymeren Studie*, VDI-Technologiezentrum Physikalische Technologien, Düsseldorf (1995), 20 pp.

[492] S. Krüger, K. Brademann-Jock, H.-J. Raabe and U. Schulz, *Abstracts 2nd Intl. Conference on Polymer Modification, Degradation and Stabilisation (MoDeSt2)*, Budapest (2002).

[493] R. Kunze, K. Brademann-Jock and D. Neubert, *Proceedings 7th European Symposium on Thermal Analysis and Calorimetry (ESTAC-7)*, Balatonfüred (1998), p. 267.

[494] I. Ericsson and R.P. Lattimer, *J. Anal. Appl. Pyrol.* **14**, 219–21 (1989).

[495] W.J. Irwin, *Analytical Pyrolysis: A Comprehensive Guide*, M. Dekker, New York, NY (1982).

[496] R.L. Levy, *Chromatogr. Revs.* **8**, 48–89 (1966).

[497] S.A. Liebman and E.J. Levy (eds.), *Pyrolysis and GC in Polymer Analysis*, M. Dekker, New York, NY (1985).

[498] T.P. Wampler (ed.), *Applied Pyrolysis Handbook*, M. Dekker, New York, NY (1995).

[499] S.C. Moldoveanu, *Analytical Pyrolysis of Natural Organic Polymers*, Elsevier, Amsterdam (1998).

[500] R.S. Lehrle and C.S. Pattenden, *Polym. Degr. Stabil.* **63**, 321–40 (1999).

[501] D.O. Hummel, *Atlas der Polymer- und Kunststoffanalyse*, Hanser Verlag, Munich (1988), Vol. 2.
[502] M. Kuch, *Determination of Additives in Polymers by the VW/Shimadzu Library, Instruction Manual*, Shimadzu, Duisburg (1998).
[503] V.G. Berezkin, *CRC Crit. Rev. Anal. Chem.* **11**, 1–48 (1981).
[504] W. Simon and G. Giacobbo, *Chem. Ing. Tech.* **37**, 709 (1965).
[505] J.P. Schmid, P.P. Schmid and W. Simon, *Chromatographia* **9**, 597 (1976).
[506] E.J. Levy, in *Analytical Pyrolysis* (C.E.R. Jones and C.A. Cramers, eds.), Elsevier, Amsterdam (1977), p. 319.
[507] R.L. Levy, *Chromatogr. Rev.* **8**, 48 (1966).
[508] S. Tsuge and T. Takeuchi, *Anal. Chem.* **49**, 348 (1977).
[509] T. Gorecki and J. Poerschmann, *Anal. Chem.* **73** (9), 2012–7 (2001).
[510] S.A. Liebman and E.J. Levy, *J. Chromatogr. Sci.* **21**, 1 (1983).
[511] M.R. Grimbley, R.S. Lehrle, R.J. Williams and D.M. Bate, *Polym. Degr. Stabil.* **48**, 143 (1995).
[512] R.W. May, E.F. Pearson and D. Scothern, *Pyrolysis Gas Chromatography*, Chemical Society, London (1977).
[513] D.J. Freed and S.A. Liebman, *Chromatogr. Sci.* **29**, 15–51 (1985).
[514] T.P. Wampler, in *Applied Pyrolysis Handbook* (T.P. Wampler, ed.), M. Dekker, New York, NY (1995), pp. 31–55.
[515] O. Watanabe, K. Teraishi, S. Tsuge, H. Ohtani and K. Hashimoto, *J. High Resolut. Chromatogr.* **14**, 269–72 (1991).
[516] J.-A. Hu, *J. Chromatogr. Sci.* **19**, 634 (1981).
[517] T. Hammond and R.S. Lehrle, *Br. Polym. J.* **21** (1), 23 (1989).
[518] R.S. Lehrle, J.C. Robb and R. Suggate, *Eur. Polym. J.* **18**, 443–61 (1982).
[519] D.M. Bate and R.S. Lehrle, *Polym. Degr. Stabil.* **53**, 39 (1996).
[520] J.W. Washall and T.P. Wampler, *Spectroscopy* **6** (4), 38–42 (1990).
[521] I. Ericsson, *J. Chromatogr. Sci.* **16**, 340–3 (1978).
[522] H. Szymanski, *Nature (London)* **188**, 403–4 (1960).
[523] H. Giacobbo and W. Simon, *Pharm. Acta Helv.* **39**, 162–7 (1964).
[524] A. Venema and J. Veurlink, *J. Anal. Appl. Spectrosc.* **7** (3), 207–14 (1985).
[525] R.P.W. Scott, *Introduction to Analytical Gas Chromatography*, M. Dekker, New York, NY (1998).
[526] K. Qian, W.E. Killinger, M. Casey and G.R. Nicol, *Anal. Chem.* **68**, 1019–27 (1996).
[527] R. Greco, in *Plastics Additives. An A–Z Reference* (G. Pritchard, ed.), Chapman & Hall, London (1998), pp. 375–85.
[528] T. Yamada, K. Okumoto, H. Ohtani and S. Tsuge, *Rubber Chem. Technol.* **63**, 191–201 (1990).
[529] M.J. Matheson, T.P. Wampler and W.J. Simonsick, *J. Anal. Appl. Pyrol.* **29**, 129–36 (1994).
[530] W.J. Irwin, *J. Anal. Appl. Pyrol.* **1**, 3–25 (1979).
[531] E.M. Andersson and I. Ericsson, *J. Anal. Appl. Pyrol.* **2**, 97–107 (1980).
[532] T.P. Wampler and E.J. Levy, *J. Anal. Appl. Pyrol.* **12**, 75–82 (1987).
[533] R.S. Lehrle and A. Shortland, *Eur. Polym. J.* **29** (10), 1277 (1993).
[534] T. Szekely and M. Blazsó, in *Analytical Pyrolysis* (C.E.R. Jones and C.A. Cramers, eds.), Elsevier, Amsterdam (1977), p. 365.
[535] S. Tsuge, T. Nagaya, Y. Sugimura and T. Kobayashi, *Toyoda Kenkyu Hokoku* (33), 50 (1980).
[536] M. Seeger and E.M. Barrall, *J. Polym. Sci., Polym. Chem. Ed.* **13** (7), 1515 (1975).
[537] E.J. Levy and J.Q. Walker, *J. Chromatogr. Sci.* **22**, 49 (1982).
[538] J.Q. Walker, *J. Chromatogr. Sci.* **15**, 267 (1977).
[539] C.E.R. Jones and C.A. Cramers (eds.), *Analytical Pyrolysis*, Elsevier, Amsterdam (1977).
[540] G.M. Brauer, *J. Polym. Sci.* **8**, 3 (1965).
[541] W.J. Irwin, *J. Anal. Appl. Pyrol.* **1**, 89 (1979).
[542] T. Hammond and R.S. Lehrle, in *Pyrolysis–Gas Chromatography* (G. Allen and J.C. Bevington, eds.), Pergamon Press, Oxford (1989), p. 589.
[543] T. Hammond and R.S. Lehrle, in *Comprehensive Polymer Science. Vol. 1: Polymer Characterization* (C. Booth and C. Price, eds.), Pergamon Press, Oxford (1989), pp. 589–612.
[544] C. Smith, in *Handbook of Instrumental Techniques for Analytical Chemistry* (F.A. Settle, ed.), Prentice Hall, Upper Saddle River, NJ (1997), pp. 893–908.
[545] T.P. Wampler, in *Applied Pyrolysis Handbook* (T.P. Wampler, ed.), M. Dekker, New York, NY (1995), pp. 1–29.
[546] *J. Anal. Appl. Pyrol.* (M. Blazsó and K.J. Voorhees, eds.), Elsevier, Amsterdam.
[547] R.P. Lattimer (ed.), Special Issue on *Analytical Pyrolysis of Synthetic and Natural Polymeric Materials, J. Anal. Appl. Pyrol.* **64** (2) (2002).
[548] M.P. Stevens, *Characterization and Analysis of Polymers by Gas Chromatography*, M. Dekker, New York, NY (1969), pp. 64.
[549] G.M. Brauer, in *Thermal Characterization Techniques* (P.E. Slade and L.T. Jenkins, eds.), M. Dekker, New York, NY (1970), Chp. 2.
[550] G.M. Brauer, in *Techniques and Methods for Polymer Evaluation* (P.E. Slade and L.T. Jenkins, eds.), M. Dekker, New York, NY (1970), Vol. 2, p. 41.

[551] C.E.R. Jones and G.E.J. Reynolds, *Rep. Progr. Appl. Chem.* **54**, 518 (1969).
[552] J.M. del Brió León and L.M. Vincente, *Rev. Plast. Mod.* **21**, 1020 (1970).
[553] V.R. Alishoev, V.G. Berezkin and A.I. Malyshev, *Zh. Anal. Khim.* **27**, 1035 (1972).
[554] P.R. De Forrest, *J. Forensic Sci.* **19**, 113 (1974).
[555] J.G. Cobler and C.D. Chow, *Anal. Chem.* **49**, 159R (1977).
[556] V.G. Berezkin, V.R. Alishoev and I.B. Nemirovskaya, *Gas Chromatography of Polymers, J. Chromatogr. Libr.* **10**, 159 (1977).
[557] D.H. Ahlstrom, *Chromatogr. Sci.* **29**, 209–76 (1985).
[558] S. Tsuge and H. Ohtani, *Asahi Garasu Zaidan Josei Kenkyu Seika Hokoku*, 29 (1993).
[559] S. Tsuge, *J. Anal. Appl. Pyrol.* **32**, 1 (1995).
[560] V.G. Zaikin, R.G. Mardanov and N.A. Plate, *Vysokomol. Soedin., Ser. A–B* **35**, B1561–78 (1993).
[561] S.A. Liebman and T.P. Wampler, in *Pyrolysis and GC in Polymer Analysis* (S.A. Liebman and E.J. Levy, eds.), M. Dekker, New York, NY (1985), pp. 53–148.
[562] J.C.J. Bart and M. Kuch, *LC.GC Europe*, 818–28 (Nov. 2000).
[563] J.C.J. Bart, *J. Anal. Appl. Pyrol.* **58–59**, 3–28 (2001).
[564] M. Blazsó, *J. Anal. Appl. Pyrol.* **39** (1), 1–25 (1997).
[565] C.G. Williams, *Phil. Trans.* **150**, 241 (1860).
[566] J.W. Washall, in *Applied Pyrolysis Handbook* (T.P. Wampler, ed.), M. Dekker, New York, NY (1995), pp. 293–310.
[567] J.M. Challinor, in *Applied Pyrolysis Handbook* (T.P. Wampler, ed.), M. Dekker, New York, NY (1995), pp. 207–41.
[568] W.H.T. Davison, S. Slaney and A.L. Wragg, *Chem. Ind.*, 1356 (1954).
[569] A.T. James and A.J.P. Martin, *Analyst* **77**, 915–32 (1952).
[570] F.C.-Y. Wang, B.B. Gerhart and C.G. Smith, *Anal. Chem.* **67**, 3681–6 (1995).
[571] R.S. Lehrle and J.C. Robb, *Nature* **183**, 1671 (1959).
[572] E.A. Radell and H.C. Strutz, *Anal. Chem.* **31**, 1890–1 (1959).
[573] S.B. Martin, *J. Chromatogr.* **2**, 272–83 (1959).
[574] B. Kolb, G. Kemmner, K.H. Kaiser, E.W. Cieplinski and L.S. Ettre, *Z. Anal. Chem.* **209**, 302 (1965).
[575] S. Tsuge and H. Ohtani, *Applied Polymer Analysis and Characterization*, Verlag Chemie, Weinheim (1986), p. 217.
[576] E. Warwick and A. Flamberg, *Chromatogr. Forum* **1**, 44 (1986).
[577] D.J. McHugh, J.D. Saxby and J.W. Tardif, *Chem. Geol.* **17**, 243 (1976).
[578] J.M. Challinor, *Forens. Sci. Intl.* **21**, 269 (1983).
[579] F.C.-Y. Wang and A.D. Burleson, *J. Chromatogr.* **A833**, 111–9 (1999).
[580] J. van Schooten, E.W. Duck and R. Berkenbosch, *Polymer (London)* **2**, 357 (1961).
[581] J. van Schooten and S. Mostert, *Polymer (London)* **4**, 135 (1963).
[582] J. van Schooten and J.K. Evenhuis, *Polymer (London)* **6**, 343 (1965).
[583] J. van Schooten and J.K. Evenhuis, *Polymer (London)* **6**, 561 (1965).
[584] N. Oguri, S. Uchino and A. Onischi, *Abstracts 11th Intl. Symposium on Analytical and Applied Pyrolysis*, Nagoya (1994), L24.
[585] N. Dadvand, R.S. Lehrle, I.W. Parsons and M. Rollinson, *Polym. Degr. Stabil.* **66**, 247–55 (1999).
[586] R.S. Lehrle, M. Rollinson, N. Dadvand and I.W. Parsons, *Polym. Degr. Stabil.* **66**, 221–31 (1999).
[587] S. Tsuge and H. Ohtani, *Pyrolysis–Gas Chromatography of Polymers – Fundamentals and Data Compilations*, Techno System Co., Tokyo (1989); Chinese translation (by X. Jin and Y.F. Luo), Chinese Science and Technology Publ. Co., Beijing (1992).
[588] T.A. Gough and C.E.R. Jones, *Chromatographia* **8**, 696 (1975).
[589] W.D. Stewart, *J. AOAC* **59**, 35 (1976).
[590] N.B. Coupe, C.E.R. Jones and S.G. Perry, *J. Chromatogr.* **47**, 291 (1970).
[591] W. Windig, P.G. Kistemaker, J. Haverkamp and H.L.C. Meuzelaar, *J. Anal. Appl. Pyrol.* **1**, 39–52 (1979).
[592] S. Tsuge and H. Ohtani, in *Analytical Pyrolysis* (K.J. Voorhees, ed.), Butterworths, London (1984), p. 407.
[593] T. Hammond, R.S. Lehrle and A. Shortland, *Polym. Degr. Stabil.* **50**, 169 (1995).
[594] L. Michajlov, P. Zugenmaier and H.-J. Cantow, *Polymer (London)* **9**, 325 (1968).
[595] K.V. Alexeeva and L.S. Solomatina, *J. Chromatogr.* **364**, 105–12 (1986).
[596] R.S. Lehrle, *Anal. Proc. (London)* **20** (11), 574 (1983).
[597] L. Dean, S. Groves, R. Hancox, G. Lamb and R.S. Lehrle, *Polym. Degr. Stabil.* **25**, 143 (1989).
[598] D.M. Bate and R.S. Lehrle, *Polym. Degr. Stabil.* **64**, 75–9 (1999).
[599] D. Gross, *Z. Anal. Chem.* **253**, 40 (1971).
[600] K.V. Mardia, J.T. Kent and J.M. Bibby, *Multivariate Analysis*, Academic Press, London (1979).
[601] M. Hida, S. Okuyama, T. Mitsui, Y. Minami and Y. Fujimura, *Chromatographia* **35**, 643–8 (1993).

[602] T. Mitsui, M. Hida and Y. Fujimura, *J. Anal. Appl. Pyrol.* **32**, 205–12 (1995).

[603] D.L. Evans, J.L. Weaver, A.K. Mukherji and C.L. Beatty, *Anal. Chem.* **50**, 857 (1978).

[604] E.A. Ney and A.B. Heath, *J. Int. Rubber Ind.* **2**, 276 (1968).

[605] A. Krishen, *Anal. Chem.* **44**, 494 (1972).

[606] B. Groten, *Anal. Chem.* **36**, 1206 (1964).

[607] T. Takeuchi, S. Tsuge and T. Okumoto, *J. Chromatogr. Sci.* **6**, 542 (1968).

[608] H. Senoo, S. Tsuge and T. Takeuchi, *J. Chromatogr. Sci.* **9**, 315 (1971).

[609] J. Zulaica and G. Guiochon, *Anal. Chem.* **35**, 1724 (1963).

[610] G. Stoev and I. Mladenov, *Dokl. Bolg. Akad. Nauk* **32**, 1385 (1979).

[611] J.M. Challinor, *J. Anal. Appl. Pyrol.* **16**, 323–35 (1989).

[612] J.P. Dworzanski, L. Berwald, W.H. McClennen and H.L.C. Meuzelaar, *J. Anal. Appl. Pyrol.* **21**, 221–32 (1991).

[613] G. Allen and J.C. Bevington (eds.), *Pyrolysis–Gas Chromatography*, Pergamon Press, Oxford (1989).

[614] W.J. Irwin and. J.A. Slak, *Analyst (London)* **103**, 63 (1978).

[615] W.J. Irwin, *J. Anal. Appl. Pyrol.* **3**, 3 (1981).

[616] S. Tsuge, *Kobunshi* **28**, 481 (1979).

[617] C.J. Wolf, M.A. Grayson and D.L. Fanter, *Anal. Chem.* **52**, 348A (1980).

[618] F.C.-Y. Wang, *J. Chromatogr.* **A843** (1/2), 413–23 (1999).

[619] Special issue on TAHM, *J. Anal. Appl. Pyrol.* (in press).

[620] J.M. Challinor, *Abstracts 14th Intl. Symposium on Analytical and Applied Pyrolysis*, Sevilla (2000), p. 61.

[621] J.C.W. Chien, J. Capistran, F.E. Karasz, M. Schen and J.L. Fan, *Polym. Prepr. (ACS, Div. Polym. Chem.)* **23** (1), 76 (1982).

[622] N. Nakagawa and S. Tsuge, *Macromolecules* **18**, 2068 (1985).

[623] P. Perlstein and P. Orme, *J. Chromatogr.* **325**, 87–93 (1985).

[624] W. Freitag, *Z. Anal. Chem.* **316**, 495 (1983).

[625] F.C.-Y. Wang and W.C. Buzanowski, *J. Chromatogr.* **A891**, 313–24 (2000).

[626] F.C.-Y. Wang, *J. Chromatogr.* **A891**, 325–36 (2000).

[627] R. Milina, *J. Anal. Appl. Pyrol.* **3**, 179–83 (1981).

[628] M.A. Roberson and G. Patonay, *J. Chromatogr.* **505**, 375–84 (1990).

[629] S.J. Shu and M.H. Chou, *Proceedings 11th Intl. Symposium on Analytical and Applied Pyrolysis (Pyrolysis '94)*, Nagoya (1994), Paper P7.

[630] S. Hirayanagi, K. Kimura, M. Sato and K. Harata, *Nippon Gomu Kyokaishi* **55** (5), 302–8 (1982).

[631] R.A. Franich, H.W. Kroese and G. Lane, *Analyst* **120**, 1927–31 (1995).

[632] F.C.-Y. Wang, *J. Chromatogr.* **A883**, 199–210 (2000).

[633] S.T. Jones, *Analyst (London)* **109** (7), 823–8 (1984).

[634] T.P. Wampler and C.P. Zawodny, *Proceedings SPE ANTEC '98*, Atlanta, GA (1998), pp. 2211–4.

[635] J.M. Challinor, *J. Anal. Appl. Pyrol.* **18**, 233 (1991).

[636] J.M. Challinor, *J. Anal. Appl. Pyrol.* **20**, 15–24 (1991).

[637] S. Mao, S. Tsuge, H. Ohtani, S. Uchijima and A. Kiyokawa, *Polymer* **39**, 143–9 (1998).

[638] I.F. Lindgren, *Proceedings 29th Natl. SAMPE (Society for the Advancement of Material and Process Engineering) Symposium*, Azusa, CA (1984), p. 1395.

[639] A.A. Faroq, D. Price, G.J. Milnes and A.R. Horrocks, *Polym. Degr. Stabil.* **33** (2), 155–70 (1991).

[640] V. Dave and S.C. Israel, *Polym. Prepr.* **31** (1), 554–5 (1990).

[641] N. Oguri, *Abstracts 12th Intl. Symposium on Analytical and Applied Pyrolysis (Pyrolysis '96)*, Venice (1996), P20.

[642] F.C.-Y. Wang, *Anal. Chem.* **71**, 2037–45 (1999).

[643] D. Machalkova, *Plasty Kauc.* **25** (4), 107–11 (1988).

[644] J. Haslam, H.A. Willis and D.C.M. Squirrell, *Identification and Analysis of Plastics*, J. Wiley & Sons, Chichester (1983).

[645] T. Yano, H. Ohtani and S. Tsuge, *Tappi J.* **74** (2), 197–201 (1991).

[646] Y. Munakata and Y. Ozaki, *Kami Parupu Kenkyu Happyokai Koen Yoshishu* **61**, 16–9 (1994).

[647] T.D. Crockett, A.A. Webb, L.G. Borchardt and D.B. Easty, *J. Chromatogr.* **407**, 330 (1987).

[648] E.J. Levy and T.P. Wampler, *J. Forens. Sci.* **31** (1), 258 (1986).

[649] T.P. Wampler and E.J. Levy, *LC.GC* **4** (11), 1112–6 (1986).

[650] J. Zimmerman, D. Mooney and M.J. Kimmet, *J. Forensic Sci.* **31** (2), 489–93 (1986).

[651] B.B. Wheals, in *Analytical Pyrolysis* (C.E.R. Jones and C.A. Cramers, eds.), Elsevier, Amsterdam (1977), p. 89.

[652] N. Sonoda, J.P. Rioux and A.R. Duval, *Stud. Conserv.* **38**, 99–127 (1993).

[653] S. Cook and R. Lehrle, *Eur. Polym. J.* **29** (1), 1–8 (1993).

[654] E.E. Sotnikov and L.A. Bubchikova, *Zh. Anal. Khim.* **42** (10), 1896–900 (1987).

[655] P. Almen and I. Ericsson, *Langmuir* **11**, 108–10 (1995).

[656] S. Muguruma, S. Uchino, N. Oguri and J. Kiji, *LC.GC Intl.* (7), 432–6 (1999).
[657] P. Perlstein and P. Orme, *J. Chromatogr.* **325**, 87–93 (1985).
[658] S. Tsuge, H. Ohtani, H. Yokoi and M.A. Becker, *Proceedings 12th Intl. Symposium on Polymer Analysis and Characterization*, La Rochelle (1999), P56.
[659] B. Adler, K. Metzner, J. Seiffarth and J. Wille, *Plaste Kautsch.* **36** (1), 8 (1989).
[660] S. Tsuge and H. Ohtani, *Pyrolysis–Gas Chromatography of Polymers – Fundamentals and Data Compilations*, Techno System Co., Tokyo (1989).
[661] V.M. Ryabikova, A.N. Zigel', T.L. Ivanova and G.S. Popova, *Polym. Sci. USSR* **31**, 237 (1989).
[662] F. Mersch and R. Zimmer, *Kautsch. Gummi Kunstst.* **39**, 427–32 (1986).
[663] M. Phair and T.P. Wampler, *Rubber World* (2), 30–4 (1997).
[664] F. Endter, *Kautsch. Gummi* **8**, WT302 (1955).
[665] G.R. Cotten and L.J. Murphy, *Kautsch. Gummi Kunstst.* **41**, 54–8 (1988).
[666] G.L. Coulter and W.C. Thompson, in *Analytical Pyrolysis* (C.E.R. Jones and C.A. Cramers, eds.), Elsevier, Amsterdam (1977), p. 1.
[667] S. Tojo, K. Nishiguchi and Y. Ishiwata, *Chiba-ken Kogyo Shikenjo Kenkyu Hokoku* (1), 21–6 (1987).
[668] S. Affolter, *Kautsch. Gummi Kunstst.* **50** (3), 216–25 (1997).
[669] A.M. Shedrinsky and N.S. Baer, in *Applied Pyrolysis Handbook* (T.P. Wampler, ed.), M. Dekker, New York, NY (1995), pp. 125–55.
[670] D. Braun and E. Čanji, *Angew. Makromol. Chem.* **29/30**, 491 (1973).
[671] S.L. Madorsky and S. Straus, *I&EC* **5**, 848–52 (1948).
[672] L.A. Wall, *J. Res. Natl. Bur. Std.* **41**, 315–22 (1948).
[673] P. Bradt, V.H. Dibeler and F.L. Mohler, *J. Res. Natl. Bur. Stds.* **50**, 201–2 (1953).
[674] R. Luyk, J. Pureveen, J.M. Commandeur and J.J. Boon, *Makromol. Chem., Macromol. Symp.* **74**, 235–51 (1993).
[675] H.-R. Schulten, B. Plage and R.P. Lattimer, *Rubber Chem. Technol.* **62**, 698 (1989).
[676] A.S. Hilton and A.G. Altenau, *Rubber Chem. Technol.* **46**, 1035–43 (1973).
[677] H.L.C. Meuzelaar, M.A. Posthumus, P.G. Kistemaker and J. Kistemaker, *Anal. Chem.* **45**, 1546 (1973).
[678] J.B. Pausch, R.P. Lattimer and H.L.C. Meuzelaar, *Rubber Chem. Technol.* **56**, 1031–44 (1983).
[679] H.-R. Schulten, H.D. Beckey, A.J.H. Boerboom and H.L.C. Meuzelaar, *Anal. Chem.* **45**, 2358 (1973).
[680] M.A. Posthumus, N.M.M. Nibbering, A.J.H. Boerboom and H.-R. Schulten, *Biomed. Mass Spectrom.* **1**, 352 (1974).
[681] R.P. Lattimer, *J. Anal. Appl. Pyrol.* **26** (2), 65–92 (1993).
[682] R.P. Lattimer, *Rubber Chem. Technol.* **68** (5), 783–93 (1995).
[683] R. Saferstein and J.J. Manura, *J. Forensic Sci.* **22**, 748–56 (1977).
[684] S. Shah, V. Dave and S.C. Israel, *ACS Symp. Ser.* **599**, 536 (1995).
[685] A.C. Tas, J. de Waart, J. Bouwman, M.C. ten Noever de Brauw and J. v.d. Greef, *J. Anal. Appl. Pyrol.* **11**, 329–40 (1987).
[686] R.J. Helleur and R. Guevremont, *J. Anal. Appl. Pyrol.* **15**, 85–95 (1989).
[687] A.C. Tas and J. van der Greef, *Trends Anal. Chem.* **12**, 60–6 (1993).
[688] A.P. Snyder, *Trends Anal. Chem.* **12**, 296–303 (1993).
[689] H.-R. Schulten, B. Plage, H. Ohtani and S. Tsuge, *Angew. Makromol. Chem.* **155**, 1 (1987).
[690] D.R. Schüddemage and D.O. Hummel, *Adv. Mass Spectrom.* **4**, 857–66 (1968).
[691] D.O. Hummel, in *Analytical Pyrolysis* (C.E.R. Jones and C.A. Cramers, eds.), Elsevier, Amsterdam (1977), p. 117.
[692] H.-R. Schulten and H. Wilcken, in *Polymeric Materials Encyclopedia* (J.C. Salomone, ed.), CRC Press, Boca Raton, FL (1996), pp. 7474–85.
[693] R.P. Lattimer, *J. Anal. Appl. Pyrol.* **31**, 203–25 (1995).
[694] R.P. Lattimer, *Abstracts 11th Intl. Symposium on Analytical and Applied Pyrolysis (Pyrolysis '94)*, Nagoya (1994).
[695] R.P. Lattimer, R.E. Harris and H.-R. Schulten, *Rubber Chem. Technol.* **58**, 577–621 (1985).
[696] H.-R. Schulten and H. Gundermann, *Neue Verpackung* **39**, 46–51 (1986).
[697] H.-R. Schulten, N. Simmleit and R. Müller, *Anal. Chem.* **59**, 2903–8 (1987).
[698] N. Simmleit and H.-R. Schulten, *Fresenius Z. Anal. Chem.* **324**, 9–12 (1986).
[699] D. Hummel, G. Czybulka and H.-J. Düssel, *Makromol. Chem., Rapid Commun.* **3**, 335–41 (1982).
[700] D.O. Hummel, H.-J. Düssel and K. Rübenacker, *Makromol. Chem.* **145**, 267–87 (1971).
[701] D.O. Hummel, F.H.J. Sadowski, D. Widdershoven and H.-J. Düssel, *Quad. Rec. Sci.* **84**, 23–41 (1973).
[702] H. Rosen, H.-J. Düssel and D.O. Hummel, *J. Appl. Polym Sci.: Appl. Polym. Symp.* **35**, 193–206 (1979).
[703] D.O. Hummel, I. Holland-Moritz, J. Kolb and H.-J. Düssel, *Acta Polym.* **32** (9), 512–23 (1981).

[704] G. Czybulka, H. Dunker, H.-J. Düssel, H. Logemann and D.O. Hummel, *Angew. Makromol. Chem.* **100**, 1–21 (1981).

[705] W. Genuit and J.J. Boon, *J. Anal. Appl. Pyrol.* **8**, 25 (1985).

[706] D.L. Zoller, S.T. Sum, M.V. Johnston, G.R. Hatfield and K. Qian, *Anal. Chem.* **71**, 866–72 (1999).

[707] D.L. Zoller and M.V. Johnston, *Anal. Chem.* **69**, 3791-5 (1997).

[708] J.J. Boon, *Intl. J. Mass Spectrom. Ion Processes* **118/119**, 755–87 (1992).

[709] T.A. Milne and M.N. Soltys, *J. Anal. Appl. Pyrol.* **5**, 93–110; 111–31 (1983).

[710] R.J. Evans, T.A. Milne and M.N. Soltys, *J. Anal. Appl. Pyrol.* **9**, 207–37 (1986).

[711] J.J. Boon, R. Heeren, J. Pureveen and E. van der Heyden, *Proceedings 11th Intl. Symposium on Analytical and Applied Pyrolysis (Pyrolysis '94)*, Nagoya (1994), Paper L6.

[712] R.M.A. Heeren and J.J. Boon, *Intl. J. Mass Spectrom. Ion Process.* **157/158**, 391–403 (1996).

[713] D. Price, G.J. Milnes, C. Lukas and T.R. Hull, *Intl. J. Mass Spectrom. Ion Process.* **60**, 225 (1984).

[714] S. Foti, A. Liquori, P. Maravigna and G. Montaudo, *Anal. Chem.* **54**, 674 (1982).

[715] R.P. Lattimer, H. Muenster and H. Budzikiewicz, *J. Anal. Appl. Pyrol.* **17**, 237 (1990).

[716] S.E. van Bramer and M.V. Johnston, *Anal. Chem.* **62**, 2643 (1990).

[717] S.E. van Bramer and M.V. Johnston, *Org. Mass Spectrom.* **27**, 949–54 (1992).

[718] S.E. van Bramer, P.L. Ross and M.V. Johnston, *J. Am. Soc. Mass Spectrom.* **4**, 65–72 (1993).

[719] P. Dworzanski and H.L.C. Meuzelaar, in *Encyclopedia of Spectroscopy and Spectrometry* (J.C. Lindon, ed.), Academic Press, San Diego, CA (2000), pp. 1906–19.

[720] S.D. Brown and A.M. Harper, in *Computer-Enhanced Analytical Spectroscopy* (C.L. Wilkins, ed.), Plenum Press, New York, NY (1993), Vol. 4, pp. 135–63.

[721] H.L.C. Meuzelaar, J. Haverkamp and F.D. Hilleman, *Pyrolysis Mass Spectrometry of Recent and Fossil Biomaterials*, Elsevier, Amsterdam (1982).

[722] J.G. Moncur and W.G. Bradshaw, *J. High Resolut. Chromatogr., Chromatogr. Commun.* **6**, 595–604 (1983).

[723] A.P. Snyder, G.A. Eiceman and W. Windig, *J. Anal. Appl. Pyrol.* **13**, 243 (1988).

[724] T.P. Wampler and C. Zawodny, *Abstracts 11th Intl. Symposium on Analytical and Applied Pyrolysis (Pyrolysis '98)*, Neuherberg/München (1998), p. 45.

[725] C.G. De Koster and J.J. Boon, *Qualitative and Quantitative Analyses of Fire Retardants in Complex Polymer Matrices by Direct Temperature Resolved Mass Spectrometry* FOM Institute of Atomic and Molecular Physics Int. Rept., Amsterdam (1995).

[725a] A.J.M. Boerboom, in *Mass Spectrometry of Large Molecules* (S. Facchetti, ed.), Elsevier Science Publishers, Amsterdam (1983), pp. 265–81.

[726] J. Maddock and T.W. Otley, in *Applied Pyrolysis Handbook* (T.P. Wampler, ed.), M. Dekker, New York, NY (1995), pp. 57–76.

[727] G. Montaudo and C. Puglisi, in *Developments in Polymer Degradation* (N. Grassie, ed.), Applied Science Publishers, London (1987), p. 35.

[728] J.B. Pausch, R.P. Lattimer and H.L.C. Meuzelaar, *Rubber Chem. Technol.* **56**, 1031–44 (1983).

[729] H.-R. Schulten and B. Plage, in *Analytiker-Taschenbuch* (R. Borsdorf, W. Fresenius, H. Günzler, W. Huber, H. Kelker, I. Lüderwald, G. Tölg and H. Wisser, eds.), Springer-Verlag, Berlin (1990), Vol. 9, p. 225.

[730] C.J. Curry, *J. Anal. Appl. Pyrol.* **11**, 213–25 (1987).

[731] S.R. Larter and A.G. Douglas, *J. Anal. Appl. Pyrol.* **4**, 1–19 (1982).

[732] D.O. Hummel, H.J. Düssel, G. Czybulka, N. Wenzel and G. Holl, *Spectrochim. Acta* **41A** (1–2), 279–90 (1985).

[733] R.P. Lattimer, *Abstracts 12th Intl. Symposium on Analytical and Applied Pyrolysis (Pyrolysis '96)*, Venice (1996), P2.

[734] H. Wilcken, *Ph.D. Thesis*, Technical University of Braunschweig, Braunschweig (1996).

[735] J.L. Savoca, R.P. Lattimer, J.M. Richards, W. Windig and H.L.C. Meuzelaar, *J. Anal. Appl. Pyrol.* **9**, 19 (1985).

[736] C. Moore, *Spectroscopy (Ottawa)* **7** (1–3), 1–10 (1989).

[737] H. Auler, *Rubber Chem. Technol.* **37**, 950 (1964).

[738] H. Auler, *Gummi Asbest Kunstst.* **14**, 1024, 1081 (1960).

[739] J.G. Kreiner, *Rubber Chem. Technol.* **44**, 381 (1971).

[740] K.T. Potts, E.G. Brügel, J.J. D'Amico and E. Morita, *Rubber Chem. Technol.* **45**, 160 (1972).

[741] Yu.G. Chikishev, N.A. Klyuev and G.A. Vakhtberg, *Khim. Geterosikl. Soedin.* **10**, 1330 (1975).

[742] J.O. Madsen, S.-O. Lawesson, A.M. Duffield and C. Djerassi, *J. Org. Chem.* **32**, 2054 (1967).

[743] Yu.G. Chikishev, N.A. Klyuev, G.A. Vakhtberg, V.P. Suboch and V.I. Vil'chinskaya, *Zh. Org. Khim.* **12**, 2222 (1976).

[744] R.P. Lattimer, H. Muenster and H. Budzikiewicz, *Rubber Chem. Technol.* **63**, 298 (1990).

[745] R.P. Lattimer and R.E. Harris, *Rubber Chem. Technol.* **62**, 548–67 (1989).

[746] J.M. Richards, W.H. McClennen, H.L.C. Meuzelaar, J.P. Shockcor and R.P. Lattimer, *J. Appl. Polym. Sci.* **34**, 1967–75 (1987).
[747] H.-R. Schulten, B. Plage and H. Gundermann, *Neue Verpackung* **41**, 114–21 (1988).
[748] J.C.-A. Hu, *Anal. Chem.* **49**, 537–40 (1977).
[749] H.-R. Schulten, *Tabakforschung International* **13**, 219–27 (1986).
[750] X. Zhu, *J. Fire Sci.* **14** (6), 443–65 (1996).
[751] M. Koel, M. Kaljurand, C.H. Lochmüller and M. Moebus, *First Intl. Chemom. InterNet Conf. (Electronic Conference)* (1994).
[752] M.H.P.M. van Lieshout, H.G. Janssen, C.A. Cramers, M.J.J. Hetem and H.J.P. Schalk, *J. High Resolut. Chromatogr.* **19**, 193–9 (1996).
[753] O. Watanabe, K. Teraishi, S. Tsuge, H. Ohtani and K. Hashimoto, *J. High Resolut. Chromatogr.* **14**, 269–72 (1991).
[754] R. Luyk, H.A.J. Govers, G.B. Eykel and J.J. Boon, *J. Anal. Appl. Pyrol.* **20**, 303–19 (1991).
[755] S. Shah, V. Dave and S. Israel, *Polym. Mat. Sci. Eng.* **71**, 231–2 (1994).
[756] R.M. Lum, *J. Polym. Sci.: Polym. Chem. Ed.* **15**, 489–97 (1977).
[757] A. Ballistreri, S. Foti, G. Montaudo, S. Pappalardo and E. Scamporrino, *J. Polym. Sci.: Polym. Chem. Ed.* **17**, 2469–75 (1979).
[758] D. Porter, E. Metcalfe and M.J.K. Thomas, *Fire & Mater.* **24**, 45–52 (2000).
[759] J.M. McGuire and C.C. Bryden, *J. Appl. Polym. Sci.* **35**, 537–48 (1988).
[760] C. Hughes, B.B. Wheals and M.J. Whitehouse, *J. Anal. Appl. Pyrol.* **103**, 482–91 (1978).
[761] W.A. Westall and A.J. Pidduck, *J. Anal. Appl. Pyrol.* **11**, 3–14 (1987).
[762] J. Vollmin, P. Kriemler, I. Omura, J. Seibl and W. Simon, *Microchem. J.* **11**, 73–86 (1966).
[763] S. Takahashi and S. Onishi, *Shimadzu Hyoron* **45** (1/2), 15–35 (1988).
[764] M. Geissler, *Kunststoffe* **87**, 194–6 (1997).
[765] F. Jungrichter, *Ph.D. Thesis*, University of Braunschweig (1997).
[766] M. Kuch, *Diplomarbeit*, University of Braunschweig (1995).
[767] J. Geyer-Lippmann, *Abstracts 2nd European Academy of Forensic Science Mtg.*, Cracow (2000), p. 183.
[768] D.E. Meier and O. Faix, in *Methods in Lignin Chemistry* (S.Y. Lin and C.W. Dence, eds.), Springer-Verlag, Berlin (1992), pp. 177–99.
[769] P. Bocchini, G.C. Calletti, S. Camarero and A.T. Martinez, *Abstracts 12th Intl. Symposium on Analytical and Applied Pyrolysis (Pyrolysis '96)*, Venice (1996), P68.
[770] K.M. Baker, G. Chantrey and D. Mauer, *Kautsch. Gummi Kunstst.* **33**, 175–81 (1980).
[771] M. Kleen, G. Lindblad and S. Backa, *J. Anal. Appl. Pyrol.* **25**, 209–27 (1993).
[772] M. Blazsó, *Macromol. Symp.* **135**, 237–45 (1998).
[773] B.A. Stankiecicz, P.F. van Bergen, M.B. Smith, J.F. Carter, D.E.G. Briggs and R.P. Evershed, *J. Anal. Appl. Pyrol.* **45** (2), 133–51 (1998).
[774] K. Vèkey, J. Tamàs, A. Somogyi, A. Bertazzo, C. Costa, G. Allegri, R. Seragli and P. Traldi, *Organic Mass Spectrom.* **27**, 1216–9 (1992).
[775] Anon., *Shimadzu News* **2**, 11 (2001).
[776] G.J.Q. van der Peyl, T.C.T. Linnartz, C.A.J.J. van Rossum and M. Zeelenberg, *J. Anal. Appl. Pyrol.* **19**, 279–83 (1991).
[777] M. Blazsó, *Rapid Commun. Mass Spectrom.* **5**, 507–11 (1991).
[778] T.P. Wampler, *Abstracts 14th Intl. Symposium on Analytical and Applied Pyrolysis*, Sevilla (2000), P22.
[779] H. Jönsson and I. Ericsson, *Abstracts 14th Intl. Symposium on Analytical and Applied Pyrolysis*, Sevilla (2000), P60.
[780] W.K. Way and C. Gloeckner, *Proceedings SPE ANTEC 2002*, San Francisco, CA (2002), pp. 2056–9.
[781] Zentralnorm PV 3935, *Kunststoffe und Polymere. Pyrolyse–Gaschromatographie/Massenspektrometrie*, Volkswagen AG, Wolfsburg (1997).
[782] L. Gruber and W. Hofman, *Anwendung der Pyrolyse–GC/MS zur Untersuchung von Kunststoffen und Kunststoffadditiven in Bedarfgegenständen*, Fraunhofer Institut f. Lebensmitteltechnologie u. Verpackung, München (n.d.).
[783] S.A. Liebmann, A.P. Snyder, D.J. Reutter, M.B. Wasserman and L.J. Schiff, *Govt. Rept. Announce* **90** (9), Abstr. No. 020,407 (1990).
[784] N. Lichtenstein and K. Quellmalz, *Fresenius Z. Anal. Chem.* **316**, 268–70 (1986).
[785] T. Meyer–Dulheuer, H. Pasch and M. Geissler, *Kautsch. Gummi Kunstst.* **53**, 574–81 (2000).
[786] T. Meyer-Dulheuer, *Proceedings 4. Würzburger Tage der instrumentellen Analytik in der Polymertechnik*, SKZ, Würzburg (1999), Paper M.
[787] N. Manabe, T. Toyoda and Y. Yokota, *Bunseki Kagaku* **48** (4), 449–56 (1999).
[788] M. Blazsó, *J. Anal. Appl. Pyrol.* **58–59**, 29–47 (2001).
[789] F.C.-Y. Wang, *J. Chromatogr.* **A886**, 225–35 (2000).
[790] E. Verdurmen, C. De Koster and J. Snijkers, *Techn. Rept. DSM Research*, Geleen (1996).
[791] B. Danzer, M. Riess, H. Thoma, O. Vierle and R. van Eldik, *Organohalogen Compd.* **31**, 108–13 (1997).
[792] F.T. Dettmer, H. Wichmann, J. de Boer and M. Bahadir, *Chemosphere* **39** (9), 1523–32 (1999).

[793] M. Riess, T. Ernst, R. Popp, Th. Schubert, H. Thoma, O. Vierle, G.W. Ehrenstein and R. van Eldik, *Organohal. Compd.* **35**, 443–6 (1998).

[794] M. Riess, H. Thoma, O. Vierle and R. van Eldik, *J. Anal. Appl. Pyrol.* **53**, 135–48 (2000).

[795] Cs. Csoma, Zs. Czégény and M. Blazsó, *Chem. Inzyn. Ekol.* **9** (2–3), 145–52 (2002).

[796] S. Sprenger and R. Utz, *Proceedings AddCon World '98*, London (1998), Paper 18.

[797] M. Kuch, in *Proceedings 4th Würzburger Tage der instrumentellen Analytik in der Polymertechnik*, Würzburg (1999), Paper K.

[798] A.A. Faroq, D. Price, G.J. Milnes and A.R. Horrocks, *Polym. Degr. Stabil.* **33** (2), 155–70 (1991).

[799] S. Zhang and X. Shao, *Fangzhi Xuebao* **11** (6), 244–8 (1990).

[800] S.M. Lomakin, S.V. Usachev, E.V. Koverzanova, L.V. Ruban, I.G. Kalinina and G.E. Zaikov, in *Proceedings Tenth Annu. BCC Conference on Flame Retardancy*, Stamford, CT (1999).

[801] J.R. Ebdon, B.J. Hunt and P. Joseph, *Abstracts First Intl. Conference on Polymer Modification, Degradation and Stabilisation (MoDeSt)*, Palermo (2000).

[802] F. Severini, R. Gallo, M. Pegoraro and G. Audisio, *Polymers Paint Colour J.* **177**, 394–401 (1987).

[803] A. Ballistreri, G. Montaudo, E. Scamporrino, C. Puglisi, D. Vitalini and S. Cucinella, *J. Polym. Sci.* **A26** (8), 2113–27 (1988).

[804] P. Carty, D. Price, J. Milnes, C. Radcliffe and S. White, *J. Fire Sci.* **17** (6), 483–93 (1999).

[805] S. Kodama, T. Tsunesada and K. Nishi, *Okayama-ken Kogyo Gijutsu Senta Hokoku* **22**, 8–10 (1996).

[806] S.W. Kim, G.S. Heo and G.H. Lee, *J. Korean Chem. Soc.* **41** (10), 524–34 (1997).

[807] M. Kuch, in *Proceedings 3rd Würzburger Tage der instrumentellen Analytik in der Polymertechnik*, Würzburg (1997).

[808] S. Köbisch, *Diplomarbeit*, Techn. Fachhochschule Berlin (1997).

[809] W. Herrmann, in *Fortbildungsseminar "Moderne Prüfmethoden – von Rohstoffen über Mischungen bis zum Produkt"*, DIK, Hannover (2000).

[810] T. Learner, in *Resins, Ancient and Modern*, SSCR Conference Preprints (J.H. Townsend and M.M. Wright, eds.), Aberdeen (1995), pp. 76–84.

[811] G. Languri, *Ph.D. Thesis*, University of Amsterdam (2004).

[812] G. Languri, J. Van der Horst and J.J. Boon, *Abstracts 14th Intl. Symposium on Analytical and Applied Pyrolysis*, Sevilla (2000), P43.

[813] A. Plum and A. Rehorek, unpubl. results (2000).

[814] H. Ohtani and S. Tsuge, *Sen'i Gakkaishi* **49** (4), P125–9 (1993).

[815] J. Odermatt, D. Meier, D. Mauler and K. Leicht, *Papier (Heidelberg)* **52** (10), 598–602 (1998).

[816] B. Schwarzinger, I. Tanczos and H. Schmidt, *Abstracts 14th Intl. Symposium on Analytical and Applied Pyrolysis*, Sevilla (2000), P70.

[817] J. Odermatt, T. Runge, D. Meier and D. Mauler, *Papier* **53** (10A), V25–8 (1999).

[818] M. Ezrin and G. Lavigne, *Proceedings SPE ANTEC 2002*, San Francisco, CA (2002), pp. 2046–50.

[819] M. Kuch, *Techn. Rept. DSM Research*, Geleen (1999).

[820] T. Meyer-Dulheuer, *Shimadzu Seminar*, Duisburg (1998).

[820a] I. Ericsson, *personal communication* (1996).

[821] H. Wilcken and H.-R. Schulten, *Fresenius J. Anal. Chem.* **355**, 157–63 (1996).

[822] G.J. Mol, R.J. Gritter and G.E. Adams, *Proceedings Applied Polymer Spectroscopy* (E.G. Brame, ed.), Academic Press, New York, NY (1978), pp. 257–77.

[823] P. Bradna and J. Zima, *J. Anal. Appl. Pyrol.* **24**, 75–86 (1992).

[824] H.-L. Hardell, *J. Anal. Appl. Pyrol.* **27**, 73–85 (1993).

[825] D. Hinzmann and M. Geissler, *Shimadzu Application Note GC-MS* 46 (1999).

[826] M. Blazsó, *Abstracts 12th Intl. Symposium on Analytical and Applied Pyrolysis (Pyrolysis '96)*, Venice (1996), L15.

[826a] M. Blazsó, Zs. Czégény and Cs. Csoma, *J. Anal. Appl. Pyrol.* **64**, 249–61 (2002).

[827] T.O. Munson, in *Applied Pyrolysis Handbook* (T.P. Wampler, ed.), M. Dekker, New York, NY (1995), pp. 157–205.

[828] S. Tsuge and H. Ohtani, in *Mass Spectrometry of Polymers* (G. Montaudo and R.P. Lattimer, eds.), CRC Press, Boca Raton, FL (2002), pp. 113–47.

[829] R.B. Barnes, R.C. Gore, R.W. Stafford and V.Z. Williams, *Anal. Chem.* **20**, 402–10 (1948).

[830] D.T. Harms, *Anal. Chem.* **25**, 1140 (1953).

[831] B.G. Achhammer, M.J. Rainey and F.W. Reinhard, *J. Res. Natl. Bur. Stds.* **47**, 116 (1951).

[832] S.A. Liebman, D.H. Ahlstrom and P.R. Griffiths, *Appl. Spectrosc.* **30**, 355 (1976).

[833] J.W. Washall and T.P. Wampler, *J. Appl. Polym. Sci.: Appl. Polym. Symp.* **45**, 377–92 (1990).

[834] J.W. Washall and T.P. Wampler, *Spectroscopy* **6** (4), 38–42 (1991).

[835] R.G. Davidson, *J. Anal. Appl. Pyrol.* **16** (2), 143–52 (1989).

[836] D. Jovanovic, *Polimeri (Zagreb)* **4** (9–10), 261–3 (1983).

[837] F. Luigart, *Kunststoffe* **70** (2), 66–7 (1980).

[838] W. Wang, D. Wang, C. Chen and Y. Liu, *J. Phys., Colloq.* **C6**, 527–32 (1983).

[839] W.L. Truett, *Am. Lab.* **9** (6), 33–8 (1977).

[840] G. Melber and H.A. Szymanski, *Progr. Infrared Spectrosc.* **3**, 25–38 (1967).

[841] C. Smith, in *Handbook of Instrumental Techniques in Analytical Chemistry* (F.A. Settle, ed.), Prentice Hall, Upper Saddle River, NJ (1997), pp. 893–908.

[842] D. Brück, *Kautsch. Gummi Kunstst.* **39** (12), 1165–74 (1986).

[843] T. Morimoto and S. Enomoto, *Natl. Techn. Rept. (Matsushita Electr. Ind. Co., Osaka)* **23** (1), 160–7 (1977).

[844] R.K. Lowry and K. Hanley, *Proceedings 24th Intl. Symposium for Testing and Failure Analysis (ISTFA '98)*, Dallas, TX (1998), pp. 399–404.

[845] A. Murase, K. Sumi, T. Kato, N. Katayama and N. Sato, *Jidosha Gijutsu* **52** (5), 25–9 (1998).

[846] M. Ishiguro, C. Matsuoka and A. Nakatsu, *Kanzei Chuo Bunsekishoho* **27**, 179–201 (1987).

[847] R.G. Davidson, *Mikrochim. Acta* **1**, 301–4 (1988).

[848] R.W. May and J. Porter, *J. Forensic Sci. Soc.* **15** (2), 137–46 (1975).

[849] W. Noble, B.B. Wheals and M.M. Whitehouse, *Forensic Sci.* **3** (2), 163–74 (1974).

[850] U. Fuchslueger, H.-J. Grether and M. Grasserbauer, *Fresenius J. Anal. Chem.* **349**, 283–8 (1994).

[851] L.R. Werner, *Adv. Tech. Mater. Invest. Fabr., Soc. Aerosp. Mater. Process Eng., Nat. Symp. Exhibit.*, 14th (1968), Western Period. Co., North Hollywood, CA (1968), I-3–5.

[852] J. Peschel, H. Vocke, A. Dostmann, F. Jenke, K.D. Ahlers, W. Kraus and R. Grabert (to VEB Chemiefaserwerk Guben), *Ger. (East) Pat.* DD 226078 (June 27, 1984).

[853] D. Price, G.J. Milnes, C. Lukas and A.M. Phillips, *J. Anal. Appl. Pyrol.* **11**, 499–510 (1987).

[854] D.O. Hummel, S. Göttgens, U. Neuhoff and H.-J. Düssel, *J. Anal. Appl. Pyrol.* **33**, 195–212 (1995).

[855] R. White, *Chromatography/Fourier Transform Infrared Spectroscopy and its Applications*, M. Dekker, New York, NY (1991).

[856] W. Herres and U. Fölster, *Farbe Lack* **91**, 6 (1985).

[857] U. Fölster and W. Herres, *Farbe Lack* **92**, 13 (1986).

[858] L.V. Azarraga and A.C. McCall, *IR Fourier Transform Spectrometry of GC Effluents*, EPA 660/2-73-034 (1974).

[859] S. Tamura, *Nippon Gomu Kyokaishi* **50** (1), 90–102 (1977).

[860] R. Oguchi, A. Shimizu, S. Yamashita, K. Yamaguchi and P. Wylie, *J. High Resolut. Chromatogr.* **14**, 412–16 (1991).

[861] C. Reddmann and H. Kurz, *LP Spec.: Chromatogr., Spectrosk.*, 166–173 (1986); *LaborPraxis* **9** (5), 454–61 (1986).

[862] D. Weber, G. Fülöp and D.O. Hummel, *Makromol. Chem., Macromol. Symp.* **52**, 151–60 (1991).

[863] E. Nishio, N. Ikuta and H. Okabayashi, *J. Anal. Appl. Pyrol.* **18**, 261–8 (1991).

[864] E. Nishio, N. Ikuta, H. Okabayashi and R.W. Hannah, *Appl. Spectrosc.* **44**, 2439 (1990).

[865] W.P. Duncan, *Am. Lab.* **20** (8), 40–6 (1988).

[866] Q. Ou, G. Wang, K. Zeng and W. Yu, *Spectrochim. Acta* **38B** (1/2), 419–25 (1983).

[867] A.B. Ross, S. Junyapoon, K.D. Bartle, J.M. Jones and A. Williams, *J. Anal. Appl. Pyrol.* **58–59**, 371–85 (2001).

[868] F.C.-Y. Wang, *J. Chromatogr.* **A743** (1), 101–8 (1996).

[869] H. Sato, K. Kondo, S. Tsuge, H. Ohtani and N. Sato, *Polym. Degr. Stabil.* **62**, 41–8 (1998).

[870] E.M. Andersson and I. Ericsson, *J. Anal. Appl. Pyrol.* **1**, 27–38 (1979).

[871] Frontier Lab., Koriyama-shi. *Application Note FL-PY 19308E* (1996).

[872] C. Watanabe, K. Sato, H. Ohtani and S. Tsuge, *Proceedings 12th Intl. Symposium on Analytical and Applied Pyrolysis (Pyrolysis '96)*, Venice (1996).

[873] M. Ezrin and G. Lavigne, *Proceedings SPE ANTEC '97*, Toronto (1997), pp. 2305–9.

[874] M.P.M. van Lieshout, H.-G. Janssen, C.A. Cramers and G.A. van den Bosch, *J. Chromatogr.* **A764**, 73–84 (1997).

[875] R. Hoogerbrugge, S.J. Willig and P.G. Kistemaker, *Anal. Chem.* **55**, 1710–2 (1983).

[876] W. Windig, J. Haverkamp and P.G. Kistemaker, *Anal. Chem.* **55**, 81–8 (1983).

[877] G.A. v.d. Doelen, K.J. v.d. Berg and J.J. Boon, *Stud. Conserv.* **43**, 249–64 (1998).

[878] R.P. Lattimer, R.E. Harris, C.K. Rhee and H.-R. Schulten, *Rubber Chem. Technol.* **61**, 639 (1988).

[879] B. Plage and H.-R. Schulten, *Macromolecules* **21**, 2018 (1988).

[880] W. Windig, E. Jacab, J.M. Richards and H.L.C. Meuzelaar, *Anal. Chem.* **59**, 317 (1987).

[881] W. Windig, T. Chakravarty, J.M. Richards and H.L.C. Meuzelaar, *Anal. Chim. Acta* **191**, 205 (1986).

[882] T.H. Risby, J.A. Yergey and J.J. Scocca, *Anal. Chem.* **54**, 2228 (1982).

[883] S.A.J. Mundy, *J. Anal. Appl. Pyrol.* **25**, 317–24 (1993).

[884] *Pyrola Application Note*, Pyrola, Lund (1996).

[885] R.A. Pesce-Rodriguez, N.F. Fell, J.M. Widder and R. Greer, *J. Energ. Mater.* **17** (2/3), 197–214 (1999).

[886] Y.J. Kwon, S.H. Nam, M.G. Kim and C. Watanabe, *Abstracts 14th Intl. Symposium on Analytical and Applied Pyrolysis*, Sevilla (2000), P8.

[887] C. Watanabe, K. Sato, A. Hosaka, H. Ohtani and S. Tsuge, *Abstracts 14th Intl. Symposium on Analytical and Applied Pyrolysis*, Sevilla (2000), P23.

[888] M. Blazsó, *Rapid Commun. Mass Spectrom.* **12**, 1–4 (1998).

[888a] J.M. Richards, W.H. McClennen and H.L.C. Meuzelaar, *J. Appl. Polym. Sci.* **40** (1–2), 1–12 (1990).

[889] H. Wilcken and H.-R. Schulten, *Anal. Chim. Acta* **336**, 201–8 (1996).

[890] J.J. Boon, J. Pureveen and D. Rainford, *Proceedings 11th Intl. Symposium on Analytical and Applied Pyrolysis (Pyrolysis '94)*, Nagoya (1994), P73.

[891] D.O. Hummel, U. Neuhoff, A. Bretz and H.-J. Düssel, *Makromol. Chem., Macromol. Symp.* **72**, 81–106 (1993).

[892] D.O. Hummel, U. Neuhoff, A. Bretz and H.-J. Düssel, *Makromol. Chem.* **194** (6), 1545–59 (1993).

[893] S. Foti, P. Maravigna and G. Montaudo, *Polym. Degr. Stabil.* **4**, 287–92 (1982).

[894] R.G. Davidson, *J. Appl. Polym. Sci.* **34** (4), 1631–44 (1987).

[895] L. Sarvaranta, M. Elomaa and E. Järvelä, *Fire & Mater.* **17**, 225–30 (1993).

[896] M. Kaljurand and E. Küllik, *Computerised Multiple Input Chromatography*, Ellis Horwood, Chichester (1989), p. 139.

[897] M. Koel and M. Kaljurand, *Crit. Rev. Anal. Chem.* **26**, 149 (1996).

[898] M. Koel, M. Kaljurand, C.H. Lochmüller and M. Moebus, *Chemom. Intell. Lab. Syst.* **30** (1), 173–8 (1995).

[899] A. Ebber, K. Ivask and M. Koel, *Eesti Tead. Akad. Toim., Keem.* **44** (1), 1–6 (1995).

[900] M. Elomaa, C.H. Lochmüller, M. Kudrjashova and M. Kaljurand, *Thermochim. Acta* **362**, 137–44 (2000).

[901] M. Elomaa, C.H. Lochmüller, M. Kaljurand and M. Koel, *Chemom. Intell. Lab. Syst.* **30** (1), 129–32 (1995).

[902] R.J. Albalek (ed.), *Polymer Devolatilization*, M. Dekker, New York, NY (1996).

[903] I.C. McNeill, *Eur. Polym. J.* **3**, 409 (1967).

[904] I.C. McNeill, *Eur. Polym. J.* **6**, 373 (1970).

[905] W.W. Wendlandt, *Thermochim. Acta* **9**, 95 (1974).

[906] F.T. Eggertsen, H.M. Joki and F.H. Stross, in *Thermal Analysis* (R.F. Schwenker and P.D. Garn, eds.), Academic Press, New York, NY (1969), Vol. 1, p. 341.

[907] D.G. Paul, D.J. Brindle and J.A. Wegener, *Pittsburgh Conf. Anal. Chem. Appl. Spectr.*, Cleveland, OH (1972).

[908] T.R.F. Fennell, G.J. Knight and W.W. Wright, in *Thermal Analysis* (H.G. Wiedemann, ed.), Birkhauser, Basel (1972), Vol. 3, p. 245.

[909] I.C. McNeill, *J. Polym. Sci.* **A1**, **4**, 2479–85 (1966).

[910] I.C. McNeill, in *Developments in Polymer Degradation* (N. Grassie, ed.), Applied Science Publishers, London (1977), Vol. 1, p. 43.

[911] T.H. Risby and A.L. Yergey, *Anal. Chem.* **50**, 326A (1978).

[912] A. Ballistreri, G. Montaudo and C.J. Puglisi, *J. Therm. Anal.* **29**, 237 (1984).

[913] J. Morelli, *J. Anal. Appl. Pyrol.* **18**, 1 (1990).

[914] J.G. Moncur, A.B. Campa and P.C. Pinoli, *J. High Resolut. Chromatogr., Chromatogr. Commun.* **5**, 322 (1982).

[915] M. Gábor and L. Pöppl, *J. Therm. Anal.* **11**, 231 (1977).

[916] H.G. Langer and R.S. Gohlke, *Anal. Chem.* **35**, 1301 (1963).

[917] H.L. Friedman, *J. Macromol. Chem.* **A1**, 57 (1967).

[918] G.P. Shulman, *J. Macromol. Chem.* **A1**, 107 (1967).

[919] G.P. Shulman and H.W. Lochte, *J. Macromol. Sci.* **A2**, 411 (1968).

[920] D. Price and. J.F.J. Todd (eds.), *Dynamic Mass Spectrometry*, Heyden & Sons, London (1978), Vol. 5.

[921] R.S. Gohlke, *Chem. Ind. (London)*, 946 (1963).

[922] S.B. Warrington, in *Thermal Analysis – Techniques and Applications* (E.L. Charsley and S.B. Warrington, eds.), The Royal Society of Chemistry, Cambridge (1992), pp. 84–107.

[923] P.S. Gill, *Du Pont Application Brief TA 43* (1974).

[924] R.L. Blaine, *Pittsburgh Conf. Anal. Chem. Appl. Spectr.*, Cleveland, OH (1974).

[925] E. Kiran and J.K. Gillham, in *Polymer Characterization by Thermal Methods of Analysis* (J. Chiu, ed.), M. Dekker, New York, NY (1974), p. 211.

[926] E. Schröder, *Pure Appl. Chem.* **36**, 233 (1973).

[927] T. Fazio, J.W. Howard and A. Sandoval, *J. Assoc. Off. Anal. Chem.* **51**, 17–21 (1968).

[928] S.S. Yushkevichute and Yu.A. Shlyapnikov, *Plast. Massy* **12**, 62–3 (1966).

[929] Yu.A. Shlyapnikov, S.G. Kiryushkin and A.P. Mar'in, *Antioxidative Stabilisation of Polymers*, Taylor & Francis, London (1996).

[930] V.G. Berezkin and A.S. Bochkov, *Quantitative Thin-Layer Chromatography: Instrumental Methods*, Nauka, Moscow (1980).

[931] S.S. Yushkevichute and Yu.A. Shlyapnikov, *Trudy AN Lit. SSR, ser. B* **3** (58), 153 (1969).

[932] K. Kurihara, F. Tanoue and T. Shoji, *Shikizai Kyokaishi* **72** (9), 546–51 (1999).

[933] M.D. Alcalá, F.J. Gotor, L.A. Pérez-Maqueda, C. Real, M.J. Dianez and J.M. Criado, *J. Therm. Anal. Calorim.* **56**, 1447–52 (1999).

[934] F. Rouquerol, J. Laureiro and J. Rouquerol, *Solid State Ionics* **63–65**, 363 (1992).

[935] K.G.H. Raemaekers and J.C.J. Bart, *Proceedings 7th European Symposium on Thermal Analysis and Calorimetry (ESTAC-7)*, Balatonfüred (1998), p. 391.

[936] D.C.M. Squirrell, *Analyst* **106**, 1042–56 (1981).

[937] C. Warner, J. Modderman, T. Fazio, M. Beroza, G. Schwartzman and K. Fominaya, *Food Additives Analytical Manual*, AOAC International, Arlington, VA (1993).

[938] I.C. McNeill, *Proceedings 6th Intl. Conference on Thermal Analysis* (H.G. Wiedeman, ed.), Birkhauser Verlag, Basel (1980), pp. 319–24.

[939] I.C. McNeill and L. Memetea, *Polym. Degr. Stabil.* **43** (1), 9–25 (1994).

[940] L. Ackerman and W.J. McGill, *J. S. Afr. Chem. Inst.* **26**, 82 (1973).

[941] W.J. McGill, in *Developments in Polymer Degradation* (N. Grassie, ed.), Applied Science Publishers, London (1984), Vol. 5, pp. 1–29.

[942] I.C. McNeill, L. Ackerman, S.N. Gupta, M. Zulfiqar and S. Zulfiqar, *J. Polym. Sci. Polym. Chem. Ed.* **15**, 2381 (1977).

[943] I.C. McNeill, *Anal. Proc. (London)* **18** (12), 540–2 (1981).

[944] X. Guo, B. Huang, T. Dyakonov, Y. Chen, L. Padron, T. Vickström, J. Kuhn, J. Hodkiewicz and W.T.K. Stevenson, *Appl. Spectrosc.* **53** (11), 1403–11 (1999).

[945] G. Camino and L. Costa, *Polym. Degr. Stabil.* **2**, 23–33 (1980).

[946] G. Camino, L. Costa and L. Trossarelli, *Polym. Degr. Stabil.* **4**, 133–44 (1982).

[947] L.J. Rigby, *Adv. Mass Spectrom.* **6**, 153–62 (1974).

[948] O. Chiantore and M. Guaita, *J. Appl. Polym. Sci.: Appl. Polym. Symp.* **48**, 431–40 (1991).

[949] I.C. McNeill, L.P. Razumovskii, V.M. Gol'dberg and G.E. Zaikov, *Polym. Degr. Stabil.* **45**, 47 (1994).

[950] I.C. McNeill and M. Bounekhel, *Polym. Degr. Stabil.* **34**, 187–204 (1991).

[951] I.C. McNeill and M.H. Mohammed, *Polym. Degr. Stabil.* **50** (3), 285–95 (1995).

[952] I.C. McNeill, *Proceedings Intl. Conference on Thermal Analysis (ICTA)*, R.F. Schwenker (ed.), Academic Press, New York, NY (1969), Vol. 1, pp. 417–33.

[953] E. Woolfenden, *J. Air Waste Managm. Assoc.* **47**, 20–36 (1997).

[954] M. Hakkarainen, A.-C. Albertsson and S. Karlsson, *J. Environm. Polym. Degrad.* **5** (2), 67–73 (1997).

[955] J.C.J. Bart, *Polym. Testing* **20**, 729–40 (2001).

[956] R.A. Wilks and S.G. Gilbert, *Food Technol.* **23**, 47 (1969).

[957] J. D. Culter, *J. Plast. Film Shtg.* **8**, 208 (1992).

[958] J.B. Pausch and I. Sockis, *Rubber Chem. Technol.* **50**, 828–34 (1977).

[959] R.P. Lattimer and J.B. Pausch, *Am. Lab.*, 80–8 (Aug. 1980).

[960] B. Kolb and L.S. Ettre, *Static Headspace – Gas Chromatography. Theory and Practice*, Wiley-VCH, New York, NY (1997).

[961] B. Kolb (ed.), *Applied Headspace GC*, Heyden, London (1980).

[962] H. Hackenberg and A.P. Schmidt, *Gas Chromatographic Headspace Analysis*, Heyden, London (1977).

[963] J.B.H. van Lierop, *De Ware(n) – Chemicus* **15**, 136–45 (1985).

[964] M. Geissler and A. Paul, *Shimadzu Application Note* SCA-180-005 (1994).

[965] S.J. Romano, J.A. Renner and P.M. Leitner, *Anal. Chem.* **45**, 2327–30 (1973).

[966] S.A. Liebman, D.H. Ahlstrom and C.I. Sanders, *Am. Lab.*, 11 (July 1975).

[967] R.A. Shanks, *Scan* **6**, 20–1 (1975).

[968] T.R. Crompton and C.W. Myers, *Plast. Polym.* (6), 205 (1968).

[969] T.R. Crompton, *Manual of Plastics Analysis*, Plenum Press, New York, NY (1998).

[970] N. Oguri, *Nippon Gomu Kyokaishi* **67** (11), 768–80 (1994).

[971] W.B. Gorman Jr. (to Bridgestone/Firestone Inc.), *U.S. Pat.* 5,191,211 (March 2, 1993).

[972] S. Tsuge and H. Ohtani, *Kagaku (Kyoto)* **44**, 212–3 (1989).

[973] S.W. Bigger, M.J. O'Connor, J. Scheirs, J.L.G.M. Janssens, J.P.H. Linssen and A. Legger-Huysman, *Adv. Chem. Ser.* **249** *(Polymer Durability)*, 249–68 (1996).

[974] M. Ezrin, E.D. Wyatt, G. Lavigne and A. Garton, *Proceedings SPE ANTEC '94*, San Francisco, CA (1994), pp. 2922–6.

[975] S. Desobry, M. Mousavi, M.C. Michalski and J. Hardy, *Packag. Technol. Sci.* **10** (4), 229–36 (1997).

[976] B. Raffael and C. Simoneau, *Abstracts 7th Intl. Symposium on Hyphenated Techniques in Chromatography and Hypenated Chromatographic Analyzers (HTC–7)*, Bruges (2002), Paper 15.3.

[977] I.P. Sizeneva, L.E. Baksanova, V.I. Budnikov and R.N. Gabdrakhmanov, *Ind. Lab. (Diagn. Mater.)* **65** (4), 216–8 (1999).

[978] N.J. Fales, L.C. Stover and J.D. Lamm, *Polym. Plast. Technol. Eng.* **21**, 111 (1983).

[979] J.P.H. Linssen, J.L.G.M. Janssens, J.P. Roozen and M. Posthumus, *J. Plast. Film Shtg.* **7**, 294 (1991).

[980] M. Ezrin and G. Lavigne, *Proceedings SPE ANTEC '91*, Montreal (1991), pp. 2230–3.

[981] A. Hagman and S. Jacobsson, *Anal. Chem.* **61**, 1202–7 (1989).

[982] A. Hagman, *Dynamic Headspace. A Versatile Method for Analysing Volatile Compounds in Polymers, Ph.D. Thesis*, University of Stockholm (1988).

[983] K. Fukui, *J. Odor Res. Eng.* **20**, 1 (1989).

[984] S.C. Hodgson, M.J. O'Connor, R.J. Casey and S.W. Bigger, *J. Agric. Food Chem.* **46** (4), 1397–1405 (1998).

[984a] S.C. Hodgson, R.J. Casey, J.D. Orbell and S.W. Bigger, *J. Chem. Ed.* **77**, 1631–3 (2000).

[985] M. Wolf, M. Riess and R. van Eldik, in *Thermische Einsatzgrenzen von Kunststoffen in Verarbeitung und Anwendung* (G.W. Ehrenstein and S. Pongratz, eds.), Springer VDI Verlag, Düsseldorf (2000), pp. 321–31.

[986] S. Tsuge, K. Kuriyama and H. Ohtani, *J. High Resolut. Chromatogr.* **12**, 727–31 (1989).

[987] M. Ezrin and G. Lavigne, *Proceedings SPE ANTEC '92*, Detroit, MI (1992), pp. 1717–9.

[988] M. Ezrin and G. Lavigne, *Proceedings SPE ANTEC '94*, San Francisco, CA (1994), pp. 3302–5.

[989] M. Ezrin, *Plastics Failure Guide*, Hanser Gardner, Cincinnati, OH (1996).

[990] W.R. Raisanen, *Proceedings SPE ANTEC '97*, Toronto, ON (1997), pp. 3603–5.

[991] J. Pawliszyn, *Solid Phase Microextraction. Theory and Practice*, J. Wiley & Sons, New York, NY (1997).

[992] Z. Zhang and J. Pawliszyn, *Anal. Chem.* **65**, 1843–52 (1993).

[993] A.J. Matich, in *Applications of Solid Phase Microextraction* (J. Pawliszyn, ed.), The Royal Society of Chemistry, Cambridge (1999), pp. 349–63.

[994] J. Ai, in *Applications of Solid Phase Microextraction* (J. Pawliszyn, ed.), The Royal Society of Chemistry, Cambridge (1999), pp. 22–37.

[995] H. Anderegg, *Schweiz. Lab.-Z.* **53**, 55–8 (1996).

[996] J.R. Dean and P. Hancock, in *Applications of Solid Phase Microextraction* (J. Pawliszyn, ed.), The Royal Society of Chemistry, Cambridge (1999), pp. 248–57.

[997] Supelco Bulletin 869, Supelco Inc., Bellefonte, PA (1995).

[998] B. MacGillavry, J. Pawliszyn, P. Fowlie and C. Sagara, *J. Chromatogr. Sci.* **32**, 317 (1994).

[999] M. Hakkarainen, *Preprints 37th Intl. Symposium on Macromolecules*, Gold Coast (1998), p. 87.

[1000] A.J. Matich, D.D. Rowan and N.H. Banks, *Anal. Chem.* **68**, 4114–8 (1996).

[1001] F. Khabbaz, A.-C. Albertsson and S. Karlsson, *Polym. Degr. Stabil.* **61**, 329–42 (1998).

[1002] Z. Penton, *SPME Application Note 7*, Varian Chromatography Systems, Walnut Creek, CA (1995).

[1003] B.D. Page, in *Applications of Solid Phase Microextraction* (J. Pawliszyn, ed.), The Royal Society of Chemistry, Cambridge (1999), pp. 423–34.

[1004] T.P. Wampler, *LC.GC Intl.*, 653–8 (Oct. 1998).

[1005] E. Woolfenden, *To-day's Chemist at Work* **7** (5), 21–5 (1988).

[1006] K. Yokoe, H. Ohtani and S. Tsuge, *Intl. J. Polym. Anal. Char.* **4**, 547–63 (1998).

[1007] T.G. Blasse, G.A. Paterson and J.H. Scrivens, *Br. Polym. J.* **21**, 37–44 (1989).

[1008] H.G. Wahl, A. Chrzanowski, N. Ottawa, H.-U. Häring and A. Hoffmann, *Gerstel Publ. 41*, Gerstel GmbH, Mülheim a.d. R. (1996).

[1009] H.G. Wahl, A. Hoffmann, H.-U. Häring and H.M. Liebich, *J. Chromatogr.* **A847**, 1–7 (1999).

[1010] DaimlerBenz, *Arbeitsvorschrift zur Bestimmung von gasförmigen und kondensierbaren Emissionen aus Fahrzeuginnenausstattungsmaterialien mit Thermodesorption, Prüfanweisung PB VWT 709* (1997).

[1010a] T.P. Wampler, *LabPlus Int.* **18** (4), 6–9 (2004).

[1011] M. Wolf, M. Riess, D. Heitmann, M. Schreiner, H. Thoma, O. Vierle and R. van Eldik, *Chemosphere* **41** (5), 693–9 (2000).

[1012] M.E. Krzymien, D.J. Carlsson, L. Deschener and M. Mercier, *Food Addit. Contam.* **18** (8), 739–49 (2001).

[1013] J.H. Scrivens and A.T. Jackson, *NATO ASI Ser., Ser.* **C521**, 201–34 (1999).

[1014] P.F. Zambetti, S.L. Baker and D.C. Kelley, *Tappi J.* **78** (4), 167–71 (1995).

[1014a] J.S. Lancaster, T.P. Lynch and P.G. McDowell, *J. High Resolut. Chromatogr.* **23** (7–8), 479–84 (2000).

[1015] T.G. Hartman, J. Lech, K. Karmas, J. Salinas, R.T. Rosen and C.-T. Ho, in *Flavor Measurement* (C.-T. Ho and C.H. Manley, eds.), M. Dekker, New York, NY (1993), pp. 37–60.

[1016] N. Manabe, T. Toyota, Y. Yokata, H. Minami and H. Takagi (to Sumitomo Denso K.K.), *Jpn. Kokai Tokkyo Koho* 2,000,146,938 (Nov. 6, 1998).

[1017] K. Kimura, T. Yoshikawa, Y. Taguchi, Y. Ishida, H. Ohtani and S. Tsuge, *Analyst* **125** (3), 465–8 (2000).

[1018] H. Knobel, *Abstracts 7th Intl. Symposium on Hyphenated Techniques in Chromatography and Hypenated Chromatographic Analyzers (HTC-7)*, Bruges (2000).

[1018a] S. Kodama, T. Tsunesada and K. Nishi, *Okayama-ken Kogyo Gijutsu Senta Hokoku* **23**, 8–11 (1997).

[1019] D. Brück, in *Fortbildungsseminar "Moderne Prüfmethoden – von Rohstoffen über Mischungen bis zum Produkt"*, Deutsches Institut f. Kautschuktechnologie, Hannover (May 2000).

[1020] *Gerstel Application Note 316*, Gerstel, Mülheim a. d. R. (n.d.).

[1021] K. Kurihara and F. Tanoue, *Bunseki Kagaku* **49** (3), 205–8 (2000).

[1022] T.P. Wampler and C.P. Zawodny, *Proceedings SPE ANTEC '98*, Atlanta, GA (1998), pp. 2211–4.

[1023] G.B. Kenion, R. Ludicky, J.H. Deitch, M. Leib and D. Doster, *Tappi J.* **77** (6), 199–205 (1994).

[1024] S.A. Buckley, A.W. Scott, B.A. Stankiewicz and R.P. Overshed, in *Abstracts 12th Intl. Symposium on Analytical and Applied Pyrolysis (Pyrolysis '96)*, Venice (1996), P73.

[1025] J.A.J. Jansen, in *Calorimetry, Thermal Analysis of Polymers* (V.B.F. Mathot, ed.), Hanser Publishers, Munich (1994), Chp. 13.

[1026] R.J. Cotter, *Anal. Chem.* **52**, 1589A–606A (1980).

[1027] T. Yoshikawa, K. Ushimi, K. Kimura and M. Tamura, *J. Appl. Polym. Sci.* **15**, 2065–72 (1971).

[1028] R.P. Lattimer, R.E. Harris, C.K. Rhee and H.-R. Schulten, *Anal. Chem.* **58**, 3188–95 (1986).

[1029] R.M.A. Heeren, C.G. De Koster and J.J. Boon, *Anal. Chem.* **67**, 3965–70 (1995).

[1030] A.I. Vit, M.N. Galbraith, J.H. Hodgkin and D. Yan, *Polym. Degr. Stabil.* **42**, 69–73 (1993).

[1031] A.P. Pleshkova and M.N. Uspenskaya, *Khim. Prom.-st. (Moscow)* (6), 321–4 (1992).

Chapter 3

Est! Est!! Est!!!

Lasers in Polymer/Additive Analysis

3.1. Lasers . . . 325
3.2. Laser Ablation . . . 331
3.2.1. Laser Ablation – Plasma Source Spectrometry . . . 335
3.3. Laser Spectroscopy . . . 341
3.3.1. Laser-induced Atomic and Molecular Fluorescence Spectrometry . . . 343
3.3.2. Laser-induced Breakdown Spectroscopy . . . 346
3.4. Laser Desorption/Ionisation Methods . . . 353
3.4.1. Laser Desorption Mass Spectrometry . . . 354
3.4.2. Laser Ionisation . . . 363
3.4.3. Decoupled Laser Desorption/Ionisation . . . 366
3.4.4. Matrix-assisted Laser Desorption/Ionisation . . . 374
3.4.5. Laser Microprobe Mass Spectrometry . . . 381
3.5. Laser Pyrolysis . . . 388
Bibliography . . . 392
Lasers . . . 392
Laser Ablation . . . 392
Laser Spectroscopy/Spectrometry . . . 392
Laser-induced Chemistry . . . 393
Laser Safety . . . 393
References . . . 393

A variety of laser techniques have been applied for direct analysis of additives in polymers. Table 3.1 shows the relation between various induced phenomena and proposed analytical methods, broadly classified as laser spectroscopy and laser mass spectrometry. Some specific features for lasers are multiphoton absorption spectroscopy (based on the frequency dependency), multiphoton ionisation spectroscopy (with resonance enhancement) and laser mass spectrometries (laser wavelength and intensity dependency; desorption/ionisation).

3.1. LASERS

Principles and Characteristics

The word "laser" is an acronym derived from *l*ight *a*mplification by the *s*timulated *e*mission of *r*adiation. Lasers are sources of radiation with unique properties, which operate by a process of induced emission. Einstein first postulated the principle of the laser in 1917. Available laser systems cover an extremely wide range, differing from each other in laser material, pump source, pumping mode, efficiency, physical size, optical, temporal and mechanical properties, such as beam divergence, pulse energy, pulse width, repetition rate, polarisation properties, stability of the beam, and vibration sensitivity related to the mechanical design, emission wavelengths and tuneability, ease of operation, maintenance costs, reliability, and safety aspects. Laser beams have a number of **unique properties** compared to other sources emitting electromagnetic radiation, such as arc lamps, which make them an almost ideal light source for use in spectroscopy. These properties are: *(i)* high degree of monochromaticity; *(ii)* high directionality (small divergence); *(iii)* high brightness; *(iv)* high degree of spatial and temporal coherence; *(v)* allowance for selective processes; *(vi)* plane polarised emission (for many types); and *(vii)* Gaussian beam profile (via special optics). Broad classifications distinguish

Table 3.1. Laser methods in polymer/additive analysis

Exploitation of photon-solid interaction	Laser spectroscopy	Laser mass spectrometry
Laser ablation	LA-AES, LA-ICP-AES	LA-MS, LA-ICP-MS
Laser excitation:		
Absorption	LAAS, UPS, RIS(MPIS)	
Emission	LIBS, PLPAS	
Fluorescence	LIFS, (ETA)-LEAFS	
Scattering	Raman, LALLS	
Laser ionisation	LEIS	
Laser desorption/(post)ionisation	UV-LDI, IR-LDI	LD-GC-MS
	SALI	LDMS (MS = FT, ToF, IT), L^2MS, LMMS, MALDI-MS, SALDI-MS, LD-RIMS, REMPI-MS
Laser photodissociation		LD/PDPI-MS
Laser pyrolysis	LPyIR	LPyMS, LPyGC-MS
Laser microscopy[a]	LCFM (CLSM)	

[a] *Cfr.* Chp. 5.3.4.

Table 3.2. Lasers

Laser medium	Laser material
Optically pumped solid-state	Ruby, Nd^{3+}:YAG, Nd^{3+}:glass, Cr^{3+}:$BeAl_2O_4$ (alexandrite), Ti^{3+}:Al_2O_3 (sapphire)
Semiconductor (diode)	GaAs, GaAlAs, InGaAsP/InP, GaInN, GaN/AlGaN, PbSnTe
Atomic and ionic gas	He/Ne, Ne^+, Ar^+, Kr^+, Xe^+
Metal vapour	Cu, Au, Sr, Mn, Ba, Pb
Molecular gas	CO_2, N_2, I_2, chemical, excimer (ArF, KrF, XeF, KrCl, *etc.*)
Dye	Rhodamine 6G
Free-electron	Free electrons

line-tuneable, continuously tuneable and discrete-wavelength or fixed-frequency lasers, as well as continuous wavelength (CW) and pulsed lasers. The output of a laser depends on the elements characterising the laser medium and construction parameters. It is beyond the scope of the present text to describe lasers in any detail. For this purpose the reader is referred to Bibliography and refs. [1–3]. Table 3.2 lists the main laser types.

Maiman [4–6] has described the first laser systems (ruby) in the 1960s. The active medium in **solid-state lasers** is generally a transparent crystal or glass into which a small amount of transition metal is doped (*e.g.* Ti/sapphire, Nd:YAG). There are two main types of **semiconductor lasers**: those operating at fixed wavelengths and those which are tuneable. Diode lasers are the most prolific of all types of laser. Most of these lasers are semiconductor compounds of Group III and V species. Diode lasers operate mostly in the near-infrared but also in the visible region. Some diode lasers are the most efficient of all lasers. Those operating in the near-IR may have an efficiency as high as 30%. The class of lasers in which the active medium is a gas covers a wide variety of devices. Generally, the gas is either monoatomic, or else it is composed of very simple molecules. The He/Ne laser is an example of an **inert gas laser** which is simple, reliable to operate and fairly inexpensive. Laser action takes place between excited energy levels of the Ne atom, the upper levels being populated partly through collisions with He atoms in metastable excited states. For an **ion laser** much higher input energy is required than for an inert gas laser in which the lasing species is a neutral atom. In this case the problem of heat dissipation is serious. The fairly expensive argon laser is the most common example of the family of ion lasers. The line-tuneable CO_2 laser is a near-IR gas laser capable of immense power with an efficiency of about 20%, comparable to an excimer laser but less efficient than a diode laser. CO_2 lasers are the most commercially successful of all lasers and are extensively used in the area of laser-induced chemical reactions. The next category of commonly available lasers consists of those in which the active medium is an excimer (*e.g.* Xe_2) or an exciplex or excited diatomic complex (*e.g.* ArF, KrF, XeF, KrCl). In spite of a clear distinction between an excimer and an exciplex, it is now common for *all* such lasers to be called **excimer lasers**. These lasers are super radiant and produce pulsed radiation with pulse durations of 10–20 ns and pulse repetition frequencies generally in the 1 to 500 Hz range. Pulse energies can be up to 1 J, with peak pulse power in the megawatt region and average power between 20 and 100 W. Excimer lasers combine high power with excellent beam quality and good spatial control The emission wavelengths lie in the short UV region where photoabsorption processes often result in rupture of chemical bonds. Often such absorption also leads to a degree of sample vaporisation; this is a process known as laser ablation. **Metal vapour lasers** are high-power efficient sources of visible light in which the monoatomic metal vapour lasing material is created by using an electric discharge to heat a plasma tube containing a small amount of metal. In **dye lasers** the medium is a liquid, *i.e.* a solution of an organic dye. Dye lasers have been demonstrated to lase in the three states of matter and their positioning between gas and solid-state is quite appropriate. A wide range of over 200 dyes can be used; the only requirements are an absorption band in the visible spectrum and a broad fluorescence spectrum. The variety of available dyes ensures that the entire spectrum from around 320 nm to beyond 1000 nm is covered by a dye laser. They have the important property of being continuously tuneable over a large wavelength range. Maeda [7] and Duarte [8] have given comprehensive listings of laser dyes and tuning ranges. The **free-electron laser** (FEL) is radically different from any of those mentioned so far. This is a laser in which the active medium consists purely of a beam of free electrons, and the optical transition on which laser action is based results from the acceleration and deceleration of these electrons in a magnetic field. FEL can produce high power UV laser light with wavelengths between 80 and 180 nm, as well as IR between 5 and 110 μm [8a]. For applications which require tuneable radiation at very high power levels, the free-electron laser is hard to beat.

Andrews [9] and others [10] have listed the **emission lines** of the most commonly available discrete-wavelength lasers (such as ruby, Nd:YAG, Er:YAG, excimer lasers) over the range 100 nm-10 μm. Molecular lasers (HF, CO, CO_2, NO) can be tuned to a large number of closely spaced but discrete wavelengths. Continuously tuneable lasers comprise some metal ion vibronic lasers (*e.g.* alexandrite and Ti:sapphire [11]), some diode and excimer lasers, dye and free-electron lasers. Tuneable sources of coherent radiation span the electromagnetic spectrum from ~300 nm to ~1 mm, with limited tuneability down to about 200 nm. Wavelength coverages of tuneable lasers have been reported [8]. In operation lasers can be either pulsed (*e.g.* various metal ion tuneable vibronic lasers, excimer and dye lasers, metal vapour) or continuous wave (major types: atomic and ionic gas lasers, dye and solid-state lasers). Most lasers with spectral output in the UV are bulky and expensive devices (especially sub 200 nm) and operate in the pulsed mode. On the contrary, many visible lasers are available which are compact, require low maintenance expenses and operate in continuous-wave (CW) mode.

General Analytical and Industrial Applications

Lasers are used both for analytical and industrial purposes. Table 3.3 summarises the main analytical fields of application. The most obvious reason to involve lasers in analytical chemistry is the directionality of the radiation (beam divergence <0.5 mrad), which implies high irradiances at remote objects (up to 10^{15} W cm^{-2}) and compatibility with miniaturised systems. Characteristics as monochromaticity and coherence are still of less importance. The monochromaticity of the laser lines is of major importance in techniques such as RS and those based on multiphoton processes. Some important analytical applications of lasers are:

(*i*) time-resolved spectroscopy with ultrashort laser pulses;

Table 3.3. Analytical applications of lasers

• Spectroscopy	• Pyrolysis
• Spectrometry	• Chromatography (detectors)
• Microscopy	• Particle-size determinations
• Desorption	• Diffraction
• Ionisation	• Process analysis
• Ablation	• Quality assurance
• Photodissociation	• Monitoring

(ii) measurements of the internal-state distribution of reaction products with LIF or REMPI;
(iii) detection of impurity atoms (ppt range) or molecules (ppb range);
(iv) chemical reaction control by selective excitation of the reactants.

Lasers are utilised as a universal tool for sampling solids. Laser desorption and volatilisation found their way to applications such as interfacing TLC to GC and to MS. In laser desorption material transport across the surface is negligible. Laser volatilisation is characterised by considerable transport of mass, momentum, and energy and occasional plasma formation.

Although the first intense monochromatic laser source was produced in 1960, the ruby laser proved to be of limited use in spectroscopy. The true dawn of modern **laser spectroscopy** arose with the development of more flexible, particularly tuneable lasers, such as the dye and diode lasers in the early 1970s. Since that time, almost all lasers have been applied in spectroscopic experiments. Lasers are now available over a wide wavelength range stretching from the microwave ($\lambda \approx 1$ cm) to VUV ($\lambda \approx 200$ nm). Each region reflects a particular type of energy level transition and hence spectroscopy; rotational in the microwave and far-IR (*ca.* 1 cm–50 μm), vibrational in the mid- and near-IR (*ca.* 50–1 μm) and electronic in the near-IR through to the VUV and beyond (*ca.* 1 μm–100 nm). A variety of coherent radiation sources in the visible and UV are available with metal ion tuneable vibronic lasers covering the 285–2500 nm range.

In contrast to the UV/VIS regions, the IR suffers from a somewhat restricted choice of tuneable coherent radiation sources. Although recent developments in laser diode and non-linear optical technology promise a potential expansion in the number of tuneable IR sources, traditionally the choice of source has been limited to molecular gas lasers, lead salt diode lasers or colour centre lasers. CO_2 lasers were amongst the earliest IR lasers to be developed; this laser is line-tuneable (between 9.6 and 10.6 μm) with CW powers of up to several 100s W and pulse energies in the multi-Joule level. Lead salt diode lasers have large bandgaps, with the result that emission occurs at wavelengths shorter than 2 μm. Continuously tuneable lasers are of particular interest and in many cases may replace wavelength-selecting elements, such as spectrometers or interferometers. Many experiments, which could not be done with conventional spectrometers because of lack of intensity or insufficient spectral resolution, are readily performed with lasers.

Short laser pulses (ranging routinely from ps to ns) often offer advantages in analysis, *e.g.* the application of time-of-flight analysers. The application of ultra-fast spectroscopy based on new fs pulsed lasers has undoubtedly been one of the most significant innovations; at the same time, the frequency stability of lasers is now in the mHz range. Demtröder [3] described time-resolved laser spectroscopy. Pulsed lasers are natural sources for time-resolved fluorimetry.

A significant drawback to continuous-wave (CW) visible lasers is that only a limited number of molecular species have absorption features in the spectral region covered by these lasers. McCoustra [11] has reviewed sources for laser spectroscopy. **Chemical spectroscopy** with lasers embraces:

- absorption spectroscopy
- emission spectroscopy (*e.g.* LIBS, laser spark emission)
- excitation spectroscopy
- ionisation spectroscopy
- photoacoustic spectroscopy (*e.g.* LPAS, PARS)
- fluorescence spectroscopy (incl. LEAFS, LIF); *cfr.* Chp. 1.4.2
- Raman spectroscopy (incl. RRS, SE(R)RS); *cfr.* Chp. 1.2.3.

For example, laser spark emission can be used in discriminating 31 paints with small differences in vehicle composition (essentially all pentaerythritol-*o*-phthalate alkyd based) [12]. In this application the technique is outstanding as compared to PyGC, PyIR, XRD or IR spectroscopy.

Many of the early laser sources such as the ruby laser and rare gas ion lasers operate at a fixed frequency, providing only a single lasing wavelength, or at most lasing on a few narrow resonances. These non-tuneable lasers are difficult to employ spectroscopically, with the obvious exception of their application in Raman spectroscopy, where excitation on a single narrow frequency band is desirable. Nitrogen lasers are used for coherent anti-Stokes Raman spectroscopy. In addition to carrying out conventional Raman experiments with laser sources new kinds of Raman experiments have become possible. Resonance Raman spectroscopy (*cfr.* Chp. 1.2.3.1.1) requires intense, monochromatic sources covering a range of wavelengths. Prior to the development of lasers, the small number of sources available limited applications. Lasers provide a much wider choice of wavelengths and are ideally suited to observing the resonance Raman effect. From an analytical point of view the arrival of Fourier transform Raman spectroscopy constitutes significant progress. Lasers do not only play a prominent role in Raman microspectroscopy (*cfr.* Chp. 5.6.3), but also in laser confocal fluorescence **microscopy** or LCFM (*cfr.* Chp. 5.3.4). One of the essential characteristics of a laser to be used in fluorescence microscopy is the wavelength of emission. The technique is particularly suited for depth profiling. A spatially resolved laser-induced ion microscope (SLIM) allows spatially resolved microanalysis. Laser profilometry and image analysis may be used for paper surface characterisation [13].

The use of lasers is proliferating in **mass spectrometry** including analysis of organic and inorganic materials, element quantisation and structural analysis of thermolabile, polar and high-MW organic compounds, trace and local analysis, microanalysis and surface studies. Lasers are used for sample volatilisation and ionisation. The great number of fundamentally different analytical applications makes complete coverage of the use of lasers in MS hard to achieve. It is well known from mass spectrometric investigations that the amount and ionisation degree of the vaporised material depend on the energy deposited into the target. Use of a laser offers better control of desorption processes than an ion beam. The different regimes of laser ionisation, desorption, vaporisation and plasma ionisation are characterised by the amount of deposited energy. With 3 mW IR laser light sufficient molecules are desorbed from a 5 to 10 μm^2 surface to record a mass spectrum. Matrix-assisted laser desorption ionisation (MALDI) employs irradiances of between 10^{10} and 10^{12} $W\,m^{-2}$; the sample desorbs as molecular or quasi-molecular ions up to high masses (10^6 Da). Even higher levels of intensity (in excess of 10^{12} $W\,m^{-2}$) are employed for microsampling in ablation processes. The mass spectrometric application is here laser microprobe mass spectrometry (LMMS). The various aspects of laser desorption/ionisation in connection with mass spectrometric detection are discussed in Chp. 3.4, with emphasis on various LD-MS modes, LMMS and MALDI-MS (*cfr.* also Chp. 9.3.1 of ref. [13a]). Related techniques, such as L-SNMS (Chp. 4.2.2), are discussed separately.

Evolved gas analysis by spectroscopy and mass spectrometry cannot only be based on volatilisation modes such as controlled direct heating (isothermal and non-isothermal) pyrolysis techniques, but also on laser plasma excitation and **laser ablation** by examination of the laser plumes. Typical laser ablation spectroscopies are LA-AES or LIBS (*cfr.* Chp. 3.3.2) and LA-ICP-AES (*cfr.* Chp. 3.3.1). Laser thermal analysis and laser pyrolysis (Chp. 3.5) are other practical analytical applications. Applications requiring vaporisation (photoablation) or the breaking of chemical bonds often employ excimer lasers. Excimer lasers have found numerous applications because of the characteristic high power and ultraviolet nature of the emission.

Various laser techniques are employed in **chromatography** to monitor eluents from GC and HPLC columns. In principle, measurements may be based on multiphoton absorption, fluorescence (LEAFS, LIF), scattering (LALLS, RALLS) and ionisation. The narrow beam width and high intensity of laser light generally means that very small detection volumes can be employed; it is usually the flow design which is the limiting factor, but volumes as small as 10^{-8} L are possible. Similarly, laser methods are also highly appropriate for "on-column" detection. Nonetheless, since very low concentrations are usually involved, the two most obvious methods of detection, namely absorption and fluorescence, are both applicable only where the substances of interest display appreciable absorption at the operating wavelength of the laser. A very sensitive photoionisation scheme, resonant multiphoton ionisation (REMPI), is based on resonant two- or three-photon ionisation of atoms and molecules in the gas phase. With this technique also liquid or solid samples can be monitored if they can be vaporised. Picogram quantities can be detected with HPLC fluorescence instrumentation. Light-scattering (LS) photometers extensively used as SEC detectors appeared in the mid-1970s and presently there are five laser-based LS photometers on the market that allow measurement of scattering intensities at different angles: low-angle laser LS (LALLS) [14], multi-angle laser LS (MALLS), triple-angle laser LS (TALLS), dual-angle laser LS (DALLS), and right-angle laser LS (RALLS) photometers. An LS detector is more sensitive to the high-MW end than a concentration detector, and less sensitive to the low-MS end. The differential laser refractometer is the preferred concentration detector for most SEC applications. Hyphenated TLC-laser desorption techniques such as TLC-LMMS, TLC-MALDI-MS and

TLC-SALDI-MS are also in use (*cfr.* Chp. 7.3.5.4 of ref. [13a]). For laser detectors in chromatography *cfr.* also Chp. 4.1 of ref. [13a] and ref. [9].

In the chemical industry product and process control by means of near-IR Raman spectroscopy also benefit from the use of lasers (*cfr.* Chps. 7.2.4 and 7.2.5). Other fields of interest are laser-induced chemical reactions (including curing and colour forming), photochemistry and laser selective chemistry (taking advantage of the tuneable wavelength characteristics) [15,16]. Particle size and shape measurements with laser diffraction is used in the toner industry for real-time **quality assurance** [17]. Circularity of the toner particles influences flow properties and charging capability. Identification of polymers and additives in recycling also benefits from lasers using thermal fingerprint techniques and LIBS (*cfr.* Chp. 3.3.2.1). For thermal applications CO_2 lasers (NIR) are used, as is the case for identification of polymeric materials. Equally worthy of notice are the particulate emissions from CO_2 laser processing of plastics. Polymers can be divided into three particulate emitter groups: high (*e.g.* PC, PA6.6, PP), medium (*e.g.* ABS) and low (*e.g.* PS, PMMA), which is essentially determined by their dominant laser processing mechanisms [18]. Analytical probing with lasers comprises pollution monitoring and aerosol analysis.

Non-contact laser scanning technology is used for advanced inspection of injection moulded plastic parts [19]. Lasers are applied for a diverse range of **materials processing** applications, such as polymer welding and cutting [20,21], surface engineering [22] and laser-marking of moulded plastic parts (PC, PC/ABS, PP, PE, PA6, PA4.6, PA6.6, PET, PBT, TPV), including automotive components, medical, specialty packaging, and a wide range of electronic devices [23]. The current understanding of laser decoration mechanisms is summarised in Table 3.4 (*cfr.* also ref. [24]).

Laser marking of a surface is effected by: *(i)* engraving (produced by local vaporisation of the surface); *(ii)* ablation (the engraving depth can be tuned by the laser intensity); and *(iii)* foaming (the material is heated locally, forming gas inclusions). On the other hand, the surface is not altered by: *(i)* carbonisation (mainly used on specially pigmented plastics); *(ii)* colour forming (as a result of a laser-induced chemical reaction); and *(iii)* selective bleaching (colorants are selectively removed by the laser beams). The introduction of dyes into a PVA matrix sensitises laser-induced degradation of polymers (514.5 and 1064 nm). Kalontarov *et al.* [25] have shown that the laser stability of polymers at the same absorbed laser power depends both on the type of incorporated dye and the type of bond between dye and polymer. The combination of pigment, polymer and laser energy must be treated as a system.

Lasers are further utilised for a wide variety of purposes such as laser imaging (particle size counting), missile aiming, light-show entertainment, cashing of consumer goods and for didactic purposes (laser pointers). The tuneable dye and semiconductor lasers are used in many diverse fields, including physics, spectroscopy, interferometry (*e.g.* materials testing), isotope separation, remote sensing and medicine (optical tomography, eye surgery, brain scanning, scalpels). Laser impulse thermography (LIT), an alternative to infrared technology, in which a CO_2 laser pulse locally heats the surface to about 200°C, allows identification of materials (recycling) with an identification rate of up to 10 objects/sec. For non-thermal applications UV lasers are employed. Laser ablation by means of UV excimer lasers finds wide application in laser diagnostics and laser cleaning (*e.g.* of art objects) [26]. Techniques utilised here include laser Raman spectroscopy, LIBS and LIF [27,28]. Real-time monitoring of pulsed excimer laser cleaning and ablation have been realised by making use of the optoacoustic effect [29]. Laser interaction affects polychromy [30,31]. A most favoured wavelength for UV laser radiation for industrial purposes is 308 nm.

The use of lasers in polymer/additive analysis is only a minor, though significant industrial application, as this chapter illustrates.

Table 3.4. Laser decoration mechanisms

Laser	Mechanisms
Excimer (193, 248, 308, 351 nm)	Ablation, colour forming
Nd:YAG (355, 532, 1064 nm)	Bleaching of colours, colour forming, foaming, carbonisation
CO_2 (10.6 μm)	Engraving, foaming, carbonisation; chemical colour change[a]

[a] DataLase (Sherwood Technology).

3.2. LASER ABLATION

Principles and Characteristics

Laser ablation is conceptually very simple, but mechanistically complicated. The process involves coupling of the photon energy of a laser pulse (typically about 20–30 ns wide, with an energy of 1–10 $J\,cm^{-2}$) into the surface of a solid, resulting in evaporation and ejection of various species from the surface (the so-called "plume") within 10^{-9} to 10^{-8} s. The first experiments were carried out in 1962 [32]. When focused to a small area, a laser beam provides enormous power densities and electromagnetic fields. The "plume", presumably a plasma, is accompanied by shock waves and electrical breakdown. The ejected material may eventually be deposited as a thin film. It is possible, by suitable selection of laser power and focus, to ablate a range of plastic materials in a controlled manner. For some matrices the polymer melts and diffuses away from the centre of the ablation site, leading to the formation of wells in the sample surface. These difficulties can be overcome by reducing the laser power.

Coupling between molecular processes and morphological changes is one of the most unique and important characteristics of laser ablation. Excitation energy relaxation dynamics and primary chemical processes of organic molecules in laser ablation have been investigated by using various time-resolved spectroscopies, such as fluorescence, absorption, Raman and IR spectroscopies. Laser ablation leads to rapid temperature elevation of the polymer matrix and thermal decomposition of the polymer. Ablation causes not only photochemical reactions but also photo-initiated thermal reactions.

The thermal effect of laser irradiation is obscure. Schawlow [33] has estimated heating rates as high as 10^{12} $K\,s^{-1}$ with temperatures up to 10^6 K, but others [34] have reported temperature assignments of 4000 to 10,000 K. Much lower estimates (650–1000°C) have also been given [35]. Anyhow, the thermal evaporation component may be deleterious for analytical purposes because elements of high

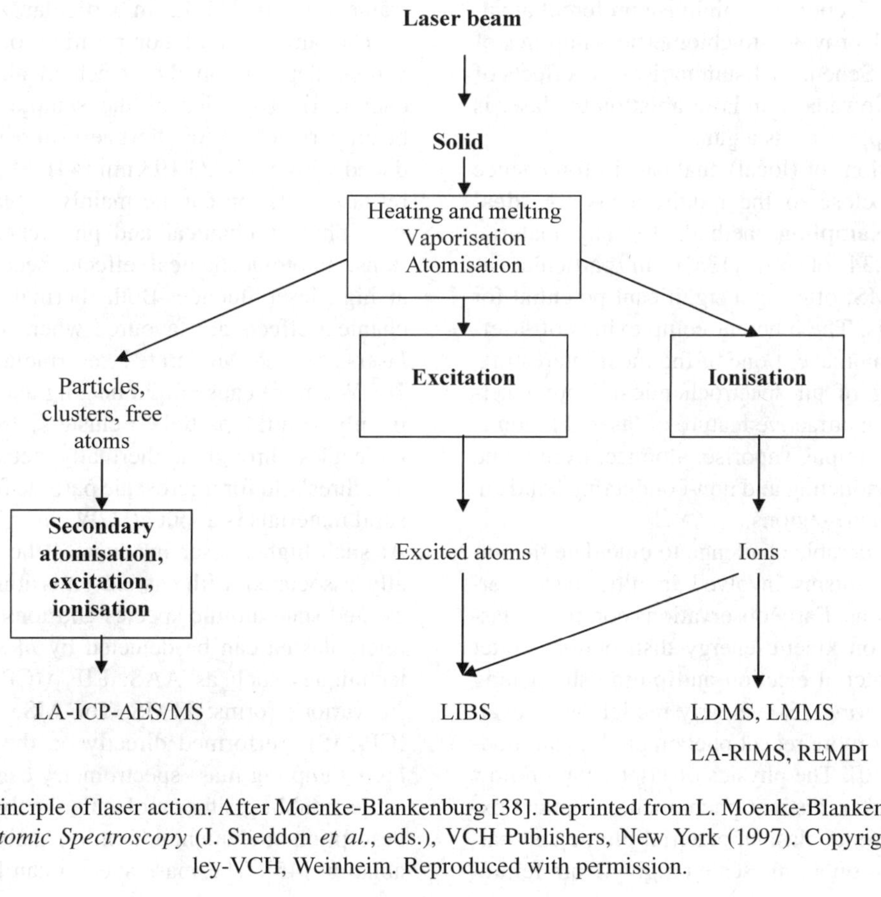

Scheme 3.1. Principle of laser action. After Moenke-Blankenburg [38]. Reprinted from L. Moenke-Blankenburg, in *Lasers in Analytical Atomic Spectroscopy* (J. Sneddon *et al.*, eds.), VCH Publishers, New York (1997). Copyright 1997 © Wiley-VCH, Weinheim. Reproduced with permission.

Table 3.5. Physics of UV laser ablation

Time scale	Observation[a]	Interpretation
fs	Crater formation: longitudinal expansion (up to 50 ps)	Absorption/electronic excitation
ps	Lateral expansion at $t > 50$ ps	Ionisation, conduction
	Laser heating of target[b]	
	Plasma ignition above target[c]	
ns	Material plume development (>500 ps)[d]	Radiation, ionisation, vaporisation, convection,
	Superheated liquid layer in solid (100 ns)	melting
μs	Microscale droplet formation	Particle ejection
	Condensation[e]	

[a] Material and fluence dependent windows.
[b] Electron heating (absorption of laser energy) and lattice heating (electron-phonon collisions).
[c] Surface electron emission and impact ionisation of gas.
[d] Semi-spherical plume clearly visible (low ns range).
[e] Most particulates (~1 nm) condense back on target (<5 μs).
After Russo *et al.* [43]. Reproduced by permission of the authors.

vapour pressure can be enriched in the vapour phase relative to the original solid sample (preferential vaporisation), rendering analysis inaccurate [36]. By using ns and shorter UV laser pulses, rapid heating and explosive ejection can minimise preferential vaporisation and provide stoichiometric sampling of the solid [37]. Scheme 3.1 summarises the effects of laser action. Nowadays, in laser ablation the laser is no longer simply used as a gun.

Laser ablation for (local) analysis, in force since 1962, comes close to the requirements for **ideal direct solid sampling** methods for any material (*cfr.* Table 8.34 of ref. [13a]), in particular as LA-ICP-ToFMS, offering a significant potential for *in situ* analysis. The inherent complexities of laser-induced plasma make it one of the most interesting, yet frustrating of all spectrochemical atom reservoirs. The main attractive feature of laser ablation is the ability to sample, vaporise, atomise, excite, and ionise both conducting and non-conducting solids in micro- and macro-regions.

It is a considerable challenge to elucidate the underlying mechanisms involved in ultra fast laser-solid interactions. Early observations comprise plasma ignition, ion kinetic energy distribution, crater formation, material ejection and plume sharpening (forward scattering). Some early models were evaluated [39]; for more recent phenomenological models, *cfr*. ref. [40]. The physics of laser ablation now distinguishes four distinct phenomena – optical excitation, electronic plasma, material plume and particle ejection – on a time scale ranging from femto- to micro-seconds. Ultra fast imaging (fs time scale) of laser ablation has allowed a description of laser-material interactions (cratering, plasma and plume formation), as shown in Table 3.5. For the ns time frame, *cfr.* refs. [41,42] in particular.

The amount and composition of the sampled vapour depends on the (mechanical, physical and chemical) properties of the sample and the laser beam parameters. Smallest aerosol particles are produced with Nd:YAG 193 nm [44]. Material removal by laser ablation can be mainly explained by thermal, photomechanical and photochemical interactions. Photomechanical effects become important at high laser fluence. Both thermal and photomechanical effects are favoured when short-pulsed IR lasers are used. Moderate laser irradiances (less than 10^7 W cm^{-2}) cause rapid heating and desorption of mainly neutral particles, clusters, free atoms and molecules through a thermally activated process. The threshold for microscale particle formation from solid materials is about 20 GW cm^{-2} laser intensity. At such higher laser irradiances the plume is usually associated with visible emission. Ground and excited state atomic species and ions formed in the microplasma can be detected by *in situ* diagnostic techniques such as AAS, LIF, (ICP)-AES or MS (in various forms: ITMS, ToF-MS, FTMS, SIMS, ICP-MS), performed directly on the plume. Thus, laser sampling mass spectrometry can be fairly versatile, enabling either molecular analysis using laser desorption or elemental (atomic) analysis using laser ablation [45]. The mass spectra can be very useful

Table 3.6. Main characteristics of laser excitation

Advantages:
- Cleanliness in depositing energy (no contamination)
- Easy tuning of delivered energy amount
- Localised sampling
- Pulsed probing
- Capability to probe insulators
- Remote and *in situ* analysis

Disadvantages:
- Need for absorption by sample of light of given wavelength
- Complexity of laser systems
- High cost, safety
- Poor understanding of laser-induced processes

for analysis of small spots if reference spectra can be obtained from known compounds. However, ionisation of the ejected material often occurs simultaneously, which complicates mass spectrometric analysis of minority species such as additives. Laser ablation gives some information about the elemental composition and functional groups of the polymer on small spots, but less structural information. If the vapour contains significant populations of excited and/or ionised atoms, direct LA-AES (LIBS) or direct laser microprobe for elemental analysis is possible. The photon beam does not induce charging of an insulating material upon irradiation unless other effects such as ion emission are involved. Table 3.6 shows the main characteristics of **laser excitation**.

The most frequently used lasers for ablation of solid samples are solid-state lasers, such as ruby (694 nm) and Nd:YAG (1064 nm, Q-switched, pulse length 5–10 ns; frequency-doubled, 532 nm, 5–8 ns; frequency-tripled, 355 nm, 4–8 ns; frequency-quadrupled, 266 nm, 3–7 ns), and also gas lasers, such as CO_2 lasers (10.6 μm) and N_2 lasers (337 nm). Also excimer lasers, such as XeF (351 nm, pulse length 20 ns), XeCl (308 nm, 20–45 ns), KrF (248 nm, 20–40 ns), KrCl (222 nm, 20–40 ns), and ArF lasers (193 nm, 20–40 ns) are applied [38]. The power delivered by normal pulsed lasers ranges from 10^3 to 10^6 W, with an irradiation time between 10^{-3} and 10^{-4} s. The introduction of the Q-switching technology allowing the use of giant pulses yields 10^6–10^8 W and a duration between 10^{-7} and 10^{-8} s. Laser characteristics such as the pulse duration, beam shape and wavelength play a significant role in the shape of the crater and the amount of ablated material. Typical crater sizes range from 10^{-2} to 10^{-5} cm^2. A comparison between Nd:YAG (1064, 532 and 355 nm) and XeCl (308 nm) lasers was reported [46]. The use of lower UV wavelengths favours better defined ablation cratering. The ablation mechanisms are totally different between NIR (1064 nm) and UV (266 nm) irradiations. For an infrared wavelength much higher thermal ablation is expected. Interaction of **IR lasers** with organic matter (IR-LA, infrared laser ablation) gives rise to multiphoton excitation over the vibrational manifolds of ground electronic states, which is then followed by thermal decomposition or pyrolysis. Ultra-fast heating permits labile, polar and high mass compounds to be released from a solid without thermal decomposition, in spite of the destructive potential of the high power density excitation. This apparent contradiction can readily be rationalised [47].

When **UV laser** radiation hits an organic polymer, the material is spontaneously etched away to a depth of 0.1 μm to several microns, apparently without thermal damage. Ablation of the surface of a polymer by a UV laser pulse is a function of the energy deposited in the solid in unit time. UV laser ablation (UV-LA) is typically carried out with a succession of pulses. Srinivasan *et al.* [48] have listed organic polymers whose ablation interactions with UV laser radiation have been studied and the analytical methods used. The products of ablation are: *(i)* volatile stable compounds (MW $<$ 200 Da); *(ii)* solid material; and *(iii)* transient species such as atoms and diatomics. Laser-induced fluorescence (LIF) spectroscopy, which combines absorption spectroscopy with fluorescence detection, is a probe for such transient species. There is evidence that the composition of the ablated products changes with UV wavelength, repetition rate and absolute fluence value. Additives *cq.* impurities may influence the laser ablation behaviour. A general mechanism that is applicable to all organic solids at all UV wavelengths does not exist. It is generally accepted that absorption of UV photons results in electronic excitation. The unique features in UV laser ablation of polymers are encountered only in the wavelength regions of electronic absorptions. The physical mechanisms involved in UV laser polymer ablation are still under discussion. With bond energies of organic polymers which are, typically, around 3–5 eV an explanation of the observed ablation rates on the basis of a purely thermal model would require very high surface temperatures, about (6–10) $\times$ 10^3 K for fluences near the ablation threshold [49]. An explanation of experimental data on the basis of purely photochemical mechanisms is equally impossible and a photophysical ablation model has been proposed [50]. At a wavelength at which a polymer has no reported absorption (*e.g.* PMMA at 308 nm) the characteristics of etching pass over from photoablation to the thermal ablation that is observed at visible and infrared wavelengths. Finally, when a UV pulse strikes a surface a loud audibly report is heard. The chemical physics of the ablation process can therefore also be studied by photoacoustics.

Rabek [51] and others [52] have described laser-induced decomposition of polymers. Comprehensive reviews have appeared on the interaction of laser radiation with solid materials and its significance in analytical chemistry [53,53a]. Various reviews cover the subjects of optical and mass spectrometry performed directly on the laser plume [54,55]. Moenke-Blankenburg [38] has described laser ablation for sample introduction. Advances in laser ablation of materials were recently reported [56,57].

Applications

Laser ablation of solids is of considerably interest in relation to chemical analysis and material fabrication. The main fields of analytical application of laser ablation are: *(i)* microanalysis; *(ii)* local analysis; *(iii)* distribution analysis with spatial resolution in microregions (migration studies); and *(iv)* bulk analysis. No firm conclusions have been obtained so far on the most suitable system for bulk analysis, localised analysis or on-line analysis, particularly regarding the different types of existing lasers. Strictly spoken, the use of lasers for sample introduction in inorganic analysis cannot be classified as laser spectroscopy.

Table 3.7 illustrates numerous possibilities in applying laser ablation-based technologies. Examples comprise semiquantitative microanalysis (screening) of polymeric compositions using LA-ICP-MS, on-line quantitative bulk analysis of polymeric compositions using LIP-AES, *etc.*

In comparison with solvent extractions and most other heat extraction methods, which are continuous processes, laser-induced ablation is a discontinuous process. **Bulk analysis**, which aims at replacing sample dissolution because sample preparation can be complex, time-consuming and costly or even impossible, can be achieved by summation of various single shots. Bulk analysis requires a fast, routine method with the same analytical performance (LODs, accuracy and precision) as that obtained by sample dissolution techniques. The key point is the availability of (certified) reference materials. This is still problematic for polymers and requires highest priority. Non-matrix-matching materials for calibration, such as the NIST 612 glass CRM or even the use of solutions, can overcome this limitation when semiquantitative results are sufficient, but not when accurate results are required. **Microanalysis** deals with very small amounts of substance (about 0.1 μg to 0.1 mg). Local analysis can be carried out by selectively ablating material from small areas on an inhomogeneous solid sample and requires a high lateral resolution, *i.e.* a small pit diameter. This diameter corresponds to a small spot size of the focused laser beam, which is mostly obtained by using optimised focusing systems, typically about 5–20 μm for polymers. Laser ablation gives compositional analysis at a spatial resolution limited only by the laser spot size and is applicable to all materials without restriction on physical properties, such as electrical conductivity. This spatial information is lost

Table 3.7. Different combinations of materials, type of analysis, dimensionality, instrument location and nature of laser-based techniques

Sample origin	Type of analysis	Dimensional considerations	Location	Technique
Material science	Quantitative	Bulk analysis	Laboratory	LA-ICP-AES
Geosciences	Semiquantitative/qualitative	Microanalysis	Remote	LA-ICP-MS
Polymers	Fingerprint isotope ratio	Local analysis	On-line	LIP-AES
Electronics		Depth profiling	Field	
Environmental science		Mapping		
Life sciences				
Forensic science				

when the solid sample is digested prior to the analysis. From NIR to UV craters get better defined (in terms of geometrical definition), which favours local analysis. Microanalysis is a quasi non-destructive method, which is crucial in some applications, such as forensic science. Laser ablation is used as a microprobe atomisation technique for resonance ionisation mass spectrometry (RIMS) of surface monolayers [58]. There is equally a need for **depth profiling**, which is performed by subsequent focusing of the laser radiation repeatedly to higher sampling depth (typically from 1 to about 100 μm). Depth profiling may be applied when a transient change in concentration occurs in the solid, which is the case with multilayers and for leaching studies in packaging in the food industry. In contrast to microanalysis, depth profiling does not require a high lateral resolution but a depth resolution of the order of 0.1 μm. Laser-based depth profiling cannot compete with conventional surface analysis methods in the nm range, but is certainly appropriate near 0.1 μm. For this application to polymers RF glow discharge is a competitive technique. **Mapping** is a growing area for laser-based technologies. Requirements are a lateral resolution of 3–50 μm, and a depth resolution of 3–10 μm for ppm to % impurity levels. Surfaces may range from mm^2 to cm^2. The maximum duration of the mapping procedure should range from min to h, depending on the scanned surface. With a 10 μm laser beam width 100–200 shots allow rapid sampling of a 1 mm $\times$ 1 mm spot, quite sufficient for the study of aggregates of additives in polymers. There is also a need for laser-based techniques in **on-line analysis** for process control, which must be fast with respect to intrinsic process time.

At present the **real strength of LA** lies in the measurement of distribution patterns of minor and trace elements in solid samples with high spatial resolution. Homogeneity testing is an application of LA-ICP-MS. There is an increasing demand for the development and validation of accurate and robust analytical technologies for the determination of the chemical characteristics of polymeric products in support of industrial needs, EC regulations (*e.g.* Directive on toy safety) or research. Needs are particularly acute for techniques able to determine trace element contents in solids with a minimum sample preparation. For this purpose, laser ablation-based methods, such as LA-ICP-AES/MS and laser-induced plasma atomic emission spectrometry (LIP-AES, LA-AES or LIBS) have already demonstrated a good potential for the determination of major, minor and trace elements with possible extension to in-line process analysis. Many of the methods based on laser ablation techniques currently being developed have not achieved satisfactory detectability, accuracy, and precision (compared to more traditional methods). The most **severe problem** with LA is the high dependence of the ablation rate upon the type of material. Polymers absorbing strongly at 10.6 μm (*e.g.* POM, PVC) show good laser ablation characteristics, as opposed to weakly absorbing materials (*e.g.* PP, PET). This makes quantitative analysis difficult. However, quantitative analysis by LA-ICP-MS is possible if appropriate standards are available.

Laser ablation technology for industrial applications has first come to prominence in 1965. For example, chlorinated rubber (CR) coatings were removed from concrete surfaces using a 60 W high power laser diode [59]. The ash particles were investigated by optical microscopy, image analysis, DTA/DTG, ESEM and EDX techniques.

3.2.1. Laser Ablation – Plasma Source Spectrometry

Principles and Characteristics

Traditional wet-chemical analysis of polymer samples is time-consuming because of the laborious sample preparation, sequential determinations, repetitive measurements and manual work in general. One of the uses of lasers is as an optical emission source. By suitable selection of laser, laser power and focus it is possible to ablate plastic materials quickly and in a controlled manner. The amount of sample being produced by the laser should be sufficient to achieve detection limits in the $ng\,g^{-1}$ range for the direct determination of trace elements in plastics. Several variations of atomic absorption using laser vaporisation have been reported [60]. However, direct analysis of the laser plume by atomic absorption is generally characterised by high % RSD due to fluctuations in laser evaporation causing inhomogeneities in the plume, non-reproducible expansion of the sample vapour, an intense continuum light emission, matrix effects, and scattering of light by particulates in the plume. From these experiments and others involving element analysis based on laser ablation and selectively excited radiation or resonance ionisation spectrometry, it has become apparent that for many analyses a clear distinction in the vaporisation and atomisation-excitation

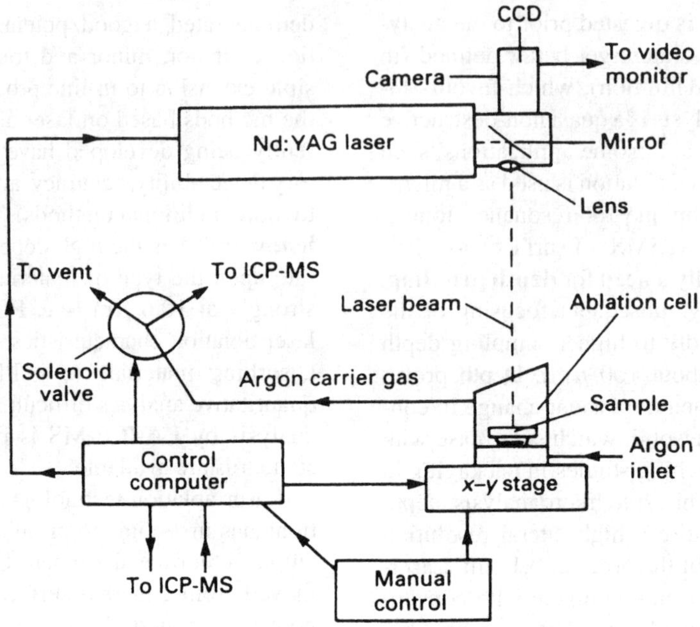

Fig. 3.1. Schematic diagram of LA-ICP-MS system. After Marshall *et al.* [69]. From J. Marshall *et al.*, *Journal of Analytical and Atomic Spectrometry* **6**, 145–150 (1991). Reproduced by permission of The Royal Society of Chemistry.

steps is desirable. Separation of the two processes allows experimental parameters to be optimised for each step. One of the earliest hybrid techniques was laser vaporisation-spark excitation (laser microprobe) [61] in which a sample was vaporised, usually with a Q-switched ruby laser, and the resulting laser plume was excited via a spark between two graphite electrodes for emission spectroscopy. This technique was not considered sufficiently accurate. Also laser ablation-assisted rf glow discharge has been reported [62].

Lasers have also been used as sample vaporisation devices for ICP [63] and microwave-induced plasma (MIP) [64]. Various types of couplings have been developed utilising both ICP-AES and direct current plasma emission spectrometry (DCP-AES) for detection [65] or ICP-MS [66]. In combination with laser ablation these methods offer direct solid sampling, rapid (semi)quantitative multi-element analysis and ease of application [67]. Using a pulsed ruby laser with low repetition frequency (<1 Hz) and high pulse energy (1 J) Gray [68] has first evaluated the potential of LA-ICP-MS. Given the sensitivity of the ICP-MS technique for liquid sample introduction, the amount of sample being produced by a laser (typically 300 μg per laser ablation event) is sufficient to achieve detection limits in the ng g^{-1} range for the direct determination of trace elements in polymeric materials. For several elements detection limits below 10 ng/g could be obtained directly on the solid sample for both conducting and non-conducting samples. Use of ruby lasers in the early systems leads again to high RSD values if heterogeneities in the sample are of greater magnitude than the laser beam spot. (Solids are defined as being analytically homogeneous if the variations in the chemical composition over the entire sample volume determined in various areas of the sample are not significantly larger than the error of the analytical procedure). Recently, the most frequently used (UV and IR) lasers for direct solid sampling have been solid-state lasers, frequency multiplied Nd:YAG, CO_2 and N_2 gas lasers, as well as excimer lasers. A typical experimental set-up for LA-ICP-MS is shown in Fig. 3.1 [69].

Solid sampling with a Nd:YAG laser for direct analysis with ICP-AES using an echelle optical system in conjunction with a solid-state detector is a straightforward technique [70].

Table 3.8 lists the **main features** of LA-ICP techniques. Because of the sensitivity of ICP-MS, for the combination of laser ablation and ICP-MS detection limits of 0.1–1 μg/g or better (down to 5–50 ng/g), have been reported [68]. Since the amount of material ablated is typically 1–10 ng/laser pulse, this

Table 3.8. Main characteristics of laser ablation-inductively coupled plasma spectroscopy

Advantages:

- *In situ* microsampling (of conducting and nonconducting samples)
- Minimal sample preparation
- Avoidance of contamination and losses
- Wavelength tuneable (VUV-IR)
- Highly controllable intensity
- Short duration (fs–ns)
- Rapid trace level multi-elemental monitoring (semiquantitative)
- Isotope ratio
- Medium-resolution spatial analysis
- Low-resolution depth profiling
- Allowance for homogeneity testing
- Spatial coherence (collimation)
- On-line control potential; remote sensing

Disadvantages:

- Difficult accurate quantitation
- Elemental fractionation
- Lack of internal standardisation and suitable matrix-matched polymeric calibration materials
- Spectral interference (LA-ICP-AES)
- No automation

represents absolute detection limits of <10 fg. The total amount ablated is usually *ca.* 500 μm^3. The method is virtually non-destructive.

Success of sample introduction methods, such as LA, ETV or LC in combination with ICP-MS, is on account of tandem source mass spectrometry [71]. Instead of one source being responsible for the vaporisation, atomisation and ionisation of the sample, these tasks are divided among two different sources: *(i)* LA/ETV: vaporisation (and atomisation); *(ii)* ICP: (atomisation and) ionisation; and *(iii)* MS: identification and quantification.

Advantages of laser microanalysis and laser ablation techniques are the minimised sample preparation and handling. As no solvent is injected, spectral interference by some polyatomic ions is greatly reduced. The technique has provision for microsampling and homogeneity testing. The ability of a laser to precisely target a small feature of interest is an important advantage compared to techniques with little or no spatial resolution such as spark source and GD-MS.

LA-ICP-MS suffers from ablation-, transport- and excitation-induced elemental fractionation. For quantitative analysis there are at least three principle problems that must be overcome: fluctuations, especially of the ablation and excitation process; matrix effects, depending on the sample composition; and chemical inhomogeneity of standard materials. Calibration is a weak point in laser ablation-plasma source spectrometry, which in turn affects accuracy. **Ablation efficiency** varies with the nature and properties of the sample (matrix effects) and laser operating parameters. Ablated mass depends on the laser characteristics (impulse energy, pulse repetition rate, wavelength, *etc.*), optical system (focusing) and materials properties (thermal conductivity, reflectivity, m.p., b.p., vaporisation enthalpy, chemical reactivity, *etc.*) [72]. Crater geometry is influenced mainly by focusing. The high power density available from fs pulsed UV lasers ($>10^{10}$ W cm^{-2}) is sufficient to cause non-thermal ablation and effectively limits melting effects and fractionation [73].

As the amount of sample volatilised is sample dependent and to compensate for pulse-to-pulse variations it is important to use an internal standard, such as a minor isotope of an element present at known concentration in sample and standard. **Quantitative analysis** on the basis of an internal standard requires that the (steady) concentration of at least one element in the sample (such as a minor isotope) is already known, as is often the case. Of course, it equally requires that the relationship between internal standard and element(s) to be analysed is known. This relationship can be established by ablating a material that contains many elements at known concentrations (*e.g.* NBS 610 glass). When an internal standard can be used for quantification the need for external standard calibration curves is obviated. The use of a single external standard (such as the NBS 610 glass material) to quantify plastic materials is highly questionable as the two samples differ in their ablation behaviour. It is commonly accepted that the best method for assessing the performance of LA-ICP-MS is the use of solid standards with identical matrix matching, especially for such complex materials as polymers. Furthermore, when using external standards in LA-ICP spectrometry, they need to be homogeneous in the microregion (the size of the laser craters produced is often about 100 μm in diameter). Calibration in LA-ICP-AES is often carried out by using solid external standards, commercially available or self-made, or by using the standard addition method, if addition is possible. Although semi-quantitative analysis by LA-ICP-AES can readily be conducted for fully unknown samples, for reliable quantitation of polymers (*e.g.* in

Table 3.9. Applications of the various laser-based techniques and types of analysis

Technique	Type of analysis				Site
	Bulk	Microanalysis	Depth profiling	Mapping	On-line
LA-ICP-AES	+				+
LA-ICP-MS	For low LODs	+	For low LODs		
LIP-AES	+	+	+	+	+

on-line control) the need for both solid standards and matrix matching of sample and standard is evident [74]. Quantitative measurements are not readily realised due to scarcity of suitable calibration materials (with the only exception of PE) and hence the LA approach should be viewed as a powerful complement to conventional ICP spectrometry. It is common practice to use previously analysed samples, or nominal formulation data for initial calibration.

In order to improve accuracy and precision of LA-ICP-MS the emission signals observed from the laser-induced plasma (LIP) during laser ablation can be used as an internal standard for correction of the ablated amount [75]. LA-ICP-ToFMS offers better future opportunities for quantitation than conventional internal standardisation methods. Because a large amount of ablated material can be injected without contamination, LA-ICP-AES is well suited to bulk analysis. In contrast, because of the very low amount of ablated sample, LA-ICP-MS is appropriate for microanalysis and has potential for generating information relating to the spatial distribution of analytes in plastic materials. LA-ICP-MS is obligatory when LODs in the order of 1–10 ppb are required. The detection limit of LA-ICP-ToFMS is 1 ppt (1 ft). LA-ICP-AES has a high analytical potential for on-line control, where usually only small changes in matrix composition occur [74]. The limits of detection of LA-ICP-AES are significantly worse than those of LA-ETAAS [76]. Selection of adequate technology is summarised in Table 3.9.

Advantages of LA-ICP-MS compared to XRF are better detection limits, simpler spectra and higher sensitivity for the light elements but application of the method is often hampered by absence of suitable standard reference materials. The method compares favourably with the electron microprobe, because of its low detection limits and wide range of elements from Li to U measurable. Very good agreement was achieved between the laser ablation results and the values determined by NAA, at the low ppm level [77]. A precision of *ca.* 10% was quoted. In comparison to GD-MS, UV-LA-ICP-MS (266 nm) is superior in terms of speed, lateral resolution and analysis of non-conductors, whereas GD-MS shows higher resolution, good depth profiling and lower matrix effects. Both techniques need reference samples/standards for calibration and highly homogeneous samples and standards due to their high spatial resolution [78]. The major problem of LA-ICP-MS is to achieve good accuracy. An absolute standardless approach might be feasible with LA-ICP-ToFMS provided that the sensitivity factors for each element are known (ionisation probability, *etc.*) [79].

Mitchell *et al.* [80] have described development of a laser ablation/direct-current argon plasma (DCP) emission spectrometry system based on a relatively low-energy, high-repetition rate Nd:YAG laser as opposed to the high-energy and low-repetition rate ruby lasers.

For laser ablation microanalysis two priorities may be defined: *(i)* comparison of existing laser technologies for a wide variety of matrices/applications (*cfr.* Table 3.7); and *(ii)* development of CRMs for bulk analysis covering a wide range of matrices.

The relation between acoustic and optical signals produced at UV-LA-ICP-AES spectrometry was discussed [81]. Moenke-Blankenburg [38] has described laser ablation for solid sample introduction for inorganic analysis. Arrowsmith [82] and Durrant [82a] have reviewed laser-assisted elemental analysis of solids by secondary plasma source mass spectrometry.

Applications

LA-ICP-AES of PVDF, PVC and PE materials containing *inorganic and organometallic additives* has been studied using a Nd:YAG laser, operated at both IR (1064 nm) and UV (355 and 266 nm) wavelengths [83]. At low energies laser-surface interaction leads to swelling of the surface. UV-LA eliminates swelling and leads to a better control of the laser ablation process. Optimum laser energy is in the range of 6 mJ (10 J/cm^2) to 10 mJ (14 J/cm^2) at 266 nm. The same experimental system ($\lambda = 1064$ and 266 nm) was used for direct determination of Ca-, Sn- and Ti-based additives in PVC and PE [74]. The signal-to-concentration ratios of the elements were strongly dependent on: *(i)* chemical structure of the elements in the additives; *(ii)* nature of the polymer matrix; and *(iii)* co-additive and additive-polymer interactions. The results confirm the need to match matrix samples and standards to achieve reliable quantitative analysis of polymers. The fact that quantitative LA-ICP-AES analysis using known prototypical glass compositions was demonstrated using Si as an internal standard [84] is in line with these observations.

Additives in PVC have been determined by UV-LA-ICP-AES ($\lambda = 266$ nm) using in-house PVC references [85]. In order to certify the amount of incorporated elements (Al, Ba, Ca, Cd, Mg, Na, Pb, Sb, Sn and Ti) as oxides (Al, Sb, Ti), hydroxides (Ca), stearates (Ba, Cd, Pb) and zeolites (Na), three alternative methods were evaluated (NAA, XRF and dissolution ICP-AES). Repeatability (1.6–5%) and reproducibility (2–5%) were satisfactory. Table 3.10 compares the LOD attainable for LA-ICP-AES for two different lasers with classical pneumatic nebulisation of a polymer solution. LOD values for LA are inferior to those of dissolution ICP-AES. However, taking into account the dilution factor in conventional analysis (typically 10 to 100) the sensitivity of both methods is rather similar [85]. LA-ICP-AES is sensitive enough to compete with some classical analytical techniques using liquid sample introduction. Poor correlations for Ba, Na and Pb were ascribed to digestion problems.

Table 3.10. Limits of detection (ppm) attainable by (LA)-ICP-AES of polymers[a]

	Nd:YAG 266 nm	XeCl 308 nm	Dissolution
Al	1.2	0.07	0.04
Ba	0.3	0.002	0.08
Ca	2.0	0.13	0.24
Cd	0.8	0.49	0.17
Mg	0.05	0.005	0.01
Sb	6.7	2.4	1.3
Sn	5.6	1.2	1.6
Ti	1.1	0.1	0.04

[a] After Hemmerlin *et al.* [85]. Reproduced from *Spectrochimica Acta* **B52**, M. Hemmerlin *et al.*, 421–430, Copyright (1997), with permission from Elsevier.

Mermet *et al.* [72] have systematically studied LA-ICP-AES of metallic additives in polymers, *i.c.* PE/(Ca, Sn, Ti) and PVC/(Ca, Sn, Ti). In this fundamental study particular attention was paid to crater and aerosol particle characteristics, LOD, matrix effects by polymer type and chemical nature of the additives, fractionation and composition of vapour fraction. The characteristics of the crater generated and the aerosol depend on the polymer nature. The ablation mechanisms for PVC and PE are quite different. LOD is related to both the polymer and additive chemical form, thus requiring an internal standard (IS). Carbon is not a good IS for some inorganic elements [72,83,85,86].

LA-ICP-AES with an Nd:YAG laser (1064 nm) has also been used for **rapid survey analysis** of polymer sheet materials (PP, PVC, rubber) and paints [87]. Although only qualitative/semiquantitative data was reported, rapid elemental monitoring capability is usually of considerable value. Measurements on both sides of PVC composite material revealed significant differences in Pb content associated with the optical properties of the surface. Such information is not available by performing a conventional bulk analysis (Pb content after acid dissolution, approximately 1000 $\mu g\, g^{-1}$).

Use of LA-ICP-MS for similar rapid semiquantitative multi-element analysis of polymeric materials containing a variety of fillers and other additives was reported for PP, PE, PVC, polyester and nylon by ICI [69] (Fig. 3.2) and for polycarbonate by DSM [88]. For semiquantitative analysis, where the ^{13}C signal was used as an internal standard in order to adjust for variations in ablation and transport of the different sample types, fair agreement ($\pm 25\%$) with XRF data was established. When compared to the external standard calibration approach, semiquantitative analysis is quicker (2 min per analysis) and provides more elemental information with only slightly less accuracy. Quantitative measurements made using LA-ICP-MS and matrix-matched calibration standards showed good agreement with nominal values (Tables 3.11 and 3.12).

Dobney *et al.* [88] have examined different sets of "in-house" polycarbonate standards (containing a selection of elements such as Ti, Sb, Cr, Co, Al, Ni, Na) to assess the feasibility of using external standard calibration and grid ablation for (bulk) quantification in LA-ICP-MS. Acceptable linear calibration

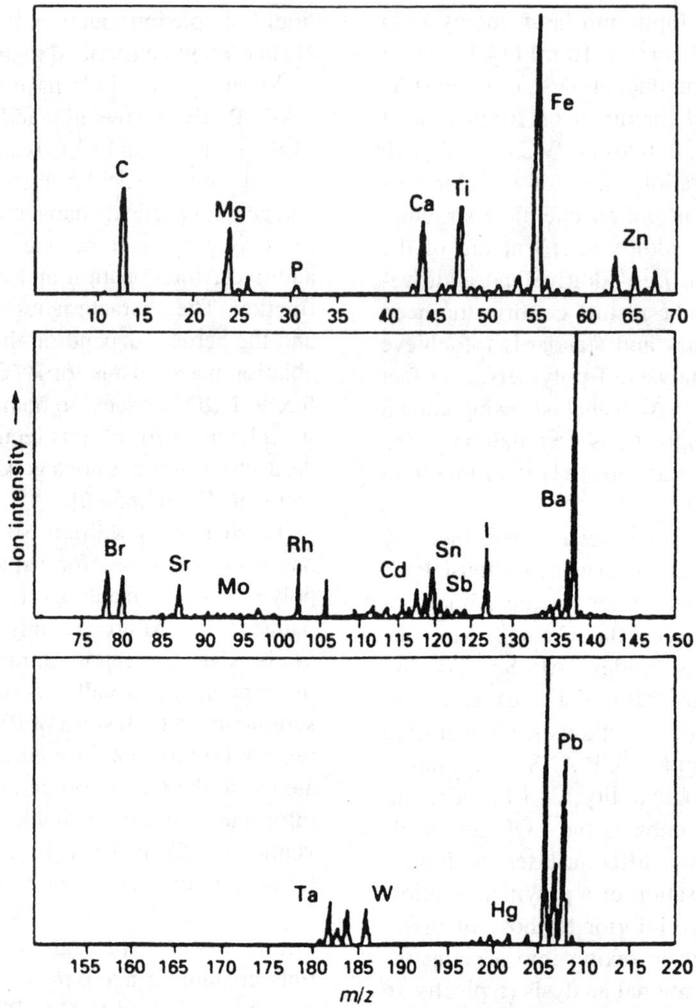

Fig. 3.2. LA-ICP-MS spectral scan of PVC material. After Marshall *et al.* [69]. From J. Marshall *et al.*, *Journal of Analytical and Atomic Spectrometry* **6**, 145–150 (1991). Reproduced by permission of The Royal Society of Chemistry.

Table 3.11. Quantitative analysis of plastics by LA-ICP-MS[a]

Element	Polyester		Polypropylene	
	LA-ICP-MS (ppm)	Nominal (ppm)	LA-ICP-MS (ppm)	Nominal (ppm)
Al	357	350	105	100
Si	720	770	600	750
P	93	105	60	45
Co	32	37	7	–
Zn	45	50	205	200
Sb	155	170	–	–

[a] After Marshall *et al.* [69]. From J. Marshall *et al.*, *Journal of Analytical and Atomic Spectrometry* **6**, 145–150 (1991). Reproduced by permission of The Royal Society of Chemistry.

Table 3.12. Quantitative determination of P, Ni and Mg in pigmented PP by LA-ICP-MS[a]

Sample	P content (ppm)		Ni content (ppm)		Mg content (ppm)	
	LA-ICP-MS	XRF	LA-ICP-MS	XRF	LA-ICP-MS	NAA
A	297	282	204	204	11	13
B	314	295	162	174	63	61
C	139	101	210	208	23	24
D	227	227	80	84	17	13
E	361	370	1	nd	29	31

nd = not detectable.

[a] After Marshall *et al.* [69]. From J. Marshall *et al.*, *Journal of Analytical and Atomic Spectrometry* **6**, 145–150 (1991). Reproduced by permission of The Royal Society of Chemistry.

curves were obtained after a relatively long analysis time per sample (*ca.* 10 min) and normalisation to the ^{13}C isotope. LA-ICP-MS has also been used to determine Pb in PE pellets [89].

LA-ICP-MS is most suitable for bulk analysis of solid samples, in particular for semiquantitative analysis of samples that are difficult to dissolve. As the size of the laser spot can be varied from 10 to 300 μm this multi-element method is also ideally suited to the study of inclusions and inhomogeneities, and in fact has recently been proposed as a tool for studying **heterogeneity** within polymers [88]. UV-LA-ICP-MS (266 nm, 2 mJ) with 40 to 90 μm craters is the more suitable tool for studying heterogeneity of polymers than the more time-consuming SS-ZAAS [88]. With LA-ICP-MS it is possible to gauge both the dimensions and scale on which the heterogeneity occurs. LA-ICP-MS can be used as a screening tool to quickly reject materials on the basis of unacceptable heterogeneity. LA-ICP-MS has also been used to study the elemental composition of gels in EPDM materials, as an alternative to PIXE [77]. The information gathered suggests how gel formation occurs in the production process. Similarly, the distribution patterns of minor constituents in solid samples have been determined [90].

LA-ICP-MS has allowed semiquantitative determination of trace elements in PVDF [91]. The technique has also been applied to investigate iron gall ink corrosion of written heritage [90a] and for **forensic discrimination** of automotive glass, as based on concentration estimates of 16 elements [92], and automotive paints [92a]. UV-LA-ICP-DRC-QMS (193 nm, 80 mJ) has been employed for routine **QC purposes** for the analysis of dopants and contaminants (1 min. simultaneous screening of 20 elements) in photo- and thermographic materials (TM) [93]. The 175 μm thick PET/Sb support layer, a 10 μm Ag-based active layer and a 1 μm Si-based protective layer of TM were easily distinguished on the basis of ^{121}Sb, ^{107}Ag and ^{30}Si. For this problem rf GD-AES is a cheaper alternative. Using LA-DCP-AES with long integration times, copper in cellulose was determined with 2–12% r.s.d. [80].

3.3. LASER SPECTROSCOPY

Principles and Characteristics

Atomic and molecular spectroscopy have been revolutionised and revitalised by laser technology. Laser spectroscopy excels in sensitivity and spectral or time resolution. Lasers enable chemists to microprobe systems of fluorophores with an exceptional degree of spectral selectivity, sensitivity, time response, and dynamic range. The four major areas of laser application in analytical atomic spectroscopy, namely LEAFS (*cfr.* Chp. 3.3.1), LA-ICP-AES/MS (*cfr.* Chp. 3.2.1), LIBS (*cfr.* Chp. 3.3.2) and LEIS, were described by Sneddon *et al.* [94]. For the types of laser and their general properties which are useful to the analytical atomic spectroscopist, *cfr.* ref. [95].

Laser-enhanced ionisation spectroscopy (LEIS) is essentially a very sensitive mono-element analysis method (as AFS or AAS) with limits of detection (LODs) often in the 1–100 pg mL^{-1} range [96,97]. LEIS is based on the measurement of the increase in ionisation of the analyte in a flame, furnace, or glow discharge by laser irradiation as a result of selective population of a level of the term diagram. In LEIS, one or two dye lasers are tuned to a wavelength characteristic of an electronic transition of the species of interest. The laser beam(s) are directed into an analytical flame, which serves as the atom reservoir. A potential is applied across the electrodes in the flame, and as the analyte atoms are selectively photoexcited and collisionally ionised, the resultant current is measured across the electrodes. The linear dynamic range of LEIS is comparable to that of OES or AFS (3–4 decades). At variance to optical emission, absorption and fluorescence spectrometry, spectral interferences in LEIS are absent as the analytical signal is not radiation but an electric current. Elements with a low ionisation potential can cause interferences, *e.g.* matrix elements (alkali and alkaline-earth metals). LEIS is a good mode of detection for LC because of the compatibility with the LC flow-rate. As the method lacks specific applications in polymer/additive analysis it is here not discussed in any detail.

In laser vaporisation experiments, generating a "plume", the laser's frequency may be synchronised with the resonance line of the element (analyte) to be analysed. The basic principles are: *(i)* absorption of the radiation by the analyte (LAAS: laser atomic absorption spectrometry); *(ii)* fluorescence (LIF, laser-induced fluorescence; LEAFS); or *(iii)* production of ionisation products (ions and electrons). LIF is an analytical method of high precision that is suitable for the measurement of diatomic species in the plume. Excitation spectroscopy or laser-excited fluorescence is not concerned with the spectral composition of the fluorescence but with how the overall intensity of emission varies with the wavelength of excitation.

The theory underlying electronic spectroscopy with lasers is essentially the theory of visible or ultraviolet photon interactions. The distinctive features that arise with the deployment of laser light in electronic spectroscopy are principally those that relate to or exploit the qualities of the electric field produced by the laser beam. Laser electronic spectroscopy is primarily based on coupling (usually of dipolar character) between the electron clouds of individual ions, atoms, chromophores or molecules of the sample with the electric field of the impinging laser radiation. The high level of monochromaticity affords the means to obtain high-resolution data.

There are several highly sensitive measurement techniques particularly suited to laser spectroscopy. These methods are all based on the monitoring of physical processes, which take place *subsequent* to absorption of radiation. Laser-based systems for absorption spectroscopy can be based either on fixed-frequency or tuneable laser sources. A special type of absorption spectroscopy, entailing multiphoton absorption, involves a particular excitation mechanism. Processes of **multiphoton spectroscopy** involve the concerted interaction of two or more photons with individual atoms or molecules. Raman scattering, which is another large application of laser spectroscopy in industrial analysis, is essentially such a process; one photon is absorbed and one is emitted in each molecular transition. However, the term "multiphoton" is generally applied to processes involving the concerted absorption of two or more photons. Single or double beam two-photon absorption requires a very intense source of light, such as a *pulsed* laser. Multiphoton studies where more than two photons are absorbed are generally based on a single beam of laser light, and transitions are subject to the condition

$$mh\nu = \Delta E \tag{3.1}$$

where m is an integer. The selection rules differ from conventional absorption. An advantage is the possibility of resonance enhancement (in analogy to that of the Raman effect). The resonance aspect of multiphoton absorption is central to a method of its detection, multiphoton ionisation spectroscopy (MPIS). Detection of ions is a highly sensitive method in absorption spectroscopy.

Strictly spoken, interactions of matter with laser light at a *fixed* wavelength are to be considered as *non*-spectroscopic chemical techniques. The advantage of studying the wavelength-dependence is the much more detailed information that is made available. In the future, hyphenated spectroscopy (F, R and LIBS in one instrument) may be envisaged.

This chapter is only concerned with spectroscopic techniques applicable to polymer/additive analysis insofar as not reported under the specific headings of laser ablation (*cfr.* Chp. 3.2) or laser pyrolysis (*cfr.* Chp. 3.5). Laser spectroscopy, which is no substitution of conventional methods but a valuable addition of the analytical toolbox, has extensively been reviewed [1,98]. Chemical spectroscopy with lasers [9] and applications of laser spectroscopy were described in monographs [1,3,99].

Applications

Laser photoacoustic spectroscopy (LPAS) can be used for selective monitoring of analytes [100]. The use of lasers for efficient ion production is becoming popular. Laser-enhanced ionisation (LEI) in flames gives detection limits down to 10^{-14} g mL^{-1} [101]. It involves multistep resonance laser excitation of atoms in flames, with their further collisional ionisation and detection of the particles formed, the number of which is proportional to the element content in a sample. Laser-enhanced electron ionisation (LEI) combined with ToFMS (GC-LEI-ToFMS) produces ag levels of sensitivity for organotin [102]. RPLC-LEIS has been used for the **speciation** of organolead compounds [103]. Sensitive detection by LEIS results in LODs comparable to those for LC-ICP-MS (0.25 ng mL^{-1} for tetraethyllead).

Time-resolved spectroscopy (*i.e.* laser flash photolysis and electron pulse radiolysis) has been used for the partial elucidation of the action of **stabilisers** by direct observation of single chemical elementary reactions and has given evidence for the formation of amine radical cations from HALS [104].

3.3.1. Laser-induced Atomic and Molecular Fluorescence Spectrometry

Principles and Characteristics

The analytical capabilities of the conventional fluorescence (CF) technique (*cfr.* Chp. 1.4.2) are enhanced by the use of lasers as excitation sources. These allow precise activation of fluorophores with finely tuned laser-induced emission. The laser provides a very selective means of populating excited states and the study of the spectra of radiation emitted as these states decay is generally known as laser-induced fluorescence (LIF, either atomic or molecular fluorescence) [105] or laser-excited atomic fluorescence spectrometry (LEAFS). In LIF an "absorption" spectrum is obtained by measuring the excitation spectrum for creating fluorescing excited state molecules. LIF is a non-destructive technique applicable *in situ* and capable of both organic and inorganic species, which exhibit fluorescence, upon irradiation with UV or visible excitation. In order to obtain detection limits in the pg/mL range LIF detection is needed. Practical advantages of **LIF detection** in comparison with CF detection are given in Table 3.13.

Table 3.13. Main characteristics of LIF detection

Advantages:
- High photon flux (improved S/N ratio, wavelength dependent)
- Extremely low detection limit
- Monochromaticity (selective excitation)
- Easy isolation from Rayleigh and Raman scattering
- Directionality (accurate focusing, spatial coherence)
- Non-invasive

Disadvantages:
- Wavelength restrictions
- Photodegradation of analytes under intense optical fields

Laser beams are particularly suited for detection of the small volumes of microbore systems (μHPLC, CE) since they can be focused to near the diffraction limit of light. LIF is the method of choice if nanoscale samples and separations are dealt with. The highly collimated beams generated from lasers allow efficient rejection of stray light during detection, consequently leading to very high mass sensitivities. LIF has the lowest limits of detection, *i.e.* in the zeptomole (10^{-21} mole) range [106].

The first applications of lasers in analytical molecular fluorescence [108] and LIF detection in combination with chromatography [109] used He–Cd lasing (325, 442 nm), *cfr.* Fig. 3.3. Introduction of diode lasers has given a major impetus to LIF. Cheap HeNe (633 nm) and diode continuous-wave lasers (635–670 nm) are appropriate excitation sources for NIR LIF detection in LC and outperform pulsed XeCl-excimer/dye and Nd:YAG/dye laser combinations [110,111]. The most important feature of fluorescence detection using red and near-infrared lasers as excitation source is reduction of the background fluorescence. A trend in LIF is the use of diode laser emitting in the blue or ultraviolet. The wavelength range of LIF detectors now extends to 248 nm. An inherent problem is the fact that most lasers only emit a few discrete wavelengths. As a result, it is difficult to achieve optimum excitation. For the laser armoury in LIF studies, *cfr.* ref. [112].

At *low* laser intensities, the processes of laser-induced excitation and subsequent fluorescence are simply related: for each laser photon absorbed an atom, ion or molecule is excited, and there is a constant probability that the subsequent decay will be via spontaneous radiation of a fluorescence photon. Provided that the sample is optically thin, the ratio of fluorescent signal to laser intensity will hence give the **absolute species concentration**. This is the basis of absolute measurements by LIF. However, at *high* laser intensities, typical of pulsed excitation, saturation occurs.

Laser-induced atomic fluorescence is increasingly being used for **trace elemental analysis**. Despite the monochromaticity of laser radiation the

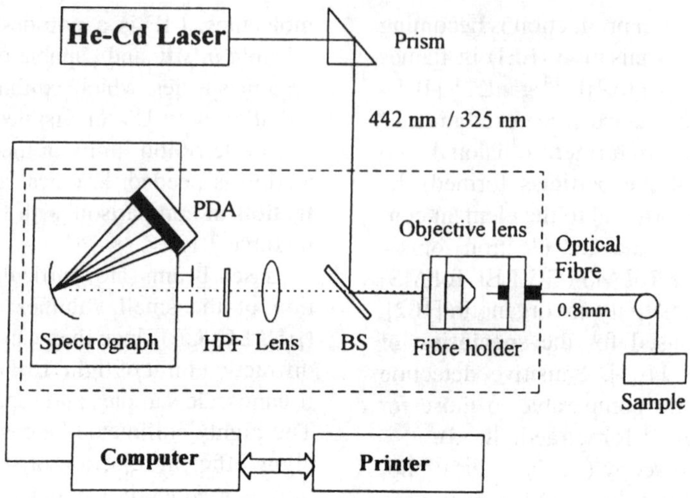

Fig. 3.3. Schematic diagram of a typical experimental module for LIF experiments. HPF: high pass filter; BS: dichroic beam splitter; PDA, photodiode array. After Anglos *et al.* [107]. Reproduced by permission of Verlag Mayer & Comp., Klosterneuburg and Vienna.

emission spectrum is complex. LEAFS or elemental LIF is a powerful technique capable of detecting ng to fg levels. This is due to the fact that when using a laser, it is possible to populate excited levels much more compared to a conventional light source such as a hollow cathode lamp [94]. Laser-induced atomic fluorescence has the ultimate capability of detecting single atoms. For atomic fluorescence measurements various lasers can be used, the provision being a high intensity of radiation in the range of absorption of the particular species of interest. The ideal laser for LEAFS is wavelength tuneable and capable of generating high peak energy and average power. Consequently, dye lasers are the most widely used, but the frequency doubled ruby laser, Nd:YAG, nitrogen or argon ion lasers, copper vapour and excimer lasers have all been successfully used in LEAFS. A detailed discussion on **basic theory on AFS** with laser excitation is available [94].

In general, the accuracy of LEAFS is to be assessed by comparison to standard reference materials (SRM®s) but is usually very good. The technique [113,114] shows freedom from spectral interferences, low matrix effects and a linear dynamic range of 5–7 orders of magnitude. With ETA-LEAFS, direct solid analysis is a fast, safe and accurate means to analyse polymers compared with methods that require difficult and time-consuming dissolution procedures. A major drawback inherent with ETA-LEAFS and ETAAS is the single-element capability.

A general comparison of various techniques for the direct determination of elements in polymers is given in Table 3.14. Rf GD-AES, rf GD-MS and laser ablation techniques are rather new, thus there are still few publications that discuss applications to the direct solid analysis of polymers (*cfr.* Chp. 3.2). The best detection limits and working range exist for ETA-LEAFS followed by techniques that involve MS. Calibration is easiest for ETA-LEAFS and ETAAS, which need simple aqueous standards, followed by NAA, which requires solid standards. Principles, instrumentation and applications of LEAFS were described recently [115]. LEAFS is extincting.

Fluorescence spectroscopy with lasers has been reviewed [3], in particular also the use of LIF for the characterisation of chemical systems [10].

Applications

Laser-induced fluorescence data provide a wide variety of detailed information about physical and chemical reactions. Laser-based time-resolved (picosecond) fluorescence spectroscopic techniques have been used to investigate the mechanism of **photostabilisation** by UVAs such as benzophenones, benzotriazoles and polymer-bound UV stabilisers [117]. Such ultrafast spectroscopic measurements can provide insight into the dynamics of the primary energy dissipation processes in polymers and polymer additives following light absorption. Excimer LIF spectra of plasticised PVC showed two distinct regions

Table 3.14. General comparison of various techniques for direct analysis of polymers[a]

Technique	Detection limit	Working range (magnitude)	Standard	Multiple elements	Analysis time	Matrix effects	Comments
ETA-LEAFS	pg g^{-1}	4–7 orders	Aqueous	No	Min	No	Laser complexity
ETAAS	μg–ng g^{-1}	1–3 orders	Aqueous	No	Min	No	Inexpensive, commercially available
NAA	μg–ng g^{-1}	3–5 orders	Solid	Yes	Min-d	No	Requires access to nuclear reactor
XRF	High μg g^{-1}	2–4 orders	Matrix matched	Yes	Min-hr	Yes	Inexpensive, commercially available
rf GD-AES	Low μg g^{-1}	2–4 orders	Matrix matched	Yes	Min-hr	Yes	Spatial information, few reported applications
rf GD-MS	High μg g^{-1}	3–6 orders	Matrix matched	Yes	Min-hr	Yes	Same as above plus high cost and isotopic analyses
LA-ICP, LA-MS	μg–ng g^{-1}	Unavailable	Matrix matched	Yes	Min-hr	Yes	Same as rf GD

[a] After Lonardo *et al.* [116]. From R.F. Lonardo *et al.*, *Journal of Analytical and Atomic Spectrometry* **11**, 279–285 (1996). Reproduced by permission of the Royal Society of Chemistry.

Table 3.15. Comparison of detection limits for phosphorous obtained with various techniques

Technique	Detection limit (based on 3σ)		Sample volume
	$\mu g\,L^{-1}$	ng	(mL)
ETA-LEAFS	0.4	0.008	0.020
ET-ICP-MS	0.30	0.015	0.050
LIF	0.7	7	10
ETAAS	550	5.5	0.010
DCP-AES	90	900*	10
ICP-MS	0.5	5*	10
ICP-AES	2	20*	10
MIP-AES	4.5	45*	10
Molybdenum Blue Colorimetry	1	10*	10
Cyclic voltammetry	40	400*	10
NAA	2.0 $\mu g\,g^{-1}$	2000	1 g

*Assuming a sample volume of 10 mL.

After Lonardo *et al.* [116]. From R.F. Lonardo *et al.*, *Journal of Analytical and Atomic Spectrometry* **11**, 279–285 (1996). Reproduced by permission of the Royal Society of Chemistry.

of emission bands, at 360 and 450 nm, attributable to the emission from the plasticiser (DOP) and carbonyl impurities, respectively. Artificial weathering quenches fluorescence from the phthalates [118]. Cullum *et al.* [119,120] have used spectrally and temporally resolved LIF to gain insight into the mechanism of **flame retardation**. They measured the effect of BFRs (HBCD and DECA) in HIPS and PVC on the concentration of OH radicals in a methane/air flame by monitoring both the LIF intensity and the fluorescence lifetime of the •OH radical. The interfacial interaction in polyimide (PI)/silica hybrid composites can be characterised by fluorescence spectroscopy [121].

LIF has gradually become the most sensitive technique for analyte detection in narrow-bore capillaries (*e.g.* in CE-LIF applications [10]). Also laser fluorometric detection for TLC was reported [122]. LIF detection often relies on derivatisation of the target molecules, as most analytes do not show native fluorescence when being excited with commercial lasers.

LIF is also used as a completely non-invasive measurement technique for determining the spatial concentration map of tracing agents, *e.g.* in pipe flows, or concentration fields in continuous stirred tank reactors.

Laser-induced fluorescence is also used in **artwork diagnostics** [27,107,123] and provides information which can be directly related to the molecular structure of pigments or other components of paintings, both inorganic and organic. Fluorescence properties of pigments, oils, and varnishes used in painting have widely been studied [124–126].

ETA-LEAFS with direct solid analysis using an excimer laser (308 nm) was used to measure phosphorous in PET over a wide range (2–3000 $\mu g\,g^{-1}$) with fg detection limits and a standard deviation of about 10% [116]; validation by ETAAS and ICP-AES. A comparison of the detection limits obtained by ETA-LEAFS with commonly used techniques for phosphorous determinations is presented in Table 3.15.

3.3.2. Laser-induced Breakdown Spectroscopy

Principles and Characteristics

Simultaneous multi-element analysis based on emission from a plasma generated by focussing a powerful laser beam on a sample (solid, liquid, or gas) is known as laser-induced breakdown spectroscopy (LIBS) and under a variety of semantic variations: time-resolved LIBS (TRELIBS), laser ablation emission spectroscopy (LAES), laser ablation atomic emission spectrometry (LA-AES), laser ablation optical emission spectrometry (LA-OES), laser plasma emission spectrometry (L-PES), laser-induced plasma spectroscopy (LIPS), laser spark spectroscopy (LSS), and laser-induced emission spectral analysis (LIESA®). Commercial LIBS analysers were already available in the 60/70s; the technique now enjoys a renaissance.

The **principles** of LIBS are similar to those of conventional plasma atomic emission spectrometry,

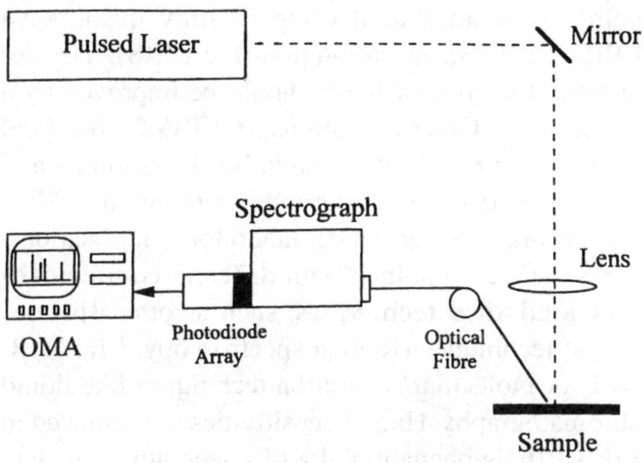

Fig. 3.4. Schematic diagram of experimental apparatus for LIBS. OMA: optical multichannel analyser. After Anglos *et al.* [107]. Reproduced by permission of Verlag Mayer & Comp., Klosterneuburg and Vienna.

such as ICP-AES, MIP-AES, DCP-AES, arc-AES and spark-AES. Characterisation of the material is based on the (atomic) emission from excited ions and atoms in the plasma. The emission lines for most elements are plentiful and because their width is in the order of several pm, there are sufficient characteristic emission lines for identification and quantitative analysis of most elements. The sample does not need to be transported to the plasma source; rather, the plasma is formed in or on the sample *in situ*. The most widely used are solid-state lasers (Nd:YAG, $\lambda = 1064$ nm, 532 nm, pulse length 5–10 ns), ruby (693 nm; 20 ns), gas lasers (CO_2: 10.6 μm, 100 ns; N_2 337 nm, 30 ps–10 ns) and excimer lasers (193 nm, 248 nm, 308 nm; 10–20 ns). Figure 3.4 shows the experimental set-up for LIBS.

Short-living plasma will be formed when the laser power density exceeds the breakdown threshold value of the solid surface. The laser-induced plasma has a life-time of approximately 20–30 μs, which is significantly longer than the laser pulse duration. The energy necessary for generation of a plasma at 15 ns laser pulses is about 100 mJ/pulse, corresponding to an intensity of some 10 MW/cm^2. Although the peak power of the laser is very high, the average power is typically less than 1 Watt. This means that the laser interaction with the sample is nearly non-thermal, *i.e.* no increase in sample temperature. The plasma is used for sampling, atomisation, excitation and ionisation in a single step. The excited evaporated components of the sample emit radiation spontaneously. For emission spectroscopy the crucial event is the transition from an excited level to an energetically lower level; the accompanying photon emission processes (bound-bound, free-bound or free-free transitions) are element specific. The emitted radiation is spectrally dispersed in a spectrometer using single- or multichannel time-resolving detectors. Usual equipment consists of a Czerny–Turner spectrograph equipped with multichannel plates (MCP) or ICCD, Paschen–Runge or Echelle spectrographs (0.02 nm spatial resolution in UV over a large spectral window). Modern equipment allows for complete elemental analysis in single shots. LIBS needs to be performed in time-resolved mode; if not gated the spectrum is dominated by continuum emission (<100 ns). Rowland spectrometers are unsuitable in time-resolved mode. The detector delay time (typically 10 μs) is very important to LIBS experiments and needs to be optimised in order to observe discrete lines. The observed emission lines are characteristic of the composition of the plasma and therefore of the atomic composition of the sample surface. The plasma emission does not contain information about the molecular structure of a polymer matrix host.

Factors influencing **laser-induced plasma (LIP) production** are:

- laser parameters: irradiance (10^6–10^7 photon/cm^2 for plasma ignition); wavelength (increasing plasma transparency in the order NIR, VIS, UV), pulse width (increasing irradiance in the order ns, ps, fs; usually ns);
- sample to laser distance;
- physical parameters of the target material;
- ambient conditions on plasma emission characteristics (quenching), mass loss and crater formation; mostly atmospheric pressure in air.

The inherent complexities of LIBS make it one of the most frustrating of all atomic reservoirs. The following experimental conditions and parameters are best suited to laser-induced emission spectral analysis: *(i)* argon as a buffer gas at reduced pressure (140 hPa); *(ii)* reduced laser irradiance; *(iii)* long delay times ($\geq$30 μs) between laser pulse and detector gate pulse; and *(iv)* use of analyte and reference spectral lines with comparable excitation energies [127].

Table 3.16 lists the **main characteristics** of LIBS. The technique can be used to analyse gases, liquids and solids directly because the plasma is produced by optical radiation. It is a simple method because it uses the single-step ability of focused laser radiation to vaporise *and* excite material. High sensitivity CCDs allow simultaneous detection of many

Table 3.16. Features of laser-induced breakdown spectroscopy

Advantages:
- Simultaneous *in situ* multi-element analysis (all elements) of arbitrary samples
- Applicable to solids, liquids, gases
- Minor or no sample preparation; no disposal of post-analytical residues
- High-speed analysis, versatile
- High selectivity
- Superb spatial resolution (1 to 100 μm)
- No restrictions on sample size
- Atmospheric conditions
- Microdestructive (2 ng–1 μg)
- Fibre-optics for remote sensing; non-invasive analysis
- Scanning microanalysis (SML)
- Mobile systems for environmental and on-line process analysis

Disadvantages:
- LOD: 0.3–100 ppm
- Lack of internal standardisation
- Qualitative or semiquantitative; calibration needed (matched standards, line ratios, normalisation)
- Relatively poor precision (5–10%)
- Matrix interference effects (particle size)
- Cost

lines, which permits overview analysis for a single shot. VUV-LIBS allows access to all elements emitting below 200 nm (*e.g.* P, S). With various laser systems, the analytical performance such as accuracy, precision, and detection limit (DL) for direct quantitative atomic spectroscopic analysis are not satisfactory. The primary reason is that the violent, nonlinear laser beam-target material interaction cannot be predicted accurately. LIBS requires calibration by means of an internal standard. Disadvantages include the difficulty in obtaining suitable standards (for this reason the technique must be regarded as semiquantitative) and a poorer sensitivity than several competing atomic spectroscopic techniques using solutions, such as ICP-AES/MS and GFAAS. Even though matrix effects are significant and detection limits not as good as for LA-ICP-MS, LIBS enjoys a major advantage. It allows fast and on-line analysis in industrial environments. When speed is more important than accuracy, LIBS can be applied to on-line analysis and product quality control. In conformity with other laser-based tools, LIBS will not be recognised as a proper analytical method as long as quantitative analysis has not improved. Ongoing work aims at developing truly quantitative LIBS (CF-LIBS or calibration-free LIBS). The detection efficiency of LIBS should be improved by a factor of ten. Other limitations for LIBS for practical application are self-absorption, line broadening, and high intensity of the background continuum.

For high performance laboratory applications, LIBS can be combined with different **complementary analytical techniques**, such as other spectroscopic techniques (Raman spectroscopy, LIF, MIR, NIR) or molecular separation techniques like liquid chromatography. Higher sensitivities are achieved in LIBS-LIF hyphenation. In this technique, analyte atoms in the plasma are selectively excited with a second laser pulse from a tuneable UV laser source.

Laser induced plasmas for analytical atomic spectroscopy have been reviewed by several authors [53, 128–131]. Time-integrated spatially resolved LIBS has been reported [132].

Applications

LIBS can be used for bulk analysis (typically: 10^2–10^3 mJ, 10^1–10^2 Hz), scanning microanalysis (10^{-1}–10^{-2} mJ, 10^2–10^3 Hz) and high speed applications (10^1–10^2 mJ laser pulse energy, 10^2–10^3 Hz repetition rate). Anderson *et al.* [133] have used LIBS to measure Ca and Sb concentrations in PVC, reporting detection limits of 0.016% and 0.04%, respectively. LIBS finds application in rapid on-line analysis of metals [134] and additive elements in polymers. In particular, LIBS is under active investigation for the analysis of heavy metals (Cd, Cr, Hg, Pb) and FRs (Sb, Br, Cl) in technical polymers, such as waste electrical and electronic equipment (WEEE), with the aim of evaluating the system's ability to rapidly identify samples on a conveyer belt [135]. Löbe *et al.* [136] have reported time-resolved LA-AES spectra of FR (4.2% Sb) PBT: 5 ns after the laser impulse mainly carbon was detected; after 250 ns also Sb was observed (Fig. 3.5). Using the 247.8 nm C line as an internal standard the Cd concentration in ABS could be determined with r.s.d. of 4.4% in the concentration range of 10^{-4} to 1 wt.%. For most elements the detection limit was about 10^{-3} to 10^{-4} wt.%.

Additives can add problems to recycling of plastics, but they can also be one of the means of identification, which might simplify the process. While analytical methods such as NIR and XRF are useful in recycling of domestic waste, LIBS finds application in case of thermoplasts (ABS, PP, PPO)

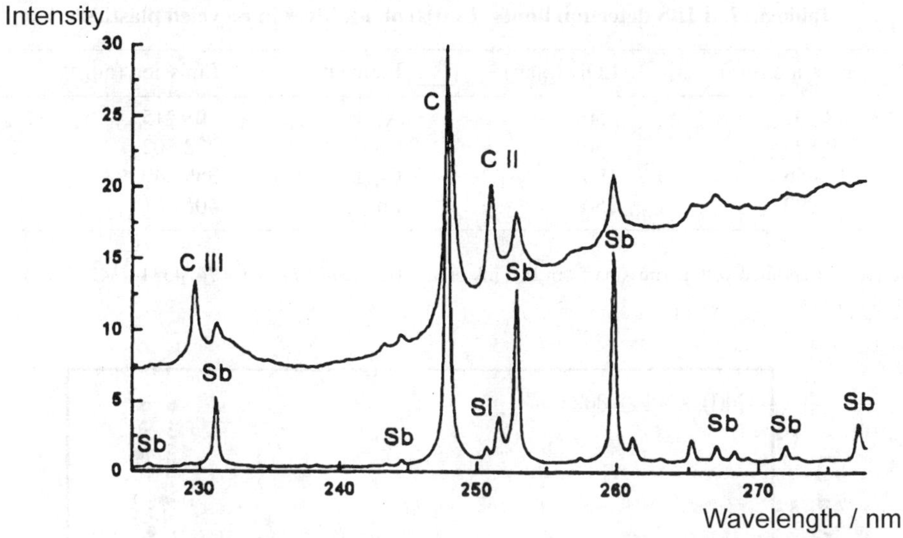

Fig. 3.5. Time-resolved LA-AES spectra of PBT/4.2% Sb flame retarded material after 5 ns (top) and 250 ns (bottom). After Löbe and Lucht [136]. Reproduced by permission of Vogel Industrie Medien, Würzburg.

from consumer electronics. Burmester [137] has examined extensively the usefulness of LIBS for the automatic qualitative identification of a variety of (additive containing) **polymer waste** components (such as PA6, PS, PC and PMMA) using a system consisting of an energetic (>0.5 J/cm^2) pulsed excimer laser ($\lambda = 248$ nm; 5–25 nm) and a diode-array spectrometer for detection of time-resolved, element specific, plasma line emissions (up to 580 nm), which secures ablation of sufficient polymeric material (*ca.* 10^{-9} g/laser pulse). The use of a UV laser ($\lambda = 248$ nm) is more advantageous than a Nd:YAG laser ($\lambda = 1064$ nm) in the NIR range because of the greater absorption in UV. Use of LIBS appears to be restricted to identification of unmodified heteroatom-containing base polymers (such as PA6, PVC) and additives: Br in PS, PC, PBT; Ca in PBT; Cd in ABS; Cl in PVC and elastomers; Si in GFR PP, PBT; and more specifically 6% Br, 4% Sb, 8% Si, 2% Al and 5% Ca in PBT. Various elements were detected by LIBS in commercial polymeric samples: PVC (Ti, Mg), PA6.6 (Ti), HDPE (Mg, Ca), LDPE (Mg, Ca) and ABS (Mg) [138].

LIBS is a suitable method for on-line analysis of the elemental composition of recycled complex thermoplasts from **consumer electronics** (ABS, PA, PC, PS, SBR, PPO, TPO, PVC, PPO/PS). The process analysis of such thermoplasts by LIBS (Nd:YAG at 266 nm) was reported [139]; data analysis consisted of multivariate methods and variable subset selection via a genetic algorithm. Reference analysis was performed by TXRF and NAA. Table 3.17 shows LIBS LODs of additives in recycled plastics using a normalisation approach based on the C(I) line at 247.856 nm. The practicability of LIBS for the elemental monitoring of recycled plastic was tested at an extruder within a recycling plant [139]. Figure 3.6 shows the extrusion of an ABS polymer with a recycled material containing 250 ppm Cd [139].

Laser-induced emission spectral analysis (LIESA®, Krupp) using a Q-switched Nd:YAG laser (1064 or 335 nm) or KrF excimer (248 nm) has been applied for **quality control** by in-stream elemental analysis of final mixes of technical rubber goods for the automotive industry (as an application of remote laser microanalysis, RELMA) [140]. Lorenzen *et al.* [141] have described a LIESA® application from 180 to 800 nm to frequent on-line homogeneity measurements of tyre rubber in the mixing shop during the forming process. To achieve accurate absolute concentration measurements the most abundant element with a high concentration should serve as the internal standard. Prominent analyte and reference lines and internal standards for LIESA® are given by ref. [141]. Analytical figures of merit are as follows: relative detection limit, 10–100 μg g^{-1}, absolute detection limit 1–100 pg; RSD value, 1–2%; relative accuracy, 3%; dynamic range, linearity over 1–3 orders of magnitude. LIESA® has been used successfully in the identification of polymers including polyamides, fluoropolymers, polycarbonates and acrylics, with detection at speeds exceeding 0.1 s

Table 3.17. LIBS detection limits (LODs) of additives in recycled plastics[a]

Element	Emission (nm)	LOD (ppm)	Element	Emission (nm)	LOD (ppm)
Ti (II)	308.802	50	Al (I)	308.215	100
Sb (I)	259.805	730	Cd (I)	228.802	30
Zn (I)	213.856	190	Cr (I)	359.349	80
Sn (I)	283.999	250	Pb (I)	405.781	70

[a] After Fink *et al.* [139]. Reprinted with permission from H. Fink *et al.*, *Analytical Chemistry* **74**, 4334–4342 (2002). Copyright (2002) American Chemical Society.

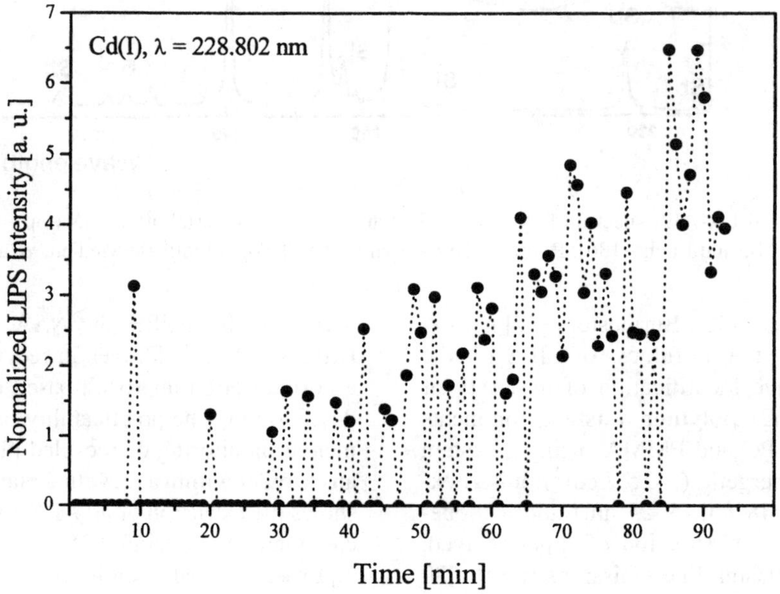

Fig. 3.6. LIBS signal of Cd in ABS during extrusion of different raw/recycled polymer mixtures. After Fink *et al.* [139]. Reprinted with permission from H. Fink *et al.*, *Analytical Chemistry* **74**, 4334–4342 (2002). Copyright (2002) American Chemical Society.

with moving objects. It is a particularly interesting technology, because the measurement error is not affected by factors such as colouring, additives such as plasticisers, or contamination on the surface.

LIBS/RELMA was tested on a series of industrial NBR compounds with various deliberate recipe errors of all components (rubber, fillers, carbon-black, plasticisers, FRs, AOs, accelerators, ZnO, sulfur, mineral oil). UV excimer laser wavelengths must be employed on polymeric surfaces since only then sharp and regular ablation patterns are produced without any thermal side-effects (at variance with IR Nd:YAG). LIBS/RELMA can be used for off-line analysis of vulcanisates, homogeneity testing and particle analysis in mixtures and element analysis of raw materials (in particular for fillers). RELMA is a QA device in the rubber mixing room allowing relative element concentrations and their distribution to be determined on a macroscopic scale in usual cycle times of internal mixers [142]. Therefore, quality statements about the compound or even end-products can be drawn very quickly. RELMA is also a QA instrument for liquids (plasticisers) and disperse solids (carbon-black) [143].

Other applications of LIBS are the identification and quantification of fillers in polymeric materials [144,145]. Hakkanen *et al.* [146] applied LIBS to measure the coating coverage, coating weight distribution, and three-dimensional distribution of elements in paper coatings. The experimental results were used to evaluate the quality of the paper coatings. LIBS fails in detecting deviations in hydrocarbon contents (mineral oil, paraffin); chlorine is also beyond reach (VUV).

Fibre-probe LIBS instruments allow *in situ* analysis of materials at a distance of 75 m [147]. LIBS is also used as a means of controlling the extent of ablation in **painted artwork conservation** [123] and *in situ* pigment identification [28,107]. Pigment analysis and depth profiling of successive paint layers is profitably carried out by LIBS (elemental information) in combination with Raman microscopy (molecular information) [148].

Industrial applications of LIBS have been described [149]. Some of the newest applications of LIBS are likely to be active topics of investigation for years to come.

3.3.2.1. *Laser-based Identification of Polymeric Materials*

Principles and Characteristics
Industry requires both technically sound and economic waste sorting processes. According to EC directives polymers containing cadmium (*e.g.* as CdS pigments) or bromine (*e.g.* PBDE) need to be eliminated from the recycling process.

A **basic requirement** for laser-based identification of polymers is detection of significant and polymer-specific differences in material behaviour. Recently an identification system for waste sorting (both recycled base polymers and additives), mainly based on thermal materials properties such as heat capacity and heat conductivity in dependency of temperature, has been proposed [137]. The thermo-optical method consists of local heating by means of a CO_2 laser (monochromatic IR irradiation), followed by evaluation in time-resolved fashion of the thermal response of matter (surface temperature distribution; thermography, IR line scanner equipped with a HgCdTe detector; emission between 3–5 and 8–12 μm). Comparison to standard samples was used for verification. In view of the carbon skeleton of thermoplasts CO_2 laser light ($\lambda = 10.6\ \mu$m, *i.e.* in the range of the stretching vibrational absorption bands of C−C bonds) is easily absorbed by these materials and thus penetrates only several hundred μm beyond the polymer surface. Polymer-specific surface temperature prophiles are thus obtained. Depending upon the absorption coefficients maximum surface temperatures of about 35°C (PE) and 160°C (PVC) are recorded and PE, PP, PTFE, POM, PA6, PC and PBT are easily distinguished from styrenics (*e.g.* ASA, SAN) and other polymeric groups (PS, ABS, PET, or PMMA, PVC).

Materials properties are greatly influenced by blending and compounding (*i.e.* addition of additives even up to 50%). For industrial use waste sorting process constraints comprise **allowance for additives** and (high concentration) fillers, in particular black pigments, a requirement not easily met. Additives, in particular black pigments, easily disturb polymer identification. Carbon-black influences the optical penetration depth of CO_2 laser light, causing a higher surface temperature. Most additives lead to an increase in absorption, especially for poorly absorbing polymers. For strongly absorbing polymers (PMMA, PVC) the optical penetration depth of CO_2 laser light is increased. However, the environmentally undesirable additives (Cd or Br containing, *etc.*) affect the thermal materials properties and absorption at $\lambda = 10.6\ \mu$m of strongly absorbing polymers (PMMA, PVC) to such a small extent that identification of such formulated additives is impossible in CO_2 laser conditions. For the purpose of recognition of additive-specific hetero-elements LIBS at 248 nm is considered to a more suitable, rapid and sensitive technique. Consequently, laser-aided polymer identification in combination with LIBS for additive identification is a powerful technical tool for recycling purposes, though not necessarily economically feasible [137].

Various other possible systems for automatic separation and identification of waste plastics (including additives) are summarised in Table 3.18. As may be seen, this table does not offer an outstanding single method for automatic identification and sorting of polymer waste, including additives, especially if one considers that methods omitted from the table are generally poor in additive recognition. In practice, a combination of techniques (XRF, XAS and MIR/NIR) was found to be industrially useful for sorting PE, PP, PVC, PET and PS, "electrotechnical scrap" and old-car plastic wastes [150], *cfr.* also Fig. 3.7.

It would appear that the use of laser-aided polymer waste sorting [137] (eventually in combination with LIBS) and fluorescent tracer systems (*cfr.* Chp. 1.4.2) [152] are technologically the most attractive, with the latter technology more geared towards the identification of polymers than additives.

Applications
Some recent variations on the aforementioned basic techniques seem to offer additional advantages.

Table 3.18. Polymer waste separation and identification techniques[a–c]

Technique	Automation	Selectivity	Confidence level	Speed	Flexibility	Polymer identification	Additive identification
MIR[d]	1	5	5	1	5	5	4
NIR[d]	5	3	3	4	3	3	2
UV/VIS[d]	2	1	2	4	1	2	2
LIBS	3	2	2	4	3	2	5
XRF	1	2	2	2	3	2	5
MS (pyrolysis)	1	3	3	1	3	2	5
Laser-aided identification[c]	4	4	4	4	5	4	2
Fluorescent tracers	2	5	5	4	5	5	1

[a] Legend: 1 (highly unsuitable) to 5 (highly suitable).
[b] Other procedures: flotation, electroseparation.
[c] CO_2 laser ($\lambda = 10.6\ \mu m$).
[d] Reflection spectroscopy.
After Burmester [137]. Reproduced by permission of the author.

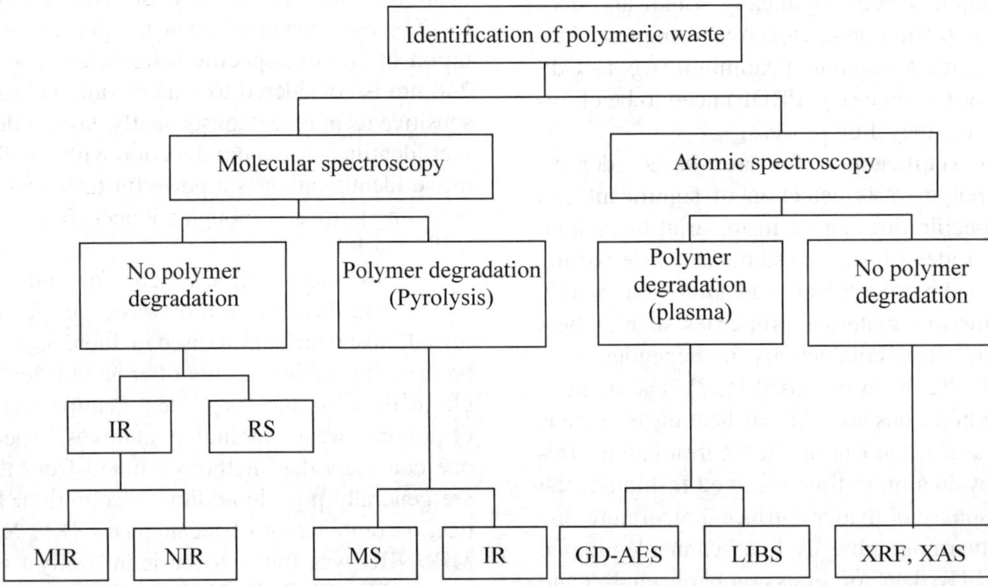

Fig. 3.7. Procedure for separation of plastics waste. After Nickel [151]. Reproduced by permission of Springer-Verlag, Copyright (1996).

NIRS (1.3–2.3 μm) in combination with **neural network analysis** applied to 300 samples of 51 different types of plastics has yielded an overall identification performance of 77% [153]. NIRS (900–1700 nm) combined to PCA and neural network analysis achieved a 97% accuracy rate for objects on a moving conveyer belt [154]. Raman spectroscopy (with PCA) has also been evaluated for plastics discrimination, with improved performance compared to NIRS [155].

Identification of different polymer types based on LIBS was reported for rather pure polymers with a negligible concentration of additives [156,157]. Discrimination between aliphatic and aromatic polymers was possible based on the C/H ratio (C(I) at 247.856 nm and H(I) at 656.285 nm). Sattmann *et al.* [156] explored the potential of LIBS for polymer identification using artificial neural networks, achieving an identification accuracy between 90 and 100% for HDPE, LDPE, PVC, PET and PP samples. Winefordner *et al.* [157] used Nd:YAG (1064 nm) LIBS to instant classification of commercial post-consumer plastic materials (PET, HDPE, LDPE, PVC, PP, PS) containing various additives by comparison to a reference library (in a 200–800 nm spectral window) based on 10 random surface spots. Using a **statistical correlation methodology** 90–99% probabilities of correct identification were achieved. The technique has excellent potential for on-line, real-time analysis of recycling materials. LIBS qualifies as an attractive methodology for rapid identification of plastics because it provides remote, *in situ* measurement capability and can be highly automated. However, further work is required here.

An overview of fast identification of plastics in recycling processes by means of spectroscopic analytical methods has recently appeared [150]. While there have been some promising results in identification and separation of mixed plastics packaging waste, there is still much work to be done in the more complex technical and engineering compounds [158].

Practical application of plastic waste sorting is greatly determined by legislation.

3.4. LASER DESORPTION/IONISATION METHODS

Principles and Characteristics

Desorption of adsorbed and surface species can be induced by photon bombardment, just as it can by electron bombardment or surface heating. The study of the phenomena involved in photon-induced desorption has taken place largely since high energy, high intensity lasers and synchrotron radiation photon sources have become available, as the cross-sections for photodesorption processes have been found to be small. It has been shown experimentally that molecules can be desorbed intact and with relatively little internal energy at high heating rates, whereas they decompose at low heating rates [159].

Laser desorption (LD) is defined as the use of short, intense pulses of laser light to induce formation of intact gaseous molecular ions or neutral molecules by direct removal of very small amounts of material (estimated at some tens of ng) from a (polymeric) matrix, including the vaporisation of thermally labile organic species without decomposition, by ultra-fast heating of a solid [160]. Laser-induced desorption techniques offer an important addition to the toolbox for the study of surface processes, as they yield information in systems that do not produce spontaneous desorption of a product species in an isothermal experiment. Laser desorption is a good approach to bring (large) molecules into the gas phase without fragmentation. Subsequent detection can be done in a variety of ways, including IR analysis after transfer (laser desorption transfer, LDT), optical spectroscopy after cooling (REMPI) or high-resolution mass spectrometry. Laser desorption is also used as a sample introduction scheme for GC analysis [161–164].

Desorption/ionisation (DI) methods may be classified as one or two-step processes with or without the use of lasers. Direct formation of ions by photon-solid interaction involves complicated energy deposition phenomena, which are largely unknown and hence completely uncontrollable. Most of the emitted particles are neutrals, not ions. Laser ionisation of the analyte isolated in the gas phase from the surrounding matrix constituents can be addressed more selectively than the solid target-matrix combination. **One-step desorption/ionisation** does not fully exploit the advantages offered by laser techniques.

Two-step desorption/ionisation methods include non-laser desorption/laser ionisation, laser desorption/non-laser post-ionisation and laser desorption/laser ionisation techniques. The main techniques of post-ionisation following non-laser desorption are discussed in Chp. 3.4.3. As to organic compounds, the direct exposure chemical ionisation (DCI) is significantly improved by the additional use of laser ionisation to make neutrals available in the gas phase. In laser desorption/non-laser post-ionisation lasers are used for desorption while secondary excitation (ionisation) of the ablated neutrals is performed by a non-laser method. Examples are LA-ICP-MS (*cfr.* Chp. 3.2.1) and laser sputtering of neutrals from the solid sample in an electric spark or glow-discharge (GD)MS [165]. The atoms are then ionised by collisions with electrons and energetic metastable atoms in the plasma. This makes a conducting sample unnecessary and may prove to permit local analysis. A particularly appealing method for surface analysis of both organic and inorganic constituents has been developed by combining laser-induced thermal desorption with EI or CI of the generated neutrals and mass analysis by FTMS [166,

167]. The application of EI or CI for post-ionisation permits interpretation of mass spectra with the aid of existing reference data. If the process of laser ablation resembles rapid thermal heating then a pyrolysis spectrum should resemble a LA/EI spectrum [168]. It is well known that pyrolysis using high heating rates gives very different results to pyrolysis with slow heating rates [169].

As in the sputtering or ablation process the ion yield is several orders of magnitude lower than the neutral atom or molecule yield, it is advantageous to post-ionise the neutral gas plume by a second laser. The processes are laser-ablation resonance ionisation spectrometry (LARIS) or laser-ablation resonance mass ionisation spectrometry (LA-RIMS), if the lasers are resonant. LA-RIMS has been proposed for the isotopic analysis of solid samples, in which an Nd:YAG laser is used for ablation and an excimer for ionisation [170]. The equivalent process for molecules rather than atoms in the plume is REMPI (*cfr.* Chp. 3.4.2.1). In these processes, the desorbing laser has ideally short pulse duration of less than 100 ns to maximise the number of neutrals within the volume irradiated by the ionising laser beam. The ionising laser should provide similar pulse durations and have adequate tuneability for various ionisation schemes.

In organic analysis a major task is the identification and characterisation of thermolabile, polar and/or high-molecular mass compounds, obtained in extremely small quantities and often as mixtures. The use of "soft" ionisation techniques, that limit the internal energy imparted by incident excitation, generally minimises fragmentation but produces low concentrations of ions. Ideally, the MS should be able to yield information on the molecular weight as well as on the functionalities present in the molecule. Hence, a soft ionisation technique alone does not provide a satisfactory answer. In fact, generation of fragments (daughter ions) always implies reduction of the parent peak intensity. A certain control over the degree of fragmentation is wanted. Moreover, MS should allow measurements in the high mass range with high mass resolution and accurate mass determination capabilities. The need for sensitivity is obvious, and a high degree of selectivity is desirable to eliminate purification steps. An ideal method would permit soft ionisation with experimental control of the degree of fragmentation. Such a technique is not yet available, but the development of pulsed laser desorption combined with multiphoton ionisation ToF-MS represents significant progress in that direction.

Laser desorption has developed into two distinct directions:

(i) laser desorption of intact *neutral molecules*, as usually achieved by means of pulsed **infrared** radiation from a CO_2 laser;

(ii) matrix-assisted laser desorption/ionisation (MALDI), in which pulsed **ultraviolet** radiation is used to produce intact gaseous *molecular ions*.

As to direct laser desorption, two major areas of laser mass spectrometry for analytical purposes have evolved: one-step laser desorption mass spectrometry (LDMS, *cfr.* Chp. 3.4.1) and laser microprobe mass spectrometry (LMMS, *cfr.* 3.4.5) with a spatial resolution of approximately 0.5 μm and ppb–ppm sensitivity. Most recently, LD has also been applied to depth profiling studies with laser microprobe instruments. LDMS offers several advantages over other desorption methods. In contrast to FAB and FD, no solvent matrix is required. Also, LD mass spectra typically give a more prominent (pseudo) molecular ion with minimal fragmentation than FAB or FD for a wide range of organic molecules. Variations in ionisation efficiencies are the limiting factor in the use of LDMS and LMMS for characterising multicomponent samples [171].

In commercial two-step laser based DI instruments [172] ultra-fast heating of solid samples by pulsed CO_2 laser irradiation, which allows desorption of labile compounds without thermal decomposition, is followed by supersonic jet cooling and REMPI. The second laser wavelength excites the analyte to an intermediate energy level and absorption of an additional photon causes ionisation.

In this Chapter the prospects of laser-based desorption/ionisation (LDI) mass spectrometric techniques for polymer/additive analysis are critically evaluated.

3.4.1. Laser Desorption Mass Spectrometry

Principles and Characteristics

Analytes with low volatility may be analysed by a variety of desorption techniques (*e.g.* FAB, FD, PD, SIMS, LD). These involve solid samples being bombarded with atoms, ions or photons. Mass spectrometric techniques such as FAB, PD and LD contribute significantly to the characterisation of complex substances [173–175]. In most cases the techniques are non-selective, and the mass spectral information from the analysis of complex mixtures is often obscured by matrix or fragmentation peaks. The main problem in obtaining direct mass spectra of solids is the vaporisation/ionisation step, which must be done while maintaining chemical integrity. Prime assets of lasers in mass spectrometry for analytical applications are selectivity as a result of the mass monitoring principle, sensitivity, ultimately down to a few ions, and the extended dynamic range. The implementation of lasers in MS experiments represents an extremely broad topic.

Desorption of surface species as a result of **photon bombardment** can arise by various processes: *(i)* thermal; *(ii)* shock-wave; and *(iii)* resonant. The former, direct heating mechanism, predominates under low-power density conditions in which desorption proceeds as a function of temperature. Laser interaction of solids enables ultra fast heating rate of the material up to extremely high temperatures. The steep gradient allows release of thermolabile compounds from the condensed phase without fragmentation. Laser-induced thermal desorption (LID or LITD) is much more likely to generate intact surface molecules, as opposed to molecular fragments commonly seen in conventional TDS experiments. In a large majority of studies involving desorption of surface species under the influence of intense infrared bombardment, desorption is due to direct heating of the substrate by absorption of the incident photons. Photon-stimulated desorption (PSD), where the excitation process leads directly to desorption of an atom or molecule, is characterised by higher power densities in which little fragmentation is observed (unlike thermal), and the sample probe or substrate does not participate in the desorption process. Indirect (or resonant) heating is associated with resonant absorption of photons causing vibrational excitation of the irradiated analyte; fragmentation is extensive near threshold, and there also appears to be no effect of the sample probe. Numerous experiments have been performed in attempts to understand the nature of the laser desorption process. The success of rapid heating approaches such as LD is due to the fact that rapid heating places non-volatile compounds within a temperature range at which the rate of vaporisation exceeds that of decomposition [176]. Difficulties arise as a result of the many experimental parameters that can be varied. Power density, wavelength, pulse duration, substrate, matrix, and sample preparation do all affect the desorption mechanisms and lead to changes in the mass spectrum. Lykke *et al.* [177] and others have studied many fundamental issues of laser desorption, such as desorption thresholds, velocity distributions, post-ionisation wavelength selectivity, *etc*.

Lasers can be used to achieve a **one-step desorption-ionisation** process (DLI, direct laser ionisation). The two physical processes involved are laser desorption and ionisation of the desorbed species. The first process depends mainly on the absorbed energy and volatility of the sample, while the second process depends principally on the ionisation potential of the desorbed species. In this case, the desorption laser also acts as the ionisation laser, limiting the ion yield and ultimately, the sensitivity.

Prompt-ion mass spectrometry, that is detection of ions produced directly by the laser desorption process [178,179], is the most widely utilised technique for monitoring laser desorbed species. **Prompt ions**, produced by laser desorption, can be detected with high sensitivity and, in many cases, can give a broad overall understanding of the composition of the surface. However, the ion yield of prompt ions from laser desorption suffers from extreme matrix effects. Thus, the measured ion abundances have little or no relation to the quantitative composition of the surface. The ions with the lowest ionisation potential (for positive ion desorption) or the highest electron affinity (for negative ion desorption) are usually the ions detected with low laser fluence. However, at higher laser fluences, many of the species present at the surface are detected. At these fluxes, however, fragmentation of desorbing molecules often becomes dominant. This will usually suffice to detect an unknown additive in a polymer.

Laser desorption (ionisation) - mass spectrometry (LDMS or one-step LDI-MS) experiments may be manipulated by varying: *(i)* sample presentation; *(ii)* power density, wavelength and pulse duration of the laser desorption technique to ablate the sample; and *(iii)* mass analyser type combined with a suitable ionisation mode. At variance to MALDI experiments in LDI-MS no matrix is required and **no special sample preparation** of any kind is necessary. The sample can be subjected to the pulsed desorption laser in a variety of ways. Most commonly, the sample is presented as a solid layer, deposited from solution, on a substrate, which absorbs at the laser wavelength. There is a well-defined threshold, above which successful desorption of neutrals can be achieved over a wide irradiance range (often several orders of magnitude). The irradiances used to achieve desorption of neutral molecules are generally in the order of 10^6–10^8 W cm^{-2}. Unlike matrix-assisted studies, laser desorption of adsorbates or surfaces produces a small population of ions. For power densities $<10^8$ W cm^{-2} the ratio of neutrals to ions is typically $\geq 10^4$:1. The formation of ions and neutrals occurs in two separate, temporally distinct, processes. Production of ions occurs instantaneously and persists for about 1 μs after onset of the laser pulse, whereas production of neutral molecules persists for about 100 μs. At high power densities ($>10^8$ W cm^{-2}) a significant yield of ions is produced in addition to neutrals. For thin sample films, laser power densities of approximately 10^8 W cm^2, and ns-μs pulse durations, it is generally accepted that laser desorption occurs via fast heating of the substrate, resulting in thermoionic emission of ions from the hot centre of the laser spot and vaporisation of intact neutral sample molecules further away at lower temperatures. As long as heating is rapid, evaporation of intact sample molecules rather than thermal degradation is favoured. Fine tuning of the laser power favours the desorption of intact neutrals (breaking intramolecular bonds) without excessive fragmentation (breaking intermolecular bonds). As pointed out in Chp. 3.2, laser desorption at moderate laser irradiances (less than 10^7 W cm^{-2}) creates suitable conditions for molecular rather than elemental analysis. For laser desorption/ablation studies typically pulsed lasers with relatively short pulse widths (<50 ns) are used because of their high peak powers and because short pulses reduce sample consumption and minimise laser-induced pyrolysis of the sample.

In this Chapter we are concerned with **laser desorption with a single wavelength**, which has achieved some success in analysis of organic polymer additives [178]. Laser desorption of neutral molecules can be achieved with pulsed IR, UV or VIS laser radiation. The choice of desorption wavelength changes the selectivity towards the materials present in the area selected for investigation and allows to study completely different desorption processes, *i.e.* electronic excitations *vs.* rovibrational excitation. **UV laser desorption** can be used for selective desorption of additives which are more susceptible to electronic excitation, whereas **IR laser desorption** can be used to thermally degrade a polymer network. Pulsed infrared laser desorption (commonly using a CO_2 laser at 10.6 μm, 100 ns) is essentially a thermal desorption process, although at present the mechanism of this process, whereby thermally labile molecules can be vaporised without decomposition, is not fully understood. Typically, the pulsed CO_2 laser is focused on the sample to a spot of diameter of approximately 100 μm–1 mm; for spatially resolved studies focal spot sizes of 20–50 μm are used. The versatility of IR laser desorption with respect to sample preparation is one of its great strengths. The sample may be present as a suspension in a viscous fluid such as glycerol, deposited on an infrared-absorbing substrate, or as a solid pellet. Typical substrates compatible with the CO_2 laser wavelength are stainless steel and glass. This technique can be used to desorb analyte molecules directly from "real" complex samples. Some workers prefer a Nd:YAG laser at 1064 nm [178] or otherwise at 532 and 266 nm [180]. Wilkins *et al.* [178] have indicated that Nd:YAG spectra generally show less fragmentation and are therefore superior for mixture analysis. A combination of different laser wavelength and laser power will allow to obtain the best desorption. There are fundamental differences with respect to power density in LDMS and LMMS.

Table 3.19 lists the **main characteristics** of LDMS. One-step desorption ionisation (DI) methods are the simplest mode of LDMS but sacrifice the selectivity that is attainable in the laser ionisation process. Despite a wide range of irradiation conditions employed (wavelengths from 10.6 μm to 266 nm, pulse widths from 5 to 150 ns or even continuous wave, power densities between 10^4 to 10^8 W cm^{-2}), these variations have only limited effect on the general features of the resulting mass spectra [181]. The same cannot be said for a change in MS type. LD spectra are distinguished by fairly high sensitivity, extended high mass range, and predominance of molecular ion species, usually even-electron ions such as $(M + H)^+$, virtually without fragmentation and decomposition, even for thermolabiles. Laser desorption allows to examine materials that are otherwise difficult or impossible to study by mass spectrometric methods, such as non-volatile organic materials. Compared to other ionisation methods, one-step laser desorption data are not only fast and easy to obtain, but they also appear to provide the highest average molecular weight. This may be explained by the harshness of the other ionisation techniques that tear apart the molecular entities during the ionisation process. For example, $\langle M_w \rangle$ of Oxypruf-20 [tetrakis(hydroxypropoxypropoxypropoxypropoxy-propyl)hydrazine] was determined as 1313.3 Da by Laser Probe® FTMS (*cfr.* Fig. 3.8), 1314.6 by LD-FTMS [182], as compared to 773.7 and 569.2 Da by SIMS and FAB ionisation techniques, respectively, which illustrates that they are not as well suited for molecular weight determinations [183].

Table 3.19. Main features of laser desorption mass spectrometry

Advantages:
- No sample preparation
- Small spot surface analysis
- Access to intractable solids (no solvent matrix required)
- Soft ionisation (alternative to FD, PD, FAB, EMD and SIMS)
- Depth profiling
- Selective process
- Identification power (molecular ion and elemental information)
- Characterisation of polar, large and involatile molecules
- Extended dynamic range
- Experimental flexibility (laser wavelength/power, ionisation mode, mass spectrometer type)

Disadvantages:
- Lack of mechanistic understanding
- Results not characteristic of the bulk (selective sputtering)
- Limited ion yield
- Limited sensitivity
- Questionable quantitation
- Extreme matrix effects
- Competition by other desorption techniques

LDMS is quite useful in many instances, but it does suffer from severe matrix effects due to the strong dependence of surface ionisation and surface neutralisation on both substrate chemistry and nature of the desorbate. The matrix effect arises from the physical processes involved: laser desorption and ionisation of the desorbed species. There can be significant selective sputtering of the surface, making the resultant mass spectrum somewhat questionable as to the relationship to the bulk structure.

In contrast to inorganic applications, organic LDMS experiments involve the use of a substantially larger range of lasers and MS analysers. In fact, almost any conceivable combination has been tried out [184]. Mass spectrometers which scan by measuring one m/z ratio at a time cannot be used effectively for laser desorption. Whereas high-resolution magnetic sector mass spectrometers are able to measure laser desorption mass spectra over a limited mass range and with relatively low resolving power [160], QIT mass spectrometers measure ions sequentially over relatively long periods of time and produce low-resolution mass spectra. Moreover, the trapped ions may undergo ion chemistry. Only mass spectrometers that can measure ions of all mass-to-charge (m/z) values near simultaneously, or mass spectrometers that can trap all the ions produced in a **single laser pulse**, are compatible with this pulsed ionisation technique. Because laser desorption is inherently a pulsed ionisation source, it should optimally be coupled to an inherently pulsed mass analyser: ToF [185], as *e.g.* in case

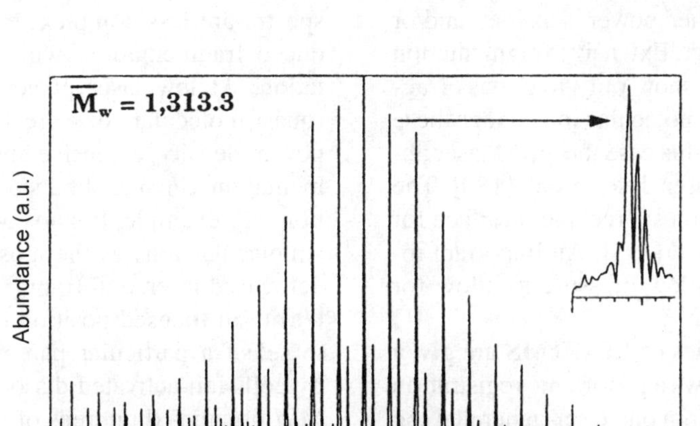

Fig. 3.8. LD-FTMS spectrum of Oxypruf-20 (MW 1313.3). After Campana *et al.* [183]. Reproduced by permission of J.E. Campana.

of MALDI, or FTICR [178,186–190], both of which provide prompt (<1 s) data acquisition of a complete mass spectrum. Both systems are commercially proven for laser desorption/ablation mass spectrometry. Whereas ToF-MS detects all ions near simultaneously, FTICR (or FTMS) is an ion trapping mass spectrometer that detects all ions simultaneously. It is obvious that the FTMS and ToF-MS technologies are much better suited than quadrupoles or magnetic sectors to study laser-produced ions. Analysis by LDMS gives even satisfactory results with amounts of material too small for a good separation by chromatographic techniques. This apparent drawback is namely fully compensated by the possibility to perform MS^n studies.

Van Vaeck *et al.* [191] have reported a comparative survey of the major characteristics of various LDI-MS instruments. Variations in laser type (TEA-CO_2, Nd:glass, CO_2, Nd:YAG, dye), mode (pulsed, continuous, Q-switch), wavelength, pulse duration, energy (W) and energy per pulse (J), spot diameter, power density (W/cm^2), geometry, mass spectrometer (B, QMS, ToF, FTMS, QITMS, *etc.*), ionisation mode and detector type determine a wide variety of experimental conditions and characteristics (mass range, mass resolution). Most devices have remained experimental in nature but various LD-FTMS systems have been commercialised. Surprisingly, the recorded mass spectra are all largely comparable in spite of the widely different experimental conditions. However, a mass discrimination effect in LD-FTMS was demonstrated [192]. Also, the characteristic ions formed on laser irradiation with a Nd:YAG laser at 266 or 532 nm differ somewhat from those for a CO_2 laser, due to higher power densities and/or the laser wavelength effect. Extensive fragmentation frequently occurs; PEGs show only low-mass fragments and no cationised molecular ions under these conditions. Therefore, in this case the term "laser ablation" is preferred to "laser desorption" [180]. The design of a laser desorption source and interface for FTMS have been described [193]. An important requirement is to maintain low pressure to allow for high resolution in FTMS.

The main characteristics of **LD-FTMS** are given in Table 3.20. FTMS allows a panoramic registration of the entire m/z range from one laser-generated ion shot only. FTMS offers potentially ultrahigh mass resolution and thus provides the additional capability for elemental composition of each ion in a single mass spectrum and the possibility of obtaining further structural information by tandem MS methods such as collisional activation, photodissociation, or selective reactions. The ability of absolute mass determination without the use of a standard, together with high resolution, permits unambiguous identification of additives (fragments) in polymers. Commercial LDMS instrumentation is available. Laser Probe® FTMS (Extrel FTMS) is capable of providing more detailed structural information compared to most other analytical techniques. For example, it is possible to measure the absolute mass of an oligomer or polymer fragment ion, and to calculate the most probable elemental composition. As an ion storage system, FTMS offers impressive capabilities for mass separation while the m/z range covered exceeds 10 kDa [191,194]. LD-FTMS spectra are less complex than EI spectra due to reduced fragmentation, which depends on fragmentation. At low laser power density predominantly (quasi)molecular ions are formed, while at higher power density extensive mass spectrometric fragmentation can be obtained for structural information. For example, Irganox MD-1024 generates quasimolecular ions as the most abundant ions, with a defocused laser, and fragment ions increasing in intensity in focused position [189]. It is also possible to select a particular parent ion for fragmentation by collision-activated dissociation (CAD). For low-MW organics (hundreds of Da or less), CAD can be used to produce structurally informative ions. Under low-energy FTMS conditions, CAD is less effective for fragmenting greater masses (several thousand Da). In these cases, photodissociation using a second laser pulse may provide an alternative.

Table 3.20. Main characteristics of LD-FTMS

Advantages:
- Ultrahigh mass resolution (accurate mass measurement)
- Panoramic positive and negative ion mass spectra (full m/z range)
- Low-pressure chemical ionisation
- Analysis of high mass materials
- Low chemical noise
- (Quasi) molecular ion spectra; minimal fragmentation
- MS^n option
- No mass discrimination, if any
- Applicability to many classes of compounds

Disadvantages:
- Sensitivity to surface characteristics
- Troublesome definition of desorption conditions for unknown substances
- Questionable reproducibility

LD gives good results in FTMS for many classes of compounds. The spectra are largely free of chemical noise. However, desorption conditions must be adjusted for different classes of compounds, and this leads to difficulties in handling unknown substances. Given that the laser spot diameter is 100 μm, the method is also well suited to identification of small polymeric contaminants, gels, *etc*. LD-FTMS spectra are sensitive to the nature of the presentation of the sample and its surface characteristics. For LD-FTMS and LD/PD-FTMS experiments finely ground sample material may be affixed directly to the probe tip. Alternatively, solid sample preparation may consist in dissolution in suitable solvents and deposition of the resulting solution onto a stainless-steel probe tip. After solvent evaporation a thin film is obtained. Salts added to enhance cationisation are either dissolved in the solvent with the sample or deposited onto the probe tip prior to or following sample deposition. Laser desorption spectra of organic molecules frequently show molecular ions produced by protonation ($[M + H]^+$) or cationisation ($[M + C]^+$ for polar compounds; C is cation) rather than electron abstraction. Some polymers (*e.g.* PDMS and PPS) produce stable molecular M^+ ions. In the negative ion mode, deprotonation (with $[M - H]^-$ formation) rather than electron attachment is usually observed. Radical ions are rarely generated. Both positive and negative ion spectra are employed in the identification process. A disadvantage of LD-FTMS is the need to improve the signal-to-noise and reproducibility of analysis. For that purpose alternative sampling/ionisation methods have been considered, such as probe introduction followed by electron impact (EI) ionisation and LD for desorbing neutrals followed by EI.

As already mentioned, while most investigations of laser desorption of organic molecules have employed pulsed CO_2 lasers, a Nd:YAG laser may be used. Adaptation of a Nd:YAG laser permits the photon energy to be multiplied by two, three, or four so that studies of MPI and **resonant laser desorption** can be done in the visible or UV. LD-FTMS experiments under resonant conditions are more reproducible than under non-resonant conditions. The use of a resonantly absorbing matrix might enable quantitative analysis by LD-FTMS.

At variance to MALDI for polymer analysis, laser desorption ionisation applied to polymer additives finds apparently fewer applications with the ToF mass analyser than with the FT mass spectrometer. **LDI-ToFMS** can produce accurate molecular weight information. Quadrupole ion trap mass spectrometers, which measure ions sequentially over relatively long periods of time and produce low-resolution mass spectra, are equally suitable for use with laser ablation and desorption. Here, no separation occurs whatsoever prior to introduction of the compounds into the mass spectrometer, at variance to on-line coupled LD-GC-MS [195]. However, MS^n experiments can by-pass the chromatographic separation. The smaller the fragmentations during desorption and ionisation, the bigger the amount of molecular ions in the trap and the better the chance of performing successful MS^n experiments. The molecular ions can easily be isolated, collisionally activated and identified through their fragmentation products. Critical advantages of **LA-ITMS** for direct solid mass spectrometry include [45]: *(i)* CID and MS^n experiments for identification of sample matrix species; *(ii)* ablated neutrals mass spectrometry by using electron ionisation; *(iii)* integration of sample ions for multiple laser pulses to increase sensitivity; and *(iv)* the possibility to employ different optical detection schemes, such as LIF, to achieve very low detection limits. Drawbacks intrinsic to LA-ITMS include: *(i)* restricted sample (pin) geometry; *(ii)* limited dynamic range resulting from space charge effects; *(iii)* possible sample matrix interferences such as the presence of easily ionised elements; and *(iv)* inability to separate sampling and ionisation since both result from the same laser pulse. Large molecules, such as proteins and synthetic materials, may be impossible to identify with LD-ITMS because their simply charged ions are beyond the ITMS molecular weight limit. ToF-MS has a higher sensitivity than ITMS with the external ion source where ions are lost during transport and trapping. Other mass spectrometric techniques, like MALDI, FD, ESI are useful for such molecules with relatively high mass.

Simpson *et al.* [138] have described polymer characterisation using **LD-IMS**, with elemental detection by means of LIBS. Ion mobility spectrometry (IMS) is a gas-phase electrophoretic technique, which uses a ^{63}Ni radioactive source (β-emitter) to ionise organic vapours by ion/molecule interactions [196]. Polymers and composites are an area of some importance for application of IMS. For polymers the laser provides a sample generation capability far superior to pyrolysis in both speed and control.

A systematic study of the **quantification** of the detection of additives in polymers by LD procedures is badly needed. If detection for direct analysis is shown to be quantifiable, then "on-line" laser desorption could be adapted to test the amount of incorporation of polymeric additives.

The current status of laser desorption has recently been reviewed [98], including LDMS [197], LD-FTMS [188,198,199] and LDMS/MS [200]. Van Vaeck *et al.* [191] have reviewed lasers in mass spectrometry, with emphasis on instrumentation.

Applications

Polymer analysis is a major application of LD-FTMS. Laser desorption allows examination of materials that are otherwise difficult or impossible to study by mass spectrometric methods. For example, LD-FTMS and LD/EI-FTMS produce good results for hydrocarbon waxes, which are generally difficult to analyse. LDMS can be used to obtain information such as: *(i)* degree of branching; *(ii)* identification of monomers in copolymers; *(iii)* $\langle M_w \rangle$; *(iv)* MWD (as an alternative to SEC, MALDI-ToFMS and HTLC); and *(v) in situ* analysis of additives in polymers/rubbers without extensive sample pretreatment. LDMS is not most suited for on-line polymer identification, as separation of the polymer fragments requires time-consuming gas chromatography.

Laser desorption MS has been used for direct analysis of **rubber additives**, *in situ* at the surface of elastomeric vulcanisates [201,202]. For example, the technique was used to analyse a sample of truck tyre that displayed premature sidewall cracking. By using a single laser pulse, the molecular ions of several intact molecular species (AOs, antiozonants and a production impurity of an additive) were observed on the rubber surface. Also McClennen *et al.* [203] have used controlled laser energy to desorb organic additives from a rubber vulcanisate.

McCrery *et al.* [204] have first used LD-FTMS to desorb and ionise organic compounds. The most widely used analytical applications of lasers in FTMS have been for desorption and ionisation of thermolabile and involatile substances and the characterisation of bulk and surface properties of (intractable) materials. LD-FTMS works especially well for polar polymers and additives with MW $<$ 10 kDa. Polymer analysis by LD-FTMS may involve direct determination of the MWDs for **oligomers** (up to 10 kDa). For high-MW **surfactants** ($>$2000 Da), which HPLC cannot resolve to individual oligomers, a single-shot LD-FTMS measurement provides a simple and accurate measure of MWD. Because this ionisation gives minimal fragmentation, it offers a reliable measure of the relative content of the oligomers, even without prior chromatographic separation of the compounds [205]. A vast improvement in resolution is noticed in detection of PEG oligomers from SEC to LDI-ToFMS and LD-FTMS [206]. LD-FTMS provides an accurate method for characterisation of polymer molecular weight averages and distributions, as illustrated for PEG, PEI, PS [207] and PPG [208]. Peaks corresponding to the K^+ or Na^+ attachment to the various oligomers dominate the spectra. Because FTMS generally does not suffer from mass discrimination it can provide accurate MWD. Polymers are now routinely used as calibration compounds for FTMS instruments [209].

Wilkins *et al.* [210] have used LD-FTMS to characterise a series of synthetic linear epoxypolyacenes, which are difficult to analyse by FAB due to their limited solubility. In this case, electron ionisation leads to extensive decomposition; LD-FTMS yields abundant molecular ions and some fragment ions in the negative ion spectra. Similarly, LD-FTMS has been used to analyse intractable solids such as poly(*p*-phenylene)s (PPP) [211] and conducting heterocyclic polymers. Kotsuka *et al.* [190] have given other examples for LD-FTMS for PDMS, Oxypruf-20, and PPG.

A large number of published applications of LD-FTMS has dealt with the analysis of polymers and **polymer additives** [188,214], *cfr.* also Table 3.21. In detailed studies, several additives in PE extracts were analysed directly by laser desorption (LD-FTMS) [178,189]. Exploitation of this technique for commercial samples has begun, with the goal of reliably detecting non-volatile additives *directly* in polymer matrices. An aspect of polymer/additive analysis that makes it a challenging problem is that the chemical nature of various additive types differs widely, ranging from salts (such as zinc or calcium stearate) to low-MW organics, oligomers and polymers. Laser desorption seems to offer an ionisation method that is effective for a wide variety of compound types, and is applicable to direct analysis of additives in the polymer, **without prior extraction**. Laser Probe® FTMS analyses using a CO_2 laser can determine many additives directly, even when the combination of laborious classical wet chemical techniques with other modern instrumental methods have proved difficult.

Table 3.21. Some additives studied by LD-FTMS[a]

Acrawax [189]
Antiozonants [201]
Carbon-black [212]
Dyes [213]
EBS wax [189]
Goodrite 3114 [178]
Irganox 245/259/1010/1035/1076/1098/3114/ MD1024 [178,189,214]
Metal stearates [178]
Naugard 76/524/BHT/DLTDP/DSTDP [178,189]
Oleamide [189]
Oxypruf-20 [190]
Pigments [212]
Polygard [189]
Sandostab PEPQ [178]
Seenox 412S [178]
Spinuvex A36 [178]
Stearamide [189]
Surfactants [212,215]
Tinuvin 144/320/326/440/622/770/900 [178,183,189,190]
Ultranox 226/236/246/626 [187,189]
Weston 618/TNNP [178,187]
Wingstay 100/300 [201]
XR250 [187]

[a] After Sheng *et al.* [212]. Reprinted with permission from L.-S. Sheng *et al.*, *ACS Symposium Series* **581**, 55 (1994). Copyright (1994) American Chemical Society.

LD-FTMS was also found to be a viable technique for identifying polymer additives in various PE extracts without prior chromatographic separation [189]. Reference spectra were obtained for 18 phenolic, thioester and phosphite AOs, 7 UVAs (Tinuvin family), and 4 amide waxes (Acrawax, EBS wax, oleamide and stearamide). The spectra contained intense quasimolecular ions (Na^+ and K^+ adducts) and also structurally informative fragments, depending on conditions used. The molecular weights of several additives in the Et_2O extracts of 3 commercial PEs were determined by using the characteristic Na^+/K^+ adduct pairs observed in the laser desorption spectra. Two additives were identified as phenolic AOs. Though MWs can easily be obtained, a database of additive molecular weights as well as spectra for confirmation is needed. This is less of a problem for a polymer chemist than for a general analytical chemist unfamiliar with additive use, since hundreds of additives are in common use and the probability of exact identification is therefore not high. While reference spectra are not needed, they can provide strong confirmation of additive identity through fragment ion information.

The exact mass capability of FTMS was illustrated by Asamoto *et al.* [189] for the case of an oxidation/hydrolysis product of Naugard 524 (473.2826 Da). While this work has not shown the feasibility of direct analysis of polymer films, the direct determination of Tinuvin 770 and Tinuvin 900 in a cross-linked polymer by means of Laser Probe® FTMS was illustrated by Campana *et al.* [183], *cfr.* Fig. 3.9. The protonated molecule was observed for Tinuvin 900 and the odd-electron molecular ion for Tinuvin 770. LD-FTMS has also provided rapid direct analysis of Irganox 1010 in dry mixes and PE films. Spectra were also obtained by extracting the additive (100 ppm) from a commercial resin [178]. Phosphite-type AOs in ABS and PVC were also examined by LD-FTMS [187]. Cody *et al.* [188] have examined various polymer additives. Salts, such as zinc or calcium stearate, pose problems for mass spectrometry because of the ionic nature of the compounds and the difficulty in dealing with doubly charged zinc and calcium cations. Nevertheless, laser desorption turns out to be a useful approach to the analysis of **stearates**. The positive ion spectrum of calcium stearate shows an ion resulting from loss of one stearate anion, leaving a net single positive charge (323 Da, or $[CaSt]^+$), and also a peak due to loss of CO at 295 Da. At high mass cluster ions are observed that are assigned to $[Ca_2St_3]^+$ and successive carbonyl losses (one per stearate). The negative ion spectrum shows the stearate anion (283 Da), and loss of a carbonyl (255 Da).

Single-step LDI/FTMS was used in the analysis of the source and chemical identity of **chromophores** present in PE lamination [216]. The technique allowed both sampling the surface and depth-profiling with high spatial resolution and accurate mass determination. Also laser-induced thermal desorption FTMS of multilayer thin films of a computer hard disk has been reported [217]. LD-FTMS could be used for the identification of fluorinated lubricants on magnetic storage media [188]. Laser Probe® FTMS analysis may also probe surface and interstitial **contaminants**.

Various studies have described a **comparison of LD-FTMS with FAB and FD** for polymer analysis [178,182,207]. Wilkins *et al.* [178] have evaluated CO_2 LD-FTMS, Nd:YAG LD-FTMS, and FAB-MS (sector and triple quadrupole) in a study of various non-volatile polymer additives

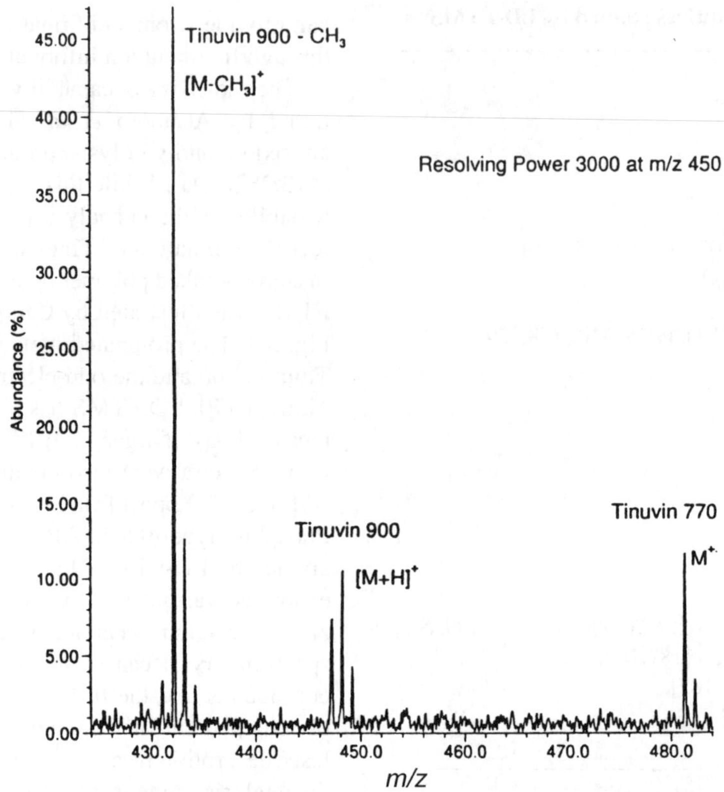

Fig. 3.9. Direct LD-FTMS of a cross-linked polymer containing a UV absorber and LM-HALS. After Campana *et al.* [183]. Reproduced by permission of J.E. Campana.

(DLTDP, DSTDP, Goodrite 3114, Seenox 412S, Irganox 1010, Spinuvex A36, Weston 618) with masses between 500 and 1300 Da. In general, FAB spectra show undesirably large amounts of fragmentation, while molecular ion spectra dominate LD-FTMS spectra. The latter spectra are also largely free of chemical noise, unlike FAB mass spectra, and detection limits are lower than those achievable by FAB. FAB yields poor quantitative information and is not recommended. LD-FTMS is clearly superior to FAB for analysis of these common polymer additives. Although both CO_2 and Nd:YAG spectra exhibit abundant cation-attached molecular ions, Nd:YAG spectra show less fragmentation, suggesting their possible superiority for mixture analysis. However, with this laser, the spectra are more dependent on a variety of factors including laser energy and power, and sample preparation. These aspects of LD-FTMS are still not fully understood. LD-FTMS spectra of alkoxylated pyrazoles (Oxypruf) and hydrazine polymers with average molecular weights between 600 and 1300 Da primarily contain K^+- and Na^+-attached intact oligomer ions and show little evidence of fragmentation [182]. In general, laser desorption spectra show more regular polymer distributions, fewer fragment ions, and less mass discrimination than SIMS and FAB spectra.

Asamoto *et al.* [185] have reported a **comparison of ToF-SIMS and LD-FTMS** for the direct identification of polymer additives. Both ToF-SIMS using keV Ar^+ sputtering and FTMS using IR laser desorption are soft desorption techniques combined with a spectrometer capable of high mass resolution, high mass range and parallel mass detection. Despite the fact that desorption and detection methods are quite different in the two approaches, the positive and negative spectra of PE extracts were surprisingly similar. Out of the seven additives found in the FTMS spectrum (as $(M+Na)^+$ and $(M+K)^+$ species) six were also found in ToF-SIMS spectra (as $(M+Ag)^+$ cationised quasimolecular ions). Both ToF-SIMS and FTMS can thus be used to obtain the molecular weight of polymer additives. The comparison suggests that IR laser desorption can be a softer vaporisation method than keV Ar^+ sputtering and produces predominantly intact molecular neutrals, which are Na or K cationised. The ToF-SIMS spectrum contains intact Ag cationised additives and more peaks due to fragmentation. The ToF-SIMS approach has inherent advantages in dynamic range, sensitivity, and reproducibility. The number of ions which can be trapped in the cell limits the dynamic range in FTMS. Irreproducibility of FTMS, which is attributed to variation in film thickness and laser power, does not affect qualitative identification of additives but would limit quantitative studies. This is not an issue in ToF-SIMS measurements.

Asamoto [199] and Cody *et al.* [188] have reviewed the use of LD-FTMS in polymer/additive deformulation. LD-FTMS is used in industrial problem solving, such as colour body analysis and in manufacturing problems.

As it turns out, a matrix is not always necessary nor even desirable to obtain a mass spectrum of a sample. In fact, to rapidly determine if an UV absorbing additive is present in a paint sample by LDI-MS, it is not convenient to alter the sample to introduce a matrix. Instead, a paint chip can directly be analysed by laser desorption with a UV or other type of laser [212].

Direct laser ablation is a widely used procedure for analysing polymers and monitoring the prompt ions produced. Lykke *et al.* [218] have used one-step LDI-ToFMS with 266 nm desorption in direct and extract analysis of vulcanisates. **UV-LDI-ToFMS** has been reported to be an appropriate technique for analysis of dyes (anthraquinones and indigo) on fibres [219]. Speciation of arsenic oxides, As_2O_3 and As_2O_5, using UV-LDI-ToFMS (266 nm) has been reported [220]. This technique can also be used to distinguish P_2O_3 and P_2O_5 in flame retarded polymers.

Gill *et al.* [45] have described **LA-ITMS** of polyimide samples. LA-ITMS has been used for the analysis of complex, non-volatile organic molecules [221] but no specific applications for additive analysis were reported.

LD-FTMS is a valuable technique for characterisation of industrial materials. Simonsick *et al.* [215] analysed novel dispersants, fluorinated surfactants, and natural oils with masses in the 500–3000 Da range. Wyplosz [219] has carried out a laser desorption mass spectrometric study of artists' organic pigments. The considerable potential of **infrared laser desorption** has yet to be fully exploited.

LD-FTMS of polymers and polymer additives was recently reviewed [222].

3.4.2. Laser Ionisation

Principles and Characteristics

Laser mass spectrometry involves sample ionisation using a laser rather than the traditional electrical methods. The use of lasers for more efficient ion production is becoming popular. The laser is unique as a means of ionising solids because by simply changing the power density in the focal spot the ionisation mechanism can be altered. However, the limited knowledge of ionisation by laser (micro) beam irradiation of solids causes intricate problems.

In **non-laser desorption/laser ionisation** techniques selective laser ionisation of gas phase neutrals (atoms and molecules), generated by evaporation, thermal heating, particle or photon bombardment, can be applied. Regarding ionisation, it is possible to differentiate between thermal impact with other particles (LEIS: laser-enhanced ionisation), ionisation due to an electric field (FILS: field ionisation laser spectrometry) and that due to photoionisation (RIS: resonance ionisation spectrometry). This type of ionisation can be combined with a mass spectrometer to detect isotopes selectively (RIMS: resonance ionisation mass spectrometry). If a polymer additive is volatile enough its presence can be ascertained via **ionisation-only** spectra. That is, the additive evaporates from the near-surface region for a certain amount of time, and the ionisation laser alone will yield a reasonable amount of signal to obtain an analysis of the additive present.

Ions may be formed by direct laser ionisation (DLI) with one laser pulse, by electron impact on pyrolysed or laser-ablated neutrals, or by laser post-ionisation. Depending on the analytical challenge, ionisation can be either as general as possible, or just the opposite, very selective for specific polymers. The most commonly used approaches are: *(i)* laser-generated cationisation; *(ii)* photoelectron attachment; *(iii)* single-photon ionisation (VUV source); *(iv)* two-photon ionisation, and; *(v)* resonance-enhanced multiphoton ionisation (REMPI) [223]. Of these, single-photon ionisation is conceptually the simplest and most general form of photoionisation by a photon with energy exceeding the ionisation potential and REMPI the most sophisticated form of photoionisation.

Major techniques of **post-ionisation** of secondary neutrals are resonance ionisation spectrometry (RIS) [224] in conjunction with MS (RIMS) [225], sputter initiated resonance ionisation spectroscopy (SIRIS), surface analysis by resonance ionisation of sputtered atoms (SARISA) and surface analysis by laser ionisation (SALI) if the laser wavelength is off-resonance and in the UV. The basic idea of RIMS concerns ion formation from atoms in the gas phase by absorption of photons that match energetically selected quantum states of these atoms [226]. Laser resonance ionisation mass spectrometry provides ultrasensitive and selective mass spectrometry of solids, but accuracy and precision are still to be assessed. Nogar *et al.* [227] have described LD/LA-RIMS. At variance to SIMS, RIMS exhibits no strong matrix effect. It is hoped that RIMS can be established as a complementary technique which can be used as a powerful aid for quantification of SIMS. The approach of ionisation of secondary neutral particles by untuned laser radiation followed by ToF-MS, SALI® [228,229], is based on non-resonant multiphoton ionisation (MUPI) of sputtered neutrals, released by primary ion bombardment. SALI detects low-mass fragments from polymers. The mass spectra are reminiscent of the lower part of static SIMS data. High-mass range signals from cationised oligomers are not present in SALI. Van Vaeck *et al.* [47] discussed these specialised techniques. Post-ionisation schemes are widely used, *e.g.* in SNMS (L-SNMS, REMPI-SNMS), GD-MS, LA-ICP-MS, TIMS, *etc.*

Laser ionisation of solids provides attractive features in both elemental and organic mass spectrometry. The photon beam is easily directed onto the sample in the confined space of an ion source and neither disturbs the electric fields nor induces sample charging of non-conducting solids. Ions produced by laser impact have been detected by various mass spectrometer types, including magnetic sector instruments [230], QMS, ToF, FTMS, ITMS, *etc.* The powerful combination of laser ionisation and MS has led to some major breakthroughs in the field of organic MS, notably ToF LMMS, FT LMMS and MALDI-ToFMS. A single ionising laser pulse is sufficient to record the entire mass spectrum.

CO_2 laser desorption usually produces mass spectra containing molecular ions with relatively few fragment ions. In forthcoming cases, CAD can sometimes be used to enhance analytical information. In an FTMS cell CAD is performed by translationally exciting ions, causing them to collide with a collision gas also present in the cell with sufficient energy to fragment. Too little excitation results in insufficient energy for fragmentation on collision and too much excitation ejects the ions from the cell. Low-energy CAD is inefficient for higher mass ions (thousands of Daltons). For such samples, photodissociation provides an attractive alternative.

The advantage of laser ionisation applied to secondary neutral atoms and molecules removed from a surface by stimulating radiation such as an ion or electron beam, as opposed to the one-step LDI, lies in the fact that most of the emitted particles are neutrals, not ions. Hence, the relative variations on the local gas phase populations are less affected by the chemical composition of the local mix. Quantitation is difficult as the total number of emitted particles and the ionisation cross-section are unknown.

Multiphoton dissociation/ionisation of polymer/additives is strongly dependent on the photon fluxes: low power accentuates only the more efficient ionisation of additives; at high power many more fragment peaks originate from cracking of the polymer backbone.

A specialised textbook is available [231].

Applications

Lykke *et al.* [218] have carried out LD, LI and two-step LDI of pure poly-TMDQ. The lack of signal from the ionisation-only experiment is indicative of the low volatility of this oligomeric HALS additive. The same authors have reported the 308 nm ionisation-only spectrum from a vulcanisate (direct analysis) showing the presence of the antioxidant di-*t*-octyldiphenylamine (DODPA, MW 393).

3.4.2.1. *Laser Multiphoton Ionisation*

Principles and Characteristics

Molecules can be ionised by simultaneous absorption of two or more photons of an intense UV laser beam. The prerequisite for ionisation is that the sum of the absorbed photon energies must exceed the ionisation potential of the target molecule. Since ionisation of most molecules requires energies in excess of that which can be supplied by a single photon, laser ionisation generally has to involve a **multiphoton excitation** process unless very short wavelengths below 150 nm are used. By absorption of the first photon the material is evaporated and brought to an excited state, and absorption of the second photon brings it to the region of the ionisation continuum, producing an ion of a given kinetic energy if the energy of the absorbed two photons exceeds the molecular ionisation potential. Hence, the term multiphoton ionisation mass spectrometry is applied to this method. Mostly the efficiencies for multiphoton processes are low, unless a vibronic intermediate state of the target compound is in resonance with the irradiated laser wavelength.

One of the most important aspects of laser mass spectrometry is the fact that the mass spectrum changes with laser wavelength. Ionisation is wavelength dependent (analyte selectivity) and 100% ionisation efficiency may be achieved. The fragmentation induced by MPI may be controlled to give mass spectra similar to those upon EI. The instrumentation is much less cumbersome than VUV sources used to generate higher energy photons for single photon ionisation. MPI is appealing as an ion source for any mass spectrometer.

In **resonance ionisation mass spectrometry** (RIMS) one or more lasers are tuned precisely to the wavelength required for the excited states and ionisation of evaporated atoms in order to get a highly selective ionisation of analyte [232]. RIMS provides ultra-high isotopic selectivity and sensitivity with detection limits up to 10^6 atoms per sample. High spectral brightness pulsed lasers are typically used to effect ionisation. RIMS has found application using a variety of atomisation sources, including filament, graphite furnace, flame, glow discharge and laser. RIMS spectra comprise merely the ions of the selected element, requiring only a simple, low-resolution mass spectrometer for analysis.

Resonant laser mass spectrometry allows two modes of operation: *(i)* wavelength selected mass spectrum (with considerably enhanced selectivity by ionising at one preselected wavelength with suppression of ion signals of interfering molecules); and *(ii)* mass selected UV spectrum (fingerprinting at one preselected mass). Some laser mass spectrometric methods using fixed wavelength excitation are already established in the field of inorganic trace analysis, and MALDI. Table 3.22 shows the **main characteristics** of resonance ionisation mass spectrometry.

Table 3.22. Main characteristics of resonance ionisation mass spectrometry

Advantages:
- Very high analyte selectivity (isomeric selectivity; mixture analysis)
- High sensitivity (sub-ppb)
- High spatial resolution ($\sim$1 μm)
- Elemental composition
- High speed analysis
- On-line capability (REMPI-ToFMS)
- Direct target-compound monitoring

Disadvantages:
- Not generally applicable (limited to UV absorbing analytes)
- Need for laser-UV spectral library
- Not suitable for multicomponent analysis of unknowns
- Not quantitative

Resonance-enhanced multiphoton ionisation (REMPI) is due to a successive absorption of two or more laser photons. The sum of the photon energies must exceed the ionisation energy. REMPI is the most sophisticated form of photoionisation, where the wavelength of an excitation laser is varied such that a second photon (from the same or another laser) ionises only when the first photon is resonant with a specific molecular level. This provides a combination of optical spectroscopy and mass spectrometry (*i.e.* a 2D analytical method). The intensity of a two-photon absorption is enhanced by several orders of magnitude if the wavelength of the laser is in resonance with the excitation energy of a UV transition of the target molecule. REMPI introduces tuneable substance selectivity into mass spectrometry. By the resonance condition, UV spectroscopy is directly involved in the ionisation process, assuring additional selectivity (wavelength selective ionisation). In REMPI spectroscopy the excitation spectra are obtained by measuring the ion yield as a function of excitation wavelength. A different mass fragmentation pattern can thus be recorded at each resonance, so providing a great deal more information on the molecular structure of the sample. Thus, with REMPI selective and soft (*i.e.* nearly fragmentationless) ionisation of many target compounds is feasible, even from complex substance mixtures. REMPI selectivity can be controlled by the selected REMPI scheme (laser wavelength(s)), laser pulse energies, laser pulse time (ns, ps or fs pulses) and temperature of the molecules to be ionised. Application of REMPI requires jet cooling [233]. The ultimate goal of laser desorption jet cooling is the ability to recognise desorbed molecules by their spectroscopy, or to selectively ionise certain molecules in a mixture by tuning to a unique resonance. Disadvantages of jet spectroscopy are that fingerprint spectra ("cold spectra") are practically unknown for identifying unknown species; moreover, not every molecule lends itself easily to REMPI. REMPI can be performed with different degrees of selectivity. Sharp UV spectra will lead to high selectivity, while broadened spectra will allow only a medium selectivity such as substance class. The lowest degree of selectivity, group selective ionisation, is still highly selective in comparison with EI ionisation. The highest degree of REMPI selectivity allows discrimination of isomers.

REMPI is not a generally applicable ionisation method, and ionisation probabilities can vary by orders of magnitude, depending on molecular electronic structure. REMPI detection is not suitable for identification of unknowns, but allows searching for specific targets with known high-resolution UV spectra (verification). Non-resonant multiphoton ionisation (NREMPI) by contrast provides essentially uniform ionisation efficiencies for a large number of elements, but requires high photon fluxes [234]. REMPI is more sensitive than LIF and can be used as a very efficient ion source for ToF-MS. The combination of laser induced resonance-enhanced multiphoton ionisation (LD-REMPI) and ToF-MS represents a highly selective and sensitive analytical technique, well suited for species-selective real-time on-line monitoring of trace products in complex gas mixtures. Fast headspace sampling is used in combination with REMPI-ToFMS. HS-REMPI-ToFMS, GC-REMPI-ToFMS [235], and HPLC-REMPI-ToFMS are all multidimensional analytical techniques.

Since multiphoton absorption generally produces molecular fragmentation patterns quite different from those of the normal mass spectrum, this method additionally provides useful insight into the dynamics of multiphoton excitation. Multiphoton ionisation and laser mass spectrometry have been reviewed [236].

Applications

For polymers with a suitable chromophore, such as aromatic groups, two-photon ionisation makes it possible to perform photoionisation at commercial laser wavelengths. This approach provides a certain degree of selectivity of detection.

High selectivity and low detection limits can be achieved by using direct ionisation of resonant analytes with a Nd:YAG laser in an MPI-FTMS experiment. Multiphoton ionisation in combination with FTMS [237,238] has been used for both surface analysis [239] and as an efficient and selective ionisation source for GC-FTMS [240]. The rugged design of REMPI-ToFMS devices allows application for target-compound analysis under industrial conditions, *e.g.* for process analytical tasks [241]. The highly selective and sensitive technique (ppb-ppt) is well-suited for real-time on-line monitoring of organic compounds in complex pyrolysis gases [242], as well as for the analysis of off-gas from an extruder. REMPI-ToFMS has been used for on-line analysis of volatiles in food [243,244]. TD/Py-REMPI-ToFMS offers a thermal preseparation, class/species selectivity and accurate mass.

3.4.3. Decoupled Laser Desorption/Ionisation

Principles and Characteristics

Direct laser desorption/ionisation mass spectrometry (LDMS) has proven to be quite useful for *in situ* analyses, but it suffers from ionisation matrix effects. Moreover, laser irradiation typically produces $\sim$1000-fold excess of neutrals over ions. To circumvent these problems, a laser post-ionisation step can be used following removal of neutral atoms and molecules by sputtering or laser desorption. It is generally useful to separate both temporally and spatially the ablation and ionisation events in order to better control these processes independently. Various approaches are possible. Examples of **laser desorption/non-laser post-ionisation** are the combination of laser ablation for sample introduction of solids in ICP-MS (*cfr.* Chp. 3.2.1) and the implementation of lasers to increase the yield and potentially the spatial resolution in glow-discharge (GD)MS. For organic applications laser-induced desorption has been used in conjunction with CI and EI in FTMS [187,245]. Laser-desorbed neutrals may be ionised by means of an electron beam passing through the trapped-ion ICR cell, in an experiment designated as LD/EI-FTMS [213]. An advantage of the use of LD/EI-FTMS is that tandem MS experiments can be performed on LD/EI generated ions because ions can be trapped for a period of several seconds or more in the FTMS instrument. A disadvantage is the low sensitivity because of low ionisation probabilities. The detection limit for LD/EI-FTMS is about an order of magnitude lower than for ATR-FTIR.

Laser desorption/laser post-ionisation (two-shot LDI) can display either non-specific detection (using non-resonant single-photon ionisation) or species-specific detection (using resonance-enhanced multiphoton ionisation). By exercising both of these options, complicated mixtures can be analysed for surface species. As efficiency of mass spectral analysis is greatly enhanced by ionising the

Table 3.23. Main features of two-step laser desorption/ionisation mass spectrometry

Advantages:
- No sample pretreatment
- Optimisation of desorption conditions
- Simultaneous vaporisation of sample components
- Minimisation of molecular fragmentation upon desorption
- Variable time delays between desorption and ionisation
- Increased ionisation efficiency with minimal fragmentation
- Minimal matrix interference
- Flexibility of choice of optically selective ionisation or non-selective photoionisation
- Wavelength-dependent photoionisation mass spectra
- Enhanced mass resolution
- Readily interpretable mass spectra (mainly molecular ion peaks)
- Spatially resolved analysis
- Wide applicability
- Direct investigation of complex mixtures
- Fast screening tool

Disadvantages:
- UV chromophore required (for R2PI)
- Difficult quantitation
- Desorption yields and ionisation cross-sections unknown
- Expensive equipment
- Experimental complexity
- Theoretically not fully understood

dominant neutral species after desorption, two-step LDI is a powerful molecular surface technique for the analysis of complex materials, such as polymer/additive systems. The main features of two-step LDI mass spectrometry are given in Table 3.23.

In LD high sensitivity is paramount since desorbed particle concentrations may be extremely low. To optimise the duty cycle of sampling and detection of a laser desorption/ionisation experiment, it is important to both temporally and spatially separate the ablation and ionisation events. By judicious selection of the variable parameters of both ablation laser and resonantly tuned ionisation laser (energy, wavelength, pulse width, respective orientation) both sensitivity and dynamic information can be maximised. Following early experiments of Schlag *et al.* [246], laser desorption of neutral molecules has been developed primarily in combination with laser photoionisation as a mass spectrometric technique. Further progress in this area was on account of refs. [247–250] giving experimental evidence that a two-laser scheme with separate desorption and ionisation steps (multiple laser shot LDI) is useful in characterising complex solid substances.

The principle of **two-step laser mass spectrometry** (L^2MS) is as follows (*cfr.* Fig. 3.10). A desorption laser in the far UV or far IR regions with short pulse widths, which is used to irradiate the polymer and cause ablation, provides the rapid heating necessary to avoid pyrolytic decomposition. During this ablation step the polymer is thermally decomposed and sample molecules are ejected from the bulk material. After a time delay, a second high-intensity pulsed laser beam (usually UV) in close proximity to the surface facilitates selective or non-selective ionisation of neutral molecules of the vaporised ejected sample. In L^2MS the pulsed ionising laser beam accomplishes ionisation over a short time interval, minimising the time for charge redistribution in the molecule, which often leads to fragmentation. The short pulse duration (typically 1–100 ns) of the desorption laser is important to maximise the spatial overlap between the ablation plume and the ionising laser field and hence the sensitivity. Short-duration pulses on the ns timescale, short-duration high-intensity pulses, monochromaticity and wavelength tuneability are the useful characteristics of laser light for this application. The two-step LDI methodology effectively reduces the likelihood of thermal decomposition and structural fragmentation of surface species. Quantitative analytical measurements and sensitive detection are significantly enhanced. The benefits of the separation of the desorption and ionisation events in laser desorption/laser photoionisation mass spectrometry (LD/PI MS or L^2MS) for *in situ* analysis of bulk polymer samples are: *(i)* desorption of neutral target molecules from the host polymer matrix with minimal decomposition; *(ii)* soft ionisation of the desorbed neutral species, resulting in readily interpretable mass spectra; *(iii)* selective ionisation of polymer additives, which have a significant one-photon absorption cross-section at the chosen ionisation wavelength; and *(iv)* highly sensitive detection of many polymer additive species. The total ionisation efficiency of L^2MS is about 100–1000 times greater than in methods where ions are directly produced on a surface (SIMS, LDMS).

The physical principles involved in L^2MS are either non-resonant single-photon ionisation or (non)-resonant multiphoton ionisation. In **two-step laser desorption/non-resonant single-photon ionisation**

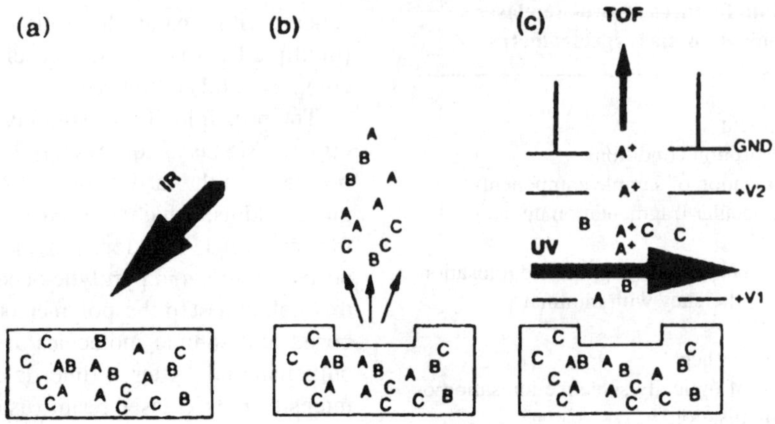

Fig. 3.10. L^2MS applied to polymer/additives (A, B, C). The polymer is irradiated by an IR laser pulse (a) and decomposes and is ejected from the bulk together with intact additives (b). Selective ionisation is carried out by a UV laser with resonant two-photon ionisation (c). The ions are mass separated and recorded in a time-of-flight mass spectrometer. After Zenobi [251]. Reproduced from R. Zenobi, *Fresenius J. Anal. Chem.* **348**, 506–509 (1994), by permission of Springer-Verlag, Copyright (1994).

for the analysis of bulk organic polymers, which is induced by a high-intensity pulsed untuned UV or VUV source rather than multiphoton ionisation, unwanted bond fragmentation is reduced. For non-selective (also called non-resonant) ionisation of desorbed species, the mass spectrometer, not the laser probe, performs the chemical differentiation of the desorbed sample. **Two-step laser desorption/non-resonant multiphoton analysis** achieves a chemically general mass spectral analysis of a surface. The powerful pulsed laser photionises all types of desorbed species with approximately equal efficiencies. This provides a uniformity of detection essential for an accurate and quantitative representation of the surface composition from a single mass spectrum. With information on the relative desorption yields of different species, an accurate quantitative assessment of surface concentrations can be attained.

The use of laser ablation coupled with RIMS detection of sputtered neutrals has a number of interesting advantages and applications. The RIMS technique offers elemental or molecular selectivity of optical spectroscopy in combination with extremely sensitive detection capabilities of an MS system. In a **two-step laser desorption/resonant multiphoton ionisation** process a pulsed CO_2 laser with emission at 10.6 μm is focused on a sample and serves as the desorption beam. The output of a frequency-quadrupled Nd:YAG laser (266 nm) softly ionises the desorbed species. The use of this high-powered, pulsed, and focused UV laser beam establishes a $1 + 1$ resonance-enhanced multiphoton ionisation (REMPI) condition. For *in situ* analysis of organic additives (with typical ionisation potentials between 7 and 10 eV) directly from polymers, absorption of two or more UV photons is required to achieve ionisation. For efficient photoionisation, a molecule must have a significant absorption cross-section at the wavelength of the ionising radiation employed, *i.e.* a suitable chromophore. If the molecule under study resonantly absorbs the first photon, greatly enhanced ionisation efficiency is found, providing a high degree of optical selectivity. Broad tunability of the ionising laser is considered essential to ensure a wide range of detectable species through various ionisation schemes whereas short pulse durations facilitate multiphoton ionisation. Resonance-enhanced laser mass spectrometry is a 2D analytical method, combining high-resolution UV spectroscopy and mass spectrometry.

In a laser desorption/photoionisation (LD/PI) experiment a number of experimental variables needs to be defined: *(i)* desorption parameters (requirement: minimal fragmentation); *(ii)* photoionisation parameters (resonant or non-resonant; near UV, far UV, VUV; single or multiple step laser shot; laser power); and *(iii)* mass analyser. By fine-tuning of the laser diminished fragmentation can be achieved by setting the laser power to produce ions near the threshold value for ionisation ("soft" ionisation).

Less internal energy is imparted to the molecular ions for each step and less fragmentation occurs. This results in readily interpretable mass spectra consisting almost exclusively of molecular ion peaks. By increasing the photoionisation power density, "hard" ionisation conditions can be produced, which induce fragmentation and provide structural information. By careful choice of the ionisation laser wavelength polymer additives with an appreciable absorption in the UV region of the spectrum can be selectively ionised in preference to the non-UV-absorbing host polymer. For selective (resonant) ionisation using a laser frequency tuned to ionise a particular molecule, the wavelength of the ionising laser contributes to the sample characterisation. By exploiting the ability of resonant two-photon ionisation (R2PI) **laser desorption/laser photoionisation (time-of-flight) mass spectrometry** (L^2MS or L^2ToFMS) stands out as a highly sensitive and optically selective method, which can serve as a very powerful tool for the direct detection of involatile and/or thermally labile organic target compounds in complex real-life sample mixtures allowing sub-femtomole (*i.e.* $<10^{-15}$ mole) identification and ps time resolution [252–254]. Near-UV wavelengths selectively ionise aromatic polymer additives, far-UV wavelengths photoionise other non-aromatic species, and VUV wavelengths provide access to all the desorbed species [177]. Coupling of FTMS or ToF-MS to selective post-ionisation wavelengths permits specificity in detecting additives in polymers. Moreover, the polymer matrix does not disturb. A complete L^2ToFMS experiment, from sample presentation to data storage, can be performed within 10 min. As the L^2ToFMS is capable of handling a large throughput of samples, it can be used as a semiquantitative screening tool.

L^2MS has several other features that are very attractive from an analytical point of view. The desorption laser can be focused to a near-diffraction limited spot on the surface (<1 μm for UV desorption, *ca.* 30 μm for CO_2 laser desorption), allowing spatially resolved analyses of heterogeneous surfaces. If the wavelength of the tuneable UV laser is optimised for REMPI ionisation of a molecular chromophore (*e.g.* aromatic ring systems, conjugated double bonds), a strong discrimination against other substances that do not carry this chromophore is usually found. This allows to solve "needle-in-a-haystack" type analytical problems, and in particular, it allows direct investigation of complex mixtures without time-consuming sample preparation or separation steps. Dependent on the absorption cross-section of the chromophores used for REMPI, detection limits can reach the attomole (10^{-18} mole) level. Disadvantages are that the analyte must contain a UV chromophore in order to achieve efficient resonance-enhanced two-photon ionisation. Knowledge of the UV spectrum of the analyte is required. Reliable calibration experiments and absolute quantitative measurements are difficult to perform. Quantitation requires standard additions and usually some assumptions with regard to the ionisation cross-sections. The absolute detection limits are both molecule- and wavelength-specific. This means that an L^2ToF mass spectrum does not give direct information on the relative concentrations of components in a mixture, unless they all have very similar photoionisation cross-sections at the wavelength used. It is possible though to determine a lower limit of detection for a particular species. Another disadvantage is the fact that the photoionisation efficiency decreases with increasing molecular size. Consequently, L^2ToFMS is limited to relatively low molecular masses, less than 10,000 Da, quite unlike MALDI which can access very high molecular masses. The mechanism of the laser-induced thermal desorption process used in L^2MS is not yet fully understood. Finally, the experimental complexity and cost of a two-laser system limit widespread use of L^2ToFMS as an analytical technique. L^2ToFMS can be used for 2D mapping of organic analytes (*cfr.* Chp. 5.9.1).

Van Bramer *et al.* [255] have advocated **photodissociation-photoionisation** (PDPI)-MS in which neutral molecules are photodissociated in the source region of a ToF-MS with a pulsed excimer laser beam (193 nm, 6.5 eV); after 1–3 μs a VUV beam (118-nm) softly ionises the dissociation products. Unlike conventional mass spectrometry, fragment ions in PDPI result predominantly from neutral rather than ionic decomposition. PDPI is highly sensitive since both dissociation and ionisation occur in the source region. Many organic compounds absorb 193-nm radiation, and virtually all absorb at 157 nm. In each case, the photon energy is more than sufficient for bond cleavage (3.6 eV for a C−C bond). The 118-nm radiation is above yet close to the ionisation potentials of most organic compounds and achieves fairly soft ionisation.

Two-step LDI-MS has recently been reviewed [223], in particular also the current status of laser desorption/laser photoionisation time-of-flight mass spectrometry and its applications [98,256–258]. Freiser [259] has reviewed the theory underlying photodissociation of ions with MS detection. Much remains to be learned about the effects of laser power density and wavelength for both desorption and photoionisation lasers.

Applications

The methodology of laser-assisted mass spectrometry is particularly a useful tool in the development and analysis of materials with **surface-related** capacities, such as slip agents, lubricants, surfactants, *etc*.

Detection and quantitation of phosphite AOs in polymers is a challenging analytical problem not easily solved by conventional methods. FTIR analysis of the intact polymer may be precluded for polymers with realistic AO concentrations (250 ppm to 2%), which show strong IR absorption of the matrix; on the other hand, the antioxidant fragments extensively in MS analysis. **LD/EI-FTMS** (1.064 μm) was used successfully to analyse Ultranox 626 diphosphite and its corresponding diphosphate oxidation product (XR 2502) in PP, ABS and PET, and Weston 618 diphosphite in ABS [187]. For each of the additives, the molecular ion M^+ was observed as the predominant species with virtually no fragmentation. Abundant molecular ions were also detected for Ultranox 626 diphosphite in a mixed polymer of PET, PP and ABS at low additive concentrations ($<0.1\%$) by direct analysis of the polymer film when the probe was heated to 200°C prior to laser desorption. LD/EI-FTMS offers a sensitive and accurate means for detecting non-volatile phosphite additives at typical concentrations in solid polymers, without the need for any chemical pretreatment. The observed sensitivity ($\sim$0.1% additive in solid polymer) is superior to FTIR detection; the method is applicable to polymers whose own IR absorption precludes direct FTIR detection of additives. LD/EI-FTMS was also used for direct determination of *ca.* 2% Tinuvin 770 and Tinuvin 900 in a cross-linked polymer using a CO_2 laser for ablation [212]. The molecular ion signals observed corresponded to detection of about 30 pmoles of additive in each laser ablation event. Exact mass measurements confirmed the identity of the two additives. Whereas direct ionisation by laser desorption of a "hard wax" sample (partially unsaturated long-chain hydrocarbon) did not provide good results, the spectrum was improved by EI ionisation of the desorbed neutrals (post-ionisation) [188].

Creasy [260] has compared thermal pyrolysis, laser microprobe and CW laser pyrolysis of a cured, fire retardant **epoxy resin** (diglycidyl ether of tetrabrominated bisphenol A). Laser ablation/electron impact (LA/EI) produced mostly low-MW fragments that gave limited information about the material. The relationships and analytical value of these techniques were discussed. The same author [167] has used LD/EI-FTMS for the analysis of oligomers.

Hsu *et al.* [213] have explored the potential of LD-FTMS for the simultaneous detection and identification of polymer and **dye components** of commercial solid polymeric materials. It was shown that LD-FTMS in LD/EI mode can detect and identify (by chemical formula) dyes in solid PMMA at concentrations at least an order of magnitude lower (0.1 *vs.* 1–2 wt.%) than those obtained by the best currently available alternative method (ATR-FTIR). The chemical formula of each dye could be established by accurate (0.5–20 ppm) mass measurement (calcd./measd. mass of 1-(methylamino)anthraquinone 238.0862 and 238.0816 Da, respectively). LD-FTMS thus offers a sensitive and accurate means for identifying dyes in solid polymers (*cfr.* Fig. 3.11).

The ultimate aim of using **L^{2}ToFMS** for polymer/additive characterisation is to provide a means for **direct detection** of these compounds in polymers. It has been demonstrated that the technique, generally using the REMPI scheme, has potential for rapid generation of readily interpretable mass spectra of polymer additives directly from their host polymer matrices. Wright *et al.* [261] have analysed phenolic **AOs** and **UVAs** (Tinuvin) by means of L^{2}ToFMS using pulsed CO_2 laser desorption and UV laser ionisation (266 or 193 nm). For all the AOs studied, the molecular ion peak dominated the 266 nm photoionisation mass spectra; very little fragmentation was observed. In contrast, at 193 nm the molecular ion peak was usually absent from the photoionisation mass spectra. Similar behaviour was exhibited by the hydroxyphenyl-benzotriazole UVAs (Tinuvin) in their photoionisation mass spectra. This wavelength-dependent fragmentation can be exploited for unambiguous identification of many polymer additives. Isomeric UVAs Tinuvin 320, Tinuvin 343, and Tinuvin 329 can be differentiated on the basis of the extent of fragmentation in their photoionisation mass spectra (Fig. 3.12).

Also several commercial polymer formulations, namely PP/(0.15 wt.% Irganox 1330, 0.5 wt.% Irgafos 168), POM/0.1 wt.% Santowhite and POM/

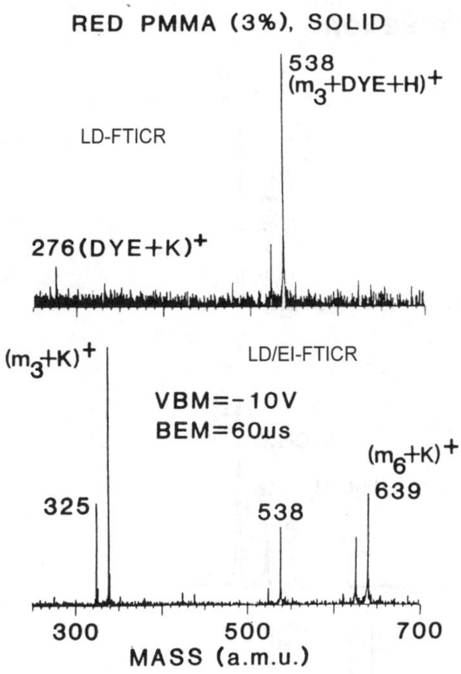

Fig. 3.11. LD-FTICR (top) and LD/EI-FTICR (bottom) mass spectra of solid PMMA/3% 1-(methylamino)anthraquinone, produced by a single laser shot. Note the enhanced signal-to-noise ratio for spectra obtained in LD/EI mode. After Hsu and Marshall [213]. Reprinted with permission from A.T. Hsu and A.G. Marshall, *Analytical Chemistry* **60**, 932–937 (1988). Copyright (1988) American Chemical Society.

0.3 wt.% Tinuvin 320, were analysed directly without any pretreatment or extraction [261]. By use of several readily available ionisation wavelengths, and with reference spectra of the pure additives, it was possible to unambiguously determine the presence of Irganox 1330 and Irgafos 168 directly from the host PP matrix. No signals were observed corresponding to the polymer matrix itself. Efficient multiphoton ionisation (MPI) requires that the target molecule has a significant absorption at the photoionisation wavelength. Neither PP nor POM possesses a UV chromophore. Thus, at low post-ionising UV laser fluences it is possible to effectively discriminate against material ablated or desorbed from the polymer matrix itself, leading to *in situ* identification of target additives. Provided sufficient material is liberated from the polymer, the limiting factor for *in situ* detection of additives at levels typically found in real polymer formulations is the efficiency with which they can be ionised. Where a suitable wavelength is available, analysis of bulk polymers with additive concentrations as low as 0.01 wt.% are possible using L^2MS. Chemical changes undergone by antioxidants, due to either processing or ageing, can be observed.

Lykke *et al.* [177,262] have used L^2MS (ToF-MS, FTMS) in resonant and non-resonant mode for the molecular analysis of complex materials, including polymer/additive systems. Different wavelengths for the post-ionisation step (near-UV, far-UV, VUV) permit selectivity that provides important additional information on the chemical constitution of these complex materials. LDI techniques render more accessible analysis of complex materials such as polymers and rubbers containing a wide variety of additives and pigments. Lykke *et al.* [218] also compared laser desorption, laser desorption/post-ionisation and laser ionisation in both direct and extract analysis of three **vulcanised rubbers** (natural rubber, SBR and poly(*cis*-butadiene)). Desorption (532, 308, 266 nm)/post-ionisation (355, 308, 266, 248, 213, 118 nm) was carried out with various lasers. Desorption (308 nm)/post-ionisation (355 nm) with REMPI detection allows preferential detection of various additives (antiozonant HPPD, m/z 268, 211, 183, 169; antioxidant poly-TMDQ, m/z 346, 311) over the ubiquitous hydrocarbons in a rubber (Fig. 3.13).

The analysis is much less matrix-dependent than prompt-ion mass spectrometry because the desorption laser provides the energy for neutral particle removal only, without the extra energy needed for direct ionisation. In addition to diminishing the matrix effect introduced by the surface, lower laser fluence is required for the removal of particles, thus reducing the chance for fragmentation. For laser desorption/laser ablation, the amount of fragmentation depends on the energy absorbed by the surface, wavelength and pulse length of the laser [48]. Operating the laser at fluences slightly above the desorption threshold for ion production, Lykke *et al.* [218] noticed no differences in product distribution by exploring various wavelengths for laser desorption (532, 308, 266 nm). In these experiments the heating rates (typically about 10^9 K/s) were much higher than in thermal-programmed desorption (TPD) experiments (1–10 K/s). Thermal desorption is relatively matrix-independent. A tuneable VUV laser allows achieving fragmentation-free photoionisation of desorbed molecules. Different post-ionisation wavelengths produce different mass spectra. Post-ionisation at longer wavelengths (>300 nm) favours

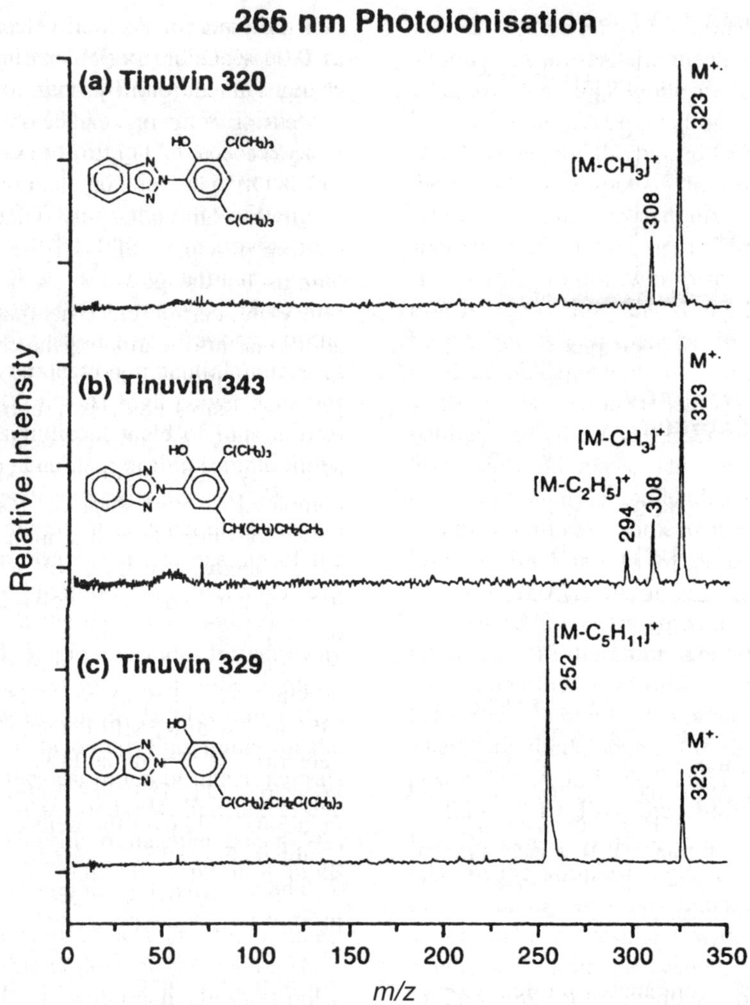

Fig. 3.12. 266 nm photoionisation mass spectra of isomeric UV stabilisers: (a) Tinuvin 320, (b) Tinuvin 343, and (c) Tinuvin 329. After Wright *et al.* [261]. Reprinted with permission from S.J. Wright *et al.*, *Analytical Chemistry* **68**, 3585–3594 (1996). Copyright (1996) American Chemical Society.

formation of ions that are indicative of additives (DODPA, m/z 393, 322, 251; DODPA impurity: tri-*t*-octyldiphenylamine, m/z 505, 448, 434), whereas with a shorter wavelength (*e.g.* 213 nm) the selectivity for additives is reduced and the spectrum is dominated by mass peaks arising from the rubber backbone (Fig. 3.14).

Vulcanisate extracts were analysed in a similar way by means of desorption (266 nm)/post-ionisation (118 nm), allowing detection of poly-TMDQ additives [218]. Higher-mass species can be observed in extracts that cannot be seen by direct analysis. It was possible to selectively detect aromatic polymer additives vaporised from rubber vulcanisates by careful choice of the ionisation wavelength. Light at 355 nm preferentially accentuates additives in the polymer (selective), 212-nm light ionises most of the other large species, while 118-nm light can be utilised to characterise the majority of the desorbed material (non-selective ionisation). This makes the combination of using 118-, 212-, and 355-nm radiation for post-ionisation an extremely powerful technique that rivals other techniques for surface and bulk analysis of polymers and their additives.

In a related L^2ToFMS study with desorption at 308 nm and photoionisation at 248 nm a dissolution of the oligomeric HALS poly-TMDQ (2,2,4-

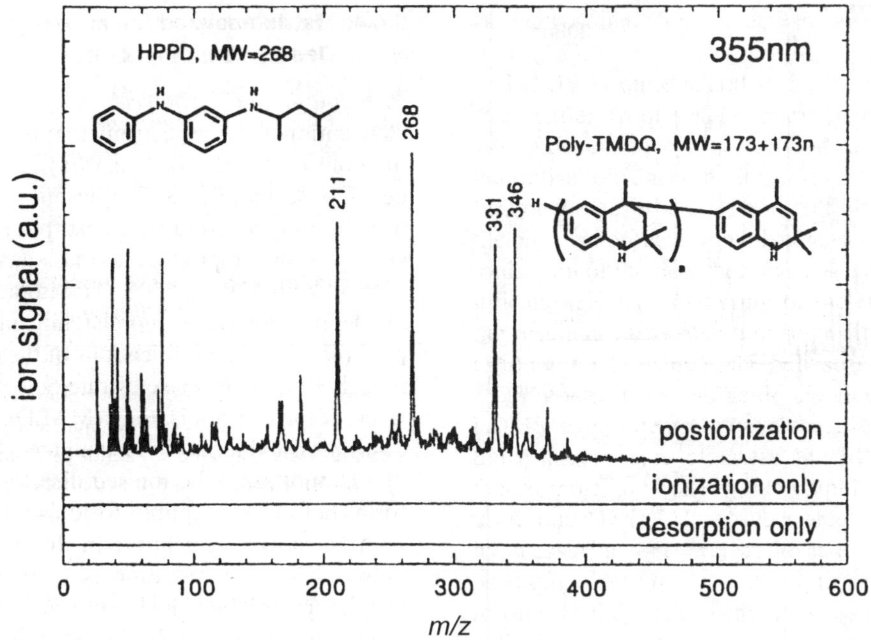

Fig. 3.13. ToF-MS spectra (desorption at 308 nm only; ionisation at 355 nm only); and L^2ToFMS spectrum (post-ionisation activated) of a rubber vulcanisate. After Lykke *et al.* [177]. Reproduced from K.R. Lykke *et al.*, *Appl. Opt.* **32** (6), 857–866 (1993), by permission of the Optical Society of America, Copyright (1993).

trimethyl-1,2-dihydroquinoline), with potential for near-resonant photon absorption at 248 nm, was examined [262]. Intense protonated molecular ions of the lower mass oligomers (m/z 692, 865, 1039, 1211, 1385 and 1558) separated by the TMDQ monomer unit (m/z 173) were observed. The most intense peak in the photo-ion mass spectrum at m/z 692 corresponds to a stable cyclic tetrameric structure of TMDQ. Poly-TMDQ is very thermally labile and undergoes demethylation and depolymerisation quite easily. In fact, other observed intense peaks correspond to demethylation and oxidation of the oligomeric unit. L^2ToFMS is thus able to detect degradation of commercially important additives. On the other hand, one-step LD-ToFMS of the ppm additive systems shows little evidence of any additive, at variance to the abundant mass signals of L^2ToFMS.

L^2MS has also potential for surface analysis and depth profiling. Zenobi *et al.* [263] have described **spatially resolved direct *in situ* analysis** of polymer additives (Tinuvin PS/234/320/326/327/328/329/343, Lowinox 22, Santowhite) in POM, PVC, PP and PET using two-step CO_2 laser ($\lambda = 10.6$ μm) mass spectrometry (L^2MS). Under usual L^2MS experimental conditions it proved difficult to observe additives directly from PP and PET samples, and higher desorption laser power was required than for POM and PVC. For the strongly absorbing polymers (*e.g.* POM and PVC), laser-induced pyroablation decomposes the host matrix but liberates intact neutral additives. Hydroxyphenylbenzotriazole UVAs and phenolic AOs were efficiently and selectively post-ionised with a UV laser pulse by two-photon REMPI. The detection limit for Santowhite powder antioxidant in a POM injection bar was 28 ppm. Weakly absorbing polymers, *e.g.* PP and PET, showed only laser melting. Additives in these polymers are not easily detectable by L^2MS. Depth profiling L^2MS experiments by stepwise CO_2 laser ablation enabled the spatial distribution of Santowhite within an injection-moulded bar of POM to be determined with μm resolution. The near-surface concentration of the antioxidant was found to differ significantly from that in the bulk.

As a slight variation on the theme laser desorption/photodissociation FTMS experiments are possible. Wilkins *et al.* [264] have investigated the utility of **LD/PD-FTMS** for the analysis of a variety of **high mass, intractable species** using 308 nm desorption. Compounds absorbing 308 nm radiation usually photodissociate as gas phase ions. Exceptionally compounds do not photodissociate, in spite

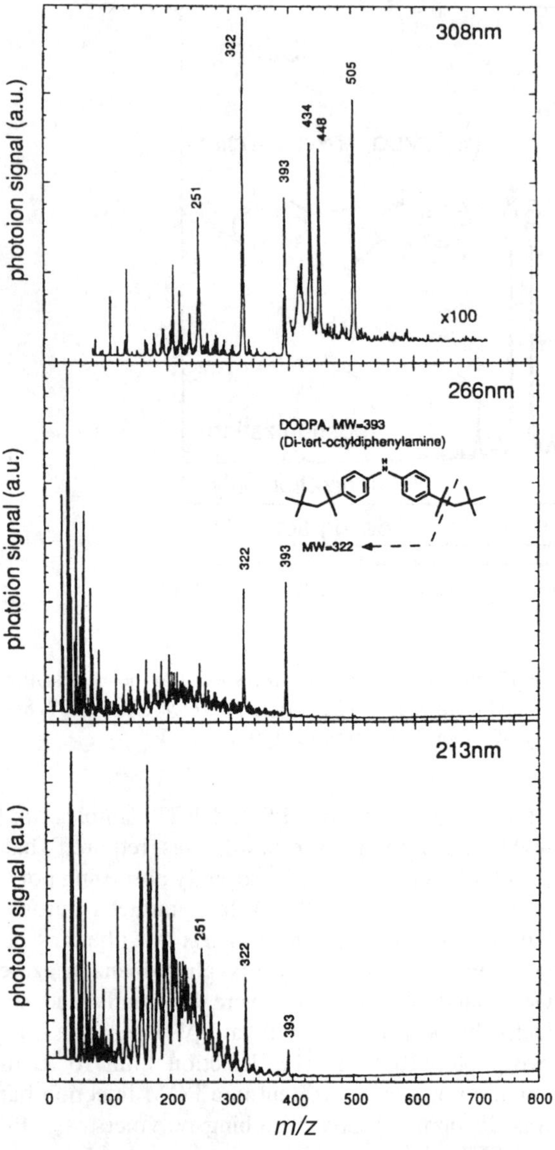

Fig. 3.14. Post-ionisation of a vulcanisate at 308, 266 and 213 nm. After Lykke *et al.* [177]. Reproduced from K.R. Lykke *et al.*, *Appl. Opt.* **32** (6), 857–866 (1993), by permission of the Optical Society of America, Copyright (1993).

of absorption bands at 308 nm in solution UV spectra. For samples not absorbing 308 nm radiation (in solution) photodissociation can be accomplished by first attaching a chromophore, either by derivatisation or by *in situ* metal ion attachment.

Becker [229] has described the analysis of (in)organic surfaces by ionisation of desorbed neutrals with untuned UV and VUV lasers.

3.4.4. Matrix-assisted Laser Desorption/Ionisation

Principles and Characteristics

Conventional mass spectrometry precludes the study of molecules with low volatility and those which decompose readily upon ionisation with an electron beam. Laser-induced desorption from a solid surface allows reproducible mass analyses for substances with MW $\leq$ 1500 Da without serious sample degradation. Even more significant advances are possible through modifications of the desorption approach. One such revised strategy is matrix-assisted laser desorption/ionisation (MALDI), a combined desorption and ionisation technique. High-MW samples cannot easily be ionised directly in their intact form, but are susceptible to ionisation once mixed with an absorbing matrix. In the MALDI process pulsed UV laser radiation is used to produce intact gaseous molecular or quasi-molecular ions from sample molecules finely dispersed in a solid matrix of small organic molecules that absorb strongly at the desorption laser wavelength in order to increase the ion yield from an analyte with unfavourable desorption characteristics. This method of laser desorption forms the basis of MALDI-MS, which is now widely used for mass analysis of polymers with molecular masses up to 1.5 10^6 Da [265]. MALDI was developed in the late 1980s [266,267].

The type of laser most commonly used for MALDI is the N_2 laser (337 nm), but Nd:YAG at 355 and 266 nm is also used. Typically, laser pulse durations in the range of 1–200 ns are used with irradiances of 10^6 to 10^7 W cm^{-2}. These power densities are achieved by focusing the laser beam to a spot of 10–100 μm diameter on the sample surface. The irradiance at the sample surface is a critical parameter for achieving successful desorption/ionisation. Ideally, the laser should deliver an efficient and controllable quantity of energy to the sample, which must be transferred quickly in order to avoid thermal decomposition.

MALDI polymer analysis consists of three steps, namely sample preparation, mass spectral recording, and data analysis. Matrices and **sample preparation** are crucial points for the applicability of MALDI-MS. The analyte molecules, often a polymer, are dispersed in a matrix of small organic molecules with strong optical absorption at the desorption laser wavelength. In this way desorption/ionisation of the analyte molecules can be achieved regardless of their absorption properties. The sample is prepared by mixing a small volume of concentrated matrix solution with a similar volume of dilute analyte solution. The molar mixing ratios between matrix and analyte are in the range of 500:1 to 10^6:1. The solvent is evaporated and the sample transferred to the vacuum chamber of the mass spectrometer. The matrix transforms the laser energy into excitation energy for the sample leading to sputtering of the analyte and matrix from the surface of the mixture.

Selection of appropriate **MALDI matrices** is based on instrumental considerations (*e.g.* the molar absorptivity at the wavelength of interest or the choice of mass analyser) and is also determined by sample considerations (*e.g.* solubilities of matrix and analyte or stability of analyte in a particular matrix). The matrix requires good solubility in organic solvents that can also dissolve the analyte, good miscibility with the analyte in the solid state and good vacuum stability. Some typical MALDI matrices for synthetic polymers are 2,5-dihydroxybenzoic acid (DHBA), dithranol (1,8,9-trihydroxyanthracene), α-cyano-4-hydroxycinnamic acid (CHCA) and 3,β-indole-acrylic acid (IAA). Räder *et al.* [268] and others [269,270] have listed commonly used matrices for many polymers. The matrices form organic crystals; incorporation of the analytes into the crystalline matrix upon evaporation of the solvent is essential for the production of intense ion signals in the MALDI process. Different matrices can result in substantial variations in analytical performance. Selection of a cationisation reagent can also be critical. The solvent system used to prepare the sample solution also affects mass spectral results. It is important to select a solvent system that will allow matrix crystallisation to take place prior to polymer precipitation. Strupat *et al.* [271] have examined the role of matrices in UV and IR laser ablation in the MALDI process. The matrix plays several roles in promoting formation of intact molecular ions from the sample:

(*i*) analyte entrapment and isolation (by acting as a solvent);
(*ii*) absorption of laser radiation (by molar absorptivity of matrix exceeding that for analyte);
(*iii*) analyte desorption (through absorbed laser energy); and
(*iv*) analyte ionisation (through active role in ion production).

For the removal of intermolecular interactions between analyte molecules a large excess of matrix molecules is required. Insoluble or poorly soluble polymers do not readily form mixed polymer/matrix crystals, a requirement for preparing good MALDI samples. Liquid matrices appear to exhibit significant advantages over solid matrices for MALDI [272]. The intimate mixing of analyte and matrix molecules and the self-healing of the sampling position through molecular diffusion permit hundreds of laser shots per sampling position with excellent shot-to-shot reproducibility. The effectiveness of a solution as a matrix for MALDI relies critically on the solubility and relative hydrophobicity of the solid matrix in the liquid support. The role of the matrix in the ionisation process is still a contentious issue. The most commonly accepted scenario involves photoionisation of the matrix molecule that can then undergo one of a number of possible reaction pathways to yield the analyte ion. An extensive study of potential matrices has been reported by Beavis [273]. Sample preparation for MALDI-MS is still a more-or-less empirical process [274]. It is important that when the laser beam hits the sample, no great variance in quality of the spectra for different areas of the sample can be observed because of heterogeneity of the matrix/analyte distribution.

The **MALDI process** formally consists of the following steps: (*i*) formation of a crystalline solid out of the solution; (*ii*) incorporation of analyte molecules into matrix crystals; (*iii*) desorption-ablation of material from crystals by the laser beam; and (*iv*) gas phase processes with transfer of charges to analyte molecules (*cfr.* Fig. 3.15). The mechanisms involved in MALDI remain poorly understood; a detailed discussion of the possible MALDI mechanisms is available [276]. Contrary to laser desorption, where the analyte is irradiated directly by laser light, in matrix-assisted laser desorption the analyte is assumed to

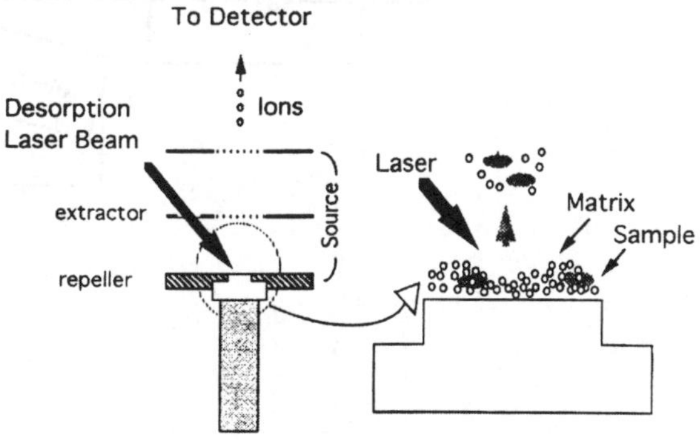

Fig. 3.15. Principle of matrix-assisted laser desorption/ionisation. After Whittal and Li [275]. Reproduced by permission of International Scientific Communications Ltd.

be homogeneously embedded in a matrix material, which transforms the laser energy into electronic excitation energy for the sample leading to sputtering of the analyte and matrix from the surface of the mixture. While this ensures efficient energy transfer, the sample molecules are spared from excessive thermal stress, which would lead to decomposition. As excitation of the internal degrees of freedom of the analyte molecules is very limited even thermally labile molecules survive desorption without structural damage. Ionisation of the analyte molecules takes place by *e.g.* proton transfer. The various mechanisms for matrix and non-matrix assisted laser desorption have been discussed [277].

The short pulse of ions produced by laser desorption is ideally suited to mass analysis by time-of-flight (ToF), which makes it possible to record a complete mass spectrum for a single laser shot. Actually, MALDI has greatly contributed to the revival of ToF-MS. The detection efficiency decreases as the ion mass increases. Key instrumental parameters of MALDI-MS are **mass resolution** and **mass accuracy**. Mass resolution of MALDI-ToFMS has increased considerably by pulsed extraction techniques. Resolutions of 30,000 or better and mass accuracy at low ppm level are achieved, which allow separation of single oligomers. The sensitivity of ToF is in the sub-fmol range. Recently also the lack of MS/MS capability has been overcome. For mass spectral recording, detection dynamic range and mass resolution must be optimised. Mass resolution can affect the mass spectral patterns and average molecular weight results. Ions produced in MALDI are generally quasi-molecular ions, rather than radical cations. These quasi-molecular ions are cationised, such as protonated $(M + H)^+$, deprotonated or alkalinated (*e.g.* $(M - Na)^+$) species. Thus, mass assignment is easy, even for mixtures containing many species. MALDI-ToFMS is to be classified as a **soft desorption/ionisation** method. Figure 3.16 shows a schematic diagram of a MALDI-ToF. In conventional MALDI-ToF instruments, the ions produced by a pulsed laser beam are immediately extracted from the source.

The important instrumental development of delayed (ion) extraction (DE) or time-lag focusing (TLF) [279] has had a great impact on MALDI development [280,281]. The major limitation to the resolution provided by MALDI-MS is the initial velocity distribution of the ions. Ions with the same mass/charge ratio but different energy distributions yield a broad peak with a decrease in resolution. Correction for this by the use of **pulsed extraction** for time-lag focusing [282] has greatly improved the quality of the mass spectra from low masses (<100 Da) up to high masses. The term **time-lag**

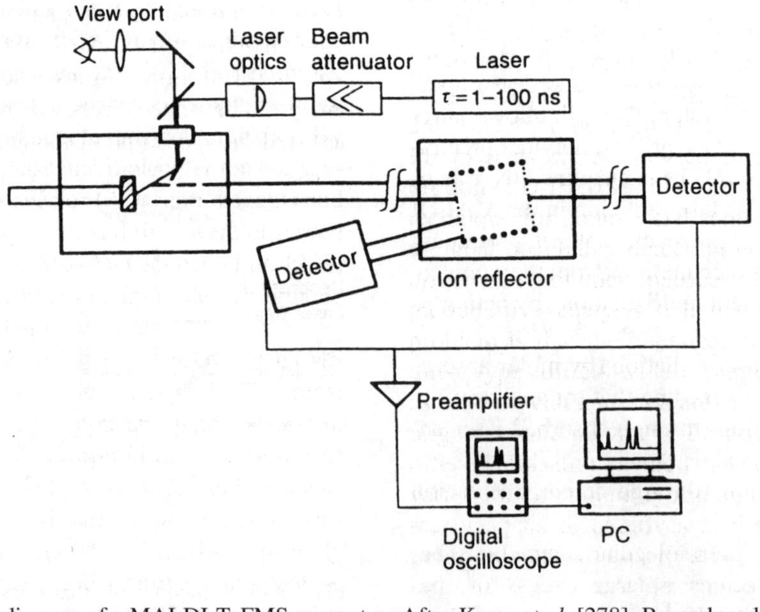

Fig. 3.16. Schematic diagram of a MALDI-ToFMS apparatus. After Karas *et al.* [278]. Reproduced form M. Karas *et al.*, *Mass Spectrom. Rev.* **10**, 335 (1991). Copyright © (1991, John Wiley & Sons, Inc.). This material is used by permission of John Wiley & Sons, Inc.

focusing MALDI has been coned to differentiate it from continuous extraction MALDI.

A linear or reflection ToF instrument equipped with TLF provides effective energy focusing of the MALDI ions, which translates into a significant improvement in resolving power. MALDI-ToFMS in pulsed ion extraction with TLF thus gives highly improved mass spectra at low m/z ratios (carbon isotope resolved peaks for molecules with MWs up to 3000 Da), as opposed to the continuous extraction mode [280,283]. TLF also provides a significant improvement in mass measurement accuracy (mass measurements with an error of less than 100 ppm using internal calibrants). The high accuracy can also be obtained over a broad mass range. Mass accuracy of better than 500 ppm can be routinely achieved for a wide range of chemicals with external standard calibration. This is comparable to or even better than the accuracy achieved using ESI with a quadrupole system. The improved mass accuracy for TLF MALDI presents an opportunity to assign mass spectral peaks based on molecular masses of the chemical species with high confidence. TLF also decouples desorption from ion extraction. Thus, laser power several-fold higher than the threshold required to generate analyte signals can be used for high-sensitivity detection while still maintaining excellent mass accuracy.

LDI-ToFMS with TLF allows analysis of small molecules directly (<100 Da) and of those which significantly absorb energy at 337 nm (with MW $<$ 1 kDa). Larger samples are analysed by mixing the sample with energy absorbing/transferring molecules (the matrix) through a MALDI process. TLF MALDI-ToFMS, in conjunction with recent advances in sample handling and sample/matrix preparation, is expected to play an increasingly important role in the analysis of industrial polymers [275,281].

Although a UV laser is commonly used to desorb/ionise matrix species in the vacuum of the mass spectrometer also IR lasers may be employed in the desorption process. IR MALDI normally uses tuneable IR (from 1.5 to 4.0 μm), Q-switched Er-YAG (2.94 μm, $t = 90$ ns) or Er-YSGG (2.79 μm, $t = 90$ ns) lasers. Smaller absorption coefficients of usable matrices in IR as compared to UV result in an at least tenfold increase in ablated material per laser exposure. With IR lasers localised thermal desorption or laser assisted pyrolysis occurs. The desorbed neutrals expand into a small ionisation chamber where they may be post-ionised by electron impact or photoionisation. The dependence of MALDI on **laser wavelength** has been studied using a tuneable free-electron laser (FEL) from 2.65 to 4.2 μm and 5.5 to 6.5 μm, covering the IR absorption band regions of O−H and C=O stretching vibrations commonly found in MALDI matrices [284]. The results indicate that different desorption/ionisation processes are operative in IR MALDI and UV MALDI. The ion desorption depends on laser *irradiance* rather than laser *fluence*. Yet, IR and UV MALDI lasers produce strikingly similar results [285]. In some cases, IR lasers are more successful at characterising higher-MW samples than are UV lasers. Nevertheless, IR MALDI is described neither by UV MALDI models nor by thermal IR LD/LITD models, but rather by explosive vaporisation or spallation processes. It has been observed that matrix melting and removal of large pieces of material dominate IR ablation, whereas the extremely shallow desorption of UV leads to the formation of very specific surface structures (trulli) [271].

Introduction of MALDI techniques has meant that the determination of the molecular masses of a very broad range of macromolecules is now commonplace and essentially routine. However, full structural characterisation remains a significant challenge. For this purpose heavy reliance is placed on the collisional activation technique. Application of ToF mass analysis for MS/MS experiments is of great interest due to its ability to provide complete product spectra from pulsed ionisation events in real time and its inherent compatibility with pulsed ion fragmentation techniques. MALDI-MS has also been demonstrated with sector mass spectrometers, FTICR-MS and QIT. Sector and FTICR mass spectrometers have higher resolution and mass accuracy than common MALDI-ToF mass spectrometers but do not have the high m/z range of MALDI-ToFMS. The mass range for an optimal added value of MALDI-FTMS is m/z 10–5000 (future: 10,000). MALDI-FTMS is particularly useful for analysing polymers less than 10 kDa in molecular weight. MALDI is implemented in FTMS using one of two approaches [286]:

(i) internal MALDI (ions are formed within or adjacent to the ICR cell);

(ii) external MALDI (ions are formed outside of the magnetic field and transported to the analyser cell).

MALDI-FTMS yields more detailed structural information (such as sequence distribution) than MALDI-ToFMS. The mass resolving power ($m/\Delta m = 120 \times$

10^6) allows base-line separation of molecular weight, chemical composition and end-group distributions. The mass accuracy of FTMS is sufficient to estimate the end-group masses with a precision of a few millimass units using exact mass measurements of individual components of a MWD. MALDI-QITMS has also been demonstrated [287,288]. Recent instrumental developments include MALDI-QIT-ToFMS for high quality MS and MS^n [289] and atmospheric MALDI (AP MALDI), which allows interchangeability with ESI-oaToFMS [290,291].

The **main characteristics** of MALDI-ToFMS are summarised in Table 3.24. A major advantage of MS is its high sensitivity, which allows mass spectra to be recorded using minute quantities of material (fmol or less) with extremely high resolution, permitting the molecular weight determination of a substance with great accuracy. However, mass measurement accuracy is still strongly dependent on the availability of internal reference peaks. MALDI-ToFMS is applicable to the measurement of polymer distributions because it mainly produces singly charged species.

Table 3.24. Characteristics of MALDI-ToFMS analysis in polymer analysis

Advantages:

- Rapid analysis (a few minutes)
- Minimum sample work-up
- Low sample consumption (less than 1 pmol)
- Ease of operation (even for polar molecules)
- Soft ionisation method (absence of fragmentation)
- High sensitivity (detection limits: pmol to amol)
- Excellent performance characteristics (broad dynamic mass range, reasonable resolving power, mass accuracy)
- Simplicity of data interpretation (mainly molecular ions)
- Measurement of complex sample mixtures
- Determination of absolute molar masses (no need for polymer standards)
- Unprecedented structural information
- Ruggedness
- At-line operation feasible

Disadvantages:

- Sample preparation more crucial than for ESI-MS
- No universally applicable standard MALDI conditions
- Method development (polymer-matrix match, sample preparation protocol)
- Different matrices for different polymers
- Selection criteria for matrices unknown
- Need for better matrices for various polymeric materials
- Exact ionisation process unknown (in particular for IR MALDI)
- Very mass sensitive (for PD > 1.5)
- Difficult quantitation
- Low shot-to-shot reproducibility
- Restricted applicability to UV-sensitive polymers
- Restricted coupling to on-line separation techniques
- Immature technique

Particular advantages of MALDI-MS for polymer analysis are the determination of absolute molar masses and molar mass distributions in a mass range (<20 kDa) that is problematic for other mass determining techniques, characterisation of chemical heterogeneity, end-group analysis and structural and compositional information via oligomer mass determination and tandem MS. Laser desorption of solid matrix-analyte deposits suffers from many disadvantages, such as low shot-to-shot reproducibility, short sample life-span and strong dependence on the sample preparation methods. These problems are generally attributed to laser-induced modifications of the surface structure of the matrix-analyte deposit and the inhomogeneity of the solid deposit. Standard MALDI ionisation conditions are not universally applicable to all classes of polymeric compounds. Each class of compounds has its own specific ionisation conditions. Development of dedicated sample preparation/matrix recipes for relevant polymers is advantageous. From a fundamental point of view it is not obvious if all polymeric materials are MALDI ionisable [292]. As there is no consensus on the mechanism of MALDI ionisation for each system the MALDI ionisation conditions (*i.e.* matrix, solvent, ratio matrix/analysis, preparative method) need to be established. In view of the analyte desorption and ionisation steps quantitation of different species, especially in broad molecular distributions, remains an important challenge for the future. The ionisation step of MALDI is not as soft as that of ESI, because the high-energy laser light can still fragment the polymer molecules. MALDI-MS results can compete with FD-MS being more generally applicable (but not for unknown systems).

MALDI-MS has mainly been used as an off-line detector for various LC separation techniques (isocratic and gradient HPLC, LCCC, SEC) of polymers. The problems of coupling of LC and MALDI-ToFMS are related to the fact that MALDI-ToF is based on the pulsed desorption of molecules from a solid surface layer and, therefore, *a priori* not compatible with liquid chromatography. Different semi-on-line interfaces have been developed for automated LC sample collection and subsequent preparation of the samples for MALDI-ToF analysis. In these interfaces, the eluate stream is deposited on the MALDI target via a spray or a drip process. The matrix required for the MALDI process is either co-added to the eluate stream or matrix-precoated MALDI targets are used. Several approaches to direct coupling have recently been described [293–295]. Continuous-flow (CF) MALDI-ToF uses a flow probe similar to a CF-FAB probe [293]; the method has not yet been used in polymer analysis. In the aerosol method for MALDI liquid introduction the matrix and analyte are dissolved in a solution that is sprayed directly into the mass spectrometer [294].

MALDI-ToFMS has also been used in connection with solid-phase microextraction (SPME) by direct introduction of the SPME fibre and desorption of the analytes into the cavity of the ion trap [296] or by loading SPE membranes coated with matrix material [287]. Internal standardisation is required for quantitative measurement because of poor reproducibility. For this purpose preferably an isotopically labelled compound should be chosen.

Trends in MALDI-ToFMS research are search for new matrices, solvent-free sample preparation, advances in analyser design (oaToF-MS, Q-ToFMS, QITMS, QIT-ToF, ToF-ToF), and (on-line) coupling to other (pre)separation techniques.

MALDI-MS has extensively been reviewed [273, 276,297–300]. Matrix and sample preparation for MALDI have been discussed [300] and advances in mass spectrometry (MALDI, ESI) of polymers have recently been reviewed [268,301]. Textbooks are available [270,302].

Applications

Chemical classes as organic polymers, additives and dyes are now all accessible by MALDI-MS on a routine basis. In the past, several approaches to the ionisation of polymer samples have been developed, including FD (which makes use of an electric field), FAB/FIB (where a beam of high-energy atoms or ions is directed at the sample), ESI (in which sample ions evaporate from small, highly charged droplets) and more recently laser desorption ionisation [299]. Until the advent of MALDI, FD was the ionisation method of choice for oligomers and polymers of moderate molecular weights which did not surpass the m/z ranges of the magnetic sector mass spectrometers. Nowadays the two favoured ionisation methods for high mass analyses, MALDI and ESI, have largely superseded the other methods for this purpose. MALDI and ESI are usually affiliated with different types of mass spectrometer. Whereas MALDI generates intact quasi-molecular ions which require the large m/z range provided by a ToF analyser, ESI produces multiply charged ions which can be measured on a typical quadrupole or magnetic sector mass spectrometer in the low m/z region (500–3000 Da). Thus the appearance of the spectra produced is quite different. There are also considerable differences in sampling method employed for the two ionisation techniques; ESI is an "on-line" technique using a continuous flow of solvent which makes it a direct coupling candidate with LC and CE, whereas MALDI handles liquid and solid matrices. ESI and MALDI are both suitable for mixture analysis, have reasonably short analysis times, fmol sensitivity, and can be automated. MALDI extends mass spectrometry to the analysis of large molecules not within reach of electron impact. MALDI deals well with thermolabile, non-volatile organic compounds, and has especially been used very successfully for analysis of biomolecules.

MALDI-MS has established itself as a powerful tool for **polymer analysis** [270,303], among other established techniques such as NMR, FTIR, SEC, DSC and end-group titrations. The interest of the polymer chemist in MALDI-ToFMS derives from the soft ionisation (no fragmentation, investigation of intact macromolecules), a large detectable mass range (analysis of monomers, oligomers and polymers), high resolution (identification of small differences in chemical structure, such as different homologous series and different end-groups) and quantification (determination of molar masses and of additives). The strength of MALDI-MS is the capability to directly study oligomers and resolve the components of the molecular weight distributions (MWD). Two features of MALDI, which are its main advantages for polymer analysis are: *(i)* analyte polarity is not critical; and *(ii)* with most matrices MALDI generates exclusively single-charged ions. Thus, in principle, MALDI mass spectra are interpretable even if a polymer sample consists of a large number of oligomers with different masses due to their chain lengths. Improvement in resolution and accuracy of mass determination by applying TLF or coupling of MALDI to FTMS allows qualitative determination of chemical composition, including end-group identity and composition, block length distribution with oligomer resolved resolution. The mass range for analysis of synthetic polymers by MALDI extends to extremely high molecular weights with virtually no fragmentation. Polar polymers such as polyesters, polyethers, polystyrenes and acrylics, which are miscible with the well-known matrices, are amenable to MALDI.

Limitations of MALDI-MS for polymer analysis are given in Table 3.25. MALDI cannot be applied to some important classes of synthetic polymers, such as polyolefins and several polymers with polar groups (*e.g.* polytetrahydrofuran oligomers). Conventional polar MALDI matrices, such as 2,5-dihydroxybenzoic acid and dithranol, are incompatible with non-polar analytes such as hydrocarbon polymers. For non-polar synthetic polymers new matrices must be devised. Another difficulty associated with MALDI analysis of non-polar polymers arises from the fact that these polymers lack an effective site for ionisation in the MALDI process. The common ionisation mechanism in MALDI-MS of polar compounds involves protonation of the analyte to form the $(M + H)^+$ ion. For synthetic polymers, formation of ions via cation attachment is much more favoured than via protonation. The fact that no MALDI spectra have been observed for saturated polyolefins has been related to the low binding energy of sodium-oligomer $(M + Na)^+$ complexes, which is insufficient to prevent dissociation of the complexes in multiple collisions occurring in the extraction section of the mass spectrometer [292]. Recently, the use of polynuclear aromatic hydrocarbons (anthracene, pyrene and acenaphthene) as charge-transfer matrices for the analysis of a variety of non-polar analytes in MALDI-MS has been described [304]. Good MALDI mass spectra were obtained for low-MW PS (MW 1940–5120 Da) and PBD (MW 760–1100 Da) using only a charge-transfer matrix without incorporating any cationisation reagent. In the absence of significant mass shifts in the presence of a cationisation reagent, metal ion-oligomer adduction apparently does not take place for these polymer samples. However, for higher molecular weights a cationisation reagent is required. Alternatively, instead of using a matrix the sample can be doped with metal salts to induce cationisation [304a]. Zenobi *et al.* [305] have recently proposed a MALDI sample preparation method suitable for insoluble polymers such as polyamides, which consists of pressing a pellet from a solid mixture of the polymer and a matrix, both in the form of finely ground powder.

Table 3.25. Limitations of MALDI-MS for polymer analysis

- Dissolution (not suspension) of the polymer is required
- Need for sample preparation protocols for POs, perfluoropolymers, and polycationic polymers
- Less mature for copolymers and blends
- Restrictions on analysis of some narrow polydispersity polymers
- Dependence on laser irradiance
- Not suitable for determinations of the degree of branching
- Need for improved overall detection sensitivity
- Overall detection sensitivity is polymer-class dependent
- No quantitative information on polymer mixtures

MALDI-ToF is more sensitive for low-MW molecules, resulting in an overestimation of these molecules compared to high-MW molecules. Desorption of higher molar masses from the matrix is more difficult than that of low molar masses. Consequently, when a polymer sample with broad MWD (high polydispersity) is analysed with MALDI-MS, the low-MW part will dominate the spectra and high-MW species will not be detected. Only polymers with a low polydispersity (<1.2) can be correctly identified with MALDI-MS. Also a dependence of MALDI spectra on laser irradiance exists: higher laser pulse energies allow desorption/ionisation of oligomers with larger m/z values. MALDI-ToFMS allows direct determination of MW and MWD distribution and the simultaneous determination of structure and end-groups in polymer samples.

Requirements for **MWD determination** by means of MS techniques are: *(i)* absence of discrimination; *(ii)* soft ionisation (DCI, FAB, FD, ESI, LD, MALDI); *(iii)* equal ionisation efficiency for high- and low-MW molecules; *(iv)* transmission efficiency (even very heavy molecules should reach the detector); *(v)* detection efficiency; and *(vi)* calibration. Soft ionisation techniques have all been applied to assorted polymers to obtain average MWDs. Of these, LDI is recommended because it is easily interfaced to both ToF-MS (capable of very high m/z detection) and FTICR (offering high mass resolution and accuracy). Sample preparation with matrix selection (out of 20–30 matrices) and the need for a great excess of matrix (10^4:1) are critical for success of MALDI-MS. Failures for correct MWD with MALDI are due to polydispersity (PD) discrimination (especially for PD > 1.1), crystallisation and ionisation processes. Mass discrimination effects observed for polymers have been attributed to sample preparation, desorption/ionisation and instrumental factors.

It is the case to notice some differences between **MALDI-MS** and **SEC** for MW and MWD determinations. Essentially, amongst these two methods SEC is the more physical and MALDI the more chemical approach. At variance to MALDI, SEC fails in determining low-MWs (<20 kDa). MALDI-ToFMS yields M_n distribution, as opposed to M_w in SEC (interconversion is not always obvious as MALDI-MS easily misses out a high-MW fraction). Compared with other traditional MW characterisation methods, MALDI uses a minimum amount of solvent. Arguably the most appropriate application of LC-MALDI-MS in the field of synthetic polymers is for calibration of a SEC system, that is, to convert SEC retention times into MW data. MALDI-MS yields inaccurate results for samples with PD > 1.3 but when combined with SEC this limitation can be overcome (*cfr.* Chp. 7.3.4.2 of ref. [13a]). On-line and off-line coupling of MALDI-ToFMS with SEC allows reliable SEC characterisation of polymers with broad MWD, where suitable standards for SEC calibration are not available. Off-line SEC-MALDI-ToFMS is used for calibration fo SEC when the universal calibration is not applicable. On-line SEC-MALDI-ToFMS is difficult (timing problem). Direct (LCCC)-MALDI-ToFMS coupling has been used for characterisation of PPO and PEO [306].

MALDI-ToFMS was used by Deckwer *et al.* [307] to study the structure of nucleating **oligomers** in emulsion polymerisation. PEG400 and PEG1000, $HO(C_2H_4O)_nH$, were analysed successfully by MALDI-ToFMS using cobalt ultrafine powder (Co-UFP) as an alternative to organic matrices to enhance the LDI process [279].

Because the mass of a polymer is much larger than the mass of a **functional group**, ultra-high mass accuracy is required for end-group determination. ToF-MS is inadequate for such determinations; MALDI-FTMS is the established method for this purpose, as illustrated for PEG1000, PEG4000 and PVP3000 by Boon *et al.* [308]. MALDI-FTMS was also used for analysis of PEG1000–6000 [286]. Quantitative determination of the end-group distribution needs calibration of the MALDI response by use of suitable standard components.

The resolution and mass accuracy advantages can be applied to more complex problems, *e.g.* the determination of **copolymer composition**. MALDI spectra allow deriving both the composition of the copolymer and the sequence distribution. Analysis of copolymer sequence by MALDI-ToFMS is a useful complement of the NMR technique.

After the initial hype of polymer mass spectrometry by MALDI a more cautious approach now focuses on issues like matrix selection, repeatability, quantitation, mass bias, *etc.*

MALDI-MS studies of **polymer additives** are rather few [309–314]. An account of MALDI-MS of polymer/additive dissolutions is given in Chp. 9.3.1 of ref. [13a]. Hsiao *et al.* [309] have reported analysis of a wide variety of additive standards (Irganox 245/259/1010/1024/1076/1098/3114, Naugard 524, Tinuvin P/144/320/326/328/440/622/770 and Chimassorb 944) and a PE/(Irganox 1010/1076, Naugard 524, erucamide) extract by MALDI-ToFMS using delayed extraction (DE), post-source decay (PSD) and collision-induced dissociation with 2,5-DHBA as the matrix. CID of the pseudomolecular ion was used to confirm the tentative assignment. Analysis of high-MW additives such as Chimassorb 944 (Fig. 3.17) and Tinuvin 622 indicated superiority of MALDI over other ionisation techniques (ESI, DCI, FAB) for this purpose. Sample preparation was more critical than DCI in the analysis of real samples. Interference of co-extracted low-MW PE molecules was noticed but could be overcome by acetone precipitation.

Following extraction/preconcentration by SPE, MALDI-ToFMS has been used for the qualitative and quantitative determination of nonylphenol ethoxylate surfactants (in surface waters) [315]. Electrospray and MALDI-MS were used for identification of spermicides in criminal investigations [316].

3.4.5. Laser Microprobe Mass Spectrometry

Principles and Characteristics

Laser microprobe mass spectrometry (LMMS, LAMMS), sometimes called laser probe microanalysis (LPA or LPMA) and often also referred to as laser microprobe mass analysis (LAMMA®, Leybold Heraeus) [317] or laser ionisation mass analysis (LIMA®, Cambridge Mass Spectrometry/Kratos) [318], both being registered trademarks, is part of the wider laser ionisation mass spectrometry (LIMS) family. In the original laser microprobe analyser, emitted light was dispersed in a polychromator. Improved sensitivity may be obtained by secondary excitation of ablated species with an electric spark. In the mass spectrometric version of the laser microprobe, ions formed in the microplasma

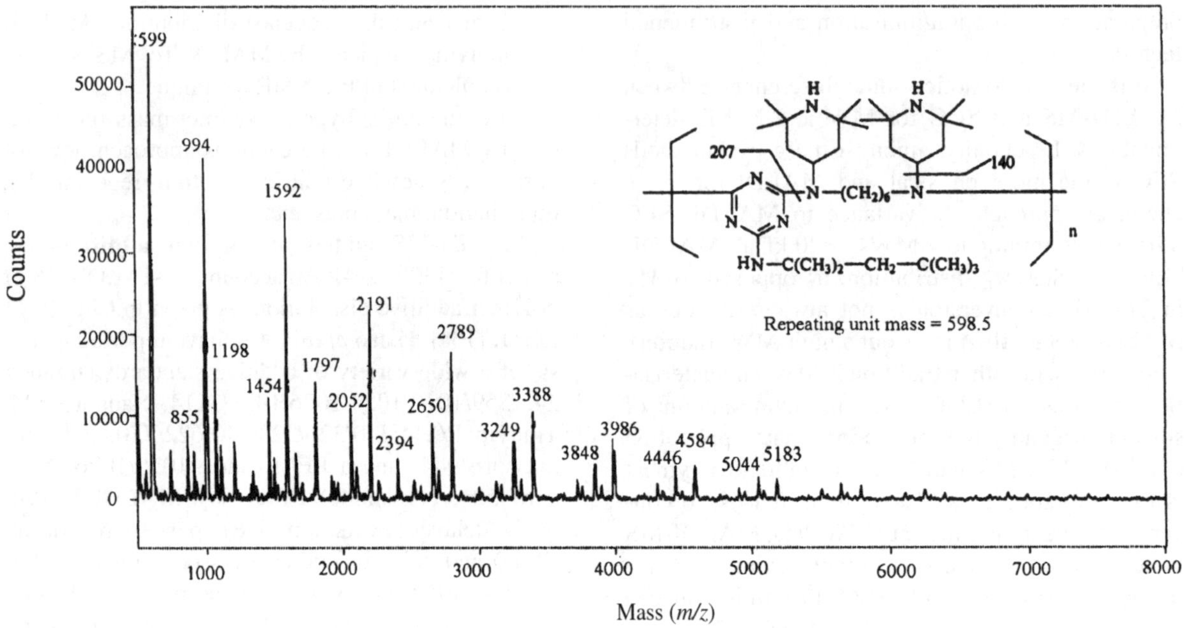

Fig. 3.17. MALDI-ToFMS spectrum of Chimassorb 944. After Hsiao *et al.* [309]. Reproduced by permission of the Chinese Chemical Society, Taipei.

are mass analysed by means of magnetic sector, ToF or FTMS instruments. The first use of a solid laser for laser microprobe analysis was reported in 1963 [319]. Growing interest in local analysis with a spatial resolution of the order of 1 μm has stimulated development of commercial laser microprobes in the 1970s.

For any true microspectrochemical analysis a microscope and a dispersing instrument are needed. An essential feature of a laser microscope is that the objectives for both the observation of the specimen and for focusing the laser radiation must be suitable for ablation, vaporisation, and excitation of the material. The defining attribute of LMMS is the use of a high power pulsed UV laser ultimately **focused** down to the diffraction-limited spot (0.5 μm at 266 nm) to vaporise, atomise, and ionise a microvolume of a solid specimen in a one-step procedure. Laser microprobe mass analysers are typically equipped with Nd:YAG lasers (1064 and 266 nm; 5–15 ns pulses) or excimer lasers (XeF, 351 nm; XeCl, 308 nm; KrF, 248 nm with about 7–30 nm pulses). Power densities of up to 10^{11} W cm^{-2} are quite common; organic compounds require attenuation to about 10^6–10^8 W cm^{-2}. By adjusting the laser power, desorption and ionisation can, to some extent, be selected over ablation and dissociation in the microplasma, enabling the technique to be used to detect molecular species in addition to elemental analysis. In modern instruments ions formed by the interaction of photons with the solid sample during about 15 nsec are continuously extracted and focused into a time-of-flight (ToF) or Fourier transform (FT) mass spectrometer.

Originally, the time definition of the ion production from the pulsed laser was matched with the registration of full mass spectra by means of a ToF analyser and the first **ToF LMMS** instrument was commercialised some 30 years ago [320]. ToF LMMS has marked a milestone in the field of microanalysis enabling element localisation, inorganic speciation, and structural characterisation of organic molecules in a given micro-object. A full mass spectrum is recorded within 0.5 ms of each laser shot, which typically evaporates about 1 μm^3 or 10^{-12} g of solid sample. The dependence of ToF LMMS data on the actual experimental conditions is a weakness, preventing routine applications. An experimental protocol has been described that provides an adequate means of reproducing a specific local regime [321]. ToF LMMS signals relate almost exclusively to ions generated during the laser pulse, *i.e.* the so-called "prompt" ions, while post-laser ion formation is important. As ToF-MS is not capable of collecting a representative fraction of the ion population the quantitation capabilities of ToF LMMS are poor. For elemental analysis, ToF LMMS cannot always compete with techniques such as SIMS or EPMA. In organic ToF LMMS soft ionisation mechanisms prevail and labile compounds are desorbed intactly; nevertheless even quite stable compounds readily undergo disintegration into carbon clusters. A desorption-ionisation (DI) model for ToF LMMS needs take into account this apparent ambiguity. ToF LMMS experiments, in which lasers are used as the combined pyrolysis and ionisation source, fail to yield informative mass spectra for polymers. In most cases polymers have given extensive fragmentation to ions of low mass and poor molecular ion yield; also virtually no additive information was reported [322,323]. Speciation by ToF LMMS relies on fingerprinting and comparison with reference spectra. Polymers can be classified by LMMS using fingerprinting and statistical approaches. The panoramic detection makes the ToF approach particularly appealing for surface and layer-by-layer analysis. However, the complex positive and negative mass spectra obtained require a significantly higher mass resolution and mass accuracy than provided by ToF LMMS. Hence a second commercial generation of LMMS, FT LMMS, was developed by combination of laser microbeam irradiation of solids with one of the most powerful MS analysers.

The ion storage principle of FT LMMS allows measuring a larger fraction of the total number of laser-generated ions. In case of inorganic compounds ion formation continues up to several hundreds of μs after the laser pulse [324]. The improved analytical utility of **FT LMMS** over ToF LMMS results from the substantially increased mass resolution and mass accuracy in FT LMMS and its ability to measure post-laser pulse ions. Van Vaeck *et al.* [325] reported an FT LMMS with external ion source and a 5-μm laser spot, suitable for microprobe analysis. In this way access is provided to the time profile of the ion production process with more degrees of freedom for optimising the optical components and ion lenses. FT LMMS is in a better position than ToF LMMS to understand the mechanisms of ion formation. Panoramic recording of a mass spectrum in FT LMMS needs accumulation of more than 100 shots, as opposed to ToF LMMS where, in principle, a full mass spectrum is available from each single laser shot. FT LMMS produces more systematically higher mass-to-charge ratio adduct ions and the deduction of molecular composition from the low and high m/z ratio signals is more obvious. The exact elemental composition of the detected ions can be derived unambiguously. FT LMMS is maximised in the high-resolution mode, permitting monitoring of one to a few selected ions. The loss in sensitivity in FT LMMS compared to ToF LMMS (a factor of 100) is compensated for by the high mass-resolution and the more accurate m/z determination, yielding more specific information. Post-laser ionisation over μs makes FT LMMS as sensitive as ToF, except for elemental ions, because of their high energy spread. With FT LMMS the analysis of less volatile molecules with MW > 500 Da is feasible, as opposed to ToF LMMS. The larger spot in FT LMMS is sufficient for many material science problems. FT LMMS costs about the same as ToF LMMS but the cost per analysis is much lower and the operational procedures are less straightforward. The technique closes the traditional gap between organic and inorganic microanalysis. The all-round nature of FT LMMS is favoured since EI/CI does not depend on the chromophoric group in the analyte, as opposed to photon ionisation. However, FT LMMS and ToF LMMS spectra are not directly comparable. In fact, the use of laser microbeam irradiation under similar conditions with respect to wavelength, pulse duration and pulse density in ToF and FT LMMS, does not ensure that the same ions are registered or that the relative intensities always closely agree.

The information contained in these spectra differs as a result of the incomparability of the time domains for sampling and analysis of the laser-generated ions. ToF LMMS registers only the prompt ions whereas FT LMMS detects a different fraction of the initial ion population, namely including the ions from post-laser DI. FT LMMS data thus give access to ions formed during *and* after the laser pulse [326, 327].

Especially for organic compounds significant differences can be noticed. FT LMMS offers superior mass resolution and mass accuracy in comparison to ToF LMMS. In FT LMMS an increased adduct ion detection for both organic and inorganic compounds is observed. ToF LMMS is often handicapped in the analysis of high-MW and polar substances. The range of compounds to which FT LMMS can be applied significantly extends towards higher molecular mass and polarity as compared to ToF LMMS. Van Vaeck *et al.* [324] have illustrated the dramatic progress of FT LMMS analysis in comparison with

Table 3.26. Analytical characteristics of LMMS

	ToF LMMS			FT LMMS
Geometry	Transmission		Reflection	Reflection
Minimum spot diameter (μm)	0.5		1	5
Information depth		10–50 nm		<50 nm
Crater depth	1 μm		100–200 nm	100 nm
Image resolution		1 μm		>1 μm
Mass accuracy		Nominal		0.1–1 ppm
Mass resolution		500[a], <850[b]		10^4–10^6
Detection limit		10^{-13}–10^{-15} g[d]		10^{-11}–10^{-12} g
Sensitivity[c]				
Element ions	10^6–10^7		Not reported	10^8–10^9
Organic ions	10^7–10^8		4×10^9	6×10^8
m/z range		H-unlimited		15–15,000
Dynamic range		10^2–10^3		10^2–10^3

[a] For organic compounds.
[b] For inorganic compounds.
[c] Neutrals in sample.
[d] Calculated according to Simons [328].
After Van Vaeck *et al.* [329]. From L. Van Vaeck *et al.*, in *Surface Characterization. A User's Sourcebook* (D. Brune *et al.*, eds.). Copyright 1997 © Wiley-VCH, Weinheim. Reproduced with permission.

ToF LMMS for PET. A clear distinction between ToF and FT LMMS is mandatory, as summarised in Table 3.26.

ToF LMMS instruments allow for analyses in transmission or in reflection mode. In the former case, the laser hits the lower surface of the sample while the upper surface faces the MS. This suits thin films or particles of about 1 μm on a polymer film. In reflection geometry the beam impinges on the sample side facing the MS and is suitable for surface analysis of bulk materials. Various FT LMMS arrangements, all using reflection geometry, have been commercialised. LMMS **samples** are preferably in the form of homogeneous thin films or coatings. Characterisation of mixtures requires suitable powder samples with uniform particle-size distributions. In microprobe applications the porosity of the matrix plays an essential role. Another requirement in LMMS is the sample stability in vacuum.

LMMS results critically depend on the procedural details. This holds true for the basic parameters, *e.g.* resolution and calibration, and the fundamental aspects, such as major ion-formation mechanism, degree of fragmentation, *etc.* Van Vaeck *et al.* [330] have presented a **standard procedure** for single-particle analysis with LMMS, which is found to determine the experimental conditions quite strictly. The proposed procedure defines the experimental conditions by directly available and generally applicable criteria.

In mass spectrometry, organic and inorganic compounds are characterised by a combination of fragments and adduct ions. As a result of the ultra-fast heating rate of the solid in LMMS thermal destruction of labile compounds is limited. **LMMS spectra** look totally different from those for conventional laser desorption (LD)MS [160]. The latter method typically yields merely cationised molecules, even from thermolabile organics, virtually without fragmentation and decomposition. There are fundamental differences between these techniques with respect to power density and ion formation. Desorption and ionisation (DI) mechanisms in LMMS are complex. Depending on the applied power density, LMMS generates elemental ions, cluster ions of higher m/z from inorganic substances, or ionised, structural elements from organic molecules. It is usually conceived that laser impact starts various processes, such as pyrolysis and thermal decomposition. A simplification is the assumption that organics are initially desorbed as neutrals, followed by electron ionisation or adduct formation by attachment of H^+ or Na^+. Organic LMMS data have a certain ambiguity. Whereas quite stable molecules are readily pyrolysed or completely disintegrated into carbon clusters, thermolabile compounds can be desorbed intact, and soft adduct ionisation processes seem to prevail; in other cases fragmentation remains abundant. Low-MW compounds show a pronounced tendency to form dimeric clusters; less volatile compounds produce fewer adducts but increased electron ionisation and prominent daughter ions. Substances that are difficult to desorb yield no adducts and only fragments. This situation is completely different from conventional MS, where each compound receives a given constant amount of internal energy. In LMMS there is a gradual transition from CI to EI conditions, depending on the physicochemical properties [330]. ToF LMMS exploits only the promptly generated ions. This time-selectivity may explain why only a fraction of the total yield for some processes is monitored in LMMS, especially cation attachment, which prevails in LDMS.

LMMS results are frequently interpreted as a yes-or-no answer about the presence of an individual target. **Qualitative identification** of local constituents in LMMS spectra can often be obtained by deductive reasoning rather than by fingerprinting. This represents a major advantage in industrial materials science applications, in particular in organic microanalysis because of the numerous structures for each molecular weight. Usually comparison with reference spectra is not even required; few reference spectra for organic molecules are available anyway as laser ionisation is a relatively recent development.

One of the basic problems in LMMS is that ion formation by focused laser irradiation of solids appears to represent extremely complicated processes, involving several competitive time- and energy-dependent mechanisms. Van Vaeck *et al.* [331] have reported a tentative model for desorption and ionisation (DI) in LMMS, allowing the rationalisation of the formation of the detected ions in terms of time, local energy and pressure. Upon ultra-fast heating rates non-volatiles get desorbed intactly (apparent contradiction). The low m/z signals give useful hints about the class and composition of the analyte. The high m/z range normally contains intense peaks from cluster ions consisting of the intact molecules. LMMS exhibits speciation capabilities. Differences in the qualitative nature of laser ionisation spectra from different matrices demands some caution to be exercised when comparing spectra.

Laser microprobes can be fitted with a **post-ablation ionisation** (PAI) capability, both resonant and non-resonant. The two-shot LMMS configuration thus consists of an ablating laser and an ionising laser (fired parallel to the sample surface). Ion mass and intensity analysis are performed in a manner similar to conventional laser microprobe analyses, except that it is usually necessary to separate the ions formed by the ablating laser pulse from those produced by the ionising laser. For detecting a specific neutral species in the ablated vapour plume the most efficient laser ionisation process is REMPI in which the ionising laser is tuned to an absorption line of the particular neutral species [332]. For survey analysis non-resonant (non-selective) multiphoton ionisation is more suitable. Since the laser PAI technique provides very high surface sensitivities, *i.e.* represents the top 5–10 Å of the sample, analysis of organic compounds adsorbed onto solid surfaces constitutes an important application of this technique. Odom [333] has described PAI for the laser microprobe.

The main features of LMMS are summarised in Table 3.27. Laser microprobe mass spectrometry is a valuable tool for **inorganic and organic analysis**. Element location and quantification on the μm scale can be achieved (spot analysis) and speciation possibilities are available, which are unsurpassed by other

Table 3.27. Strengths and weaknesses of LMMS

Advantages:
- Minimal sample preparation
- Element, inorganic and organic molecule specific information; full periodic table coverage
- Diagnostic information by deductive reasoning
- High sensitivity (10–100 ppm range for elements)
- Compound speciation
- Local and total analysis; surface analysis
- In-depth profiling
- Microprobe mapping capabilities
- Real-time analysis
- No sample charging (compatibility with non-conductive samples)
- Direct isotope information

Disadvantages:
- Difficult quantitation; problematic calibration
- Low mass resolution (in ToF LMMS)
- Vacuum stability of sample required
- UV absorbing functionality required for ionisation
- Destructive (minute material consumption)
- Spot analysis
- Relatively poor spatial resolution
- No automation

MS methods. In particular, LMMS is suitable for structural characterisation of organics by the combination of soft ionisation with extensive fragmentation. The strength of the method arises from the ease of adapting the experimental conditions, such as power density, for a given problem. A major asset of LMMS is the molecule-specific information in small volumes and mixtures (μm scale) and heterogeneities, which constitutes a major advantage with respect to most micro- and surface analysis techniques [331,334]. Microprobe analysis differs from pyrolysis measurements in that samples as small as ng of material are required as compared to 0.1–5 mg for pyrolysis. LMMS is preferentially applied to polar and ionic compounds, which are intractable by conventional techniques, and is widely appreciated as a soft ionisation method for thermolabile organics, yielding cationised molecules and virtual absence of fragmentation.

LMMS is essentially a **surface sensitive method**: structural atoms from organic samples issue from the upper 5–50 nm when a 1 μm thick sample is analysed. Detection limits for elemental ions are in the ppm range. Real depth-profiling experiments are hindered by the discrete nature of the laser interactions.

The main problem of LMMS is the **quantitation**, caused by an irreproducible sample ionisation process. The ionisation yield depends strongly on the local energy regime, *i.e.* on the applied power density and on a variety of material properties such as sample reflectivity, UV absorbance, thermal conductivity, *etc.* These features are frequently ill defined and vary from spot to spot in heterogeneous specimens. Moreover, adequate reference materials are usually not available. Another important disadvantage of the laser microprobe is the relatively poor spatial resolution, especially compared to electron microscopic methods. The resolution limit for a focussed laser is determined by the diffraction-limited beam width and is of the order of one wavelength. Finally, the equipment also suffers from poor resolution during sample observation (less than that of a high-quality optical microscope), whereas automation is problematic since the results are extremely sensitive to the laser focus adjustment in relation to the sample surface. Nevertheless, automated mapping of TLC plates without focus correction has been reported [335,336].

LMMS is a similar technique to ToF-SIMS except that the ionising primary beam is a laser; the energies involved are greatly different: eV level for LMMS, keV for SIMS. In LMMS the data are obtained from a greater depth (10–100 nm) than SIMS.

Laser microprobe mass spectrometry has been reviewed [47,328,331,334,337–339]; Van Vaeck *et al.* [330] have discussed possibilities and limitations of LMMS. Darke *et al.* [53a] have reviewed the instrumental features. A monograph is available [130].

Applications
LMMS offers a great potential for inorganic analysis and speciation as well as for organic structural characterisation. The method excels at yielding, within a relatively short period, qualitative information on local surface components from the most diverse samples, often with negligible sample preparation. LMMS allows detection of the presence of a given compound by means of structurally relevant ions, down to the 20% level in organic mixtures. This is achieved without pre-separation.

Applications are very broad. ToF LMMS has been used for fingerprinting of polymers by characteristic pyrolysis fragments [322], identification of organic residues on integrated circuits [340], of plasticisers from PE packaging material on asbestos fibres [341], and of a dibenzothiazyl disulfide agglomerate in a rubber sample [342]. Gardella *et al.* [322] have used LAMMA® (Nd:YAG, 266 nm) to vaporise/ionise various polymeric samples (*ca.* 10^{-13} g) into a ToF-MS, however without detecting additives. ToF LMMS can successfully cope with the structural characterisation of organic compounds, untractable by conventional MS techniques. A favourite field is microprobing of **local microscopic inclusions**. ToF LMMS was used to determine the element composition of impurities in PC and PA6.6 [343]. ToF LMMS has proved capable of identifying TiO_2 inclusions in PVC [344] and accelerator agglomerates in rubber [345]. Difficult problems concerning polymers are often caused by local **heterogeneities** due to poor dispersion of ingredients. LMMS does not generate extensive oligomer distributions as does static SIMS, but still has provided sufficient information to identify inadvertent PMMA inclusions in injection-moulded PC [345].

Rubber surfaces were characterised directly by ToF LMMS, LD-FTMS and TD-FTMS [201, 202]. The surface chemistry of the antiozonant *N*-(1,3-dimethylbutyl)-*N*′-phenyl-*p*-phenylenediamine (HPPD) in vulcanised natural rubber compounds was explored by ToF LMMS in order to investigate the mechanism of rubber-surface ozone ageing

(HPPD–ozone reaction) and protection [202]. Mass spectra were obtained for intact molecular ions (M^+) of organic chemical **rubber additives** such as the aromatic processing oil, and the aromatic antiozonant and AOs incorporated to protect the rubber. Molecular weight information from the molecular ions and structural information from the fragmentation ions could be obtained without interference from the fragmentation peaks of the rubber backbone. Laser and thermal desorption mass spectral techniques provide complementary structural information and can allow positive identification of various organic species present on the surface of vulcanised rubber.

UV irradiation allows selective DI of products with chromophoric groups. Various **organic pigments** (xanthones, anthraquinones, *etc.*) on a 5-μm scale in an organic matrix were analysed by means of FT LMMS [346]. Most organic components other than pigments do not have UV absorbing functionalities and hence are not ionised in LMMS. LMMS offers the possibility of analysing samples with 1–5 μm spatial resolution. Similarly, methylene blue dyed cotton fabrics were microanalysed by the same technique [347]; the adduct of indigo was detected in the single-shot mode using a 5-μm diameter spot for irradiation of jeans tissues [348] (Fig. 3.18). ToF LMMS has been applied to the study of dye spots on PS [349].

Application of FT LMMS to **inorganic compounds** offers the advantage of direct speciation [327]. FT LMMS has been applied to the analysis of a residual chromium complex on HPLC column packing material; the Cr-levels were below the detection limit in STEM EDX. The technique has also allowed to trace the origin of occasional (in)organic contaminants at the surface of microelectronic devices. Particularly interesting is a FT LMMS study of the dispersion of the magnetic elements (Fe, Mn) inside the PET matrix at the surface of faulty floppy discs [324]. Here the capability to detect both inorganic ions *and* organic structural fragments in the same spectrum was essential for troubleshooting. In the photographic industry LMMS has been involved in the study of 50 μm defects on AgBr and

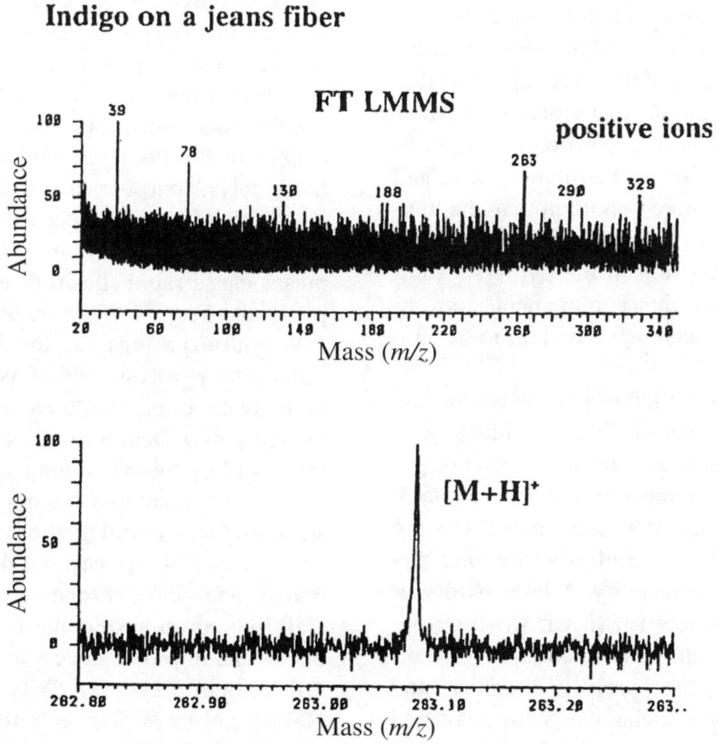

Fig. 3.18. Analysis of a commercial jeans fibre by FT LMMS. Top: positive ion mass spectrum; bottom; detection of indigo in the high-resolution mode from a single shot with a 5 μm spot. After Van Roy *et al.* [348]. Reprinted from A. Benninghoven *et al.* (eds.), *Proceedings SIMS IX*, Copyright © 1994 John Wiley & Sons, Ltd. Reproduced with permission.

PET layers [324]. These analyses were not feasible with ToF LMMS. In fact, FT LMMS allowed observing that the ion formation occurs primarily by post-laser processes, *i.e.* 100–200 μs after the pulse. ToF LMMS is essentially optimised for detection of "prompt" ions generated *within* the 15–20 ns laser pulse, rather than those formed long *after* the pulse. A variety of products with MW > 500 Da and with polar groups, which tend to produce ions after the laser pulse, cannot be analysed by ToF LMMS.

LMMS has also been used for direct analysis of normal-phase HPTLC plates (*cfr.* Chp. 7.3.5.4 of ref. [13a]) and for identification of cloth samples by means of fingerprints from dyes and fabric softeners for forensic purposes [350]. Applications of laser microprobe mass spectrometry were reviewed [53a, 328,334].

3.5. LASER PYROLYSIS

Principles and Characteristics

Lum [168] has introduced the techniques of laser probe-molecular beam sampling and mass spectrometry to the thermal analysis of polymer materials. It is now recognised that the dynamics of laser-induced thermal reactions differs greatly from that of usual thermodegradation. In pyrolysis the thermal energy deposition in a given time on the sample, which is a function of the thermal capacity and rate of heat transfer, is more important than the temperature. Laser irradiation is a suitable source for controlled energy deposition in the pyrolysis sample. Laser probe analysis thus complements conventional thermal analysis techniques and offers several unique capabilities.

The nature and distribution of pyrolysis products from a particular sample critically depend largely on the pyrolysis temperature and the specific set of pyrolysis conditions (*i.e.* temperature rise time, sample size, pyrolyser geometry). Laser pyrolysers are practically the only type of **radiative heating pyrolyser** with certain applicability. A laser pyrolyser consists of five components: *(i)* laser; *(ii)* fibre optics; *(iii)* probe for sample introduction; *(iv)* pyrolyser body, containing the pyrolysis chamber; and *(v)* heater of the pyrolysis chamber with dedicated control unit. The laser beam can be focused onto a small spot of a sample to deliver the radiative energy. This provides a special way to pyrolyse only a small portion of a sample. Only the sample itself is heated, thus eliminating interfering reactions, which could otherwise occur at the hot surfaces of the sample cell or with contaminant species liberated by cell components at elevated temperatures. The physical conditions existing at the focal point of a powerful laser beam differ drastically from those existing in any of the conventional pyrolysers. Exposure of material to an intense power concentration results in extensive degradation. Laser beam pyrolysis, therefore, offers "ultra-fast" heating ($>10^3\,°C\,s^{-1}$) to high temperatures, short temperature rise time, and rapid explosion-like expansion of the pyrolysis products into a "plume". The temperature of the plume ranges from 500°C to 2000°C. Such a unique set of degradation conditions produces product distributions which are characteristic yet different from those encountered in ordinary pyrolysis.

A variety of laser types were used for pyrolysis purposes: normal pulsed, Q-switched, or continuous wave (CW), at different energy levels (high- and low-powered). The power density of the incident irradiation on a sample has a significant influence on the composition of the vapour generated by laser pyrolysis. In general, the higher the power density incident on a polymeric material the lower the molecular weights of the species detected and the greater the number of laser produced ions. Most common are the normal pulsed high-powered lasers which give smaller, less characteristic fragments. The maximum energy of the laser can be higher than needed for the pyrolysis purposes and often needs to be attenuated. The laser can be focused to yield flux densities of the order of $10^8\,W\,cm^{-2}$ or more. Intense laser pulses cause rapid (flash) or even almost explosive pyrolysis. For a CW laser, powers between 0.5 to 5 W with exposures varying from 1 s to 5 min are utilised for pyrolysis. With CW heating using a defocused beam impinging on a relatively large area (20 to 400 μm^2), heating rates are comparatively low (but still fast compared with standard thermal methods) and the results of heating are more like those of standard fast thermal methods.

Various **laser pyrolyser designs** have been reported [351–355]. A relatively cheap (non-commercial) low-power system has been designed that simplifies and improves access to laser pyrolysis (LPy). This system uses a Nd-Cr-GGG laser that delivers 600 mJ pulses of 500 μs with a slow repeat rate of 40 s [355]. The laser energy is delivered to the pyrolysis chamber via an optical fibre.

As already pointed out in Chp. 2.2, pyrolysis techniques are characterised by a great number of

experimental variables. Obviously, in radiative-heating pyrolysis advantage may be taken of the choice in specific wavelengths corresponding to given molecular vibrations in order to break specific bonds. On the other hand, use of a laser with a broad range in characteristics (wavelength, laser fluence, pulse width, number of shots, *etc.*) adds to the complexity of the experiment. Not surprisingly, few LPyGC-MS experiments are quite comparable. Continuous and pulsed laser pyrolysis experiments have been compared [356]. In the Salford approach, a Nd:YAG laser is used for pyrolysis, and an electron beam for ionisation immediately after vapour evolution. NASA experiments were carried out with two different lasers for pyrolysis, namely pulsed Nd:YAG (1.06 μm) and CW CO_2-laser (10.6 μm), which generates much lower heating conditions. Greenwood *et al.* [357] have reported LPyMS and flash PyMS studies; also LPyGC-MS and filament PyGC-MS were compared [358]. Folmer *et al.* [359] used GC to evaluate the results of polymer pyrolysis by using filament, tube furnace and laser pyrolysis. Comparisons of analytical performance of different pyrolyser types (conventional and laser) are not straightforward because the analytical instrument at the end of the pyrolyser may influence the quality of the data. Table 2.21 lists the main characteristics of several pyrolysers.

Table 3.28 lists the **main features** of laser pyrolysis techniques. Laser pyrolysis is characterised by rapid non-linear heating, and a large ΔT. Temperature rise times and cooling times are very short, usually in the range of 100 to 300 μs [360], and quite different from conventional pyrolysis (*cfr.* Table 2.21). This contributes to the uniqueness of the degradation conditions for laser pyrolysis, which are rather different from the other pyrolysis types (*cfr.* Chp. 2.2). One of the advantages of radiant heating for flash pyrolysis is that it is "clean", that is, only the light-absorbing material is heated: the container and other surrounding objects remain cool and do not contribute to the vapour. In addition to this, the capability to pyrolyse only a very small area of the sample is characteristic for most laser pyrolysers. Inclusions, samples containing heterogeneities, *etc.*, can be successfully analysed using this technique.

Table 3.28. Characteristics of laser pyrolysis methods

Advantages:

- Very short temperature rise times/cooling times
- Unique degradation conditions
- Small area analysis (spatial resolution)
- *In situ* analytical capabilities
- High detection sensitivity

Disadvantages:

- Dependency on optical properties of sample
- Need for radiation absorbing centres
- Low reproducibility
- Variability in total pyrolysed mass
- Not quantitative
- Lack of temperature control (TRT, T_{py})
- Lack of cost-effective commercial instrumentation
- Complex laser technology

Besides formation of some unusual products due to secondary reactions, there are several problems regarding the use of lasers as an energy source for pyrolysis. A first problem is related to the direct or indirect absorption of the radiative energy into the sample. Transparent samples do not absorb the laser beam energy properly. This appears to be a serious limitation considering the broad range of wavelengths to which organic materials are transparent. CW laser pyrolysis is very dependent on the optical properties of the material. For laser power up to about 700 mW no pyrolysis was observed for a free-standing film. Therefore, this method is less general than pulsed laser ablation [260]. Efficient absorption of laser light requires addition of absorbing centres or converting the sample into absorbing derivatives. Mixing carbon powder with the sample has been used to facilitate absorption of the beam energy. However, graphite may interfere with the degradation process. Use of an inert blue cobalt glass substrate for absorption of laser energy has been described as a superior method [361]. Other disadvantages of conventional laser pyrolysis are the low reproducibility, variability in the total mass of material pyrolysed and difficulty in knowing precisely the equivalent temperature and rise time of pyrolysis. In addition, the rapid rate of heating and cooling precludes attainment of thermal equilibrium. Lack of cost-effective commercial instrumentation is another practical drawback. The cost of specialised instruments is an order of magnitude higher than that of the more usual filament and Curie-Point pyrolysers [362].

Matsuoka *et al.* [363] have reported an LPyGC system. Advantages claimed for **laser pyrolysis-gas chromatography** include relatively simple fragmentation patterns, less secondary reactions, high sensitivity, time saving and good reproducibility.

Laser pyrolysis is also used in direct coupling with an MS system. The laser can be remote from the sample, which can thus be conveniently heated inside a vacuum chamber or the ion source of a mass spectrometer. Different types of lasers (UV, IR), combinations of lasers, mass spectrometers (QMS, ToF-MS, FTICR-MS, ITMS) or other experimental set-ups were reported, as utilised in **LPyMS** [260, 364,365]. Typically, a laser power of about 100 mW is focused on a spot of 150–200 μm in diameter, and 15 eV electron impact ionisation is used. Lum [352] originally used a modest power infrared CO_2 laser focused to a fine spot size (*ca.* 100 μm) and pyrolysed the polymer samples directly in the ion source of a QMS. In LPy-ToFMS experiments the sample is located just below the ionisation chamber of the mass spectrometer, which is kept at a pressure below 10^{-4} Pa [35]. The polymer surface is then typically exposed to 0.5 ms wide laser pulses at 1.06 μm with a total energy which can be varied up to 1.0 J. The temperature rise at the surface of the sample during heating has been estimated to be within the range of 650°C to 1000°C [35]. Absorption of the energy pulses forms minute craters at the surface of the polymer sample. Species leaving this crater are analysed by ToF-MS. This technique exploits the fast scanning capability of a ToF mass spectrometer to show the evolution rates of volatilised species. The LPy-ToFMS experiment has been described in detail in ref. [366]. The fact that a ToF-MS is able to provide a complete mass spectrum per event is particularly advantageous for ionisation techniques such as plasma desorption, laser ionisation and secondary ion mass spectrometry. Comparison of LPy-ToFMS (with pattern recognition analysis, PRA) and TD-GC-MS has indicated that PyMS yields information based on structural differences in the polymer chain, where TD-GC-MS is more able to provide information on the residual volatiles and additives that are evolved below the pyrolysis temperature [367]. Laser pyrolysis methods work also well with Fourier transform mass spectrometry [260]. FTMS operates in a pulsed mode consistent with a pulsed laser. However, since ions are trapped for long times in a cell by magnetic and electric fields, they can be accumulated and detected from a low-yield continuous source. The high mass resolution capability of FTMS is useful for obtaining accurate mass measurements of ions in order to assign the chemical composition. Use of an ion trap for LPyMS allows studying the ion chemistry of the pyrolysis products and yields the possibility of relatively inexpensive MS^n type experiments for structural identification of pyrolysate fragments. LPyMS offers evolved gas analysis and information about the temporal behaviour of the evolved species. However, the processes taking place in LPyMS are not well characterised because various effects occur when the sample is irradiated with the laser beam, such as laser-induced desorption (LID), melting, pyrolysis, ionisation, *etc*. These processes depend on the laser intensity and energy (wavelength) and on the substrate and sample composition. Also, the vacuum in the MS system may play a role regarding the result of irradiation by diminishing any secondary reactions of the pyrolysate.

A simple **LPyGC-MS** instrumental arrangement has been described, based on a multipurpose pulsed Nd:YAG laser ($\lambda = 1064$ nm, pulse energies 0.1–3.0 J, pulse width 500 μs), filament pyrolyser and GC-ITMS [358]. As already noticed, in an improved design a pyrolyser for LPyGC-MS experiments was described with a simpler specialised laser system and a lower volume (μL range) pyrolysis chamber [355]. Greenwood *et al.* [368,369] also described LPyGC-MS using a CW Nd:YAG laser and a reflected light/fluorescence microscope, which enables sample viewing, selection and pyrolysis. The molecular integrity of laser pyrolysis data has been established by scrutinising against conventional pyrolysis data obtained from comparative samples. LPyGC-MS facilitates molecular analysis of microsample quantities of organic matter [370]. Ryan *et al.* [371] have described LPyGC-MS and high-resolution **laser pyrolysis IR spectroscopy**.

Zhu *et al.* [372] have reported QTLC by **laser pyrolysis scanning** (LPS). In this procedure TLC plates are placed in a chamber after development and irradiated with an IR laser to produce a high temperature at the location of the spot. The analyte is swept by a carrier gas to a GC and detected with FID or ECD. Complete analysis time is less than 20 min. The technique combines the advantage of the separation power of TLC and GC detection modes.

Moldoveanu [373] has reviewed radiative heating pyrolysis. Laser microprobe mass analysis (LMMS) has also resulted in interesting polymer pyrolysis studies (*cfr.* Chp. 3.4.5).

Applications
The specialised nature of the technique has limited applications of this pyrolysis mode. High-energy LPyGC has been investigated as a tool in analysis of polymers [361]. CO_2 laser PyGC has extensively been applied to polymers by Chinese authors [374–379]. According to this work CO_2 LPy can be

applied directly to analysis of lightly coloured or colourless transparent specimens. Ruby laser PyGC-FID was also applied to polymers [380,381]. Clear or translucent samples gave reproducible results if mixed with carbon. The carbon concentration is critical as it has a great effect on the fragmentation pattern. Fanter *et al.* [361] used a pulsed ruby laser to investigate degradation of PE and PS. The product distributions resulting from laser degradation were in general similar to those expected from high temperature (1200–1500 K) thermal pyrolysis. Laser pyrolysis of hydroxyl-terminated polybutadiene was reported [382] and IR laser pyrolysis of silane was described [383].

McClennen *et al.* [203] have investigated LPy-QMS at low electron energies (15 eV) using a CW CO_2 laser for fast, direct and reproducibly vaporising additives and pyrolysing the polymer in **solid rubber vulcanisates**. The SBR and SBR/NR formulations contained carbon-black, processing oil, stearic acid, an antiozonant (HPPD), a tackifier (*t*-octylphenol formaldehyde resin), AOs (poly-TMDQ, DODPA), ZnO, S, and an accelerator (*N*-*t*-butyl-2-benzothiazylsulfenamide). The results were compared with CuPyMS. At lower power densities, the heat absorbed by the (rubber) surface softens a significant volume of the compound to vaporise additives into the vacuum. At higher laser power densities (and shorter pulse times) the surface polymer can be flash pyrolysed to give monomers and other fragments and oligomers characteristic of the polymer composition. Less volatile additives, such as oils and large polar amines and phenols, are sensitive to exact instrument geometry for accurate detection. Large samples (>100 mg) may lead to rapid ion source contamination; instrument design must provide for control of the low volatility products, including easy removal and cleaning of ion source surfaces where condensation may occur.

Pyrolysis of a cured, fire retardant **epoxy resin** (diglycidyl ether of tetrabrominated bisphenol A) was studied by Creasy [260] under several conditions: thermal pyrolysis, laser ablation with a pulsed laser and in-source CW argon LPy-FTMS. Very different mass spectra were obtained from the brominated resin with these methods, which differ considerably in heating rate. For the pulsed laser, even the free lasing pulse gives a heating rate on the order of 10^6 K s^{-1}, higher than that for flash pyrolysis. In both cases, a range of small fragments is observed. On the other hand, a CW laser produces larger, structural molecules rather than small fragments and stable molecules as opposed to radicals. It appears that heating with a CW laser is a promising method for obtaining molecular information from small sample volumes, but this depends also strongly on the thermal and optical absorption properties of the sample.

Laser pyrolysis has found wide application for the study of **flame retarded polymers**. The conditions produced by pulsed laser heating of polymeric material are somewhat analogous to those which occur in a fire and so any information obtained regarding polymer behaviour following laser heating should give a guide to possible chemical behaviour in a fire. Lum [168] has first applied a CW argon ion laser in LPy-QMS to gain insight into polymer degradation and flame retardant behaviour of PVC and PVC/Sb_2O_3 under radiation conditions analogous to those of a real fire, however without paying specific attention to the fate of the additives. Price *et al.* [366, 367,384–387] have used LPy-EI-ToFMS to study the behaviour of FR polymers in real fire situations: plasticised PVC (Cereclor S-45, DIOP, TTP), PS/(DBDPO, Sb_2O_3, ZB 2335), PMMA/ATH, rigid PUR/DMMP foams with various isocyanate indices, and styrenic polymers. As the PVC samples absorbed the laser radiation strongly, there was no need to use extraneous material such as graphite to facilitate energy transfer to the sample. In these experiments the polymer sample was located in the ionisation chamber of the mass spectrometer so that species produced during pyrolysis were directly vaporised into a vacuum. The rationale behind the experiments is as follows. Fire behaviour of a polymer is governed by chemical reactions (pyrolysis and oxidation) in three regions, namely within the condensed phase, at the interface between the condensed phase and the gas phase, and in the gas phase. Major pyrolysis occurs at the interface region, which is also the region where condensed phase flame retardants act. An understanding of (oxidative) pyrolysis is essential to an understanding of the chemistry of polymer combustion and mechanisms of flame retardant behaviour. In a real fire situation, the heating rate at the surface of a polymer exceeds 300 K min^{-1}, which is beyond reach of conventional thermal experimental methods (<100 K min^{-1}). Consequently, conventional pyrolysis studies do not simulate real fire conditions. Rapid heating rates are easily obtainable by means of focused laser light. LPy-ToFMS is intended as a model for pyrolysis reactions, which occur in the very narrow diffusion limited region at the condensed phase-flame interface of a burning polymer. In this approach a pulsed laser is used to heat a polymer in a high-vacuum environment within the ionisation region of a ToF-MS. Continuous and pulsed laser pyrolysis experiments were compared [356]. The continuous LPy experiments ($\lambda = 514.5$ nm and 1064 nm) gave the first direct evidence for production of volatile $SbCl_3$ during decomposition of a PVC/Sb_2O_3 formulation and have provided valuable insight into the detailed mechanism of polymer pyrolysis in a controlled environment and the influence of FRs thereon. In pulsed conditions a temporal history of the pyrolysis behaviour over milliseconds is obtained. These results bear relevance to the behaviour at the condensed phase-flame interface of a burning polymer (the so-called "dark-flame" region), which is pertinent to FR behaviour. The rapid scanning capability of ToF-MS enables direct observation of the reactions occurring. LPy-ToFMS has been used to study transition metal complexes of cellulose ammonium phosphate (CAP) fabric and to monitor benzene evolution from various flame retarded PVC samples containing plasticisers: PVC/(DIOP, MoO_3), PVC/(TTP, MoO_3), PVC/(Cereclor S-45, MoO_3). Laser pyrolysis behaviour of the PMMA material has also been compared with the results of TVA-ToFMS (both under vacuum conditions) [35,388,389]. The results of LPy of PMMA/ATH are similar to those observed using conventional pyrolysis techniques. Price [390] has recently reviewed laser pyrolysis studies of flame retardance mechanisms.

LPyGC-MS was also applied to organic polymers [391]. The crucial point in pyrolysis experiments on macromolecules is represented by the reproducibility of the data. Laser micropyrolysis (*e.g.* LPyGC-MS) can be used to address the physical heterogeneity of complex organic materials [370]. Moisture analysis in glass fibre- or graphite fibre-reinforced epoxy resins and polyimides by LPyGC-MS and high-resolution LPyIR has also been reported [371].

BIBLIOGRAPHY

Lasers

P. Misra and M.A. Dubinskii, *Ultraviolet Spectroscopy and UV Lasers*, M. Dekker, New York, NY (2002).

H.-G. Rubahn, *Laser Applications in Surface Science and Technology*, John Wiley & Sons, Chichester (1999).

G.R. van Hecke and K.K. Karukstis, *A Guide to Lasers in Chemistry*, Jones and Bartlett Publ., Boston, MA (1998).

O. Svelto, *Principles of Lasers*, Plenum Press, New York, NY (1998).

D.L. Andrews, *Lasers in Chemistry*, Springer-Verlag, Berlin (1997).

W.T. Silfast, *Laser Fundamentals*, Cambridge University Press, Cambridge (1996).

F.J. Duarte (ed.), *Tunable Lasers Handbook*, Academic Press, San Diego, CA (1995).

J. Hecht, *The Laser Guidebook*, McGraw-Hill, London (1992).

F.J. Duarte (ed.), *High-Power Dye Lasers*, Springer-Verlag, Berlin (1991).

L.J. Radziemski and D.A. Cremers (eds.), *Laser-Induced Plasmas and Applications*, M. Dekker, New York, NY (1989).

Laser Ablation

J.C. Miller and R.F. Haglund (eds.), *Laser Ablation and Deposition*, Academic Press, London (1998).

R.K. Singh *et al.* (eds), *Advances in Laser Ablation of Materials, MRS Symposium Proceedings Vol. 526*, Warrendale, PA (1998).

J.C. Miller and R.F. Haglund (eds.), *Laser Ablation: Mechanisms and Applications*, Springer-Verlag, Berlin (1991).

Laser Spectroscopy/Spectrometry

H. Pasch and W. Schrepp, *MALDI-TOF Mass Spectrometry of Synthetic Polymers*, Springer-Verlag, Heidelberg (2003).

W. Demtröder, *Laser Spectroscopy, Basic Concepts and Instrumentation*, Springer-Verlag, Berlin (2003).

R.D. van Zee and J.P. Looney (eds.), *Cavity Enhanced Spectroscopies*, Academic Press, San Diego, CA (2002).

J. Sneddon, T.L. Thiem and Y.-I. Lee (eds.), *Lasers in Analytical Atomic Spectroscopy*, VCH Publishers, New York, NY (1997).

F. Hillenkamp (ed.), *Matrix-Assisted Laser Desorption Ionization Mass Spectrometry*, Elsevier, Amsterdam (1997).

D.L. Andrews and A.A. Demidov (eds.), *An Introduction to Laser Spectroscopy*, Plenum Press, New York, NY (1995).

A. Vertes, R. Gijbels and F. Adams (eds.), *Laser Ionization Mass Analysis*, J. Wiley & Sons, New York, NY (1993).

D.L. Andrews (ed.), *Applied Laser Spectroscopy: Techniques, Instrumentation and Applications*, VCH, Weinheim (1992).

Z.-G. Wang and H.-R. Xia, *Molecular and Laser Spectroscopy*, Springer-Verlag, Berlin (1991).

D.M. Lubman (ed.), *Lasers and Mass Spectrometry*, Oxford University Press, New York, NY (1990).

L. Moenke-Blankenburg, *Laser Microanalysis*, J. Wiley & Sons, New York, NY (1989).

V.S. Letokhov, *Laser Photoionization Spectroscopy*, Academic Press, Orlando, FL (1987).

Laser-induced Chemistry

D. Bäuerle, *Laser Processing and Chemistry*, Springer Verlag, Berlin (1996).

Laser Safety

D. Hughes, *Notes on Protection against Laser Radiation in the Laboratory*, HHSC, Leeds (1992).

REFERENCES

[1] J.M. Hollas, *High Resolution Spectroscopy*, J. Wiley & Sons, Chichester (1998), pp. 468–710.

[2] J. Sneddon, Y.-I. Lee, X. Hou, J.X. Zhou and R.G. Michel, in *Lasers in Analytic Atomic Spectroscopy* (J. Sneddon, T.L. Thiem and Y.-I. Lee, eds.), VCH Publishers, New York, NY (1997).

[3] W. Demtröder, *Laser Spectroscopy, Basic Concepts and Instrumentation*, Springer-Verlag, Berlin (1998).

[4] T.H. Maiman (to Hughes Aircraft Co.), *U.S. Pat.* 3,353,115 (Apr. 13, 1961).

[5] T.H. Maiman, *Phys. Rev. Lett.* **4**, 564 (1960).

[6] T.H. Maiman, *Nature* **187** (4736), 493–4 (1960).

[7] M. Maeda, *Laser Dyes*, Academic Press, New York, NY (1984).

[8] F.J. Duarte (ed.), *Tunable Lasers Handbook*, Academic Press, San Diego, CA (1995).

[8a] G. Sulmann, *Chem. Magazine*, 203–205 (May 1996).

[9] D.L. Andrews, *Lasers in Chemistry*, Springer-Verlag, Berlin (1997).

[10] G.R. van Hecke and K.K. Karukstis, *A Guide to Lasers in Chemistry*, Jones & Bartlett Publishers, Boston, MA (1998).

[11] M.R.S. McCoustra, in *An Introduction to Laser Spectroscopy* (D.L. Andrews and A.A. Demidov, eds.), Plenum Press, New York, NY (1995).

[12] R.W. May and J. Porter, *J. Forensic Sci. Soc.* **15** (2), 137–46 (1975).

[13] G. Chinga, Ø. Gregersen and B. Dougherty, *Eur. Microsc. Anal.* **84** (7), 5–7 (2003).

[13a] J.C.J. Bart, *Additives in Polymers. Industrial Analysis and Applications*, J. Wiley & Sons, Chichester (2005).

[14] A.C. Ouano, *J. Chromatogr.* **118**, 303 (1976).

[15] S.K. Sarkar and J.P. Mittal, in *Lasers in Materials Science, Mater. Sci. Forum* **301**, 1–46 (1999).

[16] C. Decker and B. Elzaouk, in *Current Trends in Polymer Photochemistry* (N.S. Allen, M. Edge, I.R. Bellobono and E. Selli, eds.), Ellis Horwood, New York, NY (1995), pp. 130–48.

[17] Anon. (Malvern), *Chem. Plants & Process.* (2), 56 (2001).

[18] R.J. Roach, E.A. Raymond, J.R. Tyrer, B.L. Sharp and D. Young, *Lasers Engng.* **6** (4), 255–71 (1998).

[19] R.P. Behm and J.J. Hahn, *Polym. Engng.*, 47–9 (Sept. 1999).

[20] T. Hoult, *Materials World*, 208–10 (Apr. 1999).

[21] H. Orlick, *DVS-Ber.* **209**, 233–7 (2000).

[22] W. Ahmed, *Chem. Br.* 40–2 (Oct. 1999).

[23] A. Burgess and K. Feng, in *Chemistry and Technology of Polymer Additives* (S. Al-Malaika, A. Golovoy and C.A. Wilkie, eds.), Blackwell Science, Oxford (1999), pp. 281–9.

[24] R. Delp and H. Kniess, *Werkstoffwoche '98, Band VIII: Symp. 10, Polym.; Symp. 14, Simul. Polym.* (W. Michaeli, ed.), Wiley-VCH Verlag, Weinheim (1998), pp. 209–14.

[25] L.I. Kalontarov and R. Marupov, *Proceedings Intl. Conference on Regulation of Polymeric Materials Stability*, Moscow (1992), p. 41.

[26] M. Cooper (ed.), *Laser Cleaning in Conservation: An Introduction*, Butterworth Heinemann, Oxford (1998).

[27] D. Anglos, M. Solomidou, I. Zergioti, V. Zafiropulos, T.G. Papazoglu and C. Fotakis, *Appl. Spectrosc.* **50**, 1331–4 (1996).

[28] D. Anglos, S. Couris and C. Fotakis, *Appl. Spectrosc.* **51**, 1025–30 (1997).

[29] Y.F. Lu, M. Meng, M.H. Hong, T.S. Low and D.S.H. Chan, *Mat. Res. Soc. Symp. Proc.* **526**, 149–54 (1998).

[30] M. Chappé, J. Hildenhagen, K. Dickmann and M. Bredo, *Restauro* (1), 27–31 (2003).

[31] M.H. Abraham, S. Scheerer, O. Madden and F. Adar, *Proceedings SPIE-Intl. Soc. Opt. Eng.* **4402**, 68–72 (2001).

[32] F. Brech and L. Cross, *Appl. Spectrosc.* **16**, 59 (1962).

[33] A.L. Schawlow, *Science* **149**, 14 (1965).

[34] R.E. Honig, in *Laser Interaction and Related Plasma Phenomena* (H.J. Schwarz and H. Hora, eds.), Plenum Press, New York, NY (1971), p. 85.

[35] F. Gao, D. Price, G.J. Milnes, B. Eling, C.I. Lindsay and P.T. McGrail, *J. Anal. Appl. Pyrol.* **40/41**, 217–31 (1997).

[36] W.T. Chan, X.L. Mao and R.E. Russo, *Appl. Spectrosc.* **46**, 1025 (1992).

[37] J.F. Ready, *Effect of High-Power Laser Radiation*, Academic Press, New York, NY (1971).

[38] L. Moenke-Blankenburg, in *Lasers in Analytical Atomic Spectroscopy* (J. Sneddon, T.L. Thiem and Y.-I. Lee, eds.), VCH Publishers, New York, NY (1997), pp. 125–95.

[39] L. Balazs, R. Gijbels and A. Vertes, *Anal. Chem.* 63, 314–20 (1991).

[40] D. Blair, M. Tillack and M. Zaghloul, unpubl. results (2003).

[41] S.S. Harilal, C.V. Bindhu, M.S. Tillack, F. Najmabadi and A.C. Gaeris, *J. Appl. Phys.* **93** (5), 2380–8 (2003).

[42] S.S. Harilal, C.V. Bindhu, M.S. Tillack, F. Najmabadi and A.C. Gaeris, *J. Phys. D.: Appl. Phys.* **35**, 2935–8 (2002).

[43] R.E. Russo, X. Mao and S.S. Mao, *Anal. Chem.* **74**, 71A–7A (2002).

[44] D. Günther and C.A. Heinrich, *J. Anal. At. Spectrom.* **14**, 1369–74 (1999).

[45] C.G. Gill and M.W. Blades, *J. Anal. Atom. Spectrom.* **8**, 261 (1993).

[46] A. Briand, P. Mauchien, F. Chartier and J.-M. Mermet, *Proceedings SPIE-Intl. Soc. Opt. Eng.* **1318**, 178–80 (1990).

[47] L. Van Vaeck, W. Van Roy and R. Gijbels, *Analusis* **21**, 53–75 (1993).

[48] R. Srinivasan and B. Braren, *Chem. Rev.* **89**, 1303–16 (1989).

[49] S.R. Cain, F.C. Burns, C.E. Otis and B. Braren, *J. Appl. Phys.* **72**, 5172 (1993).

[50] B. Luk'yanchuk, N. Bityurin, S. Anisimov, N. Arnold and D. Bauerle, *Appl. Phys.* **A62**, 397–401 (1996).

[51] J.F. Rabek, *Photodegradation of Polymers; Physical Characteristics and Applications*, Springer-Verlag, Berlin (1996), Chp. 8.

[52] A.V. Butenin and B.Ya. Kogan, *Kvantovaya Elektron. (Moscow)* **13** (10), 2149–51 (1986).

[53] R.E. Russo, X. Mao, H. Liu, J. Gonzalez and S.S. Mao, *Talanta* **57** (3), 425 (2002).

[53a] S.A. Darke and J.F. Tyson, *J. Anal. Atom. Spectrom.* **8**, 145–209 (1993).

[54] D.A. Cremers and L.J. Radziemski, in *Laser Spectroscopy and Its Applications* (L.J. Radziemski, R.W. Solarz and J.A. Paisner, eds.), M. Dekker, New York, NY (1987), Chp. 5.

[55] R.S. Houk, in *Analytical Applications of Lasers* (E.H. Piepmeier), J. Wiley & Sons, New York, NY (1986), Chp. 18.

[56] J.C. Miller and R.F. Haglund (eds.), *Laser Ablation: Mechanisms and Applications*, Springer-Verlag, Berlin (1991).

[57] R.K. Singh, D.H. Lowndes, D.B. Chrisey, E. Fogarassy and J. Narayan (eds.), *Advances in Laser Ablation of Materials*, Materials Research Society, Warrendale, PA (1998).

[58] I.S. Borthwick, K.W.D. Ledingham, C.T.J. Scott and R.P. Singhal, *Inst. Phys. Conf. Ser.* **128**, 279–82 (1992).

[59] A.A. Peligrad, M.J.J. Schmidt, L. Li and J.T. Spencer, *Opt. Laser Technol.* **32** (1), 49–57 (2000).

[60] K. Dittrich and R. Wennrich, *Progr. Anal. Atom. Spectr.* **7**, 139 (1984).

[61] W. Van Deijck, J. Balke and F.J.M.M. Maessen, *Spectrochim. Acta* **34B**, 359 (1979).

[62] H. Matsuta, K. Wagatsuma and K. Kitagawa, *Proceedings IUPAC Intl. Congress on Analytical Sciences (ICAS 2001)*, Tokyo (2001), P3006.

[63] T. Ishizaka and Y. Uwamino, *Spectrochim. Acta* **38B**, 519 (1983).

[64] F. Leis and K. Laqua, *Spectrochim. Acta* **33B**, 727 (1984).

[65] S.J. Hill, *Anal. Proc.* **29**, 399–401 (1992).

[66] H.-J. Yang, S.-J. Jiang, Y.-J. Yang and C.-J. Hwang, *Anal. Chim. Acta* **312**, 141 (1995).

[67] L. Moenke-Blankenburg, T. Schumann, D. Günther, H.-M. Kuß and M. Paul, *J. Anal. At. Spectrom.* **7**, 251–4 (1992).

[68] A.L. Gray, *Analyst* **110**, 551–6 (1985).

[69] J. Marshall, J. Franks, I. Abell and C. Tye, *J. Anal. At. Spectrom.* **6** (2), 145–50 (1991).

[70] J. Nölte, L. Moenke-Blankenburg and T. Schumann, *Fresenius J. Anal. Chem.* **349**, 131–5 (1994).

[71] G.M. Hieftje and L.A. Norman, *Intl. J. Mass Spectrom. Ion Process.* **118/119**, 519–73 (1992).

[72] J.L. Todolí and J.-M. Mermet, in *Polymer Analysis and Degradation* (A. Jimenez and G.E. Zaikov, eds.), Nova Science Publishers, Huntington, NY (2000), pp. 235–52.

[73] K.L. Mao, A.C. Ciocan and R.E. Russo, *Appl. Spectrosc.* **52**, 913 (1998).

[74] M. Hemmerlin and J.-M. Mermet, *Spectrochim. Acta* **B52**, 1687–94 (1997).

[75] M. Ohata, H. Yasuda and N. Furuta, *Proceedings IUPAC Intl. Congress on Analytical Sciences (ICAS 2001)*, Tokyo (2001), Paper 2D04.

[76] K. Dittrich, K. Niebergall and R. Wennrich, *Fresenius Z. Anal. Chem.* **328**, 330–7 (1987).

[77] A. Dobney, *Techn. Rept. DSM Research*, Geleen (1999).

[78] U. Greb, L. Rottmann and M. Hamester, *Finnigan MAT Appl. Note No. 8* (1995).

[79] G.M. Hieftje, unpubl. results (2001).

[80] P.G. Mitchell, J. Sneddon and L.J. Radziemski, *Appl. Spectrosc.* **41**, 141–8 (1987).

[81] V. Kanicky, V. Otruba and J.-M. Mermet, *Abstracts 6th Symposium on Analytical Sciences*, Deauville (1998), Paper C105.

[82] P. Arrowsmith, in *Lasers and Mass Spectrometry* (D.M. Lubman, ed.), Oxford University Press, New York, NY (1990), pp. 179–204.

[82a] S.F. Durrant, *J. Anal. At. Spectrom.* **14**, 1385–403 (1999).

[83] M. Hemmerlin and J.-M. Mermet, *Spectrochim. Acta* **B51** (6), 579–89 (1996).

[84] R.E. Russo, X.L. Mao, W.T. Chan, M.F. Bryant and W.F. Kinard, *J. Anal. At. Spectrom.* **10**, 295–301 (1995).

[85] M. Hemmerlin, J.-M. Mermet, M. Bertucci and P. Zydowicz, *Spectrochim. Acta* **B52**, 421–30 (1997).

[86] J.L. Todolí and J.-M. Mermet, *Spectrochim. Acta* **B53**, 1645–56 (1998).

[87] P.K. Booth and C.W. McLeod, *Mikrochim. Acta* **3**, 283–9 (1989).

[88] A.M. Dobney, A.J.G. Mank, K.-H. Grobecker, P. Conneely and C.G. De Koster, *Anal. Chim. Acta* **423** (1), 9–19 (2000).

[89] D. Günther, H. Cousin, B. Magyar and I. Leopold, *J. Anal. At. Spectrom.* **12**, 165–70 (1996).

[90] P. Richner, D. Evans, C. Wahrenberger and V. Dietrich, *Fresenius J. Anal. Chem.* **350**, 235–41 (1994).

[90a] B. Wagner and E. Bulska, *Abstracts Colloquium Spectroscopicum Internationale XXXIII*, Granada (2003), p. 591.

[91] M. Bertucci and P. Zydowicz, *J. Phys. IV* **6**, 853 (1996).

[92] T. Trejos, M.S. Montero and J.R. Almirall, *Anal. Bioanal. Chem.* **376**, 1255–64 (2003).

[92a] A.L. Hobbs and J.R. Almirall, *Anal. Bioanal. Chem.* **376**, 1265–71 (2003).

[93] L. Balcaen, I. Geuens, L. Moens and F. Vanhaecke, *Abstracts Laser Ablation and Related Techniques*, Zelzate (2003).

[94] J. Sneddon, T.L. Thiem and J.-I. Lee (eds.), *Lasers in Analytical Atomic Spectroscopy*, VCH Publishers, New York, NY (1997).

[95] J. Sneddon and Y.-I. Lee (eds.), *Spectrosc. Lett.* **30** (7), 1417–27 (1997).

[96] G.C. Turk, *J. Anal. At. Spectrom.* **2**, 573–7 (1987).

[97] G.J. Havrilla, S.J. Weeks and J.C. Travis, *Anal. Chem.* **54**, 2566–70 (1982).

[98] A.C. Jones, in *An Introduction to Laser Spectroscopy* (D.L. Andrews and A.A. Demidov, eds.), Plenum Press, New York, NY (1995), pp. 187–98.

[99] N.P. Fox, D.H. Nettleton and V.E. Ralph, in *Spectrophotometry, Luminescence and Colour: Science and Compliance* (C. Burgess and D.G. Jones, eds.), Elsevier, Amsterdam (1995), pp. 337–54.

[100] M.W. Sigrist, in *Progress in Photothermal and Photoacoustic Science and Technology* (A. Mandelis, ed.), Elsevier, New York, NY (1992), Vol. I, pp. 369–427.

[101] N.V. Chekalin and I.I. Vlaslov, *J. Anal. At. Spectrom.* **7**, 225–8 (1992).

[102] S.M. Colby, M. Stewart and J.P. Reilly, *Anal. Chem.* **62**, 2400 (1990).

[103] K.S. Epler, T.C. O'Haver and G.C. Turk, *J. Anal. At. Spectrom.* **9**, 79–82 (1994).

[104] O. Brede and R. Hermann, *Nachr. Chem.* **49** (10), 1162–7 (2001).

[105] J. Pfab, in *Applied Laser Spectroscopy: Techniques, Instrumentation and Applications* (D.L. Andrews, ed.), VCH Publishers, New York, NY (1992).

[106] C.A. Monnig and R.T. Kennedy, *Anal. Chem.* **66**, 280R (1994).

[107] D. Anglos, S. Couris, A. Mavromanolakis, I. Zergioti, M. Solomidou, W.-Q. Liu, T.G. Papazoglou, V. Zafiropulos, C. Fotakis, M. Doulgeridis and A. Fostiridou, in *Lasers in the Conservation of Artworks (Lacona I)* (W. Kautek and E. König, eds.), Verlag Mayer & Comp., Klosterneuburg and Vienna (1997), pp. 113–8.

[108] R. Jankow, O. Kilham, W. Renken and R. Bender, *Anal. Biochem.* **43**, 300 (1971).

[109] G.J. Diebold and R.N. Zare, *Science* **196**, 1439 (1977).

[110] A.J.G. Mank, N.H. Velthorst, U.A.Th. Brinkman and C. Gooijer, *J. Chromatogr.* **A695**, 165–74 (1995).

[111] A.J.G. Mank, N.H. Velthorst, U.A.Th. Brinkman and C. Gooijer, *J. Chromatogr.* **A695**, 175–83 (1995).

[112] T.G.M. Freegarde and G. Hancock, *J. Phys. IV* **7**, C4/15-C4/29 (1997).

[113] D.J. Butcher, J.P. Dougherty, F.R. Freli, A.P. Walton, G.T. Wei, R.L. Irwin and R.G. Michel, *J. Anal. Atom. Spectrom.* **3**, 1059 (1988).

[114] Z. Liang, R.F. Leonardo, J. Takahashi and R.G. Michel, *J. Anal. Atom. Spectrom.* **7**, 1019 (1992).

[115] X. Hou, S.-J.J. Tsai, J.X. Zhou, K.X. Yang, R.F. Lonardo and R.G. Michel, in *Lasers in Analytical Atomic Spectroscopy* (J. Sneddon, T.L. Thiem and Y.-I. Lee, eds.), VCH Publishers, New York, NY (1997), pp. 83–124.

[116] R.F. Lonardo, A.I. Yuzefovsky, K.X. Yang, R.G. Michel, E.S. Frame and J. Barren, *J. Anal. Atom. Spectrom.* **11**, 279–85 (1996).

[117] A.D. Scully and K.P. Ghiggino, in *Current Trends in Polymer Photochemistry* (N.S. Allen, M. Edge, I.R. Bellobono and E. Selli, eds.), Ellis Horwood, New York, NY (1995), Chp. 15.

[118] S.R. Ahmad and E.M. Billiet, *Polym. Degr. Stabil.* **27** (1), 99–106 (1990).
[119] B.M. Cullum, P.K. Khulbe, B.J. Marquardt and S.M. Angel, *Preprints Symp. ACS, Div. Fuel Chem.* **43** (1), 118–21 (1998).
[120] B.M. Cullum and S.M. Angel, *Abstracts 215th ACS Natl. Mtg.*, Dallas (1998), Paper FUEL-016.
[121] C.-S. Ha, H.-D. Park and C.W. Frank, *Chem. Mater.* **12** (3), 839–44 (2000).
[122] P.B. Huff and M.J. Sepaniak, *Anal. Chem.* **55**, 1992 (1983).
[123] C. Fotakis, V. Zafiropulos, D. Anglos, S. Georgiou, N. Maravelaki, A. Fostiridou and M. Doulgeridis, in *The Interface Between Science and Conservation* (S. Bradley, ed.), *British Museum Occasional Paper No. 116*, British Museum, London (1997), pp. 83–90.
[124] E.R. de la Rie, *Studies Cons.* **27**, 1, 65, 102 (1982).
[125] T. Miyoshi, *Jpn. J. Appl. Phys.* **27**, 627 (1988).
[126] T. Miyoshi, *Jpn. J. Appl. Phys.* **29**, 1727 (1990).
[127] W. Sdorra and K. Niemax, *Mikrochim. Acta* **107**, 319 (1992).
[128] Y.-I. Lee, K. Song and J. Sneddon, in *Lasers in Analytical Atomic Spectroscopy* (J. Sneddon, T.L. Thiem and Y.-I. Lee, eds.), VCH Publishers, New York, NY (1996), pp. 197–235.
[129] L.J. Radziemski and D.A. Cremers (eds.), *Laser-Induced Plasmas and Applications*, M. Dekker, New York, NY (1989).
[130] L. Moenke-Blankenburg, *Laser Microanalysis*, J. Wiley & Sons, New York, NY (1989).
[131] S. Devlin, *LabPlus Intl.* **15** (5), 16–7 (2001).
[132] Y.-I. Lee, T.L. Thiem, G.H. Kim, Y.Y. Teng and J. Sneddon, *Appl. Spectrosc.* **46** (11), 1597 (1992).
[133] D.R. Anderson, C.W. McLeod and T.A. Smith, *J. Anal. Atom. Spectrom.* **9**, 67 (1994).
[134] Ch. Carlhoff, M. Jogwich and C.-J. Lorenzen, in *Analyse von Kunststoffen durch laserinduzierte Emissionsspektralanalyse. Laser-Praxis*, C. Hanser Verlag, Munich (1994), pp. 86–8.
[135] F. Hilbk-Kortenbruck, R. Bleich, M. Stepputat and R. Noll, *LabPlus Intl.* **15** (1), 12–4 (2001).
[136] K. Löbe and H. Lucht, *Labor Praxis* **19**, 92–102 (1995).
[137] I. Burmester, *Fortschr. Ber. VDI* **5** (517), 1–115 (1998).
[138] M. Simpson, D.R. Anderson, C.W. McLeod and M. Cooke, *Analyst* **118**, 449–51 (1993).
[139] H. Fink, U. Panne and R. Niessner, *Anal. Chem.* **74**, 4334–42 (2002).
[140] T. Schneider, H.M. Ortner, C.-J. Lorenzen, M. Jogwich, W. Mertens, E. Sanzenbacher and A. Limper, *Kautsch. Gummi Kunstst.* **49** (1), 44–56 (1996).
[141] C.-J. Lorenzen, C. Carlhoff, U. Hahn and M. Jogwich, *J. Anal. At. Spectrom.* **7**, 1029–35 (1992).
[142] H. Keuter, D. Ackfeld and A. Limper, *Kautsch. Gummi Kunstst.* **53** (10), 566–73 (2000).
[143] H. Keuter, D. Ackfeld and A. Limper, *Kautsch. Gummi Kunstst.*, in press.
[144] K.M. Shafer, P.R. Griffiths and W. Shu-Qin, *Anal. Chem.* **58**, 2708 (1986).
[145] X. Chen and R.B. Smart, *Appl. Spectrosc.* **45**, 1322 (1991).
[146] H.J. Hakkanen, J.E.I. Korppi-Tommola, *Appl. Spectrosc.* **49**, 1721 (1995).
[147] A.I. Whitehouse, *Spectrosc. Europe* **12** (3), 8–12 (2000).
[148] L. Burgio, R.J.H. Clark, T. Stratoudaki, M. Doulgeridis and D. Anglos, *Appl. Spectrosc.* **54** (4), 463–9 (2000).
[149] R. Noll, A. Brysch, F. Hilbk-Kortenbruck, M. Kraushaar, I. Mönch, L. Peter and V. Sturm, *LaserOpto* **32** (6), 83–9 (2000).
[150] A.K. Błędzki and D. Kardasz, *Polimery* **43** (2), 79–86 (1998).
[151] W. Nickel, *Recycling-Handbuch*, VDI Verlag, Düsseldorf (1996), p. 224.
[152] B. Simmons, R. Ahmad and B. Overton, *Br. Plast. Rubber* (6), 4–10 (1998).
[153] T. Matsumoto, K. Tanabe, K. Saeki, T. Amano and H. Uesaka, *Bunseki Kagaku* **48**, 483 (1999).
[154] R. Feldhoff, D. Wienke, K. Cammann and H. Fuchs, *Appl. Spectrosc.* **51**, 362 (1997).
[155] V. Allen, J.H. Kalivas and R.G. Rodriguez, *Appl. Spectrosc.* **53**, 672 (1999).
[156] R. Sattmann, I. Mönch, H. Krause, R. Noll, S. Couris, A. Hatziapostolou, A. Mavromanolakis, C. Fotakis, E. Laurrauri and R. Miguel, *Appl. Spectrosc.* **52**, 456–61 (1998).
[157] J.M. Anzano, I.B. Gornushkin, B.W. Smith and J.D. Winefordner, *Polym. Eng. Sci.* **40** (11), 2423–9 (2000).
[158] A.K. Błędzki and D. Kardasz, *Intl. Polym. Sci. Technol.* **25** (4), T/90–96 (1998).
[159] D. Burgess, R. Viswanathan, I. Hussla, P.C. Stair and E. Weitz, *J. Chem. Phys.* **79**, 5200 (1983).
[160] M.A. Posthumus, P.G. Kistemaker, H.L.C. Meuzelaar and M.C. Ten Noever de Brauw, *Anal. Chem.* **50**, 985–91 (1978).
[161] J. Pawliszyn and S. Liu, *Anal. Chem.* **59**, 1475–8 (1987).
[162] A. Amirav, *Org. Mass Spectrom.* **26**, 1–17 (1991).
[163] S. Dagan and A. Amirav, *J. Am. Soc. Mass Spectrom.* **6**, 120–31 (1995).
[164] A. Amirav and S. Dagan, *Intl. Lab.*, 17A–17L (March 1996).
[165] K.R. Hess and W.W. Harrison, in *Lasers and Mass Spectrometry* (D.M. Lubman, ed.), Oxford University Press, Oxford (1990), pp. 205–22.
[166] D.M. Land, D.T.S. Wang, T.L. Tai, M.G. Sherman, J.C. Hemminger and R.T. McIver, in *Lasers and*

Mass Spectrometry (D.M. Lubman, ed.), Oxford University Press, Oxford (1990), pp. 157–78.

[167] W.R. Creasy, in *Proceedings 38th ASMS Conference on Mass Spectrometry and Allied Topics*, Tucson, AZ (1990), pp. 850–1.

[168] R.M. Lum, *Thermochim. Acta* **18**, 73 (1977).

[169] G. Montaudo, C. Puglisi, E. Scamporrino and D. Vitalini, *Macromolecules* **9**, 882 (1986).

[170] D.W. Beekman, T. Calcott, S.D. Kramer, E.T. Arakawa, G.S. Hurst and E. Nussbaum, *Int. J. Mass Spectrom. Ion Phys.* **34**, 89–97 (1980).

[171] J.A. Bennett, E.A. Schweikert, L. Van Vaeck and F.C. Adams, *J. Trace Micropr. Techn.* **7** (4), 279–92 (1989/90).

[172] J. Grotemeyer, U. Boesl, K. Walter and E.W. Schlag, *Org. Mass. Spectrom.* **21**, 645–53 (1986).

[173] M. Barber, R.S. Bordoli, G.J. Elliot, R.D. Sedgwick and A.N. Tyler, *Anal. Chem.* **54**, 652A (1982).

[174] M. Karas, U. Bahr, A. Ingendoh and F. Hillenkamp, *Angew. Chem. Int. Ed.* **101**, 805 (1989).

[175] D.F. Torgerson, R.P. Skowronski and R.D. Macfarlane, *Biochem. Biophys. Res. Commun.* **60**, 616 (1974).

[176] R.J. Buehler, E. Flanagan, L.J. Greene and L. Friedman, *J. Am. Chem. Soc.* **96**, 3390 (1974).

[177] K.R. Lykke, P. Wurz, D.H. Parker and M. Pellin, *Appl. Opt.* **32** (6), 857–66 (1993).

[178] C.L. Johlman, C.L. Wilkins, J.D. Hogan, T.L. Donovan, D.A. Laude and M.-J. Youssefi, *Anal. Chem.* **62**, 1167–72 (1990).

[179] L.Q. Huang, R.J. Conzemius, G.A. Junk and R.S. Houk, *Anal. Chem.* **60**, 1490–4 (1988).

[180] J.T. Brenna, W.R. Creasy, W. McBain and C. Soria, *Rev. Sci. Instrum.* **59**, 873 (1988).

[181] F. Hillenkamp, in *Ion Formation from Organic Solids* (A. Benninghoven, ed.), *Chem. Phys. No. 25*, Springer-Verlag, Berlin (1983), pp. 190–205.

[182] L.M. Nuwaysir and C.L. Wilkins, *Anal. Chem.* **60**, 279–82 (1988).

[183] J.E. Campana, T. Havard and A. Jarrell, *Polym. Mater. Sci. Engng.* **69**, 456–7 (1993).

[184] L. Van Vaeck and R. Gijbels, in *Microbeam Analysis-1989* (P.E. Russell, ed.), San Francisco Press Inc., San Francisco, CA (1989), XVII–XXV.

[185] B. Asamoto, S.R. Bryan, C.L. Judy, R.W. Linton, B. Hagenhoff, M. Deimal and A. Benninghoven, *Proceedings SIMS VII* (A. Benninghoven, C.A. Evans, K.D. McKeegan, H.A. Storms and H.W. Werner, eds.), J. Wiley & Sons, Chichester (1990), pp. 367–70.

[186] R.M.A. Heeren and J.J. Boon, *Intl. J. Mass Spectrom. Ion Process.* **157/158**, 391–403 (1996).

[187] X. Xiang, J. Dahlgren, W.P. Enlow and A.G. Marshall, *Anal. Chem.* **64**, 2862–5 (1992).

[188] R.B. Cody, A. Bjarnason and D.A. Weil, in *Lasers and Mass Spectrometry* (D.M. Lubman, ed.), Oxford University Press, New York, NY (1990), pp. 316–29.

[189] B. Asamoto, J.R. Young and R.J. Citerin, *Anal. Chem.* **62**, 61–70 (1990).

[190] Y. Kotsuka and S. Suzuki, *Kawasaki-shi Kogai Kenkyusho Nenpo* **22**, 19–22 (1996).

[191] L. Van Vaeck, W. Van Roy, R. Gijbels and F. Adams, *Chem. Anal. Ser. (N.Y.)* **124**, 7–126 (1993).

[192] K. Jedrzejczat and V.S. Gaind, *Analyst (London)* **118**, 149 (1993).

[193] R.B. Cody and J.A. Kinsinger, in *Fourier Transform Mass Spectrometry* (M.V. Buchanan, ed.), American Chemical Society, Washington, DC (1987), pp. 59–80.

[194] B.S. Freiser, *Talanta* **32**, 697 (1985).

[195] W. Janitschke, U. Moller, E. Uherek and K. Kleinermanns, *J. Anal. Appl. Pyrol.* **56** (1), 99–111 (2000).

[196] F.W. Karasek, *Res. Dev.* **21**, 330 (1970).

[197] D.L. Perry, *Applications of Analytical Techniques to the Characterization of Materials*, Kluwer Academic Publishers, Norwell, MA (1992).

[198] A.G. Marshall and F.R. Verdun, *Fourier Transform in NMR, Optical and MS: A User's Handbook*, Elsevier, New York, NY (1990).

[199] B. Asamoto, in *FT-ICR/MS: Applications of Fourier Transform Ion Cyclotron Resonance Mass Spectrometry* (B. Asamoto, ed.), VCH Publishers, New York, NY (1991), pp. 157–85.

[200] W.B. Emary, O.W. Hand and R.G. Cooks, in *Lasers and Mass Spectrometry* (D.M. Lubman, ed.), Oxford University Press, New York (1990), pp. 223–48.

[201] W.H. Waddell, K.A. Benzing, L.R. Evans, S.K. Mowdood, D.A. Weil, J.M. McMahon, R.H. Cody and J.A. Kinsinger, *Rubber Chem. Technol.* **64**, 622–34 (1991).

[202] W.H. Waddell, K.A. Benzing, L.R. Evans and J.M. McMahon, *Rubber Chem. Technol.* **65** (2), 411–26 (1992).

[203] W.H. McClennen, J.M. Richards, H.L.C. Meuzelaar, J.B. Pausch and R.P. Lattimer, *Polym. Mater. Sci. Eng.* **53**, 203–7 (1985).

[204] D.A. McCrery, E.B. Ledford Jr. and M.L. Gross, *Anal. Chem.* **54**, 1435–7 (1982).

[205] Z. Liang, A.G. Marshall and D.G. Westmoreland, *Anal. Chem.* **63**, 815 (1991).

[206] J.E. Campana, L.-S. Sheng, S.L. Shew and B.E. Winger, *Trends Anal. Chem.* **13**, 239 (1994).

[207] R.S. Brown, D.A. Weil and C.L. Wilkins, *Macromolecules* **19**, 1255–60 (1986).

[208] C.F. Ijames and C.L. Wilkins, *J. Am. Chem. Soc.* **110**, 2687–8 (1988).

[209] C.L. Wilkins and C.L.C. Yang, *Intl. J. Mass Spectrom. Ion Process.* **72**, 195 (1986).

[210] L.L. Miller, A.D. Thomas, C.L. Wilkins and D.A. Weil, *J. Chem. Soc., Chem. Commun.* 661 (1986).

[211] C.E. Brown, P. Kovacic, C.A. Wilkie, J.A. Kinsinger, R.E. Hein, S.I. Yaniger and R.B. Cody, *J. Polym. Sci. Part A: Polym. Chem.* **24**, 255–67 (1986).

[212] L.-S. Sheng, S.L. Shew, B.E. Winger and J.E. Campana, *ACS Symp. Ser.* **581**, 55 (1994).

[213] A.T. Hsu and A.G. Marshall, *Anal. Chem.* **60**, 932–7 (1988).

[214] C.L. Johlman and C.L. Wilkins, *Proceedings 35th ASMS Conference on Mass Spectrometry and Allied Topics*, Denver, CO (1987), pp. 785–6.

[215] W.J.J. Simonsick and C.W.I. Ross, *Int. J. Mass Spectrom. Ion Process.* **157/158**, 379–90 (1990).

[216] G.B. Kenion, R. Ludicky, J.H. Deitch, M. Leib and D. Doster, *Tappi J.* **77** (6), 199–205 (1994).

[217] D.P. Land, T.L. Tai. J.M. Lindquist, J.C. Hemminger and R.T. Melver, *Anal. Chem.* **59**, 2424–7 (1987).

[218] K.R. Lykke, D.H. Parker, P. Wurz, J.E. Hunt, M.J. Pellin, D.M. Gruen, J.C. Hemminger and R.P. Lattimer, *Anal. Chem.* **64** (22), 2797–803 (1992).

[219] N. Wyplosz, *Ph. D. Thesis*, University of Amsterdam (2003).

[220] T.M. Allen, D.Z. Bezabeh, C.H. Smith, E.M. McCauley, A.D. Jones, D.P.Y. Chang, I.M. Kennedy and P.B. Kelly, *Anal. Chem.* **68**, 4052–9 (1996).

[221] M.L. Alexander, M.E. Cisper, P.H. Hemberger, N.S. Nogar, J.D. Williams and J.E.P. Syka, *Inst. Phys. Conf. Ser.* **128**, 337–41 (1992).

[222] S.J. Pastor and C.L. Wilkins, in *Mass Spectrometry of Polymers* (G. Montaudo and R.P. Lattimer, eds.), CRC Press, Boca Raton, FL (2002), pp. 389–417.

[223] M.S. de Vries and H.E. Hunziker, in *Mass Spectrometry of Polymers* (G. Montaudo and R.P. Lattimer, eds.), CRC Press, Boca Raton, FL (2002), pp. 523–52.

[224] G.S. Hurst and M.G. Payne, *Principles and Applications of Resonance Ionization Spectroscopy*, Hilger, Bristol (1988).

[225] N.S. Nogar, R.C. Estler, B.L. Fearey, C.M. Miller and S.W. Downey, *Nucl. Instr. Meth.* **B44**, 459 (1990).

[226] G.S. Hurst, *Anal. Chem.* **53**, 1448A–56A (1981).

[227] N.S. Nogar and R.C. Estler, in *Lasers and Mass Spectrometry* (D.M. Lubman, ed.), Oxford University Press, New York, NY (1990), pp. 65–83.

[228] C.H. Becker and K.T. Gillen, *Anal. Chem.* **56**, 1671 (1984).

[229] C.H. Becker, in *Lasers and Mass Spectrometry* (D.M. Lubman, ed.), Oxford University Press, New York, NY (1990).

[230] J.L. Eloy, *Méthodes Physiques d'Analyse (GAMS)* **5**, 157–61 (1969).

[231] A. Vertes, R. Gijbels and F. Adams (eds.), *Laser Ionisation Mass Analysis*, J. Wiley & Sons, New York, NY (1993).

[232] K. Wendt, K. Blaum, B.A. Bushaw, C. Grüning, R. Horn, G. Huber, J.V. Kratz, P. Kunz, P. Müller, W. Nörtershäuser, N. Nunnemann, G. Passler, A. Schmitt, N. Trautmann and A. Waldek, *Fresenius J. Anal. Chem.* **364**, 471 (1999).

[233] M.S. de Vries. H.E. Hunziker and H.R. Wendt, in *Lasers and Mass Spectrometry* (D.M. Lubman, ed.), Oxford University Press, New York, NY (1990), pp. 383–401.

[234] J. Morellec, D. Normand and G. Petite, in *Advances in Atomic and Molecular Physics*, Academic Press, New York, NY (1982), p. 97.

[235] R. Zimmermann, U. Boesl, H.J. Heger, E.R. Rohwer, E.K. Ortner, E.W. Schlag and A. Kettrup, *J. High Resolut. Chromatogr.* **20**, 461–70 (1997).

[236] K.W.D. Ledingham, in *An Introduction to Laser Spectroscopy* (D.L. Andrews and A.A. Demidov, eds.), Plenum Press, New York, NY (1995), pp. 199–227.

[237] T.J. Carlin and B.S. Freiser, *Anal. Chem.* **55**, 955 (1983).

[238] M.P. Irion, W.D. Bowers, R.L. Hunter, F.S. Rowland and R.T. McIver, Jr., *Chem. Phys. Lett.* **93**, 375 (1982).

[239] M.G. Sherman, J.R. Kingsley, J.C. Hemminger and R.T. McIver Jr., *Anal. Chim. Acta* **178** (1), 79 (1985).

[240] T.M. Sack, D.A. McCrery and M.L. Gross, *Anal. Chem.* **57**, 1290–5 (1985).

[241] R. Zimmermann, H.J. Heger, R. Dorfner, U. Boesl and A. Kettrup, *Chem. Lab. Biotechn.* **49** (6), 210–4 (1998).

[242] R. Zimmermann, R. Dorfner and A. Kettrup, *J. Anal. Appl. Pyrol.* **49**, 257–66 (1999).

[243] R. Zimmermann, H.J. Heger, U. Boesl, R. Dorfner and A. Kettrup, *Proceedings 14th Intl. Mass Spectrometry Conference*, Tampere (1997), Paper WePo 032.

[244] R. Zimmermann, H.J. Heger, C. Yeretzian, H. Nagel and U. Boesl, *Rapid Commun. Mass Spectrom.* **10**, 1975 (1996).

[245] R.E. Shomo, A.G. Marshall and C.R. Weisenberger, *J. Am. Chem. Soc.* **57**, 2940 (1985).

[246] W.E. Henke, H.V. Weyssenhoff, H.L. Selzle and E.W. Schlag, *Verh. Dtsch. Phys. Ges.* **3**, 139 (1983).

[247] H.V. Weyssenhoff, H.L. Selzle and E.W. Schlag, *Z. Naturf.* **40A**, 674 (1985).

[248] R. Tembreull and D.M. Lubman, *Anal. Chem.* **58**, 1299 (1986).

[249] F. Engelke, J.H. Hahn, W. Henke and R.N. Zare, *Anal. Chem.* **59**, 909 (1987).

[250] J. Grotemeyer, U. Boesl, K. Walter and E.W. Schlag, *J. Am. Chem. Soc.* **108**, 4233 (1986).

[251] R. Zenobi, *Fresenius J. Anal. Chem.* **348**, 506–9 (1994).

[252] M.J. Dale, A.C. Jones, S.J.T. Pollard and P.R.R. Langridge-Smith, *Analyst* **119**, 571–8 (1994).

[253] M.J. Dale, A.C. Jones, K.F. Costello, P.G. Cummins and P.R.R. Langridge-Smith, *Anal. Chem.* **65**, 793 (1993).

[254] P. Voumard, Q. Zhan and R. Zenobi, *Rev. Sci. Instrum.* **64**, 2215 (1993).

[255] S.E. Van Bramer and M.V. Johnston, *Anal. Chem.* **62**, 2639–43 (1990).

[256] D.M. Lubman, in *Laser Ionisation Mass Analysis* (A. Vertes, R. Gijbels and F. Adams, eds.), J. Wiley & Sons, New York, NY (1993), pp. 321–68.

[257] L.J. Kovalenko, J.M. Philippoz, J.R. Bucenell, R. Zenobi and R.N. Zare, *Space Science Revs.* **56**, 191 (1991).

[258] R. Zenobi, *Chimia* **48**, 64 (1994).

[259] B.S. Freiser, *ACS Symp. Ser.* **359**, 155–74 (1987).

[260] W.R. Creasy, *Polymer* **33**, 4486–92 (1992).

[261] S.J. Wright, M.J. Dale, P.R.R. Langridge-Smith, Q. Zhan and R. Zenobi, *Anal. Chem.* **68**, 3585–94 (1996).

[262] J.E. Hunt, K.R. Lykke and M.J. Pellin, *NATO ASI Ser., Ser.* **B269**, 309–14 (1991).

[263] Q. Zhan, R. Zenobi, S.J. Scott, P.R.R. Langridge-Smith and R.R. Patrick, *Macromolecules* **29** (24), 7865–71 (1996).

[264] L.M. Nuwaysir and C.L. Wilkins, in *Lasers and Mass Spectrometry* (D.M. Lubman, ed.), Oxford University Press, New York, NY (1990), pp. 291–315.

[265] D.C. Schriemer and L. Li, *Anal. Chem.* **68**, 2721–5 (1996).

[266] M. Karas, D. Bachmann, U. Bahr and F. Hillenkamp, *Intl. J. Mass Spectrom. Ion Process.* **78**, 53–68 (1987).

[267] K. Tanaka, H. Waki, Y. Ido, S. Akita, Y. Yoshida and T. Yoshida, *Rapid Commun. Mass Spectrom.* **2**, 151–3 (1988).

[268] H.J. Räder and W. Schrepp, *Acta Polym.* **59**, 272–93 (1998).

[269] M.C. Fitzgerald, G.R. Parr and L.M. Smith, *Anal. Chem.* **65**, 3204–11 (1993).

[270] H. Pasch and W. Schrepp, *MALDI-ToF Mass Spectrometry of Synthetic Polymers*, Springer-Verlag, Heidelberg (2003).

[271] J. Kampmeier, K. Dreisewerd, M. Schürenberg and K. Strupat, *Intl. J. Mass Spectrom. Ion Process.* **169/170**, 31–41 (1997).

[272] T.P.E. Sze, G.H. Wang and T.W.D. Chan, *Proceedings 14th Intl. Mass Spectrometry Conference*, Tampere (1997), Paper TuPo 089.

[273] R.C. Beavis, *Org. Mass Spectrom.* **27**, 653 (1992).

[274] K.J. Wu, A. Steding and C.H. Becker, *Rapid Commun. Mass Spectrom.* **7**, 142 (1993).

[275] R.M. Whittal and L. Li, *Intl. Lab.* (8), 41–7 (1998).

[276] A. Vertes and R. Gijbels, in *Laser Ionisation Mass Analysis* (A. Vertes, R. Gijbels and F. Adams, eds.), J. Wiley & Sons, New York, NY (1993), pp. 127–75.

[277] F. Hillenkamp and H. Eyring, in *Mass Spectrometry in the Biological Sciences: A Tutorial* (M.L. Gross, ed.), Kluwer Academic Publishers, Dordrecht (1992), pp. 167–79.

[278] M. Karas, U. Bahr and U. Giessman, *Mass Spectrom. Rev.* **10**, 335 (1991).

[279] K. Tanaka, A. Bowdler and S. Kawabata, *Kratos Analytical Application Note MO122* (1999).

[280] R.S. Brown and J.J. Lennon, *Anal. Chem.* **67**, 1998–2003 (1995).

[281] M.L. Vestal, P. Juhasz and S.A. Martin, *Rapid Commun. Mass Spectrom.* **9**, 1044 (1995).

[282] W.C. Wiley and I.H. McLaren, *Rev. Sci. Instrum.* **26**, 1150 (1955).

[283] R.M. Whittal and L. Li, *Anal. Chem.* **67**, 1950–4 (1995).

[284] R. Cramer, R.F. Haglund and F. Hillenkamp, *Intl. J. Mass Spectrom. Ion Process.* **169/170**, 51–67 (1997).

[285] S. Niu, W. Zhang and B.T. Chait, *J. Am. Soc. Mass Spectrom.* **9**, 1–7 (1998).

[286] M.L. Easterling, T.H. Mize and I.J. Amster, *Intl. J. Mass Spectrom. Ion Process.* **169/170**, 387–400 (1997).

[287] A.W.T. Bristow, C.S. Creaser, S. Nélieu and J. Einhorn, *Analyst* **121**, 1425–8 (1996).

[288] R.M.A. Heeren, G.J. van Rooy, N. Wyplosz, P.B. O'Connor, M.C. Duursma and J.J. Boon, *Proceedings 14th Intl. Mass Spectrometry Conference*, Tampere (1997), Paper ThPo022.

[289] K. Tanaka, E. Kawatoh, L. Ding. A.J. Smith and S. Kumashiro, *Proceedings 47th ASMS Conference on Mass Spectrometry and Allied Topics* (1999), Paper TP086.

[290] V.V. Laiko, S.C. Moyer and R.J. Cotter, *Anal. Chem.* **72**, 5239–43 (2000).

[291] S.C. Moyer and R.J. Cotter, *Anal. Chem.* **74**, 469–76A (2002).

[292] R. Reinhold, R.J. Meier and C.G. De Koster, *Rapid Commun. Mass Spectrom.* **12**, 1962–6 (1998).

[293] L. Li, A.P.L. Wang and L.D. Coulson, *Anal. Chem.* **65**, 493 (1993).

[294] K.K. Murray and D.H. Russell, *Anal. Chem.* **65**, 2534 (1993).

[295] Q. Zhan, A. Gusev and D.M. Hercules, *Rapid Commun. Mass Spectrom.* **13**, 2278 (1999).

[296] M.E. Cisper, W.L. Earl, N.S. Nogar and P.H. Hemberger, *Anal. Chem.* **66**, 1897 (1994).

[297] K.O. Bornsen, *Anal. Methods Instrum.* **2** (4), 202–5 (1995).

[298] M. Karas and F. Hillenkamp, *AIP Conf. Proc.* **288**, 447–58 (1993).

[299] F. Hillenkamp, M. Karas, R.C. Beavis and B.T. Chait, *Anal. Chem.* **63**, 1193A (1991).

[300] P.A. Limbach, *Spectroscopy* **13** (10), 16–27 (1998).

[301] E. Scamporrino and D. Vitalini, in *Modern Techniques for Polymer Characterisation* (R.A. Pethrick and J.V. Dawkins, eds.), J. Wiley & Sons, Chichester (1999), pp. 233–66.

[302] F. Hillenkamp (ed.), *Matrix-Assisted Laser Desorption Ionization Mass Spectrometry*, Elsevier, Amsterdam (1997).

[303] C.L. Wilkins, *Proceedings Workshop on Mass Spectrometry of Polymers*, Catania (1999), L2.

[304] P.A. Limbach and S.F. Macha, *Polym. Prepr. Div. Polym. Chem.* **40** (2), 700–1 (1999).

[304a] I.A. Mowat and R.J. Donovan, *Rapid Commun. Mass Spectrom.* **9** (1), 82–90 (1995).

[305] R. Skelton, F. Dubois and R. Zenobi, *Anal. Chem.* **72** (7), 1707–10 (2000).

[306] C. Keil and H. Pasch, *DKI Jahresbericht 2000*, Darmstadt (2000), p. 88.

[307] R. Deckwer and K. Tauer, *Dechema Monogr.* **134**, 399–408 (1998).

[308] G.J. van Rooy, M.C. Duursma, R.M.A. Heeren, J.J. Boon and C.G. de Koster, *J. Am. Soc. Mass Spectrom.* **7**, 449–57 (1996).

[309] S.T. Hsiao, M.C. Tseng, Y.R. Chen and G.R. Her, *J. Chin. Chem. Soc. (Taipei)* **48** (6A), 1017–27 (2001).

[310] T. Meyer-Dulheuer, *Diplomarbeit*, Technical University of Darmstadt (1997).

[311] A.T. Jackson, K.R. Jennings and J.H. Scrivens, *Rapid Commun. Mass Spectrom.* **10**, 1449–58 (1996).

[312] C.G. de Koster, *Techn. Rept. DSM Research*, Geleen (1996).

[313] H. Pasch, T. Meyer-Dulheuer and M. Resch, *Kautsch. Gummi Kunstst.* **51**, 782 (1998).

[314] R.A. Kornfeld and R.D. Trengove, *Proceedings 45th ASMS Conference on Mass Spectrometry and Allied Topics* (1997), p. 419.

[315] M. Willetts, M.R. Clench, R. Greenwood, G. Mills and V. Carolan, *Rapid Commun. Mass Spectrom.* **13**, 251–5 (1999).

[316] T.P.E. Hollenbeck, G. Siuzdak and R.D. Blackledge, *J. Forensic Sci.* **44** (4), 783–8 (1999).

[317] H.J. Heinen, S. Meier, H. Vogt and R. Wechsung, *Intl. J. Mass Spectrom. Ion Phys.* **47**, 19–22 (1983).

[318] T. Dingle, *Vacuum* **31**, 571–7 (1981).

[319] R.E. Honig and J.R. Woolston, *Appl. Phys. Lett.* **2**, 138 (1963).

[320] F. Hillenkamp, E. Unsold, R. Kaufmann and R. Nitzsche, *Appl. Phys.* **8**, 341 (1975).

[321] L. Van Vaeck and R. Gijbels, *Fresenius J. Anal. Chem.* **337**, 743–54 (1990).

[322] J.A. Gardella, D.M. Hercules and H.J. Heinen, *Spectrosc. Lett.* **13** (6), 347–60 (1980).

[323] J.A. Gardella and D.M. Hercules, *Fresenius Z. Anal. Chem.* **308**, 297 (1981).

[324] H. Struyf, W. Van Roy, L. Van Vaeck, R. Van Grieken and P. Caravatti, *Rapid Commun. Mass Spectrom.* **8**, 32–9 (1994).

[325] L. Van Vaeck, W. Van Roy, H. Struyf, F. Adams and P. Caravatti, *Rapid Commun. Mass Spectrom.* **7**, 323–31 (1993).

[326] H. Struyf, W. Van Roy, L. Van Vaeck, R. Van Grieken, R. Gijbels and P. Caravatti, *Anal. Chim. Acta* **283**, 139–51 (1993).

[327] H. Struyf, L. Van Vaeck and R. Van Grieken, *Rapid Commun. Mass Spectrom.* **10**, 551–61 (1996).

[328] D.S. Simons, *Appl. Surf. Sci.* **34**, 103–17 (1988).

[329] L. Van Vaeck, W. Van Roy, H. Struyf, K. Poels and R. Gijbels, in *Surface Characterization. A User's Sourcebook* (D. Brune, R. Hellborg, H.J. Whitlow and O. Hunderi, eds.), Wiley-VCH, Weinheim (1997), pp. 354–68.

[330] L. Van Vaeck, J. Bennett, W. Lauwers, A. Vertes and R. Gijbels, *Mikrochim. Acta III*, 283–303 (1990).

[331] L. Van Vaeck, H. Struyf, W. Van Roy and F. Adams, *Mass Spectrom. Rev.* **13**, 189–208 (1994).

[332] G.S. Hurst, M.G. Payne, S.D. Kramer and J.P. Young, *Rev. Mod. Phys.* **51**, 767 (1979).

[333] R.W. Odom, *Proceedings SIMS VII* (A. Benninghoven, C.A. Evans, K.D. McKeegan, H.A. Storms and H.W. Werner, eds.), J. Wiley & Sons, Chichester (1990), pp. 185–7.

[334] L. Van Vaeck, H. Struyf, W. Van Roy and F. Adams, *Mass Spectrom. Rev.* **13**, 209–32 (1994).

[335] Z.A. Wilk and D.M. Hercules, *Anal. Chem.* **59**, 1819–25 (1987).

[336] A.J. Kubis, K.V. Somayula, A.G. Sharkey and D.M. Hercules, *Anal. Chem.* **61**, 2516–23 (1989).

[337] L. Van Vaeck and F. Adams, in *Encyclopedia of Spectroscopy and Spectrometry* (J.C. Lindon, ed.), Academic Press, San Diego, CA (2000), pp. 1141–52.

[338] A.H. Verbueken, F.J. Bruynseels, R. Van Grieken and F. Adams, in *Inorganic Mass Spectrometry* (F. Adams, R. Gijbels and R. Van Grieken, eds.), J. Wiley & Sons, New York, NY (1988), pp. 173–256.

[339] R.W. Odom and B.W. Schueler, in *Lasers and Mass Spectrometry* (D.M. Lubman, ed.), Oxford University Press, New York, NY (1990), pp. 103–37.

[340] H.J. Heinen and R. Holm, *Scann. Electron Microsc. (Pt. 3)*, 1129–38 (1984).

[341] J.K. De Waele, J.J. Gijbels, E.F. Vansant and F.C. Adams, *Anal. Chem.* **55** (14), 2255–60 (1983).

[342] R. Holm and D. Holtkamp, in *Microbeam Analysis 1989* (P.E. Russell, ed.), San Francisco Press Inc., San Francisco, CA (1989), pp. 325–9.

[343] A.P. Pleshkova and M.N. Uspenskaya, *Zavod. Lab.* **58** (11), 24–6 (1992).

[344] M.J. Southon, M.C. Witt, A. Harris, E.R. Wallach and J. Myatt, *Vacuum* **34**, 903–9 (1984).

[345] D. Holtkamp, G. Bayer and R. Holm, *Mikrochim. Acta* **1**, 245–60 (1991).

[346] W. Van Roy, A. Mathey and L. Van Vaeck, *Rapid Commun. Mass Spectrom.* **10**, 562–72 (1996).

[347] W. Van Roy, H. Struyf, P. Kennis, L. Van Vaeck, R. Van Grieken and C. Andrle, *Mikrochim. Acta* **120**, 121–37 (1995).

[348] W. Van Roy, H. Struyf, L. Van Vaeck and R. Gijbels, *Proceedings SIMS IX* (A. Benninghoven, Y. Nihei, R. Shimizu and H.W. Werner, eds.), J. Wiley & Sons, Chichester (1994), p. 463.

[349] D.M. Hercules, *Mikrochim. Acta Suppl.* **11**, 1–27 (1985).

[350] P.F. Schmidt and B. Brinkmann, in *Microbeam Analysis 1989* (P.E. Russell, ed.), San Francisco Press Inc., San Francisco, CA (1989), pp. 330–2.

[351] R.L. Hanson, N.E. Vanderborgh and D.G. Brookins, *Anal. Chem.* **49**, 390 (1977).

[352] R.M. Lum, *J. Polym. Sci., Polym. Chem. Ed.* **15**, 489 (1977).

[353] S.G. Bychkov, A.V. Desyatkov, A.A. Biketov and V.G. Kim, *Zh. Fiz. Khim.* **59** (11), 2898–900 (1985).

[354] Yu.N. Zhitnev, V.V. Timofeev and N. Yu. Ignat'eva, *Khim. Vys. Energ.* **28** (6), 543–50 (1994).

[355] W. Cecchetti, R. Polloni, G. Bergamasco, R. Seraglia, S. Catinella, F. Cecchinato and P. Traldi, *J. Anal. Appl. Pyrol.* **23** (2), 165–73 (1992).

[356] D. Price, G.J. Milnes and F. Gao, *Polym. Degr. Stabil.* **54** (2–3), 235–40 (1996).

[357] P.F. Greenwood and N. Sherwood, *APEA J. (Pt. 1)*, 633–45 (1995).

[358] A. Pozzan, F. Cecchinato, R. Seraglia, P. Traldi, W. Cecchetti and R. Polloni, *Org. Mass Spectrom.* **25** (7), 392–4 (1990).

[359] O.F. Folmer and L.V. Azarraga, in *Advances in Chromatography, 1969* (A. Zlatkis, ed.), Preston Techn. Abstr. Co., Evanston, IL (1969), pp. 216–21.

[360] D. Lichtmann and J.F. Ready, *Phys. Rev. Lett.* **10**, 342 (1963).

[361] D. Fanter, R. Levy and C. Wolf, *Anal. Chem.* **44**, 43–8 (1972).

[362] A.J.H. Boerboom, in *Mass Spectrometry of Large Molecules* (S. Facchetti, ed.), Elsevier, Amsterdam (1985), pp. 265–81.

[363] T. Matsuoka, H. Aranami, T. Funahashi, T. Ohtsuka and M. Kuzuya, *Chubu Daigaku Kogakubu Kiyo* **35**, 9–13 (1999).

[364] N. Vanderborgh and R. Jones, *Anal. Chem.* **55**, 527 (1983).

[365] H.L.C. Meuzelaar, P.G. Kistemaker and M.A. Posthumus, *Biomed. Mass Spectrom.* **4**, 139 (1975).

[366] D. Price, G.J. Milnes, C. Lukas and T.R. Hull, *Intl. J. Mass Spectrom. Ion Process.* **60**, 225–35 (1984).

[367] D. Price, G.J. Milnes, P.J. Tayler, J.H. Scrivens and T.G. Blease, *Polym. Degr. Stabil.* **25** (2–4), 307–23 (1989).

[368] P.F. Greenwood, S.C. George, M.A. Wilson and K.Hall, *J. Anal. Appl. Pyrol.* **38**, 101 (1996).

[369] P.F. Greenwood, S.C. George and K. Hall, *Org. Geochem.* **29**, 1075 (1998).

[370] P.F. Greenwood, S.C. George, W. Pickel, Y. Zhu and N. Zhong, *J. Anal. Appl. Pyrol.* **58/59**, 237–53 (2001).

[371] L.E. Ryan, M.P. Gardner and R.W. Vaughan, *Proc. 33rd Ann. Conf., Reinf. Plast./Compos. Inst.*, SPI, Washington, DC (1978), Sec. 15-A, pp. 1–6.

[372] J. Zhu and E.S. Yeung, *J. Chromatogr.* **463** (1), 139–45 (1989).

[373] S.C. Moldoveanu, *Analytical Pyrolysis of Natural Organic Polymers*, Elsevier, Amsterdam (1998).

[374] J.-Z. Shu, L.-P. Shi, L. Ang and D.-W. Zhang, *Hua Hsueh Hsueh Pao* **38** (3), 297–300 (1980).

[375] Y. Feng and J. Wu, *Lanzhou Daxue Xuebao, Ziran Kexueban* **18** (3), 69–72 (1982).

[376] W. Fu, *Hecheng Xiangjiao Gongye* **7** (2), 143–5 (1984); *C.A.* **101**, 74064 (1984).

[377] C. Hou, Y. Chen and S. Bi, *Sichuan Daxue Xuebao, Ziran Kexueban* (3), 93–7 (1986).

[378] D. Zhao, *Huaxue Shijie* **30** (11), 499–502 (1989).

[379] A. Li and X. Huang, *Huaxue Shijie* **30** (7), 305–10 (1989).

[380] O.F. Folmer, *Anal. Chem.* **43**, 1057 (1971).

[381] O.F. Folmer and. L.V. Azarraga, *J. Chromatogr. Sci.* **7**, 665–70 (1969).

[382] D.R. Esker and M.Q. Brewster, *J. Propul. Power* **12** (2), 296–301 (1996).

[383] V.V. Nosov, S.M. Repinskii and F.N. Dulchev, *Springer Ser. Chem. Phys.* **39** *(Laser Process. Diagn.)*, 205–9 (1984).

[384] D. Price, F. Gao, G.J. Milnes, B. Eling, C. Lindsay and P.T. McGrail, *Polym. Degr. Stabil.* **64** (3), 403–10 (1999).

[385] D. Price, K.A. Lincoln and G.J. Milnes, *Intl. J. Mass Spectrom. Ion Process.* **100**, 77 (1990).

[386] J.R. Ebdon, D. Price, B.J. Hunt, P. Joseph, F. Gao, G.J. Milnes and L.K. Cunliffe, *Polym. Degr. Stabil.* **69** (3), 267–77 (2000).

[387] D. Price, G.J. Milnes, C. Lukas and A.M. Phillips, *J. Anal. Appl. Pyrol.* **11**, 499–510 (1987).

[388] D. Price, J.H. Lippiatt, S.B. Howarth, G.B. Ward and P. Myres, *Intl. J. Mass Spectrom. Ion Phys.* **9**, 95 (1972).

[389] D. Price, D. Dollimore, N.S. Fatemi and R. Whitehead, *Thermochim. Acta* **42**, 323 (1980).

[390] D. Price, *Polym. News* **24** (3), 88–91 (1999).

[391] S.A. Stout and K. Hall, *J. Anal. Appl. Pyrol.* **21** (1/2), 195–205 (1991).

Chapter 4

God created solids, but surfaces were invented by the devil (W. Pauli)

Surface Analytical Techniques for Polymer/Additive Formulations

4.1. Electron Spectroscopy 408
4.1.1. Auger Electron Spectroscopy 409
4.1.2. X-ray Photoelectron Spectroscopy 411
4.2. Surface Mass Spectrometry 420
4.2.1. Secondary Ion Mass Spectrometry 422
4.2.2. Secondary Neutral Mass Spectrometry 439
4.3. Ion Scattering Techniques 441
4.3.1. Low-energy Ion Scattering 443
4.3.2. Rutherford Backscattering Spectroscopy 444
Bibliography 446
Surface Characterisation 446
Electron Spectroscopy 447
Surface Mass Spectrometry 447
Ion Scattering Techniques 447
References 447

Surface phenomena are important in numerous technological areas, such as corrosion, tribology, adhesion, catalysis, metallurgy, microelectronics, polymers and material science in general. Regions close to the surface are characterised by confined geometry and unbalanced forces. Thus, thermodynamics of these regions differs from those in the bulk. The surface chemical composition is often different from that of the bulk. Surfaces in real, industrially relevant, products are usually badly defined in terms of chemical composition, homogeneity and uniformity. Lower surface energy components tend to migrate to the surface. This process is stimulated in certain conditions; for instance, at processing temperatures above 230°C the first generation clarifying agents did "plate-out" on mould surfaces and in vents.

The very concept of "surface" has different meanings for the various characterisation methods, as will be apparent from Table 4.3. Some analytical tools essentially describe the very top layer only (*e.g.* AFM, ISS, SSIMS), other characterise the near-surface, *i.e.* several nanometers (AES, XPS, TXRF, PAS, LMMS), whereas others again have surface sensitivities in the order of micrometers (*e.g.* ATR-FTIR, PA-FTIR, Raman, UV methods), *cfr.* Chp. 1.

Table 4.1 considers the **minimum requirements** for the ideal polymer surface and interface analysis techniques. These requirements are very demanding; no single technique remotely approaches ideality (Table 4.2). Few ultra-high vacuum (UHV) surface

Table 4.1. Basic criteria for the ideal polymer surface analysis technique

- Sensitivity to top few atomic layers (depth sensitivity *ca.* 1 nm)
- Compositional information (element and chemical state differentiation, molecular speciation)
- Information on structure or local atomic arrangement
- Sampling depth variability from 0.2 to 10 nm (subsurface analysis)
- Lateral resolution of <0.1 μm
- High sensitivity
- *In situ* operation (*e.g.* in air, water)
- Insensitivity to surface roughness
- Quantitative interpretation
- Rapid turn-around time (for QC or troubleshooting purposes)
- Surface structure not affected by measurement

Table 4.2. Main features of some polymer surface analysis techniques

Feature	XPS	AES	SSIMS	RBS	LEIS	SPM	IRS
Composition	+	+	+	+	+	(+)[a]	−
Quantitation	+	+	(+)	+	(+)	−	−
Quantitative molecular speciation	−	−	(+)	−	−	−	+
Chemical state (electronic properties)	+	(+)	−	−	−	−	−
Structure	−	−	−	−	+	+	(+)
Variable sampling depth	+	+	+	+	−	(+)[a]	+
Lateral resolution < 0.1 μm	−	(+)	−	−	−	+	−
In situ capability	−	−	−	−	−	+	+
Insensitivity to surface roughness	(+)	+	+	−	−	+	−
Short analysis time, ease of use	+	+	−	−	−	+	+
Non destructive	+, −[b]	+, −[b]	+	−	+	+	+

[a] Limited, dependent on tip characteristics.
[b] With sputtering.

science techniques satisfy the essential criterion of molecular sensitivity. On the other hand, few solid-state molecular spectroscopies are surface sensitive. X-ray photoelectron spectroscopy (XPS) and static secondary ion mass spectrometry (SSIMS) stand out from the rest but lack *in situ* capability being vacuum techniques. These highly complementary techniques are increasingly being used together in both fundamental and applied investigations. Scanning probe techniques meet many of the requirements of Table 4.1 in probing topography and local material properties with close to atomic resolution and in providing chemical composition analysis in *in situ* operation. It appears that the polymer surface analysis field is now consolidating.

Surface characteristics such as morphology, structure, texture, optical properties (reflectivity, colour), chemical composition and reactivity, wettability, heat-sealing, hot-tack behaviour, polarity, adhesion, printability, lubricity, resistance to wear, friction and corrosion, environmental stability, (biological) compatibility, *etc.* are often of crucial importance in determining fundamental properties of materials. Physical and chemical modification techniques (*e.g.* flame-, corona-, plasma-, UV-treatments, surface grafting) are common processes to engineer industrial polymer surfaces. In many applications the performance of a polymeric material is greatly determined by its surface structure and interfacial interactions. For example, the adhesion of inks to packaging materials or paint to automotive bodywork depends on the interaction between two distinct phases. In some cases, *e.g.* antistatics, lubricants, antiblocking and release agents, the additive is *intended* to migrate to the surface. In other cases, *e.g.* antioxidants and plasticisers, the additive is added to modify the bulk but may, under certain circumstances, surface segregate. Surface segregation (blooming) of emulsifier and stabiliser molecules is a frequent occurrence. While migration and loss at the surface of stabilisers has a deleterious effect on product stability, complete immobilisation of the stabiliser through a graft tends to lead to deactivation. Surface stabilisation is not expected to gain practical importance [1].

The surface properties of a polymer film are greatly governed by the diffusion behaviour of surface-modifying additives. Slip agents, added to reduce the coefficient of friction (COF) of LLDPE packaging materials, migrate over time from the bulk of the film to the surface and ease handling; an optimum amount of the additive is desired to be present at the surface at all times. A relationship between additive surface concentration and COF in LLDPE/erucamide films has been reported using surface washing procedures and bulk extraction techniques [2].

Surface characterisation methods are gaining in importance also in view of the active development of surface-modification technology to render fillers of all types more acceptable to the matrix and improve interfacial bonding, *e.g.* surface-modified rubber particles as a reinforcing, elastomeric filler [3]. Other instances of deliberate surface modification are the application of sizes and finishes in glass fibre production and the spraying of moulds with release agents. Coatings of fillers and pigment particles may pass into the polymer matrix and thence to the surface. There are equally numerous possibilities for the unintended presence of additives at a surface ("surface contamination"). In many cases this is caused by external agents, such as lubricating oils, greases, hydraulic fluids, vacuum pump oils, *etc.*, used in production, all of which may end up on the article surface. Other common contaminants are polymerisation catalyst residues and dust.

Irradiation of the surface of a solid by energetic particles gives rise to various closely correlated phenomena. At the very surface, backscattering of incident particles, emission of electrons and photons, and ejection of target atoms and molecules (*i.e.* sputtering) may take place. In a near-surface region of the solid, extending to a depth which depends primarily on the incident particle's energy and the mass matching, the decelerated projectiles transfer energy and momentum to the target atoms, displacing them from their original positions. In this contact, the emission of neutral or charged atoms and molecules from that surface is of relevance. SIMS and SNMS employ these atoms and molecules sputtered from the surface to derive information on the elemental and molecular composition of the surface.

It is important to distinguish ***surface-specific* techniques**, capable of collecting information relevant to the top surface (maximum sampling depth of about 5–20 nm) and ***surface-sensitive* techniques**, whose results are not restricted to the surface and can probe up to 100 nm. Surface-specific techniques often operate under ultra-high vacuum (UHV) conditions (*e.g.* SIMS, ISS), but not always (AFM). Whereas a vacuum around 10^{-5}–10^{-6} mbar is adequate for a mean free path long enough to permit the entry of secondary particles, a vacuum lower than 10^{-8} mbar is essential in order to avoid surface contamination. Surface sensitivity can be achieved either by using a surface-sensitive method of exciting the analytical signal, or by employing a signal of high surface-sensitivity. The most popular methods are particle ejection-based spectroscopies, which rely on the use of surface sensitive signals consisting of charged particles having suitable kinetic energies. Hence, electron spectroscopic methods (XPS, AES) and analysis of secondary ions (SIMS) are presently the most common methods for chemical surface characterisations. From the viewpoint of surface-sensitivity only, SIMS and LEIS (low-energy ion scattering) are the best candidates. Methods using electromagnetic signals (X-rays, UV/VIS or IR) are less surface-specific. Table 4.3 shows surface specificity of various frequently used methods, correct at the time of writing. Instrumental performance is always likely to improve progressively with time.

Classical methods for examining surfaces are appearance, contact-angle measurement, profilometry (for surface roughness, hardness and texture), inverse gas chromatography, attenuated total reflection (ATR, *cfr.* Chp. 1.2.1.4) and microscopy (OM and SEM, *cfr.* Chps. 5.3.1 and 5.4.1). Also non-selective surface washings with a solvent, often followed by IR identification, have been practised. Similarly, thermal desorption methods (typically at 150°C), with GC-MS follow-up, are being employed for the characterisation of low-boiling species (degradation is not excluded). Recently, a new generation of optical microscopies, such as confocal techniques (providing unlimited depth of focus) and near-field scanning microscopy, coupled with image analysis methods have become available. These permit characterisation of surfaces with resolution approaching the atomic level. Also a new class of scanning probe microscopy techniques such as scanning tunnelling microscopy (STM) and atomic force microscopy (AFM) have appeared. Scanning probe spectroscopy (SPS) has become a unique surface analytical tool because it combines ultra-high spatial and energy resolution.

The **elemental surface composition** may be determined indirectly by means of many methods. They are based on effects arising from the binding energy of the electrons, like AES (Auger electron spectroscopy) and XPS [4]. The atomic mass can also be determined directly by measuring the mass of sputtered particles from a surface bombarded with ions or ionised neutrals (SIMS and SNMS) [5], or laser beams (laser microprobe mass analysis, LMMS). Ion scattering (LEIS, MEIS, RBS) is another possibility: as the scattered ions suffer energy and momentum loss, the energy and angular distribution of scattered ions can be used to determine the mass of surface atoms. The reader might wish to compare the restricted number of surface techniques for the assay of elemental composition to the large number of analytical methods for the bulk.

The choice of a surface analysis technique depends upon such important considerations as sampling depth, surface information, analysis environment, and sample suitability. Different techniques provide different, and sometimes complementary, information. Surface spectroscopy (RAIRS, ATR-FTIR, DRIFTS, SERS) is attractive in that it offers

Table 4.3. Surface specificity of various analytical characterisation techniques

Surface characterisation method	Type of information[a]	Surface specificity[b]		Resolution[c]		Other
		Information depth	Detectability	Depth	Lateral	
Auger Electron Spectroscopy (AES, SAM)	(C), D, E, M	1–3 nm	0.1 at. %	2–5 nm	5–100 nm	$Z \geq 3$
Atomic Force Microscopy (AFM)	T, elasticity, friction, *etc.*	–	–	–	0.1 nm	
Confocal Scanning Optical Microscopy (CSOM)	M (3D topography), buried interfaces	200 μm	1 μm	300–500 nm	250 nm	
Dynamic Secondary Ion Mass Spectrometry (DSIMS)	D, E, M, mass spectra	10 nm-100 μm	ppb–ppm	10 nm–5 μm	0.1–5 μm	$m/\Delta m \sim 10^4$
Energy-dispersive X-ray fluorescence (EDXRF)	E, C (bulk)	0.1–1 mm	1 μg g^{-1}	1–5 nm	0.02–1 cm	
Electron Probe Microanalysis (EPMA)	E	10 nm	0.01% (WDS) 0.1% (EDS)	1 μm	1 μm	$Z \geq 4$ (WDS) $Z \geq 11$ (EDS)
Fourier Transform Infrared Spectroscopy (ATR-FTIR, RAIRS, DRIFTS)	C, D, F	100 nm–100 μm	0.1 monolayer	0.5–5 μm[d]	15–20 μm	No depth profiling
Glow-discharge Optical Emission Spectrometry (GD-OES)	D, E	100 μm	10^{14} at. cm^{-2}	10 nm	2 mm	
Grazing-incidence X-ray Fluorescence (GIXRF)	E, layer thickness	3 nm–1 μm	10^{12} at. cm^{-2}	0.3–100 nm	1 cm	
Ion Scattering Spectroscopy (ISS)	E, S	1–2 monolayers	0.001 monol.	>2 nm	150 μm–1 mm	$Z \geq 3$
Laser Microprobe Mass Spectroscopy (LMMS)	E, point analysis	10–50 nm	10^7 at.	0.1–1 μm	3 μm	$m/\Delta m > 500$
Particle-induced Gamma-ray Emission (PIGE)	E, microprobe	50 μm	>10 ppm	n.d.	10 μm	
Particle-induced X-ray Emission (PIXE)	E, microprobe	50 μm	0.1 ppm	10 μm	1 μm	
Photoacoustic Spectroscopy (PAS)	Thermal properties, D	1–100 μm	1 monolayer	μm range	10 μm	
Rutherford Backscattering Spectroscopy (RBS)	D, S, E (quantitative)	2 μm	0.1 at. %	5–20 nm	0.5 mm	$Z \geq 10$
Scanning Electron Microscopy (SEM)	E, M, T	1 μm	0.1 at. %	1 μm	0.01–1 μm	
Scanning Tunnelling Microscopy (STM)	Surface electronic structure, T	0.3 nm	n.d.	–	Atomic	
Static Secondary Ion Mass Spectrometry (SSIMS)	C, E, M	1 monolayer	10^9 at. cm^{-2}	2–5 nm	0.1–5 μm	$m/\Delta m > 10^4$
Surface-enhanced Raman Spectroscopy (SERS)	Molecular adsorption	Few monolayers	sub-ng	n.d.	10 μm	
Total-reflection X-ray Fluorescence (TXRF)	E (quantitative)	5 nm	10^{10} at. cm^{-2}	2–7 nm	1 mm–1 cm	
Transmission Electron Microscopy (TEM)	S, morphology	1 μm	10^{-21} g	5–100 nm	0.2 nm	
X-ray Fluorescence (XRF)	E	30 nm–8 μm	ppm	1–5 nm	10 μm–0.2 mm	
X-ray Photoelectron Spectroscopy (XPS, ARXPS)	C, E, M	1–10 nm	0.1 monolayer	1–10 nm	10 μm–1 mm	$Z \geq 3$

[a] C = (molecular) chemical (surface) composition, D = depth profiling, E = elemental (surface) composition, F = functional groups, M = mapping, S = (surface) structure, T = (surface) topography.

[b] Application-dependent.

[c] Instrument-dependent.

[d] Using surface scraped, skived, or microtomed sections.

the possibility of surface analysis without the requirement for a vacuum system. Consequently, problems of sample size, vapour pressure and volatility become much less important. Lee *et al.* [6] have used UV reflection spectroscopy in surface chemical composition analysis of polymer blends. Infrared spectroscopy of polymers can also be carried out with **surface reflectance techniques** [7,8]. Surface sensitivity of flat samples can be achieved by reflecting the IR beam in glancing incidence geometry (reflection–absorption infrared spectroscopy, RAIRS). It is difficult to achieve good spatial resolution with RAIRS. For samples with rough surfaces ATR-FTIR and DRIFTS are suitable for surface analysis.

Reflectometry is a sensitive means of studying **near-surface behaviour** [9]. There are several surface sensitive techniques related to reflectometry that are performed by keeping the angle of incidence in the region of grazing incidence, such as TXRF, grazing incidence X-ray diffraction (GIXRD), and neutron reflectometry (NR). The nominal penetration depth is of the order of 10 nm for typical polymers and X-ray wavelengths. Analysing the variations in intensity with incident angle yields information on the structure of the sample. X-ray and neutron reflectrometry are conceptually closely related to other scattering techniques, but are specifically useful for studying near-surface structure due to the small incidence angles used. X-ray and neutron reflectometry may be profitably complemented by several other techniques, particularly scanning probe microscopy (SPM). SPM provides information about lateral variations in surface structure over which reflectometry averages. NR and DSIMS were used in the characterisation of polymer near-surface behaviour [9].

In-depth distribution analysis of chemical composition is a special case of local microanalysis, for which the third (axial) dimension is of primary interest. In principle, this task requires the compositional analysis of thin sections (in the ultimate dimension of monatomic layers) defined on a depth scale. It can be obtained either by non-destructive or destructive techniques. Non-destructive techniques are based on an analytical signal parameter (*e.g.* intensity and/or energy), which has a well-defined dependence on its depth of origin. For example, in electron spectroscopy, non-destructive profiling methods are based on either the energy or the emission angle dependence of the mean escape depth of the emitted electrons (*e.g.* ARXPS). Confocal microscopy is another example of optical (virtual) sectioning. Destructive techniques of depth profiling comprise the more classical methods of mechanical sectioning (by cutting or abrasion and application of laterally resolved analysis). A universally applicable "sectioning" method is surface corrosion by ion sputtering (typically at 0.5–5 keV). These methods (*e.g.* dynamic SIMS) suffer from lack of control. Hofmann [10] has reviewed depth profiling in AES and XPS.

Table 4.4. Surface-specific analytical techniques for polymer/additive analysis

Acronym	Technique	Ex/Em[a]
AES	Auger Electron Spectroscopy	E/E
DSIMS	Dynamic Secondary Ion Mass Spectrometry	I/I
EPMA	Electron Probe Microanalysis	E/P
FAB-MS	Fast Atom Bombardment Mass Spectrmetry	N/I
GD-MS	Glow-discharge Mass Spectrometry	I/I
RAIRS	Reflection Absorption Infrared Spectroscopy	P/P
ISS	Ion Scattering Spectroscopy[b]	I/I
LMMS	Laser Microprobe Mass Analysis	P/I
PIXE	Particle-induced X-ray Emission Spectroscopy	I/P
PSD	Photon Stimulated Desorption	P/I
SAM	Scanning Auger Microscopy	E/E
SERS	Surface-enhanced Raman Scattering	P/P
SEXAFS	Surface EXAFS	P/P
SNMS	Secondary Neutral Mass Spectrometry	N′/I
SSIMS	Static Secondary Ion Mass Spectrometry	I/I
XPS	X-ray Photoelectron Spectroscopy	P/E
XRF	X-ray Fluorescence Spectroscopy	P/P

[a] Excitation/emission (Ex/Em) particles: E (electrons), I (ions), N (neutrals), N′ (ionised neutrals), P (photons).
[b] LEIS, MEIS, RBS.

Surface analytical techniques can be classified in terms of the excitating and emitted probe (*cfr.* Table 4.4). The penetration of the physical probe increases from ions (ISS, RBS, SIMS) to electrons (XPS) and finally photons (UV/VIS, IR, XRF, *etc.*). Amongst the **photon beam techniques** which show some degree of surface sensitivity, in practice only XPS, total reflection X-ray fluorescence (TXRF) and laser-induced mass spectroscopic methods (LMMS), find regular application in polymer/additive analysis, as opposed to ultraviolet photoelectron spectroscopy (UPS) and (surface) extended X-ray absorption fine structure ((S)EXAFS). Of the various **electron beam techniques** only AES and electron probe microanalysis (EPMA) are applied to some extent in the field of polymer/additive analysis. Other electron beam techniques, such as appearance potential spectroscopy (APS), ionisation loss spectroscopy (ILS), high-resolution low-energy electron loss spectroscopy (HRLEELS) and techniques which analyse structure rather than chemistry, such as reflection high-energy electron diffraction (RHEED) and low-energy electron diffraction (LEED), find no application for the purpose of additive analysis. As to the **ion beam techniques**, applications in polymer/additive analyses are restricted to SSIMS, with incidental use of ion scattering techniques or PIXE. High energetic ion beam analysis techniques cause some radiation damage to polymer samples.

In surface analysis of polymeric materials it is rare that any one technique can completely characterise a surface. Each method has distinct advantages but the limitations impose restrictions on the type of surface chemistry data they can generate (*cfr.* Tables 4.2 and 4.3). Therefore it is often necessary to combine information from different techniques for consistent characterisation of complex surfaces. The benefits of a **multi-technique approach** to surface analysis have fully been recognised. Techniques employing electrons (XPS, AES) and ions (SIMS, ISS) as the detected species have proved complementary in attempts to obtain full pictures of the composition, structure and chemistry of the (near-) surface regions of the samples. If XPS is the natural partner of SSIMS, then AES has an equivalent relationship to DSIMS. AES can generally be performed at higher spatial resolution, and is quantitative if carried out with care. DSIMS has the greater sensitivity and can cope with a greater range of signal intensities, but, without standards of very similar composition to the sample under investigation, is not quantitative. Comparisons of RAIRS and XPS spectra can be very helpful in elucidating the chemical nature of an overlayer, as each technique often only gives a partial picture of the surface chemistry. Other weaponry, such as dynamic contact angle analysis, laser profilometry, reflected light microscopy, SEM and SPM, can all generate additional data which allow a complete understanding of the chemical composition and physical structure to explain a product (mis)-behaviour or resolve a production or storage problem.

For many of today's polymer and plastic products strict control of surface and interface properties is essential. Product failures will occur when the surface is out of control, *e.g.* adhesion (or release) failure, delamination, discoloration, and poor biocompatibility, printing or coating defects. Consequently, in polymer processing, product design and manufacture, surface characterisation is often of greater importance than bulk analysis. Surface analysis is frequently also an integral part of the new product development cycle from exploration through patent registration, before going to market with strong defence of product claims [11].

This Chapter mainly deals with the big four surface analysis techniques (XPS, AES, SIMS, ISS). For spatially resolved surface analytical methodologies, *cfr.* Chps. 3 and 5. Various surface analysis methods provide images of elements and other information (*cfr.* Chp. 5.9). For surface studies by means of IR spectroscopies, *cfr.* Chp. 1.

A review on the most versatile methods for studying surface properties of solids is available [12]. Takeguchi *et al.* [13] have recently reviewed progress in surface microanalysis for various polymer additives, such as stabilisers, softeners, fillers, *etc.* More extensive information on surface characterisation methods of polymers can be found in various recent books [4,5,14–16]. For quantitative surface analysis of materials, *cfr.* Chp. 6 and refs. [17,18].

4.1. ELECTRON SPECTROSCOPY

Principles and Characteristics

Photoelectron spectroscopy (PES) is a molecular spectroscopic method which is based on photoionisation. If an atom or molecule is irradiated with photons of larger than the ionisation energy of the particle ionisation may occur. In case of PES the "reaction product", an electron, is the source of analytical information. As a consequence, PES normally requires a mono-energetic radiation source of high photon intensity (10^{10}–10^{12} photons s^{-1}), a sample inlet system, a target chamber where photon-atom/molecule interaction occurs, an electron kinetic energy analyser, a detector (electron multiplier) and a recording system. Two types of photoelectron spectrometers can be distinguished on the basis of the energy of the radiation sources. In VUV

Table 4.5. Selected trace element determinations by XPS and AES

Matrix	Element(s)	Detection limit[a]	Comments	Reference
Acrylic-acid grafted PP	Pb, Ag, Cu, Fe, Ca, Cd, Hg	1 ppm	2D ion-exchange from solution	[25]
Mercury-chloride impregnated paper	As	300 ppt	Mercury arsenide deposit	[26]
Idem	Se, Sn, Sb	100 ppb	Idem; simultaneous analysis	[26]

[a] Lowest concentrations measured.

photoelectron spectrometry (UPS), which makes use of monoenergetic photons in the 10–100 eV energy range, valence shell electrons (above 6 eV) are ejected. More energetic X-ray photons, commonly in the range of 1000–2000 eV, are used for core electron ionisations. This type of photoelectron experiment is called X-ray photoelectron spectroscopy, XPS. While XPS finds regular application in polymer/additive analysis, this is not the case for UPS.

As may be seen from Table 4.4, Auger electron spectroscopy is another surface analytical technique based on the detection of emitted electrons. At variance to XPS, in this case electrons are the exciting species. For XPS and AES the depth resolution is governed by the escape depth of the emitted electrons, being the detected species. This is typically in the range of a few monolayers. The lateral resolution may be governed by the physics of the process or by the experimental arrangement, itself involving either the finite probe size, the area selected by the input optics of the analyser or the imaging properties of the system. As in any one-instrument design, the sensitivity and the spatial and energy resolutions are intrinsically linked. Theoretically, the spatial resolution of AES might be pushed to the limit of single atom analysis [19].

Prominent advantages of these methods include multi-element simultaneous analysis via commonly well-spaced spectral lines, and chemical-state information accessible via small, but characteristic and measurable shifts or shape changes in the lines. The stability of surface layers under photon or electron irradiation limits the possible duration of data acquisition time. Seah [20] has described a system for the intensity/energy calibration of electron spectrometers used in AES and XPS, necessary for quantitative analysis. Both AES and XPS may be made quantitative with reasonably good precision, although a great deal of care is necessary. Rivière [21] has compared AES and XPS to other methods of surface analysis (SIMS, ISS, EPMA, RAIRS).

Photoelectron spectrometers were recently reviewed [22]. A handbook of XPS is available [23]. Determination of trace analysis elements by electron spectroscopic methods (XPS, AES) has been reviewed [24]. XPS and AES are not outspoken trace element analysis techniques.

Applications
If a trace element is trapped as a very thin deposit (preferably in the monolayer range), XPS and AES can provide quantitative determination with high sensitivity (detection limits < ppb), as well as good accuracy and precision. The number of publications dealing with XPS/AES applications in trace element analysis is quite small. Some examples are given in Table 4.5.

4.1.1. Auger Electron Spectroscopy

Principles and Characteristics
The detection of the first "Auger" electrons was reported in 1923 [27], and the use of Auger electrons as a tool for surface analysis of solids dates from 1953 [28]. When a focused electron beam interacts with the atoms in a material, core level electrons can be ejected if the energy of the incident electrons is larger than the ionisation threshold. Auger electrons are the result of one of the decay mechanisms for the core-hole created. Auger electron spectroscopy (AES) involves more than one step. Following ejection of a core electron relaxation of the ionised atom can occur by filling the core vacancy with a less tightly bound electron from an outer shell. The relaxation energy is then dissipated in either of two ways. It can be given to a third (Auger) electron, which is emitted from the atom (Auger process), taking up the remaining excess energy as kinetic energy, or it can appear as a characteristic X-ray photon (electron X-ray microanalysis, EMA). Even though the Auger decay is conveniently described as a step process,

it is essentially a single quantum-mechanical transition, which results in a (doubly) charged ion in an excited state.

In practice, **Auger emission** prevails for holes in core levels with binding energy (*BE*) values lower than 2 keV, *i.e.* in the *BE* range explorable by XPS with the usual X-sources. More or less pronounced Auger peaks are always present in XPS spectra, where they are named XAES (X-ray excited Auger electron spectroscopy) peaks. The energy of the Auger electron depends on the chemical bonding state of the element from which it escaped. If the levels involved are of energy E_1, E_2 and E_3 respectively, then in first approximation the kinetic energy of the Auger electron is given by

$$E_K(\mathrm{A}) = E_1 - E_2 - E_3 \qquad (4.1)$$

Since all three levels are characteristic for the atom involved $E_K(\mathrm{A})$ is likewise characteristic and element specific. Auger peaks are named according to the levels involved in the electron emission process. Thus *KLL* means that the initial hole is in the *K* level and is filled by an *L* electron whose excess energy is used to eject a second *L* electron. For spectral interpretation the reader is referred to ref. [29].

AES uses a low-energy (3–10 keV) electron beam gun for surface bombardment to minimise surface heating. The maximum depth from which Auger electrons can escape is only about 0.3–6 nm for most materials. Thus, Auger spectroscopy is a technique that truly characterises the **near-surface** region of the irradiated specimen. AES can operate with spatial resolution in the 50 to 100 nm range with sensitivity down to about 0.1% of a single atom layer. AES has a high sensitivity for the light elements commonly observed in organic materials, in contrast to XRF with very small fluorescence yield for these elements.

As in case of XPS the electron energy analysis is essential and the energy analyser is therefore the central instrumental element. The electron energy analyser measures the energy distribution of Auger electrons emitted from the sample: the Auger spectrum is a plot of intensity *vs.* kinetic energy. Cylindrical mirror analysers (CMA) are particularly suited to Auger electron spectroscopy. Details of instrumental design can be found elsewhere [30].

Auger spectra contain three kinds of information. The position of the peaks along the energy axis allows **qualitative determination** of the elements present at the sample surface. AES and XPS are the most popular techniques for identifying elements and chemical states present in the outermost 5 nm at the surface of a solid sample. Because of their characteristic energies and shallow depth from which they escape without energy loss, Auger electrons are able to characterise the surface elemental composition. In some cases it is also possible to obtain **chemical state information** from the Auger peaks. With good instrumental design, one can obtain a percentage analysis of surface composition. The peak-to-peak height of the derivative feature is, to first order, proportional to the atomic concentration of the species in the near-surface region of the sample, and can be used for **quantitative information** on surface concentrations. Direct use of the Auger peak-to-peak height is not possible, since the Auger electron yield may vary considerably between different atoms. Instead of having recourse to physics to calculate the Auger electron yield, elemental sensitivity factors may be used from which the atomic concentration is derived (*cfr.* ref. [31]). It is advantageous that the sensitivity factors of elements cover a relatively small range (within one order of magnitude from one another). Either special reference materials can be used, which allow accurate measurement of sensitivity factors, or an "off-line" compensation method [20]. AES analyses accurate to within ±5% are extremely difficult, even impossible, without the use of standards with a composition very similar to that of the unknown. The irreproducibility of Auger electron spectra recorded in different laboratories arises through uncalibrated instrument functions. Areas of controversy in AES and XPS are spectral background correction in quantification. For quantitative analysis using AES and XPS the electron spectrometer requires intensity **calibration** [20]. Seah *et al.* [32] have dealt with the calibration of AES (energy and intensity scale) for valid analytical measurements. The development of a reference material and reference method to provide a calibration of the intensity scale for differential AES has been reported [33]; interlaboratory tests have been carried out [34]. When used in combination with ion sputtering to gradually remove the surface, Auger spectroscopy can determine the variation of composition of the sample with depth. Various developments in quantification of AES have been reported [35–39] and a review has appeared [40].

Table 4.6 lists the **main characteristics** of AES. It is not possible to introduce "wet" or porous materials with a high outgassing rate into the UHV chamber. The use of an electron beam for generating the

Table 4.6. Main characteristics of Auger electron spectroscopy

Advantages:
- Element specific microanalysis (all elements with $Z \geq 3$)
- Metal, semiconductor analysis; some insulators
- Good absolute sensitivity (100 ppm for most elements)
- High surface sensitivity (approximately 0.3–6 nm)
- Few spectral interferences
- Semiquantitative without standards; quantitative with standards
- 2D and 3D analysis (depth profiling, volume mapping)
- High lateral resolution (about 10 nm)
- Imaging/mapping capabilities facilities (Scanning Auger Microscopy, SAM; *cfr.* Chp. 5.4.1.3)
- Perfect correlation between the secondary electron image and the point of analysis
- Very good reproducibility
- Rapid analysis
- Operator-friendly
- Commercial equipment, databases

Disadvantages:
- Requires excellent, controlled environment (UHV)
- Vacuum-compatible materials
- Destructive to electron beam-sensitive materials
- Sample charging (especially with insulators)
- Chemical state information influenced/controlled by beam artefacts
- Not sensitive to trace or low-level concentrations (less than 0.1%)
- Beam spreading limits ultimate spatial resolutions
- Not yet fully mature
- Slow mapping due to high background signals
- Specialist user skill needed

Auger electrons usually causes charging of the surface region if the sample is insulating, as for polymers. Although lowering the primary electron beam acceleration voltage can reduce charging effects, this will limit the penetration depth of the incoming electrons. The lateral resolution in AES is limited by the diameter of the incoming electron beam. Commercial systems with spot sizes in the region of 30 nm are available.

Techniques yielding similar information to AES are XPS, SIMS, GD-AES, RBS, EDS and WDS. For a comparison of AES and XPS with alternative methods of surface analysis (SIMS, ISS, EPMA, RAIRS, DRIFTS), *cfr.* ref. [39].

AES has been reviewed [41] and several books describe this technique [42–45].

Applications

AES is widely used in the microfabrication industry as a working tool for process control and for research and troubleshooting in the entire field of materials science. For best results, Auger relies upon the sample being electrically conducting and consequently is not often used in polymer analysis. Application of AES and its imaging variant scanning Auger microscopy (SAM) to polymer-based composites is not straightforward, as the experimental parameters have to be carefully established to overcome the difficulties of the insulating matrix. This means associated degradation of spatial resolution and invariably calls for operating at a low accelerating voltage.

Lin [46] has examined the chemical constituents of dispersants, aggregation, and dispersion states of additives such as carbon-black, zinc oxide, and sulfur in vulcanised rubbers by AES combined with a SAM image analyser. A proper preparation of sample surface for accurate AES/SAM analysis was given. Although successful analysis of carbon fibre reinforced polymers (CFRP) can also be carried out with AES/SAM, the information is limited to identifying matrix and fibre specific elements that are present at an adequate concentration and using these signals to build up a chemical image of fracture surface [47]. AES excels in the analysis of metal matrix composites (MMCs), where electrostatic charging problems do not arise. Watts [48] has undertaken the analysis of interphase chemistry by examining fracture surfaces at high spatial resolution by XPS, AES/SAM and ToF-SIMS. The combination of ToF-SIMS and XPS provides a powerful means of examining the interphase region of polymer matrix composites. The role of AES in polymer/additive analysis is limited.

4.1.2. X-ray Photoelectron Spectroscopy

Principles and Characteristics

X-ray photoelectron spectroscopy (XPS), originally known as electron spectroscopy for chemical analysis (ESCA), is based on the photoelectron effect, discovered by Hertz in 1887 [49]. In this method the surface is bombarded with mono-energetic low-energy (soft) X-ray photons, which are less disruptive than an electron beam. The energy is absorbed, resulting in direct ejection of a core level electron, *i.e.* a photoelectron (*cfr.* Fig. 4.1). In the electron emission process, a singly charged ion, M^+, is produced:

$$M + h\nu \rightarrow M^+ + e^- \tag{4.2}$$

There are no specific selection rules for such a photo-ionisation process. The kinetic energy (*KE*) of the ejected electrons can be expressed as follows:

$$KE = h\nu - BE - \phi - S \tag{4.3}$$

where $h\nu$ is the X-ray energy, *BE* is the binding energy (referred to the Fermi level) of the photo-ejected electron, ϕ the characteristic work function of the spectrometer (to be measured experimentally) and *S* a correction term for surface charging (negligible in conducting, grounded samples).

The analytical system used to carry out the energy analysis is similar to that used in AES. The main components of an **XPS instrument** are the X-ray source, monochromator, sample stage, electron energy analyser, detector (all enclosed in an ultra-high-vacuum chamber), data acquisition and processing system (*cfr.* Fig. 4.2). XPS is best performed using a monochromatic X-ray source, which inflicts least damage to sensitive materials and allows chemical state sensitivity. The most commonly used X-ray target materials are Al and Mg which emit Al $K\alpha$ at 1486.6 eV (FWHM = 0.85 eV) and Mg $K\alpha$ radiation at 1253.6 eV (FWHM = 0.7 eV). With two different anode materials it is possible to resolve overlapping photoelectron and Auger electron peaks. This is because the position of the Auger

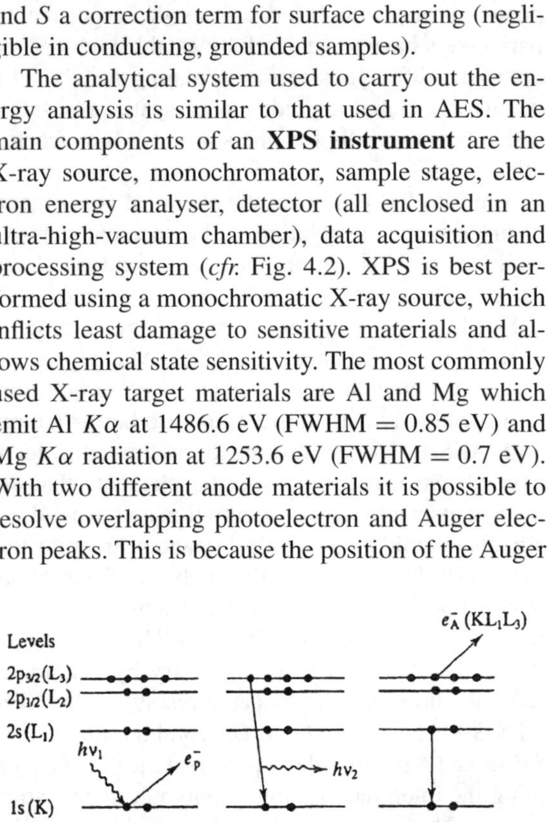

Fig. 4.1. X-ray photoemission from a 1*s* core level and subsequent relaxation processes leading to X-ray fluorescence and Auger electron emissions.

peaks varies by replacement of Al $K\alpha$ radiation by Mg $K\alpha$ radiation, but the positions of the photoelectron peaks are unaltered. The electron optical system usually consists of a concentric hemispherical analyser (CHA) and an electrostatic lens system, which focuses the electrons on the entrance of the analyser. The sensitivity of the instrument depends on the X-ray source, analysed area, geometrical factors (such as tilt-angle of the sample) and the efficiencies of lens, analyser and detector. The inherent widths of electron level and X-ray radiation and the resolving power of the spectrometer determine the energy resolution. Modern instruments combine high sensitivity with high-energy resolution [50] and allow direct imaging at <10 μm resolution [51]. An XPS instrument is usually equipped with an ion gun since ion bombardment is useful for reducing contamination on the specimen surface. For more detailed information about XPS instrumentation, *cfr.* ref. [51].

An **XPS spectrum** is a plot of the photoelectron intensity *vs.* the kinetic (or binding) energy of the photoelectrons. XPS peaks are usually named according to the photoemitting level, *e.g.* O1*s*, Fe2*p*, Au4*f*. Spectral interpretation of XPS is dealt with in refs. [29,30]. In an XPS spectrum three classes of peaks may be distinguished, namely due to photoemission from *core levels* and *valence* levels, and to X-ray excited Auger emission (*Auger series*), *cfr.* also Fig. 4.1. The major peaks reflect the electron shell structure of the surface atoms insofar as the exciting photons are capable of probing. In polymers, the valence band is 20–100 times less intense than the major core line. Auger series are the result of one of the decay mechanisms for the core hole created during photoemission. For nearly all the elements associated with polymers this mechanism dominates over X-ray fluorescence. Although XPS is less sensitive than AES and has rather poorer spatial resolution (usually about 100 μm, with developments

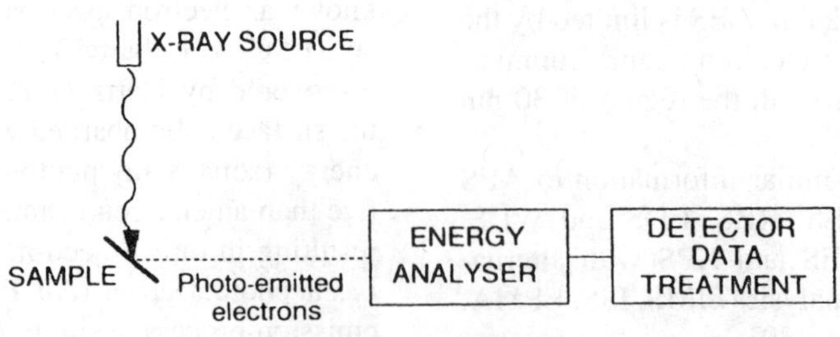

Fig. 4.2. Experimental set-up of XPS. After Garbassi *et al.* [14]. Reprinted from *Polymer Surfaces. From Physics to Technology*, F. Garbassi *et al.*, Copyright © 1998 John Wiley & Sons, Ltd. Reproduced with permission.

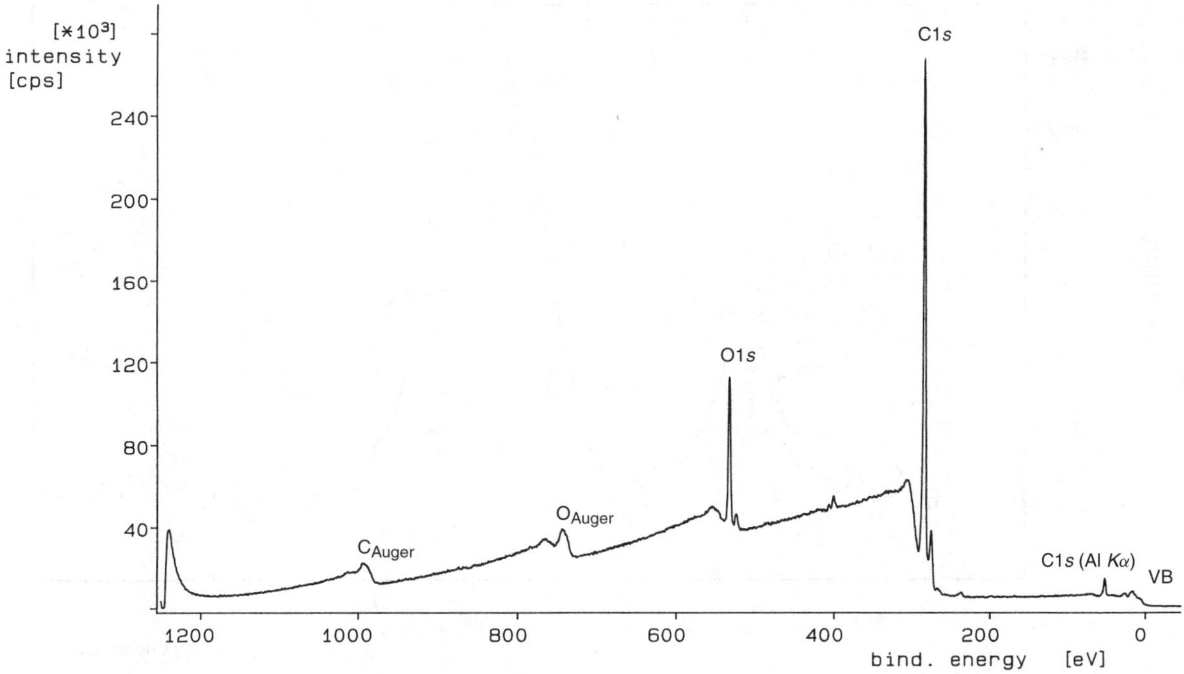

Fig. 4.3. XPS spectrum of flame-treated PP/Tinuvin 770 showing XPS, AES and valence bands. Reproduced with permission of DSM Research, Geleen.

towards 1 μm), it provides a direct measure of the binding energy of core level electrons and gives simpler spectral line shapes than AES. XPS provides three types of data: surface elemental composition, chemical state information and depth profile of elemental composition.

The basic XPS process provides a rather straightforward measurement procedure. **Qualitative analysis** is easily performed by *wide scans*, in which the full *KE* range (typically from 0–100 to 1000–1500 eV) is explored. Since the binding energy of the core electrons depends on the atomic number the elements are readily identified. Information on the elemental composition of the upper-most atomic layers (up to 5–10 nm) is obtained with a sensitivity limit of 0.1% of a monolayer. Also valence-band spectra can identify species in the surface region. XPS can be used to collect information on the chemical environment since, for a given kind of atom, the binding energy levels change with the oxidation state or net charge (valence electron density). It is possible to identify the chemical state of the elements present from small variations in the determined kinetic energies. The chemical shift between the electron energy levels varies from about 0.1 eV to 10 eV. Speciation analysis is performed by *detail scans*, where a small portion of the *KE* range (typically 20–50 eV) is acquired with a higher spectral resolution than in wide scans. Figures 4.3 and 4.4 depict an example of a wide and detail scan. As even with the use of peak-fitting routines or high-energy resolution XPS many surface functionalities cannot be identified unequivocally, due to a nearly equal chemical shift or to their low concentration, chemical derivatisation methods are in use. Functional groups on polymeric surfaces which can directly be determined using labelling experiments are epoxides, carbonyls, hydroperoxides, hydroxides, carboxyls, amines and unsaturated C−C bonds.

The **(sub)surface sensitivity** of XPS is not related to the penetration depth of the exciting X-rays (many microns), but typically to the detection of 100–1000 eV photoelectrons, for which the mean free path of travel is approximately 1 to 5 nm. The same range therefore gives the effective analytical depth sensitivity for XPS. It follows that even for an ideal sample the actual sampling depth will vary (sometimes substantially) with the energy of the source, photoelectron binding energy, and angle of emission. Thus, the depth of photoelectron detection, for a given XPS system, may vary considerably.

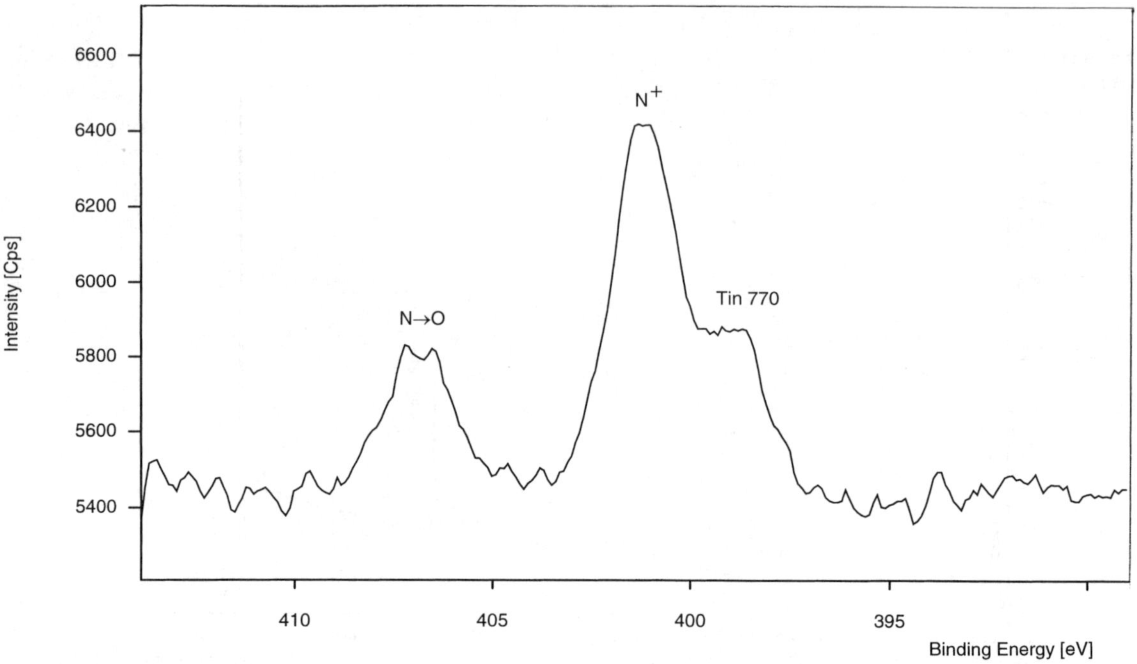

Fig. 4.4. Detail XPS N1*s* scan of flame-treated PP/Tinuvin 770 (*cfr.* Fig. 4.3) showing N−H (399 eV), N^+ (401.3 eV) and N → O (407 eV) features. Reproduced with permission of DSM Research, Geleen.

Seah *et al.* [20,32] have addressed the **calibration** of the intensity/energy response function for valid analytical measurements with electron spectrometers used in XPS and AES. Both techniques have matured to a sufficient level that calibration systems are now available which allow spectral intensities to be related from instrument to instrument. The *KE* axis is usually calibrated with reference to a signal belonging to an internal calibrant, *e.g.* the C1*s* level of adventitious carbon due to pump oil or the Au4$f7/2$ level of gold vacuum-deposited on the sample.

Quantification is in many cases *the* most important feature of XPS [52]. The X-ray photoelectron current produced by an incident X-ray photon of energy $h\nu$, which ionises core level Z in an atom of type A in a solid matrix (M), is a function of both energy- and matrix-dependent terms. Quantification of XPS spectra is thus far from simple. In fact, already the exact definition of quantification is complicated. Even for an ideal sample the actual escape depth may vary substantially. The depth of photoelectron detection, for a given XPS system, may vary dramatically from measurement to measurement, and even within the same measurement (in relation to the energy scale). After some experimental assumptions, quantification is usually performed on a relative basis, *i.e.* by selecting a particular peak as a standard and referring all measurements to it. For relative quantitative analysis one generally refers to photoelectron peaks that are fairly close in binding energy.

The development of a methodology for quantification in XPS is still a very active, necessary area of research. The theoretical approach to quantitative XPS analysis is given by ref. [53]. As absolute quantification is usually not needed, calculation of absolute intensities is generally not attempted. For a particular XPS instrument with given geometry, X-ray source and analyser, various physical parameters (cross-section, asymmetry parameter) and instrument parameters (transmission and detector efficiency) can be substituted by an experimental photoelectron yield factor. Wagner [54] and Briggs *et al.* [55] have contributed empirically derived, atomic, sensitivity factor scales. These can be used for semiquantitative results, but considerable caution is required as various assumptions regarding homogeneity and depth of analyses are being made. A comparison of experimental and theoretically derived sensitivity factors for XPS for elements from

Li to Zn has been reported [56]. In XPS, the sensitivity factors cover a relatively small scale (within one order of magnitude of one another). Either special reference materials can be used which allow accurate measurement of sensitivity factors or an "offline" compensation method is used [20]. Having a set of previously determined sensitivity factors allows one to determine the surface elemental composition of the surface layer. Measuring the relative peak intensities, and dividing them by appropriate sensitivity factors lead to the concentration of different elements on a surface. With good instrumental design and standards, one can obtain a percentage analysis of surface composition (better than $\pm 10\%$).

In analysis of technological multicomponent systems, which frequently are not very well defined, it is more convenient to work with calibration standards than with the theoretical parameters. For details on the developments in quantification in XPS the reader is referred to refs. [35,57].

A major strength of XPS is the ability to perform **depth-profiling** studies in which the composition of thin surface layers is analysed. The sampling depth of XPS analysis is dependent on the probability of electrons escaping the sample surface without interacting with atoms. The analysis can be carried out in destructive or non-destructive mode. Non-destructive profiling methods are based on either the energy or the emission angle dependence of the escape depth of the emitted electrons. The escape depth of electrons of a given kinetic energy (*KE*) varies between its full value perpendicular to the surface and a minimum at glancing emission. In **angle-resolved XPS** (ARXPS) the electron take-off angle between the specimen surface and the electron analyser optics of the XPS spectrometer is varied and a non-destructive depth-profile of the sample surface in the range of *ca.* 1 nm to 10 nm is obtained; contributions from the bulk are excluded. Depth-profiling by variation of the emission angle requires a very flat surface. A take-off angle of 45° corresponds roughly to a depth of analysis of 4.5 nm. Angle-dependent XPS (ADXPS) allows determination of the structure of the outermost molecules by following compositional changes as a function of data collection angle [58]. Data collected at angles near the surface normal provide information from the greatest depth, whereas data collected at grazing angles provides information from the top most few atom layers. In cases where the surface molecules have non-uniform atomic distributions (*e.g.* amides with C, N, O at one end and only C at the other) surface molecular orientation can be determined by ADXPS [59]. In other cases it is more convenient to use the dependence of the escape depth of the electrons on their *KE* at a constant overall take-off angle. Angular resolution studies constitute one of the principal areas for quantification in XPS. Data handling is not straightforward.

An alternative way to angle-dependent XPS analysis is profiling by ion etching. It is adopted when the layer of interest is thicker than the information depth of the XPS technique. **Sputtering** with argon ions (1–5 kV), when used to expose progressively deeper layers of the specimen (10 nm–1 μm) to XPS analysis, is a destructive mode and often gives rise to artefacts. Ion sputtering as a "sectioning" method is hardly used in the analysis of polymeric samples. Cryo-taper sectioning may be used to collect depth resolved information on additives at even greater depths (>1 μm).

XPS is essentially a large area analysis (some mm^2) with a characteristic analysis depth of several nm, as determined by the inelastic free path of the outgoing electrons (λ). The data obtained are thus the average composition in the analysed region. Normally, as large an area as possible is chosen in order to minimise the time required for the analysis. Various experimental methods are available which can increase the depth resolution to something approaching 1 μm. In the last fifteen years small area XPS (μXPS, with spatial resolution of 10 μm and depth resolution of 1 μm) has become commercially available through the use of micro-focused X-ray sources, *cfr.* Chp. 5.8.4. These systems are well suited to problem solving in industrial packaging systems and paint structures. Also chemical derivatisation of polymer surfaces is widely being practised.

Table 4.7 shows the **main characteristics** of XPS. In XPS the analysed, information containing, species (photoelectron) is generated directly. It is not surprising that XPS is the most popular surface analytical technique for providing structural, chemical bonding and compositional data for polymeric systems. XPS is capable of distinguishing different functional groups. A major advantage of XPS is that the energy resolution of practical systems is high enough to resolve the small changes in electron binding energy that accompany changes in the chemical state of the atom being excited (chemical shifts). XPS is conducted in a very high-vacuum environment ($<10^{-9}$ torr). A high vacuum is also needed to

Table 4.7. Main characteristics of X-ray photoelectron spectroscopy

Advantages:
- Determination of surface elemental ($Z \geq 3$) and chemical state compositions of solid-state compounds
- Semiquantitative without standards; quantitative with standards (about ±5% accuracy)
- Surface *cq.* subsurface sensitivity (depth probed 0.5–50 nm)
- Detectability: 0.1 monolayer
- Sensitivity approximately equal (0.1%) for all elements
- Largely non-destructive (minimal beam damage from very penetrating input beam)
- Minimal sample charging
- Rapid analysis (1–10 min typical)
- Highly reproducibible
- Depth and sputter profiling
- Imaging facilities (*cfr.* Chp. 5.8.4)
- Commercial equipment, databases

Disadvantages:
- UHV method (expensive equipment)
- Sophisticated instrument (need for skilled operators)
- Vacuum-compatible materials
- Relatively difficult interpretation
- Sample charging, mystery shifts, shake-up lines, X-ray satellites, plasmon losses, *etc.*
- Not a trace element method
- Some radiation damage to X-ray sensitive materials
- Limited lateral resolution (a few μm) by use of X-rays as primary radiation
- Poor depth resolution
- Not yet fully mature

prevent surface contamination and minimise scattering between the photoelectrons and gaseous molecules. A practical problem with using polymers in a high-vacuum environment is that low-MW components, additives, and water, may volatilise. Sample charging occurs in XPS because non-conductive samples, such as polymers or ceramic materials, do not have sufficient delocalised conduction band electrons available to neutralise charge centres that build from clustering of the positive holes created with photoelectron and/or Auger electron ejection [60]. Charging causes the associated problem of binding energy scale referencing, namely an apparent shift of the binding energy by ~2 eV to higher energy, which is more severe if a monochromatic X-ray source is used. Although XPS is referred to as a method for non-destructive analysis, which owes its origin to the fact that the XPS process only ejects electrons and does not remove nuclei as in the so-called destructive methods of analyses, polymers are somewhat unstable in the X-ray beam. Historically, obtaining useful XPS spectra from polymers has proven to be difficult. However, with care, most polymeric systems can be successfully examined in a conventional XPS and with use of a monochromator the time of successful beam exposure can be dramatically increased [61]. In order to avoid beam damage, as low an X-ray dose as possible should be used. This is also true for formulations with additives. In this regard, halogen-containing polymers are the least stable (*e.g.* PTFE). Beam damage of the chlorine-containing materials (*e.g.* PVC) seems most dramatic [61] but can largely be overcome by the use of cryoscopic techniques. Similarly, "wet" or hydrated samples can be analysed by cryo-XPS. Some problems of the past, such as low signal intensities from unstable monochromatic X-ray sources, poor spatial resolution, difficulties with charge compensation, and a lack of imaging capabilities, have now been overcome. It should be noted that, in general, the X-ray beams employed in XPS are less damaging than the electron beams of Auger or the ion beams of sputtering and the ion spectroscopies [4]. Depth profiling is difficult due to beam sizes and noise generation. Techniques yielding similar information to XPS are AES, dynamic SIMS and GD-AES. XPS and SEXAFS are complementary tools in surface analysis of polymers. Various surface problems cannot be solved with XPS and ATR-FTIR (lateral resolution, sensitivity and/or surface specificity not sufficient or not specific for certain chemical information). ToF-SIMS provides highly specific chemical information on the first surface layers on sub-μm level.

High-resolution XPS databases of organic polymers are available [59,62,63]. Various books [4,15, 42,64,65] and reviews [53] deal with XPS and surface analysis of polymers. The history of XPS is contained in refs. [15,66,67]. ARXPS has also been reviewed [68].

Applications
Areas in which X-ray photoelectron spectroscopy might be expected to perform best are the detection of surface effects (*e.g.* blooming), surface active additives (*e.g.* release and slip agents, lubricants, surfactants, *etc.*), rapidly migrating additives (*e.g.* plasticisers), or thin-film contaminants. XPS detection limits for additives (0.5 vol.%) are unfavourable for

some industrial applications (*e.g.* studies of antioxidants). However, although some of these additives are present at low bulk concentration levels (0.05–0.5%) surface enrichment may be expected. XPS is a supreme tool for problems related to migration, diffusion and orientation of additives in polymeric matrices and is profitably applied for the chemical analysis of surfaces of synthetic polymers, natural and modified textiles, wood, cellulose, fibres and paper. XPS is also a suitable means for analysing the outer surface of about 5 nm of polymers for oxidation products.

XPS is frequently called in for troubleshooting purposes. In polymer technology XPS is widely used for analysis of functional groups, often in connection with surface treatment techniques, *e.g.* high temperature (flame) or discharge (corona, plasma) techniques. Inadequate surface properties can cause problems when polymers have to be painted, dyed, printed or coated. Indeed, XPS is widely used to characterise polymeric surfaces in relation to inking and self-adhesion, phenomena which are important for the application of PE, *e.g.* in liquid food packaging. Additionally, "chemical imaging" by means of XPS allows mapping of the element distribution on the surface and distribution of chemical entities (*e.g.* C−C and C=O) with a lateral resolution of 5 μm (but in practice usually 100 μm).

Tang *et al.* [69] have used XPS analysis in the study of migration of fluorine-containing, surface-modifying **macromolecular additives** that have been evaluated for their ability to inhibit degradation of polyurethanes for medical implants. XPS was also used to examine fluorinated acrylates as modifying additives for acrylic UV-curable films based on bisphenol A-dihydroxyethyletherdiacrylate (BHEDA) [70]. Chen *et al.* [71] have reported surface analysis by XPS of paper treated with small amounts of polymeric dry strength additives, such as 0.2% PDAD-MAC and 0.5% A-PAM.

XPS confirmed that loss of **antistatic performance** of PE films for packaging electronic parts was related to the formation of antistat crystallites [72]. In order to gain information about the mechanism of antistatic action Williams *et al.* [73] have analysed glyceryl monoesters in PP by means of XPS. The approximate sampling depths (3 λ) were 20 Å for 20° and 70 Å for 80° relative to carbon electrons. As well-known, internal fatty acid ester antistats contain polar and non-polar portions. The general accepted mechanism for additive performance is migration to the polymer surface, where the polar part of the additive binds a surface layer of water to the polymer. XPS results indicate that glyceryl monooleate (GMO) has essentially no O atoms near the polymer surface. GMO does *not* migrate to the moulded PP surface and does not act as an internal antistat in PP [74]. On the other hand, glyceryl monostearate (GMS) does migrate during the moulding process and no additional migration is observed from 3 h to 7 days after moulding. Even after removal of all the GMS near the surface, there is no migration of interior GMS molecules to the surface over a period of a month. Thus GMS migration only occurs during the moulding process. XPS studies of surfaces may not only quantify the sub-monolayer but can also ascertain orientation. In the above case the orientation of GMS at the PP surface was determined; the $C_{17}H_{35}$ tail of GMS is largely at the surface of the PP plaque and is not "anchored" in the polymer matrix.

Sharma *et al.* [75] studied the mode of action of surface-active oleamide slip and stearamide **antiblocking additives**. Additives such as these are added to the bulk composition and required action relies upon migration to the surface. In order to evaluate the surface structure of the additive layers of blown PE/900 ppm Armoslip CP (oleamide) and PE/1000 ppm Armoslip 18LF (stearamide) films ARXPS analysis was carried out. The observed "bulk" concentration of oleamide at the surface is indicative of a continuous film of the additive with a thickness exceeding the XPS sampling depth (~9 nm). The stearamide surface nitrogen concentration was less than the bulk concentration of the neat material, denoting an additive layer thinner than the sampling depth of XPS and/or a discontinuous layer on the surface. In both cases no preferred orientation was observed. Migration rates of the two long chain amides were evaluated by means of quantitative analysis of 2 mm thick plaques of LDPE/0.1–0.3 wt.% Armoslip CP and LDPE/0.1–0.3 wt.% Armoslip 18LF. Oleamide appears at the surface at a considerably higher rate than stearamide and forms a continuous thick layer after ~30 days. ARXPS and surface mapping show the additive distribution to be uniform and homogeneous both in depth and laterally. The observed surface layer structure is consistent with the performance of the additives. The complete oleamide layer presents a uniform surface lubrication layer that imparts slip properties. The partial stearamide layer is sufficient to inhibit the large scale interaction of adjacent surfaces and thus imparts antiblocking properties. The observed time dependent surface oxidation of oleamide was attributed to the unsaturation in the molecule. ARXPS was also used to elucidate the distribution and molecular conformation of a perfluoropolyether lubricant on the overcoats of rigid disk storage media [76].

XPS has also been used to study the migration of **metallic components** (Sb, Sn) in heated polymeric laminates [PET(Sb)/chlorine compound.SBR/Al foil] and [PC/Sb_2O_3 or Sn chlorine compound.SBR/Al foil] [77]. Whereas the Sb catalyst residue was undetectable in bulk PET, it was found on the 100 μm thick PET film surface after thermal treatment; similarly, both Sb and Sn co-migrated to the PC film surface.

Generally, olefinic polymer films have chemically inert and non-porous surfaces with low surface tensions, causing them to be non-receptive to bonding with substrates, printing inks, coatings and adhesives. Surface treatment can be used to improve the wettability and bonding ability of virtually all plastic materials. **Corona discharge treatment** (CDT) [78] has become the primary surface treatment technology in extrusion and converting industries. Functional groups of corona-treated LDPE were studied by means of XPS, ATR-FTIR and measurement of water contact angle in relation to dyeability [79]. Similarly, the effect of corona discharge treatment and acrylic acid grafting on the dyeability of PE film were evaluated by XPS [80]. Chemical derivatisation or surface tagging can be used to quantify functional groups on polymer surfaces by means of XPS [61]. Additive loading has a significant impact on a film's ability to be treated and to retain the effect of corona treatment [78]. Higher additive loading generally reduces the ability of the film to maintain the CDT effects. Blooming or surface migration of additives masks the effect of corona treatment. This is particularly troublesome in films with high levels of additives, *e.g.* in case of high slip, with loadings from 800 to 2100 ppm. By the same token, prior to analysis of degraded or **weathered polymer surfaces** by means of XPS, it may be necessary to remove additives, as in case of plasticisers exuded on PVC surfaces [81]. XPS has been used to investigate the influence of release agents, impurities and light stabilisers on the mechanisms of flame or plasma pretreatment operations of thermoplastic materials used in the automotive industry. Jacobasch *et al.* [82] have observed migration of basic additives to the surface of PP-EPDM parts due to thermal treatment during the flame process.

Also for application in the printing and packaging industry the surface tension of inert films needs to be changed to acceptable wetting levels (*i.e.* exceed that of ink by typically 5–10 mN/m) by CDT or similar techniques. Additives, such as long chain amides (slip agents), fine silica (antiblocking agent) or siloxanes used to enhance the dispersibility of pigments, slip, mar resistance and gloss of films, may cause complications. Of the formulation variables, lubricants and stabilisers usually attract the most attention when **printability** issues are encountered. Near-surface compositional information may be gained by a variety of techniques, such as ATR-FTIR, SEM or XPS. In a typical case of poor printability of PVC films ATR indicated predominance of metal carboxylate(s) in the upper surface layers; detection of aluminium by SEM correlated with the poor printability and FTIR data [83]. XPS gave evidence for a Ba/Zn complex and hydrolysis, as follows:

$$\underset{\text{I}}{\mathrm{BaZn\,(C_{16}H_{33}CO_2)_4}} + \mathrm{H_2O} \rightarrow \underset{\text{II}}{\mathrm{BaZn\,(C_{16}H_{33}CO_2)_3OH}} + \mathrm{C_{16}H_{33}CO_2H} \tag{4.4}$$

Apparently, the mixed Ba/Zn stearate complex I migrates to the surface (as a consequence of low compatibility with the PVC matrix) and undergoes hydrolysis to give a mixture of the basic Ba/Zn stearate ($BaZn(C_{16}H_{33}CO_2)_3OH$) and free stearic acid. Since FTIR and XPS scan different depths of the surface layers, the FTIR spectrum is characteristic of I and the XPS spectrum of II.

X-ray photoelectron spectroscopy also finds wide application in the study of **adhesion** phenomena [84] and may claim a history of defining loci of failure. Adhesion failure due to moulding compound additives (such as wax, polyoxyalkylene ethers and alkylsiloxanes) in epoxy-phenolics at chip surfaces in electronic devices was studied by XPS and SAM [85]. Flame treated PP compounds have been characterised by XPS and predictive information could be obtained on paint adhesion behaviour [86]. Results are highly relevant to automotive applications replacing other time-consuming paint tests.

Critical performance applications such as aerospace **composites** but also wool require optimisation of the surface chemistry to achieve interfacial adhesion. XPS is an efficient method to analyse composition and distribution of sizing layers and interphases in glass fibre reinforcements. XPS results combined with the weight fraction (LOI) of a sizing allow to derive a quantitative value for coverage of the fibre surface by the sizing [87]. Lannon *et al.* [88] have reported various XPS studies of glass-filled PP, PA6, PA6.6, PET, PBT, PPE, PPE/PA6.6, PVC/($CaCO_3$, DOP) and PP/Mg-silicate. For compatibility issues between thermoplastic resins and glass fibres a combination of XPS, SIMS, TGA and static contact-angle measurements is frequently used [89]. Most of the understanding of the effects of the oxidative treatments necessary to improve adhesion of carbon fibre surfaces has been derived from XPS [90]. Weitzsacker *et al.* [91] used XPS to study the fibre/matrix interface in PAN based carbon fibre reinforced polyimide composites. Where XPS has greatly contributed to the surface chemistry of reinforcements for polymer matrix composites, *i.e.* carbon and glass fibres, the chemical speciation of the interphase region poses more problems and has recently been studied by SIMS and iXPS.

In relation to **heat sealability** and printability Liesegang *et al.* [92] have used XPS to study pigments and stressed polymer films and observed variations in the presence of Si on PE film surfaces with time after stretching. Siloxanes (such as PDMS, polydimethylsiloxane), which are also common additives to printing inks, trapped within a matrix of an otherwise acceptable film, are apparently induced to migrate (segregate or diffuse) to the surface causing thermal sealing inhibition and poor print adhesion. Farley *et al.* [93] studied the effect of commercial levels of slip additives on the heat-seal behaviour of LLDPE. Dillard *et al.* [94] used XPS to study the chemistry of MDI-based polyurethane hot-melt adhesive films modified by different plasticisers (as adhesion promoters).

Woods *et al.* [95] have studied the influence of the fluorocarbon-based **polymer processing additive** (PPA) Dynamar FX 9613/5920A on the surface and optical properties of polyolefin plastomer blown film by means of XPS and SSIMS. The same techniques were used to study the effect of Dynamar FX 9613 on the surface properties of HDPE [96]. Migration of fluorinated processing aids in HDPE film was also studied by XPS and ATR-FTIR [97]. Lens *et al.* [98] have reported an XPS study of the orientation of molecules of anionic **surfactants**, such as sodium dodecane sulfate (SDS), in a 10–60 Å thick coating layer onto PE; the molecules were randomly oriented as a homogeneous overlayer.

Wang [99] applied XPS to the study of the **flame retardance** mechanism of polymers. In the fluorozirconate complex Zirpro FR, which gradually loses fluorine above 350°C, two kinds of fluorine, *i.e.* covalent F−Zr and ionic F^-, were identified. The surface modification of aluminium hydroxide with various silane coupling agents was studied as well as the condensation of melamine on heating. Intumescent flame retardancy of PP/(APP/PER/ME) was studied by XPS [100]; similarly, the PP/(APP/PER) system was investigated by cone calorimetry, LOI and XPS [101]. XPS was also used to study charring in combustion of another flame retarded system, namely PVC/(Cu_2O, MoO_3) [102].

A multi-technique surface analytical study (XPS depth profiling, iSIMS, SEM/EDX, RAIRS) of automotive antiwear (zinc dialkyl dithiophosphate) films was reported [103]. XPS and SEM have been used for distribution analysis of TiO_2 and Zn phosphate in polypyrrole (PPy) [104].

XPS allows analysis of ever-smaller areas. The ability to detect small quantities of material on polymer surfaces merely through the appearance of characteristic core levels may seem to be a trivial exercise. Common polymers are comprised of a small number of elements and, furthermore, these have simple XP spectra (generally C1*s* plus one or two peaks from O1*s*, N1*s*, F1*s* and Cl2*s*, 2*p*). However, common additives or contaminants contain additional elements such as S, P, Si, Al, Na, K, Br, Sn, Cr, Ni, Ti, Zn, Ca, Sb and Ge and their presence can therefore be detected very simply, even in very low concentrations. The ability to detect such elements, especially on small or irregular "as received" samples is invaluable in **troubleshooting and QC operations** that involve surface properties (such as optical, adhesive or machine handling properties). Briggs [61] has reported a survey-scan XP spectrum from the surface of a LDPE moulding contaminated with a mould-release silicone agent which acts as a stress corrosion agent for the polymer. Siloxane contamination was also observed by XPS analysis of the surface of a carbon fibre-epoxy composite [105]. Similarly, XPS can be used as a routine method for QC monitoring of other specific hazardous compounds, such as waxes. Heat-seal failure of PP packaging film coated on both sides with vinylidenedichloride copolymer could be attributed to the

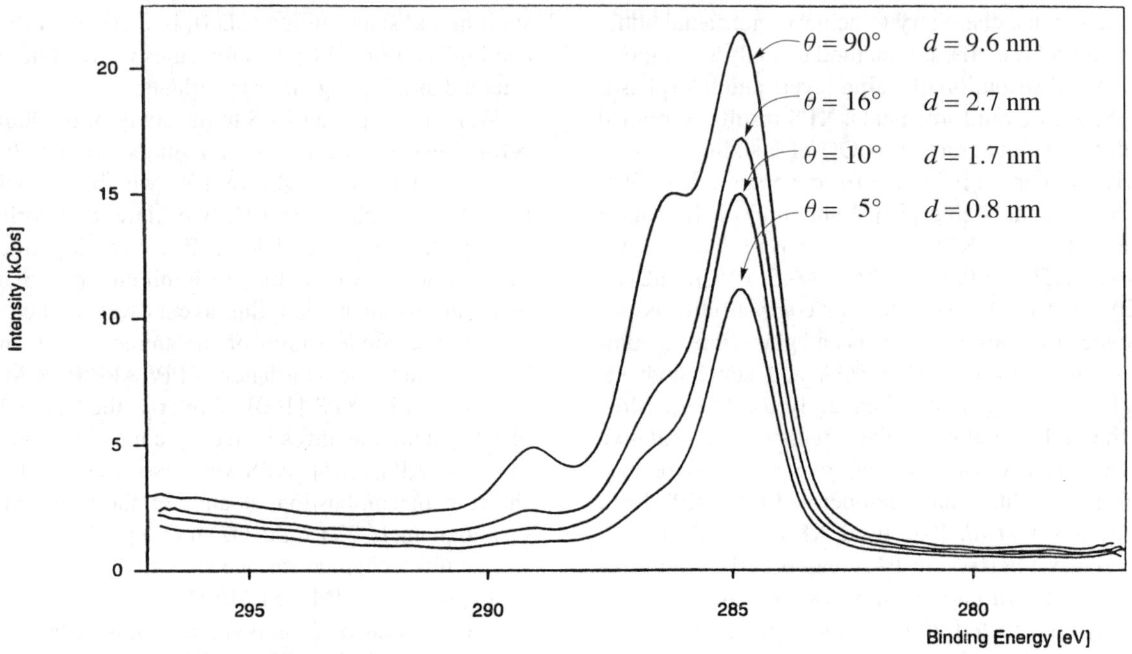

Fig. 4.5. ARXPS spectra of polymer foils coated with a Si containing additive collected at various electron take-off angles. Reproduced with permission of DSM Research, Geleen.

presence of a titanium complex, being an adhesion-promoting ink component which had migrated [61]. Non-metallised surfaces of PET packaging material have been found contaminated by aluminium-containing material. Cratering in cured paint film could be attributed to F-containing species [61].

Figure 4.5 shows a detail ARXPS scan of polymer foils coated with a Si containing additive collected at various electron take-off angles. By changing the electron take-off angle (TOA) from 90° to 5° the composition of the polymer surface changed from C:O:Si = 67:26:7 (at%) at 90° to 51:23:26 at 5°. The insert graph of the C1*s* peak shows that at very small TOAs, *i.e.* having an enhanced surface sensitivity, only very few oxygen containing carbon groups are present in the surface region. ARXPS can also profitably be applied for analysis of thin multi-layered structures.

Being essentially an element analysis technique, XPS is obviously limited in the unambiguous identification of additives on polymer surfaces, especially when complex additive packages have been used. A combination with ToF-SIMS is then an obvious choice (*cfr.* Chp. 4.2.1).

Briggs [61] has reviewed applications of XPS in polymer surface analysis problems. The substantial area of application of XPS to polymer science has been treated in several textbooks [4,15].

4.2. SURFACE MASS SPECTROMETRY

Principles and Characteristics

Surface mass spectrometry techniques measure the masses of fragment ions which are ejected from the surface of a sample to identify the atoms and molecules present. The techniques are complementary to electron spectroscopy since they provide extra absolute and surface sensitivity and give very specific molecular information. On unknown samples it is common to use a combination of electron spectroscopy *and* mass spectrometry for surface characterisation. Methods used for surface mass spectrometry are SIMS, SNMS, LDMS, LMMS, LSIMS, GD-MS and LA-ICP-MS. Of these, SIMS is by far the most important for polymer analysis.

For mass spectrometry the species of interest need to be ionised. Mass spectrometric techniques may be divided into methods with simultaneous evaporation (atomisation) and ionisation processes in the ion source (such as SSMS, ICP-MS, SIMS, LMMS) and methods with post-ionisation processes

Table 4.8. Classification of mass spectrometric techniques in respect to evaporation and ionisation processes[a]

Technique	Particles	Process
Direct ionisation methods:		
SSMS	Electrons (E /I)	Spark plasma
ICP-MS	Electrons (E /I)	Argon plasma
LMMS	Photons (E/I)	Laser plasma
SIMS	Ions (E/I)	Sputtering
Post-ionisation methods:		
GD-MS	Argon ions (E)	Argon plasma (I)
LA-ICP-MS	Photons (E)	Argon plasma (I)
TIMS	Thermal (E)	Hot filament surface (I)
SNMS	Ions (E)	EI, argon plasma, laser (I)

[a] E, evaporation; I, ionisation.

(*e.g.* SNMS, GD-MS. LA-ICP-MS and TIMS with two filaments), *cfr.* Table 4.8. With the post-ionisation methods, the processes of evaporation and atomisation or sputtering of the sample material are separated in time and space from the processes of ionising the atomic species. Because of the separation of evaporation and ionisation processes in inorganic mass spectrometry both processes can be influenced separately, which may result in easier quantification of analytical results.

Molecular surface analysis by means of mass spectrometry is only recently achieving sensitivity and selectivity comparable to elemental surface analysis. The problems to be overcome are removal of large molecular species and even thermally unstable molecular species from the surface (*e.g.* by means of an ablating laser) and ionisation of the species without alteration or fragmentation (*e.g.* by means of an ionising laser) [106]. At least two different mechanisms exist for the removal of surface molecules by laser desorption, namely thermal evaporation of species by local heating of the irradiated spot and bond rupture resulting from molecular electronic excitation, as observed for far-UV (so-called ablative photodecomposition). For a minor constituent in a sample resolution and sensitivity alone are not sufficient for species identification. Thus a preselection in the ionisation process becomes necessary. This can be achieved by varying both the wavelength and intensity of the desorption and post-ionisation laser.

An analytical technique suited for studying surface phenomena like adhesion, friction and wettability must provide molecular information with high surface sensitivity in order to identify, localise, and quantify the molecules present. SIMS and SNMS meet these requirements to an extent not provided by electron, IR-, or tunnelling spectroscopies. Secondary ion mass spectrometry (SIMS) refers generally to methods in which an energetic (primary) beam of ions is used to dislodge sample ions from a surface for mass analysis. While the term is most often used for methods employing light primary ions with kinetic energies in the kV range, most desorption techniques are in fact secondary ion mass spectrometry. Fast atom bombardment (FAB), introduced by Barber *et al.* [107], is essentially a SIMS technique that uses a liquid matrix as the sample surface. Although FAB initially distinguished itself as a method employing a neutral primary beam, energetic ion beams have proven equally effective in desorbing large molecules when used with the liquid matrix. Thus, the technique is appropriately referred to as *liquid SIMS*. The drawbacks of FAB and SIMS are that they provide no selectivity for the detection of minor sample components. Also, these techniques deposit large amounts of energy at the sample surface and typically generate predominantly fragment ions. Benninghoven [108] has compared solid and liquid SIMS.

An important difference between SIMS/SNMS and AES/PES is that the analytical information is derived from sputtered particles in the former case and from the surface in the latter case. Both laser microprobe mass spectrometry (LMMS) and static SIMS provide molecular information on local organic and inorganic compounds at variance to AES, which essentially determines only relative elemental abundancies. The primary interaction of keV ions with the sample in SSIMS, as opposed to eV-range photons in LMMS, renders the relation between detected signal and sample composition less obvious in SSIMS. LMMS allows a deductive spectral interpretation whereas SSIMS usually requires reference spectra for comparison. In LMMS ions originate from the upper 10–50 nm surface layer, although the crater depth goes up to 0.1–1 μm. Consequently, LMMS is not strictly a surface analytical technique. SSIMS generates primarily ions from the upper monolayer. For the analysis of large organic molecules lower laser powers are often used, resulting in negligible surface damage. LMMS performs spot analysis, whereas SSIMS allows imaging. The two methods are thus complementary. Depth profiling using an ion beam is inherently destructive. Furthermore, the depth resolution is often not adequate to distinguish fine detail in the compositional depth profile of the outermost few nm of the sample.

While elemental analysis of surfaces has progressed dramatically over the past two decades, **quantitative molecular surface analysis** remains difficult. This is particularly true in the analysis of complex materials such as polymers and rubbers, which contain a wide variety of additives and pigments to enhance their material characteristics. For mass spectrometric analysis the difficulty is twofold. First, desorption of surface molecules must be accomplished with minimal fragmentation and collateral surface damage. Second, the desorbed molecules must be ionised for subsequent mass analysis with high efficiency and without significant decomposition.

Current efforts are directed towards atomic-scale surface analysis by scanning probe ion mass spectrometry, combining SPM and mass analysis via field desorption and flight time determination [109]. In organic mass spectrometric methods have been reviewed [110].

Applications

While the applications of SIMS to polymer/additive analysis are quite numerous (*cfr.* Chp. 4.2.1), SNMS has not been used for this purpose. Applications of LDMS and LMMS are described in Chp. 3.4.1 and 3.4.5, respectively. For LSIMS (or FAB-MS), *cfr.* Chp. 6.2.4.1 of ref. [110a].

4.2.1. Secondary Ion Mass Spectrometry

Principles and Characteristics

As well known, transformation of atoms and high mass organic molecules from a surface-adsorbed state into the gas phase (for mass spectrometric detection) may be achieved by various methods, including field desorption, plasma desorption, laser desorption, fast atom bombardment (FAB) and ion sputtering. These techniques address different analytical problem areas. For example, the development of the laser desorption (LD) technique has been prompted by the desire to study thermally labile and high mass compounds by mass spectrometry.

Secondary ion mass spectrometry (SIMS) is the most commonly used **surface mass spectrometry** technique. SIMS analyses the secondary ions ejected from a sample following high energy (1–10 keV) bombardment with a primary ion beam as a function of their mass/electric charge (m/z) ratio (*cfr.* Fig. 4.6). The impact of the primary ion causes an atomic scale collision cascade within the surface layers of the sample and secondary ions are ejected from the surface at points remote from impact. Essentially two modes of operation are possible depending on the choice of primary ion beam dose and collimation. In *dynamic SIMS* (DSIMS), high primary ion current densities (up to a few A/cm^2) are used to allow surface erosion while secondary ions are being analysed. Early SIMS instruments [111] utilised primary ion beams with fluxes in the 1 μA/cm^2 range, which results in ablation of relatively large amounts of ions and neutrals from surface monolayers (sputtering). In *static SIMS* (SSIMS), on the other hand, the primary ion current density of the very short-pulsed (1 ns) beam is so low (*e.g.* of the order of a few nA/cm^2) that erosion effects are negligible and the chemical integrity of the sample surface during analysis is maintained. Current SIMS instrumentation provides a powerful combination of capabilities for molecular detection and trace element determination, imaging and microanalysis, and shallow depth profiling.

The relatively intense mono-energetic beams of primary ions (energy range of 0.5–50 keV) used in **dynamic SIMS** erode the sample surface at sputtering rates ranging between 0.1 and 10 nm/sec (depth probed: 2 nm–100 μm) and produce predominantly elemental ions or low-mass cluster ions. DSIMS allows mapping of elemental and molecular distributions (mapping) in all three dimensions and achieves ppm to ppb detection sensitivities for

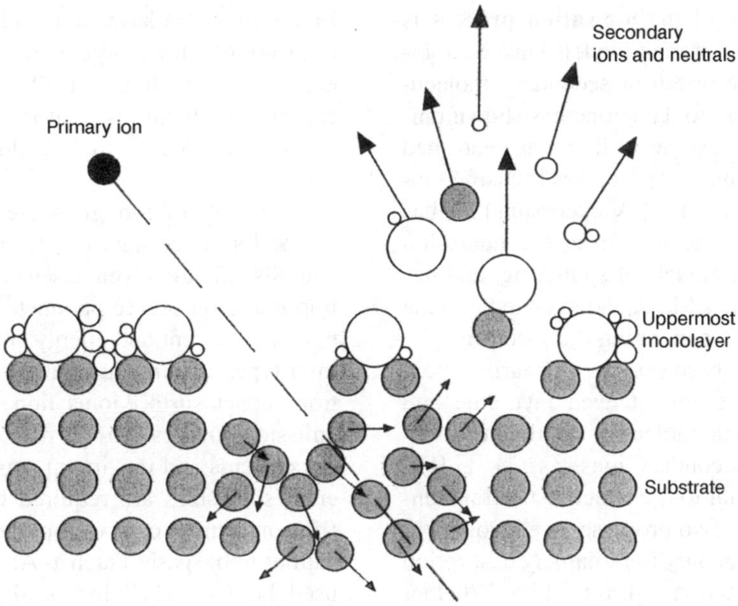

Fig. 4.6. Particle emission from a surface after excitation with primary ions of keV energy. After Benninghoven *et al.* [112]. Reprinted with permission from A. Benninghoven *et al.*, *Analytical Chemistry* **65**, 630–40A (1993). Copyright (1993) American Chemical Society.

most elements. Because of the deep sample erosion, dynamic SIMS is not a surface-specific technique. DSIMS is a frequently applied technique for chemical analysis of semiconductors and related materials [113]. Although DSIMS is expanding to a major composition analysis tool, AES is still indispensable because of the lateral resolution and good quantification. Other techniques yielding information similar to DSIMS are LMMS, PIXE, RBS, GD-AES and EDS. Dynamic SIMS, which has been described as quadrupole, magnetic sector and time-of-flight instruments, is capable of various basic modes of operation, *viz.* microvolume analysis, depth profiling, imaging and image depth profiling. The ion beam damage caused by the high primary ion current in DSIMS results in the loss of any useful structural information. Therefore, the technique hardly finds application in polymer/additive analysis at variance to static SIMS, which employs very low intensity primary ion beams.

In conditions of low primary ion fluence ($\leq 10^{12}$ ions cm^{-2}) the surface damage is low enough to ensure that the distribution of fragments ejected from the surface is not influenced by the analysis process itself (*i.e.* the surface chemistry is not changed by the analysis to a great enough extent to shift the probability of specific fragments being generated). Analysis of secondary ions emitted from a surface region unperturbed by a previous ion impact is the basis of **static SIMS** [114]. The information so derived is characteristic of the virgin or "static" surface. The lateral extension of the collision cascades initiated by each ion impact is of the order of 10 nm. SSIMS has become a key technique for surface characterisation of organic and molecular materials [115–118]. This is due to the very specific chemical information derived from characteristic secondary molecular ions.

The static mode of SIMS has many inherent features making it well suited for analysis of polymeric materials. The process may provide parent molecular secondary ions from both non-volatile and thermally labile materials, including polymers and polymer additives [119]. Large molecular ions may be detected from which chemical information about the surface can be extracted, which complements and amplifies elemental and chemical shift information from XPS and vibrational data provided by FTIR. The expansion of the technique is related to the development of high performance time-of-flight spectrometers which provide high mass resolution, unlimited mass range, high transmission, and molecular imaging capabilities in the microscope and/or microprobe modes.

The understanding of the **ionisation process** is not yet complete and different models have been involved to explain the origin of secondary molecular ions [120–122], but to date none has shown universal applicability [123]. Some light has been shed on secondary ion formation processes by using tandem mass spectrometry [124]. Vickerman [125] has recently discussed the mechanism of secondary ion generation, including models of sputtering, and has given the fundamental SIMS equation containing the parameters involved in generating the spectrum.

Organic SIMS is based on the primarily unexpected fact that sputtering of even involatile and thermally labile organic molecules results in formation of parent-like secondary ions, as $(M + H)^+$ and $(M - H)^-$. In addition, characteristic fragmentations are generated. Two processes are involved in the production of secondary ions, namely desorption (sputtering) and ionisation (*cfr.* ref. [15]). Whether these occur simultaneously or consecutively is a debated issue. There is consensus that sputtering is based primarily on the formation of a collision cascade in the target caused by the impinging primary ion. However, the ions leaving the surface region are formed initially, they may fragment during their flight to the detector. The present theories about the complex mechanism of secondary ion formation are far less advanced than the theories already available for XPS and AES, the other, most widely used surface-specific techniques.

The main components of a **secondary ion mass spectrometer** are the primary particle source (charged bombarding particles), the secondary ion source (containing the bombarded target with the sample molecules M on its surface), the *e/m* analyser, and the ion detection unit, see Scheme 4.1.

Operation in the UHV regime is even more important in SSIMS than in XPS. When studying insulating samples such as polymers it is necessary to overcome charging problems by use of an auxiliary source of electrons. Thus an electron source is an essential component of a SSIMS instrument. Lub *et al.* [126] have first applied low-energy electrons (10 eV) for charge compensation during SIMS experiments with a ToF-SIMS spectrometer. The charging problem inherent in SSIMS studies of insulators has led to the development of fast atom sources.

A variety of **ion guns** are in use [127]. Typical performance data of ion guns used in SSIMS and SNMS are given elsewhere [128]. The most important ion source parameters are brightness, extractable current and energy spread. The most common types of ion source used for SSIMS are electron impact, surface ionisation and liquid-metal field emission sources [15]. If reasonably high primary-ion currents and maximum mass resolution at moderate spot sizes are required then electron impact (EI) ion sources can be employed. For SSIMS applications noble gases (such as Ar^+, Xe^+) are generally used, but Cs^+, Ga^+, In^+ or SF_5^+ bombardment may offer specific advantages [128]. For the production of elemental and chemical surface maps with high lateral resolution (imaging) spot sizes available from EI guns are inadequate, and liquid-metal ion guns (LMIG) are used in which the primary-ion beam can be finely focused.

The low primary ion dose and very low yield of detectable secondary particles has obviously also consequences for the **mass analyser**, which is required to have a high sensitivity. Static SIMS may be carried out by a variety of analysers, *i.e.* with quadrupole mass spectrometers, either as QuadSIMS [115,129] or QQQ-SSIMS [124], time-of-flight [130,131] and double-focusing magnetic sector instruments, M-SIMS [132], or FTICR analysers [128]. QuadSIMS (or Q-SIMS) is very valuable for most inorganic and simple low mass organic analyses. Early studies on SSIMS quantification used quadrupole systems. Whilst a great deal of useful information has been obtained on polymer surface chemistry using a quadrupole mass analyser in SSIMS, the quadrupole is a low transmission device (approximately 1% of the charged fragments

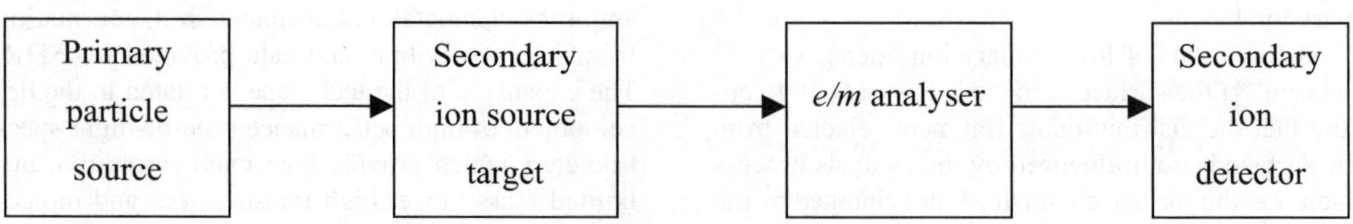

Scheme 4.1. Layout of a secondary ion mass spectrometer.

is captured for mass analysis). Furthermore, it is a scanning instrument so that it only allows the sequential transmission of ions, all other ions being discarded. The information loss is therefore very high. Since 1988 the time-of-flight spectrometer has gradually become the analyser of choice for SSIMS analysis of real problems involving complex organic and other insulating materials and high spatial resolution. Johnson *et al.* [133] have compared ToF-SIMS and Q-SIMS. The great advantage is that the ToF analyser is a non-scanning device, none of the ions are discarded in the analysis method. As a consequence, the relative sensitivity is 10^4 to 10^5 greater than a quadrupole. ToF instruments provide in theory a limitless mass range (usually in practice about 10,000 Da), as compared to about 1000 Da for Q-SIMS. ToF-SIMS thus features excellent molecular specificity through fragmentation patterns, mass spectral libraries, accurate mass analysis and lateral imaging to sub-μm scale (routinely less than 2 μm). Consequently, the ToF analyser has considerable benefits for complex organic materials analysis. The area analysed in a ToF-SIMS instrument, which mostly employs reflectron analysers (Fig. 4.7), is typically $\leq 200\ \mu m^2$ (in order to obtain high transmission), and is significantly less than in a quadrupole instrument. However, this is more than compensated for by the large gain in sensitivity (of the order of 10^4). ToF-SIMS is primarily used for analysis of static conditions. Dynamic experiments at low erosion rates may allow quantification of trace elements in the upper monolayer(s) of the sample.

Since the introduction of high performance ToF instruments various analytical advances have been made including microscope or microprobe imaging (MI) and laser post-ionisation (PI) capabilities. Sample preparation to improve ion yields has developed and statistical techniques for spectrum and image classification have been introduced. ToF-SIMS and Q-SIMS were reviewed [116]. Also M-SIMS gives relatively high mass resolution and high transmission (0.1 to 0.5), but it is still usually a scanning device. All three analyser types allow depth profiling.

Sampling for static SIMS on polymers is usually carried out in one of several modes:

- Soluble samples (*e.g.* polymer extracts) are prepared as (sub)monolayers on a noble metal substrate (usually Ag, Au or Pt).
- Insoluble samples are analysed "as received" (bulk materials) or after coating with a thin noble metal overlayer.

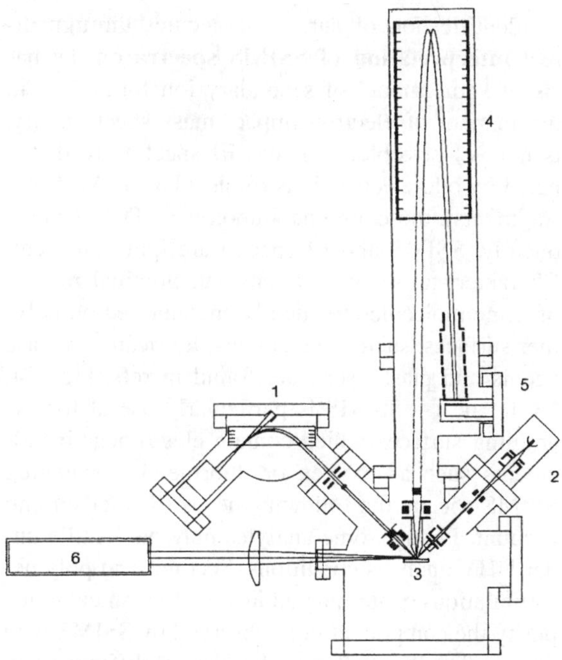

Fig. 4.7. Schematic diagram of reflectron ToF-SIMS with laser-SNMS facility. Pulsed, mass separating electron impact ion source (1), pulsed fine-focusing liquid metal ion source (2), target (3), reflectron for energy focusing the mass-separated secondary ions (4), detector (5), and laser for post-ionisation of emitted neutral particles (6). Reproduced with permission of ION-ToF GmbH, Münster.

Sample preparation is important in relation to the improved ion yield. Organic deposits on noble metal substrates lead to enhanced secondary-ion yields. The properties that make noble metals such exceptional SSIMS and SNMS substrates are: *(i)* enhanced desorption efficiency; *(ii)* cationisation; and *(iii)* stability and non-reactivity [128]. Cationisation of parent molecules is a common finding in mass spectrometry. In some techniques, like ESI-MS or MALDI-MS, the effect is widely used to identify large fragments or pseudo-molecular particles. Due to the appearance of strong quasi-molecular ion signals ($(M \pm H)^{\pm}$, $(M + Na)^{+}$, and in particular $(M + Metal)^{+}$), this kind of preparation is the preferred method for detection and identification of dissolvable organic materials and in cases where an ultimate sensitivity is required (limited amount of sample material). Detection limits in the low fmol range can be reached. In the characterisation of polymers properties like average molecular weight and oligomer weight distribution as well as the type of end-groups and additives can be determined.

Identification of surface molecules through direct **interpretation of SSIMS spectra** on the basis of some model of secondary ion formation, in the manner of electron impact mass spectrometry, is not yet possible. Whereas EI spectra are dominated by odd-electron ions (molecular ion $M^{+\bullet}$ and fragments), these are uncommon in SSIMS. Consequently, SSIMS and EI spectra are quite different. Characteristic secondary ions and nominal masses of some molecules frequently encountered on polymer surfaces, such as detergents, lubricants, release agents and plasticisers, are found in refs. [15,134, 135]. The (−)SSIMS is particularly useful for examining surfaces which contain electronegative elements such as oxygen or fluorine. In evaluating SSIMS spectra the following needs to be taken into account. Firstly, some analytes may be volatile under UHV analysis conditions. Secondly, in polymer formulations containing additives of technical grade purity the component ratio observed by SSIMS may deviate from the bulk ratio because of differences in surface activity or ion yield of the components (mass discrimination).

Hagenhoff *et al.* [128] divide SSIMS spectra in three sections: the fragment area (1–500 Da), the area of quasi-molecular ions (1150–1350 Da) and an intermediate area, where few secondary ions are detected. Quasi-molecular ions are formed either by attachment of low-MW cations and/or anions (*e.g.* salt and metal ions) to the parent mass or by the loss of small fragments (usually functional groups). Typical quasi-molecular ions are $(M + Ag)^+$, $(M + K)^+$, $(M - CH_3)^+$. Quasi-molecular ions can be desorbed intact up to 3500–15 000 Da; larger molecules tend to fragment. If no quasi-molecular ions are formed, small-fragment ions still remain visible. The observed fragment peak patterns are characteristic of the particular molecules and for the mass range up to 500 Da the term "fingerprint region" is thus used. Whereas quasi-molecular ions are comparatively easy to identify, the interpretation of the fingerprint region requires much experience or should be based on reliable libraries.

There are inherent differences between the spectral features relating to polymers *per se* and the (usually) lower molecular weight species, such as contaminants and additives, which are often detected on polymer surfaces. ToF-SIMS is highly suited to identify "small" molecules on polymer surfaces (in comparison with high-MW polymers). This is mainly due to the fact that these species usually give quasi-molecular ions, *i.e.* $(M + H)^+$, $(M - H)^-$, and characteristic fragmentation patterns. Since for most polymers the intensity in either positive or negative ion spectra falls rapidly by $m/z > 250$, surface molecules which give rise to inherently high intensity quasi-molecular ions/fragments in this region of the spectrum can easily be detected at fractional monolayer coverage [15]. Identification of surface molecules by SSIMS is currently still heavily reliant on pattern recognition.

The basis for the development of any new spectroscopic analysis technique is the creation of a **spectral database** from pure materials. This provides, progressively, the base level of spectra interpretation through matching of actual spectra with "fingerprints" from the database. As polymers are usually electrically insulating and quite sensitive to particle bombardment, rather stringent experimental restrictions are imposed on spectral data collection. The situation is complicated by the variety of instrument arrangements available using different primary ion species and impact energies (Ar^+, Xe^+, Cs^+, SF_5^+ at 8 keV, Ga^+ at 15 keV). However, SSIMS databases (comprising polymer additives) are expanding (*cfr.* ref. [134]). Various Q-SIMS libraries for standard polymers have been published [135,136]. The spectra are all perfectly understandable in terms of the known structure of the polymers. Subtle chemical state differences, which are difficult to distinguish via XPS, are clearly evident from SSIMS spectra. The Münster High Mass Resolution Static SIMS library [137] comprises 260 polymer and 170 additive ToF-SIMS spectra. Kersting *et al.* [138, 138a] have reported a ToF-SIMS library of 104 technically relevant polymer additives, based on spectra of additives in their industrially applied concentrations embedded in a host polymer (LDPE). Different primary ion bombardment conditions (monoatomic primary ions: Cs^+, Ga^+; polyatomic primary ions: SF_5^+, Au_x^+) were used to study the influence of primary ions mass and polyatomicity on the secondary ion emission. Most substances can be analysed by using Ga^+, but in some cases SF_5^+ has to be used in order to reach acceptable detection limits. Access to spectral libraries is either a "whole spectrum" approach, pattern matching or chemometric analysis.

Modern SSIM spectrometers are equipped with data acquisition/processing stations, and the programs allow control of the spectrometer during acquisition and standard spectrum interpretation and quantification (mass scale calibration, peak locating, peak intensity determination, spectrum quantification, depth profiling, spectral display, *etc.*). To make use of the abundant information generated by ToF-SIMS, powerful data handling systems, including **multivariate statistical analysis** are wanted [139–141]. This approach should point the way forward to more reliable quantitative analysis in SSIMS. A variety of data analysis methods have been developed for calibration and classification of ToF-SIMS spectra [142].

The *surface mass spectrum* characterises the surface chemical structure. The spectral intensities can be used to determine the relative surface concentrations of the different surface species. Both positive and negative ion detection modes are possible in SIMS, as in all mass spectrometry techniques. A comparison of the positive and negative ion spectra can often substantially improve the analysis of the results. In SIMS, the charged fraction of the secondary particle flux is very small (10^{-3}). Moreover, the number of sputtered ions per incident primary ion (*i.e.* the secondary-ion yield) is matrix dependent. With such yield variations direct quantification of surface species based on the number of desorbed secondary ions (*i.e.* from the SIMS data) is generally impossible [123]. Wucher *et al.* [143] have recently described a method to determine the secondary ion formation probability, *i.e.* the ionisation probability of sputtered particles in a direct and quantitative manner.

In polymer development and failure analysis quantitative information is often required, which various quantitative techniques (*e.g.* XPS, RBS, TXRF) cannot offer due to limited sensitivity. It is here that ToF-SIMS is indispensable. Although SSIMS has the reputation of being a non-quantitative method, a more subtle stand is appropriate. **Quantification** of SSIMS data is bedevilled by matrix effects (*i.e.* dependence of secondary ion yields on the chemical environment at the surface), such as the uncertainty and variations in the efficiency of fragment ejection (sputtering; damage), the influence of surface coverage, the wide range of ionisation efficiencies, and changes in surface potential caused by charge build-up due to incomplete neutralisation [123]. SSIMS spectra often contain a tremendous amount of quantitative information about surface structures, which is encoded in terms of fragment ion yields and is not readily accessible. Decoding may be based on physical principles or is empirical. A quantitative interpretation of ToF-SIMS data is still at its early stages. It is difficult to derive quantitative data from "first principles" physical models, mainly because of current lack of understanding of the underlying phenomena of secondary ion emission and of collection and detection of secondary ions. The major analytical problem facing SIMS is the development of a comprehensive understanding of the fundamental and instrumental factors contributing to the non-linear response of measured signals to concentration changes and the development of standards which allow accurate calibration of analyses, even in the presence of severe matrix effects.

It is often difficult to assess properly the accuracy of a SIMS analysis because there are no techniques capable of calibration analysis of very dilute analytes. Standardless SIMS analyses, *e.g.* using exponential ion yield relationships, are subject to sizeable errors, perhaps as much as factors of 2–3 [144]. Empirical methods of quantitative analysis can be applied to the problem of SIMS analysis to achieve useful levels of accuracy. The availability of standards is then critical. External standardisation is absolutely insufficient, and the use of internal standards is necessary. These standards must be introduced into the surface during preparation. When a suitable standards suite is available, an effective approach to quantitative SIMS analysis can be achieved through the use of the empirical method of relative elemental sensitivity factors. Detection sensitivity factors cover a wide range (six decades).

Hagenhoff [145] has given a quantitative description of organic SIMS. Various strategies for quantification have been developed; internal standards (limited applicability, because of elaborate preparational steps mostly in liquid phases), univariate and multivariate quantification. In the latter cases quantification is achieved by normalisation to uncharacteristic peaks (*e.g.* hydrocarbons), to a sum of characteristic peaks (*e.g.* in mixtures) or to the overall spectral intensity. Promising results concerning quantification of SIMS data can be obtained by evaluation of peak intensity ratios and by means of multivariate statistical methods. Using internal standards an accuracy of the quantification of better than 10% can be reached [146].

Quantification of surface coverage is within reach more often than might be expected [144]. Various areas of application have been identified where quantification is possible, *i.e.* where the change in the number of detected secondary ions truly mirrors the change in surface concentration [128]. For analytes at (sub)monolayer coverage on a substrate quantification can be achieved by normalising an analyte peak area to a substrate peak area. Absolute quantification in multilayer systems is only rarely achieved but semiquantitative information can be obtained in many cases. Galuska [147] has recently discussed quantitative ToF-SIMS methodologies for polymer analysis. Kenens *et al.* [148] have combined ToF-SIMS data with XPS measurements, which provides a standard quantification method (overlayer model) to calculate the surface coverage (atoms/cm^2).

Although the possibility of using ToF-SIMS as a quantitative method for organic materials has been demonstrated, only a few reports use ToF-SIMS in a quantitative fashion [146,149,150]. Reihs [151] has used quantitative information in industrial problem solving. An EC project has recently dealt with the quantification of additives at polymer surfaces by ToF-SIMS [152].

Standards play a critical role in particular in the realisation of quantitative analysis by SIMS. Only a limited number of materials exist that are homogeneous on the micrometer spatial scale and attain the high level of standardisation appropriate to standard reference materials. It appears that only one certified reference material (CRM) exists for SIMS analysis (for semiconductor materials) [153]. Also, standardisation of SIMS has its main focus on inorganics, such as the quantification of dynamic SIMS [154, 155]. The role of standards in SIMS has been addressed [156].

Mass calibration in ToF-SIMS is carried out *in situ*, using peaks representing secondary ions of known composition (and hence of known exact mass). As the whole spectrum is recorded simultaneously no special calibration standards are required. An accuracy of about 1 mmu can be achieved in the low mass range, in the high mass range values in the low ppm range.

Statistical process control (SPC) for SIMS was reported [157], as well as interlaboratory calibration and measurement using SIMS [158]. Recently, an **interlaboratory SSIMS study** (involving 18 laboratories and 21 SSIMS instruments) has been carried out on two PTFE and PET bulk polymers and on a thin layer of Irganox 1010 deposited on clean silver foil [159]. Reported repeatability was as good as 2%. It therefore seems that standard SSIMS spectra can be used for reference purposes independently of a particular instrument or operator.

High mass resolution of ToF-SIMS instruments (10^{-3} to 10^{-4} amu) enables **exact mass measurements**, calculation of empirical formulae and more reliable unknown peak identification [130]. Accurate mass determination by ToF-SIMS is very simple because no special calibration procedures or calibration standards are required. The accuracy is about 10 ppm for atomic species and for molecules in the low mass range. For organic molecules in the high mass range an accuracy better than 5 ppm was obtained. In cases where the mass resolution is not sufficient, mass spectral fragmentation patterns may be compared to library spectra for identification.

The SSIMS sampling depth is particularly difficult to measure experimentally and for polymer systems there is paucity of data. However, the sampling depth of SSIMS is significantly lower than that for XPS, under typical operating conditions (*i.e.* take-off angles of $>45°$, with respect to the surface). Depth profiling is usually carried out by sectioning or sputtering. Sectioning of samples is a very useful methodology for the determination of depth distributions in those cases where the collision cascade would destroy sensitive material (*e.g.* organic layer systems). Sputter depth profiling in ToF-SIMS/SNMS instruments as a sampling technique has become possible by the introduction of a second ion gun specifically designed and optimised for sputtering. The available sensitivity is still rather lower than that of dedicated DSIMS instruments. All elements of one polarity can be profiled simultaneously. In many cases no more than $\sim 10^4$–10^5 atoms need be sputtered to obtain statistically useful signals ($\sim$100 counts) so that these low detection limits can be achieved in remarkably small volumes of material. Basic aspects of sputter depth profiling were described [160].

Table 4.9 shows the **main characteristics** of SIMS. With the introduction of a charge compensation system for ToF-SIMS instruments pure bulk insulating materials have become accessible. The combined advantages of surface sensitivity, trace level detection, high mass range, and high mass resolution allow application of ToF-SIMS in the chemical industry, including characterisation and spatial distributions of surface-segregated additives. Static ToF-SIMS emerges as the method of choice for mass spectrometric investigation of polymers. To some extent, similar information can be obtained from MALDI-ToFMS. The main advantage of SSIMS resides in the capability to characterise the surface of

Table 4.9. Main characteristics of SIMS

Advantages:

- Easy sample preparation
- Suitability for insulating materials
- Mass spectral technique with very high transmission (constant over the entire mass range), quasi-simultaneous detection of all secondary ions with high mass range and high mass resolution ($m/\Delta m > 10\,000$)
- High sensitivity and dynamic range
- Detection limit: ppm to ppb for bulk analysis
- Information depth 1 nm (SSIMS); 10 nm–100 μm (DSIMS)
- Excellent spectral reproducibility
- Elemental (H to U; all isotopes) and molecular surface composition
- Analysis of mixtures
- Chemical bonding information (SSIMS)
- Very good depth resolution (>2 nm) (SSIMS)
- Inherent depth profiling (DSIMS)
- Chemically resolved imaging: fast molecular mapping capability
- Commercial equipment
- Developing databases (fingerprinting)

Disadvantages:

- UHV requirements (vacuum compatible samples needed; volatiles may be lost unless sample is cooled); *ex situ* technique
- Destructive (if sputtered long enough), limited organic imaging
- Insulator surface charging
- Low secondary ion yield (10^{-1}–10^{-9})
- Strong matrix effects
- Difficult quantification (especially in complex systems)
- Detection sensitivity factors covering wide range (six decades)
- Mass interferences of atomic and molecular ions
- Poor lateral resolution (0.1–5 μm)
- Unrelated surface contamination may complicate analysis
- Complex spectra
- Complex and expensive equipment
- Expert users needed

"as received" degraded or modified polymers. SIMS offers poor lateral resolution (1 μm) but a great deal of chemical information about adsorbates or migration of ions into deeper layers. The most important limitation of SIMS microanalysis is caused by intrinsic low ion yield for most of the elements (10^{-1}–10^{-5}).

SSIMS can be regarded as a versatile and additional, partly complementary technique to XPS, SEM-EDS, FTIR, LDMS and MALDI-MS. Yet, SSIMS should not be seen as a real competitor to established methods for elemental analysis of solids. SIMS, SNMS, LMMS and recoil spectrometry are the only techniques that can detect hydrogen. The main merit of SSIMS resides in the capability to detect molecule-specific information ("speciation"), not just element ratios. **XPS and SSIMS** are often used in combination [161–163]. This is not surprising, as these surface sensitive techniques are complementary in their information content. For example, XPS yields essentially quantitative elemental information with sensitivity to chemical functionality, whereas the strength of SIMS lies in its sensitivity to molecular structure and its power to precisely identify individual additives/components at the surface of complex industrial polymers. In case of very similar composition, XPS analysis alone is not sufficiently discriminating. Both techniques can be used for depth profiling analysis or for mapping the homogeneity of the surface condition on a micro- or macroscopic scale. The SSIMS sampling depth is difficult to measure experimentally, but is significantly lower than that for XPS.

Fast atom bombardment SSIMS (FAB-SSIMS) is not a much used version of SSIMS, which uses neutral particles (generally atoms) to excite the surface.

Several reviews [116,125,128,164–167] and textbooks [5,15,65,168] describe surface mass spectrometric techniques, with particular emphasis on SIMS and applications. Instrumentation for SIMS has been reviewed [127]. Another review deals with SIMS for the surface analysis of polymers [169]. The history of (S)SIMS has been traced by Vickerman [170] and Briggs [15]. SIMS is still an expanding and developing field. Analytical developments concentrate on the understanding of the ion formation (in thick organic layers), on imaging/molecular mapping, laser post-ionisation, sample preparation to improve ion yields, statistical techniques for spectrum and image classification and combined techniques (XPS, PDMS, MALDI, AFM).

Applications

ToF-SIMS is the most versatile of the surface analysis techniques that have been developed over the last 30 years. Analyses possible by SIMS include bulk elemental analysis in small volumes of material, in-depth analyses (depth profiles), imaging, analysis at interfaces, isotopic analysis and elemental and molecular surface analysis.

Polymer characterisation using SIMS in its various forms comprises oligomer and molecular weight distributions (M_n, M_w), characterisation of repeat units, end-groups and functional groups, copolymer and blend composition, polymer modification, general surface structural determination, surface functionalisation, diffusion/migration, surface segregation and reactivity, additives (mapping), localised microanalysis, defect and contaminant analysis. Because of the desorption process, the highest available mass is ~15 000 Da.

Two broad fields of application of SSIMS on polymers may be distinguished which relate to the sampling procedure. For soluble samples (prepared as *monolayers* on noble metal substrates) molecular information is obtained (oligomer and molecular weight distributions, end- and side-group characterisation, repeat unit, *etc.*), whereas for insoluble samples (analysis on *bulk material*) typical fragments are identified (surface functionalisation, thin film structures, surface diffusion, surface segregation and reactivity, contamination, deposition, *etc.*). ToF-SIMS can be used to analyse many types of materials, organic and inorganic, single component or mixtures, small molecules or polymers. Since organic substances are very sensitive to radiation damage, static SIMS operation (typical primary ion fluence $<10^{-13}$ cm^{-2}) is mandatory. ToF-SIMS is particularly useful in polymer surface analysis since it provides highly specific chemical information on the outermost monolayers of a surface. The main assets of the technique are the extreme surface sensitivity and the ability to generate ion images of the surface distribution of atomic and molecular species. Static SIMS applied to the **analysis of industrial polymers**, coatings, paints, inks and lacquers, covers the aforementioned areas of technological interest in relation to production processes (contamination), adhesion, failure analysis (troubleshooting), fingerprinting, chemical modification (corona treatments), migration (depth profiling), chemical imaging and general surface structural determination (surface and microphase compositions) [171]. Polymers can be fingerprinted via their characteristic fragmentation patterns. ToF-SIMS spectra of additives in polymeric systems can be delicately influenced by polymer-adsorbate interactions, as shown by a comparison of stearic acid on a variety of polymers (PE, PAA, PTFE) [172]. These matrix effects render fingerprinting difficult matter. A wide variety of additives has been studied by means of ToF-SIMS, such as plasticisers, slip agents, antiblock agents, lubricants, surfactants, mould release agents, processing aids, *etc.*, in particular in relation to: *(i)* chemical status of additives (salt formation, oxidation, ...); (ii) bulk concentration of additives; *(iii)* segregation/migration behaviour of additives; and *(iv)* imaging of the lateral additive distribution. SSIMS also plays a crucial role in defending product claims. This flexibility and range of applications make ToF-SIMS a versatile and powerful tool for polymer characterisation.

SSIMS can be used for the study of **catalyst residues** and for speciation of inorganic compounds. The fate of the weakly coordinating $[B(C_6F_5)_4]^-$ anion following commercial production of ethylene copolymers with ionic metallocene catalysts at high p, T was determined by ToF-SIMS (ion gun operating at 15 keV and 600 pA) and LDMS [173]. The ion was found to persist in the polymer product (at less than 2 ppm).

The combined use of ToF-SIMS and XPS in industrial research has been illustrated by De Lange *et al.* [161] and others [174] for problems such as surface treatment of C fibres, adhesion activation of aramid fibres, weathering and protection of wood, surfactant adsorption in pigments, grafting of PP with acrylic monomers and treatment of a perfluorinated membrane with an amphiphile. It is not surprising that XPS and SSIMS often join to tackle the same problem. As mentioned before, the two techniques are highly complementary. The combined information on molecular specificity obtained with ToF-SIMS and quantification from XPS have been used in the study of sizings of carbon fibre (CF) composites, where PDMS, dialkyl phthalates (DIBP and DIOP), phenolic antioxidants and glyceryl monostearate were identified [175]. Both techniques have also been applied for differentiation of PET originating from different production processes [163], for the characterisation of surfactant polymers [176], for surface analysis of wool [177], for carbon-black surface characterisation, as well as for the study of phosphorous-based flame retardants in acrylic polymers, *etc.* XPS and SIMS analysis are often carried out on humidity or thermally aged specimens for the safe prediction of storage time and conditions for polymeric materials. SIMS/XPS depth profiling has been reported to characterise the surface of an epoxy resin (Epikote 828) modified by the addition of a polymerisable monomeric, fluorine containing, surfactant [178]. The difference in sensitivity between the two techniques has been shown by the successful detection

Table 4.10. Examples of measured masses for known additives using ToF-SIMS. The extract results are for (M + Ag) ions while the *in situ* results are for (M + H) ions

Additive	Exact mass M	Extract mass M	Error (ppm)	*In situ* mass M	Error (ppm)
Naugard 524	646.4515	646.453	1.7	646.439	19.3
Oleamide	281.2719	281.289	60.7	281.274	7.5
Stearamide	283.2875	283.291	12.4	283.288	1.9

After Mawn *et al.* [180]. Reproduced by permission of the American Institute of Physics.

of tannic acid on PMMA by ToF-SIMS at variance to XPS [126]. Polyurethane type aircraft coating was analysed by SIMS; XPS was employed to determine pigments and extenders [179]. Watts [48] has used XPS, SIMS and AES in the study of metal matrix composites (MMC).

In **extracts** of PE/(a.o. Naugard 524, Naugard DSTDP, Irganox 3114), which consist of both additive and low-MW polymer, ToF-SIMS gives results similar to LD-FTICR analysis [181]. However, ToF-SIMS shows higher reproducibility than LD-FTICR, which facilitates quantitation with the use of standards. ToF-SIMS also has high sensitivity, and a dynamic range of $>10^6$, which makes it well suited for polymer additives identification at trace levels. An extraction procedure obviously results in bulk analysis and does not provide any information on surface-specific phenomena such as surface segregation and surface oxidation. Mawn *et al.* [180] have been the first to compare analysis of (the above) PE extracts, deposited as a submonolayer onto a roughened Ag substrate to *in situ* high-resolution ToF-SIMS analysis (using a 10 keV Ar^+ pulsed primary ion beam) of bulk PE (Table 4.10). Despite lower secondary ion intensities in the *in situ* analysis the additives detected in the extract were confirmed, except for Irganox 3114; moreover, stearamide and oleamide were identified. High mass resolution was used to assist in the assignment of molecular identities to the additives. Naugard 524 was found to be partially oxidised to a phosphate triester. Similarly, ToF-SIMS of an LLDPE/(0.15% erucamide, 0.065% Irganox 1076, 0.060% Sandostab PEPQ/Irgafos 168, 0.030% Irganox 1010) polymer extract deposited as a submonolayer on roughened silver substrate to promote silver cationisation has allowed identification of all five additives as (M + Ag)-cationised species without chromatographic separation (with evidence for Irgafos 168 and PEPQ being partially oxidised) [119]. In the untreated LLDPE polymer film only erucamide and Sandostab PEPQ were revealed by ToF-SIMS. On the other hand, ToF-SIMS of the LLDPE surface covered with a 150 nm thick Ag overlayer allowed identification of all five additives in the *in situ* surface analysis mode, including evidence for surface oxidation for Irgafos 168 and Sandostab PEPQ [119]. Figures 4.8 and 4.9 show the positive ion ToF-SIMS spectra for LLDPE sample as an extract, film, and film with silver overlayer. Briggs [15] has reported the (+)ToF-SIMS spectrum of the plasticiser di-isononyl phthalate deposited as a non-uniform film on silicon (Co^+ primary ions).

Some additives are easily measured by ToF-SIMS, such as Tinuvin 770, others are more difficult to analyse (*e.g.* a thick layer of GMS can be measured easily, a thin layer gives problems) or are even impossible (*e.g.* BHT; evaporation in UHV). Figure 4.10 shows a case of additive identification (N,N'-ethylene-bis-stearamide on PA6) by means of (+)ToF-SIMS.

For most applications, SIMS has been used in **fingerprinting** mode, *e.g.* to distinguish between polymer types and to identify additives. All substance classes are accessible either prepared as monolayers on noble metal substrates (desorption of intact molecules ($m/z < 15\,000$) and characteristic fragments) or as *in situ* bulk materials (characteristic fingerprint spectra ($m/z \leq 300$)). Several **key factors** play a role in successful detection of additives in polymeric matrices by means of SSIMS: *(i)* sensitivity (ToF-SIMS being preferred over Q-SIMS); *(ii)* nature of bombarding ions; *(iii)* characteristic mass fragment(s); and *(iv)* reference library. Benninghoven *et al.* [182] compared the characteristic molecular secondary ion emission from different polymer surfaces (PET, PP, PTFE, PS, PC, PMMA and PEG) under 10 keV Ar^+, Xe^+ and SF_5^+ bombardment and obtained high yields, in particular with SF_5^+. Figure 4.11 shows PP/Irganox 1010 under Ar^+ and SF_5^+ bombardment (intensity increase 20×). Characteristic secondary ions at m/z 219, 233 and 259 could

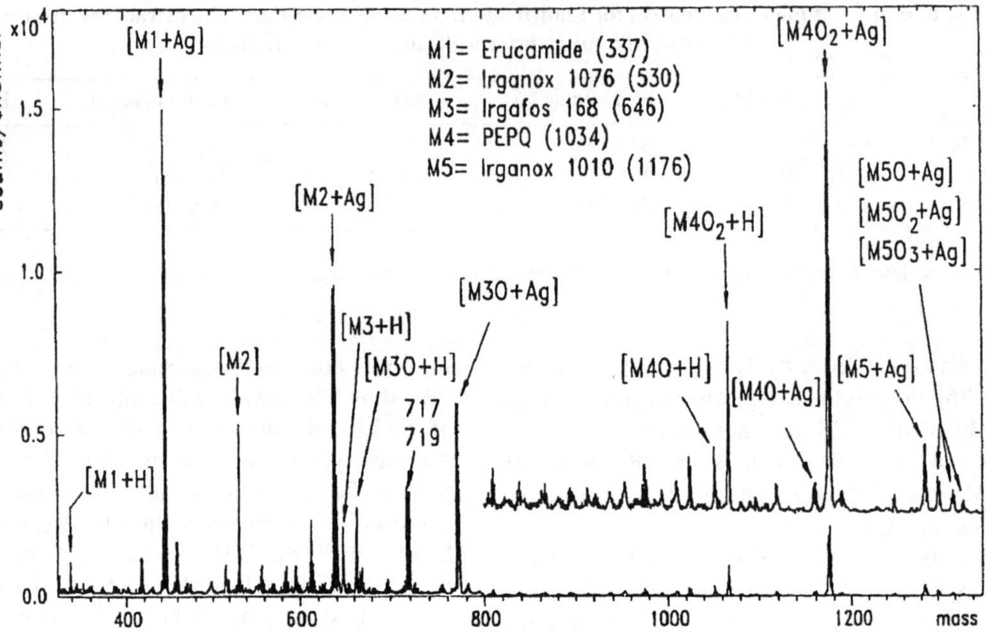

Fig. 4.8. SSIMS polymer additive extract analysis of LLDPE/(erucamide, Irganox 1010/1076, Irgafos 168, Sandostab PEPQ) showing identification of all five additives in addition to Irgafos 168 phosphate and oxidised PEPQ, forming [MO + H] and [MO_2 + H] cations. After Linton *et al.* [119]. Reprinted from R.W. Linton *et al.*, *Surface Interf. Anal.* **20** (12), 991–999 (1993). Copyright © 1993 John Wiley & Sons, Ltd. Reproduced with permission.

only be detected in the SF_5^+ generated spectrum. This pronounced enhancement of the secondary ion emission additives was also observed for other polymer/additive combinations. A variety of atomic and molecular primary ions (O^+, Ar^+, Xe^+, O_2^+, CO_2^+, SF_5^+, $C_7H_7^+$, $C_{10}H_8^+$, $C_6F_6^+$ and $C_{10}F_8^+$) with a total energy of 11 keV were compared in SSIMS analysis of PE/Irganox 1010 [183]. A strong yield enhancement was noticed with increasing mass for atomic primary ions and increasing number of constituents for molecular primary ions.

A systematic ToF-SIMS feasibility study on the determination of additives in LDPE with technically relevant concentrations has been reported using Ga^+, Ar^+, Cs^+, Au_x^+ and SF_5^+ bombardment conditions [138, 138a]. Identification and quantitative determination of polymer additives is possible with detection limits in the ppm range and a lateral distribution in the sub-μm range. Ga^+ bombardment is less performing than a liquid metal ion gun operated with Au^+ cluster ions. Irganox 565 could only be resolved using Au^+ primary ion bombardment. Low concentration Chimassorb 944FD (down to 1000 ppm) was detected and mapped in LLDPE using microfocused Ga and In beam ToF-SIMS [184]. The parent ion for the oligomer at m/z 599 is too weak for mapping the distribution of the additive in the concentration range of 0.1–0.5 wt.% in PE. Instead, imaging of the antioxidant distribution was possible to concentrations as low as 0.1% for the C_3H_8N mass fragment at m/z 58 and a linear concentration calibration curve was obtained.

Van den Berg *et al.* [185] has compiled a database of SIMS spectra of the most common inorganic pigments. ToF-SIMS offers some interesting possibilities for the spatially resolved analysis of (mixtures of) pigments in paint cross sections. However, the specific selectivity of the technique for different pigments needs to be taken into account. A further point of concern is interference with the embedding medium of the paint cross sections (matrix effect).

For **product identification** analysis of additives (as unintended markers) is a frequently used tool. Lang *et al.* [163] have examined PET samples of different suppliers by means of ToF-SIMS. Dependent on the origin, antioxidants and lubricants such as Irgafos 168, octylstearate, octylpalmitate, octylarachidate and a contaminant (polydimethylsiloxane) were found on the PET surface. Galuska [186] has developed ToF-SIMS calibration lines for PIP, PBD, PE,

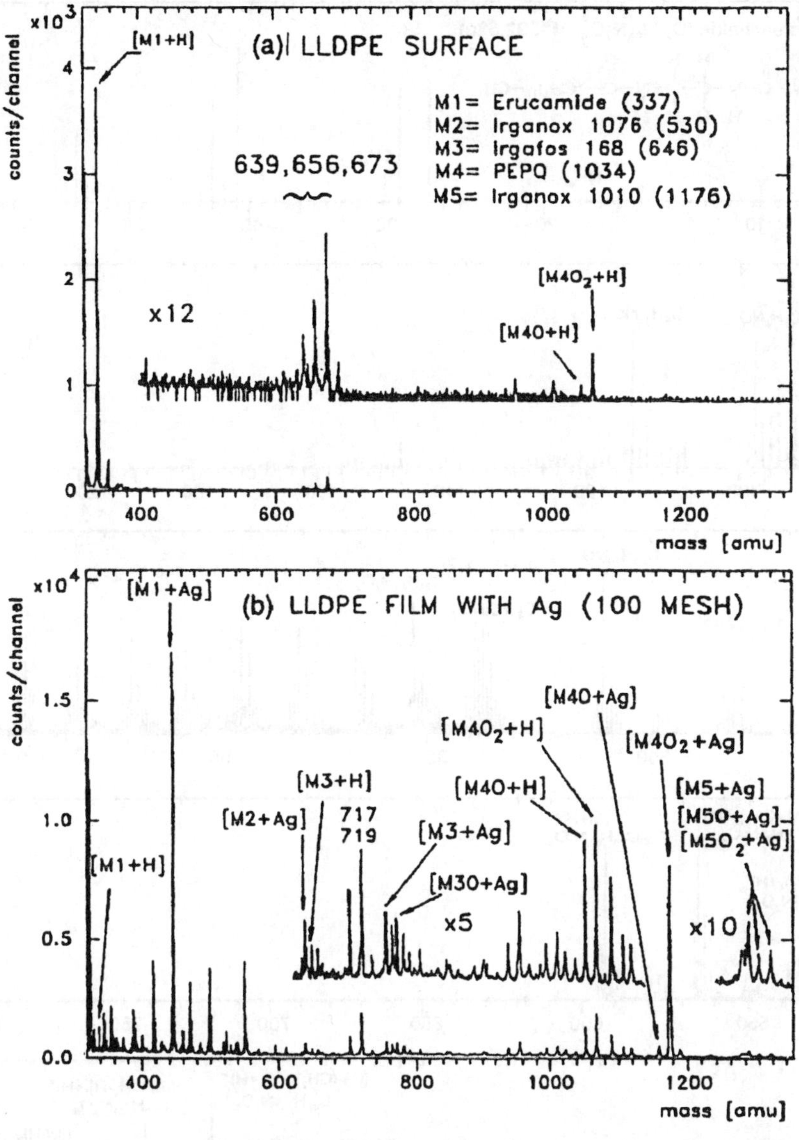

Fig. 4.9. *In situ* (+)ToF-SIMS analysis for untreated surface of LLDPE (a) and silver-patterned surface (b), showing a large improvement in additive detection with the silver overlayer. After Linton *et al.* [119]. Reprinted from R.W. Linton *et al.*, *Surface Interf. Anal.* **20** (12), 991–999 (1993). Copyright © 1993 John Wiley & Sons, Ltd. Reproduced with permission.

PS, PP and PIB low-MW standards (<20 000 Da). The MW calibrations are useful for determining the presence of unknown low-MW waxes and additives on polymeric surfaces.

Identification of low concentration (0.05–0.5%) **release agents** on EP surfaces can be carried out both by ToF-SIMS (monolayer thickness) and ATR (μm thickness). Pachuta [171] showed the use of SIMS for the identification of antioxidants, antistats, surfactants and primers on polymer surfaces and described the detection of the mould release agent N,N'-ethylene-bis-stearamide in PU parts. Lub *et al.* [126] have observed palmitic and stearic anions (m/z 255 and 283, respecively), denoting the presence of release agents in poly(bisphenol-A) carbonate (PC), as well as the i-octylphenolate anion ($m/z = 205$), originating from the end-groups of the polymer. Belu *et al.* [187] have used quantitative SIMS with PLS modeling to monitor the extent of incorporation of the **cross-linking agent** ethylene glycol dimethacrylate (EGDMA) in PMMA.

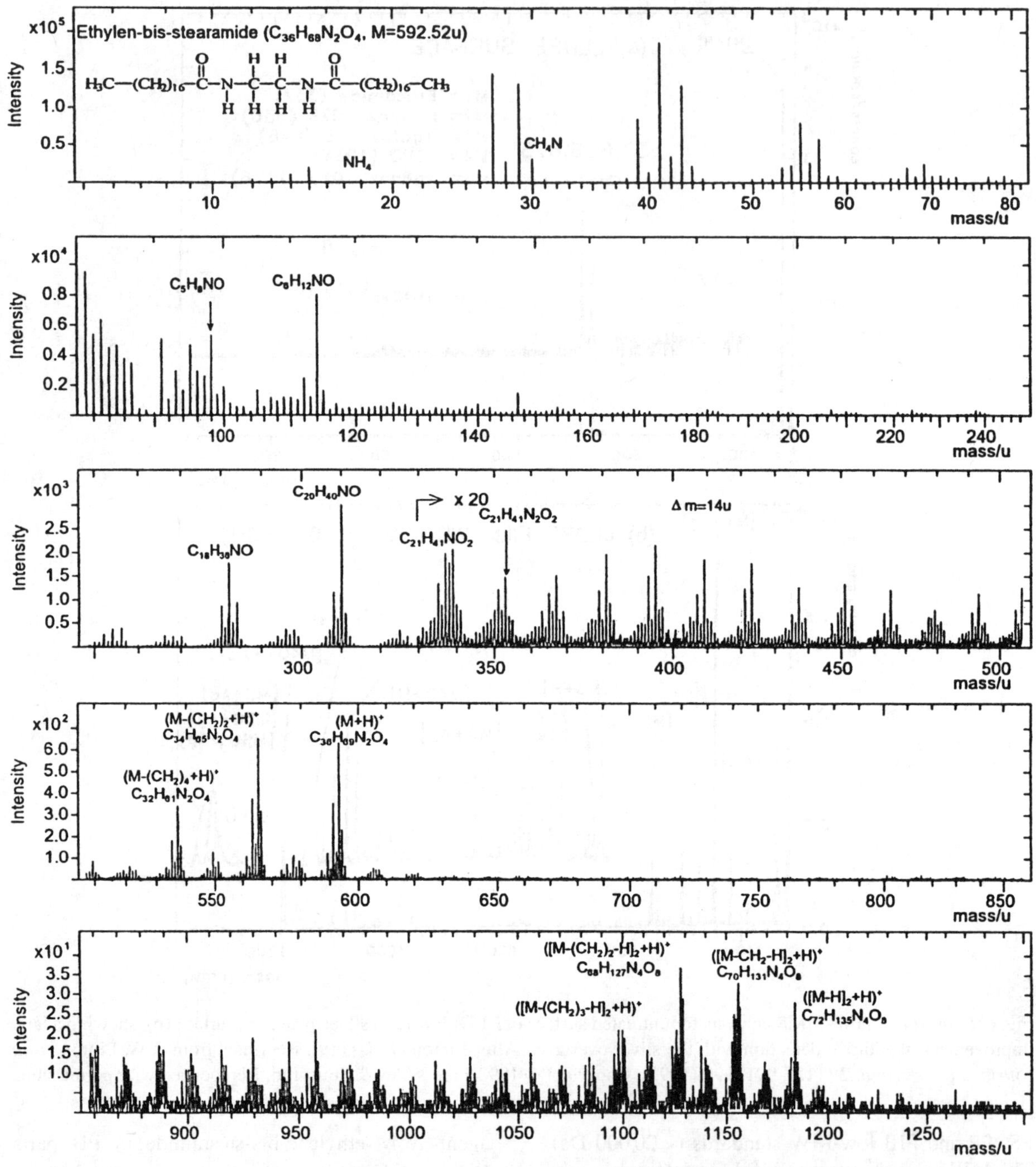

Fig. 4.10. Additive identification (EBA on PA6) by means of (+)ToF-SIMS. Reproduced with permission of TASCON, Münster.

ToF-SIMS has also been used extensively for molecular trace analysis of **antioxidants**. Boyd *et al.* [188] reported detection of Irganox 1076 (m/z 637/639 from $(M + Ag)^+$) in material extracted from a PP film surface and deposited as an Ag substrate for SSIMS. Briggs [116] has carried out a study of Biomer surfaces by SSIMS and XPS. Biomer is a commercial medical-grade poly(urethane) of undisclosed composition. Characteristic fragments of Irganox 245 (at m/z 147, 161 and 177) were detected. Literature controversy relating to the biological interactions of poly(urethanes) are likely

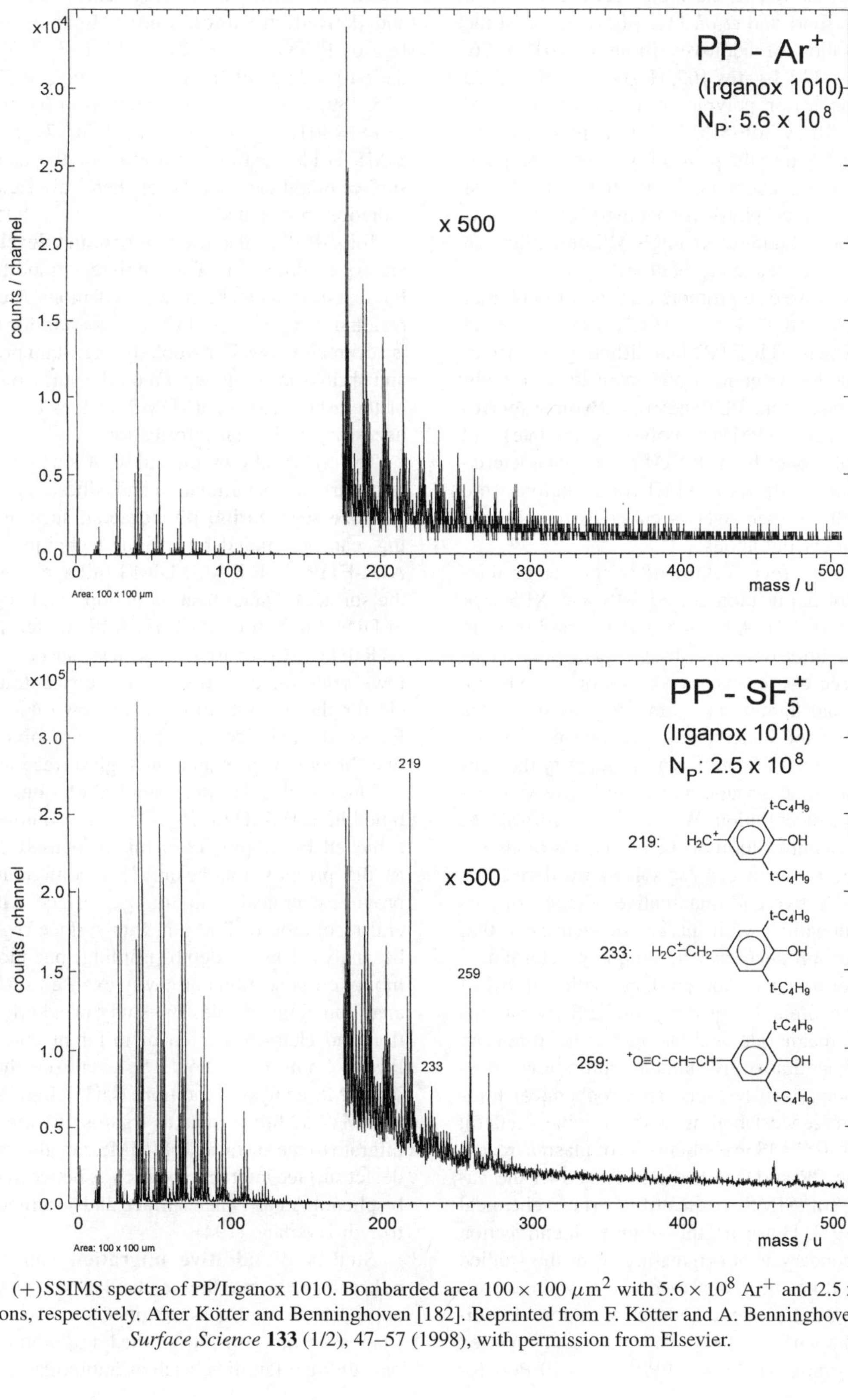

Fig. 4.11. (+)SSIMS spectra of PP/Irganox 1010. Bombarded area 100 × 100 μm^2 with 5.6 × 10^8 Ar^+ and 2.5 × 10^8 SF_5^+ primary ions, respectively. After Kötter and Benninghoven [182]. Reprinted from F. Kötter and A. Benninghoven, *Applied Surface Science* **133** (1/2), 47–57 (1998), with permission from Elsevier.

to stand in relation to the major surface chemical differences. Bertrand *et al.* [189] have reported the characterisation of additives (Irganox 1010/1076, Tinuvin 327/770, Irgafos 168, Hostavin N 30 and Cu phthalocyanine) on polymer surfaces (PS, PMMA, PET and aPP) by ToF-SIMS with the particular object of investigating the possibility of additive quantification. The authors underline the difficulty of separating surface physical-chemistry effects (surface diffusion of additives) and SIMS quantification problems in the data interpretation.

Priming is used to promote adhesion to polymer surfaces. Primed PET is a widely used industrial material which ATR-FTIR has difficulty in characterising due to the number and intensity of the absorption bands of the PET substrate, **Primer species** were identified as PMMA, poly(ethylacrylate) and poly(caprolactone) by ToF-SIMS from characteristic fragments [190]. As also PET fragment ions were observed the primer layer is either extremely thin (<10 Å) or discontinuous.

Polymer surface treatment is another productive field of application for SSIMS and XPS (*cfr.* Figs. 4.3 and 4.4). Adhesion and wettability need often to be improved in order to apply paint coatings; surface treatments such as corona or plasma discharge and flaming are commonly used for this purpose. For the understanding of plasma induced surface processes it is necessary to identify the generated functional groups and quantitatively determine their concentration. With XPS and ToF-SIMS, the unique identification of such groups is often difficult. This problem can be solved by derivatisation, *i.e.* selective and quantitative reaction of surface functionalities with marker substances so that the reaction products can be uniquely determined. Analysis of derivatisation products with ToF-SIMS extends the limited sensitivity of XPS by various orders of magnitude and, at the same time can be made quantitative by suitable calibration procedures. Comparison between spectrochemical titration (fluorescence labelling with thionine acetate), XPS and ToF-SIMS measurements of plasma treated HDPE and PP surfaces has clearly shown the capacity of ToF-SIMS to quantify surface chemical species [191]. However, this requires identification of the secondary ions originating from the studied functions. It has been observed that oxygen and nitrogen plasma treatments induce migration of additives to the surface [192]. This has been reported for PP/(Irganox 3114) and PP/(Irgafos PEPQ) for which ToF-SIMS reveals fragment-ions of additives and derived structures, namely Irganox 3114 and Irgafos PEPQ ($m/z = 219$, 647, 663, 783), degradation products of Irganox 3114 ($m/z = 278$, 349, 356, 389) and degradation products of Irgafos PEPQ ($m/z = 401, 459, 595, 611, 1051, 1067$) [193]. ToF-SIMS is also extremely useful for the analysis of surface modification by other chemical reactions and radiation treatments.

ToF-SIMS is also used for **product development** and for qualification of new polymers/modifications. For example, it can be critical to demonstrate that the residual metal content in a new coating for a rubber is acceptably low. For wool, dyeing, stain prevention and shrink-proofing depend on the surface structure of the article [61]; ToF-SIMS enables to obtain the necessary molecular information.

ToF-SIMS allows the study of surface segregation, surface contamination and adhesion properties. **Surface segregation phenomena**, such as blooming, can be studied by various techniques such as ATR-FTIR, ToF-SIMS, DIMS (after removal from the surface). Quantification is impossible by means of DIMS, difficult with ToF-SIMS and feasible with ATR-FTIR. Compared to IR mass spectrometry allows selective detection at low concentration levels for the identification of unknown components. Essential analytical capabilities for solving additive "blow-out" problems are high surface sensitivity and high molecular and spatial resolutions, all combined in ToF-SSIMS. High lateral resolution can be achieved by employing a micro-focused Ga^+ gun as the primary ion beam. This configuration can provide chemical maps of the surface with a lateral resolution of 200 nm. Sub-surface regions can be analysed using depth profiling, but more commonly cross sections or newly exposed surfaces are analysed. This enables the study of additive migration and characterisation of different coatings and layers. Some care should be exercised though, as in SSIMS analysis conditions (UHV) highly volatile low-MW additives such as Irgafos 168 are likely to migrate to the surface. ToF-SIMS can also be used to detect surface interactions, such as between the polybisphenol A carbonate surface and γ-amino propyltrihydroxysilane [194].

Studies of **additive migration** can be based either directly on the molecular ion peak or on appropriate low mass fragments of the additive. Lianos *et al.* [195] examined migration of various additive families, such as antioxidants (Irganox

1010/1076/3114, Ionox 330, Irgafos 168/PEPQ), acid acceptors (calcium stearate), UV stabilisers (Tinuvin 622), flame retardants (Adine 505, Dechlorane 505) and lubricants (erucamide) on various polymer surfaces (PP, PET) by ToF-SIMS. Additive traces could be detected even at bulk concentrations of 200 ppm. ToF-SIMS does not allow to distinguish between Ca and Na stearates although the ions are revealed. Linton *et al.* [119] have obtained information on additive surface migration and surface oxidation for LLDPE formulations. Kersting *et al.* [138] observed segregation for Irgafos 168 and Tinuvin 770, as opposed to Irganox 1076, in a static ToF-SIMS study of LDPE/Tinuvin 770 and LDPE/(Irganox 1076, Irgafos 168). Figure 4.12 shows yields of the complete molecular peak distribution of Irganox 1076 and Irgafos 168 with a leached surface as a starting point. The relative effectiveness of using Au^+ and Ga^+ ions for assessing Tinuvin 770 and Irganox 565 in LDPE was also examined. Migration of DIOP (m/z 931) in PVC was also reported [196]. Zhao *et al.* [197] studied migration of two anionic surfactants, sodium dodecyl sulfate and sodium dodecyl diphenyl ether sulfonate, in acrylic latex films both to the air–surface and the film–glass substrate interface. SSIMS and XPS data were combined.

ToF-SIMS is a powerful tool to control the surface quality of industrial polymer products, as in the polymer **coating industry** [198,199]. Strengths of ToF-SIMS for analysis of coatings are overall sensitivity (10x higher than XPS), molecular specificity, *in situ* analysis, mapping, analysis of mixtures. Each

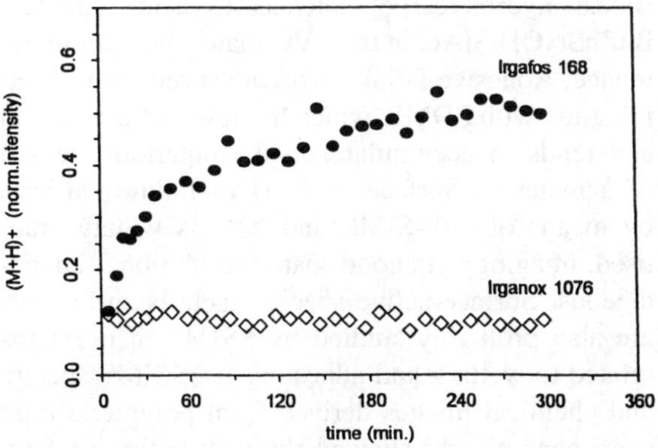

Fig. 4.12. Segregation of Irganox 1076 and Irgafos 168 to the LDPE surface as a function of time as observed by ToF-SIMS. Reproduced by permission of TASCON, Münster.

additive can be monitored through several molecular fragments. The matrix effects of the different layers affect the ion yields. Quantification requires a set of controls. The technique is extremely useful to characterise both high solids coatings, where high viscosities may limit the diffusion of materials to an interface, and waterborne coatings where there can be a large number of surface-active materials competing to be at the interface. Latex semi-gloss paints contain many components and undergo a complex series of physical changes during drying and film formation. Many of the components are surface active, and it is difficult to predict which materials will migrate and dominate the surface. Since (EVA) wax additives are commonly used to enhance surface properties, such as mar and abrasion resistance, it is of great interest to know how they are distributed. The approach in higher solids alkyd semi-gloss enamel is quite similar [200]. By removing the top layer by ablation the location of EVA can be determined as a function of depth. The finding that the wax is covered by a thin layer of resin and surfactants gives important information about the mode of action of the material, which may lead to the design of more effective mar agents. In an alkyd acrylic enamel coating containing an EVA wax also PDMS was traced. The wax additive was absent from the "as cured" paint surface, but became exposed and active when the surface had been abraded. This coating appeared to be fully top cured, but did not have complete through cure, since there are residual unsaturated acids (linolate at m/z 279 and oleate at m/z 281 Da). Examination of a peeled or delaminated surface can provide some insight into the interfacial chemistry between adjacent layers in a structure. ToF-SIMS has also been used to trace migration of additives in weathered multilayer automotive coatings, typically composed of a PU/(UVA, low-MW HALS, polymethylsiloxane) clear-coat, a PU/(UVA, high-MW HALS) basecoat, chlorinated PO primer, and rubber containing PP or PC/PBT blend substrates [201]. Similarly, Tinuvin 770 has been observed by ToF-SIMS in a cross-section of a two-layer acrylic-melamine paint system [162]. Results indicate migration of the HALS additive into the bulk of the adjacent paint layer. The ability to detect spatially the presence of additives in paint matrices by microscopic observation of the intact additive molecules is of great interest to paint research.

ToF-SIMS is typically used for **troubleshooting** purposes, such as identifying the process step

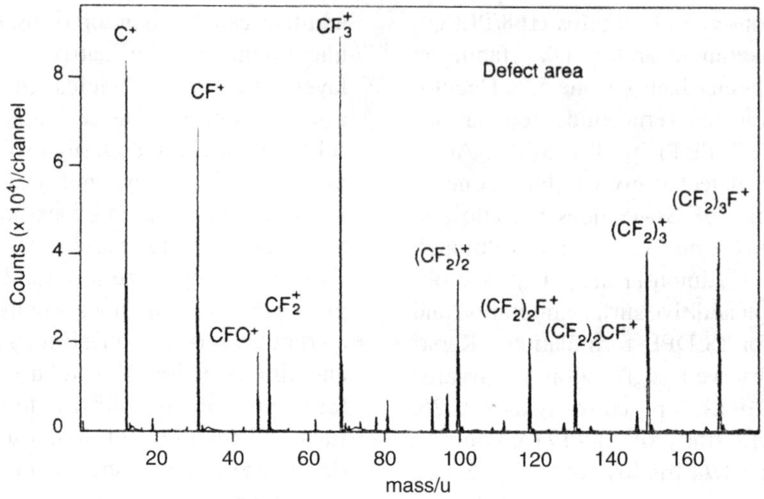

Fig. 4.13. Positive secondary ion spectrum of a defect in car paint. After Benninghoven *et al.* [112]. Reprinted with permission from A. Benninghoven *et al.*, *Analytical Chemistry* **65**, 630–40A (1993). Copyright (1993) American Chemical Society.

responsible for a defect in a final product, mostly caused by organic contaminants (silicon oils, fatty acids, perfluorinated polyethers, additives). Submonolayer quantities of lubricants and contaminants segregated to the surface usually cause heavy coating defects, which can easily be detected and identified by means of ToF-SIMS. SIMS allows the study of cratering (automotive application), which is a frequent problem in painted substrates and often caused by additives. In a typical case, fingerprint peaks in a car paint defect pointed to perfluorinated polyether (m/z 12, 31, 50, 69) (*cfr.* Fig. 4.13), which is used to lubricate assembly line components [112]. Fluorine-containing species are favoured SIMS objects because of ease of ionisation. In another case, a small paint defect of a painted PC automotive part revealed fragment ions of the polyaromatic polymer backbone of polycarbonate, Irgafos 168, and a fatty acid ester derivative of pentaerythritol, but no signals of the paint, evidence for incomplete coating [198]. A significant advantage of the method is the quasi-simultaneous detection of all chemical substances present in the uppermost monolayer of the polymer.

ToF-SIMS is a very powerful tool for identifying plasticisers and lubricants which contribute to **adhesive failure**, as in case of the interface of poly(vinylacetate-ethyl) copolymer/PVC laminations, particularly when the compounds are present at or below the detection limits of either ATR-FTIR or XPS [202]. In one case, ToF-SIMS showed fragment masses m/z 71 and 149 ($C_4H_7O^+$ and $C_8H_5O_3^+$, respectively), the latter being characteristic of dialkyl phthalate plasticiser, m/z 73, 147, 207, 221, 281 (indicative of PDMS) and m/z 268, 270, 284, 296 (from N-stearylerucamide). Negative-ion ToF-SIMS showed m/z 80, 96, 97, 265 (indicative of sodium dodecylsulfate), identifying surface species that can impact adhesive performance. ATR-FTIR was not able to spot these failures. The findings are consistent with the observation of the lubricant ethylene-bis-stearamide on sheets of PVC which demonstrated poor ink adhesion and printability [203]. ToF-SIMS and FTIR have also been used to examine the surface of other vinyl samples with poor printing properties [204]; formation of a mixed Ba/Zn hydroxycarboxylate/stearic acid complex, $BaZnSt_3OH/StAc$, in the PVC matrix was put in evidence. Adhesive failure is often caused by a contamination with PDMS, which has low surface tension and tends to accumulate on the outermost region of a material. Surface analysis of failure surfaces by means of ToF-SIMS and XPS is widely practised. Imaging with good spatial resolution is advantageous. Surface-active species such as surfactants are also profitably studied by SSIMS in problems related to wetting and adhesion. (−)SSIMS spectra and chemical images derived from polymeric fibre specimens clearly showed the distribution of fatty acids (up to C_{18}) and (+)SSIMS spectra that of fatty acid-poly(ethylene glycol)-based esters (PEG esters), $CH_3(CH_2)_nCOOCH_2CH_2O(CH_2CH_2O){\cdot}mH$ (up to $n = 16$) [196].

Dynamic SIMS has been used to measure polymer diffusion, *e.g.* interdiffusion in the PS/PPO system [205]. SIMS depth profiling enables to reveal silica present at some 10 nm below the surface. DSIMS has also been used to measure the film thickness of perfluoro-polyether lubricant coated on magnetic recording media by determining the etching time to the underlying carbon overcoat layer [206].

Foerch *et al.* [192] have observed high intensity, high mass fragments, ascribed to additive exudation, in positive **FAB-SSIMS spectra** of remote nitrogen plasma-treated LLDPE; (−)FAB-SSIMS spectra showed PO_2^- ad PO_3^- after treatment which may originate from phosphite stabilisers.

Large, intact molecules can be detected by SIMS analysis using etched Ag-substrates (Ag-SIMS) or matrix-enhanced SIMS (ME-SIMS). **ME-SIMS** [207] is a relatively new technique for the analysis of large molecules. As in MALDI-MS, in ME-SIMS the analyte molecules are prepared in a solid matrix, consisting of relatively small, organic molecules in high molar surplus. Molecular ions of organic mixtures can be detected up to masses of about 10 000 Da.

Other techniques are frequently used to confirm, complement and quantify ToF-SIMS analysis. The combination, therefore, of SSIMS with other surface spectroscopic techniques will certainly prove to be a powerful methodology. The application of SSIMS to the surface analysis of polymer materials has been reviewed [194]. Benninghoven *et al.* [112] have illustrated the application of SIMS to probing real-world samples.

4.2.2. Secondary Neutral Mass Spectrometry

Principles and Characteristics

The twin techniques of secondary neutral mass spectrometry (or sputtered neutral mass spectrometry, SNMS) and SIMS, which share bombardment of the sample surface with a focused primary ion beam (Ar^+, Cs^+, Ga^+, O_2^+) of sufficiently high ion energy (some keV), are among the most powerful surface analytical techniques for compositional characterisation of surfaces. As in SIMS, in SNMS the implanted primary ions penetrate into the solid surface to different depths (1–10 nm) and transfer their kinetic energy as a function of the sample material, primary ion energy and mass. Whereas SIMS detects the directly emitted secondary ions, in SNMS the secondary sputtered ions are suppressed by a repeller voltage and the large percentage of sputtered neutrals (>99%), whose composition ratio is almost the same as that of the sample surface, is post-ionised for mass detection. It is important to post-ionise neutrals efficiently in order to improve the detection limits (sensitivity). There do exist various other techniques that use post-ionisation sputtered particles in mass spectrometry (*cfr.* Table 4.8). Irrespective of the method effecting ionisation of the sputtered neutrals, they have in common that the emission process (*i.e.* sputtering) is *decoupled* from the ionisation step. Thus, changes in sample composition will not influence the yield of post-ionised particles in the same way as is frequently observed for the secondary ions utilised in SIMS. Furthermore, post-ionisation processes are generally far better understood than ionisation at the solid's surface during the sputtering event. The essential difference between SNMS and SIMS is therefore the separation of particle emission and ionisation in case of SNMS. Hence, the neutral-to-ion conversion factor for a specific element is independent of the chemical composition of the sample while it may influence strongly the electronic transition probability in secondary ion formation.

Post-ionisation schemes for the detection of sputtered neutral species in SNMS utilise either electron impact ionisation (e-beam SNMS), electron gas or plasma ionisation (plasma SNMS) or laser ionisation (L-SNMS). For e-beam SNMS, which is based on the use of a directed flux of essentially monoenergetic electrons towards the sputtered neutrals, high sensitivities have been obtained. Plasma or e-gas SNMS uses a low-pressure plasma (usually inert gas, *e.g.* Ar) containing ions to sputter the surface and at the same time the e-gas for ionisation of the neutrals. Although e-beam and plasma SNMS suffer from a low ionisation probability, they provide a well-established quantification scheme.

Post-ionisation using an intense pulse laser beam (usually standard pulsed excimer systems operating at 193 and 248 nm) has the advantage of a high ionisation probability. Multiphoton ionisation (MPI) processes are usually involved in L-SNMS. In SNMS two ways for laser-induced post-ionisation are being applied, which can saturate transitions from the ground or excited state of sputtered neutrals. The first method is the non-resonant approach applying single photon ionisation (SPI) or multiphoton ionisation (MPI), known as SALI® (surface analysis by laser-ionisation) [208]. The second

Table 4.11. Performance data of SSIMS and SNMS

Feature	SSIMS	SNMS
Excitation	Primary ions: Ar^+, Cs^+, O_2^+, O^-, Ga^+	Primary ions: Ar^+, Cs^+, O_2^+, Ga^+
Detection	Secondary ions	Post-ionisation of neutrals: plasma, e-beam, laser
Suitable materials	All (vacuum compatible)	All (vacuum compatible)
Sample preparation	"As received"	"As received"
Spectroscopic information	Molecular, atomic (all)	Atomic (all), (molecular)
Information depth	First monolayer	First monolayer
Detection limits	ppb (elements), fmol (molecules)	Sub-ppm
Quantification	Inherently non-quantitative	Quantitative
Matrix effect	Strong	Weak
RSC[a]	10^2–10^6	0.3–3
Calibration	10–20% RSD[b]	10–20% RSD[b]
Mass resolution	>10 000	>3000
Lateral resolution	<100 nm (LMIG)	<100 nm (atoms), μm (organics)
Depth resolution	<5 nm	<5 nm
Maximum depth	Some μm by sputtering	Some μm by sputtering
Screening for unknown elements/molecules	Yes/yes	Yes/No
Max. field of view	$100 \times 100\ \mu m^2$ (LMIG)	$700 \times 700\ \mu m^2$ (EI)
Destruction	High	High
Applications	Depth profiling, imaging, trace analysis	Depth profiling, imaging

[a] RSC, relative sensitivity coefficients.
[b] Using implantation standards.

method is the resonant approach (REMPI-SNMS). Of course, resonant schemes are extremely efficient and therefore require only moderate laser intensities. The ionisation probability for neutrals in the ionisation volume can be close to 100%. Although resonant laser post-ionisation SNMS is highly quantitative, a drawback is the loss of analytical flexibility such as the mandatory selection of a given element. When resonant transitions are involved, post-ionisation is highly element specific and, hence, a different laser set-up is generally required for every detected element or species. It is therefore ideally suited for imaging applications which rely on the efficient and laterally resolved detection of one specific atomic or molecular species at the investigated surface [209]. On the other hand, for the analysis of samples with unknown composition, non-resonant schemes must be employed to avoid element specificity of the ionisation process [208]. Non-resonant laser ionisation has a higher efficiency than electron and plasma ionisation. L-SNMS appears to be more suitable for inorganic than for organic molecules (laser damage).

In a typical arrangement used for laser post-ionisation SNMS, the plume of sputtered neutral particles is intersected by a pulsed laser, and the positively charged photo-ions generated in this way are detected, for example, by an energy-refocusing time-of-flight (ToF) mass spectrometer. Combined SIMS/SNMS instruments are available (*cfr.* Fig. 4.7).

Table 4.11 compares SIMS and SNMS (*cfr.* also Table 8.57 of ref. [110a]). Detection limits in the sub-ppm range are accessible under optimised analytical conditions. A lateral resolution of less than 100 nm and an in-depth resolution of a few nm can be achieved. One of the unique features of SNMS is the ease of analysis of insulators. This is at variance to SSMS, GD-MS and SIMS, which are handicapped by electrical charging effects. Laser SNMS is not strictly restricted to elemental analysis, but can also be applied to the characterisation of molecular surfaces. For an optimum yield of intact molecular ions and characteristic fragments it is necessary to optimise laser power density, wavelength, and pulse width [112].

In contrast to SIMS, in SNMS – with its decoupled evaporation and ionisation processes – matrix effects are significantly lower because the composition of sputtered and post-ionised neutrals corresponds more closely to that in the solid sample (compared to the sputtered secondary ions in SIMS). SNMS is useful in combination with SIMS as a quantitative tool. If it is assumed that most of the particles emitted from the sample surface are neutral, the calibration of SNMS is much simpler than SIMS. Absence of matrix effects and the more uniform ion yield across the periodic table in SNMS compensate for the higher sensitivity and dynamic range of SIMS. If quantification is not top priority, SIMS offers the highest sensitivity for spectra and dynamic range of profiles for a wide range. Combination of laser SNMS with an ion microprobe allows quantitative elemental mapping with high sensitivity. Laser SNMS imaging is not possible in the microscope mode because of the lateral dispersion of the sputtered neutrals.

In recent years, SNMS has evolved into an important tool for the characterisation of surfaces and thin films [210–217]. Mathieu *et al.* [217] have reviewed the different post-ionisation methods and the analytical use of SNMS in comparison with SIMS. A textbook is also available [5].

Applications

Industrial application of SNMS is still in its infancy. As opposed to the many SIMS applications for polymer/additive analysis (*cfr.* Chp. 4.2.1), there appear to be no records (yet) of the use of SNMS or SIMS/SNMS for this purpose despite the desirable features of the (combined) technique. However, scarcity and cost of the equipment (as well as safety aspects for L-SNMS) play a major role in application to routine problems. Also, the SIMS-XPS combination is obviously a serious proven competitor in many instances. Another drawback in many applications is of course the fact that the surface is analysed rather than the bulk. In SNMS most progress can be expected from a combination of laser post-ionisation and sputter depth profiling.

4.3. ION SCATTERING TECHNIQUES

Principles and Characteristics

Ion beam spectroscopy for polymer surface analysis comprises two general classes of experiments. One class uses a primary ion beam to generate secondary ions, which are then mass analysed. This technique, secondary ion mass spectrometry, has evolved into dynamic and static SIMS. Only the latter technique finds frequent application in polymer/additive analysis (*cfr.* Chp. 4.2.1). The second class of ion beam spectroscopy measures the energy loss of a primary ion scattered from a surface.

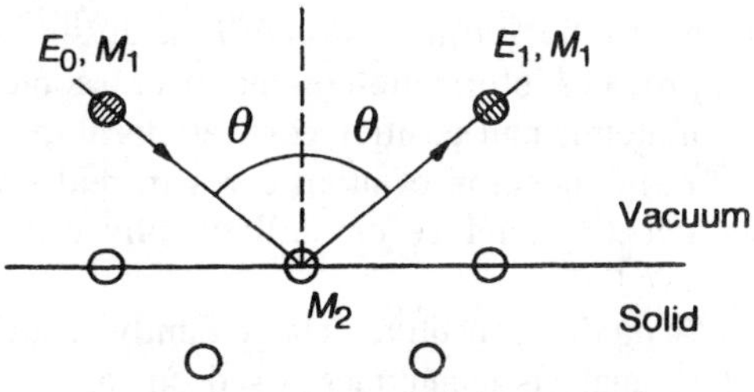

Fig. 4.14. The physical basis of ISS: an ion with energy E_0 and mass M_1 is elastically scattered by a surface atom of mass M_2. After Garbassi *et al.* [14]. Reprinted from F. Garbassi *et al.*, *Polymer Surfaces. From Physics to Technology*, Copyright © 1998, John Wiley & Sons, Ltd. Reproduced with permission.

The most important interactions between ions and atoms are as follows:

- Rutherford scattering in the Coulomb-field of the target nucleus.
- Elastic scattering caused by the nuclear forces between the two interacting nuclei (in addition to Coulomb interaction).
- Nuclear reactions with emission of γ-radiation and (or) charged particles.
- Inner shell ionisation of the atoms with subsequent emission of characteristic X-rays.

In ion scattering spectroscopy (ISS) analysis of the scattered ion energies allows **identification of surface atoms**. The interest in surface analysis by ISS is connected with the conceptional simplicity of the method, in which primary ions of a known species (usually He^+, Ne^+, Ar^+, *etc.*) and energy (keV to low MeV range) are focused on a surface; collisions of incident ions with surface atoms cause variations in their state of motion and energy, as measured under a fixed scattering angle (Fig. 4.14). The major scattering contribution in the spectra results from elastic scattering and thus the trajectories of the particles can be described by a sequence of single collisions (billiards ball game model). Hence, the elemental analysis of the surface layer and/or the determination of the geometrical position of surface atoms become straightforward and require no complicated deconvolution scheme.

ISS was first proposed for elemental identification in 1967 as a very surface-sensitive tool [218]. However, conventional noble gas ISS is matrix dependent and the exclusive first-layer specificity has been lost; multiple scattering complicates the data. ISS has thus grown from a simple "first layer" detection scheme for elemental composition to an element-specific *surface structure* analysis tool, which provides information on the macroscopic average of atomic pair position correlations at the surface, *i.e.* the position of atom cores in real space. **Mass selective surface crystallography** can thus be achieved.

Ion scattering constitutes a large family of surface structure analysis techniques based on ion scattering, which have in common the determination of nucleus positions. In ion scattering spectroscopies the primary **ion energy** utilised differentiates the techniques. Low-energy ion scattering (LEIS) uses primary ions in the keV range and Rutherford backscattering spectroscopy (RBS) employs ions in the MeV range; medium-energy ion scattering (MEIS) holds an intermediate position. It is important to notice the energy scales involved. While chemical binding effects are in the range of eVs and the "surface-sensitive" techniques are at most restricted to within a few orders of magnitude of this, the most energetic ion beam analyses are conducted in the range of a few MeV for the primary ions. This vast difference in the energy range has a number of important consequences. Methods relying on highly energetic ion beams are completely insensitive to chemical binding effects and thus offer an absolute technique, independent of chemical state effects. Consequently, XPS, AES and RBS are complementary methods. Because high energetic beams provide penetration of such solid materials to depths of several micrometres, such techniques are not strictly "surface analysis" methods so much as "near-surface analysis".

Due to the strong energy loss of the charged particles only thin layers in the range between a few nm and about 10 μm can be analysed. In contrast to activation by neutrons and photons bulk samples cannot be analysed with ion beams. However, the energy loss of the impinging ions and of the charged reaction products yields information about the distance between the target atom and the sample surface. Ion beam analysis (IBA) fills the gap between typical surface analysis methods like SIMS or EM and bulk analysis. Due to the strong deceleration of charged particles in solids, **depth profiling** with nm resolution is offered by some ion beam techniques. Depth profiles may be obtained from techniques in direct space (*e.g.* DSIMS, ISS) or by some kind of integral transform (*e.g.* neutron and X-ray reflectivity, GIXRF, *etc.*).

As the primary ion energies are so drastically different, the information content of LEIS and RBS is also very different. In RBS the primary ions are detected after scattering from depths up to several μm. By fitting the ion intensity *vs.* energy curve the atomic composition of the near-surface region is defined [219]. The greater penetration of the energetic ion beam, on the other hand, does not require removal of material to obtain a profile, as it depends only upon the relatively well-understood energy loss phenomena for ions in matter to determine the depth at which atoms of a particular species are located. In this sense, RBS is not destructive, although it may in some circumstances affect the material under study and may not always qualify as totally "non-destructive". As the energy is reduced from MeV in RBS to the low primary ion energies used in LEIS the depth resolution improves to the outermost atomic layer, the cross-sections for scattering increase and the neutralisation probability for ions scattering from below the first atomic layer increases (to nearly unity). All of these factors enhance the surface sensitivity. MEIS can probe composition and atomic structure of surface and subsurface layers non-destructively down to a few 10 nm depth with atomic layer depth resolution. LEIS has the ultimate surface sensitivity compared to other composition analysis techniques, such as AES or SIMS.

Ion beam analysis thus offers a set of fast and accurate techniques to determine concentration *vs.* depth profiles in inorganic materials and of heavy nuclei in polymer samples. The typical lateral extent of the beam is millimetres. **Features** of ion beam surface analyses, which make them unique and attractive, are the extreme surface sensitivities, low elemental detection limits, and surface mapping capabilities. Quantification is still rather arduous for noble gas ISS with ion detection only. For the general purpose of element analysis, other techniques such as AES, XPS, or SNMS seem to be superior. **Key problems** associated with the application of ion beam (SIMS, RBS, LEIS) spectroscopies on solid insulators are charging and radiation damage. This limits application of ISS for polymeric materials. Ion beam analysis of chemical composition as a function of depth can be applied to polymers if radiation damage can be minimised. The techniques require relatively smooth surface and lateral homogeneity. Ion beam analysis is not appropriate for some problems *e.g.* fibre interfaces. Competition of other techniques, in particular SSIMS, which offers a larger amount of information, also limits the applicability of LEIS. Both SIMS and LEIS are based on the interaction of low-energy ion beams with condensed phase surfaces. ISS provides lower resolution with respect to SIMS, because energies instead of masses are analysed.

Reviews on ion scattering spectroscopic techniques [220,221] are available; for textbooks, *cfr.* Bibliography.

Applications

The application of ISS to polymeric materials is not extensive, mainly because competitive techniques such as SIMS offer a larger amount of information. A combined ISS/DSIMS study of glass polymer interfaces has shown that ISS provides lower resolution with respect to SIMS, because energies instead of masses are analysed [222].

Ion beam analysis techniques form a suite of mature and well-understood techniques with potential for polymer surface and interface problems. Due to the size of the polymer chains, in addition to pure surface analysis techniques, polymer surface and interface science needs techniques that can provide depth profile information in the near-surface region (*i.e.* from 1–10 nm to 1 μm depth). This information can be provided by He ion beam techniques with energies of about 1–3 MeV. In order to avoid surface damage under the ion beam, low-damage conditions need to be developed.

The major areas of application of ion beam techniques to polymer surface and interface problems concern: *(i)* segregation at polymer surfaces and interfaces: *(ii)* polymer-polymer interfaces and diffusion; and *(iii)* transport of non-polymeric materials through thin polymer films. A prototypical case of the latter application is swelling of an amorphous polymer by a small molecule solvent. Kramer *et al.* [223,224] probed the diffusion front in swelling of PS, PMMA and other polymers by halogenated solvents. Adhesion problems, glass/polymer interfaces, and polymer surfaces have been studied.

Jones [225] has dealt with ion beam analysis of composition profiles near polymer surfaces. The application of ion beam analysis to polymer surfaces and interfaces has been reviewed [226].

4.3.1. Low-energy Ion Scattering

Principles and Characteristics

In low-energy ion scattering (LEIS), a mono-energetic ion beam, usually of $^{3}He^{+}$, $^{4}Ne^{+}$, $^{20}Ne^{+}$ or $^{40}Ar^{+}$, *etc.* is focused on the surface at low energies of the incident ion ($E_0 \approx$ 100–10000 eV). LEIS is conceptually and **theoretically simple**. LEIS is based on the measurement of energy losses of noble gas ions elastically scattered by target atoms in the solid surface. The ion scattering experiment involves analysis of that fraction of primary ions which scatter similar to billiard ball collisions. According to the laws of conservation of energy and momentum, the energy of the back-scattered ions is characteristic of the mass of the target atoms from which they are scattered. The energy spectrum obtained can thus directly be interpreted as a mass spectrum of the surface atoms. Unlike SIMS, in LEIS the ions originally in the incident beam and scattered from the surface are detected, and they are energy- rather than mass-analysed. The LEIS experiment can be carried out using a simple electrostatic analyser of the type more commonly applied to AES or XPS. An important aspect of LEIS is the extreme surface sensitivity. The information depth of LEIS is limited to one atomic layer because the low-energy noble gas ions have a high neutralisation probability. This results in a negligible scattered-ion yield from target atoms below the surface layer. From a point of view of quantitative analysis, LEIS faces a neutralisation problem. Quantitative composition determination is possible on the basis of elemental sensitivity factors provided that a calibration standard is used [227,228].

Table 4.12 lists the **main features** of LEIS. The technique excels in surface sensitivity. The principle disadvantage of LEIS is that it does not directly

Table 4.12. Main characteristics of ion scattering spectroscopy

Advantages:

- Sample material (solid, liquid, powder, insulating) only restricted as to vapour pressure
- Non-destructive measurements (for He ions dose $<10^{14}$ ions/cm^2)
- Depth resolution: outermost layer
- High sensitivity (ppm range for heavy elements)
- Lateral resolution: 100 μm
- Fast qualitative analysis
- Depth profiling by sputtering
- Quantitative analysis by comparing with reference samples (no matrix effect): $\pm$30% absolute, $\pm$10% relative

Disadvantages:

- Poor spatial resolution
- Difficult quantitation
- Subsurface analysis (depth profiles) extremely slow
- Small user base; expertise required
- Small commercial equipment base and support

give quantitative compositional information. Other limitations are the difficulty in resolving spectral peaks, which are broad and suffer from large overlap. LEIS cannot detect elements lighter than ^{3}He. When low-energy ion beams (500 eV to 1 keV) impinge on polymers, the surface builds up an electrostatic charge that rapidly impedes the acquisition of spectral information. Finally, as alluded to above, polymeric materials are quite radiation and heat sensitive. Static low-energy ion scattering using extremely low ion doses was described [229].

Applications
LEIS is not normally used solely as a means of determining surface compositions. The technique is mainly exploited in cases where it is desirable to be able to follow changes in the outermost surface layer of the sample as a function of time, or of some kind of process. An elegant use of the technique is as a highly sensitive monitor of compositional changes during ion beam depth profiling, with the same beam in use for sample erosion and LEIS measurements.

LEIS has grown to become a mature surface analysis technique. In contrast to other surface analysis techniques (XPS, AES, *etc.*) LEIS only observes the outermost atomic layer and thus is able to provide information that is very difficult to obtain otherwise. LEIS allows non-destructive and quantitative measurements on all kinds of material, including very sensitive polymer layers. However, LEIS finds most application for inorganic systems. The application of LEIS to polymeric surface analysis remains to a large extent underdeveloped. Various specific areas can be described:

(i) Detection of inorganic contamination on polymer surfaces [230].
(ii) Detection of the functional group in the side chain orientation of a polymer through the use of shadowing interactions.
(iii) Detection of surface chemical composition and reactivity due to the unique molecular orientation at the interface.
(iv) Ageing of polymer surfaces.
(v) Wettability studies.

A LEIS study of several polymer surfaces has been reported [230]. The technique is extremely sensitive to the presence of surface **impurities**. For instance, inorganic or organometallic compounds used as stabilisers in polymers tend to segregate at the surface. When examining PVC containing Sn compounds, a large amount of tin was detected at the

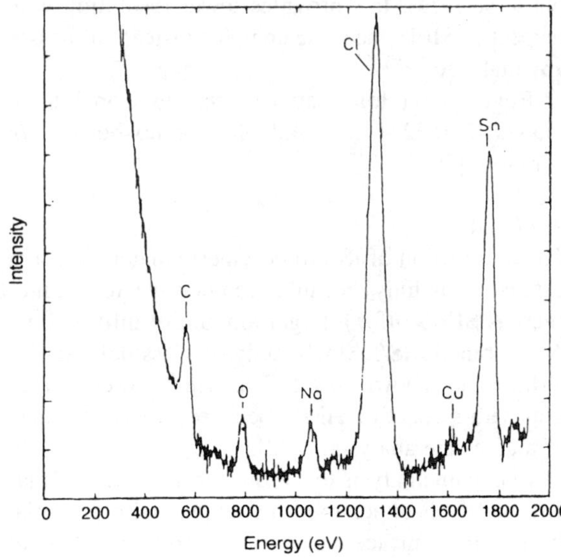

Fig. 4.15. LEIS spectrum (2 keV He$^+$) of PVC. After Thomas *et al.* [230]. Reprinted from G.E. Thomas *et al.*, *Applied Surface Science* **6**, 204–24 (1980), with permission from Elsevier.

surface [230], as shown in Fig. 4.15. Also Vargo *et al.* [231] have applied LEIS to polymer surfaces. LEIS is a very powerful tool for studying surface compositions of solid surfaces which are important in adhesion and segregation processes. When extremely low ion doses are being used, LEIS can provide very detailed information on the orientation of molecules at polymer surfaces [232].

LEIS data can be combined effectively with XPS data to provide a complementary analysis of polymer surface composition. In this approach determination of surface structure and composition of various technical polymers can be more clearly understood over a range of depths. It is to be expected that LEIS will remain only of marginal interest to polymer/additive analysis.

4.3.2. Rutherford Backscattering Spectroscopy

Principles and Characteristics
The basic and most mature method of materials analysis by **energetic ion beams** is elastic Rutherford backscattering spectroscopy (RBS). As it is practised today to interrogate a sample RBS uses typically a well collimated mono-energetic beam of α-particles from a Van der Graaf accelerator or from a variety of small accelerators, with energy $E_0 \approx 1$–5 MeV [233]. The ^{4}He^{2+} particles backscattered

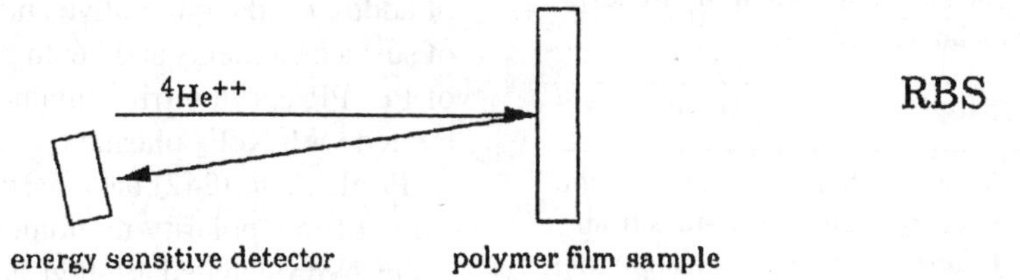

Fig. 4.16. Schematic of the geometry of Rutherford backscattering spectrometry (RBS). After Kramer [226]. Reprinted from E.J. Kramer, "*Ion-Beam Analysis of Polymer Surfaces and Interfaces*", *MRS Bulletin*, Vol. **21**, No. 1 (1996), pp. 37–42. Reproduced by permission of *MRS Bulletin*.

from heavier nuclei in the sample are detected in backward direction in a fixed solid angle with a semiconductor-diode detector that produces a current pulse proportional to the energy of the $^4He^{2+}$ particle (Fig. 4.16).

The basis of RBS is the observation that interactions between matter and charged ions of a few MeV energy fall into two distinct categories, namely collisions between ions and electrons (with continuous energy loss) and nuclear collisions. Because of the localised nature of the nucleus such collisions, though rare, lead to the elastic scattering of the ions through large angles (Rutherford scattering). The energy E of the backscattered particles encodes both the mass of the nucleus that caused the (elastic) event and its depth below the surface. The more massive the nucleus the larger the fraction E/E_0. Since the cross-section depends on the charge Z of the nucleus as Z^2, the RBS technique is much more sensitive to heavy elements than to light ones. In practice, RBS is most appropriate for elements $Z > 10$ and is of limited use for light elements in a heavy element matrix. RBS is not destructive. The energy loss of high-energy charged particles is proportional to their path length in a given material. This property is extensively used in **depth profiling** by RBS [234]. RBS is more penetrating than LEIS; the thickness of the observed layer is a function of the energy of the ion beam. Summarising, the atomic number dependent elastic scattering cross-section contains quantitative analytical information whereas the energy loss yields both elemental and depth information. Sensitivities are comparable to electron spectroscopy. At the high energies used the energy analysers are solid-state with a rather poor absolute resolution [233], leading to equally poor depth resolution. This is overcome by working at 50–500 keV energy. The resulting depth resolution (typically 20 nm) is much better than for electron-probe microanalysis (typically fractions of 1 μm).

Table 4.13. Main characteristics of Rutherford backscattering spectroscopy

Advantages:
- Well-understood theoretical foundation
- Quantitative elemental composition (down to atom-ppm level) and depth profile with nm resolution
- Reference method traceable to first principles

Disadvantages:
- Lack of sensitivity to light elements ($Z > 10$)
- Need for relatively smooth surface and lateral homogeneity
- Risk for localised radiation damage
- Measurement environment (vacuum)
- Low lateral resolution
- Limited surface specificity (information depth: 2 μm)
- Relatively complex spectrum processing
- Matrix effects
- High equipment cost (MeV ion accelerator)

Table 4.13 summarises the **main characteristics** of RBS. An advantage of RBS is that the theoretical foundation of the backscattering process is extremely straightforward. The primary mechanism is the elastic collision process. Although the principles of RBS are very simple, actual data analysis of the spectra expected for a given sample is complex [235]. A drawback of the requirement of accelerating an ion to the MeV range of energies is the need for access to expensive apparatus. Analyses require a moderately high (10^{-6} torr) vacuum. As the typical lateral extent of the beam is millimetres, the samples must be uniform laterally over at least this length scale. Radiation damage may be limited by cooling the samples to liquid-nitrogen temperature using a cold stage. A second obvious method is to

Table 4.14. Synopsis of some features of high-energy microbeam techniques

Feature	RBS	PIXE
Principles	Mass sensitivity to energetic ion beams	Characteristic X-rays from ion–atom collisions
Excitation energy	1–5 MeV	2–4 MeV
Element specificity	$Z > 10$	$Z > 14$
Limit of sensitivity	$>10^{10}$ atoms	$>10^{10}$ atoms
Surface specificity	2 μm	50 μm
Depth resolution	20 nm	1–10 μm[a]
Lateral resolution	0.5 mm	1 μm
Current density for high sensitivity[b]	$>10^{13}$	$>10^{13}$

[a] Depending on ion energy and Z of target.
[b] Particles/mm^2.

limit the ion dose on any area of the sample by frequently moving the ion beam spot.

Other techniques (AES, XPS, PIXE, GD-OES, SIMS, recoil spectrometry) yield similar information as RBS. Table 4.14 summarises some essential features of the high-energy microbeam analysis methods RBS and PIXE. RBS has the ability to analyse for the lightest elements, which PIXE cannot deal with.

Theory and instrumentation are similar to those of LEIS and have been reviewed by Gossett [235], Jones [225] and Kramer [226]. MeV ion beam profiling of polymer surfaces and interfaces has been discussed [236]. For further information on RBS the reader is referred to refs. [5,233,237]; *cfr.* also Bibliography.

Applications

Rutherford backscattering is ideal for investigations of **diffusion** near surfaces. An example is the RBS study of the diffusion of Cu(II) ions in polypropylene film oxidised between Cu foils [238]. Also the diffusion rates of copper carboxylate complexes in polymeric matrices have been studied [239] and that of iodoalkanes into polystyrene [240].

The sensitivity to heavy elements means that RBS is an excellent method for **depth profiling** of heavy elements in polymers that are usually made up of H, C, O and N. In typical polymers, the depth resolution of RBS is about 30 nm. Park *et al.* [241] have described the use of RBS and XPS for the analysis of additives, the qualitative and quantitative analysis of surface elements and depth profiling in the surface of PE, PP, acrylonitrile butadiene rubber, and paper treated with XeF_2 plasma.

Priola *et al.* [242] have used RBS and XPS in the study of low polarity monomers such as linear long chain hydrogenated n-alkyl acrylics, alkoxysilanes and fluorinated acrylates as modifying additives for highly polar acrylic UV-curable systems (Ebecryl 605 and Ebecryl 150) to show that the additives concentrate selectively at the surface. XPS yields information on the composition of the external layer of the film (about 100 Å) and RBS on the concentration profile of some elements (up to 0.3 μm in this case).

Extensive use of RBS for polymer/additive analysis cannot reasonably be expected.

BIBLIOGRAPHY

Surface Characterisation

K. Kurosaki and T. Miki (eds.), *Practical Polymer Surface Analysis*, Kohdansha Scientific, Tokyo (2001).

Y.-W. Chung, *Practical Guide to Surface Science and Spectroscopy*, Academic Press, San Diego, CA (2001).

A.J. Milling (ed.), *Surface Characterization Methods. Principles, Techniques and Applications*, M. Dekker, New York, NY (1999).

A.W. Czanderna, C.J. Powell and T.E. Madey, *Specimen Handling, Preparation, and Treatments in Surface Characterization*, Kluwer Academic Publishers, Dordrecht (1999).

F. Garbassi, M. Morra and E. Occhiello, *Polymer Surfaces. From Physics to Technology*, J. Wiley & Sons, Chichester (1998).

R.J.H. Clark and R.E. Hester, *Spectroscopy for Surface Science*, J. Wiley & Sons, Chichester (1998).

J.B. Hudson, *Surface Science. An Introduction*, J. Wiley & Sons, New York, NY (1998).

J.C. Rivière and S. Myhra (eds.), *Problem-Solving Methods for Surfaces and Interfaces*, M. Dekker, New York, NY (1998).

D. Brune, R. Hellborg, H.J. Whitlow and O. Hunderi (eds.), *Surface Characterization: A User's Sourcebook*, Wiley-VCH, Weinheim (1997).

W. Suëtaka, *Surface Infrared and Raman Spectroscopy*, Plenum Press, New York, NY (1995).

C.-M. Chan, *Polymer Surface Modification and Characterization*, Hanser Gardner, Cincinnati, OH (1993).

L. Sabbatini and P.G. Zambonin (eds.), *Surface Characterization of Advanced Polymers*, VCH, Weinheim (1993).

A.W. Adamson, *Physical Chemistry of Surfaces*, J. Wiley & Sons, Chichester (1990).

D.R. Randell and W. Neagle (eds.), *Surface Analysis Techniques and Applications*, The Royal Society of Chemistry, Cambridge (1990).

J.C. Rivière, *Surface Analytical Techniques*, Clarendon Press, Oxford (1990).

Electron Spectroscopy

D. Briggs and J.T. Grant (eds.), *Surface Analysis by Auger and X-Ray Photoelectron Spectroscopy*, IM Publications and SurfaceSpectra, Chichester (2003).

J.F. Watts and J. Wolstenholme, *An Introduction to Surface Analysis by Electron Spectroscopy*, J. Wiley & Sons, Chichester (2003).

S. Hüfner, *Photoelectron Spectroscopy*, Springer-Verlag, Berlin/Heidelberg (2003).

D. Briggs, *Surface Analysis of Polymers by XPS and Static SIMS*, Cambridge University Press, Cambridge (1998).

I. Powis, T. Baer and C.Y. Ng (eds.), *High Resolution Laser Photoionization and Photoelectron Studies*, J. Wiley & Sons, Chichester (1995).

T.L. Barr, *Modern ESCA. The Principles and Practice of X-Ray Photoelectron Spectroscopy*, CRC Press, Boca Raton, FL (1994).

C.G. Smith, *Surface Analysis by Electron Spectroscopy. Measurement and Interpretation*, Plenum Press, New York, NY (1994).

D. Briggs and M.P. Seah (eds.), *Practical Surface Analysis, Vol. 1. Auger and X-ray Photoelectron Spectroscopy*, J. Wiley & Sons, Chichester (1990).

M. Thompson, M.D. Baker, A. Christie and J.F. Tyson, *Auger Electron Spectroscopy*, J. Wiley & Sons, New York, NY (1985).

Surface Mass Spectrometry

J.C. Vickerman and D. Briggs (eds.), *ToF-SIMS: Surface Analysis by Mass Spectrometry*, IM Publications and SurfaceSpectra, Chichester (2001).

D. Briggs and M.P. Seah (eds.), *Practical Surface Analysis, Vol. 2. Secondary Ion and Secondary Neutral Mass Spectrometry*, J. Wiley & Sons, Chichester (1992).

D. Briggs, A. Brown and J.C. Vickerman, *Handbook of Static Secondary Ion Mass Spectrometry*, J. Wiley & Sons, Chichester (1989).

R.G. Wilson, F.A. Stevie and C.W. Magee, *Secondary Ion Mass Spectrometry*, J. Wiley & Sons, New York, NY (1989).

A. Benninghoven, F.G. Rüdenauer and W.H. Werner, *Secondary Ion Mass Spectrometry: Basic Concepts, Instrumental Aspects, Applications and Trends*, J. Wiley & Sons, New York, NY (1987).

Ion Scattering Techniques

H. Gnaser, *Low-Energy Ion Irradiation of Solid Surfaces*, Springer-Verlag, Berlin (1999).

J.R. Tesmer, M. Nastasi, J.-C. Barbour, C.J. Maggiore and J.W. Mayer (eds.), *Handbook of Modern Ion Beam Materials Analysis*, Materials Research Society, Pittsburgh, PA (1995).

A.W. Czanderna and D.M. Hercules (eds.), *Ion Spectroscopies for Surface Analysis*, Plenum Press, New York, NY (1991).

J.R. Tesmer, C.J. Maggiore, M. Nastasi, J.-C. Barbour and J.W. Mayer, *High Energy and Heavy Ion Beams in Materials Analysis*, Materials Research Society, Pittsburg, PA (1990).

J.R. Bird and J.S. Williams (eds.), *Ion Beams for Materials Analysis*, Academic Press, Sydney (1989).

G. Götz and K. Gärtner (eds.), *High Energy Ion Beam Analysis of Solids*, Akademie-Verlag, Berlin (1988).

F. Watt and G.W. Grime (eds.), *Principles and Applications of High-Energy Ion Microbeams*, Hilger, Bristol (1987).

W.-K. Chu, J.W. Mayer and M.-A. Nicolet, *Backscattering Spectrometry*, Academic Press, New York, NY (1978).

REFERENCES

[1] D. Munteanu, in *Developments in Polymer Stabilisation-8*, (G. Scott, ed.), Elsevier Applied Science, London (1987), pp. 179–208.

[2] M.X. Ramírez, D.E. Hirt and N. Miranda, *Proceedings SPE ANTEC 2001*, Dallas, TX (2001), pp. 2654–7.

[3] B.D. Bauman, in *Plastics Additives. An A-Z Reference* (G. Pritchard, ed.), Chapman & Hall, London (1998), pp. 584–9.

[4] D. Briggs and M.P. Seah (eds.), *Practical Surface Analysis, Vol. I, Auger and X-Ray Photoelectron Spectroscopy*, J. Wiley & Sons, Chichester (1990).

[5] D. Briggs and M.P. Seah (eds.), *Practical Surface Analysis, Vol. II, Ion and Neutral Spectroscopy*, J. Wiley & Sons, Chichester (1992).

[6] S. Lee and C.S.P. Sung, *Polymer Preprints* **40** (2), 649–50 (1999).

[7] S. Siewierska and W. Kozlowski, *Polimery* **18**, 221 (1973).

[8] J.A. Gardella and G.L. Grobe, *Anal. Chem.* **56**, 1169 (1984).

[9] M.D. Foster, *Proceedings SPE ANTEC 2000*, Orlando, FL (2000), pp. 1998–2002.

[10] S. Hofmann, in *Practical Surface Analysis* (D. Briggs and M.P. Seah, eds.), J. Wiley & Sons, Chichester (1990), Vol. 1, pp. 143–99.

[11] A.J. Swift, in *Developments in Polymer Analysis and Characterisation* (Rapra Technology, Ltd., ed.), Rapra Technology Ltd., Shawbury (1999), Paper 3.

[12] D.M. Bielinski, M. Zaborski and L. Slusarski, *Arch. Nauki Mater.* **17** (4), 307–28 (1996).

[13] H. Takeguchi and S. Takayama, *Nippon Bunseki Kagakkai* (4), 309–14 (1997).

[14] F. Garbassi, M. Morra and E. Occhiello, *Polymer Surfaces. From Physics to Technology*, J. Wiley & Sons, Chichester (1998).

[15] D. Briggs, *Surface Analysis of Polymers by XPS and Static SIMS*, Cambridge University Press, Cambridge (1998).

[16] L. Sabbatini and P.G. Zambonin (eds.), *Surface Characterization of Advanced Polymers*, VCH, Weinheim (1993).

[17] *Proceedings International Millenium Conference on Quantitative Surface Analysis (QSA-11), Surf. Interface Anal.* **31** (4), 245–342 (2001).

[18] C. Powell (ed.), *Workshop on Quantitative Surface Analysis*, NBS, Gaithersburg, MD (1986).

[19] J. Cazaux, *J. Microsc.* **145**, 257 (1987).

[20] M.P. Seah, *J. Electr. Spectrosc. Relat. Phenom.* **71**, 191–204 (1995).

[21] J.C. Rivière, *Surface Analytical Techniques*, Clarendon Press, Oxford (1990).

[22] L. Szepes and G. Tarczay, in *Encyclopedia of Spectroscopy and Spectrometry* (J.C. Lindon, ed.), Academic Press, San Siego, CA (2000), pp. 1822–30.

[23] J.F. Moulder, W.F. Stickle, P.E. Sobol and K.D. Bomben, *Handbook of X-Ray Photoelectron Spectroscopy*, Perkin-Elmer (Physical Electronics), Eden Prairie, MN (1992).

[24] M. Polak, in *Determination of Trace Elements* (Z.B. Alfassi, ed.), VCH, Weinheim (1994), pp. 359–92.

[25] M. Czuha and W.M. Riggs, *Anal. Chem.* **47**, 1836–8 (1975).

[26] M.B. Calvalho and D.M. Hercules, *Anal. Chem.* **50**, 2030–4 (1978).

[27] P.V. Auger, *Compt. Rendus* **177**, 169 (1923).

[28] J.J. Lander, *Phys. Rev.* **91**, 1382 (1953).

[29] D. Briggs and J.C. Rivière, in *Practical Surface Analysis* (D. Briggs and M.P. Seah, eds.), J. Wiley & Sons, Chichester (1990), Vol. I, pp. 85–141.

[30] J.C. Rivière, in *Practical Surface Analysis*, (D. Briggs and M.P. Seah, eds.), J. Wiley & Sons, Chichester (1990), Vol. I, pp. 19–83.

[31] C.-O.A. Olsson, S.E.Hörnström and S. Hogmark, in *Surface Characterization. A User's Sourcebook* (D. Brune, R. Hellborg, H.J. Whitlow and O. Hunderi, eds.), Wiley-VCH, Weinheim (1997), Chp. 17.

[32] P.J. Cumpson, S.J. Spencer and M.P. Seah, *Spectrosc. Europe* **10** (3), 8–14 (1998).

[33] M.P. Seah, C.P. Hunt and M. Tosa, *J. Electr. Spectrosc. Relat. Phenom.* **61**, 149 (1993).

[34] M.P. Seah, C.P. Hunt, D. Sykes, S. Valeri, R. Muller and B. Lamb, *J. Electr. Spectrosc. Relat. Phenom.* **61**, 173 (1993).

[35] M.P. Seah, in *Practical Surface Analysis* (D. Briggs and M.P. Seah, eds.), J. Wiley & Sons, Chichester (1983), Chp. 5.

[36] S.J. Mroczkowski, *J. Vac. Sci. Techn.* **A7**, 1529 (1989).

[37] M.P. Seah, *Surface Interf. Anal.* **9**, 85 (1986).

[38] P.W. Palmberg, *Anal. Chem.* **45**, 549–56A (1973).

[39] G.C. Smith, *Surface Analysis by Electron Spectroscopy. Measurement and Interpretation*, Plenum Press, New York, NY (1994).

[40] M.P. Seah, in *Practical Surface Analysis* (D. Briggs and M.P. Seah, eds.), J. Wiley & Sons, Chichester (1990), Vol. I, pp. 201–55.

[41] M. Procop, in *Werkstoffanalytische Verfahren* (H.-J. Hunger, ed.), Dtsch. Verlag f. Grundstoffindustrie, Leipzig (1995), pp. 382–408.

[42] Y.-W. Chung, *Practical Guide to Surface Science and Spectroscopy*, Academic Press, San Diego, CA (2001).

[43] M. Thompson, M.D. Baker, A. Christie and J.R. Tyson, *Auger Electron Spectroscopy*, J. Wiley & Sons, New York, NY (1985).

[44] E.H.S. Burhop, *The Auger Effect and Other Radiationless Transitions*, Cambridge University Press, London (1952).

[45] L.E. Davis, N.C. MacDonald, P.W. Palmberg, G.E. Raich and R.E. Weber, *Handbook of Auger Electron Spectroscopy*, Physical Electronics Industries, Eden Prairie, MN (1976).

[46] S.-S. Lin, *Rubber Chem. Technol.* **58**, 885–98 (1985).

[47] C. Cazeneuve, J.E. Castle and J.F. Watts, *J. Mater. Sci.* **25**, 1902–8 (1990).

[48] J.F. Watts, *Spectrosc. Europe* **6** (4), 10–15 (1994).

[49] H. Hertz, *Wiedemannsche Ann. Phys.* **31**, 982 (1887).

[50] U. Gelius, B. Wannberg, P. Baltzer, H. Fellner-Feldeg, G. Carlsson, C.-G. Johansson, J. Larsson, P. Munger and G. Vegerfors, *J. Electr. Spectrosc. Relat. Phenom.* **52**, 747 (1990).

[51] P. Coxon, J. Krizek, M. Humpherson and I.R.M. Wardell, *J. Electr. Spectrosc. Relat. Phenom.* **52**, 821 (1990).

[52] T.L. Barr, *Modern ESCA. The Principles and Practice of X-Ray Photoelectron Spectroscopy*, CRC Press, Boca Raton, FL (1994).

[53] I. Olefjord, in *Surface Characterization. A User's Sourcebook* (D. Brune, R. Hellborg, H.J. Whitlow and O. Hunderi, eds.), Wiley-VCH, Weinheim (1997), Chp. 18.

[54] C.D. Wagner, in *Quantitative Surface Analysis of Materials* (N.S. McIntyre, ed.), ASTM STP **643**, American Society for Testing and Materials, Philadelphia, PA (1978), Chp. 2.

[55] D. Briggs and M.P. Seah, *Practical Surface Analysis by Auger and Photoelectron Spectroscopy*, Wiley-Interscience, New York, NY (1983), Chp. 9.

[56] R.J. Ward and B.J. Wood, *Surf. Interface Anal.* **18**, 679 (1992).

[57] C.J. Powell, in *Quantitative Surface Analysis of Materials* (N.S. McIntyre, ed.), ASTM STP **643**, American Society for Testing and Materials, Philadelphia, PA (1978), Chp. 1.

[58] C.D. Bain, E.B. Troughton, Y.-T. Tao, J. Evall, G.M. Whitesides and R.G. Nuzzo, *J. Am. Chem. Soc.* **111**, 321 (1989).

[59] G. Beamson and D. Briggs, *High Resolution XPS of Organic Polymers*, J. Wiley & Sons, New York, NY (1992).

[60] R.T. Lewis and M.A. Kelly, *J. Electr. Spectrosc. Relat. Phenom.* **20**, 105 (1980).

[61] D. Briggs, in *Practical Surface Analysis* (D. Briggs and M.P. Seah, eds.), J. Wiley & Sons, Chichester (1990), Vol. I, pp. 437–83.

[62] B.V. Crist, *Handbook of Monochromatic XPS Spectra. Polymers & Polymers Damaged by X-Rays*, Wiley-Interscience, New York, NY (2000).

[63] G. Beamson and D. Briggs, *The High Resolution XPS of Polymers Database on CD-ROM*, Surface-Spectra/IM Publications, Chichester (2000).

[64] J.B. Hudson, *Surface Science. An Introduction*, J. Wiley & Sons, New York, NY (1998).

[65] A.J. Milling (ed.), *Surface Characterization Methods*, M. Dekker, New York, NY (1999).

[66] J.G. Jenkin, R.C.G. Leckey and J. Liesegang, *J. Electr. Spectrosc. Relat. Phenom.* **12**, 1 (1977).

[67] J.G. Jenkin, J.D. Riley, J. Liesegang and R.C.G. Leckey, *J. Electr. Spectrosc. Relat. Phenom.* **14**, 477 (1978).

[68] P.J. Cumpson, *J. Electr. Spectrosc. Relat. Phenom.* **73**, 25 (1995).

[69] Y.W. Tang, J.P. Santerre, R.S. Labow and D.G. Taylor, *J. Biomed. Mater. Sci.* **35** (3), 371–81 (1997).

[70] B. Ameduri, R. Bongiovanni, G. Malucelli, A. Pollicino and A. Priola, *J. Polym. Sci. A, Polym. Chem.* **37**, 77–87 (1999).

[71] S. Chen and H. Tanaka, *J. Wood Sci.* **44** (4), 303–9 (1998).

[72] N. Savargaonkar and P. Patel, *Proceedings Polyolefins 2000*, Houston, TX (2000), pp. 553–562.

[73] J.B. Williams, K.S. Geick, J.A. Falter and L.K. Hall, *Proceedings SPE ANTEC '95*, Boston, MA (1995), pp. 3610–14.

[74] J. Falter, K. Geick, L. Hall and J. Williams, *Proceedings SPE ANTEC '94*, San Francisco, CA (1994), p. 2804.

[75] A.H. Sharma and B.C. Beard, *J. Vinyl Addit. Technol.* **3** (4), 309–13 (1997).

[76] M.A. Karplus and D. Pocker, *J. Vac. Sci. Technol.* **A18** (4), 2033–8 (2000).

[77] T. Miki, *Anal. Sci.* **17S**, i143–6 (2001).

[78] D.A. Markgraf, *Proceedings Sixth European Polymers, Laminations & Coatings Symposium*, Copenhagen (1999), pp. 231–76.

[79] P.K. Pak, K.K. Na, H.W. Kim, S.S. Kim, G.Y Park and C.J. Park, *J. Korean Fiber Soc.* **36** (11), 847–53 (1999).

[80] S.-J. Park, J.-S. Jin and J.E. Choi, *J. Korean Fiber Soc.* **38** (5), 258–63 (2001).

[81] Y. Shichi and M. Arita, *Bunseki Kagaku* **38** (10), 463–7 (1989).

[82] H.-J. Jacobasch, K. Grundke, St. Schneider and F. Simon, *Prog. Org. Coat.* **26** (2–4), 131–43 (1995).

[83] J.W. Burley and B.C. Beard, *J. Vinyl Addit. Technol.* **5** (2), 87–93 (1999).

[84] S. Dieckhoff, in *Proceedings 3rd Würzburger Tage der instrumentellen Analytik in der Polymertechnik*, SKZ, Würzburg (1997), Paper D.

[85] A. Scandurra, R. Zafarana, Y. Tenya and S. Pignataro, *J. Adhesion Sci. Technol.* **15** (9), 1039–53 (2001).

[86] A.P. Pijpers and R.J. Meier, *J. Electr. Spectrosc. Relat. Phenom.* **121** (1–3), 299–313 (2001).

[87] J.L. Thomason and D.W.Wright, *J. Adhesion Sci. Technol.* **15** (5), 745–64 (2000).

[88] J.M. Lannon and Q. Meng, *Surf. Sci. Spectra* **6**, 87–136 (1999).

[89] P. Gao, K.B. Su, Y. Ward and L.T. Weng, *Polym. Compos.* **21** (2), 312–21 (2000).

[90] J.E.Castle and J.F.Watts, in *Interfaces in Polymer, Ceramic and Metal Matrix Composites* (H. Ishida, ed.), Elsevier, New York, NY (1988), pp. 57–72.
[91] C.L. Weitzsacker and L.T.Drzal, *Proceedings SPE ANTEC '96*, Indianapolis, IN (1996), pp. 3656–60.
[92] J. Liesegang, J. Pura and C. Panagopoulos, *Springer Proc. Phys.* **73**, 193–206 (1993).
[93] J.M. Farley and P. Meka, *J. Appl. Polym. Sci.* **51** (1), 121–31 (1994).
[94] J. Dillard, T. Ward and J. Morgan, *Proceedings Hot Melt Symposium*, Atlanta, GA (1998), pp. 21–5.
[95] S.S. Woods and A.V. Pocius, *Proceedings SPE ANTEC 2000*, Orlando, FL (2000), pp. 191–208.
[96] C.-M. Chan and S.H. Zhu, *Polym. Adv. Technol.* **8** (4), 257–60 (1997).
[97] K.B. Walters and D.E. Hirt, *Proceedings SPE ANTEC 2001*, Dallas, TX (2001), pp. 2644–8.
[98] J.P. Lens, J.G.A. Terlingen, G.H.M. Engbers and J. Feyen, *Langmuir* **14** (12), 3214–22 (1998).
[99] J. Wang, *Makromol. Chem., Macromol. Symp.* **74**, 101–10 (1993).
[100] Gy. Marosi, Gy. Bertalan, A. Balogh, A. Tohl, P. Anna and K. Szentirmay, in *Flame Retardants '96*, Interscience Communications Ltd., London (1996), pp. 115–24.
[101] J. Wang, P. Wei and J.-H. Hao, *Polym. Mater. Sci. Eng.* **83**, 61 (2000).
[102] J. Wang and B. Li, *Polym. Degr. Stabil.* **63**, 279–85 (1999).
[103] G.C. Smith and J.C. Bell, *Appl. Surf. Sci.* **144–145**, 222–7 (1999).
[104] D.M. Lenz, C.A. Ferreira and M. Delamar, *Synth. Metals* **126** (2–3), 179–82 (2002).
[105] G.A. George, H.A. Willis and R.J. Ward, *Chem. Aust.* **59** (2), 56–9 (1992).
[106] A. Benninghoven and W. Sichtermann, *Anal. Chem.* **50**, 1180 (1978).
[107] M. Barber, R.S. Bordoli, R.D. Sedgwick and A.N. Tyler, *J. Chem. Soc. Chem. Commun.*, 325 (1981).
[108] A. Benninghoven, *Proceedings SIMS IV* (A. Benninghoven, J. Okano, R. Shimizu and H.W. Werner, eds.), Springer-Verlag, Berlin (1984), pp. 342–56.
[109] M.K. Miller, *Mater. Character.* **44**, 11 (2000).
[110] J.S. Becker and H.-J. Dietze, *Intl. J. Mass Spectrom.* **197**, 1–35 (2000).
[110a] J.C.J. Bart, *Additives in Polymers. Industrial Analysis and Applications*, J. Wiley & Sons, Chichester (2005).
[111] R.S. Lehrle, J.C. Robb and D.W. Thomas, *J. Sci. Instrum.* **39**, 458–63 (1962).
[112] A. Benninghoven, B. Hagenhoff and E. Niehuis, *Anal. Chem.* **65**, 630–40A (1993).
[113] M.G. Dowsett and E.A. Clark, in *Practical Surface Analysis* (D. Briggs and M.P. Seah, eds.), J. Wiley & Sons, Chichester (1992), Vol. II, pp. 229–301.
[114] A. Benninghoven, *Phys. Status Solidi* **34**, K169 (1969).
[115] J.A. Gardella and D.M. Hercules, *Anal. Chem.* **52**, 226 (1980).
[116] D. Briggs, in *Practical Surface Analysis* (D. Briggs and M.P. Seah, eds.), J. Wiley & Sons, Chichester (1992), Vol. II, pp. 367–423.
[117] J.C. Vickerman, *Analyst* **119**, 513 (1994).
[118] A. Benninghoven, *Surf. Sci.* **299/300**, 246 (1994).
[119] R.W. Linton, M.P. Mawn, A.M. Belu, J.M. DeSimone, M.O. Hunt Jr., Y.Z. Menceloglu, H.G. Cramer and A. Benninghoven, *Surf. Interface Anal.* **20** (12), 991–9 (1993).
[120] G.J. Leggett and J.C. Vickerman, *Appl. Surf. Sci.* **55**, 105 (1992).
[121] G.J. Leggett and J.C. Vickerman, *Int. J. Mass Spectrom. Ion Proc.* **122**, 281 (1992).
[122] S.J. Pachuta and R.J. Cooks, *Chem. Rev.* **87**, 647 (1987).
[123] A. Benninghoven, F.G. Rüdenauer and H.W. Werner, *Secondary Ion Mass Spectrometry: Basic Concepts, Instrumental Aspects, Applications and Trends*, J. Wiley & Sons, Chichester (1987).
[124] G.J. Leggett and J.C. Vickerman, *Proceedings SIMS VIII* (A. Benninghoven, K.T.F. Janssen, J. Tümpner and H.W. Werner, eds.), J. Wiley & Sons, Chichester (1992), pp. 17–20.
[125] J.C. Vickerman, in *Spectroscopy for Surface Science* (R.J.H. Clark and R.E. Hester, eds.), J. Wiley & Sons, Chichester (1998), pp. 71–119.
[126] J. Lub, P.N.T. van Velzen, D. van Leijen, B. Hagenhoff and A. Benninghoven, *Surf. Interface Anal.* **12**, 53–7 (1988).
[127] R. Jede, O. Ganschow and U. Kaiser, in *Practical Surface Analysis, Vol. 2: Ion and Neutral Spectroscopy* (D. Briggs and M.P. Seah, eds.), J. Wiley & Sons, Chichester (1992), pp. 19–104.
[128] B. Hagenhoff and D. Rading, in *Handbook of Surface and Interface Analysis* (J.C. Rivière and S. Myhra, eds.), M. Dekker, New York, NY (1998).
[129] D. Briggs, *Surf. Interface Anal.* **4**, 151 (1982).
[130] E. Niehuis, *Proceedings SIMS VII* (A. Benninghoven, C.A. Evans, K.D. McKeegan, H.A. Storms and H.W. Werner, eds.), J. Wiley & Sons, Chichester (1990), pp. 299–304.
[131] P.M. Lindley, J.A. Chakel and R.W. Odom, *Proceedings SIMS VIII* (A. Benninghoven, K.T.F. Janssen, J. Tümpner and H.W. Werner, eds.), J. Wiley & Sons, Chichester (1992), pp. 219–22.
[132] S. Seki, Y. Ikebe, H. Iwamoto and H. Tamura, *Proceedings SIMS VII* (A. Benninghoven, C.A. Evans, K.D. McKeegan, H.A. Storms and H.W.

Werner eds.), J. Wiley & Sons, Chichester (1990), pp. 777–80.

[133] D. Johnson, A.J. Paul, P. Humphrey, N.M. Reed and J.C. Vickerman, *Proceedings SIMS VII* (A. Benninghoven, C.A. Evans, K.D. McKeegan, H.A. Storms and H.W. Werner, eds.), J. Wiley & Sons, Chichester (1990), pp. 911–14.

[134] D. Briggs, A. Brown and J.C. Vickerman, *Handbook of Static Secondary Ion Mass Spectrometry (SIMS)*, J. Wiley & Sons, Chichester (1989).

[135] J.C. Vickerman, D. Briggs and A. Henderson (eds.), *The Static SIMS Library*, SurfaceSpectra Ltd., Manchester (1997).

[136] J.G. Newman, B.A. Carlson, R.S. Michael, J.F. Moulder and T.A. Hohlt, *Static SIMS Handbook of Polymer Analysis*, Perkin Elmer, Eden Pairie, MN (1991).

[137] B.-C. Schwede, T. Heller, D. Rading, E. Niehuis, L. Wiedmann and A. Benninghoven, *The Münster High Mass Resolution Static SIMS Library*, ION-TOF, Münster (1999).

[138] R. Kersting, A.P. Pijpers, B. Hagenhoff, R. Verlaek, D. Stapel, A. Benninghoven and B.-C. Schwede, in *Proceedings SIMS XII* (A. Benninghoven, P. Bertrand, H.-N. Migeon and H.W. Werner, eds.), Elsevier Science Publ., Amsterdam (2000), pp. 825–8.

[138a] R. Kersting, *Ph.D. Thesis*, University of Münster (2002).

[139] A. Chilkoti, A.E. Schmierer, V.H. Perez-Luna and B.D. Ratner, *Anal. Chem.* **67** (17), 2883–91 (1995).

[140] P.J. McKeown, S.G. MacKay and M. Strobel, *Proceedings SIMS IX* (A. Benninghoven, Y. Nihei, R. Shimizu and H.W. Werner, eds.), J. Wiley & Sons (1994), Chichester, pp. 764–7.

[141] A. Belu, V.H. Perez-Luna, B.D. Ratner and J. Heller, *Polymer Prepr.* **37**, 841–2 (1996).

[142] M. Nicholas, R.W. Linton, R.M. Friedman, D. Rading and A. Benninghoven, *Proceedings SIMS X* (A. Benninghoven, B. Hagenhoff and H.W. Werner, eds.), J. Wiley & Sons, Chichester (1996), pp. 317–20.

[143] A. Wucher, R. Heinrich and C. Staudt, *Abstracts SIMS XII*, Brussels (1999), p. 333.

[144] P. Williams, in *Practical Surface Analysis* (D. Briggs and M.P. Seah, eds.), J. Wiley & Sons, Chichester (1992), Vol. II, pp. 177–228.

[145] B. Hagenhoff, *Proceedings SIMS X* (A. Benninghoven, B. Hagenhoff and H.W. Werner, eds.), J. Wiley & Sons, Chichester (1996), pp. 81–8.

[146] B. Hagenhoff, R. Kock, M. Deimel and A. Benninghoven, *Proceedings SIMS VIII* (A. Benninghoven, K.T.F. Janssen, J. Tümpner and H.W. Werner, eds.), J. Wiley & Sons, Chichester (1992), pp. 831–4.

[147] A.P. Galuska, *Abstracts SIMS XII*, Brussels (1999), p. 147.

[148] C. Kenens, T. Conard, L. Hellemans and W. Vandervorst, *Abstracts SIMS XII*, Brussels (1999), p. 162.

[149] D.M. Hercules, A.I. Gusev, A. Proctor and D.C. Muddiman, *Proceedings SIMS X* (A. Benninghoven, B. Hagenhoff and H.W. Werner, eds.), J. Wiley & Sons, Chichester (1996), pp. 637–40.

[150] K. Meyer, B. Hagenhoff, M. Deimel, A. Benninghoven and H.-J. Bauch, *Org. Mass Spectrom.* **27**, 1148 (1992).

[151] K. Reihs, *Proceedings SIMS X* (A. Benninghoven, B. Hagenhoff and H.W. Werner, eds.), J. Wiley & Sons, Chichester (1996), pp. 115–22.

[152] A. Benninghoven, *Quantification of Additives at Polymer Surfaces by ToF-SIMS*, EC Project G6RD-CT1999–0518.

[153] W.H. Gries, *Proceedings SIMS IX* (A. Benninghoven, Y. Nihei, R. Shimizu and H.W. Werner, eds.), J. Wiley & Sons, Chichester (1994), pp. 971–4.

[154] Y. Homma, *Abstracts SIMS XII*, Brussels (1999), p. 311.

[155] C.J. Powell and R. Shimizu, *Surf. Interface Anal.* **20**, 322 (1993).

[156] D.E. Newbury, in *Practical Surface Analysis* (D. Briggs and M.P. Seah, eds.), J. Wiley & Sons, Chichester (1992), Vol. II, pp. 635–55.

[157] R.S. Hockett, R. Bleiler and S. Smith, *Proceedings SIMS X* (A. Benninghoven, B. Hagenhoff and H.W. Werner, eds.), J. Wiley & Sons, Chichester (1996), pp. 645–8.

[158] V.K.F. Chia, M.J. Edgell, S. Biswas, S.D. Littlewood and I.E. Kelly, *Proceedings SIMS X* (A. Benninghoven, B. Hagenhoff and H.W. Werner, eds.), J. Wiley & Sons, Chichester (1996), pp. 693–6.

[159] I.S. Gilmore and M.P. Seah, *Proceedings SIMS XI* (G. Gillen, R. Lareau, J. Bennett and F. Stevie, eds.), J. Wiley & Sons, Chichester (1998), pp. 999–1002.

[160] K. Wittmaack, in *Practical Surface Analysis* (D. Briggs and M.P. Seah, eds.), J. Wiley & Sons, Chichester (1992), Vol. II, pp. 105–75.

[161] P.J. de Lange and J.W.G. Mahy, *Fresenius J. Anal. Chem.* **353**, 487–93 (1995).

[162] J.E. DeVries, L.P. Haack, T.J. Prater, S.L. Kaberline, J.L. Gerlock, J.W. Holubka, R.A. Dickie and J. Chakel, *Prog. Org. Coat.* **25** (1), 95–108 (1994).

[163] F.-R. Lang, Y. Pitton, H.J. Mathieu, D. Landolt and E.M. Moser, *Fresenius J. Anal. Chem.* **358** (1–2), 251–4 (1997).

[164] R.J. Cotter, *Time-of-Flight Mass Spectrometry*, ACS, Washington, DC (1997), Chp. 5.

[165] P. Bertrand and L.T. Weng, in *Surface Characterization. A User's Sourcebook* (D. Brune, R. Hellborg, H.J. Whitlow and O. Hunderi, eds.), Wiley-VCH, Weinheim (1997), Chp. 20.

[166] E. Niehuis, *Proceedings SIMS VIII* (A. Benninghoven, K.T.F. Janssen, J. Tümpner and H.W. Werner, eds.), J. Wiley & Sons, Chichester (1992), pp. 269–76.

[167] L. van Vaeck, A. Adriaens and R. Gijbels, *Mass Spectrom. Rev.* **18**, 1–47 (1999).

[168] J.C. Vickerman and D. Briggs (eds.), *ToF-SIMS: Surface Analysis by Mass Spectrometry*, IM Publications, Chichester (2001).

[169] R.W. Odom, *Microbeam Anal.* **2** (Suppl.), S99-S100 (1993).

[170] J.C. Vickerman, *Proceedings SIMS XI* (G. Gillen, R. Lareau, J. Bennett and F. Stevie, eds.), J. Wiley & Sons, Chichester (1998), pp. 19–26.

[171] S.J. Pachuta, *Proceedings Am. Chem. Soc., Div. Polym. Materials: Sci. & Engng.* **78**, 79 (1998).

[172] N. Man, A. Karen, A.Shimizu and A.Ishitani, *Proceedings SIMS XII* (A. Benninghoven, P. Bertrand, H.-N. Migeon and H.W. Werner, eds.), Elsevier Science, Amsterdam (2000), pp. 765–8.

[173] P. Brant and K.-J. Wu, *J. Mater. Sci. Lett.* **19** (3), 189–91 (2000).

[174] B.A. Keller and P. Hug, *Anal. Chim. Acta* **393**, 201–12 (1999).

[175] L.T. Weng, C. Poleunis, P. Bertrand, V. Carlier, M. Sclavons, P. Franquinet and R. Legras, *J. Adhesion Sci. Technol.* **9** (7), 859–71 (1995).

[176] K.M. Shakesheff, M.C. Davies and R. Langer, in *Surface Characterization Methods. Principles, Techniques and Applications* (A.J. Milling, ed.), M. Dekker, New York, NY (1999).

[177] R.J. Ward, H.A. Willis, G.A. George, G.B. Guise, R.J. Denning, D.J. Evans and R.D. Short, *Text. Res. J.* **63** (6), 362–8 (1993).

[178] N. Man, A. Karen and A. Ishitani, *Abstracts SIMS XII*, Brussels (1991), p. 76.

[179] A.M. Usmani and M. Donley, *Polym. News* **25** (10), 359–60 (2000).

[180] M.P. Mawn, R.W. Linton, S.R. Bryan, B. Hagenhoff, U. Jürgens and A. Benninghoven, *J. Vac. Sci. Technol.* **A9**, 1307–11 (1991).

[181] B. Asamoto, S.R. Bryan, C.L. Judy, R.W. Linton, B. Hagenhoff, M. Deimel and A. Benninghoven, *Proceedings SIMS VII* (A. Benninghoven, C.A. Evans, K.D. McKeegan, H.A. Storms and H.W. Werner, eds.), J. Wiley & Sons, New York, NY (1990), p. 367.

[182] F. Kötter and A. Benninghoven, *Appl. Surf. Sci.* **133** (1/2), 47–57 (1998).

[183] D. Stapel and A. Benninghoven, *Appl. Surf. Sci.* **174**, 261–70 (2001).

[184] M.J. Walzak, N.S. McIntyre, T. Prater, S. Kaberline and B.A. Graham, *Anal. Chem.* **71**, 1428–30 (1999).

[185] K.J. v.d. Berg and R. Huisgen, in *Proceedings Art et Chimie. La Couleur*, Louvre, Paris (1998).

[186] A. Galuska, *Surf. Interface Anal.* **25** (10), 790–8 (1997).

[187] A.M. Belu, V.H. Perez-Luna, J. Heller and B.D. Ratner, *Abstr. 211th ACS Natl. Mtg.*, New Orleans (1996).

[188] R.D. Boyd, A.M. Kenwright, J.P.S. Badyal and D. Briggs, *Macromolecules* **30**, 5429 (1997).

[189] P. Bertrand, N. Médard, C. Poleunis and X. Vanden Eynde, unpubl. results (2001).

[190] S.G. MacKay and S.J. Pachuta, *Polymer Prepr. (Am. Chem. Soc., Div. Polym. Chem.)* **37** (1), 299–300 (1996).

[191] N. Medard, M. Aouinti, F. Poncin-Epaillard and P. Bertrand, *Surf. Interface Anal.* **31** (11), 1042–7 (2001).

[192] R. Foerch and D. Johnson, *Surf. Interface Anal.* **17**, 847–54 (1991).

[193] J.C. Canry, A.D. Bass and J.C. Vickerman, *Proceedings SIMS IX* (A. Benninghoven, Y. Nihei, R. Shimizu and and H.W. Werner, eds.), J. Wiley & Sons, Chichester (1994), p. 800–803.

[194] N.M. Reed and J.C. Vickerman, in *Surface Characterization of Advanced Polymers* (L. Sabbatini and P.G. Zambonin, eds.), VCH, Weinheim (1993), pp. 83–162.

[195] L. Lianos, Tran Minh Duc, S. Reichlmaier and J. Hammond, *Proceedings SIMS IX* (A. Benninghoven, Y. Nihei, R. Shimizu and H.W. Werner, eds.), J. Wiley & Sons, Chichester (1994), p. 468–71.

[196] H.S. Munro and S. Singh, in *Polymer Characterisation* (B.J. Hunt and M.I. James, eds.), Blackie A&P, London (1993), pp. 333–56.

[197] C.L. Zhao, F. Dobler, T. Pith, Y. Holl and M. Lambla, *J. Coll. Interface Sci.* **128**, 437 (1989).

[198] R. Dietrich, *Fresenius J. Anal. Chem.* **361** (6–7), 692–4 (1998).

[199] M. Brenda, R. Doring and U. Schernau, *Progr. Org. Coat.* **35** (1–4), 183–9 (1999).

[200] E.H. Erenrich, E.A. Leone and R.D. Sedgwick, *Proceedings Intl. 23rd Waterborne, High-Solids, Powder Coat. Symp.*, New Orleans, LA (1996), pp. 327–39.

[201] A.M. Belu and J.McGinness, *Polymer Prepr. (Am. Chem. Soc., Div. Polym. Mat.: Sci. & Engng.)* **78**, 80–1 (1998).

[202] P.A. Cornelio Clark, S.A. Gardner and D. Horwat, *J. Vac. Sci. Technol.* **A13** (3, Pt. 2), 1351–8 (1995).

[203] D. Briggs, *Br. Polymer J.* **21**, 3 (1989).

[204] J.W. Burley, *Proceedings SPE ANTEC 2001*, Dallas, TX (2001), pp. 2986–90.

[205] C.C. Wang, Y.C. Ling, H.C. Lin, I.F. Tsai and A.C.-M. Yang, *Abstracts SIMS XII*, Brussels (1999), p. 69.

[206] L. Zhu, T. Liew and T.C. Chong, *Appl. Surf. Sci.* **189** (1–2), 53–8 (2002).

[207] K.J. Wu and R.W. Odom, *Anal. Chem.* **68**, 873–82 (1996).

[208] C.H. Becker and K.T. Gillen, *Anal. Chem.* **56**, 1671 (1984).

[209] H.F. Arlinghaus and X.Q. Guo, in *Proceedings SIMS XI* (G. Gillen, R. Lareau, J. Bennett and F. Stevie, eds.), J. Wiley & Sons, Chichester (1998), p. 677.

[210] W. Bock, M. Kopnarski and H. Oechsner, *Vakuum Forsch. Prax.* **1**, 21–8 (1998).

[211] H. Gnaser, *Proceedings SPIE* **4086**, 7–18 (2000).

[212] H. Oechsner, in *Thin Film and Depth Prophile Analysis* (H. Oechsner, ed.), Springer-Verlag, Berlin (1984).

[213] C.H. Becker, S.G. MacKay and D.G. Welkie, *J. Vac. Sci. Technol.* **B10**, 380 (1992).

[214] M. Wood, Y. Zhou, C.L. Brummel and N. Winograd, *Anal. Chem.* **66**, 2425 (1994).

[215] H. Oechsner, *Intl. J. Mass Spectrom. Ion Proc.* **143**, 271 (1995).

[216] Y. Hagashi, *Spectrochim. Acta* **B54**, 109–22 (1999).

[217] H.J. Mathieu and D. Léonard, *High Temp. Mater. Processes* **17** (1–2), 29–44 (1998).

[218] D.P. Smith, *J. Appl. Phys.* **18**, 340 (1967).

[219] W.A. Grant, in *Methods of Surface Analysis* (J.M. Walls, ed.), Cambridge University Press, Cambridge (1989), Chp. 9.

[220] H. Niehus, in *Practical Surface Analysis* (D. Briggs and M.P. Seah, eds.), J. Wiley & Sons, Chichester (1992), Vol. II, pp. 507–75.

[221] H.-P. Weise, W. Görner and M. Hedrich, *Fresenius J. Anal. Chem.* **369**, 8–14 (2001).

[222] A.T. Di Benedetto and D.A. Scola, *J. Colloid Interf. Sci.* **64**, 480 (1978).

[223] P.J. Mills and E.J. Kramer, *J. Mater. Sci.* **21**, 4151 (1986).

[224] P.J. Mills and E.J. Kramer, *J. Mater. Sci.* **24**, 439 (1989).

[225] R.A.L. Jones, in *Polymer Surfaces and Interfaces II* (W.J. Feast, H.S. Munro and R.W. Richards, eds.), J. Wiley & Sons, Chichester (1993), pp. 71–100.

[226] E.J. Kramer, *MRS Bull.*, 37–42 (Jan. 1996).

[227] S.N. Mikhailov, R.J.M. Elfrink, J.-P. Jacobs, L.C.A. van den Oetelaar, P.J. Scanlon and H.H. Brongersma, *Nucl. Instrum. Meth. Phys. Res.* **B93**, 149–55 (1994).

[228] H.H. Brongersma, M. Carrere-Fontaine, P. Cortenraad, A.W. Denier van der Gon, P.J. Scanlon, I. Spolveri, B. Cortigiani, U. Bardi, E. Taglauer, S. Reiter, S. Labich, P. Bertrand, L. Houssiau, S. Speller, S. Parascandola, H. Ünlü-Lachnitt and W. Heiland, *Nucl. Instrum. Meth. Phys. Res.* **B142**, 377–86 (1998).

[229] R.H. Bergmans, W.J. Huppertz, R.G. van Welzenis and H.H. Brongersma, *Nucl. Instrum. Meth. Phys. Res.* **B64**, 584–7 (1992).

[230] G.E. Thomas, G.C.J. van der Ligt, G.J.M. Lippits and G.M.M. van de Hei, *Appl. Surf. Sci.* **6**, 204–24 (1980).

[231] T.G. Vargo, J.A. Gardella, R.L. Schmitt, K.J. Hook, T.J. Hook and L. Salvati, in *Surface Characterization of Advanced Polymers* (L. Sabbatini and P.G. Zambonin, eds.), VCH, Weinheim (1993), pp. 163–80.

[232] J. Hook, R.L. Schmitt, J.A. Gardella Jr., L. Salvatti Jr. and R.L. Chin, *Anal. Chem.* **58**, 1285 (1986).

[233] W.-K. Chu, J.W. Mayer and M.-A. Nicolet, *Backscattering Spectrometry*, Academic Press, Orlando, FL (1978).

[234] J. Perreire, *Vacuum* **37**, 429 (1987).

[235] C.R. Gossett, in *Industrial Applications of Surface Analysis* (L.A. Casper and C.J. Powell, eds.), *ACS Symp. Series* **199**, 49–68 (1982).

[236] M. Geoghegan, in *Polymer Surfaces and Interfaces III* (R.W. Richards and S.K. Pearce, eds.), J. Wiley & Sons, Chichester (1999), pp. 43–74.

[237] J.W. Mayer and E. Rimini (eds.), *Ion Beam Handbook for Materials Analysis*, Academic Press, New York, NY (1977).

[238] H.H.G. Jellinek, in *Physicochemical Aspects of Polymer Surfaces* (K.L. Mittal, ed.), Plenum Press, New York, NY (1983), Vol. 1.

[239] D.L. Allara and C.W. White, *Adv. Chem. Ser.* **169**, 273 (1978).

[240] T.P. Gall, R.C. Lasky and E.J. Kramer, *Polymer* **31**, 1491 (1990).

[241] S.-W. Park, D.-H. Kim, Y.-M. Kim, B.-S. Park, W.-S. Han and B.-S. Suh, *J. Korean Soc. Anal. Sci.* **7** (3), 301–13 (1994).

[242] R. Bongiovanni, G. Malucelli and A. Priola, *Curr. Trends Polym. Sci.* **1**, 37–51 (1996).

Chapter 5

The more you look, the less you see

Microscopy and Microanalysis of Polymer/Additive Formulations

5.1. Chemical Microanalysis 458
5.2. Microscopy and Imaging Techniques 460
5.3. Light Microscopy 464
5.3.1. Conventional Optical Microscopy 466
5.3.2. Ultraviolet Microscopy 472
5.3.3. Fluorescence Microscopy 475
5.3.4. Confocal and Laser Microscopy 478
5.4. Electron Microscopy 483
5.4.1. Scanning Electron Microscopy 485
5.4.2. Transmission Electron Microscopy 494
5.4.3. Analytical Electron Microscopy 497
5.5. Scanning Probe Microscopy Techniques 501
5.5.1. Atomic Force Microscopy 504
5.5.2. Near-field Scanning Optical Microscopy 511
5.5.3. Scanning Kelvin Microscopy 514
5.6. Microspectroscopic Imaging of Additives 514
5.6.1. UV/Visible Microspectroscopy 519
5.6.2. Infrared Microspectroscopy and Imaging 521
5.6.3. Laser-Raman Microprobe and Microscopy 532
5.6.4. Fluorescence and Luminescence Imaging 541
5.7. Magnetic Resonance Imaging 546
5.7.1. Nuclear Magnetic Resonance Imaging 547
5.7.2. Electron Spin Resonance Imaging 555
5.8. X-ray Microscopy and Microspectroscopy 559
5.8.1. X-ray Microradiography 560
5.8.2. Scanning X-ray Microscopy 561
5.8.3. X-ray Microfluorescence 563
5.8.4. Micro X-ray Photoelectron Spectroscopy 564
5.9. Ion Imaging of Additives 566
5.9.1. Laser-microprobe Mapping 566
5.9.2. Imaging Secondary Ion Mass Spectrometry 567
Bibliography 573
Light Microscopy 573
Electron Microscopy 573
Scanning Probe Microscopy 574
Near-field Optics 574
Microbeam Analysis 574
Microspectroscopy 575
Imaging/Image Analysis 575
Polymer Microscopy 575
General 576
References 576

Various elemental and molecular microanalysis techniques are now available. Characterisation is usually carried out with (micro)beam techniques. By interaction of a primary beam (electrons, photons, ions), secondary signals are generated at the material (electrons, photons, ions, neutrals), which contain information on the composition and/or structure of the material. The various techniques differ in the type of information obtained, *i.e.* information depth, depth resolution, possibility to obtain depth profiles, lateral resolution, compatibility with certain types of materials (conductors *vs.* insulators), destructive or non-destructive character, and type of information (elemental, isotopic, molecular). The polymer/additive analyst's challenge is to understand and choose from the vast array of available analysis and imaging techniques.

Light microscopy was invented in the early 17th C. After a number of slow but consistent developments for just over three hundred years, the transmission electron microscope (TEM) was materialised in 1931 and progress picked up the pace afterwards. Many different techniques in light microscopy (*e.g.* phase contrast and dark field, fluorescence, confocal) as well as new non-optical microscopies were developed. Some landmarks are novel techniques such as scanning microscopies (SEM, STEM, SAM, SCAM, STM, SFM, SThM, NSOM, SNIM, STXM), tunnelling and force microscopies (SPM, STM, AFM), acoustic microscopy (SAM), near-field (NSOM, SNIM) and *in situ* microscopies (ESEM, LVESEM), as well as microbeam analysis (EDS, AEM, EPMA). These tools allow gaining a better knowledge of the actual structure of dispersed particles.

There are three basic kinds of microscopy: qualitative, quantitative and analytical. Qualitative microscopy is mostly concerned with morphology. Quantitative microscopy deals with finding out how much of a specific substance is present in a specified region of the specimen. Analytical microscopy is the characterisation of species by measurement of some physical or chemical characteristic (*e.g.* polarisation, decay time, absorbance, excitation and emission spectra, *etc.*).

In microscopy, spatial resolution is the ability to view two closely spaced objects as distinct particles. The maximum spatial resolution in a conventionally designed far-field microscope is wavelength limited. The major drawbacks of optical microscopy, namely limited resolution, poor contrast and restricted depth of field, are all improved (*cfr.* Table 5.1). Confocal microscopy was invented by Minsky [1] to improve spatial imaging and to reduce out-of-focus radiation. Near-field tunnelling and atomic force microscopes deliver high-resolution imaging, but only of near-surface features.

Considerable progress has also been due to interfacing microscopy and spectroscopy (Table 5.2). Several forms of microspectroscopy and spectromicroscopy have been developed as well as chemical imaging techniques, based on a variety of molecular spectroscopies, including infrared absorption, Raman scattering, fluorescence emission and magnetic resonance. Scanning microscopy is an example of a sequential processing system, in which the image is built up point by point. Imaging scales range from mm to nm (*cfr.* Table 5.3).

Uniting microscopy and IR spectroscopy relates microstructure and composition and marks an advance over blindly determining average composition. A single-beam dispersive transmission microscope was first introduced in 1953 [2]. Microscopes for FTIR spectrometers began appearing in 1983. State-of-the-art μFTIR spectrometers offer the confocal advantage and array detection. Microspectroscopy and spectroscopic imaging are closely related techniques. Digital image processing techniques have revolutionised the field. Non-invasive chemical imaging, or the more traditional single point-by-point mapping approach, correlates chemical information with structures in visible images. Microscopic mapping experiments generate the spectral absorbances at a given spatial position and frequency. Chemical imaging utilises state-of-the-art focal-plane-array (FPA) detector technology, which overcomes the inherent limitations of mapping. Each element of the detector array simultaneously collects spectra of individual areas of the sample. Since spectral data is acquired in a parallel, rather than sequential fashion, results are obtained at orders of magnitude faster rates. The history of uniting microscopy and spectroscopy has been traced [3].

There are many forms of microscopy that use radiation different from visible light, *i.e.* UV, X-rays, radio- and microwaves, acoustic waves, electrons, *etc.* In electron microscopes the resolution is much improved because of the shorter wavelength of the e-beam. The domain of optical microscopy is restricted; it does not include opaque materials, opalescent liquids, or media that absorb

Table 5.1. Lateral and depth resolution of different types of microscopes

Technique/operating mode	Voltage	Resolution	
		Lateral	Depth
AEM	2–5 kV	10–100 nm	0.2–2 nm
SEM	0.5–50 kV		
SE		1–10 nm	1–10 nm
BSE		0.1–0.5 μm	0.1–0.5 μm
EDS		0.1–0.5 μm	0.1–0.5 μm
WDS		0.2–1 μm	0.1–1 μm
TEM	40 kV–1.25 MV		
BF/DF		0.1–0.2 nm	5–100 nm
EDS		0.1–0.5 μm	10 nm–1 μm
EELS		2–5 nm	5–50 nm
STEM	100 kV	0.1 nm (imaging)	5–100 nm
STM	–	0.01 nm	1 pm
SOM	–	0.1–0.2 μm	0.1–10 μm

Table 5.2. Microspectroscopic and spectromicroscopic techniques

Microscopy	Spectroscopy	Effect
SEM	EDS	Microprobe
SEM	Raman	Microspectroscopy
EM	EELS	Microprobe
OM	IR	Infrared microspectroscopy
OM	Raman	Raman microspectroscopy
SFM	IR	Infrared nanospectroscopy
SPM	TA	Scanning probe based microthermal analysis

Table 5.3. Imaging scales

Scale	Distance	Technique(s)
Macroscale	>0.1 mm	VIEEW™
Mesoscale	μm	OM, μFTIR, μRS, μXRF, NMRI, LMMS
Microscale	sub-μm	SEM, iSIMS
Nanoscale	Molecular scale	STM, SFM, TEM

light completely. These restrictions are lifted with high-frequency sound waves (scanning acoustic microscopy, SAM). A microscope using sound waves with a frequency of 3000 MHz has a resolution equal to that of the optical microscope; at 16,000 MHz the resolution is 15 nm. Applications of SAM include examining the integrity of the bonds between layers in laminated and composite materials, and studying the density and distribution of glass and carbon fibres used to strengthen composite materials [4]. The acoustic microscope is very effective at probing near-surface properties. X-ray microscopy is difficult to use because of the need to generate relatively monochromatic soft X-rays that cannot be focused by Fresnel lenses. The nuclear magnetic resonance microscope (NMRI) uses radiowaves. As radiowaves are highly penetrating NMRI accepts optically opaque materials with minimal problems of transparency. NMRI is especially well suited to the study of liquid phases, a rather unusual property in the context of other microscopes. Table 5.4 classifies microscopies on the basis of some desired attribute. Some special microscopic techniques are thermomicroscopy and microthermal analysis.

Microscopic techniques require various degrees of sample preparation, from minimal (in reflection) to elaborate (in transmission). Sampling for depth profiling includes in-plane and cross-sectional microtomy techniques. In-plane microtomy is typically done with a large-scale, or slab type, microtome. Cross-sectional microtomy is readily done by large- or small-scale microtoming techniques. TEM cross-sectional sampling is unique in that cryo-microtomy is used to generate very thin sections. For depth profiling (*e.g.* of automotive coatings) on the micrometre scale in-plane microtomy may be followed by transmission mode IR microscopy, or solvent extraction for HPLC analysis [5]. In the latter case, all information on the additive's spatial distribution within the polymer is lost. Cross-section microtomy may be followed by: *(i)* optical microscopy and

Table 5.4. Classification of microscopies

Class attribute	Microscopies
High chemical information content	μFTIR, μRS, NMRI, ESRI, μNEXAFS
High spatial resolution (<100 nm)	Electron microscopies, NSOM, STM, SFM
Quantitative compositional sensitivity	μNEXAFS
Low beam damage	μNEXAFS
In situ	ESEM, LVESEM, FEG-SEM, FEG-TEM, HVEM, STM
Scanning	SEM, SAM, SCAM, STM, SFM, NSOM, STXM
Optical sectioning	CLSM, NMRI, ESRI, AFAM, SCAM

TEM (component distribution, coating layer thickness); *(ii)* Raman microscopy (microprobe) analysis (high spatial resolution chemical information); and *(iii)* ToF-SIMS (ion imaging, chemical distribution). There do exist alternatives to labour-intensive destructive sectioning. NMRI provides bladeless sectioning using a magnetic field gradient knife; similarly, CLSM allows optical sectioning using confocal laser light. AFAM allows optical sectioning using acoustics.

As to **thermal imaging**, infrared thermal wave imaging (TWI) can be used for detecting emission of energy creating images, and for mapping temperatures of a sample. TWI finds application in deformation analysis and in following stress profiles, as reported for the mechanical deformation of PP nanocomposites [6]. For microthermal analyses, *cfr.* Chp. 2.1.6.1.

A fundamental limitation of almost all microscopy investigations of materials is that the images are static and taken when the specimen is at room temperature. In some cases, as in electron microscopy (EM), the specimen is also in a high or ultra-high vacuum and under intense radiation. ***In situ* microscopy** allows observing materials dynamically under more realistic conditions approaching those of normal service life.

5.1. CHEMICAL MICROANALYSIS

Principles and Characteristics

Materials analysis can be classified as bulk, surface and small area analyses. Microanalysis contrasts to bulk analysis and is understood as being the characterisation and analysis of microscopic features using instrumental methods based on physical principles and chemical analysis of microvolumes using microscale equipment. The ultimate purpose of microanalytical techniques is to magnify the spectral information of a microsample to the size sufficient for interpretation. Microanalysis is characterised by an inhomogeneous sample, a small sample weight (up to 1 ng) and sample size (10^{-6}–10^{-12} cm^3). **Trace analysis** is usually understood as the quantitative determination of amounts of elements and compounds of the order of 10^{-6}–10^{-9} g in a relatively large amount (>1 g) or large volume (>1 cm^3) of another element, compound, or matrix (polymer). The required sample mass for various microanalysis techniques varies considerably: SEM-EDS, 10^{-12} g; TEM-EDS, 10^{-15} g; ED, 10^{-14} g; μFTIR, 10^{-4} g; XRF, 10^{-3} g; LMMS, 10^{-11} g; μDSC, 10^{-4} g; and μXRD, 10^{-4} g.

A universal microanalysis technique should be able to: *(i)* locate points and areas of interest on the surface as well as in deeper layers; *(ii)* identify elements and molecules; and *(iii)* determine the concentrations of those species with an overall sensitivity of at least ppm for surfaces and ppb for the bulk. Unfortunately, no known analytical technique can satisfy all aforementioned requirements. Nevertheless, the performance of a technique in the three areas of location, identification, and quantification can be used as a measure of its usefulness in routine analysis.

In **spatially resolved microanalysis**, measurements are obtained for many small neighbouring locations, and spatial and gradient relationships can be determined. The essence of microanalysis is the achievement of good spatial resolution. For the surface sensitive electron spectroscopies this includes both lateral resolution on the sample surface and resolution in depth. Because of the small amount of material available, some of the accuracy and precision is given up for obtaining better spatial definition.

The industrially important area of microanalysis may be divided into chemical and physical microanalysis (Table 5.5). This Chapter is mainly devoted to the methods of chemical microanalysis using optical, electron, X-ray, ion and laser beams.

Table 5.5. Chemical and physical microanalysis

Chemical:	
• Elemental	SEM-EDS, LA-ICP-MS, μXRF, μXPS, AES, SSIMS
• Molecular	μIR, μRaman, SSIMS, LMMS
• Structural	μXRD
Physical:	
• Micro-roughness	AFM/SPM
• Micro-particle analysis	SALS, PCS, PMS, CLSM
• Microthermal analysis	μDSC, μTA

Table 5.6. Tools for chemical microanalysis

Microscopic technique	Elemental composition	Molecular structure	Chemical bonding[a]/local order
μXPS	+	−	−
Imaging SSIMS	+	+	−
AFM	−	−	+
SEM-EDX	+	−	−
SAM	+	−	−
μFTIR	−	+[b]	−
μRaman	−	+[b]	−

[a]Making use of EXAFS and NEXAFS spectroscopies.
[b]Functional groups only.

Table 5.6 compares some of the tools currently in use. Since charged particles (electrons/ions), used in AES, SIMS and ISS, can be collected by electromagnetic fields into finely focused beams (1 μm to 5 nm), microanalyses are possible with such techniques (micro-AES, scanning SIMS, ion microanalysis).

Table 5.7 distinguishes four major spatially resolved analytical problem areas. Although good spectroscopic data may be collected with a few mg or less, these may not always be available, as in case of the inability to isolate or transfer a minute sample, or in failure analysis. In **microsampling**, the concern is with the area of the sample presented to the probe. The thickness of a sample needs to be the same whether the technique is macro or micro (*e.g.* consider that Beer's law is proportional to sample thickness).

Table 5.8 lists the qualifying parameters for the main microprobe techniques. Microbeam analysis usually involves depths of up to 10 μm and a surface area of less than 100 μm^2. Current highlights of **optical microanalysis** (UV, VIS, NIR, MIR) are:

- near-field microscopy (towards sub-μm or nanoworld)
- near-field microspectrometry (Raman and IR, from 1 μm down to sub-200 nm)
- surface enhanced IR analysis (10^3–10^5 increase in LOD)
- multilayer depth profiling by new ATR and PAS approaches
- mapping/imaging.

The coupling of microscopy and spectroscopy is well illustrated for XRF and SEM, resulting in two competitive techniques (μXRF and SEM-EDS):

$$\text{XRF} \xrightarrow{\textit{Chemical information}} \mu\text{XRF, SEM-EDS}$$
$$\xleftarrow{\textit{Lateral resolution}} \text{SEM}$$

Sample preparation requires auxiliaries which are not required for macrotechniques. Sample preparation is particularly important in order to reduce artefacts by handling. The need for reliable standards for both microanalysis and imaging is great. Certified PS particle size standards (100 nm–30 μm) are available for calibration (TEM, SEM, OM). For the current status on standardisation in microbeam analysis techniques (EPMA, SEM, AEM, EDS), *cfr.* ref. [8].

Table 5.7. Major micro- and nanoanalysis areas

Area	Qualifiers	Tools
Distribution	Amount (weight/volume), particle counts, detection limit, precision	Chemical microanalysis (inorganic: SEM-EDS, LA-ICP-MS, μXRF, μXRD, XPS; organic: μFTIR, μRS, XPS, ToF-SIMS, LMMS); micro-particle analysis (SALS, PCS, PMS)
Spectral	Spectral resolution, peak separation, chemical specificity	Microspectroscopy, spectromicroscopy
Surface	Lateral resolution, mapping/imaging	Far-field microscopy (reflection, transmission, polarised, fluorescence, phase-contrast, interference); near-field microscopy (AFM/SPM: morphology, micro-roughness); elemental imaging
Depth	Depth resolution, depth profiling, slicing, gradients	Confocal microscopy, μATR, SAM

As nanotechnologies are gaining in importance, so are microanalytical and microscopical techniques.

Applications

Typical microanalytical problems relate to the distribution of additives over the phases (or even interfaces) of a blend or compound, additive heterogeneity *vs.* homogeneity (as a result of industrial practices of mixing solids), surface concentration and distribution (blooming, *etc.*), gradients in the polymer (migration), foreign particles (gels, pitting, blistering, *etc.*). Detection of **contaminant species** on surfaces is similar to that of additives migrating to polymer surfaces. However, in the former case the problem is often localised and microanalytical/imaging techiques are necessary.

Depth concentration measurement is an important application of surface analytical methods. Examples are depth distribution of additives in plastics, or interface analysis where polymers are in contact with metals or ceramics. All surface methods with a good depth resolution (XPS, AES, SIMS) are suitable for depth or profile measurements. Complete multilayer coating systems require analytical methods that are applicable to small sample sizes and low concentrations. Techniques for obtaining chemical composition and component **distribution depth profiles** for automotive coating systems, both in-plane (or slab) microtomy and cross-section microtomy, include μFTIR, μRS, ToF-SIMS, optical microscopy, TEM, as well as solvent extraction followed by HPLC, as illustrated by Adamsons *et al.* [5]. Surface and interface/interphase analysis can now be done routinely on both simple monolayer coatings and complex multicomponent, multilayered coating systems. Successful tracking of UVA and HALS additives in thermoset coatings by horizontal microtoming followed by conventional microchemical analysis has been reported [9,10].

Microscopic investigations are often the most direct way of characterising the internal structure of polymeric materials and allow developing fracture hypotheses. These questions all require direct solid-state examination.

5.2. MICROSCOPY AND IMAGING TECHNIQUES

The analysis of solid matter differs from that of gases and liquids by topographical analysis. For heterogeneous systems, such as polymer/additives, it is important to simultaneously understand both the spatial distribution and the chemical behaviour of the various components within the polymer matrix. Detailed analysis over very small areas and film depths may also be needed to gain insight in the physical distribution and migration of additives in polymers. This requires analytical tools (microscopy and imaging microspectroscopy) that associate chemical and morphological information with spatial or volumetric location, ideally a "chemical microscope" that combines high spatial resolution with a highly specific chemical probe. In general, three factors are essential for successful **spatial location** of a species, which therefore govern the overall performance (*i.e.* the level of spatial resolution) of a technique. They are: *(i)* physics of the analysis process; *(ii)* lateral resolution provided by the instrumentation; and *(iii)* achievable sensitivity of the particular surface analytical technique. As shown in Fig. 5.1, there is

Table 5.8. Overview of various microanalytical techniques

Parameter	EPMA	AES	μXPS	μFTIR	μRS	SSIMS	ToF LMMS
Input probe	e (20 keV)	e (<3 keV)	ν (X)	ν (IR)	ν (VIS)	i (20 keV)	ν (UV, eV)
Output (detection)	X	e	e	ν	ν	+/− i	+/− i
Measurement	WDS, EDS	E	E	λ	λ	m/z	m/z
Resolution (detector)	20 eV, 150 eV[a]	1–15 eV	0.3–1 eV	2 cm^{-1}	0.7, 8 cm^{-1}[b]	10^3–10^4	800
Area analysed	±1 μm	0.1 μm	>5 μm	5–10 μm	1 μm	1.0 μm	1–3 μm
Information depth	<1 μm	1–3 nm	1–10 nm	10 μm	10 μm	Monolayer	0.1–1 μm
Image resolution	<1 μm	50–100 nm	Yes	mm	1 μm	0.5 μm	1 μm
Detection limit	100, 1000 ppm[a]	>1%	>1%	ppm	Major	Monolayer	ppm
Detection range	$Z > 4, \geq 11$[a]	All but H, He	All but H, He	n.a.	n.a.	All	All
Organic characterisation	No	No	No	Yes	Yes	Yes	Yes
Compound speciation	No	No	Yes	Yes	Yes	Yes	Yes
Destructiveness	(No)	No	No	No	No	Yes	Yes
In–depth profiling	No	No	No	No	No	No	Yes
Quantification	Yes	Yes	Yes	Yes	No	Difficult	Very difficult
Analysis of insulators	No	No	Yes	Yes	Yes	No	Yes
Vacuum requirements	Yes	Yes	Yes	No	No	Yes	Yes

[a] WDS, EDS.

[b] Spectrum, imaging mode; n.a. = not applicable.

After Van Vaeck and Adams [7]. Reproduced from L. Van Vaeck *et al.*, in *Encyclopedia of Spectroscopy and Spectrometry* (J.C. Lindon, ed.), Academic Press, 1141–1152, Copyright (2000), with permission from Elsevier.

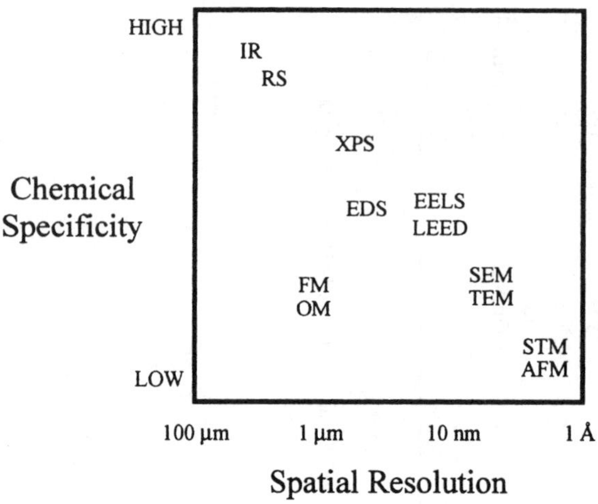

Fig. 5.1. Trade-off of chemical specificity with spatial resolution. For acronyms, *cfr.* Appendix. After Zimba *et al.* [11]. Reproduced by permission of the Society of Plastics Engineers (SPE).

typically a trade-off between spatial resolution and chemical specificity. Microscopy and imaging techniques combine physical and chemical information, respectively.

Microstructure is critical to the proper function of many products and manufacturing in a whole range of industries is forced to measure and control microstructure. Microscopic and local chemical analysis are integral parts of quality control and failure diagnosis. Although local analysis will not directly solve all the problems, it is essential to have an understanding of the potential of these technologies for the problem solving process.

With the exception of lensless microscopes, such as scanning tunnelling and near-field microscopes, the essential feature of a (far-field) microscope is the interaction with the specimen of a beam of electromagnetic waves (light), or particles that possess wave-like properties (high-energy electrons). The beam becomes modified as a result of this interaction, carrying information about the specimen, and this information is presented in the form of a magnified image with interpretable contrast related to certain surface or bulk characteristics. It is important to note that with a microscope one does not observe the specimen, but an *image* of the specimen. Table 5.9 compares the performance of the main microscopies.

Measurements may be carried out on film, plates, microtomed sections (dissecting microscopy) or replicas (not for polymers). For light microscopy studies direct examination of the specimen and indirect observation by means of replicas are used to an equal extent. Indirect examination is of special importance when using conventional transmission electron microscopy (TEM) but less so in scanning electron microscopy (SEM). Table 5.10 summarises the information content of a variety of microscopical techniques used in polymer/additive analysis. As additives are potentially heterogeneously distributed in the polymer, measurements at various positions are recommended. The considerable interest in light and electron microscopy is connected with the growing requirements regarding resolution and quantitative determination of single components in complex systems.

The increased importance in analytical science of obtaining information about the spatial distribution of specific chemical species within a system has led to strong interest in all types of **imaging** experiment, particularly those which can be described as non-destructive or non-invasive. Bulk analysis is replaced by microanalysis and chemical imaging by concentration mapping. Studies of complex materials need localised concentrations. Concentration gradients are more important than constant levels. Imaging in chemistry began with the use of the microscope [12]. There is a very extensive literature on chemical mapping in the electron microscope and on various forms of imaging optical spectroscopy using array detectors. The type of imaging experiment that can be performed depends on the source of illumination (point or single-frequency source, linear or profile source, or a multi-frequency global source), and the nature of the detector (single- or multi-channel).

Image analysis may be carried out directly from the microscope or from micrographs. Interpretation of images is not always straightforward. With an electron of optical microscope, contrast is based on complex electromagnetic diffraction effects. Determining whether a feature is protruding from the surface or recessed into it can be difficult with an image from an optical or electron microscope. Artificial contrast can occur when a sample consists of reflecting material embedded in an absorbing matrix.

Many measurement techniques (macroscopy, optical and electron microscopy, ion microscopy, tomography) can lead to multivariate images. All methods capable of giving images can give multivariate images as surfaces or volumes [13]. Univariate imaging is an extreme simplification of the

Table 5.9. Comparison of various microscopies

	Stereo-binocular	Compound	SEM	TEM	AFM
Magnification range	5–100×	30–2000×	20–1 × 10^5×	1000–2 × 10^6×	1000–2 × 10^6×
Resolution[a]	10–2 μm	1–0.2 μm	4–1 nm	1–0.1 nm	3–0.3 nm
Field of view	Very large 5 mm, 50×	Large 2 mm, 50×	Large 20 μm, 5000×	Small 2 μm, 50,000×	Small 2 μm, 50,000×
Imaging system	Light optical	Light optical	Scanning electron beam	Electron optical	Scanning solid probe
Lenses	Glass	Glass	Electromagnetic	Electromagnetic	None

[a]Typical best resolution values (depending on preparation method).

Table 5.10. Overview of microscopical techniques used in polymer/additive analysis

Feature	OM	TEM	SEM	AFM
Sample preparation	Easy to elaborate	Very difficult[a]	Easy to elaborate	Elaborate
Specimen environment	Ambient, or transparent fluid	High vacuum	High vacuum (except ESEM with FEG)	Ambient, high vacuum or fluid
Observation	Surface, transparent bulk	"Bulk" (thin slices, <0.2 μm)	Surface	Surface
Information	Structural analysis; orientation; distribution; identification of components; m.p.; contamination; birefringence; morphology at μm level; degradation	Phase/particle distribution; chemical composition; crystallography; morphology at lamellar scale (nm); defects at atomic scale (1–2 Å)	Topography; phase/particle distribution; composition	Extremely high resolution of surface structures; morphology up to atomic scale
Radiation damage	None	Severe	Severe	None
Chemical analysis	μIR, μRS	EDS, EELS	WDS, EDS, BSE	No

[a]Negative staining, metal shadowing, thin sectioning, freeze-fracturing, replication.

multivariate case. Tables 5.11 and 5.12 give some methods for obtaining 2D (surface) and 3D (volume or slice) multivariate images. Multivariate images can be obtained from one and the same technique, *e.g.* by changing the wavelength in optical microscopy (multi-wavelength or spectral imaging). A stack of congruent images measured for different wavelengths is a multivariate image. In electron microscopy, multi-wavelength X-ray detection was introduced in the 1970s. Also, many modes of operation can be used to construct multimodal images. They can also be constructed by combining images resulting from different instruments. Optical, electron and ion microscopy images of the same surface can be mixed. Correlation between images from different techniques – SIMS, AES, OM, SEM – would provide a very powerful method for surface analysis, especially if appropriate images could be acquired on a single instrument. Multivariate imaging is strongly related to spectral and spatial resolution. Images of high spectral and spatial resolution are possible. It is often necessary to balance between reduced spatial resolution and reduced spectral resolution. An intensity image may have low spectral resolution, but reasonable spatial resolution. More information is contained in a colour image.

Chemical microscopy has recently been reviewed [14], as well as the prospects of microscopic techniques with electronic imaging and digital image processing [15]. Various reviews and books deal specifically with polymer microscopy [16,17]. Vari-

Table 5.11. Methods capable of giving 2D multivariate images

Variable	Method
Electromagnetic radiation	
UV/VIS	Microscopy in general
	Fluorescence microscopy
	Macroscopy
IR, NIR	Microscopy
X-ray	X-ray microscopy
	iPIXE
Electron energy	Electron microscopy (EELS)
Ultrasound frequency	Ultrasound imaging
	Acoustic microscopy
Mass	Ion microscopy (SIMS)

Table 5.12. Methods capable of giving 3D (volume or slice) multivariate images

Variable	Method
Electromagnetic	
X-ray	X-ray tomography
Radiowaves	Magnetic resonance imaging (NMRI)
UV/VIS	Confocal microscopy
Contrast chemical	MRI

ous monographs on image analysis are available [13, 18,19], *cfr.* also Bibliography.

5.3. LIGHT MICROSCOPY

Principles and Characteristics

Many analytical techniques lend themselves to a microscopical approach so that the analysis may be applied to particles or tiny areas of larger samples. Microscopy provides information about the structure, distribution, organisation and chemical composition of objects. Some newly-developed forms of microscopy technically have little in common with traditional types of microscopy, but are nevertheless considered to be microscopy since they fulfil the basic function of a microscope, that of providing information about fine details in an object.

There are several **basic types** of light microscopes, each appropriate to a particular application. All rely on the ability of a glass lens to bend a beam of light and hence magnify an image. In different microscope types, the glass lenses are arranged in different combinations to achieve particular types of image. The basic function of a microscope to study the structure of a specimen can be extended to a range of analytical operations. These include measurement and image analysis, and the identification and localisation of specific chemical components. Routine microscopy looks at a 2D structure – a flat specimen. Computer control allows other dimensions to be explored, such as depth with optical sectioning, and variation over time. In addition, advanced systems allow controlled variations of temperature with a computer-controlled hot stage, as well as multi-fluorescence detection and analysis. A temperature-controlled stage may be used to study temperature-dependent changes such as melting or freezing, and with the addition of DSC, energy changes associated with chemical or physical changes can be directly related to the structures seen in the microscope [20]. Microscope versatility and speed of analysis are key issues. Relatively new developments have rapidly become an integrated part of light microscopes and have given rise to important techniques, such as: *(i)* confocal (laser) scanning microscopy; *(ii)* time-resolved fluorescence/luminescence imaging; *(iii)* combined use of absorption-, reflection- and luminescence-imaging at low and high magnification; *(iv)* video microscopy; and *(v)* the use of digital charged-coupled device (CCD) microscopy and subsequent image analysis. Stereomicroscopy allows investigation of surface structures (defects, weld lines) and measurement of long glass fibres in composites.

This Chapter is mainly concerned with the description and application of conventional, ultraviolet, fluorescence and confocal microscopy. A conventional microscope is a well-engineered combination of lenses and aperture diaphragms with which it is possible to study details with different light transmission properties. If the object structures do not differ in absorption, staining can be used to make structures visible. This is often sufficient to get the required information. The main advantage of UV illumination is the greater range of absorbing compounds with a high extinction coefficient, useful to give the necessary contrast. Fluorescence is not only suitable to detect the presence of certain substances; the emitted light also contains information about the specimen, which allows qualitative as well as quantitative analysis. The disadvantage of the conventional, UV and fluorescence methods is that light originating from out-of-focus planes is also collected in the imaging device. This light disturbs the image and also limits resolution. Confocal microscopy is a method developed to prevent out-of-focus light from reaching the detector. In optical microscopy, the move is towards confocal (3D) imaging.

The foundation of modern light microscopy was established more than a century ago. Abbe [21] demonstrated how diffraction of light by the specimen and the quality of the objective lens determined image resolution, defined the conditions needed to design a lens whose resolution was diffraction limited and established the role of objective lens and condenser numerical apertures on image resolution. Resolution or **resolving power** is the ability of a system to make information about fine detail distinguishable in an image. Extraordinary advances have been made starting from the invention of the first microscope by A. van Leeuwenhoek. Lateral resolution, *i.e.* resolution in the plane of focus, is defined in terms of the minimum distance d that the diffraction images of two point sources in the specimen can approach each other and still visually be distinguished as two. According to optical theory

$$d = \frac{0.61\lambda}{\mathrm{n}\sin\alpha} \tag{5.1}$$

where λ is the wavelength, α the semi-angle of the cone of rays entering the objective lens and n the refractive index of the region between specimen and the near surface of the objective. It follows that for visible light ($\lambda = 550$ nm) the theoretical resolution of the microscope is about 340 nm in air ($\mathrm{n} = 1$). Great efforts have been made to improve the resolution limits of microscopes by using other wavelengths, such as UV. In non-visible UV light at $\lambda = 220$ nm a resolution of some 100 nm may be achieved with quartz lenses. Although such methods provide some (limited) improvement in the resolution limit, great progress was achieved only when applying accelerated electrons [22]. More recently, new ways of improving and increasing the resolution of microscopes have been explored, with atomic resolution [23], as in case of TEM. Near-field scanning optical microscopy (NSOM) is an aperture size rather than diffraction limited form of microscopy with characteristics similar to scanning tunnelling microscopy (STM), *cfr.* Chp. 5.5.2.

Axial (z-axis) resolution is measured along the optical axis of the microscope, *i.e.* perpendicular to the plane of focus. Axial resolution can be defined using two criteria, either the minimum distance that the diffraction images of two points can approach each other along the axis of the microscope and still be distinguishable, or by the radius of the first minimum of the diffraction image of an infinitely small point object. The axial resolution varies with $\mathrm{NA}^{1/2}$ (NA, numerical aperture), in contrast to the lateral resolution, which rises with NA [24]. The depth of field of a microscope is the depth of the image (measured along the microscope axis translated into distances in the specimen space) that appears to be sharply in focus at one setting of the fine focus adjustment. In bright field microscopy, this depth should be approximately equal to the axial resolution, at least in theory. In all forms of light microscopy the depth of focus is limited, particularly as the magnification is increased. The result is that usually very thin samples are required for successful light microscopy so that only absorbing species with high extinction coefficients will yield acceptable contrast. Normally, a microscopist makes every effort to obtain the highest possible spatial resolution. Conventional wide-field light microscopes create images whose effective depth of field at high power is 2–3 μm; the resolving power of optical microscopy is about 0.2 μm. To avoid degradation the specimen is sometimes protected by various treatments (antibleaching agents, fade-resistant dyes). Specimen preparation for light microscopy has recently been summarised [25].

To increase our understanding of the surface morphology of multicomponent polymer systems, techniques are needed that provide two distinct types of information, namely spatial resolution on various length scales within the surface layer and sufficient depth resolution so that one can observe the transition from surface to bulk structure in the material. When the domain sizes are on the order of micrometres, they should be visible by optical microscopy.

An ordinary light microscope is not well-suited for studying the 3D structure of a specimen. It presents a 2D image consisting of a superposition of in-focus and out-of-focus regions of the specimen. Stereomicroscopes are useful in some applications requiring only low magnifications, but are not useful in high-resolution microscopy. Complicated depth structures can be studied in three dimensions by: *(i)* photographing the specimen at different focus settings; *(ii)* microtome sectioning; or *(iii)* confocal microscopy. A major development in optical microscopy is the use of video recording techniques to detect dynamic processes [26]. The dynamic video images can be manipulated using ordinary image analysis techniques and displayed as static images or video recordings as desired. **Video microscopy imaging** (VMI) and computer-assisted microscopy hold great promise indeed, and they are mainstays of some industrial and research activities.

It has been argued already that with a microscope one does not observe the specimen, but an *image* of the specimen. In microscopy, CCD cameras and image-intensified systems are appropriate imaging devices. **Quantitative microscopy** techniques have greatly improved on account of progress in imaging hardware, computing facilities and progressive evolution of software tools. Electronic imaging in light microscopy enhances the power of the technique.

The object of our interest in light microscopy is polymer/additive analysis. **Dimensions** of typical polymer-related features are to be taken into account, as follows:

- 0.001–0.01 μm, polymer molecule coils, nuclei, amorphous and crystalline domains, catalyst particles;
- 0.01–1 μm, individual domains in multicomponent systems (*e.g.* ABS), latices, pigments, fillers, inorganic nucleating agents, fibrils, lamellae;
- 1–200 μm, pigment and filler agglomerates, spherulites, morphology of fractured surfaces, fibre reinforced plastics, glass fibres; and
- 10–1000 μm, pores, foams, woven and nonwoven structures, coatings, textile fibres.

Applications

A digital image capture-process analysis system with special illumination design is used for quantitative macroscopic visualisation. The system, Video Image Enhanced Evaluation of Weathering (VIEEW™) is applied for the analysis of automotive topcoat defects created by durability tests [27], *cfr.* also Chp. 6.7. Typical image analysis applications are the determination of glass fibre length, of size, shape and distribution of particles, of the degree of dispersion of Fe particles in conducting polymer composites or of Ba titanate in PE film. A light microscopical method was described for characterising the surface, particle size, and distribution of rubber powder intended for rubber waste recycling [28].

Microtomy and transmission microscopy with or without polarised light are important means for the evaluation of polymer structures, fillers and reinforcements.

Video microscopy imaging capability has been added to thermogravimetric analysis [29,30], *cfr.* Chp. 2.1.6.

5.3.1. Conventional Optical Microscopy

Principles and Characteristics

Microscopy is the study of the fine structure, or details, of an object using a microscope. The limitations of the human eye as an instrument for the study of fine detail are overcome by three important attributes of a microscope: resolution, contrast and magnification. It is clear that **magnification** is a prime requirement, resulting in a significantly enlarged image. The optical microscope (OM) relies on spatial variations at a single wavelength of the refractive index, absorption, and reflectivity in a specimen in order to produce the modulation in light intensity necessary to form the magnified image. Theory and basic concepts of microscopy are sufficiently well known (*cfr.* refs. [31,32]). The conventional optical microscope is characterised as a parallel processing system: the whole area of the specimen is simultaneously imaged. Any light coming from out-of-focus sources cannot contribute to formation of a good quality image, but gives rise to the overall level of "noise", reducing contrast and ultimately wiping out the image. For this reason, conventional optical microscopy is most frequently applied to examination of the surface of materials, to materials which are intrinsically very thin, or to samples which have been carefully cut into thin sections using a microtome. In optical microscopy, information regarding size, shape, and relative arrangement of features is obtained.

Sampling and subsequent preparation techniques determine the nature and extent of useful information obtained [33]. Microtomy is probably the best technique. To preserve the microstructure, it is advisable to embed, grind and polish the sample. Unfilled plastic samples for optical microscopy are prepared by using a microtome to cut thin slices, typically 3–20 μm thick, from the plastic part. These slices are then placed between two glass slides and examined using transmitted polarised light. Magnifications up to 1000× are typically used. Thus, optical microscopy allows one to "see" the microstructure of the plastic.

Plastic parts containing glass or mineral fibre reinforcements generally cannot be sectioned using a microtome because the fibres tend to break or fall out of the sectioned sample. Such samples must be embedded in epoxy, polyester or acrylic, and polished using silicon carbide papers and diamond impregnated cloths. If morphology is more important than chemical analysis, it is necessary to embed; if the reverse is true, one does not embed. Some additives respond to particularly simple identification treatments on the micro scale; for example, Gale [34] has reported application of a **staining** technique (the use of hydrogen sulfide) to colour a lead stabiliser on the surface of PVC particles and thus render it identifiable using standard reflected light microscopy.

According to Abbe, the ultimate limit to the **spatial resolution** for conventional microscopy is determined by the diffraction limit of light ($\sim\lambda/2$). The resolving power of the light microscope is adequate for many areas of work in polymer science. Resolution is improved by application of confocal scanning microscopy using monochromatic light with high coherence (laser). Resolution beyond the limit imposed by the diffraction of light can be achieved by imaging with the near field of a sub-μm physical aperture.

Optical microscopists employ a variety of techniques to enhance **image contrast**, which must be adequately large. These can be non-invasive (phase contrast, polarised light, differential interference contrast), or invasive, for instance, staining the sample. Staining methods and induced fluorescence techniques should be investigated after these other non-intrusive methods have been fully exploited. The generation of contrast depends on various interactions between specimen and imaging radiation, which are usually described in terms of absorption, transmission and reflection, scattering and diffraction, interference and polarisation, phase change and fluorescence. Methods of generating contrast are bright and dark field imaging, polarised-light microscopy (based on birefringence), phase and differential interference contrast (DIC) modes (*cfr.* Scheme 5.1). These techniques will reveal much of the microstructural information of the surface.

Modern microscopes are equipped with numerous accessories to enable study of physical characteristics and chemical phenomena. The most effective microscope is one that can combine as many techniques as possible. Typical instrumental capabilities are: useful magnification, up to 2000×; transmitted light modes: bright field, dark field, phase contrast, polarised light, fluorescence; reflected light modes: bright field, dark field and Nomarski. The main methods of optical microscopy are *(i)* transmitted light, *(ii)* reflected light, *(iii)* dark field, *(iv)* polarised light, *(v)* phase contrast, and *(vi)* interference contrast microscopy. Table 5.13 gives details of these techniques.

Microscopic investigations of non-reinforced polymers are normally carried out in a transmission microscope with polarised light using microtomed sections (*ca.* 10 μm thick). For reinforced material the sample is usually embedded. **Transmitted light microscopy** (lateral resolution *ca.* 0.5 μm) reveals the internal structure of transparent material and may be used in the examination of individual fibres, particle morphology and orientation. It can also be used in the examination of thin sections of resinous coatings. **Reflected light microscopy** (bright or dark field) is used to examine surface morphology, *e.g.* paint cross-section analysis. Comparatively rough surfaces, such as those presented by many fabricated plastics products for which high gloss or transparency are not of major importance, can be successfully examined using simple reflected light methods (bright or dark field). Applications concern opaque materials, coatings, glass fibre orientation in composites, cross-sections of fibres, *etc.* Ductile fracture surfaces and the surface of powder particles may also be examined in this way.

Polarised light enhances differences between crystalline and oriented materials, and is most often used in pigment and particle identification. **Polarised light microscopy** (PLM) is distinctive in that few other techniques yield so much characterisation data useful not only for detection of trace quantities, but also for identifying small particles down to sub-ng and even sub-pg levels. Physical properties determined on such tiny samples include: size, shape (crystal system, form and habit), softness *vs.* brittleness, colour and pleochroism, transparency (translucency or opacity), crystallinity, anisotropy *vs.* isotropy, density, melting point, refractive indices, extinction angles, optical sign, solubility in multiple solvents, presence of inorganic ions or organic functional groups, phase transition temperatures, *etc.* Some of these tests require more than one single particle, although most can be done on sub-ng samples. Use of polarised light-differential interference contrast allows visualising step height differences of 1 nm. McCrone [36] regularly makes a plea for the use of polarised light microscopy.

Current conventional light microscopy allows real-time, high-resolution 3D imaging. Volume or 3D imaging can be obtained in a number of ways, such as stereo viewing, confocal microscopy (*cfr.*

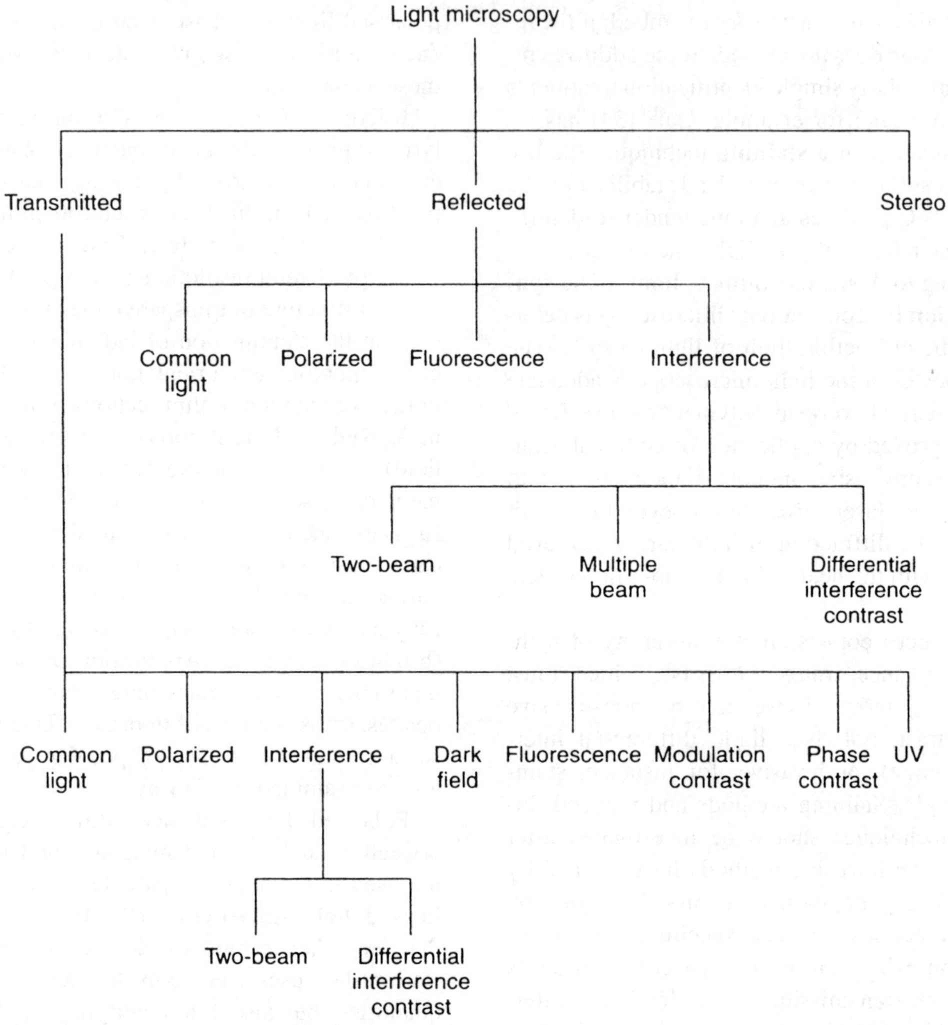

Scheme 5.1. Various types of optical microscopy techniques used to examine polymers. After Sandler *et al.* [35]. Reprinted from S.R. Sandler *et al.*, *Polymer Synthesis and Characterization, A Laboratory Manual*, Academic Press, San Diego, 1998. Copyright (1998), with permission of Elsevier.

Chp. 5.3.4) or tomography (*cfr.* Chp. 5.7). **Stereo imaging** gives greater detail than conventional microscopy. The stereomicroscope projects an image with proper orientation and depth cues. Without this tool, the general examination of objects would be incomplete. Switching from 2D to 3D brings major benefits in terms of resolution and the level of information imaged. The numerical apertures of typical stereomicroscope objectives are of the order of 0.05 to 0.2, which is very small compared to 0.1 to 1.4 for compound microscope objectives. Conventional stereomicroscopes work well, but at low magnifications (typically up to about 200×) and at low resolution. New technologies for taking 3D images, by using conventional compound microscope objectives, give significantly enhanced resolution, magnification and documentation compared to stereomicroscopes. There are two approaches to 3D technology, multiple oblique illumination (MOI) and oblique viewing [37]. Application of these techniques means that real-time 3D photography is now a reality. The new 3D microscope systems are likely to prove extremely useful for applications that require the viewing of 3D information, such as microcracks.

Images from a light microscope may be analysed (measured) and processed (by enhancement techniques) during any or all phases. Automatic analysis on the shape of features can be carried out. Mor-

Table 5.13. Optical microscopy techniques

Type	Features	Size range	Magnification
Transmitted light/bright field	Macro-, microstructures, colour, homogeneity, refractive index measurement	1 mm–0.3 μm	1–1000×[a]
Transmitted polarised light	Molecular orientation, spherulitic textures	1 mm–0.3 μm	1–1000×[a]
Transmitted light/phase contrast	Refractive index differences	1 mm–0.3 μm	100–1200×[a]
Transmitted light/dark field	Microstructures, colour	20 μm–0.3 μm	100–1000×
Transmitted light/DIC	Refractive index differences	1 mm–0.3 μm	10–500×
Reflected light/bright field	Macro-, microstructures	1 mm–0.1 μm	1–1000×
Reflected light/dark field	Macro-, microstructures	1 mm–0.3 μm	5–500×
Reflected light/DIC	Microstructures	1 mm–0.3 μm	5–500×
Fluorescence in reflected light	Macro-, microstructures	1 mm–0.3 μm	1–500×
Interference microscopy	Optical path differences	0.3–0.03 μm	50–1000×

[a] Up to 2000× with immersion oil.

phometric analysis allows **quantitative measurements** of surface, circumference, orientation, length, width, *etc.* Typical applications are the measurement of glass fibre distributions in composites and the measurement of the average size of spherulites in polymers such as PE, PP, nylon. Very fast imaging down to a few msec is now possible, enabling real-time 3D reconstructions and kinetic studies. Tanke [38] has described the impact of sophisticated image processing on light microscopy, with special emphasis on fluorescence.

The optical microscope has many **advantages** over other forms of microscopy (Table 5.14). Optical microscopy can often be carried out without the samples being "fixed", *i.e.* sectioned or stained. However, as with high-resolution techniques such as electron microscopy, optical techniques can be enhanced by staining or labelling the specimen. Visible microscopy can be made chemically specific by staining techniques that tag different materials. Modern interference light techniques allow quantitative assessments of surface roughness to a high degree of accuracy. Differential interference methods have a greater sensitivity to surface slope than SEM, so that comparatively large areas of gently sloping surface can be successfully imaged. Some **limitations** of the instrument are also listed in Table 5.14. Optical microscopy has a small depth of focus compared to SEM (*cfr.* Chp. 5.4.1). Light microscopy results ("lookology") are generally not accepted by non-microscopists with little physical background without further evidence (XRF, XRD, SEM-EDS, FTIR): seeing is not believing. In fact, conventional light microscopy does not provide chemical information. An important trend in microscopy has been towards chemical distribution and state mapping, such as provided by infrared (IR), ultraviolet (UV), Raman (and fluorescence) microscopies. All of these allow detailed viewing of a specific chemical species either by "natural contrast" or by some "labelling" mechanism. Optical nanoscopy is approaching.

In **scanning optical microscopy** (SOM), the objective lens of a light microscope focuses a parallel laser beam in a spot, the reflected light again passes the objective, and only light scattered or reflected in a limited depth of 0.1–0.2 μm and with a lateral resolution of 0.1–0.2 μm can pass a diaphragm in front of the detector (confocal mode, CSOM) [39]. SOM involves *object plane scanning*. In this, a 2D optical image of the specimen is *not* formed within the microscope, but rather the specimen itself is scanned by a focussed spot of light, and the result of the interaction of the light with successive areas of the specimen is recorded using a non-imaging photodetection device such as a photodiode or photomultiplier tube. A 2D image is built up point by point as the illuminating spot is scanned over the surface. SOM is thus the optical equivalent of scanning electron microscopy (SEM). SOM is mostly fitted with a low-power argon ion laser (*cfr.* Chp. 5.3.4).

Two-dimensional spatial images obtained by ordinary microscopy have a number of limitations. They reveal no depth profile information. Where physically slicing a fairly bulky sample into thin sections is either too time consuming or too destructive, there is a need for techniques which can be used to reject light from out-of-focus planes and allow reconstruction of clean high contrast images from

Table 5.14. Characteristics of optical microscopy

Advantages:
- Small instrumentation
- Undemanding operating conditions, services and maintenance
- No vacuum requirements; ability to operate in air or in liquids
- Frequently simple sample preparation techniques
- Relatively non-invasive observation protocols
- Non-destructive testing
- Suitability for wide variety of specimen types and size
- No need for conducting specimens
- Great versatility
- Fairly wide magnification range (up to 2000×)
- Broad range of contrast mechanisms
- No beam damage problems (non-destructive)
- Allowance for quantitative assessment
- Excellent direct visualisation (in colour; management appeal)
- Recording with photographic, video or computer techniques
- Image analysis and processing

Limitations:
- Limited resolving power (*ca.* 0.2–0.3 μm; in practice often 1–2 μm)
- Image quality degraded by out-of-focus optical information
- Finite spectral range
- Small depth of focus
- Poor images for samples with rough surface
- Restricted ability to provide chemical discrimination (staining)
- Sample preparation (when required)
- From relatively inexpensive to fairly expensive (for high quality microscopes)
- Need for highly skilled operator

within a complex material. Confocal scanning fluorescence microscopy (CSFM) is presently the most important imaging mode of the scanning optical microscope. SOM is a "far-field" imaging technique, limited by the constraints of Fraunhofer diffraction. Resolution enhancement can further be achieved by using a quite different method, **near-field scanning optical microscopy** (NSOM or SNOM). This surface probing technique is closely akin to scanning tunnelling microscopy (STM), which in principle can give spatial resolution of the order of 10 nm beyond the diffraction limit [40,41], *cfr.* Chp. 5.5.2.

For further reference the reader may consult some recent reviews on light microscopic techniques [20, 32]. Various books on optical microscopy (*cfr.* Bibliography) and videomicroscopy [26] are available.

Applications

Optical microscopy has been used in the areas of R&D and QC for many years. The conventional light microscope is a classic instrument for examination and observation of details of structures down to about 0.2 μm and is an important tool in everyday use for characterisation of structures in many areas of technology, especially those focussing on materials (including polymers) and where visualisation is a key requirement [42]. Typical applications are: morphology, particle size/shape, degree of dispersion, dynamic studies, absorption and swelling, thermal effects, identification of contaminants and failure analysis. Transparent films are usually examined by transmitted light, opaque material is viewed in reflected light.

There is a direct relationship between the microstructure of a plastic part and how it was processed. Optical microscopy makes it possible to obtain information regarding the **processing history** of moulded plastic parts. This history includes information about processing temperatures, pressures and times, gate size and location, additive distribution, contamination, *etc.* Non-uniform microstructure may have various origins: the distribution of amorphous and crystalline parts, the presence of unmelted particles, moulding defects such as voids, knit or weld lines, and the distribution of fillers and pigments. Processing defects, which are caused by incorrect processing conditions, are easily identified by optical microscopy. Optical microscopy allows a better evaluation of the quality of injection moulded thermoplastic parts. For this purpose polished samples are examined under a microscope using reflected light. Johnson [43] has evaluated processing conditions using OM indicating how processing defects can be corrected by adjusting the processing conditions. Optical microscopy can provide direct feedback relating changes in processing variables to improvements in part quality. Optical microscopy is frequently used for examining how well processing aids (with <2 μm spherical particles) or other additives are dispersed in resins and present as discrete particles [44]. The preferred equipment for this type of analysis consists of a transmitted light microscope with differential interference contrast (DIC) and a photomicrographic system. DIC gives a 3D effect to the processing aid particle, thus improving differentiability and distinguishes these particles from other materials of higher crystallinity (*e.g.* talc) that may be dispersed in the polymer [45]. Processing aid letdown level dispersions can be difficult to assess if large quantities of other particulate particles are present. Optical microscopy was also used to observe the effects of 3D chaotic mixing of melts and powder additives (*e.g.* carbon-black) in the development of electrically conducting plastic materials [46].

Various techniques may be used for measuring the **surface roughness** of extrudates. These techniques have included observations with the naked eye and light microscopy, contact profilometry, fractal analysis, digital image analysis, fibre optic surface analyser/pattern recognition, and AFM. The more popular of these techniques is profilometry. These techniques may be used to evaluate process aids for controlling surface roughness of LLDPE [47]. Optical microscopy is also a superb tool for the observation of the **morphology** of plastics. In this respect, polarisation microscopy takes a special place. Polarisation microscopes are widely employed in the assessment of spherulitic crystalline polymers as well as in the determination and measurement of orientations and internal strains in plastics. Polarised light microscopy (PLM) is used for the identification of particulates and fibres, of organic components, clays and refractories in powders and coating formulations, for defect analysis in pressure-sensitive adhesive coatings, and for the determination of particle dimensions, particle density, *etc.* using image analysis software. Magnification ranges from 100× to 1000×. Positive identification of unknown materials can be achieved as derived from measurements of refractive index, cross-polarisation, optical-retardant, optical-elongation, and birefringence behaviour. Although cross-linkers, accelerators, antioxidants and other **crystalline organic compounds** show high birefringence in the polarising microscope, this characteristic is insufficient for their unambiguous characterisation. In these cases microthermal analysis may be called in. Crystalline **inorganic fillers**, such as gypsum, kaolin and talc, but also textile and mineral fibres in rubbers exhibit birefringence and are studied conveniently by means of polarised light microscopy. PLM allows identification of mineral groups. In case of active fillers such as silica, with poor light absorbing power, phase-contrast and interference microscopy allow quantitative analysis on the basis of refractive index differences between the components. Optical microscopy is widely used for the measurement of size distribution of fillers, inorganic nucleating agents, of bubbles, voids, specks, fish eyes, pigment agglomerates in plastics. Phase contrast microscopes may aid the differentiation of layers of similar refractive index in laminates. Ahmed *et al.* [47a] have used optical microscopy in studying the morphology of water trees formed in the XLPE insulation of underground HV cables.

Various **gel types** may be detected visually or by microscopic means in a flowing melt stream. Gels due to the original high-MW polymerisation may range from tiny (seen microscopically, usually with $M_w < 10^6$; almost compatible with normal polymer) to just visible to the naked eye (*i.e.* <0.05 mm) and those which destroy the appearance of films and surfaces (0.1–0.5 mm). Gels caused by oxidation may range from small (usually <50 μm) to larger (0.025–0.5 mm; so-called "ambers", the colour being due to conjugated double bonds). Black specks are usually caused by oxidation way beyond ambers.

Light microscopy can also be used for initial characterisation of nacreous pigments; more detailed information can be obtained with a microscope spectral photometer or SEM-EDS. Poor **pigment dispersion**, or agglomeration, occurs when pigments or colorants, which are added to the moulding compound, are not well dispersed within the plastic matrix. Off-colour or non-uniform part colour is an obvious indication of poor pigment dispersion. A high accuracy, automatic image profile analysis technique for light transmission microscopy (optical sectioning) has been described for the quantitative measurement of the diffusion coefficient of pigments in plastics, as illustrated for PVC/(Pigment Red 3, DOP, Durastabe 142/2327) [48]. This technique provides a precise and quantitative assessment of the suitability of pigments as additives in polymeric systems. Pigments and dyes on textiles may be distinguished using the combination of microscopy, colour fastness and chemical analysis [49]. Visible spectromicrography and colorimetry in dye analysis and fibre comparison were reviewed [50].

Microanalytical methods are frequently used to study blooming and contamination of polymer and rubber surfaces, the formation of aggregates, particle size, shape and distribution of fillers. For this purpose high magnification (*e.g.* 150×) optical microscopy is commonly used. Success of this operation is usually dependent on generation of high quality microtome sections of the sample.

Palla [51] has worked out the applicability of light microscopic techniques for the characterisation of various components of industrial **rubbers**. Using bright-field OM or TEM various carbon-black types are often easily distinguishable. The degree of dispersion can be subjectively compared to photographs of standard examples (*e.g.* Cabot Dispersion Classification Chart for carbon-black). Optical microscopy also allows measurements on multicomponent polymers: size and size distribution as well as state of distribution of a dispersed phase, *e.g.* of **impact modifiers** (after selective contrasting or dissolution), may be established.

Light microscopy is frequently used to determine the generic class of a fibre. Microscopic examination of fibres relies on visual comparison of characteristics such as colour, diameter, cross-sectional shape, birefringence, refractive index, and fluorescence. A classical application is the determination of **glass fibre** length and diameter, distribution and orientation in reinforced plastics. For the measurement of glass fibre length in composites samples need to be taken with dimensions greatly exceeding the expected glass fibre length. The polymeric matrix is eliminated either by low temperature ashing or by selective dissolution. After dispersion of the glass fibres, image analysis may be carried out on a statistically relevant number of fibres.

Imaging of surfaces, revealed surfaces, or cross-sections has become a technique to routinely identify coating defects, degradation, morphology, or thickness, and is used in characterisation of **automotive coating systems** [5]. Lighting conditions (*i.e.*, optics configuration, masking, polarisation, and/or direction) can be adapted to optimise imaging of defects, component distribution, and interfaces.

The use of a light or optical microscope to examine **fracture surfaces** is virtually indispensable as a first step in understanding failure mechanisms of plastic parts. Visual/microscopic inspection of a failed component may assist in narrowing down the cause of failure. A specimen is commonly checked for surface imperfections, embrittlement, extent and location of cracking, nature of cracking (ductile or brittle), chalking, crazing, discoloration, contamination, *etc.*

McCrone [52] has described microscopical analysis (mainly PLM) of **contaminant particles** (such as iron oxide rust, oil soot, paper fibres) in fracture objects. In other cases, because of their visibility, contaminants must be identified in order to determine and eliminate their source. Tiny black specks in an automobile paint are a typical example. Often, the contaminant particles are common substances present generally in many locations. Because of their small particle size, their removal and identification is a problem for microscopists and referred to as a classical "needle-in-a-haystack" problem. The first step is to examine the contaminant(s) *in situ*. The second step is to postulate an identity by size, shape, and colour, combined with a knowledge of the process and possible contaminant particles. Next, the particle is removed, usually using a fine needle, and mounted for examination using a polarised light microscope.

Only a few typical examples of all the possible applications of light microscopy to the structural examination of plastics have been mentioned here; for a comprehensive review, *cfr.* ref. [53]. Hemsley [33] has described microscopy of polymer surfaces by a variety of techniques. A number of books and general references give details on the various applications of light microscopy [17,42,54,55].

5.3.2. Ultraviolet Microscopy

Principles and Characteristics

The essence of ultraviolet (UV) microscopy is that many more materials absorb radiation in the UV than in the visible region. Examination of specimens in UV light requires the use of optical components capable of transmitting such short wavelengths, which means the use of quartz rather than glass. The UV microscope is thus a quartz optics modified transmission microscope equipped with a high-pressure xenon arc or mercury lamp to provide suitable light outputs from about 400 to 250 nm [56]. Kohler [57] originally developed the UV microscope, operating in the 230–280 nm regions with the intention of taking advantage of the increased resolving power theoretically associated with shorter wavelengths. However, quartz optics are technically less advanced and neutralise this advantage in practice.

UVA light (320–400 nm) can cause reversible corneal damage on short exposure, while UVB (280–320 nm) or long exposure to UVA can produce cataracts and damage to the lens. It is most convenient to view UV microscope images with a small TV camera fitted with a UV sensitive vidicon tube. The samples of interest in UV microscopy absorb strongly in the UV but are usually transparent in the visible. In order that a UV absorbing compound may be seen in a sample it is necessary to remove that part of the spectral output of the lamp (usually visible) which is not absorbed by the sample.

With the exception of confocal methods, all transmission microscopy needs very thin samples to overcome the limited depth of focus, so that only species with very high extinction coefficients will give acceptable contrast. This is much easier to achieve in the ultraviolet. Polymer samples for UV microscopy are generally in the form of microtomed slices of 5–10 μm thickness. At the same time, many UV absorbing substances are of interest in their own right as photostabilisers for polymers in outdoor exposure. The UV microscope is inherently equipped for fluorescence, with the advantage of greater sensitivity, as the image is formed against a black background.

In order to broaden the applicability of UV microscopy several **staining procedures** have been developed to enhance contrast (but not resolution) [58]. If a small, soluble molecule with high UV absorption is uniformly distributed through the polymer the sample will appear uniformly dark in the UV microscope. Any irregularity, which leads to a variation in the solubility of the additive will disturb its distribution and give image contrast. At equilibrium, **non-reactive stains** can enhance image contrast, by their differential solubility in different regions. Applications are to be found in studies of diffusion and morphology in semi-crystalline polymers and blends. Fluorescent additives, such as Uvitex OB, are particularly suitable mobile, non-reactive stains; comparisons between fluorescence and UV absorption pictures can facilitate interpretation. An alternative to staining depending on differences in solubility is the use of **reactive stains**, which can interact with specific functional groups in the polymer. Examples are 2,4-dinitrophenylhydrazine (DNPH), which interacts with carbonyl groups, and 2,4-dinitrofluorobenzene (DNFB), which interacts with primary and secondary amino groups. Finally, polymer-bound staining reagents may be used which are covalently bound to the chain. For example, dansyl azide (*N,N*-dimethylaminonaphthyl sulfonyl azide), which combines strong UV absorption with intense fluorescence, is a suitable agent for making polymers visible in the microscope. Polymer-bound stains allow monitoring of polymer-polymer diffusion and the study of phase-separated blends.

For many applications of the UV microscope it is necessary to make **quantitative measurements** of the concentration of UV absorber as a function of position within the field of view. Billingham *et al.* [56] have described methods of making concentration measurements. For good UV microscopy, especially for quantitative work, the spectrum of the illuminating light must lie entirely within the absorption range of the absorber; light outside this range simply acts as a bright background and reduces contrast. For fluorescence, the illuminating radiation should be as close as possible to the excitation maximum.

Optical microscopy using UV light can be applied to any sample in which features of interest are, or can be made to be, UV absorbing or fluorescent. The main **advantages** of UV illumination are the greater range of absorbing compounds with a high extinction coefficient (necessary to give acceptable contrast) and its use to excite and observe fluorescing substances with high sensitivity. However, the range of suitable fluorescing compounds is rather small. Moreover, fluorescence is limited by self-absorption. The development of the electron microscope has overshadowed the small advantage in using UV light to obtain increased resolving power. Few laboratories are equipped with UV microscopes. Ultraviolet scanning microscopes using confocal imaging are in use at SR stations. At UV wavelengths the spatial resolution exceeds that of commercial confocal microscopes. Excellent references exist for UV microscopy [56,58].

Applications

Polymers containing **UV stabilisers** or fluorescent additives are an obvious target for UV microscopy, but the potential range of applications is much wider, in that UV absorbers or fluorescers can be selectively bound to specific chemical entities in the polymer or will preferentially interact with, or dissolve in, parts of the structure. A variety of applications of the UV microscope to studies of polymers has been reported (Table 5.15). Many applications of UV microscopy require quantitative analysis.

Commercially important synthetic polymers usually have no strong UV absorption in the range from 250 to 400 nm. Hence, application of the UV microscope will depend on there being added UV absorbing molecules or attached side groups whose concentration varies within the polymer. In some cases one is interested directly in the concentration or distribution of the absorbing species, as with **stabilising additives**, in other cases one might wish to observe unintended minor species, such as impurities. UV microscopy has been applied to phenolic AOs, which

Table 5.15. Applications of UV microscopy to polymers

- Qualitative and quantitative measurements of polymers with a strongly UV absorbing or fluorescent component
- Distribution analysis of additives in polymers
- Additive concentration profiling
- Additive partitioning
- Determination of the diffusion rates of UV-absorbing or fluorescent additives in solid polymers
- Polymer oxidation studies
- Morphological and structural studies
- Contaminant analysis

can be observed at their 280 nm absorption peak at concentrations down to about 1 wt.%, somewhat higher than those at which they are normally used. The fact that polyolefins are transparent to UV above 200 nm, whilst the additives absorb strongly around 320 nm, means that this system is ideal for UV microscopy. Thus **distributions of additives** (light stabilisers) in spherulitically crystalline PP have been studied [59,60]. Frank *et al.* [60] were the first to use UV microscopy to demonstrate that a UV absorbing additive (Tinuvin 328), when incorporated into polyolefins, accumulates in the interlamellar and interspherulitic regions. Thus, the technique can be used to directly observe the amorphous region of the polymer and to show that these regions are non-uniformly distributed in a solid polymer. Calvert *et al.* [61] found that during the crystallisation process UVAs are rejected from the growing spherulites, and that the ultimate distribution of these additives reflects the distribution of the amorphous content of the polymer. Billingham *et al.* [62] studied poly(4-methylpentene-1) (P4MP) and were able to confirm that UVAs are rejected from the growing crystals in the same way as in polypropylene. Also Ryan *et al.* [63] studied the behaviour of UV absorbing additives in crystallising PP in some detail using a SEM-EDS system in addition to UV and fluorescence microscopy. After sufficiently long annealing times, the distribution of the additive reflects the distribution of the crystallinity of the sample. Calvert *et al.* [64] have pointed out that this allows UV microscopy to be used as a powerful probe of spherulite structure. Spherulites appear in the UV as non-absorbing regions. Studies of the rejection of UV absorbing additives by growing spherulites during melt crystallisation also allow determination of the diffusion coefficients of additives [56].

Semi-crystalline polymers are not the only ones where differential solubility of additives is possible. Another case is a polymer blend. A good example is impact-toughened polymers, where toughness is conferred on an otherwise brittle polymer by inclusion of rubber particles. Billingham *et al.* [65] have studied **stabiliser partitioning** and oxidative degradation in rubber-toughened polypropylene. UV microscopy was also used in product development of a non-extractable stabiliser system for PE by using a masterbatch process in which a UV absorber was covalently bound to a rubber phase in high concentrations [56]. Study of the diffusion of Tinuvin 234 in different grades of a copolyester block ether by means of UV microscopy is more difficult than in polyolefin matrices due to the intrinsic absorption of the polymeric matrix and light scattering from phase separated morphology [66].

Fluorescent additives may be studied in the same way as UV absorbers. The results are very similar but slightly more care is required in quantitative interpretation since self-quenching effects can lead to non-linearity in the concentration dependence of fluorescence intensity. UV microscopy has been used to follow the distribution of fluorescent additives (such as Uvitex OB) during isothermal crystallisation and cooling of isotactic PP [64]. Billingham *et al.* [58] have observed diffusion of Uvitex OB in a PP/rubber blend using UV fluorescence microscopy. UV microscopy can be very useful in the analysis of multilayer films where one layer of polymer is intrinsically fluorescent (*e.g.* PVDC).

Quantitative analysis of the motion of an additive in a polymer can rapidly give good data on bulk diffusion rates and this is particularly useful in studies of migration and loss of stabilising additives. Molecular transport of UV absorbing or fluorescent additives can be monitored in the UV microscope by following their progress into a polymer sample [67]. For that purpose the polymer in the shape of a rod is immersed in a solution of the additive in a solvent, which does not swell the polymer. The rod is sectioned when the additive has penetrated about 100 μm and the concentration profile of the diffusant within the polymer is measured by UV microscopy of the sections. The diffusion coefficient of the additive may be determined by fitting the profile to the expected form:

$$c/c_s = 1 - \mathrm{erf}\left(x/2\sqrt{Dt}\right) \qquad (5.2)$$

where c is the measured concentration, c_s that at the surface of the polymer, x the distance from the polymer surface, t the time and D the diffusion coefficient. Understanding the diffusion of small molecules in polymers is important both for packaging and barrier materials and for controlling the migration and loss of stabilising additives. Billingham [68] used UV microscopy to study the diffusion of a benzophenone in PP samples. Concentration profiles by UV microscopy for diffusive loss of a UV stabiliser from PP were also reported [58]. Diffusion coefficients of a variety of additives (Topanol 354, CAO-5, Cyasorb UV531, Uvitex OB, Ionox 330 and Goodrite 3114) in molten PP were determined by UV microscopy [69]. Klein *et al.* [70] have described a method for measuring the diffusion coefficients of carbonyl containing compounds in PE by IR microspectroscopy. In principle, the diffusion coefficient of a small molecule is also a probe of the mobility of the polymer matrix. Spatially resolved diffusion measurements are a very interesting approach to the study of polymer structures.

Partly degraded PP contains a variety of carbonyl compounds, such as carboxylic acids, ketones and aldehydes, which absorb UV below 300 nm and so can be observed if their concentration is sufficiently high. The visibility of the oxidation of **slightly oxidised polymers** can be enhanced by means of reactive stains. Knight *et al.* [71] have used the reactive staining technique to confirm the localisation of oxidation in degraded and crystallised polypropylene. The DNPH stain shows that oxidation is extremely heterogeneous in all samples. UV microscopy has also been applied to study uniform *vs.* local distributions of oxidation products (especially carbonyl groups) produced during processing and ageing of polymers [56]. Direct evidence for heterogeneous oxidation of polypropylene has come from microscopic evidence that oxidation occurs within the amorphous region and is initiated at catalyst residues and other impurity centres [71,72]. Billingham *et al.* [58] have reported that most samples of cured epoxy (reactively stained with DNFB) show large UV-dense regions, indicative of local concentrations of unreacted hardener. Other applications of UV microscopy to studies of the oxidation and stabilisation of polymers have been reported [73–75]. UV microscopy has also been applied to thermosetting polymers [76].

5.3.3. Fluorescence Microscopy

Principles and Characteristics

Fluorescence microscopy is a technique whereby fluorescent substances are examined in a microscope. When a molecule absorbs a photon of light, it is promoted to an unstable excited state and can then release its excess energy by various pathways, amongst which **fluorescence emission** [77]. Molecules remain in the excited state for approximately 10^{-9} s before releasing their energy and returning to the ground state. The time delay between initial absorption and emission is called the "fluorescence lifetime". The advantage of fluorescence is that many lifetimes fall in the 1–20 ns range, which coincides almost perfectly with the time scale of molecular interactions. Fluorescence emission is a property of all materials. The high specificity, extreme sensitivity and excellent detection limits of fluorescence spectroscopy (*cfr.* Chp. 5.3 of ref. [77a] and Chp. 1.4.2) have made it a very popular technique for imaging. Imaging spectrometers for fluorescence microscopy have been described [78]. In microscopy, fluorescence is used to visualise either materials that have the inherent property of fluorescing, or those to which a fluorescent marker (fluorochrome) can selectively be attached. Several requirements must be met for the development of the optical arrangement for fluorescence emission detection.

Major **objectives** connected with the use of fluorescence microscopy (FM) are: *(i)* identification of a specific substance by observing its characteristic emission properties when illuminated with radiation of the appropriate absorption wavelength; *(ii)* determination of specific parameters that influence the fluorescence in a given material; *(iii)* measurement of the intensity of fluorescence; and *(iv)* localised scanning of a sample to determine the distribution of the fluorochromes. Except for this latter case, known as scanning confocal microscopy, which uses special instrumentation, the three objectives can be addressed through the use of either transmitted or reflected radiation using conventional optical microscope design. The most commonly found light (excitation) sources with fluorescence microscopes are high-pressure mercury or xenon lamps, or incandescent tungsten-halogen filament sources, which produce UV, blue, or green light. This light is passed through a monochromator or interference filter to select the excitation wavelengths that induce fluorescence in the sample being examined. The specimen is examined through a barrier filter that absorbs

Table 5.16. Applicability of fluorescence microscopy, compared with other techniques[a]

Specimen	Fluorescence	Type of microscopy		
		Absorption[b]	Polarisation, phase contrast	Reflection
Coloured	+	+	−	+
Transparent	−	−	+	−
Opaque	+	−	+	+
Dynamic	+	−	−	−
Particles below limit of resolution	+	−	−	+

[a] + Suitable, − unsuitable/impossible.
[b] Absorption microscopy is the conventional transmitted-light type.

the short-wavelength light used for illumination and transmits the fluorescence, which is therefore seen as bright against a dark background.

When UV light is used for excitation the terminology ultraviolet fluorescence microscopy is appropriate. The depth of field for a fluorescence microscope is only a few μm. Fluorescence microscopy has progressed from bright-field and dark-ground transmitted-light configurations to the now almost universal epifluorescence system. This means that the light used for excitation is reflected onto the specimen through the objective, which acts as a condenser. In essence the modern epifluorescence microscope is similar to the bright-field reflected-light instrument, but with several important differences [32]. Use of large-aperture objectives is especially desirable in fluorescence microscopy, to provide not only high resolving power, but also the brightest possible image, in order to maximise the sensitivity of the technique. Brightness in fluorescence microscopy is proportional to the fourth power of the numerical aperture (NA) of the objective. Fluorescence microscopy techniques can be added to virtually any microscope.

The chemical substances to be observed are either intrinsically fluorescent, or made so by a chemical process, or attached to a fluorescent label. Samples for fluorescence microscopy are often **stained** with a fluorescent dye (*e.g.* rhodamine), with the aim of having the physical characteristics of the probe represent the specimen characteristic. The specimen is then illuminated with light at an appropriate wavelength to excite the dye, generating fluorescent light, which is collected by the microscope. The intense fluorescence of the reactive dansyl group also determines convenient use in fluorescence microscopy, which allows the lowest concentrations

Table 5.17. Main characteristics of fluorescence microscopy

Advantages:
- High sensitivity and specificity
- Detection of particles below resolution of a light microscope
- Quantification of small amounts of fluorescent substances or small particles
- Broad applicability, including opaque or very thick objects (epi-illumination)
- Particularly well-suited for confocal microscopy

Disadvantages:
- Difficult interpretation of images
- More complex and expensive than conventional transmitted-light microscopy
- Possible photodamage

of the reagent, minimising any disturbance caused by the reagent. The development of new fluorescent probes allows a wide range of processes to be studied.

Fluorescence microscopy offers a number of **advantages** over other forms of microscopy (*cfr.* Tables 5.16 and 5.17). Its high sensitivity allows very low concentrations of specific substances to be localised. Because fluorescence is observed as luminosity on a dark background, fluorescent constituents of the specimen can be seen even in extremely small amounts. Fluorescence microscopy can also be applied to detect particles below the resolution of a light microscope. Since fluorescence involves two wavelength bands (excitation and emission), optical specificity can substantially be increased. Fluorescence microscopy, because of its complexity, gives more difficulty than usual in interpretation of the image.

Various modes of fluorescence microscopy have been developed. In the last two decades the field of **quantitative fluorescence microscopy** (QFM) has greatly advanced both on account of imaging hardware and evolution of software tools [79]. The goal of QFM is to produce a reliable quantitative estimate of some characteristic property of a specimen, such as the concentration and/or location of some molecular species. Quantitative fluorescence measurement requires a device to detect the photons emitted by the specimen [79]. For imaging applications, common detectors include silicon intensified target (SIT) cameras, charge-coupled device (CCD) cameras, and intensified CCD cameras. For non-imaging applications and for confocal microscopes, photomultiplier tubes (PMTs) are the common detectors of choice. Digitised video microscopy has been combined with fluorescence spectroscopy. Technical progress with respect to the development of high power excitation light sources (lasers) and sensitive detection devices allow the specific detection of, in principle, one fluorophore molecule. In fluorescence microscopy both some solid-state lasers and all continuous-wave (CW) gas lasers can be used.

Fluorescence is particularly suitable for confocal microscopy, which offers optical sectioning, giving very clear imaging and the possibility of building up 3D reconstructions. Scanning confocal microscopy (SCM) offers a dramatic instrumental advantage for fluorescence microscopy through discrimination against out-of-focus background interference, through inherent resolution perpendicular to the plane of focus and improved in-plane resolution (*cfr.* Chp. 5.3.4). A major improvement in the resolution of far-field fluorescence microscopy has been achieved using stimulated emission depletion (STED) microscopy [80]. The spread function of most confocal systems, which is diffraction limited, has now been reduced to 70 nm horizontal and 175 nm axial. With **laser confocal fluorescence microscopy** (LCFM) (non-destructive) depth profiling and 3D image reconstruction become possible, allowing the study of relatively thick specimens that are not accessible by conventional microscopy at all. Quantitative measurements of fluorescence intensity in LCFM images can provide precise image determinations of fluorescence marker distributions. To take advantage of the full scope of LCFM, there is a need to find dye derivatives which can be attached to a broad spectrum of polymers. The crucial limitations of SCM arise in essential features of photochemistry. A potential problem in fluorescence microscopy of organic matter is photo-damage by absorption of the radiation introduced to excite the fluorescence. In conventional fluorescence microscopes less than 5% of the fluorescence emitted from within each resolution volume is detected.

Reviews on fluorescence techniques in polymer science have been reported [58,81,82]; recent books on fluorescence imaging spectroscopy and microscopy are available [83,84]; *cfr.* also Bibliography.

Applications

Fluorescence microscopy is closely allied to transmission (absorption) microscopy in its range of application, but possesses particular advantages (Table 5.17). Because many substances are fluorescent, or can be made so, fluorescence microscopy is widely applicable to all kinds of material. In view of the more complex and expensive instrumentation than conventional transmitted-light microscopy, fluorescence microscopy is usually reserved for those applications in which its high sensitivity is of importance: *i.e.* to examine substances present in low concentrations. Fluorescence microscopy is especially a valuable tool in the biological sciences.

Fluorescence microscopy is used primarily in the examination of organic material of matrices [85,86]; the induced fluorescence can be of great value for the detection of differences in resin coatings. Several **plasticisers** fluoresce under short-wavelength excitation and allow their localisation and observation of the penetration in an elastomer matrix [51]. Alternatively, plasticisers and pores may be detected by UV fluorescence excitation of polymer samples treated with a fluorescing agent. Fluorescence microscopy has also been employed in the manifestation of hairline cracks in plastics. For that purpose, the sample was brought into contact with a fluorescent liquid (*e.g.* rhodamine), which penetrates cracks and pores and thus permits these regions to fluoresce when illuminated by UV light in the microscope.

Using UV excitation in a fluorescence microscope most white **pigments** show a characteristic fluorescence, which allows differentiating TiO_2 (anatase and rutile) and ZnO [51]. Other applications are the assessment of oxidative degradation of polymers (PE, PVC), the identification of fibres in composites and of binders/sizing on glass fibres. Tredado *et al.* [87] have performed multispectral fluorescence microscopy of 15 μm diameter polystyrene

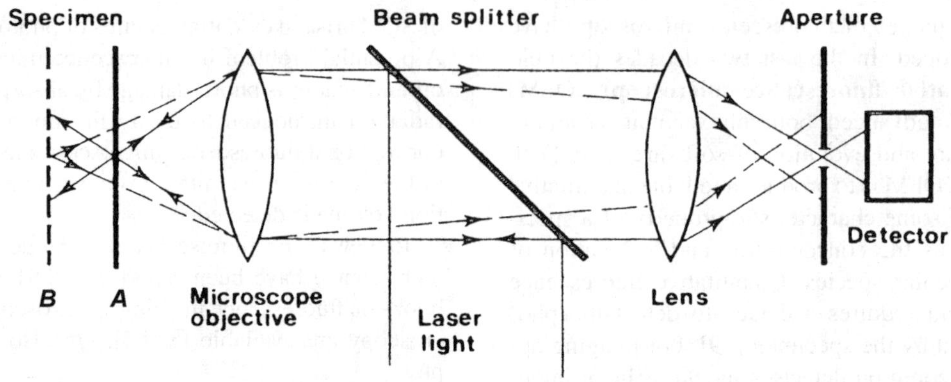

Fig. 5.2. Simplified optical beam path of a confocal microscope with incident light illumination. A focused laser beam illuminates a small specimen volume located in the focal plane of the microscope, *A*. Reflected or fluorescent light from *A* is transmitted through the detector aperture, which effectively blocks light from out-of-focus planes, *e.g. B*. By scanning either the laser beam or the specimen, an image can be recorded that represents a thin section located at *A*. Repeated scanning, using different focus settings on the microscope, results in a stack of images representing the 3D structure of the specimen. After Carlsson and Åslund [89]. Reproduced from K. Carlsson and N. Åslund, *Appl. Opt.* **26**, 3232–3238 (1987), by permission of the Optical Society of America, Copyright 1987.

microspheres tagged with different fluorescent dyes using an acousto-optical tuneable filter (AOTF).

Fluorescence microscopy has also been employed largely for 2D surface imaging. Scanning confocal fluorescence microscopy has been applied to the investigation of subsurface morphology of foams. The general knowledge on the applications of fluorescence microscopy for polymers is rather limited.

Applications of fluorescence microscopy have been reviewed [88].

5.3.4. Confocal and Laser Microscopy

Principles and Characteristics

The fundamental distinction between conventional optical microscopy and confocal optical microscopy is the manner in which the image is produced. In a normal microscope the full field of view is simultaneously and evenly illuminated, and a complete 2D image of that field of view is created by the optics which can be examined by the human eye, or projected onto film, detector or television camera. With the traditional light microscopy, the energy reaching the detector is independent of the position of the object in the object plane. In **confocal microscopy**, a point source and a point detector are used to illuminate a specific very small volume of the sample, and to reject light coming from any other part of the system. There are various technical ways to perform confocal microscopy, all with intrinsic advantages and shortcomings for defined applications. Figure 5.2 shows one way in which this can be performed.

In the confocal mode, a point light source is imaged in the object plane and a small aperture is positioned in the image plane in front of the detector, at a position confocal with the in-focus voxel. Light emanating from this in-focus voxel passes through the aperture to the detector, while that from any region above or below the focal plane is defocused at the aperture plane and is thus largely prevented from reaching the detector, thus essentially not contributing to the confocal image. The location in the image plane corresponds to the source of the point of light and the object plane. The term "confocal" relates to the fact that the image of the illuminating pinhole and the back-projection of the detection pinhole have a common focus in the object. Suppression by the pinhole of structures outside the focal plane results in a genuine resolution along the optical axis; the images produced by light arising from the in-focus specimen plane are always sharp. Images from confocal microscopy optics are produced point-by-point in the image plane from corresponding illumination points (x, y) in the specimen plane. If the illumination is focused onto a selected point in the object, then information comes from the point (x, y, z) only.

Confocal microscopy can be realised either by object scanning ("on-axis") or beam scanning ("off-axis"); a further distinction relates to single-beam and multiple-beam scanning systems. The small aperture improves the resolution and shortens the depth of focus by eliminating out-of-focus light. By using an array detector with a special configuration, the resolution of a confocal microscope can be improved.

The three characteristics associated with **confocal scanning microscopy** are point illumination, point detection and confocal imaging. The sectioning aspect plus the serial way in which the data become available make confocal scanning microscopy very suitable for coupling to a computer system. Scanning the illumination and the confocal aperture together builds up a scanned image of a selected plane. If a series of such 2D sections are taken at different depth within an intact thick sample a stack of images representing the 3D structure of the specimen can be computer constructed [90,91]. The imaging properties of confocal light microscopy are fundamentally better than in conventional light microscopy. In comparison with electron microscopy the resolution is of course considerably lower. Confocal microscopy generates thin sub-μm optical slices through thick specimens. A typical thickness of the slice being imaged is approximately 0.7 μm. Confocal microscopes [92–94] have a very high level of discrimination against light from outside the image plane, and they have shown themselves to be capable of providing high-quality images from significant depths below the surface of highly light scattering materials. Profiles can be measured with an accuracy of 0.04 μm. It is this ability to reduce out-of-focus blur, and thus permit accurate non-invasive serial optical sectioning, that makes confocal scanning microscopy so well suited for imaging and 3D tomography. The lateral resolution of a light microscope, which is limited by the wavelength of the light source, is improved to about $\lambda/2$ by application of confocal scanning microscopy. As argued before, confocal microscopes also allow for improved axial resolution. For confocal microscopy to be successfully applied, the specimen must be reasonably transparent to allow light to penetrate to regions below the surface of the specimen. If this is not the case, only microtome sectioning can solve the problem. Confocal microscopy is commonly used in Raman microscopy (*cfr.* Chp. 5.6.3).

The **advantages** of the confocal scanning optical system over conventional microscopy are several fold (Table 5.18). Pawley [93] has described the fundamental limits of confocal microscopy. Confocal designs are primarily limited to fluorescence and single wavelength bright-field images and cannot currently provide polarising, phase-contrast, or interference contrast images.

Table 5.18. Advantages of confocal scanning optics

- No time-consuming sample preparation
- Non-invasive, non-destructive optical sectioning
- No high-vacuum requirement
- Improved image contrast by reduction of out-of-focus signals
- High quality of images (sharp depth discrimination)
- Improved effective spatial resolution as compared to ordinary light microscopes
- Exceptional axial resolution
- High imaging depths (up to 200 μm)
- Unusually clear examination of thick and light scattering objects
- 3D imaging with x, y scan over wide areas of the specimen
- Allowance for quantitative studies of the optical properties of the specimen

Confocal scanning microscopy can use non-laser and laser illumination sources, but in practice only the latter provide sufficient brightness. **Confocal scanning optical microscopy** (CSOM), in which the image is built up by synchronous scanning of the source and the detector units, can operate in transmission, reflection, or fluorescence mode. The lateral resolution of CSOM is of the order of 200 nm, which is a factor 1.4 better than that of the current optical techniques. CSOM images are usually sharper than those of conventional microscopy. The technique is some twenty years old [39].

A stage-scanning confocal microscope (SSCM) was first developed by Minsky in 1957 [1,95], but its wider application had to await the arrival of low-cost reliable **lasers** and high-quality scanning systems [96]. The first operative laser was developed in 1960. In microscopy, lasers are practically used as intense, monochromatic light sources. Lasers can produce light beams with very high degree of monochromaticity, which implies a high degree of coherence. Gratton *et al.* [97] have discussed laser sources for confocal microscopy, *i.e.* lasers commonly used in fluorescence microscopy. Laser beams can be easily focused to spots of 10 μm and even down to 1 μm by means of a microscope. Davidovits *et al.* [98] have designed a **scanning laser microscope** with scanning the light beam (a He–Ne CW laser) instead of moving the object itself (as in Minsky's original microscope). It is possible to create a virtual "thin section" in the plane of interest within a thick specimen. Confocal laser microscopy has emerged from the development of various scanning methods, application of confocal optics, and the integration of lasers as light sources into optical system.

The **confocal laser scanning microscope** (CLSM), which improves resolution of a conventional reflection light microscope by replacing the light source with monochromatic light of high coherence (laser) and to a lesser extend by introducing a pinhole in the backfocal plane (confocal mode), is a powerful tool for obtaining detailed 3D information about polymer morphologies. By using the scanning technique, the lateral resolution can be improved by a factor of 2 to 3 compared with a classical light microscope. Depth resolution (focal depth) depends on the wavelength of the light source, and the pinhole size. The most common lasers are a He–Ne laser at $\lambda = 1152$ nm (IR) or $\lambda = 632.8$ nm, an Ar laser with $\lambda = 514$ nm, 488 nm, or a short wavelength He–Ne laser with $\lambda = 344$ nm, which can all be used for reflection as well as fluorescent studies. A 632.8 nm laser allows a maximum depth resolution of *ca.* 500 nm and with a 344 nm laser a depth resolution of *ca.* 300 nm can be reached by using the minimum pinhole size. Niggli *et al.* [99] evaluated the modifications necessary to affordably upgrade a commercially available CLSM for use with UV excitation. Torok *et al.* [100] have developed a new confocal scanning IR microscope. Contrast (intensity difference) observed in a CLSM is caused by different interactions of the incident beam with the sample, for example reflection, refraction, fluorescence, scattering or absorption. Both the reflection and the fluorescence modes are widely used. CLSM in reflectance mode is a very effective imaging technique [93]. Using a CLSM 3D microscopy is possible.

Table 5.19 summarises the **main characteristics** of CLSM. UV/VIS laser scanning confocal microscopes expands the range of confocal applications to include UV-excited fluorophores. Sample preparation for CLSM is comparatively simple, as long as the measurement can be done in reflection mode, and damage to the sample is negligible when compared to the influence of an electron beam. The resolution of images from the new generation of CLSMs is approaching that achieved by the microscope itself. However, the resolution limit is still rather low (about 140 nm), and structures with a size smaller than this resolution limit require electron microscopic methods. As opposed to CLSM, a normal optical microscope produces poor images when the sample surface is rough or the signal comes from a range of depths in a transparent sample. An obvious disadvantage of using focused light is the limited spatial resolution. Whereas confocal microscopes are *diffraction-limited*, the diffraction barrier can be overcome, as in stimulated emission depletion (STED) microscopy where $\lambda/20$ resolution was obtained [101,102].

Table 5.19. Main characteristics of confocal laser scanning microscopy

Advantages:
- Non-invasive serial optical sectioning ("optical microtome") rather than defocusing
- Optical rejection of out-of-focus information
- High point resolution (*ca.* 140 nm for $\lambda = 325$ and 442 nm)
- Significantly increased axial (z) resolution (*ca.* 350 nm) over conventional optical microscopy
- Extended depth of field and height profiling by axial scanning
- Good system sensitivity (high intensity)
- Wavelength-selective 3D imaging
- Improved fluorescence resolution
- Ease of operation

Disadvantages:
- Reasonably transparent specimens required
- Invisible out-of-focus structures
- Restriction on illuminating wavelength (laser dependent)
- Diffraction-limited
- Relatively expensive

CSLM is also a powerful technique for 3D imaging in fluorescence [103]. **Confocal scanning fluorescence microscopy** (CSFM) is often carried out with a dual line Ar/Kr laser (488 nm/568 nm) fitted as standard to provide a combination of blue and green excitation, while other optional lasers can be attached via fibres. Contrast is generated in CSFM by using the emission of fluorescent light after excitation with a laser wavelength of a structural unit in a sample component. Using an argon ion laser fluorescence is excited at either 488 or 514 nm and is detected at wavelengths longer than 515 or 550 nm, respectively. The use of pulsed lasers in fluorescence microscopy has been limited. As in other fluorescence microscopy techniques [104] with laser confocal fluorescence microscopy (LCFM) contrast may be enhanced by labelling or staining. A great advantage as compared with traditional methods of physical sectioning, *e.g.*, using a microtome, is that in principle the specimen is left undamaged, although photo-damage may occur. By minimising the light intensity to much less than 0.5 mW photobleaching is virtually eliminated. The same result may be achieved by image accumulation. The signal-to-noise ratio of the CSFM is significantly enhanced over a conventional fluorescence microscope. Confocal laser-scanning microscopy (commercially available as from 1987) allows slicing incredibly clean, thin optical sections out of thick fluorescent specimens, to view specimens in planes running parallel to the line of sight, to penetrate deep into light-scattering material (up to *ca.* 200 μm) for gaining impressive 3D views at very high resolution, and to improve the precision of microphotometry. Series of optical sections can be displayed as stereo pairs. The confocal, fluorescent optical sections can also be displayed side by side, with (non-confocal) bright field or phase-contrast images acquired concurrently using the transmitted, scanning laser beam. Fluorescence confocal microscopy is without doubt the major contribution to the literature dealing with confocal microscopes enabling to view weakly fluorescing domains by optical sectioning. Modern scanning fluorescence microscopes accommodate analysis techniques such as multi-photon microscopy, fluorescence correlation spectroscopy and fluorescence lifetime imaging. Some problems afflict fluorescence CLSM. As polymers usually do not show fluorescence, they have to be labelled, which sometimes makes the sample preparation complex, unless the label can simply be mixed with one of the components. Alternatively, additional synthesis steps need to be performed. The interrelations between the main expressions of light microscopy are given in Scheme 5.2. The confocal approach is also being applied in Raman microscopy.

Image processing by software enables generation of digital data sets which allow accurate quantitative measurements. Webb *et al.* [105] have dealt with **quantitative fluorescence imaging** with CLSM. As the intensity of the laser is computer-controlled, fluorescence can be measured quantitatively. Quantitative CSFM demands the highest possible efficiency of collection and detection of the fluorescence photons in order to maintain sensitivity, speed and spatial resolution. Efficiency as high as in conventional microscopy may be designed into CSFM instrumentation. Fluorescently labelled monodisperse MF particles (0.5 to 15 μm) can be used as standards for CLSM.

The foundations of confocal scanned imaging in light microscopy have been reviewed [24]. Other reviews deal with confocal microscopy [106,107] and confocal laser microscopy [108]. CLSM has also been reviewed [109], in particular also for polymer science [110]. Several (hand)books on confocal microscopy [92,93,111] and on CSFM [92,111] are available. An early report on scanning laser microscopy has appeared [98] and history has been described [95].

Applications

Some typical applications of CLSM are: observation of glass fibres, or of metal fibres in conducting polymers (distribution, length, orientation, connectivity), detection of fluorescent adhesion layers (sizings) between glass fibre and polymeric matrix, rapid assessment of surface roughness, non-destructive determination of layer thickness of multilayer laminates, *etc.* Confocal fluorescence microscopy can be used for non-destructive analysis of the 3D morphology of blends and composites [112]. A dye-labelled ethylene-butene rubber was used as a fluorescent tracer for the impact modifier phase in CSFM studies of TPO morphology [113]. Almost any fluorescently labelled specimen benefits from examination by CSFM.

CLSM can be applied as a quality control device and finds useful application for the evaluation of the degree of mixing of additives in polymers and the presence of agglomerates. CLSM can be used to identify pigment agglomerates in pellets rather

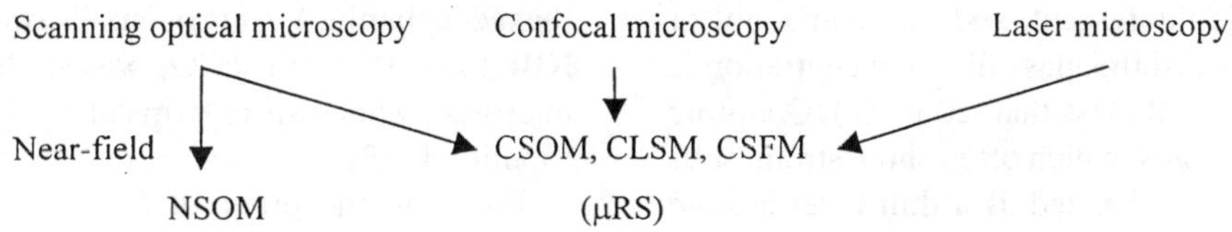

Scheme 5.2. Optical microscopy family-tree.

than diluting the material and blowing a film [114]. In fact, with these techniques it is no longer necessary to prepare thin sections to investigate the dispersion of **fillers** in transparent polymeric matrices. CLSM has been used to determine the 3D spatial distribution of silica particles used as antiblocking agents in LDPE films and masterbatch granules and to investigate the nature of silica-polymer interfacial region [115]. To obtain sufficient contrast between silica and the polymer during CLSM imaging it was necessary to fluorescently label the silicas with Rhodamine 6G dye before addition to the polymer. The performance of such porous silicas is critically dependent on the location of the particles in the film matrix. Despite the use of a two-stage mixing process prior to film blowing the distribution of 0.5 to 15 μm diameter silica particles through the film was uneven. Fluorescent contaminants in chalk and Ca- or Zn stearates can also be put to good (analytical) use in dispersion studies.

CLSM is a powerful technique for characterising **coating microstructure**, as shown for TiO_2 pigments dispersed in an acrylic urethane binder, Al flake pigments and pearlescent-pigmented coatings [116].

One of the major difficulties in any investigation of **interface regions** in heterogeneous material is the fact that the interface cannot be isolated from the sample for examination, but must be studied *in situ*. CSOM permits such non-destructive examination of interface regions (*e.g.* fibre-matrix region) or derivation of more general information on composite structure. CSOM in reflectance mode has been used in investigations of the transcrystalline interphase in fibre-reinforced thermoplastic polymer composites, such as Twaron aramid fibre in PP, and in fluorescence mode in GFR epoxy composites [117].

All commercial glass fibres contain a size, which consists of a silane, a film former and various additives, such as an emulgator, lubricant, antistatic and stabiliser. The performance of a glass fibre is strongly influenced by the size. With CSLM, the **size of glass fibres** in compounds or composites can be made visible without any sample preparation, provided the size is fluorescent, and the matrix not or hardly, and provided the glass fibre concentration is not too high (typically less than 20 wt.%). Coupling agents for glass fibres, which often show strong self-fluorescence, can be located as a thin layer around fibres and allow a detailed examination of the distribution of such coating material [117]. For glass fibres the best optical resolution is about 0.5 μm if the fibre axis is in the focal plane. For vertically oriented glass fibres the optical resolution is much worse. By means of a series of optical slices through a fibre in an injection-moulded compound, homogeneity and thickness of the size layer can be determined. Size morphology was studied by CSLM for GFR SMA and PP [118]. CLSM does not identify the whole size, but only the fluorescent component in the size.

Polymer materials studied by CLSM have included fibre-reinforced composites, where a transparent epoxy resin matrix allowed internal interfaces to be seen [117], and latex suspensions [119]. Clarke *et al.* [120] explored the maximum usable depth of scanning laser confocal imaging, comparing fibre orientation measurements of fibre-reinforced composites by both reflected light and confocal methods.

It is important to be able to monitor the **distribution** of a chemical species throughout a fibre for at least two reasons. If the material is a dye, colour fastness or durability will be affected by its distribution, surface dyeing being much more fragile than if the dye penetrates throughout the fibre. Similarly, if the material is a stain, then the ability to remove it by washing or bleaching would be affected by how deeply it penetrates the fibre and how accessible it is to cleaning systems. Moss *et al.* [90] have reported CLSM of nylon fibre dyed with Nylosan Red. Distribution profiles were measured, and as the same amount of dye was applied to each fibre, semiquantitative measurements could be made. In CSFM at least one component of a multiphase polymer system has to be labelled. As fibres show a very weak autofluorescence, fluorescence of a dye can be used to investigate its distribution in fluorescent mode. De Clerck *et al.* [121] correlated the distribution of strongly fluorescing anthraquinone or benzodifuranone dyes in polyester fibres with that of TiO_2 (present as a delustrant). In contrast to the anthraquinone dye, benzodifuranone seemed to aggregate on the TiO_2 particles, resulting in a much more heterogeneous distribution of this dye. Also impregnation of a fluorescent probe [4-(hexadecylamino)-7-nitrobenz-2-oxa-1,3-diazole, or NBD] into PP from $scCO_2$ was studied by confocal microscopy analysis in terms of partitioning and distribution [122].

Pores on the order of 5 to 15 μm are visible with confocal light microscopy. CLSM can also be used for thickness measurements of multilayer films. With confocal microscopy and digital image processing, 3D measurements of structure in phase-separated mixtures of polymers can easily be obtained. Li *et al.* [123] have been exploring the applications of CSFM to the **morphology of blends** of PS and PMMA, labelled with NBD. A requirement for application of CLSM is transparency of the matrix. This is a limiting factor for many polymers, which are slightly opaque. Yet, in order to take full advantage of the sensitivity of CLSM in those instances sections are often prepared and examined. Various applications for the CSFM are also suitable for the conventional fluorescence microscope, which however is much less sensitive. In case of a black specimen, such as aPP/bitumen (for roofing), CSFM samples the surface. Image analysis (morphology of bitumen, particle size, interparticle distances) then allows establishing the compound ratio.

Early stages of **polymer oxidation**, as for UHMWPE/HMWPE blends, can be detected by CSFM. A comparison between CSFM and chemiluminescence imaging (ICL) for this purpose is still lacking. Applications of CLSM in *in situ* (*i.e.* non-invasive) mapping have been reviewed [90].

5.4. ELECTRON MICROSCOPY

To obtain real-space information about the morphology of polymeric materials, various optical microscopic methods such as OM and CLSM are available (*cfr.* Chp. 5.3). Use of electrons as a light source for microscopy opens other perspectives [124]. Electron microscopy (EM) provides structural information in both the real and reciprocal space. Electron microscopy is used very much as an imaging method and has a great number of variants and special techniques. Electron optical techniques can be used to probe the atomic and electronic structure of materials, some with atomic resolution, both inside the materials and at their surfaces. In order to interpret the images and diffraction patterns, the elastic and inelastic scattering processes of electrons in matter must be understood. General disadvantages of EM techniques are the need for sample preparation, high vacuum conditions, radiation damage and imaging in static conditions. Table 5.20 gives an overview of the main **far-field electron microscopy techniques**. Early work on electron microscopy and significant developments for transmission electron microscopy (TEM), scanning electron microscopy (SEM) and scanning transmission electron microscopy (STEM) are due to Ruska and Knoll [125] and Von Ardenne [126,127].

TEM allows observation of structures down to sub-nanoscale but requires complicated and time consuming sample preparation. Aberration-free atomic resolution can be produced in a commercial 300 kV TEM. The TEM image is not simple to understand at high magnification. Although it is possible to calculate the image of a given structure, it is in general not possible to reverse this procedure. With SEM only the surface of a sample can be investigated. By using an etching technique, the inner structure of a specimen can be made visible.

Conventional SEM (developed originally with thermionic emitters) operates typically in high-vacuum conditions and at high accelerating voltage (*e.g.* 10–40 keV), offers an image resolution of some

Table 5.20. Electron microscopy techniques

Instrument	Conventional SEM	LVSEM[a]	TEM	STEM
Specimen type	Bulk (∞)	Thick	Ultrathin	Thin
Beam energy (kV)	10–40	1–5	80–400	80–200
Useful magnifications	20–50,000×	20–100,000×	3000–5 × 10^6	3000–300,000×
Image resolution	1 nm[b]; 4 nm	3 nm[b]; 20 nm	0.15 nm	<0.1 nm
X-ray spatial resolution	1 μm	(0.1 μm; few X-rays produced)	0.1 μm	0.1 μm
Features	Surface topography	Radiation sensitive structures	Microstructure	Microstructure
	Atomic number contrast		Microcomposition	Microcomposition
			Internal morphology	Internal morphology

[a] Low-voltage SEM.
[b] For field emission.

5 nm and is equipped with microanalytical facilities (EDS, WDS). Disadvantages of the technique are that sample preparation is usually necessary and that non-conductive samples (such as polymers) are difficult to observe. For these materials charging effects take place while observing the sample, leading to featureless image formation. This has spurred development of new techniques, such as LV-SEM (low-vacuum SEM) or environmental SEM (ESEM), LVSEM (low-voltage SEM), and more recently of FEG-SEM (field emission gun SEM using Schottky emitters). The chief disadvantage of scanning methods is that information is acquired serially, pixel by pixel, whereas in TEM all pixels are imaged simultaneously, leading to a much shorter exposure time. High-brightness field-emission guns (FEG) eliminate this difficulty. In contrast to the conventional TEM in which a large surface of the illuminated object is imaged by an objective and projected in the image plane, in a STEM the object is scanned by a focused electron beam and the electrons scattered by the transmissive object are collected by a detector and used in the modulation of intensity on the screen of a monitor. Direct imaging with TEM, or using SPM techniques, such as STM or AFM, can yield information with **molecular resolution**. The three techniques, TEM, STEM and SEM are fast approaching the limits set by basic physics. State-of-the-art instruments rely increasingly on computers needed for high performance, and for the production of images and spectra.

With the great range of TEMs, SEMs and STEMs now available, the range of acceptable specimens is extremely wide. Electron microscopy may be used to investigate polymeric materials provided that the structure, organisation of the components of the materials (additives, fillers, *etc.*) and properties of macromolecules can be preserved. In electron microscopy, **specimen preparation** (using cryo ultramicrotomy, evaporation, etching or staining) is often the essential key to good results. Most polymers show very little contrast in the electron microscope because there are generally no heavy atoms in the sample which scatter the electrons outside the objective aperture. Consequently, it is frequently necessary to introduce heavy atoms into specific parts of the specimen. In semi-crystalline polymers one generally tries to stain the amorphous regions, whereas in two-component systems it is necessary to stain one of the components. There is an increasing interest in the real surface structure of samples, *i.e.* without pretreatment, such as decoration or coating, which disturbs image formation of the surface. Sample preparation for electron microscopy was recently reviewed [128].

A fundamental limitation of almost all microscopy investigations of materials is that the images are static and taken when the specimen is at room temperature. More particularly, in EM the specimen is also in a high or ultra-high vacuum and under intense radiation. Unfortunately, all these conditions rarely represent the treatment that the material has received during its processing to final form, or the conditions it will suffer during its service life. Some of these limitations can be (partially) removed, as in case of megavolt TEMs, which accommodate heating, tensile and gas-reaction stages, and in low vacuum or variable-pressure SEMs (ESEMs). Other important improvements are the ability of recording dynamic images with modern video/CCD cameras as well as storing the images and processing them digitally. Microscopy now covers a wide field of materials characterisation, combining the abilities to heat, cool and deform bulk specimens in the SEM with the ability to image their surfaces under significant pressure of a gas that may also react chemically with the specimen. Scanning probe microscopes allow other degrees of freedom (*cfr.* Chp. 5.5). The ability to study dynamic materials directly, *in situ*, close to their natural state as they undergo reactions is a very important goal in materials research and technology. ***In situ* microscopy** under dynamic conditions with real-time monitoring of events provides information on material processes that cannot be obtained directly by other methods. Dynamic *in situ* microscopy is a rapidly expanding field [129] and comprises techniques such as HVEM, EHREM, ESEM, FESEM, FEG-ESEM and STM.

The ultimate limitation of electron beams in microscopy is, of course, the **radiation damage** they can inflict on the specimen. In EM a compromise must be sought between radiation damage, specimen thickness and the information which can be obtained. With high-resolution electron microscopy, all organic systems are generally very sensitive to electron irradiation. This is a fundamental and unavoidable problem. The atomic resolution that can be achieved in EM arises from the strong interaction between electrons and matter right down to the atomic level. In passing through matter, electrons transfer energy and cause damage by elastic scattering by excitation and ionisation of individual atoms (charging), collective (plasmon) excitation of electrons in a molecule or in a crystal lattice (atomic displacements), ejection of atoms (sputtering) and secondary effects resulting from these interactions (radiolysis). This can cause rapid degradation in many materials, but can often be alleviated by cooling the specimen to liquid-nitrogen or liquid-helium temperatures. Also low-voltage SEMs can give useful information on beam-sensitive materials. Polymers are notorious for their resistance to examination by electron microscopy in view of: *(i)* radiation damage; *(ii)* charging and heating phenomena; *(iii)* difficulty in establishing contrast; and *(iv)* difficult, detailed examination of surfaces [130]. Polymeric specimens release volatile hydrocarbons when irradiated and therefore contamination of the surface with depolymerised hydrocarbons is inevitable. Problems are alleviated (at the cost of much sample preparation) in TEM by staining (RuO_4, OsO_4) or surface replication.

When the high-resolution imaging and diffraction capabilities of TEM, SEM and STEM are combined with qualitative and quantitative **X-ray analysis**, the micro-characterisation of materials is significantly extended. Microanalysis can often be accomplished using X-rays with energies from ~1–10 keV, and this is the typical range used in the SEM. TEM has a much higher accelerating voltage, and allows to detect much higher energy X-rays. Areas from thin films as small as a few hundred Å across can be analysed in terms of their elemental nature with STEM optics (*cfr.* Chp. 5.4.3). Apart from obtaining a spectrum giving the elements in question, elemental maps can be obtained in the scanning mode from areas as small as 10^{-3} mm^2. Furthermore, the chemical composition of materials can be established with a resolution on the atomic scale, again by the use of very fine electron beams. A combination of elemental spot, line scans and elemental analysis with various topographical, structural, and crystallographic information greatly extends our knowledge of a material.

Haguenau *et al.* [131] have recently traced the history of the development of electron microscopy. Electron microscopy in polymer science was reviewed [132].

5.4.1. Scanning Electron Microscopy

Principles and Characteristics

The scanning electron microscope (SEM) is often the analytical element of choice when the light microscope no longer provides adequate resolution. In SEM an electron beam is focused into a fine probe and is subsequently raster scanned. As the beam interacts with the sample it creates various signals. By using these signals an image is formed. SEM is the most important electron-optical instrument for investigation of bulk specimens with 0.5 to 50 keV electrons, typically 20–30 keV. The original idea of scanning electron microscopy dates back to the 1930s; commercial SEMs appeared in 1965, based on the work of C.W. Oatley *et al.* [133] of Cambridge University.

The SEM consists of an electron-optical column mounted on a vacuum chamber with an electron gun placed on top of this column, a sample chamber with specimen stage and an electronic system for image display. The objective lens is used to focus the electron beam into a fine spot on the sample surface. Upon entering the sample, the electron beam interacts with the solid and a variety of signals are generated (*e.g.* secondary electrons, internal currents, photon emission, *etc.*), which are collected by dedicated detectors, amplified and displayed. The pressure in the specimen chamber (10^{-3} to 10^{-5} Pa) is much lower than the saturation vapour pressure of water, requiring special preparation of water-containing samples [134]. As the values for the depth of focus obtainable in a SEM are a factor of 100–1000 larger than in a light-optical microscope, the former is often preferred to light microscopes at low magnifications. This is particularly true when irregularly shaped specimens with large height differences are to be observed. Also better visualisation of objects such as fibres, fracture surfaces and powder particles is achieved with improved statistical interpretation. Contrast in most SEM images is largely determined by electron scattering and detector characteristics. In SEM a **topographic image** is obtained by collecting preferentially the secondary electrons (SE) which are emitted from the surface. Information can be gathered concerning size, shape and texture of many solid specimens. The practical resolution limit of SEMs is about 20 nm; FEG-SEMs allow ultra-high resolution (1.0 nm at 15 kV or 2.2 nm at 1 kV).

SEM requires little **sample preparation** for metallic and inorganic materials as the information required concerns only the surface structure and the material composition of the layer proximate to the surface. Small samples of up to several millimetres and sometimes even larger can be investigated directly in the SEM if the sample material has a sufficiently high electric conductivity to prevent charging produced by electron bombardment. However,

SEM observation of non-conductive samples, such as polymers, is not possible without a metallic or carbon conductive layer on their surface. Such a coating avoids build-up of an electric negative charge in the specimen, which would induce "imaging artefacts". Charging effects such as bright spots on the image, sample drift, or radiation damage are highly detrimental to the observation. Another possibility to avoid the charging of the specimen is to decrease the energy of the electron beam (*cfr.* LVSEM, Chp. 5.4.1.1). The different techniques of preparation of polymers for examination by SEM are given elsewhere [135]. For polymers, good cross-sections can usually be obtained by breaking the specimen at liquid-nitrogen temperatures.

In SEM various electron-specimen interactions can be used for **imaging** and **microanalysis**. The primary electrons (PE) of the electron probe produce secondary electrons (SE, exit energies <50 kV), which can escape from a small depth of about 1–10 nm. By rastering of the electron beam across the surface the secondary electron imaging (SEI) signal provides "near-surface" interpretation of sample morphology. SEI is the principal imaging mode, providing the best spatial resolution, and deriving contrast mainly from surface topography. Backscattered electrons (BSEs) are beam electrons scattered back out of the sample. BSEs with electron energies 50 eV $\leq E \leq$ eU (acceleration voltage U) cover a different information depth. These electrons penetrate much deeper into the sample than secondary electrons and still emerge from the sample to be detected. Backscattered electron imaging (BEI) derives contrast either from the mean atomic number of the substrate, or from topography – specifically line-of-sight to the detector. The percentage of beam electrons that are backscattered depends on the atomic number, hence its utility for analysing material composition. The BEI signal thus provides information regarding variability in sample composition, density, and surface geometry.

In SEM two commonly used signals for compositional investigations are X-rays and backscattered electrons. Elemental imaging provided by SEM uses high-resolution (<10 μm) electron beams for excitation. In inelastic scattering events the primary electron ionises target atoms. The ionised atoms fall back into a lower energy state with the emission of Auger electrons or X-rays. The energy of the X-rays is characteristic of the atom involved and is used in most SEMs for **elemental microanalysis**. The elemental composition can be analysed by a Si(Li) detector in EDS mode or by a wavelength-dispersive spectrometer (WDS). In combination with highly resolved images, it is possible to carry out qualitative and in part quantitative determinations of very small amounts (down to 10^{-16} g) of impurity elements by point, line or surface analysis. Microprobe analyses of plastics, filler and reinforcing substances, pigments, stabilisers and modification agents of elements $Z > 6$ can be registered quantitatively [136]. Relatively thick metal coatings must ensure electrical conductivity and thermal stability. For elemental analysis carbon-black coatings are used. The preparation is elaborate. SEM-EDS is semi-quantitative with a sensitivity of >0.1 wt.%. Average analyses corrected according to ZAF display relative errors of up to 20%. Element distributions can be registered as well. In SEM mode the lateral resolution attainable for thin sections is >5 nm. With solid samples, layers of thickness of up to *ca.* 1–10 μm can be analysed. When a high depth resolution is required for analysis, Auger spectrometers (depth resolution about 5 nm) can be coupled to SEM. XRF and AES are competitive techniques to SEM-EDS. Recently, combined electron and X-ray induced microbeam XRF in SEM has been reported [137].

Table 5.21 shows the **main characteristics** of SEM. The main advantages of SEM are the high lateral resolution (depending on the gun coherence), large depth of focus (typically 100 μm at 1000$\times$ magnification) and the numerous types of electron-specimen interactions that can be used for imaging or chemical analyses purposes. SEM has the ability to cover a wide magnification range (*e.g.* 10$\times$ to $10^5\times$) so that an area first observed at a low magnification can be studied at high magnification and resolution. The large depth of field of SEM makes it possible to image very rough surfaces with millimetres of vertical information within a single image. The principal limitations of SEM are cost and instrumental complexity because a vacuum system is required. Problems in analysis of polymers by SEM are also related to sample preparation, beam penetration effects, charging, beam damage and outgassing of low-MW components. Moreover, SEM offers only vague vertical information. Low-voltage SEM (LVSEM) offers the advantage that charging of insulating samples can be avoided.

State-of-the-art analytical capability now provides a chemical and structural analyser for SEM,

Table 5.21. Main characteristics of scanning electron microscopy (SEM)

Advantages:
- Bulk specimens; minimal sample requirements
- Non-destructive
- Established technique for large variability in magnification ($>10^5\times$), surface imaging and composition (elemental mapping)
- Lateral resolution: 1–10 nm
- Large sampling depth (few nm to few μm depending on accelerating voltage and high-contrast mode of analysis)
- Numerous types of electron–specimen interactions for imaging and analysis (SE, BSE, EBIC, CL, EDS, WDS, EBSD)
- Wide applicability

Disadvantages:
- Vacuum requirements
- Some beam damage
- Resolution limited by electron probe diameter
- No bonding information
- Only indirect depth profiling capabilities
- Relatively high instrument cost

namely a combination of SEM-EDS with micro-Raman spectroscopy. This allows unambiguous chemical and structural characterisation of a wide range of samples at the μm scale under HV, UHV, or "environmental" conditions. A considerable advantage of **SEM-Raman** is that the spectrometer explores the same area as the SEM image. The X-ray ultra microscope (XuM) is an accessory to SEMs that uses the electron beam to generate X-rays for transmission through a sample. This enables sub-μm X-ray imaging of optically opaque objects.

SEM is traditionally performed in a vacuum, with the vast majority of microscopes operating at pressures below 10^{-2} Pa. However, nowadays scanning electron microscopy can also be carried out at relatively high pressures, as in environmental SEM (ESEM) [138], low-vacuum SEM, or variable pressure SEM (VPSEM) [139]. **Variable pressure SEM** can handle large objects (up to 250 mm $\varnothing$, 70 mm in height, 2 kg). Variable pressure technology permits examination of virtually any sample without the need for traditional sample preparation techniques. Typical resolution figures are 3.0 nm at 25 kV at high vacuum and 4.5 nm at 25 kV at variable pressure. Thus it is possible to obtain ultra-high resolution imaging and analysis (VPSEI) on specimens that are completely non-conducting, moist or outgassing. High-pressure SEM (HPSEM) operates with a gas pressure in the specimen chamber in the range of 0.1 to 30 Torr. The "high pressure" in HPSEM is high only in comparison to the vacuum inside a normal SEM ($p < 10^{-6}$ Torr). High-resolution SEM (HRSEM) combines a field emission gun with a short focal length final condenser lens. HRSEM is very often operated at low beam voltage, and the technique may be referred to as "high-resolution low-voltage SEM". Thermal field emission SEM (FESEM) includes both conventional and low-vacuum instruments. A well-known technique to recover the third dimension is the usage of stereoscopic images, as in stereoscopic SEM images [140]. The latest generation of these instruments is increasingly used for process and product control.

SEM and AFM are complementary techniques for surface investigations. However, the image formation mechanisms are quite different, resulting in different types of information about the surface structure. By using two techniques which are complementary, one technique will often compensate for imaging artefacts of the other.

Detailed reviews are available [53,134,141–144]. The principles of operation of SEM are well covered in the literature (*cfr.* Bibliography). Quantitative SEM is described in ref. [145].

Applications

Typical SEM applications are:
- Analysis of variations in surface morphology *cq.* topography as related to adhesion performance
- Examination of phase distributions, rubber distributions, particulates and fibres
- Thickness measurements of polymeric coatings and films
- Determination of particle dimensions, particle density, *etc.*, using image analysis software.

The major fields of application of SEM in microanalysis are tests for pigments, colorants, fillers, flame retardants and all sorts of additives (with elements $Z > 6$) in polymers and varnishes, and the examination of metallic raw materials and catalysts. Typical applications concern the analysis of polymer surface structures in relation to technological parameters, such as gloss, strength (defects, cracks) and adhesion, fracture surface studies (*i.e.* fractography), internal resin morphology, particle size and shape, and contamination. To image the homogeneity and chemical composition of individual additives in polymers, SEM-EDS and a backscattered electron detector may be employed [146]. The depth of field and small beam size make it possible to image fibres far below the top layer.

In the determination of **filler particle size distribution** in a composite, various approaches may be used, such as removal of the polymer by solvent extraction or ashing. Both techniques preclude characterisation of filler dispersion, homogeneity and aggregation. Such detail may be derived from SEM analyses of composite fracture surfaces. However, where fracture of the composite occurs primarily through the polymer matrix fracture surfaces result in which the filler particles are embedded in the matrix material, and effectively invisible. Oxygen plasma etching for SEM observation of the structure of inhomogeneous polymers is a superior method for removing polymers from the surface by oxidising them into gases, thus avoiding a temperature increase during etching [147]. This pretreatment is effective for evaluating the distribution and orientation of fillers in filled polymers (*e.g.* PP/$CaCO_3$ [148]) and the phase structure in segmented PURs and in fluoropolymer blends. Plasma polishing of fracture surfaces enables real particle size distributions to be readily characterised without any damage to the filler particles or the location. Weale *et al.* [149] examined the **fibre orientation and distribution** (FOD) within an injection moulded nylon compound by means of SEM imaging analysis. It is well known that FOD is a function of many parameters, including component geometry, moulding conditions, matrix material, polymer melt viscosity and fibre type. Various imaging techniques for analysis of short fibre polymer composites, including assessment of fibre orientation distributions, are available [150]. SEM can be used to observe the interaction area between fibres and polymeric matrix in case of adhesive failure in fibre-reinforced composites. SEM-EDS finds application in tyre cord analysis.

Other applications are examination of porous and pigmented or filled polymers (size, shape, distribution and orientation of pores or dispersed components), and the study of multicomponent and weathered materials. Examples are the determination of the distribution of **rubber particles** in ABS, the distribution of inorganic pigments or fillers in the surface (production of specks, irregular gloss) or in the layer proximate to the surface (haze) [135]. The method is also suitable for quantitative characterisation of reinforced composites. Mills *et al.* [151] have reported novel image analysis methods for quantitative assessment of the dispersion quality of **flame retardants** in polyolefins using SEM. Owen *et al.* [152] have used SEM to study the dispersion in ABS flame retardant formulations of four Sb_2O_3 materials with average particle size ranging from 0.5 to 11.8 μm. Fan *et al.* [153] have used SEM and image analysis for the quantitative characterisation of the dispersion and distribution state of **carbon-black** in HDPE. SEM micrographs of Ketjenblack (KB) filled HDPE/iPP blends allowed the selective location of KB in the HDPE phase [154] (Fig. 5.3). The SEM micrographs also showed that KB could affect the morphology of the blends. According to ref. [155] SEM of degraded specimen surfaces of HDPE reveal a significant difference in surface topography of unfilled and $CaCO_3$ mineral filled HDPE. Imaging and quantitative image analysis have also been used to assess the homogeneity of a development product consisting of glass and PP fibres [156].

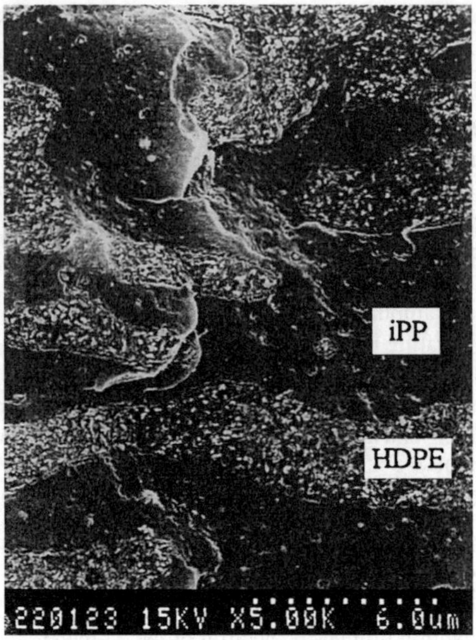

Fig. 5.3. SEM micrograph of HDPE/iPP (50/50) blend filled with 2 phr Ketjenblack moulded at 190°C for 15 min. After Zhang *et al.* [154]. Reproduced by permission of VSP-Brill.

SEM-EDS is often used to quickly identify any **additive**. An example of the value of SEM-EDS is provided by elucidation of the cause of contamination in a pigmented PVC film, where the presence of Pb points to inadequate dispersion of a Pb-based stabiliser [141]. Lead stabiliser migration from a highly plasticised PVC formulation with nonyl trimellitate was evaluated by chemical identification (SEM-EDS) of the deposit on a die insert [157]. EDS in conjunction with FTIR microscopy has been applied for examination of multilayer fragments of automobile paints [158]. Palla [51] has used SEM-EDS for the characterisation of rubber components in polymers. SEM-EDS and PyGC-MS were applied for forensic discrimination of photocopy and printer toners (typically iron oxide or carbon-black embedded in a matrix of organic binder) [159].

Stereoscopic SEM images can profitably be used for the study of the pore structure of foams. Consecutive analysis steps that can be performed are direct depth measurements, profile extraction, profile and area roughness and volumetric measurements. Partial ordering of a low-MW additive (stearyl stearamide) at LDPE foam surfaces has been confirmed by XRD, SEM and ATR-FTIR [160]. SEM backscatter imaging (BSI) and digital image analysis are reliable techniques for the characterisation of paper structure details, such as coating layer structure [161]. The suitability of using SEM for studying the structural properties of filled (dolomite, chalk) paper microstructure was described [162]. Print ink distribution details on commercially printed paper and fibre surfaces were studied using stereoscopic micrographs and SEM (BSI) [163]. The difference in atomic number between fibres and the ink pigment particles was sufficient to discern the ink by BSI.

The application of SEM to polymeric materials has been reviewed [143]. Forensic fibre analysis was reviewed [164].

5.4.1.1. *Low-voltage Scanning Electron Microscopy*

Principles and Characteristics

Scanning electron microscopy (SEM) with electron energies $E = \text{eU}$ (acceleration voltages U) in the 0.5–5 kV range is called low-voltage scanning electron microscopy (LVSEM), whereas the conventional SEM instruments work in the range of 5–30 (50) kV. Although any instrument (also the conventional SEM) can operate at low voltage, special lens arrangements are necessary to improve the resolution, which makes LVSEM *cq.* FEG-SEM or FE-SEM (field-emission gun SEM) into a unique instrument. Use of field-emission electron guns in LVSEM allows ultra-high resolution (spatial resolution of 0.5 nm at 30 kV electron beam energy, of 1.0 nm at 15 kV and of 2.2 nm at 1 kV). Ultra high-resolution FESEM is available over a wide voltage range, variable pressures and temperatures (−185°C to 200°C).

Table 5.22. Main characteristics of low-voltage scanning electron microscopy

Advantages:
- Images obtained from uncoated specimen
- Lower penetration depth (due to reduced electron range)
- Image formation limited to a near-surface layer of about 10–100 nm
- Improved topographic contrast of true surface detail
- Ultra high magnification (FESEM: 650,000×)
- Decrease of charging artefacts
- Less radiation damage
- Increased secondary electron yield
- Better application of electron spectroscopic methods

Disadvantages:
- Reduced resolution (need for correction of chromatic aberration or use of FEG)
- Higher surface contamination rate (need for ultra-high vacuum)
- Stronger sensitivity to electrostatic or magnetic stray fields
- Less effective energy-dispersive X-ray analysis
- Need for special detector strategies

Electron-specimen interactions in LVSEM are often quite different from those in conventional SEM. Reimer [165] has pointed out that the physics of the 0.5–5 kV and 5–30 kV ranges differ in many important respects. As in conventional SEM, knowledge of electron–specimen interactions is important for interpretation of image contrast in LVSEM [166].

The **main characteristics** of LVSEM/FEG-SEM (in comparison to conventional SEM) are summarised in Table 5.22. Low-voltage SEM needs an ultra high-vacuum specimen chamber. LVSEM is not a "destructive technique". No special sample preparation is needed. Image artefacts do not influence observations. Charging of insulating specimens can be avoided in many cases. LVSEMs offer dramatic improvements in image quality and resolution relative to conventional SEMs. LVSEM has the main advantage of a lower electron range, with the information more concentrated in thin surface layers. However, LVSEM is not limited to surface observation. Working at different acceleration voltages allows modification of the electron range and observation of a volume underneath the sample surface. This allows pseudo 3D imaging. For a hydrocarbon polymer (average density 0.9 g/cm^3) the electron range varies from 25 nm at 0.5 kV to 1 μm at 5.0 kV.

At the very low accelerating voltages **microanalytical (EDS) capabilities** are much restricted by lack of suitable X-ray lines for analysis; WDS is excluded altogether but EBSD facilities are available. X-ray microanalysis is still possible at low beam voltages, but is not easy. In fact, although the beam intensity is not lower, the excitation volume is smaller. Consequently, less X-rays are being produced. The low voltage allows only excitation of light elements (N, O, ...).

Using a new technique in EDS, called **position-tagged spectrometry** (PTS) [167], chemical and spatial information (<2 nm spot size) may be combined. PTS is an X-ray spectroscopic method, whereby X-ray photons generated by the scanning electron beam in SEM are tagged with the position of their origin. With PTS it is possible to construct a spectrum comprising data from all pixels belonging to a phase or chemically distinct region. The amount of each phase can be quantified by morphological image analysis [19]. FEG-SEM has been coupled with PTS to characterise the microstructure of composites at sub-μm level [168]. This technique is particularly compatible with low-voltage operation, because it minimises the dwell time at each point, thus reducing charging and specimen damage. Because microstructure is often the link between polymer processing and material properties, the extent to which microstructure can be quantified often establishes the strength of the link (SEM, X-ray microanalysis, PTS). The combination of LVSEM, light element detectors, PTS, and image processing/analysis provides the tools necessary to thoroughly characterise a material both microstructurally and chemically. Using PTS, it is possible to reconstruct maps and spectra after the fact. Because an entire X-ray spectrum is stored for every pixel, regions can be defined from which to construct spectra (*e.g.* the reinforcing component can be discriminated from the matrix phases). In fact, all the tools of image processing are available to select specific regions of the microstructure from which to perform an elemental analysis. These regions can be selected on the basis of any image-contrast mechanism, such as via secondary electrons or elemental compositions [168].

Image formation in LVSEM was discussed [165]. A special issue has appeared [169].

Applications

The application fields of conventional SEM and LVSEM are totally different. LVSEM is generally used on non-conductive samples in order to: *(i)* increase spatial resolution; *(ii)* decrease electron beam damage; or *(iii)* decrease the electron range to obtain information specific to the top surface layer. LVSEM allows investigation of polymers, and biological and insulating specimens without metal coating. The disturbing contrast by electron diffusion is strongly decreased and the information depth is reduced to a surface layer of about 10–100 nm. LVSEM (with limited resolution unless special detectors are installed) appears as a routine tool, whereas FEG-SEM is as yet mostly used for basic research.

Dudler *et al.* [170] have used LVSEM to study an ion-conductive polyamide-based antistat (Irgastat P22) in polyolefins (iPP, HDPE, LDPE). LVSEM is generally used on **insulators** either to increase the spatial resolution or to decrease any beam damage. On conductive samples, contrast due to the conduction is added to the topographical image, giving information on conduction itself. Zandbelt [171] has reported excellent 3 kV images of 50 nm latex spheres in the presence of 5 nm gold particles. Highly beam sensitive materials, such as monoglycerides, were observed. Membranes of PSU, PC, PP and teflon were also easily imaged, free of artefacts [172]. FESEM and AFM were also used to study the interphase regions in rubber-toughened epoxy polymers [173]. Although FESEM provides high-resolution micrographs, it is unable to detect the hyperfine features observed by AFM; AFM can easily distinguish the presence of rubber particles. Watkins *et al.* [174] have reported low-voltage (1 kV) secondary electron images and backscattered electron images (BSI) of Pt/PMP nanocomposite cross-sections confirming the presence of 50 nm particles throughout the thickness of the substrate.

High-magnification imaging (typically $10^5\times$) of non-conducting polymer samples in conjunction with microanalytical sampling capabilities should allow LVSEM to visualise the distribution of inorganic fillers in subsurface layers. By raising the electron beam energy in-depth filler distributions may be studied. In conventional SEM this can only be achieved by means of cross-sections.

Examination of RuO_4 stained samples in LVSEM allows the direct determination of the relative orientation of polymer phases (domain morphology) and mineral fillers, and for this reason is preferred over solvent extraction or acid etching steps that physically remove components from a blend [175].

Low-voltage X-ray microanalysis is a growth area. Position-tagged spectrometry has mainly inorganic applications, *e.g.* SiC fibres in an (in)organic matrix [168].

LVSEM rivals with AFM for high-resolution polymer morphology studies. The application of LVSEM to polymers has been reviewed [54].

5.4.1.2. *Environmental Scanning Electron Microscopy*

Principles and Characteristics

The environmental scanning electron microscope (ESEM), introduced by Danilatos [176,177], has been defined as a SEM that can operate with a specimen chamber pressure from high vacuum up to at least a pressure level that can maintain fully wet specimens, namely up to 609 Pa (or 4.6 Torr), which is the saturation water vapour pressure at 0°C. Wet and damp samples, such as paints, inks, emulsions and biological tissue, are particularly challenging for SEM. The high vacuum requirements in the sample chamber mean that lengthy specimen preparation techniques are required to remove or fix solvent or water before imaging, raising the risk of artefacts. These problems can be overcome in the ESEM, which permits **imaging of wet systems** with no prior specimen preparation. Two basic developments have made this possible. In the ESEM instrument, a series of pressure limiting apertures are placed down the column. By using a system of differential pumping, the electron gun can be maintained at high vacuum while the sample chamber can be kept at a constant pressure of 10–20 Torr. Another crucial development needed was a new type of detector that can operate at a pressure of tens of Torrs. **High-pressure SEM** (HPSEM) or low-vacuum SEM (LV-SEM) is a new family of techniques which permits imaging without sample preparation.

Table 5.23 shows the **main characteristics** of ESEM in comparison to conventional SEM. ESEM incorporates all the advantages of conventional SEM (including EDS for contaminant analysis) but without some of the disadvantages of the latter. There are some differences in operating ESEM *versus* conventional SEM, which can be either advantageous or disadvantageous depending on the situation [178]. It is generally much easier to carry out a "quick analysis" in an ESEM than a SEM, regardless of whether the sample is a powder, wet or otherwise high-vacuum incompatible. Even insulating samples do not require sample preparation. On the other hand, there are more variables to consider when optimising the conditions for imaging than in normal SEM. Charging artefacts in ESEM images are reduced. An ESEM instrument requires a higher investment than conventional SEM.

Table 5.23. Main characteristics of ESEM

Advantages:
- *In situ* microscopy tool
- No sample preparation requirements
- Suitable for "dirty" samples
- Non-destructive observation and analysis
- Allowance for large sample sizes (20 cm dia. × 8 cm height × 5 kg)
- Non-charging imaging
- EDS, WDS and EBSD analyses

Disadvantages:
- More complicated optimisation of operating conditions
- Limited to liquids with relatively low volatility
- Cost

Collins *et al.* [179] have demonstrated some unique advantages of ESEM, such as revealing of sub-surface structures of uncoated specimens. Surface chemistry and chemical reactions in general are open to ESEM. With an ESEM, practically any surface can be examined *in situ*. Wet specimens can retain their natural state because a 100% relative humidity can be maintained routinely. Dry specimens can be examined regardless of their electrical conductivity. The environmental gas becomes a good conductor because of the ionising radiation present and, thus, it substitutes the conventional conductive coatings or other treatments used to prevent specimen charging. The imaging power and information collection capability of an ESEM is not compromised over that of a conventional SEM. Detection of secondary electrons, backscattered electrons, cathodoluminescence and X-rays have been practised. Modern LV-SEMs are equipped with backscatter electron (BSE) and gaseous secondary electron (GSE) imaging facilities. X-ray microanalysis in HPSEM can be disturbed by interactions between the beam and the gas [180].

ESEM is taking a claim to be a separate technique from SEM. It would be wrong to think of ESEM as simply a "leaky" SEM. It is becoming apparent that, whereas secondary electrons in SEM basically yield topographic contrast and backscattered electrons show atomic number, ESEM may produce a contrast mechanism with no SEM equivalent.

Low-voltage (LV)ESEM is a promising new technique for polymer morphological characterisation [181,182]. Operating at low voltage has particular advantages: *(i)* high resolution; *(ii)* negligible beam damage of samples; and *(iii)* absence of the need for coating with conducting films to eliminate sample charging under a beam. Moreover, an LVESEM equipped with a field-emission gun (FEG) source, which provides high brightness, small spot size, and low energy spread in comparison to the conventional beam, can produce images of polymers at a substantially higher magnification with a better resolution than can a conventional SEM, comparable to those of TEM. From the advantages mentioned above, FEG-ESEM at low voltage offers the capability of being able to perform *in situ* deformation experiments. FEG-ESEM operates at higher pressure than ESEM.

ESEM is a substantially under-exploited technique. References [183,184] trace its developments and some recent reviews are due to refs. [185–187].

Applications

There are important applications in materials science in which the role of the environment on a sample is critical. The requirements of SEM, such as a high vacuum and the need for a thin coating if an insulator is being analysed, mean that some types of materials have always proved difficult or impossible to image straightforwardly. The possibility of examining the natural or true surface of practically any specimen has added a new dimension to electron microscopy. ESEM allows even non-conducting, outgassing, dirty, oily or wet samples to be examined non-destructively in humidified or gaseous environments: with no coatings, no cutting, no drying, no cleaning, and no manipulation. With ESEM many experiments may be performed, such as solvent action, melting and solidification cycles, paint drying, and topographic changes of polymers during curing, all in low-vacuum conditions on uncoated specimens.

ESEM was used for the morphological analysis of mixing of a low-MW additive (PDMS) in HDPE in a co-rotating twin-screw extruder with different screw geometries [188]. The structure of the extruded mixture was frozen in liquid nitrogen; PDMS was extracted in toluene, and the surface was carbon coated. A cryogenic fracture gave a surface that could be observed by ESEM. The size of the PDMS droplets in the HDPE matrix were analysed by an image processing method.

One of the early applications of ESEM was to study the surface properties of **wool fibres** by means of observing and measuring the contact angle of water and other liquids (such as detergents), or by observing water migration in fibre structures, in real-time [189]. Other applications are the study of scouring (cleaning) processes of raw wool. ESEM has been used to study wetting behaviour of PP fibres; the profiles of water droplets were clearly observed [190]. ESEM was also applied to measure oil adsorption capacities of various neutral and synthetic fibre sorbents [191]. Forensic investigations using ESEM can help in determining the cause of textile damage like tears, cuts, fibre fractures, and bites [192]. Key to forensic investigation is the validity (accuracy) of proof, non-destructiveness of the measurement method, and the speed of analysis. This calls for ESEM, as samples require no preparation and can be examined in their natural state.

Surface properties of various materials can be studied by observing the wetting properties and differential hygroscopicity on a microscale [193]. ESEM has been used for imaging both crystalline structures of polymers and ink on paper [194]. **Ink on paper** can be easily seen since the SEI is sensitive to, and allows, differentiation of surface layers (packaging application). Pressure-inked and non-inked areas of newsprint can be distinguished through brightness effects. ESEM has been used in the study of deinking of post-consumer office paper [195,196]. ESEM imaging alone does not allow unequivocal identification of mineral fillers. Knowledge of the minerals used in papermaking plus ESEM-EDS analysis do serve to identify particular mineral fillers [196]. ESEM-EDS has been used for paper pigment identification [197].

ESEM allows observations of big samples (200 mm∅ × 80 mm height × 5 kg) in high or low vacuum for non-destructive observation and analysis (*e.g.* of art objects). ESEM has been used for studying fibres and wet samples, including objects of **cultural heritage**, as the Dead Sea Scrolls.

ESEM provides opportunities for the detailed study of **dynamic phenomena** in real-time (crystallisation, corrosion, *etc.*). For example, ESEM allows visualisation of acrylic latex particles dispersed in water or the observation of a single water droplet condensing on a cellulose fibre. Images such as these may help to reveal local contact angles and also the heterogeneity of any surface treatment. The wetting properties of micro porous polymer membranes were studied in dynamic mode by ESEM [193]. In a typical example of *in situ* microscopy, FEG-ESEM with low-voltage techniques has shown to be an efficient method for studying the morphology and *in situ* micromechanical deformation processes in non-conducting polymer systems, such as ultrafine (250 nm) spherical silica-filled PE composites, as 1 μm thick microtomed sections [198]. In PE composites with 7 and 18 wt.% SiO_2, the filler particles are not finely dispersed in the PE matrix but locally form agglomerates of the order of 10–50 μm in size.

Applications of ESEM were reviewed [199], in particular also with reference to polymer science [200]. ESEM is used advantageously in failure analysis, especially if fatigue is suspected. Applications for polymer/additive analysis are less easily imagined than for paint systems. ESEM analysis of paint fragments was critically evaluated [201].

5.4.1.3. *Scanning Auger Microscopy*

Principles and Characteristics

Auger electron spectroscopy (AES) has the attributes of high lateral resolution, relatively high sensitivity, standardless semiquantitative analysis, and chemical bonding information in some cases (*cfr.* Chp. 4.1.1). AES in the form of high resolution scanning Auger microscopy (SAM) adds the surface compositional dimension to scanning electron microscopy. SAM is therefore a surface sensitive technique providing spatially resolved chemical analysis. The high spatial resolution of the electron beam and the process allows microanalysis of 3D regions of solid samples. The basic instrumentation of a scanning Auger microprobe is similar to that of SEM. The main characteristics of Auger instruments are an ultra-high vacuum (UHV) system, detector and energy analyser for electrons in the 0–2 keV range and sputter-ion gun. The first commercial Auger spectrometer appeared in 1973. Imaging in AES is now well developed.

In dedicated scanning Auger microprobes, the detector for Auger electrons is normally combined with a detector for secondary electrons. By rastering the focused primary electron beam, it is possible to get a high-resolution secondary electron image (SEI) of the surface. **Elemental maps** can be acquired from exactly the same area as seen with the secondary electron detector; Auger electron spectra and sputter-depth profiles are readily acquired from selected points or areas of the surface [202]. Scanning Auger images often give better images of the surface topography than SEI because of the effect of charging. SAM achieves a lateral resolution of up to 10 nm. Auger spot analyses are made on those sites where high contrasts are observed in the SAM images. Chemical contrast in SAM images can be enhanced by chemical treatment with a chemical oxidising reagent such as $KMnO_4$,OsO_4, Br_2 or HBr.

Table 5.24. Main characteristics of scanning Auger microscopy

Advantages:
- Non-destructive
- Elemental (surface) mapping
- High lateral resolution (>10 nm)

Disadvantages:
- Sample preparation (conductive surface)
- UHV technique
- Electron beam damage; destructive heating

Table 5.24 shows the **main characteristics** of the Auger microprobe. Heating by the primary electron beam is a serious problem in SAM analysis. The heat induced by the electron beam causes physical damage (cracking, shrinking) and unwanted chemical transformation on the surface (evaporation, charring and destruction). Thus, the image obtained at the end of the scanning may be distorted and different from that at the beginning. Scanning Auger spectroscopy is in general considered to be unsuitable for polymers because of electron beam induced degradation. Nevertheless, heating may also have positive effects, such as sputtering or spreading of volatile additives in heterogeneous materials so that detection becomes possible. Since SAM is to emphasise chemical inhomogeneities of the surface and subsurface, a mild destructive heating thus even seems to be advantageous. Requirements for successful SAM analysis are proper preparation of sample surfaces. For the analysis of additive dispersions a conductive surface is needed. Artefacts associated with the analysis, due to heating of the surface by the electron beam, and alteration of the surface topology by electron specimen interactions, should be considered.

Grahneis *et al.* [203] have compared SAM and PEEM (photo emission spectroscopy). For further information, *cfr.* ref. [202].

Applications

Scanning Auger microscopy is a versatile tool for distribution analysis [204], for example in studying the chemical constituents and homogeneity of dispersions in rubbers. The degree of **dispersion of additives** in rubber, such as sulfur and sulfur-containing species, zinc oxide and organic acids, is one of the most important factors that influence the physical properties of vulcanised rubber. AES combined with the SAM analyses produces 2D images which are useful in the elucidation of aggregations and dispersion states of rubber additives. Especially, particle inclusions and agglomerations can be studied together with the topographic features of the complementary secondary electron detector (SED) image. Additional information concerning the binding states of elastomers and additives, aggregations of excessive cross-linkages, unsaturated double bonds, and large hard segments of polymers can be imaged with the aid of chemical reactants. The information obtainable from an AES study of the dispersion of additives excluding carbon-black includes: *(i)* dimensions and distributions of additives or additive agglomerates; *(ii)* chemical nature of these particles and agglomerates; and *(iii)* chemical differences of the surface irregularities. Lin [205] has studied Auger images of commercial rubber. This author has first surveyed commercial rubber surfaces to obtain SED images at low magnification (40×) followed by SAM imaging of various elements (C, O, S, Zn, Ca, Cl) at 100× using an electron beam width of about 4 μm. In general, rubber is a material of poor heat conductance, and the dissipation of heat is quite low.

Although SAM is generally not suitable for analysing non-conducting materials such as polymers, in case of a thin enough polymer layer the carbon fibres of composite materials provide a conducting environment. Cazeneuve *et al.* [206] have used SAM to study a carbon fibre/epoxy matrix interface. Auger spectroscopy can be used to detect the presence of thin polymeric layers on the carbon fibres if a suitable, matrix-specific element is chosen to form the scanning Auger image. Auger spectroscopy enables identification of the micro-failure mechanism and of the effect of fibre surface treatment on the failure mode.

5.4.2. Transmission Electron Microscopy

Principles and Characteristics

The transmission electron microscope (TEM) basically consists of an assembly of electromagnetic lenses arranged in the form of a vertical column with a tube at the column's centre that is evacuated to 10^{-3} Pa or lower. Electrons from the electron gun mounted at the top of the tube are accelerated by a high voltage, typically between 100 and 300 kV, but up to 1.25 MV is possible (high-voltage electron microscopy). The beam is of sufficient energy to propagate through the specimen. These quasi-monoenergetic electrons are highly focused by apertures and electromagnetic lenses located below the gun. The diameter of the electron beam can typically be varied from 1 mm to about 1 nm. At the end of the column a fluorescent screen or CCD camera allows for image observation. The specimen is inserted into the column approximately midway between gun and screen. Electrons transmitted by the specimen are focused by an objective lens that surrounds the specimen. In the lower focal plane of this lens, a diffraction pattern is formed. This information is used to determine the atomic structure of the material in the sample. At a position below this plane, an image is constructed (*i.e.* at the image plane). The image can be magnified typically from 10^2 to 10^7 times. The TEM **image** is mainly formed by "absorption" of scattered electrons in the object; the contrast is primarily a function of several scattering mechanisms and mass thickness. Owing to strong interaction of electrons with matter, only a small material thickness can be penetrated; in case of high polymers, the critical thickness is about 0.1 μm if the signal-to-noise ratio of the image is to be acceptable. In solid-state physics the most popular modes are generally diffraction and phase contrast.

TEM requires sophisticated **sample preparation** involving ultramicrotoming or fracturing, in addition to chemical fixation such as staining, etching, or replication. These complex sampling techniques can severely limit the structural details present in the image. For a TEM study of thin films a classical sample preparation method is decoration, where some suitable atomic or molecular species is vapourised onto the polymer substrate and migrates to energetically favourable sites, thus revealing surface topography and crystallographic details. Techniques of TEM sample preparation have been described [135]. Table 5.25 lists TEM preparation procedures for polymers.

It must be emphasised that the spatial **resolution** of TEM is extremely high and very rarely surpassed in other instruments. The maximum resolution obtainable with this technique depends on instrument, specimen and preparation. The higher the acceleration voltage, the better the resolution. The point resolving power of commercial high performance TEMs is better than 0.2 nm. In 1993 the

Table 5.25. TEM preparation procedures of plastics

Transmission imaging	Indirect surface imaging
• Transmissive layers of objects Films, dispersions, powders, solutions • Thinning methods Ultramicrotoming Etching Peeling	• Surface impression method • Freeze etching • Surface treatment Chemical, physical, mechanical Decoration

first 1-Å TEM showed the positions of individual atoms in crystal lattices; this marks considerable progress since the first production of TEM in 1949 (50 Å). Nowadays, individual oxygen atoms can be imaged [207]. Cryo-ultramicrotomy and heavy metal staining are the primary methods for the study of polymer surfaces at resolutions of about 1 nm. The image contrast decreases with increasing energy of the electrons. Low voltages (5 kV) provide an enhanced imaging contrast nearly 20 times higher than for 100 kV, which is interesting especially for low atomic number specimens. Low-voltage TEM (LVTEM) is extremely sensitive to the thickness of the specimen.

Electron diffraction patterns (selected area and high-resolution diffraction, reflection-diffraction, *etc.*) can be obtained with a TEM by simple switching from image plane to focal plane. Selected area diffraction (SAD) combined with microscopy is an important supplementary tool to X-ray diffraction in crystal structure analysis (less so for polymers).

Various analytical techniques can be carried out in a transmission electron microscope. TEM is transformed into an analytical electron microscope (AEM) by adding an X-ray spectrometer as a detector [208]. The X-ray energy dispersive spectrometer (XEDS) is the only X-ray spectrometer currently used in TEMs. It is remarkably compact, efficient and sensitive. A combination of Si(Li) and Ge detectors can detect $K\alpha$ lines from all the elements, from B to U. XEDS is limited in terms of its need for cooling, poor energy resolution, and many spectral artefacts. The spectral resolution of EDS in a TEM is typically 120–150 eV, hence this technique is not useful in the study of fine structural detail of the electronic structure of bonds. For quantitative X-ray analysis of thin films in TEM the so-called ξ-factor method is of great use [209].

TEM with induced X-ray emission (TEM-X) and electron energy-loss spectrometry (TEM-EELS) allows local analysis for all elements for Li upwards with lateral resolution in the nm range. The local physical structure can be derived by means of electron diffraction. The EELS signal for low atomic number elements improves the X-ray signal considerably, while electron spectroscopic imaging (ESI) provides improved sensitivity for element mapping with minimum detectable mass of *ca.* 2×10^{-21} g and spatial resolution of *ca.* 0.5 nm [147]. EELS furnishes information about the energetic states of the atoms involved. In principle, EELS offers many advantages over EDS. However, for polymers specimen damage is a major problem because detection is slower than on EDS. Nevertheless, electron spectroscopic imaging (ESI) and EELS have improved considerably the capabilities of TEM. In this one instrument, a complete physical and chemical investigation of materials with nm dimensions can be undertaken, in addition to observing images with atomic resolution. The feasibility of single-atom identification in TEM, the "holy grail" of microanalysis, was demonstrated in 1991 [210].

Table 5.26 lists the **main characteristics** of TEM. Advantages of TEM are its microchemical and microstructural (electron diffraction) potential and supreme resolution. However, the technique also presents some drawbacks. One of the limitations is sample size as large samples cannot be examined. Another disadvantage of TEM is the preparation of thin specimen foils by methods such as electrolytic polishing, ion-beam sputtering, or ultramicrotomy. Organic compounds, like most polymers, cannot easily withstand irradiation in TEM or SEM, so films must be prepared on special electron microscope (metal) grids shadowed with carbon, which involves significant sample preparation effort.

Modern 200 to 300-kV FEG-TEMs equipped with an environmental cell allow relatively high sample pressure of up to 50 mbar and heating up to 1000°C for *in situ* studies.

Table 5.26. Main characteristics of transmission electron microscopy (TEM)

Advantages:

- Bulk information
- Atomic structures (by diffraction)
- Microstructural analysis (defect characterisation by image analysis)
- Light element spectroscopy (EELS)
- Indirect chemical bonding information (from diffraction and image simulation)
- High detection limits (one monolayer for relatively high-Z materials)
- High lateral resolution: 0.1–0.2 nm (BF + DF)
- Depth resolution: 5–100 nm (BF + DF)
- Imaging/mapping capabilities

Disadvantages:

- Elaborate preparation of thin specimens ($<100\ \mu$m)
- High vacuum (10^{-7} Torr)
- No specific element identification
- Destructive (specimen preparation required)
- Specialist user skill needed
- High instrumental cost; need for expensive ancillary equipment

Voigt-Martin [211] has reviewed the characterisation of polymers by TEM. The technique has recently been reviewed [212]. Various monographs deal with TEM [213–217]. For electron probe X-ray microanalysis of thin samples in TEM, *cfr.* refs. [214,217].

Applications

Transmission electron microscopy is widely applied for ultrastructural research (by diffraction and image analysis), as well as for light element spectroscopy (EELS). TEM images compliment the chemical composition, physical property, or mechanical performance information obtained by techniques such as μFTIR, μRaman, ToF-SIMS, LDMS, XPS, DSC, micro-hardness, *etc.*

During coating system development, comparisons of TEM images can effectively reveal the morphological consequences of changing the solvent mixture, matrix polymer blend, or of using additives such as pigment dispersants, adhesion promoters, levelling agents, UV absorbers, and hindered amine light stabilisers (HALS) [5].

Harper *et al.* [152] have reported a microscopical study of several brominated **fire retarded ABS/Sb_2O_3 formulations**. Whereas SEM showed good dispersion of Sb_2O_3 regardless of particle size, TEM revealed that Sb_2O_3 resided in the SAN phase of the polymer. Chemical interactions have been observed in various halogen/Sb_2O_3 flame retarded systems [218,219]. TEM analysis showed that octabromodiphenyl oxide (OBDPO), 1,2-bistribromophenoxyethane (BTBPE) and tetrabromobisphenol-A (TBBA) are more compatible in ABS than polydibromostyrene (PDBS). PDBS is easily identified in TEM and SEM because it has little affinity for the matrix and resides in large domains. The spatial distribution of additives can be evaluated by microprobe analysis. Gottlieb *et al.* [220] have used TEM/SEM/EDS in a study of the distribution of the flame retardant pentabromobenzylacrylate (PBBMA) in GFR PP/(PBBMA, Sb_2O_3, Irganox B225) that may polymerise during reactive extrusion to produce poly(pentabromobenzylacrylate) (PBBPA). Monomer and polymer were distinguished on the basis of the double bond (staining by OsO_4). Irganox B225 acts as a radical scavenger suppressing the polymerisation of PBBMA during extrusion.

Allen *et al.* [221] have used both OM and TEM in a study of lubrication of LDPE containing 5000 ppm oleamide and stearamide. The primary fatty amides migrate to the polymer surface; islands of amide gradually appear and then coalesce into a uniform coating.

Carbon-black types are differentiated on particle size and surface area measurements, using techniques such as TEM, PCS (photon correlation spectroscopy), iodine or nitrogen adsorption and mercury porosimetry. The best method for qualitative **carbon-black** determination is based on accurate measurement of the CB particle size using TEM [222,223]. Pyrolysis at 800° to 900°C followed by TEM analysis according to ASTM D 1765 allows identifying carbon-blacks used in rubber products [224]. TEM was used for identification of carbon-blacks in vulcanisates [225]. TGA has been used for quantitative determination of carbon-black [226,227]. Palla [51] used TEM for the characterisation of rubber components. Ultra-thin sections of rubbers suitable for TEM studies can be prepared by microtoming at low temperatures.

Well-dispersed **nanocomposites** with potential property improvements as to mechanical properties, heat resistance, dimensional stability, barrier and flame retardation, consist of delaminated platelets distributed homogeneously in the polymer. In these new polymer materials the silicate layers of the clay are separated at the nm scale. Degree of delamination and dispersion are commonly studied by XRD

Table 5.27. Comparison of XRD and TEM techniques for nanocomposite analysis

Technique	Advantages	Disadvantages
XRD	Ease of analysis (autosampling) Can determine *d* spacing between clay layers	Technique dependent on extent of order in clay Cannot determine difference between disordered immiscible and delaminated systems
TEM	Determines all types of clay nanostructures Analysis not dependent on order of clay Determination of *d* spacing between clay layers (calibration with internal standard) Observes orientation Image processing	Labour intensive sample preparation, analysis

and TEM [228]. Exfoliated clay is identified by disappearance of the characteristic diffraction peak (XRD) and dispersion in the polymer matrix (TEM). While XRD is the method of choice for characterising polymer-clay nanocomposites, it does have some shortcomings in that it cannot differentiate between delaminated and disordered immiscible systems. TEM is the better tool to monitor dispersion, because the clay platelets can be seen. TEM has the key advantage that it can analyse a system regardless of order or disorder in the clay, and can also determine the difference between a delaminated system and a disordered immiscible system. HRTEM is the only way to visualise the crystal lattice and to achieve information on the structure. The (crystal) structure of a nanomaterial is closely related with its physical and chemical properties. TEM can solve some of the shortcomings encountered with XRD, but it has its limitations as well; mainly, the inability to precisely determine the *d* spacing between clay layers without reference to internal standards and the very labour intensive sample preparation and analysis. Table 5.27 compares TEM and XRD techniques as complementary tools for nanocomposite analysis.

Many reports have described TEM and XRD analysis of exfoliated PP/MMT nanocomposites, prepared in various ways [229]. PP/clay nanocomposites prepared by intercalative polymerisation [230] and nanocomposites of PP-*g*-MA with organically modified clays [231] were similarly characterised by TEM and XRD. The same techniques were used in studies of delamination and clay dispersion in PAI/MMT nanocomposites [232] and in two layered organoclays (Cloisite 15A/30B) in PA6 [233]. TEM and LFM studies have indicated that organoclay (Cloisite 15A) additions in PS/PMMA migrate to the interfaces (*i.e.* act as a surfactant) and effectively compatibilise the polymer blend [234]. TEM analysis of nanostructured cured elastomeric sealant compositions prepared by sonicating a mixture of silylated apophyllite filler in PDMS showed unambiguously that silicate layers are exfoliated [235]. Watkins *et al.* [174] have studied platinum/poly(4-methyl-1-pentene) (PMP) nanocomposites by means of TEM. For nanocomposite science and technology, *cfr.* also ref. [236].

Staining techniques for detecting localised oxidation in HDPE powders and films were reviewed [237]. Optical absorbance following staining with 2,4-dinitrophenylhydrazine (DNPH) can be used as a measure of the aldehyde/ketone content in **oxidised polyolefins**. SO_2 treatment enables regions with high concentrations of hydroperoxides to be clearly distinguished.

Low-voltage TEM (LVTEM) imaging (at 5 kV) of polymer blends has been reported [238]. With LVTEM *cq.* LVSTEM it is possible to distinguish components differing very slightly in their elemental compositions, *e.g.* PE/PP, PS/PP, or PC/SAN [238]. LVTEM at 5 kV can be applied to obtain images of the phase structure of polymeric materials without any prior staining. The characterisation of polymers by TEM has been reviewed [211].

5.4.3. Analytical Electron Microscopy

Principles and Characteristics

In the late 1960s an alternative to the TEM imaging geometry was introduced by Crewe *et al.* [239], who used similar optics to produce a very small electron probe that was scanned in a raster over the area of interest. This high-resolution **scanning transmission electron microscope** (STEM) has become an

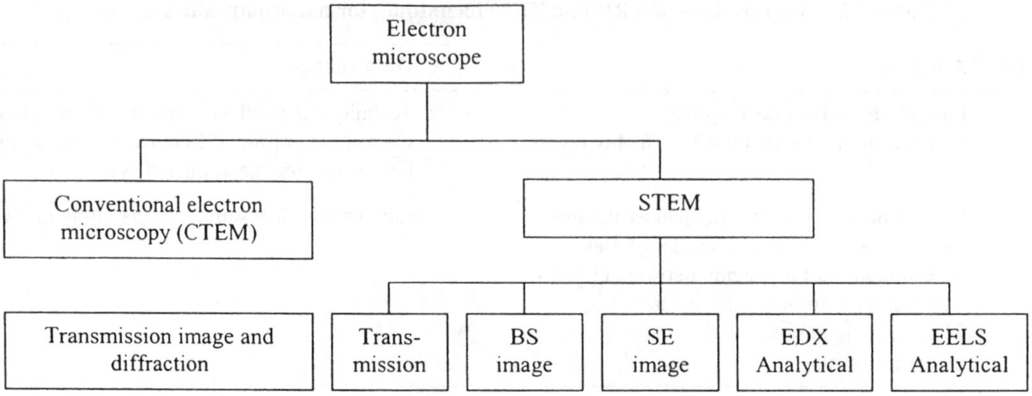

Scheme 5.3. Analytical techniques available in a modern electron microscope.

important tool for quantitative microscopy because many types of analytical signals can be used to produce an image of the scanned area (Scheme 5.3). Images of single heavy atoms were first presented in 1970 [240]. Recently, sub-Ångstrom resolution using aberration correction electron optics has been reported [241]; a world record of 0.07-nm resolution has been achieved [241a].

A dedicated STEM consists of a field-emission gun with acceleration to 100 keV and a lens for forming a mono-energetic probe of 0.2–50 nm (in diameter) that rasters over a 5–500 nm thick electron-transparent specimen in vacuum. STEM images are formed from the collected transmitted or scattered electrons. By adding scanning coils to a TEM with the ability to form a fine focused electron beam, a STEM can be realised. Most analytical TEMs with STEM attachment operate in the 80–200 kV range. STEM makes convergent beam techniques available, allows detection of secondary and backscattered electrons and chemical analysis in very small regions of the specimen, while at the same time retaining the normal microscopic facilities. Thus, unlike all the classical chemical techniques, direct correlation between chemical composition and position in the specimen is possible. This determines the unique advantage of detecting very small heterogeneities and contaminations in thin films (a few hundred Å across), which go undetected by other techniques where it is only possible to analyse larger regions. The STEM technique and its use of convergent electron beams rather than the conventional parallel beam are now considered as a powerful means of materials research. SEMs can be operated in the STEM mode. STEM provides about 100 times better spatial resolution of analysis than conventional SEM. In STEM mode transmission can occur through layers of up to 10 μm. Scanning of the object causes less damage.

Table 5.28. Main characteristics of scanning transmission electron microscopy (STEM)

Advantages:
- High versatility (record of different signals)
- Wide elemental range (Li–U)
- Fair detection limits (0.1–3.0 wt.% for EDS and EELS)
- High lateral resolution (imaging, <0.1 nm; EELS, 0.5–10 nm; EDS, 3–30 nm)
- Atomic number contrast (similar to TEM)
- Convergent beam techniques
- Microstructural, crystallographic analysis (CBED)
- Quantitative compositional analysis (EDS, EELS; 5–10% relative accuracy)
- Light element spectroscopy (EELS)
- Chemical bonding information (EELS)
- Imaging/mapping capabilities
- Low radiation doses

Disadvantages:
- Specimen preparation required (<200 nm thick for imaging and EDS; <50 nm thick for EELS)
- Long frame time
- Very high cost of instrument and ancillary equipment

Table 5.28 shows the **main features** of STEM. The technique provides a variety of facilities for bright field (BF) and dark field (DF) imaging, electron imaging (ESI, SE, BSE), elemental mapping (EDS), structural analysis (EBSD, SAD, CBED), and spectroscopy (EELS, EXELFS). A modern electron microscope uses all the signals that are generated during interaction between electron beam and specimen. Facilities are available for microscopy and elemental analysis in well-defined microscopic regions of the specimen: energy dispersive X-ray analysis (EDX) for microchemical analysis and electron energy-loss spectroscopy (EELS) for chemical and bond analysis. Elemental maps can be obtained in the scanning mode from areas as small as 10^{-3} mm^2. With EDX the minimum spatial resolution is usually about 10–20 nm with a minimum detectable mass of 5×10^{-19} g of Fe but considerably higher for Ca, S, P, O, N or C because of lower X-ray yield per scattering event and poor detector efficiency at lower energies. Compared to TEM, an advantage of STEM is that these many signals may be collected simultaneously. Taken together, these analysis techniques are termed **analytical electron microscopy** (AEM). Image processing techniques are used in analysing microstructure [242, 243]. Combination of imaging and spectrometry is most powerful and transforms a TEM to an AEM.

In **electron energy-loss spectroscopy** (EELS) a nearly monochromatic beam of low-energy electrons is directed through an ultrathin specimen, usually in a TEM or STEM [244–246]. As the electron beam propagates through the specimen, it experiences both elastic and inelastic scattering with the constituent atoms, which modifies its energy distribution. The measured signal can be used to determine quantitatively the local specimen concentration (±10–20 at.% without standards; ±1–2 at.% with standards), the electronic and chemical structure, and the nearest neighbour atomic spacings (*cfr.* analogy to EXAFS).

When the high-resolution imaging and diffraction capabilities of CTEM, STEM and SEM are combined with qualitative and quantitative X-ray analysis, micro characterisation of materials is significantly extended. Areas from thin films as small as a few hundred Å can be analysed in terms of their elemental nature with STEM optics. A combination of elemental spot, line scan and elemental analysis with various topographical, structural and crystallographic information enormously extends the knowledge about a material. **Electron probe X-ray microanalysis** (EPMA) is a well-established elemental analysis technique based upon bombarding a specimen with a focused beam of energetic electrons (beam energy 5–30 keV) to induce emission of characteristic X-rays (0.1–15 keV) [247]. The X-rays are measured in EDS (LOD: 1000 ppm) or WDS (LOD: 100 ppm) mode. The concept of the electron probe microanalyses was developed in 1951. (R. Castaing, Paris; I.B. Borovskii, Moscow) and was put into commercial production by 1956 for X-ray spectrometry. The physics of the X-ray generation process is well understood [248]. Hall [249] outlined light element detection via wavelength-dispersive X-ray analysis. EPMA is the most widely used technique in materials microanalysis. There is no essential difference between X-ray microprobes and analytical scanning electron microscopes. In practice, EDX is usually associated with imaging and EPMA with accurate analysis.

Table 5.29. Main characteristics of electron probe X-ray microanalysis

Advantages:
- Wide element range: Be to actinides
- Bulk analysis
- Non-destructive (except for beam damage)
- Quantification (standardless or pure element standards)
- Accuracy (±4% in 95% of cases; flat, polished surfaces)
- Lateral resolution: 100 nm–5 μm (energy and matrix dependent)
- Sampling depth: 100 nm–5 μm (energy and matrix dependent)
- Compositional mapping and SEM imaging
- Depth profiling (by energy variation)
- Mature technology

Disadvantages:
- Relatively poor energy resolution (WDS > EDS)
- No chemical bonding information
- Instrument cost

Table 5.29 shows the **main characteristics** of EPMA. The EDS mode provides several inherent advantages relative to WDS: *(i)* simultaneous spectral acquisition (minimising beam damage); *(ii)* allowance for quantitative microanalysis of rough surfaces or particles; and *(iii)* spectrum imaging [250]. It is waste of time to proceed with quantitative microanalysis from a XEDS spectrum without first carrying out qualitative analysis. This requires that every peak in a spectrum be identified unambiguously and with statistical certainty.

Quantitative X-ray analysis in analytical electron microscopy is now a most straightforward technique. Matrix (interelement) correction procedures based upon first principles physical models provide great flexibility in examining unknown samples of arbitrary composition. According to Castaing [251,252]:

$$C_i/C_{(i)} = kI_i/I_{(i)} \tag{5.3}$$

where C_i is the concentration of an element i in a specimen and $C_{(i)}$ of a known standard, I_i is the measured intensity emerging from the specimen (*not* generated within) and $I_{(i)}$ the measured intensity emerging from the standard, k is a sensitivity factor. The contributions to k come from three effects: Z (atomic number), A (absorption of X-rays within the specimen), and F (fluorescence of X-rays within the specimen). The correction procedure in bulk microanalysis is often referred to as the ZAF correction. For details, *cfr.* ref. [248]. If a thin electron-transparent specimen is used rather than a bulk specimen, then the correction procedure is greatly simplified because, to a first approximation, the A and F factors can be ignored and only the Z correction is necessary. In addition, if thin specimens are used, the analysed volume is substantially reduced, giving a much better spatial resolution. Consequently, **thin-foil microanalysis** has been revolutionised by using a simplification of Castaign's original ratio equation, which simply comprised the ratio of intensities gathered from two elements A and B simultaneously [253], as follows:

$$C_A/C_B = k_{AB} I_A/I_B \quad (5.4)$$

where the Cliff-Lorimer sensitivity factor k_{AB} is actually *not* a constant, but varies according to the TEM/XEDS system and the microanalysis conditions. The Cliff-Lorimer equation is the basis for quantitative microanalysis of films [254]. There are two ways to determine k factors, namely experimentally using standards (most accurate, but slow and laborious) or from first principles (less reliable, but quick), *cfr.* ref. [214]. Quantitative microanalysis of spectra from thin foils is thus essentially straightforward, so long as the k factors are determined with sufficient accuracy. For quantitation of bulk materials by EPMA the multiple reflection model is popular [255]. Also a standardless analysis method is available [256]. Low-Z element analysis ($Z < 10$) of particles is still a challenge, especially for standardless procedures, in view of the geometric effects and strong absorption effects, even within particles in the μm size range. Alternative correction methods are operative (*e.g.* Rhi Ro Z). Computer programs are available through vendors and the Microbeam Analysis Society [257]. A rigorous mathematical approach to analysing EDX spectra for improved quantification and sensitivity has recently appeared [258].

Possibilities and limitations of EPMA techniques for quantitative near-surface analysis and depth profiling were described [259]. Spatial distributions of elemental constituents can be visualised qualitatively by X-ray area scans and quantitatively by digital compositional maps. A major driving force for the development of X-ray microanalysis in AEM is the improvement in spatial resolution compared with EPMA. This improvement arises from the use of thin sections and the higher electron energy (>100–400 keV in AEM compared to 5–30 keV in EPMA).

CRMs for electron microprobe analysis of carbon and nitrogen are available [260]. A major challenge for these materials is to obtain homogeneity at the micron level. ASTM Standard E 1508 describes quantitative analysis by EDS [261]. ISO Technical Committee TC202 has launched the standardisation project No. 15632 for the specification of an EDS spectrometer.

AEM and X-ray emission spectroscopy were reviewed [262]. A monograph dealing with analytical electron microscopy has appeared [215]. Textbooks on X-ray spectrometry in electron beam instruments, particularly as it relates to the practice of EPMA and EDX, are available [263,263a].

Applications
STEMs can profitably be applied for the study of damage to polymers [264] and can produce images of polymers such as PE with exposure levels of less than 5 C/m^2 for a single image as compared to 100 C/m^2 of a conventional transmission microscope. A very useful application of micro-diffraction is the analysis of nucleation interfaces, nucleation mechanisms and the design of new nucleating compounds. EDS has severe limitations when applied to relatively low Z elements and is therefore of little use for detecting organic compounds.

Electron probe microanalysis is used for characterising surface morphology and for microanalysis of inhomogeneous samples and small volumes. Betzold [136] has discussed the use of EPMA for the determination of plastics, fillers, reinforcing materials, pigments, stabilisers, *etc.* X-ray microanalysis (Br, Mo, Sb, Ti) has been used for determining flame retardant content in ABS granules before and after extraction [265]. Also EPMA analysis of automobile paint was described [158]. EPMA of a film surface of stored PE/4,4′-thiobis(3-methyl-6-*t*-butylphenol) has revealed that only material within exuded, crystalline platelets contained sulfur [266].

One of the most important applications for EPMA is the analysis of chemical composition of **microparticles** distributed in a matrix. The microprobe can profitably be used to assess the distributions of S, ZnO, *etc.*, in (un)vulcanised rubber mixtures. Typical examples of application of EPMA analyses to polymers are inhomogeneities ("specks") in finished plastic articles or coatings (*e.g.* gels, dust, pigments, catalyst residues, metal abrasions). Recently, a new method for quantitative EPMA of individual microparticles in a matrix was proposed [267].

X-ray microanalysis was used to investigate the deterioration of organic polymers such as PE and PUR by metals [268]. Segregation of a polybromostyrene/polystyrene blend is easily studied by means of EDX mapping. The art world is a major user of microanalytical techniques [269].

5.5. SCANNING PROBE MICROSCOPY TECHNIQUES

Principles and Characteristics
The term scanning probe microscopy (SPM) encompasses a family of techniques that provides images of surface topography and, in some cases, surface properties, on the atomic scale. SPM is an imaging tool with a vast dynamic range, spanning the realms of optical and electron microscopes. The principle of SPM is very similar to profilometry, where a hard sharp tip is scanned across a surface and its vertical movements are monitored. The main SPM techniques are given in Table 5.30.

Table 5.30. Scanning probe microscopy techniques

- Tunnelling spectroscopy (STM)
- Force microscopy (AFM, CFM, EFM, LFM, MFM, PFM, FMM, UFM)
- Optical microscopy (NSOM)
- Thermomicroscopy (SThM, μTA)
- Acoustic microscopy (AFAM)

The three most important scanning probe techniques are: scanning tunnelling microscopy (STM), scanning force microscopy (SFM, also known as atomic force microscopy, AFM) and near-field scanning optical microscopy (NSOM). The three methods give different types of information (*cfr.* Table 5.31) and require correspondingly different theoretical treatments. STM probes the electronic states of a surface, SFM the force (or force gradient) between a tip and a surface, and NSOM the electromagnetic field near a surface. However, the three techniques share several common features. First, they measure local rather than average surface properties. To be useful, any theory must therefore include the local surface properties. Second, in no case it is possible to infer physical properties of the system directly from the scanning probe results. Interpretation therefore has to proceed by an indirect interpretation cycle. The family of scanning probe microscopes has revolutionary imaging capabilities at the atomic or molecular level for a wide range of materials allowing unprecedented views of surfaces and providing **local spectroscopy**. Scanning probe microscopies enable to improve our understanding of forces, dynamics, and other physical and chemical processes on the nm scale. The main feature of SPMs is that the measurements are performed with a sharp probe operating in the near-field, *i.e.* scanning over the surface while maintaining a very close spacing to the surface.

The general scheme of any SPM apparatus includes several major components which allow line-by-line scanning with an atomically sharp tip while monitoring nm scale cantilever deflections in vertical and horizontal directions. Precise 3D movements of either a sample or a cantilever (within a fraction of a nm) are provided by a tube piezoelement. The SPM

Table 5.31. Scanning probe techniques compared

Feature	STM	AFM	NSOM
Nature of probe	Electronic surface states	Local Van der Waals forces	Electromagnetic field near surface
Method	Local conductivity	Local surface hardness	Optical microscopy (not diffraction-limited)
		Surface topography of insulators	
		Surface properties	
Local spectroscopy	Most developed	Limited capabilities	Excellent capabilities
Sampling	Restricted as to conductivity and roughness	Non conducting at extremely high resolution	Specimens in air; limited spatial resolution

tip deflection is monitored by a detection scheme (*e.g.*, array of photodiodes, interferometer scheme, or piezoelectric cantilever). A microfabricated probe consists of a silicon or a silicon nitride cantilever with the integrated pyramidic tip of several microns in height with a tip end radius in the range from 5 to 200 nm.

Binnig and Rohrer [23] have first reported in 1982 the successful realisation of the **scanning tunnelling microscope** (STM). The basic working principle of the STM relies on the quantum mechanical properties of electrons. When an atomically sharp metal probe tip is brought in close proximity (within about 1 nm) of a (semi)conducting surface, electrons can tunnel through the energy barrier between probe tip and surface. The magnitude of the tunnelling current decreases exponentially as the tip-surface separation is increased. Typical conditions are a 1 nm tip-sample distance and a 1 nA tunnelling current. STMs can image the surface of the sample with sub-Ångstrom precision vertically, and atomic resolution laterally. The STM comprises a head, vibration isolation and electronics. The head contains the probe tip and its positioner. Raster scanning the tip across the surface, through the use of piezoelectric transducers while maintaining a constant tunnelling current, images a surface of constant density of electronic states. The resulting image is a convolution of topographic and electronic properties of the sample surface. Rather than measuring physical topography, STM measures a surface of constant tunnelling probability. For tunnelling to take place, both the sample and the tip must be conductors or semiconductors. Unlike AFMs, STMs cannot image insulating materials, such as polymers. In those cases a conducting surface layer (gold) is required for STM. On the other hand, AFM microscopy relies upon the effect of repulsive and attractive forces between probe and sample to bend a supporting cantilever. The bending of the cantilever, and hence the force, is extracted by monitoring the path of a laser beam reflected from the back of the cantilever. In NSOM, the sample is placed in the near-field region of a subwavelength-sized light source. The transmitted or reflected optical signal is used to form an image of the scanning sample. As the tunnelling gap can be a vacuum gap, but equally well an air or liquid gap, SPMs can be operated in a variety of environments: ultrahigh vacuum (UHV), ambient (air) or liquid.

Scanning probe microscope systems are available with multiple imaging capabilities (including contact, non-contact and intermittent contact, in-fluid, force-modulation, and others). STM has rapidly become the starting point for the development of still other microscopies, such as lateral force microscopy (LFM; measures the frictional forces between probe tip and sample surface); magnetic force microscopy (MFM; measures magnetic force gradient and distribution above the sample surface); and thermal scanning microscopy (TSM; measures the thermal conductivity of the sample surface with tip and sample not in contact). These techniques allow simultaneous acquisitions of both topographic and property data.

Fisher [270] has given an overview of the **theoretical analysis** of SPM techniques. The imaging theory for the STM technique is the best developed of all the scanning probe family [271], but much progress remains to be made in accounting correctly for the nature of the tip and for tip–sample interactions. Interpretation for STM involves a model of the atomic and electronic structure of the surface, including any adsorbates or surface defects. However, also other factors are important in STM, such as the mechanical interaction between tip and sample and the tip electric field (sometimes quite large). Distortions of both the atomic and electronic structure of the surface have been observed. The theory of NSOM is similar to that of STM, and in some ways more straightforward. The understanding of SFM data is very incomplete, particularly for experiments with resolution on the atomic scale.

Table 5.32 shows the **main characteristics** of SPM. Scanning probe microscopes are most commonly thought of as tools for generating images of a sample's surface. SPMs can also be used, however, for measuring and mapping material properties. In some cases, SPMs can measure **physical properties** such as surface conductivity, static charge distribution, localised friction, stiffness, elastic moduli, adhesion, electric or magnetic forces. Other SPM techniques include local chemical sensing and thermal modes for probing of polymer surfaces. The SPM technique is moving towards a new level of dynamical surface nanoprobing when nicely designed nanoprobes with a wide range of controllable properties will become available. This allows quantitative characterisation of polymer surface properties on a sub-μm scale and opens a door for unambiguous nanomechanical testing of surface properties.

The capabilities of SPM for imaging and manipulating surface topographies or nanostructured materials are superb, but chemical identification with SPM is limited. SPM operation in the topography

Table 5.32. Main characteristics of scanning probe microscopy

Advantages:
- Non-destructive
- No sample preparation
- Operates in various environments (UHV, air, liquid)
- Probes local geometric and electronic structure of surfaces
- Family of combined microscopic (3D imaging) and spectroscopic tools
- Measures local material properties (indirectly)
- High-resolution profilometry (STM, SFM)
- Lateral resolution: atomic (STM) to 1 nm (SFM)
- Vertical resolution: 0.01 Å (AFM) to 0.1 Å (SFM)

Disadvantages:
- Image interpretation
- Imaging artefacts
- Conductive materials (STM)
- Limited chemical identification
- Specialist skill needed

Table 5.33. Representative applications of SPM to polymeric materials

- Visualisation of surface topography and nanostructure
- Nanodomain morphology of multicomponent polymers
- Compositional mapping of heterogeneous polymer systems
- High-resolution imaging of polymer crystals (folding) and single macromolecules
- Nanoprobing of chemical composition and intermolecular interactions
- Molecular films and interfaces
- Imaging of adsorbates
- Probing of local mechanical viscoelastic and adhesive properties, incl. nanoindentation
- Monitoring of polymer structural changes induced by thermal transitions

mode only rarely provides insight into the chemical nature of a multicomponent system. The atom probe (AP) technique [272] can be used for this purpose. As the tunnelling current in STM is also a function of local electronic structure **atomic-scale spectroscopy** is possible. Fuji *et al.* [273] combined a vertical and lateral force microscope with a conventional fluorescence microscope. Even microwave frequency STM has been reported [274].

The ultra-high resolution of the SPMs has been extended with spectroscopic capabilities to elucidate local chemical and electronic information. **Scanning probe spectroscopy** (SPS) has become a unique surface analytical tool because it combines ultra-high spatial and energy resolution [275–277]. By mapping the spatial distribution of electronic states by SPS, chemically resolved information is obtained down to the atomic scale. In comparison, spectroscopies such as XPS and UPS detect and average data originating from a relatively large area (few μm–mm).

Apparently only, SPM images are among the easiest to interpret of images generated by any microscopy technique as 3D data is collected (as opposed to the case of most optical and electron microscopies). In an SPM image, a peak is unambiguously a peak, and a valley is clearly a valley so that true surface topography is involved. However, in STM the resulting image is a convolution of topographic and electronic properties of the sample surface. **Imaging artefacts** in an SPM image do arise from a phenomenon known as tip convolution or tip imaging. As long as the tip is much sharper than the feature, the true edge profile of the feature is represented. However, when the feature is sharper than the tip, the shape of the tip may dominate the image. Thus, while the periodicity of the lattice is reproduced, true atomic resolution is not necessarily achieved in all SPM techniques. In case of STM true atomic resolution is assured. In fact, because the dependence of the tunnelling current on the tip-to-sample separation is exponential, only the closest atom on a good STM tip interacts with the closest atom on the sample. True atomic/molecular resolution is tested on the ability to detect a single defect site. All scanning probe microscopies require calibration (standards: latex beads, colloidal gold particles; optical interferometry or grazing incidence X-ray reflectivity of films). STM data complements that provided by electron microscopy, XPS, SIMS and other surface analysis methods.

Scanning probe microscopy has recently been reviewed [270,278]; also several books are available (*cfr.* Bibliography). SPM of polymers has been dealt with specifically in refs. [279,280]. Meanwhile also a vast secondary and tertiary literature is available on STM: an overview [281], reviews [282,283] and books [284–286].

Applications

With a current market of over 5000 installed instruments, the breadth of applications of SPM is considerable. The atomic resolution and spectroscopic capabilities of scanning probe microscopes have enabled elucidation of the great heterogeneity of sur-

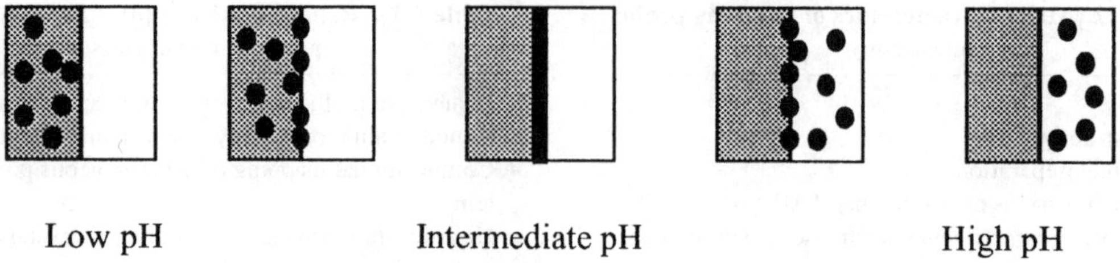

Fig. 5.4. CB localisation in a PE (white)/PS (gray)/CB (black) blend in relation to the acidity of CB. After Leclère *et al.* [287]. Reprinted with permission from Ph. Leclère *et al.*, *ACS Symposium Series* **694**, 129–140 (1998). Copyright (1998) American Chemical Society.

face sites including defects, step edges, lattice impurities, adsorbates, and grown structures. Such specific information cannot typically be acquired by spectroscopies that measure ensemble averages of the surface. The scanning tunnelling microscope is the most suited and the most developed of the various SPMs to perform local spectroscopic measurements. However, it also has the most restricted range of accessible substrates in terms of conductivity and roughness. Also other novel microscopies, such as atomic force, friction force, and magnetic force microscopy are very powerful tools for investigating supermolecular structure. Spatially resolved SPM in multiple modes has been applied to a wide variety of polymers, *cfr.* Table 5.33. Spectroscopies with SPMs have been of rapid development. The ability to study isolated or small structures of adsorbates has allowed incredible insight into the rich chemistry of surfaces, particularly in defining roles that defect-sites play.

SPM and AFM techniques are widely applied to studies of polymer materials in academia and industry. **Conducting polymer composites**, which consist of conducting filler distributed throughout an insulating polymeric material, are amenable to morphological analysis by SPM. The electrical resistivity of carbon-black (CB)-filled multiphase polymer blends depends on the CB localisation. Lateral force microscopy is a powerful tool to investigate the morphology of CB-filled polymer blends in relation to blend composition and CB loading [287]. Leclère *et al.* [287] have examined various HDPE/PS/CB blends by means of SPM (in LFM mode), in particular as to the selective localisation of CB. The CB localisation in PE/PS/CB blends stands in relation to the CB acidity (Fig. 5.4).

Kim *et al.* [288] compared the surface and structural information provided by STM and TEM/SEM images of carbon-black (high abrasion furnace, N 330, medium thermal, N 990, and graphitised MT).

In the initial stages of SPM applications simple topographical imaging prevailed. Probing of thermal, viscoelastic and near-field optical properties has been implemented. Expansion of the family of surface properties being tested by SPM is inevitable. The focus of SPM studies on polymer surfaces is now gradually changing towards **quantification** of the surface measurements and "multidimensional" characterisation of surfaces (friction properties, elastic behaviour, adhesion, magnetic and chemical composition, conductive state and thermal transformations) [289]. The very local probing capability of the SPM technique provides complementary information that is beyond the possibilities of conventional experimental techniques. Careful design of the SPM tips with controlled chemical composition of the tip end will provide improved chemical sensing capabilities. Improvements in lateral resolution can also be expected.

Applications of scanning probe microscopy have been reported in reviews [290] and books [276,291].

5.5.1. Atomic Force Microscopy

Principles and Characteristics

In 1986 Binnig *et al.* [292] have developed the atomic force microscope (AFM) which remedied a severe limitation of STM, namely imaging of conducting materials only. The first commercially available AFM was introduced in 1989. Since that time, AFM has been used with great success to study surfaces of insulators and the macromolecular architecture of polymeric materials from sub–nm to μm scale [285,286,293].

AFM is essentially a very sensitive profilometer. In AFM an atomically sharp stylus or probe (a few μm long and often less than 5 nm in diameter, made of diamond or SiN), mounted on a very soft spring or metal-coated microfabricated cantilever (length: 100–200 μm), is drawn across a surface. The tip is then raster scanned in a similar fashion as to STM, while maintaining a constant force between probe and sample, or the sample is scanned under the tip. Interatomic forces between the sharp tip and the sample surface cause the cantilever to bend, or deflect. Because the movements of the tip are very small, they need to be magnified, *e.g.* by optical systems using a laser beam. The deflections of the spring, or cantilever, are monitored either using an STM [292] or more commonly by interferometry methods [294]. As a result, the system can detect sub-Ångstrom vertical movement of the cantilever tip. AFM thus records contours of constant force due to the repulsion generated by an overlap between the electron clouds of the tip and the surface atoms. A generic AFM **instrument** comprises the following components: scanning system (*e.g.* piezoelectric tube scanner), probe (or tip), probe motion sensor and controller electronics. The microscope must be vibrationally isolated from its surroundings.

AFM is based on a **theory** that models forces between atoms and molecules. Interatomic (intermolecular) forces can be classified as short or long-range forces. Interactions for interatomic spacings smaller than 2–3 Ångstroms are always repulsive, while for larger distances interactions can either be repulsive or attractive. Short-range repulsive forces are related to overlapping of the electron clouds of two neighbouring atoms. Long-range forces include various forms of electromagnetic interactions [295]. Interatomic forces typically range from 10^{-7} N (ionic) to 10^{-11} N (Van der Waals) (*cfr.* binding energies: 10^{-8} N). Several forces typically contribute to the deflection of an AFM cantilever. The forces most commonly associated with AFM are interatomic Van der Waals forces, attractive and capillary forces, the last ones being related to a water film covering the surface under analysis.

AFM maps **surface topographic features** by monitoring the attractive and/or repulsive probe-surface interactions as the cantilever scans a surface. The technique is based on constructing digitised images from the measurement of the repulsive and attractive forces among atoms at the tip and those on the surface being analysed, allowing to obtain information related with the microstructure, superficial defects, and macroscopic properties [293,296].

Since its inception by Binnig *et al.* in 1986, AFM has become an important and widespread tool for imaging surface topography with nm resolution. By exploiting the local nature of an AFM probe and its pico-Newton force sensitivity considerably more information can be extracted from AFM than just surface topography. The magnitude of the "sticking" force and its temporal evolution can reveal details of the type and dynamics of the forces occurring between probe and surface.

There are two major categories of AFM imaging: non-oscillating and oscillating probe methods. Another way to classify the AFM imaging modes is to distinguish between contact (C) and non-contact (NC) modes depending on the forces acting on the tip [296]. Normally the force between the AFM tip and the surface is kept as small as possible, in order to prevent surface damage. In the **contact-AFM** (C-AFM) mode the AFM tip makes soft "physical contact" with the sample. The tip attached to the cantilever is scanned across the sample at a few Ångstroms from the surface, and the interaction force between the cantilever and the sample is typically in the range of repulsive interatomic forces (10^{-9} to 10^{-6} N), hence the name "atomic force microscopy". This has the advantage that one expects a large component of the force to be determined by a relatively small number of atoms near the tip apex, but the disadvantage is that the force becomes dependent on complex atomic processes involving the irreversible deformation or erosion of the tip–sample junction. Contact-mode AFM imaging requires that the molecules of interest be rigidly mounted and immobilised with well-defined orientation to avoid damage by physical contact. The inability of C-AFM to detect local molecular-scale defects at ambient conditions limits analysis of structural imperfections. Contact-mode AFM imaging of surfaces at different forces is an easy way to perform nanomechanical studies.

In the **non-contact AFM** (NC-AFM) mode the tip is kept at a constant distance from the sample (typically tens to hundreds of Ångstroms) in the attractive part of the force–distance curve (largely as a result of the long-range Van der Waals interactions). The interaction between tip and sample at large distance (non-contact interactions) can be modelled as if it were force acting between two macroscopic bodies. NC-AFM is desirable because it provides a means for measuring sample topography with little or no contact between tip and sample. The total force between tip and sample in the non-contact regime is attractive but very low, generally about 10^{-12} N.

This procedure keeps the tip in the region where the tip–sample force is (relatively) well understood, but at the price that the force is determined by the cumulative effect of a large number of atoms – hence the resolution of individual atomic-scale features is seldom possible. In the non-contact mode, the cantilever is made to vibrate at its resonant frequency, and the interaction damps the amplitude of the vibration. NC-AFM is preferable to contact AFM for measuring soft samples such as polymers, but the spatial resolution is lower.

Intermittent-contact (IC-AFM) or **tapping mode** atomic force microscopy (TM-AFM) is similar to NC-AFM, except that the vibrating cantilever tip is brought closer to the sample so that at the bottom of its oscillation it just barely hits, or "taps", the sample intermittently at kHz frequencies. This method provides resolution similar to the contact mode. Some samples, such as soft materials, are best handled using TM-AFM instead of contact or non-contact AFM. TM-AFM is less likely to damage the sample than contact AFM because it eliminates lateral forces (friction or drag) between tip and sample. Light tapping allows imaging of top surface features with lateral resolution determined by the small tip contact area (*ca.* 2–3 nm); this permits imaging of virtually any size of surface. During tapping mode imaging, the vertical force is large enough to locally deform the sample surface. Thus TM-AFM images often represent a mixture of topographic and elastic properties of the sample surface. TM-AFM can be used for imaging of polymer micro-phases and the study of micro-dispersions. Imaging with elevated forces or hard tapping allows visualisation of subsurface structures and differentiation of crystalline and amorphous regions. Clearly, technique selection (C, NC, IC) depends on sample properties and experimental objectives. AFM requires thorough approaches in order to avoid artefacts. The combined use of several modes and different experimental conditions is the basis for a comprehensive examination of polymer samples with AFM. Controlled AFM in (polymer) applications needs to consider imaging forces (from contact via tapping to non-contact), tip–surface chemistry, imaging medium, imaging temperature, imaging speed, tip radius and contact area ("the ultimate tip"), scan direction, tapping frequency, *etc.*

Because the interactions between tip and surface depend not only on the topography of the sample but also on different characteristics (such as hardness, elasticity, adhesion, or friction), the movements of the cantilever to which the tip is attached also depend on these properties. In order to interpret AFM experiments one needs to bear in mind the different types of forces that can act between tip and sample. At large distances the force most commonly present is the Van der Waals force. The slow decay of the Van der Waals force (according to z^{-2}, where z is the tip–sample separation) means that in AFM, unlike STM, the large-scale structure of the tip is important (*cfr.* Fig. 5.5).

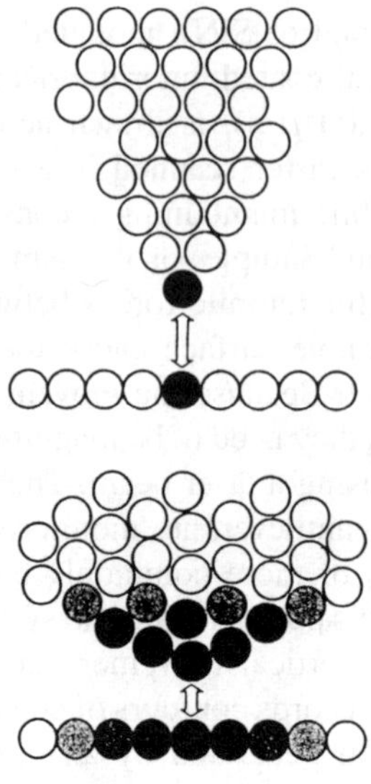

Fig. 5.5. Interatomic interaction for STM (top) and AFM (bottom). Shadowing shows interaction strength. After Howland and Benatar [296]. Reproduced by permission of Veeco Instruments Inc.

In order to obtain a true image in AFM (as well as in any other microscopy or spectroscopy experiment), the observed result and the instrumental lineshape must be deconvoluted. This is often possible in spectroscopy, but often impossible in AFM, especially for artefacts at length scales that are comparable with the tip dimensions. This tip-imaging artefact cannot be eliminated. If the sample is an insulator it may also be locally charged. **Image artefacts** are also introduced due to surface deformation.

AFM instruments offer good lateral and vertical **resolution** (about 1 nm). In principle, atomic and even subatomic resolution is possible. In practice, the probing tip limits resolution. Cantilevers and their tips are critical components of an AFM system because they determine the force applied to the sample and the ultimate lateral resolution of the system. Lüthi *et al.* [297] have discussed the resolution limits of force microscopy. The lateral resolution of an AFM image is determined by the step size of the image and the minimum radius of the tip. The lateral resolution of an AFM with the sharpest tips commercially available is 10 to 20 Å. Ideally, a tip of only a few atoms, preferably one, is necessary. It has been demonstrated that these tips can be prepared; however, they are still not available for routine analysis. Contact diameters in typical contact force microscopy are between 1–10 nm. The 10 to 20 Å resolution seems to conflict with the ubiquitous images of atomic lattices in AFM papers. The distinction between imaging atomic-scale features with accurate lattice spacing and symmetry, and true atomic resolution requires some comment. Generally, atomic-scale images have to be interpreted with care. Even when atomic-scale features are observed in normal or lateral force, the contrast originates from a multiple-atom contact. Exceptions are contact-mode imaging in liquids with ultra-low forces ($<$100 pN) and non-contact imaging in ultra-high vacuum, where true atomic-resolution can be achieved [297]. In AFM, the dependence of the force of interaction is much weaker than the exponential dependence of the tunnelling current in STM. Thus, for AFMs, each atom of the tip that participates in imaging "sees" the sample as a periodic lattice. But because the atoms of the tip are in different lateral positions, the lattice seen by each atom is shifted from the lattice seen by its neighbours. This multiple interaction affects AFM images. The AFM image observed shows the periodic features of the lattice but AFMs usually do not achieve the "true" atomic resolution needed to detect an atomic vacancy. It is also not possible to use AFM to identify unknown surface species. Table 5.34 shows the **main characteristics** of atomic force microscopy.

Table 5.34. Main characteristics of atomic force microscopy

Advantages:
- No sample preparation (no staining)
- Suitable for insulators and (semi)conducting materials
- Non-destructive
- Operates in various environments (UHV, air, liquid)
- *In situ* imaging capabilities (air, liquid)
- Multifunctional (probe of topography, nanostructure and local material properties)
- High-resolution: sub-nm (lateral), 0.1 nm (vertical)
- No radiation damage
- Easy usage
- Relatively cheap

Disadvantages:
- Image interpretation
- Imaging artefacts (quality/size of scanning tip)
- Sensitivity to vibrations
- No chemical information (but CFM)
- Surface damage (softness)
- Specialist skill needed

AFM was used initially with great success to image polymer morphology on different scales, including molecular (lattice) visualisation. The focus of AFM studies has been gradually shifted from structure visualisation to studies of **surface properties**, intermolecular and surface forces, and *in situ* monitoring of processes, often on the molecular scale. Since the advent of AFM a number of scanning force microscopy techniques have been developed utilising the principle of measuring interactions between a sharp tip and a sample. Tip-to-sample force interactions are a key issue in AFM. The surface deformation caused by the tip-to-sample force can be observed on nm scale (from conformational changes to nano indentation). SFM imaging modes can be classified according to the nature of the force between tip and sample and the type of interaction being measured. Advanced modes include magnetic force microscopy (MFM), electrostatic force microscopy (EFM), scanning thermal microscopy (SThM), pulsed force mode (PFM), force modulation microscopy (FMM) and scanning capacitance microscopy (SCM). To extract comprehensive data about surface topography, adhesion and mechanical properties, measurements should be performed at different operating conditions. Minimising the applied force helps to avoid surface deformation. This is especially important for probing surface structures with dimensions in the 1 to 100 nm range, because the estimated diameter of the tip-to-sample contact area during imaging of organic surfaces with a force of a few nN is in the range of several nanometres. The environment in which surfaces interact can play a crucial role in determining measured forces. To probe interactions determined solely by solid-surface free energies (*i.e.* base interactions), adhesion forces must be measured in ultra-high vacuum. Pulling apart the surfaces under liquid will result in their solvation upon separation. Experiments conducted under ambient conditions reflect wettability of the surfaces, since predominant interaction is the result of capillary forces.

SPM operation in the topography mode only rarely provides insight into the chemical nature of a multicomponent system. There has been considerable effort to integrate chemical analysis into the AFM's ability to image at high spatial resolution. Spectroscopy in the force microscope consists of measuring force–distance relationships on the nm scale. Such spectroscopic measurements are the equivalent of tunnelling current-distance curves in STM, which enables the barrier height to be determined. Functionalisation of an AFM tip by grafting of active molecules or deposition of a molecular layer (self-assembly monolayer, SAM) enables measurement of interaction forces between chemical groups on the probe tip and molecules present on the analysed surface. Use of functionalised AFM probes achieves chemically specific contrast and is the basis of **chemical force microscopy** (CFM), due to Frisbie *et al.* [298]. Modified tips with terminal CH_3, COOH, CH_2OH, CH_2Br, SO_3 and NH_2 groups have been reported. CFM can be used for identification of different chemical species (*i.e.* chemical sensing of surfaces on the sub-μm scale). By utilising chemically functionalised tips, force microscopy in contact mode can be used to: *(i)* probe forces between different molecular groups; *(ii)* measure surface energetics on a nm scale; *(iii)* determine pK values of the surface acid and base groups locally; and *(iv)* identify and map the spatial distribution of specific chemical species at polymer surfaces and their ionisation state (chemical sensing on the sub-μm scale). Commercial chemical microscopes are available.

Very recently, a tuneable CO_2 laser has been combined with an AFM to form an aperture-less near-field imaging system to obtain contrast in infrared absorption on a scale of about 100 nm [299]. However, the tuneable range of the CO_2 laser is limited to a region of the IR spectrum that is not particularly informative for most IR chromophores ($\sim$2300 cm^{-1}). For many applications coupling of a tuneable IR diode laser to an infrared microscope [300] is more attractive. Hammiche *et al.* [301] have used a Wollaston resistive thermometer as a photothermal probe to record IR spectra of polymers. Anderson [302] has indicated that an AFM/FTIR microscope without specialised tips can provide surface topography and chemical mapping at high spatial resolution. Direct infrared detection at a surface with the use of an AFM was tested both with filter and FTIR spectrometers (Fig. 5.6). Nowadays, IR spectroscopy at

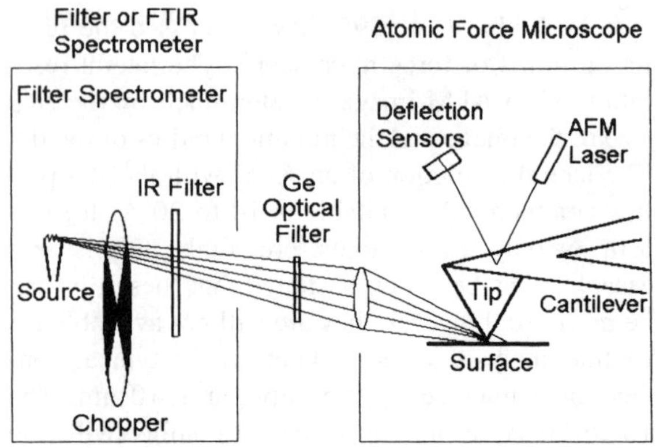

Fig. 5.6. Schematic diagram of AFM photothermal deflection test. After Anderson [302]. Reprinted with permission from M.S. Anderson, *Appl. Spectrosc.* **54**, 349–352 (2000).

high spatial resolution is quite feasible. These are relatively **new techniques**, and interpretation of the detailed local forces requires careful interpretation. **Atomic force acoustic microscopy** (AFAM) allows to probe the inner structure of a material with a lateral resolution of 1 nm using high frequency vibrations (a few MHz) of the surface. AFM is a growing family of operating and imaging modes that complement each other.

Strong interest in AFM also derives from the fact that this technique complements other analytical methods for high-resolution visualisation and mapping of heterogeneous systems. Because no current flows through the specimen, the thickness and the conductivity of the specimen are not as restrictive as for STM. High-resolution AFM imaging does not require special sample treatment, as does TEM. As opposed to some diffraction or scattering techniques (*e.g.* electron or X-ray diffraction), which result in structural information averaged over typically 10^{10}–10^{20} atoms, AFM can yield information about the local order and packing of atoms/molecules at preselected locations of the sample surface. AFM also presents surface structures in real space, whereas structural information can be deduced from diffraction data (small-angle X-ray scattering, SAXS, or small-angle neutron scattering, SANS), only in interplay with structural models. With AFM, one can observe nanoscale structural features near the surface of a thick sample that are not as accessible by TEM and SEM. In contrast to SEM, AFM allows precise determination of the depth of surface structures (*i.e.* 3D instead of 2D). Since AFM data con-

Table 5.35. Microanalytical features of AFM and complementary techniques

- Exploration of structural hierarchy (lamellar, crystalline and molecular order)
- High-resolution profiling of 3D polymer morphology and nanostructure
- Quantitative compositional mapping of heterogeneous polymers providing component concentrations, orientations, distributions, cross-linking density and properties for multicomponent systems (blends, block copolymers, and composites)
- Quantitative probing of surface topography: roughness, height, diameter, and volume distributions of grains
- Analysis of structural imperfections on the surface: depths of holes, scratches, and cracks, *etc.*
- Probing of near-surface micro structures
- Studies of local material properties: stiffness, hardness, friction, elasticity, adhesion, magnetic and electrostatic forces, hydrophilicity/hydrophobicity, thermal behaviour, *etc.*
- Understanding of structure-property relationship
- Studies of thermal phase transitions in polymer samples

tains height information, determining whether a feature is a bump or pit is straightforward, as opposed to SEM. On the other hand, while AFM can measure vertical surface variations below 0.5 Å, its ability to measure tall structures is limited (up to 10 μm). Another advantage of AFM over SEM is that the disturbing effects caused by electrostatic loading do not occur. Non-conducting coated surfaces can well be characterised. AFM is also not reliant upon high vacuum techniques such as SEM and TEM. Moreover, AFM is not only cheaper than high-resolution LVSEM, but allows *in situ* environmental studies without the problem of radiation damage of the surface. The use of AFM as small, rigid and portable devices is precluded during production runs (sensitivity to vibrations). More and more polymer scientists are solving problems with AFMs rather than with any other microscopic technique. Useful literature references are recent reviews [270,303] and various books [286,293,304]. Quantitative probing of polymer surfaces in AFM was discussed [289]. Chemical force microscopy has been reviewed [305,306].

Applications

Although STM was invented first, most progress in scanning probe microscopy of polymers has concerned atomic force microscopy. AFM is now established as an advanced microscopic tool in many academic and industrial laboratories for the study of heterogeneous surfaces. Since the first visualisation of a macromolecule, the technique has been used with great success to image polymers [307]. Nowadays, polymer scientists are solving more problems with AFMs than with any other microscopic technique. For **soft materials** with elastic moduli of a few GPa or lower, such as polymers, minimisation of force interactions between the AFM tip and the sample surface is required for non-destructive imaging of the surface. This need was met with the introduction of tapping mode imaging, which is now the predominant technique for polymer studies. In the application of AFM to polymers the force applied to the surface is usually between 1 and 10 nN. Strong interest in AFM studies of polymeric materials results from the fact that this technique substantially complements other microscopic and diffraction methods for high-resolution visualisation of polymer morphology and nanostructure, and for compositional mapping in heterogeneous systems.

The universal character of the repulsive forces between tip and sample, which are employed for surface analysis in AFM, enables examination of practically an unlimited range of materials. AFM is a multifunctional technique suitable for the characterisation of topography, and local material properties of polymer surfaces on a scale from hundreds of μm to nm that is barely accessible by other systems. Table 5.35 shows the main microanalytical features of AFM as applied to polymeric materials. **Topographic mapping** is the dominant application for AFM. In AFM imaging, tip–sample interactions can be essentially modified by surface forces. This can help to reveal spatial distributions of different components in multicomponent systems such as polymer blends and composites. AFM also allows non-destructive visualisation of subsurface structures at depths from a few to tens of nm. Local probing of mechanical response of polymer surfaces with AFM modulation techniques offers unique possibilities for conducting dynamical mechanical analysis with resolution in the nm range [308]. AFM does not probe chemical composition. Other SPM techniques have been developed which are able to provide some chemical/structural information.

Depending on the aim of AFM studies of polymer surfaces, low-force and/or force-dependent imaging should be performed. Force modulation imaging can provide measurements of surface elasticities and help identify hard and soft regions (*e.g.* in carbon fibres/epoxy composite [309]). Force modulation AFM (FM-AFM) has been used for the determination of the detailed microstructure of polymer blends, *e.g.* isobutylene-based polymers that cannot be unambiguously characterised by EM techniques because of rapid degradation under electron bombardment [310]. SFM provides a significantly more comprehensive analysis of surface structure than SEM of gold-coated polymer films. Quantitative measurements of the nanoscale surface roughness of the films have been obtained by SFM. Beake *et al.* [311] have used SFM to study bulk-filled uniaxially oriented PET films.

Addition of filler particles (China clay, silica, glass beads) during manufacture increases the final surface roughness of the films. Filler particles at or near the surface are accompanied by deep depressions aligned in the direction of draw. Negroni [312] has investigated **filler dispersion** (silica with/without silane in SBR) and rubber solubility by means of FM-AFM. Force modulation AFM is quite useful in distinguishing different types of fillers in polymeric materials [310]. AFM imaging was utilised for studying adhesion failure in an elastomer/glass system [313]. AFM combined with SAXS was used in studying the interphase properties of silica filled rubber [314]. Whereas SAXS measurements detail the morphology of the elementary silica particles (spherical, average diameter of 13 nm), AFM reveals the existence of indivisible silica aggregates and agglomerates. Coating of the fillers favours dispersion of aggregates within the elastomer matrix. Other studies have concerned carbon-black filled EPDM. AFM is an effective means for determining the real size and shape of nano-particles. Quantitative results of particle diameters, and heights, aggregate sizes and interaggregate distances can be determined reliably and directly. Also TM-AFM can be used for mapping the distribution of fillers in a polymer matrix. Comparison with TEM results has shown that TM-AFM with phase imaging is a powerful method for characterising the silica dispersion in a silicone matrix and contributes to the understanding of filled rubber reinforcement [315].

AFM provides 3D polymer **morphology/nanostructure** (high-resolution visualisation and measurement of morphology of lamellar and granular nanostructures of crystalline polymers, 2–20 nm in size) [316], quantitative compositional mapping (component concentrations, orientations, distributions of polymeric systems with fillers, oils and additives), and structural changes at thermal transitions. AFM studies of impact-modified plastics often reveal similar morphological information as TEM micrographs of ultrathin sections, however without staining and elaborate sample preparation [317]. Rodríguez *et al.* [318] have used AFM in combination with SEM and TEM for the structural and morphological study of nm-sized structural features. Shaffer *et al.* [173] have used AFM to characterise the interphase regions in rubber-toughened epoxy blends. The interface region was varied by the addition of reactive oligomers or cross-linking the sub-μm core/shell latex particles of a poly(butadiene-*co*-styrene) [P(BS)] core with a PMMA/AN shell. AFM was able to detect and quantify fracture surface features not observed with field emission SEM methods. Anderson [302] has shown how AFM/FTIR can image a plasticiser-coated PVC surface. TM-AFM has emerged as a powerful technique to provide direct spatial mapping of surface topography and surface heterogeneity with nm resolution. Phase contrast in TM-AFM often reflects differences in the properties of individual components of heterogeneous materials and is useful for compositional mapping in polymer blends, copolymers and composites. AFM allows mapping of additive dispersion on nanoscale. The technique has also been used to investigate **surface segregation** of erucamide and behenamide on LLDPE and on multilayer films with POP as the skin layer in relation to COF (coefficient of friction) studies [319,320]. AFM and OM were used in a combined study of plasticised PVC membranes used as ion-selective electrodes [321].

AFM goes far beyond high-resolution profiling by providing local properties of polymer materials, maps of sample composition, and the ability to examine underlying surface layers at μm and sub-μm scales. AFM has considerable potential in recognising structure-property relationships in advanced polymer characterisation for rubbers, paints and coatings, packaging, engineering plastics, consumer goods, and other applications.

AFM techniques are used to study technologically important parameters of coatings (gloss mechanisms, scratch resistance, film formation) and adhesives (failure mechanisms). TM-AFM provides roughness values of coated papers. The effect of plastic pigments on surface roughness was elucidated [322]. AFM provides the capability to create nano-scratches under various controllable compressive loads and rates and allows accurate determination of height and surface roughness. Gu *et al.* [323] have reported heterogeneity mapping in polymeric materials using AFM phase imaging and **nano-indentation**. Pourdeyhimi *et al.* [324] have used AFM for the evaluation of **scratch and mar resistance** in automotive coatings. Mar depth usually ranges from 50 nm to several hundred nm. Mar damage causes largely a visual effect. Scratching is a more serious form of marring. The thickness of clearcoats is about 15–20 μm, indicating that only this topcoat is affected by mar and scratch damage. The authors measured marring by nano-indentation with an SPM, optical imaging and gloss measurements. AFM is limited to micro scratch properties and cannot help with macro scale defects. For that purpose macro measurements (*e.g.* VIEEW™) are required.

Orefice *et al.* [325] have used AFM to study interactions involving polymers and silane networks. AFM results showed that phenomena such as chain penetration, entanglements, intersegment bonding, and chain conformation in the vicinity of rigid surfaces are relevant for the overall processes of adhesion and adsorption of polymeric chains within a silane network. SFM can be used to determine the maximum adhesion force with a spatial resolution of a few nm. AFM measurements of the forces on a sharp tip sliding across silicon substrates coated with perfluoropolyether polymers have provided insight into low lubricants function at the molecular level [326]. Bao *et al.* [327] have reported an AFM study of perfluoropolyether lubricants dip-coated on hard disk (amorphous C) surfaces. Nano-sampling and photothermal deformation spectroscopic analyses are both new uses of AFM.

Chemical force microscopy has been used to probe **adhesion and frictional forces** between distinct chemical groups in dry gases, organic solvents and water [306]. Schönherr *et al.* [328] have described imaging of functional group distributions in surface treated polymers by SFM using functionalised tips, reaching a lateral resolution on a sub-50 nm level.

AFM with tips chemically modified by means of methyl and hydroxyl terminated self-assembled monolayers (SAMs) of alkanethiols has been used for chemical recognition of a process stabilising agent (Irgafos 168), an antioxidant (Irganox 1010) and UV light stabilisers (Tinuvin 770, Dastib 845, Chimassorb 944 and Hostavin N 30), and for adhesion mapping of a polypropylene surface [329]. Before investigating the polymer surface, it was necessary to determine the response or fingerprint of pure spin-coated additive films toward the chemical probes. Four sets of measurements have been carried out for each compound: pull-off forces have been recorded with $-CH_3$ and $-OH$ tips, in nitrogen atmosphere and in water (Fig. 5.7). One series of data (*i.e.* one type of chemical tip in one medium) is not sufficient by itself to identify an additive, but the whole combination of the four series is necessary. A characteristic fingerprint for each additive with respect to the tip functionality and the measurement environment was put in evidence. This fingerprint enables differentiation of the additives. CFM adhesion mapping supplements information that can be obtained by ToF-SIMS analysis. ToF-SIMS data only gives information about the presence of additives on the surface, while ToF-SIMS chemical imaging of additive distribution does not provide submicroscopic lateral resolution. Laterally resolved chemical force microscopy allows studying the distribution of chemical functional groups on a surface [329]. CFM with tips functionalised with octadecanethiol or 11-mercapto-1-undecanol has also been used to study aging of PP/(Irganox 1010, Irgafos 168) and PP/(Irganox 1010, Irgafos 168, Tinuvin 770) surfaces at the submicroscopic scale [330]. Adhesion force maps with lateral resolution lower than 1 μm showed chemically heterogeneous surfaces on a sub-100 nm scale.

Magnetic force microscopy (MFM) has been used for characterising the dispersion of carbon nanotubes (CNT) in a high performance polymer matrix [331].

5.5.2. Near-field Scanning Optical Microscopy

Principles and Characteristics

Diffraction limits the spatial resolution of conventional optical microscopy instruments. In practice, the resolution limit is approximately 0.6λ, *i.e.* about 0.5 μm for optical microscopes. Resolution in "far-field" optical microscopy techniques may be improved (though slightly) by the application of UV (*cfr.* Chp. 5.3.2) or confocal laser scanning (*cfr.* Chp. 5.3.4). For confocal laser imaging with green light ($\lambda = 500$ nm), resolution is limited

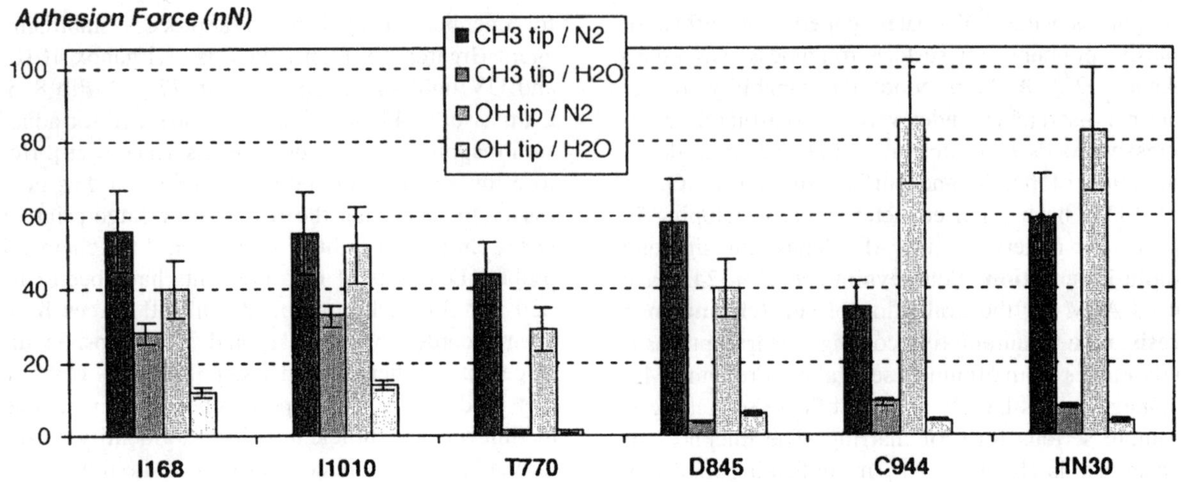

Fig. 5.7. Average adhesion forces (nN) obtained with $-CH_3$ and $-OH$ terminated tips on Irgafos 168 (I168), Irganox 1010 (I1010), Tinuvin 770 (T770), Dastib 845 (D845), Chimassorb 944 (C944) and Hostavin N 30 (HN30) spin-coated onto silicon wafers. The force-distance curves were recorded in water or in nitrogen atmosphere. After Duwez *et al.* [329]. Reprinted with permission from A.-S. Duwez *et al.*, *Langmuir* **17**, 6351–6357 (2001). Copyright (2001) American Chemical Society.

to approximately 300 nm. As already postulated by Synge [332] in 1928, the diffraction barrier to optical resolution may be circumvented by what is now called a near-field microscope.

Near-field microscopy involves illumination of a sample through an aperture significantly smaller than λ, while maintaining the sample-to-aperture separation at a distance much smaller than 0.6λ, thus overcoming the diffraction limit and enabling nanoscale optical imaging [333]. In near-field scanning optical microscopy (NSOM, also referred to as SNOM), a scanning probe microscopic technique with characteristics similar to AFM, images are produced by illuminating the specimen with a nano light source, *i.e.* light emerging from a sub-wavelength size aperture in the tip of a highly tapered probe (*e.g.* a metal-clad optical fibre) positioned in the "near-field" region [334,335]. At this proximity spatial resolution is then *not* determined by the wavelength of the light, but by the **aperture size**. Apertures smaller than 10 nm have been prepared recently. Resolution is higher than can be obtained by far-field CSOM. Raster/scanning requires accurate control of the distance of the probe to the specimen surface (*e.g.* 20 nm $\pm$ 1 nm for a 50 nm diameter aperture used with 500 nm light). The image is built up in a sequential manner point by point. The signal intensity varies exponentially with distance, as in STM. The unique advantage of this technique in comparison to other scanning probe microscopies, such as STM or AFM, is the addition of a spectral dimension which allows chemical identification of surface structures. Pohl *et al.* [333] have demonstrated the NSOM concept experimentally in 1984.

NSOM can be used to generate a visible-light image of the surface with a resolution of about 15 nm, provided that the distance between light source and sample is very short, about 5 nm. NSOM thus improves the resolution of an optical microscope by at least an order of magnitude ($<\lambda/20$) [333]. To surpass even this resolution, apertureless NSOMs have been designed with a theoretical limit of 1 nm. NSOM represents one of the most promising optical techniques that aims at overcoming the Abbe barrier, and yet retain most of the utility of a traditional optical microscopy. NSOM is capable of super-resolved transmission, reflection, polarisation, refractive index and fluorescence imaging using conventional far-field optics [336].

The **theory** of NSOM is somewhat similar to that of STM, with transport of light (or photons) replacing transport of electrical current (electrons). Instead of the Schrödinger equation, the Maxwell equations for the electromagnetic field must be solved near tip and sample, taking into account the local electromagnetic properties of each medium [270]. The resolution is lower than that attainable with STM.

Table 5.36. Main characteristics of near-field scanning optical microscopy

Advantages:
- Localised optical (spectroscopic) and 3D topographic information
- Sub-diffraction-limit spatial resolution (*ca.* $\lambda/20$)
- Single-molecule spectroscopy

Disadvantages:
- Expensive equipment
- Emerging technique

Table 5.36 shows the **main features** of NSOM. Optical fibre tips for NSOM with high light transmission permit surface analytical and spectroscopic applications (fluorescence imaging, Raman) with high spatial resolution (*ca.* 30 nm) and high chemical information content [337]. NSOM overcomes critical measurement limitations of both far-field vibrational microscopes (low spatial resolution) and scanned probe microscopes (lack of chemical specificity). NSOM offers conventional optical characterisation and contrast mechanisms with the resolution of SPM. The spatial resolution of this relatively new technique is almost competitive with that of SEM. Consequently, NSOM is expected to become a serious alternative for SEM, since it is non-destructive if visible light is used, and allows visualisation of specimens in air. The technique holds considerable promise for the future. At the present time it is still quite expensive.

There exists a remarkably wide range of different near-field techniques, which range from scattering type to aperture and wave-guide structures. The high spatial resolution of NSOM has been coupled with the chemical specificity of vibrational spectroscopy. **Near-field infrared microscopy** (IR-NSOM or SNIM) benefits from larger optical cross-sections ($\sigma = 10^{-17}$–10^{-19} cm^{-2}) than Raman scattering ($\sigma = 10^{-28}$–10^{-30} cm^{-2}). A constraint for IR-NSOM is lack of readily available bench-top light sources with adequate intensity and/or spectral tuning range. IR-NSOM is limited to cases where the instrumental sensitivity is sufficient to allow IR absorption methods of samples of sub-wavelength thickness. IR-NSOM is an *in situ*, non-destructive analysis technique for site-specific chemistry on the nm scale. Various authors [338–340] have described IR-NSOM and Raman-NSOM (R-NSOM). NSOM in combination with Raman spectroscopy and APCI-MS is used for chemical analysis [341]. With tube etching – a new method of tip fabrication – tips are available that can be used for Raman and laser desorption MS with a sensitivity below 400 molecules. Combined SFM and NSOM methods can be used in fluorescence imaging and fluorescence microscopy with a resolution of ~100 nm [342].

Since resolution in the far field is limited by wavelength, conventional optics in the radiofrequency (RF) through far-infrared (FIR) spectrum cannot resolve very small features. With the advent of near-field microscopy, RF and FIR microscopy have gained more attention [343].

Near-field scanning optical microscopy is attracting growing interest as a method for imaging [344], spectroscopy [345,346] and even material processing [347], as it can provide sub-wavelength spatial resolution over a broad range of the electromagnetic spectrum – from UV to microwave frequencies (SNMM). NSOM shows excellent **localised spectroscopic capabilities** with resolution comparable to the probe dimensions, *i.e.* tens of nm. This technique offers UV/VIS spectra of single molecules on a surface, *i.e.* chemical information together with nanometric spatial resolution. Both absorption and fluorescence spectra can be recorded and the refractive index of (sub)surface species can be measured. If IR absorption or Raman scattering is used as the contrast mechanism, vibrational spectra of samples can be obtained. Scanning near-field microwave spectroscopy (SNMM) has been used for non-contact imaging of dielectric constants [348].

Near-field microscopy (from visible to high-frequency) was recently reviewed [270,343,349–352].

Applications

While elemental analysis of surfaces is possible with a lateral resolution of a few dozen nm, analysis of *molecular* species with a resolution of better than 1 μm is very difficult. NSOM is being developed into a tool for molecular analysis with a spatial resolution in the nm range. Few surface chemical analyses in the nm region using NSOM have been described [337,353]. NSOM for fluorescence imaging of single molecules with a high signal-to-noise ratio has been reported [354,355]. The technique has been used for fluorescence imaging of dye-labelled polystyrene spheres. Recent developments in near-field imaging and fluorescence detection of individual carbocyanine dye molecules on the surface of PMMA and quartz films were reviewed [356].

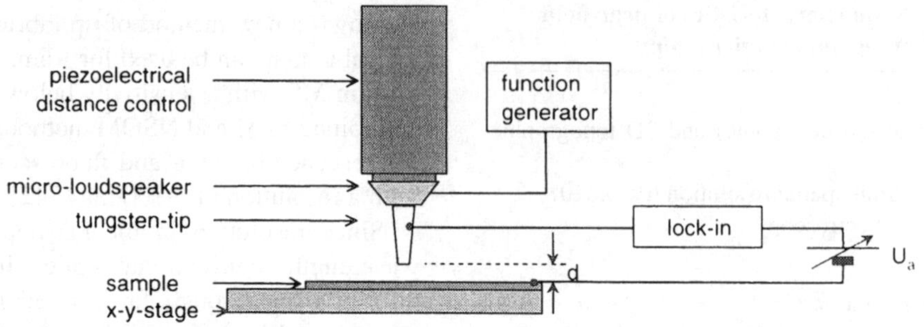

Fig. 5.8. Kelvin setup for the analysis of the conductivity distribution in heterogeneous materials. After Prasse *et al.* [367]. Reprinted from T. Prasse *et al.*, *J. Appl. Polym. Sci.* **82**, 3381–3386 (2001), John Wiley & Sons Inc., New York, NY, Copyright © (2001, John Wiley & Sons Inc.). This material is used by permission of John Wiley & Sons Inc.

Raman-NSOM has been demonstrated for a variety of samples, including adsorbed dye molecules [358–361] and polymers [357]. By combining NSOM with SERS molecular spectroscopy and imaging with a lateral resolution of 70 nm is possible [337]. Also IR-NSOM can be used for chemical imaging [338]. Thin film analysis benefits from NSOM. The technique has been used to probe the excitonic transitions in J-aggregates of 1,1′-diethyl-2,2-cyanineiodide grown in poly(vinyl sulfate) thin films [290].

5.5.3. Scanning Kelvin Microscopy

Principles and Characteristics

The vibrating capacitor or Kelvin method [362, 363] is a well-established experimental technique for measuring the contact potential difference (CPD) or work function for a variety of materials, including polymers [364] and carbon-black [365]. Here, the sensitivity of the CPD to the appearance of electronic surface states and surface charges is used. Scanning Kelvin microscopy (SKM) allows for mapping of the two-dimensional CPD distribution on sample areas of 1 cm^2 with μm resolution without extensive experimental requirements [366].

The experimental setup for SKM is shown in Fig. 5.8. The sensing element is a tungsten tip about 50 μm in diameter positioned at a distance of 10 μm above the sample surface. The sample is mounted on a x–y translation stage, above which the oscillating tungsten tip is positioned, which is leveled before the scan. An applied external voltage U_E is fixed, and the induced Kelvin current is measured at each position with a lock-in amplifier.

SKM is a non-destructive investigation method for revealing the distribution of the conductive network in a polymeric host. Filler particles participating in the percolating network are mapped selectively, whereas isolated particles or clusters not connected to the network are not resolved. This is a considerable advantage with respect to conventional transmission optical and transmission electron microscopy, which also requires greater efforts in terms of sample preparation.

Although not a scanning probe microscopy at the atomic level, scanning Kelvin microscopy shows some (macroscopic) similarities to SPM methods. At variance to STM, SKM measurements are performed with a tip operating in the far-field (μm *vs.* nm scale).

Applications

Scanning Kelvin microscopy has been used for imaging of conductive filler networks of carbon-black (CB) and carbon nanotubes (CNT) in heterogeneous bisphenol A resin materials suitable for antistatic and electromagnetic shielding [367]. SKM observes exclusively the distribution of a percolated conductive filler network. Transmission optical microscopy revealed matches between scanning Kelvin images and the sample morphologies, whereas the percolating backbone could not be distinguished in optical micrographs. Obviously, optical microscopy is not suitable for distinguishing conductive and insulating regions within the sample.

5.6. MICROSPECTROSCOPIC IMAGING OF ADDITIVES

Principles and Characteristics

Classical spectroscopic techniques generate an "average" structure over the dimensions of the sample, and no information about the distribution of the structure is obtained. Analytical techniques are moving from bulk analysis to **detailed mapping** of properties and compositions. Micro-spectroanalysers can be used for transmission spectra of dyes and pigments in paints and inks (resolution: 400 nm for UV/VIS, 4 μm in NIR). Various methods using electromagnetic signals (X-rays, or IR light) are, in their conventional form, less suitable for routine surface characterisation, although numerous variations of such methods, *e.g.*, surface sensitive XRF (TXRF) or multiple total reflectance IR spectroscopy (ATR), are now routinely used to limit response to a near-surface region with excellent sensitivity. However, these techniques sample much deeper than SIMS or XPS, on the order of 1–100 μm in the best cases. Various spectroscopic techniques (IR, Raman, UV and fluorescence) are coupled to microscopy. The technique of fluorescence is still being refined and the Fourier transform method allows IR and Raman microscopy. These techniques give information about chemical composition in 3D space.

There are four approaches to **spatially resolved spectroscopy** of samples: *(i)* microsectioning; *(ii)* microbeam methods; *(iii)* localised spectroscopic methods; and *(iv)* optical slicing. In the first case, microsections of a sample are examined individually. Microbeam techniques rely on a focused beam to probe a given sample volume (*e.g.* laser Raman spectroscopy). In a localised spectroscopic method a desired spatial region is isolated and spectroscopically excited while the remainder of the sample is unaffected. Examples are ATR, grazing angle incidence techniques; resonance techniques such as NMR and ESR use selective excitation methods. Optical slicing (or optical microtoming) involves the use of confocal microscope techniques that allow successive observations of optical sections in the axial (thickness) dimension.

Merging of spectroscopy with microscopy has generated an entirely new discipline, termed microspectroscopy, which allows measurement of the spatial distribution of chemical structures in materials. Microspectrophotometry (MSP), primarily in the UV/VIS and NIR ranges (220 to 2500 nm), has been practised in some way since the 1930s with emphasis on the microscope functionality [368, 369]. On the other hand, the recent convergence of infrared with microscopy accentuates the spectroscopic functionality. Microspectroscopy is a powerful tool for characterisation of micro samples, for examination of heterogeneous materials and for analysis of processes such as migration that involve spatial dimensions on the supramolecular scale. The **main requirements** to microspectroscopic techniques are shown in Table 5.37.

Table 5.37. Main requirements to microspectroscopic techniques

- Non-invasive, non-destructive
- Ambient conditions
- Maximum spatial and axial resolution
- High specificity
- High contrast
- High sensitivity
- Fast response times
- Experimental automation
- Robust equipment
- Wide applicability to heterogeneous materials

In imaging, it is the *spatial* distribution of the signal throughout the sample that is of principal concern. **Imaging** is primarily concerned with obtaining high spatial resolution information from a 2D sample. Imaging usually provides only limited spectral information. An imaging system is comprised of two separate functional components: *(i)* the optical component (image acquisition); and *(ii)* the computing component (image processing/analysis). During an image analysis (IA) experiment the sample must be illuminated with monochromatic energy. Various technologies have been used: grating monochromators, interference filters or acousto-optic tuneable filters (AOTFs) [370]. Multiphoton imaging facilitates data collection from samples exhibiting rapid changes.

The three primary methods of **image generation** in vibrational spectroscopy include point-by-point (or line) scanning, image reconstruction, or the use of multichannel detectors in combination with spectrometers that retain image quality. The third approach is the preferred option because the three spectral image dimensions (two spatial, one spectral) are probed directly and simultaneously. For example, Si CCD detectors have been employed with AOTFs to perform Raman imaging microscopy. For NIR and IR wavelength applications, focal-plane-array (FPA) imaging detectors have created possibilities for spectroscopy and spectral imaging [371]. Most imaging techniques are based on projection, scanning or convolution (tomography). Projection techniques require multiple sensors able to detect the radiation in each pixel almost instantaneously or in a very quick sequential way. More recent imaging techniques are based on scanning. In scanning there is usually only one detector or only one focused source of radiation (*e.g.* a laser). The sample, placed between radiation source and detector, is usually moved around. In other designs the radiation source is moved by laser scanning, or the detector can be moved. By registration of positional parameters and the measured physical or chemical property an image is constructed. Scanning techniques are also available with multiple detectors, such as diode arrays with typically 250–8000 elements for linear scanning. The scanning speed is higher than with a single detector, and resolution can become very high. Scanning is slower than projection, but produces more accurate and precise results. Some methods are hybrids between scanning and projection. Time-resolved imaging can be combined with laser scanning microscopy, either conventional or confocal. The third imaging technique is tomography. In classical tomography, the attenuation of a radiation source through a volume is measured as a line integral. There are many variants of tomography. Some are based on radiation attenuation, others on emission from inside the volume and some on magnetic field gradients in the volume (NMRI). Some techniques use a single detector, others multiple detectors, either moving or stationary. Tomography allows registration of 3D images. Tomographic reconstruction is difficult. In tomography it is not immediately clear how the image is constructed from the collected data, as in other imaging methods. In tomography, line integrals through a solid body are converted to a 3D volume or 2D slice of that body. Apart from computer-aided tomography-magnetic resonance imaging (CT-MRI) and confocal microscopy, 3D microscopy or volume visualisation is still in its infancy [372].

Key requirements for imaging experiments are an imaging method and a contrast mechanism. Various **imaging modes** may be distinguished (Table 5.38). Area excitation with area imaging detection is equivalent to the conventional microscope, telescope, and camera or other imaging optical instrument. The sample is evenly illuminated with an incoherent source. Area excitation with spot detection uses a similar illumination system, but measures the response at a specific point on or in the sample, *e.g.* imaging thermally induced distortions with a spot probe interferometer. Spot excitation with non-imaging area detection permits the widest range of spectroscopic tools to be used, including ions, electrons, photons, acoustic and thermal waves, *etc.*, and covers the widest range of imaging techniques used in analytical science. The excitation source may be scanned relative to the sample or *vice versa.* The laser has had an enormous impact on the imaging capabilities of optical spectroscopies. Spot excitation with spot detection is a mode which relies on having both a small excitation and a small detection volume. This imaging mode allows "optical" sectioning by means of confocal microscopy. Convolution imaging techniques are used when direct imaging with the desired spatial resolution is not possible. Instead, the sample is excited with a source that is temporally or spatially modulated and the response is recorded without direct imaging. The "image" is then reconstructed from the recorded data using mathematical techniques. The best-known examples are tomographic methods. Near-field techniques are used when a resolution significantly better than the wavelength of the exciting source is required. Many near-field techniques have been developed with STM being the best known example.

The second key requirement for an imaging techniques is a **contrast mechanism**, which is some property of the system that varies spatially in the same way as the sample properties to be mapped, *e.g.* a specific fragment in mass spectrometry, a specific vibrational group frequency, surface reflectivity, *etc.* In many cases, especially for polymers, samples do not have much inherent contrast under a visible light microscope. Additional contrast is desired.

Table 5.38. Imaging

Imaging mode	Representative experiment
Area excitation with area imaging detection	Conventional microscopy
Area excitation with spot detection	Spot probe interferometry
Spot excitation with non-imaging area detection	SEM, imaging MS
Spot excitation with spot detection	Confocal microscopy
Convolution techniques	Tomography, NMRI
Near-field imaging	STM, near-field spectroscopy

Table 5.39. CCD application chart

Light level	Measurement time	Typical applications
High	<3 min	Transmission, absorbance, emission, routine process
Medium	<1 h	Analytical Raman, photoluminescence, Raman process
Low	≤10 h	Research Raman, very weak light applications, astronomy

After ref. [373].

Absorption techniques offer this contrast enhancement. In fact, as both inorganic and organic compounds have strong absorbencies in the IR portion of the spectrum, there is inherent contrast in IR images for all compounds. If there is insufficient information in the absorption spectrum of the sample, the methods can be extended by using biological stains, reactive dyes, and other types of labelling. Virtually every spectroscopic technique, and many non-spectroscopic techniques can be used for imaging experiments, albeit not all with the desired resolution.

Imaging spectroscopy is the combined analysis of both spatial and spectral information so that each pixel in a 2D visualisation includes a third dimension of spectral information. Any 2D target containing spectroscopically distinguishable units is potentially a target for imaging spectroscopy. Various methods are employed for image generation: *(i)* scanning of the sample systematically through a stationary field of view defined by the collection optics and detector; *(ii)* scanning of the imaging source (or detector) in a raster pattern across the surface of the stationary sample; or *(iii)* wide field illumination and viewing (video microscopy) [26]. The imaging side of microscopy has increased considerably due to enhanced PC capability. The as yet unexplored frontier in spectroscopic imaging is the use of chemometric data analysis to identify elemental and molecular correlations.

Typical **imaging detection systems** are the human eye, a TV camera, a photographic plate, a CCD array detector or a charge induction device (CID). Two-dimensional array detectors (*e.g.* PDA, CCD) can eliminate the need for a scanning mechanism as the image is directly focused on the array detector. Spectroscopic CCD detectors operating from UV to NIR are designed to acquire data with the highest sensitivities and at the lowest possible noise (Table 5.39). Coupling of AOTF to a microscope-CCD array combination enables experiments to be carried out in which a series of images at different wavelengths is rapidly recorded, or differential images produced.

Table 5.40. Physical phenomena giving rise to imaging of materials

Type	Wavelength, frequency or energy
Electromagnetic radiation	
Gamma	0.5–140 pm, 9–2500 keV[a]
X-ray	10 pm–10 nm, 0.12–120 keV
Ultraviolet	180–380 nm
Visible	380–780 nm
Near-infrared	780–2500 nm
Mid-infrared	2500–50,000 nm
Microwave, radar	10–1000 mm, 300–40,000 MHz
Electron energy	5–500 keV[b]
Mass/charge ratio	Masses from 10 to 10,000 a.u.
Acoustic waves	20 kHz–15 MHz
Electron tunnelling	[c]
Atomic force	[c]
Gravity	[c]
Magnetism	[c]

[a] No upper limit.
[b] Typical values.
[c] Different units and ranges are in use, often not connected to wavelength or energy.
After Geladi and Grahn [13]. Reprinted from P. Geladi *et al.*, *Multivariate Image Analysis*, Copyright © 1996 John Wiley & Sons, Limited. Reproduced with permission.

Table 5.40 lists the main **physical phenomena** that give rise to imaging of materials. The ability to obtain large data sets in relatively short times means that multivariate analysis techniques come to the fore in interpreting and correlating images with product properties of heterogeneous samples such as polymers. The combination of both image and spectral data greatly assists interpretation and visualisation of product properties.

Table 5.41 lists the most current imaging modes and analytical techniques. Examples of complete imaging systems are electron microscopy and mag-

Table 5.41. Main imaging modes and techniques

Imaging mode	Technique(s)
Elemental	iXPS, SAM, iPIXE, SEM-EDS, EPMA
Chemical	ICL, fluorescence imaging
Spectroscopic	μUV, μFTIR, μRS, NMRI, NQRI, ESRI, NIVI
Functional group	μFTIR, μRS
Chemical shift	NMRI
Molecular	iSIMS
Radical	ESRI
Multispectral	UV, FTIR
Acoustic	SCAM
Macroscopic	VIEEW™, NMRI

netic resonance imaging. Sample images, spectra and spectral maps are tied up. Line maps, multiple point maps, area maps, functional group or correlation maps are easily generated. **Elemental images** offer rapid identification of areas of interest, *i.e.* differences in elemental composition, particulates, inclusions, gradients and voids. Chemical imaging couples molecular spectroscopy with digital imaging to provide a non-invasive means of spatially resolving spectral information that describes a material's chemical composition and architecture. Mesoscale chemical imaging is an invaluable tool in materials analysis, bridging the gap between high-resolution and bulk analysis [374]. Image processing, often using multivariate techniques, reveals chemical information, including molecular composition, conformation, concentration and morphology. **Chemical imaging techniques** rely on a variety of universally applicable molecular spectroscopies to generate molecular specific image contrast including IR absorption (both mid-IR and near-IR), Raman scattering and fluorescence emission. Raman imaging uses the visible region. The utility of conventional vibrational imaging techniques is limited primarily in terms of spatial resolution. The limit, imposed by diffraction effects, is of the order of 10 μm for FTIR microscopy and 1 μm for Raman microscopy. IR microscopes based on FPA detectors have recently reached 5 μm. Vibrational spectroscopic imaging techniques provide widely applicable chemical selectivity in heterogeneous materials without sample preparation. Micro-IR and micro-Raman provide complementary information. Thin specimens can be amenable to transmission IR, while thick or opaque specimens can be probed with Raman scattering. The minimum volume of sample necessary is of the order of 1000 μm^3 ($10^{-9}\mu$L) for IR spectroscopy and about 100 μm^3 for Raman spectroscopy. Developments in single reflection systems now allow FTIR viewing capability for micro-samples (typically 600 × 600 μm^2) with support for transmittance, reflectance and ATR mapping. ATR-FTIR spectroscopy has been widely applied as an analytical tool in the μm range allowing for surface characterisation and **depth profiling** of materials without the need for sectioning of the sample, and subsequent chemical analysis or surface etching such as sputtering. ATR micro-samplers (sampling area less than 250 μm) can accommodate a wide range of samples, like paint chips, single fibres, films. Except for microscopies based on electronic spectroscopy, no other imaging techniques offer such direct connections to the visible world. Multispectral imaging systems, combining imaging spectroscopy and machine vision techniques are in use for the examination of art objects (paintings).

Taylor *et al.* [375] have first reported imaging spectroscopic studies in the shortwave near-IR region (700–1100 nm) by using a CCD video camera. Robert *et al.* [376] have extended the work to the near-IR region (900–1900 nm) using a spectral approach: a sample was not characterised by a single image, but by a set of images recorded at different wavelengths. Discriminant analysis was applied for identification of the major constituents of samples. Bertrand *et al.* [377] have designed a **near-infrared imaging** system. It is possible to utilise the full spectrum in the image analysis rather than a selection of a specific frequency. This is particularly important for NIR imaging.

The relationship between an image and the chemical properties of a sample is generally complex. The power of discrimination increases when images at multiple wavelengths are recorded [378]. **Near-infrared video imaging** (NIVI) consists of illuminating samples at different wavelengths of the near-IR range and recording the corresponding sequence of images with a video camera [375,376]. NIVI is operational for practical applications [379].

Fluorescence imaging spectroscopy is an effective chemical state-imaging tool, particularly when employing highly specific fluorescent tags. **Micro fluorescence** spectrometry is an invaluable tool for trace analysis. Also mass spectrometry allows molecule-specific imaging (iSIMS), *cfr.* Chp. 5.9.2.

Acoustical micro-imaging can also be used to produce 3D images and reveals a wealth of internal features and defects. The technique provides a quick, non-destructive method of elasticity imaging and evaluating internal structure without physically sectioning the part. Van den Berg *et al.* [380] discussed the principles of **scanning acoustic microscopy** (SCAM) with respect to destructive and non-destructive analytical techniques. Macroscopic imaging techniques are X-ray tomography [381], low-frequency ultrasonic scanning [382] and NMRI [383].

This Chapter describes the use of both point mapping *and* global imaging techniques to study subtle spatial variations in polymer chemistry and morphology. Mapping and imaging of additives and crystallinity/molecular orientation in polymer articles will be illustrated [384]. Quantitative acoustic microscopy was reviewed [385] as well as scanning acoustic microscopy [386–388]. Laser ablation microanalytical techniques are discussed in Chp. 3.

Imaging techniques were recently reviewed [389]. Various monographs are available on microscopic and spectroscopic imaging [390], image processing [391,392], and image analysis applications for plastics [18]. Koenig [393] recently has described the present status of micro-spatial spectroscopic techniques (optical, IR, Raman and NMR microspectroscopic imaging) of polymers.

Applications

An expansion in image analysis applications is being witnessed. Image analysis can be used for the characterisation of industrial products. End-use performance of polymer articles such as bottles, mouldings, laminates and coated films can be critically dependent not only on additives but also on spatial variations of molecular orientation, crystallinity, colour and chemical composition, both in the bulk and near surfaces/interfaces. Coupling of optical microscopes to IR and Raman spectrometers allows such property variations to be interrogated, mapped and imaged on the μm scale, and then correlated with product performance. The study of colour or intensity of the points (pixels) in an image can be a way to obtain chemical information. For example, advantage may be taken of components fluorescing white in the visible range.

Koenig [393] in particular has examined the need for spatially resolved spectroscopic techniques for the characterisation and improvement of engineering polymers and has described the development of micro-spatial spectroscopic techniques for polymers. Near-IR spectroscopy and imaging was used for monitoring of powder blend homogeneity [379]. Lloyd *et al.* [394] have exploited the complementary nature of different surface chemical imaging techniques such as AFM, μRS and ToF-SIMS. One of the major challenges is to obtain information with these techniques from the same exact spot on the sample. **Chromographic analysis**, a method for colour coding digital images of product samples, is a valuable tool for measuring sample colour, colour homogeneity and structural integrity in a variety of applications such as establishing the nature of product deficiency [395].

Image analysis can be used for effective quality control of non-homogeneous products. Even though not trivial, imaging is less rewarding for homogeneous samples. Generic imaging applications are the determination of the composition of blends and the description of nature, size, shape and distribution of micro-particles, *etc.* A primary application of elemental mapping and FTIR or Raman microspectroscopy is the use in identification and characterisation of contaminants which contribute to material defects in industrial processes (*e.g.* dust contamination).

Applications of scanning acoustic microscopy include the detection of delaminations, grain structures, voids, microcracks, strain and surface roughness from different plastics and GFR composites. SCAM allows inspection of carbon fibre/epoxy resins and carbon fibre/PEEK composites [396]. Ultrashort probe pulses (50 MHz) provide a resolution of 100 μm and have been used for imaging of bulk microstructure of fibre-reinforced composites [397].

Analytical determinations of nitrogen are most often done using classical chemical techniques. Mitchell *et al.* [398] have applied 14-MeV NAA to characterise N distribution in polymers in a non-destructive fashion using image analysis of the proton track densities.

Multispectral imaging techniques (UV/VIS to IR, 320 to 1550 nm) are used for art conservation of paintings.

5.6.1. UV/Visible Microspectroscopy

Principles and Characteristics

Also UV spectroscopy has developed from a large area analysis method to one which has some degree of spatial resolution. There are only two ways in which such an improvement can be obtained, either operating the spectrometer in a microprobe mode in which the UV beam is reduced in dimensions (micro focus) or in microscope mode. Dudler *et al.* [399] have described a UV microspectrometer consisting of a UV microscope equipped with a monochromatic illuminating system (xenon lamp, grating monochromator) and a photomultiplier. This one-beam spectrometer allows adjusting the size of the photometric field (down to 0.5 μm) by placing a small diaphragm in the optical path. A motorised x, y stage allows line scans of the samples to be taken. Also a UV scanning microscope using confocal imaging has been developed operating at wavelengths approaching 200 nm (synchrotron radiation). At these wavelengths the spatial resolution obtainable will considerably exceed that from commercial confocal microscopes. The continuous spectrum of SR enables the selective excitation of chemical species in the sample at the maximum of their absorption bands.

The major building blocks for a visible absorbance microspectrometer are available in existing Raman microprobes [400]. Visible microspectroscopy does not supply as much structural information as IR and Raman techniques but is superior with regard to detection and quantitation in low-concentration situations. Samples as small as 2 μm in diameter can be studied.

Applications
UV microspectroscopic applications may find their origin in the fact that analytical problems in synthetics production and moulding are often associated with the determination of additives in polymers while only small sample quantities are available. Examples are the determination of stabiliser distributions or the characterisation of inhomogeneities in foils. Qualitative and quantitative characterisation of such small quantities by in-polymer UV spectrophotometry requires small pinholes (∅ 0.085 cm). Sehan *et al.* [401] have described a sample holder provision to allow for **distribution analysis** of phenolic stabilisers in polymers by direct UV analysis. For purposes of scanning (in mm steps) a specimen holder with *x, y* adjustments permits determination of the stabiliser concentration in desired increments of an industrially produced foil of relatively constant path length (200 μm). This is a way to check the distribution of substances in polymers and, if required, to optimise the manufacturing process.

Dudler *et al.* [399] have measured the **stabiliser concentration profile** of plates or films by UV microspectrometry (transmission) at the longest absorption wavelength of the UV absorbers (*ca.* 350 nm). Figure 5.9 shows line scans of an iPP (plate)/Tinuvin P sample.

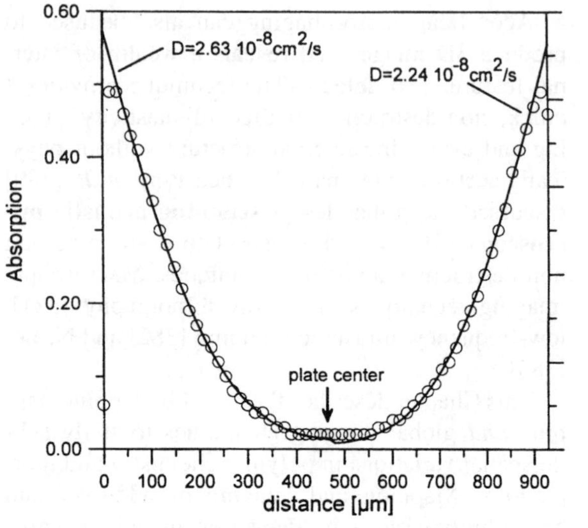

Fig. 5.9. Absorption profile of Tinuvin P in a 930 μm thick PP plate after diffusion time of 115 min at 80°C. Circles are experimental data and curves are calculated diffusion profiles for half thickness. After Dudler and Muiños [399]. Reprinted with permission from V. Dudler and C. Muiños, *Advanced Chemistry Series* **249**, 441–453 (1996). Copyright (1996) American Chemical Society.

Carter *et al.* [402] have addressed the chemical assessment of **automotive clearcoats** (usually melamine-cross-linked systems), which requires evaluation of the cross-linker type, HALS and UV absorbers. Coating systems require a variety of chemical analytical techniques for their evaluation [403], including UV microspectroscopy [402, 404], μFTIR [402,405], μRS [406,407], NMR [408], ESR [409], ToF-SIMS [410,411] and hydroperoxide titration [412]. Ideally, what is needed for industrial evaluation purposes is a set of techniques that can follow chemical changes in individual layers of a full automotive paint system, typically consisting of 45 μm clearcoat, 25 μm basecoat, 35 μm primer, and 35 μm E-coat on metal. The clearcoat must shield underlayers from UV. Unlike the case for IR radiation, examination of 30 μm and thinner clearcoat layers with 0.3 to 0.4 μm radiation lends itself quite well to the use of microscopes and microspectroscopic techniques. UV microspectroscopy of 10 μm paint system cross-sections is the method of choice (*cfr.* also Chp. 1.1). UV microspectroscopy and PA-FTIR are particular useful tools for the evaluation of chemical changes and UV protection. Normal transmission spectroscopy is used for *isolated* clearcoat samples on quartz as they are weathered. The top layer of an *intact* coating system can be examined as a function of weathering by the PA-UV techniques. Complete coating systems *removed from the surface* can best be measured in detail in microtomed cross-section by means of UV microspectroscopy. UV microspectroscopy is also used for the identification of **artist materials** in paintings.

Visible absorbance microspectroscopy has been successfully employed in a wide variety of disciplines for almost 70 years, including the analysis of organic colorants on fibres. Visible microspectroscopy has proven useful in the determination of coloured species in a single fibre [413,414]. Macrae *et al.* [414] were able to discriminate between twelve visually similar red fibres and eighteen visually blue fibres. Visible microspectroscopy is not generally successful in positive identification of pigments loaded onto PP single fibres but allows quantitation of pigment levels between 0.1 and 1.0 wt.% [415]. Visible microspectrophotometry (400–700 nm) in transmission has been used in differentiating minute smears of lipstick (composed of waxes, oils, organic dyes, and inorganic pigments) on fabric or paper-like materials for forensic purposes [416].

5.6.2. Infrared Microspectroscopy and Imaging

Principles and Characteristics

Infrared microspectroscopy can be considered as the coupling of a microscope to an infrared spectrometer. Another definition of IR microspectroscopy is the study of how infrared radiation interacts with microscopic particulates. Indeed, diffraction, refraction, reflection, and absorption effects play a much more important role in microspectroscopy than in its macroscopic counterpart. Infrared microscopy has developed rapidly and is one of the valuable and versatile tools in the analytical laboratory [390, 417–420]. The technique, which is the main development in IR spectroscopy since the advent of fast Fourier transform spectroscopy, allows spectroscopic analysis and mapping of small samples [418]. Infrared spectroscopy with the reflecting microscope was first described in 1949 [420a]. A breakthrough to high-performance FTIR microspectroscopy had to await better computing facilities in the 1980s. FTIR spectrometers offer several major advantages for microspectrophotometry over dispersive-type instrumentation, such as high-energy throughput, good spectral resolution, high S/N ratio, rapid data collection, and the ability to perform digital processing on acquired spectral data.

Essentially two FTIR microspectroscopic techniques are available – mapping and imaging. If a static sample is analysed for microscopic chemical species, **mapping** offers a convenient, fast, and cheap route for analysis. Conventional mapping equipment is the workhorse of FTIR microspectroscopy applications. For IR mapping, features must be discernable before mapping. The beam is narrowed to a very small diameter (10–20 μm) and spectra are collected *sequentially* at predefined spatial coordinates. Once a spatially related series of spectra has been acquired, a specific chemical absorption band can be selected and its magnitude or area plotted against the spatial position of each spectrum in the series. This approach allows information about the spatial distribution of the chemical species within the sample to be obtained. Table 5.42 summarises the main features of mapping and imaging μIR spectroscopy.

There are three types of IR microspectroscopy maps: point, line, and area. A point map provides several different areas of a sample to be analysed consecutively, but the spectra are not related to each

Table 5.42. Mapping and imaging μIR spectroscopy

Mapping	Imaging
• Single element detector	• Infrared array detector
• Point-by-point analysis	• 4,000–65,000 simultaneous spectra from a single scan
• Line and area infrared maps	• Visually accurate, wavelength specific images
• Auto or manual control	• Rapid data acquisition and processing (5 min)
• ATR and grazing angle	• Transmission and reflection
• Visible image capture	• Visible image capture
• 2D and 3D graphics	• 2D and 3D graphics

other spatially. An example of a point map application would be several discrete contaminants in a polymer. The line map defines a series of spectra obtained along one dimension. In line maps, chemical changes that occur along this dimension are investigated. Examples of line map applications include multilayer laminates or diffusion profiles of solvents through polymers. The area map defines a series of spectra to be collected in two dimensions (*i.e.* across a region). Mapping of large areas requires multiple positioning of a sample, spectral collection at each spatial position and much time. Area maps frequently require several hundred of thousands of spectra to be collected. Therefore, the study of dynamic processes is difficult with the mapping technique. Area maps benefit from the use of CCD array cameras as detectors.

Imaging is a technique whereby the whole area of interest is sampled *simultaneously*. Imaging spectroscopy, or hyperspectral imaging allows a large number of spectra to be acquired with fine spatial detail over an area to produce a spectral volume (x, y, λ). A spectral volume contains a spectrum for every pixel in the x, y image. Various methods have been used to measure spectral volumes [421]. The major instrumental difference between a mapping and an imaging instrument is the incorporation of a focal-plane-array (FPA) detector in the imaging microscope system. One of the major differences between IR imaging and conventional FTIR microspectroscopy is the large amount of data generated in a single imaging experiment, namely 65,536 spectra from a 256 × 256 detector array. Special data handling techniques and software are required for data analysis of files of this magnitude. The power of imaging is most apparent when small morphological features change in the course of the experiment. Infrared microscopic imaging is considered to be a specialised extension of infrared microspectroscopy. Koenig *et al.* [422] have compared FTIR mapping (using a single-element MCT detector) and imaging techniques (using an FPA detector) applied to polymeric systems.

In conventional IR microscopy, diffraction of the long wavelength radiation (5–12 μm) limits the spatial resolution to no better than a few micrometres; in practice about 10 μm. Expanding IR investigations to below the diffraction limit requires the use of more specialised approaches, such as near-field microscopy or scanning probe technology [301].

The optical requirements for an IR microscope include: *(i)* exact positioning of the sample; *(ii)* spatial isolation of the sample from a larger matrix in the IR beam; and *(iii)* capability to function in both the visible and the infrared spectral regions. For infrared microspectrometry, a thermal emission source is generally used. Fourier transform spectrometers use interferometers as an effective means to resolve photon energies. Mercury cadmium telluride (MCT) detectors have the sensitivity and speed needed for FTIR spectrometers. The use of synchrotron radiation dramatically improves infrared microspectroscopy and has the power to analyse and map samples at high resolution. SR sources have transformed the IR microspectrometer into a true IR microprobe, providing IR spectra at the diffraction limit. Optics and performance of a μFTIR interfaced with SR were described [423]. Some 15 synchrotron beam lines are equipped with IR microscopes.

Conventional microspectroscopic point-by-point mapping of a sample using only one detector may take several hours for as small as a 10 × 10 array. Application of FPA detectors with an active pixel size of about 7 μm × 7 μm in more advanced step-scan infrared imaging spectroscopy has considerably enhanced speed and allows a simpler experimental setup [424]. Imaging of large sample areas with high spatial resolution makes search for defects or specific chemical functional groups much more effective and revealing. The microscope stage, which is computer controlled, can be positioned with 0.1 μm steps. FPA systems, which consist of 4096 or 16,384 detectors (*i.e.* pixels of the array), image the sample simultaneously on the whole array and are well suited for dynamic processes (obviously at a price). This enables collection of complete hyperspectral data sets directly. Current FPA imaging technology is some 10,000 times faster than conventional microspectroscopy due to the multiplex/multichannel advantage, and the use of small detector pixels. As shown in Table 5.43 modern instruments are able to collect over 16,000 spectra in an area of 600 by 600 μm at 5 μm spatial resolution in a few minutes. The imaging technique and wavelength region employed are determined to a large degree by the sample composition. Newest methods of time-resolved FTIR hyper imaging allow the detection of dynamic processes with time scales of the order of msec [425].

State-of-the-art infrared imaging spectrometers are equipped with multidetection facilities (*e.g.* FPA

Table 5.43. FTIR microspectroscopy

Parameter	Conventional	Focal-plane-array
Analysis area	$100 \times 100\ \mu m^2$	$600 \times 600\ \mu m^2$
Pixels/analysis area	10×10	64×64, 128×128 or 256×256
Analysis time	5–9 h	1–2 min
Spatial resolution[a]	7 or 3.5 μm	7 or 3.5 μm

[a] Diffraction limit.

Table 5.44. Sampling modes in infrared microspectroscopy

Mode	Sample type
Transmission	Powders, films, laminates, fibres, crystals
Specular reflectance (Fresnel)	Highly reflective flat samples: shiny polymers, plastics, inks
Diffuse reflectance (DRIFTS)	Highly scattering samples; powders, paper products
Reflection-absorption (RA)	Thin coatings (100 nm to 20 μm) on highly reflective substrates
Grazing angle	Thin coatings on substrates
Internal reflectance (micro-ATR)	Highly absorbing samples: black rubbers, filled polymers, paper, fibre coatings; thick samples

and single detector spectroscopy), ATR imaging accessories (with a variety of crystal optics and calibrated pressure indicator) and grazing angle objectives for critical experiments, operate in an extended wavelength range (IR, NIR, VIS, UV) and allow measurements of areas as small as $10 \times 10\ \mu m^2$. Current IR step-scan imaging spectrometers are now more than just a microscope: they are a *complete* IR laboratory. Reffner [426] has described the historical development of IR microscopes.

Microspectroscopy is essentially concerned with **heterogeneous samples**. In those cases, there is an obvious need to visually discern the area of interest before starting an FTIR mapping experiment. Unfortunately, a chemically heterogeneous sample may not appear heterogeneous as such under visible light. Moreover, sampling for IR microscopy of polymers is more demanding than for the visible light microscope because of the intrinsically strong IR absorbance of polymers. Microscopy differentiates IR microspectroscopy from conventional IR microsampling techniques. Most **microsampling methods** for IR spectroscopy do not permit visual examination of the sample; they only reduce the amount of sample required for analysis. Infrared microscopes allow ng sample size as opposed to mg sample size needed for conventional FTIR analysis. With microsampling the resultant spectrum is the sum of the absorptions from all components, while in IR microspectrometry the spectra of individual phases are obtained. Sample preparation for μFTIR is sometimes difficult. Melt films are not possible for investigations of ageing phenomena, when microscopic methods are a better choice. The use of μFTIR is also limited for small dark probes. For microscopic measurements, several approaches to sampling have been reported. A complete description of some 11 different sampling techniques for IR microscopy focusing on paint samples, but useful for most polymer systems, has been reported [427].

The most common **sampling modes** for the basic IR microscope accessory are given in Table 5.44; other modes are IR emission and photoacoustic spectroscopy. Although **transmission** is the traditional mode of IR microspectroscopy, which offers the best signal-to-noise ratio, highest sensitivity, best spatial resolution, least artefacts and most sample preparation, the sample's thickness is restricted (5–10 μm). Sample preparation is a very important factor in transmission measurements. The sample should be smooth and flat. Transmission microsampling is applicable to a wide variety of physically small samples, and sensitivity as low as 1 ng has been reported [428]. Typically, to obtain microspectra of samples of limited amounts of material, the selected microsample area is masked off as desired and the beam dimensions reduced to maximise the energy passing through the sample, with apertures on the order of 10–500 μm. For larger samples, microsamples can usually be obtained by slicing with a razor blade or knife, or by microtoming. Using an IR polariser in the microscope to align the electric field of the infrared radiation can detect the orientation of molecular bonds. This is particularly useful for studying textile fibres.

As shown in Table 5.44, the FTIR microscope is not limited to transmission measurements alone. **Reflectance modes** usually require no sample preparation. Reflection-absorption is the most useful of the three external reflection forms. Microquantities of additives in solution can be cast on to an aluminium mirror and then analysed by reflectance FTIR. This method can detect and identify ng quantities of additives with surprising ease. Allen [427] has described a microscopic method for diffuse reflectance. This technique is highly effective at maximising diffusely scattered radiation, while minimising specular reflected radiation, which is a source of spectral interference. A diffuse reflectance accessory is used for the analysis of powders and rigid polymers. For example, micro diffuse reflectance is applied for detection of TLC spots [429]. For microsamples, internal reflection is limited by the size of the illuminated portion of the internal reflection element. Collecting ATR spectra with a microscope removes the barrier to analysing thick samples. Depending on the material used as IRE, the effective sample thickness varies (Table 5.45). A **microscopic ATR** probe has been developed that allows one to examine the sample optically through the probe in the microscope, position the crystal in contact with the sample, and run the spectra. Micro ATR-FTIR is becoming a well-established analytical tool due to the inherent advantages associated with its macro counterpart (*cfr.* Chp. 1.2.1.4). In addition to the ability to analyse strongly absorbing samples with little preparation, the micro adaptation has the added benefits of higher and more reproducible contact pressures applied to the sample and the ability to analyse small particulates and/or spatial domains that can be part of a much larger sample [430].

Table 5.45. ATR crystal properties

ATR material	Depth of penetration (μm)[a]	IR spot size (μm)
Ge	1.5	60
Si	2.1	70
ZnSe	4.4	100

[a]Of IR beam into a sample with refractive index of 1.5 at 1000 cm^{-1}.

The ATR accessory for FTIR microscopes is designed especially for analysis of paper products, multilayered surfaces and microcontaminations in materials. The study of soft surfaces, such as polymer laminates or tissue sections, is quite problematic. Sommer *et al.* [431] have recently reviewed ATR-FTIR microspectroscopy of soft materials, obtaining line scans or maps of pliable surfaces over an area of approximately $100 \times 100\ \mu m^2$ and overcoming many of the drawbacks of transmission analysis for these types of samples. Multilayer laminates made up of highly absorbing polymers and/or polymers that have been opacified with organic and inorganic fillers were successfully analysed. Laminates that fall into this category include packaging materials and automotive finishes. Sampling methods of IR microspectroscopy were reviewed [432].

Table 5.46 lists the **main features** of μFTIR. Conventional microscopic IR mapping experiments are only performed when there is a very specific need because the experimental set-up is complex and the measurement time is very high (typically 10 h for

Table 5.46. Main characteristics of IR microspectroscopy

Advantages:
- Non-destructive
- No sample preparation (except for transmission mode)
- Ambient operating conditions
- High speed
- Accurate positioning of IR beam
- High sensitivity
- Effective spatial selectivity
- IR spectroscopy of micro samples (*ca.* 5–10 μm in diameter) and of micro domains (5–250 μm) in macroscopic samples
- Variety of sampling modes (complete IR laboratory)
- High information content
- Imaging (functional group mapping)
- Examination of heterogeneity, profiling
- Complementary information from optical viewing
- Troubleshooting capabilities
- Growth area

Disadvantages:
- Energy-limited technique
- Relatively long data collection times (up to 24 h, unless FPA)
- Low-resolution (10 μm) in comparison to other microbeam methods (F, Raman)
- Complex experimental set-up
- Hyperspectral data analysis

mapping of rather small areas). Bailey *et al.* [300] have coupled a tuneable IR diode laser to a conventional infrared microscope. This device overcomes the low throughput problems associated with glowbar sources and provides near diffraction-limited IR images. A diode laser can be readily focused to spot size on the order of 50 μm; diodes are commercially available to cover most of the mid-IR spectrum. The improvements offered by the laser diode source over a conventional FTIR experiment are readily apparent. With the diode laser, layer structures with dimensions near that of the diffraction limit can be readily identified. Absorbance features separated by as little as 10–15 μm are resolved to near baseline. Admittedly, the diode laser gives only a single wavelength at a time, as opposed to the hyperspectral data sets obtained from FTIR-based methods; however, good chemical contrast images can be generated from only a few wavelengths.

Conventional infrared microspectroscopy shows diffraction-limited resolution. Hong *et al.* [433] have developed a **scanning near-field infrared microscope** (SNIM). This device is a new form of high-resolution IR microscopy in which the near-field images are obtained through the use of tuneable infrared transmitting fibres as scanning probe tips. SNIM allows obtaining IR spectra of localised regions at sub-μm level and acquires images at a chosen wavelength with sub-μm resolution. Since SNIM operates in transmission mode, samples must be thin. The most promising aspect of SNIM is the development of "**vibrational nanospectroscopy**". Combination of a tuneable CO_2-laser with an AFM to form an apertureless near-field imaging system can produce spatial resolution of up to $\lambda/100$ or 100 nm with high throughput and a material requirement of no more than 1 aL, but is not particularly informative for most IR chromophores ($\sim$2300 cm^{-1}) [299]. Photothermal FTIR spectroscopy by means of scanning probe microscopy is in its infancy [301].

As opposed to mid-IR microspectroscopy, **near-infrared microspectroscopy** (μNIRS, 700–2500 nm) can provide good spectra on relatively thick samples (50–500 μm) in either transmission or reflectance mode [435] and can bring useful information even if the identification of impurities or embedded particles seems to be excluded for the moment. Time-consuming microtomy, which may alter the polymer structure, is avoided. The low level of absorbance, which is due to the low absorptivity of combination and overtone bands, and the good S/N ratio of these spectra, allow the use of microspectroscopy imaging. NIR microscopes can be constructed from visible microscopes and provide nearly comparable spatial resolution [436]. Martinsen *et al.* [421] have described a NIR imaging spectrometer. Instrument developments now allow FT-NIR and NIR excited FT-Raman spectroscopy to be performed on FTIR instruments. Treado *et al.* [437] detailed the design features of an InSb FPA imaging detector for near-IR microspectrometry. FT-NIR microspectroscopy is commercially available. Non-contact NIR technologies include: interference filter-based analysers, Fourier-transform-based imaging systems and diode-array/CCD-based cameras [438]. Treado *et al.* [439] have described visible and near-infrared AOTF spectroscopic microscopy for chemical imaging allowing images at moderate spectral resolution (2 nm) and high spatial resolution (1 μm) to be collected rapidly. A CCD is used as a true imaging detector; wavelength selectivity is provided with the AOTF and a quartz halogen lamp as a tuneable source. NIR imaging systems can achieve a spatial resolution of 3–5 μm. Modern NIR microscopes complement both NIR and mid-IR analysis, combining the microscreening capabilities of μFTIR with the simplicity of NIR. It is expected that NIR imaging is going to be an important tool [440]. Advantages and drawbacks of NIR microscopy were discussed [434]. NIR microscopy has applications for thick or highly IR absorbing samples. NIR chemical imaging of heterogeneous samples provides the opportunity of combining parallel measurements and statistical approaches to both quantitative analysis and analytical sensitivity. Localised contaminants can be detected at $<0.1\%$ levels. The **main characteristics** of μFTIR and μNIRS are compared in Table 5.47.

Use of IR spectroscopy as a specific on-line detector is limited, as all common LC solvents absorb in their region, generally masking component blends. The FTIR microscope offers a solution to this problem. Infrared microspectrometers have been interfaced to various types of **chromatography** [441–443]. Chromatographic components may be deposited on suitable substrates, such as KBr disks or ZnSe plates. The solvent is removed by evaporation, leaving the component as a small spot, typically 200 μm in diameter. Using the FTIR microscope, the deposited sample may be located and

Table 5.47. Comparison of FTIR and NIR microspectroscopies[a]

Feature	μFTIR	μNIRS
Spatial resolution (theor.)	5 μm	2 μm
Sample preparation	Generally time-consuming	Often none
Sample thickness	Up to 20 μm (microtomy)	50–500 μm
Sample reuse	Occasional embedding in resin	Sample recoverable
Laboratories equipped	Numerous	Very few

[a] After Lachenal *et al.* [434]. Reproduced from *Micron* **27**, G. Lachenal *et al.*, 329–334. Copyright (1996), with permission from Elsevier.

IR spectra obtained. LC-FTIR and SFC-FTIR microscopy have been used to identify additives extracted from polymer samples (*cfr.* Chps. 7.3.3.1 and 7.3.2.1 of ref. [77a]).

Infrared microscopy, which is more widely practiced than Raman spectroscopy, is yet another opportunity to observe desired features of a sample. The technique will not replace other microscopies (such as SEM) or imaging techniques (*e.g.* iToF-SIMS), but is rather complementary.

Infrared microspectroscopy has been reviewed [436,444–447] and theory and applications have been described in several recent books [393,417–419]. An introduction to step-scan FTIR is available [448]. The role of IR and Raman microscopy/microprobe spectroscopic techniques in the characterisation of polymers, their products, and composites was reviewed [449]. McClure [450] has described NIR imaging spectroscopy and a recent review on time-resolved studies of polymers by mid- and near-infrared spectroscopy has appeared [451]. Near-infrared microspectroscopy and its applications have been reviewed [452].

Applications

The use of FTIR microspectroscopy has become commonplace in today's laboratories. The ability to quickly analyse microscopic samples has brought infrared microscopy to the top of the list of preferred analytical techniques. Table 5.48 gives a generalised view of μFTIR applications. Infrared microspectroscopy is widely used in the polymer and packaging industries. Typical sample types are solids, particles, monofilament fibres, laminates and surface coatings. μFTIR allows obtaining chemically specific data from samples including chemical composition, concentration, and molecular orientation. Infrared microscopy is of use to identify extremely small polymer samples. FTIR imaging has been applied to polystyrene microspheres [453], pre- and post-cure rubber heterogeneities [454], single solvent diffusion in a polymer film, polymer blends and semi-crystalline polymers [455]. Coupling of μFTIR and μRS (dispersive and FT) with multivariate data analysis procedures provides not only rapid, cost-effective QA methods for products, but also an efficient means to characterise physicochemical properties [456]. Imaging spectroscopy has been used for applications ranging from enhancing text on the Dead-Sea scrolls to fibre composite analysis.

Table 5.48. Typical FTIR microspectroscopy applications

- Non-destructive testing
- Surface analysis (surface treatments, coatings, oxidation, degradation, blooming)
- Identification of organic compounds (polymers, laminates, additives, intermediates)
- Chemical micro-imaging of composition (line scans, image mapping)
- Defect analysis
- Contaminant analysis
- Forensic sample analysis
- Physical distribution of additives (diffusion profiles, concentration gradients, loss)
- Oxidation profiling
- Manufacturing QC/QA
- Troubleshooting
- Customer claims
- Purity testing
- Art conservation (coloured fabrics, paintings)
- Hyphenation (HPLC-μFTIR, SFC-μFTIR)

Infrared mapping experiments have been extremely valuable since their inception [457]. Applications include obviously any work where the solid sample is very small, as often occurs in identification of contaminants, surface defects, ultrathin deposits, particulates, fibres, fillers, glass fibres, polymer mi-

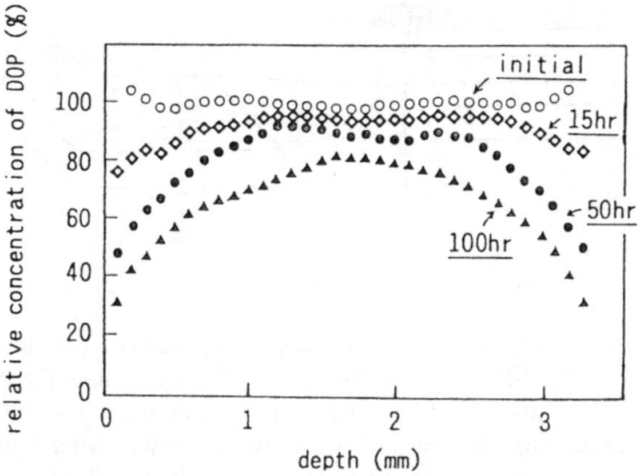

Fig. 5.10. Depth profiles of DOP in PVC/(DOP, DBTDL) heated at 120°C. After Murase *et al.* [460]. Reprinted from *Polymer Degradation and Stability* **43**, A. Murase *et al.*, 415–422, Copyright (1994), with permission of Elsevier.

crosamples, minute radioactive samples, species in forensic science or in art conservation, and in trace analysis. However, the technique is equally valuable for the examination of larger samples, such as plastics packaging material, layered polymer films, polymer blends, adhesives and inks, pharmaceuticals, electronic materials and plays a role in diffusion and ageing studies, in migration and concentration profiling of additives, surface corrosion, in new material development and failure analysis. Infrared microscopy may also be used for mapping and imaging orientation, crystallinity and chemical composition in polymer articles [384]. In particular, applications of transmission FTIR microspectroscopy have concerned: (forensic) analysis of textile fibres and fabrics for trace evidence, laminate mapping (spectra of individual layers), defect analysis (*e.g.* defect spots in HDPE film), and detection of defect homogeneity by mapping microspectroscopy.

Peitscher [458] has reported the use of (conventional) IR microspectroscopy to determine local concentration changes of **plasticisers** in polymers. Liebman *et al.* [459] reported a study of plasticiser and solvent migration in a solid propellant formulation. Migration of dioctyl phthalate in PVC/(DOP, DBTDL, DBTM) and PVC/(DOP, DBTDL) formulations was studied by depth profiling using microtoming (1–2 μm) and μFTIR (transmittance and ATR) in order to establish the effects of heating, accelerated weathering, outdoor exposure, and immersion in hot water [460]. Figure 5.10 shows the depth profiles of DOP in PVC/(DOP, DBTDL) heated at 120°C.

μFTIR can rapidly analyse samples to produce component-specific images of any organic molecular species with excellent resolution. Unknown species can be identified in seconds by searching against reference libraries of polymers and additives. Functional group images can distinguish between the various forms of silicon (silicates, siloxanes, *etc.*). The problem with direct identification of **additives** in polymeric materials is often that the concentration is low and cannot easily be observed spectroscopically. As shown in Chp. 3 of ref. [77a], this problem may be overcome by extraction procedures to concentrate a small amount of the additive. The chromatographically eluted additive sample may then be deposited on a rotating disk (typically 200 μm spot) and positioned in the IR microscope beam for analysis [461,462]. The combination of HPLC or high-resolution cSFC with the identification capabilities of μFTIR for the qualitative analysis of non-volatile chemical additives has been applied to various commercial polymeric systems [462]. μFTIR is also an effective method to identify the eluting micro components obtained from TLC [463], *cfr.* Chp. 7.3.5.2 of ref. [77a]. For oligomer identification SEC hyphenated with μFTIR may be used.

The surface of printed paper has been examined through fully automated μATR-FTIR mapping [464]. Compositional differences attributed to the printed ink, kaolinite, and cellulose distributions were revealed, which are not discernable in the visible. After spectral subtraction of the carbonate also DOP and an aromatic acrylate, both used in paper manufacturing, could be identified. Coles *et al.* [465] have compared μFTIR and ATR-FTIR in the quantitative determination of **fillers** such as kaolin clay in polyethylene/vinyl acetate. Although ATR-FTIR is not as sensitive to kaolin as μFTIR, the former provides a larger sampling area and more consistent results. ATR-FTIR is sometimes used for in-depth analysis.

Fibres are ideally suited to FTIR and Raman microscopy analysis [466]. μFTIR allows identification of the chemically similar Kevlar and Nomex aramid polymer fibres with their 1,4- and 1,3-disubstituted aromatic rings, respectively. Spectra of **fibres** with diameters down to 10 μm are suited for qualitative identification [467]. IR microscopy can also be used to identify different pigments loaded into single fibres of polypropylene. Difference micro-FTIR spectra characteristic of the interfacial region between silane-treated glass fibre and the polymeric matrix, along with contour plots of GFR epoxy composites, have indicated that specific interactions between fibre and matrix are taking place [393]. Also the effect of moisture on composite interfaces can be monitored. μFTIR in reflectance mode is more effective than DRIFT and transmission spectroscopy in analysing species on the surface of natural fibres (sisal) treated with various coupling agents such as organosilane, zirconate, titanate and N-substituted methacrylamide [468]. Sato *et al.* [469] reported the use of FTIR and ATR for direct measurement and identification of raw materials in textiles, coated and impregnated substances on paper, and surface substances on base materials, all without sample pretreatment. Non-destructive FTIR microscopy methods are extremely beneficial for analysis of samples of historical and archaeological interest such as ancient dyestuffs in coloured textile fibres [413]. Identification of a remnant dye from excavated textile is possible with μFTIR on a scale not possible by conventional extractive techniques [470].

FTIR imaging microscopy finds application in studies aiming at establishing the physical **distribution of additives**. Coleman *et al.* [471] have used both transmission IR microscopy and Raman microprobe for the analysis of compositional differences in several 10 μm thick in-plane microtomed thermoplastic olefins. No differences were observed at the surface as compared to 500 μm into the sample. The capability of examining very small sample areas (10 μm × 10 μm) makes it possible to follow concentration changes over very small distances with time. When equipped with an automated stage an IR microprobe becomes a powerful tool to obtain the **concentration profile** of a given molecular species along a given direction in a polymer. Possible applications are studies to characterise the additive/polymer interaction by probing the concentration profiles of selected species. By scanning microtomed sections this technique allows the study of the mass-transfer process from concentration profiles across the film thickness as a function of time. Bleeding of pigment additives has been studied by means of image profile analysis techniques [48], microspectrometric, microdensitometric and tracer diffusion methods. μATR-FTIR with silicon IRE of high refractive index (3.4) was applied to determine oxidation profiles of carbon-black filled nitrile and EPDM rubbers, which are usually considered to be difficult samples [472].

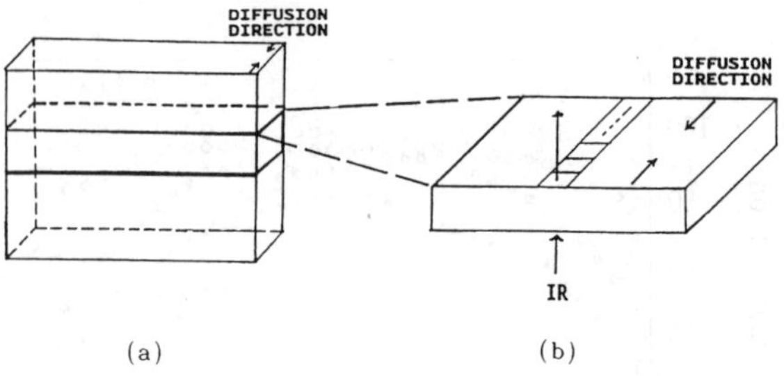

Fig. 5.11. Schematic diagram for sample preparation for IR microspectroscopic study. (a) Small piece ($0.5 \times 1.0\ cm^2$) of PP cut from the centre of a $4 \times 4\ cm^2$ plaque. (b) 250 μm-thick microtomed section. After Hsu *et al.* [473]. Reprinted with permission from S.C. Hsu *et al.*, *Appl. Spectrosc.* **46**, 225–228 (1992).

Microbeam FTIR has been used to study **diffusion** of low-MW additives in polymeric matrices, *e.g.* Cyasorb UV531 in 520 μm thickness PP plaques in a diffusion-in experiment using microtomed sections [473]. Figure 5.11 shows a schematic of the experimental method. Spectra were collected from consecutive adjacent 26 μm wide elements along the diffusion path (Fig. 5.12) and were used to derive diffusion coefficients.

The study of additive diffusion in polymers by FTIR microscopy presents several advantages: *(i)* other additives do not interfere; *(ii)* no need for a sample preparation step (minimisation of errors due to changes in crystallinity); *(iii)* the concentration profile contains more information than traditional weight sorption curves; and *(iv)* possibility of simultaneous monitoring of several additives because this method is compound/functional-group specific. Microbeam FTIR was also used to study diffusion of the antioxidant pentaerythrityl tetrabis(3,5-di-t-butyl-4-hydroxy cinnamate) in XLPE/DCP matrices [474] and of the erucamide slip agent in LLDPE films [475], both by microtome slicing techniques. Model predictions were compared against data obtained by chemical imaging. Erucamide migration in 50 μm thick LLDPE and POP single-layer and coextruded LLDPE (1%)-LLDPE (0%) bilayer films was studied by means of concentration profile mapping using SR-based μFTIR [476]. Synchrotron radiation helped to achieve a high spatial resolution (4 μm). Figure 5.13 shows a diffusion profile obtained in less than 3 min using an IR microscope equipped with a 64 × 64 element FPA detector. The normalised absorbance values of the diffusant peak are plotted against the diffusion distance and fitted to a Fickian diffusion profile.

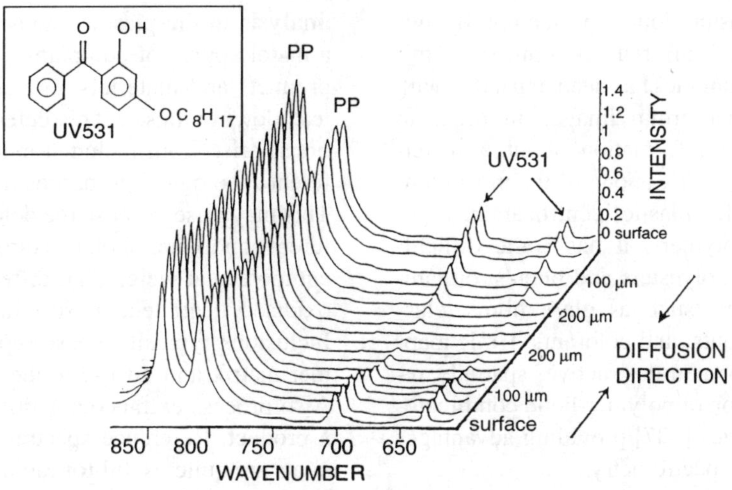

Fig. 5.12. FTIR microscopy study of diffusion-in profile of Cyasorb UV531 into PP: a 3D plot of the IR spectra of the stabiliser at different distances from the surface of the PP plaque after a diffusion experiment at 60°C. After Hsu *et al.* [473]. Reprinted with permission from S.C. Hsu *et al.*, *Appl. Spectrosc.* **46**, 225–228 (1992).

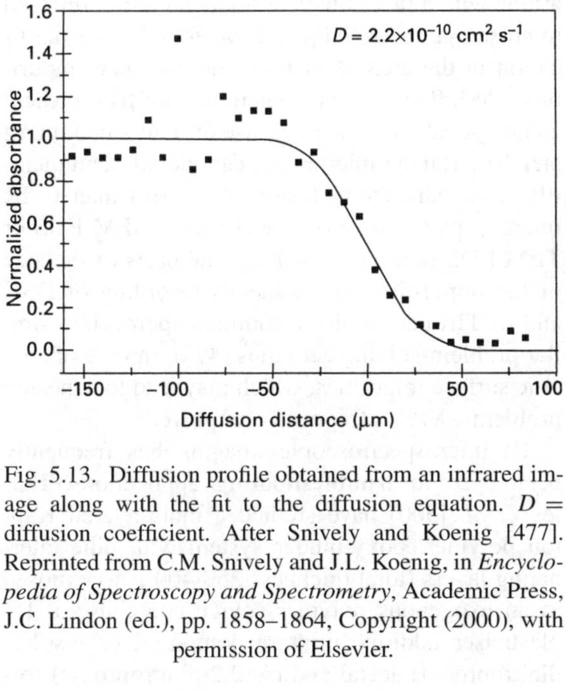

Fig. 5.13. Diffusion profile obtained from an infrared image along with the fit to the diffusion equation. $D =$ diffusion coefficient. After Snively and Koenig [477]. Reprinted from C.M. Snively and J.L. Koenig, in *Encyclopedia of Spectroscopy and Spectrometry*, Academic Press, J.C. Lindon (ed.), pp. 1858–1864, Copyright (2000), with permission of Elsevier.

Garcia *et al.* [478–480] have used FTIR microspectroscopy and mapping techniques for outdoor photodegradation of PVC siding capstock formulations as a function of exposure time and TiO_2 level. In this case advantage was taken of the complexity and specificity of the IR spectrum and the dimensional resolution of the microscope. $CaCO_3$ and acrylic impact modifier profiles for co-extruded vinyl siding samples were monitored at 1436 cm^{-1} (carboxylate) and 1733 cm^{-1} (C=O ester). The mapping approach is a much more efficient methodology to acquire a profile than microtoming consecutive slices off a surface.

A method incorporating FTIR microscopy and principal component analysis (PCA) was developed to estimate the depth of varnish penetration into paint; low-MW Dammar penetrated 18 μm into Winsor Blue paint and high-MW Paraloid B 72 only 6 μm [481].

Forensic microanalysis by FTIR comprises the characterisation of dyestuffs on wool fibres [414, 482], and the examination of household and vehicle paint fragments [158,483,484]. High quality IR spectra can be obtained in transmission from individual paint layers by using thin sections and FTIR microscopy. μFTIR may also be used to examine ink on various substrates, and to examine thin, transparent and colourless coatings. Forensics has driven development of μFTIR.

IR microscopy has been applied for the identification of oil additives (polymethacrylates, metaldithiophosphates, phenolates, sulfonates and amines) down to about 0.1 vol.% [485].

Identification of **polymer microsamples** is an important industrial problem. IR microspectroscopy allows analysis of samples in the ng range and has therefore become a most valuable tool for such identification purposes [486]. For qualitative identification it is usually possible to examine the sample directly by transmission or reflection IR microscopy. The use of infrared microprobe or microspectroscopic techniques has been expanded with the use of functional group images in order to obtain compositional information about a material [457]. Although μFTIR is one of the most common methods used for plastic identification, especially in unfilled polymers, it has some difficulties for polyamides, polyesters and blends, or samples containing fillers such as glass fibres, elastomers, flame retardants, and colorants. FTIR mapping is also used for non-destructive, spatially resolved characterisation of polymer-bond combinatorial compound libraries [487] providing advantages over the use of mass spectrometry.

Vibrational microscopy is also a powerful tool for point-mapping **orientation and crystallinity** in polymer systems [488]. FPA-FTIR [489,490] offers the potential to image properties with *ca.* 5–10 μm resolution. Infrared images taken from 10 μm sections of uniaxially drawn PET film may be used to show variations in crystallinity across the film. Apart from crystallinity, also molecular orientation may be imaged using polarised IR radiation. The very significant advantage of global IR imaging is that huge numbers of spectra are generated which form a sound basis for statistical analysis. In this way property gradients in matter are readily observed whereas such trends might easily have been missed on the basis of a few spectra only.

Analysis of **multilayer laminates** by μFTIR is one of the success stories of infrared microspectrometry. The IR microscope facilitates analysis of multilayer systems in the μm range (layers, packaging). Line maps were used in finding a thin adhesive layer and for reverse engineering of a complex multilayer laminate [491]. Figure 5.14 shows the detection of a thin adhesive layer on a labelled PP sample. When FTIR microspectroscopy in transmission is used to analyse multilayer films total superposed information is gathered, without detail about the individual layers. Christy *et al.* [492] have described multilayer laminate analysis by μFTIR in transmission mode combined with chemometrics. In studies that require non-destructive *in situ* analysis with specialty FTIR sampling capabilities microreflectance-FTIR spectra are useful. μFTIR is frequently used for mapping studies of packaging materials [491].

Vibrational microspectroscopy, in conjunction with optical microscopy and energy dispersive X-ray spectroscopy, is widely used in **product defect analysis** in the plastics industry. Contaminants are a major source of complaints concerning industrial products and materials, especially during economical downswings. Many defects (visual imperfections) arise from included material, which has different rheological properties from those of the bulk. This may arise because the defect material has a different copolymer or blend composition, is of a different molecular weight, has different end-groups, or is oxidised, degraded, or cross-linked (gels or specks). Inclusions typically phase-separate from the polymer matrix and diffuse to the surface ("fish-eyes"). Also process extraneous particles may contaminate a product. Reference spectral libraries of contaminants are quite useful for rapid fingerprinting recognition and troubleshooting. Because of the wide range of possibilities and the small amount of material, there is no universal technique for **contaminant analysis**. In some cases, it is possible to physically isolate the imperfection by simply cutting it from the article with a razor blade or diamond knife. Infrared microsampling techniques have found wide application in the area of identification of such impurities [458]. Particulates (sometimes sub-μm) present an analytical challenge because of their small diameter. The Raman microprobe can identify and quantify much smaller inclusions ($\geq$3 μm) than FTIR microscopy ($\geq$20 μm). The choice of SEM-EDS or TEM-EDS is usually made on the basis of the size of the impurities. Simultaneous recording of DSC and μFTIR data is also a common approach in similar problem-solving activities [493]. In case of specific surface impurities, which may lead to adhesion problems, XPS is frequently employed.

IR microspectroscopic imaging has frequently been used in **troubleshooting** applications. Bailey *et al.* [300] have studied a multilayered Kapton polymer/epoxy-binder system with nine alternating layers (total thickness 350-400 μm) exposed to an exogenous polymer source containing a 1:1 plasticiser additive mixture composed of bis(2,2-dinitropropyl) acetal and bis(2,2-dinitropropyl) formal (NP), using a tuneable IR diode laser at 1568 cm^{-1} (λ_{max} for NP) for the purpose of mapping the volatile contaminant (NP) distribution. It was shown that NP was confined to the epoxy-binder layer. The distribution of the additive in the layered structure did correlate with specific layers and revealed a concentration gradient suggesting a diffusive mechanism of additive migration parallel to the layered structure. Other typical cases are

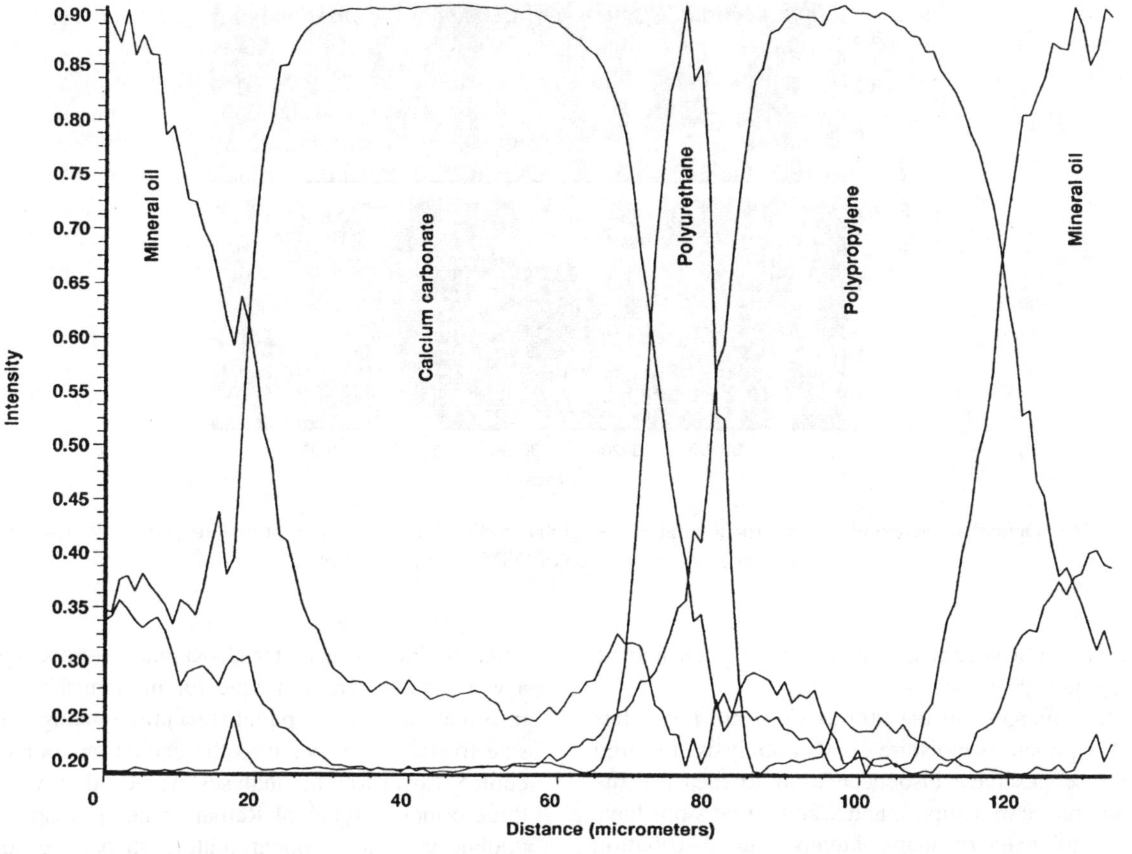

Fig. 5.14. Correlation profiles for mineral oil, calcium carbonate, polypropylene and polyurethane of a multilayer laminate using mineral oil as a mounting medium. After Martoglio Smith [491]. Reproduced from *Vibrational Spectroscopy* **24**, P.A. Martoglio Smith, 47–62 (2000), with permission of Elsevier.

the determination of local concentration changes of plasticiser or the oxygen uptake during ageing. Apart from the analysis of gel inclusions in PE [494], representative applications include the determination of Irgastab 2002 residues on PP tubing [458], of the plasticiser butylbenzylsulfonamide on PA12 tubing [458], and of a mould release agent on a polyurethane surface [486]. Ezrin [495] has shown micro IR of cellulosic film contaminant in zinc stearate. Figure 5.15 shows the globular discoloration in a PP sample, which could be ascribed to the heterogeneous distribution of talc: absent in the globule (with local poor paint adherence), higher than average in the immediate vicinity, and normal elsewhere.

In another case, chemical imaging allowed to attribute undesired inhomogeneities in high-tech foils to the additive (phosphate) distribution. Finally, a lacquer defect during car production (haze), which had caused a major production break, was rapidly identified by μATR-FTIR as being on account of 2-ethylhexyl-acrylate, a well-known softener in lacquers. In this case the lacquer producer was suffering from a dosing problem. μFTIR in combination with SEM-EDS is frequently used in the examination of **paint fragments** [497]. Application of the micro-infrared technique enables to examine the heterogeneity of such samples and to determine which components have migrated from one layer to another. McEwen *et al.* [498] compared various sample preparation methods for micro-IR analysis of a five-layer paint system for sheet moulding compound. μATR-FTIR has been applied to paint layers and coatings on glass fibres. Grazing angle microsampling has been employed for the microscopic analysis of ultra-thin (nm) sample deposits. Spatially-resolved studies of paint cross-sections are profitably carried out by combined light micro-

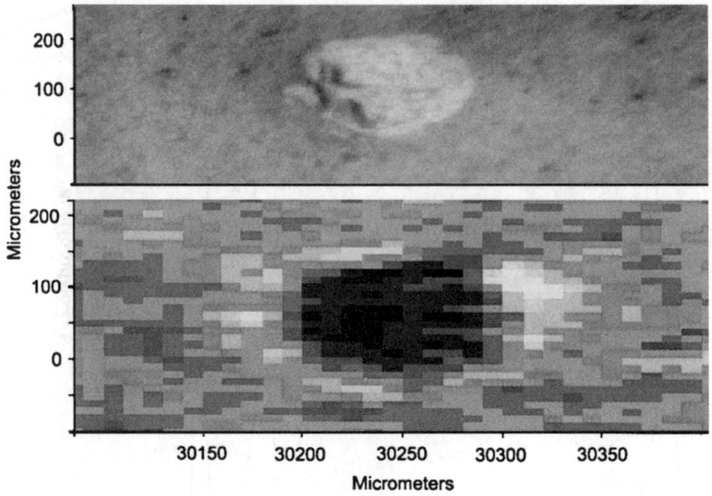

Fig. 5.15. Optical microscopy and micro-IR analysis of globular discoloration of polypropylene. After Wienke [496]. Reproduced by permission of DSM Research, Geleen.

scopic, FTIR spectroscopic and mass spectrometric imaging [499].

NIR imaging in the 900 to 1700 nm range has been applied to moisture content analysis. Infrared microscopes have also been used to measure the temperature of samples, and scanning versions have been built to enable maps of temperature distribution to be obtained.

On-line/in-line technology for monitoring extrusion processes, including FTIR microscopy, near-IR spectroscopy and optical microscopy was reviewed [500]. Several reviews describe μFTIR applications to polymers [458,501]. Line map applications of μFTIR have been discussed [491]. A recent review [502] refers to a large number of FTIR microspectroscopic studies as an important source of structural and spatial information for polymer-based articles. A monograph describes applications of FTIR microspectroscopy to polymers [393]. ASTM E 334 (1990) describes the general techniques of infrared microanalysis.

5.6.3. Laser-Raman Microprobe and Microscopy

Principles and Characteristics

Raman microscopy takes advantage of the fact that the intensity of Raman scattered light is independent of sample volume [503]. Thus, the light intensity remains essentially constant with decreasing sample size down to the dimension determined by the diffraction limit, and hence the wavelength, of the laser excitation. Raman intensity always comes from small regions of a material. Raman microscopy is a well-established technique for imaging chemical information at high spatial resolutions using a μm size focused laser beam as the excitation source of choice. Local Raman analyses are possible with all three principle types of Raman systems (dispersive double or triple monochromators, dispersive notch filter-based systems, and Fourier type systems) by coupling with an optical microscope. The microscope objective lens is used in 180° backscattering both to focus laser light onto the sample and to collect light for the Raman system. The microscope frame provides precision 3D translation of the sample relative to the objective lens. The microscope ensures that the Raman spectrum is really coming from the material of interest on a 1 μm^3 scale (**Raman microprobe**), rather than from a contaminant or non-representative part of the sample. By scanning the laser spot relative to the sample images are obtained (**Raman microscope**). The combination of a Raman spectrometer with a microscope has first been described in 1974 [504–506].

A Raman microscope consists of five **basic components**: excitation source (laser), focusing component (microscope), signal analyser (spectrometer or interferometer), photon detector (either monochannel or 2D array) and mapping unit such as a computer-controlled micromanipulator. Raman microscopes are usually equipped with low-power UV/VIS or NIR lasers, with laser spot sizes (focused laser beam) below 10 μm. Raman microscopy

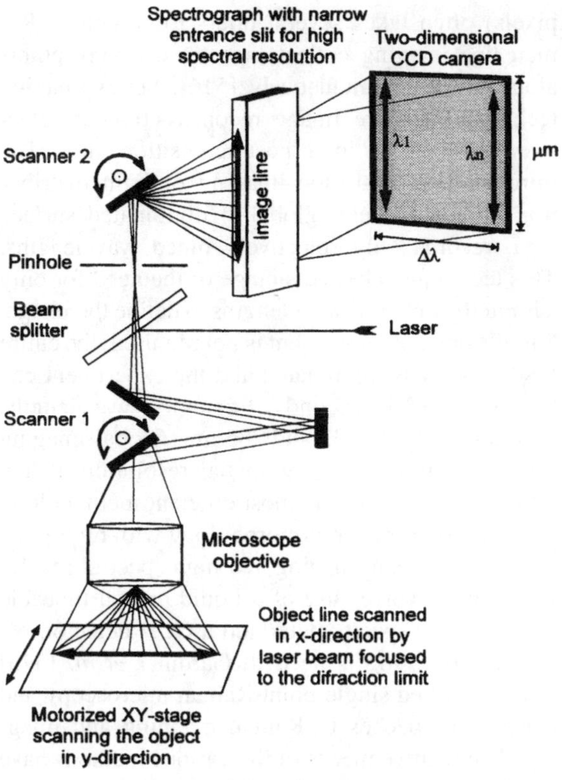

Fig. 5.16. Schematic representation of Raman imaging by confocal laser line scanning using a point focused laser for excitation of Raman scattering. After Markwort and Kip [521]. Reprinted from L. Markwort and B. Kip, *J. Appl. Polym. Sci.* **61**, 231–254 (1996), John Wiley & Sons, Inc., New York, NY, Copyright © (1996, John Wiley & Sons, Inc.). This material is used by permission of John Wiley & Sons, Inc.

is necessarily carried out in reflection mode (*cfr.* Fig. 5.16).

Both dispersive (D) and Fourier transform (FT) micro-Raman are commercially available [507–509]. The choice between D- and FT-Raman has been discussed (*cfr.* Chp. 1.2.3). However, as the intrinsic superiority of interferometers in terms of high resolution and geometrical extent cannot be exploited in micro-Raman spectroscopy, dispersive spectral analysers are preferred in combination with the microscope. FT-Raman instruments equipped with a microscope have thus far been limited to a sensitivity far lower than that of modern Raman microspectrometers equipped with multichannel detectors, which employ visible excitation. Consequently, FT-Raman microscopes often yield deceiving results with poor sensitivity and spatial resolution (from 15 to 100 μm). No fully satisfactory solution for FT-Raman microscopy is yet available. Dispersive systems offer optimum performance in terms of spatial resolution and signal-to-noise ratio.

The NIR region is a compromise between the trade-off that must be made in choosing between Raman and IR spectral imaging. Dispersive Raman microprobes using near-IR excitation beyond 1000 nm and linear array detectors with good sensitivity are useful for the investigation at the microscopic level or for remote analysis by means of optical fibres of samples which fluoresce under visible illumination. This allows manufacturing quality control.

Barbillat *et al.* [509] have developed a multichannel dispersive Raman microanalyser which features both visible and NIR technologies, two laser sources (at 532 and 1064 nm) and two detectors (InGaAs for NIR analysis, 2D CCD for VIS), and allows obtaining fluorescence-free confocal Raman spectra of microscopic samples with a spatial resolution only limited by diffraction. This dual instrument offers maximum possibilities for Raman microanalysis and remote analysis: non-fluorescent samples can be examined with visible laser excitation and CCD detection, and fluorescent samples by switching the instrument to NIR excitation and detection. The sensitivity of Ge and InGaAs detectors is fairly good and results in high S/N fluorescence-free spectra, which compare favourably with FT-Raman data obtained on much larger sample volume. The concept of dual excitation detection instrument is very powerful since it widens the field of application of Raman microanalysis.

It is possible to use an ordinary optical light microscope as the excitation beam condenser and at the same time collect very efficiently the backscattered Raman light [510]. Provided the sample under investigation is not fluorescent or light sensitive, Raman spectroscopic analysis is relatively straightforward. No particular sample preparation is necessary, and sample alignment and focusing onto microscopic features in or on the sample are easy. In many applications it is important to have good spatial resolution not only in the lateral direction, but also along the optical axis of the microscope, to provide for depth resolution. Collection of the Raman scatter can be made confocal, improving lateral and depth spatial resolution considerably [511,512]. **Confocal Raman microscopy**, which improves image contrast and reduces fluorescence (from out-of-focus planes), is now conveniently applied to point analysis and depth profiling of chemical and structural inhomogeneities. Lankers *et al.* [513] have designed an automated, point-by-point confocal Raman mapping system, whereas Brenan *et al.* [514] built a prototype instrument that combined CLSM with a FT-Raman spectrometer. Nowadays, nearly all Raman instruments based on monochromators, spectrography and interferometers are confocal, thus allowing optical sectioning. The excitation depth varies with laser wavelength. In non-resonance conditions, the sampling depth is normally of the order of the laser wavelength used; in resonance or near-resonance conditions, it can be much less. Tight control over the sampled depth is obtained via the confocal effect. In favourable samples a confocal Raman depth resolution of 1–2 μm (FWHM) can be achieved [511, 515,516]. In theory the ultimate depth resolution is about 0.3 μm [517]. When operating in confocal mode it is possible to obtain spectra with an effective volume resolution $\leq 5\ \mu m^3$. Turrell *et al.* [518] described the main features of a Raman confocal system. Everall [519,520] has critically evaluated depth resolution in confocal Raman microscopy.

Various techniques for obtaining Raman images have been pioneered and **Raman imaging** instruments have recently (1991–1993) been developed. Raman images may be formed by adding scanning optics to a micro-Raman spectrometer or by replacing a detector in a CLSM with a filter or spectrometer. Raman imaging basically involves collection of spectral data from a series of spatial points on the sample, and explicitly uses high-sensitivity 2D photoelectric detectors, *e.g.* CCD, CID or diode array detectors. Robust, low-noise CCD multichannel detectors with high quantum efficiency (up to 80%) allow imaging one or more Raman bands in the 100–1100 nm range.

Raman imaging methods are usually classified as "series-imaging or scanning" and "parallel- or direct-imaging" techniques [522]. Illumination methods of obtaining Raman maps are of the point-by-point, line-scanning and wide-field or global type. Confocal sequential points scan and sequential line scan (*cfr.* Fig. 5.16) require sample and beam movement. Raman point mapping, consisting of measuring the Raman spectrum of each pixel of the image one at a time, can yield data at high spectral resolution over a large spectral range but at fairly coarse x, y spatial resolution with the main penalty being experimental time. Coarse imaging (100×100 pixels) often takes several hours to complete. Raman line imaging collects spectra of many points along a line simultaneously [516]. Series-imaging techniques require image reconstruction at selective wavelengths in a post-acquisition step. The third method, real-time imaging of light distribution in a wide field (globally) illuminated surface area, records only selectively tuned wavelengths. This technique takes advantage of the need for only a limited number of wavelengths to define the image. Sample or beam movement is not required, the entire field of view is illuminated and the experiment can be completed in seconds. Excitation wavelengths used are typically 532 and 785 nm. Global-imaging provides fairly high x, y spatial resolution at low spectral resolution. The most efficient methodology for analysis of material morphology with high pixel definition Raman imaging and high spectral resolving power involves use of a liquid crystal tuneable filter (LCTF) spectrometer and a CCD detector. No image processing is needed. Gardiner *et al.* [523] have described single-point Raman microscopy and current approaches to Raman mapping and imaging. The relative merits of the various methods have been carefully examined [393,515,524]. The best approach to imaging depends on the application [525]. For general application in polymer science Raman imaging by confocal laser line scanning is well suited as it acquires all of the spectral and spatial data in a reasonable measurement time without sacrificing the illumination power density to the point of low Raman signal generation [515].

Raman imaging closes the gap between infrared microscopy with its comparatively poor spatial resolution, and TEM with its limited chemical information. For heterogeneities on a sub-μm scale, the value of the technique is limited to determination of average information. Table 5.49 summarises the main features of imaging Raman spectroscopy.

The minimal focal diameter of the laser beam can be in the order of the wavelength of the laser radia-

Table 5.49. Features of imaging Raman spectroscopy

- Excitation and scattered radiation (VIS or near-IR) readily guided for mapping
- Scattering technique applicable irrespective of sample form
- Compact and mechanically simple instrumentation
- Remote analysis by use of fibre-optics (visible)
- High analytical specificity due to rich spectra

tion. The use of a microscope as a sampling accessory for Raman spectroscopy allows spectral analysis on samples which are too small for conventional sampling techniques. For a given light flux of a laser source the Raman radiation flux is inversely proportional to the diameter of the focus of the laser beam at the sample. This means that an optimised Raman sample is a micro sample [526]. Typical sample sizes for μRS vary from 500 μm down to 1 μm.

The ability to combine the high sensitivity and selectivity of UV resonance Raman spectroscopy (UVRRS) with the ease of operation and spatial resolution of visible Raman microscopy or microspectroscopy is highly desirable. Many of the problems inherent in using visible excitation Raman spectroscopy for analytical applications are overcome with UV excitation. The use of UV means that most condensed-phase materials exhibit no fluorescence in the Raman region and that Raman scattering is more intense. However, until recently the excitation sources utilised for UV Raman measurements were inappropriate for **UV Raman microspectroscopy**. Many operational problems ensue from the use of low repetition rate, high peak power lasers in that sample degradation and saturation effects limit the average power. UV Raman microscopes are now available at 244 and 325 nm. Asher *et al.* [527] have used an intercavity frequency-doubled Ar^+ laser with continuous-wave (CW) excitation at 244 nm in a highly efficient UV Raman microspectrometer or UV Raman microscope with spatial lateral resolution of 3 μm $\times$ 9 μm and depth resolution of 10 μm. The ability to focus the CW laser to a spot size of a small diameter that can be efficiently imaged into the spectrometer permits high S/N ratios. The CW laser can be used to examine thermally sensitive samples including strongly absorbing solid samples.

Table 5.50 lists the **main features** of Raman microspectroscopy. Virtually any object which can be observed under a microscope can be analysed with Raman microscopy. Here, the usual constraints inherent in electron beam methods (vacuum, metallisation, *etc.*) are totally absent. Although micro-Raman spectrometers mainly use visible excitation, the confocal configuration almost eliminates fluorescence which falls outside of the focal volume. The focus area for visible lasers is ≤ 1 μm^2, whereas the focus diameter for NIR lasers is $\sim$20 μm.

Raman microscopy is not a quantitative technique as the requirement for a homogeneous sample is more demanding, if not inappropriate for a technique where spatial resolution of the sample components is its most significant feature. A useful way of normalisation of Raman spectra is use of an internal reference signal (such as the CH_2 bending at about 1450 cm^{-1}). Laser power density is limited by the thermal sensitivity of the sample (sample might melt). Combined with the very small spot size of the laser beam at the sample, limited power density means low Raman scattering intensity, hence limited sensitivity. Main limitations of micro-Raman imaging of heterogeneous polymer systems based on confocal laser line scanning are therefore sample destruction due to insufficient heat dissipation of the high-incident laser power, interferences due to fluorescence (for visible light), and instrumental instability during long collection times required for good S/N ratio spectra of weak Raman scatterers [521]. As Raman spectral lines are generally several orders of magnitude weaker than incident light, scanning a Raman image can be slow, which increases the risk of sample damage. However, with an excitation power at the sample rarely exceeding 5 mW for visible lasers, all risk of laser-induced degradation is virtually removed. Also, continuously scan-

Table 5.50. Main characteristics of Raman microspectroscopy

Advantages:
- Applicability to unprepared samples of large or non-uniform shape (*in situ*)
- Suitable for study of all types of heterogeneous material
- Non-destructiveness
- Molecular specificity
- Confocal microscopy (good depth profiling potential)
- High sensitivity (D-Raman microspectroscopy)
- High spatial (≤ 1 μm) and spectral (<1 cm^{-1}) resolution
- Multicomponent analysis
- *In situ* analysis
- Mature optical technology (visible Raman microspectroscopy)
- Relative immunity to interference
- Convenience of use, speed
- Relatively inexpensive, low-power lasers
- Portability

Disadvantages:
- Limited databases
- Qualitative only
- Sample heating
- Optical breakdown (photochemical reactions)
- Fluorescence

Table 5.51. Comparison between micro-IR and micro-Raman spectroscopy

Feature	Micro-IR spectroscopy	Micro-Raman spectroscopy
Spot size	$>10\ \mu m$	$<10\ \mu m$
Spatial resolution	$2.5–25\ \mu m^a$	$1\ \mu m$ (VIS)–$20\ \mu m$ (NIR)^b
Sampling	Thin specimens	Solids
Minimum sample mass, volume	1 pg, $1000\ \mu m^3$	0.1 pg, $100\ \mu m^3$
Minimum particle dimension	$10–20\ \mu m$	$1–2\ \mu m$
Solid-probe interaction	None	Sample degradation, burning

[a] Determined by spot size and diffraction limit.
[b] Determined by spot size only.

ning the laser beam over the sample instead of a static beam allows considerably higher laser powers without damage. Sample decomposition by the highly focused laser may also be circumvented by spinning the sample, dilution with an inert matrix, or use of a laser of different wavelength. Obviously, sample rotation leads to averaging, which limits its usefulness.

The qualifying characteristics of μRaman spectroscopy are best appreciated in comparison to μFTIR [528]. Infrared microspectroscopy is almost exclusively a Fourier transform technique, at variance to its Raman counterpart, which is mostly based on dispersive elements. Apart from the differing spectral content of the two spectroscopies, Raman has some distinct advantages for microanalysis, such as spot size limitation (*cfr.* Table 5.51). As the wavelength range of the analysed signal lies in the visible region, better spatial resolution is achieved. Raman spectroscopy simplifies microscopic and mapping analysis also for other reasons [393]: *(i)* the optimum sample for Raman spectroscopy is a microsample; *(ii)* Raman spectra are simpler than IR spectra due to a lower number of lines; *(iii)* very weak water signals, causing much less interference than in IR; *(iv)* simplified positioning of the sample in the beam (backscattering); *(v)* access to glass supported samples; and *(vi)* ease of confocal measurements (high depth and lateral resolutions), allowing analysis of embedded particles. Raman spectroscopy is more conducive to most problems of hard solid samples. Micro-IR, like IR for macro samples, works better in transmission than in reflectance. On the other hand, being a scattering phenomenon Raman is more amenable to solids in their native state. As opposed to Raman microscopy, IR imaging objectives typically employ reflective optics that do not provide high image quality. Using the rapid tuning capability of AOTF, combined with fast imaging detection allows imaging of samples as rapidly as 1 frame/s for Raman emission and at much higher rates for NIR absorption. μRS might also seem more amenable to multicomponent analyses because Raman spectra generally exhibit fewer and narrower bands over the same spectral regions. At variance to infrared, Raman microscopy may suffer from fluorescence and photoinduced damage. Also, IR spectroscopy allows a higher sensitivity than Raman methods and FTIR imaging is simpler and lower cost than Raman imaging. Finally, Raman databases are of limited size only (15,000 entries), in particular in comparison to IR tradition.

The future of Raman microspectroscopy is probably imaging and optical near-field nano-Raman spectroscopy [529], *cfr.* Chp. 5.5.2. While conventional laser Raman spectroscopy samples 10^{-3} g (mm^3), μRS handles 10^{-12} g (μm^3) and near-field Raman spectroscopy 10^{-15} g (nm^3). Mobile Raman microscopy (MRM) allows *in situ* Raman analysis [530]. One can expect **further developments** in the field of NIR multichannel Raman spectroscopy with the advent of 2D array detectors offering extended response in the NIR. With these 2D sensors it will become possible to apply in the NIR region the powerful techniques already developed in the visible, such as confocal line imaging techniques or multisite remote analysis with optical fibres.

Raman microspectroscopy and spectroscopic imaging were reviewed [393,436,488,531–533] and compared to other local-analysis techniques [534]. A review of the instrumental techniques for micro-FT-Raman indicates the power of the technique for analysis of a variety of samples [535]. For other reviews, *cfr.* refs. [536,537]. Near-IR Raman imaging microscopy (NIRIM) was recently reviewed [538]. A textbook is available [539].

Applications

Raman spectroscopy is recently gaining increased acceptance in the industrial laboratory, both as a microscopic technique [540] and in bulk analysis systems, and has begun to be used in the real world of on- or at-line process analysis and monitoring. An important advantage offered by Raman spectroscopy is flexibility in sampling for solid samples. With conventional Raman spectroscopy it was not possible to reduce and localise the analysis volume to dimensions commensurate with phase size in microstructures or with the size of the analysed object itself (*e.g.* fibres). On the other hand, Raman microscopy is powerful in polymer characterisation and competitive with TEM and ToF-SIMS, in particular at the microvolume level (a few μm^3).

Identification of micro impurities in materials and determination of chemical heterogeneity within plastics are two common applications of micro-Raman and micro-IR techniques in the chemical industry. The smaller spot sizes possible in Raman microscopy as compared to μFTIR allow detailed chemical mapping of surfaces. This is useful if there is component segregation or domain formation on a scale larger than the spot size and within the translation capability of the microscope stage. Chemically selective imaging offers a means of highlighting specific components or substructure density across a surface.

Raman imaging permits excellent molecular discrimination that is not available from many techniques. Raman microscopy has tremendous potential as a tool for mapping and imaging **chemical heterogeneity** in materials systems such as automotive coatings. μRS can be used as a technique for line profiling composition as a function of distance in one dimension or as a tool for imaging 2D chemical heterogeneity on a surface, or even 3D with confocal imaging. Image contrast is based on differences in chemistry. Raman measurements through microscopes are capable of providing information about the chemical composition within small areas through the detailed information found in vibrational spectra. This fills an important gap in the area of microanalysis since most conventional microscopic techniques do not provide detailed information about chemical structure, but are restricted to elemental or high-resolution morphological information about the areas of interest (*cfr.* Table 5.6).

Raman spectroscopy in conjunction with waveguide technology is the accepted method used in studying sub-μm thick polymer films [541,542]. Using spectral libraries Williams *et al.* [543] have identified an 80 μg EVA **inclusion** within a polymer film. This analysis would have been impossible using conventional Raman spectroscopy because of the overwhelming sample fluorescence and heating that occurs in such cases. μRS can be used for chemical composition imaging of the different phases in a multiphase polymer blend [544]. In addition, the high spatial resolution permits effective chemical composition mapping in failure analysis and cross-sectional depth profiling. Raman microscopy has been used in the diagnosis of problems in the production of plastics, where catalyst particles embedded in the polymer were identified; impurities and degradation products can also be traced. Somorjai *et al.* [545] have used UV Raman (244 nm) imaging in the study of $MgCl_2$-based PP catalysts.

The Raman microprobe has several important areas of application. Although the principal use is microspectroscopy, the microprobe is practical for rough mapping, particularly when only linear or radial distributions are needed. In such cases, 10–20 spectra are used to define the spatial features, and the microscope stage may be manually scanned.

A general strategy for analysis of **micro-impurities** (pits, gels, *etc.*) or contaminants is given in Scheme 5.4. Some measurements are destructive (DSC, PyGC, l-NMR). The Raman microprobe is used extensively for inclusion analysis. Raman can identify and quantify much smaller inclusions ($\geq$1 μm) than FTIR microscopy ($\geq$20 μm). In comparison, for μXRF a typical spot size is 300 μm. A particularly useful application of FT-Raman microprobe analysis, typical of the higher spatial resolution of Raman, is identification of the source of pinholes and craters in coating systems. In general, pinhole sizes in coatings are smaller than the spatial resolution of FTIR (*i.e.* <10 μm), making FT-Raman viable [393]. μRS of PDMS defects was reported [546].

Using confocal imaging approaches, sub-surface defects and interfaces can be analysed with minimal contribution from the matrix or overlayers, since confocality adds depth selectivity to the measurement. Confocal imaging approaches provide one strategy for non-destructive profiling changes in composition as a function of depth [511,547]. **Depth profiling** potential extends up to 200 μm in low scattering media such as unfilled polymers. Good depth resolution ($\sim$2 μm) enables study of

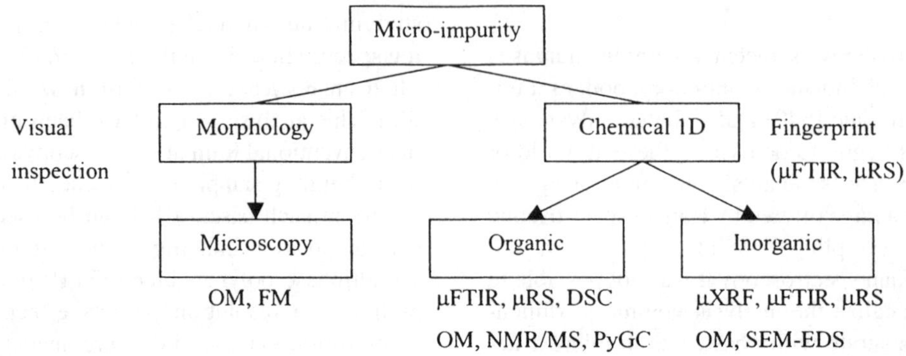

Scheme 5.4. Analysis of micro-impurities.

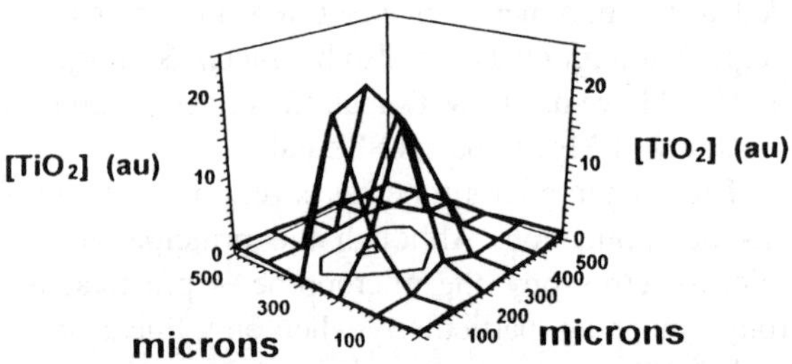

Fig. 5.17. FT-Raman map (Nd:YAG laser, 35 mW) of a white inclusion in a clear styrenated-acrylic copolymer film. After Claybourn *et al.* [546]. Reprinted with permission from M. Claybourn *et al.*, *ACS Symposium Series* **598**, 41–60 (1995). Copyright (1995) American Chemical Society.

thin (multi)layers, composite materials, impurities and embedded particles. Figure 5.17 shows a FT-Raman map of a white inclusion found in a clear styrene-acrylic copolymer film, readily identified as TiO_2.

There are specific structural and spatial problems in which Raman spectroscopy plays a dominant and important role based on higher sensitivity (due to resonance enhancement) and higher spatial resolution than FTIR. Specifically, micro-Raman spectroscopy has been applied in the analysis of (glass) fibres and their surface treatments, fibre composites, multilayer plastic films, foils and coatings, polymer blends, interfaces in composites, contaminant and paints/pigments [488].

Polymeric systems are generally weak Raman scatterers and notorious for their fluorescent capability. Interfering radiation leads to substantial loss of S/N ratio when visible excitation lasers are used. For PET and PE **domestic household plastic** samples from various sources, collected for a recycling identification test, which did not fluoresce at 1064 nm excitation, the spectra were sufficiently different to enable reliable polymer identification [509]. For mapping of **very thin films** (as thin carbon layers) a dispersive Raman spectrometer is the first choice due to the lower penetration depth of the visible excitation compared to NIR excitation. A FT-NIR Raman spectrometer should be preferred for surface mapping of samples exhibiting fluorescence contributions. This method allows determination of concentration profiles, surface impurities and surface roughness of rotating samples. Confocal Raman spectroscopy has high potential for polymer and **laminate studies** since it provides an optically slicing technique for depth profiling and fluorescence rejection [511]. No sample preparation is required, and the measurements are relatively rapid and sensitive to the polymer composition in the inner layers as well as to interactions that occur across the interfaces. Raman confocal microscopy has also been used in **fracture analysis** to differentiate a premade *vs.* *in situ* formed compatibiliser at a PS/PMMA interface [548].

Increasing amounts of radiation curable materials are being used in coating films for the surface refinement of furniture, wooden floor coverings, paper, *etc.* Confocal Raman spectroscopy of UV-cured films may be used to examine depth or lateral profiles of the cross-linking process in coatings with a resolution of approximately 1 μm^3 [549]. The chemical imaging perspective of confocal Raman microscopy addresses a variety of industrial problems including the distribution of **UV stabilisers** interacting with the UV curing process. Micro-Raman spectroscopic mapping may also be used to determine how processing affects crystallinity of a material. This may be used to detect optimally fabricated parts. Line scanning has been used to study carbon fibres and polymer degradation [550]. Polarised micro-Raman spectroscopy has been applied for quantitative analysis of the orientation of macromolecules in polycrystalline polymers. Use of NIR FT-Raman spectroscopy for a fast moving PE sample (speed up to 20 m/s) has been reported [551], which allows quality control under draw with inhibition of thermal degradation.

Commercial fibres are generally too thick optically for FTIR measurements, making it difficult to use the usual sampling methods. On the other hand, because of the simplicity of sampling, Raman microscopy has been widely used in fibre studies for many years and is an ideal tool for characterising single textile filaments and polymer fibres (viewed either across or along the fibre axis). The potential of FT-Raman microscopy to record good analytical spectra directly from single fibres with diameters as narrow as 5 μm has been demonstrated. Micro-Raman spectra of **fibres** are particularly easy to obtain by simple reflection of the focused laser beam. μRaman can identify carbon fibres in plastic material, at variance to μFTIR. In addition to identification of chemical composition of the fibre and the presence of finishes on the fibre, Raman microscopy can monitor the degree of molecular stretching (*i.e.*, the molecular elongation) and orientation. μRS of filled plastics may also be used to measure the dimensions of chopped fibres [495]. Characterisation of additives in synthetic fibres is potentially beneficial to the fields of textile, fabric, and fibre manufacturing and of interest to forensics and archaeology. μRS is a particularly exciting technique for the analysis of pigment-loaded fibres because the pigment often provides a much more intense Raman spectrum than the fibre. Confocal μRS has been used in the identification of pigments used on historic painted textiles [552]. Molecular microspectroscopy has been used to characterise different **pigments** (azoic, copper phthalocyanine and rutile TiO_2) loaded into PP fibres [415]. Single-fibre analysis by Raman, IR and visible microspectroscopies is complementary for identification and quantification of these materials. μFTIR is effective for identification and quantitation of high concentration levels (>1 wt.%). Visible microspectroscopy is not generally successful for positive identification of most pigments but is effective for quantitation of pigment levels between 0.1 and 1 wt.%. μRS is effective both for pigment identification and quantitation (0.1–10 wt.%) provided that the sample shows no signs of fluorescence or heating effects.

Small amounts of cobalt blue and red matador **dyes** (1–2 wt.%) on acrylic fibres were identified by FT-Raman after subtraction of the polymer from that of the dyed fibre [553]. Raman spectroscopy was also used for identification of dyes used in contemporary blue textiles [554]. Resonance Raman microtechniques prove very sensitive for identifying dyes on fibres [555].

Until recently, there was a lack of general applications of Raman spectroscopy in the field of **plastics additives** [556]. It would appear though that significant work is now being published. For example, micro-Raman spectroscopy was used routinely to detect an inclusion of undispersed Irganox 1076 in HDPE [557] and has revealed the antioxidant β-carotene on UHMWPE wear debris particles of orthopaedic components [558]. Raman imaging and mapping, and depth profiling may all be used to measure the distribution of small molecules in polymer matrices. As additives can easily be distributed heterogeneously in a polymer, it is of interest to examine various portions of a material. Confocal Raman spectroscopy is well suited to check the 3D **distribution profiles** or aggregation of polymeric components such as surfactants, plasticisers, monomers, *etc.*, throughout the thickness of a transparent matrix. The distribution of unreacted free melamine in a cured melamine formaldehyde resin was analysed by confocal Raman imaging using the Raman band intensity ratio of the bands at 676 and 975 cm^{-1}, which are due to vibrations in the triazine ring of melamine. Regions of high free melamine content of about 10 μm in diameter were identified in a matrix of low free melamine content [521]. Similarly, the distribution of the foaming agent azobisdicarbonamide in a PP/PE blend foil was determined based on the Raman band area ratio I_{1570}/I_{1460} and the general structure of a composite sample consisting of PE fibres in an epoxide matrix was studied [521].

Confocal Raman microscopy was used to study the distribution and redistribution (by leaching) of the fungicide Fluorfolpet FF (5%) and DOP (10–30 wt.%) in PVC films [559]. The technique was also used for depth profiling studies of small surfactant molecules (sodium dodecyl sulfate, SDS) and sulfate anions (SO_4^{2-}) in dry BuA/AA latex films [560]. Other techniques such as ATR-FTIR and step-scan PAS have extensively been used for the same purpose, but have some limitations in the depth probed and in quantitative interpretation, respectively. Confocal Raman microscopy and ATR-FTIR have been applied to examine diffusion rates and redistribution depth profiles at 70°C of various organosilane adhesion-coupling agents (Y9669, A1110, A1891) in PVC [561]. The distribution of particulate silica filler and zinc stearate curative within each phase of BIMS-BR binary polymer blends was characterised by visible (514 nm) confocal Raman micro imaging [562]. Raman microspectroscopy has also been useful in a study of minerals [563].

Micro-Raman spectroscopy has been used to study interfacial regions in fibre-epoxy **composites** [564]. The images suggest that the fibre acts as a nucleation site for areas of lower cure percentage of the epoxy. Raman imaging has also been applied to investigate chemical and physical homogeneity in interstices of glass-reinforced composites [565]. Fluorescence and Raman chemical imaging of adhesion promotion of TPO by chlorinated polyolefins (CPO) was reported [566].

As already indicated in Chp. 1.2.3 Raman spectroscopy is particularly well suited for pigment analysis and can be an effective tool for mapping pigment heterogeneity in basecoat systems [5]. Micro-Raman spectroscopy is also increasingly more important in the field of **art analysis**. It is actually contended that Raman microscopy is the ideal analytical method in art history and conservation science, in particular in relation to pigment identification [567–571]. As Raman spectroscopy is a molecular technique, art analysis is not restricted to inorganic materials, such as mineral pigments, but extends also to organic components, including natural substances, organic binding media and varnishes [572]. In this field, its speed and sensitivity are highly desirable features and small sample quantities may be examined. The technique is sufficiently sensitive to analyse pigment grains, often does not suffer from interference (from surrounding media such as binders) and is non-destructive. Other techniques used to identify pigments on manuscripts, paintings, papyri include diffuse reflection VIS and UV spectroscopy, IR spectroscopy, optical microscopy and XRD for molecular compounds, and XRF, PIXE, PIGE and SEM specifically for elements. Raman microscopy is important as a sensitive probe of pigments on manuscripts and other artefacts and can be obtained *in situ* on works of art from which samples should not be removed [573,574]. Since the probe laser wavelengths used are in the visible region, usually 0.5–0.7 μm, the spatial resolution of the experiment is of this order and therefore individual pigment grains exceeding ~0.7 μm across can be identified. The main difficulty arises from certain organic pigments which either fluoresce (or their supports or binders do), are photosensitive, or fail to yield a Raman spectrum. Numerous pigments and pigment degradation products are known to be highly sensitive to laser radiation.

The use of Raman spectroscopy in the analysis of pigments, watercolours, oil paintings, and lithographs was reported [575]. Raman microscopy and visible reflectance spectroscopy have been widely used for identification of pigments of medieval manuscripts using less than 1 μg of material [574,576]. Pigment investigation of manuscripts often avails itself of a combination of fibre optic μRS and TXRF, which yield complementary information [577–579]. TXRF provides quantitative information allowing pigment-mixing ratios to be established. Mixtures can be investigated easily, because μRS allows single particles within a size down to 1 μm in diameter to be studied. Identification of different pigment grains is based on comparison of recorded spectra with those of reference materials (fingerprinting). Other dual analytical approaches used in combination with Raman microscopy are LIBS [580,581] and PIXE [582,583], which is considered to be one of the best complementary techniques to Raman microscopy in this field. Although these studies involved the sequential use of the two methods, a tandem LIBS-Raman spectrometer has also been reported [584].

Use of Raman microscopy in art history and conservation science was reviewed [567], in particular also the application of identification of pigments on medieval manuscripts [574]. Hummel [556] has provided extensive referencing to pigment analysis (using μRS and other techniques). Raman spectra databases for historical pigments [575,585–587] and databases of Raman spectra of thousands of organic and inorganic materials are offered commercially. Algorithms for the identification of pigment groups are available [588].

Raman microscopy can be used in the characterisation of polymers and coatings before and after weathering. Defects or sites that have obviously degraded as a result of weathering can be analysed by microprobe measurements. μRaman spectrometry allows detection of oxidation phenomena. George *et al.* [589] have used resonance Raman microprobe spectral mapping in conjunction with SEM-EDS for the determination of the spatial distribution of catalyst residues and oxidation products in the early stages of photooxidation of unstabilised PP granulate and film. Laser-Raman spectroscopy has also been used in PVC **degradation studies** [590]. Degradation of PVC with different stabilisers (calcium stearate, zinc stearate and zinc chloride) was studied by μRS in the initial state taking advantage of detection of conjugated double bonds in extremely low concentrations [591]. The potential of the method lies in the possibility of revealing the working principles of stabilisers.

Although a combination of spectroscopy imaging (*e.g.* μXRF, μFTIR, μRS) would offer a powerful way to characterise materials various hurdles must be overcome to achieve the ultimate in **integrated spectroscopic imaging**. These difficulties include spatial resolution, specimen preparation, spectroscopic probe penetration depth and image integration. Same-spot (optical, μFTIR, μRS) technology is now available. The topic of Raman microscopy in combination with other microanalysis techniques (electron microscopy/X-ray microanalysis; ion microprobe mass spectrometry, and laser microprobe mass spectrometry), *i.e.* dual-use microprobe systems, has been discussed [534].

Recent reviews report many applications of Raman microscopy to polymers [488,592,593]. Applications of Raman microspectroscopy to materials science [594] and art and forensic science [595] were also reviewed.

5.6.4. Fluorescence and Luminescence Imaging

Principles and Characteristics
Different fluorescence spectroscopic techniques can be combined with various microscopic techniques (*e.g.* conventional, confocal, and near-field) for 2D and 3D measurements. Fluorescence imaging avoids the problem of looking for a low-contrast signal in the presence of a large background and therefore offers greater sensitivity. Fluorescence imaging enables the specific identification and location of trace levels of a component in a complex matrix. Modern multi-wavelength fluoroimagers equipped with CCD camera technology for highest sensitivity measure fluorescence, absorbance and luminescence of various fluorophores within one measurement and handle any fluorescent dye that is excited in the UV, visible and NIR range.

A monograph dealing with fluorescence imaging spectroscopy and microscopy is available [84].

Applications
Confocal fluorescence chemical imaging microscopy was applied in the study of an adhesion promoting primer, CPO (chlorinated polyolefin), onto a TPO surface [566,596]. By incorporating a small concentration of the fluorescent dye Nile Red in the adhesion promoter quantitative on-line monitoring of the CPO thin film uniformity, thickness and adhesion to the thermoplastic olefin surface was achieved. The solvatochromic fluorescent dye allowed monitoring of the CPO distribution within elastomer domains of TPO. Raman chemical imaging was effective for monitoring non-invasively the depth of penetration of CPO in TPO. Raman imaging requires no prior staining to obtain chemically specific information about CPO distribution. Imaging systems for fluorescence microscopy are also used for recording and analysing processes that change rapidly with time. Low-level macroscopic luminescence/fluorescence imaging is widely used for *in vivo* visualisation, imaging of micro plates and fluorescent substrate detection on TLC plates.

Fluorescence imaging and transmission UV microscopy of PP/0.5% Uvitex OB (optical brightener) were reported [63].

5.6.4.1. *Imaging Chemiluminescence*

Principles and Characteristics
Fleming and Craig [597] first pointed out that CL applications for studying autoxidation reaction mechanisms of polymers could be made more useful by adding imaging capability. The imaging chemiluminescence (ICL) technique adds spatial resolution to CL emission. In ICL the positions of the photons emitted from the specimen are registered during the measurement.

An **ICL instrument** typically consists of a temperature and atmosphere controlled oven, light-tight connected to an optical lens that projects the light emitted from the surface of a sample onto a photon counting imaging device connected with a CCD camera capable of detecting 200–1200 nm photons (*cfr.* Fig. 5.18). Various similar designs have been proposed [597–602]. By integrating the signal of the detected photons (from 2 to 10 min) an image of the CL emission is obtained as well as the integrated CL intensity. Sequential and simultaneous

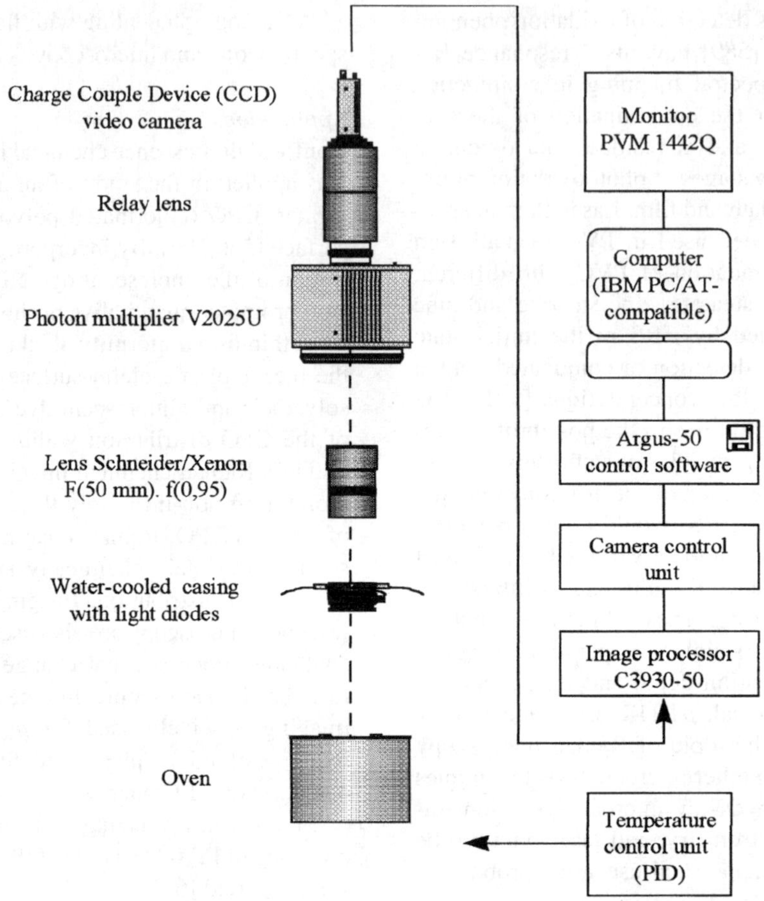

Fig. 5.18. Configuration of an imaging chemiluminescence instrument. After Ahlblad *et al.* [598]. Reprinted from *Polymer Testing* **16**, G. Ahlblad *et al.*, 59–73, Copyright (1997), with permission of Elsevier.

chemiluminescence imaging is possible. Third generation multicell ICL systems involve the possibility of simultaneous measurements of CL intensity time curves for up to 48 specimens, or monitoring spatial variations of the CL intensity for a single specimen (up to 25 × 25 cm^2) [603]. Care has been taken to avoid temperature gradients and infectious spreading to neighbouring specimens. The low quantum yield of CL (10^{-9}) and the small fraction of the polymer initially oxidising still require relatively long integration times at low temperatures at the pixel resolution necessary to resolve the oxidising centres.

In analogy with chemiluminescence (Chp. 1.4.4) various ICL experiments may be carried out in inert or oxidising atmosphere. ICL in inert atmosphere may be acquired isothermally or by linear heating. The latter approach is less satisfactory for kinetic analysis, because data are not isothermal and sample melting may cause changes in the geometric parameter G of eq. (1.12). To serve as an analytical tool in studies of heterogeneous processes during oxidation of polymers, the spatial temperature variations must be kept at a minimum.

ICL is an extremely sensitive method to study oxidative degradation of polymers and requires only very small samples (<10 μg). An image of CL emission, containing information about the rate (intensity) as well as the location of oxidation of a sample, is obtained by integrating the CL emission over a short period of time. The time to the onset of oxidation can be obtained for different samples simultaneously, or at different positions on a single sample, by integrating the CL emission at different times during the course of oxidation. ICL is particularly useful for systems that do not oxidise homogeneously. The ICL technique provides information on various types of heterogeneous oxidation of polymers, *e.g.* diffusion limited oxidation, physical spreading of oxidation and oxidation induction time distribution. Develop-

Table 5.52. Main features of imaging chemiluminescence for polymer studies

Advantages:
- No sample preparation
- Micro sample sizes (<10 μg)
- Accommodates wide range of sample geometries (film, pellet, fibre, powder, liquid)
- Highly sensitive technique
- Real-time monitoring of position and intensity of emitted photons
- Speed, simplicity
- Various experimental modes (isothermal, linear heating; oxidative, inert)
- Discrimination of low stabiliser concentrations
- Early detection of sample defects
- Applicable to volatile samples
- Acceleration *vs.* oven ageing: 10–20×
- Commercial equipment; automated testing; multisample imaging
- Applicable for industrial purposes (QC)

Disadvantages:
- No standardised testing procedures
- Relatively low resolution (20 μm/pixel)
- Not equally applicable to all polymer systems

ment of ICL has allowed improvements in the studies of polymer oxidation. ICL is now an important technique for non-destructive testing, determination of remaining useful shelf-life, and QC in manufacturing.

Table 5.52 shows the **main characteristics** of ICL for polymer degradation studies. ICL is not only a sensitive technique to study polymer degradation but can also be used in industrial research to study polymer stabilisation. Compared to the widely used oven-ageing test for assessment of the effectiveness of stabilisers, the following advantages can be noted [604]: *(i)* early detection of sample defects; *(ii)* considerable gain in speed (10–20×) without loss of comparability with oven testing; *(iii)* better discrimination between samples at low AO concentration; and *(iv)* complete automation of testing. Although further optimisation is necessary to reach the necessary confidence level required in application laboratories, CL/ICL testing is likely to catch up and even, in the near future, replace oven ageing tests for the determination of AO effectiveness. The cost of ICL is high compared to conventional oven ageing.

Recently, simultaneous ICL-DSC has been reported [605,606].

Table 5.53. Main applications of imaging chemiluminescence

- Non-destructive testing; direct imaging
- Screening of light stabilisers
- Assessment of polymer heterogeneity (*e.g.* distribution of stabilisers)
- Real-time oxidative degradation studies (in particular early stages)
- Visualisation of failure sites induced by ageing and mechanical stress
- Determination of remaining useful shelf-life
- QC in manufacturing

Applications

Table 5.53 shows the main application areas of ICL. The technique allows visualising from where CL effects originate. Isothermal ICL experiments offer a unique possibility to simultaneously measure the CL emission of different samples in a population. One of the aims of degradation studies by means of ICL has been the identification of localised zones of oxidation and an understanding of oxidation spreading, crack formation and mechanical failure.

Degradation of solid polymers is heterogeneous in nature for a variety of reasons, including a heterogeneous distribution of initiating species (*e.g.* catalyst residues, peroxides or oxygen containing groups), restricted mobility of radicals, morphological variations, and enhanced sensitivity of oxidation products to further oxidation. In addition to the morphological and chemically determined micro-scale heterogeneities, physical effects lead to macroscopic heterogeneity. One reason is the non-uniform distribution of stabilisers (apparently due to inefficient mixing during processing), also in relation to consumption, diffusivity and solubility of the stabilisers, and the tendency of many of these additives to evaporate and bloom. The other important heterogeneous effect is related to diffusion-limited oxidation (DLO), which may become significant in polymer samples during accelerated ageing in air. Heterogeneous oxidation can also occur when environmental factors interact non-uniformily with a material, *e.g.* exposure of UV light, reactive atmospheric pollutants, *etc.*

Heterogeneous oxidation of solid polymers, as apparent from mechanical failure and embrittlement very soon after the induction period when the overall damage of a material is still relatively small, is well established, especially for polyolefins, and is caused

Table 5.54. Experimental evidence for infectious spreading of PP oxidation

Technique(s)	Observation(s)	Operational level
Dynamical tests	Decrease in fracture energy with time	Macroscopic
CL	Oxidation of neighbouring particles	
SEM	Development of surface microcracks	Microscopic
Staining/UV microscopy	Non-uniform distribution of oxidation	
μFTIR	Observation of oxidation front (μm resolution)	
μRS	Chemical identification of species (1 μm resolution)	
AFM	Bubbles on photooxidised surfaces	
ICL	Non-uniform distribution of oxidation (20 μm resolution)	
SEC	Changes in molecular weight	Macromolecular
GC-MS	Formation of volatiles and sec. oxidation products in induction time	Chemical
FTIR	Oxygen uptake, formation of stable oxidation products	
ESR	Free radical formation	Reactive intermediates

by physical factors such as morphology and structure of the material, tacticity, catalyst residues, *etc.* It is often observed that oxidation starts at (undefined) edges, cracks or morphological imperfections. Heterogeneity questions the validity of applying a homogeneous kinetic analysis to solid polymer oxidation, which may be used for determining the ultimate lifetime of a material.

Various experimental techniques can monitor heterogeneous ageing, such as IR, density profiling, SEC, CL (which necessitate microtomed slices), or μFTIR, X-ray analysis and modulus profiling (more convenient direct profiling with sufficient resolution). The development of position sensitive photon detectors has provided a new possibility of direct profiling using chemiluminescence. A technique such as ICL, capable of showing a spatial distribution during *in situ* oxidation of polymers, is of great value for the understanding of the nature of physical spreading during oxidation of polymers.

ICL has been used to observe heterogeneity in **oxidation of PP** and physical spreading [600,607]. Table 5.54 collects some experimental evidence for the infectious spreading model of polymer oxidation starting from catalyst residues or other impurity centres; additional evidence comes from FTIES combined with CL and from TEM-EDS. Fayolle *et al.* [608] have studied rapid crack growth occurring soon after the end of the induction period using SEC, OM and FTIR mapping. Raman studies have indicated that catalyst residues stabilise the polymer in the immediate vicinity but generate a migratable oxidant which spreads the degradation. No correlation was established between catalyst residues and the distribution of oxidation products.

Analysis of CL and ICL data of oxidation of PP indicates that initiation occurs heterogeneously at high rates in localised zones, possibly associated with catalyst residues or other defects in the polymer, followed by spreading of the oxidation through the sample [609–611]. The induction period measures the time taken for oxidation to spread from the reactive centres. CL studies of individual PP powder particles have revealed that oxidising centres are highly active and able to initiate oxidation on neighbouring particles even if they are physically separated [607]. Apparently, physical spreading of oxidation is not limited to surface-to-surface contact of particles but may allow gas phase transport, notably involving formaldehyde and acetic acid [612]. ICL of single PP particles has shown interparticle spreading of isothermal oxidation [600]. ICL allows observing the formation of localised zones of heavy oxidation and therefore weak points in a material followed by spreading of the oxidation through the sample. A spreading model for oxidation of PP has been presented [613]. Oxidation spreads with a rate consistent with small fragment migration within the solid polymer, or by larger fragments migrating in the gas phase.

The ability to measure spatial distribution of *in situ* oxidation can also be used to study the advancement of a propagating oxidation front in polymer films under different conditions. Oxidation of stabilised PP films has been initiated by unstabilised PP particles that were kept in direct contact with the film surface [614]. Figure 5.19 shows the induction time and advancement at 150°C in air of the oxidation fronts in PP film stabilised with Irganox

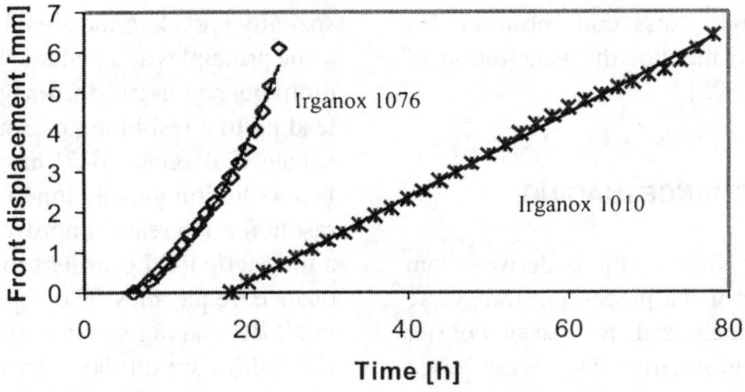

Fig. 5.19. Displacement of the oxidation fronts in PP stabilised with Irganox 1076 and Irganox 1010. After Eriksson *et al.* [614]. Reproduced by permission of G. Ahlblad, Royal Institute of Technology, Stockholm.

1076 and Irganox 1010. The performance of different stabilisers (Irganox 1010/1076/3114) in PP was assessed by ICL and μFTIR (carbonyl index I_{1710} cm^{-1}/I_{1455} cm^{-1}) from the speed of spreading of the propagating oxidation front starting from a controlled initiation of the oxidation (UV light or contacting with unstabilised material) [615]. According to George *et al.* [616] the ICL-time curve measured during thermooxidation of PP may reflect either the hydroperoxide profile or the oxidation product profile depending on the spectral wavelength analysed or the state of purity of the polymer.

Simultaneous DSC-ICL experiments (with an astronomy CCD image) for oxidative induction time (OIT) studies for PP and PVC samples have been reported [606]. Close correlations between DSC-OIT and ICL-OIT data were observed. Dudler *et al.* [604] have studied the relative **stabiliser effectiveness** of some phenolic AOs (Irganox 1010/1076/1330) by ICL and oven ageing. OIT at 150°C, measured on thin films of PP in pure oxygen, scales with the AO concentration in PP films and correlates linearly with the embrittlement time observed in the universal oven ageing test.

The thermooxidative stability of **polyamide 6** films at 100–140°C was investigated by CL/ICL and isothermal microcalorimetry (MC) techniques [617, 618]. The CL intensity in oxygen seems to be related to the content of peroxides. ICL measurements of unstabilised PA6 films denote uniform oxidation of the surface while variations in the oxidative stability of PA6/Irganox 1098 films were attributed to a non-uniform distribution of the antioxidant [617]. MC and chemiluminescence measurements were compared to oxygen uptake. All of these techniques respond to oxidation but exhibit distinctly different rate time curves. Hosoda *et al.* [599] have studied ICL of press-moulded sheets of PA6 under thermal oxidation and stress. Hydroperoxide and carbonyl index depth profiles of thick oven-aged PA6.6 were studied by ICL and FTIR [619].

Ahlblad [620] has applied ICL to the oxidation of **rubber materials** and ICL measurements in N_2 atmosphere have been used for the study of the peroxide depth profiles of HTPB rubber, pre-aged by thermal oxidation [621]. ICL has also been applied to study thermooxidative infection in populations of EPDM particles [622]. Physical spreading of oxidation by volatile oxidative species was observed in populations where the distance between particles was less than about 500 μm. The onset of spreading to adjacent particles coincides with mass loss and formation of volatiles. ICL may be of help in explaining the formation of gels. Another useful application of ICL is comparison of the efficiency of different stabiliser systems in EPDM rubbers.

Photodegradation of various **automotive coatings** (melamine cross-linked acrylic and acrylic polyurethane types) has been studied by ICL [623]. The method is sensitive enough to measure photodegradation in unstabilised and stabilised coatings after only 48 and 100–200 h of exposure, respectively. Extrapolation to the failure time of coatings by CL is not yet possible, but the technique can be used to screen rapidly the relative performance of new coatings formulations or light stabilisers added to clearcoats.

ICL was also used to monitor the spatial distribution of oxidation across **stress profiles** in pre-aged injection moulded PP, HDPE and PA6 [624]. The technique is a convenient complementary method to reveal zones of high stress concentration. Finally, ICL was used to monitor the penetration of dimethylsulfide in PP [625].

5.7. MAGNETIC RESONANCE IMAGING

The power of modern NMR methods derives from the fact that the phase of the precessing transverse magnetisation can be measured. By use of Fourier transformation phase information can be converted into probability densities of resonance or Larmor frequencies (multidimensional spectra), densities of position (NMR images), and probability densities of parameters like velocity and acceleration which quantify translational motion. The first magnetic resonance images were published in 1973 [383,626]. Magnetic resonance imaging for medical purposes uses magnetic fields up to 1.5 T (or 60 MHz proton frequency) and is used to detect proton NMR signals from body fluids. **Nuclear magnetic resonance imaging** (NMRI) at small scale and high spatial resolution, using standard NMR spectrometers designed for chemistry, has been introduced in materials science in the 1980s. It has become possible to spatially resolve virtually all magnetic resonance parameters; initially from liquid samples but increasingly from solids as well. NMR has the capability of measuring inhomogeneities in finished articles by a non-invasive and non-destructive method. Defect or non-uniform areas of the polymeric materials are clearly shown in the NMR image. NMRI may be considered as a type of chemical microscope.

Callaghan [627] has described the principles of MRI experiments in detail. The fundamental element in NMRI is the spatial dependence of the precession frequency of the spins. In imaging, the first requirement is to define the slice to be imaged. This is the basis of the **tomographic process**, which is accomplished by a method called selective excitation. Only a slice of the object is excited, and the remainder of the object does not respond. NMRI thus relies on the interaction of nuclei in only a small and controllable region of the sample by placing the sample in a spatially inhomogeneous magnetic field whose nuclear resonance frequency is matched to the rf signal in only that region. If the magnitude of the field is made non-uniform in a controlled manner, then nuclei at different points in space will precess at different frequencies, that is, the gradient system spatially encodes the spins. Other than spatially encoding the signal, imaging works on the same principles as standard NMR. The maximum radiofrequency useful for imaging is about 100 MHz, leading to a resolution of 3 m which is totally inadequate. Callaghan [627] has discussed the physical factors limiting resolution in NMR microscopy. The resolution is greatly improved by the application of a **magnetic field gradient** to disperse the NMR resonance frequencies. The highest resolution achieved in NMR imaging so far is $10 \times 10 \times 100\ \mu m^3$; typical values are on the order of 5 to 40 μm. For the reconstruction of NMR images, many different approaches are possible. It is possible to image very fast by covering reciprocal or **k**-space in a single transient, using the echo-planar imaging (EPI) technique. EPI is the only genuine real-time MRI technique, generating 2D or even 3D images in times as short as a few tens of milliseconds.

When comparing NMRI with other microscopy techniques, one has to accept that the achievable resolution is less even under optimum conditions. Although it is unlikely that the resolution of the NMR microscope will ever match that of its optical counterpart, it nevertheless has several attractive advantages: the specimen is observed in its natural state without preparation, and images can be recorded from deep within an optically opaque 3D specimen. There are many applications for which the resolution of NMRI is sufficient and the possibility to incorporate the full spectroscopic information is an invaluable advantage. It is important to concentrate on providing information that is not accessible by other techniques. In many respects, NMRI is complementary to X-ray tomography, in that X-rays mostly detect the electron-dense areas, whereas NMRI mostly traces softer areas. NQR imaging has also been reported [628].

Many substantial differences exist between ESR and NMR spectroscopies, such as the order of magnitude of frequencies used and relaxation times. Although basically similar to NMR imaging, **ESR imaging** (ESRI) has some additional requirements. ESRI is more difficult to achieve technologically than NMRI. An ESR spectrum occupies a much larger frequency range than does a NMR spectrum; moreover, the line widths of samples are considerably broader compared to those of NMR. This necessitates field gradients one or two orders of magnitude higher than are currently being used for NMRI in order to achieve practical spatial resolution. Also, the complex hyperfine structure due to electron–nucleus interactions must be removed from the projection spectra. Another problem arises due to the irradiating frequency used. Use of a conventional frequency of about 9 GHz in ESRI allows only a sample of maximum diameter of 10 mm to be examined. At frequencies as low as 200 MHz samples as large as 100 mm can be investigated. Despite various complications, which exist in the interpretation of spectra, ESR is inherently more sensitive than NMR because of the much greater electron magnetic moment and the consequently higher Larmor frequencies. However, imaging nuclear precession is considerably more advanced than ESRI. The impossibility of FT-ESR due to too short relaxation times is the greatest obstacle to shortening the acquisition time. Yet, ESRI is filled with expectation because of the selectivity and complementary properties of paramagnetic molecules compared to diamagnetic molecules detectable by NMRI.

5.7.1. Nuclear Magnetic Resonance Imaging

Principles and Characteristics

Conventional NMR spectroscopy, in which the resonance frequency of a nucleus is linearly dependent upon the strength of the static magnetic field, can be used to determine the type of chemical structure on the basis of resonance frequency, but not the spatial position of the stimulated nuclei in a heterogeneous rigid sample. NMRI is a method where the stimulating signal is spatially encoded so that an image can be reconstructed showing the **distribution of nuclei** in the sample. The magnetic resonance imaging (MRI) experiment adds a spatial dimension to the standard NMR experiment.

Most optical spectroscopic imaging techniques are based on reflection from surfaces or transmission through thin optical slices. On the other hand, nuclear magnetic resonance imaging (NMRI) or NMR microscopy is an "internal" technique, in which an internal portion of the sample to be examined is selectively excited. A 3D image is reconstructed from a collection of 2D images obtained in a series of sequential steps without destruction or extraction of the sample. 2D NMR imaging can be perceived as a particular form of 2D spectroscopy, where the frequency axes have been converted to space axes by application of magnetic field gradients. NMRI, or spatially resolved magnetic resonance, is a technique for *in situ* detecting and imaging previously **invisible internal material heterogeneities**. Because magnetic fields and radio waves both pass through the material freely, NMRI allows non-invasive visualisation of internal structures. The rf radiation in NMRI carries only low-energy quanta, and its absorption only leads to some local heating, almost always by less than 1°C.

NMR microscopy involves the acquisition of the NMR signal in the presence of a magnetic field gradient, a process known as **k**-space acquisition. A dynamic analogue of NMR imaging is the pulsed gradient spin-echo (PGSE) experiment, sometimes termed **q**-space imaging. This type of experiment can be used to study the spectrum of molecular motion as well as the morphology in porous systems (*cfr.* Chp. 1.5.1.1). While it is not customary to group together these two apparently very different applications of magnetic field gradients, there are common physical principles governing the imaging of static displacements via **k**-space and dynamic displacements via **q**-space [627]. Displacement spectroscopy (also known as **q**-space microscopy) arises when the applied gradient $\mathbf{G}(\mathbf{r})$ consists of two short pulses of duration δ separated by a phase evolution time Δ. Any net distance travelled during the diffusion time, Δ, causes attenuation of the NMR signal from the intensity acquired with the same pulse sequence but using gradients of zero strength. Using this approach a map of the diffusivity of mobile species within the material is obtained. Measurement of the nuclear spin translation via the PGSE method can achieve a spatial resolution some two orders of magnitude better than with **k**-space imaging or relaxometry. The resolution in NMRI is not particularly good by comparison with that available in optical microscopy. On the other hand, PGSE-NMR is limited, in practice, to dynamic displacements of between 100 Å and 100 μm and over time-scales of a few ms to a few seconds. In NMRI the absolute phase of the spins is measured and related to nuclear positions. For the measurement of motion phase differences are determined, for which the spin-echo (SE) is ideally suited. Both **k**-space and **q**-space NMR imaging have many potential applications in materials science opening up the study of **molecular dynamics**. It is from the range of contrast available that NMR microscopy gains its value.

NMR microscopy is limited to nuclei with a favourable sensitivity, intrinsic line width due to T_2 relaxation and repetition time allowed by spin-lattice relaxation (mainly ^1H, ^7Li, ^{13}C, ^{14}N, ^{19}F, ^{23}Na, ^{29}Si and ^{31}P). The nucleus imaged most often is the proton. Reasons are the sensitivity and the weak dipolar couplings between protons in a chemical group and between different chemical groups which dominate the signal decay by relaxation. ^{13}C is not a favourable nucleus for NMRI with its low natural abundance (1.1%) and because of low gyromagnetic ratio so that sensitivity is poor. Indirect detection techniques for ^{13}C nuclei such as cyclic J cross-polarisation (CYCLCROP) result in a significant enhancement of the NMR signal. Phosphorous is a nucleus which can readily be imaged. Fluorine micro-imaging may prove useful in materials science in applications using fluorinated solvents or polymers.

In NMR spectroscopy, nuclear spins precess about the static magnetic field, $\mathbf{B_0}$, at the Larmor frequency, ω_0, as given in eq. (5.5), where γ is the gyromagnetic ratio:

$$\omega_0 = \gamma \mathbf{B_0} \tag{5.5}$$

NMR imaging is based on the simple idea that a spatially varying magnetic field encodes the positions of the spins in their resonance frequencies, and thus the number of spins at any given location may be directly measured as the intensity of the NMR signal at the corresponding resonance frequency. In magnetic resonance imaging the recovered signal is a free induction decay recorded from the whole sample, and the excitation is a radiofrequency (rf) pulse that also interacts with the whole sample. There are several ways of spatially encoding the NMR signal. One is to apply a static magnetic field gradient along the z-axis of the sample (selection of a slice of the sample) and generally involves spin echoes or gradient echoes for refocusing nuclear magnetisation and/or avoiding artefacts due to gradient switching. When a linear magnetic field gradient, $\mathbf{G}$, is applied across the static field, the resonance frequencies of the spins become dependent on position $\boldsymbol{r}$, as given in the fundamental equation of NMRI:

$$\omega(\boldsymbol{r}) = \gamma(\mathbf{B_0} + \mathbf{G}\boldsymbol{r}) \tag{5.6}$$

By tailoring the frequency content of the rf pulse, it is possible to cause only those nuclei within a defined slice of the sample to come into resonance. For imaging in the x–y plane within this slice, a second static field gradient is imposed along the x-axis; different positions along the x-axis will experience different fields, and resonate at different frequencies. The y-axis can be added with a third field gradient. The resonance frequency is determined by the local field which varies from point to point. Following perturbation by a rf pulse the return of the spin system to equilibrium is characterised by spin–lattice (T_1) and spin–spin (T_2) relaxation time constants. Fourier transformation of the time domain data produces a frequency spectrum where the amplitude at each frequency is a measure of the number of nuclei in the corresponding region in space. In magnetic resonance imaging a multivariate slice for one pulse sequence is used if spatial resolution is most important. A full NMR spectrum can be produced for a single point inside the material if the spectral aspect is most important.

Various approaches to NMRI imaging of materials are available: liquid state and constant time methods, stray field, force detection and coherent-averaging (multiple-pulse and MAS) [629]. Wide-line methods accept the limitation to resolution imposed by the sample's line width and use strong gradients to achieve high spatial encoding. The most successful of these is stray field imaging (STRAFI), commercially available. Coherent-averaging methods have been developed to average dipolar couplings, chemical shifts, *etc.* Coherent-averaging for images is based on multiple-pulse sequences and magic-angle sample spinning.

NMR imaging methods rely on the use of static or pulsed field gradients. The image is encoded in a frequency- and phase-modulated 2D array with intensity providing the third dimension. Table 5.55 lists the main **advanced NMR imaging methods**. Spin-density imaging reveals differences in the local concentrations of hydrogen atoms of mobile regions and inhomogeneities (such as filler aggregation) throughout a sample; spin-density measurements are not very sensitive. Gradient-echo imaging shows spatial differences in the magnetic susceptibility caused by a heterogeneous distribution in the matrix. The sequence that is commonly used to measure the T_2 relaxation phenomena in images is called multiple spin-echo. T_2 images provide information on the spatial difference in cross-link density and molecular mobility ("microsoftness").

NMR usually assumes that the nuclear spins precess at the same frequency neglecting chemical shift differences arising from different chemical types of

Table 5.55. Techniques for spatial domain NMR

- Spin-density imaging
- Gradient-echo imaging
- 2D T_2 imaging
- Chemical shift imaging
- NMR-MOUSE

nuclei in substances. Much valuable information is contained in the high-resolution NMR images if the chemical shifts of the species present in a system can be sorted out correctly. It is possible to form an image from only a selected portion of the total NMR spectrum. A particular resonance peak can be selectively excited by rf irradiation and imaged to the exclusion of others in the chemical shift spectrum. This process, called **chemical shift imaging** (CSI), or spectroscopic imaging [630,631], is highly desirable. For example, separate images are due to aromatic and methyl protons, which are separated by *ca.* 4.8 ppm from each other. Chemical shift imaging techniques use pulsed magnetic field gradients. When different chemical shifts originate from different molecular species, an image taken at a specific chemical shift will provide information on the spatial distribution at the molecular level while excluding the interference of the chemical state of other components within the material. So far, however, the application of various spatially resolved NMRI techniques for the observation of high-resolution, chemically resolved spectra has been limited. Chemical shift images have been reported for two rubbery polymers, polybutadiene and polydimethylsiloxane, and also for polyether polyol with an isocyanate curing agent [632]. Little has been done in the direction of the identification of the distribution of additives in rubbers using chemical shift selective NMR microscopy. However, if we consider that spatial resolutions of 370 μm are reported for ^{13}C chemical shifts, it is concluded that the technique has as yet little to offer for the study of the distribution of additives in polymeric matrices.

Relaxation can be probed in **inhomogeneous fields**, so that the homogeneous polarisation field $\mathbf{B}_0$ is not a necessity for successful applications of soft-matter imaging. Blümich *et al.* [633] have developed a mobile NMR surface scanner of low-field (9.17 MHz), which scans with a spatial resolution of 3 mm and an adjustable depth sensitivity of 0–5 mm. The small portable NMR sensor, called NMR-MOUSE (MObile Universal Surface Explorer), provides NMR data of near-surface volume elements with the same specificity as the contrast in an NMR image [634]. Depth and size of the sensitive volume scale with the size of the coil of the NMR-MOUSE. The larger the coil, the deeper the signal-bearing volume. Scanning of depth is achieved by changing excitation and detection frequency, and lateral resolution is obtained by displacement of the scanner. Measurement times are typically of the order of 1 to 15 min. Such low-cost, mobile, sensors are suitable for investigations of arbitrarily large objects as well as for industrial process and quality control by relaxation measurements [635].

The particular utility of NMR microscopy lies in the contrasts that are available. **Image contrast** in NMRI depends on material-specific parameters (spin-density and nuclear spin relaxation times), operator-related parameters (pulse sequence, pulse delay and repetition times) and external parameters (temperature, viscosity, *etc.*). Common contrast mechanisms in solid-state NMR imaging are based on relaxation times (T_1, T_2, $T_{1\rho}$, T_{1x}) and chemical shifts. Most studies develop contrast based either on spin density or T_2 differences since these show up immediately without the need of modifying the imaging sequence. The unsurpassed soft-matter contrast of NMRI is hard to achieve with competitive methods like X-ray or computer tomography.

The major hurdle to **spatial resolution** is the poor sensitivity of NMR spectroscopy, which imposes a lower limit for the size of the sensitive volume. Spatially resolving a given volume in an NMR image is equivalent to doing NMR spectroscopy on that volume. The highest resolution reached is about $10 \times 10 \times 100\ \mu\text{m}^3$, corresponding to a voxel of 10^{-5} mm^3. In exceptional situations 5 μm resolution has been achieved. Routine measurements on liquids in solids typically have $40 \times 40 \times 100\ \mu\text{m}^3$ resolution. For this reason, many investigations of NMR imaging to material science are restricted to samples with high molecular mobility, *e.g.* the distribution of liquids in synthetic polymers. Longer measuring times and signal averaging may enhance sensitivity. Although NMRI has inferior spatial resolution compared to microscopic surface techniques and many other imaging techniques, the possibility to combine spatial features with various forms of contrast makes the method unique. Besides NMR parameters like spin density, relaxation and spectroscopic information, self-diffusion, convection and flow can be used to generate contrast due to mass transport.

Table 5.56 lists the **main features** of NMRI. As NMRI is non-invasive, multiple measurements can be made on the same sample under different conditions. The main problem with NMRI is the long data collection time, mainly due to the long spin-lattice relaxation time T_1 ($\sim$0.5 s for aqueous systems). The high cost of imaging facilities is a hindrance to exploitation of NMRI in polymer science. However,

Table 5.56. Main characteristics of NMRI

Advantages:
- Non-invasive analysis (very low rf photon energy)
- Non-destructive inspection
- No restrictions on sample geometry (except size)
- No special sample preparation required
- Absence of ionising radiation (as for imaging by UV, X-ray and higher energy electromagnetic radiation)
- Excellent power of penetration
- 3D method
- Molecular specific (chemical state imaging)
- *In situ* examination of (internal) heterogeneities in materials
- Sensitive to molecular dynamics, fluid phases
- Spatially localised diffusion measurements
- Unsurpassed soft-matter contrast

Disadvantages:
- Restrictive arrangements for sample loading
- Limited spatial resolution (10–40 μm)
- Low inherent sensitivity
- Limited applicability in materials science
- Liquid-state rather than solid-state imaging (in NMR sense)
- Need for spectroscopic and hardware inspection
- Long data acquisition times (30 min to many h)
- High complexity (need for high level of scientific expertise)
- Expensive equipment; sophisticated technique

there are some developments of low-cost low-field imagers for materials science applications.

Various distance scales may be probed by NMRI, macroscopic, microscopic and molecular, *i.e.* phenomena ranging from mapping mass transport (swelling) to molecular forces influencing diffusion contrast. At each distance scale other dynamic changes are being investigated. In NMR imaging, different regions of the sample can be made to satisfy the resonance condition at any one time by varying the intensity of the externally applied, non-uniform (gradient) magnetic field. NMR is capable of providing *in situ* information about mobility and structure, in particular the presence or absence of inhomogeneities.

In NMR terminology the words "solid" and "liquid" are used to describe the local motion of the nuclear environment. In that sense spins belonging to liquid molecules absorbed in a solid-state matrix or to mobile polymer segments in a rubbery solid are considered to be in a liquid state. **Liquid-state imaging** techniques are experimentally less demanding; the same techniques can be used for investigations of solids with narrow lines. Suitable materials are plastic solids with high molecular mobility, *e.g.* rubbers or polymers at elevated temperatures, and highly oriented rigid solids like fibres. Synthetic polymers above T_g constitute an important class of soft-matter materials. It is technically demanding to use NMRI in a dynamic mode to follow changes in polymeric materials in real-time as they undergo processing operations. It is much easier to use NMRI to observe *static* polymer structure, and the literature abounds of images of swollen polymer systems.

Elastomers are a peculiar class of solid which appear as liquids in the NMR sense. In elastomers, the macromolecular proton T_2 may be sufficiently long that it is possible to obtain an image of the polymer matrix without the need to employ special line-narrowing methods. Relaxation techniques, which probe different time regimes of molecular motion, provide the primary access to contrast in imaging of elastomers. For elastomers simple methods like Hahn-echoes and gradient echoes are useful for materials characterisation and imaging. Elastomers can be investigated by NMRI on a routine basis [636].

There are a number of limitations to the use of NMR imaging of rigid polymeric solids for non-destructive analysis, the most general being that rf fields must penetrate the material. As the solid-state line width is some 1000 times broader than its solution counterpart, with corresponding decrease in sensitivity, solid material is generally not observed in images acquired with NMRI techniques. The use of conventional imaging gradients is precluded because rigid solids have short T_2 values on the order of 100 μs [393]. These relaxation times are often too short compared with the time it takes to switch and apply field gradients for slice selection and phase and frequency encoding, while in addition extremely large gradient amplitudes are required to overcome the broad line widths. For **rigid solids**, imaging techniques relying on Hahn spin echoes or gradient echoes fail completely. Problems related with the broad lines of rigid solids can be solved using wide line methods (*e.g.* stray-field imaging, STRAFI, using strong, static gradients), phase encoding methods, and line-narrowing techniques (*e.g.* multi-pulse methods, magic-angle spinning, *etc.*). Various line-narrowing methods have been proposed or demonstrated for removing such interactions, which make solid imaging into a practical technique [637]. For the principle of the magic-echo technique and 2D-FT magic-echo imaging scheme, as applicable to drawn polymers, *cfr.* ref. [638]. Methods for imaging of rigid polymer materials have recently been developed [639,640]. NMR imaging of solids was reviewed [641,642].

For further literature the reader is referred to refs. [393,627,643–646] for NMRI principles, to refs. [393,632] for applications in polymer science and to ref. [637] for applications in food science. McBrierty [647] has given a comprehensive review of spin relaxation in solid polymers. A comparative study of various NMR imaging techniques has been given elsewhere [648,649]. A (dated) comprehensive bibliography on NMRI has also appeared [650].

Applications
In NMRI terms, samples have only two possible internal states, that is soft and hard. Most applications of NMRI are in the medical field (cellular tissue). In materials science NMRI is profitably applied only in cases where most other methods fail. Relevant applications concern imaging of heterogeneities from samples which cannot be destroyed by cutting and where the property to be imaged is altered by invasive investigation. Essentially any materials problem which can be beneficially analysed by NMR spectroscopy and in which spatial variation occurs in the specimen of interest might be a good candidate for NMRI. The potential applications of NMRI in the field of polymeric materials are many and diverse. NMRI allows imaging of various molecular and atomic properties, including the local chemical composition and molecular order, the local molecular translational motions and rotational dynamics. NMRI is a means of detecting and imaging previously invisible internal material heterogeneities. The potential applications in the field of polymeric materials are indicated in Table 5.57. Essensially two solid-state systems are of interest for which it is possible to obtain proton NMR signals with T_2 of the order of a few μs: *(i)* heterogeneous solid/liquid systems (small molecules absorbed into a solid matrix); and *(ii)* elastomers. The broad lines found in solid samples hinder the application of NMRI to solid-like materials.

Conventional NMR imaging techniques are particularly suited for the study of **dynamic processes** in polymer science, such as phenomena occurring in solid materials with mobile components, *e.g.* local motions, polymer–solvent and food–packaging interactions. NMR is sensitive to molecular mobility on a microscopic scale via the spin–spin (T_2) and spin-lattice ($T_1, T_{1\rho}$) relaxation times. The development of magnetic resonance imaging techniques means that information on molecular motion is now available with spatial resolution, albeit generally only for the most mobile components in the system, *i.e.* those with narrow NMR line widths, equivalent to long T_2 relaxation times. NMRI techniques have been used for the study of sorption and diffusion as well as desorption of multiple chemical substances in polymeric materials. Typical applications of flows of liquids in solids are swollen polymers.

Table 5.57. Potential applications of NMRI to polymeric materials

- Soft heterogeneous matter
- Liquid phase imaging: sorption, diffusion (coefficients), desorption, fluid distribution, swelling, leaching
- Monitoring of heterogeneous dynamic processes
- Determination of porosity (voids)
- Detection and imaging of subsurface defects
- Food-packaging interactions
- Local motions: differences in bulk mobility
- Chemical reactions: polymerisation, vulcanisation, curing
- Detection and characterisation of domains modified by foreign substances (additives, degradation products, contaminants)
- Non-uniform dispersion of fillers
- Cross-link heterogeneity, density and gradients ("microsoftness")
- Physical ageing
- Probing of interfacial and interphase structures
- Detection of inhomogeneities in finished articles

Little is known on the physical distribution of small molecules, including plasticisers, within polymers or on the mechanism by which desorption occurs. Such information may be collected using several techniques, including gravimetry (absorption/desorption rates), ESR (doping procedures) [651], or optical microscopy (visual observations, birefringence) [652]. Most techniques require interrupting the diffusion process and destroying the sample. The ingress of liquids in a solid can be conveniently investigated by NMRI since it allows acquiring selectively the image of the liquid. Polymer/solvent systems studied by means of NMRI comprise HDPE/toluene [653], PS/styrene [654], HIPS/blowing agents [655], nylon/H_2O [656], GFR polyesters/water [657] and water-based PVAc adhesives on wood [658]. Various other studies have employed NMRI as a tool to examine **solvents** in polymers [657–659] and motion of small molecules in swollen rubbers [660]. NMRI provides insight in the spatial distribution of solvent in the polymer and rate of desorption. Using a spin-echo imaging pulse sequence and a micro-imaging probe Weissenberger *et al.* [661] studied methanol desorption from partially swollen PMMA rods. NMRI was also used to visualise in-plane moisture transport in handsheets, heavy paperboard, and polyethylene-coated paperboard [662]. The diffusion of moisture in paper impacts many aspects of papermaking. The study of diffusion of antioxidants in polymers by means of NMRI is far more challenging than solvent ingress because of low concentrations and smaller differences in molecular mobility.

NMRI is a particularly powerful method for evaluating **diffusion processes** [663]. NMRI allows making continuous localised diffusion measurements without the need for interrupting the diffusion process or destroying the sample [627]. A true diffusion parameter image is obtained. The dynamics of the diffusion process may be followed and information on the kinetics of diffusion can be obtained. Diffusion coefficients can be quantitatively evaluated from the images recorded with different gradient field strengths. NMRI can be used to determine the mode of diffusion [661]. Diffusion of methanol into PMMA exhibits Case II diffusion and is well characterised by several techniques, including NMRI [664]. Perry *et al.* [665] have reported NMRI of the diffusion of acetone in PVC; penetrant, swollen and rigid polymer have been visualised simultaneously. NMRI has also been used to measure the penetration of acetates and alkenes into additive containing LDPE. ^{31}P NMRI can be used to gain information about the mobility of phosphorous containing additives and how this mobility changes with liquid uptake and the additive content of the polymer [666]. Motion of water in hydrogel polymers was studied by NMRI and ^{13}C l-NMR [667]. Most of the analytical methods for studying diffusion, with the exception of FTIR, cannot differentiate between two or more penetrants. NMRI allows the study of multicomponent diffusion utilising approaches based on differences in chemical shifts, relaxation times or isotopic labelling [632]. In the latter method, only one component generates a proton signal, as the other component is deuterated. Multicomponent diffusion experiments with PC/(acetone-d_6, MeOD) and PC/(acetone-MeOH-d_4) were reported [668]. The major limitation of NMRI for diffusion studies is the overall long measurement time, which restricts measurements to slowly diffusing systems. Long measurement times lead to motional artefacts and give no access to the study of fast chemical processes.

Stray-field MRI was used to measure methanol ingress into 500 μm thick PMMA pre-swollen with acetone [669]. With stray-field imaging the rigid and **swollen polymer** and the solvent are separately visualised with a resolution of the order of 20 μm. The different components are distinguished on the basis of their differing spin-spin relaxation times. For a polymer partially swollen with solvent the spatial distributions of relaxation times reveal the interactions between solvent and polymer in the diffusion process. Proton NMR images of 1,4-dioxane in swollen polybutadiene rubber were reported [393].

NMRI can potentially produce internal maps of chemical variations associated with internal homogeneities in solids: non-uniform filler dispersion, phase separation, interfaces, chemical reactions, physical ageing. Because NMRI allows obtaining the image of a slice of a polymeric sample, **internal imperfections** (voids, cracks, non-bonded regions, fibre- or resin-rich areas, resin structural defects) can be measured if they are larger than the resolution of the technique (currently >20 μm) [393]. Defects such as voids and inclusions are represented by very small image discontinuities. Swelling in a suitable solvent may enhance the visibility of defects. This approach to imaging provides the opportunity of optimising contrast in a sample-specific way by imaging the unswollen polymer, the swollen network, and different solvents with chemical-shift-selective excitation. Voids and cracks can most easily be visualised after soaking in water. Foams represent the ultimate in void content, and NMRI has been used to study the distribution of pores and their connectivity. Porosity, the volume fraction of an object that is empty space, can be determined by NMRI if the pore volume can be filled with an inert fluid that gives a strong NMRI signal. The distribution of pores in polyurethane foam has been imaged after filling the foam with water [670].

Elastomers constitute one of the industrially most relevant applications to NMRI. NMR is particularly useful for the study of elastomer networks, as the line widths of the proton resonances are narrow as the polymer is well above T_g. For elastomers, the proton-NMR line widths are not excessively broad (*ca.* 10 ms for T_2) and the resolution of the images is high (20 μm). For elastomers above T_g, where the line-width of the protons is about 2 kHz, conventional imaging methods work quite well. For more rigid samples, such as elastomers below T_g, where the linewidth is on the order of 30 kHz, more advanced techniques for imaging of rigid solids must be applied, such as imaging in combination with MAS [665]. Barth *et al.* [636] described magic-echo phase encoding solid imaging of rubber materials below T_g. Blümich *et al.* [671] have reported other applications of NMRI to elastomers.

Apart from detection and imaging of subsurface defects, NMRI also allows detection and characterisation of areas modified through introduction of foreign substances, such as additives, degradation products, and contaminants. In almost all NMRI experiments of technical elastomers inhomogeneities are detected. They are mainly due to voids, **filler agglomerations**, impurities or variations in cross-link densities. These may derive from mixing processes, vulcanisation, ageing or mechanical loading. Improper mixing of the many compounds composing technical rubbers (up to 30) leads to heterogeneities in the final product. NMRI has been useful in the determination of internal inhomogeneities arising from filler distribution, impurities, and gradients in cross-linking chemistry [672,673]. Blümich *et al.* [671] have identified filler defects derived from pressure overloading in a T_2 weighted spin-echo image of a carbon-black filled rubber gasket. PDMS reinforced by *in situ* precipitated silica was examined by NMRI [674,675]. NMRI has also been used for the study of carbon-black distribution in tyre composites [676]. The presence of carbon-black filler usually does not affect NMR analysis of the transverse relaxation decay and the longitudinal relaxation in the rotating frame for the measurement of cross-link density, in contrast to functionalised silicate filler [671]. Solvent absorption and swelling behaviour have been used to determine the cross-link density in elastomeric systems [659,677]. Blümler *et al.* [678] applied various NMRI techniques (spin-density and gradient-echo imaging, and T_2 projections) to study EPDM vulcanisates.

NMRI has also been used to study the **physical ageing** of cross-linked natural rubber filled with carbon-black [673]. The non-destructive character of NMRI allows monitoring changes in the material properties without impairing the sample during the analysis, which is of considerable importance especially for uniquely aged samples. The onset of physical ageing in natural rubber can be observed by NMRI after only two hours [632]. Physical ageing results in a change in the molecular mobility of polymer chains, and contrast is produced in the image, which increases with ageing. NMRI studies of the degradation of rubber tubing [679] and PE pipe [680] have been reported. Knörgen *et al.* [681] have described applications of NMRI to silica and carbon-black filled E-SBR (free radical polymerisation using an emulsifier) and S-SBR (polymerisation using a solvent), widely applied in tyres. Using Hahn spin-echoes the influence of filler material on the formation of ageing fronts (for ageing times of 0, 300 and 1070 min) could be visualised with a resolution of about $(100\ \mu m)^3$ voxel size. Carbon-black is more effective in preventing ageing fronts in comparison with silica.

Blümich *et al.* [671] have shown that relaxation measurements by the NMR-MOUSE are a valid alternative to relaxation measurements at homogeneous magnetic fields. The device allows scanning of the lateral surface heterogeneity of elastomeric materials [633]. Possible applications of the NMR-MOUSE for the characterisation of rubbery materials were demonstrated [682,683]. NMR-MOUSE measurements of tyre treads are non-destructive and can be carried out during tyre testing. Imaging with the NMR-MOUSE was illustrated for a rubber sheet with parallel textile fibres [671]. The NMR-MOUSE was also used for 1D imaging of stress whitening of a PS sheet. The NMR-MOUSE promises to be of use also in process and quality control of elastomers.

NMR techniques contribute to the development of numerical methods of **food packaging applications**. Greater understanding of the migration process would aid in controlling and limiting chemical contamination of food from packaging. NMRI is used to image the penetration of food or food simulant into the polymer. This provides spatially resolved quantitative measurement of the total mass uptake. The movement of the penetrant front can be followed *in situ* in real-time. PGSE-NMR can then be used to measure the steady-state self-diffusivity of the penetrating liquid within the polymer. Combination of NMRI and PGSE-NMR techniques allows imaging how the self-diffusivity of liquid within the polymer varies with liquid concentration. NMR can be used as a probe of small molecule mobility. This is particularly appropriate for the investigation of mobility and transport of species such as additives, by-products and monomers which contain phosphorous. Gladden *et al.* [684] used these techniques to identify different aspects of swelling and leaching, to probe quantitatively penetration of simulant into a polymer and migration of species from the polymer. Materials studied were HDPE/0.3% DLTDP, HDPE/1% DEHA, HDPE/0.5% Irganox 1010, HDPE/0.24% Irganox 1076 and HDPE/0.08% Irganox 1076. Molecular migration or diffusion of chemical agents in packaging materials is the rate-determining mechanism limiting the useful lifetime of the contents.

In contrast to the other spectroscopic mapping techniques such as IR and Raman spectroscopies, NMRI generates information about the relative mobility of chains in a polymer sample through contrast in the image due to contributions of T_1 and T_2. The T_2 relaxation time is sensitive to local motion of the nuclei. Generally, freely mobile molecules, with short correlation times, have long T_2 times, whereas motionally restricted or immobile molecules have long correlation times and short T_2 times. Thus, NMRI is particularly useful for processes in which large changes in molecular mobility occur. Although conventional imaging methods can be used to follow moisture migration during the early stages of drying, this is not true during the later stages, as the system becomes more solid-like and the transverse water proton relaxation times shorten to less than 1 ms. **Polymerisation** can be monitored *in situ* using relaxation times as contrast parameters. Polymerisation leads to a characteristic change in local motion, which induces a marked drop in T_2 as the polymer is formed [685]. NMRI has been used to examine benzoyl peroxide initiated methacrylic acid polymerisation [686], as well as vulcanisation processes [393]. It is also possible, using NMRI, to examine *in situ* the homogeneities and degree of curing for different **vulcanisation** formulations [687,688]. NMRI can be used to view the internal structure of adhesive bonds, which yields the potential of determining adhesion strength without destroying the bonds by testing [689]. McCarthy *et al.* [690] applied NMR imaging to the study of velocity profiles during **extrusion processing**.

Few applications of rigid-state NMRI have been reported, including the selective imaging of one component of a multicomponent blend [642]; the resolution of the image is on the order of tens of μm. Imaging of a mobile component within rigid solids, such as polymers, is relatively simple. Mobile species which can be imaged in high enough concentrations are **plasticisers**, waxes, and extrusion aids (EAs). Examples are a block copolymer of butadiene and styrene (butadiene acting as a plasticiser) and PE pipes (with large amounts of an organic lubricant as an extrusion aid). Imaging can be used as a quality check by monitoring the uniformity of EA distribution in different sections of a PE pipe [680]. Maas *et al.* [629] have described NMRI of fresh and aged solid rocket motor propellants, composed of 5 wt.% elastomeric binder material (cured hydroxy terminated polybutadiene, HTPB, plasticised with 12% DOP), highly filled with particulate oxidiser (ammonium sulfate/aluminium, 83 wt.%). NMRI is a useful tool for obtaining physical and chemical information about the binder distribution, *i.e.* that of HTPB and DOP. The technique is sufficiently mature to successfully tackle problems with length scales on the order of 10 to 100 μm. It should be realised, however, that fillers in rubbers are typically solids. Consequently, while the signal of the polymer is retained, the filler is invisible. The image intensity represents the rubber concentration, while the complement of the image represents the concentration of the filler.

Surface phenomena occurring as very thin layers (blooming, adhesion, *etc.*) cannot be observed with either T_2 or echo-imaging measurements. Similarly, since the spatial resolution of NMRI is relatively modest, the technique is not expected to be useful for studying morphologies of blends, TPVs or impact-modified thermoplastics. NMRI can be used to probe interfacial and interphase structure, *e.g.* in fibre-matrix composites [691]. NMRI also provides additional information on the microdynamic and structural properties of heterogeneous systems, such as sub-region diameters, exchange times, and phase boundary resistances [627]. The CYCLCROP imaging pulse sequence, employed in ^{13}C mapping studies, has the ability to probe the spatial distribution of one component out of a heterogeneous polymeric material (*e.g.* composed of a **blend** or mixture of polymer and additives, such as fillers). CYCLCROP imaging was first tested in a heterogeneous ^{13}C-enriched polymer system consisting of *cis*-polyisoprene (PI) and polybutadiene (PBD), as well as in a homogeneous blend of PI and polyhydroxyoctanoate (PHO). CYCLCROP has also demonstrated applicability for the acquisition of ^{13}C-edited images of natural abundance ^{13}C elastomeric materials (PI/rubber hose). Where CYCLCROP ^{13}C mapping of polymer blends has been reported no applications of polymer additives are known.

Every new NMRI application requires a significant degree of spectroscopic optimisation (*e.g.* rf and gradient pulse sequence design) and hardware optimisation (*e.g.* sample holding design). The unsurpassed soft-matter contrast of NMRI is hard to achieve with competitive methods like X-ray or computer tomography. The number of applications of NMRI to polymers is growing rapidly [627,643, 692–694].

More advantage will be taken of the ability of NMRI to investigate non-invasively the internal structure of "as is" polymer samples and to study the dynamic behaviour of such systems.

Applications of NMRI to polymer science were reviewed with examples for imaging of rigid, soft and fluid matter [635]. Further information on the applications of NMR imaging is available in several reviews [632,671] and books [393,627].

5.7.2. Electron Spin Resonance Imaging

Principles and Characteristics

ESR spectroscopy can be transformed into an imaging method for samples containing unpaired electron spins if the spectra are measured in the presence of magnetic field gradients. Herrling *et al.* [695] first devised ESR imaging (ESRI) with a modulated magnetic field gradient. This method is generally considered to be superior to the stationary field gradient method for overcoming the problem of hyperfine structure.

ESRI is a relatively new technique with unique capabilities to map the distribution of paramagnetic species in macroscopic systems. An ESR image is a representation of the spatial distribution of the ESR signal intensity in a heterogeneous sample. Various ESRI techniques have been reported such as spin-echo-detected imaging and spatial-spectral ESR imaging.

The principles of ESR and NMR imaging are similar: field gradients are used in the x-, y- and z-directions to allow a volume element to be selected. ESRI menus are 1D spatial (x, y, or z), 2D spatial-spatial (x, y plane) or 2D spatial-spectral. ESRI provides presence and concentration of a given free radical, symmetry of the electron environment, spatial mapping of free radicals and other paramagnetic species. Spatial imaging is suitable for a single component; spatial-spectral imaging is applicable to multiple components and yields spectral (lineshape) information. In **1D ESRI** experiments the concentration profile of the radicals is deduced from ESR spectra in the absence and presence of a gradient, as a function of time. Using two so-called anti-Helmotz coils with reversed currents ESRI experiments are performed by superimposing a constant gradient along the direction of the applied magnetic field of the ESR spectrometer. In the presence of such a **magnetic field gradient** the paramagnetic species placed between the coils do no longer resonate at the same value of the applied field, as in case of a conventional ESR experiment. In these conditions the resulting ESR spectrum consists of the convolution of many identical spectra having variable weight depending on the distribution of the radicals between the coils. When the ESRI experiment is performed on a sample containing one radical species showing a single narrow ESR line whose spectroscopic parameters (*i.e.* g-factor and line width) are independent on position and orientation of the radical, the integrated spectrum recorded in the presence of a field gradient provides the radical distribution. With complicated ESR spectra, as those given by nitroxyl radicals dissolved in polymeric samples, this is not true and therefore mathematical treatment of the spectrum recorded in the presence of a field gradient is requested to extract the **radical distribution function**. Simple and convenient 1D ESR imaging may provide valuable information about diffusion phenomena or kinetics of chemical reactions.

In **2D ESRI**, projections taken in a range of magnetic field gradients are used to reconstruct a 2D image that consists of the ESR spectrum along the chosen spatial coordinate. The method provides the concentration profile and the ESR line shape of the diffusant in each slice of the sample perpendicular to the direction of the gradient; determination of the translational and rotational diffusion rates in one experiment is therefore possible. Satisfactory imaging of most objects, however, requires the use of three-dimensional techniques. Lauterbur *et al.* [696] have reported practical **3D ESRI**, which allows unambiguous determination of the distribution of unpaired electrons in complex objects. The accuracy of the method is about ± 0.1 mm, which is sufficient for macroscopic samples.

Table 5.58 lists the **main characteristics** of ESRI. The basic requirement for an ESRI experiment is that a species having unpaired electrons be present in sufficient concentrations; this is in contrast to MRI where the ubiquitous proton can be used to study most materials. Fortunately, the high sensitivity of ESR compared with NMR (arising from the

Table 5.58. Main characteristics of ESRI

Advantages:
- Non-destructive (virtual) slicing
- High sensitivity
- Spatial profiling of radicals
- Unique information (diffusion and ageing phenomena)

Disadvantages:
- Need for very large magnetic field gradients
- Continuous-wave mode
- Limitations to sample size
- Limited applicability (radical species required)
- New technique, few practitioners

difference in electron and nuclear magnetogyric ratios) enables paramagnetic material to be studied in low concentrations. However, NMR has an advantage over ESR in that pulse techniques can be used to boost the signal-to-noise ratio: both the irradiating frequency and the applied magnetic field gradients can be pulsed. Most ESR experiments are carried out in the continuous-wave (CW) mode because spin-lattice relaxation times for paramagnetic materials are of the order of μsec (compared with hundreds of msec for ^{1}H NMR) and this causes considerable difficulties in achieving the short times required for the pulse experiment. Difficulties arise in ESR because the field gradients have to be much larger than in MRI (by a factor of 100 to 1000) since an ESR spectrum occupies a much larger frequency range than does an NMR spectrum. Moreover, the line widths are large (three orders of magnitude greater than in NMR). At frequencies of about 9 GHz specimens of maximum diameter of 10 mm can be examined; frequencies as low as 200 MHz permit samples as large as 100 mm to be investigated.

ESRI does not have the general applicability of NMRI because of the infrequent occurrence of unpaired electron species in useful concentrations. The stable nitroxide free radicals, however, are useful for ESRI of polymers because their distribution and kinetics and the shapes of their ESR spectra can provide information about processes in time and space, which is not easily obtainable by other techniques. ESR and 1D and 2D ESRI can be used to deduce morphology sensitive chemistry.

ESRI and its general applications (but not to polymers) were reviewed [697,698]. Two recent books deal with ESR imaging [699,700].

Applications

ESRI is important for the evaluation of transport properties of materials suitable in medical and industrial applications. Applications of ESR microscopy include the investigation of the swelling of polymers using solvents containing spin probes. In particular, the technique has been applied to investigate the reaction and diffusion of organic free radicals in polymers. For example, the diffusion into solid polymeric materials of nitroxide radicals dissolved in organic solvents has been analysed [701] and the diffusion coefficient of O_2 in PTFE (fluoroalkyl and peroxy radicals) has been determined [697]. 2D (spatial-spectral) ESRI can be used to deduce the spatial distribution and the dynamics of paramagnetic diffusants along a selected axis of the sample, as shown by Schlick *et al.* [702] in the determination of the translational **diffusion coefficient** D of various nitroxide spin probes, such as 4-trimethylamino-2,2,6,6-tetramethylpiperidine oxide iodide (TMATEMPOI).

Pedulli *et al.* [703–705] have investigated the **spatial distribution** of 2,2,6,6-tetramethyl-1-piperidinyloxyl (TEMPO) radicals in solutions and of Tinuvin 770 and other hindered amine stabilisers (HAS) in 2 mm thick PP plaques using X-band ESR-imaging. The photo-protective action of HAS involves oxidation of the amines to nitroxide radicals. Since the intermediate nitroxide radicals are very long-lived species, especially in a solid matrix such as that of the host polymer, they can be easily detected by ESR spectroscopy. 1D ESRI techniques provide information not only on the nature of the radical formed and on its concentration in the bulk but also on its distribution at various depths. For some HAS derived nitroxides a uniform radical distribution across the PP plaquette was observed after two months of UV exposure, as opposed to that after 5 months, when the nitroxide was mainly found near the external surfaces being almost absent in the centre of the plaquette (Fig. 5.20a). A strongly asymmetric **radical distribution** was also observed in irradiated samples of PP/(Tinuvin 770/328) (*cfr.* Fig. 5.20b) with nitroxide formation essentially only in proximity of the surface directly irradiated with UV light [704]. ESRI provides important information, not easily obtainable with other techniques, for a better understanding of the mechanism of protection of polymers by amine stabilisers and of their synergic interaction with other additives.

ESRI has been developed into a method for spatial and spectral profiling of radicals formed during polymer degradation. Schlick *et al.* [706–713] have reported extensive 1D (spatial) and 2D (spatial-spectral) ESRI in studies of diffusion processes in

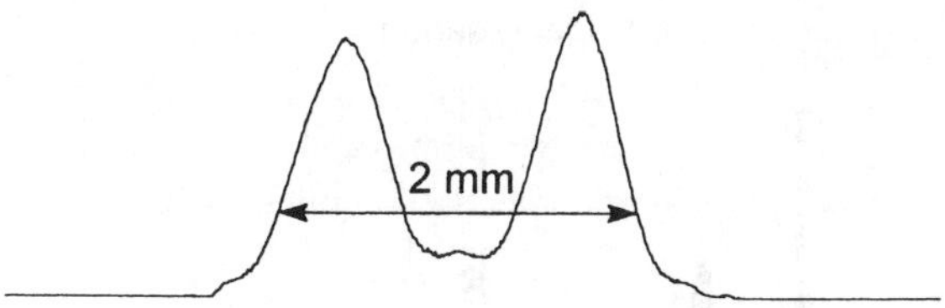

(a)

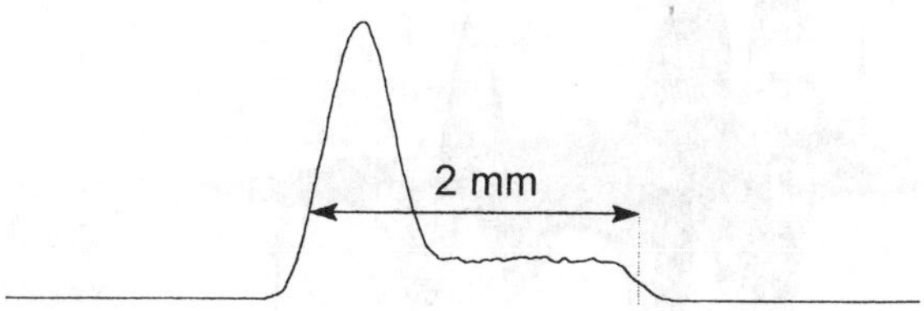

(b)

Fig. 5.20. Spatial distribution of the Tinuvin 770 nitroxyl radical across a 2 mm thick PP plaque irradiated for 5 months (a) and one-sided (left side) for 2 months (b). After Lucarini and Pedulli [704]. Reprinted from M. Lucarini and G.F. Pedulli, *Angew. Makromol. Chem.* **252**, 179–193 (1997). Copyright 1997 © Wiley-VCH. Reproduced with permission.

polymeric systems, in particular for the determination of the spatial distribution and dynamics of paramagnetic species in ion-containing polymers, polymer solutions, and cross-linked polymers swollen by solvents. In one of these ESRI investigations UV *vs.* thermal degradation of HAS (Tinuvin 770) stabilised ABS was studied by means of 1D and 2D spatial-spectral profiling of nitroxide radicals [706]. Spatial variation of the nitroxide intensity *and* of the line shapes was detected in the UV-irradiated samples. The nitroxide signal is strong on the irradiated side, increases with time on the opposite side, and is very weak in the sample interior (typical case of diffusion-limited oxidation). Nitroxides in the butadiene-rich domains are consumed rapidly on the irradiated side, and decrease to zero after 934 h of irradiation. By contrast, the radical concentration *and* the line shapes are spatially homogeneous in the polymer undergoing thermal degradation at 333 K. ESRI is capable of discriminating the early stages of thermal and UV degradation of polymers (Fig. 5.21). ESR provides details on early polymer degradation events and stabilisation that cannot be deduced from properties *averaged* over the entire sample.

1D ESRI can deduce the intensity of the HAS-derived nitroxide radicals along the UVB irradiation direction. 2D spatial-spectral ESRI can be used in order to follow non-destructively the spatial variation of the line widths and of the relative intensity of the various spectral components along the sample depth. Two spectral components in two different environments were identified. All spectra consist of a superposition of two nitroxide radicals differing in their dynamic properties: a "fast" component (F, width 32.2 G), and a "slow" component (S, width 64.2 G), assigned to the low-T_g polybutadiene-dominated domains ($T_g \approx 200$ K), and high-T_g domains dominated by polystyrene or polyacrylonitrile sequences ($T_g \approx 370$ K), respectively [710]. The intensity of the F component represents HAS-derived radicals located in *intact* (not degraded) butadiene-rich polymer domains of the polymer. In the butadiene-rich domains the nitroxides are consumed faster compared to other regions in the polymer. ESR, 1D and 2D ESRI can be used to deduce morphology-sensitive chemistry. The work demonstrates the power of spectral profiling: a non-destructive method to view both the spatial distribution of the radical intensity (by 1D and 2D ESRI), and the spatial variation of the line shapes (by 2D spatial-spectral ESRI).

In similar work the effects of UVB (290–320 nm) and a Xe arc were compared. A hierarchical variation of the HAS-derived nitroxide concentration was described: within morphological domains in ABS on the scale of a few μm and within the sample depth on the scale of mm [707]. The conclusions for ESRI on thermal degradation of ABS/Tinuvin 770 at 393K (*cfr.* Fig. 5.22) were substantiated by ATR-FTIR of the 500 μm thick outer layer of the polymer. The advantage of ESRI is the ability to provide mechanistic details on the *early* stages of the ageing process [708]. 1D ESRI has allowed visualisation of an outer layer of thickness of about 500 μm that is less degradable than the rest of the sample and is believed to be formed during sample preparation by injection moulding [709]. Because of diffusion-limited oxidation (DLO) ESRI is expected to fill a real need.

Also the spatial distribution of radicals formed in polymers after electron beam irradiation has been measured. ESRI studies of LDPE, PP and EPM oxidised in γ-irradiation conditions (2.5 Mrad) have indicated that Tinuvin 770 (HALS) is quickly oxidised by peroxy radicals and more slowly by hydroperoxides [714]. Sutcliffe [698] has reported an example of 2D spatial-spectral imaging of four specks of solid DPPH.

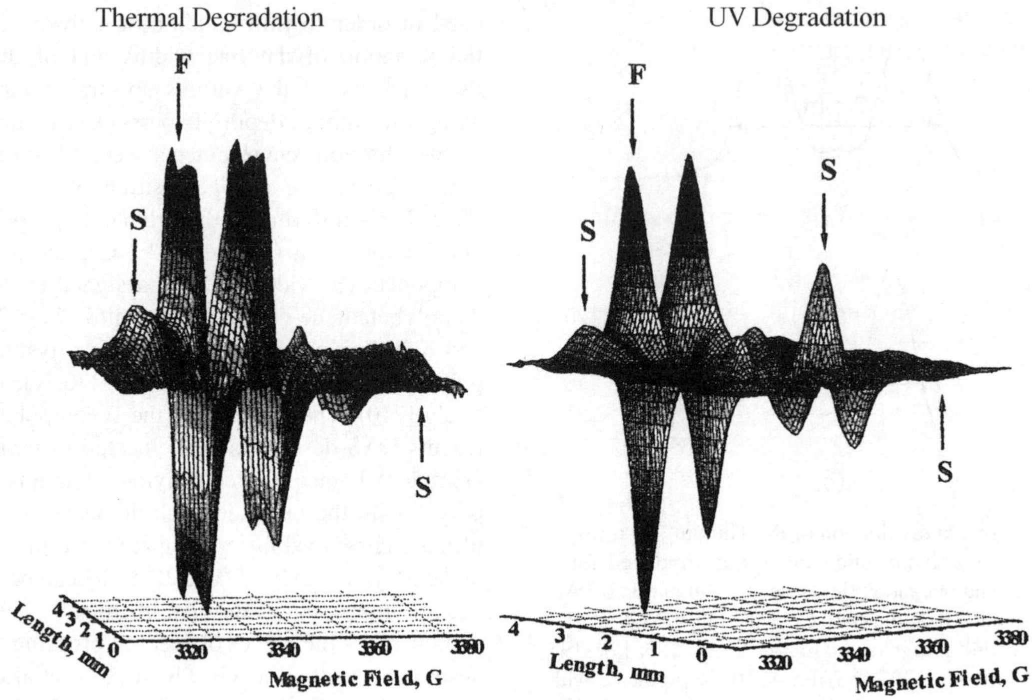

Fig. 5.21. 2D spatial-spectral ESR imaging of radicals in thermal and UV degradation of polymers. After Kruczala *et al.* [706]. Reprinted with permission from K. Kruczala *et al.*, *J. Phys. Chem.* **B104**, 3387–3392 (2000). Copyright (2000) American Chemical Society.

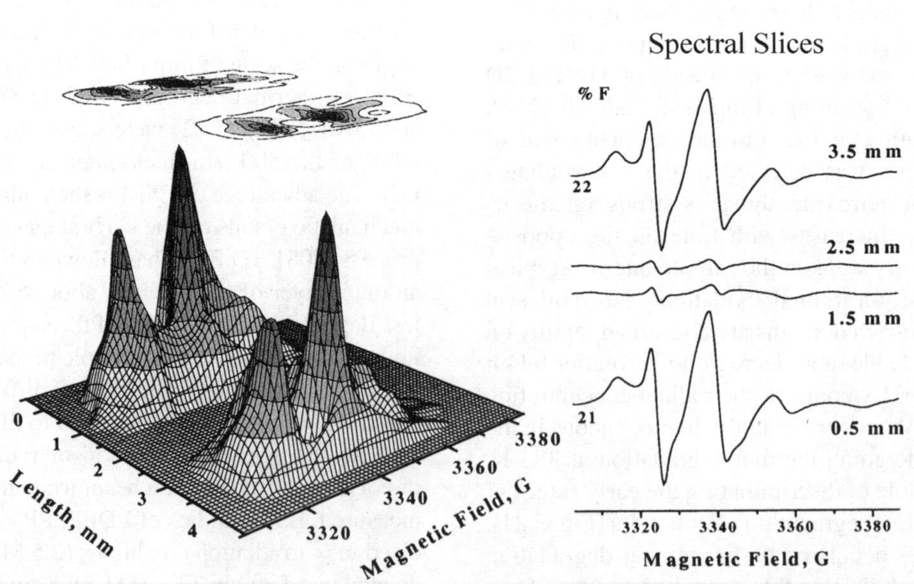

Fig. 5.22. 2D spatial-spectral contour (top) and perspective (bottom) plots for ABS/(2% Tinuvin 770) after 241 h of heat treatment at 393 K, presented in absorption. The spectral slices for the indicated depths in the perspective plot are presented in the derivative mode. Percentage nitroxides (% F) in low-T_g butadiene-rich domains are shown. After Schlick *et al.* [713]. Reproduced by permission of S. Schlick, University of Detroit.

5.8. X-RAY MICROSCOPY AND MICROSPECTROSCOPY

Principles and Characteristics

A wide variety of structural analysis tools is available employing X-rays (Table 5.59). Micro-focused X-ray sources allow small area spectroscopy, quantitative line scans and retrospective chemical state imaging based on high energy resolution spectra from user-defined areas. X-ray imaging complements electron imaging allowing to study relatively thick samples.

Spatially resolved X-ray microfluorescence (μXRF) represents **X-ray microspectroscopy**. Scanning X-ray microscopy at the sub-μm scale encompasses scanning transmission X-ray microscopy (STXM); STXM can be used to collect XAS spectra of micro domains. **Spectromicroscopy** refers to the combined use of selective energy imaging and spectroscopy at high spatial resolution. It is an integration of the spectroscopic and imaging aspects of analytical microscopy. While using many of the same concepts, spectromicroscopy is distinct from wavelength selective imaging and microprobe analysis (small spot spectroscopy) because it uses both the spatial and spectral domains to the fullest possible extent. Combination of high-resolution imaging with spectral information, such as NEXAFS spectromicroscopy, using soft X-rays, has been exploited in the characterisation of polymers [715]. Table 5.60 compares methods for X-ray analysis of small samples, up to nanoanalysis.

Problems do arise which cannot be satisfactorily solved by either light or electron microscopy, such as the measurement of the distribution of glass fibres, fillers or pigment agglomerates, and of pores in the bulk of strongly scattering or non-transparent plastics. Here it is often desirable to observe a large volume (general view) and being able to measure the additive or pore distribution at a microscopic level (detail). **X-ray microradiography** supplies such information and supplements other microscopic methods, including confocal scanning techniques. μXRD can be used to analyse very small amounts, *e.g.* of nucleating agents. Selected area diffraction (SAD) combined with microscopy is an important supplementary tool to X-ray diffraction in crystal structure analysis. SAD has the additional advantage of giving the correlation between morphology and crystal structure whenever single crystals are too small for single crystal X-ray analysis. Low amounts of crystalline components in polymer materials (notably pigments and contaminants) can be investigated by means of micro-WAXS using a Si-single crystal sample holder which is nearly background-free in the range of scattering angle $0° \leq 2\theta \leq 80°$ [716]. Qualitative analysis requires at least 0.1 mg; for quantitative analysis >1 mg is necessitated.

Table 5.59. X-ray microscopy tools

- X-ray microspectroscopy (μXRF, iXRF)
- Scanning transmission X-ray microscopy (STXM)
- X-ray spectromicroscopy (μXAS, μNEXAFS, SEM-EDS, TEM-EDS)
- X-ray microradiography (CMR, μCT)
- Microdiffraction (μXRD)
- Nuclear microscopy (μPIXE, iPIXE)
- Micro X-ray photoelectron spectroscopy (μXPS, iXPS)
- X-ray photoemission electron microscopy (XPEEM)

Although the intrinsic advantages of X-rays for elemental mapping and chemical-state imaging have long been recognised [717–719], their full potential for imaging could not be realised until (third-generation) high-brilliance synchrotron X-ray sources and high-performance X-ray microfocusing optics were developed. Synchrotron radiation scanning microprobes now achieve sub-μm spatial resolving power (focal spot of 0.25 μm, photon flux density 5×10^{10}/sec/μm^2/0.01% BW) and allow simultaneous performance of X-ray fluorescence microscopy (XFM), spectromicroscopy and 3D tomography [720]. Several X-ray microprobe SR beamlines are equipped with spatially resolved XRF (microspectroscopy) and spatially resolved XAS (spectromicroscopy) in areas as small as a few μm^2 [721,722]. **Nuclear microscopy** (or iPIXE) with its spatially resolved X-ray spectrum yields information on multi-(trace) element distribution, composition, with point- and line-scan, mapping and microtomographic analysis (*cfr.* also Chp. 8.4.2 of ref. [77a]).

Radiation damage limits probing specimens using ionising radiation and is especially significant for studies in which multiple images must be taken of the same specimen, such as for spectroscopic imaging of chemical states [715]. Structural damage caused by ionising radiation is reduced at cryo temperatures owing to reduced quantum yield for ionisation of chemical bonds.

Several reviews deal with X-ray microscopy [723, 723a] and X-ray spectromicroscopy [724]; recent books are available [725,726]. A special issue is devoted to spectromicroscopy [727].

Table 5.60. Methods for X-ray analysis of small samples

Method	Spatial resolution	Required sample mass for quantitative analysis	Required sample mass for qualitative analysis	Information
SEM-EDS	1 μm	–	10^{-12} g	Spatial distribution of elements
TEM-EDS	0.01 μm	–	10^{-15} g	Phase composition[a]
Micro-WAXS	[b]	10^{-3} g	10^{-4} g	Phase composition, crystallite size
μXRF	10 μm[c]	0.2 g	10^{-3} g	Elemental composition
μXPS	Several atomic layers	–	–	Surface composition

[a] Thin samples only.
[b] Not relevant.
[c] 1 μm for μSR-XRF.

Applications
Application of μWAXS requires separation of the additive from the polymer matrix first. μWAXS was illustrated for PBT/Fe_2O_3 (hematite) and PBT/TiO_2 (rutile and anatase) and PVC containing $Mg(OH)_2$-brucite and MgO-periclase as contaminants [716].

5.8.1. X-ray Microradiography

Principles and Characteristics
X-ray microradiography distinguishes projection microradiography and contact microradiography. In **projection microradiography** [728] X-rays produced by a 1 μm thin metal foil serving as an anticathode are transmitted by the sample placed under the metal foil and a magnified image is produced. **Contact microradiography** (CMR) employs commercial fine focusing tubes. The anticathode material and the beam voltage are chosen such that the excited X-rays are maximally absorbed by the element to be detected in the sample. The object is in direct contact with a fine-grained photographic emulsion. The advantage of contact microradiography is the possibility of investigating larger sample areas (*e.g.* up to 600 × 600 mm^2). The sample thickness has to be such that X-rays are still capable of being transmitted: it depends on the type and concentration of the incorporated additives but it can be as much as several millimetres. In the case of glass fibre reinforced plastics (30–40 wt.% glass fibre content), a sample thickness of 100–150 μm is optimal.

X-ray microscopy denotes a form of projection radiography that employs low-energy X-ray photons emitted from a point source (1 μm) to generate high-resolution images. The energy of the electron beam that is focused onto the target material to generate the X-ray source is typically <10 keV and is generally lower than is conventional either in industrial contact radiography or in microfocal radiography (15 μm X-ray source diameter).

Table 5.61. Main characteristics of X-ray microradiography

Advantages:
- No sample preparation
- Non-destructive examination of internal structures
- Fast

Disadvantages:
- No specific element imaging
- Limitations on sample thickness
- Restricted resolving power (5 μm)

Table 5.61 lists the **main features** of microfocal X-radiography. Microradiography is useful for investigating a material's interior structure that is hidden from sight, *i.e.* beneath the surface. The sample does not have to be specially prepared. Sectioning is not required. In standard X-radiographic images, the features observed are caused by absorption of X-rays by all the elements present in the sample. The inability to image specific elements is a major limitation of conventional X-radiography. Element-specific imaging can be achieved by taking "soft" and "hard" images at either side of the discontinuity in the X-ray absorption spectrum of the selected element (selective imaging). X-ray phase contrast imaging provides a particularly effective method for imaging the sample and locating suitable regions of interest for fluorescence mapping and μXANES measurements. In recent years, soft X-ray (<1 keV) microscopy has been successfully used in particular for the study of biological samples [720].

Important features in microscopy are *in situ* methods, quantitative interpretation of the object microstructure and the definition of 3D information. Of the techniques available to the microscopist today, only transmission X-ray microscopy gives nondestructive high-resolution information from the internal structure of an object under natural conditions. By combining the X-ray transmission technique with tomographical reconstruction 3D information about the internal microstructure can be derived [729]. **X-ray microtomography** (μCT) requires an X-ray microscanner (8 nm spot size), precision object manipulator, X-ray CCD camera, and microtomographical data processing [730].

X-ray projection microscopy and X-ray microtomography were reviewed [729]. The performance of the technique, with CCD detectors and hard X-rays was illustrated.

Applications

X-ray microscopy (XRM) allows non-destructive investigation of the micro-structure (fractures) of plastics, paints, adhesives, and inks. Coatings on surfaces and fibres within composite structures may be studied. Examples of X-ray micrography are the observations of inclusions in paint and ink coatings and surfaces of painted substrates.

Kämpf [135] has reported projection microradiographs of GFR PE foam (sample thickness 3 mm) and contact microradiographs of GFR PBT (sample thickness 150 μm), which give clear information concerning the local concentration, glass fibre orientation and possible glass fibre damage during compounding. Pigment size particles can be observed in paint, adhesives and inks. Similarly, in pigmented and filled plastics details regarding the coarse distribution and possible agglomeration of additives can be revealed. Test methods for assessing **pigment dispersibility** are CMR, SEM and XRF. In CMR pigmented plastic plaques are placed in contact with a film and exposed to X-rays at some distance from the source to produce microradiographs. CMR can be used to investigate the cause of poor dispersion and show whether it is due to undispersed pigment, poor distribution of masterbatch or polymer gels. However, owing to the restricted resolving power of X-ray microradiography, recognition of the primary particles (*e.g.* TiO_2 pigments of *ca.* 0.2 μm) is not possible; this is the domain of SEM. SEM is best used for systems having very good dispersion. It does not give an integral view through the sample but looks only at the surface. This necessitates taking a thin section of the sample or, with thicker sections, etching the surface with microwave-excited oxygen. A much smaller area is examined than with CMR and therefore SEM is less representative. SEM and X-ray microradiography complement each other.

Tailoring the energies of X-rays to optimise contrast in the images makes microfocal X-radiography ideal for the detection of features in a range of engineering materials applications, such as voids in plastic mouldings. Noda *et al.* [731] described the application of laser plasma soft X-ray contact imaging of ABS and PVC composites.

X-ray microtomography has been used for nondestructive imaging of defects (defectoscopy) and imaging of **composite materials** (*e.g.* fibre reinforced plastic foam) [730]. These examples are results of conventional X-ray imaging with reconstructions based on the density of the object; phase-contrast tomography offers additional possibilities.

Autoradiography was used to show non-uniform distributions of radiolabelled additives in PE, PP and PS [732]. XRM and high-resolution (to 5 μm) μCT have allowed 2D and 3D imaging of the non-uniform void and silica-supported chromium catalyst fragment distribution within PE particles [732a].

Defects and welding faults in 6 mm thick PE pipes for natural gas and water distribution may also be examined by ^{75}Se γ-ray radiography [733].

5.8.2. Scanning X-ray Microscopy

Principles and Characteristics

Images for spectromicroscopy are recorded at various energies selected in order to differentiate the chemical components of a system. In **scanning transmission X-ray microscopy** (STXM) light is focused to 20–50 nm and high-resolution images are taken by raster scanning the sample through the fixed focal spot while recording the intensity of the transmitted light. Cryo STXM using soft X-rays has been described [734]. Soft X-rays are well suited for studies of the chemical bonding state of major low-Z constituents in organic specimens. There are X-ray microscopes in both the soft X-ray ($<$1500 eV) and hard X-ray regimes ($>$1500 eV) at many synchrotron stations. Synchrotron radiation scanning X-ray microscopy allows imaging XRF, microdiffraction and micro NEXAFS (near-edge X-ray absorption fine structure) measurements [735]. SR-based X-ray microscopy in various implementations is an excellent example of spectromicroscopy.

Recently there has been considerable activity in developing inner-shell excitation spectroscopy as a high spatial resolution analytical technique, such as NEXAFS in an STXM [736] or EELS (electron energy loss spectroscopy) in a TEM [245,737]. The spatial resolution of **NEXAFS microscopy** is much lower than that obtained with TEM-EELS. EELS can be used to assess the same near-edge structures probed with NEXAFS but radiation damage in EELS is some three orders of magnitude higher than in NEXAFS spectroscopy. STXM can be used to collect NEXAFS spectra from 0.01 μm^2 regions of organic specimens [715]. In this way images with chemical specificity are obtained without aggressive staining techniques. X-ray spectromicroscopy can provide information on the spatial distribution, oxidation state, chemical environment, and chemical transformations of trace elements [717,718].

Hitchcock [738] has recently described techniques in which tuneable soft X-rays are used to provide chemical mapping via X-ray absorption spectroscopy at spatial resolution of more than 100 nm. There are a number of advantages to using soft X-rays (1–12 nm) rather than UV/VIS (180–780 nm) or hard X-rays (50–200 pm): *(i)* at the diffraction limit, short wavelengths give higher spatial resolution (50–100 nm) than longer wavelengths (1 μm for hard X-rays); *(ii)* no need for staining or fluorescent probes; *(iii)* the existence of high-contrast core edge for virtually all elements; and *(iv)* higher spectral resolution (~0.1 eV) in comparison to hard X-rays (1 eV). This enables mapping of chemical species on the basis of bonding structure rather than simply elemental content. **Soft X-ray spectromicroscopy** can distinguish very similar species, such as PE and PP, because they have small but distinct differences in their C1*s* X-ray absorption near-edge spectra (XANES). XANES or NEXAFS microscopy is a relatively new technique. Transmission NEXAFS microscopy, first shown in 1992 [715], combines a relatively high spatial resolution (about 50 nm) and low beam damage with quantitative compositional sensitivity, offered by NEXAFS spectroscopy.

Table 5.62 lists the **main characteristics** of NEXAFS microscopy. Images are obtained by rastering the sample across the X-ray focus, while keeping the high-resolution monochromator fixed. Keeping the sample stationary and tuning the monochromator across the spectral range allows obtaining NEXAFS spectra with a spectral resolution typically of about 0.1 to 0.3 eV. NEXAFS microscopy provides images with contrast that is directly proportional to the concentration of the various chemical components. NEXAFS microscopy is considered to be particularly well suited for characterisation of multicomponent polymer systems and interfaces in polymer coatings, blends and composites, and can provide spectra across interfacial regions at high spatial resolution. Principles and characteristics of X-ray absorption spectroscopy are described in ref. [717] and will not be repeated here.

Table 5.62. Main characteristics of NEXAFS microscopy

Advantages:
- Non-destructive
- No sample preparation
- Chemical sensitive X-ray imaging
- High spectral resolution (0.1 to 0.3 eV)
- Qualitative and quantitative spectromicroscopy
- Fast
- Low beam damage

Disadvantage:
- Fairly low spatial resolution (*ca.* 50 nm)

NEXAFS microscopy is complementary to high "chemical content" microscopies, such as NMRI, ESRI, μFTIR, μRaman, and high "spatial resolution" microscopies, such as various electron microscopies. While AFM is surface-sensitive, STXM images are bulk-sensitive.

Applications

Scanning transmission X-ray microscopy has been used most extensively for polymer research, *e.g.* for bulk characterisation of polymeric materials with chemical sensitivity at a spatial resolution of 50 nm [739]. STXM has also been used for the analysis (morphology, size distributions, spatial distributions and quantitative chemical compositions) of copolymer polyol-reinforcing particles in polyurethane [740]. Pitkethly [741] has reviewed the role of microscopy in the evaluation of fibre/matrix interfacial properties and micromechanical characteristics of fibre-reinforced plastic composites.

The value of direct chemical-state sensitive NEXAFS type imaging of phase distributions in polymer blends is well known [715]. NEXAFS spectroscopy has been used for blend studies [736] as well as for the quantification of composition in heterogeneous polymers [742]. Characterisation of polymer interfaces is an important analytical need in many areas of technology. STEM and NEXAFS were used in a study of the distribution of carbon-black and silica fillers in tyre compounds based on blends of a brominated isobutylene–methylstyrene copolymer and polybutadiene [743].

Ade *et al.* [11] have studied a multilayer laminate composed of PET/0.3 μm PU/1.0 μm SAN/0.8 μm CB-PVA (CB is carbon-black) by means of NEXAFS microscopy using an STXM at the National Synchrotron Light Source (NSLS) at Brookhaven National Laboratory (BNL) with the object of examining the extent of interpenetration of the SAN layer into the porous CB-PVA layer. Each of the four principle layers has a distinctly different NEXAFS spectrum (280–310 eV range).

A demonstration of speciation on a sub-μm spatial scale by NEXAFS spectromicroscopy has been presented [736]. The presence of a strong pre-peak at 5.995 keV in Cr K-edge absorption spectra allowed detection of Cr^{6+} bearing compositions using NEXAFS imaging [735]. By virtue of this effect it was possible to detect and image the heterogeneous distribution of Cr valence states by collecting image series at different strategic energies around the absorption edge.

5.8.3. X-ray Microfluorescence

Principles and Characteristics

Microfluorescence handles very small analysed areas and yields spatially resolved information of the sample composition. There are several technical solutions for obtaining a primary X-ray beam with a small diameter and sufficient intensity, from capillary optics and refractive lenses to highly sophisticated X-ray optical elements with focusing characteristics [744]. Modern compact and mobile μEDXRF spectrometers with Si drift detection for quality control, material testing, and *in situ* art and archaeometry applications nowadays offer down to 50 μm minimum focal spot size, $<$160 eV energy resolution and detection limits of the order of 20 ppm [745–747]. Using capillary optics it is possible to focus the X-ray beam of μXRF to 10 to 100 μm^2 areas.

High-density X-ray optics, CCD video imaging cameras and motorised *xyz* stage allow non-destructive, simultaneous Na through U analysis of a wide variety of solid objects, powders and liquids. **Microfluorescence imaging** using an energy-resolving detector provides a means of characterising the chemical composition of a bulk sample, offering benefits in spatial resolution and/or sensitivity compared with competitor techniques such as EDS, PIXE or SIMS. iXRF elemental maps with sub-mm spatial resolution may be obtained without moving parts using microchannel plate X-ray optics [748]. By using appropriate energy windowing around the fluorescence emission peaks it is possible to map the local distribution of several elements simultaneously.

Table 5.63. Main features of X-ray microfluorescence

Advantages:
- No sample preparation
- No limitations as to shape, diameter of the sample
- Small sample weight (mg)
- Non-destructive analysis
- Spatially-resolved qualitative and quantitative multi-element analysis (Na to U) of $>$100 μm particles or domains
- Fast
- Wide applicability range (polymers, coatings, paints, suspensions, *etc.*)
- From commercial (portable) equipment to use of synchrotron radiation

Disadvantages:
- Limited resolution (8–10 μm) for microfocus tubes
- No chemical state information

Table 5.63 shows the **main characteristics** of μXRF. Calibration standards are available for micro sample X-ray analysis (including 35 elements from Na to Bi). μXRF is competing with SEM-EDS. The primary differences of SEM-EDS with μXRF are the use of X-rays as the excitation source and large X-ray beam spot sizes, typically greater than 30 μm. The trade-off in using X-rays is greater depth penetration into a specimen compared to electron penetration. This offers greater elemental survey capability of the bulk material. On the other hand, the lower resolution provides for larger areas to be imaged than is possible with electron beams, in some cases up to 15 cm^2. The sensitivity of μXRF (down to a few ppm) is considerably better than EDS. X-ray microfluorescence (XRMF) or microscopic XRF can also be carried out in synchrotron microanalysis mode; μSR-XRF allows simultaneous mapping of multi-element distributions with high spatial resolution ($\sim$1 μm) and orders of magnitude higher sensitivity than other typical characterisation methods such as EDS, SIMS or AES; elemental LODs of 10–100 ppb are possible [720,749,750].

In comparison with electron and proton microprobes, the cross-sections of XRF excited by X-rays are typically 10 to 10^3 times higher than those excited by charged particles, and the fluorescence signal-to-background ratios are 10 to 10^5 times better for excitation by X-rays [718]. Although the spatial resolution of electron and proton microprobes can approach molecular dimensions and is better than that achievable with X-rays, the elemental sensitivity is limited to approximately 100 ppm for electron-induced X-ray microanalysis and 10 ppm for proton-induced X-ray microanalysis, considerably worse than with X-rays [719]. In addition, the energy deposition for X-rays is 10^3 to 10^5 times smaller for a given elemental detectability, resulting in substantially less radiation damage to the sample. Moreover, sample preparations for **X-ray microprobes** are far simpler than those for charged-particle microprobes [720]. Although XRF images provide detailed information on the spatial distributions of selected elements, they provide no information on the chemical states or local environments of the elements.

Advantages and pitfalls of several μXRF techniques were recently presented [751]. A review on space-resolved μXRF has appeared [744]; for further reference on micro-XRF, *cfr.* ref. [752]. A special issue has been dedicated to μXRF analysis [753].

Applications

X-ray microfluorescence can detect and identify a wide variety of flaws, as indicated in Table 5.64. Contaminants in plastic products are a common problem and one that adversely affects electrical, optical and mechanical properties of materials. Inclusion analysis comprises the distribution of catalyst residues, agglomerates in film and granulate, film contaminations, identification of pigment particles or glass fibres, metal particles in plastic granulate, inorganic or mineral inclusions that may cause holes in films or bottles. Elemental distribution and chemical state of ppm metal impurities can be measured using synchrotron-based X-ray fluorescence (μXRF) and X-ray absorption spectroscopy (μXAS), both with a few μm^2 spatial resolution [754]. Fine particle analysis (>100 μm) may allow detection of additives deposited on granule surfaces. μXRF can be used in failure and distribution analysis (linescan, mapping) of inorganic additives in plastic end-products (*e.g.* imaging of K, I, Cu and Fe in fibres). The TiO_2 pigment distribution along cross-sections of injected iPP samples aged for 515 and 3000 h has been determined using a 20 μm X-ray microbeam in order to understand the whitening process [755].

Table 5.64. Applications of μXRF

• Material testing	• Failure analysis
• Inclusion analysis, impurity mapping	• Microelectronics
• Fine particle analysis	• Forensic science
• Multi point analysis, destribution analysis	• Art and archaeology
• Quality control	

Wegrzynek *et al.* [756] have dealt with the quantitation problem when an X-ray microbeam is used to measure elemental distributions in a low-Z matrix. This μXRF technique was developed for the characterisation of sample homogeneity in a polymer matrix. Microhomogeneity studies using μSR-XRF and LA-ICP-MS on CRM BCR 680 (*cfr.* Chp. 8.3) were reported with satisfactory agreement between the sets of data [757,758].

Other application areas of non-destructive μXRF are in forensic science (spectral fingerprint), microelectronics (uniformity of deposited films – thickness and composition), and in art and archaeology (especially using handheld equipment). Historic iron gall inks used in handwritten manuscripts by Bach, Mozart and Goethe were characterised by μXRF [759]. The technique has developed into a significant analytical tool for authentication studies of artefacts.

5.8.4. Micro X-ray Photoelectron Spectroscopy

Principles and Characteristics

An imaging extension of XPS is not easily achievable or trivial, as the impinging X-rays causing photoelectrons cannot be focused or deflected by electric or magnetic fields. This imposes restrictions on the image-forming capabilities of XPS instruments, and has led to several experimental designs. As well as the imaging electron optics, other key factors in allowing progress in small spot – or imaging – XPS have been the availability of a sufficient flux of X-rays by improvements in monochromator design and the use of position sensitive detectors.

Consequently, XPS has developed from a large area analysis method to one which has some degree of spatial resolution (selected area analysis). There are essentially only two ways in which such an improvement can be obtained, operating the spectrometer in a microprobe mode, in which the X-ray beam is reduced in dimensions (the so-called defined source system), and modification of the electron collection optics (often referred to as the defined collection system). In the first category it is possible to produce a microfocus monochromator. With a highly focused electron source the X-ray spot size on the sample may be as low as 10 μm (for Al $K\alpha$ radiation). This class of spectrometer provides **selected area XPS** (SAX) spectra. Alternatively, using a conventional flood X-ray source, the analysed area is selected by limitation of the area from which the electrons are detected. This is done by an electron optical lens system [760,761]. The defined (small) area can be ~10 μm in diameter.

Various approaches to chemically specific images (intensity as a function of position) are used commercially: *(i)* scanning of a focused electron beam [762]; *(ii)* moving the defined area [763]; *(iii)* parallel imaging (VG ESCAscope) [764]; and *(iv)* operation in microanalysis mode (full spectrum from one pixel) and physically stepping the sample [765], but several other designs have been proposed [766–768]. In all cases, **chemically specific images** are acquired by tuning the analyser to pass electrons from a peak in the photoelectron spectrum, which represents the element of interest. In the parallel imaging system a spatial resolution of ~2 μm is claimed with acquisition times of a few minutes only [764]. By operating in microanalysis mode (*i.e.* full spectrum from one pixel) and physically stepping the sample, a complete image set can be acquired covering eventually the whole photoelectron spectrum. The small spot capabilities have led to μXPS, also termed **X-ray photoelectron microscopy** or **imaging XPS**.

Historically, the lack of imaging capabilities has hampered polymer characterisation by XPS. The combination of microscopy and spectroscopy has been the goal of a number of groups exploring photoelectron microscopy with X-ray or synchrotron radiation sources. The first real step towards imaging XPS (iXPS) was in 1988 (VG ESCAscope). The system allowed obtaining 2D spatial maps with a lateral resolution of <10 μm. The second generation of this instrument achieved a spatial resolution of approximately 2 μm [764,769].

Table 5.65 lists the **main characteristics** of μXPS. The lateral resolution by XPS is up to 100 times lower than the corresponding resolution obtained by AES-microprobe and SIMS. The lateral resolution of imaging XPS appears quite limited also in comparison to AFM with designed chemical tips,

Table 5.65. Main characteristics of micro XPS

Advantages:
- Selected area (local) spectroscopy
- Retrospective chemical state imaging (elemental and bonding)
- High sensitivity at small spot size
- Imaging of insulators
- Quantitative line scans, mapping
- Low sample damage
- Commercial equipment

Disadvantages:
- Limited spatial resolution (<5 μm)
- Charging effects (polymer surfaces)
- Slow process (unless parallel data acquisition)

especially on polymer surfaces because of charging effects. However, an advantage of a micro-focused X-ray source is that it facilitates positioning and collection of data from areas a small as 10 μm. Imaging XPS allows small area spectroscopy, quantitative line scans and retrospective chemical state imaging based on high energy resolution spectra from user defined areas [770].

For further information on microprobe XPS, *cfr.* refs. [768,771].

Applications

Important and straightforward industrial applications of μXPS with micro-focused X-ray sources are the determination of the diffusion characteristics of protective coatings or paint systems and the characterisation of multilayer packaging systems [770]. Characterisation of such systems by vibrational spectroscopic techniques is often made difficult by the presence of inorganic particles. μXPS has the spatial resolution needed for analysis of thin multilayer structures.

XPS techniques (surface mapping, ADXPS, quantitative XPS) were recently used for understanding slip and antiblocking action of ethoxylated oleyl amine (Armostat 710) in LDPE. Surface migration of oleamide and stearamide in LDPE/0.3% Armoslip CP (oleamide) and LDPE/0.3% Armoslip 18LF (stearamide) with time was followed by surface mapping on a 50 μm point-to-point comparison [772]. Gerlock *et al.* [773] applied XPS mapping to study the interfacial corrosion chemistry of an epoxy adhesive applied to galvanised steel.

There exists a need within polymer research, specifically with respect to coatings and adhesives, to attain molecular information at both microscopic

Table 5.66. Spatial characterisation of paint systems

Feature	Technique			
	SAM	DSIMS	SSIMS	iXPS
Lateral resolution	Sub-μm	Sub-μm	Sub-μm	10 μm
Sample charging	+	+	−	±
Beam induced damage	+	+	−	−
Molecular specificity	−	+	+	±
Atomic specificity	+	+	+	+

and macroscopic resolution. Spatial characterisation of interfaces may be carried out by SAM (scanning Auger microscopy), DSIMS, imaging SSIMS and iXPS (*cfr.* Table 5.66). Some techniques afford sub-μm lateral resolution (SAM, DSIMS, SSIMS), some pose distinct disadvantages with respect to sample charging (SAM, DSIMS) and primary beam induced damage of organic materials (SAM, DSIMS).

iXPS has been used for the differentiation of untreated and treated 6 μm carbon fibres in composites research using the chemical shift of C1*s* electrons between C−C and C=O components (about 3.6 eV) [774]. With an improved spatial resolution (<10 μm) this technique is suitable for imaging of fracture surfaces of carbon fibre reinforced polymer (CFRP) composite material.

Newer analytical tools such as iXPS and PA-FTIR can be used for surface mapping of polymers containing blooming additives [775]. These tools can be used to understand the mechanism of migration in addition to the rate of blooming.

5.9. ION IMAGING OF ADDITIVES

Characterisation of micro areas on a larger substrate commonly has been performed by electron or optical microscopy. Although these techniques provide valuable topographical and morphological information only limited chemical information is available. As shown before (Chps. 5.6–8), this can be remedied by the combination of microscopy and spectroscopy. Various microprobe techniques, including photon probes (μUV/VIS, μFTIR, μRS), X-ray probes (μXRF, μXPS, μCT), electron probes (EPMA, AES), proton probes (iPIXE) and magnetic probes (NMRI, ESRI) provide chemical information about small domains in solid structures. The spatial resolution of many of these techniques typically ranges from 1–100 μm. An alternative is use of ion probes, such as iSIMS or LMMS. Imaging MALDI-MS [775a], commercially available since 2004, shows great promise for additive distribution studies. The goal of **ion microprobe/microscopy** is the chemical analysis of unknown microstructures. In case of spatially resolved MS the limit is usually related to the diameter of the ionising beam. Quantitative evaluation of ion microscope images was discussed [776].

5.9.1. Laser-microprobe Mapping

Principles and Characteristics

Laser techniques deposit large amounts of energy at the sample surface with consequent generation of high mass organic ions characteristic of the sample. Additional benefits of laser sources are spatial resolution, surface and depth profiling capabilities and the potential for mixture analysis. Laser-based methods can be used for the direct determination of additives in a complex matrix (*cfr.* Chp. 3). It is now common to have a CCD camera and video display that gives a microscopic view of a sample when it is in the mass spectrometer. This allows contaminants, defects and areas of interest to be observed and manipulated while under the probing beam of the laser. The **spatial resolution**, defined by the spot size of the laser beam, is far higher than for dissection. Spatial resolutions of a few tens of μm have been achieved using IR laser desorption [777], while a resolution of 1 μm has been reached using a UV waveguide excimer laser for desorption [778]. This capacity for high spatial resolution gives L^2ToFMS the potential for important microanalytical tasks, such as the 2D mapping of molecular adsorbates on a wide range of substrates (*e.g.* blooming problems) and the examination of individual particles (troubleshooting applications). Moreover, the method may almost be considered "non-destructive" for the sample as only some hundreds of ng of material are taken away during the desorption.

An FTIR imaging system can be coupled to spatially resolved (UV or IR) LD-ITMS. The functional group mapping of the FTIR will serve to localise a region of interest. The LD laser beam may be targeted on it and MS analysis will confirm identity and structure of the desorbed area. These combined techniques can provide functional group distributions and detailed chemical information of selected areas. Application to additive distributions, and detection of inclusions and trace contaminants in polymers may be envisaged. However, application requires fine-tuning of the experimental conditions in order to selectively detect the additive fragments before breakdown of the (excess of) polymeric material. Migration of components within and between layers may also be researched by a combination of **LDMS and FTIR imaging**.

Applications
Detailed additive analysis over very small areas and film depths may be needed to gain insight in additive migration, *e.g.* in studies on ageing and metal-catalysed thermal degradation [779]. Apart from the direct determination of additives in a complex matrix, laser-based methods can be used to probe surface and interstitial contaminants by desorbing neutral molecules directly from the embedded pit. In the area of industrial troubleshooting the analysis of "**pitting**" is a complex problem because of the embedding of the impurity (additive or not) in the polymeric matrix. The amount of sample used and the spatial resolution of the analysis depend utterly on the physical dissection skills.

LMMS has been used to produce molecular maps with μm resolution of triphenylmethane dyes (M^+ (m/z) 385, 372, 288, 470) [780]. Similar work [781] shows that the laser microprobe is capable of providing both the chemical indentity and location of organic material. LMMS has also allowed detection of each component in multicomponent dye samples containing methylene blue, methyl red, methyl orange, phenol red and/or FD&C Yellow 5 [782].

Two-step laser mass spectrometry has been employed by Zenobi *et al.* [783] for direct **spatially resolved *in situ* analysis** of a variety of additives in different polymers (*cfr.* also Chp. 3.4.3). Sheng *et al.* [784] have proved that direct laser probe FTMS analysis of industrial samples can rapidly determine many additives directly, even when the combination of laborious classical wet chemical techniques with other modern instrumental methods is both difficult and time-consuming.

5.9.2. Imaging Secondary Ion Mass Spectrometry

Principles and Characteristics
Static secondary ion mass spectrometry (SSIMS) is extremely surface sensitive (*cfr.* Chp. 4.2.1) and has the ability to obtain ion images of the surface distribution of atomic and molecular species. Briggs [785] has first demonstrated molecular imaging SIMS with an early quadrupole instrument. The advantages provided by ToF-SIMS, namely high mass resolution, high transmission and low ion dose used, have made polymer imaging with sub-μm resolution possible using Ga^+, In^+, Ar^+, O_2^+ and SF_5^+ ion guns [786]. Rapid analytical advances in sensitivity, mass resolution and mass range have been made in particular since the introduction of high performance time-of-flight (ToF) instruments to the commercial market starting in the late 1980s, including microscope or microprobe imaging (MI) and laser post-ionisation (PI) capabilities.

The acquisition of images in SIMS may be carried out by two different means: scanned imaging in an ion microprobe or stigmatic imaging in an ion microscope. In a **microprobe** a (pulsed) microfocused primary ion beam (Cs^+, Ga^+, In^+; beam spot size $\sim$0.2 μm) is rastered across a selected portion of the surface of the specimen and the secondary ion signal intensity is recorded and displayed as a function of beam position. Under ideal conditions the image resolution in the microprobe mode is dependent on the primary ion diameter and thus the primary ion current. With a pulsed liquid metal ion gun (LMIG) an ultimate lateral resolution of approximately 20 nm can be obtained, with 100–200 nm as a more routine performance, and current densities in excess of 1 A cm^{-2}. This presents an analytical trade-off between spatial resolution and sensitivity. For sub-μm imaging a rather large primary ion pulse width in the range of 5 to 50 ns is used and correspondingly, the mass resolution in the low mass range is limited to about 150–1500 [787]. Although Ga-LMIG ensures a high lateral resolution, it does not provide the sensitivity needed to detect high-mass (molecular) ions. SF_5 primary ions give good sensitivity, but poor lateral resolution (50 to 100 μm). Top-quality SIMS images can be made with Au sources. Because it is far easier to control the position of a continuous (rather than pulsed) primary beam, quadrupole mass analysers have been used in many SIMS instruments. The development of ion microprobes capable of focusing an ion beam to a very small spot size has made SIMS perhaps the most important method for chemical imaging of surfaces.

With the **ion microscope** mode, first demonstrated by Slodzian *et al.* [788–790], the pulsed primary ion beam may illuminate the whole sample area at once and an image is displayed rapidly. The microscopic mode of imaging operates much like an optical microscope in which the spatial distributions of the secondary ions emanating from the surface are refocused onto some form of position sensitive device. The fundamental process of ion emission sets the ultimate spatial resolution limit. In practice, the lateral resolution achievable is limited by chromatic and spherical aberrations in the secondary optical system to around 0.5 μm [790a]. Generally, the improved lateral resolution in microprobe mode is achieved at the expense of poorer mass resolution. In an ion microscope, the spatial information on the region of origin of the secondary ions is preserved by the mass spectrometer and mass filtered images may be projected directly onto a screen or channel plate. As only the outer monolayer of material is being sampled, there are only a limited number of molecules available for imaging. The high primary ion beam current density necessary to obtain a reasonable image limits the utility of the ion microscope for characterisation of organic species. The signal in a ToF-SIMS image is dependent on the concentration of the surface species, the sputter yield of the particular mass species, and the probability that the sputtered species will be transformed to an ionised species. Transformation probabilities for organic materials can range from 10^{-2} to 10^{-6}, which typically limits the spatial resolution of a molecular image to 1 μm [791].

Magnetic sector field instruments can image in either the microscope and microprobe modes. For quadrupole instruments imaging can only be carried out using the microprobe approach. The mass filter is tuned to detect a signal (secondary ions of a chosen m/z), representing a species of interest, whose intensity at each pixel in the scanned array is measured. For ToF instruments both microprobe and microscope approaches are available. The latter is only possible if the spectrometer has imaging optics for imaging secondary ions so that direct images are formed at the detector position [792]. The advantage of ToF-MS is that the whole spectrum is acquired at each pixel. Multiphoton ionisation (MPI) using pulsed lasers is the most promising approach, and its use in imaging ToF mass spectrometers has been pioneered by Winograd *et al.* [793]. ToF-SIMS imaging systems are equipped with laser post-ionisation. High mass resolution of the ToF analyser ensures that peaks at the same nominal mass can be attributed with a high degree of confidence by exact mass determination. On modern ToF-SIMS instruments a separation of 0.03 Da can be resolved with ease. ToF-SIMS operated in the static regime at sub-μm resolution, with high transmission and parallel detection, ensures that the maximum information is obtained with the minimum of ion dose. The generally accepted dose necessary for static conditions is 10^{13} ions cm^{-1} (material dependent). Yet, the non-uniqueness of low-mass fragment ions, and the difficulty in obtaining unambiguous high-mass information, due to sample charging or primary ion beam damage, often results in little/no contrast in ToF-SIMS mapping studies of organic/polymeric systems. A ToF analyser allows imaging of insulating samples by use of a pulsed charge compensation method applying low energy electrons. Principal Component Analysis (PCA) has been applied to image files in order to group image pixels (*i.e.*, establish contrast) based on spectral components [794].

It is possible to generate elemental and chemical maps of the analysed surface (*i.e.*, the "imaging mode"). Imaging SIMS can be regarded as **chemical microscopy**. Eccles *et al.* [795,796] have described a chemical microscope – a low-cost automated imaging SIMS instrument (Millbrook Chemical Microscope or mini SIMS). The Millbrook Chemical Microscope is a self-contained benchtop SIMS instrument (essentially a microprobe) designed for running rapid, routine analyses and is equipped with a 300 Da QMS [797]. There are many cases where SIMS is the quickest and most direct analysis technique capable of solving a problem or monitoring a process. The chemical microscope means that SIMS analysis can now be performed in the same time and for the same overall cost as other more common analysis techniques. The chemical microscope is being used in a broad range of industrial sectors, including electronics, paints, specialist coatings and catalysts for QC and failure analysis.

ToF-SIMS used for static SIMS measurements is now also used for (rather poor) **quantification** of surface species. Quantification with SSIMS is possible with standards. Spool *et al.* [798] discussed image analysis methods for the quantification of ion images using standard tools of image processing. Quantitative 3D imaging on the ion microscope is greatly simplified using a resistive anode encoder (RAE) [799].

Table 5.67 shows the **main characteristics** of imaging SIMS. Imaging ToF-SIMS takes advantage of the high sensitivity and broad mass range in combination with good mass resolution and accuracy of ToF-SIMS. The sensitivity of high-performance imaging ToF-SIMS exceeds that of imaging AES and imaging XPS by orders of magnitude. ToF-SIMS can be used to produce chemically resolved images with a sub-μm lateral resolution of organic material by mapping the intensity of specific molecular or atomic ions while maintaining static analysis conditions. In many cases the identification of chemical species on surfaces does not provide enough information to clearly characterise all processes which influence the behaviour of a material. ToF-SIMS imaging offers the possibility of obtaining insight into the lateral distribution of atoms, molecules or functional groups on the surface. Traces of contaminants can be displayed as chemical mapping. A disadvantage is that surface damage is not improbable in imaging SIMS since the primary ion current has to be raised in order to generate sufficient secondary ions from very small areas. Care has then to be taken to obtain real surface information from imaging SIMS. If the investigated sample is non-conducting, a low-energy electron beam has to be used for efficient charge compensation.

Table 5.67. Main characteristics of imaging SIMS

Advantages:
- Lateral resolution (100 nm–1 μm)
- Good mass resolution (high resolution for ToF)
- High sensitivity (trace concentrations down to ppm level)
- Surface microanalysis; inorganic/organic chemical imaging
- Wide industrial applicability to heterogeneous specimens

Disadvantages:
- Sample charging (unless charge compensation)
- Risk of surface damage (not in case of ToF-SSIMS)
- Difficult quantification
- Expensive equipment
- Specialist use

DSIMS may be equipped with magnetic sector, quadrupole or time-of-flight mass spectrometers that are capable of imaging molecular distributions via microscope and/or microprobe modes. The use of microscopic imaging on magnetic sector instruments is ideally suited to DSIMS **image depth profiling** to depths of many μm [800]. Although information on the elemental spatial distribution is collected in real-time and 2D images are constructed during data acquisition, 3D imaging to a depth of many μm is carried out by retrospective image visualisation [771].

For a summary of instrumentation and application of the SIMS approach, *cfr.* ref. [801]; for instrumental aspects of ToF microscopes *vs.* microprobes, *cfr.* ref. [802]. Additional information on imaging SIMS techniques is available in several specialised textbooks [771,803].

Applications

Table 5.68 shows some typical applications of imaging SSIMS. ToF-SIMS is ideally suited for the analysis of surface additives and primers. The high surface sensitivity coupled with both structural and chemical information enables this method to identify additive species even within a complex matrix. The excellent reproducibility of the technique also permits semi-quantitative analysis. The capability of imaging molecular species allows determining the spatial distribution of additives in three dimensions. Briggs [785] first reported SIMS imaging from a multicomponent polymer surface (PET/DMS/PTFE) with a silicone release agent (DMS). Mawn *et al.* [804] have demonstrated the potential of molecular imaging of polymer domains as small as a few μm^2. ToF-SIMS can image electrically insulating surfaces such as paper [805] and polymers [806], *e.g.* a printed ink dot on paper [807].

One of the most important applications of ToF-SIMS imaging is in determining the distribution

Table 5.68. Typical applications of imaging SSIMS

• Surface segregation	• End-group determination
• Surface contamination	• General surface structural determination
• Adhesion properties	• Depth profiling
• Additive mapping and migration	• Quality control
• Characterisation of coatings and layers	• Failure analysis

of organic phases on surfaces with spatial resolution on the order of some μm down to 0.1 μm. Imaging SSIMS allows the study of inhomogeneous surface segregation of additives in polymer films. **Additive migration** by imaging static ToF-SIMS has been examined for erucamide in PP (following the O^-, OH^- and CN^- signals), and for glycerol monostearate (GMS) in PP and PET [808]. Walzak *et al.* [809] have measured the homogeneity of the distribution of Chimassorb 944 in LLDPE using microfocused Ga or In primary beams. Instead of the weak parent ion for the oligomer at m/z 599 imaging was carried out with the C_3H_8N mass fragment at m/z 58. Imaging of the AO distribution was possible to concentrations as low as 0.1%, and a linear concentration calibration curve was obtained. Improved imaging capabilities for a structured additive containing PP surface were reported for SF_5^+ primary ion beams in comparison to focused Ar^+ [810]. Kersting *et al.* [811,811a] have mapped LDPE/(Tinuvin 770, Irganox 565), LDPE/(Tinuvin 770, Irgafos 168), PP/EBA and PP/GMS surfaces using a recently developed polyatomic Au ion source. With this source orders of magnitude higher sensitivity for the weak parent-ion are realised and additive distribution may be mapped with the characteristic molecular ions $(M+H)^+$ with high lateral resolution (a few μm) and good intensity. The use of $(M+H)^+$ ions is preferred compared with the normally used low mass-fragments as they give higher contrast and more specific information, such as the difference between an intact and degraded additive (*e.g.* Irgafos phosphite *vs.* phosphate). It was noticed that Tinuvin 770 is poorly dispersed over a 500 × 500 μm^2 area, as opposed to Irganox 565 (Fig. 5.23). It is also possible to image cross-sections of a sample. A 3D ToF-SIMS image of LDPE/DBDPO was reported [811a]. *Cfr.* also Chp. 4.2.1 and Fig. 4.12.

Ishitani *et al.* [812] have reported ToF-SIMS imaging of Irganox 1010 ($^{163}C_{11}H_{15}O^+$, $^{203}C_{14}H_{19}O^+$, $^{219}C_{15}H_{23}O^+$, $^{259}C_{17}H_{23}O_2^-$) and stearamide ($^{284}C_{18}H_{38}NO^+$) on a PP surface. The two additives showed a different distribution. Aspects of *in situ* molecular trace analysis were pursued by the application to polymer additives, such as antioxidants [813]. At variance to UV microscopy, the application of ToF-SIMS is not limited to UV absorbers, but can also map HALS.

ToF-SIMS$^-$ imaging (m/z 392, 483) has also been used for determining the distribution of the dye Remanzol Reactive Blue 19 on the surface of used cotton fabric [814].

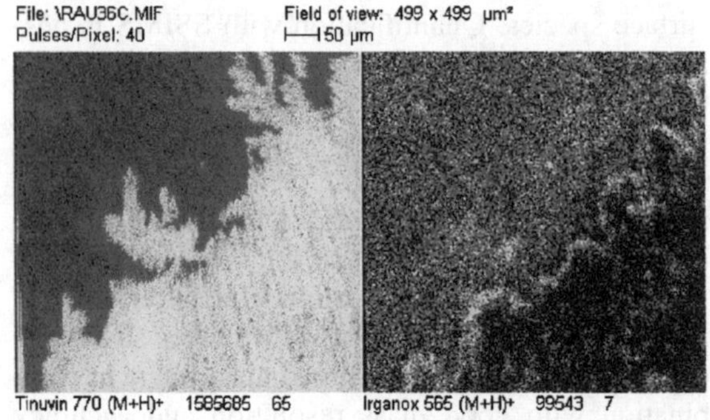

Fig. 5.23. Molecular ion surface chemical mapping by ToF-SIMS of Tinuvin 770 ($C_{28}H_{53}N_2O_2$) and Irganox 565 ($C_{33}H_{57}S_2N_4O$) on LDPE. Field of view 500 × 500 μm^2. After Kersting *et al.* [811]. Reprinted from R. Kersting *et al.*, in *Proceedings SIMS XII* (A. Benninghoven *et al.*, eds.), Elsevier Science Publishers, Amsterdam (2000), pp. 825–828, Copyright (2000), with permission from Elsevier.

Another example of ToF-SIMS analysis of polymer surfaces is the positive ion microscope analysis of a layered **automotive paint** sample, consisting of a clear melamine acrylic resin applied in two coats over an aluminum substrate [815]. Each resin layer was approximately 70 μm thick. The topcoat contained 2 wt.% of the UV photostabiliser Tinuvin 770. In order to determine the extent to which Tinuvin 770 diffused from the topcoat into the underlying coat as a function of curing cycles a paint cross-section, prepared by ultra microtoming, was examined. An image corresponding to $(M+H)^+$ of Tinuvin 770 showed rather extensively diffusion of the UV stabiliser into the undercoat layer. Similarly, ToF-SIMS operated in microscope mode has been used to characterise a cross-section of two paint layers, only one of which contained a photostabiliser additive [773]. Results indicated again possible migration of this additive into the bulk of the adjacent paint layer. Figure 5.24 shows ToF-SIMS line scans of an automotive coating system. The $^{16}O^-$ ion image delineates the primer layer; the distribution labelled "P" is the $(M-1)^-$ species of an organic red pigment at m/z 355 and the DDBSA curve is the $(M-1)^-$ species for dodecylbenzenesulfonic acid at m/z 325 [5]. Gerlock *et al.* [816] have used $^{18}O^-$ ToF-SIMS imaging of all coating layers of automotive paint systems exposed to UV light and heated in an $^{18}O_2$ atmosphere. ToF-SIMS is but one of a variety of spectroscopic techniques available for the study of chemical composition changes in

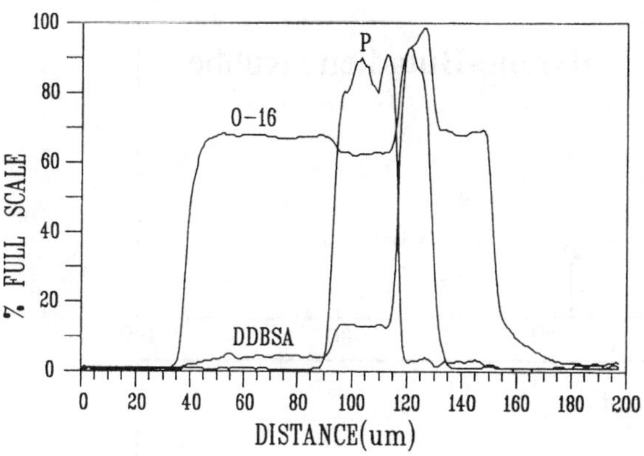

Fig. 5.24. ToF-SIMS line scans determined from mass selected ion images obtained from a cross-section of a fresh automotive coating system; line scans of ^{16}O, red pigment and DDBSA are plotted. After Adamsons *et al.* [5]. Reprinted with permission from K. Adamsons *et al.*, *ACS Symposium Series* **722**, 257–287 (1999). Copyright (1999) American Chemical Society.

complete paint systems, namely transmission FTIR, DRIFT, PA-FTIR, confocal Raman microscopy, μ-FTIR, ToF-SIMS and ESR. Hagenhoff *et al.* [816a] have shown ion-induced secondary electron images and mass resolved ion images of defects in car paint, identifying lubricants (perfluorinated polyether) and smoothing agents (silicone oil). Imaging ToF-SIMS is able to detect lubricant concentrations that are not observed with either EPMA or SAM, imaging techniques normally applied to paint cratering problems.

Fluorochemical species tend to have high ion transformation probabilities, which makes **fluorochemical additives** ideal candidates for ToF-SIMS image analysis. MacKay *et al.* [817] have reported ToF-SIMS images of the distribution of a fluorochemical additive in a cross-sectioned polymer film. The image of the CF^+ fragment ion at m/z 31 revealed concentration of the additive at the surface and a heterogeneous distribution throughout the film. Weng *et al.* [818] have reported applications of ToF-SIMS imaging in polymer processing. For the Dynamar-aided HDPE processing system, ToF-SIMS imaging clearly proves that there is a Dynamar-rich lubricant layer formed in the interface between polymer and metal die-wall. As Dynamar contains a fluoropolymer and SiO_2/MgO as a filler, the F^- and (O^-+OH^-) images were taken to represent Dynamar. A line scan on F^- and (O^-+OH^-) images showed a 3.5 μm thick lubricant layer. Hoshi *et al.* [819] have reported chemical ion images of the C_2F_4 fragment of a fluorolubricant on video tape.

Physical and chemical mapping using AFM and ToF-SIMS are appropriate means for studying surfaces of **flame retarded materials**. Various interactions may take place between ammonium polyphosphate (APP), pentaerythritol (PER) and tetraethoxysilane (TES). Whereas the reaction between APP and PER has been described [820], no direct reaction could be found between APP and TES. Marosi *et al.* [821] have reported imaging ToF-SIMS of APP-PER-ME-TES intumescent flame retarded (IFR) polypropylene. The image prepared on the basis of SiO_x ions proved the existence of a silicone layer between polymer and solid APP particles covered by PER.

ToF-SIMS chemical imaging (often in conjunction with iXPS) also plays a role in the analysis of the interphase region of fully fabricated **glass fibre composites**, particularly interaction of silane based adhesion promoters with the resin matrix. ToF-SIMS is profitably used in the packaging industry (adhesives) and food industry (contamination of contents by the packaging). The technique allows examining phase-separation of blends in the surface [822].

Samples from many industrial sources are often contaminated at the surface by processing agents or adventitious post-process **contaminants**. Detection of specific contaminant molecules on surfaces by SSIMS is, of course, an exercise similar to detection of additive molecules migrating to polymer surfaces. However, in case of contaminants the problem is often localised and the microanalytical/imaging capability of SSIMS is called upon. As the primary ion beam can be focused to a very small spot size (a few μm) even very small defects can be analysed allowing locally resolved chemical surface identification by microanalytical SSIMS. Additive mapping has been used for the identification of contaminations [823]. Lloyd *et al.* [794] have described identification of a hexamethylene tetramethyldiamine deposit on interior automotive parts. Another typical example concerns the behaviour of a non-woven PP fibre product with surface contamination by dimethylsilicone (DMS). The distribution of DMS over the 30 μm PP fibres was mapped [824]. The sensitivity of SSIMS to molecular additives and surface contamination has been illustrated by Weng *et al.* [825] who identified PDMS (m/z 28, 73, 147, 207, 221, *etc.*) and palmitic/stearic acids (m/z 239, 257, 267, 285) in polybutadiene copolymers (Fig. 5.25) by means of ToF-SIMS. SSIMS is

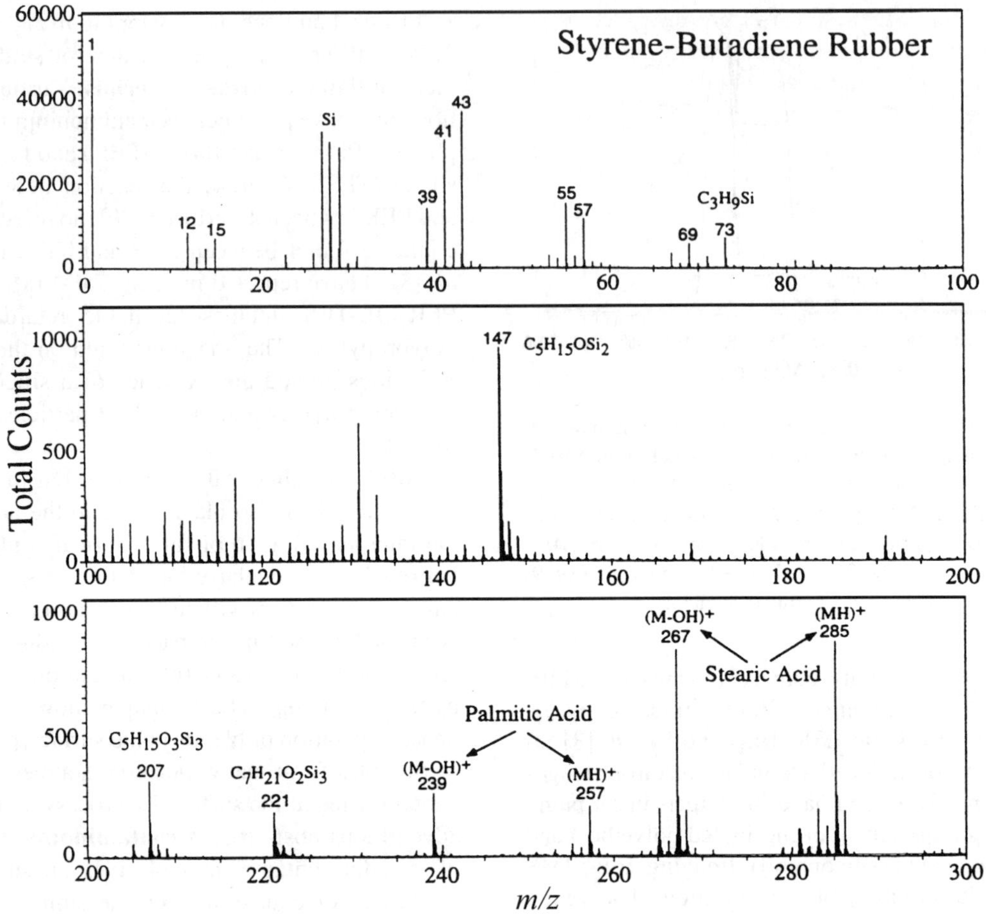

Fig. 5.25. 15 keV-Ga^+ positive-ion ToF-SIMS spectrum of styrene–butadiene rubber showing the presence of additives and PDMS contamination. After Weng *et al.* [825]. Reprinted from L.-T. Weng *et al.*, *Surface Interf. Anal.* **23**, 879–886 (1995). Copyright © 1995 John Wiley & Sons, Ltd. Reproduced with permission.

quite sensitive to silicones at very low surface coverage since the positive ion yield is high and the fragmentation pattern very distinctive. ToF-SIMS allows high precision determination of traces of fats or silicones (less than 1 ng/cm^2) [826].

ToF-SIMS is also useful in revealing **surface degradation** products [827] and in morphological studies of multiphase systems to determine the distribution (spatial partitioning) of additives within polymer blends. ToF-SIMS is equally capable of simultaneously providing quantitative trace metal element, monomer (ENB content, C_2/C_3 ratio) and oxidation analysis from microscopic polymer domains, such as full compositional analysis of EPDM gels [828].

Briggs [829] has described use of SSIMS to semi-quantitatively assess coverage of the cyclic oligomer (trimer) on a PET film surface. Lloyd *et al.* [794] have reported secondary ion maps of additives (such as low-MW PEG) on bulk polybutylene glycol (PBG). Imaging ToF-SIMS was also applied in the analysis of the chlorine distribution and migration at PP-CPO (chlorinated polyolefin) surfaces, useful for adhesion improvement [816a]. Surface analysis of LCD materials was also carried out with this technique.

Techniques such as SIMS and XPS have sufficient surface sensitivity and small area analysis capability to tackle **cratering** problems. It is often necessary to determine if the crater is truly a de-wetted area or simply a recess in the coating brought about by local contamination. The spectrum per point analysis mode enables such questions to be resolved quickly and definitively. A combination of depth profiling SIMS and post-profile crater imaging has been used to fully characterise the polymer composition and interfacial behaviour for the buried interface of a polymer/steel system [830]. Careful analysis along the crater wall from the same image enables the diffusion profile of species in the polymer to be determined. **Depth profiling** by SIMS using the ion adduct mode of analysis can reveal depth and spatial distribution of filler particles throughout a film (dispersion characteristics).

BIBLIOGRAPHY

Light Microscopy

G.C. Allen, *Modern Microscopy and Analysis of Materials*, The Royal Society of Chemistry, Cambridge (2002).

D.J. Goldstein, *Understanding the Light Microscope: A Computer-Aided Introduction*, Academic Press, London (1999).

P.M. Conn, *Confocal Microscopy*, Academic Press, San Diego, CA (1999).

J. Slavík, *Fluorescence Microscopy and Fluorescent Probes*, Plenum Press, New York, NY (1996).

F.W.D. Rost, *Fluorescence Microscopy*, Cambridge University Press, Cambridge (1995, 1996), Vol. 1.

T.R. Corle and G.S. Kino, *Confocal Scanning Optical Microscopy and Related Imaging Systems*, Academic Press, San Diego, CA (1996).

B. Herman and J.J. Lemasters (eds.), *Optical Microscopy: Emerging Methods and Applications*, Academic Press, San Diego, CA (1993).

M. Pluta, *Advanced Light Microscopy, Vol. 3. Measuring Techniques*, Elsevier/PWN-Polish Scientific Publishers, New York, NY (1993).

W.G. Hartley, *The Light Microscope: Its Use and Development*, Senecio Publ. Co., Oxford (1993).

M. Abramowitz, *Fluorescence Microscopy – The Essentials*, Olympus-America, New York, NY (1993).

R.P. Haugland, *Molecular Probes Handbook of Fluorescent Probes and Research Chemicals*, Molecular Probes, Eugene, OR (1992).

F.W.D. Rost, *Quantitative Fluorescence Microscopy*, Cambridge University Press, Cambridge (1991).

D. Simpson and W.G. Simpson, *An Introduction to Applications of Light Microscopy in Analysis*, The Royal Society of Chemistry, Cambridge (1990).

R.F. Smith, *Microscopy and Photomicrography: A Working Manual*, CRC Press, Boca Raton, FL (1990).

T. Wilson (ed.), *Confocal Microscopy*, Academic Press, London (1990).

T. Wilson and C. Sheppard, *Theory and Practice of Scanning Optical Microscopy*, Academic Press, London (1989).

J.S. Ploem and H.J. Tanke, *Introduction to Fluorescence Microscopy*, Oxford University Press, Royal Microscopical Society, London (1987).

W.C. McCrone, L.B. McCrone and J.G. Delly, *Polarized Light Microscopy*, Ann Arbor Science Publishers, Ann Arbor, MI (1978).

Electron Microscopy

M. De Graef, *Introduction to Conventional Transmission Electron Microscopy*, Cambridge University Press, Cambridge (2003).

D.J. Dyson, *X-Ray and Electron Diffraction Studies in Materials Science*, Maney Publishing, Leeds (2003).

J.C.H. Spence, *High-Resolution Electron Microscopy*, Oxford University Press, Oxford (2003).

C.C. Ahn, M.M. Disko and B. Fultz, *Transmission Electron Energy Loss Spectrometry and the EELS Atlas*, J. Wiley & Sons, Chichester (2002).

A.R. Clarke and C.N. Eberhardt, *Microscopy Techniques for Materials Science*, Woodhead Publishing Ltd., Cambridge (2002).

D. Shindo and T. Oikawa, *Analytical Electron Microscopy for Materials Science*, Springer-Verlag, Berlin (2002).

B. Fultz and J.M. Howe, *Transmission Electron Microscopy and Diffractometry of Materials*, Springer-Verlag, Berlin (2002).

M. Aindow and C.J. Kiely (eds.), *Electron Microscopy and Analysis 2001*, IOP Publishing, Bristol (2001).

P.J. Goodhew, F.J. Humphreys and R. Beanland, *Electron Microscopy and Analysis*, Taylor & Francis, London (2001).

L. Reimer, *Scanning Electron Microscopy*, Springer-Verlag, Berlin (1998).

P.L. Gai (ed.), *In Situ Microscopy in Materials Research*, Kluwer Academic Publishers, Boston, MA (1997).

D.B. Williams and C.B. Carter, *Transmission Electron Microscopy: A Textbook for Materials Science*, Plenum Press, New York, NY (1996).

L. Reimer, *Image Formation in Low-Voltage Scanning Electron Microscopy*, SPIE Press, Bellingham, WA (1993).

D. Shotton (ed.), *Electronic Light Microscopy*, J. Wiley & Sons, New York, NY (1993).

S.L. Flegler, J.W. Heckman Jr. and K.L. Klomparens, *Scanning and Transmission Electron Microscopy: An Introduction*, Oxford University Press, New York, NY (1993).

L. Reimer, *Transmission Electron Microscopy, Physics of Image Formaton and Microanalysis*, Springer-Verlag, Berlin (1989).

L. Reimer, *Scanning Electron Microscopy, Physics of Image Formation and Microanalysis*, Springer-Verlag, Berlin (1985).

D.B. Holt, M.D. Muir, P.R. Grant and I.M. Borswarva, *Quantitative Scanning Electron Microscopy*, Academic Press, London (1974).

Scanning Probe Microscopy

K.S. Birdi, *Scanning Probe Microscopes. Applications in Science and Technology*, CRC Press, Boca Raton, FL (2003).

S. Morita, R. Wiesendanger and E. Meyer (eds.), *Noncontact Atomic Force Microscopy*, Springer-Verlag, Berlin (2002).

D. Bonnell (ed.), *Scanning Probe Microscopy and Spectroscopy; Theory, Techniques and Applications*, Wiley-VCH, Weinheim (2001).

A.J. Milling (ed.), *Surface Characterization Methods*, M. Dekker, New York, NY (1998).

R. Wiesendanger (ed.), *Scanning Probe Microscopy*, Springer-Verlag, Berlin (1998).

B.D. Ratner and V.V. Tsukruk (eds.), *Scanning Probe Microscopy of Polymers*, American Chemical Society, Washington, DC (1998).

S.N. Magonov and M.-H. Whangbo, *Surface Analysis with STM and AFM, Experimental and Theoretical Aspects of Image Analysis*, VCH, Weinheim (1996).

C. Bai, *Scanning Tunneling Microscopy and Its Applications*, Springer-Verlag, Berlin (1995).

R. Wiesendanger, *Scanning Probe Microscopy and Spectroscopy. Methods and Applications*, Cambridge University Press, New York, NY (1994).

S.H. Cohen, M.T. Bray and M.L. Lightbody (eds.), *Atomic Force Microscopy/Scanning Tunneling Microscopy*, Plenum, New York, NY (1994).

D. Sarid, *Scanning Force Microscopy: With Applications to Electric, Magnetic and Atomic Forces*, Oxford University Press, New York, NY (1994).

D.A. Bonnell (ed.), *Scanning Tunneling Microscopy and Spectroscopy: Theory, Techniques and Applications*, VCH, Weinheim (1993).

R. Howland and L. Benatar, *A Practical Guide to Scanning Probe Microscopy*, Park Scientific Instruments, Sunnyvale, CA (1993).

Near-field Optics

M.A. Paesler, *Near-Field Optics: Theory, Instrumentation, and Applications*, Wiley-Interscience, New York, NY (1996).

Microbeam Analysis

D. Brandon and W.D. Kaplan, *Microstructural Characterisation of Materials*, J. Wiley & Sons, Chichester (1999).

J. Thieme, G. Schmahl, D. Rudolph and E. Umbach (eds.), *X-Ray Microscopy and Spectromicroscopy*, Springer-Verlag, Heidelberg (1998).

I. McNulty (ed.), *X-Ray Microfocusing: Applications and Techniques*, SPIE, Billingham, WA (1998).

Oxford Instruments, *The Principles and Practice of X-ray Microanalysis (CD-ROM Tutorial)*, Oxford Instruments, High Wycombe (1998).

R.F. Egerton, *Electron Energy-Loss Spectroscopy in the Electron Microscope*, Plenum Press, New York, NY (1996).

D.B. Williams, J.I. Goldstein and D.E. Newbury (eds.), *X-Ray Spectrometry in Electron Beam Instruments*, Plenum Press, New York, NY (1995).

H.J. Humecki (ed.), *Practical Guide to Infrared Microspectroscopy*, M. Dekker, New York, NY (1995).

P.C. Cheng, T.H. Lin, W.L. Wu and J.C. Wu (eds.), *Multidimensional Microscopy*, Springer-Verlag, New York, NY (1994).

B.G. Yacobi, L.L. Kazmerski and D.B. Holt (eds.), *Microanalysis of Solids*, Kluwer Academic/Plenum Publishers, New York, NY (1994).

J.I. Goldstein, D.E. Newbury, P. Echlin, D.C. Joy, A.D. Romig, C.E. Lyman, C. Fiori and E. Lifshin, *Scanning Electron Microscopy, X-Ray Microanalysis and Analytical Electron Microscopy*, Plenum Press, New York, NY (1992).

K.F.J. Heinrich and D.E. Newbury (eds.), *Electron Microprobe Quantitation*, Plenum Press, New York, NY (1991).

J.J. Hren, J.I. Goldstein and D.C. Joy (eds.), *Introduction to Analytical Electron Microscopy*, Plenum Press, New York, NY (1979).

Microspectroscopy

K. Janssens, F. Adams and A. Lindby (eds.), *Microscopic X-Ray Fluorescence Analysis*, J. Wiley & Sons, Chichester (2000).

G. Turrell and J. Corset (eds.), *Raman Microscopy. Developments and Applications*, Academic Press, London (1996).

H.J. Humecki (ed.), *Practical Guide to Infrared Microspectroscopy*, M. Dekker, New York, NY (1995).

R.G. Messerschmidt and M.A. Harthcock (eds.), *Infrared Microspectroscopy: Theory and Applications*, M. Dekker, New York, NY (1988).

P.B. Roush (ed.), *The Design, Sample Handling, and Applications of Infrared Microscopes*, ASTM Special Technical Publication **949**, ASTM, Philadelphia, PA (1987).

Imaging/Image Analysis

J.P. Hornak (ed.), *Encyclopedia of Imaging Science and Technology*, J. Wiley & Sons, Chichester (2002).

B. Blümich, *NMR Imaging of Materials*, Oxford University Press, Oxford (2000).

R. Baldock and J. Graham (eds.), *Image Processing and Analysis: A Practical Approach*, Oxford University Press, Oxford (2000).

J.J. Friel (ed.), *Practical Guide to Image Analysis*, ASM International, Materials Park, OH (2000).

J.C. Lindon, G.E. Tranter and J.L. Holmes (eds.), *Encyclopedia of Spectroscopy and Spectrometry*, Academic Press, San Diego, CA (2000).

B. Pourdeyhimi (ed.), *Imaging and Image Analysis Applications for Plastics*, SPE/Plastics Design Library, Norwich, NY (1999).

L. Wojnar, *Image Analysis*, CRC Press, Boca Raton, FL (1999).

E.M. Haacke, R.W. Brown, M.R. Thompson and R. Venkatesan, *Magnetic Resonance Imaging: Physical Principles and Sequence Design*, Wiley-Liss, New York, NY (1999).

P. Soille, *Morphological Image Analysis, Principles and Applications*, Springer-Verlag, Berlin (1999).

P. Blümler, B. Blümich, R. Botto and E. Fukushima (eds.), *Spatially Resolved Magnetic Resonance: Methods, Materials, Medicine, Biology, Rheology, Geology, Ecology, Hardware*, Wiley-VCH, Weinheim (1998).

R. Kimmich (ed.), *NMR Tomography, Diffusometry and Relaxometry*, Springer-Verlag, Berlin (1997).

B. Jähne, *Practical Handbook on Image Processing for Scientific Applications*, CRC Press, Boca Raton, FL (1997).

P. Geladi and H. Grahn, *Multivariate Image Analysis*, J. Wiley & Sons, Chichester (1996).

M. Gu, *Principles of Three-dimensional Imaging in Confocal Microscopes*, World Scientific, Singapore (1996).

X.F. Wang and B. Herman (eds.), *Fluorescence Imaging Spectroscopy and Microscopy*, Wiley-Interscience, New York, NY (1996).

K. Humphrey, *Image Analysis*, PIRA, Leatherhead (1995).

C.J. Cogswell and K. Carlsson, *Three-Dimensional Microscopy: Image Acquisition and Processing*, SPIE, Bellingham, WA (1994).

A.T. Hubbard, (ed.), *Handbook of Surface Imaging and Visualization*, CRC Press, Boca Raton, FL (1994).

M. Morris (ed.), *Microscopic and Spectroscopic Imaging of the Chemical State*, M. Dekker, New York, NY (1993).

A.M. Parikh, *Magnetic Resonance Imaging Techniques*, Elsevier, New York, NY (1992).

B. Blümich and W. Kuhn (eds.), *Magnetic Resonance Microscopy*, VCH, Weinheim (1992).

P.T. Callaghan, *Principles of Nuclear Magnetic Resonance Microscopy*, Oxford University Press, New York, NY (1991).

P. Mansfield and E.L. Hahn, *NMR Imaging*, The Royal Society of Chemistry, London (1990).

Polymer Microscopy

J.L. Koenig, *Microspectroscopic Imaging of Polymers*, American Chemical Society, Washington, DC (1997).

L.C. Sawyer and D.T. Grubb, *Polymer Microscopy*, Chapman & Hall, London (1996).

P.H. Greaves and B.P. Saville, *Microscopy of Textile Fibers*, BioScientific, Oxford (1995).

S.J. Spells (ed.), *Characterization of Solid Polymers – New Techniques and Developments*, Chapman & Hall, London (1994).

D.A. Hemsley (ed.), *Applied Polymer Light Microscopy*, Elsevier Applied Science, London (1989).

G. Kämpf, *Characterization of Plastics by Physical Methods*, C. Hanser, Munich (1986).

D.A. Hemsley, *The Light Microscopy of Synthetic Polymers*, Oxford University Press, Oxford (1984).

General

J. Heath, *Dictionary of Microscopy*, J. Wiley & Sons, Chichester (2005).

REFERENCES

[1] M. Minsky, *U.S. Pat.* 3,013,467 (July 7, 1957).
[2] V.J. Coates, A. Offner and E.H. Siegler, *J. Opt. Soc. Am.* **43**, 984 (1953).
[3] J.A. Reffner, *Intl. Lab.* **30** (5), 16–20 (2000).
[4] A. Briggs, *An Introduction to Scanning Acoustic Microscopy*, Oxford University Press, Oxford (1985).
[5] K. Adamsons, L. Litty, K. Lloyd, K. Stika, D. Swartzfager, D. Walls and B. Wood, *ACS Symp. Ser.* **722**, 257–87 (1999).
[6] A. Hernandez–Luna and N.A. D'Souza, *Proceedings SPE ANTEC 2003*, Nashville, TN (2003), pp. 1342–6.
[7] L. Van Vaeck and F. Adams, in *Encyclopedia of Spectroscopy and Spectrometry* (J.C. Lindon, ed.), Academic Press, San Diego, CA (2000), pp. 1141–52.
[8] D.J. Dyson, D.M. Poole and B.N. Shelley, *Eur. Microsc. Anal.* **64**, 13–4 (2000).
[9] G. Ligner and E. Hess, *Eur. Coat. J.* **10**, 622 (1991).
[10] G. Haacke, F.F. Andrawes and B.H. Campbell, *J. Coat. Technol.* **68**, 57 (1996).
[11] C.G. Zimba, H. Ade and A.P. Smith, *Proceedings SPE ANTEC 2000*, Orlando, FL (2000), pp. 2036–40.
[12] P. Cooke, *Anal. Chem.* **64**, 219–43R (1992).
[13] P. Geladi and H. Grahn, *Multivariate Image Analysis*, J. Wiley & Sons, Chichester (1996).
[14] P.M. Cooke, *Anal. Chem.* **68** (12), 333–78R (1996).
[15] D.M. Shotton, *Histochem. Cell Biol.* **104** (2), 97–137 (1995).
[16] A.S. Vaughan, in *Polymer Characterisation* (B.J. Hunt and M.I. James, eds.), Blackie, Glasgow (1993), pp. 297–332.
[17] D.H. Hemsley (ed.), *Applied Polymer Light Microscopy*, Elsevier Applied Science, New York, NY (1989).
[18] B. Pourdeyhimi (ed.), *Imaging and Image Analysis Applications for Plastics*, SPE/Plastics Design Library, Norwich, NY (1999).
[19] P. Soille, *Morphological Image Analysis, Principles and Applications*, Springer-Verlag, Berlin (1999).
[20] M.C. Edwards, in *Encyclopedia of Analytical Sciences* (A. Townshend, ed.), Academic Press, London (1995), pp. 3115–20.
[21] E. Abbe, *Schultzes Archiv. f. mikr. Anat.* **9**, 413–68 (1873).
[22] E. Ruska, *Z. Physik* **87**, 580–602 (1934).
[23] G. Binnig, H. Rohrer, Ch. Gerber and E. Weibel, *Appl. Phys. Lett.* **40**, 178 (1982); *Phys. Rev. Lett.* **49**, 57 (1982).
[24] S. Inoué, in *Handbook of Biological Confocal Microscopy*, Plenum Press, New York, NY (1990), pp. 1–14.
[25] P.J. Evennett and C. Hammond, in *Encyclopedia of Analytical Sciences* (A. Townshend, ed.), Academic Press, London (1995), pp. 3127–34.
[26] S. Inoué, *Video Microscopy*, Plenum Press, New York, NY (1986).
[27] F. Lee, B. Pourdeyhimi and K. Adamsons, in *Service Life Prediction of Organic Coatings* (D.R. Bauer and J.W. Martin, eds.), *ACS Symp. Ser.* **722**, 207–32 (1999).
[28] A. Spittel and R.H. Schuster, *Kautsch. Gummi Kunstst.* **47** (12), 896–902 (1994).
[29] A.V. Matzakos and K. Zygourakis, *Rev. Sci. Instrum.* **64** (6), 1541–8 (1993).
[30] G. Bongaerts, K. Rabel, K.G.H. Raemaekers and J.C.J. Bart, *Abstracts 12th Intl. Congress on Thermal Analysis and Calorimetry (ICTAC-12)*, Copenhagen (2000).
[31] S. Bradbury, *An Introduction to the Optical Microscope*, Oxford University Press and Royal Microscopical Society, Oxford (1989).
[32] P.J. Evennett, in *Encyclopedia of Analytical Sciences* (A. Townshend, ed.), Academic Press, London (1995), pp. 3106–15.
[33] D. Hemsley, in *Developments in Polymer Characterisation-1* (J.V. Dawkins, ed.), Applied Science Publishers, London (1978), p. 245.
[34] G.M. Gale, *Rapra Research Rept. No. 155*, Rapra Technology Ltd., Shawbury (1967).
[35] S.R. Sandler, W. Karo, J. Bonesteel and E.M. Pearce, *Polymer Synthesis and Characterization, A Laboratory Manual*, Academic Press, San Diego, CA (1998).
[36] W.C. McCrone, *Am. Lab.* **30** (21), 27–34 (1998).
[37] B. Foster, *Mater. World (7)*, 402–5 (1998).
[38] H.J. Tanke, *J. Microsc.* **155**, 405–18 (1989).
[39] T. Wilson and C.J.R. Sheppard, *Theory and Practice of Scanning Optical Microscopy*, Academic Press, London (1984).
[40] U. Dürig, D.W. Pohl and F. Rohner, *J. Appl. Phys.* **59**, 3318–27 (1986).
[41] D. Courjon, K. Sarayeddine and M. Spajer, *Opt. Commun.* **71**, 23–8 (1989).

[42] D. Simpson and W.G. Simpson, *An Introduction to Applications of Light Microscopy in Analysis*, The Royal Society of Chemistry, Cambridge (1990).

[43] J.A. Johnson, *Proceedings SPE ANTEC '96*, Indianapolis, IN (1996), pp. 3318–23.

[44] S.E. Amos, G.M. Giacoletto, J.H. Horns, C. Lavallée and S.S. Woods, in *Plastics Additives Handbook* (H. Zweifel, ed.), Hanser Publishers, Munich (2000), pp. 553–84.

[45] *Dynamar Polymer Processing Additive Technical Information, Optical Microscopy Method for Dispersion Analysis in Polyolefins*, Dyneon LLC, Oakdale, MN (n.d.).

[46] R.I. Danescu and D.A. Zumbrunnen, *J. Vinyl Addit. Technol.* **6** (1), 26–33 (2000).

[47] D.E. Hauenstein, D.J. Cimbalik and P.G. Pape, in *Imaging and Image Analysis Applications* (B. Pourdeyhimi, ed.), SPE/Plastics Design Library, Norwich, NY (1999), pp. 97–105.

[47a] M. Ahmed, M.A. Al-Ohali, M.A. Garwan, Z.M. Al-Hamouz and K.Y. Al-Soufi, *Europ. Microsc. Anal.* **18** (6), 21–2 (2004).

[48] C.B.C. Ohanuzue and J.H. Nobbs, *Indian J. Eng. Mater. Sci.* **6** (1), 43–7 (1999).

[49] G. Prior and H. Meyer, *GIT Fachz. Lab.* **39** (7), 633–6 (1995).

[50] H. Miyata, *Kagaku Keisatsu Kenkyusho Hokoku, Hokagaku-Hen* **47** (4), 141–7 (1994).

[51] H. Palla, in *Analyse von Kautschuken und Elastomeren* (H. Ostromow, ed.), Springer-Verlag, Berlin (1981), pp. 158–71.

[52] W.C. McCrone, *Intl. Lab.* **30** (1), 9–15 (2000).

[53] H. Orth, H. Palla and M. Polke, in *Handbuch der Mikroskopie in der Technik* (H. Freund, ed.), Umschau Verlag, Frankfurt a. Main (1958), Vol. VII.

[54] L.C. Sawyer and D.T. Grubb, *Polymer Microscopy*, Chapman & Hall, London (1996).

[55] D.A. Hemsley, *The Light Microscopy of Synthetic Polymers*, Oxford University Press, Oxford (1984).

[56] N.C. Billingham and P.D. Calvert, *Dev. Polym. Charact.* **3**, 229–59 (1982).

[57] A. Kohler, *Z. Wiss. Mikroskopie* **21**, 129, 275 (1904).

[58] N.C. Billingham and P.D. Calvert, *Inst. Phys. Conf. Ser.* **98**, 571–6 (1990).

[59] A.D. Curson, *Proc. Roy. Microsc. Soc.* **7**, 96 (1972).

[60] H.P. Frank and H. Lehner, *J. Polym. Sci., Pt. C* **31**, 193–203 (1970).

[61] P.D. Calvert and T.G. Ryan, *Polymer* **19**, 611 (1978).

[62] N.C. Billingham, P. Prentice and T.J. Walker, *J. Polym. Sci. Symp.* **57**, 287 (1976).

[63] T.G. Ryan, P.D. Calvert and N.C. Billingham, *Adv. Chem. Ser.* **169**, 261–72 (1978).

[64] P.D. Calvert and T.G. Ryan, *Polymer* **25** (7), 921–6 (1984).

[65] N.C. Billingham, O.J. Hoad, F. Chenard and D.J. Whiteman, *Macromol. Symp.* **115**, 203–14 (1997).

[66] L. Lazare, *Ph.D. Thesis*, University of Sussex, Brighton (2001).

[67] N.C. Billingham, P.D. Calvert and A. Uzuner, *Eur. Polym. J.* **25**, 839 (1989).

[68] N.C. Billingham, in *Oxidation Inhibition in Organic Materials* (J. Pospíšil and P.P. Klemchuk, eds.), CRC Press, Boca Raton, FL (1990), Vol. II, pp. 249–97.

[69] N.C. Billingham, P.D. Calvert, J.B. Knight and T.G. Ryan, *Br. Polym. J.* **11** (3), 155–9 (1979).

[70] J. Klein and B.J. Briscoe, *Polymer* **17**, 481 (1976).

[71] J.B. Knight, P.D. Calvert and N.C. Billingham, *Polymer* **26**, 1713 (1985).

[72] N.C. Billingham, *Makromol. Chem., Macromol. Symp.* **28**, 145 (1989).

[73] N.C. Billingham, P.D. Calvert and J.B. Knight, *Proceedings 28th IUPAC Makromol. Symp.*, Oxford (1982), pp. 291.

[74] N.C. Billingham and P.D. Calvert, *Pure Appl. Chem.* **57** (11), 1727–36 (1985).

[75] P.D. Calvert, N.C. Billingham, J.B. Knight and A. Uzuner, *ACS Symp. Ser.* **220**, 243–63 (1983).

[76] N.C. Billingham, P.D. Calvert and M.G. Ghaemy, *Br. Polym. J.* **15** (1), 62–5 (1983).

[77] O.S. Wolfbeis, *Fluorescence Spectroscopy*, Springer-Verlag, New York, NY (1992).

[77a] J.C.J. Bart, *Additives in Polymers. Industrial Analysis and Applications*, J. Wiley & Sons, Chichester (2005).

[78] H.R. Morris, C.C. Hoyt and P.J. Treado, *Appl. Spectrosc.* **48**, 857–66 (1994).

[79] P.L. Becker, in *Fluorescence Imaging Spectroscopy and Microscopy* (X.F. Wang and B. Herman, eds.), J. Wiley & Sons, New York, NY (1996), pp. 1–29.

[80] T.A. Klar, M. Dyba and S.W. Hell, *Appl. Phys. Lett.* **78** (4), 393–5 (2001).

[81] Y. Nishijima, *J. Polym. Sci. Symp.* **31**, 353 (1970).

[82] M.B. Rhodes, in *Encyclopedia of Analytical Sciences* (A. Townshend, ed.), Academic Press, London (1995), pp. 3134–42.

[83] E. Becker, *Fluorescence Microscopy*, Ernst Leitz, Wetzlar (1985).

[84] X.F. Wang and B. Herman, *Fluorescence Imaging Spectroscopy and Microscopy*, J. Wiley & Sons, New York, NY (1996).

[85] F.W.D. Rost, *Fluorescence Microscopy*, Cambridge University Press, Cambridge (1992), 2 vols.

[86] J.S. Ploem and H.J. Tanke, *Introduction to Fluorescence Microscopy*, Oxford University Press, New York, NY (1987).

[87] H.R. Morris, C.C. Hoyt and P.J. Treado, *Appl. Spectrosc.* **48** (7), 856–7 (1994).

[88] F. Rost, in *Encyclopedia of Spectroscopy and Spectrometry* (J.C. Lindon, ed.), Academic Press, San Diego, CA (2000), pp. 565–70.

[89] K. Carlsson and N. Åslund, *Appl. Opt.* **26**, 3232–8 (1987).

[90] M.C. Moss, J.A. Veiro, S. Singleton, D.P. Gregory, J.J. Birmingham, C.L. Jones, P.G. Cummins, D. Cummins, R.M. Miller, R.C. Sheppards, V.C. Howard and N. Bhashkar, *Analyst* **118**, 1–9 (1993).

[91] G.J. Brakenhoff, H.T.M. van der Voort, E.A. van Spronsen, W.A.M. Linnemans and N. Nanninga, *Nature* **317**, 748 (1985).

[92] P.M. Conn (ed.), *Confocal Microscopy*, Academic Press, San Diego, CA (1999).

[93] J.B. Pawley (ed.), *Handbook of Biological Confocal Microscopy*, Plenum Press, New York, NY (1990).

[94] P.C. Cheng, T.H. Lin, W.L. Wu and J.L. Wu, *Multidimensional Microscopy*, Springer-Verlag, New York, NY (1994).

[95] M. Minsky, *Scanning* **10**, 128–38 (1988).

[96] N. Åslund, A. Liljeborg, P.-O. Forsgren and S. Wahlsten, *Scanning* **9**, 227–35 (1987).

[97] E. Gratton and M.J. VandeVen, in *Handbook of Biological Confocal Microscopy*, Plenum Press, New York, NY (1990), pp. 53–67.

[98] P. Davidovits and M.D. Egger, *Nature* **223**, 831 (1969).

[99] E. Niggli, D.W. Piston, M.S. Kirby, H. Cheng, D.R. Sandison, W.W. Webb and W.J. Lederer, *Am. J. Physiol.* **266** (1), C303–10 (1994).

[100] P. Torok, G.R. Booker, Z. Laszik and R. Falster, *Inst. Phys. Conf. Ser.* **134**, 771–4 (1934).

[101] S.W. Hell and J. Wichmann, *Opt. Lett.* **19**, 11, 780 (1994).

[102] S.W. Hell, C.M. Blanco, A. Egner, M. Dyba, V. Westphal and L. Kastrup, *Proceedings Colloquium Spectroscopicum Internationale XXXIII*, Granada (2003), p. 131.

[103] G.J. Brakenhoff, H.T.M. van der Voort, E.A. van Spronsen and N. Nanninga, *J. Microsc.* **153**, 151–9 (1989).

[104] N.C. Billingham, P.D. Calvert and A. Uzuner, *Polymer* **31**, 258 (1990).

[105] K.S. Wells, D.R. Sandison, J. Strickler and W.W. Webb, in *Handbook of Biological Confocal Microscopy* (J.B. Pawley, ed.), Plenum Press, New York, NY (1990), pp. 27–39.

[106] R.G. King and P.M. Delaney, *Mater. Forum* **18**, 21–9 (1994).

[107] T. Wilson, in *Microscopy of Solids* (B.G. Yacobi, D.G. Holt and L.L. Kazmerski, eds.), Plenum Press, New York, NY (1994), pp. 219–32.

[108] H. Ishikawa, *Jikken Igaku* **12** (18), 2353–8 (1994).

[109] M.B. Rhodes, in *Encyclopedia of Analytical Sciences* (A. Townshend, ed.), Academic Press, London (1995), pp. 3120–7.

[110] A. Ribbe, *Trends Polym. Sci.* **5** (10), 333–7 (1997).

[111] M. Gu, *Principles of Three-Dimensional Imaging in Confocal Microscopes*, World Scientific, Singapore (1996).

[112] S. Tahara, S. Niiyama, H. Kumai and T. Wakabayashi, *Proceedings SPIE – Int. Soc. Opt. Eng.* **3297**, 44–51 (1998).

[113] J.-D. Tong, M. Moffitt, X. Huang, M.A. Winnik and R.A. Ryntz, *J. Polym. Sci. A: Polym. Chem.* **39** (1), 239–52 (2001).

[114] E.C. Nielsen, *Proceedings SPE ANTEC 2001*, Dallas, TX (2001), pp. 2394–8.

[115] K.R. Franklin, S. Singleton, R. Treloar and D.G. Collopy, *Proceedings AddCon Asia '97* (Rapra Technology Ltd., ed.), Shawbury (1997), Paper 16.

[116] L.-P. Sung, M.E. Nadal, P. Stutzman and M.E. McKnight, *Proceedings ACS Div. Polym. Mater. Sci. & Engng.* **83**, 343–4 (2000).

[117] J.L. Thomason and A. Knoester, *J. Mater. Sci. Lett.* **9**, 258 (1990).

[118] B.J.R. Scholtens and J.C. Brackman, *J. Adhesion* **52**, 115–29 (1995).

[119] H. Yoshida, K. Ito and N. Ise, *Phys. Rev.* **B4**, 435 (1991).

[120] A.R. Clarke, G. Archenbold, N. Davidson and N.J. Williamson, *Proceedings 9th Intl. Conf. Compos. Mater.*, Zaragossa (1993), pp. 201–8.

[121] K. De Clerck, P. Kickens and V. Oostveldt, *Proceedings 17th IFVTCC (Internationale Föderation der Vereine der Textilchemiker und Coloristen) Congress*, Vienna (1996), p. 201.

[122] Y. Wang, C. Yang and D. Tomasko, *I&EC Res.* **41** (7), 1780–6 (2002).

[123] L. Li, S. Sosnowski, C.E. Chaffey, S.T. Balke and M.A. Winnik, *Langmuir* **10** (8), 2495–7 (1994).

[124] E. Ruska, *Rev. Modern Phys.* **59** (3/1), 627–38 (1987).

[125] M. Knoll and E. Ruska, *Z. Physik* **78**, 318–39 (1932); *Ann. Physik (Leipzig)* **12**, 607–40, 641–61 (1932).

[126] M. von Ardenne, *Z. Physik* **109**, 553–72 (1938); *Z. Tech. Phys.* **19**, 407–16 (1938).

[127] M. von Ardenne, *Elektronen-Übermikroskopie*, Springer-Verlag, Berlin (1940).

[128] D. Chescoe and S. Chapman, in *Encyclopedia of Analytical Sciences* (A. Townshend, ed.), Academic Press, London (1995), pp. 3166–74.

[129] P.L. Gai (ed.), *In-situ Microscopy in Materials Research*, Kluwer Academic Publishers, Boston (1997).

[130] D.T. Grubb and G.W. Groves, *Phil. Mag.* **24**, 815–28 (1971).

[131] F. Haguenau, P. Hawkes, J. Hutchison, B. Satiat–Jeunemaître, G. Simon and D. Williams, *Microsc. Microanal.* **9** (2), 96–138 (2003).

[132] G.H. Michler, *Appl. Spectrosc. Rev.* **28** (4), 327–84 (1993).

[133] C.W. Oatley, W.C. Nixon and R.F.W. Pease, *Adv. Electron. Electron. Phys.* **21**, 181–247 (1965).

[134] A.F. Otterloo, in *Encyclopedia of Analytical Sciences* (A. Townshend, ed.), Academic Press, London (1995), pp. 3151–60.

[135] G. Kämpf, *Characterization of Plastics by Physical Methods*, Hanser Publishers, Munich (1986).

[136] J. Betzold, *Mikrochim. Acta* **2** (5–6), 391–400 (1983).

[137] B.J. Cross and K. Witherspoon, *Proceedings Colloquium Spectroscopicum Internationale XXXIII*, Granada (2003), pp. 531–2.

[138] G.D. Danilatos, *Mikrochim. Acta* **114/115**, 143–55 (1994).

[139] C. Mathieu, *Microsc. Anal.* **43**, 13 (1996).

[140] S. Scherer and O. Kolednik, *Eur. Microsc. Anal.* **70**, 15–17 (2001).

[141] R. Holm and B. Reinfandt, *Z. f. Werkstofftechn.* **9**, 153–88 (1978).

[142] G. Pfefferkorn, *Mikroscopie (Wien)* **34**, 80–112 (1978).

[143] H.-J. Purz and E. Schulz, *Acta Polym.* **30**, 377–90 (1979).

[144] L. Reimer, in *Electron Microsc. Mater. Sci., Proc. Intl. Sch.* (P.G. Merli and M.V. Antisari, eds.), World Sci., Singapore (1992), pp. 459–78.

[145] D.B. Holt, M.D. Muir, P.R. Grant and I.M. Borswarva, *Quantitative Scanning Electron Microscopy*, Academic Press, London (1974).

[146] H. Fink, U. Panne and R. Niessner, *Anal. Chem.* **74**, 4334–42 (2002).

[147] H. Nishimura, T. Yarita and M. Noshiro, *Asahi Garasu Kenkyu Hokoku* **33** (2), 151–60 (1983); *C.A.* **101**, 91757 (1984).

[148] V. Khunová, J. Hurst, I. Janigová and V. Smatko, *Polym. Testing* **18**, 501–9 (1999).

[149] D.J. Weale, J. White and D. Walton, in *Imaging and Image Analysis Applications for Plastics* (B. Pourdeyhimi, ed.), SPE/Plastic Design Library, Norwich, NY (1999), pp. 19–26.

[150] P.J. Hine, N. Davidson, R.A. Duckett and I.M. Ward, *Compos. Sci. Technol.* **53** (2), 125–31 (1995).

[151] S.L. Mills, G.C. Lees, C.M. Liauw and S. Lynch, *Proceedings Eurofillers '97* (British Plastics Federation, ed.), London (1997), pp. 259–62.

[152] S.R. Owen and J.F. Harper, *Polym. Degr. Stabil.* **64** (3), 449–55 (1999).

[153] W. Fan, R. Huang and B. Cai, *Gongneng Cailiao* **29** (Suppl.), 1145–9 (1998).

[154] C. Zhang, H.-F. Han, X.-S. Yi, S. Asai and M. Sumita, *Compos. Interf.* **6** (3), 227–36 (1999).

[155] A. Valadez-Gonzalez, J.M. Cervantes-Uc and L. Veleva, *Polym. Degr. Stabil.* **63** (2), 253–60 (1999).

[156] J.G. Koen, *Techn. Rept. DSM Research*, Geleen (1998).

[157] I. Fras, P. Cassagnau and A. Michel, *Polymer* **40** (5), 1261–9 (1999).

[158] J. Zięba-Palus, *Mikrochim. Acta (Suppl.)* **14**, 357–9 (1997).

[159] W.J. Egan, R.C. Galipo, B.K. Kochanowski, S.L. Morgan, E.G. Bartick, M.L. Miller, D.C. Ward and R.F. Mothershead II, *Anal. Bioanal.* **376**, 1286–97 (2003).

[160] R.H.B. Bouma, W.J. Nauta, J.E.F. Arnauts, Th. van den Boomgaard, J.M. Steuten and H. Strathmann, *J. Appl. Polym. Sci.* **65** (13), 2679–89 (1997).

[161] G. Chinga, T. Helle and P.O. Johnsen, *TAPPI Coat. Conf. Trade Fair*, Washington, DC (2000), pp. 309–16.

[162] A. Szymanski, B. Kaczmarek and V.N. Sokolov, *Przegl. Papier.* **49** (4), 124–9 (1993).

[163] T. Helle and P.O. Johnsen, *J. Pulp Pap. Sci.* **20** (7), 189–92 (1994).

[164] M.W. Tungol, *Diss. Abstr. Int.* **B56** (2), 789 (1995).

[165] L. Reimer, *Image Formation in Low-Voltage Scanning Electron Microscopy*, SPIE Optical Engineering Press, Bellingham, WA (1993).

[166] R. Böngeler, U. Golla, M. Kässens, L. Reimer, B. Schindler, R. Senkel and M. Spranck, *Scanning* **15**, 1 (1993).

[167] R.B. Mott, C.G. Waldman, R. Batcheler and J.J. Friel, *Proceedings Microscopy and Microanalysis 1995*, Kansas City, KS (1995), pp. 593–4.

[168] J.J. Friel and V.A. Greenhut, *J. Am. Ceram. Soc.* **80** (112), 3205–8 (1997).

[169] Special Issue on *Low Voltage SEM*, *J. Microsc.* **140**, No. 3 (1985).

[170] V. Dudler, M.C. Grob and M. Mérian, *Polym. Degr. Stabil.* **68**, 373–9 (2000).

[171] P.A. Zandbelt, unpubl. results (1998).

[172] M. Dickson, unpubl. results (1998).

[173] O.L. Shaffer, R. Bagheri, J.Y. Qian, V. Dimonie, R.A. Pearson and M.S. El-Aasser, *J. Appl. Polym. Sci.* **58**, 465–84 (1995).

[174] J.J. Watkins and T.J. McCarthy, *Polym. Mater. Sci. Engng.* **73**, 158–9 (1995).

[175] G.M. Brown and J.H. Butler, *Polymer* **38**, 3937–45 (1997).

[176] G.D. Danilatos, *Adv. Electronics Electr. Phys.* **71**, 109–250 (1988).

[177] G.D. Danilatos and V.N.E. Robinson, *Scanning* **2**, 72–82 (1979).

[178] E. Doehne, in *In-situ Microscopy in Materials Research* (P.L. Gai, ed.), Kluwer Academic Publishers, Boston (1997), pp. 45–62.

[179] S.P. Collins, R.K. Pope, R.W. Sheetz, R.I. Ray, P.A. Wagner and B.J. Little, *Microsc. Res. Techn.* **25**, 398–405 (1993).

[180] C. Mathieu, *Eur. Microsc. Anal.* (7), 5–7 (1999).

[181] J.H. Butler, D.C. Joy, G.F. Bradley and S.J. Krause, *Polymer* **36**, 1781 (1995).

[182] D.L. Vezie, E.L. Thomas and W.W. Adams, *Polymer* **36**, 1761 (1995).

[183] G.D. Danilatos, *J. Microsc.* **160**, 9–19 (1990).

[184] G.D. Danilatos, *Microsc. Res. Techn.* **25**, 354–61 (1993).

[185] R. Johnson, *Environmental Scanning Electron Microscopy*, Philips ElectroScan Corp., Wilmington, MA (1996), 55 pp.

[186] G.D. Danilatos, in *In-situ Microscopy in Materials Research* (P.L. Gai, ed.), Kluwer Academic Publishers, Boston, MA (1997), pp. 13–44.

[187] R.E. Cameron, *Microsc. Anal.* (6), 17–9 (1994).

[188] J. Cheng, R. Sanchez and D.I. Bigio, in *Imaging and Image Analysis Applications for Plastics* (B. Pourdeyhimi, ed.), SPE/Plastics Design Library, Norwich, NY (1999), pp. 79–88.

[189] G.D. Danilatos and J.H. Brooks, *Proceedings 7th Intl. Wood Textile Research Conference*, Tokyo, Vol. 1, pp. 263–72 (1985).

[190] R.R. Mather, Q.F. Wei, O.K. Risnes, J. Buckman, A.F. Fotheringham and A. Neville, *Eur. Microsc. Anal.* **83**, 23–4 (2003).

[191] H.M. Choi and J.P. Moreau, *Microsc. Res. Techn.* **25** (5–6), 447–55 (1993).

[192] W.R. Pelton, *J. Forensic Sci.* **40** (5), 874–82 (1995).

[193] R.E. De La Parra, *Microsc. Res. Techn.* **25** (5–6), 362–73 (1993).

[194] J.H. Rask, J.E. Flood, J.K. Borchardt and G.A. York, *Microsc. Res. Techn.* **25**, 384–92 (1993).

[195] J.K. Borchardt and J.H. Rask, *Tappi J.* **77** (9), 161–9 (1994).

[196] J.K. Borchardt, J.H. Rask, G.A. York and K. Cathie, *Progr. Pap. Recycl.* **4** (4), 16–27 (1995).

[197] D.A. Scott, N. Khandekar, M.R. Schilling, N. Turner, Y. Taniguchi and H. Khanjian, *Stud. Conserv.* **46**, 93–108 (2001).

[198] G.-M. Kim and D.-H. Lee, *J. Appl. Polym. Sci.* **82**, 785–9 (2001).

[199] G.D. Danilatos, *Microsc. Res. Techn.* **25**, 529–34 (1993).

[200] R.E. Cameron, *Trends Polym. Sci.* **2** (4), 116–20 (1994).

[201] N.W. Bower, D.C. Stulik and E. Doehne, *Fresenius' J. Anal. Chem.* **348** (5–6), 402–10 (1994).

[202] C.-O.A. Olsson, S.E. Hörnström and S. Hogmark, in *Surface Characterization. A User's Sourcebook* (D. Brune, R. Hellborg, H.J. Whitlow and O. Hunderi, eds.), Wiley-VCH, Weinheim (1997), Chp. 17.

[203] W. Grahneis, C. Ziethen, G.H. Fecher and G. Schonhense, *Jpn. J. Appl. Phys., Pt. 1* **38** (Suppl. 1), 317–20 (1999).

[204] S. Contarini, E. Lambers and P.H. Holloway, *Appl. Surf. Sci.* **62** (3), 181–8 (1992).

[205] S.-S. Lin, *Rubber Chem. Technol.* **58**, 885 (1985).

[206] C. Cazeneuve, J.E. Castle and J.F. Watts, *J. Mater. Sci.* **25**, 1902–8 (1990).

[207] K. Urban, *Mater. World* **11** (6), 20–1 (2003).

[208] P. Duncumb, *J. Microscopie* **7**, 581–9 (1968).

[209] M. Watanabe, Z. Horita and M. Nemoto, *Ultramicrosc.* **65**, 187–98 (1996).

[210] O.L. Krivanek, C. Mory, M. Tencé and C. Colliex, *Microsc. Microanal. Microstruct.* **2**, 257–67 (1991).

[211] I.G. Voigt-Martin, in *Applied Polymer Analysis and Characterization* (J. Mitchell, ed.), Hanser Publishers, Munich (1987), pp. 277–96.

[212] A.H. Reader, in *Encyclopedia of Analytical Sciences* (A. Townshend, ed.), Academic Press, London (1995), pp. 3160–8.

[213] B. Fultz and J.M. Howe, *Transmission Electron Microscopy and Diffractometry of Materials*, Springer-Verlag, Berlin (2001).

[214] D.B. Williams and C.B. Carter, *Transmission Electron Microscopy. A Textbook for Materials Science*, Plenum Press, New York, NY (1996).

[215] P. Goodhew, F.J. Humphreys and R. Beanland, *Electron Microscopy and Analysis*, Taylor & Francis, London (2001).

[216] L. Reimer, *Transmission Electron Microscopy, Physics of Image Formation and Microanalysis*, Springer-Verlag, Berlin (1989).

[217] B.J. Martin and A. George, *Caracterisation Experimentale des Materiaux. II. Analyse par Rayons X, Electrons et Neutrons*, Presses Polytechniques et Universitaires Romandes, Lausanne (1998).

[218] A.V. Antonov, R.M. Gitina and S.N. Novikov, *Polym. Sci. USSR* **32** (9), 1809–15 (1990).

[219] M.P. Luda di Cortemiglia, G. Camino and L. Costa, *J. Anal. Appl. Pyrol.* **11**, 511–26 (1987).

[220] H. Dvir, M. Goldraich, M. Gottlieb and S. Daren, *Proceedings 1st Intl. Conference on Modification, Degradation and Stabilisation (MoDeSt)*, Palermo (2000).

[221] D. Allen, B.J. Briscoe and D. Tabor, *Wear* **25** (3), 393–7 (1973).

[222] W.M. Hess, G.C. McDonald and E. Urban, *Mtg. ACS Rubber Division*, Cincinnati, OH (1972).

[223] D. Brück, in *Moderne Prüfmethoden – von Rohstoffen über Mischungen bis zum Produkt*, DIK, Hannover (2000).

[224] D. Coz and K. Baranwal, *Rubber World*, 30–5 (Jan. 1999).

[225] F. Mersch and R. Zimmer, *Kautsch. Gummi Kunstst.* **39**, 427 (1986).

[226] J.J. Maurer and D.W. Brazier, *Kautsch. Gummi Kunstst.* **36**, 37 (1983).

[227] P. Pautrat, B. Metivier and J. Marteau, *Rubber Chem. Technol.* **49**, 1060–7 (1976).

[228] H.R. Dennis, D.L. Hunter, D. Chang, S. Kim, J.L. White, J.W. Cho and D.R. Paul, *Plast. Engng.*, 56–60 (Jan. 2001).

[229] S.G. Lyu, Y.C. Lee and G.S. Sur, *Kongop Hwahak* **11** (5), 512–6 (2000).

[230] J. Ma, Z. Qi and Y. Hu, *J. Appl. Polym. Sci.* **82** (14), 3611–7 (2001).

[231] M. Zanetti. G. Camino, D. Canavese, A.B. Morgan, F.J. Lamelas and C.A. Wilkie, *Chem. Mater.* **14** (1), 189–93 (2002).

[232] A. Ranade, N.A. D'Souza, B. Gnade and T.D. Golden, *Proceedings SPE ANTEC 2001*, Dallas (2001), pp. 2171–5.

[233] H.R. Dennis, D.L. Hunter, D. Chang, S. Kim, J.L. White, J. Whan Cho and D.R. Paul, *Proceedings SPE ANTEC 2000*, Orlando, FL (2000), pp. 428–33.

[234] H. White and J. Fenton, *Eur. Microsc. Anal.* **84** (7), 21–3 (2003).

[235] T.C. Chao, G.T. Burns and D.E. Katsoulis, *Polym. Mater. Sci. Eng.* **82**, 266–7 (2000).

[236] P.M. Ajayan, P.V. Braun and L.S. Schadler, *Nanocomposite Science and Technology*, Wiley-VCH, Weinheim (2003).

[237] J. Scheirs, O. Delatycki, S.W. Bigger and N.C. Billingham, *Polymer Intl.* **26**, 187–93 (1991).

[238] F. Lednický, E. Coufalová, J. Hromádková, A. Delong and V. Kolařík, *Polymer* **41**, 4909–14 (2000).

[239] A.V. Crewe, J. Wall and L.M. Welter, *J. Appl. Phys.* **39**, 5861–8 (1968).

[240] A.V. Crewe, J. Wall and J. Langmore, *Science* **168**, 1338–40 (1970).

[241] P.E. Batson, N. Dellby and O.L. Krivanek, *Nature* **418**, 617–20 (2002).

[241a] P.D. Nellist, M.F. Chisholm, N. Delby, O.L. Krivanek, M.F. Murfitt, Z.S. Szilagyi, A.R. Lupini, A. Borisevich, W.H. Sides Jr. and S.J. Pennicook, *Science* **305** (5691), 1741 (2004).

[242] C.-Y. Wang. H.-X. Huang and Z.-M. Xie, *Proceedings SPE ANTEC 2003*, Nashville, TN (2003), pp. 2219–22.

[243] A. Ranade, N.A. D'Souza and B. Gnade, *Polymer* **43**, 3759 (2002).

[244] J. Hillier and R.F. Baker, *J. Appl. Phys.* **15**, 663–75 (1944).

[245] R.F. Egerton, *Electron Energy-Loss Spectroscopy in the Electron Microscope*, Plenum Press, New York, NY (1996).

[246] C.C. Ahn, M.M. Disko and B. Fultz, *Transmission Electron Energy Loss Spectrometry and the EELS Atlas*, J. Wiley & Sons, Chichester (2002).

[247] G.A. Hutchins, in *Characterizations of Solid Surfaces* (P.F. Kane and G.B. Larrabee, eds.), Plenum Press, New York, NY (1974), p. 468.

[248] K.F.J. Heinrich and D.E. Newbury (eds.), *Electron Probe Quantitation*, Plenum Press, New York, NY (1991).

[249] M.G. Hall, *Surf. Eng.* **9** (3), 205–12 (1993).

[250] I.M. Anderson, *Inst. Phys. Conf. Ser. No.* **165** (Symposium 14), 437–8 (2000).

[251] R. Castaing, *Ph.D. Thesis*, University of Paris (1951), ONERA Publication # 55.

[252] R. Castaing, *Electron Physics*, NBS Circular **527**, 305–8 (1954).

[253] G. Cliff and G.W. Lorimer, *J. Microsc.* **103**, 203 (1975).

[254] J.I. Goldstein, J.L. Costley, G.W. Lorimer and S.J.B. Reed, *Scann. Electr. Microsc.*, 315–24 (1977).

[255] M. Andrae, P. Klein, K. Röhrbacher and J. Wernisch, *X-Ray Spectrom.* **25**, 78–82 (1996).

[256] J. Wernisch, *Abstracts 5th Eur. Workshop on Modern Developments and Applications in Microbeam Analysis (EMAS '97)*, Torquay (1997), pp. 27–37.

[257] R.A. Waldo, M.C. Militello and S.W. Gaarenstroom, *Surf. Interf. Anal.* **20**, 111 (1993).

[258] J.M. Titchmarsh and S. Dumbill, *J. Microscopy* **184**, 195–207 (1996).

[259] P. Karduck, *Abstracts 5th Eur. Workshop on Modern Developments and Applications in Microbeam Analysis (EMAS '97)*, Torquay (1997), pp. 39–63.

[260] S.R.J. Saunder, P. Karduck, W.G. Sloof, M. Whitwood, P. Busch, J. Almagro, J.-F. Thiot, T. Wirth and H. Schneider, *Inst. Phys. Conf. Ser. No.* **165**, 393–4 (2000).

[261] ASTM Standard E 1508-98, *Guide for Quantitative Analysis by Energy-Dispersive Spectroscopy, ASTM Annual Book of Standards*, ASTM, West Conshohocken, PA (1998), Vol. 03.01.

[262] C.E. Lyman, *Scanning Microsc. Suppl.* **1**, 123–34 (1986).

[263] D.B. Williams, J.I. Goldstein and D.E. Newbury (eds.), *X-Ray Spectrometry in Electron Beam Instruments*, Plenum Press, New York, NY (1995).

[263a] A.J. Garratt-Reed and D.C. Bell, *Energy-Dispersive X-Ray Analysis in the Electron Microscope*, Bios Scientific, Oxford (2003).

[264] D. Vesely, *Mater. World* **10** (11), 23–4 (2002).

[265] E. Marioth, G. Bunte and Th. Härdle, *Polym. Recycl.* **2** (4), 303–8 (1996).

[266] H.E. Bair, *Polym. Eng. Sci.* **13**, 435–9 (1973).

[267] A. Berner and O. Sideleva, *Inst. Phys. Conf. Ser. No.* **165** (Symposium 9), 299–300 (2000).

[268] Y. Ohtake, *Kinzoku* **69** (5), 441–2 (1999).

[269] Special Millennium Issue on *Cultural Heritage*, in *X-Ray Spectrometry* **29** (1) (2000).
[270] A.J. Fisher, in *Encyclopedia of Spectroscopy and Spectrometry* (J.C. Lindon, G.E. Tranter and J.L. Holmes, eds.), Academic Press, San Diego, CA (2000), pp. 2060–6.
[271] W. Sacks and C. Noguera, *J. Vac. Sci. Technol.* **B9** (2), 488–91 (1991).
[272] M.K. Miller, *Mater. Character.* **44**, 11 (2000).
[273] T. Fuji, M. Suzuki, H. Yamada and K. Nakayama, *Thin Solid Films* **243** (1–2), 407–10 (1994).
[274] G.S. McCarthy and P.S. Weiss, *Proceedings SPE ANTEC 2001*, Dallas, TX (2001), pp. 1946–50.
[275] D. Bonnell (ed.), *Scanning Probe Microscopy and Spectroscopy – Theory, Techniques and Applications*, Wiley-VCH, Weinheim (2001).
[276] R. Wiesendanger, *Scanning Probe Microscopy and Spectroscopy. Methods and Applications*, Cambridge University Press, Cambridge (1994).
[277] R. Wiesendanger (ed.), *Scanning Probe Microscopy*, Springer-Verlag, Berlin (1998).
[278] J.G. Kushmerick and P.S. Weiss, in *Encyclopedia of Spectroscopy and Spectrometry* (J.C. Lindon, G.E. Tranter and J.L. Holmes, eds.), Academic Press, San Diego, CA (2000), pp. 2043–51.
[279] R.M. Overney and V.V. Tsukruk, *ACS Symp. Ser.* **694**, 2–31 (1996).
[280] B.D. Ratner and V.V. Tsukruk (eds.), *Scanning Probe Microscopy of Polymers*, American Chemical Society, Washington, DC (1998).
[281] S.P. Tear, in *Spectroscopy for Surface Science* (R.J.H. Clark and R.E. Hester, eds.), J. Wiley & Sons, Chichester (1998), pp. 121–58.
[282] F. Besenbacher, *Rept. Progr. Phys.* **59**, 1737 (1996).
[283] F. Besenbacher and I. Stensgaard, in *Encyclopedia of Analytical Sciences* (A. Townshend, ed.), Academic Press, London (1995), pp. 3196–203.
[284] C.J. Chen, *Introduction to Scanning Tunneling Microscopy*, Oxford University Press, New York, NY (1993).
[285] D.A. Bonnell (ed.), *Scanning Tunneling Microscopy and Spectroscopy: Theory, Techniques and Applications*, VCH, Weinheim (1993).
[286] V. Magonov and M.H. Whangbo, *Surface Analysis with STM/AFM, Experimental & Theoretical Aspects of Image Analysis*, J. Wiley & Sons, New York, NY (1996).
[287] Ph. Leclère, R. Lazzaroni, R. Lazzaroni, F. Gubbels, M. De Vos, R. Deltour, R. Jérôme and J.L. Bredas, *ACS Symp. Ser.* **694**, 129–40 (1998).
[288] S.J. Kim and D.H. Reneker, *Rubber Chem. Technol.* **66** (4), 559–66 (1993).
[289] V.N. Bliznyuk, J.L. Hazel, J. Wu and V.V. Tsukruk, *ACS Symp. Ser.* **694**, 252–65 (1998).
[290] C.J. Roberts, M.C. Davies, S.J.B. Tendler and P.M. Williams, in *Encyclopedia of Spectroscopy and Spectrometry* (J.C. Lindon, G.F. Tranter and J.L. Holmes), Academic Press, San Diego, CA (2000), pp. 2051–9.
[291] K.S. Birdi, *Scanning Probe Microscopes. Applications in Science and Technology*, CRC Press, Boca Raton, FL (2003).
[292] G. Binnig, C.F. Quate and Ch. Gerber, *Phys. Rev. Lett.* **56**, 930–3 (1986).
[293] D. Sarid, *Scanning Force Microscopy with Applications to Electric, Magnetic and Atomic Forces*, Oxford University Press, New York, NY (1991–94).
[294] B. Drake, C.B. Prater, A.L. Weisenhorn, S.A.C. Gould and P.K. Hansma, *Science* **244**, 1586–9 (1989).
[295] M. Rigby, E.B. Smith, W.A. Wakeham and G.C. Maitland, *The Forces Between Molecules*, Clarendon Press, Oxford (1986).
[296] R. Howland and L. Benatar, *A Practical Guide to Scanning Probe Microscopy*, Park Scientific Instruments, Sunnyvale, CA (1997).
[297] R. Lüthi, E. Meyer, M. Bammerlin, A. Baratoff, J. Lü, M. Guggisberg and H.-J. Günterodt, *ACS Symp. Ser.* **694**, 300–11 (1996).
[298] C.D. Frisbie, W. Rozsnyai, A. Noy, M.S. Wrington and C.M. Lieber, *Science* **265**, 2071 (1994).
[299] B. Knoll and F. Keilmann, *Nature* **399**, 134–7 (1999).
[300] J.A. Bailey, R.B. Dyer, D.K. Graff and J.R. Schoonover, *Appl. Spectrosc.* **54** (2), 159–63 (2000).
[301] A. Hammiche, H.M. Pollock, M. Reading, M. Claybourn, P.H. Turner and K. Jewkes, *Appl. Spectrosc.* **53**, 810–5 (1999).
[302] M.S.Anderson, *Appl. Spectrosc.* **54** (3), 349–52 (2000).
[303] S.N. Magonov and D.H. Reneker, *Annu. Rev. Mater. Sci.* **27**, 175–222 (1997).
[304] A.J. Miller (ed.), *Surface Characterization Methods*, M. Dekker, New York, NY (1999).
[305] D.V. Vezenov, A. Noy and C.M. Lieber, *ACS Symp. Ser.* **694**, 312–20 (1998).
[306] A. Noy, D.V. Vezenov and C.M. Lieber, *Annu. Rev. Mater. Sci.* **27**, 381–421 (1997).
[307] O. Marti, H.O. Ribi, B. Drake, T.R. Albrecht, C.F. Quate and P.K. Hansma, *Science* **239**, 50 (1988).
[308] M. Radmacher, R.W. Tillman, M. Fritz and H.E. Gaub, *Science* **257**, 1900–5 (1992).
[309] P. Maivald, H.-J. Butt, S.A.C. Gould, S.B. Prater, B. Drake, J.A. Gurley, V.B. Elings and P.K. Hansma, *Nanotechnol.* **2**, 103–6 (1991).
[310] A.A. Galuska, R.R. Poulter and K.O. Mcelrath, *Surf. Interf. Anal.* **25** (6), 418–29 (1997).

[311] B.D. Beake, J.S.G. Ling and G.J. Leggett, *Polymer* **41**, 2241–8 (2000).

[312] F. Negroni, in *Moderne Prüfmethoden – von Rohstoffen über Mischungen bis zum Produkt*, DIK, Hannover (2000).

[313] F. Creuzet, G. Ryschenkow and H. Arribart, *J. Adhes.* **40** (1), 15–25 (1992).

[314] P. Mele, D. Brown, S. Marceau, C. Da Silva, Y. De Puydt and N.D. Alberola, *Proceedings Eurofillers*, Łódź (2001), C-28.

[315] F. Clement, A. Lapra, L. Bokobza, L. Monnerie and P. Menez, *Polymer* **42** (14), 6259–70 (2001).

[316] S. Magonov and Y. Godovsky, *Am. Lab.* **31** (8), 52–8 (1999).

[317] S. Magonov and M.G. Heaton, *Am. Lab.* **30** (10), 10–2 (May 1998).

[318] C.S. Rodríguez, J.M. Fernández, M.A. Correa-Duarte and L.M. Liz-Marzán, *Eur. Microsc. Anal.* (65), 11–3 (2000).

[319] A.V. Janorkar, D.E. Hirt and J.J. Wooster, *Proceedings SPE ANTEC 2002*, San Francisco, CA (2002), pp. 2887–91.

[320] M.X. Ramirez, D.E. Hirt and L.L. Wright, *Nano Letters* **2** (1), 9–12 (2002).

[321] J.A. Ortuño, J. Colchero and J. Abellán, *Proceedings Colloquium Spectroscopicum Internationale XXXIII*, Granada (2003), p. 519.

[322] A. Haunschild, A. Pfau, J. Schmidt-Thümmes, K. Graf, D. Lawrenz and N. Gispert, *Proceedings TAPPI Coating Papermakers Conference 1998* (TAPPI Press), New Orleans, LA (1998), Vol. 1, p. 301.

[323] X. Gu, M. VanLandigham, D. Raghavan and T. Nguyen, *Proceedings ACS Div. Polym. Mater.: Sci. & Engng.* **82**, 50–1 (2000).

[324] B. Pourdeyhimi, X. Wang and F. Lee, in *Imaging and Image Analysis Applications for Plastics* (B. Pourdeyhimi, ed.), SPE/Plastics Design Library, Norwich, NY (1999), pp. 107–17.

[325] R. Orefice and A. Brennan, *Polim.: Cienc. Tecnol.* **8** (4), 82–9 (1998).

[326] C.M. Mate, *Phys. Rev. Lett.* **68** (22), 3323–6 (1992).

[327] G.W. Bao, M. Troemel and S.F.Y. Li, *Appl. Phys. A: Mater. Sci. (Suppl., Pt. 2)*, S 1283–8 (1998).

[328] H. Schönherr, Z. Hruska and G.J. Vancso, *Proceedings ACS Div. Polym. Mater.: Sci. & Engng.* **81**, 383–4 (1999).

[329] A.-S. Duwez, C. Poleunis, P. Bertrand and B. Nysten, *Langmuir* **17**, 6351–7 (2001).

[330] A.-S. Duwez and B. Nysten, *Langmuir* **17**, 8287–92 (2001).

[331] P.T. Lillehei, C. Park, J.H. Rouse and E.J. Siochi, *Nano Lett.* **2** (8), 827–9 (2002).

[332] E.H. Synge, *Philos. Mag.* **6**, 356 (1928).

[333] D.W. Pohl, W. Denk and M. Lanz, *Appl. Phys. Lett.* **44**, 651–3 (1984).

[334] A. Lewis, M. Isaacson, A. Harootunian and A. Muray, *Ultramicrosc.* **13**, 227–32 (1984).

[335] M. Isaacson, E. Betzig, A. Harootunian and A. Lewis, *Ann. N.Y. Acad. Sci.* **483**, 448–56 (1986).

[336] D. Higgins and P. Barbara, *J. Phys. Chem.* **99**, 3 (1995).

[337] D. Zeisel, B. Dutoit, V. Deckert, T. Roth and R. Zenobi, *Anal. Chem.* **69**, 749–54 (1997).

[338] S.J. Stranick, D.B. Chase and C.A. Michaels, *Proceedings SPE ANTEC 2001*, Dallas (2001), pp. 1961–5.

[339] C.L. Jahncke, H.D. Hallen and M.A. Paesler, *J. Raman Spectrosc.* **27**, 579 (1996).

[340] S. Webster, D.A. Smith and D.N. Batchelder, *Spectrosc. Europe* **10** (4), 22–7 (1998).

[341] R. Zenobi, *Proceedings Colloquium Spectroscopicum Internationale XXXIII*, Granada (2003), p. 133.

[342] M. Fujihira, H. Monobe, H. Muramatsu and T. Ataka, *Chem. Lett.* (3), 657–60 (1994).

[343] B. Rosner and D. van der Weide, *Rev. Sci. Instrum.* **73** (7), 2505–25 (2002).

[344] M.A. Paesler, *Near-Field Optics: Theory, Instrumentation and Applications*, Wiley-Interscience, New York, NY (1996).

[345] J.K. Trautman and J.J. Macklin, *Chem. Phys.* **205**, 221 (1996).

[346] T.D. Harris, D. Gershoni, R.D. Grober, L. Pfeiffer, K. West and N. Chand, *Appl. Phys. Lett.* **68**, 988 (1996).

[347] S. Madsen, M. Mullenborn, K. Birkelund and F. Grey, *Appl. Phys. Lett.* **69**, 544 (1996).

[348] C.P. Vlahacos, D.E. Steinhauer, S.M. Anlage, F.C. Wellstood, S.K. Dutta and J.B. Feenstra, *Eur. Microsc. Anal.*, 29–31 (Jan. 2000).

[349] H. Sakai and A. Fujishima, *Kokagaku* **18**, 39–45 (1994).

[350] H. Shiku and R.C. Dunn, *Anal. Chem.* **71**, 23–8A (1999).

[351] H. Heinzelmann and D.W. Pohl, *Appl. Phys. A: Solids Surf.* **A59** (2), 89–101 (1994).

[352] O. Marti and G. Krausch, *Phys. Bl.* **51** (6), 493–6 (1995).

[353] D.A. Smith, S. Webster, M. Ayad, S.D. Evans, D. Fogherty and D. Batchelder, *Ultramicrosc.* **61**, 247–52 (1995).

[354] G. Tarrach, M.A. Bopp, D. Zeisel and A.J. Meixner, *Rev. Sci. Instrum.* **66** (6), 3569–75 (1995).

[355] A.J. Meixner, D. Zeisel, M.A. Bopp and G. Tarrach, *Opt. Eng.* **34**, 2324–2 (1995).

[356] T. Baschke, *Angew. Chem.* **106** (17), 1805–7 (1994).

[357] Y. Narita, T. Tadokoro, T. Ikeda, T. Saiki, S. Mononobe and M. Ohtsu, *Appl. Spectrosc.* **52** (9), 1141–4 (1998).
[358] D. Zeisel, V. Deckert, R. Zenobi and T. Vo-Dinh, *Chem. Phys. Lett.* **283** (5–6), 381–5 (1998).
[359] R.M. Stockle, Y.D. Suh, V. Deckert and R. Zenobi, *Chem. Phys. Lett.* **318**, 131–6 (2000).
[360] S.R. Emory and S.M. Nie, *Anal. Chem.* **69**, 2631–5 (1997).
[361] R.M. Stockle, V. Deckert, C. Fokas, D. Zeisel and R. Zenobi, *Vibr. Spectrosc.* **22**, 39–48 (2000).
[362] W.T. Kelvin, *Phil. Mag.* **46**, 82 (1898).
[363] W.A. Zismann, *Rev. Sci. Instrum.* **3**, 367 (1932).
[364] R. Blum, A. Ivankov, St. Schwantes and M. Eich, *Appl. Phys. Lett.* **76**, 604 (2000).
[365] J.R. Harbour, M.J. Walzak, W. Limburg and J. Yanus, *Carbon* **24**, 725 (1986).
[366] R. Mäckel, H. Baumgärtner and J. Ren, *Rev. Sci. Instrum.* **64**, 694 (1993).
[367] T. Prasse, A. Ivankov, J. Sandler, K. Schulte and W. Bauhofer, *J. Appl. Polym. Sci.* **82** (13), 3381–6 (2001).
[368] T.O. Caspersson, *J. Roy. Microsc. Soc.* **60**, 8–25 (1940).
[369] T.O. Caspersson, *Z. Wiss. Mikr.* **53**, 403–9 (1936).
[370] X. Wang and E.N. Lewis, in *Fluorescence Imaging Spectroscopy and Microscopy* (X.F. Wang and B. Herman, eds.), J. Wiley & Sons, New York, NY (1996).
[371] E.N. Lewis, P.J. Treado and I.W. Levin, *Am. Lab.* (6), 16–21 (1994).
[372] S. Fang, *Intl. Lab.* **31** (5A), 6–7 (2001).
[373] S. De Goy, D. Lanzisera, K. Lopez, J. Noonan and J. Rebello, *Spectroscopy* **13** (10), 36–42 (1998).
[374] G.J. Havrilla and J.R. Schoonover, *Mater. World* **7** (10), 613–5 (1999).
[375] S.K. Taylor and W.F. McClure, *Proceedings 2nd Intl. NIRS Conference* (M. Iwamoto and S. Kawano, eds.), Korin Publishing, Tokyo (1989), p. 393.
[376] P. Robert, D. Bertrand, M.F. Devaux and A. Sire, *Anal. Chem.* **64**, 664–7 (1992).
[377] D. Bertrand, P. Robert, B. Novales and M.F. Devaux, in *Near Infrared Spectroscopy. The Future Waves* (A.M.C. Davies and P. Williams, eds.), NIR Publications, Chichester (1996), pp. 174–8.
[378] P. Geladi. S. Wold and K. Esbensen, *Anal. Chim. Acta* **191**, 473–80 (1986).
[379] A.S. El-Hagrasy, H.R. Morris, F. D'Amico, R.A. Lodder and J.K. Drennen III, *J. Pharm. Sci.* **90**, 1298, 1915 (2001).
[380] J. van den Berg, J. van Oyen and H.W. Werner, *Anal. Chim. Acta* **297** (1–2), 73–86 (1994).
[381] R.S. Ledley, G. Di Chiro, A.J. Lüssenhop and H.L. Twigg, *Science* **186**, 207 (1974).
[382] K.R. Erikson, F.J. Fry and J.P. Jones, *IEEE Trans. Sonics Ultrasonics* **SU-21**, 144 (1974).
[383] P.C. Lauterbur, *Nature* **242**, 190 (1973).
[384] N.J. Everall, J.M. Chalmers, L.H. Kidder, E.N. Lewis, M. Schaeberle and I. Levin, *Proceedings ACS Div. Polym. Mater.: Sci. & Engng.* **82**, 398–9 (2000).
[385] A. Briggs, in *Materials Science and Technology* (R.W. Cahn, P. Haasen and E.J. Kramer, eds.), VCH Publishers, New York, NY (1994), pp. 241–79.
[386] P. Mutti and G.A.D. Briggs, in *Microanalysis of Solids* (B.G. Yacobi, D.B. Holt and L.L. Kazmerski, eds.), Plenum Press, New York, NY (1994), pp. 327–55.
[387] R.J.M. De Fonseca, L. Ferdj-Allah, G. Despaux, A. Boudour, L. Robert and J. Atal, *Adv. Mater.* **5** (7–8), 508–19 (1993).
[388] B.T. Khuri-Yakub, *Ultrasonics* **31** (5), 361–72 (1993).
[389] R.M. Miller, in *Encyclopedia of Analytical Sciences* (A. Townshend, ed.), Academic Press, London (1995), pp. 2042–8.
[390] M.D. Morris (ed.), *Microscopy and Spectroscopic Imaging of the Chemical State*, M. Dekker, New York, NY (1993).
[391] B. Jähne, *Practical Handbook on Image Processing for Scientific Applications*, CRC Press, Boca Raton, FL (1997).
[392] R. Baldock and J. Graham (eds.), *Image Processing and Analysis. A Practical Approach*, Oxford University Press, Oxford (2000).
[393] J.L. Koenig, *Microspectroscopic Imaging of Polymers*, American Chemical Society, Washington, DC (1997).
[394] K.G. Lloyd, D.J. Walls, G.S. Blackman and N. Tassi, *Abstracts SIMS XII*, Brussels (1999), p. 304.
[395] C.H. Stapfer and U. Bosch, *Proceedings 3rd European Additives & Colors Conference* (J.M. Liégois, ed.), SPE Benelux, Antwerp (2003), pp. 92–7; *Proceedings SPE ANTEC 2002*, San Francisco, CA (2002), pp. 2548–51.
[396] A. Todoroki, H. Kobayashi and J.G. Lee, *Compos. Sci. Technol.* **52** (4), 551–2 (1994).
[397] V.M. Levin, S. Liu and E. Guo, *Proceedings Mater. Res. Soc. Symp.* **702** (Advanced Fibers, Plastics, Laminates and Composites), San Francisco, CA (2002), pp. 217–22.
[398] J.W. Mitchell, S. Yegnasubramanian and L. Shepherd, *J. Radioanal. Nucl. Chem.* **112**, 425 (1987).
[399] V. Dudler and C. Muiños, *Adv. Chem. Ser.* **249**, 441–53 (1996).
[400] A.J. Sommer, P.A. Martoglio and J.E. Katon, *Appl. Spectrosc.* **44**, 1471 (1990).

[401] R. Sehan and W. Kimmer, *Jena Rev.* **35** (1), 31–3 (1990).

[402] R.O. Carter III, J.L. Gerlock and C.A. Smith, *Proceedings SPE ANTEC '99*, New York, NY (1999), pp. 2480–4.

[403] D.R. Bauer, *J. Coat. Technol.* **66**, 57 (1994).

[404] J.E. Pickett and J.E. Moore, *Polym. Prepr.* **34** (2), 153 (1993).

[405] D.R. Bauer and L.M. Briggs, *ACS Symp. Ser.* **234**, 271 (1984).

[406] J.L. Dupuie, W.H. Weber, D.J. Scholl and J.L. Gerlock, *Polym. Degr. Stabil.* **57**, 339 (1997).

[407] W.H. Weber and J.T. Remillard, *Proceedings 7th Annu. ESD Advanced Coatings Technology Conf. Exhib.*, Detroit, MI (1998).

[408] K.R. Carduner, D.R. Bauer, J.L. Gerlock, M. Rokosz and D.F. Mielewski, *Polym. Degr. Stabil.* **35**, 219 (1991).

[409] J.L. Gerlock and D.R. Bauer, *J. Polym. Sci., Polym. Lett.* **22**, 477 (1984).

[410] J.L. Gerlock, T.J. Prater, S.L. Kaberline and J.E. DeVries, *Polym. Degr. Stabil.* **47**, 405 (1995).

[411] A.M. Belu and J. McGinness, *Polym. Prepr. (ACS Div. Polym. Mater. Sci. & Engng.)* **78**, 80–1 (1998).

[412] D.F. Mielewski, D.R. Bauer and J.L. Gerlock, *Polym. Degr. Stabil.* **33**, 93 (1991).

[413] P.A. Martoglio, S.P. Bouffard, A.J. Sommer, J.E. Katon and K.A. Jakes, *Anal. Chem.* **62**, 1123A (1990).

[414] R. Macrae, R.J. Dudley and K.W. Smalldon, *J. Forensic Sci.* **24**, 117–29 (1978).

[415] S.P. Bouffard, A.J. Sommer, J.E. Katon and S. Godber, *Appl. Spectrosc.* **48** (11), 1387–93 (1994).

[416] M.Y. Choudhry, *J. Forensic Sci.* **36**, 366–75 (1991).

[417] H.J. Humecki (ed.), *Practical Guide to Infrared Microspectroscopy*, M. Dekker, New York, NY (1995).

[418] R.G. Messerschmidt and M. Harthcock (eds.), *Infrared Microscopy. Theory and Applications*, M. Dekker, New York, NY (1988).

[419] P.B. Roush (ed.), *The Design, Sample Handling and Applications of Infrared Microscopes*, ASTM STP **949**, ASTM, Philadelphia, PA (1987).

[420] P. Lang, in *Handbook of Surface Imaging and Visualization* (A.T. Hubbard, ed.), CRC Press, Boca Raton, FL (1994).

[420a] R. Barer, A.R.H. Cole and H.W. Thompson, *Nature* **163**, 198–201 (1949).

[421] P. Martinsen, P. Schaare and M. Andrews, *J. Near Infrared Spectrosc.* **7**, 17–25 (1999).

[422] R. Bhargava, B.G. Wall and J.L. Koenig, *Appl. Spectrosc.* **54** (4), 470–9 (2000).

[423] J.A. Reffner, P.A. Martoglio and G.P. Williams, *Rev. Sci. Instr.* **66**, 1298–302 (1995).

[424] T. Yokoyama, *Nippon Sekigaisen Gakkaishi* **8** (2), 62–9 (1998).

[425] R. Bhargava and I. Levin, *Appl. Spectrosc.* **57** (4), 357–66 (2003).

[426] A. Reffner, *Microscopy Today* **3**, 6 (1993).

[427] T.J. Allen, *Vibr. Spectrosc.* **3**, 217 (1992).

[428] C.J. Curry, M.J. Whitehouse and J.M. Chalmers, *Appl. Spectrosc.* **39**, 174 (1985).

[429] J.M. Chalmers, M.W. Mackenzie, J.L. Sharp and R.N. Ibert, *Anal. Chem.* **59**, 415–8 (1987).

[430] J.A. Reffner and P.A. Martoglio, in *Practical Guide to Infrared Microspectroscopy* (H.J. Humecki, ed.), M. Dekker, New York, NY (1995), Chp. 2.

[431] L.L. Lewis and A.J. Sommer, *Appl. Spectrosc.* **54** (2), 324–30 (2000).

[432] P.L. Lang and L.J. Richwine, in *Practical Sampling Techniques for Infrared Analysis* (P.B. Coleman, ed.), CRC Press, Boca Raton, FL (1993), pp. 145–63.

[433] M.K. Hong, S. Erramilli, P. Huie, G. James and A. Jeung, *Proceedings SPIE – Int. Soc. Opt. Eng.* **2863**, 54–63 (1996).

[434] G. Lachenal, A. Pierre and N. Poisson, *Micron* **27**, 329–34 (1996).

[435] G. Lachenal and I. Stevenson, *NIR News* **6** (2), 10 (1995).

[436] P.J. Treado and M.D. Morris, *Appl. Spectrosc. Rev.* **29**, 1–38 (1994).

[437] P.J. Treado, I.W. Levin and E.N. Lewis, *Appl. Spectrosc.* **48** (5), 607–15 (1994).

[438] J. Workman, *NIR News* **8** (5), 13–4 (1997).

[439] P.J. Treado, I.W. Levin and E.N. Lewis, *Appl. Spectrosc.* **46**, 553–9 (1992).

[440] A.M.C. Davies, *Proc. Contr. Qual.* **9**, 139–42 (1997).

[441] S.L. Pentoney, K.H. Shafer and P.R. Griffiths, *J. Chromatogr. Sci.* **24**, 230 (1986).

[442] A.M. Haefner, K.L. Norton, P.R. Griffiths, S. Bourne and R. Curbelo, *Anal. Chem.* **60**, 2441 (1988).

[443] P.T. McKittrick, N.D. Danielson and J.E. Katon, *Microchem. J.* **44**, 1 (1991).

[444] J.A. Reffner, J.P. Coates and R.G. Messerschmidt, *Am. Lab.* **4**, 56 (1987).

[445] D.A. Clark, in *Encyclopedia of Analytical Sciences* (A. Townshend, ed.), Academic Press, London (1995), pp. 3174–82.

[446] J.E. Katon, A.J. Sommer and P.L. Lang, *Appl. Spectrosc. Revs.* **25** (3/4), 173–211 (1990).

[447] J.L. Koenig, S.-Q. Wang and R. Bhargava, *Anal. Chem.* **73**, 361–9A (2001).

[448] T.J. Johnson and G. Zachman (eds.), *Introduction to Step-Scan FTIR*, Bruker Optics Inc., Billerica, MA (2000).

[449] J.M. Chalmers and N.J. Everall, *Macromol. Symp.* **94**, 33–49 (1995).

[450] W.F. McClure, in *Making Light Work: Advances in Near Infrared Spectroscopy* (I. Murray and I.A. Cowe, eds.), VCH, Weinheim (1992), pp. 1–13.

[451] S. Okretic, N. Volkl and H.W. Siesler, *Crit. Rev. Opt. Sci. Technol.* **CR69**, 3–22 (1997).

[452] M.J. Smith and R.T. Carl, *Appl. Spectrosc.* **43**, 865 (1989).

[453] E.N. Lewis and I.W. Levin, *Appl. Spectrosc.* **49**, 672 (1995).

[454] S.J. Oh and J.L. Koenig, *Anal. Chem.* **70**, 1768 (1998).

[455] C.M. Snively and J.L. Koenig, *Macromolecules* **31**, 3753 (1998).

[456] J.M. Chalmers and N.J. Everall, *Intl. J. Polym. Anal. Charact.* **5**, 223–45 (1999).

[457] M.A. Harthcock and S.C. Atkin, *Appl. Spectrosc.* **42**, 449–55 (1989).

[458] G.W. Peitscher, *Makromol. Chem., Macromol. Symp.* **5**, 75–85 (1986).

[459] S.A. Liebman, C. Phillips, W. Fitzgerald, E.J. Levy, R.A. Pesce-Rodriguez, J.M. Morris and R.A. Fifer, *Polym. Mater. Sci. Engng.* **69**, 263 (1993).

[460] A. Murase, M. Sugiura and T. Araga, *Polym. Degr. Stabil.* **43** (3), 415–22 (1994).

[461] E.V. Misco and I.W. Guilmette, in *The Design, Sample Handling and Applications of Infrared Spectroscopes* (P.B. Roush, ed.), ASTM, Philadelphia, PA (1987), p. 97.

[462] M.W. Raynor, K.D. Bartle, H.L. Davies, A. Williams, A.A. Clifford, J.M. Chalmers and B.W. Cook, *Anal. Chem.* **60**, 427 (1988).

[463] P.R. Brown and B.T. Beauchemin, *J. Liq. Chromatogr.* **11**, 1001 (1988).

[464] P. Troost, D.W. Schiering and T.J. Tague, *Appl. Note 1/98*, Spectra-Tech Inc., Waltham, MA (1998).

[465] B.J. Coles and C.J. Hall, *Proceedings SPE ANTEC '98*, Atlanta, GA (1998), pp. 2215–8.

[466] P.L. Lang, E. Katon, J.F. O.Keefe and D.W. Schiering, *Mikrochem. J.* **34**, 319 (1986).

[467] K. Krishnan, *Polym. Prepr. (ACS Polym. Div.)* **25**, 1182 (1984).

[468] B. Singh, M. Gupta, A. Verma and O.S. Tyagi, *Polym. Intl.* **49** (11), 1444–51 (2000).

[469] S. Sato, Y. Ikehara and M. Arime, *Kanzei Chuo Bunseki Shoho* **33**, 93–105 (1994).

[470] R.D. Gillard, S.M. Hardman, R.G. Thomas and D.E. Watkinson, *Stud. Conserv.* **39** (3), 187–92 (1994).

[471] P.B. Coleman and A.C. Ramamurthy, *Appl. Spectrosc.* **53** (2), 150–6 (1999).

[472] T.T. Do, M. Celina and P.M. Fredericks, *Polym. Degr. Stabil.* **77**, 417–22 (2002).

[473] S.C. Hsu, D. Lin-Vien and R.N. French, *Appl. Spectrosc.* **46**, 225–8 (1992).

[474] K. Sheu, S.J. Huang and J.F. Johnson, *Polym. Eng. Sci.* **29**, 77 (1989).

[475] N.B. Joshi and D.E. Hirt, *Proceedings SPE ANTEC '98*, Atlanta, GA (1998), pp. 2832–6.

[476] S.Y. Sankhe and D.E. Hirt, *Proceedings SPE ANTEC 2002*, San Francisco, CA (2002), pp. 2877–81.

[477] C.M. Snively and J.L. Koenig, in *Encyclopedia of Spectroscopy and Spectrometry* (J.C. Lindon, ed.), Academic Press, San Diego, CA (2000), pp. 1858–64.

[478] D. Garcia and J. Black, *J. Vinyl Addit. Technol.* **5** (2), 81–6 (1999).

[479] D. Garcia and J. Black, *Proceedings SPE ANTEC '97*, Toronto (1997), pp. 3588–92.

[480] D. Garcia and J. Black, *Proceedings SPE ANTEC '99*, New York, NY (1999), pp. 3609–13.

[481] S, Kokot, *Preprints 37th IUPAC Intl. Symposium on Macromolecules*, Gold Coast (1998), p. 81.

[482] A.W. Hartshorne and D.K. Laing, *Forensic Sci. Int.* **34**, 107–29 (1987).

[483] J.M. Wilkinson, J. Locke and D.K. Laing, *Forensic Sci. Int.* **38**, 43–52 (1988).

[484] J.M. Wilkinson, R.A. Richard, J. Locke and D.K. Laing, *Microsc.* **35**, 233–48 (1987).

[485] H. Grupp, *Maschinenschaden* **44** (1), 28–32 (1971).

[486] M.A. Harthcock, L.A. Lentz, B.L. Davis and K. Krishnan, *Appl. Spectrosc.* **40**, 210 (1986).

[487] W.J. Haap, T.B. Walk and G. Jung, *Angew. Chem. Intl. Ed.* **37**, 3311–4 (1998).

[488] R.J. Meier and B.J. Kip, *Microbeam Anal.* **3**, 61–77 (1994).

[489] E.N. Lewis, P.J. Treado, R.C. Reeder, G.M. Story, A.E. Dowrey, C. Markott and I.W. Levin, *Anal. Chem.* **67**, 3377–81 (1995).

[490] L.H. Kidder, I.W. Levin, E.N. Lewis, V.D. Kleiman and E.J. Heilweil, *Opt. Lett.* **22**, 742–4 (1997).

[491] P.A. Martoglio Smith, *Vibr. Spectrosc.* **24** (1), 47–62 (2000).

[492] A.A. Christy, F.O. Libnau and O.M. Kvalheim, in *Surface Characterization. A User's Sourcebook* (D. Brune, R. Hellborg, H.J. Whitlow and O. Hunderi, eds.), Wiley-VCH, Weinheim (1997).

[493] F.M. Mirabella, *Appl. Spectrosc.* **40** (3), 317 (1986).

[494] M.A. Harthcock, in *The Design, Sample Handling and Applications of Infrared Microscopes* (P.B.

Roush, ed.), ASTM, Philadelphia, PA (1987), p. 85.
[495] M. Ezrin, *Plast. Engng.* **58** (2), 40–55 (2002).
[496] D. Wienke, *Techn. Rept. DSM Research*, Geleen (1999).
[497] J. Zięba-Palus, *J. Trace Microprobe Tech.* **17** (3), 299–308 (1999).
[498] D.J. McEwen and G.D. Cheever, *J. Coat. Technol.* **65**, 35–40 (1993).
[499] N. Wyplosz, G. van Rooy, M. Janson, K. Groen, R. Heeren and J. Boon, *Proceedings Art & Chemie*, Paris (1998).
[500] M. Xanthos and S.H. Patel, *Adv. Polym. Technol.* **14** (2), 151–7 (1995).
[501] J.A. Reffner, in *Microbeam Analysis* (P.E. Russell, ed.), San Francisco Press, San Francisco, CA (1989), p. 167.
[502] J.E. Katon, *Vibr. Spectrosc.* **7**, 201 (1994).
[503] M. Delhaye and M. Migeon, *C.R. Acad. Sci. Paris* **262**, 702, 1513 (1966).
[504] M. Delhaye and P. Dhamelincourt, *Proceedings 4th Intl. Conference on Raman Spectroscopy*, Brunswick, MN (1974).
[505] G.J. Rosasco, E.S. Etz and W.A. Cassat, *Proceedings 4th Intl. Conference on Raman Spectroscopy*, Brunswick, MN (1974).
[506] M. Delhaye and P. Dhamelincourt, *J. Raman Spectrosc.* **3**, 33 (1975).
[507] J. Sawatski, *Fresenius J. Anal. Chem.* **339**, 267 (1991).
[508] P.H. Turner, *Bruker Rept.* **140**, 36 (1994).
[509] J. Barbillat and E. Da Silva, *Spectrochim. Acta* **53A**, 2411–22 (1997).
[510] P. Dhamelincourt, J. Barbillat and M. Delhaye, *Spectrosc. Europe* **5** (2), 16 (1993).
[511] R. Tabaksblat, R.J. Meier and B.J. Kip, *Appl. Spectrosc.* **46**, 60–8 (1992).
[512] G. Cox, *Micron* **24** (3), 237 (1993).
[513] M. Lankers and W. Kiefer, *Fresenius J. Anal. Chem.* **349**, 224–5 (1994).
[514] C.J.H. Brenan and I.W. Hunter, *Appl. Opt.* **33** (31), 7520–8 (1994).
[515] L. Markwort, B. Kip, E. Da Silva and B. Roussel, *Appl. Spectrosc.* **49**, 1411–30 (1995).
[516] A. Feofanov, S. Sharonov, P. Valisa, E. Da Silva, I. Nabiev and M. Manfait, *Rev. Sci. Instrum.* **66**, 3146–58 (1995).
[517] T. Wilson (ed.), *Confocal Microscopy*, Academic Press, London (1990).
[518] G. Turrell, M. Delhaye and P. Dhamelincourt, in *Raman Microscopy. Developments and Applications* (G. Turrell and J. Corset, eds.), Academic Press, London (1996), pp. 27–49.
[519] N.J. Everall, *Appl. Spectrosc.* **54** (6), 773–82 (2000).
[520] N.J. Everall, *Appl. Spectrosc.* **54**, 1515–20 (2000); *Inst. Phys. Conf. Ser. No.* **165**: *Symposium 1*, pp. 33–4 (2000).
[521] L. Markwort and B.J. Kip, *J. Appl. Polym. Sci.* **61**, 231–54 (1996).
[522] G. Turrell and P. Dhamelincourt, in *Modern Techniques in Raman Spectroscopy* (J.J. Laserna, ed.), J. Wiley & Sons, Chichester (1996), pp. 109–42.
[523] D.J. Gardiner and M. Bowden, *Encyclopedia of Spectroscopy and Spectrometry* (J.C. Lindon, ed.), Academic Press, San Diego, CA (2000), pp. 3182–9.
[524] M.D. Schaeberle, H.R. Morris, J.F. Turner II and P.J. Treado, *Anal. Chem.* **71** (5), 175–81A (1999).
[525] J. Barbillat, in *Raman Microscopy. Developments and Applications* (G. Turrell and J. Corset, eds.), Academic Press, London (1996), pp. 175–200.
[526] B. Schrader, W. Niggemann, H.H. Belz and B. Schallert, *Z. Chem.* **21**, 249 (1981).
[527] V. Pajcini, C.H. Munro, R.W. Bormett, R.E. Witkowski and S.A. Asher, *Appl. Spectrosc.* **51** (1), 81–6 (1997).
[528] P. Colarusso, L.H. Kidder, I.W. Levin and E.N. Lewis, in *Encyclopedia of Spectroscopy and Spectrometry* (J.C. Lindon, ed.), Academic Press, San Diego, CA (2000), pp. 1945–54.
[529] J. Grausem, B. Humbert, M. Spajer, D. Courjon, A. Burneau and J. Ostwalt, *J. Raman Spectr.* **30**, 833–40 (1999).
[530] D.C. Smith, *Abstracts 7th Intl. Conference on Non-Destructive Testing and Microanalysis for the Diagnostics and Conservation of the Cultural and Environmental Heritage (Art 2002)*, Antwerp (2002), p. 46.
[531] K.P. Ghiggino, M.R. Harris and P.G. Spizzirri, *Rev. Sci. Instr.* **63**, 2999 (1992).
[532] P.J. Treado and M.D. Morris, *Pract. Spectrosc.* **16**, 71–108 (1993).
[533] G.J. Rosasco, in *Advances in Infrared and Raman Spectroscopy* (R.J.H. Clark and R.E. Hester, eds.), Heyden, London (1980), Vol. 7, pp. 223–82.
[534] M. Truchet, J.-C. Merlin and G. Turrell, in *Raman Microscopy. Developments and Applications* (G. Turrell and J. Corset, eds.), Academic Press, London (1996), pp. 201–42.
[535] D.B. Chase, *Appl. Spectrosc.* **48**, 14 (1994).
[536] B. Schrader, *Fresenius J. Anal. Chem.* **337** (7), 824–9 (1990).
[537] M. Delhaye, J. Barbillat, J. Aubard, M. Bridoux and E. Da Silva, in *Raman Microscopy. Developments and Applications* (G. Turrell and J. Corset, eds.), Academic Press, London (1996), pp. 51–173.
[538] B.L. McClain, H.G. Hedderich, A.D. Gift, D. Zhang, K.N. Jullad, K.S. Haber, J. Ma and D. Ben-Amotz, *Spectroscopy* **15** (9), 28 (2000).

[539] G. Turrell and J. Corset (eds.), *Raman Microscopy. Developments and Applications*, Academic Press, London (1996).

[540] G.A. Voyiatzis and K.S. Andrikopoulos, *Proceedings Spectroscopy in Process and Quality Control (SPQ)*, London (1998).

[541] C.G. Zimba, S. Turrell, J.D. Swalen, V.M. Hallmark and J.F. Rabolt, *J. Phys. Chem.* **94**, 939 (1990).

[542] R.J. Nemanich, *Microbeam Anal.*, 234 (1990).

[543] K.P.J. Williams and S.M. Mason, *Spectrochim. Acta* **46A**, 187 (1990).

[544] *LabRam Application Note R-29, Polymer Blends – Identification and Mapping of Phases*, Jobin Yvon, Edison, NJ (1998).

[545] E. Magni, F. Malizia and G.A. Somorjai, *Proceedings Polypropylene. Past, Present and Future: The Challenge Continues*, Ferrara (1998), pp. 179–210.

[546] M. Claybourn, A. Luget and K.P.J. Williams, *ACS Symp. Ser.* **598**, 41–60 (1995).

[547] S. Hajadoost and J. Yardwood, *Appl. Spectrosc.* **50**, 511 (1993).

[548] C. Koulic, Z. Yin, C. Pagnoulle, B. Gilbert and R. Jerome, *Proceedings 24th Annu. Mtg. Adhes. Soc.*, Williamsburg, VA (2001), pp. 177–9.

[549] W. Schrof, R. Schwalm and U. Meisenburg, *Abstracts Advanced Methods of Polymer Characterization: New Developments and Applications in Industry*, Mainz (1999), Paper 1.48.

[550] M. Bowden, J.W. Bradley, L.R. Dix, D.J. Gardiner, N.M. Dixon and D.L. Gerrard, *Polymer* **35**, 1654 (1994).

[551] R. Salzer, U. Roland and R. Born, in *Fourier Transform Spectroscopy* (J.A. de Haseth, ed.), Am. Inst. Physics, Vol. **430**, pp. 474–5 (1998).

[552] A.M. Macdonald, P. Wyeth and C. Rogerson, *Eur. Microsc. Anal.* **85**, 17–9 (2003).

[553] D. Bourgeois and S.P. Church, *Spectrochim. Acta* **A46**, 295–301 (1990).

[554] C. Coupry, G. Sagon and P. Gorguet-Ballestreros, *J. Raman Spectrosc.* **28**, 85–9 (1997).

[555] P.C. White, C.H. Munro and W.E. Smith, *Analyst* **12**, 835 (1996).

[556] D.O. Hummel, *Atlas of Plastics Additives: Analysis by Spectrometric Methods*, Springer-Verlag, Berlin (2002).

[557] D.L. Gerrard, in *Analytical Raman Spectroscopy* (J.G. Grasselli and B.J. Bulkin, eds.), J. Wiley & Sons, New York, NY (1991), pp. 275–324.

[558] D.W. Hahn, D.L. Wolfarth and N.L. Parks, *J. Biomed. Mat. Res.* **35** (1), 31–7 (1997).

[559] C. Mura, J. Yarwood, R. Swart and D. Hodge, *Polymer* **41**, 8659–71 (2000).

[560] F. Belaroui, Y. Grohens, H. Boyer and Y. Holl, *Polymer* **41**, 7641–5 (2000).

[561] P. Eaton, P. Holmes and J. Yarwood, *Appl. Spectrosc.* **54**, 508–16 (2000).

[562] R. Appel, T.W. Zerda and W.H. Waddell, *Appl. Spectrosc.* **54**, 1559 (2000).

[563] L. Nasdala, A. Banerjee, T. Häger and W. Hofmeister, *Eur. Microsc. Anal.* **70**, 11–3 (March 2001).

[564] M.L. Scheepers, R.J. Meier, L. Markwort, J.M. Galan, D.J. Vanderzande and B.J. Kip, *Vibr. Spectrosc.* **9**, 139 (1995).

[565] C. Stellman, K. Booksh and M. Myrick, *Appl. Spectrosc.* **50**, 552 (1996).

[566] H.R. Morris, B. Munroe, R.A. Ryntz and P.J. Treado, *Langmuir* **14**, 2426–34 (1998).

[567] G.D. Smith and R.J.H. Clark, *Rev. Conserv.* (2), 92–106 (2001).

[568] M.J. Manzaneda, J.M. Yúfera, S. Ruiz-Moreno, M.J. Soneira, P. Morillo and M. Breitman, in *Lasers in the Conservation of Artworks (Lacona I)*, W. Kautek and E. Konig (eds.), Verlag Mayer & Comp., Klosterneuburg and Vienna (1997), pp. 61–7.

[569] K. Trentelman, L. Stodulski and M. Paviosky, *Anal. Chem.* **68**, 1756 (1996).

[570] P.V. Huong, *Physica C (Amsterdam)* **180**, 128 (1991).

[571] R.J.H. Clark, C.J. Cooksey, M.A.M. Daniels and R. Withnall, *Endeavour* **17**, 191 (1993).

[572] P. Vandenabeele, B. Wehling, L. Moens, H. Edwards, M. De Reu and G. van Hooydonk, *Anal. Chim. Acta* **407**, 261–74 (2000).

[573] R.J.H. Clark, *J. Mol. Struct.* **480–81**, 15–20 (1999).

[574] R.J.H. Clark, *Chem. Soc. Rev.* **24**, 187–96 (1995).

[575] R. Davey, D.J. Gardiner, B.W. Singer and M. Spokes, *J. Raman Spectrosc.* **25** (1), 53–7 (1994).

[576] S.P. Best, R.J.H. Clark, M.A.M. Daniels, C.A. Porter and R. Withnall, *Stud. Conserv.* **40** (1), 31–40 (1995).

[577] B. Wehling, P. Vandenabeele, L. Moens, R. Klockenkämper, A. von Bohlen, G. van Hooydonk and M. De Reu, *Mikrochim. Acta* **130**, 253–60 (1999).

[578] P. Vandenabeele, B. Wehling, L. Moens, B. Dekeyzer, B. Cardon, A. von Bohlen and R. Klockenkämper, *Analyst* **124**, 169–72 (1999).

[579] A. von Bohlen, P. Vandenabeele, M. De Reu, L. Moens, R. Klockenkämper, B. Dekeyzer and B. Cardon, *Restauro* (3), 188–22 (2003).

[580] L. Burgio, R.J.H. Clark, T. Stratoudaki, M. Doulgeridis and D. Anglos, *Appl. Spectrosc.* **54**, 463–9 (2000).

[581] L. Burgio, K. Messanaki, M. Doulgeridis, R.J.H. Clark and D. Anglos, *Spectrochim. Acta* **B56**, 905–13 (2001).

[582] L. Bussotti, M.P. Carboncini, E. Castellucci, L. Giutini and P.A. Mandò, *Stud. Conserv.* **42**, 82–92 (1997).

[583] M. Christensen, T. Calligaro, S. Consigny, J.-C. Dran, J. Salomon and P. Walter, *Nucl. Inst. Methods, Phys. Res.* **B136**, 138, 869 (1998).

[584] B.J. Marquardt, D.N. Stratis, D.A. Cremers and S.M. Angel, *Appl. Spectrosc.* **52**, 1148–53 (1998).

[585] I.M. Bell, R.J.H. Clark and P.J. Gibbs, *Spectrochim. Acta* **A53**, 2159–79 (1997).

[586] L. Burgio and R.J.H. Clark, *Spectrochim. Acta* **A57**, 1491–521 (2001).

[587] R.J.H. Clark and P.J. Gibbs, *Spectrochim. Acta* **53A**, 2159 (1997).

[588] P. Vandenabeele and L. Moens, *Abstracts 7th Intl. Conference on Non-Destructive Testing and Microanalysis for the Diagnostics and Conservation of the Cultural and Environmental Heritage (Art 2002)*, Antwerp (2002).

[589] I. Blakey and G.A. George, *Polym. Degr. Stabil.* **70** (2), 269–85 (2000).

[590] H.J. Bowley, D.L. Gerrard and I.S. Biggin, *Polym. Degr. Stabil.* **20**, 257–69 (1988).

[591] A. Gupper and P. Wilhelm, *Proceedings 8th Intl. Plastics Additives and Modifiers Conference (AddCon World 2002)*, Budapest (2002), pp. 227–35.

[592] K.P.J. Williams, I.C. Wilcock, I.P. Hayward and A. Whitley, *Spectroscopy* **11**, 45 (1996).

[593] P.V. Huong, *Vibr. Spectrosc.* **11**, 17 (1996).

[594] P. Dhamelincourt and S. Nakashima, in *Raman Microscopy. Developments and Applications* (G. Turrell and J. Corset, eds.), Academic Press, London (1996), pp. 243–87.

[595] C. Coupry and D. Brissaud, in *Raman Microscopy. Developments and Applications* (G. Turrell and J. Corset, eds.), Academic Press, London (1996), pp. 421–53.

[596] H.R. Morris, J.F. Turner, B. Munroe, R.A. Ryntz and P.J. Treado, *Langmuir* **15** (8), 2961–72 (1999).

[597] R.H. Fleming and A.Y. Craig, *Polym. Degr. Stabil.* **37**, 173 (1992).

[598] G. Ahlblad, B. Stenberg, B. Terselius and T. Reitberger, *Polymer Testing* **16**, 59–73 (1997).

[599] S. Hosoda, Y. Seki and H. Kihara, *Polymer* **34**, 4602–6 (1993).

[600] M. Celina, G.A. George, D.J. Lacey and N.C. Billingham, *Polym. Degr. Stabil.* **47** (2), 311–7 (1995).

[601] D.J. Lacey and V. Dudler, *Polym. Degr. Stabil.* **51**, 115–24 (1996).

[602] D.R. Kohler and C. Kröhnke, *Polym. Degr. Stabil.* **62**, 385–93 (1998).

[603] M. Hamskog, G. Ahlblad, G. Färnert, P. Gijsman and B. Terselius, *Polymer Testing (in press).*

[604] V. Dudler, D.J. Lacey and Ch. Kröhnke, *Polym. Degr. Stabil.* **51**, 115–24 (1996).

[605] N.C. Billingham, P. Fearon, D.J. Whiteman and V. Dudler, *Proceedings AddCon '99*, Prague (1999), Paper 23.

[606] L. Woo, C.L. Sandford and H. Blom, *Proceedings SPE ANTEC 2002*, San Francisco, CA (2002), pp. 2695–7.

[607] M. Celina, G.A. George and N.C. Billingham, *Adv. Chem. Ser.* **249**, 159–74 (1996).

[608] B. Fayolle, L. Audouin, G.A. George and J. Verdu, *Polym. Degr. Stabil.* **77**, 515–22 (2002).

[609] M. Celina, G.A. George and N.C. Billingham, *Polym. Degr. Stabil.* **42**, 335–44 (1993).

[610] M. Celina and G.A. George, *Polym. Degr. Stabil.* **40**, 323–35 (1993).

[611] M. Celina, G.A. George and N.C. Billingham, *Polym. Prepr.* **34** (2), 262 (1993).

[612] P. Eriksson, T. Reitberger and B. Stenberg, *Polym. Degr. Stabil.* **78** (1), 183–9 (2002).

[613] G.A. George, M. Celina, C. Lerf, G. Cash and D. Weddell, *Macromol. Symp.* **115**, 69–92 (1997).

[614] P. Eriksson, G. Ahlblad, T. Reitberger and B. Stenberg, *Proceedings 1st Intl. Symposium on Modification, Degradation and Stabilisation (MoDeSt)*, Palermo (2000).

[615] P. Eriksson, T. Reitberger, G. Ahlblad and B. Stenberg, *Polym. Degr. Stabil.* **73** (1), 177–83 (2001).

[616] I. Blakey, G.A. George and N.C. Billingham, *Macromolecules* **34** (26), 9130–8 (2001).

[617] D. Forsström, T. Reitberger and B. Terselius, *Polym. Degr. Stabil.* **67**, 255–61 (2000).

[618] D. Forsström, L.-G. Svensson and B. Terselius, *Polym. Degr. Stabil.* **67** (2), 263–9 (2000).

[619] G. Ahlblad, D. Forsström, B. Stenberg, B. Terselius, T. Reitberger and L.-G. Svensson, *Polym. Degr. Stabil.* **55**, 287–93 (1997).

[620] G. Ahlblad, *Ph.D. Thesis*, Royal Institute of Technology, Stockholm (1998).

[621] G. Ahlblad, T. Reitberger, B. Terselius and B. Stenberg, *Polym. Degr. Stabil.* **65**, 185–91 (1999).

[622] G. Ahlblad, T. Reitberger, B. Terselius and B. Stenberg, *Polym. Degr. Stabil.* **65**, 169–77 (1999).

[623] V. Dudler, *Preprints 37th Int. IUPAC Symposium on Macromolecules*, Gold Coast (1998), p. 960.

[624] K. Jacobson, B. Stenberg, B. Terselius and T. Reitberger, *Polym. Degr. Stabil.* **68**, 53–60 (2000).

[625] A. Kron, *Ph.D. Thesis*, Royal Institute of Technology, Stockholm (1996).

[626] P. Mansfield and P.K. Grannell, *J. Phys. C* **6**, L422 (1973).

[627] P.T. Callaghan, *Principles of Nuclear Magnetic Resonance Microscopy*, Clarendon Press, Oxford (1995).

[628] E. Rommel, R. Kimmich, H. Robert and D. Pusiol, *Meas. Sci. Technol.* **3**, 446 (1992).

[629] W.E. Maas, L.H. Merwin and D.G. Cory, in *Spatially Resolved Magnetic Resonance* (P. Blümler, B. Blümich, R. Botto and E. Fukushima, eds.), Wiley-VCH, Weinheim (1998), pp. 141–61.

[630] A. Kumar, D. Welti and R.R. Ernst, *J. Mag. Reson.* **18**, 69 (1975).

[631] T.R. Brown, B. Kincaid and K. Ugurbil, *Proc. Natl. Acad. Sci. (USA)* **79**, 3523 (1982).

[632] J.L. Koenig, in *Polymer Spectroscopy* (A.H. Fawcett, ed.), J. Wiley & Sons, Chichester (1996), pp. 151–72.

[633] G. Eidmann, R. Savelsberg, P. Blümler and B. Blümich, *J. Magn. Reson.* **A122**, 104 (1996).

[634] B. Blümich, P. Blümler, A. Guthausen, R. Haken, U. Schmitz. K. Saito and G. Zimmer, *Magn. Reson. Imaging* **16**, 479 (1998).

[635] B. Blümich, P. Blümler, L. Gasper, A. Guthausen, V. Gobbels, S. Laukemper-Ostendorf, K. Unseld and G. Zimmer, *Macromol. Symp.* **14**, 83–93 (1999).

[636] P. Barth, S. Hafner and W. Kuhn, *J. Magn. Reson.* **A110**, 198–201 (1994).

[637] B. Hills, *Magnetic Resonance Imaging in Food Science*, J. Wiley & Sons, Chichester (1998).

[638] B. Traub, S. Hafner, D. Maring and H.W. Spiess, in *Spatially Resolved Magnetic Resonance* (P. Blümler, B. Blümich, R. Botto and E. Fukushima, eds.), Wiley-VCH, Weinheim (1998), pp. 179–93.

[639] C.B. Kennedy, B.J. Balcom and I.V. Mastikhin, *Can. J. Chem.* **76** (11), 1753–65 (1998).

[640] J.B. Miller and A.N. Garroway, *Rev. Progr. Quant. Non-Destructive Evaluation* **7A**, 287–94 (1988).

[641] P. Blümler and B. Blümich, in *Solid State NMR, I. Methods* (B. Blümich, ed.), Springer-Verlag, Berlin (1994), pp. 209–77.

[642] D.G. Cory, in *Encyclopedia of Spectroscopy and Spectrometry* (J.C. Lindon, ed.), Academic Press, San Diego, CA (2000), pp. 2009–17.

[643] B. Blümich and W. Kuhn (eds.), *Magnetic Resonance Microscopy*, VCH, Weinheim (1992).

[644] P. Blümler, B. Blümich, R. Botto and E. Fukushima (eds.), *Spatially Resolved Magnetic Resonance*, Wiley-VCH, Weinheim (1998).

[645] P.J. McDonald and D.M. Lane, in *Polymer Surfaces and Interfaces III* (R.W. Richards and S.K. Pearce, eds.), J. Wiley & Sons, Chichester (1999), pp. 237–67.

[646] D.M. Grant and R.K. Harris (eds.), *Encyclopedia of NMR*, J. Wiley & Sons, New York, NY (1996).

[647] V.J. McBrierty, *Magn. Reson. Rev.* **8**, 165 (1983).

[648] R.R. Ernst, G. Bodenhausen and A. Wokaun, *Principles of Nuclear Magnetic Resonance in One and Two Dimensions*, Clarendon Press, Oxford (1997).

[649] B. Brunner and R.R. Ernst, *J. Magn. Reson.* **33**, 83 (1979).

[650] J. Jaklovsky, *NMR Imaging, A Comprehensive Bibliography*, Addison Wesley, Reading, MA (1983).

[651] M. Gyor, A. Rockenbauer, L. Jokay and F. Tudos, *Polym. Bull.* **15**, 525 (1986).

[652] N.L. Thomas and A.H. Windle, *Polymer* **22**, 627 (1981).

[653] C. Malveau, F. Beaume, Y. Germain and D. Canet, *J. Polym. Sci. Pt. B: Polym. Phys.* **39** (22), 2781–92 (2001).

[654] W.P. Rothwell and P.P. Gentempo, *Bruker Rept.* **1**, 46 (1985).

[655] G. Maddinelli, P. Iwanski, A. Callaioli and G.P. Ravanetti, *Abstracts 13th Intl. Symposium on Polymer Analysis and Characterization (ISPAC-2000)*, Pittsburgh, PA (2000), Paper P-15.

[656] S. Blackband and P. Mansfield, *J. Phys. C: Solid State Phys.* **19**, L49–52 (1986).

[657] K.-P. Hoh, B.C. Perry, G. Rotter, H. Ishida and J.L. Koenig, *J. Adhesion* **27**, 245 (1989).

[658] A.O.K. Nieminen and J.L. Koenig, *J. Adhes. Sci. Technol.* **2**, 407 (1988).

[659] R.C. Clough and J.L. Koenig, *J. Polym. Sci.: Polym. Lett. Ed.* **27**, 451 (1989).

[660] A.K. Whittaker, in *Spectroscopy of Rubbers and Rubbery Materials* (V.M. Litvinov and P.P. De, eds.), Rapra Technology Ltd., Shawbury (2002), pp. 491–518.

[661] L.A. Weissenberger and J.L. Koenig, *Macromolecules* **23**, 2445–53; 2454–9 (1990).

[662] J. Leisen, H. Hojjatie, D.W. Coffin and H.W. Beckham, *Drying Technol.* **19** (1), 199–206 (2001).

[663] M. Iig, B. Pfielderer, K. Albert and E. Bayer, *Mater. Res. Soc. Symp. Proc.* **217**, 27 (1991).

[664] L.A. Weissenberger and J.L. Koenig, *Appl. Spectrosc.* **43**, 1117–26 (1989).

[665] K.L. Perry, P.J. McDonald, E.W. Randall and K. Zick, *Polymer* **35**, 2744–8 (1994).

[666] S.G. Harding, L.F. Gladden and P.J. Fryer, *Proceedings Intl. Symposium on Food Packaging*, Budapest (1996), p. 25.

[667] A. Whittaker, P. Ghi, D. Hill and M. Whittaker, *Proceedings ACS Div. Polym. Mater.: Sci. & Engg.* **79**, 467–8 (1998).

[668] R.A. Grinsted and J.L. Koenig, *Macromolecules* **25**, 1229 (1992).

[669] D.M. Lane, P.J. McDonald and J.L. Keddie, in *Spatially Resolved Magnetic Resonance* (P. Blümler, B. Blümich, R. Botto and E. Fukushima, eds.), Wiley-VCH, Weinheim (1998), pp. 242–52.

[670] B.C. Perry and J.L. Koenig, *J. Polym. Sci. A: Polym. Chem.* **27**, 3429–38 (1989).

[671] B. Blümich and D.E. Demco, in *Spectroscopy of Rubbers and Rubbery Materials* (V.M. Litvinov and P.P. De, eds.), Rapra Technology Ltd., Shawbury (2002), pp. 247–89.

[672] J.L. Koenig, *Makromol. Chem. Symp., Macromol. Chem.* **86**, 283–97 (1994).

[673] P. Blümler and B. Blümich, *Macromolecules* **24**, 2183–8 (1991).

[674] L. Garrido, J.E. Mark, C.C. Sun, J.L. Ackerman and C. Chang, *Macromolecules* **24**, 4067 (1991).

[675] L. Garrido, J.L. Ackerman, J.M. Vevea and J.E. Mark, *Polymer* **33**, 1826 (1992).

[676] S.N. Sakar and R.A. Komoroski, *Macromolecules* **25**, 1420 (1992).

[677] C. Chang and R.A. Komoroski, *Macromolecules* **22**, 600 (1989).

[678] P. Blümler, V. Litvinov, H.G. Dikland and M. v. Duin, *Kautsch. Gummi Kunstst.* **51**, 865–7 (1998).

[679] J.A. Catwalk and G.J. Hunter, *Mater. Sci. Lett.* **11**, 222 (1992).

[680] M. Sardashti, B. Baldwin and D.J. O'Donnell, *J. Polym. Sci. B: Polym. Phys.* **33**, 571–6 (1995).

[681] M. Knörgen, U. Heuert and H. Schneider, in *Spatially Resolved Magnetic Resonance* (P. Blümler, B. Blümich, R. Botto and R. Fukushima, eds.), Wiley-VCH, Weinheim (1998), pp. 211–20.

[682] G. Guthausen, A. Guthausen, F. Balibanu, R. Eymael, K. Hailu, U. Schmitz and B. Blümich, *Macromol. Mat. Eng.* **276/277**, 25 (2000).

[683] A. Wiesmath, C. Filip, D.E. Demco and B. Blümich, *J. Magn. Reson.* **149**, 258 (2001).

[684] S.G. Harding, M.L. Johns, S.R. Pugh, P.J. Fryer and L.F. Gladden, *Food Addit. Contam.* **14** (6/7), 583–9 (1997).

[685] P. Jackson, N.J. Clayden, N.J. Walton, T.A. Carpenter, L.D. Hall and P. Jerzard, *Polym. Int.* **24**, 139–43 (1990).

[686] B. Balcom, T. Carpenter and L. Hall, *Macromolecules* **25**, 6818 (1992).

[687] M.A. Rana and J.L. Koenig, *Macromolecules* **27**, 3727-34 (1994).

[688] C. Fülber, K. Unseld, V. Herrmann, K.H. Jakob and B. Blümich, *Colloid Polym. Sci.* **274**, 191 (1996).

[689] A.O.K. Nieminen and J.L. Koenig, *Intl. J. Adhes. Adhes.* **11**, 5 (1991).

[690] K.L. McCarthy, R.J. Kauten and C.K. Agemura, *Trends Food Sci. Technol.* **3**, 215–9 (1992).

[691] K.-P. Hoh, H. Ishida and J.L. Koenig, *Polym. Compos.* **11**, 192 (1990).

[692] P. Blümler and B. Blümich, *NMR* **30**, 209 (1994).

[693] W. Kuhn, *Angew. Chem. Int. Ed.* **29**, 1 (1990).

[694] K. Schmidt-Rohr and H.W. Spiess (eds.), *Multidimensional Solid-State NMR and Polymers*, Academic Press, London (1994).

[695] T. Herrling, N. Klimes, W. Karthe, U. Ewert and B. Ewert, *J. Magn. Reson.* **49**, 203 (1982).

[696] R.K. Woods, G.G. Bacic, P.C. Lauterbur and H.M. Swartz, *J. Magn. Reson.* **84**, 247–54 (1989).

[697] K. Ohno, *Appl. Spectrosc. Rev.* **22**, 1–56 (1986).

[698] L.H. Sutcliffe, in *Encyclopedia of Spectroscopy and Spectrometry* (J.C. Lindon, ed.), Academic Press, San Diego, CA (2000), pp. 437–45.

[699] G.R. Eaton, S.S. Eaton and K. Ohno (eds.), *EPR Imaging and In Vivo EPR*, CRC Press, Boca Raton, FL (1991).

[700] B. Blümich, P. Blümler, R. Botto and E. Fukushima (eds.), *Spatially Resolved Magnetic Resonance: Methods, Materials, Medicine, Biology, Rheology, Ecology, Hardware*, Wiley-VCH, Weinheim (1998).

[701] U. Ewert and K.U. Thiessenhusen, in *EPR Imaging and In Vivo EPR* (G.R. Eaton, S.S. Eaton and K. Ohno, eds.), CRC Press, Boca Raton, FL (1991), Chp. 11.

[702] S. Schlick, P. Eagle, K. Kruczala and J. Pilar, in *Spatially Resolved Magnetic Resonance* (P. Blümler, B. Blümich, R. Botto and E. Fukushima, eds.), Wiley-VCH, Weinheim (1998), pp. 221–34.

[703] M. Lucarini, G.F. Pedulli, V. Borzatta and N. Lelli, *Res. Chem. Intermed.* **22** (6), 581–91 (1996).

[704] M. Lucarini and G.F. Pedulli, *Angew. Makromol. Chem.* **252**, 179–93 (1997).

[705] M. Lucarini, G.F. Pedulli, V. Borzatta and N. Lelli, *Polym. Degr. Stabil.* **53** (1), 9–17 (1996).

[706] K. Kruczala, M.V. Motyakin and S. Schlick, *J. Phys. Chem.* **B104**, 3387–92 (2000).

[707] M.V. Motyakin and S. Schlick, *Macromolecules* **34** (9), 2854–64 (2001).

[708] M.V. Motyakin and S. Schlick, *Polym. Degr. Stabil.* **76** (1), 25–36 (2002).

[709] M.V. Motyakin and S. Schlick, *Macromolecules* **35** (10), 3984–92 (2002).

[710] M.V. Motyakin and S. Schlick, *Proceedings ACS Div. Polym. Mater.: Sci. & Engng.* **83**, 395–6 (2000).

[711] S. Schlick, K. Kruczala, M.V. Motyakin and J.L. Gerlock, *Polym. Mater.: Sci. & Engng.* **83**, 120–1 (2000).

[712] S. Schlick, K. Kruczala, M.V. Motyakin and J.L. Gerlock, *Polym. Degr. Stabil.* **73**, 471 (2001).

[713] S. Schlick, K. Kruczala and M.V. Motyakin, *Abstracts 2nd Intl. Conference on Polymer Modification, Degradation and Stabilisation (MoDeSt2)*, Budapest (2002).

[714] G. Geuskens and G. Nedelkos, *Polym. Degr. Stabil.* **19**, 365 (1987).

[715] H. Ade, X. Zhang, S. Cameron, C. Costello, J. Kirz and S. Williams, *Science* **258**, 972–5 (1992).

[716] C. Herkt-Maetzky, K. Moser and J. Gessner, *J. Appl. Polym. Sci.: Appl. Polym. Symp.* **48**, 417–29 (1991).

[717] D.C. Koningsberger and R. Prins (eds.), *X-Ray Absorption: Principles, Applications, Techniques of EXAFS, SEXAFS and XANES*, J. Wiley & Sons, New York, NY (1988).

[718] C.J. Sparks, in *Synchrotron Radiation Research* (H. Winick and S. Doniach, eds.), Plenum Press, New York, NY (1980), p. 459.

[719] B. Forslind, K.G. Malmquist and J. Pallon, *Scanning Microsc.* **5**, 877 (1991).

[720] Z. Cai, W. Yun, S.T. Pratt, R.M. Miller, E. Gluskin, D.B. Hunter, A.G. Jarstfer, K.M. Kemner, B. Lai, H.-R. Lee, D.G. Legnini, W. Rodrigues and C.I. Smith, *Adv. Photon Source Res.* **1** (1), 15–20 (1998).

[721] N. Mölders, P.J. Schilling and J.M. Schoonmaker, *Mater. Res. Soc., Symp. Proc. (Applications of Synchrotron Radiation Techniques to Material Science V)* **590**, 3–8 (2000).

[722] Y. Hwu, W.-L. Tsai, L.-W. Chang, C.H. Chen, C.C. Wu, D.Y. Noh, J.H. Je, G.H. Fecher, M. Bertolo, H. Berger and G. Margaritondo, *Jpn. J. Appl. Phys. Pt. 1* **38**, 646–9 (1999).

[723] D. Sayre and H.N. Chapman, *Acta Crystallogr.* **A51** (3), 237–52 (1995).

[723a] P.J. Duke, in *Applications of Synchrotron Radiation* (C.R.A. Catlow and G.N. Greaves, eds.), Blackie, Glasgow (1990), pp. 283–310.

[724] H. Ade, in *Experimental Methods in the Physical Sciences* (T. Lucatorto and M. de Graef, eds.), Academic Press, New York, NY (1998), Vol. 32, pp. 225–62.

[725] J. Thieme, G. Schmahl, E. Umbach and D. Rudolph (eds.), *X-Ray Microscopy and Spectromicroscopy*, Springer-Verlag, Berlin (1998).

[726] I. McNulty (ed.), *X-Ray Microfocusing: Applications and Techniques*, SPIE, Bellingham, WA (1998).

[727] Special Issue on *Spectromicroscopy* (H. Ade, ed.), *J. Electr. Spectrosc. Relat. Phenom.* **84** (1997).

[728] V.E. Cosslett and W.C. Nixon, *X-Ray Microscopy*, Cambridge University Press, London (1960).

[729] J. Cazaux, D. Erre, D. Mouze, J.M. Patat, S. Rondot, A. Sasov, P. Trebbia and A. Zolfaghari, *J. Phys. IV* **3**, 2099–104 (1993).

[730] A. Sasov and D. van Dyck, *Eur. Microsc. Anal.* **65**, 17–9 (March 1998).

[731] H. Azuma, A. Takeichi and S. Noda, *Jpn. J. Appl. Phys., Pt. 1* **33** (8), 4622–7 (1994).

[732] J.D. Moyer and R.J. Ochs, *Science* **142**, 1316 (1963).

[732a] W.C. Conner, S.W. Webb, P. Spanne and K.W. Jones, *Macromolecules* **23**, 4742–7 (1990).

[733] C. Dragnea, *Insight* **43** (6), 418–20 (2001).

[734] J. Maser, A. Osanna, Y. Wang, C. Jacobsen, J. Kirz, S. Spector, B. Winn and D. Tennant, *J. Microsc.* **197**, 68–79 (2000).

[735] R. Barrett, B. Kaulich, M. Salomé and J. Susini, *Proceedings 6th Intl. Conference on X-Ray Microscopy* (W. Meyer-Ilse, T. Warwick and D. Attwood, eds.), AIP, New York, **507**, 458–63 (2000).

[736] S.G. Urquhart, A.P. Hitchcock, A.P. Smith, H.W. Ade, W. Lidy, E.G. Rightor and G.E. Mitchell, *J. Electr. Spectr. Relat. Phenom.* **100**, 119–35 (1999).

[737] M.M. Disko, C.C. Ahn and B. Fultz (eds.), *Transmission Electron Energy-Loss Spectrometry in Materials Science*, The Minerals, Metals and Materials Society (TMS), Warrendale, PA (1992).

[738] A. Hitchcock, *Intl. Lab. News* **32** (1A), 24–5 (Feb. 2002).

[739] H. Ade, B. Hsiao, G. Mitchell, E. Rightor, A.P. Smith and R. Cieslinski, *Mater. Res. Soc. Symp. Proc.* **375**, 293–302 (1995).

[740] A.P. Hitchcock, I. Koprinarov, T. Tyliszczak, E.G. Rightor, G.E. Mitchell, M.T. Dineen, F. Hayes, W. Lidy, R.D. Priester, S.G. Urquhart, A.P. Smith and H. Ade, *Ultramicroscopy* **88** (1), 33–49 (2001).

[741] M.J. Pitkethly, *J. Microsc.* **169** (2), 183–8 (1993).

[742] H. Ade, *Inst. Phys. Conf. No.* **165**, Symp. 1, 109–10 (2000).

[743] A.J. Dias, S.G. Urquhart, H. Ade, A.P. Smith and P. Stevens, *Preprints 155th ACS Rubber Division Mtg.*, Chicago, IL (1999), Paper 30.

[744] F. Adams, K. Janssens and A. Snigirev, *J. Anal. Atom. Spectrom.* **13**, 319–31 (1998).

[745] A. Gross, *Proceedings 7th Intl. Conference on Non-Destructive Testing and Microanalysis for the Diagnostics and Conservation of the Cultural and Environmental Heritage (Art 2002)*, Antwerp (2002), p. 31.

[746] M. Schreiner, *Proceedings 7th Intl. Conference on Non-Destructive Testing and Microanalysis for the Diagnostics and Conservation of the Cultural and Environmental Heritage (Art 2002)*, Antwerp (2002), p. 127.

[747] H. Bronk, S. Röhrs, A. Bjeoumikhov, N. Langhoff, J. Schmalz, R. Wedell, H.-E. Gorny, A. Herold and U. Waldschläger, *Fresenius J. Anal. Chem.* **371**, 307–16 (2001).

[748] G.J. Price, G.W. Fraser, J.F. Pearson, I.B. Hutchinson, A.D. Holland, J. Nussey, D. Vernon, D. Pullan and K. Turner, *Spectrosc. Europe* **15** (5), 8–13 (2003).

[749] W. Yun, S.T. Pratt, R.M. Miller, Z. Cai, D.B. Hunter, A.G. Jarstfer, K.M. Kemner, B. Lai, H.-R. Lee, D.G. Legnini, W. Rodrigues and C.I. Smith, *J. Synchrotron Radiat.* **5**, 1390–5 (1998).

[750] K. Janssens, *Proceedings Colloquium Spectroscopicum Internationale XXXIII*, Granada (2003), p. 40.

[751] K. Janssens, L. Vincze, J. Rubio, F. Adams and G. Bernasconi, *J. Anal. At. Spectrom.* **9** (3), 151–8 (1994).

[752] K. Janssens, F. Adams and A. Rindby (eds.), *Microscopic X-Ray Fluorescence Analysis*, J. Wiley & Sons, Chichester (2000).

[753] Special Issue on *Microfluorescence*, in *X-Ray Spectrom.* **26** (6) (1997).

[754] S.A. McHugo, A.C. Thompson, C. Flink, E.R. Weber, G. Lamble, B. Gunion, A. MacDowell, R. Celestre, H.A. Padmore and Z. Hussain, *J. Cryst. Growth* **210** (1–3), 395–400 (2000).

[755] D.R.J. Maia, C.A. Perez and M.A. De Paoli, *J. Polym. Sci. Pt. B: Polym. Phys.* **40** (7), 657 (2002).

[756] D. Wegrzynek, B. Holynska and J. Ostachowicz, *X-Ray Spectrom.* **28**, 209–14 (1999).

[757] L. Kempenaers, C. De Koster, W. van Borm and K. Janssens, *Fresenius J. Anal. Chem.* **369** (7–8), 733 (2001).

[758] L. Kempenaers, N.H. Bings, T.E. Jeffries, B. Vekemans and K. Janssens, *J. Anal. At. Spectrom.* **16** (9), 1006 (2001).

[759] *Application Note ArtTAX 01/03*, Röntec GmbH, Berlin (2003).

[760] A.B. Christie, in *Methods of Surface Analysis* (M.J. Walls, ed.), Cambridge University Press, Cambridge (1989), p. 127.

[761] I.W. Drummond, *Microsc. Anal.* (3), 29 (1992).

[762] P.E. Larson and P.W. Palmberg (to Physical Electronics Inc.), *U.S. Pat.* 5,315,113 (Sept. 29, 1992).

[763] M.P. Seah and G.C. Smith, *Surf. Interf. Anal.* **11**, 69 (1988).

[764] P. Coxon, J. Krizek, M. Humpherson and I.R.M. Wardell, *J. Electr. Spectrosc. Relat. Phenom.* **52**, 821–36 (1990).

[765] U. Gelius, B. Wannberg, P. Baltzer, H. Fellner-Feldeg, G. Carlsson, C.-G. Johansson, J. Larsson, P. Munger and G. Vegerfors, *J. Electr. Spectrosc. Relat. Phenom.* **52**, 747 (1990).

[766] J. Cazaux, *Rev. Phys. Appl.* **10**, 263 (1975).

[767] D.J. Keast and K.S. Dowing, *Surf. Interf. Anal.* **3**, 99 (1981).

[768] N. Gurker, M.F. Ebel and H. Ebel, *Surf. Interf. Anal.* **5** (1), 13–9 (1983).

[769] G. De Stasio, W. Ng, A.K. Ray-Chaudhuri, R.K. Cole, Z.Y. Gao, J. Wallace, G. Margaritondo, F. Cerrina, J. Underwood, R. Perera, J. Kortwright, D. Mercanti and M.T. Ciotti, *Nucl. Instrum. Meth. Phys. Res.* **A294**, 351–4 (1990).

[770] J.S. Hammond, J.F. Moulder and D.J. Hook, *Abstracts 215th ACS Natl. Mtg.*, Dallas, TX (1998), PMSE-058.

[771] D. Briggs, *Surface Analysis of Polymers by XPS and Static SIMS*, Cambridge University Press, Cambridge (1998).

[772] A.H. Sharma, F. Mozayeni and J. Alberts, *Proceedings Polyolefins XI*, SPE (ed.), Houston, TX (1999), pp. 679–709.

[773] J.E. deVries, L.P. Haack, T.J. Prater, S.L. Kaberline, J.L. Gerlock, J.W. Holubka, R.A. Dickie and J. Chakel, *Progr. Org. Coat.* **25** (1), 95–108 (1994).

[774] J.F. Watts, *Vacuum* **45**, 653–71 (1994).

[775] A.H. Sharma and B.C. Beard, *Communication at SPE RETEC*, Atlantic City, NJ (1996).

[775a] R.M. Caprioli, T.B. Farmer and J. Gile, *Anal. Chem.* **69**, 4751–60 (1997).

[776] B.K. Furman and G.H. Morrison, *Anal. Chem.* **52**, 2305–10 (1980).

[777] L.J. Kovalenko, J.M. Philippoz, J.R. Bucenell, R. Zenobi and R.N. Zare, *Space Sci. Revs.* **56**, 191 (1991).

[778] M.S. de Vries, H.E. Hunziker, H.R. Wendt and D.D. Saperstein, *Anal. Chem.* **62**, 2385 (1990).

[779] A.I. Vit, M.N. Galbraith, J.H. Hodgkin and D. Yan, *Polym. Degr. Stabil.* **42** (1), 69–73 (1993).

[780] F.P. Novak, Z.A. Wilk and D.M. Hercules, *J. Trace Microprobe Tech.* **3** (3), 149–63 (1985).

[781] D.M. Hercules, *Mikrochim. Acta Suppl.* **11**, 1–27 (1985).

[782] J.A. Bennett, E.A. Schweikert, L. Van Vaeck and F.C. Adams, *J. Trace Microprobe Tech.* **7**, 279–92 (1989).

[783] Q. Zhan, R. Zenobi, S.J. Scott, P.R.R. Langridge-Smith and R.B. Patrick, *Macromolecules* **29** (24), 7865–71 (1996).

[784] L.-S. Sheng, S.L. Shew, B.E. Winger and J.E. Campana, *ACS Symp. Ser.* **581**, 55 (1994).

[785] D. Briggs, *Surf. Interf. Anal.* **5**, 113 (1983).

[786] A. Benninghoven, *Angew. Chem. Int. Ed.* **33**, 1023 (1994).

[787] E. Niehuis, *Proceedings SIMS VIII* (A. Benninghoven, K.T.F. Janssen, J. Tümpner and H.W. Werner, eds.), J. Wiley & Sons, Chichester (1992), pp. 269–76.

[788] G. Slodzian, *Ann. Phys. (Paris)* **9**, 13 (1964).

[789] R. Castaing and G. Slodzian, *J. Microsc.* **1**, 395 (1962).

[790] G.H. Morrison and G. Slodzian, *Anal. Chem.* **47** (11), 932A (1975).

[790a] M.P. Seah, *J. Electr. Spectr. Relat. Phenom.* **71**, 191–204 (1995).

[791] J. Schwieters, H.G. Cramer, T. Heller, U. Jürgens, E. Niehuis, J.F. Zehnpfenning and A. Benninghoven, *J. Vac. Sci. Technol.* **A9**, 2864 (1991).

[792] B.W. Schueler, *Microsc. Microanal. Microstruct.* **3**, 119 (1992).

[793] M. Wood, Y. Zhou, C.L. Brummel and N. Winograd, *Anal. Chem.* **66**, 2425–32 (1994).

[794] K.G. Lloyd, K.M. Stika and D.G. Swartzfager, *Proceedings SIMS XI* (G. Gillen, R. Lareau,

J. Bennett and F. Stevie, eds.), J. Wiley & Sons, Chichester (1998), pp. 555–8.

[795] A.J. Eccles and T.A. Steele, *Abstracts SIMS XII*, Brussels (1999), p. 337.

[796] A.J. Eccles, *Mater. World* **7** (10), 619–20 (1999).

[797] A.J. Eccles and T.A. Steele, *Proceedings SIMS XI* (G. Gillen, R. Lareau, J. Bennett and F. Stevie, eds.), J. Wiley & Sons, Chichester (1998), pp. 775.

[798] A. Spool and K. Kuboi, *Abstracts SIMS XII*, Brussels (1999), p. 227.

[799] G. Gillen and R.L. Myklebust, *Proceedings SIMS VIII* (A. Benninghoven, K.T.F. Janssen, J. Tümpner and H.W. Werner, eds.), J. Wiley & Sons, Chichester (1992), pp. 509–12.

[800] P.A.W. van der Heide, *Proceedings SIMS XI* (G. Gillen, R. Lareau, J. Bennett and F. Stevie, eds.), J. Wiley & Sons, Chichester (1998), pp. 821–6.

[801] R.J. Day, S.E. Unger and R.G. Cooks, *Anal. Chem.* **52**, 557A–72A (1980).

[802] R. Jede, O. Ganschow and U. Kaiser, in *Practical Surface Analysis, Vol. 2: Ion and Neutral Spectroscopy* (D. Briggs and M.P. Seah, eds.), J. Wiley & Sons (1992), pp. 19–104.

[803] P. Bertrand and L.T. Weng, in *Surface Characterization. A User's Sourcebook* (D. Brune, R. Hellborg, H.J. Whitlow and O. Hunderi, eds.), Wiley-VCH, Weinheim (1997), Chp. 20.

[804] M.P. Mawn, R.W. Linton, S.R. Bryan, B. Hagenhoff, U. Jürgens and A. Benninghoven, *J. Vac. Sci. Technol.* **A9**, 1307–11 (1991).

[805] S.J. Pachuta and J.S. Staral, *Anal. Chem.* **66**, 276–84 (1994).

[806] S. Reichlmaier, S.R. Bryan and D. Briggs, *J. Vac. Sci. Technol.* **A13**, 1217–23 (1995).

[807] J. Zehnpfenning, H.G. Cramer, T. Heller, U. Jürgens, E. Niehuis, J. Schwieters and A. Benninghoven, *Proceedings SIMS VIII* (A. Benninghoven, K.T.F. Janssen, J. Tümpner and H.W. Werner, eds.), J. Wiley & Sons, Chichester (1992), pp. 501–4.

[808] L. Lianos, Tran Minh Duc, S. Reichlmaier and J. Hammond, *Proceedings SIMS IX* (A. Benninghoven, Y. Nihei, R. Shimizu and H.W. Werner, eds.), J. Wiley & Sons, Chichester (1994), pp. 468–71.

[809] M.J. Walzak, N.S. McIntyre, T. Prater, S. Kaberline and B.A. Graham, *Anal. Chem.* **71** (7), 1428–30 (1999).

[810] D. Stapel and A. Benninghoven, *Appl. Surf. Sci.* **174**, 261–70 (2001).

[811] R. Kersting, A.P. Pijpers, B. Hagenhoff, R. Verlaek, D. Stapel, A. Benninghoven and B.-C. Schwede, in *Proceedings SIMS XII* (A. Benninghoven, P. Bertrand, H.-N. Migeon and H. Werner, eds.), Elsevier Science Publishers, Amsterdam (2000), pp. 825–8.

[811a] R. Kersting, *Ph.D. Thesis*, University of Münster (2002).

[812] A. Ishitani, A. Karen, S. Tomita and K. Okuno, *Proceedings SIMS X* (A. Benninghoven, B. Hagenhoff and H.W. Werner, eds.), J. Wiley & Sons, Chichester (1996), pp. 99–106.

[813] R.W. Linton, M.P. Mawn, A.M. Belu, J.M. DeSimone, M.O. Hunt, Y.Z. Menceloglu, H.G. Cramer and A. Benninghoven, *Surf. Interf. Anal.* **20**, 991 (1993).

[814] I. Suarez-Lopez, C.M. Carr and J.C. Vickerman, *Proceedings Colloquium Spectroscopicum Internationale XXXIII*, Granada (2003), p. 592.

[815] P.M. Lindley, J.A. Chakel and R.W. Odom, *Proceedings SIMS VIII* (A. Benninghoven, K.T.F. Janssen, J. Tümpner and H.W. Werner, eds.), J. Wiley & Sons, Chichester (1992), pp. 219–22.

[816] J.L. Gerlock, T.J. Prater, S.L. Kaberline, J.L. Dupuie, E.J. Blais and D.E. Rardon, *Polym. Degr. Stabil.* **65**, 37–45 (1999).

[816a] B. Hagenhoff and D. Rading, in *Handbook of Surface and Interface Analysis* (J.C. Rivière and S. Myhra, eds.), M. Dekker, New York, NY (1998).

[817] S.G. MacKay and S.J. Pachuta, *Polymer Prepr. (ACS Div. Polym. Chem.)* **37** (1), 299–300 (1996).

[818] L.T. Weng, J. Feng, S.H. Zhu, H.H.K. Lo and C.-M. Chan, *Proceedings SIMS XI* (G. Gillen, R. Lareau, J. Bennett and F. Stevie, eds.), J. Wiley & Sons, Chichester (1998), pp. 567–70.

[819] T. Hoshi and M. Tozu, *Proceedings SIMS X* (A. Benninghoven, B. Hagenhoff and H.W. Werner, eds.), J. Wiley & Sons, Chichester (1996), pp. 478–80.

[820] G. Camino, L. Costa, L. Trossarelli, F. Constanzi and G. Landoni, *Polym. Degr. Stabil.* **12**, 213 (1985).

[821] Gy. Marosi, I. Csontos, I. Ravadits, A. Tohl, P. Anna, F. Sommer, M. Botreau and M.D. Tran, *J. Therm. Anal. Calorim.* **56**, 1071–80 (1999).

[822] J. Feng, C.-M. Chan and L.-T. Weng, *Proceedings SPE ANTEC 2000*, Orlando, FL (2000), pp. 2434–37.

[823] S.J. Pachuta, *Proceedings ACS Div. Polym. Mater.: Sci. & Engng.* **78**, 79 (1998).

[824] J. Briggs, M.J. Hearn, I.W. Fletcher, A.R. Waugh and B.J. MacIntosh, *Surf. Interf. Anal.* **15**, 62 (1990).

[825] L.-T. Weng, P. Bertrand, W. Lauer, R. Zimmer and S. Busetti, *Surf. Interf. Anal.* **23**, 879–86 (1995).

[826] H. Feld, M. Deimel and J.-U. Riedel, *Metalloberfläche* **54** (11), 48–52 (2000).

[827] J.C. Canry, A.D. Bass and J.C. Vickerman, *Proceedings SIMS IX* (A. Benninghoven, Y. Nihei,

R. Shimizu and H.W. Werner, eds.), J. Wiley & Sons, Chichester (1994), pp. 800–3.
[828] A.A. Galuska, *Proceedings ACS Div. Polym. Mater.: Sci. & Engng.* **78**, 79 (1998).
[829] D. Briggs, *Surf. Interf. Anal.* **8**, 133 (1986).
[830] A.J. Swift, in *Developments in Polymer Analysis and Characterisation* (Rapra Technology Ltd., ed.), Rapra Technology Ltd., Shawbury (1999), Paper 3.

Chapter 6

Thou salt quantify!

Quantitative Analysis of Additives in Polymers

6.1. Sampling Procedures for Quantitative Analysis of Polymer/Additive Packages 600
6.1.1. Quantitative Analysis of Mineral Filled Engineering Plastics 605
6.1.2. Reverse Engineering of Cured Rubber Compounds 606
6.1.3. Determination of Additive Blends in Polymers 606
6.2. Quantitative Solvent and Thermal Extraction 609
6.2.1. Extraction and Quantification of Polyolefin Additives 613
6.2.2. Supercritical Fluid Extraction 614
6.2.3. Quantification of Antioxidants in Polyolefins 615
6.2.4. Determination of Plasticisers by Solvent and Thermal Extraction 619
6.2.5. Oil-extended EPDM 623
6.2.6. Migration Rates of Phthalate Esters from Soft PVC Products 624
6.3. Quantitative Chromatographic Methods 624
6.3.1. Quantitative Gas Chromatography 626
6.3.2. Quantitative Liquid Chromatography 628
6.3.3. Quantitative Supercritical Fluid Chromatography 629
6.3.4. Quantitative Thin-layer Chromatography 630
6.4. Quantitative Spectroscopic Techniques 633
6.4.1. Quantitative Ultraviolet/Visible Spectrophotometry 637
6.4.2. Quantitative Fluorescence Spectroscopy 639
6.4.3. Quantitative Infrared Spectroscopy 639
6.4.4. Quantitative Near-infrared Spectroscopy 644
6.4.5. Quantitative Raman Spectroscopy 645
6.4.6. Quantitative Nuclear Magnetic Resonance Methods 646
6.5. Quantitative Mass Spectrometric Techniques 647
6.6. Quantitative Surface Analysis Techniques 651
6.7. Quantitative Microscopy 653
Bibliography 654
General Quantitative Analysis 654
Sampling 654
Chromatography 654
Spectroscopy 655
Mass Spectrometry 655
Surface Analysis 655
Chemometric Techniques 655
References 655

Quality is more important than quantity. Yet, without proper quantitation no quality. Analytical chemists practise one of the most quantitative of all sciences. Obtaining quantitative results for any analytical method is a challenge. Quantitative chemical analysis is a highly dynamic field, as new methods are continuously being developed. In industrial research, quantitative analysis is particularly dominant. Quantitative analysis involves the whole process of analysis, from sampling to detection. The fundamental problem is to isolate a representative signal of the analyte from the matrix. This favours chemical or physical separation methods.

Quantitative analyses are described as macroanalytical (>100 mg of sample), semi-microanalytical (10–100 mg), microanalytical (1–10 mg) or ultramicro-(submicro-)analytical (<1 mg). Methods for quantification are classical chemical methods, physicochemical or instrumental methods. Important classical methods are gravimetry and titrimetry. In physicochemical methods, based on a chemical reaction of the analyte to be determined, an instrument is used to measure a physical property of the reaction product (*e.g.* potentiometry, conductimetry). Instrumental methods (optical, electroanalytical, thermal, *etc.*), which evaluate an extensive physical property, are sensitive and quite selective and can be used for automatic monitoring. When all other factors are equal a specific (or selective method) is preferable to one that is merely sensitive. In another **classification** of quantitative analytical procedures, the analytical principles, methods and techniques are considered according to the material being analysed (*e.g.* polymer additives). This has the advantage of bringing together all relevant data for a particular type of material. Different analytical procedures are often necessary for organic and inorganic analytes.

Analytical procedures may be either destructive or non-destructive. Indirect or destructive methods require a significant alteration to the sample so that the additives can be removed from the plastic material for subsequent detection. Direct or non-destructive methods involve minimal sample preparation which greatly speeds up the analytical procedure. Quantitative analytical procedures are either continuous or discontinuous. In non-continuous routine analytical procedures a small amount of substance is analysed which is considered to be representative of the whole sample. Errors are introduced by the sampling procedures and no measure of continuous changes in the composition of the material is obtained. Continuous monitoring needs more resources but can give a better and more extensive measure of the composition of the material and variations in time (*cfr.* Chp. 7).

It is fairly intuitive in which cases the analysis of additives in polymeric formulations should be extended to quantitation. Quantitative analysis may be of importance from raw material evaluation to quality control, allows to verify mass balances and is the precise goal of in-process analysis (*cfr.* Chp. 7). Legislation is an important driver in the analytical market place for support and verification of quality assurance procedures.

While industrial analysis technology has realised considerable advances, including fibre optics, filters, rugged probes, detectors, lasers, classification and multivariate calibration techniques, manufacturing practices and hardware have reduced the need for rigorous quantitative compositional analysis. The method chosen is most often a compromise between accuracy and economics (calculated risk). At the same time, pressure to reduce production costs has led to close scrutiny of quality control (QC) laboratory testing volume. Often recourse is taken to process validation without rigorous quantitation [1]. Practices of assurance of product quality with reduced testing include testing of a single control additive which represents others in a concentrate. This practice involves assumptions which are not always justified, as will be shown. Also, the concentration of additives in a polymer may vary from batch to batch and even within the same batch.

In some application areas, such as packaging for food contact, quantitative analysis of additives and of their degradation and interaction products may be more important than in others (E&E, automotive, *etc.*). However, effects of migration also occur in contact of polymeric formulations with gas (gas pipes), water (water pipes), organic liquids (fuel tank), *etc.*, affecting polymer lifetime. Depending upon the polymer application, interest in quantification of *specific* additive classes may exist. For example, in aged materials it may be of greatest interest to verify the rest stabiliser contents. In a troubleshooting laboratory the exact nature of the additive package and loadings are usually unknown, whereas in a research environment each sample may be unique as opposed to a production support laboratory.

It should be understood that the seemingly ultimate challenge of quantifying the unknown is meaningless. For example, in polymer processing 5 ppm of an unknown component is usually immaterial, whereas for odour/taste problems the same amount (and indeed often much less) is highly relevant. In the latter case even quantitation is not sufficient; identification of the odorant is often desired. Quantitative polymer/additive analysis is costly and needs to be considered carefully both in terms of time efficiency and reliability of the results. Figure 6.1 shows the steps in a typical quantitative analysis; Table 6.1 lists the **main requirements**.

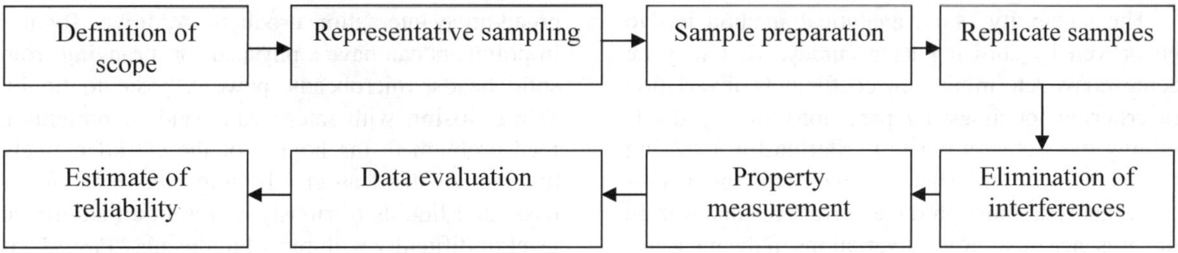

Fig. 6.1. Steps in a typical quantitative analysis.

Table 6.1. Requirements for quantitative analysis

- Scope (amount of sample, level of accuracy, choice of method, *etc.*)
- Sampling, sample preparation and dosing
- Ease of use
- Elimination of interferences (*e.g.* chromatography)
- Linearity over the range of concentrations expected (linear dynamic range)
- Measurement robustness
- Routine analysis
- Multicomponent analysis
- Experimental flexibility
- Flagging of error conditions *cq.* incorrect results
- Data processing
- Validation

Classical methods of quantitative analysis are relatively straightforward in that they use well-understood relationships between the independent data (*e.g.* spectral absorbances) and the dependent data (concentrations). Reliable quantitative method development and application require special and careful attention. In quantitative analysis reproducibility is paramount. The criteria set during method development *must be met* if quantitative results are to be meaningful. This applies whether the analysis is univariate or multivariate. Multivariate models have a distinct advantage over univariate ones in that they allow inclusion of more of the available data and thus the solutions are generally more stable. Important parameters in many solid sampling methods are particle size, and the effects of sample morphology (crystallinity or molecular orientation). Many chemical and physical properties, such as state of hydration or intermolecular interactions, may need to be considered in addition to factors such as temperature and pressure. For most quantitative analyses, in particular multivariate analyses, it is also important that the spectral ranges analysed are free from artefacts, such as atmospheric absorptions (IR: water vapour and CO_2; Raman: O_2 and N_2). Whatever method is chosen to perform the quantitation, whether a simple single wavelength assay or a complex chemometric method (usually CLS, ILS, PCR or PLS), the method should be validated to ensure it produces meaningful data. Needs expressed by European laboratories comprise CRMs/RMs of monomers, additives, trace elements and heavy metals in polymers and other synthetic materials [2]. As shown in Chp. 8.4.1, different **analytical performance parameters** are involved for validated quantitation of major components, determination of impurities or degradation products, or just determination of performance characteristics. Calibration, *i.e.* determination of the relation between instrumental response and concentration, is one of the most critical steps in quantitative analysis. In multicomponent quantitative analysis a great deal of emphasis has been placed on multivariate calibration. A quantitative method needs a representative reference standard to calibrate the detector response. In absolute methods of analysis, based on an absolute property of an analyte (*e.g.* mass), the response of the property can be used once the instrument is calibrated (*e.g.* gravimetric analysis calibrated to international standards). In standardless analyses the signal is not matrix dependent and the measured f(conc.) is stable with time; in those cases standards are infrequently required. Interesting examples of standardless analysis are XRF, IDMS and NAA. In relative analytical techniques (*e.g.* chromatography) changes in background, such as reagents or matrix, may modify the analytical performance, and frequent calibration is needed. Quantitative analyses are usually carried out by comparing the measured quantities of test samples with those of standards with known concentrations. IUPAC guidelines for calibration in analytical chemistry are available [3].

The suitability of an analytical method has to be proven by showing its accuracy. This may be achieved by determining the coefficients of variation (a criterion for assessing precision) and by determining the recovery rates (a criterion for assessing the accuracy of the mean or bias). A routine quantitative analysis must produce correct results even in the presence of significant variations in the measurement environment.

From the viewpoint of quantitative analysis, of course, any detection specificity calls for the introduction of correction factors. A number of detection techniques (ion-selective electrodes, light-scattering detection) or analytical techniques (TLC) and/or sample preparation steps (extraction) show non-linear relationships.

As to **nomenclature**, the lowest amount of analyte in a sample which can be detected but not necessarily quantitated as an exact value is termed the limit of detection (LOD). LOD is referenced to the total analytical method, including sample losses in sample preparation steps. The limit of quantitation (LOQ) is the lowest amount of analyte in a sample which can be quantitatively determined experimentally with suitable precision and accuracy. Approaches for determining LOD and LOQ may be used on visual evaluation (LOD only), S/N ratio, calibration curve or standard deviation of the blank. Standard error of prediction (SEP) is an estimate of the standard deviation.

In Chp. 8 of ref. [3a] quantitative element analysis was described; for the development of certified reference materials, *cfr.* Chp. 8.3. In this Chapter **quantitative molecular analysis**, as applied to polymer/additive formulations, is further worked out. Quantitative applications already reported will not be repeated here and reference is made to previous chapters.

6.1. SAMPLING PROCEDURES FOR QUANTITATIVE ANALYSIS OF POLYMER/ADDITIVE PACKAGES

Principles and Characteristics

In general terms, quantitative polymer/additive analysis is a **multi-analyte problem**, which is even more complicated by the heterogeneity of some formulations, the representativity of sampling of the various analytical techniques, the presence of technical ingredients, the occurrence of degradation and co-additive interaction products. Additives for use in polymers can have a physical form ranging from solid beads, microbeads, powder, paste to liquid. While **dosing** with microbeads seldom presents a feed problem in the hopper of the extruder, finely divided powders can give bridging problems, while paste and liquids obviously represent an additional level of difficulty with regard to dosing. "Pastes", as such, are difficult to dose on continuous equipment and they need to be heated to enable their dosing by special pumps. There is also a degree of uncertainty with regard to the precise dosage level being achieved owing to ambient temperature variations, which lead to changes in the viscosity of the liquid being injected. Concentrates in pellet form are very easily dosed. Concentrates can readily be mixed with virgin resins in pellet form to give a homogeneous mixture. The use of concentrates, masterbatches, one-pack additive blends and sophisticated material delivery systems can give high confidence in polymer compounding. However, even these systems have vulnerabilities (operator error, mixing equipment, failure of components, *etc.*). As to the **representativity of sampling**, it is sufficient to recall the small sample sizes needed for some techniques, such as DIMS (1 μg) and PyGC-MS (1 mg), as opposed to in-process NIRS (cm thick flow cells). Quantitative chemical analysis is a highly dynamic field. The continually increasing sensitivity of analytical instruments allows us to probe smaller samples, such as local structures. For smaller volumes, surface properties become more important.

In quantitative analysis one should be aware of the use of technical products, such as linear phthalates (some important plasticiser alcohols are mixtures of C_9–C_{11} mixtures), branched chain phthalates (C_6–C_{12} mixtures), chlorinated paraffin plasticisers (C_{10}–C_{12}, C_{12}–C_{14}, C_{14}–C_{17}, C_{18}–C_{20} fractions) [4]. Other **technical products**, such as fatty acid methyl esters (FAME), which comprise methyl laurate (C_{12}, saturated), palmitate (C_{16}, saturated), stearate (C_{18}, saturated), oleic (C_{18}:1), linoleic (C_{18}:2), linolenic (C_{18}:3), and arachidic (C_{20}:0) components, are typically accounted for by taking the total area of the methyl esters in each GC chromatogram [5]. Also lubricating agents are generally composed of primary fatty amide mixtures, such as palmitamide/stearamide/oleamide (20–25/70–80/2–5), caprylamide/capramide/lauramide/myristamide/palmitamide/stearamide/linoleamide (7–10/6–8/40–60/15–20/8–10/1–3/1–3), or stearamide/oleamide/linoleamide (4–7/85–95/2–5). Similarly, textile fibre-finishing compositions often consist of varying combinations of surfactant, C_{10}–C_{16} fatty alcohols, and caproamide or fatty methyl esters, which enhance surfactant penetration and retention in rayon fibres [6].

As pointed out by Ashton [7], quantitative relationships of additive content to polymer properties are often elusive. A reason for this fact is that the amount of additive charged to a polymer under processing is not necessarily the parameter which should be measured, but it traditionally is the easiest to measure. Ideally, the level of residual active stabiliser is of real interest. Quantitation also needs to take into account possible **degradation/oxidation products**. The analytical problem can be illustrated as follows: various methods are applicable to the regulated antioxidants butylated hydroxyanisole (BHA), butylated hydroxytoluene (BHT), dilaurylthiodipropionate (DLTDP), *n*-propylgallate (PG) and *t*-butylhydroquinone (TBHQ). As the analytical methods, described by FAAM [5], measure the presence of unoxidised substance only, the analytical results may not necessarily indicate or reflect the amount of antioxidant that was originally added to the polymer. BHA, BHT, PG en TBHQ are determined by RPLC-UV (at 280 nm) after extraction from the matrix [8]. Scheirs *et al.* [9] have used a range of techniques (GC, GC-MS, HPLC-UV, SFC, MS, ^{31}P NMR, F) to characterise and quantify the commercial AOs Irganox 1076 and Irgafos 168 and their conversion products in HDPE. Quantification of Irgafos 168 is complicated by its degradation and hydrolysis products [10].

There are three **basic alternative routes** to quantification, probably applicable to all forms of analysis: *(i)* calculation of all the relevant terms from first principles (as in primary techniques such as NAA, NMR, *etc.*); *(ii)* the use of published databases; and *(iii)* the use of locally produced standards and local databases. In practice, a combination of the three approaches may often be most effective. In secondary techniques (which are the majority of analytical techniques) the instruments are calibrated against results obtained using other analytical methods. Precision and accuracy obtained are then limited by those of the primary techniques.

For quantification of unknown (molecular) components in unknown polymeric matrices generally three functionalities are highly desirable; separation, identification and quantitation, *preferably in this order*. Sometimes it is possible to eliminate one or both of the first steps. Chapter 7 of ref. [3a] has described the hyphenation of separation and identification techniques, typically pre-chromatographic sample preparation–chromatography–spectroscopy/spectrometry.

Table 6.2. Analytical functionalities

Separation	Identification[a]	Quantitation
GC	UV/VIS	(HR)TG
HPLC	Mid-IR	NMR
SFC	MS	FID
TLC	NMR	UV
GPC	Chromatography[b]	NIRS
MS^n		
NMR		
HS		
TD		
Pyrolysis		

[a] Usually on the basis of spectral databases.
[b] Retention times.

On the basis of the preferred order for quantitation, Table 6.2 suggests only a restricted number of **(hyphenated) techniques for multi-analyte quantitation**. Of the quantitative methods the primary technique of ^{1}H l-NMR is limited to 5 ppm (matrix and analyte dependent) and FID to sub-ppm; TG allows considerably higher sensitivity and lower detection limits (100–1000 ng). By applying thermogravimetric analysis, the carbon-black content (of rubbers) can be determined quantitatively [11]. Affolter *et al.* [12,13] have shown that mixtures of plasticisers can be separated and quantified by means of TG. However, it is not common practice and indeed not very practical to have TG coupling at the end of a chain of hyphenated techniques. Rather, the following techniques allow quantitative analysis: TG, TG-MS, TG-FTIR, TG-GC-MS and TG-FTIR-MS; TG-DTA and TG-DSC are less suitable for evolved gas analysis. Having TG in pole position means that the ideal operational sequence (separation, identification, quantification) is compromised. For example, it is clear from Chp. 2.1.5.6 that even the best current TG-based method, TG-GC-MS, is not routinely used for quantitative analysis. Another possible approach is GC-MS/FID, *cfr.* Chp. 7.3.1 of ref. [3a]. For additive analysis the HPLC-NMR coupling has as yet found few adepts. It needs to be concluded that the possible approaches to hyphenated, quantitative,

Table 6.3. Characteristics of sampling procedures for quantitative analysis of polymer/additive formulations

Sampling method	Advantages	Disadvantages
Extraction	Allows access to chromatographic separations	Not generally applicable
	Homogenisation	Limited applicability for high-MW compounds
		Extraction yield analyte and matrix dependent
		Wet chemistry
Dissolution/Precipitation	Quantitative analyte transfer	Some polymeric matrices are dissolution resistant
	Access to chromatographic separations	Diluted systems
	Homogenisation	Wet chemistry
Hydrolysis	Access to analytes	Not generally applicable
	Homogenisation	Wet chemistry
In-polymer spectroscopy	Speed	Chemometrics
	Quantitative analyte sampling	Difficult multi-analyte analysis
Desorption	Speed	Limited analyte separation
		Questionable quantitative sampling
Thermal extraction	Speed	Limited separative power
	Quantitation	Difficult multi-analyte analysis
Pyrolysis	Speed	Breakdown of analytes
	General applicability	Difficult quantitation
Melt	Spectroscopic quantitation	Chemometrics
		Difficult multi-analyte analysis
		Limited use

multi-analyte analysis of polymeric formulations are actually fairly restricted.

It is useful to compare various **sampling methods** to quantitative chemical analysis and to list their respective advantages and limitations (Table 6.3). In fact, an analysis is only as good as the sample which has been introduced into the analytical instrument. The ideal way to carry out a quantitative analysis with a sampling technique is to transfer an analyte completely from the sample matrix to the analytical apparatus. This means that in principle quantitative analysis of an additive is well carried out by dissolution (100% recovery), especially when the procedure restricts additional handling (evaporation, preconcentration, redissolution, *etc.*). The routine application of μSEC-GC is a case in point. For quantitative analysis, most instruments require a solution. On-line combinations of sample treatment and analytical systems are being studied intensively. The idea behind such systems is to perform sample extraction, clean-up and concentration as an integral part of the analysis in a closed system [14].

The choice of the most appropriate analytical method is somehow also closely related to **concentration**. The development of sophisticated instrumental methods has allowed analytical chemists to probe samples for components at very low concentrations levels. The more sensitive techniques, such as chromatography and some of the spectroscopies, are therefore of increasing importance in polymer/additive analysis. Classical methods, such as titrations and precipitations, are nowadays of fairly limited use in polymer/additive analysis. They are restricted to relatively concentrated solutions (1 mM or more). In the past, for quantitative analysis of vulcanisation accelerators and their reaction products also conductometry, polarography and photometry were being considered [15,16].

One of the keys to quantitative analysis is the assumption that the concentrations of the analytes in the samples are related to the measured data. Starting from a collection of known data (the composition of standards) a calibration or training set is formed. The calibration equation will then accurately predict the quantities of the constituents of interest of "unknown" samples provided the same experimental conditions are used as in the calibration set. Some experimental methods provide "single-point" measurements for each calibration (*e.g.* single-element atomic absorption), others (*e.g.* spectroscopies) provide many data points. In the latter case many more measurements per sample are available in generating the calibration equations. The laboratory should carefully consider the calibration mode chosen: *(i)* standard additions; *(ii)* calibration curve; and *(iii)* bracketing standards. There is no calibration mode that in all cases should be recommended. All suffer from typical sources of error, *e.g.* for standard additions: non-linearity of the calibration curve, extrapolation difficulties, chemical form of calibrant added, *etc.*

In **method development** for quantitative sample analysis it is of utmost importance to optimise and show reproducibility of recovery (average ± standard deviation) from sample to sample. As long as results are reproducible and known with a high degree of certainty 100% recovery is not necessary. Wherever possible, it is helpful to determine percent recovery of a spiked, authentic standard analyte in a sample matrix that is shown to contain no such analyte. In the method of quantitation to be developed the percent recovery needs to be taken into account. Possible methods include standard additions, external/internal standard and isotopic dilution.

To obtain values for the concentration or mass of analyte in a sample, the peak height or area is compared against peak measurements of **standards** containing known amounts of analyte. The comparison is often accomplished by constructing a calibration graph showing concentration of analyte plotted against peak height or area. The goal of calibration, whether multivariate or not, is to replace a measurement of the property of interest by one that is cheaper, or faster, or better accessible, yet sufficiently accurate. In standards preparation the extreme concentrations for each component must be included, as extrapolation outside of the calibrated concentration range is dubious. There are a number of methods used to prepare the standards for calibration. Materials referred to as standards span a considerable range in quality:

(i) (Inter)national standards, such as standard reference materials (SRM®s), which have been certified through the use of either a definitive analytical technique, two or more independent techniques, or interlaboratory testing with detailed statistical evaluation of the results.

(ii) Carefully synthesised materials for which at least partial analytical characterisation has been performed, *e.g.* standard "research materials" (RM), issued by NIST.

(iii) Standards with a more modest pedigree: commercial materials or compounds synthesised locally for the purpose of solving a particular problem.

Standards with known additive loadings are required to calibrate "in-polymer" analysis techniques. In the laboratory preparation of standards, it is extremely important to refer to the actual materials used in the formulation of the final product rather than ultra pure grades of chemicals. In the selection of mixtures to be used as standards all component levels should be evenly represented in the calibration blends, in order to avoid one component from dominating the spectral information for quantitative measurements. The precision of standard mixtures generally needs to be better than that of the analytical system being developed. Preparation of good calibration standards and the choice of a suitable internal standard are of crucial importance for quantitation of polymer additives.

Standards in use for quantitation are essentially employed in three ways. With the **internal standard technique**, known quantities of a carefully selected (usually high purity) substance, the internal standard, are added to both samples and standards. The internal standard (preferably a non-commercial product) should have similar chemical and physical properties to the analyte, in particular, volatility and functional groups, in order to react in the same way to changes in the chemical environment (*e.g.* dinonyl adipate may serve as an internal standard for the determination of di(2-ethylhexyl) adipate). Solutions of pure additives used as standards may be unsatisfactory due to the difference in the evaporation profile between pure additives and those blended in the polymer samples. For example, a pure Permanax WSP sample evaporates in the ion source from about 30 to 150°C, whereas Permanax WSP blended in PE evaporates from about 120°C (m.p. of PE) to 350°C [17]. If one opts for polymer-based calibration standards the homogeneity of the samples is of crucial importance. Using internal standards in quantitative analysis is advantageous, for instance, in cases where the sample thickness cannot be determined exactly, or in gaseous samples with unknown total pressure.

Quantitative analysis using the **external standards method** is achieved by preparing a range of standards, containing known quantities of analyte, in the same matrix as the sample. The major disadvantage of this method is that the volume introduced into the GC needs to be precisely made. An internal standard is often a good alternative. The **standard addition method** is particularly useful where suitable blank material is not available to prepare external standards and matching of samples to standards is critical. Polymer analysis and foodstuffs are two areas where this method of calibration is commonly employed. Standard addition is a general method used in quantitative analytical chemistry (in particular for spectrometric analysis) to quantify the amount of an impurity element in an unknown sample. A known amount of the impurity is added to the unknown sample, and then the ratio between the known added standard and the unknown impurity is measured to establish the absolute amount of the impurity originally present.

The methods of **data treatment** used in an analysis need to be as carefully defined and understood as those used for sample handling [18]. There is a great tendency to wantonly alter spectroscopic data before performing quantitative measurements. However "good appearance" will not improve the quantitative results. Classical methods of quantitative analysis use well-understood relationships between the measurement data and concentrations. Some methods are univariate, whereas others are multivariate (solve a series of equations using many measurements per sample for one calibration value). In univariate methods there is generally one independent variable (*e.g.* spectral response) and one dependent variable (concentration). Multivariate models allow inclusion of more of the available experimental data and generally generate more stable solutions. Multivariate calibration is the collective term used for the development of a quantitative model for the reliable prediction of properties of interest ($y_1, y_2, \ldots, y_q$) from a number of predictor variables ($x_1, x_2, \ldots, x_p$). This can typically be applied in the spectroscopic analysis (*e.g.* UV, IR, NIR, XRF, NMR) of a mixture in order to measure the concentration of the various constituents. Multivariate methods may provide the means of quantifying all the components of complex mixtures [19,20]. The field of process analysis, *i.e.* the analysis of chemical systems or processes using multiple sensors, depends heavily on the applicability of multivariate calibration models for the quantitative monitoring of the systems or processes of interest (*cfr.* Chp. 7). It is not surprising that multicomponent quantitative analysis (*e.g.* of polylolefin formulations) requires extensive work in preparation of standards, calibration and maintenance. A compromise in the existing tension to assure product quality with reduced testing is provided by procedures to validate the process without rigorous quantitation. Such practices include testing of a single control additive, which represents others in a concentrate, and/or qualitative fingerprinting by some analytical technique. These practices still involve assumptions [1].

It is of course of interest to determine which of the methods of quantitation provides the most accurate and precise quantitative data. It is equally important to consider the constant trade-off for precision and sensitivity. At very low concentrations, precision often becomes limited by extraneous factors, such as wall effects. In such cases, high-precision measurements are becoming virtually unobtainable. In analogy to the Quantitative Ingredient Declarations (QUID) in food analysis, which require statements as to the uncertainty of the measurement and the variability of the results (sampling!), also for industrial polymer analysis intra- and interlaboratory variation and the meaning of average analytical results needs to be established. It is the responsibility of the analyst to adequately describe the instrumentation and **performance** to duplicate the repeatability and accuracy of the developed method.

Analytical results are required to be accompanied by the **measurement uncertainty**. The knowledge of uncertainty allows a proper evaluation of the result with regard to specifications, batch-to-batch variations, tolerances and regulatory limits. For regulatory procedures, a check analysis, preferably by a method based on different principles, should be performed to document the validity of the data. In the event that the results are required for legal proceedings, the analyst must demonstrate that the method is performing through the use of controls and recovery studies. As well known, sample preparation accounts for 60% of the total analytical time spent and therefore can contribute significantly to the overall measurement uncertainty (typically 30% of the total error bar). The measurement uncertainty of an analytical procedure depends upon many parameters, including analyte concentration, matrix properties, sample preparation technique and measurement principle. It is well known that the interlaboratory standard deviation of analytical results increases with decreasing analyte concentration [21]. Typical relative standard deviations (%) are 1.5, 4, 5, 12 and 40% for concentrations of 10%, 1%, 0.01%, 1 ppm and 1 ppb, respectively (Horwitz

Table 6.4. Feasibility conditions for quantification

- Standards must be composed of linearly independent concentration data
- In blends as standards at least two components should be at different levels
- Components should be represented evenly in mixtures comprising the calibration set
- Concentrations of components in calibration standards should evenly cover the experimental range
- The system under investigation should be overdetermined
- Precision of a quantitation cannot exceed that of the concentrations in the calibration standards
- Calibration standards should be composed of the actual materials used in the formulation of the final product (technical *vs.* pure components)
- The analysis method should not affect the sample
- Identical operating conditions should apply to calibration references and unknowns
- Calibration should be checked regularly (SPC charts)
- Any mathematical treatment should be handled with care
- Spectral manipulations must be performed in identical fashion on all spectra
- The developed method should be validated

curve). Deviations from the Horwitz curve are on account of the analytical procedure, *e.g.* a titration can be performed with higher repeatability than a chromatographic analysis. Sources of error in polymer/additive analysis may involve various analytical steps: extraction, derivatisation, separation and detection. Extraction should be done in such a way that the analyte is separated from the interfering matrix without loss or contamination, without change of speciation (if of interest) and together with the minimum of interferences. In general, the risk of producing a wrong result increases with the number of steps in a determination and with their complexity. A detailed guide is available on the determination of the measurement uncertainty of chemical analyses [22].

Compton *et al.* [18] have worded some (fairly obvious) conditions for the feasibility of a quantitative method (Table 6.4). In developing quantitative methods for polymer/additive analysis **Murphy's Law** always applies, as also apparent from some of the following Case Studies. Quantitative determinations of "new" additives should be validated, *e.g.* by calibration on the basis of more than one mother liquor, repeatability/reproducibility experiments or the use of SPC charts. The analyst must constantly be aware of unwanted interferences. For example, the presence of certain additives, particularly fillers and pigments, may cause serious interferences with the measurement of other additives present [23]. More trivially, contaminations of Chimassorb 944 residues accumulated in a rotavapour may interfere with subsequent analyses. Care should be exercised that no extractions may occur from materials composing the equipment rather than the sample.

6.1.1. Quantitative Analysis of Mineral Filled Engineering Plastics

Case Study

In order to test internal quantitative analysis procedures Nelissen [24] has used various techniques to analyse three mineral filled polyamide compounds of nominally known composition (but not to the analyst). XRF was carried out both after ashing (using borax pearl technique) and directly on a plaque; XRD was used for the identification of crystalline fillers (*e.g.* mica). Wet chemical analysis consisted of hydrolysis with 6N HCl or HF followed by identification (IR) and quantification (mass %) of the residue. Table 6.5 shows the potential of various analytical techniques for the determination of mineral fillers. The results of Table 6.6 show overall good agreement between nominal and experimental values. The accuracy of the analyses is estimated as about 5% but mineral dependent. Inaccuracies are highest for the trace components. Different techniques gave various analytical contributions: DSC (PA6, PA6.6 copolymer), FTIR (polyamide, mica, glass fibre, melamine cyanurate); ATR-FTIR (no migrating components); XRF (mineral composition, absence of bromine, glass fibre type); hydrolysis (copolymer composition, chain regulator, fatty acid; mica and glass fibres in residue); GC (antioxidants). The results of this typical analytical deformulation problem of a complex polymer composi-

Table 6.5. The potential of some analytical techniques for the determination of mineral fillers

Filler	Wet chemical	XRF	IR	XRD
Talc	2	1, 2	1	1
Clay	2	1, 2	1	0
Mica	2	1, 2	1	1
Wollastonite	2	1, 2	1	1
Chalk	2	1, 2	1	1
E-glass	2	1, 2	0	0

0 = Identification and quantification not possible.
1 = Identification possible.
2 = Quantification possible.

Table 6.6a. Quantitative analyses of mineral filled engineering plastics

Component	Composition (EP 1)		Technique(s)
	Nominal	Experimental	
PA6, PA6.6 copolymer	6.9, 62.0 wt.%	7.5, 60 wt.%	DSC, IR, 6N HCl and TLC
Mica	20.0 wt.%	19.2 wt.%	IR, XRF, hydrolysis[a]
Glass fibre	5.0 wt.%	4.7 wt.%	IR, XRF, hydrolysis[a]
TiO_2	0.4 wt.%	0.6 wt.%	XRF
Melamine cyanurate	5.0 wt.%	5.0 wt.%	IR
Ca-stearate	0.3 wt.%	0.2 wt.%	Hydrolysis, XRF
Irganox 1098	0.4 wt.%	0.23 wt.%	GC/HPLC
Chain regulator (acetic acid)	–	460 ppm	Hydrolysis
Various pigments	0.03 wt.%	–	–
Total	100 wt.%	97.4 wt.%	

[a] Insoluble.

tion illustrate the general need for the use of several experimental techniques to crack the code.

6.1.2. Reverse Engineering of Cured Rubber Compounds

Case Study
Although direct analysis of rubber compounds yields simultaneously information about polymer and additives, a quantitative determination of the additives is difficult. This requires separation of these components from the polymer and fillers by means of extraction. Following the standard procedure at Akron Rubber (*cfr.* Chp. 2.2 of ref. [3a]), Coz *et al.* [25] have examined four unknown cured rubber compounds (radial passenger tyre tread, radiator hose, oil pan seal and engine gasket). Tables 6.7 and 6.8 compare the reconstructed formulations and actual recipes for the tyre and radiator hose.

In the radiator hose it was impossible to identify tetramethyl thiuram monosulfide (TMTM). This is most probably due to the fact that during vulcanisation TMTM breaks down into fragments that are very similar to tetramethyl thiuram disulfide (TMTD) fragments. Therefore, only TMTD is reported in the reconstructed formulation.

The reconstructed formulations were very similar to the actual recipes. The examples show that by the use of proper analytical techniques rubber formulations can be successfully reconstructed from cured compounds, giving an excellent tool to rubber mixers, fabricators and end-users of rubber parts.

6.1.3. Determination of Additive Blends in Polymers

Case Study
Quantitative polymer/additive analysis is particularly troublesome in case of additive blends. For production control of Irganox B220 (a nominally 3:1 blend of Irgafos 168 and Irganox 1010) in PE in 1993 a series of samples in the concentration range of

Table 6.6b. (Continued)

Component	Composition (EP 2)		Technique(s)
	Nominal	Experimental	
PA6	52.4%	51.5 wt.%	DSC, IR
Copolyamide[a]	0.6 wt.%	–	–
Glass fibre	40 wt.%	38.8 wt.%	IR, XRF
Pigments	0.002 wt.%	0.8 wt.%	XRF (as mineral residue)
MA mod. EPR[b]	6 wt.%	6.5 wt.%	IR residue
Carbon-black	0.4 wt.%	0.5 wt.%	IR residue, TGA
Ca-stearate	–	0.05 wt.%	Hydrolysis, XRF
Chain regulator (benzoic acid)	–	20 ppm	Hydrolysis
Chimassorb 944	0.175 wt.%	–	–
Tinuvin 234	0.175 wt.%	1200 ppm	D(HFIP)/P(MTBE)[c], GC/HPLC
Irganox 1098	0.175 wt.%	820 ppm	*Idem*
Irganox 1010	–	25 ppm	*Idem*
Irgafos 168	0.175 wt.%	710 ppm	*Idem*
Irgafos 168 phosphate	–	250 ppm	*Idem*
2,4 DTBP	–	160 ppm	*Idem*
Total	100.1 wt.%	98.4 wt.%	

[a] Carrier of carbon-black masterbatch.
[b] MA maleic anhydride.
[c] Dissolution/precipitation.

Table 6.6c. (Continued)

Component	Composition (EP 3)		Technique(s)
	Nominal	Experimental	
PA6	52.3 wt.%	54 wt.%	DSC, IR
LDPE	10.4 wt.%	16 wt.%	DSC, IR, residue
MA mod. EPR[a]	6.2 wt.%		
Talc	30 wt.%	27.7 wt.%	IR, XRF, residue
TiO_2	0.8 wt.%	0.6 wt.%	XRF
Ca-montanate	0.25 wt.%	0.07 wt.%	Hydrolysis[b]
Chain regulator (benzoic acid)	–	60 ppm	Hydrolysis
Irganox 1010	0.005 wt.%	80 ppm	GC/HPLC
Irgafos 168	0.02 wt.%	190 ppm	GC/HPLC
Irgafos 168 phosphate	–	120 ppm	GC/HPLC
Pigments	0.09 wt.%	–	–
Total	100.3 wt.%	98.4 wt.%	

[a] MA maleic anhydride.
[b] Determined as Ca-stearate.

500–2500 ppm were prepared for calibration purposes using XRF (P analysis), *cfr.* Table 6.9. The results were subsequently corroborated with independent ICP measurements on different samples (Table 6.10; $R^2 = 0.9983$ for ICP/XRF relationship). A PE/Irganox B220 control sample was also measured daily for Statistical Process Control (SPC) over a 4 year period with a variation coefficient of <3%. Unexpectedly, in 1997 an interlaboratory XRF comparison of several grades showed systematic discrepancies for PE/Irganox B220 (Table 6.11), on average 8.6% between two instruments, with a small spread (2.8%) for single-sided XRF(1) and a larger spread (23.3%) for double-sided XRF(2) measurements. These results, which greatly exceeded the consistent SPC variations for XRF(1) measurements

Table 6.7. Comparison of reconstructed formulation with actual recipe for a tyre compound[a]

Ingredient	Reconstructed formulation (phr)	Actual recipe (phr)
Styrene butadiene rubber	40.00	45.00
Polybutadiene rubber	60.00	55.00
N 234 carbon-black	75.30	70.00
Zinc oxide	4.00	3.00
Aromatic oil	43.80	37.50
Stearic acid	1.00	2.00
Agerite resin D	2.00	2.00
Santoflex 13	1.00	1.00
Miscellaneous extractables	3.00	–
Santocure	1.00	1.00
Sulfur	2.00	1.75
Reogen	–	1.00
Sunolite 240	–	3.00
Total	236.10	222.25

[a] After Coz and Baranwal [25]. Reproduced by permission of *Rubber World Magazine* (Lippincott).

Table 6.8. Comparison of reconstructed formulation with actual recipe for a radiator hose[a]

Ingredient	Reconstructed formulation (phr)	Actual recipe (phr)
Ethylene propylene rubber	100.00	100.00
N 550 carbon-black	203.00	200.00
Zinc oxide	5.90	5.00
Naphthenic oil	147.80	140.00
Stearic acid	1.00	1.00
Miscellaneous extractables	2.00	–
TMTD	3.00	1.50
TMTM	–	1.50
DPTT	1.00	1.00
ZDBC	1.00	1.50
Sulfur	0.50	0.50
Paraffin wax	–	5.00
Total	465.20	457.00

[a] After Coz and Baranwal [25]. Reproduced by permission of *Rubber World Magazine* (Lippincott).

in PE/Irganox B220, were attributed to differences in sample preparation (press, temperature, migration, measurement mode, *etc.*) or could have originated from instrumental drift. By careful standardisation of the XRF measurements in 1997 a variation coefficient of 2.1% could be achieved (*cfr.* Table 6.9). Nevertheless, the internally consistent results denote a systematic deviation for all double-sided XRF measurements (1997) with respect to earlier single-sided measurements (1993), despite the use of the original granulate in all cases. On the other hand, the new ICP/XRF relationship ($R^2 = 0.9955$) did not suggest the need for revision of the calibration curve (Table 6.10). However, Table 6.9 now shows large systematic (unexplained) discrepancies between the calibration standards (XPS, 1993) and ICP (1997; +18%) and HPLC (1997: +32%) values (same granulate).

Three major error sources may be indicated: the inaccuracy in the determination of the P content, the variability in the blend composition (at macro or micro level) and the stability of the standards. Irganox B220/B225 blend analysis based merely on P analysis (with XRF) translates a 5 ppm uncertainty

Table 6.9. Determination of Irganox B220 (mg/kg) in PE

Calibration standard (1993)	Measured values (1997)					
PW 1480[a]	PW 1480[b]	PW 1404[b]	PW 2400[b]	⟨XRF⟩ (1997) / XRF (1993)	ICP	HPLC[c]
493	502 ($n = 16$)	480	490	−1%	668	645
987	1020	965	995	+1%	1114	1330
1480	1552	1488	1528	+3%	1643	1925
1973	2112	2018	2078	+5%	2228	2645
2467	2639	2574	2629	+6%	2952	3190

[a] Single-sided XRF measurements (element P).
[b] Double-sided XRF measurements (element P).
[c] Total Irgafos is Irgafos 168 + Irgafos 168 phosphate; total Irganox 1010 is Irganox 1010 + degradation products + non-extractables.
After ref. [26].

Table 6.10. Determination of phosphorous (ppm) in various PE and PP/Irganox B220 grades

Grade	ICP (1994)	XRF (1994)	XRF (1997)
PE (1)	51	52	54
PE (2)	13	16	15
PE (3)	<5	<5	<5
PP (1)	42	42	43
PP (2)	34	36	37
PP (3)	15	18	19

ICP *vs.* XRF (1994): $R^2 = 0.9983$; (1997): $R^2 = 0.9955$.

Table 6.11. Interlaboratory comparison of the P content in various PE/Irganox B220 grades measured by means of XRF

Grade[a]		Spread[c]	
	ΔXRF[b]	XRF(1)	XRF(2)
PE (4)	−13.5	4.7	26.0
PE (5)	−3.0	3.6	15.8
PE (6)	−5.4	0.0	21.1
PE (7)	−11.6	3.0	29.8

[a] Different P concentrations in various PE/Irganox B220 grades.
[b] Deviation (%) of mean values of XRF(2) with respect to XRF(1).
[c] Spread (%) in individual measurement values. For XRF(1) three different plaques were measured one-sided, for XRF(2) two different plaques were measured double-sided.

in P contents into an uncertainty in total blend content of some 200 ppm, which is clearly quite unacceptable in production control. More accurate blend analysis would require a standard error in P analysis of less than 1 ppm, which was at the limit of the then state-of-the-art. Moreover, determination of the Irganox B220 contents based merely on the contents of only one of the blend components requires a constant and exact nominal blend ratio (Irganox 1010: Irgafos 168). There are strong (HPLC) indications that this might not be the case on the analytical micro-level. Therefore, the cost-effective XRF measurement for production control may be subject to large sampling errors. Finally, the results have indicated the need for close monitoring of various PE standards (containing P, Zn, Fe, Ti, Cl, Si, Mg) by XRF, NAA, ICP and HPLC.

As an alternative tool for production control, various spectroscopic methods for blend analysis have been evaluated. Determination of Irganox B220 in PE by means of IR lacks sensitivity and NIRS is rather inaccurate at low additive levels. As shown in Chp. 6.4.1, UV spectrophotometry is the most promising method as λ_{max} for Irganox 1010, Irgafos 168 and the main degradation products are rather similar.

6.2. QUANTITATIVE SOLVENT AND THERMAL EXTRACTION

Principles and Characteristics

As quantitative analysis is best carried out with a sampling technique that assures complete transfer of an analyte from the sample matrix to the analytical instrument direct in-polymer analysis is ideal (no sample preparation). Also analyses in the polymer melt (in-process analysis, *cfr.* Chp. 7) and the dissolution/precipitation technique (*cfr.* Chp. 3.7 of ref. [3a]) qualify in this respect. Solvent or thermal extraction leads to acceptable quantitation only in case of exhaustive extraction (100% extraction yield) or reproducible recovery. In theory, partitioning of the analyte between polymer and solvent prevents complete extraction. However, as the quantity of extracting solvent is much larger than that of the polymeric material, and the partitioning coefficients usually favour the solvent, in practice at equilibrium very low levels in the polymer will result. In **exhaustive extraction**, selectivity is sacrificed to obtain a quantitative transfer of target analytes into the extracting phase; often lower oligomers are dissolved as well. One advantage of this approach is that, in principle, it does not require calibration, since all the analytes of interest are transferred to the extracting phase. On the other hand, the equilibrium approach usually requires calibration for complex samples. It is not trivial to establish that total recovery of an analyte has occurred. The problem is even more complicated in cases where the analyte has undergone some form of interaction (chemical, physical) or degradation. This may mean that: *(i)* it is no longer in its original chemical form, or *(ii)* has undergone some form of immobilisation. In order to address the problem the analyst has various ways of approach. The first is that a particular sample be extracted using two different methods or techniques and the results compared. Participation in certification or testing schemes is another approach. In practice, the course of extraction is often followed spectrophotometrically or by means of XRF in order to determine the residual amount of analyte in the polymer.

To allow for non-quantitative extraction yield, an internal standard may be introduced. In order to achieve quantification, it is necessary to choose a standard with comparable extraction behaviour (time, matrix effects). Desrosiers [23] has been unique in describing methods for preparation of **standards for extraction**. In one approach known levels of the analytes (*e.g.* a primary antioxidant and a secondary phosphite) are mixed with a known weight of resin under rather severe conditions of temperature and time. The resulting compound will be quite homogeneous but is probably degraded and oxidised. In an improved procedure a mixing head is used that is continuously purged with dry, inert gas; the head temperature should be kept as low as possible. In a further improved method an appropriate weight of the component of interest in a (volatile) solvent was added to a weighed amount of resin in a vial. Most of the solvent was later removed in a gentle stream of dry, inert gas, and ultimately in a vacuum oven. Additional mixing can be achieved by shaking action. The powder is then cast into a thin film for extraction (followed by standard HT-GC, GC-MS or HPLC analysis) or direct analysis (*e.g.* IR). Such standards have had almost "no energy" input and will be extremely close to their calculated component concentrations.

In the overall analytical process the errors involved with possible incomplete extraction of the compounds from solid samples should be taken into account. Extraction recoveries vary from one element *cq.* compound to another and are also dependent on the way the species of interest are bound in the matrix. The assumption that organometal species are weakly bound to the solid is not always warranted. A thorough evaluation of extraction implies the availability of a "real life" reference sample. It is not surprising that many intercomparison exercises demonstrated that different extraction procedures lead to a wide range of different results. In the literature considerable attention has been paid to the reduction of the overall extraction time in order to emphasise the superiority of a specific solvent extraction technique. Much less attention has been devoted to a comparison of these methods for quantitative analysis [27], *cfr.* Table 6.12. Extraction efficiencies are frequently not explicitly indicated in research papers, with some good exceptions [28,29]. For example, the additives of PP/(AO-4K, Chemantox AO-49) were extracted in reflux in high yield (98%) with an RSD of 4% for the determination of the analytes [30]. Desrosiers [23] has critically compared the influence of sample shape (pellet, thin film or ground polymer). Both ground polymer and thin film are extracted well when the extraction process imparts an additional source of energy to the sample in addition to heat, *i.e.* microwave or ultrasonic digestion. It was observed that a 1 mm thin disc gives better extraction efficiencies than ground samples. This eliminates the cost of a grinder and cleaning of the grinder between specimens. Other factors affecting the efficiency of extraction of additives from polymers are the solubility of additives in the fluid and the rate of mass transfer of additives out of the polymer matrix. For example, in SFE low trap temperature, high extraction pressure and temperature, and low fluid flow-rate with moderate extraction time result in highest extraction

Table 6.12. Comparison of extraction techniques in relation to quantitation

Technique	Advantages	Disadvantages
Soxhlet	Largest weight of sample	Unfavourable extraction efficiencies Concentration step required for low level additives
Hot block	Total extraction[a]	Traumatic for polymer[b]
Microwave	Total extraction[a]	Traumatic for polymer[b]
US	Least traumatic	Low extraction efficiencies for some additives
SFE	Not particularly traumatic	Very low, unacceptable, extraction efficiencies for some components (matrix dependency)
PFE	Total extraction	Analyte stability problems (at high T)

[a] Likelihood of wax or oligomer discharge from the polymer matrix and of oxidation/degradation reactions.
[b] Complete dissolution of polymer.

efficiency [27]. For optimisation of SFE parameters both T_g and T_m of the polymer sample need to be considered. The ideal temperature is $T_g < T < T_m$.

The high reactivity and low stability of some additives and their low concentrations make **handling of extracts** an exacting job if quantitative information is required. For example, antioxidants can be labile compounds forming complex decomposition products. This considerably complicates interpretation of analytical data, and any loss of material is liable to be significant since the quantities of AOs present are initially so low. It is generally recommended that extracts be kept in smoked glassware and used for subsequent analysis without delay. If any storage of solutions is necessary, this should be done under nitrogen, in the dark and in a refrigerator [31]. Unwanted sample changes may occur during processing of AO extracts, including losses during concentration by evaporation. One should also be aware of complications, which arise when strongly adsorbing or chemisorbing fillers or pigments are present which can invalidate quantitative extraction procedures. Of course, in cases where the stabiliser molecule has been grafted to the polymer all of the extraction methods would be less than quantitative. Ashing does not yield quantitative results, quite often, because of the decomposition of fillers.

After extraction, a clean-up step may be necessary prior to quantification of the additives. If the extraction method was ultrasonics, for example, the resulting liquid is mixed with swelled pellets, strips or ground polymer. If the method was hot block extraction or microwave digestion, the polymer may absorb most of the analyte as it reformed. In most cases, the analyte is actually held quite loosely by the polymer. In general, sample liquid must be free of particles, should be as free as practical from waxes or oligomers and must have a matrix that is acceptable to the quantifying instrument of choice. Oligomers and other high-MW waxes that are dissolved in the extract may be precipitated, *e.g.* by means of acetonitrile, which makes the system more compatible with a gradient system if a liquid chromatography analysis is to be run.

Analyte recoveries in P&T experiments can vary widely due to matrix effects, purging efficiency, volatility, purge cell design, choice of adsorbent, isolation temperature, and many other factors. Quantification with the various **headspace techniques** always requires method development in terms of extraction time and temperature in order to avoid degradation. With dynamic headspace (DHS) nearly 100% recovery of volatiles is possible provided headspace temperature is appropriate to remove most of the analyte in a reasonable time. Kolb *et al.* [32] have outlined the prospects of quantitation by means of headspace techniques.

Also in **solid-phase microextraction** (SPME) analytes are typically not extracted quantitatively from the matrix. However, when partition equilibrium is reached, the extracted amount of an analyte is proportional to its initial concentration in the sample matrix phase. As indicated by Ai [33], application of SPME for quantitative analysis is feasible also when the partition equilibrium is not attained. Pawliszyn [34] has reviewed the quantitative aspects of SPME. Provided proper calibration strategies are followed, SPME can yield quantitative data and excellent precision, reproducibility and linearity (detection limits of 15 ng/L). In terms of precision, linearity and sensitivity SPME equals HS techniques.

Relative standard deviations of highly volatile components are 1–5%, for less volatile analytes 5–15% [35].

Whereas a great deal of attention has been paid to the extraction efficiency of various solvent extraction methods, much less effort has been devoted to quantitation of **thermal desorption techniques**. Thermal desorption analyses may be successfully calibrated using an external standard method. However, for additional confidence internal standard introduction is possible via a sample valve accessory.

Quantitation of P&T-TD, DTD-GC and DTD-GC-MS analysers was addressed [36]. Well-established standard methodologies, including standard addition, internal standard, surrogate standard, and stable isotope-labelled internal standard methods are used to quantify P&T-TD and DTD analyses [37]. The stabile isotope-labelled internal standard method is the most accurate and precise means available for quantifying P&T-TD and DTD analyses. This methodology requires a mass spectrometer as detector since the isotopically labelled standards may co-elute at the same GC retention times as the target compounds. The standard addition method is particularly useful for off-flavour, contamination and packaging migration investigations.

In general terms, absolute quantification by means of TD-GC-MS or thermal volatilisation techniques is a doubtful exercise because total desorption of the analyte(s) at a given temperature is not assured, internal standards are difficult to use and mass spectrometry is not exactly well known for its quantitative excellence. No reports are available on the use of direct TD-CIS-GC-MS for quantification purposes. To that end the use of FID detection has considerable advantages (*cfr.* ref. [38]). The PTV-CT-GC-FID method is useful mainly for qualitative analysis and rapid screening. For quantitative analysis the reproducibility (now within about 20%) needs to be improved. Scrivens *et al.* [38] have proposed the use of a TD-GC-MS/FID set-up in which the concentrator gas flow is split to an FID detector for real-time monitoring of the evolution of volatiles and the measurement of the total amount of sample trapped into a sorbent trap kept at room temperature for further GC-MS processing. Quantitation is facilitated by the ability to inject a standard into the wide bore trap during volatile collection. In on-line TD-GC-FTIR-FID quantitative analysis can be made on the basis of extinction coefficients measured on standards [39].

Applications

In a fairly typical case, Zhou *et al.* [27] have compared on-line SFE-SFC, off-line SFE-HPLC and off-line ESE®-HPLC in the quantitative extraction of LDPE/(BHT, BHEB, Isonox 129, Irganox 1010/1076).

As shown in Table 6.13, the measured amount of BHT was exceedingly low via all three analyses. The difficulty in the reported BHT analyses may have various origins: evaporation of BHT during the grinding process, decomposition or dimerisation (leading to a non-extractable product). Relatively low recovery (84–87%) of BHT, due to volatility, was also observed by others [8]. The methods of Table 6.13 were certainly not optimised for BHT, which can be determined to within a few ppm [5,40]. Butylhydroxyethyl benzene (BHEB),

Table 6.13. Concentration (ppm) of the additives in LDPE with one standard deviation

Additive	Manufacturer's data	On-line SFE-SFC[a]	Off-line SFE-HPLC[b]	Off-line ESE®-HPLC[b]
BHT	875	n.d.	67 ± 1	73 ± 4
BHEB	975	900 ± 160	1020 ± 80	1010 ± 100
Isonox 129	975	780 ± 160	650 ± 7	660 ± 13
Irganox 1076	1000	830 ± 150	490 ± 10	500 ± 15
Irganox 1010	975	900 ± 110	880 ± 40	910 ± 150

[a] Sample size: 2.5 mg.

[b] Sample size: 500 mg.

n.d., not detected.

After Zhou *et al.* [27]. Reproduced from *Journal of Chromatography* **A858**, L.Y. Zhou *et al.*, 209–218. Copyright (1999), with permission from Elsevier.

Isonox 129, Irganox 1010/1076 were also quantified. Recoveries exceeding 80% for each of these additives were achieved using SFE-SFC. The concentrations of BHEB, Isonox 129 and Irganox 1010 were comparable using all three methods and consistent with the manufacturer's data. Irganox 1076 determined via on-line SFE-SFC also matched the manufacturer's data, quite at variance to both methods employing HPLC. For Isonox 129, Irganox 1010/1076 the extraction profiles were typical of analytes that are both solubility and diffusion limited. The precision of the optimised on-line SFE-SFC extraction was quite low due to the use of a very small sample size (2 mg) to avoid clogging, and possibly the heterogeneous distribution of additives in the polymer product. Lower recoveries were obtained with *off-line* SFE-HPLC because of precipitation of the co-extracted oligomer. Higher recoveries with the *on-line* SFE-SFC method were also ascribed to fewer sample handling steps. The precision of off-line SEC-HPLC was much better compared to that from on-line SFE-SFC, which undoubtedly stands in connection to the sample size difference (500 mg in SFE-HPLC). Due to clogging, with ESE® much clean-up of the system was required after extraction, as opposed to SFE. Dibutylphthalate, as well as other relatively volatile plasticisers, can be completely lost if no care is taken. The guidelines in ASTM D 3421 should be observed (extraction and analysis of plasticisers in PVC plastics) [41].

Roberson *et al.* [42] have reported an extraction recovery rate of 90–99% for one-step extraction of PP samples. Desrosiers [23] described extraction and quantification of in-polymer additives from polyolefinic materials. The problems encountered in the extraction step are release of low-MW waxes or oligomers, and formation of hydrolysis, oxidation and degradation products. Undesirable chemical reactions may also take place during the extraction process.

Quantitation limits (defined as the monomer concentration necessary to produce a peak at least three times the baseline noise or 3% of full scale) for residual monomers (such as vinylchloride, butadiene, acrylonitrile, styrene and 2-ethylhexylacrylate) as low as 0.05 ppm have been reported for solution headspace, as compared to 1 ppm for direct solution injection GC [43]. Quantification of volatiles by solid headspace sampling can be challenging. Solid headspace provides about 10-fold more sensitivity than solution headspace.

6.2.1. Extraction and Quantification of Polyolefin Additives

Case Study
Nielson [28] has compared microwave (MAE) and ultrasonic (US) extraction of HDPE/(BHT, Irganox 1010, Irganox 1076) pellets using cyclohexane/isopropyl alcohol (IPA) and methylene chloride/IPA mixtures in various ratios. Results are collected in Table 6.14.

The data reported correspond to an RSD of 1.8% for the three additives over six separate extractions or to an RSD of 1.9% for three separate extractions. The same author [28] has also compared US and MAE for several PP resins that contained a variety of additive packages. The PP formulations consisted of antistats, pigments, fillers, slip agents, and antioxidants/UV degradants. Table 6.15 shows the MAE recoveries for a selection of additives in four resins.

The results of a reproducibility study are reported in Table 6.16.

Table 6.17 compares ultrasonic extractions for 30 and 60 min. It follows that while Irgafos 168 is extracted in only 30 min, Irganox 1010 benefits from an additional 30 min to reach a 98+ % recovery.

Results were also reported for LDPE/erucamide and LDPE/(BHT, BHEB, Isonox 129, Irganox 1010/1076) [28]. Similar detailed quantitative reports are rare. The results give a good indication of the expectations of careful quantitative extraction.

Table 6.14. Microwave and ultrasonic extraction of HDPE

Additive	Microwave		Ultrasonics	
	Concentration (ppm) Cyclohexane/IPA (1:1)	Concentration (ppm) $MeCl_2$/IPA (98:2)	Concentration (ppm) Cyclohexane/IPA (1:1) (60 min)	Concentration (ppm) $MeCl_2$/cyclohexane (3:1) (30 min)
BHT	451 (90%)	455 (91%)	454 (91%)	449 (90%)
Irganox 1010	454 (91%)	459 (92%)	457 (91%)	458 (92%)
Irganox 1076	480 (96%)	474 (95%)	475 (95%)	481 (96%)

Table 6.15. Microwave extraction of PP resins with methylene chloride/isopropanol (98:2)

Resin	Additives	Amount recovered (ppm)	Amount present (ppm) ± 35 ppm	% Recovery
A	Irganox 3114	565	600	94
	Cyasorb UV531	442	500	88
B	Irgafos 168	521	500	100+
	Irganox 1010	986	1000	99
C	Ultranox 626	709	800	89
	Irganox 3114	635	800	79
D	BHT	513	500	100+
	Irganox 1010	931	1000	93

Table 6.16. Reproducibility of extraction from PP/(Irgafos 168, Irganox 1010)

Extraction	Concentration Irgafos 168 (ppm)	Concentration Irganox 1010 (ppm)
1	521	986
2	485	1034
3	509	981
4	476	1019
5	489	978
6	496	970
	(RSD = 3.3%)	(RSD = 2.6%)

Table 6.17. Ultrasonic extractions in $MeCl_2$/cyclohexane (3:1)

Additive	Concentration (ppm) 30 min	Concentration (ppm) 60 min
Irgafos 168-1	522/500	492/500
Irgafos 168-2	494/500	516/500
Irgafos 168-3	513/500	503/500
Irganox 1010-1	929/1000	982/1000
Irganox 1010-2	882/1000	998/1000
Irganox 1010-3	921/1000	980/1000

6.2.2. Supercritical Fluid Extraction

Case Studies

Salafranca *et al.* [44] have used full-factorial design for the optimisation of the extraction of virgin and recycled LDPE/(Irganox 1076, Chimassorb 81) film (200 μm) and HDPE/(Irganox 1076, Irgafos 168) pellets. A two-level design would require $2^{11} = 2048$ experiments for the extraction step and $2^7 = 128$ experiments for the collection section. In order to keep experimentation more practical the extraction step was optimised for only 4 variables (extraction temperature, system pressure, dynamic time and percentage modifier), keeping 7 parameters fixed (supercritical fluid, sample weight, extraction cell volume, sample introduction mode, supercritical fluid flow, static time and matrix modifier). Similarly, for the collection section 3 parameters were varied (restrictor, trap and elution temperatures), whereas 4 parameters were kept fixed (collecting solvent, solvent volume and flow, and number of washing steps). Consequently, optimisation of the extraction step required $2^4 = 16$ experiments by variation of pressure (200–450 atm), temperature (50–100°C), time (5–20 min) and matrix modifier concentration (0–10%). The collection step was optimised in $2^3 = 8$ steps, varying restrictor temperature (25–75°C), trap temperature (−30–0°C) and elution temperature (10–50°C). Experimental results are summarised in Table 6.18.

It was observed that the values for virgin LDPE polymers are quantitative in the optimised SFE conditions, as opposed to those for recycled polymers. This was attributed to the higher crystallinity of the recycled LDPE and to sample thickness effects. By reducing the sample thickness to 470 μm and increasing the dynamic extraction time to 30 min ("SFE drastic conditions") quantitative results were obtained. Optimum conditions for LDPE/(Irganox 1076, Chimassorb 81) were given as pressure, 450 atm; dynamic extraction time, 15 min; modifier, 10% (methanol); and extraction temperature, 75°C.

In the optimised SFE conditions of LDPE none of the additives present in HDPE (Irganox 1076, Irgafos 168) could even be detected. At this point, it may be considered that HDPE shows a more dense molecular structure than LDPE. To improve the extraction efficiency, the thickness of the HDPE sample was then reduced to 380 ± 60 μm and the dynamic extraction time was increased to 2 h. Despite

Table 6.18. Additive concentrations in polymers found by the SFE procedure (optimised and drastic conditions) and reference values obtained by total dissolution/reprecipitation

Sample	Chimassorb 81 (μg·g^{-1})			Irganox 1076 (μg·g^{-1})		
	SFE-optimised	SFE-drastic	Total dissolution[a]	SFE-optimised	SFE-drastic	Total dissolution[a]
LDPE virgin	2723 ± 148	2728 ± 148	2815 ± 122	452 ± 41	854 ± 79	876 ± 53
LDPE recycled	1023 ± 87	2204 ± 178	2245 ± 275	108 ± 15	710 ± 103	725 ± 59
EVA virgin	2460 ± 131	2460 ± 150	2540 ± 125	130 ± 11	425 ± 37	431 ± 35
EVA recycled	1377 ± 117	2015 ± 202	2031 ± 217	58 ± 7	368 ± 53	365 ± 32
Sample	Irgafos 168 (μg·g^{-1})			Irganox 1076 (μg·g^{-1})		
	SFE-optimised	SFE-drastic	Total dissolution[a]	SFE-optimised	SFE-drastic	Total dissolution[a]
HDPE pellets	n.d.	28 ± 2	471 ± 23	n.d.	59 ± 6	237 ± 11
HDPE pressed	21 ± 3	91 ± 5	473 ± 25	37 ± 3	204 ± 16	242 ± 14

[a] Toluene/methanol; Irganox 1010 as internal standard.

After Salafranca *et al.* [44]. From J. Salafranca *et al.*, *Journal of High Resolution Chromatography* **22**, 553–558 (1999). © Wiley-VCH, 1999. Reproduced by permission of Wiley-VCH.

these aggressive experimental conditions, quantitative extraction is still not possible. Besides, Irgafos 168 decomposed to the corresponding phosphate derivative (HPLC evidence). The results are to be considered as quite discouraging after so much experimental effort in comparison to the simple dissolution/precipitation procedure, taken as a reference. *Cfr.* also Chp. 3.4.2.7 of ref. [3a].

It is also to be noticed that extraction is never complete in finite time; in order to obtain the total extractable amount of a compound extrapolation procedures can be used. Clifford [45] has described an extrapolation procedure based on the initial extraction of an amount m_1 in the period of $t = 0$–t_1, and two subsequent extractions, in equal periods of time terminating at t_2 and t_3, of amounts m_2 and m_3. Algebraic manipulation leads to

$$m_0 = m_1 + m_2^2/(m_2 - m_3) \qquad (6.1)$$

where m_0 is the extrapolated value. Clifford [45] has reported the case of PP/BHT with extraction of only 57% of the additive in 8 hours and an estimate of the final amount according to eq. (6.1), which is 5.2% below the given value.

Table 6.19 shows the results of a comparison between SFE and Soxhlet extraction of vulcanised rubbers with different polarities. Analysis was carried out off-line by means of HPLC-UV. The results are comparable.

6.2.3. Quantification of Antioxidants in Polyolefins

Case Study

Recently, the Swiss Federal Laboratories for Materials Testing and Research (EMPA, St. Gallen) have promoted a series of laboratory performance studies on polymeric materials, examining the glass-transition-point by DSC (amorphous thermoplastics), antioxidant content in polyolefins, halogen concentration in plastics and rubbers, chemical resistance of elastomers (according to ISO 1817), global migration in food packaging, plasticiser content (comparative examination: TGA and extraction) and the oxidation-induction time and temperature (OIT/OIT*) of polyolefins [47]. Data evaluation was carried out according to ISO 5725 [48]. Object of one of the round-robins was the determination of the antioxidant levels in various PP/(Irganox 1010, Irgafos 168, Ca-stearate) materials; nominal composition: Table 6.22 [49]. At variance to their industrial importance only few international norms describe the determination of antioxidants, namely:

- ISO 11089 (1997): Determination of antidegradants in synthetic elastomers with HPLC.
- BS 2782 (1975): Methods of testing plastics; chemical properties; determination of antioxidants in polyolefin compounds by a spectrophotometric method.

Table 6.19. Quantitative extraction of elastomers

Matrix	Antioxidant(s)	Total extractables (%)		Antioxidant concentration (%)	
		SFE[a]	Soxhlet[b,c]	SFE[a]	Soxhlet[b,c]
EPDM	Vulkanox BKF	14.4	14.9	0.04	0.03
NR-SBR	Vulkanox BKF	4.1	4.2	0.43	0.44
EPDM	Irganox 1010/1076	6.0	6.1	<0.002, 0.02	<0.002, 0.01
NBR-SBR	Vulkanox 4020	19.2	20.2	0.84	0.54
NR	Vulkanox 4010 NA	1.9	2.3	0.31	0.29
NBR	Vulkanox HS	3.8	4.1	0.80	0.89
CR	Vulkanox 4010 NA	16.3	15.8	0.26	0.29

[a] Mean of 5–10 measurements.
[b] According to DIN ISO 1407.
[c] Mean of 3 measurements.
After Werthmann *et al.* [46]. Reproduced by permission of Hüthig GmbH.

Table 6.20. Round-robin test results for Irganox 1010 in AO-1[a]

Reference laboratory	$\langle x_i \rangle$ (ppm)	s_i (ppm)	Extraction solvent	Extraction procedure	Analytical method	Observations
8[b]	6	2.1	Acetone	Soxhlet 6 h	HPLC	
15	257	8.2	Dichloromethane	Soxhlet 8 h	HPLC	Corr. for extraction yield
26	310	26.5	Chloroform	Soxhlet 18 h	HPLC	Degrad. products ~90 ppm
41	330	30.6	Chloroform	Soxhlet 2 h	HPLC	
50	346	1.0	Xylene/acetonitrile	Diss./prec.	HPLC	Internal standard
38	425	88.1	Chloroform	Soxhlet 48 h	HPLC	
24[c]	602	16.0	Dichloromethane	Soxhlet 8 h	HPLC	Internal standard
44[b]	970	68	–	–	PyGC-MS	Non-optimised
Mean[d]	357	Table 6.23				

[a] Three independent measurements per sample.
[b] Excluded from data evaluation.
[c] $Z > 2$, *i.e.* unreliable result.
[d] Dosage 500 ppm.
After Bart *et al.* [49]. Reproduced by permission of the Japan Society for Analytical Chemistry.

- ISO 4645 (1984): Procedure for the determination of antidegradants in rubbers and rubbery products by means of TLC.

Various methods were employed by the participating laboratories: HPLC (7×), PyGC-MS (1×), for the determination of Irganox 1010; HPLC (6×), PyGC-MS, photometry, GC-MS and XRF (each 1×) for the determination of Irgafos 168. It is noteworthy that only one direct method of determination has been used, namely PyGC-MS (but no ToF-SIMS, MALDI or DIP-MS). Also only one indirect method was employed throughout (HPLC), but no HTGC, TLC, nor any spectrophotometric method apart from XRF. Also the choice in extraction procedures was limited (Soxhlet or dissolution/precipitation). Tables 6.20 and 6.21 show the results for the determination of Irganox 1010 in the two test samples (AO-1 and AO-2), available as powders (particle size < 1 mm).

For these test samples 2 h of Soxhlet extraction in $HCCl_3$ were apparently sufficient. Only one participating laboratory mentioned explicitly extraction yield correction. Dissolution/precipitation (xylene/acetonitrile) yielded acceptable results; the use of an internal standard in HPLC analysis was considered advantageous. The observed concentrations for Irganox 1010 and Irgafos 168 in the test samples are given in Table 6.22. The wide spread is disenchanting.

Table 6.21. Round-robin test results for Irganox 1010 in AO-2

Reference laboratory	$\langle x_i \rangle$ (ppm)	s_i (ppm)	Extraction solvent	Extraction procedure	Analytical method	Observations
8[b]	42	6	Acetone	Soxhlet 6 h	HPLC	
15	1019	18	Dichloromethane	Soxhlet 8 h	HPLC	Extraction yield correction
26	1180	32	Chloroform	Soxhlet 18 h	HPLC	Degrad. products ~230 ppm
50	1252	13	Xylene/acetonitrile	Diss./prec.	HPLC	Internal standard
41	1290	31	Chloroform	Soxhlet 2 h	HPLC	
38	1325	75	Chloroform	Soxhlet 48 h	HPLC	
24[c]	1615	21	Dichloromethane	Soxhlet 8 h	HPLC	Internal standard
44[b]	2770	235	–	–	PyGC-MS	Non-optimised
Mean[d]	1253	Table 6.23				

[a] Three independent measurements per sample.
[b] Excluded from data evaluation.
[c] $Z > 2$, *i.e.* unreliable result.
[d] Dosage 1500 ppm.
After Affolter *et al.* [47]. Reproduced by permission of M. Schmid, EMPA, St. Gallen.

Table 6.22. Nominal values and observed antioxidant ranges (ppm) for round-robin test samples[a]

Sample	Irganox 1010		Irgafos 168	
AO-1	500	6–970	2000	568–1826
AO-2	1500	42–2770	500	299–593

[a] After Bart *et al.* [49]. Reproduced by permission of the Japan Society for Analytical Chemistry.

Table 6.23. Statistical data of round-robin test results of Irganox 1010 in AO-1 and AO-2

Statistical value	Unity	AO-1	AO-2
Average y	ppm	357	1253
s_r	ppm	20	30
s_r, relative	%	5.5	2.4
s_R	ppm	95	155
s_R, relative	%	26.6	12.3
r	ppm	55	83
r, relative	%	15.3	6.6
R	ppm	266	433
R, relative	%	74.5	34.5
No. laboratories		6	6

Data evaluation according to ref. [48].
After Affolter *et al.* [47]. Reproduced by permission of M. Schmid, EMPA, St. Gallen.

The repeatability standard deviation s_r for the samples investigated was 5.5% (AO-1) and 2.4% (AO-2) (for statistical nomenclature, *cfr.* Table 6.24). However, the relative deviations for interlaboratory reproducibility s_R are considerable, *i.e.* 26.6% (AO-1) and 12.3% (AO-2). These results (Table 6.23) show that the determination of antioxidants in polyolefins is not a trivial matter. For cases where the interlaboratory precision is much larger than the intralaboratory precision there is obviously lack of robustness of the analytical methods used.

Surprisingly, most participants reported considerably less than the dosed Irganox 1010 content. This was generally attributed to decomposition and/or losses of Irganox 1010 during sample preparation (in atmospheric conditions as opposed to N_2 coverage in a plant) and/or analysis. Irganox 1010 is known to be stable under reflux conditions for 3 h [50]. On the other hand, even during very short microwave-assisted extractions at 140°C some degradation of Irganox 1010 has been observed [50]. However, at 125°C Marcato *et al.* [51] reported negligible degradation. Only one participating laboratory had explicitly taken into account the decomposition products of this stabiliser; overall analysis then closely conforms to dosage.

Table 6.24. Definitions of terms used in interlaboratory tests

Repeatability conditions: conditions where independent test results are obtained with the same method on identical test items in the same laboratory by the same operator using the same equipment within short intervals of time.

Reproducibility conditions: conditions where test results are obtained with the same method on identical test items in different laboratories with different operators using different equipment.

$\langle x_{\text{med},i} \rangle$	Mean value of a series of experiments in laboratory i
y	General mean value (mean value of all the test results obtained by all the laboratories at a particular level of the experiment; equivalent to y in ISO 5725 (1998))
s_i	Intralaboratory standard deviation for measurement series; measures intralaboratory spread
s_L	Interlaboratory standard deviation measuring spread of mean laboratory values
s_r	Repeatability standard deviation (standard deviation of test results obtained under repeatability conditions)
s_R	Reproducibility standard deviation (standard deviation of test results under reproducibility conditions); $s_R = \sqrt{(s_r^2 + s_L^2)}$
r	Repeatability limit $r = 2.8 s_r$ (intralaboratory 95% confidence level)
R	Reproducibility limit $R = 2.8 s_R$ (interlaboratory 95% confidence level)
Z-score	s_R-normalised deviation of a laboratory mean value from the total mean value of all laboratories participating in a round-robin

Table 6.25. Round-robin test results for Irgafos 168 in AO-1[a]

Reference laboratory	$\langle x_i \rangle$ (ppm)	s_i[e] (ppm)	Extraction solvent	Extraction procedure	Analytical method	Observations
32[c]	568	31	Schöniger digestion	–	Photometry	Phosphate determination
20[c]	835	14	Dichloromethane	Soxhlet 8 h	GC-MS	
44[b]	1380	160	–	–	PyGC-MS	Non-optimised
24	1397	106	Dichloromethane	Soxhlet 16 h	HPLC	Internal standard
15	1427	47	Dichloromethane	Soxhlet 8 h	HPLC	
38	1484	67	Chloroform	Soxhlet 48 h	HPLC	
26	1550	20	Chloroform	Soxhlet 18 h	HPLC	Irgafos 168 phosphate ~250 ppm
50	1558	27	Xylene/acetonitrile	Diss./prec.	HPLC	Internal standard
41	1573	84	Chloroform	Soxhlet 2 h	HPLC	
69	1826	33	–	–	XRF	Pressed sample
Mean[d]	1482	34[f]				

[a] Three independent measurements per sample.
[b] Excluded from data evaluation.
[c] $Z > 2$, *i.e.* unreliable result.
[d] Dosage 2000 ppm.
[e] Estimated standard deviation.
[f] Repeatability (intralaboratory) standard deviation (s_r).
After Bart *et al.* [49]. Reproduced by permission of the Japan Society for Analytical Chemistry.

Results obtained for Irgafos 168 were very similar to those of Irganox 1010 (Tables 6.25 and 6.26). Also Irgafos 168 was generally detected far below dosed quantities. Again only one laboratory had explicitly taken into account the Irgafos 168 phosphate concentration. The repeatability standard deviation s_r amounted to 2.3% (AO-1) and 4.1% (AO-2), with a poor relative interlaboratory reproducibility s_R (12.1% for AO-1 and 27.1% for AO-2), *i.e.*, far beyond acceptable values for quality assurance or accreditation. In this case Soxhlet extraction in $HCCl_3$ for 2 h was sufficient; dissolu-

Table 6.26. Round-robin test results for Irgafos 168 in AO-2[a]

Reference laboratory	$\langle x_i \rangle$ (ppm)	$s_i{}^c$ (ppm)	Extraction solvent	Extraction procedure	Analytical method	Observations
15	299	3.8	Dichloromethane	Soxhlet 8 h	HPLC	
24	331	22	Dichloromethane	Soxhlet 16 h	HPLC	Internal standard
38	356	104	Chloroform	Soxhlet 48 h	HPLC	
26	357	51	Chloroform	Soxhlet 18 h	HPLC	Irgafos 168 phosphate ∼280 ppm
41	370	10	Chloroform	Soxhlet 2 h	HPLC	
32	396	13	Schöniger digestion	–	Photometry	Phosphate determination
50	453	11	Xylene/acetonitrile	Diss./prec.	HPLC	Internal standard
20	574	75	Dichloromethane	Soxhlet 8 h	GC-MS	
69	593	3.6	–	–	XRF	Pressed sample
Mean[b]	401	16[d]				

[a] Three independent measurements per sample.
[b] Dosage 500 ppm.
[c] Estimated standard deviation.
[d] Repeatability (intralaboratory) standard deviation (s_r).
After Affolter *et al.* [47]. Reproduced by permission of M. Schmid, EMPA, St. Gallen.

tion/precipitation (xylene/acetonitrile) yielded again good results; the internal HPLC standard was advantageous; PyGC-MS was satisfactory, as opposed to photometry of phosphate according to Schöniger or GC-MS. Quantitative XRF stands out as a very rational and quick method without the need for sample preparation, but presumes absence of other phosphorous-containing analytes. From the fact that XRF detects 1826(33) ppm (AO-1) and 593(4) ppm (AO-2) of P-containing components it appears that essentially no additive is lost; at most, either polymer processing and/or the analytical procedures have degraded part of the original additive to an equal or different extent, respectively. Heterogeneities are sometimes considered to be the biggest source of error for the polymeric materials prepared on a laboratory scale (film extrusion followed by cryogenic homogenisation).

This round-robin [47] shows that still extensive use is being made of Soxhlet extraction despite allegations of much progress in modern sample preparation techniques. Dissolution/precipitation stands out invariably by very reliable results. However, interlaboratory reproducibility is quite unsatisfactory and most laboratories fail to meet the mass balance. Further interlaboratory actions are wanted. *Cfr.* also ref. [51a].

6.2.4. Determination of Plasticisers by Solvent and Thermal Extraction

Case Study

In the aforementioned round-robin [47] also solvent- and heat-extraction methods for plasticiser determinations were compared. Materials were chosen which do not exhibit polymer degradation in the range of plasticiser loss at 300°C (Table 6.27).

Table 6.28 shows the solvent-extraction norms for plasticiser content. However, the participating laboratories were actually left free in the choice of their preferred method of analysis.

As may be seen from Table 6.29, still wide use is being made of conventional exhaustive (*i.e.* non-selective) extractions, such as Soxhlet and reflux extractions. Only four laboratories exercised with the use of ASE®, none reported SFE or MAE. The table shows the need for optimisation of ASE® conditions.

Table 6.30 summarises reported weight losses in PA12-P (WM-1) by means of thermogravimetry, mainly according to ISO 9924-1 [52], corresponding to n-butylbenzenesulfonamide and (eventually) other residuals.

Table 6.31 indicates quite comparable results for solvent- and heat-extraction despite the inherent differences between these two methods. As shown in Table 6.32, in TGA the (wide) 95% confidence levels

Table 6.27. Materials for solvent- and heat-extractive determinations of plasticisers

Sample	Material	Visual appearance	Plasticiser[a]	Concentration
WM-1	PA12-P	Transparent granulate	*n*-Butylbenzenesulfonamide	*ca.* 13.5%
WM-2	NBR[b]	Black film	Di-2-ethylhexylphthalate	*ca.* 21.5%
WM-3	EPDM[b]	Green film	Mineral oil	*ca.* 11.5%
WM-4	SBR[b]	Black film	Mineral oil	*ca.* 13%

[a] Technical products.
[b] Vulcanised.

Table 6.28. Normalised methods for the determination of plasticisers from polymeric materials by means of solvent extraction

Sample	Norm	Solvent	Extraction time	Drying
WM-1	ISO 599 DIN 53738 ISO 6427	Methanol	3 h	40°C, 25 mbar
WM-2	ISO 1407	Propanol-2	16 h	2 h, 100°C, 1000 mbar
WM-3	ISO 1407	Acetone	16 h	2 h, 100°C, 1000 mbar
WM-4	ISO 1407	Ethanol–toluene azeotrope (acetone)	16 h	2 h, 100°C, 1000 mbar

are somewhat better despite the much lower sample mass probed (10 mg for TGA *vs. ca.* 3 g for solvent extraction).

The selective determination of *n*-butylbenzenesulfonamide in PA12 by means of dissolution (HFIP)/precipitation (MTBE), followed by GC analysis of the main component of this technical product, yielded 11.87 wt.% ($\sigma = 0.15$ wt.%). There is a good correlation between PyGC-MS results (13.15 wt.%, $\sigma = 0.13$ wt.%) and the mean TGA results ($\Delta w = 12.86$ wt.%), which reflect exhaustive thermal extraction in the given conditions.

In case of NBR/DEHP, EPDM/mineral oil and SBR/mineral oils, the relative intralaboratory confidence levels r were about 5% for both solvent- and heat extraction (not shown). In all cases a very large spread was noticed in the mean values $\langle x_i \rangle$ of individual laboratories, *cfr*. Table 6.32 and Fig. 6.2 for solvent extraction of NBR/DEHP and Fig. 6.3 for heat extraction of SBR/mineral oil by means of TGA.

In NBR/DEHP (WM-2) exhaustive extraction sets an upper limit for dialkylphthalates (*ca.* 22 wt.%). The two selective procedures, PyGC-MS (16.60 wt.%, $\sigma = 0.53$ wt.%) and THF extraction followed by SEC separation with RI and UV detection (17.0 and 17.25 wt.%, respectively; $\sigma = 0.22$ wt.%), are in excellent agreement. The advantage of SEC is that dialkylphthalates are determined as a group. It is quite apparent that propanol-2 extraction of NBR/DEHP according to ISO 1407 extracts other components as well.

In all four cases relative interlaboratory levels of 10–40% were observed, which were slightly better for TGA than for solvent extraction. These observations raise considerable concern. The widely scattering results are partly on account of the fact that a few selective analytical methods (PyGC-MS and GC for WM-1 and PyGC-MS and SEC for WM-2) were included, which score below average of exhaustive extractions (as expected). Nevertheless, the 95% confidence intervals stay broad. It is reassuring to notice that considerably better results are obtained if we consider only solvent extractions according to the ISO 599, DIN 53738 and ISO 6427 norms (*cfr.* Table 6.32). As shown in Table 6.33, normalised methods lead systematically to somewhat higher averages of laboratory mean values (y), but with a much reduced spread. It is to be noticed that the conditions used (mass, extraction time, drying time, *etc.*) differed even within the application of the same normalised procedure. It would appear that many in-house extraction methods qualify for incomplete extraction. It is also noticed that PA12

Table 6.29. Determination of *n*-butylbenzenesulfonamide in PA12-P (WM-1) by means of solvent extraction

Reference laboratory	n[a]	$\langle x_i \rangle$ (ppm)	s_i (ppm)	Norm	Sample preparation	Extraction method[c]	Solvent	Mass (g)
35[b]	4	9.18	0.24	–	No	ASE®	Methanol	2
57[b]	4	9.81	0.20	–	No	ASE®	Methanol	3–4
40[b]	4	10.31	0.62	–	Milling	H	Methanol	2
10	4	11.85	0.33	ISO 6427	Chopped	H	Methanol	1
26	3	11.87	0.15	–	Dissolution (HFIP)/Prec. (MTBE); GC	–	–	n.d.
50	4	11.93	0.62	–	No	H	Methanol	4–5
12	4	12.13	0.10	ISO 599	Chopped	H	Methanol	1
63	4	12.20	0.41	–	No	R	Methanol	2
53	4	13.10	0.28	–	Chopped	ASE®	Methanol	1
56	4	13.15	0.13	–	PyGC-MS	–	–	10^{-4}
24	4	13.24	0.95	ISO 6427	Milling/N_2	S	Methanol	3–5
7	4	13.48	0.63	DIN 53738	Milling/N_2	S	Methanol	3–4
25	4	13.60	0.27	ISO 599	Chopped	S	Methanol	3–4
41	4	13.60	0.21	ISO 6427	Milling	S	Methanol	2
66	3	13.71	0.52	–	No	ASE®	Methanol	1–2
39	4	13.80	0.08	ISO 6427	Chopped	S	Methanol	3–4
46	4	13.80	0.41	–	Milling	H	Methanol	3–5
20	4	13.85	0.13	ISO 6427	Chopped	S	Methanol	5
1	4	14.00	0.08	ISO 6427	No	S	Methanol	2
31	4	14.00	0.08	DIN 53738	No	S	Methanol	4
34	4	14.18	0.05	–	No	R	Methanol	2
8	4	14.33	0.05	–	Milling/N_2	S	Methanol	2–3
32	4	14.35	0.06	–	Chopped	S	Methanol	2
38	4	14.62	0.48	–	Chopped	S	Methanol	5–6
Mean		13.19	Table 6.31					

[a] Number of independent measurements per sample.
[b] $Z > 2$; *i.e.* unreliable result.
[c] H, hot extraction; R, reflux extraction; S, Soxhlet extraction.
After Bart *et al.* [49]. Reproduced by permission of the Japan Society for Analytical Chemistry.

Table 6.30. Determination of *n*-butylbenzenesulfonamide in PA 12-P (WM-1) by means of thermogravimetry

Reference laboratory	$\langle x_i \rangle$ (ppm)	s_i (ppm)	Mass (mg)	Sample preparation
27	11.48	0.41	*ca.* 10	Chopped
7	11.85	0.70	*ca.* 10	Milled
10	12.13	0.59	*ca.* 10	Chopped
24	12.50	0.29	*ca.* 10	Milled/N_2
32	12.80	0.16	*ca.* 10	Chopped
38	12.93	0.22	*ca.* 10	No
50	12.98	0.65	10–13	Chopped
26	13.09	0.14	*ca.* 10	Chopped
20	13.15	0.10	10–17	Chopped
46	13.35	0.10	7–20	Milled
41	13.47	0.06	*ca.* 10	Milled
40[a]	15.40	0.45	*ca.* 10	Chopped

[a] $Z > 2$; *i.e.* unreliable result.
After Bart *et al.* [49]. Reproduced by permission of the Japan Society for Analytical Chemistry.

Table 6.31. Comparison between the determination of *n*-butylbenzenesulfonamide in PA12-P (WM-1) by solvent and heat extraction

Statistical parameter	Extraction[a]		Extraction[b]		TGA[b]	
	Abs.	Rel. (%)	Abs.	Rel. (%)	Abs.	Rel. (%)
Average y	13.86	–	13.19	–	12.86	–
s_r	0.14	1.0	0.25	1.9	0.30	2.3
s_R	0.66	4.8	1.19	9.0	0.80	6.2
r	0.38	2.8	0.70	5.3	0.84	6.4
R	1.84	13.4	3.33	25.2	2.24	17.4
No. participants	13		24		12	

[a] Ref. [53].
[b] Ref. [47].

Table 6.32. Range of observed mean values ($\langle x_i \rangle$) (%)

Sample	Solvent extraction[a]		Heat extraction (TGA)
WM-1	9.18–14.62	11.85–14.00	11.48–15.40
WM-2	15.18–26.83	19.88–26.83	18.55–24.20
WM-3	7.46–14.85	11.33–14.28	8.83–13.68
WM-4	7.90–14.90	12.63–14.90	6.68–15.73

[a] Left: all methods (norms and in-house); right: results according to norms of Table 6.28.
After Bart *et al.* [49]. Reproduced by permission of the Japan Society for Analytical Chemistry.

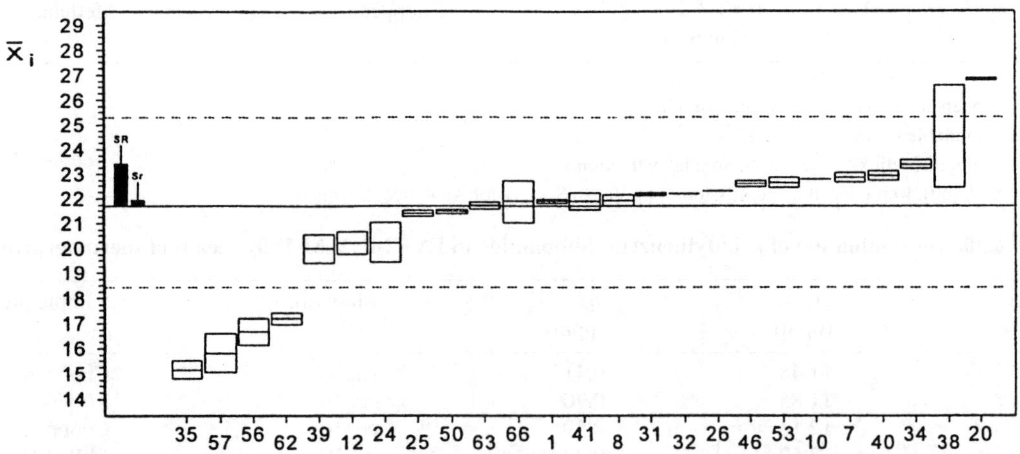

Fig. 6.2. Solvent extraction of NBR/DEHP films by 25 laboratories participating in a round-robin. Mean value: 21.72% DEHP; s_r rel. 0.97%, s_R rel. 7.87%, s_r abs. 0.210, s_R abs. 1.711 (*cfr.* Table 6.24). Intralaboratory error bars are indicated. Outlier values beyond dotted (3σ) lines. After Affolter *et al.* [47]. Reproduced by permission of M. Schmid, EMPA, St. Gallen.

was extracted by methanol in all cases (conforming to the norms), NBR mostly by propanol-2 (in accordance to ISO 1407), but also by acetone, methanol or dimethoxymethane, as well as by THF (for selective extraction purposes). For EPDM (WM-3) the recommended extraction solvent was acetone (ISO 1407), but methanol and dimethoxymethane were also used. In case of SBR (WM-4) ISO 1407 pre-

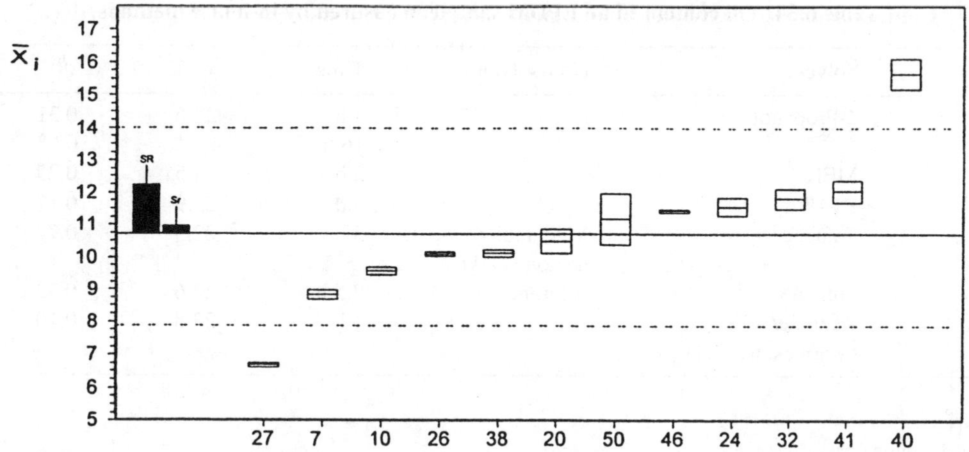

Fig. 6.3. Heat extraction of SBR/mineral oil films by means of TGA by 12 laboratories participating in a round-robin. Mean value: 10.71% mineral oil; s_r rel. 2.15%, s_R rel. 14.12%, s_r abs. 0.230, s_R abs. 1.512 (*cfr.* Table 6.24). Intralaboratory error bars are indicated. Outlier values beyond dotted (3σ) lines. After Affolter *et al.* [47]. Reproduced by permission of M. Schmid, EMPA, St. Gallen.

Table 6.33. Comparison between average values *y* (%) from solvent extraction and TGA

Sample	Material	Expected value[a]	*y* (Extraction)[b]		*y* (TGA)
WM-1	PA12-P	13.5 ± 0.5	13.19	13.36	12.86
WM-2	NBR	*ca.* 21.5	21.73	22.19	20.80
WM-3	EPDM	*ca.* 11.5	12.42	12.80	11.12
WM-4	SBR	*ca.* 13.0	13.11	13.55	10.71

[a] Dosage.
[b] Left: all methods (norms and in-house); right: according to norms of Table 6.28 only.
After Affolter *et al.* [47]. Reproduced by permission of M. Schmid, EMPA, St. Gallen.

scribed either ETA (ethanol–toluene azeotrope) or acetone (used most); again some laboratories opted for methanol and dimethoxymethane. The results do not allow suggesting unambiguously the cause of the unsatisfactory results: material heterogeneity (TGA *vs.* extraction), processing (equal for all samples) or analysis (various procedures; effect of normalisation).

The round-robin [47] also shows that TGA is a valid analytical alternative for solvent extraction, at least for the samples examined (*cfr.* Tables 6.27 and 6.33). If validated, TGA is suitable for quality assurance purposes. Advantages of TGA are: *(i)* small sample size (10 mg); *(ii)* environmental friendliness (no solvent); *(iii)* speed; and *(iv)* absence of sample preparation.

Feigenbaum *et al.* [54] have reported poor repeatability (±12%) for extraction of plasticisers from PVC by means of dissolution (THF or CH_2Cl_2)/precipitation (methanol or hexane). This was attributed to adsorption of variable amounts of plasticisers by the precipitated polymer. On the other hand, using a 6 h Soxhlet ether extraction good reproducibility was obtained. The amounts of plasticiser determined (by means of FTIR) were identical (±1%) to those indicated by the manufacturers (34.5 wt.% for DEHP and 37.5 wt.% for TEHTM), thus validating the Soxhlet extraction–FTIR determination. In cases of relatively high polymer/additive content (*e.g.* 20–30% DIOP, chlorinated polythene wax and Topanol CA, in PVC) the total material extracted can be determined by gravimetric analysis [55].

6.2.5. Oil-extended EPDM

Case Study

Determination of the oil content of oil-extended EPDM is another important analytical problem.

Table 6.34. Oil content in an EPDM sample, measured by in-house methods

Method	Solvent	Non-solvent	Time	x^a	s^b	v^c
Extraction	2-Propanol		4 h	23.6	0.31	1.3
			16 h	24.1	0.03	0.14
Extraction	MEK		2 h	24.5	0.25	1.0
Extraction	ETA[d]		n.d.	23.9	0.17	0.70
Precipitation	Toluene	Methanol/ acetone (1:1)	3 h	22.3	0.77	3.4
Precipitation	Toluene	Acetone	2.5 h	23.6	0.32	1.4
Extraction	Acetone/ cyclohexane (2:1)		1 h	23.8	0.10	0.42

[a] x = average (mass %).
[b] s = standard deviation (mass %).
[c] v = variation coefficient (%).
[d] ETA = ethanol/toluene azeotrope.
After Noordermeer [56]. Reproduced by permission of *Rubber World Magazine* (Lippincott).

Methods commonly in use are extraction and precipitation methods. The first category makes use of a solvent which dissolves the extender oil but not EPDM. Depending on the experimental configuration, extraction is achieved either by step-by-step extractions in flasks with regular renewal of the extraction medium or by means of Soxhlet apparatus. The second category uses a suitable solvent/non-solvent combination, where the oil-extended polymer is first completely dissolved; EPDM is then precipitated and separated from the liquid phase containing the extender oil. For both procedures either the EPDM moiety or the oil moiety after evaporation of the extraction medium can be used for the calculation of the oil content. Table 6.34 shows some typical results.

For QC purposes the duration of the test is an important criterion. In this respect, Soxhlet extraction performs better than the conical flask method. A systematic – unexplained – difference was noticed between Soxhlet MEK extraction and the conical flask method using a 2:1 mix of acetone and cyclohexane. The Soxhlet extraction method using MEK as the solvent has been recommended for the determination of the oil content of oil-extended EPDM [57].

6.2.6. Migration Rates of Phthalate Esters from Soft PVC Products

Case Study
Measuring the rate of release of DINP and other phthalates from toys and other children's products during child chewing activities has been severely hampered by the absence of any standard measurement procedure. As a result, published values have varied widely and have confused efforts to assess potential health risks.

For a same teether sample and using the same shaking procedure two laboratories measured a release rate differing by a factor of 200 (Table 6.35). The reason for this huge discrepancy is not clear. Actions are being undertaken in the framework of EC Measurement and Testing to increase the reliability of the migration data.

6.3. QUANTITATIVE CHROMATOGRAPHIC METHODS

Principles and Characteristics
Chromatographic methods are conceptually rather simple, and have become an ubiquitous part of quantitative chemical analysis. The resolving power and sensitivity of modern (gas and some liquid) chromatographies is amazing. As discussed previously (Chp. 6.1), for quantitative chromatographic analysis two factors are important: *(i)* preconcentration of the analyte of interest from a relatively large volume of sample to a small extract volume; and *(ii)* clean-up of the sample matrix to produce a particle-free and chromatographically clean extract. For quantitation prior separation and positive identification (*e.g.* by means of spectroscopy) are absolutely necessary. Various regulatory agencies refuse to approve analytical methods that rely only on deconvolution to obtain quantitation. Clever algorithms are

Table 6.35. Migration rates of phthalate esters from soft PVC products

Measurement type	Release rate (μg DINP/cm²/h)
In vitro	
Agitation in simulated saliva and solvent extraction	0.54–233
Idem	1.0[a]
Idem	6.7–7.5 (2.9–3.6, DEHP)
Impaction in simulated saliva and solvent extraction	0.1–4.4
Ultrasonication in simulated saliva and solvent extraction	7.9–31.4
In vivo	
Saliva extracts from adult human volunteers	10.9 (1.8–53.4)
Idem	26.0 (6.1–57.9)

[a] Reanalysis of the sample with previously reported release of 233 μg DINP/cm²/h.

After Wilkinson and Lamb [58]. Reproduced from *Regulatory Toxicology and Pharmacology* **30**, C.F. Wilkinson *et al.*, 140–155. Copyright (1999), with permission from Elsevier.

no substitute for good chromatography [59]. A chromatogram contains three quantifiable items: peak position, peak width and peak height or area. Peak width is seldom reported, as it is relatively immaterial as long as successive peaks do not overlap.

The selective quantification of components separated by chromatographic analysis is strongly dependent upon the **detector** capabilities. The response of various compounds will differ depending on their structure and the selectivity of the detector. Thus, after the detector has generated its signal, the chromatographer needs to care about data handling and calibration (conversion of area number to concentration). It is not always feasible to calibrate the detectors for their sensitivity to every component observed, and assumptions must then be made when comparing the relative yields of different products. Unless the correct assumptions are made about the significance of the area of a GC peak for each detector, the relative yields of products can appear to be widely different, *e.g.* when comparing results from FID and MS detectors. This is especially the case where the products cover a wide range of molecular weight. The lack of correspondence between relative peak areas from FID and MS data has been pointed out [60]. FID measures *weights* of components, whereas MS measures *numbers*. FID is insensitive to some small molecules. For larger molecules, it is often assumed that the sensitivity factor is proportional to the number of carbon atoms in the molecule, but deviations of this assumption occur especially for heteroatomic molecules. In mass spectrometry the overall molar sensitivity is, strictly speaking, specific for each type of molecule, because molecules have different ionisation and detection efficiencies.

The main **sources of error** in quantitation using chromatography are: *(i)* sampling technique and sample introduction; *(ii)* design of the instrument; and *(iii)* peak size measurement. The suitability of indirect methods for quantification of polymer additives depends strongly on the optimisation of the extraction methods. Only a high extraction yield can assure trustworthy quantitative results of an overall extraction-chromatography approach. The basis for all quantitative work is the fact that over the linear response range of the detector the area underneath a chromatographic peak is directly proportional to the amount of substance giving rise to the peak. This is so independently of the shape of the peak. A chromatogram is evaluated for quantitative analysis either by the measurement of peak areas or peak heights. Relative (dis)advantages are well known [59]. Peak area is the generally preferred measurement, accounting for any changes in chromatographic conditions. Integration of chromatographic peaks allows a precision of *ca.* 0.5%. Accurate quantitative analysis can only be expected in the following conditions: *(i)* adequate resolution; *(ii)* predictable detector response as a function of the concentration of the analyte; and *(iii)* correct data processing.

Quantitative methods in chromatography rely on the areas of the eluting peaks and on reference compounds to establish detector sensitivity:

$$m_i = K_i A_i \tag{6.2}$$

where m_i is the quantity of compound i injected on the column, K_i is the *absolute* response factor for compound i, and A_i is the area of the eluting peak. To calculate the response factor K_i of compound i it is essential to know precisely the injected quantity, which is difficult matter. Moreover, as K_i is not an intrinsic property of the analyte most chromatographic methods for quantitative analysis do not make use of the absolute response factor. Traditionally, there are several **standardisation techniques** employed in the practice of chromatographic analyses: external or internal standards (most commonly used), standard addition and normalisation. The external standardisation (ESTD) method is most efficient if the pure standards of the peaks of interest be available. The ESTD method is easy to use and allows measurement of the concentration of one or more components; the method employs the *absolute* response factor, K. Imprecise volume determination can lead to systematic errors. This simple method is used in industry for repetitive analysis; an autosampler is desirable. External calibration is sensitive to variations in the matrix, and therefore is unsuitable for many matrix systems. Internal standards and standard addition can overcome the matrix effect; however, a homogeneous mixture of standard and sample is difficult to obtain if solid samples are analysed [61]. The method of internal normalisation is used for mixtures in which each compound has been identified by its elution peak. In the internal standard (ISTD) method the reference or standard must be completely separated from all other components in the mixture. The major advantage of ISTD is that it is less sensitive to changes in detector response factors as most changes will affect the sample component and internal standard in the same way. The solutes of interest cannot, themselves, be selected as standards. For trace analysis, it is preferable to use a method that relies on the *relative* response factor for a compound against a reference compound (the internal standard). The (area) normalisation method requires no reference standards or calibration solutions, but in order to be applicable the detector must have the same response to all the components of the sample. An example is FID for GC analysis of high-MW paraffins.

A number of recommendations has been made in the development of quantitative chromatographic methods. ASTM disclosed substantial laboratory-to-laboratory differences in quantitative HPLC analyses [62]. A text describes the chromatographic integration methods for peak identification, validation and quantitation [63].

6.3.1. Quantitative Gas Chromatography

Principles and Characteristics

Since the introduction of GC over 40 detectors have been developed. All contemporary GC detectors that are commercially available are designed to give a linear output over a defined concentration range. The main quantitative detectors in use are FID, NPD, ECD and MS (with narrow-bore column in view of fouling). Universal or sensitive selective detection methods, *e.g.* FID or NPD, make GC techniques particularly attractive. Capillary GC-FID (temperatures up to 325°C) allows screening of the polymer extract, identification and quantification of volatile additives (up to about 700 Da) in one run. FID is nearly ideal for quantitative analysis as it has a response to solute concentration that is linear over 4 to 5 orders of magnitude. While it is often assumed in quantitative GC-FID analysis that the molar response to a component is proportional to the number of carbon atoms in that component, which renders calibration of the response unnecessary, experimentally it is better to carry out FID sensitivity calibrations using reference samples. ECD has limited use for quantitative analysis because of wide variations in electron affinity (and hence sensitivity) between different compounds; calibration of ECD is essential. Other suitable detectors for quantitative analysis are FPD, SCD and PID. GC-SCD has a linear response of 5 orders of magnitude.

In HT-GC the presence of oxidation/degradation peaks is much less of an issue than it is in HPLC. Nevertheless, some oligomers are always present and may have to be properly accounted for if they interfere with peaks representing very low levels of additives or their by-products. Additives can be quantified at very low levels with good sensitivity and confidence. High accuracies (down to about 0.1% RSD) are attainable with GC (usually GC-FID). In favourable circumstances an extractive-chromatographic determination of additives in polymers allows a relative error of determination of about 5%, with the biggest contribution to the uncertainty being on account of the extraction. Also the lack of reproducibility of HT-GC columns is of some concern. For thermally labile compounds quantitative GC analysis is difficult. When individual additives originally present generate identical degradation products derivatisation procedures may be used to prevent thermal degradation. The derivatised compounds can then be chromatographed but since the derivatisation process itself is composed of several steps this methodology is rather time-consuming.

Although it is crucial in quantitative GC to obtain a good separation of the analytes of interest, this is less critical when a mass spectrometer is used as a detector; nevertheless, it is good practice. If the GC effluent is split between the mass spectrometer and FID detector, either detector can be used for quantitation. Because the response for any individual compound will differ, it is necessary to obtain relative response factors for those compounds for which quantitation is needed.

The reproducibility of quantification of additives with GC-QMS ion scan mode is unacceptable with RSDs of 5–20%, as opposed to 0.5–2.5% for GC-FID. Various alternatives for quantitation may be considered: *(i)* simultaneous FID and QMS detection; *(ii)* separate GC-FID and GC-QMS analysis; and *(iii)* verification in scan mode and quantification in SIM mode. For quantification by means of FID peak purity is important. Quantification in SIM mode is more sensitive than by means of FID. ToF-MS detection systems are inherently efficient in the true integral of the ion intensity from GC peaks of any width, compared to scanning instruments, where the ability to characterise narrower GC peaks falls off sharply as peak widths come within an order of magnitude of scan time (post-processing required).

External standard calibration of GC-MS with scanning instruments achieves standard deviations of 5–10% [64]. Accurate quantitation in GC-MS requires addition of a known quantity of an internal standard to an accurately weighed aliquot of the mixture (matrix) being analysed. The internal standard corrects for losses during subsequent separation and concentration steps and provides a known amount of material to measure against the compound of interest. The best internal standard is one that is chemically similar to the compound to be measured, but that elutes in an empty space in the chromatogram. With MS detection the use of radioactive analogues is encouraged but is not much practised.

Quantitative GC analysis has been reviewed [65–68]. For quantitative GC-MS, *cfr.* also refs. [69–74].

Applications

Quantitative analysis of **volatile products** by TD-GC-MS has been used to evaluate the performance of flame retardants in EPs such as PC, PPE and PBT [75]. McGrattan [76] has described the quantitative analysis of volatile products of programmed degradation by trapping in a chemical adsorbent and separation by GC prior to FTIR analysis (TG-CT-GC-FTIR). The simultaneous determination of 2,4,6-trichloroanisole (2,4,6-TCA), 2,3,4,6-tetrachloroanisole (2,3,4,6-TeCA) and pentachloroanisole (PCA) extracted from packaging materials by HRGC-MIM-MS has been described using 3,5-dimethyl-2,4,6-trichloroanisole as an internal standard [29]. The mean extraction efficiency of the combined steam distillation-extraction step exceeded 90%; coefficients of variation for the GC-MS step ranged from 2 to 11%. Tinuvin 770 in PP can be determined by GC-FID (100–7000 ppm range) using an internal standard.

Isotope dilution GC-MS (IDGC-MS) procedures have been used for the evaluation of the migration of **plasticisers** from packaging materials (PE, PVC, cellulose and vinylidene chloride films) into foods [77–79]. Acetyltributyl citrate (ATBC) migration from Saran (vinylidene chloride-*co*-vinyl chloride) has been measured by means of IR spectroscopy, extraction/saponification [80] and stable isotope dilution GC-MS using [$^2H_3$] ATBC [79].

Kawamura *et al.* [81] have surveyed nonylphenol by GC-MS (with quantification by GC-SIM-MS) in 207 samples of food contact plastics and baby toys. Crompton [43] has described the quantitative GC analysis of residual vinylchloride, butadiene, acrylonitrile, styrene and 2-ethylhexylacrylate in polymers by solution headspace analysis. Considerably greater sensitivities and shorter analysis times were obtained using the headspace analysis methods than were possible by direct injection of polymer solutions into a GC. Similarly, various residual hydrocarbons (10 ppm of isobutane, *n*- and isopentane, iso- and neohexane) in expanded PS were determined by GC analysis of a solution of the sample with hydrocarbon internal standards; accuracies of 5 to 10% were reported [82]. Residual *n*- and isopentane (0.001%) in expandable and expanded PS were also determined by a solvent-free procedure consisting of heating the polymer at 240°C in a sealed tube, followed by HS-GC; calibration against known blends of *n*- and isopentane and *n*-undecane internal standard [82].

The analytical power of combining capillary PyGC with the selectivity of FID and NPD has been demonstrated for rapid quantitative and qualitative analysis of high-MW and polymer stabilisers in PP, using the standard addition method (up to 10,000 ppm) [42]. Quantitative aspects of PyGC are discussed in Chp. 2.2.1.

6.3.2. Quantitative Liquid Chromatography

Principles and Characteristics

HPLC is probably the most universal quantitative method in use at present and essentially better than GC. The three basic quantitative analytical techniques to consider in LC are again the internal and external standard methods and the normalisation method. The former is more tedious requiring a known amount of standard to be added to each sample; the standard must be completely resolved from all components of the sample. It does, however, provide the greatest accuracy (e.s.d. of 1–2% for HPLC-PDA) and precision. The external standard method requires a separate calibration chromatogram to be run, but only one standard solution needs to be made up in bulk and can be used for a large number of analyses. The method eliminates the need to determine response factors but is less precise and less accurate than the internal standard method. Normalisation is the simplest method and can provide the most accurate and precise results. Unfortunately, it can only be used in very special circumstances where the response of the detector is the same for all solutes of interest (*e.g.* RI detector for analysing polymer mixtures). The possibilities for quantitative analysis in LC are slightly different from those in GC. Liquid chromatography usually operates with concentration sensitive detectors. Consequently, the quantity is a function of flow-rate. Sensitivities in LC vary considerably for various analytes. For example, in UV detection molar extinction coefficients differ up to a factor of 10^4. Consequently, internal normalisation is not applicable to LC, as opposed to GC in which FID shows about equal response factors for the various components. In view of the precise injection volumes the external standard method is adequate; there is no need for the internal standard method. In general isocratic analyses yield more accurate results than analyses with gradient elution. It is possible to quantitate to ppm level and to detect impurities to ppb level. Major sources of error in quantitation are sample collection and preparation.

Almost all quantitative LC analyses are carried out using UV, ECD or F **detection**; also MS, CLND, ELSD and SCD play a role. The UV detector is probably the detector of choice for quantitative analysis as it combines the essential features of wide linear dynamic range with fairly high sensitivity. Most antioxidant stabilisers, whether phosphites, hindered phenols, *etc.*, exhibit UV absorptivity. Simultaneous multi-wavelength quantitation at unlimited different wavelengths is provided with HPLC-PDA. In post-run analysis, any wavelength extracted from 3D data can be used for reanalysis. In other words, the best wavelength for quantitation can be determined after data acquisition. Gradient ($CNCH_3/H_2O$) RPLC-PDA allows identification and quantification in one run. Area counts per unit quantity are normally high enough that sensitivity is quite good and levels of detection are low. The presence of oligomers with no UV absorptivity is of no concern. Both UV and MS detection are routinely used for sensitive and accurate quantitation of compounds on-line with RPLC. However, it is difficult and cumbersome to use either of these detectors for on-line compound quantitation because of the requirement for a proper calibration standard – either the authentic compound or a closely related analogue. ELSD responds similarly to compounds from the same structural class that are within a relatively narrow molecular weight range. However, for chemically dissimilar compounds, its response varies significantly, making it difficult to quantify such organic molecules. ELSD (cheap, non-volatiles, not selective, not tuneable, non-linear, ease of use) is complementary to MS (expensive, volatiles, selective, linear, complex). An alternative for universal compound detection and quantitation is CLND, which is equimolar and responds to the nitrogen content of a sample, down to low pmol levels [83] and has been adapted for use in conjunction with HPLC.

High-performance liquid chromatography is particularly useful for relatively high-MW, reactive, polar and thermolabile additives. The main assets of LC-MS (*e.g.* LC-QQQ) for quantitation are sensitivity, specificity and speed. However, disadvantages are quite numerous: *(i)* limited dynamic range (ionisation, fragmentation, clustering, multiple charging); *(ii)* high chemical background (<200 amu); *(iii)* ion suppression (co-eluting peaks, too close to solvent front); and *(iv)* expensive, complicated and "temperamental". Quantification in LC-MS requires the removal of interfering matrix compounds using suitable sample clean-up protocols. Suitable internal standards should always be used to compensate for any remaining matrix effects or other MS detector variations. HPLC may also present quantification problems in case of oxidation/degradation peaks due to total peak interference and peak overlap. In an integrated system such as LC-UV/CLND-MS quantitation of N compounds is usually performed by the CLND function. For more universal quantitation LC-UV-MS/MS is recommended. Quantitative LC analysis has been reviewed [84].

Applications

Typical industrial applications are the chromatographic determination of Irganox 1076 and Irganox 1520 in unvulcanised rubbers by RPLC-PDA at 278 nm (in 100–1500 ppm range), of Irganox 1098 in PA4.6 by RPLC-PDA at 278 nm (in 100–1000 ppm range), of Irganox 1010 and Irganox 1076 in PP by RPLC-PDA at 278 nm (in 100–2000 ppm range), of Tinuvin 622 (as diol) in HDPE by RPLC-PDA at 225 nm (in 100–5000 ppm range), of the peroxide shifter Luperco 802 in PP by RPLC-PDA at 218 nm (1–50 ppm range), and of Irganox 245 in ABS by RPLC-PDA at 276 nm (100–2500 ppm range).

From a study of the optimisation of experimental parameters for the quantification of polymer additives using SFE-HPLC it clearly emerges that it is not only important to reach 100% extraction recoveries but also to control the compounding procedure, because there could be significant losses of antioxidants during the mixing of polymer and antioxidant [85]. This total analysis procedure was illustrated for PP/(Irganox 1010, Irgafos 168). It has not been possible to recover completely the amount of antioxidants that were initially mixed into the polymer resin (*cfr.* Table 7.16 of ref. [3a]). The deviations were attributed to evaporation during mixing of the components, transformation of AOs during the mixing period, and the uniformity of distribution in the matrix. The antioxidants may also react or degrade during extraction and analysis. The reaction products are difficult to quantify.

Also a variant of liquid chromatography, SEC, has been applied for quantitative analysis, although to a much lesser extent. Jickells [86] has exploited the use of **SEC-FTIR** for quantitative additive analysis. SEC separated mixtures can also be used as a direct sample input into a mass spectrometer for mass analysis. Cortes *et al.* [87] have introduced quantitative polymer/additive analysis by multidimensional chromatography using on-line coupled microcolumn SEC as a preliminary separation. A comparative quantitative study of dissolution and dissolution/precipitation of PC/(2,4-di-*t*-butylphenol, nonylphenol isomers, Tinuvin 329, Irgafos 168) and ABS/(nonylphenol isomers, Tinuvin P, benzylbutyl phthalate, Vanox 2246, Tinuvin 328/770, Topanol CA and Acrawax) by means of μSEC-GC/LC has shown the quantitative reliability of the dissolution procedure [87]. It also appears that the precipitation technique can yield low results for additives which exhibit solubility dependence. Polymer additives may routinely be analysed using dissolution followed by μSEC-GC. SEC-GC analysis of polymer/additive dissolutions has shown the possibility of applying the absolute peak area for quantitative analysis of additives and RSD values of 0.5–2.0% for the phthalic acid esters [88].

6.3.3. Quantitative Supercritical Fluid Chromatography

Principles and Characteristics

Chester *et al.* [89] have identified some eleven essential considerations for accurate and precise trace analysis by means of **capillary SFC**, matching HPLC precision. The key to trace analysis below 1 ppm with an FID is providing an injection volume of sufficient size (with complete avoidance of splitting). By injecting volumes up to 0.5 μL relative standard deviations of less than 0.3% for the injected volume are achieved with little or no sacrifice of chromatographic performance; RSDs for solute areas of 2% are quoted. FID detection permits quantitation of well-shaped peaks as low as approximately 100 pg in mass, thus providing quantitation of sub-ppm solutes in the injection solvent. Packed column SFC, which uses standard size HPLC columns and hence standard HPLC injection systems, yields more reproducible quantitative results than cSFC. *Cfr.* also ref. [3a].

Applications

SFC-FID was used for composition analysis of the Irgafos P-EPQ mixture [90]. On-line SFE-SFC is frequently being used for quantitative analysis of polymer additives [91–93]. Bunel *et al.* [94] have developed a fast on-line SFE-SFC method to characterise and quantitatively determine PP additives. The method is better equipped than most chromatographies to monitor intact and decomposed antioxidants. On-line SFE-SFC with a C_{18} column (5 μm particle size) has been used for quantitative analysis of antioxidant blends in polyethylene (0.2–0.5 mg amounts) [95]. An overall average additive recovery of 97.6% was quoted (triplicate, *cfr.* Table 6.36). The use of UV detection rather than FID offers a distinct advantage in the analysis of polyolefin samples. The oligomeric fraction of the polymer, which is partially extracted with the additives from the polyolefin matrix, is easily detected with FID. However, only the portion of the oligomeric fraction that contains a chromophore (such as unsaturated waxes) is detected by UV detection. This results in significantly less interference in the sample chromatogram.

Table 6.36. On-line SFE-SFC analysis of antioxidant blends in PE

Base resin	Formulated antioxidant content[a] (recovery, %)
HDPE	613 ppm I-1076 (92.2–97.9), 1029 ppm E-398 (97.6–99.3)
HDPE	1545 ppm I-1010 (97.2–101.0), 1591 ppm I-168 (98.7–100.3)
LLDPE	505 ppm I-1076 (96.2–100.2), 1414 ppm W-399 (94.8–97.7)
LLDPE	959 ppm C-1790 (96.5–99.6), 1053 ppm U-626 (94.5–98.2)

[a] I-1010, Irganox 1010; I-1076, Irganox 1076; I-168, Irgafos 168; W-399, Weston 399; U-626, Ultranox 626; E-398, Ethanox 398; C-1790, Cyanox 1790.
After Tikuisis and Cossar [95]. Reproduced by permission of the authors.

On-line SFE-SFC-FID is also suited to quantitative analysis of dialkyltin additives in rigid PVC and is able to replace other tedious and time-consuming procedures. The results are highly reproducible. Table 7.14 of ref. [3a] shows other quantitative analyses of polymer additives by means of SFE-SFC.

On-line SFE-CC-pSFC-FID was used for quantitative analysis of LDPE/(BHT, BHEB, Isonox 129, Irganox 1010/1076) [27]. Results obtained for on-line SFE-SFC were comparable to those from off-line SFE-HPLC-UV and off-line ESE®-HPLC-UV (except for Irganox 1076), but of lower precision due to small sample size (2.5 mg) employed in the on-line system. On-line SFE-SFC was considered to be a reliable and robust method for application in routine quality control analysis. Bücherl *et al.* [90] consider SFE-SFC-MS with simultaneous quantitative detection by FID to be a rapid and easy analysis method for packaging materials. However, method development is not trivial.

6.3.4. Quantitative Thin-layer Chromatography

Principles and Characteristics

TLC is not considered the best chromatographic technique for quantitative analysis and, although it can provide quantitative results, the necessary procedure tends to be more cumbersome and tedious compared with other chromatographic methods. Furthermore, for accurate work, expensive scanning equipment is required. In its basic mode, classical TLC has for a long time been regarded as a semi-quantitative technique (at best). In principle, there are three approaches used in quantitative TLC: extraction of the spot and separate measurement by spectroscopic or other techniques, comparative techniques employing visual assessment, and finally optical scanning. Quantitation historically involved simply visually comparing spots of diluted and undiluted test solutions, a method with suspect accuracy and sensitivity [96]. The progress in layers, instrumentation and the development of automated scanning densitometers has led to a remarkable improvement of the method's features. Modern TLC (usually termed high-performance thin-layer chromatography, HPTLC), which started around 1975, is a fully instrumentalised version of conventional TLC, capable of providing accurate *in situ* quantitation for a wide variety of applications. Because of the involvement of optimised instrumentation with high levels of automation, HPTLC offers precise control over sample application, chromatographic development and chromatogram recording. In most cases TLC in combination with other sophisticated analytical techniques is used for quantitative analysis; prevailing techniques are conventional and video-densitometry, fluorimetry and radiometry (including NAA). Provided that suitable precautions are taken good quantitative HPTLC analysis can often be obtained. For that purpose some basic requirements need to be fulfilled, both in terms of sample, sample application and dosage [97]. Using spray-on techniques with uniform mass distribution over the full length of the bands, densitometric evaluation can be done by aliquot scanning, which ensures maximum quantitative accuracy. Quantitative results rely on the choice of spray reagent, spraying skill and other operational parameters and reproducibility is therefore poor.

The **methods** used for quantitative analysis of substances after separation on HPTLC plates include:

(i) Visual assessment: comparison of spot sizes and colour intensities between samples and standards.

(ii) Spot-size measurement: evaluation of spot areas, which are proportional to concentration of the spotted analyte.

(iii) Zone-elution: drying the layer, scraping off the appropriate region of the layer, extraction of the analyte from the adsorbent material, and separate measurement of analyte concentration by a microanalytical technique (*e.g.* titrimetry, spectrophotometry, electroanalysis, *etc.*).

(iv) *In situ* densitometry: measurement of the absorption of visible or ultraviolet light or of emitted fluorescence of resolved spots on the chromatoplates (optical scanning).

(v) *In situ* X-ray fluorescence and photothermal spectrometry.

(vi) Laser pyrolysis.

Between 1 and 10 μg of a coloured component can be estimated on a developed TLC plate by eye, but with an operator dependent low reproducibility (10–30%). Quantification by eluting the relevant band or spot from the sorbent followed by spectrophotometry (colorimetry) is widely reported but tedious and leads to poor precision. Coloured or UV absorbing substances are most readily and accurately quantified at ng levels using scanning densitometers. A significant cause for poor reproducibility of conventional TLC is the positioning error in densitometric scanning. The production of uniform TLC plates, programmable applicators and development systems, and the use of PCs, CCD cameras and colour printers, have opened new possibilities for QTLC. A real improvement in reproducibility, simplicity and speed of quantitative evaluation, has come from the use of image processing. *In situ* measurement of zones with a scanning densitometer is now the preferred technique for quantitative TLC, with RSD $<2\%$, making it a reliable quantitative tool [98].

Chromatogram evaluation can nowadays be carried out with classical and video densitometry. TLC scanners for the quantitative evaluation of thin-layer chromatograms consist of a spectral photometer, a mechanical (programmable) device for transporting the TLC plate through the light beam coming from the photometer's light source and a recorder for recording the remission-position curves. Depending on the structure of the compound (chromophore) and its concentration in the layer, substance spots absorb part of the light energy. An absorption measurement in incident light is called a remission measurement (remission = diffuse reflection). In all types of densitometry the signal from a chromatogram fraction is compared to the signal from the sample free plate background. For quantitative determination peak data of the unknowns are correlated with data from calibration standards chromatographed on the same plate. **Classical densitometry** uses a light beam of selectable length and width to scan the tracks of the chromatogram. Examination of the spectrum reveals which wavelength is suitable for quantitative analysis. When the chromatogram is scanned photometrically at the chosen wavelength a remission-position plot is obtained; the peaks and areas are a function of the amount of substance applied. For quantitation automated scanning densitometers are used equipped with D_2, W and Xe lamps, capable of measuring **UV or visible absorbance**, or fluorescence in the reflectance or transmission modes. An important practical consideration in QTLC, when choosing between quantification by either reflectance or transmission, is the nature of the support onto which the layer of sorbent has been coated. The degree of remission, the portion of the radiated light intensity which is reflected, or not absorbed, is not necessarily correlated in a linear fashion with the concentration. In many cases, there is a linear correlation using the Kubelka–Munk function, which however does not apply without qualification [99]. Evidence that the K-M approach is correct is that the logarithm of transmittance and the reciprocal value of reflectance can be plotted against the coefficient of absorption (K_a) and the coefficient of scatter (S) with near linear results. In other cases, the calibration graphs are inherently non-linear, with perhaps a pseudo-linear portion at low concentrations. Quantification is generally based on a second-order polynomial fit of the calibration graph, with the concentration of unknown samples assigned by interpolation.

Analytical methods based on **fluorescence** are preferred over absorption for TLC quantification. Fluorescent substances are excited by long-wave UV light (366 nm); the fluorescence emission in the visible region is detected. In this respect classical and video densitometry are comparable. However, video technology lacks the variable-excitation-based selectivity of classical densitometry. For the detection of fluorescent substances the intensity of the emitted fluorescence (I_{fl}) is directly proportional to the amount of substance applied. The most serious drawback of fluorescence measurements is the low quantum efficiency of the excitation process. The light intensity available for fluorescence measurements at a given intensity of illumination is, therefore, much smaller than the intensity that, under the same circumstances, would be available for direct measurement. Fluorimetry is comparable with absorption densitometry in the time of analysis (2 min) but surpasses it in the detection limit by 10^3 to 10^4 times. With fluorescence, sensitivity is often of pg levels (comparable to HPLC), calibration curves have a wider linear range (typically 10^2–10^3), and selectivity of detection is improved because of the ability to choose characteristic excitation and emission wavelengths. Fluorescence detection is, however, limited to those compounds which fluoresce or can be conveniently derivatised to become fluorescent. Fluorescent reagents and field of application are given in ref. [99].

Both sample and matrix effects influence quantitative TLC using scanning densitometry. The sorbent matrix influences the shape of calibration curves; impurity gradients, resulting from contamination of the layer, or application of derivatising reagents are common sources of baseline instability. Chemical reactions catalysed by the sorbent layer are also a source of sample instability. Fluorescence quenching and fluorescence enhancement effects can influence the reliability of quantification in the fluorescence mode [97]. Densitometry has been reviewed [100].

It is likely that densitometric evaluation of plates belongs to the past as one now moves in the direction of new video-oriented data-acquisition and processing procedures usable in QTLC. For **video densitometry** an electronic image of the plate is generated with a video or digital camera and the intensities of the image pixels are compared. Video systems with their low running costs, user friendliness, speed of evaluation, flexibility, rapid archiving on PC and availability of electronically stored chromatograms for (later) quantitative evaluation have largely replaced instant photography systems for recording and archiving thin-layer chromatograms. Image processing systems for quantitative analysis of HPTLC plates have been described [101]. Video densitometers with quantification software function only in the visible range, while classical slit-scanning densitometers can be used for quantitative analysis over the 190–800 nm wavelength range with spectral selectivity. UV-absorbing substances can be detected by video technology via the quenching of a fluorescence indicator embedded in the layer, *i.e.* detection is shifted to the visible region. Spectral selectivity, a strong point of the classical densitometer, is not accessible with the video system. The more a substance to be quantified absorbs at or near the excitation maximum of the fluorescence indicator (254 nm), the higher is sensitivity and accuracy of video quantification; it becomes comparable to that of classical densitometry. The lower the absorbance at 254 nm, the less sensitive and less accurate becomes video quantification. *In situ* spectroscopy, a feature of the TLC densitometer, is not available with video technology.

X-ray fluorescence and photothermal spectrometry are also employed for *in situ* analysis. It is possible as well to determine elements reliably and quantitatively, after removal from a plate. Laser pyrolysis scanning (LPS) may also be used as a quantification method for TLC [102]. No spray reagent is required for TLC-LPS-FID/ECD. Low ng detection for LPS-FID and pg detection for LPS-ECD is possible. The technique combines the advantage of the separation power of TLC and GC detection modes.

QTLC methods suffer from several **sources of error** inherent to the TLC procedure itself, such as the difficulty in applying a reproducible amount of sample to the layer, variations in layer thickness, the difficulty of spraying a plate uniformly and ensuring that a reproducible, quantitative reaction occurs between solute and chromogenic agent [103]. For QHPTLC fixed-volume dosimeters (100 or 200 nL) are used. The highest accuracies (0.6–1.5% RSD) in QTLC have been reported for use of fluorimetry in which a fluorescent spot on a dark background is scanned under UV, in particular for naturally fluorescing compounds that do not require treatment with a chromogenic reagent. Precision in HPTLC is 0.2% RSD for a densitometer making repetitive scans of a single sample track; for multiple applications of the same sample, it is usually 1 to 5% RSD [104]. The smallest detectable amount of substance on a TLC plate depends on the properties of the compound. For compounds with favourable absorption coefficients it is in the low ng range for absorption measurements and in the pg range for fluorescence.

Quantitative TLC has been treated in detail from theoretical and practical viewpoints, including descriptions of protocols for sample calibration, for establishing resolution, sensitivity, detectability, and optimum scan rate, and for comparing the performance characteristics of different slit-scanning densitometers [98]. **Validation** of a measurement process, such as QTLC, involves two related activities. One is quality control (QC), which develops and implements the tasks necessary to produce a measurement of requisite quality, the other is quality assessment, which verifies that the QC system is operating within acceptable limits: this latter controls the quality of measured data. Prosek *et al.* [105] have considered quality assessment in QTLC. Each laboratory should institute a quality assurance program for QTCL methods that is appropriate for the local situation. Several authors [106,107] have described validation of QTLC.

QTLC offers several **advantages**. With the provisions of automatic reproducible sample introduction onto the plates and a UV/VIS scanner the overall method error can be contained in about 2%. HPTLC features trace quantitative capabilities at low cost and is a time-effective alternative to HPLC. HPTLC is a powerful tool for quantitative analysis of complex mixtures with a high sample throughput because of parallel sample processing and can tolerate cruder samples than column methods because the stationary phase is disposable, and provides flexibility in the method and choice of detection. HPTLC is also used in process monitoring and allows quantitative results within 30 min after sampling. A **disadvantage** of HPTLC is that calibration is frequently non-linear in reflectance densitometry due to light scatter from the surface of the chromatographic medium. Calibration, linearisation and curve fitting approaches can profoundly affect method accuracy [104]. Quantitative evaluation of TLC plates has required more development work than that of chromatograms in HPLC and GC. However, provided sample application errors are eliminated and calibration is performed, now comparable accuracy and precision can be achieved in HTPLC, HPLC and GC [108]. A modified internal standard method for use in quantitative TLC was found to enhance significantly the accuracy and precision obtained [109].

Several reviews describe various aspects of QTLC [106–114] and three books [115–117] deal with this subject.

Applications

HPTLC on silica gel followed by scanning densitometry is a satisfactory method for the identification and quantitative prior analysis of dye liquors to be used for acrylic fibres and also for dyes extracted from finished products [118]. QTLC has recently been reported for a variety of applications, such as the determination of BHT in gum base (densitometry at 600 nm) [119], of free sulfur in vulcanised rubber [120], of Irganox 1330 in HDPE [121], and of organotin compounds and triphenylphosphate [122], as well as for quantitative inorganic and organometallic analysis and in radiochemical studies [123,124]. For industrial applications of QTLC the reader is referred to Treiber [125].

6.4. QUANTITATIVE SPECTROSCOPIC TECHNIQUES

Principles and Characteristics

Spectroscopic techniques (mainly UV, F, FTIR, NIRS and NMR) are widely used for quantitation because of speed and flexibility. Ideally, they require pure samples for reliable identification. Before a substance is quantified it needs to be positively identified using spectroscopy or mass spectrometry.

Spectroscopic data are often reduced to selected peak heights or areas. Molecular absorbance measurements in UV/VIS can be (though seldom are) made very precisely; however, such measurements are restricted to fairly concentrated solutions. In most other spectrometric methods, common precision is at best of the order of a few percent. The case of NIRS shows that new ways of analysing data (using matrix methods) allow high-precision measurements. Some methods, such as NMR and Raman spectroscopies, are relatively insensitive but still seem poised to become more quantitative. Peak height measurements ($A = \log_{10} I_0/I$) are used in most quantitative analyses, but tend to be prone to changes in instrumental resolution. The integrated intensity as a measure of the total intensity of a band shows less instrumental dependency.

Quantitation can be carried out using absorption or fluorescence spectroscopy. Measurements can be carried out in transmittance or reflectance mode. The basis of quantitative absorption spectroscopy in **transmission** mode (UV/VIS and FTIR) is the usual linear relationship of the Beer–Bouguer–Lambert law, which states that the absorbance A of a solute is directly proportional to its concentration c:

$$A = -\log T = \log I_0/I = \varepsilon c l \qquad (6.3)$$

where T is the transmittance, I_0 and I are the (monochromatic) incident and transmitted intensities, ε is the molar absorption or extinction coefficient, and l is the path length. The extinction coefficient is assumed constant and characteristic of a given substance under a precisely defined set of conditions, such as wavelength, solvent, and temperature. The most precise results are usually obtained for $0.8 \geq T \geq 0.1$. At concentrations above 10 mM, real deviations from Beer's law often appear. An adequate calibration and validation procedure should always be followed for all quantitative methods in order to determine the linearity of the method. This will also provide an estimate of the error in the analysis. A **calibration curve** constructed using standard solutions with known concentrations of the analyte, which bracket the concentration of the unknown sample, will yield accurate quantitative results even for very narrow bands. Once a calibration has been developed, it can then be used for the prediction of unknowns, provided two general conditions are met: *(i)* the spectra of the unknowns are recorded under the same conditions as employed in the calibration step (*i.e.*, same instrumental parameters, identical means of sample handling, *etc.*); and *(ii)* the composition of the calibration standards is representative of that of the unknowns [126]. For reliable results, the sample to be analysed must contain only the absorbing component for which the calibration has been performed. Uniform dispersion of the additives in the polymeric matrix is more important for quantitative spectroscopic analysis than for thermal techniques.

The simplest application of Beer's law is graphical. A calibration plot can be created for a single component in a simple system, such as a single component dissolved in a non-interacting solvent, by plotting chemical concentration *vs.* absorbance at a single analytical frequency. Problems in quantitative IR spectrometry arise when dealing with chemical data which show deviations from Beer's law. However, methods exist which, in many cases, can easily deal with such cases. All spectra to be used for quantitative measurements are best examined in a format where the ordinate axis is linear with sample concentration (provided Beer's law applies). The most commonly used format for the ordinate axis is "Absorbance". If the thickness of a sample is doubled, then the recorded band intensities should be doubled. In practice, the most commonly applied technique is comparison of a material containing an unknown amount of a component with standards of known composition of the component:

$$A_1/A_2 = kc_1/c_2 \tag{6.4}$$

In the simplest type of quantitative analysis, the concentration of a **single component** is measured in a solvent in a liquid cell of fixed thickness. In solid-state spectra of films, KBr discs or mulls, the thickness or concentration in the KBr or mineral oil is not known and different for different preparations. In these cases band ratios can be used since the absorbance ratio of two bands in the same spectrum should be independent of sample thickness.

Multicomponent quantitative methods are all based on the principle that the absorbance at any wavelength of a mixture is equal to the sum of the absorbance of each component in the mixture at that wavelength. Quantitation of compounds with highly overlapping spectra in a mixture is analytically difficult, especially at quite unequal analyte concentration levels. The simple approach to multicomponent analysis is based on measurements at a number of wavelengths equal to the number of components in the mixture (assuming that no interferences occur). For calibration, the absorbance of standards of known concentrations of pure components is measured to determine the extinction coefficient for each component at each wavelength selected. For many multicomponent systems, linear calibrations restrict the analyst to a narrow region of concentration of one or more of the chemical components. The various criteria and methods for the choice of the number and position of analytical wavelengths for quantitative analysis of multicomponent mixtures by least squares methods have been addressed [127]. Modern multiwavelength analysis utilises the reversed matrix representation of the Beer–Lambert law (Principal Component Analysis method, PCA). It is applicable to the simultaneous determination of a large number of components, even those with very close absorption maxima. General criteria for selecting analytical wavelengths for multicomponent mixtures by the PCA method require that, at the selected wavelength, Beer's law is obeyed and the absorbances are additive for each component. Furthermore, in an overlapping region, the selected wavelengths should be positioned at the absorption maxima of individual constituents to provide maximum sensitivity. At variance to conventional multicomponent analysis, full spectrum quantitation (FSQ) does not suffer from interactions between components which alter the absorption spectrum of an analyte.

Beer's law applies to transmission spectroscopy but has no basis for use in reflectance. Formats that may be used for quantitative measurement are Kubelka–Munk for diffuse reflectance and photoacoustic units for PAS, although the potential for

PAS quantitation is usually very limited due to severe detector saturation problems [118]. In **reflectance measurements**, $\log(1/R)$, where R is the reflectance of the sample, is proportional to concentration. The proportionality constant is not as universal as in absorption. The constant depends on factors such as particle size of the sample and moisture. The constant is thus unique for each sample and this makes quantitation using reflectance techniques very challenging. Reflectance spectra are primarily used for quantitative estimation at constant wavelength and not for taking a scan over a broader wavelength range. Reflectance measurements are commonly used in the NIR and FTIR regions.

For diffuse reflectance spectroscopy the Kubelka–Munk function, $f(R_\infty)$, is most appropriate [128, 129]. The K-M theory indicates that linear relationships of band intensity *vs.* concentration should result when intensities are plotted as the K-M function $f(R_\infty) = k/S$, where k is the absorption coefficient and S is the scattering coefficient (*cfr.* Chp. 1.2.1.3). The use of the K-M equation for quantitative analysis by diffuse reflectance spectroscopy is common for measurements in the visible, mid-IR and far-IR regions of the spectrum. Measurement of scattered light (ELSD) allows quantitative analysis.

The quantitation of compounds with highly overlapping spectra in mixture analysis is a difficult analytical problem in particular at unequal analyte concentration levels. Sampling and data acquisition in themselves are often only a half-way step to providing effective solutions to many quantitative analytical problems. Equally important, especially for more complex systems, is appropriate **data analysis**. Chemometric techniques make powerful tools for processing the vast amounts of information produced by spectroscopic techniques, the performance of which is significantly enhanced as a result. Compounds in complex mixtures may be identified by spectral subtractions, multivariate analysis over the full or limited spectral range, *etc.* In fact, the efficiency of a spectroscopic technique is currently dictated mostly by the chemometric procedure used to acquire the qualitative or quantitative information it provides. Whereas less rigorous quantitative analysis routines require that the ordinate be linear with concentration, the more advanced routines (*e.g.* PCR, PLS) can cope with data which exhibit non-linearities arising from chemical interactions in the sample.

Quantitative techniques are essentially of three kinds: single wavelength methods, multiwavelength methods and derivative spectroscopy.

Single wavelength methods. The simplest method of quantitation is to use data from a single wavelength (typically at an absorption maximum). This method of quantitation is commonly applied for absorption measurements rather than reflectance techniques, which tend to require data from more than one wavelength. Quantitation using data at a single wavelength is limited to solutions of simple compounds. The frequently applied Least Squares Regression (LSR) model (univariate method) requires isolated spectral bands that are solely related to the constituent(s) of interest and cannot be used for complex mixture analysis with overlapping spectral bands. The technique finds wide application in UV/VIS spectrophotometry.

Multiwavelength methods. Least squares curve fitting techniques may be used in the determination of multicomponent mixtures with overlapping spectral features. Two classical quantitation methods, the Classical Least Squares (CLS) mode and the Inverse Least Squares (ILS) model, are applied when wavelength selection is not a problem. CLS is based on Beer's law and uses large regions of the spectrum for calibration but cannot cope with mixtures of interacting constituents. ILS (multivariate method) can accurately build models for complex mixtures when only some of the constituent concentrations are known.

Analysis of complex mixtures with overlapping spectral bands is based on the fact that absorbances in Beer's law are additive [130]. If Beer's law holds and cell thickness and wavelength are held constant, a plot of concentration *vs.* absorbance for a single component will be a straight line. If Beer's law does not hold exactly (*e.g.* in case of hydrogen bonding), the plot will be slightly non-linear but can still be used for analyses. As eigenvector quantitation methods combine the best features of both the CLS and ILS procedures they are generally superior to the classical methods in both accuracy and robustness. These methods base the concentration predictions on changes in the data and not on absolute absorbance measurements (as used in the classical methods). The techniques are used when a single wavelength, that is specific for the analyte of interest, cannot be found. Mathematical approaches need then to be used to unravel the contribution to the spectrum of the compounds of interest and hence deduce their contribution. Different regression techniques are Multiple Linear Regression (MLR), Principal Component Regression

(PCR) and Partial Least Squares (PLS). PCR uses the model of the spectral variation to calculate the calibration equations; no wavelength selection is required. PLS is a spectral decomposition technique that is closely related to PCR. Whereas MLR requires careful choices of wavelengths to find a robust quantitation model, PCR and PLS overcome the wavelength selection problem by using the full spectrum. This is one of the best features of factor analysis-based models and allows to build models with little or no *a priori* knowledge of the spectra of the constituents of interest. The PLS and PCR factor models are able to figure out those spectral regions that are most important for calibration based on the information in the training set. A PLS calibration model is developed by compressing the spectral data for the training set into a series of mathematical "spectra", known as loading spectra or factors. When the spectrum of an unknown is analysed, PLS attempts to reconstruct the spectrum from the loading spectra, and the amounts of each loading spectrum employed in reconstructing the spectrum are then used to predict the concentration of the unknown. PLS is often considered to give superior results to PCR. The method should be validated to ensure it produces meaningful data. In particular, the parameters linearity, specificity, accuracy, precision and robustness need to be assessed.

One of the most apparent drawbacks of multivariate calibration models is the comparatively large number of training set samples required. Training set samples should be as similar as possible to the unknowns. Most samples used for factor-based multivariate quantitative spectroscopic analysis are not simple mixtures, as otherwise simpler models and calibrations could be used. Chemometrics provides the spectroscopist with many different ways to solve the calibration problem for analysing of spectral data and has found widespread use in spectroscopic quantitation using both absorption and reflectance techniques. Advantages and disadvantages of both the classical and eigenvector quantitation methods for spectroscopic application have been summarised by Duckworth [131]. For more detailed information refs. [132–135] may be consulted.

Derivative spectroscopy. A common problem in spectroscopic quantitation is that a sharp spectral feature band may overlap with a broad interfering band. Low- and high-order ($n > 2$) derivative spectroscopy is a versatile tool for quantitative estimation and analysis of substances and can be used for analysis of mixtures when two components have different bandwidths [136]. In derivative spectroscopy, the derivative of spectral absorbance is obtained as a function of wavelength, mathematically:

$$\mathrm{d}^n A/\mathrm{d}\lambda^n = cl\,\mathrm{d}^n \varepsilon/\mathrm{d}\lambda^n \qquad (6.5)$$

for the nth-order derivative. Derivative spectroscopy can be used to enhance fine structure and eliminate broad peaks. The use of derivative spectroscopy for direct quantitation is limited to compounds where the spectra show major differences, for example when a curve contains shoulders and other non-resolved regions which have their origin in overlapping signals. Quantitation can be carried out using data at single wavelengths or the whole spectrum can be used in some of the chemometric techniques.

Curve analysis by multidifferentiation is primarily employed in spectroscopic applications (notably UV/VIS, IR, F, ESR, AAS, NMR) for the enhancement of spectroscopic quantitative analysis (1 to 3 orders of magnitude more sensitive), identification by fingerprints, purity tests, signal sharpening (for separation in multicomponent analysis), *etc.* However, also various non-spectroscopic applications benefit from derivative spectroscopy, such as GC, HPLC, TLC (quantitative analysis, resolution of shoulders and inflection points) and DTA (fine resolution of temperature profiles) [136].

Fast-scanning PDA detectors and powerful signal-processing techniques, such as multiwavelength analysis and derivative spectrophotometry, greatly facilitate multicomponent determinations and eliminate or reduce interferences. However, the precision is usually degraded for mixture analysis requiring multivariate techniques when compared to analyses where interferences are not observed.

Modern quantitative software packages can easily handle up to 20 components and 500 spectra can be included in one calibration set. The success of absorption spectroscopy for routine quantitative analysis owes much to the use of double-beam systems to achieve the required measurement robustness.

Duckworth [131] and others [132] have reviewed quantitative spectroscopic analysis, including evaluations of classical quantitation methods (LSR, CLS, ILS) and eigenvector quantitation methods (PCR, PLS, factor analysis). Another useful review dealing with chemometrics (to spectra) is ref. [137].

6.4.1. Quantitative Ultraviolet/Visible Spectrophotometry

Principles and Characteristics

UV spectrophotometry is particularly useful for quantitation using data at the maximum absorbance of a chromophore. The technique is a very fast and exact tool for the quantitative determination of substances in polymers, primarily of stabilisers, directly in-polymer. However, UV/VIS spectroscopy is used primarily to measure liquids or solutions. This mode is simpler and allows more accurate quantitative analysis than do reflectance measurements on solids. The smallest sample quantity analysable amounts to about 0.1 to 0.2 mg. Such small samples permit stabiliser contents down to concentrations of 0.03% to be determined with an error of ±10% within 15 min [138]. The UV approach requires standards of measurement of extinction coefficients in order to provide a quantitative determination of additives or of the extent of their degradation.

Application of UV/VIS for the purpose of quantitative analysis requires compliance with Beer's law over the concentration range of interest. Whenever the linear dynamic range of the instrument is exceeded, and the relationship between absorbance and concentration becomes non-linear, the easiest solution is to dilute the sample to an absorbance level within the linear dynamic range. With solid samples, however, this is not possible. An alternative is to select one or more wavelengths on the side of the absorbance band, where absorptivity is lower.

Method development involves selecting the wavelength(s) that yield the best results for a particular analysis. The conventional "single measurement at a single wavelength" approach to obtaining results is insufficient for assuring optimum results. Multiple measurements at **multiple wavelengths** (simultaneous detection) or (preferably) full spectra yield the best accuracy and precision of results [139]. Multiwavelength detection with PDA allows quantitative analysis with internal or external standard, linear or polynome calibration.

Ideally, the absorbance that occurs during UV/VIS measurements should be due only to the target analyte. However, in the UV/VIS part of the spectrum both Rayleigh and Tyndall scattering may interfere. With significant scattering quantitative analysis is seriously impaired. The usefulness of UV analysis for qualitative and quantitative characterisation is also restricted by the poor selectivity of the technique (many additives show rather similar absorbance bands). UV spectrophotometry is very sensitive, but cannot easily be used to identify unknown additives or to indicate the presence of more than one antioxidant. Quantification of unknown analytes with UV detectors is difficult since UV absorption often bears no relationship to the relative masses represented by individual peaks in a chromatogram.

Previously, UV/VIS spectrophotometry was used preferably for quantitative estimations of concentrations of known substances at constant wavelength, because the fundamental spectra are mostly flat and are less characteristic than IR spectra. However, higher-order derivatives now allow for an enhancement of the sensitiveness by a factor of 10–100 or more as well as a characterisation of the substances by providing fingerprints, even in complex mixtures. This is very important for ultra microanalysis.

Joint use of UV/VIS spectrophotometry and multivariate calibration for simultaneous determinations of analytes has gained widespread acceptance in recent years as an effective alternative to sequential methods. Blanco *et al.* [140] have developed a spectrophotometric method for the simultaneous quantification of organic additives using factor design and least squares regression methods (CLS and ILS). The quality of the results obtained using CLS methodology depends greatly on the wavelength range and spectral mode used for quantification. One of the principal advantages of ILS over CLS is a high tolerance to interactions between variables. **Multicomponent UV/VIS analyses** are becoming popular with modern instruments and curve-fitting techniques.

Because many compounds exhibit either very weak or no absorbance in the UV or visible regions, a number of methods using chemical derivatisation have been developed (*cfr.* ref. [141]). Such **indirect quantification methods** usually involve adding an organic reagent, which forms a complex with strong absorptivity. The technique is considerably more sensitive and faster than NMR, but has problems of unambiguous peak assignment and quantitation.

ASTM E 169-93 describes "Practices for General Techniques of Ultraviolet-Visible Quantitative Analysis".

Applications

The simple linear relationship between absorbance and concentration and the relative ease of measurement of UV/VIS light have made UV/VIS spectroscopy the basis for thousands of quantitative analytical methods. Most UV/VIS applications are single-component quantitative analyses, including quantitative assays of **additives in solutions**. For UV/VIS spectroscopic methods, the solution requires high dilution factors and volumetric dilutions have proved to be a major source of variability in the test procedures leading to the need to investigate the potential of other techniques for these chemicals.

Rao *et al.* [142] have developed a method using UV/VIS quantification of BHT, Irganox 1076, Tinuvin 327 in PP and Irganox 1010/1076 in EP copolymers. The procedure involves an efficient solvent extraction of additives from the polymer matrix followed by estimation by UV/VIS spectrophotometry. Also the direct quantitative determination of Tinuvin 783 (a 1:1 blend of Tinuvin 622 and Chimassorb 944) in a 100 μm PE **film** has been reported [143]. The RSD value for UV measurement was 1–5% as opposed to 10–15% for IR measurements. The method is suitable for QC purposes. In IR Tinuvin 622 was determined by means of the ester carbonyl stretching vibration at 1740 cm^{-1} and an overtone or combination band in PE at 2020 cm^{-1} was used as a reference (absorbance ratios A_{1740}/A_{2020} in 0.1–0.5 wt.% range served as a calibration line). Chimassorb 944 can be measured very accurately in PE film using UV spectroscopy as opposed to IR spectroscopy [143]. The basis of the measurement is the absorbance of the triazine ring at 227 nm depending on the concentration. With reference to the Lambert–Beer law, measurement of the film thickness is needed. Calibration can be carried out on the basis of the differences in absorbances measured at 227 and 290 nm.

It has also been reported that quantitative analysis of Chimassorb 944 (λ_{max} 210–250 nm) and Irganox B 220 (λ_{max} 260–290) in HDPE/(Chimassorb 944, Irganox B220, Ca stearate) is possible using UV transmission spectroscopy of 70 μm thick films (of homogenised material) [144]. For this purpose a chemometric (PLS) model was based on the first derivative spectra of 19 samples. Typical SEP values are 36 ppm for 400 to 2700 ppm Chimassorb 944 and 46 ppm for 1000 to 2000 ppm Irganox B220. Determination of Irganox B215/220/225 blends (Irganox 1010, Irgafos 168) in PE can be based on analysis of the total amount of benzene fragments (derived degradation products included) using UV/VIS transmission spectroscopy of PE films. This compensates for variations in the nominal composition of the blends. For Irganox B900/B921 blends (Irganox 1076, Irgafos 168) direct NIRS analysis on PE granulate of additive concentrations between 300 and 3000 ppm leads to poor precision (*ca.* 100 ppm); a UV/VIS method on PE film shows much better precision. In the latter case the sample inhomogeneity is bigger than the analytical error [145].

Direct analysis of **phenolic stabilisers** (0.03–0.3%) in very small quantities of solid polymers (<1 mg) and stabiliser distribution analysis by means of UV spectrophotometry have been demonstrated for the heat stabiliser stearyl 3-(3,5-di-*tert*-butyl-4-hydroxyphenyl) propionate in polyolefins [138]. The energy of the UV spectrophotometer used was sufficient to allow quantitative analysis for a 200 μm thick foil using a pinhole of 0.085 cm diameter. Determination required 15 min with an error of $\pm 10\%$.

The application of derivative spectroscopy to the determination of polymer additives has also been reported, *cfr.* also Table 6.37. A typical case is that of the phenolic antioxidants 2,6-di-*tert*-butyl-4-methylphenol (AO-4K) and 4-substituted 2,6-xylenol (Chemantox AO-49), which exhibit virtually identical UV spectra [130]. However, the antioxidants can be distinguished in alkaline medium due to a bathochromic phenol-phenolate shift. The use of derivative spectroscopy reduces light scattering and matrix interferences when extracts from PP samples are measured. The use of derivative spectroscopy eliminates those interference phenomena which cause inaccuracies when evaluating direct absorption spectra. Shlyapnikov *et al.* [147] have used derivative ($n = 2$) UV spectrophotometry to determine antioxidants (in 0.2–2.0% concentrations) extracted from 0.02–0.1 g PE samples by distillation *in vacuo* at different temperatures with an accuracy of 1–2%. Pump *et al.* [146] used UV derivative spectroscopy for the quantitative determination of phenolic AOs in LDPE and Talsky *et al.* [148] determined the polymer/bound azo-content in PC (Fig. 6.4) by means of derivative UV/VIS.

The quantitative determination of known additives by spectroscopy, both by direct examination of polymer films and in the solvent extract, was extensively reviewed [153]. Chapter 7.2.2 describes in-process analysis by means of UV/VIS spectrophotometry.

Table 6.37. Use of derivative spectroscopy

Technique	$d^{n\,a}$	Application	Reference
UV spectroscopy	2	Pigments and phenolic AOs in PE	[146]
Idem	2	Phenolic antioxidants in PP	[30]
Idem	2	Antioxidants in PE	[147]
Idem	4	AZO-PC (*cfr.* Fig. 6.4)	[148]
FTIR spectroscopy	1	Antioxidants/antiozonants in rubber	[149]
Reflectance spectroscopy	–	Some acid dyes on wool and nylon	[150]
NIR spectroscopy	2	Additives in PP	[151]
Thermal analysis	1	Estimation of inflection points	[152]

[a] n-th derivative.

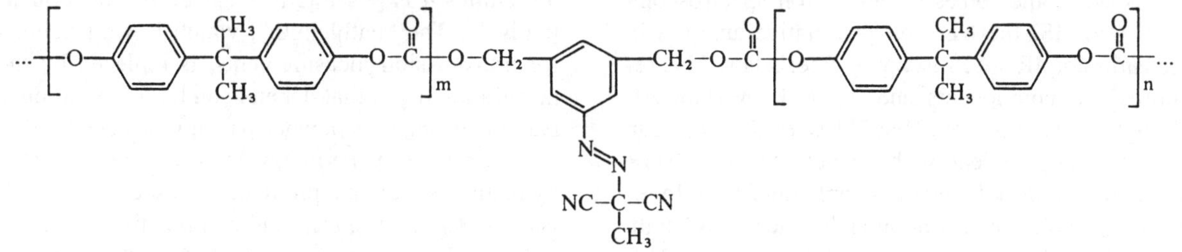

Fig. 6.4. Chemical formula of polymer-bound azo-polycarbonate. After Talsky *et al.* [148]. Reprinted from G. Talsky *et al.*, *Makromol. Chem.* **180**, 513–516 (1979). Copyright 1979 © Wiley-VCH. Reproduced with permission.

6.4.2. Quantitative Fluorescence Spectroscopy

Principles and Characteristics

Fluorescence spectroscopy can also be used for quantitation as it provides greater selectivity and sensitivity than UV spectrophotometry. Fluorescence quantitation can be described by eq. (6.6):

$$F = I_0\Phi_f(1 - e^{\varepsilon cl}) \qquad (6.6)$$

where F is the fluorescence intensity of the sample, I_0 is the intensity of the incident light and Φ_f is the fluorescence quantum yield. The quantity $1 - e^{\varepsilon cl}$ derives from Beer's law. At low concentrations with absorbance less than *ca.* 0.05 fluorescence is linearly related to concentration:

$$F = I_0\Phi_f\varepsilon cl \qquad (6.7)$$

Fluorescence has an immediate advantage over absorption in that it is not a relative technique. Provided that the sample is optically thin, the ratio of fluorescent signal to laser intensity (in LIF) gives the *absolute* species concentration. The quantitative use of fluorescence is restricted for the following reasons: *(i)* quenching (by impurities); *(ii)* temperature-dependence; *(iii)* non-linear calibration curves; and *(iv)* fluorescent impurities in solvents. Its vulnerability to the presence of fluorescence quenchers restricts its quantitative use to well defined or carefully purified samples, conditions which often apply to the effluent of a chromatographic column.

Applications

A common problem with the use of fluorescence for quantitative analysis is that many compounds can effect quenching. Adsorptive quenching is so common that it is used in TLC to identify where a thin-layer plate may contain elution bands. In this case, the TLC plate contains a chemically bond fluorescent dye. When observed under UV irradiation, the entire plate fluoresces visibly except where the plate carries adsorbates quenching the fluorescence. Polymer/additive applications are not routine.

6.4.3. Quantitative Infrared Spectroscopy

Principles and Characteristics

Various problems must be addressed when attempting to perform quantitative measurements using IR spectroscopy [18]. Important considerations concern accuracy and precision needed, concentration range, choice of spectral region, instrumental conditions, speed, pathlength determinations, transmittance measurements of optical filters, definition of appropriate standards for calibration, spectral subtraction, discriminate analysis, cost of method development, final operational cost per analysis, *etc.* A method needs only be good enough. High analytical precision and accuracy cannot easily be obtained for heterogeneous samples. The most advanced mathematical treatment for quantitative analysis is frequently not needed. Real-time analysis may only be appropriate for in-process mid-infrared polymer/additive analysis [154].

As with other types of absorption spectroscopy (*e.g.* UV/VIS) the basis of quantitative analysis in transmission IR spectroscopy is Beer's law. This requires few components and no peak overlap. Although deviations from Beer's law exist, these can usually easily be dealt with. The challenge in FTIR quantitation for polymers is sample thickness. In infrared, sample concentration and optical pathlength can seldom be controlled as tightly as in UV/VIS spectrometry. This is primarily due to the absence of suitable materials (solvents and cuvets) that are transparent over a sufficiently wide frequency range. Use of peak ratios standardises the absorbance signal and eliminates the thickness variable. Alternatively, use can be made of sealed cells with constant pathlength.

One of the difficulties associated with infrared has always been that of sampling. The most reliable technique for quantitative analysis consists in **transmission measurements** of liquid samples and is superior to reflection/transmission, ATR and fibre sampling. Quantitative measurements using IR spectroscopy are quite common for liquid solutions. Where possible, for quantitative work it is often best to dissolve the sample in a suitable solvent and subsequently treat it as a liquid. IR measurements of solids are notoriously more difficult to quantify. However, an advantage of quantitative analysis of solids is the absence of solvatochrome peak shifts. Preparing solids for transmission measurements requires some labour, except for thin films. A sample in the physical form of a film can simply be examined by standard transmission techniques. Uniformity is critical in transmission measurements, both as to sample thickness and homogeneity. For many polyaromatics the requisite thickness is much less than 100 μm. Films resulting from dissolution of a material in a solvent, followed by evaporation of the solvent (cast films), have a multitude of potential error sources, varying from solvent impurities (*e.g.* BHT in THF) to solvent volatilisation (with collateral phenomena, such as loss of volatile components, occurrence of polymorphism, formation of aggregated domains, *etc.*). Concerns regarding homogeneity apply for samples examined as alkali halide discs, including thickness, particulate distribution, air voids, pressure effects, *etc.*

The quantitative measurement of powder mixtures is at least by an order of magnitude more difficult. The measurements are classically performed in the diffuse reflectance mode. Although there are difficulties it is possible to measure powders quantitatively. The "pathlength", which is well defined for transmission measurements, is replaced by the penetration depth that depends on hard to reproduce parameters such as powder packing or density. Differences in the penetration depth are compensated by mathematical data pretreatments such as normalisation, derivatives, *etc.*, and combinations of them. For quantitative measurements to be made, powders (sample and diluent) must be carefully weighed prior to mixing so that repeatability in sample concentration can be achieved. Quantitative analysis of solids by pelleting should be avoided whenever possible. Very careful sample preparation may give results with a standard deviation of approximately $\pm 10\%$.

As a result of their total thickness and/or their embossed surfaces samples may not be amenable to direct transmission or surface reflection FTIR. **Reflectance measurements** can then be used to determine concentrations of non-absorbing samples. Reflectance spectra are primarily used for quantitative estimation at constant wavelength and not for taking a scan over a broader wavelength range. Solid sampling techniques to obtain IR spectra are the most diversified. Diffuse reflectance and photoacoustics have found limited favour as quantitative procedures, but generally are too imprecise to analyse within the bounds of stringent product specifications. For the use of DRIFTS as a research tool for quantitative analysis it is quite necessary to satisfy the basic requirement of the Kubelka–Munk (K-M) theory, namely that the scattering from the samples must be constant. This can be accomplished by careful screening samples, establishing an internal reference material (IRM) for the system, and keeping a control chart. The IRM material should not change over time. When the FTIR or DRIFT accessory alignment is changed, an IRM spectrum must be collected to determine if further adjustments are to be made to continue obtaining valid results. Good quantitative measurements require a linear calibration plot and reproducible measurements. In ATR (or internal reflection spectroscopy, IRS) the main variable encountered in quantitative measurements is uniform, repeatable contact of the sample against the IRE. This includes reproducibility in IRE area coverage and quality of contact. Use of the ATR sampling device is not recommended for the bulk quantitation of additives in polymers because it examines only the polymer surface. Moreover, ATR-FTIR is often not sensitive enough to detect low levels of additive species. HATR has become an integral component of commercial spectrometers designed as dedicated at-line process FTIR analysers. Müller *et al.* [155] have described some basic considerations concerning quantitative ATR spectroscopy.

As IR spectroscopy is a **secondary method** of analysis, the development of quantitative analysis methods requires calibration with a set of standards of known composition, prepared gravimetrically or analysed by a primary chemical method, to establish a relationship between IR band intensities and the compositional variable(s) of interest. The precision of the infrared quantitation cannot be better than the (instrumental) technique employed to provide the concentrations used for the calibration standards [156]. Mid-IR may be more accurate than near-IR if the solid sample presentation is correct.

Requirements for a **single component quantitative analysis** are: *(i)* the band should not overlap with bands of other constituents; and *(ii)* the absorbance of the chosen band should not drop below 0.2 or exceed 0.7 over the entire selected concentration range. Quantitative analysis is largely facilitated by the appearance of characteristic non-overlapping bands. In this respect mid-IR is usually more favourable than either UV or NIR spectroscopy. One of the key requirements for direct IR polymer/additive analysis is to select an absorption band of the analyte which does not interfere with or directly overlaps any of the absorption peaks of the host polymer. FTIR analysis is also sensitive to changes in the polymer matrix and its use for quantitative analysis is generally restricted to applications where the matrix is constant (such as in QC-type analyses). Also morphology may be a critical parameter to successful analysis: absorbance band intensity and shape may depend on molecular conformation, configuration and orientation.

Recently, it has become common practice to use FTIR spectra for quantitative analysis of **complex mixtures**. Prior to this, dispersion IR spectra were used primarily as a qualitative tool and for relatively simple quantitative measurements. Vibrational spectroscopy is particularly suited for **multicomponent quantitative analysis**. If intermolecular interaction between compounds can be excluded over the entire interesting concentration range, the absorbance at any given wavenumber equals the sum of the absorbances of all constituents of the sample:

$$A_i = \varepsilon_{i1} c_1 l + \varepsilon_{i2} c_2 l + \varepsilon_{i3} c_3 l + \cdots + \varepsilon_{ij} c_j l \qquad (6.8)$$

where ε_{ij} is the absorption coefficient of compound j at wavenumber ν_i. In order to determine the concentrations c_j of j components, it is obviously necessary to carry out measurements at a minimum of j wavenumbers. The system under investigation should preferably be overdetermined. More standards need to be examined than there are components in the system in order to determine error information. Although in principle there is no limitation to the number of compounds, multicomponent analysis of more than three or four constituents of a sample should be avoided. For quantifying the number of components and their relative concentrations, different numerical methods can be used, ranging from the univariate peak height or peak area method in combination with spectral subtraction or multiple linear regression (MLR) on a few wavelengths to full matrix methods, such as CLS, PCR or PLS. Maris *et al.* [157] have described non-linear multicomponent analysis by IR spectrophotometry on the basis of CLS and ILS methods and have demonstrated the applicability of "curvilinear" models to spectroscopic data. Multivariate data analysis of appropriate sets of vibrational spectra clearly enables simplified quantitative procedures to be developed for complex analysis. Fully automated analysis of compounds that can be extracted from complex matrices requires a PCA algorithm for rapid discrimination [158]. Pros and cons of performing quantitative IR analyses by various mathematical treatments have been described [159–161]. Data processing techniques for quantitative analysis include absorbance subtraction, the ratio method, factor analysis, discriminant analysis, *etc.* The combination of mid-IR spectrometry with discriminant analysis makes the tool more readily available for

QC validation by non-spectroscopists. However, this requires careful tailoring of models to articulate process and chemical information as well as close screening of training sets to insure outlier elimination. This form of validation without quantitation is directly applied to on-line data [1].

Good quantitative methods by IR spectroscopy are possible [18]. In principle, it is possible to determine compounds in the concentration range from ppm up to 100%, with a standard deviation of about 1% or less. In multicomponent analysis, the lowest concentration of each investigated component must be of the order of about 1%. The amount of substance needed for IR is in the range of a few milligrams. Analyses with an accuracy of ±1% often require preparation of a number of standards, as well as certain precautions concerning the instrumental parameters and the sample itself. In general, difference spectroscopy is capable of providing an accuracy on the order of 0.1% or even better. The feasibility of a good quantitative IR method can be assessed according to Compton *et al.* [18].

Full quantitative analysis generally requires a combination of techniques involving chromatographies (GC, HPLC) and spectroscopies (UV, IR, NMR). Vibrational spectroscopy is the technique with some of the most significant advantages and some of the most significant disadvantages. **Advantages** of vibrational spectroscopy for routine quantitative analysis are low cost and operation, and direct analysis of a wide range of sample morphologies. Sample preparation is very quick and the method is "non traumatic" to the polymer and its additives. Oligomers are not a problem except as part of the overall background. There are no dilution or concentration steps where handling errors can be made. The analysis is extremely fast and does not require high technical skills. Vibrational spectroscopy is best suited for the identification and subsequent quantification of compounds in connection with quality control. By far the most important **disadvantage** is the lower sensitivity compared to many other methods. Minimum quantifiable levels, at best, are a decade higher than those quantified by HTGC or HPLC and several decades higher than those quantified by GC-MS. Another very significant limiting factor stands in relation to sample representativity and homogeneity. Unlike chromatographic analysis, where calibration only requires the gravimetric dissolution of the additive(s) in the appropriate solvent, direct IR analysis requires the additive(s) to be homogeneously dispersed throughout the polymer matrix at the correct concentration. A sample weight of a 1.0 mm thick film is very small compared to that used in various extraction techniques. Such a film thickness is optimal in achieving a balance between lowering of minimal quantifiable levels (the greater the film thickness, the lower the minimum quantifiable level) and transmission. Some additives are quantifiable down to approximately 10 ppm while others may be difficult to quantify at all [23]. A number of inherent shortcomings of IR spectroscopy (*e.g.* extensive band overlap, failure to fulfil Beer's law over wide enough concentration ranges, irreproducible baselines, elevated instrumental noise, low sensitivity), which have previously hampered quantitative analysis, have now largely been overcome by FTIR spectrophotometers. Powerful chemometric techniques for data processing provide an effective means for tackling the analysis of complex mixtures without the need for any prior separation of their components [126,162,163]. The technique can save a great deal of time and thus lowers analytical costs. It may or may not work for certain additives groups.

An early compilation of established quantitative infrared polymer/additive methods was published [164]; no update seems to be available. Various reviews on quantitative (surface) IR analysis have appeared [18,130,159,165,166,166a]. Several textbooks discuss basic considerations concerning quantitative analysis by vibrational spectroscopy [167–169]. Data processing techniques for quantitative analysis are covered by Koenig [170], in particular regarding theory and application of FTIR to the characterisation of polymers. Hummel [171] has also discussed quantitative IR spectroscopic analysis of additives.

Various ASTM standards relate to spectroscopy, such as ASTM E 168-92 (Practices for General Techniques of Infrared Quantitative Analysis) and E 1655-97 (Standard Practices for Infrared, Multivariate, Quantitative Analysis).

Applications

Despite the fact that FTIR spectroscopy has great potential in performing quantitative analysis of polymeric materials its use for polymer/additive analysis is not really most common. An example is the determination of stearic acid in PS by dissolution (CH_2Cl_2)/precipitation (ethanol) followed by IR examination (integrated absorbance of the analytical

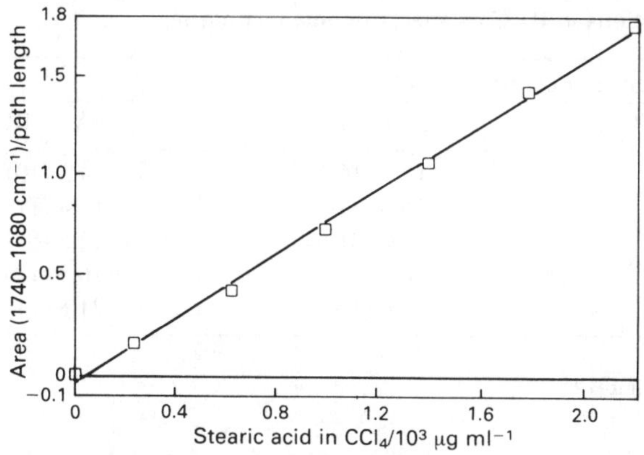

Fig. 6.5. Calibration graph for stearic acid in polystyrene resins (slope, 8×10^{-4}; intercept, 3.86×10^{-2}; R^2, 0.998). After Kumar [172]. From T. Kumar, *Analyst* **115**, 1319–1322 (1990). Reproduced by permission of The Royal Society of Chemistry.

band at 1680–1740 cm^{-1}), *cfr.* Fig. 6.5 [172]. FTIR is especially of interest to those with a very large workload or quick sample turnover, as in a **quality control** environment [23]. Quantitative methods of analysis based on IR spectroscopy have always had importance, whether for the quality assurance of end-products, determining the effects of process or fabrication variables on the polymer morphology or troubleshooting process problems.

Multivariate calibration techniques have been used for the quantitative FTIR analysis of selected additives (3400–10,000 ppm of SiO_2, erucamide and BHT) in 1 mm thick LDPE film, based on a calibration model of 60 samples [163]. The method, which is both time and cost effective, has potential for QC of polyethylene. Results (correlation coefficient R^2, standard error of estimate) were as follows: SiO_2 0.99, 30 ppm; BHT 0.84, 69 ppm; erucamide 0.91, 72 ppm. Multivariate calibration was also used in quantitative analysis of 1 mm thick HDPE/(Irganox 1010, Irgafos 168, Ca stearate) films for QC purposes [173]. Account was taken of the presence of a phosphate degradation product. Twenty different mixtures were used as calibration samples; with replicates a total of 55 samples were included in the calibration set. Concentration ranges of the additives in the calibration set were as follows: Irganox 1010 (0–700 ppm), Irgafos 168 (0–2000 ppm), Irgafos 168 phosphate (0–1000 ppm), Ca stearate (0–500 ppm). A separate test set of ten production samples (with replicates totalling 24) was used to validate the calibration. HPLC was used to determine the Irganox 1010 and Irgafos 168 concentrations. Different techniques for normalisation and calibration were compared. Another 30 samples (with replicates totalling 80 samples) of three different HDPE products were used for testing the background correction techniques. Also Leardi *et al.* [174] have applied multivariate calibration (PLS) for the prediction of additive concentrations in PE films from FTIR data (*cfr.* also Chp. 7.2.3).

Blanco *et al.* [149] have described the simultaneous determination of rubber additives (up to 5 components in solution) by FTIR spectrophotometry with multivariate calibration. The wavenumber range chosen contained appreciable absorption of all the species analysed. Use of the first derivative spectra for the additive mixtures resulted in improved quantitation; errors of prediction of 2–5% were quoted.

Although FTIR can yield helpful insights into additive degradation, interpretation may not be straightforward because of a lack of specificity, difficulty in quantitation, and possible polymer–matrix interference effects. Light stabilisers (Chimassorb 81/944; Chimassorb 81 and Tinuvin 622) were quantitatively determined in 180 μm thick **agricultural PE film** using selected absorptions (Tinuvin 622: 1740 cm^{-1}; Chimassorb 81: 1630 cm^{-1}; Chimassorb 944: 1530 cm^{-1}) [175]. The method is specific: Cyasorb UV3346 interferes with Chimassorb 944; Tinuvin 770, Irganox 1010/1076, Hostanox O3 and Plastanox STDP with Tinuvin 622, and Chimassorb 81 with other benzophenones or isocyanurate type compounds. Relative error for Chimassorb 81 amounted to 5–10%, for Chimassorb 944 30–50% in view of interference with the matrix (at 1460 cm^{-1}). Scoponi *et al.* [176] have determined the additive concentrations in LDPE/(Chimassorb 944, Tinuvin 622) films by UV at 225 nm for the absorption of the 1,3,5-triazine group of Chimassorb 944 and by FTIR at the 1734 cm^{-1} ester group absorption of Tinuvin 622. Despite reports in the literature [173,177,178] that the degree of conversion of Irgafos 168 from phosphite to phosphate can be measured by FTIR, other authors regard the method as unsuitable for such quantitative assessment. The major limitation quoted is that the phosphate P—O stretching absorption at 968 cm^{-1} is in the same region as the *trans*-vinylene group absorption in PE. The phosphate degradation product of Irgafos 168 was also observed by the m/z 662 ion by means of FTICR-MS [179] and ToF-SIMS [180] studies.

Table 6.38. Comparison of plasticiser determinations in PVC by spectroscopic techniques[a]

Sample	Plasticiser content (wt.%)			
	Theoretical	LR-NMR	ATR-FTIR	PA-FTIR
1	13.6	14.1 (2)	12.5 (10)	14.5 (10)
2	16.6	16.3 (2)	16.8 (3)	17.4 (7)
3	17.4	17.2 (2)	17.2 (3)	16.5 (6)
4	18.2	17.9 (2)	18.2 (3)	16.5 (12)
5	20.9	21.1 (2)	22.4 (10)	21.8 (7)

[a] After Herres [182]. Reproduced by permission of Carl Hanser Verlag GmbH & Co.

FTIR difference spectroscopy ($\nu = 888$, 860 and 768 cm^{-1}) was used for quantitative analysis of BHT in films of PE (500 μm) and EVA copolymer (285 μm thickness) [181].

Herres [182] has described a comparison of various spectroscopic methods (ATR-FTIR, PA-FTIR and LR-NMR) for the quantitation of **plasticiser** concentration in 0.2 mm thick PVC containing 9 wt.% TiO_2 and 15–20 wt.% plasticiser. Sample sizes were in the range of g (LR-NMR), 100 mg (ATR-FTIR) and 5 mg (PA-FTIR). Quantitative analysis in normal transmission IR is limited for highly filled materials. For those cases the surface techniques ATR-FTIR and PA-FTIR may be applied. These FTIR methods yield information about migration and accumulation of low-MW species near the surface. ATR spectroscopy is well suited for quantitative analysis provided contact between the specimen and ATR crystal is reproducible. PA-FTIR samples some 8–15 μm, *i.e.* considerably more than the ATR technique (<2 μm). The indirect detection method of PAS does not reach the sensitivity of normal FTIR measurements. Quantitative analysis of plasticisers in polymers by means of LR-NMR uses the fact that relaxation of protons in a magnetic field depends on the molecular environment. As shown in Table 6.38, LR-NMR is most precise and fast (0.5 min) in comparison to ATR-FTIR (2 min) and PA-FTIR (5 min).

Absorbance bands at approximately 1118 and 470 cm^{-1} may both be used to detect 0.01% silica in 500–1000 μm thick HDPE films [156]. Coles *et al.* [183] determined quantitatively kaolin clay in PVA-E by means of ATR-FTIR. The latter method, while providing better resolution of kaolin, does not have a large enough sampling area and is therefore subject to small shifts in concentration of filler within the sample. ATR is not as sensitive to kaolin as μFTIR, but provides a larger sampling area and more consistent results. The filler content of the polymer was confirmed by ashing. When a polymer is ashed care must be taken as the composition of the filler could change during the process. ATR-FTIR is useful also for quantitative determination of a polyacrylamide resin (PAM), which is a dry-strength additive for paper sheets [184]. The major application of DRIFTS has been the analysis of powders and the interaction of species on **fillers**. Many studies have been carried out using DRIFTS to study the interaction of silane coupling agents with fillers used to manufacture high-strength reinforced composite materials [185,186]. The K-M theory suggests that a linear relationship should exist between the concentration of the silane-coupling agent on the filler and the intensity of the reflectance spectrum for each functional group that absorbs infrared radiation.

6.4.4. Quantitative Near-infrared Spectroscopy

Principles and Characteristics
NIRS is a secondary technique requiring calibration against other techniques. The primary method therefore limits the precision and accuracy obtained using NIR. The accuracy of the NIR technique is also dependent upon the validity of the calibration data set. FT-NIR spectroscopy allows development of high-precision quantitative methods. However, the higher the precision, the more difficult will be the calibration development and instrument transfer processes. Use of NIR for quantitation is an area of great innovation.

Since the vibrational intensities of characteristic near-infrared bands are only slightly dependent on the state of the system, NIR is well suited to quantitative analysis up to the high pressures and temperatures of extruders. Moreover, in spectroscopic measurements covering an extended NIR wavenumber range, overtone and combination modes with very different molar absorption coefficients can be recorded simultaneously. This enables determination of concentrations differing by several orders of magnitude in a single experiment. Since almost all substances which are practically relevant have characteristic NIR absorption bands, quantitative analysis via NIRS is generally applicable to *on-line* concentration measurements (as in in-process conditions). Reflectance measurements (*e.g.* NIRS in the hopper) are not as accurate as transmission measurements.

The use of the Kubelka–Munk equation for quantitative analysis by diffuse reflectance spectroscopy is common for measurements in the visible, mid-IR and far-IR regions of the spectrum, but not in the near-IR region. As has been pointed out [187, 188], almost all near-IR diffuse reflectance spectra have been converted to $\log(1/R)$ (R = reflectance of the sample relative to that of a non-absorbing sample). The use of $\log(1/R)$ instead of the K-M function provides a more linear relationship between reflectance and concentration. Olinger *et al.* [189] explain this behaviour by the effective penetration depth of the beam, which is very short, when absorption is strong. For many of the algorithms developed to achieve multicomponent determinations from the diffuse reflectance spectra of powdered samples, a linear dependence of band intensity on analyte concentration is not absolutely mandatory for an analytical result to be obtained.

Jansen [190] has compared FT-NIR diffuse reflection and diffuse transmission measurements for quantitative analysis of powder mixtures. Diffuse transmission as a mode for powder measurements has been quite rarely considered in the past but appears to have some advantages, especially when it comes to low concentrations.

Applications

Hall *et al.* [151] have demonstrated the feasibility of using NIRS to measure additive levels in PP pellets obtained from two process streams. Analysis was performed directly on the polymer pellets with no sample preparation. In a rugged spectroscopic model, variations in the polymer pellet size or shape, and sample and packing density, which will affect the effective path length of the NIR radiation in the sample, should be accounted for. Even when the chemical identity of the additive is unknown, the availability of a NIR spectrum permits assignability of spectral features (*i.e.* at 2172 nm) within the matrix that can be used for quantitative purposes. Multiplicative scatter effects were compensated for by using the intensity ratio at two wavelengths, where the second-derivative intensity at 1946 nm accounts more for physical differences in polymer pellets from each extruder/pelletiser affecting the entire NIR spectrum than compositional differences. By characterising the inherent spectral variations between the two polymer pellet sub-populations, these populations could be combined indiscriminately in a single MLR spectroscopic algorithm. The scatter-corrected NIR spectroscopic model was validated by predicting the additive level for a distinct set of polymer samples obtained from both extruders.

Near-infrared spectroscopic process control is described in Chp. 7.2.4.

6.4.5. Quantitative Raman Spectroscopy

Principles and Characteristics

Raman spectroscopy is mainly utilised as a qualitative tool, but can also be employed quantitatively [191]. Quantitative analysis by Raman spectroscopy has not kept pace with the rapid growth in the use of Raman spectroscopy for structural and qualitative analysis. Quantitative FT-Raman spectroscopy can be made as routine and reliable as absorption spectroscopy [192,193].

The relation which describes the intensity of a Raman band is determined by the number of scattering molecules per unit volume N, the differential scattering cross section $d\sigma/d\Omega$, and the intensity of the incident laser beam I_0:

$$I \sim N \cdot I_0 \cdot (d\sigma/d\Omega) \qquad (6.9)$$

Hence, if the scattering cross section is independent of concentration and if the intensity of the incident laser beam I_0 remains constant, then the intensity of a band is directly proportional to the sample concentration. A less rigorous relationship is frequently used, which is analogous to the Beer–Lambert law in the case of IR intensities. The scattering intensity of the Raman band (I) can be expressed as:

$$I = k \cdot V_S \cdot c \cdot I_0 \qquad (6.10)$$

where I_0 is the intensity of the exciting radiation, V_S is the volume of the sample illuminated by the laser source, c is the concentration of the analyte, and k is a constant for each band. The concentration can be found if the absolute values of V_S and k can be defined, which is rather difficult to achieve. The single-beam Raman technique implies limitations for quantitative analysis owing to uncorrected variations in source, sample and optics.

The evaluation procedures for single as well as for multicomponent analysis are in principle identical to those used in IR spectroscopy. The intensities of Raman signals depend on a number of properties of the sample, such as refractive index and fluorescence, as well as many experimental factors. The study of Raman spectroscopy as a quantitative tool has been confined mainly to liquid mixtures in low-concentration ranges. A solid mixture is less attractive probably because the Raman intensity of a powder mixture depends not only on the concentration of each component in the mixture but also on particle size, density of packing, and homogeneity of the mixtures [194]. Quantitation of solids by Raman spectroscopy is only applicable when the experimental conditions for all materials are identical; without calibration only relative analyte contents will be derived. An internal or external standard is required. The use of an internal standard in Raman measurements provides a condition comparable to the double-beam approach for absorption measurements and can confer a comparable degree of measurement robustness. In solution measurements, the solvent is often chosen as the internal standard species. Difference in sample absorbance is the most likely cause for failure of the internal standard approach in applying quantitative Raman spectroscopy to real-life samples, particularly when using resonance enhancement. Although in principle only a single standard is necessary to convert measured intensities into concentrations, it is more advisable and usual to employ a set of standards. Robust quantitative multivariate methods need careful attention to the design of the standard set of samples, the experimental parameters, method validation and interrogation of the variances.

What makes Raman spectroscopy potentially a powerful quantitative method is that it can be used in the visible and near-ultraviolet part of the spectrum, with the optical components (such as glass and quartz) and solvents (especially water) used in visible spectrometry. Raman spectroscopy may eventually become a more useful quantitative method, depending on the ready availability of more powerful lasers. Quantitative analysis by Raman spectroscopy was reviewed [192,194a].

Applications

Applications of Raman spectroscopy for quantitative analysis have included the determination of phenols [195,196], aromatic amines [197] and azo dyes [198]. FT-Raman spectroscopy was also used for the quantitative determination of high filler ($CaCO_3$) content (up to 75 wt.%) in particulated HDPE composites as an alternative to thermogravimetric techniques [191]. In spite of the difficulties of quantitative analysis by Raman spectroscopy, copolymer composition can be carried out by the same relative band ratio method used in IR spectroscopy.

6.4.6. Quantitative Nuclear Magnetic Resonance Methods

Principles and Characteristics

Quantitative analysis is inherent in the NMR experiment. During the excitation pulse, nuclei absorb the rf energy, exciting the nuclear spins to higher energy states. Unlike any other spectroscopy, this energy absorption remains linearly quantitative across the whole NMR spectral range – absorption of rf energy depends only on the number of nuclei present, and is not enhanced or dampened by specific molecular environments or chemical functionality. The NMR spectrum intensity contains the quantitative information from the total number of nuclei, while the frequency domain contains the chemical/molecular structural information. Quantitative data from the NMR experiment can be obtained by monitoring signal intensities from experiments using both single-pulse and complex-pulse sequences. The integrated area under each peak in a given NMR spectrum is proportional to the number of emitting nuclei in the molecular structure so that the NMR signal (at least for protons) is directly proportional to concentration; Beer's law does not apply. This is the usual method for obtaining quantitative information from high-field NMR spectra. At low magnetic fields, due to the decrease in spectral resolution, the problem of overlapping resonances is accentuated. This can result in spectra that are quite complicated and difficult to interpret visually, particularly in the case of ^{1}H NMR spectra. In such cases, chemometric techniques may help in data interpretation. To quantify a single component in solution, it is not necessary to know the nature of all the other compounds that are present or to identify *all* of them in the NMR spectrum. It suffices to identify a signal generated by the analyte of interest.

Quantification in l-NMR spectroscopy necessitates optimisation of experimental conditions, such as pulse width, recycle delay and decoupler gating, as well as determination of the inherent relaxation time of the nucleus, T_1. An advantage of NMR experiments is that it is not necessary to determine an extinction coefficient to provide quantitative results. If experimental conditions are correctly set then the areas of the NMR peaks are directly proportional to the number of nuclei resonating at that frequency. NMR is a **primary analytical technique**. This is a great advantage over optical spectra where integral absorption intensities are proportional not only to the concentration but also to an absorption coefficient. To determine absolute levels of additives in samples by NMR spectroscopy, either an internal or an external calibrated standard must be used. Calibration in terms of the number of protons in a given sample peak can therefore be based on an internal proton standard of known concentration. The internal standard could be an antioxidant known to be absent from the polymer sample studied or some other high boiling compound, which does not generate conflicting NMR resonances, is stable at the temperature of the NMR experiment, and for which the proton spin–lattice relaxation times are known.

Two factors must be taken into account to obtain quantitative data. It is necessary to use a sufficient delay between rf pulses to ensure that all nuclei are fully relaxed before the next rf pulse is applied (relaxation agents may be added), and the effect of NOE must be eliminated. Consequently, ^{13}C s-NMR is normally not readily quantifiable. Precision of phosphate assays by ^{31}P NMR is consistently within 0.2–0.6%, comparable to results obtained using chromatographic methods.

NMR is occasionally used to quantify the relative ratios of the individual components in mixtures. The latter is only possible when the areas of the individual signals generated by individual components can be measured separately. Analysis can be conducted without sample preparation, without destroying the sample and, unlike many methods like chromatography, it does not require a prior standardisation step. Low-resolution FTNMR also permits quantitation, after standardisation.

Applications
In order to overcome the limitations of extraction methods, Schilling *et al.* [199] used a direct, quantitative procedure that identifies type and quantity of each additive present in stabilised polyolefin samples. High-field (11.7 T), high-resolution NMR and selective signal suppression techniques can discriminate between additives with similar molecular structure and provide a quantitative measure of each compound.

The best accuracy achievable in extensive studies of ^{1}H NMR of known mixtures is ±1%. For ^{13}C analysis the accuracy is commonly poorer, about 1–5%. In case of extruded polymer samples some precaution should be taken to minimise the event of sample to sample variations in additive concentration. Solid ^{13}C NMR spectra of filled vulcanisates allow direct quantitative analysis of the polymeric components without prior sample work-up [200].

In many cases, ^{31}P NMR has proven to be the most generally useful method for the study of phosphorous-containing antioxidants. Quantitation is straightforward once the proper experimental conditions have been defined. Typically, at least 1 g of polymer/additive material may be needed to provide enough ^{31}P nuclei to be observed.

Quantitative measurements with high precision (depending on the application and ranging from about 0.1 to 5%) and low absolute error (typically 0.5%) are also possible by means of LR-NMR.

6.5. QUANTITATIVE MASS SPECTROMETRIC TECHNIQUES

Principles and Characteristics
As already indicated by Lattimer *et al.* [201], mass spectrometry is usually employed only for qualitative analysis of additives in polymers. Quantitative analysis is frequently performed by employing techniques other than mass spectrometry. Nevertheless, a great variety of ionisation modes and mass spectrometric techniques (hyphenated or not) have been tried on polymer/additive analysis. In this Chapter we will examine the quantitative performance of some of these techniques, in particular FD-MS, FAB-MS, LSIMS, DCI-MS^n, DT-MS, MALDI-ToFMS, GC-MS, LC-MS^n, TG-MS, TPPy-MS and PyGC-MS.

Quantitation by mass spectrometry is based on the fact that the degradation is ion specific, *i.e.* a given substance always produces the same percentage of fragments. The analyte content may be determined from the total mass and the integrated fragmentation pattern. Detection should be free from mass discrimination. Quantification by MS is not straightforward because mass spectrometric measurements are not exactly reproducible. The detector response depends on parameters such as the temperature and pressure of the ion source and condition of the detector. Equimolar amounts of different compounds also do not give an equal response because the **ionisation efficiency** depends partly on molecular structure. Nevertheless, quantification of MS is possible because, when a compound is ionised, the absolute abundance of any of its ions is related to the amount of that substance, albeit by a complex relation that varies during operation of the mass spectrometer and that cannot be applied to any other compound.

Basic requirements for accurate quantification are as follows: *(i)* positive identification of the target analyte; *(ii)* sufficient linear response over the required concentration range; and *(iii)* availability of the analyte as a standard. If this is not possible, determination should be carried out using a comparable standard compound on the assumption that standard and analyte have comparable response factors in the matrix. It has frequently been confirmed that the use of isotope-labelled internal standards affords greater precision than the use of other internal standards, external standards, or the method of standard additions [202]. Calibration of the mass spectrometer with known amounts of the analyte should be performed either just before the assay is carried out on the sample or in a manner that makes the measurement independent of instrumental variability. Quantification of a compound is often brought about by monitoring just one mass peak of its mass spectrum, together with a mass peak from a chemically similar reference compound (internal standard). Although it is possible to develop a reliable mass spectrometric assay without an internal standard (IS), an IS which is a homologue, analogue, or isotopic variant of the analyte is strongly recommended to enhance assay precision, accuracy and reliability.

A typical analytical procedure for quantitative mass spectrometry regardless of particular analyte and instrumental aspects consists of: *(i)* adding an IS to the sample (with homogenisation); *(ii)* extraction; *(iii)* mass spectrometric measurement (usually GC-MS or LC-MS in SIM mode); *(iv)* determination of response ratio (compound to IS); and *(v)* quantification of the analyte by comparison with a calibration graph. One ion indicative of the analyte is sufficient for quantification by SIM monitoring. The molecular ion or its adduct/reactant ion is the most desirable ion, since it is definitely characteristic of the analyte or its derivative. The sensitivity of selected ion monitoring is some 100 to 1000 times greater than full scan mode, providing detection limits of 10^{-9} to 10^{-15} g. This limit depends on the ionisation efficiency, the fraction of the total ion current carried by the ion monitored, the contribution of background ions to the signal, mass spectrometer resolution, and the detector sensitivity. The SIM mode is able to fulfil all demands for reliable quantification. Reproducibilities of better than 2–5% can be achieved if a stable internal control standard is used.

It is assumed that calibration curves are determined from the same type of matrix in which the analyte is to be determined. Although calibration of the mass spectrometer for each assay is accepted as necessary, calibrating the analytical procedure for each assay is more controversial. A method to assess the significance of interferences is the calibration curve. If the calibration curve is linear through the lowest calibration point, then the interference is not considered to be significant. External standard methods tend to be used with non-chromatographic methods of introducing the sample into the mass spectrometer.

A modern strategy for quantitative analysis without chromatography utilises the specificity of MS/MS. Here, quantification is based on monitoring of the fragmentation of ions with a second analyser (selected reaction monitoring by MS/MS). Only few authors justify their use of MS/MS quantification by providing quantitative data.

For ionisation techniques that use a **direct insertion probe**, accurate and precise quantification is difficult to achieve, at variance to the view expressed by Millard [203]. The direct probe inlet has little potential for fractionating samples unless the components differ widely in volatility. During quantitative analysis, the components of interest in mixtures remain largely unseparated from each other and from impurities, causing many background ions. With instruments that allow the direct probe to be heated independently of the ion source, temperature programming achieves some degree of fractionation.

Fast atom bombardment (FAB) mass spectrometry has been widely used for the analysis of high-MW and/or thermally labile compounds [204]. As with all ionisation techniques that use a direct insertion probe, accurate and precise quantification is difficult to achieve here. With FAB and FIB/LSIMS the sample signal often dies away when the matrix, rather than the sample, is consumed; therefore, one cannot be sure that the ion signal obtained represents the entire sample. Quantification in FD or FAB-MS has been described as technically difficult, but possible when internal standards are used [205]. Lattimer *et al.* [206] compared FAB and FD as ionisation techniques for mass spectrometric analysis of mixtures of additives (plasticisers, antioxidants, antiozonants, oils and waxes) extracted from rubber compounds. Neither method was considered useful in quantitative analysis, in apparent contrast with the findings for FD-high energy MS [207]. Overall, FD-high energy MS is a superior ionisation technique for quantitative analysis as there are no matrix effects. According to Jackson *et al.* [207], LSIMS is not ideal for the quantitative detection of polymer additives, as matrix effects are very important.

Quantitative determination with high precision of flame retardant formulations by **in-source pyrolysis mass spectrometry** and an internal standard peak ratio method has been demonstrated [208]. In this procedure, a carefully measured quantity of an internal standard is introduced in each standard and polymer sample and the ratio of analyte and internal standard peak area (or heights) is taken as the analytical parameter for quantitation. Temperature-resolved in-source PyMS is quite suitable for the qualitative and quantitative determination of high weight percentage additives in polymeric materials (validation with XRF or NAA). In-source pyrolysis in a QMS (with a typical detection limit of 1 ng in full scan mode) is limited to additive concentration levels of at least 0.1%, *e.g.* flame retardants. The detection limit (1 ng analyte) may not easily be reached for low weight percentage additives such as stabilisers. By introduction of a larger sample size in TP-PyMS (100–200 μg) than in in-source pyrolysis (*ca.* 1 μg or 0.1 ng additive at a 100 ppm content) pyroprobe analysis achieves a lower detection limit. It is important that no material is lost in the vacuum of the ion source or during heating.

Desrosiers [23] considers **GC-MS** in selective ion monitoring (SIM) mode as being the best analytical system for the quantification of in-polymer additives in polyolefinic materials. The method is extremely sensitive and can be used for highly accurate quantitative work. Levels in the ppb range may be measured and samples in the low or sub-ppm range may be truly differentiated provided homogeneous samples are available. However, GC-SIM-MS is a most difficult system to calibrate and to maintain. In selected ion monitoring, total ion scans are obtained of each additive and byproduct of interest. A concentrated ion, specific to that additive or byproduct at the time of elution from the analytical column, is chosen as the quantitative ion. Usually, in addition, two other ions with lower abundances and specific of the component of interest, at the elution time, are selected as confirming ions. Their abundances compared to that of the quantifying ion should remain constant as concentrations change; if not, the result is questioned. The method provides absolute component identification and extremely accurate quantification. Pure separation of overlapping peaks is assured by carefully selecting different specific ions as the quantifying ion and the two confirming ions for each component. The interfering effects of oligomers and oxidation or degradation products are easily eliminated. In practice, the necessary stringent calibration procedures, strict discipline with regard to instrument and sample runs, and additional maintenance time are truly justified only if extremely low levels must be determined with extreme accuracy or if very small differences between samples, at either high or low concentrations, must be determined very accurately. It is therefore often preferred to carry out quantification with GC-FID in a second experiment following identification with GC-MS. Quantification in scan mode is not feasible. For the use of internal standards for quantitative analysis by GC and GC-MS, *cfr.* ref. [209]. The considerations made for GC-MS hold also true for PyGC-MS.

Quantitation by means of **pyrolysis techniques** has critically been considered by Bart *et al.* [210, 211]. As argued elsewhere (Chp. 2.2), PyGC and PyGC-MS do show quantitative capabilities more readily than PyMS. Table 2.38 indicates that a variety of additives (antioxidants, plasticisers) in different polymeric matrices (polyolefins, EPs, elastomers) have recently been determined quantitatively by means of PyGC-MS at concentration levels ranging from some 300 ppm to 15 wt.% with RSD values from 1–10%. It was concluded, however, that this method is not yet a routine sample preparation replacement for solvent extraction procedures aiming at quick quantitative determination of additives in solid polymeric matrices. Various factors contribute to this conclusion: *(i)* PyGC-MS requires control of a multitude of experimental parameters (even more than in the previously discussed GC-MS coupling); *(ii)* PyGC requires a dynamic flow of an inert gas; *(iii)* GC-MS coupling requires creation of vacuum conditions; *(iv)* quantitation in GC is usually based on FID rather than MS detection; and *(v)* matrix effects. An (unattractive) option would be to use PyGC-MS first for screening followed by PyGC-FID for quantitation. It should be considered that PyGC-FID-MS is technically difficult

Table 6.39. Quantitation of polymer/additive formulations by means of PyGC-MS

Sample[a]	Inhomogeneity	Matrix effect	Quantitation
Extract (A)	No	No	Feasible
Solid (P + A)	Possible	Likely	Restricted
Solid (P*g*A)	Possible	Strong	Doubtful

[a] P, polymer; A, low-MW additive; P*g*A, additive function grafted onto polymer chain.

and PyGC-FID/MS (in parallel) not attractive for the need of splitting of the analyte flow. Introduction of additional experimental parameters in PyGC-MS experiments is also not exactly wanted. Quantitation by means of PyGC-MS requires optimised operating conditions, usually different from those routinely adopted for screening (as in the VW/Shimadzu protocol [212]). As sampling in PyGC-MS is limited to *ca.* 1 mg, heterogeneity of technical materials may be reflected in the final results. Matrix effects are most likely to occur by PyGC-MS of solid samples but should not play a role in PyGC-MS on extracts. Table 6.39 casts some doubt on the general belief that PyGC-MS is an excellent tool for quantitative analysis of intractable polymer/additive formulations.

Keys to success of LC-MS in quantitative analysis are the typical detection limits in the pg (or even sub-pg) range. LC-API-MS appears to be more suited for quantification than LC-EIMS in view of better sensitivity and linearity [213]. Quantification experiments with LC-API-MS are usually collected in SIR mode, which allows maximum sensitivity of the MS detector to be obtained. Quantitative analysis of complex matrices by LC-MS is difficult and leads to high detection limits. As chromatographic resolution is higher for GC-MS the latter technique is less affected by complex matrices. Tandem LC-MS systems are well-suited for quantitative analysis of complex matrices. HPLC-MS/MS (QITMS) allows quantitation of standards and analysis of a multicomponent matrix spiked with such standards.

Multi-analyte quantification of polymer additives by means of **TG-MS** is not standard despite the fact that TG data ease quantification of MS results. Complete identification of evolved species by mass fragmentation patterns is not always possible when multiple components are present. In favourable cases and with careful calibration, however, semiquantitative compositional analysis can be made. TG-MS-PDA represents a much-needed breakthrough for this technique [214]. However, the quantitative implications of this TG-MS data evaluation method still need to be assessed.

Quantification in MS requires adequate reference samples. This is often a bottleneck. For example, for LMMS this means that not only the chemical composition but also the UV absorption, reflective and refractive properties of each microvolume must be comparable to ensure that the energy deposition and ion yield are similar. At the present time it appears that **MALDI** experiments are unsuitable for quantitative analyses of additive mixtures, (*cfr.* Chp. 3.4.4).

Quantitative mass spectrometry was reviewed [205,215] and a textbook is available [203].

Applications

Vit *et al.* [17] have reported a method for direct quantitative analysis of organic additives in very small PE samples using methane CIMS; the conditions needed for accurate and reproducible analytical results were given. Also a variety of additives in 1–2 mg PP samples were analysed qualitatively and quantitatively by means of CIMS [216]. Short-term reproducibility of peak areas of 6%, and sensitivity corresponding to 0.05% of Cyasorb UV531 in 0.3 mg samples were stated.

Relatively few studies have been made on the feasibility of quantitative FAB analysis. Riley *et al.* [217] have described a quantification procedure to monitor the paint additive Tinuvin 770 in two coating systems (acrylic melamine and a hydroxy ester melamine). Tinuvin 770 proved to be well suited for FAB analysis in coating extracts on glycerol basis using an internal standardisation procedure. Lay *et al.* [218] have developed a FAB-MS method for the quantitative analysis of **plasticisers** (DEHP, including any isomeric dioctyl phthalates) in baby PVC pacifiers that does not require sample extraction, clean-up, or chromatographic separation. A reference material, didecylphthalate (DDP), was added to a solution of the PVC sample in THF as an internal standard. Quantitation was based on the relative

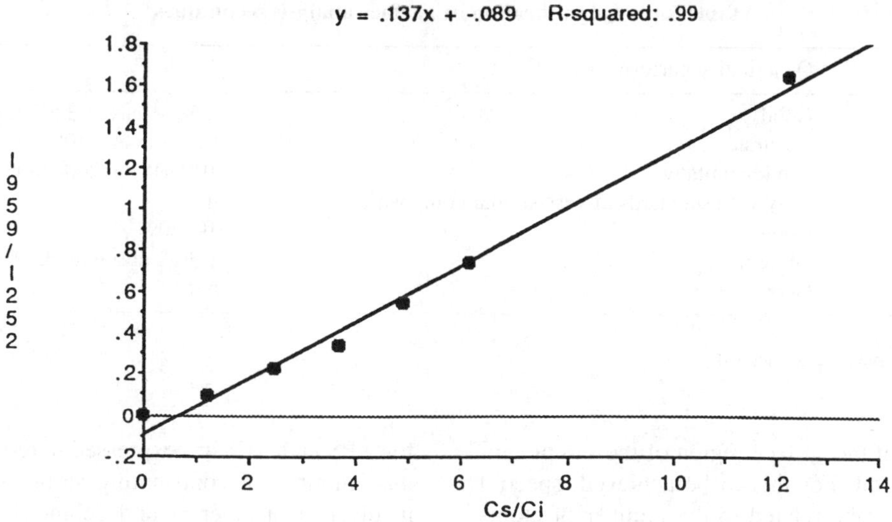

Fig. 6.6. Calibration curve for the quantitative determination of decabromodiphenyl ether ($Br_{10}DPO$). After De Koster and Boon [208]. Reproduced by permission of Consumentenbond, The Hague.

signal levels of the $[MH]^+$ ions of DEHP ($m/z = 391$) and DDP ($m/z = 447$) obtained from full-scan spectra, and use of a calibration curve. Čermák [219] has reviewed main applications of FAB-MS in qualitative and quantitative analysis of tensides, organic acids and salts, organometallic compounds, inorganic compounds, synthetic polymers and additives.

De Koster *et al.* [208] have reported direct temperature resolved (DT-MS) experiments on a double focusing (BE) mass spectrometer equipped with an in-source pyrolysis probe. For quantitative analysis of insoluble and non-homogeneous **flame retardant compositions** 1–2 g of the polymer sample were first freezer milled at LN temperature up to approximately 50–200 mesh size. An aliquot of the powder (10 mg) was then suspended in 2 mL of toluene spiked with an internal standard (perylene). A small amount (1–2 μL) of the dispersion was placed on a filament and pyrolysed at 800°C (heating rate 16.5°C/s). The commercial availability of the flame retardant to be determined, decabromodiphenyl ether ($Br_{10}DPO$), enabled development of a quantitative analysis of this analyte in the polymeric matrix by mass spectrometric techniques. The peak height intensity ratio of m/z 959 ($Br_{10}DPO$) and m/z 252 (perylene) molecular ions (I_{959}/I_{252}) was correlated with the concentration ratio of analyte to internal standard, $[c_s]/[c_i]$ (Fig. 6.6). The signal ratio I_{959}/I_{252} enabled quantitative determination of decabromodiphenyl ether in polymer matrices. Extension of this internal standard peak ratio method to the analysis of other brominated flame retardants, such as tetrabromobisphenol-A (TBBP-A), requires the availability of pure reference compounds.

Kawamura *et al.* [220] have reported the simultaneous determination method for 53 polymer additives in PE for **food packaging**. All additives were identified and quantified by GC-MS. Quantitative analysis of Irganox PS802 by LC-APCI-MS has been reported [221]. Yu *et al.* [213] described the quantification of the PP additives NC-4, Naugard-XL, 1-octadecanol and Irganox 1076 by means of LC-APCI$^+$-MS; authentic reference standards were needed. pSFC-APCI-MS can be used for quantitative analysis of a wide range of polymer additives [222].

6.6. QUANTITATIVE SURFACE ANALYSIS TECHNIQUES

Principles and Characteristics

As polymer surfaces (top 10 Å) and microscopic phases ($\leq$60 μm) influence many of today's critical technologies, their detailed quantitative characterisation is crucial. However, the spatially resolved chemical analysis of polymer surfaces and microscopic phases has historically been difficult to obtain. It is clearly the ultimate objective of surface analysis to give a quantitative description of the

Table 6.40. Quantification by surface analysis techniques

Technique	Quantitative performance	Accuracy
AES	Good	±50% (easy); ±5% (difficult)
XPS	Accurate	Better than ±10%
SSIMS	Semiquantitative	10% (IS) to 200–300% (standardless)
DSIMS	Only with standards of very similar composition	n.d.
SNMS	Good	10–20%
LEIS	Difficult[a]	±30% absolute; ±10% relative
RBS	Good	n.d.

[a] Elemental sensitivity factors needed.

composition of the surface region of the sample under investigation. For this to be achieved, spectral intensities must be related to the number of atoms in the sample emitting electrons which contribute to the spectrum. The difficulty of quantification depends on the question being asked. For example, it is relatively easy to obtain analyses using AES, which should be accurate to within ±50%, in some cases even without using standards. However, AES analyses accurate to within ±5% are extremely difficult without the use of standards with a composition very similar to that of the unknown.

Many techniques that allow the requisite spatial resolution for the characterisation of polymer interfaces provide limited quantitative chemical information. On the other hand, techniques that provide the desired level of quantitative chemical information have limited spatial resolution. The current state-of-the-art of the main surface sensitive techniques is summarised in Table 6.40. For several reasons, the quantitative capability of the techniques decreases in the order of XPS, AES and SIMS. XPS faces a background problem due to elementary excitations and hence losses of the exiting photoelectrons. However, as the physics involved is well understood, the background subtraction procedure is quantitatively established and part of commercial XPS software packages. In AES the spectral lines of the Auger electrons are in general at low kinetic energies on a background of "true" secondary electrons. The quantitative treatment of this background is more of a problem than in case of XPS.

Quantification of data from XPS or AES is complex. There appears to be no one single satisfactory method of quantification which gives reliable results in all cases. Also, the ultimate resolutions and sensitivities of the techniques are not yet totally clear. In the first-principles method the emitted intensities for XPS and AES are expressed in terms of incident flux, number of contributing atoms, cross-sections involved, instrumental and geometrical terms, and other appropriate factors depending on the fundamental physics of the process. However, these terms are not known with sufficient accuracy. First principle calculations for XPS [223] and AES [224] were reported. Instead, for both XPS and AES, it is currently routine practice in surface analytical laboratories to compare intensities in the spectrum from the unknown with reference intensities obtained by measurements of standard spectra of the elements. Observed intensities are to be normalised in some way by the use of relative empirical sensitivity factors S. It is often not feasible to compile sets of in-house sensitivity factors suitable for use with the wide range of samples and experimental conditions met in practice. There is no universal agreement on the best choice of relative sensitivity factors for a particular experimental case. Many different sets of S values are available in the literature. The XPS dataset of Wagner *et al.* [223] represents good common sense. The most popular dataset for AES is that of Davis *et al.* [225]. For accurate work it is recommended to derive experimental relative sensitivity factors under the particular instrument conditions routinely used. With these relative sensitivity factors (S_n) the relative atomic concentration of any chosen element, A, is then simply obtained from:

$$C_A = I_A S_A / \sum_n I_n/S_n \qquad (6.11)$$

where I_A is the photoelectron current for core level X of element A, S_A is the relative sensitivity factor of A (measured using level X) and C_A is usually expressed as atomic % (of all elements determined, hydrogen excluded). The significant discrepancies noted between the XPS empirical and theoretical values indicate the existence of fundamental errors in implementation of the latter approach [223]. The quantification of XPS and AES has been reviewed [226–230].

While elemental analysis of surfaces has progressed dramatically, quantitative molecular surface analysis remains difficult. This is particularly true for the analysis of complex materials such as polymers and rubbers, which contain a wide variety of additives. For mass spectrometric analysis the difficulty is twofold. First, desorption of surface molecules must be accomplished with minimal fragmentation and collateral surface damage. Second, the desorbed molecules must be ionised for subsequent mass analysis with high efficiency [231]. In recent years, the development of static SIMS techniques using time-of-flight mass spectrometers (ToF-SIMS) has provided a powerful tool for the detailed quantitative analysis of polymeric phases. A particular problem is the ion yield of the sputtered particles, which depends on the "chemical" environment of the surface (matrix effect). In general, the matrix effect prevents direct quantification of SIMS data because no direct functional correlation exists between the number of surface species and the number of detected secondary ions characteristic of those species. Quantitative interpretation of ToF-SIMS data is still at its early stages. It is difficult to derive quantitative data from first principles. It is also difficult to assess the accuracy of a SIMS analysis because there are no techniques capable of calibration analysis of very dilute analytes. Standardless SIMS analysis is subject to sizeable errors, as much as factors 2–3. The use of internal standards is crucial. By their very nature, empirical methods of quantitative analysis are dependent upon the availability of such standards. Smith *et al.* [232] have discussed several applications of the standard addition method to the quantification of SIMS depth profiles. SIMS samples are typically solids, and the standard addition is done by ion implantation. Using internal standards an accuracy of the quantification of better than 10% can be reached [233]. *Cfr.* also Chp. 4.2.1.

The principle disadvantage of ISS is that it does not directly give quantitative compositional information. Intensities can be compared with those obtained from pure element standards, but ISS would not normally be used solely as a means of determining surface compositions. In case of LEIS, quantitative analysis is of special interest because LEIS has the ultimate surface sensitivity compared to other composition analysis techniques, such as XPS, AES or SIMS. Quantitative composition determination is possible on the basis of elemental sensitivity factors provided that a calibration standard is used.

One of the most complicated problems in microprobe methods of analysis is obtaining quantitative information and checking the validity of the results. Few standard reference materials for microprobe methods of analysis are available [234].

Applications

ToF-SIMS may be used as a semi-quantitative technique if the samples are calibrated using other techniques, such as SFE-GC. This was illustrated for microtomed **automotive coatings** containing Sanduvor 3058 and Cyagard UV1164 [235]. Benninghoven [236] has addressed the quantification of additives at polymer surfaces by ToF-SIMS. In general, quantitation is rendered difficult because of matrix effects.

6.7. QUANTITATIVE MICROSCOPY

Principles and Characteristics

Imaging has become a common tool both in microscopy and spectroscopy. Imaging comprises image capture, transfer, processing, analysis and display as monochrome and colour (RGB) images (for photodocumentation or archiving digital images), which puts microscopy on a more quantitative footing. Quantitative image analysis is an analytical tool for the determination of the degree of dispersion in dispersed solid–solid systems [237]. The automatic determination of object properties is called **automatic quantitative microscopy**. There are limitations with respect to the accuracy of the object representation by a microscope image. Image analysis is now widely applied for morphological and surface analysis; picture enhancement tools have been developed. Morphometric structural analysis, in combination with advanced image analysis equipment, is of considerable importance in the quantitative statistical evaluation of light and electron micrographs, particularly for multicomponent polymers. Image analysis aims at classifying of features or objects in a 2D image and at characterising the objects by some numerical value. For an overview on image analysis, *cfr.* ref. [238]. For an in-depth presentation of the principles and applications of morphological image analysis, Soille [239] is recommended.

Applications

Imaging techniques are employed to quantify a number of important material properties, such as homogeneity, orientation, rate of growth, impurity content, void content, particle characteristics (size and shape, diameter, inter-particle distance), penetration, *etc.* Microspectroscopic imaging can be used to evaluate dimensions, measure distributions, phase area percentage, grain sizing, particle sizing and counting, coating and thickness measurement, porosity, defect counting, *etc.* Imaging provides tools to evaluate parameters that were until recently difficult to quantify. In the fields of polymers, plastics, composites and textiles, such parameters include: polymer blend morphology, texture, surface roughness, surface uniformity, fibre orientation and diameter distribution, dispersion of insoluble additives, corrosion, rate of cracking, material weathering, and many other determinants of process and/or product quality. Typical morphological image analysis problems are shape analysis and the separation of intersecting fibres [239]. Talbot *et al.* [240] have studied length and diameter estimations of **mineral fibres**. These measurements required the development of an efficient methodology for separating connected fibres in polished cross-sections or crossing fibres in SEM images. A reproducible method for the quantitative determination of the particle size distribution of additives (pigments, fillers) in plastic compounds by image analysis was described and applied to PP/$CaCO_3$ for particles with >2 μm diameter [237]. Kruse [241] has reviewed rubber microscopy (optical and electron), in particular the microstructure, qualitative and quantitative determination of carbon-blacks and light coloured fillers. Also the examination of dyed fibres by microscopy (optical, UV, fluorescence and electron) was described [242]. Spatially resolved additive analysis is a developing area, with prospects in FTIR and Raman imaging microscopy for fast images and distributions, ToF-SIMS for diffusion, migration and blooming related problems, and LMMS for impurity detection in industrial troubleshooting.

VIEEW™ (Video Image Enhanced Evaluation of Weathering) is a digital **macro scale image analysis** system allowing objective, visual evaluations for applications such as automotive clearcoat analysis, texture analysis, delamination, chalking, and defect analysis [243]. VIEEW™ allows visualisation of chromatic and geometric information, optically defines surface defects, eliminates human subjectivity, and supplies reproducible quantitative data. In VIEEW™, surface defects on all samples are detected and classified under identical test conditions. Different light geometries are used for evaluating all types of surface defects. Direct light is used to examine toplayer defects and texture of optically smooth surfaces. Diffuse light is used to examine the effects that cause changes to the surface contrast such as colour change. Image processing software allows classification of surface damages on a sample. Each sample is defined by a comprehensive statistical profile. Image analysis techniques such as VIEEW™ improve the credibility of service life prediction (SLP) methodologies. ASTM D 01.25.03 is concerned with image analysis of weathering defects.

In a different application of imaging, Figge *et al.* [244] have used direct continuous measurements in machine and cross direction of extruded films and autoradiography and liquid scintillation methods for the study of the distribution of ^{14}C-labelled additives, such as Advastab 17 MOK-^{14}C, Ionox 330-^{14}C, stearic acid [1-^{14}C] amide or *n*-butyl ester in rigid PVC, PS, HDPE and LDPE compositions.

BIBLIOGRAPHY

General Quantitative Analysis

P. De Bièvre and H. Günzler, *Measurement Uncertainty in Chemical Analysis*, Springer-Verlag, Berlin (2003).

R. de Levie, *Principles of Quantitative Chemical Analysis*, McGraw-Hill, New York, NY (1997).

E. Prichard, *Quality in the Analytical Chemistry Laboratory*, J. Wiley & Sons, Chichester (1997).

D.C. Harris, *Quantitative Chemical Analysis*, W.H. Freeman & Co., New York, NY (1995); *Lehrbuch der quantitativen Analyse*, Verlag Vieweg, Braunschweig/Wiesbaden (2002).

Sampling

P. Gy, *Sampling for Analytical Purposes*, J. Wiley & Sons, Chichester (1998).

G.E. Baiulescu, P. Dumitrescu and P.Gh. Zugrăvescu, *Sampling*, Ellis Horwood, Chichester (1991).

Chromatography

K. Grob, *Split & Splitless Injection for Quantitative Gas Chromatography*, J. Wiley & Sons, Chichester (2001).

H.P. Frey and K. Zieloff, *Qualitative und quantitative Dünnschicht-chromatographie*, VCH Verlagsgesellschaft, Weinheim (1992).

F.W. Karasek and R.E. Clement, *Basic Gas Chromatography – Mass Spectrometry Principles and Techniques*, Elsevier, New York, NY (1991).

N. Dyson, *Chromatographic Integration Methods*, The Royal Society of Chemistry, Cambridge (1990).

G. Guiochon and C.L. Guillemin, *Quantitative Gas Chromatography for Laboratory Analyses and On-line Process Control*, Elsevier, Amsterdam (1988).

J. Novák, *Quantitative Analysis by Gas Chromatography*, M. Dekker, New York, NY (1987).

E. Katz (ed.), *Quantitative Analysis Using Chromatographic Techniques*, J. Wiley & Sons, Chichester (1987).

V.G. Berezkin and A.S. Bochkov, *Quantitative Thin-Layer Chromatography*, Nauka, Moscow (1980).

Spectroscopy

B. Smith, *Quantitative Spectroscopy: Theory and Practice*, Academic Press, San Diego, CA (2004).

K. Kemsley, *Discriminant Analysis in Spectroscopy*, J. Wiley & Sons, New York, NY (1998).

H. Mark, *Principles and Practice of Spectroscopic Calibration*, Wiley-Interscience, New York, NY (1991).

Mass Spectrometry

B.J. Millard, *Quantitative Mass Spectrometry*, Heyden, London (1978).

Surface Analysis

G.C. Smith, *Surface Analysis by Electron Spectroscopy. Measurement and Interpretation*, Plenum Press, New York, NY (1994).

V.D. Scott and G. Love, *Quantitative Electron-Probe Microanalysis*, J. Wiley & Sons, New York, NY (1989).

C. Powell (ed.), *Workshop on Quantitative Surface Analysis*, NBS, Gaithersburg, MD (1986).

Chemometric Techniques

Foo-tim Chau, Yi-zeng Liang, Junbin Gao and Xue-guang Shao, *Chemometrics. From Basics to Wavelet Transform*, J. Wiley & Sons, New York, NY (2004).

T. Næs, T. Isaksson, T. Fearn and T. Davies, *Multivariate Calibration and Classification*, NIR Publications, Chichester (2002).

M. Otto, *Chemometrics, Statistics and Computer Application in Analytical Chemistry*, Wiley-VCH, Weinheim (1999).

R. Kramer, *Chemometric Techniques for Quantitative Analysis*, M. Dekker, New York, NY (1998).

D.L. Massart, B.G.M. Vandeginste, L.M.C. Buydens, S. de Jong, P.J. Lewi and J. Smeyers-Verbeke, *Handbook of Chemometrics and Qualimetrics*, Elsevier, Amsterdam (1998).

K.R. Beebe, R.J. Pell and M.B. Seasholtz, *Chemometrics. A Practical Guide*, J. Wiley & Sons, New York, NY (1998).

N.W. Barnett (ed.), *Chemometrics in Analytical Spectroscopy*, The Royal Society of Chemistry, Cambridge (1995).

M.J. Adams, *Chemometrics in Analytical Spectroscopy*, The Royal Society of Chemistry, Cambridge (1995).

R.G.Brereton, *Chemometrics Application of Mathematics and Statistics to Laboratory Systems*, Ellis Horwood, Chichester (1993).

P. Brown, *Measurement, Regression and Calibration*, Clarendon Press, Oxford (1993).

H. Martens and T. Næs, *Multivariate Calibration*, J. Wiley & Sons, New York, NY (1992).

E. Malinowski, *Factor Analysis in Chemistry*, J. Wiley & Sons, Chichester (1990).

K. Doerffel, E. Eckschlager and G. Henrion, *Chemometrische Methoden in der Analytik*, VEB Deutscher Verlag für Grundstoff-industrie, Leipzig (1990).

REFERENCES

[1] K.W. Van Every and M.J. Elder, *Proceedings SPE ANTEC '96*, Indianapolis, IN (1996), pp. 2400–5.

[2] Ph. Quevauviller, *Trends Anal. Chem.* **18** (2), 76–85 (1999).

[3] IUPAC, *Guidelines for Calibration in Analytical Chemistry. Part 1. Fundamentals and Single Component Calibration*, V.I. Document 25/91 Final draft (1996).

[3a] J.C.J. Bart, *Additives in Polymers. Industrial Analysis and Applications*, J. Wiley & Sons, Chichester (2005).

[4] A.S. Wilson, *Plasticisers – Selection, Applications and Implications, Rapra Review Report No. 88*, Rapra Technology Ltd., Shawbury (1996).

[5] C. Warner, J. Modderman, T. Fazio, M. Beroza, G. Schwartzman and K. Fominaya, *Food Additives Analytical Manual*, AOAC International, Arlington, VA (1993).

[6] B.S. Shapoval, P.A. Demchenko, Yu.V. Tanchuk, B.I. Safronov, A.S. Podrezova and G.I. Kozak, *Khim. Teknol. (Kiev)* (3), 13–4 (1973).

[7] H.C. Ashton, *Angew. Makromol. Chem.* **261–2**, 9–23 (1998).

[8] B.D. Page, *J. Assoc. Off. Anal. Chem.* **62**, 1239–46 (1979).

[9] J. Scheirs, J. Pospíšil, M.J. O'Connor and S.W. Bigger, *Adv. Chem. Ser.* **249**, 359–74 (1996).

[10] M.W. Raynor, K.D. Bartle, I.L. Davies, A. Williams, A. Clifford, J.M. Chalmers and B.W. Cook, *Anal. Chem.* **60**, 427 (1988).

[11] E.L. Charsley and D.G. Dunn, *Rubber Chem. Technol.* **55**, 382–90 (1982).

[12] S. Affolter, *Kautsch. Gummi Kunstst.* **44**, 1026 (1991).

[13] S. Affolter and S. Hofstetter, *Kautsch. Gummi Kunstst.* **48** (3), 173–9 (1995).

[14] T. Hyötyläinen and M.-L. Riekkola, *LC.GC Europe* **15** (5), 298–306 (2002).

[15] J. Heyrovsky and J. Kuta, *Grundlagen der Polarographie*, Akademie Verlag, Berlin (1965).

[16] F. Mocker, *Kautsch. Gummi* **11** (10), WT281–92 (1958).

[17] A.I. Vit, M.N. Galbraith, J.H. Hodgkin and D. Yan, *Polym. Degr. Stabil.* **42** (1), 69–73 (1993).

[18] S.V. Compton and D.A.C. Compton, in *Practical Sampling Techniques for Infrared Analysis* (P.B. Coleman, ed.), CRC Press, Boca Raton, FL (1993), pp. 217–54.

[19] D. Haaland and E. Thomas, *Anal. Chem.* **60**, 1193–202 (1988).

[20] D. Haaland and E. Thomas, *Anal. Chem.* **60**, 1202–8 (1988).

[21] W. Horwitz, L.R. Camps and K.W. Boyer, *J. Assoc. Off. Anal. Chem.* **63**, 1344–54 (1980).

[22] EURACHEM/CITAC Guide, *Quantifying Uncertainty in Analytical Measurement* (S.L.R. Ellison, M. Rösslein and A. Williams, eds.), Eurachem, Berlin (2000); *cfr.* also *http://www.measurementuncertainty.org/mu/.guide*.

[23] D. Desrosiers, *Proceedings SPE PMAD RETEC*, Atlantic City, NJ (1996), pp. 145–76.

[24] H. Nelissen, *Techn. Rept. DSM Research*, Geleen (1998).

[25] D. Coz and K. Baranwal, *Rubber World*, 30–5 (Jan. 1999).

[26] J. Rousseau, *Techn. Rept. DSM Research*, Geleen (1998).

[27] L.Y. Zhou, M. Ashraf-Khorassani and L.T. Taylor, *J. Chromatogr.* **A858**, 209–18 (1999).

[28] R.C. Nielson, *J. Liq. Chromatogr.* **14**, 503–9 (1991).

[29] F.B. Whitfield, K.J Shaw and T.H. Ly Nguyen, *J. Sci. Food Agric.* **37**, 85–96 (1986).

[30] J. Souček and E. Jelínková, *Analyst* **107**, 623–8 (1982).

[31] T.R. Crompton, *Practical Polymer Analysis*, Plenum Press, New York, NY (1993).

[32] B. Kolb and L.S. Ettre, *Static Headspace – Gas Chromatography, Theory and Practice*, Wiley-VCH, New York, NY (1997).

[33] J. Ai, in *Applications of Solid Phase Microextraction* (J. Pawliszyn, ed.), The Royal Society of Chemistry, Cambridge (1999), pp. 22–37.

[34] J. Pawliszyn, in *Applications of Solid Phase Microextraction* (J. Pawliszyn, ed.), The Royal Society of Chemistry, Cambridge (1999), pp. 3–21.

[35] H. Anderegg, *Schweiz. Lab. Z.* **53**, 55–8 (1996).

[36] T.G. Hartman, J. Lech, K. Karmas, J. Salinas, R.T. Rosen and C.-T. Ho, in *Flavour Measurement* (C.-T. Ho and C.H. Manley, eds.), M. Dekker, New York, NY (1993), pp. 37–60.

[37] T.G. Hartman, J. Lech, E. Fukuda and R.T. Rosen, *Mass Spec. Source* **XIV** (3), 18–23 (1991).

[38] T.G. Blasse, G.A. Paterson and J.H. Scrivens, *Br. Polym. J.* **21**, 37–44 (1989).

[39] J.A.J. Jansen and W.E. Haas, *Anal. Chim. Acta* **196**, 69 (1987).

[40] T. Fazio, J.W. Howard and A. Sandoval, *J. Assoc. Off. Anal. Chem.* **51**, 17–21 (1968).

[41] ASTM D 3421, *Recommended Practice for Extraction and Determination of Plasticiser Mixtures from Vinyl Chloride Plastics, Annual Book of ASTM Standards*, ASTM, Philadelphia, PA (1987).

[42] M.A. Roberson and G. Patonay, *J. Chromatogr.* **505**, 375–84 (1990).

[43] T.R. Crompton, *Analysis of Polymers. An Introduction*, Pergamon Press, Oxford (1989).

[44] J. Salafranca, J. Cacho and C. Nerín, *J. High Resolut. Chromatogr.* **22** (10), 553–8 (1999).

[45] T. Clifford, *Fundamentals of Supercritical Fluids*, Oxford University Press, Oxford (1998).

[46] B. Werthmann, V. Neyen, F. Milczewski and R. Borowski, *Kautsch. Gummi Kunstst.* **51**, 118–21 (1998).

[47] S. Affolter, A. Ritter and M. Schmid, *Ringversuche an polymeren Werkstoffen*, EMPA, St. Gallen (2000).

[48] P. Lischer, in *Schweizerisches Lebensmittelbuch: Methoden f.d. Untersuchung u. Beurteilung von Lebensmitteln u. Gebrauchsgegenständen* (Schweizer. Lebensmittelbuchkommission u.d. Eidg. Gesundheitsamt, eds.), Eidg. Drucksachen u. Materialzentrale, Bern (1964), Chp. 60b, Appendix 3.

[49] J.C.J. Bart, M. Schmid and S. Affolter, *Anal. Sci.* **17**, i729–32 (2001).

[50] H.J. Vandenburg, A.A. Clifford, K.D. Bartle, J. Carroll and I.D. Newton, *Analyst* **124**, 397–400 (1999).

[51] B. Marcato and M. Vianello, *J. Chromatogr.* **A869**, 285–300 (2000).

[51a] A. Ritter, M. Schmid and S. Affolter, in *Proceedings 6th Würzburger Tage der instrumentellen Analytik in der Polymertechnik*, Würzburg (2003).

[52] ISO 9924-1, *Determination of the composition of vulcanisates and uncured compounds by thermogravimetry. Part 1: Butadiene, ethylene-propylene copolymer and terpolymer, isobutene-isoprene, isoprene and styrene-butadiene rubbers*, International Organisation for Standardization, Geneva (1993).

[53] S. Affolter, M. Schmid and B. Wampfler, *Kautsch. Gummi Kunstst.* **52**, 519–28 (1999).

[54] E. Monroy, N. Wolff, V. Ducruet and A. Feigenbaum, *Analysis* **21** (5), 221–6 (1993).

[55] T.P. Hunt, C.J. Dowle and G. Greenway, *Analyst* **116**, 1299 (1991).

[56] J. Noordermeer, *Rubber World* **216** (4), 16–20 (1997).

[57] J. Fourreau, *Matér. Techn.* **4–5**, 80 (1992).

[58] C.F. Wilkinson and J.C. Lamb, *Regul. Toxic. Pharmacol.* **30** (2), 140–55 (1999).

[59] R.P.W. Scott, *Introduction to Analytical Gas Chromatography*, M. Dekker, New York, NY (1998).

[60] R.S. Lehrle, M. Rollinson, N. Dadvand and I.W. Parsons, *Polym. Degr. Stabil.* **66**, 221–31 (1999).

[61] M. Markelov and J.P. Guzowski, *Anal. Chim. Acta* **276**, 235 (1993).

[62] R.L. Aiken, G.T. Fritz, D.M. Marmion, K.H. Michel and T. Wolf, *J. Chromatogr. Sci.* **19**, 338 (1981).

[63] N. Dyson, *Chromatographic Integration Methods*, The Royal Society of Chemistry, Cambridge (1990).

[64] H.-J. Hübschmann, *Handbuch der GC/MS*, Wiley-VCH, Weinheim (1996).

[65] Y. Ishii, *Nippon Setchaku Kyokaishi* **16** (2), 71–8 (1980).

[66] A. Felinger, *Data Analysis and Signal Processing in Chromatography*, Elsevier, Amsterdam (1998).

[67] J. Novák, *Adv. Chromatogr.* **11**, 1–71 (1974).

[68] C.E. Reece and R.P.W. Scott, in *Quantitative Analysis Using Chromatographic Techniques* (E. Katz, ed.), J. Wiley & Sons, Chichester (1987), pp. 157–91.

[69] F.G. Kitson, B.S. Larsen and C.N. McEwen, *Gas Chromatography and Mass Spectrometry. A Practical Guide*, Academic Press, San Diego, CA (1996).

[70] M. Oehme, *Practical Introduction to GC-MS Analysis with Quadrupoles*, Hüthig Verlag, Heidelberg (1998).

[71] E.A. Bergner and W.-N.P. Lee, *J. Mass Spectrom.* **10**, 778 (1995).

[72] Z. Cai, R.V.M. Sadagopa, D.E. Giblin and M.L. Gross, *Anal. Chem.* **65**, 21 (1993).

[73] F.L. Herman, *Anal. Chem.* **65**, 1023 (1993).

[74] G. Guiochon and C.L. Guillemin, *Quantitative Gas Chromatography for Laboratory Analyses and On-line Process Control, J. Chromatogr. Libr. Ser. No.* **42**, Elsevier, Amsterdam (1988).

[75] K. Takeda and T. Nemoto, *Proceedings 10th Annual BCC Conference on Flame Retardancy*, Stamford, CT (1999).

[76] B.J. McGrattan, *Appl. Spectrosc.* **48**, 1472 (1994).

[77] L. Castle, A.J. Mercer, J.R. Startin and J. Gilbert, *Food Addit. Contam.* **5**, 9 (1988).

[78] J.R. Startin, I. Parker, M. Sharman and J. Gilbert, *J. Chromatogr.* **387**, 509 (1987).

[79] L. Castle, J. Gilbert, S.M. Jickells and J.W. Gramshaw, *J. Chromatogr.* **437**, 281–6 (1988).

[80] J.L. Heath and M. Reilly, *Poultry Sci.* **60**, 2258 (1981).

[81] Y. Kawamura, T. Maehara, H. Iijima and T. Yamada, *Shokuhin Eiseigaku Zasshi* **41** (3), 212–8 (2000).

[82] T.R. Crompton, *Manual of Plastics Analysis*, Plenum Press, New York, NY (1998).

[83] E.W. Taylor, M.G. Qian and G.D. Dollinger, *Anal. Chem.* **70**, 3339–47 (1998).

[84] R.P.W. Scott, in *Quantitative Analysis Using Chromatographic Techniques*, J. Wiley & Sons, Chichester (1987), pp. 63–98.

[85] M. Thilén and R. Shishoo, *J. Appl. Polym. Sci.* **76** (6), 938–46 (2000).

[86] S.M. Jickells, *Abstracts 4th Intl. Symposium on Hyphenated Techniques in Chromatography and Hyphenated Chromatographic Analyzers (HTC-4)*, Bruges (1996), Paper E05.

[87] H.J. Cortes, G.E. Bormett and J.D. Graham, *J. Microcol. Sep.* **4** (1), 51–7 (1992).

[88] N. Kobayashi, H. Arimoto and Y. Nishikawa, *J. Microcol. Sep.* **12** (9), 501–7 (2000).

[89] T.L. Chester and D.P. Innis, *Proceedings 5th Intl. Symposium on Supercritical Fluid Chromatography and Extraction*, Baltimore, MD (1994), pp. 7–8.

[90] T. Bücherl, A. Gruner and N. Palibroda, *Packag. Technol. Sci.* **7** (3), 139–53 (1994).

[91] L. Baner, T. Bücherl, J. Ewender and R. Franz, *J. Supercrit. Fluids* **5**, 213–9 (1992).

[92] T.W. Ryan, S.G. Yocklovich, J.C. Watkins and E.J. Levy, *J. Chromatogr.* **505**, 273–82 (1990).

[93] M. Ashraf-Khorassani and J.M. Levy, *J. High Resolut. Chromatogr.* **13**, 742–7 (1990).

[94] F. Martial, J. Huguet and C. Bunel, *Polym. Int.* **48**, 299–306 (1999).

[95] T. Tikuisis and M. Cossar, *Proceedings SPE ANTEC '93*, New Orleans, LA (1993), p. 270.

[96] A. Mohammed, M. Ajmai, S. Anwar and E. Iraqui, *J. Planar Chromatogr.* **9**, 318–60 (1996).

[97] C.F. Poole, S.K. Poole, T.A. Dean and N.M. Chirico, *J. Planar Chromatogr. – Mod. TLC* **2**, 180–9 (1989).

[98] C.F. Poole and S. Khatib, in *Quantitative Analysis Using Chromatographic Techniques* (E. Katz, ed.), J. Wiley & Sons, New York, NY (1987), pp. 193–270.

[99] H. Jork, W. Funk, W. Fischer and H. Wimmer, *Thin-Layer Chromatography*, VCH, Weinheim (1990), Vol. 1.

[100] P. Wall and I.D. Wilson, in *Encyclopedia of Analytical Science* (A. Townshend, ed.), Academic Press, London (1995), pp. 5194–203.

[101] R.M. Belchamber, J. Read and J.D.M. Roberts, *J. Chromatogr.* **395**, 47–53 (1987).

[102] J. Zhu and E.S. Yeung, *J. Chromatogr.* **463**, 139–45 (1989).

[103] R. Amos, *Talanta* **20** (12), 1231–60 (1973).

[104] M.L. Langhorst, *J. Planar Chromatogr.* **2**, 346–54 (1989).

[105] M. Prosek, M. Pukl, J. Miksa and A. Golc-Wondra, *J. Planar Chromatogr.* **6**, 62–5 (1993).

[106] M. Prosek and M. Pukl, in *Handbook of Thin-Layer Chromatography* (J. Sherma and B. Fried, eds.), M. Dekker, New York, NY (1991), Chp. 10.

[107] L. Lin and J. Zhang, *Chinese J. Chromatogr.* **15**, 310–3 (1997).

[108] S. Ebel, *J. Planar Chromatogr. – Mod. TLC* **9**, 4–15 (1996).

[109] V.N. Aginsky, in *Advances in Forensic Sciences* (B. Jacob and W. Bonte, eds.), Verlag Dr. Koester, Berlin (1995), Vol. 5, pp. 172–9.

[110] V.A. Pollak, in *Handbook of Thin-Layer Chromatography* (J. Sherma and B. Fried, eds.), M. Dekker, New York, NY (1991), pp. 249–81.

[111] G. Szepesi, in *Modern Thin-Layer Chromatography* (N. Grinberg, ed.), M. Dekker, New York, NY (1990).

[112] G. Szepesi, *Chromatogr. Sci.* **52**, 249–83 (1990).

[113] R.E. Kaiser, *J. Planar Chromatogr.* **2**, 323–6 (1989).

[114] M. Ranný, *Chem. Listy* **76** (11), 1121–46 (1982).

[115] M.P. Volynets, *Quantitative Thin-Layer Chromatography in Inorganic Analysis*, Nauka, Moscow (1993).

[116] C.F. Poole and S.K. Poole, *Chromatography Today*, Elsevier, Amsterdam (1991).

[117] V.G. Berezkin and A.S. Bochkov, *Quantitative Thin-Layer Chromatography*, Nauka, Moscow (1980).

[118] D.T. Burns and A.A. McGuigan, *Fresenius J. Anal. Chem.* **340**, 377–9 (1991).

[119] N. Marijan and M. Anzulovic, *J. Planar Chromatogr.* **10**, 463–4 (1997).

[120] C. Li, Y. Su, J. Zhai, L. Li and J. Wu, *Lanzhou Daxue Xuebao, Ziran Kexueban* **23**, 127–9 (1987).

[121] E. Dreassi, M. Bonifacio and P. Corti, *Food Addit. Contam.* **15** (4), 466–72 (1998).

[122] W. Li, T. Sun and Ch. Dong, *Hauyue Shiji* **33**, 225–6 (1992).

[123] A. Mohammed, in *Handbook of Thin-Layer Chromatography* (J. Sherma and B. Fried, eds.), M. Dekker, New York, NY (1996), Chp. 19.

[124] M.P. Volynets and B.F. Myasoedov, *Crit. Rev. Anal. Chem.* **25** (4), 247–312 (1996).

[125] Z.P. Treiber (ed.), *Quantitative Thin-Layer Chromatography and its Industrial Applications*, M. Dekker, New York, NY (1987).

[126] A.A. Ismail, F.R. van der Voort and J. Sedman, in *Instrumental Methods in Food Analysis* (J.R.J. Paré and J.M.R. Bélanger, eds.), Elsevier Science, Oxford (1997), pp. 93–139.

[127] L.L. Juhl and J.H. Kalivas, *Anal. Chim. Acta* **207**, 125 (1988).

[128] P. Kubelka and F. Munk, *Z. Tech. Phys.* **12**, 593 (1931).

[129] P. Kubelka and F. Munk, *J. Opt. Soc. Am.* **38**, 448 (1948).

[130] R.A. Crocombe, M.L. Olson and S.L. Hill, *Computerised Quantitative Infrared Analysis*, ASTM STP **934** (G.L. McClure, ed.), American Society for Testing and Materials, Philadelphia, PA (1987), pp. 95–130.

[131] J.H. Duckworth, in *Applied Spectroscopy. A Compact Reference for Practitioners* (J. Workman and A.W. Springsteen, eds.), Academic Press, San Diego, CA (1998), pp. 93–168.

[132] T. Frost, in *Encyclopedia of Spectroscopy and Spectrometry* (J.C. Lindon, G.E. Tranter and J.L. Holmes, eds.), Academic Press, San Diego, CA (2000), pp. 1931–6.

[133] P. Geladi and B.R. Kowalski, *Anal. Chim. Acta* **185**, 1–17 (1986).

[134] H. Martens and T. Næs, *Multivariate Calibrations*, J. Wiley & Sons, New York, NY (1992).

[135] J.C. Miller and J.N. Miller, *Statistics for Analytical Chemistry*, Ellis Horwood, Chichester (1993).

[136] G. Talsky, *Derivative Spectrophotometry*, VCH, Weinheim (1994).

[137] R. DiFoggio, *Appl. Spectrosc.* **54** (3), 94–113A (2000).

[138] R. Sehan and W. Kimmer, *Jena Rev.* **35** (1), 31–3 (1990).

[139] R. Łobiński and Z. Marczenko, *Crit. Rev. Anal. Chem.* **23**, 55–111 (1992).

[140] M. Blanco, J. Coello, H. Iturriaga, S. Maspoch and D. Serrano, *Fresenius J. Anal. Chem.* **363**, 364–8 (1999).

[141] T. Owen, *Fundamentals of Modern UV-Visible Spectroscopy. A Primer*, Hewlett-Packard Corporation, Publ. Nr. 12-5965-5123E (1996).

[142] P.V.C. Rao, J.V. Prasad and V.J. Koshy, *Ann. Chimica (Roma)* **85**, 171–82 (1995).

[143] G. Meszlényi, G. Körtvélyessy and M. Sipos, *ACH-Models in Chemistry* **136** (4), 407–13 (1999).

[144] G. Meyers and P. Palmen, *Techn. Rept. DSM Research*, Geleen (1996).

[145] D. Wienke, P. Knape and P. Tackx, *Techn. Rept. DSM Research*, Geleen (2000).

[146] W. Pump and D. Woltjes, *Kunststoffe* **69**, 317–20 (1979).

[147] Yu.A. Shlyapnikov, S.G. Kiryushkin and A.P. Mar'in, *Antioxidative Stabilization of Polymers*, Taylor & Francis, London (1996).

[148] G. Talsky, J. Gerum and O. Nuyken, *Makromol. Chem.* **180**, 513–6 (1979).

[149] M. Blanco, J. Coello, H. Iturriaga, S. Maspoch and E. Bertran, *Appl. Spectrosc.* **49** (6), 747–53 (1995).

[150] T.P. Bridge, R.H. Wardman and A.F. Fell, *Analyst* **110**, 1307–12 (1985).

[151] J.W. Hall, D.E. Grzybowski and S.L. Monfre, *J. Near Infrared Spectrosc.* **1**, 55–62 (1993).

[152] C. Campbell, S. Gordon and C.L. Smith, *Anal. Chem.* **31**, 1188–91 (1959).

[153] T.R. Crompton, *Chemical Analysis of Additives in Plastics*, Pergamon Press, Oxford (1977).

[154] J.C.J. Bart, *Proceedings SPE ANTEC 2000*, Orlando, FL (2000), pp. 2079–84.

[155] G. Müller, K. Abraham and M. Schaldach, *Appl. Opt.* **20**, 1182 (1981).

[156] D.R. Battiste, J.P. Butler, J.B. Cross and M.P. McDaniel, *Anal. Chem.* **53**, 2232 (1981).

[157] M.A. Maris, C.W. Brown and D.S. Lavery, *Anal. Chem.* **55**, 1694–703 (1983).

[158] J. Yang, E.J. Hasenoehrl and P.R. Griffiths, *Vib. Spectrosc.* **14**, 1 (1997).

[159] G.L. McClure, in *Laboratory Methods in Vibrational Spectroscopy* (H.A. Willis, J.H. van der Maas and R.G.J. Miller, eds.), J. Wiley & Sons, New York, NY (1987), Chp. 7.

[160] ASTM Practice E168-88, *Practices for General Techniques of Infrared Quantitative Analysis*, ASTM, Philadelphia, PA (1991).

[161] D.W. Osten and B.R. Kowalski, in *Computerised Quantitative Infrared Analysis* (G.L. McClure, ed.), ASTM, Philadelphia, PA (1987).

[162] R. Carl, *Fresenius J. Anal. Chem.* **339**, 70 (1991).

[163] B. Vigerust, K. Kolset, S. Nordenson, A. Henriksen and K. Kleveland, *Appl. Spectrosc.* **45**, 173–7 (1991).

[164] A.L. Smith and L.R. Kiley, *Appl. Spectrosc.* **18** (2), 38 (1964).

[165] T. Hirschfeld, in *Fourier Transform Infrared Spectroscopy* (J.R. Ferraro and L.J. Basile, eds.), Academic Press, New York, NY (1979), Vol. 2, Chp. 6.

[166] M. Spiekermann, in *Infrared and Raman Spectroscopy* (B. Schrader, ed.), VCH, Weinheim (1995), pp. 411–44.

[166a] H. Ishida, *Rubber Chem. Technol.* **60**, 497–554 (1987).

[167] J.M. Chalmers and G. Dent, *Industrial Analysis with Vibrational Spectroscopy*, The Royal Society of Chemistry, Cambridge (1997).

[168] B. Schrader (ed.), *Infrared and Raman Spectroscopy*, VCH, Weinheim (1995).

[169] G.L. McClure (ed.), *Computerised Quantitative Infrared Analysis*, ASTM STP **934**, American Society for Testing and Materials, Philadelphia, PA (1987).

[170] J.L. Koenig, *Fourier Transform Infrared Spectroscopy of Polymers*, Springer-Verlag, Berlin (1983).

[171] D. Hummel, *Atlas of Plastics Additives. Analysis by Spectrometric Methods*, Springer-Verlag, Berlin (2002).

[172] T. Kumar, *Analyst* **115**, 1319–22 (1990).

[173] T.V. Karstang and A. Henriksen, *Chemom. Intell. Lab. Syst.* **14** (1–3), 331–9 (1992).

[174] R. Leardi, M.B. Seasholtz and R.J. Pell, *Anal. Chim. Acta* **461** (2), 189–200 (2002).

[175] G. Meszlényi, M. Sipos, E. Juhasz and M. Lelkes, *J. Appl. Polym. Sci.: Appl. Polym. Symp.* **48**, 411 (1991).

[176] M. Scoponi, S. Cimmino and M. Kaci, *Polymer* **41**, 7969–80 (2000).

[177] M. Reeder, W. Enlow and E. Borkowski, in *Proceedings Polyolefins VI*, Houston, TX (1989), p. 181.

[178] R.C. Wieboldt, K.D. Kemfert and D.L. Dalrymple, *Appl. Spectrosc.* **44**, 1028 (1990).

[179] B. Asamoto, J.R. Young and R.J. Citerin, *Anal. Chem.* **62**, 61 (1990).

[180] M.P. Mawn, R.W. Linton, S.R. Bryan, B. Hagenhoff, U. Jurgens and A. Benninghoven, *J. Vac. Sci. Technol.* **A9**, 1307 (1991).

[181] H. Winterberg, *Plaste Kautsch.* **37**, 185 (1990).

[182] W. Herres, *Kunststoffe* **76** (7), 591–4 (1986).

[183] B.J. Coles and C.J. Hall, *Proceedings SPE ANTEC '98*, Atlanta, GA (1998), pp. 2215–8.

[184] D. Tatsumi, T. Yamauchi and K. Murakami, *Nord. Pulp Pap. Res. J.* **10**, 94 (1995).

[185] R.T. Graf, J.L. Koenig and H. Ishida, *Anal. Chem.* **56**, 773–8 (1984).

[186] L. Vågberg, P. De Potocki and P. Stenius, *Appl. Spectrosc.* **43**, 1240–8 (1989).
[187] D.L. Wetzel, *Anal. Chem.* **55**, 1165–76A (1983).
[188] L.G. Weyer, *Appl. Spectrosc. Rev.* **21**, 1–43 (1985).
[189] J.M. Olinger and P.R. Griffiths, *Anal. Chem.* **60**, 2427–35 (1988).
[190] C. Jansen, *Büchi Information Bulletin* 18 (2000), 2 pp.
[191] C. Pakjamsai and J. Suwanprateeb, *J. Appl. Polym. Sci.* **78**, 1947–54 (2000).
[192] T.J. Vickers and C.K. Mann, in *Analytical Raman Spectroscopy* (J.G. Grasselli and B.J. Bulkin, eds.), J. Wiley & Sons, New York, NY (1991), pp. 107–35.
[193] C. Petty and R. Bennett, *Spectrochim. Acta* **46A**, 331 (1990).
[194] P.D. Schweinsberg and Y.D. West, *Spectrochim. Acta, Part* **A53**, 25 (1997).
[194a] M.J. Pelletier, *Appl. Spectrosc.* **57** (1), 20A–39A (2003).
[195] N.A. Marley, C.K. Mann and T.J. Vickers, *Appl. Spectrosc.* **38**, 540–3 (1984).
[196] N.A. Marley, C.K. Mann and T.J. Vickers, *Appl. Spectrosc.* **39**, 628–33 (1985).
[197] K. Ikeda, S. Higuchi and S. Tanaka, *Bunseki Kagaku* **30**, 701–5 (1981).
[198] J.D. Womack, T.J. Vickers and C.K. Mann, *Appl. Spectrosc.* **41**, 117–9 (1987).
[199] F.C. Schilling and V.J. Kuck, *Polym. Degr. Stabil.* **31**, 141–52 (1991).
[200] R.A. Komoroski, *Rubber Chem. Technol.* **56**, 959–70 (1983).
[201] R.P. Lattimer and R.E. Harris, *Mass Spectrom. Revs.* **4**, 369–90 (1985).
[202] D.A. Schoeller, *Biomed. Mass Spectrom.* **7**, 457–63 (1980).
[203] B.J. Millard, *Quantitative Mass Spectrometry*, Heyden, London (1978), pp. 91–115.
[204] M. Barber, R.S. Bordell, G.J. Elliott, R.D. Sedgwick and A.N. Tyler, *Anal. Chem.* **54**, 645–57A (1982).
[205] S.J. Gaskell and E.M.H. Finlay, *Trends Anal. Chem.* **7**, 202–8 (1988).
[206] R.P. Lattimer, R.E. Harris, D.B. Ross and H.E. Diem, *Rubber Chem. Technol.* **57**, 1013 (1984).
[207] A.T. Jackson, K.R. Jennings and J.H. Scrivens, *Rapid Comm. Mass Spectrom.* **10**, 1449–58 (1996).
[208] C.G. de Koster and J.J. Boon, *Techn. Rept. FOM Institute of Atomic and Molecular Physics*, Amsterdam (1995).
[209] F.C. Falkner, *Biomed. Mass Spectrom.* **8**, 43–5 (1981).
[210] J.C.J. Bart and M. Kuch, *LC.GC Europe* **13** (11), 818–28 (2000).
[211] J.C.J. Bart, *J. Anal. Appl. Pyrolysis* **58–59**, 3–28 (2001).
[212] M. Geissler, *Kunststoffe* **87** (2), 194–6 (1997).
[213] K. Yu, E. Block and M. Balogh, *LC.GC Europe* **12** (9), 577–87 (1999).
[214] A. Tas, F. Wülfert, K.G.H. Raemaekers and J.C.J. Bart, unpubl. results (2000).
[215] R.A.W. Johnston and M.E. Rose, *Mass Spectrometry for Chemists and Biochemists*, Cambridge University Press, Cambridge (1996).
[216] P. Rudewicz and B. Munson, *Anal. Chem.* **58**, 358–61 (1986).
[217] T.L. Riley, T.J. Prater, J.L. Gerlock, J.E. deVries and D. Schuetzle, *Anal. Chem.* **56**, 2145 (1984).
[218] J.O. Lay and B.J. Miller, *Anal. Chem.* **59**, 1323A (1987).
[219] J. Čermák, *Chem. Listy* **83** (5), 488–505 (1989).
[220] Y. Kawamura, K. Watanabe, K. Sayama, Y. Takeda and T. Yamada, *Shokuhin Eiseigaku Zasshi* **38** (5), 307–18 (1997).
[221] N. Palibroda, O. Piringer and J. Brandsch, *Proceedings 14th Intl. Mass Spectrometry Conference*, Tampere (1997), Paper ThPo104.
[222] M.J. Carrott, D.C. Jones and G. Davidson, *Analyst (Cambridge)* **123** (9), 1827–33 (1998).
[223] C.D. Wagner, L.E. Davis, M.V. Zeller, J.A. Taylor, R.H. Raymond and L.H. Gale, *Surf. Interf. Anal.* **3**, 211 (1981).
[224] S. Mroczkowski and D. Lichtman, *Surface Sci.* **127**, 119–34 (1983).
[225] L.A. Davis, N.C. MacDonald, P.W. Palmberg, G.E. Riach and R.E. Weber (eds.), *Handbook of Auger Electron Spectroscopy*, Physical Electronics Industries Inc., Eden Prairie, MN (1976).
[226] M.P. Seah, in *Practical Surface Analysis* (D. Briggs and M.P. Seah, eds.), J. Wiley & Sons, Chichester (1990), Vol. 1, pp. 201–55.
[227] U. Gelius, B. Wannberg, P. Baltzer, H. Fellner-Feldegg, G. Carlsson, C.-G. Johansson, J. Larsson, P. Munger and G. Vegerfors, *J. Electron Spectrosc. Relat. Phenom.* **52**, 744 (1990).
[228] M.P. Seah, *J. Electron Spectrosc. Relat. Phenom.* **50**, 137 (1990).
[229] D. Briggs, *Surface Analysis of Polymers by XPS and Static SIMS*, Cambridge University Press, Cambridge (1998).
[230] G.C. Smith, *Surface Analysis by Electron Spectroscopy*, Plenum Press, New York, NY (1994).
[231] P.J. Pellin, K.R. Lykke, P. Wurz and D.H. Parker, *Inst. Phys. Conf. Ser.* **128** *(Resonance Ionization Spectroscopy)*, pp. 167–72 (1992).
[232] S.P. Smith, L. Wang, A. Saleh and J. Davies, *Abstracts SIMS XII*, Brussels (1999), p. 57.
[233] B. Hagenhoff, R. Kock, M. Deimel and A. Benninghoven, *Proceedings SIMS VIII* (A. Benninghoven, K.T.F. Janssen, J. Tümpner and H.W.

Werner, eds.), J. Wiley & Sons, Chichester (1992), pp. 831–4.

[234] A.G. Borzenko, E.V. Kletskina and F.A. Guimelfarb, *Eur. Microsc. Anal.*, 29–31 (Nov. 1999).

[235] F. Andrawes, T. Valcarcel, G. Haacke and J. Brinen, *Anal. Chem.* **70** (18), 3762–5 (1998).

[236] A. Benninghoven, unpubl. results (2000).

[237] H. Potente, J. Ansahl and A. Lappe, *Kunststoffberater* **35** (1–2), 46–9 (1990).

[238] M. Berman, L. Bischof, E. Breen and G. Peden, *Materials Forum* **18**, 1–19 (1994).

[239] P. Soille, *Morphological Image Analysis, Principles and Applications*, Springer-Verlag, Berlin (1999).

[240] H. Talbot, D. Jeulin and D. Hanton, *Microsc. Microanal. Microstruct.* **7**, 361–8 (1996).

[241] J. Kruse, *Rubber Chem. Technol.* **46**, 653–785 (1973).

[242] A. Tímár-Balázsy and D. Eastop, *Chemical Principles of Textile Conservation*, Butterworth-Heinemann, Oxford (1998), pp. 381–405.

[243] Atlas Material Testing Technology, *First VIEEW Workshop on Digital Image Acquisition, Image Storage and Image Analysis*, Frankfurt (1999).

[244] K. Figge and J. Schoene, *Dtsch. Lebensm. Rdsch.* **66** (90), 281–9 (1970).

Chapter 7

Variation is the number one enemy

Process Analytics

7.1. In-process Analysers . . . 667
7.2. Process Spectroscopy . . . 675
7.2.1. Remote Spectroscopy . . . 677
7.2.2. Process Electronic Spectroscopy . . . 679
7.2.3. Mid-infrared Process Analysis of Polymer Formulations . . . 683
7.2.4. Near-infrared Spectroscopic Process Analysis . . . 693
7.2.5. Process Raman Spectroscopy . . . 701
7.2.6. Process Nuclear Magnetic Resonance . . . 704
7.2.7. Acoustic Emission Technology . . . 716
7.2.8. Real-time Dielectric Spectroscopy . . . 719
7.3. Process Chromatography . . . 720
7.4. *In Situ* Elemental Analysis . . . 721
Bibliography . . . 722
Process Analytical Chemistry . . . 722
Process Spectroscopy . . . 722
Process Data Analysis . . . 723
References . . . 723

Traditionally the analytical chemist has provided support to an industrial process line by supplying information about the chemical composition of raw materials, intermediates and end-products. However, chemical composition information may not always fulfil the needs of the **process engineer**, who is responsible for quality management and quality assurance. The quality specifications of a product frequently use parameters other than chemical composition and the relationship between chemical composition and product quality specifications is often obscure. In a marketplace in which products are accepted on the basis of performance specifications, there is an increasing interest in on-line analytical techniques that can predict polymer product performance beyond melt index, YI, melting and crystallisation temperature.

For the polymer production industries, the competitive edge comes from the technology that excels in controlling the polymer properties in a consistent way over the entire plant and in maximising product quality and production performance while keeping safety regulations. Quality of a polymer is affected not only by the operating conditions in the reactor but also by extruding and blending operations. Polymer properties are determined both by low-order and high-order macromolecular structures. In addition, its additives set the quality of a polymeric material. Without additives there would be no polymer industry!

Although many process variables are easily measured, lack of on-line sensors for key polymer properties renders quality control of polymer plants difficult. Process control schemes based on process variables (p, T, flow-rate and feedstock compositions) alone are no longer sufficient, because these cannot reveal all material property variations. Significant efforts are being spent on improvements to process control systems, as exemplified by the numerous attempts to monitor polymer properties during processing, such as composition, density, viscosity and dispersion of a minor phase, *etc.*, all of which are somehow difficult to measure. The development of an on-line inferential system for polymer property is a very active research area of polymerisation reactor control [1]. A schematic of inferential systems is illustrated in Fig. 7.1. For highest quality

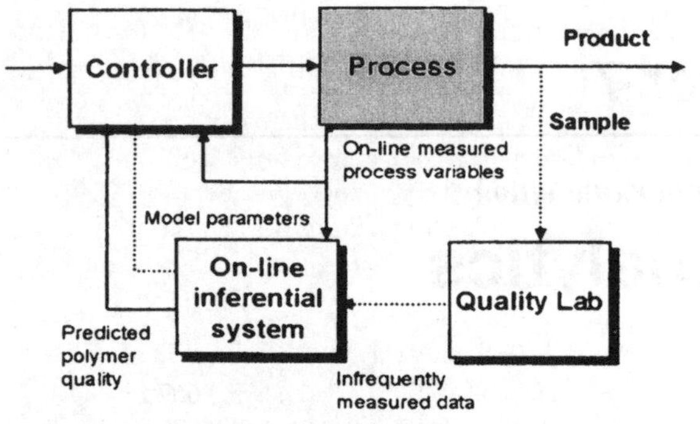

Fig. 7.1. On-line inferential system for a polymer. After Ohshima and Tanigaki [1]. Reprinted from *Journal of Process Control* **10**, M. Ohshima and M. Tanigaki, 135–148, Copyright (2000), with permission of Elsevier.

process control, sensing and optimisation should be integrated. Without quality modelling practical QC cannot be achieved. Quite clearly, process control is only part of the total quality control.

Process analysis has been identified as a **strategic research area** for industrial development. Constant pressure for increased productivity and improved product quality are forcing plastic products manufacturers to examine issues of process control (actively manipulating a process to maintain or obtain a desired situation) more closely. Efficient control schemes rely on **real-time analysis** of polymer characteristics during manufacturing, while the polymer is in the molten state, and on process streams in general. Process optimisation frequently requires feedback control based on chemical analysis. Ongoing process monitoring in production is achieved by SPC [2].

Over the last three decades, in particular gas chromatographs, electrochemical detectors and gas analysers have found their way to the process environment. Most recently, various analytical techniques that were formerly only used in the laboratory have become suitable for implementation in manufacturing. Examples are UV/VIS absorption spectroscopy, near-IR spectroscopy, refractive index measurements and more recently mid-IR spectroscopy, Raman spectroscopy, pulse NMR and mass spectrometry. In particular, the number of spectroscopic applications has increased, sometimes replacing more "established" measurement methods (like GC or gas analysers). In addition, other traditional laboratory/off-line methods are now moving towards in-process applications (*e.g.* rheometry and XRF). By an integrated approach combining process analysis, process control and process engineering more optimised plant control is achieved. Meanwhile, it should be considered that improved manufacturing processes demand rigorous quantitation in fewer cases.

The goal of every production process is an acceptable product quality in the shortest time whilst using the minimum of raw material and having the least off-spec product. Many important industrial processes are based on **feedstock material** of quite variable properties yet are expected to generate products of stable and predictable composition and characteristics. Consequently, there is scope for fast and reliable analyses, both qualitatively and quantitatively, which allow control over today's manufacturing plants. The basis of the evaluation of chemical batch processes is a comprehensive knowledge in real-time of the concentration profiles of the reactants as a function of time; in continuous processes attention is focused on end-product quality.

The **objectives of process analytical chemistry** are productivity, product quality and consistency (even tighter specifications), including prevention of off-spec material during grade changeover, process monitoring, trend or deviation spotting, process optimisation (in terms of raw materials, energy and time), documentation (for QC and management systems), legislation (environmental impact, plant effluent release, pollution prevention, crisis alerting) and safety. On-line chemical analysis integrated with sophisticated sampling and data processing capabilities is becoming critical to manufacturing and is rapidly becoming an integral part of real-time control of production processes. Complex multivariate models are necessary to relate product quality to all relevant manufacturing conditions for process control. Process chemometrics is applied to process monitoring, process control and process modelling [3,4].

Unlike laboratory samples that are stationary, process material is changing its position with solid particles or bubbles moving through the sampling point or cell. As a consequence, local optical characteristics at the analyser usually change much faster than does the average composition. The moving inhomogeneities of the sample and the finite speed of the analyser interact in a complex way. In-process analysis of process gases and fluids (feeds, reaction mixtures, product or melt streams, dispersions, emulsions, *etc.*) and solids (powders, chips, films, fibres, sheets, *etc.*) usually serves a restricted scope (very specific parameters are to be measured). For efficient control of a process it is advantageous to choose one or more critical chemical parameters, the fluctuations of which have the greatest effect on the course of the process. Obvious **requirements** for in-process analytical tools are "fit-for-purpose", rugged, robust (invariance with respect to process variations), reliable, reproducible, accurate, precise, cost- and time-effective (speed of analyses, automated, real-time feedback, low down times, little maintenance), appropriate (low sensitivity to environmental conditions: dust, temperature, season) and skill base (simple to execute, ease of use). A process analyser *must* produce the required analytical results! The acceptance and implementation of any process analysis method requires that the method be pre-programmed and made user friendly (*e.g.* menu-driven). Validation of (novel) measurement techniques for process monitoring is another critical issue [5]. Important considerations are sampling technique, minimal sample preparation (preferably none), (production) standards, calibration (instrumentation; off-line, on-line, *etc.*), method validation (may be more precise than reference, but not more accurate), location (remoteness from laboratory), long meantime between failure (MTBF) record, maintenance (serviceability) and safety. Introduction of a sample from the process stream to the analyser is critical in any on-line analytical technique. Samples can range from low- to high-viscosity liquids as well as solid materials, each of which will require different sampling systems. A low-viscosity liquid is least demanding whereas high-viscosity samples such as polymer melts are amongst the most difficult to handle. Also solid materials, ranging from fine-grained fluidised powders to large irregular pieces, are difficult to convey and analyse on-line. In process analytics the trend is towards the simplest and quickest analytical methods, such as spectroscopic techniques, physical sensors, or the use of flow-injection analysis (FIA) for liquid phases. Tables 7.1 and 7.2 compare the main qualifiers of process analytical chemistry for conventional off-line process control and *in situ* analysis, respectively.

As to **process measurements**, the intertwined questions are: *(i)* what to measure; *(ii)* how to measure (sampling problems); *(iii)* where to measure; and *(iv)* how often to measure. The quality of the analytical results, expressed in terms of speed, precision and sampling rate, defines the effectiveness

Table 7.1. Main features of off-line process control

Advantages:
- Quality assurance (specialist)
- Relatively simple equipment
- No additional investment costs

Disadvantages:
- High analysis costs
- Few data per day
- Slow process feedback
- Need for operator and transport

Table 7.2. Main characteristics of *in situ* analysis

Advantages:
- Direct analysis under actual reaction conditions
- Monitoring of process dynamics
- More analytical information
- No sample preparation (no chemical waste, no health risks)
- Elimination of maintenance overhead associated with sampling systems

Disadvantages:
- Widely different reaction environments (no unique instrumental design)
- Effect of chemistry on sensor performance/lifetime
- Cost of initial technology implementation
- Data overkill
- Culture (plant operators *vs.* lab analysts)

of process control. Various categories of process analytical tools can be distinguished: off-line, at-line, on-line, in-line and non-invasive. Each approach has its pros and cons. In the traditional **off-line** and **at-line control** situations, a sample is taken from the process flow or reaction medium and transported to the analyser, positioned in a quality laboratory or in the plant, respectively. Even in situations where the laboratory technique can report high analytical accuracies, there are several fundamental problems to this approach (*cfr.* Table 7.3). First, the entire analysis hinges on how representative the sample is which is removed from the process stream. Second, by the time the sample is analysed (typically several hours), it is too late to implement a control measurement and meanwhile a million pounds of out-of-spec product may have been generated [6]. Eventually, also the reaction equilibrium may be disturbed. These problems can be overcome by on-line analytical systems. There is a clear trend away from laboratory

Table 7.3. Main features of at-line process analysis

Advantages:
- Instruments close to process
- Relatively short response times (1 h)
- More data per day (2–6)
- Handling by plant operator
- Dedicated to particular measurements
- Limited investment

Disadvantages:
- Requires rigorous method development and testing
- Comprehensive training and troubleshooting support required
- Involves manual sampling/preparation
- May not be fast enough

Table 7.4. Main features of on-line process analysis

Advantages:
- Avoids manual sampling
- Short response times (sec to min)
- Continuous or semi-continuous monitoring ("reduced variability")
- Data used directly in control regimes
- Impurity monitoring
- Low analyst costs
- Objective

Disadvantages:
- Cost of sampling systems (true on-line)
- Dedicated equipment
- High investment costs (analyser), cost of ownership
- Operational safety requirements
- Robust calibration and validation checks
- Reliability

testing (retrospective) to qualitative and quantitative process-based testing (diagnostic).

In **on-line control**, an automated sampling system attached to a reactor or by-pass system is used to extract the sample, if needed conditioned, and presented to an analytical instrument or probe for measurement. Sampling delays can be significant in on-line installations, because of the transit line and gear pumps. On the other hand, on-line devices are isolated from the main stream by the use of a gear pump, and the temperature and pressure of the polymer sampling flow can thus be controlled. Their maintenance can therefore be done without a complete process shutdown. Many of the apparent disagreements between results from split side stream on-line analysers and results from the laboratory can be traced back to differences which occur because samples differ in acquisition time, location and stability. The main characteristics of on-line process analysis are summarised in Table 7.4.

In **in-line analysis**, the chemical analysis is done *in situ*, directly in the main process stream or reactor, using a chemically sensitive probe. A condition is that the equipment has to be placed in the plant (with consequences for maintenance and safety aspects). In this case, there still is physical contact between probe and sample. Consequently, an in-line process-monitoring device must often deal with hostile industrial processing conditions: elevated p, T, fluctuating conditions, chemically aggressive environments, electrical noise, dust, and vibrational problems. Sampling delays are very short, or non-existent for in-line devices. The feedback and control loop can be optimised in real-time manner. However, an in-line apparatus may interfere with the main process, and is also very dependent on parameters such as the temperature, and pressure of the melt, *etc.* In-line analysis offers speed often at the expense of precision, specificity and selectivity. Finally, in the **non-invasive analysis** situation, physical contact is no longer necessary.

Although process analytical chemistry has long been an important endeavour in industry, it is now receiving increased attention because of the opportunities presented by technological and methodological advances, as well as changing needs within the chemical and allied products industries. For in-process analysis specific new analytical methodologies and instruments have been developed (or are under development) for use as an integral part of manufacturing processes. Some core areas are sampling techniques, sensor and fibre optics technology, chromatography and spectroscopy (often in hyphenation), (pulse) NMR, imaging and chemometrics. The use of chemical sensors in process analytics is rapidly increasing [7]. The use of most chemical sensors is restricted to samples with a simple chemical matrix. Recently, several advances have been made in the development of on-line sensors at extruder units, *e.g.* for viscosity and yield stress measurements [8]. Fibre-optic linked NIR and Raman sensors allow monitoring of polymer properties [9, 10]. Watari *et al.* [11] utilised NIR to measure the density of PE on-line.

Significant accomplishments have been achieved in real-time measurement and data handling techniques for process monitoring and control, includ-

Table 7.5. Benefits of monitoring with process analysers

• Increased process plant operability	• Reliability
• Process and product quality improvements	• Trend and deviation spotting
• Increased manufacturing efficiency	• No sample preparation
• Reduced material wastage	• Minimised contact with hazardous materials
• Instant availability of analytical data (timeliness)	• Plant safety
• Cost benefits over conventional techniques	• Compliance with environmental regulations

ing process monitoring software – PCA, PLS, multivariate statistical process control – for evaluation of the independent variables and graphical display software. Ideally, each chemical component in a process is measured with exactly one specific (selective) sensor. Where this is not available or possible (in multicomponent process analysis) multivariate calibration may be used as a remedy. **Chemometrics** or the use of multivariate data analysis and mathematical tools to extract information from chemical data finds application in areas of pattern recognition, classification, signal resolution, instrument calibration and process analysis and control. Chemometric tools for analysis of real-time process data (*e.g.* CharmWorks™) facilitate the quantitative prediction of product quality and chemical composition as well as the identification and classification of materials. Better understanding allows better control.

Modern use of on-line and off-line monitoring with process analysers gives great **benefits** (Table 7.5). Optimal process control and real-time analysis guarantee consistent product quality. Plant capacity is maximised with less raw material usage, no off-spec products, reduced maintenance cost, energy usage and operator time. High speed and measurement precision allow rapid automatic readjustment of process parameters, thereby maintaining product quality and eliminating rejects. Modern solid-state technology with no moving optical parts ensures high reliability (automatic compensation for variations in the light source and probe contamination). Hundreds of measurements per second often allow analysis of fast moving materials and flows (*e.g.* naphtha feedstocks are now being analysed simultaneously on typically 16 properties in 1 min determining hugh savings). Measurements can be made in process and even through packaging materials and fibre optic coupled probes allow measurements in remote or hazardous environments. Plant safety is increased by reduced risks in case of fast and/or critical reactions. Optimisation of a manufacturing process has gained such a tremendous impact on the profitability of the corresponding product that it can determine the economic viability of the production plant. Modern process control systems collect and store large amounts of data. However, ultimately data is not knowledge! Data mining or multivariate feature extraction techniques may allow further optimisation of process conditions.

In the area of process analysis and control technology the balance between external publication and internal practice is obscure. Process analytical chemistry was recently reviewed [12], as well as quality control of polymer production processes [1]. Classic textbooks on process analysis are refs. [13,14].

7.1. IN-PROCESS ANALYSERS

Principles and Characteristics

The term "in-process analysers" refers to the whole range of analysers used in various processes, *i.e.* at-line, on-line, in-line or non-invasive. Development of process instrumentation is often accomplished through the transfer of well-established laboratory techniques towards the processing line, with modifications in order to ensure their robustness against the severe in-plant conditions. In addition, process control instruments are required to be routinely operated by personnel with levels of technical expertise that are different from those found in a typical laboratory. The specification for the **ideal process analyser** includes the following: safe, non-invasive (ease of sampling), non-destructive, real-time, complete sample characterisation (structure/morphology), analyses multiphase samples (liquid/solid, liquid/liquid). Other important criteria are accuracy, reliability, straightforward maintenance, wide acceptability, simplicity, linearity of response, versatility, and wide dynamic range. Most process analysers only partially fulfil these specifications.

Most mature chemical process analysis technologies, such as gas chromatography (GC), gas analysis (GA, *e.g.* FID measurements), electrochemical

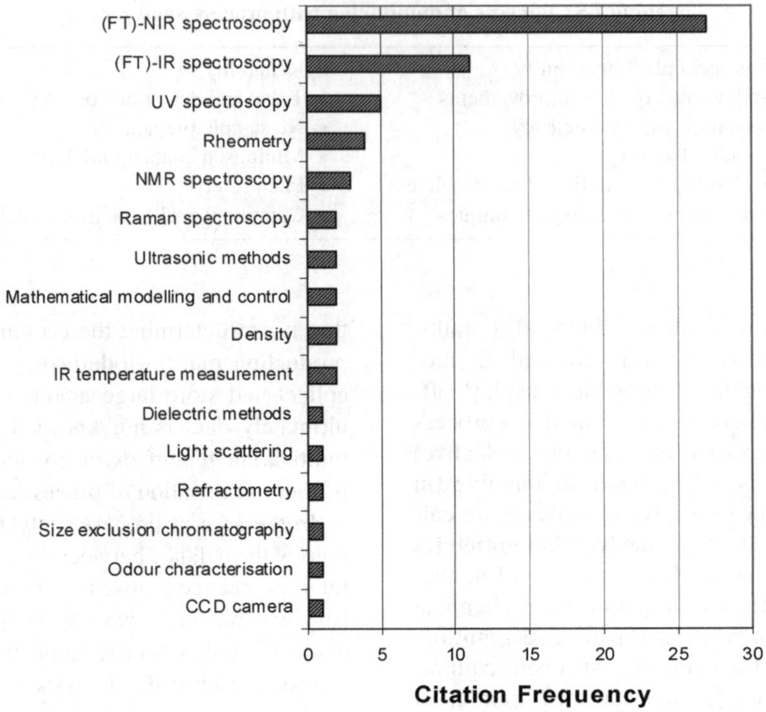

Fig. 7.2. Selected references to in-process analysis tools in polymer production. Source: Scientific and patent literature (1991–1999).

analysis (EC for pH measurements), which still account for the biggest share of installed analysers, are inherently off-line methods. In process GC only FID and TCD are in broad use. The variety of detector types available for **process chromatography** is limited because of the requirements for robustness and sensitivity. Recently, a new total concept for process GC (PGC), μGC-μTCD, has been presented [15]. Other PGC developments are multidetector technology, fast GC and hyphenation (*e.g.* GC-ICP-MS or HS-GC-ICP-MS). Other discrete process measurement systems are titrations, flow-injection analysis (FIA), *etc.* Titration is a poor analytical method in critical applications, requires carefully prepared reagents and maintenance support, and produces waste streams. While FIA is much faster than titration, the cost of a process FIA is higher than that of a process titrator while many of the disadvantages of titration remain. Tables 7.6, 7.7 and Fig. 7.2 show the growing importance of other process analytical instrumentation. GC, GA, EC, NIR, mid-IR, melt-index, determinations of humidity and oxygen are well-established in-process analytical tools, whilst UV/VIS, MS and Raman are used to a lesser extent. Spectrophotometric, mass spectrometric and electrochemical methods easily adapt for *in situ*, real-time monitoring.

Mass spectrometers have now been in use for approximately 25 years in process control applications. Nevertheless, the number of applications is still restricted because of competitive techniques, such as process spectrometry (UV/VIS, NIRS, mid-IR, Raman) and process gas chromatography (PGC). With the sample interfaces currently available for process mass spectrometers, the sample must be introduced as a gas. **Process mass spectrometers** (including IMR-MS [16]) have now replaced other gas analysers in various applications (fuel gas, liquid process streams), achieve faster control and often a reduction in process standard deviation. The proven advantages of speed of analysis, good sensitivity, high precision, excellent dynamic range and versatility outweigh, in many cases, the increased cost and complexity of mass spectrometry. Special niches in which process mass spectrometers excel are the fast analysis of light gases, environmental and ambient air monitoring, and hydrocarbons analyses. Most of the reported process-MS applications relate to sim-

Table 7.6. Status of in-process analytical techniques

Technique	Present status	New developments	Industrial applications	Citations[a]
Low-field nuclear magnetic resonance	Commercial	Improvement of sensitivity and robustness	QC food/agricultural, chemical (blends, fibres), cracker feedstock	10
Mass spectrometry	Commercial, well established alternative for process-GC	REMPI-ToFMS (selectivity; sensitivity)	Analysis of complex process gases (crackers)	31
Raman spectrometry	Commercial	Specific in-line analytical niches	(Petro)chemicals, pharmaceuticals, biomedical, catalysts, semiconductors	21
Mid-infrared	Commercial, well established	New sampling systems (cells, fibres, data handling); ATR probes	(Petro)chemical: many. Gas analysis, quantification of additives in polymers	55
Near-infrared	Commercial, well established	Rugged microspectrometers	Many: petroleum refineries, food industry, bio-technology	74
Ultraviolet/Visible	Commercial, established	Improvement of data handling	Modest; determination of stabilisers in liquid polymer melts/ granulates	17
Liquid chromatography	Commercial, limited	Short columns; HPLC chip	Limited to less time critical applications; batch-processing	29
Gas chromatography	Commercial	Reduction of analysis time and detection limit; miniaturisation	(Petro) chemical: many	40
X-ray fluorescence	Commercial, limited	Light elements; "standardless" analysis	(Petro) chemical: limited. Additives in polymer blending	31
Electrochemical	Commercial, well established	Improving life time of systems; optical sensors	(Bio) chemical: many; pH-values of liquid samples	14
Rheology, viscosity, density	Commercial, well established	Improving data analysis	Polymer processing: limited. Food industry; biotechnology	39
Dielectric spectroscopy	Commercial, established	No improvements	Moisture determinations, curing, drying	n.d.
Light scattering	In development	More robust systems	Particle size measurements: limited	n.d.
Laser absorption spectrometry	Commercial, limited	Broader use by new laser wavelengths	Ovens, burners, gases	n.d.

[a] General (review) articles (CAS, INSPEC) and patent (Derwent) references (1985–1995 period); n.d. not determined.

Table 7.7. Characteristics of in-process analytical techniques

Technique	Advantages	Disadvantages
LR-NMR	Fast response, non-invasive, linear dynamic range from detection limit to 100%	Low accuracy and precision; non specific; chemometric techniques necessary
Mass spectrometry	Fast, accurate, mass selective; large (linear) dynamic range (ppm to 100%); robust	Difficult for complex mixtures; complex calibration, difficult interfacing, cost
NIR/FT-Raman	Fundamental chemical information; ease of interpretation (no fluorescence), interfacing and sampling; cheap and long optical fibres; robust spectrometer	Relatively inaccurate; difficult quantification; spatially limited; expensive source; safety
Mid-infrared	Ease of interfacing; inexpensive; higher sensitivity than Raman; fast	Build-up of impurities; expensive optical fibres; limited penetration depth; MIR and Raman rivals
Near-infrared	Ease of interfacing, robust, fast, non-invasive, wide applicability	Robustness of calibration methods, limited information content of signal, expensive
Ultraviolet/Visible	Long path lengths, simple, low cost	Low selectivity
Liquid chromatography	Detects large molecules in complex mixtures; alternative to process-GC and -MS	Slow; relatively complicated sample handling
Gas chromatography	Simple, low cost, abundant experience, proven technology, robust	Destructive, invasive, frequent maintenance, high response times for larger molecules, discontinuous
X-ray fluorescence	Suitable for detection of trace amounts (ppm); simple quantitation; alternative to wet-chemistry analysis	Unsuitable for trace analysis of light elements; difficult interfacing
Electrochemical	Electrochemical sensors: speed, low cost	Fouling of the active surface; optical pH-sensors for niche applications
Rheology, viscosity, density	Increase in quality and reduction of cost; QA	Need for by-pass; problems for high viscous materials
Dielectric spectroscopy	Fast, ease of interpretation, rugged	Limited to conductivity
Light scattering	Use in severe conditions of T, p	Needs careful study in research environment first
Laser absorption spectrometry	Specific, ease of interfacing, extreme sampling conditions	Availability of laser wavelengths, spectral interferences

ple molecules in a simple matrix; calibration procedures are difficult for more complex systems. Solids or very high boiling liquids (b.p. $> 250°C$) generally cannot be analysed using conventional process mass spectrometer sample inlets because they are not easily vaporised. Process monitoring by means of MS can separate components in a mixture without chromatography (simultaneous monitoring of starting materials, intermediates, reaction products and impurities) and allows determination of the reaction kinetics. Problem areas are sampling of liquid and solid samples, and selectivity. Liquid sampling mass spectrometry using a GC injector port for sample volatilisation for EI ionisation in a magnetic sector mass spectrometer (negating the use of a chromatographic column) and PLS multivariate data analysis has recently been reported [17]. Reaction mixtures are often too concentrated for direct mass spectrometric analysis as the levels far exceed the working limits. Complex sample preparation and dilution procedures are often required before introduction to a mass spectrometer. A membrane-based approach alleviates this problem [18]. The selectivity problem can be taken care of by mobile on-line REMPI-ToFMS [19]. Compact mass spectrometers are more and more viewed as chemical sensors. Re-

Table 7.8. Comparison of process analysers[a]

Category	Process MS	Process IR	Process GC
Speed	Seconds to minutes	Seconds to minutes	Minutes
Maintenance	2–10 h/month	<2 h/month	2–10 h/month
Precision	0.1–1% relative	0.2–1.0% relative	0.5–2% relative
Concentration range	ppb to % depending on matrix	ppm to % depending on matrix	ppm to %
Sample size	μg–mg	g–kg	μg–mg
Safety	Pump exhaust must be safely vented	Sample cell required	Column and detectors must be safely vented
Cost	$ 100–150 K	$ 15–90 K	$ 35–60 K
Components per analyser	Usually capable of 16 or more. Average 8 in chemical process applications	Typically 1, up to 10 (for FTIR), increasing due to increased use of multicomponent photometers	Average 5
Applications	Gases or vaporisable liquids; usually mixtures	Mixtures of polar gases, liquids; unique IR peak required	Gases, vaporisable liquids; usually mixtures
Potential for interference	Peaks occupy ~1% of usable spectrum	Peaks occupy ~0.1 to 1% of usable spectrum	Minimal for 1 to 2 components

[a] After Walsh and LaPack [20]. Reproduced from *ISA Transactions* **34**, M.R. Walsh *et al.*, 67–85. Copyright (1995), with permission from Elsevier.

Table 7.9. Multicomponent capabilities of process analysers[a]

Technique	Usable spectrum	Peak width	Typical number of peaks per component	Percent of spectrum occupied	Likelihood for interferences
Mass spectrometer	2 to 200 Da	1 Da	2 to 10	2 to 5%	Likely; sample investigation needed; multivariate calibration often required
Filter IR	600 to 1800 cm^{-1}	40 cm^{-1}	2 to 10	7 to 33%	Highly likely; sample investigation essential; analysis often not feasible
FTIR	600 to 4000 cm^{-1}	2 to 8 cm^{-1}	2 to 10	0.3 to 7%	Possible; sample investigation often needed; multivariate calibrations sometimes required

[a] After Walsh and LaPack [20]. Reproduced from *ISA Transactions* **34**, M.R. Walsh *et al.*, 67–85. Copyright (1995), with permission from Elsevier.

cently, several overviews have appeared on chemical process analysis technology including mass spectrometry [20–22].

Gas chromatographs and IR analysers are common alternatives to mass spectrometers. Table 7.8 shows a comparison of process MS, IR and GC analysers [20]. As to the multicomponent capabilities of process analysers (Table 7.9), it is noticed that many process mass spectrometers require multiple bottles of calibration gases because multivariate calibration techniques are required for multicomponent analysis. It is difficult to calibrate and maintain the calibration integrity of a mass spectrometer for quantitative measurements. GC is currently still considered to be the most useful technique for process analysis of complex samples.

Table 7.10. World process analytical instrument markets

Segment	1997	2004
Spectroscopy	$ 230 × 10^6	$ 320 × 10^6
Chromatography	$ 72 × 10^6	$ 95 × 10^6

Source: Frost & Sullivan's *World Process Analytical Instrument Markets* (# 5472-30).

As to **process spectroscopy**, both low-resolution pulsed NMR [23] and high-resolution NMR find application as process analysers [23,24], where the latter is considered to be a less appropriate technique for use in a production environment. At a par with HPLC and rheometry, XRF/XRD is also little used in- and on-line. XRF and XRD are analytical methods for solids. Solid product flows usually show spatial inhomogeneity. Process analysis XRF systems range from relatively simple (yet rugged and reliable) units utilising radioisotope sources [25] with non-dispersive analysers to complex WD systems in a central location receiving samples from various process streams. In case of WDXRF both simultaneous and sequential spectrometers are used in process analytics. EDXRF is also an important method of on-line instrumentation. For process control low-cost simultaneous XRF/XRD equipment is now available. Also acoustic emission technology is attractive for in-line monitoring applications (*cfr.* Chp. 7.2.7). Key market drivers for process spectroscopy are technology improvement, a trend toward low-maintenance systems and a shift from process GC to MS, IR and NMR.

There is a more **rapid growth** of the newer types of in-process analysers with commercial potential (UV/VIS, mid-IR, NIR, Raman, LR-NMR, MS, US, rheology and viscosity analysers) compared to the more conventional ones (GC, GA, EC), *cfr.* also Table 7.10. Important developments are in the field of remote sensing (IR imaging techniques) and silicon sensors (for concentration measurements in liquid and gas streams). As the number of techniques available is increasing, so are the precision, speed, complexity and amount of information from these techniques. Gunnell [26] has recently addressed the future of process analytes. Advanced process optimisation will put strong emphasis at every stage in a processing train; cost of ownership will lead to new thinking. Demand for more difficult analytical measurements will increase. Old technologies will remain but new ones will be added such as NMR, e-nose, MS-, spectrometer- and lab-on-a-chip, nanotechnology, *etc.*

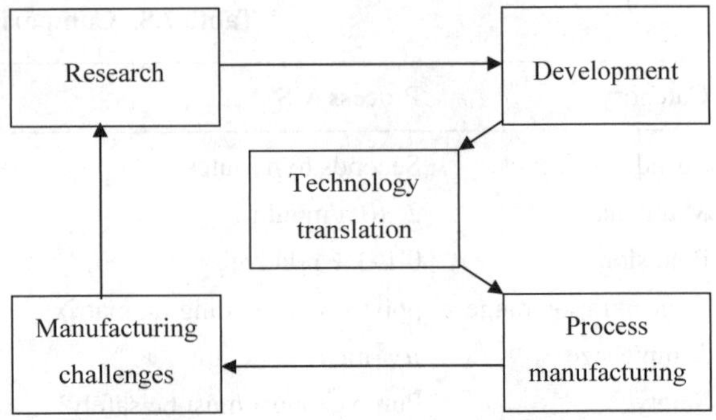

Scheme 7.1. Contributing to Manufacturing Excellence.

Consideration of elementary control concepts implies that the time difference between analysis and adjustment of the process should be as short as possible. As the time scales of chromatography and spectroscopy differ (from minutes to an hour in the former to seconds in the latter) development of in-process liquid chromatography attracts little interest. It is not surprising that more emphasis is laid on spectroscopic methods.

Research expertise required for **Manufacturing Excellence** (Scheme 7.1) requires a strong base in real-time process analytics, statistical process control and performance monitoring, chemometrics, data mining, and process modelling. There is a strong need for cross discipline appreciation. On-line analysis of process streams therefore determines the presence of **new competences** in plant environments, such as analyser technologists, spectroscopists and mathematicians. The need for a multidisciplinary approach to ensure Manufacturing Excellence requires cooperation of plant managers, chemical technologists, control systems engineers, process and chemical engineers, signal processing engineers, analytical chemists, physicists, chemometricians, statisticians, and engineering mathematicians. Only in this way advanced methods of process monitoring, control and optimisation can be implemented.

Koch [27] has discussed the process analytics of gases, liquids and solids. Thomas [28] has recently considered techniques that may be used in the production environment for process control, as opposed to spectroscopic analytical techniques that are used in the research environment.

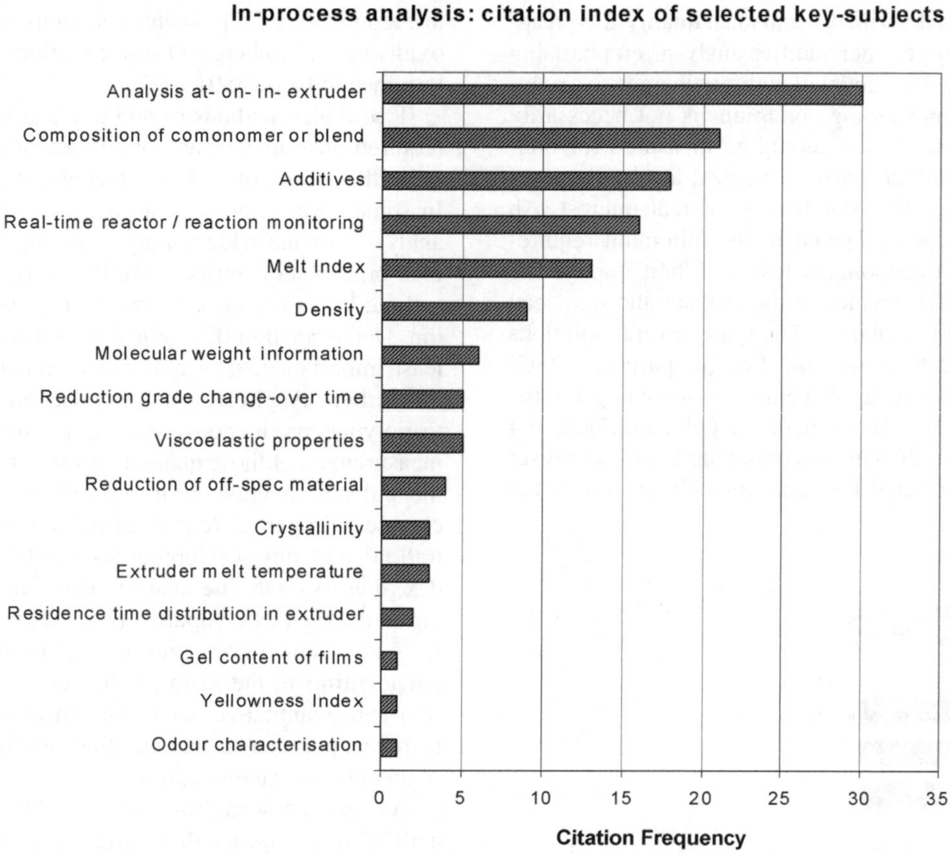

Fig. 7.3. In-process analysis in polymer production: citation index of selected key-subjects. Source: Scientific and patent literature (1991–1999).

Applications

The field of on-line monitoring of polymer processing is experiencing significant growth because of increased requirements with respect to productivity and quality. QA/QC and process measurement issues in the polymer industry are: chemical composition and sequencing; monomer residue/consumption (time to completion), end-group balance and concentration; molecular orientation (films, fibres), crystallinity, morphology; surface coating, formulated product (additive content and homogeneity). Figure 7.3 gives a broad overview of in-process analysis in **polymer production**. Some typical process analysis applications are the identification of feedstock materials, measurement of density, melt flow and tacticity in polyolefins, determination of moisture in polymers and powders, measurements in aqueous environments and at high p, T, non-contact measurements, *etc.* The number of process variables may be extraordinary high. For example, in polyester film process monitoring as many as 200 process variables may be identified [29]. Real-time monitoring of polymer processing has recently been reviewed [30].

In-line dosage of additives may be executed by injection of meltable (organic) additives into the polymer stream separately from the addition of non-meltable (inorganic) additives added as masterbatches. Static mixing ensures sufficient homogenisation (Fig. 7.4). With the high cost of additives in comparison to polymeric matrices it is not surprising that considerable savings can be gained by accurate dosage. It is desirable to be able to adjust additive dosage on-line, *i.e.* in the melt at the extruder. Similarly, there is great interest in rapid measurements of granulate and in non-contact additive analysis of moving film.

For process and quality control it is necessary to verify the additive content after processing, which might partly destroy (thermolabile) additives. Ashton [31] has described a statistical quality-driven approach to in-polymer additive analysis, emphasising the fact that the amount of additive charged to a polymer under processing conditions is not necessarily the parameter which should be measured for relationships with polymer properties; ideally the level of residual active stabiliser is of real interest. An excess dosage compared to the minimum requirements is an economical loss; a deficit impairs the technical performance of the product and may lead to justified complaints. There are several solutions to this analytical problem. For QC purposes DSC-OIT provides a rapid method of screening for the proper levels of antioxidant in a polymer. Figure 2.4 shows that a PE sample containing 0.04% stabiliser remains protected for approximately 16 min at the

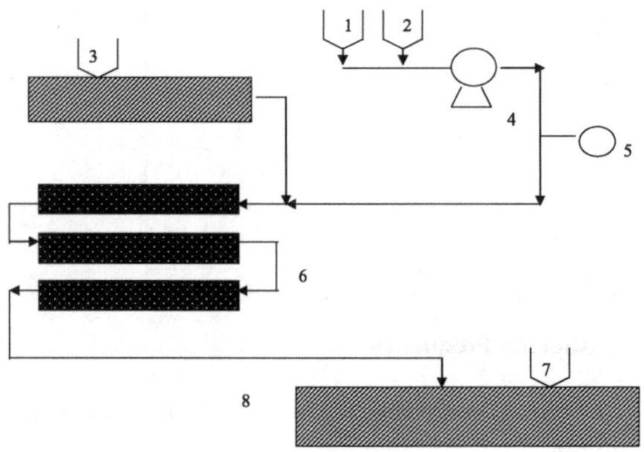

Fig. 7.4. Schematic in-line additive dosage: meltable organics (1, 2), non-meltable inorganics (3), pump (4), flow measurement (5), static mixing (6), hopper (7), extruder (8).

test temperature [32]. At elevated temperatures in an oxidising atmosphere AO concentrations can be determined to below 0.01 wt.%.

Traditional methods of additive analysis and the required instruments are often expensive and require the efforts of a skilled technician or chemist. In some cases a single instrument can not provide analyses for the wide variety of additives a particular organisation utilises. Additionally, laboratory techniques rarely provide results in a timely fashion. Determination of physical properties is not the least important if one thinks of pigments, talc and other fillers. Application of spectroscopic techniques to polymer production processes permits real-time measurement of those qualitative variables that form the polymer manufacturing specification, *i.e.* both chemical properties (composition, additive concentration) and physical properties (such as melt index, density). On-line analysis may intercept plant problems such as computer error, mechanical problems and human error with respect to **additive incorporation** in the resin production. Characterisation and quantitative determination of additives in technical polymers is an important but difficult issue in process and quality control.

As may be seen from Tables 7.11, 7.15, 7.17, and 7.21, it appears that currently in-process additive analysis in the polymer melt is most popular in UV and mid-IR applications, less so for near-IR and is not being pursued by Raman spectroscopy, most probably because the latter lacks detection power. However, the theory of application of various spectroscopies to in-line and on-line additive analysis in melts is far more advanced than

Table 7.11. Comparison of process monitoring applications by spectroscopic methods

Application	UV/VIS	Mid-IR	Near-IR	Raman	LR-NMR
Polymer/additive analysis	+	+	+	−	(+)
Polymer melt analysis	+	+	+	−	+
Additive quantitation	+	+	+	−	(+)
QA/QC purposes	+	+	+	+	+
On-line compositional polymer analysis	−	+	+	+	(+)
Reaction monitoring (*in situ* cure kinetics, polymerisation)	−	+	+	+	+
Non-contact analysis of physical parameters (density, crystallinity, orientation, cross-link density, *etc.*)	−	+	+	+	+
Colour designation	+	−	−	−	−
Molecular interactions	−	+	−	+	−
Monitoring of extrusion processes	−	+	+	−	+
Real-time measurement	+	+	+	+	+
Fibre optics	+	−/+	+	+	−

practice. Many plants still operate on the basis of gravimetry (weighing) only and take calculated risks for claims in case of off-specs instead of relying on process analytics. Simple and accurate single-component gravimetrical dosing devices are commercially available [33].

In plastics processing temperature is a decisive process control parameter. Time dependent temperatures can be measured with high accuracy and short response time (msec) using IR sensors that detect the heat radiation from molten plastic [34].

Sensor systems (e-noses) for at-line measurement, coated with various gas sensitive materials which react differently with the volatiles to be analysed, detect differences rather than absolute values. It is possible to verify quickly deviation from the standard. Process titrators are used today in many industrial fields of process control. The range of applications is extraordinary large [27,35].

7.2. PROCESS SPECTROSCOPY

Principles and Characteristics

The proper strategy for in-process analysis is selection of the application, of desired output (qualitative/quantitative), sampling, technique, technology and modelling (univariate/multivariate, outlier detection, *etc.*), in this sequence. Process spectroscopy is a series of robust analytical techniques that can deliver (near) real-time, highly reliable analyses on a manufactured product in the actual plant environment in a more timely and cost-effective way than traditional laboratory analysis in order to provide advanced process control. Small spectrometers are now available (mostly fibre-coupled) that enable direct sampling in the production line and so reduce both the cost and time required to enable real-time feedback in process control applications. In the process area spectroscopy is being used more and more, both more widely and over an increasingly wider range of spectroscopic techniques. This is not surprising as spectroscopic on-line analysis is fast, non-invasive, allows for high precision and accuracy (good for closed loop control), requires low maintenance, uses convenient fibre optic technology and permits better control of reactors and plants, leading to large savings. Spectral scans can be completed in (milli) seconds, as compared to much longer times for other traditional laboratory techniques (chromatography, titration, *etc.*). Effective use of composition monitoring requires that the measurement be fast compared to the maximum rate of composition change that might be encountered. The major difference between spectroscopy measurements and other process parameters is that spectroscopy mostly relates to material property parameters, whereas temperature, flow, pressure and humidity are all indirect process parameters.

In a broad sense, spectroscopic methods applied in process analytics comprise widely used techniques like UV/VIS, mid-IR, NIR, NMR and XRF, and less frequently used ones, such as Raman spectroscopy, fluorescence, chemiluminescence, acoustic emission and dielectric spectroscopy. Upcoming in-process analysis techniques are 2D-fluorescence, and laser absorption spectroscopy (LAS) with tuneable lasers and ppm level sensitivity. The availability of mini-spectrometers (*e.g.* UV/VIS/NIR) is not highly relevant in plant environments where safety is of primary concern.

Process analytical spectroscopy is often faced with extremely complex samples in difficult matrices under very tight time limits for delivery of the results. However, even with spectroscopy it is often necessary to collect a sample before it can be analysed. Van der Maas [36] and others [36a] have described **sampling techniques** for process spectroscopy. Sampling may be extractive (*e.g.* near-line, on-line), non-extractive (in-line) and non-invasive (*e.g.* DRIFTS, NIR, Raman). Gas analyses are best performed in a gas cell. In principle, sampling gases or vapours is rather simple, but spectral data are subject to pressure, temperature and spectral resolution. Besides, contact with the metal parts of a cell may give rise to (catalytic) reactions, while polar samples tend to adsorb onto non-coated walls. Gas spectra, which are quite unique, are ideally suited to qualitative work because of the detailed information arising from rotational contributions (monitoring air pollutants). Quantitative work is possible provided scanning conditions are well controlled (impurities in gas samples). Pure liquids and all types of solutions can be measured within 10 s in temperature controlled flow-through liquid cells. In mid-IR these cells should preferably have a path length of at most 0.1 mm and window materials have to be chosen with care (*e.g.* CaF_2), while in NIR a path length of 10 mm and glass or quartz windows may be used. Because of the fixed path length quantitative work in these cells is relatively easy and straightforward. Polymer melts, *e.g.* in extrusion processes, can be investigated in-line by transmission spectroscopy for transparent melts and diffuse

Table 7.12. Choice of spectroscopic techniques

Feature	UV	VIS	Mid-IR	NIR	Raman
Cost of analyser	\$	\$	\$\$	\$–\$\$	\$\$\$
Organic in aqueous	+	+	−	−	+
Aqueous in organic	−	+	+	+	−
Remote sampling	+	+	(+)	+	+
Calibration requirements	+	+	+	+++	+
Good sample averaging	+	+	−	+	−
Solid/slurry sampling	−	+	−	+	+
High sensitivity	+	+	+	−	−
High chemical resolution	−	−	+	−	+
General applicability of technique	−	−	+	−	−
General applicability of technology	+	+	−	+	+

Table 7.13. Comparison of NMR and NIR spectroscopy for on-line process analysis

Subject	NMR	NIR	Comments
Detection sensitivity	100 ppm	500 ppm	
Temperature	Relatively insensitive	Needs to be controlled	Small T changes affect band shapes/positions in NIR
Flow-rate	Requires control, or use of stopped-flow	Insensitive	
Calibration and modelling	Correlates NMR peaks to standards using chemometrics	Requires deconvolution of NIR peaks, and chemometrics	Distinct NMR peaks, overlapping NIR bands
Measurement response	Linear	Non-linear	Not all NIR bands have equal absorption strength

reflectance spectroscopy for opaque melts. Both experimental set-ups are suitable for quantitative in-line process analysis of multicomponent polymer mixtures [37].

Except for thin films, preparing solids for transmission measurements requires much labour (*e.g.* KBr disc technique) – quite unsuitable for process analysis. Transmission measurements on solids are less apt for monitoring or process control. In specular reflection (SR) light reflected at a flat surface at an angle θ_r, identical to the angle of incidence θ_i, is used. The intensity of the reflected beam depends on θ_i, surface roughness and absorption by the sample. In mid-IR the SR technique is useful for solids or liquids on a reflecting substrate (transflectance). Detection in the near-infrared is easier than for infrared (no reflecting support needed). Combined with chemometrics specular reflection is excellently suited for the rapid identification of materials (*cfr.* Chp. 1.2.1.2), as well as for in-process analysis. Attenuated total reflection (ATR) can be used in a flow-through way (thermo-stabilised up to 200°C and 150 psi) as ATR probes for process control allow following reactions continuously. The diffuse reflectance technique with NIR radiation has been widely and successfully used for quite some time in industries for food analysis (water, fat, sugar content), but also for additive analysis in melts.

Table 7.12 shows some parameters of judgement for allowing the choice of a spectroscopic technique, whereas Table 7.13 compares NMR and NIRS for process analysis. Process NMR has potential for the selective determination of compounds containing certain nuclei. Despite the direct nature of the chemical information contained in NMR spectra process NMR (*i.e.* LR-NMR) is not widely used for chemical composition analysis. However, NMR can be used for the measurement of opaque, viscous and optically "dirty" samples, which can cause problems in IR methods.

Most spectroscopic techniques have a number of different ways of collecting data. A most important consideration is whether a technology can supply data of sufficient quality to be used by the model. While classification techniques have not been ignored in applications like raw materials checking, unknown identification and grouping of complex materials, reported applications in process monitoring have been limited. There are many advantages of using spectroscopy as a detection technique for quality control of complex samples. It is fast, requires little or no sample preparation for most types of samples, and can be implemented at or near the source of the samples. However, often quantitative methods are being employed to simply gauge the suitability of the material being measured. In a significant number of cases, the only result that is desired is to know whether the sample falls within a defined range of allowed variability to determine if the material is of the desired quality. It is not always necessary to measure the quantities of the constituents in the sample to meet this goal [38].

Multivariate **data analysis** is quite appropriate in very common situations in which a product property depends on more parameters. Classical models, which take into account one variable at a time, are inadequate to discern the complex interplay of various factors and to extract both qualitative and quantitative information from the large (spectroscopic) process data sets which are generated routinely by in-line spectrometers. Multivariate approaches are capable to project the information into low dimensional spaces where one can easily interpret process behaviour and optimise process performance, reveal sample patterns and hidden phenomena, and variable relationships in seemingly complex data. In this way, much useful information can be extracted from the overlapping NIRS absorption bands. Infrared analysis usually requires multivariate analysis, whereas solution Raman spectroscopy can be handled with a univariate approach and is thus characterised by more robust calibration. A description of partial least-squares regression (PLS) is beyond the scope of this book, but an excellent introduction is found in the literature [39]. General guidelines to multivariate calibration have been given [40]; a tutorial will be available [41]. For multivariate quantitative spectroscopic analysis, *cfr.* also ref. [42].

Current trends in process spectroscopy are: *(i)* measurement techniques with non-invasive possibilities (acoustics, microwave, R, NMR); *(ii)* measurement techniques with high sensitivity (MS, UV, F); *(iii)* multi-probe measurements; and *(iv)* dynamic performance monitoring based on multivariate analysis. However, at plant sites there is a tendency to minimise the frequency of on/at-line analyses. It would appear that rather few new applications of on-line spectroscopic techniques (UV/VIS, mid-IR, NIR) or LR-NMR are currently under development.

UV/VIS, F and CL spectroscopies in chemical process analysis have been reviewed [43]. Siesler [44] has compared mid-IR, NIR and Raman in process monitoring, whereas Doyle [45] has made a critical comparison of near-IR and mid-IR process analysis.

Applications

As may be apparent from spectroscopic in-process additive analysis data available (*cfr.* Tables 7.15, 7.17 and 7.21) the current situation is unsettled. Actually, activities apparently are slowing down, in particular for mid-IR process analysis of additives. Yet, mid-IR is chemically the best option for on-line analysis as all additives can be measured. UV/VIS scores best in terms of technical feasibility. For applications of process spectroscopy for additive monitoring UV and mid-IR need to improve on robustness, NIR requires better detection limits and precision, Raman and acoustic spectrometry are still to be fully exploited. Polymer/additive analysis is generally considered as being difficult at the laboratory level. This is reflected in in-process trials. For real-world polymer compounds process polymer/additive analysis remains a challenge; the problem has not disappeared. A step change is needed. Practical implementation of current spectroscopic technologies in manufacturing plants is not as widespread as control via gravimetric dosing of additives (loss-in-weight feeders). Also the latter system requires external, independent validation. While bar code control may be helpful in spotting inadvertent mistakes in hopper feeding, additive decompositions go uncontrolled.

Miniature spectrometers find useful applications in plastics waste sorting. Chemiluminescence is used for the continuous determination of nitrogen oxides.

7.2.1. Remote Spectroscopy

Principles and Characteristics

In on-line spectroscopic analysis, optical probes are often used for direct interrogation of chemical processes. The function of an **optical probe** is to conduct the light into a sample medium and then allow collection of the interacted light in a precisely controlled manner. Various probe designs for (difficult) sampling tasks may be distinguished, such as: *(i)* probes for insertion in process pipelines or batch reactors; *(ii)* thin probes for insertion in cross union (facing each other, with one probe transmitting light and the other receiving light) or flow tube; and *(iii)* remote optical sensing assembly (ROSA), which allows a plant operator to draw a process sample at precisely the time that a spectroscopic analysis is made. This improves the validity of correlations made between laboratory measured "method data" and on-line spectroscopic data.

Path lengths for in-line fibre optic set-ups (flow-through cells) are 40 μm or less for mid-IR (for bulk compositions, but up to 1 mm for additives in a polyolefin) and some 40 mm for NIR. Successful on-line installation of such probes requires little engineering work and causes minimal interference to a process. Most optical probes on the market today have industrial standard diameters and are readily inserted into pipelines or vessels via standard industrial fittings [46]. Because fibre optic probes have no moving parts, are non-magnetic, and use no electricity, they can be installed in process equipment and be exposed to hazardous gases or liquids. Only the probe is exposed to the hazardous material, not a technician gathering samples, which is an important safety aspect.

To connect a guided wave spectrophotometer to a probe a single-strand **optical fibre cable**, or wave guide, is needed. The more optically efficient a guided wave probe, the more linear is the spectral response from the scanning system. Early practical fibres (with silica in the fibre core) were limited to a spectral range of 50,000 to 4500 cm^{-1} [47]. Subsequently, more IR transparent fibre optic materials have been developed but light levels are still low. The useful regions of some fibre optic materials are 4500–900 cm^{-1} (chalcogenide glass), 11,000–2100 cm^{-1} (zirconium fluoride), 25,000–3900 cm^{-1} (anhydrous quartz) or 40,000–8000 cm^{-1} (quartz). Moreover, sources for use in the mid-IR are much less bright and detectors much less sensitive than in the UV/VIS/NIR regions. Consequently, design of a practical fibre-based IR spectrophotometer is challenging. For NIR many flexible materials are available (distances up to 1000 m are feasible), as opposed to mid-IR (3 to 4 m at maximum).

Fibre optic spectroscopy offers:

- Flexibility (flexible fibre, moderate armouring) and modularity.
- Convenience (long cables): 1000 m in NIR, 3–4 m in IR.
- Real-time information from a reaction or process.
- *In situ* analysis of hazardous or air sensitive samples.
- Improved research productivity.

Two basic fibre configurations are currently in use. For longer separation distances between the optical instrument and the sensing point, only *single fibres* are practical. *Fibre bundles*, which increase the light throughput, are available only in a few metres in length, and are very expensive. In NIR, single fibre lengths of hundreds of metres can be used (attenuation at 1500 nm <1 dB/km). This enables the spectrometer to be located in a safe area remote from the measurement point where it would be impossible to install a conventional instrument, and allows a wide range of at-, in- and on-line process control applications. With monofilament fibre optics a single spectrometer can be multiplexed to a number of sample points thus further reducing costs.

Optical fibres can be used in the transmittance and ATR mode (a special ATR application is the remote sensor), and even in the reflectance mode. The development of special optical fibres for transmission, transflection or diffuse reflectance measurements favours on-line analysis of problematic product streams and reaction mixtures (solutions, suspensions, emulsions, melts, solids). Both quartz and fluoride (ZrF_4-based) glass fibres are used, with the former having poor transmission characteristics above 2000 nm.

Applications

Applications of the fibre optics transmittance or ATR probe are in quality control, reaction monitoring, skin analysis, goods-in checking, analysis at high and low temperature, radioactive or sterile conditions, and hazardous environments. Applications of the reflectance probe are for turbid liquids, powders, surface coatings, textiles, *etc.* By using an on-line remote spectrophotometer, **real-time information** is gathered about a chemical process stream (liquids, films, polymer melts, *etc.*), as often as necessary and without the need to collect samples. This determines more reliable process control. Remote spectroscopy costs less to maintain and operate than traditional techniques. Fernando *et al.* [48] have compared different types of optical fibre sensors to monitor the cure of an epoxy resin system.

Schirmer *et al.* [47] have discussed the applications of chemical sensing using fibre optics and UV/VIS/NIR spectroscopy.

7.2.2. Process Electronic Spectroscopy

Principles and Characteristics

Absorption spectroscopy of both vapour and liquid samples by wavelengths in the UV/VIS range, causing electronic transitions in the sample, can be used to quantify components in a mixture. Optical transmission measurements are preferred to diffuse reflectance, they provide higher sensitivity, more precision and enable monofilament fibre optics to be used. Spectroscopic (UV/VIS/NIR) analysis of pellets is more complicated.

UV/VIS spectrophotometry is a well developed routine technique, which is used extensively in QA/QC laboratories, but not so frequently as a process analytical technique because of lack of selectivity of the spectra (exceptions are monoaromatic hydrocarbons). Visible and near-ultraviolet spectra often do not contain so much industrially useful information. However, the method is suited for the determination of components that can be readily distinguished from the sample matrix (*e.g.* UV absorbers in polyolefins, but not in EPs; colour measurements of ABS). Provided that the analyte has a UV chromophore, UV absorbance measurements provide much greater sensitivity than NIR measurements. This enables antioxidants and other additives to be determined at the low ppm level. Polyolefins are transparent in the UV, which enables calibrations for the analysis of antioxidants to be matrix independent.

General-purpose UV/VIS fibre spectrometers allow absorbance, transmission and reflectance measurements. In-line diode array spectrophotometer systems with multi-wavelength UV/VIS monitoring in the 200–1100 nm range are now available and allow acquisition of a complete spectrum in about 0.1 s. Accurate process and **quality control** is therefore provided within milliseconds. A single photodiode array can tackle a number of applications (*e.g.* antioxidant and colour measurements) and two systems (UV/VIS and NIR) can provide full spectral coverage (up to 2200 nm). A diode array spectrometer provides high quality long-term reliable results (no moving parts). These devices can be configured to provide very rapid analysis or to gain the ultimate precision by allowing long integration of the signal.

Table 7.14. Main characteristics of in-process UV spectrophotometry

Advantages:

- Choice of optimal path length
- Long path lengths (low S/N ratios)
- Small calibration burden (univariate)
- Simple/cheap monofilament fibre optics
- Absence of disturbances due to differences in crystallinity of both additive and polymeric matrix
- Inexpensive instrumentation
- Improved consistency of product quality
- Minimisation of waste
- Tighter product quality control
- Reduced laboratory demand

Disadvantages:

- Restricted applicability
- Low selectivity
- Spectral overlap (mixtures)

Photodiode spectrometers are also less costly than scanning or FT based instruments.

As reported by Hansen *et al.* [49], UV optical fibres can be used up to 10,000 h without any transmission degradation in the 200–400 nm wavelength region. UV spectrophotometry using fibre optic probes and real-time PLS modelling performs as an ideal non-invasive, **on-line technique** in the melt with easy control of optical path length for absorbance values of additives <1 abs unit. Typically, this is in the 2.5–10 mm range, *i.e.* much longer than in films (<1 mm). UV spectroscopy thus provides much lower detection limits for additives such as antioxidants. Conditions for successful on-line UV monitoring require no p, T variations during extrusion and polymer grade specific calibration sets.

Table 7.14 shows the **main features** of UV spectrophotometry as a process analytical method for the quantitative determination of additives in **polymer melts** rather than film. Fibre optics enable measurements to be made in difficult areas where it would be impossible to install a conventional spectrometer. Depending on the application, either single strand fibre optics or fibre optic bundles are utilised to carry light between the instrument and process stream. Each is characterised by its own set of advantages with single strand systems offering flexibility, ease of installation, and performance over long distances, while bundles can provide high illumination energy and superior collection efficiency for measurements of solids or highly scattering liquids. Disadvantages are restricted applicability (only to UV active materials) and the low selectivity of the response signal (broad and overlapping absorbance bands). It is likely that UV/VIS will remain mostly an environmental/safety monitor method with limited use in some specific control applications. A separation step and multivariate analysis/calibration (spectral deconvolution), as used by refs. [50–52], are methods to increase the applicability. Collins *et al.* [53] have reported PLS model training and validation exercises for the determination of various additives in a polyethylene melt, with reference to off-line liquid chromatographic analysis. A quality control metric called "Mahalanobis distance" is often used to indicate whether predictions are within the range of the original training data [54].

Yang *et al.* [55] have published a general review on UV/VIS process analysis.

Applications

In some cases, application of spectroscopic techniques to polymer production processes permits the real-time measurement of those quality variables that form the polymer manufacturing specification, such as melt index, density and additive concentrations. This is particularly exciting as it offers the facility to effect closed-loop control of additive incorporation, and continuous **quality assurance**, resulting in substantial cost savings. Optical probes operate up to 10,000 psi (660 bar) and 250°C.

Both UV and mid-IR absorption measurements are used as some additives exhibit characteristic absorption bands only in the mid-IR region while others can be distinguished quite well in the UV. As additives exhibit specific absorption bands in certain spectral ranges the choice of the spectroscopic technique is often almost obligatory. For example, stearates cannot easily be determined by means of UV spectrophotometry due to lack of a suitable absorption band; yet, in some cases measured concentrations were quoted [56].

Table 7.15 summarises on-line multicomponent additive analyses by means of UV spectrophotometry. Schirmer [57], who appears to have generated

Table 7.15. On-line (multicomponent) additive analysis by means of UV spectrophotometry

Polymer melt	Additives *cq.* additive package	Affiliation	Reference	Year
HDPE	BHT	Guided Wave	[57]	1988
PP	BHT; Ethanox AO-330; Irganox 1010; Irganox 1076; Irganox 3114	Amoco	[58]	1991
PP	Irganox 1010, Irgafos 168, Tinuvin 770	Ciba-Geigy/ Guided Wave	[59]	1993
LDPE[a]	Chimassorb 944; Irgafos 168; Irganox 1010; Irganox 1076	DSM	[50]	1994
LDPE[b]	Chimassorb 944: Irgafos 168: Irganox 1010: Irganox 1076: [oleamide][c], [Zn-stearate][c]	DSM	[50]	1994
LDPE[a]	Irganox 1010	U Karlsruhe U Darmstadt	[56]	1995
PP	Irganox 1010, Irgafos P-EPQ, Ca-stearate, [MgAl carbonate][c]	U Karlsruhe U Darmstadt	[56]	1995
HDPE	[Ca-stearate][c], Chimassorb 944, Irganox B220	DSM	[60]	1996
PE	Three unspecified additives	BP	[53]	1998
Polyolefin	BHT, Irgafos 168, Irganox 1076	PAA	[–]	1999
LDPE	Irgafos 168, Irganox 1076, [Armostat 310/ erucamide/ Tinuvin 622][d]	U Tennessee	[49]	1999
Polyolefin	Irgafos 168, Irganox 1010, Irganox 1076, Irganox MD 1024	PAA	[54]	1999

[a] Single component analysis.
[b] Multiple component analysis (in fixed ratio).
[c] No UV chromophore.
[d] Sum of multiple additives.

the first known record, has examined HDPE/BHT melt on-line by remote optical monitoring. The butylated hydroxytoluene concentration was readily measured to 50 ppm at 280 nm with a path length of 2 mm between the probe windows. Lower concentrations could be measured by using longer optical path lengths between the probes, while higher concentrations would be best handled by measuring the absorption on the long wavelength side of the band. The addition of pigments or dyes, UV or near-IR blocking agents, and other additives can often be controlled in the same manner as antioxidant addition. UV spectrophotometry provides the sensitivity required for on-line analysis of polyolefin additions. Crompton [61] has reported the direct UV analysis of Irganox 1010 (0.01–0.1%) in molten PE (0.045–0.78 cm) and has shown the linearity of absorbance with sample thickness over the given concentration range.

Yang [58] has reported linear relationships for the UV analysis of BHT, Irganox 1010/1076/3114, and Ethanox AO-330 (up to 0.2 wt.%) in PP melts. Significant melt temperature and pressure effects on the UV absorbance were noticed. Consequently, successful on-line UV monitoring of hindered phenolic antioxidants requires that the extrusion conditions (p, T) do not vary. Spatafore *et al.* [59] have used UV spectrophotometry for on-line monitoring of hindered phenols and phosphites in PP (Himont Profax 6501). One calibration set of 19 PP samples contained Irganox 1010 and Irgafos 168 in concentrations up to 0.1 wt.%; the second calibration set of 25 PP samples was composed of PP/(0–0.3 wt.% Irganox 1010, Irgafos 168, Tinuvin 770). Tinuvin 770 does not interfere with the determination of the primary and secondary antioxidant in the melt. The technique is effective for both single component systems and those containing mixtures.

Verlaek *et al.* [50] have examined additives in LDPE by means of both UV and mid-IR absorption spectroscopy in film and melt samples. For the measurements a 14 mL-volume mini-extruder was used equipped with a melt-cell provided with optical channels for both UV and mid-IR measurements. UV measurements were carried out with fibre optic coupling; mid-IR was performed in transmission. **Single-component analysis** with UV absorption on LDPE melt gave excellent results for Chimassorb 944, Irganox 1010/1076 and Irgafos 168, all with a Standard Error of Prediction (SEP), *i.e.* 1σ of the difference between spectroscopic prediction and reference method, of *ca.* 10 ppm. For comparison, slightly higher SEP values were found for mid-IR measurements on LDPE melts. For UV measurements on films the SEP values varied from 15 to 45 ppm. **Multicomponent analysis** was performed using a complete model additive package (combination of Irganox 1010/1076, Irgafos 168, oleamide, Zn-stearate and Chimassorb 944 in a fixed ratio) in LDPE [50]. The total concentration of the additive package was varied between 100 and 1000 ppm (expressed in concentration of Irganox 1010). The SEP-value for UV and mid-IR measurements on the melt were 8 and 6 ppm, respectively (or 4 ppm averaging four measurements). It is concluded that both the UV and mid-IR methods are very promising for rapid determination of additive concentrations. The methods can be used in a laboratory situation and may be applied to an on-line mode for polymer melts. For polymer films the results are sometimes less favourable, possibly due to crystallisation effects.

Verlaek *et al.* [60] have further reported that after a careful calibration procedure (using 52 spectra) it is possible to carry out multicomponent analysis of **independently varying additive concentrations** in a melt spectrum of polyethylene. For HDPE/(Irganox B220, Chimassorb 944, Ca stearate) a typical standard error of prediction of *ca.* 75 ppm was established for both UV/VIS as compared to *ca.* 30 ppm for mid-IR measurements. It was observed that it is not possible to use the calibration set of one type of PE for another grade.

On-line photodiode array UV/VIS measurements (200–400 nm, averaged over 10 s) using fibre optic probes installed on a by-pass of an extruder of a **large-scale production plant**, followed by real-time partial least-squares (PLS) modelling, were used by BP Chemicals [53] to predict the concentrations of three (unspecified) additives in a PE melt. A self-learning software model was generated, based on 30 carefully selected samples removed from the extruder, evenly covering the concentration range, and analysed off-line using liquid chromatography. The PLS predicted data for each of the additives were validated by comparing the results with those obtained by liquid chromatography. The real-time method compares favourably with the traditional method of physical sampling analysis in the laboratory, which requires waiting several hours. It enables users to explore multivariate data sets, quantitatively predict product quality and to build **multivariate statistical process control models**. The commercially available modular AddiMet^{TM} system (Process Analysis and Automation Ltd., Farnborough, UK) is designed to perform on-line UV/VIS/NIR process monitoring of AO concentrations, such as the strong UV absorber Irganox 1076, Irgafos 168 and BHT in polyolefin melts, using UV detection and a PLS calibration set obtained from HPLC analysis of samples removed from the extruder. The concentrations of several AOs in the polyolefin melts can be measured simultaneously. Depending on the additives to be determined, analyses are made in the NIR- and UV/VIS ranges. The UV option provides the high sensitivity required for the measurement of low levels of AOs; the VIS option allows colorants and whiteness to be determined. NIRS samples physical properties and bulk composition. The use of PLS enables simultaneous on-line analysis of multiple additives. The system has been applied to on-line analysis of Irganox 1010/1076/MD 1024 and Irgafos 168 in polyolefin melts [54]. The chromophores in these compounds are all similar, which means that they have very similar UV absorption spectra. In order to quantitatively determine one component in the presence of the others some form of spectral deconvolution is required. For this purpose PLS modelling can be utilised. The technology fails in wide introduction in manufacturing plants. Possible reasons are cost, lack of robustness, immature technology or preferred alternatives.

By improved UV probe design Hansen *et al.* [49] have been able to achieve in-line monitoring of multiple additives using fibre optic UV spectroscopy; Irganox 1076, Irgafos 168 and the sum of Tinuvin 622, erucamide and Armostat 310 could be determined simultaneously in industrially used concentrations in molten LDPE using spectral data above 235 nm for analysis. In conclusion, it appears that UV and IR melt measurements achieve equal accuracy, with mid-IR being more generally applicable but UV experimentally easier to perform. UV melt measurements often outperform film measurements. The analysis time is typically 10 s.

Herman *et al.* [62] have described developments using UV methods to provide on-line data on the mixing efficiency and concentration of additives in polyolefin compounding. The application concerned development of a method for real-time evaluation of the residence time distribution in extruders without disrupting the process, providing a diagnostic for process development and troubleshooting.

The UV reflection technique can be implemented as a non-destructive, *in situ*, in-process analytical technique to continuously monitor surface chemical composition. The probed depth by UV reflection is about 500 Å for a chromophore with an extinction coefficient of 10^4 L/mol cm at 200 nm [63].

In-line colour monitoring of pigmented polymers during extrusion may be carried out using a CCD spectrometer (reflectance spectra; average values) or CCD imaging (RGB values per pixel) [63a]. The latter method shows the better precision but is more sensitive to lighting. Other reported UV/VIS applications are: colour determination of surfaces or solid materials (*e.g.* quality control of textiles or paints), colour designation (ASTM, *etc.*), colour measurement of liquids (*e.g.* end-point determination in colour or dye processes), quality control of coated glass, filters, foils, *etc.*, and control of contamination. Some applications of UV methods (fluorescence, following excitation in the UV, absorbance or reflectance) have concerned safety and environmental analysis ("summation" measurements, *e.g.* total aromatic hydrocarbons; UV/VIS is not hindered by water absorbance bands).

Some recent reviews have dealt with applications of remote chemical sensing using fibre optics and UV/VIS/NIR spectroscopy [47] and process monitoring of polymer melts using UV/VIS spectrophotometry [64].

Measurement of absolute fluorescence intensity has some significant limitations. Measuring an absolute quantity is practical only if the background is known and preferably constant in time. For this reason, absolute emission intensity is not likely to be a suitable quantity for long-term, on-line or process-control measurements. Quantitative analysis of polymer additives by process fluorescence is a doubtful exercise. Scranton *et al.* [65] have reported *in situ* cure monitoring in vinyl ester and methacrylate systems using solvatochromic probe molecules and fibre-optic **fluorescence sensors**. The solvatochromic method is based upon relative fluorescence intensity ratios avoiding problems associated with absolute fluorescence intensity measurements. Fluorescent dyes were used to measure temperature and shear gradients within flowing polymers [66]. The use of fluorescence to monitor polymer injection moulding has been demonstrated for PS and PE doped with 10–100 ppm of dye [67].

7.2.3. Mid-infrared Process Analysis of Polymer Formulations

Principles and Characteristics

Among the various spectral regions, mid-infrared offers enhanced sensitivity and selectivity because it "owns" the information of the "fingerprint" region. The mid-IR absorption bands are much sharper than NIR spectra and calibration is simpler, yet labour intensive. Consequently, the technique is in principle more suitable for on-line/*in situ* process control in the R&D stage (process development). However, IR analysis of a polymer at process stream temperature gives results which may differ considerably from the same polymer at ambient temperature. Although spectroscopy on melts is considerably different from that in the solid state, this does not limit the information content. By making measurements directly in the melt, various sources of error caused by sample preparation are evaded. Problems in reproducibility on account of sample inhomogeneity and crystallinity effects can be avoided by analysing the polymer melt directly. There exists a wide range of possible applications. Characterisation of reactions in the melt (*e.g.* reactive processing, grafting of additives, *etc.*) is an area in which mid-IR can greatly contribute to optimisation of product and process development.

The following **technological advancements** have contributed significantly to the rapid growth of FTIR analytical techniques: *(i)* flow cell design for FTIR; *(ii)* stable optical design of FTIR interferometers; *(iii)* availability of optical fibres; *(iv)* multiplexing using optical fibres; *(v)* rapid data acquisition (trade-off between speed and resolution); and *(vi)* development of statistical data analysis. Modern FTIR data systems permit real-time analysis.

Up to the early 1990s, in most plastics production plants, a small sample of plastic was taken from the compounding area and hand-carried to a QC laboratory for evaluation. The polymeric material in a particulate form was then usually hot pressed into a melt film or dissolved in a solvent for mid-IR spectrophotometry. Off-line mid-IR evaluation of the plastic's composition and concentration of constituents would take from one to two hours. **Process monitoring** using IR requires short measurement times. Analysis of IR output data is relatively rapid, lending itself to on-line processes.

In this context, the use of a dispersive IR-spectrophotometer is only recommended when one wavelength is measured. Advantages of dispersive systems are robustness and relatively low price. The development of fast Fourier transform methods has allowed the introduction of interferometry in IR spectrophotometry. FTIR spectrometers with a spectral resolution of 1 cm^{-1} in the frequency range of 510 to 14,000 cm^{-1} with 0.1 s observation times have various advantages over dispersive systems: the S/N ratio can easily be improved by increasing the number of scans, better use is made of the IR intensity and spectra are collected in seconds, allowing a quick multicomponent analysis. The AOTF technology coupled with a mercurium cadmium telluride (MCT) detector is a cost effective alternative to FTIR techniques for rapid identification of black plastics (ABS, PA, PBT, PE and PP). The system can be applied to contact-free on-line measurements. The spectrometer is mechanically very robust (no moving parts) [68]. The measurement speed of IR spectrometers meets one of the essential preconditions for using IR spectroscopy in on-line quality control, but this alone is not sufficient to achieve the short cycle times that are needed, which comprise the sampling and sample preparation stage.

The most fundamental decision to be made in planning an IR process analysis installation is whether to base the measurement on fundamental vibrational modes or on overtones and combination tones. High extinction coefficients may dictate the use of ATR and hence fundamental vibrations (as in case of QC of rubbers). On the other hand, where the need arises to use fibre optics, analysis is restricted to the high frequency overtone regions. Multicomponent in-process analysis requires **interfacing** to the process either in transmission with melt sampling using a flow-through cell or by-pass (Lambert-Beer's law applies), in transflection, or in reflection (specular, diffuse or ATR).

Sampling methods for on-line spectroscopy have recently been discussed [69]. In view of the flow problems, transmission analysis is not a widely used sample interfacing technique in IR process analysis. The fundamental absorbances corresponding to the functional groups of organic chemicals fall in the mid-IR "fingerprint" region of the spectrum and are generally very strong. On-line transmission measurements must be made on very thin samples (typically less than 100 μm), which renders analysis of molten polymers extremely difficult. For the first overtone region the optimum path length increases to some 0.5 mm, at the second and third overtones and combination tones 5 mm. The application of transmission analysis to process streams

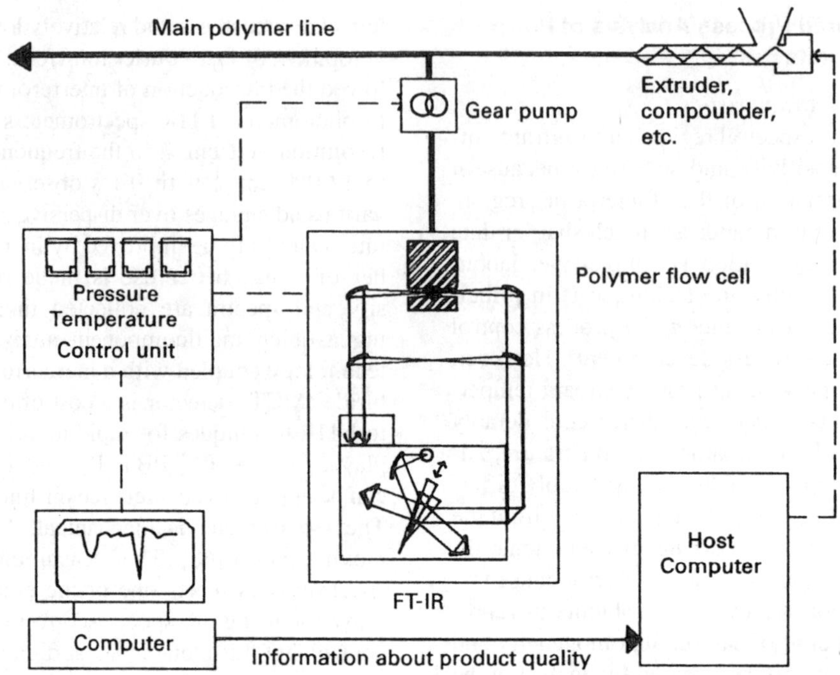

Fig. 7.5. Diagram of an on-line IR process control system for polymer production (Automatik, Germany). After Stengler and Weis [70]. Reproduced by permission of the International Society of Optical Engineering (SPIE).

requires an appropriate combination of short path length and weak absorption. The high extinction coefficients and consequently short path lengths (in the range of a few μm) of mid-IR instrumentation are often not compatible with industrial environments. While this may limit chemical composition determinations, the analysis of additives in low concentrations is less restricted. With the orders of magnitude lower absorbance of the overtones in NIR and short-wave NIR, much more robust flow cells can be used which are not susceptible to blockage.

Fast automatic sampling is essential for on-line quality control. A way of achieving this is to measure directly on the **polymer melt**. The physical problems associated with testing of materials in a moving process stream are strongly material dependent. For example, IR analysis of aqueous streams to determine the sugar and/or CO_2 content is carried out using the principle of circular internal reflection. A cylindrical crystal of an IR transmissive material is sealed into a chamber through which flows a sample stream from the beverage line. The utility of this system is limited to relatively fluid, non-viscous liquids and cannot be extended to conduct IR analysis of relatively opaque, viscous substances such as polymer melts.

Molten polymer can be pumped from the process stream to chilled calendering rolls producing a film that then passes through the mid-IR beam of the spectrophotometer. This testing technology is impractical. In later attempts at using mid-IR on-line a very viscous polymer flow was pumped at a reasonable flow-rate through an FTIR transmission flow cell [71,72]. Harvey [73] has been the first to describe a sample cell to perform IR spectrometric analysis in a non-invasive manner on a moving process stream consisting of a polymer melt. A relatively small portion of the melt flow (1 to 2 kg/hr) is diverted from the main melt channel or extruder into a melt pump by means of transfer lines heated to process temperature. This diverted melt stream is then driven through a transmission IR flow cell outfitted with special cell windows on both sides of the melt-flow channel (Fig. 7.5). The flow cell, constructed mainly of stainless steel, should typically be able to withstand pressure to 350 bar at 400°C. The path length of the flow cell can be either fixed or is (preferably) adjustable. The high mechanical specifications for the cell windows, together with the requirements that transmission in the mid-IR range (2.3 to 25 μm) should be as high as possible, severely limit the choice of window materials (ZnSe, diamond). The ability to monitor and control multiple additives directly in the polymer melt simultaneously via on-line FTIR in combination with multivariate calibration techniques opens up a new dimension in quality control possibilities in polymer manufacture and processing. The use of process FTIR spectrometers equipped with **variable-path length transmission cells** for the measurement of polymer melts on-line has been reported in the literature [71, 72,74,75]. Even in case of the bypass branch analysis of polymer melts there is still an inherent delay time (some 5 to 15 min depending on the flow-rate) between the moment a material sample leaves the process stream and actually passes through the optical beam for measurement. While this time lag is much less than the 1–2 h for off-line analysis, it still is long for effective closed-loop feedback process control.

Alternatively, it is possible to install **fibre optic probes** directly in the main stream **in-line** while the IR spectrophotometer remains remotely in a low vibration laboratory environment. In-line analysers, which do not remove any sample from the line, have the minimum possible lag time and do not change the sample physically or chemically from its nature in the process. Recently, bundles of 500 μm optic fibres have been developed for the 5000–900 cm^{-1} (2000–11,000 nm region), which permit transmission of IR energy over distances of several metres. Lowry *et al.* [76] have evaluated fibre-optic cables that might prove useful in FTIR remote sampling applications. The various optical fibres (chalcogenide, silver halide, heavy metal fluoride or sapphire) differ in their spectral window [77]. Due to the thermal stability and the spectral window, sapphire fibres are considered suitable for in-line characterisation of polymer melts in a production line (*e.g.* in an extruder head) as an alternative to discontinuously operating conventional "off-line" transmission IR spectroscopy of polymer films [78].

Quantitative information may be obtained by calculating the ratio of the peak height of CH_2 and CH_3 bands, thus compensating for different sample thickness. Quantitative determinations are always possible via FTIR but do require the building of a robust **calibration matrix**. In fact, via on-line FTIR, additives can be quantitated effectively from low ppm ranges as found in the case of most resin producers, to relatively high percentage levels as prepared by compounders and masterbatch suppliers.

The required short transmission path length due to strong IR absorption bands causes problems with the highly viscous melts in the transmission cells, as clogging can occur quite easily. Moreover, heated transfer lines to the process spectrometer are required. In order to overcome these limitations, direct measurements in the production line or in the extruder head can be performed by employing an attenuated total reflection (ATR) [77] sensing device installed directly in the production line. ATR devices are easily made into flow-through systems. ATR makes analysis of fundamental bands practical by providing the equivalent of a very thin transmission cell – typically 1 μm to 25 μm depending on ATR element material and number of reflections employed. **Process compatible ATR cells** and probes have been developed. Using a dip sensor based on conventional double reflection ZnSe ATR crystals, the polymer-analogue conversion of styrene/maleic anhydride (SMA) and a fat amine to styrene/maleimide in an extruder has been monitored [79]. However, this system suffers from relatively low sensitivity, since the signal is produced by only two internal reflections, and from inflexibility because the FTIR instrument needs to be located very close to the production line. With the use of optical fibres, which are actually elongated ATR elements, choosing an appropriate angle of light incidence can significantly increase the number of internal reflections, thus enhancing the signal intensity significantly [80]. Performance variation in ATR depends on the refractive index of medium and ATR crystal, the incident angle and number of reflections. Diamond ATR sensors are capable of dependably following even low concentrations of components participating in a reaction. For accurate quantification it is necessary to assure full optical contact of polymer melt and ATR crystal. Although the ATR method can be an effective solution for practical measurements, the "surface" nature of these probes can limit their efficiency. A disadvantage is that most suitable ATR crystal materials are toxic. Moreover, impurities may build up on the surface of crystals of ATR probes. Operating temperature is limited to about 100°C.

Specular reflectance is a much less common technique in IR analysis. It is not presently used for process analysis. It is more likely to find use in QC and product identification. It is also of interest to notice that a non-contact surface analyser uses an IR spectroscopic reflectance probe to obtain information without touching the surface. The probe is connected through mid-IR fibre-optic cables to an IR

Table 7.16. Main features of in-process FTIR spectroscopy

Advantages:
- Various sample stream interfaces
- No sample preparation (transmission)
- Real-time control
- Process and product control (*in situ* analysis)
- Relatively small calibration burden (often univariate)
- Quantitative
- Structural information (functional group analysis)
- Wide applicability

Disadvantages:
- Method development (typically 25 samples for multicomponent analysis)
- Differences with ambient temperature spectra (databases)
- Expensive thin cells (<100 μm for transmission)
- Exotic window materials (soluble, not resistant to water, acids, bases, easy to crack)
- No long distance fibre optics transmission
- Relatively poor analyte sensitivity

spectrometer and computer. The instrument's sensitivity is suitable for finding impurities or small concentrations of contaminants.

By moving the measurement from the well-controlled laboratory to the process environment, the influence of external process variables such as p, T, and flow turbulence will affect the measurements. When vibrational spectra are measured on- or in-line for process analytical and control purposes, the performance variations influence the shape of the spectra in a non-linear manner. Smilde *et al.* [81] have assessed the influence of these temperature-induced spectral variations on the predictive ability of multivariate calibration models.

Table 7.16 lists the **main characteristics** of in-process FTIR spectroscopy. Advantages realised via on-line FTIR process monitoring include the potential to reduce time-consuming lab testing and achieve real-time control rather than expensive over- and underfeeding of additives. Continuous assurance of product and process integrity enables production of more on-spec product. On-line FTIR spectroscopy is effective and advantageous compared with off-line methods (NMR, chemical determinations). When it is possible to avoid sample preparation then FTIR is usually an attractive technique. By applying FTIR to polymer melts, it is possible to introduce a well-established laboratory testing technique directly to the production floor to track levels of entire additive packs. Another advantage of FTIR is that the instrumentation is moderately inexpensive and interfacing to vapour/gas-streams is not complicated (in-line or by-pass). Moreover, mid-IR is more sensitive than NIRS. Instrument validation packages now support both the mid-IR and NIR. Wavelength validation by means of a built-in laser is secured for FTIR and better than for grating IR instruments. Mid-IR wave number and absolute transmittance standards are available for QA implementation [82]. A disadvantage is that mid-IR, at variance to NIR, transmits only through special non-glass fibres, which are brittle, (very) expensive and impractical for distances of more than a few feet. The use of optical mid-IR fibres is limited because of their lack of stability and their high loss of energy. FTIR provides highly accurate and fast concentration analysis, and suitability for in-process analysis. The high absorptivity of mid-IR requires very thin sample cells with thicknesses of only a few μm, which are of difficult construction and very expensive. Moreover, glass or quartz cells are not suitable for mid-IR because SiO_2 absorbs this light and is not transparent enough. KBr or NaCl windows are both not strong enough and soluble in water. Other mid-IR transparent materials such as crystalline CaF_2 and ZnSe are insoluble in water but quite expensive. For recent developments in mid-IR (and NIR) the reader is referred to ref. [83].

Sampling of liquids and solids/powders is more difficult in mid-IR process situations (short path length) than in NIRS; the mm cells for the analysis of low concentration additives are an exception. The conditions under which transmission mid-IR may be applied are limited to those in which thin samples are examined.

Expected **new developments** are new sample stream interfaces (*e.g.*, ATR techniques), the improvement of IR optical fibres [84] and data handling (chemometrics). A full spectrum approach provides the possibilities of multiple analysis from one measurement and correlations of the IR spectrum with other physical properties associated with composition. Discriminant analysis using principle components of mid-IR spectral data is a powerful quality identification tool where rigorous multicomponent analysis is not only costly but in many cases unwarranted. In combination with discriminant analysis, mid-IR spectroscopy becomes more readily available for QC validation by non-spectroscopists allowing validation without quantitation [85].

In situ mid-IR spectroscopy provides substantial benefits for **monitoring reactions**. It eliminates the time delay and inaccuracies associated with off-line sampling. It offers fast, sensitive, compositional information on reactants, products and by-products, thereby providing a measure of reaction progress which is more accurate than acid or hydroxyl numbers [86]. For reaction analysis [87], heat flow calorimetry has been combined with on-line FTIR spectrometry. The calorimeter provides process and safety data and FTIR supplies the on-line analysis, which allows a direct insight into the progress of a chemical reaction through the change in functional groups and chemical structures due to the reaction. It is also possible to make measurements on unstable intermediates under controlled conditions and to follow changes in the product as a function of time.

Coates *et al.* [88] have discussed the role of IR in a QC laboratory. Recently, Hansen [89] has reviewed on-line monitoring of polymeric processes by means of mid-IR and NIR discussing issues such as sample handling and residence time, multiplexing, optical path length, fibre-optic coupling, measurement sensitivity, data interpretation and practical and economical limitations on installation in a production process. Process IR has been reviewed [90], also with special reference to control systems for polymer melts and film [91].

Applications

Table 7.17 summarises on-line (multicomponent) additive analysis in **polymer melts** by means of mid-IR spectroscopy using flow cells. The transmission mid-IR method has been successfully used for various on-line determinations of the amount of a **single additive component** in a formulation. Stengler *et al.* [71] have first reported extensive on-line FTIR monitoring of additives in polymer melts using a high pressure, high temperature flow cell. Oleamide in LDPE was quantified by use of its carbonyl stretch (1715 cm^{-1}), which is characteristic of this additive [70], *cfr.* Fig. 7.6. The talc content (13 wt.%) in PA6 melts was determined by means of on-line FTIR at 1030 cm^{-1} (detection limit in ppm range) [71]. A similar application was developed for quartz in PA6.

Fidler [74,92,93] has similarly reported on-line FTIR analysis of silica antiblocking agent, erucamide and oleamide lubricant levels using a high-temperature (210°C) pressure flow cell (with 750 μm path length). High-percentage levels to low-ppm levels could be repeatedly measured to better than 5% of the silica content of known standards. Off-line measurement of silica content is typically accomplished by ashing the sample in a high-temperature oven and weighing the residual ash. This method is time-consuming and often accurate to only $\pm$10%. Moreover, error can be compounded when other fillers such as talc are present; they remain in the sample after ashing and cannot be differentiated from the silica in this methodology. The erucamide content in LDPE melt can be characterised by using amide or amide carbonyl absorption bands at 3543, 3420, or 1720 cm^{-1}. The measurement was made possible because the molten state eliminates the absorption bands of the crystalline-state polymer and sample thickness could be adjusted uniformly within the flow cell, allowing optimisation of the absorption band intensity for erucamide. Results obtained fell within 2 to 5% of the standards. Near on-line FTIR monitoring of Dowlex 2045 (LLDPE charged with five undisclosed additives) has been used for assaying all lots of resins before shipment [6]. Typical additive levels/RSDs were quoted as 600/15, 2958/48 and 1074/19 ppm. Fidler [94] has also described qualitative antioxidant analysis in molten PP by means of on-line FTIR, namely of a phosphite "U-6" (including conversion of the active phosphite to the inactive phosphate) and a blend "U-81" (hindered phenolic "I-1" and phosphite "U-6"). Primary antioxidant levels in various stabiliser blends, namely "U-81" ("I-1" and "U-6"), "U-85" ("U-2" and "U-6") and "U-87" ("U-2" and "U-6") in molten PP were compared. Rapid (near) on-line determination of polymer product changeover during extrusion via process-FTIR analysis has been reported [6]. The complete transition from HDPE to coloured TPE was observed within 5.5 min.

Verlaek *et al.* [50] have used mid-IR for the determination of the non-UV absorbers Zn-stearate, Ca-stearate and oleamide (SEP values: 29, 12 and 49 ppm, respectively in the melt as compared to 57, 37 and 34 ppm on film for typical nominal values of 450–1000 ppm oleamide and 1500 ppm stearates). It is again noticed that melt measurements using a 14 mL mini-extruder at 190°C often outperform polymer film measurements. SEP values for mid-IR measurements on LDPE melt containing Chimassorb 944, Irgafos 168, Irganox 1010/1076 were *ca.* 16 ppm; for a complete additive package in *fixed ratio* concentrations (combinations of Chimassorb

Table 7.17. On-line (multicomponent) additive analysis by means of mid-infrared spectroscopy

Polymer melt	Additive package	Reference(s)	Year(s)
HDPE	Antioxidant (0–500 ppm)	[70,71]	1989, 1990
LDPE	SiO_2 (0–1.5%) Oleamide (0–970 ppm) Irganox 1076 (0–200 ppm) Erucamide (250–1000 ppm)	[70,71]	1989, 1990
LLDPE	Irganox 1076 (0–1000 ppm) Irgafos 168 (0–1000 ppm)	[70,71]	1989, 1990
PP	Additives, lubricants, antiblocking agent, stabiliser	[70,71]	1989, 1990
EVA	$CaCO_3$ (0–3%) Erucamide (0–700 ppm) Oleamide (0–500 ppm) Irganox 1076 (0–300 ppm)	[70,71]	1989, 1990
PA6	Talc (2–40%), caprolactam (0–2%) Quartz	[70,71]	1989, 1990
PC	Slip agent (0.5–3%)	[70,71]	1989, 1990
PET	Diethylene glycol	[70,71]	1989, 1990
LDPE	Oleamide (0–1100 ppm)	[92]	1991
LDPE	Erucamide (0–1970 ppm), SiO_2 (4350–5575 ppm)	[74,93]	1992
LLDPE	Five undisclosed additives	[6]	1993
PP	Phosphite U-6 (0–0.31%) Blend U-81 (I-1 and U-6) (0–0.49%) Blend U-85 (U-2 and U-6) (0–0.49%) Blend U-87 (U-2 and U-6) (0–0.46%)	[94]	1993
LDPE	Chimassorb 944:Irgafos 168:Irganox 1010: Irganox 1076:oleamide:Zn-stearate[a] (0–1500 ppm)	[50]	1994
LDPE	Erucamide (0–1500 ppm), Sipernat (0–1500 ppm)	[60]	1996
LDPE	Oleamide (0–1500 ppm), Sipernat (0–1500 ppm)	[60]	1996
HDPE	Ca-stearate (0–2000 ppm), Chimassorb 944 (0–2000 ppm), Irganox B220 (0–2000 ppm)	[95]	1996

[a] Analysis of single and multiple components (in fixed ratio).

944, Irgafos 168, Irganox 1010/1076, oleamide, and Zn-stearate) the SEP value was 6 ppm (as compared to 8 ppm for UV measurements).

Verlaek *et al.* [60] have also carried out various **multicomponent additive analyses** on PE melt samples from a mini-extruder at 190°C by means of mid-IR and UV-methods. With careful calibration (using 36 spectra) it is possible to determine the individual concentrations of independently varying additives from a melt mid-IR spectrum for polyethylenes. This was shown for LDPE/(erucamide, Sipernat), LDPE/(oleamide, Sipernat) and for HDPE/(Irganox B220, Chimassorb 944, Ca-stearate; 1500 ppm each) in mid-IR with a

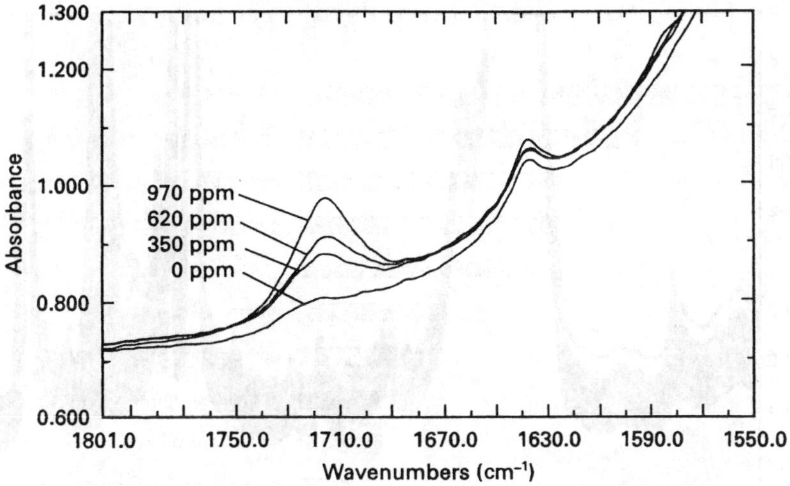

Fig. 7.6. FTIR spectra of various concentrations of oleamide in molten LDPE. After Stengler and Weis [70]. Reproduced by permission of the International Society of Optical Engineering (SPIE).

Table 7.18. Composition of B blends used in polyolefins

Code	Composition		Ratio (wt./wt.%)
Irganox B215	Irganox 1010	Irgafos 168	1:2
Irganox B220	Irganox 1010	Irgafos 168	1:3
Irganox B225	Irganox 1010	Irgafos 168	1:1
Irganox B900	Irganox 1076	Irgafos 168	1:4
Irganox B921	Irganox 1076	Irgafos 168	1:2

typical standard error of prediction of *ca.* 30 ppm. Palmen *et al.* [95] have reported scale-up experiments for molten HDPE/(Irganox B220, Chimassorb 944, Ca-stearate) using mid-IR spectroscopy. The observed poor predictive value of Ca-stearate with respect to the aforementioned mini-extruder experiments was ascribed to aggregation of this additive. As a result of clustering, absence of a (linear) correlation between concentration and extinction compromises PLS calibration. Although the polymer industry still pursues its on-line mid-IR efforts very little progress has been noted over the last few years (*cfr.* Table 7.17).

For monitoring of PE and PP production an accurate and fast analytical method is wanted for the determination of Irganox B blends, which are **blends** of Irgafos 168 with Irganox 1010 or Irganox 1076, as shown in Table 7.18. In the absence of other phosphorous containing components such blends can quantitatively be determined by means of XRF. However, strict reliance upon the determination of one element (P) only is put at risk when the blend composition shows deviations from stoichiometry. Moreover, for this purpose highly accurate phosphorous XRF analyses are required. On the other hand, extraction methods are slow, in particular for Irganox 1010. This component shows greatly varying extractability rates and speeds from various polymers. Extraction of Irganox 1010 from LDPE proceeds more rapidly than from HDPE. Therefore, development of an improved analytical method based on an on-line spectroscopic method is a useful exercise. The practical utility of the reported on-line additive analysis systems depends on the degree of control which is achieved (time delay from addition to exit). In analogy to on-line UV monitoring, it is quite noticeable that most reported efforts concern PE rather than PP.

Mid-IR spectroscopy can detect a high percentage of polyolefin additives by direct transmission measurement of **films** in a test taking less than 10 minutes (*i.e.* considerably less than extraction/chromatography) [85]. Multicomponent quantitative analysis of polyolefin formulations requires extensive work in preparation of standards, calibration and maintenance. Due to interferences from

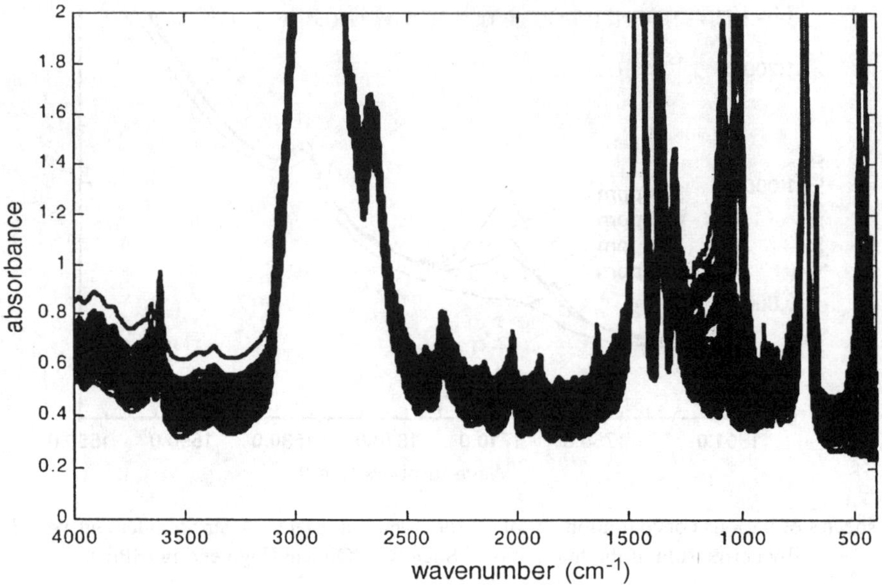

Fig. 7.7. Absorbance spectra for 109 calibration data for additive C in PE film. After Leardi *et al.* [97]. Reprinted from *Analytica Chimica Acta* **461**, R. Leardi *et al.*, 189–200, Copyright (2002), with permission of Elsevier.

similar functional groups, full quantitative analysis generally requires a combination of techniques involving spectrometries (IR), and chromatographies (GC, HPLC).

IR is limited mostly by the similarity and overlap of many additive absorption bands and by the level of sophistication required to interpret the fingerprint in detail. This presents a major opportunity for qualitative multivariate classification techniques, which can be used to recognise the many subtle details in the polyolefin formulation. Whereas classification techniques have not been ignored in applications like raw materials checking, unknown identification and grouping of complex materials, reported applications in process monitoring are limited. Van Every *et al.* [85,96] have demonstrated the application of IR spectroscopic classification techniques (Principle components/Mahalanobis distance Discriminant Analysis, PMD) for validation of polyolefin film products and have indicated the requirements for PMD in order to be a viable quality control technique. The authors have compared the ability of discriminant analysis (PMD) of mid-IR data and DSC to detect trace amounts (up to 250 ppm) of sodium benzoate (NaBz) in PP formulations. While in all cases DSC gave sensitive and consistent detection of the nucleator, PMD shows very sensitive flagging of samples down below 80 ppm level, which is the limit of detection for a univariate calibration for this compound. Discriminant analysis using principle components of mid-infrared spectral data is a powerful quality validation tool where rigorous multicomponent analysis is often costly.

Selection of variables for multivariate calibration can be considered an optimisation problem. Well performed variable selection in multivariate analysis is a very relevant step, because the removal of non-informative variables will produce better predicting and simpler models [98]. There are numerous approaches for **selection of variables**. Using FTIR spectral data Leardi *et al.* [97] have illustrated selection of variables on the basis of a genetic algorithm (GA) [99] combined with PLS for the prediction of the concentrations of three undisclosed additives (A, B and C) in PE films. The exercise aimed at developing an at-line QC tool. The entire data set consisted of 319 spectra with a significant baseline offset (Fig. 7.7). Path length correction was carried out by normalisation to a polymer peak (2662 to 2644 cm^{-1}).

Table 7.19 shows the good fit between expert-selected regions for additives B and C, based on knowledge about the spectroscopy of the additives, the polymer and the other additives in the matrix, and those derived by the genetic algorithm for automatically selecting variables for calibration without requiring spectroscopic experience from the user. In both cases, the root mean square errors of prediction

Table 7.19. Regions selected by experts and genetic algorithm

Additive	Expert-selected regions (cm^{-1})	GA regions selected[a] (window size = 19.3 cm^{-1})
B	3600–3260	**3626–3474**, 1929–1873, 1524–1506, 1234–1159, *675–580*
C	899–829	1408–1313, *1234–1121*, **906–754**

[a] The GA regions in bold agree with the expert-selected regions. Peaks in italics were also known by the experts to be related to the additive and were not included in the original expert-based model.

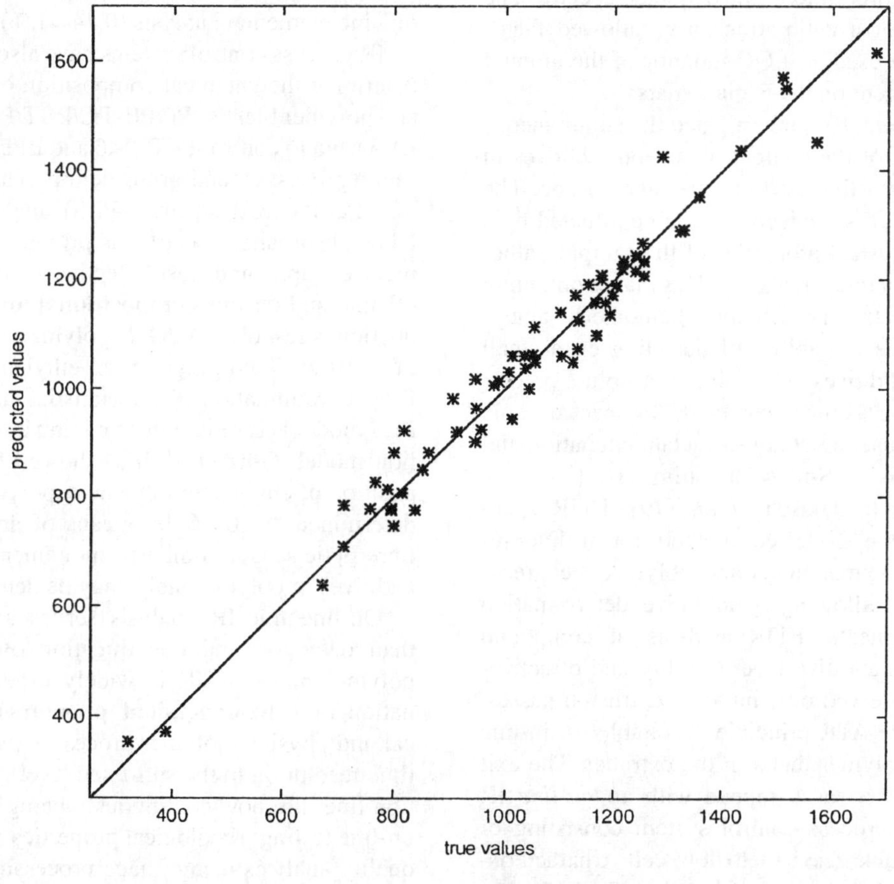

Fig. 7.8. Predicted *vs.* actual concentration plots for the validation data for additive C in PE film, ppm. After Leardi *et al.* [97]. Reprinted from *Analytica Chimica Acta* **461**, R. Leardi *et al.*, 189–200, Copyright (2002), with permission of Elsevier.

(RMSEP) were comparable, indicating that GA selected a model with equal predictive ability. Using the GA approach a non-expert will therefore be able to efficiently construct reliable calibration models with little or no intervention by an expert. Further, the approach can aid the expert with difficult calibration problems where selection of the variables is not obvious. Actually, in addition to the region(s) indicated by expert users, GA selected other regions. For additive C, the region 1234–1121 cm^{-1} was known to be related to this additive. Addition of relevant spectral regions is expected to improve the performance of the model due to a signal averaging effect. Other regions appeared to be related to the polymer. In case of additive B it is known that the catalyst "health" influences the state of this additive and the polymer produced. Therefore, it makes sense that polymer peaks would contribute to modelling this additive. Figure 7.8 shows the predicted *versus* actual concentration plots.

Factor analysis had previously been used by Culler *et al.* [100] for quality control monitoring of a polymeric composite system of γ-aminopropyltriethoxysilane (γ-APS) coupling agent on an E-glass mat substrate. In this system, factor analysis successfully indicated the number of pure components, extracted the spectra of the pure components, indicated the relative concentrations of spectra, and improved the S/N ratio in the extracted spectra. The construction of a calibration curve allowed factor analysis to be used as a QC monitor of the amount of coupling agent on the E-glass mats.

Fischer *et al.* [101] investigated the simultaneous quantification of the content of several additives in PVC with an in-line diffuse reflectance probe. The signal from diffuse reflectance can be affected by a number of physical properties of the sample, rather than just its chemical make up. This makes obtaining quantitative data very difficult. Chemometric analysis showed the possibility of detecting even small amounts of additives (3%) with an absolute prediction error of 0.3%. Step-scan PA-FTIR spectroscopic studies were used to study surfactant exudation and film formation in PS-nBA latex films [102].

According to Jakisch *et al.* [79], FTIR spectroscopy is the preferred method for in-line investigation polymer melts and polymer melt reactions/kinetics, allowing quantitative determination of all components. FTIR analysis of compound melts enables additive level stability and effectiveness to be observed over multiple extrusion passes. The use of the ATR principle is suitable for in-line analysis of polymer melts in the extruder. The exit of the extruder was equipped with an *on-line* IR transmission process control system consisting of a 150 μm thick ZnSe melt flow cell. Characteristics of such systems have been described [71,74]. Another process spectrometer with an *in situ* ZnSe-ATR dipper probe was mounted at different positions in the extruder. For in-line ATR the residence time plays no role. Only the first 5 μm (corresponding to the penetration depth of the IR radiation) are examined. Minor components are thus detected with difficulty. Jakisch *et al.* [79] monitored the conversion of styrene-maleic anhydride copolymers (SMA) with fatty amines into styrene-maleimide copolymer (SMI) during **reactive extrusion** by means of FTIR. In principle, both mid-IR and near-IR spectroscopy with ATR, transmission and diffuse reflectance probes are suitable for quantitative on- and in-line process analysis of multicomponent polymer mixtures [101]. For quantitative ATR-FTIR process spectroscopy multivariate analysis is used and calibration with a non-spectroscopic analysis technique is necessary (*e.g.* elemental analysis). On-line FTIR spectrometers can be used under real plant conditions to monitor the composition of polymer melts leaving extruders. On-line IR analysis for the major component in a methacrylate copolymer achieves a better precision (0.09 wt.%) than the corresponding off-line elemental analysis (0.44 wt.%) [103].

IR process control systems have also been used to determine the chemical composition of copolymers and polymer blends (PP/PE, PC/PBT/PET, PC/ABS, EVA) and to control PET, PA6 and EPDM polymerisation processes (end-group determination, *etc.*) [70, 92]. Partial least squares (PLS) analysis of ATR-FTIR absorbance spectra has provided an accurate, precise, rapid and cost effective method both for off-line and on-line **compositional analysis** at production sites of EO/PO copolymers in the range of 0–10 wt.% co-polymerised ethylene sites [104]. Proper examination of the statistics underlying the PLS model is essential in providing a robust calibration model. Götz *et al.* [80] showed that the composition of ethylene/propylene copolymers could be determined at 200°C by means of an IR sapphire fibre-optic sensor. Similarly, monomer residuals and additives in polymer melts may be determined.

On-line mid-IR analysis serves a wider scope than detection and quantification of additives in polymer melts. FTIR is widely used for determination of **time-dependent phenomena** in chemical and physical polymer processes (with very high time resolution to the sub-μsec level). Film production lines are now continuously being controlled by on-line testing: rheological properties and MFI, film quality analysis using image processing for continuous determination of impurities and transparency, and FTIR for layer thickness and qualitative and quantitative determination of additive content [105]. As shown in Chp. 7.2.4, NIRS is widely used for on-line reaction monitoring. Improvements in fibre optics now also enable the use of mid-IR for this purpose. *In situ* mid-IR measurements allow tracing a reaction profile by means of changes in infrared absorbance bands as a function of time. The specificity of mid-IR permits identification and tracking of intermediates, active catalytic species and by-products. On-line determination of grafting of vinyltrimethoxysilane on LLDPE by means of FTIR spectroscopy has been reported [72,75]. Kiparissides *et al.* [106] have reported on-line monitoring of emulsion co-polymerisation using an ATR probe in combination with factor analysis. Other applications concern measurement of polyethylene cross-linking and UV curing of resins. Real-time FTIR (RT FTIR) provides a unique means of following the UV cure of acrylated liquid crystals. *In situ* FTIR, ATR-FTIR and photoacoustic spectroscopic approaches have been used to study supercritical fluid impregnation, diffusion, drying, dyeing and extraction of polymeric materials. Kazarian *et al.* [107] have described *in situ* spectroscopy of CO_2-induced plasticisation of glassy polymers. *In situ* ATR-FTIR can also be used in studies of antifouling coatings. Mid-IR internal reflection spectroscopy (IRS) with reactive internal reflection elements (IREs) has also been utilised to quantitatively monitor *in situ* the adsorption of surfactant species [108].

In the past, esterifications were typically monitored and controlled by off-line determinations of the hydroxyl and acid numbers. Now utilising mid-IR based technology for the same purpose provides valuable information regarding reaction trends, kinetics and end-point determination [86]. *In situ* mid-IR spectroscopy also provides real-time monitoring of the Grignard formation in processes in which the fast reaction kinetics and overall reactivity of reactants and products precludes the removal of samples for off-line measurements [109].

Many other in-process mid-IR applications have been reported in the (petro)chemical industry, including ATR probes that work under harsh conditions. Especially physico-chemical determinations, such as the distribution of lubricating agents in transparent polymers, require the selectivity of mid-IR. *In situ* mid-IR is well established for gas analysis and gas phase applications in safety and environmental monitoring. Liquid samples in an industrial surrounding are more complicated; consequently, mid-IR is used when NIR is not informative enough, in particular for low concentrations. The spectra can be used to quantify components in mixtures or to determine product characteristics.

FTIR in combination with a cone calorimeter or in relation to various fire smoke toxicity tests has considerable potential for on-line analysis of fire gases from burning rubbers and plastics [110]. Kowol *et al.* [68] have reported the use of a FTIR-AOTF spectrometer for rapid identification of **black plastics** from automotive construction. In the wavelength region between 2.5 and 4 μm (where the fundamentals of the CH- and NH-stretching vibrations are observed), mid-IR spectroscopy allows distinction of the major types of blackened technical polymers; using 4 sec integration the detection times are worse than for NIR applications. However, as no suitable fibre optics is available which allows non-contact measurements, the object to be identified needs to be in direct contact with the sensor for about 1 s.

On-line FTIR has also been used to monitor HCl and HF emissions in fibre glass manufacture as an alternative to wet chemical techniques such as USEPA Method 26 and 26a [111].

Because of the limitations in using mid-IR fibres, "fundamental" in-process spectroscopy will remain limited and dependent on special probes (ATR, transmission, *etc.*); near-IR and Raman are mid-IR rivals. The present development of NIR spectrometers offers several advantages compared to the mid-IR technique, *e.g.* faster measurements and easier sample treatment. It appears that in-process mid-IR spectroscopy is more geared towards reaction monitoring; on the other hand, NIR spectroscopy plays a prime role for industrial on-line analysis (more general).

Calibration techniques for FTIR process monitoring have been addressed [112]. Xanthos *et al.* [83] have reported recent developments in in-line FTIR, NIR and optical microscopy for monitoring extrusion processes.

7.2.4. Near-infrared Spectroscopic Process Analysis

Principles and Characteristics

The role of quality is increasingly being recognised in chemical production. From an analytical chemical point of view, in most cases the goal in continuous processes is to keep the process composition steady at around the optimum physical and chemical conditions. Uniform quality is a requirement with many other aspects too, such as legal obligations, economical production, environmental protection, and plant safety. All of these require that the composition of various products be kept stable [113]. Consequently, reliable, selective, and sensitive process analysers are much needed.

As the purpose of the analysis is to use the data to control the process timing considerations are vital. The response time of a complete control system (to correct for any change in the concentration of the product) includes the time of sampling, analysis, calculation of the concentration and the amount

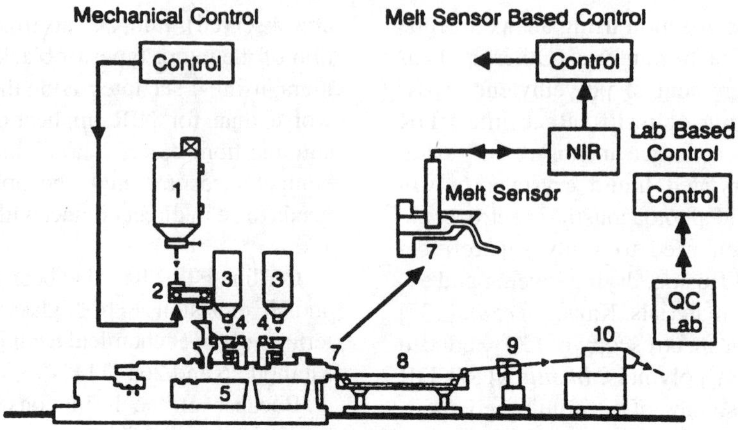

Fig. 7.9. Schematic view of feeder control, lab control and melt sensor-based control. Keys: 1, thermoplastic storage; 2, weigh feeder; 3, pigment storage; 4, loss-in-weight feeder; 5, extruder; 6, vacuum system; 7, strand die; 8, water bath; 9, air knife; 10, strand pelletiser. After Sohl [9]. Reproduced by permission of C.H. Sohl, Experimental Station, Du Pont de Nemours, Wilmington, DE.

of time required for the appropriate adjustment. The relatively long extraction times usually prohibit the use of such wet chemical methods for QC analytical applications in a plastic manufacturing plant. However, more recently, developments such as the use of microwave oven heating or accelerated solvent extraction have significantly shortened the extraction time, down to 20–60 min (*cfr.* Chp. 3 of ref. [113a]). It has also been claimed that there is considerable potential for using on-line dissolution systems, such as microwave-digestion flow systems, for process analysis [114].

Non-destructive optical analysis offers an instantaneous in-line measurement of concentration. NIR spectroscopy fits well into the list of technologies suitable for process analysis; it is fast, precise and non-destructive. When used properly it is also accurate for macro-analysis of major chemical composition parameters or contaminants. Its methodology includes the use of quantitative and qualitative chemometric techniques. In near-infrared, due to the generally much lower absorption coefficients of most combination and overtone bands with respect to the fundamental vibrations of mid-IR, undiluted materials can be analysed *in situ* in many cases through reasonable path lengths. Process control using NIRS has developed as from about 1980.

Unlike a single wavelength UV monitor, or a refractive index monitor, a NIR analyser is not a single parameter measurement device. In a process situation, NIR analysis is carried out in a manner similar to laboratory analysis in which reflectance or transmittance is measured consecutively at several wavelengths. However, process material is changing its position. Moving inhomogeneities of the sample and the finite speed of the analyser interact in a complex way [115].

NIRS allows integrated process monitoring and control and mobile product identification. NIR sensing devices are playing a significant role in on-line process control, particularly when reflectance measurements are appropriate. Sohl [9] distinguishes various **extruder control strategies**, namely feeder or product pellet stream control, lab based and melt sensor based feedback control (*cfr.* Fig. 7.9). Lab based feedback control is slow and labour intensive. Feeder control does not detect hoppers loaded with the wrong material. While NIR sensors can, and have been, placed in either the feeder bins or the product pellet streams to monitor composition, such measurements are based on NIR reflectance techniques, which generally provide analytical precisions which are a factor of 10 worse than melt based transflectance measurements. In many cases, the increased precision is well worth the minor inconveniences of melt-based sensors. Melt sensor based control offers the best of both lab and feeder control strategies. Melt measurement is delayed from the feeders by only a short time. The on-line NIR technique is especially valuable for studies under unusual or extreme conditions.

Balke *et al.* [116] have discussed the design of melt-at-die, melt-in-barrel and strand interfaces between the NIR spectrometer and the molten polymer for monitoring just before the extruder exit, in the main barrel and after the product exits from the extruder, respectively. It is important that the interfaces protect the inserted optical fibre probe from the harsh environment within the extruder (typically 200°C, 20 MPa), while permitting easy replacement of a probe without interrupting the process. The design of the interface affects multivariate analysis directed at composition prediction.

A variety of **sampling protocols** have been developed for NIRS, such as:

- direct intrusion into a liquid or vapour phase process stream using a transmission flow cell [117];
- a fast loop by-pass construction with fibre coupled interface to a remote spectrometer; or
- direct insertion fibre optic sampling probes, such as ATR or simple transmission probes.

As the NIR absorbances are typically 100–1000 times weaker than IR absorbances, NIR radiation must travel through 100–1000 times more material to obtain a useful spectrum. Thus, a greater path length (typically 0.2–5 cm) than mid-IR flow cells is to be utilised eliminating the need for a side stream of polymer melt (as in mid-IR) and resulting in more representative sampling of the melt composition. Hansen *et al.* [118] have developed various probe designs suitable for the adverse conditions typical of polymeric processes which can withstand pressures up to 1000 bar and temperatures up to 450°C and allow real-time in-line monitoring.

In some on-line NIR analysers the **sampling arrangement** is fixed; in others the sampling probe or cell is selectable. Before deciding upon the sample treatment, the basic optical arrangement has to be selected. This greatly depends on the dominating optical interaction in the sample. If the path length is adequately selected the simple **transmission** arrangement (most frequently used) works well for clear liquids. For opaque samples the near-infrared reflectance arrangement is most appropriate. For moderately opaque samples, such as liquid streams containing particulate matter, the amount of scattering is often not enough to reflect a large portion of the illuminating light; in this case transflectance provides the best quantitative results.

In NIR practice the sample may be brought to the NIR instrument (off-line), sample and NIR instruments may be coupled by fibre optics (large working distance), large instruments may be coupled optically to the sample on-line, or hand-held NIR technology may be used. Near-infrared analysis of liquids has the longest history of all sampling types. Simple two-filter, two-wavelength instruments have been used for decades, mainly for the analysis of moisture in various liquid matrices. The main advantage of the integrated liquid analysers, as opposed to the fibre optic-based ones, is the relatively higher precision of the analysis, allowed by the smaller losses in getting the light to the sampling point and by better light collection [113].

For mobile product analysis by means of NIR a reliable sampling system is a must. The various sample transport systems, namely active transport (pump), passive transport (pressure difference) and in-stream analysis (no by-pass transport), each have advantages and disadvantages. Factors influencing the precision of analyses are optimum selection of the path length and temperature control [113]. The optimum path length depends upon the wavelength and is not immediately obvious. On the short-wavelength side of the spectrum the absorption bands are so weak that in order to achieve a few tens of absorbance units the cell path length has to be increased to 10–100 mm. When the entire spectrum is used for the calibration, the optimum path length is arrived at if the mean sample absorbance approximates 0.434 [119]. Reproducibility in path length in transmission NIR analysis is equally important. Slight temperature variations change the effective path length of the cell as well as the spectral characteristics. Therefore, NIR analysers are equipped with temperature-controlled liquid cells, ensuring approximately 0.1–0.2°C temperature stability. Yet another aspect of the temperature in process analysis is that polymer samples must be hot in order to flow. One of the limitations of optical analysis of industrial samples is that the measurement is grossly affected by suspended particles and gas bubbles in the liquid and (incomplete) phase separation. An aliquot of the main stream can be continuously filtered before the analyser.

As indicated before (*cfr.* Chp. 1.2.2), five available technologies for use in **NIR process analysers** may be distinguished, namely optical filters (in photometers), scanning monochromators, PDA (photodiode array) spectrographs, interferometers and AOTF/AOTS systems [120]. In early applications of NIR for on-line measurement for process and quality control, filter-based instruments were used. Lately, the attention has shifted to scanning instruments, which allow the use of a variety of mathematical approaches for signal processing and calibration. Disadvantages of scanning instruments are expense and complexity in long-term maintenance. In process analysis, PDA offers an ideal solution, covering full range spectral data with no moving parts. High-end FT-NIR spectrometers are optimised to the special demands of on-line analysis. Due to the advantages of the FT technique, precise measurements with high sensitivity, high scan rate and high spectral resolution are possible.

Very fast, dual beam acousto-optical tuneable filter (AOTF) spectrometers have been developed primarily for multicomponent process analysis in the near-infrared [115,121]. AOTFs are opto-electronic devices that utilise the interaction of an ultrasonic and a light wavefront [122]. AOTFs are prepared from optically transparent birefrigent crystals such as TeO_2 to which an array of piezoelectric transducers are bonded. The acousto-optic effect can be used to produce tuned monochromatic radiation. A truly high-speed analyser comprises not only fast wavelength selection but also a fast detector system, fast signal processing and rapid data processing. The goal of the development was to be able to random access a selected wavelength within 100 μsec and to take a detector reading within the same time frame. AOTFs offer speed and high reliability. The wavelength repeatability, low noise, and fast wavelength access allow the acousto-optic analyser to be used for most applications that employ more conventional analysers and opens up new potential uses for fast changing samples and new process applications where compact, rugged and reliable design is of great importance. AOTF spectrometers are capable of identifying the most common plastic materials in a very short time [123].

Optical fibre sensors are particularly suitable for NIRS applications since standard silica optical fibre transmits light well over this wavelength range. Fibre optics has brought NIR to the processing line.

Moessner [124] has reviewed process NIR technology and has analysed advantages and disadvantages (in comparison with devices such as process titrators and flow-injection analysers). Table 7.20 summarises the **main features** of process NIR technology. The utility of NIR analysers for real-time process control is considerable. A NIR analyser approach yields continuous, real-time information, which has higher process control value than either slower continuous analysers requiring side-stream sampling (*e.g.* FTIR) or discrete number generating analysers (*e.g.*, GC, titration, FIA). As NIR uses low-energy radiation that does not initiate chemical reactions in the process stream fouling of flow-cells is also kept to a minimum.

Table 7.20. Advantages and disadvantages of process NIR technology

Advantages:
- Favourable hardware cost
- Gains in safety and timeliness of analysis
- High process control attainment
- Easy sampling (long distance fibre optic probing)
- Allowance for thick samples
- No reagents or waste streams
- Non-invasive, non-destructive
- High reliability
- Low maintenance costs
- Improved control capability reducing manufacturing costs
- Improved process control allowing tighter product specifications and consistent product quality

Disadvantages:
- Method development costs (model building)
- Complex method calibration
- No trace component analysis method
- Sample temperature control
- High cost of full spectrum NIR analyser

Optical analysis can be done relatively fast. NIR analysis can be performed within a few seconds, depending on the magnitude of the analytical signal, sample absorbance, and overall error of the measurement. Simple analytes like moisture require only two to three wavelengths. On the other hand, more complex analytes may be calibrated best using up to six wavelengths. Multiple linear regression is the best calibration technique [125]. Near-infrared analysis is typically a secondary analytical method, *i.e.* it has to be calibrated with several samples of known concentrations.

NIRS has been used only occasionally since the early 1950s for industrial problem solving, but a real breakthrough as a **quality- and process-control tool** occurred within the last decade following introduction of efficient chemometric evaluation techniques and development of light-fibre technology in combination with special probes. NIRS is quickly overtaking Raman and primarily mid-IR spectroscopy as a process monitoring and process control technique. The main reasons are the much easier sample presentation and the possibility to separate the sample measuring position and spectrometer by light fibres over distances of several hundred metres. Although similar arguments hold for Raman spectroscopy, interference by fluorescence and safety arguments are still limiting this industrial application on a real broad scale. As reported elsewhere [126], a few years ago about two-dozen chemical processors in refining, plastics processing, and food production used NIR spectroscopy in the closed-loop mode. NIRS can provide direct *in situ* measurements in a reaction vessel of chemical processes providing information as to reactants, intermediates, finished products and side products (reaction monitoring). Cure processes may be tracked by a variety of on-line monitoring techniques, including NIRS, dielectric analysis, ultrasonic velocity, NMR and Raman spectroscopy.

Fibre-optic diffuse reflectance probes are designed for remote monitoring of a broad spectrum of chemical processes involving **pastes, slurries, emulsions**, or other scattered media. While NIR is less sensitive than UV measurements, this makes NIR ideal for the analysis of bulk constituents and the prediction of polymer physical properties. With optical path lengths in NIRS in the order of 10 mm (as compared to ~0.1 mm for mid-IR measurements) flow cells do not block and viscous liquids may be analysed. Melts allow easy control of the optical path length and provide better sensitivity for NIR measurement. The relationship between the NIR spectrum of polyolefins and their density and melt flow index is already being exploited for on-line process analysis and control of polymer grades during polymer production. UV/VIS/NIR spectroscopic analysis of **pellets** is very difficult.

The field of process NIR is expanding rapidly due to the requirements for real-time, multi-parameter analysis for the purposes of production efficiency, waste minimisation, just-in-time manufacturing, regulatory compliance, environmental monitoring and product optimisation. The future promises to be a challenging time for NIR spectroscopists. NIR analysis will be used for many on-line applications and simultaneous, multicomponent analysis will become common. The recently developed rugged, low-cost, high-resolution NIR microspectrometers ($6 \times 4 \times 1$ in.), based on MEMS technology, represent a paradigm shift for industrial process spectroscopy [126a]. The greatest potential for near-infrared reflectance analysis (NIRA) is in the statistical process analysis of manufacturing processes. The speed and non-destructive nature of this technique make it ideally suited to continuous material control. NIR data and chemometrics can be used to study uncontrolled, ill understood but possibly systematic process fluctuations.

Regulatory issues in NIR process spectroscopy are still unsettled; few NIR methods are registered despite the hundreds of NIR spectrometers in use in industry. Apparently the field still struggles in defining **validation guidelines**.

The basis for process analytics and control strategies pertinent to multiple analytical techniques is described in several classic textbooks [13,14,127]. Workman [90,128] has published comprehensive reviews on process NIR spectroscopy; other reviews are on account of Kemeny [113,115]. A critical comparison of near-IR and mid-IR process analysis has been reported [45].

Applications
Since almost all substances which are practically relevant have characteristic NIR absorption bands, quantitative analysis via NIRS is generally applicable to *on-line* concentration measurements in connection with chemical reactions, chemical equilibria, and phase equilibria. Hansen [89] has reviewed on-line monitoring of polymeric processes by means of mid-IR and near-IR in 1991 discussing a variety of issues (*cfr.* Chp. 7.2.3). At that time, NIR (*i.e.* wavelengths of 0.8 to 2.5 μm) with low optical loss communications grade fibre-optical cables was viewed as a technological challenge. Problems related to low energy levels have been solved by multiple wavelength scans and utilisation of advanced electro-optics detectors. Multivariate calibration methods (MCM) can be used to glean quantitative information from NIR spectra of samples of essentially known composition. Only in the past decade efforts have been made to develop an **in-line monitoring** system using NIR spectroscopy for extruder control. Both FTIR and NIR spectroscopy are used for monitoring extruder processes. The harsh environmental conditions (typically 400°C and 2000 psi) provide a significant challenge in sampling. FTIR and NIR show varying degrees of applicability to polymer melt monitoring [129]. Ciurczak [130] has described mid-IR and NIR spectroscopy using flowing systems.

NIR spectroscopy has become an analytical tool frequently called upon in many production processes. Its use in polymer processing applications such as polymer extrusion [83] increases greatly product quality. The applications of non-destructive NIR methods to synthetic polymer studies have been reviewed [131–133]. Typical reported applications include process and pilot monitoring [134], real-time analysis of thermoplastic melt processes [129, 135], insoluble cross-linked systems [136], polymer flakes, fluffs and film. NIRA has controlled production in dyeing of textured PA6 carpet yarns with

C.I.Blue 127:1 [137]. De Wit [120] has reported the application of NIR reflectance spectroscopy with a multiple filter-based instrument for the on-line compositional analysis of acid-wetted cellulose chips (moisture at 1940 nm and organic acid at 1740 nm with reference regions at 1550 and 1820 nm). The analytes were monitored simultaneously in an aggressive environment with mechanical vibration, heat and acid vapours. NIR process sensing has further been used in the determination of polymer concentration in flowing solutions, and in the determination of molecular orientation of polyamide film during a drawing process [138]. A real challenge in in-process analysis is the contact-free determination of additives in running film.

Automatic **polymer waste sorting** plants based on NIR identification are operative (*cfr.* Chp. 1.2.2). For identification and sorting of carpets a portable NIR spectroscopic system – CarPID^{TM} – was developed [139]. Other reported NIRS applications are to be found in the quantitative analysis of copolymers or blends; the near-IR range allows for accurately monitoring of the monomer ratio and **residual monomer content**. Ikeda [140] used near-IR spectrochemical analysis in controlled manufacture of polyester plasticisers. Jones *et al.* [141] similarly described the use of NIR analysis for controlling plasticiser ester formation; the esterification of phthalic anhydride by isodecyl alcohol was exemplified.

Polymer melts:
Near-infrared spectroscopy in the spectral range of 4000 cm^{-1} (2.5 μm) to 12,500 cm^{-1} (0.8 μm) is an appropriate in-line method for real-time and quantitative analysis of polymer extruder processes and for monitoring components in polymer melt feeds. It is possible to separate the spectrometer from the extruder by up to 1,000 metres by application of glass fibres. Over the past decade, various applications of in-line NIR analysis of polymers in transmission have been described [9,118,124,135,142–145]. It was originally the understanding of workers in the field that only very high additive concentrations (>10,000 ppm) could be determined quantitatively by NIRS, but lower values (500 ppm) have also successfully been reported [49,146], in particular for NH- and OH-containing additives. Analysis of trace component levels is completely inappropriate for on-line NIR applications.

In-line NIR spectroscopy combined with multivariate analysis is a very powerful and simple technique, which, installed on a production line, can be very effective for quality control. The relatively weak absorptions in the NIR region allow long path lengths and enable maintaining continuous polymer flow between the probes.

Several resin producers have considered programs for the introduction of on-line NIR analysis into their production facilities, *i.e.* **process control of additive dosage**. NIRS has been used to monitor polymer melts for polymer and/or additive composition with *in situ* analysis in transmission, transflectance and reflectance modes. Research on the application of NIRS to in-line and on-line additive analysis in melts (Table 7.21) is as yet by no means as extensive as in case of UV (Table 7.15) and mid-IR (Table 7.17). Batra *et al.* [143] have applied NIRS

Table 7.21. In-line and on-line (multicomponent) additive analysis by means of near-IR spectroscopy

Polymer	Additive package	Reference	Year
PET melt	TiO_2	[143]	1994
Polyolefin melt[b]	Pigments (3.6–56 wt.%), $CaCO_3$ (up to 33.1 wt.%)	[147]	1996
PP melt[b]	Chalk (0–40 wt.%)	[148]	1997
EVA melt	Erucamide (2500 ppm), vinyl acetate (12.0 and 18.1 wt.%)	[149]	1998
PVC melt[b]	PMMA, lubricant (3 to 6%), modifier (8 to 13%)	[37]	1998
LDPE melt	Armostat 310 (0–0.2 wt.%), erucamide (0–0.5 wt.%), Irgafos 168 (0–0.1 wt.%), Irganox 1076 (0–0.1 wt.%), Tinuvin 622 (0–0.3 wt.%)	[49]	1999
Polyolefin melt	(*e.g.* BHT, Irgafos 168, Irganox 1076)	AddiMet^{TM}[a]	1999
PP melt	Irgafos 168 (up to 0.188%), Irganox 1010 (up to 0.094%), Ca-stearate (0.032–0.174%), silica (0.046–0.238%)	[146]	1999

[a] Commercial package (available from Process Analysis and Automation Ltd., Farnborough, UK).
[b] Diffuse reflectance probe.

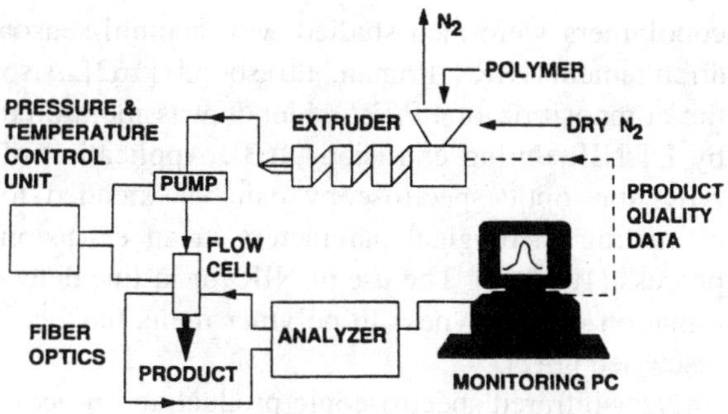

Fig. 7.10. System for in-line molten polymer analysis. After Batra *et al.* [143]. Reproduced by permission of the Society of Plactics Engineers (SPE).

to the quantitative determination of TiO_2, a white inorganic filler, in PET melt using an in-line flow cell (Fig. 7.10). The calibration and validation set was composed of 85 samples with nine TiO_2 concentrations. The observed spectral changes were essentially in the form of baseline shifts resulting from scattering due to the presence of the particulate inorganic component. Multivariate techniques were used to correlate the repeatable baseline changes to the filler content and a standard error of prediction (SEP) value of about 1% was obtained. The work demonstrates the use of NIR transmission spectroscopy for in-line composition monitoring of inorganic components in an extrusion process.

Balke *et al.* [147] reported in-line monitoring using the visible part of the spectrum of a fibre-optic-assisted VIS-NIR spectrophotometer in diffuse reflectance mode to measure the colour of "opaque", molten, pigmented polyolefins (pigment loadings from 3.6 to 56 wt.%; formulations with up to 33.1 wt.% $CaCO_3$ filler). In-line melt monitoring can distinguish within specification colour from **out-of-specification colour**. It can also be used to detect pigment degradation and to determine the upper temperature thresholds for pigmented polymer processing. Fischer *et al.* [148] have reported in-line process monitoring on polymer melts by NIRS, as applied for the quantification of filler content (pulverised chalk in PP), using a calibration model with 18 samples in three relevant spectral regions.

In the determination of three components (PMMA, lubricant, modifier) in an opaque PVC melt the results (lubricant to ±0.27%, modifier to ±0.89%) were judged sufficient for an effective process analysis [37]. Hansen *et al.* [149] used in-line fibre-optic NIRS and multivariate analysis in the 1900–2000 nm region for the simultaneous monitoring of vinyl acetate (12.04 and 18.06 wt.%) and erucamide additive concentrations (0–2500 ppm) in optically transparent EVA copolymers. Hansen *et al.* [49] have also developed and evaluated durable in-line fibre-optic probes for polymer-chemical process spectroscopy and have set up calibration models for the quantitative determination of **multiple additives** using FT-NIR and UV spectroscopy on the basis of fifteen LDPE samples. The authors have studied the feasibility of simultaneous in-line monitoring of Irganox 1076, Irgafos 168, Tinuvin 622, erucamide, and Armostat 310 in molten LDPE using a flow cell with a 7.5 mm path length. Some additives can be predicted reliably. Erucamide can be monitored (as low as 200 ppm) in the 1930–1990 nm region by fibre-optic NIRS in real-time in an extrusion process, at variance to Irganox 1076, Irgafos 168, Tinuvin 622 and Armostat 310. In another case, an extruder operated at nominal melt conditions of 270°C and 175 bar was used to analyse two (undisclosed) modifiers in molten PE using a process NIR spectrometer coupled through 120 m of fibre optical cable [150]. Fujikura Ltd. [151] has claimed an apparatus that is attached to an extruder for polyolefin extrusion with the objective of measuring the additive content.

The commercially available AddiMet™ system is designed to perform on-line measurement (NIR or UV/VIS) of antioxidant concentrations in polyolefins. The NIR option allows determination of bulk composition and prediction of physical properties of the polymer matrix. Vastenhoudt [146] reported determination of Irganox 1010, Irgafos 168, Ca-stearate and silica in PP with correlation coefficients/standard errors of 0.92/0.008%, 0.99/0.004%, 0.99/0.004 % and 0.95/0.015%, respectively. In-line NIRS was also used to monitor CO_2 dissolved in molten polymers (EPR-block-PP, LDPE, PBS) at the extrusion foaming process [152].

Sohl [9] has stressed the advantage of a "just-in-time" (JIT) compounding strategy using adequate feedback control from NIR measurements for both the polyacetal and additive feeder; the stabiliser level variations proved to be much smaller than the specification of the commercial product, even in situations of additive powder which shows clumping and bridging.

In the field of polymer processing NIRS is widely used for a variety of other applications. In particular, FT-NIR spectrophotometry can be used for on-line **chemical composition analysis** of extruded polymers and polymer blends [153–155]. Thomas *et al.* [156] have reported the in-line NIR monitoring of composition and bubble formation in expanded PS foam board containing HCFC 1426 (blowing agent) foams using probes located in the die. The talc (nucleating agent) concentration (<5 wt.%) could also be taken into account in the calibration model and be predicted accurately (<0.2 wt.%). In another application, **reactive extrusion monitoring** of a methacrylate copolymer by means of FTIR and NIR has been described; the problem of in-line NIR monitoring of a polymer melt for determination of water in the product has been addressed [129]. Parameters of interest in this application are composition of the processed polymer, moisture or reaction status in reactive polymeric systems, as well as rheological parameters such as melt flow index (MFI) or viscosity.

Rohe *et al.* [157–159] have developed a fibre optic transmission sensor for application of AOTF-NIR spectroscopy to extrusion processes, so that real *in-line* observation is possible. The parameters measured *on-line* are often not sufficient for adequate description of the polymeric melt. NIR spectroscopy can solve this lack of knowledge by *in-line* measurements of the melt. The polymer composition of a PE/PP blend during extrusion was determined with high accuracy (deviation <2%) using a polymeric melt analyser on the basis of *in-line* transmission AOTF-NIR spectroscopy and multivariate data analysis [157].

According to McPeters *et al.* [129,142] NIR spectroscopy for in-line compositional measurements on polymer blends and terpolymers inside an extruder is not as accurate as FTIR measurements. Fischer *et al.* [101] have quantified acrylic monomers in an acrylate-butadiene rubber during the mixing process in an extruder using NIRS with a transmission melt flow probe and variable optical path length or a diffuse reflectance probe. Quantitative analysis was carried out by chemometric methods (PCR and PLS). Similarly, it is possible to discriminate between EVA copolymers with different compositions, predicting the content of vinyl acetate in the copolymers and their melting points using NIR spectroscopy and chemometrics [160]. Hansen *et al.* [145,161] have studied in-line NIR analysis of molten PS/PPO blends and have reported simultaneous on-line fibre-optic NIR spectroscopy measurements of comonomer composition and rheological properties of poly(ethylene vinyl acetate). EVA copolymers were also studied with a multi-sensor arrangement (NIRS, Raman, ultrasound) [162]. Also the composition of PP/EVA blends was monitored by FT-NIR during extrusion [163]. Application of NIR fibre-optic spectroscopy can be extended to estimating rheological parameters in an extrusion process [161,164]. The use of NIR for in-line determination of yellowness in polymer melts has been discussed [142].

Near-infrared spectroscopic product and process control of polymer/additive formulations was reviewed [165]. Process control in polymer processing was discussed [166].

Polymerisation monitoring:
Near-infrared spectra can give relevant information on the chemical and physical state of polymers and polymeric composites. Degree of cure, mechanism of reaction, crystallinity/morphology, orientation, melt index/viscosity on-line, phase separation, hydroxyl number, water content and hydrogen bonding can be studied using NIR spectra without any sophisticated mathematical treatment [133]. Fibre-optic based NIR spectroscopy has also been used for monitoring polymerisation reactions [167]. Because of the high speed of NIR detector systems, it is possible to measure with much higher repetition rate compared to an interferometer FTIR spectrometer. NIR diode array spectrometers allow measurements which are 20 times faster than the normal FTIR technique. Powell *et al.* [48] have reported a comparative study for different types of optical fibre sensor developed to monitor the cure of an epoxy resin system. The optical fibre sensors were based on transmission spectroscopy, evanescent wave spectroscopy (attenuated total reflectance) and refractive index monitoring.

Typical applications include *in situ* determination of the rate or degree of cure [168], monitoring of polymerisation reactions [133,169], compositional analysis and reaction control [170]. Hartwig *et al.* [171] have reported real-time monitoring of UV induced **curing reactions** of acrylates in the region of the first C-H overtone at 1600 to 1700 nm using two NIR diode array spectrometers equipped with (extended) InGaAs array detectors (1100–2200 nm and 900–1700 nm). During UV induced polymerisation, the acrylic double bonds are converted to single bonds. As a test, the curing of lauryl acrylate with 1% Irgacure 184 photo-initiator was successfully monitored. The presence of a fibre or filler can modify the kinetics of a curing reaction or the morphology of the matrix. This should be taken into account by NIRS modelling.

Grob *et al.* [172] have reported the analysis of polyols in a pilot plant operation using a fibre optic transmission probe. Free epoxide (unreacted) and hydroxyl number were monitored to enable cost and safety control during full-scale production. The feasibility of NIR-ATR spectroscopy for the *in situ* characterisation of epoxy/amine cure reactions has been demonstrated [173].

Miscellaneous applications:
Kellar *et al.* [108] have demonstrated the potential of NIR internal reflection spectroscopy (NIR-IRS) with reactive internal reflection elements for *in situ* monitoring of surfactant adsorption. An advantage of the NIR-IRS technique over the mid-IR analogue is that a wider range of materials are transparent in the NIR region. As a result, broader ranges of substrates are available for NIR-IRS than for IR-IRS.

NIR is a widely applicable analytical technique also for the quantitative study of liquid and compressed gaseous systems, including fluid states, up to high pressures and temperatures. NIR spectroscopy measures Iodine Value more rapidly (2 min) than the traditional titration method (20 min) [174]. Other typical applications include feed gas composition monitoring and multicomponent analysis of liquids (reaction products) and solids. Applications of FT-(N)IR are reported for moisture content measurements; fermentation control; refinery analysis: distillation control, gasoline blending; hydrocarbon analysis: octane number of fuel, cetane number testing of diesel fuels; optimisation of plant operations: refining, petrochemical and polymer processes.

With an FT-NIR spectrometer and fibre optics the process operator virtually "looks" into a process stream reactor, vessel or extruder/pelletiser and can determine variations in composition of the liquid, gas or solid with **real-time feedback control** at various stages in the process. As no sample preparation or dilution is required results are generated in 30 seconds compared to several hours or more required for off-line laboratory analysis. For example, the Perkin-Elmer PIONIR 1024 process NIR analyser determines up to 20 quality parameters (of petroleum refinery products) within 15 sec, with remote signal acquisition possible via fibre optics [175].

Applications of FT-NIR spectroscopy to process monitoring were reviewed [132,176].

7.2.5. Process Raman Spectroscopy

Principles and Characteristics
As already indicated in Chp. 1.2.3, Raman scattering induced by radiation (UV/VIS/NIR lasers) in gas, liquid or solid samples contains information about molecular vibrations. Raman spectroscopy (RS) was restricted for a long time primarily to academic research and was a technique rarely used outside the research laboratory. Within an industrial spectroscopy laboratory, two of the more significant advances in recent years have been the allying of FT-Raman and FTIR capabilities, coupled with the availability of multivariate data analysis software. Raman process control (in-line, on-line, *in situ*, on-site) is now taking off with various robust commercial instrumental systems equipped with stable laser sources, stable and sensitive CCD detectors, inexpensive fibre optics, *etc.* With easy interfacing with process streams and easy multiplexing with normal (remote) spectrometers the technique is expected to have impact on product and process quality.

In situ measurements in industry must be extrapolated to on-plant monitoring. The feasibility of using fibre optic coupling between the Raman experiment and the FT interferometer has been demonstrated. For on-line use special designed probes can withstand up to 300°C and 15,000 psi. Because Raman light can remotely be focused, it is even possible to measure in a non-invasive mode (for example through a specified reactor window). A portable process Raman analyser enables both in-line and at-line measurements.

The **main features** of Raman spectroscopy for process analysis *cq.* product control are shown in Table 7.22. *In situ* real-time measurements can easily be made. Non-invasive mode measurements (*e.g.* through a reactor window or a closed sample bottle) are allowed because Raman light can remotely be focused. Also in the area of data processing on-line Raman measurements present an advantage. Many chemical systems exhibit distinct, baseline resolved (or nearly so) Raman bands which allow quantitation of important components by direct peak area or peak height measurements. Complicated chemometric methods can be avoided in these instances. On the other hand, sample information gathered by Raman comes only from a very small spot in the process or by-pass stream although there is averaging for moving samples. Because of the very low intensity and side effects (*e.g.* fluorescence, phosphorescence or

Table 7.22. Main characteristics of Raman spectroscopy for process analysis

Advantages:
- Solid, liquid or gaseous state (high pressure) samples
- No restrictions on sample shape and optical materials
- Dark-coloured samples allowed (but heating up effects)
- No sample preparation
- Non-destructive, non-intrusive
- Sampling in air, at high (800°C) and low temperatures
- Easy interfacing with process streams
- Multiplexing of several probes onto a single spectrometer
- Specially designed probes for on-line use (up to 300°C and 15,000 psi)
- Use of cheap, high efficiency fibre optics (remote probing, 100 m)
- Use of non-contact optical probes (measurements through glass, quartz, saffire flow-through cells); real-time observation
- Reflection probing
- High-resolution (1 cm^{-1}); wavenumber stability
- Fast (1 min/analysis)
- Reliable
- High information content
- Depth profiling (confocal techniques)
- Sensitive for organics in water

Disadvantages:
- Very weak phenomenon
- Limited sampling area
- Only recently useful since the development of new equipment
- Calibration systems needed (limited burden)
- Relatively inaccurate
- No laser full power (sample integrity, overheating)
- Limited use in specific application niches
- Most applications limited to the percentage range
- Expensive technique, limited lifetime of lasers
- Safety implications

band-overlap) quantitative methods are relatively inaccurate (inferior to IR). In a process environment use of high-energy lasers is an obstacle (invisible beam, safety).

Choosing a suitable Raman spectrometer for on-line process analysis requires different criteria from laboratory analysis. Some key considerations are laser safety, ruggedness, repeatability, long-term and environmental stability, high uptime, calibration transferability, ease of operation and maintenance, smart diagnostics for analyser performance, and industry-standard communication. Many processes require the analysis to be performed at multiple measurement points. A cost-effective multi-channel process Raman analyser design has been reported [177].

Fibre-optic probes simplify coupling into process streams, giving Raman an advantage over IR spectroscopy. Raman spectroscopy is well suited for determinations in aqueous solutions, in contrast to infrared. Raman spectroscopy allows greater flexibility for on-line sampling probes than does near-IR. Consequently, Raman spectroscopy is well suited to many problems involving on-line monitoring of processes in the chemical industry. It offers the potential of combining the highly specific information about molecular structure found in the mid-IR with the fibre optic sampling capability of the near-IR. Raman spectroscopy could become a competitor for mid-IR in-process analysers (at least for full spectra measurements) and near-infrared (less model maintenance). Table 7.23 shows some specific benefits and weaknesses of UV/VIS/NIR Raman spectroscopies. FT-Raman process analysis with long wavelength excitation at 1064 nm is especially useful in the following situations: *(i)* sample fluoresces when using visible excitation; *(ii)* presence of strongly scattering mixtures (*e.g.* emulsions, slurries); *(iii)* formation or consumption of symmetrical molecular homonuclear groups; or *(iv)* chemometric methods cannot be used effectively.

In process control systems it is essential to develop rapid on-line monitoring techniques to acquire structural parameters such as crystallinity, and orientation. By controlling these structural parameters the end use properties may be influenced which are essentially defined by these parameters. In order to use laser Raman spectroscopy for such purposes, calibration systems need to be developed using an independent technique.

Process monitoring using Raman spectroscopy (mainly in its NIR Fourier transform variant) is proposed for: QA/QC purposes, on-line polymer analysis, *in situ* cure kinetics, emulsion polymerisation, non-invasive analysis of physical parameters (*in situ* crystallinity determination, *etc.*) and reactor compositions, real-time measurements, molecular interactions, and components in aqueous solutions.

Applications

Raman spectroscopy is relatively new as an in-process technique, yet several applications in routine analytics, quality and process control in various branches of industry (food, pharmaceutics, mineral

Table 7.23. Strengths and weaknesses of UV/VIS/NIR Raman process spectroscopy

Strengths	Weaknesses
UV Raman:	
• High sensitivity	• Tendency to fluorescence
• Enhanced discrimination	• Expensive laser
• High spatial resolution	• Specialty fibre optics
Visible Raman:	
• Readily coupled to fibre optics	• Tendency to fluorescence
• Improved sensitivity	
• High spatial resolution	
Near-infrared Raman:	
• Reduced fluorescence	• Reduced sensitivity
• Readily coupled to fibre optics	
• High spatial resolution	
• Cheapest laser	

oil, (bio) chemical, semi- and superconductor) are now possible. A typical (new) field of application is the food industry. As Raman spectra of food supply more relevant information than NIR absorption spectroscopy, they may be employed for QC in production processes, and for the detection of preserving agents [178,179]. Successful application to food analysis has also stimulated NIR FT-Raman spectroscopy in medical diagnostics.

Many "real-world" applications reported are from an alliance of universities and chemical industries. Raman spectroscopy determines various aspects of chemical composition and physical structure (*e.g.* crystal form, polymer composition, crystallinity, molecular orientation, *etc.*). Quantitative component determination may be carried out using chemometric techniques. Many Raman applications can be handled in a more straightforward manner than infrared.

Applications of UV Raman are in the fields of biological and materials science, biochemistry, forensic sciences, *etc.*, whereas application areas for both NIR and VIS Raman are polymers, polymerisation, paints, dyestuffs, pharmaceutical materials, alkaloids, minerals, explosives, multilayer films, hard disk quality control, *etc.* Especially NIR FT Raman spectroscopy finds promising applications in various fields, from latex systems [180] to textiles [181]. Hendra *et al.* [182] and Schrader [183] have recently described application of NIR FT-Raman spectroscopy in the polymer industry.

FT-Raman is applied in QA/QC applications and in on-line polymer analysis. Process Raman spectroscopy does not excel in **polymer/additive analysis**. Reasons may be understood from the disadvantages of the technique, as listed in Table 7.22. The state and fixation of dyestuffs on cotton fabrics has been determined by vibrational spectroscopy in combination with chemometrics, PCA and PLS [184,185]. Liu *et al.* [186] have published a comparative study of dyed cotton fibres by the three most common vibrational analysis techniques, *i.e.* FT-NIR, DRIFT and FT-Raman. The results indicate that FT-Raman spectroscopy gives the best model to predict the fixation. Raman spectroscopy is also being used to control TiO_2 manufacture and ensure the correct ratio of the two crystal forms in the finished product [187].

Raman spectroscopy has also been applied as a rapid characterisation tool of ex-reactor aliphatic polyketones. Chalmers *et al.* [104] have described off-line **compositional analysis** by means of Raman and FT-Raman of EO-PO copolymers (non-ionic surfactants) for QA/QC purposes; PLS modelling can importantly decouple the spectral influences of crystallinity and orientation on Raman spectra. Simultaneous monitoring of composition and rheological properties of EVA copolymers by means of in-line fibre-optic Raman spectroscopy was reported [188,189].

Usually on-line IR spectroscopy is used to monitor the chemical evolution under UV irradiation of fast curable resins as thin films. Baillet *et al.* [190] proposed remote optical-fibre Raman spectroscopy equipped with an He–Ne 633 nm laser that allows to make local measurements within the bulk of (meth)acrylate samples (depth profiling by confocal techniques). The method is well suited to study the effect on the polymerisation rate by varying the photo-initiator amount, light intensity and film thickness. Monitoring of cure kinetics is critical to a wide range of industrial processes. In particular, the application of Raman spectroscopy to epoxy, acrylate, cyanate, bismaleimide and other systems of interest to adhesives and advanced composites is becoming easier with the use of non-visible laser sources, and confocal imaging, to reduce the effects of fluorescent interference.

Raman spectroscopy stands out for *in situ* **reaction monitoring** (reaction, intermediate and product profiles), including batch end-product determinations, aqueous emulsion polymerisations, polymorphic form identification, determination of monomer/co-monomer content, *etc.* NIR FT-Raman spectroscopy has high potentiality for fast and accurate quantitive conversion studies of latex polymerisation. *In situ* laser Raman spectroscopy is an excellent technique for the study of vinyl polymerisation kinetics. It provides accurate conversion data, since it directly probes breaking of C=C bonds and formation of C−C bonds. Moreover, it may provide evidence for molecular interactions among polymer/monomer/surfactant/co-surfactant/initiator systems. The validity of this technique has been demonstrated for bulk polymerisation of MMA and styrene, for solution polymerisation of acrylonitrile, and for styrene micro-emulsion polymerisation (*cfr.* ref. [191]). Apart from being successful in paint chemistry and technology, especially in the study of waterborne polymer latices produced by micro-emulsion polymerisation, FT-Raman spectroscopy finds application also in conventional paint technology based on oil-modified alkyd resins. Other generic examples of application are the polymerisation kinetics of styrene as $f(T)$ (1000–1700 cm^{-1}), adhesive curing (400–3400 cm^{-1}) and ageing in composite material (400–3400 cm^{-1}). Raman spectroscopy has been used in a bisphenol-A–diglycidyl ether continuous extrusion polymerisation to identify the reaction intermediates, products and contaminants [192].

On/at-line Raman spectroscopy has also scored in the determination of **physical parameters** of polymers, such as density, crystallinity and orientation. At-line FT-Raman and multivariate data analysis were used for density measurements in PET films and chip samples [104]. Laser Raman spectroscopy has been used as a non-contact method for on-line measurement of crystallinity at any time during the crystallisation process of LDPE [193]. An advantage of fibre optic laser Raman spectroscopy is that thick samples can be monitored by detection of backscattered Raman signals; dynamic structural changes occurring on the order of about 1 sec can be monitored. This time can be shortened with the use of more sensitive detectors. On-line Raman spectroscopy is a powerful tool for analysing the crystallinity, molecular orientation and composition in polymers as they are extruded, drawn and heatset. In particular, polarised μRaman spectroscopy can be used for on-line molecular orientation monitoring. Comparison of *in situ* crystallinity measurements with off-line XRD experiments was reported. As the Raman analyser is non-contacting, it does not disturb surface finish and can make measurements on much thinner samples than required for NIR analysis; in many respects it is ideal for thin-film measurements (not too thin). Determinations of commercial film properties in real-time during production allow detection of subtle changes in production-line conditions [104]. Careful application of multivariate techniques can enhance the information derived from the measurements. Raman spectroscopy is one of the optical molecular spectroscopic techniques capable of giving quantitative information about molecular orientation in gel-production lines (films/fibres) and on orientation effects in applications like mould injection.

Salzer *et al.* [194] have described FT Raman investigations of fast moving samples (20 m/s) using NIR excitation with a Nd-YAG laser and allowing quality control under draw. On-line monitoring of the molecular orientation of drawn polymers constitutes essential practical information for polymer process optimisation [195]. Obviously, a better understanding of the polymer deformation process during drawing can be obtained from on-line measurements, as opposed to off-line experiments which can only approximate true production conditions.

Chalmers *et al.* [196] have recently reviewed FTIR, FT-Raman and chemometrics in the application to polymers.

7.2.6. Process Nuclear Magnetic Resonance

Principles and Characteristics

Various NMR techniques in the frequency, time and spatial domain are useful in polymer analysis and

Table 7.24. Basic approaches to NMR in-process analysis and control

Resolution	^{1}H resonance frequency	Field heterogeneity	Response time	Operator requirements
High	200–500 MHz	<1 Hz	10 min–few h	Moderate-high
Low	<20 MHz[a]	0.5 kHz	1–10 min	Low, potentially none
High, low-field	<100 MHz	1 Hz	>5 min	Low

[a] Usually 10–30 MHz (benchtop and on-line units).

characterisation (Table 5.13 of ref. [113a]). NMR is a rapidly emerging technique focused on the acquisition of quantitative information for use in process analysis and control. Three approaches are developing: *(i)* conventional high-field spectrometers in laboratory environments used *on site* to monitor slowly changing processes; *(ii)* low-resolution instruments for a variety of QC roles; and *(iii)* high-resolution low-field spectrometers designed as *on-line* process analysers (*cfr.* Table 7.24). The first published reference to the use of NMR as a process control technique was made relatively early [197]. On-line coupling between a ^{1}H NMR spectrometer and a chemical reactor was first mentioned by BASF in 1986 [23] and is now well established in the polymer industry [198]. NMR has the potential of being a very useful tool in process environments as it is non-destructive and does not require the measurement probe to be inserted into the process liquors. Snoddy [199] examined the potential of process NMR on flowing streams. The amount of information desired from flow NMR experiments may require that the flow-rate be slow enough to allow complete relaxation of the nuclei. In instances, this renders stopped flow necessary. The use of process NMR on flowing streams is just beginning to be recognised as a powerful on-line method.

Process NMR makes frequent use of benchtop low-resolution (10–30 MHz) and low-field high-resolution (60 MHz) NMRs for lab (near-line) and off-line work (mainly QC), as well as on-line in-process units (10–30 MHz) which have a direct feed from the process. The lab or off-line units are necessary as a backup to the on-line process NMR in case of failure. The **requirements** of process NMR analysers are quite different from those of laboratory NMR instruments. For process applications, a permanent magnet or electromagnet operating at ^{1}H resonance frequencies $\leq$100 MHz is preferred over cryogenic magnets. On-line NMR analysers must have features that allow them to perform continuously in harsh environments, including automated sample introduction/removal, autotuning, autoshimming, temperature control, *etc.* The NMR probe is the conduit through which the process streams flows, and therefore must be able to withstand stream pressures and temperatures, some of which might be extreme. In a plant environment NMR instrumentation needs to be robust and possibly mobile for quality and process control at different stages of product development, manufacturing and quality control. These demands are difficult to fulfil with sophisticated pulse sequences and highly homogeneous magnetic fields $\mathbf{B_0}$. For this reason low-resolution NMR is well established in industrial laboratories.

The kind of NMR data required (*e.g.* signal amplitudes, relaxation information or chemical shift information with limited spectral resolution) plays a significant role in defining the design criteria for both hardware and software components. In common practice, in low-resolution NMR the concern is with the analysis of the NMR signal in the time domain (FID) and the characterisation of the physical structure of the bulk sample. The global characterisation of the sample in terms of molecular dynamics is key to successful use of low-field NMR. Relaxation information should provide rapid, reliable quantitative information for improved process control. The relaxation behaviour can provide extremely useful information on various aspects of mobile phases, *e.g.* moisture determination.

The intrinsic **characteristics** making NMR well suited for process analysis are given in Table 7.25. The analytical signal is directly proportional to the number of spins in the receiver coil region; the response is linear from 100% to the detection limit (in favourable cases down to *ca.* 10 ppt). Even at very low magnetic field strengths, well-resolved spectral features can be discerned. Method development is potentially simpler than for other forms of spectroscopy with greater variability in spectroscopic

Table 7.25. Main characteristics of process NMR

Advantages:
- Non-destructive, non-invasive (no sample preparation problems)
- Independent of sample aggregation state or physical condition
- Suitable for optically opaque or "dirty" samples; no granular effects
- Suitable for on/at/off-line statistical process control
- Improved product uniformity (reduced off-spec product)
- Decreased product transition times
- Increased efficiency and cost reduction
- Most informative chemical analysis technique
- Measurement flexibility
- Decreased analysis time (compared to wet analyses)
- Selective determination of compounds containing certain nuclei (^{1}H, ^{19}F, ^{31}P)
- Standardless quantitative analysis ("absolute" technique)
- High reproducibility
- Reduced solvent/waste stream (no environmental concerns)
- Simple and continuous, unattended operation
- Low maintenance, low life-cycle cost

Disadvantages:
- Method development needed (correlation installation)
- Limited robustness of calibration models for some applications
- No trace analysis (best suited for main components in process streams)
- No sulfur analysis
- Expensive (but fast payback)

transition probabilities from analyte to analyte or for the same analyte in a variable matrix. The chemically specific nature of NMR leads to simple calibration models. Most process NMR instruments operate at ^{1}H NMR frequencies, some are equipped for ^{19}F; only instruments with a relatively high magnetic field are capable of ^{31}P NMR studies. ^{13}C NMR may prove useful in selected analyses of major constituents.

Process NMR is not restricted to one method of analysis. Any NMR pulse sequence can be used but single pulse Hahn echo and solid-echo are most commonly used. Multiple-pulse sequences may be used to highlight a particular effect.

Grinsted [198] has briefly described **data treatment** from process NMR. Available methods are:

(i) R_{21} method (ratio of two FID data points – corresponding to rigid and mobile polymer components – correlating to various chemical and physical properties);

(ii) Curvefitting of FID line shapes with various functions (gaussian, lorentzian, exponential) and correlating the corresponding time constants and intensity relationships;

(iii) Use of pulse field gradients to measure molecular diffusion rates, particle and pore size distribution, and homogeneity of mixing (at low fields: 1–10 MHz; mainly for QC); and

(iv) Peak ratio in low-field solution-state NMR spectra.

Curvefitting is generally preferred over the R_{21} method. Correlations give the greatest reliability for a particular product or range of similar products. Production line-to-production line, or plant-to-plant correlations vary indicating lack of robustness of the correlation models.

The use of NMR for on-line process control and quality assurance was reviewed [200,201].

Applications

Process NMR is used for chemicals (free/bound moisture, viscosity, activity, loading efficiency in powders, catalysts, liquids, detergents, pigments) and polymers (density, crystallinity, rubber and copolymer content, dispersion of fillers, melt properties, finish content, extent of cure and cross-linking, content of solubles, plasticisers, moisture, *etc.*). Process NMR is fully operational in the polymer industry, both as on-line units [202] which provide virtually continuous process feedback control as well as off-line and laboratory units for checks of the various processes [198]. The use of NMR for advanced process control has reduced the need for frequent "wet" tests, has reduced "off-spec" materials and has improved product transition times.

7.2.6.1. *Low-field NMR*

Principles and Characteristics

Low-field NMR spectrometers (up to approximately 1.5 T or 60 MHz proton frequency) in laboratory and production environments for off-line work and on-line in-process units (5–30 MHz) are usually categorised in terms of low-, medium- and high-resolution. Recent improvements in capabilities of low-field NMR spectrometers now allow chemical shift information to be obtained from **medium-resolution** ^{1}H, ^{19}F and ^{31}P NMR spectra, and so extend the range of at-line applications of NMR in process monitoring and quality control. With low-field, medium-resolution it is not possible to obtain coupling constants or the detailed structural information provided by high-resolution NMR. Littlejohn *et al.* [203] have emphasised the role of at-line process analysis by low-field medium-resolution NMR.

Low-resolution NMR (LR-NMR) typically employs magnetic fields of 0.47 T as compared to 18.8 T for advanced 800 MHz high-resolution NMR. Modern LR-NMR spectrometers are pulsed instruments. LR-NMR instruments (first introduced in 1968) have a limited field homogeneity as they are not intended for use as true spectrometers, capable of distinguishing between protons with slightly different resonance frequencies resulting from changes in their chemical environment. They are in fact used to measure proton signal intensity as a function of time. The **main characteristics** of low-field low-resolution pulsed NMR are shown in Table 7.26.

Much of the physical and (indirectly) chemical information available through the use of NMR is associated with the **relaxation** characteristics of the nuclear magnetic moments, which can be measured using pulse NMR techniques. The energy exchange between nuclear moments and the surrounding lattice is characterised by the spin–lattice relaxation time, T_1 (commonly of the order of 1 sec), while the energy exchange among nuclear magnetic moments is described by the spin–spin relaxation time, T_2 (more commonly 10 μs–500 ms). Relaxation time methods are routine measurements. In suitable cases relaxation times (T_1 or T_2) are correlated with some bulk physical property of a sample and may yield information about the dynamic environment in which the nuclei are located (*cfr.* also Chp. 1.5.1.1). This permits studies of drying, gelatinisation and dissolution processes.

Low-resolution NMR is thus a **time domain technique**. Exactly like high-resolution spectrometers it records the decay or evolution of the magnetic resonance signal with respect to time. The time evaluation of the NMR signal after a rf pulse contains most of the information for analytical purposes: *(i)* initial signal amplitude (proportional to the total number of hydrogen nuclei in the sample volume); *(ii)* decay of NMR signals at different rates for nuclei in different phases; and *(iii)* characteristic time constant (relaxation time T_2 for each decay process). The signals from solid phases decay far more rapidly (tens of μsec) than those from soft phases (hundreds of μsec to tens of msec) or low molecular mass liquids (hundreds of msec to sec), *cfr.* Fig. 7.11. A typical time for the whole process is 5 min. Various fundamental NMR approaches form the experimental basis of the majority of applications: *(i)* simple FID measurements; *(ii)* FID and spin-echo measurements; *(iii)* solid-echo sequence; and *(iv)* relaxation time measurements.

Table 7.26. Main characteristics of low-field low-resolution ^{1}H NMR[a]

Advantages:

- No special sample requirements (suitable for solid materials, granules, powders, emulsions, suspensions, swollen gels, *etc.*)
- No sample preparation (at most weighing)
- Analysis independent of particle size and shape
- Sample volume 0.5–150 mL
- Measurement of volume-average properties
- Selective process monitoring of ^{1}H, ^{19}F and ^{31}P-containing analytes (bulk)
- Detection limit about 0.1%
- Simultaneous analysis of several sample components (high selectivity)
- No interference of mineral fillers
- Sensitivity to the physical nature of the material
- Non-invasive, non-destructive
- At-line achievable; on-line with by-pass tubes
- Rapid QC tool
- No optical fouling problems; robust
- Quantitative (0.1% up to 100%)
- Rapid (single analysis: 30 sec to few min)
- High reproducibility (operator independent); accurate and precise
- Relatively low cost
- Special probeheads up to 250°C
- Operational simplicity (fully automated; plant personnel)
- Environmental friendly (no solvents)
- Inherently safe

Disadvantages:

- Method development
- Indirect method (few calibration samples required)
- Not chemically selective
- Chemometrics and signal processing required (to restore selectivity) for some applications
- Temperature control required for some applications
- Limited field homogeneity (not intended for chemical composition analysis)

[a] Proton resonance frequencies: 10–20 MHz.

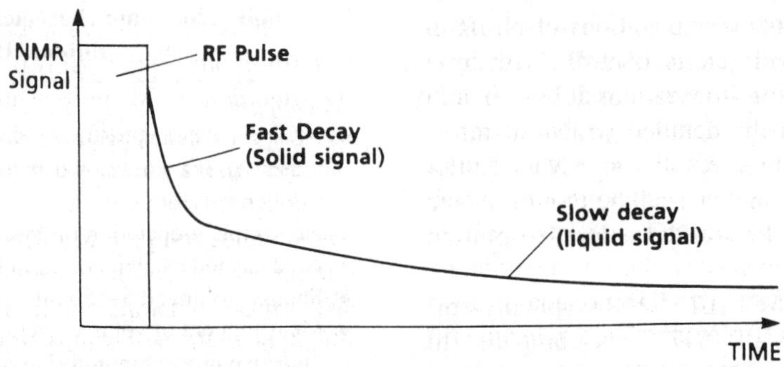

Fig. 7.11. Free-induction decay (FID) for a solid/liquid mixture.

Pulse NMR implies that a spectrum is obtained with an excitation pulse followed by detection of a free-induction decay (FID) and subsequent Fourier transformation. Pulse NMR methods are suitable for rapid and real-time measurements, which is the major requirement in manufacturing industries wishing to improve efficiency in quality and process control. The older field swept continuous-wave (CW) NMR technique – no longer used – carried considerable limitations for rapid measurements.

Three basic types of NMR experiments (single pulse, spin-echo and solid-echo) are used for measurement of T_2 **relaxation delays** [204]. In pulse NMR [204,205] the nuclei are excited by an intense pulse of rf radiation lasting only a few μs. This pulse excites all the specific nuclei of the same type, in all phases present; when the pulse is switched off, the nuclei return to their original state. The detected signal has a maximum intensity when all the nuclei have been rotated by 90° with respect to the direction of the static magnetic field. The duration of the rf pulse, variable in 100 ns steps, is adjusted to give this condition. There is a large variation in nominal 90° pulse times (typically μs). Alternatively, more than one rf pulse may be applied to give the signal to be measured. Benchtop LR-NMR analysers operate automatically using internally programmed pulse sequences. Different pulse sequences are commonly used to record the decay of the transverse magnetisation (T_2 decay) for both almost rigid (1) and mobile (2) sample fractions. (1): A solid-echo pulse sequence, $90°_x$–t_{se}–$90°_y$–t_{se}–[acquisition of the amplitude of the transverse magnetisation A(t)], to measure the T_2 free induction decay, and (2): a Hahn echo pulse sequence, $90°_x$–t_{He}–$180°_x$–t_{He}–[acquisition A(t) of the amplitude of an echo maximum], to record the slow part of the T_2 relaxation decay for the mobile fraction of samples. The second pulse in the Hahn echo pulse sequence inverts nuclear spins of mobile molecules only. It is possible to eliminate the magnetic field and chemical shift inhomogeneities, and to measure the T_2 relaxation time for mobile materials accurately.

While ^{1}H line shapes for rigid solids are very broad (tens of kHz), as a result of static dipolar couplings that are not apparent in solution due to rapid molecular motion, a material with both a rigid and a less rigid phase (the latter capable of restricted motional averaging) yields a richer time domain signal that can be fit to an appropriate model to provide a "spin count" of the protons in each component. The shape of the FID signal (Fig. 7.11) contains information about the physical nature of the sample, whereas the signal intensity gives direct quantitative information. Different phases in a sample (such as solid and liquid) give different signals and by examining the FID one can often distinguish such phases in complex samples. LR-NMR is thus a technique which entails a physical separation of rigid and mobile components in a material. If a sample consists of more than one component the signals due to each of these are superimposed. The signal due to each component decays with a characteristic time constant. By measuring the signal intensity at different points on the FID one can determine the amounts of magnetically active nuclei (usually protons) containing material in the different phases. This gives a fast, non-destructive tool for measuring the **solid/liquid ratio** S/L of a very wide variety of samples. For such composite materials the NMR signal is measured at two points after a single 90° rf pulse. The first measuring point is normally shortly after the pulse in the fast decay part of the curve of Fig. 7.11, and is in some way proportional to the total number of magnetically active nuclei in both the solid and liquid (or more mobile) phase (S+L). The second signal is measured typically at 70 μs, where there is no contribution from the solid phase protons, *i.e.* the signal only arises from liquid phase magnetically active nuclei, and is therefore proportional to the mobile phase content (L). A single pulse (90°) experiment does not provide the absolute value of the rigid/soft ratio; calibration is necessary to obtain this value. After correcting for instrumental conditions ("dead time", field homogeneity), taken into account by the calibration curve, the ratio of these two signals represents the real solid/mobile phase ratio of the sample. As there is no single experiment which provides accurate information both on hard and soft phases, a combined use of the solid-echo and spin-echo methods is often desirable. In practice, it is common use to measure L from a spin-echo intensity. This type of measurement finds wide industrial application, *e.g.* in the polymer industry, in process and quality control. For example, the method for determination of additives in polyamide copolymers makes use of the fact that the signal due to the additive decays more slowly than that of the polyamide, which decays to zero in a very short time (approximately 20 μs). The amplitude of the signal at a longer time is therefore proportional to the amount of additive present in the sample. In the application the only requirement is that the signal from additives should decay at a different rate to that due to the host polymer, *i.e.* different physical behaviour (phase state, molecular friction coefficient, viscosity) for additives and host polymer.

Pulse NMR analytical **method development** consists usually in setting up an initial calibration. Comparisons must normally be made with samples in which the quantity of interest has been determined by some other technique, usually a wet technique, as in the case of most other instrumental determinations. In a different data evaluation approach to the measurement of the solid/liquid ratio the time-domain signal or free-induction decay (FID) generated is curvefitted using two or more components depending on the composition of the polymer. The curvefits are generally a combination of gaussian and exponential functions in which rigid (crystalline) and mobile (amorphous and rubbery) components correspond. The fractions and time constants are correlated with wet test data. In process NMR chemometrics and signal processing are important. At low fields (typically 20 MHz), traditional approaches to the analysis include multivariate calibration models, which are then used to accurately determine the concentration of components in unknown samples. However, any procedure that utilises calibration models requires continual model maintenance and update. Hence, it is desirable to employ model-free procedures which do not require preparation of multiple calibration solutions. McGill *et al.* [206] have investigated potential methods for the extraction of quantitative information from low-field NMR signals in the time domain (FID), namely the continuous wavelet transform and modifications of the generalised rank annihilation method (GRAM). The ability of GRAM to resolve overlapping signals in low-field higher resolution NMR is far superior to the continuous wavelet transform.

There is considerable interest in the possibility of "absolute" type measurements which may be made independently of the weight of the sample. This eradicates a possible source of error (weight) and increases sample throughput.

Solvents, which are necessary for extraction-based analysis, are not required in the application of LR-NMR. Results are obtained in a fraction of the time taken by extraction methods and are quantitative, quite at variance to extraction methods. In comparison with GC the greatest possible advantage of LR-NMR is the analysis time [207], even not considering the extraction time. MTBE in gasoline may be analysed by LR-NMR in a few seconds, as compared to 20 minutes for GC analysis. McDonald [201] indicated speed also as the primary advantage of LR-NMR over non-NMR methods. The fact that LR-NMR is a bulk technique is considered as being an advantage over IR techniques. Carbon-black does not disturb NMR measurements, which is another asset over IR spectroscopic techniques. An additional advantage of LR-NMR compared to other techniques is the high phase/components selectivity, as apparent from the applications. The fact that a once calibrated LR-NMR instrument is suitable for operation by untrained, plant-floor personnel contrasts to most other current (wet chemistry) methods, which are generally also burdened with a variety of possible sources of error.

Homogeneous magnetic fields are not a prerequisite for imaging and relaxation measurements, and inexpensive devices like mobile low-field instruments, the NMR-MOUSE, and mobile imagers can be built for use near or in the production line and for operation by technicians. Blümich *et al.* [208] have developed an NMR **MObile Universal Surface Explorer (MOUSE)** of low-field (9.17 MHz), which scans with a spatial resolution of 3 mm and

Table 7.27. Application areas for low-resolution NMR

Area	Application
Raw material procurement	Key component evaluation
Raw materials monitoring	Ensuring consistency
Process monitoring	Monitoring feeds, products and residues
Quality assurance	Monitoring finished products
Product development and formulation	Optimisation of specifications
Research	Understanding of fundamental properties[a]

[a] For example relaxation studies of porous media.

an adjustable depth sensitivity of 0–5 mm. The NMR-MOUSE is a lightweight scanner with which NMR relaxation parameters can be acquired non-destructively from surface-near volume elements of arbitrarily large objects. Because of the simplicity of the device and the pulse sequences, it is suitable for use in a manufacturing plant and can be transported to the object of investigation for spatially resolved NMR of accessible sample regions. Because of field inhomogeneity, NMR spectroscopy of the chemical shift is not readily possible, but relaxation times and parameters of translational motion can be measured by echo techniques. These are the most important NMR parameters which are exploited for contrast in imaging.

Several monographs deal with LR-NMR spectroscopy [204,205,209].

Applications

Unlike other well-defined areas of NMR, LR-NMR applications cover all states of matter (solid state, solution, *etc.*), and all possible areas of chemistry in industry and research. Probe head size assures representative sampling for measuring inhomogeneous materials. The technique requires calibration only once, albeit with separate calibrations for different problems (*e.g.* for powder and granulate of the same material). The unique discrimination power of the technique is based on the discrimination in **mobility** between various components of a sample (*e.g.* oil in rubber, solubles, rubber content, dispersion of fillers) or between different physical structures of a molecule (*e.g.* crystallinity, density, tacticity, copolymer content, melt properties, tensile strength, *etc.*). Phase analysis by nuclear spin relaxation time measurements rests upon the assumption that each phase present will give a unique relaxation time which can be found by a multiexponential fit of a relaxation decay time. As the solid to liquid (or rigid to mobile) ratio is determined directly, no separate sample weighing is required. However, as the NMR signal is sensitive to many sample related and experimental parameters extensive method development is required. For example, NMR intensities and relaxation times are temperature sensitive. Moreover, it means that product formulation must be maintained for long periods without change if the technique is not to require regular recalibration. Under favourable circumstances the method is very accurate and moisture measurements of 0.4% absolute error have been reported. The high accuracy allows a valid statistical process evaluation. A disadvantage is that a large number of samples needs to be measured for a full statistical evaluation of the method prior to its application. However, this is true for most techniques if an established method is to be replaced. Being non-invasive and non-destructive LR-NMR is suitable for on-line analysis. Compared to high-resolution NMR, LR-NMR has the advantage of being less costly, making it more suitable for the process industry.

General application areas for LR-NMR are given in Table 7.27. Low-resolution pulsed ^{1}H NMR has found widespread application in a variety of QC laboratories and research establishments in the food industry, polymer and chemical industries, mineral oil industry, pharmaceutical and cosmetic industries, and medical research because it offers rapid analysis without the need for difficult sample preparation [30, 199,210]. Applications of LR-NMR in the food industry, *e.g.* as applied to measurement of moisture in foodstuffs, were described as long as 50 years ago [211]. Applications now include measuring oil or fat in cosmetics, oilseeds, chocolate and other foodstuffs, solid-fat content, droplet size in oil-in-water emulsions; total moisture content in seeds, milk powder, pharmaceuticals; oils in/on polymers

Table 7.28. Some typical applications of low-field low-resolution pulse NMR in the polymer industry

Product	Measurement	Analysis time	Preparation	Reference
Polyamide	Elastomer	15 s	*b*	[212]
Polyamide	Polyethylene	15 s	*b*	[212]
Polymethacrylate	Degree of polymerisation	10 s	*b*	[213]
Rubber latex	Solids content	10 s	*c*	[213]
Polyethylene	PVA	1 min	*c*	[213]
Polypropylene	Polyethylene	1 min	(*a*), *b*	[213]
HIPS	Polybutadiene	30 s	(*a*), *b*	[213]
Poly(hexene/ethylene)	Hexene	15 s	*a*, *b*	[213]
Nylon	Glass	20 s	*a*, *b*	[213]
PVC foils	Plasticiser	30 s	*a*, *b*	[213]

a = tempering; *b* = weighing; *c* = no preparation.

and fibres (spin-finish); solid and liquid phase determination in edible oils and polymers.

The polymer industry appears to be a fast growing area for analytical NMR applications. Previously, most polymers were assessed on physical properties such as melt flow, elasticity, tensile strength, or some other method using pulling, pushing, squeezing, stretching, breaking, shaking methods, *etc.*, giving very little direct evidence of chemical composition. Chemical analyses were even regarded with suspicion, since most involved the complete destruction of the sample (unlike the food industry where methods like extraction, even though indirect, at least measure the material of interest in its unmodified form). The utility of NMR must be stressed as now for the first time direct information regarding the composition of the phase components can be obtained in seconds from low-field high-resolution NMR. Moreover, it is possible to measure both mobile *and* rigid phase hydrogen content directly, whereas the continuous-wave method can only really measure liquid phase signals. This is particularly appealing as an increasingly important measurement in the polymer industry is that of **residual monomer** in finished polymer, where liquid percentages of less than 1% are often encountered, which can be handled by pulse NMR.

While many of the advantages of instrumental analysis are obvious, no matter what the technique, there are some advantages of LR-NMR over existing methods which have particular significance for process and product quality control, namely accuracy, reproducibility and speed of measurement. If we consider that the prime objective of the QC engineer is to devise a scheme which will enable a statistically valid assessment of production, these advantages of LR-NMR are crucial. In fact, any statistical scheme must be able to sample as frequently as necessary and to measure a representative sample very rapidly; the measurements must be accurate enough to allow the error limits to be within acceptable bounds. This excludes extraction methods with delivery times of over 6 h. Similarly, moisture measurements involving a simple accurate drying oven take anywhere between 4 and 12 h. On the other hand, with LR-NMR a measurement takes 20 s, and an appropriate statistical QC scheme may be set up while maintaining a high degree of accuracy. Tables 7.28 and 7.29 are illustrative of the importance of LR-NMR in **product quality control** (near-line); this type of technology makes inroads into on-line applications since the beginning of the 90 s.

Low-field low-resolution NMR is extensively being used both for process and quality control of polyolefins [198] and blends (*e.g.* ABS/PC), but less so for additive dosing and monitoring. LR-NMR has limited use for **polymer/additive analysis**. If any, given the detection limits, LR-NMR is most suited for additives present in relatively high percentage levels such as plasticisers, flame retardants, impact modifiers, fillers, or lubricants, and does not reveal antioxidants, UV stabilisers, *etc.*

Traditional techniques for measuring physical characteristics, such as flexibility or hardness, are often used as an indirect check for the **plasticiser** content of PVC. These techniques need high maintenance, skilled operators and lengthy sample prepa-

Table 7.29. Off/at/on-line LR-NMR-based product control

Material	Rigid phase[a]	Mobile phase[b]	Reference(s)
Plasticised PVC	PVC	Plasticiser	[214,215]
Flame retarded polymers	FR	Polymer	–
Filled polymers	Filler[c]	Polymer	–
Filled elastomers	Filler[c]/bound rubber	Elastomer	[216]
Polyamide/additives	Polyamide	Additives	[212]
Oil extended EPDM rubber	Rubber	Oil	[217]
PS/(oil, rubber)	PS	Oil, rubber	[202]
Finish oil/moisture on fibres	Fibre	Oil, moisture	[218,219]
Moisture in polymers	Polymer	Water	[220]
Paper	Cellulose	Water	[221]
Copolymers/blends	Variable	Variable	[202]
Impact modified polymers	Polymer	IM	–
Polypropylene	PP (hard fraction)	XS	[202]
Various polymers	Crystalline fraction	Amorphous fraction	[30,202,222]

[a] Short FID decay.
[b] Long FID decay.
[c] Not measurable by ^1H NMR.

ration. Results are obtained with poor reproducibility. Alternatively, solvent extraction may be used to measure plasticiser content. This again needs skilled operators and long analysis times. On the other hand, LR-NMR appears as an ideal, robust bench-top analysis tool for routine operation by non-specialist production workers. Consequently, LR-NMR is well established for QC in the production of flexible PVC compounds. Because of the variables involved the NMR method requires calibration using control compounds of appropriate composition. It then becomes a rapid, reliable and practical indicator of consistency. Figure 7.12 shows a calibration plot for PVC containing different levels of DIOP. LR-NMR was used to determine DIOP content of PVC/20–50 wt.% DIOP with a precision of ±0.5% on the basis of an appropriate calibration graph; for highest precision, it is essential to know the type of plasticiser present [214]. In this application LR-NMR is more accurate and faster than ATR-FTIR and PA-FTIR [215]. However, IR techniques provide additional information, *e.g.* on accumulation of plasticiser near the surface. On the other hand, NMR provides evidence about plasticiser phase separation in highly mobile domains, depending on concentration and experimental conditions.

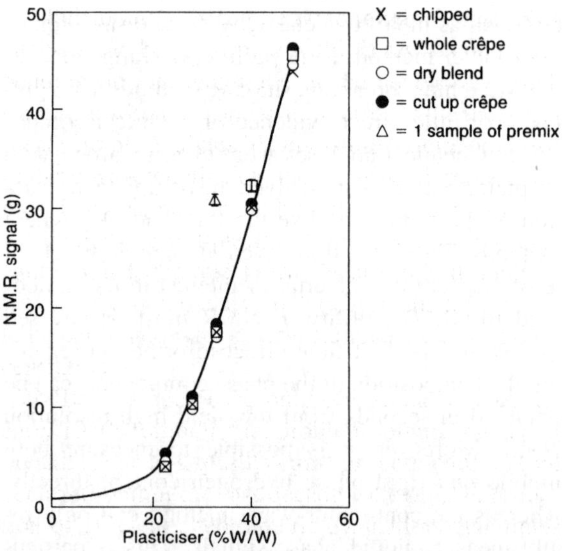

Fig. 7.12. DIOP plasticiser in PVC resin as determined by LR-NMR. Various concentrations of DIOP all milled for 10 min. 95% confidence limits of two of the points are illustrated. After Wilson [223]. Reproduced by permission of IoM Communication Ltd.

LR-NMR can also be used in FR systems if the mobilities of polymer (*e.g.* $T_{g,m} < 200°C$) and **flame retardant** (m.p. >200°C) are quite different. In that case the polymer shows the long decay at 200°C. In absolute LR-NMR analysis of GFR polymers the polymer weight fraction can be determined on the basis of the number of protons (absent in glass); the weight-normalised amplitude of solid-echo is used

Table 7.30. Main features of ^{19}F NMR

Advantages:
- Use of a fluorine probe eliminates interferences from other common additives
- No sample preparation; resin pellets can be run "neat"
- Excellent for processing aid concentrate levels (*e.g.* 3%)
- Rapid analysis (<5 min)

Disadvantages:
- Dependent on well-characterised calibration standards
- Poor sensitivity at processing aid levels below 0.1%
- Few potential users

for correlation with the percentage of glass. Other examples of problems suitable to this type of analysis include **lubricant** content in a host of materials.

Using ^{19}F NMR total fluorine in pelletised samples containing fluoropolymer **processing aids** can be analysed in a few minutes [224]. Table 7.30 shows the main features of ^{19}F NMR. LR-NMR has also been used for the determination of additive content in polyamides [215].

Although access to information on the use of NMR for QC purposes is rather limited, several applications for fast analysis of **rubbery materials** have been published, *e.g.* for determining the concentration of oil in rubbers, the distribution of carbon-black in rubber matrices, the solids content of rubber latices, *etc.* Oil in extended EPDM rubber serves as a plasticiser and softener, reducing viscosity of the rubber to that normally required in compounding. Current non-NMR methods of oil determination are quite time consuming. With off-line LR-NMR state-of-the-art instruments a coefficient of variance of 0.1–0.3% can be achieved (*cfr.* 0.3–1% for extraction methods) and no weighing of the sample (both crumb and ground) is required; the response time is about 30–40 min [217]. Since the method is based on a difference in physical behaviour of EPDM and oil, the accuracy should not be largely affected by variation in the chemical composition of EPDM and oil. Similar applications are known for waxes and paraffins. Current methods for measurement of oil content in polystyrene rely on time-consuming solvent extraction. Organic solvents used for oil extraction are costly and may be hazardous to operators and the environment. Rubber content is not normally measured directly because of the analytical difficulties. Indirect laboratory hardness testing techniques are often preferred but are very time consuming and require skilled operators for reliable results. LR-NMR offers a simple, rapid industrially applied method for determination of both oil and rubber in HIPS, eliminating the need for solvents and complex sample preparation [202]. LR-NMR has been used for determining the micro-heterogeneity of **filled elastomers** and the content of bound rubber [216]. The latter, traditionally determined by extraction, may be assessed from the change in FID. Owing to the small amounts of specimens required (0.2–0.5 g) the method can be used in evaluating the uniformity of filler dispersion in a rubber matrix. Cross-linking and interaction with filler are manifest by a shorter T_2. In silica-filled, non-vulcanised NR samples three separate regions with strongly different mobility were observed, corresponding to rubber chains tightly bound to the filler surface (lowest mobility), physically adsorbed chain portions (intermediate mobility) and free, extractable rubber chains (highest mobility) [225]. Low-field NMR has also been used for non-destructive assessment of degradation of rubbers [226] and other polymers (*e.g.* PC). The detection limit of the rubber phase in ABS/rubber is approximately 0.5%.

The traditional method of determining the **spin-finish** on fibres, be it polyester (PET) staple fibres or UHMWPE (Dyneema) fibres, is by Soxhlet extraction. Products with concentrations of 0.1% can be easily investigated by LR-NMR [218]. LR-NMR in spin-echo sequence has been used as an alternative to extraction methods for fast spin-finish determinations where the concentration of oil on the yarn is the only desired information [219]. In this procedure the signal of polymer protons and adsorbed water molecules decays during 70–100 μs after excitation, while the remaining signal is due to the oily spin-finish. Depending on proper calibration and method adjustment an accuracy of $\pm$0.02% absolute at a spin-finish level of 0.3–0.8% can be reached.

The most investigated area of non-destructive examination is detection and characterisation of **moisture** in composites and polymers [220]. In all these materials, the NMR signal amplitude was found to correlate linearly with the amount of adsorbed moisture over the range studied. Drying processes have also been analysed. In a deuterium NMR study of drawn nylon-6 fibres hydrated with D_2O the presence of three types of water and two classes of exchangeable protons has been suggested [227]. ^{2}H NMR is not used for QC. Although LR-NMR techniques might substitute classical derivatisation-GC

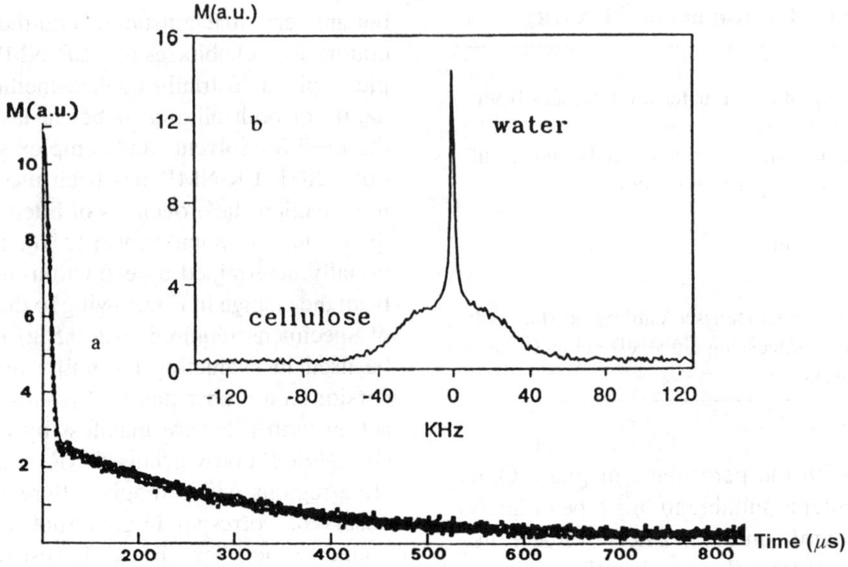

Fig. 7.13. Pulse ^{1}H LR-NMR at 57 MHz of antique paper in time domain (a) and frequency domain (b). After Attanasio *et al.* [221]. Reprinted with permission from D. Attanasio *et al.*, *ACS Symposium Series* **598**, 333–353 (1995). Copyright (1995) American Chemical Society.

methods for determination of water in polymers such as nylons, polyesters or cellulose, some inherent limitations need to be considered. In some cases the concentration is close to the detection limit (*e.g.* 50 ppm water in PET). Moreover, the NMR method is not chemically selective. Interference may occur from some other highly mobile molecules such as residual monomers.

High quality paper, a bi-component material made of cellulose, bound water and (in)organic additives and impurities, has been characterised by ^{13}C CP-MAS NMR at 100 MHz, pulse ^{1}H LR-NMR relaxation at 57 MHz, and ESR [221]. The time domain (FID) shows cellulose as a fast decaying component and water as the slowly decaying one; in the frequency domain the cellulose component is broad while the water component, which is strongly bound to cellulose, is sharp (Fig. 7.13). The state of conservation of paper correlates with the amount of paramagnetic rhombic Fe^{3+} impurities, as determined by ESR.

Apart from phase discrimination (hard *vs.* soft contents of materials), reports of **chemical composition analysis** by low-field ^{1}H NMR spectroscopy are increasing. LR-NMR allows analysis of the soft-block content of (co)polymer blends in the solid state, as *e.g.* in polyesterethers and other thermoplastic elastomers. LR-NMR can also be applied for the determination of the composition of copolymer/polymer blends, *e.g.* the ethylene content in PP copolymers, the vinyl acetate content in EVA copolymers [202]. LR-NMR is widely used in studies of the phase composition (amount) of **impact modified polymers** such as PA6.6 (Zytel ST801®, Du Pont) and for product quality control (*e.g.* SBR in Noryl®, PB in ABS and ABS/PC). The technique is a useful tool in the area of polymer blending (QC for masterbatch producers and compounders) and has found application for miniplant scaling up experiments of ethylene-octene copolymers.

LR-NMR also shows good potential in its adaptability to real **on-line measurements**. Low-field NMR spectrometers can be used for rapid determination with acceptable accuracy and precision of key quality physical parameters of polymers such as polymer content, viscosity and other rheological parameters, crystallinity, density, and tacticity that commonly constitute specifications for customer acceptance. The samples are measured "as is" without any treatment. In on-line applications a resin sample is pneumatically conveyed every few minutes from an appropriate sampling point on the process line to the measurement chamber of the spectrometer located between the poles of a permanent magnet.

It is possible to measure the **degree of polymerisation** in actual chemical processes by measuring the increasing amount of polymer in solution or in suspension as the reaction proceeds. This can be illustrated for the determination of the degree of polymerisation of styrene in industrial reaction mixtures [213]. In this case, the "spin–spin" relaxation time (T_2) changes considerably over a range of differing polymer contents.

Viscosity and other rheological parameters can be obtained from the NMR relaxation signal. Snoddy [199] has used the spectrometer to determine rapidly (within 1 min) the viscosity of polymer samples in flowing streams in a simulated production environment. The ability to continuously measure and control the **melt flow index** (MFI) is critical in order to reduce costs and maintain high quality. Unlike conventional approaches to MFI analysis, NMR technology provides the ability to directly analyse the physical and molecular structure of a solid polymer in the powder or pellet form. NMR MFI analysis is non-destructive (no melting or extruding of films) and can be performed simultaneously with a density measurement in a few minutes, making the system even very cost effective, as illustrated for PE [30]. The NMR MFI feedback results in shorter process transition times and minimises production of off-spec material.

Timely feedback of **polymer density** data for statistical process control (SPC) is equally important to minimise the production of off-grade transitional product and to maintain high quality. On-line PE density measurements using LR-NMR technology provides plant engineers with an important tool to improve process control [202,228]. Time domain signals for PE are characterised by a rapid decay rate for the crystalline segment and a slower decay for the amorphous segment of the solid polymer material. Differences in NMR signals due to density variations can be correlated to laboratory density results using modern curve fitting and statistical techniques [30]. It is possible to determine the PE density on-line with an accuracy of ± 0.0006 g/cm^3. A comparable method is used for the so-called **xylene solubles** (XS) in PP. During the first stage of the production of a polypropylene resin, the xylene soluble content of PP from the reactor is used as an important indicator of reactor efficiency. There are several different NMR methods for in/at/off-line analysis of xylene solubles in PP. All these methods are based on determination of the content of the soft/hard fractions of PP, which correlates with the fraction of xylene soluble [202]. In the LR-NMR analysis there is no need to weigh the sample or to use hazardous solvents. The LR-NMR analysis is independent of sample colour, surface, pellet size, *etc.*, thus leading to minimum calibration requirements. Accuracy is better than the standard method (extraction of PP in xylene); precision of 0.02% (σ value).

For measurement of **crystallinity** various methods exist, including the gradient column method, WAXD and DSC. The former is cumbersome and uses hazardous solvents, whilst DSC is slow and of poor reproducibility in view of small sample volume in testing (10–30 mg or a fraction of a granule). On the other hand, a 20 MHz LR-NMR analyser is highly sensitive to variations in the physical state of the phase composition and allows discrimination between the NMR signal from the crystalline fraction ("solid" part) with a short decay and that from the amorphous fraction exhibiting a longer decay. Calibration is usually carried out by comparison to a reference method (*e.g.* gradient column method). The NMR method is a reliable, non-destructive, and non-hazardous method for measuring polymer crystallinity and is almost error-source free as even no weighing is required. Results were reported for both PE [222] and polyester [30].

The **tacticity** or stereochemical configuration of polypropylene is an important product specification which is commonly measured in the laboratory by a time consuming solubility test. On-line NMR can perform this measurement directly on PP powder or pellets. The rapidly decaying region in the FID response of PP has been associated with the isotactic component and the slower decay with the atactic component. The measurement is fully automatic, highly reproducible (0.1%) and takes about three minutes. Applying advanced modelling techniques to the FID responses yields data related to PP tacticity [30].

LR-NMR also permits network structure analysis in rubbery materials [217,228a]. Cross-link densities for EPDM, PB and other rubbers can be determined. LR-NMR has also replaced the swell-index measurement of the cross-link density in the rubbery phase of ABS powder. Cross-link density of a TPE grade showed good correlation with data from swelling tests.

Benken *et al.* [222] have discussed basic principles in connection with **textile-related parameters** of NMR. Raw textile and CO_2-treated textile show only minor differences in their FID spectra, whereas the FID spectra of UV-treated textile indicate a significant change in polyester textile structure. Relaxation times (T_1) of a variety of textile samples were determined. NMR can be used for the quantitative determination of fibre species in fibre blends (*e.g.* polyamide/triacetate mixtures).

Low-field medium-resolution NMR at 29 MHz were used for QC in the determination of the average ethoxy chain length n of nonylphenol ethoxylates, $(C_9H_{18})C_6H_4O(CH_2CH_2O)_nH$; repeatability of the measurement is good (1.2% RSD) [203]. The technique has also been used for at-line determination of the ethylene oxide (EO) content of polyether polyols [229]. Direct analysis of the ^{1}H NMR FT spectra gave percentage EO concentrations of reasonable accuracy (average percentage error of 1.3%) and precision (average RSD of 1.8%), when compared with results derived from high-field ^{13}C NMR spectroscopy. Overlapping signals produced by a low-field medium-resolution instrument may be accounted for by multivariate calibration modelling or by means of the direct exponential curve resonance algorithm (DECRA) [230], which allows quantification of the NMR signal in the time domain, *i.e.* the FID signal [231].

Skloss *et al.* [24] have reported the use of a **low-field high-resolution** ^{1}H FT-NMR (42 MHz) spectrometer as a process analyser, which should combine the stability of the low-resolution type with the structure elucidating power of the high-resolution type. The instrument has been used for the determination of oxygenate additives (MTBE) in a flowing stream of gasoline. Other typical applications of high-resolution NMR (60 MHz) in simple, automated, reliable at-line and on-line analysis are also found in petroleum refinery: naphtha cracking, gasoline blending and sulfuric acid alkylation monitoring, RON, MON and benzene monitoring, *etc.*

The NMR-MOUSE technique was applied in 1D imaging of stress whitening of a sheet of polystyrene [208].

As the drive towards automation gathers momentum, more and more laboratories will adopt the low-field NMR instrument as standard. The use of NMR for on-line process control and quality assurance has been reviewed [201]. On-line analysis of polymers by means of pulse NMR was addressed by ref. [202]. Stilbs [232] has reviewed NMR methods in polymer-surfactant systems.

7.2.6.2. *High-resolution Process NMR*

Principles and Characteristics
Notably absent from the list of mature process analysis technologies is high-resolution NMR, which is rather new to the process environment [233]. As shown in Table 7.24, both low-field (<100 MHz) and high-field (>100 MHz) are in use. A high-resolution spectrum is not the most obvious measurement to make for process control applications. It requires very high magnetic field homogeneity and preferably also a high magnetic field in order to resolve resonances from different chemical groups and spin–spin couplings, which serve to fingerprint the sample. While standard high-field spectrometers may be too expensive, fragile, and operator intensive for all but the most ambitious on-line applications, they are being used in a number of sites for selected off-line process control problems that benefit from a NMR analysis even with a lag time of 1 h.

Modern FT-NMR spectroscopy offers the possibility of measuring simultaneously the time dependence of the concentrations of educts and products in a chemical reaction.

Applications
In off-line analysis, samples are taken from the reaction medium, with the disadvantage that the analysis does not occur on the reaction time scale and the reaction equilibrium itself is disturbed. For this reason, Neudert *et al.* [23] developed on-line coupling between a ^{31}P l-NMR spectrometer (360 MHz) and a chemical reactor for detection of the time dependent concentration of phosphorylated reaction components using a NMR flow tube via a by-pass. The method is a powerful tool for optimisation of chemical reactions.

7.2.7. Acoustic Emission Technology

Principles and Characteristics
The use of acoustic monitoring techniques for process analysis and control is becoming more relevant in industry. Ultrasonic signals have attributes that are well suited for characterisation of multiphase fluids and flows. The signals have the ability to interrogate fluids and dense opaque suspensions, penetrate vessel and process walls, and are not degraded by noisy process conditions because the signal frequencies differ from that of machinery.

In **passive ultrasonics**, which is usually referred to as acoustic emission, the source of the ultrasound is the process itself. Passive acoustic spectroscopy is a measure of the inherent acoustic output of a system or process. Physical processes producing acoustic emission (AE) include particle collisions, fracturing of solids, turbulent gas flow, gas evolution, fermentation, cavitation, boiling multiphase flow, and some chemical reactions. A diverse range of measurements have been made that consider ultrasonic interactions in terms of the effect of materials or processes on four ultrasound metrics: ultrasonic velocity V_{us}, attenuation α, absorption, and scattering, as functions of frequency and of composition, process reaction or phase (time/rate), and temperature. In active ultrasonics an acoustic wave is generated by means of a transducer, which propagates through the material at a characteristic velocity, is absorbed and scattered (*cfr.* Chp. 1.7).

Ultrasonic methods use piezoelectric transducers for the generation and detection of mechanical waves; pressure waves from sounds are converted into electric impulses. **Acoustic emission analysis** utilises frequencies in the range of 70 to 750 kHz (broadband transducer). In active ultrasonics the frequencies are somewhat higher, usually 1 to 200 MHz. The energy involved in ultrasonic techniques for measuring elastic or viscous polymer properties, $\sim$1.0 μW cm^{-2}, is so low that the system is not significantly perturbed. The maximum displacement of polymer molecules induced is around 0.1 Å, corresponding to small levels of stress and strain. Longitudinal waves are routinely used for polymer melts, but use of a shear wave reflection technique has also been reported [234]. Ultrasonic sensors can be designed to provide real-time, *in situ* measurement or visualisation of process characteristics; the sensors and sensing systems are compact, rugged, and inexpensive. Alig [235] has reported development of robust ultrasound sensors stable at typical polymer processing conditions (T 265°C, p 300 bar, t 2300 h).

Table 7.31 shows the **main characteristics** of passive ultrasonics. Process acoustics can be used non-invasively to study dynamic systems and provides real-time information on polymer processing suitable for process control in the noisiest industrial environments. Ultrasonic sensors can be designed to measure fluid density, viscosity, and velocity; slurry density, particle size, weight or volume percent solids concentration, stratification, and rheology; and to quantify multiphase flow interfaces, state of mixing, homogeneity, and slurry transport. Acoustic emission spectroscopy allows measurement of the degree of dispersion of a wide variety of materials, including conductive, non-conductive, transparent and opaque mixtures. AE responds to dynamic events making it suitable for process control by extracting unique, real-time information from a wide variety of processes with very high sensitivity. Ultrasonics is a powerful technique for probing molecular, chemical and physical properties such as composition, dispersion and degradation. Acoustic emission suffers from three fundamental problems: no repeatability, no comparability and no traceability. Non-invasive acoustic technology with advanced pattern recognition can be used to predict the physical properties of powders and particulates. Ultrasound facilitates polymer characterisation during processing, material identification, detection of flow instabilities during extrusion, or study of the solidification process.

Table 7.31. Main characteristics of acoustic emission spectroscopy

Advantages:
- Very fast response (real-time dynamic studies)
- Flow-rate independent
- Non-invasive
- Applicable to optical non-transparent materials
- Averaging of the response over the entire flow channel

Disadvantages:
- No repeatability, no comparability, no traceability
- Lack of robust commercial sensors for polymer processing conditions

Ultrasonic process analysis has been reviewed [12]; various monographs on ultrasonics [236,237] and on ultrasonics for process control [238,239] are available.

Applications

For decades, ultrasonic techniques have been excellent tools for non-destructive testing and imaging ultrasonic methods find application for material characterisation and process monitoring. Acoustic emission is a method of detecting discontinuities, flaws, cracks, *etc.* in plastic materials. AE operates by detecting acoustic response to applied stress. The method locates the source of emission, *i.e.* the site such as a crack or discontinuity undergoing a response to the imposition of stress. Ultrasonics have been used to characterise polymers in both the solid and molten states [240]. Acoustic emission technology (AET) is very attractive for in-line monitoring applications [30] and is used for early detection of agglomeration in fluid bed reactors (*e.g.* PE and PP production). AET has successfully been used to characterise the viscoelastic properties of polymer melts, to monitor polymer processing, chemical reactions (*e.g.* polymerisation or curing of thermosets), film formation, glue processes, or crystallisation. Ultrasound has also proved useful in discontinuous processes such as injection moulding. Acoustic emission can be used for sub-visible mechanical damage measurement. Acoustic emission is a sensitive technique for the detection of damage in fibre reinforced polymeric (FRP) components [241]. The method is used for QC in the FRP tank and pressure vessel industries.

An ultrasonic measuring device based on the measurement of the ultrasonic velocity V_{us} was introduced in the late 1970s [242] and has since been explored for diverse applications, such as control of PVC pipe extrusion [243] and monitoring of the composition of a mineral-filled polymer [244]. Ultrasonic in-line monitoring of **polymer extrusion**, with ultrasonic probes fitted to an extrusion slit die in order to generate US pulses across the flowing melt [245,246], has been exploited to control *in situ* the characteristics of the polymer being transformed in operations typically performed on twin screw extruders, such as compounding, visbreaking or reactive extrusion. Monitoring of extrusion processes by ultrasonic measurement has various advantages: *(i)* time delay free indications; *(ii)* instantaneous measurement of spatially averaged properties; and *(iii)* no disturbance of the melt flow.

Composition measurement, morphology and dispersion characterisation of multiphase systems were examined for different extrusion applications. The acoustic emission signal can be used to quantitatively infer particle size distribution, stickiness of the powder, gas flow-rate or compression properties. Application of ultrasonic techniques to polymer processing is still limited and has been used for in-line monitoring of the elastomer or **filler content** in polymer melts, and blend composition. Erwin *et al.* [247] used focused ultrasound for the measurement of mixing in polymer melts (LLDPE/20 wt.% $CaCO_3$, particle size <0.5 μm; PE/2 wt.% CB; PE-PS and PE-PP); particle agglomeration or dispersion were assessed. On-line real-time ultrasonic wave velocity measurements have been used for monitoring of the extrusion of $CaCO_3$-filled polypropylene with particle size in the 0.5 to a few μm range [244]. No evidence of agglomeration was observed up to 10 vol.% $CaCO_3$, but gross composition fluctuations and agglomeration were observed for the range

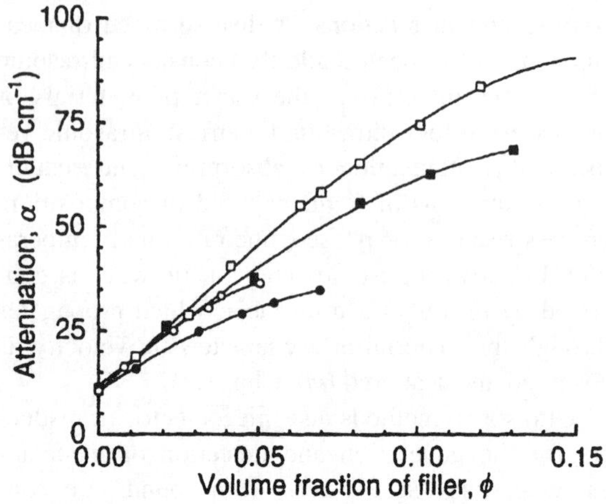

Fig. 7.14. Attenuation (α) of ultrasound for PP with different concentrations (ϕ) and grades of calcium carbonate; □, Camel-Wite, $d_p = 3.0$ μm; ■, Camel-Wite-ST, $d_p = 3.0$ μm; ○, Camel-Cal, $d_p = 0.7$ μm; ●, Camel-Cal-ST, $d_p = 0.7$ μm; "-ST" denotes stearate coated grades; d_p is the nominal mean particle diameter of calcium carbonate. After Dumoulin *et al.* [30]. Reproduced from *Trends in Polymer Science* **4**, M.M. Dumoulin *et al.*, 109–114, Copyright (1996), with permission of Elsevier.

>10 vol.%. For filled polymer systems, both V_{us} and α are sensitive to the presence of a mineral filler in a polymer matrix. Figure 7.14 shows the dependence of ultrasonic attenuation on composition for PP filled with different fillers; α depends on filler type, apparent particle size and concentration [248]. Concentration sensitivity was also used to determine residence-time distribution in an extruder, using $CaCO_3$ as a tracer in PP [249]. Continuously monitoring the ultrasonic response helps improving the compounding process by warning of variability, as small as $\pm 0.5\%$. Similar results were reported for TiO_2, and glass inclusions with sizes ranging from 0.2 to 100 μm. The feasibility of using ultrasound and neural networks together for on-line determination of filler concentration and dispersion was shown for PP/Camel-Cal and PP/Camel-Cal-ST [250]. A multi-sensor arrangement (in-line Raman, transmission NIRS and ultrasound transducer) on an extruder was recently used for real-time monitoring of EVA copolymers [162].

For optimal material properties an optimal state of mixing is required. On-line **powder blending** technology can reduce mixing times, reduce delays in processing and improve product quality. The acoustic technique may be used on any particle in almost any vessel. The acoustic signal magnitude is related to the kinetic energy of the particles; differences in shape are less detectable than density or particle size. Shape of profile and time to homogeneity are dependent on the type of particles. Passive acoustic mixing profiles were compared to simultaneously recorded profiles of the more widely accepted (equally non-invasive) technique of NIRS [251]. Homogeneity is reached when the profiles become stable. Acoustic emission spectroscopy eliminates the need for time-consuming post-processing microscopic methods for measuring the degree of **dispersion**. The increased understanding of how particle properties affect a mixing operation could lead to improved decisions when selecting materials for a formulation and potentially this could lead to improvements in scale-up of mixing processes. Results are of great relevance to masterbatch producers.

Ultrasonic sensors have also been applied in the study of physical foaming agents for foam extrusion [252]. For on-line monitoring of orientation processes birefringence, FTIR spectroscopy, fluorescence and ultrasonics are most suitable.

A comprehensive review of the applications of ultrasound to materials chemistry is available [253]. The use of ultrasonics for real-time monitoring of polymer processing was recently reviewed [30]. The necessary equipment, which is non-commercial (as opposed to the past), is relatively cheap. Only few research groups are active worldwide in this area.

7.2.8. Real-time Dielectric Spectroscopy

Principles and Characteristics

Dielectric spectroscopy (DIES) is known as a commercial in-line process technique (*cfr.* also Chp. 1.6) for measurement of chemical concentrations and physical properties, continuous quality monitoring, real-time process control and product classification. McBrearty *et al.* [254–256] have described an in-line dielectric sensor and a dielectric slit die for measuring electrical permittivities and conductivities of polymer melts and filled polymer melts over a broad range of frequencies while they are being processed through extruders or transfer lines. A microwave spectrometer gives a spectral response to the change of dielectric constant (ε') and dielectric loss (ε'') as microwave radiation passes through a sample [257]. No sample preparation is required. Dielectric analysers are among the few in-line instruments that can measure chemical concentrations in opaque as well as transparent fluids.

Applications

Permittivity measurements are potentially useful for continuous, in-line determinations of chemical composition in melts, *e.g.* co-monomer ratio in copolymers and additive concentrations in compounded products [254]. Permittivities provide a sensitive measure of chlorination level in chlorinated polyethylenes and vinyl acetate concentration in EVA copolymers [254]. In-line dielectric monitoring was used to examine the time profile of the transition of one composition to another during extrusion [255]. Processing of PP filled with Al_2O_3 and $CaCO_3$ and of EVA filled with montmorillonite clay were reported. Figure 7.15 shows permittivity *vs.* time for PS/Al_2O_3 melts. Mixing rules describe how the dielectric constant varies with concentration (*cfr.* Chp. 1.6). The dielectric slit die sensor was used for generating real-time monitoring data for compounding PA12/montmorillonite clay [256,259].

On-line real-time microdielectrometry of epoxy/fibreglass composite curing was reported [260]. DIES may be used for in-line curing or drying reactions, for the determination of water in polyamides, for (water) level indication (axiometrics) and for phase inversion detection in water/oil systems.

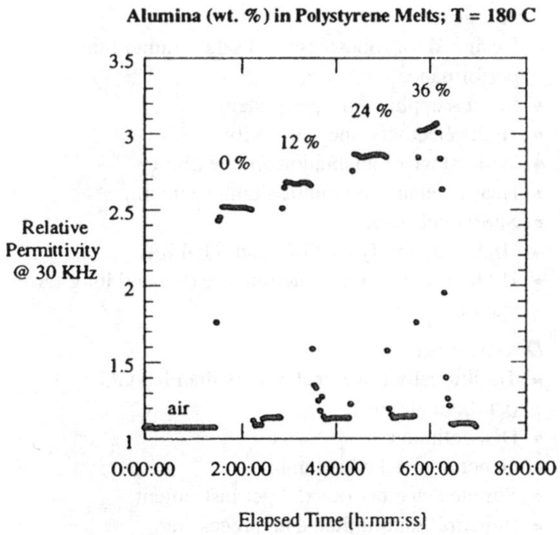

Fig. 7.15. Relative permittivity *vs.* time for extrusion of alumina-filled polystyrene. After ref. [258]. Reproduced by permission of Chemical Electrophysics Co. Inc.

7.3. PROCESS CHROMATOGRAPHY

Principles and Characteristics

Process chromatography is not the most obvious tool in relation to product quality control of polymer/additive formulations for two main reasons, namely the aggregation state of the product (melt or solid) and speed. With reference to Chp. 7.1 only those aspects of process chromatography will be outlined here which may impact additive analysis.

Process GC (PGC) dates from the late 1950s and is well established in the process environment. Table 7.32 illustrates the **main characteristics** of PGC. Various actions are possible to minimise the disadvantages: *time*: fast GC, very short narrow-bore, pressure programming, multiple detection, parallel chromatography; *auxiliary gases*: micro techniques, narrow-bore, μTCD; *cost of ownership*: micro technique, low energy; and *qualification*: modular analytics, maintenance free, remote control and maintenance, use of internet technology. Current PGC is characterised by high reliability (2 yrs.), multidetection (μTCD, FID, up to 24 on one application), capillary columns, parallel chromatography, network communication, fast GC and electronic pressure control (EPC). A new generation of GC instruments has been developed, which have been specifically designed for use in both on-line and at-line applications [261]. The simplification of multidimensional chromatography using EPC and multidetector technology can be employed to give on-line GC measurements, which are often superior to the laboratory. New requirements for process chromatographs are very short cycle times, minimum consumption of auxiliary supplies, reduced maintenance requirements, remote access for all parameters, permanent internal validation of analysis results and significant method development simplification. Key drivers for innovation in process GC are micromachining (size, weight, cost, safety), silicon technology (structure for high-resolution chromatography), valveless column switching techniques (use of HR capillary columns), improved control and greater automation, detector developments (DMD), and internet capability (remote access). On-line micro gas chromatography, which has recently been introduced, achieves analysis times of 30 s, and is therefore suitable for quality control (at a par with spectroscopic techniques). Similarly, with already available technology and a dedicated injector, MESI-SPME-fast GC enables very fast semicontinuous monitoring of both gaseous and liquid streams with separation times as short as 15 s [262].

The role of laboratory GC will decrease in favour of on-line GC. Self diagnostic fault finding and advanced calibration/validation will develop and more extensive use of multidimensional and hyphenated systems will be made. Microtechnology in process gas chromatography was recently illustrated [263]. Table 7.33 summarises the vision for PGC 2000$^+$.

As to other forms of gas chromatography, PyGC-MS is used in QC laboratories for testing of incoming materials and release of new products, as well as troubleshooting in damage cases. On-line HS-GC has been described [264].

Process HPLC, which dates from the 1970s, has more limited applicability than process GC. HPLC

Table 7.32. Main characteristics of conventional process gas chromatography

Advantages:
- Designed for robustness and safety rather than performance
- Several applications per system
- High selectivity and sensitivity
- Wide range of adaptation and flexibility
- Heavy reliance on multicolumn switching
- Short cycle time
- High availability (>98%) and reliability
- High accuracy (reproducibility ±1%) and long-term stability

Disadvantages:
- Traditionally lower technology than lab GC (isothermal only)
- Discontinuous
- Generally packed columns
- Simple detectors (max. 2 per instrument)
- Inflexible and limited data processing
- Need for auxiliary gases of high purity
- High ownership costs and investment at site
- High qualification of maintenance personnel

Table 7.33. Vision for process gas chromatography 2000$^+$

- No analytical limitations
- Nearly maintenance free
- Remote control and maintenance
- Lowest possible cost, energy consumption, size and weight
- Highest safety standards
- One sampling point per system
- One application per system

is well suited to on-line analysis for process control [265–267]. The operation of HPLC equipment in a process environment requires special considerations. As HPLC is a high-pressure technique (up to 350 bar) samples can often be transferred directly from the process to the analyser. Automation of sample processing is essential for continuous process monitoring. The analysis speed should be high enough to permit a much more rapid sampling frequency than the change of the process variable of interest. Reversed-phase chromatography (non-polar column with polar eluent) is a useful technique allowing shorter analysis times than polar columns. Microbore HPLC is useful to reduce solvent consumption, an important issue in the process environment. Barisci *et al.* [267] have described an on-line monitoring device using HPLC for unattended operation for at least a week. All analytical steps, including sample collection, pretreatment, derivatisation, injection, detection, data processing and reporting were fully automated. Use of a fully automated, on-line monitoring system based on HPLC is of great advantage for control of continuous processes. Low pressure LC, probe LC, and micro-LC are techniques important to the future of process chromatography. Process HPLC has been reviewed [268].

Requirements for SEC in process control or HTS are speed (faster than conventional SEC; <10 min/sample), less maintenance, and very high robustness [269]. Also process analysis with SFC was described [270,271]. In a moderately sized chemical plant it is often possible to conduct many more analyses per unit time by TLC-HPTLC than by GC or HPLC.

Applications
Process gas chromatography is widely used for petrochemical products (*e.g.* coke oven gas control, gasoline analysis from catalytic reforming, octane-number analysis), determination of trace organics in process streams, VOC wastewater analysis, *etc.* An at-column GC procedure has been developed for on-line determination of **polymer additives** (1200 Da; 100 ppm) in a 500 μL SEC fraction in DCM [272]. PyGC (including stepwise PyGC) is particularly amenable to product quality control [273]. PyGC-MS is used as an industrial QC tool [274], eventually in combination with PCA as in case of organic paints [275].

On-line HPLC can be used for characterisation of **raw materials**, analysis of the composition of reaction mixtures or determination of the purity and product quality of end-products [276]. Steinmüller [277] has reported on-line HPLC for the measurement of monomers in polyvinylpyrrolidone. The determination of molecular weight distributions is routine with process HPLC, *e.g.* in case of polycarbonates [268]. RPLC is also used for QC of non-vulcanised EPDM/(Irganox 1076/1520 LR) [278]. A general review on the use of on-line HPLC in the polymer industry was reported [277].

Polymer characterisation by on-line SEC has been described [279]. On-line SEC analysis of low-MW polymers and additives in a process area has been published [280]. The ability to obtain rapid status information concerning blend tank component ratios enables reduction in holding time and an increase in precision in comparison to off-line wet chemical testing procedures. For coating process control and quality evaluation of polymeric **automotive coatings** a variety of techniques (HPLC/SEC, TA, ATR-FTIR) are used [281].

7.4. *IN SITU* ELEMENTAL ANALYSIS

Principles and Characteristics
In Chp. 8 of ref. [113a] we discussed elemental analysis modes in which the sample was approached to the elemental analysis tool. Mobile spectrometers are more suitable for *in situ* or on-line monitoring. Various such tools have recently been developed, as portable XRF and LIESA®. X-ray fluorescence (XRF) analysis may be based on either electron or radio-isotope excitation. Compact, rugged, and reliable on-line XRF analysers based on radio-isotope excitation have been described [25]. On-line (micro-)XRF analysers need very little if any sample preparation compared to many other techniques. Of prime importance is that the surface at the cell window represents the whole sample stream. The instruments are capable of excellent on-line performance for many applications such as process liquids, slurries, solids (*e.g.*, polymer pellets), powders, and others. Process control XRF allows simultaneous determination of up to 32 elements.

Applications
XRF is being used increasingly in a works environment for QC type applications [282,283], for example to measure the level of additives in oil, or of metals or other elements in a polymer. The advantage is that sample preparation is minimal because

the polymer sample can be pressed into a plaque. A typical analysis would quantitate the levels of processing stabilisers containing phosphorous *vs.* a known standard. The disadvantage is that the analysis cannot differentiate between the source, such as intact phosphite and the transformation products of the stabiliser. For samples containing nitrogen, such as HALS (*e.g.* Chimassorb 944/119) nitrogen WDXRF analysis can be used to quantitate additives in polymer without extraction. Nitrogen analysis is especially valuable as a complementary technique to chromatography to confirm levels of additives. With analysis times of about 15 min to 1 h, conventional WDXRF is not as rapid as some of the on-line tests, but is considerably shorter than methods requiring extraction. In case of formulations with several nitrogen-containing additives the total level of nitrogen can be used to confirm that the additive loading process is constant.

Kalnicky *et al.* [25] have described the determination of calcium in polyolefin pellets containing Ca-stearate. The pellets (approximate dimensions $5 \times 2 \times 3$ mm^3) were measured for 600 seconds with a 20 mCi ^{55}Fe light-element on-line XRF analyser. They were poured into a slurry cell and measured "as is" without any preparation. Replicate measurements on a freshly poured sample showed reproducibility within counting statistics, indicating that packing density effects were negligible. Calibration gave $R^2 = 0.9966$ and $s = 2$ ppm for Ca from 87–139 ppm. Further improvement in the standard error (s) was achieved by including a background correction term dependent on source backscatter (BS). The statistical uncertainty of 1.3 ppm was the major source of error, not instrument stability or sample packing effects. This study showed the ability of on-line XRF instruments to analyse irregular shaped materials with minimal sample preparation/handling. On-line element analysis by EDXRF is fast.

Laser-induced emission spectral analysis (LIESA®), developed by Krupp as an in-stream elemental analysis method with broad applications (*cfr.* Chp. 3.3.2), was investigated by Schneider *et al.* [284] with respect to applicabilities and limits for QC of final mixes of **automotive rubber goods** (remote laser microanalysis, RELMA). The range of application of RELMA is identification of particles and control of incoming raw material, beside detection of mixing inhomogeneities and errors in weighing. In contrast to a tyre factory, on-line monitoring of mixes for technical rubber goods is not possible *cq.* convenient due to great matrix variety and frequent recipe changes.

Review papers of XRF/XRD as process analytical techniques are available [285,286].

BIBLIOGRAPHY

Process Analytical Chemistry

K.H. Koch, *Process Analytical Chemistry*, Springer-Verlag, Berlin (1999).

R.E. Sherman and L.J. Rhodes (eds.), *Analytical Instrumentation. Practical Guides for Measurement and Control*, Instrument Society of America, Research Triangle Park, NC (1996).

G. Oesterle, *Prozeßanalytik-Grundlagen und Praxis*, Oldenbourg, Munich (1995).

F. McLennan and B. Kowalski (eds.), *Process Analytical Chemistry*, Blackie A&P, London (1995).

K.G. Carr-Brion, J.R.P. Clarke and E.F. Harding, *Sampling Systems for Process Analysers*, Butterworth Heinemann, Oxford (1994).

Process Spectroscopy

J.M. Chalmers (ed.), *Spectroscopy in Process Analysis*, Sheffield Academic Press, Sheffield (2000).

J.C. Lindon, G.E. Tranter and J.L. Holmes (eds.), *Encyclopedia of Spectroscopy and Spectrometry*, Academic Press, San Diego, CA (2000).

J. Workman and A.W. Springsteen, *Applied Spectroscopy. A Compact Reference for Practitioners,* Academic Press, San Diego, CA (1998).

J.M. Chalmers and G. Dent, *Industrial Analysis with Vibrational Spectroscopy*, The Royal Society of Chemistry, Cambridge (1997).

R.C. Asher, *Ultrasonic Sensors for Chemical and Process Plant*, Institute of Physics Publishing, Bristol (1997).

D.M. Grant and R.K. Harris (eds.), *Encyclopedia of NMR*, J. Wiley & Sons, New York, NY (1996).

G.E. Maciel (ed.), *NMR in Industrial Process Control and Quality Control, Nuclear Magnetic Resonance in Modern Technology*, Kluwer Academic Publishers, Dordrecht (1994).

D.A. Burns and E.W. Ciurczak (eds.), *Handbook of Near Infrared Analysis*, M. Dekker, New York, NY (1992).

L.C. Lynnworth, *Ultrasonic Measurements for Process Control: Theory, Techniques, Applications*, Academic Press, Boston, MA (1989).

T.C. Farrar, *An Introduction to Pulse NMR Spectroscopy*, Farragut Press, Chicago, IL (1987).

E. Fukushima and S.B.W. Roeder, *Experimental Pulse NMR – A Nuts and Bolts Approach*, Addison-Wesley, Reading, MA (1981).

Process Data Analysis

R.G. Brereton, *Chemometrics. Data Analysis for the Laboratory and Chemical Plant*, J. Wiley & Sons, Chichester (2003).

A. Smilde, P. Geladi and R. Bro, *Multiway Analysis in Chemistry*, J. Wiley & Sons, Chichester (2003).

T. Næs, T. Isaksson, T. Fearn and T. Davies, *A User-friendly Guide to Multivariate Calibration and Classification*, NIR Publications, Chichester (2002).

R.G. Brereton, *An Introduction to Chemometrics*, Wiley-Interscience, New York, NY (2002).

C. Rauwendaal, *Statistical Process Control in Injection Molding and Extrusion*, Hanser Verlag, Munich (2000).

R. Kramer, *Chemometric Techniques for Quantitative Analysis*, M. Dekker, New York, NY (1998).

K.R. Beebe, R.J. Pell and M.B. Seasholtz, *Chemometrics: A Practical Guide*, J. Wiley & Sons, New York, NY (1998).

W.P. Gardiner, *Statistical Analysis Methods for Chemists*, The Royal Society of Chemistry, Cambridge (1997).

M.J. Adams, *Chemometrics in Analytical Spectroscopy*, The Royal Society of Chemistry, Cambridge (1995).

F. Alsup and R.M. Watson, *Practical Statistical Process Control*, Van Nostrand Reinhold, New York, NY (1993).

H. Martens and T. Næs, *Multivariate Calibration*, J. Wiley & Sons, New York, NY (1992).

E.R. Malinowski, *Factor Analysis in Chemistry*, Wiley-Interscience, New York, NY (1991).

REFERENCES

[1] M. Ohshima and M. Tanigaki, *J. Process Control* **10**, 135–48 (2000).

[2] C. Rauwendaal, *Statistical Process Control in Injection Molding and Extrusion*, Hanser Verlag, Munich (2000).

[3] S.D. Brown, S.T. Sum and F. Despagne, *Anal. Chem.* **68** (12), 21R (1996).

[4] B.K. Lavine, *Anal. Chem.* **70** (12), 209R (1998).

[5] R. Robinson, *Proceedings Advances in Process Analytics and Control Technology (APACT03)*, York (2003), Paper T29.

[6] R.A. Fidler, T. Oswald and N.W. Pewar, *Proceedings SPE ANTEC '93*, New Orleans, LA (1993), pp. 277–82.

[7] W.P. Carey, *Trends Anal. Chem.* **13**, 210–8 (1994).

[8] T.O. Broadhead, B.I. Nelson and J.M. Dealy, *Intl. Polym. Process.* **8** (2), 104–12 (1993).

[9] C.H. Sohl, *Proceedings SPE ANTEC '91*, Montréal (1991), pp. 846–9.

[10] A. Khettry and H.G. Hansen, *Proceedings SPE ANTEC '95*, Boston, MA (1995), pp. 2824–31.

[11] M. Watari, N. Mitsui, H. Higashiyama and M. Tomo, *Yokogawa Technical Information TI11VOA1-11* (1996).

[12] J. Workman Jr., D.J. Veltkamp, S. Doherty, B.B. Anderson, K.E. Creasy, M. Koch, J.F. Tatera, A.L. Robinson, L. Bond, L.W. Burgess, G.N. Bokerman, A.H. Ullman, G.P. Darsey, F. Mozayeni, J.A. Bamberger and M.S. Greenwood, *Anal. Chem.* **71**, 121–180R (1999).

[13] K.J. Clevett, *Process Analyser Technology*, J. Wiley & Sons, New York, NY (1986).

[14] G.D. Nichols, *On-line Process Analysers*, J. Wiley & Sons, New York, NY (1988).

[15] F. Müller and U. Gellert, *Abstracts 6th Intl. Symposium on Hyphenated Techniques in Chromatography and Hyphenated Chromatographic Analyzers (HTC6)*, Bruges (2000), Paper 11.2.

[16] D. Bassi, P. Tosi and R. Schögl, *J. Vac. Sci. Technol.* **A16** (1), 114–22 (1998).

[17] D. Littlejohn, S. Lancaster, A. Owen and R. Wright, *Proceedings Advances in Process Analytics and Control Technology (APACT03)*, York (2003), Poster P12.

[18] R. Clinton, C.S. Creaser and D. Bryant, *Proceedings Advances in Process Analytics and Control Technology (APACT03)*, York (2003), Paper T5.

[19] R. Zimmermann, H.J. Heger, R. Dorfner, U. Boesl and A. Kettrup, *Chem. Lab. Biotechn.* **49** (6), 210–14 (1998).

[20] M.R. Walsh and M.A. LaPack, *ISA Transactions* **34**, 67–85 (1995).

[21] J. Scrivens, *Spec. Publ.-Roy. Soc. Chem.* **154**, 145–52 (1994).

[22] T. Kotiaho, *J. Mass Spectrom.* **31** (1), 1 (1996).

[23] R. Neudert, E. Ströfer and W. Bremser, *Magn. Res. Chem.* **24**, 1089–92 (1986).

[24] T.W. Skloss, A.J. Kim and J.F. Haw, *Anal. Chem.* **66**, 536–42 (1994).

[25] D.J. Kalnicky, A.I. Ahonen, M.J. Hietala and K.J. Kauhanen, *Adv. X-Ray Anal.* **33**, 585–91 (1990).

[26] J.J. Gunnell, *Proceedings Advances in Process Analytics and Control Technology (APACT03)*, York (2003), Paper T12.
[27] K.H. Koch, *Process Analytical Chemistry*, Springer-Verlag, Berlin (1997).
[28] M.P. Thomas, *J. Vinyl Addit. Technol.* **2** (4), 330–8 (1996).
[29] S.J. Qin, *Proceedings Advances in Process Analytics and Control Technology (APACT03)*, York (2003), Paper T26.
[30] M.M. Dumoulin, R. Gendron and K.C. Cole, *Trends Polym. Sci.* **4** (4), 109–14 (1996).
[31] H.C. Ashton, *Angew. Makromol. Chem.* **261–2**, 9–23 (1998).
[32] J.J. Gibbons, *Am. Lab.* **19**, 33–9 (1987).
[33] Anon., *Additives for Polymers* (Dec. 2000), pp. 7–8.
[34] W. Obendrauf, G.R. Langecker and W. Friesenbichler, *Intl. Polym. Process.* **13** (1), 71–7 (1998).
[35] F. Müller, *Chem. Anlagen Verfahren* (6), 126–7 (1999).
[36] J. v.d. Maas, *Proceedings SPQ-98 (Spectroscopy in Process and Quality Control)*, London (1998).
[36a] J. Andrews and P. Dallin, *Spectrosc. Europe* **15** (4), 23–6 (2003).
[37] D. Fischer and K.-J. Eichhorn, *Analusis Mag.* **26** (4), M58–61 (1998).
[38] J.H. Duckworth, in *Applied Spectroscopy* (J. Workman and A.W. Springsteen), Academic Press, San Diego, CA (1994), pp. 165–90.
[39] D.M. Haaland and W.V. Thomas, *Anal. Chem.* **60**, 1193 (1988).
[40] G.R. Flåten and A.D. Walmsley, *Proceedings Advances in Process Analytics and Control Technology (APACT03)*, York (2003), Poster P4.
[41] G.R. Flåten and A.D. Walmsley, in press (2003).
[42] J.H. Duckworth, in *Applied Spectroscopy. A Compact Reference for Practitioners* (J. Workman and A.W. Springsteen, eds.), Academic Press, San Diego, CA (1998), pp. 93–168.
[43] H. Herman, in *Spectroscopy in Process Analysis* (J. Chalmers, ed.), Sheffield Academic Press, Sheffield (2000), p. 16.
[44] H.W. Siesler, in *Near Infrared Spectroscopy* (A.M.C. Davies and R. Giangiacomo, eds.), NIR Publications, Charlton (2000), pp. 331–7.
[45] W.M. Doyle, *Proceedings Annu. ISA Anal. Div. Symp.*, Toronto, **28**, 115–23 (1995).
[46] B.R. Buchanan, E.W. Richmond, Z. Lin and J. Timmermans, *Spectroscopy* **13** (10), 44–8 (1998).
[47] R.E. Schirmer and A.G. Gargus, *Am. Lab.* **18** (12), 30–9 (1986).
[48] G.R. Powell, P.A. Crosby, D.N. Waters, C.M. France, R.C. Spooncer and G.F. Fernando, *Smart Mater. Struct.* **7** (4), 557–68 (1998).
[49] J. Li and M.G. Hansen, *Proceedings SPE ANTEC '99*, New York, NY (1999), pp. 1937–41.
[50] R.L.H.M. Verlaek, E.M.F.J. Verdurmen and B.J. Kip, *Techn. Rept. DSM Research*, Geleen (1994).
[51] B. Vigerust, K. Kolset, S. Nordenson, A. Hendriksen and K. Kleveland, *Appl. Spectrosc.* **45**, 173–7 (1991).
[52] T.V. Karstang and A. Hendriksen, *Chemometr. Intell. Lab. Systems* **14**, 331–9 (1992).
[53] M. Collins and R. Hope, *Lab. Rob. Autom.* **10** (2), 77–80 (1998).
[54] M. Collins and S. Merz, *Am. Lab.* **31** (16), 22–8 (1999).
[55] J. Yang and Z. Zhao, *Jisuanji Yu Yingyong Huaxue* **5**, 270–4 (1988).
[56] R. Stengler, G. Vonderschmidt, U. Hoeppener-Kramar and G. Rinke, *Kunststoffe* **85**, 930–3 (1995).
[57] R.E. Schirmer, *Adv. Instrum.* **43**, 831–40 (1988); *ISA Transactions* **28** (2), 65–9 (1989).
[58] J.-C. Yang, *Proceedings Intl. Conf. Pet. Refin. Petrochem. Process* (X. Hou, ed.), Intl. Academic Publishers, Beijing (1991), Vol. 2, pp. 591–7.
[59] R. Spatafore, V. Catalano and K.L. Miller, *Proceedings Polyolefins VIII* (SPE, South Texas Section, ed.), Houston, TX (1993), pp. 587–601.
[60] R.L.H.M. Verlaek, G. Meyers, E. Peters and B.J. Kip, *Techn. Rept. DSM Research*, Geleen (1996).
[61] T.R. Crompton, *Practical Polymer Analysis*, Plenum Press, New York, NY (1993).
[62] H. Herman, K. Elliott, P. Hope and J. Hunter, *Proceedings SPQ-98 (Spectroscopy in Process and Quality Control)*, London (1998).
[63] A. Kaito, K. Nakayama and M. Kyotani, *J. Polym. Sci., Part B: Polym. Phys.* **29**, 1321 (1991).
[63a] F. Farahani, K. Jeong, S. Sayad, S.T. Balke, F. Calidonio and A. Rom-Roginski, *Proceedings SPE ANTEC 2003*, Nashville, TN (2003), pp. 3316–20.
[64] J.C.J. Bart, *Proceedings SPE ANTEC 2000*, Orlando, FL (2000), pp. 3741–5.
[65] J.L.P. Jessop, G.J. Blanchard and A. Scranton, *RadTech. Rept.* **12** (4), 27–32 (1998).
[66] A.J. Bur, R.E. Lowry, S.C. Roth, C.L. Thomas and F.W. Wang, *Macromolecules* **25**, 3503 (1992).
[67] A.J. Bur, F.W. Wang, C.L. Thomas and J.L. Rose, *Polym. Engng. Sci.* **34**, 671 (1994).
[68] F. Kowol, M. Oleimeulen, H. Freitag and T. Huth-Fehre, *J. Near Infrared Spectrosc.* **6**, A149–51 (1998).
[69] A. Eilert, *Proceedings 9th Intl. Diffuse Reflectance Conference*, Chambersburg, PA (1998).
[70] R.K. Stengler and G. Weis, *Proceedings SPIE-Int. Soc. Opt. Eng.* **1320**, 33–8 (1990).
[71] R.K. Stengler and G. Weis, *Kunststoffe* **79** (10), 953–60 (1989).

[72] H.-G. Fritz and S. Ultsch, *Kunststoffe* **79**, 785–90 (1989).
[73] R.J. Harvey (to Automatik Apparate-Maschinenbau), *U.S. Pat.* 4,717,827 (Jan. 5, 1988).
[74] R.A. Fidler, *Plast. Compd.* **15** (6), 47–50 (1992).
[75] B. Stöhrer, H. Kurrer, R. Stengler and G. Weis, *Kunststoffe* **79**, 1165–7 (1989).
[76] S. Lowry, T. May, A. Bornstein, Y. Weissman, R. Harman and I. Tugenthaft, *Appl. Spectrosc.* **48** (7), 852–6 (1994).
[77] N.J. Harrick, *Internal Reflection Spectroscopy*, Interscience Publishers, New York, NY (1967).
[78] P. Simak, *Kunststoffe* **78**, 234–6 (1988).
[79] L. Jakisch, D. Fischer, M. Stephan and K.-J. Eichhorn, *Kunststoffe* **85**, 1338–44 (1995).
[80] R. Götz, B. Mizaikoff and R. Kellner, *Appl. Spectrosc.* **52**, 1248–52 (1998).
[81] F. Wulfert, W.Th. Kok and A.K. Smilde, *Anal. Chem.* **70**, 1761–7 (1998).
[82] F.J.J. Clarke, *Proceedings CORM/UVSG Conference Optical Spectrometry: Applications and Instrumentation into the 21st Century*, Elsevier, Amsterdam (1998).
[83] M. Xanthos and S.H. Patel, *Adv. Polym. Technol.* **14**, 151–7 (1995).
[84] J.S. Sanghera, I.D. Aggarwal (eds.), *Infrared Fiber Optics*, CRC Press, Boca Raton, FL (1998).
[85] K.W. van Every and M.J. Elder, *Proceedings SPE ANTEC '96*, Indianapolis, IN (1996), pp. 2400–2405.
[86] *Mettler-Toledo Application Data Sheet ReactIR 7/98*, Mettler-Toledo, Schwerzenbach (1998).
[87] K. Schneider and H. Behl, *Mettler Toledo Magazine No. 2* (1997).
[88] J.P. Coates, J.M. D'Agostino and C.R. Friedman, *Am. Lab.* **18** (1), 82–6 (1986); **18** (12), 40–6 (1986).
[89] M.G. Hansen, *Proceedings SPE ANTEC '91*, Montréal (1991), pp. 840–1.
[90] J. Workman, *Appl. Spectrosc. Revs.* **34** (1/2), 1–89 (1999).
[91] J.C.J. Bart, *Proceedings SPE ANTEC 2000*, Orlando, FL (2000), pp. 2079–84.
[92] R.A. Fidler, D. Rowe and G. Weis, *Proceedings SPE ANTEC '91*, Montréal (1991), pp. 850–5.
[93] R.A. Fidler, *Proceedings SPE ANTEC '92*, Detroit, MI (1992), pp. 211–4.
[94] R.A. Fidler, *Proceedings Polyolefins VIII*, Houston, TX (1993), pp. 602–15.
[95] P. Palmen and G. Meyers, *Techn. Rept. DSM Research*, Geleen (1996).
[96] K.W. Van Every and M.J. Elder, *J. Vinyl Addit. Technol.* **2** (3), 224–8 (1996).
[97] R. Leardi, M.B. Seasholtz and R.J. Pell, *Anal. Chim. Acta* **461** (2), 189–200 (2002).
[98] E.V. Thomas, *Anal. Chem.* **66**, 795A (1994).
[99] R. Leardi, *J. Chemometrics* **14**, 643 (2000).
[100] S.R. Culler, P.C. Gillette, H. Ispida and J.L. Koenig, *Appl. Spectrosc.* **38**, 495–500 (1984).
[101] D. Fischer, E. Henßge and E. Pigorsch, *Abstracts Advanced Methods of Polymer Characterization: New Developments and Applications in Industry*, Mainz (1999), Paper O11.
[102] B.-J. Niu and M.W. Urban, *Polym. Mater. Sci. Engng.* **75**, 41 (1996).
[103] H.L. McPeters, *Anal. Chim. Acta* **238**, 83–93 (1990).
[104] J.M. Chalmers and N.J. Everall, *Proceedings SPQ-98 (Spectroscopy in Process and Quality Control)*, London (1998).
[105] A. Smith, *Plastics Additives & Compounding* **2** (5), 34–7 (2000).
[106] O. Kammona, E.G. Chatzi and C. Kiparissides, *Dechema Monographs Vol.* **134**, 365–74 (1998).
[107] S.G. Kazarian, N.H. Brantley and C.A. Eckert, *Proceedings SPE ANTEC '98*, Atlanta, GA (1998), pp. 1415–6.
[108] J.J. Kellar, W.M. Cross and J.D. Miller, *Appl. Spectrosc.* **44**, 1508–12 (1990).
[109] *Mettler-Toledo Application Data Sheet ReactIR 6/98*, Mettler-Toledo, Schwerzenbach (1998).
[110] K.T. Paul, in *Flame Retardants '94* (The British Plastics Federation, ed.), Interscience Communications Ltd., London (1994), pp. 251–62.
[111] M.D. Coleman, *Proceedings Advances in Process Analytics and Control Technology (APACT03)*, York (2003), Paper T6.
[112] D. Peters and J. Shearer, *Proceedings SPIE-Int. Soc. Opt. Eng.* **2069**, 32–41 (1993).
[113] G.J. Kemeny, in *Handbook of Near-Infrared Analysis* (D.A. Burns and E.W. Ciurczak, eds.), M. Dekker, New York, NY (1992), pp. 53–105.
[113a] J.C.J. Bart, *Additives in Polymers. Industrial Analyses and Applications*, J. Wiley & Sons, Chichester (2005).
[114] K.E. Williams, and S.J. Haswell, in *Microwave-Enhanced Chemistry* (H.M. Kingston and S.J. Haswell, eds.), American Chemical Society, Washington, DC (1997), pp. 401–22.
[115] G.J. Kemeny, in *Making Light Work: Advances in Near Infrared Spectroscopy* (I. Murray and I.A. Cowe, eds.), VCH, Weinheim (1992), pp. 14–26.
[116] R. Reshadat, S. Desa, S. Joseph, M. Mehra, N. Stoev and S.T. Balke, *Appl. Spectrosc.* **53** (11), 1412–8 (1999).
[117] D. Coombs and G. Poulter, *Spectrosc. Europe* **10** (5), 10–4 (1998).
[118] M.G. Hansen, A. Khettry, J. Batra and D.A. Stewart, *Proceedings SPE ANTEC '93*, New Orleans, LA (1993), pp. 1437–43.

[119] H. Willard, L. Merritt, J. Dean and F. Settle, *Instrumental Methods of Analysis*, Wadsworth, Belmont, CA (1981).

[120] J.S. de Wit, in *Making Light Work: Advances in Near Infrared Spectroscopy* (I. Murray and I.A. Cowe, eds.), VCH, Weinheim (1992), pp. 537–43.

[121] X. Wang, J. Soos, Q. Li and J. Crystal, *Process Control Qual.* **5**, 9–16 (1993).

[122] S.E. Harris and R.W. Wallace, *J. Opt. Soc. Amer.* **59**, 744 (1969).

[123] N. Eisenreich, J. Herz, H. Kull, W. Mayer and T. Rohe, *Proceedings SPE ANTEC '96*, Indianapolis, IN (1996), pp. 3131–5.

[124] R.C. Moessner, *Process Contr. Qual.* **2**, 237–47 (1992).

[125] W.F. McClure, in *Making Light Work: Advances in Near Infrared Spectroscopy* (I. Murray and I.A. Cowe, eds.), VCH, Weinheim (1992), pp. 1–13.

[126] R. Classon, *Chem. Engng.* (4), 102–5 (1993).

[126a] R.A. Crocombe, *Spectrosc. Europe* **16** (3), 16–9 (2004).

[127] G. Oesterle, *Proceßanalytik-Grundlagen und Praxis*, Oldenbourg, Munich (1995).

[128] J. Workman Jr., *J. Near Infrared Spectrosc.* **1**, 221–45 (1993).

[129] S. Williams and H.L. McPeters, in *Making Light Work: Advances in Near Infrared Spectroscopy* (I. Murray and I.A. Cowe, eds.), VCH, Weinheim (1992), pp. 544–53.

[130] E.W. Ciurczak, in *Applied Spectroscopy. A Compact Reference for Practitioners* (J. Workman and A.W. Springsteen, eds.), Academic Press, San Diego, CA (1998), pp. 423–35.

[131] H.W. Siesler and K. Holland-Moritz, *Infrared and Raman Spectroscopy of Polymers*, M. Dekker, New York, NY (1980), p. 320.

[132] C.E. Miller, *Appl. Spectrosc. Rev.* **26**, 277–339 (1991).

[133] G. Lachenal, *J. Near Infrared Spectrosc.* **6**, 299–306 (1998).

[134] K. Aaljoki, *Process Control Qual.* **6**, 125–31 (1994).

[135] M.G. Hansen and A. Khettry, *Polym. Engng. Sci.* **34**, 1758–66 (1994).

[136] T.G. Carswell, R.W. Garrett, D.J.T. Hill, J.H. O'Donnell, P.J. Pomery and C.L. Winzor, in *Polymer Spectroscopy* (A.H. Fawcett, ed.), J. Wiley & Sons, Chichester (1996), pp. 253–74.

[137] T. Malfait and L. Ruys, in *Near Infrared Spectroscopy* (K.I. Hildrum, T. Isaksson, T. Næs and A. Tandberg, eds.), Ellis Horwood, Chichester (1992), pp. 423–7.

[138] C.E. Miller, in *Making Light Work: Advances in Near Infrared Spectroscopy* (I. Murray and I.A. Cowe, eds.), VCH, Weinheim (1992), pp. 519–24.

[139] B. Kip, M. Huys and T. Berghmans, *Techn. Rept. DSM Research*, Geleen (1998).

[140] N. Ikeda, *Jpn. Kokai Tokkyo Koho* 95-177929 (Jan. 7, 1997).

[141] L.O. Jones, J.R. Noles Jr., R.J.L. Chimenti and H.L. Fang (to Exxon Chemical Patents Inc.), *US Pat. Appl.* 95-422561 (Apr. 14, 1995).

[142] H.L. McPeters and S.O. Williams, *Process Contr. Qual.* **3**, 75–83 (1992).

[143] J. Batra, A. Khettry and M.G. Hansen, *Polym. Engng. Sci.* **34**, 1767–72 (1994).

[144] U. Eschenauer, O. Henck, M. Hühne, P. Wu, I. Zebger and H.W. Siesler, *Near Infrared Spectroscopy*, Ellis Horwood, Chichester (1992), pp. 11–8.

[145] A. Khettry, J. Batra, D.A. Stewart and M.G. Hansen, *Proceedings SPE ANTEC '92*, Detroit, MI (1992), pp. 2674–6.

[146] T. Vastenhoudt, *SPE Benelux Techn. Mtg.*, Tilburg (1999).

[147] R. Reshadat, S.T. Balke, F. Calidonio and C.J.B. Dobbin, *Proceedings SPE ANTEC '96*, Indianapolis, IN (1996), pp. 2688–91.

[148] D. Fischer, T. Bayer, K.-J. Eichhorn and M. Otto, *Fresenius J. Anal. Chem.* **359**, 74–7 (1997).

[149] S. Vedula, M.G. Hansen and B.J. Deshpande, *Proceedings SPE ANTEC '98*, Atlanta, GA (1998), pp. 2085–9.

[150] D.C. Peters, *Proceedings SPIE-Int. Soc. Opt. Engng.* **2367**, 146–58 (1995).

[151] K. Maeda, K. Matsura, N. Natori and S. Hasegawa (to Fujikura Ltd.), *Patent* JP 07110295-A (Apr. 25, 1995).

[152] T. Nagata, M. Tanigaki and M. Ohshima, *J. Cellular Plast.* **38** (1), 11–30 (2002).

[153] S. Farquharson, *Proc. SPIE-Int. Soc. Opt. Engng.* **1172**, 164–73 (1990).

[154] Y. Thomas, K.C. Cole, E. Pellerin and M.M. Dumoulin, *Polym. Mat. Sci. Engng.* **72**, 19–20 (1995).

[155] M. Bains, R. Reshadat, R. Lew, W.R. Cluett, S.T. Balke and J.W. Hall, *Proceedings SPE ANTEC '94*, San Francisco, CA (1994), pp. 2227–31.

[156] Y. Thomas, K.C. Cole, L.E. Daigneault and L.-M. Caron, *Proceedings SPE ANTEC '96*, Indianapolis, IN (1996), pp. 1855–9.

[157] T. Rohe, W. Becker, A. Krey, H. Nägele, S. Kölle and N. Eisenreich, *J. Near Infrared Spectrosc.* **6**, 325–32 (1998).

[158] T. Rohe, W. Becker, S. Kölle, N. Eisenreich and P. Eyerer, *Proceedings SPQ (Spectroscopy in Process and Quality Control)*, London (1998).

[159] T. Rohe, S. Kölle, H. Nägele and N. Eisenreich, *Proceedings 29th Int. Annu. Conf. ICT*, Pfinztal (1998), pp. 99.1–13.

[160] K. Kitagawa, S. Hayasaki and Y. Ozaki, *Vib. Spectrosc.* **15** (1), 43–51 (1997).

[161] S. Vedula and M.G. Hansen, *Proceedings SPE ANTEC '97*, Toronto (1997), pp. 2150–4.

[162] S.E. Barnes, M.G. Sibley, E.C. Brown, H.G.M. Edwards, I.J. Scowen and P.D. Coates, *Proceedings Advances in Process Analytics and Control Technology (APACT03)*, York (2003), Paper T2; *Proceedings SPE ANTEC 2003*, Nashville, TN (2003), pp. 3311–5.

[163] Y. Kita, S. Sasao and K. Matsukawa, *Kagaku to Kogyo (Osaka)* **73** (9), 410–4 (1999).

[164] H. Lammers, M.P.B. van Uum and J.P. de Kleyn, *Macromol. Chem. Phys.* **196**, 2029–34 (1995).

[165] J.C.J. Bart, *Proceedings SPE ANTEC 2000*, Orlando, FL (2000), pp. 2074–8.

[166] D.O. Kazmer, *Proceedings SPE ANTEC 2000*, Orlando, FL (2000), pp. 3848–52.

[167] K.L. Miller, in *Making Light Work: Advances in Near Infrared Spectroscopy* (I. Murray and I.A. Cowe, eds.), VCH, Weinheim (1992), pp. 454–60.

[168] L. Xu and J.R. Schlup, *J. Appl. Polym. Sci.* **67** (5), 895–901 (1998).

[169] H. Brackemann, M. Buback and H.-P. Vögele, *Makromol. Chem.* **187**, 1977–92 (1986).

[170] M. Hühne, U. Eschenauer and H.W. Siesler, in *Making Light Work: Advances in Near Infrared Spectroscopy* (I. Murray and I.A. Cowe, eds.), VCH, Weinheim (1992), pp. 51–4.

[171] H. Hartwig and D. Wienke, *Techn. Rept. DSM Research*, Geleen (1998).

[172] R.L. Grob, D.J. Skahan, K. Dix and K. Nielsen, *Process Contr. Qual.* **2** (3), 225–35 (1992).

[173] L. Xu and J.R. Schlup, *Appl. Spectrosc.* **50**, 109 (1996).

[174] R. Cox, J. Lebrasseur, E. Michiels, H. Buys, H. Li, F.R. Van de Voort, A.A. Ishmail and J. Sedman, *J. Am. Oil Chem. Soc.* **77**, 1229 (2000).

[175] J.F. Ryan and R. Frey, *Today's Chemist at Work* (12), 36–40 (1998).

[176] D. Peters and J. Shearer, *Proceedings SPIE-Int. Soc. Opt. Eng.* **2069**, 42–9 (1993).

[177] R. Manoharan and D. Halley, *Proceedings Advances in Process Analytics and Control Technology (APACT03)*, York (2003), Paper T13.

[178] Y. Ozaki, R. Cho, K. Ikegawa, S. Muraishi and K. Kawauchi, *Appl. Spectrosc.* **46**, 1503 (1992).

[179] S. Keller, T. Löchte, B. Dippel and B. Schrader, *Fresenius J. Anal. Chem.* **346**, 863 (1993).

[180] M. Claybourn, T. Massey, J. Highcock and D. Gogna, *J. Raman Spectrosc.* **25**, 123 (1994).

[181] J.S. Church, A.S. Davie, D.W. James, W.H. Leong and D.L. Tucker, *Appl. Spectrosc.* **48**, 813 (1994).

[182] P.J. Hendra and J.K. Agbenyega, *The Raman Spectra of Polymers*, J. Wiley & Sons, Chichester (1993).

[183] B. Schrader, *Fresenius J. Anal. Chem.* **355**, 233–9 (1996).

[184] S. Kokot, N.A. Tuan and L. Rintoul, *Appl. Spectrosc.* **51**, 387 (1997).

[185] S. Kokot, K. Crawford, L. Rintoul and U. Meyer, *Vib. Spectrosc.* **15**, 103 (1997).

[186] Y. Liu, U. Meyer, A. Mohn and S. Kokot, *Preprints 37th IUPAC Intl. Symp. Macromolecules*, Gold Coast (1998), p. 130.

[187] I.R. Lewis and H.G.M. Edwards (eds.), *Handbook of Raman Spectroscopy*, M. Dekker, New York, NY (2001), pp. 936–9.

[188] B.J. Deshpande, M.S. Dhamdhere and M.G. Hansen, *Proceedings SPE ANTEC '98*, Atlanta, GA (1998), pp. 1672–5.

[189] M.S. Dhamdhere, J. Li, B.J. Deshpande, P. Patil and M.G. Hansen, *Proceedings Polymer Processing*, London (2001), pp. 46–58.

[190] G. Baillet, A. Thuillier and J.-P. Degand, *Proceedings SPQ-98 (Spectroscopy in Process and Quality Control)*, London (1998).

[191] L. Feng and K.Y.S. Ng, *Macromolecules* **23**, 1048–53 (1990).

[192] S. Farquharson and S. Simpson, *Proceedings SPIE-Int. Soc. Opt. Eng.* **1796**, 272–86 (1993).

[193] M. Cakmak, F.T. Serhatkulu, M. Graves and J. Galay, *Proceedings SPE ANTEC '97*, Toronto (1997), pp. 1794–9.

[194] R. Salzer, U. Roland and R. Born, in *Fourier Transform Spectroscopy* (J.A. de Haseth, ed.), American Institute of Physics, New York, NY (1998), Vol. 430, pp. 474–5.

[195] G.A. Voyiatzis and K.S. Andrikopoulos, *Proceedings SPQ (Spectroscopy in Process and Quality Control)*, London (1998).

[196] J.M. Chalmers and N.J. Everall, *Trends Anal. Chem.* **15** (1), 18–25 (1996).

[197] F.A.Nelson, C.A. Reilly and W.E. Savage, *I&EC* **52**, 487 (1960).

[198] R.A. Grinsted, *Proceedings SPE ANTEC 2002*, San Francisco, CA (2002), pp. 2020–4.

[199] M.L. Snoddy, *Spectroscopy* **8** (3), 41–7 (1993).

[200] A. Nordon, C.A. McGill and D. Littlejohn, *Analyst* **126**, 260–72 (2001).

[201] P.J. McDonald, in *Food Processing: Recent Developments* (A.G. Gaonkar, ed.), Elsevier Science, Amsterdam (1995), pp. 23–36.

[202] C.I. Tanzer and A.K. Roy, *Proceedings SPE ANTEC '95*, Boston, MA (1995), pp. 2700–6.

[203] A. Nordon, C.A. McGill and D. Littlejohn, *Spectrosc. Europe* **13** (3), 10–3 (2001).

[204] T.C. Farrar and E.D. Becker, *Pulse and Fourier Transform NMR. An Introduction to Theory and Methods*, Academic Press, New York, NY (1971).

[205] E. Fukushima and S.B.W. Roeder, *Experimental Pulse NMR – A Nuts and Bolts Approach*, Addison-Wesley Publishing Co., Reading, MA (1981).

[206] C. McGill, A. Nordon and K. Setarehdan, unpubl. results (2002).
[207] J.F. Haw and T.W. Skloss, *Spectroscopy* **8**, 22–7 (1993).
[208] G. Eidmann, R. Savelsberg, P. Blümler and B. Blümich, *J. Magn. Reson.* **A122**, 104 (1996).
[209] T.C. Farrar, *An Introduction to Pulse NMR Spectroscopy*, Farragut Press, Chicago, IL (1987).
[210] V. Arulmozhi and A. Srinavasa Rao, *Phys. Chem. Liq.* **27**, 219 (1994).
[211] T.F. Conway, R.F. Cohee and R.J. Smith, *Food Eng.* **29**, 80 (1957).
[212] S.A. Jones, *Minispec Applications Note 16*, Bruker Analytische Messtechnik GmbH, Karlsruhe (1983).
[213] *Minispec Applications Note 30*, Bruker Analytische Messtechnik GmbH, Karlsruhe (n.d.).
[214] P.B. Mansfield, *Chem. Ind.* **28**, 792 (1971).
[215] W. Herres, *Kunststoffe* **76** (7), 591–4 (1986).
[216] V.M. Litvinov, *Intl. Polym. Sci. Technol.* **15** (11), T28–9 (1988).
[217] V.M. Litvinov, *Techn. Report DSM Research*, Geleen (1998).
[218] W. Burk and H. Todt, *Minispec Applications Note 39*, Bruker Analytische Messtechnik GmbH, Karlsruhe (n.d.).
[219] L. König and S. Postel, *Chemical Fibers Intl.* **48**, 500–1 (1998).
[220] G.A. Matzkanin, in *Nondestructive Characterization of Materials* (P. Höller, I. Hauk, G. Dobmann, C.O. Ruud and R.E. Green, eds.), Springer-Verlag, Berlin (1989), pp. 655–69.
[221] D. Attanasio, D. Capitani, C. Federici, M. Paci and A.L. Segre, *ACS Symp. Ser.* **598**, 333–53 (1995).
[222] R. Benken, D. Knittel and E. Schollmeyer, *Tech. Mess.* **65** (1), 4–7 (1998).
[223] A.S. Wilson, *Plasticisers. Principles and Practice*, The Institute of Materials, London (1995).
[224] S.E. Amos, G.M. Giacoletto, J.H. Horns, C. Lavallée and S.S. Woods, in *Plastics Additives Handbook* (H. Zweifel, ed.), Hanser Publishers, Munich (2000), pp. 553–84.
[225] A. ten Brinke, *Ph.D. Thesis*, Twente University (2002).
[226] A.E. Somers, T.J. Bastow, M.I. Burgar, M. Forsyth and A.J. Hill, *Polym. Degr. Stabil.* **70**, 31–7 (2000).
[227] J.L. Hutchison, N.S. Murthy and E.T. Samulski, *Macromolecules* **29**, 5551–7 (1996).
[228] Oxford Instruments, *www.oxford-america.com/analytical*.
[228a] V.H. Litvinov, W. Barendswaard and M. v. Duin, *Rubber Chem. Technol.* **71**, 105 (1998).
[229] A. Nordon, C. Meunier, R.H. Carr, P.J. Gemperline and D. Littlejohn, *Anal. Chim. Acta* **472**, 133–40 (2002).
[230] W. Windig and B. Antalek, *Chemom. Intell. Lab. Syst.* **37**, 241 (1997).
[231] A. Nordon, P.J. Gemperline, C.A. McGill and D. Littlejohn, *Anal. Chem.* **73**, 4286–94 (2001).
[232] P. Stilbs, in *Polymer-Surfactant Systems* (J.C.T. Kwak, ed.), M. Dekker, New York, NY (1998), Chp. 6.
[233] R. Milward, *Proceedings Solutions for Scientist Symposium* (Advanstar, ed.), London (1999).
[234] I. Alig and D. Lellinger, *Chem. Innovation* (2), 13–8 (2000).
[235] I. Alig, *Jahresbericht DKI 2001*, DKI, Darmstadt (2002), pp. 50–4.
[236] R. Truell, C. Elbaum and B.B. Chick, *Ultrasonic Methods in Solid State Physics*, Academic Press, Boston, MA (1969).
[237] P.D. Edmonds (eds.), *Ultrasonics*, Academic Press, New York, NY (1981).
[238] L.C. Lynnworth, *Ultrasonic Measurements for Process Control: Theory, Techniques, Applications*, Academic Press, Boston, MA (1989).
[239] R.C. Asher, *Ultrasonic Sensors for Chemical and Process Plant*, Institute of Physics Publishing, Bristol (1997).
[240] R. Gendron, M.M. Dumoulin, J. Tatibouët, L. Piche and A. Hamel, *Proceedings SPE ANTEC '93*, New Orleans, LA (1993), pp. 2256–61.
[241] P.H. Ziehl and T.A. Green, *Proceedings SPE ANTEC 2003*, Nashville, TN (2003), pp. 3628–32.
[242] G. Menges and G. Wiegand, *Proceedings SPE ANTEC '78*, Washington, DC (1978), p. 557.
[243] A.D. Schiller, *Proceedings SPE ANTEC '79*, New Orleans, LA (1979), p. 378.
[244] B. Bridge and K.H. Cheng, *J. Mater. Sci. Lett.* **6**, 219–21 (1987).
[245] D.R. Franca, C.-K. Jen, K.T. Nguyen and R. Gendron, *Polym. Engng. Sci.* **40** (1), 82–94 (2000).
[246] I. Alig, D. Lellinger, R. Lamour and J. Ramthan, *Kunststoffe* **90**, 5 (2000).
[247] L. Erwin and J. Dohner, *Polym. Engng. Sci.* **24**, 1277–82 (1984).
[248] R. Gendron, J. Tatibouët, J. Guèvremont, M.M. Dumoulin and L. Piche, *Polym. Engng. Sci.* **35**, 79 (1995).
[249] R. Gendron, L.E. Daigneault, J. Tatibouët and M.M. Dumoulin, *Proceedings SPE ANTEC '94*, San Francisco, CA (1994), pp. 167–81.
[250] Z. Sun, C.V. Anghel and J. Tatibouët, *Proceedings SPE ANTEC 2003*, Nashville, TN (2003), pp. 3331–5.
[251] L. Bellamy, A. Nordon and D. Littlejohn, *Proceedings Advances in Process Analytics and Control Technology (APECT03)*, York (2003), Paper T3.

[252] A. Sahnoune, J. Tatibouët, R. Gendron, A. Hamel and L. Piche, *J. Cellular Plast.* **37** (5), 429–54 (2001).

[253] K.S. Suslick and G.J. Price, *Annu. Rev. Mater. Sci.* **29**, 295 (1999).

[254] M. McBrearty and S. Perusich, *Proceedings SPE ANTEC '98*, Atlanta, GA (1998), pp. 2080–4.

[255] A.J. Bur, S.C. Roth and M. McBrearty, *Rev. Sci. Instrum.* **73** (5), 2097–102 (2002).

[256] A.J. Bur and M. McBrearty, *Proceedings SPE ANTEC 2003*, Nashville, TN (2003), pp. 3321–5.

[257] Epsilon Industrial Inc., Austin, TX, *www.epsilon-gms.com.*

[258] Chemical ElectroPhysics, *Application Note Proceptor® In-line Dielectric Analyzer*, Hockessin, DE (n.d.).

[259] A.J. Bur, S.C. Roth and M. McBrearty, *Proceedings SPE ANTEC 2003*, Nashville, TN (2003), pp. 3326–30.

[260] F. Stephan, X. Duteurtre and A. Fit, *Polym. Engng. Sci.* **38** (9), 1566–71 (1998).

[261] T. Lynch, in *Gas Chromatographic Techniques and Applications* (A.J. Handley and E.R. Adlard, eds.), Sheffield Academic Press, Sheffield (2001), pp. 298–333.

[262] T. Górecki and J. Pawliszyn, *LC.GC Intl.* **12** (2), 123–7 (1999).

[263] T. Maurer and F. Müller, *Proceedings Advances in Process Analytics and Control Technology (APACT03)*, York (2003), Paper T19.

[264] D. Soleta, *Am. Lab.* **2**, 21–4 (1989).

[265] A.M. Bond and G.G. Wallace, *Anal. Chem.* **56**, 2085 (1984).

[266] A.M. Bond, W.N.C. Garrard, I.D. Heritage, T.P. Majewski, G.G. Wallace, M.J.P. McBurney, E.T. Crosher and L.S. McLachlan, *Anal. Chem.* **60**, 1357 (1988).

[267] J.N. Barisci and G.G. Wallace, *Anal. Chem. Acta* **310**, 79–92 (1995).

[268] J. Crandall and N. Zeug, *Messen Steuern Regeln* **33**, 305–8 (1990).

[269] H.J. Cortes and C.D. Pfeiffer, *Anal. Chem.* **65**, 1476–80 (1993).

[270] K.J. Melda, *J. Proc. Anal. Chem.* **4** (3, 4), 117 (1999).

[271] G.B. Levy, *Am. Lab.* **18** (12), 62–71 (1986).

[272] R. Perkins, D. Nicolas and R. Sasano, *www.ATAS-int.com* (2000).

[273] M.J. Matheson, T.P. Wampler and W.J. Simonsick, *J. Anal. Appl. Pyrol.* **29**, 129–36 (1994).

[274] VOLKSWAGEN AG, SEAT S.A. ŠKODA Automobilova A.S., AUDI AG, *Py-GC/MS für Kunststoffe und Elastomere, Zentralnorm PV 3935*, Wolfsburg (Dec. 1997).

[275] H. Wilcken and H.-R. Schulten, *Fresenius J. Anal. Chem.* **355**, 157–63 (1996).

[276] R. Hotop and J. König, *Automatisierungstechn. Praxis* **33**, 610–6 (1991).

[277] D. Steinmüller, *Automatisierungstechn. Praxis* **35**, 460–3 (1993)

[278] R. Scheyvens, unpubl. results (1997).

[279] L.B. Roof, G.T. Porter, E.N. Fuller and R.A. Mowery, *Proceedings* 4th *IFAC Conference*, Ghent (1980), pp. 47–53.

[280] R.L. Cotter, R.J. Limpert and C. Deluski, *Am. Lab.* **19** (12), 54–62 (1987).

[281] J. Riera Tuebols and J. Teixido Subirats, *Pint. Acabados Ind.* **31** (171), 75–81 (1989).

[282] P.L. Warren, *Anal. Proc.* **27**, 186–7 (1990).

[283] P.L. Warren, O. Farges, M. Horton and J. Humber, *J. Anal. At. Spectrom.* **2**, 245 (1987).

[284] T. Schneider, H.M. Ortner, C.-J. Lorenzen, M. Jogwich, W. Mertens, E. Sanzenbacher and A. Limper, *Kautsch. Gummi Kunstst.* **49** (1), 44–56 (1996).

[285] Y. Hasakawa, *Instrum.* **12**, 51–8 (1986).

[286] M. Hietala and D. Kalnicky, *Adv. X-Ray Anal.* **32**, 49–57 (1989).

Chapter 8

Science must be reproducible or it is not science but art (I.S. Krull, 2001)

Modern Analytical Method Development and Validation

8.1. Status of Existing Methods for Polymer/Additive Analysis . . . 732
8.2. In-polymer Additive Analysis: Method Development and Optimisation . . . 732
8.3. Certified Reference Materials . . . 736
8.4. Analytical Method Validation Approaches . . . 746
8.4.1. Analytical Performance Parameters . . . 751
8.4.2. Interlaboratory Collaborative Studies . . . 755
8.4.3. Validation of Antioxidant Migration Testing . . . 757
8.5. Total Validation Process . . . 757
8.5.1. Software/Hardware Validation/Qualification . . . 758
8.5.2. System Suitability . . . 760
8.6. Rational Step-by-step Method Development and Validation for Polymer/Additive Analysis . . . 760
Bibliography . . . 762
Method Development and Validation . . . 762
Reference Materials . . . 762
References . . . 762

Today, the vast majority of industries have ISO 9000 type accreditations, which control methods and procedures and include systems making all actions fully **traceable and auditable**. Where possible, analytical methods are closely related to national or international standards (such as DIN, EN, ASTM). Of course, it is critical to ensure that the analytical methods in use generate meaningful data. At worst, poor data can be dangerous or hazardous to health. It is also desirable to generate "equivalent" data at any location using established criteria. This is a very important aspect within global industries, where source and supply of products can be situated anywhere in the world. Finally, Operational Excellence programmes in industry require higher quality requirements for analytical measurements in terms of accuracy and precision (related to raw materials cost savings and product quality). This calls for continuous attention and search for suitable reference/calibration materials, updating of analytical methodology and round-robins (at least internally – in-company – or externally). Not surprisingly, therefore, the importance of analytical method development and validation has fully been recognised. Validation requirements for in-polymer analysis may conveniently be taken from alien areas (*e.g.* pharmaceutical drug products), despite some existing confusion at the level of various regulators (FDA, USP, and ICH). This Chapter analyses the current status and prospects for application oriented method development and validation of polymer/additive analysis. Actual validation experiences are described. It will be clear that a tremendous effort can be expended in conducting validation studies, efficiency of experimental design and documentation. In many cases retrospective validation is carried out.

8.1. STATUS OF EXISTING METHODS FOR POLYMER/ADDITIVE ANALYSIS

As shown in previous Chapters, methods for analysis of additives in polymeric materials are routinely developed, improved and applied, but rarely collaboratively studied and validated. Moreover, most instrumental approaches developed by industrial laboratories are of proprietary nature. In short, the procedures developed are of **variable rigour**. Relatively few of such methods are mentioned in large compendia, such as the Official Methods of Analysis (Association of Official Analytical Chemists, AOAC), the US Pharmacopœia (USP), or those issued by the American Society for Testing and Materials (ASTM). In comparison to the pharmaceutical industry, subject to strong regulation by various agencies, such as the US Food and Drug Administration (FDA), the chemical industry is much less regulated, although subject to similar restrictions where food-contact applications are involved, or where environmental protection is concerned. In the future greater impact of these (supra)national regulatory agencies may be expected. This will undoubtedly lead to the need for additional method development and optimisation.

Before any technique can be fully accepted it must be possible for analysts anywhere to carry out the same method and get identical results. Unfortunately, at present this is not always being implemented. However, the current trend towards globalisation in the chemical industry is stimulating method validation. Standardisation of analytical methods plays a vital role in analytical science and technologies as well as in business and trading based on analyses.

8.2. IN-POLYMER ADDITIVE ANALYSIS: METHOD DEVELOPMENT AND OPTIMISATION

Principles and Characteristics

The hierarchy of analytical methodology, proceeding from the general to the specific is given in Table 8.1 [1]. Procedures are usually considered to be standard test methods and guidelines, either from official bodies or industry specific (*e.g.* refs. [2,3]). Variations from an analysis procedure are either a documented amendment (*a priori*) or a deviation (*a posteriori*). The term "protocol" is the most specific name for a method. Protocols are typically prescribed by an official body for use in a given situation, such as regulatory processes (EPA, FDA, EC directives, GLP, *etc.*), that leave little freedom. Normalisation is the once-only drawn-up solution of a repeating problem with the scientific, technical and economic possibilities in vigour.

Capabilities of an analytical technique are sensitivity, detection limit, precision and accuracy, spatial, energy or mass resolution, amount and quality of information and also productivity (throughput). There are several valid **reasons for developing new methods** of analysis [6], such as:

- Absence of a suitable method for a particular analyte in a given sample matrix.

Table 8.1. Hierarchy of analytical methodology

Hierarchy	Definition	Example(s)
Technique	Scientific principle useful for providing compositional information	Spectrophotometry
Method	Distinct adaptation of a technique for a selected measurement process	MAE-HPLC-ELSD/UV [4][a]; strategy for control of additive packages in food industries [5]
Procedure	Written directions necessary to use a method	Standard test methods and guidelines (ASTM, AOAC, *etc.*); company standard operating procedures
Protocol	Set of definitive directions that must be followed, without exception, if the analytical results are to be accepted for a given purpose	Validation protocols; Migration Directive 82/711/EC

[a] *Cfr.* Chp. 3.4.5.1 of ref. [12a].

- Unreliability of existing methods (error-, artefact-, and/or contamination-prone, poor accuracy or precision).
- Cost (in time, energy or lack of automation, *etc.*) of existing methods.
- Inadequate sensitivity or analyte selectivity of current practice.
- Opportunities for improved methods by state-of-the-art instrumentation and techniques.
- Need for an alternative method for confirmation of analytical data obtained by other methods.

Parameters relevant to method development are accuracy, system precision, linearity, range, limit of detection (LOD), limit of quantitation (LOQ), sensitivity and robustness.

Rational method development consists of various stages: *(i)* user chooses type of method (new or existing; method transfer); *(ii)* user defines requirements (*e.g.* run time, resolution, T, pH, *etc.*); *(iii)* system suggests starting conditions (*e.g.* mobile phase, mobile phase strength, column choice, injection volume, *etc.*); *(iv)* try-out (defines sequence of experimentation); and *(v)* next experiment. Development of a new method of analysis usually starts from prior art or existing literature, and is most frequently based on already available instrumentation. Ideally, though, method development defines the method requirements first and then selects the instrumentation to utilise in complete freedom. Budgetary considerations usually limit this approach in a given laboratory. Nonetheless, analytical measurements should be made using methods and equipment which have been tested to ensure that they are fit for purpose. (It has been estimated that up to 20% of all analyses performed are not fit for purpose!) Whatever type of measurements are made, suitable, well maintained and calibrated equipment is vital to ensure success.

In most analytical R&D situations the following unit processes are distinguished: sampling; sample preparation; separation of the analyte from the matrix and enrichment; measurement; calculation and presentation of the result. The reliability of the sampling process influences the reliability of all the steps of the analytical process. The **sampling** procedure and storage conditions should be as appropriate as possible. No analysis is better than the sample itself. This is even more important for reference samples, which should normally be stored in conditions such that their retesting is possible. With samples taken for R&D purposes little might be known about their homogeneity. Any means used to homogenise the sample must not compromise the integrity of the sample. It may be convenient to have a single standard operating procedure (SOP) describing the variety of **sample treatment** methods (solvation; dissolution; digestion; extraction; surface cleaning; melting; combustion, *etc.*) used by the laboratory. **Sample size** also needs consideration [6a].

Depending on the exact analytical requirement it will normally be necessary to (quantitatively) **isolate the analyte** under study from the sample matrix. If this isolation stage is not carried out, then there is a risk that the polymer or components present in the sample matrix or in the reagent blanks interfere, in particular in the quantification process (*cfr.* Chp. 6). The lower the analyte concentration the higher the risk. The largest uncertainties in the final results usually originate from the sampling processes and the accuracy of the determination of recovery factors. It is the variability of those factors which determines the total uncertainty much more than the final (often instrumental) measuring uncertainty (signal reproducibility or repeatability)! The extraction procedures used in polymer/additive analysis are a typical case.

Method development design depends on the defined requirements (research goals), as described in Table 8.2. The impact of method development is enormous (Table 8.3).

In particular in the industrial environment, where most in-polymer analyses are being carried out, **analyst time** needs to be minimised. To this extent, autosamplers, robots, fast analysis techniques (*e.g.* ASE®, fast GC), hyphenation and standardised data output formats for further manipulation and transmission are wanted. Automation is advantageous (Table 8.4). Ideally, the whole process may be automated: analysis, data reduction and output. Unfortunately, standardisation of data handling procedures is still far off. This determines continuous, multiple efforts for training of analysts. Analytical methods should also be easy to maintain; calibration should be required at minimal levels. Sample preparation should minimise time, effort, materials and volume of sample consumed. Sample pretreatment is ideally superfluous. There should also be little inherent doubt on the representativity of the analysis (of special concern for those techniques employing minimal sample amounts: 0.1–1 μg). The method should be able to qualitatively identify the specific analyte(s) of interest, on the basis of expected behaviour (*e.g.* retention time, colour

Table 8.2. Desirable method development features for in-polymer additive analysis

• Universal nature	• Qualitative identification
• Broad screening power	• Quantitative determination
• Ease of use, simplicity	• Representativity
• Safety	• Reliability
• Application transferability (company wide)	• Minimal cost
• Standardised procedures	• Applicability of simple QA/QC procedures
• Suitable data output format	• Optimisation
• Speed (automation)	• Simplified global method validation
• Direct analysis ("as is")	• Regulatory agency guidelines
• Minimal training *cq.* method maintenance	• Spectral libraries
• Fit for purpose	• Reference materials

Table 8.3. Impact of method development

- More universally applicable methods, *i.e.* less calibration and validation
- More faster techniques (*e.g.* fast GC)
- More automated methods (autosamplers, robots)
- More instrumental methods (ISO certified)
- More sample-representative methods (0.1–1 μg to 100 mg)
- Greater use of easy accessible, well serviced equipment
- More methods suitable for R&D, *i.e.* non-routine measurements
- Fewer specific methods
- Fewer analyst-intensive methods (*e.g.* derivatisations)
- Fewer methods requiring specific library development (*e.g.* gas-phase FTIR)
- Fewer multiple hyphenated techniques (simplicity)
- Less model-building (chemometrics)

Table 8.4. Advantages of automation

- Easier sample handling
- Greater convenience: simultaneous loading of several samples
- More reproducible robotic loading: improved precision, exacting positioning
- Less chance for human error; unattended analyses
- Savings on time-consuming manual taring
- More accurate separations by use of long dwell times in overnight runs
- Multiple method storage
- Easy programming of regular calibration checks (audits)
- Increase in productivity
- Integration with existing laboratory instrumentation through instrument control

change, spectra, *etc.*) and allow quantitative determination (precision and reproducibility at the desired level). New techniques and methods normally compete with older, established ones and should present a competitive edge (*e.g.* lower investment, cheaper consumables, including solvents, less environmental damage, reduced manpower, high sample throughout, rapid sample turnaround time). Preferably, the new method developed should be extendable to simple quality assurance and quality control procedures. In method development for product quality control purposes an additional requirement is simplicity. In this case often specially designed (high safety) equipment is used. It is equally advantageous to use chemometrical evaluation methods for interpretation, because in a QC laboratory the number of samples may be so high that time-consuming visual spectra interpretation is only possible for special cases, *cfr.* Py-FIMS/PCA on industrial paints and resins [7]. Chemometrical evaluation methods constitute a powerful tool for visualisation and simplification of information.

The greater the flexibility in a method's design, the more potentially complex method development may appear. **Experimental design** may be used to identify the different factors that affect the result of an experiment, to separate the effects of the factors involved, and to minimise analytical effort.

Efficient in-polymer analysis requires **spectral libraries**, preferably from the public domain. It is in this area that much specific and proprietary library building (*e.g.* ToF-SIMS, Raman) is to be found for polymer/additive analysis. Although this is facilitated by a broad sample collection of pure additives, in many instances matrix effects are observed which complicate spectral matching.

The aforementioned considerations largely determine the selection of the instrumentation to be utilised. However, it is equally advantageous to make appropriate use of **analyte parameter values**, such as solubilities, wavelengths of detection (spectral characteristics), mass/charge ratios, *etc*.

Method development starts using only analytical standards that have been well identified and characterised, up to a preliminary evaluation of the method. Analytical figures of merit obtained, such as sensitivity, LOD, LOQ, dynamic range, linearity of calibration plots, accuracy and precision of quantitation, specificity, should then be compared to the requirements for the new method set out at the beginning. The selectivity and specificity of the method, which determines its utility, are important items to consider. At that stage it should be made clear that the new method offers advantages over existing methodology (*e.g.* in terms of sample handling, analysis time, detection limits, *etc.*) or not to warrant further optimisation. Criteria for optimisation of a new method are given in Table 8.5. Optimisation may be a limiting factor to the applicability of a specific technique, as in case of SFE (excessively great number of experimental variables) [8] or PyGC-MS (difficult quantitation) [9], as illustrated in Scheme 8.1.

Table 8.5. Criteria for optimisation of a new in-polymer additive analysis method

- Minimal sample preparation
- Adequate chromatographic and/or spectroscopic resolution
- Minimal interference and matrix effects
- LOD lower by at least one order of magnitude than needed for most samples
- Calibration plots linear over several decades, beginning with limits of quantitation
- Significantly increased sample throughput
- Documented reproducibility of analytical figures, with acceptable accuracy and precision
- Minimal cost per analysis

Sometimes, the analytical problem at hand does not require optimisation, or the cost of optimisation outweighs the benefits expected. However, in many instances a fair degree of optimisation has to be performed just to get some results at all. In any case, particular attention should be paid to the sample preparation step, *e.g.* by comparison with another technique, verification with a "known" sample, or use of a recovery SPC chart.

Characterisation of **method performance** involves a judgement as to whether the capabilities of the new method are sufficient to meet the needs of the end user (this is also known as method validation). Various options exist for characterisation of method performance. The trueness of a new method could be assessed against that of established methods, repeatability could be assessed using reference materials, and reproducibility through interlaboratory comparisons. In R&D, many of these options may not readily be available. Validation tools may be limited to the use of in-house reference materials.

Uncertainty should be estimated and quoted in a way that is widely accepted, internally consistent and easy to interpret. Where appropriate, it should be quoted with the analytical result, so that the user can be assured of the degree of confidence that can be placed on the result. Uncertainty estimations based

For an unknown sample	⇒	Specific method	⇒	Optimised method
$m \approx 100$ µg		$m < 500$ µg		$m < 500$ µg
T py = 550 °C		T py = 550 °C		T py = fitted
m/z = 33-920 amu		m/z = fitted, SIM		m/z = fitted, SIM
2 scan/sec		min. 4 scan/sec		2 scan/sec
$t \geq 120$ min		$t < 120$ min		$t < 120$ min
Targets:		*Targets* :		*Targets*:
Polymer type		Comparison		Max. sensitivity
Blends, modifiers		Quantification		Diluted sample
Memory effects		Internal standards		Optimal column
				Chemical ionisation

Scheme 8.1. Optimisation of PyGC-MS analysis of a given sample type.

on error propagation principles rely on a solid understanding of the theoretical principles of the method and the practical experience of the research workers. Optimisation with analytical standards should be followed by attempts to broaden the scope of the method to include actual sample applications. Proper analytical method development includes method validation. Validation is not optimisation, but rather a definition of the conditions under which a process is reproducible. The validity of a specific method should be demonstrated in laboratory experiments using samples or standards that are similar to the unknown samples analysed routinely.

Applications

Many of the papers published offer methods with no real advantages over most existing methods. The development of a "new" method should not be a goal in itself. Developments with high potential are rare. Some recent examples are MAE-HPLC-ELSD/UV [4] for additive analysis of polyolefins, a universal HTGC and HTGC-MS approach [10] and temperature programmed HPLC for the analysis of oligomeric additives [11,12], *cfr.* Chp. 4.4.2.2 of ref. [12a]. Useful instrumental developments are noticed for TD-GC-MS (*cfr.* Chp. 2.3.2.4); amongst in-process analysis techniques (*cfr.* Chp. 7) the application of mid-IR with PMD evaluation is of great interest [13]. Expectations for DIP-ToFMS [14], PTV-GC-ToFMS and ASE® are now high. The advantages of SFC [15–17], on-line multidimensional chromatographic techniques [18,19] and laser-based methods (*cfr.* Chp. 3) for polymer/additive analysis appear to be more limited. To ensure the **relevance** of a method, its application to real sample analysis must be demonstrated. The accuracy of an analytical method should be confirmed by an independent method or by the analysis of certified reference materials. Detailed comparative studies of the method developed with other well-established methods for polymer/additive analysis are not frequent in the analytical literature. Nevertheless, some examples may be found in Chp. 3.6 of ref. [12a] and Chp. 6. Convincing evidence of the superiority of a new method over existing ones would enhance interest.

Method development for in-polymer additive analysis in the conventional sequence of sample collection, sample preparation, extraction (polymer-analyte separation), chromatography (analyte separation), spectroscopy or spectrometry (analyte identification) and data processing requires careful planning to minimise handling, starting with the initial solvent choice. Typically, a strategy for HPLC method development may comprise a gradient-elution system, coupled to a diode-array detector. This requires defining the composition of the eluent and the detection wavelength in an efficient manner. Optimal parameters can be determined by trial-and-error using an isocratic system. Specificity and selectivity (peak purity as determined by HPLC-PDA) is a measure for success. Alternatives for components which do not absorb in UV/VIS or for degradation products are MS, refractive index detection or gas chromatography. For method development in LC, *cfr.* refs. [20,21]; for GC, *cfr.* ref. [21a].

In order to reduce the number of analyses and analysis time standardisation of existing polymer/additive analysis is important. Obvious advantages are presented by universal methods. For example, Taylor *et al.* [22] have recently proposed a hybrid SFE/ESE® technique as being a general approach to rapid sampling of both polar and non-polar analytes from polymeric matrices. Similarly, the reduction in number of various specific chromatographic methods, all based on the same column, speeds up analysis [23]. Method development needs to keep an analysis as simple as possible, such as isocratic in case of HPLC, in order to facilitate transfer to other laboratories with different equipment.

Figure 8.1 shows the proposed approach to preliminary experiments and method development in SFE and is considered the minimum effort for developing a robust, repeatable, quantitative sample preparation. Salafranca *et al.* [8] have reported full factorial design for the optimisation of supercritical fluid extraction of polymeric matrices.

A procedure for single-particle analysis with LMMS has been described which determines the experimental parameters quite strictly by directly available and generally applicable criteria [25].

8.3. CERTIFIED REFERENCE MATERIALS

Principles and Characteristics

With today's multitude of regulations, customers demand the manufacturing and testing history of formulations, *i.e.* data specific to the raw materials used in the product, the manufacturing process, and the finished product inspection or testing steps. The combination of this data is widely known as the pedigree, and must include all data specific to the traceability, characterisation, manufacturing, and inspection of a commercially produced reference material. **Traceability** is defined as: *"property of the result of a measurement or the value of a standard*

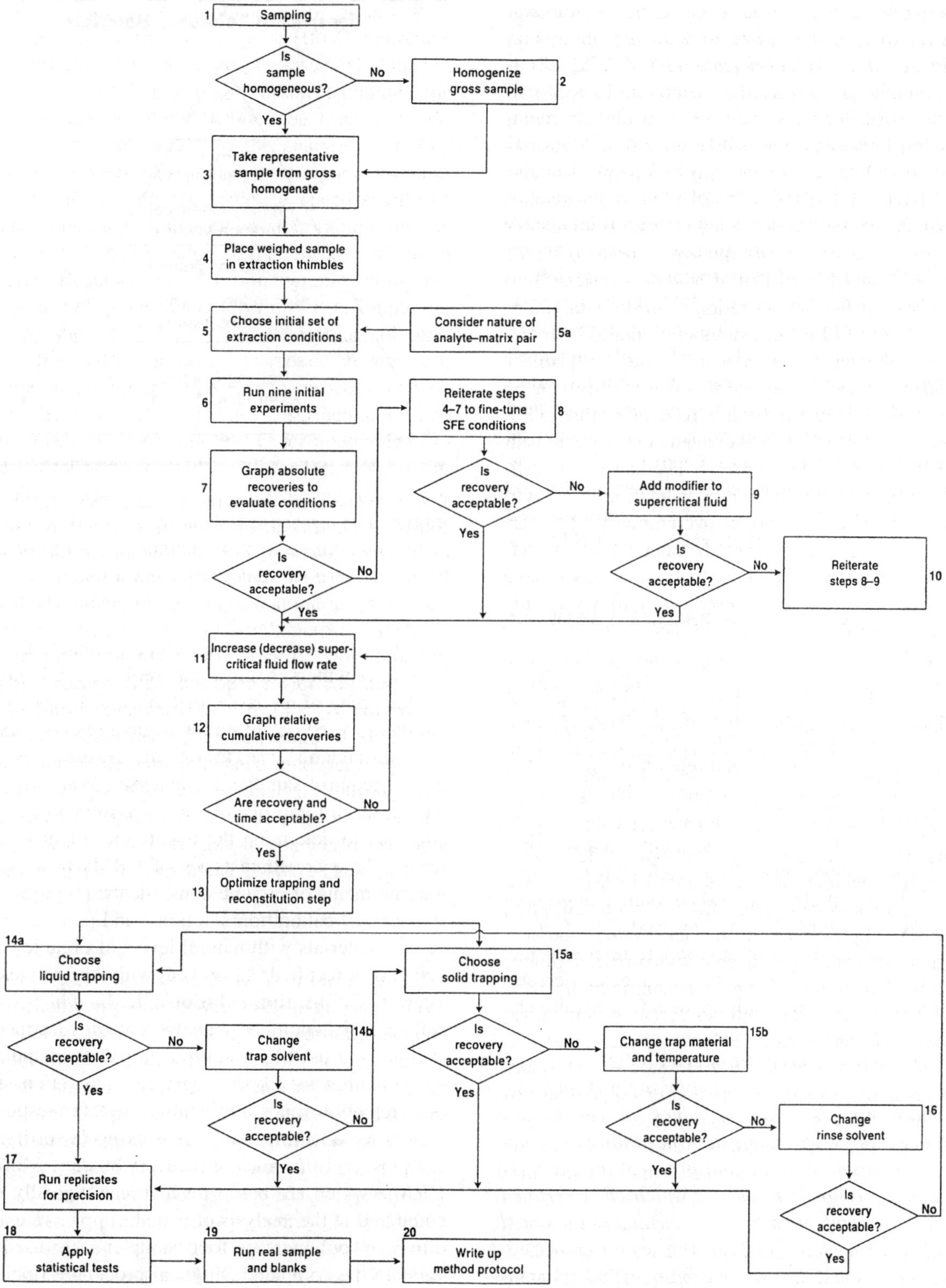

Fig. 8.1. SFE method development flow-chart. After ref. [24]. Reprinted with permission from *LC.GC Intl.*, Vol. 7, Number 7, July 1994. *LC.GC Intl.* is a copyrighted publication of Advanstar Communications Inc. All rights reserved.

whereby it can be related to stated references, usually national or international standards, through an unbroken chain of comparisons all having stated uncertainties" [26]. Thus, the term does not apply directly to laboratories, but to the results of chemical amount-of-substance measurements. Measurement principles for traceability in chemical analysis have been described [27–29]. International standards of the ISO 9000 series (Basic standards for quality management and quality assurance) or the European standards of the EN 45000 series (General criteria for the operation of testing laboratories) require that all measurements should be traceable to national or international standards (primary standards), wherever possible. Traceability can be achieved by preparing standards using a method in which the concentration is created as the direct result of fundamental measurements. The key role of reliable reference materials in the validation of analytical measurements cannot be overemphasised. Reference materials are considered strategic tools. A **reference material (RM)** is a *"material or substance one or more of whose property values are sufficiently homogeneous and well established to be used for the calibration of an apparatus, the assessment of a measurement method, or for assigning values to materials"* [26]. According to *ISO Guide for Certification of Reference Materials – General and Statistical Principles* [30], a good reference material has a number of desirable properties including a well-documented analytical value, homogeneity, stability, ready availability and traceability to a National Reference Laboratory (NRL). Useful reference materials should preferably also be similar in composition to the samples being analysed [31]. Consistent and effective use of appropriate reference materials is necessary to quality assurance and creates confidence. Such materials are therefore required by quality management systems.

A **certified reference material** (CRM) is a *"reference material, accompanied by a certificate, one or more of whose property values are certified by a procedure which establishes traceability to an accurate realisation of the unit in which the property values are expressed, and for which each certified value is accompanied by an uncertainty at a stated level of confidence"* [26,31a]. The key difference between CRM and RM is traceability. CRMs guarantee traceability of the measurement results, *i.e.* their link-up with international standards and thus ultimately with the SI base units. De Bièvre *et al.* [27] have proposed categories of reference materials, determined by their chemical nature. Figure 8.2 shows the hierarchy of reference materials. Reference materials are used: *(i)* to calibrate instruments ("calibration standards"); *(ii)* to back up measuring procedures ("control samples for analyses"); and *(iii)* to ensure the traceability of the measurement results and thus to determine the uncertainty of measurement. There are three types of CRMs to support measurements of organic constituents: *(i)* pure substances; *(ii)* calibration solutions; and *(iii)* (natural) matrix materials with natural levels of organic constituents or fortified (*i.e.* spiked) with the analytes of interest. **Calibration solution CRMs**, which typically contain a number of analytes at known concentrations, are useful for several purposes, including: *(i)* validating the chromatographic separation step (*e.g.* retention times and analyte detector response factors for quantification); *(ii)* spiking or fortifying samples; and *(iii)* analyte recovery studies. **Matrix CRMs**, which are based on matrices typically encountered in the analysis of actual samples (*e.g.* additives in polymers or food samples), are used to validate the complete analytical procedure (including solvent or thermal extraction, cleanup and isolation procedures, and chromatographic separation, detection and quantification). Little action is noticed

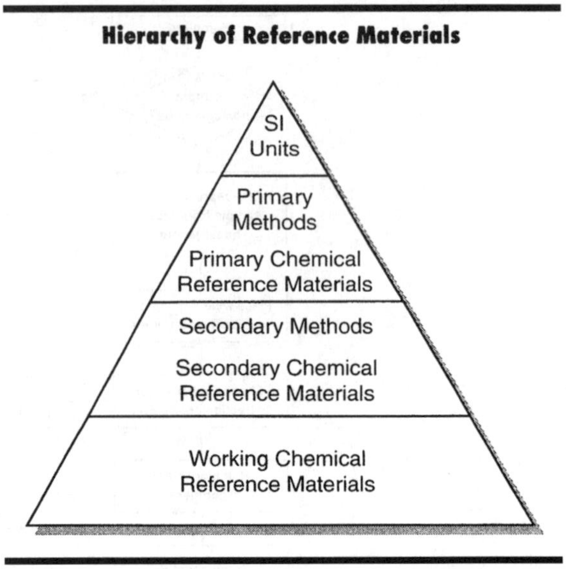

Fig. 8.2. Hierarchy of reference materials. After De Bièvre *et al.* [27]. Reproduced from P. De Bièvre *et al.*, *Accred. Qual. Ass.* **1**, 3–13 (1996), by permission of Springer-Verlag, Copyright (1996).

Table 8.6. ISO-Guides content

ISO Guide No.	Content
17025	General requirements for the competence of testing and calibration laboratories
30 (1992)	Terms and definitions used in connection with reference materials
31 (1996)	Contents of certificates of reference materials
32 (1997)	Calibration of chemical analysis and use of certified reference materials
33 (1989)	Use of certified reference materials (under revision)
34 (1996)	Quality system guidelines for the production of reference materials
35 (1989)	Certification of reference materials – General and statistical principles (under revision)
N 330	List of producers of certified reference materials

regarding the development of RMs of organic additives in polymeric matrices [32,33]. Commercial production of matrix CRMs does not make sound commercial sense [34]. The initiative should be on the side of industry.

Laboratories that are accredited to ISO 17025, rather than the well-established ISO 9000 series, are using RMs and CRMs more often and are signing up for proficiency testing (PT). ISO 17025 is the standard that provides the international aspect to any laboratory measurement process and provides the control framework to assist the production of comparable measurements. ILAC (International Laboratory Accreditation Co-operation) harmonises laboratory accreditation procedures. ISO 17025 plays an important role in international traceability and in the requirements for an internationally agreed suitable CRM.

Certified values are expected to be correct – with a probability of 95% – within the stated uncertainty intervals. However, certified data alone do not guarantee the successful, *i.e.* correct application of reference materials. Depending on the material to analyse or on the testing method to be applied, expert assessment and problem-related selection is required. The task-related application of an (instrumental) analytical method including calibration standards still demands professionally trained specialists.

Calibration, which is defined as *"the set of operations that establish, under specified conditions, the relationship between the values of quantities indicated by a measurement instrument or measuring system or values represented by a material measure of a reference material, and the corresponding values realised by standards"* [26], is one of the most critical steps in quantitative analysis. Methodological approaches to calibration were described and general classifications of RMs and their use in the calibration process were clarified [35]. Calibration serves various purposes: *(i)* quality (ISO 9000); *(ii)* reproducibility and control; *(iii)* cost saving; *(iv)* safety; and *(v)* customer requirements. Competent calibration and traceability of measurements is essential for industrial manufacture and is a major criterion in addressing product liability.

The fundamental philosophy of certification rests on the concept of independent methodology, which is the application of theoretically and experimentally different measurement techniques and procedures to generate concordant results leading to *one* reliable assigned value for the property. Such assigned values are thus *method-independent*. Extractable concentrations are generated by specific procedures and are thus *method-dependent*.

Data on RMs can be obtained from instrument manufacturers, scientific literature, data compilations [36,37], recommendations (ICTAC, GEFTA, IUPAC, *etc.*) or standardised methods (ISO, CEN, DIN, ASTM, *etc.*). A basic guide for selection and use of reference materials is readily accessible [38]. ISO has set up several rules (Table 8.6) to assure a suitable quality of reference materials, which should be clearly indicated in a certificate. The USP, NIST, national metrology institutes and many other organisations (such as BCR-IRMM, BAM, PTB, EMPA, LGC, JSAC, ICTAC) specialise in testing, preparing, guaranteeing, and marketing standard reference materials of various analyte species in different sample matrices (*cfr.* ref. [39]). Recently, a European Reference Materials initiative has been launched. Other sources of RM materials are producers of fine chemicals, *e.g.* Fluka (Sigma-Aldrich) for spectroscopy, ion chromatography [39a] and titrimetry, Starna for UV/VIS spectrophotometry (absorbance/transmission, wavelength, resolution, stray light), TA Instruments for Curie temperature

Table 8.7. Reference material manufacturing and testing

• Component identity	• Analytical verification
• Component purity	• Packaging homogeneity
• Solution preparation (mix formulation)	• Stability of mixture in storage

materials. Production of chemical reference materials is not a simple matter. RM manufacturing and testing processes typically consist of six discrete elements (Table 8.7). Organic RMs for GC-MS and isotope standards for ICP-MS are readily available [40]. Not all the produced reference materials carry traceable values [41]. Several studies at LNE (Laboratoire National d'Essais, France) and EMPA (Swiss Federal Laboratories for Materials Testing and Research) have shown that even values declared for mono-elemental standard solutions, which are used for calibration and which are relatively simple materials with respect to their matrices, are often not traceable. The need for reference materials will keep growing.

The limited number of reliable RMs that can be prepared and made available leads to the use of possibly inappropriate RMs. Typically, an OIT Reference Material is not ideal since, by its nature, it does not meet all of the criteria for a good reference material. In fact, OIT is not a thermodynamic property and is therefore not easily made traceable to a NRL. In fact, OIT is a kinetic property so its value will likely change with time and therefore lacks stability. Also the use of ferromagnetic alloys as potential standards for TG calibration is still unsatisfactory.

Major industrial areas as the cement, ferro, non-ferro, petrochemical, textile or food industry, dispose of numerous Certified Reference Materials (organic and inorganic). For example, only the ferro-industry has already more than 300 CRMs and RMs listed in COMAR, the international database (jointly operated by LNE, BAM and NPL) which lists more than 10285 RMs (as of June 1998) of more than 400 producers [42]. Notwithstanding the size of the polymer industry (total production capacity for commodity thermoplastics is equal to over 140 Mt/a, of which about 50% of polyolefinic nature) it is surprising to note the scarcity of suitable **polymer reference materials** for elemental and molecular analysis. CRMs made from a polymer material and designed for molecular analysis are lacking totally, while those for elemental analysis are rare. In fact, until quite recently, for elemental analysis of polymers, only one set of four CRMs did exist, namely for the determination of Cd in different PE parts of automobiles (VDA-CRM-001 to 004). These reference materials were certified by JRC-IRMM (Joint Research Centre – Institute for Reference Materials and Measurements) on behalf of VDA (Verband der Automobilindustrie e.V.), using thermal ionisation isotope dilution mass spectrometry with ^{111}Cd spike solutions characterised by reverse isotope dilution mass spectrometry (IDMS) based on 99.999% pure cadmium metal [43]. These CRMs, which are available on the market since 1991, were produced as a consequence of the 91/338/EC Directive on Cadmium in materials. Recently, two more multi-element CRMs have become available.

Table 8.8. Comparison of certification by IDMS with results of IMEP-2

CRM	Certified value (ppm)[a]	Round-robin value (ppm)[b]
VDA-001	40.9 ± 1.2	40.7 ± 1.2
VDA-002	75.9 ± 2.1	75.1 ± 2.1
VDA-003	197.9 ± 4.8	197.9 ± 4.9
VDA-004	407.0 ± 12	408.7 ± 8.8

[a] IDMS-based [50].
[b] IMEP-2, after ref. [51].

From an analytical point of view, the general lack of suitable CRMs is a standing problem for the effective development of analytical measurements, as it does not favour elemental analysis at the development, production and control level. To improve the accuracy and precision of the currently used analytical measurement protocols CRMs are an ideal tool. In the total validation procedure they fulfil an important role. The ultimate performance test for any calibrated analytical instrument is to analyse a CRM and obtain a result within the expected uncertainty range. Actually, if a matrix CRM is subjected to the whole analytical process then this serves to validate the entire procedure, thus saving much time and effort. More often than not, however, in R&D no CRM is available at all and it is not possible to relate a property to an existing (inter)national standard. In that case, in-house reference materials need to be used for instrument performance checks, calibration or quality control. Where R&D involves testing a large number of similar samples using a particular procedure, control samples and charts are used to monitor the continuing stability of instrument performance [44]. From a QA point of view, the limited availability of CRMs also hampers the effective introduction of quality assurance systems. Production and certification of reference materials is a costly affair and requires considerable expertise in the field of material production, material analysis and testing. An increase in the range of RMs, CRMs and Proficiency Testing (PT) programmes is not to be expected (for cost reasons). For future production of CRMs, *cfr.* also ref. [44a].

The development of SRM®s in analytical chemistry was reviewed recently [45]. The RM report reviews information on RM&PT products [46].

Development of Certified Reference Materials

Within the framework of multi-element analysis on polymer materials, a CRM would serve different purposes: *(i)* validation and intercomparison of sample destruction procedures; *(ii)* validation and intercomparison of measuring procedures; and *(iii)* validation of Quality Assurance systems.

In view of the large number of different polymeric materials and the range of elements to be determined, development of (costly) matrix CRMs for each type of polymeric material is quite impossible.

There has lately been an upsurge in interest in the determination of heavy metals in polymers and paints. Owing to the toxicity of Cd, many countries have restricted the levels of this element which may be used in plastics (*e.g.* Sweden as from 01.07.80). The European Community has also issued a cadmium Directive (18.06.91) [47]. Cadmium compounds were widely used in pigment paints and plastics and to heat-stabilise polymers containing chlorine, in particular PVC. In addition, metallic cadmium was often used to electroplate metal surfaces. In 1983 the German automobile industry decided to eliminate cadmium in their products as far as possible [48]. For direct analytical monitoring of the legal cadmium limit (75 ppm) for a great number of solid polymer samples (20,000 parts p.a.) Adam Opel AG has adopted XRF and ZAAS procedures [49].

In the absence of suitable reference materials the (plastics) industry has had to rely on in-house materials for standardisation. Later the set of four CRMs for Cd in PE (from about 40 to 400 ppm) has been certified by IRMM on behalf of VDA using isotope dilution mass spectrometry [50]. SS-ZAAS was used for homogeneity control of mg microsamples. Certified Cd values are shown in Table 8.8. The results did not show significant differences between bottles at the 300 mg level. For the sample with the highest Cd content (VDA-004; 407 ppm) the measurement reproducibility was worse, which stands in relation to heterogeneity problems, as confirmed by GF-ZAAS results.

Subsequent to their certification, the CRMs were also used in an International Measurement Evaluation Program (IMEP-2) with 23 participating laboratories using 9 different methods (XRF, ICP-AES, FAAS, GF-SS-ZAAS, GFAAS, ICP-MS, IPAA, INAA, PAA) [51–53]. The results of the program, which aimed at making a "state of the practice" overview for the determination of Cd in PE by operating under normal working conditions with free choice of measurement methods, procedures and instrumentation, are also reported in Table 8.8. The results show a reasonable spread around the certified values established by IDMS with 80 to 85% within ±10% deviation of the certified values (*cfr.* also Fig. 8.3). Nevertheless, these findings are a source of concern and stress the need for further intercomparison actions for such materials. Simmross *et al.* [54] have later used TXRF for the quantitative determination of cadmium in the same four IRMM polyethylene reference materials. Results were quite satisfactory.

Absence of suitable CRMs has limited until recently the effective implementation of EC Directive 94/62/EC (issued Dec. 20, 1994), which regulates heavy metal concentrations in plastic packaging materials. This situation was deplored by some of the largest polymer producing companies in Europe and action was taken. The **Polymeric Elemental Reference Material** (PERM) project, an SM&T initiative within the 4th Framework 1994–1998, which was a joint effort of a consortium with seventeen participants (including major European polymer manufacturers, users of packaging materials, federal institutes and public laboratories, universities, EEC institutions and commercial laboratories), aimed at designing, producing and certifying synthetic polymer reference materials (RMs) for multi-elemental analysis: Cd, Cr, Hg, Pb, As, Br, Cl, S in HDPE. A polyolefinic material was taken as the base material, as it represents a major share of the polymeric

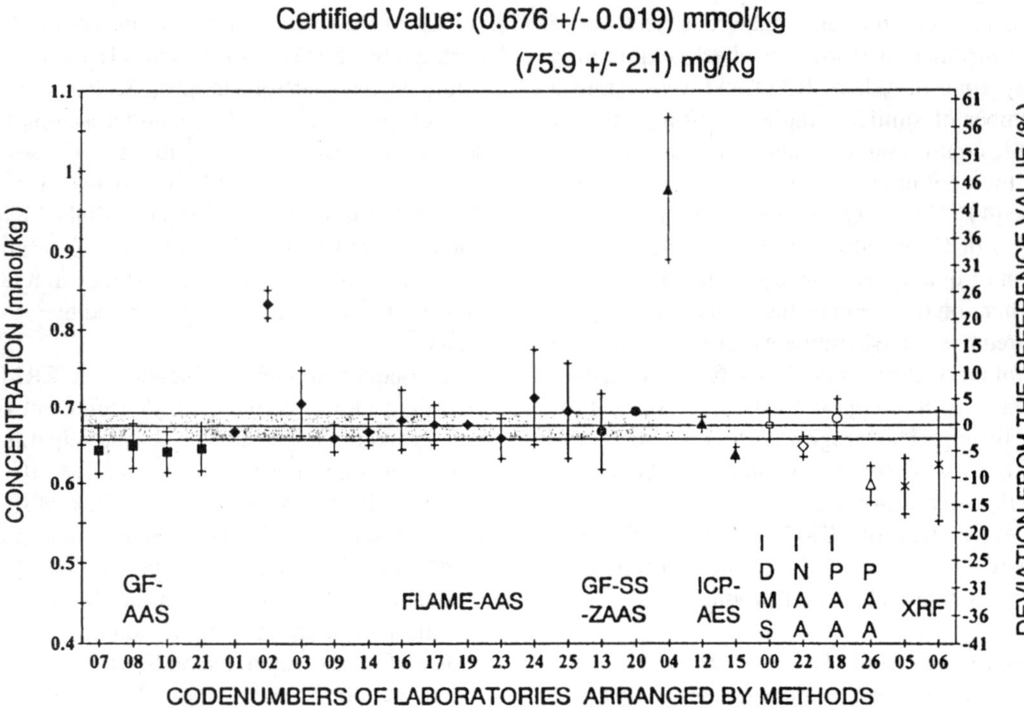

Fig. 8.3. Comparison of different methods for the determination of Cd in PE (VDA-002). After Lamberty *et al.* [51]. Reproduced from A. Lamberty *et al.*, *Fresenius J. Anal. Chem.* **345**, 310–313 (1993), by permission of Springer-Verlag, Copyright (1993).

material currently produced in Europe. The EC Directive applies to about 10,000 kt/yr of plastic packaging material in Europe alone, which is about 15% of the total packaging material used. Other expected benefits of the program are:

- the development of test methods for the assessment of the economic, technical, safety or environmental characteristics of materials, components or potential products;
- control of process wastes, particularly those which are subject to legal constraint;
- control on the environmental impact of industrial polymer processes and materials, an issue that has an increasing importance with regard to recycled polymer material. (In these materials, the heavy element entrainment is far more difficult to control and thus needs more careful monitoring);
- technical support to the achievement of total quality in measurement, through calibration or validation of chemical analyses.

The elements were chosen to meet the most typical industrial problems, being: *(i)* the determination of volatile components (Br, Cl, S); and *(ii)* the determination of heavy elements (Pb, Cd, Hg, Cr and As), as specified in EC Directive 94/62/EC. Halogen-containing additives are often used in polymers, *e.g.* in flame retardants (Br), or as stabilisers (S). The chemical form of some of these elements, in particular chromium, is a matter of debate. Some elements (notably Hg, As and Pb) are not really representative in typical additive packages.

PERM has dealt with the production and certification of a set of CRMs for use in element analysis, more specifically:

(i) the design and production of two consumable CRMs for multi-element analysis of polymer materials, consisting of a polyolefinic base material and doped with at least the heavy elements Cd, Cr, Hg and Pb at two concentration levels, namely a high level of approximately 100 mg/kg or each element (except for Hg at approximately 10 mg/kg), and at a low level of some 10 mg/kg (except for Hg at about 1 mg/kg); furthermore, the materials were doped with As, S, Cl and Br in a convenient concentration range;

(ii) the certification of the material for all elements added (*i.e.* As, Cd, Cr, Hg, Pb, Cl, Br and S),

according to the *"Guidelines for the Production and Certification of BCR Reference Materials"* (Doc. BCR/48/93, Dec. 15, 1994). The project has resulted in the availability (for each CRM) of approximately 250 kg of certified material in granular form, stored under appropriate conditions, and commercially available through BCR-IRMM.

BCR-680 and BCR-681 were prepared by mixing of finely ground (1 μm) pigments to HDPE Lupolen K 1800S (BASF, Germany), extrusion and homogenisation (*cfr.* ref. [55] for details). As shown in Table 8.9, it is apparent that other elements than the targets are also present in the materials, *e.g.* Ba, Zn and Cu (contained in phthalocyanine green). These elements were also analysed by some participants. TiO_2 powder (C.I. Pigment White 6) was added to improve the appearance; no stabilisers (UV, heat), processing aids, or other materials were added.

Development of CRMs poses problems in terms of **homogeneity** of the additives and of stability of matrix and elements. Homogeneity testing is of prime importance for the certification and use of RMs. A certification of Cd in PE [50,53] had shown that a certain risk is involved in production and certification: the production was successful though the four materials showed different homogeneities related to their concentrations in the material. Consequently, it is important that the base material does not pose severe production, stability or homogeneity problems. Criteria for sufficient element homogeneity were defined with a minimum homogeneity of 10% of a corresponding minimum mass of 50 mg for at least the elements to certify. In this way, the CRM will be adequate for elemental analysis by different analytical methods. In the case of VDA, using 60 micro-samples in the mg range, a homogeneity between 3.89% and 7.29% (95% tolerance level for 1 mg sample) was obtained for Cd-concentrations between 198 ppm and 41 ppm, respectively, with the uncertainty quoted on the reference material being applicable for minimum masses of 13 to 27 mg [53].

Homogeneity of additives *cq.* elements in a polymeric matrix is to be monitored on a micro-level (between individual pellets, mg-amounts) and a macro-level (between bottles, g-amount). The underlying reason is that CRMs must be of use for analysis techniques that use both micro and macro-amounts. For example: XRF typically uses g-amounts, whereas wet-chemical techniques that have to remove the sample matrix first are limited to mg-amounts because of pressure build-up in destruction vessels. Homogeneity studies enable estimation of the variation of average concentrations between bottles and should determine the inhomogeneity within a bottle to establish the minimum sample intake with which the certified uncertainty can be obtained [56]. An excellent means of achieving this is with solid-sampling AAS. In the PERM project ZETAAS was used to determine the minimum sample intake through a micro-homogeneity study for Cd, Hg and Pb [55]. For the other elements minimum sample intakes were derived from the certification analyses using IDMS. XRF on pressed pellets was used for the macro-homogeneity study for all certified elements. BCR-680/681 were considered sufficiently homogeneous for sample intakes of 500 and 600 mg,

Table 8.9. Additives used for preparation of BCR-680 and BCR-681[a]

Element	Target concentration		Compound added	
	BCR-680 (ppm)	BCR-681 (ppm)	Chemical name	C.I. Pigment name
Cd	140	25	(Ca, Zn)S, (Cd/Hg)S	Yellow 37
Cr	120	20	$BaCrO_4$, $PbCrO_4/PbSO_4$	Yellow 31, Yellow 34
Pd	110	15	$PbCrO_4/PbSO_4$	Yellow 34
Hg	25	5	(Cd/Hg)S	–
As	30	4	As_2O_3	–
S	650	70	$BaSO_4$, $PbCrO_4/PbSO_4$, (Cd/Hg)S	White 21, Yellow 34
Cl	800	90	Phthalocyanine green	Green 7
Br	780	100	Phthalocyanine green	Green 36

[a] After Lamberty *et al.* [55]. Reproduced from A. Lamberty *et al.*, *Fresenius J. Anal. Chem.* **370**, 811–818 (2001), by permission of Springer-Verlag, Copyright (2001).

respectively. Interlaboratory uniformity tests on the subject of metal analysis in a polymer sample are few.

Development of CRMs also requires matrix and element **stability testing** aiming at examining the long term matrix stability under various storage conditions with periodic testing and the element stability for blooming, element volatilisation, *etc*. The reason for a thorough **matrix stability** study is the fact that heavy elements in a polymer base material may degrade the polymer network. It is obviously of great importance that the physical and mechanical properties of the prospective reference material do not change over the test period such as to render the material unsuitable for processing (melting, cutting). Stability experiments with different storage temperatures, tests with artificial thermally- and UV-induced ageing were performed for BCR-681 over a period of 8 months. More precisely, the stability control was monitored by storage of the CRM in the absence of light at three different temperatures, *i.e.* +80, +20 and −18°C, reflecting two extreme and one optimum storage situation. After digestion, Cr, Cd, and Pb were measured with ID-ICPMS, Hg was measured by CV-AAS; As, S, Cl and Br were not analysed. The reason for an **element stability** study is equally obvious. Depending on their physico-chemical properties elements may migrate through the material, evaporate or change chemical form. To test element stability, aged pellets and tablets were analysed for elemental loss (by XRF). Instability turned out to be negligible for BCR-681. In fields employing CRMs which are subject to rapid deterioration, such as in clinical chemistry, it is good practice to renew CRMs on a regular basis.

General methods for the certification of RMs for elemental content are based on atomic absorption spectrometry (FAAS, ETAAS, HG-AAS, CV-AAS), atomic emission spectrometry (FAES, ICP-AES, HG-ICP-AES, DCP-AES), atomic fluorescence spectrometry (CV-AFS), mass spectrometry (IDMS, SSMS, NAMS, ICP-MS), nuclear methods (IPAA, PAA, INAA, RNAA), X-ray emission (EDXRF, WDXRF, particle induced techniques), light-absorption spectrometry (LAS, FL), electrochemistry (ASV, CSV, DPP, ISE) and other methods (Kjeldahl, combustion elemental analysis, volumetry, chromatography, gravimetry) [32]. Certification of BCR-680/681 was carried out by sixteen participating laboratories using a variety of common as well as highly specialised techniques (Table 8.10).

Table 8.10. Methods used in the certification of BCR-680/681

Elements	Final detection
As	ICP-MS, ICP-OES, IPAA, INAA, SPECT
Br	ID-TIMS, IPAA, INAA, TITR
Cd	ETAAS, ICP-OES, ICP-MS, ID-ICPMS, ID-TIMS, INAA, IPAA
Cl	IC, ID-TIMS, IPAA, INAA, TITR
Cr	ICP-MS, ICP-OES, ID-ICPMS, ID-TIMS, INAA, IPAA
Hg	CV-AAS, ICP-OES, ICP-MS, ID-ICPMS, INAA
Pb	ETAAS, ICP-OES, ICP-MS, ID-ICPMS, ID-TIMS
S	ICP-OES, ID-TIMS, INAA

CV-AAS cold vapour atomic absorption spectrometry
ETAAS electrothermal atomic absorption spectrometry
IC ion chromatography
ICP-MS inductively coupled plasma–mass spectrometry
ICP-OES inductively coupled plasma–optical emission spectrometry
ID-ICPMS isotope dilution inductively coupled plasma–mass spectrometry
ID-TIMS isotope dilution thermal ionisation mass spectrometry
INAA instrumental neutron activation analysis
IPAA instrumental photon activation analysis
SPECT spectrophotometry
TITR titration.

After Lamberty *et al.* [55]. Reproduced from A. Lamberty *et al.*, *Fresenius J. Anal. Chem.* **370**, 811–818 (2001), by permission of Springer-Verlag, Copyright (2001).

Table 8.11. Certified mass fractions in BCR-680 and BCR-681 (mg kg^{-1})[a]

Element	BCR-680			BCR-681		
	CV	U	n	CV	U	n
As	30.9	0.7	13	3.93	0.15	11
Br	808	19	9	98	5	9
Cd	140.8	2.5	25	21.7	0.7	24
Cl	810	16	10	92.9	2.8	9
Cr	114.6	2.6	23	17.7	0.6	23
Hg	25.3	1.0	16	4.50	0.15	16
Pb	107.6	2.8	13	13.8	0.7	13
S	670	70	4	78	17	4

[a] CV, certified value; U, expanded uncertainty; *n*, number of accepted sets of results.
After Lamberty *et al.* [55]. Reproduced from A. Lamberty *et al.*, *Fresenius J. Anal. Chem.* **370**, 811–818 (2001), by permission of Springer-Verlag, Copyright (2001).

Table 8.12. Additional elemental analysis (ppm) of BCR-680 and BCR-681[a]

Element	BCR-680	BCR-681
Al	51	19
Ba	2718	306
Cu	119	13.6
Sb	6.2	0.82
Ti	1174	534

[a] After Lamberty *et al.* [55]. Reproduced from A. Lamberty *et al.*, *Fresenius J. Anal. Chem.* **370**, 811–818 (2001), by permission of Springer-Verlag, Copyright (2001).

The official certified values of the PERM materials (Table 8.11) were obtained by combining the results from the different laboratories using the techniques of Table 8.10. Overall uncertainty reported comprises uncertainty resulting from the characterisation of the material, from inhomogeneity and from the stability of the material. As a bonus, additional elements (Al, Ba, Cu, Sb, Ti) were also determined, mainly by NAA (Table 8.12). The CRMs developed are expected to provide for more accurate measurements related to production, quality assurance, material research, *etc.*

Other multi-element calibration standards (TOXEL) for XRF analysis of toxic elements (Cr, Cd, Hg, Pb, As, Ni, Cu, Zn, Ba, Br) in PE are commercially available [56a], as developed in relation to RoHS compliance analysis (European legislation coming into effect on July 1, 2006). Similarly, ADPOL standards address F, Na, Mg, Al, Si, P, S, Ca, Ti, Zn in polymers [56a].

The development of control materials for polymer/additive measurements is tempered not only by lack of funding, but also by a shortage of highly skilled research expertise. In the future there will be a need for more CRMs of (other) polymer materials.

Although reliable elemental determinations are sufficiently challenging, the determination of **organics** has an additional dimension of quantitatively consistent extraction from the matrix without alteration or destruction of the organic analyte (*cfr.* Chp. 3.8 of ref. [12a]). Regrettably, with regard to low-MW *organic* additives to polymers no similar activities are foreseen, quite at difference to inter-company cross-validation exercises of additive components in drug formulations [57,58]. As shown in Chp. 6.2.3 and 6.2.4, there is a need for such actions [59]. BCR (SM&T) also aims at producing reference plastics with certified overall migration values for various food simulants (distilled water, 3% w/v acetic acid in aqueous solution, 15% v/v in aqueous solution and olive oil) [59a]; *cfr.* also ref. [60].

At the different level of **in-house reference materials**, Nagourney *et al.* [60a] have described the practical case of Witco's approach for obtaining or preparing materials to be used in quality assurance of metals (Ba, Zn, Cd) in vinyl stabilisers: *(i)* purchase of commercially available organometallic reference solutions, traceable to certified standard solutions (EMPA/BAM or NIST) [60b]; *(ii)* solubilisation of salts of known stoichiometry in small quantities of acid and 2-butoxyethanol, thereby obtaining a solution with a known quantity of metal; or *(iii)* utilisation of well-characterised *in-house* materials (either intermediates of finished products) as QA reference materials. Also other authors have shown that suitable polymer reference materials can be produced relatively easily at the laboratory level [61].

8.4. ANALYTICAL METHOD VALIDATION APPROACHES

Principles and Characteristics

Method validation in analytical chemistry is often the last step in method development. Once a candidate method has been obtained, it has to be shown to meet the requirements of the user, namely to measure a specific analyte with a given precision, accuracy, detection limit, *etc.* Method validation is carried out to ensure the quality of a method and is therefore an essential part of any quality assurance program in a laboratory. **Validation** is defined as the "*Confirmation by examination and provision of objective evidence that the particular requirements for a specified end use are fulfilled*" [62]. In practice, validation therefore means establishing *documented* evidence, which provides a *high degree of assurance* that a specific process (method, instrument or computer system) will *consistently* produce a product (or service) meeting predetermined specifications consistent with some standard quality procedure. The validation process verifies that the methodology is based on sound technical principles and that it has been reduced to practice for practical measurement purposes. Several regulatory bodies require that one must document the validation as part of a quality assurance program. Method validation requires different experimental set-ups and must always be accompanied by a statistical analysis of the data produced by the method validation experiments. Validation is one of the main fields of application of chemometrics. A method is validated when the performance characteristics are adequate and when it has been established that the measurement process is under statistical control, produces accurate results and is suitable for its intended use. Quality data and methods can only be produced using systems proven to be under control. Validation is not required by scientific journals.

As shown elsewhere (Table 1.15 of ref. [12a]), a validated method is not the only condition for analytical excellence: the sample and the experimental implementation (ISO 9000, ISO/IEC Guide 25, GLP) are equally important. The skill of the operator constitutes a critical factor in a measurement process.

In the ICH guidelines analytical validation relates to test procedures. A test procedure is a more precise and comprehensive term than analytical method as it includes the technique, the sample and standard preparation, the use of apparatus, the formulae for calculation, *etc.* (*cfr.* Table 8.1). Validation of an analytical method is primarily concerned with the identification of the sources and the subsequent quantification of the potential errors in the method. Three types of error may be encountered in an analytical measurement: gross, systematic and random; and the analyst should be able to distinguish between each type. When a new method is developed, the measurements should be compared to those of well-established existing methods to assure that the novel method is free of any systematic error. Statements of precision and accuracy are often a result of an analytical validation process, especially in case of a collaborative test exercise. Other information useful for characterising methodology or for judging its suitability for a given use includes: sensitivity to interferences, limits of detection, and useful range of measurement. What constitutes a validated method, however, is subject to analyst interpretation because there is no universally accepted industry practice for method validation.

Table 8.13 states **why, how, and when to validate and who** should take care (*cfr.* also ref. [63]). Everyone should validate/qualify each process, system and piece of equipment that is used in a laboratory. The validation process is enforced in regulated industries (FDA's and EPA's GLP) and recommended in those interested in European markets (ISO 9000). A validation program builds confidence that products are of high quality and that they can be taken as a sound basis for decision making. It is clear, however, that validation is *not* absolute proof! Also, manufacturers cannot validate users equipment.

Ideally, **validation criteria** should be compiled at different stages in the analytical procedure development, but to a different extent. It is a misconception to believe that development of an analytical procedure and validation are independent processes: they are interdependent. The actions required to ensure that valid and reliable analytical measurements are being made are not trivial. Development of a procedure and validation is an iterative process. The procedure's suitability must be studied in initial validation experiments. Pre-study validation is a formal validation protocol designed to characterise the method *before* analysis of real samples. Prerequisite is a detailed description of the method. Acceptance criteria should be established *a priori*. In the validation stage, it is necessary to demonstrate that the method works with samples of the given analyte, at the expected concentration in the anticipated

Table 8.13. Characteristics of analytical validation

- **Why validate?**
  - Ensures quality
  - Part of overall quality process
  - Demonstrates performance of method
  - Reduces number of analyses (no duplicates)
  - Good scientific practice
  - Makes good business sense
  - A regulatory requirement
- **Who validates?**
  - End-users in regulated environments
  - Those in control of quality of their manufacturing process
- **When to validate?**
  - Instruments/systems
    - Upon installation and prior to routine use
    - Post major repair
    - Post routine maintenance (on a regular basis)
  - Analytical method
    - Prior to routine use
    - Testing after changing parameters
    - Changes beyond original scope
  - Analytical systems
    - Regularly or before sample analysis
- **How to validate?**
  - Following a written, approved validation protocol, including acceptance criteria
  - Defined application and scope of the procedure
  - Defined equipment, operational conditions, accessories, reagents, solvents, standards, *etc.*
  - Qualified and calibrated equipment
  - Defined performance characteristics

matrix, with a high degree of accuracy and precision. If preliminary validation data are inappropriate, either the basic technique itself, the equipment or the acceptance criteria have to be changed. Robustness studies are part of this development phase. As pointed out elsewhere [64], it is necessary to validate the whole method (including preparatory steps) over the whole range of operating conditions and foreseeable matrices. Some parameters are quite time consuming and laborious and cannot be reported until there is a long-term experience of the overall performance (such as analyte stability studies). Consequently, it can take months to establish the basis of a test method and to validate all aspects of the measurement associated with a method. Quality control data are generated during application of the method. In the on-going validation process, suitability of the final method for the given analyte and selected sample matrices is to be demonstrated, using specified instrumentation, samples, and data handling. A method that provides all or most of the original method requirements is deemed optimised. Complete method validation can occur only after the method is developed *and* optimised. Ultimately, the method can be transferred from one instrument to another (inter instrument transfer) or from one laboratory to another. The final aim is towards validation of data at international level.

Key question regarding validation is **"How much validation"** is needed as tremendous efforts can be expended in conducting validation studies. Method validation processes easily consist of some 80–100 samples. It is particularly important in R&D that the effort put into validation balances costs, risk and technical importance, so most emphasis should be directed to those parameters which most critically affect performance. For example, when one is looking at qualitative data for a method the main thrust of method validation would be selectivity, with some attention to the limit of detection. The key to method validation is which parameters should be validated and which experimental design should be used.

Two types of method validation can be distinguished. **Full method validation**, of interest to the general scientific community, is carried out through an interlaboratory method performance study. Where a method becomes more routinely used it is reasonable to expect that the method should be fully validated. **Internal method validation** (single-laboratory method validation) is a scientific and technical alternative. It consists of validation steps carried out within one laboratory, for instance, to validate a new method that has been developed in-house or to verify that a method adopted from some other source is applied sufficiently well. A single-laboratory validation cannot assess between-laboratory variation and will provide an optimistic assessment of interlaboratory variability (*cfr.* Chp. 6.2.3 and 6.2.4). In-house method validation is described in the IUPAC, AOAC International, and ISO guidance [65,66]. There are several types of internal laboratory validation:

- Prospective validation, which is carried out when a new method is introduced.
- Suitability checks, which can be applied when transferring a method from one laboratory to another.
- Retrospective validation, where results collected over a period of time are used to determine precision of determinations over long periods.

Retrospective validation is not an acceptable substitute for formal prospective validation programs. It should only be considered for remedial action in combination with prospective validation for a non-compliant establishment. The retrospective validation process of a large multi-user chromatography data system has been described [67].

The ideal laboratory based validation technique should have broad compositional scope and sensitivity to the ingredients, be rapid for the purposes of process relevance and reduction of testing costs, and should be amenable to practice by QC lab staff without extensive training. Acceptance of any new method by others in the field will depend on the specific validation approaches used. It is the responsibility of the individual analyst to select the appropriate validation method(s). **Validation approaches** include the zero-, single- and double-blind spiking methods, comparison with a currently accepted (compendium) method and interlaboratory collaborative studies. The zero-blind approach, which might be subject to considerable analyst bias, involves a single analyst using the method with samples at known levels of analyte to demonstrate recovery, accuracy, and precision. The single-blind approach is less biased and involves one analyst preparing samples at varying levels unknown to a second analyst, who also analyses the samples. A more objective approach, however, is the double-blind approach, which involves three analysts. The first analyst prepares samples at known levels, the second does the actual analysis, and the third analyst compares both sets of data received separately from the first two analysts.

Comparison with a currently accepted compendium method is another validation approach and is frequently used in industrial research laboratories. This approach uses results from a currently accepted (analytical) method as verification of the new method's results. Agreement between results initially suggests validation. However, disagreement could cast doubts on the acceptability of the new method or may suggest that the currently accepted method is invalid. Validation of compendial methods has been addressed by the USP Chapter ⟨1225⟩ [68]. Interlaboratory collaborative studies are discussed in Chp. 8.4.2.

Which type of method validation has to be carried out depends on the application field of the laboratory. A laboratory developing its own methods largely for its own use will essentially need to carry out validation about all the analytical performance characteristics (whole method, full concentration range, all applicable matrices). It may also develop suitability checks for transfer to plant laboratories, including suitability checks for inclusion in SOPs. Validation of laboratory computer systems should be considered as a normal part of any project in laboratory automation.

A good validation is the scientific base for later adjustments without the need of **revalidation**. The maintenance of methods used in routine laboratories in a regulated environment is restricted to small adjustments of the methods. Modifications are often limited by the high cost of a revalidation of methods. It is not evident what are tolerable adjustments of methods and what are modifications with the need of a new validation and approval. If significant modifications to a method are incorporated at any time of transfer, revalidation may be necessary to ensure that the modifications have not invalidated previous, conclusive data. Revalidation is necessary whenever a method is changed and a parameter is outside the operating range; method update is also required if the sample matrix or instrument type changes. For example, substituting an alternative chromatographic column always raises the possibility of a change in specificity and resolution as well as in the quantitative aspects of the method. Therefore, such modifications require all method validation parameters to be reassessed, *i.e.* specificity (resolution), linearity, range, accuracy, precision and LOQ.

Method validation and method transfer are distinct processes. Method validation certifies that the method performs in the manner for which it was developed and is the responsibility of the method development laboratory. **Method transfer**, on the other hand, is the introduction of a validated method into a designated laboratory so that it can be used in the same capacity for which it was originally developed. Accordingly, method transfer criteria should be based on the SOPs which are unique to the designated laboratory. For the essential principles of method transfer, *cfr.* ref. [69]. Interlaboratory transfer of HPLC methods has been reported [70]. **Inter technique validation** is equally important. Both in R&D and in a production environment, the change from one technology to another must be totally transparent. A sample concentration obtained by a method in one laboratory must be the same as that obtained by another method in another laboratory.

Various branches of industry (in particular pharmaceutical manufacturing and environmental testing) are subject to very strict requirements to verify that analytical work results in reliable, valid data that help ensure that only safe and effective products reach the consumer. Method validation is receiving considerable attention from regulatory agencies, industrial committees and in the general literature. The *Guidance on the Interpretation of the EN 45000 Series of Standards and ISO/IEC Guide 25* includes a chapter on the validation of methods [71] with a list of nine validation parameters. The US FDA has proposed guidelines on submitting samples and analytical data for methods validation [72–74]. The US Pharmacopœia (USP) has published specific guidelines for method validation for compound evaluation [68,75]. This protocol specifically addresses terms and definitions, sets no official guidelines, but leaves methodology open to interpretation. This intentional omission allows flexibility in method validation. Laboratory personnel is supposed to be "skilled in the art". The International Conference on Harmonisation (ICH) has also issued a draft guideline on validation of analytical procedures with definitions and terminology [76]. The ICH Guideline on Method Validation Methodology is more explicit as to experimental design and protocol in order to improve the process of method development, optimisation and validation. Moreover, the US Environmental Protection Agency (EPA) has prepared guidance for methods development and validation for Resource Conservation and Recovery Act (RCRA), whereas the American Association of Official Analytical Chemists (AOAC), and other scientific organisations provide methods that are validated through multi-laboratory studies. AOAC has developed a peer-verified methods validation programme with detailed guidelines on what parameters should be validated [77].

Recently various papers and books have been published dealing with the validation of analytical methods in the chemical industry. Taylor [1] first and later Green [78] and Swartz *et al.* [6,78a] have given a practical guide for analytical method validation with a description of a set of **minimum requirements** and documentation for a method. Wegscheider [79] has published procedures for method validation with special focus on calibration, recovery experiments, method comparison and investigation of ruggedness. Development and validation of analytical methods (although with applications limited to the pharmaceutical product development process) have been described by Riley *et al.* [58]; *cfr.* also the general bibliography. Terminology and strategy for (internal) analytical method validation were reported elsewhere [80,81]. Statistical parameters and analytical figures of merit [82] and operational qualifications for selected equipment [83] have been reviewed. The meaning of validation and qualification applied to computer systems has been addressed [84]. An on-line resource for validation and compliance issues in analytical laboratories is available [85].

Applications

In the analysis of in-polymer additives, the entire procedure from extraction to quantitation must be validated, as safeguarding against analytical complications. The current methods for detecting and controlling additives still present some serious problems to the industrial analytical community. A first step to be validated is the (analytical) extraction in order to ensure that the additive is extracted quantitatively from the polymer. In practice, 90% or more is usually acceptable. Of course, it is important that whatever the extraction procedure used a known, constant, percentage is extracted without transformation. This requires checking samples of the polymer with known amounts of additive present. Chapters 6.2.3 and 6.2.4 give sufficient evidence that considerable improvement is required in this area.

As to inter technique validation, comparison of measurement results obtained by methods based on different physico-chemical principles is shaky ground. For example, the *generic* extraction results for plasticisers from rubbers (*cfr.* Chp. 6.2.4) are no validation for PyGC-MS, which is a selective procedure for the determination of a *specific* compound.

Lopez-Avila *et al.* [86] have considered validation of analytical **supercritical fluid extraction methods**. When developing an SFE method, various critical factors need to be considered: *(i)* solubility of the analyte in the supercritical fluid (SF); *(ii)* diffusion of the analyte from the solid matrix into the bulk fluid or displacement of the analyte by the SF that diffused into the matrix; and *(iii)* the matrix itself, which can be very adsorptive. Poor SFE recoveries may be attributed to matrix effects (*e.g.*, not all of the analyte may be extractable), to inefficient retention of the analyte in the collection solvent or on the sorbent trap, or to inefficient desorption from the sorbent trap. Once an SFE procedure has been developed and tested with real matrices, the next step

Table 8.14. Selection of typical validated company standard in-polymer additive analysis methods

Analyte(s)	Matrix	Method
EBA	PA	Wet chemical
IM	PA	Wet chemical
IM	Mineral-filled polyesters	Wet chemical
IM	GFR-PA	Wet chemical
TA, IA, TMA	Polyester TPE	LC-PDA
Diacids	Polyester TPE	LC-PDA
Nucleating agents	PE, PP	LC-PDA
Irganox 1425	PE, PP	LC-PDA
ERL 4221 (diepoxide)	PBT	GC
Stearyl stearate, nonyl stearate	PA	GC
Release agents	PA	GC
DSTDP, Santowhite powder	ABS/PA6	GC
FR 1808	PBT	HPLC
Chimassorb 944	PE	HPLC

Table 8.15. Definitions of materials components

Highly volatile matter:	Moisture, plasticiser, residual solvent or other low-boiling (at 200°C or less) components
Medium volatile matter:	Medium volatility materials degradable from 200 to 750°C (oil and polymer residues)
Combustible material:	Oxidisable material at 750°C (not volatile in unoxidised form); carbon
Ash:	Non-volatile residues in an oxidising atmosphere (metal components, fillers, inert reinforcing materials)

for full validation is an interlaboratory study. Most industrial research laboratories dispose of a standard set of validated in-polymer additive analysis methods (*e.g.* Table 8.14). For frequently recurring analyses an SPC approach is adopted.

The greatest degree of consistency often appears to be in the validation parameters applied to **chromatographic procedures**. This is particularly the case for HPLC, which is a reflection of the universal application of the technique within industry and the good agreement on the critical parameters. Validation of computerised LC systems [87] and of analysis results using HPLC-PDA [88] have been reported.

In the field of polymer/additive analysis various validated procedures (after interlaboratory tests) do exist. Various such procedures have been given in the present text (*cfr.* also Chp. 8.3 for CRM development). Here we just mention the ASTM Standard Method of Test for Carbon-Black in Polyethylene Plastics (E 1603) and the ASTM Test Method for Compositional Analysis by Thermogravimetry (ASTM Standard Method E 1131) [89], which outlines a general technique for analysis of materials by measuring mass loss through several thermal stability ranges. The general technique determines the quantity of four arbitrarily defined components, namely highly volatile matter, matter of medium volatility, combustible material, and ash. The definitions of the four components, according to Table 8.15, are based on their relative volatility. The determination of the constituents provides a compositional analysis by using an inert atmosphere for a portion of the analysis of carbon containing materials and mixtures.

The validation of an analytical method for the determination of plasticisers (DEHA, DEHP) in PVC has been reported [90], as well as that of the calibration procedure in AAS methods [91]. In an original development Van Every *et al.* [13] have applied infrared principle components/Mahalanobis distance discriminant (PMD) analysis to validate **polyolefin (film) products**. PMD is a technique designed to classify complex materials into groups or identify unknowns by using n principle components to map data characteristics into an n-space cluster [92]. An unknown material can then be assigned a distance from this cluster based on the number of standard

Table 8.16. Requirements to principle components/Mahalanobis discriminant distance analysis

- Be rapid enough to give meaningful results within the process time frame
- Have broad compositional scope and sensitivity to detect most formulation components and a wide range of possible contaminants
- Be tolerant of in-specification variations
- Be amenable to practice by QC laboratory technicians not highly skilled at interpreting IR spectra
- Insure coverage of all spectral windows for comprehensive detection
- Minimise effects of unreliable features, which could desensitise the calibration
- Avoid false flags due to unreliable features and spectral artefacts
- Monitor relationships of absorptions across separate spectral regions in addition to individual band intensities
- Be interpretive

deviations it lies in its direction from the cluster. This distance, known as the Mahalanobis distance (MD), is a sensitive qualitative parameter to measure conformance of a material to the calibration set. Table 8.16 lists the requirements to PMD for being a viable QC validation technique [13].

In case of outlier detection (such as a poorly fitting material or one having a poor quality spectrum), the operator should be guided to the most likely cause. This requires a quality training set that reflects all in-specification variations in the product, process and analysis. The combination of mid-IR spectrometry with discriminant analysis makes the tool more readily available for QC validation by non-spectroscopists. This form of **validation without quantitation** can be directly applied to on-line data, greatly reducing start-up efforts associated with quantitative method development [13]. Mid-IR classification techniques are cost-saving in comparison to more rigorous, multicomponent analyses.

In another case of on-line process control thirty samples were extracted from an extruder over a period of one week for off-line LC analysis [93]. Simultaneously, a UV spectrum of the PE melt was stored in a training model database. The training data were then modelled in a PLS regression, which allows additive concentrations to be predicted from unknown UV spectra. Comparing the results with those obtained by LC validated the PLS predicted data for each of the additives.

Calibration models in NIRS are validated in two different ways. The external prediction method requires a large and representative new set of objects which have to be kept apart from the calibration for testing purposes only. Internal validation methods such as cross-validation are based on the calibration data themselves. Cross-validation seeks to validate the calibration model with independent test data, but contrary to external validation it does not use data for testing only. The cross-validation is performed a number of times, each time with the use of only a few calibration samples as a test set.

8.4.1. Analytical Performance Parameters

Principles and Characteristics

Whereas parameters most relevant to method development are considered to be accuracy, system precision, linearity, range, LOD, LOQ, sensitivity and robustness, **method validation parameters** are mainly bias, specificity, recovery (and stability of the analyte), repeatability, intermediate precision, reproducibility and ruggedness. However, method development and validation are highly related. Also, validation characteristics are not independent: they influence each other. Acceptance criteria for validation parameters should be based on the specification limits of the test procedure. Quantitation and detection limits need a statement of the precision at their concentration levels. Procedures used for validation of qualitative methods are generally less involved than those for quantitative analytical methods. According to Riley [82], who has discussed the various parameters for validation of quantitative analytical methods, the primary statistical parameters that validate an analytical method are accuracy and precision.

According to the US Pharmacopœia (USP), method validation needs to be performed to ensure that an analytical methodology is accurate, specific, reproducible, and rugged over the specified range of analysis of an analyte. Method validation provides an assurance of reliability during normal use. Regulated laboratories are bound to perform method validation in order to comply with government regulations (*e.g.* FDA) [74]. USP [75] and ICH [76] have defined the validation parameters (also referred to as "analytical performance parameters") with some slight differences between different organisations. For a proper understanding of the impact of data elements required for assay validation, the **analytical performance parameters** are briefly described.

Accuracy is the measure of exactness of an analytical method, or the closeness of agreement between the measured value and the conventionally accepted true or reference value [94]. The true value can be obtained in several ways. Results of the method may be compared with those from an established reference method. Alternatively, accuracy can also be assessed by comparing with a sample of known concentrations, for example, a CRM. If no CRM is available, recourse can be taken to a blank sample matrix of interest spiked with a known concentration by weight or volume. The accuracy should be examined over a range that extends beyond the range of samples the method is likely to analyse. In practice, $\pm 10\%$ deviations from certified values are commonly observed, *cfr.* the VDA-001 to 004 (Cd in PE) standards. For accuracy of HPLC methods, *cfr.* ref. [95].

Precision is the measure of the degree of repeatability of an analytical method under normal operation and is usually expressed as the percent relative standard deviation for a statistically significant number of samples. Precision is considered at three levels: *repeatability* or within-run precision (refers to results of a method operating over a short time interval under the same conditions), *intermediate precision* or within-laboratory precision (refers to results from intralaboratory variations in experimental periods, operator and equipment) and *reproducibility* or between-run precision (refers to interlaboratory comparisons). Good science requires reproducible results [96]; it is irresponsible to publish data that are not reproducible! Full reproducibility can only be achieved by means of robotics. This is more difficult to achieve in analysis than in synthesis as sampling is a crucial step. Within run and between run changes in instrument response can be monitored using quality control samples and calibration standards. The objective of intermediate precision validation is to verify that in the same laboratory the method will provide the same results once the development phase is over. Validation of reproducibility is important if the method is to be used in different laboratories. Reproducibility usually means greater dispersion of measured data than repeatability as experiments are carried out in different laboratories, with different instrumentation, chemicals, or personnel (*cfr.* Chp. 6.2.3 and 6.2.4). Instrument constants in interlaboratory reproducibility studies may be eliminated by differentiation (*e.g.* DTG *vs.* TG). For compound analysis, precision is very much dependent on sample matrix, analyte concentration and analysis technique; precisions are often quoted to vary from 2% to more than 20% [80]. Random errors influence precision, repeatability or reproducibility of the determination. Precision can be improved by using an internal standard, preferably with a chemical structure similar to that of the analyte. The ability to detect stability of an analyte is dependent on the precision of the method. For further reference, *cfr.* the ASTM Standard Practice for Conducting an Interlaboratory Study to Determine the Precision of a Test Method (E 691).

Specificity is appropriately applied to analytical techniques with the ability to measure accurately and specifically only the analyte of interest in the presence of other components in the sample matrix. XRD and NAA are specific methods. Analytical procedures are usually not specific for a particular analyte. The term *specific* is generally attributed to a method that produces a response for a *single analyte* only, while the term *selective* refers to a method that provides simultaneous responses for a *number of chemical entities* in a multicomponent system that may or may not be distinguished from each other. A method is called selective if the response is distinguished from all other responses. The terms selectivity and specificity are often used interchangeably [97,98]. Kaiser [99] has given a differentiation between selectivity and specificity. The USP monograph [68] defines **selectivity** of an analytical method as its ability to measure accurately an analyte in the presence of interference, such as precursors or degradation products that may be expected in the sample matrix. The selectivity of a chromatographic method may be defined by the use of relative retention indices. Whereas the mass spectrometer provides an even higher degree of selectivity in gas chromatography, the diode-array detector (DAD) and the use of spectral deconvolution techniques are the principal tools for the determination of peak purity in liquid chromatography. This is a further step towards the evaluation of specificity (*cfr.* Fig. 8.4) [100]. Analytical procedures, especially in the food and medical applications, require that each substance be identified by at least two analytical methods, based on differing chemical or physical properties. Because of the wide variety of structures, positive structural identification is not always an easy task. Whereas selectivity can be graded (totally, highly, very, partially, *etc.*) specificity cannot be graded as it is essentially absolute. Danzer [100a] has recently proposed a quantification for these performance parameters. Specificity

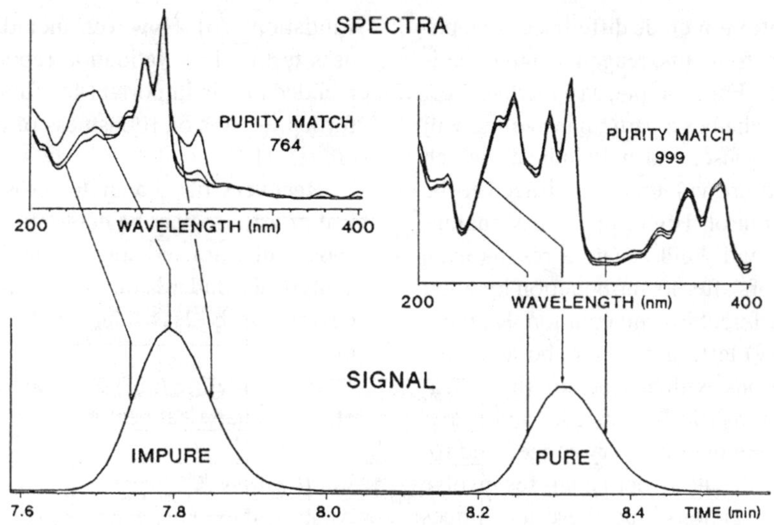

Fig. 8.4. Peak purity determination by spectral overlay. The HPLC signal does not indicate any impurity in either peak. Spectral evaluation (DAD) identifies the peak on the left as impure. After George [100]. Reprinted form S.A. George, in *Diode-Array Detection in HPLC* (L. Huber and S.A. George, eds.), Marcel Dekker Inc., New York (1993), by courtesy of Marcel Dekker Inc.

in mass spectrometry is not necessarily absolute, yet very high. Maximum specificity is afforded by full scan, but at reduced sensitivity. Maximum sensitivity with good specificity is afforded by selected monitoring techniques. Specificity can be increased by improved sample preparation, improved chromatography, alternative ionisation, tandem mass spectrometry or increased mass resolution.

The **limit of detection (LOD)** is one of the most important terms used for comparing various analytical procedures, techniques or instruments. It is defined as being the lowest concentration of the analyte that can be distinguished with reasonable confidence from the blank or background. LOD may be based on the signal-to-noise ratio, on visual non-instrumental methods (*e.g.* TLC or titrations) or computation (deviations from regression lines). The method used to determine LOD should be documented and an appropriate number of samples should be analysed at the limit to validate the level. In chromatography the detection limit is the injected amount that results in a peak with a height at least 2–3 times as high as the baseline noise level. LOD is an important parameter for quantitative measurements in trace analytical methods. Highly precise measurements are impossible at concentrations close to the LOD.

The **limits of quantitation** (LOQ) are the lowest *cq.* highest concentrations of an analyte in a sample that can be determined with acceptable precision and accuracy under given operational conditions of a method. If the required precision of the method at the limit of quantitation has been specified, the EURACHEM approach can be used [71]. In chromatography, the limit of quantitation typically requires peak heights 10–20 times higher than baseline noise. As with LOD, LOQ is relevant only in trace analytical methods when measurements are being made at concentrations close to that limit.

Linearity of an analytical method is the ability to elicit test results that are directly, or via well-defined mathematical transformations, proportional to the analyte concentration within a given range. Because deviations from linearity are sometimes difficult to detect, various graphical procedures have been proposed [80]. Linearity is generally reported as the variance of the slope of a regression line. A linear regression equation applied to the results should have an intercept not significantly different from zero. The linearity of a method should be checked over and beyond the likely operating range of the method.

The **range** of an analytical method is the (inclusive) interval between the upper and lower levels of analyte that have been demonstrated to be determined with precision, accuracy, and linearity using the method.

Ruggedness is the degree of reproducibility of results obtained by analysis under a variety of conditions, expressed as % relative standard deviation

(RSD). These conditions include differences in laboratories, analysts, instruments, reagents, and experimental periods [101]. For example, a ruggedness test will indicate firstly whether a particular method will stand up to everyday use, and will indicate which parts of the method are vulnerable and need to be subject to quality control. High ruggedness guarantees that a method will yield accurate results even when performed years after its introduction by other personnel and on different instrumentation. Ruggedness (reproducibility) testing by interlaboratory trials excludes variations with respect to time. Traditional spectrophotometric, volumetric, and gravimetric analytical techniques are very rugged and robust. However, virtually all modern analyses involve chromatographic separations, and these techniques are frequently much less rugged or robust. A ruggedness test and its application for HPLC validation has been described [102]. Those parameters related to eluent properties (composition), column temperature, pH of eluent and gradient shape have to be tested. In a reversed-phase gradient column a fair number of parameters needs to be controlled.

Robustness is the capacity of a method to remain unaffected by small deliberate variations in operational parameters (stressing the test method) and provides an indication of reliability during normal use [101]. Robustness testing covers the critical operating parameters that have the most significant effect on an analytical result (*e.g.* stability of the analyte in test, effect of temperature, sample matrix, sample preparation and pretreatment, solvent quality, injection volume, flow-rate, mobile-phase composition, column quality, detection wavelength, spectroscopic settings, *etc.*). Such data allows judging whether a method needs to be revalidated when one or more of the parameters are subject to change. A method is robust if it is student-proof. Robustness testing does not cover the effects that may show up during a transfer of the procedure to other laboratories. Notwithstanding some ambiguity in the definitions of ruggedness and robustness, it is convenient to apply the term ruggedness to the variation of errors in results arising from different operation conditions and robustness to the ease with which the critical parameters of a method may be reproduced. It also follows that ruggedness of a method is influenced by its robustness [82]. Robustness testing is part of the method (procedure) development and is not necessarily a part of the formal validation [76]. However, including data of robustness testing in a validation report is highly recommended and is important for future automation. For further aspects on robustness of analytical methods, *cfr.* ref. [103].

Recovery of an analyte across the whole analytical procedure may be determined by comparing response of extracted spikes samples and unextracted analytical standards of equivalent concentration. Recovery can be less than 100%, but must be reproducible.

For validation it is not always necessary to evaluate each analytical performance parameter.

Applications

Analytical methods have been divided into three separate categories [6]:

1. Quantitation of major components.
2. Determination of impurities or degradation products.
3. Determination of performance characteristics.

The type of method and its intended use dictates which parameters need to be evaluated, as shown in Table 8.17. In general, no quantitative parameters need to be defined for a qualitative method.

For **quality control** applications the use of SPC charts is recommended. Figure 8.5 shows 962 control measurements of P content in polypropylene by means of XRF (mean value 44.7 ppm; e.s.d. 1.2 ppm; $n = 39 > 2\sigma$, $n = 4 > 3\sigma$).

Cross-validation should be performed to compare results obtained by methods based on different techniques, *e.g.* LC-MS and HPLC-UV, or by the same method in different laboratories. Both methods should have been validated independently prior to cross-validation. Capillary electrophoresis (CE) is an alternative for HPLC for a wide range of analytical problems offering shorter analysis times. Both methods are selective and robust. Comparison of robustness implies a variation of different parameters, such as the mobile phase composition, the buffer pH and molarity, temperature, flow-rate and sample solvent [104]. Some concern has been expressed about the reproducibility of CE. Crucial parameters for robustness in CE are the mobile phase composition, which is essential for good separation, the nature of the eluents (volatility), buffer pH and concentration of the additive. Comparison of validated CE and HPLC methods shows that HPLC is about a factor of two better than CE for all quantitative parameters.

Table 8.17. USP data elements required for method validation

Analytical performance parameter	Analytical method category 1	Analytical method category 2		Analytical method category 3
		Quantitative	Limit tests	
Accuracy	Yes	Yes	*	*
Precision	Yes	Yes	No	Yes
Specificity	Yes	Yes	Yes	*
LOD	No	No	Yes	*
LOQ	No	Yes	No	*
Linearity	Yes	Yes	No	*
Range	Yes	Yes	*	*
Ruggedness	Yes	Yes	Yes	Yes

*May be required, depending on the nature of the specific test.
After Swartz and Krull [6]. Reprinted from M.E. Swartz *et al.*, *Analytical Method Development and Validation*, Marcel Dekker Inc., New York, NY (2003), by courtesy of Marcel Dekker Inc.

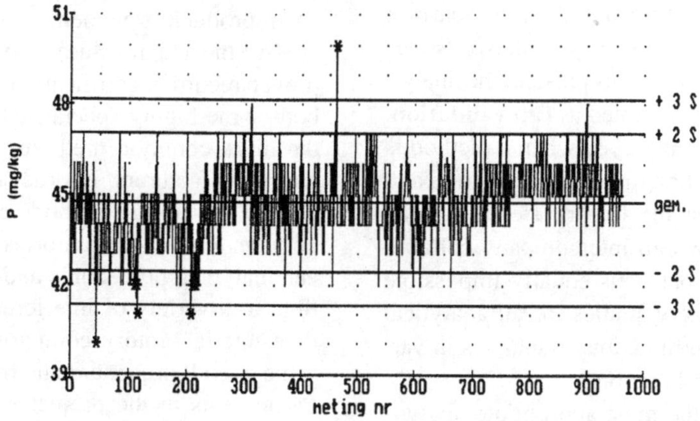

Fig. 8.5. SPC chart of XRF measurements of phosphorous in PP (period: from Febr. 23, 1994 till Mar. 26, 1998). Courtesy of DSM Plant Laboratory Services, Geleen, The Netherlands.

8.4.2. Interlaboratory Collaborative Studies

Principles and Characteristics

According to the IUPAC definition, an interlaboratory study is one in which several laboratories measure a quantity in one or more identical portions of homogeneous materials under documented conditions, the results of which are compiled into a single report. Three types of interlaboratory studies are distinguished, namely method-performance, laboratory-performance or material-certification studies. The aim of **method-performance** or collaborative studies is to assess the performance characteristics of a specific method. In **laboratory-performance** or proficiency studies a homogeneous test material is analysed of which the true concentrations are known or have been assigned in some way. The participants apply whatever method is in use in their laboratory. The results are compared to evaluate the proficiency of individual laboratories and to improve their performance. IUPAC has issued a protocol for the proficiency testing of analytical laboratories [65,66]. In **material-certification studies** a group of selected laboratories analyses, usually with different methods, a material to determine the most probable value of the concentration of a certain analyte with the smallest uncertainty possible. The objective of such a study is to provide reference materials (*cfr.* Chp. 8.3). Round-robins thus serve various purposes (Table 8.18). ISO 5725 (1994) describes the procedure for interlaboratory tests; guidelines have also been published by AOAC [105].

An interlaboratory method performance study is the ultimate procedure to validate any new analyti-

Table 8.18. Usefulness of round-robins

- Comparison of measurement results
- Method validation
- Reciprocal recognition of analytical results by industrial partners (*e.g.* supplier-customer)
- Method improvement
- Proficiency-testing
- Assessment of competence of testing laboratories in accreditation schemes
- Certification of polymeric reference materials (traceability to the SI unit system)

cal method, but suffers from several serious practical drawbacks. The collaborative approach is a limited exercise, which is costly and time consuming and can take years from start to finish. When all laboratories involved in an interlaboratory comparison have come up with overlapping quantitative values in comparison with known levels present, the analytical method is generally accepted as **full validation**. This approach is rarely employed when a method is being described for the first time in the literature and obviously loses its meaning for proprietary analytical methodology, unless an intercompany collaborative study is carried out. It is equally impossible to organise interlaboratory studies for all analytical methods in use for determination of analytes in various analyte/matrix combinations.

In the selection of the most appropriate analytical method for a standard on a specific product, interlaboratory testing of CRMs is carried out to establish the quality parameters of the method in question. CRMs are therefore important links in the chain referred to above. For **intercomparison of methods**, within a laboratory between different methods or within a company between laboratories (such as an R&D-department and production laboratory), a CRM may serve as a common reference point, with which analytical procedures can be scrutinised or adjusted.

In the future, analytical methods might be accepted as International Standards on the basis of interlaboratory tests performed on selected CRMs. ISO has observed an increasing number of calls for "ISO-certified" CRMs, CRM producers, and laboratories. In the present context, no certification or accreditation mechanisms are operated by ISO. However, ISO 9000 is a valuable tool for producers of CRMs.

Applications
As indicated in Chp. 8.3, the European polymer industry has recently taken action to improve its competitive position by promoting more accurate and reproducible analytical measurements at the R&D and production level by creating a mutual basis of recognition when it comes to interpreting analytical results between different industrial, governmental and private laboratories and universities, all of these cooperating in a project consortium. Earlier findings in an **IMEP-2 program** with the object of preparing Cd containing PE standards had already shown the usefulness and need of intercomparison actions (*cfr.* Table 8.8). In the **PERM project** (*cfr.* Chp. 8.3), which aimed at the production of well characterised CRMs (consisting of As, Cd, Cr, Hg, Pd, Br, Cl and S in PE) and the development of more accurate and reproducible elemental analysis methods (within 10% of the actual value), various laboratories with a proven record of certification have participated using both some highly specialised analysis methods and the more common methods available among polymer manufacturers, in order to favour an intercomparison of various methods currently in use among polymer analysis laboratories [55]. The use of different analytical procedures and/or techniques, susceptible to a variety of interferences, is more valuable than interlaboratory comparisons using exactly the same overall procedure and measurement technique. Participants in the project were equipped with the only adequate polymer CRM available (VDA CRM: Cd in PE), for calibration and testing of their analytical methods. Discrepancies may especially be expected for different sample destruction methods used (*e.g.* microwave destruction, ashing in an oven, acid digestion, *etc.*), in particular for the volatile elements Hg, Cl and Br. Within the frame of the PERM project [55], expertise on a number of sophisticated analysis methods for elemental polymer analysis was being shared for the first time, which has resulted in greater insight in the associated analytical difficulties and in method adjustment and improvement.

In the field of polymer/additive analysis a rather limited number of other laboratory performance studies is available. Recently, the Swiss Federal Laboratories for Materials Testing and Research (EMPA, St. Gallen) has organised a series of interlaboratory tests on polymeric materials, examining the glass transition point by DSC (amorphous thermoplastics), **antioxidant content** in polyolefins, halogen concentration in plastics and rubber, heavy metals in polymers (PVC and PUR), chemical resistance of elastomers (according to ISO 1817), **global migration in food packaging**, plasticiser content (comparative examination: TGA and extraction) and the oxidation-induction time and temperature (OIT/OIT*) of polyolefins [106]. The results of the round-robin were evaluated by means of robust statistics [107], in accordance with ISO 5725-5 (1994). Some of the results were published [59,107a, 107b] or are summarised in Chp. 6.

In a similar exercise [108] the inhomogeneity of **carbon-black** filled LDPE was quantified. Also two methods were compared for the determination of **ash content** in thermoplastic materials and cross-linked elastomers, namely: *(i)* the conventional determination of ash under air according to usual standards, *i.e.* ISO 247 (1990) (sample size: grams); and *(ii)* the thermogravimetric method similar to ISO 9924-1 (1993), optimised for the determination of ash (sample size: 10–20 mg). It was concluded that TGA is as efficient and precise as conventional standardised methods for materials with high filler contents. For materials with low contents (about 3%) the conventional determination of ash is superior (factor 5–10) to the TGA method with regard to the uncertainty of measurement. In another interlaboratory test two thermoplastics, PA12-P plasticised with a sulfonamide and PVC-P plasticised with phthalic acid esters, were examined by Soxhlet extraction and TGA [108]. It was shown that the **plasticiser content** could consistently be determined with TGA by using suitable parameters of measurement. However, it should not erroneously be concluded that TGA and Soxhlet extraction are equivalent. An essential requirement for a successful determination of plasticisers by TGA is that plasticiser evaporation is not massively interfered by degradation of other components of the material. It is possible to quantify monomeric plasticisers in PVC-P using vacuum TGA. On the other hand, polymeric plasticisers (MW > 500 to 10,000 Da) cannot be determined because the weight loss normally occurs in the region of polymer degradation [109]. Also **DSC-OIT interlaboratory tests** of HDPE and LDPE, carried out in accordance with EN 728 (1997), have been reported [108]. Recently, an interlaboratory evaluation of off-line SFE-GC-AED for the determination of organotin compounds (in soil and sediments) was reported [110]. Regrettably, interlaboratory comparisons are still lacking in many areas, *e.g.* in the application of simultaneous TG-MS in polymer analysis [111]. This is not surprising in view of the high costs involved.

8.4.3. Validation of Antioxidant Migration Testing

Case Study

In a typical experimental set-up aiming at migration testing of a PE film tests are carried out in separate cells each containing a specified amount of film. Four sets of test solutions (*e.g.* 10% ethanol) in triplicate are then analysed at various time intervals (2, 24, 96 and 240 hrs). After evaporation to dryness, the residue is dissolved in an appropriate solvent and GC analysed.

Validation experiments are normally carried out with the set of test solutions exhibiting the highest level of additive migration, typically those contacting the food simulant for the longest period (*i.c.* 240 hrs). To validate the analytical methodology, an additional three sets (in triplicate) should be run for 240 hrs. Each set of these test solutions can then be spiked with the additive at levels of 50%, 100% and 200% of the average migration value determined for the regular (unspiked) 240 hrs test solutions. Alternatively, it is also possible to carry out one large test using enough film and solvent for 12 analyses. After 240 hrs, the test solution is divided into 12 equal solutions (essentially four sets of triplicate samples). In one set (three solutions) the antioxidant content is determined. The remaining nine solutions (three sets) are spiked at concentrations corresponding to 50%, 100% and 200% of the determined additive level. Each solution is analysed as described before. Recovery calculations should be carried out. The average recovery for the various spiking levels should be within specified limits. The actual validation procedure used will, of course, depend on the particular type of analysis.

CRMs with certified $C_{p,o}$ (initial concentration of migrant in a plastic) and SM (specific migration) are in preparation [60].

8.5. TOTAL VALIDATION PROCESS

Validation is a constant, evolving process that starts before an instrument is placed on-line and continues long after method development and transfer. Validation is not a single process but a series of stages, each dependent on the integrity of the previous stage. It is broader than just instrumental standardisation, as it embraces all the regulatory aspects of documentation and control. Distinct stages in the production of valid information comprise: *(i)* a fundamental stage, dealing with the integrity of the data and integrity of the sample; *(ii)* system control (operability and GLP); *(iii)* data transformation; and *(iv)* interpretation. One of the keys to success is to ensure that the parameter space is wide enough and that the experimental design is geared to providing data embracing this parameter space. Information cannot be extracted from data which does not exist.

A well-defined and documented validation process provides regulatory agencies with evidence that the system and method is suitable for its intended use and under control. By approaching method development, optimisation, and validation in a logical, stepwise fashion, laboratory resources can be used in a more efficient and productive manner. Benefits of validated procedures are cost saving, both long term and short term (through use of vendor documentation, vendor validated systems and built-in validated system software), improved quality and reliability of data analysis, as well as an increased likelihood for successfully passing audits. The total validation process encompasses many different aspects: *(i)* software validation; *(ii)* hardware (instrumentation) validation/qualification; *(iii)* method validation; and *(iv)* system suitability.

Starting with validated software and instrument qualification a validated analytical method is developed using the qualified system. Finally, total validation is achieved by defining system suitability. The analytical chemist is mostly concerned with steps *(iii)* and *(iv)*, but he might be (rightly) suspicious regarding developments beyond his field of vision.

Software validation has been described [112] and a general proposal for instrument tests has been published [113].

8.5.1. Software/Hardware Validation/Qualification

Principles and Characteristics

In R&D it is not sufficient to adapt existing work without demonstrating that the instrumentation works properly with the new application. Care should also be exercised with novel instrumentation, where the claims of the manufacturer cannot always be made true in specific cases.

In compliance with the EURACHEM report *Guidance on Best Practice for the Equipment Qualification of Analytical Instruments* [114,115], in general terms four areas need to be addressed as to assurance of validity, namely:

(i) fitness for purpose of an instrument for the task;
(ii) compliance with the manufacturer's performance criteria;
(iii) compliance with established standards and practices; and
(iv) documented evidence for continued operability and data integrity.

Equipment Qualification (EQ) is a formal process that provides documented evidence that an instrument is fit for its intended purpose and is kept in a state of maintenance and calibration consistent with its use (Table 8.19). EQ is becoming increasingly important to demonstrate integrity of data and validity of results and is generally implemented in accordance with one of the internationally recognised quality standards: ISO 9000, Good Laboratory Practice [116] or ISO/IEC Guide 25 (ISO 17025). Design or Development Qualification (DQ) at the vendor's site covers all procedures prior to the installation of the system in a laboratory and is about what

Table 8.19. Instrument qualification terms

- EQ – Equipment Qualification
  - The overall process of equipment qualification
- DQ – Design Qualification
  - Defines functional and operational specification, selection of supplier
- IQ – Installation Qualification
  - Covers procedures relating to the installation of the instrument and its environment
- OQ – Operational Qualification
  - Determines that a laboratory instrument operates according to established specifications (before use)
- PQ – Performance Qualification
  - Demonstrates that an instrument consistently performs to specification appropriate to routine use

the instrument is required to do, and links directly to fitness for purpose. Installation Qualification (IQ) establishes that the instrument is properly installed and guarantees that the instrument works the way the manufacturer claims. The purpose of Operational Qualification (OQ) is to ensure that the instrument performs in compliance with international, national or corporate standards. Whereas DQ, IQ and OQ are designed to ensure fitness for purpose for the designated task, Performance Qualification (PQ) is intended to confirm that the instrument or analytical system continues to perform within the limits originally set (ongoing compliance) and to provide demonstrable assurance of validity of the data generated. Some accreditation schemes (*e.g.* GLP) require the performance of an instrument not only to be verified after installation but also every time it is modified, *e.g.* after repair or upgrade. Table 8.20 shows in more detail what items comprise the qualification protocols. For further guidance on equipment qualification, *cfr.* ref. [117].

Instrument qualification is an important element of laboratory validation. Suppliers's (retrospective) validation plans help with the equipment qualification process. Nowadays the regulatory compliance needs of industry on a global basis are well understood by the instrument vendors. For example, Duncan *et al.* [118] have illustrated the validation chain for benchtop LC-MS systems and Maxwell *et al.* [119] have applied the validation timeline to HPLC system validation. Both FDA and USP require that the proper operation of an HPLC system must be validated through a formal calibration program. The components of an HPLC that require calibration include: pumps, pump mixing elements, auto-injector, detector, and column heater.

The US Department of Health and Human Sciences has issued a draft guidance document (docket # 00D-1539) on the archival and maintenance of electronic records of analytical data, such as spectra and chromatograms.

Table 8.20. Items comprising the equipment qualification process

- **DQ comprises:**
  - Laboratory requirements
  - Equipment definition
  - Operational requirements
  - Purchasing policy
  - Risk analysis
  - Demonstration reports
  - Cost/benefit analysis
- **IQ comprises:**
  - Description of the instrument functionality
  - Specific instrument ID
  - Software/firmware revision
  - Instrument specifications
  - Site requirements (gases, electrical, environment, *etc.*)
  - Installation verification checklists
  - Verification of service engineer training and comprehensive qualification
  - Hazard and safety precautions
  - List of consumables
- **OQ comprises:**
  - Standard operating procedure (SOP)
  - Verification of operator training
  - Documentation listings (manuals/logs)
  - Certificate of conformity from suppliers factory
  - Functional field test/certification procedure
  - Routine maintenance procedures
- **PQ comprises:**
  - Performance monitoring that a specific process (customer methodology + samples/standards + operator) meets established specifications (ISO norms, GLP) on a consistent basis
  - Ongoing instrument performance verification
  - Regular peer review

Implementation

Validation of a **chromatographic system** is required by numerous quality assurance systems. For this purpose hardware, firmware, software and the analytical method used for analysis should be validated. Moreover, the chromatographic system needs to be tested against documented performance specifications for a given analytical method (system suitability test). Besides the prerequisites of a chromatographic separation, such as tailing factor, column plate number, range of retention factors, resolution, several analytical performance parameters, are essential. For example, the method robustness is determined with a test for the variation of parameters: for a predefined change in temperature, gradient slope or shape, pH, *etc.*, the consistency of the quantitative results is regarded. As to method ruggedness, the results of different laboratories, analysts, instruments, reagents, *etc.*, are compared by calculating the relative standard deviation of replicate measurements. Felinger [120] has reported validation of chromatographic instruments. Validation of HPLC equipment was recently discussed [121]. The implications of 21 CFR Part 11 Guidance Document (docket # 00D-1539) on chromatography data systems have been described [122].

Burgess [123] has described approaches to the validation of **spectrometers**; ASTM standards relating to spectrometry and spectrometer performance have recently been listed by the same author [124]. Where the spectrophotometer is used for regular transmittance or for absorbance measurements for quantitative purposes, the validity of the ordinate scale is of obvious relevance to the quality management system of a laboratory. FTIR spectrophotometers, which have completely displaced grating instruments for the mid-IR and far-IR spectral regions, are subject to many more possible types of systematic ordinate error than are grating instruments. Birch *et al.* [125] have discussed the sources of error in Fourier transform (FT) spectroscopy giving a structured list of 50 categories of ordinate (*i.e.* transmittance) error. Where uncertainties in transmittance and regular reflectance measurements on a grating instrument are only a few tenths of a percent, within FT spectrometers these are often over a percent, without even considering the additional errors in the reflectometer accessory. For these reasons the National Physical Laboratory (NPL) continues to use grating IR spectrophotometers for determining and supplying IR standards for the ordinate scales of various properties. This UK national measurement standards laboratory supplies an extensive range of infrared standards, such as regular reflectance, hemispherical reflectance and wavenumber calibration standards [126,127]. Also reference materials for UV, VIS and NIR spectrophotometry are available (both liquid standards and holmium glass for wavelength calibration).

In **mass spectrometry**, at the very least, daily check-ups should be made on the cleanliness of the ionisation source by devising a quickly executed sensitivity test that can be as simple as analysing a known sample and checking the absolute intensity of the ions in the mass spectrum.

8.5.2. System Suitability

The procedure known as system suitability test consists in testing an instrumental analytical system against documented performance specifications for a given analytical method. System suitability tests are based on the concept that the equipment, electronics, analytical operations and samples constitute an integral system that can be evaluated as a whole. These tests are used to make sure that the resolution and reproducibility of the system are adequate for the analysis to be performed. Documentation of system suitability can be accomplished by specifically designed software. System suitability also comprises method protection (protecting data integrity, security and traceability).

Validation requires analytical method instructions comprising a system suitability test in order to verify identical starting conditions. Part or full revalidation may be considered if system suitability tests, or the results of quality control sample analysis, are out of pre-set acceptance criteria and the source of the error cannot be tracked back to instrumental factors or anything else.

8.6. RATIONAL STEP-BY-STEP METHOD DEVELOPMENT AND VALIDATION FOR POLYMER/ADDITIVE ANALYSIS

As yet, there are no generally accepted formats for the overall method development of in-polymer additive analysis. However, one may take a lead from the work of Swartz *et al.* [6], and various other sources [20,80,87,128], who have presented a rationale for the process of successful development of (HPLC-based) analytical methods, their optimisation, and eventually validation. A sequence of steps is necessary in the development of a fully validated method for the analysis of additives in polymeric matrices, in which the user has specified validation parameters and limits, as follows:

1. *Analyte standard characterisation.*
   Aims at collecting relevant chemical and physical information about the analyte; determines the availability of standards (including degradation products) and evaluates only methods which are compatible with the sample stability.
2. *Method requirements.*
   Defines application, purpose and scope of the method as well as the analytical figures of merit (performance parameters and acceptance criteria) and practical boundary conditions (sample throughput, analysis time, equipment limitations, qualification of materials, *etc.*).
3. *Prior art.*
   Considers relevant analytical methods in the open literature and proprietary data related to analyte and matrix.
4. *Choice of an analytical method.*
   Considers adaptation, modification or extension (by analogy) of existing methods *vs.* new developments taking advantage of state-of-the-art methods and instrumentation.

5. *Preliminary experimental studies.*
Sets up the required instrumentation, prepares analyte standards, and evaluates the feasibility of the method in terms of the analytical figures of merit obtained.
6. *Optimisation.*
Uses experimental design procedures wherever possible in case of qualification taking advantage of computer-based optimisation software; definition of validation protocol and experiments.
7. *Performance of standard reference samples.*
Obtains final analytical figures of merit with standards meeting the expectations.
8. *Methods development with actual samples.*
Secures unequivocal detectability of the analyte peak, without all other potential interferences. Actual sample preparation should be compatible with the instrumental set-up. Adjustment of method parameters and/or acceptance criteria, if necessary.
9. *Validation of figures of merit.*
Evaluates precision, accuracy, linearity range, LOD, LOQ, specificity, ruggedness and robustness in pre-validation experiments.
10. *Quantitative sample analysis.*
Possible methods, which include standard additions, external/internal standard and isotopic dilution, take into account percent recovery of a spiked, authentic standard analyte into a sample matrix not containing the analyte; sample to sample reproducibility of recovery (average and standard deviation) should be determined.
11. *Method validation.*
Performs zero- and double-blind studies. Intralaboratory reproducibility (including ruggedness and robustness for real samples) should be demonstrated; additional validation using an authentic standard reference material of the analyte in the sample matrix. Definition of criteria for revalidation.
12. *Method manual.*
Prepares written protocols indicating sufficient experimental detail (equipment, suppliers, reagents, sample preparation, experimental parameters, software, spectral libraries, statistical treatment, *etc.*) as documented evidence and to facilitate method transfer. Huber [80] has detailed the contents of a validation report.
A laboratory applying a specific method should have documented evidence that the method has been appropriately validated. According to EURACHEM [71] *"The responsibility remains firmly with the user to ensure that the validation documented in the method is sufficiently complete to meet his or her needs"*. This holds for standard methods (*e.g.* from EPA, ASTM, ISO or USP) as well as for methods developed in-house. If standard methods are used, it should be verified that the scope of the method and validation comply with the laboratory's analyses requirements; otherwise, revalidation is needed. The laboratory should demonstrate the validity of the method in its own environment.
13. *Transfer of analytical method methodology.*
Continuation of method validation by (costly and lengthy) interlaboratory collaborative studies (ruggedness); statistical comparison of the validation results (*e.g.* for HPLC methods *cfr.* ref. [70]).
14. *Standard Operating Procedure.*
A summary report describes a statistical treatment of the qualitative and quantitative results. Accreditation of the method as a company standard operating procedure (SOP). Each laboratory may be expected to have SOPs in place.
15. *Routine execution.*
Based on SOPs, system-suitability tests and/or analytical quality control.
16. *Peer review.*
Preparation and acceptance of a paper describing the optimised final method and validation procedure.

The minimum requirements for validation of an experimental R&D procedure for quantification of additives in polymers may be derived by considering the three main stages of the overall process, namely:

(i) Characterisation of the calibration standard.
(ii) Isolation of the additive from the polymeric matrix (*e.g.* extraction, dissolution, destruction).
(iii) Separation and detection methods (identification, calibration, quantitation).

Characterisation of a calibration standard requires information about the concentration of the analyte (preferably to be determined by an independent absolute method), stability of the pure compound and of its solutions, mode of storage (excicator, refrigerator). It is recommended to verify the variation in time of the concentration of the analyte. The yield of methods for isolation of additives from the polymer matrix needs to be verified by an independent absolute method, analysis of a sample with known content (if available) or a recovery test with a blank polymer and a calibration standard. Actually, recovery tests are analytically suspect as "spiked samples" are more easily extractable than real samples, which have been subjected to high temperature conditions during compounding, ageing or additive-polymer interaction. Rapid extraction tests (EN 1186-15) [129] using organic solvents also need to be validated for specific migration purposes. Whatever the analysis method, it is always necessary to verify that the measured signal is fully on account of the analyte of interest (specificity). Chromatographic methods need to be calibrated (minimum/maximum concentration); options consist in external and internal methods. Repeatability and reproducibility need to be assessed (concentration, relative retention times, response factors, variation coefficients, *etc.*).

BIBLIOGRAPHY

Method Development and Validation

Chung C. Chan, H. Lam, Y.C. Lee and Xue-Ming Zhang (eds.), *Analytical Method Validation and Instrument Performance Verifications*, J. Wiley & Sons, Chichester (2004).

M.E. Swartz and I.S. Krull, *Analytical Method Development and Validation*, M. Dekker, New York, NY (2003).

A. Fajgelj and A. Ambrus, *Principles and Practices of Method Validation*, The Royal Society of Chemistry, Cambridge (2000).

C. Burgess, *Valid Analytical Methods and Procedures*, The Royal Society of Chemistry, Cambridge (2000).

L. Huber, *Good Laboratory Practice and Current Good Manufacturing Practice. A Primer*, Agilent Technologies Publ. Nr. 5968-6193E, Waldbronn (2000).

L. Huber, *Validation and Qualification in Analytical Laboratories*, Interpharm Press, Buffalo Grove, IL (1999).

W.A. Hardcastle, *Qualitative Analysis. A Guide to Best Practice*, The Royal Society of Chemistry, Cambridge (1998).

M.M.W.B. Hendriks, A.H. de Boer and A.K. Smilde (eds.), *Robustness of Analytical Methods and Pharmaceutical Technological Products*, Elsevier Science, Amsterdam (1996).

C.M. Riley and T.W. Rosanske, *Development and Validation of Analytical Methods*, Pergamon, Oxford (1996).

L. Huber, *Validation of Computerized Analytical Systems*, Interpharm Press, Buffalo Grove, IL (1995).

M.E. Double and M. McKendry, *Computer Validation Compliance: A Quality Assurance Perspective*, Interpharm Press, Buffalo Grove, IL (1994).

International Vocabulary of Basic and General Terms in Metrology (VIM), ISO, Geneva (1993).

Reference Materials

V. Barwick, S. Burke, R. Lawn, P. Roper and R. Walker, *Applications of Reference Materials in Analytical Chemistry*, The Royal Society of Chemistry, Cambridge (2001).

M. Stoeppler, W. Wolf and P.J. Jenks, *Reference Materials for Chemical Analysis*, Wiley-VCH, Weinheim (2001).

Ph. Quevauviller and E.A. Maier, *Interlaboratory Studies and Certified Reference Materials for Environmental Analysis : The BCR Approach*, Elsevier Science, Amsterdam (1999).

REFERENCES

[1] J.K. Taylor, *Anal. Chem.* **65**, 600–608A (1983).

[2] Renault Automobiles, *Methode d'Essai D40/1586. Caoutchoucs vulcanisés, Taux d'extrait acetonique ou methylique*, Guyancourt (1994).

[3] Volkswagen AG, *Prüfvorschrift PV 3935, PyGC-MS für Kunststoffe und Elastomere*, Wolfsburg (1997).

[4] M. Marcato and M. Vianello, *J. Chromatogr.* **A869**, 285–300 (2000).

[5] P. Métois, D. Scholler, J. Bouquant and A. Feigenbaum, *Food Addit. Contam.* **15** (1), 100–1 (1998).

[6] M.E. Swartz and I.S. Krull, *Analytical Method Development and Validation*, M. Dekker, New York, NY (2003).

[6a] D.L. Massart, J. Smeyers-Verbeke and Y. Vander Heyden, *LC.GC Europe* **18** (7), 390–3 (2005).

[7] H.-R. Schulten and H. Wilcken, in *Polymeric Materials Encyclopedia* (J.C. Salomone, ed.), CRC Press, Boca Raton, FL (1996), pp. 7474–85.

[8] J. Salafranca, J. Cacho and C. Nerín, *J. High Resolut. Chromatogr.* **22** (10), 553–8 (1999).

[9] J.C.J. Bart and M. Kuch, *LC.GC Europe* **13** (11), 818–28 (2000).

[10] W. Blum and R. Aichholz, in *Encyclopedia of Analytical Science* (A. Townshend, ed.), Academic Press, London (1995), pp. 1833–44.

[11] R. Trones, *Ph.D. Thesis*, University of Oslo (1999).

[12] R. Trones, T. Andersen, T. Greibrokk and D.R. Hegna, *J. Chromatogr.* **A874** (1), 65–71 (2000).

[12a] J.C.J. Bart, *Additives in Polymers. Industrial Analysis and Applications*, J. Wiley & Sons, Chichester (2005).

[13] K.W. Van Every and M.J. Elder, *Proceedings SPE ANTEC '96*, Indianapolis, IN (1996), pp. 2400–5.

[14] F. De Boever, *Micromass 2001 New Products Seminar & Users Event*, Amsterdam (2001).

[15] C. Berger, in *Supercritical Fluid Chromatography with Packed Columns* (K. Anton and C. Berger, eds.), M. Dekker, New York, NY (1997), pp. 301–48.

[16] C. Berger, *Chromatogr. Sci. Ser.* **75**, 301–48 (1998).

[17] K. Anton, M. Bach, C. Berger, F. Walch, G. Jaccard and Y. Carlier, *J. Chromatogr. Sci.* **32** (10), 430–8 (1994).

[18] H.J. Cortes, G.E. Bormett and J.D. Graham, *J. Microcol. Sep.* **4** (1), 51–7 (1992).

[19] H.J. Cortes, B.M. Bell, C.D. Pfeiffer and J.D. Graham, *J. Microcol. Sep.* **1**, 278–88 (1989).

[20] P.J. Schoenmakers and M. Mulholland, *Chromatographia* **25** (8), 737 (1988).

[21] L.R. Snyder, in *Practical HPLC Method Development* (L.R. Snyder, J. Glajch and J.J. Kirkland, eds.), Wiley-Interscience, New York, NY (1988).

[21a] J.V. Hinshaw, *LC.GC Europe* **18** (5), 276–80 (2005).

[22] M. Ashraf-Khorassani, S. Hinman and L.T. Taylor, *J. High Resolut. Chromatogr.* **22** (5), 271–5 (1999).

[23] E. Verdurmen, R. Verstappen, J. Swagten, H. Nelissen, G. Heemels and J.C.J. Bart, *Abstracts 4th Intl. Symposium on Hyphenated Techniques in Chromatography and Hyphenated Chromatographic Analyzers (HTC-4)*, Bruges (1996), Paper E38.

[24] D.R. Gere and E.M. Derrico, *LC.GC Intl.* **7** (7), 370–5 (1994).

[25] L. van Vaeck, J. Bennett, W. Lauwers, A. Vertes and R. Gijbels, *Mikrochim. Acta* **III**, 283–303 (1990).

[26] *International Vocabulary of Basic and General Terms in Metrology (VIM)*, ISO, Geneva (1993).

[27] P. De Bièvre, R. Kaarls, H.S. Peiser, S.D. Rasberry and W.P. Reed, *Accred. Qual. Assur.* **1**, 3–13 (1996).

[28] M. Valcárcel and A. Ríos, *Analyst* **120**, 2291–7 (1995).

[29] P.J. Jenks, *Spectrosc. Europe* **14** (4), 27–8 (2002).

[30] ISO Guide 35, *Certification of Reference Materials – General and Statistical Principles*, ISO, Geneva (1989), Chp. 7.

[31] J.K. Taylor, *J. Test. Eval.* **11** (6), 385–7 (1983).

[31a] P.J. Jenks, *Spectrosc. Europe* **15** (4), 27 (2003).

[32] M. Stoeppler, W.R. Wolff and P.J. Jenks, *Reference Materials for Chemical Analysis*, Wiley-VCH, Weinheim (2001).

[33] P. Quevauviller, *Trends Anal. Chem.* **18**, 76–85 (1999).

[34] P.J. Jenks, *Spectrosc. Europe* **14** (3), 29–30 (2002).

[35] L. Cudros-Rodríguez, L. Gámiz-Gracia, E.M. Almansa-López and J.M. Bosque-Sendra, *Trends Anal. Chem.* **20** (11), 620–36 (2001).

[36] K.N. Marsh (ed.), *Recommended Reference Materials for the Realisation of Physico Chemical Properties*, Blackwell Scientific Publications, Oxford (1987).

[37] COMAR Central Secretariat: Harry.Klich@BAM.de.

[38] B. King, *A Basic Guide for Laboratories and Accreditation Bodies*, *http://www.nata.asn.au/news/92.html*.

[39] A. Zschunke, *Fresenius J. Anal. Chem.* **369**, 6–7 (2001).

[39a] F. Pillong, *Analytix* (2), 10–1 (2005).

[40] Spex CertiPrep, *Total Certified Reference Materials for Inorganic and Organic Spectroscopy*, Metuchen, NJ (1999).

[41] J. McKinley and R.E. Majors, *LC.GC Europe* **13** (12), 892–901 (2000).

[42] H. Klich and R. Walker, *Fresenius' J. Anal. Chem.* **345**, 104–6 (1993).

[43] J. Pauwels, *Spec. Publ. – Roy Soc. Chem.* **238** *(Use of Matrix Reference Materials in Environmental Analytical Processes)*, 31–45 (1999).

[44] ISO 7870, *Control Charts – General Guide and Introduction*, ISO, Geneva (1993).

[44a] P.J. Jenks, *Spectrosc. Europe* **16** (3), 28 (2004); *VAM Suppl. to Spectrosc. Europe*. **17** (1) (2005).

[45] S.D. Rasberry, *Accred. Qual. Assur.* **6**, 95–9 (2001).

[46] RM Report, *www.rmreport.com*, Chichester.

[47] *Off. J. Eur. Comm.* **L186** (12.07.1991).

[48] Statement of Verband der Automobilindustrie (VDA): *Stellungnahme zum Schwedischen Cadmiumverbot* (Dec. 9. 1983).

[49] W.J. Rühl, *Fresenius Z. Anal. Chem.* **322**, 710–2 (1985).

[50] J. Pauwels, A. Lamberty, P. De Bièvre, K.-H. Grobecker and C. Bauspiess, *Fresenius J. Anal. Chem.* **349**, 409–11 (1994).

[51] A. Lamberty, P. De Bièvre and A. Götz, *Fresenius J. Anal. Chem.* **345**, 310–3 (1993).

[52] A. Götz, A. Lamberty and P. De Bièvre, *Int. J. Mass Spectrom. Ion Process.* **123**, 1–6 (1993).

[53] J. Pauwels, C. Hofmann and K.-H. Grobecker, *Fresenius J. Anal. Chem.* **345**, 475–7 (1993).

[54] U. Simmross, R. Fischer, F. Düwel and U. Müller, *Fresenius J. Anal. Chem.* **358** (4), 541–5 (1997).

[55] A. Lamberty, W. van Borm and P. Quevauviller, *Fresenius J. Anal. Chem.* **370**, 811–8 (2001).

[56] J. Pauwels, A. Lamberty and H. Schimmel, *Accred. Qual. Assur.* **3**, 51–5 (1998).

[56a] Contact PANalytical (Almelo) or DSM Resolve (Geleen).

[57] K.D. Altria, N.G. Clayton, M. Hart, R.C. Harden, J. Hezivi, J.V. Makwana and M.J. Portsmouth, *Chromatographia* **39**, 180 (1994).

[58] C.M. Riley and T.W. Rosanske (eds.), *Development and Validation of Analytical Methods*, Elsevier (Pergamon), Tarrytown, NY (1996).

[59] J.C.J. Bart, S. Affolter and M. Schmid, *Anal. Sci.* **17** (Suppl.), i729–32 (2001).

[59a] A. Stoermer, *SM&T Project G6RD-CT-2000-00411* (stoermer@ivv.fhg.de).

[60] N.H. Stoffers, *Certified Reference Materials for Food Packaging Specific Migration Tests: Development, Validation and Modelling, Ph.D. Thesis*, Wageningen University (2005).

[60a] S.J. Nagourney and R.K. Madan, *J. Test. Eval.* **19** (1), 77–82 (1991).

[60b] Fluka (Sigma-Aldrich), *AnalytixNotes* (n.d.).

[61] M. Hemmerlin, J.-M. Mermet, M. Bertucci and P. Zydowicz, *Spectrochim. Acta* **B52**, 421–30 (1997).

[62] ISO 8402, *Quality-Vocabulary*, ISO, Geneva (1994).

[63] EURACHEM Guide, *The Fitness for Purpose of Analytical Methods*, EURACHEM, Teddington (1998).

[64] B.G.M. Vandeginste, D.L. Massart, L.M.C. Buydens, S. de Jong, P.J. Lewi and J. Smeyers-Verbeke, *Handbook of Chemometrics and Qualimetrics*, Elsevier, Amsterdam (1998).

[65] W. Horwitz, *Pure Appl. Chem.* **60**, 855–67 (1988).

[66] W.D. Pocklington, *Pure Appl. Chem.* **62**, 149–62 (1990).

[67] P. Johansson, B. Wilkenstedt and R.D. McDowall, *LC.GC Intl.* **12** (2), 88–102 (1999).

[68] General Chapter ⟨1225⟩, *Validation of Compendial Methods, US Pharmacopeia XXIII, National Formulary, XVIII*, The United States Pharmacopeia Convention Inc., Rockville, MD (1995), pp. 1610–2.

[69] E. McGonigle, in *Development and Validation of Analytical Methods* (C.M. Riley and T.W. Rosanske, eds.), Pergamon, Oxford (1996), pp. 3–13.

[70] J.J. Kirschbaum, *J. Pharm. Biomed. Anal.* **7** (7), 813 (1989).

[71] EURACHEM, *Guidance Document No. WGD 2: Accreditation for Chemical Laboratories: Guidance on the Interpretation of the EN 45000 Series of Standards and ISO/IEC Guide 25*, EURACHEM, Teddington (1993).

[72] US FDA, *Technical Review Guide: Validation of Chromatographic Methods*, CDER, Rockville, MD (1993).

[73] US FDA, *General Principles of Validation*, CDER, Rockville, MD (1987).

[74] US FDA, *Guidelines for Submitting Samples and Analytical Data for Method Validation*, US Government Printing Office, Washington DC (1987).

[75] W.L. Paul, *Pharm. Technol.*, 130–41 (March 1991).

[76] International Conference on Harmonization, *Validation of Analytical Procedures*, ICH, Geneva (1996).

[77] AOAC Peer Verified Methods Program, *Manual on Policies and Procedures*, AOAC, Arlington, VA (Nov. 1993).

[78] J.M. Green, *Anal. Chem. News & Features*, 305–9A (May 1996).

[78a] M. Swartz, *Supplement to LC.GC Europe* **17** (6) (2004).

[79] W. Wegscheider, in *Accreditation and Quality Assurance in Analytical Chemistry* (H. Günzler, ed.), Springer-Verlag, Berlin (1996).

[80] L. Huber, *LC.GC Intl.* **11** (2), 96–105 (1998).

[81] W. Riepe, in *Proceedings GSF Symposium on Analytical Chemical Quality Assurance* (A. Kettrup and E. Flammenkamp, eds.), Ecomed, Landsberg (1995), pp. 107–18.

[82] C.M. Riley, in *Development and Validation of Analytical Methods* (C.M. Riley and T.W. Rosanske, eds.), Pergamon, Oxford (1996), pp. 15–71.

[83] L. Huber, *Validation and Qualification in Analytical Laboratories*, Interpharm Press, Inc., Buffalo Grove, IL (1999).

[84] J.G. Liscouski, in *Development and Validation of Analytical Methods* (C.M. Riley and T.W. Rosanske, eds.), Pergamon, Oxford (1996), pp. 303–36.

[85] L. Huber, *http://www.labcompliance.com/index.htm*.

[86] V. Lopez-Avila and W.F. Beckert, *Proceedings 5th Intl. Symposium on Supercritical Fluid Chromatography and Extraction*, Baltimore, MD (1994), pp. 8–9.

[87] W.B. Furman, T.P. Layloff and R.F. Tetzlaff, *J. AOAC Int.* **77** (5), 1314 (1994).

[88] L. Jaufmann, *Analusis Mag.* **26** (2), M22–4 (1998).

[89] D.E. Larkin, in *Compositional Analysis by Thermogravimetry* (C.M. Earnest, ed.), ASTM, Philadelphia, PA (1988), pp. 28–37.

[90] D. Spyropoulos, *Chem. Chron., Genike Ekdose* **60** (10–11), 292–9 (1998).

[91] W. Penninckx, C. Hartmann, D.L. Massart and J. Smeyers-Verbeke, *J. Anal. Atomic Spectrom.* **11**, 237–46 (1996).

[92] P.J. Gemperline and N.K. Shah, *Anal. Chem.* **62**, 465 (1990).

[93] M. Collins and P. Hope, *Lab. Rob. Autom.* **10** (2), 77–80 (1998).

[94] ISO 5725-4, *Accuracy (Trueness and Precision) of Measurement Methods and Results – Part 4. Basic Methods for the Determination of the Trueness of a Standard Measurement Method*, ISO, Geneva (1994).

[95] J. Kirschbaum, S. Perlman, J. Joseph and J. Adamovics, *J. Chromatogr. Sci.* **22**, 27 (1984).

[96] I.S. Krull, *Intl. Lab.* **31** (1), 4–5 (2001).

[97] J. Vessman, *J. Pharm. Biomed. Anal.* **14**, 867–9 (1996).

[98] B.-A. Persson, J. Vessman and R.D. McDowall, *LC.GC Intl.* **11**, 160–4 (1998).

[99] H. Kaiser, *Fresenius Z. Anal. Chem.* **260**, 252 (272).

[100] S.A. George, in *Diode-Array Detection in HPLC* (L. Huber and S.A. George, eds.), M. Dekker, New York, NY (1993), pp. 41–9.

[100a] K. Danzer, *Fresenius J. Anal. Chem.* **369**, 399–402 (2001).

[101] United States Pharmacopeial Convention, *US Pharmacopeia*, Rockville, MD (1995), pp. 1982–4.

[102] J.L. Virlichie and A. Ayache, *S.T.P. Pharma Pratiques* **5** (1), 49–60 (1995).

[103] M.M.W.B. Hendriks, A.H. de Boer and A.K. Smilde (eds.), *Robustness of Analytical Methods and Pharmaceutical Technological Products*, Elsevier Science, Amsterdam (1996).

[104] A. Kunkel, S. Günter, C. Dette and H. Wätzig, *J. Chromatogr.* **A781**, 445–55 (1997).

[105] AOAC International, *Guidelines for Collaborative Study – Procedures to Validate Characteristics of a Method of Analysis*, AOAC, Arlington, VA (1994).

[106] S. Affolter, A. Ritter and M. Schmid, *Ringversuche an polymeren Workstoffen*, EMPA, St. Gallen (2000).

[107] P. Lischer, in *Schweizerisches Lebensmittelbuch: Methoden f.d. Untersuchung u. Beurteilung von Lebensmitteln u. Gebrauchsgegenständen* (Schweizer. Lebensmittelbuch-kommission und Eidg. Gesundheidsamt, eds.), Eidg. Drucksachen- u. Materialzentrale, Bern (1964), Chp. 60b, Appendix 3.

[107a] A. Ritter, E. Michel, M. Schmid and S. Affolter, *Polymer Testing* **23**, 467–74 (2004).

[107b] M. Schmid and S. Affolter, *Polymer Testing* **22**, 419–28 (2003).

[108] S. Affolter and M. Schmid, *Int. J. Polym. Anal. Charact.* **6**, 35–57 (2000).

[109] S. Affolter and S. Hofstetter, *Kautsch. Gummi Kunstst.* **48**, 173 (1995).

[110] V. Lopez-Avila, Y. Liu and W.F. Beckert, *J. Chromatogr.* **A785**, 279–88 (1997).

[111] K.G.H. Raemaekers and J.C.J. Bart, *Thermochim. Acta* **295**, 1–58 (1997).

[112] T.A. Rooney, *Today's Chemist at Work*, 15–8 (June 1998).

[113] B. Lendi, in *Practical High Performance Liquid Chromatography* (V.R. Meyer, ed.), J. Wiley & Sons, Chichester (1999), pp. 313–8.

[114] EURACHEM, Instrumentation Working Group, *Guidance on Best Practice for the Equipment Qualification of Analytical Instruments*, EURACHEM, Teddington (1996).

[115] H. Günzler, *Accreditation and Quality Assurance in Analytical Chemistry*, Springer-Verlag, Berlin (1994).

[116] *Good Laboratory Practice – The United Kingdom Compliance Programme*, UK Dept. of Health, London (1989).

[117] P. Bedson and M. Sargent, *Accred. Qual. Assur.* **1**, 265–74 (1996).

[118] W. Duncan and D. McIntyre, *Proceedings 14th Intl. Mass Spectrometry Conference*, Tampere (1997).

[119] W. Maxwell and J. Sweeney, *LC.GC Magazine of Separation Science* **12** (9), 678–82 (1994).

[120] A. Felinger, *Data Analysis and Signal Processing in Chromatography*, Elsevier, Amsterdam (1998).

[121] S. Küppers, B. Renger and V.R. Meyer, *LC.GC Europe* **13** (5), 300 (2000).

[122] T. Meek, *LC.GC Europe* **16** (1), 36–7 (2003).

[123] C. Burgess, in *Spectrophotometry, Luminescence and Colour; Science and Compliance* (C. Burgess and D.G. Jones, eds.), Elsevier, Amsterdam (1995), pp. 21–34.

[124] C. Burgess, in *Encyclopedia of Spectroscopy and Spectrometry* (J.C. Lindon, G.E. Tranter and J.L. Holmes, eds.), Academic Press, San Diego, CA (2000), pp. 166–71.

[125] R. Birch and F.J.J. Clarke, *Spectrosc. Europe* **7** (4), 16–22 (1995).

[126] F.J.J. Clarke, *Proc. SPIE* **2775**, 6–17 (1996).

[127] Available from Division of Electrical Science, NPL. Contact: Dr. F.J.J. Clarke.

[128] T.D. Wilson, *J. Pharm. Biomed. Anal.* **8** (5), 389 (1990).

[129] EN 1186-15, *Materials and Articles in Contact with Foodstuffs – Plastics. Part 15: Alternative Test Methods to Migration into Fatty Food Simulants by Rapid Extraction into Isooctane and/or 95% Ethanol* (2004).

Appendix: List of Symbols

Acronyms of Techniques 767
Chemical Nomenclature 778
Polymers and Products 778
Additives/Chemicals 780
Physical and Mathematical Symbols 785
Physical and Mathematical Greek Symbols 789
General Abbreviations 790

ACRONYMS OF TECHNIQUES

AA(S)	Atomic absorption (spectrometry)
AC-MS	Atomic composition mass spectrometry
ADSC™	Alternating DSC
ADXPS	Angle-dependent X-ray photoelectron spectroscopy (*cfr.* ARXPS)
AE	Acoustic emission
AED	Atomic emission detection
AEM	Analytical electron microscopy
AES	(1) Atomic emission spectrometry; (2) Auger electron spectroscopy; (3) Acoustic emission spectroscopy
AET	Acoustic emission technology
AFAM	Atomic force acoustic microscopy
AFM	Atomic force microscopy
AFS	Atomic fluorescence spectrometry
AGHIS	All-glass heated inlet system
Ag-SIMS	SIMS on etched Ag substrates
AOTF	Acousto-optical tuneable filter
AOTS	Acousto-optical tuneable spectrometer/scanning
AP	Atom probe
APCI	Atmospheric pressure chemical ionisation
API	Atmospheric pressure ionisation
AP MALDI	Atmospheric pressure MALDI
APS	Appearance potential spectroscopy
ARPES	Angular resolution photoelectron spectroscopy
ARXPS	Angle-resolved X-ray photoelectron spectroscopy (*cfr.* ADXPS)
ASE®	Accelerated solvent extraction
ASV	Anodic stripping voltammetry
ATR	Attenuated total reflectance
B	Magnetic sector analyser
BEI, BSI	Backscattered electron imaging
BF	Bright field
CAD	Collision-activated dissociation
C-AFM, CM-AFM	Contact-mode atomic force microscopy
CAG	Contact angle goniometry
CARS	Coherent anti-Stokes Raman spectroscopy
CASM	Calorimetric analysis with scanning microscopy
CBED	Convergent beam electron diffraction
CC	Cryogenic collection (trap)
CCD	Charge-coupled device
CDT	Corona discharge treatment
CE	Capillary electrophoresis
CEMS	Conversion electron Mössbauer spectroscopy
CF(D)	Conventional fluorescence (detection) (*cfr.* F, FL)
CF-FAB MS	Continuous-flow fast atom bombardment mass spectrometry
CF-LIBS	Calibration-free LIBS
CFM	Chemical force microscopy

CGC	Capillary gas chromatography
CHA	Concentric hemispherical analyser
CI	Chemical ionisation
CID	(1) Collision-induced dissociation; (2) Charge-induction device
CI-MS, CIMS	Chemical ionisation mass spectrometry
CIR	Cylindrical internal reflection
CIS	Cooled injection system
CL	Chemiluminescence
CLFM	Confocal laser fluorescence microscopy (*cfr.* LCFM)
CLND	Chemiluminescent nitrogen detector
CLSM	Confocal laser scanning microscopy (*cfr.* LSCM)
CMA	Cylindrical mirror analyser
CMR	Contact microradiography
CnRTA	Constant rate thermal analysis
CP	Cross-polarisation
CPD	Contact potential difference
CPI	Correlation peak imaging
CP/MAS NMR	Cross polarisation/magic-angle spinning NMR
CrRTA	Controlled rate thermal analysis
CRTA	Controlled transformation rate thermal analysis
CRTG	Controlled rate thermogravimetry
CSFM	Confocal scanning fluorescence microscopy
CSI	Chemical shift imaging
CSOM	Confocal scanning optical microscopy
CSV	Cathodic stripping voltammetry
CT	(1) Cold trap, cryotrapping; (2) Computed X-ray tomography
CTEM	Conventional transmission electron microscopy
CuPy	Curie-point pyrolysis
CV-AAS	Cold (mercury) vapour atomic absorption spectrometry
CV-AFS	Cold vapour atomic fluorescence spectroscopy
CW-ESR	Continuous-wave electron spin resonance
CYCLCROP	Cyclic *J* cross-polarisation technique (NMR)
DAD	Diode-array detector
DALLS	Dual-angle laser light scattering
DAS	Dynamic-angle spinning (NMR)
DCI	(1) Direct chemical ionisation; (2) Desorption chemical ionisation
DCP	Direct-current (argon) plasma
DCP-AES	Direct-current plasma atomic emission spectrometry
DD	(1) Dipole-dipole interactions; (2) Decoupling/double resonance (high-power ^{1}H decoupling); (3) Direct deposition
DDSC	Derivative DSC
DDSC™	Dynamic DSC
DE	Delayed extraction
DEA	Dielectric analysis (or dielectrometry)
DEC	High-power decoupling (NMR)
DETA	Dielectric thermal analysis
DF	Dark field
DHS	Dynamic headspace
DI	(1) Desorption/ionisation; (2) Direct inlet
DIC	Differential interference contrast (Nomarski)
DIES	Dielectric spectroscopy
DI-MS, DIMS	Direct inlet mass spectrometry
DIOS	Direct ionisation on silicon
DIP	Direct inlet (insertion) probe
DLI	(1) Direct laser ionisation; (2) Direct liquid introduction; (3) Direct liquid interface
DMA	Dynamic mechanical analysis
DMD	Differential mobility detector
DMTA	Differential mechanical thermal analysis
D-NIR	Dispersive near-infrared
DOR	Double rotation (NMR)
DOSY	Diffusion ordered spectroscopy
DP	(1) Differential pressure (viscosity detector); (2) Density profiling
D/P	Dissolution/precipitation
DP-MS, DPMS	(*In vacuo*) direct probe mass spectrometry
DPP	Differential pulse polarography
D-PyGC-MS	Chemical derivatisation pyrolysis gas chromatography–mass spectrometry
DPyMS	Direct pyrolysis mass spectrometry
DR	Diffuse reflectance
DRC	(1) Dynamic rate control; (2) Dynamic reaction cell

DRIFTS	Diffuse reflectance infrared Fourier transform spectroscopy
DRS	(1) Dielectric relaxation spectroscopy; (2) Diffuse reflectance spectroscopy
DSC	Differential scanning calorimetry
DSIMS	Dynamic secondary ion mass spectrometry
DT	Differential trapping
DTA	Differential thermal analysis
DTD	Direct thermal desorption
DTG	Differential thermogravimetry
DT-MS, DTMS	Direct temperature-resolved mass spectrometry
DTPy	Direct temperature-resolved pyrolysis
E, ESA	Electric sector analyser, electrostatic analyser
EBIC	Electron-beam-induced current
EBS	Elastic backscattering
EBSD	Electron backscatter diffraction
EC	Electrochemical (analyser)
ECD	(1) Electron-capture detector; (2) Electrochemical detector
ECNI	Electron-capture negative ionisation
ECP	Enclosed Curie-point (pyrolysis)
ED	Energy dispersive
EDAX®, EDX	Energy-dispersive X-ray spectrometry
EDS	(1) Energy-dispersive spectrometry; (2) Electron diffraction spectroscopy
EDXRA	Energy-dispersive X-ray analysis (SEM)
EDXRF	Energy-dispersive X-ray fluorescence
EELS	Electron energy-loss spectroscopy
EFM	Electrostatic force microscopy
EGA	Evolved gas analysis
EGD	Evolved gas detection
EGP	Evolved gas profile
EHREM	Environmental cell high-resolution electron microscopy
EI	Electron ionisation/impact
EI-MS, EIMS	Electron impact mass spectrometry
ELS(D)	Evaporative light scattering (detector)
EM	Electron microscopy
em	Emission
EMA	Electron X-ray microanalysis
EMD	Evaporative mass detection
EMP	Electron microprobe
ENDOR	Electron nuclear double resonance
EPC	Electronic pressure control
EPI	Echo-planar imaging
EPMA	Electron-probe microanalysis
EPR	Electron paramagnetic resonance (*cfr.* ESR)
EPXMA	Electron-probe X-ray microanalysis (*cfr.* EMP, EPMA)
ERS	External reflection spectroscopy
ESCA	Electron spectroscopy for chemical analysis (*cfr.* XPS)
ESE®	Enhanced solvent extraction
E-SEM, ESEM	Environmental scanning electron microscopy
ESI	Electron spectroscopic imaging
ES(I), ESP	Electrospray (ionisation)
ESIMS, ESI-MS	Electrospray ionisation mass spectrometry
ESR	Electron spin resonance (same as EPR)
ESRI	Electron spin resonance imaging
ETAAS, ET-AAS	Electrothermal (atomisation) atomic absorption spectrometry
ETV	Electrothermal vaporisation
ex	Excitation
EXAFS	Extended X-ray absorption fine structure
EXELFS	Extended energy-loss fine structure
F, FL	Fluorescence (detector)
FAAS	Flame atomic absorption spectrometry
FAB	Fast atom bombardment
FAES	Flame atomic emission spectrometry
FD	Field desorption
FD-MS, FDMS	Field desorption mass spectrometry
FEG-ESEM	Field-emission gun environmental scanning electron microscopy
FEG-SEM	Field-emission gun scanning electron microscopy (*cfr.* FESEM)
FEL	Free-electron laser

FESEM	Field emission SEM (*cfr.* FEG-SEM, LVSEM)
FEWS	Fiberoptic evanescent wave spectroscopy
FGP	Functional group profile
FGSE	Field-gradient spin-echo
FI	(1) Field ionisation; (2) Flow injection
FIA	Flow-injection analysis
FIB	(1) Fast ion bombardment; (2) Focussed ion beam
FID	(1) Flame ionisation detector; (2) Free induction decay (NMR)
FILS	Field ionisation laser spectrometry
FIM	Field ion microscopy
FI-MS, FIMS	Field ionisation mass spectrometry
FIR	Far infrared
FL	Fluorescence, fluorometry
FLD	Fluorescence detector
FLIM	Fluorescence-lifetime imaging
FM	Fluorescence microscopy
FM-AFM	Force modulation mode AFM
FMM	Force modulation microscopy
FORS	Fibre optics reflectance spectroscopy
FPA	Focal plane array (detector)
FPD	Flame photometric detector
FRES	Forward recoil spectrometry
FRS	Forced Rayleigh scattering
FSCD	Fluorine-induced sulfur chemiluminescence detector
FT-ESR	Fourier transform electron spin resonance
FTICR	Fourier transform ion-cyclotron resonance
FTIES	Fourier transform infrared emission spectroscopy
FT-IR, FTIR	(Fourier transform) infrared spectroscopy
FTIR-μS	FTIR-microspectroscopy (*cfr.* μFTIR)
FT LMMS	Fourier transform laser-microprobe mass spectrometry
FTMS	Fourier transform mass spectrometry
FTNMR	Fourier transform NMR
FT-RS, FTRS	Fourier transform Raman spectroscopy
GA	Gas analysis
GC	Gas chromatography
GC-MS	Gas chromatography–mass spectrometry
GD-(MS)	Glow-discharge (mass spectrometry)
GD-OES	Glow-discharge optical emission spectrometry
GE	Gradient echo (NMR imaging sequence)
GFAAS	Graphite furnace atomic absorption spectrometry
GIXRD	Grazing incidence X-ray diffraction
GI-XRF, GIXF	Grazing incidence X-ray fluorescence
GPC	Gel permeation chromatography (*cfr.* SEC)
GSE	Gaseous secondary electron (imaging)
HATR	Horizontal attenuated total reflectance
hfDSC	Heat flux DSC
HG-AAS	Hydride generation AAS
HNF	Holographic notch filter
HODS	Higher-order derivative spectrophotometry ($n > 2$)
HP/DEC	High-power decoupling (NMR)
HPDSC	High-pressure DSC
HPer DSC	High-performance DSC
HPGe	High-purity germanium (detector)
HPHD	(1) High-power heteronuclear decoupling; (2) High-power proton decoupling
HPLC	High-performance liquid chromatography
HPSEM	High-pressure SEM
HPTLC	High-performance thin-layer chromatography
HRLEELS	High-resolution low-energy electron loss spectroscopy
HRMAS	High-resolution magic-angle spinning
HRMS	High-resolution mass spectrometry
HRSEM	High-resolution scanning electron microscopy
HRTEM	High-resolution transmission electron microscopy

HRTGA	High-resolution thermogravimetric analysis
HR-US	High-resolution ultrasonic (spectroscopy)
HS-GC	Headspace gas chromatography
HSSE	Headspace sorptive extraction
HS-SPME	Headspace solid-phase microextraction
HT	High temperature
HT-GC, HTGC	High-temperature gas chromatography
HT HS	High-temperature headspace
HT-PTV	High-temperature programmed thermal vaporisation
HTS	High-throughput screening
HVEM	High-voltage electron microscopy
IA	Image analysis
IC-AFM	Intermittent-contact AFM (*cfr.* TM-AFM)
ICCD	Intensified charge-coupled device
ICL	Imaging chemiluminescence
ICP(I)	Inductively coupled plasma (ionisation process)
ICP-AES	Inductively coupled plasma–atomic emission spectrometry
ICP-MS	Inductively coupled plasma–mass spectrometry
ICP-OES	Inductively coupled plasma–optical emission spectrometry
ICR	Ion-cyclotron resonance
ID(A)	Isotope dilution (analysis)
IDGC-MS	Isotope dilution gas chromatography–mass spectrometry
ID-ICPMS	Isotope dilution–inductively coupled plasma–mass spectrometry
IDMS	Isotope dilution mass spectrometry
IDP	Image depth profiling
ID-TIMS	Isotope dilution thermal ionisation mass spectrometry (*cfr.* also TI-IDMS)
IEC	Ion-exchange chromatography
IES	Infrared emission spectroscopy
ILS	Ionisation loss spectroscopy
IMA	Ion microanalysis
IMD	Ion mobility detection
IMR-MS	Ion-molecule reaction mass spectrometry
IMS	(1) Ion mobility spectrometry; (2) Infrared microspectroscopy (*cfr.* μFTIR)
INAA	Instrumental neutron activation analysis
INADEQUATE	Homonuclear J-correlated ^{13}C experiment (NMR)
IP(A)	In-process (analysis)
IPAA	Instrumental photon activation analysis
IR	Infrared
IRA	Internal reflection attachment
IRE	(1) Internal reflection element; (2) Internal reference electrode
IR-ERS	Infrared external reflection spectroscopy
IR-IRS	Infrared internal reflection spectroscopy
IR-LA	Infrared laser ablation
IR-LDI	Infrared laser desorption/ionisation
IR-NSOM	Infrared near-field scanning optical microscopy
IRRAS	Infrared reflection-absorption spectroscopy (*cfr.* RAIRS)
IRS	Internal reflectance spectroscopy (*cfr.* ATR)
ISE	Ion-selective electrode
iSIMS	Imaging secondary ion mass spectrometry
ISS	Ion scattering spectroscopy (ion surface scattering)
IT(D)	Ion trap (detector)
ITMS	Ion trap mass spectrometry
iXPS	Imaging X-ray photoelectron spectroscopy
KF	Karl Fischer (coulometry)
LA	Laser ablation
LAAS	Laser atomic absorption spectrometry
LA-AES	Laser ablation–atomic emission spectrometry (*cfr.* LIBS)
LAES	Laser ablation–emission spectrometry (*cfr.* LIBS)
LA-ICP-MS	Laser ablation–inductively coupled mass spectrometry
LA-ITMS	Laser ablation–ion trap mass spectrometry
LA(L)LS	Low-angle (laser) light scattering

LAMMA®	Laser microprobe mass analysis (*cfr.* LMMS)
LAMMS	Laser microprobe mass spectrometry (*cfr.* LMMS)
LAMS	Laser(-assisted) mass spectrometry; (resonance-enhanced) laser mass spectrometry (*cfr.* REMPI)
LA-MS	Laser ablation mass spectrometry
LA-OES	Laser ablation–optical emission spectrometry (*cfr.* LIBS)
LAP	Laser-ablated plasma
LARIS	Laser-ablation resonant ionisation spectrometry
LAS	Light absorption spectrometry
LASER	Light amplification by stimulated emission of radiation
LC	Liquid chromatography
LCCC	Liquid chromatography under critical conditions
LCD	Liquid crystal display
LCFM	Laser confocal fluorescence microscopy (*cfr.* CLFM)
LCTF	Liquid-crystal tuneable filter
LD	Laser desorption
LDI	Laser desorption/ionisation
LD-IMS	Laser desorption–ion mobility spectrometry
LDMS	Laser desorption mass spectrometry
LD/PD-ToFMS	Laser desorption/ photodissociation time-of-flight mass spectrometry
LDT	Laser desorption transfer
LE	Laser excitation
LEAFS	Laser-excited atomic fluorescence spectrometry
LED	Light emitting diode
LEED	Low-energy electron diffraction
LEI	Laser-enhanced ionisation
LEIS(S)	Low-energy ion scattering (spectroscopy)
LF	Laser flash (photolysis)
LFM	Lateral force microscopy
LI	Laser ionisation
LIAFS	Laser-induced atomic fluorescence spectrometry (*cfr.* LEAFS)
LIBS	Laser-induced breakdown spectroscopy
LID	Laser-induced desorption
LIESA®	Laser-induced emission spectral analysis (*cfr.* LIBS)
LIF(S)	Laser-induced fluorescence (spectroscopy)
LIMA®	Laser ionisation mass analyser
LIMS®	Laser ionisation mass spectrometry (*cfr.* LMMS)
LIP	Laser-induced plasma
LIP-AES	Laser-induced plasma–atomic emission spectrometry
LIPS	Laser-induced plasma spectroscopy (*cfr.* LIBS)
LIT	Laser impulse thermography
LITD	Laser-induced thermal desorption (*cfr.* LID)
LM	(1) Light microscopy; (2) Laser microanalysis
LMIG	Liquid metal ion gun
LMMS	Laser microprobe mass spectrometry
LMS	Laser mass spectrometry (*cfr.* LAMS)
L^2MS	Two-step laser mass spectrometry
l-NMR	Liquid nuclear magnetic resonance
LOES	Laser optical emission spectrometry
LPA	Laser probe microanalysis (*cfr.* LMMS)
LPAS	Laser photoacoustic spectroscopy
L-PES	Laser–plasma emission spectrometry (*cfr.* LIBS)
LPMA	Laser probe microanalysis (*cfr.* LMMS)
LPS	Laser pyrolysis scanning
LPTD	Linear programmed thermal desorption
L-Py, LPy	Laser pyrolysis
LPyMS	Laser pyrolysis mass spectrometry
LR	Laser Raman
LR-NMR	Low-resolution NMR
LRRS	Low-resolution Raman spectroscopy
LS	Light scattering
LSCM	Laser scanning confocal microscopy (*cfr.* CLSM)
LSIMS, LSI-MS	Liquid secondary ion mass spectrometry
LSM	Laser scanning microscopy
LSMS	Laser source mass spectrometry

L-SNMS Laser SNMS
LSOM Laser scanning optical microscopy
LSS Laser spark spectroscopy
LTA Local thermal analysis
L^2ToFMS Laser-desorption laser-photoionisation ToF-MS
LVEI Low-voltage electron ionisation
LVESEM Low-voltage environmental scanning electron microscopy
LVI Large-volume injection
LVSEM Low-voltage scanning electron microscopy (*cfr.* FESEM)
LV-SEM Low-vacuum scanning electron microscopy
LVTEM Low-voltage transmission electron microscopy
LW-NIR Long wavelength near-infrared spectroscopy
MAB Metastable atom bombardment
MAE Microwave-assisted extraction
MAHS Microwave-assisted headspace
MALD(I) Matrix-assisted laser desorption/ ionisation
MA(L)LS Multiple-angle (laser) light scattering
MAS (1) Magic-angle spinning; (2) Mössbauer absorption spectroscopy
MC Microcalorimetry
MCA Multichannel analyser
MCFT Multichannel Fourier transform
MCP Microchannel plate
MCT Mercury-cadmium-telluride (detector)
MDS Microwave dielectric loss spectroscopy
MDSC™ Modulated differential scanning calorimetry
MDTA Mass spectrometric differential thermal analysis
MED Microwave emission detector
MEIS Medium energy ion scattering
MEMS Micro electromechanical system
MES Mössbauer emission spectroscopy
MESI Membrane extraction with sorbent interface
ME-SIMS Matrix-enhanced SIMS
MFI Melt-flow index
MFM Magnetic force microscopy
MI Microprobe imaging
MIM Multiple ion monitoring
MIP Microwave-induced plasma
MIR (1) Multiple internal reflection; (2) Mid-infrared
MOI Multiple oblique illumination
MOUSE Mobile universal surface explorer
MP (1) Mobile phase; (2) Microplasma; (3) Modulus profiling
MPD Microwave plasma detector
MPI(S) Multiphoton ionisation (spectroscopy)
MQMAS Multiple-quantum magic-angle spinning (NMR)
MR Magnetic resonance
MRI Magnetic resonance imaging
MRM Mobile Raman microscopy
MRR Molecular rotational resonance (microwave spectroscopy)
MRS Micro Raman spectroscopy
MS Mass spectrometry
MS^n Multiple-stage mass spectrometry; tandem mass spectrometry
M-SIMS Magnetic sector type SIMS
MSP Microspectrophotometry
MSPD Matrix solid-phase dispersion
MTA Mass spectrometric thermal analysis
MTDSC Modulated temperature DSC
MTDTA Modulated temperature DTA
MTGA™ Modulated thermogravimetric analysis
MUPI Multiphoton ionisation, *cfr.* MPI
MWTA Microwave thermal analysis
MXA Microsample X-ray analysis
MXRF Micro X-ray fluorescence (*cfr.* μXRF)
μATR Micro attenuated total reflectance
μCT X-ray microtomography
μFTIR Micro Fourier transform infrared (*cfr.* FTIR-μS)
μLC Micro liquid chromatography
μRS Micro Raman spectroscopy (*cfr.* MRS)
μSEC Micro size-exclusion chromatography
μTA Micro thermal analysis
μTMA Micro thermomechanical analysis

μXAS	Micro X-ray absorption spectroscopy
μXPS	Micro X-ray photoelectron spectroscopy
μXRF	Micro X-ray fluorescence
NAA	Neutron activation analysis
NAMS	Neutron activation mass spectrometry
NC-AFM	Non-contact mode atomic force microscopy
NDE	Non-destructive evaluation
NDP	Neutral depth profiling
NDT	Non-destructive testing
NEXAFS	Near-edge X-ray absorption fine structure (*cfr.* XANES)
NFO	Near-field optics
NIR(A)	Near-infrared reflectance (analysis)
NIR-IA	Near-infrared image analysis
NIRIM	Near-IR Raman imaging microscopy
NIR-IRS	Near-infrared internal reflection spectroscopy
NIRRS	Near-infrared diffuse reflectance spectroscopy
NIRS	Near-infrared spectroscopy
NIT	Near-infrared transmittance
NIVI	Near-infrared video imaging
NMP	Nuclear microprobe
NMR	Nuclear magnetic resonance
NMRI	Nuclear magnetic resonance imaging
NOE	Nuclear Overhauser effect/enhancement
NOESY	Nuclear Overhauser and exchange spectroscopy
NPD	Nitrogen phosphorous detector or thermoionic detector
NQR	Nuclear quadrupole resonance
NQRI	Nuclear quadrupole resonance imaging
NR	Neutron reflectometry
NREMPI	Non-resonant multiphoton ionisation
NS	Neutron scattering
NSOM	Near-field scanning optical microscopy (*cfr.* also SNOM)
oaToF	Orthogonal acceleration time-of-flight
ODSC™	Oscillating DSC
OES	Optical emission spectrometry
OL	Oxyluminescence
OLM	On-line monitoring
OM	Optical microscopy
OMT	Oxidation maximum temperature
ORS	Octopole reaction system
O-SCD	Ozone-induced sulfur chemiluminescence detector
OVA	Organic vapour analyser
PA	Photoacoustics
PAC	Process analytical chemistry
PACT	Process analytics and control technology
PA-FTIR	Photoacoustic Fourier transform infrared
PAI	Post ablation ionisation
PA-NIR	Photoacoustic near-infrared spectroscopy
PARS	Photoacoustic Raman spectroscopy
PA(S)	Photoacoustic (spectroscopy)
PA-UV	Photoacoustic UV spectrophotometry
PA-VIS	Photoacoustic visible spectrophotometry
PC	(1) Paper chromatography; (2) Process control
pcDSC	Power compensation DSC
PCS	(1) Photoacoustic correlation spectroscopy; (2) Photon correlation spectroscopy
PD	^{252}Cf plasma desorption
PDA	Photodiode array (detection)
PDMS	Plasma-desorption mass spectrometry
PDPI	Photodissociation – photoionisation
PDSC	Pressure differential scanning calorimetry
PEEM	Photoemission electron microscopy
PES	Photoelectron spectroscopy
PFE	Pressurised fluid extraction
PFG-NMR	Pulsed-field gradient nuclear magnetic resonance
PFM	Pulsed force microscopy
PGC	Process gas chromatography
PGSE	Pulsed gradient spin-echo (NMR)
PI	(1) Photoionisation; (2) (Laser) post-ionisation; (3) Plasma-ionisation

PID	(1) Photon-induced dissociation; (2) Photoionisation detection
PIGE	(1) Particle-induced γ-ray emission; (2) Proton-induced γ-ray spectrometry
PIXE	(1) Particle-induced X-ray emission; (2) Proton-induced X-ray emission
PLM	Polarised light microscopy (*cfr.* PM)
PLPAS	Pyrolysis–laser photoacoustic spectroscopy
PM	Polarisation microscopy (*cfr.* PLM)
PMS	Laser particle measurement system
PMT	Photomultiplier tube
P-NMR	Pulse nuclear magnetic resonance
PR	Pulse radiolysis
PR-PAS	Phase-resolved PAS
PSD	(1) Position-sensitive detector; (2) Photon-stimulated desorption; (3) Post-source decay
pSFC	Packed column SFC
PSPD	Position-sensitive photodetector
PT, P&T	Purge-and-trap
PTA	Pulse thermal analysis
PTS	Position-tagged spectrometry
PTV	Programmed temperature vaporising (inlet)
Py	Pyrolysis
PyFTIR	Pyrolysis–Fourier transform infrared
PyGC	Pyrolysis–gas chromatography
PyGC/HRMS	Pyrolysis–gas chromatography/ high-resolution mass spectrometry
PyGC-MS	Pyrolysis–gas chromatography–mass spectrometry
PyHGC	Pyrolysis–hydrogenation gas chromatography
PyMS	Pyrolysis–mass spectrometry
QDTA	Quantitative differential thermal analysis
QFM	Quantitative fluorescence microscopy
QIA	Quasi-isothermal analysis
QIT(MS)	Quadrupole ion trap (mass spectrometer)
QMS	Quadrupole mass spectrometer
QQQ, QqQ	Triple quadrupole analyser
QSA	Quantitative surface analysis
Q-SIMS	Quadrupole type SIMS
QTLC	Quantitative thin-layer chromatography
R	(Normal) Raman
R-A	Reflection-absorption
RAE	Resistive anode encoder
RAIR(S)	Reflection-absorption IR (spectroscopy)
RALLS	Right-angle laser light scattering
RAS	Reflection-absorption spectroscopy
RBS	Rutherford backscattering spectroscopy
RCD	Redox chemiluminescence detector
RCTA	Reaction controlled thermal analysis
RELMA	Remote laser microanalysis
REMPI	Resonance enhanced multiphoton ionisation
rfGD-AES	Radiofrequency powered glow discharge–atomic emission spectrometry
RGE	Rotating wax-impregnated graphite electrode
RHEED	Reflection high-energy electron diffraction
RI(D)	Refractive index (detector)
RIMS	Resonance ionisation mass spectrometry
RIS	Resonance ionisation spectroscopy
RLIF	Remote laser-induced fluorescence
RNAA	Radiochemical neutron activation analysis
R-NSOM	Raman near-field scanning optical microscopy
ROSA	Remote optical sensing assembly
RPLC	Reversed-phase liquid chromatography
R2PI	Resonant two-photon ionisation
RRE	Resonance Raman effect
RR(S)	Resonance Raman (scattering)
RS	Raman scattering/spectroscopy
RSNOM	Raman scanning near-field optical microscopy

RTD-GC	Reactive thermal desorption gas chromatography
RT FT-IR	Real-time FT-IR
RTMS®	Real-time multiple strip (detector technology)
SAD	Selected area diffraction
SAI	Scanning Auger image/imaging
SALDI	Surface-assisted laser desorption/ionisation
SALI®	Surface analysis by laser-ionisation
SALS	Small-angle light scattering
SAM	(1) Scanning Auger (electron) microscopy/microprobe; (2) Scanning acoustic microscopy (*cfr.* SCAM); (3) Standard addition method
SANS	Small-angle neutron scattering
SARISA	Surface analysis by resonance ionisation of sputtered atoms
SATVA	Sub-ambient thermal volatilisation analysis
SAX	Selected area XPS
SAXS	Small-angle X-ray scattering
SCAM	Scanning acoustic microscopy (*cfr.* SAM)
SCD	(1) (Flame) sulfur chemiluminescence detector; (2) Segmented charged coupled device
SCM	(1) Scanning confocal microscopy; (2) Scanning capacitance microscopy
SCTA	Sample controlled thermal analysis
SDM	Selected decomposition monitoring
SE	(1) Spin echo (NMR imaging sequence); (2) Secondary electron (imaging)
SEB	Secondary electron Bremsstrahlung
SEC	Size-exclusion chromatography (*cfr.* GPC)
SED	Secondary electron detector
SEI	Secondary electron image/imaging
SEIRAS	Surface-enhanced infrared absorption spectroscopy
SELDI	Surface-enhanced laser desorption ionisation
SEM	Scanning electron microscopy
SE(R)RS	Surface-enhanced (resonance) Raman spectroscopy
SEXAFS	Surface EXAFS
SFC	Supercritical fluid chromatography
SFE	Supercritical fluid extraction
SFM	Scanning force microscopy
SGP	Specific gas profile
SGSE	Static gradient spin-echo
SHS	Static headspace
SIA	Stepwise isothermal analysis
SID	Surface-induced dissociation
SIM(-MS)	Selected-ion monitoring (single ion monitoring) mass spectrometry
SIMS	Secondary ion mass spectrometry
SIP	Solid insertion probe
SIRIS	Sputter-initiated resonance ionisation spectroscopy
SIT	Silicon intensified target (camera)
SJS	Supersonic jet spectrometry
SKM	Scanning Kelvin microscopy
SLD	Soft laser desorption
SLIM	Spatially resolved laser ion microscopy
SLP	Service life prediction
SMATCH	Simultaneous mass and temperature change
SML	Scanning microanalysis with laser spectrometry
SNIM	Scanning near-field infrared microscopy
SNMM	Scanning near-field microwave microscopy
s-NMR	Solid-state nuclear magnetic resonance
SNMS	Sputtered *cq.* secondary neutral mass spectrometry
SNOM	Scanning near-field optical microscopy (*cfr.* also NSOM)
SOM	Scanning optical microscopy
SPE	(1) Solid-phase extraction; (2) Single-pulse excitation (NMR)
SPI	Single photon ionisation
SPM	(1) Simultaneous pyrolysis methylation; (2) Scanning probe microscopy
SPME	Solid-phase microextraction
SPS	Scanning probe spectroscopy

SR	(1) Specular reflectance; (2) Synchrotron radiation
SRS	Specular reflection spectroscopy
SR-XRD	Synchrotron radiation X-ray diffraction
SR-XRF	Synchrotron radiation X-ray fluorescence
SS	Solid sampling
SSCM	Stage-scanning confocal microscope
SSIMS	Static secondary ion mass spectrometry
SSMS	(1) Spark-source mass spectrometry; (2) Solid-state mass spectrometry
SS-PAS	Step-scan photoacoustic spectroscopy
SSRS	Shifted-subtracted Raman spectroscopy
SS-ZAAS	Solid sampling Zeeman atomic absorption spectrometry
STA	Simultaneous thermal analysis
STED	Stimulated emission depletion
STEM	Scanning transmission electron microscopy
SThM	Scanning thermal microscopy
STM	Scanning tunnelling microscopy
STRAFI	Stray field imaging
STS	Scanning tunnelling spectroscopy
STXM	Scanning transmission X-ray microscopy
SW-NIR	Short wavelength NIR
SWT	Side-window tube (X-ray techniques)
TA	Thermal analysis
TAD	Thermally assisted desorption
TAHM	Thermally assisted hydrolysis and methylation (*cfr.* THM)
TALLS	Triple-angle laser light scattering
TAM	Thermal analysis microcalorimetry
TCD	Thermal conductivity detector
TD	(1) Thermal desorption; (2) Thermodilatation
TDM	Thermal desorption modulator
TD-MS	Thermal desorption mass spectrometry
TDS	Temperature-programmed desorption (*cfr.* also TPD)
TEA	(1) Thermal evolution analysis; (2) Thermoelectric analysis; (3) Thermal energy analyser
TE-GC-MS	Thermal extraction GC-MS
TEM	Transmission electron microscopy
TEM-X	TEM with induced X-ray emission
TG(A)	Thermogravimetry, thermogravimetric analysis
ThGC	Thermochromatography
THM	Thermally assisted hydrolysis and methylation
THM-GC-MS	Thermally assisted hydrolysis and methylation GC-MS
TI-IDMS	Thermal ionisation–isotope dilution mass spectrometry (*cfr.* also ID-TIMS)
TIMS	Thermal ionisation mass spectrometry
TIR	(1) Transmission infrared; (2) Thermographic infrared
TL	Thermoluminescence
TLC	Thin-layer chromatography
TLF	Time-lag focusing
TMA	Thermomechanical analysis
TM-AFM	Tapping mode AFM (*cfr.* IC-AFM)
TMBA	Thermo-molecular beam analysis
TMDSC	Temperature modulated DSC (*cfr.* MTDSC)
TMP	Thermomicrophotometry
TOA	Thermo-optical analysis
TOD	Thermo-oxidative degradation
ToF-LMMS	Time-of-flight laser-microprobe mass spectrometry
ToFMS, ToF-MS	Time-of-flight mass spectrometry
ToF-SIMS	Time-of-flight secondary ion mass spectrometry
TOL	Thermal oxyluminescence
TP	Thermal programming
TPA	Two-photon absorption spectroscopy
TPD	Thermal-programmed desorption (*cfr.* also TDS)
TPF	Temperature-programmed fractionation
TPI	Two-photon/ionisation
TPPy	Temperature-programmed pyrolysis
TPR	Thermal-programmed reduction

TRELIBS	Time-resolved LIBS
TREPR	Time-resolved ESR
TRT	Temperature-rise time
TRXRF	Total-reflection X-ray fluorescence (*cfr.* TXRF)
TSD	(1) Thermoionic specific detector; (2) Thermally stimulated discharge
TSD-GC-MS	Thermally stimulated desorption GC-MS
TSI	Thermal surface ionisation
TSL	Thermally stimulated luminescence
TSM	Thermal scanning microscopy
TTP	Temperature-time profile
TTR-PyMS	Time/temperature resolved pyrolysis mass spectrometry
TUV	Thermal ultraviolet
TVA	Thermal volatilisation analysis
TWI	Thermal wave infrared imaging
TXM	Transmission X-ray microscopy
TXRF	Total-reflection X-ray fluorescence (*cfr.* TRXRF)
UFM	Ultrasonic force microscopy
UPS	Ultraviolet photoelectron spectroscopy
US	Ultrasound
USAXS	Ultra small-angle X-ray scattering
UV	Ultraviolet
UV-LA	Ultraviolet laser ablation
UV-LDI	Ultraviolet laser desorption/ionisation
UVP	Ultraviolet photolysis
UVRRS	Ultraviolet resonance Raman scattering/spectroscopy
VIEEW™	Video Image Enhanced Evaluation of Weathering
VIS	Visible
VMI	Video microscopy imaging
VPH	Volume phase holography
VPSEI	Variable pressure secondary electron imaging
VPSEM	Variable pressure SEM
VUV	Vacuum ultraviolet
WAXD	Wide angle X-ray diffraction
WAXS	Wide angle X-ray scattering
WD	Wavelength dispersive
WDS	Wavelength dispersive spectrometry
WDXRF	Wavelength dispersive X-ray fluorescence
XAES	X-ray excited Auger electron spectroscopy
XAFS	X-ray absorption fine structure
XANES	X-ray absorption near-edge structure (*cfr.* NEXAFS)
XAS	X-ray absorption spectroscopy
XEDS	X-ray energy dispersive spectrometry
XFM	X-ray fluorescence microscopy
XPS	X-ray photoelectron spectroscopy (*cfr.* ESCA)
XRD	X-ray diffraction
XRF	X-ray fluorescence
XRM	X-ray microscopy
XRMA	X-ray microanalyser
XRMF	X-ray microfluorescence (*cfr.* MXRF)
XRR	X-ray reflectometry
XuM	X-ray ultra microscope
ZAAS	Zeeman atomic absorption spectrometry
ZETAAS	Zeeman electrothermal atomic absorption spectrometry

CHEMICAL NOMENCLATURE

Polymers and Products

ABS	Acrylonitrile–butadiene–styrene terpolymer
A-PAM	Anionic polyacrylamide
aPP	Atactic polypropylene
AS	Acrylonitrile–styrene copolymer
ASA	Acrylonitrile–styrene–acrylic ester copolymer
AU	Acrylic urethane resin
BHEDA	Bisphenol-A dihydroxyethyletherdiacrylate
BIMS	Poly(isobutylene-*co*-*p*-methylstyrene)
BMC	Bulk moulding compound
BPA-PC	Bisphenol-A polycarbonate
BPE	Branched polyethylene (*cfr.* LDPE)
bPP	Polypropylene block copolymer
BR	Butadiene rubbers, polybutadienes
Br-PC	Brominated polycarbonate
CA	Cellulose acetate
CAP	Cellulose ammonium phosphate (fabric)
CFRP	Carbon-fibre reinforced polymer

CN-PS	Poly(cyanopropyl)methylsiloxane
CPO	Chlorinated polyolefin
CPVC	Chlorinated poly(vinyl chloride), *cfr.* PVCC
CR	Polychloroprene (chloroprene rubber)
DGEBA	Diglycidyl ether of bisphenol-A (epoxy resin)
DHPVC	Dehydropoly(vinyl chloride)
DP	Degree of polymerisation
dPMMA	Deuterated poly(methyl methacrylate)
E/CO	Ethylene/carbon monoxide
EMC	Electronic moulding compounds
EO-PO	Oxyethylene–oxypropylene copolymers
EP	(1) Engineering plastic; (2) Epoxide resin
EPDM	Ethylene–propylene–diene rubber, ethylene–propylene terpolymer, poly(ethylene-*co*-propylene-*co*-3,5-ethylidene norbornene)
EPM	Ethylene–propylene copolymer
EPR	Ethylene–propylene rubber
ER	Epoxy resin
ETCL	Ethylcellulose
EVA	Ethylene–vinylacetate copolymer, poly(ethylene-*co*-vinylacetate)
FP	Functional polymer
FPO	Flexible polyolefins
FRP	Fibre reinforced polymer
GAP	Glycidylazide polymer
GFR	Glass-fibre reinforced
HDPE	High-density polyethylene
HFP-TFE	Hexafluoropropylene–tetrafluoroethylene copolymer
HIPS	High-impact polystyrene
HMW	High molecular weight
HMWPE	High molecular weight polyethylene
HPLC	Hydroxypropylcellulose
HTPB	Hydroxyl-terminated polybutadiene
IIR	Isobutylene–isopropene rubber; poly(isobutene-*co*-isoprene)
IPN	Interpenetrating network
iPP	Isotactic polypropylene
IR	Isoprene rubber; poly(*cis*-1,4-isoprene)
Kapton	Polyimide film (Du Pont)
LCP	Liquid crystalline polymer
LDPE	Low-density polyethylene
LLDPE	Linear low-density polyethylene
LPM	Low pressure melamine (prepreg)
MBS	Methylmethacrylate–butadiene–styrene terpolymer
MDPE	Medium-density polyethylene
MF	Melamine formaldehyde resin
MMC	Metal matrix composite
m-PE	Metallocene polyethylene
MPEG	Monomethoxy(polyethylene glycol)
MPW	Mixed plastic waste
Mylar	Polyethylene terephthalate film
NBR	Acrylonitrile–butadiene rubber, nitrile rubber
NR	Natural rubber; polyisoprene
OHBR	Hydroxy-terminated polybutadiene rubber
PA	Polyamide
PA6/6.6	Polyamide 6/6.6
PAA	(1) Polyalkylacrylate; (2) Poly(acrylic acid)
PAAE	Polyamide–polyamine–epichlorohydrin resin
PAE	Poly(adipic acid ester)
PAG	Poly(alkylene glycol)
PAI	Polyamidimide
Palaroid B72	Ethylmethacrylate (70%) methylacrylate (30%) copolymer (P[EMA]/[MA])
PAM	(1) Polyacrylamide; (2) Polyacrylmethacrylate
PAN	Polyacrylonitrile
PAR	Polyarylate
PAS	Polyaryl sulfone
PB, P1B	Polybutene-1
PBA	Poly(*n*-butylacrylate)
PBBPA	Poly(pentabromobenzylacrylate)
PBD	1,4-Polybutadiene
PBG	Polybutylene glycol
PBMA	Poly(butylmethacrylate)
p-Br-PS	*p*-Bromopolystyrene
PBS	Poly(butylene succinate)
P(BS)	Poly(butadiene-*co*-styrene)
PBT	Poly(butylene terephthalate)
PC	Polycarbonate
PDBS	Polydibromostyrene
PDMS	Polydimethylsiloxane
PDMT	Poly(decamethylene terephthalate)
PE	Polyethylene
PEEK	Poly(etheretherketone)
PEG	(1) Poly(ethylene glycol); (2) Polyoxyethylene lauryl ether
PEI	Polyethylene imine, polyetherimide
PEKK	Poly(ether ketone ketone)

PEMA Poly(ethylmethacrylate)
PEO Poly(ethylene oxide); α-Alkoxy-ω-hydroxy polyethylene oxide
PET, PETP Poly(ethylene terephthalate)
PEUU Poly(ether urethane urea)
PE-X Cross-linked PE (*cfr.* XPE)
PF Phenolic formaldehyde (resin)
PFC Polymerisation-filled composites
PFPAE Perfluoropolyalkyl ether
PFPE Perfluoropolyether
PFT Polymerisation-filling technique
P-*g*-A Additive-grafted polymer
PhMe-PS Poly(phenyl)methylsiloxane
PHO Polyhydroxyoctanoate
PI (1) Polyimide; (2) Polyisoprene
PIB Polyisobutylene
PK Polyketone
PKS Polyketone sulfide
PMMA Poly(methyl methacrylate)
PMP, P4MP Poly(4-methylpentene-1)
PO Polyolefins
Poly-TMDQ Poly(2,2,4-trimethyl-1,2-dihydroquinoline)
POM Poly(oxymethylene)
POP Polyolefin plastomer
PP Polypropylene
PP-*co*-PE Ethylene/propylene copolymer
PP-*g*-MA Polypropylene-graft-maleic anhydride
PPE Poly(phenylene ether)
PPG Poly(propylene glycol)
PPI Impact-modified polypropylene
PPO Poly(phenylene oxide); poly(2,6-dimethylphenylene oxide)
PPOX Polypropylene oxide
PPP Poly(*p*-phenylene)
PPS Polyphenylene sulfide
PPy Polypyrrole
PS Polystyrene
PSU Polysulfone
PTFE Poly(tetrafluoroethylene)
PTMO Poly(tetramethylene oxide)
PU(R) Poly(urethane)
PVA Poly(vinyl alcohol), *cfr.* PVAL, PVOH
PVAc Poly(vinyl acetate)
PVA-E Poly(vinylacetate–ethylene) copolymer
PVAL Poly(vinyl alcohol), *cfr.* PVA, PVOH
PVB Poly(vinylbutyral-*co*-vinylalcohol)
PVC Poly(vinyl chloride)
PVCC Chlorinated PVC, *cfr.* CPVC
PVC-NP Non-phthalate plasticised PVC
PVC-P Plasticised PVC
PVC-U Unplasticised poly(vinyl chloride)
PVDF, PVF2 Poly(vinylidene fluoride)
PVOH Poly(vinyl alcohol); *cfr.* PVA, PVAL
PVP Poly(*N*-vinyl-2-pyrrolidone)
RACO Random copolymer
RIM Reaction injection moulding
RPET Recycled PET
rPP Random polypropylene
SAN Styrene–acrylonitrile copolymer
SBR Styrene–butadiene rubber; poly(butadiene-*co*-styrene)
SMA Styrene–maleic anhydride copolymer
SMI Imidised styrene/maleic anhydride copolymer
SR Synthetic rubber
ST-DVB Cross-linked styrene-divinylbenzene
TGDDM N,N,N''-Tetraglycidyl-4,4′-diaminodiphenylmethane (epoxy resin)
TMBPA-PC Tetramethylbisphenol-A polycarbonate
TPE Thermoplastic elastomer
TPO Thermoplastic olefin
TPU Thermoplastic polyurethane
TPV Thermoplastic vulcanisate
UD-PE Ultra-drawn PE
UHMWPE Ultrahigh-molecular weight polyethylene
VC-VA Vinylchloride–vinylacetate copolymer
VLDPE Very low-density polyethylene
XLPE, XPE Cross-linked polyethylene

Additives/Chemicals

AA (1) Adipic acid; (2) Acrylic acid
ACA α-Amino caproic acid
ACN Acrylonitrile
AKD Alkenediketene
AMMO Azidomethylmethyloxetane
AN Acrylonitrile, *cfr.* ACN
AO (1) Antioxidant; (2) Active oxygen
APP Ammonium polyphosphate, $(NH_4PO_3)_n$
γ-APS γ-Aminopropyltriethoxysilane
ATBC Acetyltributyl citrate
ATH Alumina trihydrate
BA (1) Blowing agent; (2) Butylacrylate
BADGE Bisphenol-A diglycidyl ether

BAMO	Bis(azidomethyl) oxetane
BBP	Butylbenzylphthalate
BCP	Butylcyclohexyl phthalate
BEHA	(1) *N*,*N*-Bis-(2-hydroxyethyl) alkyl (C_8–C_{18}) amine; (2) Bis(2-ethylhexyl)azelate
BFR	Brominated flame retardant
BHA	Butylated hydroxyanisole; *t*-butyl-4-methoxy-phenol
BHC	*Trans*-3,5-di-*tert*-butyl-4-hydroxycinnamic acid
BHEB	Butylhydroxyethyl benzene
BHM	3,5-Di-*tert*-butyl-4-hydroxybenzylmethacrylate
BHS	3,5-Di-*tert*-butyl-4-hydroxystyrene
BHT	(1) Butylhydroxytoluene; (2) β-Hydroxytoluene
BMA	Butyl methacrylate
BOP	Benzyloctylphthalate
BP	(1) 4,4′-Bis-(2,6-di-*t*-butyl-phenol); (2) 2-Hydroxybenzophenones
BPA	Bisphenol-A
BPP	Bispyrene propane
BQM	Bis-quinonemethide
Br_xBB	Bromobiphenyl
$Br_{10}DPO$	Decabromodiphenyl ether
BSA	*N*,*O*-Bis(trimethylsilyl) acetamide
BSE	Backscattered electron
BSTFA	*N*,*O*-Bis(trimethylsilyl)trifluoro acetamide
BT	Benzothiazole
BTBP	Bis(2,4-di-*t*-butylphenyl) pentaerythritol diphosphite
BTBPE	1,2-Bistribromophenoxyethane
BTDA	Benzophenone tetracarboxylic dianhydride
BTEX	Benzene, toluene, ethylbenzene, xylenes
BuA	Butyl acrylate
BuSt	Butyl stearate
BZT	2-Hydroxybenzotriazoles
CA	Caffeic acid
CB	(1) Chain-breaker; (2) Carbon-black
CBA	Chemical blowing agent
CBS, CZ	*N*-Cyclohexyl-2-benzothiazole sulfenamide
CF	Carbon fibre
CHCA	α-Cyano-4-hydroxycinnamic acid
CHP	Cumene hydroperoxide
CLD	Caprolactamdisulfide
CNT	Carbon nanotube
COD	Cyclooctadiene
CRM	Certified reference material
CT	Charge transfer
CTP	*N*-(Cyclohexylthio)phthalimide
CVBS	Cationic vinylbenzyl silane
DAP	Diallylphthalate
DBBP	Decabromobiphenyl (*cfr.* Br_xBB)
DBBQ	2,6-Bis(1,1-dimethylethyl)-2,5-cyclohexadiene-1,4-dione
DBDPE	Decabromodiphenylether
DBDPO	Decabromodiphenyloxide (*cfr.* $Br_{10}DPO$)
DBP	Dibutylphthalate
DBS	(1) Di-*n*-butylsebacate; (2) Dibromostyrene; (3) 1,2,3,4-Di-*p*-methylbenzylidene sorbitol; (4) Sodium dodecyl benzene sulfonate; (5) Dibenzylsulfide
DBTDL	Di-*n*-butyltin dilaurate
DBTDO	Dibutyltin dioleate
DBTM	Di-*n*-butyltin maleate
DCBS	Benzothiazyl-2-dicyclohexyl sulfenamide
DCHP	Dicyclohexylphthalate
DCM	Dichloromethane
DCP	(1) Di-cresylol propane; (2) Dicumyl peroxide
DCPD	Dicyclopentadiene
DDBSA	Dodecylbenzenesulfonic acid
DDP	Didecylphthalate
DDS	4,4′-Diamino-diphenyl sulfone
DeBP	Decylbenzylphthalate
DECA	Decabromodiphenyloxide (*cfr.* DBDPO)
DEG	Diethylene glycol
DEHA	Di(2-ethylhexyl)adipate
DEHP	Di(2-ethylhexyl)phthalate
DENA	*N*,*N*-Diethyl-*p*-nitrosoaniline
DEP	Diethylphthalate
DETU	Diethylthiourea
DGE	Diglycidyl ether
DGEBA	Diglycidyl ether of bisphenol-A
DHA	Di-*n*-hexyl adipate
DHBA	2,5-Dihydroxybenzoic acid (gentisic acid)

DHDP 3,3′-Bis(1,1-dimethylethyl)-5,5-dimethoxy-1,1′-biphenyl-2,2′-diol
DHP Dihexylphthalate
DIBA Diisobutyladipate
DIBP Diisobutylphthalate
DICY Dicyanodiamide
DIDP Diisodecylphthalate
DIHP Diisoheptylphthalate
DIMP Diisopropyl methylphosphonate
DINP Diisononylphthalate
DIOA Diisooctyladipate
DIOP Diisooctylphthalate
DIPA Diisopropyladipate
DIUP Diisoundecylphthalate
DIURON 3-(3,4-Dichlorophenyl)-1,1-dimethylurea
DLO Diffusion-limited oxidation
DLTDP Dilaurylthiodipropionate
DMA (1) Dimethyladipate; (2) 1,3-Dimethyladamantane; (3) Dimethylacrylamide; (4) Dimethylacetamide
DMDTC Dimethyldithiocarbamate
DMF *N*,*N*-Dimethylformamide
DMIP Dimethylisophthalate
DMOP Dimethyl *o*-phthalate
DMP Dimethylphthalate
DMPP Dimethylpropane phosphonate
DMS (1) Dimethyl sebacate; (2) Dimethylsilicone
DNA Dinonyladipate
DNBP Di-*n*-butylphthalate
DNDP Di-*n*-decylphthalate
DNFB 2,4-Dinitrofluorobenzene
DNHP (1) Di-*n*-hexylphthalate; (2) Di-*n*-heptylphthalate
DNOP Di-*n*-octylphthalate
DNP Dinonylphthalate
DNNP Di-*n*-nonylphthalate
DNPG Dibromoneopentylglycol
DNPH 2,4-Dinitrophenylhydrazine
Dnx 2,6-Di-*tert*-butylcatechol
DOA Dioctyladipate
DODPA Di(*t*-octyl)diphenylamine
DOP Dioctylphthalate
DOPPD Dioctyl-*p*-phenylene diamine
DOS Dioctylsebacate
DOS_2 Dioctadecyldisulfide
DOTG 1,3-Di-*o*-tolylguanidine
DPB 1,3-Bis(diphenylphosphono)-benzene
DPDP Distearylpentaerythritol diphosphite
DPG 1,3-Diphenylguanidine
DPMTT Dipentamethylenethiuram-tetrasulfide
DPO Diphenylether
DPP (1) Diphenylphthalate; (2) Dipropylphthalate; (3) Diketopyrrolopyrrole
DPPD *N*,*N*′-Diphenyl-*p*-phenylene-diamine
DPPH Diphenylpicrylhydrazyl
DPTT Dipentamethylenethiuram-tetra-sulfide
DPTU Diphenylthiourea
DQ Duroquinone
DSPDP Distearyl pentaerythritol diphosphite
DSTDP Distearyl 3,3′-thiodipropionate
DTBP (1) 2,4-Di-*t*-butylphenol; (2) Di-*t*-butylperoxide
DTDM Dithiodimorpholine
DTDTDP Ditridecyl thiodipropionate
DTGS Deuterated triglycine sulfate
DTP Diethyldithiophosphate
DUP Diundecylphthalate
DVB Divinylbenzene
DZ *N*,*N*-Dicyclohexyl-2-benzo-thiazolyl sulfenamide
EA (1) Ethyl acrylate; (2) Extrusion aid
EBA *N*,*N*′-Ethylene-bis-stearamide
EBS Ethyl-bis-stearamide
EDAP Ethylene diamine phosphate
EG Ethylene glycol
EGDMA Ethylene glycol dimethacrylate
ELO Epoxidised linseed oil
EMA Ethylmethacrylate
ENB Ethylidene-norbornene (C_9)
EO Ethylene oxide, oxirane
ERM® European Reference Material
ES External standard
ETA Ethanol–toluene azeotrope
ETU Ethylene thiourea (2-mercaptoimidazoline)
FAME Fatty acid methyl esters
FEF Carbon-black, ASTM designation N 550 (S.A. 36–52 $m^2\ g^{-1}$)
FOF Finish-on-fibre
FOY Finish-on-yarn

FPA	Fluoropolymer bound processing aid
FR	Flame retardant
GAn	Ethoxylated C_{14}/C_{16} amines
GF	(1) Glass fibre; (2) Glass-filled
GMA	Glycidyl methacrylate
GMO	Glycerol monooleate
GMP	Glycerol monopalmitate
GMS	Glycerol monostearate
GR	Glass-fibre reinforced
HAF	Carbon-black, ASTM designation N 330 (S.A. 70–90 $m^2\ g^{-1}$)
HALS	Hindered amine light stabiliser
HAS	Hindered amine stabiliser
HBCD	Hexabromocyclododecane
HBHT	2,6-Di-*tert*-butyl-4-hydroperoxy-4-methylcyclohexa-2,5-dienone
HB 307	Mixture of synthetic triglycerides
HC	Hydrocarbons
HEG	Hexaethylene glycol
HET-acid	1,4,5,6,7,7-Hexachlorobicyclo-[2.2.1]hept-5-en-2,3-dicarboxylic acid
HFIP	1,1,1,3,3,3-Hexafluoroisopropanol; hexafluoropropan-2-ol
HFR	Halogenated flame retardant
HMBP	Hydroxymethoxybenzophenone
HMBT	2-(2′-Hydroxy-5′-methylphenyl)-benzotriazole
HMBTAD	*N*,*N*′-Bis(2,2,6,6-tetramethyl-4-piperidyl) 1,6-hexanediamine
HM-HALS	High molecular weight HALS
HMTA	Hexamethylenetetramine
HMW	High molecular weight
HMX	Octahydro-1,3,5,7-tetranitro-1,3,5,7-tetraazacine
HPA	3-Hydroxypicolinic acid
HPPD	*N*-(1,3-Dimethylbutyl)-*N*′-phenyl-*p*-phenylenediamine
HPVC	High production volume chemical (>1000 t/yr/producer *cq.* importer)
HRM	In-house reference material
IA	Isophthalic acid
IAA	3,*β*-Indole acrylic acid
IDBP	4,4′-Isopropylidene-bis(2,6-dibromophenol)
IFR	Intumescent flame retardant
IM	Impact modifier
IOM	Iso-octylmaleate
IOTG	Iso-octylthioglycollate
IPA	Isopropylalcohol
IPPD	*N*-Isopropyl-*N*′-phenyl-*p*-phenylene diamine
IRM	Internal reference material
IS	Internal standard
KB	Ketjenblack
KFR	Karl Fischer reagent
LMW	Low molecular weight
LPVC	Low production volume chemical (10–1000 t/yr/producer *cq.* importer)
LRM	Laboratory reference material
LS	Light stabiliser
LTTS	Long-term thermal stabiliser
MA	Methacrylic acid
MA-CY	Melamine cyanurate, *cfr.* MC
MA(H)	Maleic anhydride
MBS	Benzothiazyl-2-sulfenmorpholide
MBT	(1) 2-Mercaptobenzothiazole; (2) Monobutyltin
MBTS	Bismercaptobenzothiazole *cq.* 2,2′-dibenzothiazyl disulfide
MC	Melamine cyanurate, *cfr.* MA-CY
MDI	4,4′-Methylene bis(phenylene isocyanate); 4,4′-diphenylenemethane diisocyanate
ME	Melamine
MEK	Methylethylketone
MF	Melamine resin (fluorescently labelled microparticles)
MHCD	4-Methyl-4-hydroxy-2,6-di-*tert*-butyl-cyclohexa-2,5-dione
MMA	Methylmethacrylate
MMT	Montmorillonite
MON	Motor octane number
MOR	*N*-Oxydiethylene-2-benzothiazyl sulfenamide (morpholine derivative)
MPTD	Dimethyldiphenylthiuram-disulfide
MSMA	Trimethoxysilylpropylmethacrylate
MT	Carbon-black, ASTM designation N 990 (S.A. 6–9 $m^2\ g^{-1}$)
MTBE	Methyl-*t*-butylether
NA	(1) Nicotinic acid; (2) Norbornene dicarboxylic anhydride
NaBz	Sodium benzoate
NBD	4-(Hexyldecylamino)-7-nitro-benz-2-oxa-1,3-diazole

nBuMA	*n*-Butylmethacrylate
NDI	1,5-Naphthalene di-isocyanate
NiDRC	Nickel dialkyldithiocarbamate
NMP	1-Methyl-2-pyrrolidone
NP	(1) *p*-Nonylphenol; (2) Non-polar
NPE	Nonylphenol ethoxylates
NPEC	Nonylphenol polyethoxycarboxylate
NS	*N*-*t*-Butylbenzothiazole-2-sulfenamide
OBB	Octabromobiphenyl
OBDPO	Octabromodiphenyloxide (*cfr.* octa-BDE)
OBSH	4,4′-Oxy-bis(benzene sulfonyl hydrazide)
Octa-BDE	Octabromodiphenylether (*cfr.* also OBDPO)
ODA	Oxydiphenyldiamine
OFS	Organic formulated stabiliser
OMS	Organomodified siloxanes
OPWF	Oil-palm wood flour
OTBG	*o*-Tolyl-biguanide
OTOS	*N*-Oxydiethylenedithiocarbamyl-*N*′-oxydiethylene sulfenamide
OVI	Organic volatile impurity
PBA	Physical blowing agent
PBBMA	Pentabromobenzylacrylate
PBDD	Polybrominated dibenzo-*p*-dioxins
PBDE, PBDPE	Polybrominated diphenylethers
PBDF	Polybrominated dibenzofurans
PBN, PBNA	*N*-Phenyl-*β*-naphthyl amine
PCA	Pentachloroanisole
PCB, PCBP	Polychlorinated biphenyls
PDA	Phenylenediamine
PDAD-MAC	Poly(diallyldimethyl ammonium chloride)
PE	(1) Photoelectron; (2) Primary electron
PER	Pentaerythritol
PERM	Polymeric elemental reference material
PFA	Perfluoroalkoxy vinyl ether
PG	*n*-Propylgallate
pgm	Platinum group metals
PIC	Phenylisocyanate
PINA	Paraffins/isoparaffines/naphthenes/aromatics
Plg	Tri(mono and dinonylphenol mixture) phosphite
PM	Particulate matter
PMP	Pentamethyl piperidol
PMPME	Pentamethyl piperidol methyl ether
PP	Pentylphenol
PPA	(1) Polymer processing additive; (2) Poly(1,2-propylene adipate)
PPD	*N*-phenyl-*p*-phenylenediamine
6PPD	*N*-phenyl-*N*′-(1,3-dimethylbutyl)-*p*-phenylene diamine
PR	Primer
PROXYL	2,2,5,5-Tetramethylpyrrolidine-1-oxyl
PTR	Proton transfer
RM	Reference material
SAF	Carbon-black, ASTM designation N 110 (S.A. 125–155 $m^2\ g^{-1}$)
SAM	Self-assembled monolayer
$scCO_2$	Supercritical CO_2
SCF	Supercritical fluid (*cfr.* SF)
SDOSS	Sodium dioctylsulfosuccinate
SDS	Sodium dodecyl sulfate
SE	Secondary electron
SEX	Sodium ethyl xanthate
SF	Supercritical fluid (*cfr.* SCF)
SRF	Carbon-black, ASTM designation N 770 (S.A. 17–33 $m^2\ g^{-1}$)
SRM®	Standard Reference Material, registered trademark (NIST)
SSI	Stearyl stearamide
SSL	Sodium stearoyl-2-lactylate
St, StAc	Stearate, stearic acid
TA	(1) Terephthalic acid; (2) Triacetin
TAA	(1) Triacetoneamine; (2) 2,2,2,6-Tetra-methylpiperidin-4-one
TAAH	Tetra-alkylammonium hydroxides
TATB	1,3,5-Triamino-2,4,6-trinitrobenzene
TB	Tribromophenol
TBAC	Tributyl acetylcitrate
TBBA, TBBP-A	Tetrabromobisphenol-A
TBBP-S	Tetrabromobisphenol-*S*-bis(2,3-dibromopropyl ether)
TBBQ	2-(1,1′)-Dimethylethyl-2,5-cyclohexadiene-1,4-dione
TBBS	(1) *N*-*t*-Butyl-2-benzothiazolesulfenamide; (2) Tetrabutylbenzylsulfenamide
TBCP	*t*-Butylcumylperoxide
TBDD	Tetrabromodibenzodioxin
TBDF	Tetrabromodibenzofuran
TBE, TBPE	1,2-Bis(tribromophenoxy)ethane

TBHP	*t*-Butylhydroperoxide
TBHQ	*t*-Butylhydroquinone
TBPP	*t*-Butylperoxypivalate
TBzTD	Tetrabenzylthiuramdisulfide
TCA	Trichloroanisole
TCP	Tricresylphosphate
TeCA	Tetrachloroanisole
TEHP	Tris(2-ethylhexyl)phosphate
TEMPO	2,2,6,6-Tetramethyl-1-piperidinyl-oxyl
Tenax	Adsorbent charcoal
TEOS	Tetraethylorthosilicate
TES	Tetraethoxysilane
TET	(1) Triethyltin; (2) Tetraethyltin
TFE	Tetrafluoroethylene
TGI	Triglicidyl isocyanurate
THF	Tetrahydrofuran
TMA	Trimellitic acid
TMAH	Tetramethylammonium hydroxide
TMATEMPOI	4-Trimethylamino-2,2,6,6-tetra-methylpiperidine oxide iodide
TMDQ	2,2,4-Trimethyl-1,2-dihydro-quinoline
TMPAH	Trimethylphenylammonium hydroxide
TMQ	2,2,4-Trimethyl-1,2-dihydro-quinoline
TMS	Tetramethylsilane (internal standard)
TMSH	Trimethylsulfonium hydroxide
TMTD	Tetramethylthiuram disulfide
TMTM	Tetramethylthiuram monosulfide
TNPG	Tribromoneopentylglycol
TNPP	Tris(nonylphenyl) phosphite
TO	Thermo-oxidation
TOC	Total organic carbon
TOTM	Trioctyl trimellitate
TPC	Tri(methyl)phenylphosphate
TPP	(1) Triphenyl phosphate; (2) Triphenylphosphine
TPP-i	Intercalated/modified triphenylphosphine
TTP	Tritolyl phosphate
UDP	Undecylphthalate
UFP	Ultrafine powder
UQ	Ubiquinone
UVA	UV absorber
VA	Vinyl alcohol
VAc	Vinyl acetate
VC	Vinyl chloride
VCH	Vinylcyclohexene
VCM	Vinylchloride monomer
VOCs	Volatile organic compound(s)
VOH	Vinyl alcohol
VTMOS, VTMS	Vinyltrimethoxysilane
XS	Xylene soluble
YAG	Yttrium aluminum garnet
ZBEC	Zinc benzyldiethyldithiocarbamate
ZDBC	Zinc dibutyldithiocarbamate
ZDC	Zinc dithiocarbamate
ZDEC	Zinc-*N*-diethyldithiocarbamate
ZDMC	Zinc-*N*-dimethyldithiocarbamate
ZEPC	Zinc-*N*-ethyl-phenyl-dithiocarba-mate
ZHS	Zinc hydroxystannate
ZMBT	Zinc-2-mercaptobenzothiazole
ZnSt	Zinc stearate
ZS	Zinc stannate
Z5MC	Zinc-*N*-pentamethylenedithio-carbamate

PHYSICAL AND MATHEMATICAL SYMBOLS

A	Absorbance matrix
A	(1) Mass number of a nucleus; (2) Absorbance; (3) Area
Å	Ångstrom, unit of wavelength, $1\ \text{Å} = 10^{-8}$ cm
a, ag	Atto (10^{-18}), attogram
a	Hyperfine coupling constant (EPR)
AC	Alternating current
AU	Absorbance unit
B	(1) Minimum hole size; (2) byte
B	Magnetic field strength
B_0	Static magnetic field (flux density)
$\mathbf{B}_0$	External (applied) magnetic field amplitude (NMR)
BE	Binding energy
b.p.	Boiling point
BW	(1) Beam width; (2) Band width
C	(1) Degrees Centrigrade; (2) Coulomb
C	Concentration matrix
C, *c*	(1) Concentration or molar concentration; (2) Thermal capacity
c	Velocity of light
CA	Cluster analysis
CCM	Colour contrast matching

Ci	Curie
CLS	Classical least-squares
COF	Coefficient of friction
CP	Curie-point
c_p	(Specific) heat capacity (at constant pressure)
CV	(1) Coefficient of variation; (2) Certified value
CVA	Canonical variance analysis
CW	Continuous wave (laser)
D	(1) Debye; (2) Diffusion; (3) Dispersive; (4) Dimension
D	(1) Diffusion coefficient; (2) Distribution ratio
D_0	Self-diffusion coefficient
d	(1) Diameter, thickness; (2) Density; (3) Diffusion path length; (4) Interplanar spacing of crystal; (5) Distance
$\mathrm{d_p}$	Penetration depth
d_p	Particle diameter
Da	Dalton or atomic mass unit
dB	Decibel
DECRA	Direct exponential curve resolution algorithm
DP	Differential pressure
E	Electrical field strength
E	(1) Energy (in eV); (2) Potential; (3) Elasticity
E_{AB}	Energy of coupling interaction between nuclei A and B
E_γ	Energy of an emitted photon
E_0	Threshold energy
E_R	Recoil energy
e	Unit charge of an electron
e^-	Electron
EA	Electron affinity
EB	Electron beam
E&E	Electrical and electronic
EFA	Evolving factor analysis
EM	Electromagnetism
em	Emission wavelength used in fluorescence detection
EMI	Electromagnetic interference
EMSA	Electron microscope surface area
EOF	Electro-osmotic flow
erf(z)	Error function
ESC	Environmental stress cracking
ESD	Electrostatic discharge
ES(TD)	External standardisation (calibration)
eV	Electron volt; 23.06 kcal mol^{-1}
EVAP	Evaporative emission
ex	Excitation wavelength used in fluorescence detection
F	Fluorescence intensity
f	(1) Frequency; (2) Inhibition coefficient
f	(1) Function (general); (2) Recoil-free fraction; (3) Volume fraction
f, fg, fmol	Femto (10^{-15}); femtogram; femtomole
FA	Factor analysis
FC	Fuzzy clustering
FFT	Fast Fourier transform
FID	Free induction decay time-domain signal
FOD	Fibre orientation and distribution
FOM	Figure(s) of merit
F(r)	Interatomic/intermolecular force
$f(R_\infty)$	Reflectance function, Kubelka–Munk function
fs	Femto second (10^{-15} s)
FSQ	Full spectrum quantitation
FT	Fourier transform
FVP	Functional validation and precision
FWHH, FWHM	Full-width at half-height/maximum
G	(1) Gauss unit of magnetic field strength; (2) Giga (10^9)
G	(1) Free enthalpy (Gibbs free energy); (2) Geometric term; (3) Magnetic induction
g	(1) Gram; (2) Gradient pulse amplitude
g	Spectroscopic splitting factor (ESR)
$g(\lambda)$	Wavelength response characteristics of detector
GA	Genetic algorithm
GRAM	Generalised rank annihilation method
G(t)	Time-dependent spatially linear magnetic field gradient
Gy	Gray
H	Hamiltonian
h	Hecto
h, hr	Hour
h	Planck's constant
$\mathbf{H}_0$	Magnetic field of constant strength (ESR)

$\boldsymbol{H}_{CS}$	Chemical shift Hamiltonian
$\boldsymbol{H}_{D}$	Dipolar interaction Hamiltonian
$\boldsymbol{H}_{J}$	Nuclear-nuclear interaction Hamiltonian
$\boldsymbol{H}_{Q}$	Quadrupolar interaction Hamiltonian
$\boldsymbol{H}_{Z}$	Zeeman interaction Hamiltonian
HF	High frequency
hfs	Hyperfine splitting
HP-OIT	High-pressure oxidative induction time
HPV	High production volume
HR	High resolution
HV	(1) High voltage; (2) High vacuum
Hz	Hertz, unit of frequency (cycles per second)
$h\nu$	Photon energy in eV
I	(1) Magnetic spin of a nucleus, angular momentum quantum number (integer or half-integer); (2) Current; (3) Intensity
I_0	Intensity of incident light
I&C	Instrumentation and control
i.d.	Internal diameter
IE	Ionisation energy (formerly Ionisation potential)
ILS	Inverse least squares
IS(TD)	Internal standardisation (calibration)
J	Joule, a unit of energy
J	Mass flux
J	Spin coupling constant (NMR)
K	Kelvin
K	(1) Partition coefficient or equilibrium constant; (2) Force constant of a bond; (3) Reduced ion mobility; (4) Response factor
k	(1) Kilo (10^3); (2) Boltzmann constant
k	Wave vector
k	(1) Molar absorption coefficient; (2) Retention factor; (3) Thermal conductivity
k_{AB}	Cliff–Lorimer sensitivity factor
KE	Kinetic energy
K-M	Kubelka–Munk (theory/equation)
L	Litre
L	Length (column length)
l	Pathlength
LASER	Light Amplification by Stimulated Emission of Radiation
LN	Liquid nitrogen (temperature)
LOD	(1) Limit of detection (*cfr.* MDQ); (2) Loss on drying
LOI	Loss on ignition
LOQ	Limit of quantitation
LSR	Least-squares regression
LTHA	Long term heat ageing
M	(1) Molarity (moles/L); (2) Mega (10^6)
M	Net (macroscopic) magnetisation vector
M	(1) Atomic or molecular weight; (2) Adsorption constant at two interfaces
m	(1) Milli; (2) Metre
m	(1) Nuclear spin quantum number; (2) Mass of atom or ion
M_0, M_z	(Equilibrium) longitudinal magnetisation
$\langle M_n \rangle$	Number average molecular weight
$\langle M_w \rangle$	Weight average molecular weight
$\mathbf{M}_{x,y}$	Component of the net (macroscopic) magnetisation vector in the x, y plane
mCi	MilliCurie
MCR	Multivariate curve resolution
MD	(1) Mahalanobis distance; (2) Molecular dynamics
MDQ	Minimum detectable quantity (*cfr.* LOD)
MFI, MI	Melt flow index
mg, mmol, mL	Milligram, millimole, millilitre (10^{-3})
mil	0.001 inch
MLR	Multilinear regression
MLS	Multiple least squares
MLWR	Multilinear wavelength regression
MM	Mathematic morphology
mmu	Milli mass unit
m.p.	Melting point
MPa	Mega Pascal
MSC	Multiple scattering correction
MSPC	Multivariate statistical process control
MTBF	Mean time between failure
MVA	Multivariate analysis
MVC	Multivariate calibration
MW	Molecular weight
MWD	Molecular weight distribution

m/z	Mass-to-charge ratio
N	(1) Newton; (2) Normal
N	(1) Number of neutrons in a nucleus; (2) Noise
n	Refractive index
n_0	Refractive index of internal reflectance element
n	(1) Number of components; (2) Number of measurements; (3) Diffraction order
NA	Numerical aperture
ND	Not detectable
NEP	Noise equivalent power
ng, nm, nmol	Nanogram, nanometre, nanomole (10^{-9})
ns	Nano second (10^{-9} s)
o.d.	Outer diameter
OIT	Isothermal oxidative induction time (min)
OIT*	Oxidative induction temperature (°C)
OOS	Out-of-specification
OOT	Oxidation onset temperature (*cfr.* OIT*)
P	Calibration or regression matrix
p	Pico (10^{-12})
p	(1) Pressure; (2) Vapour pressure
p_c	Critical pressure
Pa	Pascal
PC	Personal computer
PCA	Principle component analysis
PCR	Principle component regression
PCS	Principle component score
PD	Polydispersity
PDA	Principal discriminant analysis
PFG	Pulsed field gradient
pg, pmol	Picogram, picomole (10^{-12})
phr	Parts by weight per hundred parts resin
PII	Period from injection to injection
Pixel	Picture element
PLS(R)	Partial least-squares (regression)
PM	Phase modulation
pm	Picometre
PMD	Principle components/Mahalanobis distance discriminant analysis
ppb	Parts per billion
pph	Parts per hundred
ppm	Parts per million
ppq	Parts per quadrillion
ppt	Parts per trillion
PRA	Pattern recognition analysis
ps	Pico second (10^{-12} s)
psi	Pounds per square inch
Q	(1) Quadrupolar field; (2) Electric quadrupole moment (NQR)
q	(1) Wave vector; (2) Internuclear distance; (3) Area of gradient pulse
q	Charge density
R	Isotope ratio
R	Spin displacement
R	(1) Universal gas constant; (2) Reproducibility limit ($R = 2.8 \times s_R$); (3) Reflectance; (4) Rate of luminescent reaction; (5) Resolution
R_0	Diffuse reflectivity
R_s	Resolution
R_∞	Absolute diffuse reflectance at infinite depth
R^2	Square of the multiple correlation coefficient
r	(1) Reaction rate; (2) Internuclear distance; (3) Intralaboratory 95% confidence level (repeatability limit $r = 2.8 \times s_r$); (4) Radius
RF, rf	Radio-frequency
R-G	Rosencwaig–Gersho (PAS)
RGB	Red green blue ratio
r.h.	Relative humidity
RI	Retention index
rms	Root mean square
RMSEP	Root mean square error of prediction
ROI	Residue on ignition
R&R	Reproducibility and repeatability
RRT	Relative retention time
RSC	Relative sensitivity coefficient
RSD, r.s.d.	Relative standard deviation
$R(t)$	Reaction rate
r(t)	Time-dependent spin position
r.t.	Room temperature
r_t	Retention time (*cfr.* t_R)
S	(1) Sensitivity factor; (2) Solubility
S	(1) Selectivity; (2) Solubility coefficient; (3) Scattering coefficient; (4) Surface charge (correction term)
S_f	Specific interaction factor
S_0	Electronic ground state
s	Second

s_i	Intralaboratory standard deviation for measurement series
s_L	Interlaboratory standard deviation
s_r	Repeatability standard deviation
s_R	Reproducibility standard deviation
S.A.	Surface area
SEC	Standard error of calibration
SEP	Standard error of prediction
SI	Système International d'Unités
SIMCA	Soft independent modelling of class analogies
S/N, SNR	Signal-to-noise ratio
SPC	Statistical process control
SQC	Statistical quality control
S-T	Stejskal–Tanner (NMR)
Std-OIT	Standard oxidative induction time
STP	Standard temperature and pressure
SVM	Support vector machine
T	(1) Tesla, unit of magnetic field strength (10^4 Gauss); (2) Tera (10^{12})
T	(1) Absolute temperature (K); (2) Transmittance
T_c	Critical temperature
T_{eq}	Equilibrium temperature
T_g	Glass transition temperature
T_m	Melting temperature
T_1, T_{1r}	Nuclear spin–lattice (longitudinal) relaxation time; in the rotating frame
T_2	Nuclear spin–spin (transverse) relaxation time
T_{15}	Temperature of oxidation of 15% CB
t	ton
t	(1) Time(s); (2) Layer thickness
$t_{1/2}$	Half-life time
t_1	Evolution time (NMR)
t_2	Detection time (NMR)
t_p	Pulse width
t_R	Retention time
TE	Echo time
THT	Total heating time
TIC	(1) Total ion current; (2) Total ion chromatogram
TLI	Total luminescence intensity
TOA	Take-off angle
TOIT	Temperature dependent oxidative induction time
TRT	Temperature-rise time
TTP	Temperature-time profile
TV	Television
U	(1) Acceleration voltage; (2) Expanded uncertainty
u	Unit
UHF	Ultra-high frequency
UHMW	Ultra-high molecular weight
UHV	Ultra-high vacuum
UV-A	UV wavelength range 315–380 nm
UV-B	UV wavelength range 280–315 nm
UV-C	UV wavelength range 200–280 nm
V	Volt
V	(1) Volume; molar volume; (2) Velocity
v	Recoil velocity
VI	Viscosity index
Voxel	Volume element
W	Watt, measure of RF power
w	Modulation frequency
w(r)	Interatomic/intermolecular potential energy
w/w	Weight/weight (solution concentration)
x	Crystallinity
x	Mole fraction (in general)
$x_{med,i}$	Mean value of a series of experiments in laboratory *i*
x, y, z	Cartesian co-ordinates
y	General mean
YI	Yellowing index
Z	(1) Atomic number; (2) Number of ions
z	Zepto (10^{-21})
z	(1) Axis of $\mathbf{B}_0$, the external (applied) magnetic field; (2) Number of charges on an ion; (3) Depth; (4) Tip-sample separation (SPM)
ZAF	Atomic number (Z), absorption, fluorescence correction (EPMA)
Z-score	s_r-Normalised deviation of a laboratory mean value from the total mean value

Physical and Mathematical Greek Symbols

α	(1) Orientation (with respect to $\mathbf{B}_0$) of the magnetic moment of an $I = 1/2$ nucleus; (2) Flip angle in pulsed NMR; (3) Polarisability; (4) Thermal diffusivity; (5) Attenuation (ultrasonics); (6) Angle

β Bohr magneton
γ (1) Gyromagnetic ratio of a nucleus; (2) Gamma ray
Δ (1) Shift or difference (*e.g.* ΔE, energy difference); (2) Symbol for heat; (3) Phase evolution time; (4) Duration between the gradient pulses (time over which diffusion is measured)
Δc_p Heat capacity change
Δf Line width (NQR)
ΔH_f Molar heat of fusion, J/mol
ΔH_m Melting enthalpy
ΔH_r Molar heat of reaction
δ (1) Chemical shift (ppm relative to a reference); (2) Solubility parameter; (3) Phase shift (DIES); (4) Dissipation factor; (5) Duration of the gradient pulse
ε (1) Molar extinction coefficient; (2) Dielectric constant
ε^* Complex dielectric constant
ε' Real part of complex dielectric constant
ε'' Imaginary part of complex dielectric constant (dielectric loss)
ε_o Permittivity of free space
θ (1) Angle between internuclear vector and $\mathbf{B}_0$; (2) Incident angle; (3) Bragg angle
θ_f Quantum yield of analyte molecule
λ (1) Wavelength, unit Å; (2) Decay constant
μ (1) Magnetic moment of a nucleus; (2) Dipole moment; (3) Micro (10^{-6}); (4) Reduced mass of a system; (5) Thermal diffusion length
μg, μm Microgram, micron
ν (1) Wavenumber; (2) Velocity
ρ Density; unit g cm^{-3}
σ (1) Standard deviation; (2) Nuclear shielding constant; (3) Cross-section
τ (1) Time constant (detector); (2) Lifetime; (3) Transmittance
τ_c Molecular correlation time
ϕ (1) Nuclear spin phase; (2) Volume fraction of solute (ϕ_1) and polymer (ϕ_2) in a mixture; (3) Chemiluminescence yield; (4) Spectrometer work function; (5) Diameter
ϕ_f Fluorescence quantum yield
ψ Take-off angle
ω (1) Angular velocity (rad s^{-1}); (2) Light modulation frequency; (3) Spin resonance frequency
∇ (1) Vector operator; (2) Concentration gradient

GENERAL ABBREVIATIONS

ACD Advanced Chemistry Development (Toronto, ON)
AI Artificial Intelligence
AIP American Institute of Physics (New York, NY)
AIST National Institute of Advanced Industrial Science and Technology (Tokyo, J)
AOAC Association of Official Analytical Chemists International (Arlington, VA)
AQC Analytical quality control
ASM American Society for Metals
ASME American Society of Mechanical Engineering
ASTM American Society for Testing and Materials (West Conshohocken, PA)
BAM Bundesanstalt f. Materialforschung u.-prüfung; German Federal Institute for Materials Research and Testing (Berlin, D)
BCR Bureau Communautaire de Référence; European Commission DG XII Community Bureau of Reference (Geel, B); now IRMM
BCS British Chemical Standards
BITMP Bureaux Internationaux Techniques des Matières Plastiques
BNL Brookhaven National Laboratory (USA)
BS British Standards (*cfr.* BSI)
BSI British Standards Institution (London, GB)
BTI BRG Townsend Inc. (Mt. Olive, NJ)
CAQ Computer Aided Quality Control
CAS Chemical Abstracts Service (USA)
CEC Commission of the European Communities (Brussels, B)
CEN Comité Européen de Normalisation; European Committee for Standardisation (Brussels, B)
CFR Code of Federal Regulations (USA)

CI, C.I.	Colour Index
COMAR	Code of Reference Materials
CRMMA	Chemical Reference Materials Manufacturers Association
CSBTS	China State Bureau of Technology Supervision (Beijing, PRC)
DFO	Deutsche Forschungsgesellschaft f. Oberflächenbehandlung
DIK	Deutsches Institut f. Kautschuktechnologie (Hannover, D)
DIN	(1) Deutsches Institut für Normung, German Institute on Standardisation (Berlin, D); (2) Deutsche Industrie Normen (German Industrial Standards)
DIS	Draft International Standard (ISO)
DQ	Design or Development Qualification
EC	European Community
EC DG	European Commission Directorate-General
EEC	European Economic Community
EEE, E&E	Electrical and Electronic Equipment
EFG	European Fibre Group (*cfr.* ENFSI)
EMPA	Eidgenössische Materialprüfungs- und Forschungsanstalt, Swiss Federal Laboratories for Materials Testing and Research (St. Gallen, CH)
EN	European Norm
ENFSI	European Network of Forensic Science Institutes
EPA	Environmental Protection Agency (USA)
EPG	European Paint Group (*cfr.* ENFSI)
EQ	Equipment qualification
EU	European Union
EUCAP	European Collection of Automotive Paints
EURACHEM	Association of European Chemical Laboratories (Lisbon, P)
FAAM	Food Additives Analytical Manual
FDA	Food and Drug Administration (USA)
FDIS	Final Draft International Standard (*cfr.* ISO)
GEFTA	Gesellschaft f. Thermische Analyse, German Society for Thermal Analysis
GLP	Good Laboratory Practice
GMP	Good Manufacturing Practice
ICH	International Conference on Harmonisation
ICT	Information and communication technology
ICTA(C)	International Confederation of Thermal Analysis (and Calorimetry)
ID	Identification
IEC	International Electrotechnical Commission
ILAC	International Laboratory Accreditation Co-operation
ILT	Interlaboratory test
IMEP	International Measurement Evaluation Program
INSPEC	Information Service for Physics, Electronics and Computing
IQ	Installation Qualification
IRMM	Institute for Reference Materials and Measurements (Geel, B)
ISA	Instrumentation, Systems and Automation Society (Research Triangle Park, NC)
ISO	International Organization for Standardization (Geneva, CH)
ISO-REMCO	ISO Council Committee on Reference Materials
IUPAC	International Union of Pure and Applied Chemistry
JIS	Japanese Industrial Standards (*cfr.* JISC)
JISC	Japanese Industrial Standards Committee (Tokyo, J)
JRC	Joint Research Centre
JSAC	Japan Society for Analytical Chemistry (Tokyo, J)
JSCTA	Japan Society for Calorimetry and Thermal Analysis (Tokyo, J)
JV	Joint venture
LGC	Laboratory of the Government Chemist (Teddington, UK)
LGC-ORM	LGC-Office of Reference Materials (Teddington, UK)
LNE	Laboratoire National d'Essais (Paris, F)
MQ	Maintenance Qualification
NAMAS	National Measurement and Accreditation System (UK)
NATA	National Association of Testing Authorities (AUS)
NATAS	North American Thermal Analysis Society (USA)
NBS	National Bureau of Standards (now NIST)

NEN	Netherlands Institute for Normalisation (formerly NNI) (Delft, NL)
NF	French Standards
NIST	National Institute of Standards and Technology (formerly NBS) (Gaithersburg, MD)
NMI	Nederlands Meetinstituut (Delft, NL)
NNI	Nederlands Normalisatie Instituut (now NEN)
NPL	National Physical Laboratory (Teddington, UK)
NRL	National Reference Laboratory
NSLS	National Synchrotron Light Source (USA)
OQ	Operational Qualification
PDF	(1) Portable document file; (2) Powder Diffraction File (ASTM)
PDL	Plastics Design Library (USA)
PQ	Performance Qualification
PS	Product Stewardship
PT	Proficiency Testing
PTB	Physikalisch-Technische Bundesanstalt (Braunschweig and Berlin, D)
QA	Quality Assurance
QC	Quality Control
QCAD	Quality Control of Analytical Data
QLS	Quality Assurance and Laboratory Information System
QM	Quality Management
QUID	Quantitative Ingredient Declaration
RCRA	Resource Conservation and Recovery Act (USA)
R&D	Research and Development
REMCO	Council Committee of Reference Materials (ISO, Geneva, CH)
RM&PT	Reference materials and proficiency testing schemes
RoHS	Restrictions on Hazardous Substances
SM&T	Standards, Measurements and Testing Programme, EU (formerly BCR)
SOP	Standard Operating Procedure
SPE	Society of Plastics Engineers (Brookfield, CT)
SPI	Society of the Plastics Industry (Washington, DC)
SPIE	International Society for Optical Engineering (Bellingham, WA)
STJ	SensIR ST (Japan)
TAI	TA Instruments
™	Trademark
TM	Thermographic material
TQ	Total Quality
UKAS	United Kingdom Accreditation Service (formerly NAMAS)
UL	United Laboratories
UN	United Nations
USEPA	United States Environmental Protection Agency
USP	United States Pharmacopœia
VAM	Valid Analytical Measurement
VDA	Verband der Automobilindustrie, German Federation of Car Industry (Frankfurt, D)
VDI	Verein Deutscher Ingenieure, Association of German Engineers (Düsseldorf, D)
VIM	International Vocabulary of Basic and General Terms in Metrology (ISO)
WEEE	Waste Electrical and Electronic Equipment

Subject Index

A

ABS, additives 348, 629
  Antioxidants 361, 370
  Flame retardants 25, 183 ff, 255, 271, 488, 496
  HALS 557
  Rubber distribution 488
  Volatiles 278
ABS, analysis
  EPMA 500
ABS, outgassing 288
ABS/PC, additives
  Flame retardants 197
ABS/PVC, additives
  Flame retardants 254
Accelerators, analysis
  HS-GC 285
Acid scavengers, analysis
  ToF-SIMS 437
Acoustic emission, analytical method 716 ff
  Applications 717 ff
Acrawax: trade name; lubricants
Acrylic fibres, additives
  Dyes 539, 633
Acrylics, additives 446
Actellic: trade name; pesticides
Additive blends, deformulation 606
Adekastab: trade name; nucleating agents
Adhesion, analysis
  CFM 511
  SIMS 430 ff
  XPS 418
Adhesion promoters, analysis
  ATR-FTIR 540
  Fluorescence imaging 541
  iSIMS 572
  μRS 540
Adhesives, analysis
  PyGC 230
  PyIR 263
  PyMS 240
Adine: trade name; flame retardants
AEM, analytical method 497 ff
AES, analytical method 409 ff
  Applications 411
AFM, analytical method 504 ff
  Applications 509 ff
Age Rite: trade name; aromatic amines
Alloprene: trade name; binders
Alurofen: trade name; antioxidants
Ambersorb: trade name: sorbents
Amgard: trade name; flame retardants
Analytical performance parameters 751 ff
  Accuracy 752
  Analytical range 753
  Limit of detection 753
  Limit of quantitation 753
  Linearity 753
  Precision 752
  Recovery 754
  Robustness 754
  Ruggedness 753
  Selectivity 752
  Specificity 752
Anox: trade name; phenols, phosph(on)ites
Antiblocking agents, analysis
  ATR-FTIR 31
  DIES 126
  Process IR 687
  ToF-SIMS 430
  XPS 417; iXPS 565
Antihydrolysis agents, analysis
  PyGC-FTIR 264
Antioxidants, analysis 638
  AFM 511
  CL 92; CL-OIT 88; ICL 544 ff
  DRIFTS 27
  DSC 170 ff
  DTA 174
  IR 17 ff, 21 ff; μFTIR 528; process IR 687
  iSIMS 570
  LD/EI-FTMS 370; LD-FTMS 361
  L^2ToFMS 370 ff
  MALDI-ToFMS 381
  NIRA 47; NIRS 46
  NMR 104, 647
  PyGC 229; PyGC-MS 253
  RS 646; μRS 539
  TD-GC-MS 296
  TEA-FID 278
  TGA 183
  TG-DTA 192
  TG-MS 204 ff
  ToF LMMS 387
  ToF-SIMS 431 ff
  UV 6 ff
Antioxidants, performance

DSC-OIT 170
Antiozonants, analysis
DSC-OIT 172
L^2ToFMS 371
ToF LMMS 386
Antistatic agents, analysis
TGA 183
ToF-SIMS 433
XPS 417
Antiwear agents, analysis
XPS 419
AO: trade name; phenols, amines, phosphites
aPP, additives 436
Armoslip: trade name; lubricants, slip additives
Armostat: trade name; antistatics
Art materials, diagnostics
DT-MS 274
ESEM 492
LDMS 363
LIBS 351
LIF 346
μFTIR 527
μRS 540
μUV 521
PyGC 235; PyGC-MS 257
Ash, analysis
TGA 182, 757
Atmer: trade name; antifogging additives, antistatics, slip additives, lubricants
ATR-FTIR, analytical method 28 ff
Applications 30 ff

B

BC: trade name; flame retardants
Beer–Lambert law 633, 639
Biocides, analysis
μRS 539
Biomer: trade name; PEUU grade
Blooming, analysis 213
ATR-FTIR 31
PA-FTIR 70
ToF-SIMS 436
XPS 416; iXPS 566
Blowing agents, analysis
DSC 167; PDSC 173
NMRI 551
Process NIRS 700
TG-FTIR 198
VMI-TG-MS 210
BR, additives 242, 273
Buna: trade name; rubber grade

C

CA: calcium stearates
Calibration 739
Camel: trade name; fillers
CAO: trade name; phenols, phosph(on)ites
Carbon-black, analysis 750
ATR-FTIR 33
DIES 126
LR-NMR 713; NMRI 553
μNEXAFS 563
OM 472
PA-FTIR 71
PyGC 234
SEM 488
SKM 514
SPM 504
TEM 496
TGA 186
TG-DTA 191
ToF-SIMS 430
Carbotrap: trade name; sorbents
Carbowax: trade name; sorbents
Cariflex: trade name; copolymer grade
Catalysts, analysis
EPMA 501
Fluorescence 79
ICL 544
μRS 541
μXRF 564
SSIMS 430
XPS 418
XRM 561
Cellulose acetate, additives
Plasticisers 48, 205
Cellulose, additives 341
Plasticisers 627
Wetting 492
Cellulosics, analysis
PyGC-MS 256
Cereclor: trade name; flame retardants
Chemiluminescence, elemental analysis
CLND 83
SCD 83
Chenantox: trade name; phenols
Chimassorb: trade name; HALS, UV absorbers, Ni quenchers
Chromatography, quantitative 624 ff
GC 626 ff; GC-MS 649, 651
HPLC 628 ff; RPLC 629
SFC 629
TLC 630 ff, 633
Chromosorb: trade name; sorbents
CL, analytical method 82 ff
Applications 88 ff
Russell mechanism 84
Cloisite: trade name; organoclays
CLSM, analytical method 480
Applications 481 ff
Coatings, additives 653
Binders 231
HALS 118, 520
Lubricants 571
Smoothing agents 571
UV absorbers 8, 520, 570
Coatings, analysis
AFM 510
ATR-FTIR 32
DHS-GC-FID 288
ESR 118
iSIMS 570

μUV 520
OM 472
Py-FIMS 243
PyGC 231; PyGC-MS 257
PyIR 262
ToF-SIMS 437, 653
Colorants, analysis
μVIS 521
Colour body analysis 8
Colour measurement 5 ff
Compatibilisers, analysis
μRS 538
Concentration profiling
μFTIR 528
Confocal microscopy, analytical method 478 ff
Consumer electronics, analysis
LIBS 349
Contaminants, analysis 460, 530
LEIS 444
μFTIR 526 ff
μWAXS 559
μXRF 564
OM 470; PLM 472
SIMS 430 ff; iSIMS 571
XPS 419
Controlled release 204
Corona treatment
SIMS 430
Corvic: trade name; flame retardants
Cotton, additives
Dyes 65, 703
Flame retardants 175, 256
Sizing agents 70
Coupling agents, analysis
DRIFTS 644
NIRS 44
Process IR 692
PyGC 231; PyGC-AED 265
PyGC-FTIR 264
CR, additives 242
Cratering
iSIMS 572
Cross-linking agents, analysis
ESR 115
PyGC-MS 257
SIMS 433
TG-MS 204
TVA 281
Cross-validation 754
Crystallinity 715
CSFM, analytical method 480
Applications 483
CSOM, analytical method 479
Applications 482
Curing agents, analysis
DSC 166
μRS 540
PyGC 232
Cyagard: trade name; UV absorbers
Cyanox: trade name; phenols, thiosynergists
Cyasorb: trade name; phenols, HALS, UV absorbers
Cycoloy: trade name; ABS blends

D

Dammar: natural triterpenoid resin (varnish)
Dastib: trade name; HALS
Databases
FTIR 20
MS 20
NMR 20
Raman 540
SIMS 426, 432
VW/Shimadzu, additive library 247
Dechlorane: trade name; flame retardants
Degradation products, analysis
ESR 115
FTIES 74
HS-SPME 291
ICL 542 ff
IR 23
μRS 541; RRS 63
TD-GC-MS 296
TG-FTIR 198
ToF-SIMS 436; iSIMS 572
Delamination 193
Depth profiling, analysis 335, 460
ATR-FTIR 32, 518
DRIFTS 27
L^2MS 373
μFTIR 18
μRS 537
PA-FTIR 70
PAS 68
RBS 445 ff
SIMS 428; iSIMS 573
Vibrational spectroscopy 14
XPS 415
Derivative spectroscopy, analytical method 636
Applications 638
DIES, analytical method 123 ff, 719
Applications 125 ff, 719
Diffusion, analysis 22, 105 ff
ATR-FTIR 32
DSIMS 439
ESR 116; ESRI 556
μFTIR 528
NMRI 552
RBS 446
XPS 417
Digital chromography 519
Diolpate: trade name; pesticides
Discolorants, analysis
TD-GC-MS 299
Dispersing agents, analysis
AES/SAM 411
LD-FTMS 363
Dispersion, analysis
AET 719
OM 470 ff
SAM 494
Distribution profiling, analysis

μFTIR 528
μRS 539
μUV 520; UV 7 ff
SAM 493
DOSY, analytical method 108
Doverphos: trade name; phosph(on)ites
Dowlex: trade name; LLDPE grade
DRIFTS, analytical method 25 ff
Applications 27 ff
DSC, analytical method 163 ff
Applications 165 ff
DTA, analytical method 173 ff
Applications 174 ff
DT-MS, analytical method 268
Dyeability, analysis
XPS 418
Dyes, analysis
ATR-FTIR 33
DRIFTS 27
Fluorescence 81
FTIES 75
IR 25
LD-FTMS 370
LMMS 387; LMMS mapping 567
NIRA 697; NIRS 50
NSOM 513
PA-VIS 69
Phosphorescence 82
PyGC 232; PyGC-MS 258
QTLC 633
RS 59 ff, 646; μRS 539; process RS 703
RRS 62; SERRS 65
UV 10
UV-LDI-ToFMS 363
Dynamar: trade name; processing aids
Dynamic mechanical analysis 160
Dynamic processes, analysis
NMRI 551
Dyneema: trade name; UHMWPE fibre
Dyneon: trade name; lubricants

E

Ebecryl: trade name; acrylic resin
EDS, analytical method 498
EELS, analytical method 498 ff
Elastollan: trade name; poly(ester urethane) elastomer
Elastomers, additives 79
Antioxidants 615 ff
Antiozonants 170
Ash content 757
Coupling agents 44
Cross-linking agents 257
Fillers 18, 93, 553, 713
Peroxides 167
Plasticisers 477
Vulcanisation accelerators 102, 229, 257
Elastomers, analysis
Fluorescence 79
NMR 102; NMRI 552 ff
Electron microscopy, analytical method 483 ff
Electron spectroscopy, analytical method 408 ff
Applications 409
Elemental analysis
AES 409 ff
LA-ICP-AES/MS 338 ff
LIBS 348
LMMS 385 ff
μXRF 563 ff
SIMS 422 ff
XPS 411 ff
Emission spectroscopy, analytical method 72 ff
Engineering plastics, additives
Fillers 605
EO-PO, analysis
NIRS 48
EPDM, additives
Extender oil 181, 623
Gels 341
Plasticisers 198, 205, 620 ff
Stabilisers 545
Epikote: trade name; epoxy resin
EPM, additives
HALS 557
EPMA, analytical method 499
Applications 500
Epoxy resins, additives
Flame retardants 370
Hardeners 475
Moisture 392
Epoxy resins, analysis
TPPy-MS 273
EPR, additives 82
Flame retardants 255
ERL: trade name; epoxides
ESEM, analytical method 491 ff
Applications 492 ff
ESR, analytical method 112 ff
Applications 115 ff
ESRI, analytical method 546, 555 ff
Applications 556 ff
Ethanox: trade name; phenols, phosph(on)ites
EVA, additives
Antiblocking agents 31, 126
Antioxidants 644
Fillers 33, 527
Flame retardants 105
Monomers 714
UV absorbers 8
EVA melt, additives 688, 699
Monomers 719
Evolved gas analysis 159, 192, 195 ff, 200, 227, 277
Extender oil, analysis
Extraction 623
Quantitative 623
Extenders, analysis
XPS 431
Extraction 609 ff
Extracts, analysis 240
Extrusion aids, analysis
NMRI 554
Exudation, analysis

FAB-SSIMS 439

F

F, FR: trade name; flame retardants
Failure analysis 472
DHS-GC-MS 289
DRIFTS 27
DSC 173
ESEM 493
FTIR 19 ff; μFTIR 530
ICL 543
OM 470
PyGC 234; PyGC-MS 260
PyIR 262
SIMS 430
TG-DTA 191
XPS 419
FEG-SEM, analytical method 489 ff
Fibres, additives
Colorants 521
Dyes 363
Fibres, analysis
ATR-FTIR 33
FTIES 75
μFTIR 526 ff
μRS 539
PA-FTIR 71
SEM 486, 654; ESEM 492
Fibres, identification
NIRS 51
Fillers, analysis
AET 718
AFM 510
ATR-FTIR 644
CLSM 482
CMR 561
DIES 127, 719
DRIFTS 27
DTA 175
HR-US 128
IR 18, 25; μFTIR 526 ff; process IR 687
LIBS 350
μNEXAFS 563
NIRS 52; process NIRS 699
NMR 102; NMRI 553; LR-NMR 706
PLM 471
PyGC 232
RS 59 ff; μRS 540, 646
SEM 488; ESEM-EDS 492; LVSEM 490
SKM 514
SPM 504
TGA 184 ff
TG-DSC 191
TG-FTIR 198
Finish-on-fibres, analysis
LR-NMR 706, 713
NIRS 49
PyIR 263
Firebrake: trade name; flame retardants
Firemaster: trade name; flame retardants
Fish-eyes, analysis 213
μFTIR 530
Flacavon: trade name; flame retardants
Flame retardants, analysis
DIES 126
DSC 167
DTA 175
DT-MS 651
EPMA 500
IR 18, 21, 25
iSIMS 571
LIBS 348
LIF 346
LPyMS 391
Mössbauer 123
NMR 101 ff; LR-NMR 712
NQR 112
Py-FTIR 263
PyGC 231; PyGC-AED 265; PyGC-MS 252 ff
PyMS 243
SEM 488
TD-GC-MS 627
TD-MS 300
TEM 496
TGA 183
TG-DSC 191; TG-DSC-MS 206
TG-DTA 191; TG-DTA-FTIR 207
TG-FTIR 197
TG-GC-MS 209
TG-MS 204 ff; VMI-TG-MS 210
Thermolysis-FTIR 199
ToF-SIMS 430
TPPy-MS 271
TVA 281
UV-LDI-ToFMS 363
XPS 419
Flammex: trade name; flame retardants
Flectol: trade name; aromatic amines
Fluorescence imaging, analytical method 541
Applications 541
Fluorescence microscopy, analytical method 475 ff
Applications 477 ff
Fluorescence spectroscopy, analytical method 75 ff
Applications 79 ff
Fluorescent additives, analysis
UV microscopy 473
Fluorescent pigments, use 81
Fluorfolpet: trade name; fungicides
Foaming agents, analysis
AET 719
DIES 126
DTA 175
HS-GC 285
TMA-MS 194
Food contact plastics, additives 269
Nonylphenol 627
Food contact plastics, analysis 553, 651
FTIR spectra 20
Food packaging regulations 116
Forensic science, analysis 489
ESEM 492

LA-ICP-MS 341
LMMS 388
MALDI-MS 381
μFTIR 529
μRS 539
μVIS 521
μXRF 564
PA-FTIR 71
PyGC 234; PyGC-MS 261
PyMS 243
SERRS 65
FTIES, analytical method 72 ff
Applications 74 ff
FTIR microspectroscopy, analytical method 521 ff
Applications 526 ff
FTIR spectroscopy, analytical method 14 ff

G

Geomembranes, analysis
DSC-OIT 170
Glass fibres, analysis
CLSM 481
CMR 561
iSIMS 571
μFTIR 526
μXRF 564
OM 472
TGA 185
XPS 419
Goodrite: trade name; phenols, HALS
Grafting 19, 102

H

HALS stabilisers, analysis 253, 638
ATR-FTIR 33
CL 90
ESR 117 ff; ESRI 556
IR 17; process IR 687
L^2ToFMS 372
MALDI-ToFMS 381 ff
NIRS 47; process NIRS 699
Process UV/VIS/NIR 681
PyGC 229, 231; PyGC-MS 253 ff
TD-GC 296
ToF-SIMS 431 ff; iSIMS 570
WDXRF 722
XPS 413 ff
Hardeners, analysis
PyGC 231
TPPy-MS 273
HDPE, additives 7, 22 ff, 32, 214, 492
Antioxidants 47, 92 ff, 116, 296, 539, 613 ff
Antistatic agents 183
Carbon-blacks 488
Fillers 60, 128, 488, 644, 646
Peroxides 115
Pigments 743
PPA 419
Solvents 551
Stabilisers 638
Volatiles 296
HDPE, analysis
Reference materials 741 ff
SFE 614
UV 7
HDPE melt, additives 688 ff
Stabilisers 681
Headspace sampling, analytical method 282 ff, 285 ff
Applications 284 ff, 288 ff
Heterogeneity 103, 543
GF-ZAAS 741
LA-ICP-MS 341
μFTIR 523
μRS 537
HIPS, additives
Blowing agents 551
Flame retardants 101, 112, 163, 243, 255, 271, 346
Oil 713
Rubber 713
HIPS, outgassing 288
Homogeneity testing 743
Hostanox: trade name; phenols, thiosynergists
Hostavin: trade name; HALS
HS-SPME, analytical method 289 ff
Applications 291
Hydrocarb: trade name; fillers
Hyphenated thermal analysis 192 ff
Applications 193 ff

I

ICL, analytical method 541 ff
Applications 543 ff
Image analysis 462 ff, 519
Imaging 460 ff, 514 ff, 521 ff
AFM 504 ff
SPM 501 ff
Imaging, applications 519
Imaging SIMS, analytical method 567 ff
Applications 569 ff
Impact modifiers, analysis
μFTIR 529
NMR 101
OM 472
PyGC-MS 252
SEM 488
Impurities, analysis
ICL 544
LMMS 386
μFTIR 525 ff
μRS 537
Inhibitors, analysis
Phosphorescence 82
Inks, analysis
CEMS 123
iSIMS 569
μATR-FTIR 33
μXRF 564
NIR-FTRS 65
NIRS 52
PyGC 232

SEM 489; ESEM 492
SERS 61
XPS 418 ff
Inorganics, analysis
μXRF 564
NMR 103
SEM-EDS 488
In situ analytical methods 1 ff
Instrument qualification 758 ff
Interaction products, analysis
ESR 119
Mössbauer 122 ff
Interactions
Co-additive 119, 183, 191, 198
Polymer–additives 112, 120, 196 ff
Polymer–fillers 102
Polymer–surfactants 108
Stabilisers–pesticides 8, 22
Interfaces, analysis
CSOM 482
Interlaboratory tests 755 ff
DSC-OIT 169 ff
PyGC 225; PyGC-MS 250
Pyrolysis 221
SSIMS 428
Ion imaging 566
Ion microscopy, analytical method 567
Ionol: trade name; phenols
Ionox: trade name; phenols, UV absorbers
Ion scattering, analytical method 441 ff
Applications 443
iPP, additives
Nucleating agents 167
Pigments 564
Stabilisers 92
UV absorbers 520
Whitening agents 474
Irgafos: trade name; phosph(on)ites
Irganox: trade name; phenols, thiosynergists
Irgastab: trade name; phosph(on)ites
Irgastat: trade name; antistatics
IR reflectance, analytical method 23 ff
Applications 24 ff
Isoprene rubber, additives
Stabilisers 230

K

Kane Ace: trade name; impact modifiers
Kapton: trade name; polyimide
Kemamide: trade name; slip additives
Ketjenblack: trade name; carbon-blacks
Kevlar: trade name; aromatic polyamide
Kraton: trade name; copolymer
Kubelka–Munk function 634, 645

L

Lactones, analysis
ESR 117
LA-ICP-AES, analytical method 335 ff
Applications 338 ff
LA-ICP-MS, analytical method 335 ff
Applications 338 ff
Laminates, analysis 563
μATR-FTIR 524; μFTIR 530
μRS 538
NIRS 43
PA-FTIR 70
Lankromark: trade name; PVC stabilisers
Laser ablation, analytical method 331 ff
Applications 334 ff
Laser desorption, analytical method 353 ff
Laser ionisation, analytical method 353 ff, 363 ff
Applications 364
Laser microscopy, analytical method 478 ff
Laser pyrolysis, analytical method 388 ff
Applications 390 ff
Lasers 325 ff
Applications 327 ff
Laser spectroscopy, analytical method 341 ff
Applications 342 ff
Latex films, additives
Surfactants 71
LCFM, analytical method 477, 480
LD/EI-FTMS, analytical method 366 ff
Applications 370 ff
LD-FTMS, analytical method 358 ff
Applications 360 ff
LDMS, analytical method 354 ff
LDPE, additives 8, 22 ff, 187, 191, 426, 432, 570
Accelerators 232
Antiblocking agents 31, 126, 417, 482
Antioxidants 89, 92, 170, 281, 296, 437, 612 ff, 630, 638
Carbon-black 757
Fillers 128
HALS 22, 33, 117, 171, 229, 259, 437, 557, 643
Light stabilisers 229
Lubricants 496
Release agents 419
Slip agents 90, 253, 565, 613
UV absorbers 613
Volatiles 288
LDPE, analysis
Extraction 612
SFE 614
SIMS 426
TGA 187
UV 8
LDPE melt, additives 687 ff, 699
Stabilisers 681
LEAFS, analytical method 343 ff
Applications 344 ff
LEIS, analytical method 341 ff, 443 ff
Applications 444
Leukopur: trade name; fluorescent whitening agents
LIBS, analytical method 346 ff
Applications 348 ff
LIESA®, analytical method 346 ff
Applications 348 ff
LIF, analytical method 343 ff
Applications 344 ff
Light microscopy, analytical method 464 ff

Applications 466
Light stabilisers, analysis
FTIR 643
NIRA 47
PyGC 229
UV microscopy 474
XPS 418
LLDPE, additives 32, 214, 431 ff
Antioxidants 17
HALS 432, 570
Processing aids 471
Slip agents 419, 510, 528
Stabilisers 103 ff
LLDPE, analysis
SSIMS 431
LLDPE melt, additives 687
Fillers 718
L^2MS, analytical method 367 ff
Applications 370 ff
LMMS, analytical method 381 ff
Applications 386 ff
LMMS, mapping 566 ff
Applications 567
Lotader: trade name; impact modifiers
Lowilite: trade name; UV absorbers, HALS
Lowinox: trade name; phenols, thiosynergists
Loxamid: trade name; lubricants
Loxiol: trade name; antifogging additives, lubricants
LPyMS, analytical method 390
Applications 390 ff
LR-NMR, analytical method 706 ff
Applications 710 ff
LRRS, analytical method 65
Applications 66
LS: trade name; UV absorbers, HALS
Lubricants, analysis
DSC 165
LD/EIMS 370; LD-FTMS 361
NIRA 50
NIRS 44; process NIRS 699
Process IR 687
PyGC 229; PyGC-MS 253
TD-MS 300
TGA 186
ToF-SIMS 430 ff; iSIMS 571
XPS 416
Luminescence, analytical method 75 ff
Luminor: trade name; pigments
Luperco: trade name; peroxide shifters
Luperox: trade name; peroxides
Lupolen: trade name; HDPE grade
LVSEM, analytical method 489 ff
Applications 490 ff
LV-SEM, analytical method 491 ff
Lycra Spandex: trade name; PEUU grade

M

MALDI, analytical method 374 ff
MALDI, quantitation 650
MALDI-ToFMS, analytical method 376 ff
Applications 379 ff
Mass spectrometry, quantitative 647 ff
CIMS 650
DT-MS 651
FAB-MS 648
TG-MS 650
Masterbatches, analysis 104, 198, 253
TGA 181
Medical plastics, additives 417, 434
Stabilisers 170
Melapur: trade name; flame retardants
Metal deactivators, analysis
DTA 175
Process UV/VIS/NIR 682
Metal traces, analysis
Fluorescence 79
Method development 731 ff, 760
HPLC 736
Promising approaches 736
SFE 736 ff
Method validation 731 ff, 746 ff
Antioxidant migration 757
Applications 749 ff
Polymer/additive analysis 760 ff
Microanalysis 458 ff
Applications 460
μFTIR, analytical method 521 ff
Applications 526 ff
μNIRS, analytical method 525
Microscopy 460 ff
Microscopy, quantitative 653
Mineral fibres 654
Weathering 654
Microspectroscopy 514 ff
Microthermal analysis, methods 210 ff
Applications 212 ff
μXPS, analytical method 564 ff
Applications 565 ff
μXRF, analytical method 563 ff
Applications 564
Mid-IR spectroscopy, analytical method 14 ff
Applications 16 ff
Migration, additives
Antioxidants 757
Migration, analysis 553
ATR-FTIR 32
SIMS 430, 436; iSIMS 570 ff
XPS 417; iXPS 566
Millad: trade name; nucleating agents
Mineral oils, analysis
LR-NMR 710 ff
TGA 180
Miscibility 166
Mobility 710
Modifiers, analysis
Process NIRS 699
Moisture, analysis
DHS 289
DIES 125, 719
KFR 49
LPyGC-MS 392; LPyIR 392

LR-NMR 706 ff; NMRI 552
NIRS 49; process NIRS 701
OM 471
PA-NIR 70
TGA 180
TG-DSC 191
Molecular dynamics 105 ff
Monomers, analysis
DHS-GC-MS 289
DIES 719
HS-GC 285
NMRI 551
Process IR 687
Process NIRS 698
RS 59; μRS 539; RRS 62
TD-GC-MS 296
TG-MS 202
Morphology, analysis
VMI-TG 293
Morton: trade name; antimicrobials
Mössbauer spectroscopy, analytical method 120 ff
Applications 122 ff
MRI, analytical method 546 ff

N

Nafion: trade name; fluoro-copolymer
Nanoanalysis 460
AFM 510
Nanocomposites, analysis
TEM 496
XRD 496
Naugard: trade name; aromatic amines, metal deactivators
NBR, additives 242, 350
Plasticisers 165, 180, 298, 620 ff
Neoprene: trade name; polychloroprene grade
Neozon: trade name; amines
Neviken: trade name; pesticides
NEXAFS microscopy, analytical method 561 ff
Applications 562 ff
NIRA, analytical method 35
NIRS, analytical method 34 ff
Applications 42 ff
Nitrogen, analysis
CL 81
NMR, analytical method 95 ff, 716
Applications 100 ff, 716
NMR relaxation 106
NMRI, analytical method 546 ff
Applications 551
NMR-MOUSE 549, 553, 709 ff
Nomex: trade name; aramid polymer fibre
Non-destructive analytical methods 2 ff
Noryl: trade name; PPO blends
NQR, analytical method 110 ff
Applications 112 ff
NR, additives 70, 242, 273
Antioxidants 171
Carbon-blacks 191
NSOM, analytical method 511 ff
Applications 513 ff
Nuclear Overhauser effect 97
Nuclear spectroscopy 94 ff
Nucleating agents, analysis
CL 90
DSC 167
Process NIRS 700
Nujol: trade name; mineral oil
Nylosan: trade name; dyes
Nylostab S-EED: trade name; HALS

O

Odorants, analysis
DHS-GC-MS 288
HS-GC 285
HS-SPME 291
TD-GC-MS 296 ff
Oligomers, analysis
HPLC 736
iSIMS 572
LD-FTMS 360
MALDI-ToFMS 379 ff
TD-GC-MS 298
Optical brighteners, analysis
Fluorescence 81
Fluorescence imaging 541
UV microscopy 474
Optical microscopy, analytical method 466 ff
Applications 470 ff
Outgassing, analysis
DHS-GC-MS 288
TD 295
TG-MS 205
Oxidation products, analysis
DSC-CL 93
FTIES 74
Oxyluminescence 87 ff
Oxidative induction time
DSC 168
Oxidative stability testing
DSC 165
Oxychemiluminescence, analytical method 83
Oxypruf: trade name; alkoxylated pyrazoles

P

PA4.6, additives
Heat stabilisers 126
PA6, additives 431
Antioxidants 545
Dyes 50, 697
Moisture 713
UV absorbers 253
PA6 melt, additives
Fillers 687
PA6.6, additives
Flame retardants 163, 199, 243
Impurities 386
Lubricants 186
PA12, additives
Plasticisers 531, 619 ff
Slip agents 166

PA12 melt, additives
Fillers 719
PAI, additives
Fillers 497
Palaroid: trade name; acrylic resin
Paper additives
Pigments 492
Sizing agents 269
Paper additives, analysis 270
ATR-FTIR 33, 644; μATR-FTIR 527
ESEM-EDS 492
LIBS 350
LR-NMR 714
NIRS 52
PyGC 232; PyGC-MS 258
XPS 417
Paper conservation, analysis
CL 94
PAS, analytical method 66 ff
Applications 69 ff
PB, additives
Antioxidants 171
PBMA, additives
Dyes 81
Stabilisers 122
PBT, additives
Antioxidants 296
Fillers 560
Flame retardants 101, 197, 254, 271, 300, 348, 627
Impact modifiers 102
PC, additives 300, 339, 418, 629
Flame retardants 627
Impurities 386
Release agents 433
Solvents 552
PC melt, additives
Slip agents 688
PC/PBT, additives
Antioxidants 271
Impact modifiers 271
Release agents 271
PDBS: trade name; flame retardants
PDMS, additives
Fillers 553
PE, additives 71, 269, 338 ff, 360, 381, 650 ff
Antioxidants 175, 278, 361, 431 ff, 606 ff, 630
Antistatics 417
Cadmium 741
Carbon-black 750
Catalysts 561
Extrusion aids 554
Fillers 186, 493
HALS 90, 638
Light stabilisers 21
Lubricants 229, 253
Peroxides 115
Pigments 272 ff
Slip agents 21
Stabilisers 8, 643
Volatiles 295 ff
PE, analysis
PA-FTIR 71
UV 8
Pellethane: trade name; PEUU grade
PEMA, additives
Stabilisers 122
PE melt, analysis
UV 8
Perkadox: trade name; peroxides
PERM project 741 ff
Permanax: trade name; phenols, aromatic amines
Peroxides, analysis
ESR 115
ICL 545
PET, additives 193, 346, 432
Antioxidants 370, 373
Catalysts 418
Contaminants 285
Dyes 50
Flame retardants 163
Moisture 180
Primers 436
Volatiles 285
UV absorbers 373
PET melt, additives
Fillers 699
PEUU, additives 243, 273
PFG-NMR, analytical method 108
Applications 108 ff
PGSE, analytical method 107
Phosphorescence, analytical method 81
Applications 82
Phosphorescent additives, use 82
Photo-initiators, analysis
TD-GC-MS 298
Phthalates, analysis
Migration rate 624
Pigments, analysis
CLSM 481
CMR 561
EPMA 501
Fluorescence 79; fluorescence microscopy 477
FT LMMS 387
IR 25
LA-ICP-MS 341
LDMS 363
LIBS 351
μVIS 521
μWAXS 559
OM 471
Process NIRS 699
PyGC 232; PyGC-MS 257
RS 59; μRS 539 ff
SIMS 432; iSIMS 570
TGA 185
TPPy-MS 272
UV 10; TUV 10
Plastanox: trade name; thiosynergists
Plasticisers, analysis
ATR-FTIR 32
DIES 126
DSC 165 ff

ESR 116
Extraction 757
FAB-MS 650
Fluorescence microscopy 475
FTIR 17, 644; μFTIR 527
HS-GC 285
IDGC-MS 627
LR-NMR 711 ff
NIRS 48
NMR 109; NMRI 554; process NMR 706
PA-FTIR 71
PyGC 230; PyGC-MS 253
SEC-GC 629
Solvent extraction 619 ff
TD-GC-FID 298; TD-GC-MS 298
TEA-FID 278
TGA 180, 757
TG/DTG-DTA-MS 207
TG-FTIR 198
TG-MS 205
Thermal extraction 619 ff
ToF LMMS 386
ToF-SIMS 430
TPPy-GC-MS 269
XPS 418
Plastomers, additives
PPA 419
Plate-out 184, 213
PMMA, additives 116
Antioxidants 253
Cross-linking agents 433
Dyes 370, 513
Flame retardants 391
Primers 436
Release agents 298
Solvents 552
Stabilisers 122
Polyacrylates, additives
Monomers 285
Polyamide melt, additives
Moisture 719
Polyamides, additives 339, 605, 713
Antioxidants 92
Dyes 482
Fibres 488
Flame retardants 18, 21, 104, 232, 255, 265
Optical brighteners 81
Polyamides, outgassing 288
Polybutylene glycol, additives 572
Poly(caprolactone), additives
Primers 436
Polyesters, additives 339
Dyes 482
Flame retardants 18, 232, 265
Poly(ethylacrylate), additives
Primers 436
Polygard: trade name; phosphites
Polyimides, additives 419
Moisture 392
Polymer melts, analysis
IR 23
Polymer production
In-process analysis 673
Polymers, analysis
Crystallinity 715
MALDI-MS 379
PyGC 234; PyGC-MS 251
PyMS 241
Tacticity 715
TPPy-MS 274
Polymer waste, additives
Flame retardants 206
Tracers 80
Polymer waste, analysis 351 ff
LIBS 349
NIRS 48
Polymer waste, sorting
NIRS 698
Poly(4-methylpentene-1)
UV absorbers 474
Polyolefin melt, additives 699
Polyolefins, additives 647, 650
Antioxidants 183, 474, 615 ff, 756
Antistatic agents 490
Fillers 722
Flame retardants 488
Stabilisers 47
UV stabilisers 79
Polyolefins, analysis
Extraction 613
Fluorescence 79
Polypyrrole, additives 419
Polyvinylpyrrolidone, additives
Monomers 721
POM, additives
Antioxidants 370
UV absorbers 373
Porapak: trade name; sorbents
PP, additives 46, 90, 270, 339, 437, 446, 511, 645, 650 ff
Antioxidants 22, 174, 370 ff, 373, 431, 475, 606, 613 ff
Antistatics 417
Blowing agents 167
Catalysts 541
Fibres 482
Fillers 25, 27, 59, 186, 488, 497, 654
Flame retardants 419, 496, 571
HALS 21, 90, 117 ff, 229, 259, 413 ff, 556 ff
Impurities 544
Light stabilisers 229, 474
Pigments 527
Sizings 482
Slip agents 570
Smoke suppressants 167
Stabilisers 229, 531, 544 ff, 638
UV absorbers 373, 475, 528, 613 ff
Wetting 492
PP, analysis
NIRS 645
SFE 615
PPE, additives
Flame retardants 627
PP fibres, additives

Pigments 539
PP melt, additives 687, 699
Fillers 718
Stabilisers 681
PPO, additives
Lubricants 165
PPO/PS, outgassing 288
Primers, analysis
ToF-SIMS 433; iSIMS 569
Printability, analysis
XPS 418
Proban: trade name; flame retardants
Process analysers 667 ff
Process analysis 663 ff
Process chromatography, analytical method 668, 720 ff
Applications 721
Processing aids, analysis
AFM 471
IR 18
LR-NMR 713
ToF-SIMS 430; iSIMS 571
XPS 419
Process mass spectrometry, analytical method 668
Process mid-IR spectroscopy, analytical method 683 ff
Applications 687 ff
Process NIR spectroscopy, analytical method 693 ff
Applications 697 ff
Process NMR spectroscopy, analytical method 704 ff, 716
Applications 706, 716
Process oils, analysis
LR-NMR 710 ff
NIRS 50
ToF LMMS 387
Process Raman spectroscopy, analytical method 701 ff
Applications 702 ff
Process spectroscopy, analytical method 672, 675 ff
Applications 677
Process UV/VIS spectrophotometry, analytical method 679 ff
Applications 680 ff
Process XRF, analysis 721
Applications 721 ff
Profax: trade name; PP grade
Programmed temperature vaporisation 268
PS, additives 654
Blowing agents 173, 198
Dyes 387
Fillers 127
Flame retardants 197, 271 ff, 391
Monomers 551
Volatiles 627
PS melt, additives
Blowing agents 700
Fillers 719
Nucleating agents 700
PTFE, additives 430
PUR, additives 391, 417, 431, 437
Fillers 165
Flame retardants 205, 209
Plasticisers 253
Release agents 433, 531
Smoke suppressants 197
PUR, analysis
EPMA 501
Purge-and-trap, analytical method 283, 286
PVAc, additives
Plasticisers 116, 166
PVAL, additives 265
Dyes 82
PVB, additives
Plasticisers 108
PVC, additives 34, 269, 338 ff, 348, 444, 529
Adhesion promoters 540
Antioxidants 361, 373
Coupling agents 32
Flame retardants 243, 346, 391, 419
Fungicides 539
HALS 253
Inclusions 386
Monomers 285
Pigments 119, 471
Plasticisers 17, 32 ff, 48, 60, 71, 109, 116, 126, 166 ff, 180, 197, 207, 230, 280 ff, 295 ff, 418, 510, 527, 624, 644, 650, 711 ff, 750
Stabilisers 122, 166
UV absorbers 373
PVC melt, additives 699
PVDF, additives 338, 341
Py-FIMS, analytical method 238
PyFTIR, analytical method 261 ff
Applications 262 ff
PyGC, analytical method 222 ff
Applications 228 ff
PyGC-AED, analytical method 264 ff
Applications 265
PyGC-FTIR, analytical method 263
Applications 264
PyGC-MS, analytical method 244 ff
Applications 251 ff
PyMS, analytical method 235 ff
Applications 240
Py-PIMS, analytical method 238
Pyrochek: trade name; flame retardants
Pyrolin: trade name; thermo-resistant polymer
Pyrolysers 216 ff
Pyrolysis, analytical method 214 ff
Applications 221 ff
Pyrolysis, derivatisation 228
Pyrolysis, quantitation 649
Pyrotechnics, analysis
TG-DSC 207
Pyrovatex: trade name; flame retardants

Q

Quality assurance
DSC 173
LR-NMR 710
NIRS 43
UV 680
Quality control
DSC 167 ff; DSC-OIT 168
DTA 170, 174

FTIR 19 ff, 643
LA-ICP-MS 341
LIESA® 349
LR-NMR 710 ff; NMR-MOUSE 553
NIRA 47; NIRS 45 ff, 696
PyGC 234 ff; PyGC-MS 249 ff, 260
PyGC-FTIR 264
PyIR 262
PyMS 243; Py-FIMS 238
SPC chart 754
TD-GC-MS 296
TGA 188
TPPy-MS 273
UV/VIS 679
XPS 419
XRF 721
Quantitation, additives
Antioxidants 615, 629, 638, 646 ff
Coupling agents 644
Dyes 646
Extender oil 623
Fillers 644, 646
HALS 638
Irgafos 168 616 ff; Irgafos P-EPQ 629
Irganox 1010 615 ff; Irganox B220 606
Light stabilisers 643
Paper additives 644
Plasticisers 619 ff, 644
Stabilisers 630, 638
Quantitation, analysis 597 ff
Extraction 609 ff, 619 ff
GC 626 ff; HS-GC 611
GC-MS 627, 649, 651; IDGC-MS 627
HPLC 628 ff
NMR 647
SEC-FTIR 629
SFE 614
SPME 611
TD 612; TD-GC-MS 627
TGA 619 ff
Quantitation, polyamides 605
Quantitation, polyolefins 613
Quantitation, rubbers 606

R

Radiation degradation, analysis
ESR 116
Radicals, analysis
ESR 114 ff; ESRI 556 ff
Raman microprobe, analytical method 532 ff
Raman microscopy, analytical method 532 ff
Applications 537 ff
Raman spectroscopy, analytical method 52 ff
Applications 58 ff
Raw materials, analysis
DSC 173
TGA 188
RBS, analytical method 444 ff
Applications 446
Reactive extrusion, analysis 700
Process IR 692
Recyclate, additives
Flame retardants 255
Recyclate, analysis
FTIR 19
LIBS 349
NIRS 50
PyGC-MS 255
TD-GC-MS 296
Recyclostab: trade name; recycling additives
Reference materials 736 ff
ADPOL 745
BCR 743 ff
Development 741 ff
PERM 741 ff
TOXEL 745
VDA 740 ff
Release agents, analysis
μFTIR 531
TD 298; TD/PyGC 271
ToF-SIMS 460 ff; iSIMS 569
XPS 416, 419
Remanzol: trade name; dyes
Remote spectroscopy 677 ff
REMPI, analytical method 365
Applications 366
Reofos: trade name; flame retardants
Residue analysis 192
Retarders, analysis
DHS-GC-MS 289
PyGC 234; PyGC-MS 257
RIMS, analytical method 365
Round robins 755 ff
Applications 756 ff
IMEP-2 program 756
PERM project 741 ff, 756
RRS, analytical method 61 ff
Applications 62 ff
Rubbers, additives 285, 371, 494, 606, 643
Antioxidants 245, 387
Antiozonants 386
Carbon-blacks 472, 713
Fillers 175, 510, 710 ff
Processing oils 387, 710 ff, 713
Volatiles 298
Vulcanisation accelerators 167, 206, 386
Rubbers, analysis 33, 606
Extraction 615 ff
LDMS 360
LIESA® 722
NIR-FTRS 61
PyGC 234; PyGC-FTIR 264
PyMS 242 ff; PyGC-MS 256
TD-MS 300
ToF LMMS 386 ff
TPPy-GC 270; TPPy-MS 273
Rubbers, deformulation 606

S

SAM, analytical method 493

Applications 493 ff
Sampling procedures 600 ff
SAN, additives
Flame retardants 272
Sandostab: trade name; nucleating agents, phosph(on)ites
Sanduvor: trade name; UV absorbers, HALS
Santintone: trade name; fillers
Santocure: trade name; curing agents
Santoflex: trade name; aromatic amines
Santonox: trade name; phenols, thiosynergists
Santowhite: trade name; phenols
Saytex: trade name; flame retardants
SBR, additives 70, 198, 242 ff, 273, 391
Antioxidants 256, 296
Fillers 187
Plasticisers 181, 620 ff
SBR/NR, additives 391
Sealability, analysis
XPS 419
Seenox: trade name; phenols, thiosynergists
Self-diffusion 107
SEM, analytical method 485 ff
Applications 487 ff
SERRS, analytical method 64
Applications 65
SERS, analytical method 63 ff
Applications 64
Shelf-life, analysis
DSC-OIT 172
TG-OIT 189
Silox: trade name; silanes
SIMS, analytical method 422 ff
Applications 429 ff
Simultaneous thermal analysis 189
Single-pulse excitation 97
Sipernat: trade name; antiblocking additives
Sizings, analysis
CSLM 482
PA-FTIR 70
ToF-SIMS 430
TPPy-MS 269
XPS 419
SKM, analytical method 514
Applications 514
Slip agents, analysis
AFM 510
CL 90
IR 21; μFTIR 528; process IR 687
LD/EIMS 370
Process NIRS 699
Process UV/VIS/NIR 682
TEM 496
ToF-SIMS 430; iSIMS 570
XPS 416; iXPS 565
SMA, additives
Sizings 482
Smoke suppressants, analysis
DSC 167
TG-DSC 191
TG-FTIR 197
Smoothing agents, analysis
iSIMS 571
SNMS, analytical method 439 ff
Applications 441
Softeners, analysis
LMMS 388
LR-NMR 713
Solid/liquid ratio 708
Solubles, analysis
Process NMR 706
Solvents, analysis
DHS-GC-MS 289
HS-GC 284
HS-SPME 291
NMRI 551
PyGC 232
TD 295
TGA 180
TG-FTIR 196
TG-MS 205
SOM, analytical method 469
Spaitech: trade name; PE grade
SPC chart 754
Speciation, analysis
LMMS 385 ff
μNEXAFS 563
RPLC-LEIS 343
SSIMS 429
UVRRS 63
Spectroscopy, quantitative 633 ff
Fluorescence 639
FTIR 639 ff
NIRS 644 ff
NMR 646 ff
RS 645 ff
UV/VIS 637 ff
Spermicides, analysis
MALDI-MS 381
Spinuvex: trade name; HALS
SPM, analytical method 501 ff
Applications 503 ff
Stabaxol: trade name; antihydrolysis additives
Stabilisers, analysis 630, 638
CL 92; ICL 543
DSC 166
ESR 117
FTIES 75
Luminescence 79; TSL 214
Mössbauer 122
NMR 104
Process UV/VIS/NIR 682
UV/VIS 4 ff; μUV 520
Stabilox: trade name; PVC stabilisers
Standard addition 604
Standards, quantitation 603
Internal 603
External 603
Standard test methods
DSC 169
TGA 181
Stanyl: trade name; nylon 4.6 grade
ST-DVB, additives

Cross-linking agents 231
Stearates, analysis
LD-FTMS 361
STEM, analytical method 497 ff
Applications 500 ff
STM, analytical method 501 ff
Stress cracking agents, analysis
TD-GC-FTIR-MS 299
Sulfur, analysis
Fluorescence 81
Sumilizer: trade name; phenols
Surface analysis 403 ff
ATR-FTIR 28
FTIR 23
Surface analysis, quantitative 651 ff
Surface mass spectrometry, analytical method 420 ff
Applications 422
Surface roughness
CLSM 481
OM 471
Surfactants, analysis
ATR-FTIR 32
LD/EIMS 370; LD-FTMS 360, 363
MALDI-ToFMS 381
NIRS 48
NMR 102
PA-FTIR 71
PyGC 232
RS 60; μRS 539
ToF-SIMS 430
XPS 416
Surlyn: trade name; PE ionomer grade
Swelling, analysis 34, 443, 552
ESRI 556
System suitability 760

T

Tackifiers, analysis
ATR-FTIR 32
Tacticity, analysis 715
TD-GC, analytical method 291 ff
Applications 294
TD-MS, analytical method 299
Applications 299
Tecoflex: trade name; PEUU grade
TEM, analytical method 494 ff
Applications 496 ff
Tenax: trade name; sorbents (modified PPO)
Test methods
TGA 189
Textile fibres, additives
Dyes 25
Textiles, additives
Dyes 61, 258, 387, 528
Flame retardants 167
Pigments 539
Textiles, analysis
NIRA 48
TG(A), analytical method 175 ff
Applications 179 ff
TG-DSC, analytical method 190, 206
Applications 190, 206
TG-DTA, analytical method 191, 207
Applications 191, 207
TG-FTIR, analytical method 194 ff
Applications 196 ff
TG-GC, analytical method 207 ff
Applications 209
TG-MS, analytical method 200 ff
Applications 203 ff
Thermal desorption, analytical method 275 ff
Applications 278
Thermal distillation, analytical method 279
Applications 279
Thermal evolution analysis 276
Thermal stabilisers, analysis
DIES 126
NIRA 47
NMR 101
Thermal stability 189
Thermal UV spectrometry, analytical method 10
Applications 10
Thermal volatilisation, analytical method 275 ff
Applications 278
Thermoanalytical methods 155 ff
Applications 160 ff
Thermochromatography, analytical method 274
Applications 275
Thermoluminescence, analytical method 213
Applications 214
Thermolysis-FTIR, analytical method 198 ff
Applications 199
Thermomechanical analysis 160
Thermomicroscopy, analytical method 209, 211
Applications 210, 212
Tinuvin: trade name; phenols, HALS, UV absorbers
Topanol: trade name; phenols
Toys, additives
Nonylphenol 627
TPPy, analytical method 266 ff
Applications 269 ff
Traceability 736
Trace analysis 458
SERS 64
Tracers, analysis
Fluorescence 80
IR 18
Transmission IR spectroscopy, analytical method 20 ff
Applications 21 ff
Trigonox: trade name; peroxides
Troubleshooting
FTIR 24, 643; μFTIR 530
LMMS 387; LMMS mapping 567
PyGC-MS 261
SIMS 430; ToF-SIMS 437
TD-GC-MS 298
TG 182, 198
VMI-TG-MS 210
XPS 417, 419
TVA, analytical method 280 ff
Applications 281 ff

Twaron: trade name; thermo-resistant polymer (para-aramide)
Tyres, additives 606, 608
Carbon-black 553
Fillers 563

U

UHMWPE, additives
Antioxidants 539
Ultramarine blue, analysis
ESR 119
Ultranox: trade name; phosph(on)ites
Ultrasonic spectroscopy, analytical method 127 ff
Applications 128
Urepan: trade name; poly(ester urethane) elastomer
UV absorbers, analysis
AFM 511
Fluorescence 79
IR 17
LD-FTMS 361
L^2ToFMS 370 ff
μRS 538
NIRS 48
PA-UV 69
PyGC-MS 252
ToF-SIMS 437
UV 6 ff; μUV 520; UV microscopy 473
Uvasil: trade name; HALS
Uvasorb: trade name; UV absorbers, HALS
Uvinul: trade name; UV absorbers, HALS
Uvitex: trade name; fluorescent whitening agents
UV microscopy, analytical method 472 ff
Applications 473 ff
UV microspectroscopy, analytical method 519 ff
Applications 520 ff
UV/VIS spectrophotometry, analytical method 4 ff
Applications 6 ff

V

Vacuum sublimation, analytical method 279
Applications 279
Validation 40
Criteria 746 ff
Hardware 758
Implementation 759
Software 758
Total process 757 ff
Vanox: trade name; phenols
Vapour-phase UV spectrometry, analytical method 10
Applications 10
Vapour pressure 180
Varox: trade name; thiosynergists
Vibrational spectroscopy 11 ff
VIEEW™ 466, 654
Viscosity modifiers, analysis
PyGC 232
Viton: trade name; processing aids
Volatiles, analysis
HS-GC 284
PyGC 232
TD-GC-MS 295 ff
TEA-CT-GC 278
TGA 180
TG-GC-IR-MS 209
TG-MS 202
TVA 281
Vulcanisates, additives
Antioxidants 364
Carbon-blacks 496
Vulcanisates, analysis
LDI-ToFMS 363
LPyMS 391
L^2ToFMS 372
NMR 102; NMRI 554
Vulcanisation accelerators, analysis 242
NMR 102
PyGC 229; PyGC-MS 257
PyMS 240
RS 60
TD-GC-MS 298
TG-DSC 191; TG-DSC-MS 206
TG-FTIR 198
Vulkacit: trade name; vulcanisation accelerators
Vulkanox: trade name; aromatic amines

W

Waxes, analysis
LD-FTMS 361
NIRS 52
NMRI 554
SIMS 437
Weston: trade name; phosphites
Wetting, analysis
ESEM 492
SSIMS 438
Wingstay: trade name; phenols
Wool, additives
Wetting 492

X

XLPE, additives
Antioxidants 528
Water trees 471
XPS, analytical method 411 ff
Applications 416 ff
X-ray microradiography, analytical method 560 ff
Applications 561
X-ray microscopy, analytical method 559 ff
X-ray microspectropy, analytical method 559 ff
Xylene solubles 715

Z

Zipro: trade name; flame retardants
Zytel: trade name; nylon 6.6 grade